Environmental, Health and Control Aspects of COAL CONVERSION— An Information Overview

Volume 1

Edited by

H. M. Braunstein
E. D. Copenhaver
H. A. Pfuderer

Contributors

J. T. Ensminger	N. S. Dailey	G. A. Dailey
J. K. Huffstetler	D. C. Michelson	R. F. Carrier
L. W. Rickert	R. H. Ross	D. J. Wilkes
S. S. Talmage	D. S. Harnden	

Copy editors

B. C. Winsbro
J. W. McKenna

Published 1981 by Ann Arbor Science Publishers, Inc.
230 Collingwood, P.O. Box 1425, Ann Arbor, Michigan 48106

Library of Congress Catalog Card Number 81-65350
ISBN 0-250-40445-1

Manufactured in the United States of America

First published by the Department of Energy
(ERDA), 1977. Second publication by Ann
Arbor Science Publishers, Inc.,1981.

"In this work, when it shall be found that much
is omitted, let it not be forgotten that much
likewise is performed."

Dr. Samuel Johnson
(on completion of his Dictionary, 1755)

Braunstein   **Copenhaver**   **Pfuderer**

H.M. Braunstein is a physical chemist with a broad interest in engineering and natural sciences. She is the author or coauthor of numerous publications including a recent book on biomass energy systems. During her ten years at the Oak Ridge National Laboratory (ORNL), she has held a variety of positions from her initial assignment as a predoctoral fellow in experimental thermodynamics to her present responsibility, management of studies involving energy and the environment in the Energy Division. Dr. Braunstein received her PhD from the University of Maine at Orono. Her current interests include environmental assessment of geothermal energy development and societal considerations in the hard/soft energy debate.

E.D. Copenhaver is a policy analyst in the Office of Integrated Assessments and Policy Analysis, Health and Safety Research Division of ORNL. Current research interests center around human health risk analysis and the technical and social considerations involved in siting hazardous or nuclear waste facilities. During her 18 years at ORNL, she has established two technical information centers—one in the area of toxic materials and the other dealing with environmental resources and control. She has published in the fields of environmental transport, toxic materials, and policy analysis of health and environmental legislation; has served as an editor for *Applied Physics Letters* and *NSF Trace Contaminants Abstracts*; and recently served as assistant chairman for the Third ORNL Life Sciences Symposium on *Health Risk Analysis*.

H.A. Pfuderer is currently assigned as technical assistant to ORNL's Associate Laboratory Director for Biomedical and Environmental Sciences. From 1971 to 1978 she was associated with the Ecological Sciences Information Center, four years as director. Major research interests include nuclear waste, radioecology, environmental impact modeling and coal conversion. In 1978 and 1979 she served as department head for Energy and Ecological Sciences, Information Center Complex. She has also served as Chairman of the Environmental Sciences Division of the American Nuclear Society and as editor of the *Low-Level Radioactive Waste Newsletter*. She is the author of several articles in the area of ecological effects on transuranics.

Preface

Early in the 1980s, sixteen major coal-producing and coal-consuming nations collaborated on a *World Coal Study*. Representatives of the participating countries reached the conclusion that coal resources are mankind's energy "bridge to the future" and that world coal production must increase nearly threefold by the year 2000 to keep pace with energy requirements.

The international commitment to coal is paralleled and potentially enlarged by the United States which possesses almost half of the earth's total known coal reserves. Coal makes up about 80% of the remaining fossil fuel resources in the United States, and the enormous energy potential in an estimated three trillion tons of U.S. coal offers reassurance for the future during this era of pressing energy needs and growing resource shortages. The availability of coal, however, does not guarantee untroubled decades of energy plenty. Expanded coal consumption inevitably will introduce formidable hazards and difficult problems for solution.

The United States eventually will rely heavily on coal for electricity and industrial power. The transition to coal is expected to occur slowly but steadily through the remainder of this century and into the 21st century. This trend is virtually compelled by dwindling oil and natural gas supplies, by an uncertain future for nuclear power, and by the inability of solar energy technologies collectively to meet U.S. energy needs in the foreseeable future.

Simultaneously with intensified coal use, environmental and health risks associated with coal conversion will also intensify. Thus, among the most critical energy challenges ahead is to find ways of using America's vast coal resources *safely* on a large scale over a long period.

The following *Information Overview*, originally prepared by the Information Division of Oak Ridge National Laboratory, is designed and superbly organized to provide an authoritative reference on environmental, health, and control issues that are anticipated from the use of coal as a primary energy supplier. The various effects of coal pollutants on plant, animal, and human life are analyzed in detail and thoroughly documented by this massive compilation of technical information and data.

Ann Arbor Science has published the current edition to keep it readily available for consultation by researchers, engineers, utility managers, industrialists, administrators, and any contemporary decision makers with energy responsibilities. All individuals professionally concerned with coal production or use will find the work a valuable guide to full understanding of the environmental and health consequences involved. The Coal Conversion Overview covers its subjects with a depth and scope that will keep it relevant even when later scientific developments extend the current limits of coal technology.

The team effort that went into the preparation of this Overview compares with the effort needed to compile a broad-based encyclopedia on a complex scientific topic. The contributors named as principal authors of different sections were aided by expert support staffs. Their findings were validated by many academic, industrial, and government reviewers. Only the facilities and resources of a huge national laboratory would have been sufficient to complete a study that attempts and accomplishes so much.

As additional information about the impact of coal conversion is developed, the new knowledge will complement but not replace this publication. It will remain a fundamental reference for many years. Each section contains material that must be understood and applied as coal competes with and gradually replaces oil and natural gas. The contents of the Overview will make indispensable contributions to the achievement and maintenance of an optimum balance among environmental, health, and energy factors. Neglecting to take into account the many implications of this work would hamper the development of coal and would overlook a responsibility that is a basic concern in U.S. long-range energy plans.

We believe that the continued availability of the *Information Overview* will help lead to appropriate energy choices and prudent decisions in coming years. Publication of this work is in keeping with the Ann Arbor Science policy of making established technical and reference works available as long as their publication has validity as a science service. Considering current world energy circumstances, the existence of vast coal reserves, and the known hazards of coal conversion, the present Overview clearly is among our most important current science titles.

We acknowledge the cooperative attitudes of Oak Ridge National Laboratory in agreeing that Ann Arbor Science should make this important research continually available to the scientific community.

THE PUBLISHER
Ann Arbor, Michigan
March, 1981

Section Locator
Volume 1

TABLE OF CONTENTS

G. A. Dailey and D. C. Michelson

xiii

LIST OF FIGURES

1. SUMMARY

H. M. Braunstein

It is clear that the United States in the future will come to rely much more heavily on its most
abundant fossil fuel, coal. The U.S. Geological Survey and the U.S. Bureau of Mines report that
the U.S. coal resource base is about 3 trillion tons, or about 80% of the total fossil fuel
reserves of the country. However, limited past experience with uncontrolled emissions from
pilot plant coal conversion processes indicates a pressing need to examine the possible health
and environmental consequences of full-scale industrial application of coal conversion
technologies.

This document was prepared to assist this need by attempting to identify meaningful knowledge
and, where possible, to outline gaps in the knowledge; reviewing the pertinent literature but
not necessarily all the literature; and overviewing data borrowed from kindred technologies
that bear on the question of coal conversion hazard.

In the examination of possible consequences of coal conversion releases, it becomes abundantly
clear that the areas of uncertainty are often greater than the areas of knowledge. Few studies
have addressed coal conversion effluents directly, and other bioenvironmental research, though
valuable and important, is sometimes of limited applicability. The problems are decidedly complex and resolution is difficult; nonetheless, much relevant information is evident.

Coal is a carbonaceous, nonhomogeneous, highly variable fossilized material. Formed millions
of years ago from decayed plant remains, it incorporated many constituents of its ancient swampy
environment such as clay minerals, dissolved salts, and sulfur. Under the influence of heat,
pressure, and geologic time, the plant fragments were altered, minerals were transformed, and
volatile components were driven off. Accordingly, the process of coalification was a series of
biochemical and geochemical reactions that contributed to the overall cycling of the biologically
critical element, carbon (Fig. 1.1).

Coal differences are designated by coal rank. Rank depends on the extent of coalification and
varies sequentially from low-rank lignite through subbituminous and bituminous coal to high-
rank anthracite coal. In general, carbon content increases and volatile matter decreases with
increases in rank. Ninety-seven percent of U.S. coal reserves are either bituminous (66%) or
subbituminous (31%), and the remainder is anthracite. The older Appalachian, Interior, and Gulf
region coals are of higher rank and sulfur content.

Coal is composed of minerals and macerals; minerals derive from inorganic inclusions in coal, and
macerals derive from the plant organic matter. Both fractions contribute to possible emissions
during coal utilization. Macerals decompose to desired organic fuel compounds but also to some
undesirable organics such as polycyclic aromatic hydrocarbons (PAH), whereas mineral matter is
the source of objectionable trace metals and, at times, a voluminous solid waste.

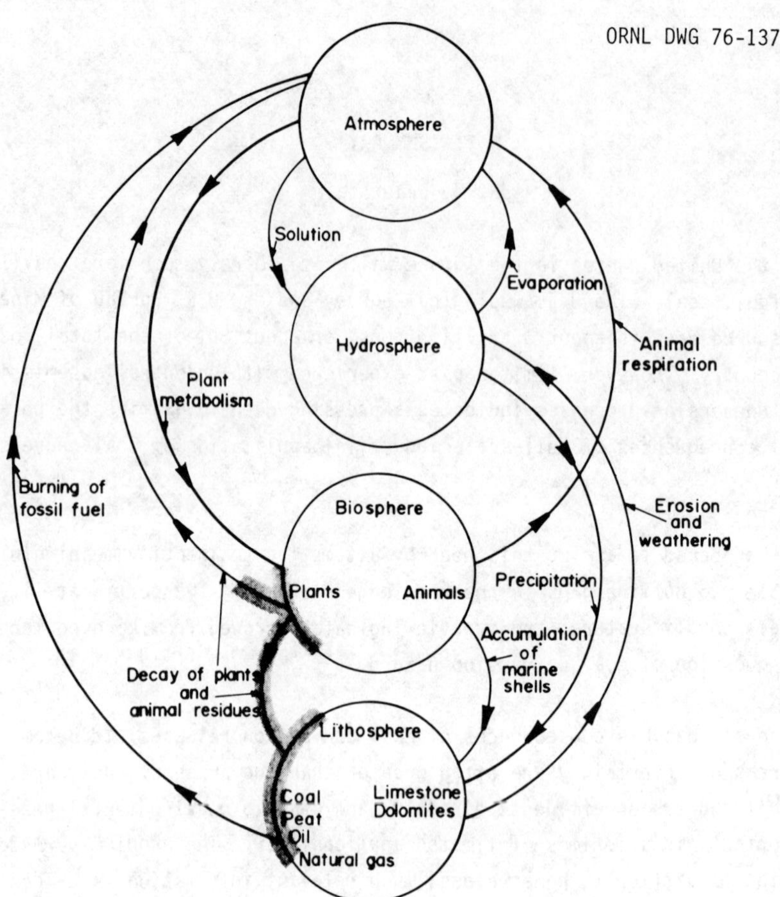

Fig. 1.1. Carbon cycle. Source: From U.S. Bureau of Mines 1974, Fig. 1, p. 3.

Sulfur in coal is environmentally significant, producing organosulfur compounds, sulfur oxides, and hydrogen sulfide. Among U.S. reserves, western coals are generally of lower rank and lower sulfur content, whereas Interior province and most Appalachian coals, which were formed under the influence of high sulfate ion concentrations in saline waters, contain greater amounts of sulfur.

Knowledge of the structure of coal is inadequate. However, serious attempts have been made to postulate a structure since the key to coal behavior is thought to lie in its structure. Coal is believed to consist largely of aromatic and hydroaromatic ring systems linked by direct carbon-carbon bonds, aliphatic groups, or ether linkages (Fig. 2.11). During coal decomposition, bonds are broken and organic fractions are released. The details of decomposition are complex and depend on many variables, including coal rank and experimental conditions. For example, during simple carbonization, volatile components pass through the hot coke and engage in a series of secondary reactions, the extent of which depends on variables such as retort temperature and contact time. The chemical composition of tar produced during carbonization is a function of these reactions and is carcinogenically important. Knowledge of coal, its structure and composition, is needed not only for identifying possible sources of undesirable effluents but also for predicting the behavior of coal as the source of gaseous and liquid fuels.

The concept of coal conversion is deceptively simple. It involves primarily two basic steps: the cracking of heavy hydrocarbons into lighter ones and the enrichment of the resultant molecules with hydrogen.

Unlike the concept, the application of coal conversion is not simple. It involves the handling of enormous amounts of a highly variable material (upwards of 25,000 tons of coal per day) often at high temperatures and pressures (greater than 1000°F and 3500 psi); it requires containment and control of both highly corrosive process materials and those that pose a possible health threat; and it calls for treatment and disposal of a voluminous solid waste and a possibly hazardous liquid or gaseous waste.

Nonetheless, coal conversion is a viable technology. Although it is currently practiced in the United States only on a small scale at pilot and demonstration plants, commercially applicable processes, capable of full exploitation of our coal resources, are being developed (Cochran 1976).

Coal can be converted to either a gas (gasification) or a liquid (liquefaction), either underground (in-situ gasification) or by mining the coal and transferring it to a conversion plant. However, because coal formations vary in size, depth, and composition, not all coals are equally suitable for a given conversion technique. Additionally, the parameters for selecting and combining a particular process and a prescribed coal seam are not yet fully delineated.

Converting coal into an environmentally acceptable gaseous fuel requires nine basic and seven auxiliary steps; these steps include coal preparation and pretreatment, gasification, shift conversion, catalytic methanation, various cleanup and disposal procedures, and final compression and dehydration of the product into pipeline gas (Fig. 3.1). The large number of named processes arise in an attempt to combine the various alternatives available in each step so that the entire procedure may be optimized. For example, a major research effort is directed toward optimizing the two key steps, gasification and methanation.

Basically, only three ingredients are needed to synthesize gas from coal — carbon, hydrogen, and oxygen. Coal provides the carbon, steam most often provides the hydrogen, and air provides the oxygen; however, the relative amounts of these elements are crucial. For example, the ratio of hydrogen to carbon largely determines the heating value of the gas.

Natural gas, which is primarily methane (CH_4), has a high hydrogen-to-carbon ratio (25% hydrogen, 75% carbon) and consequently has a high heating value (greater than 900 Btu/scf). In contrast, a typical bituminous coal contains only 5% hydrogen, 75% carbon, and 20% other, usually undesirable, constituents (sulfur and ash). Thus, when the object of coal gasification is to produce a substitute natural gas, hydrogen must be added, usually by catalytic methanation. Without added hydrogen, which is expensive, the gas will have a low or intermediate heating value, depending on whether the combustant is air or pure oxygen. Air produces a nitrogen-containing gas of low heating value (150 to 300 Btu/scf), whereas pure oxygen produces an intermediate-heating-value gas (300 to 400 Btu/scf). Low- and intermediate-Btu gases, although less expensive and simpler to produce, are considered inadequate for intermingling with natural gas for pipeline transportation and would be used at or near the production site.

Because catalytic methanation converts a low-grade gasified product to a pipeline or high-Btu gas, any low- or intermediate-Btu process becomes a high-Btu process by adding methanation. Thus, because methanation is essential to high-Btu coal gasification processes, it is the subject of considerable investigation.

Methane forms in a number of reactions but most importantly in the reaction of one part carbon monoxide with three parts hydrogen. Methanation reactions have a favorable free energy change at most temperatures (Table 3.3), but the reactions require catalytic assistance. The minerals in coal are thought to catalyze methane formation to some extent, but little is known of this catalysis and it is considered undependable.

Catalyst materials have been extensively studied. Of the metals identified as potential methanation catalysts, nickel surpasses all others as the catalyst of choice for commercial application; it is inexpensive, very active, and highly selective to methane (Table 3.4). However, nickel can be deactivated or poisoned. Sulfur compounds are the primary poison, but nickel is also deactivated by high temperatures (above 450°C), carbon deposition, nickel carbonyl formation (below 260°C), or by the deposition of iron from iron carbonyl formed in the reactor vessel. Thus, gasification processes employing catalytic methanation require rigorous cleanup of the product gas before it is allowed to contact the catalyst (Chap. 4).

Many equipment designs have been proposed for coal conversion; however, most viable gasification processes can be identified by an operating temperature, the use or absence of pressure, and the type of reactor vessel (Tables 3.9 and 3.10). Gasification processes use one of three basic reactor bed types — fixed, fluidized, or entrained bed.

A fixed-bed reactor heated internally or externally contains a grate on which coal rests. Input gases such as steam, air, or oxygen are introduced below the grate and pass through the heated coal, reacting with it to produce hot product gases which exit at the top of the reactor (Fig. 3.4). Caking coals normally cannot be used in a fixed-bed reactor, although design modifications such as stirring with water-cooled stirrers and pretreatment of the coal have attempted to loosen that constraint. Most eastern bituminous coals are caking.

The Lurgi process, a fully commercial gasification method, employs a fixed-bed reactor operating at 800 to 1000°C. It uses either air and steam or oxygen and steam as the reactants, producing low- or intermediate-Btu gas. With methanation, the product is upgraded to a high-Btu pipeline gas.

A fluidized bed is a mass of solid particles which exhibits liquidlike characteristics (it boils) when a gas flows upward through the bed. In coal conversion, gases (air or oxygen and steam) flowing through finely sized noncaking coal particles produce turbulent lifting and separation, expanding the bed and exposing a high surface area to the gaseous reactants. The Winkler gasifier (Fig. 3.6) uses a fluidized bed maintained at a sufficiently high temperature (1500 to 1800°F) to crack all the tars and heavy hydrocarbons, leaving only heavy ash particles to be removed from the bottom of the gasifier. Winkler plants, producing a low- or intermediate-Btu gas, have operated abroad since 1926.

In an entrainment system, finely sized coal particles are blown into the reactor along with the gas stream, and combustion takes place with coal particles suspended in the gas phase (Fig. 3.7). Commercial gasifiers (operating abroad) use either two or four opposing burners; the four-burner reactor can handle up to 850 tons of coal per day. The major advantage of an entrained-bed reactor, as used, for example, in the Koppers-Totzek process, is its ability to gasify any type or rank of coal, including the fines. Additionally, combustion takes place at about 3000°F, destroying tars, phenols, and light oils.

Many other techniques for gasification of coal are being explored; most employ the well-known methods with variations, but some use novel procedures. For example, the Two-Stage Combustion, PATGAS, and ATGAS processes employ an innovative means for gasification — a molten iron bath into which crushed coal is injected in the first stage. The coal's fixed carbon and sulfur dissolve, whereas the coal volatiles crack and appear as offgas. In the second stage, the dissolved carbon is gasified by hot air (1000 to 1200°F), whereas the dissolved sulfur is collected by a lime-bearing slag that floats on the molten iron. The slag is removed con-tinuously and is desulfurized and recycled (Fig. 3.9).

The Bi-Gas process uses a two-stage entrained-bed gasifier that operates at system pressures exceeding 1000 psi. When followed by a fluidized-bed methanation step, the final gas has a heating value of pipeline gas — 900 Btu/scf.

The Synthane process also involves a two-stage pressurized gasifier in which the caking proper-ties of the coal are destroyed in a fluidized-bed pretreater before carbonization.

The Hygas process avoids the problems associated with adding coal to pressurized systems through lock-hoppers. The crushed coal is slurried with a light oil and pumped under 1000 to 1500-psi pressure into the top of the gasifier (Fig. 3.21).

The CO_2 Acceptor process uses a novel technique to supply heat to the reactor. Hot, calcined dolomite (MgO·CaO) is circulated along with coal into the top of a fluidized-bed gasifier at 1500°F. Because the formation of calcium carbonate from calcium oxide is highly exothermic, the calcium oxide is allowed to react with carbon dioxide released from the coal, producing enough heat for gasification. The spent dolomite is regenerated and recycled.

Although coal is well suited to gasification, it is equally suited to liquefaction. Thus, the desire to decrease American dependence on imported crude oil and to conserve our dwindling national reserves increases the urgency for developing coal liquefaction processes. Although liquefaction implies a liquid product, processes considered as coal liquefaction can also give rise to a gaseous or solid product or, in some instances, only to a low-melting solid fuel.

Coal is liquefied primarily by either pyrolysis or dissolution. In pyrolysis, coal is heated in the absence of air to drive off volatile matter so that, in addition to a liquid product, by-product gases and char are produced (Fig. 3.28). Thus, the yield of liquid fuel is less than that from dissolution processes.

In dissolution, ground coal is dissolved in a solvent (usually a recycled solvent), the ash is filtered out, and the oil is heated by hydrocracking (Fig. 3.29). As with upgrading of gas from coal, coal liquids can be catalytically upgraded with hydrogen to produce a synthetic crude oil. Hydrogenation is needed because natural petroleum contains 11 to 15% hydrogen, 83 to 87% carbon, and up to 4% oxygen, nitrogen, and/or sulfur, whereas, as mentioned earlier, coal contains only 5% hydrogen and 75% carbon.

Dissolution processes can be distinguished as (1) those which use neither hydrogen nor a catalyst, (2) those which use hydrogen but no catalyst, and (3) those which use both hydrogen and a catalyst.

The technology for pyrolyzing coal differs little from that used for gasifying coal except that air and steam are excluded. In the COED process, for example, crushed coal is heated by stages in fluidized-bed reactors, in each of which a fraction of the coal volatiles is released and collected. Char is burned in the final stage to generate heat for the process, and the hot gases are recycled. The filtered pyrolysis oil is hydrotreated to upgrade it to a synthetic crude oil.

A unique coal liquefaction process, TOSCOAL, has borrowed technology developed for retorting oil shale. Crushed coal is contacted with heated ceramic pellets in a pyrolysis drum. Both the solid char product and pellets pass out of the drum over a trommel screen which separates the char from the pellets; the pellets are then reheated and recycled (Fig. 3.31). Pyrolysis vapors are condensed and fractionated, whereas uncondensed gases are used as pellet-heater fuel. The pyrolysis gases are free of nitrogen and carbon dioxide since only ceramic pellets are used to provide heat. The TOSCOAL process parallels that of oil shale processing except that marketable char replaces the spent shale.

The Garrett pyrolysis process uses an entrained-bed carbonizer in which coal, carried by recycled product gas, is heated at an exceedingly high rate (5000°F/sec) to prevent cracking of volatile constituents. The decomposition products flow through cyclones to separate char from the gases, whereas tars, condensed from the gases, are hydrogenated to low-sulfur fuel oil or syncrude.

Unlike the liquid fuels produced in most liquefaction processes, the Solvent-Refined Coal (SRC) process produces a low-sulfur, low-ash, low-melting solid fuel. Pulverized coal is slurried with a process-derived solvent, preheated, and sent to a reactor at 800 to 900°F and 1000 to 2000 psi where 90% of the coal organics are rapidly dissolved (Fig. 3.34). The slurry is filtered to remove inorganic ash and the small amount of undissolved carbon. The filtrate is flash-distilled to recover the solvent and to isolate the products — a light liquid by-product and a heavy residual oil called solvent-refined coal, which cools to a low-sulfur, almost ash-free pitchlike material with a melting point of 300 to 400°F and a heating value of about 16,000 Btu/lb.

The Gulf Catalytic Coal Liquids (CCL), H-Coal, and Synthoil processes all convert coal to a liquid fuel using a catalyst and hydrogen gas in contact with pulverized coal slurried in a recycle solvent. The reaction conditions, catalysts, and procedures differ in each process, but because these methods are still being developed, the parameters will be varied to optimize the conversion.

Liquid fuel can be produced by catalytic liquefaction of synthesis gas. The SASOL plant in South Africa combines the Fischer-Tropsch method for catalytically converting synthesis gas to hydrocarbons with Lurgi high-pressure steam-oxygen gasifiers, which produce a gas consisting essentially of carbon monoxide and hydrogen. Two types of reactors are used: a fixed bed of iron catalyst in vertical tubes and a fluidized bed of powdered iron catalyst (Fig. 3.43). By controlling the input to the reactors, the fixed-bed reactor produces a mixture of straight-chain, high-boiling hydrocarbons with some medium-boiling oils and diesel oil, whereas products of the fluidized-bed synthesis are mainly low-boiling hydrocarbons (C_1 to C_4) and gasoline with some oxygenated products and aromatics.

In-situ gasification of coal, which converts a coal seam into combustible gases, produces a gas rich in carbon monoxide, carbon dioxide, hydrogen, hydrocarbon gases, and other gases. Depending on the gas injected (air, oxygen, or oxygen and steam), the heating value of the product gas can range from as low as 50 Btu/scf when air alone is injected to as high as 280 Btu/scf when an intermittent air-steam sequence is used. It is thought that the carbon dioxide is formed by oxidation of the coal, whereas carbon monoxide results from the reduction of carbon dioxide.

The idea of underground gasification is very attractive. The method avoids mining and waste disposal and adds unaccessible, unworkable, or uneconomical coal deposits to the energy pool. Unfortunately, the gas has a low heating value and would either have to be used at the site or upgraded for transportation.

Starting in the 1930s, the Russians tested different methods of underground gasification, at least one of which was tried on an industrial scale. Some activity also took place outside Russia from 1945 to 1960.

In-situ gasification is not expected to replace mining, but it certainly can supplement external conversion. Two basic methods exist for underground gasification — shaft and shaftless systems. In the shaft method, large-diameter openings are cut into the seam (as in mining), whereas the shaftless method uses boreholes (Figs. 3.44 and 3.45). Russian production plants rely on the shaftless method.

Early U.S. experiments with in-situ gasification (U.S. Bureau of Mines 1946-1959) showed the process to be technically feasible (Table 3.29) but incapable of competing economically with cheap, plentiful natural gas. Studies are in progress at Hanna, Wyoming, and Pricetown, West Virginia, under ERDA sponsorship.

Conversion processes are dually oriented: (1) to convert plentiful coal into scarce liquid and gaseous fuel and (2) to remove or treat, during processing, environmentally unacceptable or health-endangering compounds. As a result, coal conversion methodology is concerned as much with processes for handling the by-products as with the products themselves.

Knowledge of the wastes and emissions from coal conversion processes is still incomplete. However, it is known that wastes will be generated during each main process stage and that these wastes are largely controllable or convertible to environmentally acceptable forms (Table 4.1, Fig. 4.1). A more serious and less tractable problem arises from possible fugitive emissions produced by inadequate containment of process streams (leaky valves) or incomplete treatment of wastes. High-temperature process streams may contain carcinogenic organic compounds similar to those implicated in an increased incidence of skin and lung cancer among coke-oven workers.

Because coal is a dirty fuel, early technologies that used coal without sulfur-emission control contributed heavily to environmental pollution. Coal conversion, however, is specifically mandated to produce clean fuel from coal; thus, a wide variety of techniques for removing sulfur are integrated into coal conversion processes. Sulfur can be removed during coal pretreatment or the conversion process and from product and waste streams.

Because organic sulfur is difficult to remove during coal pretreatment and because about 90% of the organic sulfur is converted to hydrogen sulfide in both gasification and liquefaction, this form of sulfur is estimated to be the major sulfur contaminant in gas streams. For example, of the total sulfur in a Lurgi product gas, hydrogen sulfide accounted for 95%; the remaining contaminants were carbonyl sulfide, 2.4%; carbon disulfide, 0.3%; mercaptans, 2.0%; thiophenes, 0.3%. Hydrogen sulfide is readily removed from gas streams (Table 4.6), as is sulfur dioxide, but carbonyl sulfide and carbon disulfide are more resistant to acid-gas scavenging and can pass through sulfur traps, ending up in gases emitted to the atmosphere.

Most of the sulfur in coal is expected to be recovered from conversion streams as elemental sulfur (Table 4.10), although the sulfur that reaches the stack gas as sulfur oxides and is collected by limestone scrubbing will end up as sulfate (Tables 4.11 and 4.12). Neither form of sulfur is particularly obnoxious environmentally, although the quantity to be disposed of is considerable: A 25,000-ton-per-day plant using a coal containing 4% sulfur will produce 1000 tons of sulfur daily. As opposed to throwaway disposal, new uses for elemental sulfur are being considered. Production of sulfur-based paving materials and plastics or use of sulfur in thermal and acoustical insulation are considered alternatives to landfill or minefill disposal.

Nitrogen is introduced into the conversion process not only from the inherent nitrogen in coal but also from the air used as an oxygen source. For example, a Synthane process using 3650 tons of oxygen per day separates and vents 330 million scf, or 12,200 tons, of nitrogen daily. Thus, nitrogen emissions are primarily elemental nitrogen, although nitrogen oxide levels, which can be environmentally significant, depend upon process designs and the inherent nitrogen content of coal.

No reports of nitrogen oxide emission levels or control and treatment processes designed specifically for conversion technologies were located. However, some techniques developed for nitrogen oxide control from coal combustion plants may be applicable to conversion (Sect. 4.2.1.2).

Nitrogen appears in main process streams in the form of ammonia and hydrogen cyanide gases, which are removed by water scrubbing of the gas streams (Table 4.18). Anhydrous ammonia, usually recovered for resale, can amount to 250 tons per day (El Paso Burnham I Coal Gasification Complex). Both ammonia and hydrogen cyanide retained in wastewaters are accessible to technologies used for treating industrial wastewaters (biodegradation, nitrification, and chlorination).

Disposal of solid wastes, including airborne particulates, pretreatment coal fines, and ash and char dusts, constitutes an important environmental consideration (Table 4.19). A commercial gasification plant using 8 million tons of coal per year will produce 2.3 million tons of dry refuse. Disposal problems are also encountered in coal combustion; however, the magnitude of the wastes is proportional to the amount of coal used, and the quantities of coal projected for coal conversion use far exceed those anticipated for combustion.

Most conversion processes require coal pretreatment operations such as grinding and cleansing (Chap. 3), which produce refuse such as minerals, debris, and coal fines that may constitute as much as 25 lb per 100 lb of cleaned coal. For example, the Bi-Gas process (18,267 tons/day input) produces 4567 tons of coal-washing debris per day.

Control methods exist for collecting coal dusts, fines, and airborne particulates. However, collected solid wastes stored in settling ponds may produce significant leachates about which little is known. Similarly, data have only recently surfaced describing leachate from unprotected storage areas (Table 1.1). Trace amounts of a very large number of elements appear in coal ash (Table 4.34), including some toxic elements such as mercury, cadmium, arsenic, selenium, and fluorine, although these elements or their compounds are volatile and more likely to be associated with fly ash or airborne particulates (Table 4.33).

Table 1.1. Characteristics of coal pile drainage

Constituents	Concentration (mg/liter)	
	Plant J	Plant L
Acidity (total), as $CaCO_3$	1700	270
Calcium	240	350
Chemical oxygen demand	9	
Chloride	0	
Conductance, μmho/cm	2400	2100
Dissolved solids (total)	3200	1500
Hardness, as $CaCO_3$	600	980
Magnesium	1.2	0.023
pH, unit	2.9	2.9
Potassium		0.5
Silicon (dissolved)	91	
Sodium		4.1
Sulfate	2600	
Suspended solids (total)	550	810
Turbidity, JTU	300	
Aluminum	190	
Arsenic	0.01	0.009
Barium		0.1
Beryllium		<0.01
Cadmium	<0.001	0.006
Chromium	<0.005	<0.005
Copper	0.56	0.18
Iron	510	830
Lead	<0.01	0.023
Manganese	27	110
Mercury	<0.0002	0.027
Nickel	1.7	0.32
Selenium	0.03	0.003
Titanium	<1	
Zinc	3.7	1.0

Source: Chu, Ruane, and Steiner 1976, Table 14, p. 34. Reprinted by permission of the publisher.

The fates of trace elements during the coal conversion process are not yet completely known; however, the majority of elements are expected to remain in the solid by-products such as ash, chars, and filter cake. Nonetheless, material balance studies are indicated to trace the pathway of hazardous elements during conversion and to determine the concentrations and forms of the elements in gas streams, condensates, and waste streams.

The release of conversion process wastewater into natural waters also constitutes a possible source of environmental pollution. The aqueous streams, which arise when process gases are condensed or scrubbed, may contain components of the product gas — carbon monoxide, carbon dioxide, hydrogen, hydrogen cyanide, and methane — along with contaminants such as sulfur and nitrogen compounds, ash (dust and particulates), phenols, emulsified tar, and oil. Fugitive tar collected in wastewater contains a wide variety of organic compounds (Table 4.53), and wastewater particulates may collect and adsorb PAH. Of the water-soluble organics, phenols appear to be important (58% in SRC process effluent water), along with alcohols such as catechol, resorcinol, and their methylated derivatives (200 to 1000 µg/ml in COED separator liquor). Basic components such as pyridines, quinolines, and indoles have also been identified in product wastewater (Table 4.71).

Ammonia in the cleanup water enhances the solubility of hydrogen sulfide and carbonyl sulfide and increases the uptake of hydrogen cyanide. Miscellaneous materials in wastewater may include hydrogen fluoride, fatty acids, other trace organics, and trace elements.

Because waste ammonia liquor from coking plants is so similar to by-product water from conversion processes such as gasification (Table 4.39), water treatment methods developed for the coking and petroleum industries may be adapted for coal conversion. Wastewater treatment for removal of organic compounds is usually carried out in three stages: separation of the separable tars and oils; flotation, chemical flocculation, and sedimentation of the unseparated organics; and chlorination, carbon absorption, filtration, and biological treatment of the residuals. Collected organic waste is most effectively, but also most expensively, disposed of by high-temperature incineration. Alternatively, the recycling of organic waste through the conversion process may be environmentally acceptable.

Information concerning atmospheric emission of PAH from coal conversion processes is scarce. However, coal burning studies indicate that the majority of PAH is adsorbed on particulate matter, and, although aerosol formation has been observed industrially in the presence of very high airborne PAH concentrations, control of air emissions by particulate collection would be expected to remove most PAH.

Although knowledge of coal-derived organic compounds that can be emitted from conversion process streams is vital for assessing possible health impacts, these materials have not yet been completely characterized. In fact, even the identity of the organic products and by-products is uncertain, partly because their amounts and characteristics are variable, influenced by both the coal characteristics and the process operating conditions (Table 4.60). Additionally, it has been estimated that, of the 10,000 compounds in coal tar and hydrogenation products, only 1000 have been identified (Sects. 4.5.1.1 and 4.5.1.2).

Coal-derived liquids contain a high percentage of aromatic and polyaromatic compounds, although, as in petroleum, saturated paraffins dominate the low-boiling fractions (Tables 4.56 and 4.57).

However, in contrast to petroleum, synthetic crude oils can possess higher levels of acidic components such as phenols and cresols (about 1.6 wt % in synthoil), some of which appear in the low-boiling distillate.

With only slight modification, standard oil-refining techniques should be applicable to synthetic crude oils. Although the composition of refined products from syncrudes will undoubtedly depend on the starting materials, the concentration of PAH in the heavier fractions is expected to exceed that of petroleum products.

Knowledge of the environmental interactions of possible coal conversion pollutants — hydrocarbons, trace elements, sulfur oxides, and nitrogen oxides — will help to assess the environmental stress presented by a coal conversion facility, either during normal operation or following an accidental effluent release.

The behavior of a chemical pollutant introduced into the environment depends largely upon two factors — the nature of its surroundings and the physical-chemical properties of the pollutant. For example, transformations can occur in all environmental media — the hydrosphere, lithosphere, atmosphere, and biotic component of the environment. Likewise, in the medium, the behavior of a chemical is determined by its solubility, vapor pressure, adsorption ability, and degradability.

PAH, some of which are potential health hazards, are thought to have been in man's environment throughout man's history. They are found in small but detectable concentrations (typically from 0.001 to 10 µg/liter) in air and water as well as in soil samples of all types, including those distant from highways and industries (Table 6.1). PAH in soil are thought to arise from natural sources such as pyrolytic decomposition of wood, transformation of plant organic matter, and activity of soil microorganisms. Thus, although PAH may be produced and released by man's activities, not all environmental concentrations can be assumed to have originated from anthropogenic sources; certain bacteria and plants synthesize PAH during normal development and thus provide a natural background level of PAH as high as 100 to 1000 µg/kg in the upper layers of the earth. For example, researchers have found that benzpyrene is synthesized by wheat and rye seedlings (10 to 20 µg per kilogram of dried material) both in the presence and absence of light, and benzo[*a*]pyrene (BaP) could be formed from phytoplanktonic lipids under the influence of anaerobic bacteria, particularly *Clostridium putride*.

Estimates for total BaP emissions in the United States vary between 500 and 1300 tons annually (Tables 6.6 and 6.7). By far the greatest contribution is from both residential and industrial coal use for heating and power generation, although, surprisingly, forest fires account for more than 10% of the total.

Annual average airborne BaP concentrations for 1966-1970 in U.S. nonurban areas are generally an order of magnitude less than those in urban areas (Tables 6.8 and 6.9). The country's highest recorded BaP value for 1966-1970 (29.5 ng/m^3 of air in 1967 for Altoona, Pennsylvania), which exceeds most other urban sites by an order of magnitude, is sevenfold less than the proposed industrial standard (200 ng/m^3) and fourfold less than a proposed atmospheric standard (120 ng/m^3). PAH have been detected in wastewater samples; in many instances, domestic effluents have higher concentrations than do factory effluents, particularly during periods of heavy rain (Table 6.18).

PAH entering natural water systems will suffer decomposition, storage, or removal. In aquatic environments, decomposition is accomplished primarily through photooxidation, with biodegradation assuming a lesser role; storage is through consumption by biota (Chap. 9) or incorporation into sediments; and removal by physical transport may occur in the absence of degradation and bioaccumulation. Solubility is important in determining which aquatic interaction will prevail — whether a hydrocarbon will go into solution, become suspended in the water column, or adsorb onto the sediments. However, mechanisms for hydrocarbon solubility, particularly PAH, are poorly understood, and data are incomplete and discordant.

The fate of hydrocarbons in soil results primarily from biodegradation, with leaching also being important. Soil leaching rates determine how long a chemical is retained in the top soil, where it is most subject to degradation or dissipation. Four major elements of leaching are (1) soil adsorption, which determines an underlying pattern; (2) porous flow and diffusion, which disperse the chemical; (3) adsorption dynamics, which introduces a factor of hysteresis; and (4) water infiltration and evaporation, which determine the actual amount of water movement and, hence, the observed amount of chemical movement. BaP has been found to withstand degradation by at least 99.5%, but also to be accumulated and converted by special strains of microorganisms.

PAH activity in the atmosphere is largely determined by particulate matter, as most PAH are associated with particulates and undergo aerial transport and deposition as particulates. Particle size is the physical property having the greatest influence on the bioenvironmental behavior of aerosols containing polycyclic organic matter; most PAH associate with particles less than 3.0 μm in diameter. The major mode of PAH removal from the atmosphere is through chemical reactions, probably the most important being photooxidation, which leads to the formation of quinones and hydroxyquinones. A recent report suggests that present atmospheric PAH collection methods may not yield totally accurate determinations of PAH levels.

Like PAH, trace elements also have two modes of entry into the environment, natural and anthropogenic. The rates of emission of many elements from gasification processes were estimated in one study to exceed those from combustion (Tables 6.29 and 6.31).

Some of the more hazardous trace elements present in coal (e.g., arsenic, mercury, and cadmium) are present naturally in soils, but in small concentrations. The background concentrations of trace elements in river water are low when compared with those in soil; for example, arsenic has a concentration of 6 ppm in soil and 0.0004 ppm in river water, and mercury has a concentration of 0.03 ppm in soil and 0.00008 ppm in river water. A summary of elemental concentrations in rocks, soils, waters, marine and land plants and animals, and air, plus some information on the relative toxicity of each element, is presented in Table 6.37.

In aquatic environments, most trace elements are distributed between the water column and the sediment; few are found in the biota, although certain metals bioaccumulate (Chap. 9). Arsenic, cobalt, copper, iron, lead, and zinc, for example, concentrate in the feces of a primary consumer, adult kelp crabs, *Pugettia producta*.

Trace element behavior and transport in aquatic systems result from specific processes. In rivers — the dominant source for transport of metals — trace elements can be transported either in the dissolved state or associated with particulate matter. In estuaries, where the primary trace metal input is from rivers, five major categories of reactions influence trace metal behavior: flocculation and sedimentation, mineral-water interaction, adsorption-desorption, diagenesis and remobilization of trace metals in sediments, and biological processes. Estuarine processes in turn influence the amount and form of trace elements transported to the ocean, where the primary transport processes are precipitation and concentration by marine organisms.

Chemical and biological transformations of trace elements are equally important in determining their fate in natural waters. In sediments, for example, heavy metals such as mercury, lead, cadmium, and zinc are chemically bound but can be converted to soluble forms at low pH (acidic) conditions. An example of trace element biotransformation is the microbial methylation of inorganic mercury to form monomethylmercury compounds and the highly volatile dimethylmercury.

Trace elements enter the soil by one or more of the following mechanisms: weathering by parent rock, aerial fallout, rainout, solid waste application, chemical fertilizer application, irrigation, and trace-element-containing insecticide application. Although trace elements, particularly trace metals, often have long retention times in soils, some elements may undergo chemical or biological changes that increase their mobility. For example, at a soil pH of 6.0, manganese is reduced to the soluble divalent form; when the pH reaches 6.5, the element undergoes autooxidation to the tetravalent state, which if hydrated can be reduced both bacterially and chemically.

Trace element fate in air is determined initially by dispersion, which is controlled primarily by meteorological factors; secondly, by chemical reactions, such as the reaction of aerosols of manganese dioxide with sulfur dioxide to form manganous sulfate; and finally, by fallout, after which the fate of trace elements is determined by processes in water and soil.

Sulfur dioxide as a pollutant has been the subject of more investigation than any other single pollutant. Annual anthropogenic sulfur dioxide emissions are estimated at 100 million tons, whereas natural sources, mainly volcanoes, contribute about 1.5 million tons per year. The pollutant is capable of traveling great distances in the atmosphere, thereby increasing its effect as an air contaminant. The majority of sulfur dioxide is removed from the atmosphere via rainout and washout and is deposited on soil, water, and vegetation. However, while in the atmosphere, sulfur dioxide may undergo significant chemical reactions such as the gas phase oxidation to sulfur trioxide and the subsequent dissolution in water droplets to form sulfuric acid.

In contrast to sulfur dioxide, most global nitrogen oxide (NO_x) emissions originate from natural sources, although a major buildup of NO_x occurs in urban areas as a result of man's activities. The chemical reactions that NO_x enter are important in evaluating their impact as pollutants. Hydrocarbon intervention in the atmospheric nitrogen dioxide photolytic cycle (Sect. 6.4.3), for example, leads to ozone accumulation. Ozone is a serious atmospheric contaminant; in addition to its neurotoxicity and phytotoxicity, it contributes to photochemical smog. Peroxyacyl nitrate (PAN), another photochemical oxidant, is also thought to be formed from reactions that derive from hydrocarbon intervention in the nitrogen dioxide photolytic cycle.

Microorganisms are ecologically important because they are producing, consuming, and transforming members of the ecosystem. Their functional versatility makes them sensitive indicators of environmental alteration and gives them the ability to use as their sole carbon source a wide variety of compounds, including contaminants in most environmental media.

The principal source of water pollution in coal conversion is process waters, the composition of which is expected to be similar to that of ammonia liquor from coking industries. These polluted waters will be primarily biotreated. However, some chemicals in ammonia liquor may be resistant to bacterial degradation, whereas other chemicals — including heavy metals, cyanides, halogenated compounds, phenols, mercaptans, guanidines, and thiourea — may be detrimental to bacterial survival. However, a bacterial mutant of *Pseudomonas* sp. capable of growth on ammonia wastes (e.g., thiocyanate) in the presence of heavy metals has been discovered.

Studies have shown nitrification and denitrification to be common, natural processes. Bacteria use ammonia and oxygen in an oxidative reaction that transforms ammonia to nitrite. The nitrite produced by this bacterial action generally is further oxidized to nitrate by nitrifying bacteria. This two-step process is called nitrification. A sludge system has been developed using *Nitrosomonas* bacteria that convert ammonia to nitrate. Nitrification occurs in an activated sludge system only when conditions are suitable for the retention and accumulation of nitrifying bacteria. A sludge age adequate to retain and prevent the washout of the slower-growing nitrifying bacteria is the factor upon which successful nitrification depends. Combined and separate sludge systems are illustrated in Fig. 7.1.

Denitrification is an anaerobic nitrogen reaction carried out by anaerobic bacteria; it is the dissimilatory reduction of nitrate, in the presence of organic hydrogen donors, to nitrite and the subsequent reduction of nitrite to nitric oxide, nitrous oxide, and molecular nitrogen.

Coal contains environmentally significant sulfur compounds which are accessible to microbes during outdoor coal storage or as part of coal conversion solid or liquid wastes. For example, low-pH waters from coal washing promote proliferation of *Thiobacillus ferrooxidans*, a well-known sulfur-oxidizing bacterium. This species can remove up to 90% of pyritic iron and sulfur in coal during simple storage and can oxidize ferrous iron, thiosulfate, and sulfur. Hypothetical oxidation pathways are shown in Fig. 7.8. Sulfur bacteria are also important parts of the sulfur cycle and perform both assimilatory and dissimilatory sulfate reduction.

Phenol and/or phenolic compounds, organic pollutants in coal conversion effluent, are commonly found and treated in industrial wastewaters. Phenol is produced from the carbonization of coal. Work with the alga *Selenastrum capricornutum* showed that phenol toxicity increased with an increase in temperature. This finding is significant because many effluents from coal conversion processes are formed and released at elevated temperatures. Phenol is degraded readily to resorcinol and pyrocatechol by bacteria of the *Pseudomonas* genus. Successful treatment of phenolic wastes now employs vigorous aeration methods to supply molecular oxygen (O_2), which is necessary for hydroxylation and ring cleavage reactions. If O_2 is not present, phenols will be converted to objectionable chlorophenols. Some heterocyclic ring systems, however, undergo hydroxylation reactions by aerobic bacteria in which water, not molecular oxygen, serves as the source of the hydroxyl group. Figure 7.11 depicts the formation of catechol (a well-established example of an orthodihydric phenol that undergoes ring cleavage) by bacterial degradation of phenol and a variety of aromatic hydrocarbons and carboxylic acids.

A medium for detecting phenol-degrading bacteria that represents an improvement in efficiency, savings in cost and labor, and a decrease in the time required for results has been developed (Sect. 7.2.1.6). The assay medium isolates bacteria growing in water contaminated with phenol, cresol, mono- to pentachlorophenol, and other substituted phenols. One microorganism responsible for destruction of the major constituents of coal carbonization waste liquors is similar to *Thiobacillus thioparus*. Two other bacteria strains that can metabolize phenols in liquors are *Comamonas* sp., which can grow on a wide range of phenols, including cresols and xylenols, and a member of the *Moraxella-Acinetobacter* group, which does not grow on cresols or xylenols.

Normal alkanes, produced as part of the coal carbonization process, are potential pollutants and possess an organic structure accessible to microbial attack. Microbial degradation that begins with alkanes, for example, is a major means of oil spillage cleanup.

Hydrocarbon uptake by microbes is explained by several theories, the most plausible of which is the "micelle" theory. Micelles, which are thought to be made up of surface-active agents produced by microorganisms, solubilize the hydrocarbon drops into the aqueous medium to a size optimal for microbial pinocytosis. This theory explains the high rate of hydrocarbon transfer despite the fact that hydrocarbons have very low solubility in water. The increased solubility provides the large driving force necessary for high bacterial growth rates. A lag phase in bacterial growth is predicted until sufficient surface-active compound is produced by the cells. Exponential growth occurs as long as submicron hydrocarbon droplets, with adjoining nutrient salt ions, bring sufficient substrate to the cells. If the cell concentration becomes high enough that the rate of consumption exceeds the rate of supply, the growth rate becomes linear and is proportional to the rate of hydrocarbon supply.

PAH, another coal conversion waste, have been found to be degraded by common soil bacteria. However, BaP, a carcinogen, has given inconsistent results; some investigators have reported up to 85% degradation, whereas others have reported almost no degradation under similar circumstances. As demonstrated in one series of tests using cultures of either *Bacillus sphaericus*, *Bacillus megaterium* mutilate, or *Pseudomonas*-146, soils containing a greater quantity of aromatic compounds apparently provide more favorable conditions for promoting the processes of PAH metabolism; however, the mechanism of this phenomenon (perhaps induction) remains obscure.

Activated sludges from municipal waste treatment systems have been shown to be mostly ineffective in the treatment of aromatic, organic industrial wastes. Poor oxidation performance of activated sludge is probably due to the refractory chemical structure of the carcinogenic compounds. Likewise, saprophytic microorganisms were used to break down organic compounds, including BaP; no degradation of the BaP was observed to at least 99.5%.

Microbial PAH synthesis has been investigated. Biosynthesis of BaP in nature, a process that produces background levels of BaP and thus complicates pollution studies, has been shown. In one investigation, for example, forest soil collected far from large industrial centers contained 1 to 2 μg of BaP per 100 g of soil. After the soil was set to culture under anaerobic conditions with *Clostridum putride* for six months, the BaP content had increased appreciably to 4 µg per 100 g of soil. Significantly, an experiment using *Escherichia coli*, a facultative aerobic-anaerobic microbe common in the mammalian gastrointestinal tract, produced comparable results.

Pyridine compounds are released into the environment by industrial and domestic use of coal; also, pyridines used as organic solvents inevitably find their way into effluents. Because pyridine concentrations do not increase substantially in the soil environment, many of these compounds, both natural and synthetic, are thought to be ultimately degraded by actinomycetes soil bacteria. Microbial metabolism of simple pyridine compounds produces dihydroxypyridine compounds.

Pollution of soils and groundwaters by mixed hydrocarbons (e.g., oil) occurs frequently; only microbial metabolic reactions can effect complete cleanup of an oil spill. Members of the genus *Pseudomonas* play an important role in oil degradation. Representative biodegradation pathways of alkane and aromatic compounds are given in Figs. 7.28 and 7.29. Specificity by some microorganisms has been noted; for example, *Aspergillus athecius* grows on n-tetradecane, but not on n-tridecane. Cometabolism is an important microbial process in oil degradation; cyclohexane, for example, is broken down after cometabolic transformation by one of two *Pseudomonas* strains, each of which is unable to metabolize it independently.

Oil in sediments is decomposed by microbial action under anaerobic conditions at the expense of oxygen in nitrates and sulfates, leading to the formation of nitrites, free nitrogen, and hydrogen sulfide. The bacteria most commonly reported as active in oily soils are members of the coryneform group and include the genera *Corynebacterium, Brevibacterium, Arthrobacter, Mycobacterium,* and *Nocardia*. Under varying saline concentrations representative of an estuarine environment, *Pseudomonas* was the only genus found to grow on naphthenic crude oil at the end of the test's incubation period. Also, *Pseudomonas jianni* and *Pseudomonas abikonensis* were shown to decompose an organic sulfur compound, dibenzothiophene, to five water-soluble organic compounds.

Trace metals released during coal conversion are transformed by microorganisms and lead to an overall metal content rise in the microbial biomass. For example, the marine bacterium *Leucothrix* was shown to concentrate metals added to an experimental medium; periwinkles feeding on the bacteria subsequently exhibited an increase in metal content. Trace elements deposited on soil due to ash or tars are likely to undergo microbial degradation. *Thiobacillus ferrooxidans* has been shown, under low pH conditions, to grow on chunks of coal. Microorganisms in lake sediments can transform certain inorganic and organic lead compounds (e.g., lead nitrate and lead chloride) into volatile tetramethyl lead. Mercury, another volatile element present in trace amounts in coal, is highly susceptible to microbial transformations that can lead to more toxic forms such as methylmercury.

Plants affect man: They are a part of his diet, his aesthetic environment, and his economic situation; they form the basis of the environmental food chain; and they play a role in the cycling of atmospheric and aquatic pollutants. Plant-pollutant interactions are complex and intermingled; the precise determination of cause and effect is difficult. The interdependence of a variety of factors ultimately determines the effects of pollutants on vegetation:

environmental influences (relative humidity, temperature, edaphic conditions, climate, light conditions, ambient levels of other pollutants) and characteristics of the plant itself (species, variety, age, disease, nutritional condition). Although abundant descriptive data regarding the physiological effects of some pollutants are available, the primary sites of action in the plant largely remain to be defined, as do tolerance and detoxification mechanisms (Sect. 8.2.1).

Such interactions are not limited to pollutant effects on plants. Plants can also contribute to atmospheric pollution by emitting volatile organics containing reactive hydrocarbons in a quantity six times greater than the amount produced globally by man's activities (i.e., 175×10^9 kg/year natural vs 27×10^9 kg/year anthropogenic). Calculations of inputs and outputs of atmospheric trace gases on a global scale indicated that the major source of hydrogen sulfide, ammonia, nitrous oxide, and hydrocarbons was natural emissions, including those from plant foliage. Thus, natural sources produce larger quantities of air pollutants, except for sulfur dioxide, than do anthropogenic sources.

Also, the endogenous biosynthesis of PAH in some plants has been confirmed during growth of rye, wheat, lentils, tobacco, lettuce, kohlrabi, and algae. However, careful research with higher plants indicates that definitive proof of PAH biosynthesis is yet to be obtained. Benzo[*e*]pyrene, BaP, perylene, anthanthrene, benzo[*ghi*]perylene, dibenzo[*a,h*]anthracene, and coronene were not detected in various crop plants (rye, tobacco, soybeans) grown in filtered air in special rooms with air locks, whereas those grown in open air in ordinary greenhouses contained all these hydrocarbons (Table 8.26).

Natural emissions (and possible biosynthesis) of hydrocarbons increase the difficulties of (1) sorting out effects data, (2) correctly labeling acute chronic effects, and (3) assessing the additive, synergistic, and antagonistic effects of toxicant mixtures.

Well-studied air pollutants that are known to have a major impact on plants include ozone, sulfur dioxide, other oxidants, fluorine, and ethylene, each of which may be important when considering potential pollution from the coal conversion industry. Other possible coal conversion atmospheric pollutants such as ammonia, chlorine, volatile trace metals, phenols, other organic compounds, and particulates may also be implicated.

Results of phytotoxicity research may be difficult to interpret. The effects of acute exposure can sometimes be positively identified; for example, acute exposure to high-level atmospheric pollutants have caused visible plant injuries such as sclerosis, necrosis, abscission of plant parts, and disturbance of the pigment system (Sect. 8.2.5.1). However, injury from long-term exposure to low levels of gaseous pollutants is more difficult to document; recent studies involving careful measurements of crop plants have begun to reveal subtle reductions in growth rate, plant height, root and top weight, and leaf areas, as well as retardation in floral initiation, smaller and fewer flowers, a measurable reduction in yield, and changes in community composition. Ultrastructural damage in the absence of macroscopically visible injury has been observed.

Acid rain, largely the result of sulfur dioxide pollution, may damage plants at distances of up to 1000 km from the pollutant source. Indirect evidence suggests that it has caused a reduction in forest productivity in certain areas, and it has also been shown to cause plant abnormalities and to have a detrimental effect on plant nutrition. Acid rain also can destory the protective waxy layers of foliage, cause enhancement of disease, and decrease plant productivity. Because plant uptake of heavy metals is increased by lower than normal pH values, a synergistic effect can reasonably be expected where fallout of heavy metals and acid rain appear simultaneously.

In aquatic ecosystems, pollution by hydrocarbons, heated effluents, trace metals, ammonia, and various inorganic compounds such as sulfates and sulfides have caused effects ranging from plant growth inhibition to growth stimulation, reproductive alterations, toxic reactions, and changes in species diversity.

Trace elements and heavy metals may be taken up and accumulated by both aquatic and terrestrial plants, with or without injury to the plant itself. Humans have suffered toxic effects from the ingestion of selenium-bearing plants, whereas livestock have incurred molybdenosis and fluorosis from the ingestion of plants containing high concentrations of molybdenum and fluorine.

Human exposure to PAH includes ingestion of those found in food plants (Sect. 8.3.1). The origin of these PAH has been correlated with uptake from polluted air and soil. Also, plants possibly biosynthesize these compounds, which may act as natural hormones. PAH have been observed to stimulate normal plant growth as well as the development of abnormal tissue on algae, fungi, and culture-grown higher plant structures (Sect. 8.3.1.9).

Injury to plants by pollutants may result in economic loss, ecosystem disturbance (and destruction, in some cases), and possible health effects for man and animals through contamination of food, air, and water. However, the interactions between plants and pollutants can sometimes be used for monitoring and purification. Certain plants, particularly the lichens and bryophytes, are used as inexpensive and sensitive monitors of pollutants (Table 8.3). The factor indicating the presence of a particular pollutant may be (1) the presence or absence of a certain species or (2) a high pollutant accumulation by the plant. Other factors to take into consideration when using plants as indicators include establishing a true cause and effect relationship; recognizing possible variations in the sensitivity of foliage due to the amount of waxy deposit on leaf surfaces; and determining the ecological range of plants being considered.

Plants can play a part in the cycling of materials through the ecosystem, in some cases converting the materials to other forms and thereby cleansing water and air. The water hyacinth, for example, is capable of removing phenols from polluted wastewaters with apparent metabolic breakdown. Removal capabilities of up to 160 kg of phenol per hectare of plants per 72 hr at any phenol concentration less than toxic for the plants have been projected (Sect. 8.3.2.2). Vegetation could be an important sink for such gaseous pollutants as hydrogen fluoride, sulfur dioxide, nitrogen dioxide, ozone, chlorine gas, ammonia, and to a lesser extent, PAN. Important pollutants which plants are unable to take up effectively are carbon monoxide and nitric acid.

Effluents from coal conversion processing, through interactions with natural biological systems, have the potential not only for disturbing individual organisms, but also, in so doing, for altering the complex and delicate balance that characterizes natural ecosystem stability.

Aquatic systems, as major recipients of industrial pollutants, have been studied much more extensively than others, providing some general insights into the bioenvironmental fate of anthropogenic emissions. Of the possible coal conversion emissions that differ from or greatly exceed those from existing sources, PAH, heterocyclic compounds, and trace elements stand out as a potential ecosystem hazard worthy of considerable attention.

Because marine waters may be the ultimate depositional sink for PAH and because crude oil spills have directed attention to the problem of hydrocarbon release to the marine environment, a disproportionate, but nonetheless valuable effort has been expended toward examining the uptake, accumulation, and metabolism of PAH in marine organisms. BaP, the PAH most often monitored, has been detected in a wide variety of marine organisms from remote as well as heavily industrialized areas (Tables 9.3 and 9.4). BaP concentrations ranging between 2 and 540 ppb (oysters) have been reported with some organisms (mussels) possessing as much as 3400 ppb total PAH; marine plankton were found to contain BaP concentrations as high as 400 ppb dry weight.

Aquatic invertebrates tend to accumulate PAH from their surroundings and to release them unchanged in a relatively PAH-free environment. Bioaccumulation factors ranging between 2 and 1100 have been observed; bioaccumulation appears to correlate directly with the PAH molecular weight and the organism lipid content and inversely with rates of excretion (Fig. 9.8). Like other hydrocarbons, PAH storage in tissues is directly related to the tissue lipid content. Thus, most PAH are localized in the high-lipid viscera, the hepatopancreas being the most probable final PAH storage site.

The rate and extent of PAH metabolism in organisms bear on the extent of its trophic level biomagnification. Unfortunately, information about PAH levels and metabolism in organisms from exposure to concentrations normally found in the environment is scarce. However, the inability of some organisms at the base level in the food chain (e.g., tubifex worms) to degrade PAH may have far-reaching effects throughout the food web. Thus, in the absence of definitive information, two nonconflicting mechanisms may be operating to explain the observed tendency of PAH concentrations to increase in organisms with their position in the food chain: Magnification may be due to contributions from both food chain transfer and individual species accumulation without degradation. For example, in one study, PAH content of fish livers and benthos appear to be more dependent on background levels in the organism's immediate environment than on its trophic position. Nonetheless, the influence on the ecosystem of the high PAH level in zooplankton, an initial link in many aquatic food chains, should not be ignored. Unfortunately, data necessary for evaluating the importance of the various contributions to uptake, accumulation, and magnification are not yet available.

The metabolism and elimination of PAH is strongly species dependent; among experimental laboratory mammals, even the strain of animal is important (Chap. 10). Aquatic vertebrates such as fish readily metabolize PAH to the water-soluble metabolites typical of mammalian systems (Table 9.11); aquatic invertebrates, however, generally are incapable of PAH metabolism, although at least two species (spider crabs and daphnia) have recently been found capable of metabolizing small amounts of administered PAH (naphthalene and anthracene).

The lack of PAH metabolism by edible shellfish such as oysters, clams, and mussels is perceived by some as a potential health threat to humans. Bivalves take up and retain PAH by equilibrating rapidly with their immediate environment (Tables 9.13 and 9.14), accumulating PAH in polluted waters and releasing them when transferred to PAH-free waters. Additionally, mussels release PAH in their eggs, which solubilize PAH readily in their fatty substance.

Although PAH may produce carcinogenic, teratogenic, or mutagenic responses in exposed organisms, no such effects have been directly correlated with the low concentrations normally found in the environment. Few studies have been directed at examining these effects on organisms in their natural habitat. In one study, however, increased fish tumor incidence was related to waters heavily polluted with aromatics (Table 9.27).

As noted by Auerbach (1975), the relationship of potential effects of compounds to their environmental fate is complicated. For example, a potent carcinogen may be taken up by sediment where it never encounters a target organism, whereas a weak carcinogen present in the environment at very low levels may bioaccumulate and be transferred to higher organisms in the food chain with serious consequences. In some cases even minute amounts may have unsuspected secondary effects. Reports of feeding and reproductive behavioral modifications in marine organisms, elicited by 10 to 100 ppb of soluble aromatics, led investigators to speculate on the possibility that aromatics interfere with chemical communication by altering pheromone structure.

Limited information is available concerning the bioenvironmental effects of heterocyclic compounds; however, the behavior of heterocyclics is probably similar to that of PAH.

Trace elements from coal use — including those from catalysts as well as coal itself — pose at least as serious a potential bioenvironmental hazard as do organic effluents. Because some trace elements are resistant to metabolic detoxification and all are nondegradable, they can accumulate in biota. Also, once released, they persist in the environment indefinitely. Some trace metals already exist in the environment at maximum or near-maximum permissible levels.

Of the coal-associated trace elements selected by Comar (1975) as particularly important (lead, mercury, arsenic, vanadium, selenium, nickel, and possibly cadmium and fluoride), those capable of long-distance transport (lead, mercury, and cadmium) were isolated by Auerbach (1975) as major pollutants capable of bioaccumulation. Catalysts can contribute nickel, cobalt, molybdenum, chromium, vanadium, and zinc (Table 4.73).

Because no two elements are alike, the interaction of a trace element with the bioenvironment is rarely predictable. As examples, (1) chromium is most toxic in its most oxidized state (hexavalent), whereas arsenic is most toxic in its most reduced state (as arsine); (2) of the multitude of trace elements in the environment, only mercury, arsenic, and possibly lead are biotransformed by soil microbes into highly toxic, volatile, methylated forms; (3) some elements, such as copper and zinc, that are essential to life are not only toxic at elevated concentrations, but the level of toxicity of one may also be related to the concentration level of the other; and (4) a trace amount of selenium, itself a toxic element, reduces the toxicity of cadmium and mercury. Similar examples abound; elements elicit a wide variety of physiological responses in a wide variety of organisms and situations. Nonetheless, efforts are in progress to elucidate the parameters necessary for developing hypotheses concerning the impact of trace elements on ecosystems.

The human health impact of coal conversion can be considerable. Sulfur oxides, nitrogen
oxides, heavy metals, and PAH, released from coal during conversion, can lead to a wide range
of effects, from acute toxicity to mild irritation; effects depend on many parameters, the
most important of which is the level of exposure. Decisions about industry standards and the
concomitant control technologies will need to lean heavily on health effects data; unfortunately,
the available data are hardly definitive.

Problems of interpreting health data are common to the evaluation of all chemicals introduced
into the environment. In the absence of man as an experimental subject, animal species are a
primary basis for predicting health effects.

Extrapolating animal data to man requires great care. Susceptibility of animals to toxic agents
may vary with the species of animal and sometimes with strains of the same species; experimental
outcomes may depend upon dosing routes and regimes; and the statistical significance of animal
experiments may be doubtful because of the necessary limitations in the size of test groups.

Similarly, epidemiological study, although our most valuable source of direct evidence, has
serious limitations. Accidentally exposed populations are heterogeneous, uncontrolled, and
poorly defined. For planned populations, periods of study are of necessity often too long for
resolving near-term problems.

However, despite the deficiencies, the animal and epidemiological studies are our best and
most reliable tools for at least focusing on the pertinent questions, if not providing firm
answers.

Possibly the most health-endangering coal-derived compounds are PAH; many are known animal
carcinogens and suspected human carcinogens. Workers exposed to high levels of PAH-containing
coal volatiles show an increased incidence of skin and lung cancers. BaP, a component of coal
tar and an identified carcinogen, is the most tested PAH and also the compound often chosen as
the indicator compound for monitoring PAH. It has been observed in synthetic oil samples
(41 ppm), coal and petroleum tars (2000 and 200 ppm), polluted air (38 ppb), surface water
(0.01 to 0.1 ppb), groundwater (0.001 to 0.010 ppb), soil (40 ppb), plants (1 to 10 ppb), and
sediments undisturbed since the year 3 AD (19.5 ppb). BaP has produced tumors in mice, rats,
hamsters, guinea pigs, rabbits, ducks, and monkeys by oral, skin, and intratracheal administration.

Humans are most likely to encounter PAH by accidental ingestion, inhalation, or skin contact.
From laboratory animal studies we know that many PAH ingested with food or water are largely
eliminated in feces, indicating poor absorption from the gastro-intestinal tract (Table 10.1).
Repeated application to the skin of mice, however, caused pathological changes in the blood,
spleen, lymph nodes, and bone marrow, effects which suggest that PAH can be absorbed
percutaneously.

Knowledge of the metabolism of PAH is crucial to an evaluation of their hazard; it is clear
from experiments that the metabolites are often more toxic or carcinogenic than the parent
compounds. In general, PAH such as naphthalene, anthracene, and phenanthrene, which are
lipophilic molecules and thus more soluble in fatty tissue than in the body's aqueous systems,

are metabolized in the liver to quinones and diols, water-soluble forms that can be eliminated in the urine more readily than the parent compounds. However, independent of the mode of administration, carcinogenic PAH such as BaP and its metabolites are excreted largely in the feces through the biliary system and appear only sparingly in the urine. In studies using mice, the rate of excretion of PAH varied according to the compound; for example, $t_{1/2}$ was 1-3/4 weeks for BaP and 12 weeks for dibenzo[a,h]anthracene, and carcinogenicity of the compound was directly proportional to the retention time.

Although not isolated, the intermediate compound in PAH metabolism is believed to be an epoxide or diol-epoxide, thought by many to be the active PAH carcinogen. Epoxides are highly reactive and can rearrange by a number of pathways (Fig. 10.2).

It is well established that the enzymes responsible for PAH metabolism, of which aryl hydrocarbon hydroxylase (AHH) is the most important, may either inactivate PAH or activate them to more carcinogenic forms. Because the AHH enzyme system may be a key factor in chemical carcinogenesis, it is significant that AHH in human lymphocytes is similar to that found in rat liver in culture. AHH is inducible by PAH, whole cigarette smoke, steroid hormones, barbiturates, and insecticides; the inducibility is genetically determined in laboratory animals. AHH is found in 90% of the tissues of mouse, rat, hamster, and monkey. Administration of a PAH induces the hydroxylase to higher levels in the liver, lungs, gastrointestinal tract, and kidneys, although the magnitude of induction varies greatly with the animal species and the tissue (Table 10.7).

Induced AHH activity is lower in the monkey than in other experimental animals; also, the monkey is more resistant to PAH carcinogenesis than are other experimental animals.

Variations in AHH inducibility also exist among humans, suggesting that greater cancer susceptibility may be associated with greater inducible AHH activity. Studies have shown that AHH induction in cultured human leukocytes is under genetic control. Likewise, studies of 85 healthy unrelated adults showed that the normal white population in the United States could be divided into three distinct groups of low, intermediate, and high degrees of inducibility. Bronchogenic carcinoma in humans was associated with higher levels of inducible AHH activity.

Tumorogenic activity of PAH can be altered significantly by association with other materials. Unfortunately, there is a serious deficiency in data on these combined effects. For example, PAH carcinogenesis is enhanced in animals by exposure to PAH in the presence of chemicals such as iron oxides and long-chain aliphatic hydrocarbons or is inhibited by materials such as Vitamin A and selenium.

PAH are thought to be carried into the lung primarily by particulate matter. PAH are associated with atmospheric particulates, especially those less than 5 μm in diameter. This size fraction is most readily deposited in the pulmonary portion of the respiratory tract (lung alveoli) and least readily cleared from it (Fig. 10.5). Iron oxide particulate especially appears to enhance the formation of bronchogenic carcinoma by known PAH carcinogens, although it exhibits no irritating or toxic effect itself.

N-dodecane, a long-chain aliphatic compound produced in coal carbonization, was found to enhance the cutaneous carcinogenic potential of BaP and BaA 1000-fold when used as the diluent for the carcinogens.

Some natural compounds suppress the activity of PAH carcinogens. For example, vitamin A was shown to reduce markedly the incidence of respiratory tract tumors in hamsters when the animals were injected intratracheally with a mixture of BaP and ferric oxide (a cancer promoter). Vitamin A is thought to act by inhibiting BaP metabolic pathways, especially those proceeding through an epoxide intermediate. Like vitamin A, selenium combined with vitamin E significantly reduced the number of PAH-induced skin tumors in mice. Selenium may act as an antioxidant. A number of antioxidants as well as some weakly carcinogenic hydrocarbons have been shown to exhibit anticarcinogenic properties when applied with potent carcinogens. Unfortunately, too little of the carcinogenesis literature is concerned with cancer inhibition and preventive effects.

Many substances which exhibit carcinogenic properties are also teratogens. PAH were shown to permeate the placental barrier in experimental animals, suggesting the need to study their teratogenic properties. The compounds were taken up by fetal tissue (Table 10.10), stimulating AHH enzyme activity which was three- to eight-fold higher in the fetal membranes than in the placenta (Table 10.11). In general, the distribution of AHH in human placentas shows a positive relationship between a high level of AHH activity and a history of cigarette smoking (cigarettes contain PAH). However, AHH activity is found in placentas of some women who do not smoke and is not found in some women who do smoke, supporting the contention that genetic differences exist in human AHH inducibility.

PAH have been shown to produce tumors in animal offspring of mothers treated during pregnancy; pulmonary adenomas and skin papillomas have been produced in mice in utero.

Epidemiology studies provide the strongest evidence for human susceptibility to PAH carcinogenesis. The best-documented study involved a long-term examination of 58,828 steel workers, 3530 of whom had worked in the coke plant. Data on coke-oven workers indicate that they experience an increased risk of lung cancer over total steelworkers; however, the rate is related to the work area. Full topside jobs (working above the slot ovens with a maximum exposure to coal-tar-pitch volatiles and particulates) accounted for all the excess mortality from cancer of the lung (Table 10.17). The atmosphere in an uncontrolled coke-oven area has been compared to the inside of a cigarette smoker's lungs immediately after inhaling — thick with particulates and vapors. Workers' lungs and skin are bathed continuously by chemicals and periodically by clouds of coal dust.

In a more recent study, risk to workers exposed to coal-tar-pitch volatiles (topside of the ovens averaged 3.15 mg/m^3) was a function of both exposure level and length of exposure. Mortality from lung cancer varied from 10 to 40 years, with an average of 25 years. A threshold limit of 0.2 mg/m^3 for a period of 30 years did not increase the carcinogenic risk.

Possible health effects from sulfur oxide emissions have been studied extensively in connection with coal-fired power plants. Sulfur dioxide, the major emitted oxide, is classified as a mild irritant; workers occupationally exposed daily to levels up to 2 ppm for several years showed no loss of health or pulmonary function. More than 90% of inhaled sulfur dioxide is absorbed in the airways above the larynx before the air reaches the bronchi or lungs. However, adsorbed onto the surface of particulate matter, sulfur dioxide can penetrate deep into the lungs; where uncleared, it may interfere with proper lung function. Particulate matter capable of oxidizing sulfur dioxide to sulfuric acid can enhance pulmonary irritation, causing bronchoconstriction. Episodes of high sulfur dioxide and smoke and fog pollution have resulted in increased mortality and morbidity. Adverse effects are greatest on elderly subjects, those suffering from cardio-vascular disease, and asthmatics.

Unlike sulfur dioxide, inhaled nitrogen dioxide is only slightly water soluble and passes through the trachea and bronchi into the moist alveoli of the lungs where it forms corrosive nitrous acid. In epidemiologic studies, high nitrogen dioxide levels were found to be correlated with increased childhood respiratory illness and decreased ventilatory performance.

Although there is a considerable body of knowledge of acute health effects from trace elements, little is known about low-level chronic effects which are apt to be produced by the amounts and forms of these elements as they may be emitted from coal conversion operations. From the aspect of human health, trace elements of importance are thought to be lead, chromium, manganese, arsenic, fluoride, nickel, cadmium, zinc, and copper. Threshold limit values have been established for airborne concentrations of trace substances, including metals, to which nearly all workers may be repeatedly exposed without adverse effects (Table 10.32).

LITERATURE CITED

Auerbach, S. I. 1975. Testimony presented to the subcommittee on environment and the atmosphere. U.S. House of Representatives, Hearings on costs and effects of chronic low-level environmental pollution, Nov. 12, 1975.

Chu, T. J.; Ruane, R. J.; and Steiner, G. R. 1976. Characteristics of wastewater discharges from coal-fired power plants. In *31st annual Purdue industrial waste conference, May 4-6, 1976*. West Lafayette, Indiana: Purdue University.

Cochran, N. P. 1976. Oil and gas from coal. *Sci. Am.* 234(5): 24-29.

Comar, C. L. 1975. *Conference proceedings: Workshop on health effects of fossil fuel combustion products*. Palo Alto, Calif.: Electric Power Research Institute. PB-242 418.

U.S. Bureau of Mines. 1974. *Low-temperature evolution of hydrocarbon gases from coal*. RI-7965.

2. COAL: ORIGIN, CLASSIFICATION, AND PHYSICAL AND CHEMICAL PROPERTIES

J. T. Ensminger

ABSTRACT

Fossilized material from ancient plants, converted by exposure to high temperatures in the
earth's crust over long periods of time, makes up the complex carbonaceous substance known as
coal. Coal was formed through a continuous series of alterations: living plants > peat >
lignite > subbituminous coal > bituminous coal > anthracite. This process is referred to as
metamorphosis; the degree of metamorphosis is called rank. Carbon content increases, and
volatile matter decreases, with increase in rank.

The United States contains about half of the total world reserves of coal. Coals of the
Appalachian and Interior regions are about 150 million years older than western coals, and thus
are of relatively higher rank. Having been formed under the influence of seawater, eastern
coals are usually characterized by higher sulfur contents than those of western coals.

Coal is composed basically of two types of materials: (1) inorganic, crystalline minerals from
surrounding geological formations and (2) macerals, the fragmentary organic remains of plants.
An elemental analysis of coal reveals the presence of many trace elements, some of which may
be environmentally hazardous. Sulfur in coal occurs in three forms: in organic combination
with the coal material, as the minerals pyrite or marcasite, or as inorganic sulfate. No coal
is uniform throughout in rank or in content of minerals and macerals. Coals of the same rank
from different basins may vary widely because of different environmental influences.

The extremely complicated and variable nature of coal has made impossible any efforts to
describe accurately its structure and composition. However, most workers believe that coal
is an essentially aromatic material. Tars and pitches, aromatic materials encountered in coal
conversion processes, are sources of a seemingly endless list of chemical compounds.

The physical characteristics of coal — moisture content and porosity, for example — may
significantly affect their suitability for conversion processes. Specific gravity, hardness,
strength, friability, grindability, and moisture content are all at least generally related
to coal rank.

2.0 INTRODUCTION

The increasing demand of the United States and the world for clean energy in the form of oil and gas, together with the ever decreasing natural reserves of these products, has made essential the study of alternative energy sources. One of these alternatives is fuel from the conversion of coal to liquid and gaseous forms.

The conversion of coal has been studied intensively for a relatively short time, but a large amount of data has already been produced, and new information is constantly being published. There are, however, still many unknowns and controversies associated with coal chemistry and structure.

The purpose of this chapter is to present an overview of the present knowledge of coal, including its origin, classification, and chemical and physical characteristics, from the existing literature.

2.1 ORIGIN OF COAL

Coal is fossilized plant material formed from the remains of plants that flourished millions of years ago. The partially decayed plant material was buried and exposed to high temperatures seldom exceeding 200°C (Teichmuller and Teichmuller 1966) through geologic time, thus being altered to its present complex form. In the past, pressure was considered to have an accelerating effect on the coalification process; however, more recent information (Teichmuller and Teichmuller 1966) indicates that, although pressure does bring about some physical changes in coal, it does not accelerate chemical coalification (Sect. 2.3.1). Coal consists mostly of carbon compounds derived directly from plant organic matter, but it also contains minerals and elements from the earth's crust that were included in the coal both during and after completion of the coalification process.

2.1.1 Influence of age

The major worldwide coal deposits, laid down in limited periods of time, are shown in Table 2.1. Coals of the Appalachian and Interior regions of the United States were formed during the Carboniferous period 300 million years ago, whereas the coals of the western United States were deposited much later during the Cretaceous and Tertiary periods 60 to 120 million years ago. The table shows Gulf Province coals as having been formed during the Carboniferous period; however, Texas coals are now classified as Interior Province coals, leaving only lignites of Tertiary age in the Gulf Province (Given 1977).

The age of a coal seam is important because coal constituents vary according to the different materials found in plant matter, and these change during the evolution of the plants. Spackman (1973) cites as an example the difference in exinite content of coals formed before and after the Tertiary period. During the Jurassic and Cretaceous periods, the dominant plant group was the gymnosperms, which are today represented by pines, fir, spruce, etc. In their reproductive cycle, gymnosperms produced and released large numbers of spores into the air. These spores were in such abundance during the dominant period of the gymnosperms that they constituted a large proportion of a hydroaromatic plant component of coal called exinite. Angiosperms are plants that appeared during the Cretaceous period and are today the dominant plant group.

Table 2.1. Classification of geological strata and the age of coals

| (Millions of years) | | | | Principal folding periods | Principal species | | Principal coal formations |
Age	Mean age of coals	Duration of period	Stratigraphic period		Animals	Plants	
					Cenozoic		
1	<1	1	Quaternary — Holocene Pleistocene		Mammals, man; insects	Angio- and gymnosperms	Moorland peats in Europe, asia (Siberia), N. and S. America; Peat swamps in Indonesia, Africa; Forest peats in N. America (Florida, Wisconsin, Alaska, Br. Columbia)
12	2	11	Pliocene		Domination of mammals; fishes, birds	Angiosperms (herbaceous palms, many trees), gymnosperms (Taxodium, Sequoia)	Lignites in Austria (Linz), Hungary (Plattensee, Zala), Yugoslavia (NE Bosnia), Rumania, Bulgaria, Italy (Toscana), Japan, N. Zealand, Tasmania, Alaska
23	15	11	Tertiary — Miocene	Alpinic			Lignites in Germany (Cassel, Ober- and Nieder-Lausitz in Silesia), Netherlands (S. Limburg), Denmark (Jutland), Austria (Mirz, Graz), Czechoslovakia (N. Bohemia), Hungary (Matra), Yugoslavia (NE Bosnia), S. America (Patagonia), China, Japan, Indonesia, E. and S.E. Australia, New Zealand, Tasmania, W. Indies, Venezuela, Mexico, Greenland
35	28	12	Oligocene				Lignites in England (Bovey Tracey), France (Provence), Yugoslavia (Ljubljana), Rumania (Petroszeny, Klausenburg), Spitzbergen, Br. Columbia
55	60	20 15	Eocene Paleocene				Lignites and subbituminous coal in Germany (Thüringia, Saxony, Harz), Hungary, Yugoslavia (Istria, Dalmatia, NE Bosnia), Greece, Sardinia, N. America (N. Dakota, Montana, Texas), Alaska, Arctic Islands, Burma
70					*Mesozoic*		
135	120	65	Cretaceous — Upper Lower (Weald)		Dwindling of ammonites, belemnites and sauria; origin of mammals and birds	Gymnosperms (Cordaitales, Ginkgoales, Coniferales)	Lignites, subbituminous and bituminous coals in N. America (Western and Rocky Mountains area of U.S.A. and Canada, Alaska), Mexico, Chile, Peru, France (Fuveau, Estavar), Germany (Hannover), Austria (Gisau), Yugoslavia, Bulgaria, Sachalin, Japan, Assam, S Nigeria, Queensland, N. Zealand
180	160	45	Jurassic — Upper Middle Lower		Domination of reptiles and amphibians (Dinosauria and Ichthyosauria); molluscs (ammonites and belemnites)	Gymnosperms; (Cordaitales, Ginkgoales, Coniferales)	Coals in Sweden (Schonen), Hungary (Pecs), Rumania (Resicza and Anina), Russia, Siberia (Irkutsk, Vladivostok), China (Szechwan, Chungking), Queensland, NS Wales, S. Australia, S. Victoria, Tasmania, N. Zealand
	200	40	Triassic — Upper Middle Lower		Crustaceans; first reptiles and mammals	Pteridophyta, Cordaitales, Coniferales	Coals in Poland (Dombrova), Austria (Linz), N. America (N. Carolina, Virginia), S. Mexico, Queensland

Table 2.1. (continued)

(Millions of years)			Stratigraphic period	Principal folding periods	Principal species		Principal coal formations
Age	Mean age of coals	Duration of period			Animals	Plants	
220					*Paleozoic*		
	245	50	Permian (Permo-Carboniferous)		Protozoans (Foraminifera), fishes, molluscs, arthropods (insects); first amphibians	Pteridophyta, Cordaitales, first Coniferales (pines and firs)	Coals in France (Central Plateau), Saxony, Russia (Urals), Siberia (Kuznetsk), Manchuria, China (NE and E), India, Africa (Rhodesia, Transvaal, Cape, Natal), W Australia, Queensland, NS Wales, Tasmania, Brazil, Argentina, U.S.A. (Eastern Province)
270		50	Upper carboniferous	Hercynic		Thallophyta (algae), Phycophyta (fungi), Bryophyta (mosses), Pteridophyta (ferns), Articulatae, Lepidonhyta, (lepidodendron, sigillaria), Cordaitales	Coals in France (Commentry, Autun, St. Etienne, Le Creusot), Gr. Britain (Scotland, Cumberland, Durham, Lancashire, Yorkshire, Derbyshire, N and S Wales, Staffordshire, Bristol, Kent). France (Nord, Pas de Calais), Belgium (Borinage, Charleroi, Liege, Campines), Netherlands (S. Limburg, Peel, Achterhoek), Germany (Aachen, Ruhr, Osnabrück) Saar, Poland (Silesia), Russia (Donetz Basin), Turkey, Spain (Santander, Granada), S. America (Brazil), N. America (Appalachian Region, Interior and Gulf Provinces)
300			Carboniferous				
320		30	Lower carboniferous				Edge coals in Scotland, anthracites of Pennsylvania, coals of Arctic Islands, Spitzbergen, Nova Scotia, New Brunswick, Russia (Moscow and Ural basins)
350	380	50	Devonian		Domination of coelenterates; molluscs	First Pteridophyta, Phycophyta	Coals in Bear Island, cannel coals in Arctic Islands, Halserite coal (Eiffel)
400		90	Silurian	Caledonic	Protozoans (Infusoria, Radiolaria, Foraminifera); first fishes	Thallophyta (algae)	
490		110	Cambrian		Protozoans, coelenterates, first molluscs		Shungite in NW Russia (Olonetz)
600		>1000	Algonkian / Archaean	Huronic	*Proterozoic*		

Source: van Krevelen 1961, Table III, pp. 36, 37. Reprinted by permission of the publisher.

These plants evolved an embryo sac retained in an ovule on the sporophyte generation of the plant, and spores were no longer produced in such great abundance. Thus, one of the coal types common in the older coals is relatively rare in tertiary coals. It is apparent that the evolutionary changes that occurred in plants greatly influenced the composition of coals and presently affect the suitability of different coals for various uses.

2.1.2 Influence of geology

Geological and geochemical differences also influenced the production and composition of coal seams. For example, Illinois seams developed in close association with marine water, producing large concentrations of pyritic and organic sulfur in the coal, whereas coals that developed in parts of the Dakotas experienced minimal marine water influence and contain small concentrations of sulfur.

Differences in the rate and extent of pressure and temperature buildup within various geological formations with which the coal seams were associated also resulted in different compositional reactions. Some of the different characteristics of Colorado and Appalachian coals of the same rank are the result of different rates of heating. The Colorado coals were subjected to rapid heating by igneous intrusions, whereas the Appalachian coals were exposed to slower increases in temperature associated with orogenic forces in the earth's crust (Spackman 1973).

Thiessen (1947) suggested that coal formation was affected also by the size and form of the topographic depression involved, the nature of the surrounding rock and soil, and the prevailing climate and physiography of the area.

2.1.3 Influence of microbes

On the basis of information available at this time, it seems likely that microorganisms capable of carrying out the same activities as the present soil and peat microorganisms were present in every period of coal formation. Therefore, the physiology and ecology of modern microorganisms should be a reliable guide to their geochemical role in the past (Given 1971). Microorganisms were essential to the early stages of the coalification process because they performed the function of converting dead plant material to the partially decayed organic matter called peat. Reasons for partial rather than total decay of the plant material are still somewhat open to debate; however, studies of present peat deposits in Florida indicate two important factors — the inherent resistance of certain plant materials such as lignins to decay and the chemical condition of the swamp water. For example, the pH, oxygen content, and mineral matter would have determined whether the dominant bacterial population of the water was aerobic or anaerobic and thus would have controlled the rate of decomposition (Parks 1963).

Information on specific organisms involved in the formation of peat is sparse at this time. Dickinson (1971), as cited in Given (1971), studied the peat deposits of Florida, south of Lake Okeechobee. Peat samples were taken, and populations of microorganisms in surface layers (0- to 8-cm depth) were determined from plate counts. Investigation of the peat fragments showed that those cultivated for sugar cane crops were colonized by hyphae of fungi and/or streptomycetes and that few were sterile. Conversely, 40 to 90% of the undisturbed peats were sterile, and observed growth occurred largely in streptomycetes rather than fungi. This finding indicates that streptomycetes were probably the more important agents for decomposing the plant material (Given 1971).

2.2 FORMATION OF COAL

Swamps that contained luxuriant plant growth are believed to have constituted the major environment supporting the decay of plant material necessary for coal formation. Thiessen (1947) describes peat beds in the Dismal Swamp of Virginia, which is flooded by Lake Drummond. The existence of such peat beds supports the autochthonus theory of coal formation, which states that peat is formed from material originating from the area of deposit. Conversely, the allochthonus theory states that peat is formed from plant debris that has been carried by wind or water and deposited in enclosed basins. As with most natural phenomena, there is no sharp dividing line, and both processes are believed to be involved, though the former is probably more important in the peak forming areas.

Van Krevelen (1961) gives the following account of the processes involved in coal bed formation: During the period of coal formation lasting from the lower Carboniferous to the Permian, uninterrupted swamps covered large areas of the northern hemisphere. Due to the great depth of Carboniferous coal formations, it seems likely that these swamps underwent a process of subsidence that was largely balanced by sedimentation. Such sunken areas, called geosynclines, are usually formed by strong lateral, compressive forces acting on the earth's crust. The geosynclines probably did not sink at a constant rate, but alternated between periods of rapid and slow subsidence. During the periods of slow subsidence, shallow lagoons formed and sedimentation occurred at a more rapid rate than did the sinking, resulting in luxuriant growths of aquatic plants. The plant debris settled to the bottom and was acted upon by microorganisms to form peat deposits, which grew until the rate of subsidence increased again. Eventually, the swamp was submerged and covered by sediment, and a coal seam was formed. This pattern was probably the one most often followed in formation of coal deposits; however, some coal seams were probably formed by the buildup of silt in deeper waters (van Krevelen 1961).

It is generally accepted that there were at least two stages of coal formation from plant material: (1) the biochemical period of accumulation and preservation of plant material as peat and (2) the geochemical period of conversion of peat to coal. Although most coal researchers accept this theory, agreement on the details of the actual chemical and physical changes that occurred in the process is not so widespread (Parks 1963).

2.2.1 <u>Lithotypes</u>

Spackman (1973) states that coal seams are laminar, sedimentary rock bodies ranging in thickness from a fraction of an inch to several hundred feet. Most of the economically significant seams of the United States are 14 in. to 20 ft thick, a few being as thick as 75 ft. A seam may consist of a number of layers (lithobodies), and these layers may contain different mixtures of macerals (Sect. 2.5.1) and minerals (Sect. 2.5.2), giving them different characteristics. The physical and chemical properties and behaviors of these coal types (lithotypes) differ during the liquefaction and gasification processes. Van Krevelen's (1961) description of the formation of major lithotypes is briefly summarized below.

2.2.1.1 Fusain

Fusain was formed rapidly under very dry conditions. Many researchers feel that, because of its similarity to charcoal, fusain may have formed as the result of forest fires caused by lightning. Another theory suggests that it was produced by some exothermic microbial process.

2.2.1.2 Vitrain

Vitrain was formed more slowly under semidry conditions, as in a swamp where the groundwater level was just under the surface. The dead plant material sunk into the wet humus soil and was preserved. Because the stagnant water contained so little oxygen, aerobic decomposition was inhibited, and the lignified material, which was also relatively resistant to anaerobic decay, was preserved.

2.2.1.3 Durain

Under true wet swamp conditions, decay of plant material was carried out to a much higher degree, resulting in only the most resistant fragments being left, thus forming durain.

2.2.1.4 Clarain

Clarain is believed to have been formed under conditions between the relatively dry environment of vitrain and fusain and the true wet swamp environment of durain. It is characterized by a relatively high vitrinite (Sect. 2.5.1.1) content. The woody tissue was evidently macerated before burial because clarains contain spores and other materials imbedded in a matrix of vitrinite (Gwen 1977).

2.2.1.5 Cannel coal

Cannel coal was probably formed in lakes and pools where spores accumulated by wind and water were deposited in the muck on the bottom.

2.2.1.6 Boghead coal

Boghead coal is very similar to cannel coal, but it originated from different material. It consists mainly of algae that died and settled into the mud (van Krevelen 1961).

2.2.1.7 Other terms

The previous six terms are used in the international system of classification of coal types. The U.S. Bureau of Mines system also uses the names cannel and boghead coal; however, different terms — bright, semisplint, and splint coal — are used for the other types. The two systems of classification are compared in Table 2.2. Generally, bright coal consists predominately of vitrain and clarain, semisplint coal consists mostly of clarain with some vitrain and durain, and splint coal is mostly durain with some vitrain and clarain (Yancey and Geer 1968). The U.S. Bureau of Mines system is no longer widely used.

During the 1930s it was observed that certain ranks of bright coal (vitrain and clarain) possessed very good coking properties, whereas dull coal produced very unsatisfactory coking results. Fusain and durain were for years believed to inhibit coke formation because they do

Table 2.2. Comparison of International and U.S. Bureau of Mines
terminology used for basic coal types

International	U.S. Bureau of Mines
Humic coals	Banded coals
Fusain (charcoallike)	Bright (<20% opaque matter)
Vitrain (black, vitreous)	Semisplint (20 to 30% opaque matter)
Clarain (striated, glossy)	Splint (>30% opaque matter)
Durain (nonstriated, matte)	
Liptobiolithic coals	Nonbanded coals
Cannel	Cannel (abundant spore remains)
Boghead	Boghead (abundant algal remains)

Source: Spackman 1973, Fig. 4.

not soften and fuse upon heating as do vitrain and clarain. Furthermore, pockets of fusain, which could not be wetted by the reactive components, produced weak spots in the coke. It has since been demonstrated that the addition of inert components such as fusain, evenly dispersed throughout the coke charge, actually increases the thickness and strength of coke cell walls, resulting in a stronger coke (Harrison and Latimer 1968).

2.3 COAL RANK

Coal is formed through an apparently continuous series of alterations: living material > peat > lignite > subbituminous coal > anthracite. This succession of changes in the properties and structure of coal is called metamorphism. The degree of metamorphism is called "rank" (Given 1973*a*).

2.3.1 Influence of time

Teichmuller and Teichmuller (1966) believe that time was an important factor in bringing about advances in the rank of coals. As an example, they compare coals from the U.S. Gulf Coast and the Lower Elbe Trough in northwestern Germany; the Gulf Coast coals taken from a depth of 6000 m are of much lower rank than are coals taken from a depth of 5100 m at the German site. They conclude that the advanced age of the German coal (about 250 million years) accounts for the major difference in rank.

2.3.2 Influence of pressure

According to Teichmuller and Teichmuller, the major effects of pressure upon coal formation are related to coal's physical properties. Notably, the porosity and moisture content of coal are decreased by pressure. The authors cite experiments by Huck and Patteisky (1964) as proof that static pressure actually retards the chemical process of coalification. Chemical changes of high-volatile bituminous coal containing 33% volatile matter at a temperature of 350°C increased with decreasing pressure.

2.3.3 Influence of temperature

Temperature, more than any other factor, causes the metamorphism of peat to coal (Teichmuller and Teichmuller 1966). The temperature gradient of the earth's crust is not constant; however, as an average, the earth's temperature increases by 2 to 5°C per 100 m of depth; temperatures of about 200°C would have been sufficient for anthracite formation (van Krevelen 1961). In some cases frictional heat produced by tectonic movements such as shearing have brought about increases in coalification, but such instances cannot account for increases in coal rank with depth (Teichmuller and Teichmuller 1966).

2.3.4 Influence of metamorphism

Figure 2.1 shows the linear progression of coal through the stages of development and some of the associated properties. The carbon content of the material increases from 60% at the peat stage to 90% at the low-volatile bituminous stage.

Reflectance, which is used in identifying coal constituents, also increases with increase in rank and carbon content. The molecular structure of coal is directly related to molar refraction and absorption of light; therefore, optical properties of coals can be important indicators of their constitution. Of the optical properties, reflectance is the most conspicuous and easiest to determine (van Krevelen 1961).

Volatile matter, which decreases with increasing rank, consists of gaseous substances which are driven off by heat during the coalification process.

The surface area of coal is related to the plasticity and agglomerating characteristics of the coal on heating, an important factor in both coking and conversion processes. When exposed to rising temperatures, coals give off gases and condensable vapors, leaving a residue consisting mostly of carbon. Concurrently, the coal then softens and becomes plastic, and the particles cake to form a compact mass, which swells and resolidifies to form the porous coherent substance called coke (Loison et al. 1963). Figure 2.1 shows that minimum surface area and maximum plasticity are usually attained by high-volatile A or medium-volatile bituminous coals and that only bituminous coals are suitable for coking. Van Krevelen (1961) states that virtually no plastic behavior occurs at carbon contents lower than 80.5% or higher than 91%.

The dashed vertical line drawn through the region of minimum surface area indicates that coals consisting of about 85% carbon and 28% volatile matter should be very plastic and highly expanding. Van Krevelen indicates that coals containing about 87% carbon and 29% volatile matter are most plastic and subject to expansion. Figure 2.1 also indicates that organic acid groups are essentially absent in bituminous coals and that phenol groups are rare in anthracite.

Actually, the relationship of a coal property to the degree of metamorphism cannot be realistically illustrated by a line; a band would be more appropriate. No coal bed is uniform in properties throughout; vertical samples taken from a single coal deposit will vary somewhat in analysis from all the others because of the increase in temperature with depth. Coals of the same rank from different basins would produce even wider bands.

ORNL-DWG 76-7651

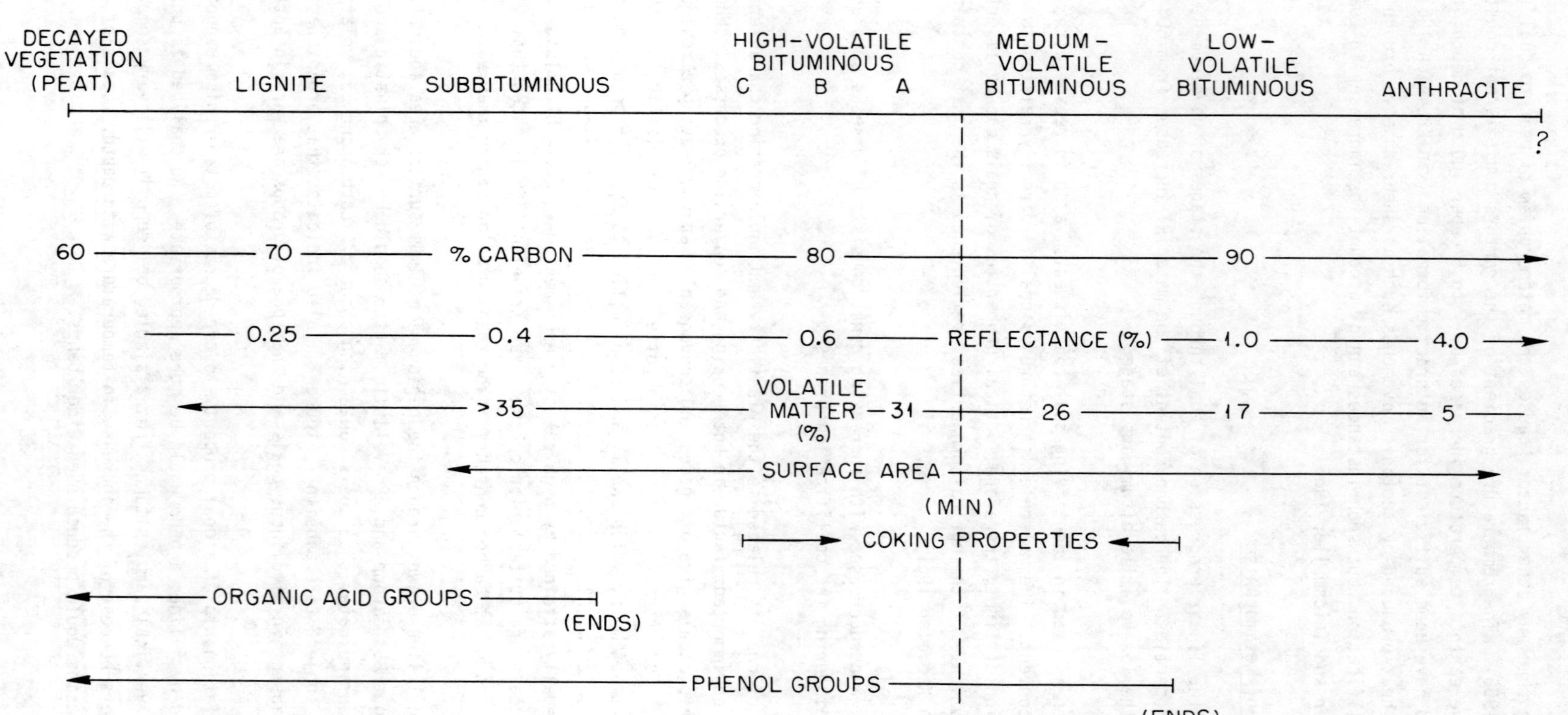

Fig. 2.1. The coal series. Source: From Given 1973*a*, Fig. 1. Reprinted by permission of the publisher.

2.3.4.1 <u>Mechanism of metamorphism</u>

Existing studies of the metamorphism of coal indicate that, in some coal deposits, advances of rank are the result of thermal distillation of volatile substances, including oxygen, caused by the intrusion of igneous bodies into, or in close proximity to, the sedimentary strata. However, the extensive deposits of high-rank coals known today cannot be attributed solely to heat from magmatic rocks. Coals of high rank, especially anthracite, are also sometimes associated with sedimentary rock strata that have undergone extensive folding, faulting, and fracturing. Tremendous orogenic and tectonic forces produced these disturbances, which were characterized by differential movement of the strata in several different planes under tremendous pressure and frictional heat (Parks 1963).

According to Teichmuller and Teichmuller (1966), increases in the rank of coal by magmatic intrusions and by frictional heat produced by tectonic movements of the rock strata are less common than coal rank increases brought about by increased temperatures associated with increased depth of the coalbed in the earth's crust.

Mott (1942) gives a five-stage description of the chemical changes associated with the metamorphosis of peat to coal: Peats of woody origin not containing large concentrations of resins and waxes were altered under acid conditions to an ulmic material containing about 65% carbon, 4% hydrogen, and 30% oxygen. This alteration was accompanied by dehydrogenation and some deoxygenation. Subsequently, the material was covered by sediment and underwent a base exchange with sodium chloride to create alkaline conditions, resulting in anaerobic decomposition which caused carbon dioxide to be given off with deoxygenation of ulmins and which increased hydrogen and carbon contents. This second, or lignite, stage of coalification usually resulted in low-rank bituminous coals containing about 79% carbon, 5.5% hydrogen, and 14.1% oxygen.

The third and fourth stages of coal development consisted of the conversion of low-rank bituminous to medium- and high-rank bituminous and then to anthracite. During the third stage, the hydrogen content remained fairly constant and the oxygen content decreased. Hydrogen content then began to decline, and oxygen was reduced more slowly. The major by-products of these two stages were probably CH_4, CO_2, and H_2O. In the fifth stage, oxygen content remained constant and hydrogen fell more rapidly than before. The major by-products of this stage were CH_4, a little H_2O, but no CO_2 (Mott 1942).

2.3.5 <u>Classification by rank</u>

For those involved in the use of coal, information on rank is vital because rank indicates coal composition and how the coal will behave in processing. The American Society for Testing and Materials (ASTM) classification of coals by rank is shown in Table 2.3, which details the parameters used to determine the rank of coals. According to this system, there are two ranks of lignite, three ranks of subbituminous coals, and five ranks of bituminous coals. The bituminous ranks include three levels of high-volatile matter content indicated by A, B, and C, with A being the highest rank and containing the least volatile matter. These ranks are followed by medium- and low-volatile-matter bituminous coals. Anthracite coals are of higher rank than bituminous and are also separated into three levels. Only the bituminous coals are indicated to be agglomerating. The ASTM system, as well as the International System of Europe, uses volatile matter content to classify high-rank coals and calorific values for lower-rank coals.

Table 2.3. Classification of coals by rank[a] (ASTM D 388)

Group	Fixed carbon limits[b] (%)		Volatile matter limits[b] (%)		Calorific value limits[c] (Btu/lb)		Agglomerating character
	Equal or greater than	Less than	Greater than	Equal or less than	Equal or greater than	Less than	
Anthracite							
Metaanthracite	98			2			
Anthracite	92	98	2	8			Nonagglomerating
Semianthracite[d]	86	92	8	14			
Bituminous							
Low-volatile bituminous	78	86	14	22			
Medium-volatile bituminous	69	78	22	31			Commonly agglomerating[f]
High-volatile A bituminous		69	31		14,000[e]		
High-volatile B bituminous					13,000[e]	14,000	
High-volatile C bituminous					11,500	13,000	
					10,500[d]	11,500	Agglomerating
Subbituminous							
Subbituminous A					10,500	11,500	
Subbituminous B					9,500	10,500	
Subbituminous C					8,300	9,500	
Lignite							Nonagglomerating
Lignite A					6,300	8,300	
Lignite B						6,300	

[a] This classification does not include a few coals, principally nonbanded varieties, which have unusual physical and chemical properties and which come within the limits of fixed carbon or calorific value of the high-volatile bituminous and subbituminous ranks. All of these coals either contain less than 48% dry, mineral-matter-free fixed carbon or have more than 15,500 moist, mineral-matter-free Btu/lb.
[b] Dry, mineral-matter-free basis.
[c] Moist, mineral-matter-free basis; moist refers to coal containing its natural inherent moisture but not including visible water on the surface of the coal.
[d] If agglomerating, classify in low-volatile group of the bituminous class.
[e] Coals having 69% or more fixed carbon on the dry, mineral-matter-free basis shall be classified according to fixed carbon, regardless of calorific value.
[f] It is recognized that there may be nonagglomerating varieties in these groups of the bituminous class, and there are notable exceptions in high-volatile C bituminous group.

Industrial classification systems were designed primarily as a guide for the coking industry (Given 1973*b*). As previously mentioned, only bituminous coals are used for coking because only they possess the softening and swelling characteristics.

The International system assigns coal a three-digit number. The first digit indicates the rank or class of the coal, the second indicates the behavior of the coal when it is heated rapidly, and the third indicates the behavior of the coal when it is heated slowly, as in a coke oven. Table 2.4 shows a comparison of the International and ASTM systems of ranking coal.

Of the coal components known, only vitrinite, a phytogenic component (maceral), shows the characteristics necessary for coking in bituminous coals. Changes in the other macerals in relation to the degree of metamorphism are less well defined, and with some macerals are relatively small (Given 1973*a*). It is possible for coals having identical vitrinites at the same stage of

Table 2.4. Parameters of International coal classification
compared with ASTM rank[a]

International class	ASTM rank	Volatile matter
		Dry, ash-free (%)
1A	Anthracite	3 - 6.5
1B	Anthracite	6.5 - 10
2	Semianthracite	10 - 14
3	Low-volatile bituminous	14 - 20
4	Medium-volatile bituminous	20 - 28
5	Medium-volatile bituminous	28 - 33
		Moist, ash-free (Btu)
6	High-volatile A bituminous	> 13,950
7	High-volatile B bituminous	12,960 - 13,950
8	High-volatile C bituminous or subbituminous A	10,980 - 12,960
9	Subbituminous B	10,260 - 10,980

[a]For exact information on International classification, consult Bureau of
Mines Report of Investigations 5435.

Source: Yancey and Geer 1968, Table 1-2, p. 1-6. Reprinted by permission
of the publisher.

metamorphism to have different properties because of variations in other components of the coal.
Nevertheless, according to Given (1973a), the rank of coal ideally should be based upon the
vitrinite it contains if the rank is to be used for scientific purposes.

In Table 2.5, Given (1974) summarizes some important correlations of the various ranks of coal
with aromatic carbon and the number of benzene rings. He notes that the aromaticity of coal
increases with increasing rank and that much of the nonaromatic carbon is in hydroaromatic rings.

A geological classification of U.S. coals by rank is illustrated in Table 2.6. Coals of the
western and central states were formed about 150 million years later than those of the eastern
and southeastern states; therefore, the plants that provided the organic material for these
coals were different. Because the locations were different, the coals were also exposed to
greatly differing geological conditions. Thus, it is not surprising that the coals of Appalachia
differ greatly from those of the Dakotas. Basically, there are three differences between the
coals of the eastern and western United States: (1) The coals of the western regions were formed
much more recently than the eastern coals and are thus generally of much lower rank; (2) the
difference in geological location resulted in different mineral distribution between coals of
the two regions (e.g., eastern coals generally have a higher sulfur content); and (3) both the
difference in geologic time and location resulted in the appearance of different plants so that
different plant materials were incorporated into each of the coals.

Table 2.5. Approximate values of some coal properties in different rank ranges

| Component | Lignite | Subbituminous | Bituminous | | | Medium-volatile | Low-volatile | Anthracite |
| | | | High-volatile | | | | | |
			C	B	A			
Carbon (mineral-matter-free), %	65 - 72	72 - 76	76 - 78	78 - 80	80 - 87	89	90	93
Oxygen, %	30	18	13	10	10 - 4	3 - 4	3	2
Oxygen, as COOH, %	13 - 10	5 - 2	0	0	0	0	0	0
Oxygen, as OH, %	15 - 10	12 - 10	9	?	7 - 3	1 - 2	0 - 1	0
Aromatic carbon atoms, % of total carbon	50	65	?	?	75	80 - 85	85 - 90	90 - 95
Average number of benzene rings per layer	1 - 2	?	2 - 3	2 - 3	2 - 3	2 - 3	5?	>25?
Volatile matter, %	40 - 50	35 - 50	35 - 45	?	31 - 40	31 - 20	20 - 10	<10
Reflectance, % (vitrinite)	0.2 - 0.3	0.3 - 0.4	0.5	0.6	0.6 - 1.0	1.4	1.8	4
Density	←			minimum				→

Source: Given 1974, Table 1, p. 91. Reprinted by permission of the publisher.

Table 2.6. Distribution of coals by province in the United States

	Appalachian	Interior	North Great Plains	Rocky Mountain	Pacific	Gulf	Alaska
States	Pennsylvania, Ohio, Virginia, West Virginia, E. Kentucky, Alabama, Tennessee	East coal region Indiana, Illinois, W. Kentucky West coal region Oklahoma, Missouri, Kansas, Arkansas	N and S Dakota, Montana, NE Wyoming	Many distinct basins of different geological history in SW Wyoming, Colorado, Utah, New Mexico, Arizona	Washington, Oregon, California	Parts of Arkansas, Texas, Louisiana, Mississippi, W Alabama	Alaska
Age, million years	Carboniferous (300)	Carboniferous	Some Cretaceous, mostly early Tertiary (100 - 50)	Mostly Cretaceous, some early Tertiary (130 - 60)	Tertiary (60 - 15)	Tertiary (70 - 30)	Cretaceous and early Tertiary (100 - 50)
ASTM rank	Principal deposits of high-volatile A and B, medium-volatile, and low-volatile bituminous, anthracite	Principal deposits of subbituminous, high-volatile A, B, and C; minor deposits of low-volatile bituminous and anthracite	Principal deposits of lignite and subbituminous	Principal deposits of subbituminous and high-volatile A, B, and C bituminous; minor deposits of medium-volatile bituminous and anthracite	Principal deposits of lignite, subbituminous and high-volatile A, B, and C bituminous; minor deposits of anthracite	Principal deposits of lignite	Principal deposits of subbituminous and high-volatile A, B, and C bituminous

Source: Given 1973*a*, Fig. 7. Reprinted by permission of the publisher.

2.4 COAL DISTRIBUTION IN THE UNITED STATES

The United States is believed to possess 48.2% of the total world coal reserves within seven major coal provinces (Table 2.7). The Great Plains province contains about 45% of the country's coal reserve, much of which consists of lignite. The east-west distribution of U.S. coal is given in Table 2.8; two thirds of the nation's coal reserve, most of which is Cretaceous or Tertiary, is located west of the Mississippi River (Spackman 1973). A geographic distribution of U.S. coals according to rank is given in Fig. 2.2.

Table 2.7. Coal reserves of the major coal provinces
in the United States

Province	Approximate reserve estimate (in millions of tons)		
	Proven (with 0 - 3000-ft cover)	Total	Age
Appalachian	286,907	382,485	Paleozoic
Eastern Interior	194,740	317,240	Paleozoic
Western Interior	59,981	133,209	Paleozoic
Rocky Mountain	176,444	598,444	Cretaceous - Tertiary
Great Plains	695,122	1,458,122	Cretaceous - Tertiary
Pacific Coast	136,604	316,704	Cretaceous - Tertiary
Gulf Coast	7,248+	7,248+	Cretaceous - Tertiary
Total	1,557,046	3,213,452	

Source: Spackman 1973*a*, Table 4, p. 12.

Table 2.8. Age and distribution of U.S. coal reserves

	Approximate reserve estimate (in millions of tons)	
	Proven (with 0 - 3000-ft cover)	Total
Distribution		
East of Mississippi River	483,647	701,725
West of Mississippi River	1,073,399	2,511,727
Total	1,557,046	3,213,452
Age		
Paleozoic	541,628	832,934
Cretaceous - Tertiary	1,015,418	2,380,518
Total	1,557,046	3,213,452

Source: Spackman 1973[a], Table 3, p. 11.

2.5 COAL PETROGRAPHY

Coal is composed basically of two types of material — inorganic crystalline minerals and phytogenic noncrystalline macerals. Both occur in coals as grains, particles, and fragments ranging in size from 2 μm^3 to several cubic centimeters and larger (Spackman 1973).

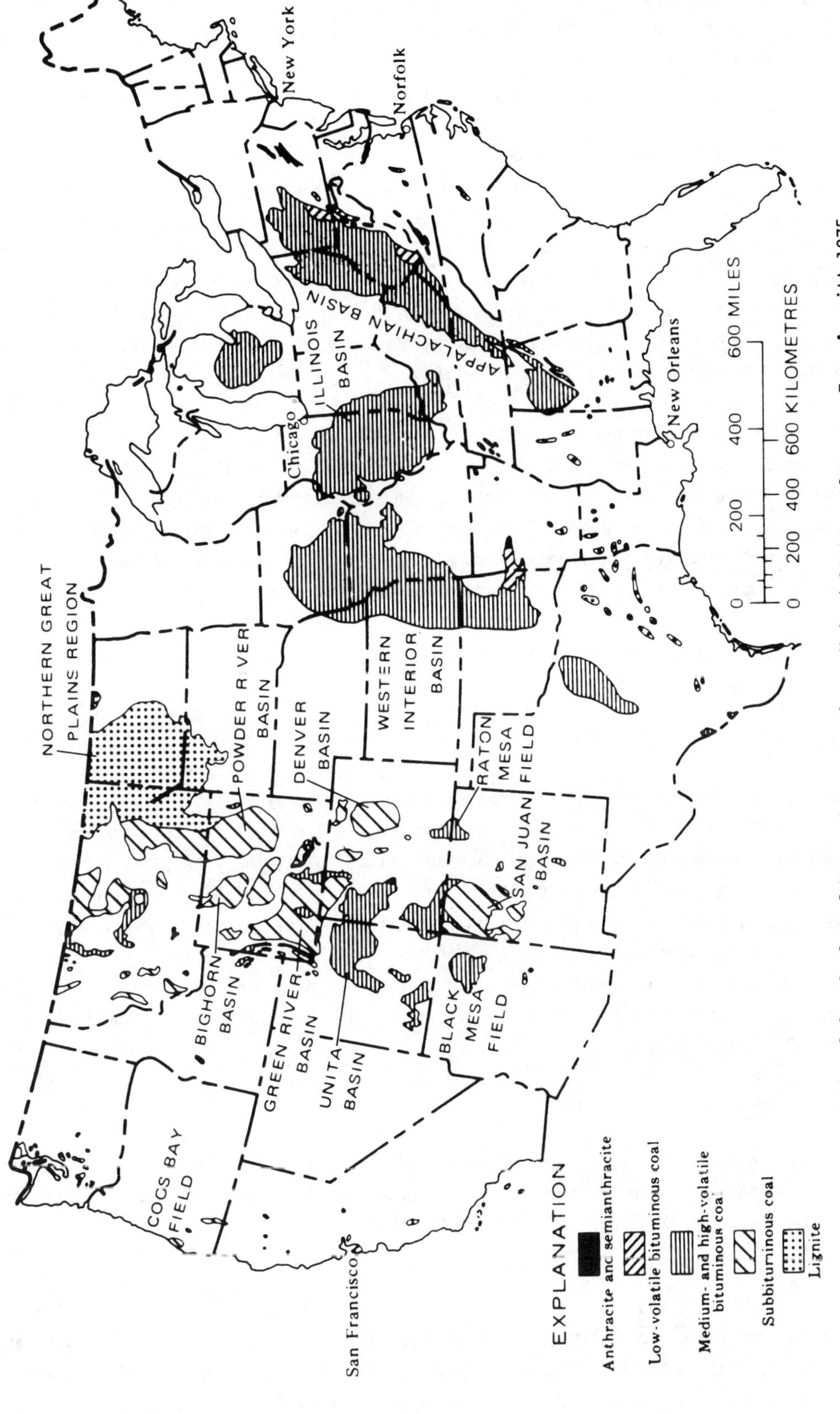

Fig. 2.2. Coal fields of the conterminous United States. Source: From Averitt 1975, Fig. 1, p. 5.

2.5.1 Macerals

Macerals are the fragmentary organic remains of plants that died, were altered to peat by partial decay, and, because of exposure to heat in the earth's crust through time, were converted to their present state in coal. The carbonaceous, combustible fraction of coal is made up of macerals; thus, by definition, macerals comprise more than half of the coal mass (Spackman 1973).

Probably the most widely used method of determining the maceral content of coal is by measurement of reflected light, a technique developed to a point of usefulness by workers of the Prussian Geological Survey (Parks 1963). Each maceral reflects characteristic amounts of light when direct light is applied to polished coal surfaces. It is generally agreed that the reflectance of the macerals increases with the increasing rank of coal, but there has been some controversy as to whether the increase occurs in a series of sudden jumps at certain ranks, or is a continuous change process. In any case, reflectance measurements offer an objective method of classifying the petrographic constituents of coals (Parks 1963).

Another method of determining the petrographic composition of coals is the transmitted light technique developed by Thiessen (1920), as cited by Parks (1963), at the U.S. Bureau of Mines. This method is frequently referred to as the American system of coal petrography, in contrast to the German system mentioned above. The transmitted light technique has not been as widely adopted as the reflected light technique, probably because of the difficulties encountered in performing the required thin-section preparation and analysis (Parks 1963).

Described in the following subsections are the common coal macerals and their distinguishing characteristics according to Spackman (1973).

2.5.1.1 Vitrinite

These macerals form the humic fraction of coal and are produced by relatively slow alteration of plant cell wall substances. A woody texture and brown color characterize vitrinites in coals of early metamorphic stages. In bituminous and anthracite coals, the vitrinites appear texture-less and black with a vitreous luster to the unaided eye. In transmitted light through a micro-scope, low-rank vitrinites show colors from buff and cream to yellow, tan, and pale orange red. In vertically incident light, they appear grey and have less than 5% reflectance. Vitrinites of intermediate rank are tan, orange red, reddish brown, and deep red, and they reflect 0.5 to 2.5% of vertically incident light. High-rank vitrinites in anthracites are opaque and have a reflectance of 2.5 to 6.0%.

2.5.1.2 Pseudovitrinite

Pseudovitrinites generally have the same characteristics as vitrinites, but show greater reflectance. They may have a remnant cell structure, serrate margins, and greater relief than do vitrinite. Unlike vitrinite, they may be inert in coking.

2.5.1.3 Fusinite

These macerals form the charcoal-like fraction of coal and are produced by rapid charring and alteration of cell wall material. They are opaque in thin section and appear white in vertically incident light. They frequently include residual cell wall structures.

2.5.1.4 <u>Semifusinite</u>

These macerals are characterized by optical properties intermediate between those of vitrinite and fusinite.

2.5.1.5 <u>Micrinite</u>

Micrinite macerals consist of attritus and are formed by granulation and metamorphism of material from cell walls. They appear translucent and yellowish brown to brown in lignites when observed in thin section and dark gray in incident light. In high-rank coals they are opaque and white. They range from 1 to 6 μm in diameter.

2.5.1.6 <u>Macrinite</u>

These macerals appear white in incident light, but are not as reflective as fusinite. They never contain remnant cell walls and range from ten to hundreds of micrometers in their greatest dimension.

2.5.1.7 <u>Exinite</u>

Exinite macerals are derived from waxy secretions such as plant cuticles and spore and pollen exines. In low- to intermediate-rank coals, they are yellow in thin section and dark gray to black in the medium range. In very high-rank coals, either they have the same optical properties as vitrinite, or they disappear.

2.5.1.8 <u>Resinite</u>

These are hydrogen-rich macerals formed from resinous secretions and excretions. They usually appear yellow in thin section and are oval, irregular, or rod-shaped. Like exinite, they either take on the optical properties of vitrinite or disappear in high-rank coals.

2.5.1.9 <u>Sporinite, cutinite, and alginite</u>

These macerals are easily identified by their shape. They are composed of the material that their name indicates — spores, cutin, or algae. The term exinite includes both sporinite and cutinite.

2.5.1.10 <u>Sclerotinite</u>

As described by van Krevelen (1961), these macerals consist of fossilized remains of fungal sclerotia. Sclerotinites are opaque and highly reflective.

Figure 2.3 illustrates the international terminology for common macerals. The U.S. Bureau of Mines terminology, which differs from the international method, is shown in Fig. 2.4.

2.5.1.11 <u>Maceral classification</u>

Van Krevelen (1961) groups the macerals into three major categories: (1) macerals whose origin is definitely due to woody and cortical tissues (vitrinite, fusinite, semifusinite); (2) macerals whose origin is definitely due to plant material other than woody tissues (exinite, sporinite,

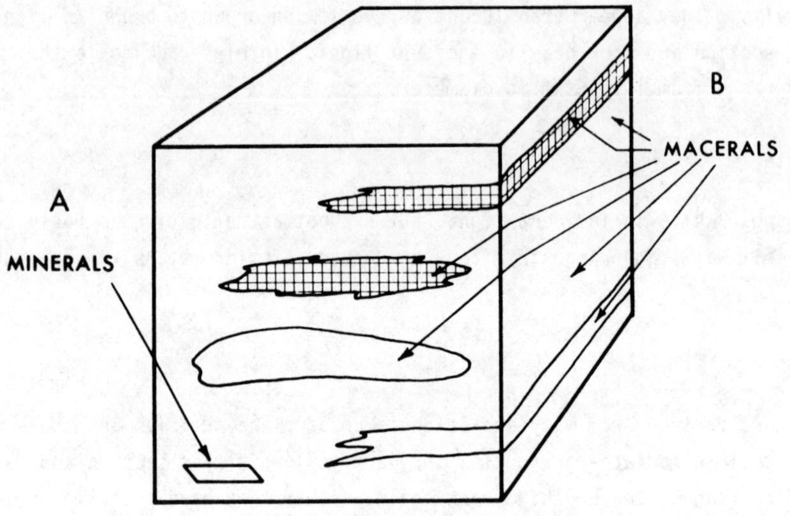

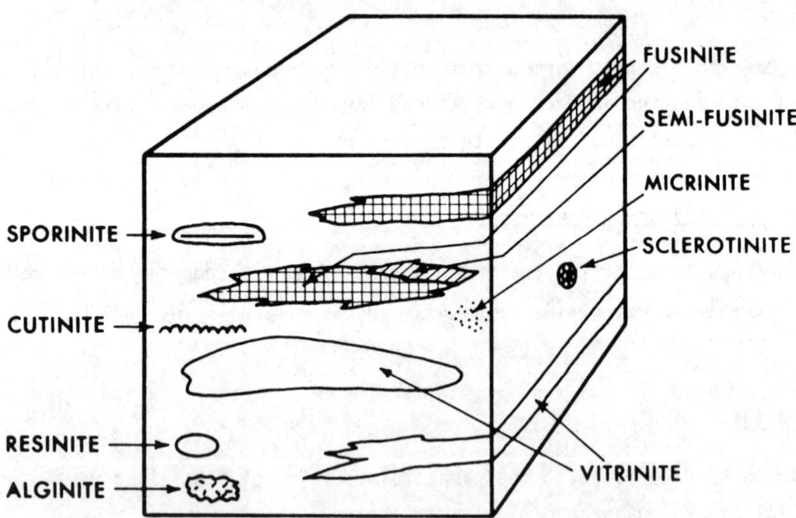

Fig. 2.3. International terminology. Source: From Spackman 1973, Fig. 1.

cutinite, resinite, alginite); and (3) a maceral whose origin has not yet been traced to a specific vegetable tissue (micrinite).

THE BASIC CONCEPT

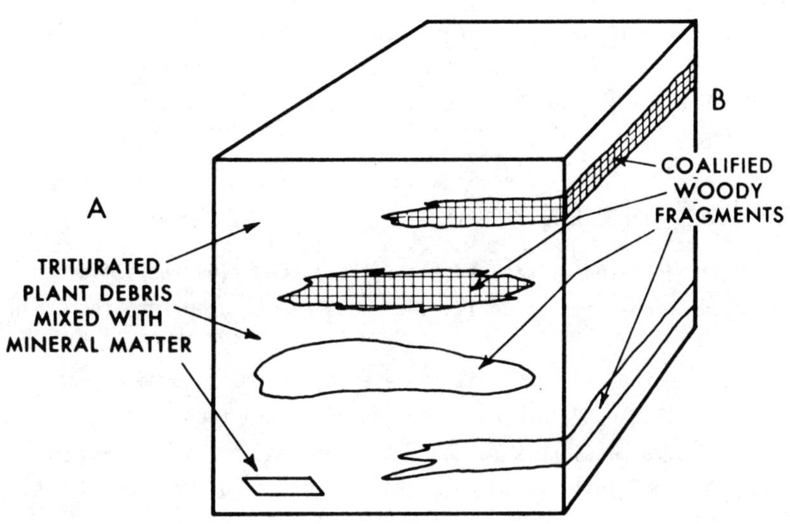

THE BASIC WORDS

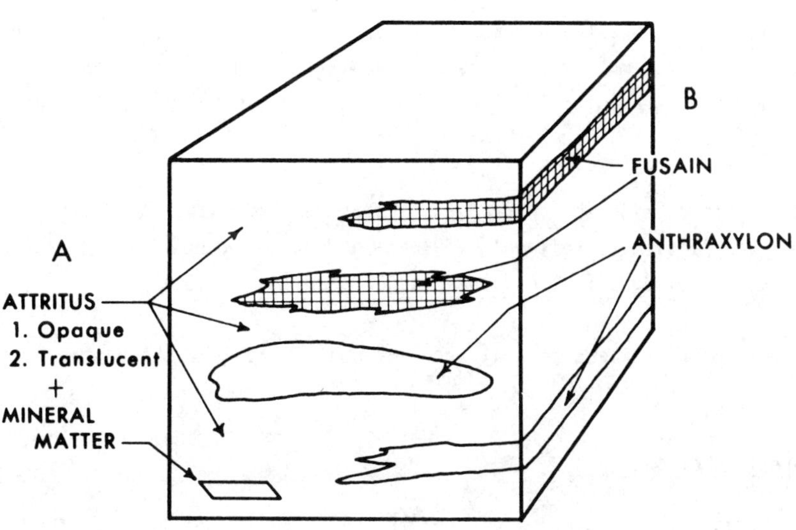

Fig. 2.4. Terminology employed in U.S. Bureau of Mines publications. <u>Source</u>: From Spackman 1973, Fig. 2.

Fusinite, semifusinite, sclerotinite, and micrinite behave as inert substances on heating and are collectively referred to as "inertinite." Sporinite, cutinite, resinite, and alginite are rich in hydrogen and become very plastic on heating. They may be included in the exinites (van Krevelen 1961).

Vitrinite is the best coke-forming maceral. In anthracite it is relatively inert, and in low-rank, high-volatile coals it will neither cake nor soften. However, if the volatile matter content of the coal is from about 19 to 33%, vitrinite is responsible for the actual coking properties. Coking coals usually have characteristic contents of vitrinite, inertinite, and exinite (van Krevelen 1961).

2.5.1.12 Influence of macerals on coal properties

Hoffman and Hoehne (1954), as cited by van Krevelen (1961), describe the influence of petrographic composition on the properties of coal. The coals are grouped into four main types:

1. Euplastic coals include the clarains, vitrinite being the most abundant maceral. During heating, they first undergo a small period of contraction, followed by rapid expansion; once maximum expansion is reached, they remain constant in volume. The coal rank and exinite-inertinite ratio determine the amount of increase in volume.

2. Subplastic coals undergo only contraction, which is characteristic of low-rank coals and those having an excess of inertinite.

3. Fluidoplastic coals like the euplastics first contract, then expand, upon heating; however, when the maximum dilation is attained, the volume drops drastically so that it is smaller than the initial volume. This behavior is usually exhibited by exinite-rich coals.

4. Perplastic coals combine the characteristics of euplastic and fluidoplastic types. An initial contraction is followed by dilation and then a limited contraction, followed by a leveling off at a volume larger than the original.

Coke manufacturers usually blend several types of coal to achieve the desired product (van Krevelen 1961).

2.5.1.13 Properties of macerals

Given (1973b) suggests that macerals must be physically separated to study their properties. Because vitrinites sometimes occur in pure bands, they are the easiest to study. Lithotypes containing 60 to 70% sporinite can be found in European coals; however, North American coals containing sporinite concentrations above 40% are rare. Higher concentrations of the macerals can be obtained by gravitational separation of appropriate lithotypes.

Given (1973b) gives the order of maceral density as fusinite > micrinite > vitrinite > sporinite. Relatively few European coals and even fewer American coals have been separated into their component macerals. It appears that only one American coal (Reggel, Wender, and Raymond 1970) has been found to contain all four macerals. Thus, little is known of the properties of macerals and their behavior in processes other than coking.

Given, Peover, and Wyss (1960) separated vitrinites, exinites, and fusinites from the same coals and analyzed them for their content of carbon, hydrogen, oxygen, ash, and petrographic purity. The results of these tests are given in Table 2.9; hydroxyl contents of vitrinite and exinite, given in Table 2.10, are compared with those of some exinite-rich duraines. Given, Peover, and Wyss (1960) note that exinites closely resemble vitrinites in many of their chemical character- istics; however, in this study three significant differences appeared: (1) Exinite is relatively insoluble in pyridine, as compared with vitrinite, whereas nitrobenzene and trichlorobenzene appear to be better solvents for exinite than for vitrinite (Table 2.11); (2) exinite did not yield humic-acid-like products on oxidation as do vitrinites; and (3) the extent of reduction of exinite with lithium, as well as the variation of the extent with rank, was less in all cases than the reduction of vitrinites (Table 2.12).

It is suggested that the solubility and oxidation differences between exinite and vitrinite are due to different intermolecular forces and molecular weights. Exinites are thought to have a more open hydroaromatic character. (Interlayer spacing for exinites has been shown to be 4.7 Å, whereas the spacing in vitrinites is 3.5 Å.) Because of the closer molecular arrange- ment of vitrinites, different intermolecular forces from those in exinite are believed to be in effect. No explanation is given for the difference in reducibility (Given, Peover, and Wyss 1960).

Reggel, Wender, and Raymond (1970) studied vitrinite, exinite, micrinite, and fusinite from Hernshaw bed high-volatile A bituminous coal, Boone County, West Virginia. By comparing the analytical and dehydrogenation data shown in Tables 2.13 and 2.14, the researchers formed the following conclusions. All four macerals are similar in sulfur content, but fusinite has the lowest nitrogen content. Exinite and vitrinite contain similar amounts of carbon, but the hydrogen-carbon ratio is much higher in exinite; in fact, exinite generally shows a higher con- tent of hydrogen and volatile matter and a higher hydrogen-carbon ratio than any other maceral. Vitrinite and micrinite are similar in volatile matter, fixed carbon (incorporated into organic compounds), carbon, and hydrogen contents, but the hydrogen-carbon ratio for vitrinites is normally a little higher. The hydrogen-carbon ratio of fusinite is very low, and fusinite has a much lower volatile matter and hydrogen content than does vitrinite. It is concluded from the low hydrogen-carbon ratio of fusinite that this maceral must be more aromatic than vitrinite. Vitrinite and exinite are hydroaromatic, whereas micrinite, having a hydrogen-carbon ratio close to that of vitrinite, contains some structure that is not aromatic but also does not give off hydrogen during dehydrogenation (Reggel, Wender, and Raymond 1970).

Boyer et al. (1961) conclude that straight-chain paraffins and olefins produced during coking are products almost exclusively of exinite. Their data indicate that a relatively large part of exinite has a paraffinic structure with long chains of about 30 carbon atoms.

Spackman (1973) describes another, more recently recognized, maceral material called bituminite, which is believed to yield hydrocarbons when heated. Bituminite can be detected only by fluorescence microscopy, and because this procedure has only recently been employed in analyses by American petrographers, it has not yet been described in American coals. Bituminite is not expected to be present in the relatively high-rank Appalachian coals.

Table 2.9. Analyses of purified maceral concentrates[a]

Source (Colliery and seam)	Element[a] (wt %)						Petrographic purity (%)	Ash (%)
	C	H	N	S	O Direct	O By diff.		
Vitrinites								
V1 Cannock Wood, Shallow	79.6	4.9				12.8	98	0.8
2 Teversal, Dunsil	81.5	5.1				10.8	98	0.3
3 Markham Main, Barnsley	82.3	5.5				9.3	98	1.0
4 Dinnington Main, Barnsley	85.1	5.3				6.9	96	0.6
5 Aldwarke Main, Silkstone	86.9	5.4				4.9	98	0.8
6 Chislet, No. 5	88.6	5.3				3.9	96	0.3
7 Roddymoor, Ballarat	88.8	5.3				3.6	96	0.4
8 Coegnant, Gellideg	91.4	4.6				1.8	89	0.7
9 Blaenhirwaun, Pumpquart	93.8	3.1				1.2	98	0.5
Semifusinites and fusinites								
F1 Cannock Wood, Shallow	90.3	2.8				5.9	96	2.6
2 Teversal, Dunsil	87.7	3.9				6.9	93	2.9
3 Markham Main, Barnsley	91.6	3.6				3.4	92	4.8
4 Dinnington Main, Barnsley	92.4	3.1				3.5	93	12.3
5 Aldwarke Main, Silkstone	92.1	3.7				3.0	96	4.6
6 Chislet, No. 5	92.2	3.8				2.7	93	5.2
Exinites								
E1 Cannock Wood, Shallow	81.0	7.0				9.9	90	1.6
3 Markham Main, Barnsley	82.6	7.4				7.0	88	0.7
	81.6	7.6	1.3	1.0	8.3	8.5		
4 Dinnington Main, Barnsley	84.8	7.1				6.1	91	1.7
	84.3	7.6	1.2	1.0	5.6	5.9		
5 Aldwarke Main, Silkstone	87.2	7.4				3.5	95	0.2
	87.1	7.4	1.2	0.4	4.5	3.9		
6 Chislet, No. 5	89.8	6.2				2.4	85	1.0
	89.1	6.2	1.3	0.2	4.0	3.2		

[a]Dry, mineral-matter-free basis.

[b]Where two analyses are given, the first is dry, mineral-matter-free and the second is dry, ash-free.

Source: Given, Peover, and Wyss 1960, Table 1, p. 324. Reprinted from *Fuel*: "Chemical properties of coal macerals" by permission of the publishers, IPC Business Press Ltd.

Table 2.10. Hydroxyl contents of vitrinites,
exinites, and durains

Sample	Carbon (%)	$Oxygen_{OH}$ (%)	$Oxygen_{OH}$/total oxygen (%)
Vitrinite			
V1	79.6	5.3	41
V3	82.3	5.7	61
V4	85.1	5.1	74
V5	86.9	3.9	80
V6	88.6	2.8	72
Exinite			
E1	81.0	3.2	32
E3	81.6	3.7	45
E4	84.3	2.9	52
E5	87.1	1.4	31
E6	89.1	0.8	20
Durain pyridine extract of			
BD1	85.5	2.3	40
BD1	83.5	2.3	40
residue from extraction of			
BD1	83.4	2.5	40
BD2	82.8	3.5	45
BD3	86.2	2.5	53

Source: Given, Peover, and Wyss 1960, Table 4, p. 328.
Reprinted from *Fuel*: "Chemical properties of coal macerals"
by permission of the publishers, IPC Business Press Ltd.

2.5.2 Minerals

Gluskoter (1975) states that the term "mineral matter in coal" includes both the mineral species and the elements generally considered to be inorganic. By this definition, all elements except carbon, hydrogen, oxygen, nitrogen, and sulfur are included in mineral matter; however, each of these elements except nitrogen may be found in inorganic combination in coals; therefore, they may also be considered part of the mineral matter.

2.5.2.1 Identification and quantitation

Minerals found in coal are classified by Nelson (1953) as syngenetic or epigenetic. Syngenetic minerals were included in the coal during the biochemical changes associated with coalification, whereas epigenetic minerals were deposited in the cleats and cracks of the coal after the coalification process was complete.

The total amount of mineral matter in coals varies considerably, but it is usually present in sufficient concentrations to be significant in any coal use (Gluskoter 1975). O'Gorman and Walker (1972) found mineral matter contents ranging from 9.05 to 32.6% in 16 coal samples from various locations in North America.

The study of mineral matter in coal has intensified since the 1960s because several elements and minerals, such as beryllium, mercury, and asbestos, have been recognized as hazardous to health, and others have been suspected of being detrimental to the environment. More recently, there has been concern about the effects of minerals on coal conversion processes (Gluskoter 1975).

Table 2.11. Relative solubilities of the exinites[a]

Sample	Cellosolve	Dimethylformamide	Pyridine	Ethylene carbonate	Chlorobenzene	Nitrobenzene	1,2,4-Trichlorobenzene
Solvent							
Exinite							
E3	<1	2	2	<1	<1	3	3
E5			3			7	20
E6	<1	1	5	<1	1	20	15
Vitrain, 82% C			10			4	3[b]

[a]Numbers assigned on the basis of color intensity.
[b]Another vitrain (84% carbon) gave 7 with this solvent.

Source: Given, Peover, and Wyss 1960, Table 3, p. 326. Reprinted from *Fuel*: "Chemical properties of coal macerals" by permission of the publishers, IPC Business Press Ltd.

Table 2.12. Reduction cf exinites with lithium[a]

| Exinite sample | Solubility of product in pyridine (%) | Atoms of hydrogen added per 100 atoms of carbon | | Atoms of oxygen added per 100 atoms of carbon to exinite |
		Exinite	Vitrain of same % carbon (estimated)	
E1	27	26	6[b]	−5
3	7	3	14	0
4		6	23	0
5	9	4	32	−1
6	28.5	23	37	1
BD1	35	24	27	0

[a]Analytical results are expressed as atoms added per 100 carbon atoms, as this shows most clearly the effect of reactions; however, the procedure does magnify the effect of analytical errors and is only really justified if no carbon is lost in the reaction by oxidation or other side effects.
[b]Extrapolated.

Source: Given, Peover, and Wyss 1960, Table 5, p. 329. Reprinted from *Fuel*: "Chemical properties of coal macerals" by permission of the publishers, IPC Business Press Ltd.

Table 2.13. Analyses of macerals used in dehydrogenation
(moisture- and ash-free basis)

Material	C	H	N	S	O (diff.)	Volatile matter	Fixed carbon
Vitrinite (Hernshaw)	84.56	5.46	1.55	0.71	7.72	33.7	66.3
Exinite (Hernshaw)	84.63	6.47	1.09	0.70	7.11	55.3	44.7
Micrinite (Hernshaw)	83.94	5.07	1.31	0.63	9.05	31.4	68.6
Fusinite (Hernshaw)	89.97	3.36	0.53	0.41	5.73	13.4	86.6
Vitrinite (Herrin)	78.09	5.56	1.89	2.73	11.73		
Fusain (Herrin)	89.07	3.38	0.21	2.04	5.30		
Fusinite (Herrin) (purified)[a]	90.40	3.50	0.20	0.73	5.17		
Vitrain (Hartshorne)	88.66	4.87	1.71	0.66	4.10	33.3	66.7
Fusain (Hartshorne)	84.36	4.76	1.56	2.03	7.29	24.8	75.2
Vitrinite (No. 2)	83.93	5.22	1.61	0.60	8.64	31.97	
Exinite (No. 1)	84.73	6.38	1.33	0.44	7.12	59.81	
Micrinite (No. 3)	85.88	4.24	1.18	0.38	8.32	23.37	
Vitrinite (No. 3E)	83.73	5.26	1.75	0.83	8.43	31.0	
Exinite (No. 5C)	85.65	6.70	1.17	0.68	5.80	55.7	
Inertinite (No. IIIB)	86.70	4.23	1.15	0.68	7.24	21.0	

[a]Moisture-, ash-, and chlorine-free basis.

Source: Reggel, Wender, and Raymond 1970, Table 3, p. 283.
Reprinted from *Fuel*: "Catalytic dehydrogenation of coal" by permission of the publishers, IPC Business Press Ltd.

Table 2.14. Results of dehydrogenation of macerals[a]

Material	Hydrogen evolved			Hydrogen evolved (H/100C): corresponding vitrain as comparison	H/C atomic	O/C atomic
	(%)[b]	(H/100C)[c]	(mg/g)[d]			
Vitrinite (Hernshaw)	33.0	25.4	201	100	0.770	0.069
Exinite (Hernshaw)	33.8	30.8	243	121	0.911	0.063
Micrinite (Hernshaw)	24.3	17.5	137	69	0.719	0.081
Fusinite (Hernshaw)	11.8	5.0	43	20	0.445	0.048
Vitrinite (Herrin)	42.4	36.0	260	100	0.848	0.113
Fusain (Herrin)	6.5	2.9	24.4	8	0.453	0.0446
Fusinite (Herrin) (purified)	6.9	3.2	26.7	9	0.462	0.0429
Vitrain (Hartshorne)	21.5	14.1	116	100	0.655	0.0347
Fusain (Hartshorne)	9.0	6.1	48	43	0.673	0.0649
Vitrinite (No. 2)	25.1	18.7	146	100	0.742	0.771
Exinite (No. 1)	26.7	23.9	189	128	0.897	0.631
Micrinite (No. 3)	20.8	12.2	98.1	65	0.588	0.727
Vitrinite (No. 3E)	30.4	22.8	178	100	0.749	0.755
Exinite (No. 5C)	33.4	31.2	249	137	0.932	0.508
Inertinite (No. IIIB)	19.1	11.1	89.7	49	0.582	0.626

[a]These results are for the actual samples as used; they have not been corrected back to the theoretical 'pure' macerals.
[b]Percent of hydrogen in starting material which is evolved as H_2 gas.
[c]Atoms of hydrogen evolved as H_2 gas per 100 C atoms in starting material.
[d]Milliliters of hydrogen gas evolved per gram of moisture- and ash-free starting material.

Source: Reggel, Wender, and Raymond 1970, Table 4, p. 284.
Reprinted from *Fuel*: "Catalytic dehydrogenation of coal" by permission of the publishers, IPC Business Press Ltd.

The amounts and kinds of minerals found in coal vary widely and undoubtedly depend on the conditions of formation. Table 2.15 lists some of the common coal minerals. The most abundant minerals are the alumino silicates (clay minerals), and the most abundant of these are illite, kaolinite, and mixed-layer montmorillonite. Pyrite is the most abundant sulfide. Sulfates are relatively rare in unweathered coals but increase easily with weathering. Table 2.15 shows a gypsum concentration in sulfates of up to 20% of the mineral matter. Pyrite is easily oxidized to iron sulfate at room temperature, as shown in Table 2.16, which lists iron sulfate minerals found in weathered Illinois coals (Gluskoter 1975). Carbonates form readily in nonacid seas. Of the carbonates, dolomite and ankerite are probably encountered most often, and quartz, which is found in almost all coals, sometimes occurs in concentrations as high as 20%.

2.5.2.2 Minerals in ash

Ash is the residue remaining after complete incineration of coal. It is related to mineral matter, but is different in chemical composition from, and found in less quantity than, the original mineral matter. The heat of the combustion process induces changes in the mineral portion of coal such as loss of water of constitution by silicate minerals, loss of carbon dioxide from carbonate minerals, oxidation of iron pyrites to iron oxide, and fixation of oxides of sulfur by bases such as calcium and magnesium. The conditions of incineration determine the extent to which these changes take place; therefore, standard procedures have been prescribed for determining ash content. These vary somewhat between countries, but generally consist of burning a 1- to 2-g sample of coal in a well-ventilated muffle furnace at temperatures of 700 to 850°C (Ode 1963).

Ash consists mainly of compounds of silicon, aluminum, iron, and calcium with measurable contributions from the compounds of magnesium, titanium, sodium, and potassium. Analyses of ash are commonly reported as oxides; however, the constituents occur in ash predominantly as a mixture of silicates, oxides, and sulfates, with smaller quantities of other compounds. The silicon and aluminum oxides are derived primarily from the silicates, whereas iron oxide comes from pyrite, which forms ferric oxide and sulfur oxides when burned. Calcium and magnesium oxides are formed from the decomposition of carbonate minerals, and sulfates are produced by reactions among carbonates, pyrite, and oxygen (Ode 1963).

2.5.3 Trace elements

According to Hall, Varga, and Magee (1974), the term "trace elements" is applied to elements that are present in the earth's crust in concentrations of 0.1% (1000 ppm) or less. Trace element concentrations are usually somewhat enriched in coal ash so that crustal abundance usually falls between the amount in coal and the amount in ash. Elements present in amounts greater than 0.1% are often called minor and major elements as concentrations increase (Hall, Varga, and Magee 1974).

Table 2.15. Semiquantitative analysis of the mineral composition of some coals[a] (% of total mineral matter)

Classification	Mineral constituents	Chemical formula	Elkhorn No. 3 seam, Ky.		Hartshorne seam, Ky.
			PSOC 3	PSOC 4	PSOC 142
Silicates	Kaolinite	$Al_2Si_2O_5(OH)_4$	3 - 40	1 - 10	1 - 10
	Illite	$(OH)_4K_y(Si_{8-y} \cdot Al_y)Al_4O_{20}$	Trace	1 - 10	1 - 10
	Muscovite	$KAl_2(AlSi_3)O_{10}(OH)_2$	b	b	b
	Chlorite	$(Mg,Fe)_6(AlSi)_4O_{10}(OH)_8$	Trace	1 - 10	1 - 10
	Montmorillonite	$Na_y(Al_{2-y}Mg_y)Si_4O_{10}(OH)_2$	b	Trace	b
	Mixed-layer illite - montmorillonite		Trace	1 - 10	b
	Plagioclase	$(Na,Ca)Al(SiAl)Si_2O_8$	b	b	b
Carbonates	Calcite	$CaCO_3$	b	b	b
	Aragonite	$CaCO_3$	b	b	b
	Dolomite	$CaMg(CO_3)_2$	b	b	b
	Ankerite	$CaCO_3 \cdot (Mg,Fe,Mn)CO_3$	b	b	b
	Siderite	$FeCO_3$	b	1 - 10	30 - 40
Oxides	Quartz	SiO_2	40 - 50	1 - 10	1 - 10
	Hematite	Fe_2O_3	b	10 - 20	b
	Rutile	TiO_2	1 - 10	b	b
Sulfates	Gypsum	$CaSO_4 \cdot 2H_2O$	1 - 10	10 - 20	1 - 10
	Jarosite	$KFe_3(OH)_6(SO_4)_2$	b	b	b
	Thenardite	Na_2SO_4	b	1 - 10	b
Sulfides	Pyrite	FeS_2	1 - 10	10 - 20	1 - 10
	Marcasite	FeS_2	b	b	b

[a]Analysis by infrared spectrometry.
[b]No data available.

Source: Modified from Given 1973, Table 1, p. 13, from data of O'Gorman and Walker 1972. Reprinted by permission of the publisher.

Table 2.16. Iron sulfate minerals in
weathered Illinois coals

Szomoluokite	$FeSO_4 \cdot H_2O$
Rozenite	$FeSO \cdot 4H_2O$
Melanterite	$FeSO_4 \cdot 7H_2O$
Coquimbite	$Fe(SO_4) \cdot 9H_2O$
Rosmerite	$Fe_2(SO_4) \cdot Fe_2(SO_4)_3 \cdot 12H_2O$
Jarosite	$(Na,K)Fe_3(SO_4)_2(OH)_6$

Source: Gluskoter 1975, p. 3.
Reprinted by permission of the publisher.

2.5.3.1 Identification and quantitation

Considerable data have been compiled on the occurrence of minor elements in U.S. coals (Ode 1963).
Data in Table 2.17 were obtained by Nunn, Lovell, and Wright (1953) by investigating the minor
and trace elements in Pennsylvania anthracites. Studies of low-rank coals from Texas, Colorado,
North Dakota, and South Dakota by Devl and Arnnel (1956), as cited by Ode (1963), produced the
data in Table 2.18. Some data compiled by Ode on elements in Canadian, British, and German
coals are shown in Table 2.19. It is apparent that a fair portion of the periodic table of the
elements is represented by an elemental analysis of coal.

Table 2.17. Range of concentration of
minor and trace elements in anthracite
ash and burning bank samples (% of ash)[a]

Titanium oxide (TiO_2)	1.5 - 2.0
Magnesium oxide (MgO)	0.4 - 0.9
Calcium oxide (CaO)	0.2 - 0.4
Vanadium oxide (V_2O_5)	0.02 - 0.04
Germanium oxide (GeO_2)	Trace - 0.02
Manganese oxide (MnO)	0.006 - 0.008
Arsenic	Trace - 0.01
Copper	0.001 - 0.01
Chromium	0.001 - 0.01
Lead	0.001 - 0.01
Lithium	Trace - 0.01
Phosphorus	$0.x$[b]
Gallium, nickel, tin, zinc	$0.0x$
Beryllium, cobalt, tungsten, molybdenum	$0.00x$

[a]Some trace elements (Ag, B, Bi, Cd, Hg, La,
Sb, Te, Zr) reported by other investigators
were not detected in these preliminary
studies. More specific examination of the
samples for those particular elements may
show their presence, but in general the
concentration will probably not exceed $0.00x$.
[b]x denotes order of magnitude of the element
concentration in cases where the spectrographic
lines from the unknown were not within the
range of the standard currently available.

Source: Nunn, Lovell, and Wright 1953,
Table 1, p. 56.

Table 2.18. Maximum concentration of minor elements in ash of western coals (%)

Number of samples	Locality	Concentration				
		10 - 1	1.0 - 0.1	0.1 - 0.01	0.01 - 0.001	0.001 - 0:0001
151	Harding County, S. Dakota		B, Ba, Sr, Mn, Ti, Mo, Co, Ni, Zn, As, Pb, Zr, Be, U	Cu, Cr, V, Li, Y, La, Ga, Ge, Sc	Sn	Yb, Ag
59	Perkins County, S. Dakota		B, Ba, Sr, P, Ti, Mo, Zr, U	Cu, Cr, V, Ni, Co, Pb, Mn, Ga, Ge, Zn, La	Y, Sc, Sn	Be, Yb Ag
26	Bowman County, N. Dakota		B, Ba, Sr, Mn, Ti, As, Pb, V Zr	Cu, Cr, Mo, Ni, Co, Ga, Ge, U	Y, Sc, Sn	Be, Yb
4	McKenzie County, N. Dakota		B, Ba, Sr, Mn, Pb	Ti, V, Cu, Li Sn, La, Zr, Co	Y, Mo, Ga, Ge, Ni, Cr	Be, Yb
35	Jefferson County, Colorado		B, Ba, Sr, Mn, Ti, Mo, Y	Cu, Cr, V, Ni, Co, Zr, Pb	Sn, Sc, Ga, U Ge, La, Yb, Be	
48	Milam County Texas	P	B, Ba, Sr, Mn, Ti, Ni, Sn	Cu, Cr, V, Co, Pb, Zr, Zn, Y	Mo, Sc, Ga, Ge, Be, Yb	

Source: Ode 1963, Table 5, p. 227. Reprinted by permission of the publisher.

The average concentrations of some important trace elements in coal from four areas of the United States are given in Table 2.20. Six of the elements are found in greatest concentrations in the Eastern Interior region, five in the Northern Plains, three in the Appalachian, and one in the Western Interior region. The geographic distribution of some more volatile and thus potentially hazardous trace elements in coal is given in Table 2.21. Concentrations of mercury, lead, and zinc are highest in the Eastern Interior region. Concentrations of antimony, arsenic, beryllium, cadmium, and selenium are highest in the Western Interior; Appalachian coal is also high in beryllium. The lowest concentrations for all of these elements is found in Powder River Basin coals.

2.5.3.2 Correlation

Only recently has it been recognized that knowledge of the trace elements in coal is important for a thorough understanding of the behavior of coals during conversion processes. Some elements may act as catalysts in promoting the conversion reactions (Ruch, Gluskoter, and Shimp 1974). Thus, it would be useful to be able to correlate the concentrations of coal elements with each other and with coal properties.

Ruch, Gluskoter, and Shimp (1974) did an extensive study of 101 coals; 82 were taken from the Illinois Basin, and 19 were collected from other states. Mean values resulting from their analyses, including trace, minor, and major elements, are shown in Table 2.22. A total of 23 trace and 9 minor and major elements were identified.

The 101 coals were separated into three groups according to the geographic region from which they were collected — 82 samples from the Illinois Basin, 11 from the eastern United States, and 8 from the western United States. Statistical parameters determined for the trace and major elements, high- and low-temperature ashes, and the proximate and ultimate analyses for each group include arithmetic mean, standard deviation ranges, and the linear correlation

Table 2.19. Minor elements in Canadian, British, and German coals and ashes

| | In ash (ppm) | | | | | In coal (ppm) | | | | |
| Element | Sydney, Nova Scotia | | Barnsley Vitrain, Great Britain | Germany | Ruhr, Germany, | Sydney, Nova Scotia | | Germany | | Rhur, Germany |
	Av	Range		max	max	Av	Range	Av	Max	max
Antimony			100-200	>1000	3000			10-30	≥10	17
Arsenic	900	280-2300		10,000		100	33-270	100	500	
Barium	300	18-2200		>1000	a	35	2-257	a	100	a
Beryllium	14		50-100	4000	1000	2	1-2	13	40	20
Bismuth				2000					100	
Boron	148	52-220	200-3000	>1000		17	6-25		≥10	
Chromium	45	18-79	100-1000	a	5000	5	2-9	a	a	50
Cobalt	87	26-196	100-300	2000	2000	10	3-34	14	30	12
Copper			800-1000	≥10,000	4000			≥25	10,000	50
Gallium			80-300	>3000	1000			30	100	20
Germanium	44	9-70	300-1000	5000	1000	5	1-8	19	50	20
Lead	572		200-800	31,000	3000	66	25-120	140	3000	30
Manganese	1200	165-2200	100-2000	≥10,000	22,000	140	9-254	a	≥5000	700
Molybdenum	60	18-105	80-200	1000	6000	7	2-12	21	200	50
Nickel	131	52-645	500-3000	≥3000	16,000	15	6-74	24	≥60	30
Silver				60	a			0.3	3	a
Strontium	560	225-750		>1000	a	65	76-87	a	100	a
Tir	9	4-18	50-200	1000	6000	1		3	300	120
Titanium			3000-8000	12,000	30,000			700	1000	1500
Varadium	120	61-244	400-5000	>1000	11,000	14	7-28	18	>100	20
Zirc	218	115-550	500-700	21,000	8000	25	13-64	170	2000	100
Zirconium			<100-500	a	7000			a	a	140

[a]Nct significant.

Source: Ode 1963, Table 6, p. 228. Reprinted by permission of the publisher.

Table 2.20. Trace element content of American coals (ppm in coal)

| | Region | | | |
Element	Northern Great Plains	Western Interior	Eastern Interior	Appalachian
Beryllium	1.5	1.1	2.5	2.5
Boron	116	33	96	25
Titanium	591	250	450	340
Vanadium	16	18	35	21
Chromium	7	13	20	13
Cobalt	2.7	4.6	3.8	5.1
Nickel	7.2	14	15	14
Copper	15	11	11	15
Zinc	59	108	44	7.6
Gallium	5.5	2.0	4.1	4.9
Germanium	1.6	5.9	13	5.8
Molybdenum	1.7	3.1	4.3	3.5
Tin	0.9	1.3	1.5	0.4
Yttrium	13	7.4	7.7	14
Lanthanum	9.5	6.5	5.1	9.4

Source: Zubovic 1975, Table 2, p. 11 A.

Table 2.21. Distribution of environmentally hazardous
trace elements (ppm in coal)

| | Region | | | |
Element	Powder River Basin	Western Interior	Eastern Interior	Appalachian
Antimony	0.67	3.5	1.3	1.2
Arsenic	3	16	14	18
Beryllium	0.7	2	1.8	2.0
Cadmium	2.1	20	2.3	0.2
Mercury	0.1	0.13	0.19	0.16
Lead	7.2		34	12
Selenium	0.73	5.7	2.5	5.1
Zinc	33		250	13

Source: Zubovic 1975, Table 3, p. 12 A.

Table 2.22. Mean analytical values for 101 coals

Constituent	Mean	Standard deviation	Min	Max
Arsenic, ppm	14.02	17.70	0.50	93.00
Boron, ppm	102.21	54.65	5.00	224.00
Beryllium, ppm	1.61	0.82	0.20	4.00
Bromine, ppm	15.42	5.92	4.00	52.00
Cadmium, ppm	2.52	7.60	0.10	65.00
Cobalt, ppm	9.57	7.26	1.00	43.00
Chromium, ppm	13.75	7.26	4.00	54.00
Copper, ppm	15.76	8.12	5.00	61.00
Fluorine, ppm	60.94	20.99	25.00	143.00
Gallium, ppm	3.12	1.06	1.10	7.50
Germanium, ppm	6.59	6.71	1.00	43.00
Mercury, ppm	0.20	0.20	0.02	1.60
Manganese, ppm	49.40	40.15	6.00	181.00
Molybdenum, ppm	7.54	5.96	1.00	30.00
Nickel, ppm	21.07	12.35	3.00	80.00
Phosphorus, ppm	71.10	72.81	5.00	400.00
Lead, ppm	34.78	43.69	4.00	218.00
Antimony, ppm	1.26	1.32	0.20	8.90
Selenium, ppm	2.08	1.10	0.45	7.70
Tin, ppm	4.79	6.15	1.00	51.00
Vanadium, ppm	32.71	12.03	11.00	78.00
Zinc, ppm	272.29	694.23	6.00	5,350.00
Zirconium, ppm	72.46	57.78	8.00	133.00
Aluminum, %	1.29	0.45	0.43	3.04
Calcium, %	0.77	0.55	0.05	2.67
Chlorine, %	0.14	0.14	0.01	0.54
Iron, %	1.92	0.79	0.34	4.32
Potassium, %	0.16	0.06	0.02	0.43
Magnesium, %	0.05	0.04	0.01	0.25
Sodium, %	0.05	0.04	0.00	0.20
Silicon, %	2.49	0.80	0.58	6.09
Titanium, %	0.07	0.02	0.02	0.15
Organic sulfur, %	1.41	0.65	0.31	3.09
Pyritic sulfur, %	1.76	0.86	0.06	3.78
Sulfate sulfur, %	0.10	0.19	0.01	1.06
Total sulfur, %	3.27	1.35	0.42	6.47
Sulfur by X-ray fluorescence, %	2.91	1.24	0.54	5.40
Air-dry loss, %	7.70	3.47	1.40	16.70
Moisture, %	9.05	5.05	0.01	20.70
Volatile matter, %	39.70	4.27	18.90	52.70
Fixed carbon, %	48.82	4.95	34.60	65.40
Ash, %	11.44	2.89	2.20	25.80
Btu/lb	12,748.91	464.50	11,562.00	14,362.00
Carbon, %	70.28	3.87	55.23	80.14
Hydrogen, %	4.95	0.31	4.03	5.79
Nitrogen, %	1.30	0.22	0.78	1.84
Oxygen, %	8.68	2.44	4.15	16.03
High-temperature ash, %	11.41	2.95	3.28	25.85
Low-temperature ash, %	15.28	4.04	3.82	31.70

Source: Ruch, Gluskoter, and Shimp 1974, Table 5, p. 18.

coefficients determined by the missing-data correlation analysis. The following geochemical associations were determined from the researchers' correlation data:

1. A positive correlation occurred between zinc and cadmium.
2. Arsenic, cobalt, nickel, lead, and antimony showed positive correlations with each other and are commonly encountered in nature as sulfides; thus, they are included in the chalcophile element group. Germanium also showed a positive correlation with many of the chalcophile elements in the coals studied.
3. Positive correlations were observed between potassium, titanium, aluminum, and tin. These elements are commonly encountered in nature as silicates and are thus included in the lithophile element group. In coals, they usually are found in clay minerals and quartz.
4. A positive correlation was found between manganese and calcium in the Illinois Basin. Manganese can replace calcium in calcite ($CaCO_3$) and was probably in that form.
5. A positive correlation was found between sodium and chlorine in all the samples and might be attributed to their deposition in coal from saline groundwater.

Samples from the Illinois Basin were stratigraphically subdivided into four groups: coals from below the Harrisburg-Springfield (No. 5) coal member, from the Harrisburg-Springfield (No. 5) coal member, from the Herrin (No. 6) coal member, and from above the Herrin (No. 6) coal member. Using data from calculation of the means, standard deviations, and correlation coefficients for these groups, Ruch, Gluskoter, and Shimp (1974) observed the following relationships:

1. Lower concentrations of arsenic, copper, lead, and aluminum were found in younger coals of the Illinois Basin.
2. By using boron as an indication of paleosalinity, it was determined that the Illinois Basin became more saline during the time between the deposition of the younger and older coals.
3. The correlation between sodium and chlorine increased in the younger coals, possibly because of the increased salinity of the Illinois Basin (Ruch, Gluskoter, and Shimp 1974).

The ratio of the mean concentration for a trace element in the coal to its "Clarke value" (the average percentage of an element in the earth's crust) gives an "enrichment factor." Table 2.23 shows the elements of the Illinois Basin, eastern United States, and western United States having enrichment factors of 10 or 0.1; that is, elements that exist in coal by an order of magnitude greater or less than their Clarke values. Only six elements were found to be enriched or depleted in coal by one order of magnitude or more. Selenium was enriched 30- to 60-fold in all coals although its concentration ranged only between 1.6 and 3.4 ppm in coal (Ruch, Gluskoter, and Shimp 1974).

2.5.3.3 Mercury

In the past few years interest has greatly increased in mercury as an air pollutant. Studies of the mercury content of commercial coals in 1971 and 1972 revealed concentrations of less than 1 ppm. Table 2.24 contains a survey of the geographical distribution of mercury by producing region. Most coals have mercury contents of less than 0.2 ppm, but contents of 1 ppm have been found in a few specific locations. A good average for U.S. coals might be around 0.15 ppm (Hall, Varga, and Magee 1974).

Table 2.23. Enrichment factors of chemical elements in coal[a]

Region	Element	Enrichment factor	Mean value in coal (ppm)	
			Author	Different source[b]
Illinois Basin	Beryllium	11.38	113.79	10.0
(81 samples)	Cadmium	14.4	2.89	0.2
	Fluorine	0.09	59.30	625.0
	Manganese	0.06	53.16	950.0
	Plutonium	0.06	62.77	1050.0
	Selenium	39.80	1.99	0.05
Eastern United States	Fluorine	0.10	62.5	625.0
(9 samples)	Manganese	0.03	28.5	950.0
	Plutonium	0.09	94.5	1050.0
	Selenium	67.33	3.37	0.05
Western United States	Chromium	0.09	9.0	100.0
(8 samples)	Manganese	0.04	38.0	950.0
	Selenium	31.31	1.57	0.05

[a]Only those enriched or depleted by one order of magnitude or more are listed.
[b]Clarke and Washington 1974 (as cited in Ruch; Gluskoter, and Shimp 1974).

Source: Ruch, Gluskoter, and Shimp 1974, Table 9, p. 32.

2.5.3.4 Sulfur

Because of its widespread environmental significance, sulfur is probably the most widely publicized and one of the most studied elements associated with coal. It may occur in three forms: in organic combination with the coal material, as the mineral pyrite or marcasite, or as sulfate (van Krevelen 1961). At sulfur contents above 0.6%, pyritic and organic forms are often roughly equal, whereas below 0.6% about 70 to 100% is in the organic form (Given 1973a).

Although sulfur is usually encountered in the three forms mentioned above, elemental sulfur concentrations as high as 15% have been found by some workers (Ode 1963).

The Interior province coals and most Appalachian coals were formed under the influence of high sulfate ion concentrations in saline water. As a result, much pyritic sulfur is included in these coals. Coals of other provinces formed under freshwater conditions contain lower sulfur concentrations. Figures 2.5 through 2.7 show the sulfur content of bituminous, subbituminous, and lignite coals from each of the U.S. provinces. Large deposits of low-sulfur coals are located in the West, whereas ample reserves of medium- to high-sulfur bituminous coals are located in Illinois, Indiana, Missouri, Kansas, and West Kentucky (Given 1973a).

Concentrations of the various forms of sulfur and the total sulfur content in 101 coals are given in Table 2.25. For most samples, sulfate makes up only a small percentage of the total sulfur; of the remaining sulfur, pyrite is more prevalent than the organic form, although in about 25% of the samples organic sulfur is equal to or greater than the inorganic pyrite.

Table 2.24. Geographical distribution of mercury, 1971 - 72 (ppm in coal)

Region and state	Analysis by				Total number of samples
	Neutron activation[a]	Neutron activation + atomic absorption[b]	Flameless - atomic absorption[c]	Average[d]	
Appalachian					
Pennsylvania	0.16, 0.28	0.15		0.20(2.0)	3
Ohio	0.10, 0.13, 0.15	0.14, 0.28, 0.49		0.21	6
West Virginia		0.07, 0.18		0.12(6.6)	2
East Kentucky				(0.25)	
Eastern Interior					
Illinois	0.04, 0.49, 0.60, 1.15			0.18(0.19)	53
Indiana		0.08		0.08(0.31)	
Western Interior					
Missouri		0.19		0.19	
Northern Great Plains					
Montana	0.06	0.07, 0.09		0.07(33.0)	
Western United States					
Utah	0.04		0.03 - 0.08	0.05	15
Colorado	0.02, 0.02	0.05	0.03 - 0.06	0.04(0.22)	3
Wyoming			0.03 - 0.06	0.05(18.6)	6
Arizona	0.02	0.06	0.04 - 0.08	0.05	6
Nevada			0.04 - 0.05	0.05	7
New Mexico			0.05 - 0.29	0.15	37

[a]Illinois State Geological Survey, bulletin EGN-43, 1971 (as cited in Hall 1974).
[b]National Bureau of Standards, 1972 (as cited in Hall 1974).
[c]Southwest Energy Study, App. J, draft, January 1972 (as cited in Hall 1974).
[d]Values from Joensuu (1971) (as cited in Hall 1974) shown in parentheses for comparison, including lithotypes; extremes show no relationship to more representative average samples.

Source: Hall 1974, Table 3, p. 46.

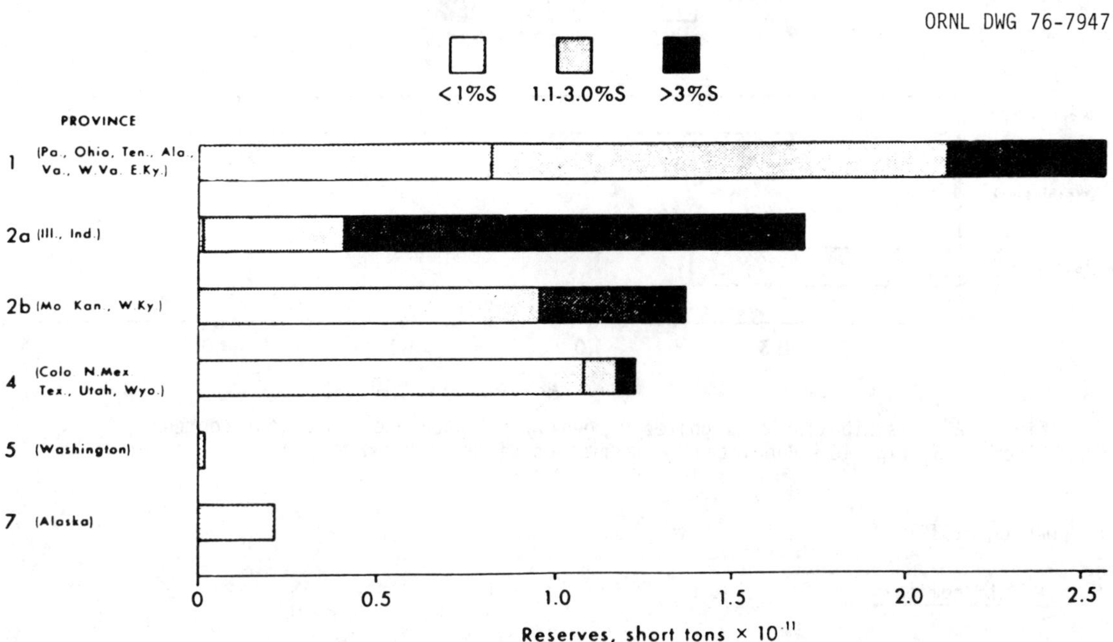

Fig. 2.5. Distribution of U.S. bituminous coals by geological province and sulfur content. Source: From Given 1973a, Fig. 8. Reprinted by permission of the publisher.

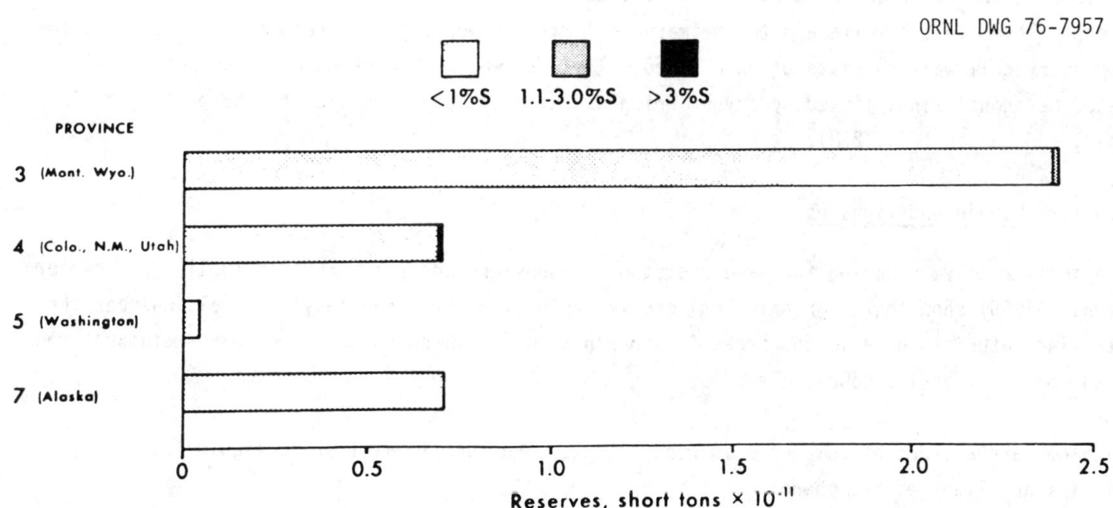

Fig. 2.6. Distribution of U.S. subbituminous coals by geological province and sulfur content. Source: From Given 1973, Fig. 9. Reprinted by permission of the publisher.

ORNL DWG 76-7946

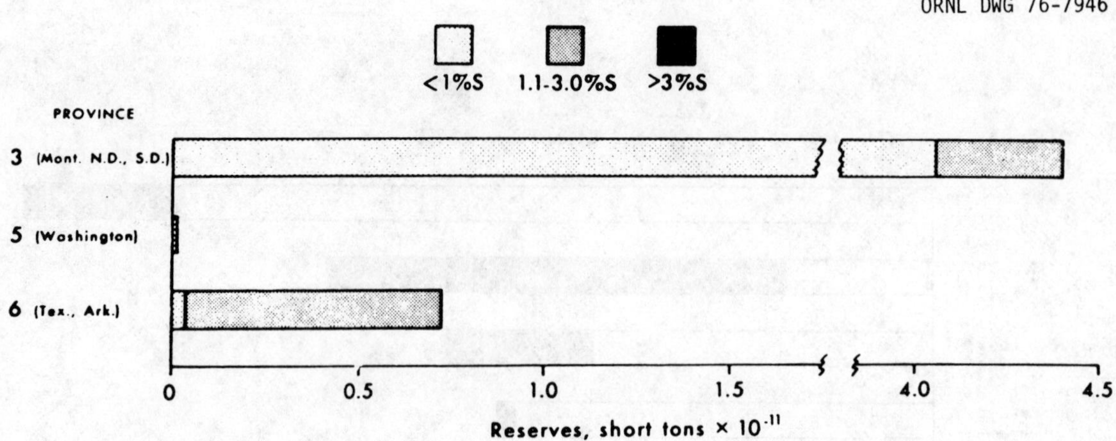

Fig. 2.7. Distribution of lignites by geological province and sulfur content. <u>Source:</u> From Given 1973, Fig. 10. Reprinted by permission of the publisher.

2.6 COAL CHEMISTRY

2.6.1 <u>Plant chemicals</u>

To understand the chemistry of peat and coal, it is helpful first to consider the chemicals found in plants. Van Krevelen (1961) describes the major classes of plant chemicals — carbohydrates, lignins, proteins, waxes, and resins.

2.6.1.1 <u>Carbohydrates and allied components</u>

Cellulose, the major constituent of the cell wall, is the most abundant member of the carbohydrate group. The middle lamella and the primary cell wall in woody plants become lignified, but the secondary cell wall consists of unmodified cellulose, which is made up of a long chain of saturated hexagonal rings linked by oxygen bridges. Related compounds are starch, pectin, alginic acid, and chitin (Fig. 2.8).

2.6.1.2 <u>Lignin and lignanes</u>

Lignin is a polymer having the same skeleton as phenylpropane (van Krevelen 1961). Freudenberg et al. (1955) show that lignin in conifers is synthesized from coniferyl alcohol, whereas the starting material in deciduous trees is sinapin alcohol, which contains one more methoxyl group than does coniferyl alcohol.

Lignanes are a group of compounds in woody plants structurally similar to lignin (Fig. 2.9). Lignins and lignanes are phenolic.

2.6.1.3 <u>Proteins and other nitrogen compounds</u>

Proteins are polymers of amino acids joined by peptide (amide) bonds. Most amino acids are related to aliphatic compounds; a few contain aromatic or heterocyclic rings. Plant proteins are among the compounds least resistant to decomposition processes, but amino acids are involved to some extent in coalification.

Table 2.25. Sulfur analyses (%, moisture-free coal)

Sample number	Organic	Pyritic	Sulfate	Total	By x-ray fluorescence
C-13854	0.66	1.27	0.0	1.93	2.18
C-17089	0.0	0.0	0.0	0.56	0.88
C-16787	0.54	1.10	0.03	1.66	1.60
C-15678	2.10	3.21	0.05	5.36	4.83
C-16919	0.37	0.76	0.10	1.23	1.05
C-16408	1.07	3.78	0.05	4.90	3.24
C-15943	1.26	3.02	0.05	4.33	2.99
C-17601	1.28	2.35	0.18	3.81	3.18
C-15944	0.91	2.27	0.02	3.20	3.48
C-13039	1.60	2.41	0.01	4.02	3.96
C-17304	2.09	1.52	0.51	4.13	3.83
C-14646	2.07	2.72	0.04	4.83	4.34
C-14650	1.32	3.38	0.11	4.81	4.27
C-15263	0.85	2.27	0.04	3.16	2.58
C-15566	1.42	3.38	0.05	4.85	4.70
C-15331	1.75	3.78	0.06	5.59	4.23
C-15496	2.34	1.28	0.05	3.67	3.94
C-16564	2.10	1.67	0.03	3.80	4.48
C-12495	1.62	2.67	0.06	4.35	4.47
C-13046	1.66	3.02	0.01	4.69	4.23
C-13983	1.78	2.47	0.02	4.27	4.27
C-14194	1.29	2.26	0.08	3.63	3.18
C-14609	1.07	1.81	0.04	2.92	2.35
C-14735	1.65	2.34	0.02	4.01	3.43
C-14774	2.24	1.42	0.02	3.68	3.74
C-14796	0.58	0.74	0.02	1.34	1.44
C-15012	1.11	2.04	0.02	3.17	2.59
C-15208	2.03	1.96	0.07	4.06	3.66
C-15384	1.63	2.13	0.14	3.90	3.63
C-15448	2.26	2.56	0.12	4.94	4.43
C-16264	2.14	2.33	0.05	4.52	4.47
C-16729	0.83	2.30	0.04	3.17	2.48
C-17001	1.51	2.62	0.02	4.14	3.49
C-17305	1.44	2.00	1.06	4.50	3.61
C-17721	0.88	1.37	0.01	2.26	2.01
C-17984	0.57	1.50	0.01	2.09	1.92
C-17988	0.80	1.35	0.01	2.17	1.95
C-18040	2.11	1.74	0.01	3.86	2.85
C-12059				3.14	3.55
C-12831	1.28	1.29	0.03	2.60	2.70
C-12942	1.38	2.42	0.12	3.92	3.87
C-13324	1.59	2.69	0.02	4.30	3.37
C-13433	0.85	2.17	0.01	3.03	1.87
C-13464	1.75	1.63	0.67	4.05	4.08
C-13895	2.12	2.43	0.02	4.57	4.55
C-13975	0.71	1.82	0.01	2.54	2.34
C-14574	1.20	0.94	0.04	2.18	2.18
C-14613	0.77	0.65	0.01	1.43	1.35
C-14630	0.63	0.57	0.02	1.22	1.23
C-14654	1.33	1.44	0.02	2.79	2.46
C-14721	1.94	1.76	0.03	3.73	3.24
C-14538	2.56	1.66	0.03	4.25	3.96
C-14970	2.19	2.02	0.04	4.25	4.13
C-14982	2.12	1.57	0.01	3.70	3.37
C-15038	0.53	0.99	0.01	1.53	1.32

Table 2.25 (continued)

Sample number	Sulfur				By x-ray fluorescence
	Organic	Pyritic	Sulfate	Total	
C-15079	1.78	2.13	0.07	3.98	3.15
C-15117	1.86	2.26	0.08	4.20	3.74
C-15125	2.02	1.42	0.01	3.45	3.40
C-15231	2.59	1.69	0.03	4.31	4.12
C-15432	0.71	0.98	0.05	1.74	1.34
C-15436	1.89	1.37	0.07	3.33	4.63
C-15456	2.03	2.36	0.06	4.45	3.26
C-15717	2.56	1.59	0.04	4.19	4.11
C-15791	0.72	1.14	0.02	1.88	1.49
C-15868	0.56	0.29	0.0	0.85	0.88
C-15872	1.82	1.81	0.05	3.68	3.29
C-15999	1.44	1.79	0.08	3.31	3.04
C-16030	1.60	1.87	0.04	3.51	3.20
C-16139	2.46	2.27	0.11	4.84	4.01
C-16265	1.94	1.22	0.04	3.20	3.53
C-16317	1.95	0.97	0.33	3.25	3.22
C-16501	1.15	1.21	0.01	2.37	2.79
C-16543	1.91	1.20	0.04	3.15	3.32
C-16741	1.96	1.54	0.05	3.55	3.63
C-16993	1.50	1.75	0.90	4.15	3.07
C-17016	2.59	2.87	0.10	5.56	5.40
C-17278	3.20	1.47	0.18	4.85	5.03
C-18044	1.85	1.85	0.01	3.71	3.29
C-15278	1.65	1.60	0.10	3.35	3.27
C-15418	0.42	0.54	0.02	0.98	0.79
C-17053	1.78	1.97	0.02	3.77	4.00
C-17215	1.41	2.67	0.10	4.18	2.88
C-17045	0.31	0.10	0.02	0.44	0.54
C-17046	0.56	0.53	0.02	1.11	0.80
C-17047	0.70	1.16	0.02	1.88	0.98
C-17054	0.49	0.31	0.02	0.83	0.87
C-17096	0.40	0.24	0.04	0.69	0.68
C-17097	0.46	0.07	0.02	0.55	0.57
C-17307	1.84	3.65	0.98	6.47	4.02
C-17309	0.34	0.06	0.01	0.42	0.54
C-17092	1.41	2.37	0.16	3.94	3.96
C-17095	1.42	2.59	0.19	4.20	2.94
C-17098	0.46	1.01	0.05	1.53	0.92
C-17099	1.01	1.82	0.21	3.04	2.33
C-17243	1.32	2.41	0.42	4.15	3.21
C-17244	0.69	1.85	0.12	2.66	1.02
C-17205	1.05	2.39	0.24	3.68	2.65
C-17246	0.74	0.19	0.02	0.96	1.23
C-17303	0.75	0.48	0.06	1.29	1.46
C-17970	0.0	0.0	0.0	0.0	1.25
C-16009	0.72	0.27	0.08	1.07	1.37

Source: Ruch, Gluskoter, and Shimp 1974, Table 4, p. 15.

ORNL DWG 76-7941

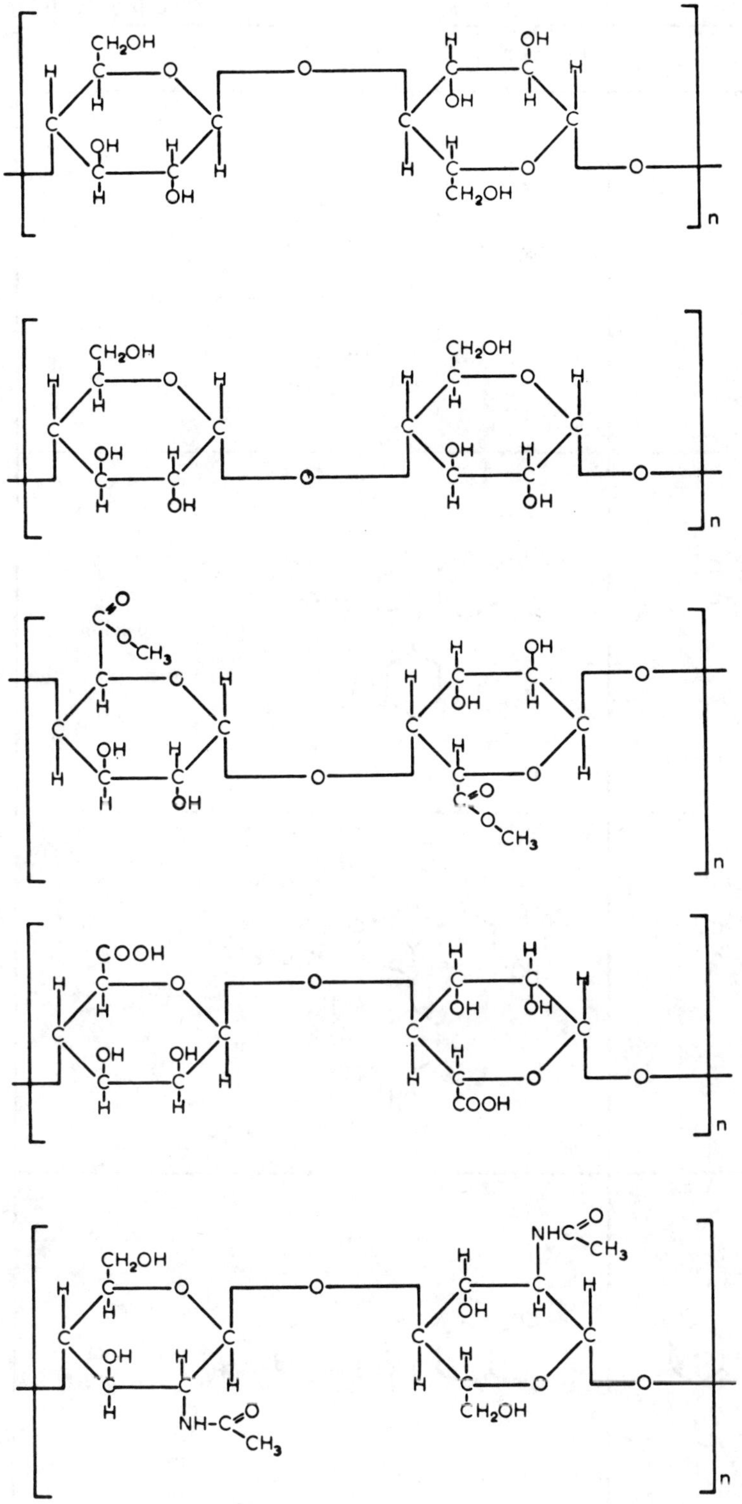

Fig. 2.8. Polymeric carbohydrates and related compounds. Source: From van Krevelen 1961, Diagram V.1, p. 90. Reprinted by permission of the publisher.

ORNL DWG 76-7942

Skeleton	Example	
I		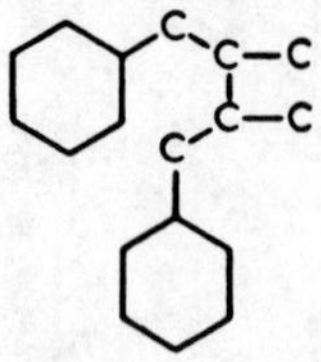 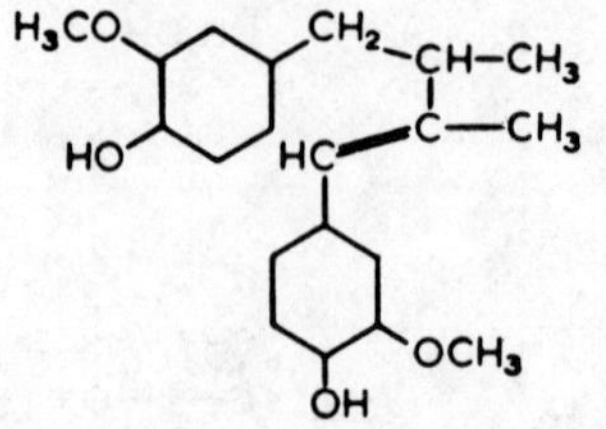 guaiaretic acid
II		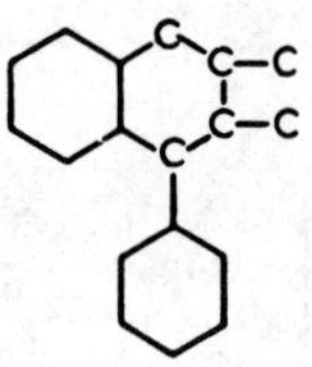 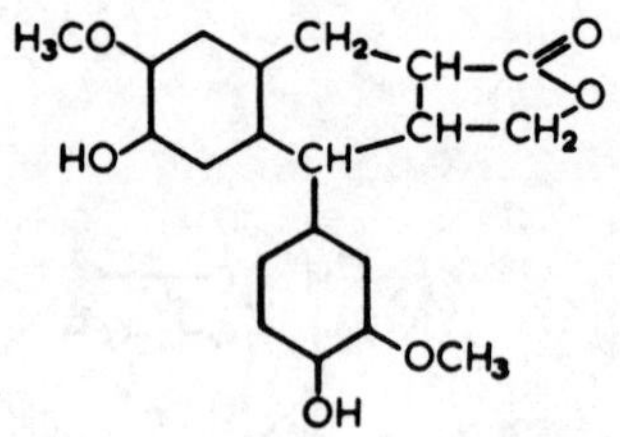 conidendrin
III		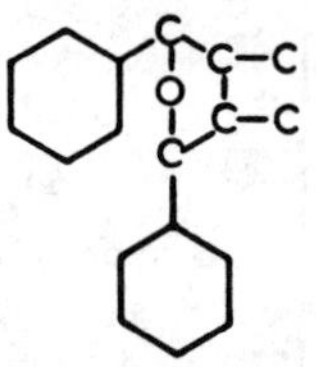 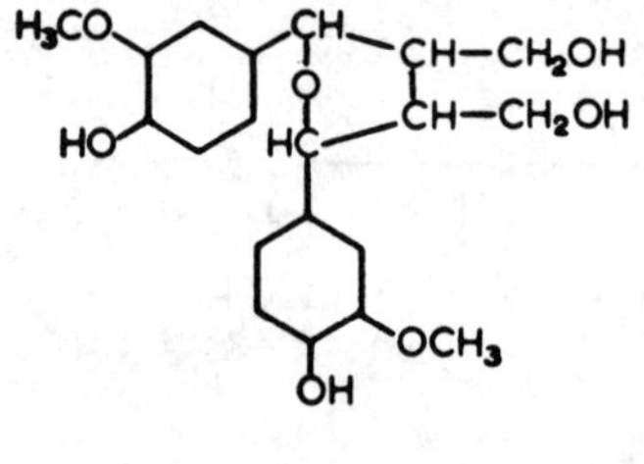 olivil
IV		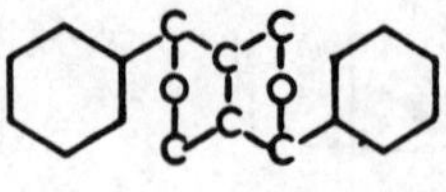 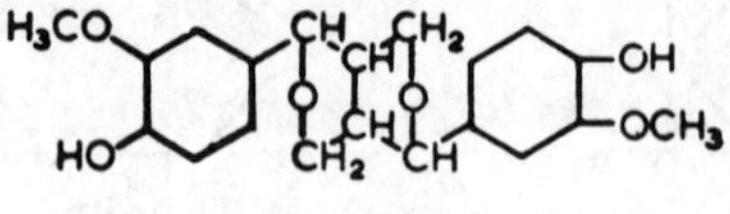 pinoresinol

Fig. 2.9. Summary of the principal lignanes. <u>Source</u>: From van Krevelen 1961, Diagram V.3, p. 94. Reprinted by permission of the publisher.

2.6.1.4 Fats and waxes

Fats and waxes are esters derived from fatty acids and glycerine and from fatty acids and wax
alcohols respectively (Table 2.26). Fats and sterols are especially involved in the formation
of boghead coal. The products of wax alcohols and waxy acids are found in cuticles, spore
exines, and cork along with lignin and tannins.

Table 2.26. Nomenclature of wax ingredients

Compound	Name	Number of carbon atoms per molecule	Empirical formula
Monohydroxy alcohols (saturated)	Ceryl	26	$C_{26}H_{53}OH$
	Octacosanyl	28	$C_{28}H_{57}OH$
	Montanyl	29	$C_{29}H_{59}OH$
	Myricyl	30	$C_{30}H_{61}OH$
	Melissyl	31	$C_{31}H_{63}OH$
	Lacceryl	32	$C_{32}H_{65}OH$
Unsaturated alcohols (branched)	Phytol	20	$C_{20}H_{39}OH$
Ketones	Montanone	30	$C_{29}H_{59}CO$
Monobasic acids	Lignoceric	24	$C_{23}H_{47}COOH$
	Cerotic	26	$C_{25}H_{51}COOH$
	Octacosanic	28	$C_{27}H_{55}COOH$
	Montanic	29	$C_{28}H_{57}COOH$
	Myricinic	30	$C_{29}H_{59}COOH$
	Melissic	31	$C_{30}H_{61}COOH$
	Lacceric	32	$C_{31}H_{63}COOH$
Hydroxy acids and more complicated acids	Phloionolic	18	$C_{17}H_{32}(OH)_3COOH$
	Phloionic	18	$C_{16}H_{30}(OH)_2(COOH)_2$
	Phellonic	22	$C_{21}H_{42}(OH)COOH$
	Phellogenic	22	$C_{20}H_{40}(COOH)_2$
	Cutic	26	$C_{26}H_{50}O_6$
	Cutinic	26	$C_{26}H_{44}O_6$
	Hydroxy montanic	29	$C_{28}H_{56}(OH)COOH$

Source: van Krevelen 1961, Table V.1, p. 98. Reprinted by permission of the publisher.

2.6.1.5 Resins

According to Shell (1976), typical resin compounds are diterpene resin acids such as alicetic
and pimaric acids which are "isoprene" derivatives structurally related to steroids. A variety
of other terpenes (both acyclic and cyclic) derived from two or more isoprene units are frequently
associated with resinous plant exudates. Many of these materials are susceptible to a variety of
acid-catalyzed transformations. Rearranged as well as polymerized products would be expected to

form during coalification. Rubber is a natural polymeric isoprene derivative (Shell 1976).
Polymeric components of plant material are classified in Table 2.27.

Table 2.27. Classification of polymeric plant constituents

Polymer	Unit	Dimer	Related to
Cellulose and other polysaccharides	Hexose	Disaccharides (cellobiose, etc.)	Plant gums, reserve carbohydrates
Lignins	Phenylpropane unit	Lignanes	Anthocyans, catechins
Proteins	Amino acids	Dipeptides	Alkaloids, nucleic acids
Cutin, exine	Waxy alcohols and waxy acids	Esters of low-molecular weight	Cork
Polyterpenes (rubber)	Isoprene	Terpenes	Carotinoids, sterols

Source: van Krevelen 1961, Table V.2, p. 99. Reprinted by permission of the publisher.

2.6.1.6 <u>Role of plant chemicals in coal formation</u>

Cohen (1968) reports that tissue fragments of the mangrove, *Rhizophora mangle*, found in peat
still contained the cell fillings and that such cells constituted a large part of the organic
matter observed under the microscope. Spackman and Barghoorn (1966) report that cell walls of
woody tissue found in coal are highly birefringent, indicating that the cell structure is well
preserved. Cellulose in cell walls is somewhat crystalline and must be responsible for the
birefringence (Given 1971).

In the past there has been some controversy concerning whether cellulose or lignin plays the
major role in coal formation. According to van Krevelen (1961), such controversy is pointless;
the reaction mechanism of coalification is exceedingly complicated, and it is not unexpected
that the product of the process is a complicated and chemically heterogeneous substance. Lignin
has, however, been observed by Huck and Karweil (1953) to be more resistant to decay than is
cellulose.

Phenolic substances may be derived from lignin by cleavage or oxidation reactions. They consist
of aromatic compounds with side chains of one to three carbon atoms (Flaig 1966). These substances are highly resistant to decay and are believed to make up a large portion of the humic
materials formed from rotting plant material (Flaig 1966) as in peat.

Figure 2.10 shows some of the lignin decomposition products. Guaiacylglycerol-β-coniferylether
is a dimeric product of coniferous lignin, and dehydrovanillin may be formed by dimerization of
vanillin, a product of lignin. Various mono-, di-, and triphenols may be formed by decomposition
of lignins found in conifers, deciduous trees, and grasses (Flaig 1966).

ORNL DWG 76-7944

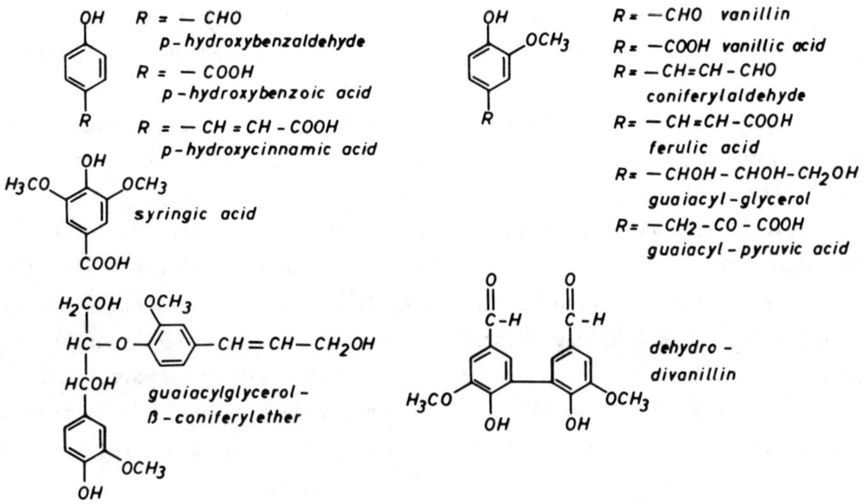

Fig. 2.10. Structure of identified lignin decomposition products. Source: From Fla[...] 1966, Fig. 1, p. 60. Reprinted by permission of the publisher.

2.6.2 Coal structure

Coal is a complex mixture of various macromolecules; thus, to establish a reasonable chemical characterization of coal, certain essentials must be known (Larsen 1975):

1. *The molecular weight distribution of the macromolecules.* This parameter should correlate with reactivity; however, little work has been done in this field.

2. *The nature of the constituent molecules.* Data to date indicate that the constituent molecules are polycyclic aromatic units of various sizes containing a variety of functional groups; however, the manner of linkage and the number found in an average molecule are not known with certainty.

3. *The nature of the cross links.* Much evidence indicating the presence of short methylene bridges between aromatic units has been collected. Gaps in the knowledge of cross links and their relative reactivities make designing coal conversion systems very difficult. Many other variables such as rank, ash content, and reaction conditions affect the reactivity of coal and must also be considered in designing such systems.

4. *The functional groups present.* Some current research, involving direct determination of aromatic and aliphatic carbon concentrations in coal by solid-phase ^{13}C nuclear magnetic resonance spectroscopy, supports data previously obtained by indirect techniques concerning the groups; however, some more recent work suggests that functional groups in coal are not randomly distributed, as is assumed in all current attempts at forming an idea of coal structure. Also, in regard to sulfur, much is yet unknown about the different forms of organic sulfur in coal and the manner in which these forms are distributed. Additionally, almost nothing is known about the organometallic compounds in coal (Larsen 1975).

The true structure of coal is presently unknown; most workers agree that the structure is so complicated and variable that it is impossible at this time to form an accurate model of it. Despite, however, the present inadequate knowledge of coal constituents and their conformation, many significant attempts have been made to gain insight into the structure of coal, and through these attempts the knowledge of its general character has been greatly improved.

A variety of methods have been used in the study of coal structure. Van Krevelan and Chermin (1954) presented a formula for statistical structural analysis of coal based on its elementary composition and density. Schuyer and van Krevelen (1954) used molar refraction values calculated from the density, refractive index, and molecular weight of organic compounds to determine the number of carbon atoms per average structural unit in coals. Montgomery and Holly (1957), in their oxidation studies of bituminous coals, concluded that such coals are comprised largely of eight aromatic nuclei (methylnaphthalene, benzene, biphenyl, naphthalene, phenanthrene, C_5 benzene, benzophenone, and toluene) which are probably linked together by more readily oxidizable structures. Heréday, Kostyo, and Neuworth (1965), by depolymerizing coals of different rank with phenol-boron trifluoride, studied the bridges that link together the structural units of coal.

Cartz and Hirsch (1960), in their x-ray diffraction studies of coal structure, found that variations in the molecular layers make the construction of an unambiguous coal model impossible. They conclude that the structural units of bituminous coals consist of aromatic ring systems probably containing one to three rings in low-rank vitrains and two to five rings in vitrains containing 90% carbon; the aromatic layers are believed to be linked by direct C-C linkages, by aliphatic groups, and by ether linkages. The layers are thought to form large sheets, which are either buckled by hydroaromatic groups, or in which adjacent aromatic ring systems are rotated relative to each other about an axis normal to the plane of the sheets. The sheets are also believed to contain side groups consisting of aliphatic chains and hydroxyl and quinone groups. In vitrains containing more than 90% carbon, the condensed aromatic layers appear to increase rapidly in size with increasing rank (Cartz and Hirsch 1960). Cartz and Hirsch feel that the results of their x-ray diffraction studies are reasonably reconcilable with the results obtained from some of the other investigations of coal structure. They state, however, that each of the methods used is subject to errors and uncertainties which leave the true structure still in doubt. They further suggest that the problem of coal structure can only be solved by synthesizing molecules which have all the properties of coal.

The diagram shown in Fig. 2.11 is an attempt by Wiser (1973) to use the present knowledge of coal to represent an average bituminous coal in a single plane, showing the aromatic and hydro-aromatic components. He suggests that perhaps coal should be thought of as being composed of stacked layers rather than as being three-dimensional. The layers may then consist of several aromatic clusters, each perhaps randomly oriented in a plane different from the surrounding clusters. Wiser's diagram does not rule out the possibility of three-dimensional molecules in coal. It is a theoretical model, but it is representative of the current attempts to improve the understanding of coal structure.

2.6.3 Functional groups

To understand the organic chemistry of coal, it is helpful to become familiar with the "functional groups." Such groups are often considered as reactive oxygen, nitrogen, and sulfur groups; however,

ORNL DWG 76-7955

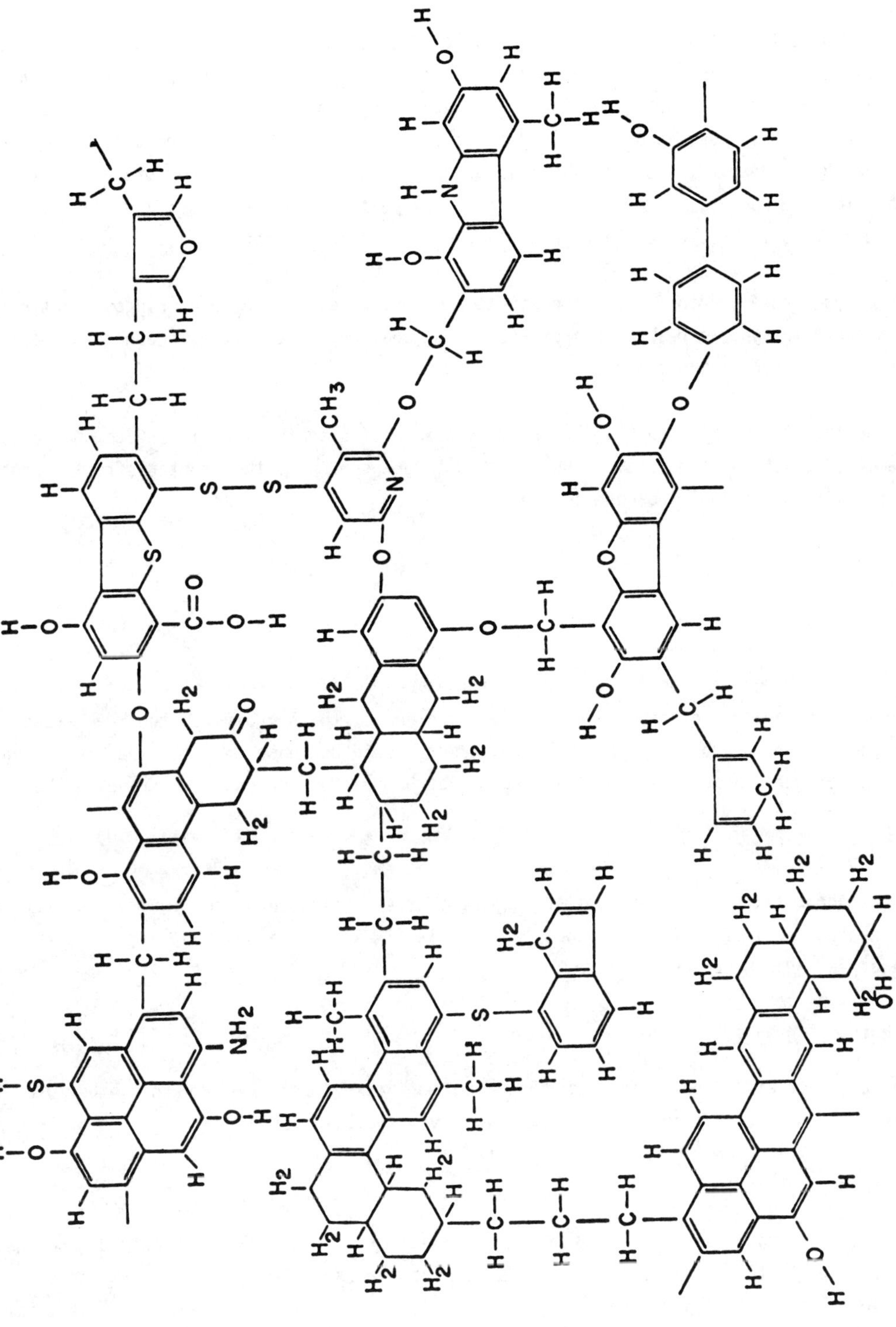

Fig. 2.11. Schematic representation of structural groups and connecting bridges in bituminous coal.
Source: From Wiser 1973, Fig. 5. Reprinted by permission of the publisher.

for the purpose of this paper, the definition is extended, as in Tingey and Morrey (1973), to include all organic structural elements.

2.6.3.1 Bridge groups

Given (1960) suggests that aromatic and hydroaromatic compounds, when joined together, form the basic coal structure. This structure cannot be accurately described at the present time; however, it has been observed that the rings are joined together in a variety of patterns — single rings, condensed double rings, condensed triple rings, etc. Clusters thus formed are believed to be held together by bridges: (1) short aliphatic groups, principally methylene, probably not many longer than four carbon atoms; (2) ether linkages; (3) sulfide and disulfide; and (4) biphenyl types. The short aliphatic groups are probably the most common (Wiser 1973).

Hereday, Kostyo, and Neuworth (1965) demonstrated the presence of methylene bridges by high-resolution nuclear magnetic resonance spectroscopy of coal depolymerization products of coals ranging from 70 to 90% carbon content.

According to Mazumdar et al. (1966), the methyl contents of bituminous coals (80 to 90% carbon) range from 4.0 to 4.6% of the total carbon. At 80 to 86% carbon, methyl contents remain constant, but decrease with additional increase in rank.

2.6.3.2 Hydroxyl groups

On the basis of available experimental data, hydroxyl groups in coals are mainly phenolic or acidic; there is little evidence to indicate the presence of alcoholic or weakly acidic hydroxyl groups. Hydroxyl oxygen may be present in concentrations of 9% in brown coals, but is usually about 8% with 65% carbon and 27% oxygen. As rank increases, the hydroxyl content decreases slightly until, at 80% carbon and 12.5% oxygen, a more rapid decrease is initiated, resulting in a hydroxyl oxygen concentration of less than 1% at 90% carbon content (van Krevelen 1961).

2.6.3.3 Carboxyl groups

Research has shown that carboxyl groups are absent in most coals above the lignite stage of development, but are present in brown coals and lignites. At 83% carbon content, all carboxyl groups disappear (van Krevelen 1961).

2.6.3.4 Methoxyl groups

Methoxyl groups are among the first lost during metamorphism of coal; as a result, they usually do not appear in significant amounts in true coals (van Krevelen 1961).

2.6.3.5 Carbonyl groups

Carbonyl groups are found at all levels of coalification; however, analytical methods and their results have been quite variable. Blom, Edelhausen, and van Krevelen (1957) report carbonyl oxygen contents of 0.2 to 1.9% of total oxygen at 85% carbon content and higher. These data were obtained by oximation (the formation of oximes by the reaction of carbonyl with hydroxyl-amine) and consecutive hydrolysis, and were expressed as minimum values. Given and Peover (1960), using reductive acetylation, obtained data for solvent extracts indicating that the carbonyl contents (quinone carbonyl) were higher, comprising 4.4 to 9.0% of the total oxygen at an 85% carbon content and higher.

2.6.3.6 Unreactive oxygen

At this time, the amount of oxygen present in "unreactive groups" in coal is uncertain. Halleux, Delavarenne, and Tschamler (1961) found about 2.5 wt % of unreactive oxygen in hard coals of all ranks.

2.6.3.7 Oxygen-containing groups

Some data on oxygen-containing groups by Blom (1960), as cited by van Krevelen (1961), are summarized in Table 2.28 and are shown in graphical form in Fig. 2.12. Van Krevelen concludes from this information that, during the metamorphic process, in coals containing up to 70 to 80% carbon, the methoxyl groups are the first lost; then the carboxyl and the carbonyl groups decrease rapidly. At 81 to 89% carbon content, the hydroxyl groups decrease rapidly. At greater than 92% carbon content, almost all oxygen is in nonreactive, stable forms. Data on hydroxyl and carbonyl oxygen concentrations of various coals (Tables 2.29 and 2.30) compiled by Dryden (1963) support van Krevelen's conclusion. Dryden further suggests that hydroxyl and carbonyl groups may account for 70 to 90% of the oxygen in bituminous coals. Given and Peover (1960) found that the two groups may comprise as much as 95% of the oxygen in such coals (Table 2.31). The structures of some oxygen-containing compounds that may occur in coals are shown in Fig. 2.13.

Table 2.28. Functional group analysis of coal (wt %)

Coal	Carbon	Hydrogen	Nitrogen	Sulfur	Oxygen					
					Total	COOH	OCH_3	OH	$C{=}O^a$	ΣO
Peat	60.0	4.9	0.9	0.05	33.9	5.1	0.9	12.3	11.6	29.9
Brown coal	65.5	5.1	0.9	0.2	27.8	8.0	1.1	7.2	6.5	22.7
Lignites	69.6	4.6	0.3	3.1	21.9	3.9		12.5	7.2	23.6
	71.7	4.9	0.8	0.4	22.6	5.1	0.4	7.8	9.3	22.6
	75.9	5.3	1.6	1.1	16.2	1.6	0.3	7.5	7.4	16.8
Hard coals	79.5	5.5	3.2	0.8	11.1	0.3	0	6.1	4.9	11.3
	80.2	4.9	1.2	0.6	13.4	1.0	0	8.3	3.1	12.4
	85.5	5.3	0.6	0.5	7.9	0.0	0	5.6	1.6	7.2
	87.0	5.3	1.9	0.7	5.1	0	0	3.1	1.7	4.8
	88.6	5.0	2.3	0.8	3.0	0	0	1.9	1.8	3.7
	90.3	4.7	1.7	1.0	2.9	0	0	0.5	2.0	2.5
	90.9	4.1	1.7	0.9	2.0	0	0	0.6	1.9	2.5

[a] Half of this $O_{C{=}O}$ content may be considered as certain; a possibility remains that the other half (or part of it) in reality consists of nonacetylatable hydroxyl or ether groups.

Source: van Krevelen 1961, Table IX,2, p. 172. Reprinted by permission of the publisher.

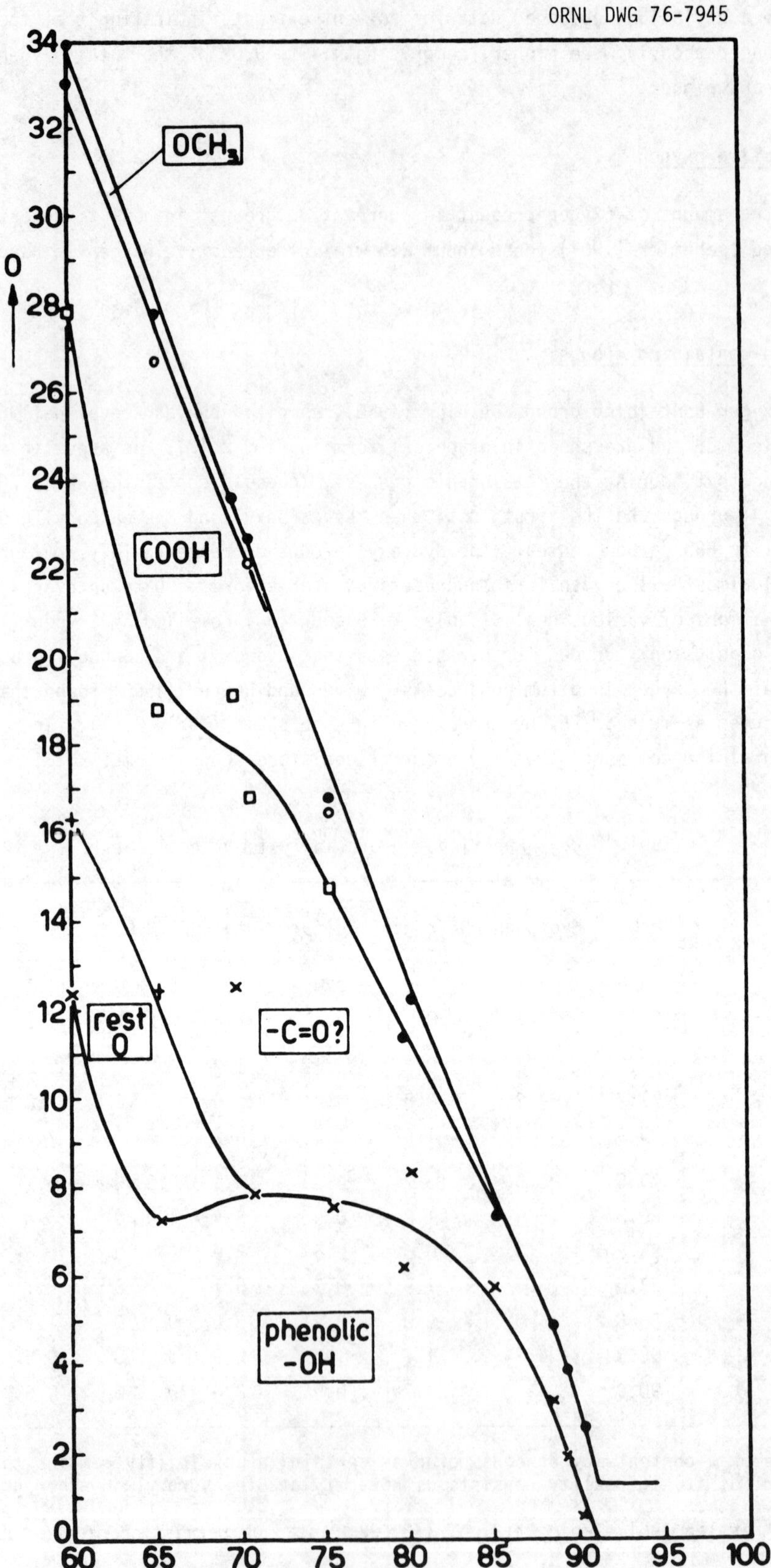

Fig. 2.12. Functional oxygen groups in coals. <u>Source</u>: From van Krevelen 1961, Fig. IX.4, p. 173. Reprinted by permission of the publisher.

Table 2.29. Determinations of hydroxyl groups in coal

Method	Medium	Temperature, time	Coal or extract	Range of rank (% carbon)	$Oxygen_{OH}$ (% of coal or extract)	$Oxygen_{OH}$/oxygen (%)
Sorption and titration	Water	Room, 4 days	Coal residues after extraction			
Sorption and titration	Water	Room, days	Coals	65 - 93	7.1 - 0.1	53 - 4
Neutralization	40% aqueous alcohol	Room, 20 hr	Coals (including macerals)	67.5 - 85.9	2.4 - 0.0	
Neutralization	Ethanol, 85% ethanol	Reflux, room, 1 hr	Coals	70.5 - 89.6	4.0 - 0.9	
Titration	Ethylenediamine	Room, 1 - 2 hr	Coals	67.5 - 94.3	7.8 - 0.2	67 - 14
			Vitrains	71.3 - 92.1	7.3 - 0.4	64 - 17
Methylation	Water	100°C, 4 hr	Coals	72.4 - 94.9	3.1 - 0	16 - 0
Methylation	Ether	−5°C, 1 week	Vitrains	65.5 - 90.9	5.4 - 0.7	38 - 15
Acetylation	Pyridine	100°C, 5 hr	Coals	77.7 - 91.8	5.6 - 0.2	44 - 6
			Extracts	81.8 - 86.5	4.5 - 2.6	46 - 31
Acetylation	None	Room to 110°C, 35 - 85 hr	Coals	81.9 - 93.0	6.4 - 0.6	100 - 42
Acetylation	Acetic anhydride	Reflux, 3 hr	Coals	65.5 - 91.1	11.4 - 0	49 - 0
			Vitrains	79.2 - 90.0	7.2 - 0.4	55 - 13
Acetylation	Pyridine	90°C, days	Vitrains	65.5 - 90.9	7.8 - 0.5	71 - 17
Etherification	Water plus dimethylformamide	20°C, days	Vitrains	79.5 - 90.9	3.7 - 0.3	33 - 10.3
Etherification	Pyridine	Reflux, 4 hr	Coals	66.9 - 93.3	7.8 - 0.2	56 - 14
			Vitrains	68.5 - 91.2	8.8 - 0.5	52 - 13
Acetylation	Acetic anhydride	Reflux, 3 hr	Coals	70.5 - 89.6	9.5 - 0.9	

Source: Dryden 1963, Table 4, p. 264. Reprinted by permission of the publisher.

Table 2.30. Determinations of carbonyl groups in coals

Method	Medium	Temperature, time	Coal or extract	Range of rank, (% carbon in coal)	Oxygen$_{CO}$ (% of coal or extract)	Oxygen$_{CO}$/oxygen (%)
Controlled potential electrolysis	Dimethylformamide	Room, >1 hr	Extract	81.7 - 89.2	(2.1-3.3) - 2.0	(20-32) - 54
Controlled potential electrolysis plus acetyl determination	Dimethylformamide	Room, >1 hr	Acetylated extract	78 - 89	(2-2.5) - 0.9	(14-18) - 20[a]
Formation of hydrazone	Water	90°C, 0.5 hr	Coal	65 - 93	1.25 - 0.4	9 - 20
Formation of hydrazone	Water	100°C, 0.5 hr	Coal	Brown, long flame	0.4 - 0.8	
Formation of oxime	Pyridine	90°C, 20 hr	Vitrain	65.5 - 91	0.5 - 0.2	5 - 10

[a]Based on oxygen in extracts.

Source: Dryden 1963, Table 5, p. 266. Reprinted by permission of the publisher.

Table 2.31. Distribution of oxygen in solvent extracts of coals
(carbonyl being assumed as chelated to unreactive hydroxyl)

Coal[a]	Carbon (% in coal)	Extracting solvent[b]	Oxygen (%) as C=O	OH (total)[c]	Oxygen (%) accounted for as C=O + OH
DO	77.7	DMF	2.25	8.75	78
Scottish vitrain		DMF-naphthol	2.0	>5.0	>61
DIII	81.1	DMF-naphthol	1.4	>5.2	>66
		Pyridine	2.1	6.3	93
			2.0	6.7	95
		DMF	2.0	6.9	93
		Butanone	2.05	>5.65	>84
DVI	83.9	DMF	1.6	>4.1	>77
DIX	86.0	DMF-naphthol	1.1	>2.9	>53
		DMF-pyridine	1.4	4.0	90
		DMF	1.65	4.45	94
DXIII	89.2	DMF	0.9	2.3	73
			0.9	2.1	91

[a]DO, DIII, DVI, DIX, and DXII are designations for five highly vitrainous low-ash coals (see Given and Peover 1958, as cited in Given and Peover 1960, for full details).

[b]DMF = dimethylformamide.

[c]Total of directly determined OH plus chelated OH equal to C=O content.

Source: Given and Peover 1960, Table 3, p. 399. Reprinted by permission of the publisher.

ORNL DWG 76-7960

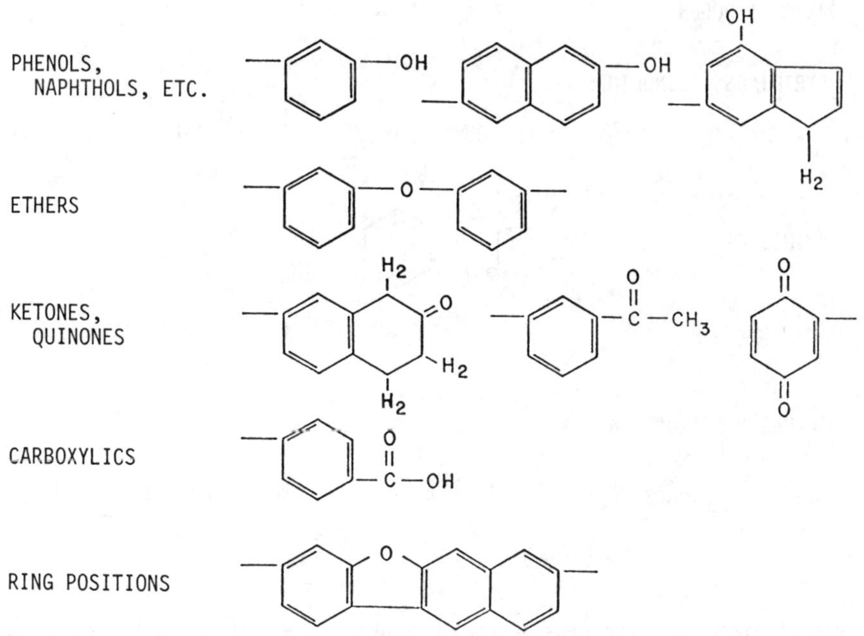

Fig. 2.13. Types of oxygen structures in coal. Source: From Wiser 1973, Fig. 3. Reprinted by permission of the publisher.

2.6.3.8 <u>Nitrogen-containing groups</u>

According to Horton and Randall (1947), the nitrogen content of most coals is generally below 2%; however, samples taken from adjacent positions may vary by as much as 20%. Coal fractions obtained from a 2-ton batch of "bright" coal by extensive float and sink separations yielded the nitrogen and sulfur values shown in Table 2.32. It may be noted that nitrogen content increases with a rise in specific gravity to a maximum of 1.88% in fraction 4, which also has the highest sulfur content — 1.47% (Horton and Randall 1947).

Table 2.32. Nitrogen and sulfur composition (%)
of fractions of different density

Fraction	Specific gravity	Approx. % of whole	Nitrogen	Sulfur
1	< 1.250	2	1.39	0.84
2	1.250 - 1.275	4	1.75	1.07
3	1.275 - 1.300	47	1.85	0.91
4	1.300 - 1.325	5	1.88	1.47
5	1.325 - 1.375	36	1.76	0.98
6	> 1.375	6	0.96	5.66

Source: Horton and Randall 1947, p. 129. Reprinted by permission of the publisher.

Francis and Wheeler (1925) observe that nitrogen in coal is almost completely in cyclic structures. Some nitrogen structures that may be present in coals are shown in Fig. 2.14.

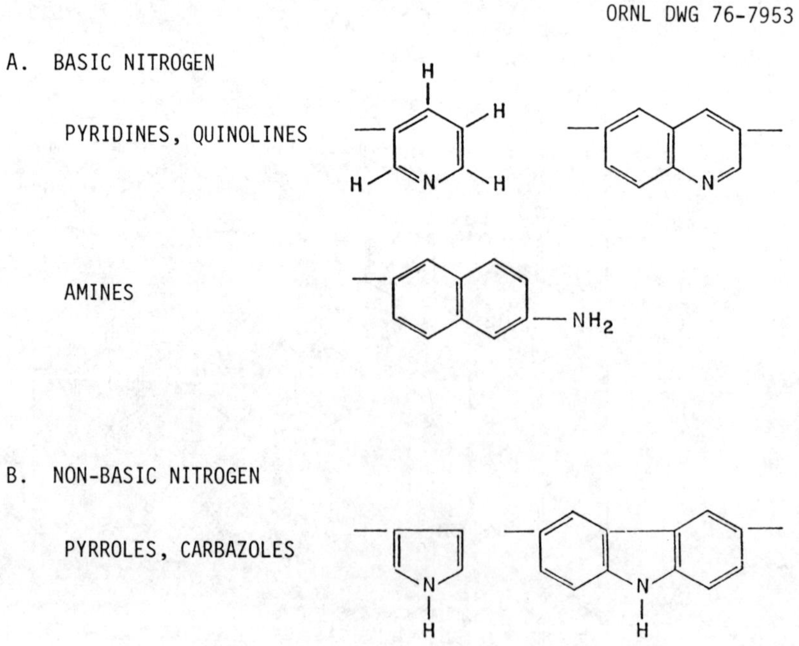

ORNL DWG 76-7953

A. BASIC NITROGEN

PYRIDINES, QUINOLINES

AMINES

B. NON-BASIC NITROGEN

PYRROLES, CARBAZOLES

Fig. 2.14. Types of nitrogen structures in coal. <u>Source</u>: From Wiser 1973, Fig. 4, p. 24. Reprinted by permission of the publisher.

Huck and Karweil (1953) estimate the nitrogen content of the vegetable matter that was converted to coal to be between 0.6 and 1.0%. Because of the coalification process, the nitrogen content of the material is believed to increase up to 45% volatile matter content, then decrease slowly with increase in rank. The researchers believe that the polar character of nitrogen is important in the formation of coal and that both nitrogen and oxygen influence the swelling and softening capability of the coal by affecting the balance of cohesive forces.

2.6.3.9 Aromatic groups

Hayatsu et al. (1975) report that their results obtained by aqueous $Na_2Cr_2O_7$ oxidation of bituminous coal support the idea that "coal is predominantly an aromatic material." Wiser (1973) suggests that, in bituminous coal, 70 to 75% of the carbon present is in the aromatic form. Six-membered rings are in the majority, but there are some five-membered ones. Fifteen to twenty-five percent of the carbon is involved with hydroaromatic structures (Wiser 1973).

The distribution of carbon in aromatic, hydroaromatic, and methyl forms in coal determined from pyrolysis by Mazumdar et al. (1966) is presented in Table 2.33. The three forms are shown to comprise 97.5 to 98.0% of the total carbon in low-rank (<86% carbon) bituminous coals and 99.4% in high-rank (90.4% carbon) bituminous coals. The addition of carboxyl groups to the amounts shown for low-rank bituminous brings the total up to the maximum of 98%. The remainder of the carbon in such coals may be accounted for by formation of CO and CO_2. The formation of CO and CO_2, other than the amounts formed by decarboxylation of carboxyl groups, appears to result from carbonyl groups, the true nature of which is presently unknown (Mazumdar et al. 1966). According to Given and Peover (1960), carbonyl may be either quinonoid or ketonic. In the first case, carbonyl should be included in aromatic carbon, but in the latter case, it should be grouped with aliphatic carbon. In any case, practically all the carbon in coal may be accounted for in these groups (Mazumdar et al. 1966).

Almost all the work up until the present has been performed under the widely accepted theory that coal is a largely aromatic structure. However, this view is not shared by everyone. Chakrabartty and Berkowitz (1974) state that the skeletal carbon arrangements of coal appear to be made up largely of nonaromatic structures. They conclude from their research on oxidation of coal with sodium hypohalite that these structures are modified bridged tricycloalkane systems or polyamantanes. Chakrabartty and Berkowitz (1974) hold that the characterization of coal as a nonaromatic structure is not inconsistent with existing experimental evidence, but that it is a matter of interpretation of the existing data.

Ghosh, Banerjee, and Mazumdar (1975) and Aczel et al. (1975) have presented rebuttals to the concept of the nonaromatic structure of coal. Also, a variety of oxidation studies such as performed by Hayatsu (1975) supports the aromatic concept of coal structure.

Thirty-five aromatic carboxylic acids identified by Hayatsu et al. (1975) from the oxidation of bituminous coal with aqueous $Na_2Cr_2O_7$ are listed in Table 2.34. Several series of aromatic derivatives are found in the chemically neutral fraction: dibenzo[*p*]dioxins, benzoxanthanes, benzanthraquinones, and acridones.

Table 2.33. Self-consistency of the maximum estimate of C-methyl
(assessed from pyrolysis) with other forms of carbon in coal

Coal number	Rank (% carbon[a])	$f_a + f_{har}$[b] (% of total carbon)	C-methyl (% of total carbon)	Carboxylic carbon (% of total carbon)	Sum of aromatic hydroaromatic, C-methyl, and carboxylic carbon (% of total carbon)	Carbonyl oxygen content vis-a-vis unaccounted carbon		
						Probable O_{CO} content[c] (% coal[a])	Oxygen appearing as CO and CO_2 during pyrolysis[d] (% coal[a])	Carbonyl carbon (% of total carbon)
2	82.4	93.0	4.6	0.3	97.9	3.1	2.6	2.8
3	83.6	93.0	4.6	0.2	97.8	3.0	2.3	2.7
4	85.5	93.1	4.6	Nil	97.7	2.9	2.3	2.5
7	90.4	96.0	3.4	Nil	99.4	1.0	1.4	1.2
8	69.7	86.0	4.0	3.3	93.3	7.2	9.7	7.7

[a]Dry, mineral-matter-free.

[b]This is derived from the retention of carbon in char consequent upon prior dehydrogenation and is considered to be the maximum estimate of the sum of the aromatic and hydroaromatic carbon. Individual determinations of aromatic and hydroaromatic carbon also lead to a similar assessment.

[c]These values of carbonyl oxygen are not determined by the present authors but are taken rankwise from Blom's determination by hydroxylamine method (Blom 1960, as cited in Mazumdar et al. 1966). These values are considered to be the maximum estimate for carbonyl oxygen and are about double the estimate obtainable by phenylhydrazine method. The latter method is suspect owing to various complications. In spite of uncertainties still remaining about the most probable estimate of carbonyl oxygen in coal, it is of interest to note that it corresponds largely to the amount of carbon monoxide and carbon dioxide formed during pyrolysis. Further, the "carbonyl" carbon is also roundly consistent with the unaccounted carbon left over after accounting for all other known forms of carbon. The coincidence is possibly more than fortuitous.

[d]This is corrected for the carbon dioxide that would form during pyrolysis owing to carboxyl group in coal.

Source: Mazumdar et al. 1966, Table VIII, pp. 490-491. Reprinted by permission of the publisher.

Table 2.34. Aromatic acids found in the aqueous $Na_2Cr_2O_7$ oxidation of coal and identified as methyl esters[a]

Nucleus	Number of -COOCH$_3$	Precise mass $(M-OCH_3)^+$ Elemental composition	Observed	Deviation X 10^3	Relative[b] abundance (±15 - 20%)
Benzene	2	$C_9H_7O_3$	163.0400	0.6	38
	3	$C_{11}H_9O_5$	221.0463	1.4	100
	4	$C_{13}H_{11}O_7$	279.0516	1.2	64
	5	$C_{15}H_{13}O_9$	337.0559	0.1	11
OCH$_3$-benzene[c]	1	$C_8H_7O_2$	135.0452	0.6	17
	2	$C_{10}H_9O_4$	193.0518	1.8	17
Benzophenone	1	$C_{14}H_9O_2$	209.0582	-1.9	8
	2	$C_{16}H_{11}O_4$	267.0677	2.0	9
Naphthalene	1	$C_{11}H_7O$	155.0488	-0.8	1
	2	$C_{13}H_9O_3$	213.0547	-0.4	9
	3	$C_{15}H_{11}O_5$	271.0610	0.6	27
	4	$C_{17}H_{13}O_7$	329.0654	-0.6	8
CH$_3$-naphthalene	1	$C_{12}H_9O$	169.0618	-3.4	1
	2	$C_{14}H_{11}O_3$	227.0711	0.3	3
	3	$C_{16}H_{13}O_5$	285.0737	2.5	5
Phenanthrene[d]	2	$C_{17}H_{11}O_3$	263.0699	-0.8	20
	3	$C_{19}H_{13}O_5$	321.0756	-0.6	17
	4	$C_{21}H_{15}O_7$	379.0813	-0.4	8
OCH$_3$-phen- anthrene[c]	2	$C_{18}H_{13}O_4$	293.0790	-2.3	8
Fluoranthene[e]	3	$C_{21}H_{13}O_5$	345.0744	-1.8	4
	4	$C_{23}H_{15}O_7$	403.0842	2.5	3
Fluorenone[f]	2	$C_{16}H_9O_4$	265.0517	1.8	5
OCH$_3$-fluor- enone[f]	1	$C_{15}H_9O_3$	237.0568	1.7	1
Anthra- quinone[f]	2	$C_{17}H_9O_5$	293.0455	0.6	8
Dibenzofuran[g] and/or naphthofuran	1	$C_{13}H_7O_2$	195.0456	1.1	18
	2	$C_{15}H_9O_4$	253.0506	0.6	35
	3	$C_{17}H_{11}O_6$	311.0576	2.1	28
	4	$C_{19}H_{13}O_8$	369.0688	2.0	8

Table 2.34 (continued)

Nucleus	Number of -COOCH$_3$	Precise mass (M-OCH$_3$+)			Relative[b] abundance (±15 - 20%)
		Elemental composition	Observed	Deviation X 10$_3$	
Benzonaphtho-furan	2	$C_{19}H_{11}O_4$	303.0692	3.6	5
	3	$C_{21}H_{13}O_6$	361.0738	2.7	7
Xanthone	1	$C_{14}H_7O_3$	223.0397	0.3	33
	2	$C_{16}H_9O_5$	281.0433	-1.6	32
Benzothiophene	2	$C_{11}H_7O_3S$	219.0122	0.6	9
	3	$C_{13}H_9O_5S$	277.0134	-3.5	1
Dibenzothiophene and/or naphthothio-phene	2	$C_{15}H_9O_3S$	269.0244	-2.7	5

[a]Illinois bituminous coal was finely powdered, heated with 12 N HCl and washed; the same procedure was repeated three times, and the coal dried at 120°C.   The sample contained 71.33% carbon.
[b]Benzene tricarboxylic acid methyl ester is normalized to 100.
[c]Methoxy compounds may result from the reaction of phenolic compounds with diazomethane; methoxyphenanthrene may be derived from the reaction of fluorenone with diazomethane.
[d]Although the phenanthrene ring is not affected during oxidation, anthracene is oxidized to anthraquinone.
[e]Methyl esters of pyrene carboxylic acids show very similar mass spectra; however, the pyrene ring is extensively degraded by present procedure.
[f]These compounds might be oxidized further to benzene-carboxylic acids in our experimental conditions.
[g]It is probable that our estimates of the first two dibenzofurans are somewhat high because of contribution from fragments of xanthones which lose both CO and OCH$_3$ to give the corresponding dibenzofuran ions.

Source: Hayatsu 1975, Table 1, p. 378. Reprinted by permission of the publisher.

2.6.3.10 <u>Heterocyclic groups</u>

Table 2.35 lists seven nitrogen-containing aromatic acids produced by photochemical oxidation of coal in alkali solution, identified as methylesters of carboxylic acids. They have pyridine, quinoline, or carbazole nuclei. Some structures believed by Hayatsu et al. (1975) to be indigenous to coal are given in Fig. 2.15. It is not known whether some of the aromatic units identified are indigenous to the coal or are produced in the processing. Hayatsu et al. (1975) note, however, that benzo- and dibenzothiophene were probably not formed by the low temperatures used in their aqueous oxidation experiments; therefore, thiophene derivatives are probably indigenous. Hayatsu agrees with Wiser (1973) that the complexity of the aromatics indicates a high degree of aliphatic or alicyclic linkage.

Heterocyclic units having oxygen, sulfur, and nitrogen rings can present difficulties in conversion processes. While studying the use of catalysts to remove organic and inorganic sulfur from Indiana No. 5 and Homestead Mine, Kentucky coals, Akhtar et al. (1974) found that, without the use of catalysts, the liquid products from the conversion contained 14 organic sulfur compounds, all of which were thiophene derivatives except diallylsulfide (Table 2.36). Organic

Table 2.35. N-Heterocyclics found in the photochemical oxidation of coal in alkali solution and identified as methyl esters of carboxylic acids[a]

Nucleus	Number of $-COOCH_3$	Precise mass $(M-OCH_3)^+$			Relative[b] abundance (±15 - 20%)
		Elemental composition	Observed	Deviation X 10^3	
Pyridine	3	$C_{10}H_8O_5N$	222.0408	0.5	3.0
	4	$C_{12}H_{10}O_7N$	280.0488	3.2	3.5
Quinoline and/or isoquinalone	1	$C_{10}H_6ON$	156.0412	-3.7	0.5
	2	$C_{12}H_8O_3N$	214.0468	-3.5	
Carbazole[c]	1	$C_{13}H_8ON$	194.0598	-0.7	10
	2	$C_{15}H_{10}O_3N$	252.0660	0.0	9
	3	$C_{17}H_{12}O_5N$	310.0708	-0.6	5

[a]Demineralized fine powdered coal was suspended in either 10% HCl or 5% KOH aqueous solution and oxidized with air by the radiation of a high-pressure mercury lamp for 7 to 10 days at 40 - 50°C.

[b]Benzene tricarboxylic acid, which is the most abundant aromatic acid in this oxidation product, is normalized to 100.

[c]Carbazole is sensitive to oxidation with oxidizing agents; however, this nucleus is stable during photochemical oxidation as shown by our own experiments with authentic carbazole and N-ethyl carbazole.

Source: Hayatsu 1975, Table 2, p. 379. Reprinted by permission of the publisher.

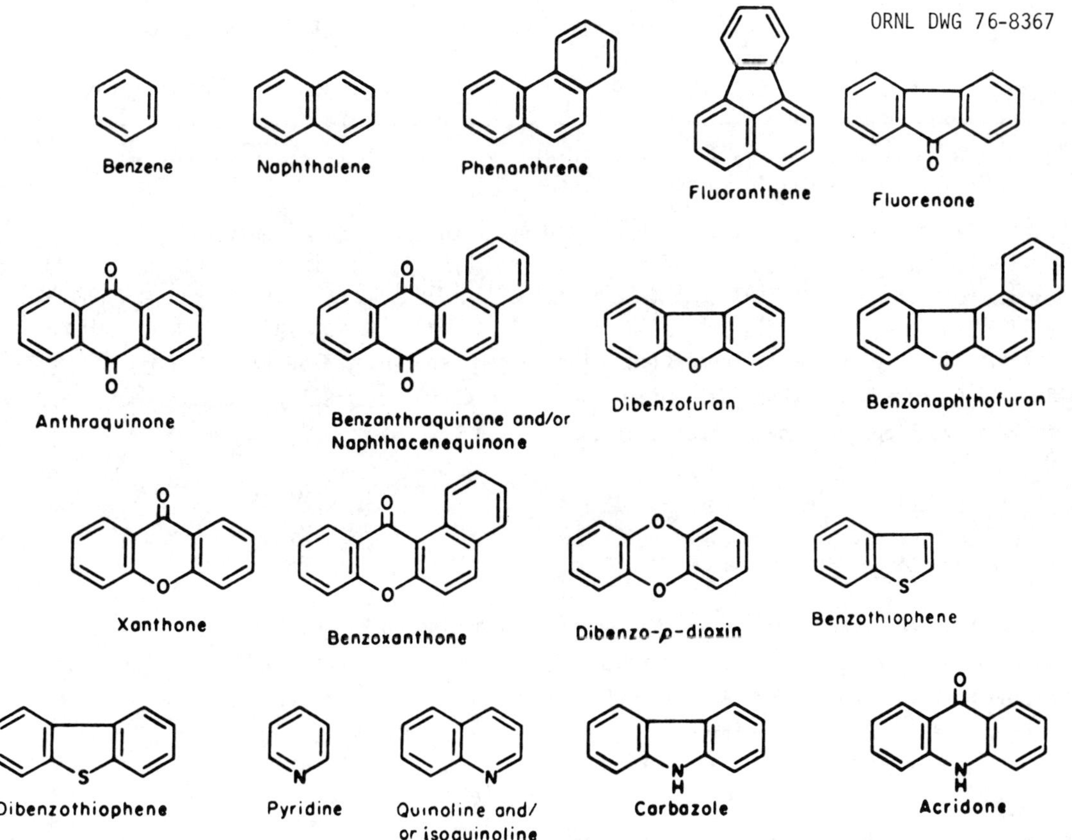

Fig. 2.15. Aromatic units indigenous to coal. Source: From Hayatsu 1975, Fig. 1. p. 379. Reprinted by permission of the publisher.

Table 2.36. Organic sulfur compounds in the products
of coal hydrogenation

| Product | Indiana No. 5 coal | | Kentucky coal | |
	Molecular formula	Identification[a]	Molecular formula	Identification[a]
Light oil	C_8H_6S	Benzothiophene	C_5H_6S	Methylthiophene
	C_9H_8S	Methylbenzothiophene	$C_6H_{10}S$	Diallylsulfide
	$C_{10}H_{10}S$	Dimethylbenzothiophene		
Heavy oil	C_5H_6S	Methylthiophene	$C_{14}H_8S$	Benzo[*def*]dibenzo-thiophene
	$C_8H_{10}S$	Tetrahydrobenzo-thiophene	$C_{16}H_{10}S$	Naphthobenzothio-phene
	$C_{11}H_{10}S$	Benzylthiophene	$C_{17}H_{12}S$	Methylnaphtho-benzothiophene
	$C_{12}H_8S$	Dibenzothiophene	$C_{20}H_{12}S$	Dinaphthothiophene
	$C_{13}H_{10}S$	Methyldibenzothiophene	$C_{21}H_{14}S$	Methyldinaphtho-thiophene
	$C_{14}H_8S$	Benzo[*def*]dibenzo-thiophene		
	$C_{16}H_{10}S$	Naphthobenzothiophene		
	$C_{17}H_{12}S$	Methylnaphthobenzo-thiophene		
	$C_{20}H_{12}S$	Dinaphthothiophene		
Asphaltenes	Spectrum not well resolved		$C_{16}H_{10}S$	Naphthobenzothio-phene
			$C_{20}H_{12}S$	Dinaphthothiophene

[a]Based upon molecular formula determined by high-resolution mass spectrometry; other
isomeric forms are possible in some instances.

Source: Akhtar 1974, Table 3, p. 6. Reprinted by permission of the publisher.

sulfur can also be present in coal in the form of thiol, thioether, disulfide, and γ-thiopyrone.
Because these forms were not detected by Akhtar et al. (1974), it was suggested that, being
very unstable at the experimental conditions, they decomposed spontaneously. By using a
$Co-Mo-SiO_2-Al_2O_3$ catalyst, Akhtar determined that the order of difficulty of removal is dibenzothiophene > benzothiophene > naphthobenzothiophene.

2.6.4 Coal chemicals

2.6.4.1 Tars and pitches

Tars and pitches are aromatic materials encountered in coking and in coal conversion processes.
Because of the different conditions to which coals are exposed in these two types of processes,
the tars and pitches produced by each may differ significantly in composition; however, some
insight into their complex composition may be gained from studies of the materials produced by
coking.

Pitches are obtained through distillation of tars and represent from 30 to 60% of the tar
components. Pitches have been estimated by McNeil (1966) to contain about 5000 compounds.

McNeil describes the formation of pitch as follows. During carbonization, coal passes through two stages of decomposition: At 400 to 500°C, plasticity begins to develop, and at 650 to 750°C, more advanced decomposition occurs. During these changes, the volatile products of the coal are released, and as they pass through the hot coke, the volatile products become involved in a series of secondary reactions. They emerge from the retort and are separated by fractional condensation or absorption into tar, ammonia liquor, benzole, and illuminating or heating gas (McNeil 1966). The major reactions involved in the conversion of primary carbonization products into tars according to McNeil 1966 are (1) the cracking of higher-molecular-weight paraffins to gaseous paraffins and olefins; (2) dehydrogenation of cyclohexane derivatives to aromatic hydrocarbons and phenols; (3) dealkylation of alkyl aromatic hydrocarbons, alkyl pyridines, and alkyl phenols; (4) dehydroxylation of phenols; (5) synthesis of polynuclear aromatic hydrocarbons by condensation of simpler structures with olefins and diolefins; and (6) disproportionation of aromatic hydrocarbons to both simpler and more complex structures.

The extent of these reactions is affected by the temperature of carbonization and the time of contact with the hot coke bed and the heated walls of the retort. Thus, tars that have been exposed to different carbonization processes vary in chemical composition. Distillation of crude tar yields a series of primary oil fractions and a residue of refined tar or pitch, depending on the extent of the distillation process. Distillate oils obtained by steam or vacuum distillation of pitch or pitch crystalloids or from the coking of pitch are the only fractions from which pure chemical compounds can normally be isolated.

Tars produced at temperatures up to 700°C have been referred to as low-temperature tars. However, because aromatization reactions begin making significant contributions at about 600°C, the maximum temperature for a low-temperature tar should be about 500°C (Karr 1963). High-temperature tars are the decomposition products of coal produced at temperatures in excess of 800°C, such as in coke ovens (Weiler 1963).

Kruber and coworkers (as cited by McNeil 1966) identified 127 compounds in tars and pitches (Table 2.37) (A more comprehensive list of polycyclic aromatic hydrocarbons is presented in Appendix C). A boiling point of 300°C was arbitrarily chosen as the starting point. Although a wide variety of compounds are listed, they are mostly condensed polynuclear aromatic hydrocarbons or heterocyclic compounds having three to six rings. Methyl and hydroxyl were the only substituent groups found, and occasionally saturation of one ring was noted. Only single-ring compounds contained side chains longer than methyl; usually hydroxyl and longer alkyl side chains are found together. Polyphenyl compounds rarely occurred (McNeil 1966).

Shultz et al. (1972) used high-resolution mass spectrometry to investigate coal carbonization products. Analysis of an 80 to 85°C softening point pitch yielded molecular formulas for 239 compounds of 108 structural types, including their alkyl derivatives. Some of these species are listed in Table 2.38 and summarized in Table 2.39. Table 2.40 shows elemental combinations that may or may not represent molecular ions. Shultz et al. (1972) believe that these combinations probably result from rearrangement ions produced by fragmentation of complex heterocyclic hydrocarbons because they occur at relatively low molecular weights, and that they indicate the presence of other material previously undetected in pitch. In addition to the organic compounds in Table 2.38, empirical formulas, which may represent either molecular entities or rearrangement ions, are listed. For example, the 23 CHON structures could contain a

Table 2.37. Compounds of boiling point above 300°C identified in high-temperature tars and pitches

Compound		Formula	Compound		Formula
Acridine		$C_{13}H_9N$	1-Azapyrene		$C_{15}H_9N$
Anthracene		$C_{14}H_{10}$			
1-Azacarbazole		$C_{11}H_8N_2$	1,2-Benzacridine		$C_{17}H_{11}N$
2-Azafluoranthene		$C_{15}H_9N$			
			3,4-Benzacridine		$C_{17}H_{11}N$
13-Azafluoranthene		$C_{15}H_9N$	1,2-Benzanthracene		$C_{18}H_{12}$
4-Azafluorene		$C_{12}H_9N$			

Table 2.37 (continued)

Compound		Formula
Benzanthrone	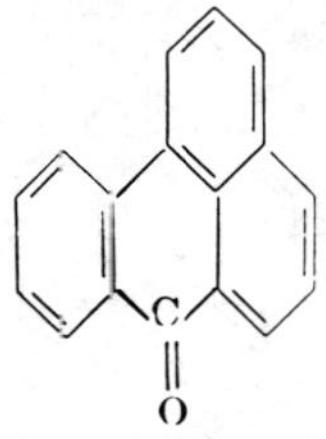	$C_{17}H_{10}O$
2,3-Benz-1-aza-carbazole	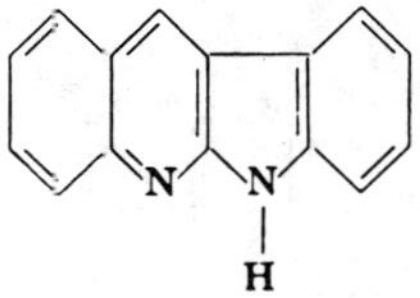	$C_{15}H_{10}N_2$
2,3-Benz-4-aza-fluorene	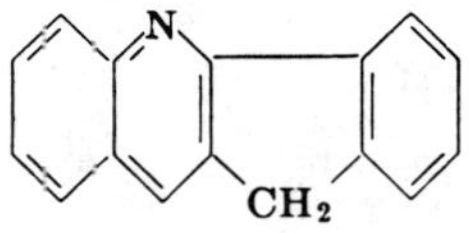	$C_{16}H_{11}N$
1,2-Benzcarbazole	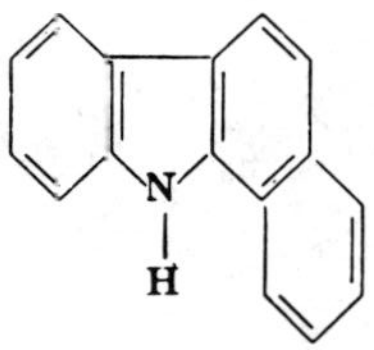	$C_{16}H_{11}N$
2,3-Benzcarbazole	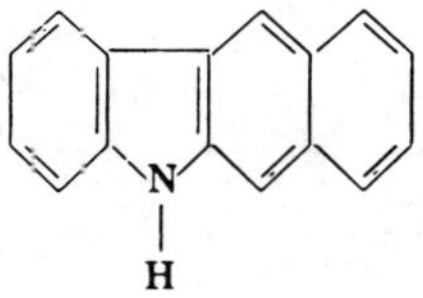	$C_{16}H_{11}N$
3,4-Benzcarbazole	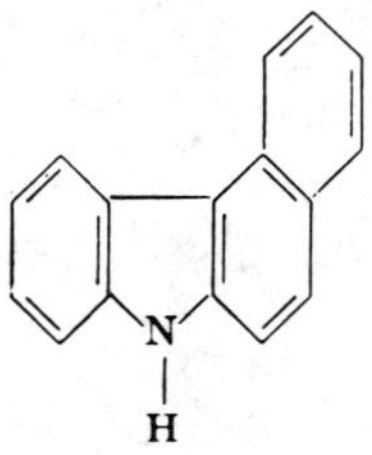	$C_{16}H_{11}N$
2,3-Benzchrysene	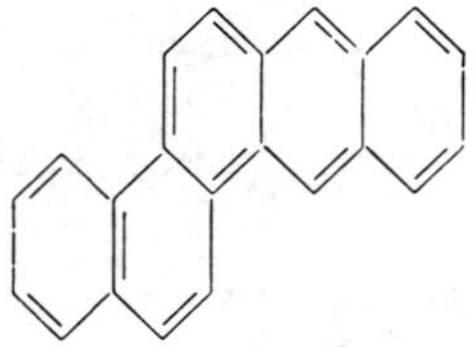	$C_{22}H_{14}$
1,2-Benzdiphen-ylene oxide	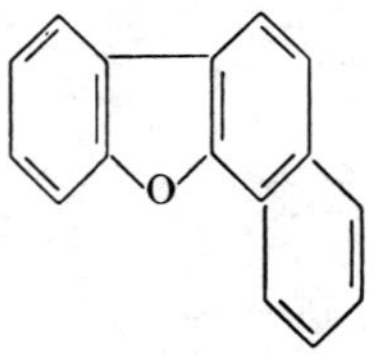	$C_{16}H_{10}O$
1,2-Benzdiphen-ylene sulphide	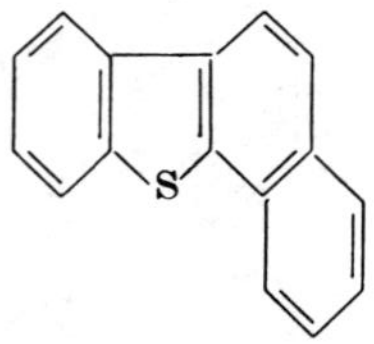	$C_{16}H_{10}S$
1,2-Benzfluorene	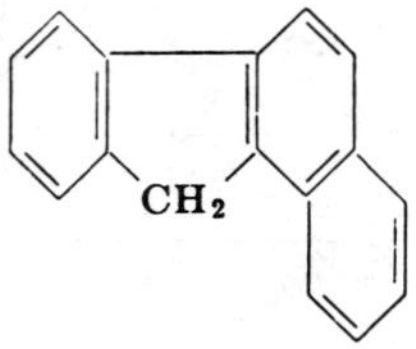	$C_{17}H_{12}$

Table 2.37 (continued)

Compound	Formula	Compound	Formula
2,3-Benzfluorene	$C_{17}H_{12}$	10,11-Benzfluo-ranthene	$C_{20}H_{12}$
3,4-Benzfluorene	$C_{17}H_{12}$	11,12-Benzfluo-ranthene	$C_{20}H_{12}$
2,3-Benzfluorene nitriles	$C_{18}H_{11}N$	1,2-Benznaphtha-cene	$C_{22}H_{14}$
Benz(m,n,o)fluor-anthene	$C_{18}H_{10}$	1,9-Benzoxan-thene	$C_{16}H_{10}O$
3,4-Benzfluoran-thene	$C_{20}H_{12}$	1,2-Benzpenta-cene	$C_{26}H_{16}$

Table 2.37 (continued)

Compound	Formula	Compound	Formula
1,12-Benzper-ylene	$C_{22}H_{12}$	4,5-Benzpyrene	$C_{20}H_{12}$
1,2-Benzpicene	$C_{26}H_{16}$	5,6-Benzquino-line	$C_{13}H_9N$
1,2-Benzpyrene	$C_{20}H_{12}$	7,8-Benzquinoline	$C_{13}H_9N$
		4,5-Benzthio-naphthene	$C_{12}H_8S$
		5,6-Benzthio-naphthene	$C_{12}H_8S$

Table 2.37 (continued)

Compound		Formula	Compound		Formula
6,7-Benzthio- naphthene		$C_{12}H_8S$	1,2,3,4-Dibenz- anthracene		$C_{22}H_{14}$
Brazan		$C_{16}H_{10}O$	1,2,5,6-Dibenz- anthracene		$C_{22}H_{14}$
Carbazole		$C_{12}H_9N$	1,2,7,8-Dibenz- anthracene		$C_{22}H_{14}$
Chrysene		$C_{18}H_{12}$			
Coronene		$C_{24}H_{12}$			

Table 2.37 (continued)

Compound		Formula	Compound		Formula
3,4,8,9-Dibenz-pyrene	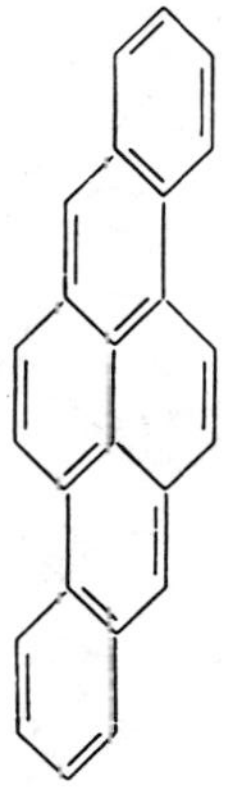	$C_{24}H_{14}$	2,3,5,6-Dibenz-thionaphthene		$C_{16}H_{10}S$
3,4,9,10-Dibenz-pyrene	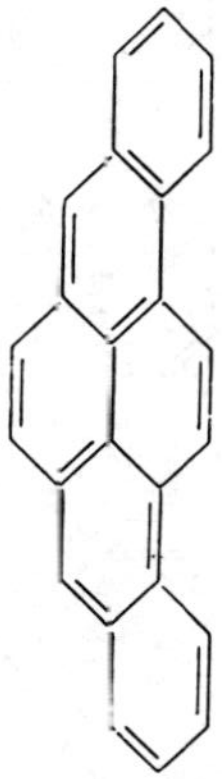	$C_{24}H_{14}$	9,10-Dihydroacri-dine	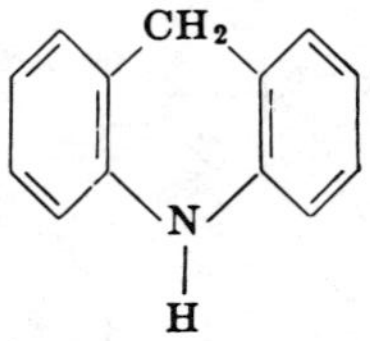	$C_{13}H_{11}N$
3,4,8,9-Dibenz-tetraphene		$C_{26}H_{16}$	9,10-Dihydro-anthracene	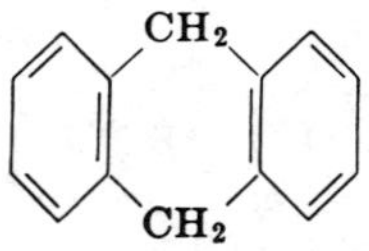	$C_{14}H_{12}$
			5,12-Dihydro-naphthacene	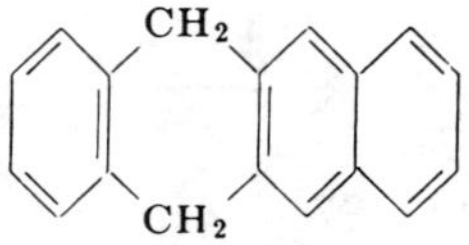	$C_{18}H_{14}$
			2,3-Dimethyl anthracene	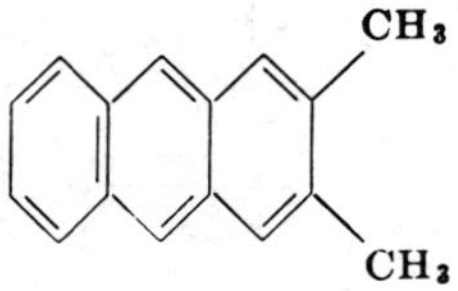	$C_{16}H_{14}$
			(?)-Dimethyl diphenylene oxide	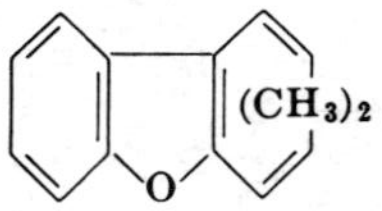	$C_{14}H_{12}O$
			1,8-Dimethyl diphenylene oxide	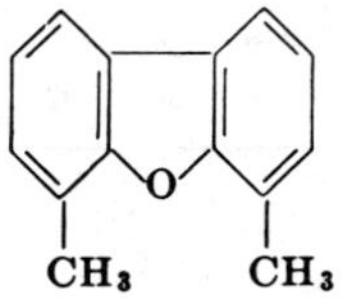	$C_{14}H_{12}O$

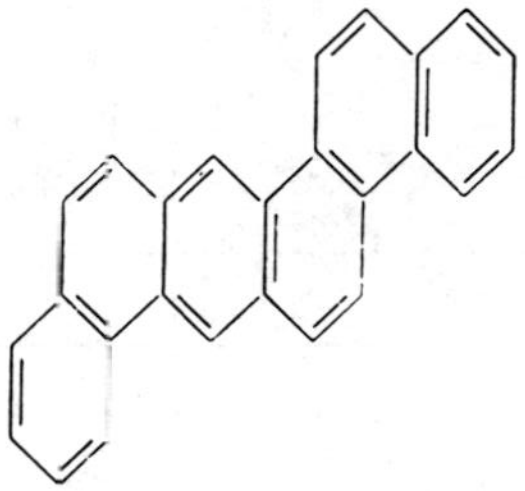

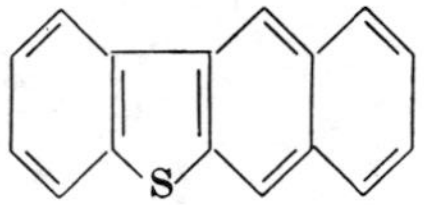

Table 2.37 (continued)

Compound		Formula	Compound		Formula
1,8-Dimethyl-diphenylene sulphide	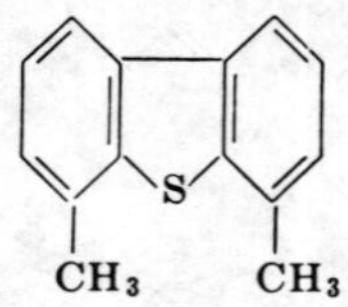	$C_{14}H_{12}S$	5-Diphenyl-γ-pyrone	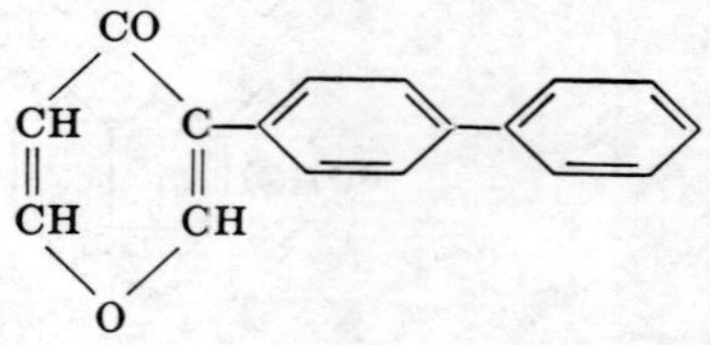	$C_{17}H_{12}O_2$
3,6-Dimethyl phenanthrene	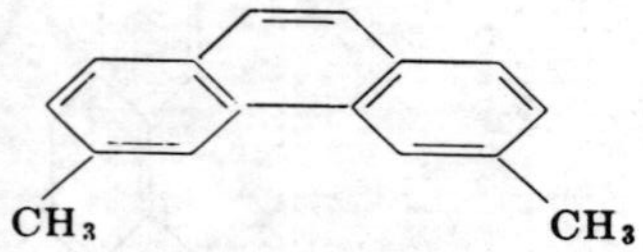	$C_{16}H_{14}$	Fluoranthene	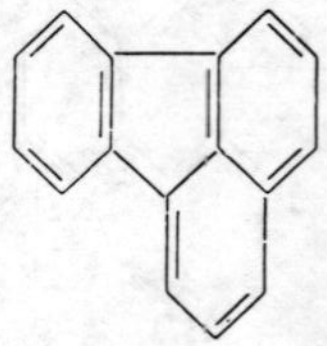	$C_{16}H_{10}$
2,2-Dinaphthyl	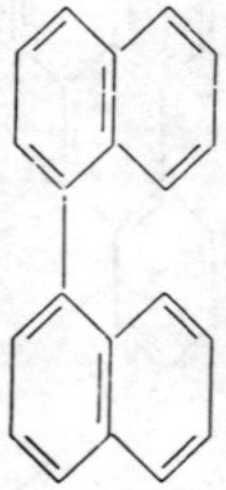	$C_{20}H_{14}$	Fluorene nitriles	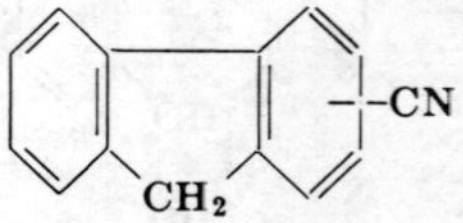	$C_{14}H_9N$
			n-Heptadecane	$CH_3(CH_2)_{15}CH_3$	$C_{17}H_{36}$
o,o'-Diphenol	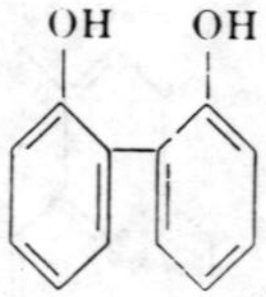	$C_{12}H_{10}O_2$	Hydroxyanthracenes	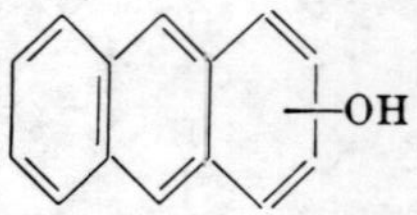	$C_{14}H_{10}O$
Diphensuccindan	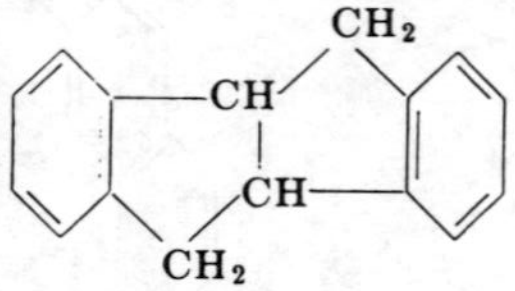	$C_{16}H_{14}$	4-Hydroxydiphenyl	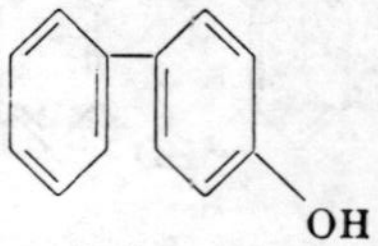	$C_{12}H_{10}O$
Diphenylene sulphide	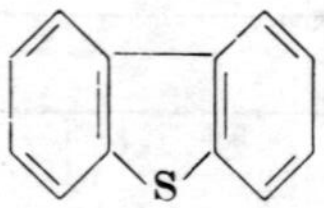	$C_{12}H_8S$	2-Hydroxy-diphenylene oxide	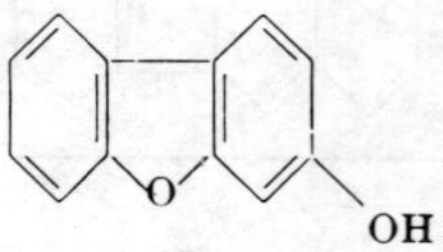	$C_{12}H_8O_2$

Table 2.37 (continued)

Compound		Formula
3-Hydroxy-diphenylene oxide	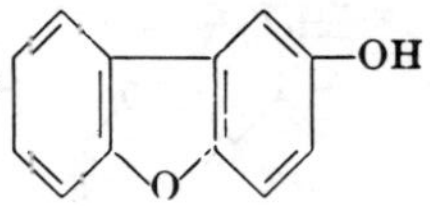	$C_{12}H_8O_2$
2-Hydroxyfluorene	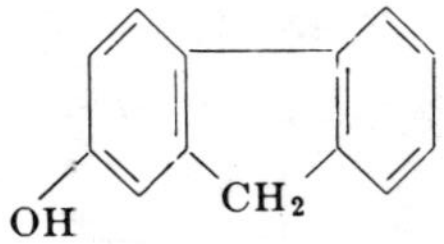	$C_{13}H_{10}O$
9-Hydroxy-4-methyl fluorene	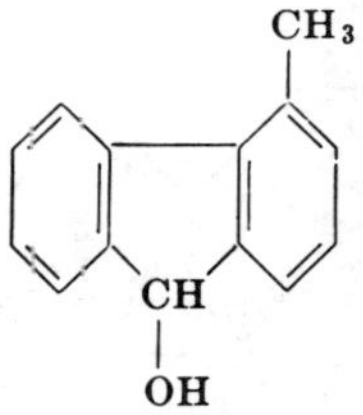	$C_{14}H_{12}O$
4,5-Iminophenanthrene	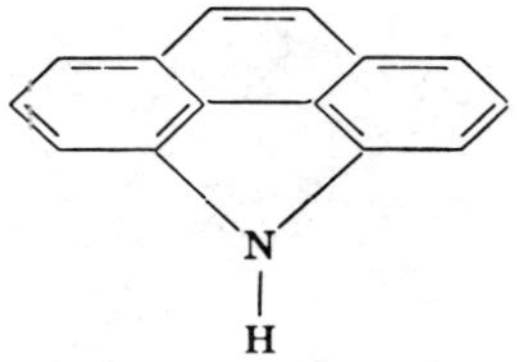	$C_{14}H_9N$
p-Methoxybenzophenone	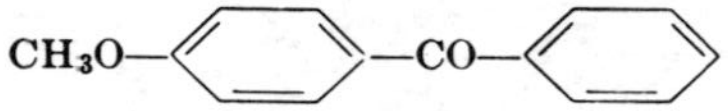	$C_{14}H_{12}O_2$
2-Methyl acridine	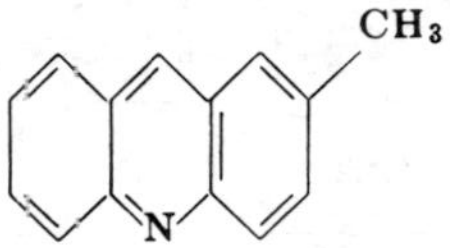	$C_{14}H_{11}N$
1-Methyl anthracene	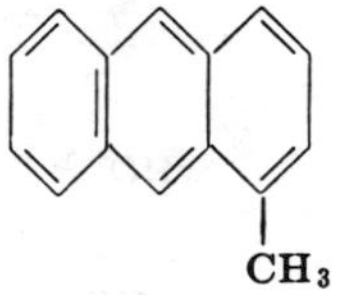	$C_{15}H_{12}$
2-Methyl anthracene	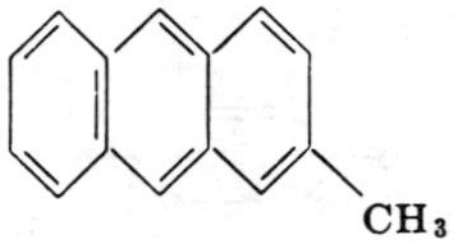	$C_{15}H_{12}$
2-Methyl-5,6-benzquinoline	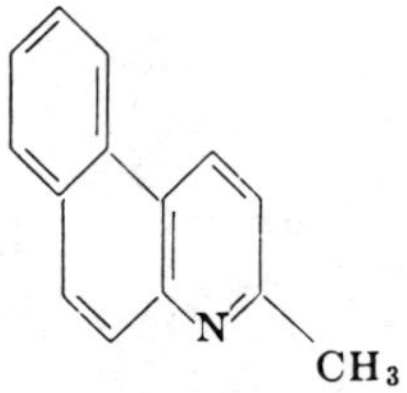	$C_{14}H_{11}N$
?-Methyl brazan	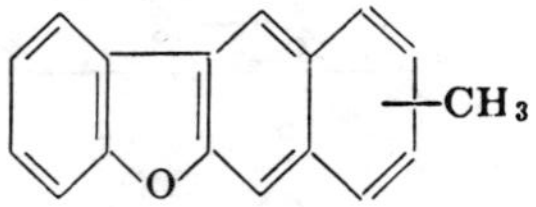	$C_{17}H_{12}O$
2-Methyl carbazole	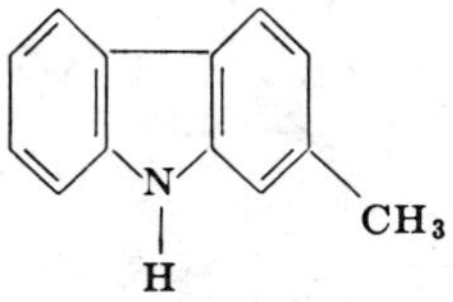	$C_{13}H_{11}N$

Table 2.37 (continued)

Compound	Formula	Compound	Formula
3-Methyl carba-zole 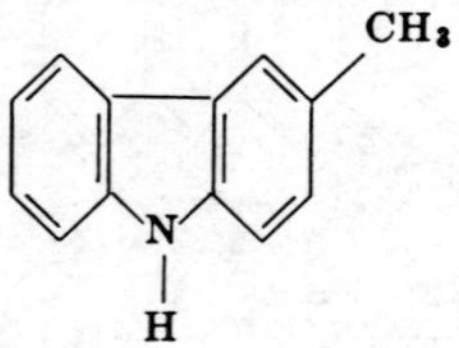	$C_{13}H_{11}N$	2-Methyl fluorene 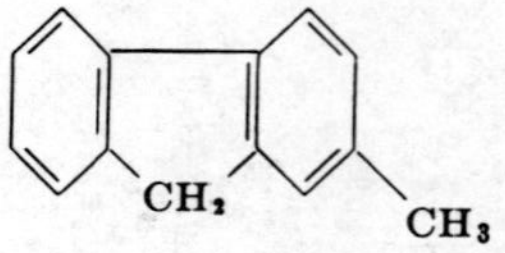	$C_{14}H_{12}$
1-Methyl chry-sene 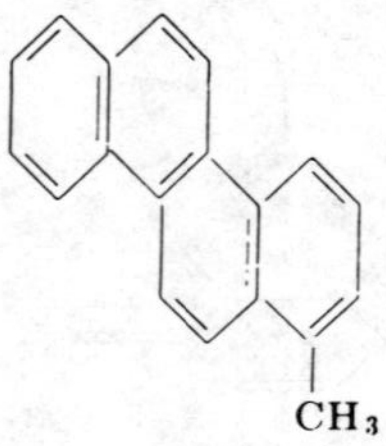	$C_{19}H_{14}$	3-Methyl fluorene 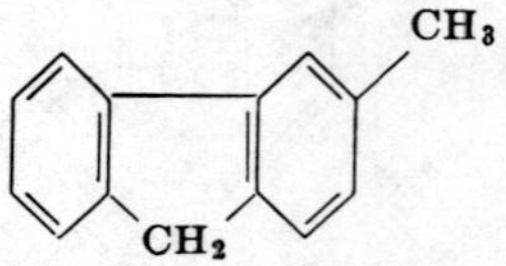	$C_{14}H_{12}$
2-Methyl diphenylene oxide 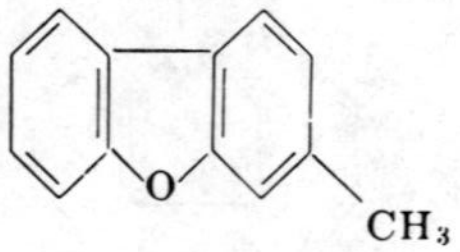	$C_{13}H_{10}O$	1-Methyl phenan-threne 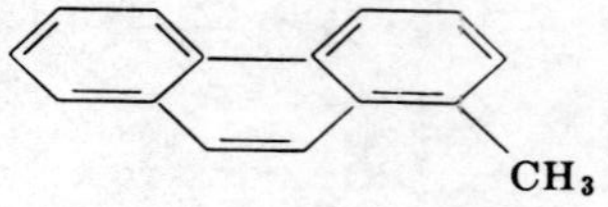	$C_{15}H_{12}$
3-Methyl diphenylene oxide 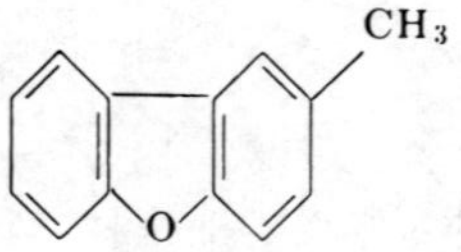	$C_{13}H_{10}O$	2-Methyl phenan-threne 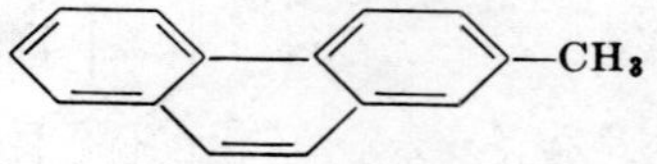	$C_{15}H_{12}$
1-Methyl fluorene 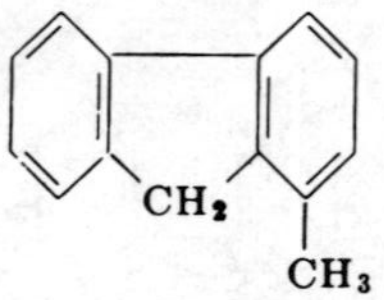	$C_{14}H_{12}$	3-Methyl phenan-threne 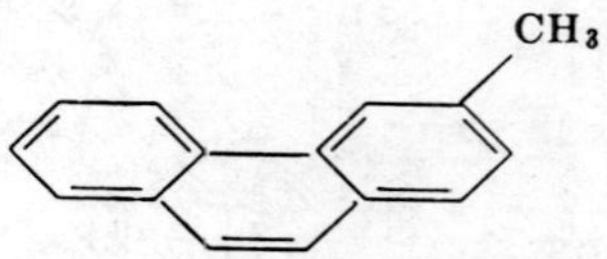	$C_{15}H_{12}$
		9-Methyl phenan-threne 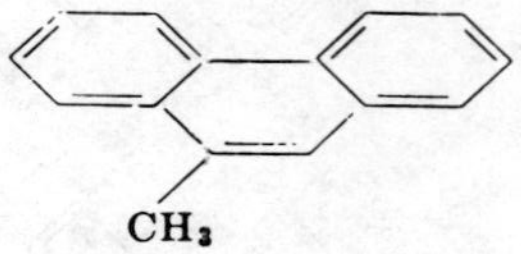	$C_{15}H_{12}$

Table 2.37 (continued)

Compound		Formula	Compound		Formula
1-Methyl pyrene	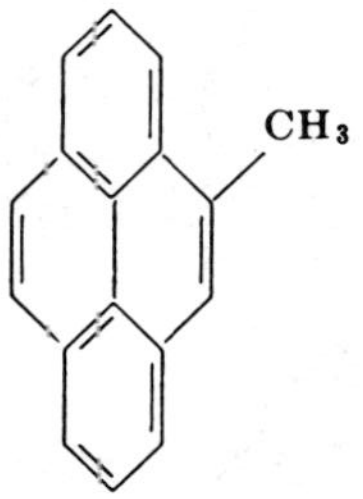	$C_{17}H_{12}$	Naphtho-2′,3′-1,2-anthracene	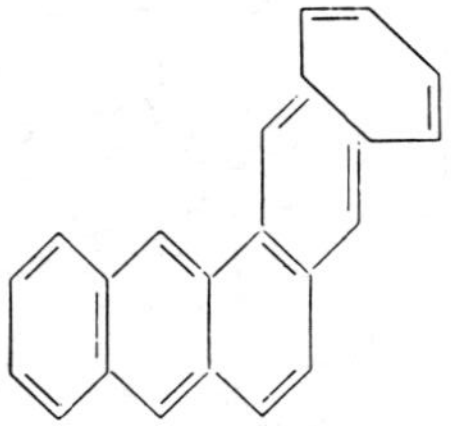	$C_{22}H_{14}$
3-Methyl pyrene	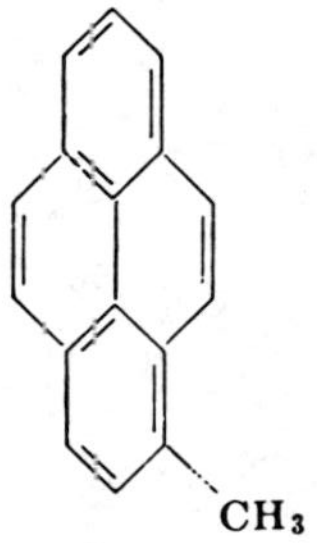	$C_{17}H_{12}$	(?)-Naphthofluorene		$C_{21}H_{14}$
			2-Naphthonitrile	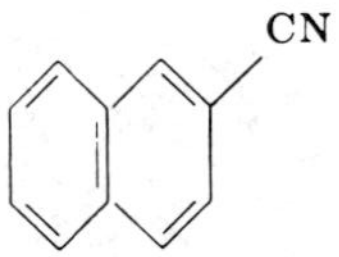	$C_{11}H_7N$
4-Methyl pyrene	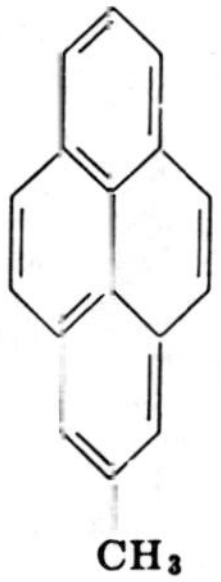	$C_{17}H_{12}$	*peri*-Naphthoxanthene	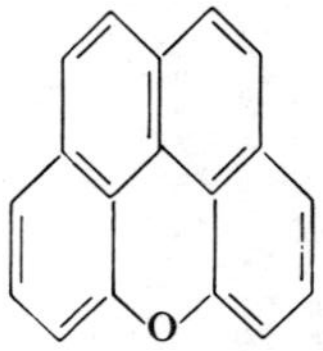	$C_{18}H_{10}O$
			1-Naphthylamine	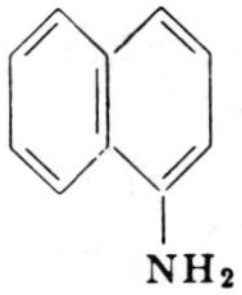	$C_{10}H_9N$
Naphthacene	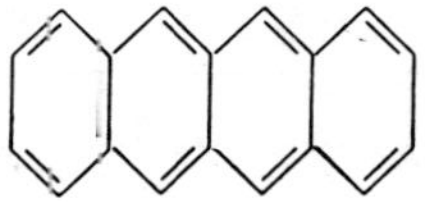	$C_{18}H_{12}$	*o*(β-Naphthyl)-phenol	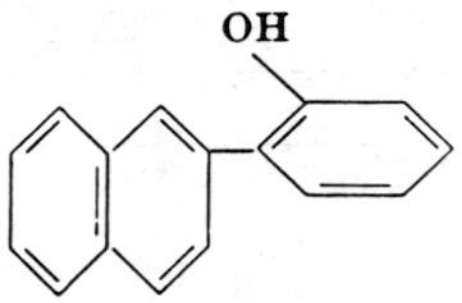	$C_{16}H_{12}O$

Table 2.37 (continued)

Compound		Formula	Compound	Formula
Nonadecane	$CH_3(CH_2)_{17}CH_3$	$C_{19}H_{40}$	Phenanthridone	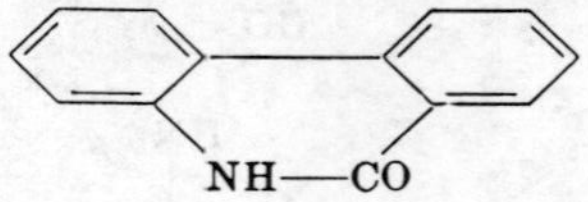 $C_{13}H_9ON$
1-Oxo-1,2-dihy-dro-2-azapyrene	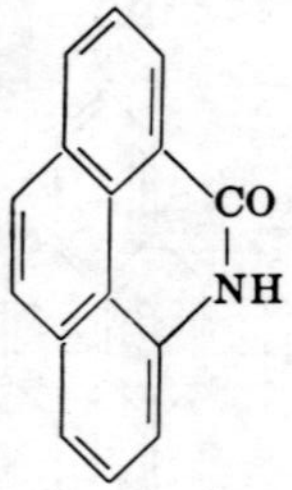	$C_{15}H_9ON$	2-Phenanthrol	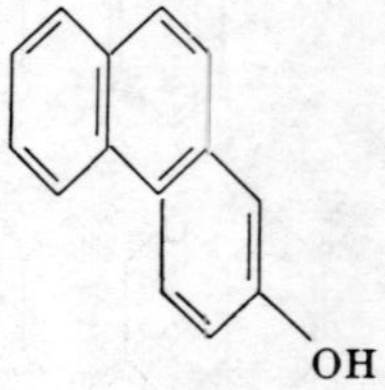 $C_{14}H_{10}O$
Perylene	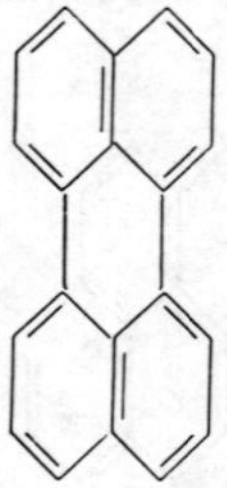	$C_{20}H_{12}$	4-Phenanthrol	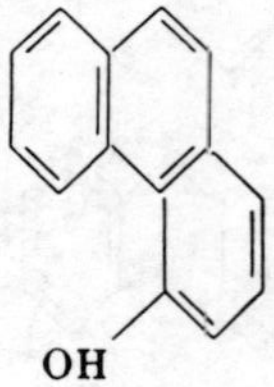 $C_{14}H_{10}O$
Phenanthrene	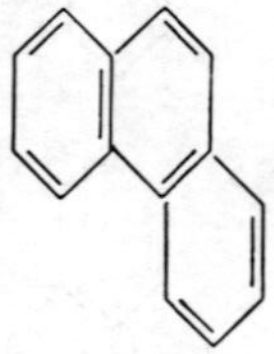	$C_{14}H_{10}$	4,5-Phenanthry-lene methane	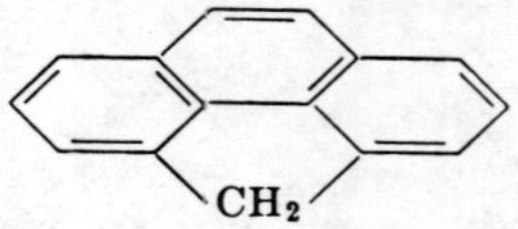 $C_{15}H_{10}$
Phenanthridine	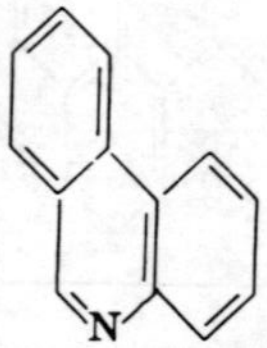	$C_{13}H_9N$	1-Phenyl benzan-throne	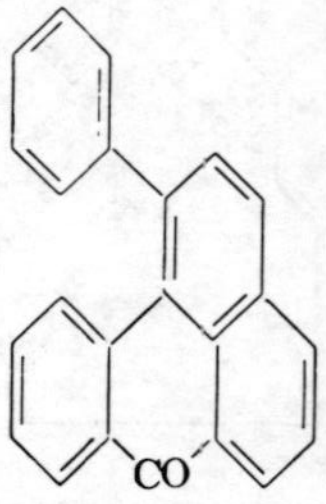 $C_{23}H_{14}O$

Table 2.37 (continued)

Compound	Formula	Compound	Formula
2,3-*o*-Phenylene pyrene	$C_{22}H_{12}$	Pyrene	$C_{16}H_{10}$
1-Phenylnaph-thalene	$C_{16}H_{12}$	1,2,3,4-Tetrahy-droacridine	$C_{13}H_{13}N$
2-Phenyl naph-thalene	$C_{16}H_{12}$	1,2,3,4-Tetrahy-droanthracene	$C_{14}H_{14}$
2-Phenyl phenan-threne	$C_{20}H_{14}$	1,2,3,4-Tetrahy-drofluoranthene	$C_{16}H_{14}$
Picene	$C_{22}H_{14}$	Tetramethyl di-phenol	$C_{16}H_{18}O_2$

Table 2.37 (continued)

Compound		Formula	Compound		Formula
2,3,6,7-Tetra-methyl naph-thalene	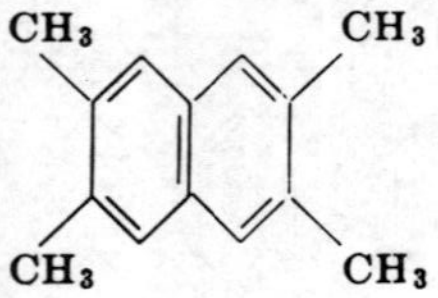	$C_{14}H_{16}$	Triphenylene	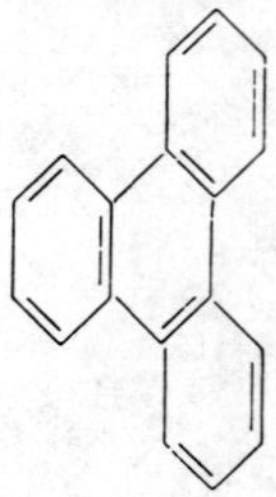	$C_{18}H_{12}$
1,3,7-Trimethyl-naphthalene	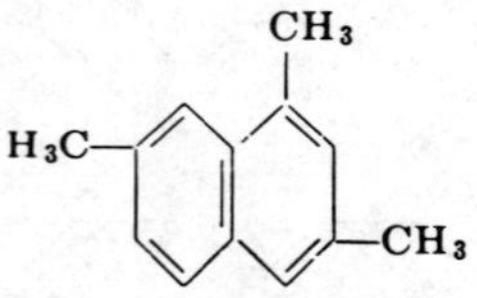	$C_{13}H_{14}$			
1,2,8-Trimethyl phenanthrene	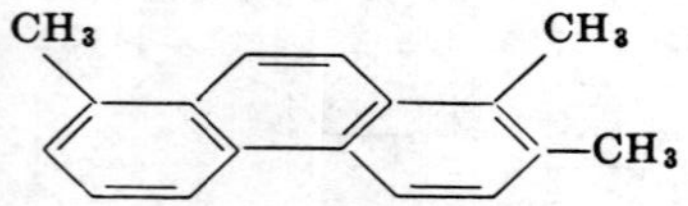	$C_{17}H_{16}$	Xanthene	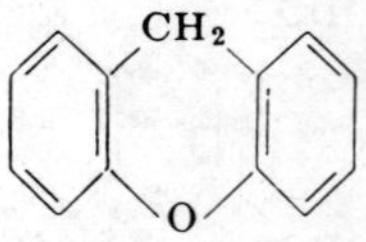	$C_{13}H_{10}O$

Source: McNeil 1956, Table 5-2, pp. 159-92. Reprinted by permission of the publishers.

Table 2.38. Organic species detected in an 80 - 85°C softening point coal-tar pitch by high-resolution mass spectrometry

Hydrocarbon class	Example of structural type
CH	Acenaphthene Acenaphthylene Anthracene Paracyclene Pyrene Chrysene Benzo[*ghi*]fluoranthene Perylene Benzochrysene Anthranthrene Benzoperylene Coronene Dibenzochrysene Peropyrene Benzocoronene
CHO	Phenylfuran Acenaphthenol Dibenzofuran 9-Fluorenone 9-Anthraldehyde Benzonaphthofuran Benzanthrone Naphthoxanthene Dinaphthofuran Coeranthrone Benzonaphthoxanthene Benzodinaphthofuran Hydroxycoronene
CHO_2	Phenylbenzoquinone Xanthene-9-one Anthraquinone Phenylnaphthoquinone Hydroxybenzanthrone Benzanthraquinone Naphthylnaphthoquinone Naphthoanthraquinone

Hydrocarbon class	Example of structural type
CHO_2N_2	Methylnitroaniline Nitroquinoline Methylnitrocarbazole
CHN	Methylquinoline Dimethylphenylpyridine Naphthonitrile Acridine Benzo[*def*]carbazole Azapyrene Benzacridine Azabenzofluoranthene Azabenzopyrene Dibenzacridine Azabenzoperylene
CHN_2	Dimethylazaindole Azacarbazole Azabenz[*def*]carbazole
CHS	Diphenylsulfide Dibenzothiophene Naphthylthiophene Thiophenanthrene Benzonaphthothiophene Thionobenzofluorene Thiobenzophenanthrene Dinaphthothiophene Dinaphthylthiophene Thiodibenzophenanthrene
CHON	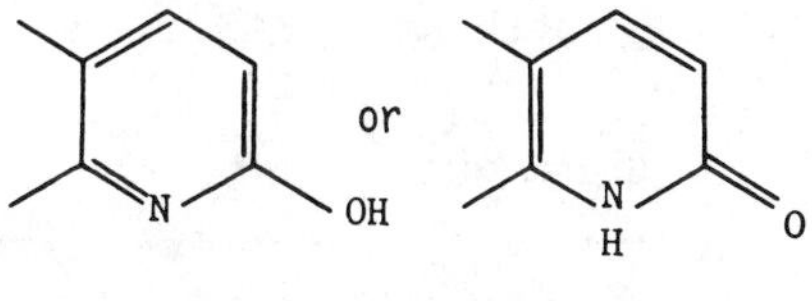

Source: Shultz et al. 1972, Table 2, pp. 244-45. Reprinted from *Fuel*: "High-resolution mass spectrometric investigation of heteroatom species in coal-carbonization products" by permission of the publishers, IPC Business Press Ltd.

Table 2.39. Summary of combinations of elements detected in coal-tar pitch by high-resolution mass spectrometry (softening point 80 - 85°C)

Elemental combination[b]	Author's work		Previously identified[a]	
	No. of structural types	No. of molecular formulas[c]	No. of structural types	No. of molecular formulas
Part A				
CH	15	40	14	30
CHO	13	41	9	15
CHO_2	8	14	3	4
CHN	11	44	9	19
CHN_2	4	6	2	2
CHS	10	17	2	3
CHON	8	23	3	3
CHO_2N_2	7	13		
Total	76	198	42	76
Part B[d]				
CHO_2N_2	7	7		
CHO_3	11	14		
CHO_4	5	8		
CHN_3	1	1		
CHOS	2	3		
CHO_2N	2	2		
CHN_2S	1	1		
CHO_2S	2	2		
CHO_6	1	3		
Total	32	41		
Grand total	108	239		

[a]Data for compounds with molecular weights >170 summarized with same restrictions as mass spectrometric data (Anderson and Wu 1963).
[b]Only the number of heteroatoms indicated.
[c]Includes alkyl derivatives.
[d]Elemental combinations indicate either molecular entities or rearrangement ions.

Source: Shultz et al. 1972, Table 1, p. 243. Reprinted from *Fuel*: "High-resolution mass spectrometric investigation of heteroatom species in coal-carbonization products" by permission of the publishers, IPC Business Press Ltd.

Table 2.40. Empirical formulas detected in an 80 - 85°C coal-tar pitch that may represent molecular entities or rearrangement ions formed from larger molecules

Nominal mass	Atomic carbon/hydrogen composition in nominal mass for								
	CHO_2N_2	CHO_3	CHO_4	CHN_3	CHOS	CHO_2N	CHN_2S	CHO_2S	CHO_6
144			6/8						
154		8/10							
168		9/12							
170		9/14	8/10						
180		10/12							
184			9/12						
196			10/12						
199				12/13		12/9			
200					12/8				
204		12/12							
208		12/16							
210			11/14						
214					13/10				
220		13/16							
230		14/14			14/14				
242		15/14							
244	14/16	15/16							
246	14/18							14/14	
248			14/16						
254		16/14							
256							15/16		
258		16/18							
260									13/8
268		17/16							
270	16/18		16/14						
274								16/18	14/10
277						18/15			
278		18/14							
280	17/16								
284			17/16						
288									15/12
292	18/16								
310	19/22								
316	20/16								

Source: Schultz et al. 1972, Table 3, p. 245. Reprinted from *Fuel*: "High-resolution mass spectrometric investigation of heteroatom species in coal-carbonization products" by permission of the publishers, IPC Business Press Ltd.

hydroxyl or a carbonyl group, assuming ring nitrogen, and of the 20 CHO_2N_2 structures, at least 13 could contain one or both of these functional groups. Most unidentified constituents of pitch are thought to be higher-molecular-weight homologues of the known compounds (Shultz et al. 1972).

2.6.4.2 Commercial hydrocarbons

Fieser and Fieser (1961) state that over 90% of the benzene and toluene obtained from coal is recovered by the scrubbing of coke-oven gas; however, a small amount of light oil also results from distillation of tar. Coal normally yields about 3% of its weight as tar. The major commercial hydrocarbons obtained from coal tar by distillation are shown in Fig. 2.16. Coal is also a source of cyclopentadiene, which is found in light oil containing small amounts of several paraffinic and olefinic hydrocarbons. Unless specially processed, coal-tar benzene contains very small amounts of thiophene, which is more stable and chemically inert than are the olefins. Because of the presence of trace amounts of naphthacene, anthracene has a yellowish tinge, and due to the presence of carbazole (Fig. 2.17), it gives a positive test for nitrogen. Naphthalene has an average yield of about 11% and is the most abundant constituent.

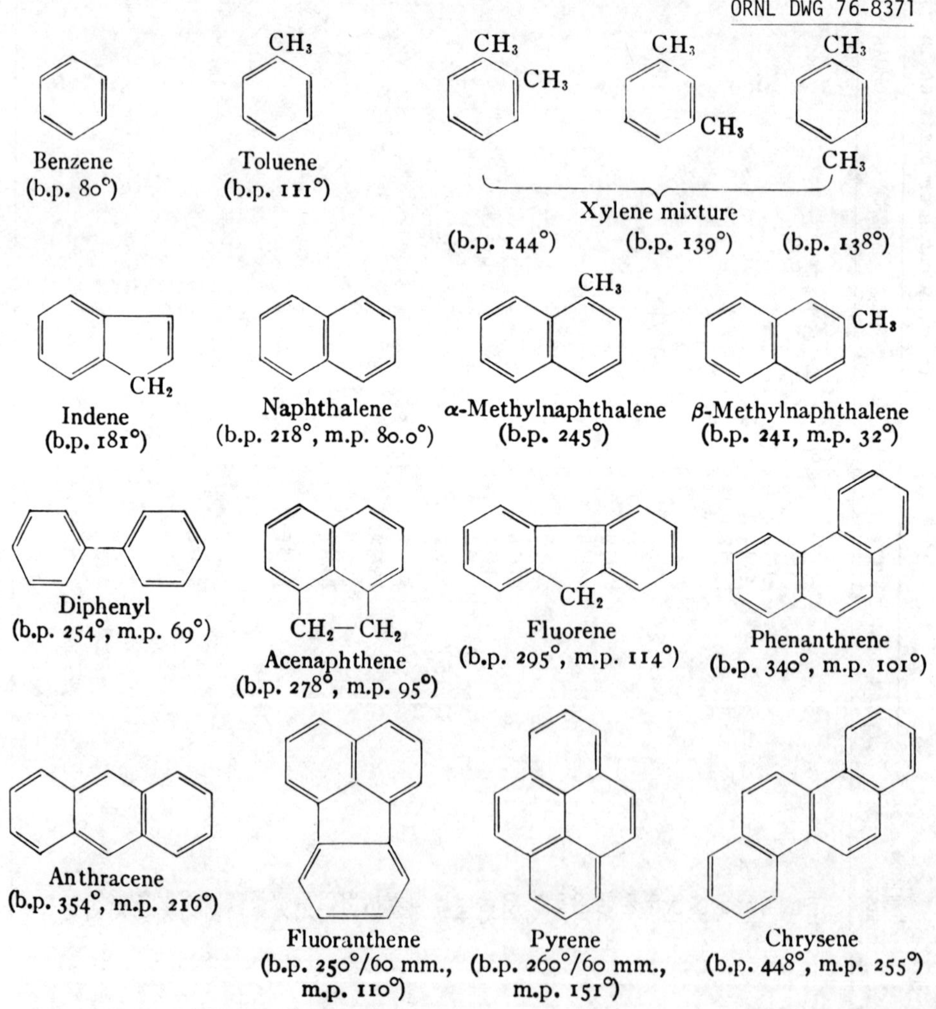

Fig. 2.16. Hydrocarbons available from coal tar. Source: From Fieser 1961, Chart 1, p. 641. Reprinted by permission of the publisher.

ORNL DWG 76-8369

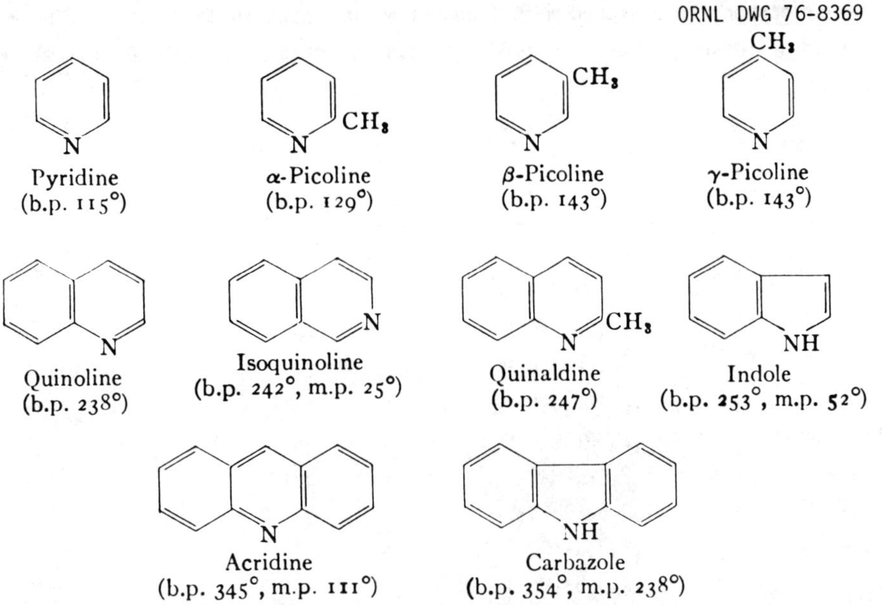

Fig. 2.17. Principal nitrogen compounds from coal tar. <u>Source</u>: From Fieser 1961, Chart 2, p. 643. Reprinted by permission of the publisher.

Fieser and Fieser's (1961) list of major commercial nitrogen components of coal tar is shown in Fig. 2.17; all except indole and carbazole are weakly basic. The oxygen-containing compounds are included in Fig. 2.18. The phenolic constituents most often obtained from coal tar are o-cresol and a mixture of m- and p-cresol.

ORNL DWG 76-8370

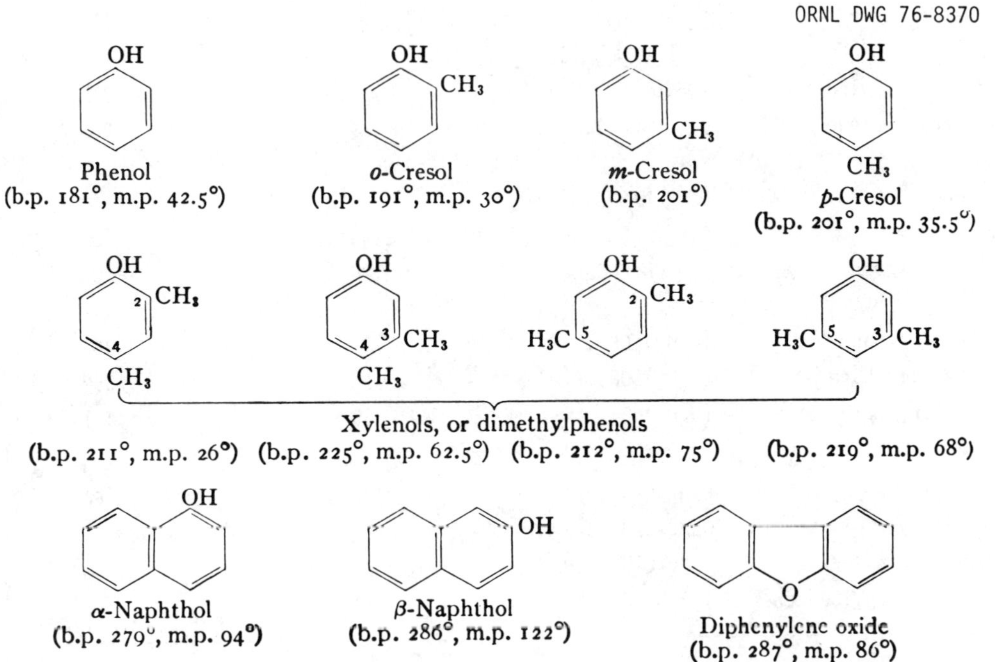

Fig. 2.18. Principal oxygen-containing compounds from coal tar. <u>Source</u>: From Fieser 1961, Chart 3, p. 643. Reprinted by permission of the publisher.

The benzenoid hydrocarbons encountered most frequently are shown in Table 2.41. The average heat of combustion of these hydrocarbons is 781 kcal/mole, or 8.79 kcal(34.8 Btu)/ml (Fieser and Fieser 1961).

Table 2.41. Benzenoid hydrocarbons[a]

Name	Formula	Melting point (°C)	Boiling point (°C)
Benzene	C_6H_6	5.4	80.1
Toluene	$C_6H_5CH_3$	−93	110.6
o-Xylene	$1,2\text{-}(CH_3)_2C_6H_4$	−28	144
m-Xylene	$1,3\text{-}(CH_3)_2C_6H_4$	−54	139
p-Xylene	$1,4\text{-}(CH_3)_2C_6H_4$	13	138
Hemimellitene	$1,2,3\text{-}(CH_3)_3C_6H_3$	Liquid	176
Pseudocumene	$1,2,4\text{-}(CH_3)_3C_6H_3$	Liquid	169
Mesitylene	$1,3,5\text{-}(CH_3)_3C_6H_3$	−57	165
Prehnitene	$1,2,3,4\text{-}(CH_3)_4C_6H_2$	−4	205
Isodurene	$1,2,3,5\text{-}(CH_3)_4C_6H_2$	Liquid	196
Durene	$1,2,4,5\text{-}(CH_3)_4C_6H_2$	80	195
Pentamethylbenzene	$C_6H(CH_3)_5$	53	231
Hexamethylbenzene	$C_6(CH_3)_6$	166	265
Ethylbenzene	$C_6H_5CH_2CH_3$	−93	136
n-Propylbenzene	$C_6H_5CH_2CH_2CH_3$	Liquid	159.5
Cumene	$C_6H_5CH(CH_3)_2$	Liquid	152
n-Butylbenzene	$C_6H_5CH_2CH_2CH_2CH_3$	Liquid	180
t-Butylbenzene	$C_6H_5C(CH_3)_3$	Liquid	168
p-Cymene	$p\text{-}CH_3C_6H_4CH(CH_3)_2$	−73.5	177
1,3,5-Triethylbenzene	$1,3,5\text{-}(CH_3CH_2)_3C_6H_3$	Liquid	215
Hexaethylbenzene	$C_6(CH_2CH_3)_6$	129	305
Styrene	$C_6H_5CH\!=\!CH_2$	Liquid	146
Allylbenzene	$C_6H_5CH_2CH\!=\!CH_2$	Liquid	156
Stilbene ($trans$)	$C_6H_5CH\!=\!CHC_6H_5$	124	307
Diphenylmethane	$(C_6H_5)_2CH_2$	27	262
Triphenylmethane	$(C_6H_5)_3CH$	92.5	359
Tetraphenylmethane	$(C_6H_4)_4C$	285	431
Diphenyl	$C_6H_5\cdot C_6H_5$	70.5	254
p-Terphenyl	$C_6H_5\cdot C_6H_4\cdot C_6H_5$	171	
p-Quaterphenyl	$C_6H_5\cdot C_6H_4\cdot C_6H_4\cdot C_6H_5$	320	428^{18mm}
1,3,5-Triphenylbenzene	$1,3,5\text{-}(C_6H_5)_3C_6H_3$	174.5	

[a]The hydrocarbons listed that contain only one benzene ring have specific gravities in the range 0.86 - 0.90.

Source: Fieser and Fieser 1961, Table 18.1, p. 645. Reprinted by permission of the publisher.

Refined coke-oven gas consists mainly of hydrogen (52%) and methane (32%) (Fieser and Fieser 1961), accompanied by smaller amounts of carbon monoxide (4 to 9%), carbon dioxide (2%), nitrogen (4 to 5%), and ethylene and other olefins (3 to 4%). About 11,200 ft^3 of these gases are produced per ton of coal, giving an average calorific value of 570 Btu/ft^3. Through the refining process (about 3.2 gal per ton of coal), light oil is recovered. It contains 60% benzene, 15% toluenes, and some xylenes and naphthalene (Fieser and Fieser 1961).

2.7 COAL CHARACTERISTICS

2.7.1 Specific gravity

Because coal is a microporous solid (Sect. 2.7.8), the value obtained for the density or specific gravity depends on the dilatometric fluid used. The specific gravities of the pure organic substance of coals, measured in water, covers the range 1.23 to 1.72. Values for actual coals depend on the nature and content of their associated minerals, and also on the method of measurement. Nevertheless, it can be said that the specific gravities of coals in water pass through a minimum (Franklin 1949, as cited in van Krevelen 1961); the rank level at which the minimum occurs cannot be exactly specified, but is probably in the high-volatile A bituminous class (Franklin 1949). The close relationship between specific gravity and ash content has been used to determine ash content of coals from the anthracite region of Pennsylvania; the percentage of float or sink associated with a selected specific gravity bath has been used to control ash content of cleaned coal shipments (Yancey and Geer 1968).

2.7.2 Hardness

The hardness of coals is related to rank. Hardness varies with rank by decreasing from a maximum in coals containing about 40% volatile matter to a minimum in coals containing about 15% volatile matter, then increasing again in coals containing about 5% volatile matter. The minimum hardness value is attained at about 85 to 90% carbon content (Yancey and Geer 1968).

2.7.3 Coal strength

The strength of coal in crushing and grinding is a highly variable characteristic. Laboratory studies have shown that coals from the same horizon or from different horizons in the same bed give widespread results, probably due to different manipulative techniques and the presence of random cracks. Generally, however, the compressive strength is related to rank, and minimum strength is reached at 20 to 25% volatile matter content (Fig. 2.19) (Yancey and Geer 1968). The compressibility of the lower-rank coals increases as volatile matter is lost during metamorphism, then decreases as the higher-rank coals become more hardened and compact.

2.7.4 Friability

Friability, which refers to the amount of breakage that results from the handling of coal, is affected by several coal characteristics such as toughness, elasticity, fracture characteristics, and strength. Highly friable coals result in increased surface area upon handling, allowing more rapid oxidation, making conditions more favorable for spontaneous combustion, and lowering coking quality (Yancey and Geer 1968).

ORNL DWG 76-7943

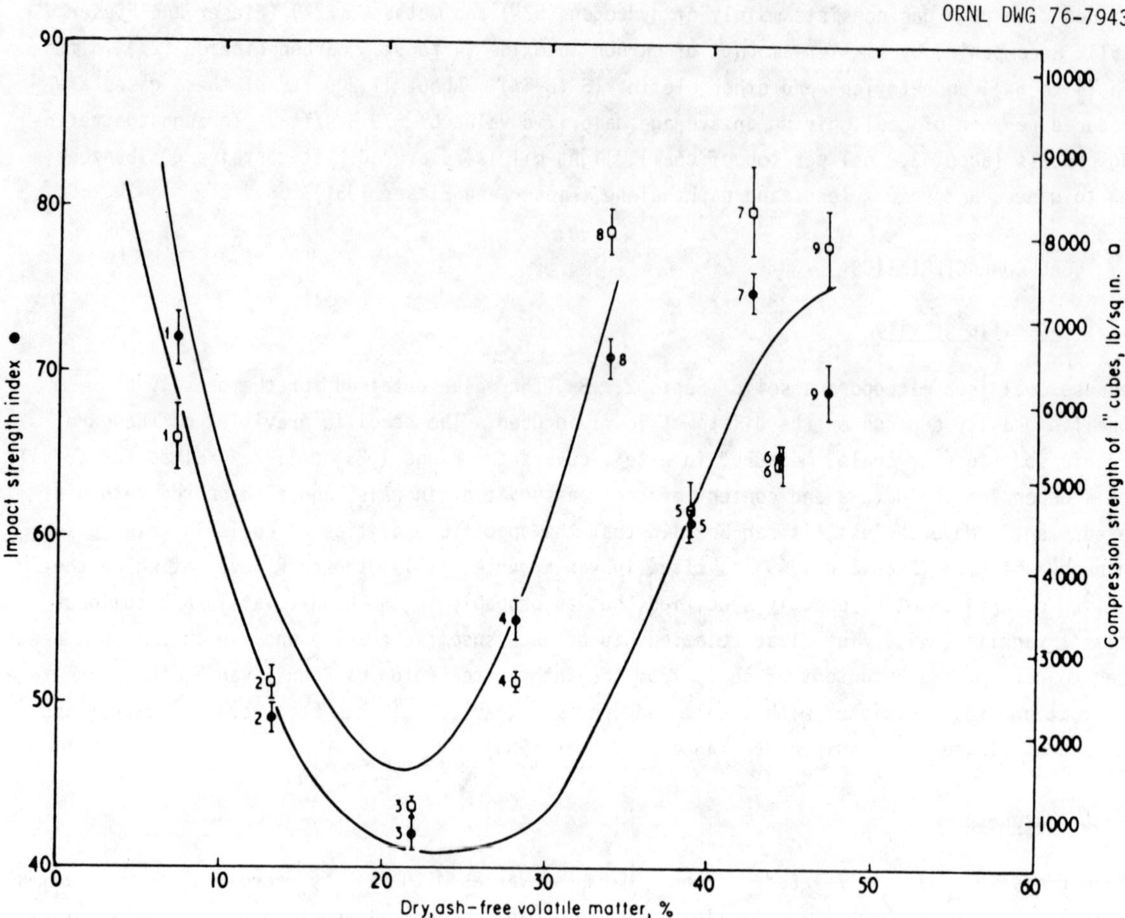

Fig. 2.19. Strength and rank. <u>Source</u>: From Brown and Hiorns 1963, Fig. 4, p. 131. Reprinted by permission of the publisher.

Laboratory tests have shown that friability and rank of coals are related (Table 2.42): Lignites are the least friable, and friability increases with rank up to the low-volatile bituminous coals, then decreases in anthracite. Although a relationship does exist between rank and friability, it is only a general one. Friability of coals of the same rank may vary widely.

Table 2.42. Average friability values of different ranks of coal

Rank of coal	Number of tests	Friability (%)
Anthracite	36	33
Low-volatile bituminous	27	70
Medium-volatile bituminous	87	43
Subbituminous A	40	30
Subbituminous B	29	20
Lignite	16	12

Source: Yancey and Geer, p. 1-31. Reprinted by permission of the publisher.

2.7.5 Grindability

A general relationship also exists between grindability and rank: Medium- and low-volatile groups are much easier to grind than are either high-volatile bituminous and subbituminous coals or anthracites (Yancey and Geer 1968). The ASTM uses the Hardgrove test of grindability, in which a 50-g portion of closely sized coal is ground for 60 revolutions in a ring-and-ball type mill. A grindability index is calculated by a formula that compares the percent of <200-mesh material obtained from the test with a base coal chosen as 100 grindability (Yancey and Geer 1968). Table 2.43 indicates the wide range of grindability values of some representative American coals.

Table 2.43. Grindability indexes of some American coals

State	County	Bed	Hardgrove grindability index	Rank[a]
Pennsylvania	Cambria	Lower Kittanning	109	lv Bituminous
Pennsylvania	Indiana	Lower Freeport	92	mv Bituminous
Pennsylvania	Washington	Pittsburgh	55	hv Bituminous
Pennsylvania	Westmoreland	Upper Freeport	65	hv Bituminous
West Virginia	Fayette	Sewell	86	mv Bituminous
West Virginia	McDowell	Pocahontas No.3	96	mv Bituminous
West Virginia	Wyoming	Powellton	58	hv Bituminous
West Virginia	Wyoming	No.2 Gas	70	mv Bituminous
Virginia	Wise	Morris	43	hv Bituminous
Virginia	Wise	Taggart	62	hv Bituminous
Virginia	Dickenson	Upper Banner	84	hv Bituminous
Virginia	Buchanan	Raven	98	mv Bituminous
Illinois	Sangamon	No.6	55	hv Bituminous
Illinois	Williamson	No.6	57	hv Bituminous
Illinois	Fulton	No.5	63	hv Bituminous
Illinois	Vermillion	No.7	56	hv Bituminous
Kentucky	Pike	Elkhorn Nos.1 & 2	42	hv Bituminous
Kentucky	Bell	Hight Splint	40	hv Bituminous
Kentucky	Muhlenburg	No.12	55	hv Bituminous
Ohio	Harrison	No.8	51	hv Bituminous
Ohio	Belmont	No.9	50	hv Bituminous
Indiana	Sullivan	No.V	55	hv Bituminous
Alabama	Walker	Black Creek	44	hv Bituminous
Utah	Carbon	Castle Gate	47	hv Bituminous
Pennsylvania	Schuylkill	Various	38	Anthracite

[a]According to the 1976 Keystone Coal Industry Manual. hv = high-volatile; mv = medium-volatile; lv = low-volatile.

Source: Yancey and Geer 1968, Table 1-7, p. 1-33. Reprinted by permission of the publisher.

2.7.6 Abrasiveness

Abrasiveness can be an important coal characteristic for the industrial consumer as well as for miners because of the wear it produces on coal-handling equipment. Studies of coal abrasiveness have demonstrated that it is caused more by the impurities in coal than by the coal material itself. Therefore, a well-cleaned and -prepared coal will be much less abrasive than unwashed coal (Yancey and Geer 1968).

2.7.7 <u>Moisture content</u>

Generally, coal freshly removed from even a relatively dry area in a mine is saturated with moisture. The amount of moisture varies with rank, ranging from about 45% in lignite, to 1 to 3% in bituminous coals, and to still less in anthracite. Moisture can be considered an impurity because it acts as a diluent and reduces the available energy in relation to the amount of coal present (Yancey and Geer 1968).

2.7.8 <u>Porosity</u>

As a result of the manner in which the coal molecules are linked together, an extensive internal pore system exists in coal. This system is extremely important in coal conversion because the conversion processes require rapid and efficient contact with the coal surface by both reacting gases and catalysts (Wiser 1973).

2.7.8.1 <u>Micropores and macropores</u>

Bangham (1943), as cited by Maggs (1943), put forth the hypothesis that coal is a dried-out colloid consisting of innumerable minute particles called micelles which are held together by cohesive forces less powerful than chemical bonds.

Van Krevelen (1961) also characterizes coal as a solid colloid having a certain porosity, to which it owes some of its properties — the ability to absorb gases and vapors, to swell in vapors and liquids, and to develop heat on wetting. The porosity, or volume percentage occupied by pores, may be determined from density measurements by using helium and mercury as displacement liquids. Zwietering and van Krevelen (1954) conclude from such experiments under pressure that coal contains two pore systems — a macropore system, which will allow entry of mercury under pressure, and a micropore system, which cannot be permeated by mercury even at high pressures. At normal pressure, mercury will not penetrate into the pore system if the pores are no larger than 10 μm in diameter.

The internal surface of coal resulting from the pores can also be determined by measuring the temperature change, due to the "heat of wetting," when coal is exposed to a liquid. Penetration of liquids such as methanol into a coal causes it to start swelling. During this process, more energy is liberated by the interaction of the coal surface and the liquid than is required for the swelling to occur; the heat given off is the "heat of wetting." The energy is believed to be due mainly to van der Waals and dipole forces. Methanol has been used widely for this purpose because of the speed with which the heat of wetting is released (van Krevelen 1961). Maggs (1943), using methanol as the wetting agent, determined the "heat of wetting" of coal to be about 1 cal/10 m^2 of internal surface. According to Maggs (1943), D. H. Bangham directed the research in which the "heat of wetting" property of coal was first used to determine the internal surface area. The results of this research indicated that a 1-lb lump of coal could have an internal surface area of 100,000 to 1,000,000 ft^2. Surface area values obtained by heat of wetting and gas sorption techniques differ considerably in low-rank coals; however, true surface area values for such coals probably lie between those obtained by the two techniques (van Krevelen 1961).

Van Krevelen states that the micropores of coal are unlike a normal porous system because they have a cagelike structure (which has been observed in other organic compounds), which traps gases. The macropores, he continues, result from cracks.

Gan, Nandi, and Walker (1973) worked with 27 coals, including samples representing all ranks and all major coalfields of the United States. Tests were performed on particles of 40 x 70 mesh size; therefore, they do not necessarily represent the whole coal (Table 2.44). The pore structure of the coals was studied by determining (1) the total volume from helium and mercury densities, (2) the pore surface areas from absorption of nitrogen at 77°K and carbon dioxide at 298°K, (3) the macropore size distribution from mercury porosimetry, and (4) size distribution of pores below 300 Å from nitrogen isotherms measured at 77°K.

Because helium, as a small atom, can penetrate almost all the pores of the coal, tests were performed at normal pressure at 32.5°C. Application of some pressure, however, is required for penetration by mercury. Gan, Nandi, and Walker (1973) determined that negligible penetration of coal by mercury occurs above 60 psia; therefore, mercury density values were calculated from measurements of mercury displaced by the samples at 60 psia. Helium and mercury densities, along with the total open pore volumes of 12 coals, are given in Table 2.45. Both helium and mercury densities decreased gradually, reaching a minimum at about 80% carbon, then rose rapidly as the carbon content increased to 90%. Total open pore volumes were calculated from the differences of the reciprocals of mercury and helium densities. Gan, Nandi, and Walker (1973) measured the surface area of several American coals by nitrogen adsorption at 77°K and carbon dioxide adsorption at 298°K. The results are summarized in Fig. 2.20. Walker (1973a) suggests that the differences in values thus obtained were caused by micropores of the 4- to 5-Å range. Low- and high-rank coals were observed to be very porous, whereas the intermediate-rank coals (78% carbon) were observed to exhibit relatively little porosity. Table 2.46 gives the total porosity of several American coals and the distribution of pore sizes.

There seems to be no apparent correlation between total pore volume and rank; however, low-rank coals usually have a greater number of large pores than do high-rank coals (Walker 1973a). The pore and capillary structure of coal is important because the materials involved in chemical reaction with the coal must pass through these spaces (Dryden 1963).

2.8 EFFECTS OF COAL COMPOSITION ON COAL CONVERSION

2.8.1 <u>Liquefaction</u>

Given et al. (1975a) studied the conversion of coal to oil in order to relate the yield, viscosity, and composition of the product oil to characteristics of the coals.

The petrographic and chemical analyses of some samples are shown in Table 2.47; results of the experiments are given in Table 2.48.

The data obtained by a Gulf procedure involving catalytic hydrogenation in batch autoclaves indicate that lignites and subbituminous coals yield less oil than do coals of higher rank; however, in coals having carbon contents greater than 90%, the amount of oil yielded drops drastically. A more precise description of this drop in yield is not given because coals containing 86 to 89% carbon seem to be rare in North America (Given et al. 1975a). High-volatile

Table 2.44. Analyses of coals

Sample	Location	Seam	Carbon	Hydrogen	Nitrogen	Sulfur	Oxygen (by diff.)	Vitrinite content (vol. %)[b]
			Ultimate analysis (%)[a]					
PSOC-85	PA	No. 8 Leader	91.2	3.8	0.60	1.1	3.1	97.5
PSOC-80	PA	Buck Mountain	90.8	2.6	0.76	0.64	5.1	96.3
PSOC-130	WV	Pocahontas No. 3	90.5	4.3	1.1	0.55	3.4	72.7
PSOC-127	PA	Lwr. Kittanning	89.5	5.0	1.0	0.83	3.6	77.7
PSOC-134	AL	Pratt	89.3	4.7	1.5	1.0	3.3	75.5
PSOC-135	AL	Pratt	88.3	4.9	0.25	0.65	5.7	83.1
PSOC-4	KY	Elkhorn No. 3	83.8	5.7	1.5	0.88	7.9	67.3
PSOC-102	PA	Pittsburgh	83.4	5.2	1.1	1.4	8.7	72.8
PSOC-106	IN	Indiana No. 1 Block	82.7	5.5	0.55	0.69	10.4	26.2
PSOC-95	WA	Queen or No. 4	81.6	6.0	1.0	1.7	9.4	89.6
PSOC-105A	IN	Indiana No. 1 Block	81.3	5.7	1.0	1.8	9.9	62.5
PSOC-215	KY	Kentucky No. 9	81.3	5.7	1.6	4.5	6.7	79.4
PSOC-24	IL	No. 2 Colchester	80.0	5.5	1.0	4.5	8.8	88.1
Rand	IN		79.9	5.2	1.5	4.3	8.9	
PSOC-213	KY	Kentucky No. 9	78.8	5.6	1.7	4.2	9.4	73.4
PSOC-22	IL	Illinois No. 6	78.7	5.8	1.5	2.8	11.0	88.3
PSOC-26	IL	Illinois No. 6	77.2	5.6	1.1	7.4	8.3	88.5
POC-197	PA	'B' Seam (Lwr. Kittanning)	76.5	5.6	1.2	4.5	12.2	85.3
PSOC-189	IL	Illinois No. 1	75.9	5.6	1.0	5.5	11.8	78.0
PSOC-190	IL	Illinois No. 6	75.5	5.3	1.0	3.3	14.6	88.9
PSOC-97	WY	No. 80	75.0	5.5	0.58	1.3	17.5	86.7
PSOC-138	TX	Darco	74.3	4.9	0.37	0.75	19.6	75.1
PSOC-100	WY	Wyodak-Roland	72.0	5.2	0.81	0.55	21.2	86.0
PSOC-141	TX	Darco	71.7	5.2	1.3	0.90	20.8	75.3
PSOC-90	MT	Lower Lignite Seam - Tongue River Member	71.5	4.8	0.83	0.50	22.2	60.9
PSOC-87	ND	Zap	71.2	5.2	0.56	0.69	22.2	64.7
PSOC-89	ND	Harmon	63.3	4.6	0.47	1.5	29.9	70.3

[a]Dry, ash-free basis.
[b]Mineral-matter-containing basis.

Source: Gan, Nandi, and Walker 1973, Table 1, p. 273. Reprinted from *Fuel*: "Nature of porosity in American coals" by permission of the publishers, IPC Business Press Ltd.

Table 2.45. Helium and mercury densities and total
open pore volume of coals

Sample	Carbon (%)[a]	LTA[b] mineral matter (%)	Helium density (g/cm³)[c]	Helium density (g/cm³)[d]	Mercury density (g/cm³)[c]	Mercury density (g/cm³)[d]	Total open pore volume (cm³/g)[d]	Open porosity (%)
PSOC-80	90.8	19.2	1.67	1.53	1.51	1.37	0.076	10.4
PSOC-127	89.5	6.6	1.38	1.33	1.29	1.25	0.052	6.5
PSOC-135	88.3	5.4	1.36	1.32	1.29	1.25	0.042	5.3
PSOC-4	83.8	2.5	1.30	1.28	1.25	1.23	0.033	4.1
PSOC-105A	81.3	10.3	1.35	1.27	1.15	1.07	0.144	15.5
Rand	79.9	14.0	1.36	1.25	1.24	1.14	0.083	9.5
PSOC-26	77.2	18.2	1.41	1.27	1.19	1.06	0.158	16.7
POC-197	76.5	25.3	1.49	1.29	1.33	1.13	0.105	11.9
PSOC-190	75.5	10.9	1.38	1.30	1.07	1.00	0.232	23.2
PSOC-141	71.7	9.9	1.42	1.35	1.24	1.17	0.114	13.3
PSOC-87	71.2	9.0	1.46	1.40	1.28	1.22	0.105	12.8
PSOC-89	63.3	13.7	1.55	1.45	1.41	1.31	0.073	9.6

[a]Dry, ash-free basis.
[b]Remaining after low-temperature ashing of coal in an oxygen plasma at about 150°C.
[c]Mineral-matter-containing basis.
[d]Mineral-matter-free basis.

Source: Gan, Nandi, and Walker 1973, Table 3, p. 275. Reprinted from *Fuel*: "Dependence of coal liquefaction on coal characteristics" by permission of the publisher, IPC Business Press Ltd.

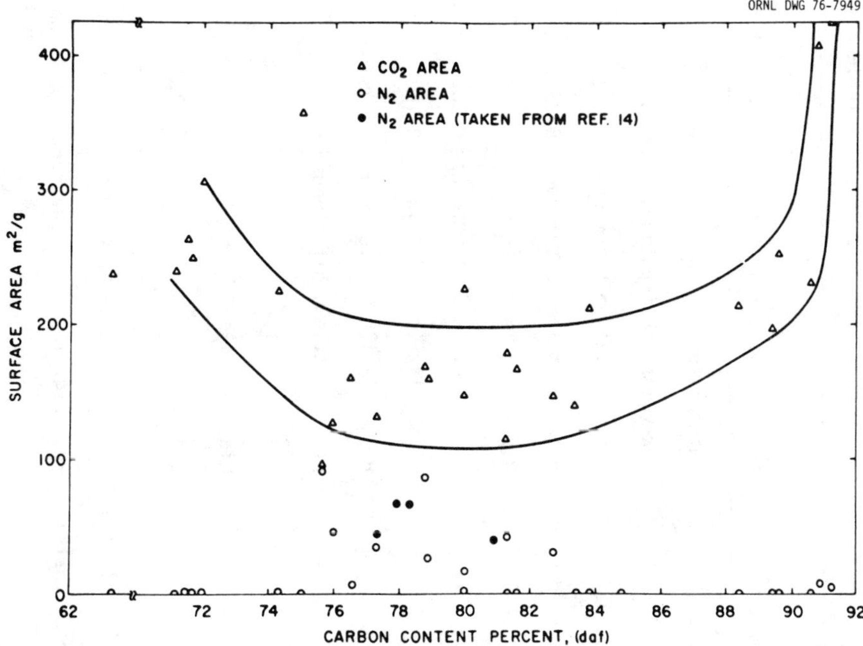

Fig. 2.20. Surface area of several American coals, as measured by nitrogen adsorption at 77°K and carbon dioxide adsorption at 298°K. <u>Source</u>: Walker 1973*a*, Fig. 11. Reprinted by permission of the publisher.

Table 2.46. Gross pore distributions in coals

Sample	Rank	Pore size[a] (cm^3/g)				Pore distribution (%)		
		V_t	V_1	V_2	V_3	V_3	V_2	V_1
PSOC-80	Anthracite	0.076	0.009	0.010	0.057	75.0	13.1	11.9
PSOC-127	Low-volatile bituminous	0.052	0.014	0.000	0.038	73.0	Nil	27.0
PSOC-135	Medium-volatile bituminous	0.042	0.016	0.000	0.026	61.9	Nil	38.1
PSOC-4	High-volatile A bituminous	0.033	0.017	0.000	0.016	48.5	Nil	51.5
PSOC-105A	High-volatile B bituminous	0.144	0.036	0.065	0.043	29.9	45.1	25.0
Rand	High-volatile C bituminous	0.083	0.017	0.027	0.039	47.0	32.5	20.5
PSOC-26	High-volatile C bituminous	0.158	0.031	0.061	0.066	41.8	38.6	19.6
POC-197	High-volatile B bituminous	0.105	0.022	0.013	0.070	66.7	12.4	20.9
PSOC-190	High-volatile C bituminous	0.232	0.043	0.132	0.070	24.6	56.9	18.5
PSOC-141	Lignite	0.114	0.088	0.004	0.022	19.3	3.5	77.2
PSOC-87	Lignite	0.105	0.062	0.000	0.043	40.9	Nil	59.1
PSOC-89	Lignite	0.073	0.064	0.000	0.009	12.3	Nil	87.7

[a] V_t (total porosity) = 4 - 30,000 Å; V_1 (macropores) = 300 - 30,000 Å; V_2 (transitional) = 12 - 300 Å; V_3 (micropores) = 4 - 12 Å.

Source: Walker 1973a, Fig. 12, p. 18. Reprinted by permission of the publisher.

Table 2.47. Analytical data for coals studied

Sample (PSOC number)	Macerals[a] (vol %)[b]								Mineral matter (%)[c]	Elements (%)							Volatile matter (%)[d]
	V	PV	F	SF	MM	GM	S	R		Sulfur			C^d	H^d	N^d	O(diff.)[d]	
										Total[c]	Pyritic[c]	Organic[d]					
141	75	0	7	3	4	7	3	1	9.5	0.82	0.15	0.75	72.6	5.2	1.4	20.1	48.4
190	84	0	3	1	1	4	6	1	10.1	3.1	1.1	2.1	76.8	5.4	1.1	14.6	36.7
26	86	0	2	1	1	3	6	1	14.2	6.7	4.2	2.4	80.2	5.8	1.2	10.5	44.6
149	67	5	4	6	2	6	7	3	20.0	6.8	3.9	1.9	81.8	5.3	0.2	10.8	41.8
68	69	11	2	2	10	3	3	0	6.0	0.85	0.27	0.55	82.0	5.7	1.5	10.3	41.3
70	83	7	0	0	0	2	7	1	29.4	0.35	0.04	0.4	82.4	4.7	0.8	10.7	40.4
111	74	12	1	1	2	2	8	0	8.1	1.6	0.6	1.1	83.4	5.6	1.7	8.3	38.3
104	73	12	2	2	3	2	5	1	12.1	3.4	2.0	1.6	84.8	5.6	1.4	6.7	39.5
95	80	7	0	0	1	2	7	3	23.8	1.4	0.6	1.0	84.4	5.9	1.1	7.6	39.9
257	67	22	3	2	0	3	1	2	10.1	0.8							29.3
87	65	0	5	7	3	5	7	8	10.6	0.6	0.2	0.5	72.0	5.2	0.6	21.7	53.6
99	61	23	2	2	3	3	6	0	27.3	0.7	0.3	0.6	72.5	5.2	0.8	20.9	67.1
151	79	0	4	2	2	2	2	9	5.7	0.45	0.05	0.4	78.3	5.8	1.2	14.3	45.4
185	79	6	3	1	1	2	8	0	16.1	3.2	1.6	1.7	81.8	5.8	1.1	9.5	42.4
187	75	6	3	2	2	3	9	0	7.5	1.1	0.35	0.8	80.5	5.6	1.0	12.2	39.0
105	63	0	6	3	3	4	20	1[e]	13.4	2.2	1.2	1.1	81.9	5.6	0.4	11.0	37.6
160A	77	5	4	4	2	2	5	1	6.2	0.5	0.02	0.5	82.5	5.4	1.8	9.8	41.1
110	65	12	2	1	3	3	13	1	7.4	1.2	0.6	0.7	85.8	5.7	1.3	6.6	37.6
129	84	1	5	2	3	4	0	1	7.5	0.7	0.3	0.5	91.0	5.0	1.0	2.6	19.2

[a] V = vitrinite, PV = pseudovitrinite, F = fusinite, MM = massive micrinite, GM = granular micrinite, S = sporinite, R = resinite.

[b] Mineral-matter-free basis.

[c] Dry basis.

[d] Dry mineral-matter-free basis.

[e] This sample also contained 0.2% alginite and 2% cutinite.

Source: Given et al. 1975a, Table 1, p. 35. Reprinted from *Fuel*: "Dependence of coal liquefaction on coal characteristics" by permission of the publisher, IPC Business Press Ltd.

Table 2.48. Results of autoclave experiments (all runs at 400°C)

Sample (PSOC number)	Seam, Mine, State	Province	ASTM rank	R_o[a] (%)	C (% dmmf)	CV[b] (MJ/kg)	CV[b] (Btu/lb)	Solvatn.	Asph.	Oil	Gas + H_2O	Asph./oil	Benzene insol.	Sats.	Viscosity (mm^2/sec)
141	Darco, Texas	Pacific	Lignite	0.35	72.6	18.9	8,120	66	35.2	23.7	7.4	1.5			
190	Ill. No. 6, Mecco, Ill.	East Interior	Subbituminous	0.45	76.8	23.5	10,910	75	39.1	26.4	7.9	1.5			
26	Illinois No. 6 Carrier	East Interior	High-volatile C bituminous	0.54	80.2	29.35	12,605	83	45.6	28.4	8.4	1.55			
149	Outcrop, unnamed seam, Floyd French Farm Okla.	West Interior	High-volatile B bituminous	0.54	81.8	32.25	13,860	80	41.3	28.8	9.5	1.44			
68	Lower Sunnyside, Geneva, Utah	Rocky Mountain	High-volatile B bituminous	0.64	82.0	32.45	13,950	77	42.1	26.1	8.9	1.6			
70	No. 3, Big Seam Strip Roslyn, Wash.	Pacific	High-volatile B bituminous	0.77	82.4	33.15	14,220	100	48.8	34.9	16.7	1.4			
111	Pittsburgh, Marian, Pa.	Appalachian	High-volatile A bituminous	0.79	83.4	35.05	15,060	84	42.8	28.9	11.5	1.5			
104	Pittsburgh, Mathies, Pa.	Appalachian	High-volatile A bituminous	0.89	84.8	34.95	15,030	82	41.1	29.5	11.1	1.4			
95	Queen, Queen, Wash.	Pacific	High-volatile A bituminous	0.84	84.4	35.05	15,060	94	48.6	32.4	13.0	1.5			
257	Upper Freeport, Simca, Pa.	Appalachian	Medium-volatile bituminous	1.22		36.2	15,575	56	31.5	14.3	8.8	2.1			
87A	Zap, Indian Head, N.Dak.	Interior	Lignite	0.30	72.0	20.0[c]	8,600	95.9	26.3				29.9	1.1	7.7
87B								97.1	31.4				56.8	1.1	27
99	School, Dave Johnston, Wyo.	North Great Plains	Lignite	0.31	72.5	20.75	8,900	78.8	43.8				19.2	2.6	3.9
151	Blue, McKinley, N.Mex.	Rocky Mountain	Subbituminous	0.40	78.3	27.6	11,845	72.1	23.9				46.9	1.5	3.9
185	Ill. No. 6, Minnehaha, Ind.	East Interior	High-volatile B bituminous	0.55	81.8	30.5	13,100	88.2	39.2				41.7	1.3	8.5
187	Ill. No. 6, Orient 3, Ill.	East Interior	High-volatile B bituminous	0.64	80.5	30.85	13,235	88.6	33.9				53.3	0.9	25
68	Lower Sunnyside, Geneva, Utah	Rocky Mountain	High-volatile B bituminous	0.64	82.0	32.45	13,950	86.4	33.3				41.5	1.7	13
105	Indiana No. 1, Old Glory, Ind.	East Interior	High-volatile B bituminous	0.73	81.9	31.8	13,690	80.8	38.2				24.7	0.9	8.1
70	No. 3, Big, Seam Strip Roslyn, Wash.	Pacific	High-volatile B bituminous	0.77	82.4	31.15	14,220	90.7	43.4				63.1	1.4	10
160A	Unnamed Seam, Big, Colorado	Rocky Mountain	High-volatile A bituminous	0.75	82.5	33.0	14,180	92.4	34.3				52.5	1.8	15
95[d]	Queen, Queen, Wash.	Pacific	High-volatile A bituminous	0.84	84.4	35.05	15,060	97.6	29.6				29.2		9.4
110	Pittsburgh, Marian, Pa.	Appalachian	High-volatile A bituminous	0.93	85.8	35.05	15,065	87.8	27.5				53.9	0.8	20
129	Lower Kittanning, Cambria 33, Pa.	Appalachian	Low-volatile bituminous	1.68	91.0	36.4	15,675	?	(Tar)						550[e]

[a] R_o = mean maximum reflectance of vitrinite under oil immersion.
[b] CV = calorific value, moist mineral-matter-free basis; 1 MJ/kg - 430 Btu/lb.
[c] On untreated coal.
[d] Solvation data obtained in run at 400°C; at 385°C, the product was not filterable.
[e] At 150°C.

Source: Given et al. 1975a, Tables 3 and 4, pp. 37-38.

Reprinted from *Fuel*: "Dependence of coal liquefaction on coal characteristics" by permission of the publisher, IPC Business Press Ltd.

bituminous coals seem to produce the best conversion results, but no trends were observed in rank parameters of such coals (Given et al. 1975a).

The molecular structure parameters of asphaltenes from sodium-rich and sodium-free lignites (87A and B in Table 2.49) were found to be very different, indicating a considerable difference in the mechanism of the liquefaction process. Although not shown in the data, resin contents were determined. They were mainly phenols and were found to make up 20% of the total weight of filtrate from the sodium-rich sample and 11.3% of the filtrate weight from the sodium-free sample. These values include some phenols from the anthracene oil that was used as a vehicle in the analysis. More phenolic material seems to result from hydrogenation when the carboxylic acids are in the form of sodium salts rather than free acids (Given et al. 1975a).

Table 2.49. Molecular parameter analyses of asphaltenes from coal hydrogenation runs, obtained by proton nuclear magnetic resonance

Sample (PSOC number)	Seam	Fractionated aromatic carbon	Aromatic rings/ average molecule	Alkyl substituent/ average molecule	Carbon atoms/alkyl substituent	Group types[a]		
						mono-	di-	tri-
87A	Zap	0.74	1.5	1.6	1.8	59.5	28.2	12.3
87B		0.50	1.6	3.9	2.2	38.1	61.9	0.0
99	School	0.60	1.8	3.2	1.9	51.4	17.2	31.4
151	Blue	0.54	1.5	3.1	2.2	70.1	14.3	15.6
185	Illinois No. 6	0.69	1.7	2.3	1.8	54.0	18.8	27.2
187	Illinois No. 3	0.66	1.5	2.2	1.9	66.3	17.2	16.5
68	Lower Sunnyside	0.59	1.4	2.7	2.0	71.2	14.1	14.7
105	Indiana No. 1 Block	0.62	1.4	2.4	2.0	74.5	7.7	17.8
70	No. 3 Big	0.57	1.7	3.7	1.8	56.7	21.2	22.0
160A	Big	0.61	1.5	2.5	2.0	54.6	45.4	0.0
110	Pittsburgh	0.65	1.8	2.8	1.7	46.2	31.8	22.0

[a]Percentage of aromatic systems with 1, 2, and 3 fused rings.

Source: Given et al. 1975a, Table 5, p. 38. Reprinted from *Fuel*: "Dependence of coal liquefaction on coal characteristics" by permission of the publisher, IPC Business Press Ltd.

Given et al. (1975b) also attempted to determine by catalytic hydrogenation the role of the petrographic composition of coals upon the behavior of the coal during the liquefaction process. A high-volatile A bituminous rank coal containing 84% carbon and having a mean maximum reflectance between 0.8 and 0.9% was liquefied. The conversion yield, as plotted against the sum of the reactive macerals (vitrinite, pseudovitrinite, and sporinite) in Fig. 2.21, correlates with the maceral content. It is suggested that sporinite may be more reactive than vitrinite (Given et al. 1975b).

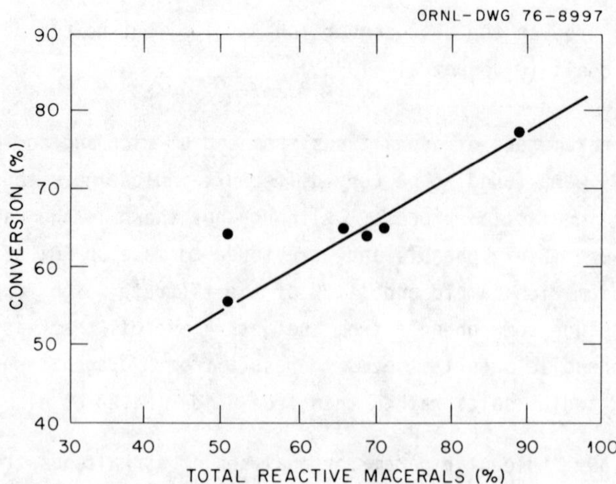

Fig. 2.21. Solvation of lithotypes from the Elkhorn No. 3 seam, Kentucky. <u>Source</u>: From Given et al. 1975*b*, Fig. 2, p. 45. Reprinted from *Fuel*: "Dependence of coal liquefaction on coal characteristics" by permission of the publisher, IPC Business Press Ltd.

In another study, a pseudovitrinite-rich sample produced a higher conversion than did a sample containing a greater amount of vitrinite, indicating that pseudovitrinite is reactive in liquefaction although it is unreactive in coking (Given et al. 1975*b*).

Cutinite, resinite, alginite, and sporinite have characteristically high hydrogen contents and volatile matter yields at low decomposition temperatures and a high fluidity during pyrolysis, which should make them desirable for liquefaction processes (van Krevelen 1961); however, one such coal sample was tested by two groups (Given et al. 1975*b*) with almost no oil being produced in either test. In one of the tests, all of the vehicle was absorbed and a tarry solid was produced. It was suggested that, because the sample was a "block coal" containing 7.4% alginite and 40% exinite, its algal materials may have caused cross-linking of the coal molecules, thus altering the mechanical properties. Conversely, another block coal had a much better conversion and asphaltene-to-oil ratio. A coal sample from Utah (PSOC-155) resembled a block coal in containing 9.0% alginite, but also contained 74% vitrinite; good yields of oil were produced by this lithotype (Given 1975*b*).

2.8.2 Gasification

Walker (1973*b*) lists the gasification reactions, shown in Table 2.50, that are important in coal conversion processes. The oxygen reactions are important for the production of heat and low-Btu gas, the carbon–carbon dioxide and carbon–steam reactions are important for production of both low- and high-Btu gas, and the carbon–hydrogen reactions are important for production of high-Btu gas. Rates of the reactions are very different. However, impurities such as iron, cobalt, and nickel catalyze the gasification process as illustrated in Fig. 2.22. The effect of adding 50 ppm of each of these three metals to pure graphite on the rate of the carbon–carbon dioxide reaction is dramatic.

In their review of the structure and reactivity of coal, Tingey and Morrey (1973) conclude that a large number of compounds act as catalysts in the reactions of coal with hydrogen, oxygen, carbon dioxide, and water vapor, including many of the minerals in coal which act as internal catalysts.

Table 2.50. Coal gasification reactions

Reactions	Relative rates[a]	Enthalpy change (kcal/mole)
C-O$_2$	1×10^5	
$C(\beta) + O_2(g) = CO_2(g)$		-94.03
$C(\beta) + 1/2\ O_2(g) = CO_2(g)$		-26.62
$CO(g) + 1/2\ O_2(g) = CO_2(g)$		-67.41
C-CO$_2$	1	
$C(\beta) + CO_2(g) = 2CO(g)$		+40.79
C-H$_2$O	3	
$C(\beta) + H_2O(g) = CO(g) + H_2(g)$		+31.14
$CO(g) + H_2O(g) = CO_2(g) + H_2(g)$		- 9.65
$C(\beta) + CO_2(g) = 2CO(g)$		+40.79
$C(\beta) + 2H_2(g) = CH_4(g)$		-17.87
C-H$_2$	3×10^{-3}	
$C(\beta) + 2H_2(g) = CH_4(g)$		-17.87

[a]At 800°C and 0.1 atm pressure.

Source: Walker 1973*a*, Figs. 1 and 5, pp. 7, 11. Reprinted by permission of the publisher.

The chemical form of the catalyst is important in determining its activity. As shown in Fig. 2.23, when 50 ppm of iron was added to the carbon—carbon dioxide reaction at 850°C, gasification was at first measurable, but the catalyst was rapidly poisoned and the reaction stopped. Susceptibility measurements were taken, and it was determined that the iron had been oxidized to iron oxide. Carbon monoxide was added at the points indicated by the letters until the iron oxide was reduced to elemental iron, and the gasification reaction increased again (Walker 1973*b*).

The amount of the catalyst and its particle size are also important; this fact is illustrated in Fig. 2.24, where the rate of the reaction at 1000°C is plotted against the amount of iron. It is apparent that small amounts of iron increased the reaction more than did larger additions. Also, iron added in the form of iron oxalate catalyzed the reaction better than did the micron-sized powder form, especially the larger particles (Walker 1973*b*).

Walker cautions that difficulty may be encountered in predicting results on mixed gases from results on individual gases. Results of the reaction between carbon dioxide and pure carbon at 850°C are plotted in Fig. 2.25. When 10 mtorr of carbon dioxide remained, a small amount of hydrogen was added and the carbon—carbon dioxide reaction rate was greatly inhibited.

Interest has been shown in the use of coal chars for the gasification process. To produce the char, coals are heated in helium to 1000°C, then exposed to 1 atm of air at 500°C. Sometimes the coals are acid-washed in hydrochloric acid and demineralized in hydrofluoric acid prior to charring.

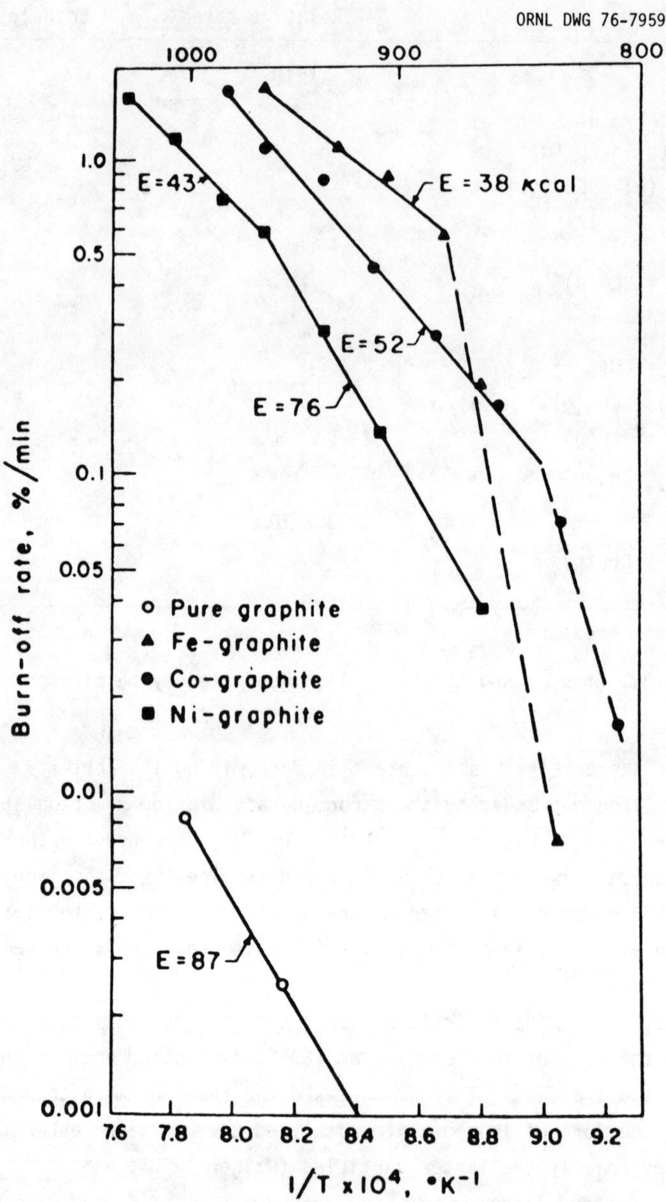

Fig. 2.22. Catalysis of gasification process by such impurities as iron, cobalt, and nickel. _Source_: From Walker 1973*a*, Fig. 6, p. 12. Reprinted by permission of the publisher.

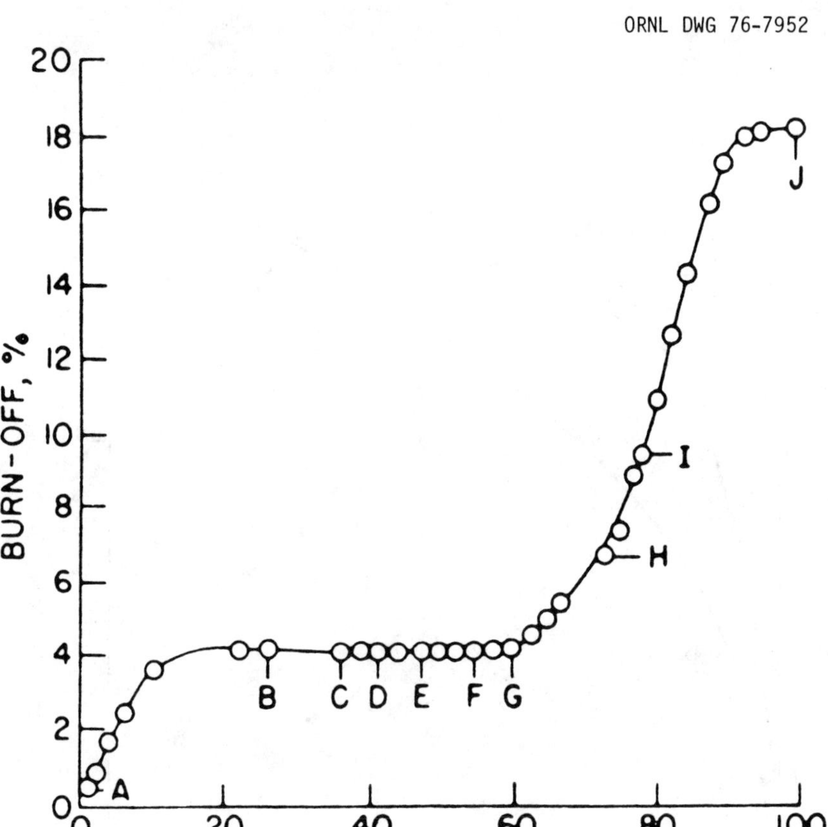

Fig. 2.23. Graphite reactivity at 850°C showing effect of CO_2–CO mixtures on catalyst activity. <u>Source:</u> From Walker 1973*a*, Fig. 7, p. 13. Reprinted by permission of the publisher.

The reactivity of chars is dependent upon the rank of the coal, as illustrated in Fig. 2.26. Some lignite chars are more than 100 times more reactive than high-rank coal chars (Walker 1973*b*).

Reactivity of low-volatile coal chars is increased by acid demineralization. Walker attributes this behavior to the removal of mineral matter and to the opening up of the macropores, which are less abundant in high-rank coals. The reactivities of chars from some acid-treated coals are listed in Table 2.51. The effect of increasing the heat treatment of chars is to decrease reactivity; therefore, heat should be held to a minimum. Walker concluded by stating that more work concerning the effects of structure, of mass transfer resistance, and of inherent chemical resistance is required before the role of catalysis in the gasification process can be fully understood.

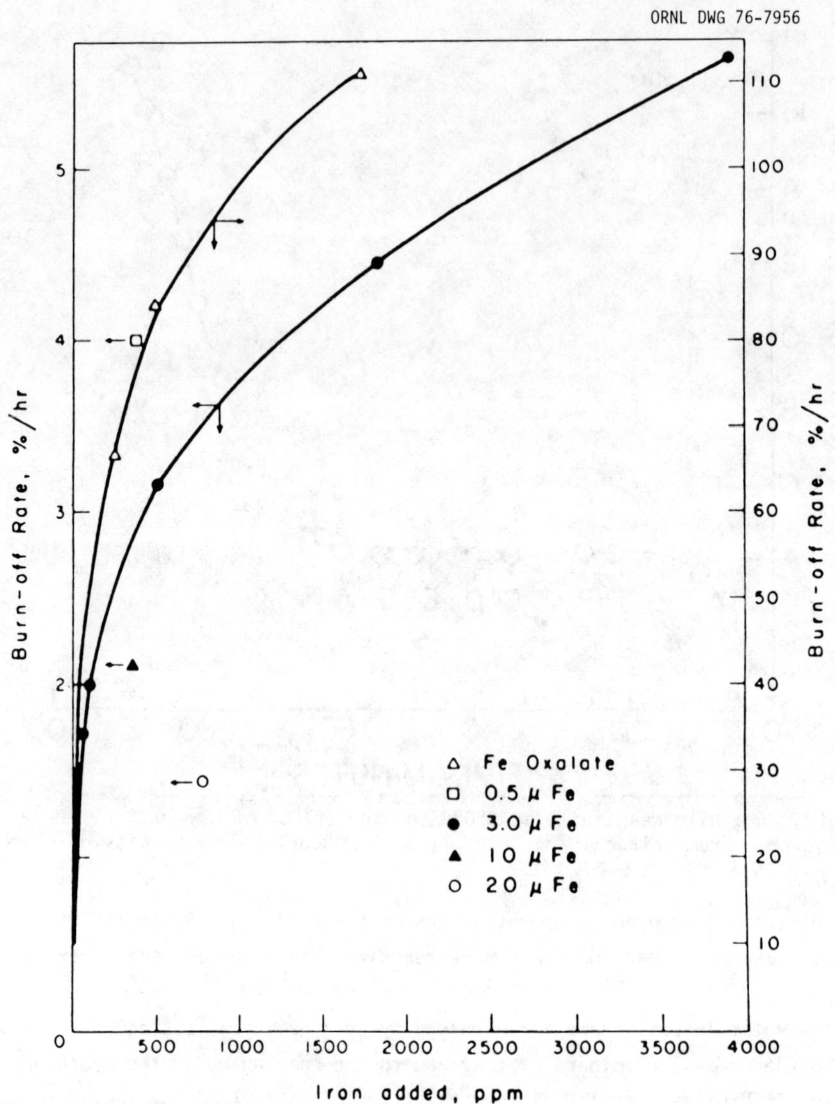

Fig. 2.24. Effect of amount and particle size of catalyst on gasification reaction.
<u>Source</u>: From Walker 1973*a*, Fig. 8, p. 14. Reprinted by permission of the publisher.

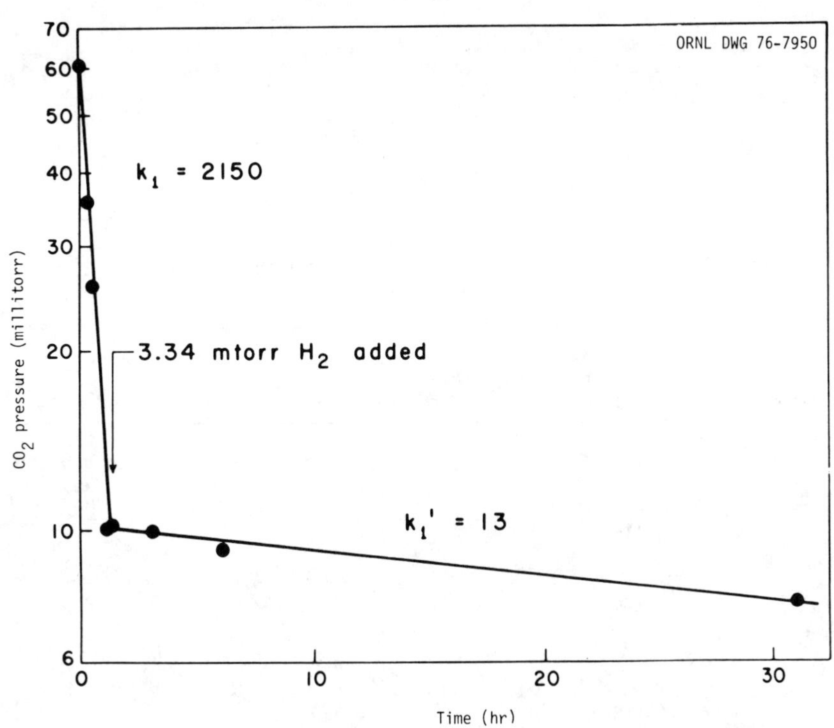

Fig. 2.25. Results of reaction between carbon monoxide and pure carbon at 850°C. Source: From Walker 1973*a*, Fig. 9, p. 15. Reprinted by permission of the publisher.

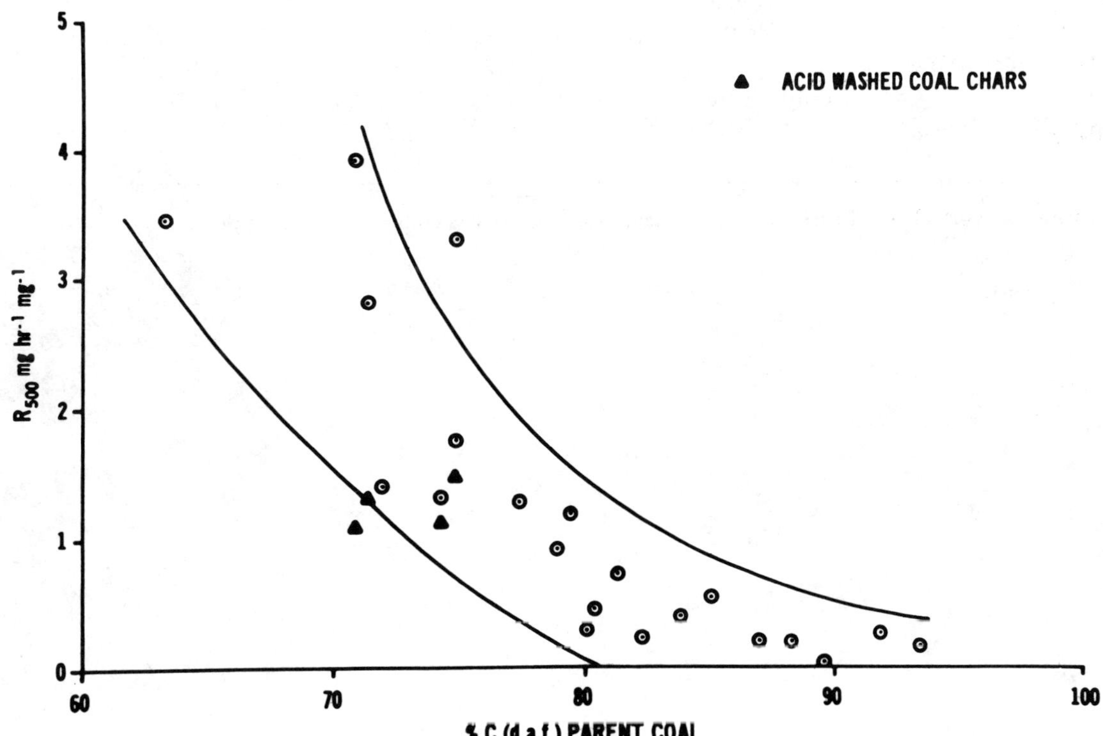

Fig. 2.26. Dependence of char reactivity on coal rank. Source: From Walker 1973*a*, Fig. 13, p. 19. Reprinted by permission of the publisher.

Table 2.51. Reactivities in air of chars prepared at 1000°C
from acid-treated coals

Coal number	Type	Ash in coal (%)	R_{500} (mg hr^{-1} mg^{-1})
89	Lignite	11.4	3.5
89, acid-washed		7.5	3.4
91	Lignite	7.9	4.0
91, acid-washed		2.7	1.1
87	Lignite	7.0	2.9
87, acid-washed		3.2	1.3
87, demineralized		<1	0.27
138	Lignite	8.5	1.3
138, acid-washed		5.1	1.1
101	Subbituminous	6.2	3.4
101, acid-washed		2.5	1.5
127	Low-volatile bituminous	5.0	0.04
127, demineralized		1.4	0.12
81	Anthracite	9.7	0.30
81, demineralized		<1	0.23

Source: Walker 1973*a*, Fig. 14, p. 20. Reprinted by permission of the publisher.

LITERATURE CITED

Aczel, T.; Gorbaty, M. L.; Maa, P. S.; and Schlosberg, R. H. 1975. Stability of adamantane to donor liquefaction conditions: Implications toward the structure of coal. *Fuel* 54: 295.

Akhtar, S.; Sharkey, A. G., Jr.; Shultz, J. L.; and Yavorsky, P. M. 1974. Organic sulfur compounds in coal hydrogenation products. Paper presented at 167th National Meeting of the American Chemical Society, Mar. 31 — Apr. 5, 1974, Los Angeles, California.

Anderson, H. C., and Wu, W.R.K. 1963. Properties of compounds in coal carbonization products. U.S. Bureau of Mines Bull. 606. Washington, D.C.: U.S. Government Printing Office.

Averitt, P. 1975. *Coal resources of the United States, January 1, 1974.* U.S. Geol. Surv. Bull. 1412.

Blom, L.; Edelhausen, L.; and van Krevelen, D. W. 1957. Chemical structure and properties of coal. XVIII — Oxygen groups in coal and related products. *Fuel* 36: 135-53.

Boyer, A. F.; Ferrand, R.; Ladam, A.; and Payen, P. 1961. Some indices on the structure of coals after the analysis of their degradation products. *Chim. Ind.* 86(5): 523-30.

Brown, R. L., and Hiorns, F. J. 1963. Mechanical properties. In *Chemistry of coal utilization,* supplementary volume, ed. H. H. Lowry, pp. 119-49. New York: John Wiley & Sons, Inc.

Cartz, S., and Hirsch, P. B. 1960. A contribution to the structure of coals from x-ray diffraction studies. *Philos. Trans. Royal Soc. Lond.* 252(1019): 558-602.

Chakrabartty, S. K. 1975. Organic chemistry of coal and chemical oxidation. In *NSF workshop on the fundamental organic chemistry of coal, July 19, 1975, University of Tennessee, Knoxville, Tennessee,* pp. 89-104.

Chakrabartty, S. K., and Berkowitz, N. 1974. Studies on the structure of coals. 3. Some inferences about skeletal structures. *Fuel* 53(4): 240-45.

Cohen, A. D. 1968. *The petrology of some peats of southern Florida.* Ph.D. Thesis, Pennsylvania State University.

DeCarlo, J. A.; Sheridan, E. T.; and Murphy, Z. E. 1966. *Sulfur content of United States coals.* U.S. Bureau of Mines, IC 8312.

Dryden, I.G.C. 1963. Coal constitution and reactions of coal. In *Chemistry of coal utilization,* supplementary volume, ed. H. H. Lowry, pp. 232-95. New York: John Wiley & Sons, Inc.

Fieser, L. F., and Fieser, M. 1961. Aromatic hydrocarbons. In *Advanced organic chemistry.* New York: Reinhold Publishing Company.

Flaig, W. 1966. Chemistry of humic substances in relation to coalification. In *Coal science,* Advances in Chemistry Series 55, ed. R. F. Gould, pp. 58-68. Washington, D.C.: American Chemical Society.

Francis, W., and Wheeler, R. V. 1925. The properties and constitution of coal ulmins. Studies in the composition of coal. *J. Chem. Soc.* 127: 2236-45.

Freudenberg, V. K.; Reznik, H.; Fuchs, W.; and Reichert, M. 1955. Experiments concerning the investigation of formation of lignins and wood. *Naturwissenschaften* 42(2): 29-35.

Gan, H.; Nandi, S. P.; and Walker, P. L., Jr. 1972. Nature of porosity in American coals. *Fuel* 51(4): 272-76.

Given, P. H. 1960. The distribution of hydrogen in coals and its relation to coal structure. *Fuel* 39: 147-53.

Given, P. H. 1971. Biological aspects of the geochemistry of coal. In *Advances in organic geochemistry, proceedings of the 5th international meeting on organic geochemistry,* ed. H. R. v. Gaertner and H. Wehner, pp. 69-92. New York: Pergamon Press.

Given, P. H. 1973*a*. How may coals be characterized for practical use? Paper presented at Short Course on Coal Characteristics and Coal Conversion Processes, Oct. 29 — Nov. 2, 1973, Pennsylvania State University, University Park, Penn.

Given, P. H. 1973*b*. Organic chemistry of coals. Paper presented at Short Course on Coal Characteristics and Coal Conversion Processes, Oct. 29 — Nov. 2, 1973, Pennsylvania State University, University Park, Pennsylvania.

Given, P. H. 1974. The chemistry of coals as it may relate to the materials technology of conversion processes. In *Workshop on materials problems and research opportunities in coal conversion, Apr. 16-18, 1974, Columbus, Ohio,* vol. 2.

Given, P. H. 1977. Personal communication.

Given, P. H.; Cronauer, D. C.; Spackman, W.; Lovell, H. L.; Davis, A.; and Biswas, B. 1975*a*. Dependence of coal liquefaction on coal characteristics. I. Vitrinite-rich samples. *Fuel* 54: 34-39.

Given, P. H.; Cronauer, D. C.; Spackman, W.; Lovell, H. L.; Davis, A.; and Biswas, B. 1975*b*. Dependence of coal liquefaction on coal characteristics. II. Role of petrographic composition. *Fuel* 54: 40-49.

Given, P. H., and Peover, M. E. 1960. Investigation of carbonyl groups in solvent extracts of coals. *J. Chem. Soc. (Lond.),* Part 1: 394-400.

Given, P. H.; Peover, M. E.; and Wyss, W. F. 1960. Chemical properties of coal macerals. I — Introductory survey, and some properties of exinites. *Fuel* 39: 323-40.

Ghosh, G.; Banerjee, A.; and Mazumdar, B. K. 1975. Skeletal structure of coal. *Fuel* 54: 294-95.

Gluskoter, H. J. 1975. Mineral matter and trace elements in coal. In *Trace elements in fuel,* Advances in Chemistry Series 141, ed. S. P. Babu, pp. 1-22. Washington, D.C.: American Chemical Society.

Hall, H. J.; Varga, G. M.; and Magee, E. M. 1974. Trace elements and potential pollutant effects in fossil fuels. In *Symposium proceedings: Environmental aspects of fuel conversion technology, May 1974,* pp. 35-47. EPA-R2-73-188. PB-220 376.

Halleux, A.; Delavarenne, S.; and Tschamler, H. 1961. Change of oxygen distribution with rank in vitrains and their extracts. *Fuel* 40: 74-76.

Harrison, J. A., and Latimer, I. S. 1968. Petrographic composition and nomenclature; preparation and carbonization. In *Coal preparation,* ed. J. W. Leonard and D. R. Mitchell, pp. 7-17. 3rd ed. New York: The American Institute of Mining, Metallurgical, and Petroleum Engineers, Inc.

Hayatsu, R.; Scott, R. G.; Moore, L. P.; and Studier, M. H. 1975. Aromatic units in coal. *Nature* 257: 378-80.

Headlee, A.J.W., and Hunter, R. G. 1955. The inorganic elements in coals. *W. Va. Geol. Surv. Bull.* XIII(A): 36-122.

Hereday, L. A.; Kostyo, A. E.; and Neuworth, M. B. 1965. Studies on the structure of coals of different rank. *Fuel (Lond.)* 44: 125-33.

Horton, L., and Randall, R. B. 1947. The occurrence of sulfur and nitrogen in coal. *Fuel* 26(5): 127-32.

Huck, V. G., and Karweil, J. 1953. Attempt at modeling the fine structure of coal. *Brennstoff-Chemie* 34: 97-102, 129-35.

Huck, V. G., and Karweil, J. 1955. Physical-chemical problems of carbonization. *Brennstoff-Chemie* 36: 1-11.

Karr, C., Jr. 1963. Low-temperature tar. In *Chemistry of coal utilization,* supplementary volume, ed. H. H. Lowry, pp. 539-79. New York: John Wiley & Sons, Inc.

Larsen, J. W. 1975. Summary — an evaluation of the current level of understanding of the fundamental organic chemistry of coal. In *NSF workshop on the fundamental organic chemistry of coal, July 17-19, 1975, University of Tennessee, Knoxville, Tennessee,* pp. 1-9.

Loison, R.; Peytavy, A.; Boyer, A. F.; and Grillet, R. 1963. The plastic properties of coal. In *Chemistry of coal utilization,* supplementary volume, ed. H. H. Lowry, pp. 150-201. New York: John Wiley & Sons, Inc.

Maggs, F.A.P. 1943. The relation between heat of wetting and the absolute value of the surface area of coals. *Inst. Fuel J.* 17: 49-54.

Mazumdar, B. K.; Ganguly, S.; Sanyal, P. K.; and Lahiri, A. 1966. Aliphatic structures in coal. In *Coal science,* Advances in Chemistry Series 55, ed. R. F. Gould, pp. 475-92. Washington, D.C.: American Chemical Society.

McNeil, D. 1966. The physical properties and chemical structure of coal tar pitch. In *Bituminous materials: Asphalts, tars, and pitches,* ed. A. J. Hoiberg, pp. 139-216. New York: Interscience Publishers.

Montgomery, R. S., and Holly, E. D. 1957. Decarboxylation studies of the structures of the acids obtained by oxidation of bituminous coal. *Fuel (Lond.)* 36: 63-75.

Mott, R. A. 1942. The origin and composition of coals. *Fuel (Lond.)* 21(6): 129-35.

Nelson, J. B. 1953. Assessment of the mineral species associated with coal. *Br. Coal Util. Res. Assoc.* 17(2): 41-55.

Nielson, G. F., and Bel, L. C., eds. 1976. *1976 Keystone coal industry manual.* New York: McGraw-Hill.

Nunn, R. C.; Lovell, H. L.; and Wright, C. C. 1953. Spectrographic analysis of trace elements in anthracite. In *Trans. Ann. Anthracite Conf., Lehigh Univ.* 11: 51-65.

Ode, W. H. 1963. Coal analysis and mineral matter. In *Chemistry of coal utilization,* supplementary volume, ed. H. H. Lowry, pp. 202-31. New York: John Wiley & Sons, Inc.

O'Gorman, J. V., and Walker, P. L., Jr. 1972. *Mineral matter and trace elements in U.S. coals.* OCR-RDR-61-IR-2. Washington, D.C.: U.S. Government Printing Office.

Parks, B. C. 1963. Origin, petrography, and classification of coal. In *Chemistry of coal utilization,* supplementary volume, ed. H. H. Lowry, pp. 1-34, New York: John Wiley & Sons, Inc.

Reggel, L.; Wender, I., and Raymond, R. 1970. Catalytic dehydrogenation of coal: Part 7. A comparison of exinite, micrinite and fusinite with vitrinite. *Fuel (Lond.)* 49: 281-86.

Ruch, R. R.; Gluskoter, H. J.; and Shimp, N. F. 1974. *Occurrence and distribution of potentially volatile trace elements in coal: A final report.* Environ. Geol. Notes 72, Illinois State Geological Survey.

Shell, F. 1976. Personal communication.

Shultz, J. L.; Friedel, R. A.; and Sharkey, A. G., Jr. 1972. *Detection of organic compounds in respiratory coal dust by high-resolution mass spectrometry.* U.S. Bureau of Mines Technical Prog. Rept. 61.

Shultz, J. L.; Kessler, T.; Friedel, R. A.; and Sharkey, A. G., Jr. 1972. High-resolution mass spectrometric investigation of heteroatom species in coal-carbonization products. *Fuel* 51: 242-46.

Schuyer, J., and van Krevelen, D. W. 1954. Chemical structure and properties of coal III — molar refraction. *Fuel (Lond.)* 33: 176-83.

Spackman, W. 1973. What is coal? The range of U.S. coals available. Paper presented at Short Course on Coal Characteristics and Coal Conversion Processes, Oct. 29–Nov. 2, Pennsylvania State University, University Park, Pennsylvania.

Spackman, W., and Barghoorn, E. S. 1966. Coalification of woody tissue as deduced from a petrographic study of Brandon lignite. In *Coal science,* ed. R. F. Gould, pp. 695-707. Washington, D.C.: American Chemical Society.

Teichmuller, M., and Teichmuller, R. 1966. Geological causes of coalification. In *Coal science,* ed. R. F. Gould, pp. 133-55. Washington, D.C.: American Chemical Society.

Tingey, G. L., and Morrey, J. R. 1973. *Coal structure and reactivity.* A Battelle Energy Program Report. Battelle Pacific Northwest Laboratories, Richland, Washington.

Thiessen, R. 1947. *What is coal?* U.S. Bureau of Mines Inf. Circ. 7397.

van Krevelen, D. W. 1961. *Coal.* New York: Elsevier Publishing Company.

van Krevelen, D. W., and Chermin, H. A. G. 1954. Chemical structure and properties of coal I — elementary composition and density. *Fuel (Lond.)* 33: 79-87.

Vorres, K. S., and Sage, W. L. 1974. *American Chemical Society short courses: Cleaner fuels from coal.* Washington, D.C.: American Chemical Society.

Walker, P. L., Jr. 1973*a*. Coal characteristics and gasification behavior. Paper presented at Short Course on Coal Characteristics and Coal Conversion Processes, Oct. 29–Nov. 2, Pennsylvania State University, University Park, Pennsylvania.

Walker, P. L., Jr. 1973*b*. Physical structure of coals. Paper presented at Short Course on Coal Characteristics and Coal Conversion Processes, Oct. 29–Nov. 2, Pennsylvania State University, University Park, Pennsylvania.

Weiler, J. F. 1963. High-temperature tar. In *Chemistry of coal utilization,* supplementary volume, ed. H. H. Lowry, pp. 580-628. New York: John Wiley & Sons, Inc.

Wiser, W. H. 1973. Some chemical aspects of coal liquefaction. Paper presented at Short Course on Coal Characteristics and Coal Conversion Processes, Oct. 29–Nov. 2, Pennsylvania State University, University Park, Pennsylvania.

Yancey, H. F., and Geer, M. R. 1968. Properties of coal and impurities in relation to preparation. In *Coal preparation,* ed. J. W. Leonard and D. R. Mitchell, pp. 3-56. 3rd ed. New York: The American Institute of Mining, Metallurgical and Petroleum Engineers, Inc.

Zubovic, P. 1975. *Geochemistry of trace elements in coal.* Reston, Va.: U.S. Geological Survey.

Zwietering, P., and van Krevelen, D. W. 1954. Chemical structure and properties of coal. IV — Pore structure. *Fuel* 33: 331-37.

3. CONVERSION PROCESSES

J. K. Huffstetler   L. W. Rickert

ABSTRACT

Coal is a versatile energy source; it can be burned to produce heat, gasified and upgraded to possibly produce a useful pipeline gas, or liquefied to produce a fuel that can be treated like a crude petroleum and used as a substitute for it.

In converting coal to a gaseous fuel, sufficient hydrogen must be added to the approximately 75% carbon–5% hydrogen content ratio in coal to create a mixture that will burn with an acceptable level of conversion efficiency. The problems of sulfur removal, corrosion and erosion of materials, and disposal of wastes must also be resolved.

Low- and medium-Btu gasification systems, classified according to heating value, vary widely in engineering design, and a single system that clearly demonstrates superiority has yet to appear. High-Btu systems (catalytically methanating lower heating value gas to hydrogenate available carbon) aim at producing a high-methane gas which can be intermingled with or directly substituted for natural gas. Although successfully demonstrated in pilot plants, a catalytic methanation system has not been used in a large-scale commercial facility. Plans have been announced for six demonstration–commercial-scale coal gasification plants, all based on technology developed by Germany during the two world wars.

Interest also has been revived in in-situ gasification of coal, where coal is ignited underground under pressurized air and steam. Although this concept is a century old, it has been practiced successfully on a commercial scale only in the Soviet Union.

In addition to liquid products, coal liquefaction can also produce a combustible gas and a solid fuel, both of which can either be sold or returned to the process to generate needed heat or hydrogen.

Current coal liquefaction investigations are based on either pyrolysis or dissolution of the coal. Pyrolysis involves decomposition by heat of the organic substances in coal; volatile matter is collected and condensed, whereas the solid char residue can be gasified to produce low-Btu fuel gas. The dissolution process involves dissolving coal in a suitable solvent under high temperature and pressure.

Hydrogen treatment of coal reduces the organic sulfur content of the final products, which can be treated by petroleum refining methods to yield such products as fuel oil, gasoline, and chemicals. Liquefaction processes are not limited to the use of a specific type of coal, and they currently range in size from laboratory scale through preliminary demonstration plant scale.

3.0 INTRODUCTION

The techniques for conversion of coal to gas and liquid hydrocarbons are very old. In theory, at least, only two basic steps are involved: the breaking, or "cracking," of the heavy hydrocarbon molecules into lighter molecules and the simultaneous enrichment of the molecules with hydrogen. Natural gas is about 75% carbon and 25% hydrogen with only traces of undesirable constituents (Linden et al. 1976). Petroleum, consisting of 83 to 87% carbon, 11 to 15% hydrogen, and up to 4% oxygen, nitrogen, and/or sulfur, is more variable (Ellison 1966). Coal has about 75% carbon, only 5% hydrogen, and 20% undesirable constituents, including pyritic and organic sulfur (Linden et al. 1976). Thus, coal conversion processes must deal with the chemical and thermal operations designed to increase the ratio of hydrogen to carbon and to remove undesirable components in an environmentally acceptable manner.

All modern industrial techniques for the conversion of coal had their origins in Germany (Faucounau 1974). Two world wars forced the Germans to invest heavily in developing alternative sources of liquid and gaseous fuel; petroleum and natural gas had to be imported and thus were vulnerable to blockades or bombings, whereas coal was a readily available natural resource. Beginning in 1915 when the German chemist Bergius built the first pilot plant manufacturing synthetic gasoline, most of the significant technical advances were made in Germany. Bergius's process consisted of a first phase in which finely ground coal was hydrogenated by amalgamation with tar oils. The product oil was fractionated by distillation: The light fraction was subjected to cracking with hydrogen-enriched steam, providing a liquid rich in the aromatics (benzene, toluene, and xylenes), whereas the heavy fraction provided the tar oils for the first hydrogenation phase.

Between 1915 and 1944, the Bergius process was greatly improved. Investigators concentrated on seeking (1) better operating conditions to avoid the production of gaseous light hydrocarbons, such as methane and ethane, which decreased the yield of gasoline; (2) techniques permitting the elimination of sulfur because it shortened the life of the hydrogenation catalysts; and (3) more economical processes for producing the necessary hydrogen. At first hydrogen was produced by reaction of the light hydrocarbons (methane and ethane) with steam in the presence of refractory materials at very high temperatures (1250°C):

$$CH_4 + 2H_2O = CO_2 + 4H_2 \quad . \tag{1}$$

Because the reaction was endothermic (i.e., requiring a supply of heat), it was necessary to work discontinuously, alternating periods of heating by combustion of methane with periods of conversion. Numerous variations were made by using other sources of hydrogen such as hydrogen from coke oven gases or hydrogen from the reaction of steam with iron.

In 1926 Fischer and Tropsch introduced a novel technique when they produced "water gas" as a source of hydrogen in a process originally developed to convert coal to a heating gas. Steam was reacted with red-hot coke to produce carbon monoxide and hydrogen:

$$C + H_2O = CO + H_2 \quad . \tag{2}$$

The product gases were reacted over a cobalt catalyst on a thoria support to form liquid hydrocarbons.

By 1939, on the eve of World War II, Germany had eight synthetic fuel plants based on coal and lignite. By the beginning of 1944, there were 18 plants in operation having a production capacity of close to 5 million tons of various fuels for automobiles and airplanes per year and a production of about 700,000 tons of lubricating oils per year (Faucounau 1974).

While liquefaction techniques were being developed, a parallel effort was taking place in techniques for gasification of coal. In 1930 the Lurgi process was developed for gasifying noncaking coals. This process, which was not fundamentally different from the Fischer process, involved three basic reactions:

$$C + H_2O = CO + H_2 \quad , \tag{3}$$

$$CO + H_2O = CO_2 + H_2 \quad , \tag{4}$$

$$CO + 3H_2 = CH_4 + H_2O \quad . \tag{5}$$

Because this scheme was overall endothermic, it was necessary to supply heat for continuous operation. Thus, in the Lurgi process, heat was supplied by partial combustion of the coal with air at the top of the reactor, or "gasifier." The fixed bed required preparation of the coal (grinding, preheating, etc.). The product gas required purification and acid gas scrubbing, but it was then suitable for use as a fuel source for heating boilers. It has a low (180 Btu/scf) or intermediate (300 Btu/scf) heating value. Later, in the 1960s, work was begun on upgrading the Lurgi product gas, by catalytic methanation, to a high-Btu (900+ Btu/scf) gas.

Prior to development of the Lurgi process, in 1922, Dr. Fritz Winkler discovered a fluid-bed process for gasification of coal. The first commercial unit was installed in Leuna, Germany, in 1926 (Banchik 1974). This gas producer operated at atmospheric pressure and moderate temperature (1500 to 1800°F) to produce a gas of 287 Btu/scf (Katz et al. 1974).

Both gasification and liquefaction techniques were used in Germany during the two world wars; after the 1920s, both techniques were also used in areas of the world in which the scarcity and cost of other fossil fuels made these techniques economically competitive. There were and are, however, many technological and environmental considerations in the development of conversion processes as viable alternatives to the use of petroleum and natural gas. Coal is an exceedingly complex material that contains varying amounts of undesirable constituents: Tars and coke plug lines and corrode valves; ash must be disposed of in an environmentally acceptable manner; sulfur poisons catalysts; pressure and heat requirements stretch the limits of materials technology; and water requirements put great strains on arid ecosystems. It is little wonder that liquefaction and gasification were not usually competitive in the days of cheap oil and abundant natural gas.

3.1 GASIFICATION

Acceleration of coal gasification development is important in solving the problem of decreasing natural gas supplies in the United States. Table 3.1 lists some planned commercial and demonstration coal gasification plants, whereas Table 3.2 summarizes and briefly describes the high-Btu coal gasification research and development projects as of May 1, 1975.

Table 3.1. Announced commercial and demonstration coal gasification plants (as of May 1, 1975)

Controlling company[a]	Site	Process	Coal feed (tons/day)	Plant output (million cfd)	Status
High-Btu gas projects					
El Paso Natural Gas Co.	Four Corners Area, New Mexico	Lurgi gasification	28,250	288	El Paso Natural Gas Co. plans to construct and operate the Burnham Coal Gasification Complex at the Navajo Indian Reservation. During the week of April 20, 1975, the FPC granted the request by El Paso to defer decision on the proposal until water and coal supply contracts are resolved.
WESCO, Texas Eastern Transmission Corp., and Pacific Lighting Corp. (Utah International Corp.)	Four Corners Area, New Mexico	Lurgi gasification with methanation	102,500	1000 (4 plants)	The firms plan to construct and operate four plants on the Navajo Indian Reservation near Farmington, New Mexico. Utah International will supply the coal and water. First plant estimated project cost is $853 million. The FPC has allowed a price of $1.38 per 1000 ft^3 of gas for the 6-month startup period.
Panhandle Eastern Pipe Line Co. (Peabody Coal Co.)	Eastern Wyoming	Lurgi gasification with methanation	25,000	270	Plant operation is anticipated in the 1978-1980 period, assuming timely receipt of all required governmental authorizations.
Natural Gas Pipeline Co. of America	Dunn County, North Dakota	Lurgi gasification with methanation	108,500 (4 mines)	1000 (4 plants)	The company plans to build at least four and possibly eight gasification plants. Coal will be mined from 110,000 acres of leases in North Dakota. The first plant is scheduled to go on line in 1982.
American Natural Gas Co. (North American Coal Corp.)	Beulah-Hazen Area, North Dakota	Lurgi gasification with methanation		275	Initial gas production is scheduled for late 1981 if FPC authorization is received in first quarter of 1976. Recent plant cost estimated at $778 million; mine cost, $126 million; and gas cost, $2.50 per 1000 ft^3. Interest during construction will be provided by a surcharge on Michigan-Wisconsin gas sales.
Northern Natural Gas Co. Cities Service Gas Co. (Peabody Coal Co.)	Powder River Basin, Montana	Lurgi gasification with methanation		1000 (4 plants)	Northern Natural and Cities Service plan to construct four 250-million-cfd coal gasification plants. Peabody Coal has agreed to supply about 500 million tons of coal, and the gas companies are negotiating for another like amount. Through 1975, $10 to $11 million will be spent for preliminary development. Construction of the first plant could start in 1976-1977 with operation in 1979-1980.
Texas Gas Transmission Corp. (Consolidation Coal Co.)	Western Kentucky			80	Texas Gas Transmission and the State of Kentucky have signed an agreement to build an 80-million-cfd gasification plant, expandable to 250 million cfd. A Federal Government commitment to fund a portion of capital and operating costs is being sought and is an essential condition for construction.
Colorado Interstate Gas Corp. (Westmoreland Coal Co.)	Southeast Montana		25,000	250	Colorado Interstate has an option on 300 million tons of coal and 10,000 acre-ft of water per year to be supplied by Westmoreland for development of a coal gasification project.

Table 3.1 (continued)

Controlling company[a]	Site	Process	Coal feed (tons/day)	Plant output (million cfd)	Status
The Columbia Gas System, Inc.	Illinois			300	Columbia Gas has agreed to exchange a 50% interest in 43,400 acres of its 300,000 acres of W. Virginia coal lands for a 50% interest in 35,000 acres of Illinois coal lands held by Exxon's Carter Oil Co. The Illinois coal will be held by Columbia for coal gasification pending development of an economically and technically sound coal gasification process.
Consolidated Natural Gas Co.	Southwest Pennsylvania				The company has purchased about 450 million tons of recoverable coal reserves in southwest Pennsylvania for eventual use as feedstock for coal gasification.
Pennsylvania Gas and Water Co.	Pennsylvania	HYGAS or similar one	5,000	80	The company has proposed to the Office of Coal Research a plan for financing and operating a demonstration plant.
Southern Natural Gas Co.	Illinois			250	Southern Natural Gas Co. has acquired an option to purchase coal reserves in the Illinois Basin from Consolidation Coal Co. The option will not be exercised unless the FPC allows inclusion of the cost of the coal reserves in Southern's rate base.
Texas Eastern Transmission Corp. (Peabody Coal Co.)	Southern Illinois			250	Texas Eastern has obtained tentative dedication of a large reserve of Peabody's southern Illinois coal while a feasibility study of a gasification plant is being made.
Panhandle Eastern Pipe Line Co. (Peabody Coal Co.)	Southern Illinois	Lurgi gasification with methanation			Feasibility studies under way.
Exxon Corp (Carter Oil)	Northern Wyoming				Feasibility study started in late 1973. State and Federal coal leases committed.
El Paso Natural Gas Co.	Southwestern North Dakota				Four plants are projected with the first to be in operation by 1981. Reserves of 2 billion tons of coal are under lease.
Island Creek Coal Co.	Kentucky				Island Creek Coal Co. is negotiating with an undisclosed utility.
Texaco, Inc.	Wyoming				2 billion tons of coal reserves obtained from Reynolds Metals. Plant may be for either gas or liquids.
Combination gas and liquid project					
Coalcon Co. (Chemico and Union Carbide)		Union Carbide	2,400	22.4 (plus 3900 bbl/day liquid fuel)	Plant cost will be $237.2 million. ERDA will fund an initial $21 million for plant design. The balance will be shared equally by the Federal Government and industry.

[a]Mining partners in parentheses.

Source: Linden et al. 1976, Table 1, pp. 66-67. Reprinted by permission of the publisher.

Table 3.2. Summary of high-Btu coal gasification R & D projects
(as of May 1, 1975)

| Name of process | Owner or contractor (and site) | Description of process | Coal | | Plant output (million cfd) | Status and funding |
			Type	Consumption (tons/day)		
Lurgi pressure gasification (Lurgi Gesellschaft fur Warme und Chemotechnik m.b.H.)	El Paso Natural Gas Co. (Four Corners Area, N.M.)	A bed of crushed coal is introduced to the gasifier vessel through lock hoppers and travels downward as a moving bed. The process operates at pressures up to 450 psi. Steam and oxygen are introduced through a revolving steam-cooled grate that also removes ash at the bottom of the gasifier. The hydrogen-rich gas passes up through the coal bed, producing some methane by hydrogenation of coal. The product gas is purified and methanated to produce 972 Btu/scf gas.	Subbituminous	800	35 (Raw synthesis gas)	A single Lurgi gasifier will be installed to test improvements in the process. Goals are operation at 20% above design capacity and 30% above design pressure, gasifier mechanical improvements, coal fines gasification, and improved pollution control. Capital investment will be $19 million; first-year operating cost will be $5 million. Construction will require 13 months after necessary approvals have been obtained. The FPC has granted intermediate approval for the project.
Lurgi pressure gasification	Conoco Methanation Co. and Scottish Gas Board (Westfield, Scotland)	Methanation of purified low-Btu product gas from a Lurgi pressure gasifier to demonstrate the commercial feasibility of the combination of Lurgi gasification and methanation to produce a 950+ Btu/scf gas.			2.5	Conoco Methanation Co. designed and constructed facilities to carry out a 1-year test at a total cost of $6 million. Fourteen U.S. companies sponsored the test, which was successfully completed in September 1974. The construction contractor was Woodall-Duckham Ltd.
Lurgi pressure gasification	South African Coal, Oil, and Gas Corp. and Lurgi (Sasolburg, South Africa)	Methanation of the product gas from a Lurgi gasifier to produce 950+ Btu/scf gas.				An experimental program has been completed to demonstrate commercial feasibility of methanation. A slipstream from the SASOL plant will be used as feed to the methanator.
COGAS	COGAS Development Co. (FMC Corp., Panhandle Eastern Pipe Line Co., Tenneco Inc., Consolidated Natural Gas Service Co., Republic Steel Corp.) (Princeton, N.J.)	Coal is charged to a series of fluidized-bed reactors with increasing stage temperatures to pyrolyze the coal and drive off volatile fractions in each stage. After separation of the oils and gas produced in the initial processing, the residual char reacts with steam (heat supplied indirectly by combustion of char with air) to produce a synthesis gas that is purified and methanated to a 950+ Btu/scf gas. The operating pressure is 50 psig. The oil by-product is hydrotreated to make a synthetic crude oil.	Subbituminous, bituminous			A 100-ton/day pilot plant built at Leatherhead, England, uses a process in which char is gasified and ash is removed as slag. The $7 million Princeton, N.J. facility, originally a $4.5 million COED pilot plant completed in 1970, removed residue as fly ash and has been shut down; the Leatherhead process is indicated as superior. COGAS Development Co. is evaluating process alternatives and economics for an 800 - 1000-ton/day coal gasification-liquefaction facility.

Table 3.2 (continued)

| Name of process | Owner or contractor (and site) | Description of process | Coal | | Plant output (million cfd) | Status and funding |
			Type	Consumption (tons/day)		
Kellogg molten salt process	M.W. Kellogg Company	In the latest version of this process, oxygen, steam, and coal are injected into a high-purity alumina reaction vessel, where molten sodium carbonate catalyzes the coal for complete gasification. The product gas can be purified and methanated to produce 950+ Btu/scf gas.	All U.S. coal types			OCR funded a bench-scale program from 1964 to 1967. Total expenditures were $1.7 million. Major difficulties were experienced with materials of construction. OCR ceased sponsorship because of this problem, budgetary restrictions, and assignment of higher priorities to other coal gasification processes. M.W. Kellogg has carried out additional bench-scale development work since 1967, but support has not yet been obtained for construction of a large-scale pilot plant.
Simplified coal gasification process	Garrett Research and Development Co. (Sponsored by Colorado Interstate Gas Co.)	Coal is introduced into a single reactor at atmospheric pressure, where gas is produced by rapid devolatilization of coal. Large amounts of excess char are produced.	All U.S. coal types			Early stage of development. Tested in small pilot plant in La Verne, Calif.
None	Gulf General Atomic and Stone & Webster	Coal is dissolved and the solution hydrogasified to produce gas that is purified and methanated to 950+ Btu/scf gas. Heat supply from nuclear reactor.	All U.S. coal types			Feasibility study completed by State of Oklahoma under $300,000 funding ERDA. Support being sought for 2-year R&D program to cost $650,000 for first phase.
ATGAS	Applied Technology Corp.	Coal is injected into a molten-iron bath, where steam and oxygen react with the carbon to produce hydrogen and carbon monoxide. The gases are then methanated to produce 950+ Btu/scf gas.	All U.S. coal types			Emphasis now being placed on low-Btu gas production.
Synthane	ERDA (Pilot plant to be operated by Lummus)	Coal is introduced into a single reactor after pretreatment with oxygen at pressure. Hydrogen-rich gas for the reaction is produced by use of oxygen and steam in the reactor. The process operates at 500 - 1000 psia. The product gas can be purified and methanated to produce 950+ Btu/scf gas.	All U.S. coal types	70 (Pilot plant)	1.4	Construction by Rust Engineering Co. is essentially complete and operation is starting. Estimated cost is $10 million.

Table 3.2 (continued)

| Name of process | Owner or contractor (and site) | Description of process | Coal | | Plant output (million cfd) | Status and funding |
			Type	Consumption (tons/day)		
HYGAS	Institute of Gas Technology (Chicago, Ill.)	Ground, dried coal is pre-treated with air, slurried with by-product oil, and fed to a two-stage fluidized-bed hydrogasifier operating at 1000 - 1500 psia; hydrogen-rich gas for the reaction can be furnished by processes using electric energy or oxygen, or by the steam-iron process. Gas from the reactor is purified and methanated to produce 950+ Btu/scf gas.	All U.S. coal types	75 (Pilot plant)	1.5	Pilot plant in operation. Preliminary demonstration, plant design complete. Original cost of pilot plant was about $9.5 million. Project funded under joint ERDA/American Gas Association program. A run of 27 days completed in March 1974, using lignite and externally supplied hydrogen to produce high-Btu gas. Steam-oxygen gasification section to produce hydrogen has been added, and a fully integrated operation successfully performed in April 1975.
Steam-iron process	Institute of Gas Technology (Chicago, Ill.)	A fluidized-bed process in which producer gas, made from steam, air, and hydrogasifier char is used to reduce a circulating stream of iron oxide. The reduced iron oxide is subsequently oxidized with steam to produce a hydrogen-rich gas for the hydrogasifier.	Hydrogasified char	36 (Pilot plant)	2.0 (44% hydrogen)	Pilot plant under construction for completion near end of 1975. Project is funded at $18.2 million under the joint ERDA/American Gas Association program.
CO_2 Acceptor	Conoco Coal Development Co. (Rapid City, S.D.) (Pilot plant constructed and operated by Stearns-Roger Gorp.)	Coal is charged to a gasifier vessel at 150 psi. Dolomite (the "acceptor") is added to the gasifier, where it reacts with carbon dioxide. Acceptor regeneration and ash removal are carried out in an air-blown regenerator vessel where spent char is combusted. The product gas can be purified and methanated to product a 950+ Btu/scf gas.	Lignite, subbituminous	30 (Pilot plant)	Up to 2 (375 Btu/ft^3)	Original cost about $9.3 million. Ten-day run completed in September 1974. Methanation stage recently added now in initial startup.
Bi-Gas	Bituminous Coal Research, Inc. (Homer City, Pa.) (Pilot plant to be managed by Phillips Petroleum Co. and operated by Stearns-Roger Corp.)	Coal is introduced into a reactor, where it is contacted and partially gasified with hydrogen-rich gas produced in the lower section of the reactor by the slagging gasification of recycle char with oxygen and steam. The pressure of operation is 1000 psia. The product gas can be purified and methanated to produce 950+ Btu/scf gas.	All U.S. coal types	120 (Pilot plant)	2.3	Construction by Stearns-Roger Corp. scheduled for completion in fall 1975. Plant cost will be about $18 million. Sponsored under joint ERDA/American Gas Association program.

Table 3.2 (continued)

| Name of process | Owner or contractor (and site) | Description of process | Coal | | Plant output (million cfd) | Status and funding |
			Type	Consumption (tons/day)		
Hydrane	ERDA (Pittsburgh, Pa.)	Raw coal reacts directly with hydrogen in a free-fall zone, and the intermediate char formed reacts further with hydrogen in a fluidized-bed. Residual char gasified with steam and oxygen to produce syngas for hydrogen. Research goals are a reactor product stream with 70% methane content.	All U.S. coal types			A 10 L/hr integrated pilot plant is in operation. Scale-up to a 24-ton/day pilot plant to be built at Morgantown, W.Va., is planned to cost $25 million.
None	Exxon Corp. (Baytown, Texas)	Company states that process does not use oxygen. No other details are available.				Exxon has spent $20 million on coal gasification and liquefaction since 1966 and is committing $10 million for additional R&D. First phase of gasification experiments finished in 1974. Next phase, construction of $75- to $80-million pilot plant charging 500 tons/day of coal postponed indefinitely.
Agglomerating ash gasification process	Battelle-Columbus Laboratories (West Jefferson, Ohio) (Process developed by Union Carbide)	Coal or char is combusted with air in one fluidized bed and is steam-gasified in a separate fluidized bed. The heat for the gasification reaction is provided by circulation of hot ash agglomerates from the burner to the gasifier. The gasification section will operate at 100 psi. The synthesis gas can be purified and methanated to produce 950+ Btu/scf gas.	Bituminous	25 (Pilot plant)	0.8 (Synthesis gas)	Pilot plant under construction by Chemico is scheduled for completion in fall 1975. Funded in joint ERDA/ American Gas Association program.
Liquid phase methanation	Chem Systems Inc. (Hackensack, N.J.)	The synthesis gas flows through the reactor with an inert liquid that picks up heat in the fluidized catalyst bed. The vaporized inert fluid is condensed and recycled. Methanation takes place in a single pass. The catalyst life is estimated to be longer than 1 year.				Chem Systems was awarded a $1.9-million contract under the joint ERDA/American Gas Association program. During the past 2-1/2 years the liquid phase methanation process has been under development. A skid-mounted transportable pilot plant is planned for operation by June 1975.
Lurgi gasification	American Natural Gas Co. and Natural Gas Pipeline Co. at South Africa Coal, Oil, and Gas Corp. with Lurgi (Sasolburg, South Africa)	Lurgi pressure gasification plus methanation of the product gas from a Lurgi gasifier to produce 950+ Btu/scf gas using the Lurgi hot-gas recycle process.	North Dakota lignite			An experimental program to demonstrate commercial feasibility of methanation. A slipstream from the SASOL plant was used as feed to the methanator. 12,000 tons of North Dakota lignite was transported to South Africa for testing.
Catalytic methanation	Catalyst & Chemicals Inc. with El Paso Natural Gas, COGAS Development Co., and WESCO (Louisville, Ky.)	The adiabatic fixed-bed process utilizes 3 beds and recycles product gas as inert diluent.				A 3-year pilot plant program recently completed. 4000-hr life tests were run on two different catalysts. Results indicate catalyst life of over 2 years.

Table 3.2 (continued)

Name of process	Owner or contractor (and site)	Description of process	Coal		Plant output (million cfd)	Status and funding
			Type	Consumption (tons/day)		
Lurgi slagging gasifier	Conoco Methanation Co. and British Gas Corp. (Westfield, Scotland)	Modified Lurgi gasifier operates at temperatures to cause slagging of coal ash. This allows operation at lower steam rates, higher throughput, and higher thermal efficiency.				Conoco will coordinate project, British Gas Corp. will be project operator, and Lurgi will provide technical assistance. 3-yr test period. The process was tested on a pilot-plant scale during the 1962-64 period by BGC. The project cost estimated at $10 million.
GE fixed-bed gasifier	General Electric Co., Schenectady, N.Y.	Unit is a fixed-bed reactor using extrusion for injection of coal to reactor. Uses inert diluting agents such as silicon carbide and coal ash to reduce swelling and caking of coal.				Work to date has been done in a feasibility unit operating at atmospheric pressure and consuming about 500 lb of coal per day. Successful runs, using low-grade coals from Illinois and Missouri, have been made. Feasibility studies have been carried out at atmospheric pressure for production of SNG from caking coals. A demonstration gasifier operating at 300 psig is planned within 1 year.
Modified Lurgi gasification	American Gas Association by British Gas Corporation (Westfield, Scotland)	Lurgi gasifier is modified to include rabble-arm stirrer that makes it applicable to use with caking coals.				Test concluded in April 1974, successfully gasified U.S. caking coals (Illinois and Pittsburgh). Caking coals required substantially more steam and oxygen than noncaking. Fines in all coal types reduced gasifier capacity.

Source: Linden et al. 1976, Table 2, pp. 68-71. Reprinted by permission of the publisher.

Gasification is only one of nine basic and seven auxiliary steps in the generalized scheme **for** converting coal into an environmentally acceptable gaseous fuel (Fig. 3.1) (Haynes and Forney 1976). Although most parts of the overall scheme are still developing, major effort is being directed toward pretreatment, gasification, and catalytic methanation.

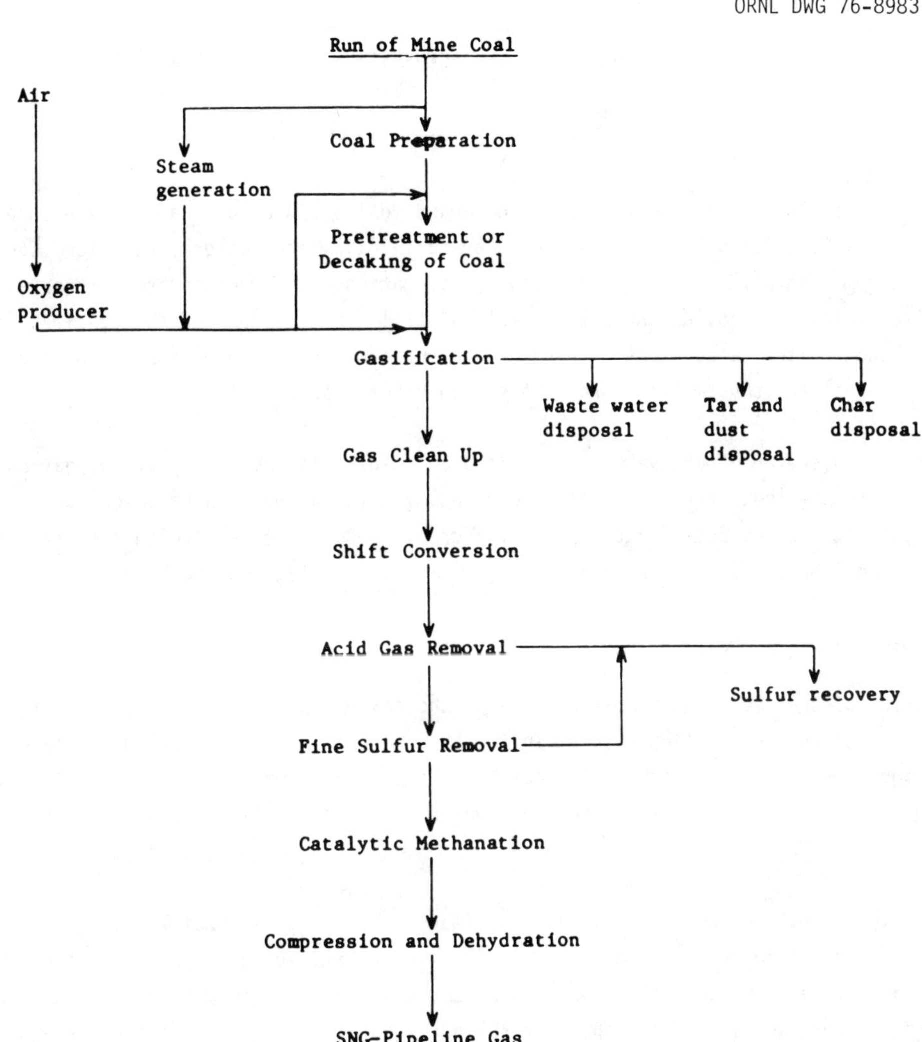

Fig. 3.1. Generalized process scheme for gasifying coal. Source: From Haynes and Forney 1976, Fig. 1, p. 3.

In many ways natural gas is the most satisfactory of all fossil fuels: It is thermally efficient to transport and use; it contains no residue such as ash; its production makes little aesthetic impact on the landscape; it is almost pollutant-free, only carbon dioxide and water being vented to the atmosphere; and its supply can be controlled at the point of use to ensure efficiency (Rooke and Hebden 1974). Thus, in recent years natural gas use has grown to supply about one-third of the total energy requirement of the United States (Linden et al. 1976).

3.1.1 <u>Need</u>

Since 1966 gas use has exceeded domestic production of natural gas, and the gap between yearly additions to reserves and consumption is growing rapidly. According to a study quoted by the National Academy of Engineering (1974*a*), deficiency in supply will increase as follows (in trillions of cubic feet):

Year	Demand	Production	Deficiency
1980	30.5	22.7	7.8
1985	34.6	21.4	13.2
1990	38.1	19.8	18.3

This forecast includes gas imported from Canada and Mexico and assumes a pipeline supply from Alaska after 1980 (National Academy of Engineering 1974*a*). The National Petroleum Council (1972) presents a somewhat different point of view by estimating that "undiscovered" reserves of natural gas are sufficient to provide three times the 393 trillion ft^3 produced through 1970. The authors concede, however, that a substantial part of the undiscovered gas deposits may be located in less accessible areas and thus may be more costly than prior discoveries.

Until these undiscovered reserves are exploited and made available, the natural-gas shortage will continue. For the short term, liquefied petroleum gas (LPG) and naphtha-based substitute natural gas (SNG) plants can be built, but they only increase dependence on foreign imports. For the long term, coal-based gaseous fuels are being actively investigated (Rooke and Hebden 1974).

3.1.2 <u>Chemical considerations</u>

Gasification of coal became a viable process in the early nineteenth century when gas was produced by the destructive distillation of coal in the absence of air. Gasification can be defined as any process for converting solid or liquid fuel to a gaseous product suitable for use as (1) a direct source of energy or (2) a feedstock for the synthesis of chemicals, liquid fuels, or other gaseous fuels.

Although highly heterogeneous, a typical bituminous coal contains about 75% carbon, 5% hydrogen, and 20% less desirable constituents (i.e., primarily organic sulfur, pyritic sulfur, and ash). Because natural gas contains about 75% carbon, 25% hydrogen, and only traces of undesirable constituents, converting coal to a usable natural gas substitute requires either the addition of a large amount of hydrogen or the rejection of a large amount of carbon. It is more effective to add hydrogen (Linden et al. 1976).

3.1.2.1 <u>Low- and intermediate-Btu gas</u>

Three ingredients are needed to synthesize gas chemically from coal — carbon, hydrogen, and oxygen. Coal provides the carbon; steam most often provides the hydrogen (although hydrogen is sometimes introduced from an external source); and air provides the oxygen (unless oxygen is needed in the pure form) (University of Oklahoma 1975). The reactions involved are

$$Coal + H_2 = CH_4 + C \ , \qquad\qquad (6)$$

$$C + 2H_2 = CH_4 \ , \qquad\qquad (7)$$

$$C + H_2O = CO + H_2 \quad , \qquad (8)$$

$$C + O_2 = CO_2 \quad . \qquad (9)$$

Because Reaction (8) is highly endothermic and will not occur unless sufficient heat is present (1900 to 2500°F), some of the coal is usually reacted with oxygen or air to provide a heat source [Reaction (9)]. If air is used as a combustant, the raw gas will have a low heating value (roughly 150 to 300 Btu/scf, depending upon design characteristics) and will contain undesirable constituents such as carbon dioxide, hydrogen sulfide, and nitrogen. The use of pure oxygen, although expensive, results in an intermediate gas (300 to 400 Btu/scf) with carbon dioxide and hydrogen sulfide as by-products (National Academy of Engineering 1974*b*). Both carbon dioxide and hydrogen sulfide can be removed from cooled low- or intermediate-Btu gas by any of several available processes (see Table 4.6); most of these processes are detailed in Chap. 4.

3.1.2.2 High-Btu gas

In 1965, the Office of Coal Research issued guidelines for coal gas, which may be substituted directly for, or intermingled with, natural gas. The gas should be pumpable at 1000 psig and should have the following desirable characteristics (Strakey, Forney, and Haynes 1975):

- Heating value > 900 Btu/scf (higher heating value)
- CO < 0.1%
- H_2S < 0.25 grains/100 scf
- Total sulfur < 10 grains/100 scf
- Inerts < 5%
- CO_2 < 3%
- Water < 7 lb/million scf
- Specific gravity of 0.59 to 0.62 recommended
- Hydrocarbon dewpoint < -40° at 1000 psig
- No poisonous compounds or gum formers

If a high-Btu gas of "pipeline" quality (900 to 1000 Btu/scf) is desired, steps must be taken to increase the methane content of the gas. This enrichment requires hydrogen, which is produced by reaction of steam with carbon or carbon monoxide:

$$CO + H_2O = CO_2 + H_2 \quad ; \qquad (10)$$

$$CO + 3H_2 = CH_4 + H_2O \quad . \qquad (11)$$

Reaction (10), the water-gas shift reaction, is performed to adjust the hydrogen–carbon monoxide ratio to 3:1, the required ratio for carrying out the methanation step [Reaction (11)]. At high temperatures the reaction goes to equilibrium, probably catalyzed by ash and reactor surfaces at temperatures greater than about 1200°F; at lower temperatures the reaction is carried out over a catalyst (Schora 1974):

$$2H_2 + 2CO = CH_4 + CO_2 \quad . \qquad (12)$$

Reaction (12) is a combination of Reactions (10) and (11).

Normally, after the water-gas shift conversion, the carbon dioxide created during earlier steps and the sulfur pollutants are removed, leaving a product gas containing only carbon monoxide and hydrogen in the proper ratio for catalytic methanation.

However, Moeller, Roberts, and Britz (1974) have suggested that a product of higher calorific value can be obtained if the synthesis gas contains, instead of a stoichiometric composition, a surplus of carbon dioxide that can be used to reduce the hydrogen content to less than 1 vol % by carbon dioxide methanation:

$$CO_2 + 4H_2 = CH_4 + 2H_2O \quad . \tag{13}$$

Whereas Reaction (10) is only mildly exothermic (heat-producing), Reactions (11) through (13) are highly exothermic and have high negative ΔH values (Table 3.3). The free energy change is also favorable at the lower range of temperatures. However, the reactions are relatively slow, and catalysts are needed to accelerate the reactions to acceptable commercial rates.

Table 3.3. Thermodynamic values for methanation reactions

Temperature		Reaction				
$^\circ$K	$^\circ$C	(10)	(11)	(12)	(13)	(14)
		Heat of reaction, ΔH_f° (kcal)				
300	27	-9.838	-49.298	-39.460	-59.136	-41.227
400	127	-9.710	-50.360	-40.650	-60.070	-41.434
500	227	-9.518	-51.297	-41.779	-60.815	-41.499
600	327	-9.292	-52.084	-42.792	-61.376	-41.460
700	427	-9.050	-52.730	-43.680	-61.780	-41.350
800	527	-8.799	-53.248	-44.449	-62.047	-41.190
900	627	-8.549	-53.654	-45.105	-62.203	-40.996
1000	727	-8.304	-53.957	-45.653	-62.261	-40.729
		Free energy of reaction, ΔF° (kcal)				
300	27	-6.827	-33.904	-27.077	-40.731	-28.621
400	127	-5.841	-28.610	-22.769	-34.451	-24.385
500	227	-4.894	-23.062	-18.168	-27.954	-20.111
600	327	-3.991	-17.338	-13.347	-21.329	-15.836
700	427	-3.127	-11.493	-8.366	-14.620	-11.574
800	527	-2.298	-5.567	-3.269	-7.865	-7.332
900	627	-1.500	+0.594	+1.921	-1.079	-3.108
1000	727	-0.729	-6.444	+7.173	+5.715	+1.090
		Equilibrium constant, $\log K_p$				
300	27	4.973	24.698	19.724	29.670	20.849
400	127	3.191	15.630	12.44	18.822	13.322
500	227	2.139	10.080	7.940	12.219	8.790
600	327	1.453	6.314	4.861	7.768	5.768
700	427	0.976	3.588	2.611	4.564	3.613
800	527	0.628	1.521	0.893	2.148	2.003
900	627	0.364	-0.144	-0.466	0.261	0.755
1000	727	0.159	-1.408	-1.568	-1.248	-0.238

Source: Mills and Steffgen 1973, Table 1, p. 165. Reprinted by permission of the publisher.

3.1.2.3 Catalytic methanation

Catalytic methanation is one of the most critical, most complex, and least understood step in the production of high-Btu gas.

For example, one problem of catalytic methanation can be the Boudouard reaction, in which elemental carbon is formed by heterogeneous decomposition during methane synthesis (Moeller, Roberts, and Britz 1974):

$$2CO = CO_2 + C \ .$$ (14)

Also, heterogeneous decomposition of methane as the reverse reaction of methane formation from the elements can cause carbon formation:

$$CH_4 = C + 2H_2 \ .$$ (15)

Either reaction can cause carbon deposits which create equipment problems and reduce catalyst activity. Moeller, Roberts, and Britz (1974) suggest that

> Carbon formation occurs only when a certain limiting value for the driving force for the Boudouard reaction is exceeded. This critical limiting value for carbon formation is a specific parameter for each catalyst and depends to a certain extent on reaction conditions, such as pressure and temperature. This limiting value can be determined only by experience.

Mills and Steffgen (1973) state that Reaction (15) and related reactions, such as the steam-carbon reaction, do not occur significantly at temperatures used in catalytic methanation (250 to 450°C). However, the high heat release of Reactions (11) through (13) makes it difficult to prevent overheating and inactivation of the catalyst. Catalytic methanation, whether carbon monoxide or carbon dioxide, is strongly exothermic. Moeller, Roberts, and Britz (1974) predict a temperature rise (from heat release) of about 450°C (842°F), but this heat is not enough to be of value for the rest of the process. It cannot, for example, be used in support of Reaction (8) because Reaction (8) occurs at a much higher temperature. Some of the heat can be used to raise steam, but much of it is discarded, constituting a process inefficiency (Linden et al. 1976).

Although the heats of reaction are not influenced greatly by temperature, changes in free energy and equilibrium constants for methanation reactions are quite sensitive to temperature (Table 3.3). Thus, equilibrium methane yields are reduced critically at high temperature, requiring catalyst beds to be operated at the lowest temperatures that are consistent with acceptable catalyst activity.

According to Mills and Steffgen (1973), pressure does not appreciably affect methane yield until the temperature exceeds 450°C (723°K). However, for temperatures above 700°K, increasing pressure tends to decrease the minimum hydrogen–carbon monoxide ratio required to prevent carbon deposition (Fig. 3.2).

Considerable work has been done on possible catalyst materials. Only five — ruthenium, nickel, cobalt, iron, and molybdenum — have been identified as having commercial importance (Strakey, Forney, and Haynes 1975). Ruthenium is very active, but because it is relatively rare, there is a question as to whether the tonnages required for large-scale use would be available. Certainly, the price of ruthenium would be a limiting factor in its large-scale use; Mills and Steffgen (1975) have estimated that a methanation reactor employing a 0.5 wt % ruthenium catalyst would require $750 to $1000 worth of ruthenium for each million cubic feet of daily gas capacity.

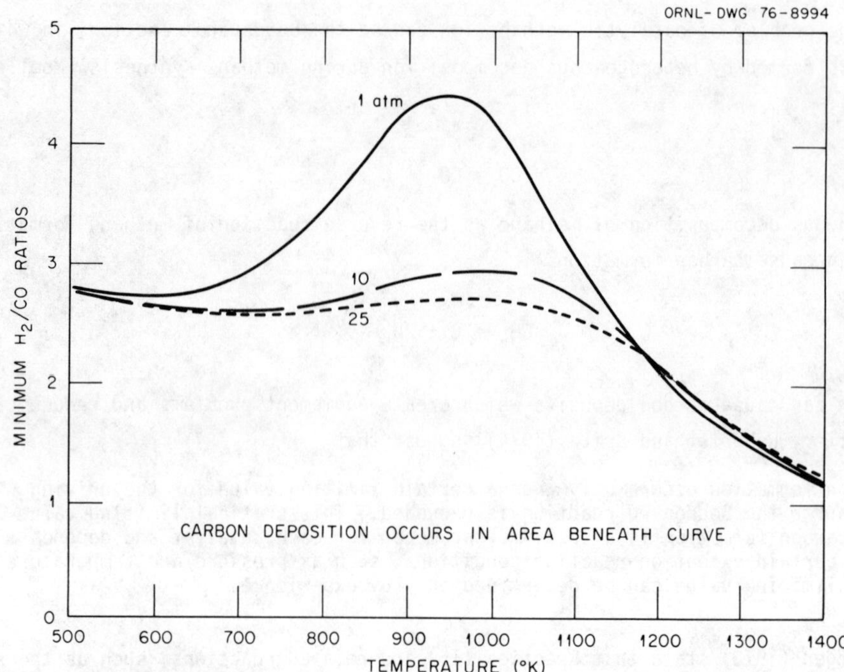

Fig. 3.2. Carbon deposition boundaries. <u>Source</u>: From Strakey, Forney, and Haynes 1975, Fig. 1, p. 6.

Nickel is inexpensive, very active, and highly selective to methane; thus, it is the catalyst of choice for most commercial operations. Cobalt is less active and less selective than nickel; iron is less active than cobalt and catalyzes carbon formation; molybdenum is of lower activity than iron and fairly selective, but it has the advantage of being resistant to sulfur.

Commercial nickel catalysts consist of 25 to 77 wt % nickel on a high-surface-area, refractory support such as kieselguhr or alumina (Table 3.4). A widely used catalyst is Raney nickel, an alloy composed of 42 wt % nickel and 58 wt % aluminum, which is leached with sodium hydroxide to create a spongy skeletal catalyst that is very active in methanation.

Catalyst deactivation, a severe problem in catalytic methanation, can occur through any of several mechanisms (Strakey, Forney, and Haynes 1975):

(a) $Ni + H_2S = NiS + H_2$;

(b) $Ni\ (75\ Å) = Ni\ (1000\ Å)$;

(c) $2CO = C + CO_2$;
$CH_4 = C + 2H_2$;

(d) $3Ni + 2CO = Ni_3C + CO_2$;
$3Ni + CH_4 = Ni_3C + 2H_2$;

(e) $Ni + 4CO = Ni(CO)_4$;

(f) $Fe(CO)_5 = Fe + 5CO$.

Table 3.4. Composition of nickel methanation catalysts

Type	Composition	Proportions	Comments
Alloy	Ni-Al		Fluid bed, fused
	Ni-Al (Raney)	42:58	Caustic leached
	Ni-Al-Al$_2$O$_3$	30-35% Ni, 5-18% Al, 52-60% Al$_2$O$_3$·3H$_2$O	Leached Raney nickel
	Ni-Cu	<4% Cu	>4% Cu reduces CH$_4$ yield
	Ni-Min-Al	10:2:1	
Nickel compound	2NiO·Al$_2$O$_3$·4.7H$_2$O		Dehydrated in vacuo at 500°C, then H$_2$ treated at 500°C; 203 m^2/g
	Ni$_2$B		Precipitated on silica gel from NiCl$_2$ using NaBH$_4$
Alumina-supported	Ni-Al$_2$O$_3$ and	23% Ni	
	Ni-η-Al$_2$O$_3$	12% Ni	Reduced at 420°C
	Ni-Ca aluminate on β-Al$_2$O$_3$ hydrate	27:25:48	High activity, CCI catalyst $\left(K_w \sim 30{,}000\right)$
	Ni-K$_2$O-η-Al$_2$O$_3$	11.5:2.5:86	Reduced at 450°C
	NiO-MgO-Al$_2$O$_3$	20:55:25	Reduced at >600°C
	Ni-Mn on Al$_2$O$_3$ and diatomaceous earth	100:20:350:100	Nitrates coprecipitated
Spinel-supported	NiO-MgAl$_2$O$_4$	9-10% Ni	Reduced at 420°C
	NiO-K$_2$O-MgAl$_2$O$_4$	9-10% Ni and 2-3% K$_2$O	Reduced at 570°C
Chromia-supported	NiO-Cr$_2$O$_3$	4.2:1M	Reduced at 300°C
	Ni-Cr$_2$O$_3$		Use Fe catalyst in first stage
	Ni-Cr$_2$O$_3$	50:50	Reduced at 350°C
	NiO-Cr$_2$O$_3$	1:0, 1.8:1, and 9:1	Optimum reduction temperatures differ
	Ni-Cr$_2$O$_3$	23% Ni	Less active than kieselguhr-based catalyst
Kieselguhr-supported	Ni-Kslgr	59% Ni	CH$_4$ and CO from CO$_2$
	Ni-Kslgr		Comparison with Raney nickel
	Ni-Kslgr	58% Ni	Harshaw catalyst used
	Ni-Kslgr	23% Ni	Compares favorably with other compositions
	Ni-Cr$_2$O$_3$-Kslgr	29.2:4.1:66.7	
	Ni-MgO-Kslgr		Organic sulfur compounds removed first using sulfurized nickel
	Ni-MgO-Kslgr	60:5:35	Methanates at 165 - 205°C
	Ni-MgO-Kslgr	60:10:30	Methanates >15% CO$_2$ at 180 - 250°C in two stages in 98 - 99% yield

Source: Mills and Steffgen 1973, Table 5, pp. 190-91. Reprinted by permission of the publisher.

Reaction (a), the poisoning of the catalyst by sulfur, is common to all metallic catalysts except molybdenum. Strakey, Forney, and Haynes (1975) try to keep sulfur concentration below 0.1 ppm to avoid irreversible contamination.

The sintering effect is represented by Reaction (b). It is generally known that, at temperatures above 450°C, nickel crystallites grow in size and a loss of surface area leads to reduced catalytic activity. Reaction (c) is the Boudouard reaction previously discussed. Nickel carbide forms as a result of Reaction (d). This formation deactivates the catalyst material, but the reaction is reversible; treatment with hydrogen at >250°C will reactivate a catalyst poisoned by nickel carbide (Strakey, Forney, and Haynes 1975). A potentially serious reaction is the one that forms nickel carbonyl [Reaction (e)]. The Occupational Safety and Health Administration has set the maximum allowable exposure to nickel carbonyl at 1 ppb (Mills and Steffgen 1974). Because this reaction occurs only at low temperatures, it is avoided by contacting the catalyst with synthesis gas at temperatures above 260°C and maintaining that temperature until all carbon monoxide has been purged from the system.

The final Reaction (f) is one that has its beginnings outside the reactor vessel. Iron carbonyl can form when carbon monoxide reacts, at high pressure and low temperature (100 to 200°C), with carbon steel piping. The carbonyl is carried into the reactor, where it decomposes according to Reaction (f) and deposits iron on the nickel catalyst, forming an iron catalyst. At methanation temperatures, carbon is deposited on the iron and poisons the catalyst. Choosing less reactive piping material, such as stainless steel, will prevent the reaction from occurring (Strakey, Forney, and Haynes 1975).

3.1.2.4 Hydrogasification

Not all high-Btu gasification technologies depend entirely upon catalytic methanation. A number of gasification processes use hydrogasification, involving direct addition of hydrogen to coal under pressure to form methane directly.

The incoming coal reacts with a hydrogen-rich gas to form methane:

$$Coal + H_2 = CH_4 + C \quad . \tag{16}$$

The hydrogen-rich gas for hydrogasification can be manufactured from steam by using the char that leaves the hydrogasifier. Appreciable quantities of methane are formed directly in the primary gasifier. Because the heat released by methane formation is at a sufficiently high temperature to be used in the steam-carbon reaction to produce hydrogen, less oxygen is used to produce heat for the steam-carbon reaction and less heat is lost in the low-temperature methanation step. These factors lead to a higher overall process efficiency.

In theory, hydrogasification has an ideal thermal efficiency of about 90% because it operates at lower temperatures and uses methane-forming reactions of relatively less exothermicity. As in catalytic methanation, the reason for temperature control in hydrogasification is thermodynamic: Lower reaction temperature increases the equilibrium yield of methane, as does higher pressure.

Two practical approaches to hydrogasification have been (1) addition of steam to the hydrogen introduced into the reactor and (2) substitution of raw synthesis gas for the hydrogen. The first of these approaches uses steam as a heat moderator because the steam will decompose at an increasing rate as the temperature rises above 1500°F, thereby absorbing some of the excess heat. Steam also generates hydrogen internally, thereby reducing the need for externally produced hydrogen. However, the use of steam adds to the carbon oxide content of the product gas.

Raw synthesis gas also adds carbon oxides to the product gas, which becomes a mixture of methane, carbon oxides, and unreacted hydrogen unsuitable for direct use as a pipeline gas unless catalytic methanation is applied.

Usually, in hydrogasification, only one third of the methane is formed by catalytic methanation, the rest being formed directly by reaction of hydrogen and coal. Greater overall plant efficiency is thus achieved — 60 to 70% as opposed to 50 to 55% for traditional synthesis-gas methanation (Linden et al. 1976).

3.1.2.5 Coal caking behavior

All gasification technologies require some initial processing of the coal feedstock. The type and degree of pretreatment is a function of the specific process and/or type of coal used (University of Oklahoma 1975). For example, the Lurgi process will accept "lump" coal (1 in. to 28 mesh), but it must be noncaking coal with the fines removed (National Academy of Engineering 1974*b*). Caking coals are defined as those coals that stick together, or agglomerate, when heated to the temperature at which they become plastic, particularly in the presence of hydrogen. Characteristically, they contain more than 24 to 26% volatile matter by weight and have a free-swelling index of more than 2.0 (Kavlick and Lee 1967). Most bituminous coals found in the eastern United States are classified as caking, whereas most western coals are noncaking.

Because agglomerating coals tend to form a plastic mass in the bottom of a gasifier, they plug up the system and drastically reduce process efficiency. Thus, some step to reduce caking tendencies must be included in some process designs. Oxidation of the coal particle surfaces helps destroy this agglomerating tendency. Kavlick and Lee (1967) found that the minimum pretreatment requires an oxygen consumption of 1.0 to 1.5 scf per pound of coal feed and that the practical pretreatment temperature range is quite narrow — between 700 and 750°F. Their studies were conducted only on Pittsburgh No. 8 coal from the Ireland mine, on Ohio No. 6 seam coal from the Broken Arrow mine, and on West Virginia No. 5 block coal; coals of different composition might require other pretreatment conditions.

3.1.3 Design considerations

All gasification processes can be classified according to the type of bed used, the type of reactor vessel, and whether or not the system operates under pressure.

3.1.3.1 Reactor bed

There are primarily three basic types of beds, that is, schemes for the physical placement of coal within the reactor vessel, which differ in their ability to use caking coals.

In a fixed-bed process the coal is supported by a grate. Gases (steam, air, oxygen, etc.) pass through the supported coal, and the hot product gas exits from the top of the reactor. Heat can be supplied internally or from an outside source. Caking coals cannot be used in an unmodified fixed-bed reactor (University of Oklahoma 1975).

The fluidized-bed system uses finely sized coal particles. A fluidized bed is a mass of suitably sized solid particles which exhibits liquidlike characteristics such as boiling when a gas flows upward through the bed. Gas flowing through the coal produces turbulent lifting and separation of particles. The result is an expanded bed having greater coal surface area to promote the chemical reaction. Fluidized-bed systems have only a limited ability to handle caking coals.

An entrainment system uses finely sized coal particles blown into the gas steam prior to entry into the reactor. Combustion takes place with the coal particles suspended in the gas phase; the product gas and residual ash are drawn off separately. Both caking and noncaking coals can be used in an entrainment system (Smoot and Hanks 1975).

3.1.3.2 Reactor type

Reactor vessels are designed to operate at either high or atmospheric pressure. High pressure maximizes the hydrogasification reaction, improves the quality of the product gas, reduces the required size of the reactor vessel, and eliminates the need to pressurize the gas before it is introduced into a pipeline if a high-Btu gas is to be the final product (University of Oklahoma 1975). However, working at high pressure creates problems involving the method of introducing the coal feedstock into the reactor. Also, design and materials requirements are much more stringent for pressurized systems. Three basic reactor types — a gasifier reactor, a hydrogasifier, and a devolatilizer — are shown along with their principal gasification reactions in Fig. 3.3. The choice of reactor depends on the product gas desired and on temperature and pressure considerations (Fig. 3.3).

ORNL DWG 76-8976

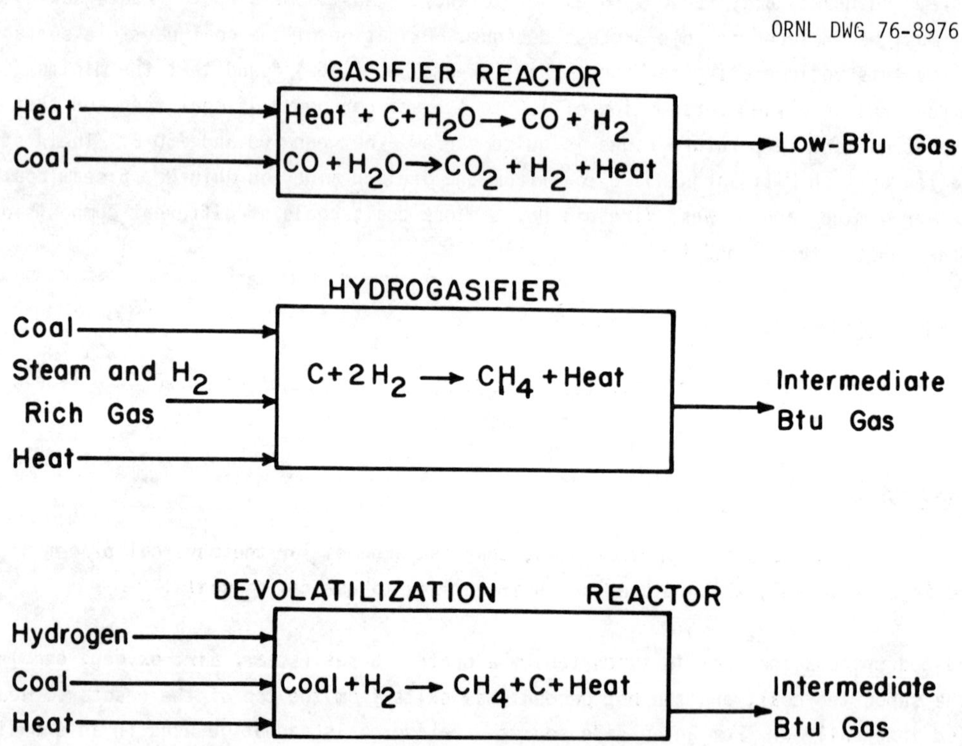

Fig. 3.3. Principal coal gasification reactions and reactor types.
Source: From University of Oklahoma 1975, Fig. 1-27, p. 1-71.

3.1.3.3 Advantages and disadvantages of low- and high-Btu gasification

The principal advantage of low- or intermediate-Btu gasification processes is their lower initial cost. After gasification, purification is required only before the synthetic gas is ready for use; no shift conversion is needed, and gas cleanup is relatively simple because no protection of a methanation catalyst is required. Purification of the product gas is less expensive than processes for purification of stack gases. Sulfur in the gas appears as hydrogen sulfide rather than sulfur dioxide, and effective processes exist for removing the hydrogen sulfide and recovering it as elemental sulfur in either a liquid or solid form (National Academy of Engineering 1974b). However, the ratio of heating value to volume makes low- and intermediate-Btu gas uneconomical for use in existing city gas systems, residences, or present-day pipelines. Carbon monoxide levels are also unacceptably high.

High-Btu gas, however, makes use of existing storage, transportation, and utilization facilities made available by the decline in natural gas. By 1968 the natural-gas pipeline system had grown to a total of 862,000 miles in the United States (Erickson and Sherman 1975). High-Btu gas can be fed into this already existing network and delivered to the point of use without equipment modification of any kind. Additionally, both industrial and residential consumers are equipped to use a product gas with a high heating value. No expensive retrofitting would be required. High-Btu gas also shares with natural gas the advantage of being almost pollutant-free at the point of use.

The most obvious constraint to the development and use of high-Btu gasification facilities is the fact that catalytic methanation, the final processing operation that increases the heating value of the product gas to pipeline quality, is expensive and has not yet been fully proved on a commercial scale. Lurgi has done considerable research at a plant in Westfield, Scotland (Hatten 1975), and the Institute of Gas Technology has produced up to a million cubic feet of pipeline-quality gas per day at its HYGAS pilot plant in Chicago (Linden 1974), but a fully commercial plant to produce high-Btu gas has yet to be built and operated successfully.

3.1.3.4 Uses of low- and high-Btu gas

Scenarios for the development and use of low- or intermediate-Btu systems assume that the gases will be used in (1) industrial or utility boilers that have been retrofitted specifically to accept gas of a lower heating value; (2) an energy park where the gasification facility is surrounded by an industrial complex attracted by the availability of an uninterrupted supply of fuel; or (3) an electricity generating facility based on the combined-cycle concept.

Some research has been done on the cost and technical feasibility of adapting existing industrial or utility boilers to burn low- or intermediate-Btu gas (Squires 1974). Burner ports and feed apparatus would have to be changed, but adjustments could be made with some sacrifice in boiler efficiency (Tillman 1976).

The energy park concept, sometimes called a grass roots facility, postulates that a low- or intermediate-Btu gasification plant, built within a reasonable distance of an assured supply of coal, could attract customers to an industrial park. Certain industries are heavily dependent on an uninterrupted source of fuel, and these industries might be willing to relocate in exchange for a guaranteed supply of fuel. There seems to be general agreement that the energy park concept is economically feasible if developed on a sufficiently large scale (Squires 1974; Linden 1974).

The most discussed idea for the use of low- or intermediate-Btu gas is that of the combined-cycle electric generating facility, which means that gas and steam turbines will be used in combination to generate electricity. The gas turbine, or first, cycle

> ...would consist of burning low- or intermediate-Btu gas at pressure with
> air. Pressure letdown of the hot combustion gas would be used to drive a
> turbine compressor (to compress the air for combustion of the gas) and then
> would go to a turbine driving an electric generator. Heat in the expanded
> turbine exhaust gases would produce steam for a normal steam cycle for addi-
> tional power generation (National Academy of Engineering 1974*b*).

Although the net efficiency of the combined-cycle facility would be somewhat less than that of a new coal-fueled electricity generating plant (33% net efficiency for a combined-cycle facility if assuming an 80% efficiency in gas production, as opposed to 39% otherwise), it is reasonable to assume that further research could lead to an improvement in the gas turbine efficiency. To this end, ERDA's Fossil Energy Division has recently announced the awarding of contracts for a six-year multiphase program to develop a more efficient high-temperature gas turbine system (Energy Research and Development Administration 1976).

Possible use of pressurized and nonpressurized low- and intermediate-Btu gas by various industries has recently been studied by Abrams, Ahn, and Burton (1976) (Tables 3.5 to 3.8). Uses of high-Btu gas are obvious; it will serve the same purposes as natural gas, that is, as pipeline gas or substitute natural gas (SNG) (sometimes referred to as synthetic natural gas). As evaluated by Abrams, Ahn, and Burton (1976) (Tables 3.5 to 3.8), high cost is the major disadvantage of SNG.

3.1.4 Processes

The principal gasification processes and their characteristics are given in Tables 3.9 and 3.10. Manufacture of low-Btu gas is similar to that of intermediate-Btu gas, except that air is substituted for oxygen; intermediate-Btu gas is similar to high-Btu gas, except that methanation is deleted. The high-Btu commercial processes differ primarily in reactor bed type and thus in usable coal type; however, studies are under way to make the processes accessible to all coal types.

3.1.4.1 Lurgi low- and intermediate-Btu process

The Lurgi gasification process, based on work begun in Germany in the early 1930s, is a currently available, fully commercial coal gasification technique. The system employs fixed-bed, pressurized, nonslagging gasifiers, which operate with either air or oxygen and steam (Fig. 3.4) (Katz et al. 1974). The countercurrent flow of reactants in this fixed-bed reactor promotes the transfer of heat by hot gases from the hot coal near the base of the gasifier to the cooler incoming coal.

Table 3.5. Application matrix — chemical and allied products

| Form of coal utilization | Steam generation | | | | | | | |
| | Process and space heat | | Electricity generation on site | | Fuel for direct heat | | Feedstock | |
	New	Retrofit	New	Retrofit	New	Retrofit	New	Retrofit
Low-Btu gas Atmospheric	Yes; large capacity	Unlikely; derating	Unlikely; high cost	Unlikely; derating	Yes; large capacity	Likely; large capacity for space available	No; too high inert content	No; too high inert content
Pressurized	No; high cost	No; equipment not pressurized	Yes; large capacity combined-cycle	No, steam system Likely, pressurized gas turbine system if space available	No; high cost	No; equipment not pressurized	No; too high inert content	No; too high inert content
Medium-Btu gas Atmospheric	No; high cost	Likely; minimum derating, large capacity if space available	No; high cost	Likely, low-pressure system if space available No, gas turbine system; high cost	Likely; special flame characteristics required	Likely; special flame characteristics required	Unlikely; high compression cost	Unlikely; high compression cost
Pressurized	No; high cost	No; high cost	No; high cost	No, low-pressure steam system Unlikely, gas turbine system; high cost	No; high cost	No; high cost	Yes; best substitute for natural gas	Yes; best substitute for natural gas
SNG	No; high cost	No; high cost	No; high cost	No; high cost	No; high cost	No; high cost	No; high cost	No; high cost

Source: After Abrams, Ahn, and Burton 1976, Table II, p. 11.

Table 3.6. Application matrix — coal gasification

Form of coal utilization	Coke oven firing		Sintering of iron ore		Blast furnace smelting	
	New	Retrofit	New	Retrofit	New	Retrofit
Low-Btu gas Atmospheric	Yes, if large capacity	Yes; least costly substitute	Likely; may be temperature limit	Likely; may be temperature limit	Unlikely; high compression cost	Unlikely; high compression cost
Pressurized	No; high cost	No; high cost	No; high cost	No; high cost	Yes; all cases	No; derate problem
Medium-Btu gas Atmospheric	Yes if large plant; multiple use	Yes, if large plant; multiple use	Unlikely; more costly than low-Btu	Unlikely; more costly than low-Btu	Unlikely; high compression cost	Unlikely; high compression cost
Pressurized	No; high cost	No; high cost	No; high cost	No; high cost	Yes, all cases; similar to present use	Yes, all cases; similar to present use

Blast furnace steelmaking		Annealing		Heat treating		Nonferrous smelting	
New	Retrofit	New	Retrofit	New	Retrofit	New	Retrofit
Unlikely; high compression cost	Unlikely; high compression cost	Unlikely; temperature and composition control problems	Unlikely; temperature and composition control problems	Yes, similar to 90 Btu/scf gas being used	Yes, similar to 90 Btu/scf gas being used	Yes, all cases	Unlikely; derate problem
Insufficient information	Insufficient information	No; high cost	No; high cost	No; high cost	No; high cost	No; high cost	No; high cost
Research needed	Research needed	Yes, similar to present use	Yes, similar to present use	Unlikely; more expensive than low-Btu	Unlikely; more expensive than low-Btu	Unlikely; more costly than low-Btu	Yes, all cases
Research needed	Research needed	No; high cost	No; high cost	No; high cost	No; high cost	No; high cost	No; high cost

Source: Abrams, Ahn, and Burton 1976, Table III, pp. 12 - 13.

Table 3.7. Application matrix — petroleum refining

| Form of coal utilization | Steam generation | | | | Feedstock to refinery | |
| | Process and space heat | | Electricity generation on site | | | |
	New	Retrofit	New	Retrofit	New	Retrofit
Low-Btu gas						
Atmospheric	Yes; applicable for all cases	Unlikely; derating problems	Unlikely; high cost	Unlikely; derating problems	Not applicable	Not applicable
Pressurized	No; high cost	No; high cost	Yes; combined-cycle plant application	No; for low-pressure steam system Likely, for gas turbine if space available	Not applicable	Not applicable
Medium-Btu gas						
Atmospheric	No; high cost	Likely; minimum derating if space is available	No; high cost	Likely; for low-pressure steam system if space available No, for gas turbine	Not applicable	Not applicable
Pressurized	No; high cost	No; high cost	No; high cost	No, for low-pressure steam system Unlikely, for gas turbine; high cost	Not applicable	Not applicable
SNG	No; high cost	No; high cost	No; high cost	No; high cost	Not applicable	Not applicable

Source: Abrams, Ahn, and Burton 1976, Table IV, p. 14.

Table 3.8. Application matrix — glass, stone, and clay products

Form of coal utilization	Glass		Stones (cement, lime, and gypsum)		Clays (bricks and structural tile, pottery and related)	
	New	Retrofit	New	Retrofit	New	Retrofit
Low-Btu gas						
Atmospheric	Yes, if large plant	Unlikely; derating	Likely; research needed in application	Unlikely; derating	Likely; research needed in application	Unlikely; derating
Pressurized	No; high cost	Likely; minimizes derating	No; high cost	No; high cost	No; high cost	No; high cost, derating
Medium-Btu gas						
Atmospheric	No; high cost	Likely; eliminate derating	Yes, all cases	Yes; minimizes derating	Yes, all cases	Yes; minimizes derating
Pressurized	No; high cost	No; high cost	No; high cost	No; high cost	No; high cost	No; high cost
SNG	No; high cost	No; high cost	No; high cost	No; high cost	No; high cost	No; high cost

Source: Abrams, Ahn, and Burton 1976, Table V, p. 15.

Table 3.9. Coal gasification processes for production of low-Btu gas

Process	Reactor bed type	Nature of residue	Pressure (atm)	Temperature ($^\circ$F)	Capacity (tons/day)
Commercial					
Winkler	Entrained	Dry ash	1	1500	2000
Demonstration					
Lurgi	Fixed	Dry ash	20	1000	2000
Pilot plants					
Stirred fixed producer					
(U.S. Bureau of Mines)	Fixed	Dry ash	20	1000	20
General Electric fixed-					
bed	Fixed	Dry ash	8	1000	0.25
Combustion Engineering –					
Consolidated Edison					
gasifier	Entrained	Slag	1	>2100	180
Westinghouse Electric				1300 – 1700	
Corp. gasifier	Multiple fluid beds	Dry ash	10 – 16	and 2000	15
Pittsburg-Midway					
gasifier	Entrained (two-stage)	Slag	4 – 35	>2100	1200
IGT U-Gas	Fluid bed	Dry ash	20	1900	30 – 50

Source: Perry 1975, Table 4-28, p. 408. Reprinted by permission of the publisher.

Table 3.10. Coal gasification processes for production of medium- and high-Btu gas

Process	Reactor bed type	Gasifying medium	Nature of residue	Pressure (atms)	Temperature (°F)	Capacity (tons/day)
Commercial						
Lurgi gasifier	Fixed	Oxygen-steam (air being tested)	Dry ash	30 - 35	500 - 800 (top) <2000 (bottom)	1000
Koppers-Totzek	Entrained	Oxygen or limited oxygen-steam (air being tested)	Dry ash	1	1750 - 2350	850
Winkler	Fluidized	Oxygen-steam and air-steam	Dry ash	1	1500 - 1800	100
Small demonstration						
HYGAS[a] (IGT)	Fluid	Hydrogen in hydro-gasifier	Dry char	75 - 100	1200 - 1500 (first stage) 1700 - 1800 (second stage)	80
CO_2 Acceptor	Fluid	Air in regenerator, steam in the gasifier	Dry ash	10 - 20	1575	30
Pilot plants						
Synthane	Fluid	Oxygen-steam	Dry char	40 - 70	1100 - 1450 (first stage) 1750 - 1850 (second stage)	75
Bi-Gas	Entrained	Oxygen-steam	Slag	70	2700 (first stage) 1700 (second stage)	120
Co-Gas	Fluid	Steam in gasifier	Dry ash	1 - 3	1600 - 1700	50
Carbide-Battelle	Fluid	Steam in gasifier	Dry ash	6	1600 - 1800	25
IGT — electrothermal	Fluid	Steam in gasifier	Dry char	75 - 100	1900	25
IGT — steam-oxygen	Fluid	Steam-oxygen	Dry ash	100	1500	25
IGT — steam-iron	Fluid	Steam in oxidizer	Dry ash	100	1500	25
Hydrane	Fluid	Hydrogen in hydro-gasifier	Dry char	70	1600 - 1700	0.25

[a]The HYGAS process is a hydrogasification route. Three alternative methods to produce the hydrogen required have been proposed: electrothermal, steam-oxygen, and steam-iron processes; they are described under "Pilot plants."

Source: Perry 1975, Tabel 4-27, p. 400. Reprinted by permission of the publisher.

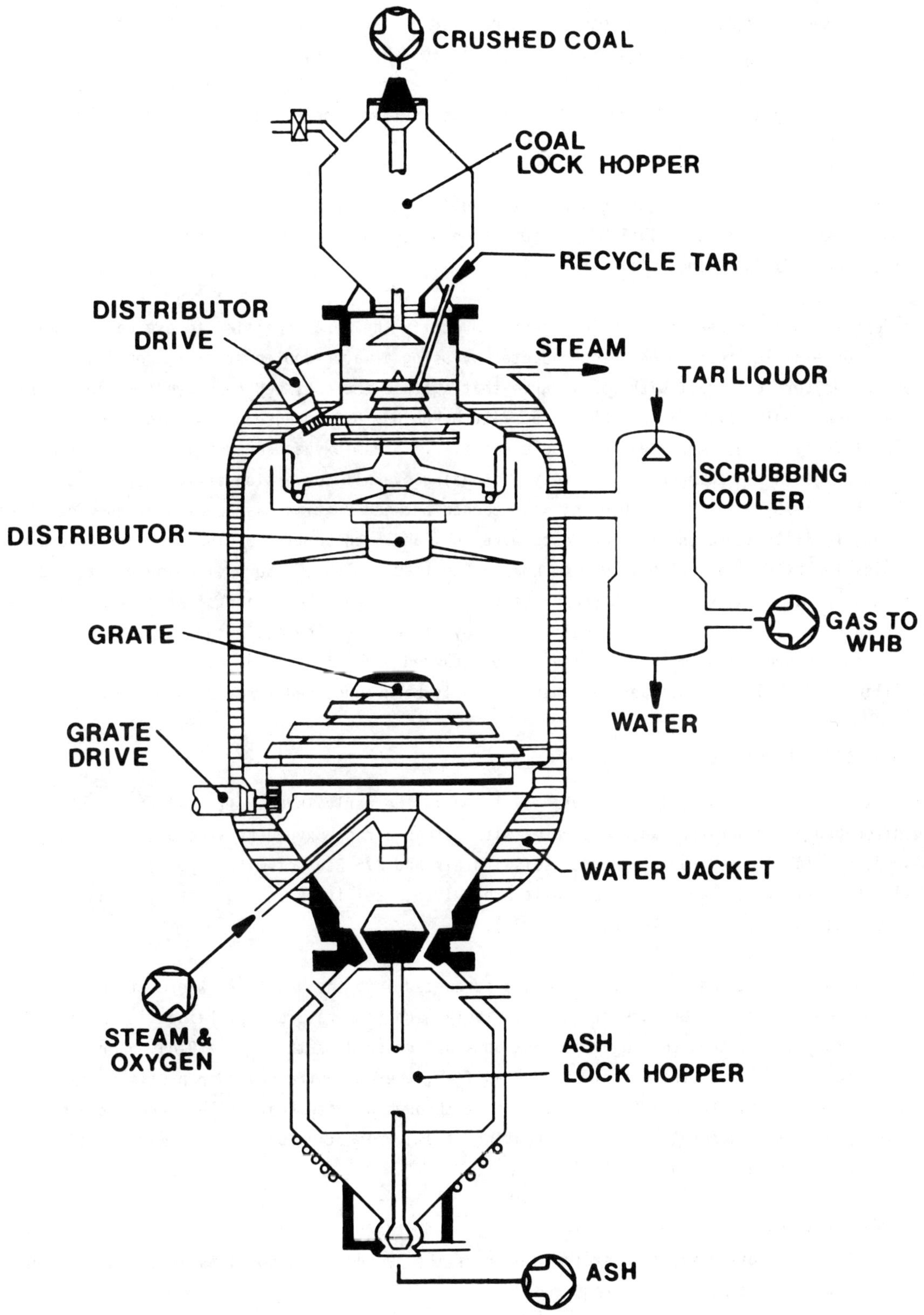

Fig. 3.4. Lurgi pressure gasifier. <u>Source</u>: Dravo Corporation 1976, Fig. 1.18-1, p. 81.

Noncaking or very weakly caking coal is sized from 3/16 to 1-3/4 in. without pretreatment and is introduced through the lock hopper into the top of the water-jacketed gasifier, where the coal falls onto the rotating grate. Steam and either air or oxygen are introduced beneath the grate. Ash falls out through the grate and is removed via an ash hopper, whereas the hot product gas (800 to 1000°F) exits from the top of the reactor to a scrubber that removes particulate matter. Ultimately, the gas is desulfurized and the hydrogen sulfide converted to elemental sulfur (Fig. 3.5).

A comparison of the Lurgi product gas when the coal is treated with either air and steam or oxygen and steam is given in Table 3.11. Dilution of the air-produced gas by nitrogen accounts for its lower heating value.

Considerable work has been done in an attempt to modify the Lurgi gasifier to use caking American coals. For example, tests have been conducted by using a water-cooled stirrer, consisting of a power-driven vertical shaft with rabble arms that agitate the coal bed to prevent agglomeration. Four American coals (Illinois Nos. 5 and 6, Pittsburgh No. 8, and Rosebud, Wyoming) were tested in 1973-1974 by using a modified full-scale gasifier. Steam-oxygen consumption was substantially higher for the three caking coals, thereby decreasing overall conversion efficiency (Woodall-Duckham Ltd. 1975). In 1975, agreement was reached between British Gas (owners of the Westfield, Scotland, facility where the earlier tests were run) and a consortium of 15 American oil and gas companies led by Conoco Coal Development Company to test a high-pressure slagging gasifier developed by British Gas's Midland Research Center. The new gasifier rejects ash as a molten slag rather than as a solid. Higher thermal efficiency, four times the gas output, and fewer environmental and effluent problems are claimed. Two extended runs of seven days, producing about 25 million ft^3 of low-Btu gas per day, have recently been announced (*Weekly Energy Report* 1976).

3.1.4.2 Winkler gasifier

The Winkler coal gasifier (Fig. 3.6) was developed in the early 1920s. It is a single-stage, fluidized-bed gas generator, which operates with either air or oxygen at atmospheric pressure to produce a low-Btu gas (about 120 Btu/scf with air and 275 Btu/scf with oxygen) (Katz et al. 1974). A total of 16 plants has been built since 1926, facilities in Yugoslavia, Turkey, and India operating as late as 1974 (Banchik 1974).

Noncaking coal is crushed to 3/8-in. size and screw-fed into the gasifier, where it reacts with air or oxygen and steam. Because of the relatively high temperatures maintained in the gasifier (1500 to 1800°F), all tars and heavy hydrocarbons are cracked (Katz et al. 1974). Larger, heavier particles of ash fall out the bottom of the gasifier, whereas entrained char particles exit in the high-velocity gas stream. The hot gas passes through a waste heat boiler, where the sensible heat is recovered by generating and superheating steam, preheating boiler feed water, and preheating combustion water (Banchik 1974).

Char particles are removed from the gas stream in two stages. Bulk removal is accomplished dry in the waste heat boiler and in a cyclone, whereas the second stage consists of a wet scrubber and an electrostatic precipitator, if required.

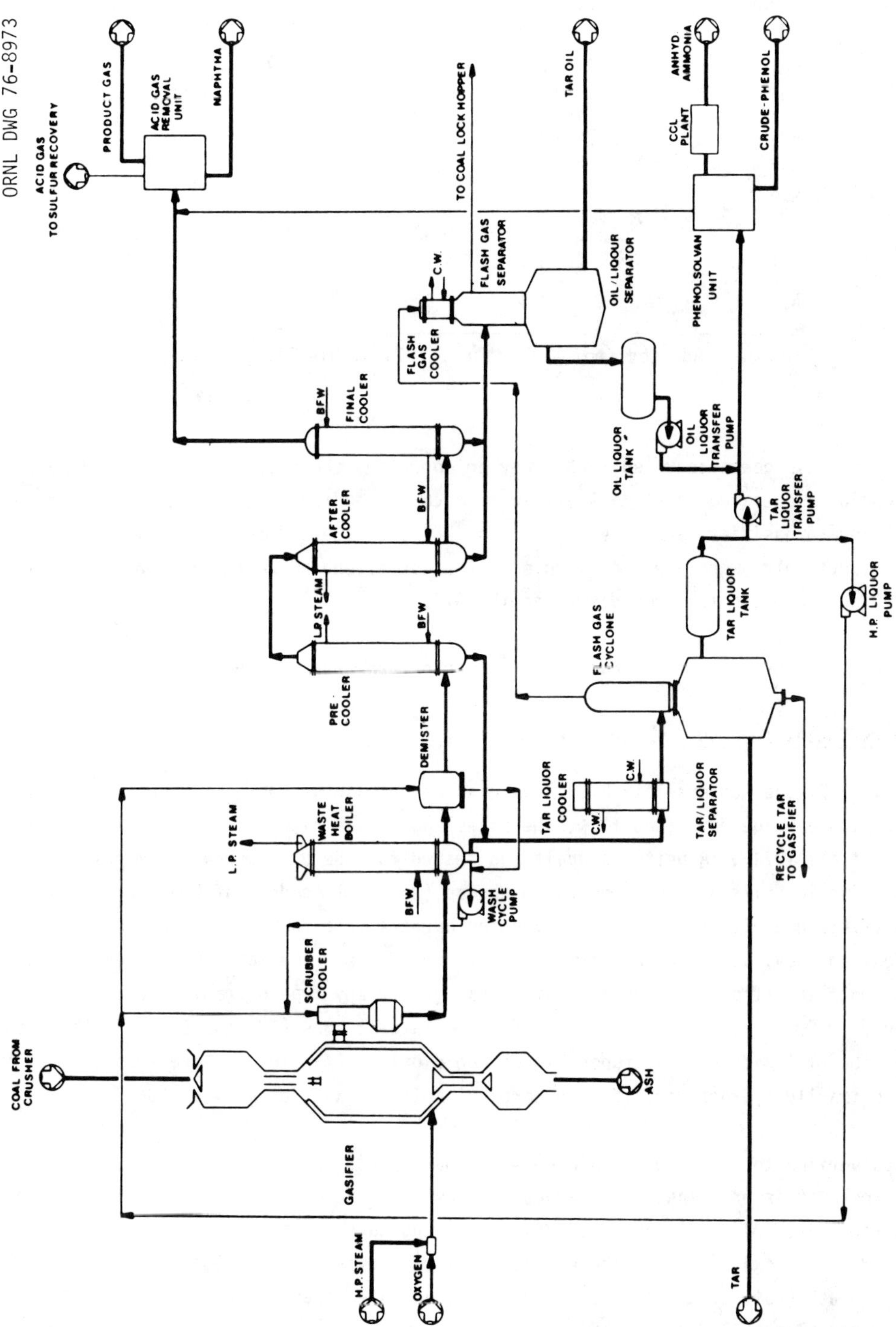

Fig. 3.5. Lurgi coal gasification system. Source: From Dravo Corporation 1976, Fig. 1.18-2, p. 82.

Table 3.11. Lurgi low-Btu product gas composition (%)

Component	Air and steam	Oxygen and steam
Btu/scf	180	300
CO	13.3	9.2
CO_2	13.3	14.7
H_2	19.6	20.1
CH_4	5.5	4.7
C_2H_6		0.5
H_2O	10.1	50.2
H_2S	0.6	0.6
COS	0.1	
N_2	37.5	

Source: Adapted from Katz et al. 1974, Table III, p. 63.

Laboratory studies have been conducted in Germany on operating the gasifier at higher pressures. The gasifier could be operated at up to 9 atm, with a corresponding increase in throughput and production. Standard gasifiers are rated to process 700 and 1100 std tons of coal per day at 1 and 3 atm, respectively, when operated with air; the corresponding rates for operation with oxygen are 1000 and 1500 std tons per day (Banchik 1974).

The Winkler air-blower product gas is compared with an oxygen-treated product in Table 3.12.

3.1.4.3 Koppers-Totzek process

As shown in Fig. 3.7, the Koppers-Totzek process uses an entrained-bed reactor having a commercially proved, single-stage entrained flow; the ash-slagging gasifier operates at atmospheric pressure (Katz et al. 1974). A unit was built and tested for the U.S. Bureau of Mines in 1949, and since that time 16 units have been operated abroad (National Academy of Engineering 1974*b*). Commercial gasifiers use either two or four opposing burners; the four burners can handle up to 850 tons of coal per day, achieving a thermal efficiency of 77%. The major advantage of the Koppers-Totzek entrained-flow gasifier is that it can gasify any type or rank of coal, including the fines (Perry 1975). The product gas is an intermediate-Btu gas (290 to 295 Btu/scf) composed of about 56% carbon monoxide, 35% hydrogen, 7% carbon dioxide, 1% nitrogen, and less than 1% trace constituents, primarily hydrogen sulfide and carbonyl sulfide (Katz et al. 1974).

Feed coal is pulverized to 70% <200 mesh (0.003-in.) particles, dried, and screw-fed to the gasifier nozzles, where it is entrained with oxygen and low-pressure steam (University of Oklahoma 1975). Combustion takes place at about 3000°F, destroying tars, phenols, and light oils (Perry 1975). The product gas exits through the central vertical outlet. About 50% of the coal ash exits as fine fly ash (Katz et al. 1974). Product gas is waterspray-cooled to 2200 to 2400°F and sent to a waste heat boiler (Fig. 3.8).

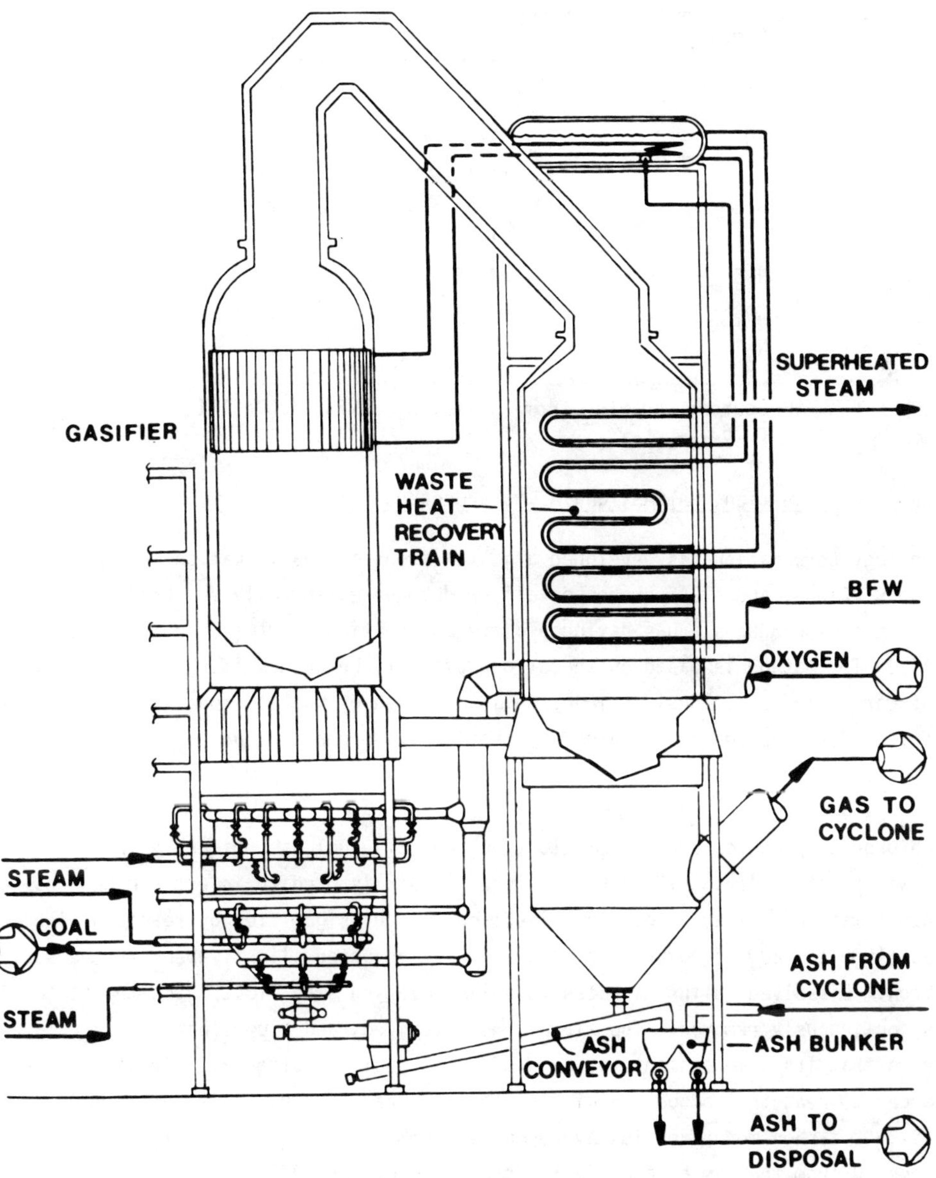

Fig. 3.6. Winkler gasifier. <u>Source</u>: From Dravo Corporation 1976, Fig. 1.16-1, p. 71.

In the process of cooling the gas to about 350°F, enough superheated, high-pressure steam is created to power the air compressors for the oxygen plant and supply all needed process steam. The gas is further cooled in a high-energy scrubbing system, which reduces the entrained solids to 0.002 to 0.005 grain/scf (Katz et al. 1974). Particulate-laden water from the cooldown processes is sent to a clarifier, whereas sludge is filtered or sent directly to a disposal area. The clean water circulates through a cooling tower and then reenters the process system. In areas where water availability is a problem, air cooling can be used to some degree. Oxygen consumption is high (0.5 to 0.8 lb per pound of dry coal), but the process itself is simple and easily controlled (National Academy of Engineering 1974*b*).

Table 3.12. Winkler product gas composition (%)

Component	Air	Oxygen
CO	19	25.7
CO_2	6.2	15.8
H_2	11.7	32.2
CH_4	0.5	2.4
H_2O	11.5	23.1
H_2S	0.13	0.25
COS	0.02	0.04
N_2	51.1	0.8

Source: Adapted from Katz et al. 1974, Table III,
p. 63.

3.1.4.4 Two-Stage Combustion, PATGAS, and ATGAS processes

Applied Technology Corporation has developed a molten iron coal gasification method. The Two-Stage
Coal Combustion, PATGAS, and ATGAS processes all produce an essentially sulfur-free fuel gas;
however, each process produces a gas having a different heating value (Table 3.13) (LaRosa and
McGarvey 1975). The method is based on the strong affinity between sulfur and iron and on the
solubility of carbon in iron. Coal is injected into a bath of molten iron; the coal's fixed
carbon and sulfur dissolve and are retained by the iron, whereas the coal volatiles crack and
appear as offgas.

The Two-Stage process produces a hot, low-Btu (190 Btu/scf) fuel gas primarily intended for combustion in power plant boilers (Fig. 3.9). Compressed air is used to inject crushed coal and
limestone into a mass of molten iron. The dissolved carbon is gasified by reaction with additional heated (1000 to 1200°F) combustion air, which is injected slightly beneath the surface of
the molten iron. Dissolved sulfur migrates to a lime-bearing slag floating on top of the iron.
This slag is continuously removed. The slag is desulfurized, and a portion is recirculated to
the gasifier so that its lime content may be used. The gasifier offgas is circulated through
a heat exchanger to heat the combustion air and then is ready for use as a boiler feed gas.
Composition is 30% carbon monoxide, 15% hydrogen, and 55% nitrogen. Particulate matter would
have to be removed from the gas before use (LaRosa and McGarvey 1975).

The PATGAS process produces an intermediate-Btu gas (315 Btu/scf), which can be used in utility
or industrial boilers, energy parks, or combined-cycle power stations (Fig. 3.10) (Katz et al.
1974). The basic technology is the same as that for the Two-Stage Combustion process, except
that oxygen, rather than compressed air, is used.

The ATGAS process is the most elaborate of the three coal gasification schemes developed by
Applied Technology Corporation (Fig. 3.11). The gasification step is identical to that used in
the PATGAS process. However, after scrubbing and cooling, the product gas is compressed, sent
through shift conversion, purified, and then methanated to produce a gas of 940 Btu/scf and composed of 93% methane, 4% hydrogen, and 3% inerts. A process efficiency of 57% is claimed (LaRosa
and McGarvey 1975).

ORNL DWG 76-8972

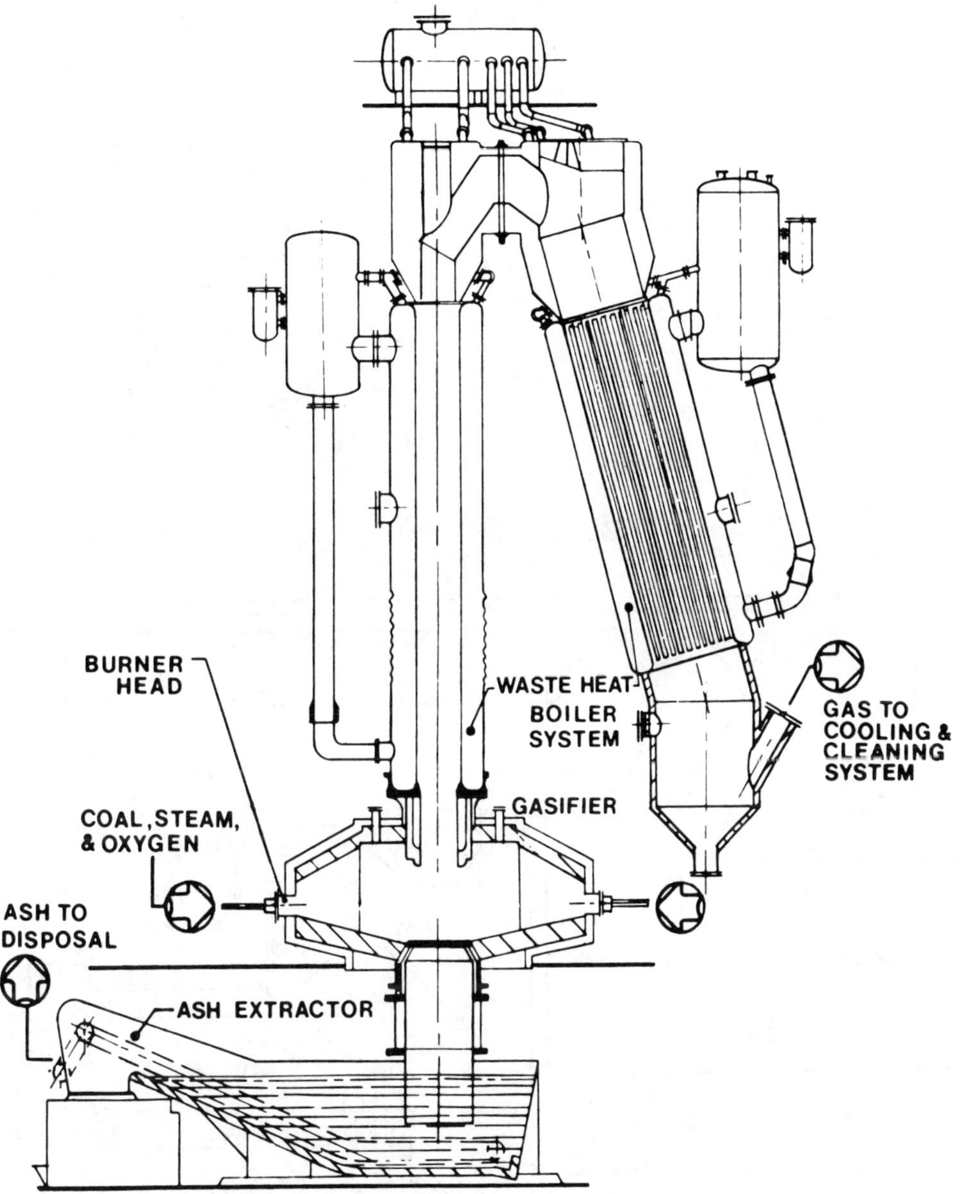

Fig. 3.7. Koppers-Totzek two-headed reactor. <u>Source:</u> From Dravo Corporation 1976, Fig. 1.5-1, p. 23.

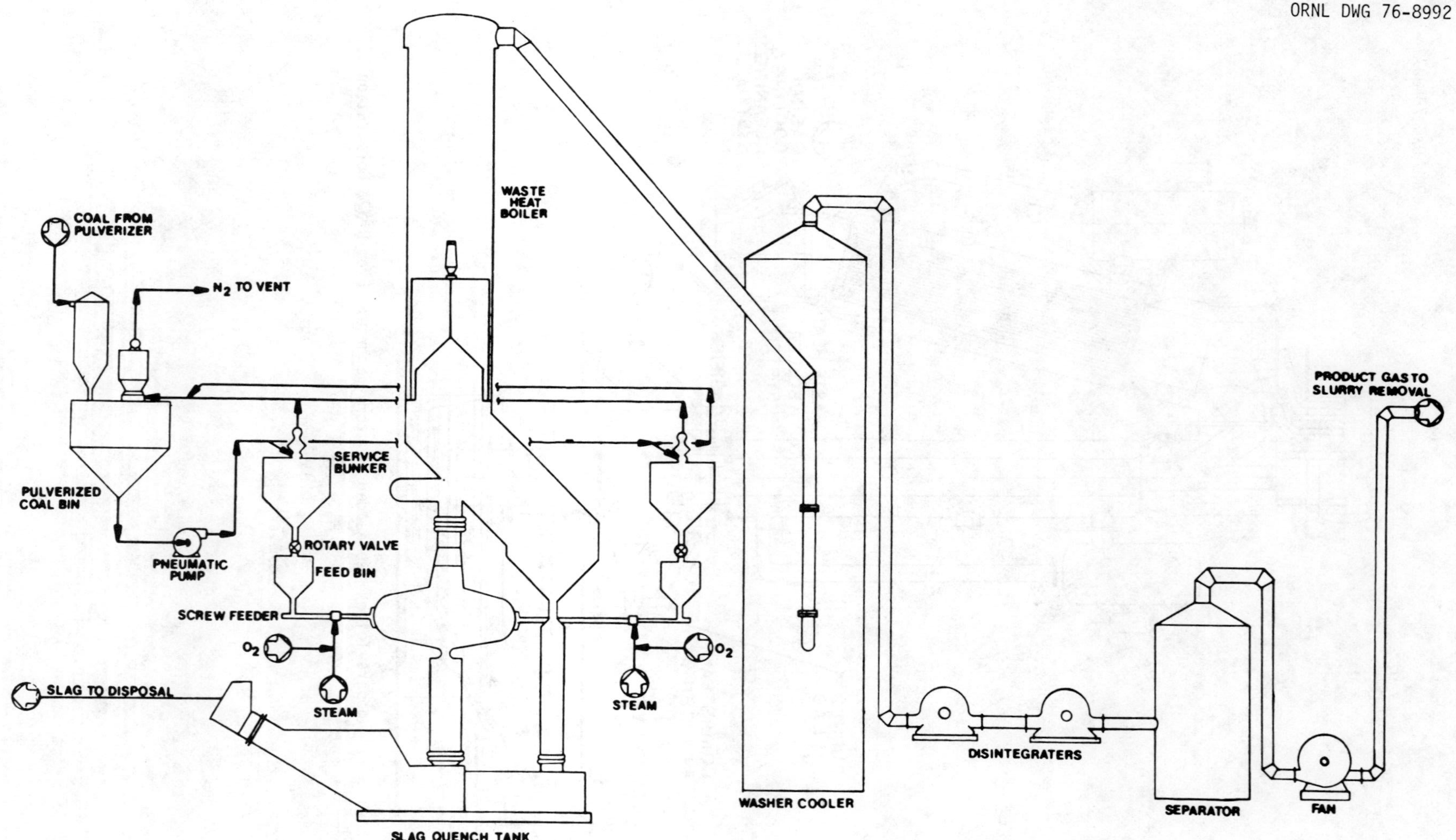

Fig. 3.8. Koopers-Totzek coal gasification system. Source: From Dravo Corporation 1976, Fig. 1.5-2, p. 24.

Table 3.13. Process description — Two-Stage, PATGAS, and ATGAS

Process[a]	Raw materials	Product gas		Pressure (psig)	Heating value (Btu/scf)
		Composition	Temperature ($^\circ$F)		
Two-Stage	Coal	30% CO	2040	2 - 5	190
	Air	15% H_2			
	Limestone	55% N_2			
PATGAS	Coal	63.5% CO	100	2/1000	315
	Oxygen	36% H_2			
	Steam	0.5% N_2			
	Limestone				
ATGAS	Coal	93% CH_4	100	1000	940
	Oxygen	4% H_2			
	Steam	3% inerts			
	Limestone				

[a]All processes yield as by-products granulated iron, elemental sulfur, and desulfurized slag.

Source: LaRosa and McGarvey 1975, Table I, p. 234. Reprinted by permission of the publisher.

3.1.4.5 Bi-Gas process

Bituminous Coal Research, Inc. began work in 1965 on the Bi-Gas process (Katz et al. 1974), a process involving a two-stage, entrained, super-pressure gasifier, which accepts both caking and noncaking coals and can produce pipeline gas (Fig. 3.12). Pulverized coal (70% through 200 mesh) is fed into the upper section of the gasifier, where hot gases (carbon monoxide, hydrogen, and water, at 2700°F) from the lower section provide heat for devolatilization of the coal and for thermal cracking of oils and tars. Because the coal is blown into the gasifier in a dilute, entrained state, caking coal can be used without pretreatment; the coal is slurried in water, pressurized to over 1500 psi, spray-dried, and separated in a cyclone separator before it is added to the gasifier along with steam (Figs. 3.13 and 3.14) (Hayne and Forney 1976).

Ash flows down the walls of the gasifier and is withdrawn at the bottom as molten slag, which is water-quenched and periodically removed from slag hoppers. The raw product gas and char exit from the top of the gasifier, where the char is separated by a cyclone separator and reintroduced into the lower section. A 380-Btu/scf gas is produced when the system is operated with oxygen at system pressures exceeding 1000 psig, whereas a 175-Btu/scf gas is produced at 300 psig with air.

Hydrogen, produced in the shift converter by reaction of carbon monoxide with steam, is used in the methanation step. The combined hydrogen, carbon monoxide, and methane are cleaned in the scrubber for removal of acid gases, which are potential sources of marketable sulfur and ammonia (Wall 1975). Following fluidized-bed methanation, the final product gas has a heating value of over 900 Btu/scf. An overall thermal efficiency of 63.2% for the Bi-Gas coal-to-methane system has been estimated (Katz et al. 1974).

ORNL DWG 76-8975

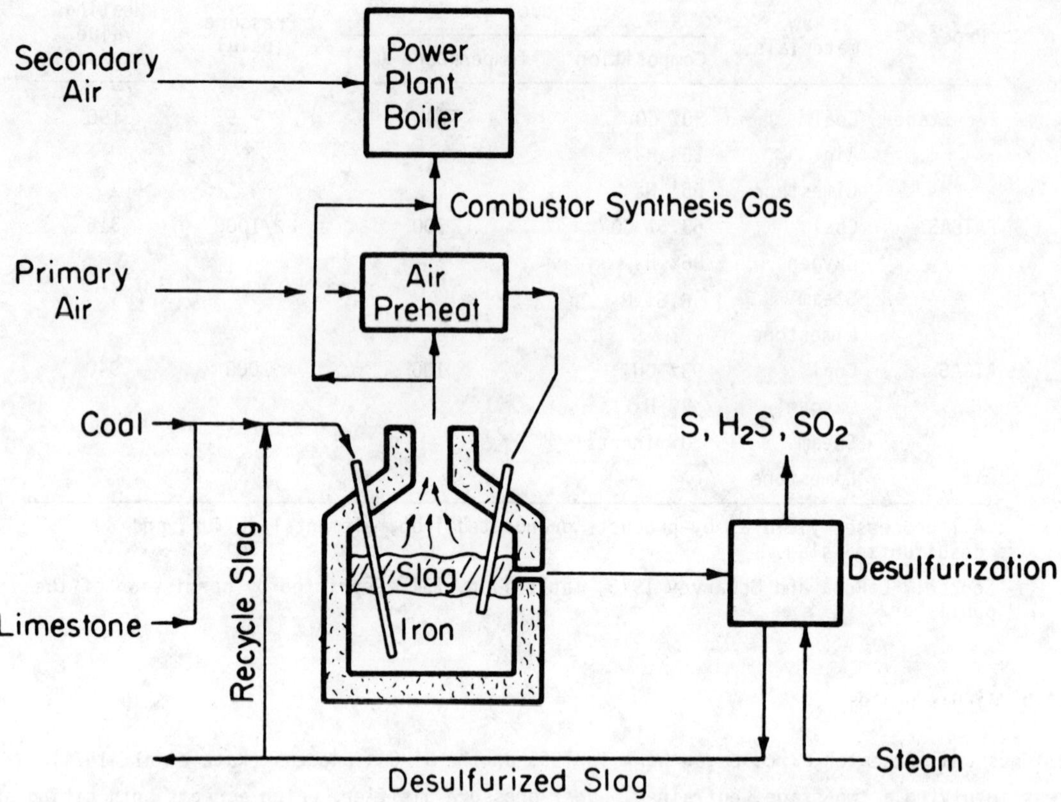

Fig. 3.9. Applied Technology Corporation's two-step gasification system. Source: From Katz et al. 1974, Fig. 31, p. 205. Reprinted by permission of the publisher.

A 120-ton/day pilot plant to produce 2.4 million cfd is currently under construction at Homer City, Pennsylvania (Wall 1975).

3.1.4.6 U-Gas process

The U-Gas process, under development by the Institute of Gas Technology (IGT), uses a fluidized-bed, pressurized reactor to convert any type of coal to a low-Btu gas suitable for utility or industrial applications (Wall 1975) (Fig. 3.15).

Coal is crushed to 1/4-in. particles. (Caking coals are first introduced into a fluidized bed operating at 700 to 800°F, where they are reacted with air to destroy their agglomerating tendencies.) The sized coal is fed into a single-stage fluidized-bed gasifier at 1900°F and 300 to 350 psig. Residence time is about 45 min (Patel and Loeding 1975). Patel and Loeding (1975) describe the process:

> ...the ash is agglomerated into larger and heavier particles for selective separation from the bed. Part of the fluidizing steam and air mixture enters the gasifier through a grid that is sloped toward one or more inverted cones contained in the grid. The remaining fluidizing gas flows upward at a high velocity through the throat at the cone apex, creating a submerged spout

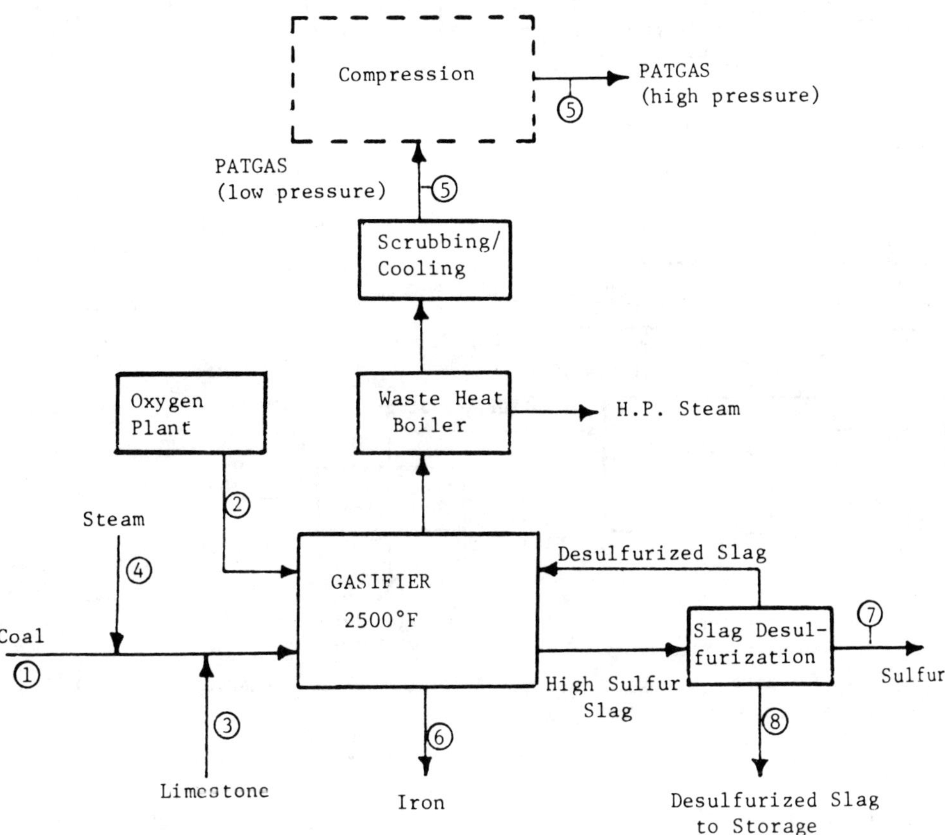

Fig. 3.10. PATGAS process. <u>Source</u>: After LaRosa and McGarvey 1975, Fig. 4, p. 238. Reprinted by permission of the publisher.

> within the fluidized bed.... Ash in gasified coal particles is heated to near its softening point, and the sticky surface of the particles makes them stick together; ash agglomerates grow in size in the jet until they are heavy enough to counter the force of the high-velocity gas stream from the throat of the inverted cone, at which time they fall out of the fluidized bed.

Fines are returned to the gasifier through cyclones, where they are instantly gasified and agglomerate with the normal bed-produced ash. The effluent gas is purified by low-temperature processes (Wall 1975).

The IGT has prepared three economic analyses of the U-Gas process: (1) as a retrofit system for existing oil- or gas-fired boilers, (2) as a combined-cycle process operation, and (3) as a completely new separate facility for industrial use. All three schemes are based on the production of a gas having a heating value of 155 Btu/scf. No shift conversion or methanation step is planned (Patel and Loeding 1975).

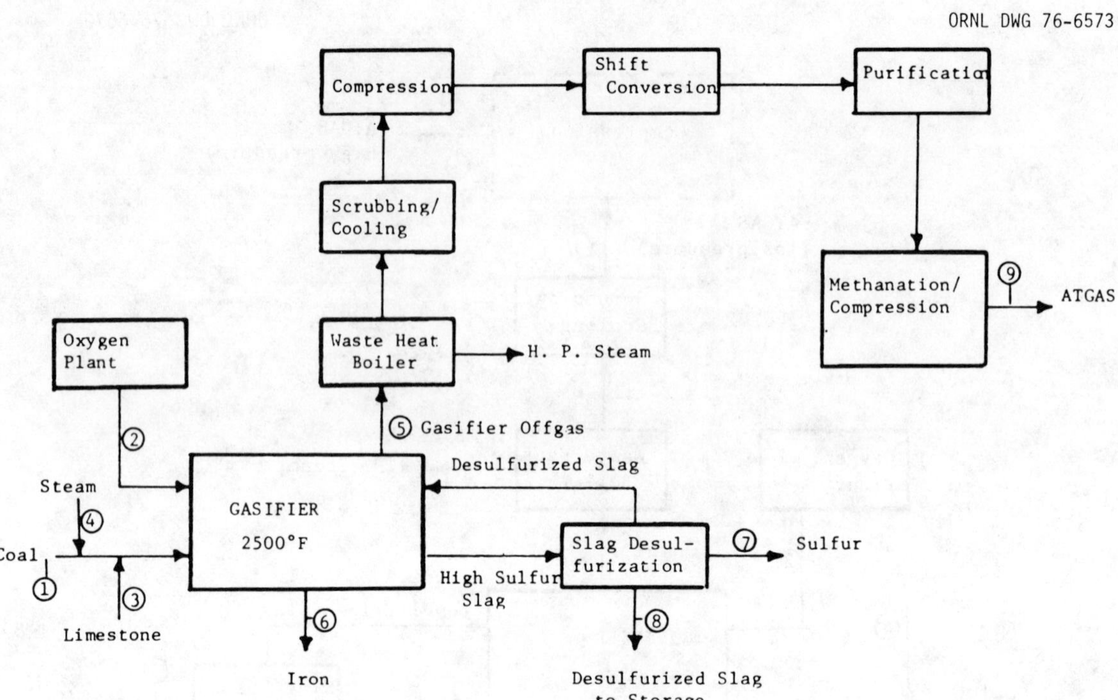

Fig. 3.11. ATGAS process. <u>Source</u>: After LaRosa and McGarvey 1975, Fig. 5, p. 240. Reprinted by permission of the publisher.

3.1.4.7 <u>Wellman-Galusha gasifier</u>

The Wellman-Galusha gasifier is a commercially available unit offered as a standard producer or as an agitator type, which has a slowly revolving horizontal arm (Fig. 3.16). Its interior dimensions range from 18 in. (brick-lined) to 10 ft (water-jacketed). Although usually operated with air and steam, the Wellman-Galusha gasifier can be adapted to use oxygen–steam, oxygen–carbon dioxide, or air–carbon dioxide (Hamilton 1963). Crushed and screened, mildly caking or noncaking bituminous coal or anthracite (charcoal or coke can also be used) is fed continuously from an overhead bunker to a moving-bed gasifier, which operates at atmospheric pressure. Capacities range from about 500,000 Btu/hr for an 18-in. unit to 88 million Btu/hr for a 10-ft unit. For the 10-ft unit, this is a throughput of about 7000 lb/hr (National Academy of Engineering 1974*b*).

As described by Waitzman et al. (1975), air and steam are admitted into the bottom of the gasifier, where (1) the carbon in the coal is oxidized to carbon monoxide and carbon dioxide and (2) carbon and carbon monoxide react with the steam to form carbon monoxide, carbon dioxide, and hydrogen. Methane and heavier hydrocarbons are formed at the top of the gasifier. A large quantity of the heavier hydrocarbons are tarry in nature. In the reducing atmosphere present in the gasifier, sulfur in the coal joins with hydrogen to form hydrogen sulfide. Ash is removed at the bottom with a rotating eccentric grate.

The hot, raw gas is scrubbed and purified to yield a gas of 168 Btu/scf that is composed of 28.6% carbon monoxide, 15% hydrogen, 2.7% methane, 3.4% carbon dioxide, and 50.3% nitrogen.

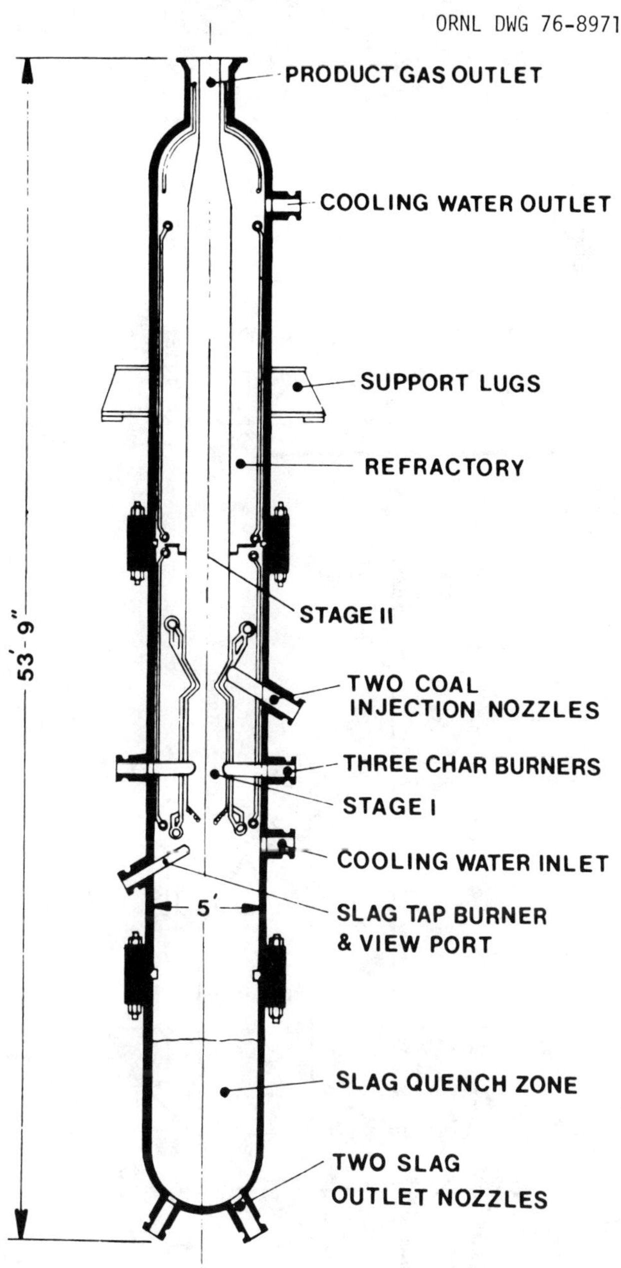

Fig. 3.12. Bi-Gas gasifier. <u>Source:</u> From Dravo Corporation 1976, Fig. 1.2-1, p. 10.

The Morgantown Energy Research Center has conducted pilot-plant tests using a 42-in. Wellman-Galusha gasifier at pressures up to 300 psia (Waitzman et al. 1975).

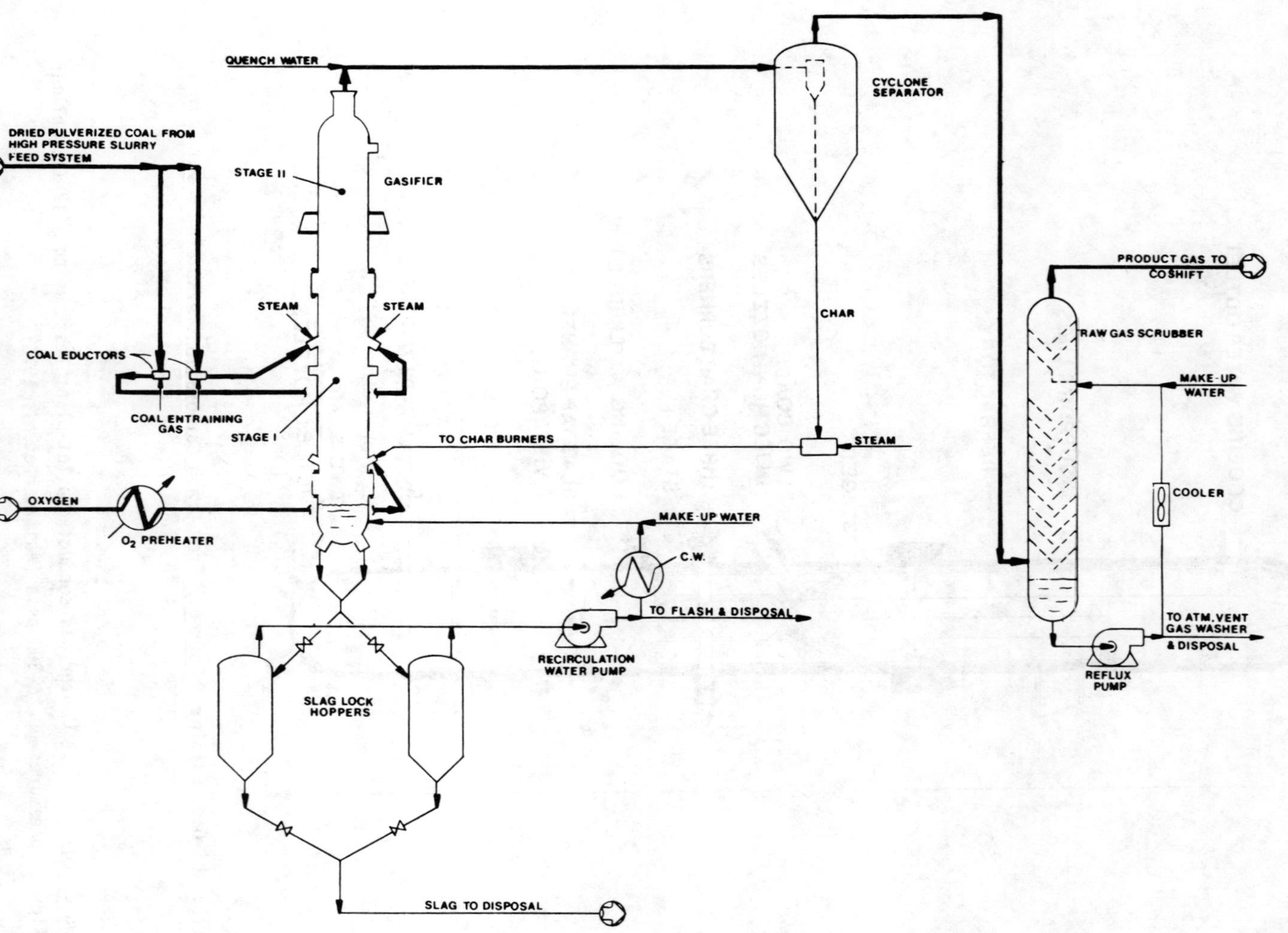

Fig. 3.13. Bi-Gas coal gasification process. Source: From Dravo Corporation 1976, Fig. 1.2-3, p. 12.

ORNL DWG 76-8979

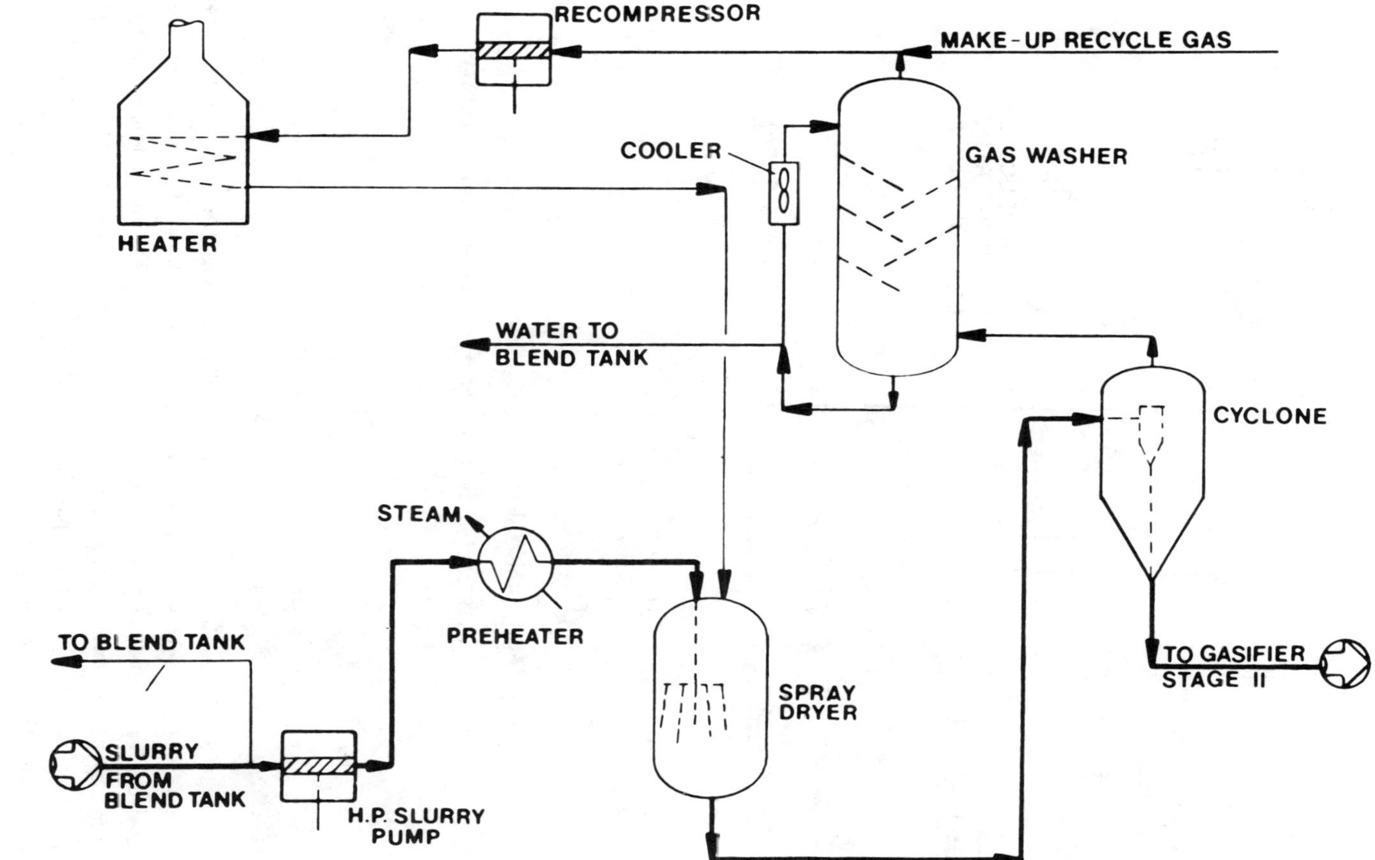

Fig. 3.14. Bi-Gas high-pressure slurry feed system. _Source:_ From Dravo Corporation 1976, Fig. 1.2-2, p. 11.

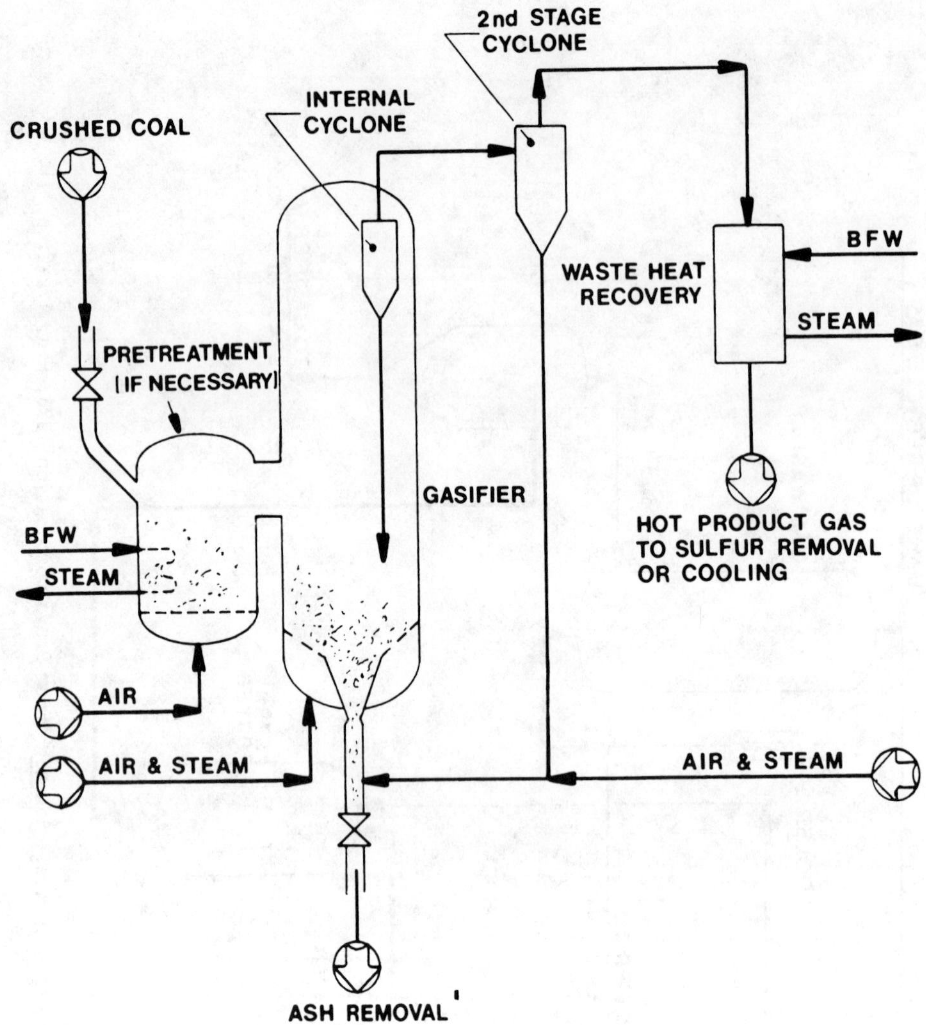

Fig. 3.15. U-Gas coal gasification system. Source: From Dravo Corporation 1976, Fig. 1.14-1, p. 64.

3.1.4.8 Synthane process

The Synthane process (Fig. 3.17), under development by the Pittsburgh Energy Research Center, involves a two-stage pressurized gasifier in which the coking properties of the coal are destroyed in a fluidized-bed pretreater before carbonization (Fig. 3.17). A 72-ton/day pilot plant has recently been completed and is undergoing commission tests (Haynes and Forney 1975; Nakles et al. 1975). Crushed coal (0.003-in. particles) is pressurized with carbon dioxide in lock hoppers and fed to the fluidized-bed pretreatment reactor, where it is contacted with steam and oxygen at 800°F and 1000 psi (Fig. 3.18). The decaked and partially devolatilized coal, along with volatile matter, gases, and excess steam, enters the top of the two-stage, fluidized-bed gasifier, where the free-falling coal particles are countercurrent-contacted with additional oxygen and steam. Char, containing about 30% of the carbon from the original coal, and ash are

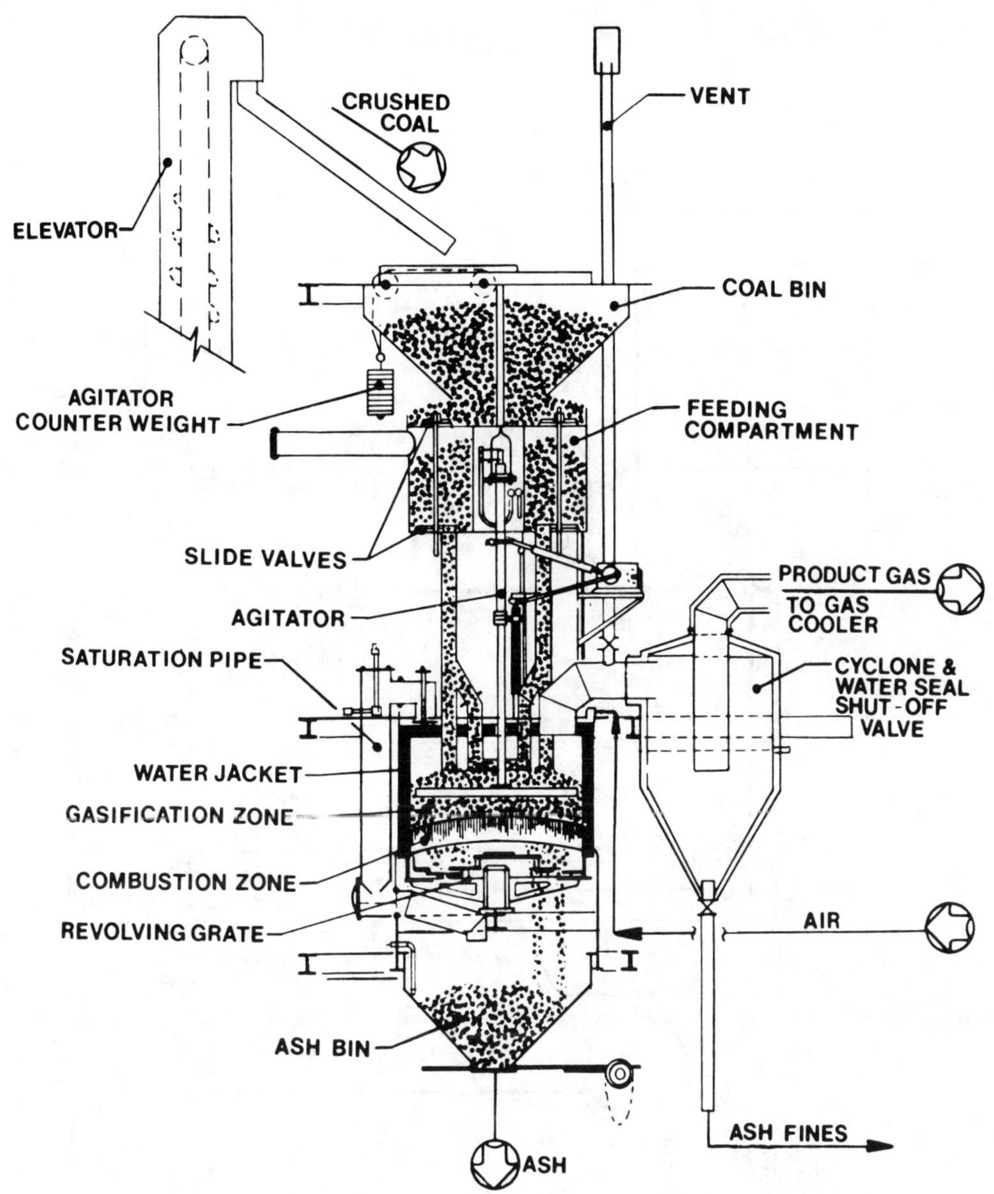

Fig. 3.16. Wellman-Galusha agitated gasifier. <u>Source:</u> From Dravo Corporation 1976, Fig. 1.19-1, p. 85.

discharged from the bottom of the gasifier through a char cooler. Hot 1800°F char from the gasifier is cooled by vaporization of water spray, which generates steam for the shift reactor. The char, together with the dust and tar recovered from the product gas, is burned to provide steam and power (Haynes and Forney 1975). The hot product gas leaves the top of the gasifier via an internal cyclone, which removes most of the dust and returns it to the reactor vessel. After shift conversion to increase the hydrogen–carbon monoxide ratio from 1.7 to 3.1, the gas is scrubbed with hot carbonate to remove carbon dioxide and hydrogen sulfide, further purified

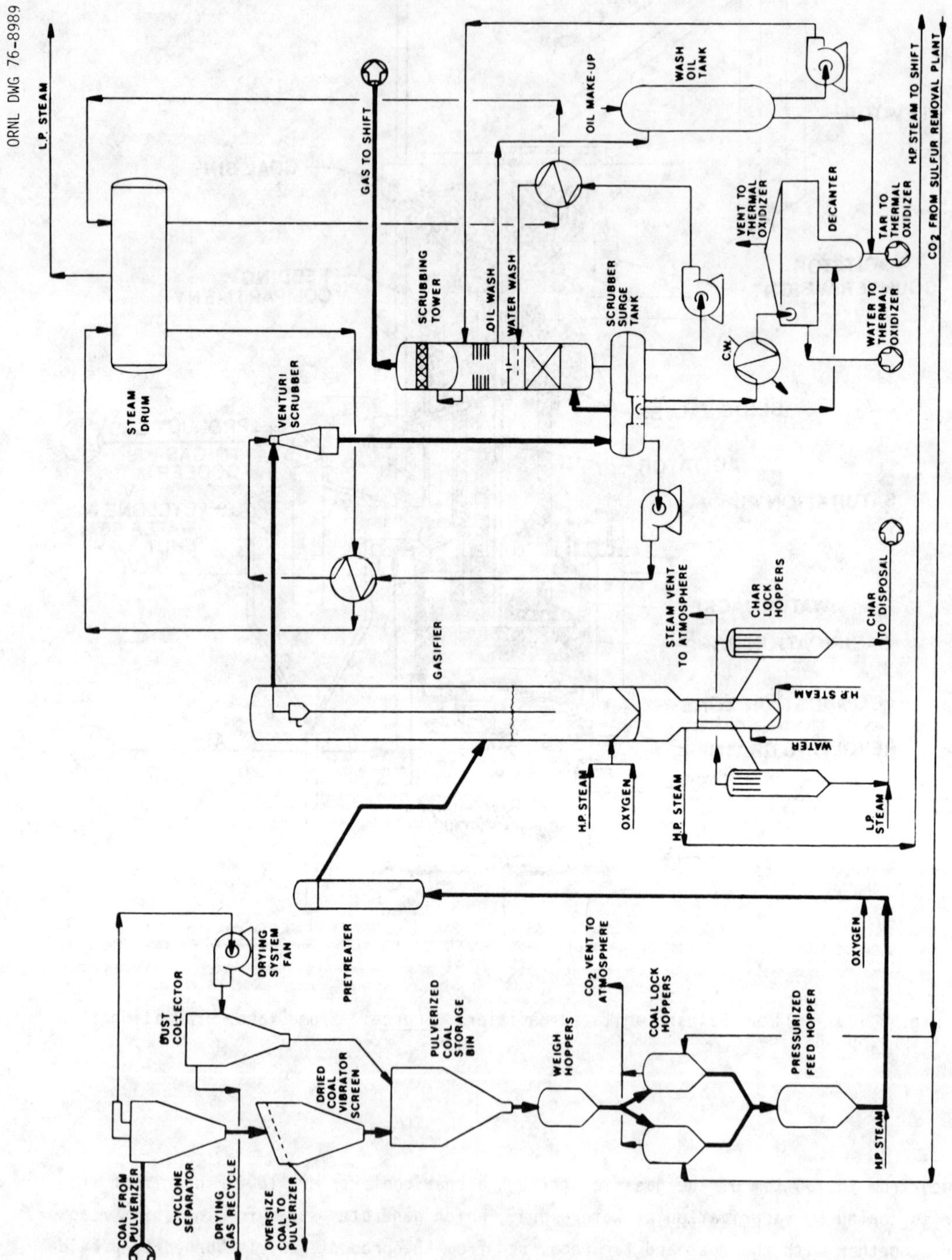

Fig. 3.17. Synthane coal gasification process. Source: From Dravo Corporation 1976, Fig. 1.13-2, p. 61.

ORNL DWG 76-8990

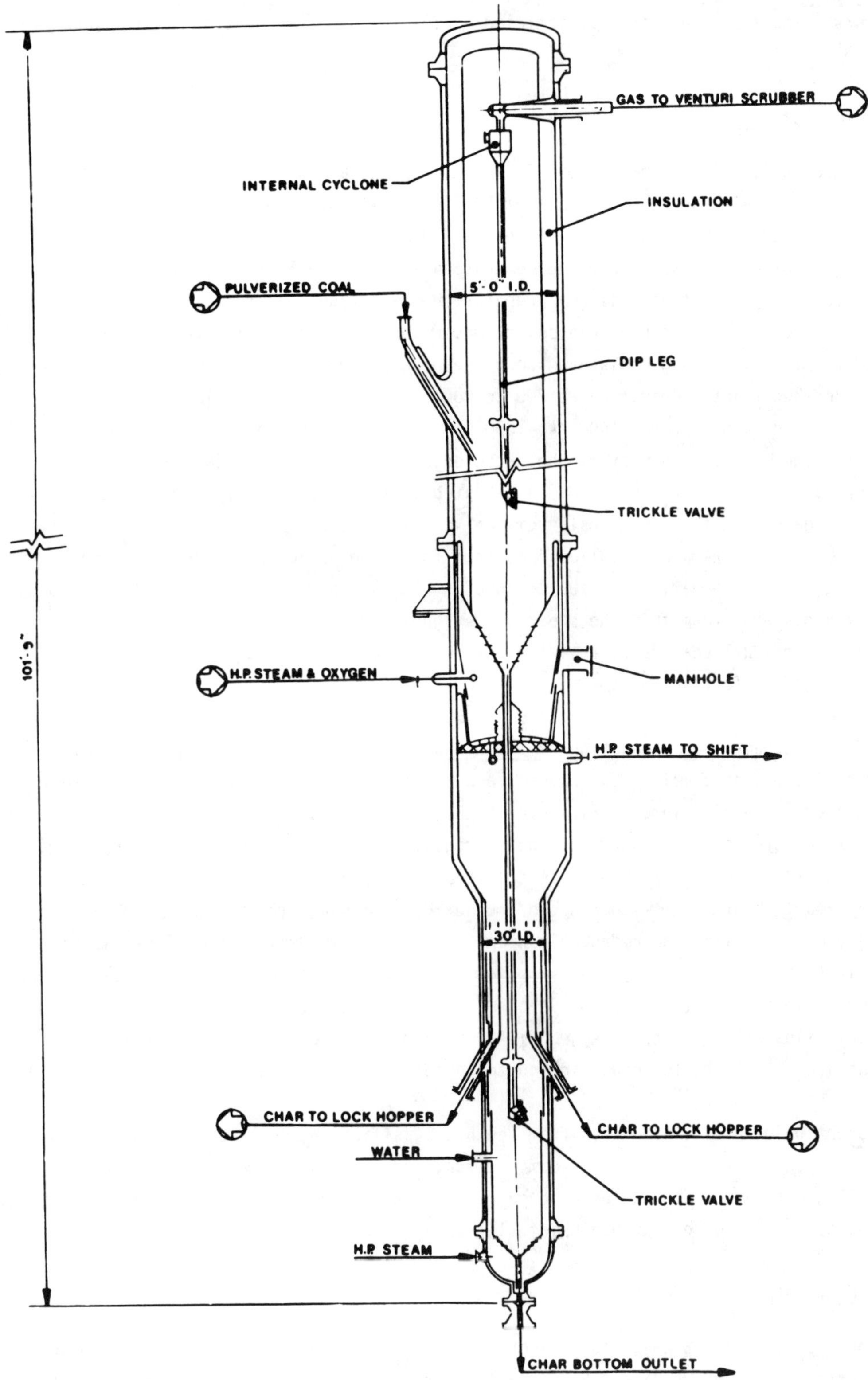

Fig. 3.18. Synthane gasifier. <u>Source</u>: From Dravo Corporation 1976, Fig. 1.13-1, p. 60.

with activated carbon to reduce sulfur to below 0.1 ppm, and then methanated. A heating value
of 927 Btu/scf and an overall thermal efficiency of about 60% is reported (Haynes and Forney
1975).

3.1.4.9 HYGAS process

The HYGAS process is being developed by IGT with sponsorship from the American Gas Association
and the Office of Coal Research, now part of ERDA's Fossil Energy Division. The work is based
on gasification studies begun at IGT in 1944 (IGT 1975a); an 80-ton/day pilot plant was started
in August 1969, with fully self-sustained operation in April 1975 (IGT 1975b). The process is
complex, involving pretreatment, slurry preparation, two-stage coal-to-methane hydrogasification,
fluidized-bed operation, high-pressure shift conversion, and methanation (Fig. 3.19). Coal is
crushed to 1/8 in.; if caking coals are used, they are pretreated with air at atmospheric pressure
in a fluidized-bed reactor operating at 700 to 800°F. The crushed coal is slurried with a light
by-product oil and pumped, under 1000- to 1500-psi pressure, into the top of the hydrogasifier
(Fig. 3.20). The oil is driven off to be recycled, and the dried coal particles fall into stage
1 of the hydrogasifier, where they are flash-heated to the reaction temperature (1200 to 1300°F)
by rising hot gases. Methane conversion of the volatile matter and the active carbon takes place
in a few seconds. The remaining solids fall into the second stage of the hydrogasifier, a
fluidized-bed reactor operating at 1700 to 1800°F. In this stage methane formation, fed by
hydrogen-rich gas and steam from the steam-oxygen gasifier (Fig. 3.21) continues (Haynes and
Forney 1976). Residual char falls from the bottom of the reactor and is used to provide carbon
for the water-gas shift reaction.

Three different schemes for production of the hydrogen needed by the process (Fig. 3.22) have been
investigated: electrothermal, steam-oxygen, and steam-iron. The electrothermal scheme has been
abandoned because the rise in electricity cost has made it uneconomical (Linden et al. 1976). In
the steam-oxygen process, hot char from the hydrogasifier is reacted with steam and oxygen in a
high-pressure, fluidized-bed reactor. In the steam-iron process, hydrogen is produced in an
oxidizer by reaction of steam with reduced iron oxide. Although the steam-oxygen process is
mechanically much simpler, the hydrogen content of its raw synthesis gas is lower than that from
the steam-iron process. Presently, a new pilot plant, which will develop the steam-iron process
as an alternative method of producing hydrogen, is being completed. Although sharing certain
support facilities with the steam-oxygen HYGAS pilot plant, the new pilot plant will be operated
independently so that data for economic and technical comparisons can be generated (IGT 1975-76).

In the present pilot plant, hot gases from the hydrogasifier are methanated, and the liquid
aromatics, carbon dioxide, unreacted steam, sulfur, and ammonia are removed. During a 20-hr run
in April 1975, catalytic methanation produced a product gas of 96.5% methane and 3.5% hydrogen
having a heating value of over 970 Btu/scf (IGT 1975b).

3.1.4.10 Hydrane process

The Hydrane process (Fig. 3.23), under laboratory-scale development by the Pittsburgh Energy
Research Center, is based on direct hydrogenation of coal in a two-zone hydrogasifier without
coal pretreatment. It has been called the most thermally efficient high-Btu process (Gray and
Yavorsky 1975). Pulverized raw coal, caking or noncaking, is fed into the upper zone of the
hydrogasifier (Fig. 3.24). The freely falling coal devolatilizes while flowing in dilute phase

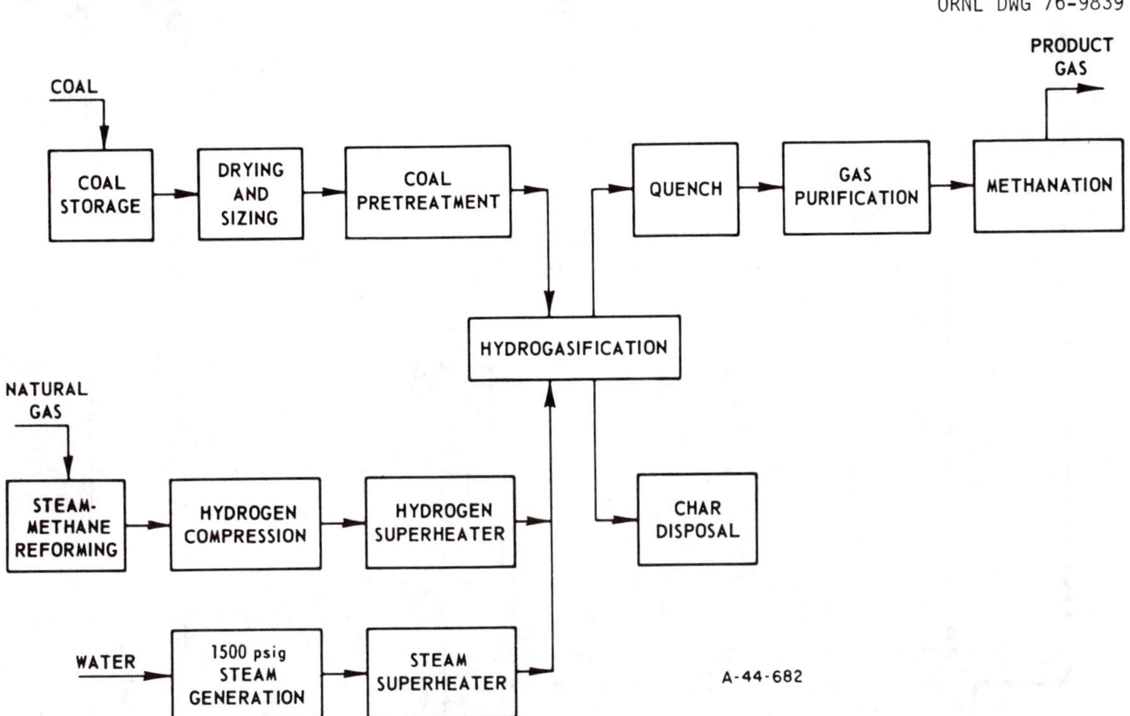

Fig. 3.19. Block flow diagram of the HYGAS pilot plant. <u>Source</u>: From Schora 1974, Fig. 3. p. 1010. Reprinted by permission of the publisher.

suspension through hot methane and hydrogen-rich gas from the lower zone. About 20% of the carbon in the raw coal is converted to methane at this stage. The remaining coal forms very porous, reactive char particles, which fall into the lower zone of the hydrogasifier, operated at 70 atm and 900 to 980°C. Hydrogen feed gas maintains the particles in a fluidized state and reacts with an additional 25% of the carbon to produce methane (Gray and Yavorsky 1975).

The product gas is purified to remove carbon dioxide, hydrogen sulfide, and water. Light methanation is required to produce a high-Btu gas of about 954 Btu/scf. Residual char from the hydrogasifier will be reacted with steam and oxygen in a separate facility to provide hydrogen. Remaining char from the hydrogen plant (calculated at about 0.137 lb per pound of dry coal) could be used as fuel for steam and power generation. Development work has proceeded slowly on this process although announced plans include the design and construction of a large (20 ton/day) process development unit at Morgantown Energy Research Center (Gray and Yavorsky 1975).

ORNL DWG 76-8987

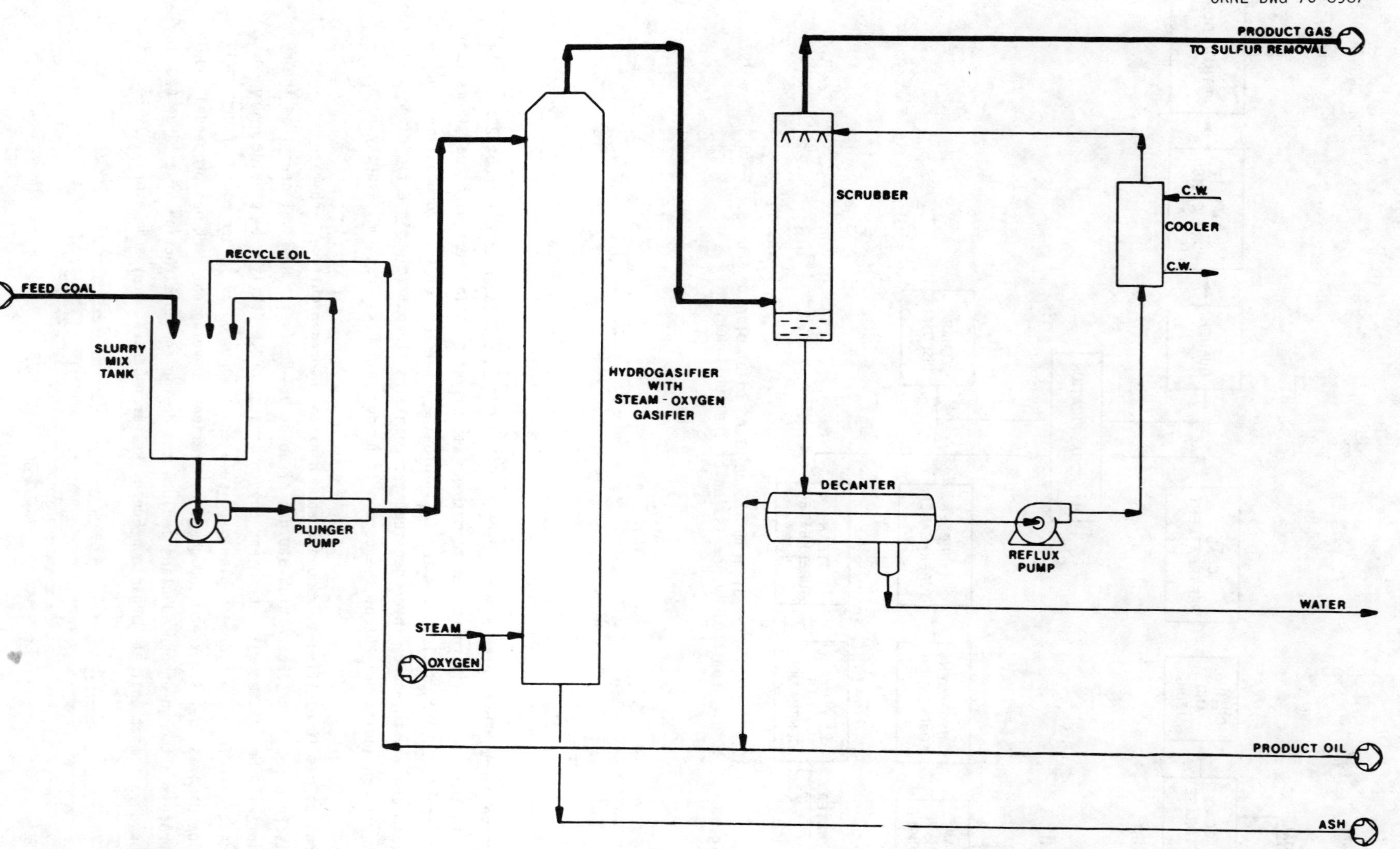

Fig. 3.20. HYGAS coal gasification process (pilot plant setup). Source: From Dravo Corporation 1976, Fig. 1.12-3, p. 56.

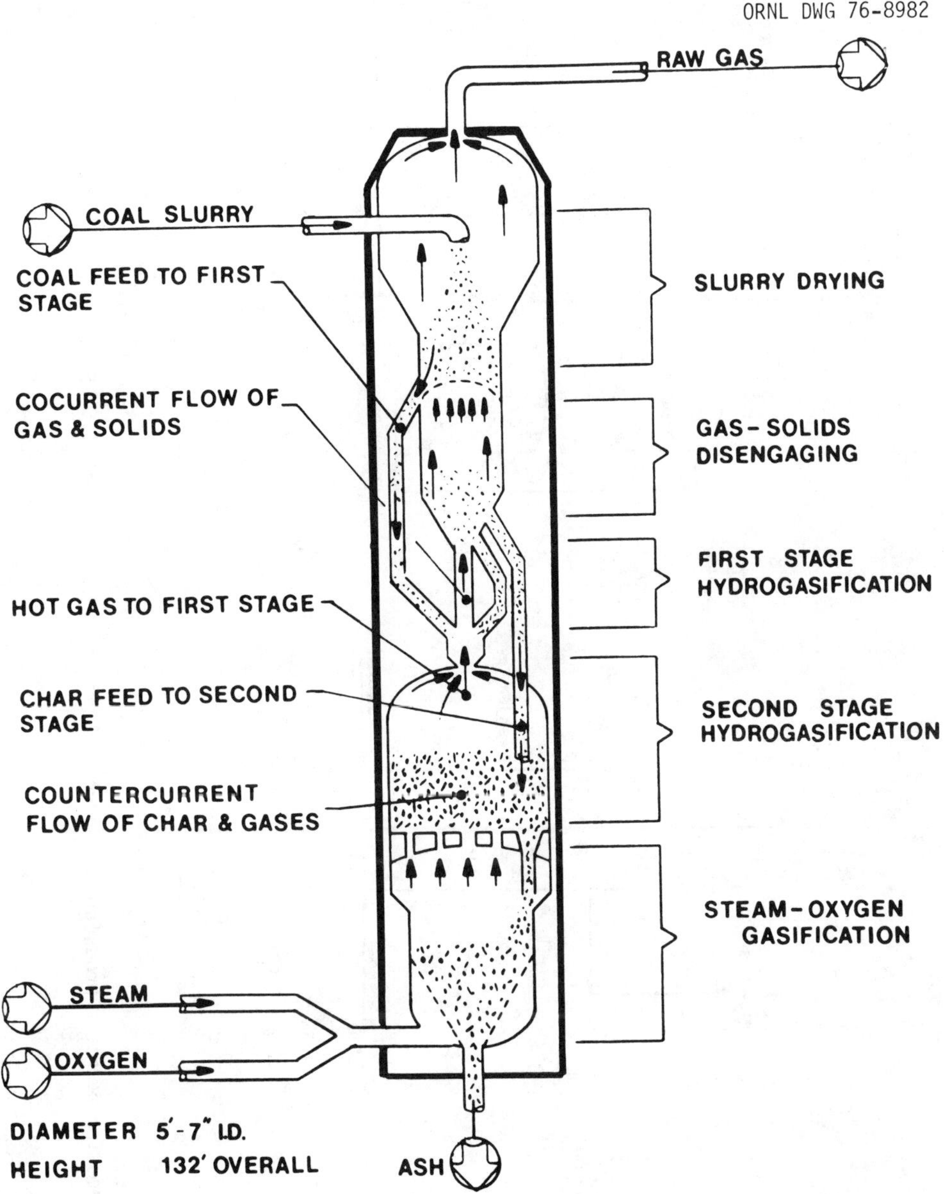

Fig. 3.21. HYGAS gasifier with steam-oxygen gasification. <u>Source</u>: From Dravo Corporation 1976, Fig. 1.12-1, p. 54.

ORNL DWG 76-8986

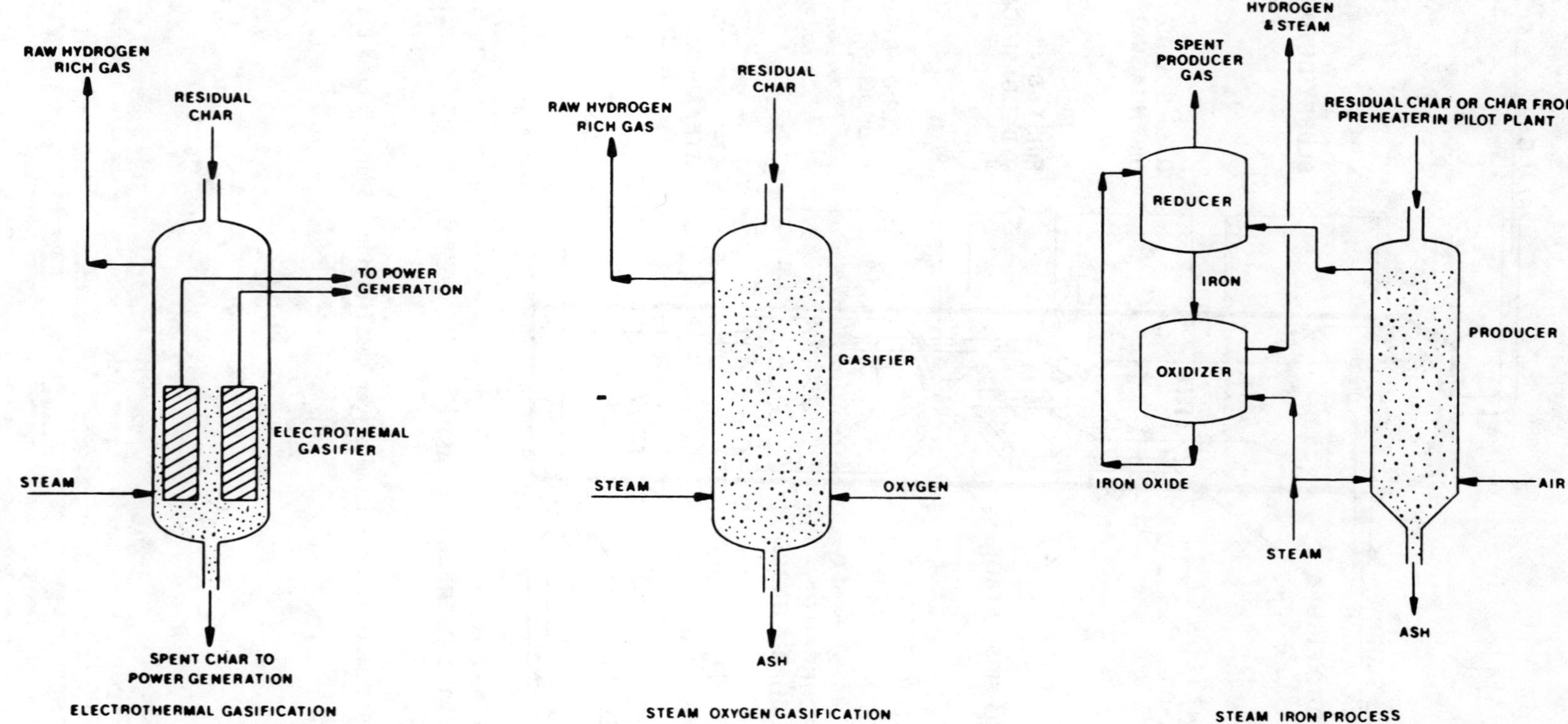

Fig. 3.22. HYGAS systems for hydrogen production. Source: From Dravo Corporation 1976, Fig. 1.12-2, p. 55.

ORNL DWG 76-8985

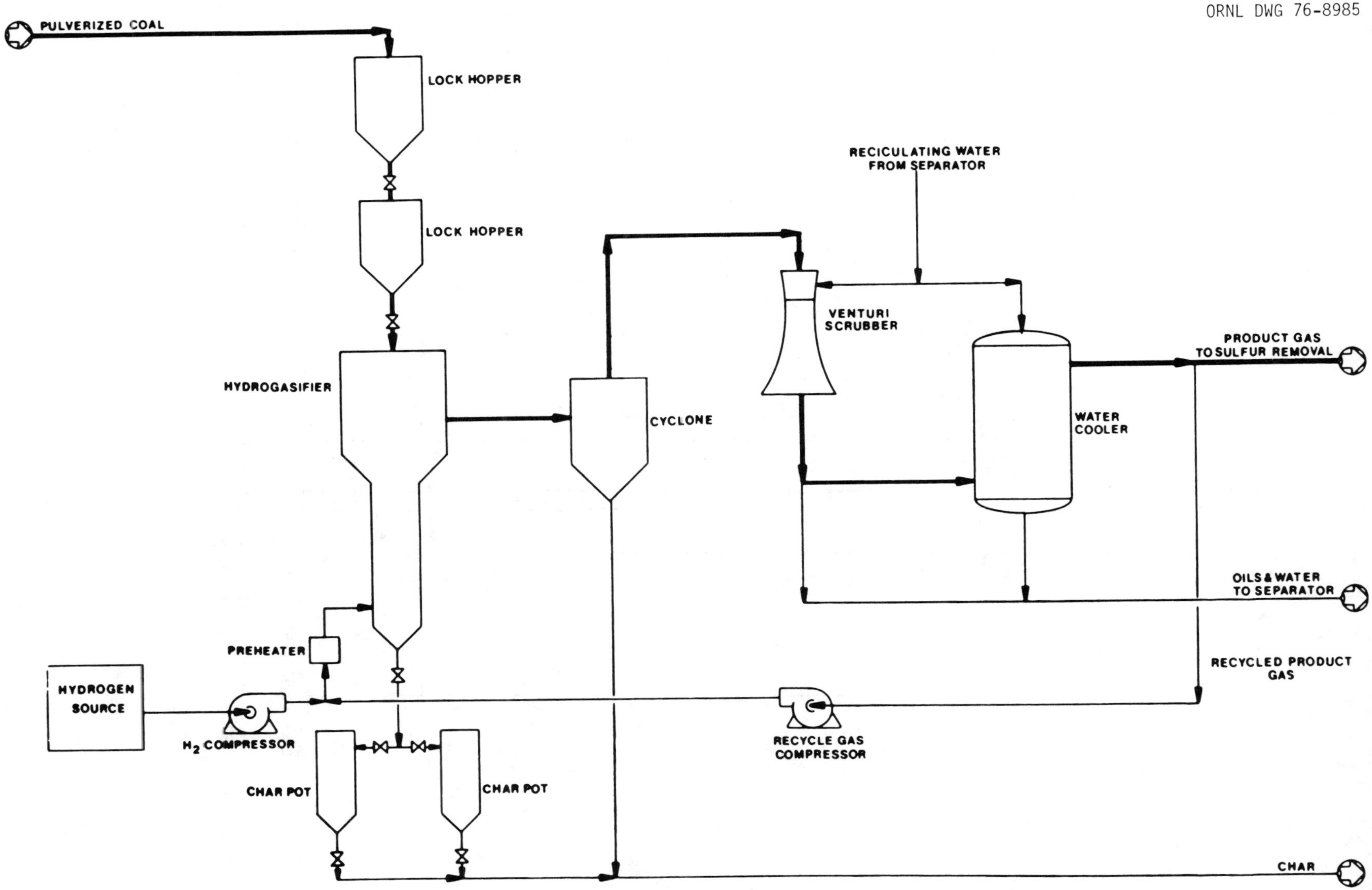

Fig. 3.23. Hydrane coal gasification process. Source: From Dravo Corporation 1976, Fig. 1.11-2, p. 49.

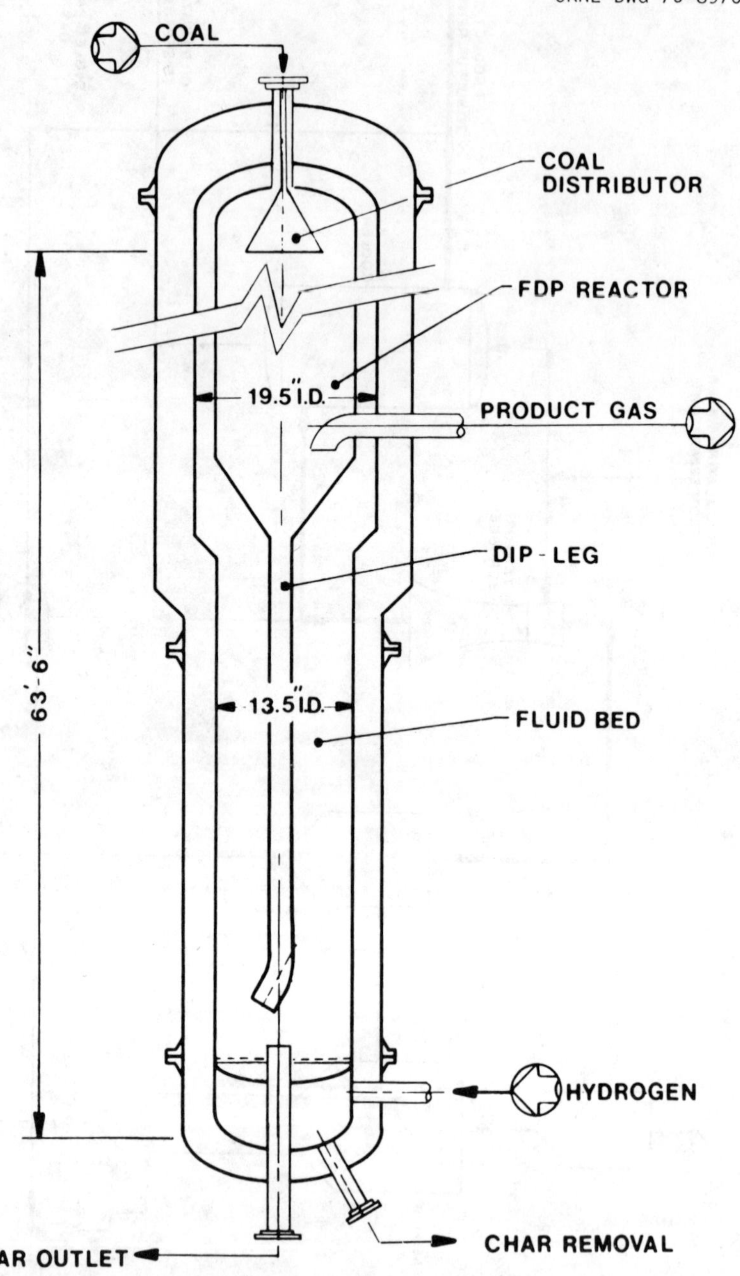

Fig. 3.24. Process development unit (PDU) Hydrane gasifier. <u>Source</u>: From Dravo Corporation 1976, Fig. 1.11-1, p. 48.

3.1.4.11 Lurgi high-Btu gasification process

The Lurgi scheme for production of a high-Btu gas is the same as that for production of an intermediate-Btu gas through the gasification stage (Fig. 3.25).

ORNL DWG 76-8978

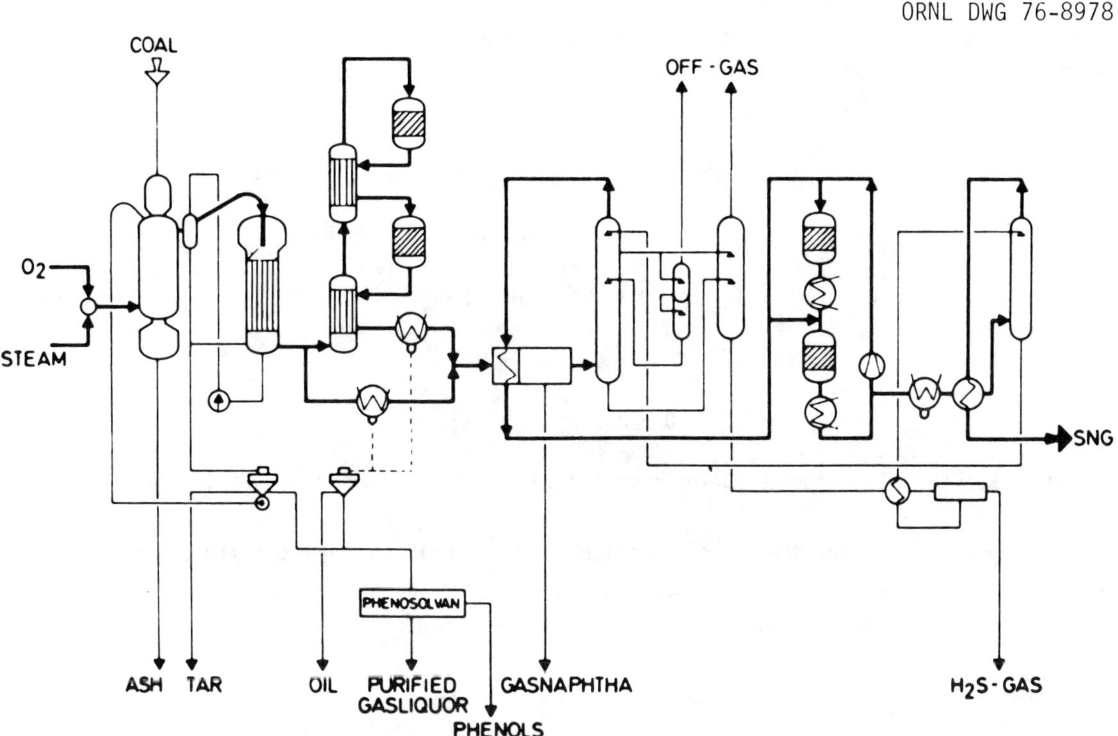

Fig. 3.25. Lurgi process to produce SNG from coal. Source: From Haynes and Forney 1976, Fig. 3, p. 5.

Pilot plant research has been carried out at the South African Coal, Oil and Gas Corporation facility in Sasolburg, South Africa (Moeller, Roberts, and Britz 1974), and at the Lurgi plant in Westfield, Scotland (Linden et al. 1976). In a year-long test completed in September 1974, the Westfield unit produced an average of 2.1 million cfd of 979-Btu/scf gas, proving the feasibility of catalytic methanation of Lurgi feed gas. Many details of the test remain proprietary, particularly the methanation catalyst, which is described only as being of a high nickel content and supported on a carrier (Moeller, Roberts, and Britz 1974).

Noncaking coal is sized and fed to a fixed-bed gasifier at 700 to 1100°F and 350 to 450 psi. The crude gas is quenched with oil to remove tar oil and dust, leaving naphtha, phenols, and ammonia to pass into the shift converter with the gas. A rectisol wash precedes the methanation step, which requires the total sulfur content of the gas to be less than 0.1 ppm. The methane synthesis employs two adiabatic reactors (Haynes and Forney 1976).

As of May 1, 1975, announcement had been made of six demonstration or commercial-scale conversion plants to use Lurgi gasification plus methanation. All will be built west of the Mississippi River and will use noncaking western coals as feedstock (Linden et al. 1976).

3.1.4.12 CO_2 Acceptor process

The CO_2 Acceptor process, under development by Conoco Coal Development Company, uses a novel technique to supply heat to the reactor. Hot, calcined dolomite ($MgO \cdot CaO$), the acceptor, is circulated into the top of a fluidized-bed gasifier operating at about 1500°F and 150 psi along with lignite or subbituminous coal that has been sized to 1/8 in. (Fig. 3.26). Steam enters from the bottom of the reactor. Heat is released when the calcium oxide forms calcium carbonate by combining with the carbon dioxide released in both the steam-carbon and the carbon monoxide shift reactions (Curran, Fink, and Gorin 1967):

$$CaO + CO_2 = CaCO_3 \ , \quad \Delta H = -76,200 \ Btu/lb\text{-}mole \ at \ 77°F \tag{17}$$

$$C + H_2O = CO + H_2 \ , \tag{18}$$

$$C + 2H_2 = CH_4 \ , \tag{19}$$

$$CO + H_2O = CO_2 + H_2 \ , \tag{20}$$

$$CaO + H_2S = CaS + H_2O \ . \tag{21}$$

As a bonus, Reaction (21) removes sulfur from the product gas before it leaves the gasifier.

The partially combusted coal leaves the gasifier as char. Both the char and the spent dolomite ($MgO \cdot CaCO_3$) enter the dolomite regenerator vessel, where the char is combusted with air to heat the dolomite and drive off the carbon dioxide:

$$Heat + CaCO_3 = CaO + CO_2 \ . \tag{22}$$

The hot regenerator effluent gas is used for power recovery and steam production after cyclone removal of char, ash, and dolomite fines.

The CO_2 Acceptor process (Fig. 3.27) generates a raw synthesis gas, which is low in carbon monoxide, carbon dioxide, and sulfur. Because the hydrogen–carbon monoxide ratio is about 3.2, there is no need for shift conversion. The raw gas is purified to remove the remaining carbon dioxide and sulfur, and then it undergoes catalytic methanation. A 30-ton/day pilot plant test operation at 150 psi gave 45.6 million scf per ton of coal and a thermal efficiency of 77% (Haynes and Forney 1976).

3.2 COAL LIQUEFACTION

The goal of coal liquefaction is to supplement dwindling national reserves of petroleum. Demand for petroleum products is increasing at the rate of 3.5% per year; by 1980 demand for crude petroleum is expected to reach 18 million bbl/day. The ratio of proved reserves to annual consumption in the United States has decreased from 11 years in 1950, to 8 in 1965, to about 7 in 1970 (Seglin and Eddinger 1971). President Ford's State of the Union message suggests that we produce 1 million bbl of synthetic fuels per day by 1985, 70% of which would be high-Btu gas (Linden 1975). Thus, to decrease dependence upon imported oil and to extend the ratio of reserves to consumption, production of liquid or solid fuels, as well as gaseous products, from coal is advantageous.

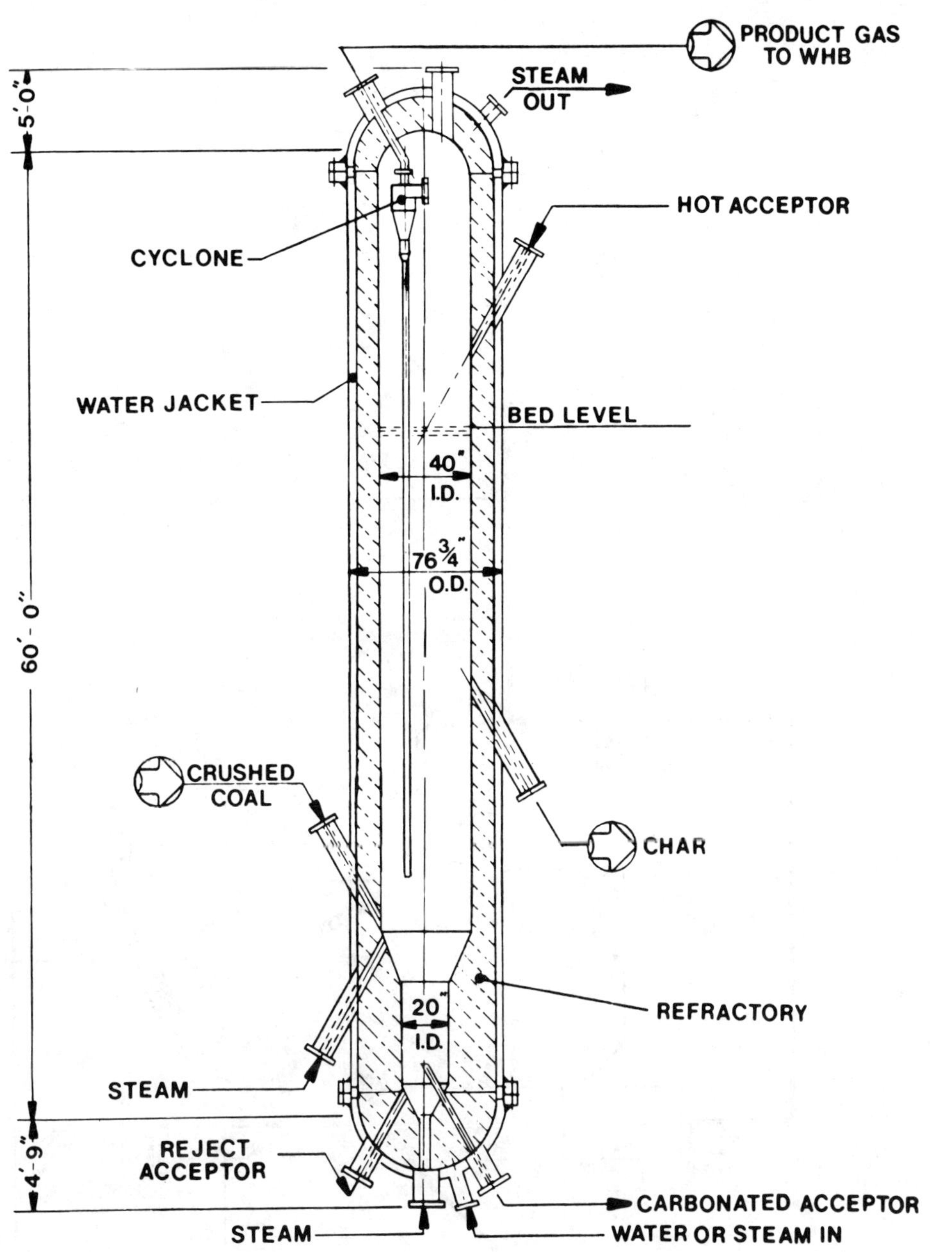

Fig. 3.26. CO_2 Acceptor gasifier. <u>Source</u>: From Dravo Corporation 1976, Fig. 1.9-1, p. 37.

Although liquefaction implies a liquid product, processes considered as coal liquefaction can give rise also to a gaseous or solid product or, in some instances, only to a low-melting solid fuel. For example, solvent-refined coal is a coal-derived solid fuel having properties that make it attractive for power plants (Booz-Allen Applied Research 1974). It contains substantially less sulfur than the original feed coal and little or no ash; it has a sufficiently low melting point to be handled as a fluid; and it can readily be ground into a fine powder, whose heating value is 16,000 Btu/lb, regardless of the coal source from which it is produced (Jimeson 1966).

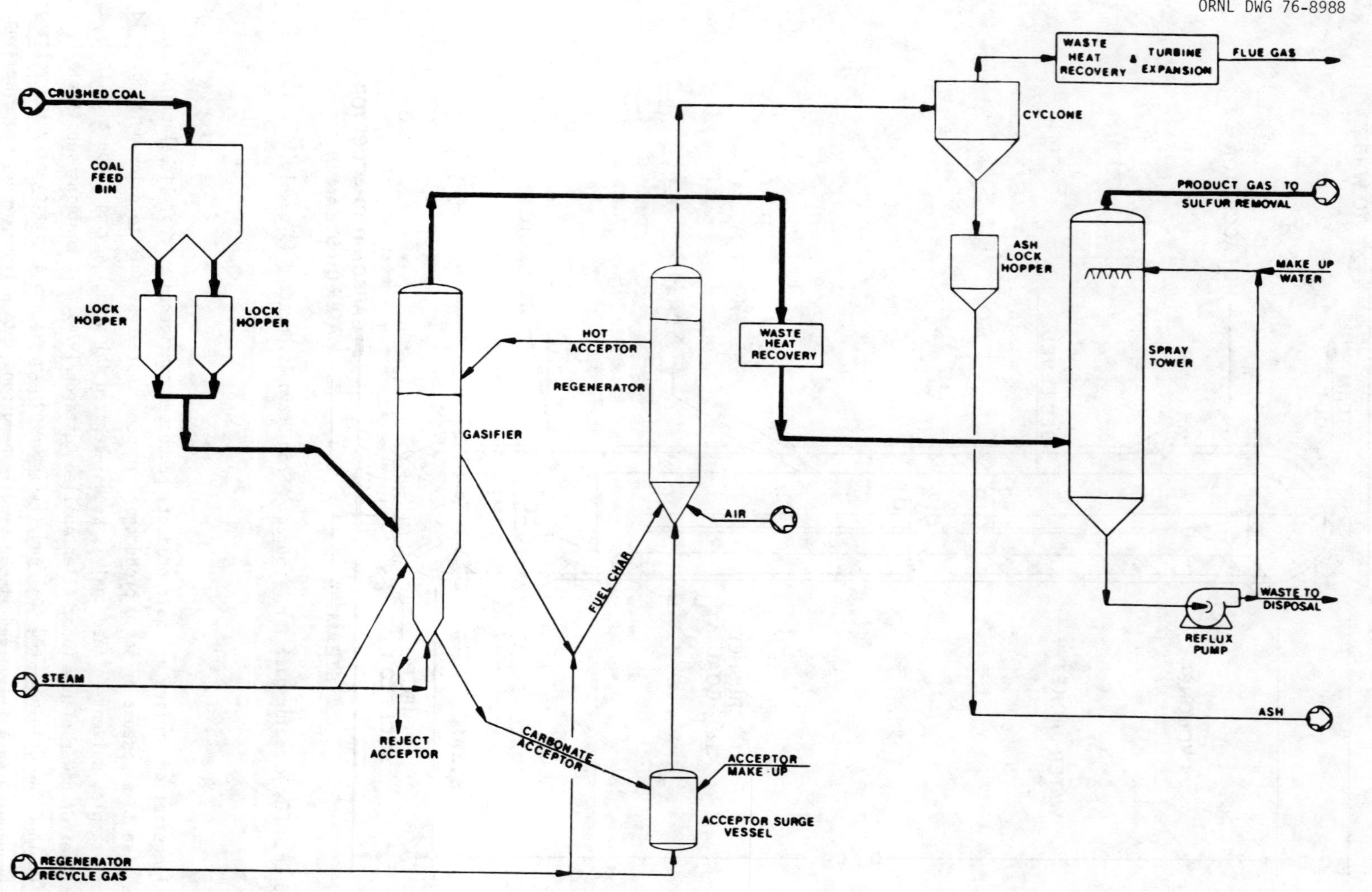

Fig. 3.27. CO$_2$ Acceptor coal gasification process. <u>Source</u>: From Dravo Corporation 1976, Fig. 1.9-2, p. 38.

3.2.1 Considerations of liquefaction

The renewed interest in coal conversion processes is the result of (1) dwindling domestic supplies of petroleum and natural gas and (2) the higher cost of foreign imports with which synthetic
fuels are now economically competitive. Many new processes for coal conversion are still at the
bench-scale stage; relatively few have been developed to the pilot-plant stage.

Commercialization of coal liquefaction will depend on many considerations. From the process
development unit and the pilot plant, the next stage is construction of a large pilot plant.
According to Epperly and Siegel (1974), it is estimated that the time involved in design, construction, and operation of a large pilot plant would be about five years. An additional four
years might be necessary for design and construction of a commercial-size plant.

In 1973 a workshop sponsored by the National Science Foundation—Research Applied to National
Needs (NSF/RANN) identified the areas in coal utilization where additional work appeared to be
needed. The six areas, as summarized by Rubin (1974), are listed in Table 3.14.

3.2.1.1 Pyrolysis (carbonization)

Pyrolysis causes the thermal decomposition of coal by heating it in the absence of oxygen to drive
off volatile matter; pyrolysis produces a liquid product as well as by-product gas and char (Fig.
3.28). After separation into gaseous and liquid products, the gases are freed from unwanted
substances (sulfur compounds, water, etc.), and a usable gas is obtained in addition to the
sought liquids (Epperly and Siegel 1974). The gas and char may be used to produce hydrogen for
hydrotreating the liquid reaction product to lower its sulfur content and to produce a high-Btu
synthetic crude oil. Char may also be burned with air or oxygen to produce the necessary heat
for pyrolysis.

Although the solids (char) contain less sulfur than does the initial feed coal, char produced
from eastern high-sulfur coals would not be suitable for direct use as fuel without preliminary
treatment for sulfur removal.

Because pyrolysis processes yield low volumes of liquids and require an outlet for the char produced, pyrolysis methods will probably not be a major source of synthetic petroleum (Epperly
1974). Three current pyrolysis processes under study are the COED (Char-Oil-Energy Development
of FMC Corporation), TOSCOAL (The Oil Shale Company), and Garrett processes (Garrett Research &
Development Company, subsidiary of Occidental Petroleum Company).

3.2.1.2 Dissolution

Dissolution is a more severe method of liquefying coal than is the pyrolysis method. The crushed
or ground coal is dissolved in a selective solvent, the ash is filtered out, and the oil is
treated by hydrocracking. Pyritic sulfur, together with the insoluble ash, is removed from the
solvent; however, organic sulfur in the coal is carried along with the liquid fuels. After
separation from the solids, the liquids can be catalytically upgraded to produce a synthetic
crude oil. The solids may be sold as fuel or used for process heat or generation of the hydrogen
used in the process. Figure 3.29 represents the steps involved in dissolution and variations
thereon. Treatment with hydrogen, with or without a catalyst, improves the product quality and
removes the organic sulfur.

Table 3.14. Areas in coal utilization requiring additional work

---

Fundamentals of chemical-physical processes and properties

1. Investigate mechanisms of coal reactions, including (a) dissolution, (b) pyrolysis, (c) hydrogenation, and (d) other reactants.
2. Investigate fundamentals of solids-gas contacting in (a) fluidized beds, (b) dilute phase, and (c) fixed beds.
3. Investigate the plastic properties of coal as controlled by the processing environment.

Each of the above research areas is needed to increase fundamental knowledge and understanding, provide building blocks for new processes, and improve upon present concepts and processes.

Materials and mechanical systems

1. New concepts in solids feeding and removal. Need for economic, reliable systems yielding continuous feed and removal from high-pressure processes.
2. Improved management of solids movement and monitoring in high-temperature-pressure (HTP) systems. Need for improved sealing and corrosion-resistant components (valves, etc.).
3. Develop a sulfur-resistant, high-temperature, strong metal alloy (e.g., a "sulfur-resistant Incoloy 800").
4. Investigate interactions of refractory with (a) coal-ash materials and (b) process environment in order to better define and understand problems such as fluxing of coal ash and plugging of reactors.

Gas-solid-liquid separation systems

1. Removal of particulate matter from coal-derived liquids. Need for efficient, reliable, and economic systems to simplify downstream processing.
2. Removal of particulate matter from HTP gases. Need for operation at high temperatures with good removal of fine particulates.
3. Chemical purification of HTP gases. Need for more efficient, reliable, and economic systems.
4. Improved gas-solids stripping techniques to avoid product gas contamination.
5. Improved liquid-solids stripping techniques to recover valuable liquids from wastes.
6. Improved effluent control techniques for coal conversion processes. Need for reliable, economic systems to meet effluent limitations on liquid wastes.

Measurement and instrumentation techniques

1. Develop improved process instrumentation for control and safety of HTP systems. Need for improved sensitivity and repeatability of measurements of (a) solids sampling, (b) fluidized-bed or dilute-phase-bed density, and (c) dilute oxygen measurement.
2. Develop analytical techniques with improved sensitivity for determining the structure of coal-derived products.

Table 3.14 (continued)

Catalyst development

1. Develop methanation catalysts and systems that are stable at high temperatures and resistant to sulfur.
2. Develop coal liquefaction catalysts for primary conversion. Need for catalysts that are long-lived, active, and selective.

New and innovative processes

1. Investigate innovative approaches to coal conversion. Need for new, improved, and more economical techniques.
2. Investigate innovative approaches to hydrogen production, including improved and more economical techniques.

Source: Rubin 1974, p. 1001.

ORNL DWG 76-9840

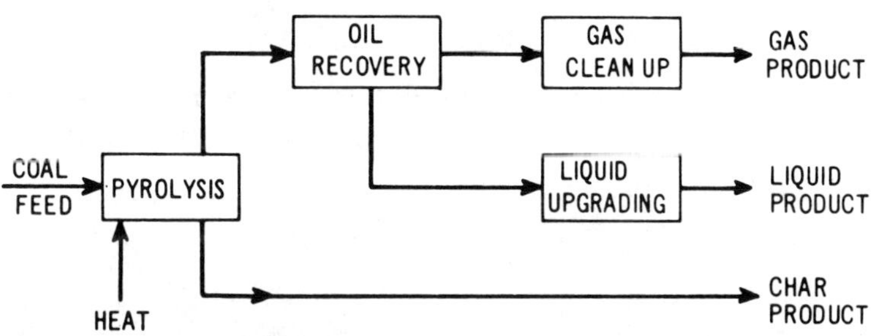

Fig. 3.28. Schematic flow plan of coal pyrolysis. <u>Source</u>: From Epperly 1974, Fig. 1, p. 1021. Reprinted by permission of the publisher.

Three types of dissolution processes can be distinguished: (1) those that use no catalyst and no hydrogen, (2) those that use hydrogen, but no catalyst, and (3) those that use both catalyst and hydrogen. A method using neither catalyst nor hydrogen currently under study is the Consol Synthetic Fuel (CSF) process of Consolidation Coal Company, a subsidiary of Continental Oil Company. The Solvent-Refined Coal (SRC) process of Pittsburgh & Midway Coal Company uses hydrogen, but no catalyst. Three processes using both catalyst and hydrogen that are currently under study are the Synthoil (U.S. Bureau of Mines), H-Coal (Hydrocarbon Research, Incorporated), and Catalytic Coal Liquids (CCL) (Gulf Research & Development) processes.

ORNL DWG 76-9838

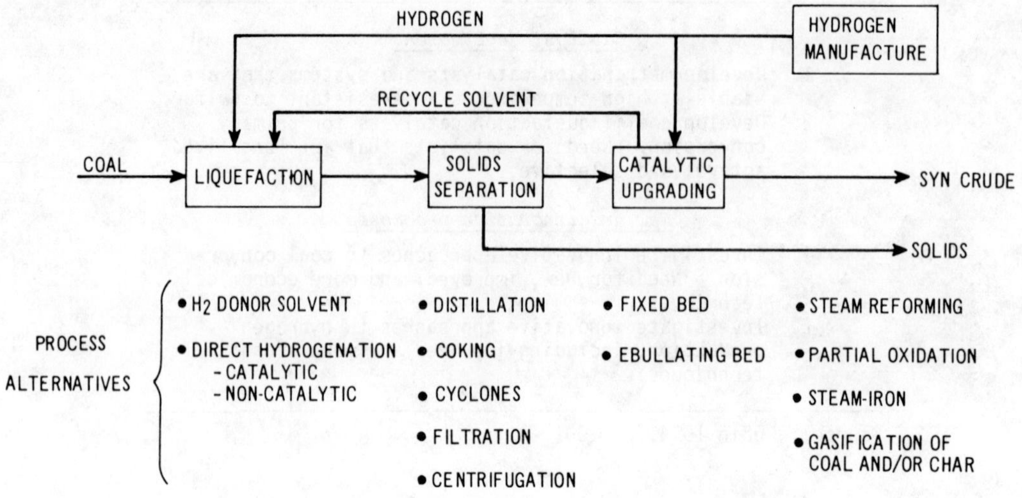

Fig. 3.29. Dissolution process scheme. Source: From Epperly 1974, Fig. 2, p. 1022. Reprinted by permission of the publisher.

Table 3.15 lists these dissolution processes and gives a brief summary of their products (Beychok 1975). A more detailed description of the operating conditions for the dissolution processes is given in Table 3.16 (University of Oklahoma 1975).

3.2.2 Process descriptions

3.2.2.1 Char-Oil-Energy Development (COED)

Project COED, sponsored by the Office of Coal Research (OCR), Department of the Interior, has been under development by FMC Corporation since 1962. Using pyrolysis, the process converts coal into cleaner and more valuable products — synthetic crude oil, gas, and char. The gas can be sold as fuel gas or converted to pipeline gas, and the char can be gasified or used as power plant fuel.

Coal, crushed to <1/8-in. particles, is dried and heated to successively higher temperatures by stages in a number of fluidized-bed reactors. In each reactor, a fraction of the volatile matter of the coal is released. The maximum temperature of each stage is chosen to be just below the temperature at which the coal will agglomerate or cake, defluidizing the bed. Pressure of the operation is between 6 and 10 psig. Typically, the temperatures and number of stages will depend on the agglomerating characteristics of the coal used. In the final stage char is burned to generate heat for the process, and the hot gas and hot char are transferred from the final stage to the other vessels (Scotti, Jones, Ford, and McMunn 1975).

Table 3.15. Coal liquefaction processes

| Developer | Process name | Fuel products | Dissolution requires | | Catalytic hydrotreater for oil |
			Catalyst	H_2	
Pyrolysis processes					
FMC Corp.	COED	Gas, oil, and char[a]			Yes
Garrett R&D	Coal pyrolysis	Gas, oil, and char[a]			Yes
The Oil Shale Corp.	TOSCOAL	Gas, oil, and char[a]			Yes
Dissolution processes					
Consolidation Coal Co.	CSF	Gas and oil[b]	No	No	Yes
Exxon Res. & Eng.[c]	Exxon	Gas and oil[b]	No	No	Yes
Pittsburgh & Midway Mining Co.	SRC	Gas, oil, and refined coal	No	Yes	Yes
Hydrocarbon Research	H-Coal	Gas, oil, and coker feed[d]	Yes	Yes	No
Bureau of Mines	Synthoil	Gas, oil, and coker feed[d]	Yes	Yes	No
Gulf R&D	CCL	Gas, oil, and coker feed[d]	Yes	Yes	No

[a]Char most probably gasified to supply low-Btu fuel gas.
[b]Produces char that will most probably be converted to supply hydrogen for hydrotreater.
[c]The technical literature indicates this process to be similar to the CSF process.
[d]The product coke might possibly be converted to supply hydrogen to dissolution reactor.

Source: Beychok 1975, Table 15, p. 129.

Figure 3.30 is a block diagram of the COED process, showing four stages of pyrolysis. Hot gas from stage IV is recycled back through stage III to stage II. Pyrolysis gas from stage II is quenched with water to condense the entrained oil, which is separated from the water and filtered. Gas from the oil recovery section is scrubbed to remove ammonia, carbon dioxide, and hydrogen sulfide and then steam-reformed to produce hydrogen

$$CH_4 + H_2O \overset{>500°C}{=} 3H_2 + CO \ . \tag{23}$$

The filtered crude oil, containing less than 0.1 wt % solids, is hydrotreated at about 750°C in a fixed-bed reactor to produce a synthetic crude oil of about 25°API gravity.[*]

Through March of 1974 the COED plant in Princeton, New Jersey, processed over 18,000 tons of coal. Table 3.17, which includes the American Society of Testing Materials (ASTM) rank of the coals processed, summarizes the sources of six different coals evaluated over the operating

---

[*]API gravity is a factor governing the quality of crude petroelum; crude petroleum prices are most frequently posted against values in degrees API. It can be represented by the equation

$$(141.5/\text{sp gr } 60/60°F) - 131.5,$$

where sp gr 60/60°F is the ratio of the mass of a given volume of the liquid at 60°F to the mass of an equal volume of pure water at the same temperature (American Society of Testing Materials 1967).

Table 3.16. Characteristics of coal liquefaction technologies

Process	Coal feedstock	Reactor temperature (°F)	Reactor pressure (psi)	Product oil grade (°API)	Heating value of oil (Btu/lb)	Yield (bbl/ton of coal)
		Hydrogenation				
Synthoil	Pulverized, dried, caking or noncaking	850	2000 - 4000	Not applicable	17,700	3.0
H-Coal	Pulverized, dried, caking or noncaking	Nominal 850	2700	5 - 17	*a*	4.4
SRC	Pulverized, dried, caking or noncaking	800	1000	*b*	*b*	*b*
CSF	Pulverized, dried, caking or noncaking	800	1000	*b*	*b*	*b*
		Pyrolysis				
COED	Pulverized, dried, caking or noncaking[c]	600 - 1600 in four reactors	6 - 10	25	*b*	1.0 - 1.5
TOSCOAL	Pulverized, dried, caking or noncaking	970	Atmospheric	6 - 13	16,000	0.5
		Catalytic conversion				
Fischer-Tropsch	Depends on gasification process	Two reactors at different temperatures	330 - 360	Various	Various	*b*

[a]Different coal feedstocks may require a greater number of pyrolysis feedstocks.
[b]Unknown.
[c]One reported run; lower temperatures give lower yield of liquids.

Source: University of Oklahoma 1975, Table 1-40, p. 1-96.

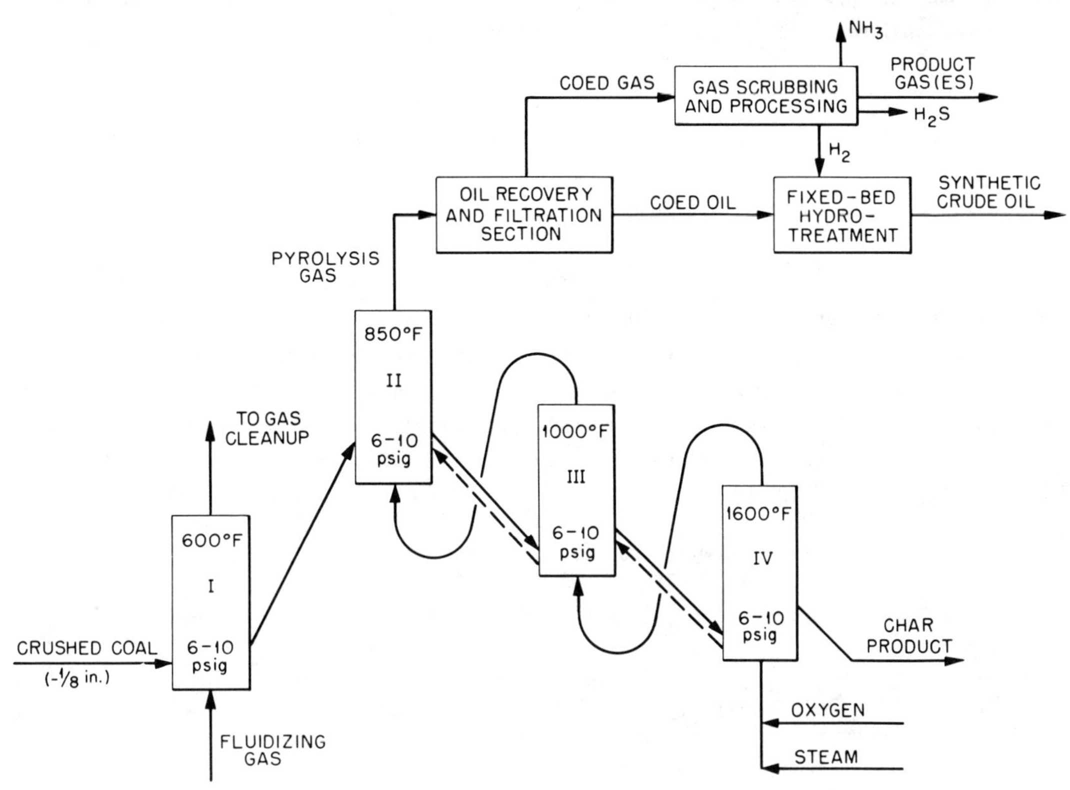

Fig. 3.30. Schematic of the COED process. <u>Source</u>: After Scotti et al. 1975a, Fig. 1, p. 61. Reprinted by permission of the publisher.

Table 3.17. Coals processed in COED pilot plant

Coal source	Seam	Mine	Rank (ASTM D388-38)
Colorado	Somerset C	Bear	High-volatile B bituminous
Wyoming	Monarch	Big Horn	Sub-bituminous B
Illinois	No. 6	Sahara No. 6	High-volatile B bituminous
Illinois	No. 6	Peabody No. 10	High-volatile C bituminous
North Dakota		Glen Harold	Lignite
Utah	A	King	High-volatile B bituminous
Western Kentucky	Nos. 9 and 14	Colonial	High-volatile B bituminous

Source: Scotti, Jones, Ford, and McMunn 1975, Table 1, p. 62. Reprinted by permission of the publisher.

period. The oil yield (21.5 wt %, dry-coal-feed basis) from the Utah A seam coal was the highest obtained from any coal feed.

An oil absorber tower recovery system was installed downstream as an alternative to the aqueous quench recovery system in stage II. Comparison of the total oil yields from the absorber with the yields from the conventional recovery system (55.6 vs 41.3%) indicates a potentially improved pyrolysis product for hydrotreating from the absorber system. There is also a decreased content of resins and asphaltenes and a lower Conradson carbon[*] residue in the absorber product.

After filtration, the COED oil is hydrotreated until more than 90% of heteroatom impurities (sulfur, nitrogen, oxygen) are removed. Hydrotreatment at pressures of 1750 to 2500 psig and temperatures of 700 to 800°F produces a syncrude of 25 to 30°API. All the coals processed in the COED plant (Table 3.17) have undergone successful hydrotreating operations.

Filtered pyrolysis oil has been evaluated for conversion by hydrotreating to COED syncrude, then to a refinery feedstock and/or a low-sulfur fuel. The COED syncrude can be used as feedstock to a petroleum refinery or fractionated into naphtha and low-sulfur fuel oil fractions (No. 4 fuel oil equivalent). Table 3.18 shows typical distillation data and hydrocarbon-type analysis for syncrude produced from two types of coal (Jones 1975). Also, a U.S. Navy destroyer using COED syncrude is reported to have operated successfully, giving performance similar to standard Naval distillate fuel (Scotti et al. 1975).

Table 3.18. COED syncrudes: typical distillation data and hydrocarbon-type analysis

	Coal	
Property	Illinois No. 6	Utah King
Hydrocarbon-type analysis, vol %		
Paraffins	10.4	23.7
Olefins		
Naphthenes	41.4	42.2
Aromatics	48.2	34.1
API gravity, 60°F	28.6	28.5
ASTM distillation, °F		
Initial boiling point (IBP)	108	260
50% distilled	465	562
End point (EP)^{a}	746	868
Fractionation yields, wt %		
IBP - 180°F	2.5	
180 - 390°F	30.2	5.0
390 - 525°F	26.7	35.0
390 - 650°F	51.0	65.0
650 - EP	16.3	30.0
390 - EP	67.3	95.0

^{a}95%, except for Illinois No. 6, which is 98%.

Source: Jones 1975, Table 10, p. 188. Reprinted by permission of the publisher.

Since the char product from the COED process contains the same percentage of sulfur as the coal from which it is derived, combustion of the char would be environmentally unacceptable without stack gas cleaning. Therefore, three low-Btu gasification methods were reviewed for gasifying COED char to clean fuel gas: the Rummel single-shaft, Koppers-Totzek, and Winkler gasifying systems. All three operate at atmospheric pressure and would be suitable for addition to a

---

[*] A Conradson carbon residue value is an indication of the carbonaceous residue formed after evaporation and pyrolysis of a petroleum product. The residue does not consist completely of carbon, but is a coke which can be further changed by pyrolysis. Low values would indicate a low content of coking material (American Society of Testing Materials 1965).

COED unit. From a literature survey, heat and material balances were drawn up for a gasifying
unit that receives 13,000 tons of char per day from a 25,000-ton/day COED plant. Because the
fluid bed of the Winkler gasifier operates using air and allows hot, unpulverized char to be fed
directly, it was recommended for further evaluation in a commercial COED facility. However, because of the large amount of commercial experience with Koppers-Totzek gasifiers, addition of
such a unit would probably be the quickest way to build a COED plant having added char gasification. Tests with an existing Koppers-Totzek unit were planned for 1975 to obtain commercial
design data (Scotti et al. 1975).

3.2.2.2 TOSCOAL

The Oil Shale Corporation has developed a process for retorting oil shale and has applied the
technology to the processing of Gillette, Wyoming (Wyodak) subbituminous coals in a 25-ton/day
pilot plant. The principal goals were (1) to carbonize the coal to upgrade the heating value of
subbituminous coal, thus decreasing the transport cost per Btu and (2) to recover liquid hydrocarbons.

Figure 3.31 is a diagram of the TOSCOAL process. Crushed coal is fed to a surge hopper and then
preheated by hot flue gas in a fluid bed; from there, the coal is sent to a pyrolysis drum, where
it is contacted with heated ceramic pellets. The solid product char leaves the pyrolysis drum
and passes through a trommel screen, which separates the ceramic pellets from the char. The char
is cooled and sent to storage, and the cooled ceramic pellets are returned to the pellet heater
by an elevator. Pyrolysis vapors are condensed and fractionated, whereas uncondensed gas is used
as a pellet-heater fuel. The use of heated ceramic pellets to provide the required retorting heat
is unique and results in a retort gas not diluted by nitrogen and carbon dioxide.

The handling of solid and liquid products of the TOSCOAL process is similar to that of oil shale
except that marketable char replaces the spent shale. Unfortunately, the char product is dustier
and more pyrophoric than coal so that conventional coal handling methods are not suitable for
this product. An economic evaluation showed that the TOSCOAL process is of considerable interest
but that solutions must be found for the problem of char transport before the process is commercially practicable.

Analyses of the Wyodak coal and its resultant char are given in Tables 3.19 and 3.20 respectively.
Over half the coal is converted to char, which has a gross heating value of 11,826 to 12,963 Btu/
lb, as compared with the original coal's gross heating value of 8139 Btu/lb. Additionally, the
low sulfur content of the char makes it an environmentally acceptable solid fuel.

The product yields are given in Table 3.21. As the retorting temperature was raised from 800 to
970°F, the amount of char produced decreased, whereas the amount of oil increased. Oil produced
rose from 13.2 to 21.7 gal per ton of coal as the retorting temperature rose from 800 to 970°F.
As with the char, a low-sulfur oil (0.2 to 0.4 wt %) results (Carlson, Yardumian, and Atwood
1975).

The pilot plant chars have been evaluated by several boiler manufacturers, who indicate that the
chars could be used satisfactorily as a boiler fuel. The char produced at 970°F, which had a
volatile content of 15.9%, ignited readily, and combustion was smooth and complete (Carlson,
Yardumian, and Atwood 1974).

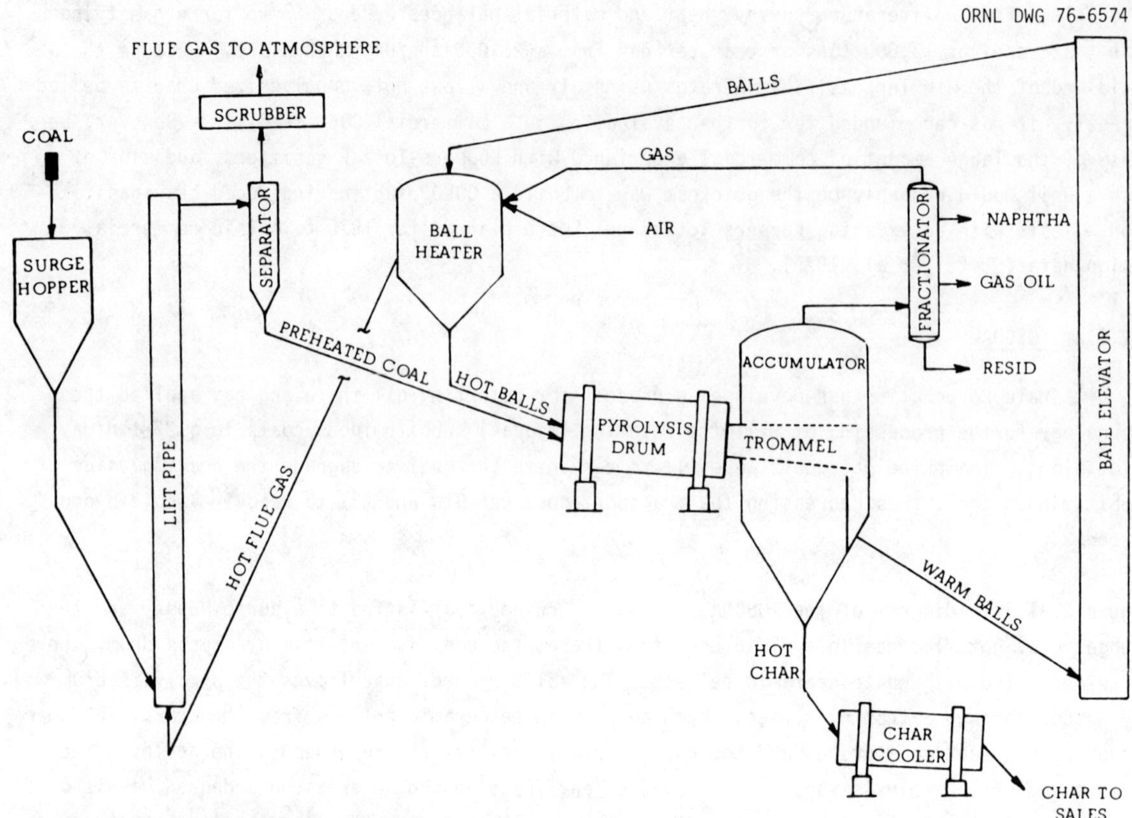

Fig. 3.31. TOSCOAL process. Source: From Carlson, Yardumian, and Atwood 1975, Fig. 1, p. 497. Reprinted by permission of the publisher.

3.2.2.3 Garrett pyrolysis

In developing its coal pyrolysis process, Garrett Research and Development Company recognized that certain criteria would have to be satisfied for a successful process. There had to be high heat transfer rates to accommodate the endothermic decomposition of the coal, and oxygen had to be excluded to maximize production of the most valuable products — the liquids. In addition, a short residence time in the reactor would provide high liquid yields, prevent further cracking of volatile constituents, and inhibit carbon monoxide formation.

The Garrett process is illustrated by the flow chart in Fig. 3.32. Coal, dried and pulverized, is carried by recycled product gas into an entrained-bed carbonizer, where the coal is heated extremely rapidly (>5000°F/sec) by the recycled char. The product stream flows through cyclones to separate the gases from the char. Part of the char is cooled and stored; the remainder is heated with air to 1200 to 1600°F in a char heater and recycled to the pyrolysis reactor.

Gas from the reactor is cooled and scrubbed to remove the tar. A portion of the uncondensable gas is used for conveying the coal and char to the reactor. The remainder can be upgraded to pipeline gas or used as a hydrogen source in the plant.

Table 3.19. Wyodak coal assay
(samples 120-8, 120-10)

Component	Amount
Proximate, wt %	
Moisture	30.0
Ash	5.3
Volatile matter	30.7
Fixed carbon	33.9
Total	99.9
Heating values	
Gross, Btu/lb	8139
Net, Btu/lb	7570
Ultimate, wt %	
Carbon	46.4
Hydrogen	2.8
Oxygen	14.7
Nitrogen	0.7
Sulfur	0.3
Chlorine	0.0
Moisture	30.0
Ash	5.3
Total	100.2
Equilibrium moisture, wt %[a]	27.6
Hardgrove grindability	56.0
Bulk density (>3/4 in.), lb/ft^3	
Poured	42.4
Packed	47.7
Sulfur forms (DCB)	
Pyrite	0.21
Sulfate	0.01
Organic	0.19
Total	0.41
Ash fusion temperatures ($^{\circ}$F)	
Reducing	
Initial deformation	2200
Softening (H=W)	2220
Softening (H=1/2W)	2230
Fluid	2255

[a]Moisture content at 97% relative humidity.

Source: Carlson, Yardumian, and Atwood
1975, Table 2, p. 503.
Reprinted by permission of the publisher.

Condensed tars are hydrogenated to produce either a low-sulfur fuel oil or a syncrude. Unfortunately, the char produced from most eastern coals contains sulfur in excess of proposed Environmental Protection Agency standards; however, it has been possible to desulfurize the char experimentally because of its high reactivity and porosity.

Tests conducted with a high-volatile eastern United States bituminous coal showed a maximum yield of 35% conversion of coal to liquids at a temperature of 1075°F, a pressure of 25 psig, and a residence time of 0.03 min (Epperly and Siegel 1974). A material balance for the process indicated 56.7% char, 35.0% tar, 6.6% gas, and 1.7% water. Although the ash content of the char is higher than that of the original coal, the heating value of the char is essentially equal to that of a bituminous coal. The pyrolysis gas has a heating value of about 700 Btu/ft^3 on a dry,

Table 3.20. TOSCOAL char properties

Char properties	Retort temperature		
	800°F	900°F	970°F
TOSCO II run number	C-8[a]	C-2	C-3
Proximate, wt %			
Moisture	0.0	0.0	0.0
Ash	12.4	10.0	9.8
Volatile matter	25.3	19.7	15.9
Fixed carbon	62.3	70.3	74.3
Total	100.0	100.0	100.0
Ultimate, wt %			
Carbon	68.8	74.7	77.5
Hydrogen	3.4	3.0	2.9
Oxygen	13.3	11.8	8.3
Nitrogen	1.0	1.2	1.3
Sulfur	0.5	0.2	0.3
Chlorine	0.0	0.0	0.0
Moisture	0.0	0.0	0.0
Ash	12.4	10.0	9.8
Total	99.4	100.9	100.1
Equilibrium moisture, wt %	10.0	10.8	9.9
Hardgrove grindability	78.2	49.1	43.6
Heating values, Btu/lb			
Gross	11,826	12,560	12,963
Net	11,516	12,280	12,693
Bulk density, (1/4" x 0), lb/ft^3			
Packed	51.2	48.8	47.8

[a]Feed Wyodak coal was different from that used in runs C-2 and C-3.

Source: Carlson, Yardumian, and Atwood 1975, Table 4, p. 505.
Reprinted by permission of the publisher.

Table 3.21. TOSCOAL retorting of Wyodak coal — product yields
(lb/ton of as-mined coal)[a]

	Retort temperature		
	800°F	900°F	970°F
TOSCOAL run number	C-8	C-2	C-3
Char	1049.0	1011.7	968.7
Gas (C$_3$ and lighter) (scf/ton)	119.0 (1250.0)	156.7 (1777.0)	126.0 (1624.9)
Oil (C$_4$ and heavier) (gal/ton)	114.0 (13.2)	143.0 (17.4)	186.2 (21.7)
Water[b]	702.0	702.0	702.0
Total	1984.0	2013.4	1982.9
Recovery (%)	99.2	100.7	99.1

[a]Numbers in parentheses provide different units of measure as marked.

[b]Value assumed from Fischer assay and moisture content; addition of
steam to the process prevented accurate measurement of water
produced in retorting.

Source: Carlson, Yardumian, and Atwood 1975, Table 3, p. 504.
Reprinted by permission of the publisher.

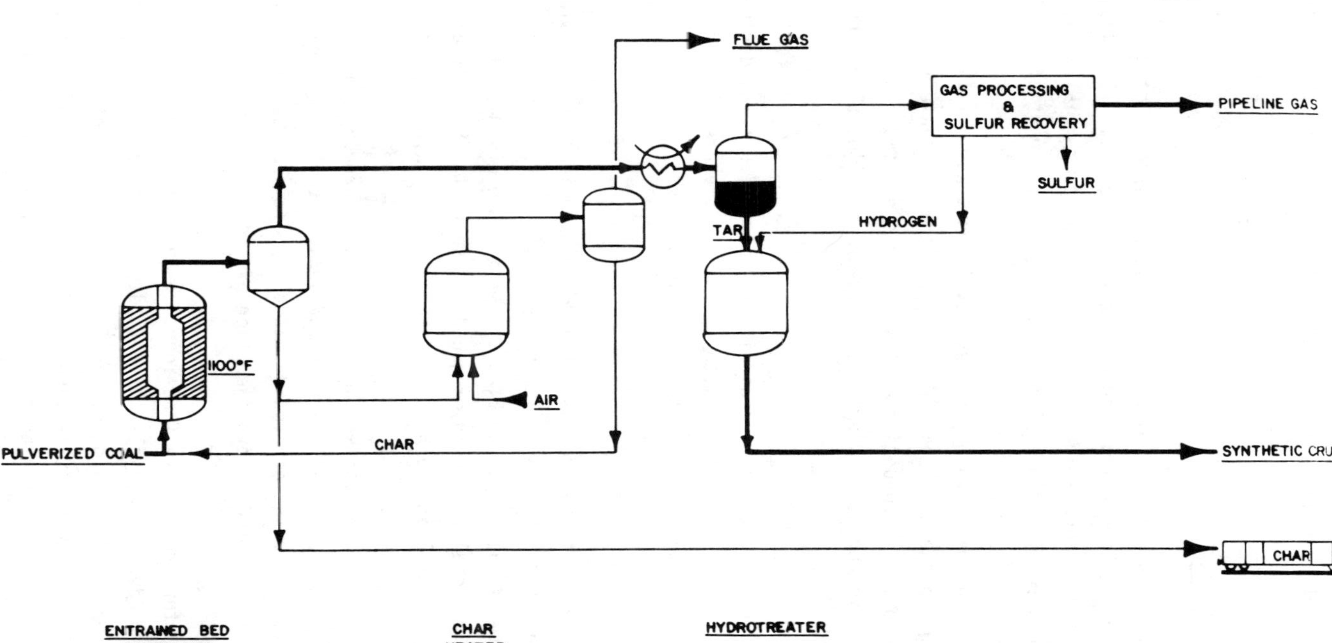

Fig. 3.32. Schematic flowsheet of Garrett's coal pyrolysis process. Source: After Sass 1974, Fig. 1, p. 72.

hydrogen-sulfide-free basis and is high in hydrocarbons. Because the tar undergoes little cracking, its hydrogen content is lower than that of most other low-temperature tars (Sass 1974),
requiring hydrotreatment to produce a synthetic crude oil. By using the product char as heat
carrier in the entrained-bed reactor, oxygen could be excluded from the reactor and a rapid rate
of heat transfer would be assured in a turbulent solid–gas system.

A pilot plant was constructed at LaVerne, California, and is being operated for further development of the process (Sass 1974).

3.2.2.4 Consol Synthetic Fuel (CSF)

According to Consolidation Coal Company (1973), the CSF process consists of a partial conversion
of high-sulfur caking coal to a solid residue and a synthetic crude oil; the oil is further converted to marketable liquid fuels, such as gasoline, by means of commercially available refining
procedures. Hydrogen is produced by commercial methods and is not a part of the CSF process.

Sponsorship of the Consolidation Coal Company process development was initiated by the OCR in
1963, when the project was given the code name "Project Gasoline," despite the fact that experimental development was aimed only at production of low-sulfur synthetic crude oil distillate.
Support by OCR was terminated in August 1972 after three years of pilot plant operation (November
1966 to February 1970).

The process is based on solvent extraction of coal with a "natural" solvent derived from hydrogenation of the extract. By starting initially with synthetic solvents, an inventory of extract
was built up in successive cycles and hydrogenated to yield a solvent for the next extraction
cycle.

Figure 3.33 shows a block diagram of the process, as incorporated in the pilot plant at Cresap,
West Virginia. Coal, crushed to 3/8-in. particles and partially dried, is first sent into a preheater, where its temperature is raised to 450°F; the coal is then discharged into slurry mixing
tanks, where it is mixed with a coal-derived hydrogen donor solvent. Solvent extraction of the
coal takes place in a stirred vessel at 765°F and 150 to 400 psig. Liquid products and solid
residue, consisting of unreacted coal and minerals, are separated in hydroclones. Ash content
of the extract is about 0.90%. The ash content was lowered to 0.35% by using filters designed
to operate at 600°F and 150 psig, but the filters proved incapable of leak-free operation. On
the basis of bench-scale results, it was concluded that the additional ash present would not
detrimentally affect the catalysts used in hydrogenation of the extract.

Vapors produced in the extractor are sent to fractionation and solvent recovery units, where they
are treated with the liquid products. Recycle solvent, distillate, and liquid product for the
hydrotreatment are produced here. The solid residue is sent in a concentrated slurry to a low-
temperature carbonization reactor, where it is pyrolyzed in a fluid bed of char at about 925°F
and 15 psig. Recovered solvent is sent to the slurry system. The char obtained is used in a
second carbonization unit to carbonize unconverted residue from the extract hydrogenation operation and to recover additional distillate.

ORNL DWG 76-9841

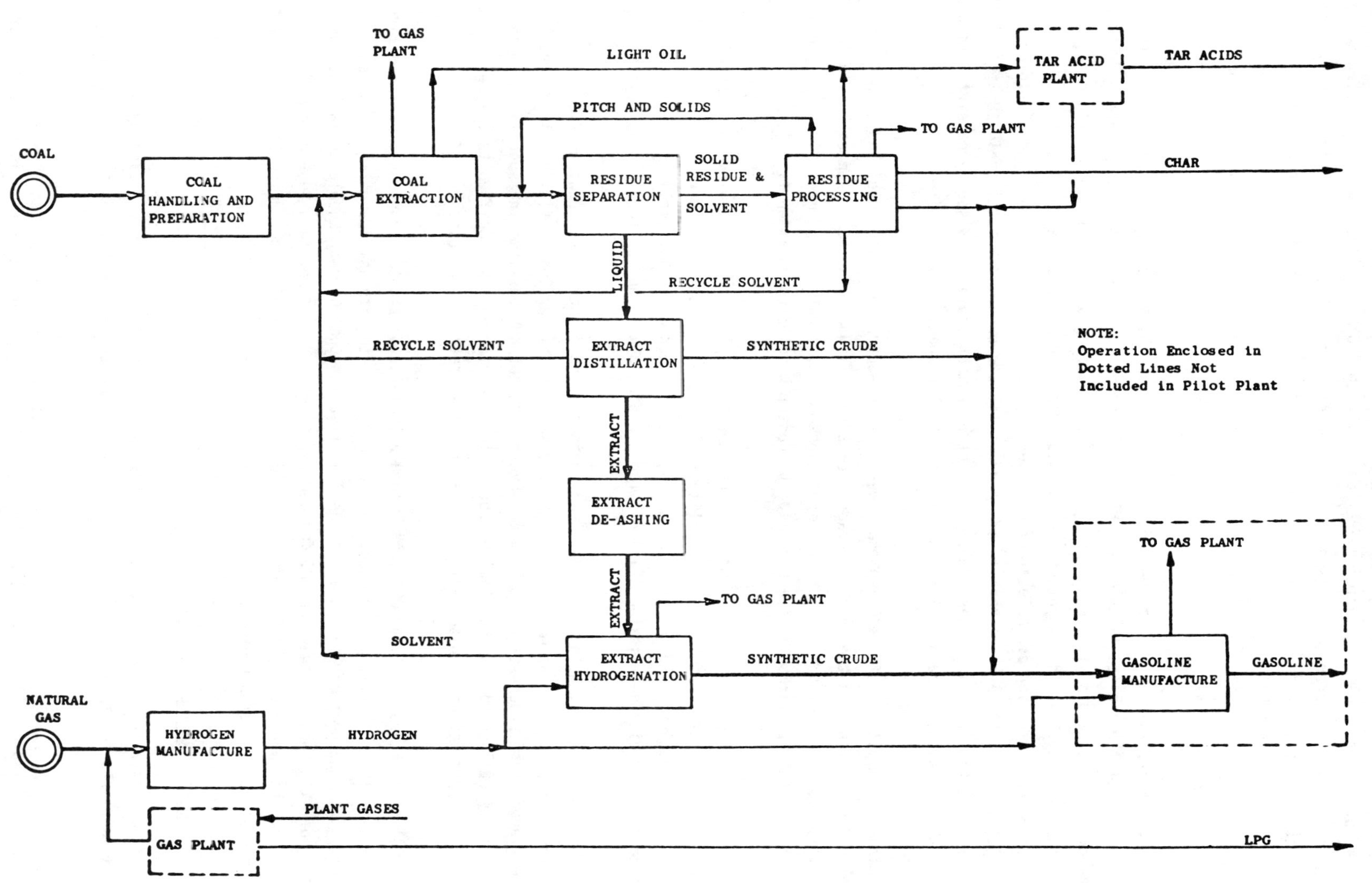

Fig. 3.33. Schematic flow diagram — Consol Synthetic Fuel process. <u>Source:</u> From Consolidation Coal Company 1973, Fig. 2, p. 5.

A distillate oil product is taken off the fractionation unit overhead. The bottoms are treated with hydrogen in a catalytic reactor operating at 775 to 850°F and 4200 psig. This step converts the heavy liquids to donor solvent and naphtha product. The pilot plant was capable of processing 20 tons of coal per day to yield about 60 bbl of oil (C_{5+} synthetic crude) per day from highly caking Eastern bituminous coal. By-products were high-ash char, light hydrocarbon gases, hydrogen sulfide, ammonia, and water. Hydrogen for the plant was generated from methane by a steam reforming process.

Samples of Cresap extract were processed under subcontract by Hydrocarbon Research, Incorporated in 1970 to test the applicability of the H-Coal process to conversion of the extract. Hydrocarbon Research, Incorporated's yield data, in agreement with Cresap and Consol bench-scale data, indicated that a 90 wt % yield of synthetic crude liquid averaging less than 0.2% sulfur could be produced by hydrogenation of the extract. Universal Oil Products found that hydrorefining of light distillate and two-stage hydrocracking of heavy distillate produced a total of 101.8 wt % yield of clear 91 octane (Research) gasoline.[*] The octane number could be raised by catalytic reforming at the expense of liquid yield (Consolidation Coal Company 1973).

Independent evaluations of the CSF process were made by Schroeder, Hiteschue, and Conn, Foster Wheeler Corporation, and the National Academy of Engineering and judged economically feasible (Consolidation Coal Company 1973). Recommendations of modifications and mechanical revisions to achieve better performance have been made by Foster Wheeler Corporation (1972).

Consolidation Coal Company has recommended that the pilot plant be reactivated and renovated to optimize the production of low-sulfur utility fuel from highly caking Eastern coals. The technology of the process would permit using high-sulfur Eastern coals in compliance with 1978 EPA and state sulfur dioxide emission standards for new stationary sources (about 0.9% sulfur extract).

In 1975, under contract with ERDA, the Fluor Engineers and Constructors (Lewton 1975) undertook a survey of filtration equipment to evaluate abilities to filter Consol fluids. In at least three liquefaction processes (CSF, SRC, and Synthoil), the current limiting specification is 0.1 wt % residual solids in the final product. This requirement is specified to eliminate the necessity of precipitator installation on the flue gas stacks of the users.

A subsidiary of Standard Oil of Ohio, the Old Ben Coal Corporation, announced plans for a five-year, $73-million, multicompany project for refining coal to both solid and liquid fuels. The technology of the project would draw on the Consolidation Coal Company knowledge and the SRC process of Pittsburg & Midway Coal Mining Company. The aim is to build a 900-ton/day demonstration plant that would produce 515 tons of solid fuel (16,000 Btu/lb) per day plus 600 bbl of low-sulfur distillate fuel per day (Chementator 1973).

3.2.2.5 <u>Solvent-Refined Coal (SRC)</u>

In 1962 OCR contracted with the Spencer Chemical Company to develop a coal deashing process. In 1965 a 100-lb/hr continuous flow unit successfully demonstrated the feasibility of the process

---

[*]Research octane number is a value determined by a standard method that measures antiknock performance of a gasoline under mild operating conditions in an engine (American Society for Testing Materials 1971).

now known as the Solvent-Refined Coal process. During the term of the contract, Gulf Oil Corporation had acquired Spencer Chemical Company and assigned the project to Pittsburg & Midway Coal Mining Company, another subsidiary. In 1966 an OCR contract was awarded to Pittsburg & Midway to develop the process to a pilot plant capable of processing 50 tons of coal per day. Two years after design of the plant had been completed, funds became available for construction, and Rust Engineering Company was selected to construct the pilot plant at Fort Lewis, Washington. Construction was started in 1972 and was essentially completed by April 1974. Total cost was estimated to be about $20 million (Hinderliter 1974).

In the SRC process (Fig. 3.34), coal is pulverized and mixed with a solvent, later to be derived from the process, to form a slurry containing 25 to 35 wt % coal. The slurry is mixed with hydrogen and sent through a slurry-preheater to a "dissolver," or reaction zone, maintained at about 800 to 900°F and 1000 to 2000 psig. Here about 90% of the organic material in the coal is dissolved in about 20 min. The remaining solids consist of the ash-forming inorganic material and a small amount of undissolved carbon. The product stream is separated to remove excess hydrogen, and the slurry of liquid and undissolved solids is filtered. The excess hydrogen is freed from hydrogen sulfide by acid gas absorption and then recycled to the system. The undissolved solids are washed with light solvent after filtration and dried to remove adhering wash solvent. The mineral residue is not processed further in the pilot plant.

The liquid filtrate is flash-distilled in vacuum to recover solvent. The overhead from this distillation is fractionated into three fractions: a light liquid by-product, a wash solvent fraction for the filters, and a process solvent for recycle to the slurry preparation area. The small amount of solvent converted to light liquids and gases during each pass through the reactor is more than balanced by the amount of coal converted to solvent, with the result being that the process requires solvent only during initial startup.

The bottoms product from the flash distillation is a heavy residual oil called solvent-refined coal that contains less than 0.1% ash and less than 0.8% sulfur. On cooling, it is a pitchlike material having a melting point of 300 to 400°F and a heating value of about 16,000 Btu/lb, regardless of the quality of the coal feedstock (Schmid 1975).

The composition of the recycle solvent is of vital importance because the solvent plays an integral part in the coal dissolution process, which occurs by reaction with about 2 wt % hydrogen (Anderson 1975). Much of the hydrogenation occurs through transfer of hydrogen from the solvent to the coal. Thermal decomposition of the coal is believed to lead to free radicals that may be capable of removing a hydrogen atom from a donor solvent molecule. Known hydrogen donors include hydroaromatic materials, for example, tetralins, hydrophenanthrenes, and hydroanthracenes. Rehydrogenation of the solvent may occur in the presence of coal minerals and hydrogen.

Further developmental work on the SRC process has also been carried out at Wilsonville, Alabama, by Southern Services, Incorporated and the Electric Power Research Institute. The Wilsonville plant, having a 6-ton/day feed rate, has operated successfully for about one year. The product generally contains less than 0.8% sulfur and less than 0.15% ash (Schmid 1975).

ORNL DWG 73-7747

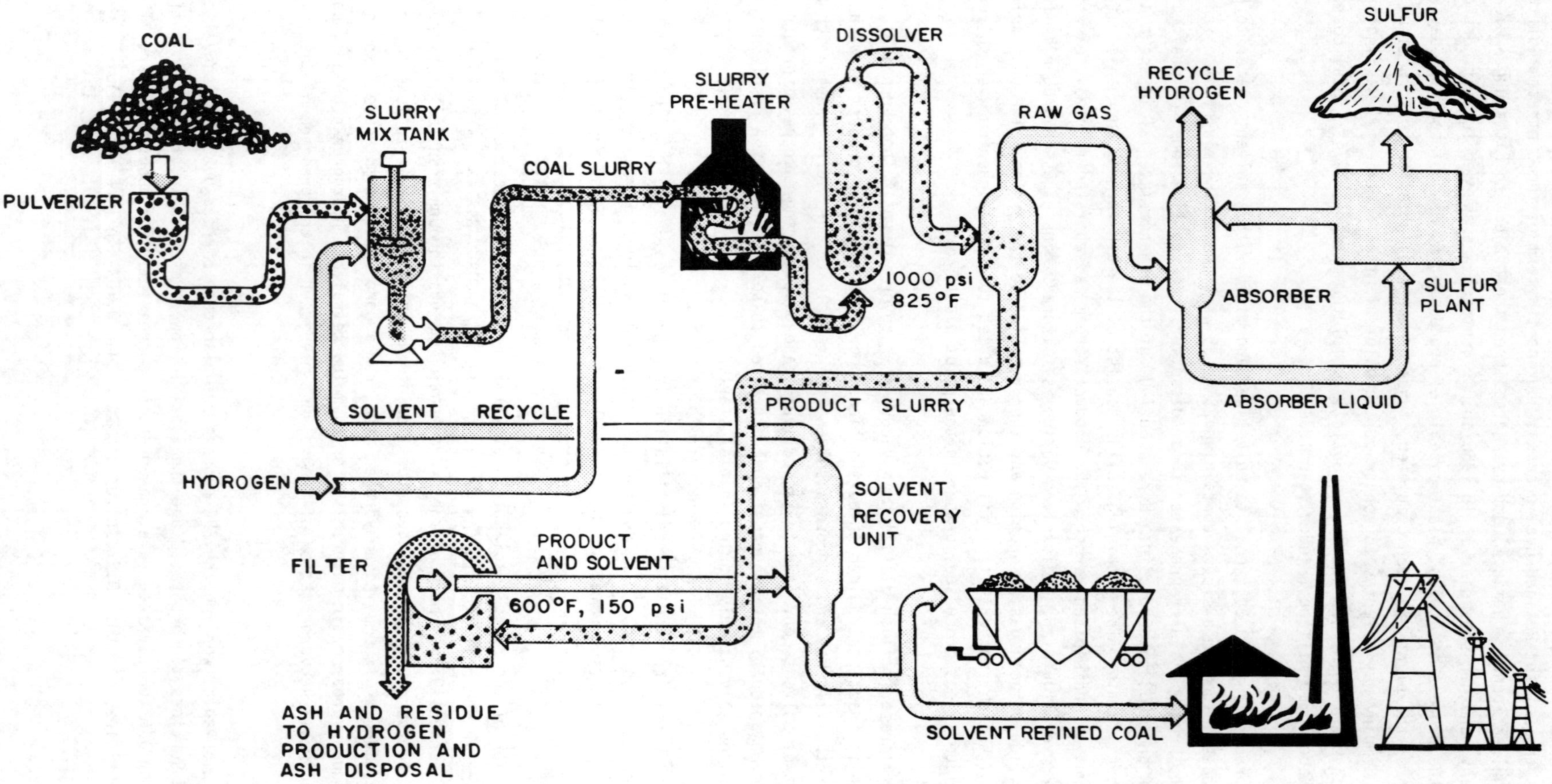

Fig. 3.34. Solvent Refined Coal process. Source: From Hinderliter 1974, Fig. 1, p. 161.

The SRC process was taken one step further in the preliminary design for an environmentally acceptable demonstration-scale plant by the Ralph M. Parsons Company for OCR. The plant, for producing clean boiler fuels from coal, consists of a coal preparation section, a coal liquefaction section, and a gasification section. It is designed to produce two low-sulfur liquid fuels sufficient to supply a 600-MW power plant. Some naphtha and by-product sulfur are also formed. The light hydrocarbons formed are burned as plant fuels (O'Hara et al. 1974).

The plant is designed to convert 10,000 tons of coal feed (Illinois No. 6) per day into five products: (1) liquid fuel oil having 0.2% sulfur, (2) heavy liquid fuel having 0.5% sulfur, (3) naphtha having 1 ppm sulfur, (4) elemental sulfur, and (5) fuel gas burned for plant operation (Fig. 3.35). Major process waste streams include waste gases (19,430 tons/day), wastewater (6390 tons/day), and a solids waste stream (710 tons/day) consisting primarily of gasifier slag (O'Hara et al. 1974).

Liquefaction feed, consisting of <1/8-in. coal particles as a 50 wt % slurry in recycle solvent, is contacted in reactors with reducing gases at 850°C and 1000 psig. The gas phase of the discharge is largely recycled; the solid phase is filtered from the liquid phase. The resulting filter cake serves as feed to the gasification section, whereas the liquid filtrate produced in the filtration operation is fractionated into an overhead naphtha stream, a distillate light boiler fuel, and a residual fuel oil. Low-sulfur fuel suitable for boiler firing results from further hydrogenation of the distillate fuel.

Table 3.22 shows the composition, output, and contaminants of the principal streams.

Wet filter cake from the liquefaction process is sent to a slagging, suspension-type gasifier, where the cake is reacted with steam and oxygen at 3000°F and 200 psig to produce synthesis gas (carbon monoxide and hydrogen) principally. Hydrogen sulfide is removed from most of the synthesis gas, which is fed to the liquefaction process. The remainder of the synthesis gas is subjected to shift conversion, carbon dioxide removal, and methanation; the resulting hydrogen-gas stream is used to hydrogenate the light distillate and naphtha streams from the liquefaction section.

Actual samples of process solvent, mineral residue, process effluent water (before treatment), process "naphtha," and fuel product were provided to Battelle Pacific Northwest Laboratories by Pittsburg & Midway from its Fort Lewis operation. A partial characterization of the organic constituents of the samples has been completed, and a list of compounds identified is included in Table 3.23.

The same major organic components appear in all the samples investigated, but their distributions vary with the type of sample. In addition, the compounds identified are not necessarily those eventually escaping into the atmosphere. Future analyses will be made of the actual plant effluents, including process water after treatment, gaseous emissions, and any particulates (Petersen 1975).

ORNL DWG 76-6581

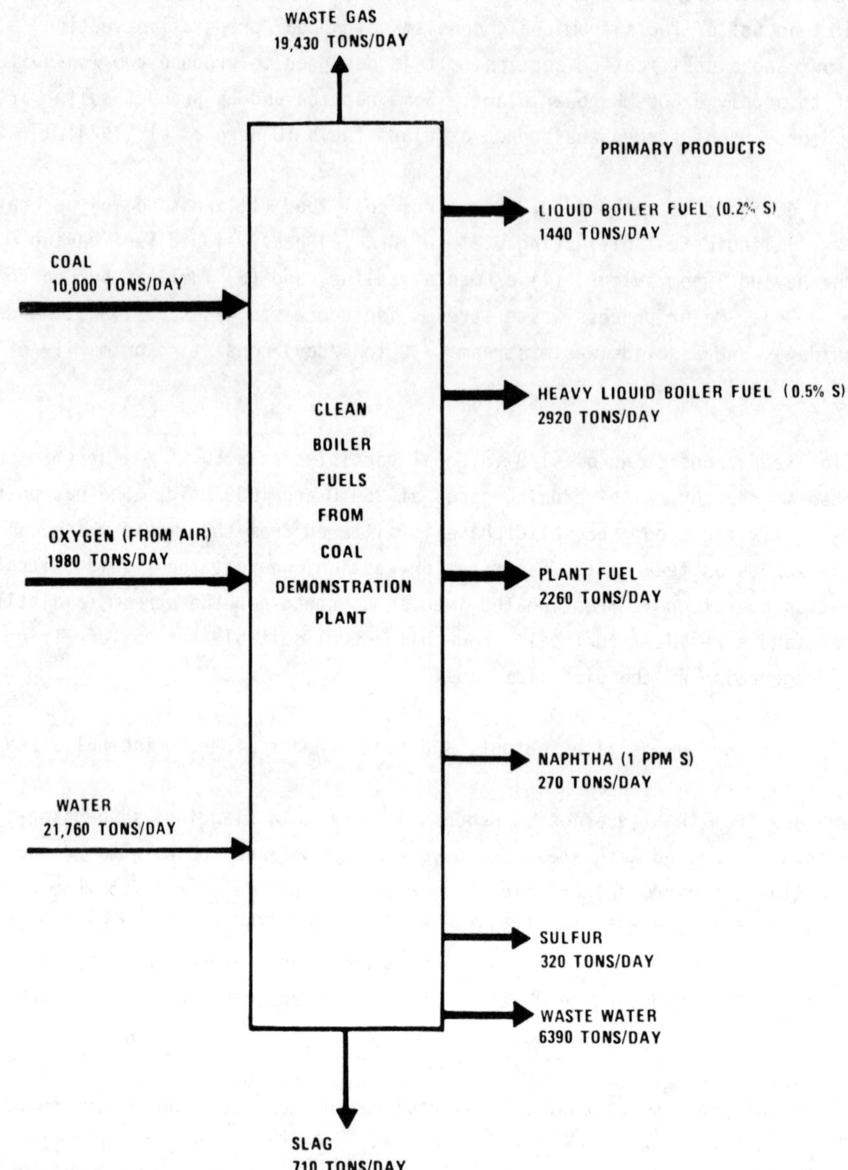

Fig. 3.35. Material balance for proposed Solvent-Refined Coal plant.
<u>Source:</u> From O'Hara et al. 1974, Fig. 2, p. 172.

3.2.2.6 <u>Gulf Catalytic Coal Liquids (CCL) process</u>

Gulf Research and Development Company has developed a bench-scale coal liquefaction unit (0.5 to
5 lb of coal per hour), which can be operated at pressures up to 10,000 psig and at temperatures
up to 1000°F.

According to Chun (1974), research has been carried out on coals from six states — Pennsylvania,
Kentucky, Colorado, Montana, Wyoming, and North Dakota — with most of the work being done on subbituminous and bituminous coal. A low-sulfur liquid (0.05 to 0.2% sulfur) was produced.

Table 3.22. Gaseous waste process streams

Waste gas stream	Composition	Output (million cfd)	Content (million Btu/day)	Major contaminant
Combustion gases				
Fuel gas	32% O_2, 9.4% CO_2, 15.6% H_2O, 71.3% N_2	935	14,150	4 - 15 ppm SO_2 50 - 100 ppm NO_x
Fuel oil	3.3% O_2, 10.8% CO_2, 12.7% H_2O, 73.2% N_2	54	850	80 - 100 ppm SO_2 100 - 150 ppm NO_x 35 ppm particulates
Oxygen plant	N_2 and rare gases	64	Nil; ambient conditions	None
Sulfur removal plant tail gas	72% N_2, remainder CO_2	32.4	Nil; ambient conditions	10 ppm max H_2S
CO_2 removal	99% CO_2	14.3	Nil; ambient conditions	None

Source: O'Hara 1974, Fig. 4, p. 174.

The catalyst, which is proprietary, has been thoroughly tested and was developed through 20 years of process and catalyst research. It is described as being coke-resistant, metal-tolerant, and highly active.

The flow diagram for the process (Fig. 3.36) shows that coal, ground and slurried in a recycle solvent, is sent together with hydrogen through a catalytic, fixed-bed reactor at 900°F and 2000 psi. The products from the reactor are flashed to separate gases and liquids; hydrogen, separated from the other gases, is compressed and returned to the process, whereas the liquids, subjected to a low-pressure flash separation, yield water and recycle solvent. The remaining material is filtered; the filtrate is distilled to yield the product, a low-sulfur synthetic crude oil, and the solids are coked to produce gaseous and liquid products, which are combined with the filtrate to be distilled.

A 1-ton/day pilot plant located in Harmarville, Pennsylvania, began operation in January 1975. The plant is designed to produce 3 bbl of oil per ton of moisture- and ash-free coal and to provide design data for a larger demonstration plant (Howard-Smith and Werner 1975).

3.2.2.7 H-Coal

The H-Coal process for catalytic conversion of coal to liquid fuels uses the ebullated bed reactor system, which was originally developed by Hydrocarbon Research, Incorporated, to upgrade petroleum residual oils (Johnson et al. 1973).

Crushed coal is slurried with a heavy recycle oil product, mixed with hydrogen gas, heated to about 850°F, and reacted at about 2500 psig in an 8-in.-diam reactor handling 200 lb of coal per hour (Fig. 3.37). A bed of cobalt molybdate catalyst particles in the reactor is kept in an ebullated state by the upward flow of the coal suspension and rising hydrogen bubbles. Catalyst can be added or removed continuously as required. Because of its high cost, attrited catalyst must be recovered from the unconverted coal or coal ash (Perry 1974).

Table 3.23. Compounds identified in Solvent-Refined Coal process

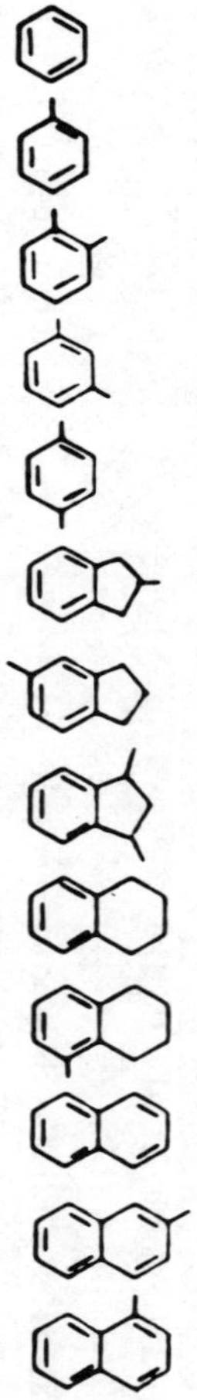

Benzene

Toluene

o-Xylene

m-Xylene

p-Xylene

2-Methylindan

5-Methylindan

1,3-Dimethylindan

1,2,3,4-Tetrahydronaphthalene

5-Methyl-1,2,3,4-tetrahydronaphthalene

Naphthalene

2-Methylnaphthalene

1-Methylnaphthalene

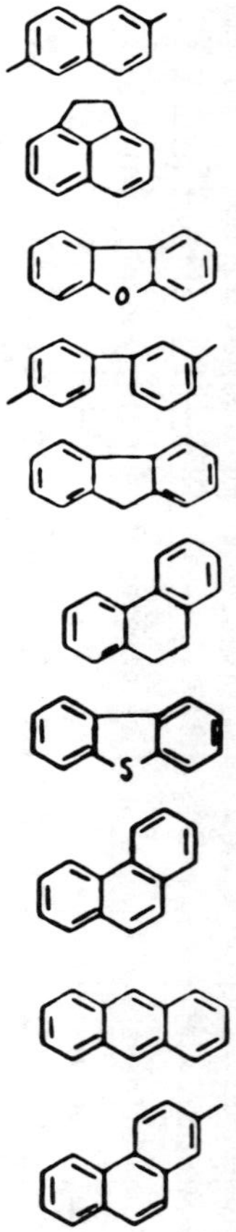

2,6-Dimethylnaphthalene

Acenaphthene

Dibenzofuran

3,4'-Dimethylbiphenyl

Fluorene

9,10-Dihydrophenanthrene

Dibenzothiophene

Phenanthrene

Anthracene

2-Methylphenanthrene

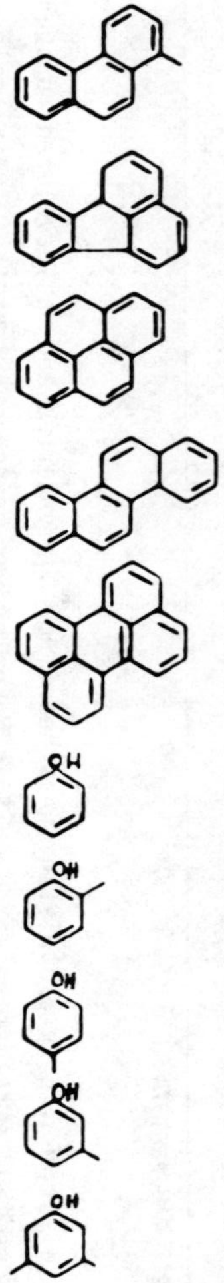

1-Methylphenanthrene

Fluoranthene

Pyrene

Chrysene

Perylene

Phenol

o-Cresol

p-Cresol

m-Cresol

3,5-Xylenol

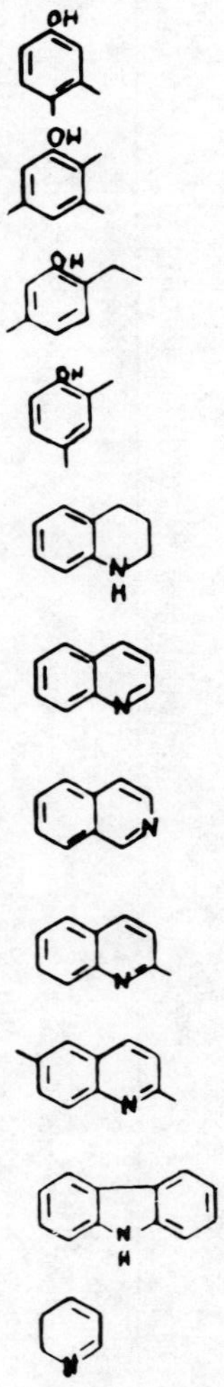

3,4-Xylenol

2,3,5-Trimethylphenol

2-Ethyl-5-methylphenol

2,4-Xylenol

1,2,3,4-Tetrahydroquinoline

Quinoline

Isoquinoline

2-Methylquinoline

2,6-Dimethylquinoline

Carbazole

Pyridine

Source: Petersen 1975, Appendix, p. 9.

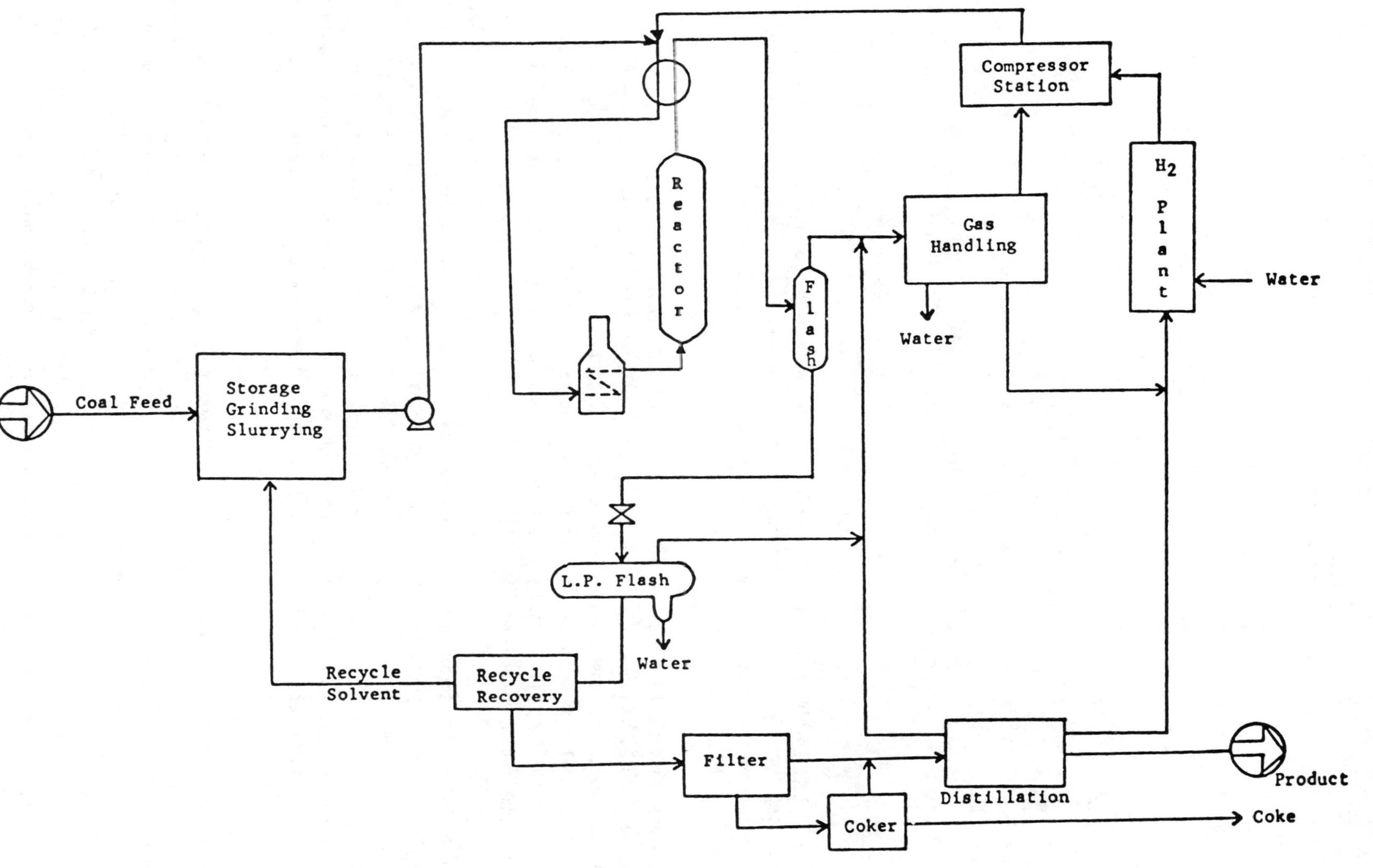

Fig. 3.36. Conceptual flow diagram — Gulf CCL process. Source: Chun 1974, p. 267.

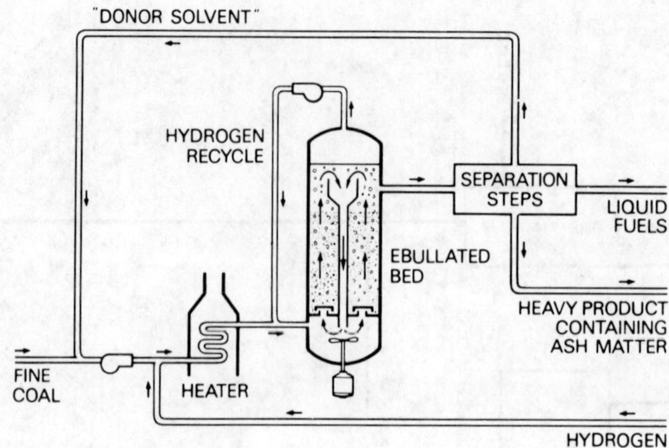

Fig. 3.37. Schematic diagram of the H-Coal process of Hydrocarbon Research, Inc., Trenton, N.J. <u>Source</u>: From Squires 1974, Fig. 12, p. 11. Reprinted by permission of the publisher.

Oil used for the slurry donates hydrogen to the coal within the ebullated bed. The oil is recirculated from above the catalyst suspension to the bottom of the reactor and thus keeps the temperature of the reactor uniform. Spent solvent oil is continuously hydrogenated in the reactor to restore its hydrogen-donating capacity (Squires 1974).

Gas taken off from the top of the reactor is treated for removal and recovery of sulfur and ammonia. Unused hydrogen is also separated and returned to the dissolution process. Raw oil from the reactor is flashed down to low pressure; the flashed vapors are distilled in an atmospheric distillation unit, whereas the flashed liquids are distilled under vacuum. Products are low-sulfur fuel oils, a heavy recycle oil to the coal slurry preparation, and a residuum slurry that can be coked (Beychok 1975).

If operated at 200 atm and 455°C, the process can yield a light synthetic crude oil. At 100 atm and about 430°C, a synthetic heavy fuel oil can be made. One of the major problems, however, has been the separation of solids: Hydroclones remove only about two thirds of the solids, centrifuges are too expensive, and magnetic separation results are not good. Filtration is less effective than vacuum distillation of the solid-liquid mixture followed by coking (Perry 1974). Because a good solids separation technique is not available, the process is more suited for production of a synthetic crude (Squires 1974).

H-Coal process development has proceeded from conceptual studies through a 3-ton/day process demonstration unit. Plans for an $80-million, 600-ton/day demonstration plant at Catlettsburg, Kentucky, are being drawn up by Hydrocarbon Research, Incorporated. A consortium composed of Hydrocarbon Research, the State of Kentucky, Ashland Oil Company, Standard Oil Company (Indiana), and the Electric Power Research Institute is negotiating with ERDA for the contract to build the plant, construction of which will take three years. The final phase will consist of operating the commercial demonstration plant (H-Coal process nears commercialization 1976).

3.2.2.8 Synthoil

The Bureau of Mines Synthoil process produces low-sulfur fuel oil by catalytic hydrodesulfurization of coal in a fixed-bed reactor with turbulent flow of hydrogen through the bed at elevated
temperatures and pressures. Hydrogenation conditions are controlled to promote desulfurization
of the coal and to minimize additional hydrogenation of products from the primary liquefaction.
The liquid fuel can be used as a low-sulfur utility fuel either before or after removal of ash
and unreacted coal.

Akhtar, Friedman, and Yavorsky (1974) of the Pittsburgh Energy Research Center report work with
three different feed coals. The coals, ground to 70% <200 mesh and slurried initially in a high-
temperature tar derived from coking metallurgical-grade bituminous coal, were sent with hydrogen
through a preheater before entering a 5/16-in.-ID by 68-ft tube reactor folded inside a 10-ft-
long, 10-in.-diam cylindrical furnace (Fig. 3.38). The reactor was charged with a Co-Mo–SiO$_2$-
Al$_2$O$_3$ catalyst. Reactor conditions were 450°C and 2000 psi with a short residence time for the
feed. The product stream leaving at the top of the reactor was sent through a water-cooled condenser to a high-pressure receiver for liquids, whereas the gases were fed through a second high-
pressure condenser and receiver, then scrubbed with an alkali solution to remove hydrogen sulfide
and ammonia.

ORNL DWG 76-6578

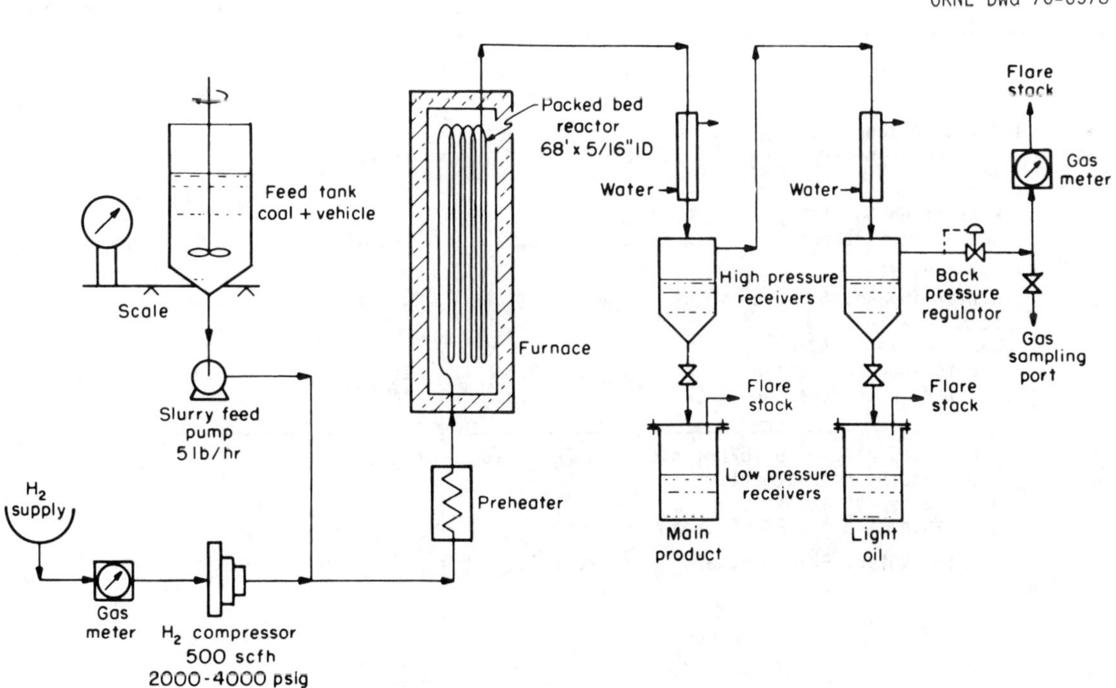

Fig. 3.38. Simplified schematic diagram of Synthoil 100-lb/day hydrodesulfurization
pilot plant. Source: From Akhtar, Friedman, and Yavorsky 1974, Fig. 1, p. 426. Reprinted
by permission of the publisher.

By this method Pittsburgh seam coal containing 1.3% sulfur was desulfurized to a fuel containing 0.3% sulfur, an Indiana No. 6 coal containing 3.4% sulfur was desulfurized to a fuel containing 0.42% sulfur, and a Middle Kittanning No. 6 coal containing 3.0% sulfur was desulfurized to a fuel containing 0.36% sulfur. The heating value of the three fuels was about the same — 16,400 Btu/lb.

In anticipation of future commercial operations, coal was also hydrodesulfurized in a coal-derived liquid. The liquid was prepared by hydrodesulfurizing a 30 wt % slurry of Middle Kittanning No. 6 coal in tar and repeating the operation three times, each time using a slurry of 30 wt % coal in the centrifuged liquid products from the preceding cycle. The fourth-cycle liquid products were estimated to contain about 80% coal-derived liquids (recycle oil). Table 3.24 shows the sulfur, ash, heating value, and hydrogen consumption for the coals. The hydrogen consumption drops considerably when recycle oil is used for slurrying the coal. Centrifugation of the liquid product gave a low-ash, low-sulfur fuel (0.31 wt % sulfur and 1.3 wt % ash).

Table 3.24. Hydrodesulfurization of coal slurries at 450°C and 2000 psi[a]

Feed	Gross oil product analysis			
	S (wt %)	Ash (wt %)	Calorific value (Btu/lb)	H_2 consumed (scf/100 lb of feed)
30 wt % Pittsburgh seam coal in tar[b]	0.3	1.5	16,450	284
30 wt % Indiana No. 5 seam coal in tar[b]	0.42	2.7	16,415	304
30 wt % Middle Kittanning No. 6, seam coal in tar[b]	0.36	2.5	16,310	330
30 wt % Middle Kittanning No. 6 seam coal in coal-derived liquid[c]				
Before centrifugation	0.48	3.5	16,200	139
After centrifugation[d]	0.31	1.3	16,840	

[a] Slurry feed rate — 5 lb/hr; hydrogen feed rate — 500 scf/hr.
[b] Three-pass hydrotreatment.
[c] One-pass hydrotreatment.
[d] Centrifuge cake = 9.5% of gross product.
Source: Akhtar, Friedman, and Yavorsky 1974, Table 2, p. 429.

Yavorsky et al. (1975) have successfully processed five different coals in a 100-lb/day unit and in a 0.5-ton/day unit (reactor unit has a 1-in. ID and is 14 ft long). Typically, a Kentucky coal containing 5.5% sulfur and 16% ash can be converted to oil with 0.2% sulfur and 0.2% ash. Oil yields of over 3 bbl/ton are reported for process conditions of 2000 or 4000 psig, 2-min reaction time, and 450°C. In runs of 500 hr no serious catalyst deactivation was reported. Data obtained from the 500-hr runs permitted calculating the mass balance for the process (Fig. 3.39). The hydrogen sulfide is converted into sulfur by known methods, and the ammonia can be sold.

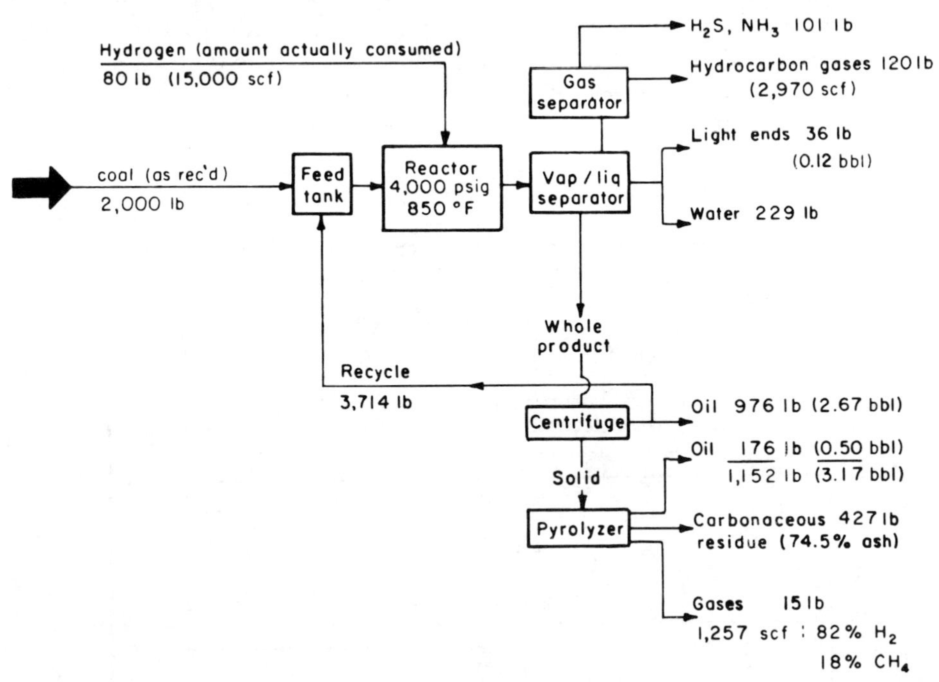

Fig. 3.39. Material balance for a Synthoil run with Western Kentucky coal. Source: From Yavorsky et al. 1975, Fig. 1, p. 80. Reprinted by permission of the publisher.

Hydrocarbon gases can be reformed to supplement the hydrogen needed for the process, and the pyrolyzed residue can be added with the coal to the steam-oxygen gasification to produce the bulk of the hydrogen needed. The product oil has a heating value of 17,400 Btu/lb.

A 10-ton/day pilot plant is being designed, with construction to start in the summer of 1975; startup is expected in the summer of 1976.

3.2.2.9 University of Utah—Intermediate Coal Hydrogenation Process

The University of Utah process is a two-stage process for producing aromatic hydrocarbons from coal. In the first step the coal is converted to a synthetic oil by catalytic hydrogenation. In the second step this oil is converted to benzene-toluene-xylene (BTX) by a combination of hydrorefining, hydrocracking, and solvent extraction.

In bench-scale tests a high-volatile coal was ground to <100 mesh, impregnated with catalyst, dried, and fed to a dilute-phase free-fall reactor by a screw feeder (Fig. 3.40). Hydrogenation was carried out under various conditions of temperature, pressure, and coal feed rate and with different catalysts to optimize operating conditions. Temperature was varied from 450 to 650°C, pressure from 1000 to 4000 psi, coal feed rate from 10 to 100 g/min, and catalyst concentrations from 5 to 20 wt % of the coal. Catalysts investigated were $SnCl_2$, $ZnCl_2$, and $(NH_4)_2MoO_4$.

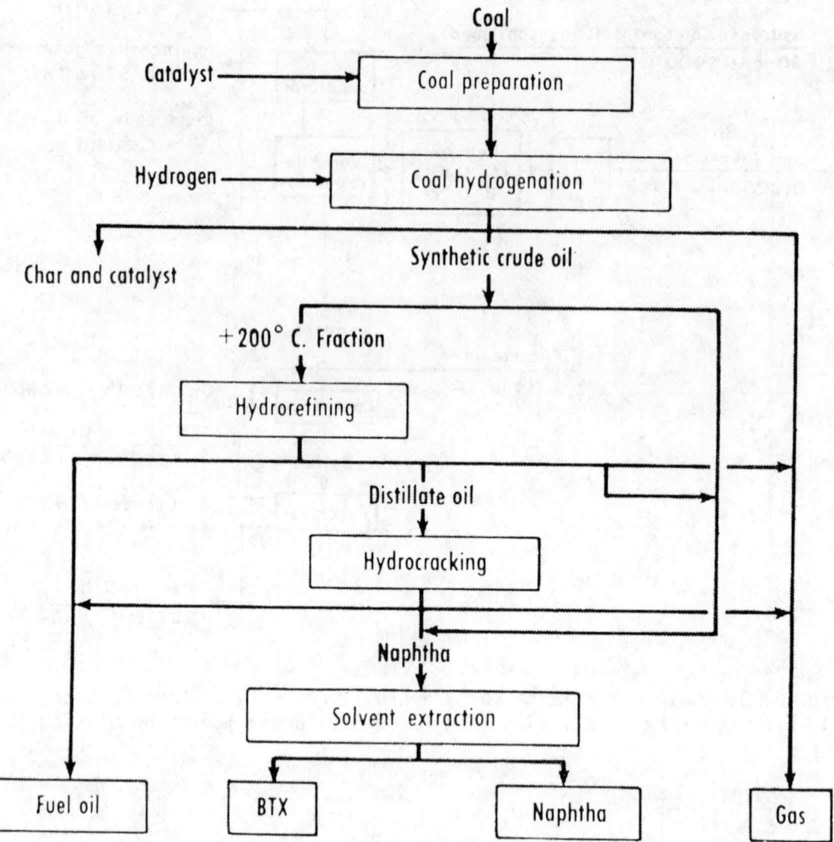

Fig. 3.40. University of Utah Intermediate Coal Hydrogenation process for production of aromatics from coal. <u>Source</u>: From Qader 1974, Fig. 1, p. 58. Reprinted by permission of the publisher.

Based on data from several runs, the process parameters were optimized at 500°C, 3000 psi, 25-g/min feed, and a catalyst concentration of 10%. Synthetic oil from the first stage was separated from unreacted coal by washing with acetone. The oil was then distilled to separate fractions boiling below and above 200°C.

The fraction boiling above 200°C was hydrorefined and hydrocracked in a continuous bench-scale fixed-bed reactor by using commercial catalysts. Hydrorefining was carried out over a cobalt-molybdate catalyst (hydrodesulfurization catalyst), reducing sulfur and nitrogen content to less than 0.1%. The hydrorefined product was distilled to separate three fractions: (1) 200°C, (2) 200 to 360°C, and (3) >360°C. The 200 to 360°C fraction was then hydrocracked over a $NiS\text{-}WS_2\text{-}SiO_2\text{-}Al_2O_3$ catalyst for conversion to naphtha. The naphtha contained 80% aromatics and 20% saturates.

An overall material balance (Table 3.25) for the process shows that about 33% of the coal is converted to BTX, and 30% to gas. The catalyst is expected to come out of the process in mixture with char and to be recoverable. The C_1-C_4 gases will be a good feedstock for SNG.

Table 3.25. Material balance — University of Utah process[a]

Component	Amount	Wt % of coal
Input, lb		
Coal[b]	2000	100.00
Hydrogen	91.37	4.57
Catalyst	200	10.0
Total	2291.37	
Output		
C_1 - C_4 gas, scf	9611	30.15
Fuel oil, gal	25.54	15.83
Naphtha, gal	24.09	3.35
BTX, gal	90.76	33.08
Char, lb	200	10.00
H_2S, scf	201	0.90
NH_3, scf	730	1.64
H_2O, gal	23	9.62
Catalyst, lb	200	

[a]Conditions: 500°C, 3000 psi, catalyst zinc chloride (10%).
[b]Moisture- and ash-free.

Source: Qader 1974, Table 5, p. 59.
Reprinted by permission of the publisher.

3.2.2.10 Coalcon

The Coalcon Company of New York was formed in a joint venture by Union Carbide Corporation and Chemical Construction Company in May 1974 to demonstrate the commercial feasibility of Union Carbide's hydrocarbonization technology in converting high-sulfur bituminous coal to clean liquid boiler fuel and gaseous fuels. Coalcon was awarded a $237-million contract on January 17, 1975, by ERDA for a demonstration plant to convert three types of Eastern coals to clean liquid boiler fuel and gaseous fuels. A site near Belleville, Illinois, was chosen from among 16 proposed locations, each of which had been evaluated separately by Coalcon and ERDA. Construction of the plant was to begin in 1977 and be completed in 1980. It was to process 2600 tons of a high-sulfur coal per day into 3900 bbl of liquid fuels for boilers per day and 22 million ft^3 of high-Btu pipeline gas per day (Chironis 1975).

The technology is based on the extensive pilot plant program undertaken by Union Carbide in the early 1960s to produce chemicals from coal by using a hydrocarbonization process. The coal used was a low-sulfur Western subbituminous coal (Lake de Smet coal). The process, which was scaled up from a 1-lb/hr bench-scale unit to 10 lb/hr and then to 20 tons/day, operated successfully. Further work was suspended in 1964, when the assessment was made that chemicals from coal could not compete economically with petroleum-derived products.

Figure 3.41 is a simplified block diagram of the Coalcon process showing the treatment and separation of the reaction products. The hydrocarbonization process differs from other processes in using a dry, noncatalytic, fluid bed of coal particles suspended in hydrogen gas instead of a fixed-bed or liquid-phase process to produce char, gas, and liquid fuels. The process also converts the organic sulfur compounds in the coal into hydrogen sulfide gas, which can be removed and further processed to produce sulfur (Chironis 1975).

ORNL DWG 76-6580

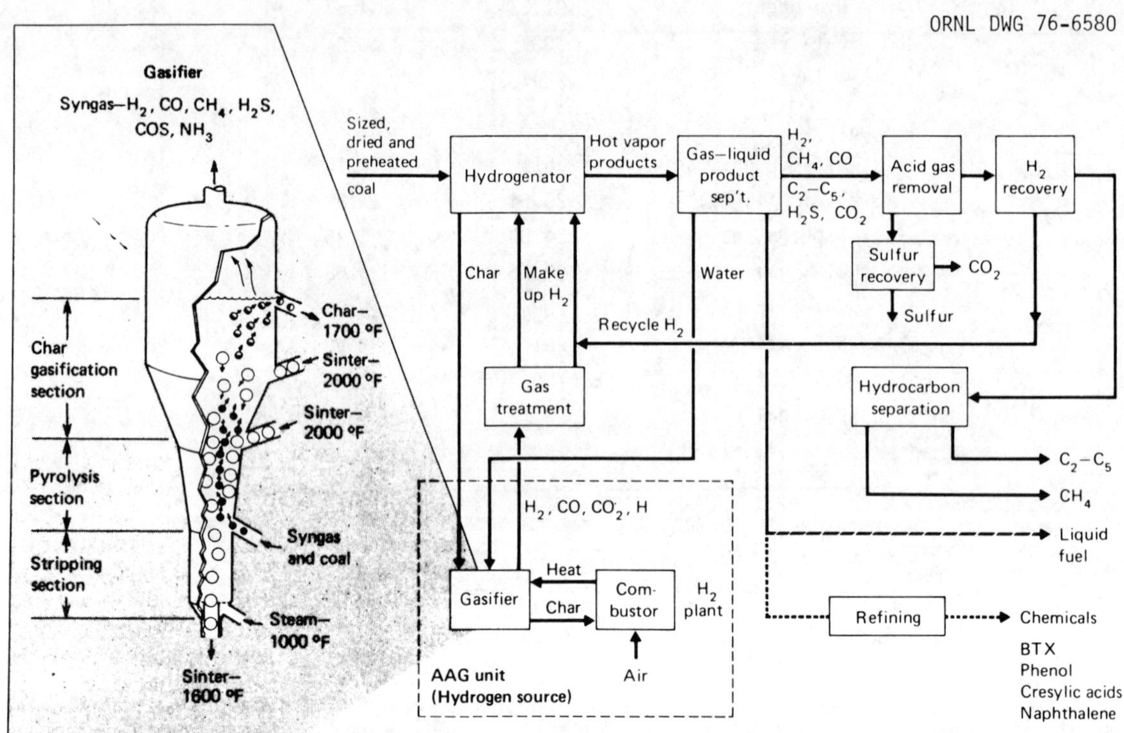

Fig. 3.41. Coalcon process. <u>Source</u>: From Chironis 1975, p. 77. Reprinted by permission of the publisher.

The three coals chosen for testing in the demonstration plant are Pittsburgh No. 8, Illinois No. 6, and Western Kentucky No. 11. Coal to be used in the hydrocarbonization reactor (hydrogenator) is ground to a size range of 60 to 325 mesh and dried by warm flue gas to about 1% moisture. The sized coal, held temporarily in a hopper, is fed to a coal heater, which keeps the heat balance of the reactor within bounds. Without heating, more fluidization gas would be necessary to maintain the reaction temperature in the reactor, whereas too much heating causes the coal to become sticky and agglomerate. A hot oxygen-free flue gas is used to heat the coal to about 325°C and carry it to a coal feed hopper. The coal is then pressurized with steam in a jacketed lock-hopper system before being fed by gravity on a 20- to 40-min cycle at 617°F to the coal injection vessel, where the coal is fluidized with hydrogen from the bottom at 260 psi above the reactor pressure.

The reactor will operate under the following tentative conditions: temperature, 560°C; pressure, 555 psia; solids residence time, 25 min; gas residence time in the bed, 25 sec; hydrogen partial pressure in the bed, 419 psia; and superficial velocity, 1.0 fps. Gases leaving the bed surface are charged with solids, which are removed by two cyclones. An internal cyclone returns the solids to the bed, and an external cyclone extracts the final solids from the gas and feeds them to the char surge hopper. The cyclones are designed to remove enough solids to meet specifications for solids content of the heavy oil product.

Most of the char is removed from the bottom of the reactor vessel, quenched with water and steam, depressurized to about 60 psia, cooled in a series of fluid beds by nitrogen, ground in a mill, and fed by a screw feeder to a Koppers-Totzek gasifier, where it is reacted with oxygen and steam to produce hydrogen for the hydrogenation process. Oxygen will be generated by a conventional cryogenic air separation plant. About 680 tons of oxygen per day are needed for the gasifier. Steam is produced by vaporizing process wastewater from various parts of the plant.

The gas coming from the external cycle of the hydrogenator is hot (1040°F) and under high pressure (555 psia), and it may contain some entrained solids. A fractionator is used to split the gas into four streams: (1) overhead gas (H_2, CO, CO_2, CH_4); (2) light liquid (e.g., No. 2 fuel oil); (3) heavy liquid (e.g., No. 5 or 6 fuel oil); and (4) wastewater.

The heavy oil is cooled and flashed to atmospheric pressure and pumped to storage. The fractionator overheads are sent to a decanter after cooling against boiler feed water and cooling water. The light fuel, overhead gas, and wastewater are separated. Some of the light fuel oil is used as reflux at the top of the fractionator, and some is pumped to storage.

Ammonia generated by the process is contained in the decanter product gas and wastewater. These streams are fed to a Phosam unit designed by U.S. Steel, where ammonia is stripped from the liquid and scrubbed from the gas. The acid gases (CO_2, H_2S) are to be removed either by a Sulfinol absorption process or a cold methanol wash. The gas is then sent directly to the cryogenic processing unit if a cold methanol wash has been used. Otherwise, an adsorption bed containing carbon and molecular sieves is used to separate possible aromatics, carbon dioxide, and hydrogen sulfide.

The cryogenic processing unit divides the gaseous feed into three streams: (1) purified hydrogen (93% H_2, 6% CO, 1% CH_4); (2) C_{2+} product (40% CH_4, 42% C_2, 13% C_3, 4% C_4, 1% C_5); and (3) methanation synthesis gas (48% H_2, 3% N_2, 16% CO, 35% CH_4, 1% C_2H_5). The C_{2+} stream is to be split in a deethanizer into CH_4-C_2H_6 and C_{3+} products. The C_{3+} components are liquid and will be sold as liquified petroleum gas. The methane and ethane are to be added to the methanator product to form the SNG stream.

The hydrogen–carbon monoxide ratio of the methanation feed stream is adjusted to the correct stoichiometric ratio for reacting in the methanator ($3H_2 + CO = CH_4 + H_2O$). After methanation, the product is dried and the gas from the deethanizer is added to give a high-Btu (about 1030 Btu/ft^3) SNG product.

Figure 3.42 shows the anticipated yields of products from operation of the 2790-ton/day coal hydrocarbonization demonstration plant (Morgan 1975).

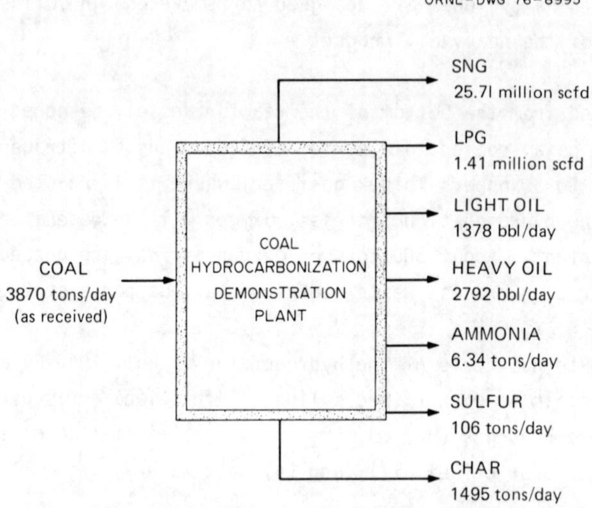

Fig. 3.42. Coalcon demonstration plant feed-product distribution.
Source: From Morgan 1975, Fig. 1-B.

Because of funding problems for FY77, the future of the plant is currently undecided.

3.2.2.11 Fischer-Tropsch process

The Fischer-Tropsch process is a method for indirect conversion of coal to liquid products: Coal is gasified to a synthesis gas, which is then converted to a liquid by catalytic liquefaction.

F. Fischer and H. Tropsch at the Kaiser Wilhelm Institute in Germany extensively studied the catalytic reduction of carbon monoxide with hydrogen in the 1920s and developed a normal-pressure process for production of liquid hydrocarbons. In 1934, Ruhrchemie, having purchased rights to the process, erected and successfully operated a semicommercial plant. In the following year four full-scale plants were erected. The main steps of the process, as practiced in Germany, were synthesis gas manufacture, gas purification, synthesis of hydrocarbons, condensation of liquid products, and recovery of gasoline from product gas (Pichler and Nector 1964). In 1936 Fischer and Pichler determined the optimum conditions for carrying out the medium-pressure hydrogenation of carbon monoxide with a cobalt-kieselguhr catalyst. Yields were 10 to 15% higher than with the normal-pressure process. Later, yields were increased with the use of iron catalysts, but this method was not used commercially until after World War II. Fischer-Tropsch synthesis of hydrocarbons from carbon monoxide contributed greatly to the energy requirements of Germany during the war by producing gasoline, gas oil, and paraffin. The gas oil was excellent diesel fuel; the greater part of German production was used for this purpose. A total of nine plants were put into operation in Germany during 1936-1939, reaching peak production in 1943 at 585,000 tonnes of synthetic oil.

At present there is only one commercial plant, SASOL (Suid-Afrikaanse Steenkool-, Olie- en Gaskorporasie), operating in the Republic of South Africa, for production of liquid hydrocarbons from coal by the Fischer-Tropsch process. The following description of the Fischer-Tropsch process is based on this plant.

The Republic of South Africa, possessing some 90% of the proven coal reserves of the African Continent, is richly endowed with bituminous coal, but economically recoverable petroleum deposits have not been found. The coal is a low-grade bituminous containing between 25 and 30% ash, 42% fixed carbon, and 8% moisture and having a gross heating value of 8700 Btu/lb.

The coal charged to the SASOL plant is used for two purposes. About 40% goes to a boiler house, where steam is produced for power and for production of synthesis gas. The balance of the non-caking coal, crushed to 3/8- to 1-1/2-in. particles and dried, is converted in a battery of Lurgi high-pressure, steam-oxygen gasifiers at 350 to 450 psi to produce a gas consisting essentially of carbon monoxide and hydrogen; tar and oil are removed by quenching. To prevent poisoning of the Fischer-Tropsch catalysts, the crude synthesis gas is purified by a low-temperature scrubbing with methanol (Rectisol process), which removes the last traces of tar and oil, as well as 98% of the carbon dioxide, all the hydrogen sulfide and organic sulfur, and ammonia and phenol. Hydrocarbons and entrained solids are also removed (Rousseau 1962; Hoogendoorn and Salomon 1957a).

The two chemical reactions typifying the Fischer-Tropsch formation of hydrocarbons are

$$n\text{CO} + 2n\text{H}_2 = (\text{CH}_2)_n + n\text{H}_2\text{O} \quad (\text{H}_2 \text{ to CO ratio of 2:1}) \quad ; \tag{24}$$

$$2n\text{CO} + n\text{H}_2 = (\text{CH}_2)_n + n\text{CO}_2 \quad (\text{H}_2 \text{ to CO ratio of 1:2}) \quad . \tag{25}$$

These two reactions are connected by the water-gas shift reaction:

$$\text{CO} + \text{H}_2\text{O} = \text{H}_2 + \text{CO}_2 \quad . \tag{26}$$

The SASOL plant uses two types of reactors to carry out these reactions: One is of German design (ARGE reactor) and works with a fixed bed of iron catalyst in vertical tubes, and the other is of American design (M. W. Kellogg reactor) and operates with a fluidized bed of powdered iron catalyst. The catalyst is highly pyrophoric and is based on a special type of iron ore whose activity is enhanced by a small percentage of metals of group I or III of the Periodic Table of the Elements (Hoogendoorn and Salomon 1957b).

A major portion of the purified synthesis gas is sent to the ARGE reactor, where the hydrogen and carbon monoxide (in a 2:1 ratio) are converted at 220 to 255°C and 250 kg/cm^2 to a mixture of straight-chain, high-boiling hydrocarbons, with some medium-boiling oils, diesel oil, liquified petroleum gas, and oxygenated compounds such as alcohols and acids.

The balance of the synthesis gas, together with tail gases from the ARGE and Kellogg reactors, is sent to a methane-reforming unit before being sent to the Kellogg reactor, where a higher ratio of hydrogen to carbon monoxide (3:1) is needed for the reaction occurring at about 320°C and 230 kg/cm^2. Product gas and catalyst leaving the reactor are separated by cyclones, and the catalyst is recycled to the reactor. Products of the fluidized-bed synthesis are mainly low-boiling hydrocarbons (C_1-C_4) and gasoline, with substantial amounts of oxygenated products and aromatics.

Table 3.26 lists the products of the process, as percentage of total volume production, from the fixed-bed and the fluid-bed sections and includes a breakdown of the liquid products in volume percent from the two beds.

Products from the two synthesis reactors differ greatly. The ARGE process produces less of the oxygenated hydrocarbons and more of the heavier hydrocarbons such as diesel oil and paraffin waxes, essentially straight-chain products. The Kellogg process yields higher amounts of oxygenated hydrocarbons and lower-molecular-weight hydrocarbons such as liquefied petroleum gas and gasoline. The hydrocarbons are more branched, and a significant amount of aromatics are present. Figure 3.43 shows the overall scheme of the SASOL plant and the products formed by working the ARGE and Kellogg reactor products (Rousseau 1962).

Table 3.26. Typical products — Fischer-Tropsch process

Component	Composition (vol %)	
	Fixed-bed process	Fluid-bed process
Liquified petroleum gas (C_3-C_4)	5.6	7.7
Petrol (C_5-C_{11})	33.4	72.3
Middle oils (diesel, furnace, etc.)	16.6	3.4
Waxy oil or gatsch	10.3	3.0
Medium wax, mp 203-206°F	11.8	
Hard wax, mp 203-206°F	18.0	
Alcohols and ketones	4.3	12.6
Organic acids	Traces	1.0

Liquid products Fraction	C_5-C_{10}	C_{11}-C_{18}	C_5-C_{10}	C_{11}-C_{14}
Paraffins	45	55	13	15
Olefins	50	40	70	60
Aromatics	0	0	5	15
Alcohols	5	5	6	5
Carbonyls	Traces	Traces	6	5

Source: Bodle and Vyas 1974, p. 78. Reprinted by permission of the publisher.

In his paper dealing with the SASOL operation, Rousseau (1962) points out that the Fischer-Tropsch process could make economic headway only in an area in which there is a large differential between the cost of coal and price of oil.

ORNL DWG 76-6581

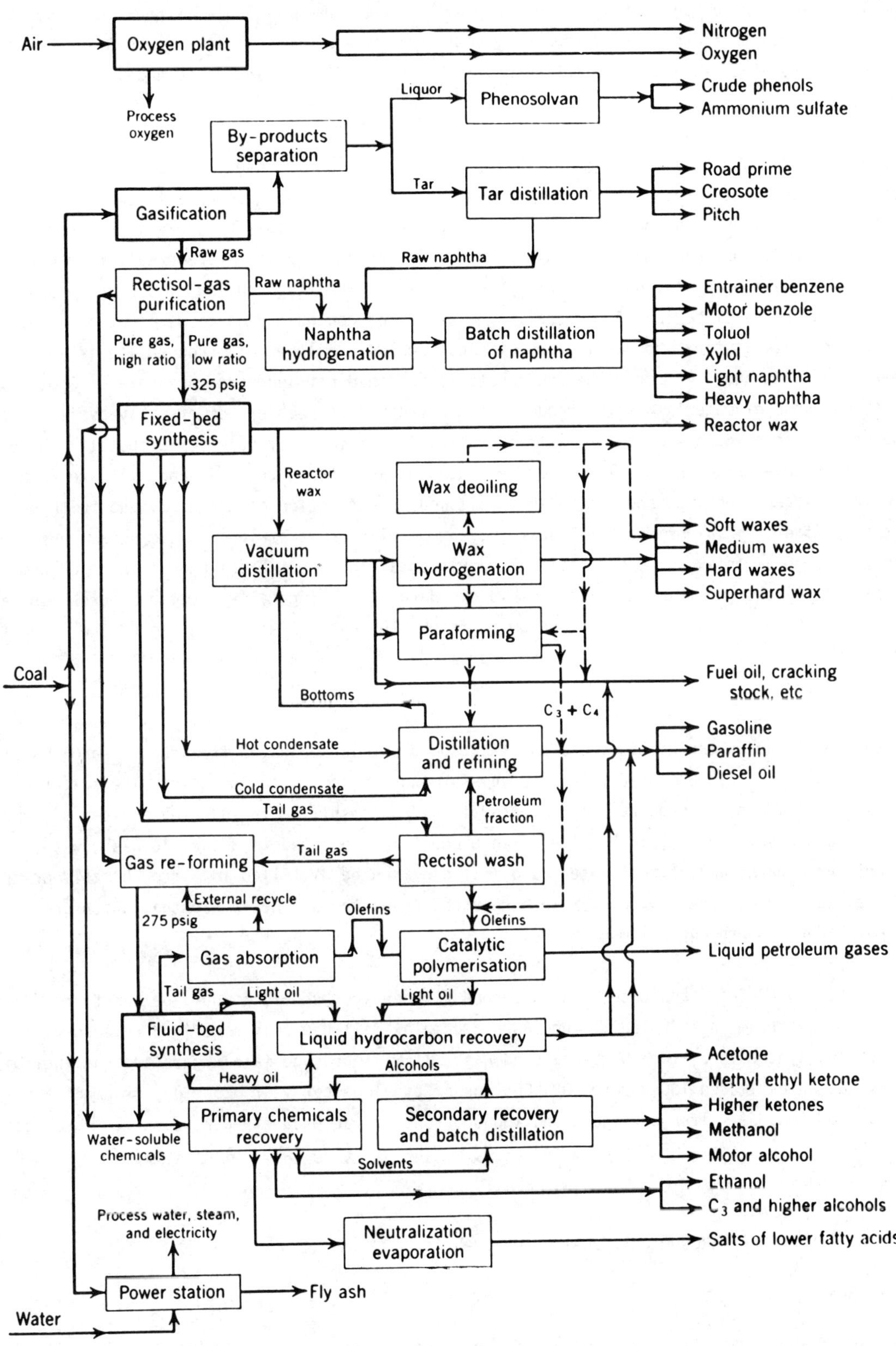

Fig. 3.43. Overall scheme of the SASOL plant. <u>Source</u>: From Hoogendoorn and Salomon 1957*a*, Fig. 3, p. 242. Reprinted by permission of the publisher.

Hydrocarbon products from the two synthesis units are separated in petroleum refinery units especially adapted to handle the Fischer-Tropsch product. Water-soluble oxygenated products from the two reactors are treated together; unsaturated gases (primarily propylene and butylenes) are treated together in a polymerization unit and converted into high-grade gasoline (Hoogendoorn and Salomon 1957*a*).

According to Hoogendoorn and Salomon (1957*a*), "an interesting feature [of the process] is the fact that although the recovery of ammonia and tar acids from the aqueous effluent is essential before this [effluent] can be discharged into the Vaal [River], this recovery is so organized that an erstwhile nuisance has been turned into a source of revenue." The greatest amounts of potentially environmentally unacceptable products result from the cooling of the gas produced in the pressure gasification of the coal. The stream of condensate contains light oils (almost entirely aromatic), tar, and gas liquor. Typical analysis of the total condensate is 98.3% water, 0.3% oil, and 1.4% tar. The condensate is obtained hot and under pressure of 300 psig. The three types of material in the condensate are treated to recover commercially available products and to produce a wastewater stream, which is sufficiently pure for disposal. Phenols are first removed from the gas liquor by the Phenosolvan method; then tar and oils are removed by gravity separation and sand filtration, followed by extraction of the gas liquor with butyl acetate. After removal of the butyl acetate, the gas liquor is stripped of ammonia, present mainly as free ammonia, and the purified liquor, containing about 2 ppm phenols and 300 ppm ammonia, is discharged from the plant. This discharge is biologically treated before disposal. The ammonia vapor is worked up to ammonium sulfate. The butyl acetate containing the extracted tar acids is distilled first at atmospheric pressure and then in a vacuum distillation tower. The butyl acetate is recovered and reused. Recovered phenols are sold in crude form.

Naphtha resulting in the Rectisol gas purification plant is separated from the methanol used in the cleanup and is hydrogenated to remove gum-forming substances and phenols. The sulfur compounds contained in the naphtha are converted to hydrogen sulfide in so doing and can be removed by an alkali wash. The liquid hydrogenation product is subjected to a caustic wash, water wash, sulfuric acid wash, and a final water wash. It can then be distilled to "motor benzol" or fractionated to give benzene, toluene, xylene, neutral creosote, and heavy naphtha, which are salable solvents (Hoogendoorn and Salomon 1957*d*).

In December 1974, SASOL announced that a second Fischer-Tropsch plant would be constructed and would come onstream in 1979-1981. The SASOL 2 complex will use only a refined Kellogg Synthoil process for synthesis of the Lurgi-produced gas. It will produce mainly gasoline and fuel oil, as well as an estimated 150,000 to 200,000 tons of ethylene per year, 250 tons of ammonia per day, 92,000 tons of sulfur per year, and 190,000 tons of tar products per year. The projected plant represents a scale-up, by a factor of 8, of the existing SASOL plant (Howard-Smith and Werner 1975).

3.3 IN-SITU OR UNDERGROUND GASIFICATION OF COAL

3.3.1 Goal

The goal of in-situ gasification of coal is to convert coal, without mining, into combustible gases by initiating combustion of a coal seam and injecting a current of air, oxygen, or oxygen and steam. In-situ combustion results in the production of an artificial or synthetic gas rich in carbon monoxide, carbon dioxide, hydrogen, hydrocarbon gases, and other gases, depending on the current injected and the method of combustion.

It is believed that carbon dioxide is formed by partial oxidation of the coal, whereas carbon monoxide is produced by reduction of the carbon dioxide. Hydrogen results from the water-gas reaction ($C + H_2O = CO + H_2$), and hydrocarbon gases are formed by carbonization of the coal (Magnani and Farouq Ali 1975).

Table 3.27 indicates the heating values of gas obtained from underground gasification as a function of the current injected for the combustion (Nadkarni, Bliss, and Watson 1974).

Table 3.27. Gas heating value from underground
coal gasification

Gasification agents	Heating value of product gas (Btu/scf)
Air	50 - 140
Oxygen	180 - 250
Steam cycle when using intermittent air-steam sequence	250 - 280
High-hydrogen gas produced for short periods when operating stream method without airflow	200 - 235

Source: Nadkarni, Bliss, and Watson 1974, Table 3, p. 234. Reprinted by permission of the publisher.

3.3.2 Background

As long ago as 1868, Siemens in Germany suggested gasification of slack and waste coal in the mine. In Russia Mendeleev suggested true gasification in 1888. A British patent granted to Betts in 1909 proposed gasifying coal underground by igniting coal at the base of one shaft or borehole, supplying it with air and steam, and withdrawing the gas formed through the same or a different borehole. In 1912 Sir William Ramsay in England suggested that coal be retorted underground to produce a gas containing hydrogen and carbon monoxide. This gas product was to be used in gas engines at the pit mouth to produce electricity (Elder 1963).

With government subsidization in the 1930s, the Russians tested different methods of in-situ gasification. At least one method was tried on an industrial scale. Outside Russia, most of the activity in underground gasification of thin seams occurred from 1945 to 1960. During this period work was carried out in England, Morocco, Belgium, Italy, Czechoslovakia, and the United States (Nadkarni, Bliss, and Watson 1975).

3.3.3 Advantages and disadvantages

The advantages of gasifying coal underground would be manifold. Not only could mining and its dangers be partially or fully eliminated, but the method would also be cheaper. Coal reserves would be increased because coal seams that were formerly inaccessible, unworkable, or uneconomical to mine could be put to profit; this potential is particularly great for low-grade coals such as lignites. Strip mining, with its accompanying environmental impacts, as well as the problems of spoil banks and acid mine drainage could be avoided. High-ash coal could be worked without great disturbance of the surface because burnt ash would remain as backfill.

Because of its low heating value, however, the gas would have to be either used directly at the production site for generation of electricity or upgraded to pipeline quality for economical transmission by pipeline over long distances.

3.3.4 Methods of in-situ gasification

Shaft systems and shaftless systems constitute the basic methods for underground gasification. Combinations of these systems are also possible. Selection of the method to be used depends on the permeability of the coal seam and the geology of the coal deposit; on the seam's thickness, depth, inclination, and tectonics; and on the amount of mining desired.

The shaft method involves driving large-diameter openings into the seam, a process requiring underground labor. Shaftless methods use boreholes for gaining access to the coal and do not require mining work.

3.3.4.1 Shaft methods

Chamber or warehouse method

Figure 3.44 shows the principles of the chamber method, in which coal panels are isolated with brickwork and underground galleries. This method, used by the Russians, relies on natural porosity of the coal for flow through the system. The panel was ignited at one side, and the gas produced was removed at the opposite side of the panel. Gasification and combustion rates were low, and the product gas had a variable composition, sometimes containing unconsumed oxygen. Variations of this method consisted of breaking the coal by hand or drilling a number of holes in the seam and charging them with dynamite. As gasification of the coal progressed through the seam, it was hoped that the coal would be crushed in advance of the reaction zone by a series of explosions. In practice, however, the Russians found that the explosions and coal breakage were irregular and that gas was obtained only intermittently. The gas contained unconsumed oxygen, large amounts of carbon dioxide, and little of the desired products (carbon monoxide and hydrogen). Almost complete combustion of the products occurred before they could be withdrawn. At times, however, a gas having a heating value of 145 to 200 Btu/ft^3 could be obtained (Elder 1963).

Borehole producer method

Further studies by the Russians led to the development of the borehole producer method (Fig. 3.45), in which parallel galleries, used for air inlet and product gas withdrawal, were constructed in the coal bed. Boreholes were drilled from one gallery to another about 5 yards

ORNL DWG 76-10174

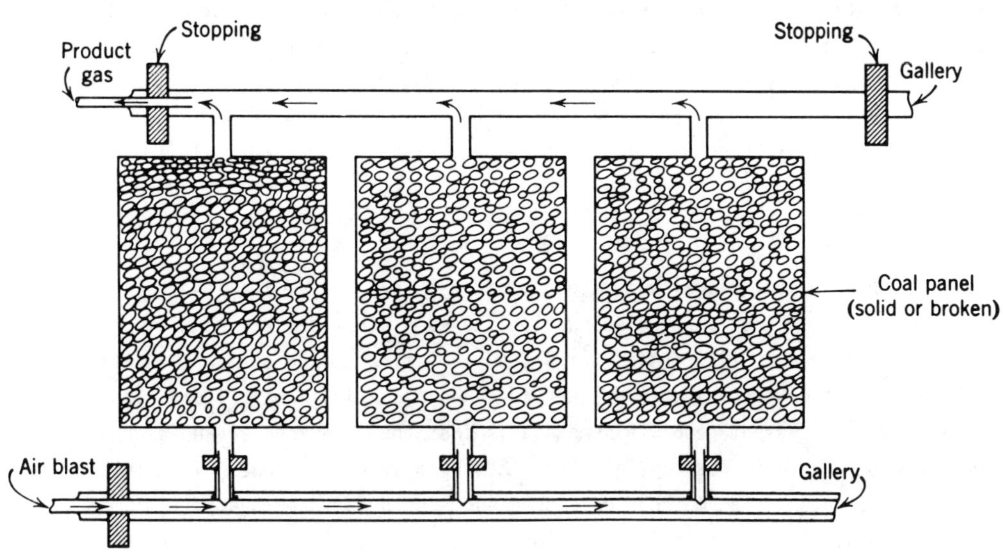

Fig. 3.44. Chamber method. <u>Source:</u> From Elder 1963, Fig. 1, p. 1024. Reprinted by permission of the publisher.

ORNL DWG 76-10172

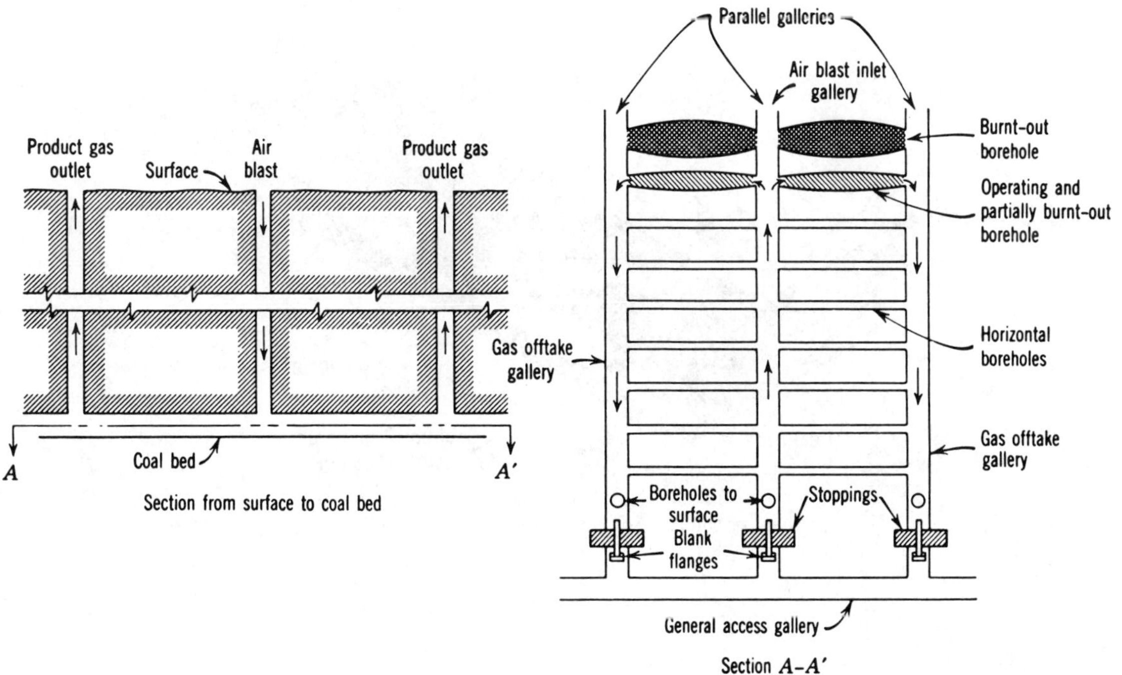

Fig. 3.45. Borehole producer method. <u>Source:</u> From Elder 1963, Fig. 2, p. 1025. Reprinted by permission of the publisher.

apart and were fitted with valves at their inlets and with iron seatings at their outlets. Electric ignition of the coal in each borehole was provided by remote control, and the process was managed so that the maximum amount of coal was burned from each hole and an approximately constant gas production rate was maintained (Elder 1963). This method is applicable to relatively horizontal coal beds, but because a large amount of underground labor is required, it is limited in application. The Russians used this method successfully in 1935.

<u>Stream method</u>

This method can be applied generally to steeply pitched coal beds. Inclined galleries following the dip of the coal seam are constructed parallel to each other and are connected at the bottom by a horizontal gallery or "fire drift" (Fig. 3.46). A fire in the horizontal gallery starts the gasification, which proceeds upward. Air comes down one inclined gallery, and gas departs through the other. An advantage of this method is that the fire zone advances upward, and the ash, together with any roof that may fall, collects below the fire zone. This method was the most successful of the Russian methods developed and continues to be one of the important systems for underground gasification of coal (Elder 1963).

ORNL DWG 76-10173

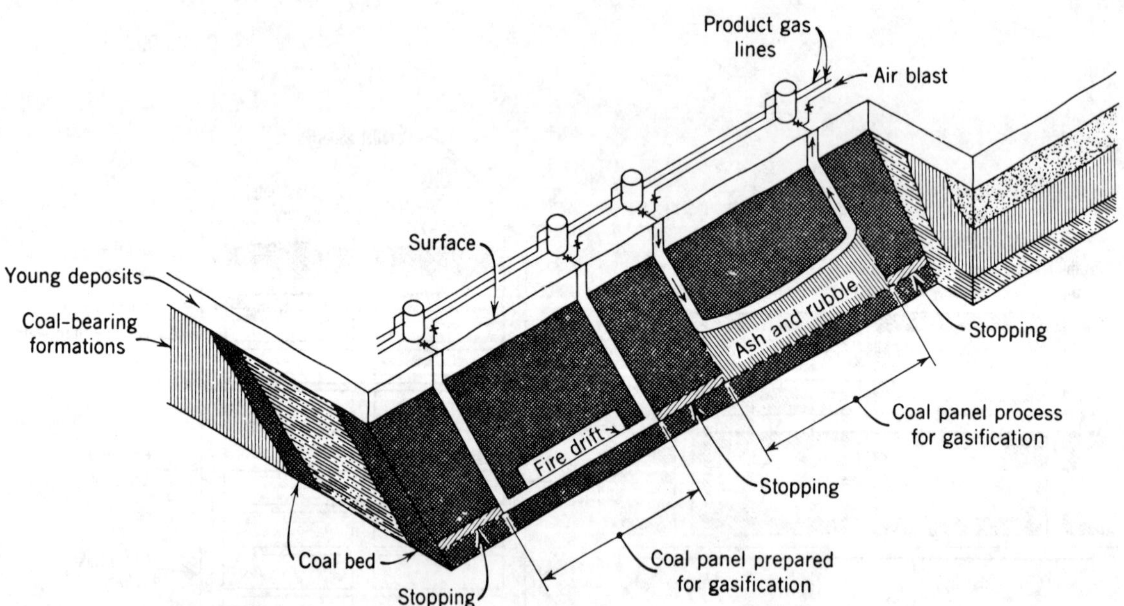

Fig. 3.46. Stream method. <u>Source:</u> From Elder 1963, Fig. 3, p. 1026. Reprinted by permission of the publisher.

3.3.4.2 <u>Shaftless methods</u>

<u>Percolation or filtration method</u>

One example of a shaftless system for underground gasification of coal is the percolation or filtration method, in which two boreholes are drilled from the surface, preferably through the coal seam. The distance between boreholes depends on the seam permeability. Air or air and steam are blown through one hole, and gas is removed from the second. Forward and reverse combustion are possible. According to Fischer, Brandenberg, and Schrider (1975), "Forward combustion is defined as combustion front and injected air movement in the same direction, whereas reverse combustion is defined as combustion front movement countercurrent to injected air movement." Figure 3.47 illustrates the percolation method.

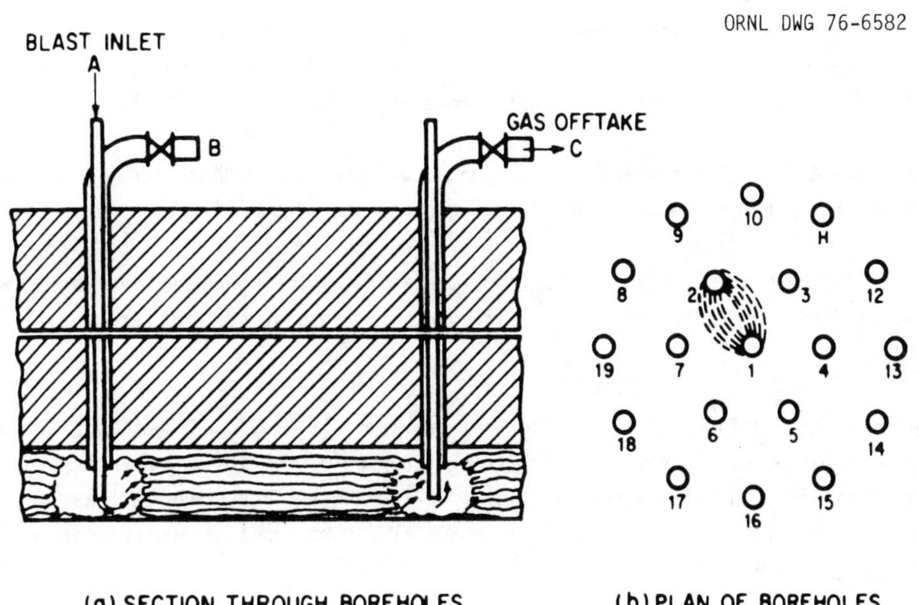

Fig. 3.47. Percolation method. Source: From Nadkarni, Bliss, and Watson 1975, Fig. 5, p. 635. Reprinted by permission of the publisher.

As burning progresses, the permeability of the seam increases. Compressed air blown through the seam also enlarges cracks in the seam. As combustion of the zone nears completion, the process is transferred to another pair of boreholes. Subsidence of the roof may occur and eventually reach the surface. To maintain a continuous gas supply, it is necessary to divide the area to be gasified into zones of different borehole pairs. Oxygen and steam may be added to fuel the fire and to improve the quality of the gas (Hucka and Das 1973).

Most lower-rank coals such as lignites have a considerable natural permeability and can be gasified without linking the boreholes underground; with higher-rank coals this linkage becomes necessary. When multiple boreholes are present, the inability to control gasification in a particular segment becomes a problem because of the parallel flows involved (Nadkarni, Bliss, and Watson 1975).

3.3.5 Current status of Russian work

The number of Russian papers appearing in the literature of in-situ gasification decreased after 1965. Until recently, it was thought that work in this area had been abandoned, possibly owing

to discovery of natural gas deposits. However, Russian work had continued, and several gasification plants are now commercial. Russian technology has been licensed by Texas Utilities
Services, Incorporated and by Resource Sciences Corporation (Arscott et al. 1975).

The licensed Russian technology may prove invaluable in future developments of underground gasification undertaken by Texas Utilities; it points out both successful and unsuccessful techniques.
The currently operating commercial installations in Russia, which are producing a low-Btu gas by
injection of air, are located in Angren, near Tashkent in the Uzbek Republic, and in Yuzhno-Abinsk
in Siberia. A former in-situ gasification site in Tula near Moscow now serves as a research and
development laboratory (R. Z. Mason, personal communication 1976; Gregg, Hill, and Olness 1976).

As noted in Table 3.28, which summarizes underground gasification methods, the Russian production
plants rely on the shaftless method, that is, no underground mining (Little 1972).

An overview of the Russian in-situ gasification literature by the Lawrence Livermore Laboratory
(LLL) has identified the significant factors in successful Russian systems (Gregg, Hill, and
Olness 1976).

Gas leakage from the burning zone is an important problem: It was found that operation of the
system at the lowest possible pressure and with a higher seam permeability between the injection
and exhaust boreholes than between the reaction zone and the surface prevented gas loss. The
Russians established highly permeable paths in the coal before gasification and sealed possible
surface cracks by filling with mud.

A high gas flow was found necessary both to compensate for the intrusion rate of water into the
seam and to achieve an economically high production rate. This factor is closely related to
the permeability between boreholes. Gas flow rates of 3000 to 10,000 m^3/hr at pressures not exceeding about 2.5 atm proved successful.

Directional control of the gas flow was achieved by using large, highly permeable linkage paths
at the bottom of the coal seams. These paths made the process independent of natural permeabil-
ities in the coal seams and permitted design of a predictable and reproducible system.

Because a high surface area is needed for the gasification reaction, the Soviet horizontal coal
seams were undercut by the flame front, allowing coal to fall into the void as rubble. In steeply
dipping beds, the coal falls into the cavity at the bottom where combustion was initiated.

Intruding water and pyrolysis tars produced by combustion of the coal can be controlled by forming
a channel at the bottom of the coal seam during the initial linking step. This channel collects
the liquids, which migrate to the bottom, and permits their removal by vaporization. In the
absence of a bottom channel, the liquids would collect at the bottom of the seam and cause the
flame front to move upward and across the top of the seam, forming a channel there. Coal
ash would then seal the bottom of the channel and hinder further combustion of the coal. Removal
of the liquids as vapor also prevents contamination of aquifers by the phenols formed in the
process; in addition, the recovered coal tars have an economic value.

Table 3.28. Underground gasification methods

Country	Location	Coal seam Type	Thickness (in.)	Depth (ft.)	Dip	CV (Btu/lb)	Technique	Type operation	Linkage	Pattern	Blast	CV, product gas (Btu/scf)	Remarks
							Shaft (underground development)						
Russia	Lisichansk	Bituminous	30	100	40	11,600	Streaming	Commercial	Gallery	Panels	30% O$_2$	86	
Russia	Gorlovka	Bituminous	72	200	75	10,000	Streaming	Commercial	Gallery	Panels	30% O$_2$	130	
Poland			39				Producer	Experimental	Drill holes	Parallel	Air	80	6% heat loss
U.S.	Gorgas	Bituminous	35	25	2	14,000	Streaming	Experimental	Gallery	U-shape	Air	47	High gas loss
Belgium	Bois-la-Dame	Semianthracene	36	525	87	14,000	Streaming	Experimental	Gallery	37-ft. panel	Air	56	35 - 45% thermal efficiency
							Shaftless (boreholes)						
U.S.	Gorgas	Bituminous	37	180	2	15,000	Percolation	Experimental	Electro	Square	Air	93	30 - 50% gas loss
											Oxygen	195	
											Water gas	279	
U.S.	Gorgas	Bituminous	37	180	2	15,000	Percolation	Experimental	Hydraulic	X pattern	Air	84	44% thermal efficiency
U.S.	Gorgas	Bituminous	37	180	2	15,000	Percolation	Experimental	Hydraulic	Straight line	Oxygen	124	
U.S.	Gorgas	Bituminous	37	180	2	12,000	Percolation	Experimental	Hydraulic	50-ft. circle	Air	90	40% gas loss
U.K.	Newman-Spinney	Bituminous	36	75	8	12,800	Streaming	Experimental	Boreholes	Straight line	Air	85	
U.K.	Newman-Spinney	Bituminous	36	100	8	12,800	Percolation	Experimental	Pneumatic	Rectangular	Oxygen	100	
Russia	Yushno Abinsk	Bituminous	276		75	12,000	Percolation	Commercial	Pneumatic	Inclined	Air	130	65% heat recovery
Russia	Lisichansk	Bituminous	30	100	40	11,600	Percolation	Commercial	Pneumatic	Inclined	Air	100	High pressure used
Russia	Moscow Field	Lignite	72	65	0	4,900	Percolation	Commercial	Pneumatic	75-ft. square	Air	85	
Russia	Shatsky	Lignite	120	150	0	4,900	Percolation	Commercial	Pneumatic	25-m grid	Air		
Russia	Stalinsk	Bituminous	98	1500	80	14,000	Streaming	Commercial	Pneumatic	Inclined	Air		
Russia	Tula	Lignite	390	180	0	4,900	Percolation	Commercial	Pneumatic	25-m grid	Air	105	
							Combination (shafts plus boreholes)						
U.S.	Gorgas	Bituminous	42	125	2	14,000	Streaming	Experimental	Gallery	Straight line	Air	70 - 90	4 - 40% gas loss
U.K.	Newman-Spinney	Bituminous	36	240	8	12,800	Streaming	Commercial	Boreholes	Parallel unit	Air	57	84% coal recovery
Morocco	Djerada			160	90		Streaming	Experimental	Gallery	Panels	Air		Combustible gas produced
Russia	Lisichansk	Bituminous	30	100	40	11,600	Streaming	Commercial	Boreholes	Panels	43% O$_2$	100	47% efficiency

[c]Proposed.

Source: Little 1972, Table 4, p. 38.

The Russian systems protect the major investment in the access pipes by placing them so that they are removed from the subsiding zone until they are no longer needed. This procedure permits pulling and reusing casings.

There appears to be no maximum limit in the thickness of the coal seams for which the processes are applicable. There does appear to be a lower limit of 3 to 4 ft, however, below which the heating value of the gas becomes too low. The method can also be applied to multiple layers of coal by working downward.

Under some circumstances the Russians found that the process could be shut down for several hours and started up again, an advantage for meeting peak load demands for power. A product gas composition having a constant heating value could be maintained.

The method has been applied in lignite, subbituminous, and bituminous coal seams and gives reproducible and predictable results within reasonable limits.

3.3.6 U.S. experiments in in-situ gasification

3.3.6.1 U.S. Bureau of Mines

The first experiments in the United States were conducted jointly by the U.S. Bureau of Mines and the Alabama Power Company from 1946 to 1959 in Gorgas, Alabama. The test area comprised about 200 acres and included three coalbeds of high-volatile A bituminous rank, whose thickness varied from 50 to 62 in. The overlying strata ranged from 128 to 178 ft in thickness.

According to Capp et al. (1961),

> the purpose of the ... experiments was to investigate the feasibility of the
> underground process and explore some of the techniques or methods of making
> a path through the coalbed for the introduction of combustion air and removal
> of product gases. Efforts to open a path or to increase the air acceptance
> of the coal included the use of the stream method — which is simply a passage-
> way mined in the coal — and the electrolinking-carbonization technique, in
> which an electric current carbonizes a path through the coalbed. Coal also
> was hydraulically fractured with a sand-oil suspension in an effort to in-
> crease its air acceptance.... In this hydraulic fracturing experiment the
> air-acceptance of the coalbed was increased to a maximum of 650 std. c.f.m.
> from prefracture values of 6 to 13 std. c.f.m. Moreover, the increased air
> acceptance lasted for at least two years and the effects of the fracture
> extended as far as 600 ft. away from the point of application. Successful
> gasification of the fractured coal demonstrated the feasibility of the
> method and the recovery of energy in the coal (as represented by the heat-
> ing value and quantities of the product gas) ranged between 5000 and 5800
> B.t.u. per pound of coal affected.

Sand suspended in kerosene was used for the first fracturing test; larger-grained sand suspended in a water-starch gel was used for the second. Results were comparable for the two fracturing media. These tests were conducted to determine (1) the extent of the area fractured by the fluid, (2) the air acceptance of the coal (air acceptance is defined as flow rate of air that can be injected into the coalbed at 65 psig), and (3) the rate at which a gasification path could be linked through the coal.

On the average, linkage rates of 5 ft per day were attained in attempts to link two boreholes with one another underground by reverse burning (Capp et al. 1961). During the gasification period, air was injected at about 1000 scfm, and temperature in the borehole was kept below 1350°F to prevent formation of slag in the borehole. Table 3.29 shows the average operating data for a period extending from 50 to 94.3 days after an ignition.

Table 3.29. Average operating data — U.S.
Bureau of Mines process

Component	Amount
Air input, thousand scfd	1440.0
Air injection pressure, psig	43
Gas produced, thousand scfd	1460.0
Recovery of air as gas, %	84
Analysis of product gas, %	
Carbon dioxide	12.4
Oxygen	0.5
Illuminants	0.1
Hydrogen	9.0
Carbon monoxide	10.3
Methane	1.9
Nitrogen	65.8
Specific gravity	0.951
Heating value, Btu/scfd	84
Coal utilization, tons/day^a	
Completely gasified	5.88
Carbonized only	4.38
Total affected	10.26
Energy recovered in product gas, million Btu/ft of path	27.1

^{a}Moisture- and ash-free coal.

Source: Capp et al. 1961, p. 21.

Although the process was found technically feasible, the products could not compete economically with natural gas at the time. The experiments were also hindered by high gas leakage rates and the availability of only thin seams. The program was eventually cancelled (Fischer, Brandenburg, and Schrider 1975).

Hanna, Wyoming test

Because of the increased interest in producing energy from sources other than petroleum and petroleum-derived products and also because the concept had been appraised as viable (Nadkarni, Bliss, and Watson 1974), the U.S. Bureau of Mines[*] revived its experimental work in underground gasification near Hanna in Carbon County, Wyoming, in 1972.

The Hanna No. 1 seam in the Hanna Formation was chosen for the field test. It is subbituminous in rank, lies 350 to 400 ft below the surface, and has a thickness of 26 to 30 ft. The Hanna Formation forms the floor of the Hanna Basin and is a near-surface aquifer. The seam itself lies about 180 to 200 ft below the water table; for this reason, high injection pressures were initially necessary to free the area from water (Campbell, Brandenburg, and Boyd 1974).

[*]The Hanna, Wyoming, work came under ERDA sponsorship in 1975.

Analysis of oriented core samples showed a permeability trend in the seam. A pattern of wells was drilled to maximize product gas recovery by intersecting this trend. Sixteen wells were drilled and cased with line pipe. The casing extended 4 ft into the coal seam and was cemented to the surface with heat-resistant cement. A smaller-diameter hole was drilled from the bottom of the casing to within 4 ft of the bottom of the coal seam, as shown in Fig. 3.48. The ignition injection wells had a similar construction (Fig. 3.49).

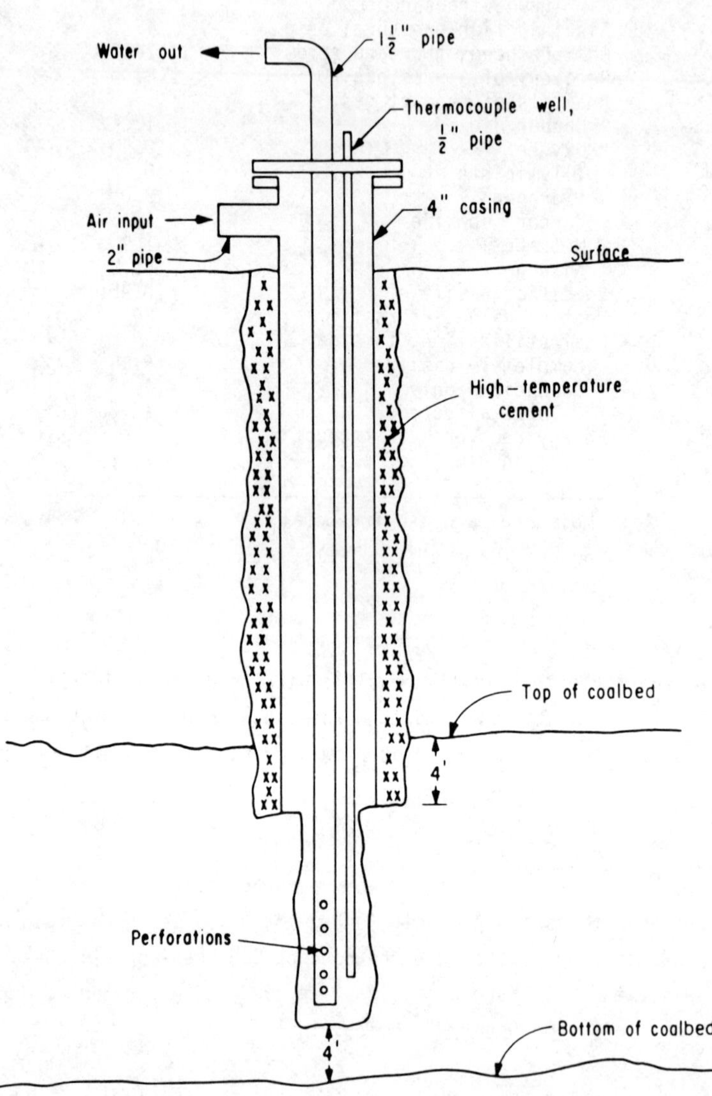

Fig. 3.48. Schematic of production wells. <u>Source:</u> From Campbell, Brandenburg, and Boyd 1974, Fig. 5, p. 7.

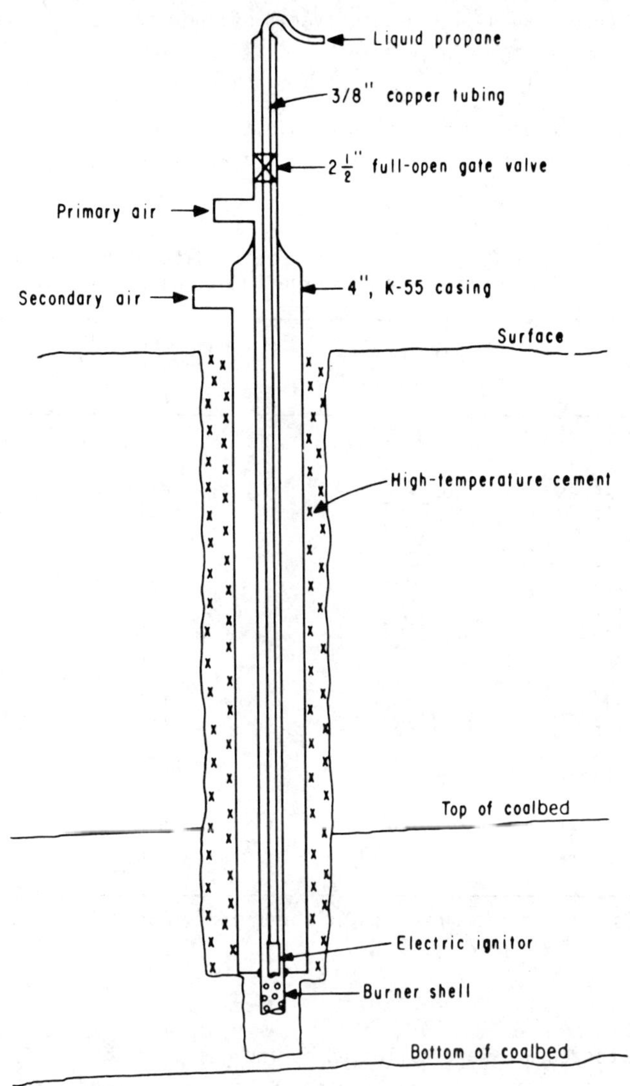

Fig. 3.49. Schematic of ignition injection well. <u>Source</u>: From Campbell, Brandenburg, and Boyd 1974, Fig. 6, p. 7.

Low air acceptance and high injection pressure of two wells made it necessary to fracture one of them hydraulically. After injection of water, followed by injection of gelled water containing sand, the well was flushed again with water. Air acceptance was increased fivefold.

Forward combustion, that is, combustion front movement and gas flow proceeding in the same direction, encountered difficulties. Presumably, coal tars produced in the combustion zone plugged the fracture system in the coal seam.

Reverse combustion (i.e., combustion front movement and gas flow moving in opposite directions) succeeded in creating underground linkage of several of the wells and produced gas having a heating value between 109 and 166 Btu/ft^3.

Table 3.30 indicates the average amounts of air injected and gas produced and the heating values of the gas during the 5-1/2-month operating period. Typical composition of the dry gas produced is listed in Table 3.31.

Table 3.30. Average operating data of underground
coal gasification experiment, Hanna, Wyoming

Period	Air injected (thousand scfd)	Gas produced (thousand scfd dry gas)	Heating value (Btu/scf)
1973			
September 16-30	907	1215	111
October 1-15	1054	1721	155
October 16-31	1055	1639	141
November 1-15	871	1367	135
November 16-30	1265	1824	124
December 1-15	1162	1724	124
December 16-31	1242	1798	113
1974			
January 1-15	902	1120	65
January 16-31	765	1010	96
February 1-15	1224	1981	165
February 16-28	1394	2224	157
Average value	1076	1602	126
	(30,450 Nm^3/day)	(45,340 Nm^3/day)	(1122 kcal/Nm^3)

Source: Fischer, Brandenburg, and Schrider 1975, Table 1, p. 285. Reprinted by permission of the publisher.

Samples of the liquid condensate were collected during gas production. This condensate consisted of about 90% water and 10% coal tars. Table 3.32 shows the gas composition of samples taken during periods of low and high gas production. The higher methane content of sample 1 indicated that carbonization was occurring, whereas the high content of carbon monoxide of sample 2 indicated that gasification conditions prevailed. Table 3.33 compares the boiling range distribution of the coal tars from the two samples with that for coal tar produced by laboratory carbonization of a Hanna coal sample at 500°C. Concerning the lack of residue, King, Brandenburg, and Lanum (1975) remark that

> while the underground gasification may be similar to aboveground gasification, the coal tars produced in underground gasification are carried to the surface as a steam distillate. Since very little heavy tar will reach the surface, under these conditions, very little if any residue is expected, and the coal tar will not be entirely representative of the total tar generated in the combustion zone.

Table 3.31. Typical produced dry gas composition,[a] underground coal gasification experiment, Hanna, Wyoming

Constituent	Mole percent
H_2	15.96
Argon	0.76
N_2	53.18
CH_4	3.91
CO	6.33
C_2H_6	0.39
CO_2	19.22
C_3H_8	0.13
C_3H_6	0.04
iso-C_4H_{10}	0.01
H_2S	0.07

[a]Heating value, 124 Btu/scf (1104 kcal/Nm^3).

Source: Fischer, Brandenburg, and Schrider 1975, Table 2, p. 285. Reprinted by permission of the publisher.

Table 3.32. Gas analyses on collection dates (mole %)[a]

Component	Sample 1 (Aug. 4, 1973)	Sample 2 (Dec. 10, 1973)
Hydrogen	9.5	16.14
Argon	1.17	1.01
Nitrogen	55.62	53.84
Methane	9.62	3.74
Carbon monoxide	0.80	6.94
Ethane	0.81	0.28
Carbon dioxide	21.90	17.91
Propane	0.16	0.08
Propene	0.12	0
n-Butane	0.01	0
iso-Butane	0.05	0
Hydrogen Sulfide	0.23	0.05
Heating value, Btu/scf	154	119

[a]Values are expressed in mole %, except for heating value.

Source: King, Brandenburg, and Lanum 1975, Table 1, p. 132. Reprinted by permission of the publisher.

Table 3.33. Boiling range distribution (%)

Sample	100 – 400°F	400 – 500°F	500 – 600°F	600 – 700°F	700 – 800°F	800 – 900°F	900 – 1000°F	Residue
1	30.0	46.7	22.0	1.3	0	0	0	0
2	6.2	16.9	25.6	28.2	16.0	5.3	1.8	0
Carbonized	0	11.3	16.3	13.1	15.2	12.4	7.5	24.2

Source: King, Brandenburg, and Lanum 1975, Table 2, p. 133. Reprinted by permission of the publisher.

The coal tars were separated into tar bases, tar acids, and neutral fractions. Samples 1 and 2 were composed of the following weight percentages:

	Sample 1	Sample 2
Tar base	4.0	8.7
Tar acid	42.1	14.5
Neutral	53.9	76.8

There was 90.9% recovery of sample 1 and 96.4% recovery of sample 2. Mass spectral analysis of the sample 1 tar acid fraction showed a high content of substituted phenols (King, Brandenburg, and Lanum 1975).

Further analysis of the fractions by coulometric determination and nonaqueous titration showed that the various fractions contained total nitrogen, total sulfur, and titratable nitrogen as noted in Table 3.34 (King, Brandenburg, and Lanum 1975). Examples of weak bases would be pyridines or quinolines. Amides and primary and secondary anilines titrate as very weak bases. No further analysis of the acid and neutral fractions of sample 2 was reported by King, Brandenburg, and Lanum (1975).

Table 3.34. Total nitrogen, total sulfur, and titratable
nitrogen for acid, base, and neutral fractions (%)

Component	Total sulfur	Total nitrogen	Titratable nitrogen[a]
Sample 1			
Tar base	0.00	9.64	8.10 (WB), 1.36 (VWB)
Tar acid	0.33	0.35	0.264 (WB), 0.081 (VWB)
Neutral	0.02	0.08	0.035 (VWB)
Sample 2			
Tar base	0.23	4.82	4.45 (WB), 0.36 (VWB)
Tar acid	0.14	0.09	0.058 (VWB)
Neutral	0.20	0.37	0.037 (WB), 0.145 (VWB)

[a]WB and VWB refer to weak and very weak base.

Source: King, Brandenburg, and Lanum 1975, Table 7, p. 135. Reprinted by permission of the publisher.

By means of nuclear magnetic resonance, mass spectrometry, ultraviolet, and gas-liquid chromatography data, specific compounds or compound types were identified from the tar base fraction
of sample 1 (Table 3.35); 46.6% were substituted pyridines, 13.4% were anilines, 9.1% were
quinolines, and the remainder were probably substituted pyridines.

Table 3.35. Compounds identified from the
tar base fraction of sample 1

Component	Percent total of base fraction	Compound type or compound
1	1.18	Pyridine
2	6.40	2-Picoline
3	6.20	2,6-Lutidine
4	9.61	3-Picoline, 4-picoline, and 2-ethyl pyridine
5	12.90	2,4-Lutidine, 2,5-lutidine, plus some methyl ethyl pyridine
6	3.22	2,3-Lutidine
7	7.11	Trimethyl pyridine and an ethyl pyridine
8	5.98	Aniline
9	4.83	2-Methylaniline
10	2.54	A dimethyl aniline and an ethylaniline
11	4.08	Quinoline, a dimethyl aniline, and a trimethyl aniline
12	5.02	A methyl quinoline, a C_3 aniline; quinoline, a dimethylquinoline

Source: King, Brandenburg, and Lanum 1975,
Table 8, pp. 136-37. Reprinted by
permission of the publisher.

Analysis of the base fraction of sample 2 suggested that it contained alkylated anilines or
quinolines and substituted quinolines.

Because of the difference in the compounds formed under gasification and carbonization conditions (and therefore under different temperature conditions), future temperature monitoring of
the combustion zone in the Hanna experiments should allow more definite conclusions to be drawn
about the conditions under which the coal tars are formed.

King, Brandenburg, and Lanum (1975) note that "possible herbicidal properties of components identified would require precautions against spillage to minimize environmental impacts. Monitoring underground migration of these organic fluids to determine what effect they may have on water supplies is being studied."

According to calculations by Fischer, Brandenburg, and Schrider (1975), the average dry gas production of 45,300 Nm^3/day with an average heating value of 1122 kcal/Nm^3 would have been sufficient to generate about 1 MW of power at 40% efficiency.

The portion of liquid organic materials carried along with the product gas stream was condensed with an air condenser. An estimated 8300 liters of liquids having an estimated heating value of 9300 kcal/liter were produced per month.

Fischer, Brandenburg, and Schrider (1975) calculate an energy return rate of 3.5 to 3.6 on the basis of energy input and output (Fig. 3.50). Use of more efficient air compressors and a flaring system designed for low-Btu gas combustion could increase energy return to 8 and raise the overall energy efficiency of the process from 50 to 55%.

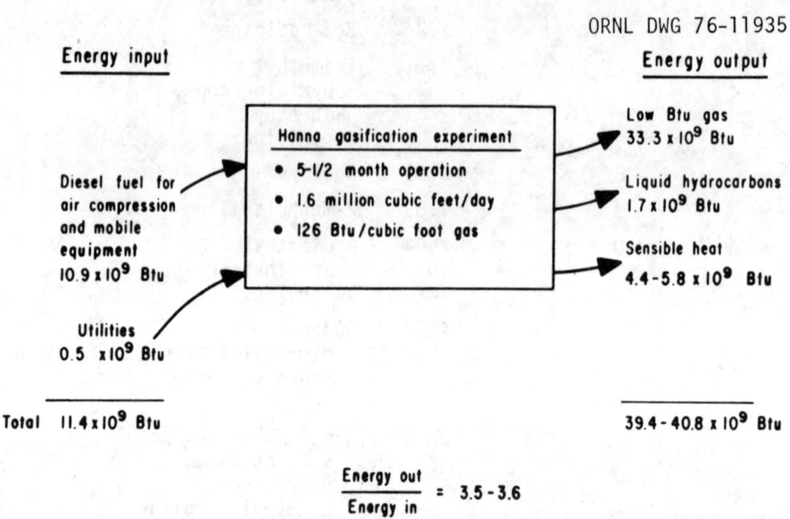

$$\frac{\text{Energy out}}{\text{Energy in}} = 3.5 - 3.6$$

Fig. 3.50. Energy return ratio. Source: From Fischer, Brandenburg, and Schrider 1975, Fig. 6, p. 288. Reprinted by permission of the publisher.

This first experiment showed that sustained production of gas having a stable heating value can be achieved.

The second Hanna experiment is to be conducted in three phases. The first phase is designed to determine (1) whether the natural permeability directions of the coal have a major influence on the preferred flow direction, (2) whether permeability could be increased by simply injecting high-pressure air rather than burning a pathway between wells, and (3) whether changes in process design based on experience from the first Hanna experiment would bring improved results. This phase was recently completed (Schrider and Fischer 1976).

Oriented core studies from an area located about 700 ft west of the Hanna I experiment indicated a preferred flow direction, as shown in Fig. 3.51, from well 1 to 2 and a minor flow from well 1 to 3. However, high-pressure air injected at well 1 produced the greatest air production at well 3. Well 1 was subsequently used as the ignition well, and air was injected to ensure sustained combustion. Air injection was then carried out at well 3 to achieve reverse combustion between wells 1 and 3. Linkage was achieved, and gasification was conducted for 38 days. During this period, gas having an average 152 Btu/scf (1353 kcal/Nm^3) was produced at an average rate of 2.7 million scfd (76,500 Nm^3/day). Air injection rates averaged 1.9 million scfd (53,800 Nm^3/day).

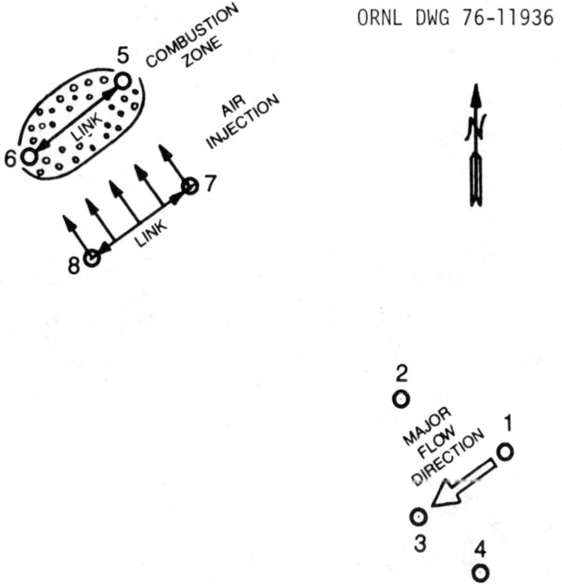

Fig. 3.51. Well pattern for Hanna experiment; wells are about 60 ft (18 m) apart. Source: From Schrider and Fischer 1976, Fig. 4, p. 23. Reprinted by permission of the publisher.

Phases 2 and 3 of the second Hanna experiment will use wells 5 through 8 (Fig. 3.51). Wells 7 and 8 will be linked by reverse combustion; wells 5 and 6 will be linked by reverse combustion and gasified. Then air will be injected simultaneously into wells 7 and 8 in an attempt to produce a broad linkage between the 5-6 and 7-8 well pairs. This is known as a "line-drive" system. If this method is successful, it will provide the basis for further testing through pilot and demonstration plant phases to yield an evaluation of underground coal gasification feasibility.

Already plans are being drawn up for FY 1977 for a further underground design test about 14 times larger, Hanna III, which is to gasify about 60,000 tons of coal in a 6-month period and which should produce about 45 million scfd (L. Schrider, personal communication, April 20, 1976).

Sandia Laboratories has recently received $676,000 from ERDA to develop instrumentation and process controls for in-situ processing of coal, oil shale, and other fossil fuels. Sandia will monitor the experiments at the Hanna test site and also will support the LLL project. The monitoring is expected to result in a process control system (Sandia Labs gets ERDA coal monitoring project 1975).

Morgantown, West Virginia Energy Research Center[*]

The Morgantown Energy Research Center (MERC) is investigating the "longwall generator" in-situ coal gasification concept. Directional holes are drilled from and back to the surface horizontally through the coal seam and perpendicular to the maximum permeability direction. The reaction zone moves horizontally through the seam between horizontal bore holes. It is hoped that the method will produce better gas flow, possibly without the need for preliminary fracturing of the seam. The tests are to be conducted in a 6-ft-thick Pittsburg seam at a depth of 800 to 1200 ft (Arscott et al. 1975). An idealized view of the process is shown in Fig. 3.52 (Komar, Overbey, and Pasini 1973).

ORNL DWG 76-10170

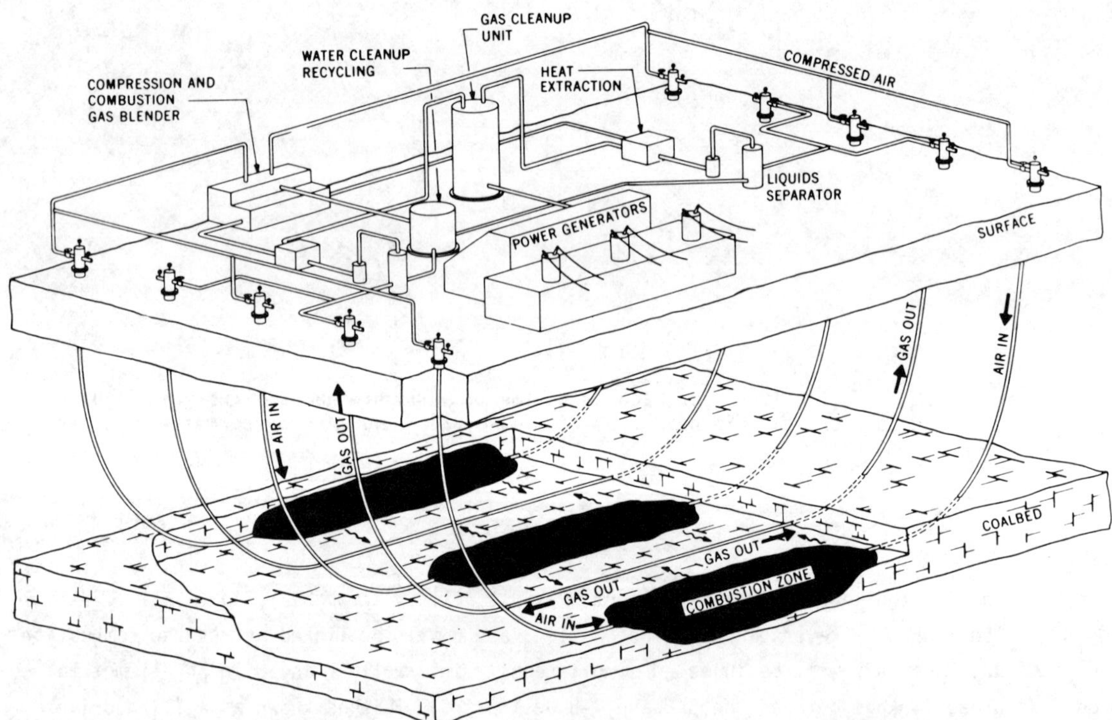

Fig. 3.52. Proposed longwall generator concept for underground gasification of coal. Source: From Komar, Overbey, and Pasini 1973, Fig. 7, p. 7.

---

[*]Although the Morgantown experimentation was initiated by the U.S. Bureau of Mines, it came under ERDA sponsorship in 1975.

A comprehensive analytical scheme for characterization of coal core samples, which also includes determination of the properties important in the burning of coal, has been developed by MERC. This scheme will aid in evaluating sites for underground gasification. Coal cores from the Hanna, Wyoming, underground gasification site, in addition to cores from an eastern site near Pricetown, West Virginia, were analyzed. The lateral continuity in chemical and mineralogical composition for both sites indicates that the stratigraphic variations are systematic and thus predictable (Estep-Barnes and Kovach 1975).

3.3.6.2 Lawrence Livermore Laboratory, University of California

The Lawrence Livermore Laboratory proposal for in-situ gasification of coal is aimed at the production of high-Btu gas from deep thick coal seams, located between 600 and 3000 ft below the surface (Fig. 3.53).

ORNL DWG 76-10169

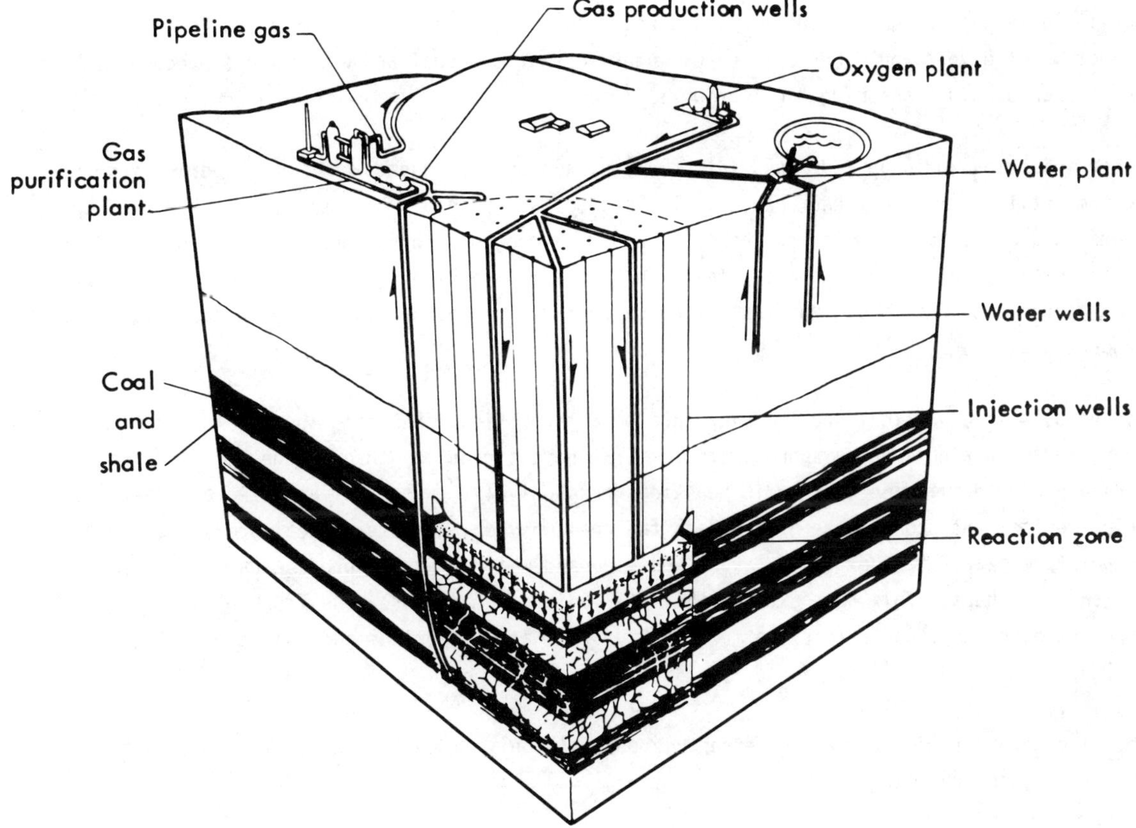

Fig. 3.53. In-situ coal gasification concept. Source: Arscott et al. 1975, Fig. 1, p. 3.

According to calculations made by Gulf Research & Development Company in evaluating the LLL method, there are about 84×10^9 tons of coal in the United States which could possibly be gasified by the packed-bed concept. Assuming that 30% of the coal is left to form the walls of the underground gasifier and assuming a sweep efficiency within the gasifier of 50%, Arscott et al. (1975) predict that, if the combined efficiency of the in-situ and surface upgrading methods is 40%, as in the Lurgi process, then the gas produced would be about equal to the present proven reserves of natural gas, or an 11-year supply at the current consumption rate.

In the LLL method, the coalbed would first be fractured in place by using explosives, such as an ammonium nitrate–aluminum–diesel oil mixture. Only the region to be processed would be broken; thus, the fractured coal seam would resemble a packed-bed reactor. Subsequent blasting would be carried out to ensure that fractures did not reach a previously processed region. After fracturing, the next step would be to drill and case access holes to the top and bottom of the fractured seam. Casing the holes would prevent entry of water. The third step would be to inject oxygen and start combustion at the top of the broken zone. By restricting outlet flow until pressure in the zone is equal to the hydrostatic potential in the coal-bearing zones, there should be little leakage of gas or entry of unwanted water. At a temperature of about 700°K (427°C), injection of oxygen would be replaced by water so that the coal and water would produce methane and carbon dioxide, assuming equilibrium is reached.

Water injection would be continued into the top, and product gases would be withdrawn from the bottom until all the coal had reacted. The product gases should be a mixture of methane, carbon dioxide, and water vapor with traces of nitrogen, hydrogen, and carbon monoxide. After cleanup and removal of water and carbon dioxide, a high-quality pipeline gas should remain. If, however, underground methanation did not take place, a surface methanator could be used to upgrade the product gas.

Figure 3.54 is a vertical section view through a theoretical coal zone showing the chemical reactions occurring and the approximate temperature distribution through the bed; $+\Delta H$ and $-\Delta H$ indicate endothermic and exothermic reactions respectively. Water is vaporized and heated in the inlet region. As soon as the gases reach the coal, oxygen and water react, producing the high-temperature peak. Farther down, carbon monoxide and water react to produce carbon dioxide, methane, and heat. This heat extends the intermediate temperature zone farther and farther down the high-temperature reaction front. Finally, water vaporizes and condenses (Higgins (1972).

The direct conversion of coal to methane is possible only if enough oxygen is supplied to provide the necessary heat by way of the reaction,

$$\text{Coal} + O_2 = H_2O + CO_2 \quad , \tag{27}$$

which is a highly exothermic reaction.

Certain problem areas of the process must be considered. There could possibly be uncontrollable entry of water into the coal zones, or the coal might not be sufficiently permeable to gases after fracturing. There could also be surface subsidence on the order of 75 to 90% of the thickness of the coal bed. Considerable amounts of carbon dioxide would be released.

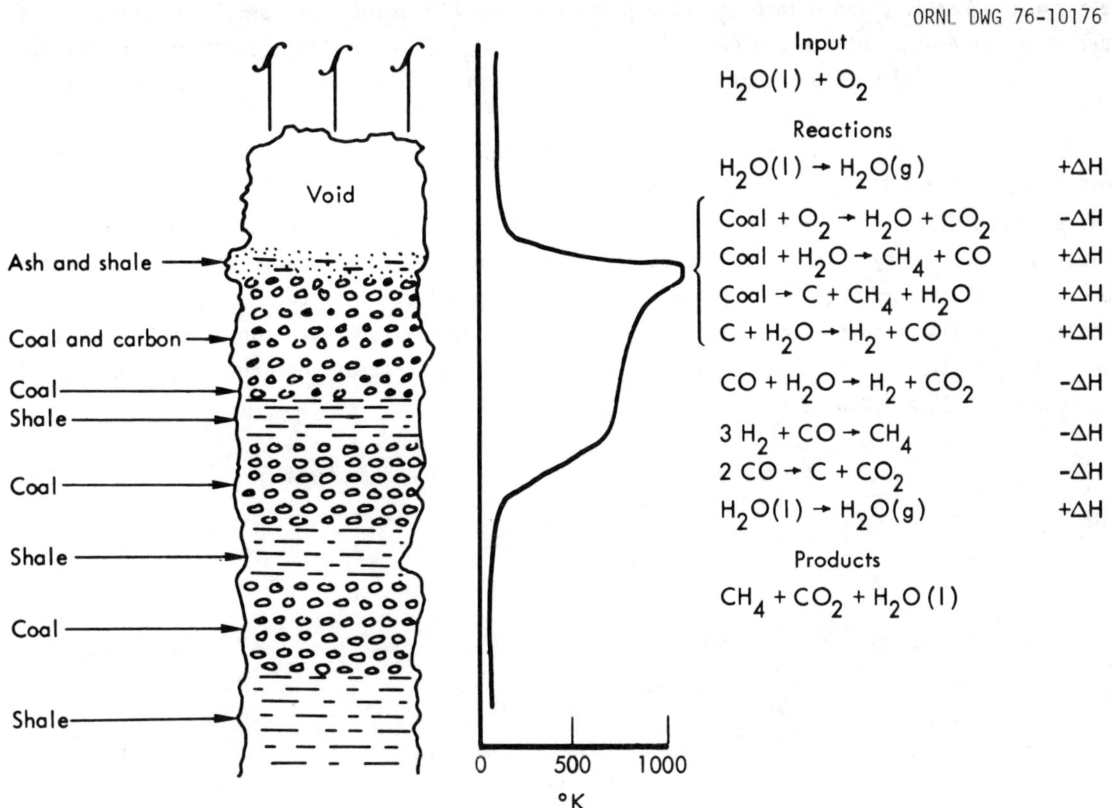

Fig. 3.54. Chemical reactions and temperature gradient along a vertical section of the reacting coalbed. <u>Source</u>: From Higgins 1972, Fig. 4, p. 12.

On the other hand, the method would not release sulfur-containing gases or fly ash. Waste heat would remain underground. Little or no high-quality water would be needed because water appearing as liquid condensate in the bottom of the broken zone could be recycled through the process (Higgins 1972).

Since 1972, when the LLL proposal was made, a research and development program has been directed at developing the packed-bed method for in-situ high-Btu gas production. Laboratory and field work is intended to determine (1) whether suitable deposits exist for in-situ gasification, (2) whether adequate permeability can be produced by explosive fracturing and maintained during gasification, (3) whether the reaction zone can be maintained and controlled, and (4) whether the process can be applied commercially.

The research and development work consists of a laboratory coal fracturing program, an experimental coal processing program, and a pilot plant field gasification program. The coal fracturing program aims at developing a method for predicting the extent and distribution of explosive-induced fractures in coal and at successfully demonstrating the method. The coal processing program is intended to develop sufficient understanding of in-situ coal gasification to be able to control the reaction and analyze and interpret results of in-situ gasification experiments. Field gasification tests are included in this program.

The field program is divided into six sequential phases: (1) exploratory drilling, (2) site
characterization, (3) shallow and deep site characterization, (4) multiple explosive fracturing
and gasification, (5) multiple shot fracturing, and (6) in-situ gasification and demonstration.

The site chosen for the first shallow field fracture experiment was at Hoe Creek, Campbell County,
Wyoming, where the Felix No. 2 coal seam is about 25 ft thick at a depth of about 118 ft.
Original plans called for a five-hole pattern, consisting of a central injection well and four
collection wells arranged in a square pattern as shown in Fig. 3.55. The intention was to fracture
the central well, along with each of four other explosive holes, with about 1500 lb of chemical
explosives. Coal was to be ignited in the center explosive well and burn concurrently downward
with an oxygen-steam blast. Dewatering of the coal would be necessary before and during gasification (Gregg, Hill, and Olness 1976).

ORNL DWG 76-6583

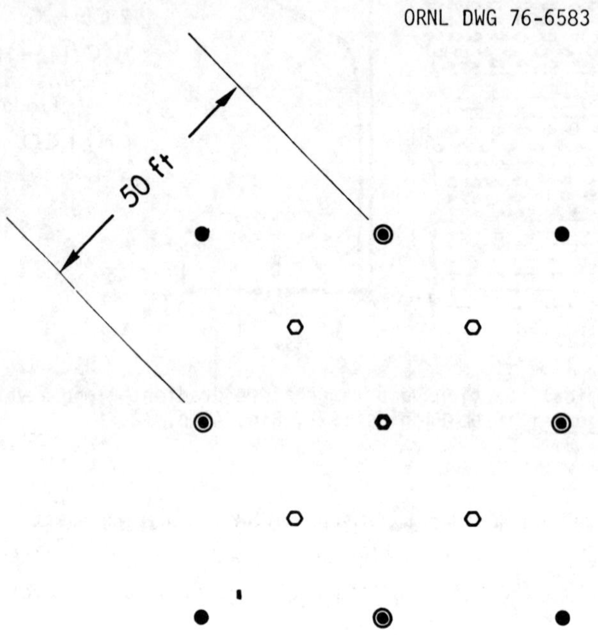

Fig. 3.55. Plan view of the packed-bed experiment. Source: From Stephens and Lentzner
1975, Fig. 4, p. 6.

In September 1975 it was decided that a simple fracturing and postshot characterization should
be performed before fracturing the five-well pattern. This was done by detonating 750 lb of
DuPont EL-836 explosive (a heavily aluminized water slurry based on ammonium nitrate) in the
injection hole (INJ) and in the high explosive hole (HE) (Fig. 3.56). Postshot hydrological and
gas flow data were inconsistent with the predicted calculated permeability. Gas flows were
lower than predicted. Gas flow from the injection well to the production well was essentially
zero. Drilling operations for Hoe Creek Experiment No. 2 were suspended December 2, 1975, due
to the unexpected permeability data obtained from Experiment No. 1. This data will be analyzed
before Experiment No. 2 is carried out (Gregg, Hill, and Olness 1976).

ORNL DWG 76-10175

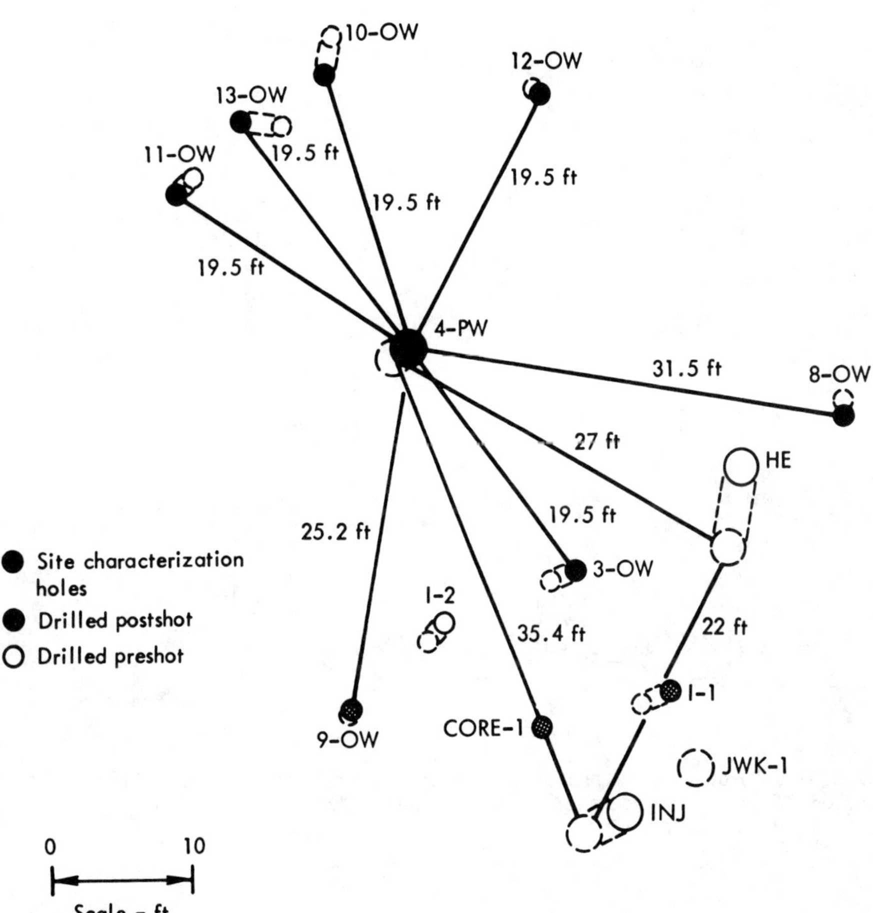

Fig. 3.56. Hole locations in Experiment No. 1. Dashed circles indicate positions of hole
bottoms. All wells drilled preshot, except 4-PW, had 0.10-m(4-in.)-diam PVC casing. Well 4-PW
was modified preshot by adding a 0.18-m(7-in.)-diam casing from the surface to within 1.5 m (5 ft)
of the bottom of the Felix No. 2 coal seam. Source: From Stephens and Lentzner 1976, Fig. 1,
p. 4.

3.3.6.3 Gulf Research & Development Company

In 1968-1969 the Gulf Research & Development Company carried out an underground combustion of coal
in a seam being strip-mined in Western Kentucky. The coal was a 9-ft-thick seam of No. 14 bituminous coal at a depth of 107 ft. Each of two test patterns located 100 ft apart contained
an injection well, a temperature observation well, and a sampling well perpendicular to the strip
mining wall (Fig. 3.57). The intention, after ignition of the coal, was to inject air into the
first pattern and to inject air and water simultaneously into the second to gasify the coke.

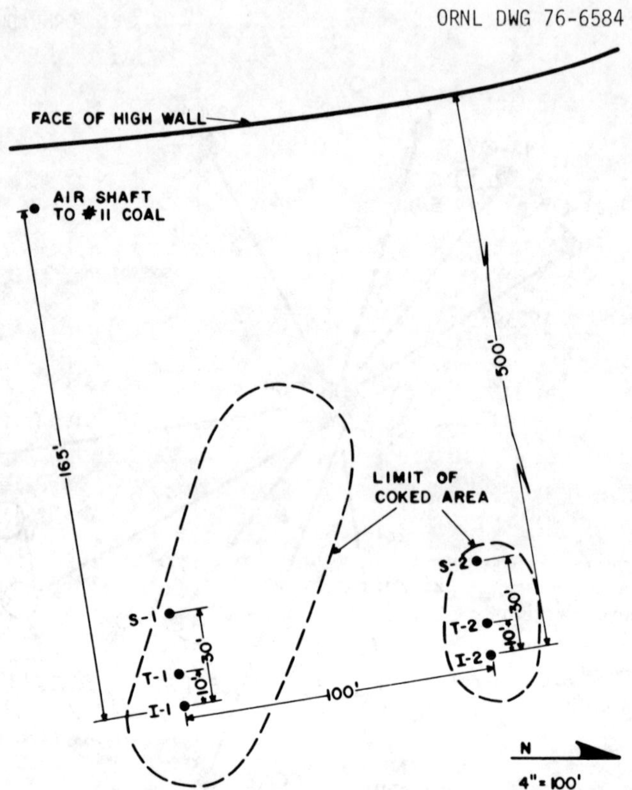

Fig. 3.57. Colonial mine project. Source: From Raimondi, Terwilliger, and Wilson 1975,
Fig. 1, p. 36. Reprinted by permission of the publisher.

After ignition of well I-1, the air injection rate gradually rose to 155,000 cfd under a
constant pressure of 310 psig. A rate of 1.2 million cfd obtained at 350 psig could be maintained while the pressure gradually decreased to 130 psig. Similar conditions prevailed in
well I-2. When the experiment began to interfere with mining operations, air injection into the
wells was stopped. The injection of air and water together could not be carried out. Injection
of cooling water began after air injection stopped. The shape of the burned-out void in the
seam was preserved by injection of cement so that the pattern could be retained and observed by
uncovering the test site. On removal of the overburden, a swelling above the test sites, amounting to about 3 ft near the injection wells and then tapering off, was noted.

Examination of the coke and coal showed that the volatile matter content in the coke or char was less than 10% and that of the coal next to the coke was 30 to 35%. About 50 ft downstream from the coked area, however, the coal had a volatile matter content about 5% higher than that of the original coal. It was thought that the additional volatile matter could have come from the retorting zone.

Flammable gases, the average heat content of which was 270 Btu/ft^3, were produced throughout the test. Gas samples were also analyzed for their composition during the experiment. Table 3.36 shows the analytical results (Raimondi, Terwilliger, and Wilson 1975).

The considerably higher heat content of the gas produced, in comparison with that from other shaftless systems, was explained by the fact that a frontal-like displacement had occurred; that is, the oxygen of the injected air was completely consumed at the rear of the front. Hot combustion gases swept through and devolatilized the coal by retorting.

Conclusions drawn from this experiment were that an injection pressure of about 30% above overburden pressure made it possible to maintain the desired air injection rate and sustain combustion in a coal seam. It was also speculated that the hot tar and hydrogen produced aided in solubilizing the coal ahead of the combustion front and increasing the yield of liquid products.

When invitations were sought for industry participation in the LLL coal gasification program, Gulf Research & Oil Company cooperated in an interchange of technical data with LLL. The review of the LLL in-situ gasification program that resulted recommended that the program be continued, with emphasis on enhancing permeability of the coal seam and on studying the problems related to subsidence. Use of hydraulic and pneumatic fracturing techniques, as well as explosive fracturing, was also suggested.

Table 3.36. Composition and gross heat content of produced gas[a]

Identification															
Date, 1968	10/14	10/14	10/15		10/16		10/17		10/28		10/31	11/11		11/21	
Cumulative air injected in I-1, million ft^3	0.07	0.14	0.17		0.20		0.23		1.37		1.97	8.85		19.46	
Well number	S-1	S-1	S-1	S-2	S-1	S-2	S-1	S-2	S-1	S-2	S-1	S-1	S-2	S-1	S-2
Component, vol %															
Benzene	0.14	0.00	0.12	0.00	0.24	0.00	0.24	0.00	0.19	0.00	0.12	0.31	0.00	1.62	0.17
Hydrogen	13.23	11.16	8.66	8.74	8.74	8.48	9.61	8.56	9.40	9.82	9.03	15.68	11.8	18.61	10.90
Methane	24.21	16.22	19.52	17.16	25.53	28.59	20.59	27.42	14.75	15.6	13.25	17.56	4.08	23.62	6.71
Ethane	1.32	0.71	0.95	0.64	1.26	0.81	0.89	0.96	0.68	0.78	0.63	0.52	0.26	0.86	0.31
Propane	0.27	0.17	0.26	0.18	0.27	0.17	0.22	0.25	0.21	0.14	0.22	0.14	0.16	0.22	0.17
Butanes +	0.28	0.35	0.11	0.14	0.56	0.16	0.38	0.23	0.11	0.75	0.14	0.00	0.00	0.47	0.00
Carbon monoxide	7.62	8.33	6.51	7.66	4.16	4.84	6.32	5.47	7.05	7.16	7.38	15.13	13.26	19.8	6.93
Carbon dioxide	10.43	11.76	12.85	1.94	13.22	3.67	12.99	5.04	14.34	12.83	15.02	7.43	12.88	13.65	19.08
Nitrogen	41.86	50.65	50.44	62.7	45.55	52.55	48.15	51.38	52.36	52.26	53.28	41.38	56.44	18.69	59.89
Argon	0.54	0.64	0.63	0.83	0.58	0.73	0.61	0.68	0.65	0.66	0.67	0.53	0.74	0.19	0.72
Oxygen	0.11	0.00	0.00	0.00	0.00	0.00	0.00	0.00	0.10	0.11	0.15	0.00	0.00	0.00	0.00
Ammonia	0.00	0.00	0.00	0.00	0.00	0.00	0.00	0.00	0.00	0.00	0.00	0.27	0.20	0.00	0.00
Hydrogen sulfide	0.00	0.00	0.00	0.00	0.00	0.00	0.00	0.00	0.00	0.00	0.00	1.04	0.14	2.64	0.00
Other sulfur compounds	0.00	0.00	0.00	0.00	0.00	0.00	0.00	0.00	0.16	0.00	0.00	0.00	0.00	0.00	0.00
Heat content, Btu/ft^3	355	253	274	244	356	351	301	350	229	257	212	299	130	465	140

[a]Water-vapor-free.

Source: Raimondi, Terwilliger, and Wilson 1975, Table 2, p. 37. Reprinted by permission of the publisher.

LITERATURE CITED

Abrams, R. N.; Ahn, Y. K.; and Burton, D. C. 1976. The why, when and how of coal gasification. Paper presented at the American Nuclear Society Topical Meeting, Environmental Aspects of Non-Conventional Energy Resources, Denver, Colorado, Feb. 29-Mar. 3. Reading, Pa.: Gilbert/Commonwealth.

Akhtar, S.; Friedman, S.; and Yavorsky, P. M. 1974. Low-sulfur liquid fuels from coal. *Energy Sources* 1(4): 423-34.

American Society of Testing Materials. 1965. *Standard method of test for Conradson carbon residue of petroleum products.* ASTM D 189-65.

American Society for Testing Materials. 1967. *Standard method of test for density, specific gravity or API gravity of crude petroleum and liquid petroleum products by hydrometer method.* ASTM D 1298-67.

American Society for Testing Materials. 1971. *Standard specifications for gasoline.* ASTM D 439-71.

Anderson, R. P. 1975. Recycling solvent techniques for the SRC process. *Chem. Eng. Prog.* 71(4): 72-74.

Arscott, R. L.; Fair, J. C.; Garon, A. M.; Raimondi, P.; Trump, R. P.; Bidlack, D. L.; Adelmann, C. R.; Cramer, J. L.; Emerson, D. O.; Maimoni, A.; Pasternak, A. D.; Stephens, D. R.; and Thompson, D. S. 1975. *The LLL in-situ coal gasification research program in perspective — a joint GR&DC/LLL report.* TID-26825.

Banchik, I. N. 1974. Power gas from coal via the Winkler process. Paper presented at the Symposium on Coal Gasification & Liquefaction, University of Pittsburgh, Aug. 6-8. Lakeland, Fla.: Davey Powergas, Inc.

Beychok, M. R. 1975. *Process and environmental technology for producing SNG and liquid fuels.* EPA-660/2-75-011.

Bodle, W. W., and Vyas, K. C. 1974. Clean fuels from coal. *Oil Gas J.* 72(34): 73-88.

Booz-Allen Applied Research. 1974. *Emissions from processes producing clean fuels.* Bethesda, Maryland: Booz-Allen Applied Research.

Capp, J. P.; Plants, K. D.; Fies, M. H.; Pears, C. D.; and Hirst, L. L. 1961. *Underground gasification of coal: Second experiment in preparing a path through a coalbed by hydraulic fracturing.* U.S. Bureau of Mines R.I. 5808.

Campbell, G. G.; Brandenburg, C. F.; and Boyd, R. M. 1974. *Preliminary evaluation of underground coal gasification at Hanna, Wyo.* U.S. Bureau of Mines TPR 82.

Carlson, F. B.; Yardumian, L. H.; and Atwood, M. T. 1975. The TOSCOAL process: Coal liquefaction and char production. In *Papers — clean fuels from coals symposium II,* pp. 495-509. Chicago, Ill.: Institute of Gas Technology.

Carlson, F. B.; Yardumian, L. H.; and Atwood, M. T. 1974. TOSCOAL process for low temperature pyrolysis of coal. *Soc. Mining Eng., AIME, Trans.* 256: 128-31.

Chementator. 1973. *Chem. Eng.* 80(Dec. 24): 18.

Chironis, N. P. 1975. New Athens, Illinois ... Winning site for Coalcon's winning technique of converting coal to liquids and gases. *Coal Age* 80(13): 74-77.

Chun, S. W. 1974. Gulf Catalytic Coal Liquids (CCL) process. In *Materials problems and research opportunities in coal conversion,* vol. II, pp. 263-76. Columbus, Ohio: Ohio State University, Corrosion Center.

Consolidation Coal Company. 1973. *Final report: Development of CSF coal liquefaction process,* vol. V. OCR-RDR-39.

Curran, G. P.; Fink, C. E.; and Gorin, E. 1967. CO_2 acceptor gasification process: Studies of acceptor properties. In *Fuel gasification*, Advances in Chemistry Series 69, ed. R. F. Gould, pp. 141-65. Washington, D.C.: American Chemical Society.

Dravo Corporation. 1976. *Handbook of gasifiers and gas treatment systems*. FE-1772-11. Pittsburgh, Pennsylvania: Dravo Corporation.

Elder, J. W. 1963. The underground gasification of coal. In *Chemistry of coal utilization*, supplementary volume, ed. H. H. Lowry, pp. 1023-40. New York: John Wiley & Sons, Inc.

Elliott, M. A., and Linden, H. R. 1966. Manufactured gas. In *Kirk-Othmer encyclopedia of chemical technology*, vol. 10, pp. 353-442, 2nd ed. New York: John Wiley & Sons, Inc.

Ellison, S. P., Jr. 1967. Petroleum (origin). In *Kirk-Othmer encyclopedia of chemical technology*, vol. 14, pp. 838-45, 2nd ed. New York: John Wiley & Sons, Inc.

Energy Research and Development Administration (ERDA). 1976. ERDA selects four manufacturers for program using coal derived fuels in high temperature gas turbines. ERDA press release No. 76-78, Mar. 22, 1976.

Epperly, W. R., and Siegel, H. M. 1974. Status of coal liquefaction technology. In *Proceedings of the 9th intersociety energy conversion engineering conference, San Francisco, Calif., Aug. 26-30*, Paper 749131, pp. 1020-26. New York: American Society of Mechanical Engineers.

Erickson, K., and Sherman, D. 1975. *Glossary of energy terminology*. VEO-75/01. Richmond, Va.: Virginia Energy Office.

Estep-Barnes, P. A., and Kovach, J. J. 1975. *Chemical and mineralogical characterization of core samples from underground coal gasification sites in Wyoming and West Virginia*. MERC/RI-75/2. Morgantown, W. Va.: Morgantown Energy Research Center.

Faucounau, J. 1974. Le charbon de l'an 2000. *La Recherche* 5(51): 1062-71.

Fischer, D. D.; Brandenburg, C. F.; and Schrider, L. A. 1975. Energy recovery from in situ coal gasification. *Energy Sources* 2(3): 281-92.

Foster Wheeler Corporation. 1972. *Engineering evaluation and review of Consol synthetic fuel process*. OCR-70.

Gray, J. A., and Yavorsky, P. M. 1975. The hydrane process. In *Papers — clean fuels from coal symposium II, June 23-27, 1975, Chicago, Ill.*, pp. 159-76. Chicago, Ill.: Institute of Gas Technology.

Gregg, D. W.; Hill, R. W.; and Olness, D. U. 1976. An overview of the Soviet effort in underground gasification of coal. UCRL-52004. Lawrence Livermore Laboratory.

Hamilton, G. W. 1963. Gasification of solid fuels. *Cost Eng.* 8(3): 4-11.

Hatten, J. L. 1975. Pipeline quality gas from coal. *Mech. Eng.* 97(7): 31-35.

Haynes, W. P., and Forney, A. J. 1975. The synthane process. In *Papers — clean fuels from coal symposium II, June 23-27, 1975, Chicago, Ill.*, pp. 150-57. Chicago, Ill.: Institute of Gas Technology.

Haynes, W. P., and Forney, A. J. 1976. *High Btu gas from coal: Status and prospects*. PERC/IC-76-1. Pittsburgh, Pa.: Pittsburgh Energy Research Center.

H-Coal process nears commercialization. 1976. *Chem. Eng. News* 54(12): 8.

Higgins, G. H. 1972. *A new concept for in situ coal gasification*. UCRL-51217, rev. 1. Lawrence Livermore Laboratory.

Hinderliter, C. R. 1974. Environmental aspects of the SRC process. In *Symposium proceedings: Environmental aspects of fuel conversion technology (May 1974, St. Louis, Mo.)*, pp. 159-64. EPA-650/2-74-118.

Hoogendoorn, J. C., and Salomon, J. M. 1957*a*. SASOL: World's largest oil-from-coal plant. *Brit. Chem. Eng.* 2(5): 238-44.

Hoogendoorn, J. C., and Salomon, J. M. 1957*b*. SASOL: World's largest oil-from-coal plant. *Brit. Chem. Eng.* 2(6): 308-12.

Hoogendoorn, J. C., and Salomon, J. M. 1957*c*. SASOL: World's largest oil-from-coal plant. *Brit. Chem. Eng.* 2(7): 368-73.

Hoogendoorn, J. C., and Salomon, J. H. 1957*d*. SASOL: World's largest oil-from-coal plant. *Brit. Chem. Eng.* 2(8): 418-19.

Howard-Smith, I., and Werner, G. J. 1975. CCL process. In *Coal conversion technology. A review*, pp. 1-2. Brisbane, Australia: Millmerran Coal Pty. Ltd.

Hucka, V., and Das, B. 1973. In situ gasification of coal: Solving the energy crisis. *Mining Eng.* 25(8): 49-50.

Institute of Gas Technology. 1974. HYGAS: A status report. *Gas Scope* 28: 1-3.

Institute of Gas Technology. 1975*a*. *HYGAS: 1972 to 1974, Pipeline gas from coal — hydrogenation (IGT hydrogasification process)*. OCR-RDR-110, Int. Rept. 1.

Institute of Gas Technology. 1975*b*. HYGAS passes another milestone. *Gas Scope* 31: 7.

Institute of Gas Technology. 1975-76. IGR updates an old hydrogen production method. The steam-iron process. *Gas Scope* 34: 1-4.

Interagency Task Force on Synthetic Fuels from Coal. 1974. *Project independence.* U.S. Dept. of the Interior. Washington, D.C.: U.S. Government Printing Office.

Jimeson, R. M. 1966. Utilizing solvent refined coal in power plants. *Chem. Eng. Prog.* 62(10): 53-60.

Johnson, C. A.; Hellwig, K. C.; Johanson, E. S.; Wolk, R. H.; and Stotler, H. H. 1973. H-Coal: Conversion of western coals. *Soc. Mining Eng., AIME, Trans.* 254: 235-38.

Jones, J. F. 1975. Clean fuels from coal power generation. *Energy Sources* 2(2): 179-202.

Katz, D. L.; Briggs, D. E.; Lady, E. R.; Powers, J. E.; Tek, M. R.; Williams, B.; and Lobo, W. E. 1974. *Evaluation of coal conversion processes to provide clean fuels.* EPRI 206-0-0, Parts II and III. Ann Arbor, Mich.: University of Michigan.

Kavlick, V. J., and Lee, B. S. 1967. Coal pretreatment in fluidized bed. *Fuel gasification,* Advances in Chemistry Series 69, ed. R. F. Gould, pp. 8-17. Washington, D.C.: Chemical Society.

King, S. B.; Brandenburg, C. F.; and Lanum, W. J. 1975. Characterization of nitrogen compounds in tar produced from underground coal gasification. *ACS Fuel Chem. Div. Preprints* 20(2): 131-39.

Komar, C. A.; Overbey, W. K., Jr.; and Pasini, J., III. 1973. *Directional properties of coal and their utilization in underground gasification experiments.* Bureau of Mines TPR 73.

LaRosa, P., and McGarvey, R. J. 1975. Fuel gas from molten iron coal gasification. In *Papers — clean fuels from coal symposium II, June 23-27, 1975, Chicago, Ill.,* pp. 228-42. Chicago, Ill.: Institute of Gas Technology.

Lewton, L. C. 1975. *Evaluation of filtration equipment for CRESAP testing, special report.* FE-1517-16. Los Angeles, Calif.: Fluor Engineers and Constructors, Inc.

Linden, H. R. 1974. Prospects and problems of coal gasification. Paper presented at public hearings on "Western Regional Resource Development, including coal, oil shale, and synthetic fuels" of the Federal Energy Administration, August 6-9, Denver, Colorado. IGT-4293. Chicago, Ill.: Institute of Gas Technology.

Linden, H. R. 1975. *Statement to subcommittee on energy research and water resources, U.S. Senate, March 3, 1975.* IGR-4275. Chicago, Ill.: Institute of Gas Technology.

Linden, H. R.; Bodle, W. W.; Lee, B. S.; and Vyas, K. C. 1976. Production of high-Btu gas from coal. In *Annual review of energy,* vol. 1, ed. J. M. Hollander and M. K. Simmons, pp. 65-85. Palo Alto, Calif.: Annual Reviews, Inc.

Little, Arthur D., Incorporated. 1972. *A current appraisal of underground coal gasification.* PB-209 274. Cambridge, Mass.: Arthur D. Little, Inc.

Magnani, C. F., and Farouq Ali, S. M. 1975. Mathematical modeling of the stream method of underground coal gasification. *Soc. Pet. Eng. J.* 15(10): 425-36.

Mills, G. A., and Steffgen, F. W. 1973. Catalytic methanation. *Catal. Rev.* 8(2): 159-210.

Moeller, F. W.; Roberts, H.; and Britz, B. 1974. Methanation of coal gas for SNG. *Hydrocarbon Process.* 53(4): 69-70.

Morgan, W. D. 1975. Coalcon's clean boiler fuels from coal demonstration plant. Paper 59C, presented at the AIChE Convention, Los Angeles, Calif., Nov. 19, 1975.

Nadkarni, R. M.; Bliss, C.; and Watson, W. I. 1974. Underground gasification of coal. *Chemtech* 1974: 230-37.

Nadkarni, R. M.; Bliss, C.; and Watson, W. I. 1975. Underground gasification of coal. In *Papers — clean fuels from coal symposium II, June 23-27, 1975*, pp. 625-51.

Nakles, D. V.; Massey, M. J.; Forney, A. J.; and Haynes, W. P. 1975. *Influence of Synthane gasifier conditions on effluent and product gas production.* PERC/RI-75/6. Pittsburgh, Pa.: Pittsburgh Energy Research Center.

National Academy of Engineering. 1974*a*. *Evaluation of coal-gasification technology. Part I — Pipeline-quality gas.* OCR-RDR-74; Int. Rept. 1.

National Academy of Engineering. 1974*b*. *Evaluation of coal-gasification technology. Part II — Low- and intermediate-Btu fuel gases.* OCR-RDR-74, Int. Rept. 2.

National Petroleum Council. 1972. *U.S. energy outlook, a summary report of the National Petroleum Council.* Washington, D.C.: National Petroleum Council.

O'Hara, J. B.; Rippee, S. N.; Loran, B. I.; and Mindheim, W. J. 1974. Environmental factors in coal liquefaction plant design. In *Symposium proceedings: Environmental aspects of fuel conversion technology (May 1974, St. Louis, Mo.)*, pp. 169-80. EPA-650-/2-74-118.

Parson, R. M., Company. 1970. *1970 final report — Consol synthetic fuel process.* OCR-RDR-45.

Patel, J. G., and Loeding, J. W. 1975. Institute of Gas Technology U-gas process. In *Papers — clean fuels from coal symposium II, June 23-27, 1975, Chicago, Ill.*, pp. 193-206. Chicago, Ill.: Institute of Gas Technology.

Perry, H. 1974. The gasification of coal. *Sci. Am.* 230(3): 19-25.

Perry, H. 1975. Coal conversion technology. In *Perspectives on energy: Issues, ideas, and environmental dilemmas*, ed. L. C. Ruedisili and M. W. Firebaugh, pp. 397-420. New York: Oxford University Press.

Peterson, M. R. 1975. *Progress report: Organic constituents in process streams of a solvent-refining coal plant.* BNWL-B-470. Richland, Wash.: Battelle Pacific Northwest Laboratories.

Pichler, H., and Nector, A. 1964. Carbon monoxide–hydrogen reactions. *Kirk-Othmer's encyclopedia of chemical technology*, vol. 4, pp. 446-89, 2nd ed.

Qader, S. A. 1974. Process yields BTX from coal for 25¢/gal. *Oil Gas J.* 72(29): 58-59.

Raimondi, P.; Terwilliger, P. L.; and Wilson, L. A., Jr. 1975. A field test of underground combustion of coal. *J. Pet. Technol.* XXVII: 35-41.

Rooke, D. E., and Hebden, D. 1974. Natural gas substitutes from fossil fuels. In *9th world energy conference*, vol. IV, pp. 273-96.

Rousseau, P. E. 1962. Organic chemicals and the Fischer-Tropsch synthesis in South Africa. *Chem. Ind. (Lond.)* 1962: 1958-66.

Rubin, E. S. 1974. Research and development needs for enhancing U.S. coal utilization. In *9th Intersociety energy conversion engineering conference proceedings, San Francisco, Calif., Aug. 26-30, 1974*, Paper 749160, pp. 997-1005. New York: American Society of Mechanical Engineers.

Sandia Labs gets ERDA coal monitoring project. 1975. *Coal Age* 80(5): 30.

Sass, A. 1974. Garrett's coal pyrolysis process. *Chem. Eng. Prog.* 70(1): 72-73.

Schmid, B. K. 1975. Status of the SRC project. *Chem. Eng. Prog.* 71(4): 75-78.

Schora, F. C., Jr. 1974. Clean energy from coal. In *Proceedings of the 9th intersociety energy conversion engineering conference, San Francisco, Calif., August 26-30*, Paper 749107, pp. 1006-14. New York: American Society of Mechanical Engineers.

Schrider, L. A., and Fischer, D. D. 1976. In situ coal gasification, a unique means of recovery. *Mech. Eng.* 98(3): 20-24.

Scotti, L. J.; Jones, J. F.; Ford, L.; and McMunn, B. D. 1975. The project COED pilot plant. *Chem. Eng. Prog.* 71(4): 61-62.

Scotti, L. J.; Merrill, R. C.; McMunn, B. D.; Romelczyk, S. J.; Domina, D. J.; Ford, L.; Terzian, H. D.; and Jones, J. F. 1975. *Char Oil Energy Development*. FE-1212-5, Int. Rept. 5. Princeton, N.J.: FMC Corporation, Chemical Research and Development Center.

Seglin, L., and Eddinger, R. T. 1971. Synthetic crude oil from coal. In *Kirk-Othmer's encyclopedia of chemical technology*, supplementary volume, pp. 177-98. 2nd ed.

Smoot, L. D., and Hanks, R. W. 1975. *Mixing and gasification of coal in entrained flow systems*. FE-1767-T1. Provo, Utah: Brigham Young University.

Squires, A. M. 1974. Clean fuels from coal gasification. *Science* 184: 340-46.

Squires, A. M. 1974. Coal: A past and future king. *AMBIO* 3(1): 1-14.

Stephens, D. R., and Lentzner, H. L., eds. 1975. *LLL in situ coal gasification program, quarterly progress report, April through June 1975*. UCRL-50026-75-2. Lawrence Livermore Laboratory.

Stephens, D. R., and Lentzner, H. L., eds. 1975. *LLL in situ coal gasification program, quarterly progress report, October through December 1975*. UCRL-50026-75-4. Lawrence Livermore Laboratory.

Strakey, J. P.; Forney, A. J.; and Haynes, W. P. 1975. *Methanation in coal gasification processes*. PERC/IC-75/1. Pittsburgh, Pa.: Pittsburgh Energy Research Center.

Tillman, D. A. 1976. Status of coal gasification. *Environ. Sci. Technol.* 10: 34-38.

University of Oklahoma. 1975. *Energy alternatives: A comparative analysis*. Washington, D.C.: U.S. Government Printing Office.

Waitzman, D. A.; Faucett, H. L.; Kindahl, E. E.; Tomlinson, S. V.; Nichols, D. E.; and Broadfoot, W. J. 1975. *Evaluation of fixed-bed low-BTU coal gasification systems for retrofitting power plants*. TVA Bull. Y-91. Muscle Shoals, Ala.: Tennessee Valley Authority.

Wall, J. M., ed. 1975. Gas processing handbook. *Hydrocarbon Process.* 54(4): 79-138.

Weekly Energy Report, 4(13): 13. 1976.

Woodall-Duckham, Ltd. 1975. *Trials of American coals in a Lurgi gasifier at Westfield, Scotland*. FE-105. Crawley, Sussex, England: Woodall-Duckham, Ltd.

Yavorsky, P. M., and Akhtar, S. 1974. Environmental aspects of coal liquefaction. In *Symposium proceedings: Environmental aspects of fuel conversion technology (May 1974, St. Louis, Mo.)*, pp. 325-30.

Yavorsky, P. M.; Akhtar, S.; Lacey, J. J.; Weintraub, M.; and Reznik, A. A. 1975. The Synthoil process. *Chem. Eng. Prog.* 71(4): 79-80.

4. PROCESS EFFLUENTS: QUANTITIES AND CONTROL TECHNOLOGIES

N. S. Dailey

ABSTRACT

Wastes are generated during each main coal conversion stage: feedstock storage and preparation, conversion operations, and waste stream treatment. Any material entering the conversion facility is a potential pollutant. For most pollutants, processes exist for controlling emissions or converting them to environmentally acceptable forms. Information is needed on the specific nature of effluents so that satisfactory controls can be applied.

Through technologies applied before, during, and after conversion, the largest portion (50 to 98%) of the sulfur in coal can be recovered. Lesser amounts are retained in the ash, tars, chars, or liquefaction solids, and only a small fraction is emitted to the environment. Nitrogen emissions (primarily nitrogen oxides, ammonia, and hydrogen cyanide) may be reduced by appropriate combustion techniques, operating modifications, or removal processes. Particulates and solid residues from coal conversion activities, as with any other industry using coal, pose a potentially serious emission problem and require treatment for control. After removal, they are disposed of as landfill or minefill, recycled as fuel, or in some instances sold as by-products.

Fates of trace elements in coal conversion are not completely known. Trace element pathways may include adsorption on particulate matter, inclusion in condensates, deposition on equipment surfaces, inclusion in by-products and final products, and emission as fugitive pollutants, thus requiring development of a variety of suitable control technologies.

Process wastewaters arise from moisture in the coal, water of constitution or decomposition, or water added for stoichiometric process requirements, by-product recovery, or gas scrubbing. Gas liquor, the sum of the aqueous streams condensed in the processing units, can contain components of the product gas (carbon monoxide, carbon dioxide, hydrogen, hydrogen cyanide, methane, and nitrogen) and contaminants such as sulfur and nitrogen compounds, particulates, phenols, soluble salts, and emulsified tars and oil. Water recycling or reuse processes are proposed for most conversion designs. However, problems from contaminant buildup or corrosion are not completely known, and some aqueous discharges are probably inevitable. Wastewater treatment facilities include mechanical, chemical, and biological removal processes similar to those for other industries.

Typical organic compounds that may be present in the gas, liquid, and solid waste streams from coal conversion activities include naphtha, coal tar, light and heavy oil, BTX (benzene, toluene, and xylene), solvent-refined coal, and product gases. The fate of polycyclic aromatic hydrocarbons (PAH) and heterocyclic hydrocarbons is not yet known. Atmospheric and wastewater emissions of organics are expected, and possible means of control are being studied.

Spent catalysts, radioactive emissions, carbon dioxide, thermal effluents, and noise are additional coal conversion waste products, all of which require environmentally acceptable controls or disposal. Possible problems from odors and fugitive emissions arising from startups and shutdowns may require special treatment.

4.0 INTRODUCTION

The primary objectives of this chapter are to (1) identify the potential sources and general nature of coal conversion wastes and (2) present the possible methods of waste management that are accessible to coal conversion technologies. A few conversion technologies within the United States have progressed to the bench-scale or pilot-plant stage so that preliminary studies and engineering reports are beginning to identify sources of emissions and the nature of their effluents. Although the liquefaction and gasification technologies have different processing units, similarities of their waste streams and waste management practices allow them to be discussed concurrently (Table 4.1). However, the exact nature and quantity of wastes vary according to coal type, operational design, and plant size. The chemical and physical properties of coal (Chap. 2) are important in the performance of processing equipment; they influence the size and cost of sidestream processing trains and affect the magnitude of environmental discharges (Hittman Associates 1975a). Influential operative design features include temperature, process steps, and process sequencing. Because of the wide variance of these features among plants, base-size plant conversion information has been compiled in an appendix at the end of this chapter (Appendix 4A, p. 4-140) to enable general comparisons between material inputs, product yields, and generated wastes. Process emissions are significantly affected by design, coal feed, and plant size; thus, strict comparisons made between different processes may not be completely valid.

4.0.1 Pollution standards

Federal environmental pollution standards for coal conversion processes have not yet been established. However, Federal performance standards for related industries (petroleum refineries, coal carbonization industries, and fossil-fueled power plants) are relevant to conversion waste management. Federal wastewater standards for oils, phenolics, sulfide, and ammonia in petroleum and coking industries are given in Table 4.2; Table 4.3 compares them on equivalent process feedstock energy input. Federal air standards controlling petroleum- and fossil-fueled power plant emissions are compared with New Mexico gasification plant air standards in Table 4.4.

4.0.2 Sources of waste

Three main sources of waste from conversion processes (Glaser, Hershaft, and Shaw 1974) are (1) storage and preparation of feedstocks, water, hydrogen, and oxygen; (2) conversion operations; and (3) waste stream control and treatment processes (Fig. 4.1).

In the preparatory stages of coal conversion (Fig. 4.1a), coal is stockpiled, cleaned, dried, and crushed; water is pretreated; and steam, oxygen, or hydrogen is generated. Emissions at this stage are characteristic of those from other large coal-using industries such as power plants; thus, pollutants specific to coal preparative operations are known, and effective control equipment is available (Interagency Task Force on Synthetic Fuels from Coal 1974).

Table 4.1. Potential coal conversion pollutants

Potential pollutants	Process source	Treatment processes
Runoff water	Coal storage	Collection, neutralization, demineralization, ion exchange, settling
Coal-washing aqueous wastes	Coal processing	Collection, neutralization, demineralization, ion exchange, settling
Coal dusts and fines	Coal storage and processing	Cyclone separators, fabric filters, negative pressure grinders, wet scrubbers, electrostatic precipitators
Mineral debris	Coal processing	Landfill, minefill
Mineral residues	Water processing	Landfill, minefill
Water treating chemicals	Water processing	Landfill, minefill
N_2	Oxygen generation	Venting
Stack gas	Utility and steam production	Limestone scrubbing, electrostatic precipitators, Wellman-Lord, Cation-Oxidation, MgO scrubbing
H_2S	Conversion	Acid gas scrubbed, Stretford, Giammarco Vetrocoke, and Takahax (1-step removal); Amine, Economine, Alkazid, Benfield, Catacarb CO_2 removal, Purisol, Adip, Fluor, Selexol, Rectisol, Sulfinol, SNPA-DEA processes (2-step removal requiring Claus-based sulfur recovery units)
NO_x	Conversion	Venting, Continuous Catalytic Absorption, Modified Alkalized Alumina, equipment design modification
CO_2	Conversion	Stack gas cleanup of sulfur and combustibles, venting

Table 4.1 (continued)

Potential pollutants	Process source	Treatment processes
COS, CS_2	Conversion	Recovered in some scrubbing processes for H_2S; fugitive emission from sulfur-recovery unit tail gas
NH_3	Conversion	Stripped by Phosam or Chevron unit for resale recovery; residual NH_3 treated in wastewater
HCN	Conversion	Follows ammonia to wastewater treatment
Miscellaneous gases (HF, HCl, CO)	Conversion	Wet-scrubbed from product gas stream and transferred to wastewater treatment processes such as biological oxidation ponding or lime neutralization for removal
Volatile trace metals (Hg, As, Se, Sb, Tl)	Conversion	Stack gas cleaning: modified limestone scrubbing, modified Double Alkali, Cation-Oxidation, DAP-Mn (wet), and Chiyoda processes
Trace metals	Conversion	Recovered in ash, tars, chars, and wastewater
Ash	Conversion	Electrostatic precipitators, minefill, landfill, possible resale options
Particulate matter	Conversion	Cyclone separators, fabric filters, negative pressure grinders, wet scrubbers, electrostatic precipitators
Char	Conversion	Recycled as gasifier, steam or utility fuels; requires sulfur recovery before or after combustion; wastewater treating potential
Tar	Conversion	Hydrodesulfurized for fuel oil, gasified or combusted, residual treated in wastewater
Oils	Conversion	Hydrodesulfurized for fuel oil, residual treated in wastewater

Table 4.1 (continued)

Potential pollutants	Process source	Treatment processes
Gaseous hydrocarbons	Conversion	May be scrubbed into wastewater for activated carbon or biological oxidation pond removal
Phenols	Conversion	Stripped by Phenosolvan for recovery and resale; residual removal by active carbon absorption, biological oxidation ponding, or solar evaporation
Process wastewater	Conversion	NH_3 and phenol stripping (Chevron, Phenosolvan, Phosam), lime neutralization, active carbon absorption, biological oxidation ponding, residual tar and oil removal, evaporation, zeolite treatment
Spent catalysts	Conversion, control treatment	Regeneration, disposal in landfill or minefill
Sulfur recovery tail gas (H_2S, SO_2, CO_2, CS_2)	Control and treatment	Beavon, Cleanair, Chiyoda, IFP (Institut Francais du Petrole) Aquaclaus, SCOT (Shell Claus Offgas Treating), venting, or combustion
Slag	Control treatment generated (limestone processes)	Granulated for landfill, minefill, or resale

Table 4.2. Wastewater effluent limitations for petroleum
refineries and by-product coking

Pollutant	Maximum 30-day average[a]	
	Petroleum refineries (lb/1000 bbl feedstock)	By-product coking (lb/1000 lb coke)
5-day biochemical oxygen demand	1.35 - 18.85	b
Total suspended solids	0.95 - 12.47	0.0104
Chemical oxygen demand	6.85 - 130.5	b
Oil and grease	0.43 - 5.80	0.0042
Phenolics	0.0098 - 0.1247	0.0002
Ammonia (as N)	0.28 - 11.02	0.0042
Sulfide	0.0073 - 0.1015	0.0001
Total chromium	0.0226 - 0.3045	b
Hexavalent chromium	0.00038 - 0.0052	b
pH (for all categories)	6.0 - 9.0	6.0 - 9.0
Cyanides amenable to chlorination	b	0.0001

[a]Daily limits not included here; refinery ranges reflect EPA promulgations as of May 1974.
[b]Not applicable.

Source: Rubin and McMichael 1975, Table 4, p. 114. Reprinted with permission from
Environmental Science and Technology. Copyright by the American Chemical Society.

Table 4.3. Adjusted new source performance standards for by-product
coke making and petroleum refining, 30-day maximum

Pollutant	Petroleum refineries	By-product coke making
	(pounds of pollutant/10^{12} Btu feedstock[a])	
5-day biochemical oxygen demand	210 - 2900	b
Total suspended solids	140 - 1920	600
Chemical oxygen demand	1050 - 20,000	b
Oil and grease	70 - 890	240
Phenolics	1.5 - 19	12
Ammonia (as N)	40 - 1700	240
Sulfide	1.1 - 16	5.8
Total chromium	3.5 - 47	b
Hexavalent chromium	0.06 - 0.80	b
Cyanides amenable to chlorination	b	5.8

[a]Assumes heating values of 6.5 million Btu/bbl for crude oil and 12,000 Btu/lb for coal, with
a coke yield of 0.69 lb coke/lb coal.
[b]Not applicable.

Source: Rubin and McMichael 1975, Table 6, p. 115. Reprinted with permission from
Environmental Science and Technology. Copyright by the American Chemical Society.

Table 4.4. Existing State (New Mexico) and Federal emission standards
having implications for future coal refineries

Pollutant	Federal		New Mexico	
	Petroleum refineries	Fossil-fueled steam generators	Gas-fired power plant associated with coal gasification plants	Gasification plants
Sulfur dioxide	0.10 grain H_2S/scf (dry) in fuel gas (plant gas fuel combustion)[a]	1.2 lb/10^6 Btu (solid) 0.8 lb/10^6 Btu (liquid)	0.16 lb/10^6 Btu	b
Particulate matter	0.27 grain/scf + 0.10 lb/10^6 Btu aux. fuel; 30% opacity, except for 3 min/hr (catalytic cracking unit catalyst regenerator)	0.1 lb/10^6 Btu 20% opacity	0.03 lb/10^6 Btu	0.03 grain/scf
Hydrocarbons	Specified vapor pressure limits and required control devices (petroleum storage vessels)	b	b	b
Nitrogen oxides	b	0.7 lb/10^6 Btu (solid) 0.3 lb/10^6 Btu (liquid) 0.2 lb/10^6 Btu (gas)	0.20 lb/10^6 Btu	b
Sulfur (total)	b	b	b	0.008 lb/10^6 Btu
Reduced sulfur (sum of H_2S, CS_2, and COS)	b	b	b	100 ppm
Hydrogen sulfide	b	b	b	10 ppm
Hydrogen cyanide	b	b	b	10 ppm
Hydrogen chloride and hydrochloric acid	b	b	b	5 ppm
Ammonia	b	b	b	25 ppm

[a]Equivalent to 0.1 lb/10^6 Btu for gas with 250 Btu/ft^3.
[b]Not applicable.

Source: Rubin and McMichael 1975, Table 5, p. 115. Reprinted with permission from _Environmental Science and Technology_. Copyright by American Chemical Society.

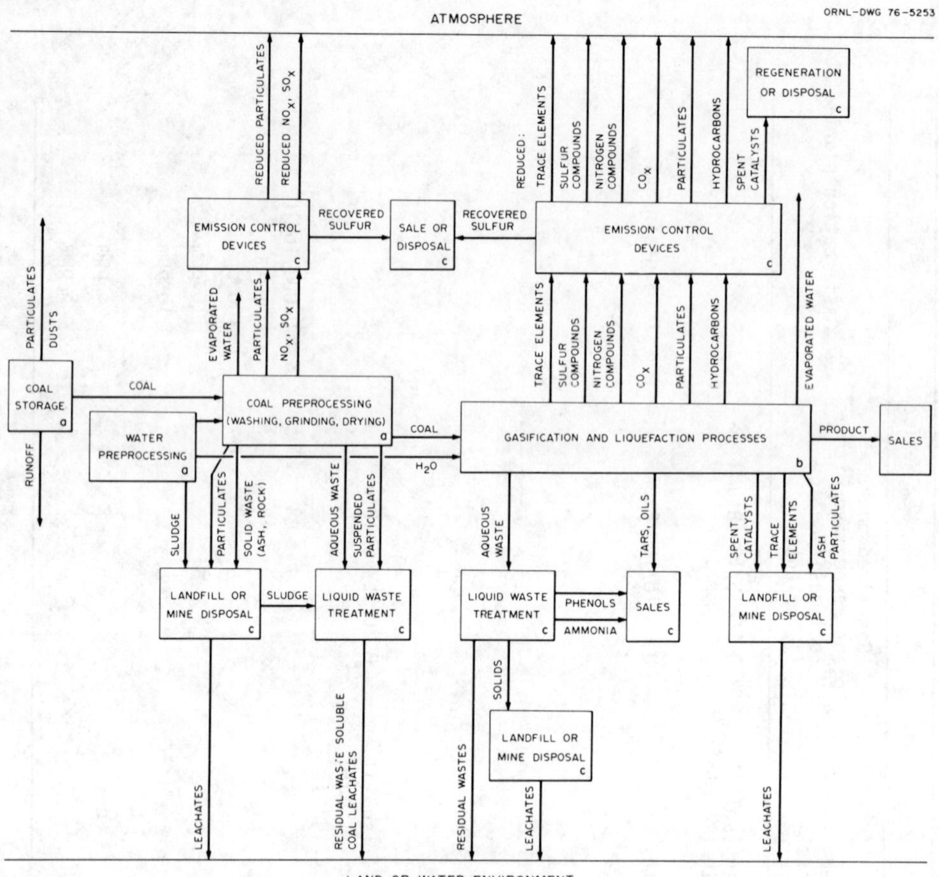

Fig. 4.1. Emission fates for coal conversion process. <u>Source</u>: Adapted from Interagency Task Force on Synthetic Fuels in Coals 1974, Fig. 6-1, p. 91.

Utility production systems include facilities for steam, power, and air (O$_2$ or H$_2$) generation. Where coal is used in boilers, stack gas cleaning facilities for sulfur and particulates will be applicable. Thermal efficiency is lower, however, for conversion facilities using some of the low-sulfur, low-Btu gas for utility requirements (Jahnig 1975*b*).

Wastes from coal conversion units (Fig. 4.1*b*) are now being identified through engineering reports, bench-scale studies, and pilot plant experience. Numerous steps in coal conversion produce emissions that are potential pollutants. Coal gasification steps include gasification, tar separation, shift conversion, quenching, gas purification, methanation, compression, and dehydration; each step produces effluents requiring treatment. Steps in coal liquefaction processes that can give rise to emissions are hydrogenation, pyrolysis, gas purification, hydrotreating, mixing, and solids separation. Effluents of particular importance to conversion units appear to be acid gas, pyrolysis gas, ammonia stripping gas, ash, tars, chars, filter cakes, particulates, spent catalysts, process wastewater, cooling water, and blowdown water.

Process streams (including products and byproducts) contain all the coal-derived compounds plus added materials. All the process streams are capable of producing emissions through incomplete recovery, evaporation of constituents, spills, or leaks. Because coal-derived compounds produced at high temperatures can be carcinogenic, containment of any fugitive emissions is important.

Control and treatment processes, applied to minimize pollutants produced in the conversion areas, include particulate control, stack gas control, acid gas treatment, wastewater treatment, char treatment, tar treatment, heat control, and noise control (Fig. 4.1c). The task of control and treatment processes is not to eliminate the pollutants, but to convert them to less offensive forms. Cleanup processes can generate their own pollutants requiring treatment; thus, achievement of zero effluent, although desirable, can become an infinite process.

Emissions from pollution treatment facilities include (1) converted and fugitive pollutants derived from process emission streams and (2) constituents used within the treatment systems. Emission disposal options include the resale of by-products to other industries, use as landfill or minefill, or return for further treatment.

4.0.3 Factors influencing control technologies

Technical, operational, and economic factors are important in selecting treatment facilities. Understandably, complete material balances will be needed for control systems planning, optimal by-product recovery, lower treatment costs, and acceptable discharges. Factors affecting the cost, performance, and selection of environmental control systems are listed in Table 4.5.

Economically and environmentally acceptable control of the wastes produced in coal conversion requires knowledge of the potential pollutants. As the research literature expands, pollutant pathways through the conversion units, possible means of pollutant escape to the environment, pollutant emission form, ultimate fate of the pollutant within the process units, and effective waste treatment processes are being identified.

Emissions from the pretreatment facilities, coal conversion units, and associated treatment facilities include sulfur (S_2, S_4, H_2S, SO_x, COS, CS_2, mercaptans, thiophenes), nitrogen (NO_x, NH_3, HCN, thiocyanates), particulate matter (debris, fines, ash, chars), trace elements, wastewaters, organics, spent catalysts, thermal effluents, noise, carbon dioxide, and radioactivity.

4.1 SULFUR

The principal contaminant in fossil fuels is sulfur. Various forms of sulfur in coal are reflected in the possible effluents: elemental sulfur (S_x), hydrogen sulfide, sulfur dioxide, carbonyl sulfide, carbon disulfide, sulfides, mercaptans, thiophenes, and thiocyanate.

4.1.1 Pretreatment operations

Chemical pretreatment of coal to reduce sulfur content has been studied by the U.S. Bureau of Mines and the TRW Systems Group (Powell and Ulmer 1974; Meyers et al. 1972). The U.S. Bureau of Mines process uses a 10% aqueous sodium hydroxide at 440°F (227°C) to remove inorganic sulfur from coal and acid washing to remove the mineral matter (Powell and Ulmer 1974). The TRW Systems Group's Meyers process removes 84 to 98% of the pyritic sulfur in coal by using aqueous ferric

Table 4.5. Factors influencing control
system selection

Stream to be treated

Components
Concentration
Quantity
Pressure and temperature
Contaminants to systems
Recoverable values
Level of abatement required
Toxicity

Ultimate discharges

Saleable by-products
Hazardous effluents
Further treatment of effluents
Ultimate disposal
Innocuous materials

Control system

Input stream volume
Input stream concentrations
Operating temperature and pressure ranges
Flexibility toward input changes
Raw materials and chemicals
Catalysts
Energy required
Space requirements
Capital cost
Operating costs
Discharges produced
Efficiency
Reliability and compatibility
Manpower required

Source: Adapted from Hittman Associates
1975*b*, Table VI-3. Reprinted by permission of the publisher.

iron salts (ferric sulfate or chloride) (Meyers et al. 1972; Hamersma et al. 1972; Hall et al. 1975). Total sulfur reduction is from 40 to 82%, and reaction products are dissolved ferrous sulfate, sulfuric acid, and precipitated free sulfur. Regeneration of the leach solution is by oxidation. The precipitated sulfur is recovered by solvents vaporization or condensation (Hall et al. 1975). Neither of these processes treats organic sulfur.

The TRW Systems Group studied the chemical removal of nitrogen and organic sulfur from four coals and found that phenol, *o*-chlorophenol, cresol, nitrobenzene, and sodium or potassium hydroxide (aqueous or alcoholic) were effective for removal of organic sulfur, but nitrogen removal effectiveness was questionable (Meyers, Land, and Flegal 1971).

Battelle Memorial Institute (proprietary) is developing a hydrothermal process that uses a water slurry of coal and sodium hydroxide (430 to 650°F at 350 to 2500 psi) to convert sulfur (pyritic, and 70% of the organic) and part of the ash to soluble forms for removal (Worthy 1975; Hall et al. 1975).

Organically bound sulfur (mercaptans, organic sulfides, organic disulfides, and thiophenes) cannot be removed as easily from coal by mechanical, physical, or wet cleaning methods as can pyritic sulfur (Hittman Associates 1975a). However, physical methods can remove from 20 to 80% of the iron sulfides in most coals. Density, or specific gravity, is the most commonly employed property in physical cleaning processes: "Pure" bituminous coal has a specific gravity of about 1.3; "rock," about 2.0 to 2.7; and pyrite, about 4.8 to 5.0 (Hall et al. 1975). Froth flotation processes, unlike washing, enable separation of fine coal, but require drying. Froth flotation processes, like the Bureau of Mines process, use frothing agents that float the coal from the mineral matter; a second stage removes the fine pyrite particles, yielding a cleaner coal (Hall et al. 1975).

Coal drying and preheating units require options for lowering emissions. Coal(or fines)-fired boilers, using 8 to 9% of total fuel requirement to supply heat, require stack gas treating processes. If product gas is used, overall thermal efficiency is lowered. Therefore, a number of processes will use a combination of coal (or fines) and product gas to lower sulfur dioxide emissions to acceptable levels. For example, the CO_2 Acceptor process, using lignite fines for coal drying and preheating, emits 1.6 lb of sulfur dioxide per million Btu fired (1.2 lb/million Btu is allowed for large stationary power generation); but using 75% lignite fines and 25% product gas emits about 1.2 lb of sulfur dioxide per million Btu (Jahnig and Magee 1974). Use of bituminous coal in the Koppers-Totzek process would amount to 1.4 lb of sulfur dioxide per million Btu so that a coal-gas drier feed may be used to meet sulfur emission limits (Magee, Jahnig, and Shaw 1974).

4.1.2 Conversion

Sulfur control in conversion processes primarily involves purification of the product gas and oil and sulfur removal from residues such as tars, chars, and ash. Reduced sulfur compounds, primarily hydrogen sulfide and to a lesser extent carbonyl sulfide and carbon disulfide, are produced during both gasification and liquefaction and are found in the process streams, especially during the initial stages (Glaser, Hershaft, and Shaw 1974). About 90% of the organic sulfur in coal is converted to hydrogen sulfide during gasification (Williams and Dressel 1973). Sulfur compounds in the product gas stream of the El Paso Burnham I Coal Gasification Complex (Lurgi technology) are estimated to be 95% hydrogen sulfide, 2.4% carbonyl sulfide, 0.3% carbon disulfide, 2.0% mercaptans, and 0.3% thiophenes (Gibson, Hammons, and Cameron 1974). Sulfur compounds from the Char Oil Energy Development (COED) process streams were reported by Shults (1976):

Compounds	Pyrolysis gas (ng/ml)	Stack gas (ng/ml)
COS	6.4	11.0
H_2S	68.0	12.8
Thiophen[e]	16.0	0.9
$(CH_3S)_2$	0.8	

However, these gas samples were received in August 1974, and the tabulated results were obtained in October 1974. Earlier quantitative analysis showed higher concentrations of the components, particularly hydrogen sulfide; such discrepancy indicates the need to analyze fresh samples (Shults 1976).

Liquefaction acid gas streams resulting from the treatment of raw pyrolysis gas and the hydro-treating bleed gas stream from the COED process contain about 7% hydrogen sulfide (Kalfadelis and Magee 1975).

Sulfur compounds must be removed from the gas streams prior to methanation to prevent fouling of the methanation catalysts or the nonsulfur-resistant shift conversion catalysts used in hydrogen manufacture (Glaser, Hershaft, and Shaw 1974).

Raw gasifier gas processing routes include the following:

1. Hot gas goes to the shift conversion using a sulfur-resistant catalyst, followed by acid gas removal and methanation.
2. Raw gas is cooled and scrubbed for acid gas removal and then goes to shift conversion, carbon dioxide removal, and methanation.
3. Gas is desulfurized at high temperature (amine or hot carbonate), then methanated with steam, followed by carbon dioxide removal (Jahnig 1975*b*).

Most acid gas removal systems require cooling of the gas stream (routes 1 and 2); a few methods are being studied to remove the hydrogen sulfide from the hot gases (Panel on Coal Gasification Technology 1974), enabling higher thermal efficiencies. The Bureau of Mines is developing a process using iron (Bureau of Mines 1974), and CONSOL is studying the use of half-calcined dolomite (Jahnig 1975*b*).

4.1.2.1 Acid gas removal

A resume for processes for removing acid gas (primarily hydrogen sulfide and carbon dioxide) is given in Table 4.6. Acid gas removal, when used industrially, employs at least one of three possible techniques (Katz et al. 1974): (1) absorption into a solvent, (2) chemical conversion into another compound, (3) adsorption on solids. Solvents for the first category include amines, methanol, or hot carbonate (Magee and Shaw 1974). Reagents for chemical conversion include zinc oxide, ferric oxide, calcium oxide (Katz et al. 1974), and vanadium pentoxide (Princiotta 1972). Absorbents include iron sponge, charcoal (Katz et al. 1974), and activated carbon (Glaser, Hershaft, and Shaw 1974).

The Giammarco-Vetrocoke (Fig. 4.2) (using arsenic-activated alkali), Stretford [using Na_2CO_3 and anthraquinone disulfonic acid (ADA)] (Fig. 4.3), and Takahax (Fig. 4.4) processes remove hydrogen sulfide from the product gas and convert it to solid elemental sulfur in one step (Panel on Coal Gasification Technology 1974). Other acid gas stripping processes require the use of a recovery plant to produce elemental sulfur or sulfuric acid.

The Giammarco-Vetrocoke process uses an alkali (sodium carbonate) arsenate and arsenite solution to scrub hydrogen sulfide from the synthesis gas (Wall 1973). Process reactions are

$$Na_3AsO_3 + 3H_2S = Na_3AsS_3 + 3H_2O \; , \tag{1}$$

Table 4.6. Processes for removing H_2S and CO_2 from low- and intermediate-Btu gases

Process	Sorbent	Removes
Amine	Monethanolamine, 15% in water	CO_2, H_2S
Economine	Diglycolamine, 50-70% in water	CO_2, H_2S
Alkazid	Solution M or DIK (potassium salt of dimethylamine acetic acid), 25% in water	H_2S, small amount of CO_2
Benfield, Catacarb	Hot potassium carbonate, 20-30% in water (also contains catalyst)	CO_2, H_2S; selective to H_2S
Purisol (Lurgi)	n-Methyl-2-pyrolidone	H_2S, CO_2
Fluor	Propylene carbonate	H_2S, CO_2
Selexol (Allied)	Dimethyl ether polyethylene glycol	H_2S, CO_2
Rectisol (Lurgi)	Methanol	H_2S, CO_2
Sulfinol (Shell)	Tetrahydrothiophene dioxide (sulfolane) plus diisopropanol amine	H_2S, CO_2; selective to H_2S
Giammarco-Vetrocoke	K_3AsO_3 activated with arsenic	H_2S
Stretford	Water solution of Na_2CO_3 and anthraquinone disulfonic acid with activator of sodiummeta vanadate	H_2S
Activated carbon	Carbon	H_2S
Iron sponge	Iron oxide	H_2S
Adip	Alkanolamine solution	H_2S, some COS, CO_2, and mercaptans
SNPA-DEA	Diethanolamine solution	H_2S, CO_2
Takahax	Sodium, 1,4-napthoquinone, 2-sulfonate	H_2S

Source: Adapted from Panel on Coal Gasification Technology 1974, Table I-C, p. 62.

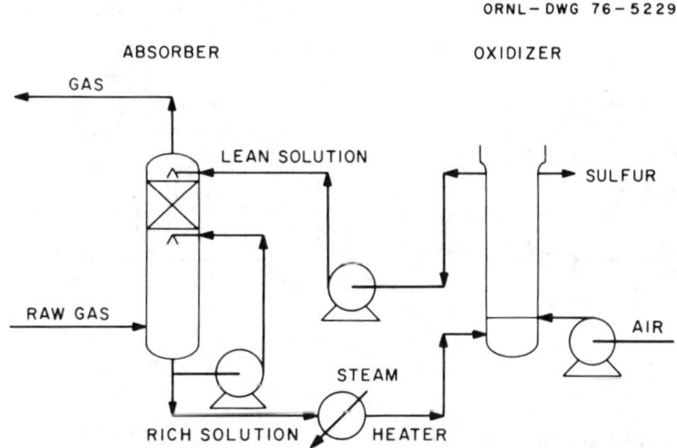

Fig. 4.2. Giammarco-Vetrocoke process flow chart. Source: After Wall 1973, p. 108. Reprinted by permission of the publisher.

$$Na_3AsS_3 + 3Na_3AsO_4 = 3Na_3AsO_3S + Na_3AsO_3 \quad , \tag{2}$$

$$Na_3AsO_3S = Na_3AsO_3 + S \quad , \tag{3}$$

$$Na_3AsO_3 + 1/2\ O_2 = Na_3AsO_4 \quad . \tag{4}$$

In the oxidizer, elemental sulfur is removed by froth flotation, vacuum-filtrated, and washed; the oxidizing reaction (Eq. 4) reestablishes the arsenate–arsenite solution by oxidizing some arsenite to arsenate (Wall 1973).

The Stretford process removes acid gas by washing with an aqueous solution of sodium carbonate, sodium vanadate, and anthraquinone disulfonic acid (Wall 1973):

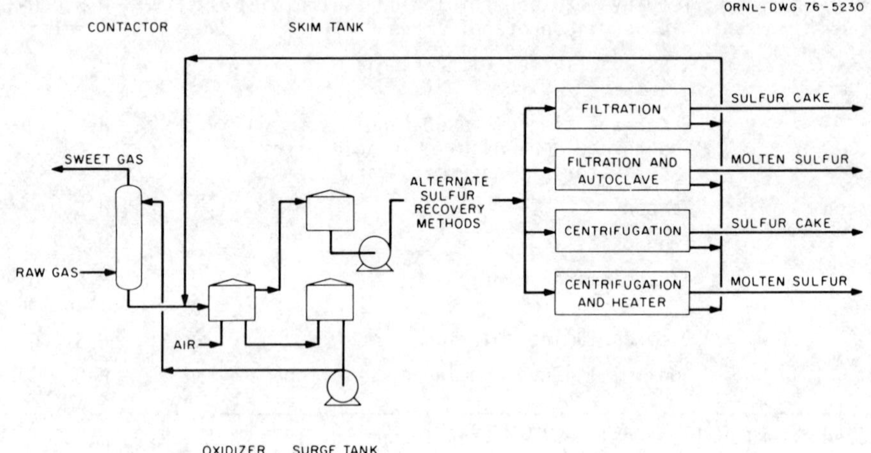

Fig. 4.3. Stretford process flow chart. <u>Source</u>: After Wall 1973, p. 109. Reprinted by permission of the publisher.

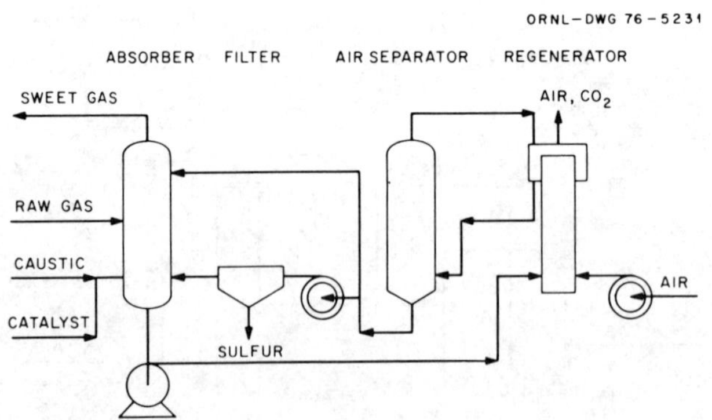

Fig. 4.4. Takahax process flow chart. <u>Source</u>: After Wall 1973, p. 110. Reprinted by permission of the publisher.

$$H_2S + Na_2CO_3 = NaHS + NaHCO_3 \quad \text{(Beers 1973)}. \tag{5}$$

The hydrosulfide produced reacts with vanadium to form elemental sulfur:

$$HS + V^{5+} = S + V^{4+} \quad \text{(Beers 1973)}. \tag{6}$$

Scrubbing liquor is regenerated by air blowing, and the vanadium is restored by oxygen transfer of the ADA:

$$V^{4+} + ADA = V^{5+} + \text{reduced ADA} \quad \text{(Beers 1973)}. \tag{7}$$

Stretford developers have developed processes that convert organic sulfur compounds in hydrocarbon streams to hydrogen sulfide by treatment with steam over a U_3O_8-containing catalyst, enabling the resulting hydrogen sulfide to be removed in the Stretford system (Kalfadelis and Magee 1974). The Synthane process, using 1.6%-sulfur-content feed coal and using the Stretford process, produces about 140 tons of elemental sulfur per day (Kalfadelis and Magee 1974).

The Takahax process (Fig. 4.4) uses an alkaline solution containing sodium 1,4-naphthoquinone-2-sulfonate as the reduction-oxidation catalyst to remove hydrogen sulfide. The hydrogen sulfide reacts with sodium carbonate to form sodium bisulfide and sodium bicarbonate. The bisulfide is oxidized by the catalyst, precipitating solid sulfur. Reduced naphthoquinone sulfonate (naphtho-hydroquinone sulfate) is oxidized in the regenerator for recycle (Wall 1973). Hydrogen cyanide (if present in feed) is absorbed by the alkali and oxidized to sodium thiocyanate, which eventually must be purged from the system (Wall 1973).

Two-step acid gas removal processes include Amine, Econamine, Alkazid, Benfield-Catacarb, Purisol, Fluor, Selexol, Rectisol, Sulfinol, Adip, and SNPA-DEA. Processes such as the Amine, Econamine (Fig. 4.5), and Alkazid (Fig. 4.6) scrub the acid gas with a water-solvent solution. The hydrogen sulfide and carbon dioxide in the acid gas react to form loosely bound salts. Steam stripping and pressure reduction release the hydrogen sulfide and carbon dioxide and enable recycling of the absorber solution (Panel on Coal Gasification Technology 1974). Monoethanolamine (MEA) scrubbing (Amine process) cannot effectively remove carbonyl sulfide or carbon disulfide because of degradation of the MEA solvent (Glaser, Hershaft, and Shaw 1974).

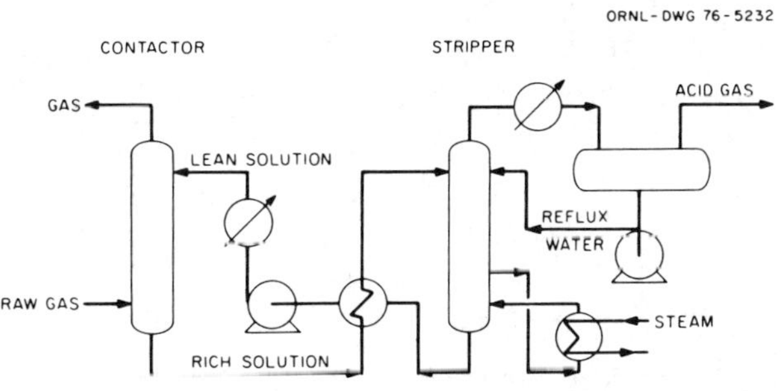

Fig. 4.5. Econamine process flow chart. Source: After Wall 1973, p. 94. Reprinted by permission of the publisher.

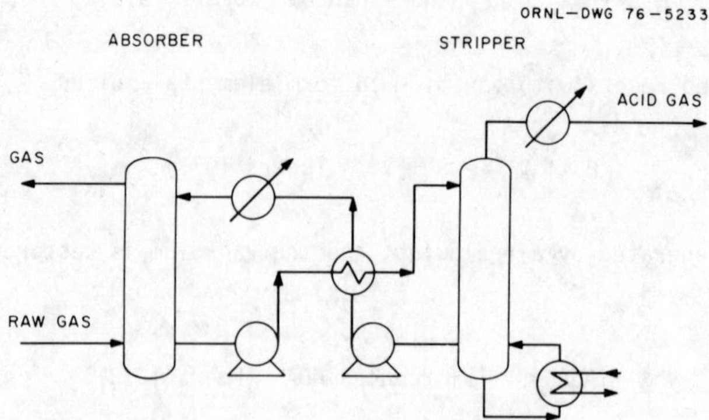

Fig. 4.6. Alkazid process flow chart. Source: After Wall 1973, p. 91. Reprinted by permission of the publisher.

The Benfield (Fig. 4.7) and Catacarb (Fig. 4.8) processes use hot potassium carbonate to remove acid gas. Steam stripping and pressure reduction release the hydrogen sulfide and carbon dioxide. Unfortunately, the high carbon dioxide removal by the carbonate reduces the efficiency of the Claus sulfur recovery plant (Panel on Coal Gasification Technology 1974).

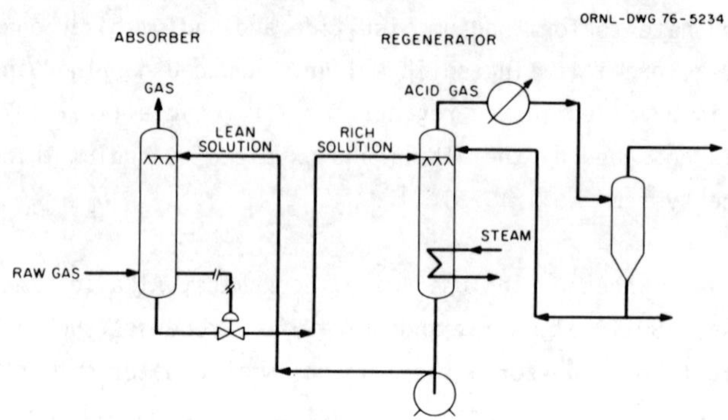

Fig. 4.7. Benfield process flow chart. Source: After Wall 1973, p. 92. Reprinted by permission of the publisher.

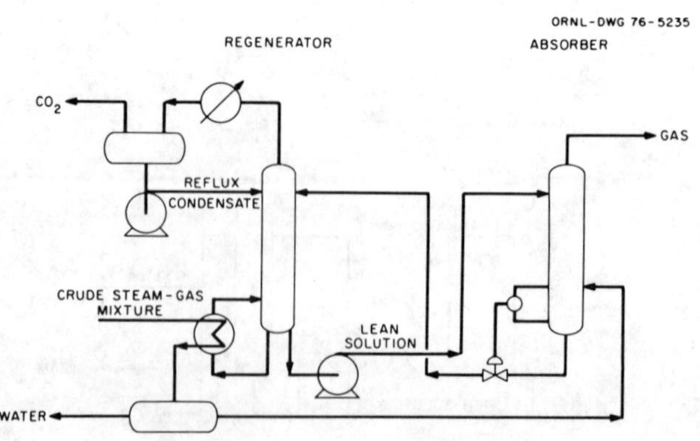

Fig. 4.8. Catacarb CO_2 Removal process flow chart. Source: After Wall 1973, p. 93. Reprinted by permission of the publisher.

The Purisol (Fig. 4.9), Fluor (Fig. 4.10), and Selexol (Fig. 4.11) processes absorb acid gas
(and some methane) at elevated pressure and release them at reduced pressure. Because these
processes use expensive solvents, solvent losses must be minimized (Panel on Coal Gasification
Technology 1974).

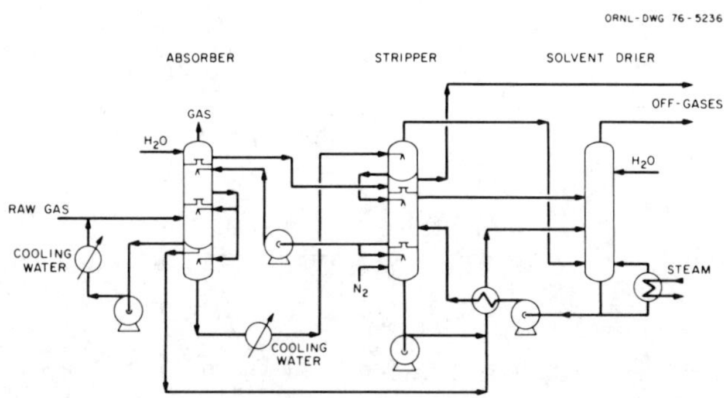

Fig. 4.9. Purisol process flow chart. <u>Source:</u> After Wall 1973, p. 98.
Reprinted by permission of the publisher.

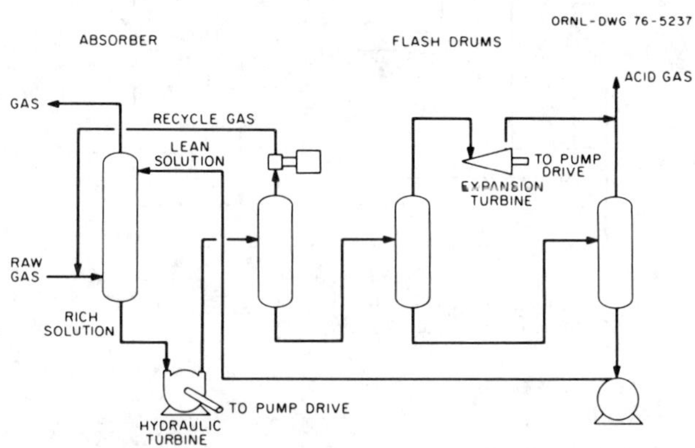

Fig. 4.10. Fluor process flow chart. <u>Source:</u> After Wall 1973, p. 95.
Reprinted by permission of the publisher.

Methyl alcohol is the solvent for the Rectisol process (Fig. 4.12); however, the process operates
at temperatures requiring refrigeration (Panel on Coal Gasification Technology 1974). The
Rectisol process used by the El Paso Burnham I Gasification (Lurgi) Complex removes sulfur to
0.2 vol ppm as well as hydrogen cyanide. Unlike most sulfur recovery units, Rectisol removes
saturated and unsaturated hydrocarbons without contaminating the methanol solvent beyond regeneration (Gibson, Hammons, and Cameron 1974). Shown below is the approximate composition of the
offgas from the Rectisol solvent regeneration section for the El Paso Burnham I Coal Gasification
(Lurgi) Complex; the offgas may be sent to a Stretford unit for sulfur recovery (Gibson, Hammons,
and Cameron 1974).

Component	Mole %
CO_2	97.63
H_2S	0.75
C_2H_4	0.24
CO	0.07
H_2	0.43
CH_4	0.56
C_2H_6	0.32
N_2 + A	
Total	100.00

ORNL-DWG 76-5238

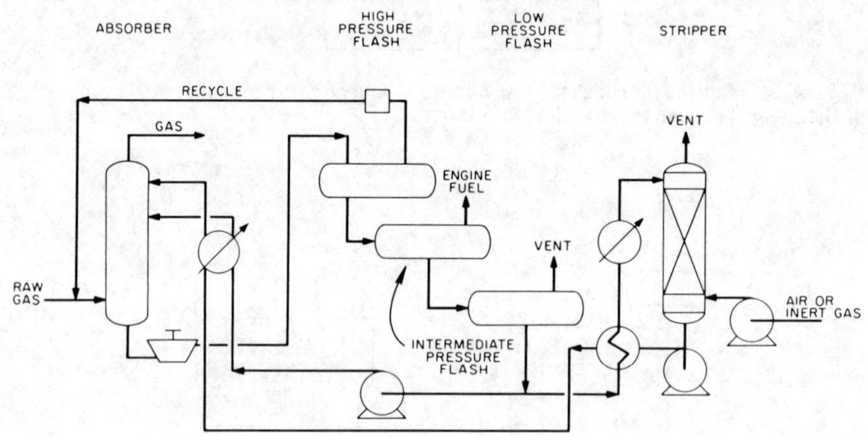

Fig. 4.11. Selexol process flow chart. <u>Source</u>: After Wall 1973, p. 100. Reprinted by permission of the publisher.

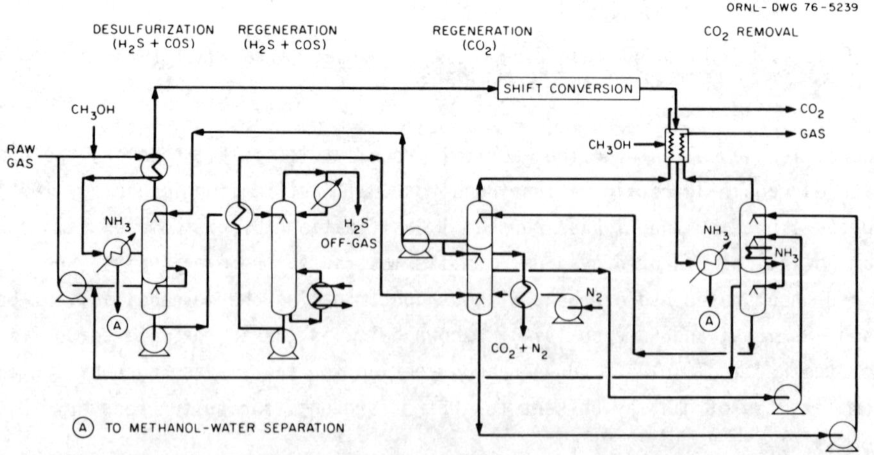

Fig. 4.12. Rectisol process flow chart. <u>Source</u>: After Wall 1973, p. 99. Reprinted by permission of the publisher.

The vented Rectisol carbon dioxide stream (284 million scf per day) for the Wesco Coal Gasification (Lurgi) Plant is composed of carbon dioxide (98.3%), hydrocarbons (CH_4, C_2H_6, C_2H_4), and 35 ppm carbonyl sulfide (Berty and Moe 1974). A Rectisol unit at South African Coal, Oil and Gas Corporation (SASOL) has achieved sulfur recovery removal to 0.007 vol ppm (Gibson, Hammons, and Cameron 1974).

The Sulfinol process (Fig. 4.13) operates similarly to amine processes and is selective for hydrogen sulfide (Panel on Coal Gasification Technology 1974), carbon disulfide, and mercaptans (Glaser, Hershaft, and Shaw 1974).

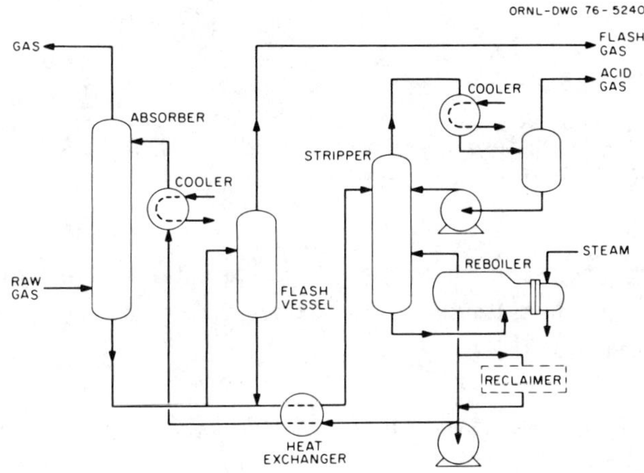

Fig. 4.13. Sulfinol process flow chart. Source: After Wall 1973, p. 102. Reprinted by permission of the publisher.

The Adip process (Fig. 4.14) uses an aqueous solution of an alkanolamine to remove the acid gas. In the absorber, the hydrogen sulfide feed is contacted countercurrently with the Adip solution. In the regenerator, acid gases are stripped. The absorbent is then cooled and recycled (Wall 1974).

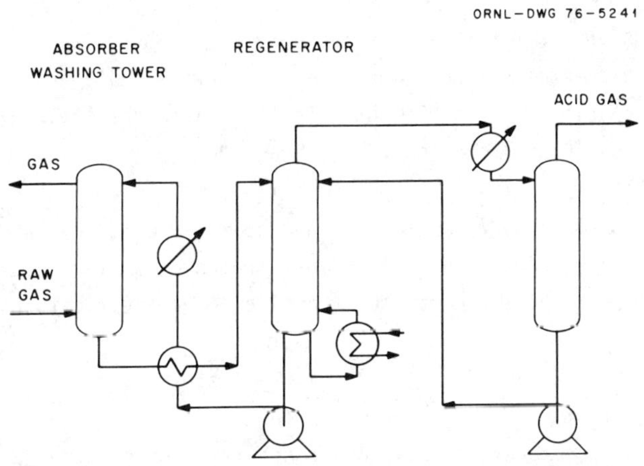

Fig. 4.14. Adip process flow chart. Source: After Wall 1973, p. 90. Reprinted by permission of the publisher.

The SNPA-DEA (Société National des Petroles d'Aquitaine-Diethanolamine) process uses an aqueous solution of DEA to remove the hydrogen sulfide and carbon dioxide (Fig. 4.15). In the regenerator the acid gases are stripped from the DEA, cooled, and routed to a sulfur recovery unit (Wall 1974).

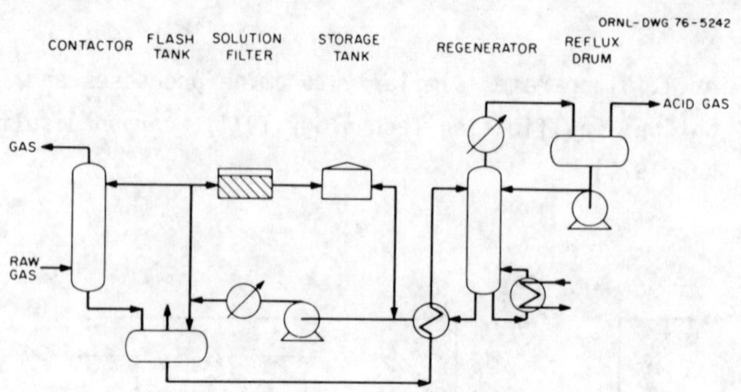

Fig. 4.15. SNPA-DEA process flow chart. Source: After Wall 1973, p. 101. Reprinted by permission of the publisher.

Fugitive sulfur, trace amounts remaining in the product gas stream after acid gas removal, is adsorbed by sulfur guards (or drums) such as iron oxide, zinc oxide, or carbon prior to methanation (Glaser, Hershaft, and Shaw 1974).

Iron sponge (wood shavings impregnated with hydrated iron oxide), which selectively absorbs hydrogen sulfide, is economical for gases containing up to 300 ppm of hydrogen sulfide. Regeneration is by air oxidation (Panel on Coal Gasification Technology 1974).

Activated carbon absorbs hydrogen sulfide, but minimal carbon dioxide and methane must be present in the gas stream for proper operation. Regeneration is by steam and air. This process is best for removing low levels of sulfur because the carbon is easily overloaded, deactivated, and abraded (Panel on Coal Gasification Technology 1974).

4.1.2.2 Sulfur recovery

Stripped sulfur compounds are sent to a sulfur recovery unit, where they are converted to either dilute sulfuric acid or elemental sulfur. Most sulfur recovery units are designed to use Claus technology and to recover sulfur in its elemental form; environmentally, this form is the most compact and least obnoxious (Glaser, Hershaft, and Shaw 1974).

Claus plants produce sulfur from hydrogen sulfide and sulfur dioxide by gas phase reactions near atmospheric pressure; catalysts are used in the final stages of sulfur recovery. The following reaction takes place in the catalytic converter (Panel on Coal Gasification Technology 1974):

$$2H_2S + SO_2 \xrightarrow[\text{catalyst}]{\text{bauxite}} 1.5S_2 + \text{steam} \quad . \tag{8}$$

Claus reaction variations accommodate the various concentrations of acid gas feeds; variations include direct oxidation, split-flow, straight-through, and sulfur recycle processes (Fig. 4.16). The optimum Claus process depends primarily on the hydrogen sulfide concentration in the feed.

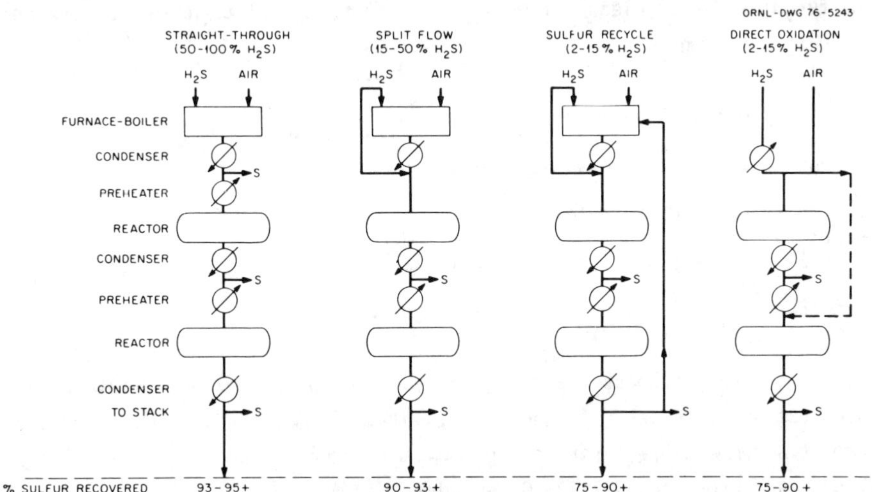

Fig. 4.16. Claus plant gas phase reaction variations. <u>Source</u>: After Grekel, Palm, and Kilmer 1968, Fig. 1. Reprinted by permission of the publisher.

The direct oxidation and sulfur recycle variations are designed for dilute acid gas feeds (low H_2S, 15 mole %). The split-flow variation is suited for intermediate hydrogen sulfide concentrations (about 50 mole %), whereas high hydrogen sulfide concentrations (90 mole %) are required for the straight-through variation. Reheat variations within the Claus variations have little effect on the sulfur recovery efficiency, but are important in adapting the processes to the acid gas producing plant. Conversion percentages are affected by the number of catalytic stages: The higher the number of catalytic stages, the higher the sulfur recovery (Beers 1973).

Hydrocarbons in the feed to Claus units cause an increase in undesirable side reaction products such as carbonyl sulfide and carbon disulfide (Barry 1972). The presence of hydrocarbons in the feed gas to the straight-through process requires increased equipment capacity and lowers sulfur recovery, whereas hydrocarbons in the feed gas to the split-flow process help to maintain the flame stability. Bypassed hydrocarbons in the reactor crack over the bauxite catalyst and contaminate the product sulfur for the split-flow, direct oxidation, and sulfur recycle processes (Grekel, Palm, and Kilmer 1968).

4.1.2.3 <u>Tail gas cleanup</u>

Claus technology units, themselves, must have tail gas treatment units to reduce sulfur emissions (Beers 1973). Unrecovered sulfur in Claus plant tail gas includes primarily hydrogen sulfide, elemental sulfur, sulfur dioxide, and lesser amounts of other sulfur compounds (Beers 1973). Stretford units (using Claus technology) are not capable of recovering sulfur from carbonyl sulfide or carbon disulfide, which as a result, are presently vented to the atmosphere with fugitive hydrogen sulfide (Glaser, Hershaft, and Shaw 1974). Before incineration, tail gas from Amoco Production Company sulfur recovery plants (using Claus technology) is composed predominately of nitrogen, carbon dioxide, and water vapor with small quantities of hydrogen sulfide (0.5 to 1.5 mole %), sulfur dioxide (0.25 to 0.75 mole %), carbonyl sulfide (0.01 to 0.04 mole %), carbon disulfide (0.01 to 0.05 mole %), plus fractions of a percent of hydrogen, carbon monoxide, and elemental sulfur (vapor and entrained) (Beers 1973).

Options for Claus tail gas disposal include venting, incineration, which converts the sulfur compounds to sulfur dioxide, or cleanup (Beers 1973). A number of technologies are being developed to reduce Claus unit emissions:

SCOT (Shell Claus Offgas Treating)
Aquaclaus
Chiyoda Thoroughbred 101 flue gas desulfurization
Cleanair
IFP (Institute Francais du Petrole)
Beavon sulfur removal
Wellman-Lord SO_2 recovery or flue gas desulfurization
Sulfreen
Topsoe-SNPA (Société National des Petroles d'Aquitaine)

Claus plant tail gas cleanup processes are of two types — low-temperature Claus processes and conversion–concentration processes. In the first, sulfur formation is promoted by operating a catalytic system at a temperature favoring thermodynamic equilibrium: Topsoe-SNPA and Sulfreen (Naber, Wesselingh, and Groenendaal 1973; Groenendaal and Meurs 1972). In the second, sulfur compounds are converted to hydrogen sulfide or sulfur dioxide, which is then converted to elemental sulfur, sulfuric acid, or gypsum. The SCOT, Aquaclaus, Chiyoda Thoroughbred, Cleanair, IFP, Beavon, and Wellman-Lord processes are of the second type.

The Topsoe-SNPA process uses a dry catalysis method (catalytic oxidation and reduction) for treating Claus tail gas (Fig. 4.17). Entering tail gas is heated, converting hydrogen sulfide, sulfur, hydrogen, and carbon monoxide to their oxidized forms. Cooled gas from the converter passes to the two-stage absorption–acid concentration system. Cooling and some acid contact absorption occurs in the first stage. In the second stage, absorption is completed. Gas from the absorber contains about 25 ppm sulfur trioxide and less than 400 ppm sulfur dioxide and passes through a mist eliminator before being emitted (Slack and Hollinden 1975).

ORNL DWG 76-11474

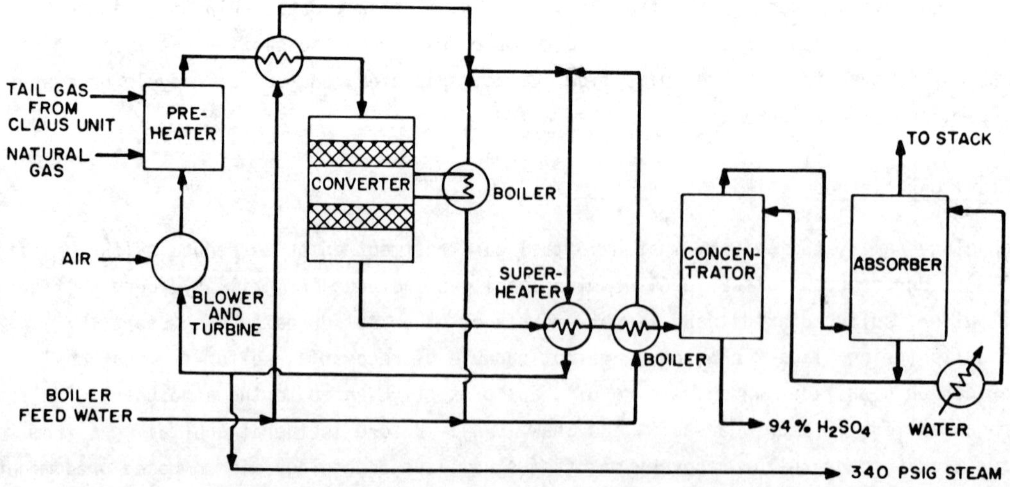

Fig. 4.17. Topsoe-SNPA process for treating Claus plant tail gas. Source: From Slack and Hollinden 1975, Fig. 11.4, p. 286. Reprinted by permission of the publisher.

In the Sulfreen process (Fig. 4.18), hydrogen sulfide and sulfur dioxide from Claus tail gas react at temperatures below the sulfur dew point by means of an alumina or activated carbon catalyst:

$$2H_2S + SO_2 = 3S + H_2O + 35 \text{ kcal} \qquad (9)$$

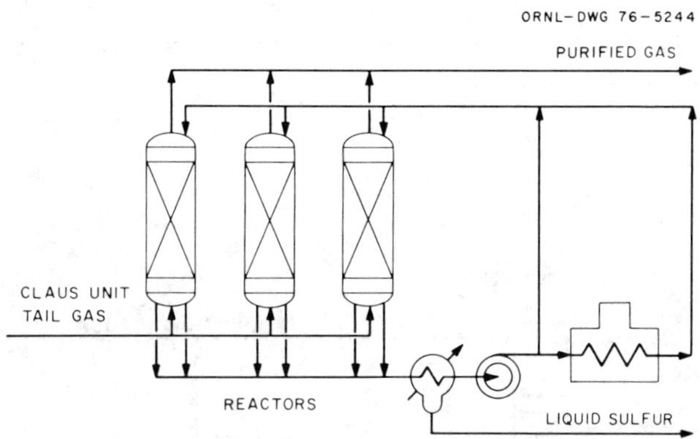

Fig. 4.18. Sulfreen process reactors. Source: After Wall 1973, p. 115. Reprinted by permission of the publisher.

The catalyst adsorbs the sulfur and releases it upon regeneration (Wall 1973). Because only solid adsorbents are used, no liquid wastes (other than product sulfur) are produced (Wall 1973). The number of reactors is determined by economic considerations (Wall 1973). The Sulfreen process removes up to 90% of the sulfur dioxide from the Claus tail gas (Barry 1972).

The SCOT process (Fig. 4.19) removes residual sulfur from Claus tail gas streams in two stages — reduction and absorption. In the reduction stage, sulfur compounds and elemental sulfur in the offgas are reduced to hydrogen sulfide. In the absorption stage, water is removed by condensation, and hydrogen sulfide, removed by alkanolamine absorption-regeneration, is recycled to the Claus unit. The catalyst is cobalt-molybdenum-on-alumina (Naber, Wesselingh, and Groenendaal 1973).

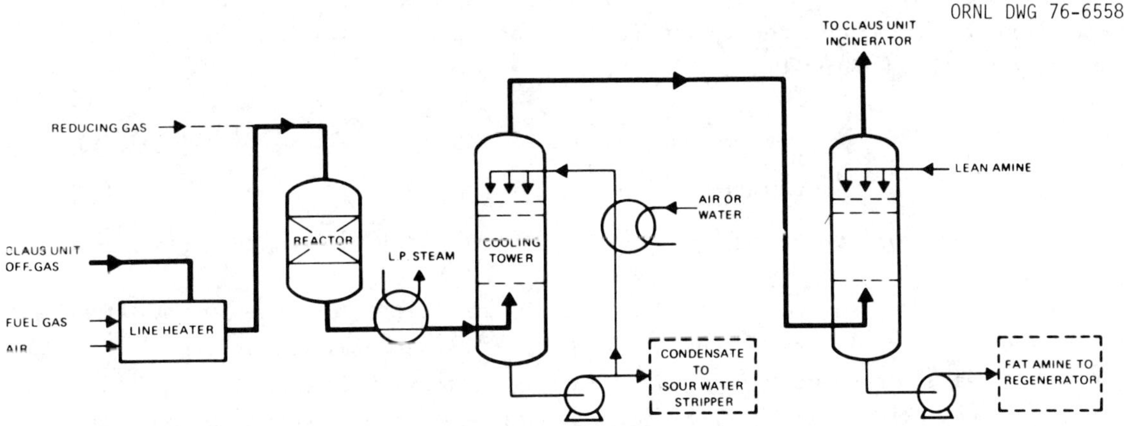

Fig. 4.19. Shell Claus Offgas Treating (SCOT) process flow diagram. Source: From Naber, Wesselingh, and Groenendaal 1974, Fig. 1, p. 29. Reprinted by permission of the publisher.

The Aquaclaus process was developed as a Claus plant tail gas cleanup, but it has a wider applicability for other sulfur-producing industries and power plants (Fig. 4.20). In the Aquaclaus
absorber, sulfur dioxide is removed by an absorbent solution (phosphate base) (Slack and Hollinden
1975) so that gaseous effluents from the absorber vented to the atmosphere contain less than
100 ppm of sulfur dioxide. The absorber solvent, rich in sulfur dioxide, then reacts with hydrogen
sulfide, causing sulfur to be precipitated and the solvent to be regenerated. The sulfur slurry
is processed for sulfur recovery in the elemental form. Sodium sulfate from side reactions (2
to 3% of the sulfur dioxide in the feed gas) must be removed; sulfate is capable of being processed
for resale or disposal (Hayford, VanBrocklin, and Kuck 1973; 1975).

ORNL DWG 76-6559

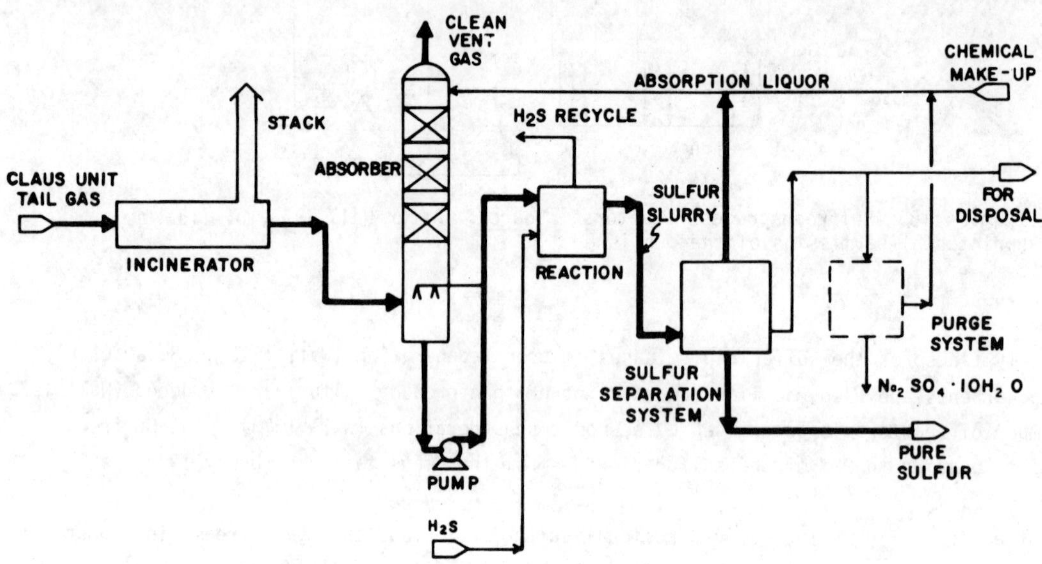

Fig. 4.20. Conceptual flow diagram of the new Aquaclaus process system. Source: From
Hayford, VanBrocklin, and Kuck 1973, Fig. 1, p. 54. Reprinted by permission of the publisher.

The Chiyoda Thoroughbred process is a wet desulfurization process capable of high sulfur (SO_x)
removal from flue gases, producing gypsum for resale or diposal (Fig. 4.21). The process reaction
sequence is as follows (Beers 1973):

$$\text{Absorption:} \quad SO_2 + H_2O = H_2SO_3 \; ; \qquad (10)$$

$$\text{Oxidation:} \quad H_2SO_4 + 1/2 \; O_2 = H_2SO_4 \; ; \qquad (11)$$

$$\text{Crystallization:} \quad H_2SO_4 + CaCO_3 + H_2O = CaSO_4 \cdot 2H_2O \; (s) + CO_2 \; (g) \; . \qquad (12)$$

In the absorber, sulfur dioxide and sulfur trioxide in the flue gas (precooled and dust-free) are
absorbed and removed by dilute sulfuric acid at 50 to 70°C. Sulfurous acid in the dilute sulfuric
acid is oxidized into sulfuric acid by using a catalyst. The sulfuric acid (from the flue gas SO_2
and SO_3) is sent to the crystallizer, whereas the remainder is recirculated to the absorber. In
the crystallizer, the sulfuric acid is neutralized with calcium compounds (natural limestone,
quicklime, or calcium oxide, slaked lime, calcium hydroxide, and carbide residue), ultimately
producing gypsum (Beers 1973).

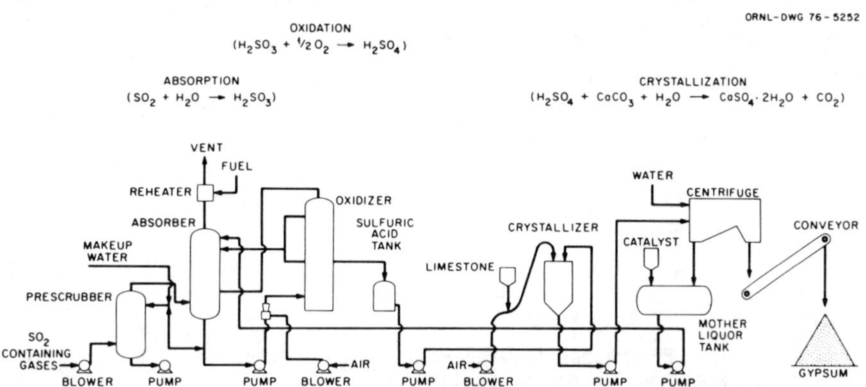

Fig. 4.21. Flow diagram of Chiyoda Thoroughbred process. <u>Source</u>: After Beers 1973, pp. 134-35. Reprinted by permission of the publisher.

The Cleanair process was developed as an addition to existing Claus plants or a part of future Claus plants (Fig. 4.22). In the first stage, sulfur dioxide in the Claus tail gas stream is converted to elemental sulfur, which is recovered for resale or disposal. Stage 2 employs the Stretford process, which uses a gas washing system in which the gas is contacted countercurrently with an alkaline washing solution, ultimately producing elemental sulfur. In stage 3, carbonyl sulfide and carbon dioxide are removed (Beers 1973). The process is capable of removing sulfur dioxide to 250 ppm or less (Barry 1972).

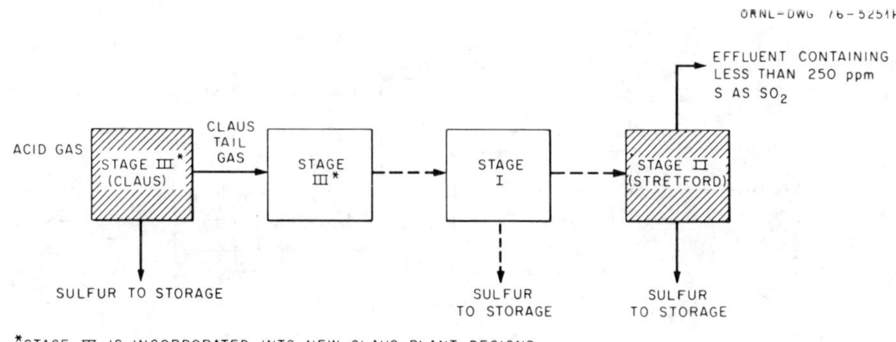

Fig. 4.22. Cleanair sulfur process scheme. <u>Source</u>: After Beers 1973, p. 112. Reprinted by permission of the publisher.

The reaction for the IFP process is

$$2H_2S + SO_2 = 3S + 2H_2O \tag{13}$$

in a catalyst-containing solvent (Fig. 4.23). The catalyst combines with the sulfur dioxide, hydrogen sulfide, and sulfur to yield an active complex, which in turn reacts with additional feed hydrogen sulfide and sulfur dioxide to produce elemental sulfur. For maximum conversion, the ratio of hydrogen sulfide to sulfur dioxide should be between 2.0 and 2.4. The IFP process does not recover carbonyl sulfide and carbon disulfide, but requires their minimized production

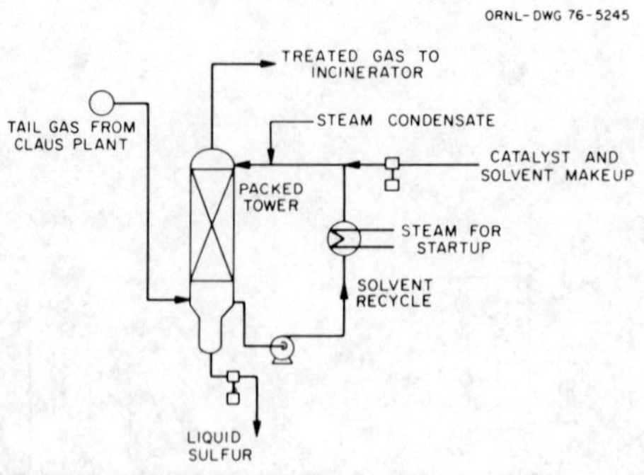

Fig. 4.23. IFP process flow chart. <u>Source</u>: After Beers 1973, p. 143. Reprinted by permission of the publisher.

from the Claus plant (Beers 1973). The system reduces Claus unit tail gas to 1500 to 2500 ppm after incineration (Barthel et al. 1971) or to 1000 ppm, as reported by Barry (1972). Both the Sulfreen (vapor phase) and IFP (liquid phase) processes merely use different catalysts to extend the Claus reaction (Barry 1972).

In the first step of the Beavon process (Fig. 4.24), Claus tail gas sulfur compounds (SO_2, S_x, COS, and CS_2) are converted to hydrogen sulfide by hydrogenation and hydrolysis in the presence of a catalyst. The stream is cooled by contact with a slightly alkaline buffer solution. A Stretford process absorber using an oxidizing alkaline solution produces elemental sulfur (Wall 1973). Sulfur dioxide removal to 250 ppm or less is achieved (Barry 1972).

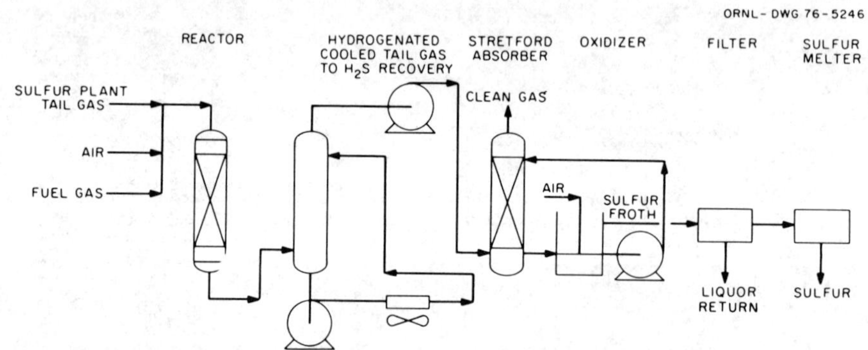

Fig. 4.24. Beavon process flow chart. <u>Source</u>: After Wall 1973, p. 111. Reprinted by permission of the publisher.

The Wellman-Lord process (Fig. 4.25) is a wet regenerative sulfur dioxide, sulfur trioxide, acid mist, and fly ash removal system producing sulfur dioxide in pure gaseous or liquid forms suitable for the manufacture of sulfuric acid or elemental sulfur. It uses sodium sulfite–bisulfite solution based absorption, recovering sulfur dioxide for reuse in sulfur recovery plants (Beers 1974). The sulfur dioxide is removed by converting the absorbent to the bisulfite form. The sulfite is regenerated in the evaporator–crystallizer section by indirect heating (Wall 1973), producing sulfur dioxide, water, and sodium sulfite crystals (Craig 1972). Exit gas containing levels less than 100 vol ppm of sulfur dioxide (Wall 1973) [90% SO_2 removal efficiency (Hall et al. 1975)] is incinerated to convert trace hydrogen sulfide before exhaustion (Cleaning up SO_2 1973). Absorbent makeup rate is about 0.45 moles of sodium hydroxide per mole of sulfur dioxide removed

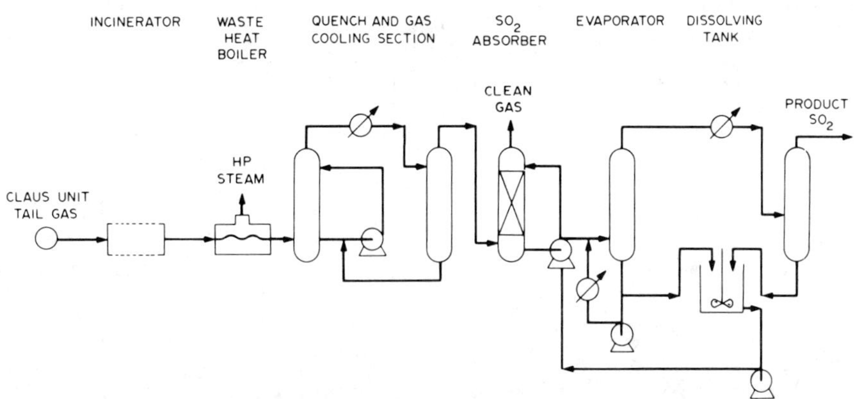

Fig. 4.25. Wellman-Lord SO$_2$ Recovery flow chart. <u>Source</u>: After Wall 1973, p. 116. Reprinted by permission of the publisher.

(10% of the sodium content) (Hall et al. 1975); however, process improvements should reduce the makeup to 0.10 moles of sodium hydroxide per mole of sulfur dioxide removed (Beers 1973). Sodium sulfate (formed by SO$_3$ absorption, disproportionation, and sulfite oxidation) and other sodium salts must be purged from the system (Princiotta 1974; Davis 1971).

For the Bi-Gas process, sulfur dioxide in the Claus tail gas and stack gas is scrubbed with sodium sulfite to form sodium bisulfite. The sulfite is regenerated by thermal decomposition, and the liberated sulfur dioxide is returned to the Claus plant for retreatment. Sodium sulfate, also formed during scrubbing, must be continually disposed of, thus necessitating makeup of sodium hydroxide to the scrubber. Estimates of sodium sulfate quantities produced range from 12 to 114 tons per day for the Solvent-Refined Coal (SRC) process and 66 to 315 tons per day for the Bi-Gas process, or 0.25 ton of Na$_2$SO$_4$ per ton of sulfur dioxide feed (Hittman Associates 1975a). Sodium sulfate and other soluble salts that have accumulated in the cooling water may have to be separated out and sold, stored, or disposed of in the ocean (Shaw 1976, personal communication). Sodium sulfate from sulfur recovery is water-soluble and is a possible leachate and groundwater contaminant if released in ash or other land-disposal form (Hittman Associates 1975a).

The sulfur dioxide stack gas pollution treatment technologies developed for industries such as sulfuric acid plants, smelters, and power plants may be immediately transferrable to Claus emissions (Barry 1972). Sulfur dioxide removal technologies can be grouped according to whether the trapped sulfur is contained in a "throwaway" product or recovered as elemental sulfur or sulfuric acid (Slack and Hollinden 1975):

Throwaway	Recovery
Dry process (lime)	Alkali absorbents
Wet processes	Alkaline earth absorbents
Direct lime-limestone scrubbing	Metal oxides
Indirect lime-stone scrubbing	Adsorption
	Catalytic oxidation and reduction

Throwaway operations, using lime and/or limestone, are of two types — wet or dry. Dry systems use boiler-injected limestone (as is or calcined) to recover sulfur oxides from the gas stream (Fig. 4.26). However, the disadvantages of the system (low SO$_2$ absorption, excess limestone

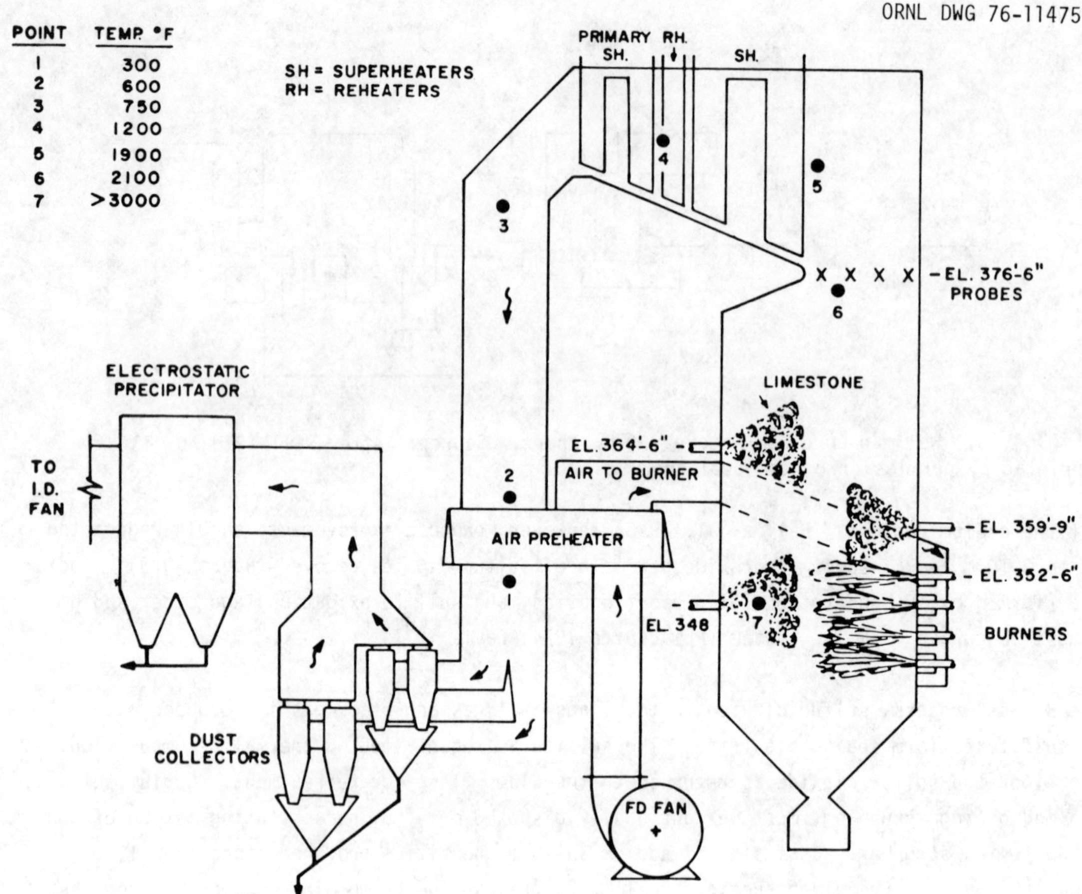

Fig. 4.26. Dry system for removing SO_2 from stack gas by sorption in limestone. Source: From Slack and Hollinden 1975, Fig. 2.1, p. 29. Reprinted by permission of the publisher.

requirements, fouling, lowering of electrostatic precipitator efficiency) make the process inadequate for sulfur dioxide emission control (Slack and Hollinden 1975).

Wet systems are of two types — direct and indirect lime-limestone scrubbers (Slack and Hollinden 1975). Direct processes use limestone ($CaCO_3$) or hydrated lime [$Ca(OH)_2$] added to the scrubber circuit or limestone injected into the boiler, causing calcination to lime, followed by lime slurry scrubbing (Fig. 4.27). The discharge stream from these processes, containing the throw-away reaction products, fly ash, and unreacted alkali, is dewatered and sent to a disposal pond or landfill site. The scrubbing reactions involved in these systems are:

$$\text{Limestone: } CaCO_3 + SO_2 + 1/2\ H_2O = CaSO_3 \cdot 1/2\ H_2O + CO_2 \ ; \qquad (14)$$

$$\text{Lime: } Ca(OH)_2 + SO_2 = CaSO_3 \cdot 1/2\ H_2O + 1/2\ H_2O \ . \qquad (15)$$

The Commonwealth Edison Will County Station achieves a 75 to 85% sulfur dioxide removal efficiency; major problems are demister pluggage, mechanical problems, and waste gypsum sludge disposal. Coal-fired power plants using carbide sludge [$Ca(OH)_2$] remove up to 80 to 95% of the sulfur dioxide in the flue gases (Princiotta 1974).

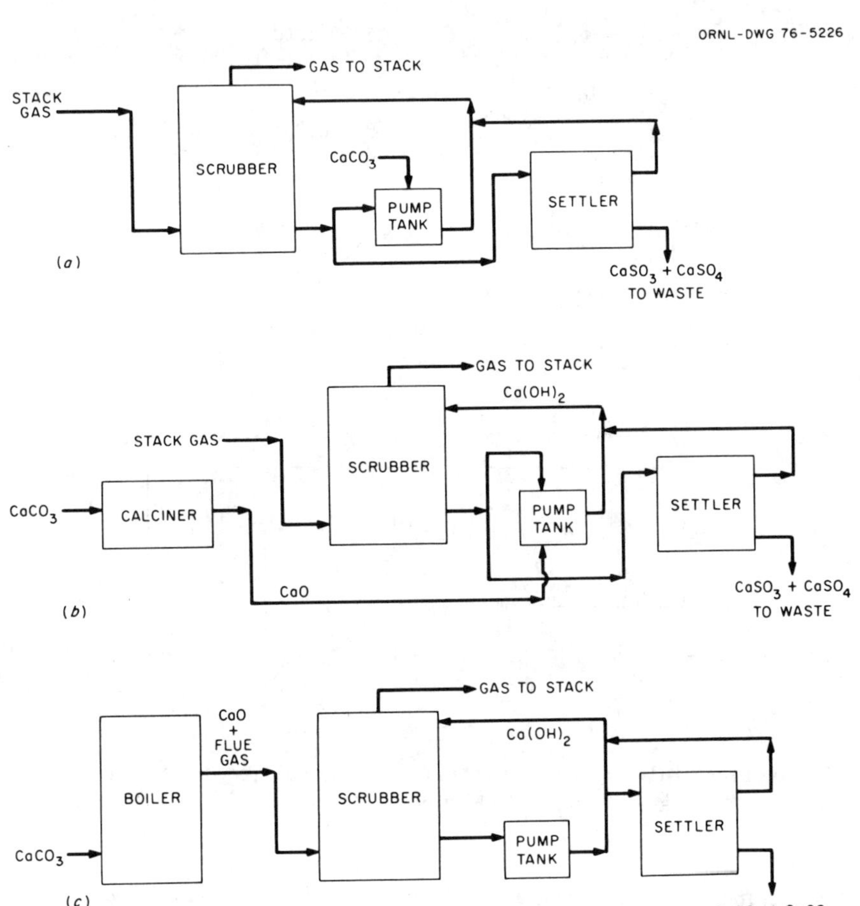

Fig. 4.27. Major process variations for use of lime or limestone for removal of SO$_2$ from stack gases: (*a*) scrubber addition of limestone; (*b*) scrubber addition of lime; (*c*) boiler injection of limestone. <u>Source</u>: After Princiotta 1974, Fig. A-1, p. 286.

Indirect lime-limestone scrubbing uses a soluble medium to absorb the sulfur dioxide, followed by the scrubber effluent solution of lime or limestone for medium regeneration or neutralization (Slack and Hollinden 1975). Absorbents used in this system include sodium salts, ammonium salts, water, ammonium sulfate, or carbon; the system is thus termed Double Alkali (Slack and Hollinden 1975). Double Alkali flue gas desulfurization systems (Fig. 4.28) scrub flue gases with sodium or ammonium salts in solution. The resulting liquor is then treated with lime or limestone, producing a throwaway sludge. Reactions for this system are:

$$\text{Scrubber:} \quad Na_2SO_3 + SO_2 + H_2O = 2NaHSO_3 \quad ; \tag{16}$$

$$\text{Regenerator:} \quad 2NaHSO_3 + Ca(OH)_2 = Na_2SO_3 + 3/2\ H_2O + CaSO_3 \cdot 1/2\ H_2O \quad . \tag{17}$$

The gypsum solids require disposal. A sulfur dioxide removal efficiency of 90 to 99% is expected on the basis of pilot-scale results (Princiotta 1974; Cornell and Dahlstrom 1975; LaMantia, Lunt, and Shah 1974; 1975). Performance results (Cornell and Dahlstrom 1975) from the Envirotech Double-Alkali process using sodium hydroxide (SO$_2$ sorbent) and lime (alkali regenerator) are listed below:

Inlet gas volume   3000 actual cfm from
                                                      fossil-fueled boiler

Particulate removal

    Inlet   0.2–0.3 grain/scf, dry
    Exit   0.01–0.02 grain/scf, dry

SO_2 removal

    Inlet   250–400 ppm
    Exit   15–40 ppm

Lime consumption   1.0–1.1 stoichiometric
                                            level based on SO_2 removal

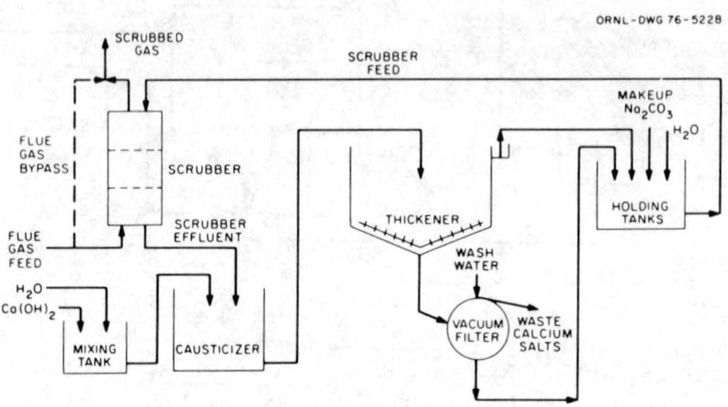

Fig. 4.28. Double Alkali process variation — sodium sulfite scrubbing with regeneration of the sulfite with lime. Source: After Princiotta 1974, Fig. A-2, p. 288.

Lime-limestone scrubbing sludge will contain calcium hydroxide, calcium carbonate, calcium sulfate, and calcium sulfite (Hittman Associates 1975*b*). In a power plant, sulfur residues are produced at an amount approximately 50% greater (dry weight) than fly ash without scrubbing; thus, the sludge plus ash throwaway is about 2.5 times the normal coal ash disposal tonnage (Hittman Associates 1975*b*).

Variables affecting sludge composition include composition of the coal, the alkali added, the scrubber process operation, oxidation, and alkali utilization efficiency (Selmeczi and Knight 1973). Selmeczi and Knight (1973) analyzed lime process sludges from various power plant sources and showed their compositions to be highly variable (Table 4.7).

Landreth and Mahloch (1975), in assessing the pollution potential and leachability of waste SO_x sludges (from the Double Alkali and limestone processes when using eastern and western coal and from the lime process when using eastern coal), have identified sludge and elutriate composition (Table 4.8). Sulfur dioxide, chlorine, and calcium are found to be the primary sludge and elutriate components. However, elutriate metals most frequently exceeding the water quality standards are selenium, mercury, manganese, and arsenic (Landreth and Mahloch 1975). Water layers equilibrated with sludge have shown 8000 ppm or more of total dissolved solids, thus necessitating leakproof ponding (Shaw 1976, personal communication).

Disposal options for gypsum (hydrated calcium sulfate) include adding it to the ash for landfill or minefill; thus, the term "throwaway" is used. Recent literature has detailed possible new uses for gypsum, such as in wallboard manufacture (Beers 1973; Slack and Hollinden 1975). One

Table 4.7. Chemical analysis of lime process sludges
on dry solid basis (%)

Component	Sample[a]					
	A	B	C	D	E	F
CaO	18.1	43.2	40.7	43.4	25.6	43.8
MgO	2.4	0.2	b	0.001	1.2	b
Total sulfur	7.2	18.9	18.1	20.0	10.9	22.9
SO_2	12.1	33.0	32.9	29.2	10.8	45.8
SO_3	2.9	5.9	4.8	13.6	13.6	c
CO_2	3.2	6.7	2.3	7.1	2.2	1.0
Free carbon	b	b	b	2.8	0.14	c
SiO_2	31.6	4.9	3.76	0.58	21.3	0.18
Al_2O_3	18.3	3.4	1.71	1.21	11.3	0.39
Fe_2O_3	4.3	0.6	0.86	0.39	5.6	0.29
Na_2O	b	b	b	0.35	0.76	0.09
K_2O	b	b	b	0.03	0.98	0.01
Free base as CaO	0.3	1.3	7.9	0.06	0.06	c

[a]A — power station prior to fly ash collections; B — power station after fly ash collection; C — Chemico using carbide lime; D — power plant using proprietary scrubbing; E — wet limestone pilot plant scrubber; F — molybdenum sulfide pilot plant scrubbing effluent.
[b]Not determined.
[c]Not detected.

Source: Selmeczi and Knight 1973, Table 1, p. 124.

SO_x removal process waste composed of CaO and $CaSO_4$ is being tested for wastewater treatment potential; it appears to be effective in removing turbidity and orthophosphate from wastewater (Lin 1975). Disposal options for the ash and sludge combination wastes from wet lime-limestone scrubbing currently under study are diagrammed in Figs. 4.29 and 4.30; however, use as a mineral aggregate or pozzolanic base appears to be the most promising option (Gogineni et al. 1972).

The sulfur dioxide scrubbing systems that yield a recoverable product use alkali absorbents, alkaline earth absorbents, metal oxides, adsorption, or catalytic oxidation and reduction. Alkali absorbents used include ammonium, sodium, potassium, lithium, hydrogen, or aluminum (Slack and Hollinden 1975). Processes using alkali absorbents are IFP, Wellman-Lord, Alkalized Alumina, Citric Acid, Aquaclaus, and Molten Carbonate. Descriptions of the IFP, Wellman-Lord, and Aquaclaus processes are presented earlier in this section.

Flue gas to the Alkalized Alumina process (Fig. 4.31) is fed to the alkalized alumina absorber, where the sulfur dioxide and sulfur trioxide are absorbed. The spent absorbent is transported to the regenerator, heated to 1200°F, and treated with producer gas. Absorbed sulfur dioxide reacts with the hydrogen and carbon monoxide of the producer gas to yield hydrogen sulfide, carbon dioxide, and water. One third of the hydrogen sulfide is oxidized to sulfur dioxide in a Claus unit; the gas streams are mixed and passed over the bauxite catalyst to recover elemental sulfur (Reid and Streebin 1972; Katell 1966).

Table 4.8. Chemical analysis of flue gas
desulfurization sludges

Parameter	Sludge	Elutriate/leachate
Total organic carbon (elutriate only)		X
Total solids	X	
Dissolved solids		X
Suspended solids (leachate only)		X
pH		X
Hardness (elutriate only)		X
Conductivity		X
Arsenic	X	X
Beryllium	X	X
Cadmium	X	X
Calcium	X	X
Chromium	X	X
Copper	X	X
Lead	X	X
Magnesium	X	X
Manganese	X	X
Mercury	X	X
Nickel	X	X
Selenium	X	X
Zinc	X	X
Chloride		X
Cyanide	X	X
Fluoride (calcium fluoride sludge only)		
Nitrate	X	X
Nitrite	X	X
Sulfate	X	X
Sulfite	X	X

Source: Adapted from Landreth and Mahlock 1975, Table 2, p. 217.
Reprinted by permission of the publisher.

ORNL DWG 76-6205

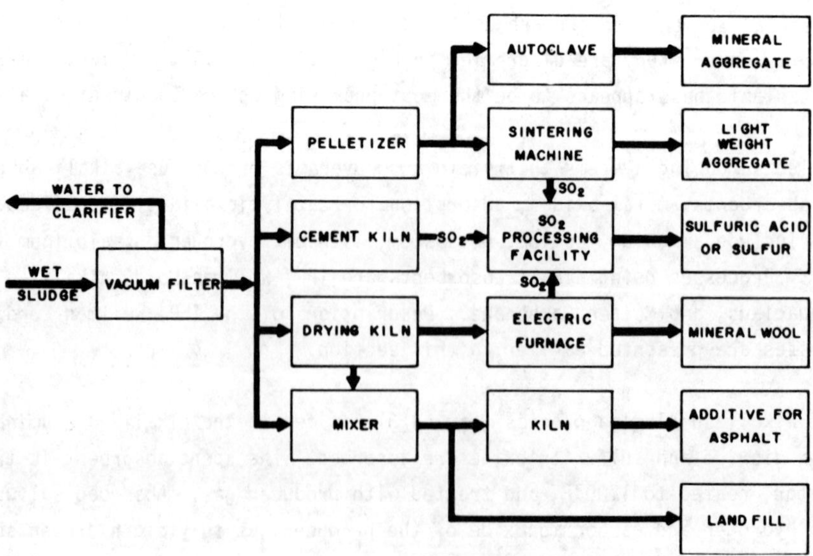

Fig. 4.29. Sludge disposal schemes for a single-scrubber system. Source: From Gogineni
et al. 1972, Fig. 7, p. 149.

ORNL DWG 76-6206

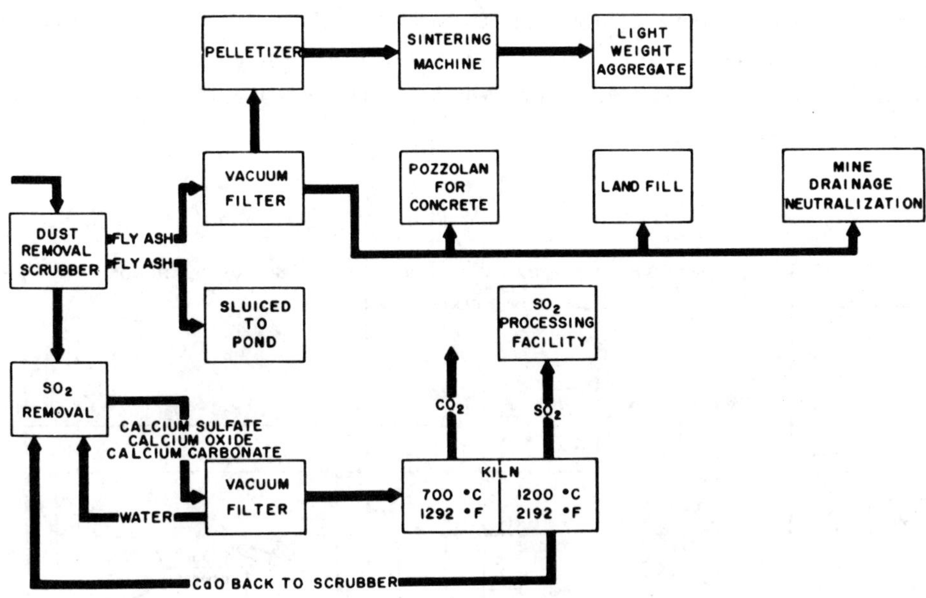

Fig. 4.30. Sludge disposal schemes for a dual-scrubber system. <u>Source</u>: From Gogineni et al. 1972, Fig. 8, p. 149.

ORNL DWG 76-6560

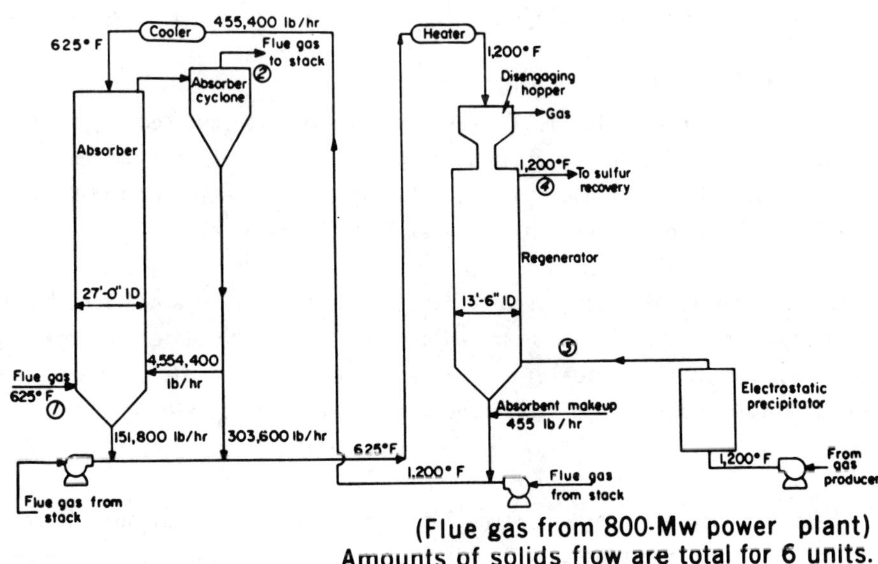

Fig. 4.31. Flow diagram of Alkalized Alumina process. <u>Source</u>: From Katell 1966, Fig. 3, p. 69. Reprinted by permission of the publisher.

The Citrate process (Fig. 4.32) of stack gas scrubbing has been used to treat lean smelter and coal-fired generating plant gases (Cleaning up SO_2 1973). The gases are cleaned of particulate matter and sulfur trioxide. In a packed absorption tower, the gas is scrubbed with a partially neutralized solution of citric acid and sodium carbonate. A sulfur dioxide removal efficiency of 90 to 95% is possible (Cleaning up SO_2 1973). The sulfur-dioxide-laden stream reacts with hydrogen sulfide to yield elemental sulfur, which is filtered from the recyclable sorbent (Cleaning up SO_2 1973).

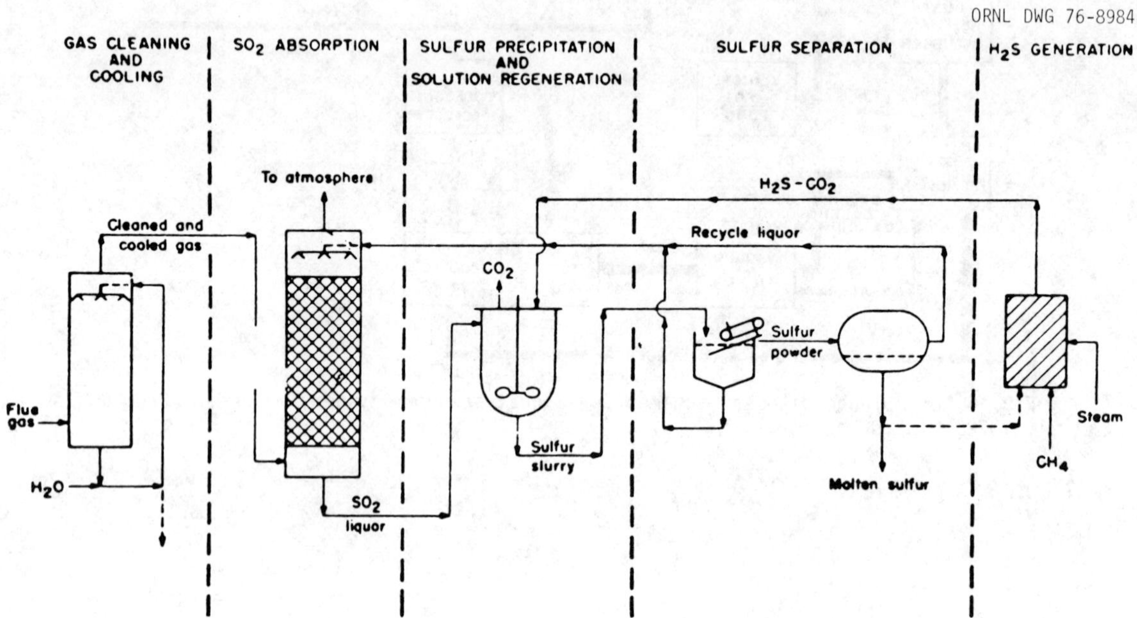

Fig. 4.32. Citrate process flow diagram. <u>Source</u>: From Hall et al. 1975, Fig. 8.25, p. 8-29.

The Molten Carbonate process (Fig. 4.33) uses a mixture of lithium, sodium, and potassium carbonates to remove SO_x from power plant stack gas. The product is a molten alkali (metal sulfites, sulfates, and unreacted carbonate) solution. Regeneration of the sorbent produces a hydrogen sulfide stream for recovery by a Claus plant (Yosim et al. 1972).

Alkaline earth adsorbents are being used for their easy thermal regeneration; however, they are in pilot or prototype stages. Alkaline earth adsorbents presently studied include magnesium (MgO Scrubbing process) primarily and calcium. Processes using calcium have not been developed as far as the magnesium-based methods; sulfur recovery from the product $CaSO_3$ or $CaSO_4$ is still being studied.

An aqueous slurry of magnesium oxide, magnesium sulfite, and magnesium sulfate removes sulfur dioxide in the MgO Scrubbing process (Fig. 4.34). The reaction involves the formation of additional magnesium sulfite by combining sulfur dioxide and magnesium oxide:

$$MgO(aq) + SO_2 = MgSO_3 \text{ (Koehler 1975)} \quad . \tag{18}$$

ORNL DWG 76-6203

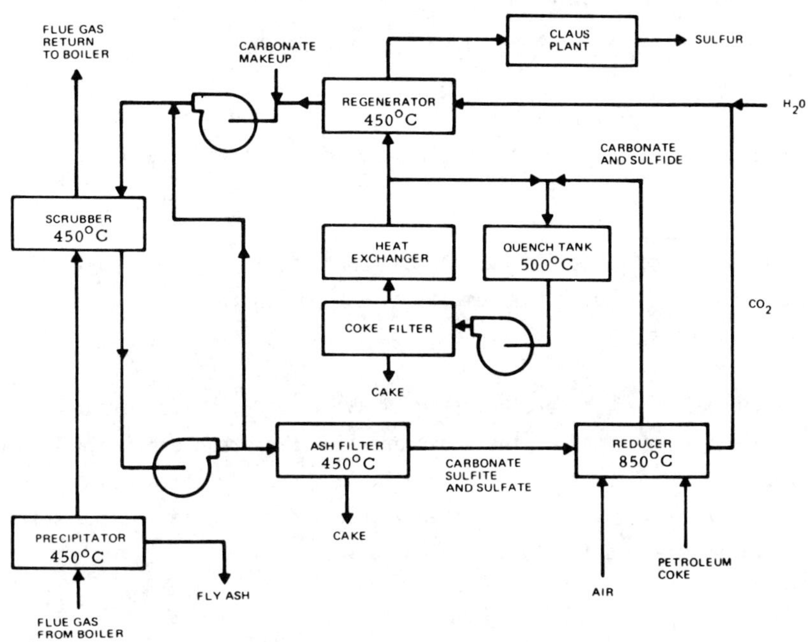

Fig. 4.33. Molten Carbonate process flow diagram. <u>Source</u>: From Yosim et al. 1972, Fig. 1, p. 175.

ORNL-DWG 76-5227

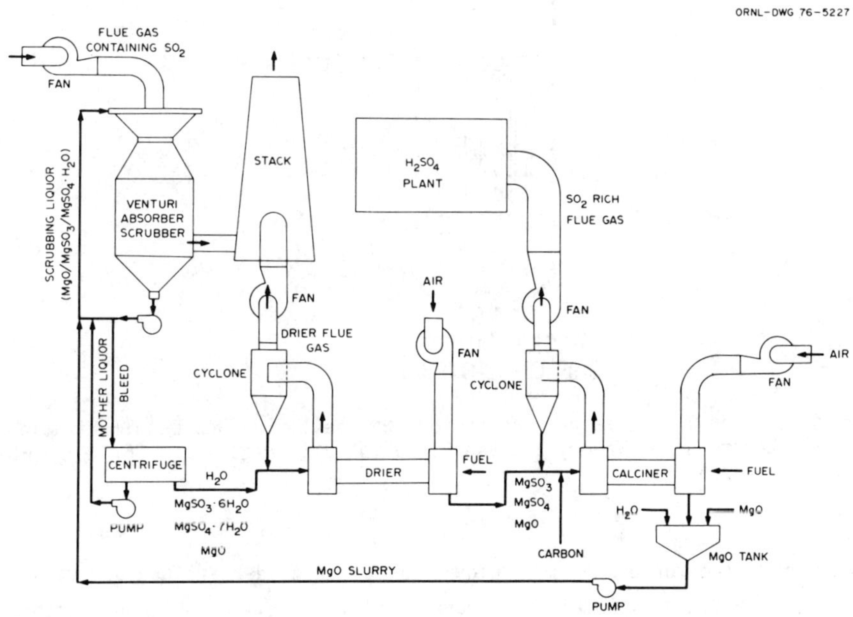

Fig. 4.34. MgO Scrubbing process flow chart. <u>Source</u>: After Princiotta 1974, Fig. A-3, p. 290.

Removed magnesium sulfite is dried and calcined to remove the sulfur dioxide and regenerate magnesium oxide through thermal decomposition:

$$MgSO_3 = MgO + SO_2 \text{ (Koehler 1975)} \tag{19}$$

The sulfur dioxide can be converted to sulfuric acid or to elemental sulfur. The MgO Scrubbing process is capable of removing 90% of the sulfur dioxide for a wide range of flue gas concentrations (Princiotta 1974; Koehler 1975).

Metal oxide processes, using primarily CuO and MnO_2 and some FeO, ZnO, V_2O_3, and V_2O_4, use the adsorbent in the solid form in a dry system (Slack and Hollinden 1975) such as the Shell Flue Gas Desulfurization process.

The Shell Flue Gas Desulfurization process (Fig. 4.35) is a dry sulfur dioxide removal process, which uses a copper-on-alumina bed as the acceptor. In swing reactors, metallic copper is oxidized to copper oxide. A fraction of the available copper oxide reacts with sulfur dioxide and oxygen to form copper sulfate. The sulfate is reduced to the metal, and sulfur dioxide is liberated for acceptor regeneration (Dautzenberg et al. 1971).

ORNL DWG 76-6561

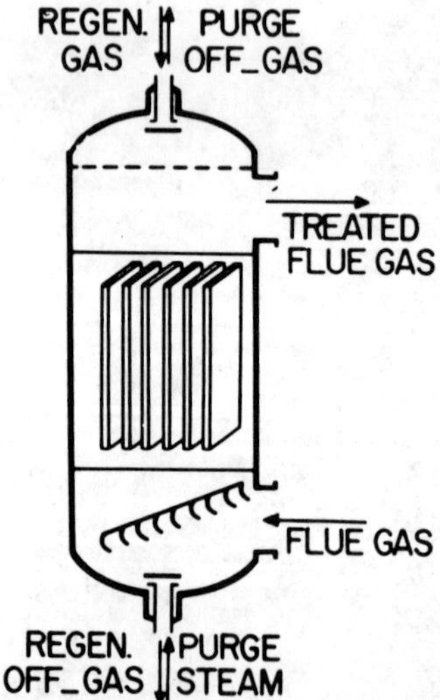

Fig. 4.35. Parallel passage reactor for the Shell Flue Gas Desulfurization process. Source: From Dautzenberg, Nader, and Van Ginnekin 1971, Fig. 1, p. 86. Reprinted by permission of the publisher.

Adsorbents used or tested for use in the removal of sulfur oxides include activated carbon (primarily), molecular sieves (containing natural zeolites), silica gel, and bentonites (Slack and Hollinden 1975). Adsorbent processes discussed here include the Reinluft, Westvaco, and the PuraSiv S.

The Westvaco sulfur dioxide recovery process uses a dry, fluidized activated carbon process to recover elemental sulfur (Fig. 4.36). The sulfur dioxide (as H_2SO_4) is sorbed on carbon and converted to molten sulfur by reaction with hydrogen sulfide. This process may prove applicable to clean up Claus tail gas (Ball et al. 1972). Pilot plant sulfur dioxide removal efficiencies averaged 94% from an inlet concentration of 2000 ppm (Slack and Hollinden 1975).

ORNL DWG 76-6204

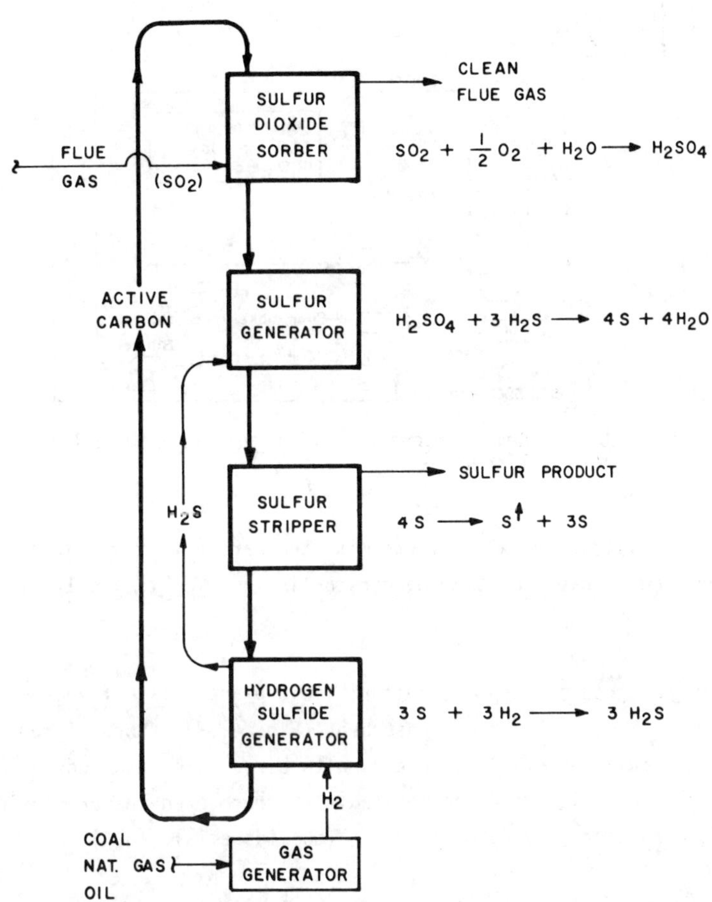

Fig. 4.36. Westvaco SO_2 Recovery process using dry, fluidized activated carbon. <u>Source</u>: From Ball et al. 1972, Fig. 1, p. 184.

The Reinluft process (Fig. 4.37) uses a fixed absorbent on a slowly moving bed of an activated charcoal. The feed flue gas, at an elevated temperature, enters the bottom of the absorber, where sulfur trioxide is removed. The gas is drawn off, cooled to 220°F, and returned to the absorber, where the sulfur dioxide in the gas is oxidized to sulfur trioxide and absorbed with water on the char to form sulfuric acid (Reid and Streebin 1972; Katell 1966). In the regenerating section, the temperature is raised to 700°F, causing the sulfuric acid to dissociate into sulfur

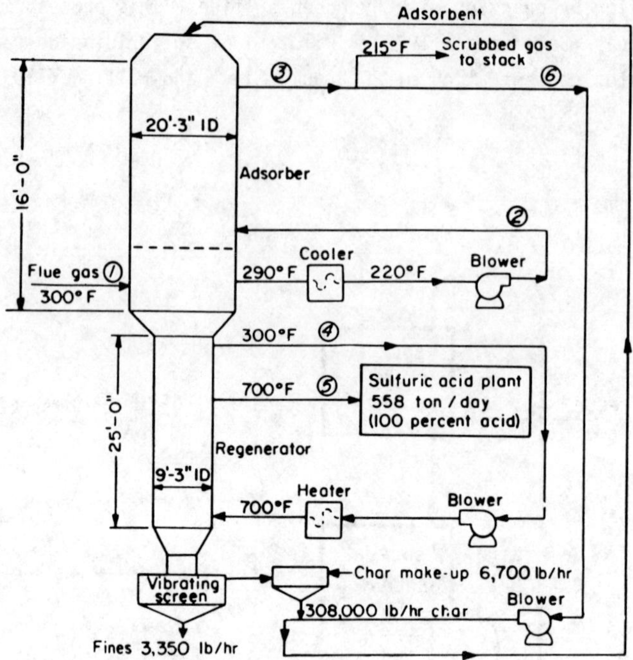

Fig. 4.37. Flow diagram of the Reinluft process. <u>Source:</u> From Katell 1966, Fig. 1, p. 67. Reprinted by permission of the publisher.

trioxide and water. The sulfur trioxide reacts with the carbon to form carbon dioxide and sulfur dioxide. The product gas from the regenerator is then fed to a sulfuric acid plant (Katell 1966).

The PuraSiv S system (designed for contact sulfuric acid plants) is a fixed-bed sulfur dioxide adsorption process using a molecular sieve adsorbent (Fig. 4.38). Sorbent regeneration is by heating and purging the sulfur-dioxide-laden beds with air. Using two beds allows the first to be adsorbing while the second is being regenerated. The recovered sulfur dioxide is recycled to the sulfuric acid plant for conversion to H_2SO_4 (Collins et al. 1974).

Catalytic oxidation and reduction sulfur dioxide removal processes convert the sulfur dioxide to recoverable compounds within the main gas stream. The sulfur dioxide is either oxidized to sulfur trioxide or reduced to sulfur. Sulfur is recovered, but sulfur trioxide is further converted to H_2SO_4 or is absorbed by another material to produce a useful product (Slack and Hollinden 1975). The Topsoe-SNPA, described in Sect. 4.1.2.3, and the Catalytic Oxidation processes are examples of this type.

The Catalytic Oxidation process (Fig. 4.39) uses a fixed catalyst bed of vanadium pentoxide to remove the sulfur from flue gases. The flue gases pass through a high-temperature electrostatic precipitator, where fly ash is removed; the gases are then sent to the catalyst, where the sulfur dioxide is oxidized to sulfur trioxide. Cooling of the exit gas to 200°F causes formation

ORNL DWG 76-6563

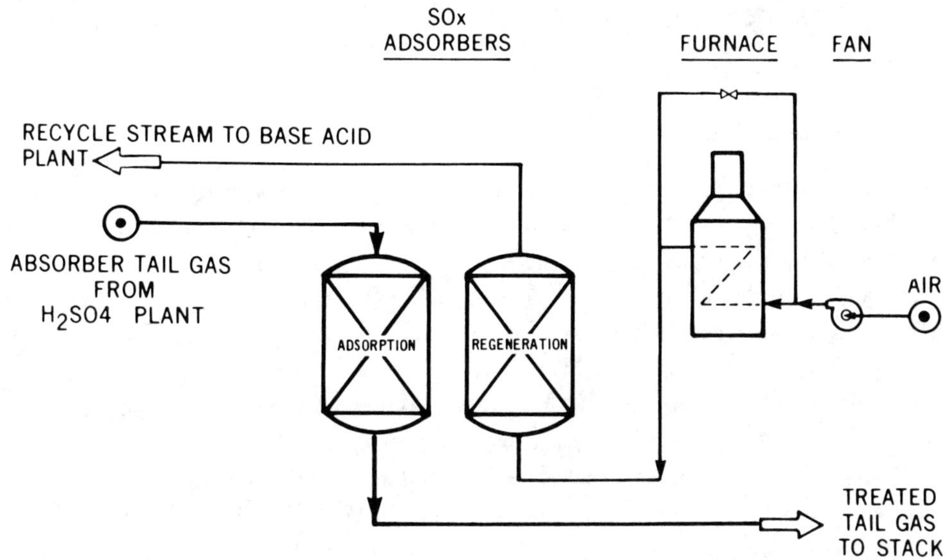

Fig. 4.38. Schematic flow diagram for the molecular sieve PuraSiv S unit. Source: From Collins et al. 1974, Fig. 2, p. 59. Reprinted by permission of the publisher.

ORNL DWG 76-6564

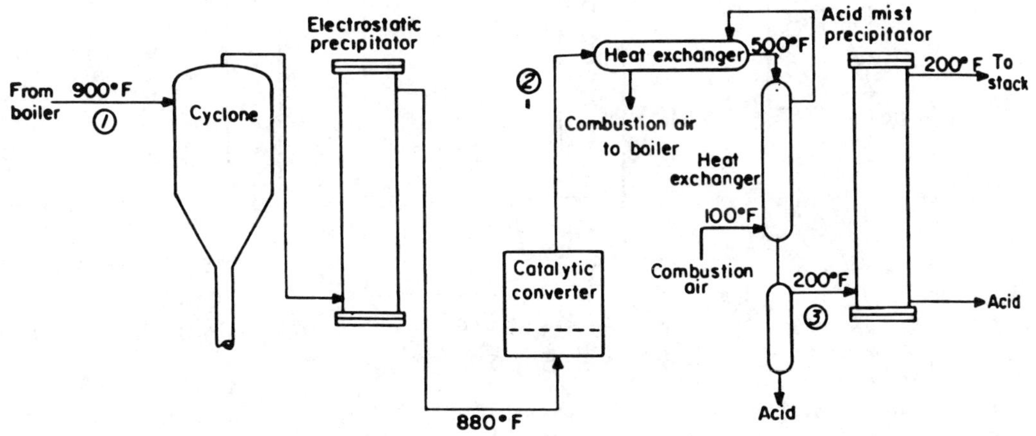

Fig. 4.39. Flow diagram of a catalytic-oxidation process. Source: From Katell 1966, Fig. 2, p. 68. Reprinted by permission of the publisher.

of sulfuric acid mist. The acid mist and droplets of condensed acid are removed by an electrostatic precipitator (Reid and Streebin 1972). The process removes 99% of the fly ash and 85%
of the sulfur dioxide and produces 78% sulfuric acid for resale (Miller 1974; 1975; Katell 1966;
Princiotta 1974).

4.1.2.4 Hydrodesulfurization

Hydrotreating of synthetic crude oils is required for pyrolysis and noncatalytic dissolution
processes. Hydrotreating desulfurizes the oil and provides a hydrogenated solvent oil (Beychok
1975). For the COED process (Kalfadelis and Magee 1975), hydrotreating upgrades the heavy
pyrolysis oil and converts sulfur to hydrogen sulfide, nitrogen to ammonia, and oxygen to water.
Hydrodesulfurization involves the general reaction (Schuit and Gates 1973):

$$\text{organic sulfur compound} + H_2 = H_2S + \text{desulfurized organic compound} \ \ .$$

Specific hydrodesulfurization processes have not been located for liquefaction processes. However, a simplified hydrodesulfurization process scheme is presented in Fig. 4.40.

ORNL DWG NO 76-11848

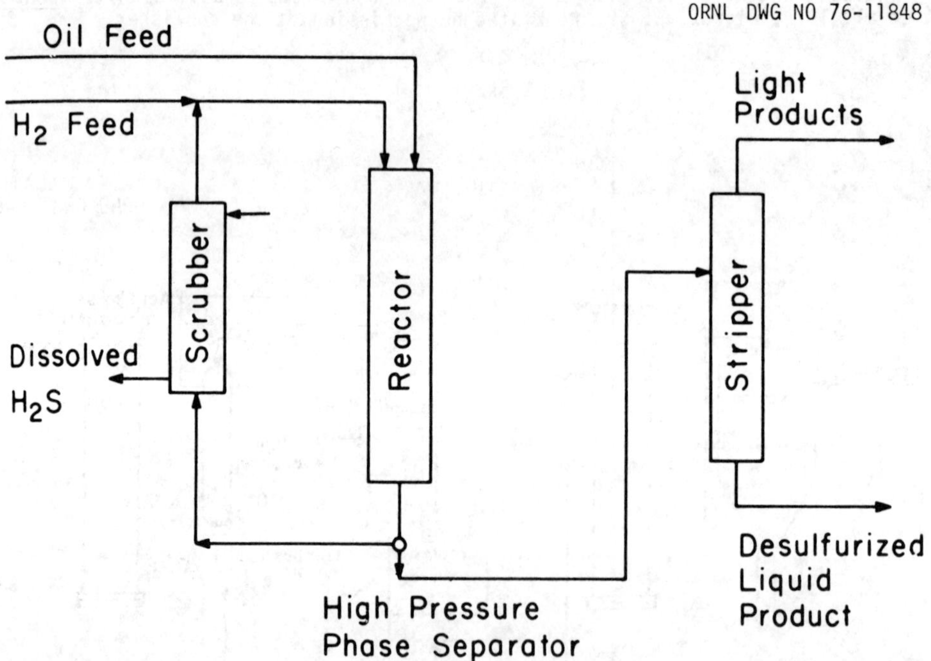

Fig. 4.40. Simplified diagram of hydrodesulfurization action process. <u>Source</u>: From
Schuit and Gates 1973, Fig. 1, p. 419. Reprinted by permission of the publisher.

Hydrodesulfurization uses fixed-, fluidized-, trickle-, or slurry-bed reactors, depending on the
type of feed (Schuit and Gates 1973). For crude oil hydrodesulfurization, process selection
varies according to the boiling ranges of the feed oil (Glaser et al. 1974). In general, lower-
boiling oils contain less sulfur and more easily treated forms than do higher-boiling materials
from the same crude oil (Glaser et al. 1974). Residual hydrodesulfurization, its residuum being
dirtier and more complex, is more difficult than distillate desulfurization (Glaser et al. 1974).

Katzer et al. (1975) are presently studying reaction networks and reaction kinetics for hydrodesulfurization and hydrodenitrogenation of compounds found in coal-derived liquids. Chirakaparambil (1974) studied the noncatalytic hydrodesulfurization of coal-derived liquids and found
the desulfurization of the feed oils to be relatively low over the range of variables (temperature, pressure, and space time) studied.

4.1.3 Sulfur balances

Balances comparing amounts of sulfur put in, recovered, and contained in products or by-products
(gas, oils, tars, chars) with amounts of fugitive emissions escaping to the environment through
atmospheric or wastewater releases indicate that the largest portion of the sulfur can be
recovered and that fugitive environmental emissions are a fraction of the total input sulfur for
conversion processes as a whole (Tables 4.9 through 4.16).

4.1.3.1 Ultimate disposal and/or recovery of sulfur

The largest amounts of sulfur (ranging from 50 to 98% of the total input) are expected to be
recovered by the recovery facilities and will require resale or environmentally safe disposal
as land- or minefill. Options for elemental sulfur disposal now include resale, use as landfill,
or return to coal mines; sulfuric acid allows only resale. Routes for elemental sulfur resale
include production of sulfur-containing fertilizers to slow the release of urea, sulfur-based
paving material as a bulk aggregate limestone substitute, sulfur foams for pavement subsurfaces,
and thermal and acoustical insulation (Hittman Associates 1975*b*). New uses for sulfur include
plastics, surface bonding structural materials, mortar and brick panels, asphalt paving mixes
and binders, soil stabilizers, and electrodes for lithium batteries (West 1975). Potential
agricultural uses of sulfuric acid include reclamation of sodic soils to increase plant nutrient
availability; treatment of alkaline or ammoniated irrigation water; control of certain weeds
and pathogens; and improvement of range grass establishment (Miyamoto, Ryan, and Stroehlin 1975).

Finding new routes for sulfur disposal may be important; the amounts of sulfur anticipated from
coal use may overload the present sulfur resale market, requiring development of either new
sulfur uses or acceptable methods of environmental sulfur disposal.

Significantly less amounts will be retained in the ash, tars, chars, and liquefaction solids
(filter cake). Sulfur in the ash is fixed to the carbon lattice and should be stable (Glaser,
Hershaft, and Shaw 1974). The sulfur content of tars (organic sulfides, mercaptans, and
thiophenes) (Glaser, Hershaft, and Shaw 1974) produced from the Synthane process ranges from
0.5 to 2.7 wt % (Forney et al. 1974). According to Yavorsky (1976), chars produced from pyrolysis
contain the highest amounts of sulfur. Liquefaction solids from the filtration of product oils
vary in sulfur composition with process types (Bazelmans et al. 1973). For the SRC process, the
mineral phase contains primarily ferrous sulfide reduced from the original pyrites (Bazelmans
et al. 1973). Consol process solids contain inorganic sulfur as ferrous sulfide and pyrite
combined as pyrrhotite. Sulfur levels, averaged from a number of Consol tests, are:

Type of sulfur compound	Percent sulfur in filter cake solids
Organic	44
Ferrous sulfides	46
Pyrites	10

Conversion processes designed to use their tars, chars, or filter cake as utility or steam production boiler fuel require stack gas cleaning units to treat the released sulfur dioxide emissions.

Even smaller amounts will be retained in the conversion gas or oil products. Estimated sulfur levels in various liquefaction solid and liquid streams are detailed in Table 4.9. Organic sulfur compounds identified in the Synthoil products formed by hydrogenation of Kentucky and Indiana No. 5 coals are listed in Table 2.36.

Table 4.9. Estimate of sulfur levels in combined liquid and solid products from liquefaction processes (%)

Process	Original sulfur content of coal[a]	Sulfur in liquid product	Sulfur in solids	Estimated sulfur in combined product
Solvent refined coal	3.5 - 4.3	1.0	4.6 - 7.0	1.7 - 2.3
H-coal	5.0	0.20	7.2	1.7
Consol	4.2	0.30	5.2	2.6
Bureau of Mines hydrodesulfurization	3.0	0.31	2.1	0.8

[a]For bituminous coals.

Source: Bazelmans et al. 1973, Table 7.18, p. 7-57.

For the Synthoil process (50,000 bbl of oil per day), sulfur recovery includes production of ammonium sulfate in addition to the elemental sulfur and sulfuric acid. The ammonia and hydrogen sulfide collected as aqueous solutions of ammonium sulfides in the coal hydrogenation unit are stripped from the water solutions and passed through a saturator to produce ammonium sulfate crystals (Process Evaluation Group 1975).

Synthoil ammonium sulfate production (tons per day)

Product	Coal type	
	Wyodak (0.7% S)	Western Kentucky (4.7% S)
Elemental sulfur	40	564.6
Ammonium sulfate	182.9	514
Sulfuric acid	136	380

4.1.3.2 Fugitive sulfur

Sulfur compounds (hydrogen sulfide, carbonyl sulfide, carbon disulfide) are released to the atmosphere due to incomplete stripping and recovery, leakage from equipment, evaporation from wastewater streams, reformation during quenching, or use of sulfur-bearing residues or slags (Rubin and McMichael 1975). In comparison with the total input sulfur, only minor amounts of sulfur are released to the atmosphere in offgases from the recovery and treatment facilities. Atmospheric sulfur dioxide emissions for the coal conversion processes (Synthane, CO_2 Acceptor, Lurgi, SRC, COED, and HYGAS) studied by Glaser, Hershaft, and Shaw (1974) (Table 4.10) ranged from 0.14 to 1.13 tonnes/hr, or 0.1 to 15% of the input sulfur.

Table 4.10. Estimated sulfur balance (in tonnes/hr)[a]

	Process and coal type										
	Synthane				CO_2 Acceptor	Lurgi	SRC	COED[c]		HYGAS	
	Pittsburgh[b]		Illinois No. 6[b]		North Dakota lignite	Navajo	Kentucky No. 11	Utah A seam	Illinois No. 6	Montana subbituminous (g-moles/sec)	Illinois No. 6 (g-moles/sec)
	A	B	A	B							
Input	8.62	8.62	18.30	18.30	6.66	7.36	12.8	1.89	13.98	40.2	225.5
Tar	1.03	1.03	0.25	0.25		0.1					
Ash					0.43		0.14			0.4	2.1
Char								1.312	1.976		
Product						0.06[d]	2.6	0.019	0.132	38.2	221.2
Elemental sulfur	6.42	7.38	16.79	17.73	5.13	7.06	9.9	0.556	11.757	38.2	
Recovered sulfides	0.10	0.12	0.21	0.23	0.07						
Released SO_2	1.07	0.09	1.13	0.09	1.03	0.14	0.1			1.1	1.9
Output	8.62	8.62	18.30	18.30	6.66	7.36	12.74[e]	1.89	13.98	39.7[f]	225.2[g]

[a]All measurements in tonnes/hr, except for those for HYGAS, which are in g-moles/sec.
[b]A — direct char combustion; B — indirect char combustion.
[c]Both COED processes are gas reforming.
[d]Includes tar oil, naphtha, and ammonia solution.
[e]Liquid by-products contain 0.06 tonnes of sulfur per hour.
[f]Includes 0.5 g-moles/sec in BTX oil and 0.1 g-moles/sec in wastewater.
[g]Includes 0.3 g-moles/sec in BTX oil and 0.1 g-moles/sec in wastewater.

Source: Glaser, Hershaft, and Shaw 1974, Tables IV-4, V-4, VI-6, VII-5, VIII-3, IX-6, pp. IV-10, V-13, VI-16, VII-20, VIII-13, IX-19.

Sulfur input varies by a factor of from 2 to 4 for the various coals outlined for the SRC process (833,333 lb/hr of coal) (Table 4.11); the estimated atmospheric discharge, when using tail gas recycle, is constant at 2 tons per day (Hittman Associates 1975a). Atmospheric sulfur dioxide discharge for the Bi-Gas process (Table 4.11) exceeds that of the SRC process by at least an order of magnitude, but is still only a small fraction, about one-tenth, of the sulfur input (Hittman Associates 1975a). Boiler stack gas, after scrubbing and removal of about 80% of the sulfur dioxide as calcium sulfate, releases about 4.7 tons of sulfur to the atmosphere per day; fugitive Claus tail gas emissions amount to 1.4 tons of sulfur per day; and the superheater releases about 0.9 ton of sulfur per day for the Lurgi process (Table 4.12) (Beychok 1975). Thus, the Lurgi process emits a total of 14 tons of sulfur dioxide (7 tons of sulfur) per day, which is less than 3% of the 233 tons of sulfur input to the gasifier and boiler plant per day (Beychok 1975).

The COED and CSF (Consol Synthetic Fuel) processes (Table 4.13) emit totals of 13 and 20 tons of sulfur per day when using Illinois and Pennsylvania coals; these emissions amount to 1.5 and 1.1% of the input sulfur respectively (Beychok 1975). The COED process (Table 4.14) sulfur dioxide emissions from sulfur recovery plants are estimated to be 0.1 ton per hour when using Illinois No. 6 coal (Kalfadelis and Magee 1975). The CO_2 Acceptor process (Table 4.15 and Fig. 4.41) recovers 72.4% of the total sulfur input as by-product and 2.3% as solid residues; 25.3% is discharged to the air (Jahnig and Magee 1975). Flue gas sulfur dioxide emissions estimated for the Synthane process (Table 4.16) total 2290 lb/hr, or 12% of the input sulfur (Kalfadelis and Magee 1974).

Gas quenching prior to acid gas removal causes hydrogen sulfide, carbonyl sulfide, and some sulfides, mercaptans, thiophenes, and thiocyanate to be dissolved in the quench water (Glaser, Hershaft, and Shaw 1974). Incomplete stripping of gas liquors also causes some sulfur compounds to be present in wastewater streams. The amount of hydrogen sulfide remaining in the wastewater after stripping has been estimated to be 44 to 210 ppm, or about 2 to 9 tons per day for the Bi-Gas process (Hittman Associates 1975). Wastewater components are subsequently treated in the wastewater treatment section. Synthane process atmospheric emissions of sulfur dioxide from aqueous condensates are estimated to be 100 lb/hr, or 0.5% of the input sulfur (Kalfadelis and Magee 1974). Yearly estimated sulfur emissions for the proposed Wesco facility (Lurgi technology), when using pollution controls, are shown in Table 4.17. Without pollution control, the sulfur emissions, based on 226 tons of sulfur entering the gasifiers per day, would produce 83,400 tons of hydrogen sulfide per year, and steam superheaters and boiler offgases would carry 21,600 tons of sulfur dioxide per year (Bennett, Kerr, and Kolstad 1976), an amount 32 times greater than that produced with pollution control.

4.2 NITROGEN

4.2.1 Preparatory operations

4.2.1.1 Elemental nitrogen

Oxygen is used in some gasification processes to pretreat the coal and in some conversion processes when oxidation or pyrolysis is required. Because air is the principal source of oxygen, oxygen plant emissions are composed of the other components of air, principally elemental nitrogen, which is not a pollutant. Shaw and Magee (1974) report the composition of the oxygen plant effluent stream for the Lurgi process to be 98.9% nitrogen, 0.9% oxygen, 0.2% water, and 42.9 ppm carbon dioxide. The Lurgi process uses three oxygen plants to produce 6000 tons of

Table 4.11. Estimates of sulfur recovery and discharges for the Bi-Gas and SRC processes
(tons per day)

Description	Appalachian region	Eastern interior region	Fort Union region	Powder River region	Four Corners region
Bi-Gas					
Input sulfur					
To refinery	879	1985	578	408	458
To boilers	167	378	110	77	87
Total	1046	2363	688	485	545
Sulfur discharge					
H_2S in wastewater	4	9	3	2	2
SO_2 from boilers to atmosphere	67	151	44	32	35
95% tail gas recycle					
Recovered elemental sulfur	997	2255	656	464	523
Total SO_2 to atmosphere	68	152	44	32	35
Total Na_2SO_4 produced	140	315	92	66	73
No tail gas recycle					
Recovered elemental sulfur	991	2244	652	461	520
Total SO_2 to atmosphere	78	176	51	37	41
Total Na_2SO_4 produced	139	315	92	65	73
SRC					
Sulfur input, total	1046	2363	687	485	546
Sulfur discharges					
Sulfur in products	267	254	250	267	252
H_2S in wastewater	2	6	Trace	Trace	1
95% tail gas recycle					
Recovered elemental sulfur	770	2030	430	210	288
SO_2 to atmosphere	2	2	2	2	2
Na_2SO_4 produced	43	114	24	12	16
No tail gas recycle					
Recovered elemental sulfur	766	2020	429	210	286
SO_2 to atmosphere	10	22	6	2	4
Na_2SO_4 produced	43	114	24	12	16

Source: Hittman Associates 1975*a*, Tables IV-32 and IV-33, pp. 222 and 225. Reprinted by permission of the publisher.

Table 4.12. Sulfur balance for Lurgi gasifier and boiler plant

Component	Sulfur (tons/day)	Actual form
Gasifier		
Input		
In gasification coal	200	Sulfur compounds
Output		
In ash	10.0	Sulfur compounds
In tars, oils, naphtha	5.8	Organic sulfur
In CO_2 vent	0.5	Carbonyl sulfide
By-product	174.5	Sulfur
Calcium sulfate solids	7.8	$CaSO_4$ (gypsum)
SO_2 emissions to air from tail gas theater	1.4	SO_2
Total	200.0	
Boiler		
Input		
In coal fines	32.7	Sulfur compounds
In by-product oil fuel	0.9	Organic sulfur
Total	33.6	
Output		
In ash	1.6	Sulfur compounds
Calcium sulfate solids	26.4	$CaSO_4$ (gypsum)
SO_2 emissions to air		
From boiler stack gas scrubbers	4.7	SO_2
From superheater	0.9	SO_2
Total	33.6	

Source: Beychok 1975, pp. 40-41.

98% pure oxygen per day (Shaw and Magee 1974), whereas the Koppers-Totzek process uses 4000 tons per day (Magee, Jahnig, and Shaw 1974).

The Synthane process, using 3650 tons of oxygen per day, results in the separation of 330 million scf of nitrogen per day (Kalfadelis and Magee 1974). The COED process, using 3760 tons of oxygen per day to the last pyrolysis stage, emits about 340 million scf of nitrogen per day (Kalfadelis and Magee 1975). The nitrogen vent for the Lurgi process produces a stack gas of 502 million scfd, or 18,500 tons per day, of almost pure nitrogen (Beychok 1975). Nitrogen may be used to serve as a transport medium for combustible powders or dusts, to act as an inert stripping agent in regeneration or distillation, to dilute other effluent gas streams, or to pressurize raw oil filters (Kalfadelis and Magee 1974; 1975).

4.2.1.2 Nitrogen oxides

Emissions from utility boilers (steam, hydrogen, oxygen, or electric power plants) may require control and treatment for nitrogen oxides. Nitrogen oxide emission levels vary with process

Table 4.13. Estimated COED and CSF sulfur balances

Component	Sulfur (tons/day)
COED (Illinois coal)	
Input	
25,000 tons/day coal (3.5 wt % S)	875
Output	
26,000 bbl/day product oil (0.3 wt % S)	12
1330 million scf/day product gas (10 grains H_2S/100 scf)	10
18,900 bbl/day plant fuel oil (0.3 wt % S)	9[a]
By-product sulfur (@ 99.5% recovery)	840
Sulfur recovery tail gas	4[a]
Total	875
CSF (Pennsylvania coal)	
Input	
23,360 tons/day coal (4.0 wt % S)	934.4
Output	
47,600 bbl/day product oil (0.3 wt % S)	22.0
251 million scfd net product gas (10 grains H_2S/100 scf)	1.8
12,700 bbl/day residuum fuel oil (0.7 wt % S)	15.3[b]
98 million scfd fuel gas (10 grains H_2S/100 scf)	0.7[b]
By-product sulfur (@ 99.5% recovery)	890.1
Sulfur recovery tail gas	4.5[b]
Total	934.4

[a]Total sulfur emissions = 13 tons/day.
[b]Total sulfur emissions = 20.5 tons/day.

Source: Beychok 1975, p. 138.

Table 4.14. COED sulfur balance (tons/hr)

Illinois No. 6 seam	Reported by FMC[a]	Estimated per design
Coal (total input)	41.0	41.0
Syncrude	0.2	0.2
Elemental sulfur	22.4	20.8
SO_2 emissions[b]	0.1	0.1
Char (to gasifier)	18.3	19.9

[a]As cited in Kalfadelis and Magee 1975.
[b]From sulfur recovery plants. Sulfur emission may be mostly in the
form of carbonyl sulfide if Beavon tail gas treatment is used. This
balance assumes no sulfur emission in the purge gas stream from stage
1 pyrolysis and recycle-to-extinction of aqueous process condensates.
Additional sulfur emission approximately equal to the SO_2 value given
above may be expected from the auxiliary gas cleaning facility of the
char gasification plant

Source: Kalfadelis and Magee 1975, Table 4.14, p. 43. .

Table 4.15. Sulfur balance
for CO_2 acceptor process (%)

By-product sulfur	72.4
Reject acceptor	1.0
Spent ash	1.3
Regenerator flue gas	16.9
Drier vent gas	6.2
Claus tail gas	2.2
Total	100.0

Source: Jahnig and Magee 1975,
Table 4, p. 32.

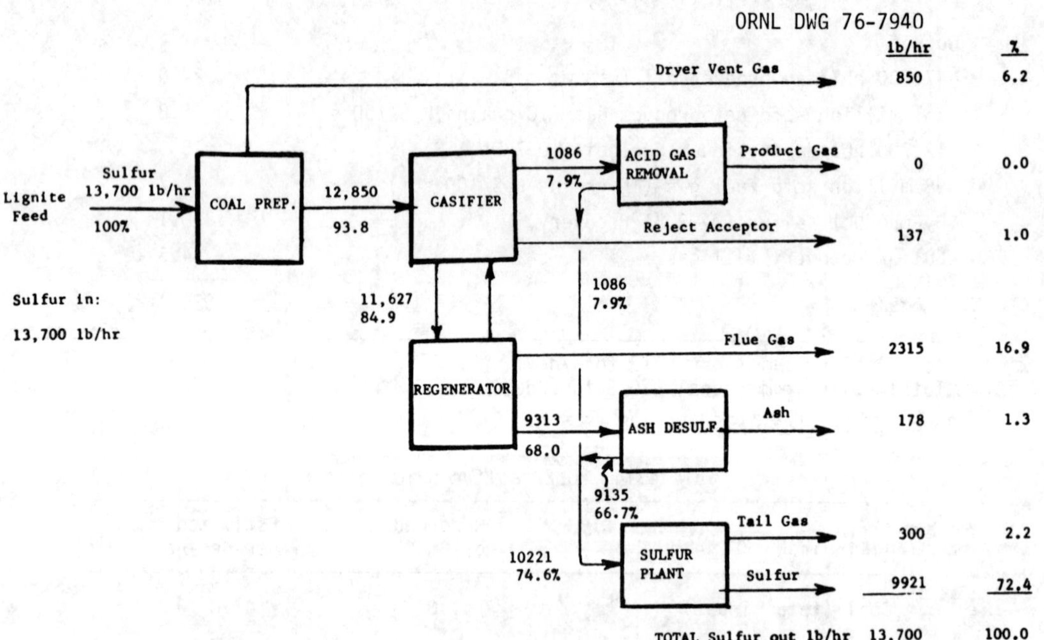

Fig. 4.41. Sulfur balance for CO_2 Acceptor process. Source: From Jahnig and Magee 1975, Fig. 7, p. 33.

4-49

Table 4.16. Estimates of sulfur balance for the Synthane process

Description	Quantity (lb/hr)	(%)
In feed coal (Pittsburgh seam coal, 1.6% S)	19,000	
H_2S converted to sulfur product	11,400	60.0
Char from gasifier	5,435	28.6
Final ash	210	1.1
As SO_2 in flue gas	2,290	12.1
As sulfated residual from flue-gas scrubber	2,935	15.4
Organic sulfur in raw gas[a] (thiophene, mercaptan, sulfides)	165	0.9
Hydrolyzed in shift converter	80	0.4
Hydrolyzed in Benfield	55	0.3
Sulfur guard	30	0.2
Total sulfur in aqueous condensate[a]	935	4.9
Recovered as H_2S	270	1.4
Oxidized, released to atmosphere	100	0.5
As sludge to incineration	565	3.0
Tar (by difference)	1,065	5.6

[a]Estimated from Forney et al. 1974.

Source: Kalfadelis and Magee 1974*b*, Table 7, p. 57.

Table 4.17. Yearly estimated emissions (tons/year) from Wesco Coal Gasification (Lurgi) Plant with pollution control[a]

Item	Pollution control	SO_2	Particu-lates	COS	H_2S
Coal lock fan	Wet cyclone		0.4		
Ash lock fan	Wet cyclone		0.8		
Local vent[b]	None	7.8			Trace
Flare[b]	None	20.7			
Coal feed bin	Baghouse		3.9		
Coal lock	None			24	Trace
Boiler feed bin	Baghouse		1.0		
Steam boilers	Hot electrostatic precipitator and scrubber	2145	272		
Steam superheater	None	133	6.4		
Coal handling	Dust suppression		Trace		
Claus sulfur plant	Incineration and scrubber	480			
Stretford plant	Incineration and scrubber	63			
Gas liquor expansion gas	Incineration and scrubber	126			
Limestone handling	Baghouse		5.6		
Gas liquor vent gas	Incineration and scrubber	66			
Rectisol vent gas	Incineration	140			
Shift catalyst regeneration[c]	Scrubber	7.3			
Recovered hydrocarbons	Incineration and scrubber	73			
Total		3261.8	290.1	24	Trace

[a]Based on 7970 hr/year of full-capacity operation (N.B. -- 24 hr/day x 365 days/yr = 8760 hr/yr).
[b]During startup after periodic shutdowns.
[c]During catalyst regeneration only.

Source: Bennett, Kerr, and Kolstad, Table IV, p. 14.

designs, combustion parameters (temperature and oxygen availability), and inherent nitrogen content of coal (Rubin and McMichael 1975; Van der Kooij and Elshout 1974). NO_x control techniques and processes are being developed; however, the major emission, nitrogen oxide, is more difficult to control.

About 176 lb/hr of NO_x is emitted from the Lurgi superheater-incinerator; this amount meets the EPA regulation for new fossil-fuel-fired steam generating units of more than 250 million Btu/hr (Shaw and Magee 1974).

The proposed Wesco Coal Gasification (Lurgi) Plant, using electrostatic precipitators and scrubbers, expects NO_x emissions to be 5805 tons per year from the steam boiler and 385 tons per year from the superheater without pollution controls (Bennett, Kerr, and Kolstad 1976).

Combustion techniques designed to reduce NO_x production for boilers, as reported by Bartok, Crawford, and Skopp (1971), may be applicable to conversion designs. Proposed techniques consist of modifying the operating and design features that control NO_x production: combustion temperature; availability of combustion air; the mixing of fuel, air, and combustion products; heat release and removal; and fuel type (Bartok, Crawford, and Skopp 1971; Van der Kooij and Elshout 1974). Operating modifications defined for these purposes are low excess air firing, two-stage combustion, flue gas recirculation, water injection, and combinations of these techniques. Design modifications include burner configuration, location and spacing, types of firing, and combustion methods (Bartok, Crawford, and Skopp 1971; Van der Kooij and Elshout 1974). However, the Pittsburgh Energy Research Center (1975) is modifying the Alkalized Alumina process to remove both sulfur and nitrogen oxides from power plant flue gas (Fig. 4.42). Tests showed that, by lowering the absorption temperature (from 330 to 130°C or lower), sorbent attrition is reduced, and nitrogen oxides as well as sulfur dioxide are removed with 88 and 98% removal efficiency respectively (Pittsburgh Energy Research Center 1975). The recovered NO_x, about two-thirds of which is decomposed to nitrogen and oxygen, can be used to manufacture nitric acid, or it can be returned to the boiler for flame destruction. Regeneration of the sorbent produces a hydrogen-sulfide-rich offgas suitable for Claus plant conversion to elemental sulfur. Regenerated sorbent is cooled and recycled (Pittsburgh Energy Research Center 1975).

Processes such as Continuous Catalytic Absorption have been developed for emission control for other NO_x-producing industries (Fig. 4.43). It uses a stripped nitric acid solution as the absorption medium. The conversion oxidation reaction is:

$$2NO_x + H_2O + (2.5x)O_2 = 2HNO_3 \tag{20}$$

Fugitive HNO_3 vapor losses can be recovered by addition of water trays before the gas leaves the unit (Mayland and Heinze 1973).

4.2.2 Conversion operations

In fossil fuel conversion, nitrogen inherent in the organic coal materials forms primarily nitrogen oxides (NO, NO_2) during combustion and ammonia (NH_3) during hydrogenation. Reducing conditions, as in gasification, produce ammonia and hydrogen cyanide, whereas in liquefaction the amine or ring compounds may remain intact (Hittman Associates 1975a). Organic nitrogen compounds

ORNL DWG 76-6208

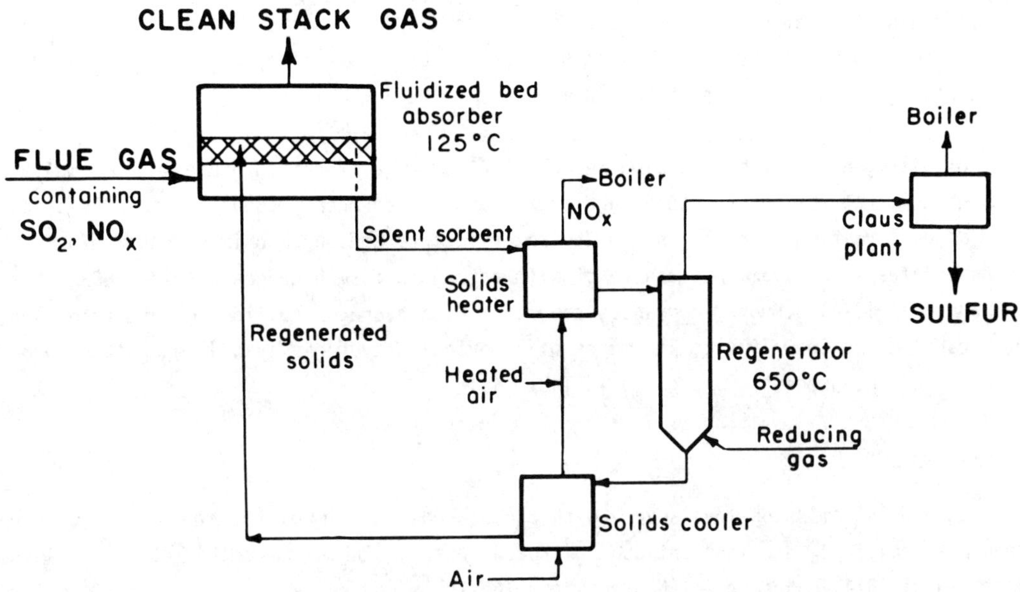

Fig. 4.42. Pittsburgh Energy Research Center stack gas cleaning process (simultaneous removal of SO_2 and NO_x). <u>Source</u>: From Pittsburgh Energy Research Center 1975, p. 23.

ORNL DWG 76-6565

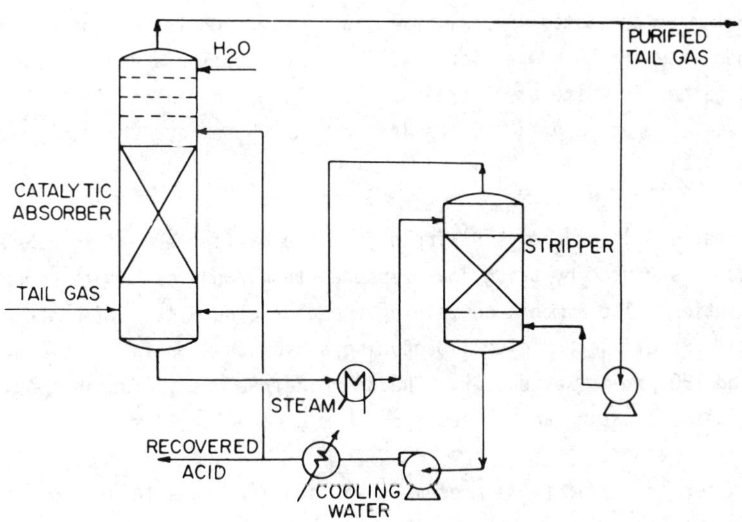

Fig. 4.43. Continuous catalytic absorption of nitrogen oxides. <u>Source</u>: From Mayland and Heinze 1973, p. 75. Reprinted by permission of the publisher.

such as amines (Hittman Associates 1975*b*) may exist as such in coal (and are released intact) or may be formed during hydrogenation. Jahnig (1975) reports a high (>1%) nitrogen content in the heavy SRC product. Shults (1976) reports indole at 210 ppm and skatole at 128 ppm and expects pyridines and chinolines at much higher levels in the synthoil. Heavier organic nitrogen compounds have been reported in coal carbonization products by Anderson and Wu (1973). Shearer (1973) reports the nitrogen content of the COED pipeline gas to be 1.6 mole %.

The extent of nitrogen conversion to ammonia (about 60% for gasification processes) is influenced by processing temperatures and pressures and the amount of hydrogen present (Hittman Associates 1975*a*). Cyanide formation is dependent on the above factors and on ammonia concentration (Hittman Associates 1975*a*); ammonia can react with carbon to form hydrogen cyanide (HCN) (Glaser, Hershaft, and Shaw 1974). Hydrogen cyanide can react with hydrogen sulfide to form thiocyanates (Glaser, Hershaft, and Shaw 1974). Estimates of ammonia and hydrogen cyanide quantities produced by the SRC and Bi-Gas processes are given in Table 4.18.

4.2.2.1 Nitrogen oxides

Again, nitrogen oxide emission levels vary with process design, combustion parameters (temperature and oxygen availability), and inherent coal nitrogen content (Rubin and McMichael 1975; Van der Kooij and Elshout 1974). Reports of NO_x emission levels or control and treatment designs for coal conversion technologies have not been located. However, combustion techniques and control processes for fossil fuel boilers reported earlier may prove applicable.

4.2.2.2 Ammonia and hydrogen cyanide

Ammonia in fuel gas process streams is removed by water scrubbing. Initially, ammonia and cyanide from the main process stream are expected to occur as gases dissolved in condensates from the gas streams (Hittman Associates 1975*b*). Ammonia in solution favors the absorption of hydrogen cyanide in the condensate (Hittman Associates 1975*a*). Thus, ammonia and hydrogen cyanide are likely to appear together in wastewater streams. Dissolved ammonia is either treated for removal or recovered for resale. Recovery facilities include the Phosam and Chevron processes (Glaser, Hershaft, and Shaw 1974).

In a Phosam unit (Fig. 4.44) ammonia is stripped from the wastewater and scrubbed from the gas streams; the liquid is stripped by using low-pressure steam, and the gas is scrubbed by using a phosphoric acid solution. The ammonia-containing phosphoric acid is detarred and fed to a stripper column and a fractionator column, producing anhydrous ammonia. Residual ammonia is about 7 ppm in the gas and 150 ppm in wastewater. The ammonia-free gas is further purified of other contaminants, and the wastewater can be used for steam generation (Morgan 1975).

The Chevron Waste Water Treatment (WWT) process (Fig. 4.45) is capable of processing sour water to separate (1) high-purity hydrogen sulfide (99.9%), high-purity ammonia (99.9%), and clean water (99.9+%) or (2) pure sulfide and water, but a lower-quality ammonia for incineration; or the WWT process may be added to the existing foul-water treatment process for additional recovery. Ammonia can be recovered as anhydrous liquid or as 25% aqueous solution for fertilizer sales (Annessen and Gould 1971). The wastewater stream is fed to a degassing unit to remove dissolved hydrogen, methane, and other light compounds and to a surge tank to remove floating hydrocarbons by skimming. Hydrogen sulfide is then stripped and sent to a recovery plant. The

Table 4.18. Estimates of ammonia and hydrogen cyanide from coal refining

Description	Appalachian region	Eastern interior region	Fort Union region	Powder River region	Four Corners region
Bi-Gas process, tons per day					
Nitrogen to plant	493	726	779	883	864
Ammonia produced	180	249	275	306	298
Wastewater stream	49,709	48,789	48,889	47,459	51,500
Ammonia in wastewater @ 50 ppm	2.5	2.4	2.4	2.4	2.6
Recoverable ammonia	178	247	273	304	295
Solvent-Refined process, tons per day					
Nitrogen to SRC (coal only)	517	721	722	105	876
Nitrogen to gasifier (coal only)	70	142	206	a	152
Nitrogen in gasifier effluent	183	236	313	291	730
Ammonia in gasifier effluent	56	72	95	88	222
Separator water	3,242	8,957	44,590	20,348	14,458
Ammonia remaining in separator water @ 50 ppm	0.16	0.45	2.2	1.0	0.72
Recoverable ammonia	56	72	93	87	221
HCN estimates (basis 4% of NH$_3$)					
Bi-Gas, tons per day	11	16	17	19	19
Bi-Gas, ppm	221	328	348	400	369
SRC, tons per day	4	5	6	6	14
SRC, ppm	1,230	558	135	295	968

aNo coal to gasifier in this case.

Source: Hittman Associates 1975a, Table IV-35, p. 230. Reprinted by permission of the publisher.

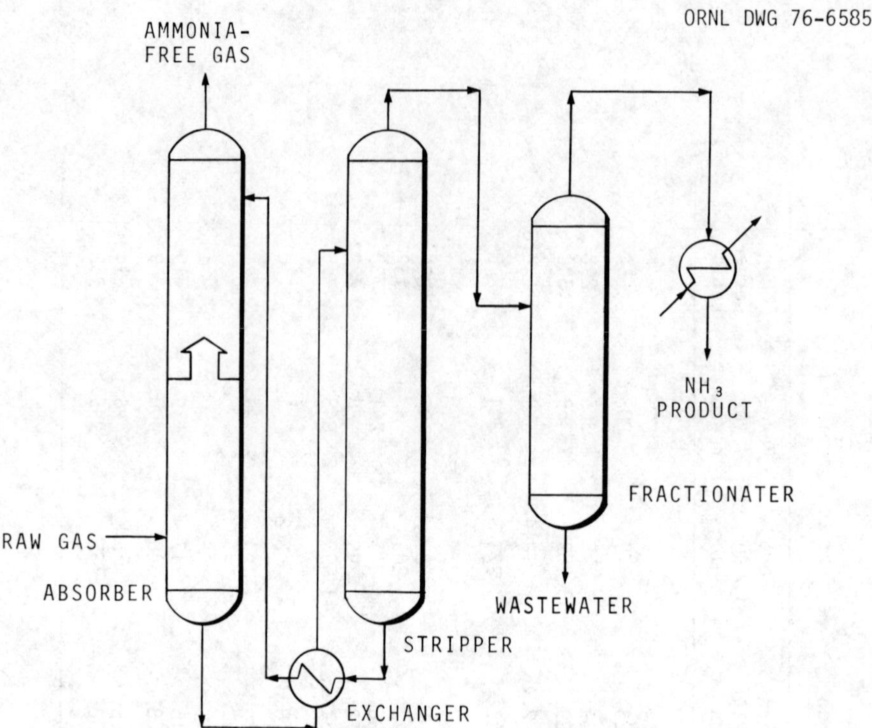

Fig. 4.44. Ammonia recovery (COALCON). Source: Adapted from Morgan 1975, p. 14.

ammonia is stripped from the water in a fractionator. Ammonia is about 98% pure when it leaves the condenser; after leaving the scrubbing system, where trace hydrogen sulfide and water is removed, the ammonia is of high purity (99.9%) (Annessen and Gould 1971).

Anhydrous ammonia recovery amounts to 122 tonnes per day from Illinois coal and 38 tonnes per day from Montana coal in the Hygas process, 790 kg/hr in the SRC process (Glaser, Hershaft, and Shaw 1974), and 250 tons per day in the El Paso Burnham I Coal Gasification Complex (Gibson, Hammons, and Cameron 1974). The Synthoil process recovers ammonium sulfate (Bureau of Mines 1975); other processes may recover ammonia in solution.

Practical technologies for removing ammonia from industrial wastewaters include biological synthesis, nitrification, ion exchange, air and steam stripping, and chlorination (Adams 1973). Residual ammonia (amounts too low to allow recovery and fugitive emissions from the recovery unit) is usually sent to biological oxidation units or aeration ponds for degradation (Chap. 7). Coking plants using biological oxidation report 90% ammonia removal, 57% cyanide removal, and 17% thiocyanate removal, the latter being more resistant to biodegradation (Jahnig 1975). Activated carbon can be used as a final water treatment following the biological oxidation to remove remaining trace contaminants (Jahnig 1975).

Biological nitrification (oxidation to nitrites or nitrates) by an activated-sludge-type process removed ammonia at concentrations as high as 500 mg/liter with 90% efficiency when using a single-stage system and 97% efficiency when using a two-stage system (Adams 1973). In the two-stage system for industrial wastewaters, biochemical oxygen demand is reduced in a separate aeration

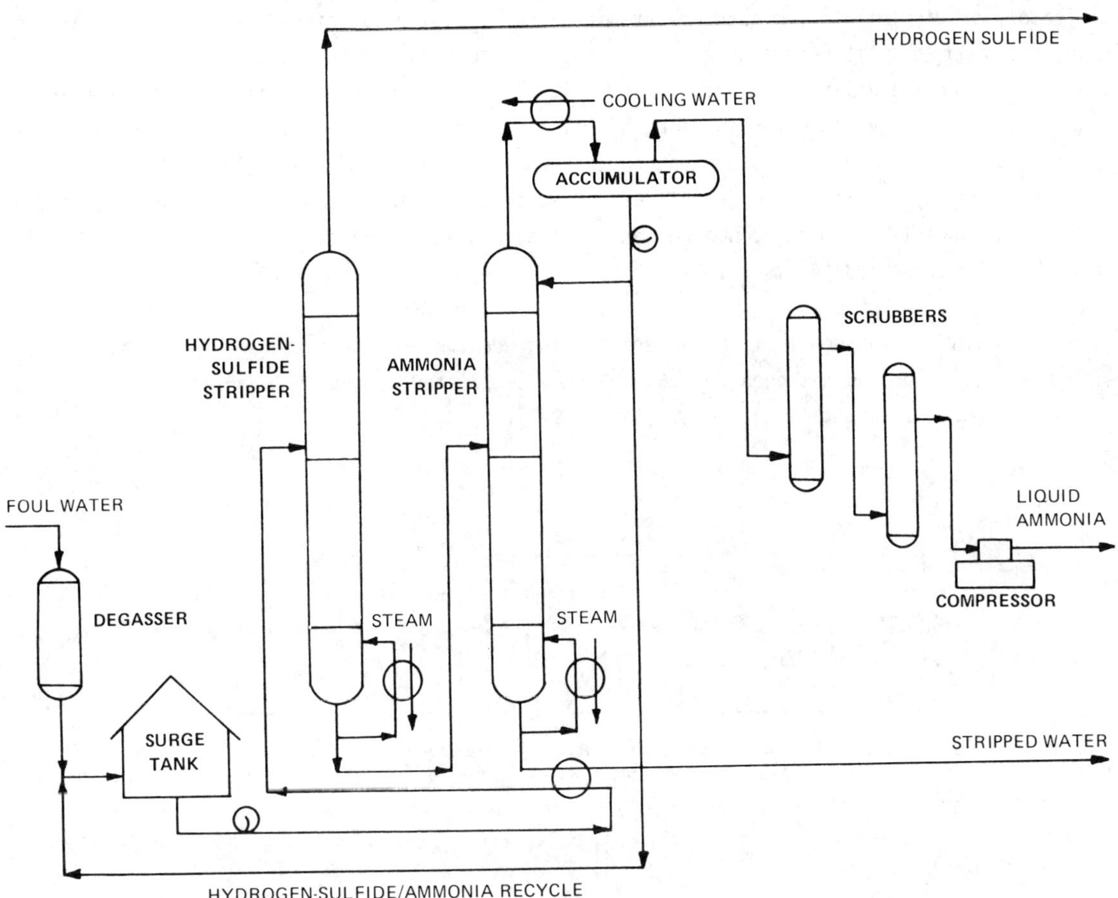

Fig. 4.45. Chevron Waste Water Treatment process — hydrogen sulfide and recovery process flowsheet. Source: After Annessen and Gould 1971, p. 68. Excerpted by special permission from *Chem. Eng.* Copyright 1971 by McGraw-Hill, Inc., New York, N.Y. 10020.

basin-clarifier system, followed by nitrification in an aeration basin-clarifier system. If inhibitors of biodegradation are present or if effluent levels less than 5 mg/liter of ammonia-nitrogen are required, a two-stage system will probably be required (Adams 1973). Limiting factors for nitrification are temperature (optimum, 28 to 32°C), shock loading, pH (7.8 to 8.3), dissolved oxygen (below 3.0 mg/liter for larger flocs), and toxicity of other compounds in the wastewater (heavy metals, cyanides, halogenated compounds, phenols, mercaptans, quinidines, and thiourea) to the organisms (Adams 1973). The presence of even trace amounts of these toxic compounds may lower the nitrification rate (Adams 1973). For consistent nitrification of higher-strength wastes, the activated sludge process is the most dependable (Adams 1973) (Chap. 7).

Laboratory and pilot plant studies of the biological treatment of concentrated ammonia wastewaters (725 mg/liter) from the manufacture of ammonium nitrate and urea fertilizers show that the system should achieve about a 90% reduction of the ammonia-nitrogen and should produce an effluent having 75 mg/liter of ammonia-nitrogen, 75 mg/liter of nitrate-nitrogen, and 575 mg/liter nitrite-nitrogen (Hutton and LaRocca 1975).

Biological mechanisms are the most practical and dependable methods for removal of nitrate and nitrated compounds from industrial wastewaters where ion exchange is inapplicable (Adams 1973). Assimilatory or dissimilatory mechanisms are involved in biological nitrate reduction. Nitrate-nitrogen is used as a nitrogen source by conversion to ammonia and is subsequently incorporated into the cell for assimilatory reduction (Adams 1973). In nitrogen dissimilation, nitrate serves as the terminal hydrogen acceptor in place of molecular oxygen, producing nitrite, ammonia, nitrous oxide, or nitrogen gas, depending on the organisms and pH of the system (Adams 1973). Denitrification systems include anaerobic activated sludge, anaerobic ponds, and upflow anaerobic filters. Methanol is the most economical external carbon source on the basis of cost and available carbon when required (Adams 1973).

Biological nitrification-denitrification systems include two- or three-stage sludge systems (Figs. 4.46 and 4.47). Improved nitrogen removal may be accomplished by the three-stage system, but it involves increased capital investment (Adams 1973).

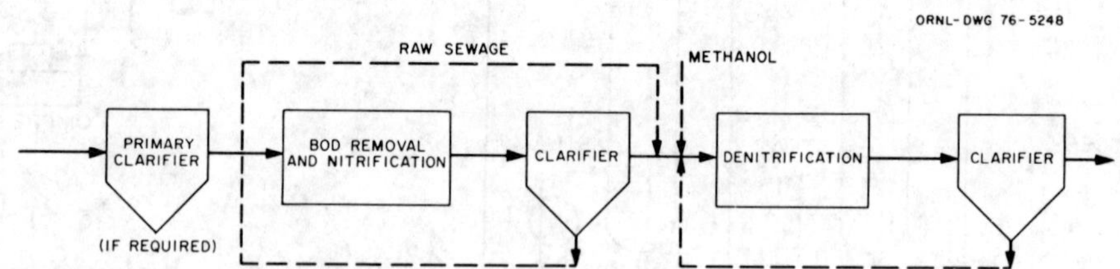

Fig. 4.46. Two-stage sludge system for nitrogen removal. Source: After Adams 1973, p. 698. Reprinted with permission from *Environ. Sci. Technol.* Copyright by the American Chemical Society.

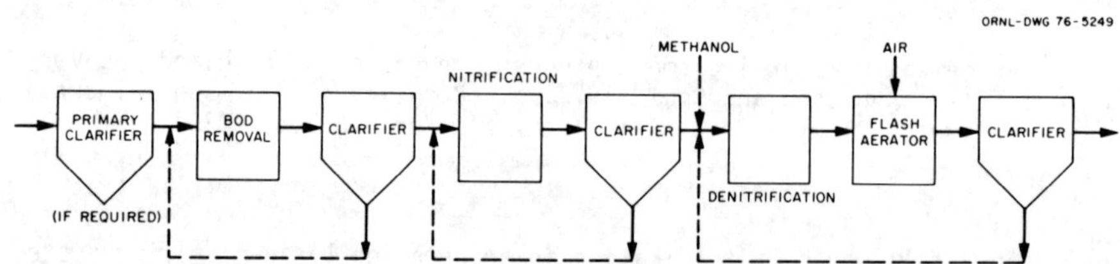

Fig. 4.47. Three-stage sludge system for nitrogen removal. Source: After Adams 1973, p. 699. Reprinted with permission from *Environ. Sci. Technol.* Copyright by the American Chemical Society.

Removal of the ammonia in wastewater by ion exchange uses the strong, selective affinity of clinoptilolite, a natural zeolite, for the ammonium ion in solution (Fig. 4.48) (Adams 1973; Mercer et al. 1970; Koon and Kaufman 1975). The clinoptilolite can be regenerated with sodium or calcium salts (sodium chloride, calcium chloride, or lime) at high pH, air-stripped to remove the ammonia, and recycled. Clinoptilolite use has been limited to low-strength municipal wastewater, and applicability to industrial wastewater is limited (Adams 1973; Jorgensen 1975). However, synthetic ion exchange resins have been used for industrial wastewater treatment. At Farmers Chemical Association, Incorporated, Harrison, Tennessee, a cation resin was used to remove the ammonium ion, and an anion resin was used to remove the nitrate ion. Ion regenerants

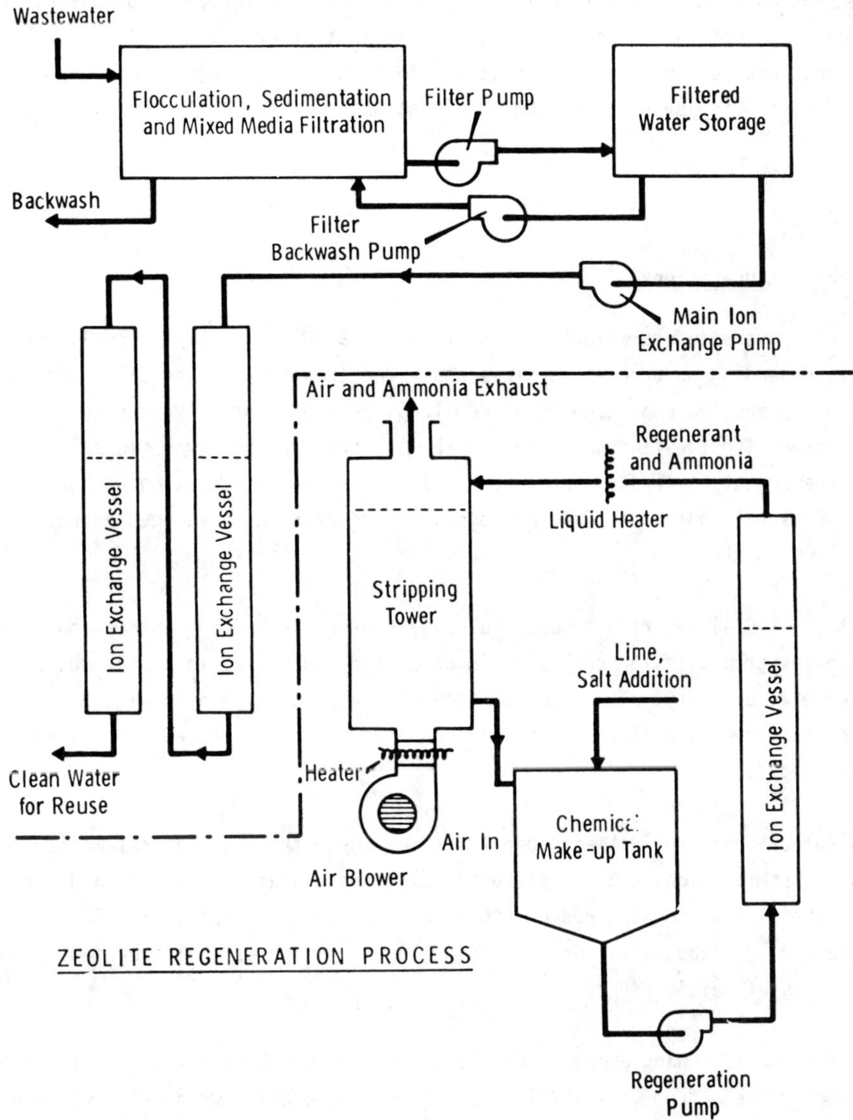

Fig. 4.48. Flowsheet for ammonia-selective ion exchange process — NH_3 selective ion exchange (top) and zeolite regeneration (bottom). Source: From Mercer et al. 1970, Fig. 7, p. R101. Reprinted by permission of the publisher.

were nitric acid (cation) and aqua ammonia (anion), forming ammonium nitrate as the spent regenerant. Regenerant solutions were blended and concentrated by evaporation to produce a marketable by-product fertilizer (Adams 1973). The presence of organic material may severely retard the effectiveness and useful life of the resins; thus, synthetic ion exchange is presently limited to wastewaters of primarily inorganic nature (Adams 1973). Mercer et al. (1970) report 93 and 97% ammonia-nitrogen removal efficiencies using clinoptilolite columns containing 15 mg ammonia-nitrogen per liter of unclarified secondary effluent and 16 mg ammonia-nitrogen per liter of clarified secondary effluent respectively (Fig. 4.48).

Chlorination to convert ammonia in wastewaters produces monochloramine, trichloramine, and dichloramine (Stasiuk, Hetling, and Shuster 1974). Disadvantages in using breakpoint chlorination include high cost, neutralization if hydrochloric acid is formed, and possible necessity for dechlorination. However, experiments have shown that chlorination followed by activated carbon treatment removed ammonia nitrogen at the 100% level, although some residual nitrate-nitrogen was formed. An advantage is the short detention time (Stasiuk, Hetling, and Shuster 1974).

4.3 PARTICULATES

4.3.1 <u>Pretreatment operations</u>

The majority of conversion operations will store about a 30-day supply of coal, requiring 30 to 50 acres depending on process requirements. One exception is the SRC process, which will store only a 3-day supply, amounting to 37,500 tons, including utility feedstocks (Jahnig 1975*b*). As a result, weathering (dusting and runoff) from the area may be environmentally significant. Coal storage areas must be treated to prevent dusting or runoff. Coating methods such as spraying oil or asphalt on the surface or covering with plastics have been proposed (Jahnig 1975*b*).

Farnsworth et al. (1974) describe a dead (reserve) storage technique for coal supplies used in the Koppers-Totzek process. The coal pile is compacted and sealed to reduce susceptibility to dusting during wind activity. The outer surface of the pile is sprayed with an organic polymer crusting agent to prevent dusting or rain erosion. The pile is located on a waterproof base to prevent water seepage.

Coal-derived solids from pretreatment operations include extraneous mineral matter and other debris produced from physical coal grinding and cleaning operations. Grace and Diehl (1974) estimate that the Bi-Gas process produces 4567 tons of coal preparation refuse (mineral matter and debris) requiring disposal per day. Jahnig (1975*b*) estimates that the Bi-Gas process produces 4804 tons of refuse per day.

Solid wastes from coal washing amount to 12 to 25 lb per 100 lb of clean coal (Interagency Task Force on Synthetic Fuels from Coal 1974). For the Bi-Gas process when using western Kentucky coal, this amounts to 3904 tons of coal washing debris per day (Jahnig 1975*b*). Coal washing wastes may be used as a cement additive (Berestovoi, Obruchev, and Rosov 1953), sinter enhancer (Deutsch Eisenwerke 1952), or lightweight aggregate (Myers, Pfeiffer, and Ornig 1964).

Rejected solids from the SRC process amounts to about 2500 tons per day. Fine solids accumulate in the coal washing water tailing pond and amount to 52 tons/hr for the SRC process. For example, a tailing pond having a 1-acre surface area (when assuming fine solids at 100 lb/ft^3) will sediment at a rate of 180 ft per year (Jahnig 1975*b*). Leaching of pyrites, iron, or other materials from washing refuse piles may be significant.

Coal dusts and fines from the drying and grinding operations are controlled by a variety of control units: cyclone separators, bag filters, negative pressure grinders, wet scrubbers, or electrostatic precipitators. Selection and use of a control method depends on the temperature, moisture content, and contaminant content of the particulate stream. Dust loading and particulate characteristics (specific gravity, shape, size, abrasiveness, resistivity, viscidity, flammability, solubility) are equally important (Tomany 1975).

Mechanical cyclone collectors use centrifugal, inertial, and gravitational forces to remove particles (Fig. 4.49). Removal efficiencies equal to or greater than 90% are typical for particles over 3 μ; however, lower efficiencies occur with smaller particulates (Power from Coal 1974).

ORNL DWG 76-6549

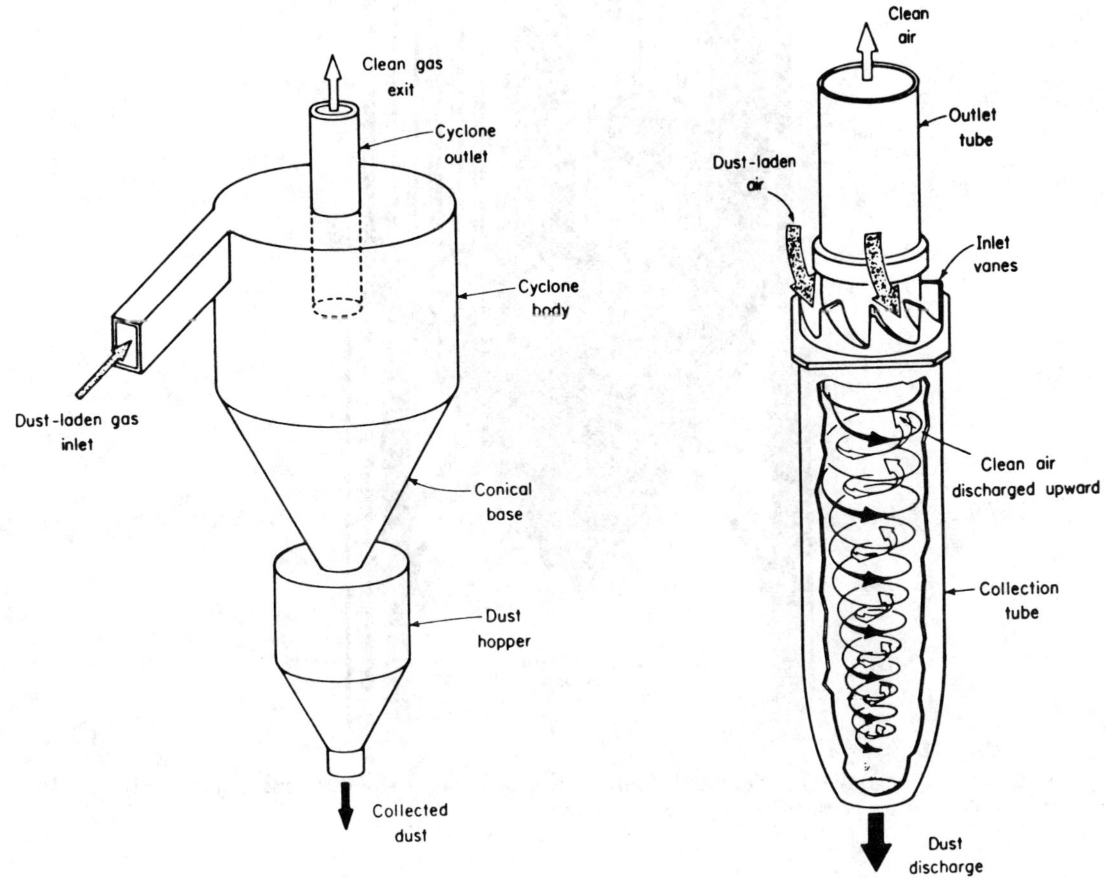

Fig. 4.49. Cyclonic separators. Source: From Tomany 1975, Figs. 5-7 and 5-8, pp. 111 and 113. Reprinted by permission of the publisher.

Removal efficiencies as high as 99% are observed with fabric filters, which trap dust from a flow-through gas stream. However, filter surfaces must be maintained at minimal moisture content to obtain these high efficiencies. Particles as small as 0.5 μ are trapped by fabric filters although some 0.01-μ particles are also removed from the stream (Fig. 4.50) (Tomany 1975).

ORNL DWG 76-8368

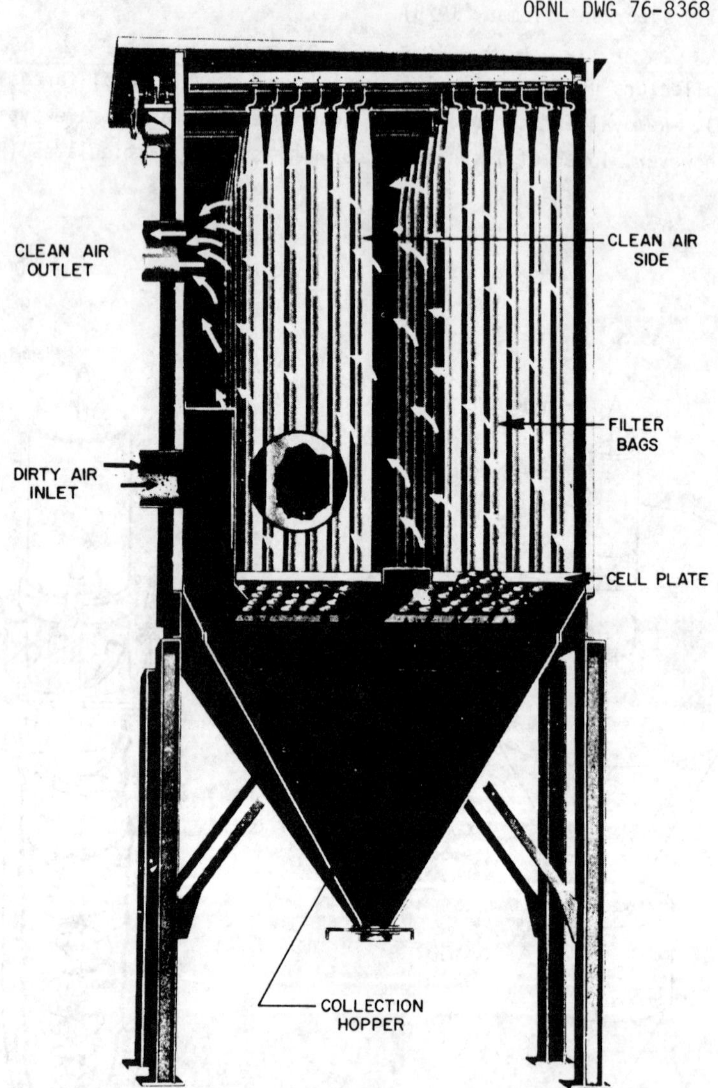

Fig. 4.50. Tubular-type fabric filter baghouse. <u>Source</u>: From Tomany 1975, Fig. 5-31, p. 157. Reprinted by permission of the publisher.

Wet scrubbers use water or scrubber liquor to mix with a particulate-laden gas for particle removal (Figs. 4.51 through 4.56). Commonly, scrubbers are divided into two parts, a contacting section and a deentrainment section. Particulate-containing gas and scrubbing liquor mix in the contact compartment, and particles are separated in the deentrainment compartment, which contains inertial devices that promote agglomeration of particulate matter (Tomany 1975). Selection of scrubber depends on particulate size rather than on solubility (Tomany 1975).

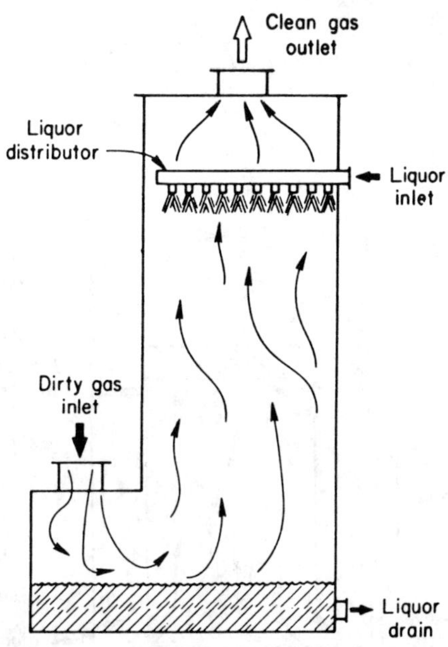

Fig. 4.51. Open spray tower. Source: From Tomany 1975, Fig. 5-44, p. 191. Reprinted by permission of the publisher.

Electrostatic precipitators use static electricity to remove solid or liquid particulates (Fig. 4.57). Charged particles are drawn to oppositely charged collecting plates, which vibrate periodically, so that the dust can be collected in storage hoppers (Tomany 1975).

In general, for all particulate control processes, care in sizing, temperature control, equipment maintenance, and loading contributes to high removal efficiencies. Combinations of the above techniques ensure maximum removal rates (Power from Coal 1974). Fines collected in cyclones and bag filters from coal grinding for the Synthane process may amount to 60 tons per day (Kalfadelis and Magee 1975b); these fines may be recycled to the coal feed. Synthoil dust from crushing and transfer operations will amount to about 1 ton/hr (Akhtar, Friedman, and Yavorsky 1975). Yearly particulate emissions for the proposed Wesco Coal Gasification (Lurgi) Plant amount to 290.1 tons per year (Table 4.17), but without pollution control, emissions could amount to as much as 152,000 tons per year (Bennett, Kerr, and Kolstad 1976).

Options for disposal or use of coal fines are minefill, briquette to Lurgi feed size, power plant or utility feed, recycle to gasification or liquefaction section, or resale (Williams and Dressel 1973; Jahnig 1975).

4.3.2 Conversion operations

Dust removal from the gas streams is accomplished by scrubbing or cycloning; however, the Bi-Gas process developers are studying the use of sand filters for efficient dust removal at high temperature. When sand filters are used, particulates are not carried in the wastewater stream, but they may complicate the pathways of trace elements that volatilize (Jahnig 1975b).

ORNL DWG 76-6556

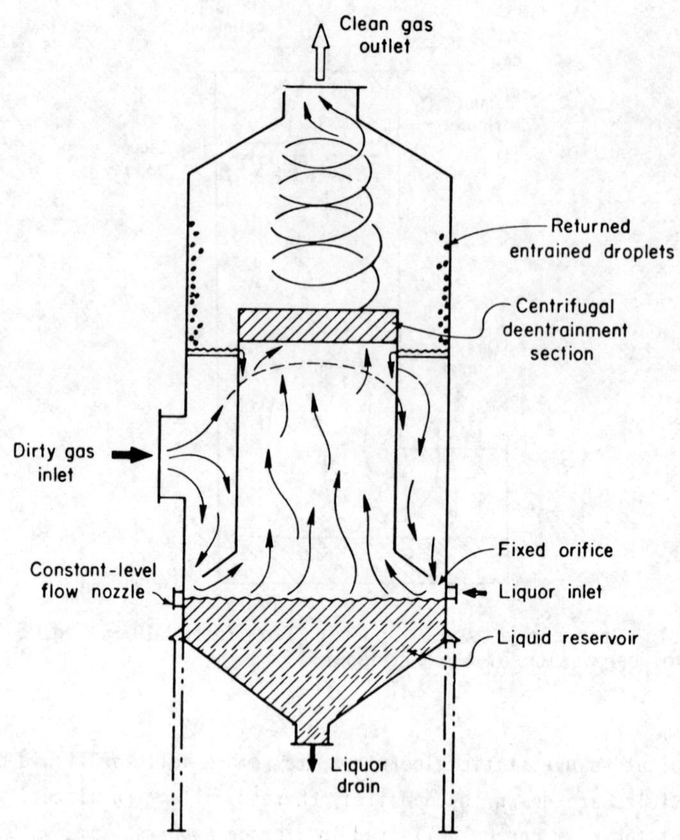

Fig. 4.52. Liquor impingement scrubber. Source: From Tomany 1975, Fig. 5-45, p. 192. Reprinted by permission of the publisher.

Solid residues produced in the coal conversion processes are primarily ash or char. Ash, which is mainly mineral matter, has little or no heating value, whereas char can be recycled as a fuel for either the originator or another industry. The amount and kind of solid waste produced vary according to the type of coal and the processing steps used; however, both ash and char require environmentally acceptable means of disposal.

4.3.2.1 Ash

Ash from the combustion units, utility (energy) production, or conversion operations is produced in dry (including fly ash and particulate matter), melted (slag), or softened (self-agglomerating) forms (Katz et al. 1974). Fly ash or particulate matter in the fuel gas is collected by cyclones, electrostatic precipitators, or wet scrubbers prior to acid gas cleanup. Melted ash is quenched with water, forming glasslike particles. The slurry produced from the wet scrubbers and quenching operations are sent to clarifiers and/or thickeners to remove excess moisture.

Ash and char discharge estimates for the Bi-Gas and SRC processes are given in Table 4.19. A commercial gasification plant (250 million cfd) using 8 million tons of Illinois coal per year will generate 2.3 million tons of ash and dry refuse per year (Sather et al. 1975). A yearly

ORNL DWG 76-6554

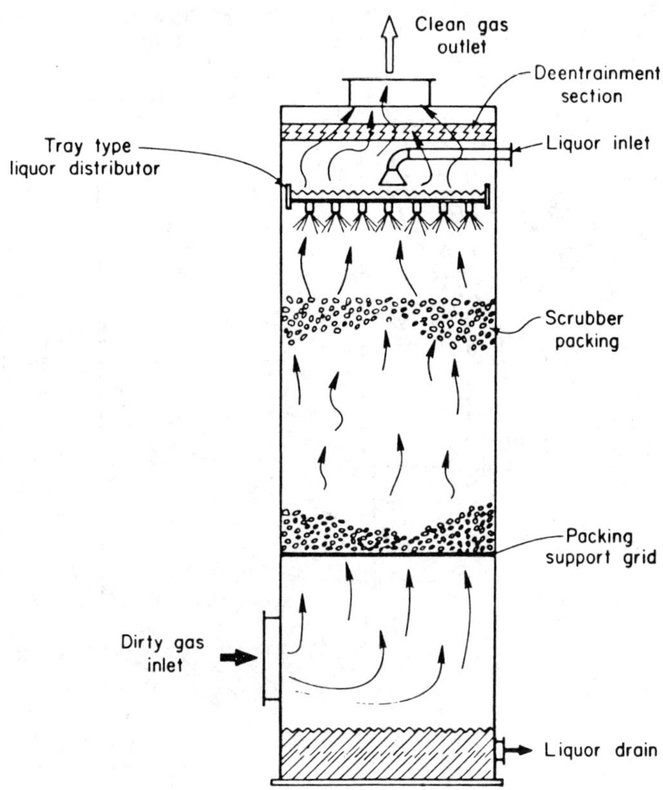

Fig. 4.53. Countercurrent-flow packed scrubber. Source: From Tomany 1975, Fig. 5-47, p. 195. Reprinted by permission of the publisher.

estimate of 400,000 tons of slag or ash for a 250-million-scfd gasification plant is reported by Van Meter and Erickson (1975). About 2,806,700 tons of bottom ash recovered from the coal-fired boilers and fly ash collected by the precipitators are expected for the Wesco Coal Gasification (Lurgi) Complex and are expected to be used for minefill (Bennett, Kerr, and Kolstad 1976). Slag from the Bi-Gas process amounts to 68,391 lb/hr (820 tons per day) when using western Kentucky coal; for disposal as landfill, this corresponds to about 90 acre-ft per year (Jahnig 1975*b*). The disposal problems of coal ash from coal conversion are similar to those from coal combustion.

Gasifier ash, which is similar to boiler bottom ash, is difficult to handle because of its size and abrasive characteristics (Williams and Dressel 1973). The major constituents of bituminous coal bottom ash are SiO_2, Fe_2O_3, and Al_2O_3; lesser constituents are CaO, MgO, Na_2O, K_2O, and SO_3 (Table 4.20).

Ash composition varies with the coal's origin, rank, and location within a bed (Chap. 2). However, nearly all trace elements show an enrichment in ash relative to their crustal abundance (Abernethy, Peterson, and Gibson 1969*b*). The three main constituents of coal ash are SiO_2, Al_2O_3, and Fe_2O_3. Abernethy, Peterson, and Gibson (1969*a*) report that, for their samples of

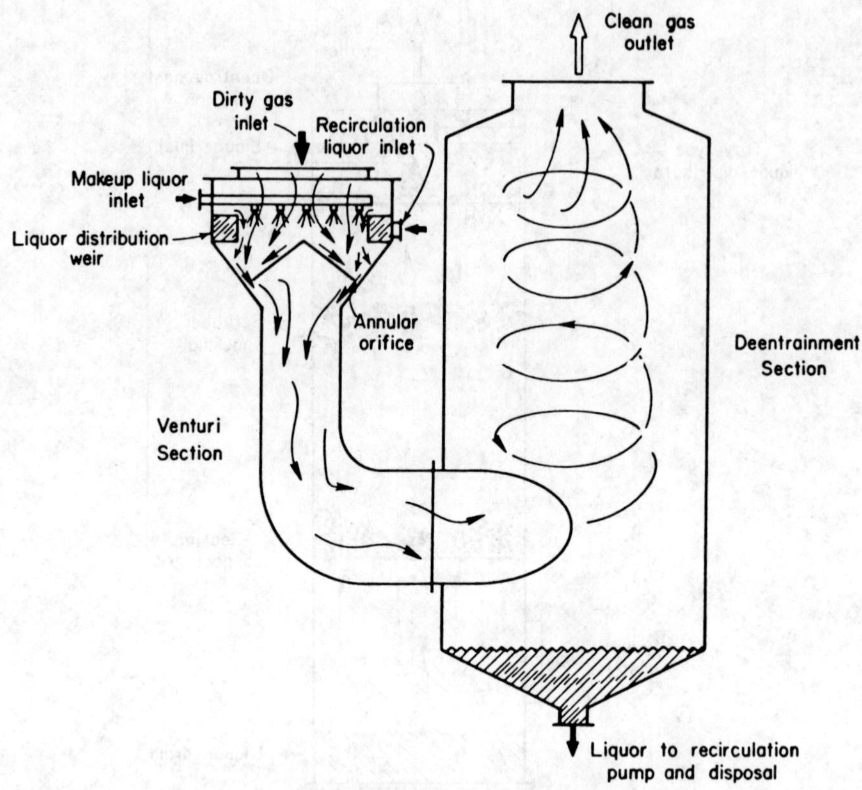

Fig. 4.54. Venturi scrubber. <u>Source</u>: From Tomany 1975, Fig. 5-50, p. 199. Reprinted by permission of the publisher.

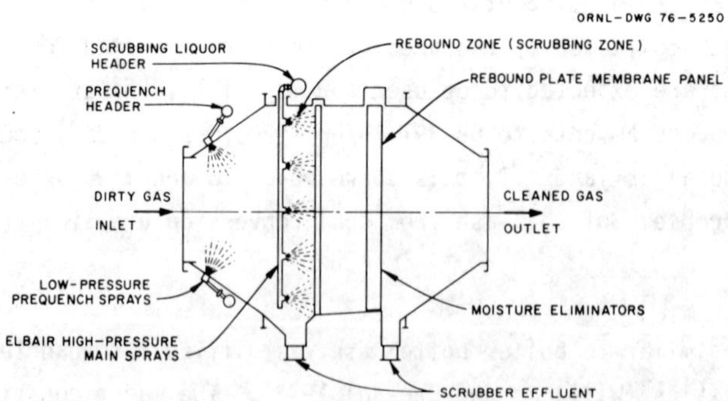

Fig. 4.55. High-pressure spray scrubber. <u>Source</u>: After Tomany 1975, Fig. 5-49, p. 197. Reprinted by permission of the publisher.

ORNL DWG 76-6552

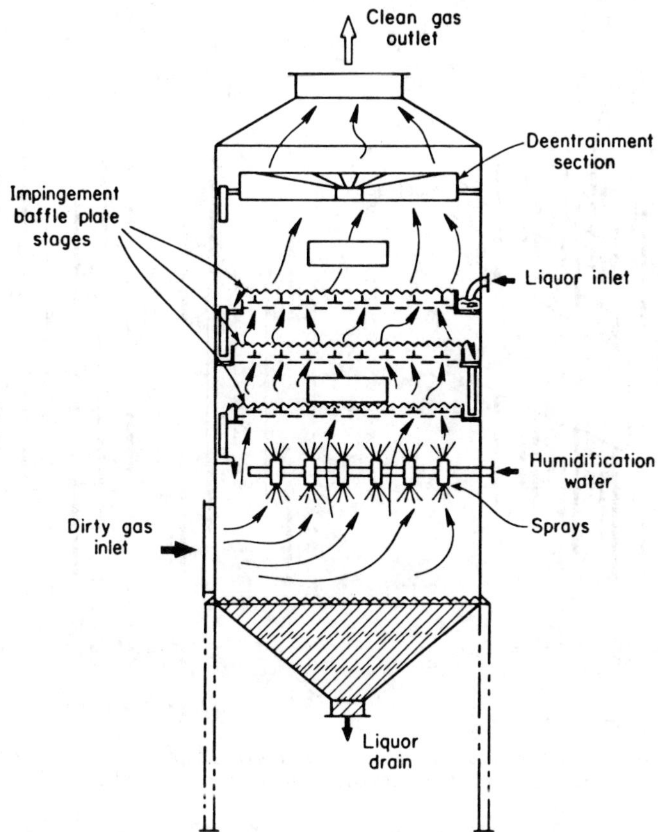

Fig. 4.56. Baffle impingement scrubber. Source: From Tomany 1975, Fig. 5-48, p. 196. Reprinted by permission of the publisher.

bituminous coals, the average percentage for each constituent was 45.7, 26.0, and 18.1% respectively. The chemical analyses of subbituminous and bituminous coal ash are compared in Table 4.21.

Sodium, potassium, calcium, and magnesium were reported in fairly high concentrations in ash by von Lehmden, Jungers, and Lee (1974). Sulfur has been reported at the 0.05 wt % level in ash from coal conversion (Shaw and Magee 1974). Sulfur and nitrogen emissions in the ash are low due to their containment in the raw product gas as ammonia and sulfides (Table 4.22) (Sather et al. 1975).

Trace amounts of a very large number of elements appear in ash. Abernethy, Peterson, and Gibson (1969b) found barium, beryllium, boron, chromium, cobalt, copper, gallium, germanium, lanthanum, lead, lithium, manganese, molybdenum, silver, nickel, scandium, strontium, tin, vanadium, ytterbium, zinc, and zirconium. Also, arsenic, bismuth, cerium, neodymium, niobium, rubidium, and thallium were detected in many of their ash samples. The following elements also have been reported in ash: mercury, cadmium, antimony, selenium, fluorine (von Lehmden, Jungers, and Lee 1974), rubidium, tungsten (Headlee and Hunter 1953), and phosphorus (Manz 1973). Maximum, minimum, and average concentrations of 21 elements found in coal ashes from anthracite and sub-bituminous coals are given in Table 4.23.

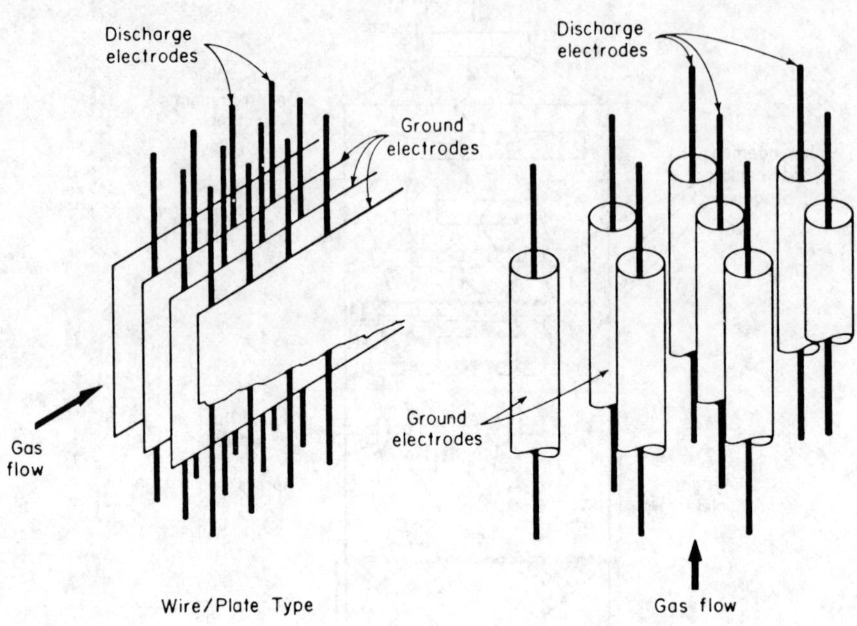

Fig. 4.57. Schematic arrangement of wire/plate and wire/tube precipitators. Source: From Tomany 1975, Fig. 5-20, p. 135. Reprinted by permission of the publisher.

Amounts of trace elements in ash produced by a coal gasification plant processing 20,000 tons of Illinois No. 5 coal per day may be as follows:

Element	Quantity (lb/day)
Arsenic	1-2
Beryllium	75-80
Fluorine	10-20
Lead	750-2700
Vanadium	600-750

Greater amounts of chromium, copper, manganese, and nickel were found in the ash than in the original coal (possibly from equipment wear) (Sather et al. 1975). The processing of 20,000 tons per day of coal that contains 1 ppm of trace elements leaves 40 lb per day of element in the plant streams to be accounted for (Magee 1975).

Iron oxide and alumina extraction processes from power plant fly ash have been studied on pilot- and demonstration-plant scale with reliable and economical results (Nowak 1973).

Van Meter and Erickson (1975), in a preliminary report on the characterization of coal conversion pilot plant solid wastes and their leachates, found the effluent from the CO_2 Acceptor process

Table 4.19. Estimates of slag, char, and ash discharges, Bi-gas and SRC processes (tons per day)

Description	Appalachian region	Easterr interior region	Fort Union region	Powder River region	Four Corners region
		Bi-Gas			
Slag from gasifier	3,701	5,245	5,860	3,535	9,075
Char from scrubber	342	397	689	476	523
Refinery ash	715	1,013	1,137	686	1,761
Power plant ash	361	931	1,023	594	1,509
Slag quench water	14,798	20,979	23,493	14,149	36,286
Char scrubber water	1,368	1,588	2,756	1,903	2,092
Makeup brine to slag quench	652	1,233	524	103	2,705
Dry ash disposal					
Wet slag (10% H_2O)	4,112	5,828	6,511	3,928	10,083
Wet char (10% H_2O)	380	441	766	529	581
Dry ash	1,346	1,944	2,160	1,280	3,270
Total solids discharge	5,838	8,213	9,437	5,737	13,934
Brine to evaporation	1,572	2,194	2,552	1,561	3,731
Wet ash disposal					
Ash slurry water	5,384	7,776	8,640	5,120	13,080
Makeup brine to slag quench and ash slurry	1,326	2,205	1,604	743	4,340
Wet ash (10% H_2O)	1,496	2,160	2,400	1,422	3,633
Total solics discharge	5,988	8,429	9,677	5,879	14,297
Brine to evaporation	2,095	2,950	3,392	2,059	5,003

Table 4.19. (continued)

Description	Appalachian region	Eastern interior region	Fort Union region	Powder River region	Four Corners region
		SRC			
Slag from gasifier	4,439	6,311	7,448	4,255	10,953
Power plant ash	631	931	1,023	594	1,509
Slag quench water	17,756	25,244	29,792	17,020	43,812
Makeup brine to slag quench	2,219	3,155	3,724	2,128	5,477
Dry ash disposal					
Wet slag (10% H_2O)	4,932	7,012	8,276	4,728	12,170
Dry ash	631	931	1,023	594	1,509
Total solids discharge	5,563	7,943	9,299	5,322	13,679
Brine to evaporation	1,726	2,454	2,896	1,655	4,260
Wet ash disposal					
Ash slurry water	2,524	3,724	4,092	2,376	6,036
Makeup brine to slag quench and ash slurry	2,535	3,620	4,236	2,425	6,231
Wet ash (10% H_2O)	701	1,034	1,137	660	1,677
Total solids discharge	5,633	8,046	9,413	5,388	13,847
Brine to evaporation	1,972	2,816	3,294	1,886	4,846

Source: Hittman Associates 1975*a*, Tables IV-26 and IV-27, pp. 201 and 204. Reprinted by permission of the publisher.

Table 4.20. Chemical analysis of selected bottom ashes (%)[a]

Source	SiO_2	Al_2O_3	Fe_2O_3	CaO	MgO	Na_2O	K_2O
Kammer	48.9	21.9	14.3	1.4	5.2	0.7	0.1
Kanawha River	53.6	28.3	5.8	0.4	4.2	1.0	0.3
Mitchell	45.9	25.1	14.3	1.4	5.2	0.7	0.3
Muskingham	47.1	28.3	10.7	0.4	5.2	0.8	0.4
Willow Island	53.6	22.7	10.3	1.4	5.2	1.2	0.1
Typical limits[b]	20 - 60	10 - 35	5 - 35	1 - 20	0.3 - 4	1 - 4	1 - 4

[a]No SO_3 detected.
[b]From Melvig, W. A., and Gibson, F. H. 1956. *Analysis of ash from United States coals.* Bu. Mines Bull. 567.
Source: Moulton 1973, Table 2, p. 153.

Table 4.21. Chemical analysis of fly ash (%)

Chemical components	Subbituminous coal	Bituminous coal
SiO_2	50.50	43.5 - 57.0
Al_2O_3	33.70	18.0 - 28.0
Fe_2O_3	6.85	7.9 - 16.0
CaO	2.80	4.0 - 10.0
MgO	1.50	1.0 - 5.5
SO_3	0.75	0.9 - 3.3
Combustion losses	2.85	($Na_2O + K_2O$)

Source: Nowak 1973, Table 1, p. 225.

residue to be highly alkaline. The potential of the ash for leaching trace elements may differ significantly from that of the ash found in conventional power plants because portions of the ash may be formed by different routes (e.g., at lower temperature in gasification reactions) (Fulkerson et al. 1974).

Ash disposal options include return to the coal mine, ponding, use as landfill, or resale to industries. When burial, ponding, or landfilling is chosen, either the ash or the site must be treated so that no trace metals can leach from the ash into groundwater systems. Little is known about the environmental destiny of ash leachates (calcium chloride, magnesium sulfate, fluorides). In particular, ash leachates may constitute a significant source of trace metal environmental pollution.

In addition to heavy metal content, fugitive ash emissions may be environmentally significant as adsorption media for polynuclear aromatic hydrocarbons. In fact, polycyclic organic matter in the atmosphere may be associated exclusively with particulate matter, especially soot (National Academy of Sciences 1972). This subject is discussed further in Sect. 4.6.

Once ash composition can be controlled for resale purposes, ash may be considered a by-product rather than a waste, and only fugitive emissions will be considered pollutants. Ash from coal-fired industries is presently being used, or tested for use, as a soil stabilizer, concrete

Table 4.22. Moisture, carbon, sulfur, nitrogen, and ash content of Illinois coal and ash samples[a] (wt %)

Component	No. 5 coal		No. 5 ash		No. 6 coal		No. 6 ash	
	As received	Dry	As received	Dry	As received	Dry	As received	Dry
Moisture	8.49		0.45		8.49		0.35	
Carbon[b]	62.5	68.3	3.8	3.8	59.3	64.8	5.4	5.4
Sulfur[c,d]	3.03	3.31 (3.61)	0.47	0.47	3.00	3.28 (3.43)	0.58	0.58
Nitrogen	1.17	1.28	0.04	0.04	1.12	1.22	0.05	0.05
Ash[d]	8.34	9.11 (9.05)	96.19	96.62	9.67	10.57 (10.53)	94.55	94.88

[a]Argonne analyses.
[b]Estimated accuracy $\pm$ 5% (relative).
[c]Estimated accuracy $\pm$ 10% (relative).
[d]Values in parentheses obtained at Peabody Coal Company's Central Laboratory.

Source: Sather et al. 1975, Table 1, p. 8.

Table 4.23. Range in amount of trace elements
present in coal ashes (ppm)

Element	Anthracites			Lignites and Subbituminous		
	Max	Min	Average(5)	Max	Min	Average(13)
Ag	1	1	b	50	1	b
B	130	63	90	1,900	320	1020
Ba	1340	540	866	13,900	550	5027
Be	11	6	9	28	1	6
Co	165	10	81(4)	310	11	45
Cr	395	210	304	140	11	54
Cu	540	96	405	3,020	58	655
Ga	71	30	42	30	10	23(12)
Ge	20	20	b	100	20	b
La	220	115	142	90	34	62
Mn	365	58	270	1,030	310	688
Ni	320	125	220	420	20	129(8)
Pb	120	41	81	165	20	60
Sc	82	50	61	58	2	18(10)
Sn	4250	19	962	660	10	156
Sr	340	80	177	8,000	230	4660
V	310	210	248	250	20	125
Y	120	70	106	120	21	51
Yb	12	5	8	10	2	4
Zn	350	155	b	320	50	b
Zr	1200	370	688	490	100	245

[a]Figures in parentheses indicate the number of samples used to
compute average values.
[b]Insufficient figures to compute an average value.

Source: O'Gorman and Walker 1972, Table 18, pp. 99-101.

aggregate, paving material, fire-control or -proofing material, acoustical insulation, ingredient
of cement blocks, face brick, and paving tile, and in tana grass manufacture (Browning 1973; Manz
1973; Pedlow 1973; and Bracket 1973). Pulp and paper mills have investigated the use of fly ash
in water pollution control: Ashes having a high lime content appeared to have the greatest
potential for treating highly colored industrial wastes (e.g., those from pulp and paper mills),
but by chemical precipitation rather than adsorption (Whittemore 1973). High-temperature gasification processes (Koppers-Totzek) produce a slag that may be suitable for fill material, road
aggregate, or cinder block material (Magee, Jahnig, and Shaw 1974). Tests have shown that gasifier
coal ashes having significant iron content are suitable sorbents for desulfurization of low-
calorific-value fuel gas from coal (Schrodt, Hilton, and Rogge 1975).

4.3.2.2 Char

Char is produced in some conversion processes (Synthane, COED, Garrett Pyrolysis) when carbon
removal from the coal is incomplete (Glaser, Hershaft, and Shaw 1974). If significant carbon is
present, char can be a useful source of energy for either the conversion process or another
industry. Conversion processes may use chars for gasifier feedstock, for utility fuel production,
or as a hydrogen source.

Amounts of net char produced from coal conversion processing vary from 10 to 55% of the coal
feed weight depending on the process steps. The amount of char produced by a single conversion

plant could be as much as 4 to 5 million tons annually (Nelson et al. 1969). As a fuel, char
is a potential source of pollution, requiring treatment of sulfur dioxide emissions.

The COED process produces a char of 50 to 60% of the original coal feed, amounting to 12,512 tons
per day (Kalfadelis and Magee 1975). A feed of 25,000 tons of dry coal per day to the modified
COED process plant will produce 13,440 tons of char per day (Strom and Eddinger 1971). Table
4.24 is a COED char analysis by Hamshar, Terzian, and Scotti (1974).

Table 4.24. Properties of COED char product

	Utah	Illinois No. 6
Proximate analysis, wt %, dry		
Volatile matter	6.1	2.7
Fixed carbon	80.2	77.0
Ash	13.7	20.3
Ultimate analysis, wt %, dry		
Carbon	81.5	73.4
Hydrogen	1.3	0.8
Nitrogen	1.5	1.0
Sulfur	0.5	3.4
Oxygen	1.5	1.0
Ash	13.7	20.3
Chlorine	0.006	0.1
Iron[a]	0.28	
Higher heating value, Btu/lb, dry	12,310	11,040

[a]Included in ash above.

Source: Hamshar, Terzian, and Scotti 1974, Table 6, p. 154.

Char properties depend on rank of the coal feed, coal characteristics (ash, sulfur), and type of
treatment (severity, temperature, reactive atmosphere). In general, ash levels in char are higher
for gasification or solvent processes than for processes using thermal treatment (COED) (Nelson
et al. 1969). The ultimate and proximate analyses of chars from the Synthane process are compared with the coals from which they are derived (Table 4.25).

Analysis of the char produced in the Garrett Pyrolysis process is given below:

Element	Wt %
Carbon	74.0
Hydrogen	1.9
Nitrogen	1.0
Sulfur	.6
Oxygen	3.9
Ash	18.6

Table 4.25. Representative analyses of coals and chars (wt %)

Component	Illinois No. 6	Western Kentucky	Wyoming subbituminous	North Dakota lignite	Pittsburgh seam
Coals					
Moisture	8.3	4.3	18.1	20.6	2.5
Volatile matter	37.5	34.6	31.9	32.9	30.9
Fixed carbon	43.0	44.5	32.0	38.2	51.5
Ash	11.2	16.6	18.0	8.3	15.1
Hydrogen	5.3	4.7	5.4	5.7	4.7
Oxygen	15.9	10.9	30.3	32.6	9.3
Carbon	63.0	62.7	45.2	51.5	68.4
Nitrogen	1.1	1.2	0.6	0.7	1.2
Sulfur	3.5	3.9	0.5	1.2	1.3
Chars					
Moisture	0.8	1.2	0.5	1.2	1.4
Volatile matter	4.0	4.8	5.1	10.0	1.6
Fixed carbon	69.9	63.3	38.1	50.2	69.3
Ash	25.3	30.7	56.3	38.6	27.7
Hydrogen	1.0	1.0	1.0	0.9	1.0
Oxygen	1.3	1.1	1.2	0.0	1.7
Carbon	70.4	64.5	40.6	58.9	68.9
Nitrogen	0.6	0.7	0.4	0.2	0.5
Sulfur	1.4	2.0	0.5	2.0	0.2

Source: Forney et al. 1974, Table 6, p. 7.

Chars show considerable variation in ash, sulfur, and volatile matter content (Nelson et al. 1969). Total sulfur in char is lower than that of the original coal, the most pronounced reduction occurring in chars from gasification processes. These constituents are compared for various coals and their chars for different conversion processes (Table 4.26).

Analyses for the trace elements in chars have not been located, but a percentage is expected from the ash content in the char. Neither char combustion nor gasification emissions have been located.

Char use options reported for the Bi-Gas process (Hittman Associates 1975) include (1) dewater and use as fuel for steam generation, (2) dewater and recycle to gasifier, (3) use in applications in which the carbon content may exhibit chemical or physical absorption properties, or (4) dispose of as a solid waste.

Char use considerations for the COED process include onsite utility combustion or gasification in the Koppers-Totzek process or COGAS system (Kalfadelis and Magee 1975).

Nelson et al. (1969) found the prime char market (other than in-plant transfers) to be power plants for combustion use. Properties determining char performance as boiler fuel are volatile matter, porosity, grindability, and ash content (Nelson et al. 1969). However, use of char as a fuel will require treatment of residual sulfur content; it can be removed either before or after combustion. Synthane's utility boiler, when combusting 4350 tons of char per day, would emit 10,900 lb of sulfur dioxide per hour, requiring about 115 tons of limestone for treatment per day (Kalfadelis and Magee 1975*b*).

Table 4.26. Examples of analysis of coal feeds and resulting chars for the various processes (dry basis) (%)

Coal type	Coal				Char			
	Volatile matter	Fixed carbon	Ash	(Sulfur)	Volatile matter	Fixed carbon	Ash	(Sulfur)
FMC (COED)								
Pittsburgh-Federal	36.8	57.0	6.2	(2.9)	3.7	86.8	9.5	(1.9)
Illinois No. 6	38.6	50.0	11.4	(3.8)	3.5	76.4	20.1	(3.1)
McKinley	41.8	49.6	8.6	(0.7)	4.4	79.5	15.4	(0.7)
Atlantic-Richfield (Seacoke)								
Illinois No. 6	41.4	48.2	10.4	(4.0)	4.2	76.9	17.7	(3.1)
McKinley	40.3	51.0	8.7	(0.6)	1.7	83.9	13.0	(0.5)
Hydrocarbon Research, Inc. (H-Coal)								
Illinois No. 6	38.1	50.3	11.6	(3.1)	3-5	*a*	44.0	*a*
Consolidation Coal Company (Liquefaction)								
Pittsburgh No. 8 (Ireland)								
80% depth of conversion	39.4	45.8	14.8	(4.3)	9.4	46.2	44.4	(5.1)
50% depth of conversion	39.4	45.8	14.8	(4.3)	12.9	62.8	24.3	(4.6)
Western Hi-Vol B								
80% depth of conversion	41.2	51.3	7.6	(1.0)	12.1	59.0	28.9	(0.9)
50% depth of conversion	41.2	51.3	7.6	(1.0)	14.6	71.4	14.0	(0.9)
Consolidation Coal Company (Gasification)								
Pittsburgh No. 8 (Ireland)								
80% depth of conversion	39.4	45.8	14.8	(4.3)	1.0	56.4	42.6	(1.0)
50% depth of conversion	39.4	45.8	14.8	(4.3)	1.5	75.4	23.1	(0.9)
Western Hi-Vol B								
80% depth of conversion	41.2	51.3	7.6	(1.0)	1.5	69.7	28.8	(0.2)
50% depth of conversion	41.2	51.3	7.6	(1.0)	2.0	84.1	13.9	(0.2)
Institute of Gas Technology								
Pittsburgh No. 8 (Ireland)	32.7	52.3	14.1	(4.3)	1.2	77.5	21.3	(1.7)

[a]Not available.

Source: Nelson et al. 1969, Table 3, p. 31.

Coal desulfurization processes that remove sulfur before combustion, such as the TRW Leaching process and the Institute of Gas Technology (IGT) Flash Desulfurization process described in Sect. 4.1.1, are being considered for char desulfurization. The FMC Corporation is designing a char desulfurization process that involves contacting a mixture of char and an acceptor, dolomite, with hydrogen in a fluidized or rotary bed. Dolomite regeneration is required, and elemental sulfur is recovered (Nelson et al. 1969).

Processes that remove sulfur during or after coal combustion (Limestone Injection, Lime Scrubbing, atalytic Oxidation, Wellman-Lord, Double Alkali, and Citrate) may prove applicable for use with chars.

Nandi and Walker (1972) have studied adsorption properties of coal chars from several conversion processes: two chars from IGT (hydrogen and steam partial gasification), one from FMC (coal pyrolysis and partial combustion in air), and one from Consolidation Coal Company (CCC) (CO_2 acceptor steam gasification). Adsorption values varied from 25 and 101 mg/g (IGT) to 121 mg/g (CCC) and 31 mg/g (FMC); thus, some of the chars showed significant adsorption — over 25% of that shown by commercial activated carbons (Nandi and Walker 1972). The adsorption capacity of chars depends upon extent of gasification of the coal particle, coal type, and uniformity of gasification through the coal particle (Nandi and Walker 1972). Although the chars were not premium-quality active carbons, it may be possible to use them in areas where high activity is not required, for example, in sewage treatment or removal of sulfur dioxide from fuel gases (Stacy and Walker 1972).

4.3.2.3 Filter cake

A filter cake similar to char, produced by the filtration of the liquefaction products, is composed of most of the coal mineral matter. In the COED process, hot filter cake (38% oil, 52% char, 10% filter; 350°F) is discharged at the rate of 15 tons/hr (1.5 tons/hr filter oil, 5.8 tons/hr raw oil, and 7.9 tons/hr char fines) and is recycled to the coal feed (Kalfadelis and Magee 1975). The SRC filter cake will be recycled to the gasification unit (Hittman Associates 1975a). The SRC filter cake slurry is composed of 713 tons of ash per day, 818 tons of unreacted char per day, and 1530 tons of oil per day. Gasification of the filter cake slurry produces a slag (requiring land- or minefill disposal) and a synthesis (low-Btu) gas for utility production (Jahnig 1975). Slag withdrawn from the gasifier requires disposal. Dry char, tars, or oils in the synthesis gas are removed by cycloning or scrubbing (Jahnig 1975).

4.3.2.4 Miscellaneous solids

The CO_2 Acceptor gasification process will require land- or minefill of spent dolomite acceptor solids at a rate of about 2% per day, or about 162,810 lb/hr of the input 7,164,000 lb/hr (Jahnig and Magee 1974).

4.3.2.5 Trace elements

Numerous studies list elements found in coal; Table 4.27 lists 65 elements that have been reported in the references cited. A two-volume bibliography containing 2450 citations was published by the Toxic Materials Information Center in 1975.

The fates of trace elements in coal conversion are not yet completely known; full material balances are still needed for each conversion process. Factors governing trace metal distribution

Table 4.27. Coal elements

Aluminum	Gallium	Samarium
Antimony	Germanium	Scandium
Arsenic	Hafnium	Selenium
Barium	Holmium	Silicon
Beryllium	Indium	Silver
Bismuth	Iodine	Sodium
Boron	Iron	Strontium
Bromine	Lanthanum	Tantalum
Cadmium	Lead	Tellurium
Calcium	Lithium	Terbium
Cerium	Lutetium	Thallium
Cesium	Magnesium	Thorium
Chlorine	Manganese	Tin
Chromium	Mercury	Titanium
Cobalt	Molybdenum	Tungsten
Copper	Neodymium	Uranium
Dysprosium	Nickel	Vanadium
Erbium	Niobium	Ytterbium
Europium	Potassium	Yttrium
Fluorine	Praseodymium	Zinc
Gadolinium	Rubidium	Zirconium

Sources: Kessler, Sharkey, and Friedel
1973; Ondov et al. 1975; Magee,
Hall, and Varga 1973; Ruch,
Gluskoter, and Shimp 1973;
Von Lehmden, Jungers, and Lee
1974.

are process operating conditions and interaction of coal constituents (Shaw and Magee 1974).
Trace element pathways within conversion processes may include adsorption on particulate matter,
inclusion in condensates, deposition on equipment surfaces (scales, etc.), inclusion in by-
products, inclusion in final products, and emission as fugitive pollutants.

Elements adsorbed on particulate matter may be removed by electrostatic precipitators or cyclones
during stack gas and product gas cleaning. Most of the particulate matter mass from fossil
fuel combustion sources is composed of sodium, potassium, magnesium, aluminum, silicon, calcium,
titanium, iron, sulfur, and presumably oxygen; other constituents are cobalt, copper, arsenic,
strontium, silver, molybdenum, cadmium, tin, antimony, barium, mercury, lead, lithium, beryllium,
boron, vanadium, chromium, manganese, nickel, and zirconium (Hillenbrand, Engdahl, and Barrett
1973).

Recovery of elements in the acid gas scrubbing or condensing operations may result in side
reactions with other substances. For example, conversion processes may produce organometallic
compounds of environmental concern (Fulkerson et al. 1974b). Mercury, cadmium, and zinc are
removed from solution by acclimated activated sludge in respectively decreasing amounts (Newfeld
and Hermann 1975). On the other hand, chromium present in the wastewater is detrimental to
biological oxidation units (Jahnig 1975). Table 4.28 gives the trace element composition of
aqueous condensates from the Synthane (laboratory-scale) process (Forney et al. 1974).

Kalfadelis and Magee (1975b), using values given in Forney et al. (1974) (Table 4.29), calculated
the quantity of major coal elements (excluding carbon, hydrogen, oxygen, and nitrogen) that would

Table 4.28. Synthane process: trace elements in condensate from an
Illinois No. 6 coal gasification test

Component	No. 1	No. 2	Average (by weight)
Major constituents, ppm			
Calcium	4.4	3.6	4
Iron	2.6	2.9	3
Magnesium	1.5	1.8	2
Aluminum	0.8	0.7	0.8
Minor constituents, ppb			
Selenium	401	323	360
Potassium	117	204	160
Barium	109	155	130
Phosphorus	82	92	90
Zinc	44	83	60
Manganese	36	38	40
Germanium	32	61	40
Arsenic	44	28	30
Nickel	23	34	30
Strontium	33	24	30
Tin	25	26	20
Copper	16	20	20
Columbium	7	5	6
Chromium	4	8	6
Vanadium	4	2	3
Cobalt	1	2	2

Source: Forney et al. 1974, Table 2, p. 4.

Table 4.29. Major elements in coal

Element	Concentration in feed coal (wt %)	Concentration in feed to gasifier (lb/day)	Concentration in condensate[a] (ppm)	Discharge to water treatment (lb/day)
Silicon	4.0	1,140,000		
Iron	1.5	428,000	3	50
Sulfur	1.6	456,000	1400	22,500
Aluminum	1.0	285,000	0.8	13
Calcium	0.5	142,500	4	65
Potassium	0.22	62,700	0.16	3
Magnesium	0.17	48,500	2	32
Sodium	0.069	19,700		

[a]From Forney et al. 1974.

Source: Kalfadelis and Magee 1975b, Table 10, p. 62.

enter the Synthane gasifier and the quantity that would enter the raw gas condensate. The bulk
of the major elements is expected to be contained in the ash. Table 4.30 similarly lists some
of the minor (or trace) elements in coal, whereas Table 4.31 lists the hazardous trace elements.
However, original feed coal trace element concentrations vary significantly, as do their
solubility and reactivity.

Table 4.30. Minor trace elements

Element	In feed coal		In condensate	
	(ppm)	(lb/day)	(ppm)[a]	(lb/day)
Boron	165	4700		
Barium	100	2900	0.13	0.2
Zinc	44	1250	0.06	1
Vanadium	35	1000	0.003	0.05
Lithium	25	710		
Chromium	22	620	0.006	0.1
Nickel	14	400	0.03	0.5
Germanium	12	340	0.04	0.6
Copper	12	340	0.02	0.3
Manganese	10	290	0.04	0.6
Molybdenum	8.2	230		
Yttrium	7.7	220		
Lanthanum	5.1	150		
Uranium	5	140		
Cobalt	3.8	110	0.002	0.03
Tin	1.5	43	0.02	0.3

[a]From Forney et al. 1974.

Source: Kalfadelis and Magee 1975b, Table 11, p. 63.

Elements that are deposited on the conversion equipment can be potential hazards and may require
removal to prevent interference with normal operations. For example, titanium tends to accumulate
on the surface of liquefaction catalysts and deactivate them (Given 1974), or it can appear in the
product oils (Jahnig 1975). Vanadium may react with alkali in the conversion reducing atmosphere
to form a vanadate slag that can cause corrosion within the units (Magee, Hall, and Varga 1973).
Corrosion from chloride-formed acids may also require treatment. The more refractory trace
elements will be contained predominately in heavy by-product residues such as tars. If tars are
coked and gasified by reaction with steam and air, most of the refractory elements will be
discharged in the ash (Fulkerson et al. 1974b).

The majority of elements are expected to remain in the solid by-products such as ash, chars, and
filter cakes although other by-products such as tars may also contain small amounts of them.

Elements remaining in the product streams must be removed to prevent fouling of catalysts in the
final conversion stages and to reduce pollution levels of the product fuel. Trace constituents

Table 4.31. Potentially hazardous trace elements

Element	In feed coal (ppm)	To gasifier (lb/day)	Condensate[a] (ppm)	Raw gas[a,b] (ppm)	Tar[a] (ppm)
Beryllium	2	60			
Fluorine	85	2430			
Arsenic	31	880	0.03 (0.5)		0.7 (0.7)
Selenium	2.2	65	0.36 (6)		
Cadmium	0.14	4			
Mercury	0.2	6		0.0001[c] (0.04)	0.003 (0.003)
Lead	7.7	220			
HCN				20[c] (1000)	

[a]Numbers in parentheses indicate quantities in pounds per day appearing in design streams.
[b]No mercury or HCN detected in product gas.
[c]Units as given in Forney and Magee 1972.

Source: Kalfadelis and Magee 1975b, Table 12, p. 64.

in the SRC product are compared with those in the coal feed in Table 4.32. Jahnig (1973) reports that the titanium content is high and that beryllium (at 0.7 ppm), copper, cobalt, and lead levels in the SRC product oil can be significant and might pose a pollution problem when burned. Product streams may contain significant amounts of particulate material, organometallic compounds, and soluble compounds such as the hydrides of arsenic, antimony, boron, fluorine, or tellurium (Fulkerson et al 1975).

On the other hand, there are strong indications that some of the mineral constituents (sodium and pyrite) of coals may actually be beneficial, acting as additional catalysts for the liquefaction process (Given et al. 1974). They may increase yields or alter the characteristics of the oil. Sodium ions associated with acid groups have a marked effect on the oil viscosity produced by hydrogenation (Given et al. 1974). Pyrite, reduced to pyrrhotite or to metallic iron under liquefaction conditions, may be an important liquefaction cocatalyst (Given et al. 1974).

Trace metals in Synthoil (Yavorsky and Akhtar 1974) are as shown below.

Metal	Concentration in oil (ppm)
Silica	135.5
Iron	67.4
Aluminum	29.1
Potassium	5.0
Sodium	2.9
Calcium	2.3
Magnesium	2.2
Molybdenum	<0.2
Cobalt	<0.1

Table 4.32. SRC process: analysis of coal and product samples composition (ppm)

Element	Atomic absorption			Neutron activation			Air emission		
	Sample 1 (1000 psig)	Sample 2 (2000 psig)	Sample 3 (feed coal)	Sample 1 (1000 psig)	Sample 2 (2000 psig)	Sample 3 (feed coal)	Sample 1 (1000 psig)	Sample 2 (2000 psig)	Sample 3 (feed coal)
Aluminum							96	130	12,000
Antimony	<4	<4	<4	0.25	0.30	10.6			
Arsenic				1.4	0.5	19			
Barium							0.22	0.55	50
Beryllium	0.7	0.4	0.9				<0.4	<0.2	<10
Bismuth							<0.2	<0.1	<5
Boron	100	51	94				15	36	200
Bromine							3.8	3.9	7.2
Cadmium	<0.1	<0.1	<1.5				<1	<0.7	<33
Calcium	180	70	3,400				100	84	4,800
Cobalt	2.2	0.8	17	0.35	0.23	6	<0.4	<0.2	<10
Copper	3.7	2.5	6				1.1	3.9	8.6
Chromium	<2	<0.2	31	1.3	0.88	38	1.1	1.5	78
Fluorine				<100	<100	300			
Germanium							<0.4	<0.2	<10
Gold							<0.4	<0.2	<10
Iron	98	161	24,000				30	160	20,000
Lead	<2	0.4	8				<0.4	<0.2	<10
Lithium	<0.02	<0.02	7.4						
Magnesium	23	9	550				20	15	890
Manganese	3	1	39				2.8	2.7	75
Mercury	0.05	0.01	0.05						
Molybdenum	<50	<50	<50				1.7	1.4	49
Niobium							<2	<1	<44
Nickel	2.5	4	29				3.6	9.8	120
Potassium	<2	6	1,300	5	17	1,790			
Samarium				0.36	0.16	1.9			
Selenium				2	<1	7			
Silicon				600	900	18,000			
Silver	<2	<2	<2				0.2	0.9	0.8
Sodium	45	25	166	30	21	367	10	20	320
Strontium							<0.8	<0.5	<20
Tantalum							<2	<0.9	<50
Tellurium				1.4	0.6	5.8			
Thorium							<0.8	<0.5	<20
Tin				31	9	104	2.9	2.0	40
Titanium	300	74	460				260	160	600
Tungsten							<1	<0.7	<30
Uranium							<4	<2	<100
Vanadium	17	16	175				17	12	200
Ytterbium				0.39	0.25	0.51			
Zinc	6	3	39	4.5	0.8	42			
Zirconium				2.0	1.1	6.3	5.4	3.9	35

Source: Jahnig 1975, Table 7, pp. 50 and 51.

Arsenic, beryllium, mercury, selenium, cadmium, fluorine, and lead are toxic and volatile (Magee, Hall, and Varga 1973). All are expected to volatilize during coal conversion (Interagency Task Force on Synthetic Fuels from Coal 1974), resulting in possible fugitive emissions (Jahnig, Magee, and Kalfadelis 1974). The fate of trace elements through coal conversion may be different from that evidenced by coal combustion. Conversion operates in reducing atmospheres, whereas combustion uses oxidation. In coal conversion, the elements may form compounds such as hydrides, carbonyls, or sulfides, which may be more volatile than the oxidized compounds produced by combustion (Jahnig 1975*b*).

A Hygas (bench-scale) process study of the fate of 11 elements indicates substantial loss of mercury, selenium, arsenic, tellurium, lead, and cadmium from the ash (Table 4.33). However, Shaw and Magee (1974) report that these results were based on a limited number of samples and that further tests are necessary. Coal and ash trace element compositions from the Lurgi process are compared in Table 4.34. On the basis of the above results, calculated amounts of trace elements that may be contained in the ash for the Lurgi plant processing 20,000 tons of coal per day are

Element	Amount (lb/hr)
Arsenic	1-2
Beryllium	75-80
Fluorine	10-20
Lead	750-2700
Vanadium	600-750

However, these values pertain only to the particular coal and ash samples analyzed in the study. Trace elements lost (i.e., higher values present in coal than in ash) are expected to be in the raw product gas stream. Trace element (chromium, copper, manganese, nickel) increases may be due to contamination or wear of equipment surfaces (Sather et al. 1975).

Table 4.33. HYGAS: percent disappearance of trace elements

Element	Preheater (430°C, 1 atm)	Hydrogasifier (650°C, 74 atm)	Electrothermal (1000°C, 74 atm)	Sum
Hg	30	48	19	97
Se	41	21	12	74
As	22	25	18	65
Te	36	18	9	63
Pb	25	19	19	63
Cd	24	23	14	61
Sb	13	7	13	33
V	-9[a]	18	21	30
Ni	8	8	8	24
Be	-9[a]	7	21	19
Cr	-13[a]	7	7	1

[a]The minus sign indicates that more element was recovered than was put in.

Source: Attari 1973, as cited in Shaw et al. 1974, Table 10, p. 36.

Table 4.34. Lurgi gasification: trace element concentrations (ppm) in Illinois coal and in the unquenched gasifier ash obtained from it

Element	Coal		Ash	
	Peabody[a]	Argonne[b]	Peabody[a]	Argonne[b]
Illinois No. 5 coal				
Ag	0.3		3.0	
As	1.6	1.9 ± 1.0 (B)	0.3	
B	307		673	
Ba				490 ± 250 (B)
Be	2.2	2.0 ± 0.1	22	19.8 ± 1.0
Br		6.6 ± 1.0 (A)		
Cd	<0.3		<0.3	
Ce				41 ± 4 (AA)
Co	3.7	3.8 ± 0.6 (A)		38 ± 4 (AA)
Cr	15	15 ± 2 (A)	551	592 ± 59 (AA)
Cs				11 ± 2 (A)
Cu	10		273	
Dy				
Eu				
F	59	55 ± 11		4.6 ± 0.9
Fe ($\times 10^4$)		1.3 ± 0.1 (AA)		15 ± 2 (AA)
Hg	0.20	0.17 ± 0.02	0.01	0.016 ± 0.002
K ($\times 10^3$)		1.3 ± 0.1 (AA)		14 ± 1 (AA)
La		3.6 ± 0.5 (A)		42 ± 4 (AA)
Li	5.5		54	
Mn	21	23 ± 2 (AA)	338	305 ± 30 (AA)
Mo	7		8	
Na ($\times 10^2$)		2.8 ± 0.3 (AA)		29 ± 3 (AA)
Ni	32		462	
Pb	30	28.1 ± 2.8	219	200 ± 10
Sb	0.3	0.1 ± 0.02 (A)	0.3	19 ± 2 (AA)
Sc		1.6 ± 0.2 (AA)		
Se		9 ± 5 (B)		
Sm				
Ta				1.3 ± 0.7 (B)
Tb				(C)
V	21		181	
Yb				11 ± 2 (B)
Zn ($\times 10^2$)	1.82	2.4 ± 1.2 (B)	15.8	16 ± 2 (A)
Illinois No. 6 coal				
Ag	0.3		3.8	
As	1.0	2.1 ± 1.0 (B)	0.1	
B	132		622	
Ba				
Be	1.8	1.55 ± 0.08	14	13.4 ± 0.7
Br		4.1 ± 0.7 (A)		
Cd	<0.3		<0.3	
Ce				38 ± 4 (AA)
Co	4.3	3.2 ± 0.5 (A)	40	34 ± 3 (AA)
Cr	22	18.3 ± 2.7 (A)	705	806 ± 81 (AA)
Cs				
Cu	12		239	
Dy				8.5 ± 1.3 (A)
En		0.2 ± 0.1 (B)		

Table 4.34   (continued)

Element	Coal		Ash	
	Peabody[a]	Argonne[b]	Peabody[a]	Argonne[b]
F	79	79 $\pm$ 16		5.2 $\pm$ 1.0
Fe $(\times 10^4)$		1.2 $\pm$ 0.1 (AA)		13 $\pm$ 1 (AA)
Hg	1.00[c]	1.18 $\pm$ 0.12	0.04	0.007 $\pm$ 0.001
K $(\times 10^3)$		1.5 $\pm$ 0.1 (AA)		16 $\pm$ 2 (AA)
La		3.9 $\pm$ 0.4 (AA)		40 $\pm$ 20 (B)
Li	9.2		74	
Mn	20	18.6 $\pm$ 1.9 (AA)	243	156 $\pm$ 16 (AA)
Mo	7		6	
Na $(\times 10^2)$		3.0 $\pm$ 0.3 (AA)		27 $\pm$ 3 (AA)
Ni	14		456	
Pb	12	8.0 $\pm$ 0.8	96	46.0 $\pm$ 2.3
Sb	0.1	(C)	0.2	
Sc		2.1 $\pm$ 0.2 (AA)		24 $\pm$ 2 (AA)
Se		(C)		
Sm		0.005 $\pm$ 0.003 (B)		
Ta				
Tb				3.1 $\pm$ 1.6 (B)
V	29		301	
Yb		1.4 $\pm$ 0.7 (B)		11 $\pm$ 2 (A)
Zn $(\times 10^2)$	0.43	(C)	4.69	(C)

[a]The precision of the Peabody results is estimated to be $\pm$10% in all cases.

[b]The accuracy of the Argonne results for mercury is estimated to be $\pm$10%; the precision of the Argonne results for beryllium is +5%, for lead +5 to 10%, for fluorine +20%. The confidence ratings shown for the Argonne results obtained by neutron activation analysis correspond to the following accuracy levels: AA, +10%, A, +15%; B, +50%, and C, identification only.

[c]Not representative of seam; contamination suspected.

Source: Sather et al. 1975, Tables 2 and 3, pp. 10 - 13.

Forney et al. (1975), in a study of the trace element and major component balances around the Synthane PDU (pilot demonstration unit) gasifier, found that the largest percentage of the 16 elements considered most important were contained primarily in the chars. These elements included arsenic, boron, beryllium, cadmium, chlorine, chromium, fluorine, mercury, manganese, nickel, phosphorus, lead, uranium, selenium, vanadium, and zinc. Almost all of the chlorine was found in the water; significant percentages of mercury, boron, selenium, and fluorine were also found in the water. Tars contained some arsenic, lead, cadmium, and mercury. Most of the mercury appeared in the tar and water, but analyses were incomplete. However, some of the recovery percentages for the 65 elements in the study were inaccurate, requiring further analysis and investigation of better methods (Forney et al. 1975).

Results for 47 trace elements in COED process coal, coal product, and waste stream samples are listed in Table 4.35 and are expected to be within a factor or two of the actual concentrations (Shults 1976).

Jahnig (1975b), using the approximate degree of volatilization shown for various elements and their corresponding coal concentrations (hypothetical as typical), has estimated the amount (in

Table 4.35. Trace element results (ppm) for the COED process

Element	Feed	Raw oil	Filter cake	Syncrude	2nd-stage liquor	1st-stage liquor	Char	Fines, oil	Fines, cyclone	Filtrate
Al	5,000	5,000	10,000	90	70	4	10,000	100	30	2,000
As	3	3	7	2	0.7	0.01	10	2	1	2
B	200	200	200	4	10	3	300	3	20	100
Ba	5	1	10	1	3	<0.5	50	2	0.1	1
Bi	<0.3	<0.3	<0.3	<0.3	<0.3	0.3	3	<0.3	<0.3	<0.3
Br	<0.2	0.7	<0.2	<0.1	<0.3	<0.3	<4	<0.3	3	0.1
Ca	1,000	3,000	3,000	30	100	500	10,000	100	200	1,000
Cd	<0.3	<0.3	<0.3	<0.3	<0.3	<0.3	<2	<0.3	<0.3	<0.3
Cl	20	100	20	3	1	10	300	30	10	20
Co	1	0.2	2	0.1	0.1	1	5	0.4	<0.1	0.3
Cr	5	2	2	0.7	1	1	300	5	2	2
Cs	<0.1	1	0.5	0.2	0.2	5	1	<0.5	<0.2	0.6
Cu	3	2	10	0.3	0.3	5	150	1	0.4	2
F	4	2	10	1	2	<0.5	50	3	0.2	2
Fe	10,000	7,000	10,000	400	500	300	20,000	3,000	20	15,000
Ga	1	0.5	2	0.2	0.1	<0.1	4	0.3	<0.1	0.2
Ge	1	0.8	1	<0.7	<0.5	<0.5	3	<0.5	<0.5	1
K	200	250	500	10	100	4	3,200	100	10	200
La	10	1	5	1	1	0.1	10	5	0.1	2
Mg	80	200	500	20	20	80	4,000	50	20	500
Mn	5	5	10	0.4	0.6	7	100	5	0.4	2
Mo	5	0.5	20	0.7	<0.5	0.8	7	<0.5	5	<0.5
Na	50	100	500	10	30	10	<500	50	300	50
Nb	2	0.1	0.4	0.3	0.3	<0.1	3	<0.5	0.5	0.2
Nd	10	2	5	1	2	1	5	3	0.5	3
Ni	30	30	50	4	10	100	40	50	10	7
P	20	10	70	3	5	0.2	200	20	2	10
Pb	3	1	10	1	2	0.1	20	2	<0.1	1
Pr	2	0.5	3	0.4	0.3	0.4	3	1	0.1	1
Rb	1	0.5	2	0.2	0.2	0.02	10	0.5	<0.1	0.3
Sb	0.5	<0.2	0.1	<0.1	<0.1	<0.1	<2	0.1	<0.2	<0.2
Sc	3	1	1	0.4	0.5	<0.5	10	1	<0.5	1
Se	0.3	<0.1	0.4	<0.1	<0.1	0.2	3	1	2	0.1
Si	3,000	7,000	15,000	1,000	2,000	30	30,000	8,000	1,000	15,000
Sn	<0.3	<0.3	<0.3	<0.3	<0.1	<0.1	<3	<0.5	0.8	<0.3
Sr	10	5	30	10	4	0.3	60	10	<0.5	10
Ta	<1	<1	<1	<0.5	<0.5	<1	<3	<1	<3	<1
Te	<0.5	<0.5	<0.5	<0.5	<0.5	<0.5	<3	<0.5	<0.5	<0.5
Th	0.7	0.2	0.6	0.8	0.4	<0.1	4	0.7	<0.1	0.2
Ti	200	50	100	20	20	<1	2,500	70	<1	50
Tl	2	0.5	3	0.5	1	0.3	<3	0.3	<0.5	0.5
U	2	0.7	2	1	0.5	<0.1	3	<0.6	<0.5	0.6
V	50	5	30	2	2	0.1	300	5	0.1	15
W	3	<0.5	2	0.4	0.4	<0.5	<5	<0.5	<0.5	<0.5
Y	7	1	5	1	1	0.5	25	3	<0.3	2
Zn	10	3	10	2	1	3	400	1	1	4
Zr	10	3	20	2	3	<0.5	30	10	3	1

Source: Shults 1976, Table 4.9, pp. 38 - 39.

pounds per day) that might be carried out with the hot gases leaving the gasifier (Table 4.36).
The degree of volatilization for zinc, boron, and fluorine has not been determined (Jahnig
1975b).

Table 4.36. Example of trace elements that may appear in gas cleaning section

Element	Possible ppm in coal[a]	Percent volatile for example[b]	In gas[c] (lb/day)
Chlorine	1500	>90+	32,400
Mercury	0.2	90+	5
Selenium	2.2	74	39
Arsenic	31	65	484
Lead	7.7	63	116
Cadmium	0.14	62	2
Antimony	0.15	33	1
Vanadium	35	30	252
Nickel	14	24	81
Beryllium	2	18	9
Zinc	44	(10)	106
Boron	165	(10)	396
Fluorine	85	(10)	204
Titantium	340	(10)	816
Chromium	22	Nil	Nil

[a]Mainly based on Pittsburgh seam coal (Kalfadelis and Magee 1974).
[b]Mainly based on lower-temperature gasifier (Attari 1973) and indicated at 10% for zinc, boron, and fluorine in absence of data.
[c]For 12,000 tons of coal feed per day.

Source: Jahnig 1975b, Table 7, p. 42.

However, in power plant flue gas, mercury, arsenic, selenium, thallium, and antimony from coal
combustion are likely to exist in vapor form as Hg, As_4O_6, Sb_4O_6, Tl_2O, and SeO_2 (Gorman et al.
1974). Studies of a coal-fired power plant showed that 80% of the mercury and most of the arsenic
and selenium vaporized during combustion (Bolton et al. 1973). In a 25,000-ton/day plant, coal
containing 0.2 ppm of mercury releases 10 lb of mercury in vapor form per day (Fulkerson et al.
1974b). Cadmium and lead also vaporize in coal combustion, but Schultz, Hattman, and Booher
(1973) indicate that these would be retained with the fly ash. Toca, Cheever, and Berry (1973),
in a study of the fugitive particulate effluent from a coal-fired boiler, conclude that the
highest concentrations and amounts of cadmium and lead are found in the smaller-sized,
respirable-range (<5 μm) fractions of the particulates and that there was an increase in
concentrations with decreasing particle size.

Phillips (1973) could account for only 16% of the beryllium in the feed coal in ash in his study
of the Hayden Power Plant; this report suggests that beryllium may volatilize.

In a study of the trace element emissions of the coal-fired Chalk Point Electric Generating
Station, Gladney (1974) estimated the mass flow of elements and found that only 6 of the 35

elements present (bromine, iodine, gallium, selenium, mercury, and lead) were thought to have significant gas phase components.

Klein et al. (1975) studied the pathways of 37 trace elements through the TVA Allen Steam Plant in Memphis, Tennessee. Their results showed that most of the mercury, some of the selenium, and probably most of the chlorine and bromine were emitted as gases. Arsenic, cadmium, copper, gallium, molybdenum, lead, antimony, selenium, and zinc were more heavily concentrated in the fly ash than in the slag, and primarily in the fly ash discharged from the stack rather than that collected by the precipitator. Aluminum, barium, calcium, cerium, cobalt, europium, iron, hafnium, potassium, lanthanum, magnesium, manganese, rubidium, scandium, silicon, samarium, strontium, tantalum, thorium, and titanium were found in the slag and in collected and emitted fly ash (Klein et al. 1975).

In a semiquantitative study of power plant effluents, Gorman et al. (1974) found that the modified Limestone Scrubbing, the modified Double Alkali, Catalytic Oxidation, DAP-Mn (wet), and Chiyoda processes used for sulfur dioxide removal also have the potential to remove other vapor phase pollutants such as mercury, arsenic, selenium, antimony, and thallium.

The distribution of trace elements in the various conversion process streams needs further study before actual control measures can be proposed. Hazardous elements possibly may be concentrated in a process area and then enter the environment at a concentration higher than that of the original coal (Jahnig 1975).

4.4 WASTEWATER

4.4.1 Water-treating chemicals

Water is used in large quantity in many parts of the conversion process — in the pretreatment of coal, as feed for the conversion process, and ultimately for cooling or quenching. Most water supplies must be at least demineralized and clarified by water-treating chemicals. The Lurgi process, for example, involves the disposal of about 2 tons of water-treating chemicals per hour, of which 1000 lb/hr are associated with demineralization (caustic, sulfuric acid, and resins) (Shaw and Magee 1974). Cooling tower water is chemically treated to control corrosion, scale formation, plant growth, and pH (Shaw and Magee 1974). Of the chemicals used (alum, chlorine, chromium, sulfuric acid, sodium hydroxide, ferric chloride, and calcium carbonate), those having potential air pollutants by evaporation or water pollutants from blowdown discharges will require control.

4.4.2 Pretreatment operations wastewater

Runoff from an untreated coal storage area (20 to 25 acres) could amount to 5000 gpm during major precipitation (Kalfadelis and Magee 1975b). Liquid runoff wastes from the storage and preprocessing of coal are potential environmental pollutants; they contain suspended coal particulates and leachates due to rainfall or washing. Water is a polar solvent; thus, it can react with coal to form acids, extract polar organics, dissolve sulfur and metal compounds, and suspend coal particulates. The exact nature of the aqueous leachate has not been thoroughly investigated; however, it may be similar to acid mine water (Magee, Jahnig, and Shaw 1974). Methods for treating acid mine water include reverse osmosis, demineralization, ion exchange, active biochemical sludge followed by limestone neutralization, and hydrated lime neutralization. Hydrated lime neutralization is effective and is the only extensively used method of treating acid mine drainage (Reid and Streebin 1972).

Effluent limitation guidelines by EPA for the coal mining industry under the Refuse Act Permit Program may apply to the coal storage facilities (Environmental Protection Agency 1972). As a result, impounding and settling facilities will be required. However, as with acid mine water treatment, runoff from coal storage will require treatment designed for the exact water characteristics.

Coal washing requires large amounts of water. For example, jigging operations require 1500 to 2000 gal of water per ton of coal processed. More than 15% of the wash water, containing about 1% dissolved solids, is discharged (Interagency Task Force on Synthetic Fuels in Coal 1974). Refuse (mineral matter, debris, rock) from coal washing may be a source of water pollution (e.g., by oxidation of pyrite to sulfuric acid in the presence of oxygen and water). Wastewaters from the coal washing operations are sent to settling ponds for fines removal. Leaching or seepage from the settling ponds must be controlled.

4.4.3 Conversion operations wastewater

4.4.3.1 Process wastewater

Wastewaters arise from moisture in the coal, water of constitution or decomposition, water added for stoichiometric process requirements, or water introduced for by-product recovery or gas scrubbing (Rubin and McMichael 1975).

Process wastewaters are produced in the coal conversion processes (1) when gases are scrubbed to remove soluble contaminants, quenched to control operating temperatures, compressed, or dehydrated; (2) when steam used in the units cools and condenses or is reformed during methanation or hydrotreating; or (3) when slags and ash are water-quenched for slurry removal. Wastewater contaminant content varies according to feed coal characteristics, conversion process steps (types and severity), and upstream subsidiary treatments. The composition of process water, which is the principal source of water pollutants, is expected to be analogous to the ammonia liquor produced in coking industries (Rubin and McMichael 1975; Glaser, Hershaft, and Shaw 1974).

Gas liquor, or sour water, is the sum of the aqueous streams from condensing or scrubbing in the processing units. Components of the product gas (carbon monoxide, carbon dioxide, hydrogen, hydrogen cyanide, and methane) will be found in various quantities in the gas liquor (Hittman Associates 1975a) along with contaminants such as sulfur and nitrogen compounds, ash (dust and particulates), phenols, emulsified tar and oils, and soluble salts. Soluble salts may accumulate in aqueous streams in amounts as high as 3000 ppm (Shaw 1976, personal communication).

Water-soluble hydrogen sulfide and carbonyl sulfide can be picked up during gas scrubbing (producing sour water), especially if ammonia is present; ammonia increases sulfur solubility (Glaser, Hershaft, and Shaw 1974) and also favors absorption of hydrogen cyanide in the condensate (Hittman Associates 1975a). Ammonia, along with cyanide and thiocyanates, can also enter the gas liquor during gas scrubbing. Phenols and cresols, water-soluble aromatic compounds, may be carried into the water from the gasification units. Particulate matter, tars, and oils originating from coal combustion can be suspended in wastewater from gas scrubbing. Donaldson (1974) reports that all organic compounds identified in the tars from conversion technologies [i.e., Synthane process (Forney et al. 1974)] must be considered during wastewater analysis. Rubin and McMichael (1975) conclude that the conversion wastewaters can contain practically all the organic compounds found in coal. Miscellaneous materials in the wastewater may include hydrogen fluoride, fatty acids, pyridines, trace organics, and trace elements. Hydrogen fluoride can follow ammonia into the aqueous waste stream, but will probably be neutralized with the basic materials present in the water (Shaw and Magee 1974).

Methanation reformed-water contaminants should be minimal (Jahnig 1975*b*) because the gases have been cleaned prior to this step. Hydrotreating produces a sour water stream containing hydrogen sulfide, ammonia, oils, char, and other coal-derived materials. Slag or ash quenching, a process that involves adding about 50 wt % water to ease slurrying and then dewatering for disposal, produces a wastewater containing any water-soluble components of the slag or ash.

Table 4.37 shows the composition of five water streams within the Koppers-Totzek gasification plant (commercial); because of the high gasification temperature (3000 to 3500°F), phenols, pyridines, and other organics are not found (Glaser, Hershaft, and Shaw 1974).

Table 4.38 compares the gasifier condensate from the Synthane (laboratory-scale) process when using various coals with a coking-plant weak ammonia liquor. Depending on the type of coal used, the cyanide and thiocyanate waste is decreased one to two orders of magnitude, whereas suspended solids, phenols, and ammonia are about the same from Synthane gasification as from coking plant by-product water. Trace element components of the Synthane (laboratory-scale) process condensate are given in Table 4.28. Because by-product water from the Synthane process and the waste ammonia liquor from coking plants are so similar (Table 4.39), water treatment methods developed for other industries, especially coking, should be adaptable for coal conversion.

Characteristics of the aqueous condensate streams from the COED process are identified in Table 4.40. The FMC Corporation indicates that these condensates, following ammonia and hydrogen sulfide removal, will be recycled to the pyrolyzer or char gasifier (Kalfadelis and Magee 1975).

Sour water from the gasifier scrubbing and shift condensates may contain amounts of volatile trace metals, which, after recycling, may concentrate and require recovery or treatment (Jahnig 1975*b*).

Glaser, Hershaft, and Shaw (1974) report only trace amounts of aromatics and pyridines in the gasifier scrubber water from the Synthane process (Table 4.41).

4.4.3.2 Cooling water

Conversion operations use large amounts of water for cooling purposes. Leakage of process stream components into the cooling circuit should be minimized through mechanical design. However, inevitable process leakage may input organic and inorganic process stream components into the system (Magee 1975). Cooling tower circuit losses include evaporation, drift, and blowdown; replacement of the losses is termed makeup water. Drift loss may be significant because it may contain solids and result in environmental deposits (Jahnig 1975). Evaporation of cooling tower chemicals or coal-derived materials may cause significant releases of contaminants. Blowdown can be treated and should not be environmentally significant.

Table 4.37. Koppers-Totzek coal gasification water analyses,
Kutahya, Turkey

Component	Amount (mg/liter)[a] at sample location[b]				
	I	II	III	IV	V
pH^a	8.8	8.8	8.9	8.8	8.9
CaO	78	101	78	135	179
MgO	97	161	194	145	113
Na	17.5	17.5	17.5	17.5	17.5
K	5.6	8.8	10.0	8.0	8.0
Zn	0.01	0.03	0.02	0.02	0.02
Fe	0.05	0.22	1.95	0.20	0.64
NH_4	0.32	157	184	137	122
NO_2	0.02	0.13	4.47	0.24	4.37
NO_3	58.2	3.32	13.7	24.7	22.9
PO_4, total	1.89	0.81	1.21	0.81	2.70
Cl	18	85	96	57	46
SO_4	42	216	155	255	109
CN	0.26	0.52	12.5	1.4	14.0
H_2S^c					
$KMnO_4$, consumed	8	9	400	11	145
Chemical oxygen demand	14	18	128	16	63
SiO_2	14.8	16.0	14.8	19.8	42.6
Suspended solids	14	4612	5084	3072	50
Cu	0.01	0.01	0.01	0.01	0.06

[a]All measurements in mg/liter except for pH.
[b]I — cooling water to slag quench tank (stream 5); II — water from
slag quench tank (stream 14); III — wash water from gas cooler
(stream 17); IV — total water to clarifier; V — water out of
clarifier.
[c]Not detected.

Source: From Farnsworth, Mitsak, and Kamody 1974, Table 4, p. 126.

4.4.4 Water balances

Water balances (Tables 4.42-4.48) illustrate estimated amounts of water that are input into
process streams and cooling towers, lost through evaporation, drift, or blowdown, required as
makeup, treated, and discharged to the environment. Liquefaction and gasification processes
differ with respect to water use requirements; liquefaction processes require less overall water
(Hittman Associates 1975a).

About 1.5 tons of water are needed to gasify each ton of coal (Williams and Dressel 1973). The
Wesco Coal Gasification (Lurgi) Plant uses 1.4 lb of water per pound of coal, which is about
5100 gpm raw water intake for 2800 tons of coal (Berty and Moe 1974).

Table 4.38. By-product water analysis from Synthane gasification of various coals (in mg/liter[a])

Component	Coke plant	Illinois No. 6	Wyoming subbituminous	Illinois char	North Dakota lignite	Western Kentucky	Pittsburgh seam
pH	9	8.6	8.7	7.9	9.2	8.9	9.3
Suspended solids	50	600	140	24	64	55	23
Phenol	2000	2,600	6,000	200	6,600	3,700	1,700
Chemical oxygen demand	7000	15,000	43,000	1700	38,000	19,000	19,000
Thiocyanate	1000	152	23	21	22	200	188
Cyanide	100	0.6	0.23	0.1	0.1	0.5	0.6
NH_3	5000	8,100[b]	9,520	2500	7,200	10,000	11,000
Chloride		500		31			
Carbonate		6,000[c]					
Bicarbonate		11,000[c]					
Total sulfur		1,400[d]					

[a]Except for pH.
[b]85% free NH_3.
[c]Not from same analysis.
[d]S^{2-} = 400; $SO_3{}^{2-}$ = 300; $SO_4{}^{2-}$ = 1400; S_2O_3 = 1000.

Source: Forney et al. 1974, Table 1, p. 3.

Table 4.39. Composition of wastewaters
(in mg/liter^a)

Pollutant	Coke plant waste ammonia liquor	Synthane process by-product water
pH	8.3 - 9.1	7.9 - 9.3
Chemical oxygen demand	2,500 - 10,000	1,700 - 43,000
Ammonia	1,800 - 4,300	2,500 - 11,000
Cyanide	10 - 37	0.1 - 0.6
Thiocyanate	100 - 1,500	21 - 200
Phenols	410 - 2,400	200 - 6,600
Sulfide	0 - 50	b
Alkalinity (as $CaCO_3$)	1,200 - 2,700	b
Specific conductance, mho/cm	11,000 - 32,000	b

^{a}Except for pH and specific conductance.
^{b}Not determined.

Source: Rubin and McMichael 1975, Table 3, p. 113. Reprinted with
permission from *Environmental Science and Technology*.
Copyright by American Chemical Society. Reprinted by permission
of the publisher.

Table 4.40. Properties of COED process liquors (wt %)

Component	Illinois No. 6	Utah
First-stage pyrolysis liquor		
Carbon		0.17
Nitrogen	0.05	0.01
Sulfur	0.07	0.01
Phenol	0.00	9.2 ppm
Entrained oil		0.01
Suspended solids	0.49	0.03
pH	3.6	8.1
Second-stage pyrolysis liquor		
Carbon		2.0
Nitrogen	0.93	0.88
Sulfur	0.18	0.13
Phenol	0.38	0.40
Entrained oil	0.0 - 0.5	0.2
Suspended solids	1.09	0.64
pH	8.8	9.0
Hydrotreating liquor		
Carbon	0.8	1.7
Nitrogen	5.0	5.2
Sulfur	8.7	4.2
pH	9.3	11.5

Source: Hamshar, Terzian, and Scotti 1974,
Table 9, p. 155.

Table 4.41. Composition of wastewater
from Synthane gasifier

Component	Quantity	
	(mg/liter[a])	(lb/hr)
Suspended solids	23	15
Chemical oxygen demand	19,000	12,710
Carbon dioxide	13,000	8,680
Ammonia	11,000	7,360
Phenol	1,700	1,140
Cyanide	0.6	<1
Thiocyanate	188	130
Sulfur	300 ppm	200
Aromatics	Trace	
Pyridines	Trace	

[a]Except for sulfur.

Source: Glaser, Hershaft, and Shaw 1974,
Table IV-2, p. IV-6.

Table 4.42. SRC process: treated and wastewater balances

Use	Quantity (lb/hr)
Treated water	
Process (net)	121,183
Boiler feed makeup	213,738
Potable	149,909
Cooling tower makeup	1,328,378
Total	1,813,208
	(3,626 gpm)
Wastewater	
Boiler feed blowdown	61,123
Cooling tower blowdown	301,555
Cooling tower evaporation and drift	1,026,823
Sour water to biological oxidation pond	19,606
Water to sanitary sewer	149,909
Total	1,559,016
Actual wastewater discharge	532,193
	(1,064 gpm)

Source: Jahnig 1975, Table 13, p. 68.

Table 4.43. SRC process: cooling water required

Description	Cooling water circulated (gpm)
Coal preparation	2,000
Coal slurrying and pumping	1,760
Coal liquefaction and filtration	32,676
Dissolver acid gas removal	37,500
Coal liquefaction product distillation	410
Fuel oil hydrogenation	2,259
Naphtha hydrogenation	45
Fuel gas sulfur removal	3,100
Gasification	6,209
Acid gas removal	17,220
Shift conversion	422
CO_2 removal	
Methanation	80
Sulfur plant	
Oxygen plant	17,400
Instrument and plant air	90
Raw water treatment	
Process wastewater treatment	
Power generation	
Product storage	
Slag removal system	
Steam generation	
Total	121,171
Raw water makeup	2,666

Source: Jahnig 1975, Table 12, p. 67.

4.4.5 Wastewater treatment

Separate collection facilities for the various wastewater streams facilitate treatment of process area specific contaminants. For example, dissolved sulfur compounds and carbon dioxide in conversion plant wastewaters are released by flashing or stripping and are sent to the sulfur recovery unit. Suspended particulates are removed in settling ponds or by clarifiers, thickeners, filters, or API separators, which can remove floating oil and particulates simultaneously.

Separation of ammonia and/or phenol as by-products may be economical for some conversion processes: The Lurgi process can recover both ammonia and phenol for resale (Shaw and Magee 1974), the Hygas and SRC recover ammonia (Glaser, Hershaft, and Shaw 1974), and the Synthoil process recovers ammonium sulfate (Process Evaluation Group 1975). Ammonia in wastewater is removed by steam stripping; the stripped ammonia can be recovered as anhydrous ammonia or in solution by the Phosam and Chevron processes (described in Sect. 4.2.2.2).

The Phenolsolvan (Lurgi proprietary) process recovers crude soluble phenols from process waters (described in Sect. 4.5.6.1).

Table 4.44. CO_2 acceptor process: water requirements

Description	Water (lb/hr)
Consumed	
By reaction in gasifier	1,053,000
In wet ash rejected	233,000
To ash desulfurizer	15,800
Evaporation in cooling tower	457,000
Drift loss in cooling tower	43,000
Handling loss on condensate	68,000
Total	1,869,800
Available	
From methanator	483,000
From gas compressor	3,700
Total	486,700
Net makeup required	1,383,100 (2766 gpm)[a]

[a]If the moisture in the vent gas were recovered, the makeup would be about half as much (i.e., 1400 gpm).

Source: Jahnig and Magee 1974, Table 12, p. 49.

Table 4.45. Koppers-Totzek process: water balance

Description	Water (lb/hr)
Consumed	
Ash cooling tower	
Evaporation	218,598
Drift loss	16,220
In wet slag	4,865
In wet ash	62,802
Utility cooling tower	
Evaporation	1,050,000
Drift loss	147,528
Consumed in gasifier	11,208
Handling loss on condensate	63,890
Available	
In coal @ 2%	9,585
From O_2 plant	8,926
Net water required	1,556,600 (3113 gpm)

Source: Magee, Jahnig, and Shaw 1974, Table 9, p. 37.

Table 4.46. COED process: plant water requirements

Description	Water (gpm)
Users	
Cooling tower makeup	9,000
Treated boiler feedwater (includes H_2 plant requirement	1,550
Raw process water	-40^a
Potable water	70
Total	10,580
Streams to waste treating	
Cooling tower blowdown	2,400
Boiler blowdown	270
Oily process water	300
Sanitary	20
Total	2,990
Raw water makeup (assumes 100% reuse)	7,590

^{a}Net water makeup.

Source: Kalfadelis and Magee 1975, Table II, p. 53.

Table 4.47. Lurgi process: water balance

Description	Water (gpm)	Water (%)
Consumed		
Reaction	1971	31.7
Evaporation	3543	57.0
Vent	79	1.3
Drift	260	4.2
Ammonia by-product	160	1.7
Wet ash	145	2.3
Fuel and incineration	108	1.7
Total	6212	100
Supplied		
Raw river water reservoir	4908	79.0
Coal	713	11.5
Produced in methanation	591	9.5
Oxygen plant condensate	0 - 19	
Total	6212 - 6231	100

Source: Shaw and Magee 1974, Table 8, p. 25.

Table 4.48. Water requirements for
Bi-Gas base case (gpm)

Cooling water circulation	
Coal preparation	0
Gasification	450
Quench and dust removal	0
Shift conversion	0
Acid gas removal	114,000
Methanation and drying	1,930
Oxygen plant	117,600
Sulfur plant	1,100
Power generation	27,500
Total	262,580
Cooling tower makeup	
Drift loss	526
Evaporation	5,252
Blowdown (net)	1,200
Total	6,978

Source: Jahnig 1975*b*, Table 11, p. 56.

Residual contaminants (fugitive sulfur compounds, ammonia, phenols, cyanides, and thiocyanates, or amounts too low for economical recovery) and oxygen demand may be treated in a retention pond or in biological oxidation units using trickling filters or activated sludge. Biological treatment consists of feeding the contaminants in the wastewater to active bacteria, which are able to use the contaminant as food and, in doing so, produce inoffensive by-products. Trickling filters have bacteria living on the surface of stones or other media in the bed. The waste is distributed over the media, and the bacteria absorb the contaminants. Activated sludges use suspended bacteria in the tank. Biological floc suspensions are then separated by settling, a portion of the thickened sludge is recirculated to the aeration basin for contact with additional waste, and excessive sludge accumulation is disposed of (Sittig 1974). In biological treatment, however, the pollutants are incorporated into the cellular material or sludge, which then requires disposal. Since such sludge usually contains 80 to 95% water, disposal is difficult (Shaw 1976, personal communication).

Biological oxidation processes give a high degree of contaminant reduction. Optimum operating conditions include adequate oxygen supply, correct temperature and pH, adequate nutrients, suitable physical conditions, and absence of inhibitory substances (Nebolsine 1957; Adams 1973). Thus, biological treatment necessitates several coordinated and ordered steps. For example, hydrogen cyanide, which follows ammonia in wastewater, is toxic to the biological oxidation unit bacteria; therefore, its removal must be complete before ammonia biological oxidation treatment (Shaw and Magee 1974). Phenol is toxic to nitrifying bacteria; and ammonia, in high concentrations, severely inhibits the phenol-degrading bacteria (Fulkerson et al. 1974*b*).

Activated carbon may also be used for treating conversion wastewater contaminants and oxygen demand (Kalfadelis and Magee 1975).

Electrolytic decomposition of cyanides has been shown to be effective for coking industries under certain conditions; however, the effectiveness of electrolytic decomposition of phenol and thiocyanate is disputed (Kuhn 1970).

However, coal conversion operations may also produce aromatic compounds in the wastewater that resist biodegradation (Jahnig 1975). Oily wastes can be separated in gravity settling facilities such as API separators, skim ponds, or parallel plate separators. Secondary treatment facilities for oily and chemical wastes include dissolved air flotation units, granular-media filtration, or chemical flocculation units (Kalfadelis and Magee 1975*b*).

Fulkerson et al. (1974*b*) envision a coal conversion wastewater processing scheme that would include (1) physical ammonia removal, preferably through ion exchange, leaving a sufficient concentration of ammonia to serve the nutritional requirements for subsequent biological operations; (2) phenol removal via biological oxidation in a biochemical reactor; (3) nitrification (if required) of the remaining ammonia to nitrate in a second biochemical reactor; (4) denitrification (if required) of the nitrate to nitrogen gas in the third and final reactor; and (5) polycyclic compound removal by carbon adsorption.

4.4.6 <u>Water reuse</u>

Water reuse or recycle designs use one conversion area's waste stream as feed for another area. El Paso's Burnham I Gasification (Lurgi) Complex reuse design (Fig. 4.58) includes the (1) use of treated process condensate (containing only trace amounts of hydrogen sulfide, ammonia, or phenols after their treatment and removal) and boiler blowdown as cooling tower makeup, (2) use of methanation water-of-reaction as boiler feedwater, and (3) reuse of plant blowdown and treated condensates in the gasifier ash removal system (Milios 1975). Of the total water input (including river water and cooling water), 67% will be returned to the atmosphere as pure water vapor by evaporation, 24% will be consumed by chemical reactions to produce the product gas, 2.5% will be contained in ammonia by-product, and 6.5% will be contained in the plant flue gas, ash, and cooling tower drift (Milios 1975).

Other conversion process recycle schemes include (1) recycling the (filtered) slag quenching water to the scrubbing system, (2) recycling treated (H_2S and ammonia removed) sour water stream to the last pyrolysis stage for destruction (COED process) (Kalfadelis and Magee 1975), (3) use of treated (H_2S and ammonia removed) sour water stream to mix with the ash or slag for slurry removal, (4) recycling sour water to the gasifier (Bi-Gas process) (Jahnig 1975*b*), steam production area, or the coal slurry section ahead of the reactor (SRC process) (Jahnig 1975). Water from the biological oxidation units may be recycled as makeup water for the cooling towers. However, complications from the buildup, transfer, or destruction of process contaminants through fouling or corrosion in the systems are not completely known.

Most conversion designs propose closed-loop water systems; however, some aqueous discharges due to ash-wetting and evaporation will be inevitable. Water treatment facilities will have to be designed for each coal conversion process to best treat the various wastes produced by each. The overall water disposition scheme for the proposed Wesco Coal Gasification (Lurgi) Complex is given in Fig. 4.59.

4.5 ORGANICS

Coal conversion technologies produce gas, liquid, and solid waste streams, all of which may contain coal-derived organic compounds derived from the organic constituents of the feed coal. Product streams, as well as process streams, are important because their components can be

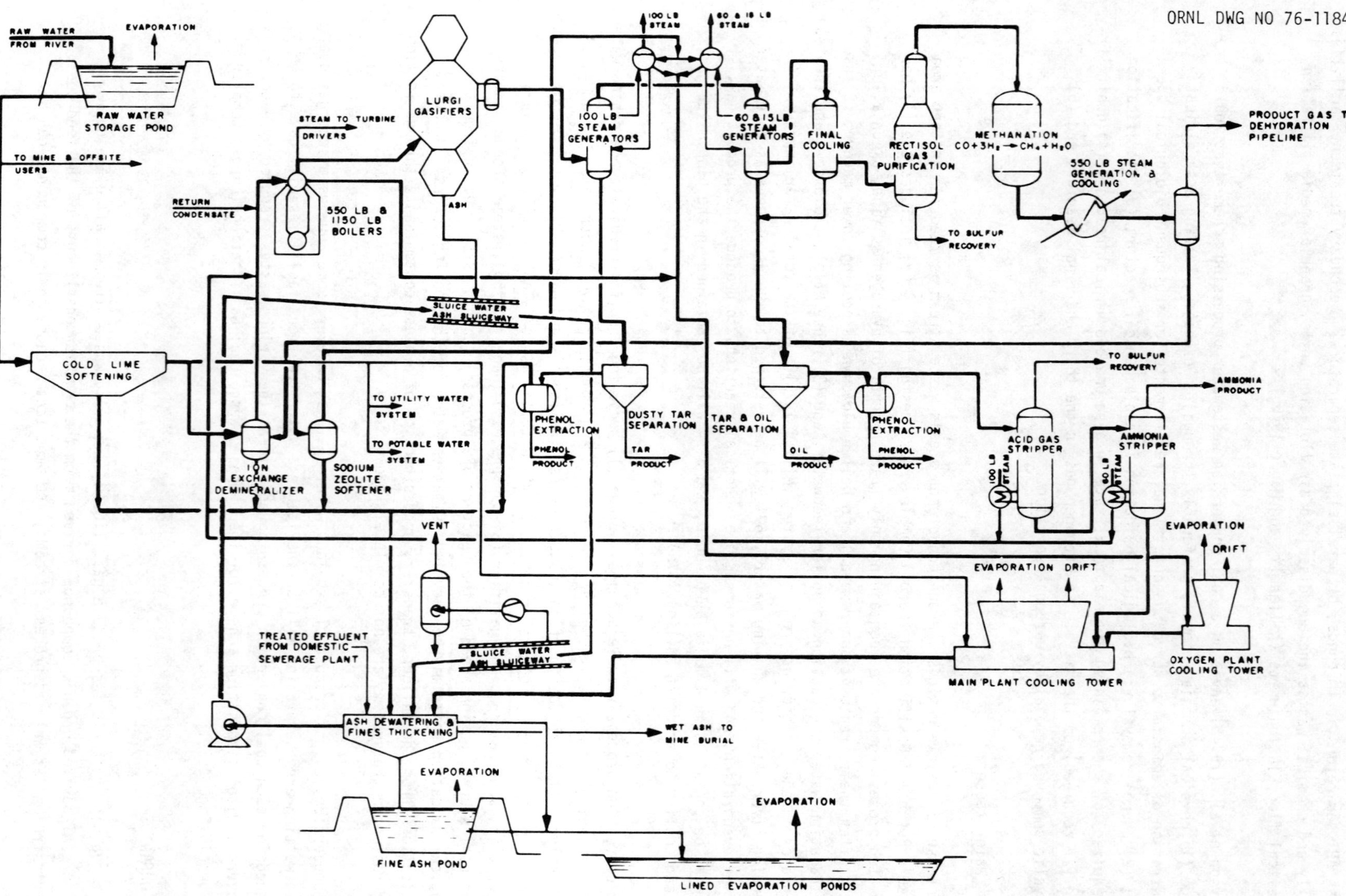

Fig. 4.58. El Paso Burnham I Coal Gasification (Lurgi) Complex water reuse system. From Milios 1975, Fig. 2, p. 100. Reprinted by permission of the publisher.

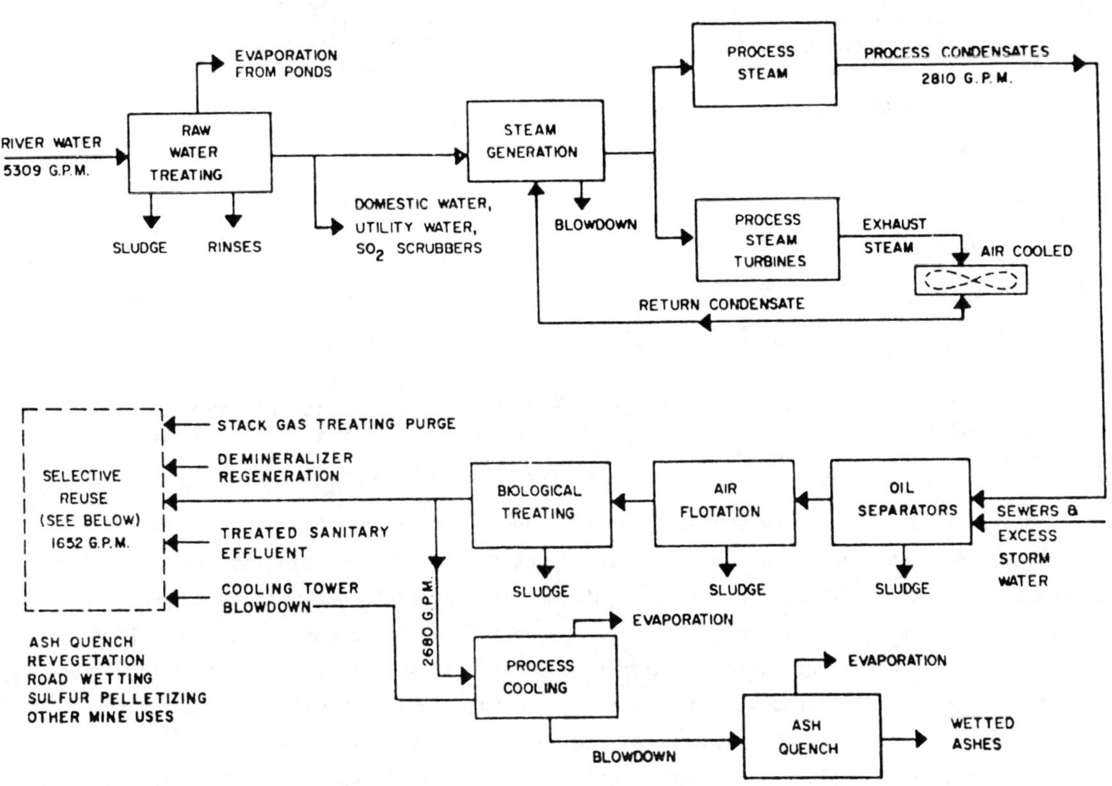

Fig. 4.59. Water disposition scheme for the proposed Wesco Coal Gasification (Lurgi) Complex. <u>Source</u>: From Bennett, Kerr, and Kolstad 1976, Fig. 2, p. 19. Reprinted by permission of the publisher.

potential environmental pollutants. Components of the products (gases, tars, and oils) may appear in the waste streams from incomplete product or by-product recovery. Hydrocarbons may be emitted to the atmosphere by incomplete combustion, leaks in hydrocarbon by-product transfer, evaporation from aqueous effluent of cooling streams during slurry mixing, or emission of flue gases from coal, char, and oil combustion.

Vapor leaks from pump seals are of primary importance in the liquefaction slurry preparation and transfer area. The concentration of organic vapors in the air depends on the size of the leak, slurry temperature, and air circulation (Akhtar, Freidman, and Yavorsky 1975). For the petroleum refining industry, the cost of controlling hydrocarbon losses may be offset by the saving of valuable products that otherwise escape into the plant atmosphere (Laster 1973).

Product oil filtration is a procedure having potential for leaks, spills, and maintenance complications; it involves filter precoating, filter cake scraping, and reslurrying with oil for transport to the gasifier. The level of hydrocarbons emitted from incomplete combustion is not expected to exceed 0.007 lb/million Btu, measured as methane to the atmosphere (Hangebrauck, von Lehmden, and Meeker 1964; Shaw and Magee 1974). Estimates of evaporative hydrocarbon emissions from leakage and storage of by-products for the Lurgi design (using API suggested methods) are given in Table 4.49. Startups, shutdowns, maintenance, and upsets can produce environmentally unacceptable emissions (Jahnig 1975a).

Table 4.49. Estimated evaporative hydrocarbon emmissions
from leakage and storage of by-products

Hydrocarbon	Emission rate (lb/hr)
Crude phenol	1.3
Tar oil naphtha	2.3
Tar	3.5
Naphtha	1.8
Methanol	1.4
Product gases	2.8

Source: Shaw et al. 1974, p. 33.   Reprinted by permission of the
publisher.

The exact nature of the organic emissions from coal conversion has not been characterized;
however, compounds identified in various coal fractions from other technologies may be similar
to conversion streams. In fact, streams from intermediate-temperature gasification processes
such as the Synthane process may contain all the products found in carbonization and coking
streams. However, processes operating at higher temperatures, such as the Koppers-Totzek process,
do not produce tar, oil, or phenols (Magee, Jahnig, and Shaw 1974). Similarities among the
Synthoil synthetic crude oil, other syncrudes, and petroleum crude oils indicate that much of
the present petroleum technology (with some modifications) is applicable to coal-derived syncrude
processing (Woodward et al. 1976).

Amounts and characteristics of the typical organics [product gas, naphtha, coal tars, light oils,
BTX (benzene, toluene, xylene), heavy oils, solvent-refined coal, gases] in the products, by-
products, and emissions depend on the conversion operating conditions and the constituents of
the coal feed. However, Peterson (1975) found that each of the SRC process samples (process
solvent, mineral residue, effluent water, naphtha, and product fuel) contained the same major
organic components, but differed with respect to their relative concentrations.

Coal-associated organics are potential environmental pollutants, but are also valuable products.
For example, desulfurized tars, liquid products, and SRC are valuable fuels, recovery of which is
essential to conversion efficiencies. The BTX stream is valuable as a motor fuel component
because of its high octane rating (Glaser, Hershaft, and Shaw 1974).

Miscellaneous hydrocarbons, including PAH, are environmentally significant because of the
possible health hazards they present.

4.5.1 Products

Tars and oils are differentiated primarily by their boiling points: Tars have a higher boiling
point. One of the primary contaminants in tars is sulfur, in the form of organic sulfides,
mercaptans, and thiophenes. Oils may also contain similar but lower-molecular-weight sulfur
compounds, in addition to dissolved hydrogen sulfide and carbonyl sulfide (Glaser, Hershaft, and
Shaw 1974), the amounts depending on the quenching and condensing sequences. Tars and oils may

be amenable to hydrodesulfurization techniques (Glaser, Hershaft, and Shaw 1974); however, should they contain particulates, residual hydrodesulfurization techniques may be more applicable.

4.5.1.1 Tar

The Synthoil (liquefaction) process yields 60 to 80 lb of tar per ton of coal (Yavorsky and Akhtar 1974) having a sulfur content of 1.1 to 2.8% (Forney et al. 1974). The sulfur content of tars produced from the Synthane gasification process ranged from 0.5 to 2.7 wt % (Forney et al. 1974). Forney et al. (1974) also found that the Synthane tars from Illinois No. 6 coal contained 0.003 ppm mercury and 0.7 ppm arsenic, which, if the tars were burned, would probably end up in the stack gas. Tars from the Lurgi process contain 0.27% sulfur (Shaw and Magee 1974).

Major components of German high-temperature conversion-process coal tar are listed in Table 4.50; naphthalene, phenanthrene, and fluoranthene show the highest concentrations. Analysis from coal gasification by combined gas chromatography-mass spectrometry also indicates a high relative concentration of naphthalene and phenanthrene, as well as fluorene and phenols (Sharkey et al. 1975). Concentrations of the major structural compound types in tar from the gasification of Illinois No. 6 coal show high relative concentrations of acenaphthenes, 3-ring aromatics, and naphthalenes (Table 4.51).

Representative coal carbonization compounds selected by the Coal Tar Research Association are listed in Table 4.52 and are grouped, according to chemical constitution, as paraffins, olefins, monocyclic aromatic hydrocarbons, polycyclic aromatic hydrocarbons (PAH), oxygen-containing compounds, nitrogen-containing compounds, and sulfur-containing compounds. The Bureau of Mines, in 1963, published a list of 800 organic compounds and 32 inorganic compounds derived from coal carbonization, which could be included in the above categories (Anderson and Wu 1963). Freudenthal, Lutz, and Mitchell (1975) prepared a partial list of the coal tar constituents grouped according to boiling point (Table 4.53). However, it has been estimated that only 1000 of the estimated 10,000 components have been identified in coal tar and hydrogenation products (Peterson 1975).

4.5.1.2 Liquid products

H-Coal process oil compositions of various distillation fractions from Illinois No. 6 coal and Wyoming subbituminous coals include paraffins, olefins, monocyclic and polycyclic aromatic compounds, and phenolic compounds (Tables 4.54 and 4.55) (Hydrocarbon Research Inc. 1968). Overall, aromatic compounds constitute the largest percentage.

The major gas chromatogram peaks of process naphtha (liquid volatile organics) (Fig. 4.60) from a 50-ton/day SRC pilot plant using high-sulfur (4%) eastern coal were PAH (naphthalenes and phenanthrenes) and simple phenolic compounds (Petersen 1975). Basic process naphtha fractions consisted primarily of quinoline-related compounds (Petersen 1975).

Of the various Synthoil acidic components, phenol and cresol concentrations were about 1.6 wt % of the sample. The nitrogen compounds indole and skatole were measured at 210 and 128 ppm, respectively, in the Synthoil; the amounts of pyridines and chinolines are expected to be much higher (Shults 1976).

Table 4.50. Major components of German high-temperature conversion process coal tar

Component	Boiling point (°C)	Melting point (°C)	Average weight percent
Naphthalene	217.9	80.3	10.0
Phenanthrene	338.4	100.0	5.0
Fluoranthene	383.5	111.0	3.3
Pyrene	393.5	150.0	2.1
Acenaphthylene	270.0	93.0	2.0
Fluorene	297.9	115.0	2.0
Chrysene	441.0	256.0	2.0
Anthracene	340.0	218.0	1.8
Carbazole	354.8	244.4	1.5
2-Methylnaphthalene	241.1	34.6	1.5
Diphenylenoxide	285.1	85.0	1.0
Indene	182.4	-1.5	1.0
Acridine	343.9	111.0	0.6
1-Methylnaphthalene	244.7	-30.5	0.5
Phenol	181.8	40.9	0.4
m-Cresol	202.2	12.2	0.4
Benzene	80.1	5.5	0.4
Diphenyl	255.0	69.2	0.4
Acenaphthene	227.5	95.0	0.3
2-Phenylnaphthalene	359.8	101.0	0.3
Toluene	110.6	-95.0	0.3
Chinoline	237.1	-14.2	0.3
Diphenylensulfide	331.4	97	0.3
Thionaphthene	219.9	31.3	0.3
m-Xylene	139.1	-47.9	0.2
o-Cresol	191.1	31.0	0.2
p-Cresol	201.9	34.7	0.2
Isochinoline	243.3	26.5	0.2
Chinaldine	247.6	-1.0	0.2
Phenanthridine	349.5	107.0	0.2
7,8-Benzochinoline	340.2	52.0	0.2
2,3-Benzodiphenylenoxide	394.5	208.0	0.2
Indole	254.7	52.5	0.2
3,5-Dimethylphenol	221.7	63.3	0.1
2,4-Dimethylphenol	210.9	24.5	0.1
Pyridine	115.3	-41.8	0.02
α-Picoline	129.4	-66.7	0.02
β-Picoline	144.1	-18.3	0.01
γ-Picoline	145.4	3.7	0.01
2,6-Intidine	144.0	-6.1	0.01
2,4-Intidine	158.4	-64.0	0.01

Source: Shults 1976, Table 2.1, pp. 4, 5. Reprinted by permission of the publisher.

Table 4.51. Mass spectrometric analyses of water contaminants and
benzene-soluble tar from Synthane (laboratory-scale)
gasification (vol %)

Structural type[a]	Pittsburg[b]	Illinois No. 6[b]
Benzenes	1.9	2.1
Indenes	6.1[c]	8.6[c]
Indanes	2.1	1.9
Naphthalenes	16.5	11.6
Fluorenes	10.7	9.6
Acenaphthenes	15.8	13.5
3-Ring aromatics	14.8	13.8
Phenylnaphthalenes	7.6	9.8
4-Ring, peri-condensed	7.6	7.2
4-Ring, cata-condensed	4.1	4.0
Phenols	3.0	2.8
Naphthols	[c]	[c]
Indanols	0.7	0.9
Acenaphthenols	2.0	
Phenanthrols		2.7
Dibenzofurans	4.7	6.3
Dibenzothiophenes	2.4	3.5
Benzonaphthothiophenes		1.7
n-Heterocyclics	(8.8)	(10.8)
Average molecular weight	202	212

[a] Includes alkyl derivatives.
[b] Spectra indicate traces of 5-ring aromatics.
[c] Includes any naphthol present (not resolved in these spectra).
[d] Data on n-free basis since isotope corrections were estimated.

Source: Forney et al. 1974, Table 4.

The <207°C Synthoil syncrude distillate showed that the base fraction (about 0.7%) was mostly analines with some pyridines, the acidic fraction (35.3%) was almost entirely phenols, and the hydrocarbon (neutral) portion (61%) was made up of paraffins, cycloparaffins, aliphatic olefins, monoaromatics, and diaromatics (Woodward et al. 1976). Saturate concentrates from the higher-temperature distillates (Table 4.56) showed approximately equal concentrations (16% total) of the saturate compounds (paraffins) in the 207 to 363°C distillate and lower paraffin concentrations (9% total) in the 363 to 531°C distillate (Woodward et al. 1976). Monoaromatic, diaromatic, and polyaromatic percentages for the 207 to 303°C and 363 to 531°C fractions were 27 and 5%, 22% (both), and 30 and 59% respectively (Woodward et al. 1976). Distillate compositions of the original Synthoil were about 9.5% saturates, 13% monoaromatic hydrocarbons, 15.4% diaromatics, 14.6% polyaromatic hydrocarbons, and 17.4% other compound types (heteroatomic types, weak acids, strong acids, bases, unknowns, and unanalyzed material) (Woodward et al. 1976).

The major structural compound types detected in the heavy oil and asphaltene fractions from Synthoil liquefaction of Kentucky coal (Table 4.57) include alkylbenzenes, indenes, indans, acenaphthylenes, biphenols, anthracenes, phenanthrenes, phenylnaphthalenes, 4-, 5-, and 6-membered rings, and phenols. Hydrocarbons to C_{28} have been detected for the components with the following combinations of atoms: C_xH_y, C_xH_yO, C_xH_yN, C_xH_yON, $C_xH_yO_2$, and $C_xH_yO_3$ (Sharkey et al. 1975).

Table 4.52. Representative coal carbonization compounds

Constituent	Formula
Paraffins	
Pentane	C_5H_{12}
Hexane	C_6H_{14}
Olefins	
1-Pentene	C_5H_{10}
1-Hexene	C_6H_{12}
1,3-Cyclopentadiene	C_5H_6
Dicyclopentadiene	$C_{10}H_{12}$
Monocyclic aromatic hydrocarbons	
Benzene	C_6H_6
Methyl benzene (toluene)	$C_6H_5(CH_3)$
Vinyl benzene (styrene)	$C_6H_5CH{=}CH_2$
Ethyl benzene	$C_6H_5(C_2H_5)$
1,2-Dimethyl benzene (*o*-xylene)	$C_6H_4(CH_3)_2$
1,3-Dimethyl benzene (*m*-xylene)	$C_6H_4(CH_3)_2$
1,4-Dimethyl benzene (*p*-xylene)	$C_6H_4(CH_3)_2$
Isopropyl benzene (cumene)	$C_6H_5CH(CH_3)_2$
1,2,3-Trimethyl benzene (hemimellitene)	$C_6H_3(CH_3)_3$
1,2,4-Trimethyl benzene (pseudocumene)	$C_6H_3(CH_3)_3$
1,3,5-Trimethyl benzene (mesitylene)	$C_6H_3(CH_3)_3$
Polycyclic aromatic hydrocarbons	
Indene	C_9H_8
Naphthalene	$C_{10}H_8$
Tetrahydronaphthalene (tetralin)	$C_{10}H_{12}$
Decahydronaphthalene-*cis* (decalin-*cis*)	$C_{10}H_{18}$
Decahydronaphthalene-*trans* (decalin-*trans*)	$C_{10}H_{18}$
1-Methyl naphthalene	$C_{10}H_7(CH_3)$
2-Methyl naphthalene	$C_{10}H_7(CH_3)$
Acenaphthene	$C_{12}H_{10}$
Diphenyl	$C_6H_5 \cdot C_6H_5$
1,6-Dimethyl naphthalene	$C_{10}H_6(CH_3)_2$
1,7-Dimethyl naphthalene	$C_{10}H_6(CH_3)_2$
2,3-Dimethyl naphthalene	$C_{10}H_6(CH_3)_2$
2,6-Dimethyl naphthalene	$C_{10}H_6(CH_3)_2$
2,7-Dimethyl naphthalene	$C_{10}H_6(CH_3)_2$
Fluorene	$C_{13}H_{10}$

Table 4.52 (continued)

Constituent	Formula
Anthracene	$C_{14}H_{10}$
Phenanthrene	$C_{14}H_{10}$
Fluoranthene	$C_{16}H_{10}$
Pyrene	$C_{16}H_{10}$
Chrysene	$C_{18}H_{12}$

Oxygen-containing compounds

Constituent	Formula
Acetone	$(CH_3)_2CO$
Phenol	C_6H_5OH
2-Methyl phenol (*o*-cresol)	$C_6H_4(CH_3)(OH)$
3-Methyl phenol (*m*-cresol)	$C_6H_4(CH_3)(OH)$
4-Methyl phenol (*p*-cresol)	$C_6H_4(CH_3)(OH)$
1,2-Dihydroxybenzene (catechol)	$C_6H_4(OH)_2$
1,3-Dihydroxybenzene (resorcinol)	$C_6H_4(OH)_2$
1,4-Dihydroxybenzene (hydroquinone)	$C_6H_4(OH)_2$
Benzofuran (coumarone)	C_8H_6O
2,3-Dimethyl phenol (2,3-xylenol)	$C_6H_3(CH_3)_2(OH)$
2,4-Dimethyl phenol (2,4-xylenol)	$C_6H_3(CH_3)_2(OH)$
2,5-Dimethyl phenol (2,5-xylenol)	$C_6H_3(CH_3)_2(OH)$
2,6-Dimethyl phenol (2,6-xylenol)	$C_6H_3(CH_3)_2(OH)$
3,4-Dimethyl phenol (3,4-xylenol)	$C_6H_3(CH_3)_2(OH)$
3,5-Dimethyl phenol (3,5-xylenol)	$C_6H_3(CH_3)_2(OH)$
2-Ethyl phenol	$C_6H_4(C_2H_5)(OH)$
3-Ethyl phenol	$C_6H_4(C_2H_5)(OH)$
4-Ethyl phenol	$C_6H_4(C_2H_5)(OH)$
3-Methyl-5-ethyl phenol	$C_6H_3(CH_3)(C_2H_5)(OH)$
1-Naphthol	$C_{10}H_7OH$
2-Naphthol	$C_{10}H_7OH$
Dibenzofuran (diphenylene oxide)	$C_{12}H_8O$

Nitrogen-containing compounds

Constituent	Formula
Acetonitrile	CH_3CN
Pyrrole	C_4H_5N
Pyridine	C_5H_5N
Aniline	$C_6H_5NH_2$
2-Methyl pyridine (2-picoline)	$(CH_3)C_5H_4N$
3-Methyl pyridine (3-picoline)	$(CH_3)C_5H_4N$
4-Methyl pyridine (4-picoline)	$(CH_3)C_5H_4N$
2-Methyl aniline (*o*-toluidine)	$(CH_3)C_6H_4NH_2$
3-Methyl aniline (*m*-toluidine)	$(CH_3)C_6H_4NH_2$

Table 4.52 (continued)

Constituent	Formula
4-Methyl aniline (*p*-toluidine)	$(CH_3)C_6H_4NH_2$
2,4-Dimethyl pyridine (2,4-lutidine)	$(CH_3)_2C_5H_3N$
2,6-Dimethyl pyridine (2,6-lutidine)	$(CH_3)_2C_5H_3N$
Indole	C_8H_7N
Quinoline	C_9H_7N
Isoquinoline	C_9H_7N
2-Methyl indole	$(CH_3)C_8H_6N$
3-Methyl indole (skatole)	$(CH_3)C_8H_6N$
7-Methyl indole	$(CH_3)C_8H_6N$
2-Methyl quinoline (quinaldine)	$(CH_3)C_9H_6N$
4-Methyl quinoline (lepidine)	$(CH_3)C_9H_6N$
Carbazole	$C_{12}H_9N$
Acridine	$C_{13}H_9N$

Sulfur-containing compounds

Constituent	Formula
Carbon disulfide	CS_2
Thiophene	C_4H_4S
Thionaphthene	C_8H_6S

Source: Coal Tar Research Association 1958, Table A-1.

Table 4.53. Coal tar constituents by boiling point

≤100°C	*200 - 250°C*	*Above 300°C*

≤100°C

Pentane
Hexane
Heptane
Cyclohexane
Butylene
Amylene
Hexene
Heptene
Cyclohexene
1,5-Butadiene
Crotonylene
Cyclopentadiene
Cyclohexadiene
Benzene
Acetone
Methylethyl ketone
Acetonitrile
Carbon disulfide
Methylmercaptan
Ethylmercaptan
Dimethyl sulfide
Diethyl sulfide
Thiophene

100 - 150°C

Octane
Methylcyclohexane
Dimethylcyclohexane
Nonaphthene
Toluene
o-Xylene
m-Xylene
p-Xylene
Ethylbenzene
Styrene
Pyrrole
Pyridine
2-Methylpyridine
3-Methylpyridine
4-Methylpyridine
2,6-Dimethylpyridine
Thiotoluene
Thioxene

150 - 200°C

Decane
Dicyclopentadiene
Hydrindene
Isopropylbenzene
o-Ethyltoluene
m-Ethyltoluene
p-Ethyltoluene
n-Propylbenzene
Mesitylene
Pseudocumene
Hemellitene
Cymene
Durene
Isodurene
Indene
2,7-Dimethylindene
3,6-Dimethylindene
Phenol
m-Cresol
Coumarone
Dimethylcoumarone

Aniline
Toluidine
2,3-Dimethylpyridine
2,4-Dimethylpyridine
2,5-Dimethylpyridine
3,4-Dimethylpyridine
Dimethylaniline
2,4,5-Trimethylpyridine
2,4,6-Trimethylpyridine
Benzonitrile

200 - 250°C

Naphthalene
Dihydronaphthalene
α-Methylnaphthalene
β-Methylnaphthalene
4,6-Dimethylindene
5,7-Dimethylindene
m-Cresol
p-Cresol
1,2,3-Xylenol
1,2,4-Xylenol
1,3,4-Xylenol
1,3,5-Xylenol
1,4,2-Xylenol
o-Ethylphenol
m-Ethylphenol
p-Ethylphenol
3-Methyl-5-ethylphenol
Isopseudocumenol
Durenol
2,2'-Dihydroxydiphenyl
Acetophenone
Benzoic acid
Dimethyl coumarone
1,2,3,4-Tetramethylpyridine
Quinoline
Isoquinoline
2-Methylquinoline
Thionaphthene

250 - 300°C

Diphenyl
3-Methyldiphenyl
2-Methyldiphenyl
4-Methyldiphenyl
4,4'-dimethyldiphenyl
3,4-Dimethyldiphenyl
Acenaphthene
1,2-Dimethylnaphthalene
1,6-Dimethylnaphthalene
1,7-Dimethylnaphthalene
2,6-Dimethylnaphthalene
2,7-Dimethylnaphthalene
2,3-Dimethylnaphthalene
2-Ethylnaphthalene
1-Ethylnaphthalene
1,2-Cyclopentanonaphthalene
Fluorene
α-Naphthol
β-Naphthol
Diphenylene oxide
α-Naphthofurane
β-Naphthofurane
1-Methyldiphenylene oxide
Indole
2-Methylindole
3-Methylindole

4-Methylindole
5-Methylindole
7-Methylindole
3-Methylquinoline
4-Methylquinoline
5-Methylquinoline
6-Methylquinoline
7-Methylquinoline
1-Methylisoquinoline
3-Methylisoquinoline
1,3-Dimethylisoquinoline
2,8-Dimethylquinoline
5,8-Dimethylquinoline
1-Naphthonitrile

Above 300°C

Nonadecane
Phenanthrene
Anthracene
2-Methylfluorene
3-Methylfluorene
Dihydroanthracene
4,5-Phenanthrylene methane
Chrysogen
Methylanthracene
Fluoranthene
Pyrene
Tetrahydrofluoranthrene
2,7-Dimethylanthracene
1,2-Benzfluorene
2,3-Benzfluorene
Chrysene
1,2-Benzanthracene
Naphthacene
Triphenylene
Perylene
1,2-Benzpyrene
Crackene
4,5-Benzpyrene
Picene
1,2-Benznaphthacene
Naphtho-2,3-1,2-anthracene
Acridine
Diphenylene sulfide
Benzerythrene
1-Methylphenanthrene
3-Methylphenanthrene
9-Methylphenanthrene
Dibenzocoumarone
Carbazole
Benzacarbazole
Truxene
2-Methyldiphenylene oxide
2-Hydroxyfluorene
1,9-Benzoanthrene
2-Naphthylamine
Hydroacridine
2-Methylcarbazole
3-Methylcarbazole
2-Phenylnaphthalene
p-Phenylphenol
2-Hydroxydiphenylene oxide
Phenanthridene
2-Phenanthrol
1-Naphthylamine
2-Naphthonitrile

Source: Adapted from Freudenthal, Lutz, and Mitchell 1975, Fig. 1, pp. 6-10.

Table 4.54. Inspection of products from Illinois No. 6 coal

Compound	Component	Weight percent	Total weight percent
	C_4 - 400°F fraction		
Saturated compounds	nC_4	0.10	
	iC_5	0.20	
	nC_5	0.69	
	C_6 - C_{12}	11	
			11.99
Alkyl benzenes	C_6 - C_{12}	17.55	
			17.55
Saturated naphthenes	Monocycloparaffins	42.64	
	Dicycloparaffins	8.50	
	Tricycloparaffins	0.19	
			51.33
Unsaturated naphthenes	Monocycloparaffins	5.32	
	Dicycloparaffins	4.98	
	Tricycloparaffins	0.90	
			11.20
Other compounds	Indans	6.44	
	Naphthalenes	0.59	
	Phenols (mol. wt) - 108, 122, 136, 150	1.0	
			7.93
			100.00
	400 - 650°F fraction		
Saturated compounds	n-paraffins	4.8	
	i-paraffins	1.7	
	Monocycloparaffins	14.0	
	Dicycloparaffins	7.9	
	Tricycloparaffins	2.6	
			31.0
Unsaturated nonaromatic	Monocycloparaffins	4.3	
			4.3
Aromatic compounds	Alkyl benzenes	12.6	
	Indans & tetralins	30.8	
	Indenes	5.7	
	Naphthalenes	0.2	
	Naphthalenes	3.5	
	Acenaphthenes (C_nH_{2n-14})	4.0	
	Acenaphthenes (C_nH_{2n-16})	2.2	
	Tricyclics (C_nH_{2n-18})	0.4	
			59.6
Other compounds	Phenols (mol. wt) - 108, 122, 136, 150, 164, 178	2.0	
	Other nonhydrocarbons	3.10	
			5.10
			100.00

Table 4.54. (continued)

Compound	Component	Weight percent	Total weight percent
	650 - 919°F fraction		
Saturated compounds	Paraffins	1.4	
	Monocycloparaffins	3.1	
	Bicycloparaffins	0.6	
	Tricycloparaffins	0.7	
	Tetracycloparaffins	0.4	
	Pentacycloparaffins	0.2	
	Hexacycloparaffins	0.1	
	Phenyls	0.3	
			6.8
Unsaturated nonaromatic	Paraffins	0.0	
	Monocycloparaffins	0.5	
	Bicycloparaffins	0.3	
	Tricycloparaffins	0.2	
	Tetracycloparaffins	0.2	
	Pentacycloparaffins	0.1	
	Hexacycloparaffins	0.1	
	Phenyls	0.2	
			1.6
Other compounds	Alkyl benzenes	3.0	
	Indans and/or tetralins	0.5	
	Other aromatics[a]	72.8	
	Phenolic compounds	1.5	
	Other nonhydrocarbons	13.8	
			91.6
			100.00

[a]An approximate breakdown of aromatic-type compounds is:

Component type	Millimoles per 100 g
Naphthalenes	93.4
Phenanthrenes	91.1
Chrysenes	21.9
1-2 Benzanthracenes 3-4 Benzphenanthrenes	14.6
Pyrenes	15.4
5-Ring compounds	5.1

Source: Hydrocarbon Research Inc. 1968, Table 27, pp. 88-90.

Of the COED process samples studied by Shults (1976) (unfiltered raw oil, filter cake, filtered raw oil, syncrude, product separator liquor, drier-stage-1 liquor, char, drier-stage-1 fines, and product separator fines), the highest phenol and cresol concentrations occurred in the unfiltered and filtered raw oil. Organosulfur compounds have been detected in COED aqueous and syncrude samples.

Medium COED oils have been fractionated into saturate (55%), mono-, di-, and triaromatic (30%), and polyaromatic (11%) fractions (Nichols 1975). Light-oil COED sample components include (in order of decreasing concentrations) methylcyclohexane, cyclohexane, ethylcyclohexane, and toluene. Other components in lower concentrations include n-hexane, methylcyclopentane, 1,3-dimethylcyclohexane, o-, m-, and p-xylene, ethyl benzene, ethyl toluene, and propylcyclohexane.

Table 4.55. Inspection of products from Wyoming subbituminous coal

Compound	Component	Weight percent	Total weight percent
	C₄ - 400°F fraction		
Saturated paraffins	C_4	0.20	
	iC_5	0.20	
	nC_5	0.88	
	$C_6 - C_{12}$	13.42	
			14.70
Olefins, diolefins, etc.	C_6	0.02	
	$C_8 - C_{11}$	4.22	
			4.24
Saturated naphthenes	Monocycloparaffins	36.82	
	Dicycloparaffins	6.09	
			42.91
Unsaturated naphthenes	Monocycloparaffins	7.71	
	Dicycloparaffins	7.28	
	Tricycloparaffins	1.67	
			16.66
Alkyl benzenes	$C_6 - C_{11}$	14.16	
			14.16
Other compounds	Indans	5.11	
	Naphthalenes	0.79	
	Phenols (mol. wt) - 108, 122, 136, 150	1.71	
			7.61
			100.3
	400 - 650°F fraction		
Saturated compounds	n-Paraffins	8.8	
	i-Paraffins	2.1	
	Monocycloparaffins	7.5	
	Dicycloparaffins	2.7	
	Tricycloparaffins	1.1	
			22.2
Unsaturated nonaromatic	Paraffins	1.4	
	Monocycloparaffins	1.9	
	Dicycloparaffins	0.8	
	Tricycloparaffins	0.4	
			4.5
Aromatic compounds	Alkyl benzenes	6.7	
	Indans and tetralins	23.2	
	Indenes	8.7	
	Naphthalenes	12.3	
	Acenaphthenes (C_nH_{2n-14})	12.7	
	Acenaphthenes (C_nH_{2n-16})	1.7	
	Tricyclics (C_nH_{2n-18})	0.5	
			65.8
Other compounds	Phenols (mol. wt) - 108, 122, 136, 150	2.8	
	Other nonhydrocarbons	4.7	
			7.5
			100.0

Table 4.55. (continued)

| | | Weight | Total weight |
Compound	Component	percent	percent
	650 - 900°F fraction		
Saturated compounds	Paraffins	7.2	
	Monocycloparaffins	1.4	
	Dicycloparaffins	0.4	
	Tricycloparaffins	0.6	
	Tetracycloparaffins	0.5	
	Pentacycloparaffins	0.3	
	Hexacycloparaffins	0.3	
			11.1
Unsaturated nonaromatic	Paraffins	0.6	
	Monocycloparaffins	0.7	
	Dicycloparaffins	0.1	
	Hexacycloparaffins	0.1	
			1.5
Other compounds	Alkyl benzenes	0.2	
	Other aromatics[a]	74.0	
	Phenolic compounds	0.6	
	Other nonhydrocarbons	12.6	
			87.4
			100.0

[a]An approximate breakdown of aromatic-type compounds is:

Component type	Millimoles per 100 g
Naphthalenes	91.7
Phenanthrenes	116.7
Chrysenes	26.1
1-2 Benzanthracenes 3-4 Benzphenanthrenes	15.2
Pyrenes	19.2
5-Ring compounds	5.1

Source: Hydrocarbon Research Inc. 1968, Table 28, pp. 91-93. Reprinted by permission of the Battelle Energy Program.

Table 4.56. Composite semiquantitative analysis of the 207° to 363°C and the 363° to 531°C distillates of Synthoil from West Virginia, Pittsburgh seam coal

Mass Z series	Distillate (wt %)		Crude (wt %)		Probable compound types
	207° to 363°C	363° to 531°C	207° to 363°C	363° to 531°C	
			Saturates		
	2.549	2.849	1.087	0.777	0-ring paraffins
	4.013	3.322	1.711	0.906	1-ring paraffins
	3.733	0.634	1.592	0.173	2-ring paraffins
	3.569	0.835	1.522	0.228	3-ring paraffins
	2.089	0.946	0.891	0.258	4-ring paraffins
		1.137		0.310	5-ring paraffins
	0.493	0.342	0.210	0.093	Monoaromatics (not classified as specific types)
			Monoaromatics		
-6	2.104	0.173	0.898	0.047	Alkylbenzenes
-8	11.144	0.331	4.753	0.090	Alkylnaphthenobenzenes
-10	10.828	0.837	4.618	0.228	Alkyldinaphthenobenzenes
-12	2.698	1.536	1.151	0.419	Alkyltrinaphthenobenzenes
-14	0.045	1.287	0.019	0.351	Alkyltetranaphthenobenzenes
-16		0.357		0.097	Alkylpentanaphthenobenzenes
-18		0.162		0.044	Alkylhexanaphthenobenzenes
			Diaromatics		
-12	7.531	0.188	3.211	0.051	Alkylnaphthalenes
-14	9.330	2.347	3.979	0.640	Alkylnaphthenonaphthalenes and/or alkyldiphenylalkanes
-16	3.949	4.596	1.684	1.254	Alkyldinaphthenonaphthalenes and/or alkylnaphthenodiphenylalkanes
		0.951		0.259	Alkyldinaphthenonaphthalenes and/or alkylfluorenes
-18	0.803	4.479	0.343	1.221	Alkyltrinaphthenonaphthalenes and/or alkyldinaphthenodiphenylalkanes
-20	0.023	4.297	0.010	1.172	Alkyltetranaphthenonaphthalenes and/or alkyltrinaphthenodiphenylalkanes
-22		3.628		0.989	Alkylpentanaphthenonaphthalenes and/or alkyltetranaphthenodiphenylalkanes
-24		1.308		0.357	Alkylhexanaphthenonaphthalenes and/or alkylpentanaphthenodiphenylalkanes
-26		0.580		0.158	Alkylheptanaphthenonaphthalenes and/or alkylhexanaphthenodiphenylalkanes
-28		0.196		0.053	Alkyloctanaphthenonaphthalenes and/or alkylheptanaphthenodiphenylalkanes
-30		0.063		0.017	Alkylnonanaphthenonaphthalenes and/or alkyloctanaphthenodiphenylalkanes

Table 4.56. (continued)

Mass Z series	Distillate (wt %)		Crude (wt %)		Probable compound types
	207° to 363°C	363° to 531°C	207° to 363°C	363° to 531°C	
				Polyaromatics	
-16	1.538	0.663	0.656	0.181	Alkylacenaphthalenes
-18	0.406	0.006	0.173	0.002	Alkylphenanthrenes/anthracenes
	0.274		0.117		Alkylnaphthenoacenaphthalenes
	1.328	2.728	0.566	0.744	Alkylnaphthenoacenaphthalenes and/or alkylphenanthrenes/anthracenes
-20	0.312		0.133		Alkyldiacenaphthalenes
		0.047		0.013	Alkylphenylnaphthalenes
	0.106		0.045		Alkyldinaphthenoacenaphthalenes
	1.264	0.163	0.539	0.044	Alkylnaphthenophenanthrenes/anthracenes
	0.165	5.140	0.071	1.401	Alkyldinaphthenoacenaphthalenes and/or alkylnaphthenophenanthrenes/anthracenes
	0.067		0.028		Alkyldinaphthenoacenaphthalenes and/or alkylnaphthenaphenanthrenes/anthracenes and/or alkyldiacenaphthalenes
-22	0.184	2.399	0.078	0.654	Alkylpyrenes
		0.036		0.010	Alkylnaphthenophenylnaphthalenes
	0.284		0.121		Alkyldinaphthenophenanthrenes/anthracenes
		6.135		1.672	Alkyltrinaphthenoacenaphthalenes and/or alkyldinaphthenophenanthrenes/anthracenes
	0.084		0.036		Alkyltrinaphthenoacenaphthalenes and/or alkyldinaphthenophenanthrenes/anthracenes and/or alkylnaphthenodiacenaphthalenes
	0.330		0.141		Alkylnaphthenodiacenaphthalenes and/or alkyldinaphthenophenanthrenes/anthracenes
	0.973		0.415		Alkyldinaphthenophenanthrenes/anthracenes and/or alkylpyrenes
-24	0.026	0.041	0.011	0.011	Alkylnaphthenopyrenes
		0.022		0.006	Alkyldinaphthenophenylnaphthalenes
	0.080		0.034		Alkyltrinaphthenophenanthrenes/anthracenes
	0.060	1.207	0.026	0.329	Alkyltrinaphthenophenanthrenes/anthracenes and/or alkylchrysenes
	0.119		0.051		Alkyldinaphthenodiacenaphthalenes and/or alkyltrinaphthenophenanthrenes/anthracenes
		4.445		1.212	Alkyltetranaphthenoacenaphthalenes and/or alkyltrinaphthenophenanthrenes/anthracenes
		2.503		0.682	Alkyltetranaphthenoacenaphthalenes and/or alkyltrinaphthenophenanthrenes/anthracenes and/or alkylchrysenes
	0.046		0.020		Alkyltetranaphthenoacenaphthalenes and/or alkyltrinaphthenophenanthrenes/anthracenes and/or alkyldinaphthenodiacenaphthalenes

Table 4.56. (continued)

Mass Z series	Distillate (wt %) 207° to 363°C	Distillate (wt %) 363° to 531°C	Crude (wt %) 207° to 363°C	Crude (wt %) 363° to 531°C	Probable compound types
-26		0.073		0.020	Alkyldinaphthenopyrenes
	0.005		0.002		Alkylnaphthenochrysenes
		0.017		0.005	Alkyltrinaphthenophenylnaphthalenes
	0.019		0.008		Alkyltrinaphthenodiacenaphthalenes
	0.092		0.039		Alkyltetranaphthenophenanthrenes/anthracenes
	0.018		0.008		Alkyltrinaphthenodiacenaphthalenes and/or alkyltetranaphthenophenanthrenes/anthracenes
		0.209		0.057	Alkyltetranaphthenophenanthrenes/anthracenes and/or alkylnaphthenochrysenes
		3.364		0.917	Alkylpentanaphthenoacenaphthalenes and/or alkyltetranaphthenophenanthrenes/anthracenes
		1.772		0.483	Alkylpentanaphthenoacenaphthalenes and/or alkyltetranaphthenophenanthrenes/anthracenes and/or alkylnaphthenochrysenes
-28	0.003		0.001		Alkylbenzopyrenes
		0.340		0.093	Alkyltrinaphthenopyrenes and/or alkylperylenes
	0.057		0.025		Alkylpentanaphthenophenanthrenes/anthracenes
		1.895		0.517	Alkylpentanaphthenophenanthrenes/anthracenes and/or alkyldinaphthenochrysenes
		2.494		0.680	Alkylhexanaphthenoacenaphthalenes and/or alkylpentanaphthenophenanthrenes/anthracenes
-30		0.219		0.060	Alkyltetranaphthenopyrenes and/or alkylnaphthenoperylenes
		1.660		0.453	Alkylheptanaphthenoacenaphthalenes and/or alkylhexanaphthenophenanthrenes/anthracenes
	0.037	1.289	0.016	0.351	Alkylhexanaphthenophenanthrenes/anthracenes and/or alkyltrinaphthenochrysenes
-32		0.577		0.157	Alkyldinaphthenoperylenes
	0.017	0.242	0.007	0.066	Alkylheptonaphthenophenanthrenes/anthracenes
		0.621		0.169	Alkylheptanaphthenophenanthrenes/anthracenes and/or alkyltetranaphthenochrysenes
		0.555		0.151	Alkyloctanaphthenoacenaphthalenes and/or alkylheptanaphthenophenanthrenes/anthracenes
-34		0.101		0.028	Alkyltrinaphthenoperylenes
	0.013		0.005		Alkyloctanaphthenophenanthrenes/anthracenes
		0.097		0.026	Alkyloctanaphthenophenanthrenes/anthracenes and/or alkylpentanaphthenochrysenes
-36		0.009		0.002	Alkyltetranaphthenoperylenes

Table 4.56. (continued)

Mass Z series	Distillate (wt %) 207° to 363°C	Distillate (wt %) 363° to 531°C	Crude (wt %) 207° to 363°C	Crude (wt %) 363° to 531°C	Probable compound types
				Others	
-4	0.005	0.009	0.002	0.003	Alkyltriolefins and/or alkylcyclodiolefins and/or dicycloolefins
-10S	0.078		0.033		Alkylbenzothiophenes
-12S	0.038		0.016		Alkylnaphthenobenzothiophenes
-16S	0.142	0.010	0.061	0.003	Alkyldibenzothiophenes
-18S	0.064	0.007	0.027	0.002	Alkylnaphthenodibenzothiophenes
-22S		0.379		0.103	Alkylphenanthrenothiophenes
-24S		0.398		0.108	Alkylnaphthenophenanthrenothiophenes
-26S		0.514		0.140	Alkyldinaphthenophenanthrenothiophenes
-28S		0.046		0.012	Alkyltrinaphthenophenanthrenothiophenes
		0.008		0.002	Alkylthienoperylenes
-16(O)	1.143		0.488		Alkyldibenzofurans
-18(O)	0.188		0.080		Alkylnaphthenodibenzofurans
-7N	0.011		0.005		Alkylpyrindans
-9N	0.121		0.052		Alkylindoles
-11N	0.083		0.035		Alkylnaphthenoindoles
-15N	0.353	0.341	0.150	0.093	Alkylcarbazoles
-17N	0.229	0.583	0.098	0.159	Alkylnaphthenocarbazoles
	0.014		0.006		Alkylacridines
-19N		0.138		0.038	Alkyldinaphthenocarbazoles
-21N		0.644		0.176	Alkylbenzocarbazoles
-23N		0.500		0.136	Alkylnaphthenobenzocarbazoles
-25N		0.287		0.078	Alkyldinaphthenobenzocarbazoles
-27N		0.115		0.031	Alkyltrinaphthenobenzocarbazoles
-29N		0.014		0.004	Alkyltetranaphthenobenzocarbazoles
0, -14	0.014		0.006		Unknown
-2, -16		0.017		0.005	Unknown
-4, -18	0.052	0.018	0.022	0.005	Unknown
-6, -20	0.047		0.020		Unknown
-8, -22	0.081	0.015	0.035	0.004	Unknown
-10, -24	0.148	0.004	0.063	0.001	Unknown
-12, -26	0.008		0.003		Unknown
-5N, -19N		0.022		0.006	Unknown
-7N, -21N	0.003	0.035	0.001	0.009	Unknown
	7.220		3.079		Weak acids (not separated from 363° to 531°C PAP fraction)
	9.138	6.845	3.898	1.866	Strong acids
	3.356	4.729	1.431	1.289	Bases
	3.255	5.079	1.388	1.385	Material not analyzed from GPC separation
	1.337	0.794	0.570	0.217	Material not analyzed from silica-alumina separation

Source: Woodward et al. 1975, Table 3, pp. 11-14.

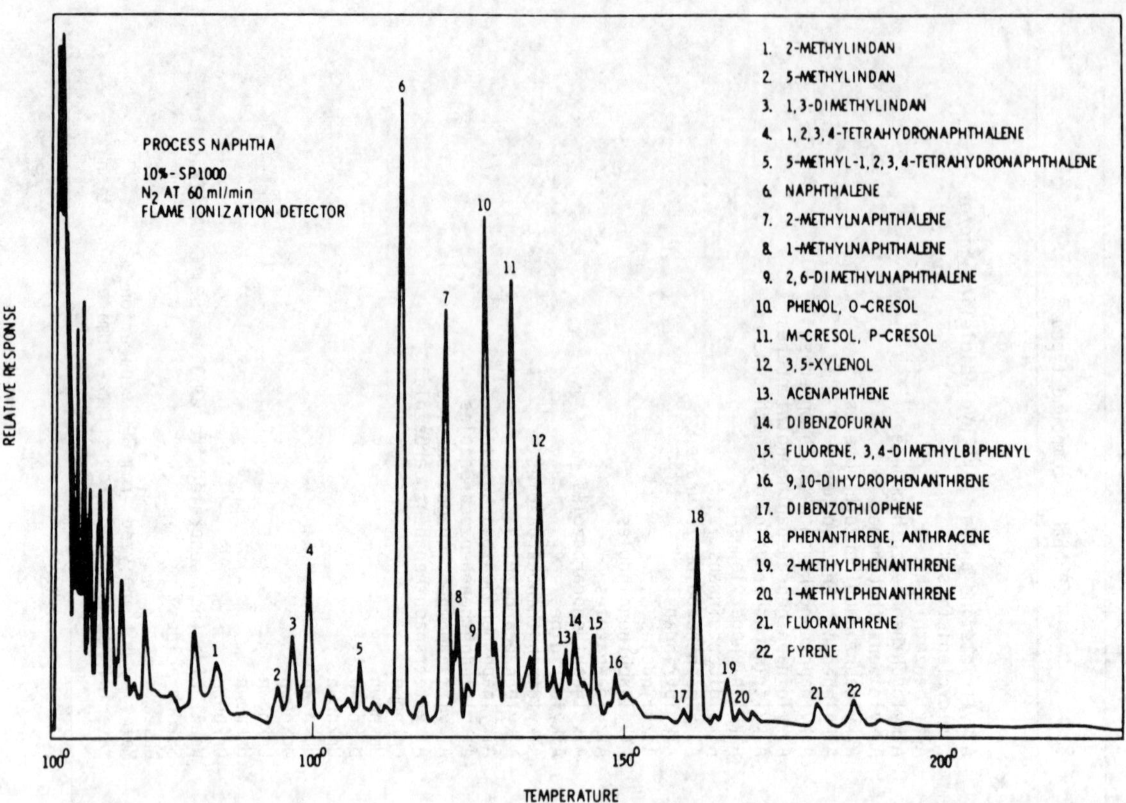

Fig. 4.60. Gas chromatogram of process naphtha from a 50-ton/day SRC pilot plant using high-sulfur eastern coal. <u>Source</u>: From Petersen 1975, Fig. 1, p. 5. Reprinted by permission of the publisher.

Greater than 95% of the material was eluted when the gas chromatography temperature reached 145°C, corresponding to a boiling range of 25 to 333°C, and was predominantly saturated hydro-carbons (Nichols 1975).

The major organic sulfur compounds detected in the liquid product from the hydrodesulfurization of Kentucky coal are thiophenes listed in Table 2.36.

Various syncrude distillates are compared in Table 4.58. Saturate concentrations were lower in Synthoil for both boiling ranges. Monoaromatic concentration in the 363 to 531°C distillate of Synthoil was lower than the other concentrates shown, which indicates a sample having more aromaticity, as evidenced by the high concentrations of polyaromatics. Heteroatomic types were more concentrated in both distillates of Synthoil. The Synthoil distillate (363 to 531°C) contained 67% aromatics having four or more rings; Utah syncrude contained 40%; western Kentucky syncrude contained 49%; and comparable petroleum distillates contained 17 to 21%. The <207°C Synthoil distillate was only 4.4% of the original synthetic crude, as compared with 21% for western Kentucky syncrude and 13% for Utah syncrude (Woodward et al. 1976).

4.5.1.3 <u>BTX</u>

Streams containing BTX have been identified in the Lurgi and HYGAS flow charts (Glaser, Hershaft, and Shaw 1974). Removed from the product gas streams in the first gas treatment stage by solvent-based acid-gas removal processes, the BTX stream, according to data from the HYGAS pilot plant,

Table 4.57. Major structural types and percent of total ionization
in heavy oil and asphaltene fractions
from Synthoil product

Structural types[a]	Percent of total ionization	
	Heavy oil	Asphaltene
Alkylbenzene	11	9
Indenes	7	4
Indans	9	2
Naphthalenes	6	2
Acenaphthylenes	12	8
Biphenyls	21	11
Anthracenes; phenanthrenes	6	4
Phenylnaphthalenes	5	6
4-ring, peri-condensed	5	11
4-ring, cata-condensed	3	9
5-ring, peri-condensed	4	15
5-ring, cata-condensed	1	5
6-ring, peri-condensed	1	10
Phenols	9	4

[a]Including alkyl derivatives.

Source: Sharkey et al. 1975, Table 4, p. 7.

is composed of 50% very light oil (similar to kerosene or No. 1 fuel oil), minor amounts of
hydrogen sulfide, and BTX (Glaser, Hershaft, and Shaw 1974). A BTX and oil production rate of
484,480 liters/day (128,000 gpd) is predicted for Illinois No. 6 coal, whereas 862,980 liters/day
(228,000 gpd) is predicted for Montana subbituminous coal, when using the HYGAS process (Glaser,
Hershaft, and Shaw 1974). Estimates of 20,000 to 25,000 gpd of BTX have been reported for the
Synthane process (Kalfadelis and Magee 1975b). Forney et al. (1974) report that Synthane
(laboratory-scale) gasifier gas contained benzene (10 to 1730 ppm) and toluene (3 to 185 ppm),
the amount depending on the coal type.

4.5.2 Polycyclic aromatic hydrocarbons and heterocyclic aromatic compounds

The three principal carcinogenic classes of compounds which may be produced in coal conversion
are PAH, polyheterocyclic aromatic compounds (including "azo" derivatives), and aromatic amines.

4.5.2.1 Liquid products

The PAH content in Synthoil oil and COED process streams is listed in Tables 4.59 and 4.60
respectively (Shults 1976). Amounts of PAH present in a coal-derived liquefaction product are
listed in Table 4.61.

Table 4.58. Comparison of distillates of similar boiling ranges from three syncrudes on a weight-percent-of-distillate basis

	West Virginia Synthoil, 207° to 363°C	Western Kentucky COED, 205° to 380°C	Utah COED, 204° to 381°C	West Virginia Synthoil, 363° to 531°C	Western Kentucky COED,[a] >380°C	Utah COED,[a] >381°C
Saturates	16.0	25.0	27.8	9.7	23.8	25.8
Monoaromatics	27.3	42.0	25.1	4.7	25.1	14.4
Diaromatics	21.6	13.0	17.5	22.6	24.3	18.4
Polyaromatics	7.9	5.4	7.1	41.1	20.0	25.1
Heteroatomics	22.2	4.4	15.2	15.6	4.5	7.4
Distillate-weight-percent of syncrude	42.6	54.2	45.4	27.3	24.2	40.3

[a]Neither Western Kentucky nor Utah COED syncrudes contained appreciable material boiling above 530°C.

Source: Woodward et al. 1976, Table 4, p. 18.

Table 4.59. PAH compounds in Synthoil

Compound	Result (ppm)
Phenanthrene	413
Benzo[*a*]anthracene	18
Benzo[*a*]pyrene	41

Source: Shults 1976, Table 3.4, p. 22.

Table 4.60. PAH in COED samples

Sample type	Results (ppm)		
	BaP	BaA	Phenol
Product separator liquor	8.0		
Drier liquor	9.3		
Syncrude	51	Trace	Trace
Unfiltered raw oil	107	42	270
Filtered raw oil	96	52	400

Source: Shults 1976, Table 4.5, p. 32.

Table 4.61. Quantities of PAH in a coal-derived
liquefaction product

PAH	Quantity (ppm)
Naphthalene	347
2-Methylnaphthalene	1325
1-Methylnaphthalene	383
Biphenyl	89
2,6-Dimethylnaphthalene	328
1,3-Dimethylnaphthalene	181
1,5- and/or 2,3-Dimethylnaphthalene	202
Benzo[*a*]pyrene	41
Benzo[*a*]anthracene	18
Phenanthrene	413
Acenaphthalene	20
Acenaphthene	61
Fluorene	205
1-Methylfluorene	152
2-Methylfluorene	245
1-Methylphenanthrene	107
9-Methylanthracene	42
Chrysene	98
2-Methylchrysene	102
6-Methylchrysene	64
3-Methylchrysene	106

Source: Guerin et al. 1975, Table 3, p. 670.

4.5.2.2 Coal tar

Coal-associated PAH and heterocylic emissions from sources other than coal conversion have been reported. Matsushita et al. (1972) identified 24 polynuclear hydrocarbons in coal tar samples. Carcinogens, or suspected carcinogens, identified in the Matsushita et al. study were chrysene, benzo[a]anthracene (BaA), benzo[a]pyrene (BaP), benzo[b]fluoranthene, benzo[j]fluoranthene, benzo[e]pyrene, indol[1,2,3-cd]pyrene, benzo[ghi]pyrene, dibenzo[a,h]pyrene, and dibenzo[a,i]pyrene. Benzo[e]pyrene and perylene were also identified in that study (Matsushita et al. 1972). Badger (1962), in addition to most of the above, identified dibenzo[a,h]anthracene in coal tar. Kornreich (1974) reports dibenzo[a,h]anthracene and benzo[c]acridine in coal tar. The PAH in coal tar, as identified by Combes (1954), include benzo[a]anthracene and its derivatives: benzo[a]pyrene, 9,10-dimethyl-1,2-anthracene, 1,2,5,6-dibenzanthracene, 20-methylcholanthrene, 5,9,10-trimethyl, benzo[a]anthracene, and phenanthrene. Azo compounds in coal tar, as identified by Combes (1954), include 4-amino-2,3-azotoluene, 2,3-azotoluene, and p-dimethylamino-azobenzene. Buu-Hoi (1958) reports small amounts of dibenzo[a,h]pyrene present in coal-tar pitch, and Schoental (1957) reports dibenzo[a,i]pyrene in coal tar.

Lijinsky et al. (1963) noted higher concentrations of less-volatile pentacyclic hydrocarbons present in coal tar than in cresote (Table 4.62). Average concentrations of PAH measured in asphalt and coal-tar pitches are compared in Table 4.63; other compounds found in low concentrations but not listed are benzonaphthothiophenes, benzonaphthofuranes, benzacridines, and benzo[ghi]fluoranthene (Wallcave et al. 1971). The PAH content of the coal-tar pitches was several orders of magnitude higher than that of the petroleum asphalts examined. It is suspected that these differences may be caused by differences in thermal histories (Wallcave et al. 1971). Thus, a prominent difference may be expected between the amounts of chemical carcinogens present in heavier and more aromatic fractions of coal-derived fuels and those associated with other fuel production processes such as petroleum processing (Fulkerson et al. 1974).

Amounts of PAH and heterocyclic compounds present in an air sample polluted by coal-tar pitch are given in Table 4.64 (Sawicki, Meeker, and Morgan 1965).

4.5.2.3 Atmospheric emissions

At the Union Carbide Institute West Virginia Plant, high concentrations of BaP in aerosols (as high as 19,000 ng/m^3) were measured near the solids removal operations and pitch treatment section (Ketcham and Norton 1960). The lowest concentrations were measured around the paste processing section and coal storage area (Ketcham and Norton 1960).

In a petroleum processing plant, atmospheric values of PAH, BaP, and 1,2-benzoperylene were lowest in the area of the atmospheric-vacuum and cracking equipment, and higher around the equipment for processing the high-boiling products (fuel oil, tar, and cracking residues) (Kireeva and Yanysheva 1970). Especially high discharges of PAH (including 1,2,5,6-dibenzanthracene, 1,2-benzanthracene, 1,2,3,4-dibenzanthracene, and the noncarcinogenic anthracene) were found in the air at the coking equipment, where the processing temperatures were high (500°C) and processes were carried out in equipment that was not hermetically sealed (Kireeva and Yanysheva 1970).

In the petroleum refining industry, it is in the regeneration of the cracking catalyst, through the combustion of coke on the catalyst surface, that BaP and other polycyclic organic matter are

Table 4.62. PAH in creosote and coal tar

PNA	Concentration in creosote (g/kg)		Concentration (g/liter)	Concentration in coal tar (g/kg)	
	(1)	(2)		(1)	(2)
Anthracene	12.1	12.0	6.2	2.88	4.35
Benz[a]anthracene	2.77	2.94	2.75	6.24	6.98
Benzo[b]chrysene	0.03	0.06		0.93	0.80
Benzo[j]fluoranthene	0.29	0.29		0.63	0.45
Benzo[k]fluoranthene	0.30	0.11		1.08	1.07
Benzo[g,h,i]perylene				1.23	1.89
Benzo[a]pyrene	0.14	0.22	0.12	2.08	1.76
Benzo[e]pyrene	0.18	0.15		1.85	1.88
Carbazole	2.20	1.42	2.75	1.32	1.27
Chrysene	1.34	0.94	1.27	2.13	2.86
Dibenz[a,h]anthracene				0.30	0.23
Fluoranthene	24.8	22.2	7.8	17.7	17.8
Perylene	0.04	0.04	0.04	0.70	0.76
Phenanthrene	39.9	33.3	47.9	13.6	17.5
Pyrene	9.1	6.8	4.2	7.95	10.6

Source: Lijinsky et al. 1963, Table II, p. 954.

formed (National Academy of Sciences 1972). However, organic pollution arising from coal conversion may be different from that associated with other fuel production technologies such as petroleum refining. Most significant is the greater concentration of chemical carcinogens in the heavier and more aromatic fractions of the coal conversion streams (Fulkerson et al. 1974b).

PAH emissions from coal-burning units vary widely (Table 4.65), depending on the quality of combustion achieved (Hangebrauck, von Lehmden, and Meeker 1964). Factors contributing to lower PAH emission rates from larger coal-fired units include more efficient combustion of fuel attainable with closely regulated air-fuel ratios, uniformly high combustion-chamber temperatures, and relatively long retention times in the high-temperature zone (Hangebrauck, von Lehmden, and Meeker 1964). Comparing PAH emissions with other products of incomplete combustion, Hangebrauck, von Lehmden, and Meeker (1964) indicate that PAH emission rates are generally high when carbon monoxide and total gaseous hydrocarbons are high.

Diehl, du Breuil, and Glenn (1967) (see Table 4.66), considering the PAH emissions per unit of time to be the important factor (rather than concentration per volume of flue gas as above), feel that milligram-per-hour values are more significant, and that, in these terms, larger installations emit as much or more PAH than do smaller installations. They also suggest that the manner in which a coal is burned has more influence on PAH emissions than does the coal type.

Coal-fired furnace emissions of PAH and heterocyclics are listed in Table 4.67 (Sawicki, Meeker, and Morgan 1965).

Table 4.63. Concentration averages of PAH in asphalt and coal-tar pitches

Pitch		Anthracene	Phenanthrene	Pyrene	Fluoranthene	Benzofluorenes	Benz[a]anthracene	Triphenylene	Chrysene	Benzo[a]pyrene
Asphalt	A	—[b]	2.3 (4.5)	0.6 (1.2)	+[b]	+	0.15 (1.1)	0.25 (2.4)	0.2 (4.0)	0.5 (2.9)
	B	—	0.4 (7.5)	1.8 (18)	+	+	2.1 (46)	6.1 (31)	8.9 (101)	1.7 (12)
	C	—	3.5 (24)	4.0 (8.6)	2.0 (+)	+	1.1 (6.9)	3.1 (6.9)	2.3 (12)	1.3 (3.8)
	D	—	1.3 (62)	8.3 (34)	+	+	0.7 (9.7)	3.4 (8.5)	3.9 (31)	2.5 (7.2)
	E	—	0.6 (6.5)	0.9 (6.7)	+	+	0.9 (13)	3.8 (12)	3.2 (25)	1.6 (8.4)
	F	—	35t (89)	38 (89)	5 (+)	+	35 (109)	7.6 (43)	34 (158)	27 (69)
	G	—	1.1 (6.3)	0.3 (1.9)	—	+	0.2 (1.2)	1.0 (3.2)	0.7 (6.2)	0.1 (1.0)
	H	—	2.3t (0.8)	0.08 (0.8)	—	—	— (0.05)	0.3 (0.7)	0.04 (0.4)	—
Coal-tar pitch	A	0.86t[b]	3.1	2.0 (0.2)	4.0 (0.15)	0.73t	0.89 (0.15)	0.15 (0.04)	0.74 (0.09)	0.84 (0.15)
	B	1.0t	2.9t	2.9 (0.24)	4.3 (0.14)	0.51t	1.25 (0.22)	0.11 (0.06)	1.0 (0.27)	1.25 (0.11)

Pitch		Benzo[e]pyrene	Benzo[k]fluoranthene	Perylene	Anthanthrene	Benzo[ghi]perylene	Indeno[1,2,3-cd]pyrene	Picene	Coronene
Asphalt	A	3.8 (11)	+[c] (—)[d]	—	—	2.1 (7.4)	Tr	+	1.9 (—)
	B	13 (30)	—[d]	39 (9.7)	Tr[b]	4.6 (9.6)	—	+	0.8 (0.5)
	C	2.9 (5.5)	+ (—)	2.2 (0.8)	Tr	1.0 (1.6)	Tr	+	0.5 (—)
	D	3.2 (6.8)	+ (—)	6.1 (2.0)	Tr	1.7 (2.9)	Tr	+	0.2 (—)
	E	6.5 (24)	+ (—)	2.9 (1.4)	+ (—)	2.7 (4.5)	Tr	+	0.9 (0.5)
	F	52 (141)	—	3.0 (—)	1.8 (—)	15 (41)	1.0 (—)	1.0 (+)	2.8 (1.9)
	G	1.6 (4.1)	—	0.1 (—)	—	0.6 (0.8)	—	+	0.9 (—)
	H	0.03 (0.06)	—	—	—	Tr	—	—	—
Coal-tar pitch	A	0.54 (0.06)	0.71 (+)	0.20 (+)	0.13 (+)	0.32 (+)	0.73 (+)	NE[b]	(0.07) (+)
	B	0.70 (0.08)	0.90 (+)	0.33 (+)	0.21 (+)	0.33 (0.05)	0.93 (+)	0.20 (+)	(0.07) (+)

[a]Asphalt values are given in parts per million; coal-tar pitches in percent; and alkyl derivatives in parentheses.
[b]Abbreviation: —, not present; t, estimate includes alkyl derivatives; + , not estimated but present in small amount; Tr, trace; NE, not estimated but present in substantial amount.
[c]Benzo[b]fluoranthene usually associated with benzo[k]fluoranthene.
[d]Contains benzo[b]fluoranthene.

Source: Wallcave et al. 1971, Table 1, p. 46.

Table 4.64. Amounts of polycyclic and heterocyclic compounds in an air
sample polluted by coal tar pitch

Compounds	Milligram of compound per gram of:		Milligram of compound per 1000 m^3 air
	Benzene-sol. fraction	Airborne particulates	
Acridine	7.800	4.400	0.870
Benzo[*f*]quinoline	3.800	2.100	0.420
Benzo[*h*]quinoline	2.300	1.300	0.260
Phenanthridine	0.200	0.100	0.020
Benz[*a*]acridine	1.800	1.000	0.200
Benz[*c*]acridine	1.100	0.600	0.120
Indenol[*1,2,3-ij*]isoquinoline	0.240	0.140	0.030
11H-Indeno[*1,2-b*]quinoline	1.700	0.950	0.190
Dibenz[*ah*]acridine	0.090	0.050	0.010
Dibenz[*aj*]acridine	0.010	0.006	0.001
Anthracene	7.000	3.800	0.800
Phenanthrene	36.000	19.000	4.000
Benz[*a*]anthracene	∿6.000	∿3.300	∿0.700
Fluoroanthene	∿80.000	∿43.000	∿9.000
Pyrene	72.000	39.000	8.300
Benzo[*a*]pyrene	3.300	1.800	0.400
Benzo[*e*]pyrene	2.100	1.100	0.240
Perylene	0.450	0.240	0.050
Benzo[*ghi*]perylene	0.500	0.270	0.060
Anthanthrene	0.044	0.024	0.005
Coronene	0.030	0.016	0.003
Carbazole	20.000	11.000	2.300
Total weight, mg	246.000	133.000	28.000
ACR/PHD	39.000		
BeACR/IND[*1,2-b*]Q	0.640		
BaP/BhQ	1.400		
BaP/DB[*ah*]ACR	37.000		

Source: Sawicki, Meeker, and Morgan 1965, Table 3, p. 296. Reprinted by permission of the publisher.

4.5.2.4 Particulate matter

The majority of PAH compounds are adsorbed on particulate matter; also, polycyclic organic matter
in the atmosphere is exclusively associated with particulate matter, especially soot (National
Academy of Sciences 1972). Particulate matter emissions of PAH by municipal and commercial
incinerators and open-burning sources are given in Table 4.68 (Hangebrauck, von Lehmden, and
Meeker 1964). Water-spray flue-gas scrubbing for fly-ash control reduced the level of PAH
atmospheric emissions significantly (Hangebrauck, von Lehmden, and Meeker 1964).

4.5.3 Process solids

Solvent-refined (pilot-plant) fuel-product neutral components using high-sulfur (4% S) eastern
coal are mainly aromatic compounds, anthracene, fluoranthene, and pyrene (Petersen 1975). Components identified from the concentrated benzene extract of the SRC (pilot-plant) mineral residue
included naphthalene, 2-methylnaphthalene, and less abundant phenolic compounds (Petersen 1975).

4.5.4 Process solvent

The most predominant gas chromatogram peak of the SRC (pilot-plant) process solvent was anthracene,
with only a small amount of phenolics. Greater than 95% of the material was aromatic hydrocarbons
(phenanthrene, pyrene, fluoranthene, fluorene, dibenzofuran, and naphthalenes) (Petersen 1975).

Table 4.65. Polynuclear hydrocarbon emission summary — heat-generation sources

Group 1 (Pyrene through Coronene) and Group 2 (Anthracene through BaA) are in µg/million Btu heat input.

Source number	Fuel used	Firing method	BaP (µg/1000 m³)	BaP (µg/lb fuel)	Pyrene	Pyrene	BeP	Perylene	BghiP	Anthanthrene	Coronene	Anthracene	Phenanthrene	Fluoranthene	BaA
1	Coal	Pulverized	42	0.22	19	150								180	19
2			75	0.43	32	240	92					370		550	
3		Chain grate stoker	71	0.44	37	390	130							680	
4		Spreader stoker	49	0.35	26	590	350				26			360	
5		Underfeed stokers	7,900	140	10,000	16,000	7,900	1,600	4,500	290	330	850	10,000	38,000	3,900
6			61	1.6	120	1,700	230						1,000	3,200	
7			3,400	52	3,800	7,700	5,400		580		1,200		29,000	47,000	560
8		Hand-stoked	340,000	6,000	400,000	600,000	100,000	60,000	300,000	90,000	30,000	400,000	1,000,000	1,000,000	
9	Oil	Steam-atomized	<38	<0.3	<20	49								56	
10			40	0.89	47	300							1,800	270	27
11		Low-pressure air-atomized	1,900	18	900	6,100			300		2,100	3,900	3,500	1,900	
12		Centrifugal-atomized	<26	<0.9	<40	1,800							8,900	5,000	
13			<27	<1	<60	15								76	
14		Vaporized	<34	<2	<100	1,200								15,000	
15	Gas	Premix burners	<29	<0.4	<20	160					14			100	
16			350	4.6	200	18,000	490		1,800	200	5,300			2,900	
17			<23	<0.5	<20	170								320	
18			<30	<0.6	<20	120	18						77	110	

[a]"Less than" values for BaP were calculated for those samples having concentrations below the limit of quantitative determination (about 0.6µg per sample. Similar calculations were not included for the other PNA (indicated by blanks in the table).
[b]Micrograms per 1000 m³ of flue gas at standard conditions (70°F, 1 atm).

Source: Hangebrauck, von Lehmden, and Mecker 1964, Table IV, p. 270. Reprinted by permission of the publisher.

Table 4.66. Summary of data from field installations

Code	Type of burner	Capacity (lb steam/hr)	Load factor sampled	BaP			BeP			BaA		
				(μg/1000 m^3)	(μg/10^6 Btu)	(mg/hr)	(μg/1000 m^3)	(μg/10^6 Btu)	(mg/hr)	(μg/1000 m^3)	(μg/10^6 Btu)	(mg/hr)
1	Chain grate	33,000	0.61				250	119	3.5			
2	Chain grate	35,000	0.45	9	4	0.1	91	42	1.0	44	20	0.4
3	Chain grate	40,000	0.47	110	75	2.8	440	300	11			
4	Chain grate	52,000	0.41	850	500	20	530	310	12	1,400	820	32
5	Chain grate	55,000	0.81[a]									
6	Underfeed	160,000	0.74	240	120	13	140	72	7.7	380	200	21
7	Underfeed	25,000	0.40[a]	53	31	1.1	590	350	12			
8	Spreader	24,000	0.75	120	68	1.3	820	470	9.1			
9	Pulsating grate	2,470	1.00	650	330	1.3	1,300	670	2.7	1,300	670	2.7
10	Pulverized coal	40,000	0.80	420	230	9.1	860	470	19	400	220	8.7
11	Pulverized coal	240,000	0.88	180	120	45	1,300	880	330			
12	Pulverized coal	600,000	1.10				160	64	0.06			
13	Pulverized coal	830,000	1.05	66	21	31	110	34	50			
14	Pulverized coal	1,060,000	1.00	18	6	10	380	130	220			
15	Pulverized coal	1,250,000	1.02	80	32	59	68	27	50			
16	Pulverized coal	2,030,000	0.78	59	20	54	130	44	120			
17	Pulverized coal	2,100,000	0.85	56	17	28	120	37	62			
18	Cyclone	2,200,000	0.99	40	16	49	140	55	170			

[a]Fluctuating load.

Source: Diehl, du Brevil, and Glenn 1967, Table 3, p. 281. Reprinted by permission of the publisher.

Table 4.67. Amounts of polynuclear compounds
in air pollution source effluents

Compound	Coal heating	
	Benzene solute (mg/g)	Gas (mg/1000 m^3)
Acridine	1.10	111
Benzo[*f*]quinoline	0.55	57
Benzo[*h*]quinoline	0.36	38
Phenanthridine	0.31	32
Benz[*a*]acridine	0.25	26
Benz[*c*]acridine	0.14	15
Benzo[*lmn*]phenanthridine		
Indeno[1,2,3-*ij*]isoquinoline	∿0.16	∿17
11H-Indeno[1,2-*b*]quinoline	0.23	24
Dibenz[*a,h*]acridine	0.16	17
Dibenz[*a,j*]acridine	0.02	2
Anthracene	7.60	780
Phenanthrene	18.00	1800
Benz[*a*]anthracene	∿13.00	∿1300
Chrysene	∿7.00	∿720
Fluoroanthene	∿29.00	∿2900
Pyrene	21.00	2200
Benzo[*a*]pyrene	9.90	1000
Benzo[*e*]pyrene	4.80	500
Perylene	1.20	120
Benzo[*ghi*]perylene	7.40	760
Anthanthrene	1.80	190
Coronene	0.30	30
Total weight, mg	124.00	13200
ACR/PHD	3.50	
BcACR/IND[1,2-*b*]Q	0.60	
BaP/BhQ	28.00	
BaP/DB[*a,h*]ACR	62.00	

[a]Taken from stack of residential coal furnaces.

Source: Sawicki, Meeker, and Morgan 1965, Table 2, p. 295.
Reprinted by permission of the publisher.

Table 4.68. Polynuclear hydrocarbon content of particulate matter emitted by incineration and open-burning sources[a]

Source Number	Type of unit	Sampling point	Group 1							Group 2			
			BaP	Pyrene	BeP	Perylene	BghiP	Anthanthrene	Coronene	Anthracene	Phenanthrene	Fluoranthene	BaA
			(μg/gram of particulate)										
20	Municipal incinerators											2.2	0.09
	250-ton/day	Breeching (before	0.016	1.9	0.08				0.06				
	Multiple chamber	settling chamber)									9.8	2.5	
21	50-ton/day multiple	Breeching (before	3.3	28	6.5		19		8.2				
	chamber	scrubber)										5.5	0.26
		Stack (after scrubber)	0.15	3.6	0.97		1.1		1.1				
	Commerical incinerators												
22	5.3-ton/day single chamber	Stack	58	350	49	3.3	98	7.1	23	51	150	240	5.0
23	3-ton/day multiple chamber	Stack	180	2600	180	36	540	45	130	53	62	2400	210
	Open-Burning												
24	Municipal refuse	In smoke plume	11	29	4.5					4.7		13	
25	Automobile tires		1100	1300	450	72	660	53	81	110	450	470	560
26	Grass clippings, leaves, branches		35	120	21		5.4			4.7		110	25
27	Automobile bodies		270	670	120	33	150	12	15	220	160	450	40

[a]A blank in the table for a particular compound indicates that it was not detected in the sample.

Source: Hangebrauck, von Lehmden, and Meeker 1964, Table VI, p. 271.

4.5.5 Organic wastewater contaminants

Water scrubbing and condensing systems used throughout conversion operations to remove contaminants from gas streams may pick up tar, oil, and other organic contaminants in addition to the contaminants described earlier. Tars separated from the Synthane (laboratory-scale gasifier using Pittsburgh and Illinois coals) aqueous condensates were partially identified by Forney et al. (1974) (see Table 4.51).

The SRC process [pilot-plant, using high-sulfur (4%) eastern coal] effluent water was alkaline (pH, 10) and contained sulfide ions (Petersen 1975). Neutral components carried over in the process water may be composed of a mixture of the process naphtha (primarily naphthalenes) and process solvent (containing anthracene, fluorene, and dibenzofuran). Fifty-eight percent of the organics in the water appeared to be phenol, and there were some cresols present in the acidic fraction. The basic components (primarily quinolines) were higher in relative concentrations and more numerous than were those from the other samples in this test (Petersen 1975).

Compound types, primarily phenolics, detected in product water from coal gasification are given in Fig. 4.61 and Table 4.69.

ORNL DWG 76-11472

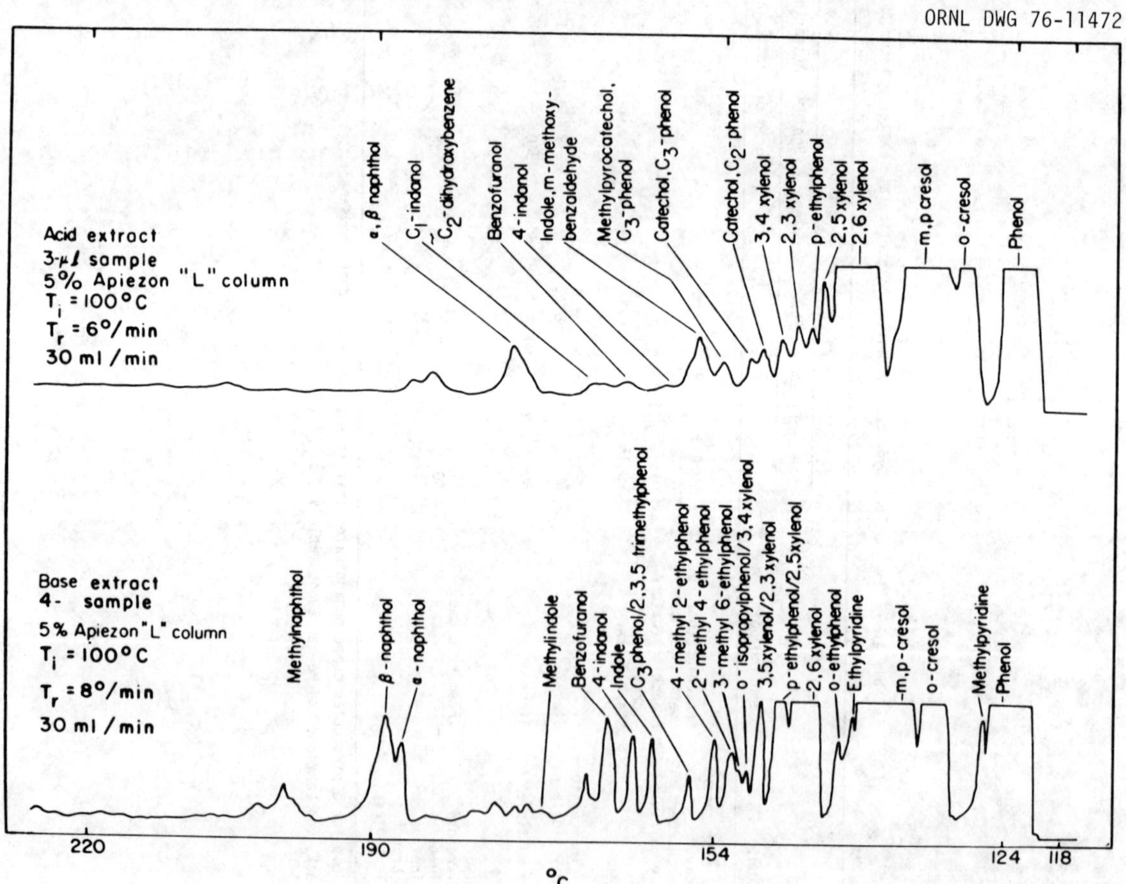

Fig. 4.61. Contaminants in product water from coal gasification: gas chromatography-mass spectrometry data. <u>Source</u>: From Sharkey et al. 1975, Fig. 8, p. 19. Reprinted by permission of the publisher.

Table 4.69. Contaminants in product water gasification
of Illinois No. 6 coal[a]

Compound	Quantity (ppm)
Phenol	3400
Cresols	2840
C_2-Phenols	1090
C_3-Phenols	110
Dihydrics	250
Benzofuranols	70
Indanols; acetophenones	150
Hydroxybenzaldehyde; benzoic acids	60
Naphthols	160
Indenols	90
Benzofurans	
Dibenzofurans	
Biphenols	40
Benzothiophenols	110
Pyridines	
Quinolines	
Indoles	

[a]Low-voltage mass spectrometry data.

Source: Sharkey et al. 1975, Table 4, p. 695.
Reprinted by permission of the publisher.

Organic compounds identified in COED pilot-plant product-separator liquor and their approximate concentrations include: catechol, ~200 µg/ml; resorcinol, ~1000 µg/ml; 2-methylresorcinol; orcinol, ~500 µg/ml; 4-methylcatechol, 10 to 15 µg/ml; and 3-methylcatechol (Shults 1976). Hydroquinone and a hydroxyacetophenone-natured compound are possible contaminants. Phenols, cresols, xylenols, and organosulfur compounds have also been identified in COED product-separator liquor (Shults 1976). COED drier scrubber samples contained dioctylphthalate and organosulfur compounds (Shults 1976).

Investigations of petrochemical industrial effluents showed that the highest concentrations of BaP (up to 27 mg/liter) and BaA (up to 2.8 mg/liter) were found in the aqueous distillate from installations involved in tar distillation and pyrolysis (Ershova 1971).

The presence of BaP in aqueous streams from the COED process (Table 4.60) and the reduced PAH emissions in coal combustion flue gas attributed to waste spray scrubbing (Table 4.68) are significant. PAH may be present in the conversion aqueous streams in a dissolved state or adsorbed onto suspended particulate matter. PAH have a low solubility in water, but the presence of surfactants may increase solubility by a factor of 2 to 7 (Andelman and Snodgrass 1974). The presence of PAH in COED aqueous liquors suggests the need for analysis and identification of trace organics in the aqueous liquor of other conversion processes.

4.5.6 Organic treatment and control

4.5.6.1 Tar, oil, and phenol

Tar and oil from processing leaks, spills, and wastewater treatment separators will be either incinerated or recycled and consumed in the process. Recycling to the gasifier section appears to be the most environmentally acceptable method (Hittman Associates 1975).

Contaminant hydrocarbons (tar, tar-oil, naphtha, PAH, and others) in the crude gas streams are separated by scrubbing, cooling, and condensing them in the water stream (Berty and Moe 1974). In the wastewater treatment facility, tar and oil separators utilize the different specific gravities of tar, oil, and water to effect separation. Tar organics are heavier than water, and oil organics are lighter than water (Gibson, Hammons, and Cameron 1974). Separators include API separators, parallel plate separators, and skim ponds. Removal of oils and tars is usually required previous to most secondary and tertiary wastewater treatment processes because of the deleterious effect of oils and tars on them (Sittig 1974). However, gravity separators neither separate soluble organics nor break emulsions (Sittig 1974).

Recovery of phenol may prove to be essential for pollution control and economical for some coal conversion processes. The Sasol process operates a parallel, two-stream Phenolsolvan process (Fig. 4.62) prior to ammonia recovery (Hoogendoorn and Salomon 1957). Residual tar and oil is removed from the gas liquor by gravity separation and sand filtration. Filtered liquor is then extracted with butyl acetate in a seven-stage countercurrent horizontal extractor. The liquor pH is maintained at about 8.5 to avoid hydrolysis of the butyl acetate by saturating the gas liquor with carbon dioxide prior to extraction. The extracted gas liquor is stripped of dissolved butyl acetate with live steam in a Raschig-ring-packed stripping tower. The butyl acetate (containing the extracted tar acids) is treated first in an atmospheric distillation tower and then in a vacuum distillation tower. The bottom product of the vacuum tower is crude phenol, consisting of the following approximate composition: 40% phenol (carbolic acid), 30% cresols, 7% xylenols, and 23% higher-boiling tar acids. Butyl acetate is recycled. Subsequent to the removal of butyl acetate from the gas liquor, the gas liquor is stripped of ammonia. Purified liquor (containing about 2 ppm phenols and 300 ppm ammonia) is biologically treated. Ammonia vapor (about 6% ammonia) from the ammonia stripper is converted to ammonium sulfate (Hoogendoorn and Salomon 1957). Shaw and Magee (1974) similarly describe the Phenolsolvan process used for the Lurgi process using isopropyl ether. Glaser, Hershaft, and Shaw (1974) report the residual phenol concentration to be as low as 20 ppm and Rudolph (1974) reports the residual ammonia concentration to be 60 ppm.

Lauer, Littlewood, and Butler (1969) (Fig. 4.63) describe the development and operation of a similar solvent extraction dephenolization process for coke plant aqueous wastes which has a removal efficiency of 99+% and wastewater residuals of 1 to 4 ppm.

Crook, McDonnell, and McNulty (1975) describe a promising phenol removal and recovery process from industrial waste effluents using Amberlite XAD polymeric adsorbents.

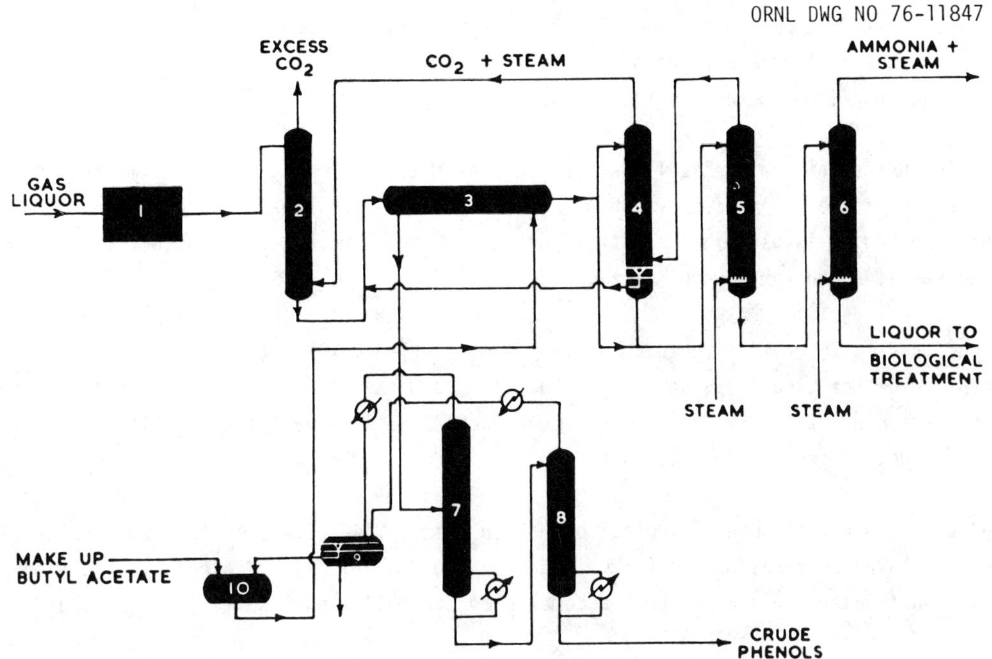

Fig. 4.62. Phenolsolvan plant (only one stream shown). <u>Source</u>: From Hoogendoorn and Salomon 1957, Fig. 1, p. 419.

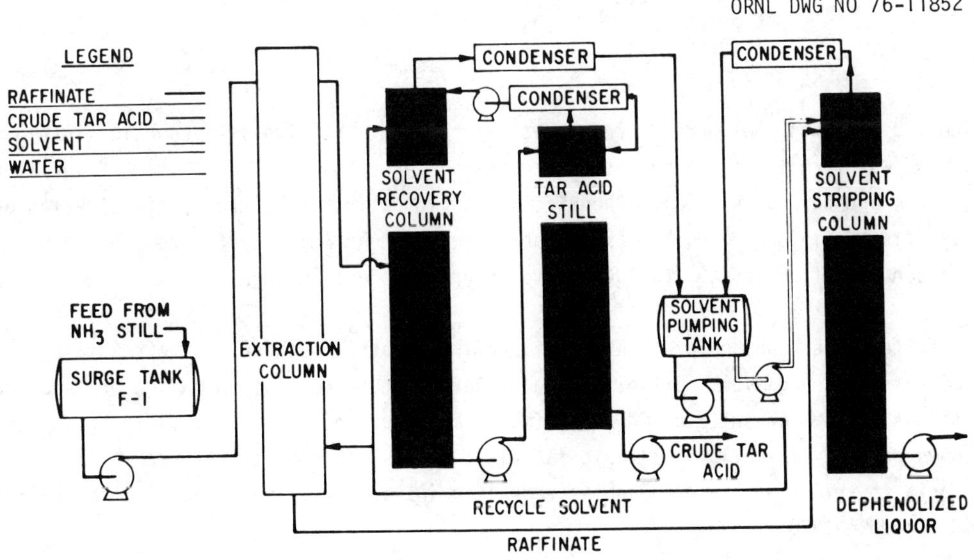

Fig. 4.63. Plant flow diagram. <u>Source</u>: From Lauer, Littlewood, and Butler 1969, Fig. 4, p. 102.

Intermediate organic wastewater treatment facilities include flotation, chemical flocculation, and biological treatment units. In the flotation process, separation of the phases is enhanced by pressurization of the incoming or recycled wastewater. Pressure reduction provides minute bubbles which attach themselves to, or become enmeshed in, the suspended oil particles. The air-solid particles rise to the surface of the unit and are removed by a sludge collection mechanism. The resulting underflow represents the clarifier effluent (Sittig 1974).

Flocculation units combine chemical coagulation-precipitation with gravity sedimentation. The addition of chemicals significantly enhances the removal of colloidal and suspended materials. A scraping mechanism is used in the bottom portion of the quiescent zone to collect and concentrate the resulting bottom sludge (Sittig 1974).

Biological treatment involves the elimination of most organic constituents by bacteria. Organic compounds are either used for synthesis of bacterial substances or oxidized for energy production (Reid and Streebin 1972). Biological treatment facilities include trickling filters and activated sludge (described in Sect. 4.4.5).

Coke and steel industries use biological oxidation (Fig. 4.64) with satisfactory results (Reid and Streebin 1972; Ludberg and Nicks 1969; Jüntgen and Klein 1974). Biological oxidation of coke plant wastewaters in a 24-hr test accomplished the following results (Jüntgen and Klein 1974):

Compound	Percent removal
Phenols	99.9
Ammonia	90
Chemical oxygen demand	80
Cyanides	57
Thiocyanates	17

At the Shell Oil Company refinery in Anacortes, Washington, the five-day reduction in biological oxygen demand was from 175 to 25 mg/liter, and phenol reduction was from 30 to 0.6 mg/liter (Newman, Reno, and Burroughs 1958). Phenolic waste liquor from an ammonia still treated by a three-stage (two activated sludge tanks and one percolating filter) system removes input phenols at concentrations of 400 to 700 ppm to 30 ppm and less (Clough 1961).

Trickling filters are popular for treating oil refinery wastes because of their ability to resist shock loads of toxic organics. Oil-oxidizing bacteria require dissolved oxygen or an oxygen source such as nitrate or sulfate and an optimum temperature of 25 to 37°C (Reid and Streebin 1972). However, they may allow the toxic materials to pass through the filters. The use of plastic media in trickling filters enables reduction in land area and higher loading capacities (Reid and Streebin 1972).

Activated sludges using complete-mixing systems can absorb shock loads and handle higher organic concentration than can conventional activated sludges. Again, adequate microbial oxygen transfer is vital. Complete-mixing activated sludge can be used to produce any degree of phenol reduction required (up to 99.9%) (Reid and Streebin 1972).

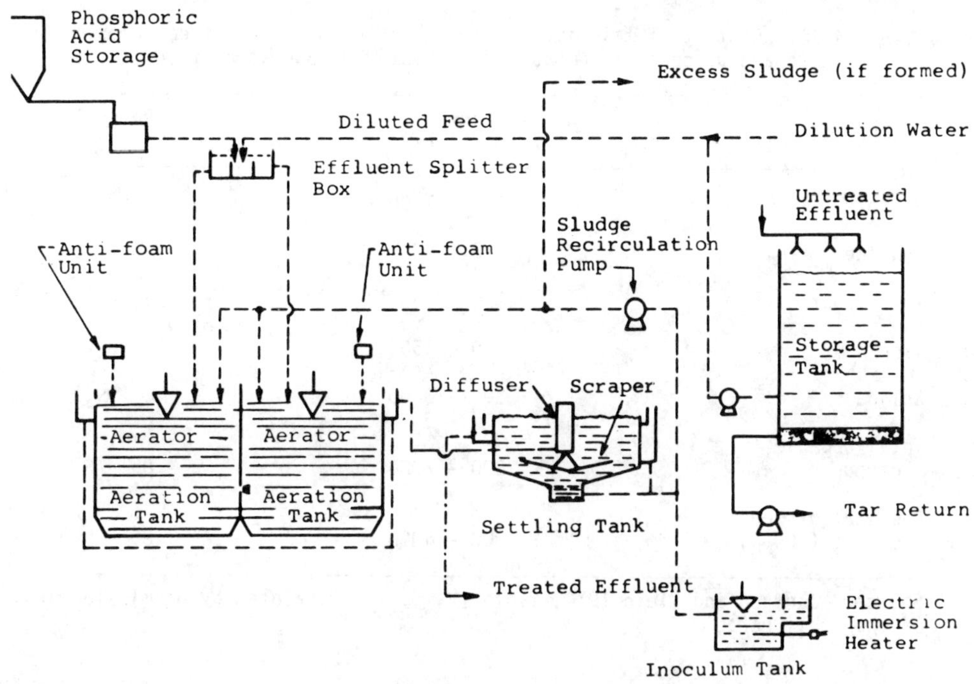

Fig. 4.64. Flow diagram of phenol biological processing plant at Dominion Foundries and Steel of Hamilton, Ontario. Source: From Ludberg and Nicks 1969, Fig. 2, p. 11.

Tertiary (final) organic wastewater facilities may include chemical treatment, carbon absorption, granular media filtration, or biological treatment. Chemical agents such as chlorine, chlorine dioxide, or ozone, which are used for destruction of waste phenols from iron plants, may be used by conversion plants. Chlorination uses the strong oxidizing power of chlorine for treatment of residual organics (Sittig 1974). However, the use of chlorine on phenols and ammonia can produce chloro-compounds (Nebolsine 1957; Stasiuk, Hetling, and Shuster 1974), and chlorine dioxide and ozone are expensive.

The activated carbon process uses the adsorptive affinity of carbon for the removal of contaminants (Sittig 1974). Preliminary tests on a semitechnical coal gasification plant show that removal of organic impurities (phenolic wastes) by adsorption on activated carbon is possible (Jüntgen and Klein 1974). Activated carbon is effective in removing some wastewater impurities in coking plant wastes (Tables 4.70 and 4.71). However, Jüntgen and Klein (1974) report a 2.0 to 2.5% carbon loss due to inactivation. Char from some of the conversion processes may have the attributes of activated carbon for removal of impurities (Nandi and Walker 1972).

Granular media filtration, such as rapid sand filtration, can be used to polish wastewater effluents; using gravity and pressure, rapid sand filtration removes suspended solids (Sittig 1974). Biological treatment facilities for the treatment of residual organics include oxidation ponds and aerated lagoons. Oxidation ponds require long retention times (60 days) but little operational control. However, effluent quality fluctuates radically from summer to winter, corresponding to temperature fluctuations. Aerated lagoons permit additional organic loading over that for oxidation ponds because of the use of aeration units to supply adequate oxygen

Table 4.70. Content of main impurities in coking pilot plant decanter waste and percent of removal by activated carbon adsorption

Impurities	Content (mg/liter)	Removal (%)
Phenols	650 - 1400	>99
Cyanide	5 - 35	45 - 70
Thiocyanate	120 - 450	30 - 80
Iron	40 - 150	30
Solids	300 - 3000	>99
Total organic carbon	800 - 2000	85 - 95
Chemical oxygen demand (COD_{Mn})	2000 - 4000	80 - 99

Source: Jüntgen and Klein 1974, Table IV, p. 72. Reprinted by permission of the publisher.

Table 4.71. Treatment of coke plant waste effluents with activated carbon

Component	Quantity (mg/liter[a])		
	Raw wastewater	Wastewater after clarification - filtration	Wastewater after adsorption
pH	7.2	8.0	8.0
Suspended solids	50	<5	<5
Color, APHA units	70	70	0
Chemical oxygen demand	6600	6340	1260
Total organic carbon	2100	1750	156
Soluble organic carbon	1900	1750	156
Phenol	2235	1950	<0.1
Cyanide	0.01	0.01	0.01
Ammonia	4000	4000	4000
Thiocyanate	700	700	<700

[a]Except for pH and color.

Source: van Stone 1972, Tables I, II, and III, pp. 63-4.

transfer, thus providing a high degree of waste stabilization. The use of three aerated lagoons removes 98% of phenols and 100% of sulfides (Reid and Streebin 1972).

The use of water hyacinth (*Eichhorinia crassipes*) sewage filtration to absorb water pollutants is being evaluated (Wolverton 1975). Water hyacinth roots absorb lead, mercury, strontium-90, phenols, and some organic compounds believed to be carcinogenic (Water Hyacinths Soak up Pollution 1976; Wolverton 1975). For example, water hyacinths removed 36 mg of phenol per gram dry weight of plant material in 72 hr. Thus, one hectare of water could remove 160 kg of phenol in a 72-hr period (Wolverton 1975). Plants could then be converted to methane and used for fertilizer, poultry feed or cattle-feed additive, or as plant mulch (Water Hyacinths Soak up Pollution 1976).

Serial treatment processes using combinations of those described for wastewater may prove to be the most efficient. When using biological oxidation followed by activated carbon treatment, 99.97% phenol removal efficiency was reached, whereas efficiencies of 99.6 and 99.1%, respectively, were accomplished by the single processes (Short, DePrater, and Myers 1974).

4.5.6.2 PAH wastewater treatment

Although PAH have a low solubility in water, they may occur in conversion streams as aqueous solutions associated with or adsorbed onto a variety of colloidal materials and thereby be transported through the water environment (Andelman and Suess 1970). The efficacy of the various wastewater treatment technologies for the removal of PAH in wastewater depends primarily on the state (dissolved or suspended) of PAH in the system (Andelman and Snodgrass 1974).

Separation processes, such as sedimentation and filtration with sand or activated carbon, can be effective for portions of the PAH adsorbed on particles (Andelman and Snodgrass 1974). Activated carbon adsorbs up to 99% of the PAH in the influent, but can leave a residual in the effluent (McGinnes and Snoeyink 1974). PAH reduction of 80 to 90% was accomplished by primary and secondary sedimentation of sewage (Reichert et al. 1971, as cited in Andelman and Snodgrass 1974). Sedimentation (60 min) and sand filtration tests by Il'nitskii (1969) showed 57 and 99% BaP removal efficiencies for high initial concentrations and 54 ± 16% for sedimentation of low initial concentrations.

For dissolved PAH, however, mechanical separation processes have little effect (Andelman and Snodgrass 1974), and further processing may be required.

Although bacterial cells are capable of converting various PAH (Chap. 7), biological treatment of PAH appears to be ineffective (Andelman and Snodgrass 1974). Petrochemical effluents containing BaP (up to 27 mg/liter) and BaA (up to 2.8 mg/liter) were subjected to biochemical purification in aeration tanks prior to discharge, but still contained up to 1.7 μg/liter of BaP and up to 0.6 μg/liter of BaA after discharge (Ershova 1971). Activated sludge treatment of domestic sewage from Nashville, Tennessee, was unable to effect significant removal of PAH by oxidation within normal detention times (Malaney et al. 1968). Reduction of PAH in the sewage was attributed to adsorption by the activated sludge (Andelman and Snodgrass 1974).

Chlorination of water lowered, but did not completely eliminate, the concentration of BaP (Trakhtman and Manita 1965). After chlorination, the water contained the products of the reaction, assumably 5-mono-chloro-3,4-benzpyrene and 3,4-benzpyrenequinone (Trakhtman and Manita 1965). Sensitivities to chlorine treatment varied with respect to PAH type and various water types (deionized water and doubly distilled water) (Sforzolini et al. 1974). Average PAH removal efficiencies reported in the two water types, when using chlorine (2 ± 0.25 mg/liter) for 30 min, showed BaP to be the compound most sensitive to treatment (Table 4.72) (Sforzolini et al. 1974).

Table 4.72. Average PAH removal efficiencies, when using chlorine, in deionized and doubly distilled water

| Compound | Decrease (%) | |
	Doubly distilled water	Deionized water
Pyrene	37.34	22.30
Benzo[*a*]anthracene	82.65	75.62
Benzo[*a*]pyrene	100	100
Benzo[*b*]fluoranthene	15.18	26.25
Benzo[*k*]fluoranthene	22.71	38.69

Source: Modified from Sforzolini et al. 1974.

Free chlorine was more efficient than chloramines in removing BaP, and its efficiency was only slightly improved by increasing the chlorine doses (Gabovich, Vrochinskii, and Kurinnyi 1969). Similarly, Il'nitskii et al. (1971) showed that chlorination of aqueous BaP solution (initial concentration 1×10^{-7} g/liter), after 30-min contact with 0.5 mg/liter residual chlorine, left 81% of the BaP; contact with 1 mg/liter residual chlorine left 76%.

Ultraviolet radiation tests also showed that PAH compounds differed in their stabilities. Decreasing stability was noted for the following compounds: BaP, pyrene, dibenzo[*a,h*]anthracene, perylene, benz[*a*]anthracene, and 7,12-dimethylbenz[*a*]anthracene (Il'nitskii et al. 1971). Reduction of aqueous BaP in solution by ultraviolet radiation approximating conditions found in waterworks was fairly effective; irradiation for 2.5 min destroyed 60 to 80% of the BaP, and for 15 min, destroyed 80 to 90% (Il'nitskii et al. 1971).

Ozonation tests also showed different resistances with respect to PAH type and, to a lesser extent, to water type. Ozone resistance diminished along the series BaP > BaA > 9,10-dimethyl-1,2-benzanthracene (Gabovich, Vrochinskii, and Kurinnyi 1969). Solutions of BaP, 9,10-dimethyl-1,2-benzanthracene, 1,2-benzanthracene, 1,2,5,6-dibenzanthracene, and pyrene in water studies showed that all were inactivated by contact with ozone for 2.5 min, but that resistances varied, BaP being the most stable and DMBA being the least resistant (Il'nitskii et al. 1968). Additionally, BaP that is sorbed on soil particles is more stable, with respect to ozone resistance, than BaP in solution (Il'nitskii et al. 1968). Ozone at a concentration of 0.4 ± 0.05 mg/liter acted effectively on PAH in both doubly distilled and deionized water (Sforzolini et al. 1974).

The PAH most sensitive to ozone action were BaP and benzo[k]fluoroanthene, each having 100% removal efficiencies in both water types. The BaA and pyrene decreased 100% in doubly distilled water and 87.72 and 74.89%, respectively, in deionized water; benzo[b]fluoranthene decreased 80.81% in doubly distilled water and 70.56% in deionized water (Sforzolini et al. 1974).

Ozonation of BaP proved to be more effective than chlorination and ultraviolet irradiation (Il'nitskii et al. 1971).

Ozone may be the most effective treatment process for the reduction of PAH in water for drinking purposes (Andelman and Snodgrass 1974). However, the efficacy of ozone treatment may be reduced by the presence of other organic compounds (Sforzolini et al. 1974; Andelman and Snodgrass 1974). The efficacy results of chlorination, ultraviolet irradiation, and ozonation may be variable with respect to PAH and water type. Lower removal efficiencies than those required for drinking water may be sufficient for conversion of streams contained in-house.

Serial treatments recommended for removal of PAH in wastewater include (1) sand filtration, activated carbon treatment, and chlorination for the removal of carcinogenic PAH compounds from river water with small loads, and (2) flocculation and precipitation, sand filtration, activated carbon treatment, and chlorination for larger PAH loads (Il'nitskii 1969). Flocculation, using flocculant VA-2 in a concentration of 2 mg/liter, followed by sand filtration, eliminated practically all (99%) of the suspended BaP in wastewaters having initially high BaP concentrations (Il'nitskii 1969). Dissolved PAH, however, will pass through the stages of flocculation and filtration without any considerable decrease (Il'nitskii 1969). A flocculation plant removed 98.5% of PAH when using ferrous sulfate and chlorine followed by ozone and ferric chloride, and then sand filtration (Borneff 1969, as cited in Andelman and Snodgrass 1974).

4.6 SPENT CATALYSTS

A variety of catalysts will be used widely in coal conversion technologies (Table 4.73). As chemical reactants or purification sorbents, catalysts deactivate with use and require regeneration or disposal. Regeneration frequently involves return to the supplier; disposal frequently involves return to the mine with the ash. Catalysts can be removed as spent catalysts, lost through abrasion and escape into the ash, or incorporated into product streams as ultrafine particulate matter (Fulkerson et al. 1974b).

Because of contamination, catalyst activity varies inversely with increasing operating time. Free carbon, sulfur, and chlorine can poison catalysts; additionally, in the hydrodesulfurization feed streams, organometallic compounds are a primary cause of catalyst deactivation. The rate of deactivation is influenced by feedstock characteristics, catalyst characteristics, and operating severity (Gregoli and Hartos 1972). Preliminary data using spark-source mass spectrometry analysis indicate that many of the elements detected in the original coal are found on used-catalyst surfaces (Sharkey et al. 1975). The removal and inerting of the COED hydrotreating catalyst (cobalt-molybdenum and nickel-tungsten) may result in ammonium sulfide and metal carbonyl emissions, which require treatment in gas-cleanup sections (Kalfadelis and Magee 1975).

Table 4.73. Conversion catalysts

Catalysts	Use
Activated carbon	Purification
Iron oxide	Purification
Methanol	Purification
Propylene carbonate	Purification
Sodium carbonate	Purification
Potassium carbonate	Purification
Amines	
Monethanolamine	Purification
Diethanolamine	Purification
Diglycolamine	Purification
Zinc oxide	Purification
Cobalt-molybdenum	Shift conversion, liquefaction (hydrotreating), purification
Limestone-dolomite	Sulfur recovery
Molten salt	
Nickel	Methanation or liquefaction
Vanadium	
Dolomite	Purification
Bauxite	Sulfur recovery
Iron	Shift conversion or liquefaction
Isopropyl ether	Phenol recovery
n-Methyl-2-pyrrolidine	Purification
Dimethyl ether polyethylene glycol	Purification
K_3AsO_3	Liquefaction
Tungsten	Liquefaction
Zinc chloride	Liquefaction
Sodium sulfite	Purification
$Co-Mo/SiO_2-Al_2O_3$	Liquefaction
Sulfoxide	Sulfur recovery
Chelated iron salt	Sulfur recovery
Nickel-tungsten	Liquefaction (hydrotreating)
Ruthenium	Methanation

In the petroleum industry, the formation of BaP and other polycyclic organic matter has been noted (National Academy of Sciences 1972). Trace elements may also be released by volatilization during catalyst regeneration (Fulkerson et al. 1974b).

Environmental problems can arise by unintentional loss of desulfurization or hydrogenation catalysts. For example, catalyst addition for the H-Coal process pilot plant was reported to be about 1 lb per ton of coal processed (Fulkerson et al. 1975b). Catalyst losses would amount to about 700 lb of cobalt and 1 ton of molybdenum for the cobalt-molybdenum catalyst (Schuit and Gates 1973). The losses of these elements, if primarily nonvolatile, would increase their concentrations in the ash. Dolomite losses by the CO_2 Acceptor processes amount to 162.810 lb/hr, or about 2% of the input (7,164,000 lb/hr) (Jahnig and Magee 1974).

4.7 RADIOACTIVITY

The natural radioactivity of coal varies and is mainly due to uranium and thorium in concentrations usually no greater than 400 to 500 ppm. Most of the radioactivity is contained in the escaping fly ash or in the ash collected by electrostatic precipitators. Radium fly-ash concentrations vary from 0.4 to 14 pCi/g, a value about twice that found in collected ash. The potential radionuclide hazard of fly ash is insignificant because of high dilution factors.

Modern air cleaning operations at coal-fired plants reduce the radioactivity emitted in stack effluents by high fly-ash removal efficiencies. The recovered ash contains large quantities of alpha-emitting nuclides. Disposal or use of this ash could have significant radiological consequences (Kerr et al. 1975). Unknown concentrations of radium will be contained in the conversion process streams (Kerr et al. 1975). Bayliss and Whaite (1966), in their study of a Sydney power station, concluded that the alpha activity (4.2×10^{-17} μCi/cm^3 of air) associated with the fly ash represented a negligible radiological health hazard.

Bedrosian's (1970) radiological study of the Widow's Creek Steam Station (coal-fired) determined that the hourly lung dose emitted was from 0.28 to 1.23 μrem, 17 μrem/hr being the radiation guide value for the general population.

4.8 CARBON DIOXIDE

Carbon dioxide may be the largest single waste material (by weight) produced by high-Btu gasification, and, although the compound is not toxic, continuous discharge under stagnant atmospheric conditions could cause the formation of a carbon-dioxide-enriched air layer (Hittman Associates 1975*b*). Concern has been voiced over the theories of a long-term climatic effect known as the greenhouse effect, from increased carbon dioxide concentrations.

Using feed gas to produce hydrogen for hydrotreating of the SRC product will result in about 809 tons of carbon dioxide being vented to the atmosphere per day (Jahnig 1975*a*).

Manufacture of hydrogen may include steam reforming, cryogenic separation, or partial oxidation. The COED process uses steam reforming of the process gases to produce 914 tons/hr of hydrogen with major gaseous effluents being the products of combustion from the product-gas-fired heaters and the carbon dioxide stream (60 tons/hr) removed from the process gas after reforming (Kalfadelis and Magee 1975). The Wesco Coal Gasification Plant has a Rectisol offgas carbon dioxide stream amounting to 284 million scfd (Berty and Moe 1974).

4.9 THERMAL EFFLUENTS

Like any other manufacturing industry, coal conversion will probably produce thermal effluents, which can be given off through heat from cooling water and product streams. Heat can also be lost in the extraction and purification of the conversion products. Lost energy is dissipated to the environment.

The level of thermal pollution is inversely proportional to the process efficiency (Glaser, Hershaft, and Shaw 1974).

Thermal efficiency values for various conversion processes are given in Table 4.74 as ranges because of the variability of process designs concerning the inclusion by-products. Factors affecting thermal efficiency include the heating value of the feed coal and of the products, marketability, and heating value of the by-products, conversion process design relating to input power requirements (coal drying, process reheat and gas compression requirements, etc.), and utility requirements.

Table 4.74. Thermal efficiency for various processes

Process	Thermal efficiency (%)	Reference
Lurgi	52.9 - 66.6	Shaw et al. 1974
Lurgi	68 - 72	Rudolph 1974
COED	57.8 - 72.2	Kalfadelis and Magee 1975
CO_2 Acceptor	60.2 - 76	Jahnig and Magee 1974
Koppers-Totzek	53 - 69	Magee, Jahnig, and Shaw
SRC	60.3 - 65.5	Jahnig 1975*a*
Bi-Gas	65 - 70 (77% if low-Btu gas is produced)	Hittman Associates 1975; Jahnig 1975*b*
Bi-Gas	60 - 66.0	Kalfadelis and Magee 1975

The thermal efficiencies (the combustion heat in the product gas divided by the combustion heat quantity contained in the raw coal feed) determined for gasification processes range from 46.4 to 77.8% (Wen 1975). Greater thermal efficiencies can be attained by the direct feeding of raw coal to the gasifiers, thereby eliminating coal pretreating. However, associated mechanical and design problems may affect the higher thermal efficiencies (Wen 1975). Optimization studies showed higher efficiencies for processes using methane formation in the gasification subsystem rather than in the methanation units (Wen 1975).

Thermal discharges amount to about 0.4 to 0.8 units of energy for each unit of clean energy produced (Glaser, Hershaft, and Shaw 1974). Nuclear power plants operate at 30 to 33% efficiency, and fossil fuel power plants at about 38% (Glaser, Hershaft, and Shaw 1974).

4.10 NOISE

Noise pollution associated with coal conversion will probably originate primarily from coal preparation (grinding and drying operations). Coal gasification facilities will consume up to 20,000 tons of coal per day, requiring the use of multiple grinding units (the current capacity is 200 tons/hr). Noises can also be generated in conversion areas using pumps, fans, compressors, gasifiers, pressure relief valves, fired heaters, gas turbines, and other mechanical equipment. Noise from individual equipment components may not represent the total noise level, which includes all equipment items, motor drives, piping, ductwork, reverberations from adjacent equipment and buildings, and sound interferences from different sound sources. Noise sources from a 5000-ton/day coal-burning power plant are shown in Fig. 4.65. The study of this facility indicated that over one-third of the working population received unacceptably high exposures (Broderson, Edwards, and Green 1975). Excessive noise from an SNG-producing plant has been reported (Anderson 1974).

Noise control at the source is an engineering problem which requires modification or redesigning of the source. Noise reduction along the path which the sound travels can be accomplished by shielding or enclosing the source and the receiver (Cohen 1976, personal communication). Standard engineering practices such as sound-absorbing insulation, walls, and mufflers are expected to be

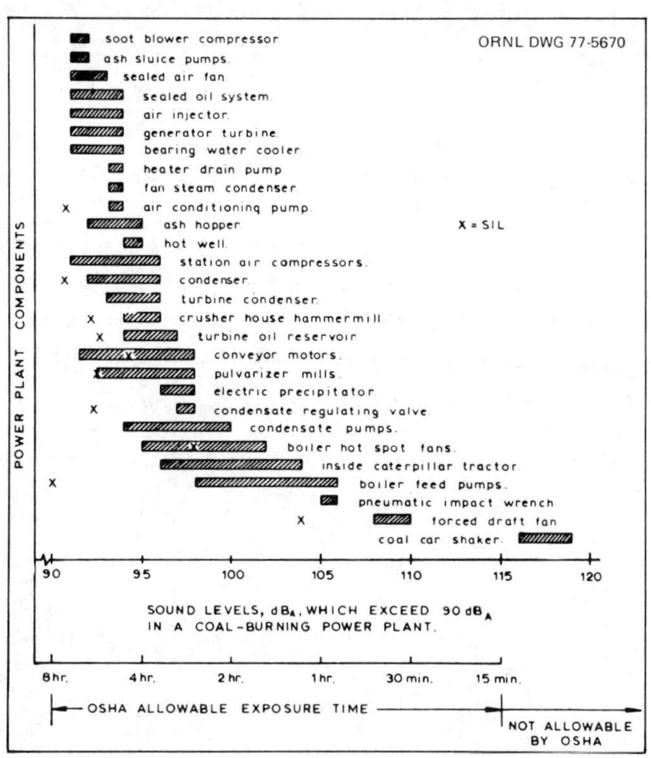

Fig. 4.65. Noise sources in the power plant. <u>Source</u>: From Broderson, Richards, and Green 1975, Fig. 7, p. 24. Reprinted by permission of the publisher.

included in conversion designs. Additional personnel precautions may be needed in the area with excessive noise; ear plugs, booth enclosure, other personal equipment, and regulation of exposure time are expected to be provided.

APPENDIX 4A

Conversion Process Size Descriptions from Literature

To enable general comparisons among material inputs, product yields, and generated wastes associated with coal conversion activities, this appendix is provided for use with Chap. 4. Given in tabular data are the various conversion processes cited in the text and their related literature sources, the type, characteristics, and amount of coal entering the conversion operation, and the type and amount of final product.

Process	Source	Coal type	Coal characteristics	Input	Output
Bi-Gas	1	Appalachian	3.4% moisture 1.8% sulfur 7.7% ash	66,300 tons/day	1.146×10^9 scfd SNG
		Eastern Interior	11.20% moisture 3.5% sulfur 9.4 ash	77,400 tons/day	1.108×10^9 scfd SNG
		Fort Union	37.50% moisture 0.6% sulfur 6.2% ash	131,100 tons/day	1.103×10^9 scfd SNG
		Powder River	19.10% moisture 0.6% sulfur 5.3% ash	92,100 tons/day	1.124×10^9 scfd SNG
		Four Corners	11.40% moisture 0.7% sulfur 14.10% ash	88,600 tons/day	1.113×10^9 scfd SNG

Bi-Gas	2	Western Kentucky	12,330 Btu/lb HHV[a]
			8.4% moisture
			3.48% sulfur
			6.7% ash
			dry analysis
			  13,285 Btu/lb HHV
			  1.3% moisture
			  3.75% sulfur
			  7.2% ash
				23,243 tons/day to gasifiers (14,535 tons/day after cleaning)	250×10^6 scfd pipeline gas
					943 Btu/ft^3

Bi-Gas	3	Pittsburgh seam	7.03% ash 2.46% sulfur	18,267 tons/day (gasifiers: 12,000; auxiliary purposes: 1700; washing refuse: 4567 tons/day)	250×10^6 scfd SNG

CO_2 Acceptor	4	North Dakota lignite	3927 kcal/kg (7068 Btu/lb)
			33.67% moisture
			0.59% sulfur
			7.21% ash
			dry analysis
			  0.89% sulfur
			  10.87% ash
				1129 tonnes/hr (1244 tons/hr)	260×10^6 cfd high-Btu gas
					8474 kcal/m^3 (952.3 Btu/ft^3)

Process	Source	Coal type	Coal characteristics	Input	Output

CO$_2$ Acceptor	5	Lignite	33.67% moisture
			dry analysis
			  0.89% sulfur
			  10.86% ash
			preheated lignite feed
			  11,120 Btu/lb HHV
			  0.90% sulfur
			  11.45% ash
				28,517 tons/day	262.2 × 10^6 scfd SNG
					  952 Btu/scf HHV

COED	4	Utah A	7517 kcal/kg 6.0% moisture 0.5% sulfur 6.1% ash	402.5 tonnes/hr (443.7 tons/hr)	591.7 bbl/day Syncrude oil 820,800 ft^3/hr low-Btu fuel gas 187.1 tonnes/hr char (206.2 tons/hr)
		Illinois No. 6	6956.1 kcal/kg 11.0% moisture 3.7% sulfur 12.4% ash	419.6 tonnes/hr (462.5 tons/hr)	377.9 bbl/day Syncrude oil 562,500 ft^3/hr low-Btu fuel gas 208.8 tonnes/hr char (230.2 tons/hr)

COED	6	Illinois	12,500 Btu/lb
			3.5% sulfur
				25,000 tons/day	26,000 bbl/day oil
					  5.9 × 10^6 Btu/bbl
					  0.3% sulfur
					1330 × 10^6 scfd gas
					  215 Btu/scf
					18,900 bbl/day plant fuel oil
					  0.3% sulfur

COED	7	Illinois No. 6	12,420 Btu/lb HHV
			5.9% moisture
			3.8% sulfur
			10.6% ash
			dry analysis
			  4.1% sulfur
			  11.3% ash
				25,512 tons/day	3945 tons/day hydrotreated oil
					(24,925 bbl/day)
					  19,100 Btu/lb
					123 tons/hr product gas
					  505 Btu/scf
					12,512 tons/day product char
					  11,040 to 11,700 Btu/lb

COED	8	Utah	13,900 Btu/lb HHV
			6.4% moisture
			6.6% sulfur
			6.4% ash
				24,000 tons/day	31,700 bbl/day syncrude
					113 million scfd net
					  pyrolysis gas[b]
					  510 Btu/scf HHV
					13,200 tons/day char

Process	Source	Coal type	Coal characteristics	Input	Output

		Illinois No. 6	12,150 Btu/lb HHV
			14% moisture
			4.1% sulfur
			12.1% ash
				24,000 tons/day	26,000 bbl/day syncrude
					112 million scfd net
					  pyrolysis gas[b]
					  510 Btu/scf HHV
					12,950 tons/day char

CSF	6	Pennsylvania	10,830 Btu/lb 4.0% sulfur	23,360 tons/day	47,600 bbl/day oil 6×10^6 Btu/bbl 251×10^6 scfd net product gas 12,700 bbl/day residuum fuel oil 98×10^6 scfd fuel gas

Hygas	4	Montana subbituminous	$13,332 \times 10^6$ Btu/hr
			22% moisture
			0.51% sulfur
			dry analysis
			  0.66% sulfur
			  7.16% ash
				21,481 tonnes/day
				(23,678 tons/day)
					250×10^6 cfd pipeline gas
					966 Btu/ft^3

Hygas	4	Illinois No. 6	13,332 Btu/hr
			6.5% moisture
			3.93% sulfur
			dry analysis
			  4.20% sulfur
			  11.54% ash
				15,891 tonnes/day
				(17,517 tons/day)
					250×10^6 cfd pipeline gas
					963 Btu/ft^3

Koppers-Totzek	9	Bituminous	8830 Btu/lb HHV
			16.5% moisture
			0.63% sulfur
			17.3% ash
				8110 tons/day
				  (gasifier: 6750;
				  boiler: 1360 tons/day)
					290×10^6 cfd product gas
					303 Btu/ft^3 HHV
					300 ppm sulfur

Lurgi	4	Navajo	4813 kcal/kg (8664 Btu/lb) 16.25% moisture 0.69% sulfur dry analysis 0.82% sulfur	25,626 tonnes/day (28,249 tons/day)	288×10^6 cfd pipeline-quality gas 8489 kcal/m^3 (954 Btu/ft^3)

Process	Source	Coal type	Coal characteristics	Input	Output
El Paso Burnham I Gasification (Lurgi) Complex	10			28,254 tons/day	288×10^6 scfd SNG
Wesco Coal Gasification (Lurgi) Plant	11			21,800 tons/day to gasifiers	250×10^6 scfd SNG
Lurgi	6	New Mexico	0.91% sulfur	25,620 tons/day (gasifiers: 21,860; boilers: 3760 tons/day)	250×10^6 scfd SNG 60×10^9 Btu/day by-product fuel
Lurgi	12	Navajo	7500 to 10,250 Btu/lb HHV 16.5% moisture 0.95% sulfur 17.3% ash	26,000 tons/day	250×10^6 scfd SNG
SRC	4	Kentucky No. 11	6624 kcal/kg (11,924 Btu/lb) 9.51% moisture 3.14% sulfur 6.63% ash	9754 tonnes/day (10,752 tons/day)	5316 tonnes/day SRC (5860 tons/day) 8760 kcal/kg (15,768 Btu/lb) 1.17% sulfur 0.48% ash
SRC	1	Appalachian	3.4% moisture 1.8% sulfur 7.7% ash	66,300 tons/day	31,560 tons/day SRC 5909 tons/day light oils
		Eastern Interior	11.20% moisture 3.5% sulfur 9.4% ash	77,400 tons/day	30,240 tons/day SRC 5651 tons/day light oils
		Fort Union	37.50% moisture 0.6% sulfur 6.2% ash	131,100 tons/day	29,624 tons/day SRC 5547 tons/day light oils
		Powder River	19.10% moisture 0.6% sulfur 5.3% ash	92,100 tons/day	32,289 tons/day SRC 6044 tons/day light oils

Process	Source	Coal type	Coal characteristics	Input	Output
		Four Corners	11.40% moisture 0.7% sulfur 14.10% ash	88,600 tons/day	30,638 tons/day SRC 5726 tons/day light oils

SRC	2	Illinois type bituminous	12,821 Btu/lb HHV
			dry analysis
			  2.7% moisture
			  3.38% sulfur
			  7.13% ash
				12,500 tons/day
				(converts 10,000 tons/day)
					25,000 bbl/day heavy clean
					  liquid fuel
					    0.5 to 0.2% sulfur
					2915 tons/day heavy liquid
					  16,660 Btu/lb HHV
					1442 tons/day hydrotreated liquid
					  18,330 Btu/lb HHV
					272 tons/day hydrogenated light
					  oils

Synthane	4	Pittsburgh	2.5% moisture 1.6% sulfur 7.4% ash	12,930 tonnes/day (14,250 tons/day) 91.22 $\times$ 10^9 kcal (362.0 $\times$ 10^9 Btu)	250 $\times$ 10^6 cfd SNG 231.8 $\times$ 10^9 Btu
Synthane	4	Illinois No. 6	2.9% sulfur		

Synthane	13	Pittsburgh seam	13,701 Btu/lb
			2.5% moisture
			1.6% sulfur
			7.4% ash
				14,250 tons/day to
				  gasifiers
					250 $\times$ 10^6 scfd product gas
					  927 Btu/scf HHV
					4350 tons/day char
					  3980 $\times$ 10^6 Btu/hr

Synthoil	14	Wyodak	29.0% moisture
			0.7% sulfur
			6.6% ash
			dry analysis
			  0.5% moisture
			  1.0% sulfur
			  9.2% ash
				1,336,400 lb/hr	50,000 bbl/stream-day

[a]HHV = higher heating value (also gross heating value); the total heat released when a fuel is burned.
[b]Gas yield after supplying plant fuel gas and feedstock to the hydrogen plant.

Source:
1. Hittman Associates 1975a
2. Jahnig 1975
3. Grace and Diehl 1974
4. Glaser, Hershaft, and Shaw 1974
5. Jahnig, Magee, and Kalfadelis 1974
6. Beychok 1975
7. Kalfadelis and Magee 1975
8. Hamshar, Terzian, and Scotti 1974
9. Magee, Jahnig, and Shaw 1974
10. Gibson, Hammons, and Cameron 1974
11. Berty and Moe 1974
12. Shaw and Magee 1974
13. Kalfadelis and Magee 1974b
14. Bureau of Mines 1975

LITERATURE CITED

Abernethy, R. F.; Peerson, M. J.; and Gibson, F. H. 1969*a*. *Major ash constituents in U.S. coals*. U.S. Bureau of Mines Report of Investigations 7420.

Abernethy, R. F.; Peterson, M. J.; and Gibson, F. H. 1969*b*. *Spectrochemical analysis of coal ash for trace elements*. U.S. Bureau of Mines Report of Investigations 7281.

Adams, C. E., Jr. 1973. Removing nitrogen from wastewater. *Environ. Sci. Technol.* 7(8): 696-701.

Akhtar, S.; Freidman, S.; and Yavorsky, P. M. 1975. Environmental aspects of Synthoil process for converting coal to liquid fuels. In *EPA symposium, environmental aspects of fuel conversion technology II, Hollywood, Fla., Dec. 15-18, 1975.*

Andelman, J. B., and Snodgrass, J. E. 1974. Incidence and significance of polynuclear aromatic hydrocarbons in the water environment. *CRC Crit. Rev. Environ. Control* 4(1): 69-83.

Andelman, J. B., and Suess, M. J. 1970. Polynuclear aromatic hydrocarbons in water environment. *Bull. World Health Organization* 43: 479-508.

Anderson, D. E. 1974. First large scale SNG plant. *Oil Gas J.* 72(3): 74-6.

Anderson, H. C., and Wu, W.R.K. 1963. *Properties of compounds in coal carbonization products.* U.S. Bureau of Mines Bull. 606. Washington, D.C.: U.S. Government Printing Office.

Annessen, R. J., and Gould, G. D. 1971. Sour-water processing turns problem into payout. *Chem. Eng.* 78: 67-69.

Attari, A. 1973. *Fate of trace constituents of coal during gasification.* EPA-650/2-73-004. PB-223 001.

Badger, G. M. 1962. Mode of formation of carcinogens in humans. Environment. National Cancer Institute Monograph 9, 1. In IARC Monographs.

Ball, F. J.; Brown, G. N.; Davis, J. E.; Repik, H. J.; and Torrence, S. L. 1972. Recovery of sulfur dioxide from stack gases as elemental sulfur by a dry fluidized activated carbon process. In *Pollution control and energy needs, Adv. Chem. Ser.* 127, ed. R. M. Jimeson and R. S. Spindt, pp. 183-94. Washington, D.C.: American Chemical Society.

Barry, C. B. 1972. Reduce Claus sulfur emission. *Hydrocarbon Process.* 51(4): 102-6.

Barthel, Y.; Bistri, Y.; Deschamps, A.; Renault, P.; Simadoux, J. C.; and Dutriau, R. 1971. Treat Claus tail gas. *Hydrocarbon Process.* 50: 89-91.

Bartok, W.; Crawford, A. R.; and Skopp, A. 1971. Control of NO_x emissions from stationary sources. *Chem. Eng. Prog.* 67(2): 64-76.

Bayliss, R. I., and Whaite, H. M. 1966. A study of the radium alpha activity of coal, ash, and particulate emission at a Sydney power station. *Air Water Pollut.* 10: 813-19.

Bazelmans, C. L.; Birford, F. T.; Dunley, R. D.; Haas, P. A.; Hightower, J. R.; Holmes, J. M.; Morris, E. B.; Nichols, J. P.; Ruggeri, S.; and Solmon, R. 1973. *Study of options for control of emissions from an existing coal-fired electric power station.* ORNL/TM-4298. Oak Ridge, Tenn.: Oak Ridge National Laboratory.

Bedrosian, P. H.; Easterly, D. G.; and Cummings, S. L. 1970. *Radiological survey around power plants using fossil fuel.* EERL 71-3. PB-20 2414.

Beers, W. D. 1973. *Characterization of Claus plant emissions.* EPA-R2-73-188. PB-220 376.

Bennett, W. S.; Kerr, D. M.; and Kolstad, C. D. 1976. *Wesco coal gasification plans: Navajo considerations.* LA-6247-MS.

Berty, T. E., and Moe, J. M. 1974. Environmental aspects of the Wesco coal gasification plant. In *Symposium proceedings: environmental aspects of fuel conversion technology (May 1974, St. Louis, Missouri)*, pp. 101-06. EPA-650/2-74-118.

Beychok, M. R. 1975. *Process and environmental technology for producing SNG and liquid fuels.* EPA-660/2-75-011.

Berestovoi, P. G.; Obruchev, I. S.; and Rosov, M. N. 1953. Use of coal-washing wastes in raw mixes for cement. *Tsement* 19(1): 27-29.

Bolton, N. E.; Carter, J. A.; Emery, J. F.; Feldman, C.; Fulkerson, W.; Hulett, L. D.; and Lyon, W. S. 1973. Trace element mass balance around a coal fired steam plant. *ACS Div. Fuel Chem.* 18(4): 114-23.

Borneff, J. 1969. Elimination of carcinogenic polycyclic aromatic compounds during water purification. *Gas-Wasserfach* 110(2): 29. In Andelman and Snodgrass 1974.

Boyland, E.; Dentwill, W.; Falk, H. L.; Gori, B. G.; Kipling, M. D.; Miller, R. W.; Terracini, B.; and Cooper, J. 1972. *IARC Monographs on the evaluation of the carcinogenic risk of chemicals to man: certain polycyclic aromatic hydrocarbons and heterocyclic compounds*, vol. 3. IARC working group on the evaluation of carcinogenic risk of chemicals to man, Lyon, Dec. 5-11.

Bracket, C. E. 1973. Production and utilization of ash in the United States. In *Ash utilization proceedings: 3rd international ash utilization symposium, Mar. 13-14, 1973*, ed. J. H. Faber, W. E. Eckard, and J. D. Spencer, pp. 12-18. U.S. Bureau of Mines Inf. Circ. 8640.

Broderson, A. B.; Edwards, R. G.; and Green, W. W. 1975. Noise dose and hearing loss in a coal-burning power plant. *Sound Vib.* 9: 22-30.

Browning, J. E. 1973. Ash — the usable waste. *Chem. Eng.* 80: 68-70.

Bureau of Mines. 1974. *Removal of hydrogen sulfide from hot producer gas by solid absorbents.* Report of Investigations 7947.

Buu-Hoi, N. P. 1958. Presence of 3:4-8:9-dibenzpyrene in coal-tar. *Nature (London)* 182: 1158-59.

Cavanaugh, G.; Burklin, C. E.; Dickerson, J. C.; Lebowitz, H. E.; Tam, S. S.; Smithson, G. R., Jr.; Nack, H.; and Oxley, J. H. 1975. *Potentially hazardous emissions from the extraction and processing of coal and oil.* EPA-650/2-75-008. PB-241 803.

Chirakaparambil, F. G. 1974. *Development of a process for producing an ashless, low-sulfur fuel from coal. Vol. IV. Product studies. Part 8. A preliminary of non-catalysed hydrodesulfurization of coal-derived liquids.* Res. Dev. Rept. 53; Int. Rept. 17. FE-496-T8.

Cleaning up SO_2. 1973. *Chem. Eng.* 80(9): 16.

Clough, G.F.G. 1961. Biological oxidation of phenolic waste liquor. *Chem. Proc. Eng.* 42: 11-14.

Coal Tar Research Association. 1958. *The coal tar data book* (rev. ed.). Oxford Road, Gomersal, Leeds.

Cohen, Murray, personal communication to E. D. Copenhaver 1976.

Collins, J. J.; Fornoff, L. L.; Manchanda, K. D.; Miller, W. C.; and Lovell, D. C. The PuraSiv S process for removing acid plant tail gas. *Chem. Eng. Prog.* 70(6): 58-62.

Combes, F. C. 1954. In *Coal tar and cutaneous carcinogenesis in industry.* Springfield, Ohio: Charles C. Thomas. In Freudenthal et al. 1975.

Cornell, C. F., and Dahlstrom, D. A. 1975. Performance results on a 2500-ACT ft^3/min double-alkali plant for SO_2 removal. In *Air. II. Control of NO_x and SO_x emissions*, AiChe Symp. Ser. 148, vol. 71, ed. C. Rai and R. D. Siegel, pp. 272-82. New York: American Institute of Chemical Engineers.

Craig, T. L. 1972. Tail gas desulfurization operations successful. *Oil Gas J.* 6: 65-66.

Crook, E. H.; McDonnell, R. P.; and NcNulty, J. T. 1975. Removal and recovery of phenols from industrial waste effluents with Amberlite XAD polymeric adsorbents. *Ind. Eng. Chem., Prod. Res. Dev.* 14(2): 113-18.

Dautzenberg, F. M.; Nader, J. E.; and Van Ginnekin, A.J.J. 1971. Shell's flue gas desulfurization process. *Chem. Eng. Prog.* 67(8): 86-91.

Davis, J. C. 1971. SO_2 absorbed from tail gas with sodium sulfite. *Chem. Eng.* 78: 43-45.

Deutsche Eisenwerke, A. G. 1952. Ore sinter. *Brit.* 668; 838.

Diehl, E. K.; du Breuil, F.; and Glenn, R. A. 1967. Polynuclear hydrocarbon emission from coal-fired installations. *J. Eng. Power* 89(A1): 276-82.

Donaldson, W. T. 1974. Preliminary analysis of aqueous wastes from coal conversion plants (a recommended approach). In *Symposium proceedings: environmental aspects of fuel conversion technology (May 1974, St. Louis, Missouri).* EPA-650/2-74-118.

Environmental Protection Agency. 1975. *Coal mining industry-effluent limitation guideline.*

Ershova, K. P. 1971. Studies of the content of polycyclic hydrocarbons in effluents of petrochemical industry and surface waters. *Hyg. Sanit.* 36(6): 474.

Farmer, M. H., and Bertrand, R. R. 1971. *Long range sulfur supply and demand model.* PB-208 993.

Farnsworth, J. F.; Mitsak, D. M.; and Kamody, J. F. 1974. Clean environment with Koppers-Totzek process. In *Symposium proceedings: environmental aspects of fuel conversion technology (May 1974, St. Louis, Missouri),* pp. 115-30. EPA-650/2-74-118.

Fink, C. E., and Vardaman, M. H. 1973. CO_2 acceptor process: status of development. In *Technology and use of lignite proceedings: Bureau of Mines-University of North Dakota Symposium, Grand Forks, North Dakota, May 9-10,* ed. G. H. Gronhovd and W. R. Kube, pp. 229-35. Inf. Circ. 8650.

Forney, A. J.; Haynes, W. P.; Gasior, S. J.; Kornosky, R. M.; Schmidt, C. E.; and Sharkey, A. G. 1975. *Trace element and major component balances around the Synthane PDU gasifier.* PERC/TPR-75/1.

Forney, A. J.; Haynes, W. P.; Gasior, S. T.; Johnson, G. E.; and Strakey, J. P., Jr. 1974. *Analyses of tars, chars, gases, and water found in effluents from the Synthane process.* U.S. Bureau of Mines Technical Progress Report TPR 76.

Freudenthal, R. I.; Lutz, G. A.; and Mitchell, R. I. 1975. *Carcinogenic potential of coal and coal conversion products.* Batelle Energy Program Report. Columbus, Ohio: Batelle Columbus Laboratories.

Fulkerson, W.; Bolton, N. E.; Cresia, D. A.; Endelman, F. J.; Gehrs, C. W.; Guerin, M. R.; Holmes, J. M.; Meyer, A. S.; and Shults, W. D. 1974a. *A proposal to study the health and environmental problems of coal conversion technologies: liquefaction and gasification.* CTL-560. Oak Ridge, Tenn.: Oak Ridge National Laboratory.

Fulkerson, W.; Jolley, R. L.; Kimball, R. F.; Lincoln, T. A.; Moore, G. E.; Nettesheim, P.; Russell, W. L.; Shults, W. D.; and Turner, J. E. 1974b. *Biomedical and environmental research in support of coal conversion technology.* Oak Ridge, Tenn.: Oak Ridge National Laboratory.

Fulkerson, W., et al. 1975. *Biomedical and environmental research for support and evaluation of coal conversion technology.* Oak Ridge, Tenn.: Oak Ridge National Laboratory.

Gabovich, R. D.; Vrochinskii, K. K.; and Kurinnyi, I. L. 1969. Decolorization, deodorization and decontamination of drinking water by ozonization. *Hyg. Sanit.* 34(6): 336-41.

Gibson, C. R.; Hammons, G. A.; and Cameron, D. S. 1974. Environmental aspects of El Paso's Burnham I Coal Gasification Complex. In *Symposium proceedings: environmental aspects of fuel conversion technology (May 1974, St. Louis, Missouri),* pp. 91-100. EPA-650/2-74-118.

Given, P. H. 1974. Problems in the chemistry and structure of coals as related to pollutants from conversion processes. In *Symposium proceedings: environmental aspects of fuel conversion technology (May 1974, St. Louis, Missouri),* pp. 27-34. EPA-650/2-74-118.

Given, P. H.; Cronauer, D. C.; Spackman, W.; Lovell, H. L.; Davis, A.; and Biswas, B. 1974. *Dependence of coal liquefaction behavior on coal characteristics.* R&D Rept. 61; Int. Rept. 9. Pennsylvania State University.

Gladney, E. S. 1974. *Trace element emissions of coal-fired power plants: A study of the Chalk Point Electric Generating Station.* Ph.D. thesis, 75-15, 774. College Park, Md.: University of Maryland.

Glaser, F.; Hershaft, A.; and Shaw, R. 1974. *Emissions from processes producing clean fuels.* BA 9075-015. Bethesda, Maryland: Booz-Allen Applied Research.

Gorman, P. G.; Nebgen, J.; Smith, I.; and Trompeter, E. 1974. *Assessment and development of control technology applicable to removal of mercury and other potentially hazardous pollutant vapors from SO_2 bearing waste gases (power plants).* Midwest Research Institute Int. Rept. 2, draft.

Gogineni, M. R.; Taylor, W. C.; Plumley, A. L.; and Jonakin, J. 1972. Wet scrubbing of sulfur oxides from flue gases. In *Pollution control and energy needs, Adv. Chem. Ser.* 127, ed. R. M. Jimeson and R. S. Spindt, pp. 135-51. Washington D.C.: American Chemical Society.

Grace, R. J., and Diehl, E. K. 1974. Environmental aspects of the Bi-Gas process. In *Symposium proceedings: Environmental aspects of fuel conversion technology (May 1974, St. Louis, Missouri)*, pp. 131-34. EPA-650/2-74-118.

Gregoli, A. A., and Hartos, G. R. 1972. Hydrodesulfurization of residuals. In *Pollution control and energy needs, Adv. Chem. Ser.* 127, ed. R. M. Jimeson and R. S. Spindt, pp. 98-104. Washington, D.C.: American Chemical Society.

Grekel, H.; Palm, J. W.; and Kilmer, J. W. 1968. Why recover S from H_2S? *Oil Gas J.* In Beers 1973.

Groenendaal, W., and van Meurs, H.C.A. 1972. *Petro. Petrochem. Int.* 12(9): 54-58.

Guerin, M. R.; Griest, W. H.; Ho, C.-H.; and Shults, W. D. 1975. Chemical characterization of coal conversion pilot plant materials. In *Proceedings of the 3rd environmental protection conference, Chicago, Ill., Sept. 23-26, 1975*, pp. 661-85. ERDA-92.

Hall, H. J., and Bartok, W. 1971. NO_x control from stationary sources. *Environ. Sci. Technol.* 5(4): 320-26.

Hall, E. H.; Peterson, D. B.; Foster, J. F.; Kiang, K. D.; and Ellzey, V. W. 1975. *Fuels technology: a state-of-the-art review.* EPA-650/2-75-034. Columbus, Ohio: Batelle Columbus Laboratories. PB-245 535.

Hamersma, J. W.; Kraft, M. L.; Koutsoukas, E. P.; and Meyers, R. A. 1972. Chemical removal of pyritic sulfur from coal. In *Pollution control and energy needs, Adv. Chem. Ser.* 127, ed. R. M. Jimeson and R. S. Spindt, pp. 69-79. Washington D.C.: American Chemical Society.

Hamshar, J. A.; Terzian, H. D.: and Scotti, L. J. 1974. Clean fuels from coal by the COED process. In *Symposium proceedings: environment aspects of fuel conversion technology (May 1974, St. Louis, Missouri)*, pp. 147-57. EPA-650/2-74-118.

Hangebrauck, R. P.; von Lehmden, D. J.; and Meeker, J. E. 1964. Emissions of polynuclear hydrocarbons and other pollutants from heat generation and incineration processes. *J. Air Pollut. Control Assn.* 14(7): 267-78.

Hayford, J. S.; VanBrocklin, L. P.; and Kuck, M. A. 1973. Sulfur developments: Stauffer's Aquaclaus process. *Chem. Eng. Prog.* 69(12): 54-55.

Hayford, J. S.; VanBrocklin, L. P.; and Kuck, M. A. 1975. Stauffer Aquaclaus system. In *Air. II. Control of NO_x and SO_x emissions*, AiChE Symp. Ser. 148, ed. C. Rai and R. D. Siegel, vol. 71, pp. 299-303. New York: American Institute of Chemical Engineers.

Headlee, A.J.W., and Hunter, R. G. 1953. Elements in coal ash and their industrial significance. *Ind. Eng. Chem.* 45: 548-51.

Hillenbrand, L. J.; Engdahl, R. B.; and Barrett, R. E. 1973. *Chemical composition of particulate air pollutants from fossil-fueled combustion sources.* PB-219 009. Columbus, Ohio: Battelle Columbus Laboratories.

Hittman Associates. 1975a. *Baseline data environmental assessment of a large coal conversion complex.* R&D Rept. 101; Int. Rept. 1, vol. II, June 1973-August 1974.

Hittman Associates. 1975b. *Impacts and issues related to large scale coal refining complexes.* R&D Rept. 101; Int. Rept. 2.

Hoogendoorn, J. C., and Salomon, J. H. 1957. SASOL: World's largest oil-from-coal plant. *Brit. Chem. Eng.* 2(8): 418-19.

Hutton, W. C., and LaRocca, S. A. 1975. Biological treatment of concentrated ammonia wastewaters. *J. Water Pollut. Control Fed.* 47(5): 989-97.

Hydrocarbon Research, Incorporated. 1968. *Project H-Coal report on process development.* PB-234 579. OCR/RDR 26/Final.

Il'nitskii, A. P. 1966. Control of the pollution of water basins by carcinogenic hydrocarbons. *Hyg. Sanit.* 31(12): 386-92.

Il'nitskii, A. P. 1969. Experimental investigation of the elimination of carcinogenic hydrocarbons from water during its clarification and disinfection. *Hyg. Sanit.* 34(9): 317-21.

Il'nitskii, A. P.; Ershova, K. P.; Khesina, A. Ya.; Rozhkova, L. G.; Klubkov, V. G.; and Koroleve, A. A. 1971. Stability of carcinogens in water and efficacy of methods of decontamination. *Hyg. Sanit.* 36(4): 9-13.

Il'nitskii, A. P.; Khesina, A. Ya.; Cherkinskii, S. N.; and Shabad, L. M. 1968. Effect of ozonation upon aromatic hydrocarbons, including carcinogens. *Hyg. Sanit.* 33(3): 323-27.

Interagency Task Force on Synthetic Fuels from Coal. 1974. *Project independence.* U.S. Dept. of Interior. Washington, D.C.: U.S. Government Printing Office.

Jahnig, C. E. 1975a. *Evaluation of pollution control in fossil fuel conversion processes. Liquefaction: Sect. 2. SRC process.* EPA-650/2-74-009-f. PB-241 792.

Jahnig, C. E. 1975b. *Evaluation of control in fossil fuel conversion processes. Gasification: Sect. 5. Bi-Gas process.* EPA-650/2-74-009-g. PB-243 694.

Jahnig, C. E., and Magee, E. M. 1974. *Evaluation of pollution control in fossil fuel conversion processes. Gasification. Sect. I. CO_2 Acceptor process.* PB-241 141. EPA-650/2-74-009-d.

Jahnig, C. E.; Magee, E. M.; and Kalfadelis, C. D. 1974. Overall environmental considerations of conversion technology. In *Symposium proceedings: environmental aspects of fuel conversion technology (May 1974, St. Louis, Missouri)*, pp. 197-201. EPA-650/2-74-118.

Jorgensen, S. E. 1975. Recovery of ammonia from industrial wastewaters. *Water Res.* 9(12): 1187-91.

Jüntgen, H., and Klein, J. 1974. Purification of wastewater from coking and coal gasification plants using activated carbon. *ACS Div. Fuel Chem. Preprint* 19(5): 67-84.

Kalfadelis, C. D., and Magee, E. M. 1974. *Evaluation of pollution control in fossil fuel conversion processes. Gasification. Sect. 1. Synthane process.* EPA-650/2-74-009b. PB-237 113.

Kalfadelis, C. D., and Magee, E. M. 1975. *Evaluation of pollution control in fossil fuel conversion processes. Liquefaction: Sect. 1. COED process.* EPA-650/2-74-009-e. PB-240 371.

Katell, S. 1966. Removing sulfur dioxide from flue gases. *Chem. Eng. Prog.* 62(10): 67-73.

Katz, D. L.; Briggs, D. E.; Lady, E. R.; Powers, J. E.; Tek, M. R.; Williams, B.; and Lobo, W. E. 1974. *Evaluation of coal conversion processes to provide clean fuels. Part II. Final Report.* EPRI 206-0-0. PB-234 203.

Katzer, J. R.; Gates, B. C.; Olson, J. H.; Kwart, H.; and Stiles, A. B. 1975. *Kinetics and mechanism of desulfurization and denitrogenation of coal-derived liquids. First quarterly report for period June 20, 1975-September 20, 1975.* FE-2028-1.

Kerr, G. D.; Haywood, F. F.; Thorngate, J. H.; and Jones, T. D. 1975. Radiological impact of the utilization and conversion of coal. In Fulkerson et al. 1975.

Kessler, T.; Sharkey, A. G., Jr.; and Friedel, R. A. 1973. *Analysis of trace elements in coal by spark-source mass spectrometry.* Pittsburgh, Penn.: Pittsburgh Energy Research Center.

Ketcham, N. H., and Norton, R. W. 1960. The hazards to health in the hydrogenation of coal. III. The industrial hygiene studies. *Arch. Environ. Health* 1: 194-207.

Kireeva, I. S., and Yanysheva, N. Ya. 1970. Loading of the atmospheric air with carcinogenic polycyclic aromatic hydrocarbons from petroleum processing installations. *Gig. Naselennyka Mest. Resp. Mezhved Sb.* 9: 113-17.

Klein, D. H.; Andren, A. W.; Carter, J. A.; Emery, J. F.; Feldman, C.; Fulkerson, W.; Lyon, W. S.; Ogle, J. C.; Talmi, Y.; Van Hook, R. I.; and Bolton, N. 1975. Pathways of thirty-seven trace elements through coal-fired power plant. *Environ. Sci. Technol.* 9(10): 973-79.

Koehler, G. R. 1975. New England SO_2 recovery project system performance. In *Air. II. Control of NO_x and SO_x emissions*, AiChE Symp. Ser. 148, ed. C. Rai and R. D. Siegel, vol. 71, pp. 283-92. New York: American Institute of Chemical Engineers.

Koon, J. H., and Kaufman, W. J. 1975. Ammonia removal from municipal wastewaters by ion exchange. *J. Water Pollut. Control Fed.* 47(3): 448-65.

Kornreich, M. R. 1974. *A preliminary assessment of the problem of carcinogens in the atmosphere.* MTRE Tech. Rept. MTR-6874.

Kuhn, A. T. 1971. Electrolytic decomposition of cyanides, phenols and thiocyanates in effluent streams — a literature review. *J. Appl. Chem. Biotechnol.* 21(2): 29-34.

LaMantia, C. R.; Lunt, R. R.; and Shah, I. S. 1974. SO_2 processing: dual alkali process for sulfur dioxide removal. *Chem. Eng. Prog.* 70(6): 66-67.

LaMantia, C. R.; Lunt, R. R.; and Shah, I. S. 1975. Dual alkali process for SO_2 control. In *Air. II. Control of NO_x and SO_x emissions*, AiChE Symp. Ser. 148, ed. C. Rai and R. D. Siegel, vol. 71, pp. 324-29. New York: American Institute of Chemical Engineers.

Landreth, R. E., and Mahloch, J. L. 1975. Stabilization of hazardous waste SO_x sludges. In *Proceedings of the national conference on management and disposal of residues from the treatment of industrial wastewaters, Feb. 3-5, 1975, Washington, D.C.*, pp. 215-24.

Laster, L. L. 1973. Atmospheric emissions from the petroleum refining industry. EPA-650/2-73-017. PB-225 040.

Lauer, F. C.; Littlewood, E. J.; and Butler, J. J. 1969. Solvent extraction process for phenols recovery from coke plant aqueous waste. *Iron Steel Eng.* 99-102.

Lijinsky, W.; Singer, G.; and Taylor, W. 1975. Carcinogenesis by environmental pollutants related to energy production. In Fulkerson et al. 1975.

Lijinsky, W.; Domsky, I.; Mason, G.; Ramahi, H. Y.; and Safavi, T. 1963. The chromatographic determination of trace amounts of polynuclear hydrocarbons in petrolatum, mineral oil, and coal-tar. *Anal. Chem.* 35(8): 952-56.

Lin, P. W. 1975. Wastewater treatment with an SO_2-removal by-product. *J. Water Pollut. Control Fed.* 47(9): 2271-80.

Ludberg, J. E., and Nicks, G. D. 1969. Phenols and thiocyanate removed from coke plants effluents. *Industrial Wastes, A Water and Sewage Works Supplement* 116(11): 10-13.

Magee, E. M. 1975. Environmental impact and R and D needs. In *EPA Symposium, environmental aspects of fuel conversion technology II, Hollywood, Fla., Dec. 15-18, 1975.*

Magee, E. M.; Hall, H. J.; and Varga, G. M., Jr. 1973. *Potential pollutants in fossil fuels.* EPA-R2-73-240. PB-225 039.

Magee, E. M.; Jahnig, C. E.; and Shaw, H. 1974. *Evaluation of pollution control on fossil fuel conversion processes. Gasification: Sect. I. Koppers-Totzek process.* EPA-650/2-74-009-a. PB-231 675.

Magee, E. M., and Shaw, H. 1974. Technology needs for pollution abatement in fossil fuel conversion processes. In *Symposium proceedings: environmental aspects of fuel conversion technology (May 1974, St. Louis, Missouri)*, pp. 309-13. EPA-650/2-74-118.

Malaney, G. W.; Lutin, P. A.; Cibulka, J. J.; and Hickerson, L. H. 1968. Resistance of carcinogenic organic compounds to oxidation by activated sludge. *J. Water Pollut. Control Fed.* 39(12): 2020. In Andelman and Snodgrass 1974.

Manz, O. 1973. Utilization of lignite and subbituminous ash. In *Technology and use of lignite proceedings: Bureau of Mines-University of North Dakota Symposium*, ed. G. H. Gronhovd and W. R. Kube. Inf. Circ. 8650.

Masek, V. 1971. Benzo(a)pyrene in the workplace atmosphere of coal and pitch coking plants. *J. Occup. Med.* 13: 193.

Matsushita, H.; Esumi, Y.; Susuki, A.; and Handa, T. 1972. An analytical method for polynuclear hydrocarbons in coal tar. *Bunseki Kagaku (Tokyo)* 21(11): 1471-78.

Mayland, B. J., and Heinze, R. C. 1973. Continuous catalytic absorption for NO_x emission control. *Chem. Eng. Prog.* 69(5): 75-76.

McGinnes, P. R., and Snoeyink, V. L. 1974. *Determination of the fate of polynuclear aromatic hydrocarbons in natural water systems.* Water Resources Council Rept. 80. PB-232 168.

Mercer, B. W.; Ames, L. L.; Touhill, C. J.; Van Slyke, W. J.; and Dean, R. B. 1970. Ammonia removal from secondary effluents by selective ion exchange. *J. Water Pollut. Control Fed.* 42(2): R95-R107.

Meyers, R. A.; Hamersma, J. W.; Land, J. S.; and Kraft, M. L. 1972. Desulfurization of coal. *Science* 177: 1187-88.

Meyers, R. A.; Land, J. S.; and Flegal, C. A. 1971. Chemical removal of nitrogen and organic sulfur from coal. PB-204 863. ADTD 0845.

Milios, P. 1975. Water reuse at a coal gasification plant. *Chem. Eng. Prog.* 71(6): 99-104.

Miller, W. E. 1974. SO_2 processing: the cat-ox process at Illinois Power. *Chem. Eng. Prog.* 70(6): 49-52.

Miller, W. E. 1975. The cat-ox project at Illinois Power. In *Air. II. Control of NO_x and SO_x emissions,* AiChE Symp. Ser. 148, ed. C. Rai and R. D. Siegel, vol. 71, p. 293. New York: American Institute of Chemical Engineers.

Miyamoto, S.; Ryan, J.; and Stroehlein, J. L. 1975. Potentially beneficial uses of sulfuric acid in southwestern agriculture. *J. Environ. Qual.* 4(4): 431-37.

Morgan, W. D. 1975. Coalcon's clean boiler fuels from coal demonstration plant. Paper 59c, presented at the AiChE Convention, Los Angeles, Calif., Nov. 19, 1975.

Moulton, L. K. 1973. Bottom ash and boiler slags. In *Ash utilization proceedings: 3rd international ash utilization symposium, Mar. 13-14, 1973,* pp. 148-69. U.S. Bureau of Mines Inf. Circ. 8640.

Myers, J. W.; Pfeiffer, J. J.; and Orning, A. A. 1964. *Production of lightweight aggregate from washery refuse.* U.S. Bureau of Mines Report of Investigations 6449.

Naber, J. E.; Wesselingh, J. A.; and Groenendaal, W. 1973. New Shell process treats Claus off-gas. *Chem. Eng. Prog.* 69(12): 29-34.

Naber, J. E.; Wesselingh, J. A.; and Groenendaal, W. 1975. Shell Claus off-gas treating process. In *Air. II. Control of NO_x and SO_x emissions,* AiChE Symp. Ser. 148, ed. C. Rai and R. D. Siegel, vol. 71, p. 304. New York: American Institute of Chemical Engineers.

Nandi, S. P., and Walker, P. L., Jr. 1972. *Adsorption characteristics of coals and chars.* Office of Coal Research, U.S. Dept. of Interior, R&D Rept. 61; Int. Rept. 1. Pennsylvania State University.

National Academy of Sciences, Committee on Biologic Effects of Atmospheric Pollutants. 1972. *Particulate polycyclic organic matter.* Washington, D.C.

Nebolsine, R. 1957. The treatment of water-borne wastes from steel plants. *Iron Steel Eng.* 34(12): 125-51.

Nelson, H. W.; Schuler, R. E.; Shilhan, M. J.; and Engdahl, R. B. 1969. *Study of the identification and assessment of potential markets for chars from coal processing systems.* PB-182 966.

Newfeld, R. D., and Hermann, E. R. 1975. Heavy metal removal by acclimated activated sludge. *J. Water Pollut. Control Fed.* 47(2): 310-29.

Newman, E. D.; Reno, C. J.; and Burroughs, L. C. 1958. Waste disposal at Anacortes. *Oil Gas J.* 56(20): 124.

Nichols, J. P. 1975. *Coal technology program progress report for October 1974.* ORNL/TM-5046.
CT-74-3. Oak Ridge, Tenn.: Oak Ridge National Laboratory.

Nowak, Z. 1973. Iron and alumina extraction from power plant fly ash in Poland. In *Ash utilization proceedings: 3rd international ash utilization symposium, Mar. 13-14, 1973,* pp. 224-30.
U.S. Bureau of Mines Inf. Circ. 8640.

Oak Ridge National Laboratory. 1975. *Coal technology program quarterly progress report for the
period ending Sept. 30, 1975.* ORNL-5093. Oak Ridge, Tenn.: Oak Ridge National Laboratory.

O'Gorman, J. V., and Walker, P. L., Jr. 1972. *Mineral matter and trace elements in U.S. Coals.*
OCR-RDR-61-IR-2. Washington, D.C.: U.S. Government Printing Office.

O'Hara, J. B.; Rippee, S. N.; Loran, B. I.; and Mindheim, W. J. 1974. *Environmental factors in
coal liquefaction plant design.* Office of Coal Research, R&D Rept. 82; Int. Rept. 3.
PB-235 802.

Ondov, J. M.; Zoller, W. H.; Olmez, I.; Aras, N. K.; Gordon, G. E.; Rancitelli, L. A.; Abel, K. H.;
Filby, R. H.; Shah, K. R.; and Ragaini, R. C. 1975. Elemental concentrations in the
National Bureau of Standards' environmental coal and fly ash standard reference materials.
Anal. Chem. 47(7): 1102-9.

Ad Hoc Panel on Coal Gasification Technology. 1974. *Evaluation of coal gasification technology.*
Part II. Low and intermediate-Btu fuel gases. R&D Rept. 74; Int. Rept. 2. Office of
Coal Research, U.S. Dept. of Interior. PB-234 042.

Pedlow, J. W. 1973. Cenospheres. In *Ash utilization proceedings: 3rd international ash
utilization symposium, Mar. 13-14, 1973,* pp. 33-43. U.S. Bureau of Mines Inf. Circ. 8640.

Perry, H., and Berkson, H. 1971. Must fossil fuels pollute? *Technol. Rev.* 74(2): 34-43.

Petersen, M. R. 1975. *Progress report: organic constituents in process streams of a solvent
refining coal plant.* BNWL-B-470; UC-90a. Battelle Pacific Northwest Laboratories.

Pittsburgh Energy Research Center. 1975. *Processes under development by the Pittsburgh Energy
Research Center for clean energy from coal.* PERC 75/1.

Phillips, M. A. 1973. Investigations into levels of both airborne beryllium and beryllium in
coal at the Hayden Power Plant near Hayden, Colorado. *Environ. Lett.* 5(3): 183-88.

Powell, E. M., and Ulmer, R. C. 1974. Controlling emissions from fossil-fueled power plants.
Combustion 45: 23-34.

Power from coal. Part III. Combustion, pollution controls. 1974. *Power* 118(4): S49-S64.

Princiotta, F. T. 1972. *Control of sulfur oxide pollution from power plants.* PB-228 706.

Princiotta, F. T. 1974. Status of flue gas desulfurization technology. In *Symposium proceedings:
environmental aspects of fuel conversion technology (Mar. 1974, St. Louis, Missouri).*
EPA-650/2-74-118.

Process Evaluation Group, Morgantown, West Virginia. 1975. *Economic analysis of Synthoil plant
producing 50,000 barrels per day of liquid fuels from two coal seams: Wyodak and western
Kentucky.* U.S. Dept. of the Interior, Bureau of Mines. ERDA 76-35. FE-2083-1.

Reichert, J.; Kunte, H.; Borneff, J.; and Engelhardt, K. 1971. Carcinogenic substances occurring
in water and soil. XXVII. Further studies on the elimination from waste of certain
polycyclic aromatic hydrocarbons. *Zentralbl. Bakteriol. (Naturwiss)* 155(1): 18. Also in
Andelman and Snodgrass 1974.

Reid, G. W., and Streebin, L. E. 1972. *Evaluation of wastewaters from petroleum and coal
processing.* EPA-R2-72-001. PB-214 610.

Rubin, E. S., and McMichael, F. C. 1975. Impact of regulations on coal conversion plants.
Environ. Sci. Technol. 9(2): 112-17.

Ruch, R. R.; Gluskoter, H. J.; and Shimp, N. F. 1973. Distribution of trace elements in coal.
In *Symposium proceedings: environmental aspects of fuel conversion technology (May 1974,
St. Louis, Missouri),* pp. 49-53. EPA-650/2-74-118.

Rudolph, P.F.H. 1974. The Lurgi process route to substitute natural gas (SNG) from coal.
Chem. Age India 25(5): 289-99.

Sather, N. F.; Swift, W. M.; Jones, J. R.; Beckner, J. L.; Addington, J. H.; and Wilburn, R. L. 1975. *Potential trace element emissions from the gasification of Illinois coal.* PB-241 220. IIEQ 75-08.

Sawicki, E.; Meeker, J. E.; and Morgan, M. J. 1965. The quantitative composition of air pollution source effluents in terms of aza heterocyclic compounds and polynuclear aromatic hydrocarbons. *Int. J. Air Water Pollut.* 9: 291-98.

Schoental, R. 1957. Isolation of 3:4-9:10-dibenzopyrene from coal-tar. *Nature (London)* 180: 606.

Schrodt, J. T.; Hilton, G. B.; and Rogge, C. A. 1975. High temperature desulfurization of low-CV fuel gas. *Fuel* 54(4): 269-72.

Schuit, G.C.A., and Gates B. C. 1973. Chemistry and engineering of catalytic hydrodesulfurization. *AiChE J.* 19: 417-38.

Schultz, H.; Hattman, E. A.; and Booher, W. B. 1973. The fate of some trace elements during coal pretreatment and combustion. *ACS Div. Fuel Chem.* 18(4): 108-13.

Selmeczi, J. G., and Knight, R. G. 1973. Properties of power plant waste sludges. *Ash utilization proceedings: 3rd international ash utilization symposium, Mar. 13-14, 1973*, pp. 123-38. U.S. Bureau of Mines Inf. Circ. 8640.

Sforzolini, G. S.; Savino, A.; Monarca, S.; and Conti, P. 1974. Decontamination of water contaminated with polycyclic aromatic hydrocarbons (PAH). II. Action of chlorine and ozone on PAH dissolved in drinking and river water. *Igiene Moderna* 66(6): 595-619.

Sharkey, A. G., Jr.; Shultz, J. L.; Schmidt, C. E.; and Friedel, R. A. 1975*a*. Mass spectrometric analysis of coal derived liquids. In *Proceedings of the 3rd environmental protection conference, Chicago, Ill., Sept. 23-26, 1975*, pp. 686-701. ERDA-92.

Sharkey, A. G., Jr.; Shultz, J. L.; Schmidt, C. E.; and Friedel, R. A. 1975*b*. *Mass spectrometric analysis of streams from the coal gasification and liquefaction processes.* PERC/RI-75-r.

Shaw, H. 1976. Personal communication to E. D. Copenhaver.

Shaw, H., and Magee, E. M. 1974. *Evaluation of pollution control in fossil fuel conversion processes. Gasification: Sect. I. Lurgi process.* EPA-650/2-74-009-c. PB-237 694.

Shearer, H. A. 1973. The COED process plus char gasification. *Chem. Eng. Prog.* 69(3): 43-49.

Short, T. E., Jr.; DePrater, B. L.; and Myers, L. H. 1974. Controlling phenols in refinery wastewaters. *Oil Gas J.* 72(47): 119-24.

Shults, W. D., ed. 1976. *Preliminary results: chemical and biological examination of coal-derived materials.* ORNL/NSF/EATC-18. Oak Ridge, Tenn.: Oak Ridge National Laboratory.

Sittig, M. 1974. *Pollution control in the organic chemical industry.* Park Ridge, New Jersey: Noyes Data Corporation.

Slack, A. V., and Hollinden, G. A. 1975. Sulfur dioxide removal from waste gases. *Pollution Technology Review* 21 (2nd ed.). Park Ridge, New Jersey: Noyes Data Corporation.

Smith, W. M. 1970. Evaluation of coke oven emissions. *Yearb. Am. Iron Steel Inst.* 163.

Stasiuk, W. N., Jr.; Hetling, L. J.; and Shuster, W. W. 1974. Nitrogen removal by catalyst-aided breakpoint chlorination. *J. Water Pollut. Control Fed.* 46(8): 1974-83.

Strom, A. H., and Eddinger, R. T. 1971. COED plant for coal conversion. *Chem. Eng. Prog.* 67(3): 75-80.

Stacy, W. O., and Walker, P. L., Jr. 1972. *Structure and properties of various coal chars.* Office of Coal Research, U.S. Dept. of Interior, R&D Rept. 61; Int. Rept. 3. Pennsylvania State University.

Toca, F. M.; Cheever, C. L.; and Berry, C. M. 1973. Lead and cadmium distribution in the particulate effluent from a coal-fired boiler. *Am. Ind. Hygiene Assn. J.* 34: 396-403.

Tomany, J. P. 1975. *Air pollution: the emissions, the regulations, and the controls.* New York: American Elsevier Pub. Co., Inc.

Toxic Materials Information Center. 1975. *Trace elements in coals and other fuels. Parts I and II.* Oak Ridge, Tenn.: Oak Ridge National Laboratory.

Trakhtman, N. N., and Manita, M. D. 1966. Evaluation effect of chlorination of water on pollution of 3,4-benzpyrene. *Hyg. Sanit.* 31(3): 316-20.

Van der Kooij, J., and Elshout, A. L. 1974. The emission of nitrogen oxides by power stations. *Het. Ingenicursblad* 43e(10): 311-17.

Van Meter, W. P., and Erickson, R. E. 1975. *Environmental effects from leaching of coal conversion by-products.* Int. Rept., June-September 1975. FE-2019-1.

van Stone, G. R. 1972. Treatment of coke plant waste effluent. *Iron Steel Eng.* 49(4): 63-66.

von Lehmden, D. J.; Jungers, R. H.; and Lee, R. E., Jr. 1974. Determination of trace elements in coal, fly ash, fuel oil, and gasoline — a preliminary comparison of selected analytical techniques. *Anal. Chem.* 46(2): 239-45.

Wall, J. M., ed. 1973. NC/LNG/SNG handbook. *Hydrocarbon Process.* LII: 88-132.

Wallcave, L.; Garcia. H.; Feldman, R.; Lijinsky, W.; and Shubik, P. 1971. Skin tumorigenesis in mice by petroleum asphalts and coal tar pitches of known polynuclear aromatic hydrocarbon content. *Toxicol. Appl. Pharmacol.* 18: 41-52.

Water hyacinths soak up pollution. 1976. *Bio. Sci.* 26(3): 224.

Wen, C. Y. 1975. *Optimization of coal gasification processes.* OCR-R&D Rept. 66; Int. Rept. 2.

Williams, J. E., and Dressel, J. H. 1973. Coal gasification, the new energy source. *Can. Mining Met. Bull.* 66: 72-77.

Whittemore, R. C. 1973. An evaluation of the adsorptive properties of fly ash with reference to a pulp and paper mill waste effluent. In *Ash utilization proceedings: 3rd international ash utilization symposium, Mar. 13-14, 1973,* pp. 296-317. U.S. Bureau of Mines Inf. Circ. 8640.

Wolverton, B. C. 1975. *Water hyacinths for removal of phenols from polluted waters.* NASA Tech. Memo. TM-X-72722.

Woodward, P. W.; Sturm, G. P., Jr.; Vogh. J. W.; Holmes, S. A.; and Dooley, J. E. 1976. *Compositional analyses of Synthoil from West Virginia coal.* BERC/RI-76/2.

Worthy, W. 1975. Hydrothermal process cleans up coal. *Chem. Eng. News* 53(27): 24-25.

Yavorsky, P. M. 1976. ORNL seminar, Oak Ridge National Laboratory, Oak Ridge, Tenn., Jan. 20, 1976.

Yavorsky, P. M., and Akhtar, S. 1974. Environmental aspects of coal liquefaction. In *Symposium proceedings: environmental aspects of fuel conversion technology (May 1974, St. Louis, Missouri),* pp. 325-30. EPA-650/2-74-118.

Yosim, S. J.; Grantham, L. F.; McKenzie, D. E.; and Stegman, G. C. 1972. The chemistry of the molten carbonate process for sulfur oxides removal from stack gases. In *Pollution control and energy needs, Adv. Chem. Ser.* 127, ed. R. M. Jimeson and R. S. Spindt, pp. 174-82. Washington, D.C.: American Chemical Society.

Younger, A. H. 1975. Process for sour natural gas treating. In *Air. II. Control of NO_x and SO_x emissions,* AiChE Symp. Ser. 148, ed. C. Rai and R. D. Siegel, vol. 71, pp. 216-19. New York: American Institute of Chemical Engineers.

5. ANALYSIS OF COAL AND COAL PRODUCTS

G. A. Dailey   D. C. Michelson

ABSTRACT

Analysis of coals for determination of chemical and physical properties has been a concern of
the Bureau of Mines for many years. The increasing interest in coal conversion and use is
leading to improved methods of coal analysis which provide greater efficiency and more accurate
results.

Proximate analysis of coal determines four products by heating under standard conditions:
(1) water or moisture, (2) volatile matter consisting of gases and vapors driven off during
pyrolysis, (3) fixed carbon, remaining as the nonvolatile fraction of the pyrolyzed coal, and
(4) ash, derived from mineral matter impurities.

Ultimate analysis expresses the composition of coal in percentages of carbon, hydrogen,
nitrogen, sulfur, oxygen, and ash. This form of analysis does not show the origin of the
elements and includes mineral carbonates as well as organic coal substance under the single
carbon percentage.

Coal mineral matter content ranges from 9.4 to 22.3% and totals nearly 8.9×10^6 tons per year
of coal mined. Most minerals in coal are either aluminosilicates, carbonates, sulfides, or
silica. Of these, sulfides, present mostly as pyrite, pose the greatest environmental pollution problem. Determination of minerals in coal with respect to oxidation states and compound
types is difficult, but recent success has been achieved using an electronic (radiofrequency)
low-temperature ashing technique.

Spectrophotometric techniques are basic tools for many coal analysis methods. One of the most
widely used techniques is atomic absorption spectrometry (AAS), which gives satisfactory
results for many metals and is especially suitable for particulates collected in air samples.
Interferences occur less often with AAS than with many other methods although the major disadvantage of AAS is the relatively large sample required.

Mine and coal dusts in the respirable size range have been examined for trace elements by
spark-source mass spectrometry (SSMS). SSMS and thermal emission mass spectrometry (TEMS)
have been used for trace element determination in solid coal samples.

Sulfur compounds are major coal emissions which pose special problems in sampling and
determination. Sulfur dioxide determinations are difficult due to the presence of oxygen and
other pollutants that are oxidants or reductants. Sulfur trioxide and sulfuric acid determination methods range from acid-base titrations to spectrophotometric measurements. Hydrogen
sulfide emissions are collected as samples by aspirating air through an alkaline cadmium

hydroxide suspension and precipitating out the sulfide for spectrophotometric measurement. Sulfate determinations are made by air particulate collection on standard high-volume sampler filters and testing by either a turbidimetric (sulfaver) method or a methylthymol blue method.

Polycyclic aromatic hydrocarbon (PAH) determination and identification methods are based on available pure samples in most cases, with some applications to analysis of air particulate samples. Analytical methods include computerized gas chromatographic-mass spectrometric (GC-MS) systems and liquid chromatography. Gas-liquid chromatographic columns are limited to low-molecular-weight PAH because the liquid stationary phase "bleeds" during high-temperature separation of four- and five-ring PAH compounds.

Benzo[*a*]pyrene (BaP) is of special concern as it is a known carcinogen. Quantitative analysis of BaP is difficult since it usually occurs in a complex hydrocarbon mixture and is often accompanied by an isomer, benzo[*e*]pyrene (BeP). BaP and BeP have not been separated routinely by gas chromatography.

Phenols in low-temperature bituminous coal tar distillates have been determined by gas-liquid chromatography. This technique is simpler and more direct than infrared spectrometry. This class of compounds is analyzed fairly routinely to monitor water quality.

Finally, the development and adaptation of separations methodologies are of prime importance in testing and monitoring coal conversion operations and the products and by-products. This is an area of increasing activity.

5.0 INTRODUCTION

Chapter 5 presents a compilation of significant techniques for the analysis of coal and coal emissions. The arrangement of information is from broadest analytical categories to smaller divisions such as individual elements or compounds. The chapter is not all-inclusive; it attempts to touch on those areas of greatest current concern and having the greatest collection of information. Not all elements present in coal are discussed because many are found in such minute amounts or are so easily detected that their inclusion would be pointless. Also, not all methods of analysis are included since many are variations of current techniques and others have not been widely accepted by those involved in the field.

A discussion of general coal analyses and, in particular, proximate, ultimate, and trace element analysis is provided. Sulfur emissions, including sulfur dioxide, hydrogen sulfide, sulfuric acid, and particulate sulfur, are discussed since they are present in quantitatively large amounts. PAH emissions, particularly benzo[a]pyrene, are discussed due to their carcinogenicity. The discussion of PAH includes problems in separation, detection, identification, quantification, and sampling. Phenolic by-products and general separation approaches are also discussed.

Some knowledge of the chemical and physical properties of coal is essential in the selection of suitable methods for the chemical analysis of coal. Coal conversion processes and the coals used in these processes have been studied by the U.S. Bureau of Mines for many years. These methods and other more detailed methods of coal analysis are described by Ode (1963). Many new methods for the identification and quantification of organic components of coal and coal-derived products are being developed rapidly.

5.1 STANDARD METHODS

Coal that is to be used in a conversion process is sampled and analyzed to establish its utility with respect to a particular process. Some standard methods are applied to all coals. As a first step, the samples are air-dried at 30 to 55°C to approximate equilibrium with air in the laboratory. To obtain a representative sample for analysis, the air-dried sample is pulverized to pass through a No. 60 sieve (250 μ), reduced to about 50 g, placed in a 4-oz stoppered bottle, and mixed thoroughly on a mixing wheel. About 1 g of coal is removed for each analysis.

5.1.1 Proximate analysis

Proximate analysis of coal is a first-order determination of four products obtained during heating under standard conditions: (1) water or moisture; (2) volatile matter, consisting of gases and vapors driven off during pyrolysis; (3) fixed carbon, remaining as the nonvolatile fraction of the pyrolyzed coal; and (4) ash, derived from the mineral matter impurities (Ode 1963).

5.1.1.1 Moisture

Moisture content in coal is determined by passing dried air or nitrogen over a weighed coal sample that is held at a temperature of 105°C for 1 hr. The dried air or nitrogen takes up moisture from the sample, and the loss in sample weight is calculated as the percent moisture. Errors inherent in this and other techniques include driving off gases such as methane and other volatile components that are commonly contained in coal (Ode 1963). This loss would contribute to weight loss in the coal sample, thus presenting an erroneous moisture content. The effect of carrier gas (oxygen or nitrogen) moisture, if present, must be considered because this influence was found not negligible by Mukherjee, Basak, and Lahiri (1953).

5.1.1.2 Volatile matter

Determinations of volatile matter do not identify particular compounds in coal; rather, these are empirical procedures that result in a weight loss by the sample. In one standard procedure, a 1-g, air-dried coal sample is heated in a platinum crucible in a 950°C (±20°C) oven for exactly 7 min. The crucible is removed and air-cooled on a steel plate. The cooled sample is weighed; the loss of weight minus the weight of moisture (determined at 105°C) times 100 equals the percentage of volatile matter (Bureau of Mines 1967). Results depend directly on experimental conditions, which must be followed rigidly for valid comparisons among samples. Coals vary widely in volatile matter content according to rank (lignite, anthracite, etc.).

5.1.1.3 Fixed carbon

The percentage of fixed carbon, the nonvolatile fraction of pyrolyzed coal, is determined by subtracting the sum of the percentages of moisture, ash, and volatile matter from 100. The remainder is the percentage of nonvolatile carbon.

5.1.1.4 Ash

All elements in coal other than carbon, hydrogen, oxygen, nitrogen, and sulfur are referred to as "mineral matter in coal." This mineral matter cannot be accurately determined by normal high-temperature ashing at 750°C. Most major minerals, except quartz, are altered: hydrates lose water of hydration, pyrite minerals are oxidized to ferric oxide and sulfur dioxide, and calcium carbonate is calcined to its oxide (Gluskoter 1975). Possible analytical errors due to identifying ash with mineral matter in coal are given in Table 5.1. By determining mineral-derived carbon dioxide, corrections can be applied to the volatile matter, carbon, and oxygen contents of coal (Given 1973).

Ash content is determined by placing a sample of the coal in a cold muffle furnace and heating it gradually to 500°C in 1 hr and to 750°C in 2 hr. The sample is burned to completion, which is shown by a constant weight. Ash has a different composition from coal mineral matter, which must be determined if coal properties are tabulated on a mineral-free basis. Coal ash compositions vary directly with the coal mineral compositions. Typical ash constituents are shown in Table 5.2.

5.1.2 Ultimate analysis

Ultimate analysis provides information on the elemental composition of coal in terms of carbon, hydrogen, nitrogen, sulfur, and oxygen. Results are given in percentages.

Carbon and hydrogen contents are determined by burning coal in pure dry oxygen in a combustion tube and collecting the gaseous combustion products in an absorption train. Anhydrous magnesium perchlorate traps the hydrogen as water, and sodium hydroxide-asbestos absorbs the evolved

Table 5.1. Possible analytical errors due to
mineral matter in coals

Source of error	Nature of error
Water in clay minerals	Ash weight less than original mineral matter; organic H and volatile matter too high
Carbonates	Loss of CO_2, ash less than original mineral matter; residual oxides fix S as SO_4; organic C and volatile matter too high
Pyrite	Burns to Fe_2O_3; ash less than original mineral matter; partially decomposed in volatile matter test; contributes to calorific value
Chlorine	Partially lost on ashing and in volatile matter test
Cations in low-rank coals	On ashing, fix S as SO_4

Source: Given 1973, Fig. 11. Reprinted by permission of the publisher.

Table 5.2. Typical limits of ash composition of
bituminous coals

Constituent	USA	England	Germany
SiO_2	20 - 60	25 - 50	25 - 45
Al_2O_3	10 - 35	20 - 40	15 - 21
Fe_2O_3	5 - 35	0 - 30	20 - 45
CaO	1 - 20	1 - 10	2 - 4
MgO	0.3 - 4	0.5 - 5	0.5 - 1
TiO_2	0.5 - 2.5	0 - 3	
$Na_2O + K_2O$	1 - 4	1 - 6	
SO_3	0.1 - 12	1 - 12	4 - 10

Source: Ode 1963, Table 1, p. 209.
Reprinted by permission of the publisher.

carbon dioxide. Multiplication of weight differences in the collection tubes (a Marchand tube
for hydrogen and a Nesbitt bulb for carbon) by gravimetric factors gives the percentages of
carbon and hydrogen in the coal.

There is no simple direct method for determination of oxygen content in coal (Bureau of Mines
1967; Given and Yarzab 1975). Oxygen content determinations are estimated by subtracting the
sum of carbon, hydrogen, nitrogen, sulfur, and ash from 100. The remaining figure is the
percentage of oxygen content. All errors accumulated in the determinations of carbon, nitrogen,
hydrogen, sulfur, and ash will affect the oxygen content figure.

5.1.3 Incorporation of mineral matter into coals

Mineral matter is incorporated into coal in two ways: (1) deposition from water percolating
through decaying vegetation in early stages of coal formation and (2) infiltration of cracks
and fissures in the formed coal during subsequent stages. Mineral matter in coal is commonly

classified as either "inherent mineral matter" — material that is too closely associated with the coal substance to be readily separated from it by physical methods — and "adventitious mineral matter" — material that is less intimately associated with the coal and that can be readily separated from it (O'Gorman and Walker 1972). Additionally, some minerals (inorganic elements) are an integral part of the original vegetation composing the coal, and they define the lower limit to which a coal can be separated physically from its mineral impurities (Ode 1963). The identification and estimation of the types of mineral matter are important for assessing the performance of the coal in conversion processes and for evaluating possible process effluents.

5.1.3.1 <u>Ranges of mineral matter content in coal</u>

Mineral matter content of coal ranged from 9.4 to 22.3% in a recent study of 65 Illinois coals (Gluskoter 1975), and since about 590×10^6 tons of coal are produced yearly, about 89×10^6 tons of mineral matter is unwanted. Most minerals in coal are in one of four groups: aluminosilicates (clay minerals), carbonates, sulfides, and silica. Of these, the sulfides, present mostly as pyrite, pose the greatest environmental pollution problem. The carbonate minerals in most coals consist of dolomite ($CaCO_3$-$MgCO_3$) and ankerite ($2CaCO_3 \cdot FeCO_3$). Silica and the aluminosilicate clays often make up more than 50% of the coal mineral matter.

5.1.3.2 <u>Mineral matter analysis</u>

The separation of minerals from the parent coal for identification and quantitation is difficult. Even low-temperature ashing in an oxygen stream provides limited information because many of the minerals are oxidized (Gluskoter 1975). The recent technique of electronic (radiofrequency) low-temperature ashing has been more successful and consists of forming activated oxygen by passing oxygen first through a radiofrequency field and then over a coal sample, oxidizing the organic matter at low temperatures, usually between 149 and 163°C (O'Gorman and Walker 1972). Most major minerals are unaffected by this type of ashing (Gluskoter 1975). The minerals in the ash are identified by well-established analytical methods: x-ray diffraction, which is the most common method; infrared absorption; differential thermal analysis; and electron microscopy. Infrared and x-ray diffraction techniques are considered as complementary to each other. Infrared analysis is extremely sensitive to short-range ordering, whereas x-ray diffraction analysis methods are sensitive to longer-range order and to a periodic arrangement of atoms in a crystalline structure (O'Gorman and Walker 1972). Pyrite has long been studied directly during petrographic analysis of coal by optical microscopy (Bureau of Mines 1967).

5.1.3.3 <u>Carbonate analysis</u>

Carbonate minerals are so common in coal that methods for determining carbonate carbon dioxide content have been established for decades (Pringle 1963; Bureau of Mines 1967). Only recently, however, has an effort been made to improve the analytical accuracy, sensitivity, and speed — factors important in routine analyses of coals containing small quantities of carbonates. The standard method consists of boiling coal in dilue hydrochloric acid and absorbing the liberated carbon dioxide on a sodium hydroxide-asbestos matrix (Bureau of Mines 1967). A titrimetric method, in which carbon dioxide is absorbed in a benzylamine solution, was compared with and rated as equal or superior to both gravimetric and manometric standard methods (Allan et al. 1966). Knott and Belcher (1975) report a semimicro simple, sensitive, rapid gravimetric method that features a soda-asbestos absorber preceded by an efficient vertical scavenging train.

5.1.3.4 Ashing coal for mineral matter analysis

Ashing coal in air at 600°C yields an ash product that can be analyzed for both major and minor trace elements. Amounts of the major metallic elements — silicon, aluminum, calcium, iron, magnesium, titanium, sodium, and potassium — can vary widely (Table 5.3); the variability is important in determining the fusion point of the ash and the resultant behavior of the coal in a high-temperature conversion or combustion environment. Atomic absorption and emission spectroscopic methods are used to analyze for the major elements in coal ash. An atomic absorption procedure was developed by Medlin, Suhr, and Bodkin (1969). The method requires 80.0 mg of 200 mesh sample mixed with 400 mg of lithium metaborate in a plastic vial on a dentist's amalgamator for about 30 sec. The mixture is placed in a graphite fusion crucible and fused for 10 min at 1000°C in a muffle furnace. The molten bead is poured directly into a Teflon beaker containing 40.0 ml of a 3% nitric acid solution. The solution is stirred magnetically until completely dissolved (about 10 min) and is used directly or diluted with a 1% lanthanum nitrate solution. The lanthanum was used as a releasing agent for chemical inter-ferences (O'Gorman and Walker 1972). The dc arc technique has been used for emission spec-troscopic analysis. The sample is mixed with an internal standard and graphite in the ratio of 1:1:2 (by weight) respectively. The internal standard is composed of a fused mixture of 0.1% lutetium oxide (Lu_2O_3) plus 0.2% cadmium oxide (CdO) in sodium borate ground to a fine powder. The mixtures (sample plus standard plus graphite) are packed into electrode cups and burned in a Stallwood jet with an argon-oxygen atmosphere at 14 A. Photographic plates are used to record the spectral lines. An NSL Comparator-Densitometer is used for intensity measurements. Results are converted from density to parts per million, producing working curves of parts per million vs relative intensity (O'Gorman and Walker 1972).

5.2 TRACE ELEMENT ANALYSIS

As mentioned above, trace elements and other noncombustible components of coal typically com-prise 10 to 20% of the total mass. The trace elements occur as an integral portion of the organic coal structure and are not released until the coal is destroyed, as by combustion (O'Gorman and Walker 1972). The rise in public awareness of trace elements in the environment resulting from coal and other sources has placed a greater emphasis on analytical methods that are economical, efficient, and quick. Analytical methods that are reliable in the fractional parts-per-million range of contaminant in the condensed phase or micrograms per cubic meter of air are necessary for reliable results (Ahuja et al. 1973). Trace elements released by coal may be identified by a variety of methods.

5.2.1 Spectrophotometric techniques

Spectrophotometric techniques (Kolthoff and Elving 1964) have been the basis of many coal analysis methods. Spectrophotometry is based on measuring light absorbance by a particular sample at a particular wavelength. Analysis by these methods normally involves the reaction of a coloring reagent with the element of interest, addition of needed masking or buffering agents, and measurement of absorbance by the element-reagent complex in a spectrophotometer. For example, an absolute sensitivity of 0.5 to 20 µg has been reported for spectrophotometric measurements of arsenic in biological and environmental samples (Talmi and Bostick 1975) with complexes of silver-diethyldithiocarbamate, ammonium molybdate, and 8-mercapto-quinoline. A widespread method for cadmium determination is the extraction of cadmium with dithiazone followed by spectrophotometric reading. This method has a sensitivity of 50 µg/liter (Drury 1975).

Table 5.3. Variations in coal ash composition (%) with rank

Rank	SiO_2	Al_2O_3	Fe_2O_3	TiO_2	CaO	MgO	Na_2O	K_2O	SO_3
Anthracite	48 - 68	25 - 44	2 - 10	1.0 - 2	0.2 - 4	0.2 - 1			0.1 - 1
Bituminous	7 - 68	4 - 39	2 - 44	0.5 - 4	0.7 - 36	0.1 - 4	0.2 - 3	0.2 - 4	0.1 - 32
Subbituminous	17 - 58	4 - 35	3 - 19	0.6 - 2	2.2 - 52	0.5 - 8			3.0 - 16
Lignite	6 - 40	4 - 26	1 - 34	0.0 - 0.8	12.4 - 52	2.8 - 14	0.2 - 28	0.1 - 1.3	8.3 - 32

Source: O'Gorman and Walker 1972, Table 13, p. 74. Reprinted by permission of the publisher.

5.2.2 Atomic absorption spectrometry

One of the most widely used techniques for analysis of trace elements is atomic absorption spectrometry (AAS) (Tables 5.4 and 5.5), in which samples and standards are aspirated into an AAS flame. A hollow cathode lamp provides a source of radiation that is characteristic for the element of interest. The absorption of characteristic energy by atoms of the metal in the flame can be related to the metal concentration in the aspirated sample. Sensitivity is adequate for all metals collected in air samples, but not satisfactory for beryllium, cadmium, calcium, chromium, manganese, molybdenum, nickel, or tin in biological samples (Kneip et al. 1975). The relative standard deviation of AAS measurements is about 3% in the ranges listed in Table 5.4.

Interferences occur less often with AAS than with many other methods (Jones and Manahan 1975). Potential interferences include (1) background or nonspecific absorption from particles produced in the flame, (2) spectral interferences caused by the adsorption of radiation by atoms other than those of the element being measured, (3) ionization interferences, (4) chemical interferences, and (5) physical interferences due to changes in viscosity or surface tension of the sample. Corrections for many of the interferences are given in Table 5.5.

In some instances, increased sensitivity can be obtained by replacing the AAS flame with a graphite, carbon, or tantalum-ribbon furnace in which the sample is heated to incandescence (flameless spectrometry). Talmi and Morrison (1972) and Talmi (1974) describe a radiofrequency induction furnace in which an induction-heated graphite crucible produces vaporization and atomization of the sample (Fig. 5.1). Flameless AAS may be used on all types of samples, but it works best on direct analyses of solid biological and environmental samples without previous processing (Drury 1975). Absolute sensitivities with the radiofrequency induction furnace are generally inferior to other nonflame systems, but matrix interferences are substantially lower (Talmi and Crosmun 1974). The relative sensitivity of the method is 10^{-4} to 10^{-6}% for zinc, silver, lead, cadmium, indium, and magnesium, with an average standard deviation of 8%. Talmi and Crosmun (1974) report detection limits of 5 pg for lead and copper. Interferences may result from smoke or salt particles produced during heating which cause light scattering (Drury 1975). Flameless techniques usually produce an absorbance peak of shorter duration than do flame techniques.

A major disadvantage of AAS is that at least 1 to 2 ml of solution is needed for determination of each element. Thus, for small samples, the necessary dilution will increase sensitivity requirements.

5.2.3 Spark-source mass spectrometry

Spark-source mass spectrometry involves high-energy excitation of solid samples by a radio-frequency spark, followed by spectrometric separation of the ionic fragments from the sample (Laitinen 1973). Detection limits for this method are given in Table 5.6. Generally, SSMS is capable of extremely high sensitivity and is applicable to about 70 elements (Laitinen 1973). Accuracy is usually 20 to 30%, but Carter et al. (1976) report that SSMS with isotope dilution is capable of trace element determination with a precision of 10%. Limitations of this method are its high cost, limited general accuracy, and the fact that most samples must be in solid form.

Table 5.4. Sensitivity, detection limit, and optimum working range for elements

Element	Sensitivity (μg/ml)	Range (μg/ml)	Detection limits		Minimum threshold limit value[a] (μg/m³)
			(μg/ml)	(μg/m³)	
Ag	0.036	0.5 - 5.0	0.003	0.1	10 (metal and soluble compounds)
Al	0.76	5 - 50	0.04	2	NL
Ba	0.20	1 - 10	0.01	0.4	500 (soluble compounds)
Be	0.017	0.1 - 1.0	0.002	0.08	2
Bi	0.22	1 - 10	0.06	3	NL
Ca	0.021	0.1 - 1.0	0.0005	0.02	5,000 (CaO)
Cd	0.011	0.1 - 1.0	0.0006	0.03	200 (metal dust and soluble salts)
					100 (cadmium oxide fume)
Co	0.066	0.5 - 5.0	0.007	0.3	100 (metal fume and dust)
Cr	0.055	0.5 - 5.0	0.005	0.2	100 (chromic acid and chromates, as CrO_3)
					500 (soluble chromic, chromous salts)
					1,000 (metal and insoluble salts)
Cu	0.040	0.5 - 5.0	0.003	0.1	100 (fume)
					1,000 (dusts and mists)
Fe	0.062	0.5 - 5.0	0.005	0.2	10,000 (iron oxide fume, as iron oxide)
					1,000 (soluble compounds)
In	0.38	5 - 50	0.05	2	100 (metal and compounds)
K	0.010	0.1 - 1.0	0.003	0.1	NL
Li	0.017	0.1 - 1.0	0.002	0.08	25 (as lithium hydride)
Mg	0.003	0.05 - 0.50	0.0003	0.01	10,000 (as magnesium oxide fume)
Mn	0.026	0.5 - 5.0	0.003	0.1	5,000 (metal and compounds)
Na	0.003	0.05 - 0.50	0.0003	0.01	2,000 (as sodium hydroxide)
Ni	0.066	0.5 - 5.0	0.008	0.3	1,000 (metal and soluble compounds)
Pb	0.11	1 - 10	0.02	0.8	150 (inorganic compounds, fumes, and dusts)
Rb	0.042	0.5 - 5.0	0.003	0.1	NL
Sr	0.044	0.5 - 5.0	0.004	0.2	NL
Tl	0.28	5 - 50	0.02	0.8	100 (soluble compounds)
V	0.88	10 - 100	0.1	4	500 (V_2O_5 dust)
					100 (V_2O_5 fume)
Zn	0.009	0.1 - 1.0	0.002	0.08	1,000 ($ZnCl_2$ fume)
					5,000 (ZnO fume)

[a]NL signifies no limit expressed for this element or its compounds.

Source: Kneip et al. 1975, Table 2, p. 385. Reprinted by permission of the publisher.

Table 5.5. Flame and operating conditions for elements

Element	Type of flame	Analytical wavelength (nm)	Interferences[a]	Remedy[a]
Ag	Air-C_2H_2 (oxidizing)	328.1	IO_3^-, WO_4^{2-}, MnO_4^{2-}	b
Al[c]	N_2O-C_2H_2 (reducing)	309.3	Ionization, SO_4^{2-}, V	b, d, e
Ba	N_2O-C_2H_2 (reducing)	553.6	Ionization, large concentration of Ca	d, f
Be[c]	N_2O-C_2H_2 (reducing)	234.9	Al, Si, Mn	b
Bi	Air-C_2H_2 (oxidizing)	223.1	None known	
Ca	Air-C_2H_2 (reducing) N_2O-C_2H_2	422.7	Ionization and chemical	d, e
Cd	Air-C_2H_2 (oxidizing)	228.8	None known	
Co[c]	Air-C_2H_2 (oxidizing)	240.7	None known	
Cr[c]	Air-C_2H_2 (oxidizing)	357.9	Fe, Ni	b
Cu	Air-C_2H_2 (oxidizing)	324.8	None known	
Fe	Air-C_2H_2 (oxidizing)	248.3	High Ni concentration, Si	b
In	Air-C_2H_2 (oxidizing)	303.9	Al, Mg, Cu, Zn, $H_xPO^x_4{}^{3-}$	b
K	Air-C_2H_2 (oxidizing)	766.5	Ionization	d
Li	Air-C_2H_2 (oxidizing)	670.8	Ionization	d
Mg	Air-C_2H_2 (oxidizing) N_2O-C_2H_2 (oxidizing)	285.2	Ionization and chemical	d, e
Mn	Air-C_2H_2 (oxidizing)	279.5	None known	
Na	Air-C_2H_2 (oxidizing)	589.6	Ionization	e
Ni	Air-C_2H_2 (oxidizing)	232.0	None known	
Pb	Air-C_2H_2 (oxidizing)	217.0 283.3	Ca, high concentration SO_4^{2-}	b
Rb	Air-C_2H_2 (oxidizing)	780.0	Ionization	d
Sr	Air-C_2H_2 (reducing) N_2O-C_2H_2 (reducing)	460.7	Ionization and chemical	d, e
Tl	Air-C_2H_2 (oxidizing)	276.8	None known	
V[c]	N_2O-C_2H_2 (reducing)	318.4	None known in N_2O-C_2H_2 flame	
Zn	Air-C_2H_2 (oxidizing)	213.9	None known	

[a]High concentrations of Si in the sample can cause an interference for many of the elements in this table and may cause aspiration problems. No matter what elements are being measured, if large amounts of silica are extracted from the samples the samples should be allowed to stand for several hours and centrifuged or filtered to remove the silica.

[b]Samples are periodically analyzed by the method of additions to check for chemical interferences. If interferences are encountered, determinations must be made by the standard additions method, or, if the interferent is identified, it may be added to the standards.

[c]Some compounds of these elements will not be dissolved by the procedure described here. When determining these elements, one should verify that the types of compounds suspected in the sample will dissolve if using this procedure.

[d]Ionization interferences are controlled by bringing all solutions to 1000 ppm Cs (samples and standards).

[e]1000 ppm solution of La as a releasing agent is added to all samples and standards.

[f]In the presence of very large Ca concentrations (greater than 0.1%), a molecular absorption from $Ca(OH)_2$ may be observed. This interference may be overcome by using a background correction when analyzing for Ba.

Source: Kneip et al. 1975, Table 1, p. 384. Reprinted by permission of the publisher.

5.2.4 <u>Microwave plasma emission</u>

Microwave plasma emission spectrometry has been used to analyze organomercury compounds (Talmi and Mesmer 1975), alkyl-arsenic acids (Talmi and Bostick 1975), and selenium (Talmi and Andren 1974). An intense argon or helium discharge, powered by a microwave generator, is produced in a capillary tube, and the sample (previously separated by gas chromatography) is placed inside. The sample is instantly fragmented into molecular species and free atoms. The plasma is an efficient spectroscopic excitation source and produces a characteristic spectrum which can be monitored. The detection limits are 20 and 40 pg for arsenic and selenium, respectively, and 0.002 to 0.0005 ng for mercury.

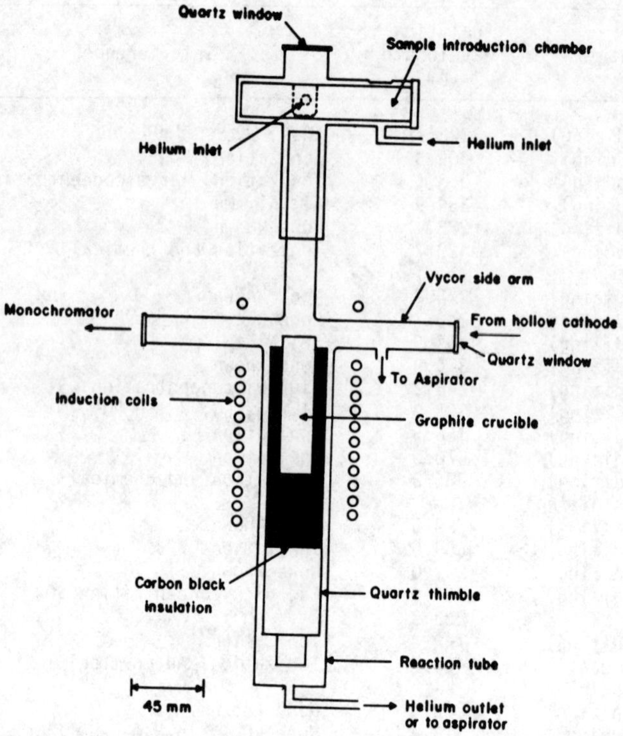

Fig. 5.1. Radiofrequency source chamber. Source: From Talmi and Morrison 1972, Fig. 2, p. 1456. Reprinted with permission from *Anal. Chem.* Copyright by the American Chemical Society.

Table 5.6. Detection limits (ng) for elements by spark source mass spectrometry

Ag, 0.2	Er, 0.5	Mn, 0.05	Sm, 0.5
Al, 0.02	Eu, 0.2	Mo, 0.3	Sn, 0.3
As, 0.06	Fe, 0.05	Na, 0.02	Sr, 0.09
Au, 0.2	Ga, 0.09	Nb, 0.08	Ta, 0.2
Ba, 0.2	Gd, 0.5	Nd, 0.4	Tb, 0.1
Be, 0.008	Hg, 0.6	Os, 0.4	Th, 0.2
Bi, 0.2	Ho, 0.1	Pb, 0.3	Ti, 0.05
Ca, 0.03	In, 0.1	Pd, 0.3	Tl, 0.2
Cd, 0.3	Ir, 0.3	Pr, 0.1	Tm, 0.1
Ce, 0.1	K, 0.03	Pt, 0.5	V, 0.04
Co, 0.05	La, 0.1	Re, 0.2	W, 0.5
Cr, 0.05	Li, 0.0006	Rh, 0.09	Yb, 0.5
Cs, 0.1	Lu, 0.1	Ru, 0.03	Zn, 0.1
Cu, 0.08	Mg, 0.03	Sc, 0.04	Zr, 0.1
Dy, 0.5			

Source: Dulka and Risby 1976, Table VII, p. 649 A.
Reprinted by permission of the publisher.

5.2.5 X-ray fluorescence

X-ray fluorescence (XRF) is a cheap and rapid quantitative analysis tool, requiring no time-consuming chemical concentration or alteration steps (Sparks et al. 1974). With XRF, K or L orbital electrons are ejected from atoms of the element when the sample is irradiated by an x-ray source. A series of x-ray lines are emitted when the ejected electrons are replaced by those from outer orbits; the x-ray spectrum given off is then read by a detector (Drury 1975).

Major limitations of the method are lack of sensitivity, dependence on sample thickness, and application restricted to the heavier elements (Laitinen 1973). Detection limits for several elements are given in Table 5.7. Sparks et al. (1974) describe a system that uses a mono-chromatic x-ray diffractor and a solid-state-detector energy-analyzing system (Fig. 5.2). The monochromation by diffraction improves detectability as compared with systems using white radiation or transmission target x-ray tubes. This system is capable of detection limits below 1 ppm by weight for heavy elements to 10 ppm for lighter elements in a variety of samples. The accuracy of the monochromatic method compares favorably with NBS standard reference materials for many elements (Table 5.8).

Table 5.7. Detection limits (μg) for x-ray fluorescence spectrometry[a]

Ag, 1.2	Cs, 0.15	Nd, 0.30	Sr, 0.00007
Al, 5.0	Cu, 0.00002	Ni, 0.06	Tb, 159/ml
As, 0.11	Eu, 0.66	P, 0.001	Te, 0.12
Au, 0.001/cm^2	Fe, 0.0085	Pb, 0.0003	Th, 6.5/ml
Ba, 0.12	Ga, 0.01	Rb, 0.0075	Ti, 0.001
Bi, 0.61	Hg, 0.24	Rh, 103/ml	U (as UO$_2$), 0.72
Ca, 0.100	In, 1.1	Sc, 0.38	U, 0.00002
Cd, 0.40	K, 0.52	Se, 0.020/cm^2	Y, 0.22
Ce, 0.17	La, 0.12	Si, 170 ppm	Yb, 6.8/ml
Co, 0.05	Mn, 0.00015	Sm, 4.1/ml	Zn, 0.00004
Cr, 0.00006	Mo, 0.072	Sn, 3.9 ppm	Zr, 0.00002

[a]Elements not available: Be, Dy, Er, Gd, Ge, Hf, Ir, Li, Lu, Mg, Na, Nb, Np, Os, Pa, Pd, Pr, Pu, Re, Sb, Ta, Tm, and W.

Source: Dulka and Risby 1976, Table X, p. 651 A. Reprinted by permission of the publisher.

5.2.6 Neutron activation analysis

Neutron activation analysis (NAA) is a highly sensitive (Table 5.9), nondestructive method for analysis of many elements (Lyon and Emery 1975). Basically, samples are irradiated in a nuclear reactor, which produces radioisotopes of the element of interest. Appropriate measurement of daughter activities affords a means for calculating the parent isotope (element) concentrations (Drury 1975). Disadvantages of this method are the expense of facilities and the slowness of analysis, especially for elements yielding nuclei having long half-lives (Laitinen 1973). As a result, NAA is usually employed only in multielement analysis, where cost and time are not of great importance.

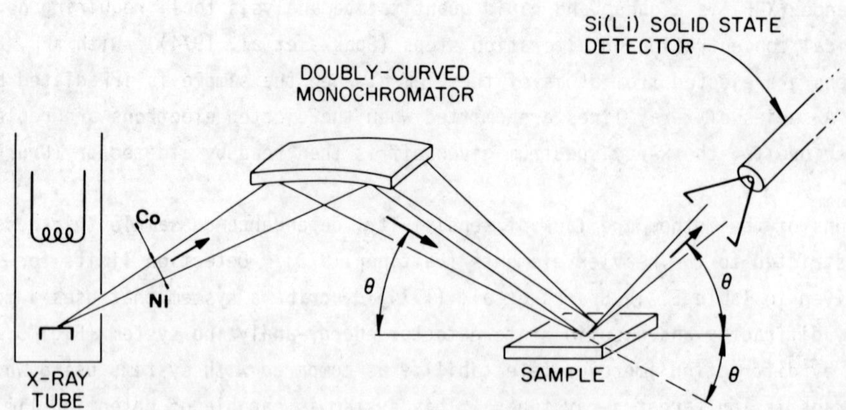

Fig. 5.2. Arrangement using a primary beam monochromator to select a monoenergetic x-ray source to excite characteristic x rays for fluorescent analysis. Solid state detector provides the energy resolution for identifying the fluorescent energies from the sample. Source: From Sparks et al. 1974, Fig. 1, p. 362. Reprinted by permission of the publisher.

Table 5.8. Comparison of concentrations determined by XRF
with NBS standard reference materials

	Orchard leaves		Coal	
Element	NBS 1571 reference standard (ppm)	XRF (ppm)	NBS 1632 reference standard (ppm)	XRF (ppm)
Fe	300 ± 20	301 ± 2.5	(8600)	7150 ± 800
Cu	12 ± 1	9.6 ± 1.7	18 ± 2	22.6 ± 3
Zn	25.3 ± 3	25 ± 1.6	37 ± 4	36.6 ± 1.4
As	14 ± 2	15.3 ± 0.5	5.9 ± 0.4	3 ± 2
Se			2.8 ± 0.2	5.5 ± 0.5
Br	(10)	6.6 ± 0.4		23.7 ± 3.2
Rb	12 ± 1	11.2 ± 0.4		28.6 ± 3.2
Sr	(37)	34.3 ± 0.5		155 ± 15
Pb	45 ± 3	40 ± 2	30 ± 9	23 ± 0.9

Source: Sparks et al. 1974, Table I, p. 365. Reprinted by permission of the publisher.

Table 5.9. Sensitivities[a] for elements by neutron activation analysis[b]

Sensitivity (g)	Elements
$10^{-13} - 10^{-12}$	Dy, Eu
$10^{-12} - 10^{-11}$	Au, In, Mn
$10^{-11} - 10^{-10}$	Hf, Ho, Ir, La, Re, Rh, Sm, V
$10^{-10} - 10^{-9}$	Ag, Al, As, Ba, Co, Cu, Er, Ga, Hg, Lu, Na, Pd, Pr, Sb, Sc, U, W, Yb
$10^{-9} - 10^{-8}$	Cd, Ce, Cs, Gd, Ge, Mo, Nd, Os, Pt, Ru, Sr, Ta, Tb, Th, Tm
$10^{-8} - 10^{-7}$	Bi, Ca, Cr, Mg, Ni, Rb, Se, Te, Ti, Tl, Zn, Zr
$10^{-7} - 10^{-6}$	Pb
$10^{-6} - 10^{-5}$	Fe

[a]Sensitivity based on irradiation period of 0.5 T ½ or 10 h; whichever
is less at a flux of 10^{13} neutrons cm^{-2} s^{-1}. Activity measured by NaI
(Ti) gamma spectrometry.
[b]Sensitivities for activation analysis are dependent on the thermal
neutron fluxes at the irradiation facility, time of irradiation, and
half lives of the radionuclides measured, among other factors. Hence,
the numbers quoted here are only representative of the high sensitivities that can be obtained by neutron activation analysis.

Source: Dulka and Risby 1976, Table VI, p. 649 A. Reprinted by permission
of the publisher.

5.3 TRACE ELEMENT QUANTIFICATION

Quantification of trace elements in coal is a first step in appraising the potential for trace
element catalysis of coal conversion reactions and the possible environmental impact of the
conversion process. Trace elements have been determined by a wide variety of techniques, some
of which have been discussed in more detail in Sect. 5.2. Pollock (1975) used dry ashing and
acid dissolution followed by conventional AAS to determine lithium, beryllium, vanadium,
chromium, manganese, cobalt, nickel, copper, zinc, silver, cadmium, and lead. Flameless AAS
determinations of bismuth, selenium, tin, tellurium, beryllium, lead, arsenic, cadmium,
chromium, antimony, and germanium were conducted with the aid of a graphite furnace accessory.
Mercury can be determined by this technique after oxygen bomb combustion.

5.3.1 Optical emission spectroscopic techniques

Dreher and Schleicher (1975) discuss optical emission spectroscopic techniques used to analyze
16 trace elements in coal ash produced at high temperatures. Coal samples are ashed in a high-
temperature muffle furnace, which mainly produces oxides of the elements measured in this
method. Two kinds of instruments for emission spectrometric measurements are discussed: (1)
a direct-reading spectrometer, used to analyze tin, boron, lead, copper, cobalt, nickel,
beryllium, chromium, vanadium, molybdenum, and germanium, and (2) a photographic plate recording
spectrograph for determination of boron, manganese, chromium, lead, beryllium, vanadium, silver,
copper, zinc, zirconium, cobalt, and nickel. Standard operating conditions for both instruments are shown in Table 5.10. Major drawbacks in the use of emission techniques include the
great care that must be exercised in the ashing procedure to guard against contamination or
loss of certain elements and the sample inhomogeneities that cause variations in the emission
intensities.

Table 5.10. Standard operating conditions

	Spectrometric	Spectrographic
Instrument	Jarrell-Ash Model 750 Atomcounter	Jarrell-Ash 3.4-m Ebert with 3-lens collimating system
Arc current, A	15 (short circuit)	11 (true)
Arc voltage, V, dc		250
Electrode gap, mm	6	4
Exposure time, sec	65	Completion + 10 (usually about 150)
Electrodes		
Sample	National L-3979	National L-3903 (ASTM C-13)
Counter	National L-4036 (ASTM C-1)	National L-4037 (ASTM C-1 with radial tip)
Electrode atmosphere	80% argon, 20% oxygen at 10 scfh	80% argon, 20% oxygen at 14 scfh
Sample charge, mg	15	20
Attenuation[a]		Neutral density filter, 6% T
Step sector		1:1.585, 8 steps
Entrance slit width, μm	10	25
Exit slit width, μm	50	
Development		3.0 min in Eastman D-19, 30 sec in 2.5% acetic acid, 4.0 min in Eastman fixer

[a]Because the concentrations of most trace elements in the ash were too high to be recorded on a total energy ignition, it was necessary to insert a neutral density filter in the light path after the second lens of the three-lens collimating system.

Source: Dreher and Schleicher 1975, Table II, p. 39. Reprinted by permission of the publisher.

5.3.2 SSMS in mine and coal dust analysis

Sharkey, Kessler, and Friedel (1975) have used spark-source mass spectrometry to determine trace elements in respirable-range mine dusts and prepared coal dusts. The instrument used was a commercial Mattaugh-Herzog mass spectrometer equipped with photographic and electrical detection systems and a radiofrequency spark source; instrument resolution is 1 part in 5000. Electrodes are prepared by mixing the samples with equal parts of pure graphite plus 50 ppm indium to ensure homogeneous mixing and to determine the plate sensitivity. This mixture, one for each sample, is pressed into electrodes in polyethylene slugs in a commercial isostatic die. Analysis of samples is conducted by making graded photoplate exposures ranging from 1×10^{-12} to 3×10^{-7} coulombs.

5.3.3 AAS analysis of siliceous materials

Muter and Nice (1975) have used atomic absorption spectroscopy to analyze coal samples for major and minor constituents in siliceous materials. A lithium tetraborate fusion-acid dissolution technique was applied to coal ash for the determination of silicon, aluminum, iron, titanium, calcium, magnesium, sodium, potassium, manganese, nickel, barium, silver, gold, chromium, copper, gallium, indium, molybdenum, antimony, and zinc. The technique used

high-temperature ashing to produce samples for analysis. Standards for ash analyses are
prepared in aqueous solution from commercially available pure salts, with appropriate acid
additions, where necessary, to match acid concentrations in the samples and to hold materials
in solution. The lithium tetraborate fusion technique makes determinations of mercury, and
possibly lead and tin, questionable because of the volatility of the elements. Relative error
values for the elements ranged from 0% to a high of 18%, with most values being 5% or less.
The 18% error was for indium and was attributed to the low concentration of the element.

5.3.4   XRF analysis of whole coal

Kuhn, Harfst, and Shimp (1975) have used x-ray fluorescence for the analysis of whole coal.
High- and low-temperature ashes were analyzed prior to whole coal samples. The first phase
of the work was to prepare calibration curves by independent methods and use the curves as
standards. The accuracy of the x-ray fluorescence method is dependent on the accuracy of the
methods used to prepare the calibrating standards. The comparative accuracy and detection
limits of this method are shown in Table 5.11. Coal samples are ground with a binder and
pressed into a 1-1/8-in.-diam disc under 40,000 psi. Samples showed the greatest precision
when ground to -325 mesh and failed to yield acceptable results when left as large as -60 mesh.
This method yields reasonably accurate analyses of whole coals with a simplicity and speed
that makes it important in large-scale surveys of coal resources (Kuhn, Harfst, and Shimp 1975).

Table 5.11. Comparative accuracy for whole coal and limits
of detection based on 50 samples

Element	Accuracy (%)	Limit of detection (%)
Al	±0.08	0.012
Si	±0.10	0.016
S	±0.04	0.003
Cl	±0.01	0.0015
K	±0.02	0.003
Ca	±0.04	0.0005
Mg	±0.010	0.015
Fe	±0.10	0.005

Element	Accuracy (ppm)	Limit of detection (ppm)
Ti	±6.3	7.5
V	±3.1	2.5
Ni	±1.9	3.5
Cu	±2.5	1.0
Zn	±23.0	2.0
As	±4.3	3.2
Pb	±7.7	1.8
Br	±1.0	0.5
P	±15.0	15.0
Co	±1.3	2.5
Mn	±3.4	4.5
Cr	±2.1	1.5
Mo	±5.2	5.0

Source: Kuhn, Harfst, and Shimp 1975, Table II, p. 69.
Reprinted by permission of the publisher.

5.3.5 SSMS and TEMS in coal analysis

Carter, Walker, and Sites (1975) have used spark-source mass spectrometry (SSMS) and thermal emission mass spectrometry (TEMS) to determine trace elements in coal. Conducting electrodes for general scan analysis are prepared by mixing the pulverized coal or ash sample with an equal amount of pure silver powder (99.999% silver). The mixture is then pressed into polyethylene slugs in an isostatic electrode die at 25,000 psi for 1 min.

5.3.6 TEMS in lead and uranium determination

Thermal emission mass spectrometry has been used to determine lead and uranium in coal and fly ash. A three-stage thermal emission mass spectrometer is used in this work. Lead analyses are conducted by thermally producing lead ions at rhenium filament temperatures of 1100 to 1300°C and producing uranium ions at rhenium filament temperatures of 1700 to 1825°C (Cameron, Smith, and Walker 1969). Small quantities of uranium (10 to 100 ng) in the form of uranium nitrate are loaded onto rhenium V-type canoe filaments, which produce ions for analysis. Pulse counting is used for ion detection. The loaded filaments are loaded into the instrument, and the source region is pumped down to a pressure of 1×10^{-5} torr. The filaments were carburized in situ, with benzene vapor introduced through a Granville-Phillips variable leak. The resulting carburized filaments reduce the UO^+ and UO_2^+ ion signals to near zero and enhance the U^+ signal. The precision of an isotope dilution analysis, when using this technique, where reagent blank contributions are kept insignificant, is usually within ±1% (Carter, Walker, and Sites 1975).

5.3.7 Rapid coal analysis by NAA

Frost et al. (1975) have used NAA for the determination of antimony, arsenic, bromine, cadmium, cesium, gallium, mercury, rubidium, selenium, uranium, and zinc. The technique involves irradiating the coal sample with thermal neutrons and determining the elements with an NaI (TI) detector and a 400-channel, pulse-height analyzer. The use of a solid-state Ge (Li) detector and a 1000- to 4000-channel analyzer can extend instrumental NAA to as many as 43 major, minor, and trace elements. The range and scope of the NAA methods make it a good procedure for rapidly monitoring the composition of many coal samples.

5.4 ANALYSIS OF SULFUR EMISSIONS

5.4.1 Sulfur dioxide

Atmospheric sulfur dioxide determination is complicated by the presence of oxygen and the coexistence of other pollutants such as oxidants (ozone and nitrogen oxides) or reductants (hydrogen sulfide and disulfide) that either react with the sulfur dioxide itself during sampling and storage or interfere with subsequent analytical processes (SCOPE 1973). Acidic and basic constituents in air also contribute to the problem of accurate sulfur dioxide measurement. A variety of techniques are employed for atmospheric sulfur dioxide determination; conductometry, polarography, iodometry, acidimetry, gas chromatography, flame photometry, coulometry, and spectrophotometry. Of these methods, the spectrophotometric method introduced by West and Gaeke in 1956 appears to be the best, especially for critical studies, because this technique is sensitive, accurate, and relatively free from interferences.

The basis of the West-Gaeke method is the absorption and stabilization of sulfur dioxide from air by a sodium (or potassium) tetrachloromercurate(II) solution to form the dichlorosulfito-mercurate(II) complex (SCOPE 1975; Chess 1974). This complex resists oxidation by atmospheric oxygen and is stable in the presence of small amounts of oxidants. The addition of acid-bleached pararosaniline hydrochloride and formaldehyde to the sulfito complex and measurement of the resulting colored pararosaniline methylsulfonic acid permit quantitative determinations of the collected sulfur dioxide. When high levels (2 ppm or above) of nitrogen oxides are expected, 1 ml of sulfamic acid can be added to the samples 10 min prior to analysis to counteract the high concentrations. The interference of large amounts of heavy metals may be eliminated by the addition of 0.079 of EDTA (ethylenediaminetetraacetate, tetrasodium salt) per liter of tetrachloromercury(II) ion prior to sampling.

The SCOPE (1973) report also describes the collection and determination of sulfur dioxide by a method incorporating permeation through a silicon membrane with subsequent determination by the West-Gaeke method. This incorporation allows sampling and determining of the average sulfur dioxide concentration in the ambient atmosphere for periods of from 6 hr to one week. The advantages of this procedure are that it is selective, sensitive, reproducible, and suitable for field studies. During a 24-hr sampling period, the lower limit of detection is 5 $\mu g/m^3$ and the analytical range is from 5 to 1000 $\mu g/m^3$. Under laboratory conditions, the precision is 2 to 3% when repetitive samples are run by the same operator, whereas in field studies, the accuracy is primarily limited by variations of ambient temperature; variations of $\pm 10°C$ will cause errors no greater than 5%. Particulates do not interfere with determination because they cannot penetrate the membrane; nor is there interference by hydrogen sulfide at a concentration of 600 $\mu g/m^3$, ozone at 500 $\mu g/m^3$, or nitrogen dioxide at 300 $\mu g/m^3$.

Stauff and Jaeschke (1975) report an analytical procedure that is a modification of the West-Gaeke method for sulfur dioxide determination. This chemiluminescence method stabilizes the sulfur dioxide by tetrachloromercury(II), as in the West-Gaeke method, but instead of reacting the sulfite with the pararosaniline to form pararosaniline methylsulfonic acid (a compound which is bleached by ozone), the sulfite is oxidized by potassium permanganate with the accompanying chemiluminescence being monitored by photomultiplier tubes. The lower detection limit is 0.5 ng/cm^3, and the concentration at which the error is less than 10% is 20 ng/cm^3. The new chemiluminescence method (Fig. 5.3) does not have the inhibition of dye development by atmospheric trace substances, as does the West-Gaeke method, and it permits measurement of sulfur dioxide traces in unpolluted air where the pararosaniline method fails to give meaningful results.

Jacobs (1960) describes six wet chemistry methods for the determination of sulfur dioxide in air, one of which is the previously described West-Gaeke method. The hydrogen peroxide method operates by trapping the sulfur dioxide in an impinger or a bubbler containing a dilute solution of hydrogen peroxide, thus forming sulfuric acid, which can be titrated to estimate the amount of sulfur dioxide present. Since this method titrates all acidic components, it can be used to determine total acid in the air with no interference from the hydrogen peroxide that decomposes to water.

The second wet chemistry method described by Jacobs is the iodine method, which works by trapping sulfur dioxide in a standard impinger containing a standard sodium hydroxide solution, which is subsequently acidified, forming sulfurous acid; the sulfurous acid is liberated and

ORNL DWG 76-14193

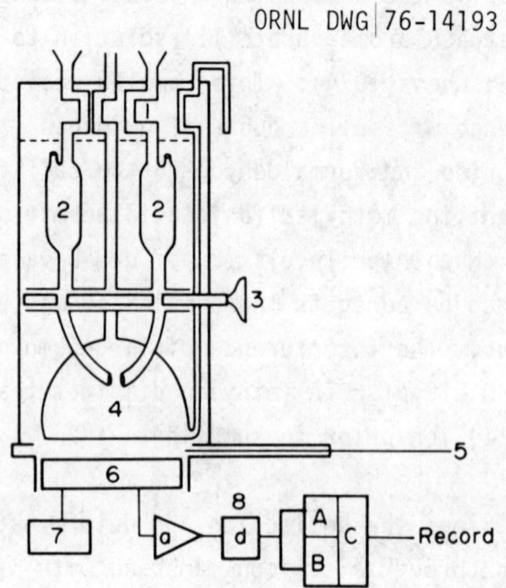

Fig. 5.3. Schematic of the apparatus used to measure chemiluminescence. Source: From Stauf and Jaeschke 1975, Fig. 1, p. 1038. Reprinted by permission of the publisher.

titrated with a standard iodine solution. The hydrogen peroxide method is more suitable for field work since the sulfuric acid sample can be titrated long after sampling, whereas with the iodine method, titration must be done immediately after sampling.

The fuchsine or rosaniline method uses sulfur dioxide trapped in a standard solution of sodium hydroxide in a microimpinger. The range of sulfur dioxide concentration is obtained by iodine titration, and then the trapping solution is used to develop a color in a decolorized fuchsine-formaldehyde solution. This method is relatively specific for sulfur dioxide.

The fourth, or iodine-thiosulfate, method is a common sulfur dioxide determining method, which functions by drawing the air through a standard solution of iodine in potassium iodide. The remaining iodine is subsequently estimated by titration with standard sodium thiosulfate solution.

The disulfitomercurate method or West-Gaeke method has been previously described.

The sixth method described by Jacobs for sulfur dioxide measurement is the lead peroxide cylinder method, which was designed in Great Britain to provide an index of the activity of sulfur dioxide in the atmosphere as a measure of its attack rate on buildings and other permanent fixtures constantly exposed to the slight concentrations of sulfur dioxide in the atmosphere. Lead peroxide reacts with sulfur dioxide,

$$PbO_2 + SO_2 = PbSO_4 , \qquad\qquad (1)$$

with the yield of lead sulfate being proportional to the concentration of sulfur dioxide for all concentrations up to one part per thousand. Jacobs (1960) cites Wilsdon and McConnell (1934) in stating that (1) this reaction proceeds at a uniform rate until more than 15% of the lead peroxide has been converted; (2) the reaction was nearly independent of air speed; and (3) a 10°C rise in temperature increased the reactivity by 4%.

Jacobs reports that Hochheiser, Braverman, and Jacobs (1955) have shown that the hydrogen peroxide, iodine, and fuchsine methods can be used interchangeably. The hydrogen peroxide method is more suitable for field work than the iodine method because the sample can be kept for long periods without titration, whereas titration must be conducted immediately after collection for the iodine method. The hydrogen peroxode is also easily adopted for laboratory use because (1) it is simple in operation; (2) it is fast; (3) the samples, as mentioned, have good stability; and (4) it is useful for determining the total acid content of air.

Table 5.12 indicates the variety of physical methods of analysis, and Table 5.13 shows commercial instruments available for sulfur dioxide detection.

Table 5.12. Physical methods of analysis

Method	Remarks
Flame ionization analyzer	Responds in proportion to number of carbon atoms in gas sample.
Mass spectrometry	Determines charge-mass ratio of organic fragments or ions.
Infrared absorption	Sample absorbs radiation in infrared region of spectrum; differences in absorption measured. Other regions of spectrum used (e.g., UV).
Emission spectroscopy	Excited solid provides spectra of metals present.
Flame photometry	Flame excites samples with low excitation potentials.
Atomic absorption	Sample absorbs radiation; emitted radiation is a function of atoms present.
Fluorescence spectroscopy	Excited sample may reemit excess of excited energy.

Source: Stern et al. 1973, Table 15-4, p. 180. Reprinted by permission of the publisher.

Table 5.13. Continuous automatic gaseous pollutant analyzers for sulfur dioxide

Method	Reaction	Comment
Electrolytic conductivity (pure water)	$SO_2 + H_2O = H_2SO_3$	Good sensitivity, possible CO_2 interference, not specific
Electrolytic conductivity (H_2O_2 and H_2SO_4)	$SO_2 + H_2O_2 = H_2SO_4$	Good sensitivity, not specific, most commonly used method, most experience
Amperometric	$SO_2 + I_2 + 2H_2O = 2I^- + SO_4^{2-} + 4H^+$	Not specific
Coulometric	$SO_2 + I_2 + 2H_2O = 2I^- + SO_4^{2-} + 4H^+$ $SO_2 + Br_2 + 2H_2O = 2Br^- + SO_4^{2-} + 4H^+$	Not specific
Colorimetric	West and Gaeke	Slow response, specific
Ultraviolet	Correlation spectrometry Flame emission	Used for long path measurement; measures all sulfur gases

Source: Stern et al. 1973, Table H.13, p. 186. Reprinted by permission of the publisher.

5.4.2 Sulfur trioxide and sulfuric acid

The analysis of sulfur trioxide and sulfuric acid emissions or any pollutant-containing emission from stationary sources and from ambient air involves two distinct operations: (1) collection of the sample by a variety of techniques such as absorption, condensation, or filtration and (2) subsequent analysis of the sample (Urone 1974). Urone reports that the greatest weakness in existing methods for analyzing emission sources is the sampling procedure.

5.4.2.1 Sampling

When sampling for sulfur trioxide and sulfuric acid, a serious problem arises because the sulfur dioxide oxidizes in solution, leading to high sulfate results. Depending on the sampling method used, other interferences may come from particulates, hydrochloric acid, nitrogen oxides (or nitric acid), various sulfate compounds, and heavy metal ions, which catalyze the oxidation of sulfur dioxide in solution by dissolved atmospheric oxygen (Driscoll 1972; Berger, Driscoll, and Morgenstern 1970; Johnstone and Coughanowr 1958; and Junge and Ryan 1958, as cited by Urone 1974).

Urone (1974) states that the selective removal of sulfur trioxide and sulfuric acid from a gas sample involves some variation or combination of two basic techniques — absorption or condensation followed by filtration.

Absorption in isopropyl alcohol solutions

Figure 5.4 illustrates the EPA acid mist analytical method, which uses a heated sampling probe fitted with a stainless steel nozzle to obtain the sample. The sample gas is then passed through a standard Greenburg-Smith impinger containing 100 ml of 80% isopropyl alcohol cooled in an ice bath. Urone (1974) cites Seidman (1958) and Fielder and Morgan (1960) as stating that the isopropyl alcohol presumably absorbs the sulfur trioxide and sulfuric acid and prevents the oxidation of sulfur dioxide. Any sulfur trioxide that escapes the system is assumed to be hydrolyzed to sulfuric acid mist, which is then collected with any mist not captured by the impinger by a glass-fiber filter. Sulfur dioxide, which is not highly soluble in the isopropyl alcohol solution, is absorbed in the following two impingers (Fig. 5.5), which contain 100 ml of 3% hydrogen peroxide. Water and any residual polar gases or vapors are stripped off by the fourth impinger, which contains 200 g of silica gel. One serious difficulty with this method is that, when sampling dry stack gases under field conditions, it is extremely difficult to prevent partial evaporation of the isopropyl alcohol absorbing solution.

Absorption in alkaline solutions

Sulfur trioxide and sulfuric acid measurements are generally high and erratic when sampled with alkaline solutions (Driscoll 1972, as cited by Urone 1974). The reason for such measurements is that sulfur dioxide is absorbed by alkaline solutions, and since it oxidizes somewhat readily, the determination of sulfuric acid and sulfur trioxide by the analysis of sulfur dioxide is difficult, if not impossible.

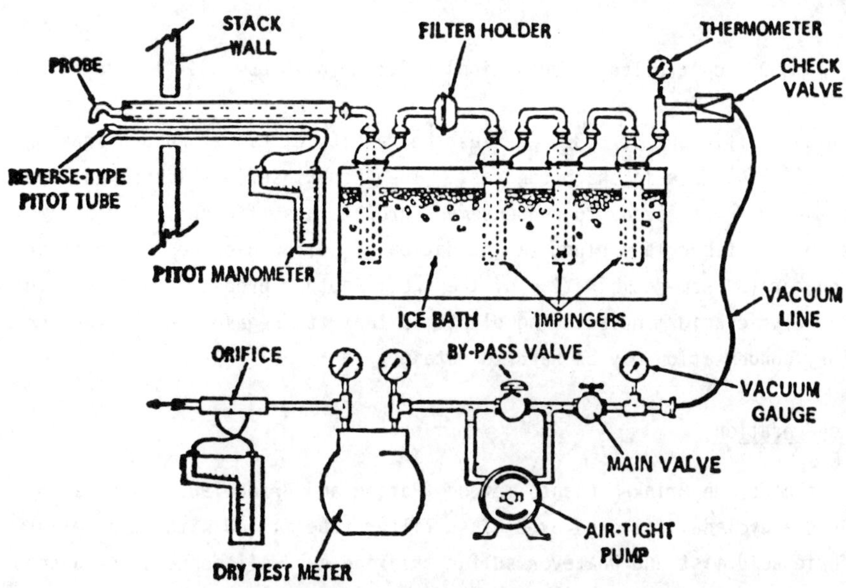

Fig. 5.4. EPA sulfuric acid mist sampling train. <u>Source</u>: From Urone 1974, Fig. 1, p. 247. Reprinted by permission of the publisher.

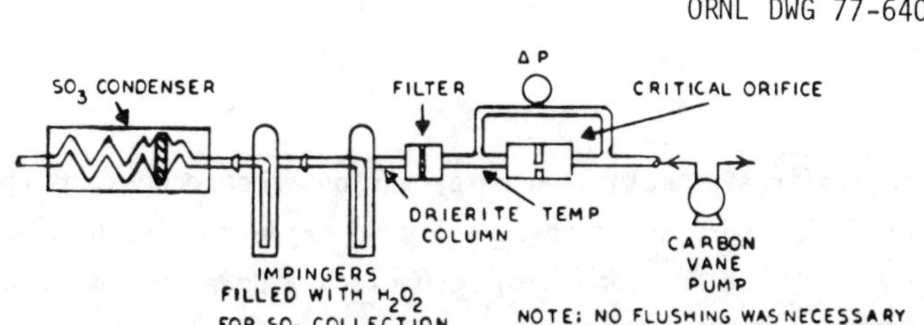

Fig. 5.5. Walden Research train "A" controlled condensation. <u>Source</u>: From Urone 1974, Fig. 2, p. 247. Reprinted by permission of the publisher.

Condensation

In reviewing papers by a number of authors, Urone (1974) finds that a number of sampling methods are based on the condensation of sulfur trioxide as sulfuric acid mist, followed by filtration, impingement, etc.

Controlled condensation

In the controlled condensation method (Fig. 5.5), the sample air stream passes through the coiled condenser before passing through a fritted glass disc built into the condenser. Urone (1974) notes that different researchers report contrasting results with the use of this method. For example, Walden Research Corporation found that, for a contact sulfuric acid manufacturing plant, the controlled condensation method gives results of less than 50% accuracy by the relation,

$$\text{ppm } SO_3 \text{ (controlled condensation)} = 0.14 + 0.45 \text{ ppm } SO_3 \text{ (IPA method)} , \qquad (2)$$

with a correlation coefficient of 0.80, whereas the results of Lisle and Sensenbaugh (1965), as cited by Urone (1974), are in sharp contrast. They report serious losses as a result of (1) failure of the sulfur trioxide to condense and grow to a sufficiently large particle size to be collected by the fiberglass plug, (2) oxidation of sulfur dioxide in the isopropyl alcohol solution, or (3) some combination of the two. Another problem that might influence the controlled sulfuric acid manufacturing plants is that stack gases can be very dry, and thus satisfactory condensation may be hard to obtain.

Uncontrolled condensation

Figure 5.6 illustrates the Brink-Monsanto method (Patton and Brink 1963, as cited by Urone 1974), which uses a cyclone, followed by a glass filter tube filled with pyrex glass wool to catch the sulfuric acid mist and whatever sulfur trioxide gas will condense or adsorb on the glass wool and walls of the sampling train up to the filter. Although the efficiency of isopropyl alcohol solutions for capturing all sulfur trioxide and sulfuric acid and preventing the oxidation of sulfur dioxide is in doubt, some methods (Fielder and Morgan 1960, as cited by Urone 1974) do pass the flue gases directly into impingers containing isopropyl alcohol solutions.

Filtration

Urone (1974), in reviewing filtration methods used by various researchers, finds that fiberglass and other types of filters can adsorb and oxidize sulfur dioxide to give high sulfate and low sulfur dioxide results, a fact that has long been recognized. However, the stability of sulfuric acid mist on a filtering medium is questionable. For example, Urone (1974) cites Richards (1973) as successfully volatilizing most of the sulfuric acid from a Teflon filter with warm dry air at 50°C; acid concentrations ranged from 0.08 to 16 μg. Although the purpose of Richard's experiment was to volatilize the sulfuric acid for flame photofluorescent measurement without decomposition of other sulfate compounds, including ammonium sulfate, it does indicate that the volatility of sulfuric acid is such that it could occur at temperatures encountered under field sampling conditions.

5.4.2.2 Analysis

Although most of the problem lies in the collection rather than the analysis of sulfur trioxide and sulfuric acid, the analytical procedures are discussed here to provide an understanding of the total process of sulfur trioxide and sulfuric acid monitoring. These analytical methods range from acid-base titrations to spectrophotometric measurements. Some of these, reviewed by Urone (1974), are described in the following subsections.

Sodium hydroxide titration

This method is simple and is used by many laboratories; however, it is not specific and depends on the assumption that sulfuric acid is the only acid present and that there are no basic substances present to interfere.

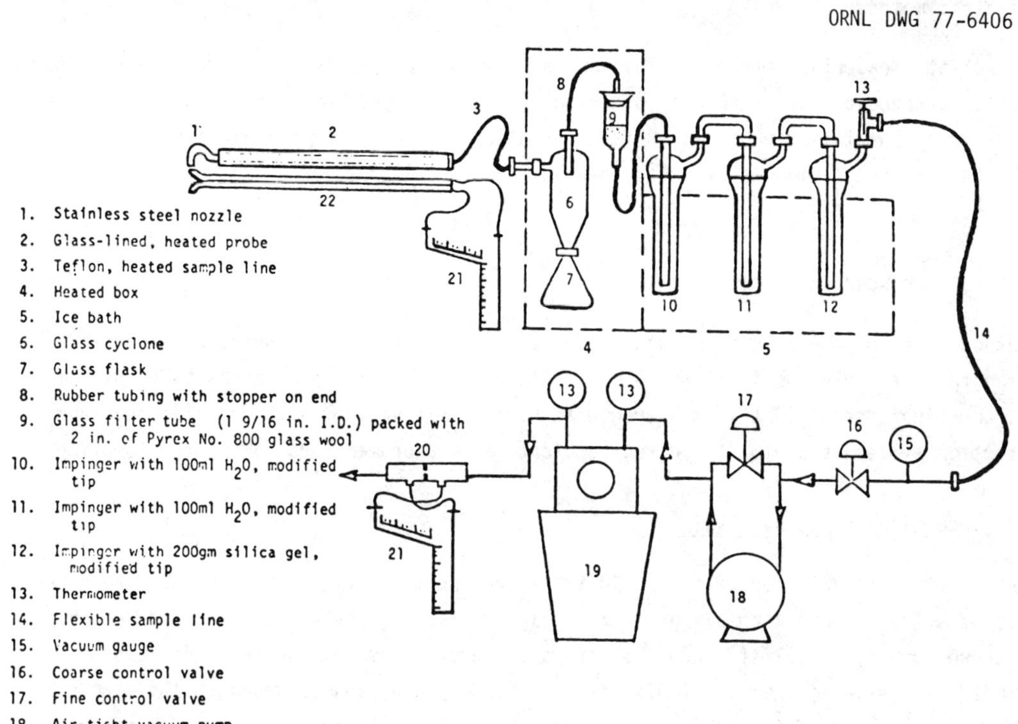

Fig. 5.6. Brink-Monsanto mist sampling train by Environmental Engineering, Inc. <u>Source</u>: From Urone 1974, Fig. 5, p. 249. Reprinted by permission of the publisher.

Barium ion titration

Thorin and nitrochromazo are the indicators for this method, which is largely specific for the sulfate ion. This titration is similar to, if not the same as, the barium sulfate method described by Jacobs (1960). Jacobs' procedure involves passing sulfuric acid mist and sulfur trioxide through an absorber of the bubbler type or other trapping device containing a known quantity of standard sodium hydroxide solution (usually 0.02 N). After sampling, the amount of alkali that reacted with the sulfuric acid and sulfur trioxide may be determined by titrating the excess sodium hydroxide with a solution of hydrochloric acid. The sulfate can then be precipitated with barium chloride, and the barium sulfate determined by estimation.

Chloranilate method

The release of the colored chloranilate ion from the insoluble barium chloranilate salt is the basis of this method. The chloranilate ion, released from the reaction of sulfate with barium chloranilate (which forms insoluble barium sulfate), colors the solution in proportion to the amount of sulfate originally present. This method is sensitive enough to be used for ambient air sulfuric acid measurements and is reliable as an analytical method for source emissions. The interference by heavy metal ions may be nullified by passing the adjusted sampling solution through an ion exchange column.

Turbidimetric method

In this sulfate measuring method, barium ion is used to precipitate the sulfate as a uniformly distributed suspension, the turbidity of which is measured spectrophotometrically. Jacobs (1960) reports that this method, which is a variation of the barium sulfate method, is preferred when very small amounts of sulfuric acid, sulfur trioxide, or sulfates are being determined.

Flame photoluminescent detector

This method works on the principle that sulfur compounds fluoresce in hydrogen-rich flames at a wavelength relatively specific for sulfur. After passing through a narrow-band optical filter, the fluorescent light is measured with high sensitivity by a photomultiplier tube. The detection system is useful for direct ambient air measurement and for source sampling.

5.4.2.3 Determination by pyrolysis

Dharmarajan et al. (1975) report that acid titration and turbidimetry methods for the determination of sulfuric acid aerosols are unreliable and lack sensitivity. Other methods based on spectrophotometry or sulfate reduction are also reported as inadequate. The method recommended by Dharmarajan et al. (1975) is developed by them and is based on the pyrolysis of an organic sulfate precipitate. This pyrolysis involves the thermal reduction of perimidylammonium sulfate $[(PDA)_2SO_4]$ (Fig. 5.7). The pyrolyzed effluent is then bubbled into sodium tetrachloromercurate(II), and the amount of sulfur dioxide is determined by the West-Gaeke method (Maddalone, Thomas, and West 1976). The most significant fact of the pyrolysis of perimidylammonium sulfate is the conversion efficiency of sulfate to sulfur dioxide; from 1 to 50 μg of the sulfate, as $(PDA)_2SO_4$, was converted to sulfur dioxide with 100% efficiency. Also important is the sensitivity of the pyrolytic method, which was calculated by using the West-Gaeke procedure at 0.1 μg of sulfate.

Thomas et al. (1976) also describe the measurement of sulfuric acid aerosol using the compound perimidylammonium. In the work by Thomas et al. (1976), the perimidylammonium, in association with bromide (PDA-Br), is impregnated on glass-fiber filters, whereas fluoropore filters were used by Dharmarajan et al. (1975). Table 5.14 illustrates that the impregnated glass-fiber filters are 90% more efficient than the fluoropore filters.

Thomas et al. (1976) point out that previous methods have suffered from sampling errors when coexisting pollutants such as Fe_2O_3, Al_2O_3, and PbO react with the sulfuric acid, thus neutralizing part of the acid to form salts. These sampling errors are eliminated by the new method because the reaction of the acid with PDA-Br on the filter surface forms stable perimidylammonium sulfate instantaneously, thus inhibiting any reaction between the acid aerosol and collected copollutants.

Glass-fiber filters impregnated with PDA-Br were shown to be 25% more efficient in the collection of acid aerosol than nonimpregnated filters. This increased efficiency is attributed to increased stabilization of the sulfuric acid aerosol, reduction in pore size, and coating of the alkaline sites on the glass-fiber filter. Direct pyrolysis of the impregnated filters with 0.5 to 78 μg of sulfuric acid results in an average recovery of 96 ± 5% without interference from salt-derived sulfate.

ORNL DWG 76-13713

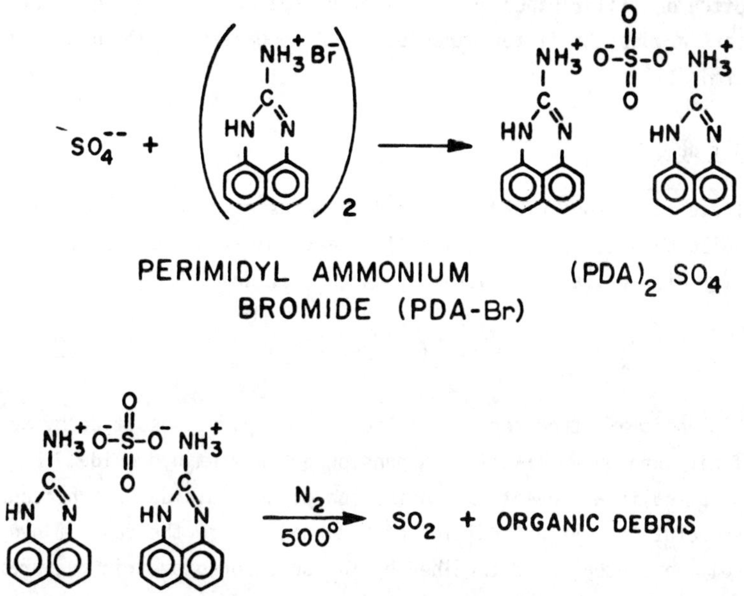

$^-SO_4^{--}$ + (PDA-Br)$_2$ → (PDA)$_2$ SO$_4$

PERIMIDYL AMMONIUM
BROMIDE (PDA-Br)
(PDA)$_2$ SO$_4$

$\xrightarrow[500°]{N_2}$ SO$_2$ + ORGANIC DEBRIS

Fig. 5.7. Pyrolytic sulfate method. <u>Source</u>: From Dharmarajan et al. 1975, Fig. 6, p. 291. Reprinted by permission of the publisher.

Table 5.14. Filter efficiency of impregnated glass fiber filters as compared with fluoropore

| Group | H$_2$SO$_4$ collected (µg) | | Relative increased efficiency (%) |
	Fluoropore	Impregnated glass fiber filters	
1	3.0	6.3	110
2	4.4	7.6	73
3	4.4	7.9	80
4	3.6	6.9	91

Source: Thomas et al. 1976, Table I, p. 640.
Reprinted by permission of the publisher.

One method that can be readily used for field studies as well as laboratory samples of sulfuric acid or total acidity is the ring-oven method (SCOPE 1973; Dharmarajan 1975). Described as sensitive, rapid, and easy to perform, the ring-oven method is based on the measurement of dissociated protons. The stoichiometric release by protons of bromine from a mixture of bromide and bromate is the basis for the determination:

$$5Br^- + BrO_3^- + 6H^+ = 3Br_2 + 3H_2O \ . \tag{3}$$

The liberated bromine is reacted with fluorescein to produce tetrabromafluorescein. The sample is then spotted on filter paper and the various reagents added sequentially so that a ring of color appears around the filter paper edge. A comparison is then made with standards for the final determination.

5.4.3 Hydrogen sulfide

The iodine-thiosulfate method described earlier as a means of determination for sulfur dioxide (Sect. 5.4.1) can also be used for hydrogen sulfide determination (Jacobs 1960). The iodine, which oxidizes sulfur dioxide, also oxidizes hydrogen sulfide:

$$H_2S + I_2 = 2HI + S \ . \tag{4}$$

Adams et al. (1975) report a method for the collection of hydrogen sulfide by aspirating a measured volume of air through an alkaline suspension of cadmium hydroxide. Precipitation of the sulfide as cadmium sulfide prevents air oxidation of the sulfide, which occurs rapidly in an aqueous alkaline solution. After addition of STRactan 10[®] to the cadmium hydroxide slurry precipitate, the sulfide content is determined by the spectrophotometric measurement of the methylene blue that is produced by the reaction of the sulfide with a strongly acid solution of N,N-dimethyl-p-phenylenediamine and ferric chloride. Completion time for the analysis is 24 to 26 hr, with the collection efficiency becoming variable below 10 $\mu g/m^3$.

Some interferences of this hydrogen sulfide collection method include the following:

1. Strong reducing agents such as sulfur dioxide can inhibit color formation.
2. The presence of 0.5 $\mu g/ml$ or more nitrogen dioxide gives a pale yellow color with the sulfide reagents.
3. Ozone at 57 ppb reduces recovery of sulfide precipitated as cadmium sulfide by 15%.
4. Atmospheric oxygen will oxidize sulfides in solution unless inhibitors such as cadmium and STRactan 10[®] are present.
5. The photodecomposition of cadmium sulfide will occur unless 1% STRactan 10[®] is added to the absorbing solution prior to sampling.
6. If the sulfide solution is oxidized to free sulfur by atmospheric oxygen, impingers or bubblers with fritted-end gas delivery tubes should not be used because the sulfur collects on the fritted glass membrane and may significantly change the flow rate of the air sample through the system.

5.4.4 Sulfate

The Environmental Protection Agency (1974) describes two methods for the analysis of suspended sulfate — the turbidimetric (sulfaver) method and the methylthymol blue method. The same sample collection and extraction procedure is used for both methods. This involves the collection of suspended sulfate from ambient particulate matter by the standard high-volume sampler with glass-fiber or other filter media. After a 24-hr sampling period, the exposed filter is equilibrated and weighed; a 2- by 20-cm (3/4- by 8-in.) strip of it is placed in an Erlenmeyer flask, 25 ml of distilled water is added, and the mixture is refluxed for 30 min. When the solution has cooled, it is filtered through acid-washed Whatman No. 42 filter paper and then diluted to 50 ml with distilled water.

In the turbidimetric method of analysis, barium chloride is used to treat an aqueous extract of the sample in the presence of sulfaver, a proprietary stabilizing agent, forming barium sulfate crystals of uniform size. The absorbance of the barium sulfate suspension is measured by a spectrophotometer, with the sulfate ion concentration being determined by comparison with a previously determined standard. Provided that uniformly good technique is used and interferences are absent, sulfate ion concentrations are reproducible within ±5.8% at a 95% confidence level.

In the methylthymol blue method, the analysis is automated. The Environmental Protection Agency (1974) indicates that the methylthymol blue method has replaced the turbidimetric method because it makes possible a higher rate of analysis and reduces the chance for operator error.

> The filter extract sample is first passed through a cation-exchange column to remove
> interferences. The sample containing sulfate is then reacted with barium chloride
> at a pH of 2.5 to 3.0 to form barium sulfate. Excess barium reacts with methyl-
> thymol blue to form a blue-colored chelate at a pH of 12.5 to 13.0. The uncomplexed
> methylthymol blue color is gray; if it is all chelated with barium, the color is blue.
> Initially, the barium chloride and methylthymol blue are equimolar and equivalent to
> the highest concentration of sulfate ion expected; thus the amount of uncomplexed
> methylthymol blue, measured at 460 nm, is equal to the sulfate present.

The methylthymol blue method has a 95% confidence level at 200 ppm and a detection limit of 10 ppm.

5.4.5 Organic sulfur compounds

Rasmussen (1974) reports that trace organic sulfur compounds such as methyl mercaptan, dimethyl-sulfide, and dimethyldisulfide have been neglected or underestimated in connection with atmospheric chemistry. The lack of comprehensive analyses of these trace compounds and other trace sulfur gases probably results from the difficulty involved in the gas chromatographic analysis of ambient air for low (parts per billion) concentrations. Rasmussen lists four major technical problem areas that have hindered a successful solution of the analysis of trace sulfur gases in air:

> 1) sulfur compounds at ppb concentrations are lost through adsorptive site scavenging
> on the column and in the sampling-injection system of the instrument; 2) selective
> detection of small concentrations of sulfur gases in the presence of many times
> greater concentrations of other air constituents is difficult; 3) it is necessary
> to optimize the flow characteristics of the hydrogen-rich flame and to optimize the
> photomultiplier bias voltage with suppression of the high-frequency, noise-output
> signal from the photomultiplier tube prior to amplification by the electrometer;
> 4) a routine high-ratio enrichment freeze-out loop capable of handling large volumes
> of air for subsequent slug-injection onto the head of the column is also necessary.

Rasmussen (1974) describes an instrumental analysis that has been successfully employed in the study of trace quantities of organic sulfur compounds.

> The Brody-Chaney Flame Photometric Detector (FPD) is designed to provide a semi-
> specific response to volatile sulfur compounds at 394 nm when burned in a hydrogen-
> rich flame; this effect provides a two orders-of-magnitude increase in sensitivity
> for sulfur compounds compared to the detection limits of the conventional flame
> ionization detector for organic sulfur compounds, thus making it possible to detect
> 0.1 ng of sulfur.

To obtain optimum sensitivity, there must be proper selection of the bias voltage (-1000 to -750 V) and the temperature of the photomultiplier tube. Also, the column support phase and the FEP Teflon components must be properly conditioned and equilibrated to the trace sulfur gases to obtain surface adsorption stabilization. For parts-per-billion analyses, the Teflon tubing is packed with 1/4- to 1/2-mm-diam Teflon chips to provide a sampling system compatible with the needed stabilized sulfur adsorptive losses.

Rasmussen (1974) also describes the gas chromatographic analysis, which is made by using packed columns in a Varian-Aerograph 1490-5 sulfur gas analyzer.

> The operating conditions were optimized as follows: column — 10-ft x 1/8-in. FEP Teflon packed with 5% Triton X-305 coated on 100- to 120-mesh Chromosorb G; column temperature: 50°C isothermal, with a programmed temperature rate of 12°C/min to 80°C and hold. Column adsorption stabilization was obtained by conditioning the system to 63 ppb H_2S, 40 ppb MeS, and 75 ppb DMS at 50°C for 72 hr. No ghosting or memory peaks were obtained after 2 hr of secondary conditioning with SCOT® "zero" air. Conditions were as follows: carrier gas flow: air — 78 ml/min, hydrogen — 64 ml/min, make-up air — 38 ml/min; attenuation: 2 x 10^{-9} AFS; sample size: variable from 25 to 1000 ml. All samples were injected via the freeze-out loop.

5.4.6 Particulate sulfur

One of the promising techniques for accurate determination of submicrogram quantities of particulate sulfur, sulfur compounds on particles, is the thermal decomposition of particulate sulfur compounds to gaseous products and subsequent detection by a flame photometric detector (Fig. 5.8) (Husar, Husar, and Stubits 1975). Samples of ambient particulate matter were collected on glass-fiber tapes, and the filter deposits were extracted in 0.25 to 1 ml of double-distilled, deionized water at 25°C. Sample extracts and solutions were transferred to a tungsten boat and heated at 60°C for 30 sec until dry. The residue was then vaporized by capacitor discharge. Vaporized gaseous decomposition products of the sulfur compounds were carried to a flame photometric detector by a stream of clean charcoal-filtered air. The lower detection limit of this method is controlled by the purity of the distilled water; a 5-μl sample of distilled water gives a signal equivalent to 0.3 ± 0.05 ng sulfur, corresponding to a solution concentration of 0.06 μg/ml.

The precision inferred from three consecutive analyses performed at concentrations of 0.6, 1.6, 3.2, and 6.4 μg/ml sulfur as sulfuric acid is in the range of 2.5%, exluding extraction errors. When the extraction step is included, the precision of the flash vaporization-flame photometric detector is about half as good. Table 5.15 shows that the percent of sample recovery is between 88% (sample No. 40077B) and 10% (sample No. 40076F).

Ten samples of size-classified ambient aerosol were analyzed by x-ray fluorescence and then reanalyzed by the vaporization-flame photometric detector technique. A comparison indicates that the sulfur concentrations obtained by x-ray fluorescence are, on the average, 23% higher than those obtained by vaporization-flame photometric detection.

5.5 POLYCYCLIC AROMATIC HYDROCARBON EMISSIONS

Studies for more efficient determination and identification methods of polycyclic aromatic hydrocarbons (PAH) have been under way for more than a decade. Development of analytical

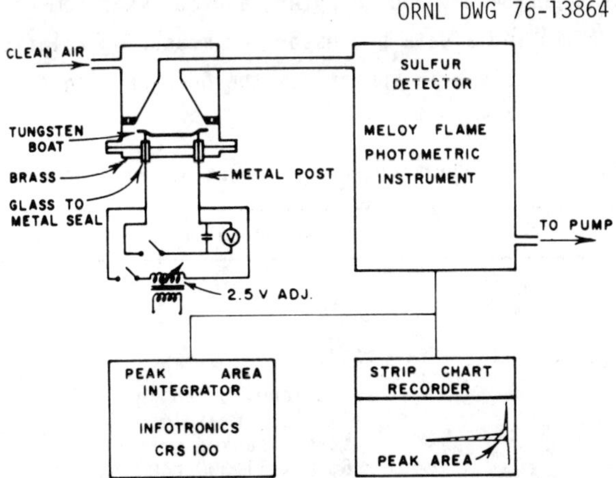

Fig. 5.8. Schematic of the sulfur detection system consisting of the flash vaporizer, Meloy SA-160 flame photometric detector, strip chart recorder, and the peak area integrator. Source: From Husar, Husar, and Stubits 1975, Fig. 1, p. 2063. Reprinted by permission of the publisher.

Table 5.15. Recovery of known amounts of sulfur as $(NH_4)_2SO_4$ added to samples[a]

Sample number	Sulfur (µg/ml)				Recovered sulfur (%)
	Sample	Added	Total	Recovered	
40075 F	4.06	3.09 ± 0.07	7.04	2.98	96.3
40075 B	4.00	3.09 ± 0.07	6.84	2.84	91.9
40076 F	3.38	3.09 ± 0.07	6.77	3.39	109.7
40076 B	3.23	3.09 ± 0.07	6.59	3.36	108.7
40077 F	3.94	3.09 ± 0.07	6.96	3.02	97.7
40077 B	3.22	3.07 ± 0.07	5.95	2.73	88.3

[a]Recovered sulfur mean 3.05 ± 0.27 with percent standard deviation from the mean 8.8%

Source: Husar, Husar, and Stubits 1975, Table II, p. 2064. Reprinted by permission of the publisher.

methods is based on commercially available pure samples in most cases, with some applications to the analyses of particulate samples from air over cities and coal-burning plants, air samples from inside coking ovens, petroleum products, and other fossil-derived materials (e.g., coal conversion liquid products).

5.5.1 Identification and analysis of PAH compounds

Lao et al. (1975) have applied a computerized gas chromatographic-mass spectrometric (GC-MS) system to the rapid quantitative analysis of PAH in environmental samples. The system collects mass spectra, total ion current, and GC data and performs data reduction — all under complete computer control. The method consists of three steps: (1) preliminary separation of PAH by solvent extraction and/or column chromatography; (2) identification of PAH by a combination

of GC with quadrupole mass spectrometry and computer; and (3) measurement of PAH by GC with a
flame ionization detector (FID) and a data processor for measuring relative retention times and
calculating response factors. Table 5.16 summarizes the GC operating conditions.

Table 5.16. Chromatographic operating conditions

	Packed column	SCOT column
Detector	FID	FID
Detector temperature, °C	300	300
Liquid sample volume, µl	0.5	0.2
Sample injector temperature, °C	325	325
Column	12 ft long, 0.125 in. in diam, stainless steel, packed with 6% Dexsil 300, 400 or 410 on 80/100 mesh Chromosorb W (HP)	35 ft long, 0.02 in. in diam, stainless steel coated with 2% Dexsil 300
Carrier gas (helium) flow, ml/min	40	4
GC/MS split ratio	9:1	1:1
Initial temperature, °C	165, held for 2 min	165, held for 2 min
Programmed temperature, °C/min	4	4
Final temperature, °C	295, held for 50 min	295, held for 80 min
Recorder attenuation	160 and 640	160
GC/MS interface temperature, °C	325	325

Source: Lao et al. 1973, Table 1, p. 909. Reprinted by permission of the publisher.

A quadrupole mass spectrometer has several advantages over the magnetic scanning equipment
used in an earlier study by Lao et al. (1973): for example, relatively low initial cost yet
reasonable resolution; simple design and low volume, permitting high pumping speeds and
straightforward maintenance and repair; and extremely high speed mass scans, which are linear
with mass and which greatly simplify system control, data logging, and spectra interpretation
(Lao, Oja, Thomas, and Monkman 1973). The principal disadvantages are low resolution and
sensitivity at higher masses, which may be largely overcome by using computer data processing
methods.

The mass spectrometer was calibrated with perfluorobutylamine (FC-43). Once the spectrometer
was properly calibrated, the instrument response for mass range did not change over the
experimental period. The instrument sensitivity was checked each day with a sample of
decafluorotriphenylphosphine. This compound has a base peak of m/e 198 and key fragments of
m/e 275, 442, and 443 in its spectrum; therefore, it is convenient to use for an ion abundance
calibration of the high-mass region, where the PAH have their molecular and fragmentation ions.

A computerized data processing system is necessary for acquiring and storing the great quantity
of data produced by the MS. The computer controls the MS scan, and in addition to reducing
the burden of analyzing the data, it produces final results much more quickly and accurately
than would be possible by manual data handling.

Samples included an airborne particulate sample, a coal-tar sample, a filter sample on which
coal-tar volatile material had been collected, and four samples from coke-oven sources in the
steel industry. Two of the coke-oven samples were collected on glass filters, and two were

collected on silver membrane filters of 0.8-μ porosity. The filters were extracted in a Soxhlet extractor for 24 hr with spectrograde cyclohexane; the extracts were concentrated and injected into the GC-MS and GC-FID systems.

Airborne particulates were collected on glass-fiber filters with a commercial high-volume sampler, which pulled 1500 m^3 of air through each preweighed filter. After collection, the filter was allowed to equilibrate; weighed and aliquot portions of the filter were extracted in a Soxhlet extractor for about 8 hr with cyclohexane. The extract was put through a Rosen separation, which uses adsorption chromatography, to isolate the PAH fraction from the aliphatic and heterocyclic compounds (Rosen and Middleton 1955). The concentrated PAH fraction was injected into the GC-MS system, and mass spectra were obtained for each eluted GC peak. These spectra were compared with spectra obtained from over 40 primary PAH standards of known purity. The standard or reference solutions were prepared by combining known amounts of pure PAH. These synthetic mixtures were run on the GC-MS system to obtain GC retention times and response factors and MS spectra. Table 5.17 lists the GC parameters for the synthetic mixture of PAH on a Dexsil-300 packed column.

Fluoranthene, which has been found in all air samples and is eluted from the GC column as a pure isolated compound, was selected as the internal reference for both relative retention times and response factors.

Gas chromatographic parameters were obtained on both packed (Dexsil-300) and surface coated open tubular (SCOT) columns. The columns were preconditioned for 24 hr at 350°C. There was no apparent column bleeding during the GC runs. The final temperature of 295°C was maintained for 50 min to ensure the complete elution of the seven-ring compounds. Concentrations of the PAH in the air sample were calculated by using the response factors relative to fluoranthene obtained from gas chromatograms of the standard PAH. The Dexsil-300 packed column chromatogram is shown in Fig. 5.9. Results of the analyses of this particulate air sample on the Dexsil-300 packed column and the SCOT column are shown in Table 5.18.

The packed column yielded better GC results than did the SCOT column; coronene, the highest-molecular-weight PAH generally found in air, was completely eluted only on the packed column at the final temperature of 295°C. No further compounds were detected, even with a 60-min extended final temperature program. The calculations of relative percentages were based on the assumption that all components had been eluted. Since coronene was not eluted from the SCOT column, the total integrated area of the chromatogram was less than that of the packed column. The percentage composition of the individual components thus had a calculated difference.

One of the problems inherent in gas-liquid chromatography schemes for separating four- and five-ring PAH is bleeding of the liquid stationary phase from the column packing (support) at elevated temperatures. Janini, Muschik, and Zielinski (1976) have synthesized a high-temperature nematogenic crystal, N,N'-bis(p-butoxybenzylidene)-α,α'-bi-p-toluidine (BBBT), which greatly reduces the problem of bleeding. An 8% loss of liquid phase was reported from a BBBT column operated at a column temperature of 285°C for 100 hr. The resolution and separation factor of chrysene-benz[a]anthracene on BBBT remained constant within 2% over a continuous 146-hr operation.

Table 5.17. Retention time and response factor data of primary PAH standards on Dexsil-300 packed column

Compound name	Relative retention time[a]	Response factor[a]
Biphenyl	0.211	0.751
1,2,3,4,5,6,7,8-Octahydro-anthracene	0.250	0.776
Benzindene	0.388	0.810
Fluorene	0.412	0.864
9,10-Dihydro-phenanthrene	0.473	0.827
9,10-Dihydro-anthracene	0.499	0.803
2-Methyl-fluorene	0.546	0.916
1-Methyl-fluorene	0.555	0.916
9-Methyl-fluorene	0.603	0.916
Phenanthrene	0.648	0.920
Anthracene	0.664	0.880
Benzoquinoline	0.688	0.926
Acridine	0.708	0.918
2-Fluorene carbonitrile	0.733	0.926
3-Methyl-phenanthrene	0.790	1.042
2-Methyl-phenanthrene	0.795	1.042
2-Methyl-anthracene	0.806	0.998
Dihydro-pyrene	0.962	0.962
Fluoranthene	1.000	1.000
Pyrene	1.064	1.067
Benzo[a]fluorene or 1,2-benzofluorene	1.153	1.137
Benzo[b]fluorene or 2,3-benzofluorene	1.172	1.137
Benzo[c]fluorene or 3,4-benzofluorene	1.177	1.137
2-Methyl-fluoranthene	1.193	1.070
4-Methyl-pyrene	1.196	1.144
3-Methyl-pyrene	1.235	1.142
1-Methyl-pyrene	1.233	1.138
Benzo[c]phenanthrene	1.342	1.238
Benzo[ghi]fluoranthene	1.406	1.250
Benzo[a]anthracene	1.418	1.245
Chrysene	1.428	1.239
Triphenylene	1.432	1.242
4-Methyl-benzo[a]anthracene	1.545	1.328
1-Methyl-chrysene	1.553	1.334
6-Methyl-chrysene	1.565	1.321
β,β'-Binaphthyl	1.582	1.267
7,12-Dimethyl-benzo[a]anthracene	1.636	1.424
9,10-Dimethyl-benzo[a]anthracene	1.631	1.436
Benzo[j]fluoranthene	1.698	1.293
Benzo[k]fluoranthene	1.743	1.426
Benzo[b]fluoranthene	1.748	1.330
Benzo[a]pyrene	1.853	1.322
Benzo[e]pyrene	1.851	1.331
Perylene	1.892	1.332
3-Methyl-cholanthrene	1.985	1.337
1,2,3,4-Dibenzanthracene	2.295	1.342
2,3,6,7-Dibenzanthracene	2.304	1.342
Benzo[b]chrysene	2.440	1.348
o-Phenylenepyrene or 2,3-phenylenepyrene	2.418	1.354
Picene	2.457	1.354
Benzo[c]tetraphene	2.499	1.351
Benzo[ghi]perylene	2.595	1.356
Anthanthrene	2.574	1.350
Coronene	3.523	1.483
1,2,3,4-Dibenzpyrene	3.592	1.485
1,2,4,5-Dibenzpyrene	3.592	1.481

[a]Relative to fluoranthene.

Source: Lao et al. 1973, Table II, p. 909. Reprinted by permission of the publisher.

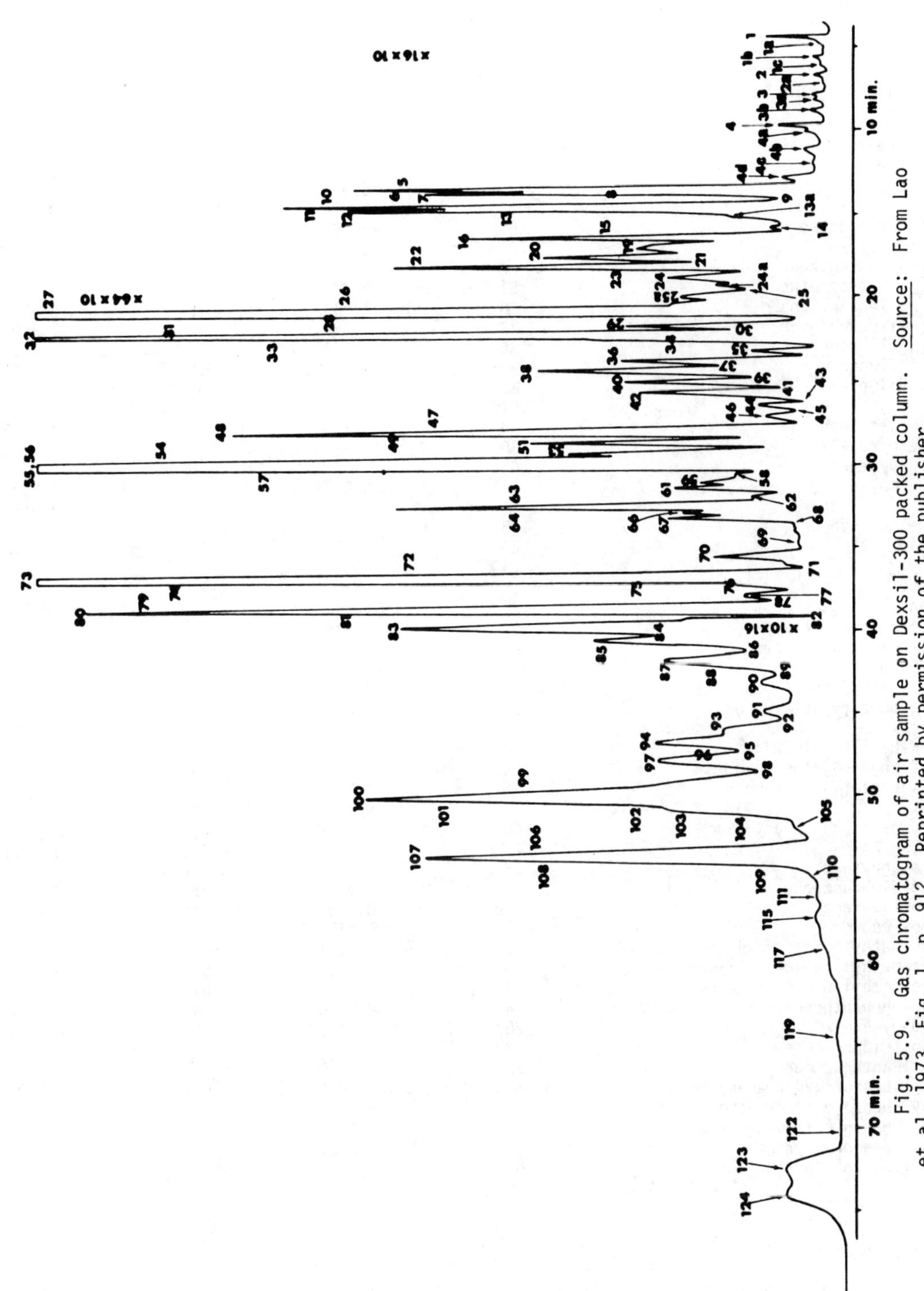

Fig. 5.9. Gas chromatogram of air sample on Dexsil-300 packed column. Source: From Lao et al. 1973, Fig. 1, p. 912. Reprinted by permission of the publisher.

Table 5.18. PAH concentrations in airborne particulate sample

Compound	Dexsil-300 packed column		Dexsil-300 SCOT column	
	Peak number	Concentration[a] (µg)	Peak number	Concentration[a] (µg)
Acridine[b]	12	0.6745	12	0.1907
Anthanthrene[b]	107		107	
Anthracene[b]	7	0.5252	7	0.1058
Benzindene[b]	3	0.0131	3	0.0011
	3a	0.0147	3a	0.0262
Benzo[a]anthracene[b]	55	44.6912	55	13.1333
Benzo[b]chrysene[b]	100	6.5796	100	1.9482
Benzo[b]fluoranthene[b]	72		72	
Benzo[ghi]fluoranthene[b]	51	3.7276	51	0.8372
Benzo[j]fluoranthene[b]	70	1.9437	70	0.4158
Benzo[k]fluoranthene[b]	72	32.8984	72	11.5132
Benzo[a]fluorene[b]	36	2.5566	36	0.5937
Benzo[b]fluorene[b]	38	4.5780	38	1.3954
Benzo[c]fluorene[b]	38		38	
Benzo[ghi]perylene[b]	107	3.8119	107	1.1328
Benzo[c]phenanthrene[b]	48	9.7808	48	3.1300
Benzo[a]pyrene[b]	80	14.9268	80	5.1223
Benzo[e]pyrene[b]	80		80	
Benzoquinoline[b]	11	0.6899	11	0.0609
			11a	0.0159
Benzo[c]tetraphene[b]	105			
β,β'-Binaphthyl[b]	66	1.1838	66	0.7974
Biphenyl[b]	1	0.0941	1	0.0013
Chrysene[c]	55, 58		55, 58	
Coronene[b]	123	1.2522		Not eluted
1,2,3,4-Dibenzanthracene[b]	94	1.3348	94	0.4477
2,3,6,7-Dibenzanthracene[b]	97	1.4697	97	0.3397
Dibenzpyrene[b]	124	1.3621		Not eluted
Dihydro-anthracene[b,c]	4a	0.0305	4a	0.0023
Dihydro-benzo[a]anthracene[c]	53	2.4037	53	0.2029
			53a	0.3694
Dihydro-benzo[a]fluorene[c]	29	2.3624	29	0.5384
Dihydro-benzo[b]fluorene[c]	29		29	
Dihydro-benzo[c]fluorene[c]	35	0.5813	35	0.4057
			35a	0.0332
Dihydro-benzo[c]phenanthrene[c]	46	0.5166	46	0.0138
			46a	0.0508
Dihydro-chrysene[c]	53		53, 53a	
Dihydro-fluoranthene[c]	25	0.1826	25	0.0394
Dihydro-fluorene[c]	1b	0.0330	1b	0.0046
Dihydro-fluorene[c]	1c	0.0171	1c	0.0003
Dihydro-methyl-benzo[k&b]fluoranthenes[c]	67	1.6251	67	0.5141
Dihydro-methyl-benzo[ghi]fluoranthene[c]	59	1.0593	59	0.2162
Dihydro-methyl-benzo[a&e]pyrenes[c]	67		67	
Dihydro-phenanthrene[b,c]	4	0.0556	4	0.0102
Dihydro-pyrene[b,c]	25a	0.1168	25a	0.0633
Dihydro-triphenylene[c]	53		53, 53a	
Dimethyl-anthracene[c]	22		22	
Dimethyl-benzo[a]anthracene[b,c]	68a	0.0966	68a	0.979
Dimethyl-benzo[b]fluoranthene[c]	91		91, 93	
Dimethyl-benzo[k]fluoranthene[c]	91	0.3264	91, 93	0.1409
Dimethyl-benzo[a]pyrene[c]	93	0.6661	91, 93	
Dimethyl-benzo[e]pyrene[c]	93		91, 93	

Table 5.18 (continued)

Compound	Dexsil-300 packed column		Dexsil-300 SCOT column	
	Peak number	Concentration[a] (μg)	Peak number	Concentration[a] (μg)
Dimethyl-chrysene[c]	69	0.1136	69	0.0423
Dimethyl-phenanthrene[c]	20		20	0.1424
Ethyl-anthracene[c]	22	1.0451	22, 20a	0.1272
Ethyl-phenanthrene[c]	20	0.8691	20a, 20	0.0199
Fluoranthene[b]	27	15.7200	27	4.9620
Fluorene[b]	3b	0.0242	3b	0.0044
Fluorene carbonitrile[b,c]	13a	0.1082	13a, 14	0.0297
Fluorene carbonitrile[b,c]	14	0.0270		
Hexahydro-chrysene[c]	48		48	
Methyl-anthanthrene[c]	119		119	
Methyl-anthracene[b]	19	0.5985	19	0.1048
Methyl-benzo[a]anthracene[b]	61	1.4318	61	0.0253
Methyl-benzo[b]fluoranthene	77		77	
Methyl-benzo[k]fluoranthene	77	1.7520	77	0.4858
Methyl-benzo[b]chrysene[c]	117	0.0791	117	0.0207
Methyl-benzo[ghi]perylene[c]	122	0.1748		Not eluted
Methyl-benzo[a]pyrene	87	0.8756	87	0.1155
Methyl-benzo[e]pyrene	87		87	
Methyl-benzo[c]tetraphene[c]	117		117	
Methyl-β,β'-binaphthyl[c]	68	0.9169	68	0.1600
Methyl-biphenyl	2	0.0245	2	0.0008
Methyl-biphenyl	2a	0.0098	2a	0.0088
3-Methyl-cholanthrene[b]	85	0.8912	85	0.1072
Methyl-chrysene[b]	64	6.1948	64	1.4590
Methyl-dibenzanthracene[c]	115	0.1587		
Methyl-fluoranthene[b]	40	2.1762	40	0.5211
1-Methyl-fluorene[b]	4c	0.0136	4c	0.0005
2-Methyl-fluorene[b]	4b	0.0405	4b	0.0027
9-Methyl-fluorene[b]	4d	0.0576	4d	0.0070
Methyl-phenanthrene[c]	16	0.5825	16	0.2065
Methyl-o-phenylene-fluoranthene[c]	111	0.0415		
Methyl-o-phenylene-pyrene[c]	119	0.1270	119	0.0111
Methyl-picene[c]	119			
Methyl-pyrene[b]	42	2.6224	42	0.5634
			43	0.0308
Methyl-triphenylene	62	0.3963	62	0.3446
Octahydro-anthracene[b,c]	1a		1a	
Octahydro-fluoranthene[c]	24	0.2949	24	0.0385
Octahydro-phenanthrene[c]	1a	0.0163	1a	0.0018
Octahydro-pyrene[c]	24a	0.0832	24a	0.0299
Perylene[b]	83	2.1666	83	0.6619
Phenanthrene[b]	5	0.6102	5	0.2199
o-Phenylene-fluoranthene	90	0.1566	90	0.0092
o-Phenylenepyrene[b]	100		100	
Picene[b]	105	0.0984		
Pyrene[b]	32	16.8904	32	4.8121
Tetrahydro-methyl-benzo[a]anthracene[c]	58	0.9814	58	0.0988
Trimethyl-fluoranthene[c]	44	0.6456	44	0.0628
Trimethyl-pyrene[b,c]			44a	0.0258
Triphenylene[b,c]	55, 58, 68a		55, 58, 68a	

[a]Calibrated with fluoranthene the specific responses 0.3455 μg/unit area.
[b]Determined by comparison with a primary standard compound of known purity with respect to relative retention time and mass spectrum.
[c]Compounds found in the air sample which have not been previously reported.

Source: Lao et al. 1973, Tables III and IV, pp. 910, 911, 914, and 915. Reprinted by permission of the publisher.

A differential scanning calorimetry (DSC) scan of BBBT showed phase transition temperatures of 159°C (solid-smectic), 188°C (smectic-nematic), and 303°C (nematic-isotropic). Plots of the retention times of anthracene and phenanthrene on BBBT vs reciprocal absolute temperature show that the retention times decrease with increasing column temperature until the smectic-nematic transition is approached. After the transition temperature is passed, the retention times again increase with temperature increase until the solute volatility overrides the increasing temperature-raising retention effect.

Chromatograms of pentacyclic aromatic hydrocarbons indicate that benzo[a]pyrene (BaP) can be separated from benzo[e]pyrene as well as from other members of the benzpyrene fraction. In addition, a fairly complex mixture of three- to five-ring PAH isomers can be separated using BBBT. Chromatograms of progammed-temperature gas-liquid chromatographic analyses of a synthetic mixture containing 13 PAH show the retention profiles and elution orders of these compounds. The PAH in this synthetic mixture, in order of increasing retention time, are phenanthrene, anthracene, fluoranthene, pyrene, benzo[mno]fluoranthene, triphenylene, benz[a]anthracene, chrysene, naphthacene, benzo[k]fluoranthene, benzo[e]pyrene, perylene, and BaP (Janini, Muschik, Zielinski 1976). Liquid crystals that have nematic-isotropic transition temperatures above 200°C and have been used as high-temperature gas-liquid chromatographic liquid phases are listed in Table 5.19.

Sawicki, Hauser, and Stanley (1960) performed some of the earliest studies of the ultraviolet, visible, and fluorescence spectral analysis methods for PAH. In these studies, they separated and identified a mixture of 50 PAH, including benzene, naphthalene, anthracene, phenanthrane, benz[a]anthracene, benzo[e]pyrene, perylene, benzo[ghi]perylene, anthanthrene, and coronene.

Commercial preparations of hydrocarbons were purchased, and mixtures were prepared in a pentane solvent. The total concentration in pentane was 10^{-5} moles per hydrocarbon. Analysis was accomplished by using a Cary model II recording quartz spectrophotometer to determine absorption wavelengths and an Aminco-Bowman spectrophotofluorometer to determine activation and fluorescence spectra [sensitivity was set at 50, slit arrangement No. 2 (narrowest slit width 1/32 in.), and RCA 1P21 phototube].

Results showed pentane to be the ideal solvent for analysis of PAH mixtures (two to six fused rings). The authors found that in many instances it was possible to obtain the pure fluorescence spectrum of a PAH in a mixture by determining the fluorescence spectrum of the mixture at an appropriate activating wavelength maximum of a particular PAH. This analytical method was developed as a reinforcement for ultraviolet-visible analytical procedures. This method potentially simplifies the determination and identification of many hydrocarbons.

Sawicki, Johnson, and Kosinski (1966) studied methods of thin-layer and column chromatographic separation and photometric and fluorometric analysis for aromatic amines and heterocyclic imines. All aromatic amines and imino-heterocyclic compounds studied were obtained commercially or synthesized by reported procedures. Compounds subjected to column chromatography were adsorbed onto 1 g of alumina, which was then carefully dried. This material was placed in a 0.5- by 15-in. column containing 9 g of alumina and 0.5 g of silica gel, neither containing eluting solvent. The column was then eluted with successive 100-ml volumes of pentane solutions containing various percentages of ether, acetone, or methanol; column elution required about 8 hr. The column was protected from light while 15-ml fractions were collected.

Table 5.19. Liquid crystals having nematic-isotropic transition temperatures above 200°C

Code	Structure	Mol wt	Smectic (°C)	Nematic (°C)	ΔT nematic
I	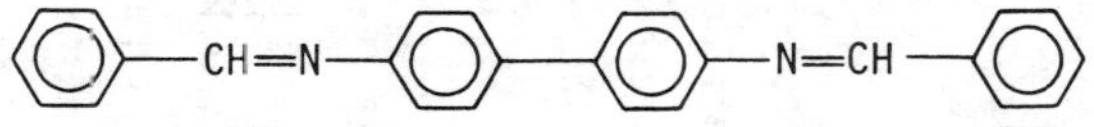 phenyl—CH=N—⬡—⬡—N=CH—phenyl	358		239 – 265	26
II	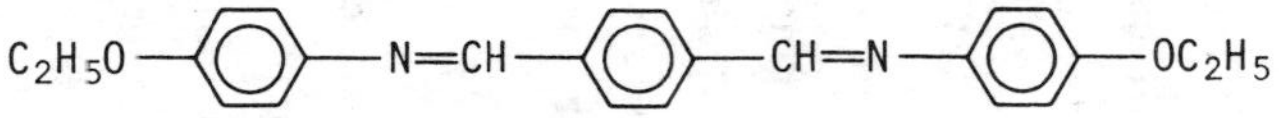 C_2H_5O—⬡—N=CH—⬡—CH=N—⬡—OC_2H_5	372		200 – 320	120
III	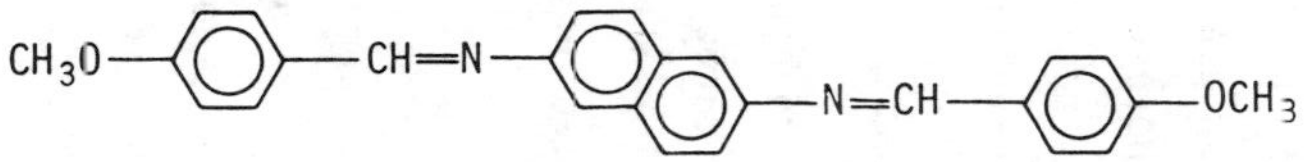 CH_3O—⬡—CH=N—⬡—⬡—N=CH—⬡—OCH_3	394		189 – 356	167
IV	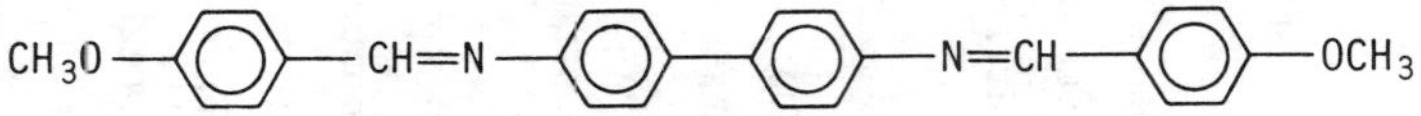 CH_3O—⬡—CH=N—⬡—⬡—N=CH—⬡—OCH_3	420		266 – 390	>124
V	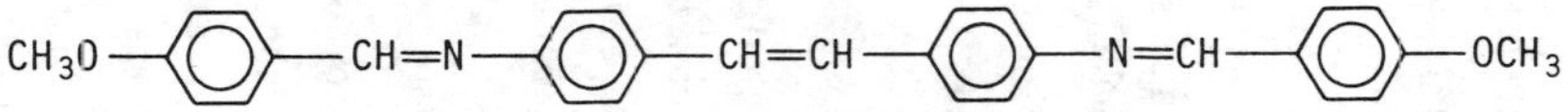 CH_3O—⬡—CH=N—⬡—CH=CH—⬡—N=CH—⬡—OCH_3	446		274 – 340	>66
BMBT	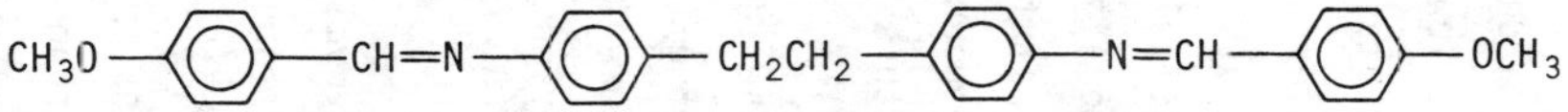 CH_3O—⬡—CH=N—⬡—CH_2CH_2—⬡—N=CH—⬡—OCH_3	448		181 – 320	139
VI	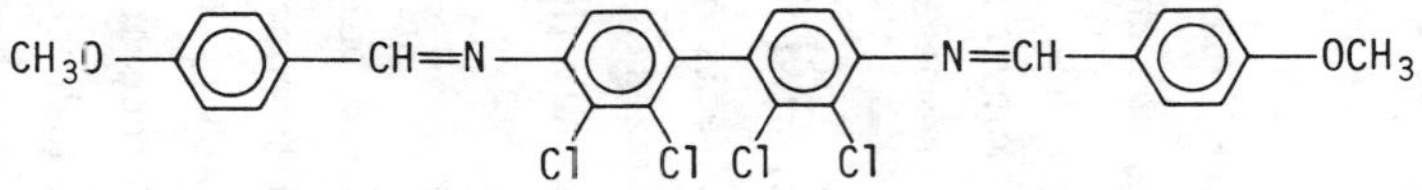 CH_3O—⬡—CH=N—⬡(Cl Cl)—⬡(Cl Cl)—N=CH—⬡—OCH_3	489		154 – 344[a]	190
BBBT	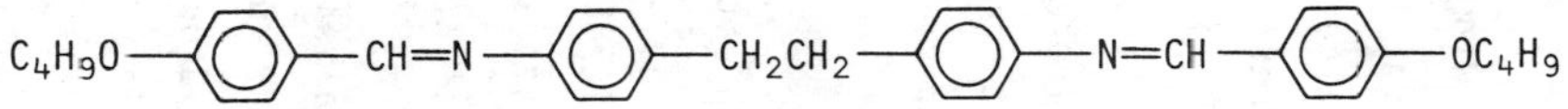 C_4H_9O—⬡—CH=N—⬡—CH_2CH_2—⬡—N=CH—⬡—OC_4H_9	532	159 – 188	188 – 303	115

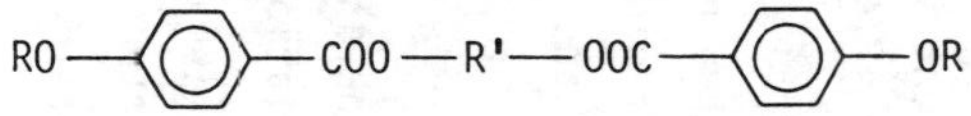

RO—⬡—COO—R'—OOC—⬡—OR

Code	R	R'	Mol wt	Smectic (°C)	Nematic (°C)	ΔT nematic
VII	n-C_4H_3 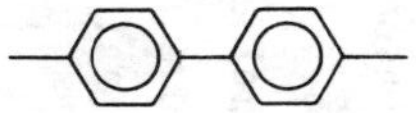	—⬡—⬡—	538	171 – 184	184 – 358[a]	174
VIII	n-C_7H_{15} 	—⬡—	546	83 – 125	125 – 206	81
IX	n-C_7H_{15} 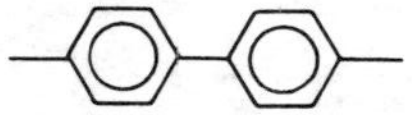	—⬡—⬡—	622	150 – 211	211 – 316	105

[a]Decomposes.

Source: Janini, Muschik, and Zielinski 1976, Table II, p. 813. Reprinted by permission of the publisher.

Thin-layer chromatography was applied as another separation technique. The absorbent was placed on glass plates to a thickness of 2 mm. The plates were cut into strips for analysis with a glass cutter.

Results showed that cellulose thin-layer chromatography was excellent for separating mixtures of aromatic amines. The use of direct fluorometric examination of thin-layer plates for aromatic amines was useful in the nanogram-to-microgram range. Fluorescence and absorption spectra for compounds studied are shown (Table 5.20).

Sawicki and Johnson (1964) characterized the fluorescence and phosphorescence spectra of some aromatic compounds at low temperatures. Compounds studied were subjected to paper chromatography and then lowered to the temperature of liquid nitrogen for fluorescence identification.

Samples were separated by paper or thin-layer chromatography. The solvent for the paper chromatography procedure was trifluoroacetic acid. The spots were dried, lowered to liquid nitrogen temperature, and examined for fluorescence and phosphorescence. After preliminary examination, each spot was treated with either a drop of concentrated sulfuric acid or 29% methanolic tetraethylammonium hydroxide and reinspected for fluorescence and phosphorescence.

Alumina thin-layer plates were sprayed with Neatan, a developer; the appropriate part of the alumina was then removed with transparent tape. The low-temperature fluorescence, room-temperature fluorescence, and phosphorescence were obtained with neutral trifluoroacetic acid and tetraethylammonium hydroxide treatment. The material of interest was examined in a Chromato-Vue cabinet, which is equipped with 3660-A, 2537-A, and white-light lamps. Results showed that the typical aliphatic fraction of airborne particulate samples gave a light blue fluorescent color at room temperature and a blue fluorescent and a green phosphorescent color in liquid nitrogen. The results also showed that compounds that fluoresce at room temperature usually fluoresce at liquid-nitrogen temperatures. Many compounds that are nonfluorescent at room temperature are strongly fluorescent at low temperatures.

Sawicki, Stanley, and Johnson (1965) investigated the technique of quenchofluorometric analysis for polynuclear compounds. The quenching effects of nitromethane, carbon disulfide, acetophenone, 2,3-butanedione, o-nitromethane, m-dinitrobenzene, pieric acid, chloranil, o-cresol, azobenzene, tetranitromethane nitrosobenzene, and tetracyanoethylene were studied on a variety of polynuclear compounds (Tables 5.21-5.23).

Quenchofluorometry is useful in the analysis of polycyclic compounds because of its high selectivity caused by solvent variation. Determinations should be made using the following solvent types: (1) aza-heterocyclic hydrocarbons in the presence of other polynuclear compounds and vice versa, (2) fluoranthenic hydrocarbons in the presence of other types of aromatic hydrocarbons, (3) aza-heterocyclic hydrocarbons in the presence of some aromatic amines and vice versa, and (4) aza-heterocyclic hydrocarbons in the presence of imino-heterocyclic hydrocarbons and vice versa.

Mariich and Lenkevich (1973) developed a method of capillary chromatography for rapid determination of main components in coal tar. This method uses steel columns that are 50 m long with an inside diameter of 0.25 mm. The column was treated by purging with hydrogen for 24 hr, washing three times each with benzene, ether, ethanol, water, weak ammonia solution, and then

Table 5.20. Fluorescence and absorption spectra of some aromatic amines in neutral, acid, and alkaline solutions on a thin-layer plate

Compound	Spectra	Media[a]	Solvent	Detection limit[b] (ng)	Total analysis [μg or (molarity)]	Absorption or excitation spectra[c] λ(mμ)	Absorption or excitation spectra[c] 100 MM.T[d]	Emission spectra[c] λ(mμ)	Emission spectra[c] 100 MM.T
1-Naphthylamine	F	Soln.	DMF[e]	5	(10^{-6})	333	28	420	28
	A	Soln.	DMF	14,300	(10^{-4})	334	68		
	F	Soln.	DMF-TEA[f]	30	(10^{-5})	330[g]	114	540	130
	F	Soln.	DMF-HCl[h]	250	(5×10^{-5})	275	28	340	32
	A	Soln.	DMF-HCl	2,400	(10^{-4})	270 280 290s[i]	54 62 44		
	F	C	DMF	10	0.02	262 330	50 200	420	250
2-Naphthylamine	F	Soln.	DMF	7	(5×10^{-6})	293 350	84 90	403	89
	F	C	DMF	30	0.2	258 300 370	780 1100 1100	405	1300
1-Anthramine	F	Soln.	DMF	60	(10^{-5})	278 367 408	55 44 55	528	55
	F	C	DMF	40	0.2	280 340s 375 410	140 110 130	528	160
2-Anthramine	F	Soln.	DMF	30	(10^{-6})	282 320s 334 350 415	47 9 15 10 23	500	47
	F	C	DMF	50	0.2	282 330s 345 361 422	510 160 240 200 400	500	520
9 Anthramine	F	Soln.	DMF	6	(5×10^{-6})	370 375 425	90 42 36	510	86
	F	C	DMF	200	0.2	275 385 425		510	
8-Aminofluoranthene	F	Soln.	DMF	10	(10^{-6})	282s 308 333 371 415	8 25 14 14 25	514	27
	F	C	DMF	20	0.2	290s 320 380s 425	100 170 170 250	510	330
	A	Soln.	DMF	4,300	(4×10^{-5})	300s 310 332 346s 370 422	48 75 27 19 19 29		
	F	Soln.	DMF-TEA[j]	70	(10^{-5})	340 412 540s 555	23 37 48 53	590	59
	A	Soln.	DMF-TEA[k]	1,500	(5×10^{-5})	327 421 540s 565	145 68 55 70		
	F	Soln.	DMF-HCl (1:1)	13	(2.5×10^{-6})	265s 285s 290 312s 330 350 363	29 54 61 12 37 52 54	515[l]	66

Table 5.20 (continued)

Compound	Spectra	Media[a]	Solvent	Detection limit[b] (ng)	Total analysis [μg or (molarity)]	Absorption or excitation spectra[c] λ(mμ)	100 MM.T[d]	Emission spectra[c] λ(mμ)	100 MM.T
	A	Soln.	DMF-HCl (1:1)	3,600	(5×10^{-5})	270	160		
						281	190		
						291	108		
						310	22		
						330	32		
						348	37		
						366	42		
6-Aminochrysene	F	Soln.	DMF	5	(10^{-6})	280	41	*433*	51
						350	49		
	A	Soln.	DMF	1,300	(2×10^{-5})	280	110		
						351	28		
	F	C	DMF	7	0.2	285	240	*445*	33
						360	310		
	F[m]	Soln.	DMF-TEA[k]	6	(10^{-6})	283	32	*460*	40
						335	41	480s	28
						395s	15		
						415	30		
						452	47		
	A[m]	Soln.	DMF-TEA[k]	2,000	(5×10^{-5})	275	152		
						340	70		
						395	21		
						418	30		
						445	46		
	F	Soln.	DMF-HCl (1:1)	60	(5×10^{-6})	*270*	42	370	38
						295	19	*385*	45
						310	24	405	24
						320	26	430	19
	A	Soln.	DMF-HCl	1,200	(10^{-5})	261	68		
						270	110		
						299	13		
						312	15		
						325	15		
1-Aminopyrene	F	Soln.	DMF	0.2	(2.5×10^{-8})	287	25	*432*	67
						373s	64		
						380	66		
						407	64		
	F	C	DMF		0.2	293[n]		*443*	
	A	Soln.	DMF	4,300	(2×10^{-5})	288	53		
						369	33		
						386	33		
						403	11		
7-Aminobenz[*a*]anthracene	F	Soln.	DMF	30	(10^{-5})	*305*	94	*475*	95
						395s	70		
						410	75		
	F	C	DMF		0.2	*309*	100	*480*	78
						410	90		

[a]C = cellulose thin-layer chromatogram.
[b]Absorption spectra, ng/ml; fluorescence spectra, ng/0.1 ml; in both cases, entire spectrum obtained at detection limit.
[c]Italicized values are emission (excitation) wavelength maxima at which excitation (emission) spectra are obtained.
[d]For absorption spectra, value is 100 x absorbance.
[e]DMF = dimethylformamide.
[f]DMF — 40% aqueous tetraethylammonium hydroxide (1:1).
[g]Excitation wavelength maximum of neutral compound.
[h]DMF — concentrated hydrochloric acid (1:1).
[i]s = shoulder.
[j]DMF containing 2% of 29% methanolic tetraethylammonium hydroxide. With 0.3% of 40% aqueous tetraethylammonium hydroxide in DMF, the long-wavelength excitation maximum and the emission maximum are 565 and 600 mμ respectively.
[k]DMF·containing 0.3% of 40% aqueous TEA.
[l]Emission spectrum of neutral compound.
[m]Unstable solution.
[n]Longer wavelength bands hidden.

Source: Sawicki, Johnson, and Kosinski 1966, Table 5, pp. 88-93. Reprinted by permission of the publisher.

Table 5.21. Fluorescence spectra of aza-heterocyclic hydrocarbons
in nitromethane containing 1% trifluoroacetic acid

Compound	Detection limit (ng)	Molar conc.	Emission spectra[a]		Excitation spectra[b]	
			λ max (nm)	MM.T	λ max (nm)	MM.T
Benzo[f]quinoline[c]	14	10^{-5}	437	1.26	381	1.26
Benzo[h]quinoline[c]	39	10^{-5}	439	0.46	380	0.46
Phenanthridine[c]	120	10^{-4}	408	1.53	378	1.53
Acridine[c]	4	10^{-6}	480	0.46	406	0.46
Indeno[1,2,3-ij]iso-quinoline[d,e]	2000	10^{-4}	470	0.07	379	0.07
Acenaphtho[1,2-b]-pyridine[c]	180	10^{-4}	450	1.11	380	1.11
Benzo[lmn]phen-anthridine[c]	2	10^{-6}	487	1.20	408	1.20
Benz[a]acridine[c]	1	10^{-6}	480	2.19	401	2.19
Benz[c]acridine	2.5	10^{-6}	487	0.92	390	0.85
					410	0.90
Benz[c]acridine[d]	100	10^{-5}	386	0.22	386	0.26
			409	0.26		
Dibenz[a,h]acridine	0.7	10^{-7}	460	0.45	410	0.39
					430	0.45
Dibenz[a,h]acridine[d]	2.8	10^{-7}	391	0.57	391	0.25
			417	0.39		
			440	0.11		
Dibenz[a,j]acridine	0.3	10^{-7}	450	0.94	*409*[f]	0.60
					424	0.94
Dibenz[a,j]acridine[d]	20	10^{-6}	319	0.60	391	0.58
			417	0.45		
			440	0.14		
14-Phenyldibenz[a,j]-acridine	24	10^{-6}	462	0.22	409	0.15
					430	0.22
14-Phenyldibenz[a,j]-acridine[d]	20	10^{-6}	400	0.25	*380*	0.06
			428	0.26	398	0.25
			450	0.08		
7-Phenyldibenz[c,h]-acridine	2.4	10^{-6}	481	1.86	409	1.50
					421	1.80
7-Phenyldibenz[c,h]-acridine[d]	36	10^{-6}	392	0.15	371	0.04
			420	0.18	391	0.20
			449	0.06		
Pyrenoline[c]	250	10^{-5}	550	0.33	391	0.08
					468	0.33

[a]Instrument set at most intense excitation wavelength maximum.
[b]Instrument set at most intense emission wavelength maximum.
[c]Nonfluorescent in nitromethane containing 1% triethylamine.
[d]In nitromethane containing 1% triethylamine.
[e]Nonfluorescent in nitromethane containing 1% trifluoroacetic acid.
[f]Italicized values are shoulders.

Source: Sawicki, Stanley, and Johnson 1965, Table I, p. 179. Reprinted by permission of the publisher.

repeating the process in reverse order. Apiezon L was used as the stationary phase and was
applied to the cleaned column from a 5% solution in benzene under a hydrogen pressure of 2 atm.
Column conditioning involves a 24-hr purging with hydrogen under 1 atm pressure and successive
stepwise heating (100-150-200-250°C).

The capillary column was fitted to a KhPOM-Z chromatograph with a flame-ionization detector.
The operating conditions were 265°C column temperature, argon carrier gas, and 1.12-atm pressure.
A splitter was included in the apparatus and set to a ratio of 1:200 for a 0.7-µliter sample.
The fractions were saved for further analysis.

Table 5.22. Compounds that are nonfluorescent and colored in nitromethane-aluminum chloride

Compound	Visible color[a]	Compound	Visible color[a]
Anthracene	B	Dibenzothiophene	Pi
Phenanthrene	*b*	Dibenzofuran	R
Fluorene	G	Benzo[*b*]naphtho[2,3-*d*]furan	R
Fluoranthene	R	Dinaphtho[2,1-*b*:1',2'-*d*]furan	R
Triphenylene		Benzo[*kl*]xanthene	Bk
Benzfluorenes		Naphtho[2,1,8,7-*klmn*]xanthene	G
Benzo[*c*]phenanthrene	RBr	Carbazole	B
Chrysene	R	11 H-Benzo[*a*]carbazole	B
Naphthacene	G	5 H-Benzo[*b*]carbazole	BG
Pyrene	G	7 H-Benzo[*c*]carbazole	
Benzo[*a*]pyrene	P	4 H-Benzo[*def*]carbazole	B
Benzo[*e*]pyrene	G	5 H-Naphtho[2,3-*c*]carbazole	R
Benzo[*b*]fluoranthene		N-Phenylcarbazole	BG
Benzo[*k*]fluoranthene	G	N-Phenylphenothiazine	RBr
Benzo[*ghi*]fluoranthene	Br	2-Aminoanthracene	
Perylene	BV	4-Aminofluoranthene	
Benzo[*b*]chrysene	RBr	6-Aminobenzo[*a*]pyrene	
Dibenz[*a,c*]anthracene	B	3-Nitro-9-ethylcarbazole	R
Dibenz[*a,h*]anthracene		6-Nitrochrysene	V
Dibenz[*a,j*]anthracene		1-Nitropyrene	P
Picene		1-Naphthol	G
Dibenz[*a,e*]pyrene	Br	2-Naphthol	G
Naphtho[1,2,3,4-*def*]chrysene	R	6-Hydroxychrysene	
Anthanthrene	Pi	9-Carboxyanthracene	
Indeno[1,2,3-*cd*]pyrene	GrG	1-Carboxypyrene	
Coronene	lG	Fluorenone	
Benzo[*a*]coronene		Phenanthridone	
Indeno[3,2-*j*]acenaphtho-[1,2-*k*]-fluoranthene		Dibenzo[*f,h*]quinoxaline	
Diacenaphtho[1,2-*j*:1',2'-*l*]-fluoranthene		Acenaphtho[1,2-*b*]-quinoxaline	
Phenanthraquinone		Acenaphtho[1,2-*b*]benzo[*f*]-quinoxaline	R
Aceanthrene-1,2-dione		Acenaphtho[1,2-*b*]benzo[*g*]-quinoxaline	
Dibenzo[*def,mno*]chrysene-6,12-dione	B	Diacenaphtho[1,2-*b*:1',2'-*d*]-thiophene	Br
Naphtho[2,3-*g*]chrysene-11,16-dione			

[a]B = blue, Bk = black, Br = brown, G = green, Gr = grey, P = purple, Pi = pink, R = red, V = violet, l = light.
[b]A shade of yellow or a very light color probably due to impurities.

Source: Sawicki, Stanley, and Johnson 1965, Table II, p. 180. Reprinted by permission of the publisher.

Results of the separation and identification recorded 242 individual compounds in coal tar, of which 184 were identified; the main components are listed in Table 5.24.

5.5.2 Separation of PAH compounds

Sawicki, Guyer, and Engel (1967) studied paper and thin-layer electrophoretic separations of polynuclear aza-heterocyclic compounds. These compounds have been found in air pollution resulting from coal-burning operations.

Table 5.23. Fluorescence spectra of aza-heterocyclic and
other hydrocarbons

Compound	Detection limit (ng)	Molar conc.	Emission spectra[a]		Excitation spectra[b]	
			λmax, nm	MM.T	λmax, nm	MM.T
Acetophenone-trifluoroacetic acid (99:1)						
Acridine	7	10^{-6}	480	0.25	400	0.25
Benz[*a*]acridine	0.7	2×10^{-7}	480	0.67	407	0.67
Benz[*a*]acridine[c]	30	10^{-5}	410	0.74	384	0.74
Benz[*c*]acridine	4	10^{-6}	490	0.60	390	0.59
					419	0.51
					430	0.43
Benz[*c*]acridine[c]	19	10^{-6}	409	0.12	388	0.12
Fluoranthene	560	10^{-4}	452	0.36	382	0.36
Benzo[*k*]fluoranthene	8	10^{-6}	410	0.32	380	0.32
			430	0.27	401	0.53
			460	0.11		
Perylene	17	10^{-6}	442	0.77	400	0.25
			472	0.54	412	0.52
			500	0.15	440	0.67
Benzo[*a*]pyrene	30	2.5×10^{-6}	405	1.20	390	1.20
			430	0.75		
			456	0.16		
Carbon disulfide-trifluoroacetic acid (99:1)						
Acridine	90	10^{-5}	*457d*	0.18	408	0.23
			480	0.22		
Benz[*a*]acridine	14	10^{-6}	460	0.16	410	0.16
Benz[*c*]acridine	70	10^{-5}	478	0.43	390	0.44
					410	0.38
					437	0.33
Benz[*c*]acridine[c]	230	10^{-5}	412	0.14	390	0.14
			437	0.05		
Dibenz[*a,h*]acridine[c]	28	10^{-6}	420	0.16	400	0.15
			448	0.02		
Dibenz[*a,j*]acridine[c]	19	10^{-6}	410	0.23	*390*	0.04
			430	0.15	400	0.15
			452	0.05		
Fluoranthene[e]	330	10^{-4}	410	0.60	387	0.78
			438	0.76	408	1.20
			462	0.56		
Benzo[*k*]fluoranthene[c]	25	10^{-6}	410	0.14	390	0.13
			437	0.13	410	0.37
			464	0.05		
Benzo[*k*]fluoranthene	14	10^{-6}	410	0.54	390	0.45
			437	0.43	410	0.88
			464	0.18		
Perylene	2	10^{-7}	452	0.53	400	0.10
			482	0.32	420	0.30
			512	0.08	450	0.50
1% Trifluoroacetic acid in o-cresol						
Fluoranthene	30	10^{-5}	465	0.74	*330*	0.35
					350	0.63
					360	0.76
Benzo[*k*]fluoranthene	2	10^{-6}	415	0.60	310	0.58
			435	0.60	330	0.20
					340	0.15
					365	0.20
					380	0.42
					405	0.48
Perylene	1	2×10^{-7}	445	1.0	395	0.46
			470	0.70	412	0.90
			500	0.28	440	1.15
Pyrene	10	2×10^{-6}	375	0.30	325	0.15
			390	0.34	340	0.35
Benzo[*a*]pyrene	3	5×10^{-7}	410	0.94	*355*	0.40
			430	0.68	370	0.80
			455	0.28	390	0.90
Benzo[*def*]carbazole	20	10^{-5}	380	0.77	338	0.80
					345	0.80
1-Aminopyrene	10	10^{-6}	430	0.20	345	0.20
					370	0.12
					395	0.08

Table 5.23 (continued)

Compound	Detection limit (ng)	Molar conc.	Emission spectra[a]		Excitation spectra[b]	
			λmax, nm	MM.T	λmax, nm	MM.T

			N,N-Dimethylaniline			
Benzo[k]fluoranthene	80	10^{-5}	525	0.30	375	0.25
					390	0.30
					408	0.20
Perylene	500	10^{-4}	550	0.49	*380*	0.32
					398	0.44
					427	0.45
					450	0.48
Pyrene	400	10^{-4}	490	0.40	355	0.40
Benzo[a]pyrene	500	10^{-4}	530	0.45	375	0.40
					390	0.45
4-H-Benzo[def]carbazole	45	10^{-5}	445	0.42	355	0.42
1-Aminopyrene	30	10^{-5}	478	0.85	360	0.44
					410	0.85
			5% m-Dinitrobenzene + 1% trifluoroacetic acid in chloroform			
Acridine	40	10^{-5}	450	0.35	425	0.38
			480	0.43	435	0.40
Benz[a]acridine	15	2×10^{-6}	458	0.34	423	0.32
Benz[c]acridine	4	10^{-6}	470	0.60	435	0.60
Perylene	2500	10^{-4}	450	0.07	430	0.10
			470	0.10	448	0.10
			68% Nitrogen dioxide in trifluoroacetic acid			
Acridine	20	10^{-5}	450	0.70	380	0.90
			470	1.00	390	1.20
			500	0.60	404	1.32
					425	0.90
Benz[a]acridine	3	10^{-6}	460	0.77	405	0.76
Benz[c]acridine	15	10^{-6}	472	0.15	382	0.15
					418	0.14
					432	0.13
			10% Aluminum chloride in nitromethane			
Acridine	220	10^{-4}	480	0.80	410	0.80
Phenanthridine	900	10^{-4}	410	0.20	387	0.20
Benz[a]acridine	20	5×10^{-6}	480	0.60	410	0.60
Benz[c]acridine	330	10^{-4}	490	0.70	420	0.70
Benzo[lmn]phenanthridine	70	10^{-5}	490	0.30	410	0.30
			10% o-Nitrotoluene and 1% trifluoroacetic acid in chloroform			
Acridine	15	5×10^{-6}	*454*[c]	0.64	429	0.80
			470	0.79		
Benz[a]acridine	10	2×10^{-6}	460	0.45	424	0.45
Benz[c]acridine	10	2×10^{-6}	470	0.48	430	0.48
Benzo[lmn]phenanthridine	10	2×10^{-6}	470	0.39	420	0.39
Dibenz[a,h]acridine	20	10^{-6}	454	0.14	435	0.14
			478	0.11		
Dibenz[a,j]acridine	8	10^{-6}	445	0.34	429	0.34
14-Phenyldibenz[c,h]-acridine	4	10^{-6}	450	0.92	430	0.90
7-Phenyldibenz[c,h]-acridine	5	10^{-6}	475	0.78	430	0.78
Pyrenoline	63	10^{-5}	550	0.38	464	0.40
					484	0.30
					440	0.19
Acenaphtho[1,2-b]-benzo[f]quinoxaline	8	5×10^{-7}	490	0.21	447	0.21
Acenaphtho[1,2-b]-benzo[g]quinoxaline	2300	5×10^{-4}	500	0.65	434	0.67
Indeno[1,2,3-cd]pyrene	2800	5×10^{-4}	*480*	0.42	*450*	0.45
			500	0.53	460	0.53
Indeno[3,2-j]-acenaphtho[1,2-l]-fluoranthene	830	10^{-4}	475	0.46	440	0.50
			501	0.40		

[a]Instrument set at most intense excitation wavelength maximum that can be used.
[b]Instrument set at most intense emission wavelength maximum that can be used.
[c]1% Triethylamine instead of trifluoroacetic acid.
[d]Italicized values are shoulders.
[e]Almost identical spectra in carbon disulfide or alkaline carbon disulfide.

Source: Sawicki, Stanley, and Johnson 1965, Table III, p. 181. Reprinted by permission of the publisher.

Table 5.24. Qualitative and quantitative chromatographic characteristics
of the main components of coal tar

Components	Relative retention time at 250°C	Correction factor		
		Calculated	Experimental	
			1	2
Naphthalene	1.00	1.00	1.00	1.00
Indole	1.33	1.14	1.72	1.82
Quinoline	1.92	1.12	2.50	2.65
β-Methyl-naphthalene	2.00	1.04	1.18	1.12
α-Methyl-naphthalene	2.33	1.04	1.18	1.22
Diphenyl	2.67	1.00	1.20	1.22
Acenaphthene	5.25	1.00	1.17	1.21
Diphenylene oxide	5.83	1.09	1.42	1.46
Fluorene	7.67	1.00	1.47	1.44
Diphenylene sulfide	15.67	1.20	1.50	1.58
Carbazole	16.83	1.09	2.00	2.10
Phenanthrene	17.67	0.99	1.25	1.37
Anthracene	18.33	0.99	1.41	1.54
Fluoranthene	45.33	0.98	1.28	1.38
Pyrene	55.83	0.98	1.23	1.38
2,3-Benzofluorene	68.83	0.95	1.25	1.30
1,2-Benzofluorene	71.00	0.95	1.25	1.30
1,2-Benzanthracene	144.00	0.99	1.32	1.58
Chrysene	149.33	0.99	1.30	1.10

Source: Mariich and Lenkevich 1973, Table 1, p. 1058. Reprinted by permission
of the publisher.

The aza-heterocyclic compounds were obtained commercially in a pure form and, in the paper
electrophoresis, were spotted in 25 ng of sample on Whatman No. 1 paper about 2 in. from the
edge. The cathode end of the paper was placed in the buffer solution and wetted, and the
buffer was allowed to diffuse toward the spots. When the entire paper was wet, the mixtures
were separated at 500 V for 75 min. The pherogram was examined under ultraviolet light.

The thin-layer technique used 8- by 8-in. dry-cellulose glass plates spotted with the samples.
The cathode and anode edges of the plate were connected to the cold buffer solution by means
of wet paper strips. Except for the spot area, the whole plate was wetted with a piece of
filter paper wet with buffer. The mixtures were separated at 500 V for 75 min. The thin-
layer pherogram was examined under ultraviolet light. A pH 2.0 buffer was made by mixing
formic acid (31.2 ml) and glacier acetic acid (59.2 ml) made up to 1 liter with distilled water.

The thin-layer and paper electrophoretic processes proved to be much more rapid than comparable
chromatographic methods. Resolution in separation proved to be good due to small sample size
and 3- to 4-mm spot size.

Kogan and Sorokina (1974) studied gas chromatographic methods for the determination of
anthracene, phenanthrene, and carbazole in coal-tar fractions. This work involved gas
separation using a temperature lower than the normal 350°C, which is hard to regulate.

The columns used were stainless steel, 3 mm wide and 1 m long. Test column 1, operating temperature of 180°C, was packed with Chromosorb W (80 to 100 mesh) impregnated with 2 wt %
polyethyleneglycol maleinate saturated with anthracene. This stationary liquid phase gives

two peaks — one for anthracene plus phenanthrene and another for carbazole. The carrier gas was helium at a flow rate of 120 ml/min with an evaporator temperature of 310°C and an analysis temperature of 180°C. A flame-ionization detector was used. Column 2 had the same column dimensions and packing as column 1, but was impregnated with 10% polyethyleneglycol maleinate in equal parts of benzene and acetone. All other conditions were the same. The use of polyethyleneglycol maleinate was to permanently retain the anthracene in the stationary liquid phase and produce a peak for phenanthrene alone. Anthracene content in the mixture is measured by the difference in peak size in the chromatograms from columns 1 and 2. The sample size of the coal-tar mixture was 1 to 4 µl.

Results showed that obtaining a pair of chromatograms took 45 min. The lower-temperature chromatograph of less than 350° worked well and was easy to regulate, and it can be applied to other commercial mixtures for separation.

Pierce and Katz (1975) developed a chromatographic procedure to separate PAH having the same molecular weight and possessing similar or isomeric structures. This process uses adsorption thin-layer chromatography for group separation of the isomeric arenes followed by partition thin-layer chromatography for the resolution of individual isomers. A total of 12 PAH — 5 pentacyclics and 7 hexacyclics — having similar or isomeric configuration were separated in this work (Table 5.25 and Fig. 5.10).

The PAH investigated in this work were pericondensed, and all contained a pyrene nucleus. All solvents discussed were of spectrograde quality. The isomeric groups of the PAH mixture were separated by preliminary thin-layer adsorption chromatography on neutral aluminum oxide using hexane and ether (19:1, by volume) as the mobile phase. The plates were prepared by making a slurry of 25 g of aluminum oxide G, type E, 10 to 40-µm minimum grain size, in 50 ml of distilled water and mixing in a blender for 5 min at high speed. The slurry was spread on 20- by 20-cm clean glass plates with a Desaga spreader. The wet plates were air-dried for 1/2 hr, then vertically dried in an oven for 30 min at 110°C. The activated plates were stored at a constant relative humidity of 50 ± 2% and had a uniform thickness of 250 µm.

Aliquots of the PAH were added to the plates 1.5 cm from the bottom by use of clean, disposable capillary pippettes of 2.0-µl capacity (0.1 to 0.5 mg of PAH per spot). The solvent was added, and the front allowed to progress 15 cm. The chromatograms were removed from the development chamber, dried, and viewed under nondestructive ultraviolet light. The perimeters of the desired PAH were scribed, and the adsorbent, containing the extract, was removed from the plate. The PAH were eluted with 20 to 30 ml of hot spectrograde dichloromethane for 10 to 15 min, and the solvent was evaporated under a stream of dry nitrogen. The residues were taken up in a specific volume of benzene.

Resolution of the PAH present in each isomeric group was accomplished by alumina thin-layer chromatography using thin-layer chromatographic glass plates prepared with cellulose of varying degrees of acetylation (20, 30, and 40%) with n-propanol, acetone, and water (2:1:1, by volume) as the developing phase. Acetylated cellulose plates were prepared by making a slurry of 15 g of 20, 30, or 40% acetylated cellulose in 75 ml of absolute ethanol. The slurry was blended and applied in the same manner and thickness as above. The plates were air-dried for 1/2 hr and then dried in an oven at 40°C for 1/2 hr. Storage, spotting, and eluting were done as described above. The eluting agent was anhydrous diethyl ether (20 to 30 ml for 30 min)

Table 5.25. Rb values and percent recovery of isomeric arenes separated by thin-layer chromatography

Compound number	Compound name	Acetylated cellulose (n-propanol:acetone:water)(2:1:1)			Aluminum oxide (hexane:ether) (19:1)	Percent recovery[a]
		20%	30%	40%		
I	Benzo[e]pyrene	3.93	4.76	5.93	1.02	86.0 ± 5.3
II	Benzo[a]pyrene	1.00	1.00	1.00	1.00	85.2 ± 5.1
III	Benzo[b]fluoranthene	1.17	1.24	1.37	0.99	92.1 ± 3.5
IV	Benzo[k]fluoranthene	2.20	2.65	3.15	0.99	91.2 ± 3.7
V	Perylene	3.17	3.89	4.71	0.97	90.4 ± 4.6
VI	Dibenzo[def,mno]chrysene	1.17	1.32	1.67	0.90	90.9 ± 3.9
VII	Benzo[ghi]perylene	3.27	5.03	6.57	0.91	88.7 ± 4.0
VIII	Naphtho[1,2,3,4-def]chrysene	2.74	5.33	7.01	0.74	87.9 ± 5.5
IX	Benzo[rst]pentaphene	1.14	1.68	1.89	0.73	91.8 ± 4.2
X	Dibenzo[b,def]chrysene	0.27	0.33	0.40	0.73	89.8 ± 3.9
XI	Naphtho[2,1,8-gra]naphthacene	1.23	4.11	4.90	0.73	90.7 ± 5.1
XII	Dibenzo[def,p]chrysene	2.77	5.95	7.99	0.76	84.1 ± 6.2

[a]Percent recovery ± reproducibility for two-step preliminary aluminum oxide-final cellulose acetate TLC.

ORNL DWG 77-6407

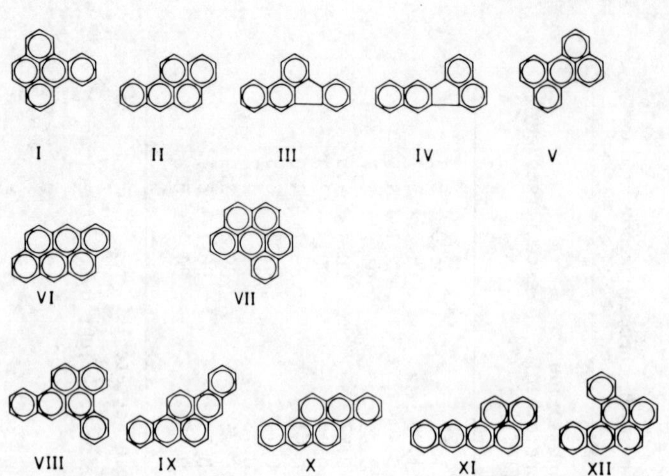

Fig. 5.10. Chemical structures of isomeric polycyclic arenes. Names are given in Table 5.25. Source: From Pierce and Katz 1975, Fig. 1, p. 1744. Reprinted with permission from *Anal. Chem.* Copyright by the American Chemical Society.

rather than dichloromethane because dichloromethane dissolves cellulose. Each residue, after solvent evaporation, was recovered from the cellulose plate and added to a specific volume of spectrograde hexane for fluorescence analysis.

Soedigdo, Angus, and Flesher (1975) studied the use of high-pressure liquid chromatography using three constant solvent compositions rather than gradients to purify and identify polycyclic hydrocarbons and their derivatives. The chromatograph consisted of a Waters high-pressure pump set to maintain a pressure of 250 kg/cm^2 at the inlet, a stainless steel column (1 m by 2.0 mm) packed with Permaphase ODS. The temperature of the column was regulated by circulating temperature-controlled water through a jacket around the column. An ultraviolet detector set to 254 nm and connected to a recorder was used to produce the chromatogram. The sample consisted of known PAH dissolved in thanol, ethoxyethanol, or ethyl acetate at a concentration of 1 μmole/ml. The eluting solvent was a methanol-water mixture, which was varied for each of three runs (75:25, 65:35, 55:45, methanol to water).

Results, given in retention times and millimeters, are shown in Tables 5.26 and 5.27. The 75:25 eluant of methanol-water proved to be the most convenient as it gave fairly short retention times and sharp peaks for most compounds tested.

5.5.3 Sampling and analysis of airborne PAH compounds

Sawicki et al. (1960) describe a simplified procedure for the separation and characterization of PAH in urban airborne particulates. This method involves one pass of benzene-soluble compounds through a column chromatograph and the subsequent ultraviolet, visible, and fluorescence studies on the fractions. The final step involves destructive analysis by a color test or spectral analysis in sulfuric acid.

Chromatographic analysis used Merck acid-washed aluminum oxide, which was washed with ether, dried, and heated in an oven for 30 min at 130°C. The alumina was determined to contain 12%

Table 5.26. Retention of BaP and derivatives

Compound	Retention					
	(mm)			(min)		
	75:25	65:35	55:45	75:25	65:35	55:45
4,5-Dihydro-4,5-dihydroxybenzo[a]pyrene	13.0	15.5	20.0	1.0	1.2	1.6
Benzo[a]pyrene-3,6-dione	18.5	23.5	52.5	1.5	1.9	4.1
6-Hydroxymethylbenzo[a]pyrene	19.0	32.0	76.0	1.5	2.5	6.0
4,5-Epoxy-4,5-dihydrobenzo[a]pyrene	24.0	44.5	112.0	1.9	3.5	8.8
6-Acetoxybenzo[a]pyrene	24.5	54.0	171.0	1.9	4.3	13.5
6-Methoxymethylbenzo[a]pyrene	34.0	68.5	226.5	2.7	5.4	17.8
6-Acetoxymethylbenzo[a]pyrene	36.0	83.5		2.8	6.6	
6-Bromomethylbenzo[a]pyrene	40.0	88.5		5.2	7.0	
6-Formylbenzo[a]pyrene	45.4	98.0		3.6	7.8	
Benzo[a]pyrene	74.5	194.0		5.9	15.3	
6-Methylbenzo[a]pyrene	99.0	309.0		7.8	24.3	
6-Benzoyloxymethylbenzo[a]pyrene	135.0			10.6		

Source: Soedigdo, Angus, and Flesher 1975, Table 1, p. 665. Reprinted by permission of the publisher.

water, requiring the addition of water until a final concentration of 13.7% was achieved. The
mixture was stirred well and allowed to stand for 12 hr in a sealed container. One gram of the
treated alumina was added to a small volume of a chloroform solution containing 50 to 150 mg
of the benzene-soluble fraction of the air particulate sample. The chloroform was evaporated
so that the organic material was homogeneously dispersed in the alumina. The material was
dispersed in the alumina prior to chromatography because it is only slightly soluble in the
primary eluting solvent. The mixture was added to a 0.5- by 15-in. column containing a lower
layer of 9 g of treated alumina and an upper layer of silica gel, neither containing eluting
solvent.

The column was eluted with successive 100-ml volumes of pentane containing 0, 3, 6, 9, and
12% of ether respectively. The chromatographic run took 2 to 3 hr, during which the column
was protected from light. Fractions of 15 to 20 ml were collected and evaporated under
vacuum at room temperature and in the dark. Residues were dissolved in pentane and tested
for ultraviolet-visible absorption spectra from 220 to 450 mμ.

Results showed that ultraviolet-visible absorption spectra of analogous fractions obtained
from different communities are similar. Airborne particulates from over 100 communities consistently contained pyrene, fluoranthene, benzo[a]fluorene, benzo[b]fluorene, chrysene,
benzo[a]anthracene, BaP, benzo[e]pyrene, benzo[k]fluoranthene, perylene, benzo[ghi]perylene,
anthanthrene, and coronene.

PAH in suspended particulate matter from New York City air were separated by Dong, Locke, and
Ferrand (1976), using a high-pressure liquid chromatographic technique. Particulate matter
was collected in high-volume air samplers set at 40 to 50 ft^3/min and run for 24 hr. The
samples were collected on glass-fiber filters and weighed between 90 and 200 mg, depending on
the sampling site. Cyclohexane in a Soxhlet extractor was used to extract the PAH from the
glass filters. The extract, a clear yellow solution, was concentrated to 5 ml in a rotary
evaporator.

Table 5.27. Retention of DMBA and derivatives

Compound	Retention					
	(mm)			(min)		
	75:25	65:35	55:45	75:25	65:35	55:45
5,6-Dihydro-5,6-dihydroxy-7,12-dimethylbenz[a]anthracene	14.0	15.5	19.0	1.1	1.22	1.5
7,12-Dimethylbenz[a]anthracene-5,6-dione	15.0		35.0	1.2		2.8
7-Hydroxymethyl-12-methylbenz[a]anthracene	16.0	24.0	49.0	1.3	1.9	3.9
5,6-Epoxy-5,6-dihydro-7,12-dimethylbenz[a]anthracene	22.0	39.0	93.5	1.7	3.1	7.4
7-Methoxymethyl-12-methylbenz[a]anthracene	25.0	51.5	161.0	2.0	4.1	12.7
7-Acetoxymethyl-12-methylbenz[a]anthracene	25.0	56.5	194.5	2.0	4.5	15.3
7-Formyl-12-methylbenz[a]anthracene	29.0	64.5		2.3	5.1	
7-Chloromethyl-12-methylbenz[a]anthracene	27.5	70.5		2.1	5.6	
7-Bromomethyl-12-methylbenz[a]anthracene	29.0	70.0		2.3	5.5	
7,12-Dimethylbenz[a]anthracene	69.5	228.0		5.5	18.0	
7-Benzoyloxymethyl-12-methylbenz[a]anthracene	86.5			6.8		

Source: Soedigdo, Angus, and Flesher 1975, Table 2, p. 666. Reprinted by permission of the publisher.

PAH were separated from olefins, paraffins, heterocyclics, etc., also present in the extract
by thin-layer chromatography. The plates were coated to a depth of 500 μ with silica gel G;
no fluorescence indicator was included in the coating. The solvent for the thin-layer
chromatographic separation was cyclohexane-benzene (1.5:1.0).

Liquid chromatographic separation of individual PAH compounds was carried out in a 25-cm long
by 2.1-mm-ID stainless steel column with OBS Zorbax packing. The extract sample size was
30 to 50 μliters when the detector was a Cary 14 YV-VIS spectrophotometer, and it was 5 μliters
when a Perkin-Elmer MFA-ZA spectrofluorometer was used. Both detectors were tried with the
ultraviolet preference stated. The chromatograph conditions were a temperature of 60°C, an
inlet pressure of 1200 psi, a methanol-water (65:35) solvent, and a flow rate of 0.21 ml/min.

The high-pressure liquid chromatography with an ultraviolet detector gave the most complete
separation with the least interference. The PAH that were separated are shown in Table 5.28.

Pellizzari et al. (1976) developed a technique to determine trace hazardous organic vapor
pollutants in ambient air samples collected in different cities around the United States.
Samples were collected in Tenax GC cartridges, thermally desorbed, and analyzed by capillary
gas-liquid chromatographic column coupled to a mass spectrometer. A computer, which was
connected on-line, recorded data on magnetic tape and generated normalized mass spectra and
mass fragmentograms. This process identified 21 halogenated hydrocarbons, including vinyl
chloride and trichloroethylene. The process also separated automobile exhaust hydrocarbons
and numerous oxygen, sulfur, nitrogen, and silicon compounds.

Tenax GC (2,6-diphenyl-p-phenyleneoxide polymers) was purified by 18-hr extraction with methanol
in a Soxhlet apparatus. The sorbent was put in Pyrex glass tubes (1.5-cm-ID by 10-cm-long)
to a depth of 6 cm (36 to 60 mesh) and held in place by 1-cm glass wool plugs. The Tenax GC
sampling cartridges were conditioned for 20 min at 270°C under a helium flow (50 ml/min) in
a thermal desorption chamber to remove background vapors. The cartridges were transferred to
Corex tubes with Teflon-lined screw caps (all organic material had been removed from the
tubes by heating them in an oven for 2 hr at 500°C) and carried to and from field sample sites.

Collected vapors were recovered by a thermal desorption inlet manifold interfaced to a Varian
MAT CH-7 gas chromatograph with three columns. Inlet temperature was 180°C, and the carrier
gas was helium, flowing at 4 ml/min into SCOT columns. The SCOT capillary columns (200 ft
and packed with OV-17, 200 ft and packed with OV-101, and 400 ft and packed with OV-101) were
programmed from 20 to 200°C at 4°C/min. The detector was a mass spectrometer which allowed the
carrier gas to enter through a single-stage glass jet separater. Mass spectra were continuously
obtained and automatically accumulated along with retention time data on magnetic tape by an
on-line Varian 620i computer.

Results are shown in a total ion current chromatogram for one sample (Fig. 5.11) to demonstrate
capabilities; peaks showed 119 compounds in the sample, many of which were identified. The
more interesting identified compounds include the toxic bromoform and tetrachloroethylene.

Mulik et al. (1975) developed a gas chromatographic—gas phase fluorescent detection system for
the assay of BaP from 24-hr particulate samples. This system can assay ambient air concentrations as low as 20 pg per cubic meter of air. The gas chromatograph partially separates the

Table 5.28. Comparison of quantitation methods on air sample S303

Compound names	Concentration (μg/1000 m^3)			
	HPLC/UV	HPLC peak height	GC peak area	Av U.S. urban concentration
Phenanthrene	0.36 ⎱ 0.41	0.71 ⎱ 0.80	0.5 ⎱ 0.6	
Anthracene	0.05 ⎰	0.09 ⎰	0.1 ⎰	
Fluoranthene	1.1	1.8	1.9	4
Pyrene	1.05	2.7	2.0	1.3 - 19
Triphenylene	0.58	0.95		
Benz[*a*]anthracene	1.4 ⎱ 3.0 ⎱ 3.6	3.9[a]	4.3	4
Chrysene	1.6 ⎰			1.3 - 11.6
Benzo[*ghi*]fluoranthene	0.19	0.3	1.0	
Benzo[*c*]phenanthrene	0.04		0.2	
Benzo[*j*]fluoranthene	0.3			
Benzo[*b*]fluoranthene	0.8	⎱ 4.0	3.2 for BaP	
Benzo[*e*]pyrene	1.4		and BeP	5
Perylene	0.1	0.2	0.2	0.7
Benzo[*k*]fluoranthene	0.6	0.6	3.0 for BbF	0.5 - 2.0
Benzo[*a*]pyrene	1.15	1.3	and BkF	5.7
Benzo[*ghi*]perylene	0.9	⎱ 2.0[a]	2.3	8
Indeno[1,2,3-*cd*]pyrene	0.3		1.4 ⎰	
Anthanthrene	0.1	0.18		0.26
Coronene	0.4	0.5	1.5	2
Dibenzopyrene				

[a]Using area.

Source: Dong and Locke 1976, Table I, p. 370. Reprinted with permission from *Anal. Chem.*
Copyright by the American Chemical Society.

BaP from other PAH compounds, and then the BaP is completely resolved optically from difficult
compounds (e.g., benzo[*e*]pyrene, perylene, benzo[*k*]fluoranthene) by use of a gas phase
fluorescent detector. The assay of BaP can be done in about 1 hr by this method.

High-volume field samples on filters were collected and returned for analysis. Each filter was
halved and shredded, then added to 250 ml of cyclohexane and homogenized in a tall beaker with
a high-speed mixer. The fiber pulp and solution were separated by filtration through a medium-
grade, sintered-glass filter. The solution was concentrated by evaporation. The concentrate
was placed by drops on a 4-cm-square piece of fiberglass filter mat, and the solvent allowed
to evaporate. The mat squares were then placed in a 1/4- by 5-in. glass tube and inserted
into the loop position of an eight-port high-temperature valve attached to the injection part
of the gas chromatograph.

The injection temperature was 420°C (after preheating to 300°C). The column was a 6-ft-long by
1/4-in.-OD glass column packed with a mixed stationary phase. The support was high-performance
Chromosorb W of 100 to 200 mesh coated with 6 wt % Dexsil 300 and 6 wt % Dexsil 400. The
column was run isothermally at 280°C, with the connector from column to gas phase fluorescent
detector held at 350°C. The detector was an Aminco-Bowman spectrophotofluorometer equipped
with a quartz flow-through cell and heated to 350°C. The fluorescent emission signal was
recorded by use of a strip chart recorder. Optimum response of BaP with the detector was
found at an excitation wavelength of 360 nm and an emission wavelength of 404 nm.

The method was found to be highly selective and sensitive for the analysis of BaP from
atmospheric samples. The minimum detectable quantity of BaP was found to be about 50 ng.

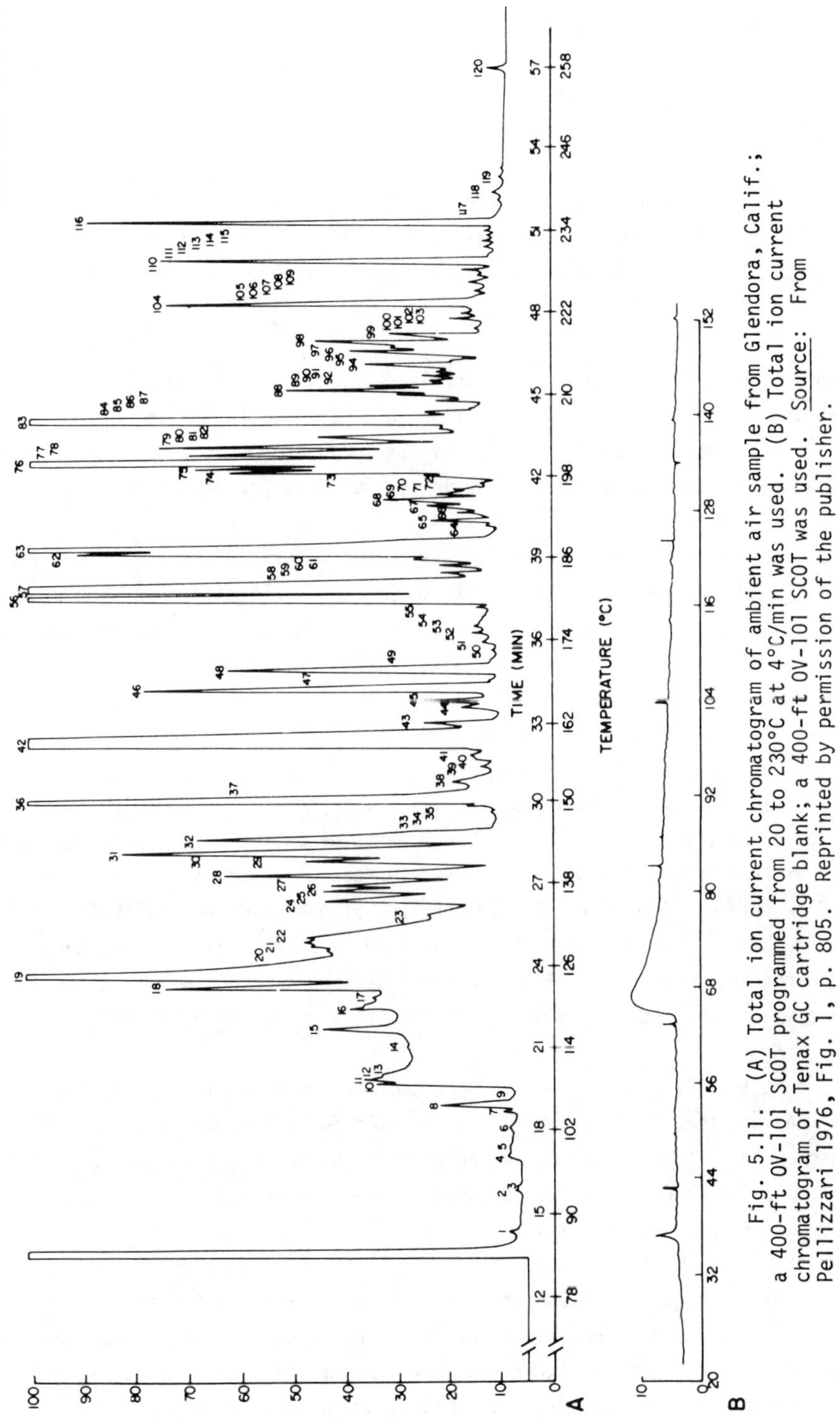

Fig. 5.11. (A) Total ion current chromatogram of ambient air sample from Glendora, Calif.; a 400-ft OV-101 SCOT programmed from 20 to 230°C at 4°C/min was used. (B) Total ion current chromatogram of Tenax GC cartridge blank; a 400-ft OV-101 SCOT was used. Source: From Pellizzari 1976, Fig. 1, p. 805. Reprinted by permission of the publisher.

Sawicki, Meeker, and Morgan (1965) studied the quantitative composition of coal combustion air pollution particulates in terms of aza-heterocyclic compounds and polynuclear aromatic hydrocarbons.

Coal combustion particulates were collected by use of a water bubbler, a freeze-out trap, and a high-efficiency glass-fiber filter. The aromatics were extracted from the filter by 4 hr of Soxhlet extraction with chloroform, the addition of a drop of triethylamine, and extraction for an additional 4 hr.

Column chromatography was used to separate components of the extract. The column was packed with alumina, and the solvent was varying percentages of ether or acetone in pentane. The detector was an ultraviolet absorption device, and each sample was a milliliter solution with pentane.

Results showed extremely high amounts of polynuclear aromatic and aza-heterocyclic hydrocarbons (246 mg of benzene-soluble PAH in 133 g of particulate matter) present in the coal combustion particulates. Carcinogens isolated from the samples include dibenzo[*ah*]acridine, dibenzo-[*aj*]acridine, alkyl derivatives of benzo[*c*]acridine, BaP, and benzo[*a*]anthracene.

Variation in concentration of tar constituents in coke-oven gas was studied by Girling and Ormerod (1964). Coke-oven gas was collected from the oven's ram side ascension pipe at a point about 18 in. above the top of the oven. The brown fog was drawn into a partially evacuated glass flask by opening a stop to a 3/16-in.-OD stainless steel tube protruding into the ascension pipe. One milliliter of toluene was added to the flask and moved around by swirling; the tar readily dissolved into the solvent.

The toluene-tar solutions were examined on two gas-liquid chromatograph columns. Both columns were constructed of 20-ft-long by 3/16-in.-OD copper tubes packed with 100 to 200 mesh Celite carrying a 4.3 wt % of Apiezon L grease. The carrier gas was argon at an inlet pressure of 60 psi and a flow rate of 30 ml/min. The temperature of operation was 140°C for one column and 200°C for the other. Temperatures were maintained by electrically heated thermostated jackets. The recorder, which was a micro-argon detector containing radium-226 foil as the ionizing source, was connected to a 500-μV potentiometric graph-making recorder.

Results showed that the 140°C column was best suited for the analysis of compounds up to the methylnaphthalenes and the 200°C column was best suited for compounds in the naphthalene to anthracene range. Forty-four peaks were detected in chromatograms from the two columns. Eleven peaks, ten of which were identified, were categorized as major; the rest, of which only three were identified, were categorized as minor.

Schulte et al. (1975) conducted a NIOSH study on analytical methods for determination of coke-oven emissions. Eight sampling procedures and seven analytical techniques were used to test samples collected for this study. Six of the analytical methods were discussed in the paper. The techniques were mostly concerned with the detection and measurement of PAH.

Sampling techniques were of either area or personal types. The area samplers were operated at a given station in the coke plant and consisted of the following seven types.

1. A Dorr-Oliver 10-ml cyclone, operating at 1.7 liters/min and exhausting into a filter held by a 37-ml millipore field monitor cassette, was chosen for a respirable sample with a widely used device.

2. A British Cast Iron Research Association aluminum alloy cyclone, operating at 2.0 liters/min and exhausting into a filter held by a 37-ml millipore field monitor cassette, was chosen for a respirable sample to compare with the one collected by the above sampler (No. 1).

3. A five-plate horizontal elutriator, operating at 5.0 liters/min and exhausting into a filter held by a 37-ml millipore field monitor cassette, was used because it is a well-established method for sampling respirable particulates.

4. A filter held in a 37-ml millipore field monitor cassette is operated with the face cap removed at 5.0 liters/min and the inlet facing down; this method was chosen to collect a gross sample at a flow rate near the peak efficiency rate for the filter.

5. This involves the same method as No. 4, but was a different type of filter for comparison.

6. A charcoal tube is sampled at 0.5 liter/min to collect any possible organic solvents.

7. A general Metal Works high-volume air sampler with an 8- by 10-in. glass filter without an organic binder, operated at 20 ft^3/min, was chosen so as to have a sample large enough to analyze by all the chemical methods used.

The personal sampler was used to collect samples in the breathing zone of the worker. A personal unit consisted of a Mine Safety Appliance, model G, battery-operated vacuum pump with pulsation dampener operated at 1.7 liters/min with a Dorr-Oliver 10-ml nylon cyclone, which exhausted into a filter held by a 37-ml millipore field monitor cassette. Analytical methods were modifications of published methods and included:

1. _Benzene-soluble extract method._ This method uses filter extraction with hot benzene using a Soxhlet extractor. The weight of the benzene-extractable material is determined by the different filter weights before and after the extraction.

2. _Gas chromatography for PAH._ This method employs the use of a commercially purchased 3% Dexsil 300 GC column fitted with a trapping device. Trapped eluant is tested with either an ultraviolet-visible spectrophotometer or a spectrophotofluorometer. This method of separation is used for fluoranthene, pyrene, benzo[c]acridine, chrysene, benzo[a]anthracene, benzo[a]anthrene, benzo[b]fluoranthene, benzo[j]fluoranthene, benzo[k]fluoranthene, benzo[e]pyrene, BaP, dibenzo[ah]anthracene, benzo[ghi]perylene, and anthanthrene.

3. _Thin-layer chromatography for BaP._ This method is used to separate BaP from other PAH. The developing mixture is a pentane-ether mixture. The BaP is extracted from the plate with diethyl ether and is dissolved in sulfuric acid; BaP content is measured fluorometrically at 470-nm excitation and 540-nm emission wavelengths.

4. _Liquid column chromatography for PAH._ The column is packed with alumina and the solvent is an n-pentane—ether mixture. The fractions are collected and scanned by absorption spectroscopy, with specific fractions examined by fluorometry to quantitate the amounts present. The PAH separated were pyrene, fluoranthene, benzo[a]anthracene, chrysene, benzo[e]pyrene, perylene, benzo[k]fluoranthene, and anthanthrene.

5. _Emission index._ The sampler filters are extracted with cyclohexane, partitioned with a dimethylsulfoxide-phosphoric acid mixture, and then again partitioned with iso-octane. The iso-octane extract is used to obtain the ultraviolet-visible spectrum from 200 to 600 nm, the fluorometric emission spectra from 300 to 500 nm with excitation at 350, 400,

and 450 nm, and the fluorometric excitation spectra from 300 to 500 nm with emission at 350, 400, and 450 nm.

6. <u>Atomic absorption for trace metals</u>. The sample filters are wet-ashed with ultra-pure acids and the ash taken into solution and diluted to 25 ml. The solutions are analyzed with an atomic absorption spectrophotometer for cobalt, chromium, copper, manganese, nickel, and zinc.

7. <u>Gas chromatography for organic solvents</u>. Organic solvents are collected on activated charcoal and then desorbed by elution with carbon disulfide. An aliquot of the carbon-disulfide solution is injected into a gas chromatograph using an 8% SE-30 packing to attain separation of benzene, toluene, xylene, and naphthalene.

The high-volume filter sampling methods showed consistent results between specific brands. The charcoal tubes were found to be unreliable sampling devices at high temperatures. No analytical method could be singled out as the best method, but the gas and liquid high-pressure techniques hold promise because of their specificity and ease of operation.

A combined gas chromatograph-ultraviolet absorption spectrometric method for monitoring petroleum pitch volatiles was developed by Greinke and Lewis (1975). Many of the major volatiles in petroleum pitch were methyl-substituted aromatics and include methylchrysene, dimethylchrysene, methylphenanthrene, and methylpyrene.

The gas chromatograph was composed of a 10-ft-long by 1/8-in.-OD stainless steel column, packed with a 3 wt % Dexsil 300 GC on Chromosorb G (AW, DMCS treated 80 to 100 mesh). The temperature of the chromatograph was programmed at a 6°C/min rise from 175 to 350°C. The injection port temperature was held at 340°C since temperatures above this point could cause some of the polynuclear compounds to decompose. The effluent from the column was passed through a 10:1 splitter, which allowed 9% of the effluent to pass to the flame ionization detector and 91% to flow to a trapping device. The eluant collected by the trapping device was held for subsequent ultraviolet measurement. The ultraviolet detector was a Cary model 14 spectro-photometer. The trapped samples were washed into the ultraviolet measuring cell with spectro-grade cyclohexane. This analytical method showed a detection limit of about 5 µg per compound.

5.5.4 Benzo[a]pyrene

The ubiquitous compound BaP is of special concern as it has been proved to be one of the most carcinogenic PAH known to man (Hornreich 1975; Mahar and Zimmerman 1975; International Agency for Research on Cancer 1972; National Academy of Sciences 1972). Quantitative analysis of BaP is difficult since it usually occurs in a complex mixture of hydrocarbons. In addition, BaP often is accompanied by an isomer, benzo[e]pyrene (BeP). BaP and BeP have not been routinely separated by gas chromatography.

Mulik et al. (1975) have developed a routine method for the selective identification and determination of BaP in atmospheric samples. This technique uses a gas chromatograph for the partial separation of the cyclohexane-extracted PAH and a spectrophotofluorometric detector (SPFD) for the selective detection and measurement of BaP. An excitation wavelength of 360 nm and an emission wavelength of 404 nm gave a strong response for BaP, but no observable response for BeP concentrations as high as 1 µg. The general procedure used for measuring ambient levels of BaP included five steps: (1) collection of 24-hr particulate air samples;

(2) extraction of the sample with cyclohexane; (3) concentration of the extract; (4) evaporation of an aliquot of concentrate onto a 4- by 4-cm piece of fiberglass filter; and (5) sublimation of the sample into the gas chromatograph.

Blank filters spiked with known amounts of BaP standard were carried through the procedure as a check for possible loss of BaP. A recovery value of 100.5% with a standard deviation of 2.4% at the 95% confidence level was obtained. When the solution was allowed to go to dryness during the concentration step, a severe loss of BaP occurred. The use of vacuum distillation did improve BaP recovery in this step.

Field samples were collected using a high-volume particulate sampling technique. As a measure of collection efficiency, control filters, one half of which were spiked with a known concentration of BaP and the other half unspiked, were run at the same time. The blank section of the control filters showed no BaP when run through the procedure.

Extraction of the particulate samples was accomplished by homogenizing the filter with cyclohexane in a high-speed mixer. Cyclohexane is nonpolar and renders a cleaner extract than the more polar compounds do. After filtering, the extract was concentrated, and an aliquot was introduced to the injection system of a gas chromatograph fitted with a 6-ft-long by 1/4-in.-OD glass column. A support of Chromosorb W (100 to 120 mesh) coated with 6 wt % Dexsil 300 and 6 wt % Dexsil 400 was used. The column effluent was then transferred to the SPFD, and the resulting fluorescent emission signal was recorded by the strip chart recorder. The weight of BaP in the sample was calculated with the aid of a fluorescent emission response vs nanograms of BaP curve derived from GC-SPFD measurements of a standard BaP solution. The GC-SPFD method was used to determine ambient levels of BaP in samples from four western sampling stations; results are shown in Table 5.29.

Table 5.29. Ambient BaP concentrations in 24-hr particulate samples from western smelter sites

Site	Date	BaP (μg/1000 m^3)
Douglas, Arizona	4/24/74	0.122
Kellogg, Idaho	4/23/74	0.519
McGill, Nevada	4/10/74	1.76
Salt Lake City, Utah	1/16/74	1.05

Source: Mulik et al. 1975, Table I, p. 522. Reprinted by permission of the publisher.

In actual atmospheric particulate samples, there may be residual organic constituents that give an SPFD response; however, duplicate spiked and unspiked filter samples indicated that this procedure is a very specific technique for measuring low to high levels of atmospheric BaP; the limit of detection for BaP was about 50 ng. A relative standard deviation of 2.3% at the 95% confidence level was achieved for five replicate injections. The assay of BaP from a high-volume filter can be done in about 1 hr; there are no apparent interferences.

Bunn et al. (1975) have used three different sampling techniques in their characterization of organic compounds found in Kansas City air. One method employs high-volume samplers for collecting air particulates on filters, and two methods accomplished the collection of samples by drawing air through columns containing XAD-2 resin and Tenax GC material respectively. All three methods may be used for remote field sampling if battery-powered pumps are used.

Identification of the compounds in the extracts, fractionated extracts, and in the case of the Tenax column, the total desorbed components, was accomplished with a computerized GC-MS system. The computerized system used a data acquisition program called IFSS (integration time as a function of signal strength), which resulted in a nearly constant signal-to-noise ratio. Weak signals are automatically monitored for long periods of time; strong signals are monitored for relatively shorter periods. At the conclusion of each gas-liquid chromatographic (GLC) run, ion abundance data were summed, normalized, and plotted as a function of spectrum index number by a digital incremental plotter. GC-MS separations were done with a 6-ft coiled glass column (0.078-in.-ID) packed with 3% OV-1 on 80 to 100 mesh Chromosorb W. Carrier gas was helium, at a flow rate of 30 ml/min; temperature at the injection port was 230°C. Standard samples, when available, were run on the GC-MS system to help identify the sample components.

Commercial high-volume samplers, using 8- by 10-in. flash-fired and organic binder free filters, were used to collect the airborne particulate samples. About 2800 m^3 of air were drawn through each filter during a 24-hr period. Each filter was extracted in a Soxhlet extractor with methylene chloride for about 8 hr. The concentrated extract (about 1 ml) was put through a Rosen separation, using activated silica gel to isolate the extract components into aliphatic, aromatic, and oxygenated fractions. Each of these fractions was concentrated to about 1 to 3 ml before injection into the GC-MS system.

Ambient air samples were collected by pulling a metered amount of air through a bed of 20 to 60 mesh purified XAD-2 resin. The resin bed was 8 cm long and was contained in a 2-cm-diam glass column. The flow rate through the bed was about 2.8 liters/min. Preextracted glass wool plugs held the resin in place in the column. Purification of the resin was accomplished by sequential Soxhlet extractions with methanol, acetonitrile, and methylene chloride. The purified resin was stored under methanol.

After sampling, the resin columns were eluted with 200 ml of dichloromethane; the extracts were concentrated and fractionated by Rosen separation, as were the extracts from the air particulate samples. Concentrated fractions were injected into the GC-MS system. Unused filters for the air particulate samplers and unused resin for the resin samplers were subjected to extraction, fractionation, and analysis of concentrates to check for background contamination.

The third type of sampling was done by drawing air at a rate of 500 ml/min for 8 to 24 hr through a column containing a 45-mm-thick bed of 60 to 80 mesh Tenax GC adsorption material. The Tenax was confined in the column by means of preextracted glass wool plugs. Before actual sampling was begun, the Tenax columns were preconditioned by passing nitrogen through them at a flow rate of 30 ml/min at a temperature of 300°C for about 8 hr.

The design of the column containing Tenax allowed the collected sample to be desorbed directly onto the analytical GLC column. The Tenax columns were fabricated from 6-mm-OD (2.5-mm-ID) Pyrex tubing. The tubes were 200 mm long and had a 22-gauge injection syringe needle embedded in one end of each tube. By means of the needle, the column could be connected to the analytical GLC column by piercing the rubber septum of the injection port. The sorbed components were desorbed onto the GLC analytical column by heating the Tenax column at 250°C and flowing helium through it at 10 ml/min. The desorption oven was a 17/64-in.-ID stainless steel tubing wrapped with heating tape. Oven temperatures were monitored by thermocouple. Desorption was continued for 8 min with the GLC column held at 40°C. Helium flow through the column was adjusted to 30 ml/min, and a GLC temperature-programmed run from 40 to 250°C at 8°C/min was started. Background contamination was checked by subjecting unused Tenax columns to the same procedures as the columns containing the collected samples.

Desorbing collected samples directly onto the analytical GLC column increases the analytical sensitivity by about 1000 times the sensitivity derived from injection of concentrated extract into the instrumental system. In addition, direct desorption prevents loss of volatile sample components.

Table 5.30 summarizes some of the results of analysis of samples collected at an EPA National Air Surveillance Network station located in Kansas City. It is interesting to note that measured concentrations of phenol from air samples collected by resin and impinger collectors were 30 and 28 $\mu g/m^3$, whereas high-volume air particulate samplers indicated only a trace of phenol present.

Table 5.30. Some organic compounds found in air samples collected at Kansas City

Compound type	Compound	Tenax[a]	Hi-vol aromatic fraction	XAD-2 resin aromatic fraction
Alkyl benzenes	Toluene, C_7H_8	X		
	C_8H_{10}	X		X
	C_9H_{12}	X		X
	$C_{10}H_{14}$	X		X
	$C_{11}H_{16}$	X		X
Naphthalenes	Naphthalene, $C_{10}H_8$	X		X
	Methylnaphthalene, $C_{11}H_{10}$	X		X
	$C_{12}H_{12}$	X		X
Other aromatic hydrocarbons	Biphenyl, $C_{12}H_{10}$			
	Anthracene, $C_{14}H_{10}$		X	
	Fluoranthene, $C_{16}H_{10}$		X	
	Pyrene, $C_{16}H_{10}$		X	
	Benzofluorene, $C_{17}H_{12}$		X	
	$C_{18}H_{12}$		X	
	Methylchrysene, $C_{19}H_{14}$		X	
	Benzofluoranthene, $C_{20}H_{12}$		X	
	Benzpyrene, $C_{20}H_{12}$		X	

[a]No fractionating procedure was performed prior to injection into the GC-MS system.

Source: Bunn et al. 1975, Table 1, pp. 370-71. Reprinted by permission of the publisher.

5.6 ANALYSIS OF PHENOLIC COMPOUNDS

Karr, Brown, and Estep (1958) used gas-liquid partition chromatography for identification and quantitative determination of low-temperature bituminous coal-tar distillates. The phenols identified were phenol, cresols, xylenols, and ethylphenols. The purpose of the experiment was to compare the gas-liquid partition chromatographic method with the conventional fractional distillation followed by infrared analysis.

The distillate chromatographed was from a sample of low-temperature (500°C) bituminous coal tar. The sample was fractionated in a vapor-phase chromatographic apparatus. The column was a 12-ft-long by 1/4-in.-diam metal tube packed with C-22 firebrick of 30 to 60 mesh that contained 31 wt % of di-n-octyl phthalate. The temperature was 160°C, and the carrier gas was helium flowing at a rate of 15 cm^3/min (15 gauge p. 5.1 at inlet). The sample charge was 250 μl; fractions were trapped in short glass vials (12-mm ID) fitted with rubber caps and containing 1 ml of spectrograde cyclohexane. Each sample trapped was submitted to infrared analysis. The method of detection and production of the chromatogram were not discussed by the author.

The authors stated, on the basis of their results, that the chromatographic method was simpler and more direct than infrared analysis and that the results were limited only by the difficulty of resolving certain of the peak-producing compounds that have similar partition coefficients. This is not too serious a problem with modern GC techniques.

Karr, Estep, and Hirst (1960) reported a fractionation of high-boiling phenols from a low-temperature coal tar by countercurrent distribution. The presence of 84 individual compounds was demonstrated; 24 compounds were identified with respect to specific isomers, and 36 with respect to structural type. The remainder of the compounds were acids of unknown type (Table 5.31).

A 200-g portion of tar acids, 3.54 wt % of a low-temperature coal tar, was distilled at 20 mm of mercury through a column filled with glass helices. All material boiling up to 118°C head temperature, equivalent to 232°C at 760 mm of mercury, was removed as a single fraction, leaving a residue of 50 g of high-boiling phenols, or 25 wt % of the original mixture. A charge of 41.68 g of high-boiling tar acids was distilled at 2.9 mm and a reflux ratio of 20 to 1 in a spinning band still. Infrared spectra were obtained on all fractions, which were then combined on the basis of qualitative similarity to give 10 samples. All samples except the lowest-boiling one were fractionated by countercurrent distribution.

The countercurrent distribution instrument used was a 60-tube all-glass model, with 200 tubes in the fraction collector and an automatic robot mechanism. The tube capacity for each phase was 40 ml. The instrument was operated to give 100 to 105 transfers or plates. The average sample size was 185 mg; the upper phase consisted of spectrograde cyclohexane, and the lower phase was a phosphate buffer. After completion of each fraction, 8 ml of 1:1 hydrochloric acid was added to each tube to neutralize the buffer and dissolve the phenols in the cyclohexane. Infrared spectra were obtained on each cyclohexane solution.

Table 5.31. Countercurrent distribution of phenols

	Constituent	Observed analytical wave lengths (UV in mμ, IR inμ)	Partition coefficient	Peak tube	
Number	Identity			Calculated	Experimental
		Boiling at 238 - 251° C, 100 transfers, pH 11.58			
1	Unknown I	285.2, 275.2, 252.5, 245.5 mμ	0		0
2	3,5-Xylenol	281.2, 273.8 mμ	0.14[a]	12	12
3	3,4-Xylenol	285.0, 278.7 mμ, 13.72, 12.52, 12.32, 11.85, 11.58, 10.57, 9.95, 9.80, 8.97, 8.66, 8.42, 7.90, 7.73 μ	0.16[a]	13	14
4	4-Indanol	276.4, 271.0, 268.7 mμ, 14.25, 13.07, 12.98, 10.18, 10.00, 9.52, 8.72, 8.32, 7.85, 7.62 μ	0.31[a]	23	21
5	5-Indanol	289.6, 283.3, 280.2 mμ, 14.45, 13.49, 12.52, 12.32, 11.91, 11.65, 10.68, 9.23, 8.82, 8.52, 7.92 μ	0.35[a]	26	26
6	3-Ethyl-5-methylphenol	280.9, 276.5, 273.3, 271.0, 267.0, 264.0 mμ, 14.47, 12.02, 11.86, 11.67, 11.07, 10.40, 8.68, 8.58 μ	0.59[a]	37	37
7	3-Ethyl-4-methylphenol	12.40, 11.49, 10.81, 8.61, 8.31, 7.95, 7.70 μ	0.64		39
8	4-*n*-Propylphenol	285.4, 278.9, 276.0, 273.0, 270.0, 267.0, 264.0, 261.0 mμ, 14.20, 13.08, 12.62, 12.20, 9.02, 8.56, 6.19 μ	0.87[a]	46	41
9	2,3,5-Trimethylphenol	282.4, 277.9, 273.4 mμ, 12.06, 11.90, 10.30, 9.27, 8.77, 8.62, 8.17, 7.88, 7.62 μ	2.6[a]	23FC[b]	25FC
10	2,6-Dialkylphenol (2-ethyl-6-*n*-propylphenol)	285.6, 277.5 mμ, 13.83, 13.08 μ	9.8		6FC
11	2,3,5,6-Tetramethylphenol	13.08, 12.95, 12.06, 9.17, 8.46 μ	59[a]	1FC	1FC
		Boiling at 251 - 258° C, 100 transfers, pH 11.20			
12	4-Indenol	—, 299.5, 287.8, 252.5 mμ, 14.45, 13.45, 13.42, 12.20, 11.10, 10.98, 10.60, 10.00, 6.42 μ	0.13[a]	11	13
13	5-Indenol	307.0, 290, 268, 258 mμ, 14.60, 14.48, 13.65, 13.00, 10.52, 9.40, 9.00, 8.25, 6.42, 6.29, 6.27 μ	0.31		24
4	4-Indanol	276.4, 271.0, 268.7 mμ	0.68[a]	41	39
5	5-Indanol	289.6, 283.3, 280.2 mμ	0.79[a]	45	45
14	Unknown II	No distinctive bands	0.90		50
15	3,4,5-Trimethylphenol	284.7, 280.0, 276.0 mμ, 14.33, 13.52, 12.04, 8.82, 8.44, 8.06 μ	1.2[a]	55	56
16	1-, 2-, or 3-methyl-5-indanol (1-methyl-5-indanol?)	288.8, 284.0, 280 mμ, 12.54, 12.32, 11.95, 10.68, 9.24, 8.55 μ	3.3		18FC[b]
17	1-, 2-, or 3-methyl-4-indanol (3-methyl-4-indanol?)	276.5, 271.5, 269.0 mμ, 14.27, 13.12, 12.98, 10.21, 8.73, 8.34 μ	3.8		16FC
18	3,5-Dialkylphenol (3-methyl-5-*n*-propylphenol)	280.3, 276.6, 273.4 mμ, 14.47, 12.04, 11.95, 10.38, 8.72 μ	6.0		10FC

Table 5.31 (continued)

Constituent		Observed analytical wave lengths (UV in mμ, IR in μ)	Partition coefficient	Peak tube	
Number	Identity			Calculated	Experimental
19	3,4-Dialkylphenol (4-ethyl-3-*n*-propylphenol)	285, 278.3 mμ, 14.34, 13.60, 12.84, 12.45, 12.35, 11.87, 10.53, 10.37, 8.70 μ	15		4FC
		Boiling at 258 - 260° C, 100 transfers, pH 10.49			
20	Methyl-1-indanone (4-methyl-1-indanone?)	314.0, 302.5, 296.0, 278.0, 257.4, 253.5, 249.5, 245.6, 242.0 mμ, 12.95, 8.65, 5.92 μ	0.35		26
21	Methyl-1-indanone (6-methyl-1-indanone?)	314.0, 278.0, 271.0, 257.5, 253.5, 249.7, 242.0 mμ, 5.93 μ	1.0		56
13	5-Indenol	307.0, 290.0, 280.0, 268.0, 258 mμ, 14.60, 13.00, 10.52, 10.38, 8.25, 6.27 μ	2.0		30FC[b]
5	5-Indanol	289.5, 284.0, 280.0 mμ, 14.45, 13.47, 12.55, 12.32, 10.68, 9.25, 8.82, 8.55 μ	2.9[a]	20FC	24FC
22	2,3- or 2,6-alkylalken-1-ylphenol (3-methyl-2-propen-1-ylphenol)	296.4 mμ, 14.10, 12.88, 10.20μ	3.0		20FC
16	1-, 2-, or 3-methyl-5-indanol (1-methyl-5-indanol?)	288.5, 282.9, 279.4, 274.5 mμ, 14.45, 12.07, 12.30, 9.24, 8.55 μ	13		5FC
23	3,4-Dialkylphenol (3-ethyl-4-*n*-propylphenol)	280.5, 273.5 mμ, 14.35, 13.61, 12.37, 11.88, 9.21, 9.02, 8.90, 8.68 μ	13		5FC
24	2-Phenylphenol	14.25, 13.90, 13.70, 13.32, 13.02, 12.04 μ	34[a]	2FC	4FC
		Boiling at 260 - 270° C, 101 transfers, pH 11.86			
20	Methyl-1-indanone (4-methyl-1-indanone)	314.0, 302.0, 257.0, 253.0, 249.0, 246.0 mμ	0.020		2
25	3,4-Dinuclearphenol	284.8, 280.7, 275.5 mμ	0.020		2
26	Alkenylphenol I	292.0 mμ	0.086		8
27	1-, 2-, or 3-methyl-4-indenol	309.6, 299.7, 288.6, 265.0, 260.7, 254.5, 249.3 mμ, 14.14, 13.15, 13.00, 12.89, 12.77, 12.42, 10.72, 10.51, 9.85, 9.15, 7.25, 6.28, 6.17, 6.14 μ	0.19		16
28	Unknown III	303.0, 240.0 mμ	0.38		28
29	Cycloalkenylphenol I?	253.0 mμ	0.51		34
30	Methylindenol	296.5, 261.0, 254.5, 249.3 mμ, 10.42, 8.71, 8.45 μ	0.60		38
31	Alkenylphenol II	308.0 mμ	0.71		42
32	7-Methyl-5-indanol	287.8, 283.0, 279.5 mμ, 14.47, 13.10, 12.07 10.53, 10.20. 9.87, 8.92, 8.47, 8.07, 7.58, 7.52, 7.26, 6.22 μ	1.2		55
33	Methyl-5-indanol (6-methyl-5-indanol)	287.8, 283.0, 279.5 mμ, 14.45, 13.60, 12.60, 12.50, 12.10, 11.97, 10.52, 10.05 μ	1.4		58
34	2,4- or 3,4-dialkylphenol?	287.5, 281.5, 277.0, 272.5 mμ	1.6		37FC[b]

Table 5.31 (continued)

Constituent		Observed analytical wave lengths (UV in mμ, IR inμ)	Partition coefficient	Peak tube	
Number	Identity			Calculated	Experimental
35	2,4- or 3,4-dialkylphenol (4-isopropyl-3-n-propyl-phenol)	287.5, 280.5, 273.4 mμ	4.5		13FC
24	2-Phenylphenol	283.0, 245.5 mμ, 14.25, 13.90, 13.70, 13.32, 13.02, 12.04, 8.49 μ	4.5[a]	13FC	13FC
36	2,4- or 3,4-dialkylphenol (2,4-di-n-propylphenol)	285.0, 279.0, 273.0 mμ, 12.45 μ	59		1FC
		Boiling at 270 - 280° C, 100 transfers, pH 10.28			
37	Unknown IV	248.0 mμ	0.15		13
38	Unknown V	307.5, 301.0, 296.0, 286.0, 265.4, 256.5 mμ, 14.57, 13.50, 13.00, 12.50, 10.84, 9.24, 9.15, 9.07, 9.02, 8.78, 8.52, 8.32, 8.12, 6.34 μ	0.22		18
39	Unknown VI	243.5 mμ, 14.69, 12.74 μ	0.33		25
40	2-Naphthol	328.6, 324.0, 321.0, 314.0, 310.0, 307.0, 301.0, 285.4, 273.8, 263.5, 254.0 mμ, 13.47, 12.52, 12.41, 11.97, 10.42, 8.95, 8.75, 8.60, 6.60, 6.22, 6.12 μ	0.59[a]	37	37
41	Unknown VII	No distinctive bands	0.79		44
42	Ketone (alkyl indanone)	314, 263.8, 259.0, 255.0, 248.0 mμ, 5.92 μ	1.0		50
43	Alkenylphenol III	301.0 mμ	1.2		56
27	1-, 2-, or 3-methyl-4-indenol	309.6, 299.7, 288.6, 265.0, 254.5, 249.3 mμ, 14.14, 13.15, 13.00, 12.89, 12.77, 12.42, 10.72, 10.51, 9.85, 9.15, 6.28, 6.17, 6.14 μ	2.2		27FC
28	Unknown III	303.0, 240.0 mμ, 14.55, 10.87, 10.10, 8.54 μ	3.9		15FC
30	Methylindenol	296.5, 254.5, 249.0 mμ	5.4		11FC
44	5,6,7,8-Tetrahydro-2-naphthol	287.8, 279.5 mμ, 14.45, 13.60, 12.60, 12.50, 12.10, 11.97, 10.52 μ	6.7[a]	9FC	9FC
45	Unknown VIII	256.0 mμ, 10.40, 10.25, 8.92, 8.70, 8.50 μ	9.8		6FC
46	5,6,7,8-Tetrahydro-1-naphthol	279.0, 273.0 mμ, 14.10, 13.04, 12.07 μ	25	2FC	1FC
24	2-Phenylphenol	283.5, 245.5 mμ, 14.25, 13.90, 13.70, 13.32, 13.02, 12.04, 8.49 μ	51[a]	1FC	1FC
		Boiling at 280 - 297° C, 101 transfers, pH 9.94			
38	Unknown V	307.5, 296.3, 265.0, 255.8, 243.5 mμ	0.31		24
47	Cycloalkenylphenol II?	304.5, 292.5, 273.7, 264.5, 256.5 mμ, 13.30 μ	0.53		35
40	2-Naphthol	328.6, 324.0, 321.0, 314.0, 310.0, 307.0, 301.0, 285.4, 273.8, 263.5, 254.0 mμ, 13.47, 12.52, 12.42, 11.98, 10.42, 8.94, 8.80, 8.60, 6.60, 6.24, 6.12 μ	0.91[a]	48	48
48	Unknown IX	No distinctive bands	1.2		55

Table 5.31 (continued)

Constituent		Observed analytical wave lengths (UV in mμ, IR inμ)	Partition coefficient	Peak tube	
Number	Identity			Calculated	Experimental
49	1-, 2-, or 3-polyalkyl-4-indenol	312 (sh), 307.5, 299.5, 296.3, 265.0, 256.0 mμ, 14.65, 13.12, 12.40, 9.13, 6.33, 6.14 μ	1.8		32FC
50	Methyl-2-naphthol (4-methyl-2-naphthol?)	331.5, 327.0, 324.2, 317.0, 290.5, 276.0, 267.0, 258.0 mμ	2.5		24FC
51	Dimethyl-1-naphthol (5,7-dimethyl-1-naphthol?)	327.4, 320.0, 312.0, 306.0, 303.0, 265.0 mμ	3.7		16FC
52	4-Methyl-1-naphthol	326.0, 319.0, 312.0, 303.0, 290.0 mμ, 13.15, 12.40, 6.30 μ	5.5[a]	11FC	10FC
53	Methyl-2-naphthol (3-methyl-2-naphthol?)	331.5, 317.0, 290.0, 278.0 mμ	6.6		9FC
54	1-Methyl-2-naphthol	335.0, 320.0, 304.0, 265.0 mμ, 13.48, 13.72, 12.45, 12.34, 0.55, 10.10, 9.83, 9.45, 8.82, 7.45, 6.63, 6.24, 6.15 μ	7.4		8FC
55	Alkyl-1-naphthol (2-ethyl-1-naphthol?)	328.2, 321.0, 314.0, 300.0, 290.0 mμ, 12.52 μ	30		2FC
56	2-Cyclohexylphenol	279 mμ, 13.30 μ	2400	0 - 1FC	1FC

Boiling at 297 - 300° C, 100 transfers, pH 10.10

Constituent		Observed analytical wave lengths (UV in mμ, IR inμ)	Partition coefficient	Peak tube	
Number	Identity			Calculated	Experimental
38	Unknown V	307.5, 296.3, 265.0, 255.8, 243.5 mμ	0.25		20
47	Cycloalkenylphenol II?	304.5, 292.5, 273.7, 264.5, 256.5 mμ, 13.30 μ	0.43		30
40	2-Naphthol	328.6, 324.0, 321.0, 314.0, 310.0, 307.0, 301.0, 285.4, 273.8, 263.5, 254.0 mμ, 13.48, 12.52, 12.42, 11.98, 10.42, 8.60 μ	0.72[a]		42
49	1-, 2-, or 3-polyalkyl-4-indenol	312 (sh), 307.5, 299.5, 296.3, 265.0, 256.5 mμ, 14.65, 13.12, 12.40, 9.13, 6.33, 6.14 μ	1.4		58
57	Unknown X	No distinctive bands	1.6		38
58	Unknown XI	309.0, 303.0, 297.0, 285.0, 276.0 mμ	1.9		31FC[b]
50	Methyl-2-naphthol (4-methyl-2-naphthol?)	331.5, 327.0, 324.2, 317.0, 290.5, 276.0, 267.0, 258.0 mμ, 13.40 μ	2.5		24FC
59	Unknown XII	No distinctive bands	2.9		21FC
60	4-Phenylphenol	255.6 mμ, 14.60, 14.05, 13.20, 12.05 μ	3.7[a]	16FC	16FC
54	1-Methyl-2-naphthol	335.0, 320.0, 304.0, 265.0 mμ	5.5		11FC
61	Dimethyl-1-naphthol (2,5- or 2.7-dimethyl-1-naphthol?)	327.0, 311.0, 303.0, 290.0, 280.0 mμ	7.9		7FC
62	Dimethyl-1-naphthol (2,6-dimethyl-1-naphthol?)	331.5 mμ	9.3		6FC
55	Alkyl-1-naphthol (2-ethyl-1-naphthol?)	328.2, 321.0, 314.0, 300.0, 290.0 mμ, 12.52 μ	20		3FC
56	2-Cyclohexylphenol	279 mμ, 13.30 μ	2100	0 - 1FC	2FC

Boiling at 300 - 325° C, 105 transfers, pH 11.85

Constituent		Observed analytical wave lengths (UV in mμ, IR inμ)	Partition coefficient	Peak tube	
Number	Identity			Calculated	Experimental
40	2-Naphthol	328.6, 314.0, 285.0, 273.8 mμ, 13.48 μ	0.040[a]	5	4
49	1-, 2-, or 3-polyalkyl-4-indenol	307.5, 299.5, 296.3, 246.5 mμ	0.050		5

Table 5.31 (continued)

Constituent		Observed analytical wave lengths (UV in mμ, IR in μ)	Partition coefficient	Peak tube	
Number	Identity			Calculated	Experimental
50	Methyl-2-naphthol (4-methyl-2-naphthol?)	331.0, 317.0, 290.5, 276.0, 267.0, 258.0 mμ, 13.40 μ	0.094		9
60	4-Phenylphenol	255.6 mμ, 14.60, 14.05, 13.20, 12.05, 11.02, 9.30, 8.97, 7.95 μ	0.11[a]	10	10
63	Unknown XIII	303.0, 296.0, 290.0 mμ	0.44		32
64	Methyl-2-naphthol (7-methyl-2-naphthol?)	331.0, 317.0, 275.0, 266.3 mμ	0.59		39
65	Unknown XIV	258.5 mμ	0.72		44
66	Unknown XV	305.5 mμ	0.84		48
67	Unknown XVI	304.0, 294.0 mμ	1.3		45FC[b]
68	Unknown XVII	261.5, 255.0 mμ	1.9		31FC
69	Unknown XVIII	260 mμ	3.0		20FC
55	Alkyl-1-naphthol (2-ethyl-1-naphthol?)	328.2, 321.0, 314.0, 300.0, 290.0 mμ, 14.53, 13.48, 12.52, 12.12, 10.07 μ	15		4FC
70	2-Alkylcycloalkylphenol I	287.0 mμ, 13.30 μ	15		4FC

Boiling at 325 - 331° C, 100 transfers, pH 11.85

Number	Identity	Observed analytical wave lengths	Partition coefficient	Calculated	Experimental
40	2-Naphthol	328.6, 286.0, 279.5, 273.0 mμ, 14.00, 13.48, 12.52, 12.42, 11.98, 10.42 μ	0.042[a]	5	4
71	Unknown XIX	306.0, 299.0, 294.0, 275.2 mμ	0.042		4
60	4-Phenylphenol	256 mμ, 14.60, 14.05, 13.20, 12.05, 9.30, 8.97, 7.95 μ	0.11[a]	10	10
72	Methyl-2-naphthol (6-methyl-2-naphthol?)	331.0 mμ	0.14		12
73	3-Phenylphenol	281.0, 249.5 mμ, 14.40, 13.25, 8.67, 8.56 μ	0.16[a]	15	14
74	Methyl-2-, -3-, or -4-fluorenol	311.0, 298.5, 282.5, 277.0, 258.0 mμ	0.20		16
75	Unknown XX	278.5 mμ	0.27		21
76	Unknown XXI	280.0, 260.7, 254.8 mμ	0.54		35
77	Unknown XXII	302.0, 282.0, 268.0, 260.5, 255.0, 249.0 mμ	0.89		47
78	8-Methyl-2-naphthol?	14.30, 13.76, 13.16 μ	2.3		26FC[b]
79	Unknown XXIII	No distinctive bands	3.5		17FC
80	Unknown XXIV	No distinctive bands	6.6		9FC
55	Alkyl-1-naphthol (2-ethyl-1-naphthol?)	328.2, 321.0, 314.0, 300.0, 290.0 mμ, 14.53, 13.48, 12.52, 12.12, 10.07 μ	29.5		2FC
81	2-Alkylcycloalkylphenol II	13.27 μ	59		1FC

[a]Partition coefficient determined from authentic specimen; other values calculated from experimental peak tube.
[b]Tube in fraction collector.

Source: Karr, Estep, and Hirst 1960, Table II, pp. 464-67. Reprinted by permission of the publisher.

Results showed that constituents fell into one of three groups. The first consisted of compounds for which authentic specimens were available. The second group consisted of compounds for which ultraviolet and/or infrared spectra were available from the literature and, in a few instances, partition coefficients. The third group was composed of compounds for which only the literature boiling point was available. Compounds in groups 1 and 2 were nearly always identified with complete certainty; members of group 3 were almost never identified with certainty as to the isomer. Compounds isolated and identified included alkylphenols, alkenylphenols, cycloalkylphenols, indanenes, indanols, indenols, and naphthols.

Junk et al. (1974) has studied the use of macroreticular resins for the analysis of water contaminated with phenols. The most widely used resin has been Rohm and Haas Amberlite XAD-2. This resin is a low-polarity styrenedivinylbenzene copolymer which possesses the macroreticular characteristics essential for high sorptive capacity (Junk et al. 1974).

The macroreticular resin was purified by sequential solvent extractions with methanol, acetonitrile, and diethyl ether in a Soxhlet extractor for 8 hr per solvent. The purified resins were stored under methanol in glass-stoppered bottles. The purified resin was packed as a methanol slurry in a 0.6-cm-ID by 10-cm-long glass tube stoppered with silanized glass wool plugs both above and below the slurry. Samples were introduced by a 2-liter reservoir attached to the top of the column and allowed to pass through the column in a gravity flow. The resins should isolate any organic impurities in the water by sorption.

Elution of the resin column was accomplished by washing with 25-ml diethyl ether. Residual water was removed from the diethyl ether by immersing the receiver in liquid nitrogen for two 10-sec intervals. The ether was decanted into a concentration vessel, a boiling chip was added, and a Snyder column was attached. The ether was boiled over a hot plate with a solvent evaporation rate of 0.5 to 2.0 ml/min. The ether was boiled until 1.00 ml remained in the concentration vessel. Compounds in the ether were subjected to gas chromatography on a single-column Varian 1200 gas chromatograph equipped with a linear temperature programmer and a flame ionization detector. The column was 6 ft x 1/8 in. OD stainless steel packed with 80 to 100 mesh acid-washed DMCS-treated Chromosorb W coated with 5 wt % OV-1 liquid phase. A DuPont 21-490-1 combination gas chromatograph-mass spectrometer was used to identify organics eluted from the XAD resins.

5.7 APPROACH TO SEPARATIONS PROBLEMS — CHROMATOGRAPHY

On March 21, 1903, Mikhail Semenovich Tsvet, a Russian botanist, presented his paper entitled "On a New Category of Adsorption Phenomena and Their Application to Biochemical Analysis" before the Warsaw Society of Natural Scientists (Heftmann 1975). In this paper, Tsvet described his investigations of the adsorption of plant pigments and other substances on over 100 different adsorbents. He called the technique "chromatography" after the Greek *chromatus* (color) and *graphein* (to write) because many of the extracts of plant pigments resolved into colored bands when passed through a column containing an adsorbent (Grob 1975).

Since the time of Tsvet's experiments, a wide variety of techniques that use the word "chromatography" have been developed, but they have very little to do with color. Today the term is applied to any of a variety of techniques which effect a separation through the distribution of sample between two immiscible phases. A feature that distinguishes chromatography

from other separation techniques such as extraction is that one of the phases must be
stationary, whereas the other is mobile and percolates through the stationary phase. Usually,
the mobile phase is a gas or a liquid, and the stationary phase is a liquid or a solid.
Separation of a sample into its component parts is possible because of differences in their
adsorption, solution, or reaction rates with the mobile and stationary phases.

Chromatography is divided into four major types, depending on the physical states of the two
phases: gas-liquid chromatography; gas-solid chromatography; liquid-liquid chromatography;
and liquid-solid chromatography. Table 5.32 shows the composition of the different chromatographic systems and the method of separation employed by each.

Table 5.32. Chromatographic systems

System	Mobile phase	Stationary phase	Configuration	Separation
Gas	Gas	Liquid	Column	Partition
	Gas	Solid	Column	Adsorption
Liquid	Liquid	Liquid	Column	Partition
	Liquid	Solid	Column'	Adsorption
Paper	Liquid	Paper	Sheet or strip	Partition or adsorption
Thin-layer	Liquid	Solid	Thin film	Adsorption
Ion exchange	Liquid	Solid	Column	Ionic replacement reactions

Source: Grob 1975, Table 2, p. 6. Reprinted by permission of the publisher.

5.8 CHEMICAL CLASS FRACTIONATIONS

When analyzing a mixed sample, it is often important to obtain first a general or qualitative
idea of the characteristics of the components and to approximate their quantity and concentration. Class separations are used for general qualitative and quantitative analysis and can be
accomplished by a number of different methods, used either singly or in combination. Wilson
and Wilson (1968) describe a number of class separations, including liquid chromatography in
columns, gas chromatography, gas-liquid chromatography (GLC), ion exchange chromatography,
and distillation. Other methods include liquid-liquid chromatography, liquid-solid chromatography, gel chromatography, paper chromatography, thin-layer chromatography, solvent extraction,
filtration, precipitation, centrifugation, volatilization, electrophoresis, and electrodeposition.

Sample components may be separated into groups according to their chemical behavior or physical
properties. Groups may be precipitated from a solution by varying the pH or adding precipitating
reagents; or by using boiling points, they may be selectively distilled. Chromatographic
characteristics may allow column separation of groups, or groups may be separated by the most
powerful separatory technique, solvent extraction. Some of the simplest techniques for
separating large particles from small particles employ sieving or sizing with a series of
calibrated screens. Other simple methods separate particles or flocs from solution by filtration or centrifugation. In most gross separations, conditions can be altered to increase or
decrease the number of components in each separated group. The addition of certain compounds

to a solution containing several components will selectively complex or "tie up" ion groups in such a way as to alter their solubility, adsorptivity, volatility, or electroactivity.

5.8.1 PAH separations

Methods for group separation of the elements have been well developed and routinely used for a long time. However, organic analyses are not as well mapped; this is especially true for the PAH. These compounds are ubiquitous, formed in the high-temperature pyrolysis of organic materials such as coal, petroleum, wood, and plants (Andelman and Snodgrass 1974). Although PAH concentrations in air, water, soil, and plants are usually higher in industrialized urban areas than in rural areas, environmental samples, on the whole, contain very small quantities of PAH. Some PAH concentrations found in environmental samples are listed in Table 5.33.

The occurrence of PAH in the parts-per-million and parts-per-billion range increases the need for high sensitivity in separation. Additionally, PAH occur in very complex combinations, including difficult-to-separate isomers such as BaP and BeP. Quantitative separation of these isomers is necessary to isolate carcinogenically active BaP from inactive BeP. Dimethyl-benz[a]anthracene, chrysene, and triphenylene are also difficult to separate; dimethyl-benz[a]anthracene is considered one of the most carcinogenic compounds, whereas the carcino-genicity of chrysene is debatable (International Agency for Research on Cancer 1972).

5.8.2 Whole sample class fractionations

Complex samples are usually subjected to a procedure which divides the components into groups. One method for group separation of organic compounds, developed by Swain, Cooper, and Stedman (1969) for large-scale fractionation of cigarette smoke condensate, separates the components into acidic, basic, and neutral fractions. Since many of the organic compounds in cigarette smoke condensate also occur in environmental samples and in process and waste streams from coal conversion processes, the method can be applied to samples from these sources.

To prevent loss or change of sample components, the cigarette smoke condensate used in this study was collected in dry ice–acetone cold tapes from which the condensate was removed by dissolution in acetone. After removal of the acetone by vacuum at a low temperature, the cigarette smoke condensate was shipped under nitrogen at -20°C and stored in a deep-freeze box until fractionation was started. Elapsed time between sample receipt and start of fractiona-tion was usually no more than two days.

Analysis of cigarette smoke condensate used 720 g of the material warmed to room temperature and transferred to a 6-liter separatory funnel with the aid of several small portions of ether and 1.0 N sodium hydroxide. Weights of the insoluble fractions were determined by vacuum filtration on a Buchner funnel, followed by air-drying and weighing. Drying of solutions was done by evaporation in vacuo in tared flasks using rotating evaporators at a bath temperature of 30°C or less. Evaporation was continued until successive weighing produced no evidence of further weight loss. All solutions were dried over sodium sulfate before evaporation, which reduced the variability of weights due to retaining water compared to air-dried solids. A more drastic drying procedure was avoided because of the possibility of losing or altering constituents.

Table 5.33. Typical concentration ranges of BP and PAH
in various fresh surface waters

Source	Concentration (µg/liter)		
	BP	Carcinogenic PAH	Total PAH
Rhine River at various points	0.05 - 0.11	0.01 - 0.73	0.73 - 1.50
Various German rivers	0.001 - 0.04	0.04 - 1.30	0.12 - 3.1
One American river	0.078 - 0.150		
Rivers receiving effluents from industries that are sources of PAH	0.0001 - 12		

Concentration of BP in surface waters (µg/liter)

Source	BP
Moscow reservoirs, slightly polluted	4 - 13
Volga River below discharge site of oil refinery	0.0001
Pskov region, remote from exogenous sources of BP	$10^{-5} - 10^{-4}$
Sunzha River, 3 - 4 km downstream from discharge sites of petroleum refinery (23 samples)	0.05 - 3.5
Sunzha River, 25 km downstream from lowest discharge site of petroleum refinery	0.07 - 1.06
Oyster River, Conn., U.S.	
Sample 1	0.078
Sample 2	0.125
Sample 3	0.150

Concentration of BP in surface water environment (µg/kg dry material)

Source	BP
Rublevskoye Reservoir	
Plankton	0.7 - 1.8
Bottom sand	44
Pond weeds	0.6 - 2.7
Khimkinskoe Reservoir	
Plankton	7.3 - 8.8
Bottom sand	390 - 500
Pond weeds	1.7 - 37.8
Bottom sediments of Sunzha River, 3 - 4 km downstream from discharge site of refinery	9.2 - 19.0
Bottom sediments of Sunzha River 25 km downstream from discharge site of refinery	Trace - 3.1
Silt in Moscow Reservoir	1.8 - 5.0
Pskov region, remote from exogenous sources of BP	
Algae	5
Bottom sand	1 - 2

Source: Andelman and Snodgrass 1974, Table 3, pp. 76-77. Reprinted by permission of the publisher.

To check the reproducibility of this fractionation procedure, nine different runs were made. A reconstituted cigarette smoke condensate was prepared for each run by combining one-third of each of the 12 fractions obtained in a run; the remaining two-thirds of each fraction was evaporated and weighed. Table 5.34 summarizes the results of nine runs, and Table 5.35 summarizes the variability in cigarette smoke condensate fraction yields.

Table 5.34. Results of nine repeated cigarette smoke condensate fractionations

| Fraction | | Run number | | | | | | | | | Average |
Name[a]	Number	1	2	3	4	5	6	7	8	9	$\pm$S.E.[b]
Reconstituted CSC	2	219.6[c]	208.1	212.0	184.8	208.1	227.8	192.2	199.4	227.9	208.9 $\pm$ 5.0
$B_I a$	3	8.3	6.8	6.6	4.9	4.6	3.0	5.5	6.0	5.5	5.7 $\pm$ 0.5
$B_I b$	4	4.0	3.2	1.6	2.1	3.8	2.4	6.3	2.7	3.5	3.3 $\pm$ 0.5
B_F	5	28.9	28.2	23.0	16.8	43.8	25.3	33.4	33.9	31.5	29.4 $\pm$ 2.5
B_W	6	15.3	11.9	8.7	11.8	15.9	6.9	8.2	13.0	13.8	11.7 $\pm$ 1.1
WA_I	7	28.8	40.4	64.1	31.9	26.6	44.7	27.7	33.5	43.0	37.9 $\pm$ 4.0
WA_E	8	46.6	52.8	33.4	45.9	57.6	48.5	45.4	57.0	45.5	48.1 $\pm$ 2.4
SA_I	9	4.5	8.3	10.1	10.9	7.5	6.1	11.8	7.2	8.3	8.3 $\pm$ 0.8
SA_E	10	8.8	9.3	11.7	12.9	17.0	12.8	13.0	21.1	17.2	13.8 $\pm$ 1.3
SA_W	11	160.6	193.9	184.5	195.7	150.9	186.4	163.6	133.6	166.9	170.7 $\pm$ 7.0
N_{MeOH}	12	29.7	26.3	31.3	26.4	15.8	21.3	20.5	22.0	24.8	24.2 $\pm$ 1.6
N_{CH}	13	82.2	72.0	67.0	60.0	77.4	92.8	73.5	99.0	77.6	77.9 $\pm$ 4.1
N_{NM}	14	14.3	12.6	14.6	14.2	12.4	10.5	9.3	19.0	18.4	13.9 $\pm$ 1.1
Total	2 - 14	651.6	673.8	668.6	618.3	641.4	688.5	610.4	647.4	683.9	653.8 $\pm$ 9.2

[a]CSC, cigarette smoke condensate; A, acids; B, bases; N, neutrals; S, strong; W, weak. Subscripts: I, insoluble; E, ether-soluble; W, water-soluble; MeOH, methanol-soluble; CH, cyclohexane-soluble; NM, nitromethane-soluble.
[b]S.E. = standard error of the mean, $\sqrt{V/9}$; V, variance, $\Sigma(X_i-\overline{X})^2/8$. One-third of the isolated amounts of fractions 3 - 14 were pooled to give fraction 2. The values for fractions 3 - 14 in the table represent the yields obtained from the remaining two thirds.
[c]From 720 g CSC.

Source: Swain, Cooper, and Stedman 1969, Table 1, p. 582. Reprinted by permission of the publisher.

Table 5.35. Variability of cigarette smoke condensate fraction yields[a]

Fraction		Mean (g)	Average deviation	V	S.D.	Coefficient of variation
Name	No.					
$B_I b$	4	3.3	0.99	1.9	1.38	42.0
WA_I	7	37.9	9.05	143.1	11.95	31.0
SA_E	10	13.8	3.13	15.9	3.99	29.0
SA_I	9	8.3	1.80	5.4	2.33	28.0
BW	6	11.7	2.50	10.1	3.37	28.0
B_E	5	29.4	5.50	58.3	7.63	26.0
$B_I a$	3	5.7	1.10	2.3	1.50	26.0
N_{NM}	14	13.9	2.40	10.4	3.22	23.0
N_{MeOH}	12	24.2	3.90	23.5	4.85	20.0
N_{CH}	13	77.9	8.90	147.4	12.14	15.0
WA_E	8	48.1	5.20	53.6	7.32	15.0
SA_W	11	170.7	17.30	441.9	21.02	12.0
Reconstituted CSC	2	208.9	11.50	224.1	14.97	7.1
Total	2 - 14	653.8	22.20	759.2	27.55	4.2

[a]Mean, $\Sigma X_i/9$; average deviation, $\Sigma(X_i-\overline{X})/9$; $V = \Sigma(X_i-\overline{X})^2/8$; S.D., standard deviation, $\sqrt{V}$; coefficient of variation, 100 S.D./mean. See Table 5.31 for abbreviations.

Source: Swain, Cooper, and Stedman 1969, Table 2, p. 583. Reprinted by permission of the publisher.

The greatest variabilities were caused by (1) the difficulty of quantitatively removing and recovering the four insoluble fractions and (2) losses that occurred during solvent removal by evaporation in vacuo. The cigarette smoke condensate samples themselves probably contributed to the variability of results as there is considerable variability of component concentrations between condensates. Still, an average recovery of 90.8% for the sums of the 12 subfractions and the reconstituted condensates was achieved (Swain, Cooper, and Stedman 1969).

LITERATURE CITED

Adams, D. F.; Frohliger, J. O.; Falgout, D.; Hartley, A. M.; Pate, J. E.; Plumley, A. L.; Scaringelli, F. P.; and Urone, P. 1975. Hydrogen sulfide in air analytical method. *Health Lab. Sci.* 12(4): 362-68.

Ahuja, S.; Cohen, E. M.; Kneip, T. J.; Lambert, J. L.; and Zweig, G., eds. 1973. *Chemical analysis of the environment and other methods.* Progress in Analytical Chemistry Series 5. New York: Plenum Press.

Allan, F. J.; Lambie, D. A.; McKissock, M. C.; and Shaw, T. C. 1966. Determination of carbon dioxide in Scottish coals. *Fuel* 45(1): 93-95.

Andelman, J. B., and Snodgrass, J. E. 1974. Incidence and significance of polynuclear aromatic hydrocarbons in the water environment. In *CRC Crit. Rev. Environ. Control* 4(1): 69-83.

Bunn, W. W.; Dean, E. R.; Klein, D. W.; and Kleopper, R. D. 1975. Sampling and characterization of air for organic compounds. *Water Air Soil Pollut.* 4(3-4): 367-80.

Cameron, A. E.; Smith, D. H.; and Walker, R. L. 1969. Mass spectrometry of nanogram-size samples of lead. *Anal. Chem.* 41: 525-26.

Carter, J. A.; Walker, R. L.; and Sites, J. R. 1975. Trace impurities in fuels by isotope dilution mass spectrometry. In *Trace Elements in Fuel,* Advances in Chemistry Series 141, ed. S. P. Babu, pp. 74-83. Washington, D.C.: American Chemical Society.

Carter, J. A.; Christie, W. H.; Donohree, D. L.; Franklin, J. C.; Smith, D. H.; and Stelzner, R. H. 1976. Development and application of multi-element isotope dilution, spark-source mass spectrometry, and ion microprobe analysis of trace contaminants. In *Progress report October 1974 - December 1975,* pp. 117-24. ORNL/NSF-EATC-22. Oak Ridge, Tenn.: Oak Ridge National Laboratory.

Dharmarajan, V.; Thomas, R. L.; Maddalone, R. F.; and West, P. W. 1975. Sulfuric acid aerosol. *Sci. Total Environ.* 4: 279-98.

Dong, M.; Locke, D. C.; and Ferrand, E. 1976. High pressure liquid chromatographic method for routine analysis of major parent polycyclic aromatic hydrocarbons in suspended particulate matter. *Anal. Chem.* 48(2): 368-72.

Dreher, G. B., and Schleicher, J. A. 1975. Trace elements in coal by optical emission spectroscopy. In *Trace elements in fuel,* Advances in Chemistry Series 141, ed. S. P. Babu, pp. 35-47. Washington, D.C.: American Chemical Society.

Drury, J. S. 1975. Analysis for cadmium. In *The environmental impact of cadmium* (working draft), pp. 9-104. ORNL/EIS-75-76. Oak Ridge, Tenn.: Oak Ridge National Laboratory.

Environmental Protection Agency. 1974. *Health consequences of sulfur oxides: A report from CHESS, 1970-1971.* EPA 650/1-74-004.

Frost, J. K.; Santoliquido, P. M.; Camp, L. R.; and Ruch, R. R. 1975. Trace elements in coal by neutron activation analysis with radiochemical separations. In *Trace elements in fuel,* Advances in Chemistry Series 141, ed. S. P. Babu, pp. 84-97. Washington, D.C.: American Chemical Society.

Girling, G. W., and Ormerod, E. C. 1964. Variation in concentration of some tar constituents in coke-oven gas. *J. Appl. Chem.* 14: 36-40.

Given, P. H. 1973. How may coals be characterized for practical use? Paper presented at Short Course on Coal Characteristics and Coal Conversion Processes, Oct. 29-Nov. 2, 1973, Pennsylvania State University, University Park.

Given, P. H., and Yarzab, R. F. 1975. *Problems and solutions in the use of coal analysis.*
PSU-TR-1; FE-0390-1. University Park: Pennsylvania State University, Coal Research
Section.

Gluskoter, H. J. 1975. Mineral matter and trace elements in coal. In *Trace elements in fuel,*
Advances in Chemistry Series 141, ed. S. P. Babu, pp. 1-22. Washington, D.C.: American
Chemical Society.

Greinke, R. A., and Lewis, I. C. 1975. Development of a gas chromatographic-ultraviolet
absorption spectrometric method for monitoring petroleum pitch volatiles in the environment. *Anal. Chem.* 47(13): 2151-55.

Grob, R. L., ed. 1975. *Chromatographic analysis of the environment.* New York: Marcel Dekker,
Inc.

Heftmann, E., ed. 1975. *Chromatography: A laboratory handbook of chromatographic and
electrophoretic methods.* New York: Van Nostrand Reinhold Co.

Hornreich, M. R. 1975. *A preliminary assessment of the problem of carcinogens in the
atmosphere.* MTR-6874. Mitre Corporation.

Husar, J. D.; Husar, R. B.; and Stubits, P. K. 1975. Determination of submicrogram amounts
of atmospheric particulate sulfur. *Anal. Chem.* 47(12): 2062-65.

International Agency for Research on Cancer. 1972. *IARC monographs on the evaluation of the
carcinogenic risk of chemicals to man: Certain polycyclic aromatic hydrocarbons and
heterocyclic compounds,* vol. 3. From the Meeting of the IARC Working Group on the
Evaluation of the Carcinogenic Risk of Chemicals to Man, Dec. 5-11, 1972, Lyon, France.

Jacobs, M. B. 1960. *The chemical analysis of air pollutants.* New York: Interscience
Publishers, Inc.

Janini, G. M.; Muschik, G. M.; and Zielinski, W. L. 1976. N,N'-Bis (p-butoxybenzylidene)-
α,α'-bis-p-toluidine: Thermally stable liquid crystal for unique gas-liquid
chromatography separations of polycyclic aromatic hydrocarbons. *Anal. Chem.* 48(6):
809-13.

Jones, D. R., IV, and Manahan, S. E. 1975. Atomic absorption detector for chromium
organometallic compounds separated by high speed liquid chromatography. *Anal. Lett.*
8: 569-74.

Junk, G. A.; Richard, J. J.; Grieser, M. D.; Witiak, D.; Witiak, J. L.; Arguello, M. D.;
Vick, R.; Svec, H. J.; Fritz, J. S.; and Calder, G. V. 1974. Use of macroreticular resins
in the analysis of water for trace organic contaminants. *J. Chromatog.* 99: 745-62.

Karr, C., Jr.; Brown, P. M.; and Estep, P. A. 1958. Identification and determination of
low-boiling phenols in low temperature coal tar. *Anal. Chem.* 30(8): 1413-16.

Karr, C., Jr.; Estep, P. A.; and Hirst, L. L., Jr. 1960. Countercurrent distribution of
high-boiling phenols from a low-temperature coal tar. *Anal. Chem.* 32(4): 463-75.

Kneip, T. J.; Ajemian, R. S.; Driscoll, J. N.; Grunder, F. I.; Kornreich, L.; Loveland, J. W.;
Moyers, J. L.; and Thompson, R. J. 1975. General atomic absorption procedure for trace
metals in airborne material collected on membrane filters. *Health Lab. Sci.* 12: 383-93.

Knott, A. C., and Belcher, C. B. 1975. Determination of carbon dioxide in coal and minerals.
Talanta 22(9): 751-53.

Kogan, L. A., and Sorokina, N. L. 1974. Gas-chromatographic determination of anthracene,
phenanthrene, and carbazole in coal-tar fractions at low temperatures. *Coke Chem.*
1974(3): 32-34.

Kolthoff, I. M., and Elving, P. J., eds. 1964. *Treatise on analytical chemistry.* Part I,
vol. 5, pp. 2707-3346. New York: Wiley Co.

Kuhn, J. K.; Harfst, W. F.; and Shimp, N. F. 1975. X-ray fluorescence analysis of whole
coal. In *Trace elements in fuel,* Advances in Chemistry Series 141, ed. S. P. Babu,
pp. 66-73. Washington, D.C.: American Chemical Society.

Laitinen, H. A. 1973. Analytical techniques for heavy metals other than mercury. Presented
at the Heavy Metals in the Aquatic Environment Conference, Dec. 4-7, 1973, Vanderbilt
University, Nashville, Tenn.

Lao, R. C.; Oja, H.; Thomas, R. S.; and Monkman, J. L. 1973. Assessment of environmental problems using the combination of gas chromatography and quadrupole mass spectrometry. *Sci. Total Environ.* 2(3): 223-33.

Lao, R. C.; Thomas, R. S.; and Monkman, J. L. 1975. Computerized gas chromatographic-mass spectrometric analysis of polycyclic hydrocarbons in environmental samples. *J. Chromatogr.* 112: 681-700.

Lao, R. C.; Thomas, R. S.; Oja, H.; and Dubois, L. 1973. Application of a gas chromatograph-mass spectrometer-data processor combination to the analysis of the polycyclic aromatic hydrocarbon content of airborne pollutants. *Anal. Chem.* 45(6): 908-15.

Luther, G. W., and Meyerson, A. L. 1975. Polarographic analysis of sulfate ion in seawater samples. *Anal. Chem.* 47(12): 2058-59.

Lyon, W. S., and Emery, J. F. 1975. Neutron activation analysis applied to the study of elements entering and leaving a coal-fired steam plant. *Int. J. Environ. Anal. Chem.* 4: 125-33.

Maddalone, R. F.; Thomas, R. L.; and West, P. W. 1976. Measurement of sulfuric acid aerosol and total sulfate content of ambient air. *Environ. Sci. Technol.* 10(2): 162-68.

Mahar, H., and Zimmerman, N. 1975. *Evaluation of selected methods to assess the potential hazards associated with industrial particulate emissions: interim report.* M74-70. Mitre Corporation. Prepared for Process Measurements Branch, Industrial Environmental Research Lab., EPA, Research Triangle Park, North Carolina.

Mariich, L. I., and Lenkevich, Zh. K. 1973. Capillary chromatography as a method for the rapid determination of the main components in coal tar and coal tar fractions. *Zh. Anal. Khim.* 28(6): 1193-98.

Medlin, J. H.; Suhr, N. H.; and Bodkin, J. B. 1969. Atomic absorption analysis of silicates employing $LiBO_2$ fusion. *At. Absorpt. Newsl.* 8(2): 25-29.

Mukherjee, P.; Basak, N. G.; and Lahiri, A. 1953. Moisture in coal: Part I. *J. Sci. Res.* 12B: 15-24.

Mulik, J.; Cooke, M.; Guyer, M. F.; Semeniuk, G. M.; and Sawicki, E. 1975. A gas liquid chromatographic fluorescent procedure for the analysis of benzo(a)pyrene in 24 hour atmospheric particulate samples. *Anal. Lett.* 8(8): 511-24.

Muter, R. B., and Nice, L. L. 1975. Major and minor constituents in siliceous materials by atomic absorption spectroscopy. In *Trace Elements in Fuel*, Advances in Chemistry Series 141, ed. S. P. Babu, pp. 57-65. Washington, D.C.: American Chemical Society.

National Academy of Sciences (NAS). 1972. *Particulate polycyclic organic matter*. Washington, D.C.: National Academy of Sciences.

Ode, W. H. 1963. Coal analysis and mineral matter. In *Chemistry of coal utilization,* supplementary volume, ed. H. H. Lowry, pp. 202-31. New York: John Wiley & Sons, Inc.

O'Gorman, J. V., and Walker, P. L. 1972. *Mineral matter and trace elements in U.S. coals.* R&D Rept. 61; Int. Rept. 2. Office of Coal Research, U.S. Dept. of Interior.

Pellizzari, E. D.; Bunch, J. E.; Berkeley, R. E.; and McRae, J. 1976. Determination of trace hazardous organic vapor pollutants in ambient atmospheres by gas chromatography/ mass spectrometry/computer. *Anal. Chem.* 48(6): 803-07.

Pierce, R. C., and Katz, M. 1975. Determination of atmospheric isomeric polycyclic arenes by thin-layer chromatography and fluorescence spectrophotometry. *Anal. Chem.* 47(11): 1743-48.

Pollock, E. N. 1975. Trace impurities in coal by wet chemical methods. In *Trace elements in fuel*, Advances in Chemistry Series 141, ed. S. P. Babu, pp. 23-24. Washington, D.C.: American Chemical Society.

Pringle, W.J.S. 1963. A rapid titrimetric method for the determination of carbonate carbon dioxide in coal. *Fuel* 42: 63-68.

Rasmussen, R. A. 1974. Analysis of trace organic sulfur compounds in air. In *Laboratory instrumentation: Chromatography*, series I, vol. I, pp. 293-97. Greens Farms, Conn.: International Scientific Communications, Inc.

Rosen, A. A., and Middleton, F. M. 1955. Identification of petroleum refinery wastes in surface waters. *Anal. Chem.* 27(5): 790-94.

Sawicki, E.; Elbert, W.; Stanley, T. W.; Hauser, T. R.; and Fox, F. T. 1960. Separation and characterization of polynuclear aromatic hydrocarbons in urban air-borne particulates. *Anal. Chem.* 32(7): 810-15.

Sawicki, E.; Guyer, M.; and Engel, C. R. 1967. Paper and thin-layer electrophoretic separations of polynuclear aza heterocyclic compounds. *J. Chromatogr.* 30: 522-27.

Sawicki, E.; Hauser, T. R.; Elbert, W. C.; Fox, F. T.; and Meeker, J. E. 1962. Polynuclear aromatic hydrocarbon composition of the atmosphere in some large American cities. *Am. Ind. Hyg. Assoc. J.* 23: 137-44.

Sawicki, E.; Hauser, T. R.; and Stanley, T. W. 1960. Ultraviolet, visible and fluorescence spectral analysis of polynuclear hydrocarbons. *Int. J. Air Pollut.* 2: 253-72.

Sawicki, E., and Johnson, H. 1964. Characterization of aromatic compounds by low-temperature fluorescence and phosphorescence: Application to air pollution studies. *Microchem. J.* 8: 85-101.

Sawicki, E.; Johnson, H.; and Kosinski, K. 1966. Chromatographic separation and spectral analysis of polynuclear aromatic amines and heterocyclic imines. *Microchem. J.* 10: 72-102.

Sawicki, E.; Meeker, J. E.; and Morgan, M. J. 1965. The quantitative composition of air pollution source effluents in terms of aza heterocyclic compounds and polynuclear aromatic hydrocarbons. *Int. J. Air Water Pollut.* 9: 291-98.

Sawicki, E.; Stanley, T. W.; and Johnson, H. 1965. Quenchofluorometric analysis for polynuclear compounds. *Mikrochem. Acta* 1965: 178-92.

Sawicki, E.; Stanley, T. W.; McPherson, S.; and Morgan, M. 1966. Use of gas-liquid and thin-layer chromatography in characterizing air pollutants by fluorometry. *Talanta* 13: 619-29.

Scientific Committee on Problems of the Environment (SCOPE). 1975. *Environmental pollutants selected analytical methods*, SCOPE 6. Ann Arbor, Mich.: Ann Arbor Science Publishers, Inc.

Sharkey, A. G., Jr., Kessler, T.; and Friedel, R. A. 1975. Trace elements in coal dust by spark-source spectrometry. In *Trace elements in fuel*, Advances in Chemistry Series 141, ed. S. P. Babu, pp. 48-56. Washington, D.C.: American Chemical Society.

Shulte, K. A.; Larsen, D. J.; Hornung, R. W.; and Crable, J. V. 1975. Analytical methods used in a study of coke oven effluent. *Am. Ind. Hyg. Assoc. J.* 26(2): 131-39.

Soedigdo, S.; Angus, W. W.; and Flesher, J. W. 1975. High-pressure liquid chromatography of polycyclic aromatic hydrocarbons and some of their derivatives. *Anal. Biochem.* 67: 664-668.

Sparks, C. J., Jr.; Carrn, O. B.; Harris, L. A.; and Ogle, J. C. 1974. Simple, quantitative x-ray fluorescent analysis for trace elements. In *Trace substances in environmental health VII. University of Missouri, Columbia, Mo., June 12-14, 1974*, pp. 361-68.

Stauff, J., and Jaeschke, W. 1975. A chemiluminescence technique for measuring atmospheric trace concentrations of sulfur dioxide. *Atmospher. Environ.* 9(11): 1038-39.

Swain, A. P.; Cooper, J. E.; and Stedman, R. L. 1969. Large-scale fractionation of cigarette smoke condensate for chemical and biologic investigations. *Cancer Res.* 29: 579-83.

Talmi, Y., and Morrison, F. H. 1972. Induction furnace method in atomic spectrometry. *Anal. Chem.* 44: 1455-66.

Talmi, Y. 1974. Determination of zinc and cadmium in environmentally based samples by the radiofrequency spectrometric source. *Anal. Chem.* 46: 1005-10.

Talmi, Y., and Andren, A. W. 1974. Determination of selenium in environmental samples using gas chromatography with a microwave emission spectrometric detection system. *Anal. Chem.* 46: 2122-26.

Talmi, Y., and Crosmun, R. 1974. Applicability of the RF-furnace technique for AA and AE analysis of trace elements in environmental samples. In *Trace substances in environmental health VII. University of Missouri, Columbia, Mo., June 12-14, 1974*, pp. 379-85.

Talmi, Y., and Bostick, D. T. 1975. The determination of arsenic and arsenicals. *J. Chromatogr. Sci.* 13: 231-37.

Talmi, Y., and Mesmer, R. E. 1975. Studies on vaporization and halogen decomposition of methyl mercury compounds using GC with a microwave detector. *Water Res.* 9: 547-552.

Thomas, R. L.; Dharmarajan, V.; Lundquist, G. L.; and West, P. W. 1976. Measurement of sulfuric acid aerosol, sulfur trioxide, and the total sulfate content of the ambient air. *Anal. Chem.* 48(4): 639-42.

Urone, P. 1974. Source and ambient air analysis for sulfur trioxide and sulfuric acid: A state-of-the-art report. *Health Lab. Sci.* 11(3): 246-54.

U.S. Bureau of Mines, 1967. *Methods of analyzing and testing coal and coke.* Bull. 638. Washington, D.C.: U.S. Dept. of the Interior, U.S. Office of Coal Research.

West, W. P., and Gaeke, G. C. 1956. Fixation of sulfur dioxide as disulfitomercurate(II) and subsequent colorimetric estimation. *Anal. Chem.* 28: 1816-19.

White, F. A., and Collins, T. L. 1954. A two stage magnetic analyzer for isotopic ratio determinations of 10^4 to 1 or greater. *Appl. Spectrosc.* 9: 169-73.

Wilson, C. L., and Wilson, D. W., eds. 1968. *Comprehensive analytical chemistry. Vol. IIB Physical Separation Methods.* Amsterdam: Elsevier Publishing Co.

Zubovic, P. 1975. *Geochemistry of trace elements in coal.* Reston, Va.: U.S. Geological Survey.

Environmental, Health and Control Aspects of COAL CONVERSION— An Information Overview

Volume 2

Edited by

H. M. Braunstein
E. D. Copenhaver
H. A. Pfuderer

Contributors

J. T. Ensminger	N. S. Dailey	G. A. Dailey
J. K. Huffstetler	D. C. Michelson	R. F. Carrier
L. W. Rickert	R. H. Ross	D. J. Wilkes
S. S. Talmage	D. S. Harnden	

Copy editors

B. C. Winsbro
J. W. McKenna

Published 1981 by Ann Arbor Science Publishers, Inc.
230 Collingwood, P.O. Box 1425, Ann Arbor, Michigan 48106

Library of Congress Catalog Card Number 81-65350
ISBN 0-250-40445-1

Manufactured in the United States of America
All Rights Reserved

First published by the Department of Energy
(ERDA), 1977. Second publication by Ann
Arbor Science Publishers, Inc.,1981.

Section Locator
Volume 2

S. S. Talmage

xx

6. ENVIRONMENTAL INTERACTIONS

R. H. Ross

ABSTRACT

Because pollution-abatement measures that are implemented at coal conversion plants may not be 100% effective, a knowledge of the environmental interactions of possible coal conversion pollutants — hydrocarbons, trace elements, sulfur oxides, and nitrogen oxides — will help to assess the environmental stress presented by the introduction of these pollutants.

The presence of polycyclic aromatic hydrocarbons (PAH) in the environment cannot be attributed solely to man's activities; certain bacteria and plants synthesize PAH during normal development, and naturally caused forest and prairie fires can add significant amounts. PAH entering natural water systems will undergo decomposition, storage, or removal, with PAH solubility playing an important role in determining which specific aquatic interactions will prevail. The fate of hydrocarbons in soil is primarily influenced by biodegradation, while PAH activity in the air is largely determined by association with particulate matter.

Like PAH, trace elements also are introduced into the environment both by natural and anthropogenic activities. In aquatic systems, most trace elements are found either in the water column or sediment, with little in the biota; behavior and transport result from specific processes such as association with particulate matter in rivers, flocculation in estuaries, and precipitation in oceans. In soils, trace elements often have long retention times but may undergo chemical or biological changes that increase their mobility. The fate of trace elements in air is determined by dispersion, chemical reaction, and fallout.

Anthropogenic sulfur dioxide (SO_2) emissions far exceed natural SO_2 background levels. SO_2 can travel great distances in the atmosphere; most of the pollutant is removed via rainout and washout and is deposited on soil, water, and vegetation. However, while in the atmosphere, SO_2 may undergo significant chemical reactions such as conversion to sulfuric acid.

In contrast to SO_2, most global nitrogen oxide (NO_x) emissions originate from natural sources, although major man-caused buildups of NO_x occur in urban areas. The chemical reactions that NO_x enter into are important in evaluating NO_x as pollutants; hydrocarbon intervention in the nitrogen dioxide photolytic cycle, for example, leads to ozone accumulation and probably peroxyacyl nitrate production.

6.0 INTRODUCTION

The future of coal conversion cannot be adequately assessed without considering the environmental
impact of possible effluents from conversion processes. This chapter will attempt to explore the
fate of potential coal conversion pollutants in the environment by considering such factors as
mobility, transport, bioaccumulation, retention, and degradation. Four principal classes of
pollutants will be reviewed: hydrocarbons (particularly polycyclic aromatic hydrocarbons), trace
elements, sulfur oxides, and nitrogen oxides.

The behavior of a chemical pollutant introduced into the environment depends largely upon two
factors — the nature of its surroundings and the physicochemical properties of the pollutant. For
example, transformations can occur in all environmental media — the hydrosphere, lithosphere,
atmosphere, and biotic component of the environment. Likewise, in the medium, the behavior of a
chemical is determined by its solubility, vapor pressure, adsorption ability, and degradability
(Haque and Ash 1973).

6.1 POLYCYCLIC AROMATIC HYDROCARBONS

Most of the current information concerning polycyclic aromatic hydrocarbons (PAH) has been generated not from coal conversion processes, but from related technologies. However, as coal conversion pilot plants continue to operate and full-scale plants are built, the literature should
begin to address more of the environmental aspects of coal conversion, such as the effects of
associated effluents. For example, Herbes, Southworth, and Gehrs (1976) have compiled an extensive critical literature review of the environmental hazards of anticipated organic components
of aqueous coal liquefaction products, presented at the recent Trace Substances in Environmental
Health Conference (University of Missouri 1976).

The environmental role of PAH is of particular concern because some PAH are potential health
hazards. PAH are found in small but detectable concentrations in air, water, and soil samples of.
all types. The concentrations found are typically small, ranging from 0.001 to 10 µg/liter, but
they accumulate in organic fatty material and can thus concentrate in the food chain (McGinnes
and Snoeyink 1974).

6.1.1 Background concentrations

Not all PAH are the result of man's activities. Biosynthesis of PAH by plants and microorganisms
is well known; the National Academy of Sciences (NAS) (1972) emphasizes the importance of examining the possible contribution of vegetation to the total carcinogenic burden in the environment.

Andelman and Suess (1970) cite the research of Gräf (1964 and 1965), Gräf and Diehl (1966), and
Gräf and Nowak (1966) on the synthesis and physiological functions of PAH in plants (Chap. 8).
Wheat and rye seeds containing only a trace of PAH were grown in PAH-free solutions in both the
presence and absence of light. Five- to ten-day-old seedlings contained 10 to 20 µg of benzpyrene
(BP) per kilogram of dried material; BP was the only PAH analyzed for in the experiments. The
researchers conclude that BP is synthesized by the plants in both the presence and absence of
light and thus that PAH likely are synthesized worldwide and have always existed in man's environment. Andelman and Suess agree that the evidence indicates that PAH and other hydrocarbons are
naturally ubiquitous. However, Grimmer and Duvel (1970) were unable to detect PAH in higher
plants grown in PAH-free environments (Chap. 8).

A planktonic member of the plant kingdom, the alga *Chlorella vulgaris*, is known to synthesize several PAH (Borneff et al. 1968); algal extractions contained 10 to 50 µg per kg of PAH (Borneff 1964). Phytoplankton play an important role in the annual production of hydrocarbons (paraffin-naphthene and aromatic); Smith (1954) calculates a production rate of 13 tons (87 bbl) of hydrocarbons per square mile of ocean surface. Smith used a planktonic hydrocarbon content of 2000 ppm of the dry weight (furnished by an analysis of phytoplankton by B. H. Ketchum of the Woods Hole Oceanographic Laboratory) and a value of 6400 tons per square mile of ocean surface for the annual production rate of phytoplankton to determine the annual hydrocarbon output by oceanic phytoplankton.

Blumer (1961) studied the BP content of rural soils in Massachusetts and Connecticut (Table 6.1). He suggests that the BP in these soils is indigenous because the rural areas sampled are distant from major highways and industries. The natural sources are thought to be pyrolytic decomposition of wood, transformation of plant organic matter to peat and lignite, and activity of soil microorganisms. These studies prompted Blumer (1961) to postulate, as did Gräf and coworkers, that hydrocarbons, including carcinogenic species, have been in the environment throughout man's entire history.

Table 6.1. BaP in soils

Origin and type of soil	Concentration (µg/kg)
Oak forest, West Falmouth, Cape Cod, Mass.	40
Pine forest, West Falmouth, Cape Cod, Mass.	40
Mixed forest, West Falmouth, Cape Cod, Mass.	1300
Mixed forest, eastern Conn.	240
Garden soil, West Falmouth, Cape Cod, Mass.	90
Plowed field, eastern Conn.	900

Source: Blumer 1961, Table 1, p. 474.

Andelman and Suess (1970) report the synthesis of BP by soil bacteria in citing the research of Knorr and Schenk (1968) (see Chap. 7). Laboratory-cultured bacteria synthesized and accumulated BP in amounts of 2 to 6 µg per kilogram of dried material. Suess (1975) reports the existence of bacterially produced carcinogenic PAH in the upper layers of the earth in concentrations of 100 to 1000 µg/kg; these PAH leach through the soil, giving groundwater PAH concentrations of 0.001 to 0.010 µg/liter in the absence of anthropogenic sources.

Mallet (1972) studied old sediments and found 19.5 µg per kilogram of benzo[*a*]pyrene (BaP) in a fragment from a 50-m depth. He concludes that this hydrocarbon could be formed from phytoplank-tonic lipids under the influence of anaerobic bacteria, particularly *Clostridium putride*. Mallet and Tissier (1972) further investigated the biosynthesis of BP by bacteria in a series of laboratory experiments using *C. putride* and *Escherichia coli*. Their initial experiment used humus from a forest soil near an industrial area. This soil was sterilized, analyzed for BP, inoculated with *C. putride*, and stored for four months at laboratory temperature. The initial concentration, 80 to 100 µg of BP per 100 g of soil, increased to 160 µg at the end of four months. A second experiment involved forest soils extracted from rural areas having an initial BP concentration of 1 to 2 µg per 100 µg of soil. One batch of this rural soil was ground, sterilized dry at 170°C, and incubated for six months anaerobically after inoculation with *C. putride*. The same experiment was also performed with *E. coli*. Table 6.2 shows that after six months about 4 µg of BaP was detected in soils containing both *C. putride* and *E. coli*. In another experiment, Mallet and Tissier extracted from the soils what they believe to be the active substances for the formation of BaP, namely, fatty acids, and added them to soft agar inoculated with *C. putride*. After four months of incubation at 31°C, no trace of BaP was detected in the control tubes; however, significant but small amounts of the hydrocarbon were found in the samples inoculated with the *C. putride*.

Table 6.2. Soil sterilized in the Pasteur oven at 170°C

Amount of BP without addition of bacteria (µg)	Bacteria added	Amount of BP after experiment (µg/100 g)
Possible traces	*Clostridium putride*	2.6
Traces	*Clostridium putride*	4.2
1.40	*Clostridium putride*	3.95
Traces	*Clostridium putride*	1.95
2	*Clostridium putride*	2.6
0.75	*Escherichia coli*	2.25
Traces	*Escherichia coli*	2.20
1	*Escherichia coli*	4.75
	Escherichia coli	5.1
	Escherichia coli	5

Source: Mallet and Tissier 1972, Table 1, p. 106. Reprinted by permission of the publisher.

These investigations corroborate the results observed in marine muds, where the amounts of BP are generally considerable; BP is capable of being synthesized by anaerobic microorganisms or reduced by certain aerobic bacteria of the *Bacillus* type (Mallet and Tissier 1972).

Other papers by Mallet (Mallet, Zanghi, and Brisou 1972; Mallet, Heros, and Brisou 1972) also suggest the biosynthesis of BaP by anaerobic bacteria at the expense of lipids. A coworker of Mallet, Brisou (1972) reports that it now seems possible to accept that (1) the biosynthesis of polybenzene hydrocarbons, including BaP for certain anaerobic bacteria occurs; (2) this biosynthesis depends on only certain bacterial strains and is not a property vested in all the anaerobes; and (3) this biosynthesis has for a starting point not only lipids (purified fatty acids), but other substrates (particularly aliphatic terpenes). However, in a recent paper, Hase and Hites (1976) found no biosynthesis of PAH by anaerobic bacteria present in New England river sediments, but they did find that the bacteria accumulated PAH from the sediments.

The PAH compounds also occur naturally as minerals (Blumer 1975). These relatively rare minerals — curtisite, idrialite, and pendletonite — are found in association with mercury ores at scattered locations such as Shaggs Springs, Sonoma County, California; and Ordejov, Moldavia, Czechoslovakia.

Geissman, Sim, and Murdoch (1967) are cited by Blumer (1975) as postulating that these minerals originated from organic matter by pyrolysis and were deposited by distillation or from solution. A pyrolytic origin is consistent with the fact that the minerals were found in association with bituminous substances and inflammable gases. The PAH and sulfur-containing composition of curtisite and idrialite is shown in Table 6.3. In contrast to the variety of PAH in these two minerals, pendletonite is almost pure coronene with only a small alkyl-derivative contribution.

Table 6.3. PAH and sulfur-containing series in curtisite and idrialite

First member mass (amu)	Occurrence[a]	Suggested structure	Chromatographic position
216	CU	Benzofluorene (+ thienologs?)	With chrysene
228	CU	Chrysene	(UV evidence)
234	(CU)	Tribenzothiophene	With chrysene
252	(CU)	4,5-10,11-dimethylenechrysene	With chrysene
254	CU	Naphthenochrysene (2 alicyclic carbons)	With alkylchrysenes
260	(CU)	Naphthenotribenzothiophene (2 alicyclic carbons)	With chrysene
266	CU, ID	Dibenzofluorene (+ thienologs?)	With picene
276	(CU)	Anthanthrene	After picene (UV: not benzoperylene)
278	CU ID	Picene	(UV evidence)
284	CU, ID	Tetrabenzothiophene	Before picene
296	CU	Dinaphthenobenzofluorene (3 alicyclic C/ring)	With benzofluorene
300	(CU)	Coronene	With anthanthrene
302	(CU)	Dibenzofluoranthene, dimethylenepicene	After picene
304	ID, CU	Naphthenopicene (2 alicyclic C)	With picene
310	(ID)	Naphthenotetrabenzothiophene (2 alicyclic C)	With tetrabenzothiophene
316	ID, (CU)	Tribenzofluorene	After picene
322	ID	Dinaphthenotetrabenzothiophene (3 alicyclic C)	With picene/tetrabenzothiophene
328	ID	Benzopicene	After picene, with tribenzofluorene
334	ID	Pentabenzothiophene	With alkylpicenes
340	(ID)	Tetranaphthenopicene (5 alicyclic C)	With alkylpicenes
354	ID	Naphthenobenzopicene (2 alicyclic C)	With benzopicene
366	(ID)	Tetrabenzofluorene	Last PAH fraction
372	(ID)	Dinaphthenopentabenzothiophene (3 alicyclic C)	After picene
378	(ID)	Dibenzopicene	Last PAH fraction

[a]CU = curtisite; ID = idrialite. Major series underlined; minor series in parentheses.

Source: Blumer 1975, Table II, p. 251. Reprinted by permission of the publisher.

In a recent article, Blumer (1976) proposes mechanisms for the geosynthesis of aromatic hydrocarbon minerals such as idrialite, curtisite, and pendletonite. They arise from sediments containing organic compounds that are located deeper in the earth than those regions in which petroleum is formed and in which pyrolytic temperatures may reach 500°C. Although Blumer's mineral specimens come from museum collections, he believes that they are representative of the original composition in situ since no contamination peaks were detected by mass spectrometry (Blumer 1975).

Volcanic activity and forest fires are also sources of background PAH (Suess 1975). Forest fires especially can be important in determining background levels, particularly in the atmosphere. Also, Blumer and Youngblood (1975) and Youngblood and Blumer (1975) suggest that PAH compounds in soils and recent sediments are formed largely from natural forest and prairie firest. This suggestion is based on the high degree of similarity in the molecular weight distribution of the many series of alkyl homologs in the PAH compounds, and the fact that this distribution exhibits little variance over a wide range of depositional environments. Table 6.6 indicates that 140 tons of BP are emitted each year as a result of the open burning of forest and agricultural land as compared with the 192 tons emitted per year by coke production.

Hydrocarbons, including PAH, are ubiquitous in the environment, and their presence cannot be attributed exclusively to man's activities.

6.1.2 Existing concentrations

Existing concentrations of PAH include those from natural and anthropogenic sources. These sources release PAH compounds into the water environment (Table 6.4) and atmosphere (Table 6.5). Table 6.4 indicates the PAH compounds commonly found in water along with information about their relative carcinogenicity; Table 6.5 lists polycyclic compounds found in air — those that occur in cigarette smoke and exhaust gas and those that are associated with coal.

6.1.2.1 Air

The yearly global emission of BP, from data for 1966-1969, is estimated by Suess (1975) to be about 5000 tons, the greatest contribution coming from coal combustion (Table 6.6); he estimates U.S. yearly emission of BP to be about 25% of the worldwide total, or 1283 tons. In comparison, Olsen and Haynes (1969) estimate the U.S. annual BaP emissions to be 481 tons (Table 6.7) by assuming an annual consumption and emission rate that was derived from a large number of sources.

Measured average annual airborne BaP concentrations for 1966-1970 in urban areas in the United States are reported in Table 6.8. Altoona, Pennsylvania, appears to have the highest recorded value (29.5 ng/m^3 in 1967), followed by Chattanooga, Tennessee (22.9 ng/m^3 in 1967). The U.S. average BaP concentration for the five-year period is about 2.0 ng/m^3.

Table 6.9 lists nonurban annual average BaP concentrations for 1966-1970. As expected, these values are generally an order of magnitude less than those for the urban concentration in Table 6.8. Clarion County, Pennsylvania, which had the highest values for each of the five years (1.5, 2.1, 1.0, 1.2, and 1.2 ng/m^3), compares favorably with most urban sites, but the values are only a fraction of the highest values recorded at some of the Pennsylvania urban sampling stations (e.g., Pittsburgh, 13.8 ng/m^3 in 1969).

Table 6.10 demonstrates the seasonal effect on BaP concentrations in urban air; concentrations are higher in winter than in summer, the difference often being an order of magnitude. Urban concentrations of other PAH are compared with those of BaP in Table 6.11. These values are for the summer of 1958 and winter of 1959.

Shabad (1971) cites Yanysheva and Balenko (1966) in reporting the safe lifetime BP dose for the human lungs as 4.3 mg. On the basis of this value, the concentration of atmospheric BP should

Table 6.4. PAH commonly found in water

Structure	IUPAC name[a]	Earlier name[b]	Mol wt	Relative carcino-genicity[c]	Abbreviation
	Anthracene	Anthracene	178	?	An
	Benzo[a]anthracene	1,2-Benzanthracene	228	+	BaA
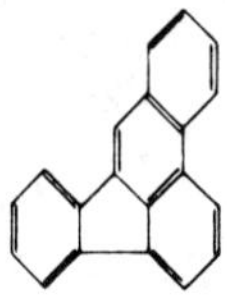	Benzo[b]fluoranthene	3,4-Benzfluoranthene	252	++	BbF
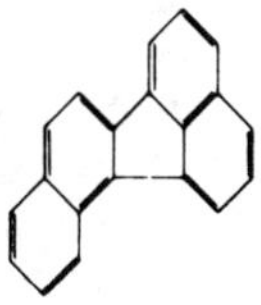	Benzo[j]fluoranthene	10,11-Benzfluoranthene	252	++	BjF
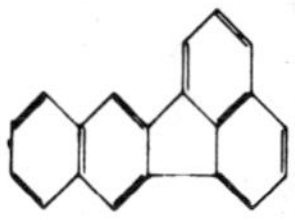	Benzo[k]fluoranthene	11,12-Benzfluoranthene	252	-	BkF
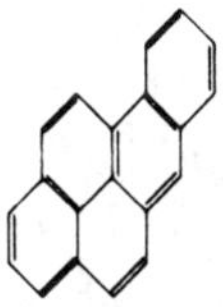	Benzo[a]pyrene	3,4-Benzpyrene	252	+++	BaP
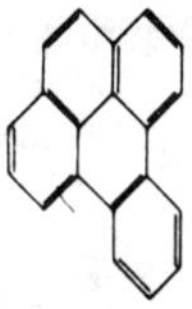	Benzo[e]pyrene	1,2-Benzypyrene	252	+	BeP
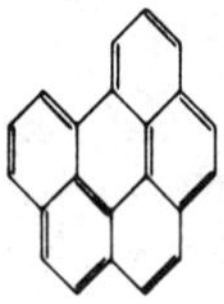	Benzo[ghi]perylene	1,12-Benzperylene	276	-	BghiP
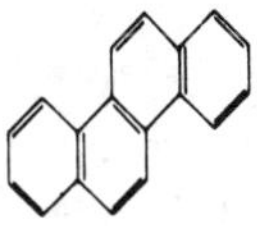	Chrysene	Chrysene	228	+	Ch
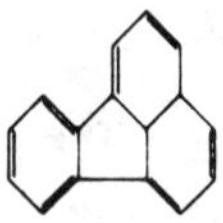	Fluoranthene	Fluoranthene	202	-	Fl

Table 6.4. (continued)

Structure	IUPAC name[a]	Earlier name[b]	Mol wt	Relative carcino-genicity[c]	Abbreviation
	Indeno[1,2,3-c,d]pyrene	2,3-o-Phenylenepyrene	276	+	IP
	Phenanthrene	Phenanthrene	178	?	Ph
	Perylene	Perylene	252	-	Per
	Pyrene	Pyrene	202	-	Pyr

[a] IUPAC 1957 rules. 1960. *J. Am. Chem. Soc.* 82:5545-84.
[b] Pre-1957 naming.
[c] Relative activity on mouse epidermis: +++, active; ++, moderate; +, weak; ?, unknown; -, inactive.

Source: Harrison, Perry, and Wellings 1975, Table 1, pp. 332-33. Reprinted by permission of the publisher.

not exceed 120 ng/m^3. A recent standard of BaP concentration (Standards Advisory Committee on Coke Oven Emissions 1975) for industrial workers was determined to be 200 ng/m^3, about twice as great as the suggested atmospheric value. The highest recorded BaP value in the United States for 1966-1970 (29.5 ng/m^3 in 1967 for Altoona, Pennsylvania, Table 6.8) is almost an order of magnitude less than the proposed industrial standard and fourfold less than the atmospheric standard.

6.1.2.2 Water

According to the NAS report (1975), about 6 million tons of petroleum hydrocarbons enter the oceans annually; the major contributors are marine transportation and runoff (urban and river), yielding 2.1 and 1.9 million tons per year respectively (Table 6.12). Coastal refineries and industrial and domestic waste together were estimated to contribute 0.8 million tons per year; natural seeps and atmospheric fallout each add another 0.6 million tons per year.

Table 6.13 shows the concentration of BP in marine sediments, primarily off the Mediterranean coast. These sediments are muds, sands, and shells; there is no easily discernible correlation between sediment, depth of sample, or BP concentration.

Concentrations of PAH in surface waters, mainly German rivers, are compiled in Table 6.14 (multiple values correspond to different sources of the same information). The average total PAH concentration is about 1.2 µg/liter; the highest values (12 µg/liter) are recorded in samples taken at the discharge site of shale oil and coke by-product effluent. However, 500 to 3500 m

Table 6.5. Polycyclic compounds found in air, cigarette smoke, and exhaust gases

Compound[a]	Occurrence[b]
Naphthalene	A T
2-Methylnaphthalene	A T
Alkylnaphthalenes	A
Azulene	A T
Acenaphthene	A T
Acenaphthylene	A T
Dibenzofuran	A
Carbazole	A
Dibenzothiophene	A
Fluorene	A T
Anthracene (A)	A T
Phenanthrene	A T
2-Methylanthracene	T
11H-Benzo[b]fluorene	A
11H-Benzo[a]fluorene	A T
7H-Benzo[c]fluorene	A
Fluoranthene (Fluor)	A T
8-Methylfluoranthene	T
Alkyl fluoranthene	T
Naphthacene	G
Benz[a]anthracene (BaA)[c,d]	A T
Chrysene (Ch)[c,d]	A T
Alkyl chrysene	T
Benzo[c]phenanthrene[c]	T
Pyrene (P)	A T G D
1-Methylpyrene	A T
4-Methylpyrene	T
2,7-Dimethylpyrene	A
Naphtho[2,1,8,7-klmn]xanthene	T
10,11-Dihydro-9-H-benzo[a]cyclopent[i]anthracene	T
2,3-Dihydro-1-H-benzo[a]cyclopent[h]anthracene	T
7H-Dibenzo[c,g]carbazole[c]	T
Benzo[b]fluoranthene[c]	A T G D
Benzo[ghi]fluoranthene	A T
Benzo[j]fluoranthene[c,d]	A T
Benzo[k]fluoranthene (BkF)	A T G D
2-Methylfluoranthene[c]	T
Methylfluoranthene	T
Benzo[a]naphthacene	T
Dibenzo[b,h]phenanthrene	T G D
Dibenz[a,h]anthracene[c,d]	T
Benzo[a]pyrene (BaP)[c,d]	A T G D
Methylbenzo[a]pyrene	T
Hydroxybenzo[a]pyrene	T
Benzo[a]pyrenequinone	A
Benzo[e]pyrene (BeP)[c,d]	A T
Perylene (Per)	A T
Dibenzo[a,l]naphthacene	A G D
Dibenzo[a,j]naphthacene	T
Naphtho[2,1,8-qra]naphthacene (Naphtho[2,3-a]pyrene)	A
Phenalen-1-one	A
Dibenzo[a,i]pyrene[c,d]	A T
Dibenzo[a,e]pyrene	T G D
Dibenzo[cd,jk]pyrene (anthanthrene)	A T G D
Dibenzo[cd,jk]pyrene-6,12-dione (anthanthrone)	A
Benzo[ghi]perylene (BghiP)	A T G D
Dibenzo[b,pqr]perylene	G
Coronene (Cor)	A T
Dibenzo[a,h]pyrene[c,d]	A T
Tribenzo[a,i]fluorene	T
13H-Dibenzo[a,i]fluorene	T
Dibenzo[a,c]naphthacene	T
Benzo[h]quinoline	A
Ra-Benzo[h]quinoline[e]	A
Rb-Benzo[h]quinoline[e]	A

Table 6.5 (continued)

Compound[a]	Occurrence[b]
Benz[c]acridine[f]	A G
Ra-Benz[c]acridine[e]	A
Rb-Benz[c]acridine[e]	A
Dibenz[a,h]acridine[c,f]	A T
Indeno[1,2,3-ij]isoquinoline	A G
Phenanthridine	A
11H-Indeno[1,2-b]quinoline	A
Acridine	A
Benzo[f]quinoline[e]	A
Ra-Benzo[f]quinoline[e]	A
Rb-Benzo[f]quinoline[e]	A
Benz[a]acridine[e]	A
Rb-Benz[a]acridine[e]	A
Dibenz[a,j]acridine[c,f]	A T
Rb-Dibenz[a,j]acridine	A
7H-Benz[de]anthracen-7-one	A
Indeno[1,2,3-cd]pyrene[c,d]	A
Dibenz[e,1]pyrene[c]	A
Xanthene-9-one	A
Dibenz[a,i]acridine	A

[a] Abbreviations used for most common compounds in parentheses.
[b] A — air; T — tobacco smoke; G — gasoline exhaust; D — diesel exhaust.
[c] Reported by Public Health Services to háve carcinogenic activity.
[d] Coal-associated PAH
[e] R-alkyl groups — substituted alkyl groups; Ra and Rb — various substitutes.
[f] Coal-associated heterocyclics

Source: EPA 1975, Table 5-1, pp. 5-1 - 5-3.

downstream of the effluent discharge, concentrations are reduced to 1 to 2 µg/liter. The PAH levels in Thames River water (Table 6.15) are comparable with those in the German rivers. The distribution of PAH compounds in German rivers is given in Table 6.16. No trend is immediately apparent. This is generally the case and precludes the measurement of any one PAH to estimate the concentration of all the others.

Concentrations of PAH have been detected in wastewater samples (Table 6.17). In many instances, domestic effluents have higher PAH concentrations than do factory effluents. The PAH levels in domestic sewage increase by 50- to 1000-fold during periods of heavy rain (Table 6.18) due to increased runoff, primarily from highways (Borneff and Kunte 1965). Table 6.19 shows the PAH content of road dust and a comparison of it with the atmospheric dust of urban air in Dortmund, a large German city. Many of the PAH concentrations in road dust are more than 1000-fold those levels found in soil (Table 6.20). Also, the PAH concentrations in road dust are shown to be several times greater than those in the atmospheric dust of Dortmund. Therefore, it is evident why runoff from highways causes a dramatic increase in the PAH concentration found in sewage during a heavy rain as opposed to PAH sewage levels during dry weather (Table 6.18).

6.1.2.3 Soil

Concentrations of PAH in forest soil are listed in Table 6.20. The soil samples were taken from beech, oak, and mixed woods; in one location, soils from beech woods had the highest PAH concentrations, but in the other, soils from mixed woods had the highest levels. Thus, no apparent correlation between PAH concentration in soil and ground cover can be detected from this table.

Table 6.6. Estimated BaP emissions to the atmosphere

Sources of formation	BaP emission (tons/year)		
	USA	Worldwide (excluding USA)	Worldwide
Heating and power generation, using			
Coal	431	1945	2376
Oil	2	3	5
Gas	2	1	3
Wood	40	180	220
Subtotal	475	2129	2604
Industrial processes			
Coke production	192	841	1033
Catalytic cracking	6	6	12
Subtotal	198	847	1045
Refuse and open burning			
Enclosed incineration			
Commercial and industrial	23	46	69
Other	11	22	33
Open burning			
Coal refuse fires	340	340	680
Forest and agriculture	140	280	420
Other	74	74	148
Subtotal	588	762	1350
Vehicles			
Trucks and buses	12	17	29
Automobiles	10	6	16
Subtotal	22	23	45
Grand total	1283	3751	5044

Source: Suess 1975, Table 1, p. 3. Reprinted by permission of the publisher.

6.1.3 Fate in natural water

In natural stream systems, PAH will be transported to the ocean, decomposed, or removed. Removal can consist of consumption by biota, adsorption on biological material, and/or adsorption on suspended or bottom mineral matter (McGinnes and Snoeyink 1974). Decomposition depends on biological metabolism or chemical oxidation, whereas physical transport usually occurs in the absence of the other mechanisms. Transient storage compartments for hydrocarbons include water and organisms, whereas sediments probably constitute the most significant long-term storage site (Fig. 6.1). Losses from an aqueous system include evaporation, tidal and riverine flushing, and microbiological decomposition (DiSalvo, Guard, and Hunter 1975).

6.1.3.1 Marine

In marine environments, pollutants are transported and distributed by currents, waves, and turbulent mixing. However, present information is insufficient to predict accurately either the complete physical behavior or the entire biological fate of a pollutant in oceanic deep waters (Ruivo 1970). For example, the fate of a pollutant may be determined as often by the microbial

Table 6.7. Estimated annual BaP emissions for the United States

Source	Estimated BaP emission rate	Estimated annual consumption or production	Estimated annual BaP emission (tons)
	Heat generation		
	$(\mu g/10^6$ Btu$)$	$(10^{15}$ Btu$)$	
Coal			
Residential			
Hand-stoked	1,400,000	0.26	400
Underfeed	44,000	0.20	9.7
Commercial	5,000	0.51	2.8
Industrial	2,700	1.95	5.8
Electrical	90	6.19	0.6
Oil	200	6.79	1.5
Gas	100	10.57	1.2
Total			421.6
	Refuse burning		
	$(\mu g/$ton$)$	$(10^6$ tons$)$	
Incineration			
Municipal	5,300	18	0.1
Commercial	310,000	14	4.8
Open burning			
Municipal refuse	310,000	14	4.8
Grass, leaves	310,000	14	4.8
Auto components	26,000,000	0.20	5.7
Total			20.2
	Industry		
	$(\mu g/$bbl$)$	$(10^6$ bbl$)$	
Petroleum catalytic			
Cracking (catalyst regeneration)			
Fluid catalytic cracker			
No CO boiler[a]	240	790	0.21
With CO boiler	14	790	0.012
Houdriflow catalytic cracker			
No CO boiler	218,000	23.3	5.6
With CO boiler	45	43.3	0.0024
Thermo for catalytic cracker (air lift)			
No CO boiler	90,000	131	13.0
With CO boiler	<45	59	<0.0029
Thermo for catalytic cracker (bucket lift)			
No CO boiler		119	0.0041
With CO boiler	<31	0	0
Asphalt road mix	50 µg/ton	187,000 tons	0.000010
Asphalt air blowing	<10,000 µg/ton	4,400 tons	<0.000048
Carbon-black manufacturing	Atmospheric samples indicate that BaP emissions		
Steel and coke manufacturing	from these processes are not		
Chemical complex	extremely high		
Total			18.8
	Motor vehicles		
	$(\mu g/$gal$)$	$(10^{10}$ gal$)$	
Gasoline			
Automobiles	170	4.61	8.6
Trucks	>460	2.01	>10
Diesel	690	0.257	2.0
Total			>20.6
Total (all sources tested)			481

[a]CO boiler — carbon monoxide waste heat boiler.

Source: Olsen and Haynes 1969, Table 6, pp. 104-6.

Table 6.8. Annual average ambient BaP concentrations
at National Air Sampling Network urban stations
(ng/m^3)

Station	1966	1967	1968	1969	1970
Alabama					
Birmingham	18.5				
Gadsden	3.5		2.4	1.8	2.5
Huntsville		3.1	2.7	1.8	1.6
Mobile	6.5		4.2	2.6	
Montgomery		2.3	2.9	2.0	1.3
Alaska					
Anchorage	2.3	1.9	1.7	1.3	0.8
Arizona					
Phoenix	1.7	2.5	2.1	2.2	
Tucson	0.6	0.7	0.7	0.5	0.4
Arkansas					
Little Rock	1.2	0.9	0.9	1.1	0.7
West Memphis	1.1	2.2	2.2	2.4	0.6
California					
Burbank	2.5			2.9	1.9
Glendale		1.0	1.6	1.6	1.0
Long Beach		2.1	2.1	2.3	1.0
Los Angeles	2.1	1.3	1.8	1.9	1.2
Oakland	2.7	1.7	1.6	1.6	1.0
Ontario			0.9	0.6	0.6
Pasadena	1.8		2.3		0.7
Riverside			1.3	0.8	0.7
Sacramento			1.4	1.8	0.7
San Bernardino			1.0	0.9	0.8
San Diego	1.7	1.6	1.2	1.4	0.7
San Francisco	1.1	1.5	1.8	1.2	0.6
Colorado					
Denver	2.3	2.4	2.3	2.5	2.2
Connecticut					
Hartford	2.3	2.1	1.4	2.0	1.4
New Haven	3.5	1.9	1.4	2.1	1.2
Delaware					
Newark	1.0	1.4	0.9		0.4
Wilmington	2.2	2.7	1.9	1.7	1.1
District of					
Columbia	2.4	1.9	1.9	4.3	
Florida					
Jacksonville			2.9	2.3	1.4
Tampa			1.5	1.0	0.5
Georgia					
Atlanta	1.4	3.0	1.8	1.9	0.9
Hawaii					
Honolulu	0.2	0.5	0.6	0.6	0.2
Idaho					
Boise City	3.5	2.4	2.0	6.0	1.1
Illinois					
Chicago	3.3	3.0	3.1	3.9	2.0
Springfield			1.1	1.3	0.9

Table 6.8 (continued)

Station	1966	1967	1968	1969	1970
Indiana					
East Chicago	6.8	5.7	1.9	6.8	5.3
Hammond	3.9	2.5	2.1	3.3	1.7
Indianapolis	10.4	5.7	4.1	5.2	2.3
Muncie	2.4	1.6			
New Albany	5.4			4.3	3.7
South Bend	2.2	2.6	3.7	3.7	2.4
Terre Haute		3.7		4.0	2.8
Iowa					
Davenport	3.2			1.7	0.9
Des Moines	2.5	2.7	1.1	0.9	0.7
Cedar Rapids		0.8	0.7		0.3
Kansas					
Kansas City	1.2		1.2	1.1	2.4
Topeka		0.5	0.7	0.4	0.3
Wichita	0.8	0.5	1.0	0.7	0.5
Kentucky					
Ashland	10.5		9.3	10.9	6.7
Covington	3.1	1.9	3.6	4.1	4.4
Lexington		1.8	3.0		1.6
Louisville	2.5	2.1	2.7	1.9	
Louisiana					
New Orleans	2.3	1.8	1.6	1.5	1.1
Maine					
Portland			2.3		1.1
Maryland					
Baltimore	2.8	3.8	2.3	2.8	2.1
Massachusetts					
Worcester			1.7	1.5	1.6
Michigan					
Detroit	4.7	5.4	5.1	3.9	2.6
Flint		1.4	0.8	1.7	1.5
Grand Rapids		2.8	3.4	1.7	0.9
Trenton			1.4	1.6	0.8
Minnesota					
Duluth	2.2		2.7	2.1	1.1
Minneapolis	1.6	1.3	1.1	1.4	0.6
Moorhead	0.7		0.9	1.0	1.6
St. Paul	1.8	2.3	1.8	1.8	1.0
Missouri					
Kansas City			1.8	1.6	1.1
St. Louis		2.3		3.3	
Montana					
Helena		0.8	0.9	0.5	
Nebraska					
Omaha	2.7	1.3	1.9	1.6	1.0
Nevada					
Las Vegas	1.3	1.1	1.4		
Reno		4.6	3.1		
New Hampshire					
Concord	0.6	1.5	1.0	0.7	0.6

Table 6.8 (continued)

Station	1966	1967	1968	1969	1970
New Jersey					
Camden	3.0		1.6	2.4	1.9
Glassboro	0.7	0.8	1.2	1.1	1.2
Jersey City	4.2	3.5	2.3	2.7	4.7
Marlton	1.2	1.6	1.3		1.4
Newark	2.1	3.3	2.1	1.8	1.5
Patterson		1.9	2.0	1.2	1.2
Perth Amboy	2.1	2.1	1.2	1.2	1.0
Trenton	2.2		1.0	1.5	1.1
New Mexico					
Albuquerque	2.0	1.9	1.8	1.1	1.1
New York					
New York City	4.1	3.9		3.6	3.0
North Carolina					
Charlotte	5.7	6.3	5.6	4.9	1.9
Durham			8.0	3.4	3.9
North Dakota					
Bismarck			0.9	1.0	0.4
Ohio					
Akron	4.1	3.7	3.0		
Cincinnati	3.6	1.9	1.8	2.9	2.6
Cleveland	3.1	2.9	3.0	3.8	2.8
Columbus	2.9	1.7	2.2	2.7	1.6
Dayton	2.7	3.7	2.4	1.9	1.5
Toledo	1.8	1.9	1.8	1.5	1.4
Youngstown	7.3	8.2	5.6	9.9	7.1
Oklahoma					
Oklahoma City	1.5	0.7	0.7	0.7	0.9
Tulsa	0.7	0.6	0.8	0.5	0.8
Oregon					
Eugene		2.4			
Medford		4.8	8.2	4.1	
Portland	3.3	3.5	4.1	2.6	2.3
Pennsylvania					
Allentown		1.8	1.2	1.9	2.4
Altoona		29.5	18.0	22.3	19.3
Bethlehem		2.9	2.1	2.0	2.7
Harrisburg			1.3	1.5	1.5
Lancaster	2.3				
Philadelphia	3.8	5.9	2.9	4.0	2.4
Pittsburgh	4.9	7.0	6.3	13.8	5.9
Reading	2.3	2.9	2.4	1.8	1.6
Scranton		5.2	6.1	7.7	2.9
Warminster	0.9	2.2	0.9	1.0	
West Chester		1.1	1.0	1.3	
Wilkes Barre			1.6	1.5	1.3
York		1.8	1.9	2.0	1.2
Rhode Island					
East Providence		1.6	1.2	1.2	1.2
Providence	3.6	2.8	2.0	2.2	2.1

Table 6.8 (continued)

Station	1966	1967	1968	1969	1970
South Carolina					
Columbia		4.2	6.2	1.3	
Greenville	5.0		18.6	7.0	3.4
Tennessee					
Chattanooga	8.4	22.9	7.4	4.2	5.5
Knoxville		7.0	9.8	4.7	
Memphis	1.7	1.6	1.3	0.7	1.4
Nashville	5.5	7.0	6.0	2.8	3.6
Texas					
Dallas	1.4			2.0	1.9
Houston	0.9				1.2
San Antonio	0.6	1.4	0.9	0.6	1.0
Utah					
Ogden	0.5		0.8	0.7	2.5
Salt Lake City	1.2	0.7	1.0	0.7	1.4
Vermont					
Burlington	0.8		0.7	0.5	0.7
Virginia					
Danville	3.2		2.5	1.8	2.7
Hampton		2.2	1.5	0.9	1.1
Lynchburg		9.2	8.7	6.3	4.5
Norfolk	2.8	3.5	4.9	3.9	1.8
Portsmouth		7.7	10.2	3.4	4.9
Richmond		5.2		2.2	2.1
Roanoke		7.5	7.7	5.3	6.2
Washington					
Seattle	2.7	1.8	2.0	1.6	1.5
West Virginia					
Charleston	3.4		4.6	2.6	2.1
Wisconsin					
Kenosha			1.4	1.7	1.3
Madison			1.3		1.1
Milwaukee	4.1		4.7	4.0	2.5
Superior			3.3	1.6	1.5
Wyoming					
Casper			0.9	0.6	0.4
Cheyenne	0.5		0.6	0.5	0.4

Source: EPA 1975, Table 5-4, pp. 5-6 - 5-10.

environments as by the chemical and physical surroundings. Many pollutants, including the components of crude oil, may be degraded by the action of bacterial yeasts and other microflora in either water or bottom sediments (Chap. 7). In the event that a benthic population is altered or destroyed, the loss of microbial degradation may cause harmful chemicals to be released (these same chemicals ordinarily could be degraded by a healthy benthic population).

Most chemical reactions occur in the ocean at the air-sea interface, the sediment-water interface, or the boundaries between the water and the particles that are suspended in the water (Horne 1969, as cited in Gross 1970). Materials that concentrate at the air-water interface are subject to oxidation, ultraviolet radiation, evaporation of volatile constituents, and polymerization

Table 6.9. Annual average ambient BaP concentrations
at National Air Sampling Network nonurban stations
(ng/m^3)

Station	1966	1967	1968	1969	1970
Arizona					
Grand Canyon	0.3	0.2	0.2	0.2	0.1
Maricopa County		0.2	0.5	0.3	0.3
Arkansas					
Montgomery County	0.3	0.1	0.2	0.2	0.1
California					
Humboldt County	0.4	0.4	0.3	0:5	0.1
Idaho					
Butte County			0.2	0.1	0.1
Indiana					
Monroe County	0.5		0.5	0.3	0.2
Parke County	0.9		0.4	0.3	0.4
Maine					
Acadia National Park	0.2		0.3	0.1	0.2
Missouri					
Shannon County		0.2	0.2	0.2	0.2
Montana					
Glacier National Park		0.3	0.4	0.4	
Nebraska					
Thomas County	0.2		0.2	0.1	0.1
Nevada					
White Pine County	0.1		0.1	0.1	0.1
New Hampshire					
Coos County	0.2	0.2	0.2	0.1	0.1
New York					
Jefferson County	0.2		0.2	0.3	0.2
North Carolina					
Cape Hatteras	0.2		0.2	0.1	0.2
Oklahoma					
Cherokee County	0.2	0.2	0.2	0.2	0.2
Oregon					
Curry County	0.1	1.1	0.1	0.1	0.1
Pennsylvania					
Clarion County	1.5	2.1	1.0	1.2	1.2
Texas					
Matagorda County	0.3	0.1	0.2	0.1	0.3
Vermont					
Orange County	0.9		0.3	0.3	0.2
Virginia					
Shenandoah National Park	0.9	0.3	0.3	0.3	0.2

Source: EPA 1975, Table 5-5, pp. 5-11 and 5-12.

Table 6.10. Seasonal effect on the BaP concentrations
of various urban atmospheres

Location	Year	Concentrations (μg/m^3 of air)	
		Summer (low)	Winter (high)
Atlanta, Ga.	1958	0.0016	0.015
	1960	0.0009	0.014
Birmingham, Ala.	1958	0.006	0.074
	1960	0.003	0.062
Cincinnati, Ohio	1958	0.002	0.026
	1960	0.0012	0.018
Detroit, Mich.	1958	0.0034	0.031
Location I	1961	0.0072	0.017
Location II	1961	0.0036	0.0137
Location III	1963	0.0002	0.0018
Los Angeles, Calif.	1958	0.0004	0.013
Nashville, Tenn.	1958	0.0014	0.055
New Orleans, La.	1958	0.0020	0.006
	1960	0.0006	0.007
New York, N.Y.			
Herald Square	1963 - 64	0.0005	0.0094
Columbus Circle	1964	0.0007	0.0026
Philadelphia, Pa.	1958	0.0025	0.019
San Francisco, Calif.	1958	0.0003	0.0075
South Charleston, W. Va.	1960	0.0006	0.012

Source: Olsen and Haynes 1969, Table 9, p. 111.

(Gross 1970). When considering the rate of sediment supply to the oceans and the apparent rate
of accumulation in deep ocean basins, it is likely that 90% or more of river-borne particles are
deposited on continental shelves. The particles that are greater than 0.5 μ in diameter have
residence times in the oceanic water column of less than 100 years. For smaller particles the
residence may be between 200 and 600 years (Arrhenius 1963, as cited in Gross 1970). There is
also rapid removal of particles by filter feeders that concentrate the particles in fecal pellets,
which drop to the ocean bottom (Gross 1970).

6.1.3.2 Accumulation in biota

Upon entering the aquatic environment, hydrocarbons are available for uptake by the biota.
Farrington and Quinn (1973), in studying marine mollusks, report that marine organisms can
synthesize some low-boiling hydrocarbons, but state that synthesis of naphthenic and aromatic
hydrocarbons is too complex for those marine organisms. Thus, when they extracted hydrocarbons
from clams collected in Narragansett Bay, they were able to identify them as petroleum hydrocarbons
because they included aromatic species. Marine mussels have also been recognized as good indicators of petroleum hydrocarbons. They appear to take up and release the hydrocarbons in

Table 6.11. Polynuclear hydrocarbon content of particulate matter for selected cities
(μg/g benzene soluble fraction)

City	Month	Compound[a]								Total
		BghiP	BaP	BeP	BkF	P	Cor	Per	Anth	
Winter 1959										
Atlanta	Feb	830	690	440	560	560	400	100	48	3628
Birmingham	Feb	1090	1500	610	800	1000	210	330	130	5670
Detroit	Feb	2100	2000	1500	1300	1600	410	390	130	9430
Los Angeles	Feb	510	150	230	160	170	330	44	4.5	1599
Nashville	Jan	880	1300	710	790	1400	240	230	92	5642
New Orleans	Feb	760	430	670	410	240	280	84	11	2885
San Francisco	Jan	590	180	230	130	150	380	27	8.1	1695
Seattle	Jan – Mar	1200	790		710	590	1300	210	91	4891
Sioux Falls	Jan – Mar	1000	480		330	480	440	99	27	2856
South Bend	Jan – Mar	1500	2000	1700	1300	3900	480	370	190	11440
Wheeling	Jan – Mar	1100	1600		990	1700	360	200	110	6060
Youngstown	Jan – Mar	1600	2000		1300	2400	240	530	240	8310
Summer 1958										
Atlanta	July	510	160	150	130	73	250	40	20	1333
Birmingham	July	950	730	670	520	240	270	240	29	3649
Cincinnati	July	600	390	400	350	170	280	93	6.5	2290
Detroit	July	1500	950	840	770	440	290	270	61	5121
Los Angeles	July	200	43	54	39	23	190	29	2.4	580.4
Nashville	July	440	180	150	130	75	160	27	7.7	1170
New Orleans	July	530	230	360	210	39	290	45	12	1716
Philadelphia	July	960	480	350	350	310	410	110	17	2987
San Francisco	July	720	69	150	67	25	450	12	6.1	1499

[a]BghiP = benzo[g h i]perylene; BaP = benzo[a]pyrene; BeP = benzo[e]pyrene; BkF = benzo[k]fluoranthene; P = pyrene;
Cor = coronene; Per = perylene; Anth = anthanthrene.

Source: Olsen and Haynes 1969, Table 8, p. 110.

Table 6.12. Comparison of estimates for petroleum
hydrocarbons annually entering the ocean, ca.
1969 - 1971

| | Authority (millions of tons per year) | | |
Source	MIT SCEP report[a] (1970)	USCG impact statement[b] (1973)	NAS workshop (1973)
Marine transportation	1.13	1.72	2.133
Offshore oil production	0.20	0.12	0.08
Coastal oil refineries	0.30		0.2
Industrial waste		1.98	0.3
Municipal waste	0.45		0.3
Urban runoff			0.3
River runoff[c]			1.6
Subtotal	2.08	3.82	4.913
Natural seeps			0.6
Atmospheric rainout	9.0[d]		0.6
Total	11.08		6.113

[a]From Study of Critical Environmental Probe (SCEP). 1970.
Man's impact on the global environment. Assessment and recommendations for action. Cambridge, Mass.: MIT Press.
[b]From U.S. Coast Guard. 1973. *Draft environmental impact statement.*
For International Convention for the Prevention of Pollution from
Ships, 1973.
[c]Input from recreational boating assumed to be incorporated in the
river runoff value.
[d]Based upon assumed 10% return from the atmosphere.

Source: NAS 1975*a*, Table 1-6, p. 6.

proportion to the hydrocarbon concentrations in surrounding sea water (DiSalvo, Guard, and Hunter 1975). (The uptake and metabolism of aromatic hydrocarbons are discussed in detail in Chap. 9.) Lee, Sauerheber, and Benson (1972) also report the presence of hydrocarbons in mussels. Their experiments were performed in the laboratory on the marine mussel *Mytilus edulis*. The mussel rapidly filtered hydrocarbons across its gills, but seemed to lack the ability to metabolize any of the hydrocarbons. The nontoxic paraffinic hydrocarbons were taken up to a greater extent (10 mg per mussel) than were the aromatics (2 to 20 µg per mussel). When the polluted mussels were placed in uncontaminated seawater, they discharged the hydrocarbons: 90% of the heptadecane, 80% of the terralin (1,2,3,4-tetrahydronaphthalene), and most of the BaP; naphthalene and toluene were also discharged.

Other experiments by Lee and co-workers (Lee 1975; Lee, Ryan, and Neuhauser 1976) indicates that unlike mussels, marine copepods and the blue crab, *Callinectes sapidus,* have the ability to metabolize paraffinic and aromatic hydrocarbons. When transferred to fresh seawater, both copepod and blue crab released most of the accumulated radiolabeled hydrocarbons. After 17 days the temperate water copepod *Calanus plumchrus* discharged all but 1×10^{-5} µg of benzypyrene of the original ingested amount, 22×10^{-4} µg. The hepatopancreas, assumed to be the main site of hydrocarbon metabolism, was the only organ in the blue crab where an appreciable buildup of radioactivity occurred. In yet unpublished data, Lee and co-workers report that similar rates of hydrocarbon uptake, metabolism, and discharge have been observed in the California spiny lobster *Panulirus interruptus,* the spider crab *Libinia emarginata,* and the American lobster *Homarus Americanus,* leading Lee and co-workers to speculate that all marine crustaceans have the ability to metabolize hydrocarbons.

Table 6.13. Concentration of BP in marine sediments

Source	Sample	Depth (m)	BP concentration (µg/kg of dry sample)
Greenland, west coast	Sand	0.20	5
Italy, Bay of Naples	Mud, sand, shell[a]	15 - 45	1,000 - 3,000
(6 locations)	Sand, shell	13	7.5
	Mud, sand	2 - 65	10 - 530
	Mud, sand, shell[b]	55	260 - 960
	Muddy sand	120	1.4
	Mud, sand[c]		100 - 560
French Mediterranean coast		0 - 0.03	1,800
		0.03 - 0.08	3,600
		0.08 - 0.13	5,000
		0.13 - 0.18	2,500
		0.23 - 0.28	2,200
		0.33 - 0.38	730
		0.48 - 0.53	420
		1.00	26
		2.00	16
	Sand	14	400
	Black mud	16	1,500
	Sand	48	75
	Sand	53	Traces
	Beige mud	82	400
Estuary, French Mediterranean	Sand		34
	Sand	1	20
	Sand	4	15
	Sand	5	25
	Mud	102	Not detected
French Mediterranean coast			50
French Mediterranean coast			20
French Channel coast	Mud		15,000
French Channel and Atlantic coasts (11 locations)	Mud, sand		Not detected to 1,700

[a]Five samples were collected 300 m from shore; area is highly industrialized.
[b]Four samples were collected in the vicinity of volcanic pollution.
[c]Four samples; this island is affected by pollution.

Source: Andelman and Suess 1970, Table 8, p. 489. Reprinted by permission of the publisher.

Table 6.14. Concentration of PAH in surface waters

Source	Concentration (µg/liter)		
	BP	Carcinogenic PAH	Total PAH
Bodensee		0.003	
		0.0004[a]	
	0.0013	0.030	0.065
Alprhine		0.005[a]	
River Rhine			Present
River Rhine		0.050 to 0.500[b]	
River Rhine at Mainz		0.080[a]	
River Rhine at Mainz (Mar. 1964)	0.049	0.240	0.73
(Mar. 1964)	0.114	0.730	1.50
River Main, at Seligenstadt	0.0024	0.155	0.48
River Danube, at Ulm (Apr. 1964)	0.0006	0.055	0.28
(May 1964)		0.078	0.20
River Gersprenz, at Munster			
(Jan. 1964)	0.0096	0.055	0.16
(Apr. 1964)		0.038	0.12
River Aach, at Stockach[a]		0.50	
River Aach, at Stockach[a]			
Sample I	0.043	1.30	3.0
Sample II	0.016	0.90	2.5
Sample III	0.004	0.50	1.4
Sample IV	0.005	1.10	3.1
River Schussen (Bodensee)		0.50[a]	
River Schussen	0.01	0.20	1.0
River Argen (Bodensee)		0.07[a]	
River Seine	Considerable amount		
River Plyussa			
At discharge site of shale-oil effluent	12		
3500 m downstream	1		
At the water intake of Narvy	0.1		
A river			
15 m below discharge of coke by-product effluent	8 - 12		
500 m downstream	2 - 3		
Peat (turf) water	0.05		

[a]Extrapolated from centrifugate fractions.
[b]Extrapolated from activated-carbon adsorption analyses.

Source: Andelman and Suess 1970, Table 9, p. 490. Reprinted by permission of the publisher.

The oyster, a marine counterpart of the mussel, also has the ability to take up hydrocarbons. In one study (Ehrhardt 1972), oysters from Galveston Bay were found to contain hydrocarbons, 56% of which were aromatic. Ehrhardt considers the stability of hydrocarbons, as they pass through the food chain, responsible for uptake of the chemical by the oyster. He cites a study by Blumer (1967), in which hydrocarbons in the liver of the basking shark were found to reflect the hydrocarbon content of its planktonic food. On the basis of this supporting evidence, Ehrhardt concludes that even highly organized animals accumulate the entire range of hydrocarbons to which they are exposed.

Table 6.15. PAH levels in Thames River water (ng/liter)

PAH	Kew Bridge	Albert Bridge	Tower Bridge
	Location		
Fluoranthene	140	200	360
Benzo[k]fluoranthene	80	40	120
Benzo[a]pyrene	130	160	350
Perylene	40	70	130
Indeno[1,2,3-cd]pyrene	50	110	210
Benzo[ghi]perylene	60	110	160

Source: Acheson et al. 1975, as cited in Harrison, Perry, and Wellings 1975, Table 4, p. 334. Reprinted by permission of the publisher.

From laboratory studies of North Sea plankton, Whittle (1974) proposed that plankton concentrate hydrocarbons from seawater, possibly via surface absorption; this further supports the accumulation of hydrocarbons by lower members of the aquatic food chain. Another indication that highly organized animals can take up hydrocarbons is from research by Ogata and Miyake (1973). They demonstrated that aromatic hydrocarbons — benzene, xylene, and toluene — rapidly infiltrate fish and possibly impart an offensive odor. They conclude that aromatic hydrocarbons can accumulate in fish and that the hydrocarbons might be transferred to man through his consumption of fish. Stegeman and Teal (1973) also found hydrocarbon uptake in their laboratory study of the oyster, *Crassostrea virginica*. They note that the hydrocarbon concentration in the seawater, as well as the lipid content of the oyster, apparently affects the rate and extent of accumulation. The oysters having a higher lipid content accumulated more hydrocarbons. When oysters were transferred to uncontaminated seawater, there was a rapid discharge of about 90% of the hydrocarbons taken in. However, the amount retained was over 30 times that found in the oysters prior to exposure to the hydrocarbons. The hydrocarbons discharged by the oysters were greater in aromatic content and seemed to be somewhat degraded relative to the hydrocarbons present at the time of exposure; this prompted Stegeman and Teal to suggest the possibility that oysters modify the hydrocarbons.

6.1.3.3 Accumulation in sediments

Shelton and Hunter (1975) document the existence of hydrocarbons in bottom sediments. They cite the American Petroleum Institute's estimate of the national pollution potential from oil as about 450 million gal annually, some of which reaches aquatic environments. Shelton and Hunter state that a significant portion of the aquatic input is incorporated into bottom sediments of receiving waters. Much of the PAH found in bottom sediments may be the result of natural forest and prairie fires (Sect. 6.1.1). The PAH, while adsorbed on particles or in solution, may eventually reach the beds of the various water bodies and be taken up by bottom organisms or, in the absence of light and under anaerobic conditions, remain in a stable condition for long periods (Suess 1975). Additional discussion of hydrocarbons in sediments is presented in the following sections.

Table 6.16. PAH levels in German Rivers (ng/liter)

PAH	Location												
	River Gersprenz at Munster		River Danube at Ulm		River Main at Seligenstadt				River Aach at Stockach				River Schussen
Fluoranthene	38.5	71.3	94	61	107	21.3	128	192	694	474	379	761	358
Benzo[a]anthracene	18.8	4.3	11	14	7.4	7.0	14.4	16.2	385	199	101	128	57
Chrysene					11.8	26.4	38.2						
Benzo[b]fluoranthene	10.4	13.2	24.2	23.9	12.2	19.9	32.1	67.0	362	177	76	332	41
Benzo[j]fluoranthene	4.6	14.7	10.1	23.4	14.2	21.5	35.7	75.5	337	394	144	420	53
Benzo[k]fluoranthene	9.6	4.8	7.7	14.1	4.2	6.4	10.6	21.6	130	136	132	173	33
Benzo[a]pyrene	9.6		0.6		1.3	1.1	2.4	6.5	43	16	4	5	10
Benzo[ghi]perylene	12.9	1.6	9.5	9.5	8.0	13.2	21.2	25.9	84	105	42	46	46
Indeno[1,2,3-cd]pyrene	12	5.4	9.5	16.4	12.5	19.5	32.0	23.7	217	116	144	188	45

Source: Borneff and Kunte 1964 and 1965, as cited in Harrison, Perry, and Wellings 1975, Table 5, p. 334. Reprinted by permission of the publisher.

Table 6.17. PAH in wastewater samples (ng/liter)

PAH	Domestic effluents	Factory effluents	Domestic effluents	Sewage (high percentage industry)	
Fluoranthene	2416	2198	273	3420	2660
Pyrene	1763	1957		3120	2560
Benzo[a]anthracene	319	167	191	1360	343
Benzo[b]fluoranthene	202	114	36	870	525
Benzo[j]fluoranthene	205	45	37	1740	1100
Benzo[k]fluoranthene	193	32	31	460	336
Benzo[a]pyrene	74	100	38	100	368
Benzo[ghi]perylene	219	73	40	480	120
Indeno[1,2,3-cd]pyrene	238	57	22	930	476

Source: Borneff and Kunte 1965, as cited in Harrison, Perry, and Wellings 1975, Table 9, p. 336. Reprinted by permission of the publisher.

Table 6.18. PAH in domestic sewage during dry and
wet weather (ng/liter)

PAH	Concentration in sewage during	
	Dry weather	Heavy rain
Fluoranthene	352	16,350
Pyrene	254	16,050
Benzo[a]anthracene	25	10,360
Benzo[b]fluoranthene	39	9,910
Benzo[j]fluoranthene	57	10,790
Benzo[k]fluoranthene	22	4,180
Benzo[a]pyrene	1	1,840
Benzo[ghi]perylene	4	3,840
Indeno[1,2,3-cd]pyrene	17	4,980

Source: Borneff and Kunte 1965, as cited in Harrison, Perry,
and Wellings 1975, Table 10, p. 336.
Reprinted by permission of the publisher.

Table 6.19. PAH in road dust (mg/kg)

Substances detected	Dust from Federal Road 31					Atmospheric dust Dortmund
	1	2	3	4	5	
Fluoranthene	126	166	78	37	32	3.8
Pyrene	114	147	62		27	2.77
Benzo[a]anthracene	21	59	19	17	16	0.53
Benzo[j]fluoranthene	60	97	36	14	6	0.62
Benzo[b]fluoranthene	26	62	19	18	15	0.16
Chrysene	22	64				
Benzo[a]pyrene	9	17	12	2	3	0.05
Benzo[ghi]perylene	22	47	21	6	12	0.51
Benzo[k]fluoranthene	13	26	13	6	8	0.48
Indeno[1,2,3-cd]pyrene	24	61	20	8	9	0.96
Amount of known compounds	437	746	280	108	128	9.88
Amount of unidentified compounds	400	800	22	17	18	0.20
Total amount of PAH	837	1546	302	125	146	10.08
Amount of carcinogenic PAH	162	360	106	59	49	2.31
% fraction of carcinogens of the total amount	19	23	35	47	33	23
Number of substances isolated	32	32	22	20	20	18

Source: Borneff and Kunte 1965, Table 2, p. 230. Reprinted by permission of the publisher.

Table 6.20. Concentration of PAH found in forest soil (µg/kg)

PAH	Location					
	South of Darmstadt			Near Lake Constance		
	Mixed woods	Beech woods	Oak woods	Spruce woods	Mixed woods	Beech woods
Benzo[*a*]pyrene	2.5	4.0	1.5	1.5	2.5	1.5
Benzo[*ghi*]perylene	10	70	20	10	20	10
Benzo[*b*]fluoranthene and benzo[*j*]fluoranthene	30	110	50	35	25	25

Source: Borneff and Kunte 1963, Table 2, p. 333. Reprinted by permission of the publisher.

ORNL DWG 76-7302

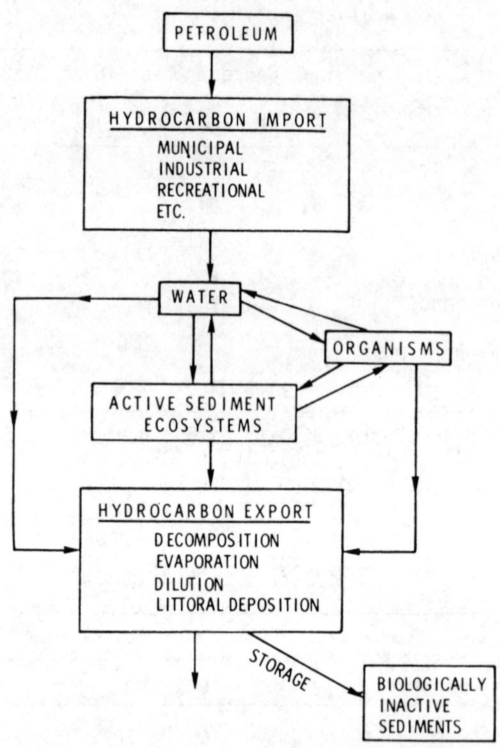

Fig. 6.1. Conceptual model of probable flows of hydrocarbons through major compartments of an estuarine ecosystem. Source: From DiSalvo, Guard, and Hunter 1975, Fig. 1, p. 247. Reprinted with permission from *Environ. Sci. Technol.* Copyright by the American Chemical Society.

6.1.3.4 Solubility

The fate of hydrocarbons in natural water systems is influenced by their solubilities: The solubility of a hydrocarbon determines whether it will go into solution, become suspended in the water column, or adsorb onto the sediments.

In discussing hydrocarbon solubility, it is important to consider the possible incorporation of hydrocarbons into micelles, a process thought to be responsible for some hydrocarbon solubilization in natural systems (Elworthy, Florence, and MacFarlane 1968; Boehm and Quinn 1973). A micelle is composed of an aggregate of single-surface-active molecules (surfactants), each possessing a hydrophobic hydrocarbon chain and an ionizable or water-soluble group, hydrophyllic in nature (e.g., sodium lauryl sulphate) (Elworthy, Florence, and Macfarlane 1968):

$$CH_3CH_2CH_2CH_2CH_2CH_2CH_2CH_2CH_2CH_2CH_2CH_2-SO_4^-Na^+$$

Hydrophobic   Hydrophyllic

The solubilization of a hydrocarbon is thought to occur by association of the hydrocarbon (1) with the hydrocarbon core of a micelle, (2) with the polar surface of a micelle, or (3) with both the interior and the surface (Fig. 6.2); a hydrocarbon having low solubility in pure water (e.g., decane) can be incorporated into the micellear structure and become solubilized. The sodium lauryl sulphate monomer is a surfactant.

ORNL DWG 76-7326

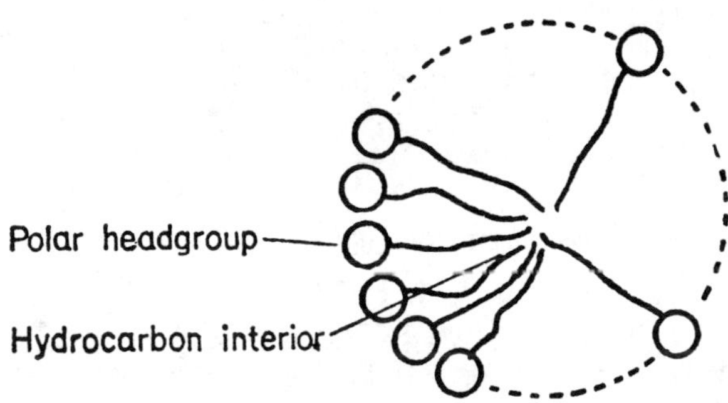

Fig. 6.2. Suggested structure of a cross-section of a spherical micelle; only a few of the molecules forming the micelle are shown. Source: From Elworthy, Florence, and Macfarlane 1968, Fig. *i*, p. 11. Reprinted by permission of the publisher.

Boehm and Quinn (1973) list the following natural sources for surfactant humic materials in seawater: (1) dissolved organic matter in seawater (2) fulvic acid extracted from a marine sediment, (3) organic matter extracted by seawater from a marine sediment, (4) material isolated from seawater at a chloroform-seawater interface, and (5) organic matter contributed by a municipal sewage effluent. Removal of this dissolved organic matter from natural samples resulted in a 50 to 99% decrease in the amounts of hydrocarbons (primarily n-alkanes) solubilized. Solubilities of phenanthrene and anthracene were unaffected by the removal. However, high concentrations of surfactants (10 to 50 mg/liter) can increase the solubility of BP by a factor of 2 to 10 (Il'nitskii, Klubkov, and Shabad 1971, as cited in Andelman and Snodgrass 1974).

Another source of surfactants is microbiological: Microorganisms are thought to produce surfactants as metabolites to solubilize hydrocarbon droplets prior to assimilation (Barnett, Velankar, and Houston; Goma, Pareilleux, and Durand 1973) (see Chap. 7).

McAuliffe (1966) has performed extensive determinations of the solubility of hydrocarbons (Tables 6.21 through 6.24). His studies indicate that branching increases the solubility of

Table 6.21. Solubility in water at room temperature of paraffin and
branched-chain paraffin hydrocarbons

Hydrocarbon	Solubility (g of hydrocarbon/10^6 g of water		Molar vol (ml/mole at 20°C)
	This work[a]	Literature[b]	
Methane	24.4 ± 1.0	21.7, 22.8, 21.5	39
Ethane	60.4 ± 1.3	56.6, 58.3, 51.6	55
Propane	62.4 ± 2.1	67.0, 65.6	88.1
n-Butane	(136) 61.4 ± 2.6	72.7, 67.2	100.4
Isobutane	(165) 48.9 ± 2.1		104.3
n-Pentane	38.5 ± 2.0	360	115.2
Isopentane	47.8 ± 1.6		116.4
2,2-Dimethylpropane	(54) 33.2 ± 1.0		122.1
n-Hexane	9.5 ± 1.3	140, 36	130.7
2-Methylpentane	13.8 ± 0.9		131.9
3-Methylpentane	12.8 ± 0.6		129.7
2,2-Dimethylbutane	18.4 ± 1.3		132.7
n-Heptane	2.93 ± 0.20	50, 10	146.5
2,4-Dimethylpentane	4.06 ± 0.29		148.9
n-Octane	0.66 ± 0.06	14	162.6
2,2,4-Trimethylpentane	2.44 ± 0.12		165.1
2,2,5-Trimethylhexane	1.15 ± 0.08		181.3

[a]Numbers following ± symbol indicate standard deviation from mean; numbers in
parentheses indicate calculated solubility of liquid hydrocarbon at 25°C.
[b]See McAuliffe 1966 for sources of the values given.

Source: McAuliffe 1966, Table 1, p. 1270. Reprinted with permission from *J. Phys. Chem.*
Copyright by the American Chemical Society.

paraffin, olefin, and acetylene hydrocarbons, but not of cycloparaffin, cycloolefin, and aromatic
hydrocarbons. For each homologous series of hydrocarbons, the logarithm of the solubility in
water was found to be a linear function of the hydrocarbon molar volume. McAuliffe determined his
solubilities in distilled water with measurements by a gas-liquid partition chromatographic
technique. The chromatograph indicated slight impurity peaks, but McAuliffe assumed that these
impurities did not interact with the hydrocarbon to alter the solubility. Thus, the solubility
values in Tables 6.21 through 6.24 are assumed to be in pure water without the interference of
such impurities as surfactant humic material.

Solubility of PAH in natural water systems seems to be poorly defined. Boehm and Quinn (1973)
report the PAH, phenanthrene, to be quite water-soluble. Three concentrations of phenanthrene
(1.0, 0.5, and 0.1 mg per 100 ml) added to Narragansett Bay water yielded solubility values
of 60 to 70 µg per 100 ml for all three concentrations. In two samples of Providence River water,
Boehm and Quinn found phenanthrene solubilities of 24.6 and 15.5 µg per 100 ml. The authors
give no explanation for the higher solubility of phenanthrene in Narragansett Bay water than in
Providence River water. They define soluble hydrocarbons as "those hydrocarbons passing a 0.5-µm
filter" and solubilization as "the process resulting in passage of the hydrocarbons by this
filter." These researchers also found that neither salinity nor dissolved organic carbon had an
effect on the solubility of phenanthrene, implying that phenanthrene is not solubilized by
micelles.

McGinnes and Snoeyink (1974) report that PAH are not soluble in water, but are present either as
particulate material or as material adsorbed onto solid surfaces in natural water systems; the
authors provide no numbers, but they cite Borneff and Knerr (1960) for support. Andelman and
Suess (1970) report that the solubility of PAH is very low in pure water, but that "the increase

Table 6.22. Solubility in water at room temperature of olefin hydrocarbons

Hydrocarbon	Solubility (g of hydrocarbon/10^6 g of water)		Molar vol. (ml/mole at 20°C)
	This work[a]	Literature[b]	
Olefins			
Ethene	131 + 10	134, 131	54.5
Propene	(2040) 200 + 27	183 at 30°C	81.9
1-Butene	(615) 222 + 10		94.3
2-Methylpropene	(670) 263 + 23	289,314	94.4
1-Pentene	148 + 7		109.5
2-Pentene	203 + 8		107.0 - 108.2[c]
3-Methyl-1-butene	(156) 130 + 14		111.8
1-Hexene	50 + 1.2		125.0
2-Methyl-1-pentene	78 + 3.2		123.4
4-Methyl-1-pentene	48 + 2.6		126.7
2-Heptene	15 + 1.4		138.7 - 140.0[c]
1-Octene	2.7 + 0.2		157.0
Diolefins			
1,3-Butadiene	(1980) 735 + 20		87.1
2-Methyl-1,3-butadiene	642 + 10		100.0
1,4-Pentadiene	558 + 27		103.1
1,5-Hexadiene	169 + 6		118.7
1,6-Heptadiene	44 + 3		134.0

[a]Numbers following + symbol indicate standard deviation from mean; numbers in parentheses indicate calculated solubility of liquid hydrocarbon at 25°C.
[b]See McAuliffe 1966 for sources of the values given.
[c]Molar volume for *cis-trans* forms.

Source: McAuliffe 1966, Table II, p. 1271. Reprinted with permission from *J. Phys. Chem.* Copyright be the American Chemical Society.

of solubility of PAH by the addition of water soluble organic compounds is a phenomenon that is important in the passage of these compounds into and through environmental waters." Thus, the solubilization of PAH is enhanced by the presence of micelles; this conclusion seems to contradict the results of Boehm and Quinn (1973), who found dissolved organic carbon to have no influence on phenanthrene and anthracene solubility. However, since no two PAH compounds can be assumed to behave similarly with regard to solubility or any other property without sufficient experimental data, the results of Boehm and Quinn (1973) cannot definitely be reported as contradictory to the work of Andelman and Suess (1970), who do not identify the specific PAH compounds about which they are writing. Grasselli (1973) records phenanthrene as insoluble, but conditions for determination were not stated.

Clearly, the literature indicates that the solubilities of hydrocarbons, particularly PAH, in natural waters are poorly understood, presumably because of insufficient experimental data.

Table 6.23. Solubility in water at room temperature
of acetylene hydrocarbons

Hydrocarbon	Solubility (g of hydrocarbon/ 10^6 g of water)[a]	Molar vol (ml/mole at 20°C)
Acetylenes		
Propyne	3640 $\pm$ 125	60
1-Butyne	(5150) 2870 $\pm$ 101	83
1-Pentyne	1570 $\pm$ 33	98.7
1-Hexyne	360 $\pm$ 17	114.8
1-Heptyne	94 $\pm$ 3	131.2
1-Octyne	24 $\pm$ 0.8	147.7
1-Nonyne	7.2 $\pm$ 0.5	164.1
Diacetylenes		
1,6-Heptadiyne	1650 $\pm$ 25	112
1,8-Nonadiyne	125 $\pm$ 3	147

[a]Numbers following $\pm$ symbol indicate standard deviation
from mean; number in parentheses indicate calculated
solubility of liquid hydrocarbon at 25°C.

Source: McAuliffe 1966, Table III, p. 1271. Reprinted
with permission from *J. Phys. Chem.* Copyright by the
American Chemical Society.

6.1.3.5 Degradation — Photooxidation

Photooxidation of hydrocarbons, particularly PAH, is the most important process in hydrocarbon
degradation in air and water (Suess 1975). Suess refers to results from his other published
data (Suess 1972; Andelman and Suess 1971) and predicts that BP degradation in natural waters
will be higher in the upper strata because the factors that promote degradation — illumination,
temperature, and oxygen concentration — are high; degradation will decrease with depth due to a
reduction in water clarity, illumination, temperature, and oxygen concentration. He also predicts that degradation by photooxidation of PAH in river, lake, and sea deposits will be minimal
due to the lack of penetrating radiation and oxygen.

Feldman (1973) found that toluene and related aromatics having side chains can be photooxidized
in the presence of a variety of materials such as H_2O, H_2O_2, and H_2O + $FeCl_3$; tetrahydro-
naphthalene, for example, photooxidized in the presence of H_2O and O_2, gives polymeric materials.
Feldman suggests that hydrocarbon photooxidation is a precursor to the formation of tar in marine
waters.

McGinnes and Snoeyink (1974) studied the photodecomposition of particulate PAH in water using
benzpyrene (BP) and 1,2-benzanthracene (BA). The BP particles, 1.5 μm in diameter, irradiated for
four days with ultraviolet light of varying intensities, decomposed (as shown in Fig. 6.3) from
1 mg/liter to a nominal concentration of 0.3 to 0.45 mg/liter; 55 to 65% of the original BP in
suspension was photodecomposed. The residual BP did not further decompose, regardless of the
length of irradiation, because the decomposed outer layers (about 0.2 μm thick) functioned as a
shield, protecting the inner BP from photodecomposition. This implies that a residual BP core
will remain in natural systems for all photodecomposed BP particles having a diameter greater than
0.4 μm.

Table 6.24. Solubility in water at room temperature of cycloparaffin, cycloolefin, and aromatic hydrocarbons

| Hydrocarbon | Solubility (g of hydrocarbon/10^6 g of water) | | Molar vol. (ml/mole at 20°C) |
	This work[a]	Literature[b]	
Cycloparaffins			
Cyclopenthane	156 + 9		94.1
Cyclohexane	55 + 2.3	63	108.1
Cycloheptane	30 + 1.0		121
Cyclooctane	7.9 + 1.8		135
Methylcyclopentane	42 + 1.6		112.4
Methylcyclohexane	14.0 + 1.2		127.6
1-*cis*-2-Dimethylcyclohexane	6.0 + 0.8		140.9
Cycloolefins			
Cyclopentene	535 + 20		88.2
Cyclohexene	213 + 10		101.3
Cycloheptene	66 + 4		116
1-Methylcyclohexene	52 + 2		118.7
1,4-Cyclohexadiene	700 + 16		93.6
4-Vinylcyclohexene	50 + 5		130
Cycloheptatriene	620 + 20		103
Aromatics			
Benzene	1780 + 45	1740, 1790, 1775, 1740 1730, 1720, 1450	88.7
Toluene	515 + 17	530, 627, 470, 536	106.3
o-Xylene	175 + 8	204	120.6
Ethylbenzene	152 + 8	168, 208, 140, 165	122.4
1,2,4-Trimethylbenzene	57 + 4		137.2
Isopropylbenzene	50 + 5	73	139.5

[a]Numbers following + symbol indicate standard deviation from mean.
[b]See McAuliffe 1966 for sources of the values given.

Source: McAuliffe 1966, Table IV, p. 1272. Reprinted with permission from *J. Phys. Chem.* Copyright by the American Chemical Society.

The primary BP photodecomposition products are identified as benzpyrene quinones, probably benzpyrene-6,12-quinone and benzpyrene-1,6-quinone:

ORNL DWG 76-7318

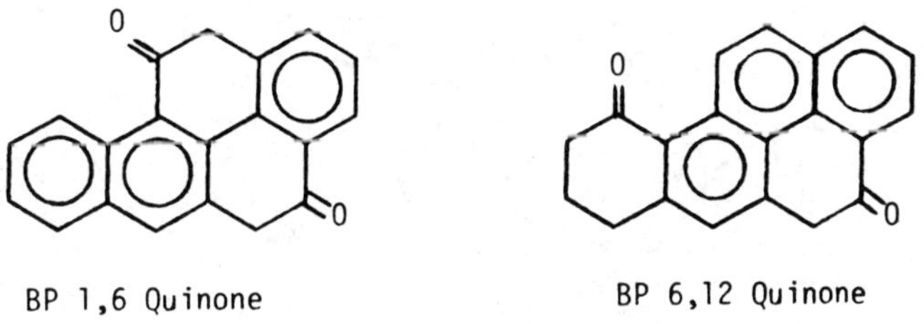

BP 1,6 Quinone

BP 6,12 Quinone

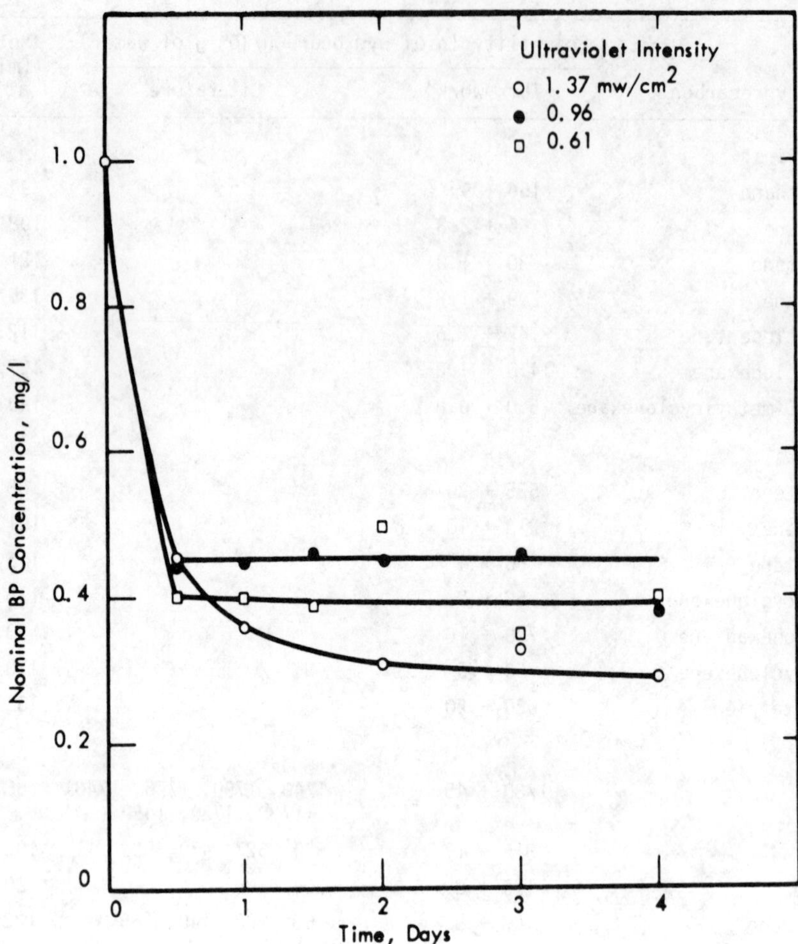

Fig. 6.3. Decomposition of particulate suspensions of BP under ultraviolet light of various intensities. <u>Source</u>: From McGinnes and Snoeyink 1974, Fig. VI, p. 29.

The decomposition products are unstable in the presence of ultraviolet light and are assumed by McGinnes and Snoeyink to decompose rapidly to secondary and tertiary decomposition products.

McGinnes and Snoeyink (1974) also irradiated BA particles having a diameter of 1.25 μm in a manner similar to that used for BP. Figure 6.4 shows that, in contrast to BP, BA continues to decompose with time. BA photocomposes to a benzanthracene-7,12-quinone:

BA 7,12 Quinone

ORNL DWG 76-7316

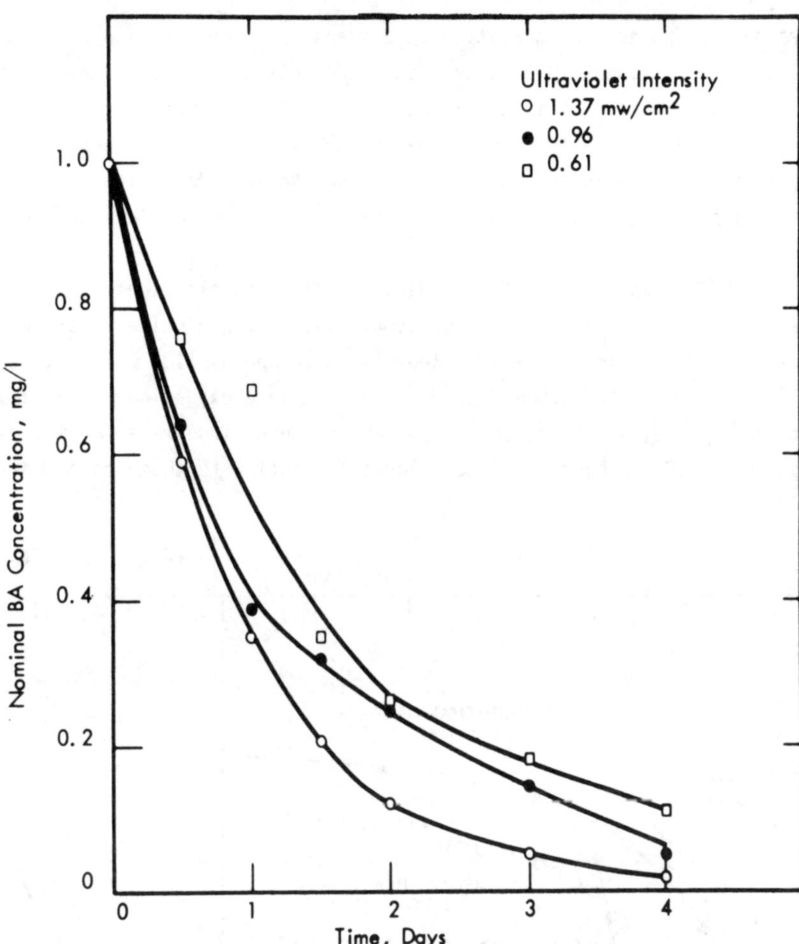

Fig. 6.4. Decomposition of particulate suspensions of BA under ultraviolet light of various intensities. <u>Source</u>: From McGinnes and Snoeyink 1974, Fig. VII, p. 31.

A secondary decomposition product is speculated. The BA photodecomposition products were assumed to be either soluble in water or transparent to ultraviolet radiation, thus explaining the complete decomposition of BA. Although these are laboratory test results, the researchers feel confident that the results can be extended to natural water systems.

The effect of particle size on decomposition of particulate PAH was also investigated by McGinnes and Snoeyink. Samples were irradiated at an ultraviolet intensity of 0.96 milliwatts (mW) per cm^2; BP samples underwent irradiation until a steady state was reached while BA samples were irradiated for a 24-hr period. This investigation showed an increase in percent decomposition for all samples when particle size was decreased, thereby increasing total surface area available for photodecomposition. This implies surface decomposition of both BA and BP.

Armstrong, Williams, and Strickland (1966) describe the photooxidation of pyridine and other organic compounds in artificial sea water as going to completion within a 2- to 3-hr irradiation period.

6.1.3.6 Freshwater biodegradation

Several researchers have proposed biodegradation as a means for decomposition of hydrocarbons in the environment (Chap. 7). Feldman (1973) states that the interactions or pathways of such materials as trace or heavy metals, vitamins, amino acids, EDTA, nutrient forms of phosphorus and nitrogen, DDT, and PCB influence the biodegradation of hydrocarbons.* These trace materials and compounds tend to concentrate in the surface layer and can interfere with the succession of biota, each of which uses a particular petroleum fraction, or an intermediate product.

Ludzack and Ettinger (1963) measured the biodegradability of naphthalene, ethyl benzene, and tetralin; the data were expressed in terms of the cumulative carbon dioxide recovered as a percent of theoretical recovery vs time in days. Theoretical carbon dioxide was obtained from the stoichiometric ratio of chemical to carbon dioxide after complete oxidation. The results (Fig. 6.5) indicate that naphthalene, tetralin, and ethyl benzene were biodegraded at a moderate rate (more than 15 days for 50% CO_2 recovery); microbiological adaption to these compounds was poor.

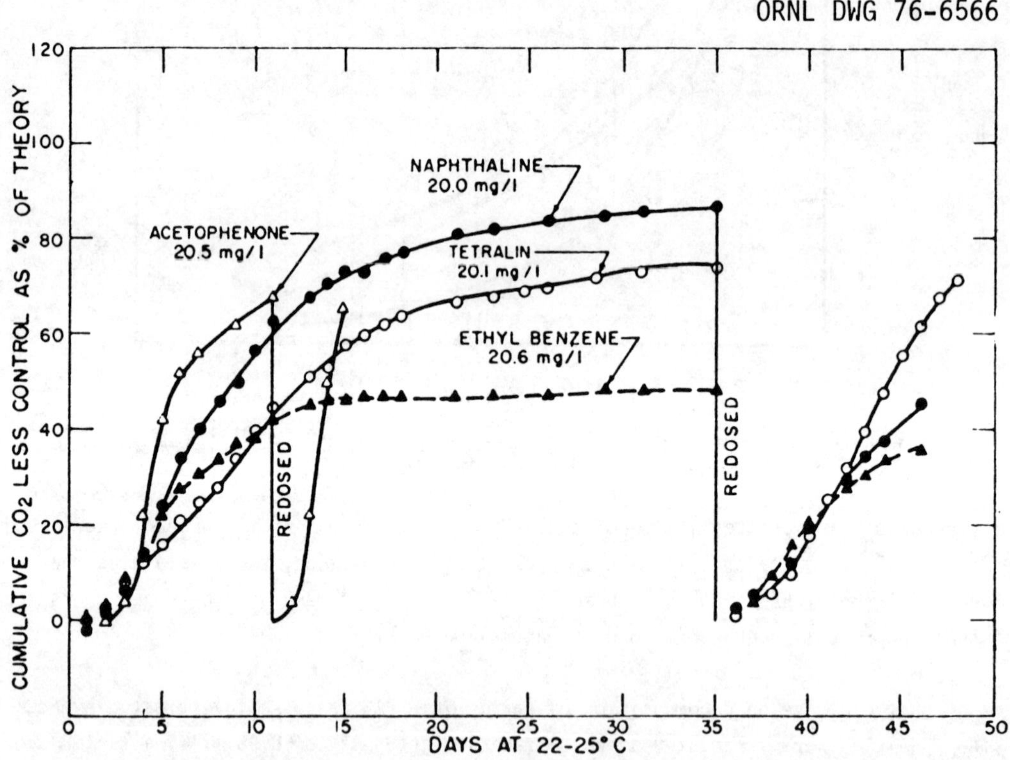

Fig. 6.5. Biodegradability of acetophenone, naphthaline, tetralin, and ethyl benzene in Ohio River water. <u>Source</u>: From Ludzack and Ettinger, Fig. 1, p. 279.

Kostyaev (1973) states that the activity of phenol-oxidizing bacteria, which exist in air, water, and soil, depends primarily on the temperature of the surrounding medium. He conducted experiments at 10°C and found that 5% of the original phenol decomposed, whereas at 30°C, 65% of the original

---

*EDTA = ethylenediaminetetraacetic acid; DDT = dichloro-diphenyl-trichloro-ethane; PCB = polychlorinated biphenyl.

phenol decomposed. Jeffery (1961) found that the rate of phenol oxidation in his studies increased with increased air flow through the medium.

6.1.3.7 Marine biodegradation

Suess (1975) points out the existence of some controversy concerning the ability of marine organisms to degrade PAH. He cites the works of Reichart et al. (1971) and Malaney et al. (1968) as two examples of biological treatment processes having little or no efficiency in removing PAH and the work of Poglazova et al. (1972) as an example of successful degradation of PAH by a mixed microbial culture (Chap. 7). According to an NAS report (1975a),

> microorganisms capable of oxidizing constituents under the right conditions have been found in virtually all parts of the marine environment that have been examined....
> Both laboratory experiments and some field observations have shown that microorganisms consume the least toxic fraction of petroleum (normal alkanes) in a few days or months, depending on temperature and nutrient supply. The fraction containing aromatics and naphthenes is more toxic than the alkanes and also degrades more slowly.

The degradation of PAH may also be assisted by more highly developed aquatic fauna (Suess 1975).

6.1.3.8 Sedimentary biodegradation

A significant portion of the oil pollutants discharged into aquatic environments is incorporated into bottom sediments of receiving waters. Shelton and Hunter (1975) cite Tauson (1934) in discussing one mechanism of the anaerobic decomposition of hydrocarbons in freshwater sediments. Tauson's calculations indicate that the lower-molecular-weight saturated hydrocarbons may be oxidized microbially to CO_2 and H_2O at the expense of sulfate oxygen, whereas higher members of the series, starting with nonane (C_9H_{20}), may be transformed into naphthenes and polynaphthenes that are of no use to sulfur-reducing bacteria due to the negative heat effect accompanying their oxidation with sulfate. Shelton and Hunter (1975) report that there will be a gradual loss of the low-boiling petroleum fractions and an accumulation of naphthenic and polynaphthenic acids in sediments if the transformation of oil under natural anaerobic conditions occurs as Tauson indicates. These acids can be expelled to the surface and completely oxidized if, for example, there is gas formation on a river bottom to act as a carrier.

Baker (1960) suggests that the fate of hydrocarbons in marine sediments is to form crude oil. The oil-forming mechanism is as follows:

1. Recent sediments and crude oils contain some of the same kinds of hydrocarbons that appear to be derived from the organic compounds produced by living organisms.
2. Because the majority of the hydrocarbons, which may later become collected to form crude oil deposits, appear to be present in the sediments relatively soon after decomposition, no geologic time factor seems to be involved in hydrocarbon formation.
3. Formation waters may contain natural solubilizers, which would enhance the solubility of hydrocarbons. Such waters could selectively dissolve and release sediment hydrocarbons in proportions that correspond to hydrocarbon occurrence in crude oils.
4. The solubilizing ability of the formation waters is thought to be sensitive to both salt concentration and dilution. Consequently, hydrocarbons could be abruptly released as the interstitial waters pass from shale to sand or as the micelles diffuse throughout the sand.

5. The hydrocarbons that are released would appear for the first time as tiny oil droplets
    (oil shows). They could coalesce and collect in the highest parts of the reservoir
    rock to form regions of high oil saturation (oil pools). The principles involved here
    are the classical ones of buoyancy and capillary-pressure–entry-pressure mechanisms.

In summary, the ultimate fate of hydrocarbons in the water environment is (1) dispersal in the
water column, (2) incorporation into sediments, or (3) oxidation by chemical or biological means
to CO_2 (NAS 1975a).

6.1.4 Fate in soils

Soils exhibit differences in pH, texture, structure, and chemical composition (Ahlrichs 1972).
Thus, movement of additives within the soils is affected by the interrelation of the macrofeatures
(texture, structure, pore space, and density), the humic nature of the surface horizons, and the
four components of soil (solid, liquid, gaseous, and living). The solid component has a surface
microcharacter that adsorbs, reacts with, and catalyzes reactions of both added and indigenous
substances. This microcharacter is generated by crystalline mineral colloids, amorphous mineral
colloids, organic colloids, and their surface character. The soil liquid component has special
acidity-alkalinity properties due to the variable salt solution concentration, which may be
buffered and which change with changing soil moisture. Since the water phase is strongly
adsorbed, it competes for solid surfaces. The soil atmosphere (gaseous component) is a modification of normal air; microbial activity and limited rates of diffusion cause a lower oxygen content and a higher carbon dioxide content than in air. The living component is the most complex:
Organisms introduce into the soil many enzymes capable of metabolizing both natural organic
substances and synthetic organic additives.

6.1.4.1 Leaching

Contamination of groundwater by leaching of organic chemicals through soil is an environmental
concern (Hamaker 1975). The ability of a chemical to reach groundwater depends upon not only
its movement through, but also its disappearance from, the soil; that is, if the rate of
degradation is sufficiently rapid, as compared with the rate of leaching, the chemical will
disappear before it can reach the groundwater. Determination of soil leaching rates indicates
how long a chemical is retained in the top soil, where it is most subject to degradation or
dissipation. Four major elements of leaching (Hamaker 1975) are (1) soil adsorption, which
determines an underlying pattern; (2) porous flow and diffusion, which disperse the chemical;
(3) adsorption dynamics, which introduce a factor of hysteresis; and (4) water infiltration and
evaporation, which determine the actual amount of water movement and, hence, the observed amount
of chemical movement.

6.1.4.2 Adsorption

Many different intermolecular interactions are involved in adsorption of organic chemicals by
soils (Table 6.25) (Hamaker and Thompson 1972). Clay minerals (mainly kaolinite and montmorillonite), organic matter (humin, humic acids, and fulvic acids), and hydrated metal oxides
(aluminum and iron oxides), the three major soil components for adsorption, can each contribute
to or compete for the adsorption of a soil contaminant, depending on the interactions available
to the system. Hamaker and Thompson conclude that adsorption in soil is largely due to small
molecules interacting with macromolecules such as the soil organic matter.

Table 6.25. Intermolecular interactions

Type of interaction	Forces involved
Van der Waals-London	Electrostatic
Hydrophobic bonding	Entropy generation
Charge transfer	Electrostatic
Hydrogen bonding	Electrostatic
Ligand exchange	Electrostatic
Ion exchange	Electrostatic
Direct and induced ion-dipole and dipole-dipole forces	Electrostatic
Chemisorption	Electrostatic
Magnetic bonding	Magnetic

Source: Hamaker and Thompson 1972, Table 1, p. 52.
Reprinted by permission of the publisher.

Baker and Luh (1971) attribute pyridine sorption onto sodium kaolinite and sodium montmorillonite from aqueous solution to cationic exchange of pyridinium ions with sodium ions. The extent of sorption depends upon the pH of the solution: Maximum pyridine sorption for sodium kaolinite occurs at a pH of 5.5, and maximum pyridine sorption for sodium montmorillonite occurs at a pH of 4.0. Sorption does not occur if the pH is greater than 6.0.

Roberts, Street, and White (1964) describe the physical processes of phenol adsorption by amine-activated montmorillonite. Wetting with water usually results in a considerable basal spacing expansion, whereas the adsorption of the phenol causes a progressive contraction of the mineral lattice. Formation of a quaternary ammonium chloride montmorillonite results in a much greater basal expansion and a correspondingly greater adsorption of phenol. The amount of phenol adsorbed increases as the number of milliequivalents of quaternary ammonium base in the clay increases.

6.1.4.3 Porous flow, adsorption dynamics, and water infiltration

A chemical flowing through a porous medium is either dispersed or retarded, depending on the porosity of the medium (Hamaker 1975). This is a dynamic effect, which depends upon the movement of the liquid through pores. Diffusion allows the chemical to flow into stagnant pores to be released slowly when the main body of chemical has passed. A "frictional" effect caused by adsorption results in slow release of a chemical as soil readjustment takes place.

The pattern of rainfall and evapotranspiration determines the infiltration and movement of water into and through soil. Chemicals are leached downward following a rain, but as evaporation dries out the upper layers of soil, there is a compensating upward movement of water and chemicals. Letey and Oddson (1972) observed that energy from the sun causes evaporation at a soil surface, resulting in the soil becoming drier; consequently, soil suction increases and a hydraulic gradient causes upward water flow, which could transport organic chemicals back to the surface.

Hamaker (1975) concludes, on the basis of published information, that little, if any, organic chemical ever leaches to groundwater (see Suess 1975, Sect. 6.1.1).

6.1.4.4 <u>Decomposition</u>

The decomposition of hydrocarbons by soil microorganisms is the most important process in hydrocarbon degradation (Chap. 7). Meikle (1972) discusses the decomposition of saturated, unsaturated, alicyclic aliphatic, and aromatic hydrocarbons. Saturated aliphatic hydrocarbons undergo three types of reaction in soil and in cultures of microorganisms — oxidation of the terminal carbon atom, oxidation of the C_2-carbon atom, and dehydrogenation:

$$CH_3[CH_2]_n CH_3 = CH_3[CH_2]_n CH_2OH \quad , \tag{1}$$

$$CH_3[CH_2]_n CH_2CH_3 = CH_3[CH_2]_n \overset{\overset{\textstyle OH}{\textstyle |}}{C}HCH_3 \quad , \tag{2}$$

$$CH_3[CH_2]_n CH_2CH_2[CH_2]_m CH_3 = CH_3[CH_2]_n CH{=}CH[CH_2]_m CH_3 \quad . \tag{3}$$

According to Meikle, the predominant soil reaction is Reaction (1). However, the usual isolated product of the microbiological oxidation of saturated aliphatic hydrocarbon chains is a carboxylic acid (Stewart et al. 1959, as cited by Meikle 1972) or glycolipid (Jones and Howe 1968; Makula and Finnerty 1968, as cited by Meikle 1972) rather than the hydroxyl compound implied in Reaction (1). Alcohols are thought to be the intermediates in the sequence that produces the carboxylic acid. Meikle (1972) believes Reaction (2) to be of relatively minor importance in the oxidation of saturated aliphatics. As evidence of Reaction (3), Meikle cites the work of Tamura and Manire (1968), in which hexadecane was converted to a mixture of internal monohexadecanes.

Microbial decomposition of alicyclic aliphatic hydrocarbons is discussed by Meikle (1972). Reaction (4) illustrates the hydroxylation of an unactivated methylene group, and Reaction (5) shows a steroid aromatization reaction.

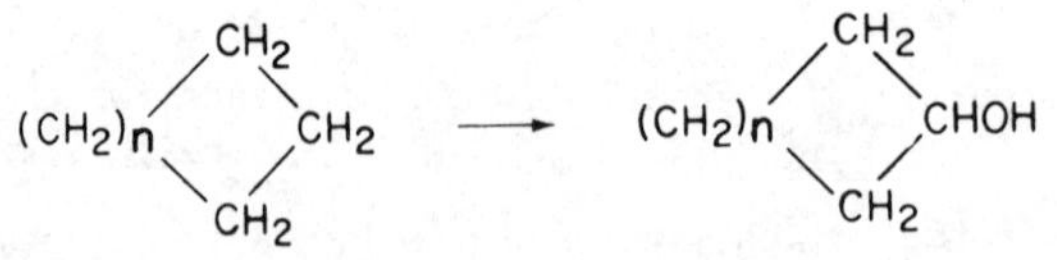

ORNL DWG 76-7298

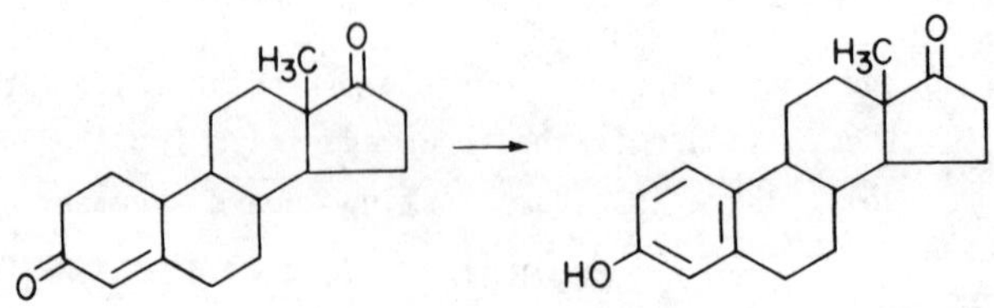

Microbiological degradation of the aromatic nucleus is accomplished with relative ease, resulting in destruction of the ring and leading to the ultimate use of the fragments by microorganisms as a source of carbon for growth (Chap. 7). The oxidative dissimilation of the aromatic ring is by two distinct mechanisms: (1) cleavage of a catechol between adjacent carbon atoms bearing hydroxyl groups (orthocleavage) (Fig. 6.6) and (2) cleavage of the ring between a carbon atom bearing a hydroxylated carbon atom and the adjacent nonhydroxylated carbon atom (metacleavage) (Fig. 6.7) (Meikle 1972).

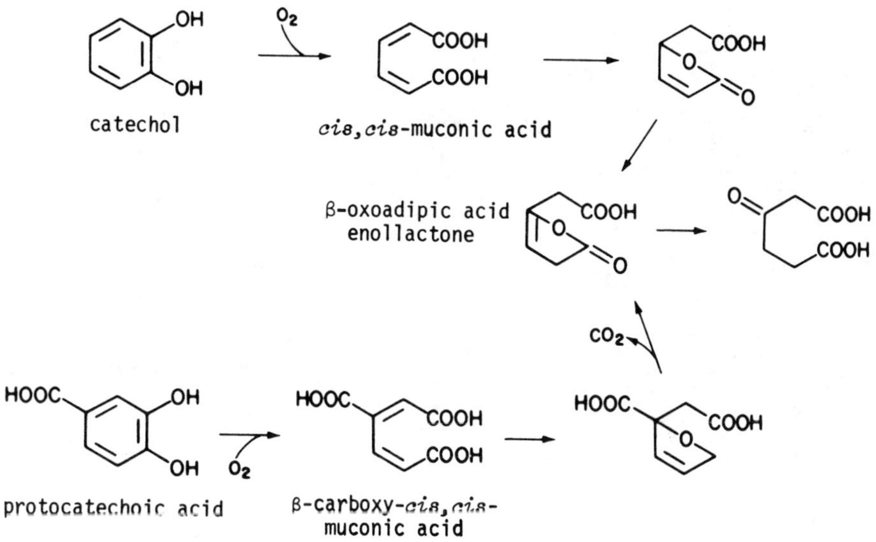

Fig. 6.6. Orthocleavage. Source: From Meikle 1972, p. 166. Reprinted by permission of the publisher.

Fernley and Evans (1958) and Davies and Evans (1964) both present the degradative pathways of the PAH naphthalene by soil pseudomonads (Figs. 6.8 through 6.10). Both groups of researchers agree that naphthalene is oxidatively metabolized by soil pseudomonads through D-*trans*-1,2-dihydroxy-naphthalene to 1,2-dihydroxynaphthalene, which then undergoes ring cleavage. At this point, however, disagreement arises. Fernley and Evans say that the ring fission product is *O*-carboxy-*cis*-cinnamic acid (Fig. 6.8), whereas Davies and Evans believe that the ring fission product is *cis-O*-hydroxybenzalpyruvate, isolated as the crystalline perchlorate (Fig. 6.10). The two groups of researchers do agree that naphthalene is completely dissimilated by soil pseudomonads.

Phenanthrene and anthracene can be completely dissimilated by soil pseudomonads (Evans, Fernley, and Griffiths 1965) (Fig. 6.11). As with naphthalene metabolism, the ring fission mechanism with ring cleavage occurs at the 3,4-dihydroxyphenanthrene (Fig. 6.11). The final metabolism is accomplished through the naphthalene pathway (Fig. 6.10). Anthracene metabolism by ring fission (Fig. 6.12) is similar to that for phenanthrene, although the final metabolism is conducted by unknown pathways. Evans, Fernley, and Griffiths (1965) state that PAH, occurring as various derivatives, can be completely dissimilated by soil pseudomonads (Chap. 7).

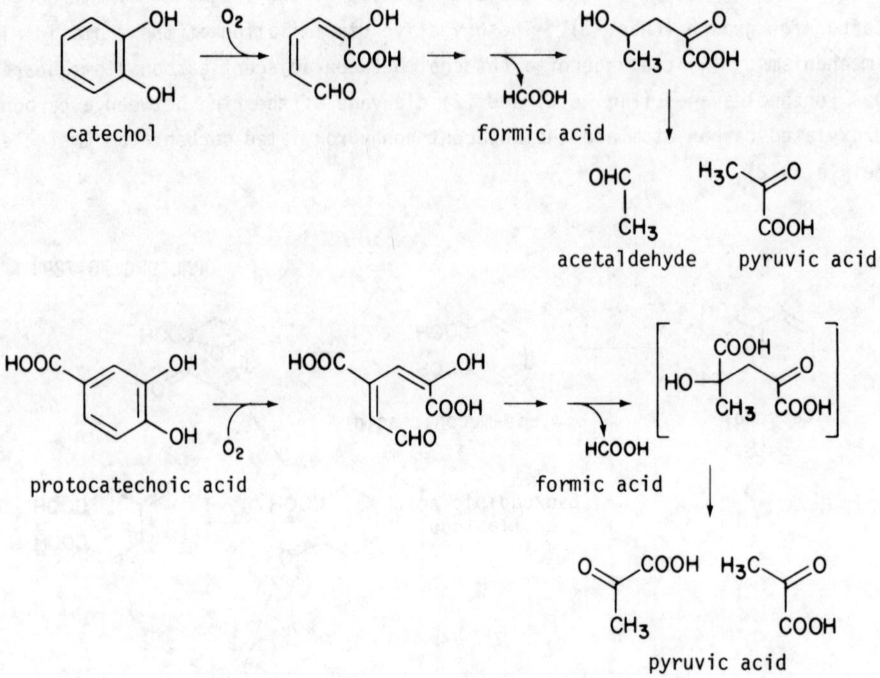

Fig. 6.7. Metacleavage. Source: From Meikle 1972, p. 167.
Reprinted by permission of the publisher.

Muller and Korte (1975) studied the microbial degradation of BaP. The chemical was sprayed on
ground waste, and after three weeks of composting, the digested material was extracted by organic
solvents. They found that BaP withstood degradation by at least 99.5%; however, the researchers
indicate that Lorbacher, Püls, and Schlipköter (1971) found that BaP can be accumulated and converted by special strains of microorganisms.

Hussien, Tewfik, and Hamdi (1974) demonstrated the ability of 22 strains of *Rhizobium* to degrade
the aromatic compounds, catechol, protocatechuic acid, p-hydroxybenzoic acid, and salicylic acid.
At 1 mM concentration and in the presence of 4.8 mM sodium-glutamate, all rhizobia tested
degraded catechol (99 to 100%), p-hydroxybenzoic acid (79 to 99%), protocatechuic acid (81 to
97%), and salicylic acid (20 to 83%). Increased glutamate concentration favored degradation of
p-hydroxybenzoic and salicylic acids, had little observable effect on catechol, and inhibited
the degradation of protocatechuic acid. The researchers found that most strains of rhizobia
tested readily used catechol and protocatechuic acid without detectable accumulation of intermediate phenolic compounds, but p-hydroxybenzoic acid apparently was converted to protocatechuic
acid before ring cleavage. Also, they found salicylic acid to be slowly degraded by different
strains of rhizobia in liquid cultures, being first converted to another compound identified as
gentisic acid.

In conclusion, an understanding of simultaneous leaching and degradation is needed to properly
understand the movement of chemicals through the soil and into the groundwater (Hamaker 1975).

ORNL DWG 76-7308

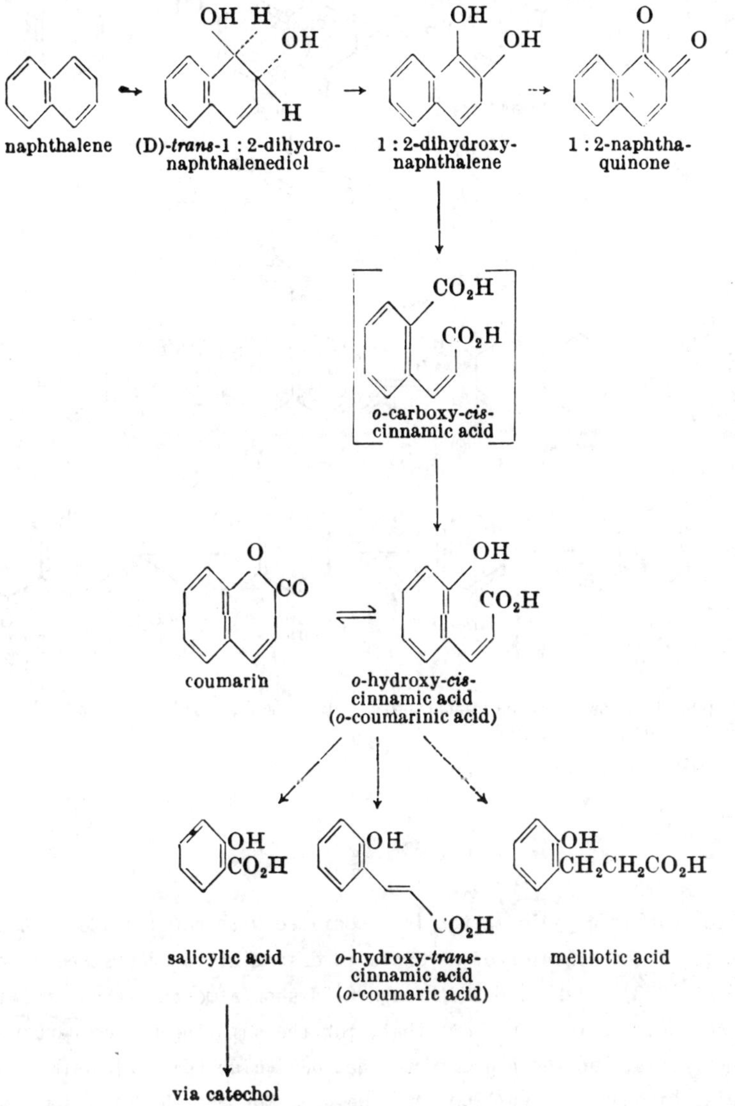

Fig. 6.8. Suggested pathway for the end-ring attack on naphthalene by soil pseudomonads.
Source: From Fernley and Evans 1958, Scheme 1, p. 374.

Reprinted by permission of the publisher.

ORNL DWG 76-7311

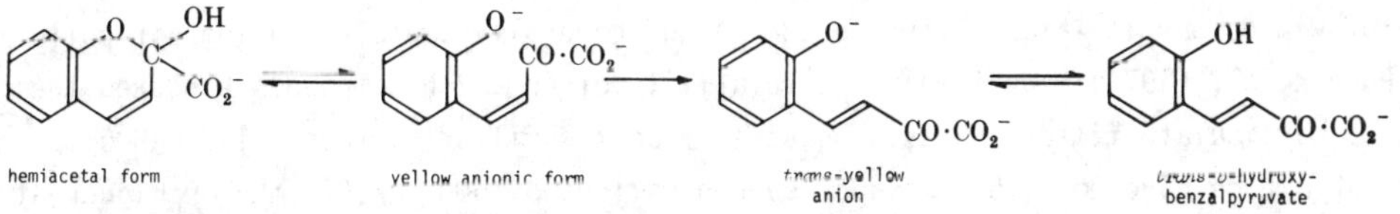

Fig. 6.9. Possible transformations of cis-o-hydroxybenzalpyruvate. Source: From Davies and Evans 1964, Scheme 1, p. 257.

Reprinted by permission of the publisher.

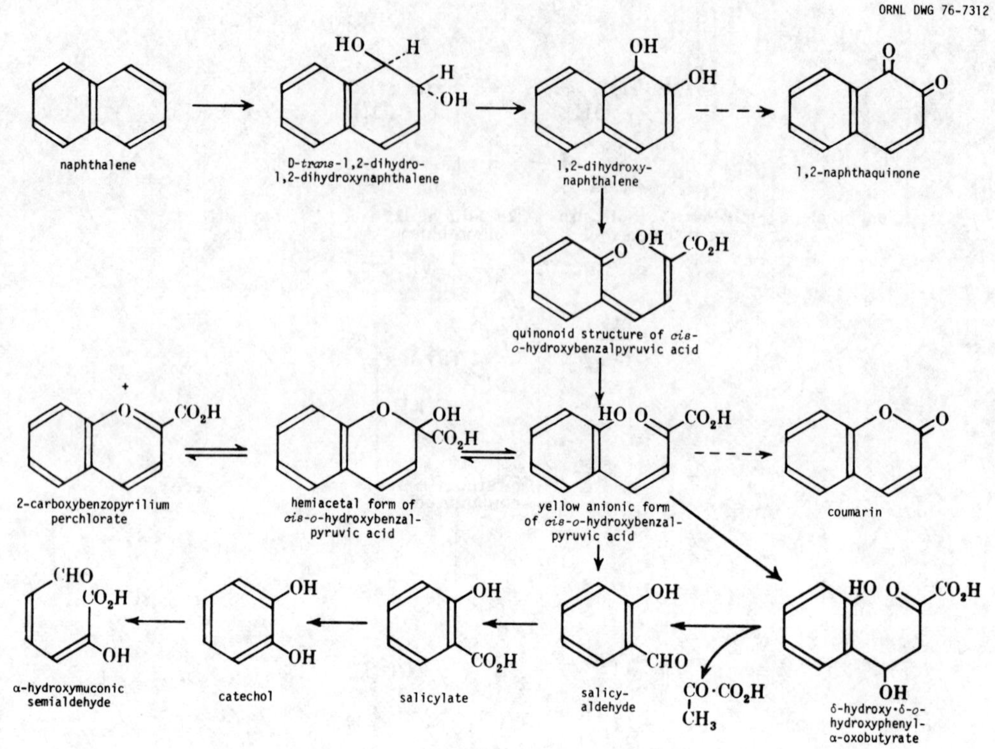

ORNL DWG 76-7312

Fig. 6.10. Proposed pathway of naphthalene metabolism by soil pseudomonads. _Source_: From Davies and Evans 1964, Scheme 2, p. 259.
Reprinted by permission of the publisher.

6.1.5 Fate in air

In the atmosphere, polycyclic organic matter is associated with particulate matter, especially soot. Thus, knowledge of the behavior of aerosols is essential to understanding the fate of PAH in the atmosphere (NAS 1972). Tables 6.26 through 6.28 show atmospheric concentrations of a commonly measured PAH, BP. Table 6.26 shows that, for the sampling period January through March 1959, Altoona, Pennsylvania, had the highest recorded BaP value (61 μg/1000 mg^3) for urban areas, followed by St. Louis, Missouri (54 μg/1000 m^3). Helena, Montana, having a BaP concentration of 0.11 μg/m^3, had the lowest urban concentration. Most nonurban sampling sites had BaP concentrations of less than 1.0 μg/1000 m^3 (Table 6.27). Table 6.28 lists BaP levels in the air of various parts of the world. Some of the higher concentrations were recorded in London, England, and in Milan, Italy. Figure 6.13 illustrates some carcinogenic polycyclics that have been identified in urban air.

Collection of PAH on particulate filters is the method generally employed during air pollution surveys. Pupp et al. (1974) report that the volatility of some PAH compounds hindered their collection on particulate filters; losses occurred from the filter surface. The PAH compounds, pyrene and BaP, which have equilibrium vapor concentrations of 500 μg/10^{-3} m^{-3} or higher at ambient temperatures, were reported to have considerable losses from filters during air sampling. These losses (plus losses by sublimation) and the fact, reported by Thomas, Mukai, and Tebbens (1968), that not all PAH are absorbed even in the presence of excess soot led Pupp et al. to question the accuracy of present collection methods. If the present collection methods are not totally representative of ambient PAH concentrations, then the discrepancy would most likely be

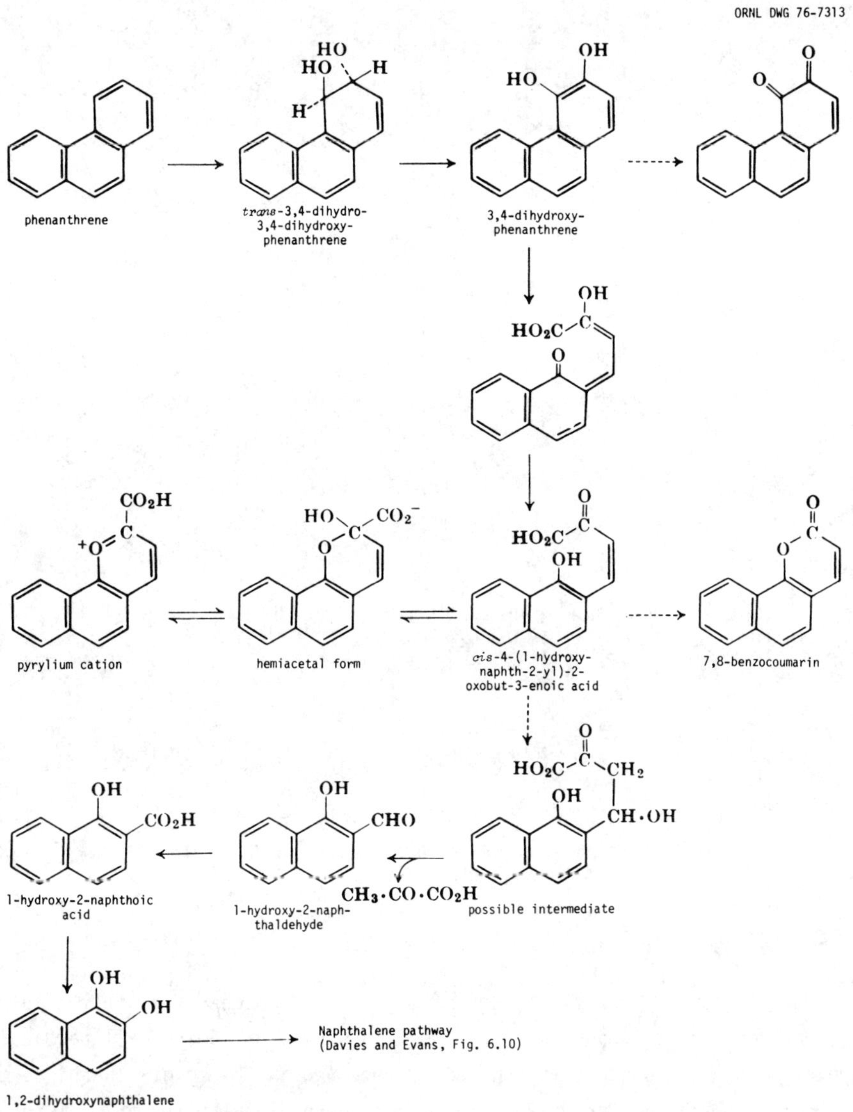

Fig. 6.11. Proposed pathway of phenanthrene metabolism by soil pseudomonads. Source: From Evans, Fernley, and Griffiths 1965, Fig. 1, p. 829.

Reprinted by permission of the publisher.

in such a direction that the reported PAH atmospheric concentrations would be lower than the true value due to losses and unmeasured PAH.

6.1.5.1 Physical reactivity

Particle size is the physical property having the greatest influence on the bioenvironmental behavior of aerosols containing polycyclic organic matter (NAS 1972). A seasonal study of PAH association with particulate matter found that most PAH were associated with particles less than 3.0 μm in diameter (Pierce and Katz 1975). During the summer months, 6 to 23% of the total atmospheric PAH were associated with particles having diameters ≤1.0 μm, whereas particles having a diameter ≤3.0 μm contained 56 to 70% of the total PAH content. During the colder winter months (this study was conducted in Toronto), the PAH content of the smaller respirable particles (<1.0 μm) increased. Pierce and Katz (1975) note that particles having a diameter ≤5.0 μm are in the

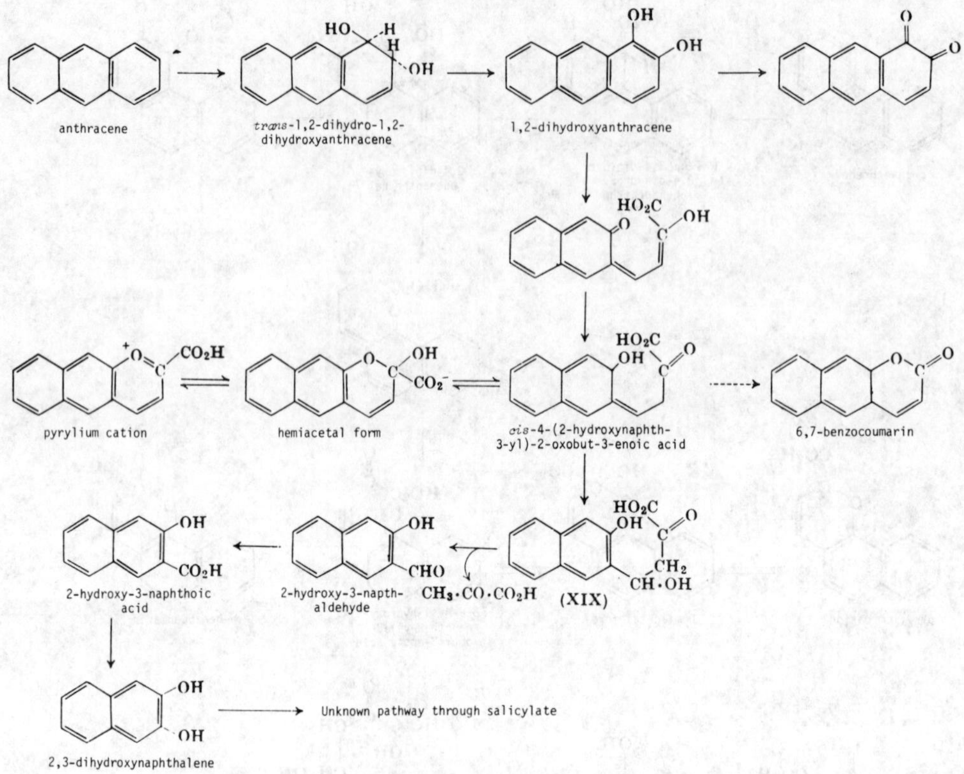

Fig. 6.12. Proposed pathway of anthracene metabolism by soil pseudomonads. <u>Source</u>: From Evans, Fernley, and Griffiths 1965, Fig. 2, p. 830.

Reprinted by permission of the publisher.

"respirable size range," that is, the size range most likely to be deposited in the pulmonary portion of the respiratory tract (Chap. 10). Figure 6.14 shows that 60% of all particulate matter in the size range of 0.5 to 2.0 μm that is inhaled is retained in the lung. The highest retention, 77%, occurs at 1.0 μm. DeMaio and Corn (1966) found that aerosol particles <5.0 μm in diameter contained more than 75% of the weight of selected polycyclic hydrocarbons.

Atmospheric suspended particles evolve by the following mechanisms (NAS 1972): (1) growth or change in particles by homogeneous or heterogeneous chemical reactions of gases on the surface of particles; (2) change in particles by attachment and adsorption of trace gases and vapors to aerosol particles; (3) net change by collision between particles undergoing Brownian motion or differential gravitational settling; (4) net change by collision between particles in the presence of turbulence in the suspending gases; (5) gain or loss in concentration by diffusion or convection from neighboring air volumes; (6) loss by gravitational settling; (7) removal at the earth's surface on obstacles by impaction, interception, Brownian motion, and turbulent diffusion; (8) loss or modification by rainout within clouds; and (9) loss by washout below cloud level. These mechanisms can influence aerosol chemical composition, as a function of size, and can affect the observed size distribution of aerosols (NAS 1972).

The NAS report states that the lifetime of polycyclic organic matter in air depends on the carrier aerosol and on the chemical alteration of the matter itself. In dry atmospheric conditions, residence times of particles <5 μm in diameter exceed 100 hr, but in the presence of

Table 6.26. BP concentrations in urban sampling sites for January through March 1959

State	City	μg BaP/gm part	μg BaP/gm benzene soluble fraction	μg BaP/1000 m³ air	State	City	μg BaP/gm part	μg BaP/gm benzene soluble fraction	μg BaP/1000 m³ air
Alabama	Montgomery	340	2000	24	Michigan	Dearborn	110	960	9.0
Alaska	Anchorage	64	540	3.8		Flint	140	1400	15
Arizona	Phoenix	15	160	5.0		Grand Rapids	91	1400	15
Arkansas	Little Rock	20	230	1.5	Minnesota	Duluth[b]	110	1500	12
California	Berkeley	41	260	2.9		Minneapolis	73	1600	14
	Glendale	5.3	38	0.8	Mississippi	Jackson	24	230	1.2
	San Bernar-dino	13	140	2.3	Missouri	Kansas City	46	540	6.5
						St. Louis[a]	200	1800	54
	San Diego	20	150	2.1	Montana	Helena	2.4	51	0.11
	San Jose	7.2	91	0.56	Nebraska	Omaha	33	460	3.5
Colorado	Denver	51	290	6.9	Nevada	Las Vegas	16	160	1.4
Connecticut	Hartford	68	730	6.5	New Hampshire	Manchester	53	600	6.0
	New Britain	50	450	5.5	New Jersey	Bayonne	33	410	5.5
	New Haven	53	580	5.3		Jersey City	33	440	6.0
Delaware	Wilmington	55	650	10		Newark	46	500	4.5
District of Columbia	Washington	71	59	9.3		Paterson	51	610	6.3
					New Mexico	Albuquerque	15	460	6.3
Florida	Miami	28	250	1.9	North Carolina	Charlotte	290	2100	39
	Orlando	110	810	11		Raleigh	180	1300	14
	Tampa	140	1200	15	North Dakota	Bismarck	5.8	130	0.44
Georgia	Savannah	>49	>480	>4.3	Ohio	Cleveland	110	1200	24
Illinois	Chicago	74	950	15		Columbus	70	930	9.5
	Rockford	63	660	7.3		Dayton	78	760	7.9
Indiana	East Chicago[a]	34	710	11.1		Hamilton	83	600	14
	Indianapolis	120	1100	26		Toledo	100	1200	11
	Hammond	280	2600	39		Youngstown	190	2000	28
	South Bend	91	200	16	Oklahoma	Tulsa	13	180	1.0
Iowa	Des Moines	160	1600	23	Oregon	Portland	96	730	8.0
Kansas	Topeka	40	510	3.1	Pennsylvania	Allentown	26	440	3.4
	Wichita	19	310	2.3		Altoona[a]	280	1400	61
Kentucky	Louisville	70	860	16		Erie	70	1000	9.5
Louisiana	Shreveport	6.4	150	0.65		Johnstown	58	660	16
Maine	Portland	180	2100	21		Pittsburgh	16	200	5.1
Maryland	Baltimore	64	650	14		Scranton	33	360	6.1
Massachusetts	Boston	45	730	9.6		York	31	510	5.6
	Lowell	40	410	3.1	Rhode Island	Providence	24	240	2.9
	New Bedford	81	1000	4.4	South Carolina	Charleston	68	530	5.6
	Worcester	66	700	14		Columbia[a]	120	750	24

Table 6.26 (continued)

State	City	μg BaP/gm part	μg BaP/gm benzene soluble fraction	μg BaP/1000 m³ air	State	City	μg BaP/gm part	μg BaP/gm benzene soluble fraction	μg BaP/100 m³ air
South Dakota	Sioux Falls	31	480	4.0	Virginia	Norfolk	59	580	8.4
Tennessee	Chattanooga	120	1000	31		Richmond	410	1900	45
	Knoxville	210	1900	24		Roanoke	160	1100	18
Texas	Beaumont	13	200	0.82	Washington	Seattle	81	790	9.0
	Dallas	6.1	160	1.4	West Virginia	Charleston[a]	40	900	14
	Galveston	3.4	50	0.16		Wheeling	140	1600	21
	Houston	12	210	1.6	Wisconsin	Madison	80	830	4.9
	San Antonio	5.8	110	0.86		Milwaukee	60	730	8.5
Utah	Salt Lake City	5.4	54	0.52	Wyoming	Cheyenne	36	340	1.2
Vermont	Burlington	28	39	1.0	Puerto Rico	San Juan	15	160	1.2

[a]In respect to the cities with BP levels greater than 11 μg/1000 m³ of air, five cities had a concentration of particulates in the air 1.5 to 2 times higher than in the corresponding January to March period in 1958.
[b]The concentration of particulates in the air for this city in January to March 1959 was one half that found in the corresponding period of 1958.

Source: Sawicki et al. 1960, Table IV, p. 447.

Reprinted by permission of the publisher.

Table 6.27. BP concentrations in nonurban sampling sites

State	County	Location	μg BaP/gm part	μg BaP/gm benzene soluble fraction	μg BaP/1000 m³ air
Alabama	Baldwin	Near Ft. Morgan--Gulf State Park	2.9	63	0.076
Arizona	Coconino	Grand Canyon Park	3.0	31	0.041
Arkansas	Montgomery	Quachita National Forest	6.0	79	0.23
California	Humboldt	Trinidad C. G. Station	3.9	100	0.21
Connecticut	Lichtfield	Wigwam Water Reservoir	19	180	0.69
Delaware	Kent	Bombay Hook Wildlife Refuge	8.9	180	0.69
Hawaii	Oahu Island	Near Barbers Point	0.39	11	0.030
Indiana	Montgomery	In Turkey Run State Park	41	430	1.8
Iowa	Clayton	Backbone State Park	9.5	150	0.54
Kentucky	Pulaski	In Cumberland Falls State Park	14	240	0.58
Louisiana	St. Tammany	Near Slidell	4.1	73	0.15
Maine	Hancock	Acadia National Park	4.4	70	0.18
Maryland	Calvert	On Solomon's Island	26	330	0.70
Massachusetts	Nantucket	On Nantucket Island	1.2	25	0.43
Michigan	Huron	Sleeper State Park	40	490	1.2
Minnesota	Cook	Superior National Forest	25	350	0.66
Missouri	Shannon	Missouri St.	0.76	120	0.025
Nebraska	Thomas	Nebraska National Forest	5.4	56	0.10
New Jersey	Cape May	On Cape May	15	230	0.95
North Carolina	Hyde	Near Cape Hatteras	7.9	140	0.25
Ohio	South Bass Island	Perry National Monument	18	400	0.90
Oregon	Curry	Near Cape Blanco	~ 0.15	~ 9.3	~ 0.01
Pennsylvania	Clarion	In Cook Forest	51	730	1.9
Puerto Rico		Loquillo Mountains Park	1.8	23	0.031
Rhode Island	Washington	In Pine Hill Forest	8.3	120	1.9
South Carolina	Richland	Near Pontiac	36	280	1.1
West Virginia	Webster	In Holly Run State Park	33	530	0.89
Wisconsin	Door	Peninsula State Park	25	410	0.72

Table 6.28. Variation in BP in various parts of the world
(μg BP/1000 m^3 air)

Month	London, England County Hall	Milan, Italy	Sheffield, England	Cannock, England	Copenhagen, Denmark	Oslo, Norway
January	147 (1950)	231 (1958)	78 (1950)	32 (1950)		
February	95 (1950)	127 (1958)	64 (1950)	27 (1950)	15.4 (1956)	
March	101 (1950)	94 (1958)	65 (1950)	27 (1950)		3.45 (1955)
April	48 (1950)	26.6 (1958)	44 (1950)	11 (1950)		
May	25 (1950)	3.9 (1958)		16 (1950)	6.2 (1956)	
June	26 (1949)	5.2 (1958)	20 (1949)	4 (1950)		0.86 (1955)
July	12 (1949)	3.8 (1958)	24 (1949)	6 (1949)		
August	21 (1949)	2.9 (1958)	21 (1949)	6 (1949)	5.4 (1956)	
September	14 (1949)	11.3 (1958)	33 (1949)	11 (1949)		
October	44 (1949)	38.3 (1958)	58 (1949)	27 (1949)		12.4 (1955)
November	80 (1949)		56 (1949)	31 (1949)	14.1 (1956)	
December	122 (1949)		63 (1949)	27 (1949)		15.2 (1955)

Source: Sawicki et al. 1960, Table VI, p. 449.

Reprinted by permission of the publisher.

ORNL DWG 76-6571

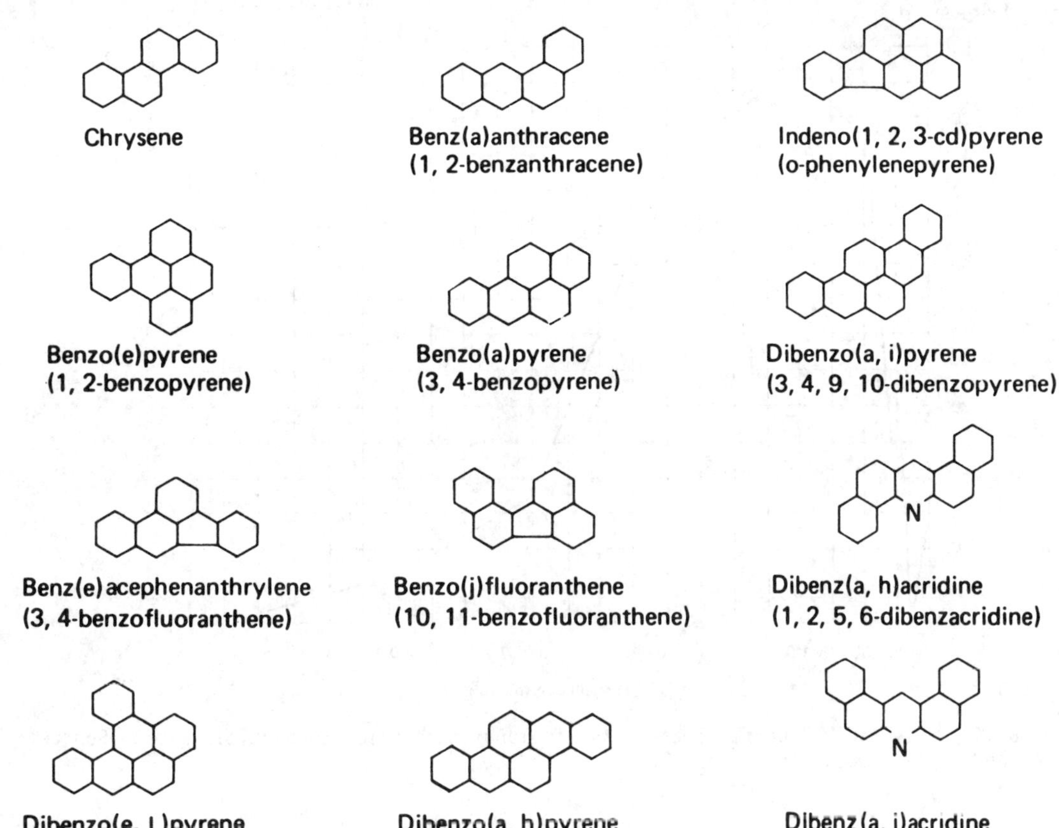

Fig. 6.13. Carcinogenic polycyclics identified in urban air. <u>Source</u>: From Olsen and Haynes 1969, Fig. 4, p. 98.

sunlight, chemical reactivity may lead to transition of polycyclic organic matter adsorbed on soot to other material in several hours.

6.1.5.2 Chemical reactivity

The major mode of polynuclear compound removal from the atmosphere is through chemical reactions; for example, photooxidation is probably one of the most important processes in atmospheric removal of PAH (NAS 1972). Thomas, Mukai, and Tebbens (1968) indicate that BaP and other arenes are primarily adsorbed on the surface of soot through hydrogen bonding. Chemisorption is apparently not feasible because of the rapid rate of photomodification, indicating that BaP must be primarily in an exposed position on the surface.

Tebbens, Thomas, and Mukai (1966) demonstrated a 15 to 50% loss of BaP in smoke on filters with 6 hr of sunlight exposure. Another study (Falk, Markul, and Kotin 1956) shows a 10% destruction of BaP adsorbed on soot or a filter after a 48-hr exposure to light of unstated intensity and a 50% destruction after a 1-hr exposure to light and synthetic smog.

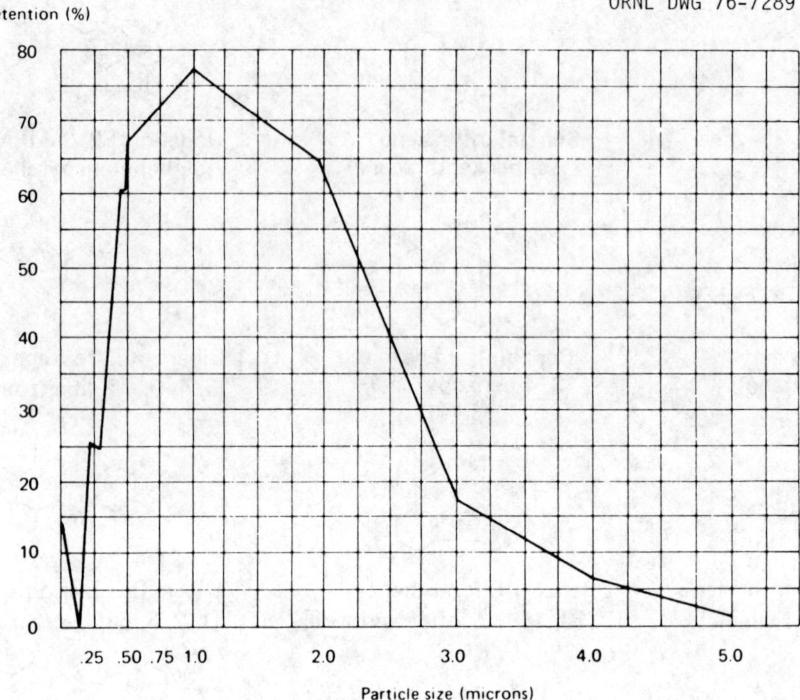

Fig. 6.14. Retention of particulate matter in lung in relation to particle size. <u>Source</u>: From Olsen and Haynes 1969, Fig. 1, p. 23.

Products of photooxidation of adsorbed aromatic hydrocarbons (anthracene and pyrene adsorbed on silica gel and alumina) are given in the 1972 NAS report:

The oxidation of both anthracene and pyrene took place in the presence of ultraviolet light and oxygen, the degree of anthracene oxidation being somewhat dependent on the adsorbent. The pyrene oxidation was conducted on thin-layer chromatography plates.

PAH also react with ozone, peroxides, nitrogen oxides, and sulfur oxides and undergo one-electron oxidation (NAS 1972). They react with ozone in one of four ways: (1) cleavage of phenanthrenelike double bonds, which results in diacid formation under oxidative conditions, (2) oxidation at anthracene-9,10-like positions to yield quinones; (3) a more complex nuclear oxidation, and (4) a side-chain oxidation. A typical example is the ozonolysis of benz[a]anthracene, which undergoes cleavage of the phenanthrenelike double bond to yield a cleavage product, a quinone, and a phthalic acid:

ORNL DWG 76-7325

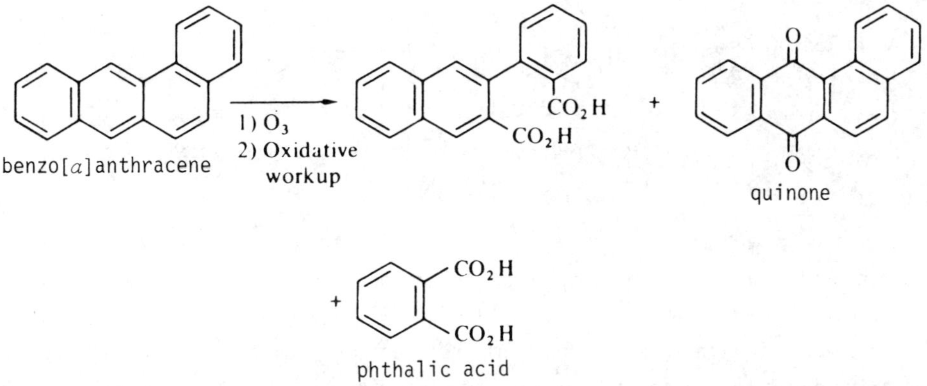

The peroxide reaction involves the interaction of benzoyl peroxide [$(C_6H_5CO_2)_2$] with the PAH, in this case BaP, to yield the 6-benzoyloxy derivative:

ORNL DWG 76-7319

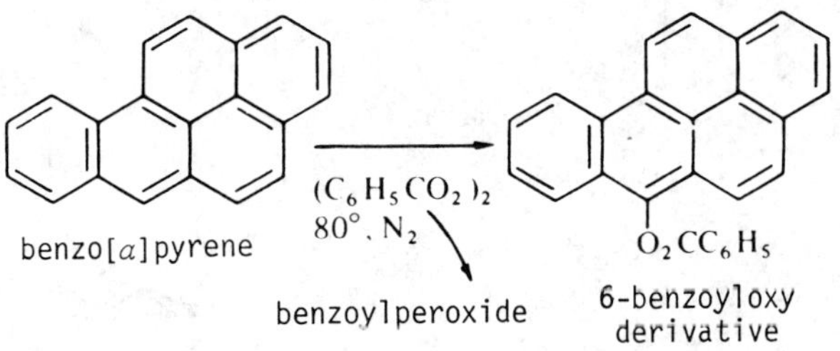

In the presence of oxygen, this peroxide reaction could lead to further oxidation.

Nitrogen oxides or dilute aqueous HNO_3 oxidizes anthracene to anthraquinone:

ORNL DWG 76-7320

Anthracene → (dil. aq. HNO₃ or NO$_x$) → Anthraquinone

Specifically, the addition and substitution of NO_2 in anthracene yields an oxidation product by loss of HNO_2 from the intermediate compound:

ORNL DWG 76-7321

Anthracene → (NO_2) → intermediate → ($-HNO_2$) → substitution product

The one-electron oxidation of PAH results in the formation of radical cations as primary products; these react rapidly with water or other nucleophiles or with oxygen. The complete reaction leads to formation of quinones (with water and oxygen), nucleophile adducts, or dimeric hydrocarbons. One-electron oxidation of BaP also illustrates the formation of quinones:

ORNL DWG 76-7323

Benzo[a]pyrene → ($-e$) → Radical cation → (H_2O, [O]) → 1,6-Dione and other quinones

Dimer

PAH also react with sulfur oxides, yielding sulfuric and sulfonic acids as products; these sulfur compounds will no longer appear in the benzene-soluble fraction because they are water-soluble.

6.1.6 Environmental transport summary

The environmental transport of hydrocarbons is summarized by Suess (1975):

> ... the transport pattern of PAH in the environment appears to be relatively simple. The background PAH, which are formed by biosynthesis, are quite static and, obviously, remain in the plants and microorganisms in which they were formed and, more generally seen, stay within their own ecosystems. However, it appears probable that PAH in ground-water are leached out from the soil.
>
> In contrast, PAH formed by high temperature processes, whether resulting from natural open burning and volcanic eruptions or from man-induced combustion reactions, are all emitted into the atmosphere, and thus are subject to the same dynamic forces which govern the movement, transport and fallout of aerosols generally. Because a significant portion of PAH, adsorbed onto the aerosols, will decompose by photo-oxidation while still in the atmosphere, either stationary or in motion, their fallout at greater distances from the source (delayed fallout) will be relatively very limited. However, where fallout of PAH occurs, it will contaminate the upper layers of the earth, including vegetation and forests, as well as rivers, reservoirs and lakes, and some of them will also reach the oceans. Runoff and the rivers will carry eventually some of this fallout to the open seas and oceans. As waste treatment plants do not remove all PAH, the coastal waters will receive an additional load from domestic and industrial waste effluents either directly, or indirectly through the rivers....
>
> The degradation of PAH in the atmosphere by photooxidation will also continue to some extent when they have settled back on earth and water surfaces, for as long as they are exposed to sunlight. However, some PAH will be degraded by soil bacteria and aquatic organisms. PAH, while adsorbed on particles or in solution, may eventually reach the bed of the various water bodies and be taken up by bottom organisms, or remain there stable for extremely long periods, given absence of light and anaerobic conditions.

6.2 TRACE ELEMENTS

A maximum credible estimate of the trace element emission rate from a 10,000 ton/day coal combustion plant was determined by Vaughan et al. (1975) (Table 6.29). The 30 listed trace elements were assumed to adsorb on particulate matter of the indicated size distribution. In most cases, one third to one half of the elements were adsorbed on respirable particles (<6 μm). Vaughan et al. (1975) point out the deficiency of knowledge about the fate of trace elements during gasification. However, on the basis of elemental loss during processing (Table 6.30), they estimated emission rates for gasification (Table 6.31). The rates of emission from gasification processes are estimated to exceed those of combustion for most elements.

Many trace elements present in fly ash, including Be, C, Ca, Cr, K, Li, Mn, Na, P, Pb, S, Tl, V, and Zn, are shown to be concentrated on the particle surfaces (Natusch et al. 1976). The particle surface is limited in this study to the layer 1000Å in depth from the external surface. Natusch et al. found that for large particles (75 to 100 μm diameter) only a small fraction of a given element was present in this 1000Å layer, whereas for smaller particles (1 μm diameter) this surface layer contained as much as 80% of the mass of a given trace element. The concentration of many trace elements, particularly metals such as lead and vanadium, is known to increase with decreasing particle size in fly ash derived from coal combustion (Lee and von Lehmden 1973). Therefore, because these smaller respirable particles (<5 μm) contain significant trace element concentrations, and because particle surface deposition might occur as Natusch et al. indicate,

Table 6.29. Maximum credible estimate of emission rate vs particle size for trace elements from coal combustion (g/sec)

Element	Particle size (in μm)						Total
	0 - 1	0 - 3	3 - 6	6 - 9	9 - 12	12 - 15	
Ag[a]	5.4×10^{-5}	2.3×10^{-4}	2.4×10^{-4}	1.8×10^{-4}	1.5×10^{-4}	9.9×10^{-5}	9.0×10^{-4}
As	1.5×10^{-3}	7.2×10^{-3}	1.6×10^{-2}	8.0×10^{-3}	4.6×10^{-3}	2.3×10^{-3}	3.8×10^{-2}
Ba[a]	5.4×10^{-4}	2.3×10^{-3}	2.4×10^{-3}	1.8×10^{-3}	1.5×10^{-3}	9.9×10^{-4}	9.0×10^{-3}
Bi[b]	1.5×10^{-3}	7.2×10^{-3}	1.6×10^{-2}	8.0×10^{-3}	4.6×10^{-3}	2.3×10^{-3}	3.8×10^{-2}
Cd[b]	1.5×10^{-4}	7.2×10^{-4}	1.6×10^{-3}	8.0×10^{-4}	4.6×10^{-4}	2.3×10^{-4}	3.8×10^{-3}
Co[c]	3.0×10^{-4}	1.2×10^{-3}	9.5×10^{-4}	6.5×10^{-4}	6.1×10^{-4}	4.2×10^{-4}	3.8×10^{-3}
Cr[c]	1.8×10^{-4}	7.1×10^{-4}	5.8×10^{-4}	3.9×10^{-4}	3.7×10^{-4}	2.5×10^{-4}	2.3×10^{-3}
Cu[c]	3.0×10^{-2}	1.2×10^{-1}	9.5×10^{-2}	6.5×10^{-2}	6.1×10^{-2}	4.2×10^{-2}	3.8×10^{-1}
Fe	9.6×10^{-1}	3.7×10^{0}	3.0×10^{0}	2.0×10^{0}	1.9×10^{0}	1.3×10^{0}	1.2×10^{1}
Ga[a]	5.7×10^{-3}	2.4×10^{-2}	2.6×10^{-2}	1.9×10^{-2}	1.6×10^{-2}	1.0×10^{-2}	9.5×10^{-2}
Ge[a]	1.4×10^{-2}	5.8×10^{-2}	6.2×10^{-2}	4.6×10^{-2}	3.9×10^{-2}	2.5×10^{-2}	2.3×10^{-1}
Hg[d]	3.0×10^{-4}	1.4×10^{-3}	3.1×10^{-3}	1.6×10^{-3}	8.9×10^{-4}	4.4×10^{-4}	7.4×10^{-3}
Mo[c]	6.0×10^{-3}	2.3×10^{-2}	1.9×10^{-2}	1.3×10^{-2}	1.2×10^{-2}	8.3×10^{-3}	7.5×10^{-2}
Mn[c]	3.0×10^{-2}	1.2×10^{-1}	9.5×10^{-2}	6.5×10^{-2}	6.1×10^{-2}	4.2×10^{-2}	3.8×10^{-1}
Na[a]	7.2×10^{-2}	3.0×10^{-1}	3.2×10^{-1}	2.4×10^{-1}	2.0×10^{-1}	1.3×10^{-1}	1.2×10^{0}
Ni[c]	1.2×10^{-2}	4.7×10^{-2}	3.8×10^{-1}	2.6×10^{-2}	2.4×10^{-2}	1.7×10^{-2}	1.5×10^{-1}
Pb[b]	3.0×10^{-3}	1.4×10^{-2}	3.2×10^{-2}	1.6×10^{-2}	9.0×10^{-3}	4.5×10^{-3}	7.5×10^{-2}
Rb[a]	4.5×10^{-4}	1.9×10^{-3}	2.0×10^{-3}	1.5×10^{-3}	1.3×10^{-3}	8.3×10^{-4}	7.5×10^{-3}
Sb[b]	3.0×10^{-4}	1.4×10^{-3}	3.2×10^{-3}	1.6×10^{-3}	9.0×10^{-4}	4.6×10^{-4}	7.5×10^{-3}
Sc	4.5×10^{-4}	1.9×10^{-3}	2.0×10^{-3}	1.5×10^{-3}	1.3×10^{-3}	8.3×10^{-4}	7.5×10^{-3}
Se[b]	3.0×10^{-4}	1.4×10^{-3}	3.2×10^{-3}	1.6×10^{-3}	9.0×10^{-4}	4.5×10^{-4}	7.5×10^{-3}
Sn[c]	1.2×10^{-2}	4.6×10^{-2}	3.8×10^{-2}	2.5×10^{-2}	2.4×10^{-2}	1.6×10^{-2}	1.5×10^{-1}
Ta[a]	1.1×10^{-4}	4.8×10^{-4}	5.1×10^{-4}	3.8×10^{-4}	3.2×10^{-4}	2.1×10^{-4}	1.9×10^{-3}
Te[b]	1.5×10^{-4}	7.2×10^{-4}	1.6×10^{-3}	8.0×10^{-4}	4.6×10^{-4}	2.3×10^{-4}	3.8×10^{-3}
Tl[b]	3.8×10^{-5}	1.8×10^{-4}	4.0×10^{-4}	2.0×10^{-4}	1.1×10^{-4}	5.7×10^{-5}	9.5×10^{-4}
Th[a]	7.2×10^{-4}	3.0×10^{-3}	3.2×10^{-3}	2.4×10^{-3}	2.0×10^{-3}	1.3×10^{-3}	1.2×10^{-2}
Ra[a]	5.0×10^{-11}	1.8×10^{-10}	2.0×10^{-10}	1.4×10^{-10}	1.2×10^{-10}	0.8×10^{-10}	7.2×10^{-10}
V[b]	3.0×10^{-3}	1.4×10^{-2}	3.2×10^{-2}	1.6×10^{-2}	9.0×10^{-3}	4.5×10^{-3}	7.5×10^{-2}
W[a]	5.7×10^{-3}	2.4×10^{-2}	2.6×10^{-2}	1.9×10^{-2}	1.6×10^{-2}	1.0×10^{-2}	9.5×10^{-2}
Zn[c]	3.0×10^{-2}	1.2×10^{-1}	9.5×10^{-2}	6.5×10^{-2}	6.1×10^{-2}	4.2×10^{-2}	3.8×10^{-1}

[a]Distribution similar to Sc.
[b]Distribution similar to As.
[c]Distribution similar to Fe.
[d]Distribution similar to As; 95% volatilized during combustion with recondensation upon particles beyond precipitation.

Source: Vaughan et al. 1975, Table 6, p. 13.

Reprinted by permission of the publisher.

Table 6.30. Elemental losses during the gasification process

Element	% Loss	Element	% Loss
Ag[a]	<10	Na[a]	<10
As	65	Ni	25
Be	20	Pb	60
Bi[a]	50	Rb[a]	<10
Cd	60	Sb	30
Co[a]	<10	Sc[a]	<10
Cr[a]	20	Se	75
Cu[a]	20	Ta[a]	<10
Fe[a]	<10	Te	65
Ga[a]	<10	Th[a]	<10
Ge[a]	<10	Tl[a]	50
Hg	99	V	30
Mo[a]	50	W[a]	<10
Mn[a]	<10	Zn[a]	25

[a]No previous literature estimate available; behavior estimated by comparison with other elements of similar volatility.

Source: Vaughan et al. 1975, Table 7, p. 14.

Reprinted by permission of the publisher.

Table 6.31. Estimated trace element emissions
from the gasification process

Element	Emission (g/sec)	Element	Emission (g/sec)
Ag	2.1×10^{-4}	Na	2.6×10^{-1}
As	5.6×10^{-1}	Ni	8.7×10^{-1}
Be	4.2×10^{-2}	Pb	1.0×10^{0}
Bi	4.3×10^{-1}	Rb	1.7×10^{-3}
Cd	5.2×10^{-2}	Sb	5.2×10^{-2}
Co	8.7×10^{-4}	Sc	1.7×10^{-3}
Cs	1.0×10^{-1}	Se	1.3×10^{-1}
Cu	1.7×10^{0}	Ta	4.3×10^{-4}
Fe	2.6×10^{0}	Te	5.6×10^{-2}
Ga	1.6×10^{0}	Th	2.6×10^{-3}
Ge	6.5×10^{0}	Tl	1.1×10^{-2}
Hg	8.6×10^{-2}	V	5.2×10^{-1}
Mo	8.7×10^{-1}	W	1.1×10^{0}
Mn	8.7×10^{-2}	Zn	6.5×10^{-1}

Source: Vaughan et al. 1975, Table 8, p. 15.

Reprinted by permission of the publisher.

the toxic material present at the point of contact of a particle with living tissues may be higher than previously supposed on the basis of bulk analyses (Natusch et al. 1976).

Vaughan et al. (1975) cite ways in which a metal interacting with green plants or aquatic organisms may become insignificant at the receptor level: (1) low emission rate, (2) dilution in a large (available) endogenous pool in soil, (3) low bioaccumulation rates by either plants or aquatic organisms, and (4) its negligible effect as a potential toxicant. After considering these variables, Vaughan listed only 8 of the possible 30 trace elements as requiring concern (Table 6.32). The comparison between combustion and gasification of the eight metals shows about a fivefold increase in gasification emission rates for cadmium, copper, and nickel and a tenfold

Table 6.32. Factors for possible increases in biologically available metals[a,b]
in the vicinity of a coal combustion and coal gasification facility
after 40 years of operation

Element	Average emission rate (g/sec)	Ratio of emission rates gas./comb.	Factor of increase in			
			Runoff water, soil		Green plant tissue	
			Combust.	Gas.	Combust.	Gas.
Cd	3.8×10^{-3}	7.3	Negligible	Negligible	4.8	28.0
Co	3.8×10^{-3}	0.2	Negligible	Negligible	1.3	1.1
Cu	3.8×10^{-1}	4.5	1.6	3.8	1.1	1.4
Hg	7.4×10^{-3}	11.6	1.7	8.9	1.4	4.1
Mo	7.5×10^{-2}	11.6	3.5	29.6	3.4	29.3
Ni	1.5×10^{-1}	5.8	Negligible	1.2	1.6	4.3
Sn	1.5×10^{-1}	?	1.1	?	?	?
W	9.5×10^{-2}	11.5	Negligible	Negligible	2.1	13.6

[a]Provided emission rates are no more than ten times those listed in Table 6.29, other metals will contribute less than 10% to the endogenous concentrations of metals which are biologically available; gasification emission rates are listed in Table 6.31; 95% ash recovery is assumed for combustion (1400 MWe) facility.
[b]Data are inadequate to evaluate exogenous increases for Ga, Ge, Ra, Te, Th, Tl, and V.

Source: Vaughan et al. 1975, Table A, p. iv.

Reprinted by permission of the publisher.

increase for mercury, molybdenum, and tungsten. This leads to almost a thirtyfold increase in the biological availability to plant tissue of cadmium and molybdenum and a thirteenfold increase of tungsten availability.

6.2.1 Background concentration

Knowledge of the background concentrations of trace elements is important for assessing the possible threat from anthropogenic sources. Natural levels are found in the soil, water, and atmosphere, but the forms and concentration in each medium are different.

6.2.1.1 Soils and rocks

Some of the more hazardous trace elements present in coal (e.g., arsenic, mercury, and cadmium) are present in soils, but in small concentrations (Table 6.29). Of the elements common to soil (Table 6.33) and coal (Table 4.27), silicon has the highest concentration (330,000 ppm), and cadmium the smallest (0.06 ppm).

Table 6.34 indicates the background concentrations of trace elements in igneous rocks, shales, sandstones, limestones, and coal. Not all elements are present in coal. However, many of those that are present are greatly enriched relative to the accompanying rock layers (Chap. 2).

Table 6.33. Composition of soils[a]

Element	Mean ppm in dry soil (range)
Ag	0.1 (0.01 - 5)
Al	71,000 (10,000 - 300,000)
As	6 (0.1 - 40)
B	10 (2.0 - 100)
Ba	500 (100 - 3,000)
Be	6 (0.1 - 40)
Br	5 (1 - 10)
C	200,000
Ca	137,000 (7,000 - 500,000)
Cd	0.06 (0.01 - 0.7)
Ce	50
Cl	100
Co	8 (1 - 40)
Cr	100 (5 - 3,000)
Cs	6 (0.3 - 25)
Cu	20 (2 - 100)
F	200 (30 - 300)
Fe	38,000 (7,000 - 550,000)
Ga	30 (0.4 - 300)
Ge	1 (1 - 50)
Hf	6
Hg	0.03 (0.01 - 0.3)
I	5
K	14,000 (400 - 30,000)
La	30 (1 - 5,000)
Li	30 (7 - 200)
Mg	5,000 (600 - 6,000)
Mn	850 (100 - 4,000)
Mo	2 (0.2 - 5)
N	1,000 (200 - 2,500)
Na	6,300 (750 - 7,500)
Ni	40 (10 - 1,000)
O	490,000
P	650
Pb	10 (2 - 200)
Ra	8×10^{-7} ($3 - 20 \times 10^{-7}$)
Rb	100 (20 - 600)
S	700 (30 - 900)
Sb	(2 - 10?)

Table 6.33 (continued)

Element	Mean ppm in dry soil (range)
Sc	7 (10 - 25)
Se	0.2 (0.01 - 2)
Si	330,000 (250,000 - 350,000)
Sn	10 (2 - 200)
Sr	300 (50 - 1,000)
Th	5 (0.1 - 12)
Ti	5,000 (1,000 - 10,000)
Tl	0.1
U	1 (0.9 - 9)
V	100 (20 - 500)
Y	50 (25 - 250)
Zn	50 (10 - 300)
Zr	300 (60 - 2,000)

[a]The figures refer to oven-dried soils. Soils near mineral deposits have been omitted in computing ranges. Insufficient data are available for Ag, Be, Cd, Ce, Cs, Ge, Hf, Hg, La, Sb, Sn, Tl, and U, and the values quoted for these elements may require revision.

Source: Bowen 1966b, Table 3.4, pp. 39 - 40.

Reprinted by permission of the publisher.

6.2.1.2 Fresh and salt water

The background concentrations of trace elements in river water are low (Table 6.35) when compared with those in soil (Table 6.33). For example, arsenic has a concentration of 6 ppm in soil and 0.0004 ppm in river water, and mercury has a concentration of 0.03 ppm in soil and 0.00008 ppm in river water. Also, river water contains most of the same elements found in the soil because river waters represent the average composition of the soil solutions in the regions they drain (Livingstone 1963, as cited in Bowen 1966b).

Seawater contains more elements than does fresh water. Table 6.36 shows the elemental concentrations of trace elements in seawater, their probable chemical forms, their oceanic residence times, and their percentage retention by seawater.

6.2.1.3 Summary

A summary of elemental concentrations in rocks, soils, and waters is presented as a part of Table 6.37. Also included in Table 6.37 are concentrations in marine and land plants and animals, concentrations in air, and some information on the relative toxicity of each element. The table is a modification of data given by Bowen (1966c). Bowen's data were compiled from the available data on biogeochemistry of the elements, with concentrations in ppm dry weight. The elements listed are those from Table 4.27.

Table 6.34. Elementary composition of igneous and sedimentary rocks (ppm)

Element	Igneous rocks[a]	Shales[b]	Sand-stones[b]	Lime-stones[b]	Coal[c]
Ac	5.5×10^{-10}				
Ag	0.07	0.07	0.05	0.05	0.1
Al	82,300	80,000	25,000	4,200	
Ar	3.5				
As	1.8	13	1	1	25
Au	0.004	0.005	0.005	0.005	<0.125
B	10	100	35	20	100
Ba	425	580	50	120	1 - 3,000
Be	2.8	3	<1	<1	0.1 - 1,000
Bi	0.17	1	0.3		1
Br	2.5	4	1	6.2	
C	200	15,300	13,800	113,500	800,000
Ca	41,500	22,100	39,100	302,000	
Cd	0.2	0.3	0.05	0.035	0.25
Ce	60	59	92	12	
Cl	130	180	10	150	3,000
Co	25	19	0.3	0.1	15
Cr	100	90	35	11	60
Cs	1	5	0.5	0.5	1.3
Cu	55	45	5	4	300
Dy	3	4.6	7.2	0.9	
Er	2.8	1.9	1	0.36	
Eu	1.2	1.1	0.55	0.2	
F	625	740	270	330	80
Fe	56,300	47,200	9,800	3,800	
Ga	15	19	12	4	5.5
Gd	5.4	4.3	2.6	0.7	
Ge	5.4	1.6	0.8	0.2	25 - 3,000
H	1,400	5,600	1,800	860	50,000
He	0.008				
Hf	3	2.8	3.9	0.3	
Hg	0.08	0.4	0.03	0.04	
Ho	1.2	0.61	0.51	0.17	
I	0.5	2.2	1.7	1.2	6
In	0.1	0.1	0.05	0.05	<0.1
Ir	0.001				
K	20,900	26,600	10,700	2,700	
Kr	0.0001				
La	30	20	7.5	6.2	<1,000
Li	20	66	15	5	<25
Lu	0.5	0.33	0.096	0.067	
Mg	23,300	15,000	7,000	47,000	
Mn	950	850	50	1,100	
Mo	1.5	2.6	0.2	0.4	10
N	20				15,000
Na	23,600	9,600	3,300	400	480
Nb	20	11	0.05	0.3	
Nd	28	16	11	4.3	
Ne	0.005				
Ni	75	68	2	20	35

Table 6.34 (continued)

Element	Igneous rocks[a]	Shales[b]	Sand-stones[b]	Lime-stones[b]	Coal[c]
O	464,000	482,600	491,700	496,800	50,000
Os	0.0015				
P	1050	700	170	400	
Pa	1.4×10^{-6}				
Pb	12.5	20	7	9	5
Pd	0.01				0.005
Po	2×10^{-10}				
Pr	3.2	6	2.8	1.4	
Pt	0.005				≤ 0.035
Ra	9×10^{-7}	11×10^{-7}	7×10^{-7}	4×10^{-7}	2×10^{-7}
Rb	90	140	60	3	15
Re	0.005				
Rh	0.001				0.001
Rn	4×10^{-13}				
Ru	0.001				
S	260	2,400	240	1,200	10,000
Sb	0.2	1.5	0.05	0.2	
Sc	22	13	1	1	3
Se	0.05	0.6	0.05	0.08	≤ 7
Si	281,500	73,000	368,000	24,000	
Sm	6	5.6	2.7	0.8	
Sn	2	6	0.5	0.5	10
Sr	375	300	20	610	1,000
Ta	2	0.8	0.05	0.05	
Tb	0.9	0.58	0.41	0.071	
Te	0.001				
Th	9.6	12	1.7	1.7	
Ti	5,700	4,600	1,500	400	$\leq 20,000$
Tl	0.45	1.4	0.82	0.05	0.05 - 10
Tm	0.48	0.28	0.3	0.065	
U	2.7	3.7	0.45	2.2	0.005 - 200
V	135	130	20	20	40
W	1.5	1.8	1.6	0.6	
Xe	0.00003				
Y	33	18	9.1	4.3	5
Yb	3	1.8	1.3	0.43	
Zn	70	95	16	20	40
Zr	165	160	220	19	≤ 250

[a] Taylor, S.R., 1964 (as cited in Bowen 1966a).
[b] Turekian and Wedepohl 1961; Haskin and Gehl 1962 for lanthanides (as cited in Bowen 1966a).
[c] Rankama and Sahama (1950), Stutzer (1940) and Bethell (1962); Balashov et al. 1964 for additional data on lanthanides (as cited in Bowen 1966a).

Source: Bowen 1966a, Table 2.1, pp. 16-17.

Reprinted by permission of the publisher.

6.2.2 Existing concentrations

This section contains information concerning concentrations of some trace elements found in the air, water, and soil. Existing concentrations constitute both background and anthropogenic levels.

6.2.2.1 Air

The amounts of elements mobilized into the atmosphere as a result of weathering and the combustion of fossil fuels are shown in Table 6.38. Some of the more toxic elements are shown to be mobilized into the atmosphere in greater amounts from weathering than from fossil fuel (e.g.,

Table 6.35. Elementary composition of soil solutions and river waters (ppm)[a]

Element	Soil solution[b]		River water[c]	
	Median	Range	Median	Range
Ag			0.00013	0.00001 - 0.0035
Al			0.24	0.01 - 2.5
Ar			0.6	
As			0.0004	< 0.0004 - 0.23
Au			< 0.00006	
B		0.03 - 10	0.013	0.01 - 1
Ba			0.054	0.009 - 0.15
Be			0.001	0.0001 - 0.001
Br		< 0.001 - 0.01	0.021	0.005 - 140
C (HCO_3-)	4	2 - 7	11	6 - 19
Ca	32	1 - 60 (- 1500)	15	4 - 120
Cd			0.08	
Cl	10	7 - 50 (- 8000)	7.8	5 - 35
Co			0.0009	< 0.006
Cr			0.00018	0.0001 - 0.08
Cs			0.0002	0.00005 - 0.0002
Cu		0.01 - 0.06	0.01	0.0006 - 0.4
F			0.09	
Fe		0.1 - 0.25 (- 25)	0.67	0.01 - 1.4
Ga			< 0.001	
Hg			0.00008	
I		0.01	0.002	
K	3.5	1 - 11 (- 400)	2.3	1.4 - 10
Li			0.0011	0.00007 - 0.04
Mg	25	0.7 - 100 (- 2400)	4.1	1.5 - 6
Mn		0.02 - 2 (- 800)	0.012	0.00002 - 0.13
Mo		< 0.001	0.000035	< 0.007
N (NO_3-)		2 - 800	0.23	0.01 - 0.8
Na	15	9 - 30 (- 3500)	6.3	3 - 25
Ni			0.01	0.0002 - 0.02
P	0.005	0.001 - 30	0.005	0.001 - 0.3 (- 12)
Pb			0.005	0.0006 - 0.12
Ra			3.9×10^{-10}	
Rb			0.0015	< 0.008
Rn			1.7×10^{-15}	
S	5	< 3 - 5000	3.7	0.9 - 30
Se		0.001 - 3	< 0.02	
Si		0.5 - 12	6.5	2 - 12
Sn			0.00004	
Sr		< 0.1	0.08	0.003 - 0.8
Th			0.00002	
Ti		< 0.07	0.0086	< 0.11
U			0.001	0.00002 - 0.05
V			0.001	< 0.007
Zn		0.1 - 0.3	0.01	0.0002 - 1
Zr			0.0026	0.00005 - 0.022

[a]Many elements vary in abundance in river water seasonally (e.g., Si) or even from day to day (e.g., Mn); abnormal ranges are given in parentheses.
[b]Swaine 1955, Wiklander 1958, Vinogradov 1959, Fried and Shapiro 1961, Barber et al. 1963, Bowen and Cawse 1965 (as cited in Bowen 1966b).
[c]Livingstone 1963a, Durrum and Haffty 1963 (as cited in Bowen 1966b).

Source: Bowen 1966b, Table 3.3, pp. 31-32.

Reprinted by permission of the publisher.

Table 6.36. Elements in seawater[a]

Element	Chemical form	Amount (ppm)	Residence time (years x 1000)	Percentage retention
Ag	$AgCl_2^-$	0.0003	2,100	0.4
Al		0.01	0.1	0.00001
Ar	Ar	0.6		140
As	AsO_4H^{2-}	0.003		0.15
Au	$AuCl_4^-$	0.000011	560	0.25
B	$B(OH)_3$	4.6		46
Ba	Ba^{++}	0.03	84	0.006
Be		0.0000006	0.15	0.00002
Bi		0.000017	45	0.01
Br	Br^-	65		2300
C	CO_3H^-, organic C	28		12
Ca	Ca^{2+}	400	8,000	0.9
Cd	Cd^{2+}	0.00011	500	0.05
Ce		0.0004	6.1	0.0006
Cl	Cl^-	19,000		13,000
Co	Co^{2+}	0.00027	18	0.001
Cr		0.00005	0.35	0.00004
Cs	Cs^+	0.0005	40	0.04
Cu	Cu^{2+}	0.003	50	0.005
F	F	1.3		0.2
Fe	$Fe(OH)_3$	0.01	0.14	0.000016
Ga		0.00003	1.4	0.0002
Ge	$Ge(OH)_4$	0.00007	7	0.004
H	H_2O	108,000		6,850
He	He	0.0000069	20,000	0.05
Hf		< 0.000008		< 0.00024
Hg	$HgCl_4^{2-}$	0.00003	42	0.03
I	I^-, IO_3^- ?	0.06		1.1
In		< 0.02		
K	K^+	380	11,000	1.6
Kr	Kr	0.0025		130
La		0.000012	0.44	0.0009
Li	Li^+	0.18	20,000	0.75
Mg	Mg^{2+}	1,350	45,000	5.1
Mn	Mn^{2+}	0.002	1.4	0.0002
Mo	MoO_4^{2-}	0.01	500	0.6
N	Organic N, NO_3^-, NH_4^+	0.5	2.5	220
Na	Na^+	10,500	260,000	40
Nb		0.00001	0.3	0.00004
Ne	Ne	0.00014		2
Ni	Ni^{2+}	0.0054	18	0.005
O	OH_2, O_2, SO_4^{2-}	857,000		164
P	PO_4H^2	0.07		0.006
Pa		2×10^{-9}		0.13
Pb	Pb^{2+}	0.00003	2	0.000002
Ra		6×10^{-11}		0.006
Rb	Rb^+	0.12	270	0.12
Rn	Rn	6×10^{-16}		0.1
S	SO_4^{2-}	885		300
Sb		0.00033	350	0.15
Sc		< 0.000004	5.6	< 0.00002
Se		0.00009		0.16
Si	$Si(OH)_4$	3	8	0.001
Sn		0.003	100	0.1
Sr	Sr^{2+}	8.1	19,000	1.9
Ta		< 0.0000025		< 0.0001
Th		0.00005	0.35	0.0005
Ti		0.001	0.16	0.00002
Tl	Tl^+	< 0.00001		< 0.002

Table 6.36   (continued)

Element	Chemical form	Amount (ppm)	Residence time (years x 1000)	Percentage retention
U	$UO_2(CO_3)_3^{4-}$	0.003	500	0.1
V	$VO_5H_3^{2-}$	0.002	10	0.001
W	WO_4^{2-}	0.0001	1	0.006
Xe	Xe	0.000052		300
Y		0.0003	7.5	0.0008
Zn	Zn^{2+}	0.01	180	0.01
Zr		0.000022		0.000012

[a]Goldberg 1963, Goldberg 1965, Shultz and Turekian 1965 (as cited in Bowen 1966*a*).

Source: Bowen 1966*a*, Table 2.2, pp. 19-20. Reprinted by permission of the publisher.

beryllium is mobilized by fossil fuel, primarily coal, at an annual rate of 0.41×10^9 g compared with 5.6×10^9 g per year from sediment weathering, and lead is mobilized from fossil fuel at 3.6×10^9 g compared with 110×10^9 g per year from river flow weathering).

Table 6.39 shows the levels of cadmium and zinc in the air of cities and nonurban areas. Mean cadmium concentrations range from 0.050 $\mu g/m^3$ in Covington, Kentucky, to 0 $\mu g/m^3$ in Houston, Texas. The highest maximum recorded is 0.370 $\mu g/m^3$ in Youngstown, Ohio.

Changes in cadmium and zinc concentrations for the period 1954-1964 are shown in Table 6.40. The amount of cadmium in the air of 13 cities decreased; however, there was a considerable increase in Chicago, Baltimore, St. Louis, and Philadelphia. In the remainder, there was little change. Significant zinc concentration increases occurred in Detroit, Philadelphia, and Pittsburgh, whereas concentrations in St. Louis and Cincinnati decreased significantly from 1959 to 1964.

The concentrations of nickel in urban areas are shown to be greater for the Atlantic and Pacific coastal states (0.037 $\mu g/m^3$, mean) than in the southern (0.004 $\mu g/m^3$, mean) or midwestern (0.003 $\mu g/m^3$, mean) states for 1966 (Table 6.41).

Table 6.42 indicates the same trend for nickel in nonurban areas as is shown for urban areas (Table 6.41). The coastal states had a mean concentration of 0.0044 $\mu g/m^3$, almost three times as great as the mean value for interior states, 0.0015 $\mu g/m^3$.

Lead is present in the urban air of many countries (Table 6.43). Paris has the highest concentration of those listed, followed by Los Angeles with 5 to 9.8 $\mu g/m^3$ and 4.0 $\mu g/m^3$ respectively.

6.2.2.2 <u>Water</u>

A summary of trace metals in drainage basins of the United States for a five-year period (Oct. 1, 1962 through Sept. 30, 1967) is shown in Table 6.44. This table is a modification of data found in Kopp and Kroner (1967). Kopp and Kroner also list the number of positive occurrences and the frequency of detection of each trace element along with the trace metal concentrations in individual rivers and lakes of each drainage basin in Table 6.44. The blanks for an element indicate that no positive occurrences were found (e.g., cadmium in the Tennessee River basin). If the

Table 6.37. Elemental concentrations in the environment (ppm)[a]

| Element | Rocks | | | | | | | Water | | | Plants | | Animals | | Soils | Toxicity |
	Igneous	Shale	Sandstone	Limestone	Shale, igneous	Sandstone, limestone	Igneous sedimentary	Fresh	Sea	Air (μg/m)	Marine	Land	Marine	Land		
Ag							0.07	0.00013	0.0003		0.25	0.06	3 - 11	0.006	0.1	Highly toxic to fungal spores, plants, and animals.
Al			25,000	4,200	82,000			0.24	0.01	<3	60	500	10 - 50	4 - 100	71,000	Moderately toxic to most plants; slightly toxic to mammals.
As	1.8	13				1.0		0.0004	0.003	<0.01	30	0.2	0.005 - 0.3	≤0.2	6.0	Moderately toxic to plants although tolerated by some fungi; highly toxic to mammals, especially as AsH_4; carcinogenic.
B	10	100	35	20				0.013	4.6		120	50	20 - 50	0.5	10 (2 - 100)	Moderately toxic to plants; slightly toxic to mammals.
Ba	425	580	50	120				0.054	0.03		30	14	0.2 - 3.0	0.75	500	Moderately toxic to plants; slightly toxic to mammals (or scarcely toxic as $BaSO_4$).
Be	2.8	3.0				<1.0		<0.001	0.0000006	<0.0001	0.001	<0.1		<0.0003 - 0.002	6 (0.1 - 40)	Very toxic to plants; very toxic to mammals if injected intravenously; probably carcinogenic
Bi	0.17	1.0	0.3						0.000017			0.06	0.04 - 0.3	<0.004		Moderately toxic to plants; moderately toxic to mammals if taken in soluble form.
Br	2.5	4.0	1.0	6.2				0.2	65		740	15	60 - 1,000	6.0	5.0	Br_2 is very toxic; Br^- relatively harmless to organisms.
Cd	0.2	0.3	0.05	0.035				<0.08	0.00011		0.4	0.6	0.15 - 3.0	≤0.5	0.06	Moderately toxic to all organisms; a cumulative poison in mammals.
Ce	60	59	92	12					0.0004		320			≤0.03	50	Moderately toxic to mammals when injected intravenously.
Cl	130	180	10	150				7.8	19,000	1.2	4,700	2,000	5,000 - 90,000	2,800	100	Relatively harmless as Cl^-; highly toxic as Cl_2, ClO^-, or ClO_3^-.

Table 6.37. (continued)

| | Rocks | | | | | | | Water | | Air (µg/m) | Plants | | Animals | | | Toxicity |
Element	Igneous	Shale	Sandstone	Limestone	Shale, igneous	Sandstone, limestone	Igneous sedimentary	Fresh	Sea		Marine	Land	Marine	Land	Soils	
Co	25	19	0.3	0.1				0.0009	0.00027	<0.0007	0.7	0.5	0.5 - 5.0	0.03	8.0	Very toxic to plants, but tolerated by one strain of yeast (*Saccharomyces cerevisiae*); moderately toxic to mammals when injected intravenously.
Cr	100	90	35	11				0.00018	0.00005	<0.002	1.0	0.23	0.2 - 1.0	0.075	100 (5 - 3,000)	Cr(III) is moderately toxic; Cr(IV) is highly toxic to organisms; probably carcinogenic.
Cs	1.0	5.0				0.5		0.0002	0.00005		0.07	0.2		0.064	6.0 (0.3 - 25)	Relatively harmless to all organisms.
Cu	55	45	5.0	4.0				0.01	0.003	<0.02	11	14	4 - 50	2.4	20 (2 - 100)	Very toxic to algae, fungi, and seed plants, but two species of fungi can grow in saturated CuSO$_4$ solution; highly toxic to invertebrates; moderately toxic to mammals.
Dy	3.0	4.6	7.2	0.9								<0.02		<0.01		Scarcely toxic
Er	2.8	1.9	1.0	0.36								≤46	0.02 - 0.04	2.0		Scarcely toxic
Eu	1.2	1.1	0.55	0.2								0.021	0.01 - 0.06	0.00012		Scarcely toxic
F	625	740	270	330				0.09	1.3	<0.01	4.5	0.5 - 40	2.0	150 - 500	200	Moderately toxic to all organisms.
Gd	5.4	4.3	2.6	0.7								≤70	0.06			Scarcely toxic
Ge	5.4	1.6	0.8	0.2					0.00007				0.3 ?		1.0	Scarcely toxic, except as GeH$_4$, which is highly toxic to mammals.
Hf	3.0	2.8	3.9	0.3					<0.000008	<0.4	0.01 ?			0.04	3.0 ?	Scarcely toxic
Hg	0.08	0.4	0.03	0.04				0.00008	0.00003	Probably present	0.03	0.015		0.046	0.03 - 0.8	Very toxic to fungi and green plants; highly toxic to mammals if injected intravenously or by mouth as HgCl$_2$; a cumulative poison in mammals.

Table 6.37. (continued)

Element	Rocks							Water		Air (µg/m)	Plants		Animals			Toxicity
	Igneous	Shale	Sandstone	Limestone	Shale, igneous	Sandstone, limestone	Igneous, sedimentary	Fresh	Sea		Marine	Land	Marine	Land	Soils	
Ho	1.2	0.6	0.51	0.17							$\leq$16	0.005 - 0.01	0.5			Scarcely toxic
Fe	56,300	47,200	9,800	3,800			0.67	0.01	<3.0	700	140	400	160	38,000		Slightly toxic to organisms.
Ca	41,500	22,100	39,100	302,000			15	400	<2.0	10,000	18,000	1,500 - 20,000	200 - 85,000	13,700 (7,000 - 500,000)		Relatively harmless to all organisms.
Ga	15	19	12	4.0			<0.001	0.00003		0.5	0.06	0.5	<0.006	0.4 - 6.0		Moderately toxic to mammals by intravenous injection.
I	0.5	2.2	1.7	1.2			0.002	0.06		30 - 1,500	0.42	1.0 - 150	0.43	5.0		Scarcely toxic
K	20,900	26,600	10,700	2,700			2.3	380		52,000	14,000	5,000 - 30,000	7,400	14,000		Moderately toxic to mammals when injected intravenously; otherwise relatively harmless
La	30	20	7.5	6.2				0.000012		10	0.085	0.1	0.0001	30		Slightly toxic
Li	20	66	15	5.0			0.0011	0.18		5.0	0.1	1.0	<0.02	30		Slightly toxic; causes abnormal development of embryos; some strains of yeast are adapted to Li.
Lu	0.5	0.33	0.096	0.067						$\leq$4.5	0.003	0.00012				Slightly toxic
Mg	23,300	15,000	10,700	2,700			4.1	1,350	<1.0	5,200	3,200	5,000	1,000	5,000		Moderately toxic when injected intravenously into mammals; otherwise relatively harmless.
Mn	950	850	50	1,100			0.012	0.002	<0.01	53	630	1 - 60	0.2	850		Moderately toxic; a few fungi are adapted to high concentrations of Mn.
Mo	1.5	2.6	0.2	0.4			0.00035	0.01	<0.0005	0.45	0.9	0.6 - 2.5	<0.2	2.0		Moderately toxic; excessive Mo inhibits reduction of sulfate.
Na	23,600	9,600	3,300	400			6.3	10,500	1.1	33,000	1,200	4,000 - 48,000	4,000	6,300		Relatively harmless
Nb	20	11	0.05	0.3				0.00001			0.3	<0.001				Moderately toxic to mammals by intravenous injection.
Nd	28	16	11	4.3						5.0	$\leq$460	0.5				Scarcely toxic.

Table 6.37. (continued)

| | Rocks | | | | | | | Water | | | Plants | | Animals | | | Toxicity |
Element	Igneous	Shale	Sandstone	Limestone	Shale, Igneous	Sandstone, limestone	Igneous, sedimentary	Fresh	Sea	Air (μg/m)	Marine	Land	Marine	Land	Soils	
Ni	75	68	2.0	20				0.01	0.0054	<0.002	3.0	3.0	0.04 - 25	0.8	40	Very toxic to most plants and to fungi; moderately toxic to mammals, but Ni(CO)$_4$ is highly toxic; probably carcinogenic.
P	1,050	700	170	400				0.005	0.07		3,500	2,300	4,000 - 18,000	17,000 - 44,000	650	White P$_4$ and PH$_3$ are highly toxic to mammals; phosphates are relatively harmless.
Pb	12.5	20	7.0	9.0				0.005	0.00003	<0.2	8.4	2.7	0.5	2.0	10	Very toxic to most plants; moderately toxic to mammals, where it acts as a cumulative poison.
Pr	8.2	6.0	2.8	1.4							5.0	≤46	0.5			Slightly toxic
Rb	90	140	60	3.0				0.0015	0.12		7.4	20	20	17	100	Scarcely toxic in the presence of K.
In	0.05 - 1.0	0.1				0.05								0.016		Slightly toxic to plants ard animals.
S	260	2,400	240	1,200				3.7	885	3 - 50	12,000	3,400	5,000 - 19,000	5,000	700	S$_8$ is highly toxic to most bacteria and fungi, relatively harmless to green algae, seed plants, and mammals; S^{2-} is moderately toxic; H$_2$S is highly toxic to mammals although tolerated by many bacteria; SO$_2$ is moderately to highly toxic; SO$_4^{2-}$ is relatively harmless.
Sb	0.2	1.5	0.05	0.2					0.00033	<0.004		0.06	0.2	0.006	2 - 10 ?	Moderately toxic to all organisms
Sc	22	13				1.0			<0.000004			0.008		0.00006	7.0	Scarcely toxic
Si	281,500	73,000	368,000	24,000				6.5	3.0	<4.0	1,500 - 20,000	200 - 5,000	70 - 1,000	120 - 6,000	330,000	Scarcely toxic, but large amounts in mammalian lungs are harmful.
Se	0.05	0.6	0.05	0.08				<0.02	0.00009		0.8	0.2		1.7	0.2	Moderately toxic to plants; highly toxic to mammals.
Sm	6.0	5.6	2.7	0.8								0.0055	0.04 - 0.08	0.01		Slightly toxic to plants

Table 6.37. (continued)

Element	Rocks							Water		Air (µg/m)	Plants		Animals		Soils	Toxicity
	Igneous	Shale	Sandstone	Limestone	Shale, igneous	Sandstone, limestone	Igneous, sedimentary	Fresh	Sea		Marine	Land	Marine	Land		
Sn	2.0	6.0			0.5			0.00004	0.003	<0.01	1.0	<0.3	0.2 - 20	<0.15	10 (2 - 200)	Very toxic to plants and green algae; moderately toxic to mammals by intravenous injection; very toxic as SnH_4.
Sr	375	300	20	610				0.08	8.1		260 - 1,400	26	20 - 500	14	300	Scarcely toxic unless Ca is absent.
Ta	2.0	0.8			0.05				<0.0000025				≤410			Moderately toxic to mammals after intravenous injection; otherwise scarcely toxic to organisms.
Tb	0.9	0.58	0.41	0.071								<0.0015	0.006 - 0.01			Slightly toxic
Te	0.001											2 - 25		0.02		Moderately toxic to plants; highly toxic to mammals.
Th	9.6	12			1.7			0.00002	0.00005				0.003 - 0.03	0.003 - 0.1	5.0	Moderately toxic to mammals if injected intravenously; otherwise slightly toxic.
Tl	0.45	1.4	0.82	0.05					<0.00001					<0.4	0.1	Moderately toxic to plants; highly toxic to mammals.
Ti	5,700	4,600	1,500	400				0.0086	0.001	<0.01	12 - 80	1.0	0.2 - 20	<0.2	5,000	Relatively harmless
U	2.7	3.7	0.45	2.2				0.001	0.003			0.038	0.004 - 3.2	0.013	1.0	Highly toxic to mammals if injected intravenously; moderately toxic to all organisms.
V	135	130	20	20				0.001	0.002	<0.001	2.0	1.6	0.14 - 2.0	0.15	100	Highly toxic to mammals if injected intravenously; moderately toxic to all organisms.
W	1.5	1.8	1.6	0.6					0.0001			0.07	0.0005 - 0.05	0.005	1.0	Moderately toxic to plants; slightly toxic to mammals.
Y	33	18	9.1	4.3					0.0003			<0.6	0.1 - 0.2	0.04	50 ?	Slightly toxic

Table 6.37. (continued)

Element	Rocks							Water		Air (µg/m)	Plants		Animals		Soils	Toxicity
	Igneous	Shale	Sandstone	Limestone	Shale, igneous	Sandstone, limestone	Igneous, sedimentary	Fresh	Sea		Marine	Land	Marine	Land		
Yb	3.0	1.8	1.3	0.43							<0.0015	0.02		0.0012		Slightly toxic
Zn	70	95	16	20				0.01	0.01	<0.07	150	100	6.0 - 1,500	160	50	Moderately toxic to plants, but tolerated by a few fungi; slightly toxic to mammals.
Zr	165	160	220	19				0.0026	0.000022		≤20	0.64	0.1 - 1.0	<0.3	300	Moderately toxic to plants; slightly toxic to mammals.

^{a}All measurements in ppm except for concentrations in air, which are in µg/m.

Source: Bowen 1966c, pp 173 - 209.

Reprinted by permission of the publisher.

Table 6.38. Amounts of elements mobilized into the atmosphere as a result of weathering processes and the combustion of fossil fuels

Element	Fossil fuel concentration[a,b] (ppm)		Fossil fuel mobilization[b,c] (X 10^9 g/year)			Weathering mobilization (X 10^9 g/year)	
	Coal	Oil	Coal	Oil	Total	River flow[d]	Sediments[e]
Li	65		9			110	12
Be	3	0.0004	0.41	0.00006	0.41		5.6
B	75	0.002	10.5	0.0003	10.5	360	
Na	2,000	2	280	0.33	280	230,000	57,000
Mg	2,000	0.1	280	0.02	280	148,000	42,000
Al	10,000	0.5	1,400	0.08	1,400	14,000	140,000
P	500		70			720	
S	20,000	3,400	2,800	550	3,400	140,000	
Cl	1,000		140			280,000	
K	1,000		140			83,000	48,000
Ca	10,000	5	1,400	0.82	1,400	540,000	70,000
Sc	5	0.001	0.7	0.0002	0.7	0.14	10
Ti	500	0.1	70	0.02	70	108	9,000
V	25	50	3.5	8.2	12	32	280
Cr	10	0.3	1.4	0.05	1.5	36	200
Mn	50	0.1	7	0.02	7	250	2,000
Fe	10,000	2.5	1,400	0.41	1,400	24,000	100,000
Co	5	0.2	0.7	0.03	0.7	7.2	8
Ni	15	10	2.1	1.6	3.7	11	160
Cu	15	0.14	2.1	0.023	2.1	250	80
Zn	50	0.25	7	0.04	7	720	80
Ga	7	0.01	1	0.002	1	3	30
Ge	5	0.001	0.7	0.0002	0.7		12
As	5	0.01	0.7	0.002	0.7	72	
Se	3	0.17	0.42	0.03	0.45	7.2	
Rb	100		14			36	600
Sr	500	0.1	70	0.02	70	1,800	600
Y	10	0.001	1.4	0.0002	1.4	25	60
Mo	5	10	0.7	1.6	2.3	36	28
Ag	0.5	0.0001	0.07	0.00002	0.07	11	0.03
Cd		0.01		0.002			0.5
Sn	2	0.01	0.28	0.002	0.28		11
Ba	500	0.1	70	0.02	70	360	500
La	10	0.005	1.4	0.0008	1.4	7.2	40
Ce	11.5	0.01	1.6	0.002	1.6	2.2	90
Pr	2.2		0.31			1.1	11
Nd	4.7		0.65			7.2	50
Sm	1.6		0.22			1.1	13
Eu	0.7		0.1			0.25	2.1
Gd	1.6		0.22			1.4	13
Tb	0.3		0.042			0.29	
Ho	0.3		0.042			0.36	2.3
Er	0.6	0.001	0.085	0.0002	0.085	1.8	5.0
Tm	0.1		0.014			0.32	0.4
Yb	0.5		0.07			1.8	5.3
Lu	0.07		0.01			0.29	1.5
Re	0.05		0.007				0.001
Hg	0.012	10^f	0.0017	1.6	1.6	2.5	1.0
Pb	25	0.3	3.5	0.05	3.6	110	21
Bi	5.5		0.75				0.6
U	1.0	0.001	0.14	0.001	0.14	11	8

[a]These average values were obtained from a review of published values for the abundances of elements in coal and fuel oils. Wide variations exist for a given element and the figures are presented as reasonable estimates from available data.
[b]Coal and lignite are combined in this representation.
[c]Based upon 1967 production figures.
[d]In the main, calculated from the data of Turekian (as cited in Sax 1974).
[e]In the main, calculated from the data of Goldberg (as cited in Sax 1974).
[f]This Hg value may be unrepresentative of petroleums in general. It is the only published value for Hg in fuel oils of which we are aware. The samples came from the Cymric oil fields of California, an area near known Hg deposits.

Source: Sax 1974, Table 14.13, p. 373. Copyright 1974 by Litton Educational Publishing, Inc. Reprinted by permission of Van Nostrand-Reinhold Books.

Table 6.39. Cadmium and zinc in the air of cities
and nonurban areas

City or area	Cd (μg/m^3)	Zn (μg/m^3)	City or area	Cd (μg/m^3)	Zn (μg/m^3)
Youngstown, Ohio	0.370[a]	0.75	Wichita, Kan.	0.009[a]	0.18
Nashville, Tenn.	0.094[a]	0.13	Birmingham, Ala.	0.008	1.09
New York, N.Y.	0.069[a]	0.89	Bridgeton, N.J.	0.007	0.08
Portland, Ore.	0.064[a]	0.84	Akron, Ohio	0.007	0.48
Hammond, Ind.	0.055[a]	0.84	Rock Island, Ill.	0.006	0.13
Covington, Ky.	0.050	1.70	Somerville, Mass.	0.005	0.21
Salt Lake City, Utah	0.048[a]	0.04	Altoona, Pa.	0.005	0.27
Ashland, Ky.	0.041	0.54	Beverly Shores, Ind.	0.005	0.20
Muskegan, Mich.	0.035	1.30	Burbank, Calif.	0.004	0.20
Rochester, N.Y.	0.034[a]	0.31	W. Lafayette, Ind.	0.004	0.04
Tampa, Fla.	0.028[a]	0.29	Lincoln, Neb.	0.004	0.50
San Bernardino, Calif.	0.023[a]	0.05	Portsmouth, Va.	0.004	0.08
Atlanta, Ga.	0.017	0.52	Moline, Ill.	0.003	0.00
Bethlehem, Pa.	0.016	0.67	Baltimore, Md.	0.003	0.34
Johnstown, Pa.	0.014[a]	0.01	Springfield, Mass.	0.003	0.07
Indianapolis, Ind.	0.013[a]	0.42	Kalamazoo, Mich.	0.003	0.04
Bridgeport, Conn.	0.013	1.60	Paradise Valley, Ariz.	0.002	0.10
Mean of maximum values	0.058	0.63	Houston, Tex.	0	0.28
Mean of average values	0.025	0.90	Mean	0.005	0.24
Ratio, maximum values	10.86[a]		Ratio	48.0	
Ratio, average values	36.8				

[a]Maximum values; means not available.

Source: Schroeder 1971*b*, Table 2, p. 4.

Reprinted by permission of the publisher.

minima, maxima, and mean are the same (e.g., cadmium in the Southeast basin) or if only a mean value is listed (e.g., beryllium in the Western Great Lakes drainage basin), only one positive occurrence was observed.

Nonradioactive inorganic pollutants found in fresh water are listed in Table 6.45. These pollutants exist as ions, salts, and organometallic compounds.

6.2.2.3 Soil

The greater the distance from a point source of pollution, the less contaminated the soil becomes (Table 6.46). This table shows the decreasing concentrations of lead, zinc, copper, and cadmium found at distances south of a lead smelter. A significant decrease was detected 1 mile south of the smelter with a manyfold decrease found at 5 miles.

The geometric mean compositions and geometric deviations of samples of soils and other surficial materials in the conterminous United States are shown in Table 6.47. These samples were taken at an approximate depth of 8 in. from locations 50 miles apart. Aluminum has the highest geometric mean (45,000 ppb), and beryllium the lowest (0.6 ppb).

Table 6.40. Changes in airborne cadmium and zinc, 1954-59 to 1964,
maximum values (μg/m^3)

City	Cadmium		Zinc	
	1959	1964	1959	1964
Birmingham, Ala.	0.018	0.029	8.40	8.60
Phoenix, Ariz.	0.068	0.020	1.60	1.50
Los Angeles, Calif.[a]	0.020	0.022	2.00	1.80
Denver, Colo.	0.087	0.030	1.40	1.90
Atlanta, Ga.[a]	0.044	0.020	7.90	6.40
Chicago, Ill.[a]	0.017	0.049	2.40	3.80
Des Moines, Iowa	0.009	0.000	0.079	0.000
New Orleans, La.	0.019	0.000	0.065	0.000
Baltimore, Md.	0.034	0.110	2.00	2.50
Detroit, Mich.[a]	0.051	0.000	1.80	4.30
St. Louis, Mo.[a]	0.150	0.290	8.20	4.10
Newark, N.J.[a]	0.310	0.350	4.50	3.30
New York, N.Y.[a]	0.210	0.040	2.80	1.80
Cincinnati, Ohio[a]	0.093	0.110	17.00	14.00
Philadelphia, Pa.[a]	0.016	0.110	1.40	6.10
Pittsburgh, Pa.[a]	0.067	0.028	4.50	8.00
Chattanooga, Tenn.	0.039	0.014		2.10
El Paso, Tex.	0.310	0.300	2.90	3.60
Seattle, Wash.	0.011	0.000	1.70	0.71
Tacoma, Wash.	0.170	0.000	0.75	0.47
Charleston, W. Va.	0.071	0.000		0.62
Milwaukee, Wis.[a]	0.043	0.000	1.40	1.00
Mean	0.084	0.069	4.05	3.48
Cities increasing	4		3	
Cities decreasing	13		3	
Little or no change	5		14	

[a]Zinc slab manufacturer in city or nearby.

Source: Schroeder 1971*b*, Table 3, p. 5.

Reprinted by permission of the publisher.

6.2.3 Fate in natural water

According to Vaughan et al. (1975), metals enter the aquatic environment in small soil particles;
these are redistributed during runoff as suspended particles in streams, lakes, and coastal marine
locations. Trace metals in natural waters are important; they are toxic to man and aquatic life,
and they accumulate in aquatic food chains (Singer 1973). Natural interactions of trace metals
with water, sediments, and the atmosphere determine trace metal fate in natural environments.
Trace metal activity in aquatic systems depends on the metal species formed by natural hydro-
dynamic, chemical, and biological forces. As indicated by Singer (1973), the ability of the
metal species to associate with other dissolved and suspended components of the aquatic system
is of major importance.

6.2.3.1 Bioaccumulation

Most of the metallic elemental content of aquatic ecosystems is present in the sediments and in
the water; only a small fraction resides in the biota. Plankton (zooplankton and phytoplankton)
constitute the most important living elemental reservoir in terms of turnover and total physical
transport and redistribution processes (Wolfe and Rice 1972). However, in contaminated environments, commercially harvested species higher in the trophic scheme may concentrate certain
elements to potentially harmful levels for people who consume large quantities of seafood.

Table 6.41. Nickel in air by geographical area, urban,
means by state, yearly averages

State	Number of cities	Amount (µg/m^3)	State	Number of cities	Amount (µg/m^3)
Massachusetts	4	0.042	Minnesota	2	0.002
Connecticut	3	0.034	Michigan	2	0.011
New York	2	0.085	Indiana	7	0.008
New Jersey	2	0.055	Illinois	2	0.010
Maryland	1	0.034	Ohio	2	0.015
Virginia	2	0.021	Pennsylvania	4	0.014
Kentucky	2	0.015	Total and mean	19	0.010
Florida	1	0.014			
Puerto Rico	1	0.021	Idaho	1	0.000
Oregon	1	0.041	Nebraska	1	0.005
California	4	0.023	Kansas	1	0.003
Total and mean	23	0.037^a	Iowa	1	0.003
			Utah	1	0.004
Georgia	1	0.007	Arizona	1	0.002
Alabama	2	0.004	New Mexico	1	0.002
Texas	1	0.006	Total and mean	7	0.003
Tennessee	2	0.003			
Arkansas	1	0.001			
Total and mean	7	0.004			

^{a}Differs from other means, P < 0.01 - 0.001.

Source: Schroeder 1970a, Table 9. p. 11.

Reprinted by permission of the publisher.

Table 6.42. Nickel in air (µg/m^3) according to geographical area,
by state, yearly averages, nonurban areas (1966)

State	Amount (µg/m^3)	State	Amount (µg/m^3)
Maine	0.0110	Montana	0.0033
New Hampshire	0.0021	Wyoming	0.0000
Vermont	0.0045	Colorado	0.0012
Rhode Island	0.0083	New Mexico	0.0024
New York	0.0033	Arizona	0.0014
Pennsylvania	0.0039	Nevada	0.0008
Maryland	0.0032	South Dakota	0.0007
Virginia	0.0024	Nebraska	0.0011
North Carolina	0.0026	Iowa	0.0015
South Carolina	0.0013	Missouri	0.0017
Mississippi	0.0120	Wisconsin	0.0017
Oregon	0.0017	Indiana	0.0030
California	0.0013	Texas	0.0018
Mean	0.0044^a	Arkansas	0.0012
		Oklahoma	0.0013
		Mean	0.0015^a

^{a}Means differ by Student's t, P <0.001.

Source: Schroeder 1970a, Table 10, p. 12.

Reprinted by permission of the publisher.

Table 6.43. Ambient air lead levels

City and country	Average lead level ($\mu g/m^3$)
Zurich, Switzerland	2.8
London, England	3.2
Paris, France	5 - 9.8
Cincinnati, U.S.A.	1.6
Los Angeles, U.S.A.	4.0
Philadelphia, U.S.A.	1.9

Source: Haley 1969, Table II, p. 26.

Reprinted by permission of the publisher.

Table 6.48 indicates the concentration ratios of trace metals in some aquatic organisms; the concentration ratio is defined as "the ratio of the concentration of an element or radionuclide in an aquatic organism or its tissues to that in the surrounding water under equilibrium or steady-state conditions" (Thompson et al. 1972). Examination of the table reveals that (1) beryllium, bismuth, molybdenum, nickel, antimony, and radium have low concentration ratios; (2) cadmium is readily accumulated in marine invertebrates; (3) mercury levels are high in freshwater invertebrates; (4) silver has a low concentration ratio in freshwater fish; and (5) manganese levels are high in freshwater and marine invertebrates. Because aquatic invertebrates are largely filter feeders, Vaughan et al. (1975) believe that the high manganese levels indicate the tendency for manganese to sorb onto particulates. Marine zooplankton, which are invertebrates, have three possible means of concentrating metals: (1) particulate ingestion of suspended matter from seawater, (2) ingestion of elements via their preconcentration in food material, and (3) complexing of metals by coordinate linkages with appropriate organic molecules (Brooks and Rumsby 1965).

Ion exchange is probably the major mechanism for concentration of alkali metals, and perhaps the heavier alkaline earths, by marine plankton (Leland, Shukla, and Shimp 1973). Bowen (1966*d*) lists the order of affinities in marine organisms as follows: plankton — zinc > lead > copper > manganese > cobalt > nickel > cadmium; brown algae — lead > manganese > zinc > copper, cadmium > cobalt > nickel.

Leland, Shukla, and Shimp (1973) report that marine organisms appear to accumulate the heavier divalent metals to a greater degree than the lighter ones, an occurrence that may be related to the greater polarizability of the heavier divalent metals. They believe, however, that it is doubtful that any single factor is responsible for the order of concentration observed. In support of this statement, they cite Pringle et al. (1968) as stating that apparent affinities for trace elements among various mollusks depend upon (1) environmental concentrations of an element or elements, (2) physiochemical properties of each element, (3) availability of organic ligands for chelation, (4) stability of metal-organic ligands formed, and (5) processes of transport and storage.

Table 6.44. Summary of trace metals in drainage basin rivers and lakes of the United States, 1962 - 1967 (μg/liter)[a]

Element	North-east	North Atlantic	South-east	Tennessee River	Ohio River	Lake Erie	Upper Mississippi	Western Great Lakes
Zinc	4 - 697 (96)	3 - 504 (49)	4 - 495 (52)	5 - 172 (28)	3 - 787 (81)	10 - 1183 (205)	3 - 380 (45)	2 - 406 (24)
Cadmium	1 - 12 (5)	1 - 7 (3)	5 - 5 (5)		2 - 11 (7)	6 - 120 (50)	4 - 8 (6)	3 - 8 (5)
Arsenic	15 - 58 (34)	10 - 97 (47)	5 - 109 (35)	50 - 50 (50)	26 - 128 (66)	281 - 336 (308)	16 - 210 (69)	13 - 83 (37)
Boron	6 - 160 (32)	2 - 221 (42)	6 - 120 (29)	1 - 66 (24)	1 - 752 (67)	28 - 700 (210)	6 - 626 (105)	2 - 49 (19)
Phosphorus	4 - 232 (44)	4 - 288 (48)	4 - 120 (43)	6 - 120 (42)	8 - 725 (130)	16 - 350 (153)	9 - 1750 (243)	5 - 104 (31)
Iron	3 - 520 (51)	1 - 195 (19)	4 - 660 (120)	2 - 686 (37)	1 - 223 (28)	6 - 312 (35)	7 - 363 (35)	1 - 168 (22)
Molybdenum	4 - 61 (25)	4 - 168 (33)	2 - 53 (15)	3 - 67 (25)	6 - 473 (70)	21 - 108 (68)	4 - 360 (88)	4 - 129 (28)
Manganese	0.3 - 40 (3.5)	0.3 - 17 (2.7)	0.4 - 37.4 (2.8)	0.4 - 11 (3.7)	0.4 - 3230 (232)	1.6 - 900 (138)	0.5 - 257 (9.8)	0.3 - 7.4 (2.3)
Aluminum	1 - 148 (28)	2 - 136 (22)	4 - 1050 (117)	3 - 92 (30)	7 - 1430 (141)	18 - 138 (56)	2 - 130 (18)	1 - 71 (17)
Beryllium	0.02 - 0.02 (0.02)	0.02 - 0.4 (0.12)	0.05 - 0.05 (0.05)	0.16 - 0.16 (0.16)	0.05 - 1.22 (0.28)	0.16 - 0.19 (0.17)		(0.05)
Copper	2 - 100 (15)	2 - 155 (17)	3 - 110 (14)	2 - 33 (11)	2 - 280 (23)	5 - 39 (11)	2 - 112 (14)	1 - 34 (7)
Silver	0.1 - 6.0 (1.9)	0.3 - 2.5 (0.9)	0.1 - 0.7 (0.4)		0.4 - 8.2 (2.1)	1.1 - 9.0 (5.3)	0.9 - 6.0 (3.4)	0.1 - 3.8 (1.4)
Nickel	1 - 21 (8)	1 - 49 (8)	2 - 17 (4)	4 - 6 (4)	2 - 114 (31)	9 - 130 (56)	1 - 42 (15)	2 - 28 (10)
Cobalt	9 - 20 (14)	6 - 13 (9)	1 - 1 (1)	0 (0)	3 - 48 (19)	20 - 46 (33)	(18)	2 - 21 (11)
Lead	4 - 48 (17)	3 - 72 (14)	3 - 21 (8)	5 - 38 (17)	10 - 140 (30)	16 - 90 (39)	5 - 119 (33)	3 - 55 (14)
Chromium	1 - 112 (14)	1 - 29 (6)	1 - 22 (4)	2 - 20 (6)	1 - 36 (7)	6 - 25 (12)	1 - 20 (7)	1 - 20 (6)
Vanadium	8 - 10 (9)	5 - 40 (12)	10 - 10 (10)		2 - 54 (22)	45 - 63 (54)	(20)	
Barium	7 - 52 (21)	3 - 64 (25)	3 - 340 (26)	2 - 52 (25)	4 - 195 (43)	10 - 140 (42)	6 - 110 (39)	5 - 41 (15)
Strontium	19 - 224 (76)	4 - 270 (62)	7 - 77 (26)	9 - 118 (47)	43 - 520 (130)	82 - 960 (260)	7 - 310 (105)	9 - 108 (44)

Table 6.44 (continued)

Element	Missouri River	Southwest Lower Mississippi	Colorado River	Western Gulf	Pacific Northwest	California	Great Basin	Alaska
Zinc	4 - 572 (39)	3 - 1080 (85)	3 - 312 (51)	6 - 405 (92)	2 - 330 (40)	3 - 47 (16)	8 - 159 (44)	5 - 88 (28)
Cadmium			2 - 2 (2)	10 - 10 (10)	1 - 13 (5)		1 - 1 (1)	
Arsenic	106 - 150 (123)	56 - 126 (91)	26 - 80 (53)	20 - 24 (22)	16 - 300 (68)		20 - 20 (20)	32 - 36 (34)
Boron	48 - 600 (154)	9 - 1020 (131)	11 - 1800 (179)	34 - 1726 (289)	4 - 148 (30)	31 - 391 (143)	13 - 330 (84)	10 - 120 (28)
Phosphorus	23 - 5040 (353)	5 - 329 (81)	10 - 580 (121)	18 - 570 (173)	2 - 183 (47)	10 - 189 (83)	8 - 104 (37)	10 - 82 (40)
Iron	3 - 248 (37)	3 - 837 (69)	2 - 251 (40)	3 - 952 (173)	2 - 256 (32)	3 - 227 (46)	3 - 557 (70)	2 - 85 (25)
Molybdenum	8 - 354 (83)	11 - 1100 (95)	10 - 444 (130)	4 - 59 (24)	2 - 128 (30)	14 - 124 (45)	3 - 338 (145)	2 - 54 (17)
Manganese	1.0 - 414 (13.8)	0.6 - 50 (9.0)	1.3 - 49 (12)	0.6 - 42 (10)	0.4 - 28 (2.8)	0.5 - 7.8 (2.8)	0.4 - 23 (7.8)	0.5 - 163 (18)
Aluminum	10 - 2760 (213)	7 - 588 (68)	7 - 200 (50)	5 - 924 (335)	3 - 179 (30)	10 - 232 (63)	2 - 27 (15)	2 - 24 (11)
Beryllium	0.01 - 0.56 (0.23)				0.02 - 0.03 (0.02)			
Copper	3 - 133 (17)	2 - 250 (19)	1 - 35 (10)	2 - 38 (11)	1 - 37 (9)	3 - 45 (12)	2 - 70 (12.6)	2 - 20 (9)
Silver	0.8 - 1.5 (1.2)	1.1 - 9.0 (4.3)	0.4 - 38 (5.8)	0.4 - 6.6 (3.5)	0.1 - 3.7 (0.9)		0.3 - 0.3 (0.3)	(1.1)
Nickel	1 - 10 (5)	5 - 50 (17)	3 - 26 (12)	3 - 3 (3)	1 - 30 (10)	2 - 34 (10)	2 - 7 (4)	3 - 7 (5)
Cobalt	8 - 8 (8)	36 - 36 (36)	10 - 11 (11)		1 - 17 (8)			
Lead	23 - 57 (39)	9 - 75 (37)	14 - 64 (32)	4 - 4 (4)	4 - 79 (15)	2 - 6 (4)	2 - 66 (18)	2 - 35 (12)
Chromium	9 - 27 (17)	2 - 90 (16)	3 - 63 (16)	5 - 56 (25)	1 - 36 (6)	2 - 45 (15)	3 - 6 (4)	3 - 19 (9)
Vanadium	158 - 184 (171)	4 - 67 (25)	7 - 300 (105)	9 - 9 (9)	3 - 20 (13)	21 - 40 (30)		(32)
Barium	8 - 192 (63)	13 - 262 (90)	3 - 232 (60)	10 - 174 (67)	2 - 100 (27)	6 - 163 (47)	10 - 113 (41)	4 - 41 (17)
Strontium	81 - 1000 (342)	78 - 5000 (540)	115 - 5500 (697)	21 - 2720 (652)	3 - 334 (68)	34 - 560 (153)	30 - 331 (152)	14 - 341 (81)

[a]Range of values is given (minimum - maximum); mean is given in parentheses.

Source: Modified from Kopp and Kroner 1967.

Table 6.45. Listing of inorganic pollutants in fresh water

Element	Symbol	Element	Symbol
Aluminum[a]	Al	Bromine	Br
Aluminum ion	Al^{3+}	Bromide	Br^-
Antimony	Sb	Bromate	BrO_3^-
Stibous	Sb^{3+}	Hypobromate	BrO^-
Stibic	Sb^{5+}	Bromaurate	$AuBr_4^-$
Triethylantimony	$Sb(C_2H_5)_3$	Bromoplatinate	$PtBr_6^{2-}$
		Bromoselenate	$SeBr_6^{2-}$
Argon	A	Bromostannate	$SnBr_6^{2-}$
Arsenic[a]	As	Liquid bromine	Br_2
Arsenous	As^{3+}	Cadmium	Cd
Arsenic	As^{5+}	Cadmium ion	Cd^{2+}
Arsenide	As^{3-}		
Metaarsenite	AsO_2^-	Calcium	Ca
Arsenite	AsO_3	Calcium ion	Ca^{2+}
Metaarsenate	AsO_3^-	Carbon	C
Arsenate	AsO_4^{3-}	Carbonate	CO_3^{2-}
Cacodylate	$As(CH_3)_2^{2+}$	Bicarbonate	HCO_3^-
Methylarsenate	$HAsCH_3$	Carbon monoxide	CO
Phenylarsenate	$C_6H_5AsO_3H^-$	Carbon dioxide	CO_2
Arsine	H_3As	Carbonyl sulfide	COS
Trimethylarsine	$As(CH_3)_3$	Carbon disulfide	CS_2
Tetramethyldiarsyl	$As_2(CH_3)_4$	Cerium	Ce
Barium[a]	Ba	Cerrous	Ce^{3+}
Barium ion	Ba^{2+}	Cerric	Ce^{4+}
Beryllium	Be	Cesium	Cs
Beryllium ion	Be^{2+}	Cesium ion	Cs^+
Bismuth	Bi	Chlorine	Cl
Bismuthous	Bi^{2+}	Chloride	Cl^-
Bismuthic	Bi^{3+}	Hypochlorite	ClO^-
		Chlorate	ClO_3^-
Boron	B	Perchlorate	ClO_4^-
Boron ion	B^{3+}	Chloraurate	$AuCl_4^-$
Metaborate	BO_2^-	Chloroplatinate	$PtCl_6^{2-}$
Peroxyborate	BO_3^-	Chlorostannate	$SnCl_6^{2-}$
Borate	BO_3^{3-}	Chloroplumbate	$PbCl_6^{2-}$
Diborate	$B_2O_5^{4-}$	Chlorine gas	Cl_2
Tetraborate	$B_4O_7^{2-}$	Chromium[a]	Cr
Pentaborate	$B_5O_8^-$	Chromous	Cr^{2+}
Borotungstate	$BW_{12}O_{40}^{5-}$	Chromic	Cr^{3+}
Organoborate	RBO_2H	Chromate	CrO_4^{2-}
Trimethylboron	$B(CH_3)_3$	Peroxychromate	CrO_8^{3-}
Decaborane	$B_{10}H_{14}$		

Table 6.45 (continued)

Element	Symbol	Element	Symbol
Dichromate	$Cr_2O_7^{2-}$	Periodate	IO_4^-
Cobalt[a]	Co	Paraperiodate	HIO_5^{2-}
Cobaltous	Co^{2+}	Iodine solid	I_2
Cobaltic	Co^{3+}	Iridium	Ir
Copper[a]	Cu	Iron[a]	Fe
Cuprous	Cu^+	Ferrous	Fe^{2+}
Cupric	Cu^{2+}	Ferric	Fe^{3+}
Dysprosium	Dy	Ferrocyanide	$Fe(CN)_6^{3-}$
Erbium	Er	Ferricyanide	$Fe(CN)_6^{4-}$
Europium	Eu	Krypton	Kr
Fluorine	F	Lanthanum	La
Fluoride	F^-	Lead	Pb
Fluoborate	BF_4^-	Plumbic	Pb^{2+}
Hexafluophosphate	PF_6^-	Plumbate	PbO_4^{4-}
Fluorophosphate	PO_3F^{2-}	Lithium	Li
Difluorophosphate	$PO_2F_2^-$	Lithium ion	Li^+
Fluosilicate	SiF_6^{2-}	Lutetium	Lu
Fluoaluminate	AlF_6^{3-}	Magnesium	Mg
Fluosulfonate	SO_3F^-	Magnesium ion	Mg^{2+}
Fluorine gas	F_2	Manganese	Mn
Gadolinium	Gd	Manganous	Mn^{2+}
Gallium	Ga	Manganic	Mn^{3+}
Germanium	Ge	Permanganate	MnO_4^-
Ethylgermanium oxide	$(C_2H_5GeO)_2O$	Manganate	MnO_4^{2-}
Gold	Au	Mercury	Hg
Aurous	Au^+	Mercurous	Hg^+
Auric	Au^{3+}	Mercuric	Hg^{2+}
Hafnium	Hf	Ammonomercuric	$Hg(NH_2)^+$
Helium	He	Organomercuric	RHg^+
Holmium	Ho	Diorganomercury	HgR_2
Hydrogen	H	Molybdenum	Mo
Hydrogen ion	H^+	Molybdic	Mo^{6+}
Hydrogen gas	H_2	Molybdate	MoO_4^{2-}
Deuterium	D	Paramolybdate	$Mo_7O_{24}^{6-}$
Indium	In	Phosphomolybdate	$PMo_{12}O_{40}^{3-}$
Iodine	I	Neodymium	Nd
Iodide	I^-	Neon	Ne
Iodate	IO_3^-	Nickel	Ni
		Nickelous	Ni^{2+}

Table 6.45 (continued)

Element	Symbol	Element	Symbol
Hexaminenickel	$Ni(NH_3)_6^{2-}$	Rhenium	Re
Niobium	Nb	Perrhenate	ReO_4^-
Nitrogen	N	Rhodium	Rh
Azide	N_3^-	Rubidium	Rb
Nitrite	NO_2^-	Rubidium ion	Rb^+
Nitrate	NO_3^-	Ruthenium	Ru
Cyanide	CN^-	Samarium	Sm
Cyanate	OCN^-	Scandium	Sc
Thiocyanate	SCN^-	Selenium[a]	Se
Ammonium	NH_4^+	Selenium ion	Se^{4+}
Carbamate	$NH_2CO_2^-$	Selenide	Se^{2-}
Cyanaurate	$Au(CN)_4^-$	Selenite	SeO_3^{2-}
Cyanocobaltate	$Co(CN)_6^{3-}$	Selenate	SeO_4^{2-}
Nitrogen gas	N_2	Silicon	Si
Nitrous oxide	N_2O	Metasilicate	SiO_3^{2-}
Nitric oxide	NO	Silicate	SiO_4^{4-}
Hydrazine	NH_2NH_2	Disilicate	$Si_2O_5^{2-}$
Hydroxylamine	NH_2OH	Silver	Ag
Osmium	Os	Argentous	Ag^+
Oxygen	O	Argentic	Ag^{2+}
Peroxide	O_2^{2-}	Sodium	Na
Hydroxide	OH^-	Sodium ion	Na^+
Oxygen gas	O_2	Strontium	Sr
Ozone	O_3	Strontium ion	Sr^{2+}
Palladium	Pd	Sulfur	S
Phosphorus[a]	P	Sulfide	S^{2-}
Metaphosphate	PO_3^-	Sulfite	SO_3^{2-}
Hypophosphate	PO_3^{2-}	Sulfate	SO_4^{2-}
Phosphite	PO_3^{3-}	Thiosulfate	$S_2O_3^{2-}$
Phosphate	PO_4^{3-}	Hydrosulfite	$S_2O_4^{2-}$
Pyrophosphate	$P_2O_7^{4-}$	Pyrosulfite	$S_2O_5^{2-}$
Hypophosphite	$H_2PO_2^-$	Dithionate	$S_2O_6^{2-}$
Phosphine	H_3P	Pyrosulfate	$S_2O_7^{2-}$
Platinum	Pt	Peroxydisulfate	$S_2O_8^{2-}$
Platinic	Pt^{4+}	Trithionate	$S_3O_6^{2-}$
Potassium[a]	K	Tetrathionate	$S_4O_6^{2-}$
Potassium ion	K^+	Pentathionate	$S_5O_6^{2-}$
Praseodymium	Pr	Sulfur dioxide	SO_2
Radium	Ra	Hydrogen sulfide	H_2S

Table 6.45 (continued)

Element	Symbol	Element	Symbol
Tantalum	Ta	Titanium	Ti
Tellurium	Te	Tungsten	W
Tellurate	TeO_4^{2-}	Tungsten ion	W^{4+}
Diethyl telluride	$(C_2H_5)_2Te$	Tungstate	WO_4^{2-}
Dimethyl telluronium-dichloride	$C_2H_6Cl_2Te$	Phosphotungstate	$P(W_3O_{10})_4^{3-}$
Terbium	Tb	Uranium	U
		Uranous	U^{4+}
Thallium	Tl	Uranic	U^{6+}
Thallium ion	Tl^+	Vanadium[a]	V
Thorium	Th	Vanadate	VO_3^-
Thulium	Tm	Xenon	Xe
Tin[a]	Sn	Ytterbium	Yb
Stannous	Sn^{2+}	Yttrium	Y
Stannic	Sn^{4+}	Zinc[a]	Zn
Organostannic	RSn^{3+}	Zinc ion	Zn^{2+}
Diorganostannic	R_2Sn^{2+}	Zirconium	Zr

[a]Elements found in the condensate of an Illinois No. 6 coal gasification test.

Source: A. D. Little, Inc. 1971, Table I, pp. 13-17.

Boothe and Knauer (1972) report that arsenic, cobalt, copper, iron, lead, and zinc are definitely concentrated in the feces of a primary consumer, adult kelp crabs, *Pugettia producta*. A comparison of the amount of elements in algae with that in the feces of the kelp crab shows that primary consumers accumulate only a small portion of their elemental content from their food. Frankenberg and Smith (1967) report that fecal material appears to play a significant role in the trophic relationships of coastal benthic communities. Thus, the fecal concentrating mechanism may influence the levels of trace and heavy metals in coprophagous organisms and other members of detrital food webs.

Burton (1966) reports that vanadium is concentrated by marine organisms, particularly by tunicates, who concentrate more than 1000 ppm by uptake from solution or possibly from particulate material. Hem (1972) also reports the uptake of metals (cadmium and zinc) by marine organisms.

6.2.3.2 Interactions in fresh water

A source, a means of transport (wind, water, or ice), and a sink (freshwater lakes, rivers, or oceans) are the three elements in the geochemical cycle of a trace metal (Barnes and Schell 1973). Table 6.49 illustrates the geochemical cycle: (1) a trace element source, mining production; (2) a means of trace element transport, rivers and atmospheric washout; and (3) a trace element sink, oceans.

Table 6.46. Heavy metal content of soils, south of smelter

Distance from smelter (miles)	Horizon[a]	Metal (μg/g dried at 100°C)			
		Pb	Zn	Cu	Cd
0.25	01	2,760	230	34	22
	Oh	15,250	590	240	60
	Soil	217	45	11.3	3.7
0.5	01	2,760	200	34	17
	Oh	16,500	775	270	74
	Soil	53	23	6.2	<0.33
1	01	690	88	13	4.5
	Oh	1,120	149	26	7.2
	Soil	38	28	7.3	0.43
2	01	470	80	11	3.2
	Oh	2,080	266	47	12
	Soil	57	15	3.4	<0.33
3	01	400	68	9.6	2.4
	Oh	810	152	26	4.5
	Soil	31	21	3.3	<0.33
4	01	164	50	6.6	1.2
	Oh	360	92	12	2.8
	Soil	23	26	4.8	<0.33
5	01	203	56	8	1.6
	Oh	360	110	14	2.8
	Soil	17	29	4.3	0.35

[a]01 = slightly decomposed leaf litter; Oh = well decomposed leaf litter;
Soil = <80 mesh fraction, 0 - 1 in. depth.

Source: Bolter et al. 1974, Table 5.

Reprinted by permission of the publisher.

Mechanisms of transport

Troup and Bricker (1975) report that rivers are a dominant source for transport of metals; the metals can be transported in dissolved or particulate form. Dissolved metals are transported organically or inorganically: The organic form involves complexes of trace elements and dissolved organic matter, whereas the inorganic form involves the "free" hydrated ion and inorganic ion pairs (e.g., $Fe(OH)_2$, $FeCl_2$).

Transport of metals in association with particulates involves the following mechanisms:

1. Metals are adsorbed on the surface of particles — most suspended particles in natural waters are negatively charged, and cations are attracted to their surfaces. Additionally, trace elements are adsorbed on ion exchange positions of clay minerals and on hydrous oxides of iron and manganese. Both kinds of adsorbed ions are easily desorbed from particles when the chlorinity of the surrounding solution changes.

2. Metals are coprecipitated with iron and manganese in hydrous oxide coatings; during weathering, iron is released to solution and immediately precipitates (at a pH greater than 4 to 4.5) as a hydrated oxide. This material readily scavenges trace metals by either coprecipitation or sorption. Manganese in solution commonly coprecipitates with iron. These fine-grained high-surface-area hydrous oxides coat detrital particles

Table 6.47. Geometric mean compositions, and geometric deviations, of samples of soils and other surficial materials in the conterminous United States[a]

Element	The conterminous United States N ≈ 863		Western United States (west of 97th meridian) N ≈ 492		Eastern United States (east of 97th meridian) N ≈ 371	
	Geometric mean (ppm)	Geometric deviation	Geometric mean (ppm)	Geometric deviation	Geometric mean (ppm)	Geometric deviation
Al	45,000	2.41	54,000	2.02	33,000	2.70
B	26	2.05	22	2.09	32	1.92
Ba	430	2.06	560	1.80	300	2.19
Be	0.6	2.49	0.6	2.47	0.6	2.53
Ca	8,800	3.92	18,000	2.93	3,200	2.87
Ce	75	1.67	74	1.64	78	1.70
Co	7	2.21	8	2.01	7	2.55
Cr	37	2.32	38	2.16	36	2.52
Cu	18	2.28	21	2.00	14	2.54
Fe	18,000	2.30	20,000	1.90	15,000	2.76
Ga	14	2.11	18	1.71	10	2.53
K	12,000	2.71	17,000	1.60	7,400	3.56
La	34	1.85	35	1.81	33	1.90
Mg	4,700	3.19	7,800	2.21	2,300	3.39
Mn	340	2.70	389	1.94	285	3.65
Na	4,000	4.11	10,200	1.98	2,600	4.11
Nb	12	1.66	11	1.74	13	1.54
Nd	39	1.72	36	1.81	44	1.61
Ni	14	2.26	16	2.03	13	2.60
P	250	2.74	320	2.33	180	3.03
Pb	16	1.96	18	1.93	14	1.96
Sc	8	1.79	9	1.74	7	1.85
Sr	120	3.39	210	2.12	51	3.56
Ti	2,500	1.87	2,100	1.82	3,000	1.84
V	56	2.16	66	1.91	46	2.41
Y	24	1.77	25	1.66	23	1.93
Yb	3	1.81	3	1.67	3	2.03
Zn	44	1.86	51	1.78	36	1.89
Zr	200	1.90	170	1.78	250	1.95

[a]Too few molybdenum values were available to make a statistical evaluation.

Source: Shacklette et al. 1971, Table 3, p. D7.

Table 6.48. Concentration ratios, based on stable element data,
of edible portions of aquatic and marine species

Element	Freshwater invertebrates	Freshwater fish	Marine invertebrates	Marine fish
Ag	769	2	3,330	3,330
As	333	333	333	333
Be	10	2	200	200
Bi		15		15
Cd	2,000	200	250,000	3,000
Co	200	20	1,000	100
Cr	20	40	2,000	400
Cu	1,000	200	1,670	667
Ge	33	3,300	15,700	3,300
Hg	100,000	1,000	33,300	1,670
Mn	40,000	100	10,000	600
Mo	10	10	10	10
Ni	100	100	250	100
Pb	100	300	1,000	300
Sb	10	1	5	40
Se	167	167	1,000	4,000
Sn	1,000	3,000	1,000	3,000
Te	75	400	100,000	
Tl	15,000	10,000	15,000	10,000
Th	500	30	2,000	10,000
Ra	250	50	100	50
V	3,000	10	50	10
W	10	1,200	30	30
Zn	10,000	1,000	100,000	2,000

Source: Vaughan et al. 1975, Table 21, p. 44.

Reprinted by permission of the publisher.

Table 6.49. Global trace metal production
and potential ocean input

Element	Mining production (10^6 tons/year)	Transport by rivers to oceans (10^6 tons/year)	Atmospheric washout (10^6 tons/year)	Ratio aeolian/ fluvial
Pb	3	0.1	0.3	3.0
Cu	6	0.25	0.2	0.8
V	0.02	0.03	0.02	0.7
Ni	0.5	0.01	0.03	3.0
Cr	2	0.04	0.02	0.5
Sn	0.2	0.002	0.03	15
Cd	0.01	0.0005	0.01	20
Hg	0.009	0.003	0.08	27

Source: Barnes and Schell 1973, Table 1, p. 45.

and are transported with them by rivers. The coatings are stable in highly oxic
surface waters, but are unstable under anoxic conditions and quickly dissolve,
releasing the contained trace metals.

3. Metals are precipitated on the surface of detrital particles; some trace metals can
precipitate in the form of pure phases on surfaces of clay minerals or other detrital
particles at low temperatures and pressures. These precipitates can dissolve under
varying conditions of pH and pE.

4. Metals are bound within the crystalline lattice of sediment particles; these trace
   metals can only be released to solution under harsh chemical conditions rarely
   encountered in natural waters.
5. Metals are bound within organic particulates; they are stable in oxic environments and
   unstable in anoxic waters and sediments. Rate of decomposition and subsequent release
   of trace metals depends on the composition of the organic matter and the intensity of
   bacterial activity.

Troup and Bricker (1975) point out the difference in chemical reactivity of trace elements in
each of the fractions mentioned above; consequently, the elemental behavior in natural waters is
influenced by elemental distribution among these fractions.

Knowledge of the mechanisms of trace metal transport in rivers is necessary for an understanding
of trace metal chemical cycles in nature. Like Troup and Bricker, Gibbs (1973) identifies some
chemical mechanisms to consider when studying the transport of trace metals in rivers: (1) (a)
dissolving of ionic species and inorganic associations, (b) complexing with organic molecules in
solution, (2) adsorption on solids, (3) precipitation and coprecipitation on solids (metallic
coatings), (4) incorporation in solid biological materials, (5) incorporation in crystalline
structures. Figure 6.15 and Table 6.50 show the relative contributions of these five mechanisms
in transport of several trace metals in the Amazon and Yukon Rivers. Incorporation into
crystalline sediment structures is the major transport mechanism for copper, cobalt, and chromium
and the second most important mechanism for iron, manganese, and nickel.

<u>Forms and transformations</u>

In sediments, most heavy metals (mercury, lead, cadmium, zinc, copper, nickel, cobalt, manganese,
chromium, and silver) are chemically bound and require both heat and a low pH to convert them to
soluble form (Hutchinson and Fitchko 1974). Various physical, chemical, and biological factors
(such as the scouring action of water currents, dredging, changes in chemical parameters and in
microbial activity) can bring about the redistribution and partial solution of heavy metals from
sediments. Sediments constitute a major sink for metals, and those brought back into the water
column can enter into food webs. In their study, which was conducted in the Great Lakes,
Hutchinson and Fitchko (1974) list the pathways of heavy metals to the Great Lakes as (1) runoff
of surface waters, (2) aerial fallout [especially zinc and nickel into Lake Michigan (Winchester
and Nifong 1971)], (3) direct contamination, and (4) lotic inputs. Factors responsible for the
concentration of heavy metals in outlet sediments of Great Lake tributaries can be divided into
first- and second-order factors (Hutchinson and Fitchko 1974). First-order factors are (1)
magnitude and proximity of industrial and agricultural sources, (2) geology of area, (3) erosion
in drainage basins, (4) efficacy of pathways, and (5) geomorphological characteristics in the water
course (e.g., the occurrence of swamps and lakes). Second-order factors represent binding and
retention mechanisms such as the quantity of hydrated iron oxides and iron sulfides, the amount
of organic matter, and the percentage of fine sediment (e.g., clay and fine silt). According to
Hutchinson and Fitchko (1974), the heavy metals, due to their different chemistries, are influenced
by first- and second-order factors in varying degrees; however, in heavily polluted areas, first-
order factors become dominant.

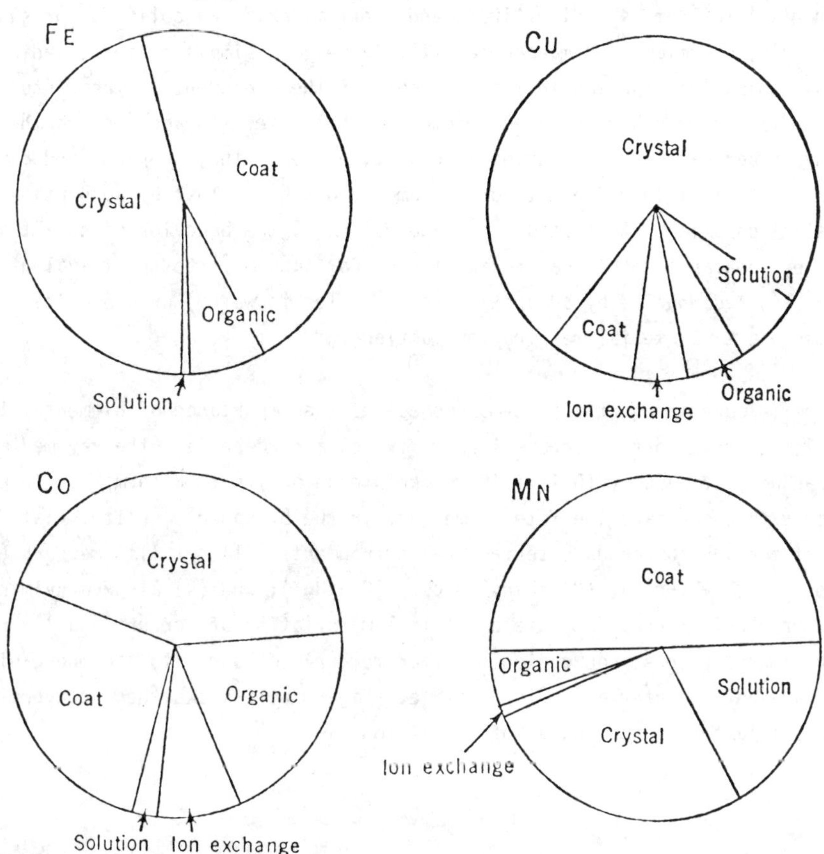

Fig. 6.15. Relative contributions of five mechanisms involved in transporting iron, copper, cobalt, and manganese in the Amazon River. Source: From Gibbs 1973, Fig. 1, p. 71. Reprinted with permission from *Science*. Copyright 1973 by the American Association for the Advancement of Science.

Table 6.50. Relative contributions of five mechanisms in transport of trace metals in the Yukon and Amazon Rivers

Mechanism	Fe	Ni	Co	Cr	Cu	Mn
Amazon River						
In solution and organic complexes	0.7	2.7	1.6	10.4	6.9	17.3
Adsorbed	0.02	2.7	8.0	3.5	4.9	0.7
Precipitated and coprecipitated	47.2	44.1	27.3	2.9	8.1	50
In organic solids	6.5	12.7	19.3	7.6	5.8	4.7
In crystalline sediments	45.5	37.7	43.9	75.6	74.3	27.2
Yukon River						
In solution and organic complexes	0.05	2.2	1.7	12.6	3.3	10.1
Adsorbed	0.01	3.1	4.7	2.3	2.3	0.5
Precipitated and coprecipitated	40.6	47.8	29.2	7.2	3.8	45.7
In organic solids	11.0	16.0	12.9	13.2	3.3	6.6
In crystalline sediments	48.2	31.0	51.4	64.5	87.3	37.1

Source: Gibbs 1973, Fig. 1, p. 71. Reprinted with permission from *Science*. Copyright 1973 by the American Association for the Advancement of Science.

Benes and Steinnes (1974) found that, 2 hr after sampling, the state of trace elements in river water is (1) alkali and alkaline earth elements exist as simple cations; (2) trivalent and tetravalent elements (aluminum, scandium, iron, and thorium) exist as colloidal or suspended particles; (3) cobalt, chromium, and manganese exist largely as simple cations, indicating the importance of truly dissolved species in the transport of these elements. Their studies also indicate that iron may carry other trivalent elements; that water storage (one month) has little effect on the physiochemical state of rubidium, cobalt, barium, zinc, thorium, and barium; and that iron and chromium exist in different forms, some of which are lost by adsorption, leaving the anions of hexavalent chromium. As a result of water storage, the behavior of cobalt and manganese could be changed by spontaneous chemical (oxidation of Co^{2+} and Mn^{2+} to the trivalent state and subsequent hydrolysis accompanied by adsorption to particles in water) or biological (precipitation of carbonates due to bacterial respiration) actions.

The methylation of mercury in aquatic systems emphasizes the importance of elemental transformations in trace element transport. Because lakes, rivers, and oceans are the key media for environmental transport of mercury (D'Itri 1972), knowledge of its transformation in natural waters is necessary to understand the fate of mercury in the biosphere. D'Itri lists the following forms of mercury compounds entering the environment: (1) metallic mercury (Hg^0); (2) inorganic divalent mercury (Hg^{++}); (3) phenylmercury ($C_6H_5Hg^+$); and (4) alkoxyalkylmercury ($CH_3OCH_2CH_2Hg^+$), or alkylmercury. Because biological methylation of mercury usually requires the inorganic divalent mercury ions, conversion of other mercury compounds to the inorganic divalent form must take place before methylation is possible. Figure 6.16 shows these conversions in natural waters and illustrates the routes of conversion.

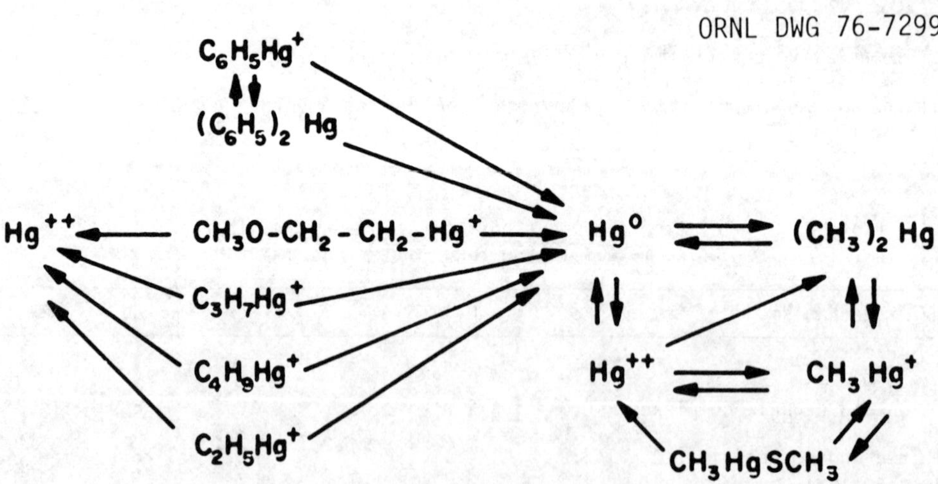

Fig. 6.16. Hypothetical model of the conversion of mercury in the aquatic environment. <u>Source</u>: From D'Itri 1972, Fig. 4, p. 64. From *The environmental mercury problem*, copyright The Chemical Rubber Co., 1972. Used by permission of The Chemical Rubber Company.

Microbial activity is thought to be primarily responsible for the formation of the highly toxic mono- and dimethylmercury compounds:

$$Hg^{++} + 2R-CH_3 = CH_3HgCH_3 = CH_3Hg^+ \tag{4}$$

$$Hg^{++} + R-CH_3 = CH_3H_9^+ \overset{R-CH_3}{=} CH_3HgCH_3 \tag{5}$$

Dimethylmercury is more volatile than the monomethylmercury ion and thus would be released into the atmosphere, decreasing the extent of bioaccumulation by aquatic organisms. In alkaline water having a high pH, the formation of dimethylmercury is favored over monomethylmercury; the reverse is true in acidic water. Therefore, at a low pH (acidic waters), more of the inorganic divalent mercury (Hg^{++}) will be converted to monomethylmercury and remain in the aquatic system (Jernelöv 1972; D'Itri 1972).

In the laboratory, bacterial methylation activity is frequently higher under anaerobic conditions than under aerobic conditions; however, in natural fresh water the reverse seems to be true (Jernelöv 1972). In sediments, where anaerobic conditions are more likely to occur, hydrogen sulfide, ubiquitous in the natural environment under anaerobic conditions, combines with mercury to form insoluble mercuric sulfide (Jernelöv 1972; Vostal 1972). Thus, under aerobic conditions in fresh water, methylation rates are likely to be higher than under anaerobic conditions, where the formation of mercuric sulfide binds mercury and hinders methylation.

Chappell and coworkers at the University of Colorado have extensively studied the transport of molybdenum in the environment. Molybdenum, described as being relatively mobile as compared with other heavy metals (Chappell 1973), is an important coal conversion catalyst. Lake cycling of molybdenum is probably both physical and biological; it is possibly taken up by algae, where it seems to be essential for nitrogen fixation (Chappell 1973). In waters having a pH of 8 to 8.5, molybdenum exists in solution as the dissolved molybdate anion (MoO_4^{2-}), but at a pH of 5 or 6, the ion $HMoO_4^-$ becomes dominant; in very acidic (low-pH) water containing large amounts of ferric ions, a significant amount of molybdenum is transported as ions adsorbed on colloidal particles of iron hydroxide. With some exceptions, molybdenum in the waters of the Southwest is probably carried almost entirely in dissolved form (Chappell 1973). Fletcher and Doyle (1974) found that concentrations of dissolved molybdenum are greatest under alkaline conditions. Also, cobalt, manganese, nickel, and zinc concentrations were found to be greater in sediments and less in neutral or weakly alkaline streams than in acid streams; copper distribution appears to be the element least affected by pH.

Pita and Hyne (1975) report that zinc and lead occur in the 2.0 to 2.9 specific gravity portions of the reservoir they studied; this observation suggests that zinc and lead are deposited as either ions or organometallic complexes adhering to clay surfaces. The adsorption of zinc and lead ions or organometallic complexes to clay mineral surfaces could cause flocculation and deposition in bottom sediments, thus removing zinc and lead from the aqueous system. Pita and Hyne (1975) postulate that reservoirs having large water residency times could act as sinks for certain heavy metals.

Perhac (1972) took samples from four streams in northeast Tennessee and found heavy metals partitioned between a dissolved phase, a colloid fraction, and a coarse suspended particulate fraction. The coarse particulate and colloid fractions were separated by centrifugation, whereas the dissolved solids were obtained by evaporation of the water. Although the heavy metals were more concentrated in the colloids (Table 6.51) relative to the coarse suspended material, by far the greatest total quantity of each element occurred in the dissolved state (Table 6.52). Only a few percent of the elements were associated with coarse particulates, and less than 1% with colloids. However, despite their low total metal content, colloids are capable of carrying large amounts of metal on a weight basis. Therefore, if sufficient colloids are present, these can carry considerable quantities of an otherwise immobile element (one with low water solubility).

Table 6.51. Metal content of coarse and colloidal
particulates (ppm metal in solid)

Element	Sample			
	1	2	3	4
	Coarse			
Cd	15	21	24	13
Co	40	45	61	72
Cu	85	153	647	9120
Fe	1.46%	2.36%	2.39%	2.39%
Mn	1115	1270	2375	1750
Ni	46	73	69	71
Pb	124	213	653	123
Zn	228	2480	823	256
	Colloidal			
Cd	227	167	69	1615
Co	182	288	167	159
Cu	2520	4750	1575	4650
Fe	2.38%	2.53%	2.96%	2.39%
Mn	469	940	2945	468
Ni	222	321	394	481
Pb	62	2820	2820	<850
Zn	<50	1840	1360	875

Source: Perhac 1972, Tables 3 and 4, pp. 182 and 183. Reprinted by permission of the publisher.

Bourg and Filby (1974) report on the binding and retention mechanism of fine sediment; their efforts were directed toward study of the adsorption of zinc by clays found in bottom sediments of streams and lakes. These researchers indicate that an ion exchange mechanism rather than physical adsorption must be taking place due to the rapidity of the adsorption reaction. There are two possible explanations for zinc adsorption: (1) Zn^{2+} ions are adsorbed between the layers of clay minerals to neutralize the excess negative charge created because of the isomorphous substitution of Al^{3+} for Si^{4+} or by the substitution of Mg^{2+} for Al^{3+} (ion exchange); or (2) Zn^{2+} ions may diffuse into the clay mineral lattice. Zinc adsorption may occur either through the first explanation or possibly through a combination of explanations (1) and (2). Bourg and Filby (1972) found that, whereas Zn^{2+} between layers may reach a saturation value, diffusion of Zn^{2+} into the mineral lattice results in no observable maximum value for adsorption of zinc on clays.

Turekian and Scott (1967) report the concentrations of trace elements in suspended material in streams. They cite Kennedy (1965) as stating that the cation exchange capacity of the clay fraction from streams of the eastern United States is 14 to 28 mg per 100 g, whereas that from streams of central and west central United States is 25 to 65 mg per 100 g. As listed in Table 6.53, Turekian and Scott (1967) show that, even though the cation exchange capacities of the

Table 6.52. Percentage of element occurring in
dissolved and particulate solids

Element	Sample			
	1	2	3	4
Cadmium				
Dissolved solid	95.4	95.3	95.8	85.1
Coarse particulate	3.9	4.2	3.5	8.9
Colloid	0.8	0.5	0.7	6.0
Cobalt				
Dissolved solid	95.9	93.2	95.9	82.3
Coarse particulate	3.9	6.2	3.5	17.5
Colloid	0.2	0.6	0.7	0.2
Copper				
Dissolved solid	95.0	94.4	90.4	93.0
Coarse particulate	3.6	3.8	8.2	5.8
Colloid	1.4	1.8	1.4	1.2
Iron				
Dissolved solid	18.8	12.5	26.9	20.4
Coarse particulate	79.5	86.0	67.1	75.5
Colloid	1.7	1.4	6.0	4.1
Manganese				
Dissolved solid	23.2	41.7	18.5	10.4
Coarse particulate	76.4	57.6	74.8	89.5
Colloid	0.4	0.7	6.7	0.1
Nickel				
Dissolved solid	96.5	95.2	96.6	84.7
Coarse particulate	3.4	4.5	2.4	14.8
Colloid	0.2	0.3	1.0	0.5
Lead				
Dissolved solid	95.0	90.9	87.9	89.5
Coarse particulate	5.0	7.6	9.3	10.1
Colloid	Trace	1.6	2.9	0.4
Zinc				
Dissolved solid	85.0	53.3	91.9	81.1
Coarse particulate	15.0	46.2	7.3	18.5
Colloid	Trace	0.5	0.9	0.4

Source: Perhac 1972, Table 5, p. 185.

Reprinted by permission of the publisher.

eastern U.S. stream sediments are the lowest of the rivers sampled, these streams have a higher
concentration of most elements than do the western U.S. streams (Table 6.53). They believe these
data indicate that the major trace element transport mechanism cannot be simple cation exchange.
These researchers found a strong correlation of the trace elements with manganese, possibly indicating (1) coprecipitation of trace elements with iron and manganese oxides in weathering profiles, (2) association with organic material from soils, or (3) industrial contamination. They
also note the excellent ability of trace element scavenging shown by fresh precipitates of iron
oxide (hydroxide). They conclude by suggesting that the trace element transport by some rivers
such as the Susquehanna (Table 6.54) may be large enough for economic trace element recovery.

In another study of the Susquehanna River, McDuffie et al. (1976) conclude that a large fraction
of the trace metal load of a river is carried by suspended solids, and that at times of low
river flow, bottom sediments are good scavengers of trace metals released from urban pollution
sources.

Table 6.53. Trace element composition of suspended material in rivers

River and state	Suspended load (mg/liter)	Cr	Ag	Mo	Ni	Co	Mn
		Element (ppm)					
Brazos, Tex.	954	100	0.4	11	30	20	690
Colorado, Tex.	150	82	0.6	10	40	17	780
Red, La.	436	37	0.3	5	6	7	320
Mississippi, Ark.	185	150	0.7	18	100	33	2,300
Tombigbee, Ala.	25	220	1.0	22	200	31	5,900
Alabama, Ala.	54	150	4.0	19	100	34	3,700
Chattahoochie, Ga.	71	190	7.0	20	100	35	2,400
Flint, Ga.	12	210	1.0	28	100	39	5,100
Savannah, S. C.	30	460	2.0	35	250	36	4,400
Wateree, S. C.	37	200	1.5	24	100	34	7,000
Pee Dee, S. C.	188	150	0.4	15	100	23	1,300
Cape Fear, N. C.	61	130	0.7	16	70	21	1,700
Neuse, N. C.	36	380	4.0	22	70	30	3,000
Roanoke, N. C.	33	240	4.9	21	100	45	7,900
James, Va.	41	290	7.0	29	300	60	15,000
Rappahannock, Va.	28	140	1.0	31	80	46	2,200
Potomac, Va.	34	170	1.5	23	400	94	7,700
Susquehanna, Pa.	54	290	15.0	32	>1000	>500	12,000
Rhone, France Avignon, June 1966	296	150	0.7	14	60	29	820
Rio Maipo, Chile Puente Alto, S of Santiago, September 1966	41	68	1.0	44	40	76	2,400
Shales[a]		90	0.1	2.6	68	19	850

[a]Turekian and Wedepohl 1961 (as cited in Turekian and Scott 1967).

Source: Turekian and Scott 1967, Table I, p. 942. Reprinted with permission from *Environ. Sci. Technol.* Copyright by the American Chemical Society.

Table 6.54. Trace element transport by the Susquehanna River

Element	Tons transported per year[a]
Chromium	870
Silver	45
Molybdenum	97
Nickel	3,000
Cobalt	1,500
Manganese	120,000

[a]Assuming 300×10^3 tons of suspended sediment per year (measured for the Juniata River).

Source: Turekian and Scott 1967, Table II, p. 942. Reprinted with permission from *Environ. Sci. Technol.* Copyright by the American Chemical Society.

6.2.3.3 Interactions in estuaries

Rivers are the dominant source of trace metals entering estuaries although some estuarine trace metal input comes from rainfall and particulate fallout (Troup and Bricker 1975). Troup and Bricker (1975) characterize trace metals as either conservative or nonconservative, depending on their behavior in estuaries. They cite major ions as being conservative because they exhibit simple physical mixing between river and seawater. Most trace elements such as iron are nonconservative, meaning that physical mixing alone cannot account for their estuarine distribution.

The amount and form of trace elements transported to the ocean are affected by estuarine processes; five major categories of reactions influence trace metal behavior in estuaries (Troup and Bricker 1975).

1. <u>Flocculation and sedimentation</u>. These processes are encouraged by the increased salinity of estuarine waters. As a result of the salt wedge type of circulation commonly observed (Fig. 6.17), sedimentation is concentrated in the upper reaches of estuaries. Suspended sediment discharged by rivers is carried toward the ocean by fresh water at the surface of the estuary. As the solid particles settle, they fall into the salt wedge layer moving toward the river mouth. The particles continue to settle, and sedimentation is concentrated in the low-velocity region near the tip of the salt wedge.

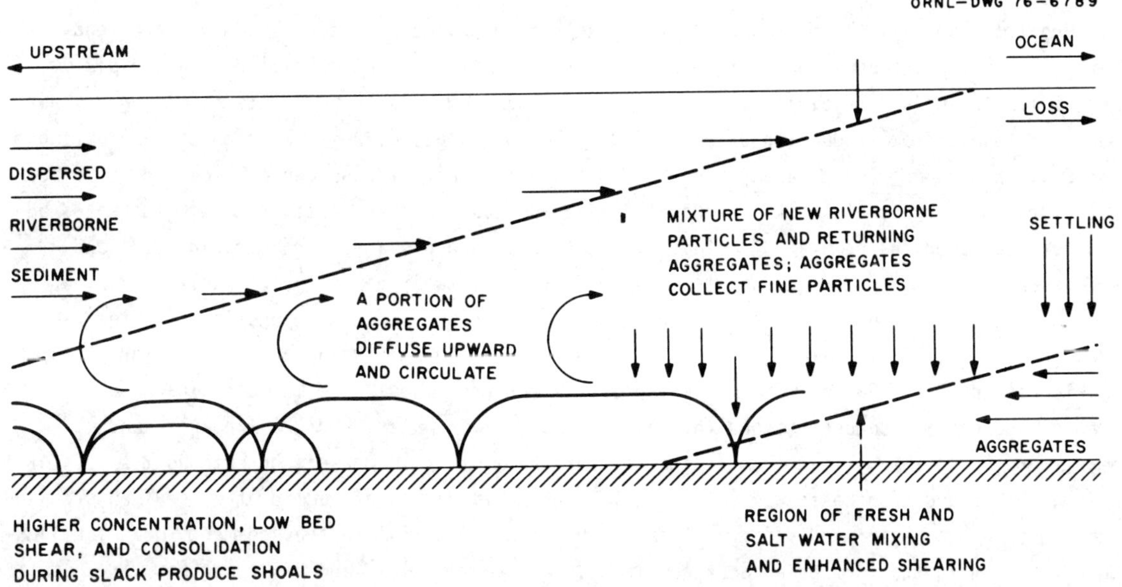

Fig. 6.17. Diagrammatic sketch of an estuarine salt wedge. <u>Source</u>: From Troup and Bricker 1975, Fig. 11.

2. <u>Mineral-water interaction</u>. The rapid changes in salinity, pH, and oxygen content of estuarine waters can affect the precipitation or dissolution of minerals (for instance, the dissolution of the hydrous oxide coatings under low oxygen conditions in water or sediments).

3. <u>Adsorption-desorption</u>. Due to the high concentrations of sodium and manganese in seawater, ion exchange reactions occur rapidly in estuarine waters as sediment particles adsorb the seawater cations and desorb trace elements.

4. <u>Diagenesis and remobilization of trace metals in sediments</u>. The decreased pH and strong pH gradient in sediment interstitial waters brought about by decomposition of organic matter significantly affects the solubility of trace element solid phases and ion exchange equilibria. Trace metal concentrations in sediment pore waters are

typically orders of magnitude greater than in the overlying water, and the flux of trace elements across the sediment-water interface may constitute an important mass transfer process in estuaries.

5. <u>Biological processes</u>. Interactions with aquatic organisms may strongly influence the behavior of trace elements during their passage through estuaries.

The mobilization of metals is related to the existence of tidal effects in fluvial and marine regions, especially in river estuaries. From lab experiments, de Groot and Allersma (1973) found that the decomposition products of organic matter form soluble organometallic complexes with metals from suspended matter and that the degree of mobilization depends on the stability constant of the metal with the organic ligand. For some metals, copper for example, mobilization can be promoted by the ability to form stable complexes with both positively and negatively charged organic ligands.

Helz, Huggett, and Hill (1975) report that concentrations of copper, zinc, cadmium, and lead entering the estuarine environment from a wastewater treatment plant outfall decrease rapidly downstream. An active immobilization process, which was removing these metals to sediments, was thought to result from production of a pH of 9 by intense algal blooms. With an average effluent pH of 6.9, the sudden rise to a daytime pH of 9 could induce rapid inorganic deposition of trace metal hydroxides, carbonates, or phosphates. Helz and his coworkers felt that this increased pH may have encouraged the removal of metals by sorption on organic and inorganic particles and that, because the algae can concentrate many trace metals, their death could result in further sediment concentration of trace metals. Manganese and cadmium concentrations in the water decreased at the outfall, but began to increase downstream toward the ocean. They offered two possible reasons: (1) Saline water entering Back River from Chesapeake Bay contained higher manganese and cadmium concentrations than did the Back River water, and (2) manganese and cadmium were being remobilized from sediments near the mouth. The researchers thought that, for cadmium, the second hypothesis was valid, that is, remobilization of cadmium from sediments caused by cation exchange, aging and decay of metal-rich organic material in the sediment, or solubilization of cadmium in the form of aqueous complexes. Apparently, Helz and his coworkers were not sure about manganese, as no explanation was stated concerning the increase in manganese concentration near the river's mouth.

6.2.3.4 <u>Interactions in oceans</u>

Spencer, Brewer, and Sachs (1972) studied trace element composition in the Black Sea. They list four processes that influence the vertical profiles of the elements studied: (1) the presence of detrital silicates, (2) precipitation as sulfides in deep water, (3) coprecipitation with or adsorption by MnO_2 that is precipitated just above the oxygen zero boundary (due to the upward flux of dissolved Mn(II) by advection and diffusion), (4) concentration by marine organisms in the surface waters. Their results indicate that the vertical zinc profile was found to be dominated by process (2) and the manganese profile by process (3); the vertical profiles of antimony, cobalt, and mercury were influenced by processes (1), (2), and (3) respectively.

Table 6.55 shows the concentrations of trace metals in the ocean, with the sample locations given as degrees of longitude and latitude (Fabricand et al. 1962). Table 6.56 indicates the concentrations of some heavy metals in the coastal seawater (filtrate) of the British Isles, and Table

Table 6.55. Concentrations of trace metals in the ocean (µg-atom/liter)

Position	Depth (ft)	Metal					Date sample taken
		Cu	Fe	Mn	Ni	Zn	
13°17´N, 59°20´W	1600	0.025	0.10	0.054	0.0063	0.0097	5/21/61
13°17´N, 59°20´W	1100	0.025	0.12	0.054	0.0054	0.0095	5/21/61
13°17´N, 59°20´W	700	0.028	0.10	0.076	0.0051	0.010	5/21/61
14°21´N, 58°03´W	1000	0.033	0.12	0.069	0.0054	0.010	5/28/61
14°21´N, 58°03´W	500	0.025	0.11	0.063	0.0054	0.011	5/28/61
14°21´N, 58°03´W	300	0.030	0.10	0.066	0.0037	0.011	5/28/61
15°41´N, 61°08´W	1000	0.033	0.13	0.070	0.0027	0.0094	6/4/61
15°41´N, 61°08´W	500	0.030	0.10	0.063	0.0045	0.010	6/4/61
16°13´N, 62°28´W	1000	0.028	0.12	0.073	0.0044	0.0099	6/4/61
16°13´N, 62°28´W	500	0.030	0.11	0.070	0.0024	0.011	6/4/61
17°08´N, 64°02´W	1000	0.033	0.12	0.063	0.0031	0.0099	6/4/61
17°08´N, 64°02´W	500	0.030	0.12	0.063	0.0022	0.011	6/4/61
18°34´N, 65°40´W	1000	0.033	0.11	0.063	0.0038	0.010	6/5/61
18°34´N, 65°40´W	500	0.025	0.10	0.063	0.0034	0.011	6/5/61
19°40´N, 64°16´W	1000	0.035	0.11	0.069	0.0029	0.0091	6/8/61
19°40´N, 64°16´W	500	0.025	0.12	0.073	0.0044	0.011	6/8/61
21°18´N, 60°44´W	1000	0.028	0.11	0.061	0.0038	0.010	6/9/61
21°18´N, 60°44´W	500	0.025	0.11	0.063	0.0031	0.0099	6/9/61
23°31´N, 55°20´W	1000	0.025	0.11	0.080	0.0060	0.0099	6/10/61
23°31´N, 55°20´W	500	0.025	0.11	0.061	0.0031	0.010	6/10/61
25°05´N, 49°15´W	1000	0.018	0.11	0.073	0.0038	0.0099	6/11/61
25°05´N, 49°15´W	500	0.022	0.11	0.066	0.0032	0.011	6/11/61
28°16´N, 56°46´W	1000	0.024	0.10	0.066	0.0034	0.0095	6/16/61
28°16´N, 56°46´W	500	0.028	0.10	0.054	0.0041	0.011	6/16/61
31°47´N, 65°11´W	1000	0.028	0.11	0.066	0.0027	0.0092	6/17/61
31°47´N, 65°11´W	500	0.028	0.12	0.073	0.0041	0.0092	6/17/61
32°40´N, 117°26´W	Surface	0.031	0.11	0.072	0.0049	0.0092	7/21/61
Spread		0.018	0.10	0.054	0.0022	0.0091	
		-0.035	-0.13	-0.080	-0.0063	-0.011	

Source: Fabricand et al. 1962, Table 1, p. 1026.

6.57 shows some heavy metal concentrations in shoreline seawater (filtrate) from the Irish Sea (Preston et al. 1972).

Another possible transport process is suggested by Martin (1970), who indicates that trace metals may be transported by moulted copepod exoskeletons. This process occurs by adsorption of the metals to the exoskeletons for each adult copepod. Martin postulates that there are more trace elements adsorbed on exoskeletons at great depths (≥100 m) than on surface inhabitants. Food is scarcer than at the surface; because moulting is food-dependent, the exoskeletons would remain attached to the copepod for longer periods, thus enhancing adsorption of trace elements. Martin (1970) believes that, if this hypothesis is correct, moulted copepod exoskeletons may be important in biogeochemical cycles.

Kharkar, Turekian, and Bertine (1968), in order to estimate the flux of trace elements flowing from streams to the ocean, identify the forms in which trace elements are transported to the ocean: (1) in solution, (2) in the lattice of minerals of the suspended load, (3) adsorbed on the minerals of the suspended load, and (4) associated with organic material. Of these transport mechanisms, similar to those cited by Troup and Bricker (1975) and Gibbs (1973), the last two are critical to trace metal interaction in oceans; the adsorption-deposition behavior of trace

Table 6.56. Concentrations of selected heavy metals in British Isles coastal seawater (filtrate)
(μg/liter)

Element	Area					
	1	2	3	4	4A	5
1969						
Zinc						
Geometric mean	6.4	7.1	6.5			4.6
Range	4.3 - 9.0	3.0 - 20.0	1.8 - 14.8			
Number of observations	4	16	22			1
Iron						
Geometric mean	0.7	0.1	0.5			2.0
Range	0.2 - 2.3	0.1 - 0.6	0.1 - 1.1			
Number of observations	4	16	22			1
Manganese						
Geometric mean	0.4	0.5	0.6			0.4
Range	0.3 - 0.5	0.3 - 0.8	0.2 - 2.2			
Number of observations	4	16	22			1
Copper						
Geometric mean	1.3	1.3	1.3			1.4
Range	1.1 - 1.6	0.7 - 2.8	0.8 - 1.9			
Number of observations	4	16	22			1
Nickel						
Geometric mean	0.7	0.5	0.7			0.5
Range	0.5 - 1.3	0.3 - 1.2	0.3 - 1.4			
Number of observations	4	16	22			1
1970						
Zinc						
Geometric mean	2.0	3.0	4.2	2.0	3.0	3.1
Range	1.2 - 3.8	0.8 - 9.0	2.3 - 7.5	1.3 - 3.4	1.4 - 7.0	1.5 - 6.9
Number of observations	4	10	21	8	5	4
Iron						
Geometric mean	0.3	0.09	0.18			0.06
Range	0.06 - 1.3	0.03 - 0.6	0.06 - 1.9			0.06 - 1.5
Number of observations	4	10	21	8	5	4
Manganese						
Geometric mean	0.32	0.53	1.95	0.18	0.06	0.34
Range	0.10 - 0.49	0.15 - 2.6	0.22 - 14.6	0.02 - 0.49	0.03 - 0.09	0.24 - 0.54
Number of observations	4	10	21	8	5	4
Copper						
Geometric mean	0.46	0.59	0.66	0.34	0.26	0.48
Range	0.23 - 1.29	0.18 - 3.75	0.28 - 0.98	0.19 - 0.62	0.05 - 0.80	0.24 - 1.70
Number of observations	4	10	21	8	5	4
Nickel						
Geometric mean	0.38	0.38	0.71	0.53	0.43	0.31
Range	0.22 - 0.95	0.22 - 0.55	0.32 - 22.9	0.36 - 0.79	0.29 - 0.66	0.16 - 0.51
Number of observations	4	10	21	8	5	4
Lead						
Geometric mean	0.17	0.19	0.11	<0.05		0.21
Range	<0.05 - 1.1	<0.05 - 1.2	<0.05 - 1.0			<0.05 - 0.8
Number of observations	4	10	21	8	5	4
Cadmium						
Geometric mean	0.06	0.11	0.04	<0.01	0.04	0.41
Range	<0.01 - 0.38	<0.01 - 0.52	<0.01 - 0.62	<0.01 - 0.18	<0.01 - 0.41	0.29 - 0.60
Number of observations	4	10	21	8	5	4

Source: Preston et al. 1972, Table 3, p. 75.

Reprinted by permission of the publisher.

elements determines their fate in seawater. Kharkar, Turekian, and Bertine (1968) report that
adsorbed trace elements are always released, although to a greater or lesser extent, on contact
with seawater; this is due to displacement of trace metal ions by magnesium and sodium ions
present in seawater in high concentration.

Table 6.58 shows the results of a laboratory study in which cobalt, silver, molybdenum, and
selenium were adsorbed on three types of clay (montmorillonite, illite, and kaolinite) as well
as on ferric oxide, manganese oxide, hydrated ferric oxide, and peat in distilled water; the four
elements were desorbed in natural seawater. Under conditions resembling those in streams,
montmorillonite and illite adsorb about 90% of cobalt, 20 to 30% of silver, and 30 to 50% of
selenium present in solution, whereas ferric hydroxide adsorbs 95% of cobalt, 60% of silver, and
90% of selenium. Between 30 and 70% of the adsorbed metals were released in seawater. It was
found that Cr^{6+} and Mo^{6+} were not adsorbed to any significant degree by any of the solids
(Kharkar, Turekian, and Bertine 1968).

Table 6.57. Heavy metal concentrations in
shoreline seawater (filtrate) from the
Irish Sea, 1970 (µg/liter)

Element	Area	
	2	3
Zinc		
Geometric mean	6.6	6.8
Range	4.9-11.1	3.8-49.1
Number of observations	9	11
Iron		
Geometric mean	6.1	11.9
Range	2.5-23.9	5.3-24.7
Number of observations	9	11
Manganese		
Geometric mean	2.2	6.1
Range	0.7-10.8	2.5-25.5
Number of observations	9	11
Copper		
Geometric mean	1.4	1.7
Range	0.9-2.7	1.1-3.1
Number of observations	9	11
Nickel		
Geometric mean	1.4	2.6
Range	0.9-3.1	1.3-9.8
Number of observations	9	11
Lead		
Geometric mean	1.6	1.3
Range	0.9-2.9	0.6-2.4
Number of observations	7	8
Silver		
Geometric mean	0.08	0.04
Range	0.02-0.24	0.03-0.16
Number of observations	9	11
Cadmium		
Geometric mean	0.41	0.46
Range	0.03-1.43	0.15-1.14
Number of observations	9	11

Source: Preston et al. 1972, Table 4, p. 76.
Reprinted by permission of the publisher.

Using known data on the clay content of streams and results of their adsorption-desorption experiments, Kharkar, Turekian, and Bertine (1968) estimate the total flux of soluble trace elements supplied to the oceans by streams (Table 6.59). As a rough estimate, the authors conclude that the worldwide contribution of desorbable cobalt to the oceans from streams would be about twice the dissolved load. For silver and selenium, the desorbable component adds an additional 10% to the dissolved load.

Elements solubilized in the oceans would be expected to remain as such for a long time. Average residence times for some elements are given in Table 6.60.

Manganese, one of the trace elements in coal, was found in the condensate from a coal gasification test at about 40 ppb, a concentration similar to that of zinc, arsenic, and nickel (Table

Table 6.58. Adsorption and desorption experiments

Concentration (mg/liter)	Percent Co-60 adsorbed			Percent Ag-110 adsorbed		Percent Mo adsorbed from distilled water	Percent Se-75 adsorbed		
	From distilled water	Desorbed in seawater	From seawater	From distilled water	Desorbed in seawater		From distilled water	Desorbed in seawater	From seawater
				Montmorillonite 22A, Miss.					
200							53.7	27.6	4.8
250						7.5			
1,000	86.3	67.7	24.2						
10,000	89.4	99.5		33	26.5				
				60	30.7				
				59.5	35.0				
100,000	96.6	65.9							
				Illite 35, Fithian, Ill.					
200							32.0	55.5	12.3
1,000	90.2	48.1		13.9					
				28.0					
10,000	93.9	27.0		74.6	25.9				
				83.0					
				Kaolinite 7, S.C.					
200							13.4	70.5	5.0
1,000	No detectable adsorption			13.0					
10,000	41.5	56.5		13.0					
				Ferric oxide					
200							86.5	28.0	40.0
1,000	No detectable adsorption			5.0					
				Manganese dioxide					
200							85.0	60.0	11.5
250						5.1			
1,000	25.0	No detectable desorption		81.4	96.1				
				89.8					
				Ferric hydroxide, freshly precipitated at pH ~7.0					
200							90.0	18.3	68.0
250						10.1			
1,000	86.1	15.0	94.6	59.0	17.0				
	99.2	12.6							
				Peat, oyster pond, Mass.					
200							14.6	60.0	No adsorption
250									
1,000						6.6			

Source: Kharkar, Turekian, and Bertine 1967, Table 4, p. 292.

Reprinted by permission of the publisher.

Table 6.59. Dissolved, desorbable, and total supply of cobalt, silver, and selenium to the oceans (μg/liter)[a]

River	Cobalt			Silver			Selenium		
	Dissolved in stream	Adsorbed in stream and desorbed in contact with seawater	Total soluble load	Dissolved in stream	Adsorbed in stream and desorbed in contact with seawater	Total soluble load	Dissolved in stream	Adsorbed in stream and desorbed in contact with seawater	Total soluble load
Mississippi	0.11	0.41	0.52	0.24	0.02	0.26	0.114	0.03	0.14
Susquehanna	0.35	0.71	1.06	0.37	0.02	0.39	0.325	0.96	0.39
Mad	0.14	0.59	0.73	0.26	0.02	0.28			
Neuse	0.073	0.080	0.16	0.36	0.02	0.38			
Rhone	0.10	0.37	0.47	0.38	0.03	0.41	0.154	0.03	0.18
Amazon	0.115	0.17	0.29	0.23	0.01	0.24	0.21	0.04	0.25

[a]The following average adsorption and desorption data were used:

Clay	Cobalt		Silver		Selenium	
	Adsorbed (%)	Desorbed (%)	Adsorbed (%)	Desorbed (%)	Adsorbed (%)	Desorbed (%)
Montmorillonite	90.0	70.0	30.0	30.0	50.0	50.0
Illite	90.0	20.0	20.0	20.0	30.0	50.0
Kaolinite (upper limit)	40.0	60.0	10.0		10.0	70.0

Source: Kharkar, Turekian, and Bertine 1967, Table 6, p. 294.

Reprinted by permission of the publisher.

Table 6.60. Mean oceanic residence times of
the soluble trace element supply

Element	Mean residence time (years)
Silver	40,000
Antimony	7,000
Chromium	6,000
Cobalt	30,000
Rubidium	4,000,000
Cesium	600,000
Selenium	20,000
Molybdenum	200,000

Source: Kharkar, Turekian, and Bertine 1967,
Table 7, p. 295.

Reprinted by permission of the publisher.

4.28). In seawater, soluble divalent manganese from land is oxidized to insoluble tetravalent manganese, partly by bacteria (*Leptothrix, Crenothrix*) and partly by chemical reactions (Schroeder 1971*a*). With plankton as a nucleus, manganese and iron form nodules with thin structures, incorporating other elements such as nickel (2%), cobalt (0.2 to 2%), and copper (2 to 3%) as the nodule enlarges. An estimated 1.5 trillion tons of nodules, containing 16 to 50% manganese, lie on the bottom of the Pacific Ocean, representing a possible source for mining (Schroeder 1971*a*). On the basis of research on the marine geochemistry of manganese in the Gulf of Aden, Glasby, Tooms, and Cann (1971) conclude:

1. Two types of manganese-iron encrustations are forming in the Gulf of Aden.
2. The trace-element-enriched massive manganese-iron encrustations forming on weathering basalts on the ridges away from the median valley are deposited from seawater.
3. The trace-element-impoverished, friable manganese deposits found on the underside of basaltic flows are being precipitated on the basalt from interstitial waters at or near the sediment surface.
4. Under reducing conditions within the organically rich sediments, manganese, nickel, and molybdenum are being leached from the sediments, and the interstitial water is being enriched with these elements.

Schroeder (1971) notes that the bioaccumulation of manganese by marine plants is 100 to 100,000 times that in seawater; bioaccumulation by simple marine animals is 100 to 10,000 times that in seawater; and bioaccumulation by fish is 100 times that in seawater. Bioaccumulation by man is 3 to 4 times the amount present in food. In contrast, Hartung (1974), on the basis of studies in the lower Mississippi River, indicates that aquatic organisms do not readily concentrate manganese. This contradiction may be due to the difference in the study medium: Schroeder's study was in seawater, and Hartung's in fresh or brackish water.

6.2.4 Fate in soils

Trace element entry into the soil environment can be accomplished by one or more of the following mechanisms: (1) weathering by parent rock, (2) aerial fallout (dry deposition), (3) rainout (e.g., deposition of fly ash by scavenging), (4) solid waste application (e.g., sewage sludge and landfill), (5) chemical fertilizer application, (6) irrigation water, and (7) trace-element-containing insecticides (e.g., arsenicals). The fate of trace elements in soil is determined by such parameters as persistence, mobility, and soil pH.

6.2.4.1 Persistence

Purves (1972) wrote about the contamination of soils by several elements. Copper, lead, and zinc contamination appears to be largely irreversible. Because plants take up only a small percentage of the available trace metals, the potential for retention of trace metals by soils is great (Mitchell 1945). Long-term leaching removes only a small fraction of the soil content of such trace metals as zinc and copper. Roberts and Goodman (1973) determined the persistence of heavy metals in soils. On bare ground the half-life of zinc, copper, and nickel was 2 to 5 years, but as grass grew, the retention time increased the half-life to 10 to 30 years for the same soil. Wentink and Etzel (1972) recognize the importance of soil ion exchange capacity in removing metals from wastewater applied to soil. They report that, as the soils become finer (sands → loams → clay loams), the ion exchange capacity increases because the organic content usually increases as the clay content increases. Wentink and Etzel (1972) also report that:

1. Soil completely removed Cu^{2+} from a solution dripping onto a soil column after an initial conditioning period.
2. Zn^{2+} was removed with an efficiency as great as 99.7%, even with applications as high as 300 mg/liter.
3. Cr^{3+}, in concentrations up to 300 mg/liter, was completely removed from solution to the soil. Leachability of the exchanged ions was essentially nonexistent with either tap water or deionized water.

This low leachability seems to confirm the results of Purves (1972) and of Roberts and Goodman (1973): Metals have long retention times in soils.

6.2.4.2 Mobility

Tiller (1958) found that, during weathering, elements in the basalt-covering soils of South Australia exhibit the following order of mobility: calcium, phosphorus > sodium, magnesium, cobalt, zinc > copper, manganese > nickel > potassium, vanadium, gallium, molybdenum, iron, aluminum, titanium. The mobility of zinc seems to be determined by the soil type (Barrows, Neff, and Gammon 1960). Zinc movement through four soils for one year was found to be in the following order: Lakeland fine sand > Red Bay fine sandy loams > Savannah fine sandy loams > Arredondo loamy fine sand. Zinc movement can be explained not on the basis of any one chemical or physical property of soils, but on a combination of such factors as organic matter, phosphates, pH, and clay content. Similar findings were expressed by Singh (1974), who reports the movement of zinc in soils to be dependent on cation exchange, water flow rate, pH, initial zinc content of soils, and migration of other soil fractions.

In flooded soil or sediment in which reducing conditions are intense and S^{2-} is present, the metal ions Fe^{2+}, Mn^{2+}, Hg^{2+}, and Cu^{2+} may form insoluble and relatively stable sulfides (Connell and Patrick 1968, as cited in Engler and Patrick 1975). Engler and Patrick (1975) state that the precipitation of certain metal ions as sulfides is an important mechanism for regulating the solution concentration of both S^{2-} and the metal ions, either of which can be toxic. If a flooded soil or sediment is drained and subsequently aerated, the sulfides may be transformed to the more soluble oxidized salts of the metal. The solubilities of the metal sulfides, in decreasing order of solubility, are MnS, FeS, ZnS, CuS, and HgS (Engler and Patrick 1975).

6.2.4.3 Effect of pH

Nichol, Horsnail, and Webb (1967) studied manganese and iron found in the Moorland soils of England and Wales. When compared with freely drained soils, the Moorland soils were found to have a low pH and Eh. The researchers conclude that, in the acidic waterlogged Moorland soils, manganese and iron readily pass into solution and migrate with circulating groundwaters, but are precipitated when they enter the drainage channel due to the increase of pH and Eh. They also note the role of freshly precipitated iron and manganese oxides as adsorbing and scavenging agents for various trace metals. The equilibrium of the various forms of manganese in soils has been expressed as water-soluble $Mn^{++} \rightleftharpoons$ exchangeable $Mn^{++} \rightleftharpoons$ colloidal hydrated (reducible) $MnO_{1-2} \rightleftharpoons$ relatively inert MnO_2. This process involves the conversion of bivalent to tetravalent manganese due to an increase in pH and Eh and possibly due to bacteria as well (Sherman, McHarque, and Hodgkiss 1942).

Schroeder (1971a) reports the occurrence of manganese in soils in both trivalent and tetravalent forms. In acid soils (pH 6), trivalent manganese is reduced to the soluble divalent form, and at pH 6.5 autooxidation takes place to the tetravalent state, which is hydrated and can be reduced both bacterially and chemically. However, in alkaline soils trivalent manganese can be oxidized to the tetravalent form (MnO_2 or pyrolusite), which is insoluble in natural solutions. Bohn and Aba-Husayn (1971) studied the uptake in alkaline soils (pH 7.5 to 9.5) of manganese, iron, copper, and zinc by sacaton grass (*Sporobulus wrightii*). They found that the concentrations of copper and zinc in the grass decreased with increasing soil pH, whereas the concentrations of manganese and iron increased with increasing pH. This was thought to be due possibly to reducing conditions that result from the low air and water permeability of the sodic soils during the growing season. Acidic soils apparently yield nickel to plants more so than do alkaline soils (Painter, Toth, and Bear 1953). At a concentration of about 2 ppm, alfalfa from New Jersey soils having a pH greater than 6.8 removed a total of only 4.8 to 5.6 g of nickel from the soil, whereas cuttings from soils having a pH of 5.9 to 6.8 removed 5.2 to 7.2 g of nickel. The highest plant concentration of nickel was 7.8 ppm in ragweed grown in soil having a pH of 4.4 (Painter, Toth, and Bear 1953). Chappell (1973) explored the relationship of metals with soil pH: Molybdenum is generally less available to plants at low pH, high sulfate, and low soil moisture levels.

6.2.5 Fate in air

Dispersion, chemical reactivity, and fallout are important phenomena for evaluation of trace element fate. Dispersion and fallout rates determine where and to what extent deposition will occur, whereas the chemical reactivity of trace elements determines the form in which an element may exist, thus influencing the method of deposition.

6.2.5.1 Dispersion

The most important meteorological factors influencing the concentration of trace metal pollutants in air are the stability of the atmosphere, the height of the mixing column (or mixing depth), and the wind speed and direction (Kleinman, Kneip, and Eisenbud 1973). These factors determine emission dispersal rate. In New York City in 1972, for example, a maximum dispersion occurred in the spring months, resulting in a minimum concentration of airborne trace metal pollutants; a minimum dispersion occurred in the summer, resulting in a maximum pollution concentration. The trace metals were collected as airborne particulate matter. The monthly variation in the concentration of total suspended particles is shown in Fig. 6.18.

6.2.5.2 Chemical reactions

Although the fate of air pollutants is poorly understood (Bohn 1972), Tullar and Suffet (1975) present a rather detailed picture of the fate of vanadium in an urban air shed. The lower vanadium oxides oxidize to V_2O_5 through the following two-step reaction:

$$V_2O_3 + 1/2 \ O_2 = V_2O_4 \ , \tag{6}$$

$$V_2O_4 + 1/2 \ O_2 = V_2O_5 \ . \tag{7}$$

Thus, the major part of atmospheric vanadium oxides consists of V_2O_5 in particulate combination with fly ash and alumina silica catalyst emissions. These researchers believe that, because V_2O_5 and SO_2 coexist in stack emissions and because the catalytic vapor phase reaction,

$$SO_2 + 1/2 \ O_2 \ \underset{\text{3rd body}}{\overset{V_2O_5}{=\!=\!=}} \ SO_3 \ , \tag{8}$$

is a commercial way of making sulfuric acid (Shreve 1945, as cited by Tullar and Suffet 1975), there is the possibility of such a reaction occurring in stack emissions containing V_2O_5. Creation of an appreciable quantity of SO_3 could set the stage for formulation of particulate H_2SO_4 aerosols, which would increase the rate of deposition for vanadium in the atmosphere. Deposition is the only discernible sink for vanadium. Using Stokes' law, Tullar and Suffet (1975) calculated a settling velocity of 12 m per day for a 2-μ-diam particle having a specific gravity of 1.0 (roughly corresponds to the specific gravity of V_2O_5-H_2SO_4 aerosols) falling in 70°F air, with velocity varying as the square of the diameter and as the first power of density. They report that most V_2O_5 emissions are removed by deposition of large aerosols near industrial sources, leaving the fine particles, which can represent a serious inhalation threat (Lee et al. 1972), to account for the low observed atmospheric concentration. According to Tullar and Suffet (1975), V_2O_5 in a raindrop may be ionized to $H_3V_2O_7^-$ when the pH is between 4 and 4.5 or to $H_2VO_4^-$ at a higher pH; and because V_2O_5 can be dissolved and ionized, it could run off when it reaches land and enter natural waters. Figure 6.19, which diagrams the fate of vanadium in air, indicates a residence time of 1 day or less.

Schroeder (1971a) studied the fate of manganese in the atmosphere. Aerosols of manganese dioxide react with SO_2 to form manganous sulfate or dithionate and with nitrogen dioxide to form manganous nitrate. The manganous sulfate catalyzes the oxidation of sulfur dioxide (SO_2) to sulfuric acid (H_2SO_4); the reaction rate is slow in sunlight, fairly rapid in fog, and dependent on the concentration of manganese.

ORNL DWG 76-6570

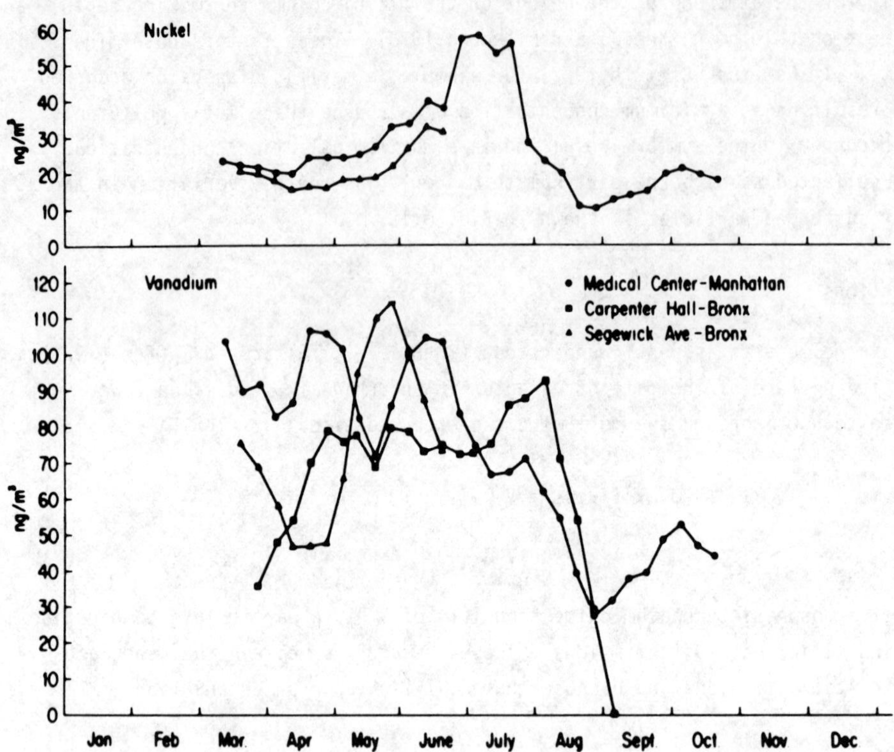

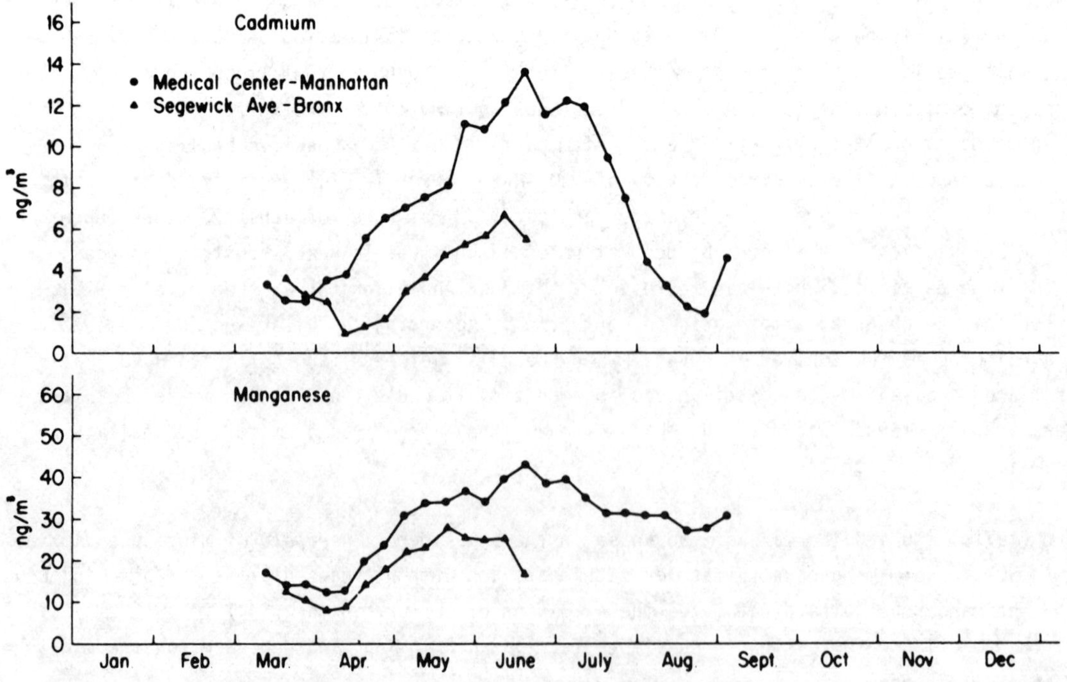

Fig. 6.18. Concentrations of particulate matter, iron, copper, zinc, nickel, vanadium, cadmium, and manganese in New York City air. Source: From Kleinman, Kneip, and Eisenbud 1973, Figs. 2, 3, 5, and 6, pp. 163-65.

Reprinted by permission of the publisher.

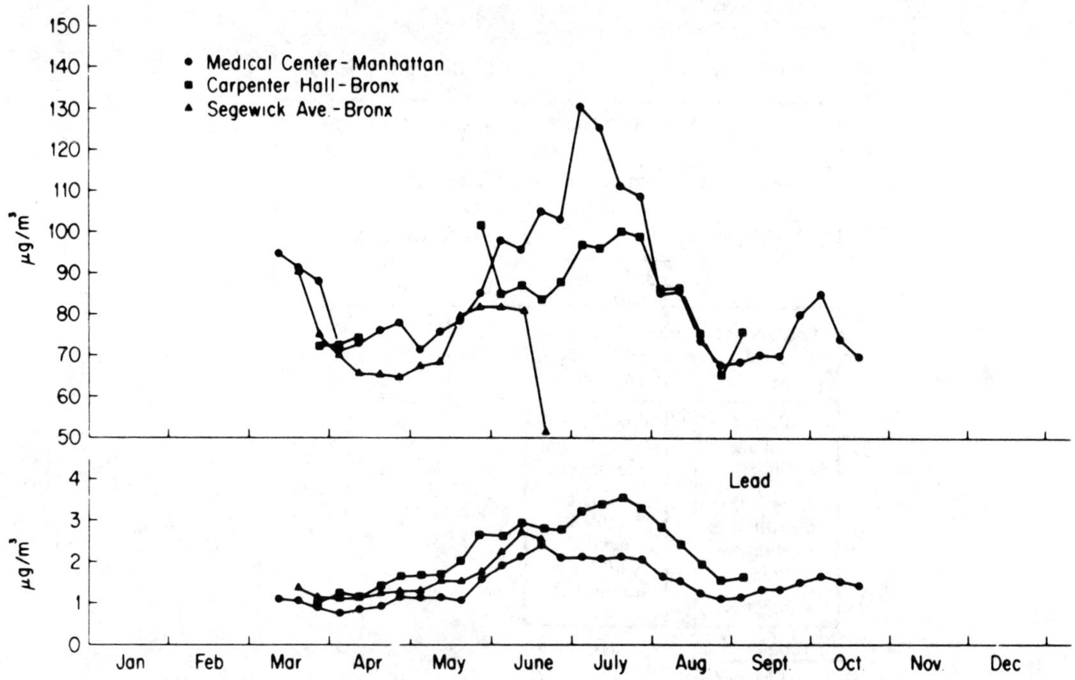

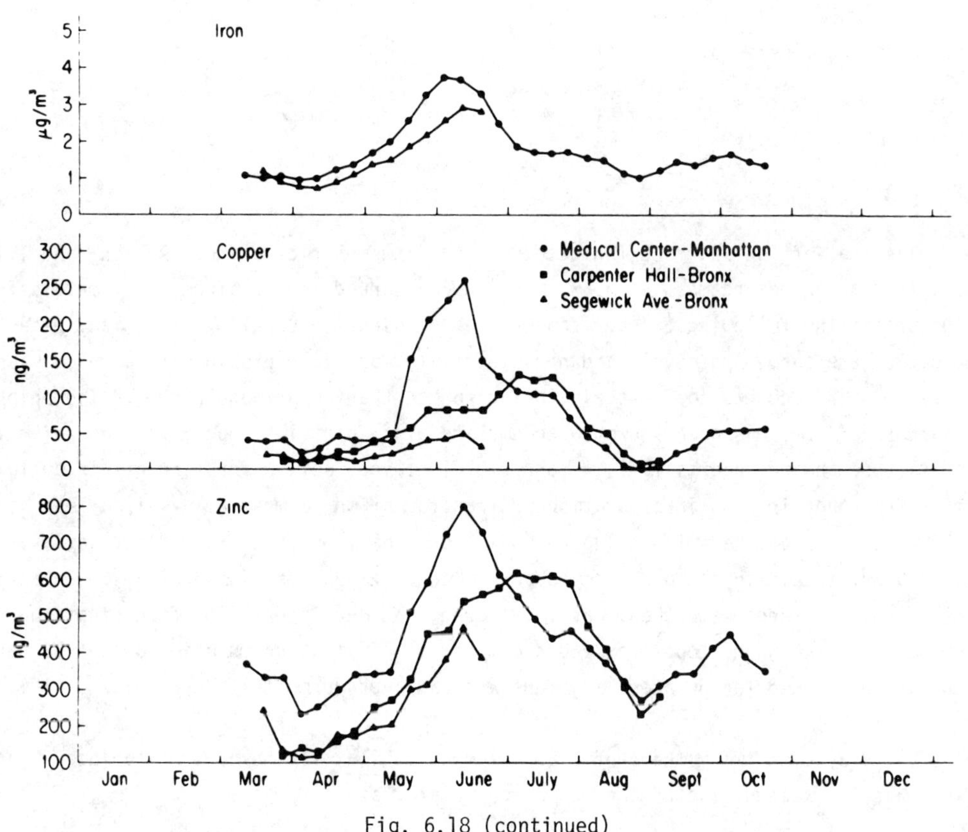

Fig. 6.18 (continued)

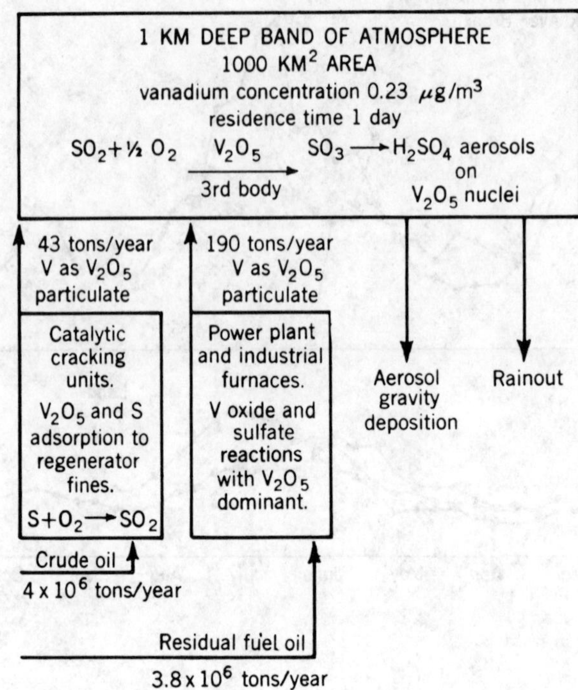

Fig. 6.19. Flow diagram of the fate of vanadium in air. <u>Source:</u> From Tullar and Suffet 1975, Fig. 3, p. 286.
Reprinted by permission of the publisher.

6.2.5.3 Fallout

Plants and soils are, of course, good indicators of the atmospheric makeup. Ruhling and Tyler (1970) found that the woodland moss *Hylocomium splendens* sorbed and retained heavy metals from dilute solutions in the following order: copper, lead > nickel > cobalt > zinc, manganese. Copper and nickel are sorbed passively and nonselectively, and zinc probably is sorbed by an active process, which accounts for less zinc being sorbed than copper and nickel. The epiphytic moss *Hypnum cupressiforme* has been shown to accumulate zinc, nickel, copper, and magnesium via air transport (Goodman and Roberts 1971). Bohn (1972) defines soil absorption of air pollutants as adsorption from the air plus incorporation or transformation, almost always in a less harmful form, by all components of the soil including soil microorganisms. He states that soils are able to inactivate heavy metals in the order of copper > lead > zinc > cadmium > nickel. In a six-month study of Walker Branch watershed, Harris, Andren, and Harrison (1975) found that the input of zinc, chromium, nickel, and copper exceeded output, signifying accumulation on the landscape. In this study, zinc showed the greatest movement with an average monthly input flux of 18%.

Peirson et al. (1973) cite two methods of trace element fallout to the earth's surface. The first is known as the washout factor and can be calculated as

$$W = \frac{\text{concentration in rain } (\mu g\ kg^{-1})}{\text{concentration in air } (\mu g\ kg^{-1})} \tag{9}$$

Particles greater than 10 μm settle promptly by sedimentation; smaller particles are dispersed by updrafts and turbulence. Mean monthly concentrations of 29 trace elements, including the washout factor, W, and dry deposition, D, for a relatively unpolluted site near Lake Windermere, England, in 1971 are given in Table 6.61. Although the average dry deposition usually amounted to no more than about 15% of the total, on occasion and for short periods, it accounted for more than half the total (Peirson et al. 1973).

The variation of the washout factor according to element is plotted in Fig. 6.20. For all the elements except sodium, chlorine, and selenium, the washout factors lie between 500 and 2000 despite a 10,000-fold variation in concentration. According to Peirson et al. (1973), trace elements having higher values of W have higher concentrations at rain-forming altitudes (3 km) than do trace elements having lower values of W. Therefore, a high W is associated with a greater dispersion in altitude (transport from a distant source), whereas a low W is associated with a more local source of trace elements. An analysis of the washout factor vs the dry deposition velocity allowed the authors to determine which of three possible sources (sea, soil, or industry) was the origin of each of the trace elements (Fig. 6.21). Part of the consideration was particle size; it was suggested that elements associated with larger aerosol particles, such as iron and aluminum, had wind-eroded soil dust as their origin, whereas industrial processes yielded smaller particles.

Hardy (1976) agrees with Peirson et al. (1973) in reporting on the size of trace-element-associated particles. Hardy found that, in urban areas, high concentrations of trace elements in large-particle-size fractions (>4 μm) were attributed to soil dust and that trace element emissions from combustion sources might be enhanced in the smaller respirable-particle-size range (0.25 to 4 μm). These small-particle-associated elements include sulfur, lead, bromine, chlorine, nickel, zinc, manganese, and vanadium.

6.3 SULFUR DIOXIDE

Sulfur dioxide as a pollutant has been the subject of more investigation than any other single pollutant (Rasmussen, Kabel, and Taheri 1974). This section will only briefly treat some of the research devoted to SO_2. Kellogg et al. (1972) cite anthropogenic SO_2 emissions as 100 million tons per year. Natural sources, mainly volcanoes, contribute about 1.5 million tons per year.

6.3.1 Physical transport

Smith and Jeffrey (1975) report that SO_2 is capable of traveling considerable distances in the atmosphere. The role of wind in SO_2 dispersal is noted by Padmanabhamurty (1975) and by Zeedijk and Velds (1973). The conditions favoring long-distance transport of SO_2 are little disturbance of the atmosphere and simultaneous fast movement of the air by wind. Worley (1971) (cited in Zeedijk and Velds 1973) lists four interactions that SO_2 may undergo during its long-distance transport: (1) reaction with other gases, with or without photochemical effects, (2) absorption by water droplets, (3) chemisorption onto dry metal oxides, or (4) absorption by aerosols containing solutions of metal salts that convert SO_2 to a sulfate.

Table 6.61. Annual summary of Wraymires results (January - December 1971)[a]

| | Concentration | | | | | | |
Element	In rain (μg/liter)	In air (ng/kg)	Washout factor, W	Total deposition/month, T (μg/cm)	Dry deposition/month, D (μg/cm)	D/T ratio	Dry deposition velocity, V_g (cm/sec)
Na	2,300	800	2,900	27.5	1.35	0.05	0.53
Al	160	260	620	1.90	0.36	0.19	0.44
Cl	4,100	1,750	2,300	48.5	2.50	0.05	0.44
Ca	<1,100	520	<2,100	<13	0.95	0.06	0.57
Sc	0.042	0.059	710	0.0005	0.00019	0.38	1.01
V	4.1	8.0	510	0.049	0.0075	0.15	0.29
Cr	2.9	2.4	1,200	0.035	0.0022	0.06	0.29
Mn	8.1	10.6	760	0.096	0.008	0.08	0.24
Fe	200	230	890	2.35	0.41	0.17	0.57
Co	0.25	0.32	780	0.003	0.0003	0.10	0.29
Ni[b]	<6	5.5	<1,100	<0.07	<0.02		<1.1
Cu[b]	23	26	880	0.27	0.006	0.03	0.07
Zn	85	80	1,050	1.00	0.051	0.05	0.20
As	1.6	2.5	640	0.019	0.0019	0.10	0.24
Se[b]	0.34	0.90	380	0.004	0.00026	0.07	0.09
Br	17.0	27.1	620	0.20	0.017	0.09	0.20
Rb[b]	0.67	1.04	640	0.008	<0.00012	<0.15	<0.04
Cd[b]	<17.7	<3		<0.21	<0.02	<0.1	
In[b]	<0.59	0.3	<2,000	<0.007	<0.0007		<0.73
Sb	1.8	1.6	1,150	0.021	0.00027	0.01	0.05
I	<2.5	<2.2		<0.03	<0.0004		
Cs[b]	0.17	0.16	1,050	0.002	0.00012	0.06	0.23
La[b]	<0.3	<0.5		<0.004	<0.0005	<0.3	
Ce[b]	0.42	0.30	1,400	0.005	<0.0005	<0.1	<0.52
W[b]	0.35	<0.86		<0.004	<0.0009		
Au	0.01	<0.01		0.00011	<0.00001		
Hg	<0.2	<0.17		<0.002	<0.0002	<0.1	
Pb[b]	39	87	450	0.46	<0.05	<0.1	<0.18
Th[b]	0.12	0.13	920	0.001	0.0002	0.10	0.48

[a]Annual rainfall 142.7 cm.
[b]Incomplete data.

Source: Peirson et al. 1973, Table 1, p. 253.

Reprinted by permission of the publisher.

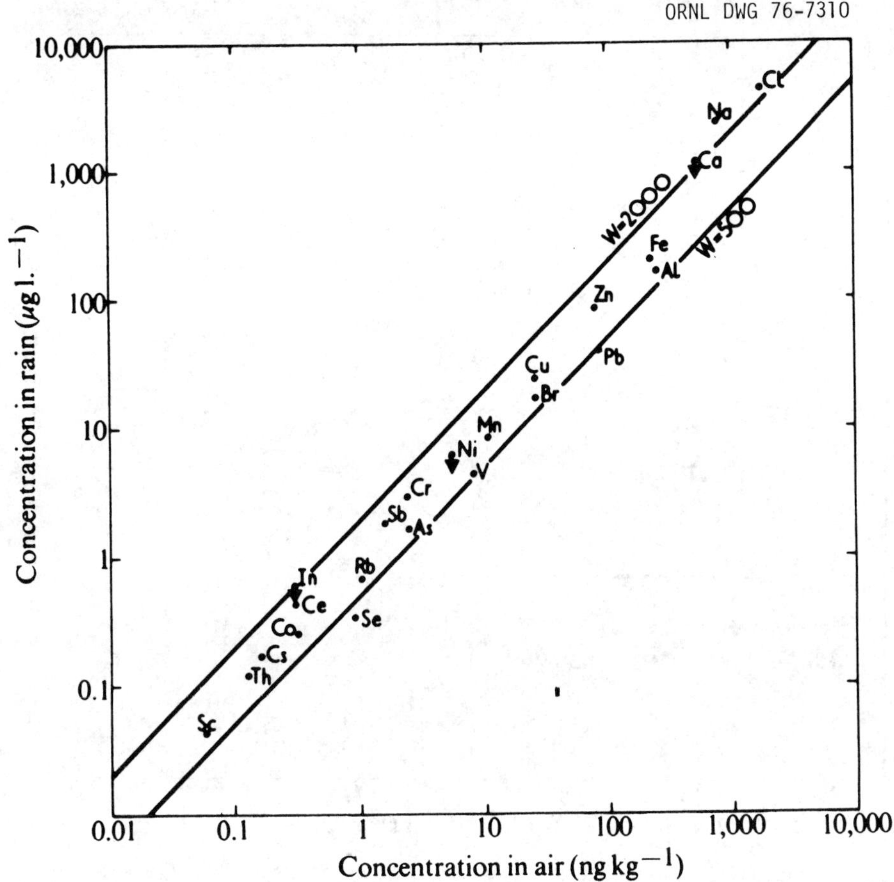

Fig. 6.20. Elemental concentrations in air and rain at Wraymires (1971). <u>Source</u>: From Peirson et al. 1973, Fig. 3, p. 254.
Reprinted by permission of the publisher.

6.3.2 Atmospheric chemical reactions

Two chemical reactions that SO_2 may enter into in the atmosphere are the gas phase oxidation of SO_2 to form sulfate (Harker 1975) and the oxidation of SO_2 in water droplets in the presence of ammonia to form ammonium sulfate (McKay 1971). In fog or cloud droplets, SO_2 becomes sulfurous acid (H_2SO_3), which is rapidly oxidized to H_2SO_4 by dissolved O_2 (Kellogg et al. 1972). Kellogg et al. (1972) report a three-body reaction of SO_2 with atomic oxygen,

$$SO_2 + O + M = SO_3 + M \quad , \tag{10}$$

where M is a molecule of O_2 or N_2. The third-body M, by carrying off excess reaction energy, prevents prompt reversal of the reaction. Kellogg et al. (1972) schematically summarize the chemical processes involving environmental sulfur and indicate the mean lifetime of the sulfur species (Fig. 6.22).

6.3.3 Deposition

The major portion of atmospheric SO_2 is probably removed by rainout and washout (Rasmussen, Kabel, and Taheri 1974). Rainout is the removal of SO_2 within clouds, and washout is removal by

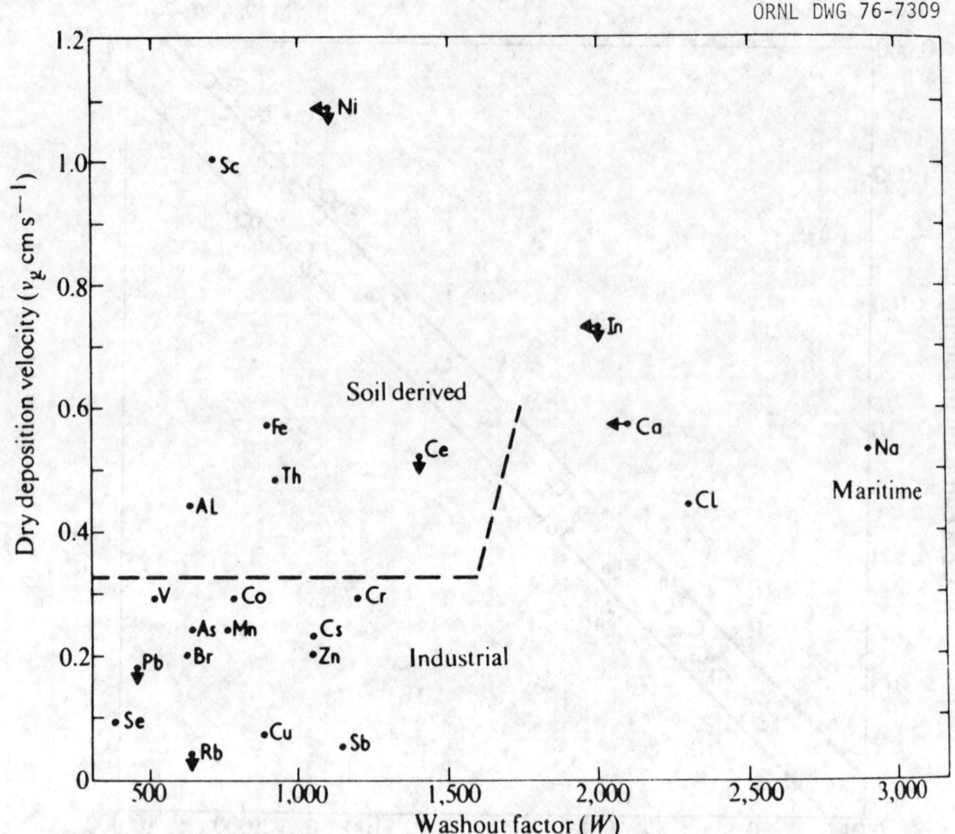

Fig. 6.21. Dry deposition velocity and washout factors at Wraymires (1971). <u>Source</u>: From Peirson et al. 1973, Fig. 4, p. 254.

Reprinted by permission of the publisher.

precipitation below the clouds (Kellogg et al. 1972). The deposition rate of SO_2 to land and sea from the atmosphere is calculated as $V_d \times C$, where C = low-level concentrations and V_d = a velocity of deposition (1 cm/sec on land and 0.8 cm/sec over the sea) (Smith and Jeffrey 1975).

6.3.3.1 <u>On soil</u>

Yee, Bohn, and Miyamoto (1975) report on the ability of calcareous soils to sorb SO_2. Sorption of SO_2 from dry air by dry soil reached saturation in 10 to 15 min, regardless of soil type, soil aggregate size, or SO_2 concentration, whereas in moist soils, SO_2 sorption from a dry gas stream was complete after 15 to 20 min. The absorption capacities for both dry and initially moist soils in dry gas streams fit the equation

$$W_c = \alpha C^\beta \, , \tag{11}$$

where

W_c = sorption capacity,
C = concentration of SO_2 in dry air streams,
α, β = constants determined by plotting W_c vs C on a log-log scale.

When moisture was added to the gas stream, Yee and his coworkers found that the soil absorption capacity for SO_2 increased.

ORNL DWG 76-7303

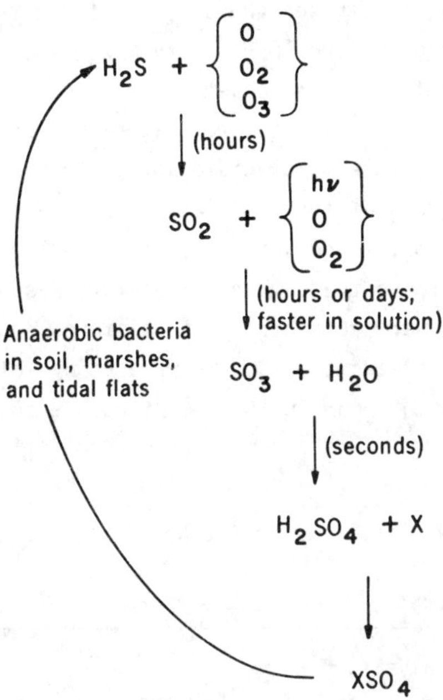

Fig. 6.22. Schematic representation of the chemical processes involving environmental sulfur. *Source*: From Kellogg et al. 1972, Fig. 1, p. 588. Reprinted with permission from *Science*. Copyright 1972 by the American Association for the Advancement of Science.

In summary, Yee, Bohn, and Miyamoto (1975) indicate that SO_2 sorption increases with the presence of moisture in either the soil or the air stream; they conclude that the effect of water is probably to hasten SO_2 oxidation and the reactions of its oxidation products with soil bases.

Payrissat and Beilke (1975), having studied the uptake of SO_2 by different European soils, report that uptake is dependent on meteorological and soil parameters. They found that the deposition velocity of SO_2 is dependent on such parameters as wind velocity and on such physicochemical soil surface properties as pH and buffer capacity for water surfaces and pH, moisture content, vegetation type, and effective surface area for land surfaces. In regard to pH, Payrissat and Beilke found that SO_2 uptake increased strongly with increasing pH of the soil between 4.5 and 7.6, corresponding to a tenfold decrease in soil surface resistance (r_{soil}) or to a threefold increase in deposition velocity. Acidic soils absorbed less SO_2 per injection than did alkaline soils, and after a 24-hr recovery period, all soils recovered their initial SO_2 absorption capability. Payrissat and Beilke (1975) explain the decreased SO_2 sorption by soils with continued exposure by suggesting that SO_2 molecules that are deposited on the soil surface might impede the deposition of additional SO_2 from the atmosphere. They note that, when the relative humidity is high, SO_2 oxidation in the surface water film is probably important to the absorption capacity, especially when the soil pH is also high. The subsequent increase in soil resistance is thought to be caused by a fall in pH at the surface due to the increased formation of sulfate from the oxidation of SO_2. They conclude that soil surfaces without vegetation are not perfect sinks for SO_2.

Experiments by Terraglio and Manganelli (1966) indicate that soil moisture has an influence on the adsorption of atmospheric sulfur dioxide by soil. Table 6.62 shows the characteristics of the two New Jersey soils studied by these researchers. They found that when 600 liters of air, containing 6.03 mg SO_2/m^3, was passed through 10 g of dry Nixon soil, 1134 µg of SO_2 were retained. With 2% soil moisture, retention increased to 1908 µg; with 5% soil moisture, retention increased to 2888 µg; and with 10% soil moisture, retention increased to 2909 µg. The retention of SO_2 by dry Lakewood soil (the more acidic soil), when 600 liters of air, containing 7.27 mg SO_2/m^3, was passed through 25 g of the soil, was zero. This result supports the data of Payrissat and Beilke (1975); they, too, found acid soils to have low SO_2 adsorption capabilities. Terraglio and Manganelli, however, found that, when 2% moisture was present in the Lakewood soil, 285 µg of SO_2 were retained; 383 µg were retained at 5% moisture, and 567 µg were retained at 10% moisture. Abeles et al. (1971) states that SO_2 is removed by soil primarily through some chemical reaction (reduction of 100 to 8 ppm in 15 min) with some contribution from microbes (Chap. 7).

Table 6.62. Analysis of two New Jersey soils

Constituent	Lakewood soil	Nixon soil
Typical mechanical analysis, %		
Sand	95	65
Silt	3	20
Clay	2	15
Volatile matter, %	0.96	2.49
Moisture, %	0.053	0.52
Cation exchange capacity, (meq/100 g)	2.5	7.5
Density, g/ml	2.53	2.46
Bulk density, g/ml	1.55	1.39
pH	4.0	5.5

Source: Terraglio and Manganelli 1966, Table 1, p. 784.
Reprinted by permission of the publisher.

6.3.3.2 On vegetation

Knowing that SO_2 may be removed from the atmosphere in rain by oxidation to particulate material and by direct interaction of the gas molecules with the lower boundary of the atmosphere (dry deposition) (Garland et al. 1974), one can logically assume that SO_2 may adsorb onto the surface of vegetation since most of the land area in the world is covered by vegetation of one form or another.

Shepherd (1974) reports that the deposition velocity of SO_2 onto grass in a polluted area is about 8 mm/sec in the summer months and 3 mm/sec in autumn, with surface resistances of about 80 and 300 sec/m respectively. Surface resistance is normally the limiting factor for deposition. Garland et al. (1974) found a mean SO_2 deposition value of 0.0055 m/sec (5.5 mm/sec) onto short grass, and they cite Meetham (1950), who calculated the deposition rate for a hectare as 70 kg of SO_2. Hill (1971) showed that the wind velocity above plants, the height of the plant canopy, and the light intensity all affect the removal rate of SO_2 by plants. Linzon, McIlveen, and Temple (1973) indicate that SO_2 can be injurious to vegetation; Table 6.63 classifies herbaceous and woody plants as sensitive, intermediate, or resistant to SO_2.

Table 6.63. Observed sensitivity of herbaceous and woody plants to SO_2 in vicinity of sulphite pulp and paper plant in Ontario, August 1971

I. Herbaceous plants

A. Sensitive

Alfalfa	*Medicago sativa* L.
Beggars' ticks	*Bidens vulgata* Greene
Bindweed	*Convolvulus arvensis* L.
Burdock	*Arctium minus* (Hill) Bernh.
Carrot	*Daucus Carota* L. var. *sativa* DC.
Chickory	*Cichorium Intybus* L.
Cocklebur	*Xanthium strumarium* L.
Dahlia	*Dahlia variabilis* (Willd) Desf.
Daisy, fleabane	*Erigeron annuus* (L) Pers.
Dandelion	*Taraxacum officinale* Weber
Delphinium	*Delphinium* sp.
Goldenrod	*Solidago* sp.
Hollyhock	*Althaea rosea* Cav.
Lady's-thumb	*Polygonum* sp.
Oxeye daisy	*Chrysanthemum Leucanthemum* L.
Peony	*Paeonia lactiflora* Pall
Prickly lettuce	*Lactuca Scariola* L.
Ragweed	*Ambrosia artemisiifolia* L.
Rhubarb	*Rheum Rhaponticum* L.
Sunflower	*Helianthus annuus* L.
Thistle	*Cirsium* sp.
Tomato	*Lycopersicon esculentum* Mill.
Yarrow	*Achillea Millefolium* L.
Zinnia	*Zinnia elegans* Jacq.

B. Intermediate

Bean, garden	*Phaseolus* sp.
Bouncing Bet	*Saponaria officinalis* L.
Clover	*Trifolium* sp.
Dayflower	*Commelina communis* L.
Geranium, garden	*Pelargonium* sp.
Ground ivy	*Glechoma hederacea* L.
Japanese knotweed	*Polygonum cuspidatum* Sieb. & Zucc.
Lamb's-quarters	*Chenopodium album* L.
Milkweed	*Asclepias syriaca* L.
Oxalis	*Oxalis europaea* Jord.
Pigweed	*Amaranthus retroflexus* L.
Silverweed	*Potentilla anserina* L.

C. Resistant

Birds'-foot trefoil	*Lotus corniculatus* L.
Cyprus spurge	*Euphorbia Cyparissias* L.
Phlox	*Phlox* sp.
Sweet corn	*Zea Mays* L.

II. Woody plants

A. Sensitive

Ash, red	*Fraxinus pennsylvanica* Marsh.
Birch, European	*Betula pendula* Roth
Birch, white	*Betula papyrifera* Marsh.
Birch, yellow	*Betula lutea* Michx.
Cottonwood	*Populus deltoides* Marsh.
Mahonia	*Mahonia* sp.
Maple, Manitoba	*Acer Negundo* L.
Raspberry, red	*Rubus idaeus* L.
Rose, tea	*Rosa* sp.
Virginia creeper	*Parthenocissus quinquefolia* (L.) Planch.

Table 6.63 (continued)

B. Intermediate

Apple	*Pyrus Malus* L.
Basswood	*Tilia americana* L.
Catalpa	*Catalpa speciosa* Warder
Currant, red	*Ribes* sp.
Dogwood, red osier	*Cornus stolonifera* Michx.
Elm, Chinese	*Ulmus parvifolia* Jacq.
Grape, wild	*Vitis riparia* Michx.
Honeysuckle, tatarian	*Lonicera tatarica* L.
Hydrangea	*Hydrangea paniculata* Sieb.
Lilac	*Syringa vulgaris* L.
Mock orange	*Philadelphus coronarius* L.
Mock orange	*Philadelphus virginalis* Rehd.
Mountain ash	*Pyrus (Sorbus) Aucuparia* (L.) Gaertn.
Oak, white	*Quercus alba* L.
Spiraea	*Spiraea Vanhouttei* Zabel.
Spruce, white	*Picea glauca* (Moench.) Voss.
Weigela	*Weigela* sp.

C. Resistant

Forsythia	*Forsythia viridissima* Lindl.
Linden, little-leaved	*Tilia cordata* Mill.
Maple, Norway	*Acer platanoides* L.
Maple, silver	*Acer saccharinum* L.
Maple, sugar	*Acer saccharum* Marsh.
Spruce, blue	*Picea pungens* Engelm.

Source: Linzon, McIlveen, and Temple 1973, Table III, p. 133.
Reprinted by permission of the publisher.

6.3.3.3 On water

The solubility of sulfur dioxide [10.8 g per 100 g H_2O at 20°C and 1 atm (Rasmussen, Kabel, and Taheri 1974)] would indicate that the earth's water bodies are possible SO_2 sinks. Spedding (1972), in studying the SO_2 adsorption by seawater, found a linear relationship between the flow rate of an SO_2-air mixture over seawater and the velocity of deposition (Fig. 6.23); the implication is that the deposition is controlled largely by gas phase diffusion. When seawater was exposed by Spedding (1972) to 1445 µg/m SO_2 for 4 hr, no change in pH was observed, but when Terraglio and Manganelli (1967) exposed distilled water at pH 9 to 5540 µg/m SO_2, the pH dropped to 4 in 1 hr due to the formation of sulfurous acid upon contact of the SO_2 with water. Spedding (1972) believes that this difference illustrates the buffer capacity of the ocean, which, along with its alkaline pH, makes it a potentially important SO_2 sink. The high solubility of SO_2 in water makes rain a very effective means of SO_2 removal from the atmosphere (Terraglio and Manganelli 1967).

In summary, SO_2 is removed by several mechanisms, but mainly by rainout and washout. Both land and water serve as SO_2 sinks although the form depends on the presence of moisture, oxygen, and sulfur-metabolizing microorganisms. Figure 6.24 (Kellogg et al. 1972) illustrates a global sulfur cycle.

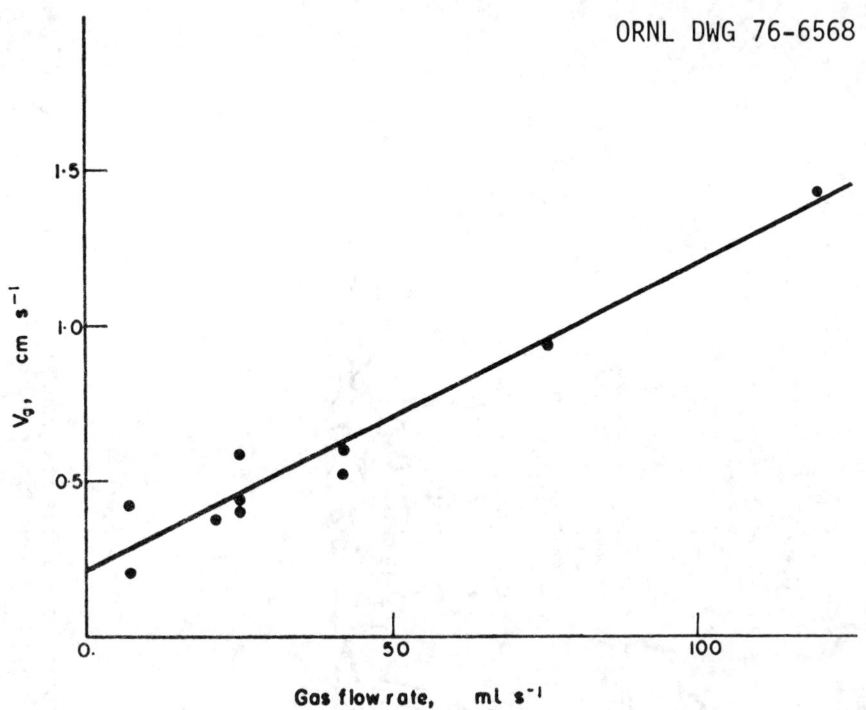

Fig. 6.23. Velocity of deposition of SO_2 into sea water vs flow rate of SO_2-air mixture over the sea water. <u>Source</u>: From Spedding 1972, Fig. 2, p. 585.
Reprinted by permission of the publisher.

6.4 NITROGEN OXIDES

Nitrogen and oxygen combine to form several nitrogen oxide compounds: nitric oxide (NO), nitrogen dioxide (NO_2), nitrous oxide (N_2O), nitrogen sesquioxide (N_2O_3), nitrogen tetroxide (N_2O_4), and nitrogen pentoxide (N_2O_5), and nitrogen trioxide (NO_3) [Environmental Protection Agency (EPA) 1971]. The term NO_x is used to indicate the sum of NO and NO_2, the only significant nitrogen oxide air pollutants. The other nitrogen oxide compounds are either inert or present in such small concentrations that they contribute little to air pollution. Therefore, the following discussion will be limited to NO and NO_2 (NO_x).

6.4.1 <u>Background concentrations</u>

Background levels of NO_x have been reported in widely scattered nonurban areas. For example, the Piedmont section of North Carolina was found to have NO_2 concentrations of 10.6 $\mu g/m^3$ at 4 ft above the land surface, 14.3 $\mu g/m^3$ at 30 ft, 11.5 $\mu g/m^3$ at 120 ft, and 6.4 $\mu g/m^3$ at 5120 ft. Nitric oxide concentrations were found to be 2.34 $\mu g/m^3$ at 4 ft and 2.72 $\mu g/m^3$ at 5120 ft (Ripperton, Kornreich, and Worth 1970). During the dry season in Panama, the atmospheric NO_2 level was 0.9 ppb (1.8 $\mu g/m^3$), whereas during the rainy season the NO_2 value quadrupled — 3.6 ppb (7.1 $\mu g/m^3$) (Lodge and Pate 1966, EPA 1971). Nitrogen dioxide concentrations of 8.0 $\mu g/m^3$ were reported at Pikes Peak in Colorado (Hamilton et al. 1968, as cited in EPA 1971). Average values in North America were 4 ppb (8 $\mu g/m^3$) for NO_2 and 2 ppb (2 $\mu g/m^3$) for NO (Robinson and Robbins 1970; EPA 1971).

ORNL DWG 76-7305

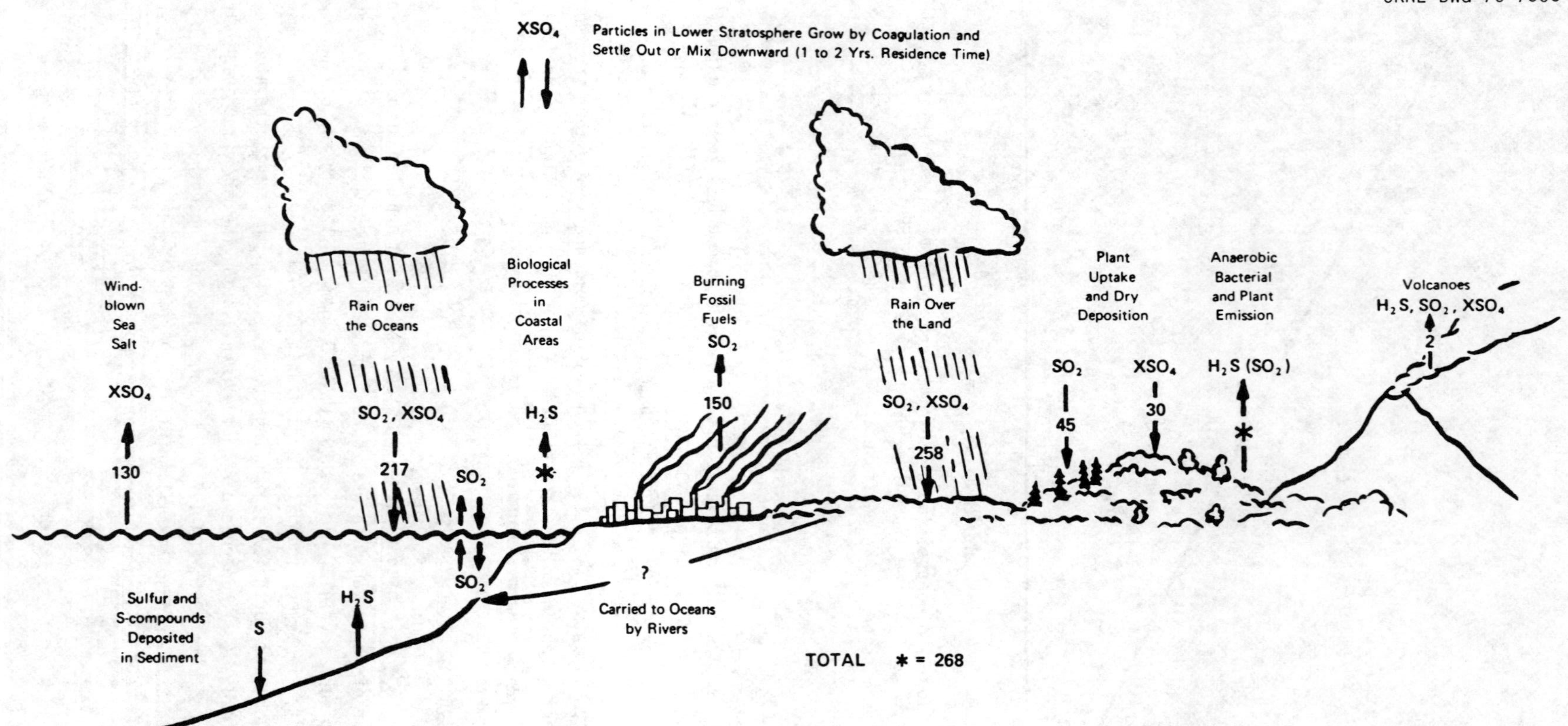

Fig. 6.24. Sources and sinks of atmospheric sulfur compounds; units are 10⁶ tons calculated as sulfate per year. Source: From Kellogg et al. 1972, Fig. 2, p. 594. Reprinted with permission from *Science*. Copyright 1972 by the American Association for the Advancement of Science.

Bacterial action is responsible for the major proportion of global NO_x, which is mostly NO (EPA 1971). This biological action produces at least 10 times more NO_x than is formed as a result of pollutant emissions (Table 6.64) (Robinson and Robbins 1970).

Table 6.64. Estimated annual global emissions of nitrogen compounds (tons/year)[a]

Compound	Source	Quantity of source	Estimated emissions	Emissions as nitrogen
NO_2	Coal combustion	3074×10^6	26.9×10^6	8.2×10^6
	Petroleum refining	$11,317 \times 10^6$ (bbl)	0.7×10^6	0.2×10^6
	Gasoline combustion	379×10^6	7.5×10^6	2.3×10^6
	Other oil combustion	894×10^6	14.1×10^6	4.3×10^6
	Natural gas combustion	20.56×10^{12} (ft^3)	2.1×10^6	0.6×10^6
	Other combustion	1290×10^6	1.6×10^6	0.5×10^6
Total			52.9×10^6	16.1×10^6
NH_3	Combustion		4.2×10^6	3.5×10^6
NO	Biological action		501×10^6	234×10^6
NH_3	Biological action		1160×10^6	957×10^6
N_2O	Biological action		592×10^6	378×10^6

[a]All values in tons/year except for quantity of petroleum refining (bbl) and natural gas combustion (ft^3) sources.

Source: Robinson and Robbins 1970, Table 1, p. 304.

Reprinted by permission of the publisher.

6.4.2 Man-made concentrations

The major source of anthropogenic NO_x emissions in the United States is fuel combustion (10 million tons per year), followed by transportation (8.1 million tons per year). These values represent 87.8% of the total nationwide nitrogen oxide emissions from man-made sources in 1968 (Table 6.65).

Table 6.66 shows NO_2 concentrations measured at National Air Sampling Network sites throughout the United States and Puerto Rico from 1967 through 1969. New York City had the highest 1967 concentration (0.179 ppm), and Long Beach, California, had the highest values for 1968 and 1969 (0.216 and 0.182 ppm respectively). In 1967 the lowest recorded value was 0.027 ppm at Portland, Oregon; the minimum value in 1968 was 0.024 ppm at Acadia National Park; and in 1969 the lowest concentration (0.012 ppm) was shared by Grand Canyon National Park, Little Rock, Arkansas, Humboldt County, California, and Glacier National Park.

Table 6.65. Summary of nationwide nitrogen oxides emissions, 1968

Source category	Emissions			
	10^6 tons/year		Percent	
Transportation	8.1		39.3	
Motor vehicles		7.2		34.9
Gasoline			6.6		32.0
Diesel			0.6		2.9
Aircraft[a]		b		b
Railroads		0.4		1.9
Vessels		0.2		1.0
Nonhighway		0.3		1.5
Fuel combustion in stationary sources	10.0		48.5	
Coal		4.0		19.4
Fuel oil		1.0		4.8
Natural gas[c]		4.8		23.3
Wood		0.2		1.0
Industrial processes	0.2		1.0	
Solid waste disposal	0.6		2.9	
Miscellaneous	1.7		8.3	
Forest fires		1.2		5.8
Structural fires		b		b
Coal refuse		0.2		1.0
Agricultural		0.3		1.5
	20.6		100.0	

[a]Emissions below 3,000 ft.
[b]Not reported; estimated less than 0.05×10^6 tons/year.
[c]Includes LPG and kerosene.

Source: EPA 1971, Table 3-1, p. 3-2.

6.4.3 <u>Chemical reactions</u>

Nitric oxide is a primary contaminant; nitrogen dioxide is considered a secondary contaminant because it is a conversion product of the reaction:

$$NO + O_3 = NO_2 + O_2 \quad . \tag{12}$$

This reaction is strongly influenced by sunlight, but will not completely stop in its absence as long as the ozone (O_3) supply is not depleted (EPA 1971). Thus, during the hours just after dawn (6 to 8 AM), with the advent of automobile traffic, NO levels increase. As ultraviolet light becomes available, NO_2 concentrations increase at the expense of O_3 (Reaction 12) until most of the NO is converted to NO_2. The atmospheric NO_2 photolytic cycle is shown in Fig. 6.25. Ultraviolet light energy breaks the N–O bonds, releasing O, which rapidly joins with atmospheric O_2 to form O_3; the O_3 then reacts with NO to form NO_2.

6-117

Table 6.66. Average 24-hr NO_2 concentration[a] at National Air Sampling
Network sites, 1967 through 1969 (ppm)

State	City	Station number	1967	1968	1969
Alabama	Birmingham	3			0.093[b]
	Mobile	1			0.025
	Montgomery	1			0.016
Alaska	Fairbanks	1	0.046	0.052	0.045
Arizona	Grand Canyon National Park	1			0.012[b]
	Phoenix	2		0.117[b]	0.089
	Tucson	1			0.028[b]
Arkansas	El Dorado	1		0.057	0.047
	Little Rock	1			0.012
California	Anaheim	1		0.159	0.148
	Berkeley	1			0.027[b]
	Fresno	1			0.042
	Glendale	1			0.081[b]
	Humboldt County	1			0.012
	Long Beach	1		0.216[b]	0.182[b]
	Oakland	1			0.053[b]
	Sacramento	1			0.019[b]
	San Bernardino	1		0.131	0.106
	San Diego	1		0.113	0.106
	San Francisco	1		0.110	0.095
	San Jose	2		0.123	0.116
	Santa Ana	1			0.059
Colorado	Denver	1			0.032
	Denver CAMP	2	0.070	0.092[b]	0.076[b]
Connecticut	Bridgeport	1		0.112	0.106[b]
	Hartford	1	0.109	0.071	0.077
	New Haven		0.094	0.107	0.072
	Waterbury	1			0.037
Delaware	Kent County	1		0.047[b]	0.042
	Newark	1	0.045	0.075	0.054[b]
	Wilmington	1	0.096	0.104	0.071[b]
District of Columbia	Washington	1			0.040
	Washington CAMP	3			0.069
	Washington CAMP	2	0.132	0.101	
Florida	Jacksonville	2			0.038
	Miami	2		0.072	0.061
	St. Petersburg	2			0.020[b]
	Tampa	2		0.080	0.079
Georgia	Atlanta	1		0.118	0.096
	Columbus	1			0.025[b]
	Savannah	1			0.031
Illinois	Chicago	1			0.054
	Chicago CAMP	2	0.144	0.155	0.160
	Peoria				0.050[b]
	Rockford	1			0.036
Indiana	East Chicago	1	0.081	0.081	0.086
	Evansville	1	0.044	0.032	0.031
	Gary	1			0.045
	Hammond	1			0.054
	Indianapolis	1	0.051	0.099	0.079
	Monroe	1	0.028[b]	0.032	0.034
	New Albany	1	0.086[b]		
	New Albany	2		0.081[b]	0.072[b]
	South Bend	2			0.031
Iowa	Des Moines	1	0.051	0.070	0.033
	Dubuque	1	0.030	0.091	0.072
Kansas	Topeka	1			0.023
	Wichita	1		0.057	0.064
Kentucky	Covington	1	0.088	0.110	0.090
	Lexington	1		0.077	0.062[b]
	Louisville	1		0.115	0.096
Louisiana	Carville	1		0.053	0.043
	New Orleans	2		0.080[b]	0.061

Table 6.66 (continued)

State	City	Station number	1967	1968	1969
Maine	Acadia National Park	1		0.024[b]	0.020
Maryland	Baltimore	1		0.095[b]	0.099
Massachusetts	Boston	1	0.063[b]	0.055[b]	0.040[b]
	Springfield	2		0.115[b]	0.086
	Worcester	1		0.090[b]	0.087
Michigan	Detroit	1	0.100	0.130	0.119
	Flint	1		0.093	0.085
	Grand Rapids	1		0.093	0.090
	Lansing	1			0.038
	Saginaw	1			0.031
Minnesota	Minneapolis	1	0.071	0.076	0.076
Missouri	Kansas City	1	0.076	0.081[b]	0.045
	St. Louis	1	0.094	0.116[b]	0.135
	St. Louis CAMP	2	0.116	0.115	0.108
Montana	Glacier National Park	1			0.012[b]
Nebraska	Omaha	1	0.076	0.072	0.075
New Jersey	Burlington County	2	0.097	0.091	0.062
	Camden	1	0.106	0.121	0.128
	Glassboro	1	0.071	0.059	0.020
	Jersey City	1	0.131	0.066[b]	0.065
	Newark	1	0.116	0.137	0.092
	Patterson	1		0.129	0.099
New Mexico	Albuquerque	1		0.060	0.048
New York	Albany	1		0.084[b]	0.071
	Buffalo	1		0.079	0.029
	Buffalo	3	0.082[b]		
	New York City	1	0.179	0.148[b]	0.142[b]
	Rochester	1		0.096[b]	0.086
	Syracuse	1		0.093[b]	0.074
	Utica	1		0.062	0.070
North Carolina	Durham	1		0.102	0.076[b]
	Greensboro	1			0.076
	Greensboro	2		0.099	
Ohio	Akron	1		0.107	0.038
	Canton	1		0.103	0.092
	Cincinnati	1		0.102	0.099
	Cincinnati CAMP	3	0.087	0.096	0.091
	Cleveland	1	0.072	0.119	0.099
	Columbus	1		0.105	0.087
	Dayton	1	0.032	0.043	0.057[b]
	Toledo	1		0.093	0.096[b]
	Youngstown	1	0.088	0.096	0.083[b]
Oklahoma	Oklahoma City	1	0.074	0.090	0.048[b]
	Tulsa	1	0.039	0.056	0.033
Oregon	Portland	1	0.027[b]	0.074	0.056[b]
Pennsylvania	Allentown	1		0.077	0.090
	Clearfield County	1			0.033
	Indiana County				0.040
	Johnstown				0.080
	Lancaster			0.101[b]	
	Philadelphia	1		0.088[b]	0.024
	Philadelphia	2	0.125		
	Pittsburgh	1	0.079	0.132	0.113
	Reading	1		0.112	0.081
	Warminster	1	0.075	0.066	0.054
	West Chester	1	0.049	0.066	0.042
	York	2		0.096	0.076
Puerto Rico	Bayamon	2		0.047	0.041
	Bayamon	1	0.040[b]		
	Guayanilla	1	0.029	0.031[b]	
	Guayanilla	2			0.034
Rhode Island	Providence	1	0.051	0.100	0.087
South Dakota	Custer	1		0.025	0.015[b]

Table 6.66 (continued)

State	City	Station number	1967	1968	1969
Tennessee	Chattanooga	1	0.076	0.089[b]	0.042
	Memphis	1		0.090[b]	0.078[b]
	Nashville	1	0.084	0.101	0.068[b]
Texas	Austin	2			0.030[b]
	Beaumont	1			0.042[b]
	Corpus Christi	1			0.026[b]
	Dallas	2		0.100	0.074
	El Paso	1	0.043	0.071	
	El Paso	2			0.048[b]
	Fort Worth	1		0.083[b]	0.071
	Houston	1		0.116[b]	0.105[b]
	Lubbock	1			0.021[b]
	Pasadena	1	0.035	0.074[b]	
	Pasadena	2			0.050
	San Antonio	1		0.068	0.065
	Tom Green County	1			0.015
Utah	Salt Lake City	1	0.048	0.084	0.060
Virginia	Norfolk	1		0.087	0.077
	Page	1		0.035	
	Shenandoah National Park	1			0.015
	Richmond	2		0.102	0.088[b]
Washington	Seattle	1	0.052	0.096	0.095[b]
	Tacoma	1			0.023
West Virginia	Charleston	1	0.076	0.109	0.092
Wisconsin	Milwaukee	1	0.093	0.109	0.090
Wyoming	Casper	1	0.035	0.032	0.025

[a]Determined by integrated Jacobs-Hochheiser method.
[b]The number and distribution of individual values making up this average do not meet the NASN criteria for calculating a yearly average; nevertheless, it is felt that the average is adequate for drawing the general relationships.

Source: EPA 1971, Table 6-10, pp. 6-33 — 6-36.

In equation form, the photolytic cycle would be

$$NO_2 \xrightarrow{\text{uv}} NO + O \, , \tag{13}$$

$$O + O_2 + M = O_3 + M \, , \tag{14}$$

$$O_3 + NO = NO_2 + O_2 \, . \tag{15}$$

The symbol M represents a collision molecule, which is necessary for the transfer of excess energy from the O_3 molecule. Without this transfer of energy, the O_3 molecule will decompose to O_2 and O (U.S. Department of Health, Education, and Welfare 1970; Davis, Smith, and Klauber 1974).

6.4.3.1 Ozone formation

According to NAS (1975*b*), ozone is not formed in situ at low altitudes. This is because the radiation that dissociates O_2, thus producing atomic oxygen, which in turn leads to ozone formation, does not penetrate to low altitudes in significant amounts; its level of maximum absorption is in the upper stratosphere. Therefore, any appreciable ozone concentrations found at low altitudes must result from processes other than in situ formation. The photolytic cycle (Fig. 6.25) and Reaction (14) show ozone formation, but this O_3 quickly breaks down as it reacts with NO to form NO_2 (Reaction 15) during photolysis, leaving no excess O_3.

ORNL DWG 76-7317

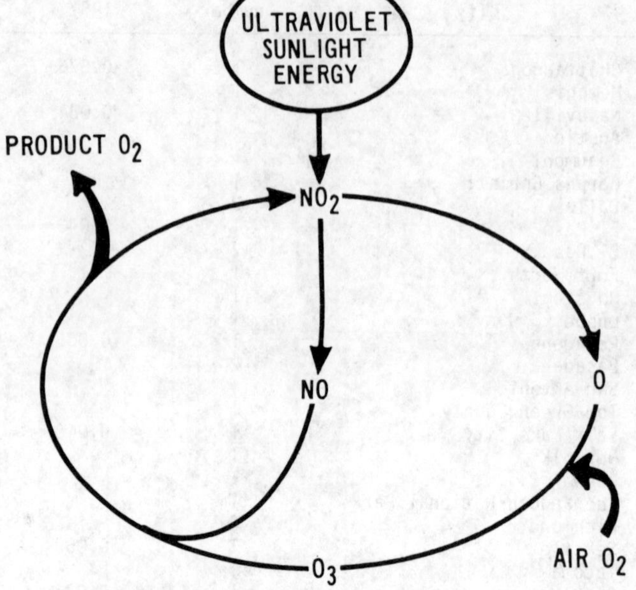

Fig. 6.25. Atmospheric NO_2 photolytic cycle. <u>Source</u>: From EPA 1971, Fig. 2-1, p. 2-3.

However, ozone accumulation is enhanced at low altitudes when hydrocarbons (notably olefins and substituted aromatics) enter the NO_2 photolytic cycle (Fig. 6.26) (U.S. Department of Health, Education, and Welfare 1970). The role of hydrocarbons in the formation of ozone photochemical smog has been examined in detail (Robinson and Robbins 1970; Schuck and Stephens 1969; Schuck, Pitts, and Wan 1966; Romanovsky, Ingels, and Gordon 1967).

Hydrocarbons are thought to interact through the following steps: (1) Hydrocarbons are oxidized by O atoms to form a free radical (Fig. 6.26); (2) the free radical, being very reactive, undergoes a series of changes in which it reacts with O_2 and oxidizes NO to NO_2; (3) the free radical oxidizes more than one molecule of NO to NO_2; (4) there is a rapid buildup of NO_2 and an accumulation of O_3.

The accumulation of O_3 via the photolytic cycle is possible because the hydrocarbon free radical that is formed substitutes for O_3 in the oxidation of NO to NO_2 (U.S. Department of Health, Education, and Welfare 1970; Chameides and Stedman 1976; Calvert 1976a; Calvert 1976b). Figure 6.27 shows an urban diurnal variation of NO, NO_2, and O_3. Nitric oxide levels are shown to rise in early morning, then decrease as the NO is converted to NO_2. The figure shows an accumulation of O_3, implying the intervention of hydrocarbons into the NO_2 photolytic cycle. This accumulation of O_3 may result in ozone air pollution levels in rural areas downwind of urban pollution sources (Chameides and Stedman 1976).

6.4.3.2 <u>Peroxyacyl nitrate formation</u>

Peroxyacyl nitrate (PAN), like ozone, is a photochemical eye irritant and phytotoxicant (Chap. 8) (Schuck, Pitts, and Wan 1966). Two possible modes of PAN formation have been suggested. The

ORNL DWG 76-7295

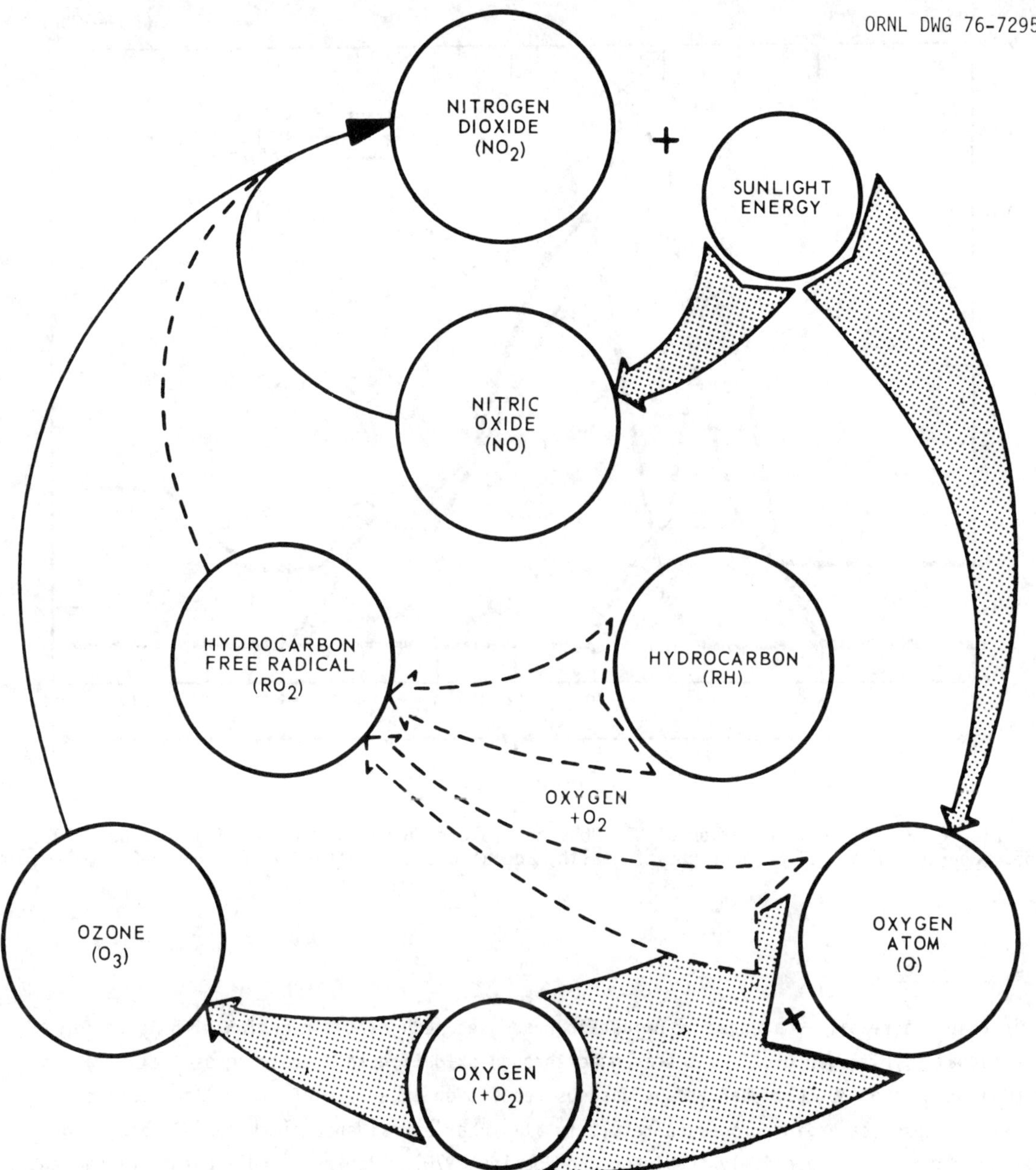

Fig. 6.26. Interaction of hydrocarbons with atmospheric nitrogen dioxide photolytic cycle.
Source: From U.S. Department of Health, Education, and Welfare 1970, Fig. 2-3, p. 2-7.

first mechanism requires the presence of atmospheric nitrites which has not been demonstrated
conclusively (U.S. Department of Health, Education, and Welfare 1970). If present, however,
irradiation of the parent nitrite (RONO) would result in PAN formation. A second method involves
the reaction of a photolytic free radical (Fig. 6.26) with O$_2$ to form a peroxyacyl radical
($RC\overset{\text{O}}{\underset{\|}{O}}O$), which would react with NO$_2$ to form peroxyacyl nitrate ($RCOON\overset{\text{O}}{\underset{\|}{O}}_2$) (U.S. Department of
Health, Education, and Welfare 1970; Stephens 1969).

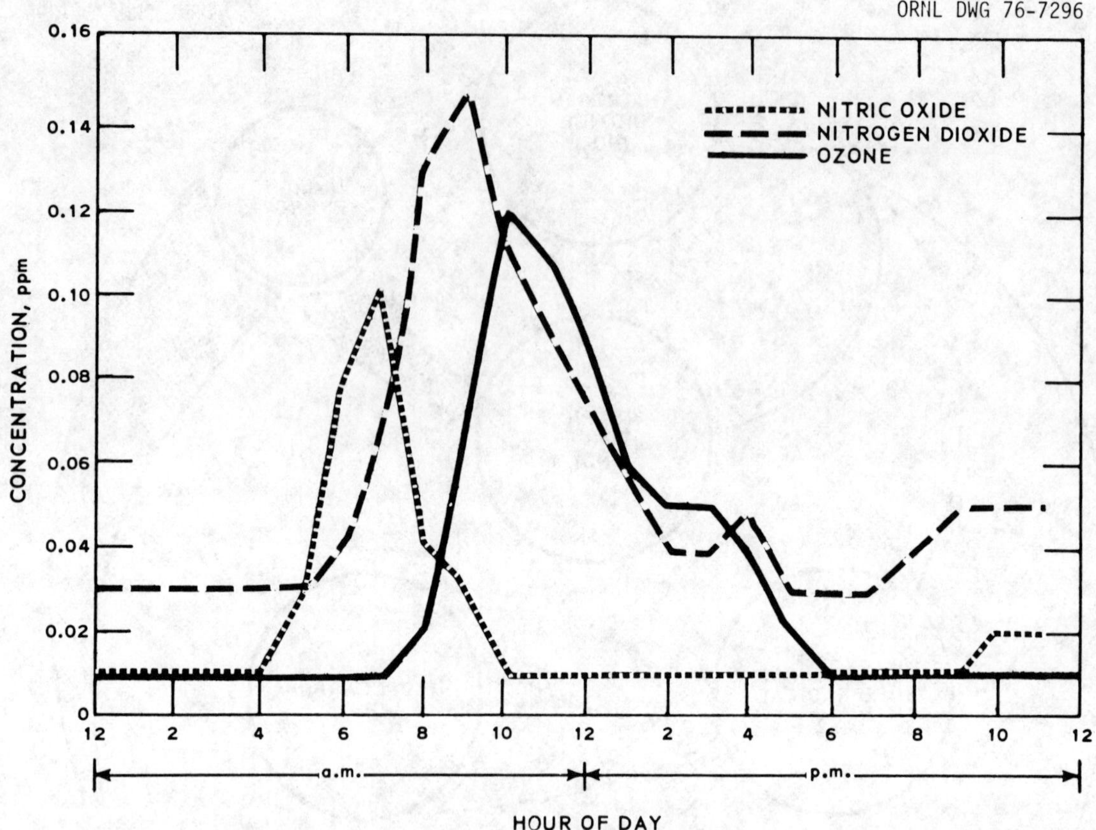

Fig. 6.27. Diurnal variation of NO, NO$_2$, and O$_3$ concentrations in Los Angeles, July 19, 1965. <u>Source</u>: From U.S. Department of Health, Education, and Welfare 1970, Fig. 2-2, p. 2-5.

6.4.4 <u>Residence</u>

Nitric oxide is readily scavenged by oxidation processes such as the reaction with O$_3$ to form NO$_2$. Globally, nitrogen dioxide is scavenged through oxidation to nitrate aerosol at a rate of 462 million tons of NO$_3$ per year and by gaseous deposition at a rate of 478 million tons of NO$_2$ per year. These scavenging processes result in atmospheric residence times for NO$_2$ and NO of three and four days respectively (Robinson and Robbins 1970). Robinson and Robbins believe that, although these scavenging processes do not proceed rapidly enough to affect urban hourly concentrations, there appears to be little reason to expect any long-term global buildup of nitrogen oxides.

LITERATURE CITED

Abeles, F. B.; Craker, L. E.; Forrence, L. E.; and Leather, G. R. 1971. Fate of air pollutants: removal of ethylene, sulfur dioxide, and nitrogen dioxide by soil. *Science* 173: 914-16.

Ahlrichs, J. L. 1972. The soil environment. In *Organic chemicals in the soil environment*, ed. C.A.I. Goring and J. W. Hamaker, pp. 3-43. New York: Marcel Dekker, Inc.

Andelman, J. B., and Snodgrass, J. E. 1974. Incidence and significance of polynuclear aromatic hydrocarbons in the water environment. *CRC Crit. Rev. Environ. Control* 4(1): 69-83.

Andelman, J. B., and Suess, M. J. 1970. Polynuclear aromatic hydrocarbons in the water environment. *Bull. W. H. O.* 43: 479-508.

Angino, E. E. 1969. Iron, manganese, nickel, cobalt, strontium, lithium, zinc, and silicon dioxide in streams of the Lower Kansas River Basin. *Water Resour. Res.* 5(3): 698-705.

Armstrong, F.A.J.; Williams, P. M.; and Strickland, J.D.H. 1966. Photo-oxidation of organic matter in sea water by ultra-violet radiation, analytical and other applications. *Nature* 211: 481-83.

Baker, E. G. 1960. A hypothesis concerning the accumulation of sediment hydrocarbons to form crude oil. *Geochim. Cosmochim. Acta.* 19: 309-17.

Baker, R. A., and Luh, M.-D. 1971. Pyridine sorption from aqueous solution by montmorillonite and kaolinite. *Water Res.* 5: 839-48.

Barnes, R. S., and Schell, W. R. 1973. Physical transport of trace metals in the Lake Washington watershed. In *Cycling and control of metals*, ed. M. G. Curry and G. M. Gigliotti, pp. 45-53. Cincinnati, Ohio: EPA National Environmental Research Center.

Barnett, S. M.; Velankar, S. K.; and Houston, C. W. 1974. Mechanism of hydrocarbon uptake by microorganisms. *Biotechnol. Bioeng.* XVI: 863-65.

Barrows, H. L.; Neff, M. S.; and Gammon, N., Jr. 1960. Effect of soil type on mobility of zinc in the soil and on its availability from zinc sulfate to tung. In *Soil science society proceedings 1960*, pp. 367-72.

Benes, P., and Steinnes, E. 1974. Migration forms of trace elements in natural fresh waters and the effect of the water storage. *Water Res.* 9: 741-49.

Blumer, M. 1961. Benzpyrenes in soil. *Science* 134: 474-75.

Blumer, M. 1975. Curtisite, idrialite and pendletonite, polycyclic aromatic hydrocarbon minerals: Their composition and origin. *Chem. Geol.* 16: 245-56.

Blumer, M. 1976. Polycyclic aromatic hydrocarbons in nature. *Sci. Am.* 234(3): 35-45.

Blumer, M., and Youngblood, W. W. 1975. Polycyclic aromatic hydrocarbons in soils and recent sediments. *Science* 188: 53-55.

Boehm, P. D., and Quinn, J. G. 1973. Solubilization of hydrocarbons by the dissolved organic matter in sea water. *Geochim. Cosmochim. Acta* 37: 2459-77.

Bohn, H. L. 1972. Soil absorption of air pollutants. *J. Environ. Qual.* 1(4): 372-77.

Bohn, H. L., and Aba-Husayn, M. M. 1971. Manganese, iron, copper and zinc concentrations of *Sporobolus wrightii* in alkaline soils. *Soil Sci.* 112(5): 348-50.

Bolter, E.; Wixson, B. G.; Butherus, D. L.; and Jennett, J. C. 1974. *Distribution of heavy metals in soils near an active lead smelter*. Rolla, Missouri: Dept. of Civil Engineering, Environmental Research Center, University of Missouri.

Boothe, P. N., and Knauer, G. A. 1972. The possible importance of fecal material in the biological amplification of trace and heavy metals. *Limnol. Oceanogr.* 17(2): 270-74.

Borneff, J., and Kunte, H. 1963. Carcinogenic substances in water and soil. XIV. Further investigation concerning polycyclic aromatic hydrocarbons in soil samples. *Arch. Hyg.* 147: 401-9.

Borneff, J., and Kunte, H. 1965. Carcinogenic substances in water and soil. XVII. Concerning the origin and estimation of the polycyclic aromatic hydrocarbons in water. *Arch. Hyg.* 149: 226-43.

Borneff, J.; Selenka, F.; Kunte, H.; and Maximos, A. 1968. Experimental studies on the formation of polycyclic aromatic hydrocarbons in plants. *Environ. Res.* 2: 22-29.

Bourg, A.C.M., and Filby, R. H. 1974. Adsorption isotherms for the uptake of Zn^{2+} by clay minerals in a fresh-water medium. In *Proceedings of the international conference on transport of persistent chemicals in aquatic ecosystems,* ed. A.S.W. de Freitas, D. J. Kushner, and S. U. Qadri, pp. II-1-II-5. Ottawa: National Research Council of Canada.

Bowen, H. J. 1966a. Outline of terrestrial geochemistry. In *Trace elements in biogeochemistry,* pp. 8-24. New York: Academic Press.

Bowen, H. J. 1966b. The composition of the soil. In *Trace elements in biogeochemistry,* pp. 25-41. New York: Academic Press.

Bowen, H. J. 1966c. The biochemistry of the elements. In *Trace elements in biogeochemistry,* pp. 173-210. New York: Academic Press.

Bowen, H. J. 1966d. The uptake and excretion of elements by organisms. In *Trace elements in biogeochemistry,* pp. 85-101. New York: Academic Press.

Brisou, J. 1972. Lipids, sterols, terpenes and bacterial biosynthesis of 3,4-benzopyrene. In *Pollution des milieux vitaux par les hydrocarbures polybenzeniques du type benzo-3,4 pyrene,* ed. L. Mallet, pp. 181-83. Paris: Librairie Maloine.

Brooks, R. R., and Rumsby, M. G. 1965. The biogeochemistry of trace element uptake by some New Zealand bivalves. *Limnol. Oceanogr.* 10: 521-27.

Burton, J. D. 1966. The marine geochemistry of vanadium. *Nature* 212(5066): 976-78.

Calvert, J. G. 1976a. Test of the theory of ozone generation in Los Angeles atmosphere. *Environ. Sci. Technol.* 10(3): 248-56.

Calvert, J. G. 1976b. Hydrocarbon involvement in photochemical smog formation in Los Angeles atmosphere. *Environ. Sci. Technol.* 10(3): 256-62.

Chameides, W. L., and Stedman, D. H. 1976. Ozone formation from NO_X in "clean air." *Environ. Sci. Technol.* 10(2): 150-53.

Chappell, W. R. 1973. Transport and the biological effects of molybdenum in the environment. Paper presented at International Conference on Heavy Metals in the Aquatic Environment, Nashville, Tenn., Dec. 4-7.

Davies, J. I., and Evans, W. C. 1964. Oxidative metabolism of naphthalene by soil pseudomonads. *Biochem. J.* 91: 251-61.

Davis, D. D.; Smith, G.; and Klauber, G. 1974. Trace gas analysis of power plant plumes via aircraft measurement: O_3, NO_X, and SO_2 chemistry. *Science* 186: 733-36.

de Groot, A. J., and Allersma, E. 1973. Field observations on the transport of heavy metals in sediments. Paper presented at International Conference on Heavy Metals in the Aquatic Environment, Nashville, Tenn.

DeMaio, L., and Corn, M. 1966. Polynuclear aromatic hydrocarbons associated with particulates in Pittsburgh air. *J. Air Pollut. Control Assoc.* 16: 67-71.

DiSalvo, L. H.; Guard, H. E.; and Hunter, L. 1975. Tissue hydrocarbon burden of mussels as potential monitor of environmental hydrocarbon insult. *Environ. Sci. Technol.* 9(3): 247-51.

D'Itri, F. M. 1972. *The environmental mercury problem.* Cleveland, Ohio: CRC Press.

Ehrhardt, M. 1972. Petroleum hydrocarbons in oysters from Galveston Bay. *Environ. Pollut.* 3: 257-71.

Elworthy, P. H.; Florence, A. T.; and Macfarlane, C. B. 1968. Solubilization by surface-active agents. London: Chapman and Hall Ltd.

Engler, R. M., and Patrick, W. H., Jr. 1975. Stability of sulfides of manganese, iron, zinc, copper, and mercury in flooded and non-flooded soil. *Soil Sci.* 119(3): 217-21.

Environmental Protection Agency (EPA). 1971. *Air quality criteria for nitrogen oxides.* Air Pollution Control Office Publication AP-84.

Environmental Protection Agency (EPA). 1975. *Scientific and technical assessment report on particulate polycyclic organic matter (PPOM).* EPA-60016-75-001.

Evans, W. C.; Fernley, H. N.; and Griffiths, E. 1965. Oxidative metabolism of phenanthrene and anthracene by soil pseudomonads — the ring-fission mechanism. *Biochem. J.* 95: 819-31.

Fabricand, B. P.; Sawyer, R. R.; Ungar, S. G.; and Alder, S. 1962. Trace metal concentrations in the ocean by atomic absorption spectroscopy. *Geochim. Cosmochim. Acta* 26: 1023-27.

Falk, H. L.; Markul, I.; and Kotin, P. 1956. Aromatic hydrocarbons. IV. Their fate following emission into the atmosphere and experimental exposure to washed air and synthetic smog. *A. M. A. Arch. Ind. Health* 13: 13-17.

Farrington, J. W., and Quinn, J. G. 1973. Petroleum hydrocarbons in Narragansett Bay. I. Survey of hydrocarbons in sediments and clams (*Mercenavia mercenaria*). *Estuarine Coastal Mar. Sci.* 1: 71-79.

Feldman, M. H. 1973. *Petroleum weathering: Some pathways, fate and disposition on marine waters.* Corvallis, Oregon: Pacific Northwest Environmental Research Laboratory, Program Element 1BA025. EPA 660/3-73-013. PB-227 278.

Fernley, H. N., and Evans, W. C. 1958. Oxidative metabolism of polycyclic hydrocarbons by soil pseudomonads. *Nature* 182: 373-75.

Fletcher, K., and Doyle, P. 1974. Factors influencing trace element distribution in the eastern Yukon. *CIM Bull.* 67: 61-65.

Frankenberg, D., and Smith, K. L., Jr. 1967. Coprophagy in marine animals. *Limnol. Oceanogr.* 12: 443-50.

Garland, J. A.; Atkins, D.H.F.; Readings, C. J.; and Caughey, S. J. 1974. Deposition of gaseous sulphur dioxide to the ground. *Atmos. Environ.* 8: 75-79.

Gibbs, R. J. 1973. Mechanisms of trace metal transport in rivers. *Science* 180(4081): 71-73.

Glasby, G. P.; Tooms, J. S.; and Cann, J. R. 1971. The geochemistry of manganese encrustations from the Gulf of Aden. *Deep-Sea Res.* 18: 1179-87.

Goma, G.; Pareilleux, A.; and Durand, G. 1973. Kinetics of degradation of hydrocarbons by *Candida lipolyfica. Arch. Mikrobiol.* 88: 97-109.

Goodman, G. T., and Roberts, T. M. 1971. Plants and soils as indicators of metals in the air. *Nature* 231: 287-92.

Grasselli, J. G. 1973. *Atlas of spectral data and physical constants for organic compounds.* Cleveland, Ohio: CRC Press.

Grimmer, G., and Duvel, D. 1970. Investigations of biosynthetic formation of polycyclic hydrocarbons in higher plants. *Z. Naturforsch.* 25b: 1171-75.

Gross, M. G. 1970. Waste removal and recycling by sedimentary processes. In *Marine pollution and sea life, Food and Agriculture Organization of the United Nations Conference, Rome,* ed. M. Ruivo, pp. 152-58. London: Fishing News (Books) Ltd.

Haley, T. J. 1969. *Air quality monographs: A review of the toxicology of lead.* Monogr. 69-7, American Petroleum Institute, Washington, D.C.

Hamaker, J. W. 1975. The interpretation of soil leaching experiments. In *Environmental dynamics of pesticides,* ed. R. Haque and V. H. Freed, pp. 115-33. New York: Plenum Press.

Hamaker, J. W., and Thompson, J. M. 1972. Adsorption. In *Organic chemicals in the soil environment,* ed. C.A.I. Goring and J. W. Hamaker, pp. 49-143. New York: Marcel Dekker, Inc.

Haque, R., and Ash, N. 1973. Factors affecting the behavior of chemicals in the environment. In *Survival in toxic environments,* ed. M.A.Q. Khan and J. P. Bederka, Jr., pp. 357-71. New York: Academic Press, Inc.

Hardy, K. A. 1976. Elemental constituents of Miami aerosol as function of particle size. *Environ. Sci. Technol.* 10(2): 176-82.

Harker, A. B. 1975. The formation of sulfate in the stratosphere through the gas phase oxidation of sulfur dioxide. *J. Geophys. Res.* 80(24): 3399-3401.

Harris, F.; Andren, A.; and Henderson, G. 1975. Trace elements from coal-fired plants. *Review* 8(2): 20-24. Oak Ridge, Tenn.: Oak Ridge National Laboratory.

Harrison, R. M.; Perry, R.; and Wellings, R. A. 1975. Polynuclear aromatic hydrocarbons in raw, potable and waste waters. *Water Res.* 9: 331-46.

Hartung, R. 1974. Heavy metals in the lower Mississippi. In *Proceedings of the international conference on transport of persistent chemicals in aquatic ecosystems*, ed. A.S.W. de Freitas, D. J. Kushner, and S. U. Qadri, pp. I-93–I-98. Ottawa: National Research Council of Canada.

Hase, A., and Hites, R. A. 1976. On the origin of polycyclic aromatic hydrocarbons in recent sediments: biosynthesis by anaerobic bacteria. *Geochem. Cosmochim. Acta* 40: 1141-43.

Helz, G. R.; Huggett, R. J.; and Hill, J. M. 1975. Behavior of Mn, Fe, Cu, Zn, Cd, and Pb discharged from a wastewater treatment plant into an estuarine environment. *Water Res.* 9: 631-36.

Hem, J. D. 1972. Chemistry and occurrence of cadmium and zinc in surface water and groundwater. *Water Resour. Res.* 8(3): 661-79.

Herbes, S. E.; Southworth, G. R.; and Gehrs, C. W. 1976. Organic contaminants in aqueous coal conversion effluents: environmental consequences and research priorities. In *Tenth annual conference on trace substances in environmental health, Univ. of Missouri, June 8-10* (to be published).

Hill, A. C. 1971. Vegetation: A sink for atmospheric pollutants. *J. Air Pollut. Control Assoc.* 21(6): 341-46.

Hussien, Y. A.; Tewfik, M. S.; and Hamdi, Y. A. 1974. Degradation of certain aromatic compounds by rhizobia. *Soil Biol. Biochem.* 6: 377-81.

Hutchinson, T. C., and Fitchko, J. 1974. Heavy metal concentrations and distributions in river mouth sediments around the Great Lakes. In *Proceedings of the international conference on transport of persistent chemicals in aquatic ecosystems*, ed. A.S.W. de Freitas, D. J. Kushner, and S. U. Qadri, pp. I-69–I-77. Ottawa: National Research Council of Canada.

Jeffery, J. 1961. Experiments on the biological treatment of phenolic effluents. *Gas World — Coking* 153(3994): 23-25.

Jernelöv, Å. 1972. Factors in the transformations of mercury to methylmercury. In *Environmental mercury contamination*, ed. R. Hartung and B. D. Dinman, pp. 167-72. Ann Arbor, Mich.: Ann Arbor Science Publishers, Inc.

Kellogg, W. W.; Cadle, R. D.; Allen, E. R.; Lazrus, A. L.; and Martell, E. A. 1972. The sulfur cycle. *Science* 175(4022): 587-96.

Kharkar, D. P.; Turekian, K. K.; and Bertine, K. K. 1968. Stream supply of dissolved silver, molybdenum, antimony, selenium, chromium, cobalt, rubidium, and cesium to the oceans. *Geochim. Cosmochim. Acta* 32(3): 285-98.

Kleinman, M. T.; Kneip, T. J.; and Eisenbud, M. 1973. Meteorological influences on airborne trace metals and suspended particulates. In *Seventh annual conference on trace substances in environmental health, Univ. of Missouri, June 12-13*, ed. D. D. Hemphill, pp. 161-166.

Kopp, J. F., and Kroner, R. C. 1967. *Trace metals in waters of the United States: A five-year summary of trace metals in rivers and lakes of the United States (Oct. 1, 1962–Sept. 30, 1967).* U.S. Dept. of the Interior, Federal Water Pollution Control Administration, Cincinnati, Ohio.

Kostyaev, V. Ya. 1973. Effect of phenol on the hydrochemical regime, phytoplankton, and plant overgrowth in artificial bodies of water. *Tr. Inst. Biol. Vnutr. Vod. Akad. Nauk. SSR* 24: 119-51.

Lee, R. E., Jr.; Goranson, S. S.; Enrione, R. E.; and Morgan, G. B. 1972. National air surveillance cascade impactor network. II. size distribution measurements of trace metal components. *Environ. Sci. Technol.* 6(12): 1025-30.

Lee, R. E., Jr., and von Lehmden, D. J. 1973. Trace metal pollution in the environment. *J. Air Pollut. Control Assoc.* 23(10): 853-57.

Lee, R. F. 1975. Fate of petroleum hydrocarbons in marine zooplankton, pp. 549-53 in *Proceedings of the conference on prevention and control of oil pollution.* American Petroleum Institute, Washington, D.C.

Lee, R. F.; Ryan, C.; and Neuhauser, M. L. 1976. Fate of petroleum hydrocarbons taken up from food and water by the blue crab *Callinectes sapidus. Marine Biol.* 37: 363-70.

Lee, R. F.; Sauerheber, R.; and Benson, A. A. 1972. Petroleum hydrocarbons: Uptake and discharge by the marine mussel *Mytilus edulis. Science* 177: 344-46.

Leland, H. V.; Shukla, S. S.; and Shimp, N. F. 1973. Factors affecting distribution of lead and other trace elements in sediments of southern Lake Michigan. In *Trace metals and metal-organic interactions in natural waters,* ed. P. C. Singer, pp. 89-129. Ann Arbor, Michigan: Ann Arbor Science Publishers, Inc.

Letey, J., and Oddson, J. K. 1972. Mass transfer. In *Organic chemicals in the soil environment,* ed. C.A.I. Goring and J. W. Hamaker, pp. 399-440. New York: Marcel Dekker, Inc.

Linton, R. W.; Natusch, D.F.S.; Loh, A.; Evans, C. A., Jr.; and Williams, P. 1976. Surface predominance of trace elements in airborne particles. *Science* 191: 852-54.

Linzon, S. N.; McIlveen, W. D.; and Temple, P. J. 1973. Sulphur dioxide injury to vegetation in the vicinity of a sulphite pulp and paper mill. *Water Air Soil Pollut.* 2: 129-34.

A. D. Little, Inc. 1971. Inorganic chemical pollution of freshwater. *Water quality criteria data book,* 2: 11-17. Washington, D.C.: U.S. Government Printing Office.

Lodge, J. P., and Pate, J. B. 1966. Atmospheric gases and particulates in Panama. *Science* 153: 408-10.

Ludzack, F. J., and Ettinger, M. B. 1963. Biodegradability of organic chemicals isolated from rivers. In *Industrial waste conference, Purdue Univ.,* pp. 278-82.

Mallet, L. 1972. Investigation of polybenzene hydrocarbons in oil sediments. In *Pollution des milieux vitaux par les hydrocarbures polybenzeniques du type benzo-3,4 pyrene,* ed. L. Mallet, pp. 111-13. Paris: Librairie Maloine.

Mallet, L., and Tissier, M. 1972. Experimental biosynthesis of polybenzene hydrocarbons of the benzo-3,4-pyrene type at the expense of forest soils. In *Pollution des milieux vitaux par les hydrocarbures polybenzeniques du type benzo-3,4 pyrene,* ed. L. Mallet, pp. 105-107. Paris: Librairie Maloine.

Mallet, L.; Heros, M.; and Brisou, J. 1972. Biosynthesis and bioregression of the carcinogenic polybenzene hydrocarbons of the benzo-3,4-pyrene type, taken as controls, at the expense of lipids. In *Pollution des milieux vitaux par les hydrocarbures polybenzeniques du type benzo-3,4 pyrene,* ed. L. Mallet, pp. 177-80. Paris: Librairie Maloine.

Mallet, L.; Zanghi, L.; and Brisou, J. 1972. Investigations of the possibilities of biosynthesis of polybenzene hydrocarbons of the benzo-3,4-pyrene type by a *Clostridium putride* in the presence of marine plankton lipids. In *Pollution des milieux vitaux par les hydrocarbures polybenzeniques du type benzo-3,4 pyrene,* ed. L. Mallet, pp. 153-57. Paris: Librairie Maloine.

Martin, J. H. 1970. The possible transport of trace metals via moulted copepod exoskeletons. *Limnol. Oceanogr.* 15(5): 756-61.

McAuliffe, C. 1966. Solubility in water of paraffin, cycloparaffin, olefin, acetylene, cycloolefin, and aromatic hydrocarbons. *J. Phys. Chem.* 70(4): 1267-75.

McDuffie, B.; El-Barbary, I.; Hollod, G. J.; and Tiberio, R. D. 1976. Trace metals in rivers — speciation, transport, and role of sediments. In *Tenth annual conference on trace substances in environmental health, Univ. of Missouri, June 8-10* (to be published).

McGinnes, P. R., and Snoeyink, V. L. 1974. *Determination of the fate of polynuclear aromatic hydrocarbons in natural water systems.* University of Illinois at Urbana-Champaign, Water Resources Center, WRC Res. Rept. 80, UILU-WRC-74-0080. PB-232 168.

McHardy, W. J.; Thompson, A. P.; and Goodman, B. A. 1974. Formation of iron oxides by decomposition of iron-phenolic chelates. *J. Soil Sci.* 25(4): 471-82.

McKay, H.A.C. 1971. The atmospheric oxidation of sulfur dioxide in water droplets in presence of ammonia. *Atmos. Environ.* 5: 7-14.

Meikle, R. W. 1972. Decomposition: Qualitative relationships. In *Organic chemicals in the soil environment,* ed. C.A.I. Goring and J. W. Hamaker, pp. 145-251. New York: Marcel Dekker, Inc.

Mitchell, R. L. 1945. Cobalt and nickel in soils and plants. *Soil Sci.* 60: 63-70.

Muller, W. P., and Korte, F. 1975. Microbial degradation of benzo-[a]-pyrene, monolinuron, and dieldrin in waste composting. *Chemosphere* 3: 195-98.

National Academy of Sciences (NAS). 1972. *Particulate polycyclic organic matter.* Washington, D.C.: National Academy of Sciences.

National Academy of Sciences (NAS). 1975a. *Petroleum in the marine environment.* Washington, D.C.: National Academy of Sciences.

National Academy of Sciences (NAS). 1975b. *Atmospheric chemistry: Problems and scope.* Washington, D.C.: National Academy of Sciences.

Nichol, I.; Horsnail, R. F.; and Webb, J. S. 1967. Geochemical patterns in stream sediment related to precipitation of manganese oxides. *Inst. Min. Metall., Trans., Sect. B* 76(726): B113-15.

Ogata, M., and Miyake, Y. 1973. Identification of substances in petroleum causing objectionable odour in fish. *Water Res.* 7: 1493-1504.

Olsen, D., and Haynes, J. L. 1969. *Preliminary air pollution survey of organic carcinogens. A literature review.* APTD 69-43. National Air Pollution Control Administration, Raleigh, N.C.

Padmanabhamurty, B. 1975. The role of wind in pollution dispersion. *J. Air Pollut. Control Assoc.* 25(9): 956-57.

Painter, L. I.; Toth, S. J.; and Bear, F. E. 1953. Nickel status of New Jersey soils. *Soil Sci.* 76: 421-29.

Payrissat, M., and Beilke, S. 1975. Laboratory measurements of the uptake of sulfur dioxide by different European soils. *Atmos. Environ.* 9: 211-17.

Peirson, D. H.; Cawse, P. A.; Salmon, L.; and Cambray, R. S. 1973. Trace elements in the atmospheric environment. *Nature* 241: 252-56.

Perhac, R. M. 1972. Distribution of Cd, Co, Cu, Fe, Mn, Ni, Pb, and Zn in dissolved and particulate solids from two streams in Tennessee. *J. Hydrol.* 15: 177-86.

Pierce, R. C., and Katz, M. 1975. Dependency of polynuclear aromatic hydrocarbon content on size distribution of atmospheric aerosols. *Environ. Sci. Technol.* 9(4): 347-53.

Pita, F. W., and Hyne, N. J. 1975. The depositional environment of zinc, lead and cadmium in reservoir sediments. *Water Res.* 9: 701-6.

Preston, A.; Jefferies, D. F.; Dutton, J.W.R.; Harvey, B. R.; and Steele, A. K. 1972. British Isles coastal waters: The concentrations of selected heavy metals in sea water, suspended matter and biological indicators — a pilot survey. *Environ. Pollut.* 3: 69-82.

Pupp, C.; Lao, R. C.; Murray, J. J.; and Pottie, R. F. 1974. Equilibrium vapour concentrations of some polycyclic aromatic hydrocarbons, As_4O_6 and SeO_2 and the collection efficiencies of these air pollutants. *Atmos. Environ.* 8: 915-25.

Purves, D. 1972. Consequences of trace-element contamination of soils. *Environ. Pollut.* 3: 17-24.

Rasmussen, K. H.; Kabel, R. L.; and Taheri, M. 1974. *Sources and natural removal processes for some atmospheric pollutants.* EPA-650/4-74-032. PB-237 168.

Ripperton, L. A.; Kornreich, L.; and Worth, J.J.B. 1970. Nitrogen dioxide and nitric oxide in non-urban air. *J. Air Pollut. Control Assoc.* 20(9): 589-92.

Roberts, A. L.; Street, G. B.; and White, D. 1964. The mechanism of phenol adsorption by organo-clay derivatives. *J. Appl. Chem.* 14: 261-65.

Roberts, T. M., and Goodman, G. T. 1973. The persistence of heavy metals in soils and natural vegetation following closure of a smelter. In *Seventh annual conference on trace substances in environmental health, Univ. of Missouri, June 12-14,* ed. D. D. Hemphill, pp. 117-25.

Robinson, E., and Robbins, R. C. 1970. Gaseous nitrogen compound pollutants from urban and natural sources. *J. Air Pollut. Control Assoc.* 20(5): 303-6.

Romanovsky, J. C.; Ingels, R. M.; and Gordon, R. J. 1967. Estimation of smog effects in the hydrocarbon–nitric oxide system. *J. Air Pollut. Control Assoc.* 17(7): 454-59.

Ruhling, A., and Tyler, G. 1970. Sorption and retention of heavy metals in the woodland moss *Hylocomium splendens* (Hedw.). *Br. Sch. Oikos* 21: 92-97.

Ruivo, M. 1970. Summary of discussion. In *Marine pollution and sea life, Food and Agriculture Organization of the United Nations Conference, Rome,* ed. M. Ruivo, pp. 113-14. London: Fishing News (Books) Ltd.

Rust, B. R., and Waslenchuk, D. G. 1974. The distribution and transport of bed sediments and persistent pollutants in the Ottawa River, Canada. In *Proceedings of the international conference on transport of persistent chemicals in aquatic ecosystems,* ed. A.S.W. de Freitas, D. J. Kushner, and S. U. Qadri, pp. I-25–I-40. Ottawa: National Research Council of Canada.

Sawicki, E.; Elbert, W. C.; Hauser, T. R.; Fox, F. T.; and Stanley, T. W. 1960. Benzo[a]pyrene content of the air of American communities. *J. Amer. Ind. Hyg. Assoc.* 21: 443-51.

Sax, N. I. 1974. *Industrial pollution.* New York: Van Nostrand Reinhold Co.

Schroeder, H. A. 1970a. *Air quality monographs: Nickel.* Monogr. 70-14, American Petroleum Institute, Washington, D.C.

Schroeder, H. A. 1971a. *Air quality monographs: Manganese.* Monogr. 70-17, American Petroleum Institute, Washington, D.C.

Schroeder, H. A. 1971b. *Air quality monographs: Cadmium, zinc, and mercury.* Monogr. 70-16, American Petroleum Institute, Washington, D.C.

Schuck, E. A.; Altschuller, A. P.; Barth, D. S.; and Morgan, G. B. 1970. Relationship of hydrocarbons to oxidants in ambient atmospheres. *J. Air Pollut. Control Assoc.* 20(5): 297-302.

Schuck, E. A.; Pitts, J. N.; and Wan, J.K.S. 1966. Relationships between certain meteorological factors and photochemical smog. *Air Water Pollut. Int. J.* 10: 689-711.

Schuck, E. A., and Stephens, E. R. 1969. Oxides of nitrogen. In *Advances in environmental sciences,* vol. 1, ed. J. N. Pitts and R. L. Metcalf, pp. 119-46. New York: Wiley-Interscience.

Shabad, L. M. 1971. Concerning the possibility of establishing maximum permissible doses and concentrations (MPDs and MPCs) for carcinogens. *Hygiene Sanitation* 36(10-12): 115-20.

Shacklette, H. T.; Hamilton, J. C.; Boerngen, J. G.; and Bowles, J. M. 1971. *Elemental composition of surficial materials in the conterminous United States.* Geol. Surv. Prof. Paper 574-D, Washington, D.C.: U.S. Government Printing Office.

Shelton, T. B., and Hunter, J. V. 1975. Anaerobic decomposition of oil in bottom sediments. *J. Water Pollut. Control Fed.* 47(9): 2256-70.

Shepherd, J. G. 1974. Measurements of the direct deposition of sulfur dioxide onto grass and water by the profile method. *Atmos. Environ.* 8: 69-74.

Sherman, G. D.; McHarque, J. S.; and Hodgkiss, W. S. 1942. Determination of active manganese in soil. *Soil Sci.* 54: 253-57.

Singer, P. C. 1973. Preface. In *Trace metals and metal-organic interactions in natural waters*, ed. P. C. Singer, pp. v-vi. Ann Arbor, Mich.: Ann Arbor Science Publishers, Inc.

Singh, B. R. 1974. Migration of ions in soils. *Plant Soil* 41(3): 619-28.

Smith, F. B., and Jeffrey, G. H. 1975. Airborne transport of sulfur dioxide from the U.K. *Atmos. Environ.* 9: 643-59.

Smith, P. V., Jr. 1954. Studies of the origin of petroleum: Occurrence of hydrocarbons in recent sediments. *Bull. Am. Assoc. Petrol. Geol.* 38(3): 377-404.

Spedding, D. J. 1972. Sulphur dioxide absorption by sea water. *Atmos. Environ.* 6: 583-86.

Spencer, D. W.; Brewer, P. G.; and Sachs, P. L. 1972. Aspects of the distribution and trace element composition of suspended matter in the Black Sea. *Geochem. Cosmochim. Acta* 36: 71-86.

Standards Advisory Committee on Coke Oven Emissions. 1971. Advisory panel recommends adoption of benzo(a)pyrene as hazard indicator. *Occupational Safety and Health Reporter* 4(50): 1639-40.

Stegeman, J. J., and Teal, J. M. 1973. Accumulation, release and retention of petroleum hydrocarbons by the oyster *Crassostrea virginica*. *Mar. Biol.* 22: 37-44.

Stephens, E. R. 1969. The formation, reactions, and properties of peroxyacyl nitrates (PANs) in photochemical air pollution. In *Advances in environmental sciences*, vol. 1, ed. J. N. Pitts and R. L. Metcalf, pp. 119-46. New York: Wiley-Interscience.

Suess, M. J. 1975. The environmental load and cycle of polycyclic aromatic hydrocarbons. Presented at The International Conference on Environmental Sensing and Assessment, Las Vegas, Sept. 14-19.

Tebbens, B. D.; Thomas, J. F.; and Mukai, M. 1966. Fate of arenes incorporated with airborne soot. *J. Am. Ind. Hyg. Assoc.* 27: 415-22.

Terraglio, F. P., and Manganelli, R. M. 1966. The influence of moisture on the adsorption of atmospheric sulfur dioxide by soil. *Air Water Pollut. Int. J.* 10: 783-91.

Terraglio, F. P., and Manganelli, R. M. 1967. The absorption of atmospheric sulfur dioxide by water solutions. *J. Air Pollut. Control Assoc.* 17(6): 403-6.

Thomas, J. F.; Mukai, M.; and Tebbens, B. D. 1968. Fate of airborne benzo[a]pyrene. *Curr. Res.* 2(1): 33-39.

Thompson, S. E.; Burton, C. A.; Quinn, D. J.; and Ng, Y.C. 1972. *Concentration factors of chemical effluents in edible aquatic organisms.* UCRL-50564, Rev. 1. University of California, Riverside: Lawrence Livermore Laboratory.

Tiller, K. G. 1958. The geochemistry of basaltic materials and associated soils of south-eastern South Australia. *J. Soil Sci.* 9: 225-41.

Troup, B. N., and Bricker, O. P. 1975. *Processes affecting the transport of materials from continents to oceans.* Microfiche No. COO-3279-17.

Tullar, I. V., and Suffet, I. H. 1975. The fate of vanadium in an urban air shed: The Lower Delaware River Valley. *J. Air Pollut. Control Assoc.* 25(3): 282-86.

Turekian, K. K., and Scott, M. R. 1967. Concentrations of Cr, Ag, Mo, Ni, Co, and Mn in suspended material in streams. *Environ. Sci. Technol.* 1(11): 940-42.

U.S. Dept. of Health, Education, and Welfare. 1970. *Air quality criteria for photochemical oxidants.* National Air Pollution Control Administration Publication AP-63.

Vaughan, B. E.; Abel, K. H.; Cataldo, D. A.; Hales, J. M.; Hane, C. E.; Rancitelli, L. A.; Roufson, R. C.; Wildung, R. E.; and Wolf, E. G. 1975. *Review of potential impact on health and environmental quality from metals entering the environment as a result of coal utilization.* Battelle Energy Program Report, Pacific Northwest Laboratory.

Vostal, J. 1972. Transport and transformation of mercury in nature and possible routes of exposure. In *Mercury in the environment*, pp. 15-27. Cleveland, Ohio: CRC Press.

Wentink, G. R., and Etzel, J. E. 1972. Removal of metal ions by soil. *J. Water Pollut. Control Fed.* 44(8): 1561-74.

Whittle, K.; Mackie, P. R.; and Hardy, R. 1974. Hydrocarbons in the marine ecosystem. *S. Afr. J. Sci.* 70(5): 141-44.

Winchester, J. W., and Nifong, G. D. 1971. Water pollution in Lake Michigan by trace elements from pollution aerosol fallout. *Water Air Soil Pollut.* I(1): 50-64.

Wolfe, D. A., and Rice, T. R. 1972. Cycling of elements in estuaries. *Fish. Bull.* 70(3): 959-72.

Yee, M. S.; Bohn, H. L.; and Miyamoto, S. 1975. Sorption of sulfur dioxide by calcareous soils. *Soil Sci. Soc. Am., Proc.* 39(2): 268-70.

Youngblood, W. W., and Blumer, M. 1975. Polycyclic aromatic hydrocarbons in the environment: homologous series in soils and recent marine sediments. *Geochem. Cosmochim. Acta* 39: 1303-14.

Zeedijk, H., and Velds, C. A. 1973. The transport of sulfur dioxide over a long distance. *Atmos. Environ.* 7: 849-62.

7. MICROBIAL INTERACTIONS

G. A. Dailey

ABSTRACT

Microorganisms are important ecologically in the degradation of most coal conversion wastes.
Pollutants, both elemental and organic, can be used in microbial metabolic processes. Activated
sludge has been shown to be an effective utilizer of pollutants confined to the waste streams
from coal conversion processes, while common soil and water bacteria have exhibited a similar
utilization of pollutants escaping to the environment.

Organic wastes produced by coal conversion processes are readily degraded by microbes. Process
waters are the principal sources of these organic wastes, the composition of which is expected
to be similar to ammonia liquor from coking industries. Phenols and ammonia are two common
pollutants found in the waste streams, which are primarily biotreated by activated sludge
systems. Ammonia wastes (e.g., thiocyanate) are not greatly reduced by sludge due to microbial
poisoning by trace quantities of heavy metals which are also present in the wastes. A
Pseudomonas mutant capable of growth on these nitrogen compounds in the presence of heavy metals
has been discovered. Studies have shown nitrification and denitrification to be common, natural
processes.

Phenol, the major waste discharge toxicant, is degraded readily to resorcinol and pyrocatechol
by bacteria of the *Pseudomonas* genus. The reaction requires molecular oxygen for ring cleavage
and thus demands vigorous aeration during waste treatment.

Polycyclic aromatic hydrocarbons (PAH), another coal conversion waste, have been found to be
degraded by common soil bacteria. However, benzo[*a*]pyrene (BaP), a carcinogen, has given inconsistent results; some investigators have reported up to 85% degradation, whereas others have
reported almost no degradation under similar circumstances. Microbial PAH biosynthesis has also
been investigated. Biosynthesis of BaP in nature, a process which produces background levels
of BaP and thus complicates pollution studies, has been shown.

Hydrocarbon uptake by microbes is explained by several theories, the most plausible of which is
the "micelle" theory. Micelles, which are thought to be surface-active agents produced by
microorganisms, solubilize the hydrocarbon drops to an optimal size for microbial pinocytosis.
Mixed hydrocarbons (e.g., oil) are first attacked in the *n*-alkane range. Such selective attacks
may cause PAH enrichment of aged mixture samples.

Pyridine compounds, released into the environment by the industrial and domestic use of coal,
are degraded by actinomycetes soil bacteria. Microbial metabolism of simple pyridine compounds
produces dihydroxypyridine compounds.

Elements released into the environment by coal conversion processes are commonly transformed by microorganisms. *Thiobacillus* species are able to remove up to 90% of pyritic iron and sulfur in coal during simple storage. These bacteria also oxidize ferrous iron, thiosulfate, and sulfur. Sulfur bacteria are also important parts of the sulfur cycle and perform both assimilatory and dissimilatory sulfate reduction.

Trace metals released during coal conversion are transformed by microorganisms and can lead to an overall metal content rise in the microbial biomass. Lead and mercury are biomethylated, forming the more toxic and volatile dimethylmercury and tetramethyl lead.

7.0 INTRODUCTION

Microorganisms are ecologically important because they are producing, consuming, and transforming members of the ecosystem (Buck 1974). Their functional versatility makes them sensitive indicators of environmental alteration and gives them the ability to use as their sole carbon sources a wide variety of compounds, including contaminants in most environmental media.

Interaction with environmental contaminants makes microbes particularly useful as a part of the waste treatment process for coal conversion systems. Organic wastes such as phenol are readily degraded by most microbes, whereas polycyclic aromatic hydrocarbons (PAH) can be degraded up to 85% by common soil bacteria.

PAH are also microbially synthesized. Some anaerobic soil bacteria biosynthesize PAH compounds (e.g., benzpyrene and benzo[a]pyrene) to the extent that they contribute to PAH background concentrations in the environment.

Elements released into the environment are transformed by microorganisms. Microbes are essential components of the environmental cycle for organic and inorganic sulfur, nitrogen, and a variety of heavy metals; lead and mercury are biomethylated, forming the more toxic and volatile dimethylmercury and tetramethyl lead.

Oil in spills is susceptible to degradation by bacteria; however, degradation time is greatly affected by environmental conditions such as aeration, fertilization, and temperature. Bacterial attack on oil usually begins with the n-alkanes, which can, theoretically, concentrate the more complex organics, such as PAH, in the oil.

7.1 INORGANIC COMPOUNDS

7.1.1 Nitrogen

7.1.1.1 Ammonia

Process waters are the principal sources of water pollution in coal conversion; their composition is expected to be similar to that of ammonia liquor from coking industries (Rubin and McMichael 1975). Biological treatment is a primary method of treating these polluted waters. However, some chemicals in ammonia liquor may be resistant to bacterial degradation, whereas other chemicals may be detrimental to bacteria survival. Ammonia liquor constituents detrimental to nitrifiers include heavy metals, cyanides, halogenated compounds, phenols, mercaptans, guanidines, and thiourea (Adams 1973).

Ammonia is not removed by activated sludge in the treatment of coking plant effluent. Nitrification by nitrifying bacteria is common in soil biology, but no reduction of ammonium ion concentration may be attributed to biological oxidation of coke plant effluent during activated sludge treatment (Kostenbader and Flecksteiner 1969). This lack of nitrification may be due to the presence of detrimental compounds poisoning the bacteria; however, the Canadian Centre for Inland Waters discovered a bacterial mutant of *Pseudomonas* sp. capable of growth on nitrogen compounds and survival in wastewaters (Denitrification 1973).

7.1.1.2 Ammonia nitrification

Sutton et al. (1975) developed a sludge system using *Nitrosomonas* bacteria that converted ammonia to nitrate. It was known that nitrification occurs in an activated sludge system only when conditions are suitable for the retention and accumulation of nitrifying bacteria. A sludge age adequate to retain and prevent the washout of the slower-growing nitrifying bacteria is the factor upon which successful nitrification depends. A carefully monitored sludge wasting program, which determines the solids retention time (SRT), or sludge age, will determine the ability of the nitrifying bacteria to grow.

Two basic sludge system schemes are discussed by Sutton et al. (1975): a combined sludge system, which may be a single- or multistage system, and a separate sludge system, which is normally a two-stage system. These systems are shown in Fig. 7.1. The combined system employs simultaneous carbon removal and nitrification. Nitrification may be maintained if the growth rate of the nitrifying bacteria is rapid enough to compensate for the organisms lost through sludge wasting. Thus, nitrification depends on the interaction of the growth rate of the nitrifying bacteria with the net solids production rate of the process (Sutton et al. 1975). In the separate sludge system, carbon is removed by heterotrophic microorganisms in one complete step, which is separate from the subsequent nitrification carried out by autotrophic bacteria in a second step. This study compared pilot-plant carbon removal and nitrification efficiencies from three process configurations: single- and two-stage combined sludge systems and a two-stage separate sludge system. The performances were compared under a range of temperatures and SRTs. Temperature changes were studied because the *Nitrosomonas* growth rate and subsequent substrate utilization rate are a function of temperature.

ORNL DWG NO 76-7958

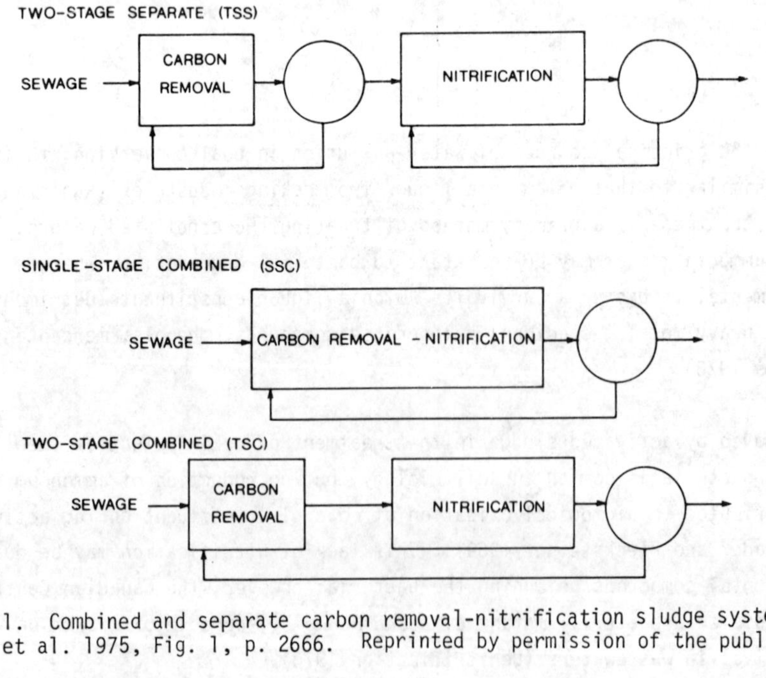

TWO-STAGE SEPARATE (TSS)

SEWAGE | CARBON REMOVAL | NITRIFICATION

SINGLE-STAGE COMBINED (SSC)

SEWAGE | CARBON REMOVAL - NITRIFICATION

TWO-STAGE COMBINED (TSC)

SEWAGE | CARBON REMOVAL | NITRIFICATION

Fig. 7.1. Combined and separate carbon removal-nitrification sludge systems. Source: From Sutton et al. 1975, Fig. 1, p. 2666. Reprinted by permission of the publisher.

Results of the comparison by Sutton et al. (1975) are shown in Table 7.1. Variations due to
temperature change are defined by an Arrhenius relationship,

$$K = K^{*}e^{-E/R\left(\frac{1}{T} - \frac{1}{T_0}\right)} , \qquad (1)$$

in which

$K^{*} = Ae^{-A/RT_0}$;

K = reaction rate constant, per hour;

A = frequency factor;

E = activation energy, cal/g-mole;

R = universal gas constant, cal/g-mole/°K;

T = temperature, °K;

T_0 = median of the temperature range, °K.

Changes in the parameters expressed above minimize the interaction between the frequency factor,
A, and the activation energy, E, which makes the Arrhenius equation a difficult expression to fit.
Rate data of the separate or combined sludge systems indicated no lack of fit by analysis of
variance when applied to this modeling procedure.

Table 7.1. Results of analysis of variance for rate variation with temperature[a]

Reactor configuration	E (cal/g-mole)	A	$\dfrac{\text{Mean square lack of fit}}{\text{Mean square pure error}}$	F α = 0.99	F α = 0.95
Separate sludge					
4-day SRT	25,900	2.15×10^{18}	6.74	7.0	3.8
7-day SRT	20,000	6.60×10^{13}	1.19	3.6	2.7
10-day SRT	15,050	1.51×10^{10}	6.39	99.3	19.3
4-, 7-, and 10-day SRTs			13.33	2.8	2.0
4-, 4-, and 10-day SRTs			11.91	2.7	2.0
(pooled data with K:K*)					
Combined sludge					
4-day SRT	25,100	3.08×10^{17}	17.58	6.6	3.7
7-day SRT	21,250	2.89×10^{14}	1.41	4.0	2.7
10-day SRT	11,900	3.18×10^{17}	3.84	99.3	19.3
4-, 7-, and 10-day SRTs			4.65	2.7	2.0
4-, 7-, and 10-day SRTs			4.19	2.6	2.0
(pooled data with K:K*)					
Combined and separate:					
4-day SRT			40.61	3.6	2.5
7-day SRT			7.07	2.7	2.0
10-day SRT			24.07	14.6	6.0
Combined and separate					
(pooled data with K:K*)					
4-day SRT			22.60	3.5	2.4
7-day SRT			4.37	2.6	2.0
10-day SRT	13,000		9.02	14.5	5.9

[a]E = activation energy; A = frequency factor; F = F-variable statistic test.

Source: Sutton et al. 1975, Table 4, p. 2670. Reprinted by permission of the publisher.

7.1.1.3 <u>Denitrification</u>

Nitrate and other nitrogen compounds cause an oxygen depletion in natural bodies of water and act as inorganic nutrients for algal growth. Jeris and Owens (1975) developed a high-rate biological denitrification unit for removal of nitrates from water. The unit (Fig. 7.2) employs a biological process, using activated carbon as a biological growth support medium, in which bacteria reduce nitrite or nitrate in the influent waste stream to a harmless nonpolluting end-product of nitrogen gas.

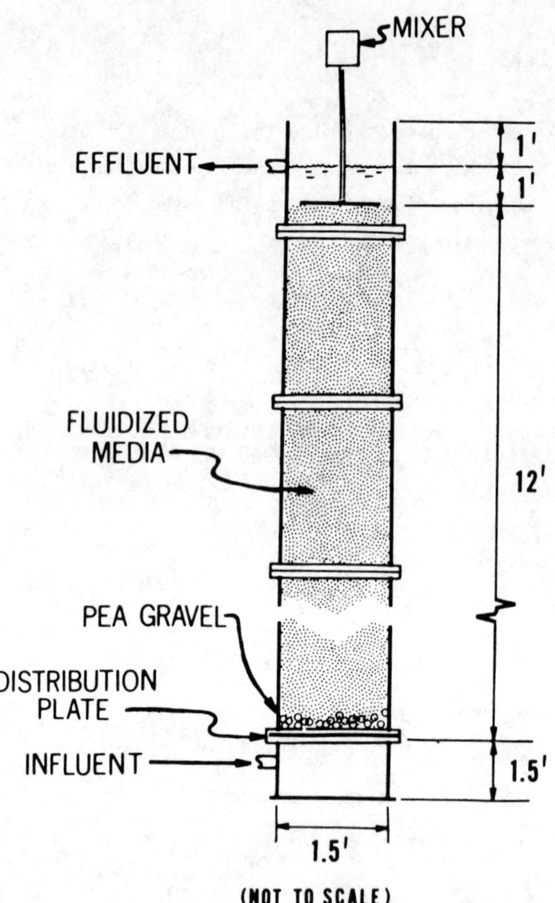

Fig. 7.2. Denitrification reactor. <u>Source</u>: From Jeris and Owens, Fig. 1, p. 2044. Reprinted by permission of the publisher.

The Jeris and Owens (1975) denitrification reactor passes wastewater upward in a cylinder through a bed of small particles, such as activated carbon or sand, at a velocity sufficient to cause motion or fluidization of all the medium. The use of small particles as a support growth medium vastly increases the amount of bacteria able to grow due to the increase in surface area. The increase in bacteria allows a concomitant increase in nitrite or nitrate removed in a given volume of fluid in the reactor. The use of medium fluidization increases effective surface area as compared with that of a packed bed and minimizes operational problems such as clogging.

The bacteria cultivated in this denitrification reactor are heterotrophic and require an organic carbon source to maintain metabolic processes. Because secondary effluents usually contain insufficient carbon, an external source of carbon must be introduced into the treatment system. Methanol was used by Jeris and Owens in their work because of its availability, low cost, and low bacterial cell yield. Most bacteria used in this work were facultative and could function as either aerobes or anaerobes. The bacteria preferred elemental oxygen to nitrate as their energy source, meaning that anaerobic conditions must be maintained to obtain the greatest levels of nitrate reduction. The nitrogen reduction reactions are part of bacterial anaerobic respiration, which produces a rather small cell yield, and proceed as follows:

$$NO_3^- + 1/3\ CH_3OH = NO_2^- + 1/3\ CO_2 + 2/3\ H_2O\ ; \qquad (2)$$

$$NO_2^- + 1/2\ CH_3OH = 1/2\ N_2 + 1/2\ CO_2 + 1/2\ H_2O + OH^-\ ; \qquad (3)$$

$$\text{Net:}\ \ NO_3^- + 5/6\ CH_3OH = 1/2\ N_2 + 5/6\ CO_2 + 7/6\ H_2O + OH^-\ . \qquad (4)$$

Results of the denitrification reactor operation for a typical period of time and with ample amounts of methanol added are shown in Table 7.2. Jeris and Owens (1975) found that methanol control was a very important consideration in the reactor operation. Excess methanol in the reactor was economically undesirable and left a residual biochemical oxygen demand in the effluent. A methanol shortage in the reactor slowed cell metabolism and decreased performance of the system.

Table 7.2. Summary of operation, Oct. 3 - Dec. 12, 1973

Parameter	Value
Flow (gpm)	26.0
Flux (gpm/ft^2)	15.6
Empty bed detention time (min)	6.5
Temperature (°F)	68
Influent pH	6.6
Effluent pH	7.5
Influent DO (mg/liter)	1.5
$CH_3OH:NO_3$-N (mg/liter)	4.2
Influent $(NO_3 + NO_2)$-N (mg/liter)	21.5
Effluent $(NO_3 + NO_2)$-N (mg/liter)	0.2
Percentage $(NO_3 + NO_2)$-N removed	99.0
$(NO_3 + NO_2)$-N removed (lb/day/1000 ft^3)	335.0

Source: Jeris and Owens 1975, Table 1, p. 2047.
Reprinted by permission of the publisher.

Jeris and Owens (1975) tested their denitrification reactor at the highest flow rate attainable (40 gpm = 151 liters/min) to determine the effect of hydraulic load. The reactor, less 75 lb of sand-biological mass to allow for bed expansion at the higher flow rate, still removed an excess of 95% of the nitrogen compounds (Table 7.3 and Fig. 7.3). The effects of diurnal flow

Table 7.3. High-flux study

Time (days)	Flow (gpm)	Flux (gpm/ft^2)	NO_3-N (mg/liter)		NO_2-N (mg/liter)		$NO_2 + NO_3$)-N (mg/liter)		Removal (%)
			Influent	Effluent	Influent	Effluent	Influent	Effluent	
0.0	40	24							
0.8	40	24	16.0	0.2	0.3	0.2	16.3	0.4	97.7
1.1			16.3	0.2	0.4	0.1	16.7	0.3	98.3
1.7	39	23	17.0	0.2	0.4	0.1	17.4	0.3	98.3
2.0			19.0	0.0	0.2	0.1	19.2	0.1	99.8
2.9	39	23	18.0	0.0	0.2	0.1	18.2	0.1	99.8
3.1			21.5	0.0	0.2	0.1	21.5	0.1	99.8
4.0	37	22	21.0	0.1	0.2	0.1	21.2	0.1	99.3
4.9	38	23	16.0	0.0	0.2	0.1	16.2	0.1	99.1
5.8	37	22	15.0	0.1	0.1	0.1	15.1	0.1	99.2
7.1			24.0	0.0	0.1	0.1	24.1	0.1	99.8

Source: Jeris and Owens 1975, Table 3, p. 2051. Reprinted by permission of the publisher.

ORNL DWG NO 76-6591

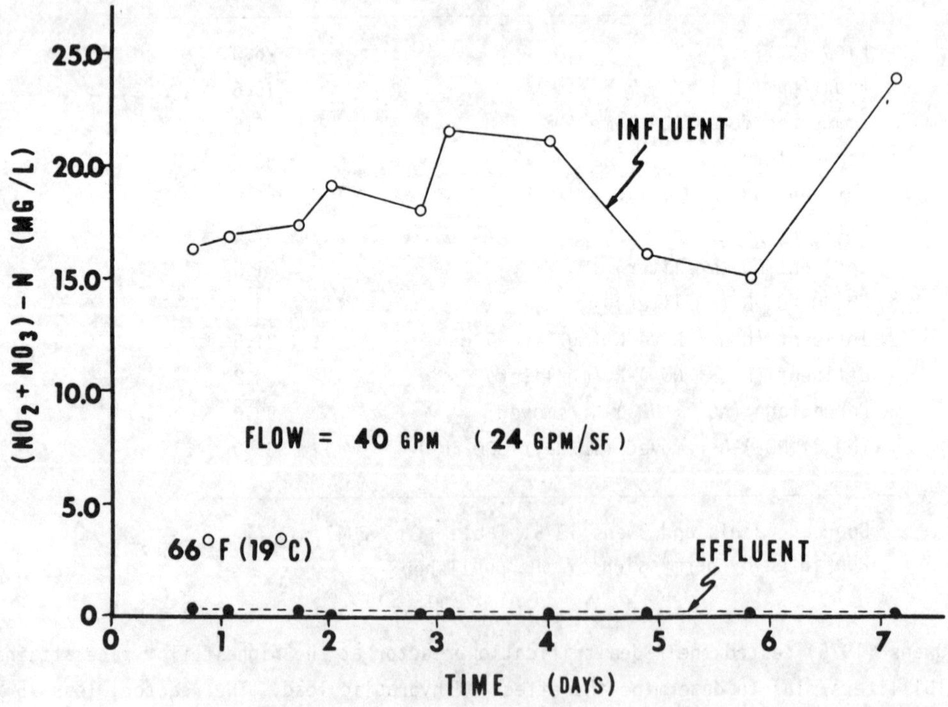

Fig. 7.3. Results of high-flux study. Source: From Jeris and Owens 1975, Fig. 11, p. 2052. Reprinted by permission of the publisher.

variation were also studied by manually increasing the flow rate during the day and reducing the
rate at night. Again, there were no detrimental effects on system performance (Fig. 7.4).

ORNL DWG NO 76-6592

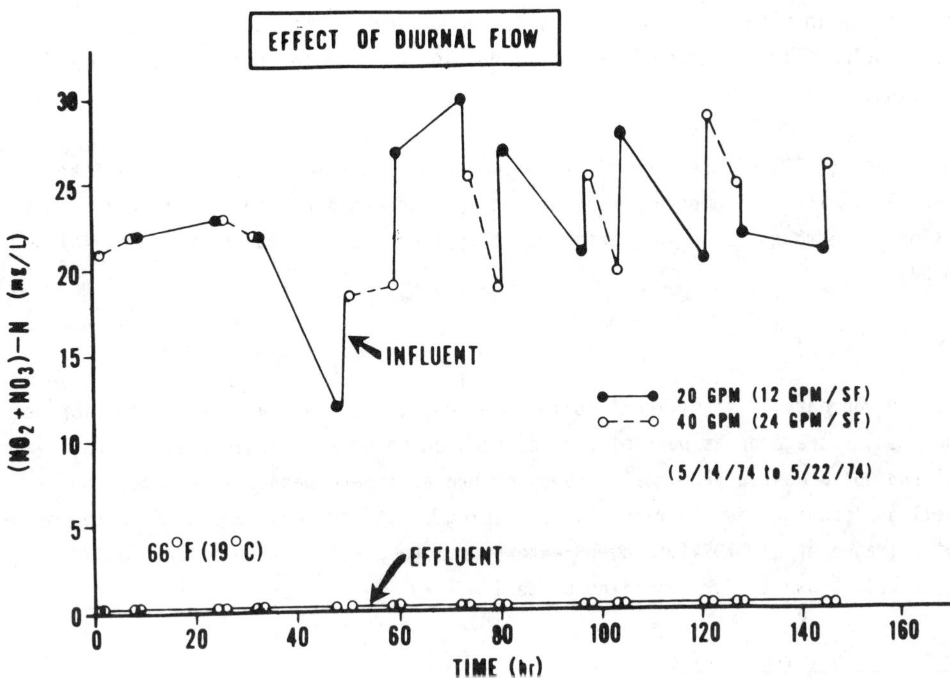

Fig. 7.4. Effects of diurnal flow variation on nitrite and nitrate concentration (top)
and nitrite and nitrate quantities (bottom). <u>Source</u>: From Jeris and Owens 1975, Fig. 13,
p. 2053. Reprinted by permission of the publisher.

The use of bacteria for nitrification and denitrification stems from the fact that both processes
are routinely carried out by common soil bacteria. Ammonia liberated during protein breakdown
or use serves as an energy source for nitrite bacteria. The bacteria use ammonia and oxygen in
an oxidation reaction that transforms ammonia to nitrite. The nitrite produced by this bacterial
action generally is further oxidized to nitrate by nitrifying bacteria. This two-step process
is called nitrification (Rheinheimer 1974). The first reaction involved,

$$NH_4^+ + 1\text{-}1/2\ O_2 = NO_2^- + H_2O + 2H^+ + 76\ \text{kcal}\quad, \qquad (5)$$

is called the nitritation step. This step is carried out by nitrite bacteria such as *Nitrosomonas
europea*. The second nitrification step is called nitratation:

$$NO_2^- + 1/2\ O_2 - NO_3^- + 24\ \text{kcal} \qquad\qquad (6)$$

Nitratation is carried out by nitrate bacteria such as *Nitrobacter winogradskyi* or *Nitrospina
gracilis* and *Nitrococcus mobilis* in the oceans. Formation of nitrate is the final step of the
mineralization of organic nitrogen compounds (Rheinheimer 1974). Nitrification is strictly
aerobic even though a very small concentration of oxygen will suffice for the reactions.

Denitrification is an anaerobic nitrogen reaction carried out by anaerobic bacteria; it is the dissimilatory reduction of nitrate, in the presence of organic hydrogen donors, to nitrite and the subsequent reduction of nitrite to nitric oxide (NO), nitrous oxide (N_2O), and molecular nitrogen (N_2). Under anaerobic conditions, nitrate, and then nitrite, acts as a hydrogen acceptor; this process represents anaerobic respiration, also called nitrate respiration. Actual anaerobic respiration in nature involves mainly the reduction of nitrate to nitrite because more organisms are capable of this reaction than the reduction of nitrite to free nitrogen (Rheinheimer 1974).

Robinson and Robbins (1970) discuss the nitrogen compounds present in the environment and their sources. Generally, nitrogen compounds released as a consequence of biological action, mainly bacterial action, were found to be in greater quantity than those released by industrial sources (see Table 6.64).

7.1.2 <u>Sulfur</u>

Coal contains environmentally significant sulfur compounds, which are accessible to microbes during outdoor coal storage or as part of coal conversion solid and liquid wastes. For example, coal and solid wastes from coal conversion processes are washed with water; these waters, as well as leachings due to rainfall or washing of stored coal, are similar to low-pH acid mine water (Magee et al. 1974). Low-pH waters promote proliferation of *Thiobacillus ferrooxidans*, a well-known sulfur-oxidizing bacterium.

7.1.2.1 <u>Pyritic sulfur, iron pyrite</u>

Iron- and sulfur-reducing bacteria have been studied since the beginning of this century, and work on the identification of iron-reducing bacteria was recorded as early as 1927 (Starkey and Halvorson 1927). Studies of the physiology of iron bacteria were first recorded in the early 1930s and were continued through the 1940s. The studies involved *Gallionella* species, which were thought to be strict autotrophs, and species of *Leptothrix* and *Crenothrix*, which were thought to be facultative autotrophs (Starkey 1945). The study of *Thiobacillus* species for the oxidation of sulfur compounds was under way in 1935 (Starkey 1935).

Bacteria may reduce the pyritic sulfur content up to 80 to 90% (Capes et al. 1973). *Thiobacillus ferrooxidans* is commonly found in coal being weathered before processing (Silverman et al. 1963). Low-pH treatment streams may promote the growth of both *Thiobacillus thiooxidans*, which is a well-known sulfur oxidizer, or *Chlamydobacteriales* members, which are iron-bearing bacteria (Colmer and Hinkle 1947). Iron content in a low-pH treatment stream may be decreased by the iron-oxidizing *Metallogenium*, which is often found in acid mine drainage (Walsh and Mitchell 1975). Acidophilic bacteria in acid streams have been shown to readily oxidize ferrous iron at a rate of up to 600 mg/liter/hr if adequate amounts of oxygen, carbon dioxide, nitrogen, and phosphorus are present (Whitesell et al. 1971).

Kelly and Tuovinen (1975) studied *Thiobacillus ferrooxidans*, best known for its ability to oxidize ferrous iron at a pH of 1 to 4 and for its ability to oxidize sulfur and thiosulfate. All strains studied could oxidize sulfur and thiosulfate and, when cultures were transferred, they could oxidize iron again (Kelly and Tuovinen 1975). *T. ferrooxidans* could be adapted to grow on thiosulfate or tetrathionate, but after growth on iron, very few of the organisms could grow on tetrathionate (Kelly and Tuovinen 1975). Table 7.4 shows the development of the organism and the readaptation of all *T. ferrooxidans* cultures to iron oxidation after growth on sulfur compounds. Oxidation rates for sulfur compounds by *T. ferrooxidans* are shown in Fig. 7.5.

Table 7.4. Development of colonies of *Thiobacillus ferrooxidans* on membrane filters on agar media after prior culture in liquid media with various substrates

Liquid culture substrate	Percentage of organisms producing colonies on agar containing		
	$FeSO_4$	$Na_2S_2O_3$	$K_2S_4O_6$
$FeSO_4$	100	$<10^{-6}$	66 - 82
Thiosulfate	8 - 48	100^a	97 - 100
Tetrathionate	~40	100	100
Sulfur	100 (max)		72 (max)

[a]All these colonies oxidized $FeSO_4$ when the membranes were transferred to $FeSO_4$ agar.

Source: Kelly and Tuovinen 1975, Table 1, p. 80.

ORNL DWG NO 76-11833

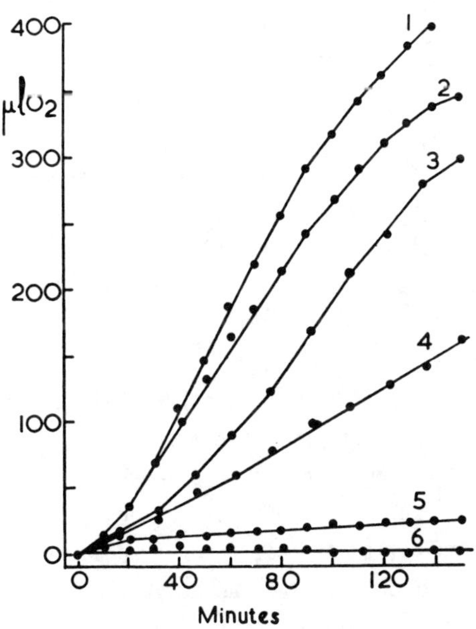

Fig. 7.5. Oxidation of sulfur compounds by *Thiobacillus ferrooxidans* (2.86 mg/Warburg flask) previously grown on thiosulfate at pH 3.9 to 4.0. Flasks received: (1) $K_2S_3O_6$— 10 μmoles; (2) $K_2S_4O_6$ — 5 μmoles; (3) sulfur — 30 mg and $Na_2S_2O_3$ — 10 μmoles; (4) sulfur — 30 mg; (5) $Na_2S_2O_3$ — 10 μmoles; (6) $Na_2S_2O_6$ — 10 μmoles. <u>Source</u>: From Kelly and Tuovinen 1975, Fig. 1, p. 80.

Process regulation by microorganisms responsible for degradation of waste streams or waters containing iron and/or sulfur has drawbacks worth considering. The iron and sulfur bacteria responsible for degradation in water produce odors and unsavory tastes. Early experiments found that superchlorination and dechlorination were effective methods of reducing bacterial populations to a desirable level (Alexander 1941).

7.1.2.2 Inorganic sulfur

Bacterial types

Colorless sulfur bacteria belonging to the families *Thiobacteriaceae*, *Beggiatoaceae*, and
Achromatiaceae have been identified as oxidizers of reduced inorganic sulfur compounds (Kuenen
1975; Buchanan and Gibbson 1974). Genera of the three families are shown in Table 7.5.
Thiobacillus is the best known genus of the colorless sulfur bacteria. The listing of microorganisms as *Thiobacilli* is based on physiological rather than morphological criteria. *Thiobacilli* have in common their ability to derive energy from the oxidation of reduced inorganic
sulfur compounds.

Table 7.5. The colorless sulfur bacteria

Thiobacteriaceae	Beggiatoaceae	Achromatiaceae
Thiobacterium	*Beggiatoa*	*Achromatium*
Macromonas	*Thiospirillopsis*	
Thiovulum	*Thioploca*	
Thiospira	*Thiothrix*	
Thiobacillus	*Thiodendron*	
Thiomicrospira		
Sulfolobus		

Source: Kuenen 1975, Table 1, p. 51.
Reprinted by permission of the publisher.

Colorless bacteria can be found almost everywhere in nature; different habitats, along with the
name of the colorless sulfur bacteria that would occur in the habitat, are shown in Fig. 7.6.
All colorless sulfur bacteria are not equally versatile; there appears to be a complete spectrum
from obligate chemolithotrophs to heterotrophs, as shown in Fig. 7.7.

Obligate chemotrophs are shown at the top, and the mixotrophs, which have a flexible metabolism
allowing them to grow either heterotrophically or autotrophically, are shown at the right.
Mixotrophic organisms that are not colorless sulfur bacteria can oxidize thiosulfate (aerobically)
and grow autotrophically (in the dark) on some sulfur compounds (Kuenen 1975). *Thiobacillus*
perfometabolis can derive energy by sulfur compound oxidation, but cannot grow autotrophically.
This organism is classified as a chemolithotrophic heterotroph. Heterotrophs such as *Beggiatoa*
sp. cannot oxidize sulfur, but others are catalase-negative and dispose of toxic intracellular
hydrogen peroxide by reaction of this compound with hydrogen sulfide. Heterotrophs that can
oxidize sulfur compounds but do not seem to benefit from the oxidation also exist. Some
heterotrophs may live in association with obligate chemolithotrophs and stimulate their growth.
They live on metabolic waste products of the obligate chemolighotrophs and, in this way, contribute indirectly to the oxidation of sulfur compounds.

Oxidation reactions

Two major physiological groups exist within the genus *Thiobacillus*. The first group is made
up of obligate chemotrophs, which depend on the oxidation of reduced inorganic sulfur compounds
to obtain energy and use carbon dioxide as the major carbon source under all growth conditions.

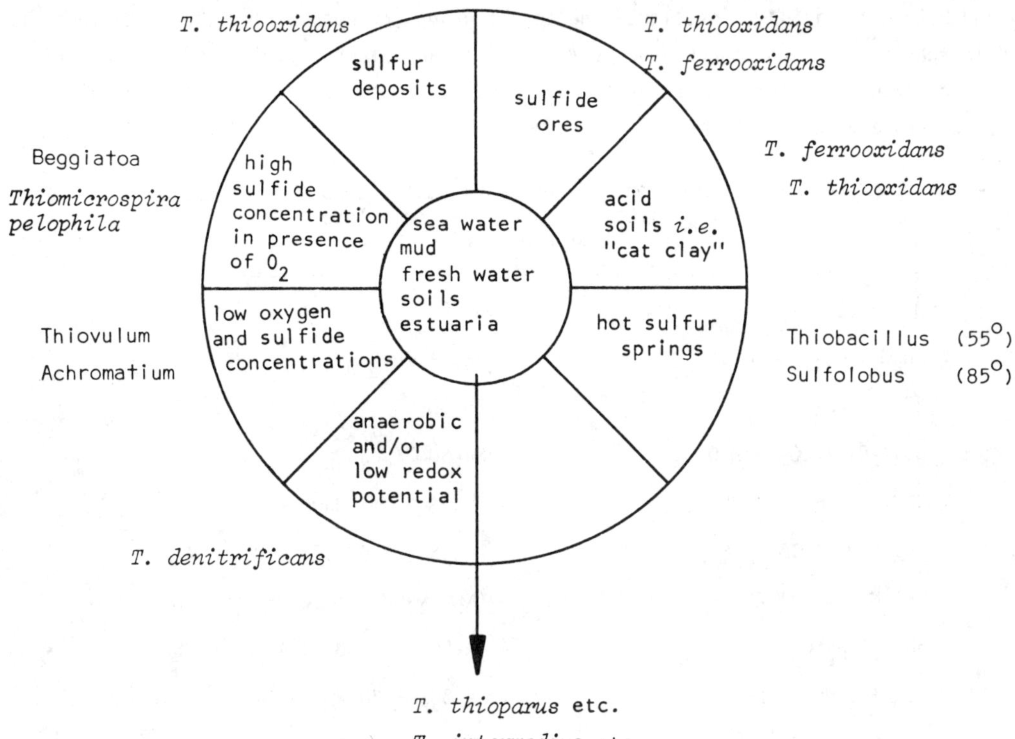

Fig. 7.6. Occurrence of colorless sulfur bacteria in exceptional (outer circle) and moderate (inner circle) environments. <u>Source</u>: From Kuenen 1975, Fig. 3, p. 58. Reprinted by permission of the publisher.

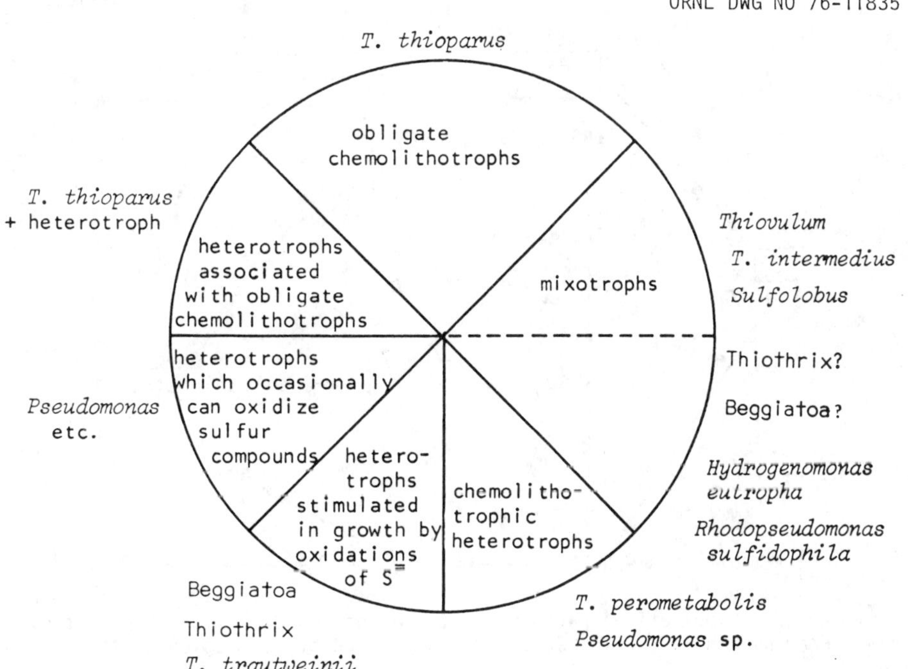

Fig. 7.7. Spectrum of organisms involved in the aerobic metabolism of inorganic reduced sulfur compounds. <u>Source</u>: From Kuenen 1975, Fig. 6, p. 69. Reprinted by permission of the publisher.

Mixotrophs (facultative autotrophs), which can grow not only autotrophically, but also mixotroph-
ically and heterotrophically, constitute the second group (Kuenen 1975). The oxidations of
reduced inorganic sulfur compounds used by *Thiobacilli* and other colorless sulfur bacteria are
given in Table 7.6. A brief summary of the hypothetical pathways for the oxidation of thiosulfate, sulfur, and sulfide by members of thiobacilli is shown in Fig. 7.8.

Table 7.6. Oxidations of inorganic sulfur compounds effected
by thiobacilli and other colorless
sulfur bacteria

$H_2S + 2O_2$	$\rightarrow H_2SO_4$
$2H_2S + O_2$	$\rightarrow 2S^0 + 2H_2O$
$2S^0 + 3O_2 + 2H_2O$	$\rightarrow 2H_2SO_4$
$Na_2S_2O_3 + 2O_2 + H_2O$	$\rightarrow Na_2SO_4 + H_2SO_4$
$4Na_2S_2O_3 + O_2 + 2H_2O$	$\rightarrow 2Na_2S_4O_6 + 4NaOH$
$2Na_2S_4O_6 + 7O_2 + 6H_2O$	$\rightarrow 2Na_2SO_4 + 6H_2SO_4$
$2KSCN + 4O_2 + 4H_2O$	$\rightarrow (NH_4)_2SO_4 + K_2SO_4 + 2CO_2$
$5H_2S + 8KNO_3$	$\rightarrow 4K_2SO_4 + H_2SO_4 + 4N_2 + 4H_2O$
$5S^0 + 6KNO_3 + 2H_2O$	$\rightarrow 3K_2SO_4 + 2H_2SO_4 + 3N_2$
$5Na_2S_2O_3 + 8NaNO_3 + H_2O$	$\rightarrow 9Na_2SO_4 + H_2SO_4 + 4N_2$

Source: Kuenen 1975, Table 2, p. 55. Reprinted by permission of the
publisher.

ORNL DWG NO 76-11836

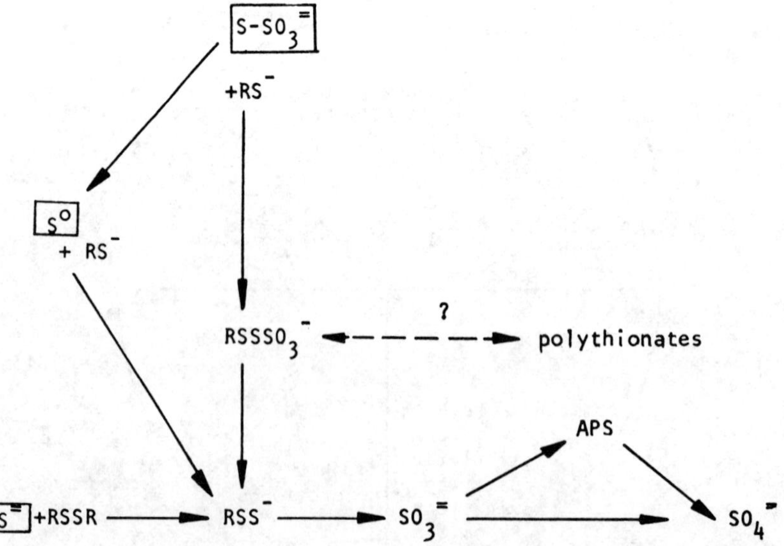

Fig. 7.8. Hypothetical pathways of the oxidation of sulfur compounds by thiobacilli.
Source: From Kuenen 1975, Fig. 2, p. 57. Reprinted by permission of the publisher.

Oxidation-reduction cycle

Pfennig (1975) reports that members of the photosynthetic purple and green sulfur bacteria, as
well as members of the purple nonsulfur bacteria, have the unique characteristic of oxidizing
hydrogen sulfide to sulfate under anaerobic conditions in the course of their anoxigenic photosynthesis. Sulfate, one pole of the sulfur cycle, is stable under both aerobic and anaerobic
conditions, whereas hydrogen sulfide, the second pole of the sulfur cycle, and compounds containing a sulfhydryl group persist only under anaerobic conditions because of autooxidation
with molecular oxygen (Fig. 7.9). Assimilatory sulfate reduction forms reduced sulfur compounds
that are cell constituents that are protected within the cells against reaction with hydrogen.
However, degradation of cell material under anaerobic conditions produces hydrogen sulfide.
Dissimilatory sulfate reduction, the other biological sulfate-reducing process, which is
quantitatively more important in nature, is carried out under anaerobic conditions by sulfate-
reducing bacteria that accumulate hydrogen sulfide in their environment. Assimilatory and
dissimilatory sulfate reduction form a sharp discontinuity in the redox potentials occurring
between the aerobic and hydrogen-sulfide-containing habitats. In this environment, elemental
sulfur (S^0) is formed as the result of an inorganic oxidation of hydrogen sulfide with
molecular oxygen (Pfennig 1975).

ORNL DWG NO 76-11837

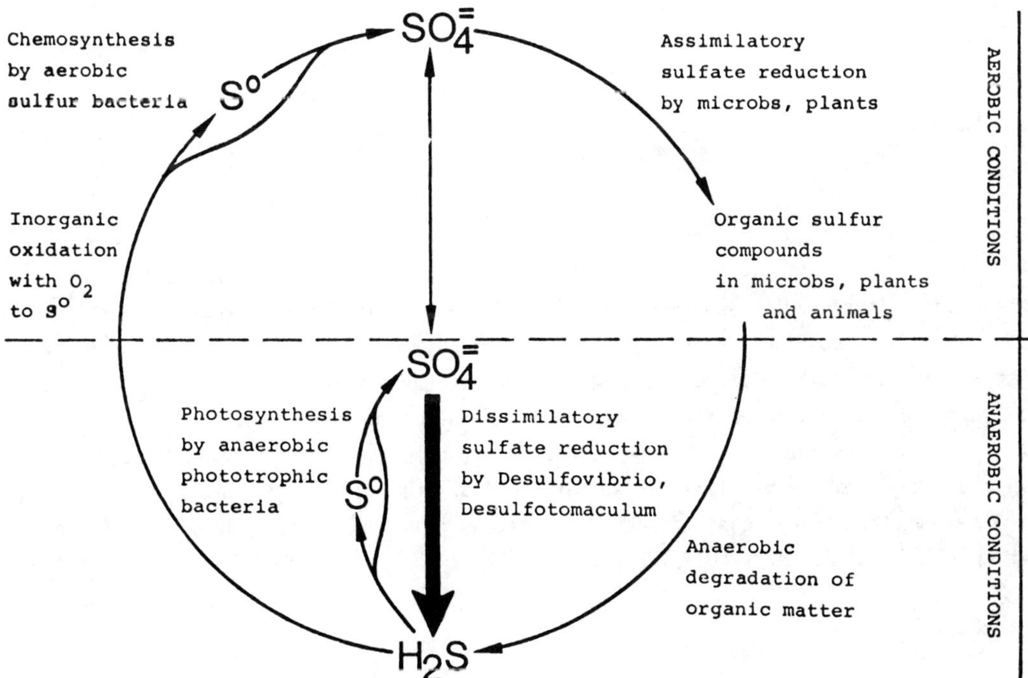

Fig. 7.9. Sulfur cycle. Source: From Pfennig 1975, Fig. 1, p. 2. Reprinted
by permission of the publisher.

Light-dependent reactions

A detailed study of pure cultures of phototrophic bacteria differentiated five major types of
light-dependent hydrogen sulfide and elemental sulfur utilization (Table 7.7) (Pfennig 1975).
Types 1, 2, and 3 are characteristic for the traditional purple and green sulfur bacteria,

Table 7.7. Types of light-dependent H_2S and S^o utilization in the phototrophic bacteria

Type	Extra-cellular substrate	Intermediary S^0 formation	Extra-cellular end-product	Organisms
1a	H_2S	At low H_2S concentration ⟶	SO_4-	Purple sulfur bacteria
b	H_2S	→ S^0, intracellular ⟶	SO_4-	Chromatium, Thiocystis
c	S^0	⟶	SO_4-	Thiocapsa, Thiospirillum
				Thiosarcina, Lamprocystis
				Thiodictyon, Amoebobacter
				Thiopedia
2a	H_2S	At low H_2S concentration ⟶	SO_4-	Purple sulfur bacterium
b	H_2S	→ S^0, extracellular ⟶	SO_4-	Ectothiorhodospira
c	S^0	⟶	SO_4-	Green sulfur bacteria
				Chlorobium, Prostheco-chloris, Pelodictyon
3a	H_2S	At low H_2S concentration ⟶	SO_4-	Green sulfur bacteria
		incomplete		Marine strains of
				Chlorobium vibrioforme
b	H_2S	→ S^0, extracellular ----⟶ incomplete	SO_4-	
c	S^0	-----------------------⟶	SO_4-	
4	H_2S	⟶	SO_4-	Purple nonsulfur bacteria
				Rps. palustris
				Rps. sulfidophila
5	H_2S	⟶	$-S^0$	Purple nonsulfur bacteria
				Rps. sphaeroides
				Rps. capsulata, Rps. rubrum

Source: Pfennig 1975, Table 1, p. 5. Reprinted by permission of the publisher.

which can oxidize hydrogen sulfide and intra- and extracellular elemental sulfur (Table 7.7, Types 1c, 2c, and 3c). Types 1a, 2a, and 3a (Table 7.7) show that no elemental sulfur is formed when homocontinuous cultures are grown with hydrogen sulfide as the growth-limiting factor and the hydrogen sulfide concentration in the growth medium kept at very low concentrations. Types 4 and 5 (Table 7.7) are characteristic nonsulfur purple bacteria that are unable to metabolize sulfur as well as do the purple and green sulfur bacteria. *Rhodopseudomonas palustris* and *Rhodopseudomonas sulfidophila* metabolize hydrogen sulfide (or thiosulfate) to sulfate without the formation of elemental sulfur, even at high sulfide concentrations (Table 7.7, type 4). The other purple nonsulfur bacteria metabolize hydrogen sulfide only to elemental sulfur, which is accumulated extracellularly (Table 7.7, type 5) (Pfennig 1975).

7.1.2.3 Tetrathionate and thiosulfate

Members of the Enterobacteriaceae family that belong to the genera *Proteus*, *Citrobacter*, and *Salmonella* can reduce tetrathionate to thiosulfate, and thiosulfate to sulfite and sulfide (Oltmann et al. 1975). Three representatives of these genera and the specific enzyme activities involving inorganic sulfur are listed in Table 7.8. It was concluded that the reductases involved in reduction of tetrathionate and thiosulfate are formed constitutively because their specific activities after growth without an inorganic electron acceptor are essentially similar to those found after inorganic growth in the presence of substrates (Oltmann et al. 1975).

Table 7.8. Specific reductase activities in some Enterobacteriaceae after various growth conditions

Organism	Inorganic electron acceptor during growth	Specific reductase activities (μmoles substrate/mg protein/hr) for the reduction of			
		NO_3^-	ClO_3^-	$S_2O_3^-$	$S_4O_6^-$
Proteus mirabilis	None	2.8	23.6	32.0	14.7
	NO_3-	30.8	38.0	0.0	0.0
	S_2O_3-	1.7	16.1	29.6	17.3
	S_4O_6-	2.3	14.3	21.5	13.5
	O_2	0.4	0.7	0.0	0.0
Salmonella typhimurium	None	5.1	12.5	1.2	3.2
	NO_3-	117.7	126.0	0.0	0.0
	S_2O_3-	2.9	10.6	3.4	
Citrobacter freundii	None	4.4	10.1		
	NO_3-	52.3	77.3		
	S_2O_3-			2.1	2.3

Source: Oltmann et al. 1975, Table 1, p. 154. Reprinted by permission of the publisher.

Thiobacillus ferrooxidans has been shown to produce thiosulfate from tetrathionate anaerobically (Table 7.9) (Kelly and Tuovinen 1975). A ratio of 1:1 is formed between the amount of tetrathionate consumed and the amount of thiosulfate produced. Anaerobic thiosulfate production by *T. ferrooxidans*, the rapid anaerobic production of thiosulfate from trithionate by *T. neapolitanus* strain C, and the inability of *T. A2* to anaerobically degrade tetrathionate are compared in Table 7.9.

Table 7.9. Anaerobic metabolism of tetrathionate or trithionate by three thiobacilli[a]

Length of incubation (min)	*Thiobacillus species*					
	T. ferrooxidans		*T. A2*		*T. neapolitanus*	
	$S_4O_6^{2-}$	$S_2O_3^{2-}$	$S_4O_6^{2-}$	$S_2O_3^{2-}$	$S_4O_6^{2-}$	$S_2O_3^{2-}$
0	1.88	0.08	4.78	0	8.00	0
5			4.86	0		
10						0.51
20	1.28	0.70	5.04	0		1.11
40	0.86		4.94	0		2.14
60	0.76	1.18				3.55
80	0.00	1.70	4.86	0		
90						5.13
135						6.05

[a] *Thiobacillus ferrooxidans* (3 mg/ml) and *Thiobacillus A2* (5 mg/ml) received $K_2S_4O_6$ (2 and 5 μmoles/ml respectively), and *Thiobacillus neapolitanus* C (4 mg/ml) received $K_2S_3O_6$ (8 μmoles/ml). The suspensions were incubated at 30°C under 100% nitrogen, and samples periodically removed for analysis. The concentration of sulfur compounds is expressed as μmoles/ml.

Source: Kelly and Tuovinen 1975, Table 2, p. 82. Reprinted by permission of the publisher.

7.1.2.4 <u>Elemental sulfur</u>

Sulfur scavenged from the process, product, and cleanup coal conversion streams may be converted to elemental sulfur for disposal.

A study by Adamczyk-Winiarska, Król, and Kobus (1975) of the effect of elemental sulfur on soil showed that sulfur microorganisms can oxidize the sulfur to sulfate, which leads to an acidic condition in the soil. However, the addition of $CaCO_3$ to the soil neutralizes the acid. Figure 7.10 shows the formation of sulfuric acid as a result of elemental sulfur oxidation by soil microorganisms. The soil pH fell to 2 at high sulfur application rates without calcium carbonate. Addition of $CaCO_3$ protected the soil from pH change and thus provided better developmental conditions for sulfur-oxidizing microorganisms. Solids treated with $CaCO_3$ contained three times more sulfate than untreated soils (Adamczyk-Winiarska, Król, and Kobus 1975).

ORNL DWG NO 76-11838

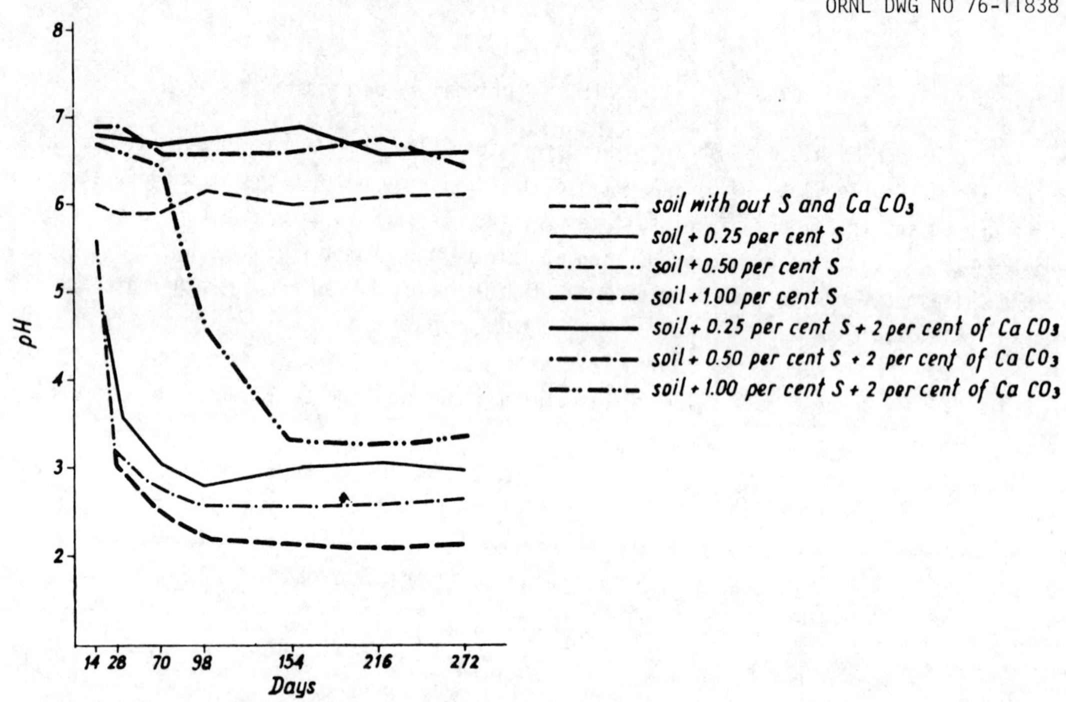

Fig. 7.10. Changes in pH as a result of sulfur oxidation in the soil. <u>Source</u>: From Adamczyk-Winiarska, Król, and Kobus 1975, Fig. 1, p. 97. Reprinted by permission of the publisher.

7.2 ORGANIC COMPOUNDS

Organic wastes are produced during the coal conversion processes. Process waters are the principal sources of these organic wastes, and their composition is expected to be similar to ammonia liquor from coking industries (Rubin and McMichael 1975). Biological treatment is a primary method of treating organically polluted waters; however, some chemicals in ammonia liquor may be resistant to bacterial degradation, whereas others may be detrimental to the survival of bacteria. For example, thiocyanate, a major constituent of ammonia liquor, was

not greatly removed by bacterial action (Baker and Thompson 1973) because conditions were not
conducive to nitrifiers; however, the oxidizing organisms are sensitive, and other chemical
compounds in ammonia liquor may poison them.

7.2.1 Phenols and xylenols

An organic pollutant found commonly in wastewaters is phenol and/or phenolic compounds. Carbonization of coal produces benzene, which is the principal source of phenol.

7.2.1.1 Toxicity of phenols

Baker and Weston (1975) and Reynolds et al. (1974) estimate phenol to be the major toxicant in
waste discharges and believe high temperature may have a significant effect on the toxicity of
the wastes. Experiments by Tomcsik (1955) showed that phenol denatured the cytoplasmic membrane
in *Bacillus megatherium*. Reynolds et al. (1974), working with the alga *Selenastrum capricornutum*, determined that phenol toxicity increased with an increase in temperature. This
finding is significant because many effluents from coal conversion processes are formed and
released at elevated temperatures.

7.2.1.2 Utilization of phenols

Methyl homologs of phenol, cresols, and xylenols result directly from high-temperature carbonization of coal. Coal tar may have as many as 40 different mono- and dihydric phenols (Chapman
1971). Bacterial use of nonheated phenolic compounds as carbon sources is known to require
molecular oxygen (O_2) for hydroxylation and ring cleavage reactions. If molecular oxygen is not
present, phenols will be converted to objectionable chlorophenols. Successful treatment of
phenolic wastes now employs vigorous aeration methods to supply the necessary molecular oxygen.

In wastewaters

Phenolic wastewaters have also been investigated by Semov and Andreeva (1972), who reported
that the bacteria *Pseudomonas crucivae* and *Pseudomonas phenolis putilina* oxidize phenol and,
without preliminary adaptation, they oxidize phenol to resorcinol and pyrocatechol. Work with
an activated sludge showed that both mono- and divalent phenols may be decomposed. The phenols
tested were from coking plant wastewaters that were simultaneously purified with domestic
wastes by sewage sludge (Semov and Andreeva 1972).

As cresol

Degradation of cresol, a monomethylated phenol, has been accomplished by many *Pseudomonas* species.
At least one strain has been shown to thrive on cresol concentrations of 1450 ppm, which is about
three times greater than any concentrations previously reported (Cobb et al. 1974). It was shown
that the ability to degrade cresol may be episomal in nature and that the genetic trait could be
easily lost. To produce a stable, reliable system for the treatment of cresylic acid wastes,
organisms with stable genetic characteristics must be available. Fortunately, techniques for
producing that stability are at hand. It is possible to insert the extra chromosomal DNA back
into the chromosome of the original organism (Cobb et al. 1974); this technique may produce the
stability needed for successful use of the *Pseudomonas* species.

Chlorinated phenols

Bacterial isolates that metabolize chlorinated phenols and phenolic compounds have been found.
Chlorinated phenols are formed when waters are treated with chlorine for purification. A bacterial culture that metabolizes pentachlorophenol was isolated from a heterogeneous population
of microorganisms used to show the feasibility of biological treatment for industrial wastewater
containing penta and other phenolics (Chu and Kirsch 1972). Continuous culture technique was
used to provide additional enrichment of the penta-degrading microorganisms in the mixed culture.
An isolate that removed pentachlorophenol was repeatedly reisolated by dilution plating to effect
purification. This bacterium was designated KC-3. A compilation of taxonomic observations is
given in Table 7.10. The morphological and physiological characteristics of KC-3 suggest a relationship to the saprophytic coryneform bacteria (Chu and Kirsch 1972).

Table 7.10. Taxonomic characteristics of bacterium KC-3

Criteria	Description of organism
Cell morphology	Slender bacilli, 2 to 4 µm long, long, nonmotile, nonspore-forming, frequent V-shapes, occasional palisade arrangements, and Chinese character aggregates.
Staining	Gram-variable, but predominantly gram-negative with metachromatic granules; non-acid-fast; granules stain with Sudan Black B.
Colonial morphology	Circular, convex, entire, lemon-yellow colonies on complex organic media, including Trypticase soy agar and nutrient agar; small, pale yellow colonies on cellobiose-mineral salts agar; and tiny opaque colonies on pentachlorophenol-mineral salts agar.
Physiological and cultural	Galatin not liquified; nitrate not reduced; glucose oxidized aerobically, but not fermented; catalase positive; cytochrome-oxidase negative; no growth in lauryl tryptose broth; no growth in brilliant green bile broth; cellobiose, glucose, acetate, and selected chlorinated phenols used as sole carbon sources; sucrose, lactose, citrate, glutamic acid, xylose, pyruvate, amlate, and cellulose do not support growth in a mineral salts medium.

Source: Chu and Kirsch 1972, Table 1, p. 1034. Reprinted by permission of
the publisher.

The isolation of a bacteria able to dissimilate and assimilate pentachlorophenol, considered one
of the most refractory compounds of the chlorophenol family, shows the fallacy of the belief that
microorganisms are unable to cope with unusual "man-made" chemicals (Chu and Kirsch 1972).
Chlorinated phenols may be degraded by certain bacteria even when some compounds contain up to
six chlorine atoms per molecule (Baxter et al. 1975). *Nocardia* sp. (NCIB 10603) and *Pseudomonas*

sp. (NCIB 10643) were investigated by Baxter et al. (1975) for their ability to degrade a variety
of polychlorinated biphenyls. Results, shown in Table 7.11, indicate the degree of biodegradation
of the compounds tested.

Table 7.11. Biodegradation of polychlorinated biphenyls over a period of time

Compound	Percent degradation/days	
	NCIB 10603	NCIB 10643
2,4'-Dichlorobiphenyl	70/7	60/73
2,4'-Dichlorobiphenyl + biphenyl	100/7	100/73
4,4'-Dichlorobiphenyl	Nil/121	50/15
4,4'-Dichlorobiphenyl + biphenyl	Nil/121	50/10
2,3-Dichlorobiphenyl	67/8	
2,3-Dichlorobiphenyl + biphenyl	64/8	
3,4-Dichlorobiphenyl	80/8	
3,4-Dichlorobiphenyl + biphenyl	100/8	
2,3,2'-Trichlorobiphenyl	50/7	
2,3,2'-Trichlorobiphenyl + biphenyl	95/7	
2,3,4'-Trichlorobiphenyl	94/7	
2,3,4'-Trichlorobiphenyl + biphenyl	100/7	
2,5,4'-Trichlorobiphenyl	Nil/73	15/73
2,5,4'-Trichlorobiphenyl + biphenyl	60/73	60/73
3,4,3'-Trichlorobiphenyl	76/12	
3,4,3'-Trichlorobiphenyl + biphenyl	70/12	
2,4,6-Trichlorophenol	Nil/12	Nil/84
2,4,6-Trichlorophenol + biphenyl	Nil/12	Nil/84
2,4,2',4-Tetrachlorobiphenyl	Nil/9	
2,4,2',4-Tetrachlorobiphenyl + biphenyl	Nil/9	
2,4,6,2'-Tetrachlorobiphenyl	Nil/9	
2,4,6,2'-Tetrachlorobiphenyl + biphenyl	Nil/9	
2,3,4,5,2',3'-Hexachlorobiphenyl + 2,3,2'- and 2,3',4'-trichlorobiphenyl	Nil	
2,3,4,5,2',3'-Hexachlorobiphenyl + 2,3,2'- and 2,3'4'-trichlorobiphenyl + biphenyl	50/11	

Source: Baxter et al. 1975, Table 2, p. 58. Reprinted by permission of the publisher.

Preliminary experiments showed that the rates of degradation were increased both by the initial
addition of biphenyl and by subsequent additions at intervals during extended periods of exposure
(Baxter et al. 1975).

7.2.1.3 Degradation of phenols

Extensive work has been done to determine the reactions used by various bacterial species in the
degradation of phenolic compounds. Research since 1932 on phenol-reducing microorganisms has
shown that the metabolism of benzenoid compounds is dependent on the presence of molecular oxygen
for such reactions as hydroxylation and ring cleavage (Chapman 1971). Some heterocyclic ring
systems, however, undergo hydroxylation reactions by aerobic bacteria in which water, not molecular
oxygen, serves as the source of the hydroxyl group. Water also supplies the hydroxyl group for
hydroxylation of 2-hydroxy-cyclohexane carboxylic acid by *Rhodopseudomonas palustris* (Chapman

1971). Figure 7.11 depicts the formation of catechol (a well-established example of an ortho-dihydric phenol that undergoes ring cleavage) by bacterial degradation of phenol and a variety of aromatic hydrocarbons and carboxylic acids. Hussien et al. (1974) worked with three strains each of *Rhizobium meliloti*, *R. trifolii*, *R. leguminosarum*, *R. phaseoli*, *R. japonicum*, *R. lupini*, and cowpea rhizobia to show the ability of *Rhizobium* to degrade phenolic compounds. The *Rhizobium* cultures were grown on a medium containing only catechol, protocatechuic acid, *p*-hydroxybenzoic acid, or salicylic acid at a concentration of 1 millimole as the sole carbon source. The growth rate of the cultures, measured by optical density, and the percentage of degradation of phenols are shown in Table 7.12. To determine whether these phenolic compounds are cleaved as such or converted to other metabolites before the ring is opened, ether extracts of cultures of *Rhizobium phaseoli* 405 grown in the presence of 1 millimole of the phenolic compounds and 8.0 millimoles of sodium glutamate were concentrated and analyzed by thin-layer chromatography. Extraction of cultures grown with protocatechuic acid and catechol showed no spots other than those corresponding to traces of the original compounds.

Gibson (1968) studied the metabolic pathways involved in microbial degradation of benzenoid compounds. *Pseudomonas putida* was isolated as an organism that could use toluene as its sole source of carbon. This organism was also able to metabolize benzene. Cell extract examinations and radioisotope-trapping experiments implicated *cis*-benzeneglycol as an intermediate in the formation of catechol by benzene metabolism (Fig. 7.11). Washed cell suspensions oxidized benzene, catechol, and *cis*-benzenediol at approximately equal rates, whereas phenol and *trans*-benzenediol were only slowly metabolized.

The identification of *cis*-dihydrodiols as metabolic intermediates in microbial oxidation of benzene, toluene, ethylbenzene, and naphthalene suggests that dioxetanes may be formed although the mechanism of formation is obscure. Dioxetane, a hypothetical cyclic peroxide, formed by enzymes in the microbial oxidation of aromatic hydrocarbons can be postulated. The *cis*-stereochemistry of the isolated dihydrodiols must be taken into account as well as the fact that both atoms of oxygen in the dihydroxylated intermediates are supplied by molecular oxygen (Gibson 1971; Gibson and Yeh 1973).

7.2.1.4 Degradation of xylenols

Further examples of additional pathways available for bacterial use of substituted phenols are provided by a series of studies using xylenols. The structures of the isomeric xylenols are shown in Fig. 7.12. Interestingly, the degradation of xylenol is strongly dependent on the isomeric form.

Liquid enrichment cultures using each xylenol as the sole carbon source in a mineral salts medium inoculated with polluted estuarine water yielded bacterial isolates with five of the xylenol isomers. Repeated attempts to isolate a 2,6-xylenol-utilizing organism were unsuccessful. All organisms isolated by this technique were members of the genus *Pseudomonas* (Chapman 1971). The degradation of 2,4-xylenol by *P. putida*, a fluorescent *Pseudomonas* strain, was studied by identifying compounds accumulated under conditions of limited aeration and in the presence of inhibitors. This data, together with oxidation studies by intact cells and cell extracts, helped allow the formation of a reaction pathway as shown in Fig. 7.13 (Chapman 1971).

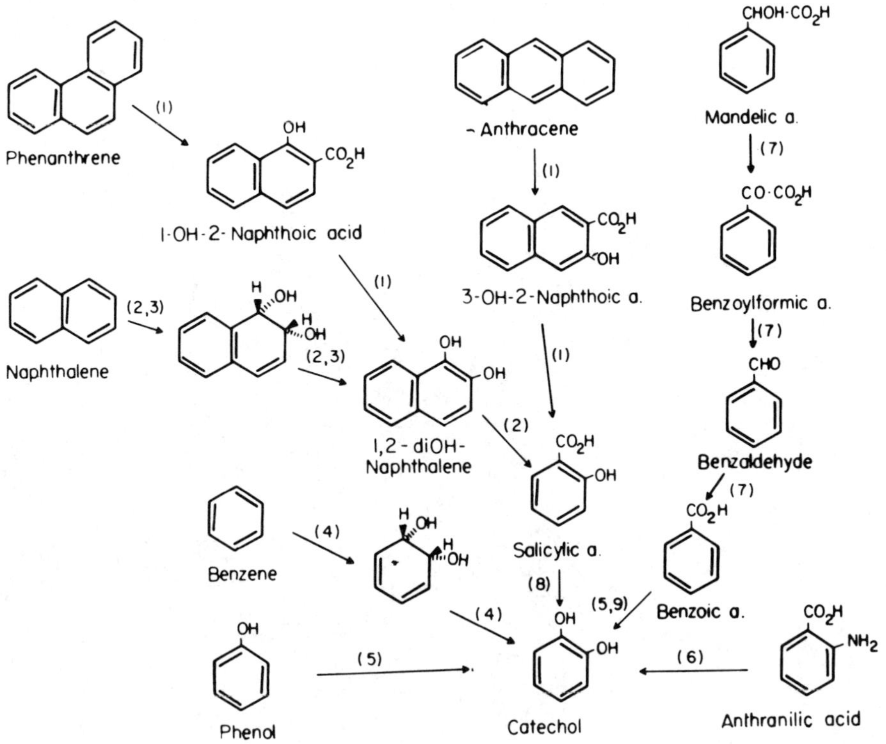

Fig. 7.11. Role of catechol as a central metabolite in the bacterial degradation of phenol and other benzenoid components. Source: From Chapman 1971, Fig. 1, p. 20. Reprinted by permission of the publisher.

The 2,4-xylenol is first converted to 4-hydroxy-3-methylbenzoic acid by oxidation of the paramethyl substituent. Further oxidation of 4-hydroxy-3-methylbenzoic acid forms 4-hydroxy-isophthalic acid, which in turn forms protocatechuic acid after oxidative decarboxylation. Protocatechuic acid is then metabolized into β-ketoadipic acid in a reaction catalyzed by protocatechuic acid-3,4-dioxygenase. The metabolism of 2,4-xylenol is in marked contrast to the metabolism of 2,3-xylenol and 3,4-xylenol when metabolized by *Pseudomonas putida*. Metabolism of the 2,3- and 3,4-xylenol isomers leaves the methyl substituents intact, whereas both methyl groups of 2,4-xylenol undergo oxidation in succession, and the original orthomethyl group is eliminated and replaced by hydroxyl to yield protocatechuic acid (Chapman 1971).

7.2.1.5 Degradation of other alkylphenols

Bacterial isolates showing capacity for degrading alkylphenols were discovered by enrichment culture technique. A nonfluorescent *Pseudomonas* sp. was isolated by enrichment culture on 2,5-xylenol and was found to also readily metabolize *m*-cresol, 3,5-xylenol, 3-ethyl-5-methylphenol, and 2,3,5-trimethyl-phenol as carbon sources. A fluorescent organism, identified as *P. putida*, was isolated by enrichment culture on 3,5-xylenol and was found to also metabolize *m*-cresol and 3-ethyl-5-methylphenol (Chapman 1971). The use of inhibitors subsequently showed that both the

Table 7.12. Growth of *Rhizobium* strains and the amount of phenolic compounds (millimoles) degraded in the presence of 4.0 millimoles sodium glutamate within 15 days[a]

Rhizobium strain	Catechol		*p*-Hydroxybenzoic acid		Protocatechuic acid		Salicylic acid	
	G	D	G	D	G	D	G	D
Rhizobium meliloti								
D71	18.0	99.0	18.0	38.5	17.0	90.4	1.5	81.2
M-160	20.0	98.8	16.5	98.2	17.0	95.9	1.5	21.5
Rhizobium trifolii								
TR3	17.0	99.6	16.5	98.7	20.0	97.2	1.9	83.1
204	18.0	99.2	16.8	98.7	16.0	81.0	1.1	76.6
Rhizobium leguminosarum								
329	22.0	98.9	15.0	98.7	17.0	98.3	1.5	52.4
307	18.0	99.1	17.0	98.7	18.0	95.1	1.9	63.1
M-360	20.0	99.0	14.0	99.8	16.0	97.3	1.7	52.2
Rhizobium phaseoli								
405	21.0	99.1	17.0	97.1	15.0	90.0	1.8	52.4
407	20.0	99.0	17.0	98.7	16.0	89.0	1.9	23.6
Rhizobium japonicum								
138	19.0	99.0	18.0	98.5	16.0	89.0	1.9	70.6
516	20.0	100.0	18.0	98.5	16.0	97.3	1.9	54.3
E41	18.0	99.1	15.9	99.2	17.0	94.5	1.7	44.8
Cowpea rhizobia								
603	19.0	99.7	17.8	98.7	16.0	93.2	1.8	63.9
605	21.0	99.2	20.0	99.7	17.0	93.2	1.5	51.2
Rhizobium lupini								
802	20.0	99.0	14.0	98.5	16.0	87.6	1.5	20.5
807	22.0	99.2	19.0	98.0	17.0	93.2	1.7	38.2
850	22.0	99.8	16.0	98.7	17.0	97.3	1.5	60.5

[a]G = growth (optical density x 100); D = degraded (%).

Source: Hussien, Tewfik, and Hamdi 1974, Table 2, p. 379. Reprinted by permission of the publisher.

ORNL DWG NO 76-6594

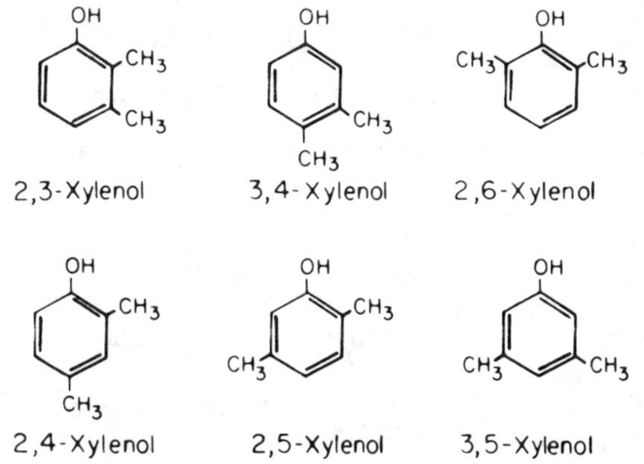

Fig. 7.12. Different xylenol isomers. <u>Source:</u> From Chapman 1971, Fig. 12, p. 38. Reprinted by permission of the publisher.

ORNL DWG NO 76-6595

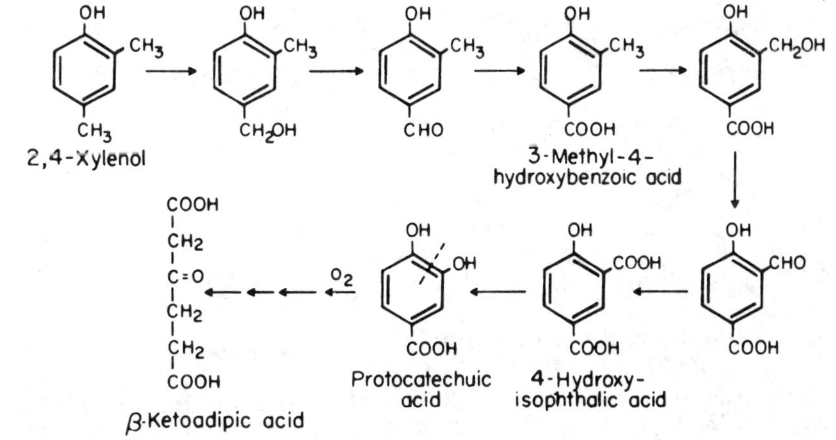

Fig. 7.13. Bacterial metabolism of 2,4-xylenol. <u>Source:</u> From Chapman 1971, Fig. 13, p. 40. Reprinted by permission of the publisher.

2,5- and 3,5-xylenol isolates were able to oxidize all usable alkylphenols to alkyl-substituted 3-hydroxybenzoic acids and alkyl-substituted gentisic acids (Fig. 7.14) (Chapman 1971).

7.2.1.6 <u>Detection of phenol-degrading bacteria</u>

A medium for detection of phenol-degrading bacteria that represents an improvement in efficiency, savings in cost and labor, and a decrease in the time required for results has been developed. The assay medium isolates bacteria growing in water contaminated with phenol, cresol, mono- to pentachlorophenol, and other substituted phenols (Ralston and Vela 1974).

ORNL DWG NO 76-11806

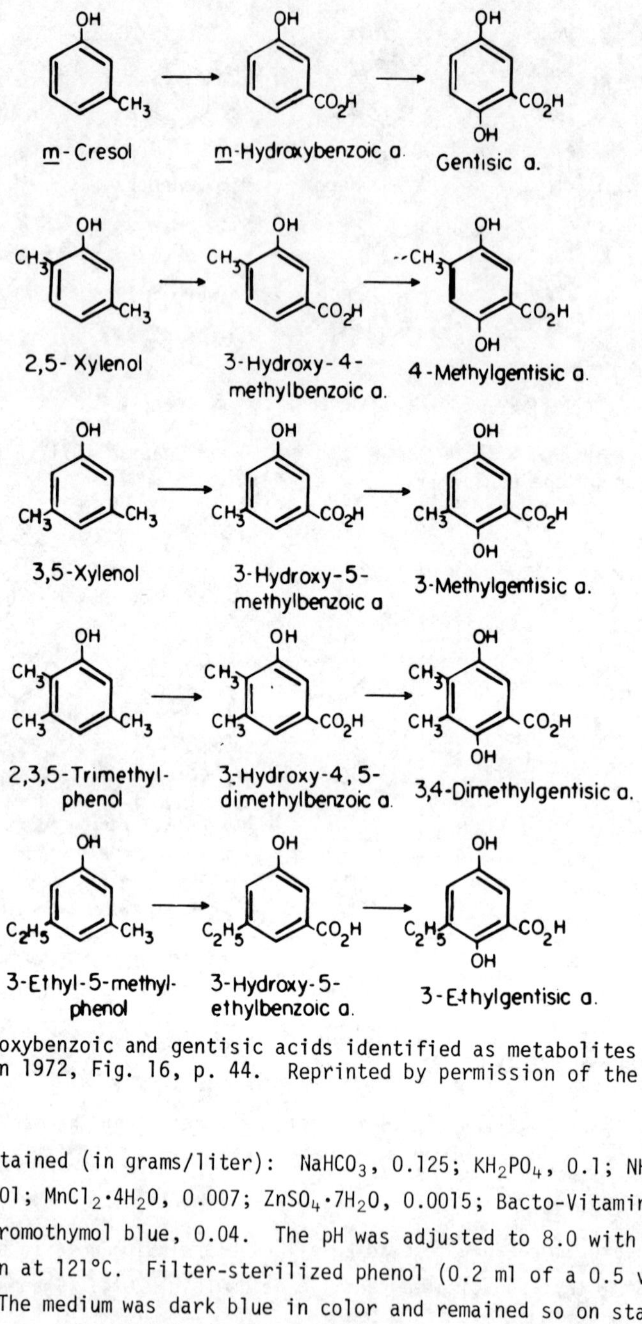

Fig. 7.14. Hydroxybenzoic and gentisic acids identified as metabolites of different phenols. <u>Source</u>: From Chapman 1972, Fig. 16, p. 44. Reprinted by permission of the publisher.

The assay medium contained (in grams/liter): $NaHCO_3$, 0.125; KH_2PO_4, 0.1; NH_4Cl, 0.07; Na_2SiO, 0.02; $FeSO_4 \cdot 7H_2O$, 0.01; $MnCl_2 \cdot 4H_2O$, 0.007; $ZnSO_4 \cdot 7H_2O$, 0.0015; Bacto-Vitamin-Free Casamino Acids (Difco), 0.01; and bromothymol blue, 0.04. The pH was adjusted to 8.0 with 1 N NaOH and was autoclaved for 15 min at 121°C. Filter-sterilized phenol (0.2 ml of a 0.5 vol % solution) was added aseptically. The medium was dark blue in color and remained so on standing (Ralston and Vela 1974). A change in color from dark blue to yellowish green after inoculation with suspected bacteria and incubation at 35°C on a reciprocal shaker for 24 to 48 hr is considered indicative of phenol-degrading bacteria (Ralston and Vela 1974).

<u>In coal carbonization waste liquor</u>

Jones and Carrington (1972) studied microorganisms isolated from a sludge that were successful in treating the waste liquors of a coal carbonization plant. The liquors consisted mainly of solutions of phenols and thiocyanate. Eight strains of bacteria were isolated from the sludge, three of which were found to be responsible for destruction of the major constituents of the waste liquor. The three strains of bacteria were studied further for identification purposes. One organism, which was grown on thiosulphate or thiocyanate, was autotrophic, did not produce polythionate, and lowered the pH of the medium to about 5.0. This organism was found to be similar to *Thiobacillus thioparus*. Bacterial degradation of thiocyanate was increased in a study by Catchpole and Cooper (1972) when they added small amounts of para-aminobenzoic acid to liquors to observe the changes in the sludge process. The two bacterial strains found to grow on phenols were *Comamonas* sp., which could grow on a wide range of phenols, including cresols and xylenols, and a member of the *Moraxella-Acinetobacter* group that was taxonomically close to *Achromobacter* sp. NCIB 8250. This bacteria strain was restricted in the range of phenols it could metabolize because it could not grow on cresols or xylenols (Jones and Carrington 1972).

<u>In coke-oven effluent</u>

Stafford and Callely (1973) report the presence of bacteria in a tip-lagoon used in purifying coke-oven effluents that contained phenols and thiocyanate. The bacteria were aerobic, gram-negative bacilli that could metabolize thiocyanate and phenols. The organisms isolated in this experiment were not used to a high level of efficiency due to the lack of aeration in the lagoon.

7.2.2 Alkanes

Normal alkanes have been shown to be produced as part of the coal carbonization process and are potential pollutants. They possess an organic structure accessible to microbial attack. Oil spills are a major problem to ecosystems, and microbial degradation which begins with alkanes is a major means of oil spillage cleanup. The normal alkane has been extensively studied in both the laboratory and the field.

7.2.2.1 Mechanisms of uptake

Finnerty et al. (1973) examined ultrastructure of *Acinetobacter* sp. (formerly named *Micrococcus cerificans*) grown on paraffinic and olefinic hydrocarbons. The purpose of the experiment was to keep a pure culture of hydrocarbon using bacteria grown on defined and chemically pure hydrocarbons and to examine fixed whole cells and cell sections by electron microscopy. Cells grown on hexadecane indicated that the hydrocarbon was dispersed as an oil in water emulsion in the culture medium, and further work showed that bacteria preinduced to growth on hydrocarbons were essential for formation of the emulsion. Examination of the culture by light and electron microscopes showed a physical relationship between preinduced bacteria and microdroplets of hexadecane. The bacteria had adhered to the surface of the microdroplet and covered it uniformly. This finding suggested a mechanism for active transport of hydrocarbons. Analysis of cellular and extracellular lipids showed that a number of specific lipids which could act as a biodetergent and promote pseudo-solubilization of hydrocarbons were present. The specific role of these lipids was not shown, but their known physical properties would suggest an influential role in maintaining a finite level of hydrocarbon in aqueous solution as macro- or micro-emulsions.

Thin-section electron microscope analysis showed that hydrocarbon-grown cells exhibited hydrocarbon inclusion bodies that are unique to hydrocarbon-grown cells. An inexplicable growth phenomenon was also discovered in that hydrocarbon-grown cells transformed into giant cells at irregular intervals to become four to ten times as large as normal. The enlargement is due to extensive intracytoplasmic membrane development and occurs only when the bacteria are grown on hydrocarbons.

A theory to identify the mechanism of hydrocarbon uptake by microorganisms has been proposed by Barnett and Velankar (1974). The hydrocarbons are thought to be present in submicron-size drops, which would be the "edible" size for microorganisms (e.g., pinocytosis). These small hydrocarbon drops are used by the microorganisms as though the hydrocarbon were in the dissolved state. This theory explains the high rate of hydrocarbon transfer despite the fact that hydrocarbons have very low solubility in water. It is thought that the hydrocarbon drops are solubilized into the aqueous medium by the presence of surface-active compounds produced by the microorganisms. The increased solubility provides the large driving force necessary for the high bacterial growth rate (Barnett and Velankar 1974).

Micelle formation

The hydrocarbon microdroplet may have an ionic hydrophilic surface, which would make it possible for several layers of nutrient salt ions to accumulate near the surface of the microdroplet, much as a double layer forms adjacent to a charged surface. These microdroplets are called "micelles" (Fig. 7.15) (Velankar et al. 1975). This multiple layer of ion salts in turn would cause the microbial cells to migrate preferentially to areas of high nutrient concentrations. The microdroplet with the ions of nutrient salts can thus serve as a carbon and nutrient source (Barnett and Velankar 1974).

ORNL DWG NO 76-11807

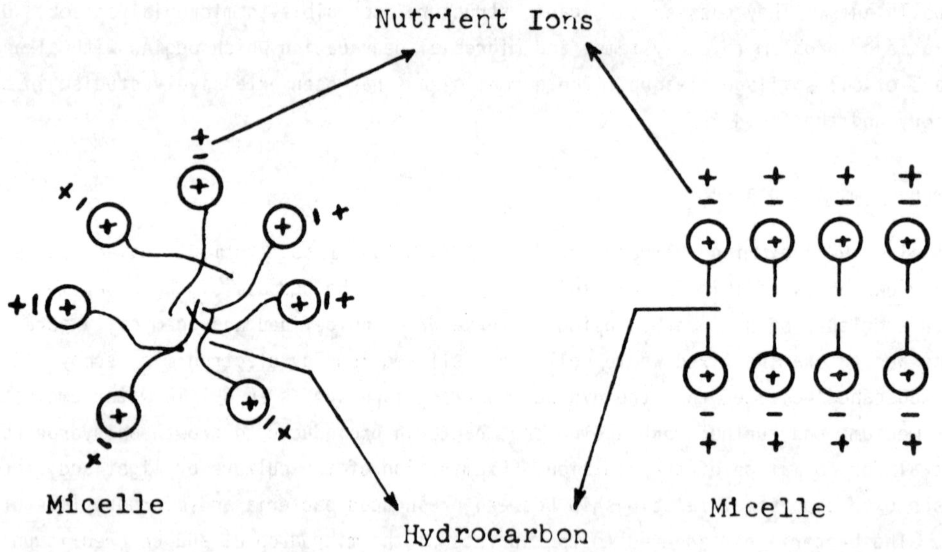

Fig. 7.15. Two possible structures of micelles with hydrocarbon inside and nutrient ions near the surface (ionic layers shown in the idealized form). Source: From Velankar et al. 1975, Fig. 3, p. 243. Reprinted by permission of the publisher.

Velankar et al. (1975), while working on the transport of hydrocarbons in a fermentation system, proposed that micelles would be formed by surface-active agents added to the medium or perhaps be produced by the microorganism itself. They proposed that the micelles act as "transport packets" for hydrocarbons. When the micelles are immediately adjacent to the cells, hydrocarbon is transferred more quickly to the cell surface because diffusion through the aqueous phase is eliminated. Thus, a high hydrocarbon transport rate could be achieved. The rate of hydrocarbon

transfer would depend on both the number of transport packets and the amount of hydrocarbon that
each carries. The proposed mechanism of transport of hydrocarbon in a fermentation system is
shown in Fig. 7.16. As hydrocarbons are reacted or depleted in the micelles, they are replaced
by hydrocarbons diffusing from large droplets to micelles. The hydrocarbon droplet is about
10 times larger than the bacterial cell, which in turn is 10 times larger than the micelle.

ORNL DWG NO 76-11308

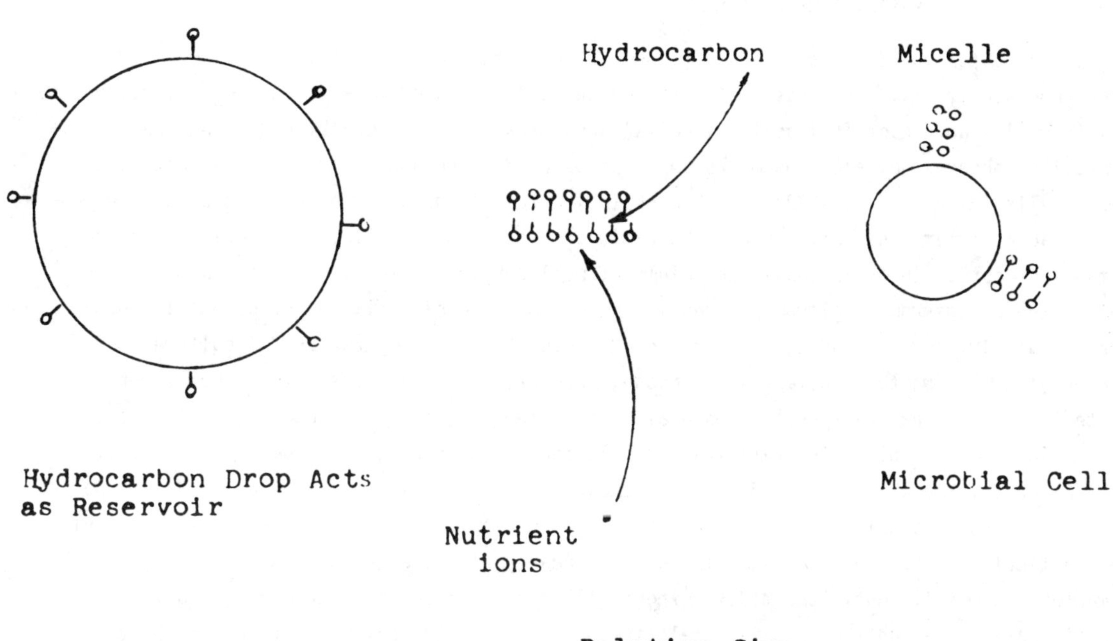

Fig. 7.16. Mechanism of hydrocarbon transport in fermentation system. Source: From
Velankar et al. 1975, Fig. 2, p. 243. Reprinted by permission of the publisher.

Solubility and cell growth

The theory proposed by Barnett and Velankar (1974) predicts a lag phase in bacterial growth until
sufficient surface-active compound is produced by the cells. Exponential growth occurs as long
as submicron hydrocarbon droplets, with adjoining nutrient salt ions, bring sufficient substrate
to the cells. If the cell concentration becomes high enough that the rate of consumption exceeds
the rate of supply, the growth rate becomes linear and is proportional to the rate of hydrocarbon
supply.

Work by Yoshida and Yamane (1971) with *Candida tropicalis* demonstrated that the yeast cells
accumulated hydrocarbons existing in the aqueous phase. This work was concerned with the uptake
of n-hexadecane, the solubility of which was determined by gas chromatography to be 5.57 x 10^{-9} g
per cubic centimeter of water. When the rate of solubility was compared with the rate of growth,
the yeast absorption rate was found to be higher than the rate of solubility. Measurement of the

saturation concentration of n-hexadecane in water gave 11 x 10^{-8} g/ml, which is twenty times larger than the measured solubility of n-hexadecane in water. The results show that the n-hexadecane was absorbed into the water, not as individual molecules but as aggregates of molecules or extremely fine droplets. To interpret the results of the experiment on a culture of *Candida tropicalis* in a rotating disk fermentor with the aqueous culture medium, one must infer that the hydrocarbon uptake by direct contact with the liquid hydrocarbon phase is unimportant (Yoshida and Yamane 1971). The results recorded in this experiment closely resemble those of Barnett and Velankar (1974).

7.2.2.2 Effect on bacterial morphology

Candida tropicalis pK 233 yeast cells have been studied for morphological changes when grown on hydrocarbon substrates. The yeast cells formed cultures composed of a mixture of filamentous form cells (F-cells) and yeast form cells (Y-cells) when grown on a substrate containing n-alkanes. The F-cells, observed by electron microscope, were found to be a hypha divided by speta into several cells (Hirai et al. 1972). Carbon chain lengths of the n-alkanes tested as growth substrates had a significant influence on the ratio of F-cells to Y-cells. The cells grown on n-alkanes of carbon 10 to 12 consisted mainly of well-developed F-cells with a small portion of Y-cells, whereas pseudo-F-cells were predominant in the case of n-alkanes of carbon 13 to 16 (Hirai et al. 1972). F-cells have never been detected in *C. tropicalis* pK 233 cultures when grown on glucose, fructose, galactose, mannose, saccharose, and acetate. The F-cells were detected only when the yeast cells used n-alkanes as their sole carbon source (Hirai et al. 1972). Figure 7.17 depicts the relative morphological changes of *C. tropicalis* pK 233 cells as to the F- or Y-forms when subjected to various carbon sources. Teranishi et al. (1974) report that F-cells show a significantly higher n-alkane-oxidizing capacity. The respiratory activity of a yeast culture grown on n-alkanes increased linearly with the growth and development of the filamentous-form cells until the middle exponential growth phase; it decreased gradually toward the stationary phase and then remained relatively constant at an oxygen quotient of about 200 µl/hr/mg, dry weight.

Teranishi et al. (1974) report that yeast organisms grown on hydrocarbon substrates (*C. tropicalis* pK 233 cultures as well as cultures of *C. albicans* IFO 0587, *C. guilliermondii* IFO 0566, *C. intermedia* NRRL Y-6328-1, and *C. lipolytica* NRRL Y-6795) produced microbody-like granules, which could be observed by electron microscope. The hydrocarbon substrate consisted of a mixture of n-alkanes, including n-decane, n-undecane, n-dodecane, and n-tridecane. These granules were profuse and thought to be the same as "glyoxysomes" or "peroxisomes," which were detected and named in higher plants, a species of protozoa, and mammalian cells.

7.2.2.3 Effect on enzyme activity

The enzyme activities of the *Candida* yeasts grown on hydrocarbons were found to be much higher than those of yeasts grown on glucose, ethanol, or lauryl alcohol (Table 7.13). The increase in enzyme activity seems to be consistent with the increase in microbodies in yeast cultures grown on hydrocarbon substrates. *C. tropicalis* pK 233 was used by Mishina et al. (1973) to show a possible relationship between the large numbers of microbodies and a markedly high catalase activity. Lipid contents of *Candida* cells were observed to be higher when hydrocarbons were used as the carbon source. Work with *C. lipolytica* grown on pure n-hexadecane by Chenouda and

ORNL DWG NO 76-11809

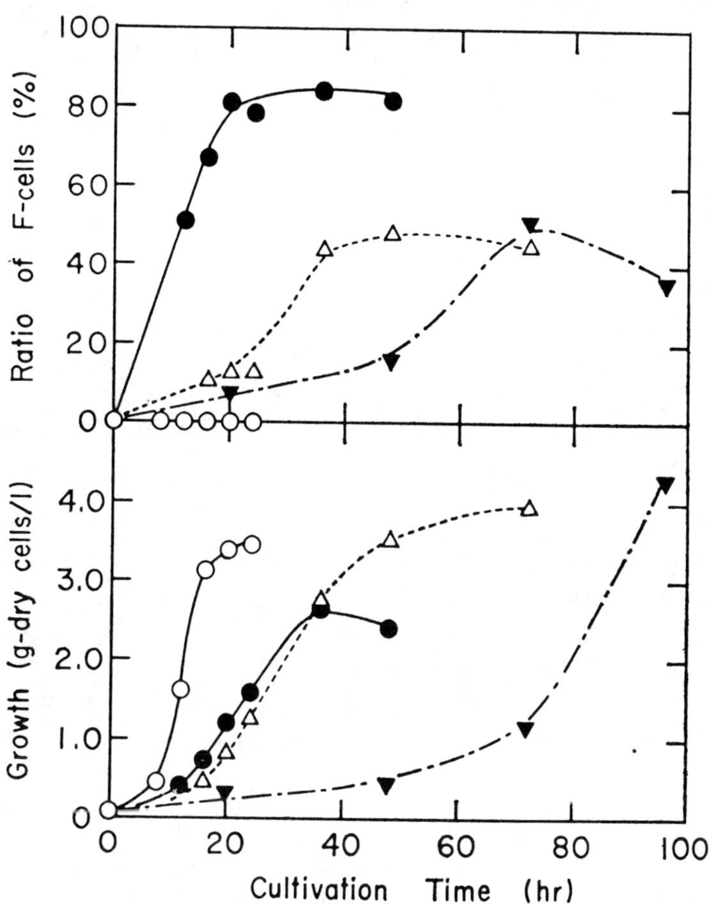

Fig. 7.17. Growth and morphological changes of *Candida tropicalis* pK 233 cultivated on various carbon sources. The yeast was cultured at 30°C on glucose (○—○); *n*-decane (●—●); *n*-tridecane (△--△); or *n*-hexadecane (▼---▼). <u>Source</u>: Hirai et al. 1972, Fig. 2, p. 2337. Reprinted by permission of the publisher.

Jwanny (1972) also showed the cellular lipid content changes. Teranishi et al. (1974) demonstrated the inducibility of catalase activity and concomitant microbody increase when *C. tropicalis* pK 233 was cultured on hydrocarbons. An increase in catalase activity and microbody production when the yeast was cultured on hydrocarbons was shown, but the information on the physiological role of microbodies and also of catalase in the yeasts was not discussed. Very little in general is known about the physiological actions of catalase and microbody activity in yeasts grown on hydrocarbons.

7.2.2.4 Growth kinetics

A model was developed by Moo-Young and Shimizu (1971) to present the growth kinetics of *Candida* yeasts on hydrocarbon substrates. *C. lipolytica* was grown on a dodecane substrate to determine the quantitative effects of the hydrocarbon-fermentation operating conditions on experimental cell growth. The rate data can be accounted for by a simple mechanistic model, which is based on the assumption that the cell growth kinetics are controlled by the extent of attachment between cells and oil droplets of much smaller size. The effects may be expressed in terms of an

Table 7.13. Catalase activities (μmoles/min/mg cells) of *Candida* yeasts grown on various substrates[a]

Yeast	Glucose		Ethanol		Lauryl alc.		Hydrocarbons	
	E	S	E	S	E	S	E	S
Candida albicans	2	6	3	11			145	28
Candida guilliermondii	8	4	8	5			143	46
Candida intermedia	7	7	3	2			169	92
Candida lipolytica	8	16	3	2			56	14
Candida tropicalis	7	10	25	4	47	10	370	72

[a]The yeasts were harvested at the exponential (E) and the stationary (S) phases.

Source: Teranishi et al. 1974*a*, Table III, p. 1215. Reprinted by permission of the publisher.

impeller Weber number or agitation power input, which is shown to be of practical importance to scale-up considerations (Moo-Young and Shimizu 1971).

Yoshida, Yamane, and Nakamoto (1973) cultured the yeast *C. tropicalis* on *n*-hexadecane dispersed in water as submicron droplets. The droplets of *n*-hexadecane were supplied continuously into a fermentor containing an aqueous medium. This type of operation, in which only the growth-limiting substrate is fed continuously into a batch fermentor, is sometimes called the fed-batch (or continuous) culture. Theoretical analysis of the fed-batch culture would be to consider the balance for the substrate,

$$\frac{d(VS)}{dt} = FS_F - \frac{1}{Y}\frac{d(VX)}{dt} \; , \qquad (7)$$

where

 V = volume of the medium in the fermentor,
 S = the concentration of limiting substrate, subscript F designating the feed,
 F = the volumetric rate of the feed containing the limiting substrate,
 Y = the cell/substrate yield constant,
 X = the cell mass concentration.

The rate of growth of organisms is represented by

$$\frac{d(VX)}{dt} = \mu VX \; , \qquad (8)$$

where μ is the specific growth rate and is generally a function of the substrate concentration. It is unknown whether the Monod-type equation holds for the relationship between the specific growth rate and the concentration of the substrate that exists as a fine dispersed phase. The limiting substrate could be fed to the fermentor in two different ways: (1) at a constant feed rate or (2) at feed rates increasing exponentially in proportion to the growth of organisms (exponential fed-batch culture).

With a constant feed rate, the volume of the culture medium should increase linearly with time, that is,

$$V = V_0 + Ft \; , \qquad (9)$$

where V_0 is the initial volume of the medium. Furthermore, when the volume of the medium is very large relative to the feed rate and can be regarded as practically constant, Eqs. (7) and (8) reduce to Eqs. (10) and (11) respectively:

$$V \frac{ds}{dt} = FS_F - \frac{V}{Y} \frac{dX}{dt} \quad ; \tag{10}$$

$$\frac{dX}{dt} = \mu X \quad . \tag{11}$$

Integration of Eq. (10) with the condition that $S_0 = 0$ gives

$$VS = FS_F t - \frac{V}{Y} (X - X_0) \quad , \tag{12}$$

where subscript 0 designates initial values. In most of the experiments, the feed rate F was constant, and the volume of the medium in the fermentor was kept relatively constant. Because the substrate n-hexadecane was practically insoluble in water, Eqs. (10) and (12), respectively, can be written as

$$\frac{dS}{dt} = \frac{F\phi\rho}{V} - \frac{1}{Y} \frac{dX}{dt} \quad , \tag{13}$$

and

$$VS = F\phi\rho t - \frac{V}{Y} (X - X_0) \quad , \tag{14}$$

where ϕ is the volume fraction of hydrocarbon in the feed, and ρ is the density of liquid hydrocarbon.

The yield of yeast with respect to hydrocarbon consumed was calculated from the experimental data for both the exponential and linear growth phases. If the substrate concentration in the fermentor at t_1 is extremely low, as will be shown later, the left-hand side of Eq. (14) is nearly zero. Hence, the yield for the exponential growth phase Y_E is given by

$$Y_E \cong \frac{(X_1 - X_0)\, V}{F\phi\rho t_1} \quad , \tag{15}$$

where X_0 and X_1 are the cell mass concentrations at the beginning and end of the exponential growth phase respectively; t_1 is the culture time at the end of the exponential growth phase. X_1 and t_1 were determined from the intersection of the extrapolated growth curves for the exponential and linear growth phases. The fact that the dissolved oxygen concentration increased in the cell population indicates that the oxygen consumption per unit mass of cells decreased with time during the linear growth phase.

7.2.2.5 <u>Assimilation</u>

The assimilation of hydrocarbons has been shown to be of value in the classification of yeasts for some time. Scheda and Bos (1966) tested over 1200 yeast strains of the genera *Saccharomyces*, *Hensenula*, *Pichia*, *Debaryomyces*, *Candida*, *Torulopsis*, *Brettanomyces*, *Kloeckera*, *Crytococcus*, and *Rhodotorula* for this ability. Yeast cultures were tested for assimilation of n-decane and n-hexadecane and positive growth cultures were tested to determine whether growth on these substrates corresponds with growth on kerosene. The strains of yeast found to assimilate hydrocarbons are given in Table 7.14.

Table 7.14. Number of strains assimilating *n*-hexadecane, *n*-decane, and kerosene in some species of *Pichia*, *Debaryomyces*, *Torulopsis*, and *Candida*

Species	Number of strains tested	Number of strains growing on		
		n-Hexadecane	*n*-Decane	Kerosene
Pichia farinosa	22	22	22	19
Pichia guilliermondii	37	37	37	33
Pichia haplophila	2	2	2	2
Pichia pastoris	2	2	2	2
Pichia polymorpha	3	3	3	2
Pichia robertsii	2	2	2	2
Pichia scolyti	5	1	1	1
Pichia vini	5	5	5	5
Debaryomyces hansenii	45	45	45	45
Debaryomyces vanriji	4	4	4	4
Torulopsis dattila	2	2	1	1
Torulopsis famata	12	12	12	12
Torulopsis haemulonii	2	2	2	2
Torulopsis sake	2	1	1	1
Candida brumptii	6	6	5	4
Candida catenulata	5	5	5	1
Candida intermedia	10	10	10	6
Candida lipolytica	12	12	12	8
Candida melinii	4	1	1	1
Candida parapsilosis	11	11	11	9
Candida pulcherrima	17	16	16	11
Candida reukaufii	7	7	5	2
Candida rhagii	4	4	3	4
Candida tenusis	8	4	3	2
Candida tropicalis	24	24	24	24

Source: Scheda and Bos 1966, Table 1, p. 660. Reprinted by permission of the publisher.

Makula and Finnerty (1971) worked with *Micrococcus cerificans* grown on hexadecane. This work studied the assimilation of the hexadecane carbon into the cellular lipid. The bacteria was grown in the presence of 49.4 mCi of hexadecane-1-^{14}C. Samples (25 ml) were removed at 2-, 4-, 8-, and 12-hr intervals, and a sodium oxide (1 ml of 10% solution) was added immediately to each sample to kill the bacteria. The cells were collected by centrifugation, washed with basal medium, lyophilized, and extracted for total lipid. The crude lipid was fractionated by silicic acid chromatography into nonmetabolized hexadecane, neutral lipid, and phospholipid by hexane, chloroform, and methanol elution to obtain the respective lipid fractions. Distribution of the products is shown in Table 7.15.

Table 7.15. Distribution of hexadecane-1-^{14}C assimilation products in the lipids of *Micrococcus cerificans*

Time (hr)	Cell mass (mg, dry wt)	A	B	C	D	E	F
		Total^a (10^3 counts/min)	Chloroform-soluble (10^3 counts/min)	Hexadecane (10^3 counts/min)	Neutrals (10^3 counts/min)	Phospholipids (10^3 counts/min)	Recovery (%)
2	150	250.0(0.83)^b	100.0(40)^b	80.0(80)^b	7.5(37.5)^b	2.2(11)^b	90^c
4	150	407.0(1.36)	210.0(52)	172.0(82)	10.8(28.4)	6.4(17)	90
8	150	975.0(3.25)	485.0(50)	338.0(70)	18.0(12.2)	31.0(21)	80
12	150	2378.0(7.92)	665.0(28)	394.0(59)	52.0(19)	104.0(38.4)	82

^{a}Total counts per minute were determined by solubilizing 5 mg of dry cells in hyamine hydroxide.

^{b}A = Total counts per minute/counts per minute of hexadecane-1-^{14}C added × 100; B = B/A × 100; C = C/B × 100; D = D/B-C × 100; E = E/B-C × 100.

^{c}F = C + D + E/B × 100.

Source: Makula and Finnerty 1971, Table 2, p. 810. Reprinted by permission of the publisher.

7.2.2.5 <u>Respiration</u>

Le Petit et al. (1975) studied the bacterial strains classified as similar to the *Achromobacter*, *Alcaligenes*, *Pseudomonas*, *Acinetobacter*, and *Arthrobacter* genera and the respiratory activity of each when using hexadecane as their carbon source. The bacteria were isolated from the sea in a region near Marseilles that contained hydrocarbon pollutants (station I) and another region that had no hydrocarbon pollution (station II). The bacterial isolates were grown on media by using either hexadecane or acetate as the carbon source. Respiratory activity (as O_2 consumption) was measured by the use of a Warburg apparatus. Chloramphenicol was added to the system at a rate of 50 µg/ml to inhibit protein synthesis. The chloramphenicol helped separate the bacterial isolates into two groups: Group 1 showed immediate respiratory activity on hexadecane that was not inhibited by chloramphenicol; group 2 showed either an immediate or delayed respiratory activity that was always inhibited by chloramphenicol. Maintenance or suppression of respiration by chloramphenicol proved to be a characteristic that was homogeneously distributed among the species. The respiratory activities of the two groups, which indicate metabolic activity, are shown in Tables 7.16 and 7.17.

Table 7.16. Respiratory activities of bacterial strains cultivated on acetate (group 1)[a]

| | | Substrate | | |
| | | Hexadecane[b] | | |
Bacterial strain[b]	Acetate (QO_2)	A	B	C
Alcaligenes sp. 2				
L-Jan. 16	47	55	51	35
L-Feb. 7	95.5	73	106	63
L-March 3	111	87.5	70	33.5
L-April 3	64	55	64.5	31
L-May 4	224	171	118	144
L-June 1	104.5	56	51.5	32
L- Aug. 6	112	102	104.5	73
L-Sept. 3	133	105	111	83
L-Oct. 6	91.5	92	86	74
L-Nov. 1	45	40	54	38
C-Feb. 5	113	61	53	47
C-May 1	129	52.5	78	26.5
C-July 2	67	67.5	69.5	52.5
C-Aug. 2	114	46.5	88	27
C-Nov. 2	136	113	110.5	27
C-Jan. 1	67	82.5	63	50

[a]Each value is expressed, the endogenous O_2 being deducted.
[b]Designation of the strains: L and C correspond respectively to stations I and II; all dates are in 1973, except for C-Jan. 1, which is in 1974.
[c]A — QO_2 (oxidation quotient) measured between 0 and 1 hr after contacting the cells with hexadecane; B — QO_2 measured between 2 and 3 hr of hexadecane contact; C — QO_2 measured in the presence of chloramphenicol.

Source: Le Petit et al. 1975, Table 2, p. 375. Reprinted by permission of the publisher.

Table 7.17. Respiratory activities of bacterial strains
cultivated on acetate (group 2)[a]

Bacterial strain[b]	Acetate (QO_2)	Hexadecane[c]		
		A	B	C
Achromobacter sp. 1	33.5	2.75	12.5	2.5
Alcaligenes sp. 4				
L-Jan. 8	50	16.5	19	2
L-Feb. 3	43	21	37	6
L-March 1	66	25	60	6.5
L-April 7	65	12	19	2.5
L-Sept. 1	96	18.5	37.5	1
C-Feb. 1	37	19.5	37	4.75
C-April 1	115.5	53	74.5	12.5
C-Oct. 2	97.5	32	69	11
C-Nov. 1	53.5	41.5	48.5	10
C-Dec. 1	80	38	52	5
C-Jan. 4	100	54	53 (78)[d]	14.5
Alcaligenes sp. 5				
L-June 6	67	38.5	46	4
L-June 1	40	35	41	6
Acinetobacter sp.	55	9	30	0
Arthrobacter sp.	71.2	0	11 (28)[e]	0

[a]Each value is expressed, the endogenous O_2 being deducted.
[b]Designation of the strains: L and C correspond respectively to stations I and II; all dates are in 1973, except for C-Jan. 4 and L-Jan. 1, which are in 1974.
[c]A — QO_2 (oxidation quotient) measured between 0 and 1 hr after contacting the cells with hexadecane; B — QO_2 measured between 2 and 3 hr of hexadecane contact; C — QO_2 measured in the presence of chloramphenicol.
[d]QO_2 between 1 and 2 hr.
[e]QO_2 between 6 and 7 hr.

Source: Le Petit et al. 1975, Table 3, p. 376. Reprinted by permission of the publisher.

7.2.2.7 Metabolism

McKenna (1971) studied the microbial metabolism of hexadecane and other alkanes. Some 20 bacterial species from several different genera were used to study growth on hexadecane, and all strains were found to grow profusely on the hydrocarbon. The 20 bacterial strains found to assimilate hexadecane were *Micrococcus cerificans* H.O. I-0, *M. cerificans* H.O. 3, *M. cerificans* H.O. 4, *M. cerificans* S-18.2, *M. cerificans* S-14.1; *Pseudomonas aeruginosa* Davis, *P. aeruginosa* 115 JWF, *P. aeruginosa* 191 JWF, *P. aeruginosa* Sol 20 JS, *P. fluorescens* JWF; *Mycobacterium phlei* No. 451, *M. fortuitum* No. 389, *M. rhodochrous* No. 382, *M. smegmatis* No. 422; *Nocardia opaca*, *N. rubra*, *N. erythropolis*, *N. polychromogenes*, *N. corallina*; Isolate RTMP.

Makula and Finnerty (1972) studied the distribution of cellular fatty acids in defined lipid classes in *M. cerificans* grown on specified hydrocarbons. Carbon sources used in this study were nutrient broth, acetate, and hexadecane. The cultures were grown on the specified hydrocarbon substrate, and then the cells were extracted with chloroform-methanol (2:1 v/v) and centrifuged. The cell residue remaining after chloroform-methanol extraction was saponified for 8 hr with 20%

methanolic KOH. The fatty acids were partitioned into chloroform after acidification and reduced to dryness in vacuo. These procedures define three derived lipid classes: (1) neutral lipid fraction, (2) phospholipid fraction, and (3) the cell residue fraction. The neutral lipid fraction obtained from hexadecane-grown cells was analyzed before and after saponification. A four-fold increase in palmitate and hexadecan-1-ol and a doubling of palmitoleic acid were measured after saponification. The increase in palmitic acid and hexadecan-1-ol was due to the presence of cetyl palmitate in the neutral lipid fraction. Cetyl palmitate was isolated from the neutral lipid fraction by thin-layer chromatography employing Silica Gel G. The presence of bound fatty alcohol in the neutral lipid fraction of cells grown on hexadecane indicates that wax esters occur within the cell as well as being extracellular products of hydrocarbons found in the growth medium. The percent composition of fatty acid with the same chain length as the alkane substrate was 98% for hexadecane. The membrane phospholipids reflect a fatty acid pattern related to the carbon number of the alkane substrate. The ratio of unsaturated to saturated fatty acid corresponding to the same chain length as the hydrocarbon substrate was 2.31 for hexadecane. This increase in unsaturated fatty acid content of cells when grown on hydrocarbons indicates a physiological response that manifests itself in phospholipid fatty acid. Because fatty acids are direct products of hydrocarbon oxidation, it was of interest to determine the free fatty acid (FFA) content of cells using hexadecane in comparison to nonhydrocarbon growth substrates. Figure 7.18 shows the data obtained from growth of *M. cerificans* on specified carbon sources as measured by culture optical density (left ordinate). Cultures from nutrient broth and acetate exhibited 0.4 µmole of FFA per milliliter of culture when entering the stationary growth phase. Samplings of the culture show FFA increases as indicated by the right-hand ordinate (Fig. 7.18). A sevenfold increase was determined for the FFA content of hexadecane-grown cultures, and the increase was parallel with cellular growth. In all cases, the FFA reached a maximal concentration during the early stationary phase.

Cooney and Walker (1973) studied *Cladosporium (Amorphotheca) resinae,* a filamentous fungus, which is one of the most prominent hydrocarbon-using fungi. Two strains of *C. resinae* were isolated from contaminated jet fuel; each was tested for growth ability on 55 individual hydrocarbons as its sole source of organic carbon with NH_4NO_3 as the nitrogen source. Growth results were best on *n*-alkanes of intermediate length, but growth also occurred on alkanoic alcohols and acids, *n*-alkenes, and some cyclic and aromatic compounds. Compounds supporting growth included dodecane, hexadecane, 1-dodecanol, 1-hexadecanol, dodecanoic acid, hexadecanoic acid, dodecane-1, cyclohexane, benzene, and *o*-xylene. The *C. resinae* strains did resemble other hydrocarbon-using bacteria and yeasts in that they grew more slowly on hydrocarbons than on substrates of glucose.

Comparison of *C. resinae* cells grown on glucose with those grown on *n*-alkanes indicated cellular differences. Hydrocarbon-grown cells contained more total lipids than did glucose-grown cells. Growth on *n*-alkenes does not greatly affect total cellular fatty acids. The major phospholipids are the same in glucose- or *n*-alkane-grown cells, but fatty acids varied from one phospholipid to another and were influenced by substrate (Cooney and Walker 1973).

Metabolic pathway

Carnes (1972) studied the growth of *M. cerificans* on a hexadecane medium. The purpose of the study was to determine metabolic pathways in the use of hexadecane by the microorganism. The

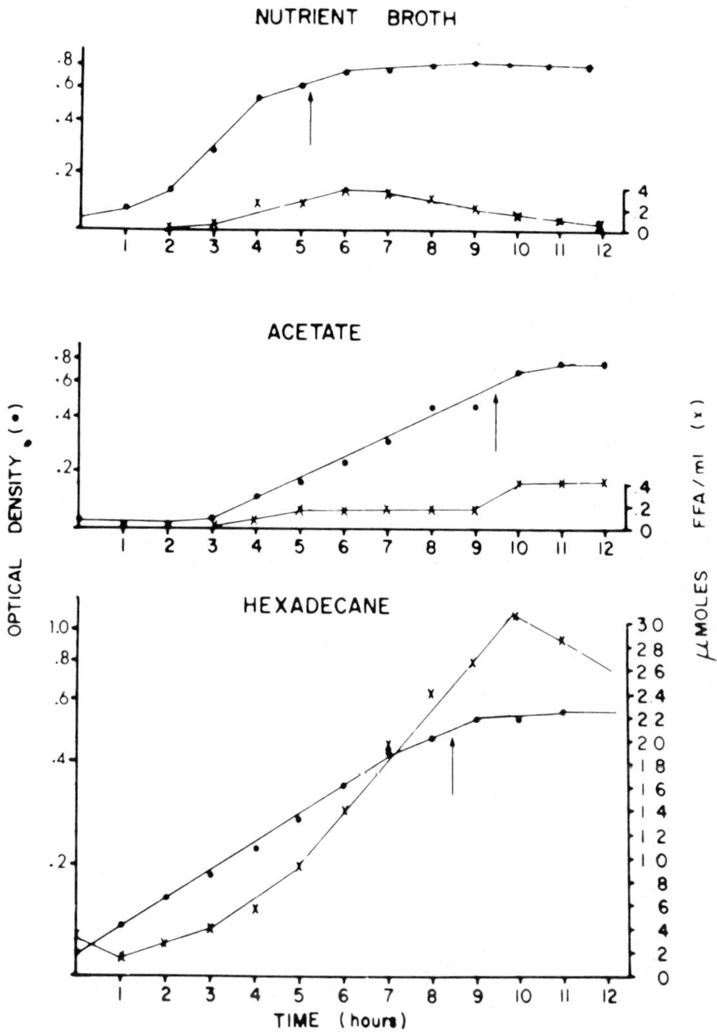

Fig. 7.18. Free fatty acid production by *Micrococcus cerificans* grown on different carbon sources. Arrows indicate the point in the growth cycle at which cultures were harvested for lipid analysis. <u>Source:</u> From Makula and Finnerty 1972, Fig. 5, p. 403. Reprinted by permission of the publisher.

proposed metabolic pathway is shown in Fig. 7.19. This study showed that *M. cerificans* has a functioning glyoxylate bypass, forms pyruvate rapidly, fixes little carbon dioxide, and has an active tricarboxylic acid cycle when grown on hexadecane.

7.2.2.8 <u>Effect of pressure (marine organisms)</u>

Microorganisms present in Atlantic Ocean sediment samples collected at a depth of 4940 m were studied by Schwarz, Walker, and Colwell (1974) for their capability of using *n*-hexadecane under both ambient (1 atm) and in-situ (500 atm) pressures. The bacterial isolates obtained in this study and examined in both mixed and pure culture may represent those bacteria capable of surviving the decompression and increase in temperature associated with sample retrieval; hence, they

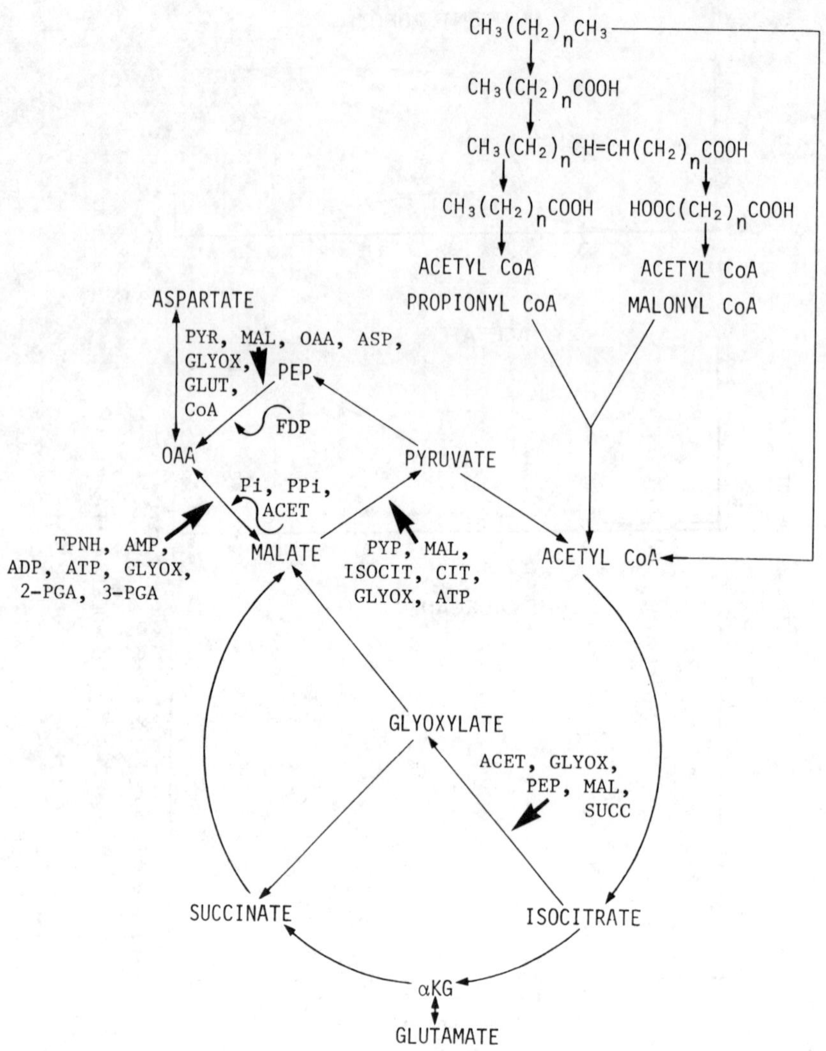

Fig. 7.19. A proposed metabolic pathway for hexadecane metabolism by *Micrococcus cerificans*. Large arrows indicate inhibition; small arrows indicate stimulation. <u>Source</u>: After Carnes 1972, Fig. 34, p. 117.

may represent only a small fraction of the deep-sea sediment population. This study yielded seven isolates presumptively identified as *Aeromonas*, *Pseudomonas*, and *Vibrio* species. The bacterial strains were grown in a salts medium containing 0.37 vol % n-hexadecane, which provided a two-phase system. The n-hexadecane enrichment procedure included three serial transfers, accomplished by aseptic removal of 1 ml of stationary phase culture and transfer of this portion to fresh salts medium containing 0.37 vol % n-hexadecane. The percentage of n-hexadecane used at stationary phase was determined by gas-liquid chromatography (GLC) for each of three enrichments (Table 7.18). The times required for the mixed culture to reach the stationary phase decreased from seven to five days, from the first to the third serial transfer. The lag phase associated with each serial transfer (i.e., enrichment) decreased from three days

Table 7.18. Growth of indigenous sediment bacteria
inoculated into n-hexadecane medium with
subsequent transfer of the original enrichment into
fresh n-hexadecane medium[a]

Enrichment on n-hexa-decane	Lag phase (days)	Time required to reach stationary phase (days)	Viable count at stationary phase (cells/ml)	Percent n-hexadecane utilized at stationary phase
Original	3	7	2.05×10^{8}	38.6
Second	2	6	4.22×10^{8}	68.7
Third	1	5	1.03×10^{9}	99.7

[a]The salts medium used contained 0.37% n-hexadecane (vol/vol).

Source: Schwarz, Walker, and Colwell 1974, Table 1, p. 983.
Reprinted by permission of the publisher.

to one day for the first and third transfer respectively. At the same time, the total viable
count (TVC) measured at the stationary phase for each enrichment increased about fivefold.
The increase in TVC was accompanied by an increase in n-hexadecane utilization (38.6 to 99.7%).
Additional enrichments on n-hexadecane did not increase either the TVC at stationary phase or
the extent of n-hexadecane utilization determined by GLC analysis.

Growth characteristics of the mixed culture at 1 and 500 atm are shown in Fig. 7.20. The
curve representing the culture under atmospheric pressure exhibited an increase in TVC, which
reached a maximum at 14 days. The curve representing growth on n-hexadecane at 500 atm showed
an initial decrease in TVC, followed by a sharp increase paralleling results at 1 atm. The
TVC reached a maximum at 21 days and decreased thereafter. The GLC analyses of n-hexadecane use
are summarized in Table 7.19. The culture held at 1 atm showed about 50% degradation of the
available n-hexadecane within the first 7 days and about 92% at 14 days. Hydrocarbon utilization at 1 atm paralleled results for growth shown in Fig. 7.20. The GLC analysis of samples
of the culture incubated at 500 atm revealed very slight hydrocarbon utilization at 7 days;
utilization increased rapidly after the seventh day and peaked at 21 days, also paralleling
results for growth shown in Fig. 7.20. The GLC tracings did not indicate accumulation of
smaller chain breakdown products of the n-hexadecane.

7.2.3 Cyclic alkanes

Beam and Perry (1973) searched for soil microorganisms that use unsubstituted cycloparaffinic
hydrocarbons (e.g., cyclohexane) as their sole carbon source, but were unsuccessful. Earlier
work had shown that nonproliferating cell suspensions of *Mycobacterium vaccae* strain JOB 5 and
cultures of *Pseudomonas fluorescens* grew with the cycloalkane as their sole substrate and
accumulated cyclohexanone when sodium bisulfite was added to the growth medium. The results
suggest that cycloparaffinic hydrocarbons might be biodegraded in nature by a cometabolism
mechanism through the less recalcitrant cycloalkanone. The inability to isolate organisms from
soil that can use cycloparaffinic hydrocarbons as their sole substrate cannot be taken as proof
that such organisms are not present in the environment. The relative abundance of cycloalkanone

ORNL DWG NO 76-11811

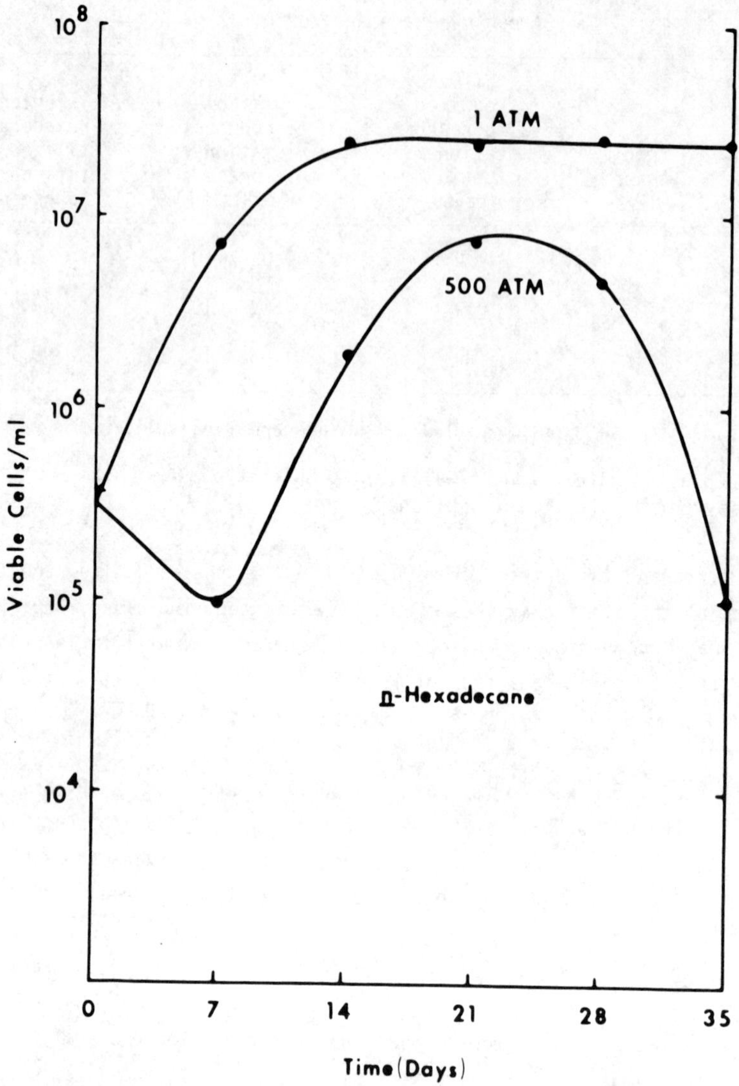

Fig. 7.20. Growth of a mixed culture of strains of deep-sea-sediment bacteria on n-hexadecane at 20°C and at 1 and 500 atm expressed as total viable counts after enumeration on modified seawater yeast extract agar plates. <u>Source</u>: From Schwarz, Walker, and Colwell 1974, Fig. 1, p. 983. Reprinted by permission of the publisher.

users and the capacity for oxidation of cycloalkanes to the homologous alkanone is strongly suggestive that cometabolism does play a role in the degradation of these compounds in nature. This would be similar to those cometabolic pathways suggested for other environmentally important molecules such as DDT and other chlorinated hydrocarbons.

7.2.4 Polycyclic aromatic hydrocarbons

Industrial waste such as that expected from coal conversion processes contains polycyclic aromatic hydrocarbons (PAH). It is important to draw on related industry experience with PAH to assess the role of microorganisms in determining the fate of these compounds in the environment.

Table 7.19. Bacterial utilization (%) of n-hexadecane under conditions of ambient and in-situ pressure[a]

Pressure (atm)	Time (days)				
	7	14	21	28	35
1	49.20	91.55	95.35	96.23	96.18
	49.60	91.85	94.25	95.37	96.02
Mean	49.4 ± 0.28	91.7 ± 0.21	94.8 ± 0.77	95.8 ± 0.60	96.1 ± 0.11
500	14.63	80.19	95.26	95.14	96.15
	10.77	73.81	93.54	95.06	95.85
Mean	12.7 ± 2.72	77.0 ± 4.51	94.4 ± 1.21	95.1 ± 0.05	96.0 ± 0.21

[a]Samples were processed in duplicate. Percent utilization was calculated relative to peak area from sterile and poisoned controls with OV-1 column programmed from 60 to 200°C at 5°C/min.

Source: Schwarz, Walker, and Colwell 1974, Table 2, p. 984. Reprinted by permission of the publisher.

The literature is inconsistent in defining the fate of benzo[a]pyrene (BaP). Authors in both laboratory and field degradation experiments report highly variable information on degradation. Benzpyrene (BP) has been found almost universally to be degraded by soil microorganisms.

7.2.4.1 Degradation by soil microorganisms

Soil microflora can destroy PAH. BP-contaminated soil was collected and analyzed by Khesina et al. (1969), who showed that soil microflora can destroy up to 70% of the BP contained in the soil. A mutant strain of the soil bacteria *Bacillus megaterium* was isolated and found capable of metabolizing (apparently by an oxidation reaction) BP and several other PAH. The bacteria metabolized the PAH compounds both in culture media and in natural soil contaminated with industrial waste products. The bacteria could degrade 80 to 90% of the BP originally present in 8 days of cultivation (Khesina et al. 1969).

Benzo[a]pyrene

Poglazova et al. (1966, 1967, 1968, and 1971) reported a series of experiments in which bacteria were isolated from BaP-polluted and nonpolluted soils collected in various areas of the U.S.S.R. Non-spore-forming bacteria were isolated from BaP-contaminated soil (100 mg of BaP per kilogram of soil) surrounding an oil refinery. Pure cultures were isolated by repeated inoculation of soil samples onto meat-peptone agar. Seventeen bacterial strains of different morphological and cultural characteristics were isolated. All the isolates assimilated BaP to some degree, all were spore-forming, and many belonged to the genus *Megaterium*. Four of the seventeen cultures (PBK No. 2/I, PBK No. 2/II, PBK No. 5, and PBK No. 13) were selected for quantitative investigations. The selected bacteria strains were cultured on nutrient agar to which BaP had been added. BaP decomposition or modification was determined by measuring the difference between the BaP introduced into the medium and the amount in the medium and microorganisms after incubation. The experiment was run four times, and the results (Table 7.20) were averaged. Cultures of bacterial strains No. 13 and 2/II showed a gradual disappearance of BaP; cultures containing bacteria strains No. 2/I and No. 5 completely retained the BaP in the medium. These results indicate that BaP-polluted soils do contain bacteria capable of modifying the BaP.

Using non-spore-forming bacteria isolated from his 1966 work, Poglazova et al. (1967) used members of the *Mycobacterium* genus and the mutant *Bacillus megaterium* mutilate for measurement of BaP assimilation. Results (Table 7.21) showed that *Mycobacterium rubrum* and a *M. flavum* strain exhibited the highest BaP degradation — 54 and 48% respectively. All the *Mycobacteria* could degrade BaP to at least 14%. Table 7.22 shows BaP degradation potentials of *Bacillus megaterium* mutilate and isolates PBK No. 5 and PBK No. 13. The cultures were subjected to both solid and aerated liquid media. The bacteria showed a greater BaP dissimilatory ability in the liquid media than on the solid media; this tendency is possibly due to closer contact of the bacteria cells with the hydrocarbon in liquid media (Poglazova et al. 1967).

Poglazova et al. (1968) again used *Bacterium megaterium* mutilate and isolates PBK 5 and PBK 13 in a study of BaP degradation. Test soil contaminated with BaP was moistened with meat-peptone broth and sterilized. Test cultures were inoculated onto the soil and incubated. Results (Table 7.23) show that all cultures could oxidize BaP (up to 85%) in the contaminated and sterilized soil. Continued studies with incubation of unsterilized soil contaminated with BaP showed that natural soil microorganisms could also degrade the hydrocarbon when it was present in high concentrations (Table 7.23).

Table 7.20. Change in the total content of BaP in the medium and in cells of soil microorganisms during their culturing

Bacterial cultures	Duration of culturing (days)									
	2				3		4				
	BaP extracted (μg)			Untreated with bacteria (%)	BaP extracted (μg) Expt. 4	Treated with bacteria (%)	BaP extracted (μg)				Treated with bacteria (%)
	Expt. 1	Expt. 2	Expt. 3				Expt. 1	Expt. 2	Expt. 3	Expt. 4	
PBK No. 13	176	174	175	13	114	43	63	84	117	118	53
PBK No. 2/II	176	192	165	11	154	23		132	118	150	33
PBK No. 5	201	201	188	0	204	0	199	195	207		0
PBK No. 2/I	198	188	182	0	195	0	210	180	195		0
Control	208	201	198		203		210	198			

Source: Poglazova et al. 1966, p. 539. Reprinted by permission of the publisher.

Table 7.21. BaP content change due to dissimilation by *Mycobacteria*

Culture	BaP introduced (μg)	BaP remaining after four days of culturing (μg)	Deficit (%)
Mycobacterium flavum var. *methanicum* strain B_4b	280	200	29
M. flavum var. *methanicum* strain C_1	280	145	48
M. lacticolum strain B_3	280	188	33
M. lacticolum strain D_5	280	237	15
M. rubrum var. *propanicum* strain B_4a	280	241	14
M. rubrum var. *propanicum* strain D_2	280	219	22
M. rubrum	280	128	54
M. smegmatis	280	180	35

Source: Poglazova et al. 1967, Table 1, p. 650. Reprinted by permission of the publisher.

Poglazova et al. (1971) again compared BaP degradation between natural soil bacteria and selected species. Natural bacteria cultures were isolated from soils containing varying amounts of BaP. Results (Table 7.24) show that natural soil microorganisms in all the soil samples decomposed significant (20 to 40%) amounts of BaP. Poglazova et al. (1971) also sterilized the soil samples and inoculated them with either *Bacillus sphaericus, Bacillus megaterium* mutilate, or *Pseudomonas*-146. The bacterial cultures were previously shown to be BaP metabolizers and, after incubation, showed degradation of 70 to 80% in the most heavily polluted soils (Table 7.25). Apparently, of the tested soils, those containing a greater quantity of aromatic compounds provide more favorable conditions for promoting the processes of PAH metabolism; however, the mechanism of this phenomenon (perhaps induction) remains obscure (Poglazova et al. 1971).

Muller and Korte (1975) experimented with BaP degradation by natural soil microorganisms. BaP in an acetone-water solution was sprayed on ground waste (10 to 20 μg/g) and composted for three weeks. Results showed that BaP was able to withstand the composting process to at least 99.5%. No degradation by microorganisms was observed in the extractable matter.

Other PAH

Fedoseeva et al. (1968) cultured *Bacillus megaterium* mutilate with a variety of PAH compounds. The *B. megaterium* mutilate was known to be a BaP-degrading bacteria from the earlier work by Poglazova; the work by Fedoseeva et al. (1968), however, examined the ability of the bacteria strain to degrade PAH compounds other than BaP. The compounds used in this study were 9,10-dimethyl-1,2-benzanthracene (DMBA), 1,2,5,6-dibenzanthracene (DBA), 1,2-benzanthracene (BaA), pyrene (P), 1,2-benzpyrene (BeP), BaP, 1,12-benzperylene (1,12-BPL), and perylene (PL). One of the above hydrocarbons was added to each growth medium to a concentration of 10 mg per milliliter of medium. Each hydrocarbon-containing medium preparation was then inoculated with the strain of bacteria. Cultures were incubated for four days. After incubation, the remaining undegraded hydrocarbon was removed from the medium along with the microorganisms growing in it by extraction with *n*-octane accompanied by mechanical shaking for 24 hr.

Table 7.22. Comparison of BaP dissimilation by microorganisms in solid and liquid media

Culture	Solid medium			Liquid medium		
	BaP introduced (μg)	BaP remaining after four days of culturing (μg)	Deficit (%)	BaP introduced (μg)	BaP remaining after four days of culturing (μg)	Deficit (%)
B. megaterium mut.	200	170	15	570	370	36
PBK No. 13	200	166	17	570	535	6
PBK No. 5	200	160	20	520	240	54

Source: Poglazova et al. 1967, Table 2, p. 650. Reprinted by permission of the publisher.

Table 7.23. Destruction of BaP by soil microorganisms[a]

Culture	BaP remaining after eight days of culturing (μg/10 g)	Deficit (%)
Sterilized soil		
No. 13	270	79
No. 5	200	85
B. megaterium mutilate	234	82
Unsterilized soil		
Natural soil microflora	525	60
Natural microflora + No. 13	370	71
Natural microflora + No. 5	350	73
Natural microflora + *B. megaterium* mutilate	290	78

[a]Soil naturally contaminated with BaP; original BaP concentration 1306 μg per 10 g of soil.

Source: Poglazova et al. 1968, Table 7.16, p. 201. Reprinted by permission of the publisher.

Table 7.24. Decomposition of BaP by the characteristic microflora of various soils

Location of the soil sample	Initial concentration of BaP (μg/ml)	Concentration of BaP after 5 days of cultivation (μg/ml)	Deficit (%)
Territory of the Klyaz'menskii reservoir	5	4	20
Territory of the MONIKI hospital	5	3	40
Territory of the "Neftegaz" factory	5	3	40

Source: Poglazova et al. 1971, Table 1, p. 349. Reprinted by permission of the publisher.

Results of the Fedoseeva et al. (1968) work (Table 7.26) shows that *B. megaterium* mutilate could oxidize all the PAH compounds tested and was not BaP-specific. The percentage of the hydrocarbon oxidized was found to be relatively the same for each compound, regardless of the compound's solubility or its ability to act as an enzyme inducer.

Fedoseeva et al. (1968) compared the oxidizing ability of *B. megaterium* mutilate with high and low concentrations of perylene, which has a low solubility, and BaP, which has a high solubility. The results (Table 7.27) show that the bacteria could oxidize both hydrocarbons over a wide range of concentrations with practically no percentage change. These results seem to indicate that microorganisms can oxidize PAH compounds in their surroundings to an extent that is practically independent of compound concentration (Fedoseeva et al. 1968).

Table 7.25. Decomposition of BaP in the soils by several microorganisms

Location and strain	Initial concentration of BaP in 10 g of soil (μg)	Concentration of BaP in 10 g of soil after 5 days of cultivation (μg)	Deficit (%)
Territory of the Klyaz'menskii reservoir			
B. sphaericus		0.007 ± 0.0005	48
B. megaterium mutilate	0.0135 ± 0.0015	0.0055 ± 0.0015	59
Pseudomonas - 146		0.006 ± 0.0001	55
Territory of the MONIKI hospital			
B. sphaericus		0.9 ± 0.2	82
B. megaterium mutilate	5.15 ± 0.55	1.3 ± 0.1	75
Pseudomonas - 146		0.7 ± 0.1	86
Territory of the "Neftegaz" plant			
B. sphaericus		2.4 ± 1.0	76
B. megaterium mutilate	10.0 ± 2.0	2.0 ± 1.0	80
Pseudomonas - 146		2.0 ± 1.2	80

Source: Poglazova et al. 1971, Table 2, p. 350. Reprinted by permission of the publisher.

Degradative pathway

Phenanthrene decomposition was studied by Rogoff and Wender (1957) to identify intermediate transformation products in the oxidative pathway used by soil microorganisms. The microorganisms were isolated from mixed cultures grown from soil samples. The isolate was a gram-negative motile bacillus that produced small, circular, granular colonies and liquefaction on gelatin plates, and saccate liquefaction with a whitish sediment in gelatin stabs. Further biochemical tests showed the isolate to be a member of the genus *Pseudomonas* and resembled the type species *P. aeruginosa*. The isolate also showed some resemblance to the naphthalene-attacking *P. boreopolis*.

The experiment was carried out in 500-ml Erlenmeyer flasks, with 96-hr incubation at 37°C before determinations of residual phenanthrene were taken. Decomposition varied from 46 to 78% with the different isolate cultures tested. The *Pseudomonas* isolate, which consistently decomposed from 65 to 78% of the phenanthrene, was used in the study. Interestingly, after 36 to 48 hr of incubation, the fermentation liquor became a deep buff color and gave a positive bluish-green ferric chloride test, which indicates the presence of a phenolic compound, possibly a hydroxynaphthoic acid (Rogoff and Wender 1957).

The oxidation of phenanthrene by soil isolates was also studied much earlier by Tausson (1928). The soil isolate was able to decompose relatively large amounts of phenanthrene in a fairly short time. Growth experiments were with *o*-hydroxybenzyl alcohol, *o*-hydroxybenzaldehyde, salicylic acid, and catechol as substrates. Tausson postulated, on the basis of the rapidity with which the compounds were assimilated, that the primary attack on the phenanthrene molecule occurred at the 9,10-position and resulted in the formation of the above-named compounds as intermediates (pathway I, Fig. 7.21). Rogoff and Wender (1957) isolated 1-hydroxy-2-naphthoic acid from

Table 7.26. Oxidation of aromatic hydrocarbons by microorganisms

Compound investigated	Concentration (μg/ml nut. med.) (control)	Concentration (μg/ml med.) after 4 days incub.	Hydrocarbon metabolized (% of control)
DMBA	10.5	7.3	
	10.8	6.8	37
	9.5	5.2	
BaP	8.4	5.8	
	8.0	5.2	35
	8.8	5.4	
	9.55	5.8	
DBA	9.45	5.0	42
	9.9	6.1	
	8.6	6.3	
BaA	9.7	6.1	32
	8.6	6.0	
	7.9	4.2	
P	7.8	3.2	55
	9.0	3.8	
	10.0	7.9	
BeP	10.5	6.7	23
	9.5	8.4	
1,12-BPL	9.0	5.8	
	10.1	5.9	40
	9.5	5.3	
	12.6	9.4	
PL	13.0	7.4	33
	12.0	9.5	

Source: Fedoseeva et al. 1968, Table 2, p. 685. Reprinted by permission of the publisher.

their cultures and found that the phenanthrene-grown cells could oxidize salicylic acid and catechol. Based on their results, Rogoff and Wender (1957) postulated that the attack on the phenanthrene molecule occurred via the end-ring (pathway II, Fig. 7.21). The end-ring attack first occurs at the 3,4-bond of the phenanthrene molecule (Rogoff 1962). The possibility of differences between points of attack in the Tausson postulate and the Rogoff and Wender postulate due to the metabolic pattern differences of the organisms used cannot be eliminated (Rogoff and Wender 1957).

7.2.4.2 Degradation by marine organisms

Sisler and ZoBell (1947) describe the microbial use of aromatic hydrocarbons (naphthalene, anthracene, phenanthrene, BaA, and DBA) in laboratory cultures. The bacteria were mixed cultures of marine origin. This work involved the dispersal of the aromatic hydrocarbon on ignited sand and the immersion of the sand and substrate into seawater. Seawater is a physiologically balanced solution and contains all the essential salts necessary to satisfy the mineral requirements of marine microorganisms (Sisler and ZoBell 1947). The amount of hydrocarbon used was measured by the amount of carbon dioxide evolved from oxidation of the

Table 7.27. Oxidation of aromatic hydrocarbons in relation
to their concentration in the culture medium

Compound investigated	Concentration (μg/ml)		Hydrogen metabolized (% of control)
	Before incubation (control)	After incubation	
PL	12.6	9.4	
	13.0	7.4	33
	12.0	9.5	
The same	2.04	1.55	
	1.98	1.43	31
	2.3	1.39	
BP	9.7	4.8	
	10.25	6.3	46
	10.0	5.0	
The same	1.9	1.1	
	2.1	1.05	45
	2.0	1.1	

Source: Fedoseeva et al. 1968, Table 3, p. 685. Reprinted by permission of the publisher.

substrate. The carbon dioxide produced by the control was subtracted from the total carbon dioxide produced by the culture acting on the substrate to give the net amount of hydrocarbon oxidized by the microorganisms (Table 7.28). The marine bacteria assimilated phenanthrene and anthracene more rapidly than naphthalene, or the higher homologues (BaA and DBA). No intermediates in the oxidation processes were isolated.

The degradation of naphthols by marine organisms has received little attention. Sikka, Miyazaki, and Lynch (1975) studied the ability of selected species of marine bacteria, yeasts, and filamentous fungi to degrade 1-naphthol. The compound used in the study was 1-naphthol-1-^{14}C having a specific activity of 15.2 mCi/millimole. Organisms isolated from a marine environment and used in this study were *Brevibacterium* sp., *Flavobacterium* sp., *Serratia marina*, and *Spirrillum* sp. (bacteria); *Candida parapsilosis*, *Rhodotorula glutinus*, and *Trichosporon fermentans* (yeasts); *Aspergillus fumagutus*, *Culcitalna achraspora*, *Halosphaeria mediosetigera*, and *Humicola alopallonella* (filamentous fungi). The bacteria were grown on Difco marine broth, whereas the yeasts and filamentous fungi were grown in a medium containing 5 g of glucose, 2.4 g of NH_4NO_3, and 1.0 g of yeast extract per liter of artificial seawater prepared from Rila marine mix (Rila Products, Teaneck, N.J.). The organisms were grown on their respective nutrient media containing 1.0 ppm of 1-naphthol. The microorganisms showed the ability to degrade 1-naphthol and also produced different types and metabolites of 1-naphthol (Table 7.29). The cultures of *Brevibacterium*, *Candida*, *Rhodotorula*, and *Trichosporon* incubated with ^{14}C-1-naphthol produced only small amounts of water-soluble metabolites from 1-naphthol. The remaining organisms were able to convert 1-naphthol to both ether- and water-soluble metabolites. *Culcitalna*, *Halosphaeria*, *Humicola*, and *Aspergillus* were more effective than *Flavobacterium*, *Spirrillum*, and *Serratia* in converting 1-naphthol to its water-soluble metabolites. In the

ORNL DWG NO 76-11812

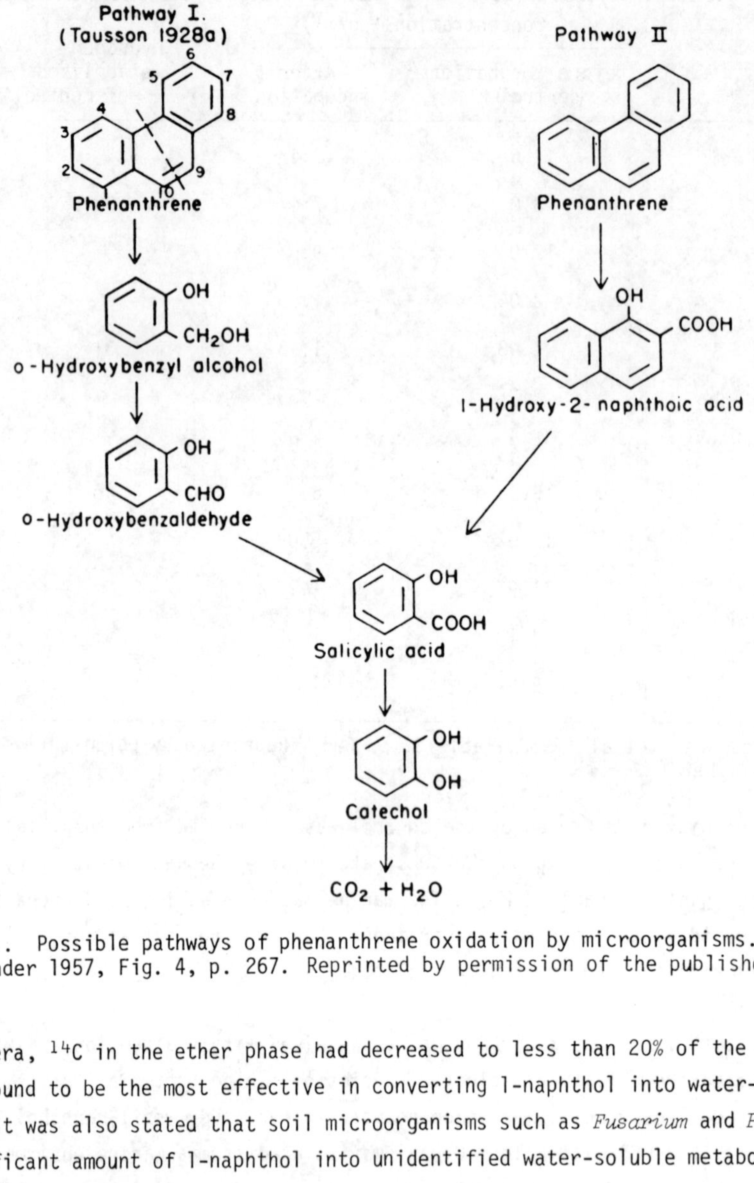

Fig. 7.21. Possible pathways of phenanthrene oxidation by microorganisms. Source: From Rogoff and Wender 1957, Fig. 4, p. 267. Reprinted by permission of the publisher.

first four genera, ^{14}C in the ether phase had decreased to less than 20% of the original amount. *Humicola* was found to be the most effective in converting 1-naphthol into water-soluble metabolites. It was also stated that soil microorganisms such as *Fusarium* and *Pseudomonas* converted a significant amount of 1-naphthol into unidentified water-soluble metabolites (Sikka et al. 1975).

7.2.4.3 Resistance to degradation

Activated sludges from municipal waste treatment systems have been shown to be mostly ineffective in the treatment of aromatic, organic industrial wastes. The inability of treatment systems to degrade the hydrocarbons could result in introduction of the hydrocarbons into local water tables as pollutants.

Oxidated sludge

Malaney et al. (1967) studied the resistance of carcinogens to oxidation by oxidated sludge. The compounds studied (Fig. 7.22) were mostly aromatic, organic industrial wastes. Sludge samples were taken from municipal treatment plants at Nashville, Franklin, and Ashland City, Tennessee.

Table 7.28. Amount of carbon dioxide produced by action of bacteria on 25 mg of hydrocarbons in four days at 32°C

Hydrocarbon	CO_2 produced (mg)	Amount oxidized (%)
Naphthalene	44.2	51
Anthracene	53.5	64
Phenanthrene	58.5	68
Diaminobenzene	14.0	23
1,2-Benzanthracene	44.2	47
1,2,5,6-Dibenzanthracene	11.6	13
None (control)[a]	0	

[a]The controls used as checks against adventitious organic matter produced less carbon dioxide than the endogenous controls, hence are condensed as one control shown here.

Source: Sisler and ZoBell 1947, Table 1, p. 522. Reprinted with permission from *Science*.

Aeration-tank-mixed liquors were collected on the day of the experiment, blended for 10 sec, and analyzed for suspended solids concentration (MLSS) by the membrane filter technique. The MLSS concentration was then adjusted to 2500 or 5000 mg/liter. Twenty milliliters of the sludge, which was neither washed nor treated with mineral salts, was pipetted into 125-ml Warburg flasks containing the individual organic compounds. Substrate concentration in the Warburg flask reaction compartment was 500 mg/liter. Warburg apparatus water bath temperature was 20°C. Readings were made every 2 hr for the first 24 hr and then every 4 hr for the remaining 120 hr of the experimental run. Results are shown in Tables 7.30, 7.31, and 7.32. Results rapidly indicated that most of the compounds tested were markedly resistant to biological oxidation (Malaney et al. 1967).

It could be concluded that the poor oxidation performance of the activated sludge is due to the refractory chemical structure of the carcinogenic compounds. These compounds are condensed nuclear hydrocarbons whose structures are relatively strain-free. The stable structures result from carbon-carbon bonds being in the staggered configuration (the same configuration as methane) and all bond angles being tetrahedral, which produces the least strain (Malaney et al. 1967).

Waste stream microorganisms

Saprophytic microorganisms were used in a waste compost to break down organic compounds, including BaP. No degradation of the BaP was observed to at least 99.5% (Muller and Forte 1975).

Fedorenko (1964) studied BaP-containing waste streams from a by-product coke plant. The waste streams were subjected to biochemical treatment for purification before being released into a local river. Fedorenko (1964) tested the waste stream for BaP content before and after biochemical treatment. Results showed that the treatment, including passage through a biological basin, decreased the BaP content only slightly; the result was contamination of the river at the site of the waste stream discharge. Gibson and Jarina (1975) created a mutant strain of *Beijerinckia*

Table 7.29. Distribution of ^{14}C in the culture solution of various microorganisms incubated with ^{14}C-1-naphthol

Organism	Percent of initial ^{14}C						
	Ether-soluble products						Water-soluble products[a]
	Rf-values						
	0	0.09	0.31	0.39 (1-naphthol)	0.64	Total	
Brevibacterium sp.		0	0	77.8	0	77.8	3.6
Flavobacterium sp.	10.6	0	0	72.2	0	76.8	4.7
Serratia marina	5.7	0	0	59.4	0	65.1	15.8
Spirrillum sp.	7.9	0	0	57.1	0	65.0	14.3
Candida parapsilosis	0	0	0	74.7	0	74.7	7.5
Rhodotorula glutinis	0	0	0	73.0	0	73.0	8.0
Trichosporon ferments	0	0	0	71.8	0	71.8	9.6
Aspergillus fumigatus	2.4	2.4	6.9	0	1.1	12.8	74.8
Culcitalna achraspora[b]	4.0	2.1	3.2	0.3	3.1	12.7	70.7
Halosphaeria mediosetigera[b]	1.7	3.0	11.6	2.8	0	19.1	70.7
Humicola alopallonella[b]	0	2.1	2.8	0	0.9	5.8	90.2

[a]Corrected for nonbiological conversion.
[b]These organisms were incubated with 1-naphthol for 7 days, whereas the others were incubated for 3 days.

Source: Sikka, Miyazaki, and Lynch 1975, Table 2, p. 669. Reprinted by permission of the publisher.

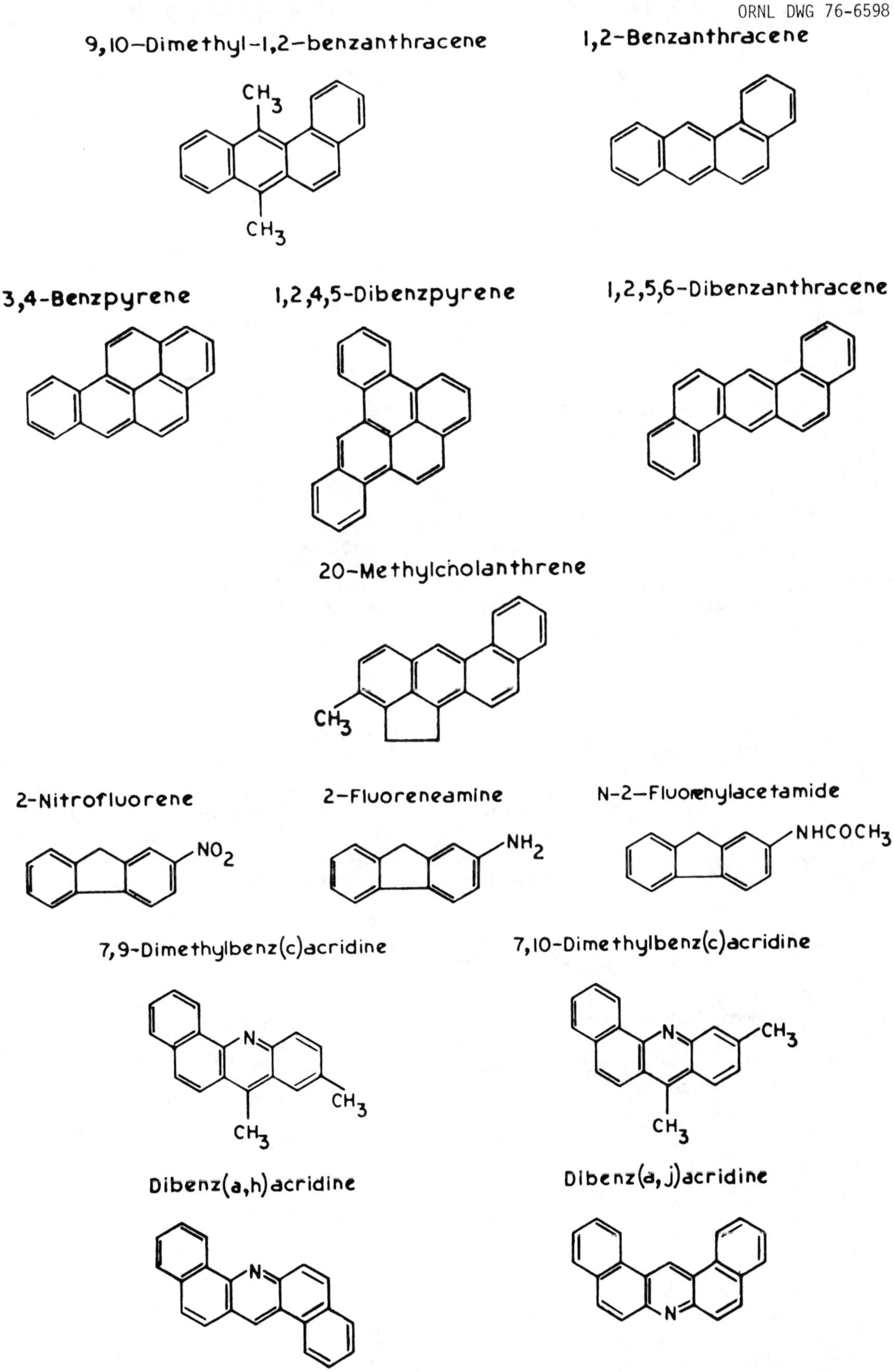

ORNL DWG 76-6598
9,10-Dimethyl-1,2-benzanthracene
CH_3
CH_3
1,2-Benzanthracene
3,4-Benzpyrene
1,2,4,5-Dibenzpyrene
1,2,5,6-Dibenzanthracene
20-Methylcholanthrene
CH_3
2-Nitrofluorene
NO_2
2-Fluoreneamine
NH_2
N-2-Fluorenylacetamide
$NHCOCH_3$
7,9-Dimethylbenz(c)acridine
N
CH_3
CH_3
7,10-Dimethylbenz(c)acridine
N
CH_3
CH_3
Dibenz(a,h)acridine
N
Dibenz(a,j)acridine
N

Fig. 7.22. Carcinogens and their molecular structures that were studied for biologic oxidation. Source: From Malaney et al. 1967, Figs. 1 through 4, p. 2021. Reprinted by permission of the publisher.

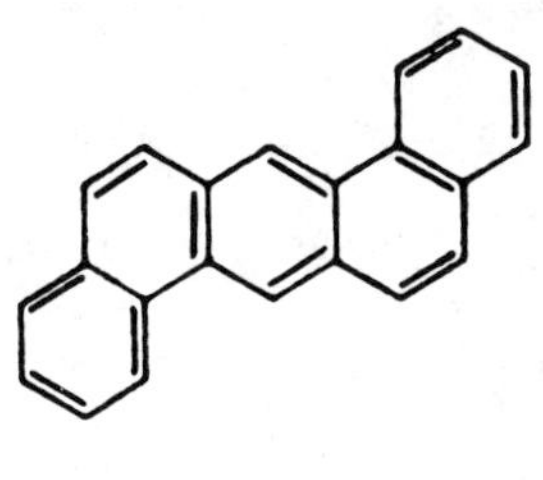

2,3-Butylene oxide

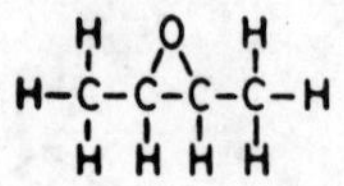

beta-Propiolactone

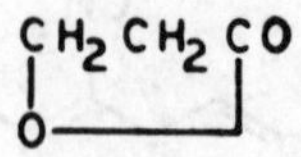

Thiourea

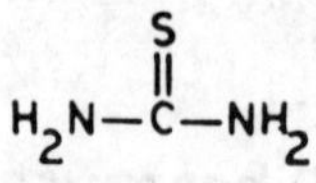

Ethyl carbamate

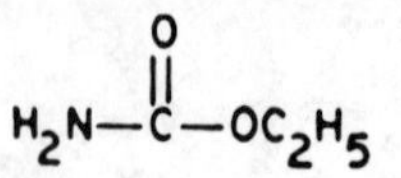

2-Thiouracil

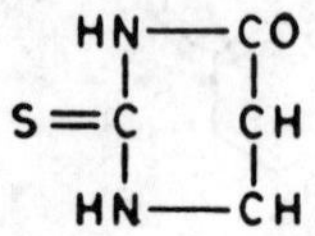

4-Ethoxyphenylurea

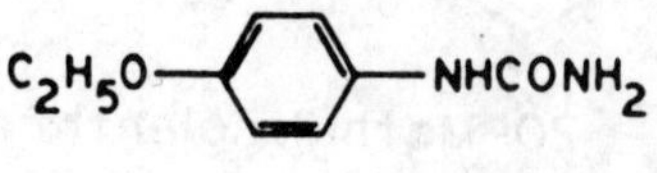

Benzidine

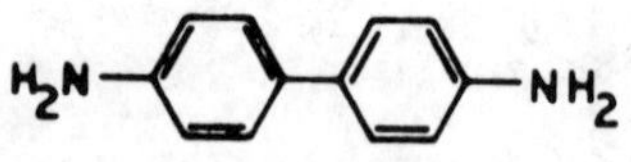

4,4'-Dihydroxy-a,b-diethylstilbene

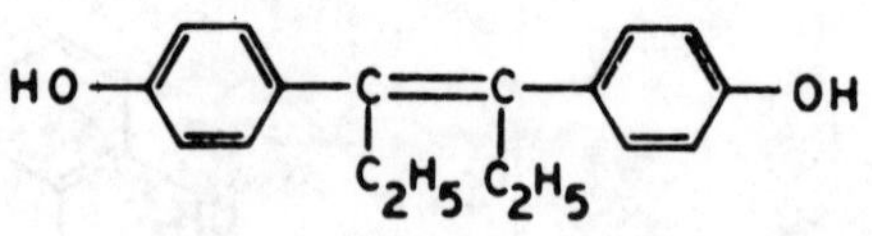

p-Phenylazophenol

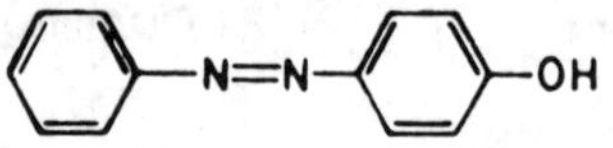

p-Phenylazoaniline

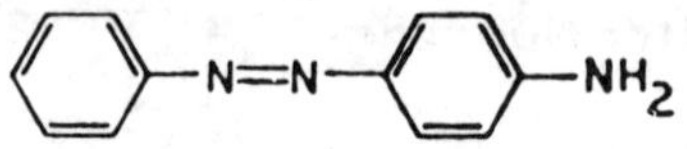

4,4-Bis(dimethylamino)benzophenone

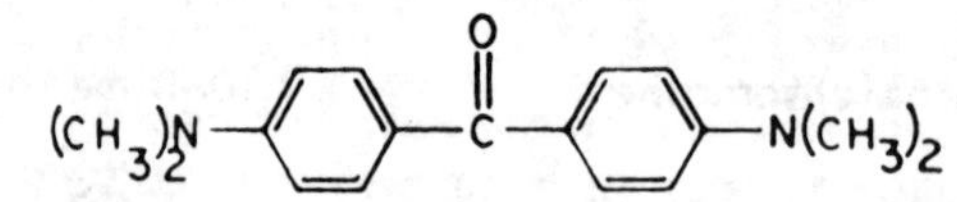

beta-Naphthylamine

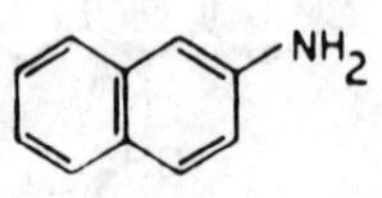

9,10-Dimethylanthracene

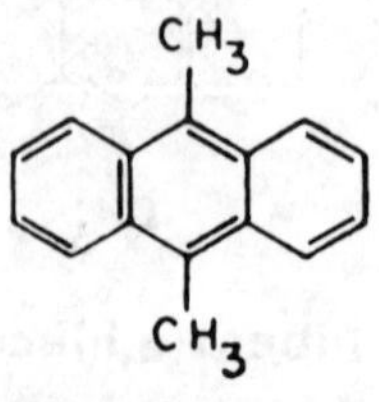

7-Methyl-1,2-benzanthracene

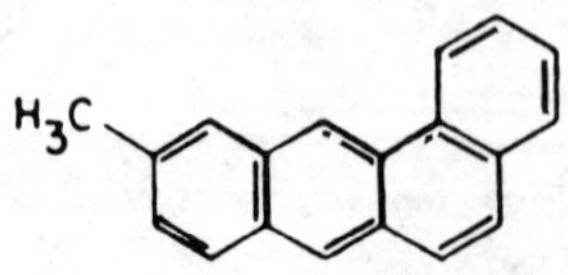

Fig. 7.22 (continued)

Table 7.30. Oxidation of carcinogens by Ashland City activated sludge at 2500 mg/liter[a]

Compound	Percentage of theoretical oxygen demand[b]			
	6 hr	24 hr	72 hr	144 hr
2,3-Butylene oxide	0.3	T	T	2.4
β-Propiolactone	T	T	30.3	55.9
Thiourea	T	T	T	T
Ethylcarbamate	T	T	T	T
2-Thiouracil	T	T	T	T
4-Ethoxyphenylurea	T	T	T	T
Benzidine	T	T	T	T
4,4'-Dihydroxy[a,b]diethylstilbene	T	T	T	T
2-Naphthylamine	0.4	T	T	T
4,4'-Bis(dimethylamino)benzophenone	T	T	0.9	0.6
p-Phenylazophenol	T	T	T	T
p-Phenylazoaniline	T	T	T	T
9,10-Dimethylanthracene	0.1	0.2	0.0	0.0
1,2-Benzanthracene	0.1	0.0	0.0	0.3
7-Methyl[1,2]benzanthracene	0.1	T	0.0	0.4
9,10-Dimethyl[1,2]benzanthracene	0.1	0.2	0.3	0.5
1,2,5,6-Dibenzanthracene	0.1	T	0.0	T
3,4-Benzpyrene	0.2	1.5	1.6	1.7
1,2,4,5-Dibenzpyrene	0.0	T	T	T
20-Methylcholanthrene	0.1	0.1	0.0	T
2-Nitrofluorene	T	T	T	T
2-Fluoreneamine	T	T	T	T
N-2-Fluorenylacetamide	T	T	T	T
7,9-Dimethylbenz[c]acridine	0.1	0.1	0.7	T
7,10-Dimethylbenz[c]acridine	T	0.0	0.0	T
Dibenz[a,h]acridine	0.3	0.3	0.5	0.7
Dibenz[a,j]acridine	T	0.1	0.0	T

[a]Figures in tables obtained using following calculation:

$$\% \text{ TOD} = \frac{O_2 \text{ uptake substrate} - O_2 \text{ uptake control}}{\text{TOD}} \times 100 .$$

[b]T = toxic, i.e., O_2 uptake control was greater than O_2 uptake substrate, indicating that the substrate was inhibitory or lethal to the sludge organisms.

Source: Malaney et al. 1967, Table I, p. 2022. Reprinted by permission of the publisher.

capable of oxidizing BaP. The original *Beijerinckia* species was isolated from a polluted stream by virtue of its ability to grow with biphenyl as its sole source of carbon and energy. Treatment of this organism with N-methyl-N'nitro-N-nitrosoguanidine led to the isolation of the mutant strain *Beijerinckia* B-836. This organism was then grown on succinate in the presence of biphenyl and, after this procedure, it oxidized BaP and BaA to polar compounds. The compounds were identified as *cis*-dihydrodiols, which suggests that the intermediates are not formed from arene oxide precursors because *trans* configuration would then occur. It is possible that

Table 7.31. Oxidation of carcinogens by Nashville activated sludge at 2500 mg/liter

Compound	Percentage of theoretical oxygen demand			
	6 hr	24 hr	72 hr	144 hr
2,3-Butylene oxide	0.5	2.0	4.1	9.6
β-Propiolactone	T	T	T	T
Thiourea	T	T	T	T
Ethylcarbamate	0.1	1.6	T[a]	T
2-Thiouracil	0.8	2.4	12.2	12.8
4-Ethoxyphenylurea	1.9	6.3	13.0	9.4
Benzidine	1.0	4.1	T	T
4,4'-Dihydroxy[a,b]diethylstilbene	T	T	T	T
2-Naphthylamine	0.3	2.8	T[a]	T
4,4'-Bis(dimethylamino)benzophenone	T	0.2	3.7	4.9
p-Phenylazophenol	0.1	4.7	T[a]	T
p-Phenylazoaniline	1.5	4.3	T[a]	T
9,10-Dimethylanthracene	0.5	1.6	3.2	19.5
1,2-Benzanthracene	0.5	1.1	1.6	2.1
7-Methyl[1,2]benzanthracene	T	T	1.6	3.1
9,10-Dimethyl[1,2]benzanthracene	0.5	1.4	9.4	12.7
1,2,5,6-Dibenzanthracene	0.8	1.8	5.5	7.9
3,4-Benzpyrene	0.3	1.4	3.5	2.7
1,2,4,5-Dibenzpyrene	0.3	0.9	1.0	1.8
20-Methylcholanthrene	0.7	1.7	6.1	9.3
2-Nitrofluorene	1.4	2.8	10.3	13.7
2-Fluoreneamine	0.9	4.2	T[a]	T
N-2-Fluorenylacetamide	1.3	3.7	3.2	6.3
7,9-Dimethylbenz[c]acridine	0.3	0.9	1.2	4.1
7,10-Dimethylbenz[c]acridine	0.0	0.1	1.6	4.3
Dibenz[a,h]acridine	0.6	0.9	3.1	4.6
Dibenz[a,j]acridine	0.0	0.6	3.3	4.7

[a]Result is consequence of sudden increase in oxygen uptake by sludge alone, perhaps onset of nitrification.

Source: Malaney et al. 1967, Table II, p. 2023. Reprinted by permission of the publisher.

eukaryotic microorganisms use a monooxygenase enzyme system for the oxidation of aromatic hydrocarbons (Gibson and Jerina 1975). Barnsley (1975) studied cultures of four bacteria strains and their metabolism of BaP solutions. The rate of disappearance of BaP increased as a culture entered the stationary phase of growth. The addition of cyanide did not increase the rate of disappearance. The rates observed with the four species during 6-hr incubations in the stationary phase were *Pseudomonas* NCIB 9816, 0.13 μmoles/hr; *Pseudomonas* ATcc 17483, 0.05 μmoles/hr; *Pseudomonas putida* PpG7, 0.05 μmoles/hr; and *Escherichia coli*, not detected.

Table 7.32. Oxidation of carcinogens by Franklin sludge at 2500 mg/liter

Compound	Percentage of theoretical oxygen demand			
	6 hr	24 hr	72 hr	144 hr
2,3-Butylene oxide	1.8	T	T	T
β-Propiolactone	T	T	T	T
Thiourea	0.5	3.5	1.4	5.7
Ethylcarbamate	T	T	T	T
2-Thiouracil	1.7	2.7	5.5	6.2
4-Ethoxyphenylurea	T	5.8	24.2	20.6
Benzidine	T	T	1.9	1.9
4,4'-Dihydroxy[a,b]diethylstilbene	0.4	6.0	T[a]	5.0
2-Naphthylamine	T	T	T	T
4,4'-Bis(dimethylamino)benzophenone	0.6	2.1	4.1	4.5
p-Phenylazophenol	T	6.4	4.2	T[b]
p-Phenylazoaniline	T	9.8	4.2	T[b]
9,10-Dimethylanthracene	0.1	0.2	0.6	1.6
1,2-Benzanthracene	0.5	0.7	1.2	1.6
7-Methyl[1,2]benzanthracene	T	T	T	T
9,10-Dimethyl[1,2]benzanthracene	0.1	0.2	0.5	0.5
1,2,5,6-Dibenzanthracene	0.1	0.1	0.3	0.9
3,4-Benzpyrene	1.2	1.8	3.8	6.1
1,2,4,5-Dibenzpyrene	0.1	0.1	0.0	T
20-Methylcholanthrene	0.8	1.1	2.1	4.4
2-Nitrofluorene	1.2	3.6	7.2	12.4
2-Fluoreneamine	1.1	9.7	8.0	T[a]
N-2-Fluorenylacetamide	2.6	7.5	10.8	12.3
7,9-Dimethylbenz[c]acridine	0.1	0.4	T	T
7,10-Dimethylbenz[c]acridine	0.1	0.4	0.7	1.1
Dibenz[a,h]acridine	0.1	0.2	T	T
Dibenz[a,j]acridine	T	T	T	T

[a]Result is consequence of sudden increase in oxygen uptake by sludge alone, perhaps onset of nitrification.
[b]Result is consequence of unexplained decrease in oxygen uptake by sludge-substrate mixture.
Source: Malaney et al. 1967, Table III, p. 2024. Reprinted by permission of the publisher.

7.2.4.4 Synthesis by microorganisms

Biosynthesis of polybenzene compounds by microorganisms was proposed by Mallet et al. (1964). Several studies indicate increases in PAH concentrations that could only be attributed to biosynthesis; other investigations using bacterial and plankton extracts showed the synthesis of PAH when fatty acids were the initial substrates. Brisou (1969) showed that at least part of the BaP, a carcinogen, found in nature could have a biological origin. Several common anaerobic soil bacteria isolated from soil samples demonstrated the ability to produce BaP under laboratory conditions. However, a study by Hase and Hites (1976) disputed the biosynthesis studies and stated that microorganisms actually act as bioaccumulators.

<u>Detection of PAH in soils and deep sediments</u>

The presence of polybenzene hydrocarbons in plant soils and marine sediments was demonstrated by Mallet and Schneider (1964). Samples were taken at different depths below the surface; often samples below the surface had higher BaP concentrations than did surface samples.

Calcareous sediments were studied to determine whether the increased hydrocarbon content was from alteration with air and light. In Gouviex, France, a sample was mechanically cut at a depth of 50 m within an underground quarry with a very homogeneous calcareous base. This sample, which had had no contact with an industrial environment, yielded a BaP content of 1.95 µg per 100 grams of material, whereas surface soil samples taken in the vicinity of Molineaux, France, yielded only 1.5 µg per 100 grams of material (Mallet and Schneider 1964). The investigations showed that the PAH did not, as might be expected, have an external origin. Mallet and Schneider (1964) and Mallet (1969) considered the hydrocarbons detected in sediments to have endogenous biosynthetic origins.

Mallet (1969) studied both planktonic calcareous sediments, which contain decomposition products of phytoplankton, and boghead coal, which is rich in algae. Transformation of these animal and plant materials into hydrocarbon products, amino acids, fatty acids, and waxes may result in aromatic hydrocarbon production under the influence of bacteria. Mallet et al. (1972) proposed that the starting substrates for polycyclic hydrocarbon biosynthesis may be more varied and include sterols, terpenes, and lipids as well as those hydrocarbons previously mentioned. Brisou (1972) also proposed that BaP biosynthesis by bacteria has for a starting point lipids, purified fatty acids, and other substances including aliphatic terpenes. Mallet (1969) hypothesized that anaerobic microorganisms form polycyclic hydrocarbons at the expense of fatty acids present in sediments. Identification tests showed that *Clostridium putride* was particularly important in the biosynthesis of polycyclic hydrocarbons.

Hase and Hites (1976) collected sediment from the Charles River Basin and examined it for PAH and bacterial types. The sediment contained at least eleven different ring structures and abundant alkyl-substituted derivatives of the ring structures. They found indications that at least some of the PAH compounds could originate from natural sources such as biosynthesis by the mixed anaerobic bacteria present in the sediment. Mixed anaerobic bacteria cultures were prepared from the sediment samples and grown on a chemically defined medium. All equipment, solvents, and glassware were shown to be free of PAH by gas chromatographic mass spectrometry. The media were prepared by mixing with 40 liters of tap water and autoclaving at 120°C. Examination (by gas chromatographic mass spectrometry) of extracts from the mixed anaerobic bacteria cultures and blanks showed the presence of several PAH compounds. The PAH in the medium blank could originate from the tap water or the chemicals or from autoclaving the dextrose at 120°C. High-resolution mass spectrometry was used to confirm the presence of PAH and their many alkyl homologs in the bacterial and blank extracts.

Results of the high-resolution mass spectrometry showed that the PAH composition and distribution in the bacteria were about the same as in the media in which they were grown. This fact shows that the bacteria bioaccumulated the PAH from 200 liters of total medium to about 400 ml in the total wet weight of the bacteria (Hase and Hites 1976).

Most details of biosynthesis experiments — especially the care with which the blank measurements are taken — are not described. The sensitivity of blank measurements is particularly important when considering the bioaccumulation effect and must be much greater than that used for analysis (Hase and Hites 1976).

Culture studies

Mallet et al. (1967), working with planktonic sediments, studied the production of polybenzene hydrocarbons with *Clostridium*-type microorganisms grown in a lipid medium produced from marine plankton. They placed 2 to 4 mg of total lipids extracted from marine plankton in a sterile Erlenmeyer flask, added 10 ml of nutritive broth with seawater, and inoculated the flask with the bacterial culture. Ten flasks were prepared as described; three were inoculated with strict anaerobic bacteria, to which were added 5 ml of thioglycollate medium with 0.5% agar, whereas the seven remaining flasks were inoculated with aerobic bacteria. Incubation of the groups was carried out by placing the anaerobic cultures in a vacuum incubator at 37°C for 15 days and placing the aerobic cultures in a shaker at 20°C for 15 days. Results of the experiment are shown in Table 7.33.

Table 7.33. Biosynthesis of BaP by microorganisms
on a plankton lipid extract

Plankton	Lipid extract	BaP for the whole of the samples (µg)
Plankton 148		None detected
Plankton 148	Anaerobic 57 (*Clostridium*)	0.02
Plankton 185		Traces
Plankton 185	Anaerobic 481 (*Clostridium*)	0.05
Plankton 115		
Plankton 115	Anaerobic 11 (*Clostridium*)	0.04

Source: Modified from Mallet, Zanghi, and Brisou 1967, Table 1,
p. 154. Reprinted by permission of the publisher.

Results of the Mallet et al. (1967) work emphasized the possibility of biosynthesis of BaP by anaerobic bacteria. The anaerobic bacteria used in this work were *Clostridium* types isolated from different coastal plankton. The aerobic bacteria in culture Nos. 3598 and 3599 were *Flavobacterium*, those in culture Nos. 3490, 3491, 3547, and 3600 were *Phytobacterium*, and the bacteria in culture No. 3493 was a *Pseudomonas*.

The results seem to confirm the biosynthesis of BaP from plankton lipids by anaerobes of the *Clostridium* genera (Mallet et al. 1967).

Knorr and Schenk (1968) measured the BaP-synthesizing capability of various bacteria. Their work concentrated on strains of *Mycobacterium smegmatis*, which were believed to convert cholesterol into one of the structurally related carcinogens. Results, given in Table 7.34, show the BaP produced by strains of *Mycobacterium smegmatis* grown on agar. The same or higher

Table 7.34. BaP biosynthesis by strains of *Mycobacterium smegmatis*[a]

Strain	Dry weight (g)	BaP (μg)	Corrected to 100 g dry mass
SN 1	50	0.5	2.8
SN 2	33	0.4	3.4
SN 7	40	0.7	4.8
SN 10	25	0.3	3.4
1675	45 (25)	0.4 (0.3)	2.4 (2.8)
1970	35	0.25	2.0
170	70	0.2	2.0
1619	30 (40)	0.3 (0.5)	2.8 (3.6)

[a]Values in parentheses are BaP-free control values.

Source: Knorr and Schenk, Table 1, p. 283. Reprinted by permission of the publisher.

yields of BaP were obtained with bacterial types other than *Mycobacteria* such as *E. coli*, *Proteus*, *Pseudomonas*, *Serratia*, and *Myxobacteria* (Table 7.35).

Table 7.35. BaP formation by bacterial types other than *Mycobacterium*

Strain	Dry weight (g)	BaP (μg)	Corrected to 100 g dry mass
E. coli 91	37	0.5	4.6
E. coli 92	27	0.5	5.0
Proteus	37	0.75	5.6
Pseudomonas fluorescens	45	0.5	3.0
Serratia marcescens	67	0.5	2.0
Myxobacteria Z6	40	0.9	6.0

Source: Knorr and Schenk, Table 2, p. 284. Reprinted by permission of the publisher.

Brisou (1969) used bacteria of the genera *Welchia* and *Clostridium* to identify the substrate that would be most favorable for biosynthesis of BaP by the bacteria. The substrates studied and the amounts of BaP detected are shown in Table 7.36. The bacterial cultures were grown at 37°C either in a reducing medium with thioglycollate or in brain-heart broth maintained in a vacuum oven. The results varied from one bacterial strain to the other for the same substrate.

Brisou (1969) showed that part of the carcinogenic BaP found in nature might have a biological origin. The production of polybenzenes seems dependent on anaerobic microbes, which are commonly found in soils and sediments (Brisou 1969). The results obtained with pure cultures are clearly

Table 7.36. Biosynthesis of BaP on different substrates

Substrate added	BaP detected (μg/100 g dry medium)
Pentadecanoic acid	Traces
Margaric acid	Traces
Leneicosanoic acid	Traces
Nonadecanoic acid	0.0025
Sodium oleate	0.02
Thioglycollate broth	0
Cholesterol	0
Equigyne (mixture of estrogens)	0.06
p-Aminobenzoic acid + succinic anhydride	0

Source: Brisou 1969, p. 773. Reprinted by permission of the publisher.

lower than those recorded with mixtures of lipids of plankton, which permits the assumption of the intervention of several substrates, and of complex organic materials (Brisou 1969).

Soil studies

Mallet and Tissier (1969) used *Clostridium putride* isolated from compost soils to show the bio-synthesis of BaP in sterilized soils inoculated with the bacteria. The oven-sterilized humus was inoculated with *C. putride* and left for four months at ambient laboratory temperatures. The BaP was measured at the start of the experiment and found to be 80 to 100 μg per 100 grams of soil; the concentration at the end of the four-month period was 160 μg per 100 g of soil.

The large initial amount of polybenzene hydrocarbons lead Brisou (1969) to collect some forest soil far from large industrial centers. The BaP measurements at the time of collection were 1 to 2 μg per 100 g of soil. The soil was ground, dry-sterilized at 170°C, and set to culture under anaerobic conditions with *C. putride* at laboratory temperatures for six months. At the end of the six-month period, the BaP content was 4 μg per 100 g of soil, which is an appreciable increase. Significantly, an experiment using *Escherichia coli*, a facultative aerobic-anaerobic common in the mammalian gastrointestinal tract, produced comparable results.

7.2.4.5 Effect on microorganism growth

Hass and Applegate (1975) studied the effect of 3-, 4-, and 5-ring unsubstituted PAH on the growth of *Escherichia coli* (ATCC No. 26922). The PAH studied were BaA, 1,2,5,6-dibenzanthracene (DBA), BaP, tetracene, pyrene, anthracene, phenanthrene, chrysene, 1,2,3,4-DBA, and pentacene (Fig. 7.23). The rate of bacterial growth on nutrient broth medium (with added PAH) was measured by determining the logarithmic phase of the growth curve (k_{PAH}) and comparing it with the rate of growth in flasks without PAH ($k_{CONTROL}$). The comparison indicated growth enhancement (K > 1.0) or growth retardation (K < 1.0). The results were plotted as graphs and are presented in Figs. 7.24, 7.25,

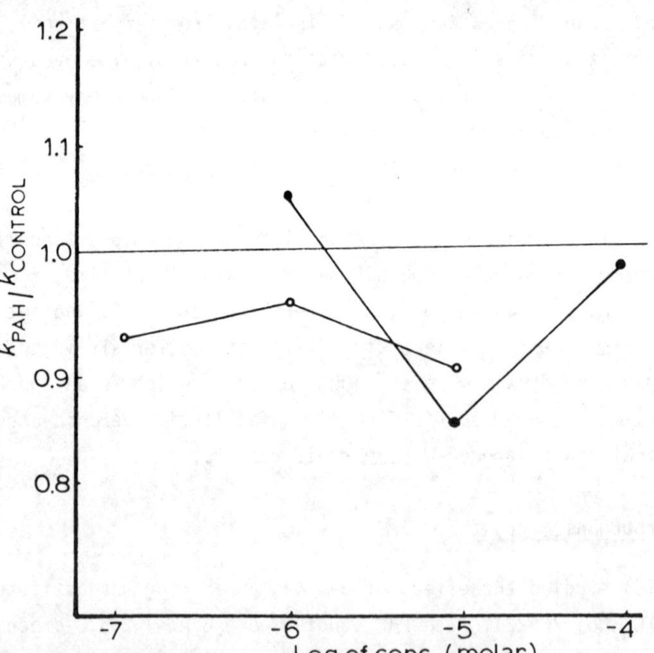

ORNL DWG 76-6590

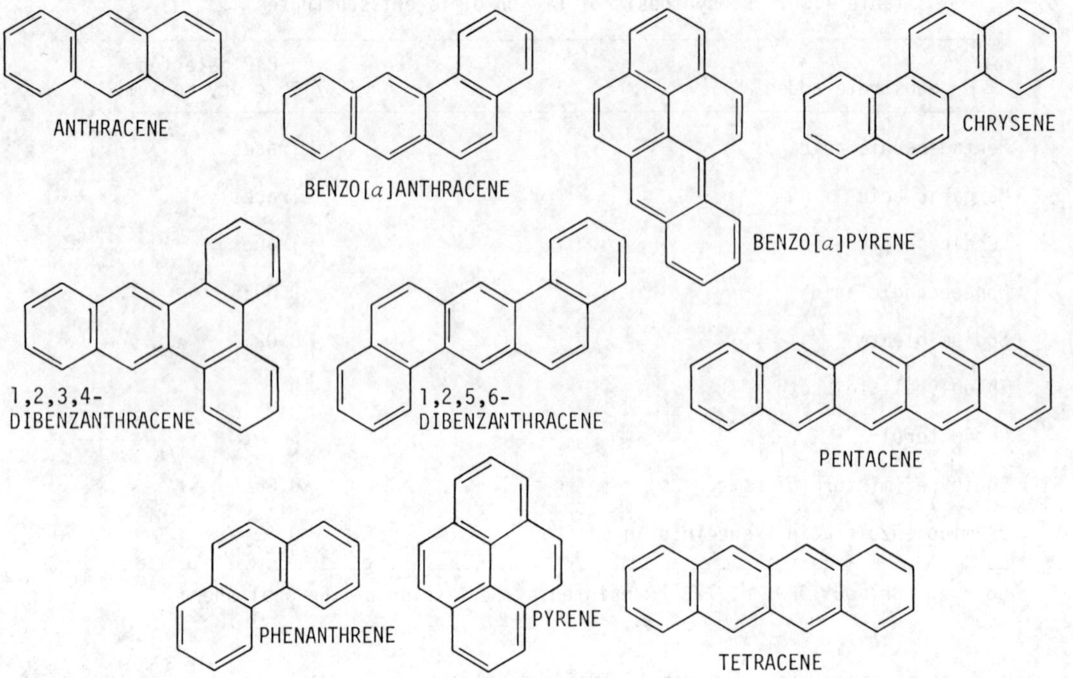

Fig. 7.23. Structures of organic compounds studied by Hass and Applegate (1975).

ORNL DWG NO 76-11813

Fig. 7.24. Effect of phenanthrene (●) and anthracene (○) on growth of *E. coli*. Source:
From Hass and Applegate 1975, Fig. 1, p. 266. Reprinted by permission of the publisher.

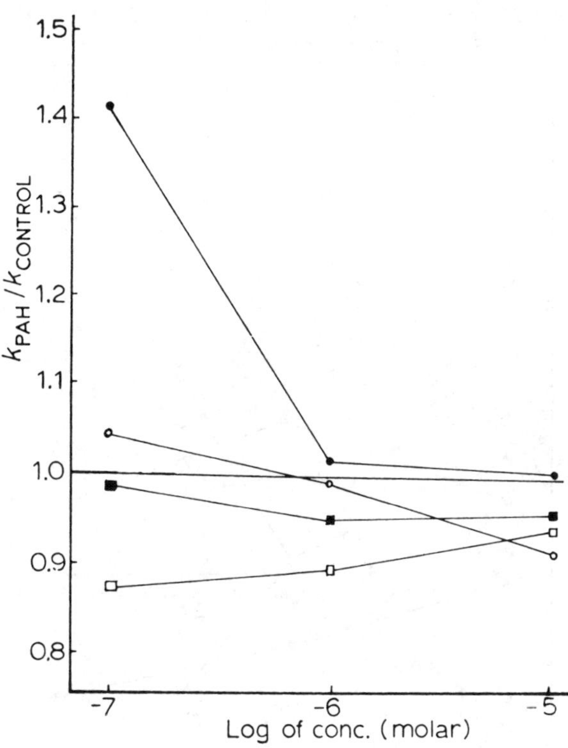

Fig. 7.25. Effect of BaA (●), pyrene (○), tetracene (■), and chrysene (□) on growth of *E. coli*. <u>Source</u>: From Hass and Applegate 1975, Fig. 2, p. 266. Reprinted by permission of the publisher.

and 7.26. The angular hydrocarbon configurations of BaA, 1,2,5,6-DBA, and BaP promote growth on *E. coli*; tetracene and pyrene had little or no effect on growth; anthracene, phenanthrene, chrysene, 1,2,3,4-DBA, and pentacene inhibited cell growth at most concentrations. Apparently, on the basis of these results, angular configuration is necessary in the PAH for *E. coli* growth stimulation (Hass and Applegate 1975).

7.2.5 Heterocyclic aromatic hydrocarbons

7.2.5.1 Pyridine compounds

Pyridine compounds are released into the environment by industrial and domestic use of coal; also, pyridines used as organic solvents inevitably find their way into effluents (Houghton and Cain 1972). Because their concentrations do not increase substantially in the soil environment, many of these pyridine compounds, both natural and synthetic, are thought to be ultimately degraded. The carbon-nitrogen skeleton of pyridine is thus mineralized, and the elements recycled. Microbial metabolism of simple pyridine compounds produces dihydroxypyridine compounds.

Utilization

Microorganisms having the ability to use pyridine or its hydroxy derivatives as their sole source of carbon, nitrogen, and energy were obtained by elective culture of a few crumbs of soil (0.5 g) or sewage from an activated-sludge plant (Houghton and Cain 1972). The bacteria were grown on a

ORNL DWG NO 76-11815

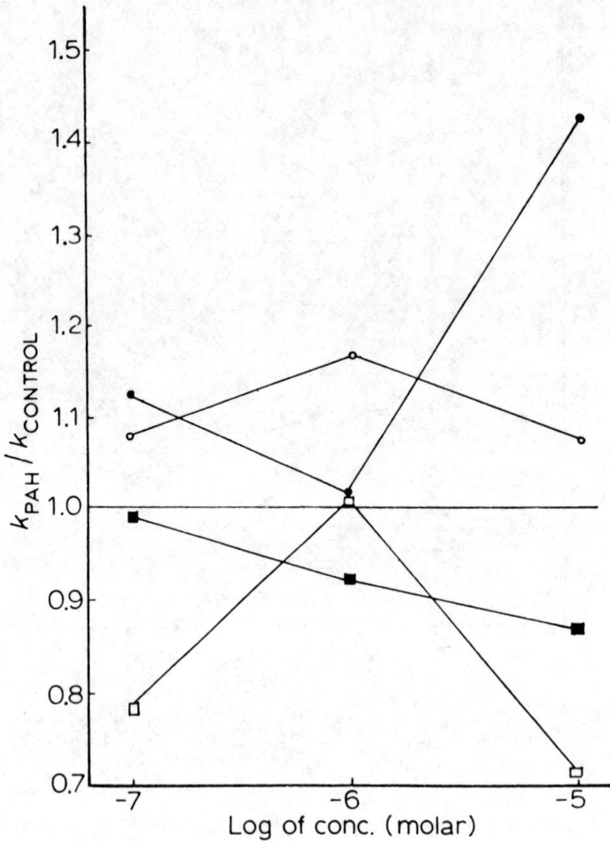

Fig. 7.26. Effect of 1,2,5,6-DBA (●), BaP (○), 1,2,3,4-DBA (■), and pentacene (□) on growth of *E. coli*. From Hass and Applegate 1975, Fig. 3, p. 267. Reprinted by permission of the publisher.

basal medium, which was supplemented with yeast extract (0.5 g/liter) where necessary. The bacterial strains and their optimum growth conditions for degrading pyridine are shown in Table 7.37.

Contrary to general experience, the bacteria were not hard to isolate. Several strains of bacteria, isolated by elective culture, were capable of degrading this heterocyclic substrate and have been identified as actinomycetes. The bacteria isolates that grew well on the 2- and 3- isomers of hydroxypyridine were 7N and 2L (Table 7.37) and resembled *Achromobacter cycloclastes* in morphology and in most physiological and biochemical characters. Strain G2, also tentatively identified as an *Achromobacter* grew on 2-hydroxypyridine. Strain 35S, identified by the National Collection of Industrial Bacteria (NCIB), was *Agrobacterium*, distinct from *Agrobacterium radiodurans* and *Agrobacterium radiobacter*. Strain 35S, deposited with the NCIB under accession number 10413, is an aerobic, gram-negative, peritrichously flagellated organism with a thich (1 µg) capsule. Strain Z1 was a *Nocardia* species that could metabolize pyridine as the sole carbon, nitrogen, and energy source, but otherwise had a very limited heterocyclic nutritional spectrum.

Table 7.37. Optimum conditions for growth and yield of cells from organisms degrading pyridine compounds[a]

Organism	Substrate added	Optimum substrate concentrations (%, w/v)	Yeast extract (%, w/v)	Optimum pH	Average cell yield (g wet wt/liter of medium)
Nocardia Z1	Pyridine	0.1	None	8.0	
Achromobacter G2	2-Hydroxypyridine	0.1	None	7.0	3
Achromobacter 7N	3-Hydroxypyridine	0.1	0.05	6.0	3
Achromobacter 2L	3-Hydroxypyridine	0.1	0.05	7.0	3
Agrobacterium 35S	4-Hydroxypyridine	0.025	0.015	8.0	0.75

[a]The optimum substrate concentrations, the optimum pH for growth, and the effects of added yeast extract were determined in growth experiments in 250-ml flasks with side arms suitable for nephelometry and containing 75 ml of the appropriate medium and 0.1% (v/v) pyridine as the only source of carbon and nitrogen, or the same medium supplemented with 0.01% (w/v) yeast extract. Flasks were shaken at 30°C, and growth followed nephelometrically. Cell-yield values were obtained from large batch cultures.

Source: Houghton and Cain 1972, Table 1, p. 881. Reprinted by permission of the publisher.

Metabolism

Watson and Cain (1975) worked with the *Nocardia* strain Z1. The strain of *Nocardia* metabolized pyridine to glutarate semialdehyde. Labelled glutarate semialdehyde 2,4-dinitrophenyl-hydrazone was isolated from the products of metabolism of $2,4-^{14}C$-pyridine (Table 7.38). Radioactivity was incorporated into glutarate semialdehyde with no decrease in specific radioactivity, showing that the carbon skeleton of glutarate semialdehyde was probably derived from the pyridine ring. The results suggest that glutarate semialdehyde was a metabolite of pyridine by *Nocardia* Z1. The amount of glutaric semialdehyde 2,4-dinitrophenylhydrazone formed in replicate experiments was calculated from its molar extinction in 10% (w/v) Na_2CO_3 solution that 10 to 15% of the stoichiometric amount was accumulated from the available pyridine. About a fiftyfold increase of activity was found in extracts of cells grown on pyridine as compared with that present in cells grown on succinate. The activity of cells grown on glutarate itself was also increased, but was still twentyfold lower than that in fully induced extracts (Watson and Cain 1975). These data are shown in Table 7.39.

Wright and Cain (1972) used *Achromobacter* D to study the microbial metabolism of the pyridinium compound, 4-carboxy-1-methylpyridinium chloride. The heterocyclic ring of *N*-methylisonicotinic acid (4-carboxy-1-methylpyridinium chloride) is broken down to formate, succinate, and methylamine; the 4-carboxyl group was released as carbon dioxide. The only precursor of these endproducts so far implicated by either chemical identification or enzyme pattern in *N*-methylisonicotinate-grown cells was succinic semialdehyde. The biodegradation could be affected in several ways, but the characterization of a C-4 and C-1 product from *N*-methylisonicotinate suggested a cleavage between C-2 and C-3 of the pyridine ring. To confirm the possible site of cleavage of the heterocyclic ring, as well as the origin of some of the other metabolic products in the *N*-methylisonicotinate molecule, the radioactive products were examined. In the case of succinate, the four carbon atoms of the pyridine ring form the symmetrical molecule (Fig. 7.27) (Wright and Cain 1972).

Table 7.38. Isolation of radioactive glutarate semialdehyde 2,4-dinitrophenylhydrazone after metabolism of [2,6-^{14}C]pyridine by *Nocardia* Z1[a]

Sample	Radioactivity (dpm/mg)
Reisolated glutarate semialdehyde 2,4-dinitrophenylhydrazone	
1st recrystallization	567
2nd recrystallization	612 } mean value = 608
3rd recrystallization	604

Specific radioactivity

Specific radioactivity of the available [2,6-^{14}C]pyridine was 6750 dpm/μmole.

Total label in the reisolated derivative (608 x 30) was 18,240 dpm.

Therefore, specific radioactivity in the original 2.2 μmole of glutarate semialdehyde 2,4-dinitrophenylhydrazone was 8290 dpm/μmole.

[a]The incubation mixture contained in a total volume of 10 ml: pyridine-grown *Nocardia* Z1 cells (equivalent to 35.3 mg dry wt); potassium phosphate buffer, pH 6.0, 500 μmole; semi-carbazide-HCl adjusted to pH 6.0, 150 μmole; [2,6-^{14}C]pyridine, 20 μmole (135,000 dpm). A control flask contained no pyridine. After shaking for 2 hr at 30°C, the cells were removed by centrifugation, and 0.5 ml of 2.0% (w/v) 2,4-dinitrophenylhydrazone in 2 M HCl was added to the supernatants. After 12 hr at room temperature, the mixtures were extracted with ethyl acetate and the acidic 2,4-dinitrophenylhydrazones, then re-extracted back into 50 ml of 10% (w/v) Na$_2$CO$_3$ solution. The E$_{375}$ of this solution was 0.69, corresponding to the formation of 2.2 μmole of [^{14}C]glutarate semialdehyde 2,4-dinitrophenyl-hydrazone, a yield of 11% from the available pyridine. Glutarate semialdehyde 2,4-dinitro-phenylhydrazone (30 mg) was added to the carbonate solutions, which were acidified with concentrated HCl, and the derivatives extracted into ethyl acetate. The ethyl acetate was removed in vacuo, and the residues recrystallized three times from water. After each recrystallization 0.5 to 1.0 mg of the derivative was dissolved in toluene (2 ml), and 10 ml of Insta-Gel scintillator was added; the counting efficiency of the colored solutions was 70 to 80%. The counts were corrected for background (28 dpm/mg) from the material isolated from the control incubation without pyridine.

Source: Watson and Cain 1975, Table 2, p. 163. Reprinted by permission of the publisher.

Table 7.39. Glutarate semialdehyde dehydrogenase and glutaric dialdehyde dehydrogenase in extracts of *Nocardia* Z1[a]

Growth substrate	Nicotinamide nucleotide used	Glutaric semialdehyde dehydrogenase (μmole NAD(P)H/min/mg of protein)	Glutaric dialdehyde dehydrogenase (μmole NAD(P)H/min/mg of protein)
Pyridine	NADP+	0.34	0.046
	NAD+	0.275	0.185
Glutarate	NADP+	0.017	[b]
	NAD+	0.015	[b]
Succinate	NADP+	0.006	0.009
	NAD+	0.008	0.034

[a]*Nocardia* Z1 was grown on the substrates indicated and harvested during exponential growth; activity of glutarate semialdehyde dehydrogenase and glutaric dialdehyde dehydrogenase determined as units (μmole of substrate disappeared or product formed/min) per mg of protein.
[b]Not determined.

Source: Watson and Cain 1975, Table 3, p. 164. Reprinted by permission of the publisher.

ORNL DWG NO 76-11816

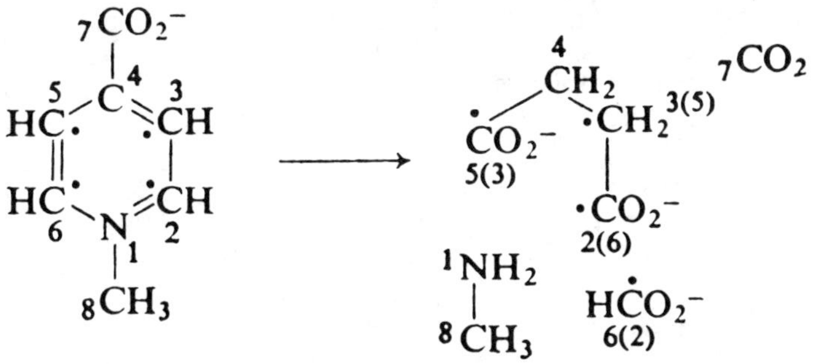

Fig. 7.27. Distribution of radioactivity in the products of N-methyl $[2,3-^{14}C_2]$ isonicotinate metabolism by extracts of *Achromobacter* D. <u>Source</u>: From Wright and Cain 1972, Scheme 1, p. 565. Reprinted by permission of the publisher.

Skryabin et al. (1972) worked with the phenomenon of coupled oxidation to show the oxidation of 3-methylpyridine, a coal carbonization product, into nicotinic acid. The reactions were obtained from microbial cultures grown on media using n-alkanes as the growth nutrient. It was shown that 80% yields of nicotinic acid could be obtained in cultures to which 3-methylpyridine had been added; this coupled oxidation proceeded not only with n-alkanes as the growth substrate, but also with isoalkanes and some amino acids as nutrients. The possibility of oxidizing 3-methylpyridine under continuous fermentation conditions was studied, and results showed that, under chemostatic conditions and at a flow rate of 0.025 to 0.05 liter/hr, 3-methylpyridine could be oxidized to nicotinic acid. It was found that the greatest number (over 50%) of active microbial strains capable of oxidizing 3-methylpyridine to nicotinic acid under coupled oxidation conditions were species of *Nocardia* and *Arthrobacter*, as shown in Table 7.40.

Table 7.40. Comparative evaluation of methods of selection of cultures achieving oxidation of alkylpyridines under coupled oxidation conditions[a]

Method of selection	Number of cultures selected	Taxonomic classification	Cultures giving oxidation of alkylpyridines (%)
Cumulative cultures containing n-alkanes as the carbon source	150	*Mycobacterium, Nocardia*	52
Cumulative cultures containing n-alkanes and alkylpyridines as the carbon source	120	*Nocardia, Arthrobacter, Proactinomyces*	16

[a]From Skryabin et al. (1972).

Source: Skryabin et al. 1972, Table 1, p. 107. Reprinted by permission of the publisher.

7.2.6 <u>Crude oil</u>

Pollution of soils and groundwaters by mixed hydrocarbons occurs frequently (Vanloocke et al. 1975). Accidents during surface transport of oil products are reported regularly, and leaks in pipelines are not uncommon. Table 7.41 shows the number and importance of pipeline and pumping station leaks in Western Europe. The consequences of oil spills can be serious because they can

Table 7.41. Spillages from oil industry cross-country pipelines in W. Europe

| Year | Number of reported incidents on | | Volume of spilled oil (m^3) | Total transport by pipeline $(10^6 \ m^3)$ |
	Pipelines	Pump stations		
1967	3	2	33.2	224
1968	2	0	1.8	236
1969	3	3	61.5	248
1970	9	2	719.1	250
1971	7	4	2720.5	310
1972	21	1	2720.0	433
1973	15	5	1154.0	558

Source: Vanloocke et al. 1975, Table 1, p. 99. Reprinted by permission of the publisher.

radically disturb both macro- and microecologies of ecosystems. Microbial metabolic reactions theoretically can effect complete cleanup of an oil spill (Vanloocke et al. 1975).

7.2.6.1 Degradation

Members of the genus *Pseudomonas* play an important role in oil degradation. Representative bio-degradation pathways of alkane and aromatic compounds are given in Figs. 7.28 and 7.29. Specificity by some microorganisms has been noted; a one-atom change in some n-alkanes can be the difference between total or no bacterial degradation. For example, *Penicillium spinulosum*, *Fusarium oxysporum*, and *Aspergillus niger* all grow on n-undecane, but not on n-decane, whereas *Aspergillus athecius* grows on n-tetradecane, but not on n-tridecane (Vanloocke et al. 1975).

7.2.6.2 Cometabolism

Cometabolism is an important process used by microorganisms for degradation of oil hydrocarbons. Cyclohexane is broken down after cometabolic transformation by one of two *Pseudomonas* strains, each of which is unable to metabolize it independently. Alkylbenzenes having a methyl- or ethyl-side chain can only be metabolized by cometabolism (Vanloocke et al. 1975).

7.2.6.3 Benthal decomposition

Decomposition of oils in aquatic sediments occurs aerobically and anaerobically and is called benthal decomposition (Shelton and Hunter 1975). Aerobic conditions prevail at the surface of oil-bearing sediments due to the dissolved oxygen (DO) present in overlying groundwater. Oxygen will penetrate into the sediment only as far as the balance between oxygen diffusion and oxygen consumption permits. This is usually not below the superficial sediment layers. Below this level, anaerobic oil decomposition takes place.

ORNL DWG NO 76-11817

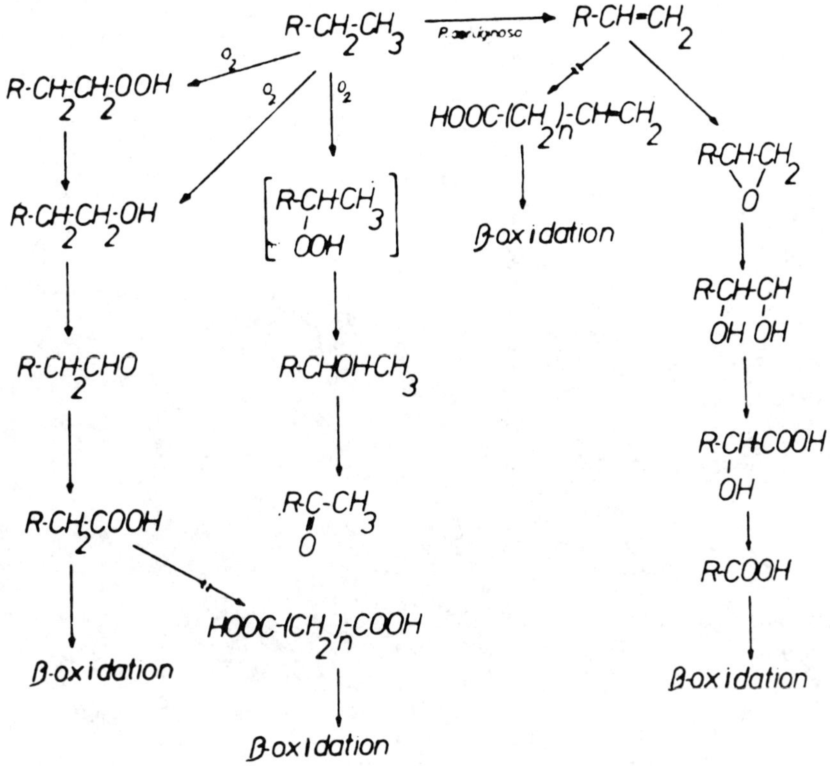

Fig. 7.28. Biodegradation pathways of aliphatic hydrocarbons. <u>Source:</u> From Vanloocke et al. 1975, Fig. 3, p. 106. Reprinted by permission of the publisher.

It is now widely accepted that oil in sediments is decomposed by microbial action under anaerobic conditions at the expense of oxygen in nitrates and sulfates, leading to the formation of nitrites, free nitrogen, and hydrogen sulfide (Shelton and Hunter 1975). Assimilation of aliphatic hydrocarbons by *Desulfovibrio* sp. (obligate anaerobes; curved rods; move by single polar flagellum; no growth in pH of 5.5 or less) following the pattern described above has been observed. The sulfate reducers were found to catalyze the anaerobic oxidation of a considerable number of mixed hydrocarbons with insoluble fatty acids as intermediate metabolites that underwent further degradation. *Desulfovibrio* sp. is also thought to use a dehydrogenase system in association with hydrocarbon use (Shelton and Hunter 1975).

7.2.6.4 Microorganism growth

Walker, Seesman, and Colwell (1975) studied the effect of spilled south Louisiana crude oil and No. 2 fuel oil on microorganisms isolated from creek water. The area from which the microorganisms were taken had not previously been contaminated with oil. The work showed that both south Louisiana crude oil and No. 2 fuel oil had little influence or effect on the growth of yeasts and fungi (Figs. 7.30 and 7.31). The heterotrophic bacteria used in this investigation showed significant growth on south Louisiana crude oil; however, No. 2 fuel oil seemed to limit bacterial growth, as shown in Fig. 7.32. The crude oil also supported a greater growth of proteolytic, lipolytic, chitinolytic, and cellulolytic bacteria (Figs. 7.33 through 7.36) than did fuel oil. The amounts of lytic bacteria were found to be lower in the oil cultures than in

ORNL DWG NO 76-11818

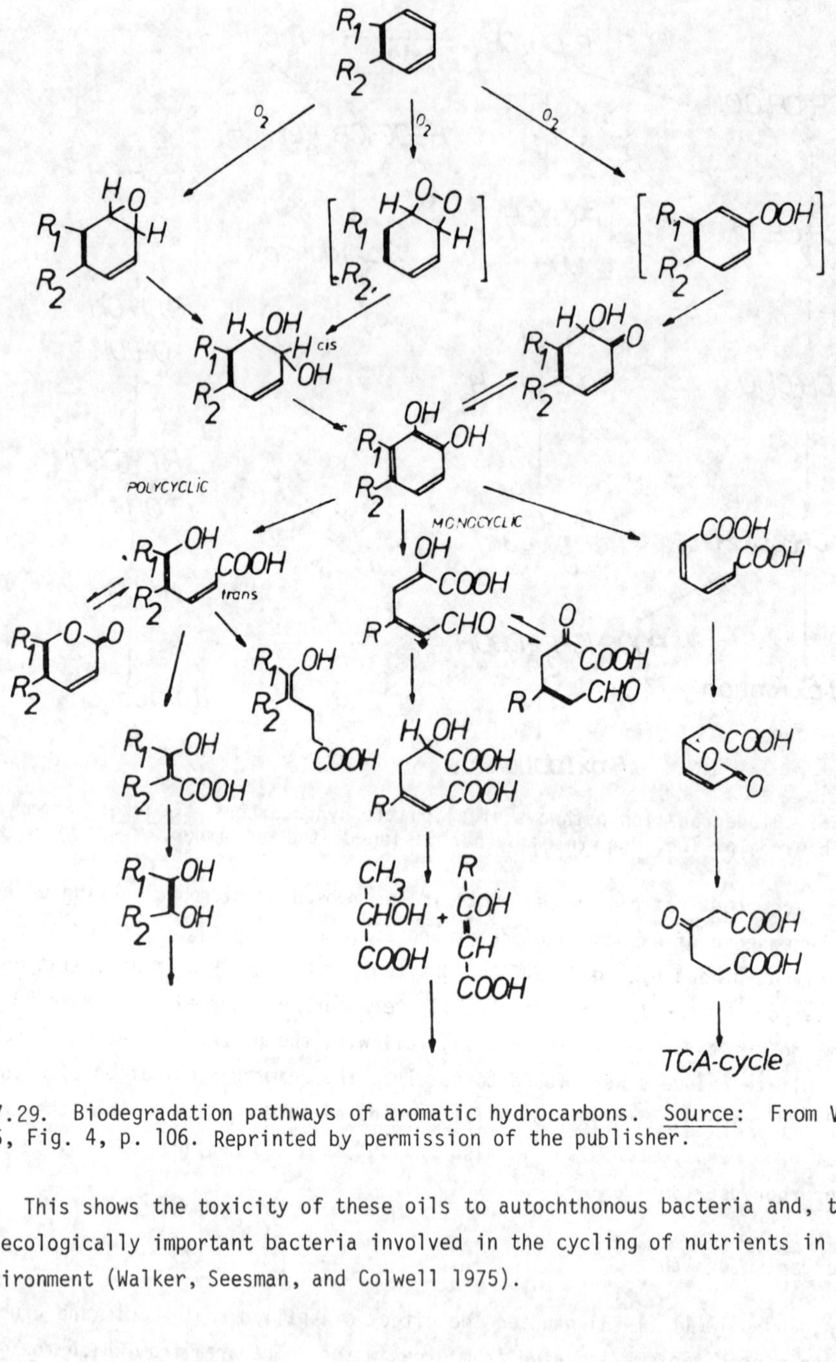

Fig. 7.29. Biodegradation pathways of aromatic hydrocarbons. <u>Source:</u> From Vanloocke et al. 1975, Fig. 4, p. 106. Reprinted by permission of the publisher.

the controls. This shows the toxicity of these oils to autochthonous bacteria and, thus, the toxicity for ecologically important bacteria involved in the cycling of nutrients in the estuarine environment (Walker, Seesman, and Colwell 1975).

<u>In soils</u>

Jensen (1975) treated soil with oily waste and recorded increasing bacterial numbers as well as increased microbial activity. The bacteria most commonly reported as active in oily soils are members of the coryneform group and include the genera *Corynebacterium, Brevibacterium, Arthrobacter, Mycobacterium,* and *Nocardia.* Jensen (1975) believes that much confusion exists in the definition of these genera and that strains defined as *Corynebacterium,* and possibly *Mycobacterium,* might well be referred to as either *Arthrobacter* or *Nocardia.* Jensen (1975) states

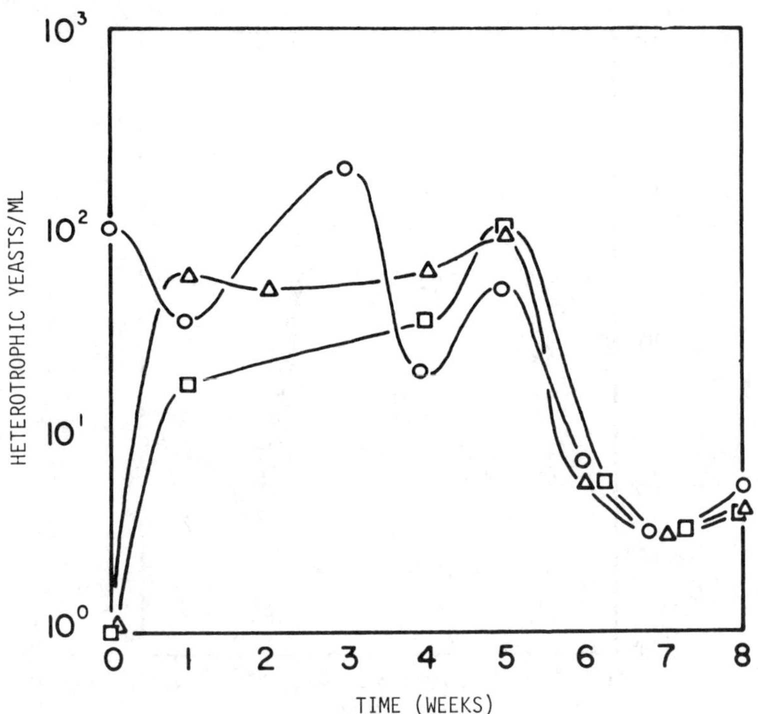

Fig. 7.30. Growth of heterotrophic yeasts without added substrate (○); with fuel oil No. 2 (□); and with south Louisiana crude oil (△). <u>Source</u>: From Walker, Seesman, and Colwell 1975, Fig. 6, p. 23. Reprinted by permission of the publisher.

that, in general, *Corynebacterium* members are not normally proper soil residents; therefore, strains isolated in his work were designated as *Arthrobacter*- or *Nocardia*-type rods. Isolations, by dilution plates and enrichment cultures, showed that *Arthrobacter* members were the most active and most important hydrocarbon-decomposing bacteria in the experiment. Fluorescent pseudomonads were also isolated in this work. Most cultures were obtained by enrichment cultures, which indicate that low numbers of bacteria were present. Data for the percentage of bacterial morphological types taken in the experiment are shown in Table 7.42. The number of bacteria in each percentage group is shown in Table 7.43. The *Arthrobacter-Nocardia* group had the highest percentages in the control soil, but the numbers per gram of soil were considerably higher in the oil-treated soil (Jensen 1975). The difference was mainly caused by the dominance of a single *Arthrobacter* species in the oil soil that formed large, slimy, easily recognizable colonies. The *Arthrobacter-Nocardia* group in the control soil was composed of many bacterial types.

Mulkins-Phillips and Stewart (1974) studied naturally occurring bacterial populations taken from Chedabucto Bay, Nova Scotia, Canada. The area in and around Chedabucto Bay had been heavily polluted with oil in February of 1970 when the oil tanker *Arrow*, carrying 108,000 bbl of Bunker C fuel oil, ran aground and ruptured. This study allowed close observation of bacterial populations that had been enriched in situ with a Bunker C substrate. Mixed bacterial cultures taken from beach and water samples were found to grow on Bunker C fuel oil at temperatures as low as 5°C. Decreased temperatures did extend the lag phases of both the growth curve and the Bunker C fuel oil degradation (Figs. 7.37 and 7.38). A pure culture was isolated from one of the mixed

ORNL DWG NO 76-11820

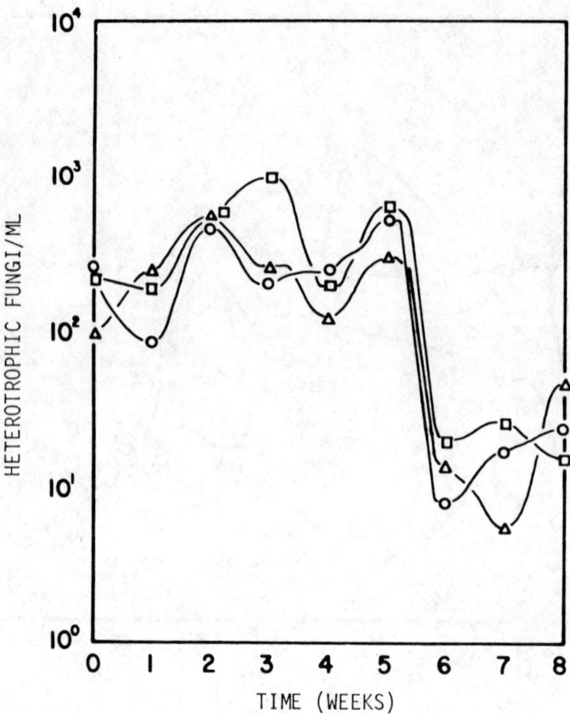

Fig. 7.31. Growth of heterotrophic fungi in the absence of added substrate (○); with fuel oil No. 2 (□); and with south Louisiana crude oil (△). <u>Source</u>: From Walker, Seesman, and Colwell 1975, Fig. 7, p. 24. Reprinted by permission of the publisher.

culture samples. The isolate appeared as a long, slender Gram-positive bacillus when first grown on Trypticase soy agar (TSA) media. The isolate had changed morphologically by day 6 and had assumed a gram-positive coccobacillus-like form. The isolate was not acid-fast and produced filiform cream-colored growth on TSA. Further testing led to the classification of the isolate as a *Nocardia* species (Mulkins-Phillips and Stewart 1974). The *Nocardia* species grew on hexadecane, a mixture of hydrocarbons (consisting of naphthalene, anthracene, dibenzothiophene, decalin, hexadecane, hexadecane-1, octadecane, dodecane, and iso-octane), glucose, and Venezuelan crude oil at rates similar to those shown in Table 7.44. The organism grew equally well on hexadecane at initial pH values between 7.0 and 8.5, but a lowering of pH produced a concomitant decrease in organism growth (Fig. 7.39) (Mulkins-Phillips and Stewart 1974).

<u>In saline environments</u>

Brown, Phillips, and Tennyson (1970) studied bacteria able to metabolize naphthenic crude oil and refined motor oil under varying saline concentrations representative of an estuarine environment. Microorganisms were incubated at various salinities in 10% oil enrichment water samples. Bacteria were collected from seawater and bottom sediments, inoculated into the test media, and incubated at 30°C. In this work, the bottom sediment bacteria had a faster oil degradation rate than did the overlying water bacteria. Salinities ranging from 17 to 34 ppt produced the highest degradation rates. The naphthenic crude oil proved to be more specific than the crude oil for the growth

ORNL DWG NO 76-11821

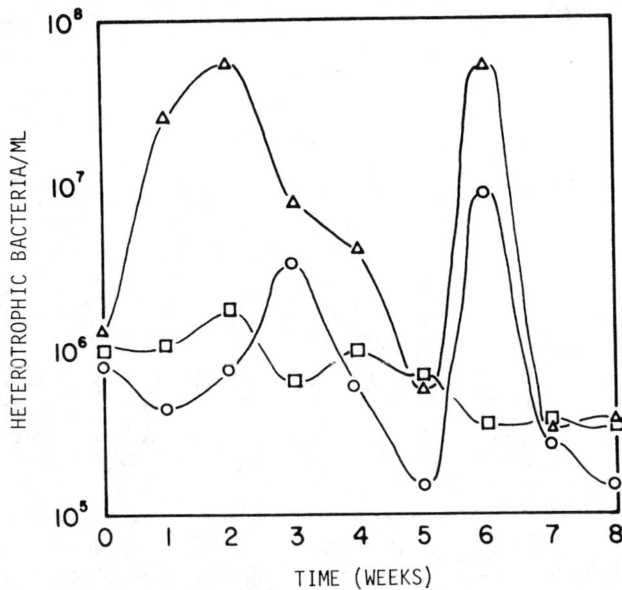

Fig. 7.32. Growth of heterotrophic bacteria in the absence of added substrate (○); with fuel oil No. 2 (□); and with south Louisiana crude oil (△). <u>Source:</u> From Walker, Seesman, and Colwell 1975, Fig. 8, p. 24. Reprinted by permission of the publisher.

ORNL DWG NO 76-11822

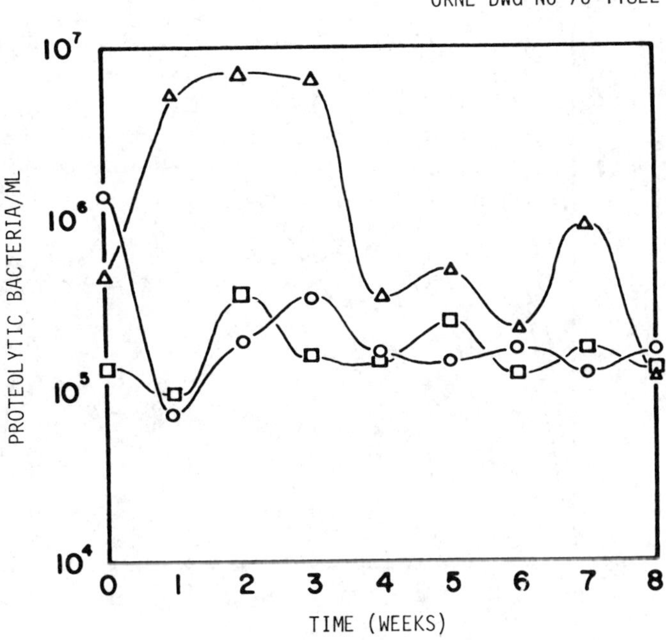

Fig. 7.33. Growth of proteolytic bacteria in the absence of added substrate (○); with fuel oil No. 2 (□); and with south Louisiana crude oil (△). <u>Source:</u> From Walker, Seesman, and Colwell 1975, Fig. 9, p. 25. Reprinted by permission of the publisher.

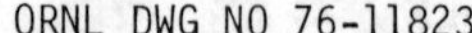

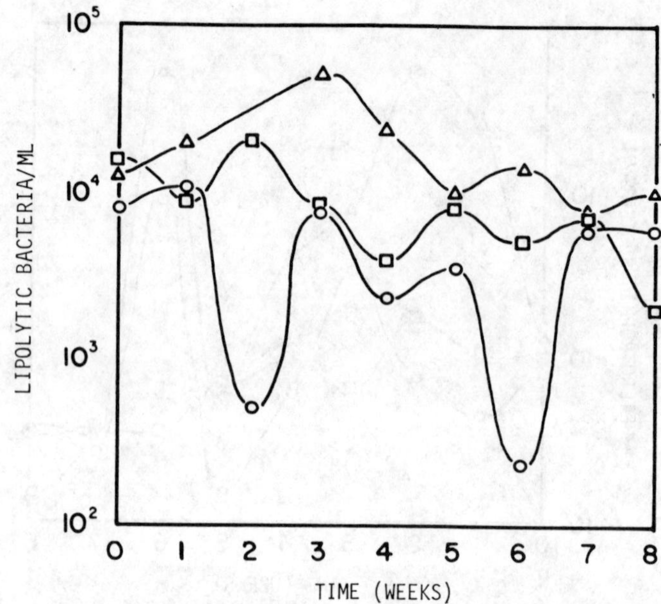

Fig. 7.34. Growth of lipolytic bacteria in the absence of added substrate (○); with fuel oil No. 2 (□); and with south Louisiana crude oil (△). Source: From Walker, Seesman, and Colwell 1975, Fig. 10, p. 25. Reprinted by permission of the publisher.

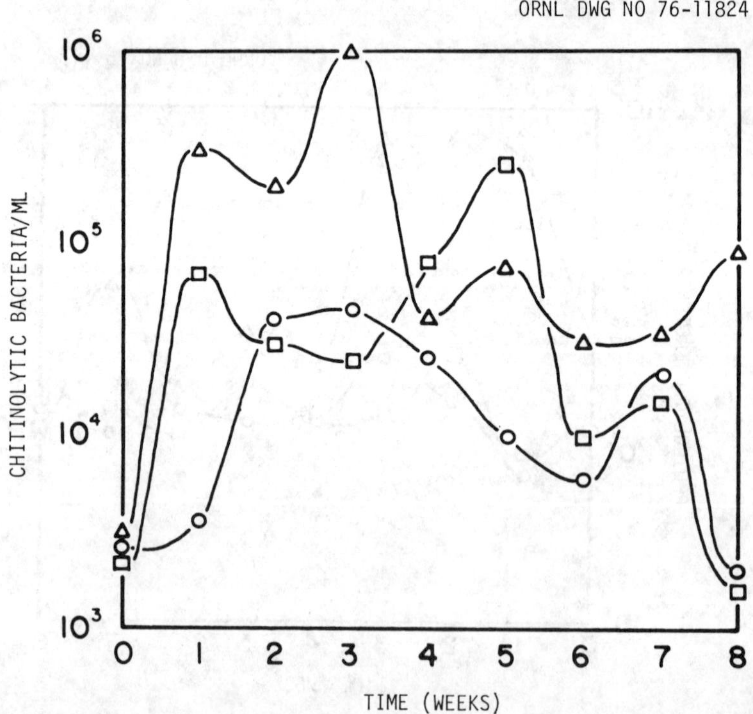

Fig. 7.35. Growth of chitinolytic bacteria in the absence of added substrate (○); with fuel oil No. 2 (□); and with south Louisiana crude oil (△). Source: From Walker, Seesman, and Colwell 1975, Fig. 11, p. 26. Reprinted by permission of the publisher.

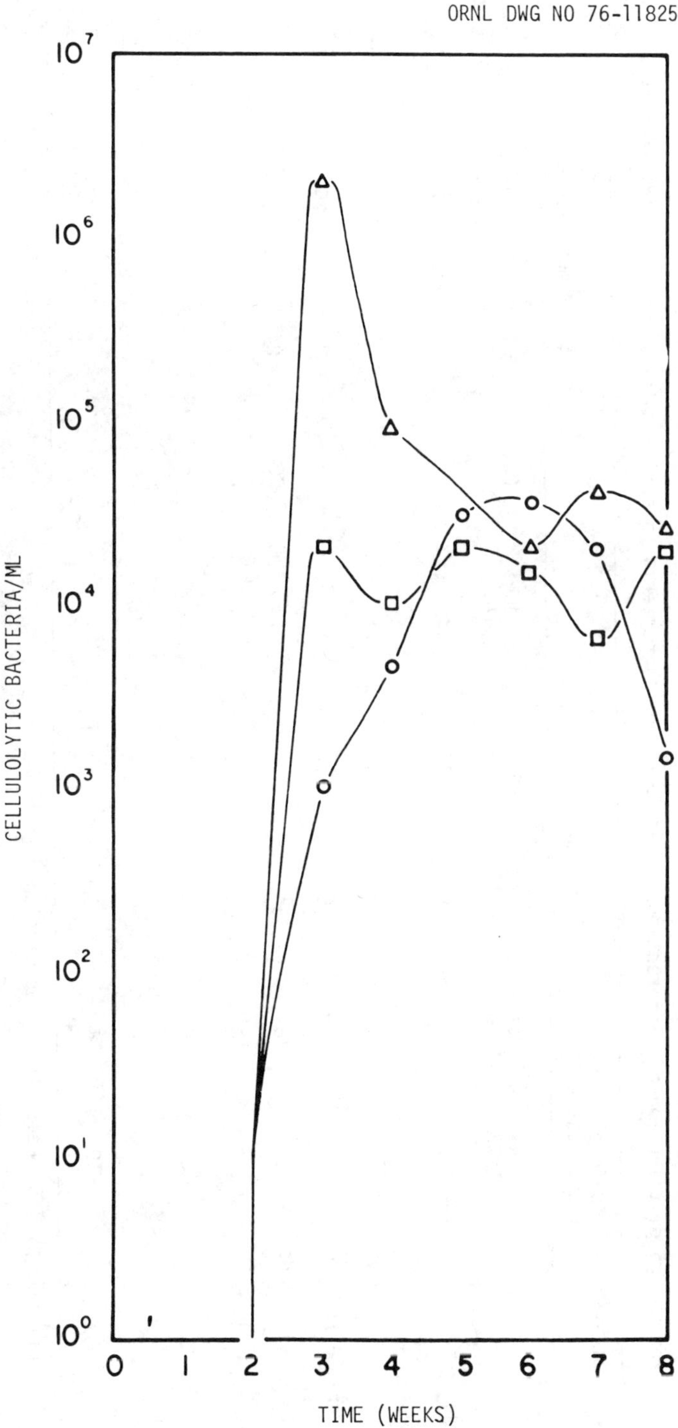

Fig. 7.36. Growth of cellulolytic bacteria in the absence of added substrate (○); with fuel oil No. 2 (□); and with south Louisiana crude oil (△). Source: From Walker, Seesman, and Colwell 1975, Fig. 12, p. 27. Reprinted by permission of the publisher.

Table 7.42. Percentage distribution of different morphological types of bacteria in soil samples from the field experiment

Sampling date	Before oil application		Control area				Experimental area		
	4 May 1973	1 Oct. 1973	18 Dec. 1974	28 Mar. 1974	20 June 1974	1 Oct. 1973	18 Dec. 1973	28 Mar. 1974	20 June 1974
Spore-forming rods	9	8	8	19	40	2	0	2	11
Nonspore-forming gram + or ± rods	27	11	10	17	14	9	30	33	57
Nonspore-forming gram − rods	5	7	4	2	6	9	25	35	10
Arthrobacter–Nocardia-like rods	21	28	34	25	19	20	24	13	11
Slow-growing pleomorphic rods	6	24	24	13	3	51	17	14	6
Streptomycetes	20	18	18	21	16	2	0	2	2
Others	2	0	0	0	0	0	3	0	2
No growth after transfer	10	4	2	3	2	7	1	1	1

Source: Jensen 1975, Table 5, p. 155. Reprinted by permission of the publisher.

Table 7.43. Numbers (million g^{-1} dry soil) of different morphological types of bacteria in soil samples from the field experiment

Sampling date	Before oil application		Control area			Experimental area			
	4 May 1973	1 Oct. 1973	18 Dec. 1973	28 Mar. 1974	20 June 1974	1 Oct. 1973	18 Dec. 1973	28 Mar. 1974	20 June 1974
Spore-forming rods	6.2	32.5	9.3	15.4	9.5	42.6	0	7.1	11.9
Nonspore-forming gram + or ± rods	18.5	44.7	11.6	13.8	3.4	191.9	214.5	116.3	61.7
Nonspore-forming gram − rods	3.4	28.4	4.6	1.6	1.4	191.9	178.7	123.4	10.8
Arthrobacter–Nocardia-like rods	14.4	113.7	39.5	20.2	4.5	426.3	171.6	45.8	11.9
Slow-growing pleomorphic rods	4.1	97.5	27.8	10.5	0.7	1087.2	121.5	49.4	6.5
Streptomycetes	13.7	73.1	20.9	17.0	3.8	42.6	0	7.1	2.2
Others	1.4	0	0	0	0	0	21.5	0	2.2
No growth after transfer	6.8	16.2	2.3	2.4	0.5	149.2	7.2	3.5	1.1
Total	68.5	406.1	116.0	80.9	23.8	2131.7	715.0	352.6	108.3

Source: Jensen 1975, Table 6, p. 155. Reprinted by permission of the publisher.

ORNL DWG NO 76-11826

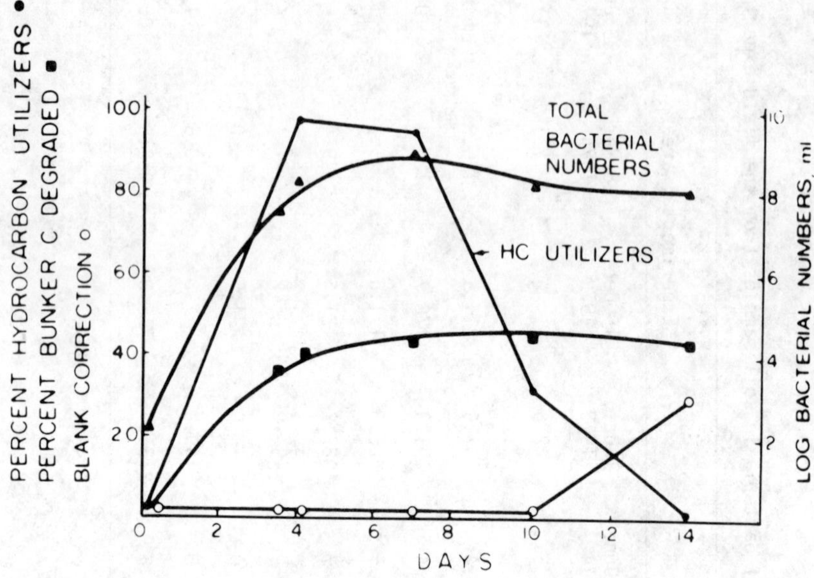

Fig. 7.37. Enrichment of sample MA36 at 28°C with 0.125% Bunker C in the minimal medium (duplicate trials). <u>Source</u>: From Mulkins-Phillips and Stewart 1974, Fig. 1, p. 918. Reprinted by permission of the publisher.

ORNL DWG NO 76-11827

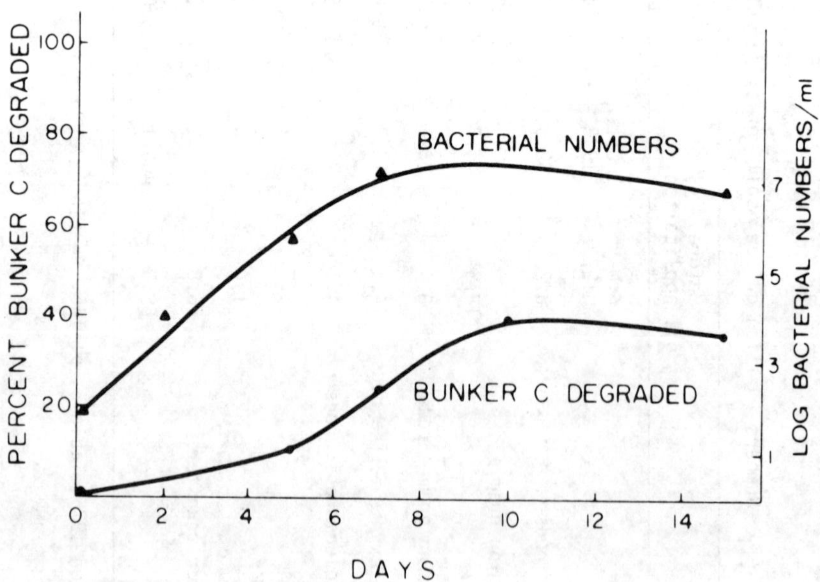

Fig. 7.38. Enrichment of sample MA36 at 15°C with 0.125% Bunker C in the minimal medium (duplicate trials). <u>Source</u>: From Mulkins-Phillips and Stewart 1974, Fig. 2, p. 918. Reprinted by permission of the publisher.

Table 7.44. Effect of temperature on generation time of
the *Nocardia* sp. on different
hydrocarbon substrates

Substrate	Generation time (hr)		
	15°C	5°C	C_5/C_{15}
Hexadecane	11.3	24.0	2.1
Hydrocarbon mixture	12.0	29.3	2.4
Glucose	12.0	29.3	2.4
Crude oil	12.0	27.7	2.3
Bunker C	16.0	37.7	2.35
Naphthalene	132.0	224.0	1.7
Control	No growth	No growth	

Source: Mulkins-Phillips and Stewart 1974, Table 3, p. 918. Reprinted by permission of the publisher.

ORNL DWG NO 76-11828

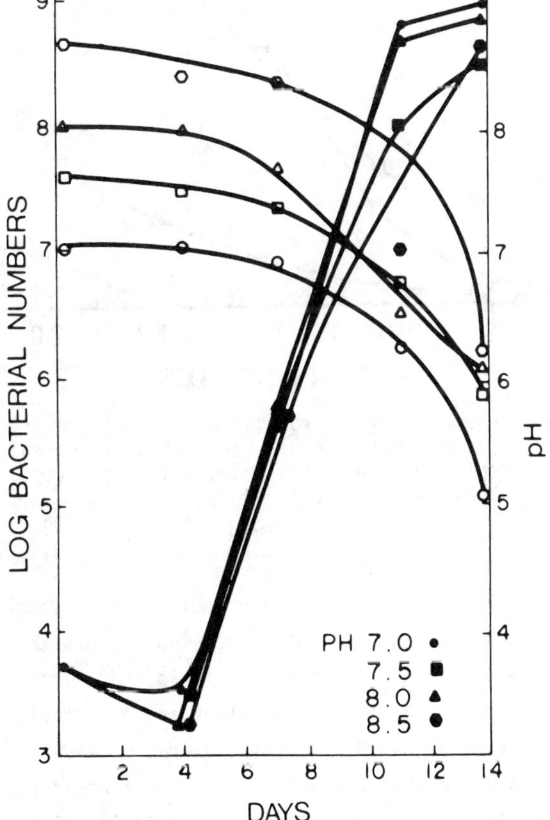

Fig. 7.39. Effect of pH on the growth of *Nocardia* sp. at 15°C. Open symbols represent pH measurements, closed symbols represent the corresponding bacterial counts (duplicate trials). Source: From Mulkins-Phillips and Stewart 1974, Fig. 3, p. 919. Reprinted by permission of the publisher.

of microorganisms. *Pseudomonas* was the only genus found to grow on naphthenic crude oil at the end of the incubation period. Comparative degradation rates, as determined by oxygen consumption, for the two oil types are shown in Figs. 7.40 and 7.41.

ORNL DWG NO 76-11829

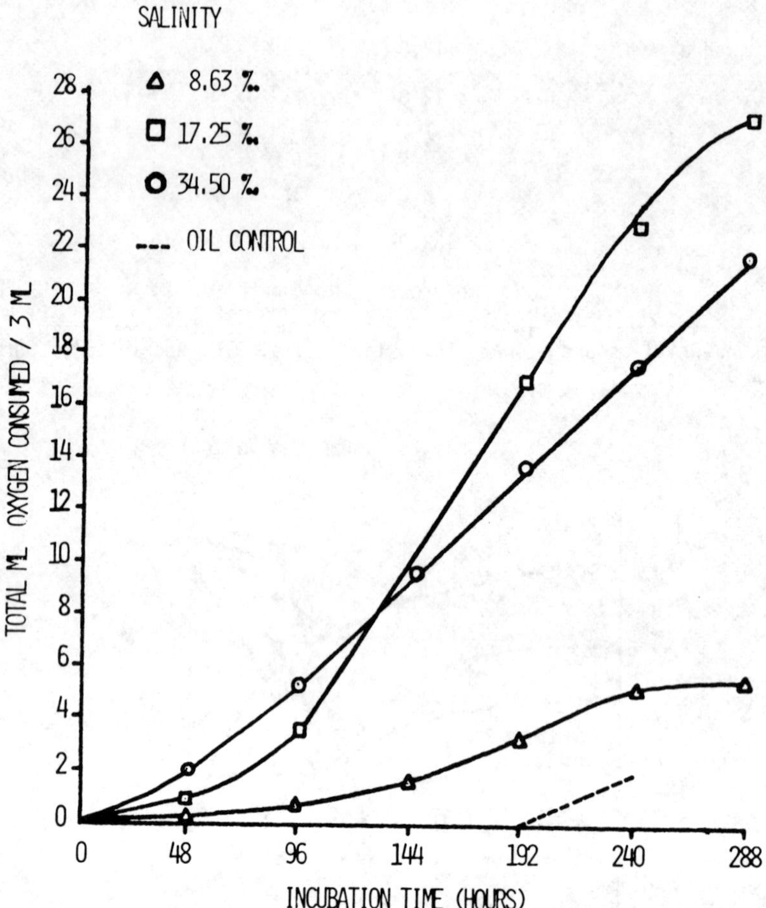

Fig. 7.40. Total oxygen consumption by microorganisms during the degradation of naphthenic crude oil at varying salinities. Conditions: 900 ml of triton marine salts at varying salinities and 100 ml of oil with added inoculum and incubated under shake conditions at 30°C. <u>Source</u>: From Brown, Phillips, and Tennyson 1970, Fig. 7, p. 23.

Bartha and Atlas (1973) studied the limiting factors of petroleum biodegradation in seawater. A *Flavobacterium* and a *Brevibacterium* received intensive study by growth on an artificial seawater medium with 1% filter-sterilized Sweden crude oil as the only carbon source. Biodegradation of petroleum was measured by carbon dioxide evolution and quantitative gas chromatography. Results showed that low water temperatures predictably lowered degradation rates, but lag periods preceded the onset of any measurable degradation. The largest portion of this lag period was due to the time required for evaporation of volatile inhibitors present in crude petroleum. Addition of fertilizer was found to speed degradation by as much as tenfold. A combination of octylphosphate and a slow-release paraffinized urea gave the best results over nitrogen and phosphorus salts.

ORNL DWG NO 76-11830

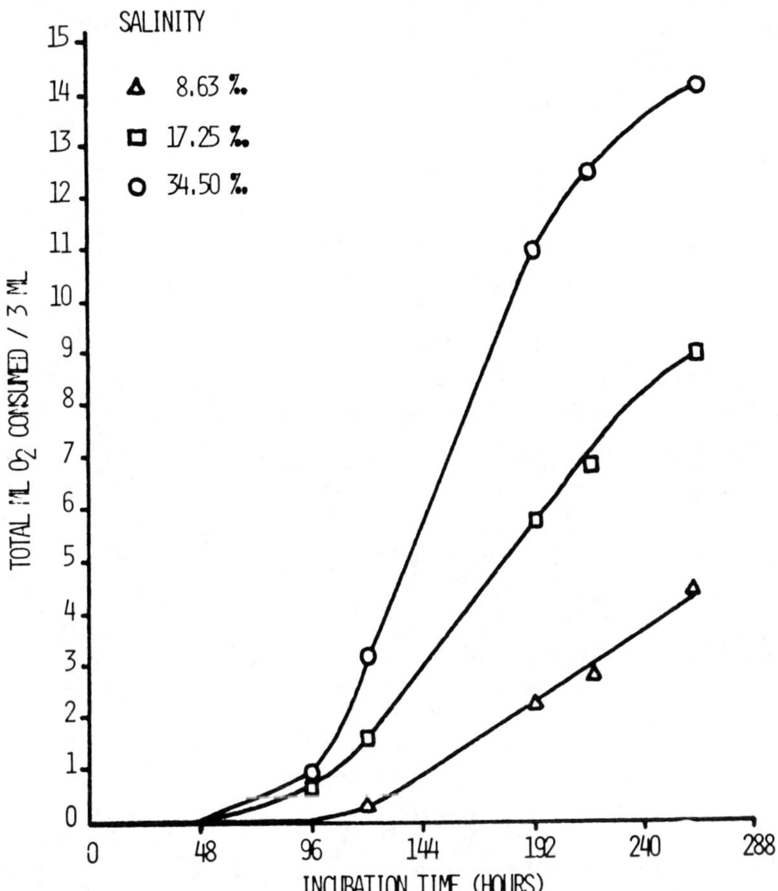

Fig. 7.41. Total oxygen consumption by microorganisms during the degradation of refined motor oil at varying salinities. Conditions: 900 ml triton marine salts at varying salinities and 100 ml of oil with added inoculum and incubated under shake conditions at 30°C. Source: From Brown, Phillips, and Tennyson 1970, Fig. 9, p. 25.

Atlas and Bartha (1973) studied the effect of commercial oil herders, dispersants, and bacterial inocula on biodegradation of oil in seawater. The oil herders and dispersants significantly increased the mineralization rate, but this effect is apparently due to the increased oil droplet surfaces and lowered levels of microorganism toxicants. Phosphorus and nitrogen supplements were needed for increased mineralization as none of the oil herders or dispersants increased carbon dioxide production without supplementation. All commercial bacterial inocula tested were totally ineffective in stimulating biodegradation. Measured biodegradation rates of inoculated and uninoculated seawater containing crude oil were identical.

7.2.6.5 Field plot studies

An experiment to test the oil degradation ability of naturally occurring heterotrophic bacteria was carried out by Lehtomäki and Niemelä (1975). Light fuel oil was poured onto test field plots (2 x 2 m) and subjected to three treatments (irrigation, fertilization, and aeration) given at random and in all combinations. Control plots with no oil were subjected to the same treatment as were the oily soil plots. Total colony counts of heterotrophic bacteria in the oily plots were found to be consistently higher than those in the control plots, which indicates participation of heterotrophic bacteria in oil degradation (Lehtomäki and Niemelä 1975). The soil treatments also changed the bacterial profile as shown in Table 7.45. The treatment having the greatest effect on oil degradation was aeration. Organic fertilizers, added to the test areas, speeded degradation of the light fuel oil and accomplished a more complete degradation. The organic fertilizers having the greatest effect were waste brewery yeast and peat as shown in Fig. 7.42. Brewery yeast had the most favorable effect on the bacterial degradation of oil and was tested more extensively (Fig. 7.43) to determine whether the yeast acted only as a fertilizer or participated in the degradation. Tests showed that brewery yeast in sterilized soil could not bring about oil decomposition (Lehtomäki and Niemelä 1975).

Table 7.45. Amount of oil consumed in 2 months in 8 experimental field plots[a]

Other treatments	Aerated	Nonaerated	Difference
None	42.1	37.9	4.2
Irrigation	32.5	19.5	13.0
Fertilization	23.4	31.5	8.1
Irrigation and fertilization	33.9	28.1	5.8
Mean	33.0	29.3	3.7

[a]Irrigation = watering three times a week with 2 liters except in the case of rain; fertilization = inorganic fertilizer (N-P-K:15-8.7-12.4, 500 kg/ha); aeration = boring and keeping open holes of about 1 cm in diameter and 10-cm spacing to a depth of 25 cm.

Source: Lehtomäki and Niemelä 1975, Table 1, p. 126. Reprinted by permission of the publisher.

7.2.6.6 Soil rehabilitation studies

Gudin and Syratt (1975) conducted work on the rehabilitation of land sites contaminated with hydrocarbon mixtures such as oil. The work showed that treatment should include

1. Improved aeration, such as disc harrowing or, if the soil is saturated with oil, mixing the contaminated soil with fresh soil.

2. Addition of a nitrogenous fertilizer to restore the carbon-nitrogen balance and to increase the rate of microbial degradation.

3. (a) Covering the contaminated area with black polyethylene sheeting during winter to increase soil temperature, thus shortening the lag phase for the microorganisms or (b) covering the contaminated area with transparent polyethylene in summer if the area is in a dry climate, thus reducing water evaporation.

4. Establishment of a vegetative cover to improve the rhizosphere; leguminous plants appear to be very suitable for this purpose.

The use of leguminous plants as a vegetative cover is important due to the symbiotic relationship of the plant with the nitrogen-fixing *Rhizobium* species of bacteria. This practice allows the

ORNL DWG NO 76-11831

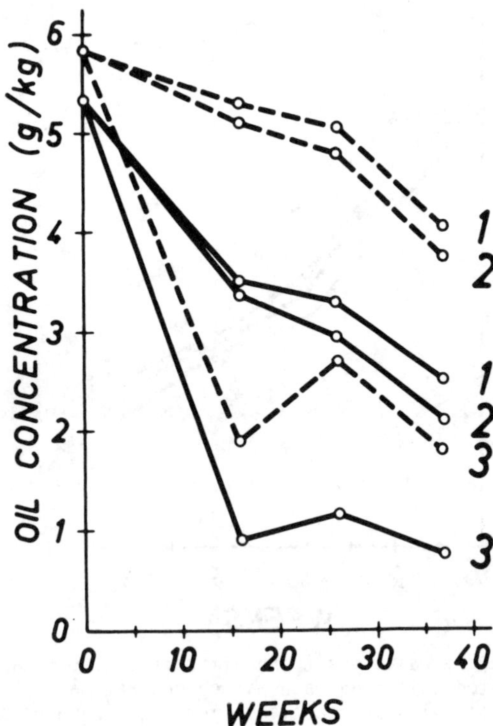

Fig. 7.42. Effects of organic fertilizers on oil decomposition in unsterilized soil. Means of 10 pots kept in the laboratory at 20°C. 1 = no fertilizer, 2 = addition of 0.75% fertilized peat, 3 = addition of 0.75% fresh brewery waste yeast; solid line indicates light fuel oil, and broken line indicates refinery waste oil. Source: From Lehtomäki and Niemelä 1975, Fig. 1, p. 127. Reprinted by permission of the publisher.

leguminous plants to grow without dependence on soil nitrogen, for which the soil microorganisms are already strongly competitive.

7.2.7 Organic sulfur

Organic sulfur compounds (e.g., thiophene) can be degraded by soil bacteria. Kurita et al. (1971) isolated four different bacterial strains from oil well sediments and crude oil reservoir bottom samples. All isolates were found to be gram-negative bacilli.

Sediment isolates were grown on culture media (Table 7.46), designated A in this work, and tested for sulfur decomposition in the apparatus shown in Fig. 7.44. Cultures showing large hydrogen sulfide production were isolated and anaerobically cultured on medium B, which contained 1% of a selected organic sulfur compound (thiophene, dimethyl sulfide, 1-butanethiol, or polysulfides). Colonies appearing to have a black color, and producing a dark brown ferrous sulfide precipitate, were isolated and inoculated onto medium C. Cultures exhibiting sulfur-reducing properties on medium C were retained for further thiophene degradation experiments.

Thiophene degradation was measured by hydrogen sulfide formation. Figure 7.45 shows the growth of bacteria in thiophene medium C and the concomitant formation of hydrogen sulfide. The lag period is shown to be about 15 hr with the maximum hydrogen sulfide formation rate observed

ORNL DWG NO 76-11832

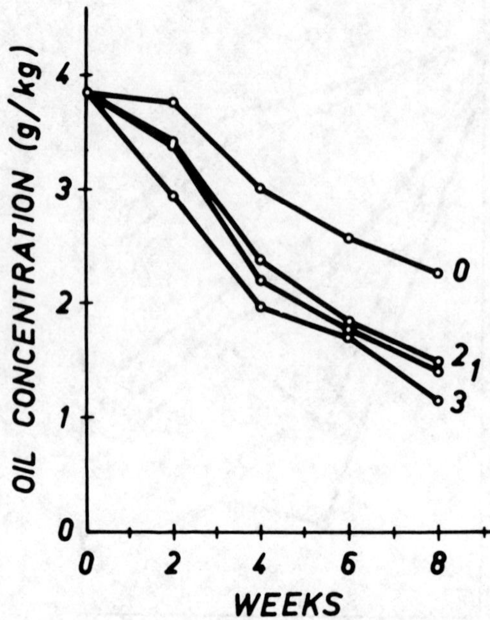

Fig. 7.43. Effect of brewery waste yeast in different states of activity on decomposition of light fuel oil in unsterilized soil. Average oil concentrations in duplicate pots kept in the laboratory at 20°C. 0 = soil without yeast addition, 1 = fresh yeast, 2 = fresh yeast killed by exposure to 70°C for 5 min, 3 = yeast stored for 6 months at 4°C. Source: From Lehtomäki and Niemelä 1975, Fig. 2, p. 128. Reprinted by permission of the publisher.

Table 7.46. Composition of culture media

	Medium (g)[a]			
	A	B	C	D
Polypeptone	5.0	5.0	0.2	0.2
Lactic acid	4.0	4.0	0.4	0.25
Glucose	0.1			
Meat extract	0.1			
Yeast extract	0.5	0.5		
Urea	0.1	0.1	0.02	0.02
KH_2PO_4	0.1	0.1	0.04	0.05
$CaCl_2$	0.1	0.1	0.04	0.05
$(NH_4)_3PO_4$	0.1		0.04	0.05
Na_2SO_4	0.1			
$Fe(NH_4)(SO_4)_2 \cdot 6H_2O$		(0.05)		
Organic sulfur compounds		1.0	0.2 - 3.0[b]	1[c] (50 ml)[d]
Tap water (liter)	1	1	1	1
pH (adjusted with NaOH solution)	7.2	7.2	7.2 - 7.8	7.2

[a]All units in grams except for tap water and pH.
[b]Thiophene.
[c]Asphaltene.
[d]Hydrogenated residue oil or crude oil.

Source: Kurita et al. 1971, Table 1, p. 186. Reprinted by permission of the publisher.

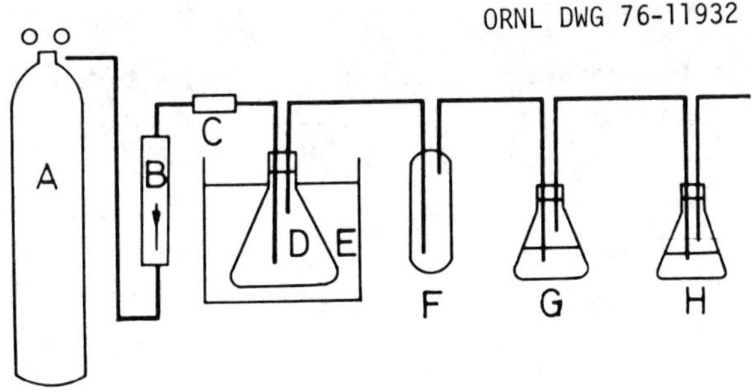

Fig. 7.44. Apparatus for the measurement of H_2S production from organic sulfur compounds by anaerobic "organic sulfur compound decomposing bacteria." A, cylinder; B, flow meter; C, cotton-wood column; D, reaction vessel; E, thermostat; F, fluid trap; G, H_2S trap; H. thiophene trap. <u>Source</u>: From Kurita et al. 1971, Fig. 1, p. 186. Reprinted by permission of the publisher.

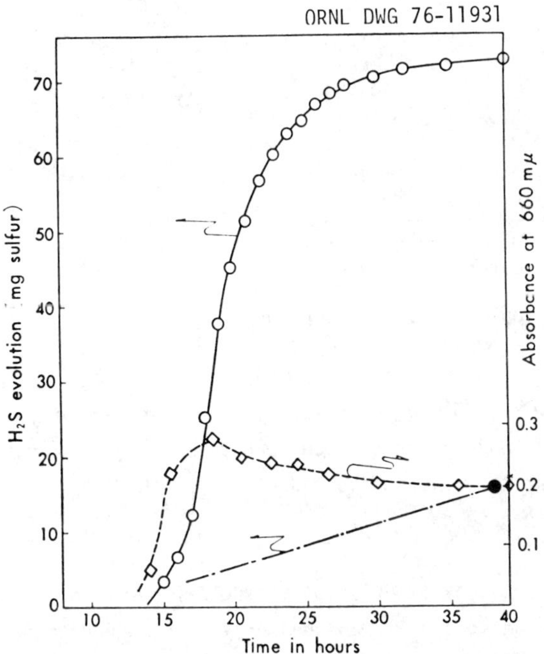

Fig. 7.45. Growth and H_2S evolution curves of thiophene-decomposing bacteria. —◇— Absorbance of the culture at 660 mµ; —○— evolution of H_2S from 0.2 ml of thiophene in medium C (5 liters); —●— evolution of H_2S from medium C (5 liters) in the absence of thiophene. <u>Source</u>: From Kurita et al. 1971, Fig. 3, p. 194. Reprinted by permission of the publisher.

about 18 hr after inoculation. The pH value did not change during the experiment. The effects of temperature and pH changes during growth were studied and are shown in Tables 7.47 and 7.48. Optimum growing temperatures at a constant pH were shown to be 38 to 40°C, and the optimum pH range was 7.0 to 9.0.

Cell-free extracts of the thiophene-degrading isolate were tested in a Warburg apparatus to determine hydrogen sulfide production (Kurita et al. 1971). The reaction mixture contained 0.31 mg cell-free extract, 2 µmoles methyl viologen (a dye), and 12.7 µmoles thiophene in 0.02 M phosphate buffer of pH 7.2 (3 ml). The gas phase was H_2. The reaction was started, and

Table 7.47. Effect of temperature on the growth of thiophene-decomposing bacteria[a]

Substrate	Thiophene			Hydrogenated residue oil			Asphaltene		
Period of growth, hr	12	24	48	12	24	48	12	24	48
Temperature, °C 20	−	−	−	−	−	−			
30	−	+	++[b]	−	++	++[b]			
38	+	+++	+++[b]	+	+++	+++[b]	+	+++	+++[b]
40	+	+++	+++[b]	+	+++	+++[b]	+	+++	+++[b]
60	−	−	−	−	−	−	−	−	−
Composition of culture media	5 liters of culture medium C containing thiophene (0.2ml)			200ml of culture medium D (poly-peptone 2g/liter) containing hydrogenated residue oil (50ml)			200ml of culture medium D (poly-peptone 1g/liter) containing asphaltene (1.0g)		
Inoculation	50ml of culture (medium C) containing thiophene-decomposing bacteria			10ml of culture (medium C) containing thiophene-decomposing bacteria			10ml of culture (medium C) containing thiophene-decomposing bacteria		
Gas phase	N_2			N_2			N_2		
pH	7.8			7.2			7.2		

[a]Symbols: −, no growth; +, ++, medium growth; +++ heavy growth.
[b]H_2S formation proportional to the growth was observed.

Source: Kurita et al. 1971, Table 2, p. 191. Reprinted by permission of the publisher.

hydrogen sulfide was produced within 5 min as the methyl viologen was reduced (indicated by a blue color). Methyl viologen and hydrogen were required for hydrogen sulfide production (Table 7.49), a fact that suggests that the reaction is reductive and that the cell-free extract has hydrogenase and lactate dehydrogenase capabilities.

Kodama et al. (1970) reported the identification of products from microbial degradation of dibenzothiophene. *Pseudomonas jianni* and *Pseudomonas abikonensis* decomposed dibenzothiophene (DBT) to five water-soluble organic compounds. The incubation was aerobic, at 28°C on a 110-rpm shaker for seven days, in a liquid metal extract medium containing 1.0 g DBT per liter of medium, at an adjusted pH of 7.3, and with an inoculum size of 5%. The fermentation broth was filtered and chromatographed after fermentation, and five organic compounds were isolated. The organic compounds were subjected to different identification procedures such as infrared and nuclear magnetic resonance spectra. These tests identified three of the compounds as 3-hydroxy-2-formyl-benzothiophene, dibenzothiophene-5-oxide, and 3-oxo[3'-hydroxy-thionaphtheryl-(2)-methylene]-dihydrothionaphthene (Fig. 7.46). Two other products could not be identified.

7.3 TRACE ELEMENTS

Trace elements, primarily trace metals, are released during coal conversion. Some of these may enter the environment through volatilization and dissolution or by incorporation into solid wastes.

Table 7.48. Effect of pH on the growth of thiophene-decomposing bacteria[a]

Substrate	Thiophene			Hydrogenated residue oil			Asphaltene		
Period of growth (hr)	12	24	48	12	24	48	12	24	48
pH 6.4 - 6.6	—	—	—	—	—	—	—	—	—
7.0 - 7.2	+	+++	+++[b]	++	+++	+++[b]	+	+++	+++[b]
7.4 - 7.8	+	+++	+++[b]	++	+++	+++[b]	+	+++	+++[b]
9.0	+	++	+++[b]	+	++	++[b]	+	++	++[b]
Composition of culture media	5 liters of culture medium C containing thiophene (0.2ml)			200ml of culture medium D (poly-peptone 2g/liter) containing hydrogenated residue oil (50ml)			200ml of culture medium D (poly-peptone 1g/liter) containing asphaltene (1.0g)		
Inoculation	50ml of culture (medium C) containing thio-phene-decomposing bacteria			10ml of culture (medium C) con-taining thio-phene-decomposing bacteria			10ml of culture (medium C) con-taining thio-phene-decomposing bacteria		
Gas phase	N_2			N_2			N_2		
Temperature	38°			38°			38°		

[a]Symbols: —, no growth; +, ++, medium growth; +++, heavy growth
[b]H_2S formation proportional to the growth was observed.

Source: Kurita et al. 1971, Table 3, p. 192. Reprinted by permission of the publisher.

Table 7.49. H_2S production from thiophene by cell-free extract of thiophene-decomposing bacteria

	H_2S produced (μg sulfur)
Complete system	3.9
Without methyl viologen	0.0
Without cell-free extract	0.0
Without thiophene	0.4

Source: Kurita et al. 1971, Table 6, p. 196. Reprinted by permission of the publisher.

ORNL DWG 76-6588

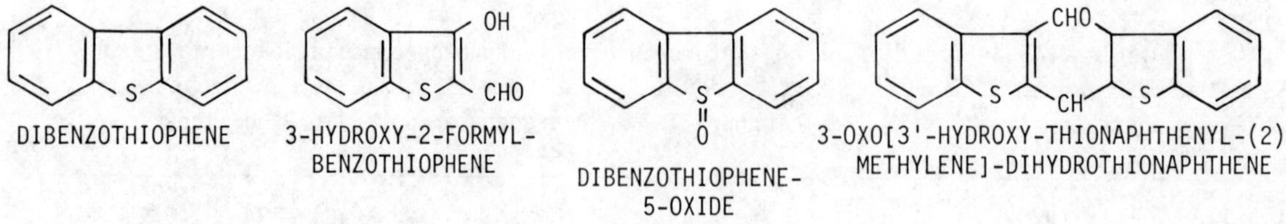

OH
CHO
DIBENZOTHIOPHENE
3-HYDROXY-2-FORMYL-
BENZOTHIOPHENE
S
O
DIBENZOTHIOPHENE-
5-OXIDE
CHO
CH
CH
3-OXO[3'-HYDROXY-THIONAPHTHENYL-(2)-
METHYLENE]-DIHYDROTHIONAPHTHENE

Fig. 7.46. Structures of dibenzothiophene and its microbial products. <u>Source</u>: Adapted from Kodama et al. 1970, Fig. 7, p. 1324.

7.3.1 Ashes and tars

Ashes and tars produced by coal conversion processes are known to contain trace elements; thus, washings or rain on stored ash or tar deposits can leach the soluble components. Ash deposits in stream systems have been found to reduce slightly the total number of bacteria present, but not to reduce the number of bacterial types (Cherry and Guthrie 1975). Trace elements that may be deposited on soil due to ash or tars are likely to undergo microbial degradation. *T. ferrooxidans* has been shown to grow on chunks of coal if the pH is low. Gleen (1950) has shown that iron deposited on soil undergoes biological oxidation to over 50% of the divalent ions.

7.3.2 Heavy metals

7.3.2.1 Rotifer assay

The bdelloid rotifer *Philodina acuticornis* was used by Buikema, Cairns, and Sullivan (1974) as a bioassay organism to identify varying concentrations of heavy metals in fresh water. *P. acuticornis* was chosen for its wide distribution in nature and because it is possible to obtain a large number of uniform test animals. The heavy metals studied were chromium, cadmium, cobalt, copper, lead, mercury, nickel, silver, and zinc. These metals have been shown to be in coal conversion effluents that enter freshwater streams. The rotifer cultures were maintained in a soft artificial dilution water made with doubly distilled or deionized water. They were tested in similar soft water and, in some instances, hard water. The soft water was about 25 ppm hardness with calcium and magnesium each contributing 17 and 18 ppm hardness respectively. Alkalinity was 24 ppm and pH varied from 7.4 to 7.9. The hard water had 53 ppm calcium and 28 ppm magnesium for a total hardness of 81 ppm. Alkalinity varied from 54 to 67 ppm, and pH varied from 7.4 to 7.8. The toxicant solutions were prepared at a concentration five times greater than that desired in the final test samples just prior to experimentation. Three 4-ml samples of the culture containing rotifers were pipetted into a small plastic culture dish. The population sizes ranged from 20 to 450 rotifers per sample, averaging about 70/sample. No effect of sample size on toxicity was noted. One milliliter of a specific toxicant solution (5 x concentration) was added to each sample. At least seven toxicant concentrations, which spanned the EC_{50} range, were used for each bioassay. (EC_{50} was defined as the estimated concentration of toxicant that would affect 50% of the organisms after a specified length of time.) A separate bioassay was conducted for each EC_{50} determination. To help ensure uniformity of toxicant composition and uniformity in culturing, the EC_{50} values for a specific toxicant were obtained simultaneously whenever feasible. The toxicant levels used for EC_{50} calculations were the calculated or introduced amounts.

To determine the percentage of affected organisms at the end of the assay period, the number of affected organisms in each sample was counted, formalin was added to the sample, and the total number in each of the triplicate samples was counted. Results for the triplicates for each test concentration were averaged for the EC_{50} determination. The EC_{50} values for experiments were calculated by probit analysis. Some EC_{50} values were determined by the straight-line estimation method because there were not enough x-values to use the probit analysis. These values are given in Table 7.50 for soft water and Table 7.51 for hard water (Buikema, Cairns, and Sullivan 1974).

Table 7.50. Toxicity of heavy metals, salts, ammonium chloride, and phenol to the rotifer *Philodina acuticornis* in soft water[a]

Chemical	24 hr		48 hr		96 hr	
	EC_{50} from data	EC_{50} from probit	EC_{50} from data	EC_{50} from probit	EC_{50} from data	EC_{50} from probit
Cadmium chloride	11.5	6.2	1.4	1.5	0.5	0.5
Cadmium sulfate	8.2	5.2	2.0	0.5	0.2	0.1
Cobalt chloride	32.0	27.8	*b*	183.0	*b*	59.0
Cupric sulfate	1.5	1.9	0.8	1.0	0.6	0.7
Lead chloride	50.3	56.2	50.5	47.4	50.4	40.8
Mercuric chloride	2.0	1.0	1.4	1.3	0.8	0.7
Nickelous chloride	7.2	7.6	4.0	4.1	2.9	2.6
Nickelous sulfate	7.2	7.1[c]	7.2	7.1[c]	7.4	7.2[c]
Potassium dichromate	42.0	31.6	50.0	31.2	3.1	3.1
Silver nitrate	3.8	5.3	*b*	15.7	1.4	1.7[c]
Zinc chloride	4.2	3.7	3.0	2.4	1.5	1.3
Zinc sulfate	1.4	1.2	1.4	0.9	1.2	1.2
Ammonium chloride	1175.0	1140.0	1150.0	1110.0	980.0	934.0
Phenol	325.0	282.0	205.0	202.0	260.0	248.0[c]

[a]All data are in ppm of the metal ion.
[b]Survival was less than 50% on all concentrations tested.
[c]Standard methods — straight-line graphical interpolation.

Source: Buikema, Cairns, and Sullivan 1974, Table 1, p. 651. Reprinted by permission of the publisher.

In comparing EC_{50} values read from the actual data and computed from the probit analysis (Tables 7.50 and 7.51), there were some discrepancies. Further, when one examines the variations in the slopes of the data curves (Fig. 7.47), it is evident that the effects of reproduction, and possibly other factors, modify the results (Buikema, Cairns, and Sullivan 1974).

7.3.2.2 In marine environment

Metals in the marine environment were studied by Lee, Patrick, and Loutit (1975). Test organisms for the demonstration of metal concentration during food chain passage were the marine bacterium *Leucothrix* and a periwinkle *Melarapha*. The *Leucothrix* bacterium was feeding on the surface of sea lettuce *Ulva* in Brock's medium (BM), to which six metals were added. Periwinkles were added to the medium to feed on the bacteria, and the cultures were left for one week. The *Leucothrix*

Table 7.51. Toxicity of heavy metals, salts, ammonium chloride, and
phenol to the rotifer *Philodina acuticornis* in hard water[a]

Chemical	24 hr		48 hr		96 hr	
	EC_{50} from data	EC_{50} from probit	EC_{50} from data	EC_{50} from probit	EC_{50} from data	EC_{50} from probit
Cadmium chloride						
Cadmium sulfate	4.5	4.8	0.8	1.4	0.3	0.3
Cobalt chloride						
Cupric sulfate	5.8	6.4	5.8	5.4	1.1	1.1
Lead chloride						
Mercuric chloride	2.8	2.3	2.0	1.7	2.1	1.6
Nickelous chloride						
Nickelous sulfate						
Potassium dichromate	23.0	22.0[b]	19.0	21.0[b]	15.0	15.0[b]
Silver nitrate						
Zinc chloride						
Zinc sulfate	4.1	2.6	2.4	2.0		
Ammonium chloride						
Phenol						

[a]All data are in ppm of the metal ion.
[b]Standard methods — straight-line graphical interpolation.

Source: Buikema, Cairns, and Sullivan 1974, Table 1, p. 652. Reprinted by permission of the publisher.

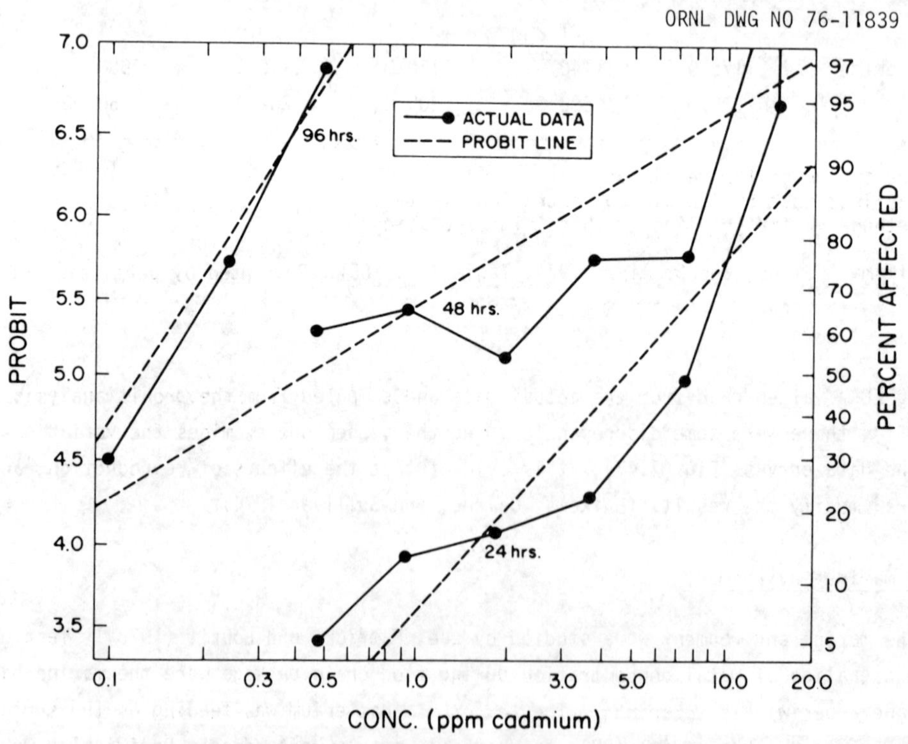

Fig. 7.47. Toxicity of cadmium sulfate to *Philodina acuticornis* tested in soft water.
<u>Source</u>: From Buikema et al. 1974, Fig. 1, p. 653. Reprinted by permission of the publisher.

strain was able to concentrate the metals added to the medium (Lee, Patrick, and Loutit 1975). When the periwinkles fed on the bacteria, there was an increase in their metal content. This level increased with the metal level of the *Leucothrix*. The concentration of metals through the *Leucothrix* and *Melarapha* is shown in Table 7.52.

Table 7.52. Concentration of metals in a marine bacterium and a periwinkle[a]

Medium and species[b]	Cr	Cu	Mn	Fe	Pb	Zn
BM	c	0.34	0.06	0.16	0.06	0.27
BM plus 1 µg/ml of each metal	0.64	0.94	1.08	1.11	1.02	1.23
BM plus 2 µg/ml of each metal	1.74	1.48	2.10	1.73	1.88	2.30
Seawater	c	0.03	0.02	0.06	0.03	0.01
Leucothrix in BM	6.53	641.18	6.60	556.66	94.70	523.48
Leucothrix in BM plus 1 µg/ml of each metal	439.84	1320.26	44.68	976.73	953.02	1187.07
Leucothrix in BM plus 2 µg/ml of each metal	1075.58	2068.31	145.65	1539.03	1279.07	2023.01
M. cincta fed *Leucothrix* no metal added	2.99	104.56	66.47	151.10	37.42	83.02
M. cincta fed *Leucothrix* 1 µg/ml added	4.89	115.83	91.02	168.83	144.81	105.36
M. cincta fed *Leucothrix* 2 µg/ml added	6.53	159.55	100.70	193.27	157.87	121.44

[a]Concentrations in media and seawater are given in µg/ml, and concentrations in bacteria and periwinkles are given in µg/g dry wt of cells.
[b]BM — Brock's medium.
[c]Not detectable at a sensitivity of 0.02 µg/ml.

Source: Lee et al. 1975, p. 18. Reprinted by permission of the publisher.

7.3.2.3 Thermal effect

The chemical pollution of streams by metallic wastes similar to those released by coal conversion processes was shown to have, overall, less effect on bacterial community structures than thermal pollution (3 to 5°C above the normal temperature for the season). The bacterial density and diversity were changed only by a lowering of chromogenic bacterial counts, whereas thermal pollution significantly lowered both bacterial numbers and diversity (Cherry and Guthrie 1974). Little is known about the actual behavior of the *Thiobacillus* species in relation to degradation due to the diversity of conflicting data (Boyer and Gleason 1975).

7.3.2.4 Lead

Lead in fresh water was studied by Wong, Chau, and Luxon (1975), who found that microorganisms in lake sediments could transform certain inorganic and organic lead compounds into volatile tetramethyl lead (Me_4Pb). Lake sediment and water were collected and put in a filtering flask with a side arm. Nutrient broth (0.5%) and glucose (0.1%) were added to stimulate microbial growth, and the flask was capped and sealed. Nitrogen was used to create anaerobic conditions. After incubation for 2 weeks at 20°C, the gas phase in the incubation flasks was withdrawn through the side arm by means of a peristaltic pump and transferred to a U-tube containing 3% OV-1 at -70°C. The sample trapped in the U-tube was swept into a gas chromatograph-atomic absorption spectrophotometer system for the separation and analysis of the volatile lead compound. Tetramethyl lead was detected in the air sample with reference to the retention time of a synthetic compound. Identity of the Me_4Pb peak in the air sample was further confirmed by gas chromatography and mass spectrometry. Addition of inorganic lead nitrate or organic trimethyl lead acetate (Me_3PbOAc) at 5 mg (expressed as lead) per liter of sample greatly increased the Me_4Pb production. Not all sediment samples examined, however, produced Me_4Pb in conditions identical to those described.

In some cases, the addition of inorganic lead nitrate or lead chloride failed to affect the transformation, but in all cases the transformation of Me_3PbOAc to Me_4Pb was observed. There was no transformation if the systems were autoclaved. Ultraviolet light caused no conversion of Me_3PbOAc to Me_4Pb in the absence of microorganisms, ruling out the possibility of chemical disproportionation reactions activated by ultraviolet light (Wong, Chau, and Luxon 1975).

The conversion of lead nitrate and lead chloride to Me_4Pb was observed on several occasions during the experiments, but no transformations were observed from lead hydroxide, lead cyanide, lead oxide, lead bromide, or lead palmitate. Lake sediment isolates of *Pseudomonas*, *Alcaligenes*, *Acinetobacter*, *Flavobacterium*, and *Aeromonas* could readily conduct biological methylation from Me_3Pb^+ to Me_4Pb, but the conversion of inorganic lead to organic lead was a difficult process, probably requiring specific physical, chemical, and biological conditions (Wong, Chau, and Luxon 1975).

7.3.2.5 Mercury

Mercury is a volatile element present in trace amounts in coal. It is highly susceptible to microbial transformations that can lead to more toxic forms such as methylmercury.

Resistance to mercury

Some microorganisms possess mercury-resistance genes for a naturally occurring resistance transfer factor (RTF). A strain of *Escherichia coli* can convert 95% of 10^{-5} M Hg^{2+} (chloride) to metallic mercury at a rate of 4 to 5 nmoles of Hg^{2+} per minute per 10^8 cells (Summers and Silver 1972). Mercury-resistant cells volatilize added $HgCl_2$, whereas mercury-sensitive cells do not (Fig. 7.48). There is a slight difference in the rapidity of mercury volatilization between resistant cells pregrown with 10^{-5} M $HgCl_2$ (i.e., induced) and uninduced resistant cells (Summers and Silver 1972). The rate of volatilization of mercuric chloride is temperature-dependent, as shown in Fig. 7.49, where uninduced cells do not volatilize ^{203}Hg at 2.5°C (Fig. 7.49A), and volatilization activity is found only after an induction lag of about 10 min at 26°C (Fig. 7.49A) or 5 min at 37°C (Fig. 7.49A). Mercury volatilization by induced cells (Fig. 7.49B) is barely measurable at 2.5°C, but begins without a measurable lag at temperatures above 15°C (Summers and Silver 1972). The optimum temperatures for volatilization occur between 15 and 25°C (Sommers and Floyd 1974). The rate of volatilization is temperature-dependent (Fig. 7.49, inset). The rate of volatilization of mercuric chloride is not only temperature-dependent, but also varies as a

ORNL DWG NO 76-11840

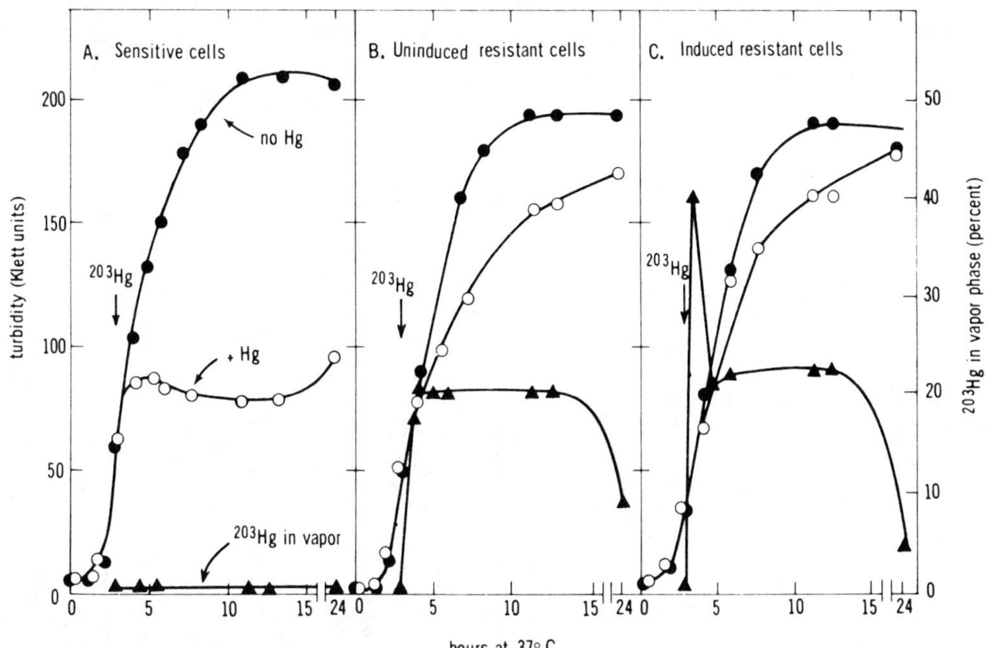

Fig. 7.48.   Volatilization of $^{203}HgCl_2$ by mercury-resistant cells. The bacterial cells used in parts A and B were grown overnight in unsupplemented broth. In part C, the cells were grown overnight in broth supplemented with 10 μmoles of $HgCl_2$. Cells were diluted 100-fold into fresh broth at the beginning of the experiment. Turbidity was followed with the Klett colorimeter by using the side-arm flasks. After 3-hr growth at 37°C (i.e., in late exponential phase), $^{203}HgCl_2$ (10 μmoles; 14 μCi/μmole) was added to one of each pair of flasks. Monitoring of turbidity continued (open and closed circles), and periodically 3-ml samples of the vapor phases from the flasks with ^{203}Hg were removed and injected in scintillation fluid (closed triangles). Source: From Summers and Silver 1972, Fig. 1, p. 1229. Reprinted by permission of the publisher.

function of substrate concentration (Fig. 7.50). There is a maximum initial rate of mercury loss of 4 to 5 nmoles/min/10^8 cells at 37°C and a substrate concentration at half-maximum rate of 10^{-5} to 2 x 10^{-5} M $HgCl_2$ (Fig. 7.50) (Summers and Silver 1972).

Summers and Lewis (1973) report the mercury resistance of *Escherichia coli*, *Staphylococcus aureus*, and *Pseudomonas aeruginosa*. These bacterial strains have independently isolated plasmids with genes determining the resistance to mercury and can convert $HgCl_2$ to a volatile form of mercury that is soluble in organic solvents. This form of mercury is very likely metallic mercury rather than an alkyl mercury compound (Summers and Lewis 1973). Volatilization and extraction data are shown in Table 7.53. Only the resistant strains could volatilize mercury, and the rate of loss of ^{203}Hg was much higher in all cases when the cells had been induced by growth in 10^{-5} M $HgCl_2$. Resistant cells also converted added mercury chloride into a form of mercury that was soluble in chloroform.

Colwell and Nelson (1975) studied bay water and sediment bacteria as part of the mobilization of mercury in sediments and water. Samples of water and sediments were collected, isolated, and tested by the ability of the population to form colonies on a simple solid growth medium supplemented with selected organic and inorganic mercury compounds. The preparation and incubation of the medium was varied. The bacterial concentration in each test sample that could

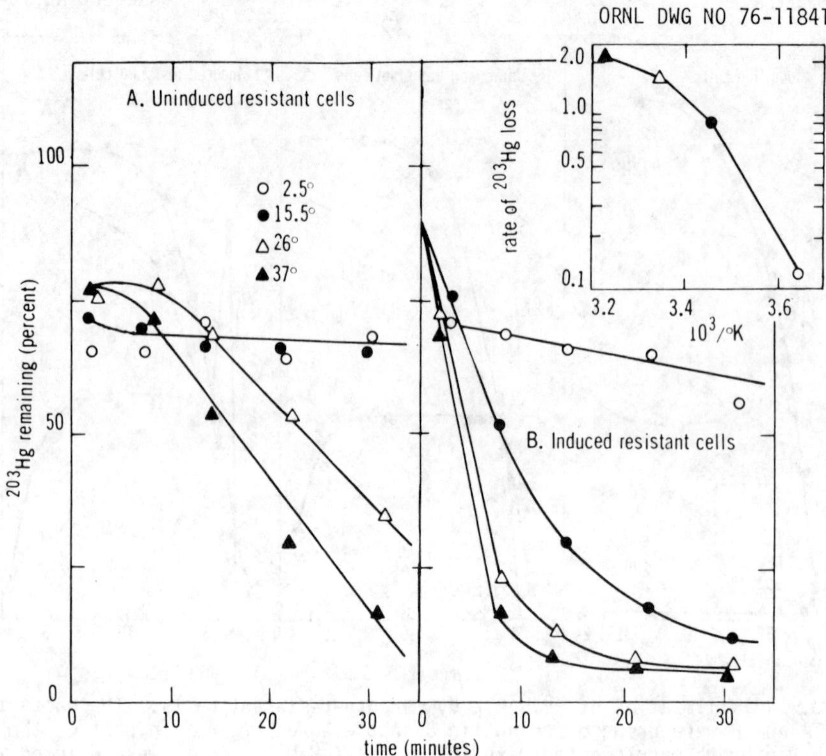

Fig. 7.49. Temperature dependence of (A) the induction of the system which volatilizes ^{203}Hg and (B) the volatilization of ^{203}Hg. The induced cells were grown overnight in 10 μmole HgCl$_2$, then into late lag phase with added HgCl$_2$ and then reinduced by adding 10 μmole HgCl$_2$ 30 min before beginning the experiment. For each reaction, 2 ml of culture in a glass vial, fitted with an injection-ported screw cap, was preincubated at the indicated temperature for 10 min, then 20 μmoles of ^{203}HgCl$_2$ was added, and 0.1-ml aqueous samples were removed at various times and counted. Source: From Summers and Silver 1972, Fig. 2, p. 1230. Reprinted by permission of the publisher.

grow on the mercury medium was expressed as the "total viable, aerobic, heterotrophic bacterial count" (Colwell and Nelson 1975). The total viable count (TVC) was found to be lower for three samples incubated anaerobically than for those incubated aerobically (Table 7.54) (Colwell and Nelson 1975). This result agreed with the isolation and identification of the mercuric-ion-reducing bacteria. The variety of bacterial genera isolated and characterized during this investigation is shown in Fig. 7.51. Also shown in Fig. 7.51 are (1) the average comparative population distributions of HgCl$_2$-resistant and total bacterial populations and (2) the greater diversity of the total population with respect to the generic categories characterized.

Mercury resistance in bacteria is often accompanied by resistance to other heavy metals and to drugs (Colwell and Nelson 1975). A group of bacterial isolates, which were representative mercury-resistant strains, were tested for other heavy metal resistance. Varied resistance was observed (Table 7.55), suggesting that resistance was not generalized and that determinants of resistance are independent (Colwell and Nelson 1975). Resistance is believed to be conferred by factors borne by transferrable extrachromosomal fragments (plasmids) (Colwell and Nelson 1975).

Mercury-resistant microorganisms have been shown to degrade oil (Walker and Colwell 1974). The mercury-resistant bacteria were examined for petroleum-degrading capability (Table 7.56). The use of petroleum by *Pseudomonas* sp. may be important because it was also the most resistant to mercuric chloride.

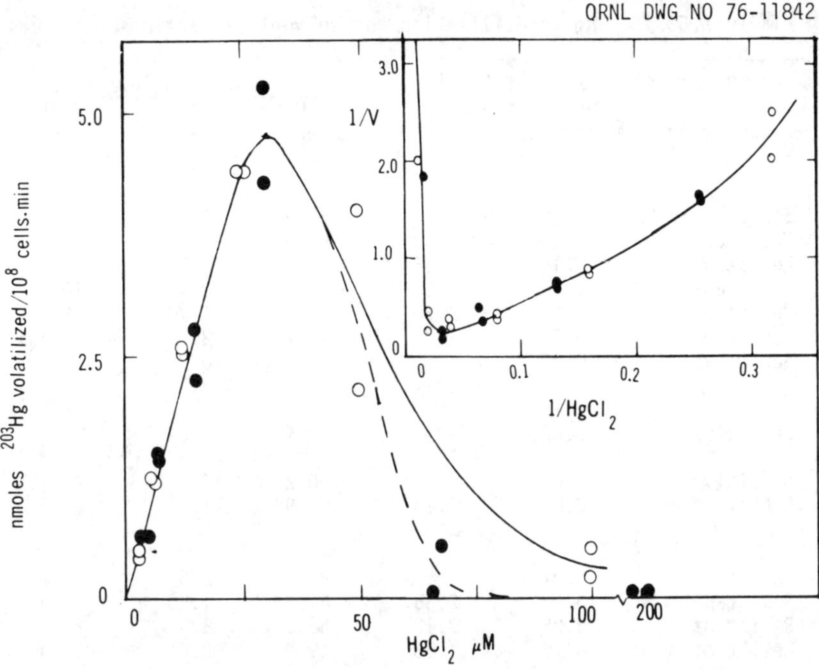

Fig. 7.50. Initial rate of volatilization of ^{203}HgCl$_2$ as a function of HgCl$_2$ concentration at 37°C. ^{203}HgCl$_2$ was added at the indicated concentrations to 2.0 ml of induced, lag-phase resistant cells in tryptone broth at 37°C, and the liquid medium was sampled repeatedly for ^{203}Hg to determine the initial rate of volatilization. After 20 min, another addition of the same initial amount of ^{203}HgCl$_2$ was made to each assay vial, and the rate of volatilization was determined again. All concentrations were done in duplicate. INSET: Lineweaver-Burk plot of 1/initial rate of reaction vs 1/substrate concentration. Source: From Summers and Silver 1972, Fig. 4, p. 1231. Reprinted by permission of the publisher.

Methylation

River sediment bacteria were studied by Spangler et al. (1973) in a long-term incubation investigation to show the biomethylation of Hg^{2+}. Various combinations of growth media and gaseous atmospheres were used to identify the lake sediment organisms capable of methylating Hg^{2+}. Cultures were purified by streaking, and pure cultures were tested for methylmercury degradation. Four pure cultures degraded methylmercury (Fig. 7.52) and were found to produce elemental mercury (Hg^0) and methane in the head gas of the closed flasks. These species have not been characterized; all were short, gram-negative rods and appeared to be *Pseudomonas* species (Spangler et al. 1973).

Bisogni and Lawrence (1975) presented experimentally derived relationships describing the rates of mercury methylation as a function of such parameters as inorganic mercury concentration, availability of inorganic mercury, pH, microbial activity, temperature, and redox potential. This could serve as the basis of a semiempirical model of mercury methylation in an aquatic system. All methylation schemes present an alkylcorrinoid, an enzyme, or an alkylcorrinoid plus an enzyme as a precursor to the reaction. Bisogni and Lawrence (1975) postulate that "the rate of methylmercury production is a function of free or available mercuric ion concentration and either the concentration or production rate of akylcorrinoids (methylcorrinoids) or enzymes involved in methylation." A generalized kinetic equation to give the kinetics of enzyme-catalyzed systems (Bisogni and Lawrence 1975) is of the form

Table 7.53. ^{203}Hg volatilization and solvent extraction

| Strain | *Mer* phenotype | Rate of ^{203}Hg volatilization[a] | | Percent ^{203}Hg extracted[b] | | | |
| | | Uninduced | Induced | Uninduced | | Induced | |
				Aqueous	Organic[c]	Aqueous	Organic[c]
E. coli							
AB1932-1	Sensitive	0.05		59.3	0.95		
AB1932-1/JJ1	Resistant	0.10	1.90	64.8	9.0	16.2	33.6
AB1932-1/U150	Resistant	0.15	2.55	64.3	7.3	12.0	26.8
AB1932-1/U305	Resistant	0.15	2.3	65.8	0.90	18.5	19.1
AB1932-1/U480	Resistant	0.60	2.50	62.5	1.7	8.9	39.3
09-37/R-ASKCTSuM	Resistant	0.70	1.75	59.7	2.4	15.1	25.2
K10C4	Sensitive	0.03		67.0	10.7[d]		
S. aureus							
8325-4	Sensitive	0.02		90.6	0.84		
8325-4/P1258*mer*-14	Sensitive	0.00		100.2	1.5		
8325-4/P1258*mer*	Resistant	0.05	0.15	94.9	1.3	67.3	3.3
P. aeruginosa							
PU21	Sensitive	0.01		83.5	2.5		
PU21/FP$^+$	Resistant	0.22	2.35	61.3	1.96	16.6	18.7
PU21/Stone	Resistant	0.85	2.10	63.5	9.9	20.8	25.4
PU21/PS18	Resistant	0.70	1.90	49.8	2.6	16.4	14.3

[a]Data expressed as the initial rate of volatilization: n-moles/min/ml of cells.
[b]Data expressed as percent of the initial radioactivity added.
[c]Extractions with chloroform.
[d]This extraction only with diethyl ether.

Source: Summers and Lewis 1973, Table 2, p. 1071. Reprinted by permission of the publisher.

Table 7.54. Comparative effects of aerobic and anaerobic incubation on bacterial resistance to $HgCl_2$

| Sample (date) | Percent resistant[a] | |
	Aerobic incubation	Anaerobic incubation[b]
B2 (5/24/73)	7.0	1.0
A2 (5/24/73)	1.3	0.3
EB1 (5/25/73)	1.8	0.3

[a]Samples of sediment were diluted and spread on basal medium agar, with and without 6 ppm $HgCl_2$, and incubated for 7 days at 25°C.
[b]Incubated in BioQuest (Cockeysville, Maryland) anaerobic jars containing CO_2-enriched anaerobic atmosphere produced by the Gas Pak (BioQuest).

Source: Colwell and Nelson 1975, Table 3, p. 7.

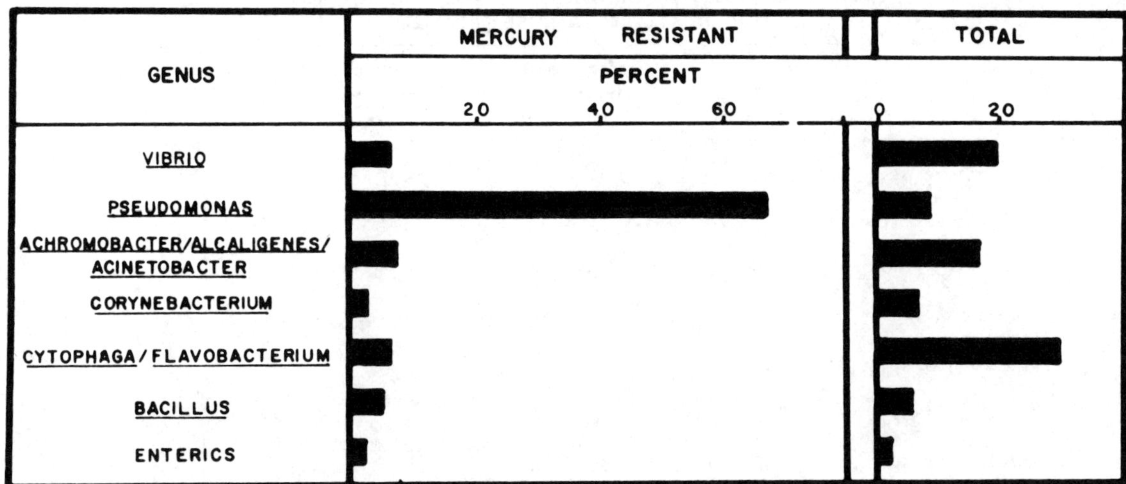

Fig. 7.51. Average distribution of genera in total and $HgCl_2$-resistant populations.
Source: From Colwell and Nelson 1975, Fig. 1, p. 9.

Table 7.55. Heavy metal resistance of mercury-resistant bacteria[a]

Culture number	Maximum concentration (ppm)[b] of metal ion[c]									
	Al^{3+}	Pb^{2+}	Ag^{+1}	As^{+5}	Co^{2+}	Cu^{2+}[d]	Zn^{2+}	Cd^{2+}	Cr^{+6}	Hg^{2+}
72	167.0[d]	167.0	16.7	501.0	334.0	334.0	55.1	100.2	668.0	4.0
119	167.0	167.0	16.7	501.0	55.1	334.0	167.0	167.0	1.7	20.0
187	16.7	167.0	1.7	501.0	55.1	334.0	110.2	50.1	668.0	16.0
132	16.7	167.0	1.7	501.0	167.0	334.0	167.0	167.0	16.7	40.0
85	167.0	167.0	1.7	501.0	167.0	334.0	55.1	50.1	100.2	12.0
639	16.7	167.0	1.7	501.0	110.2	334.0	55.1	100.2	50.1	12.0
94	16.7	167.0	1.7	501.0	55.1	334.0	55.1	16.7	334.0	24.0
244	16.7	167.0	1.7	501.0	55.1	334.0	55.1	50.1	100.2	50.0
127	16.7	167.0	1.7	501.0	18.4	334.0	18.4	16.7	16.7	24.0

[a]One drop of culture was added to 3 ml of basal broth containing dilutions of filter-sterilized solution of heavy metal, and the tubes were incubated, without agitation, at 25°C for 7 or more days. Tubes showing turbidity were scored positive. Uninoculated and inoculated controls, with and without metals, were included in the assay.

[b]Maximum concentrations for all cultures tested were: Al — 167.0, Pb — 167.0, Ag — 167.0, As — 501.0, Co — 668.0, Cu — 334.0, Zn — 334.0, Cd — 334.0, Cr — 668.0, Hg — 100.0.

[c]Metal ion salts added were: $AlCl_3 \cdot 6H_2O$, $Pb(C_2H_3O_2)_2 \cdot 3H_2O$, $AgNo_3$, $Na_2AsO_4 \cdot 7H_2O$, $CoCl_2 \cdot 6H_2O$, $CuSO_4 \cdot 5H_2O$, $ZnSO_4 \cdot 7H_2O$, $(3CdSO_4) \cdot 8H_2O$, $(NH_4)_2CrO_4$, and $HgCl_2$.

[d]Broth with salt added was adjusted to pH 7 with 1 N NaOH and filter-sterilized.

Source: Colwell and Nelson 1975, Table 7, p. 13.

Table 7.56. Mercury-resistant strains of bacteria isolated from
samples collected in the Chesapeake Bay and tested
for ability to degrade petroleum

Genus	Strain number	Maximum HgCl tolerance (mg/liter)	Petroleum utilization		
			7 days	14 days	21 days
Pseudomonas sp.	94	24 - 40	+		
Arthrobacter sp.	72	4 - 5	+		
Flavobacterium sp.	119	20 - 24	+		
Vibrio sp.	639	12 - 16	+		
Pseudomonas sp.	244	50 - 60	+		
Pseudomonas sp.	127	20 - 24		+	
Citrobacter sp.	132	10 - 50		+	
Pseudomonas sp.	187	16 - 20		+	
Enterobacter sp.	85	12 - 14			+

Source: Walker and Colwell 1974, Table 3, p. 286. Reprinted by permission of the publisher.

ORNL DWG NO 76-11844

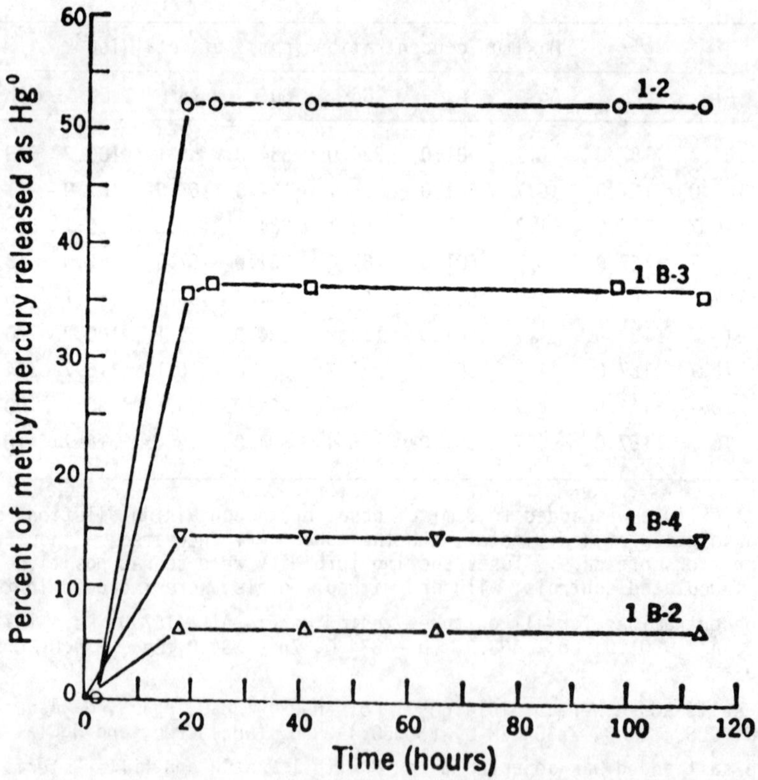

Fig. 7.52. Degradation of methylmercury by pure cultures isolated from sediment. Volatile mercury was monitored in a trap containing $HgBr_2$ and KBr attached to the flask. Curves are shown only for cultures that degraded methylmercury. Source: From Spangler et al. 1973, Fig. 3, p. 193. Reprinted with permission from *Science*. Copyright 1973 by the American Association for the Advancement of Science.

$$K_R = [S]^n \left(\frac{\bar{V}_{max}}{\bar{V}} - 1 \right) , \tag{16}$$

where

K_R = equilibrium constant for the reaction,

$[S]$ = substrate of effector concentration, moles/liter,

$\bar{V}_{max}$ = maximum reaction velocity, $(time)^{-1}$,

$\bar{V}$ = reaction velocity, $(time)^{-1}$,

n = order of reaction.

Equation (16) may be arranged and written as follows:

$$\log \bar{V} = n \log[S] - \log K_R + \log (\bar{V}_{max} - \bar{V}) . \tag{17}$$

The basic parameters used to describe the methylation rates for this laboratory work are present in natural systems. The parameters are shown in the equation describing the overall net production of methylmercury, which is

$$NSMR = \gamma(Hg^{2+})^n , \tag{18}$$

where

$$NSMR = \text{net specific methylation rate}, \frac{\mu g \text{ of } (CH_3)_2Hg \text{ or } CH_3HG^+ \text{ as Hg,}}{g \text{ of VSS-day}}$$

γ = coefficient determined by the microbial growth rate of the system,

(Hg^{2+}) = concentration of free mercuric ions, mg/liter,

n = pseudo-order of the reaction,

VSS = concentration of volatile suspended solids (a measure of biomass), g/liter.

To make Eq. (18) more general, the term β is defined as the ratio of free mercuric ions to total inorganic mercury. Substitution of β into Eq. (18) yields

$$NSMR = \gamma(\beta Hg_{total})^n , \tag{19}$$

where Hg_{total} = concentration of all forms of inorganic mercury, mg/liter (Bisogni and Lawrence 1975).

The four system parameters may be extended to a real aquatic system by proper evaluation of the necessary coefficients. Equation (19) predicts that the methylmercury production rate for benthic regions in natural water systems depends on (1) sediment mercury complexing characteristics, (2) metabolic activity of sediment microorganisms, and (3) the total inorganic mercury concentration of the sediment (Bisogni and Lawrence 1975). The methylation rate also depends upon the aerobic or anaerobic nature of the system, because the exponential n seems to depend on the redox potential of the sediment (Bisogni and Lawrence 1975).

LITERATURE CITED

Adams, C. E. 1973. Nitrogen from wastewater. *Environ. Sci. Technol.* 7(8): 696-701.

Adamczyk-Winiarska, Z.; Krol, M.; and Kobus, J. 1975. Microbial oxidation of elemental sulphur in brown soil. *Plant Soil* 43(1): 95-100.

Alexander, J. A. 1941. Control of iron and sulfur organisms by super-chlorination and dechlorination. *J. Amer. Water Works Assoc.* 32(7): 1137-46.

Atlas, R. M., and Bartha, R. 1973. Effects of some commercial oil herders, dispersants and bacterial inocula on biodegradation of oil in seawater. In *The microbial degradation of oil pollutants*, pp. 283-89. LSU-SG-73-01. Louisiana State University, Baton Rouge.

Baker, D. A., and Weston, R. F. 1975. Petroleum processing wastes. *J. Water Pollut. Control Fed.* 47(6): 1476-79.

Barker, J. E., and Thompson, R. J. 1973. *Biological removal of carbon and nitrogen compounds from coke plant wastes.* EPA-R2/73-167.

Barnett, S. M., and Velankar, S. K. 1974. Mechanism of hydrocarbon uptake by microorganisms. *Biotech. Bioeng.* 16(6): 863-65.

Barnsley, E. A. 1975. The bacterial degradation of fluoranthene and benzo[*a*]pyrene. *Canadian J. Microbiol.* 21(7): 1004-8.

Bartha, R., and Atlas, R. M. 1973. Biodegradation of oil in seawater: Limiting factors and artificial stimulation. In *The microbial degradation of oil pollutants*, pp. 147-152. LSU-SG-73-01. Louisiana State University, Baton Rouge.

Baxter, R. A.; Gilbert, P. E.; Lidgett, R. A.; Mainprize, J. H.; and Vodden, H. A. 1975. The degradation of polychlorinated biphenyls by microorganisms. *Sci. Total Environ.* 4(1): 53-61.

Beam, H. W., and Perry, J. J. 1973. Co-metabolism as a factor in microbial degradation of cycloparaffinic hydrocarbons. *Appl. Mikrobiol.* 91: 87-90.

Bisogni, J. J., and Lawrence, A. W. 1975. Kinetics of mercury methylation in aerobic and anaerobic aquatic environments. *J. Water Pollut. Control Fed.* 47(1): 135-52.

Boyer, J. F., and Gleason, V. E. 1975. Coal and coal mine drainage. *J. Water Pollut. Control Fed.* 47(6): 1466-73.

Brisou, J. 1969. Biosynthesis of 3,4-benzopyrene and anaerobiosis. 1969. *C. R. Soc. Biol.* 163(3): 772-74.

Brisou, J. 1972. Lipids, sterols, terpenes, and bacterial biosynthesis of 3,4-benzopyrene. In *Pollutions des Milleux vitraux par les hydrocarbures cancerigenes*, ed. L. Mallet, pp. 181-83. Paris: Librairie Maloine.

Brown, L. R.; Phillips, W. E.; and Tennyson, J. M. 1970. *The effect of salinity on the oxidation of hydrocarbons in estuarine environments.* PB-193 706. Water Resources Research Institute, Mississippi State University, State College, Miss.

Buchanan, R. E., and Gibbson, N. E., eds. 1974. *Bergey's manual of determinative bacteriology*, 8th ed. Baltimore: Williams and Wilkins Co.

Buck, J. D. 1974. A survey for monitoring ecologically important microorganisms in the marine environment. In *Proceedings of seminar on methodology for monitoring the marine environment, Seattle, Wash., October 1973*, pp. 385-406. EPA-600/4-74-004.

Buikema, A. L.; Cairns, J.; and Sullivan, G. W. 1974. Evaluation of *Philodina acuticornis* (Rotifera) as a bioassay for heavy metals. *Water Res. Bull.* 10(4): 648-61.

Capes, C. E.; McIlhinney, A. E.; Sirianna, A. F.; and Puddington, I. E. 1973. Bacterial oxidation in upgrading pyritic coals. *CIM Bull.* 66: 88-91.

Carnes, D. 1973. *Microbial assimilation of hydrocarbons: Characterization of enzymes involved in intermediary metabolism.* Ph.D. dissertation. University Microfilms 73-5663. Ann Arbor, Mich.: University Microfilms.

Catchpole, J. R., and Cooper, R. L. 1972. The biological treatment of carbonization effluents. III. New advances in the biochemical oxidation of liquid wastes. *Water Res.* 6(12): 1459-74.

Chapman, P. J. 1971. An outline of reaction sequences used for the bacterial degradation of phenolic compounds. In *Conference proceedings on the degradation of synthetic organic molecules in the biosphere, National Academy of Sciences, June 12-13,* pp. 17-55.

Chenouda, M. S., and Jwanny, E. W. 1972. Utilization of hydrocarbons by microorganisms: Lipids and phospholipids of *Candida lipolytica* grown on hexadecane and on glucose media. *J. Gen. Appl. Microbiol.* 18(3): 181-88.

Cherry, D. S.; Guthrie, R. K.; and Harvey, R. S. 1974. Bacterial populations of aquatic systems receiving different types of stress. *Water Resour. Bull.* 10(5): 1009-16.

Cherry, D. S., and Guthrie, R. K. 1975. The significance of ash discharged into aquatic drainage systems. *Aware* 56: 12-14.

Chu, J. P., and Kirsch, E. J. 1972. Metabolism of pentachlorophenol by an axenic bacterial culture. *Appl. Microbiol.* 23(5): 1033-35.

Cobb, H. D.; Atherton, R.; and Olive, W. 1974. *An ecological approach to the problem of biodegradation of phenolic wastes.* AD/A-004 517. Air Force Office of Scientific Research.

Colmer, A. R., and Hinkle, N. E. 1947. The role of microorganisms in acid mine drainage: A preliminary report. *Science* 108: 253-56.

Colwell, R. R., and Nelson, J. D. 1975. *Metabolism of mercury compounds in microorganisms.* EPA-600/3-75-007.

Continental Oil Company (CORPAUTH). 1971. *Microbiological treatment of acid mine drainage waters.* 14010 ENW 09/71. Environmental Protection Agency, Water Pollution Control Series.

Cooney, J. J., and Walker, J. D. 1973. Hydrocarbon utilization by *Cladosporium resinae.* In *The microbial degradation of oil pollutants,* pp. 25-32. LSU-SG-73-01. Louisiana State University, Baton Rouge.

Denitrification. 1973. *Effluent Water Treatment J.* 13(12): 791.

Fedorenko, Z. P. 1964. The effect of biochemical treatment of waste water of a by-product coke plant on the 3,4-benzpyrene content. *Hyg. San.* 29(3): 19-21.

Fedoseeva, G. E.; Khesina, A. Ya.; Poglazova, M. N.; Shabad, L. M.; and Meisel', M. N. 1968. The oxidation of aromatic polycyclic hydrocarbons by microorganisms. *Dokl. Akad. Nauk SSSR* 183(1): 208-11.

Ferguson, J., and Bubela, B. 1974. The concentration of Cu(II) and Zn(II) from aqueous solutions by particulate algal matter. *Chem. Geol.* 13: 163-86.

Finnerty, W. R.; Kennedy, R. S.; Lockwood, P.; Spurlock, B. O.; and Young, R. A. 1973. Microbes and petroleum: Perspectives and implications. In *The microbial degradation of oil pollutants,* pp. 105-25. LSU-SG-73-01, Louisiana State University, Baton Rouge.

Gibson, D. T. 1968. Microbial degradation of aromatic compounds. *Science* 161: 1093-97.

Gibson, D. T. 1971. The microbial oxidation of aromatic hydrocarbons. *CRC Crit. Rev. Microbiol.,* pp. 199-223.

Gibson, D. T., and Jerina, D. M. 1975. Oxidation of the carcinogens benzo(a)pyrene and benzo(a)anthracene to dihydrodiols by a bacterium. *Science* 189: 295-97.

Gibson, D. T., and Yeh, W. K. 1973. Microbial degradation of aromatic hydrocarbons. In *The microbial degradation of oil pollutants,* pp. 33-38. LSU-SG-73-01. Louisiana State University, Baton Rouge.

Gleen, H. 1950. Biological oxidation of iron in soil. *Nature* 166(4229): 871-72.

Gudin, C., and Syratt, W. J. 1975. Biological aspects of land rehabilitation following hydrocarbon contamination. *Environ. Pollut.* 8(2): 107-12.

Hase, A., and Hites, R. A. 1976. On the origin of polycyclic aromatic hydrocarbons in recent sediments: Biosynthesis by anaerobic bacteria. *Geochem. Cosmochem. Acta* 40(9): 1141-43.

Hass, B. S., and Applegate, H. G. 1975. The effects of unsubstituted polycyclic aromatic hydrocarbons on the growth of *Escherichia coli*. *Chem.-Biol. Interactions* 10: 265-68.

Hirai, M.; Shimizu, S.; Teranishi, Y.; Tanaka, A.; and Fukui, S. 1972. Effects of hydrocarbons on the morphology of *Candida tropicalis* pK 233. *Agr. Biol. Chem.* 36(13): 2335-43.

Houghton, C., and Cain, R. B. 1972. Microbial metabolism of the pyridine ring: Formation of pyridinediols (dihydroxypyridines) as intermediates in the degradation of pyridine compounds by microorganisms. *Biochem. J.* 130: 879-93.

Hussien, Y. A.; Tewfik, M. S.; and Hamdi, Y. A. 1974. Degradation of certain aromatic compounds by rhizobia. *Soil Biol. Biochem.* 6: 377-81.

Jahnig, C. E. 1975. *Evaluation of pollution control in fossil fuel conversion processes. Liquefaction: Sect. 2. SRC process.* PB-241 792.

Jensen, V. 1975. Bacterial flora of soil after application of oily waste. *OIKOS* 26(2): 152-58.

Jerina, D. M.; Yagi, H.; and Yeh, H.J.C. 1975. Oxidation of the carcinogens benzo(a)pyrene and benzo(a)anthracene to dihydrodiols by a bacterium. *Science* 189: 295-97.

Jeris, J. S., and Owens, R. W. 1975. Pilot-scale, high rate biological denitrification. *J. Water Pollut. Control Fed.* 47(8): 2043-57.

Jones, G. L., and Carrington, E. G. 1972. Growth of pure and mixed cultures of microorganisms concerned in the treatment of carbonization waste liquors. *J. Appl. Bact.* 35(2): 395-404.

Kelly, D. P., and Tuovinen, O. H. 1975. Metabolism of inorganic sulphur compounds by *Thiobacillus ferrooxidans* and some comparative studies on *Thiobacillus* A2 and *Thiobacillus neapolitanus*. *Plant Soil* 43(1): 77-93.

Khesina, A. Ya.; Shcherbak, N. P.; Shabad, L. M.; and Vostrov, I. S. 1969. Benzpyrene breakdown by soil microflora. *Bull. Exper. Biol. Med.* 68(7): 1139-41.

Knorr, M., and Schenk, D. 1968. The question of the synthesis of polycyclic aromatics by bacteria. *Arch. Hyg. Bakteriol.* 152(3): 282-85.

Kostenbader, P. D., and Flecksteiner, J. W. 1969. Biological oxidation of coke plant weak ammonia liquor. *J. Water Pollut. Control Fed.* 41: 199-207.

Kuenen, J. G. 1975. Colourless sulfur bacteria and their role in the sulfur cycle. *Plant Soil* 43(1): 49-76.

Lee, Y. L.; Patrick, F. M., and Loutit, M. 1975. Concentration of metals by a marine bacterium *Leucothrix* and a periwinkle *Melarapha*. *Dunedid, New Zealand, Univ. of Otagi Medical School Proceedings* 53(1): 17-18.

Lehtomäki, M., and Niemelä, S. 1975. Improving microbial degradation of oil in soil. *AMBIO* 4(3): 126-29.

Le Petit, J.; Bertrand, J. C.; N'Guyen, M. H.; and Tagger, S. 1975. On the taxonomy and physiology of bacteria utilizing hydrocarbons in the sea. *Ann. Microbiol. (Inst. Pasteur)* 126A(3): 367-80.

Magee, E. M.; Jahnig, C. E.; and Shaw, H. 1974. *Evaluation of pollution control in fossil fuel conversion processes.* PB-231 675.

Makula, R. A., and Finnerty, W. R. 1972. Microbial assimilation of hydrocarbons: Cellular distribution of fatty acids. *J. Bacteriol.* 112(1): 398-407.

Makula, R. A., and Finnerty, W. R. 1971. Microbial assimilation of hydrocarbons: Phospholipid metabolism. *J. Bacteriol.* 107(3): 806-14.

Malaney, G. W.; Lutin, P. A.; Cibulka, J. J.; and Hickerson, L. H. 1967. Resistance of carcinogenic organic compounds to oxidation by activated sludge. *J. Water Pollut. Control Fed.* 39(12): 2020-29.

Mallet, L., and Scheider, C. 1964. Presence of polybenzene hydrocarbons of the benzo-3,4-pyrene type in geological and archeological levels. *Compt. Rend.* 259(3): 675-76.

Mallet, L.; Zanghi, J.; and Brisou, J. 1967. Investigations of the possibilities of biosynthesis of polybenzene hydrocarbons of the benzo-3,4-pyrene type by a *Clostridium putride* in the presence of marine plankton lipids. *Acad. Sci., Paris Ser. D.* 264(11): 1534-37.

Mallet, L., and Tissier, M. 1969. Experimental biosynthesis of polybenzene hydrocarbons of the benzo-3,4-pyrene type at the expense of forest soils. *C. R. Soc. Biol.* 163(1): 63-65.

Mallet, L. 1969. Investigation of polybenzene hydrocarbons in ancient sediments. *C. R. Soc. Biol.* 163(2): 319-20.

Mallet, L.; Heros, M.; and Brisou, J. 1972. Biosynthesis and bioregression of the carcinogenic polybenzene hydrocarbons of the benz-3,4-pyrene type, taken as controls, at the expense of lipids. In *Pollutions des milleux vitaux par les hydrocarbures cancerigenes*, pp. 177-80. Paris: Librairie Maloine.

McKenna, E. J. 1971. Microbial metabolism of normal and branched chain alkanes. In *Conference proceedings on degradation of synthetic organic molecules in the biosphere, National Acad. of Science*, June 12-13, pp. 73-97.

Mishina, M.; Yanagawa, S.; Tanaka, A.; and Fukui, S. 1973. Effects of chain-length of alkane substrate on fatty acid composition and biosynthetic pathway in some *Candida* yeasts. *Agr. Biol. Chem.* 37(4): 863-70.

Moo-Young, M., and Shimizu, T. 1971. Hydrocarbon fermentations using *Candida lipolytica*. II. A Model for cell growth kinetics. *Biotech. Bioeng.* 13(6): 761-78.

Mulkins-Phillips, G. J., and Stewart, J. E. 1974. Effect of environmental parameters on bacterial degradation of Bunker C oil, crude oils, and hydrocarbons. *Appl. Microbiol.* 28(6): 915-22.

Muller, W. P., and Korte, F. 1975. Microbial degradation of benzo-a-pyrene, monolinuron and dieldrin in waste composting. *Chemosphere* 4(3): 195-98.

Ohmori, T.; Ikai, T.; Minoda, Y.; and Yamada, K. 1973. Utilization of polyphenyl and polyphenyl related compounds by microorganisms. Part I. *Agr. Biol. Chem.* 37(7): 1599-1605.

Oltmann, L. F.; Van Der Beek, E. G.; and Stouthamer, A. H. 1975. Reduction of inorganic sulfur compounds by facultatively aerobic bacteria. *Plant Soil* 43(1): 153-69.

Pfennig, N. 1975. The phototrophic bacteria and their role in the sulfur cycle. *Plant Soil* 43(1): 1-16.

Poglazova, M. N.; Fedoseeva, G. E.; Khesina, A. Ya.; Meisel, M. N.; and Shabad, L. M. 1966. On the possible modification of benz(a)pyrene by soil microorganisms. *Dokl. Akad. Nauk SSSR* 169(5): 1174-77.

Poglazova, M. N.; Fedoseeva, G. E.; Kesina, A. Ya.; Meisel, M. N.; and Shabad, L. M. 1967. Further investigations of the decomposition of benz(a)pyrene by soil bacteria. *Dokl. Akad. Nauk SSSR* 176(5): 1165-67.

Poglazova, M. N.; Fedoseeva, G. E.; Khesina, A. Ya.; Meisel, M. N.; and Shabad, L. M. 1968. The oxidation of benz(a)pyrene by microorganisms in relation to its concentration in the medium. *Dokl. Akad. Nauk SSSR* 179(6): 1460-62.

Poglazova, M. N.; Fedoseeva, G. E.; Khesina, A. Ya.; Meisel, M. N.; and Shabad, L. M. 1971. Metabolism of benzo(a)pyrene by various soil microflora and isolated microorganisms. *Dokl. Akad. Nauk SSSR* 198(5): 1211-13.

Ralston, J. R., and Vela, G. R. 1974. A medium for detecting phenol-degrading bacteria. *J. Appl. Bact.* 37(3): 347-51.

Reynolds, J. H.; Middlebrooks, E. J.; and Procella, D. B. 1974. Temperature-toxicity model for oil refinery waste. *J. Environ. Eng. Div., Proc. Amer. Soc. Civil Engineers* 100: 557-77.

Rheinheimer, G. 1974. *Aquatic microbiology*. London: John Wiley & Sons.

Robinson, F., and Robbins, R. C. 1970. Gaseous nitrogen compound pollutants from urban and natural sources. *J. Air Pollut. Control Assoc.* 20(5): 303-06.

Rogoff, M. H., and Wender, I. 1957. The microbiology of coal: The bacterial oxidation of phenanthrene. *J. Bacteriol.* 73: 264-68.

Rogoff, M. H. 1962. Chemistry of oxidation of polycyclic aromatic hydrocarbons by soil pseudomonads. *J. Bacteriol.* 83: 998-1004.

Rubin, E. S., and McMichael, F. C. 1975. Impact of regulations on coal conversion plants. *Environ. Sci. Technol.* 9(2): 112-17.

Scheda, R., and Bos, P. 1966. Hydrocarbons as substrates for yeasts. *Nature* 211: 660.

Schwarz, J. R.; Walker, J. D.; and Colwell, R. R. 1974. Deep sea bacteria: Growth and utilization of hydrocarbons at ambient and in situ pressure. *Appl. Microbiol.* 28(6): 982-86.

Semov, V., and Andreeva, L. *Microbial degradation of phenols in the purification of phenolic waste waters with activated sludge.* ORNL/tr-2983.

Shelton, T. B., and Hunter, J. V. 1975. Anaerobic decomposition of oil in bottom sediments. *J. Water Pollut. Control Fed.* 47(9): 2256-70.

Sikka, H. C.; Miyazaki, S.; and Lynch, R. S. 1975. Degradation of carbaryl and 1-naphthol by marine organisms. *Bull. Environ. Contam. Toxicol.* 13(6): 666-72.

Silverman, M. P.; Rogoff, M. H.; and Wender, I. 1963. Removal of pyritic sulphur from coal by bacterial action. *Fuel* 42: 113-24.

Sisler, F. D., and ZoBell, C. E. 1947. Microbial utilization of carcinogenic hydrocarbons. *Science* 106: 521-22.

Skryabin, G. K.; Golovleva, L. A.; Andreev, L. V.; Finkel'shtein, Z. I.; and Novik, S. N. Enzymatic oxidation of heterocyclic compounds in microbial cultures. *Dokl. Akad. Nauk SSSR* 203(2): 483-85.

Sommers, L. E., and Floyd, M. 1974. Microbial transformations of mercury in aquatic environments. PB-241 486. Tech. Rept. 54. Purdue: Water Resources Research Center.

Spangler, W. J.; Spigarelli, J. L.; Rose, J. M.; and Miller, H. M. 1973. Methylmercury: Bacterial degradation in lake sediments. *Science* 180: 192-93.

Stafford, D. A., and Callely, A. G. 1973. The role of microorganisms in waste tip-lagoon systems purifying coke oven effluents. *J. Appl. Bact.* 36(1): 77-87.

Starkey, R. L., and Halvorson, H. O. 1927. Studies on the transformations of iron in nature. II. Concerning the importance of microorganisms in the solution and precipitation of iron. *Soil Sci.* 24(6): 381-402.

Starkey, R. L. 1935. Isolation of some bacteria which oxidize thiosulfate. *Soil Sci.* 39(3): 197-219.

Starkey, R. L. 1945. Precipitation of ferric hydrate by iron bacteria. *Science* 102(2656): 532-33.

Summers, A. O., and Silver, S. 1972. Mercury resistance in a plasmid-bearing strain of *Escherichia coli.* *J. Bacteriol.* 112(2): 1228-36.

Summers, A. O., and Lewis, E. 1973. Volatilization of mercuric chloride by mercury-resistant plasmid-bearing strains of *Escherichia coli, Staphylococcus aureus,* and *Pseudomonas aeruginosa.* *J. Bacteriol.* 113(2): 1070-72.

Sutton, P. M.; Murphy, K. L.; Jank, B. E.; and Monaghan, B. A. 1975. Efficacy of biological nitrification. *J. Water Pollut. Control Fed.* 47(11): 2665-73.

Tausson, W. O. 1927. Die oxydation des phenanthrens durch bakterien. *Planta* 5: 239-73.

Teranishi, Y.; Tanaka, A.; Osumi, M.; and Fukui, S. 1974. Catalase activities of hydrocarbon-utilizing *Candida* yeasts. *Agr. Bio. Chem.* 38(6): 1213-20.

Teranishi, Y.; Kawamoto, S.; Tanaka, A.; Osumi, M.; and Fukui, S. 1974. Induction of catalase activity by hydrocarbons in *Candida tropicalis* pK 233. *Agr. Biol. Chem.* 38(6): 1221-25.

Teranishi, Y.; Kawamoto, S.; Tanaka, A.; Osumi, M.; and Fukui, S. 1974. Induction of catalase activity by hydrocarbons in *Candida tropicalis* pK 233 grown on hydrocarbon and glucose. *Agr. Biol. Chem.* 38(9): 1581-87.

Tomcsik, H. 1955. Effect of disinfectants and of surface-active agents on bacterial protoplasts. *Proceedings Soc. Exper. Biol. Med.* 89: 459-63.

Vanloocke, R.; DeBorger, R.; Voets, J. P.; and Westraete, W. 1975. Soil and groundwater contamination by oil spills: Problems and remedies. *Intern. J. Environ. Stud.* 8: 99-111.

Velankar, S. K.; Barnett, S. M.; Houston, C. W.; and Thompson. A. R. 1975. Microbial growth on hydrocarbons — some experimental results. *Biotech. Bioeng.* 17(2): 241-51.

Walker, J. D., and Colwell, R. R. 1974. Mercury-resistant bacteria and petroleum degradation. *Appl. Microbiol.* 27(1): 285-87.

Walker, J. D.; Seesman, P. A.; and Colwell, R. R. 1975. Effect of south Louisiana crude oil and No. 2 fuel oil on growth of heterotrophic microorganisms, including proteolytic, lipolytic, chitinolytic, and cellulolytic bacteria. *Environ. Pollut.* 9: 13-33.

Walsh, F., and Mitchell, R. 1975. Mine drainage pollution reduction by inhibition of iron bacteria. *Water Res.* 9: 525-28.

Watson, G. K., and Cain, R. B. 1975. Microbial metabolism of the pyridine ring. *Biochem. J.* 146(1): 157-72.

Wong, P.T.S.; Chau, Y. K.; and Luxon, P. L. 1975. Methylation of lead in the environment. *Nature* 253: 263-64.

Wood, D. K., and G. Tchobanoglous. 1975. Trace elements in biological waste treatment. *J. Water Pollut. Control Fed.* 47(7): 1933-45.

Wright, K. A., and Cain, R. B. 1972. Microbial metabolism of pyridinium compounds. *Biochem. J.* 128(3): 561-68.

Yoshida, F.; Yamane, T.; and Yagi, H. 1971. Mechanism of uptake of liquid hydrocarbons by microorganisms. *Biotech. Bioeng.* 13(2): 215-28.

Yoshida, F., and Yamane, T. 1971. Hydrocarbon uptake by microorganisms — a supplementary study. *Biotech. Bioeng.* 15(2): 257-70.

Yoshida, F.; Yamane, T.; and Nakamato, K. 1973. Fed-batch hydrocarbon fermentation with colloidal emulsion feed. *Biotech. Bioeng.* 15(2): 257-70.

8. PLANT INTERACTIONS

R. F. Carrier

ABSTRACT

The interactions of environmental pollutants with vegetation are as complex as they are important.
Acute exposure to atmospheric pollutants such as sulfur dioxide, fluorides, ozone, oxides of
nitrogen and other oxidants, particulates, hydrochloric acid, ammonia, ethylene, and chlorides
and chlorine has caused visible plant injuries such as chlorosis, necrosis, abscission of plant
parts, and disturbance of pigment systems. These symptoms have been used as indicators of
environmental pollutants and to estimate agricultural damage.

Although injury from long-term exposure to low levels of gaseous pollutants (chronic exposures)
is more difficult to document, recent studies involving careful measurements of crop plants have
begun to reveal subtle reductions in growth rate, plant height, root and top weight, and leaf
area, as well as retardation in floral initiation, smaller and fewer flowers, a measurable reduction in yield, and changes in community composition. Ultrastructural damage in the absence of
macroscopically visible injury has been observed.

The interdependence of a variety of factors ultimately determines the effect of pollutants on
vegetation. These factors include environmental influences (relative humidity, temperature,
edaphic conditions, climate, light conditions, ambient levels of other pollutants) and characteristics of the plant itself (species, variety, age, disease, nutritional condition). As examples,
plant injury from peroxyacetylnitrate (PAN) probably will not occur if a polluted air mass
reaches vegetation in the late evening when there is insufficient light for PAN to cause damage;
and when non-injurious concentrations of sulfur dioxide and ozone are combined, injury to
sensitive plants may result.

Acid rain, largely the result of sulfur dioxide pollution, may damage plants at distances of up
to 1000 km from the pollutant source. Indirect evidence suggests that it has caused a reduction
in forest productivity in certain areas, and it has been shown to cause plant abnormalities and
to have detrimental effects on plant nutrition. Acid rain also can destroy the protective waxy
layers of foliage, cause enhancement of disease, and decrease plant productivity. Because the
uptake of heavy metals is increased by lower than normal pH values, a synergistic effect can
reasonably be expected where fallout of heavy metals and acid rain appear simultaneously. In
some types of ecosystems, however, the beneficial effects of increased nitrogen inputs from acid
rain may outweigh the damage caused by increases in hydrogen ion input. Also, an increase in
soil acidity may not occur because of the balancing effect of plant sulfur metabolism.

In the aquatic ecosystem, pollution by hydrocarbons, heated effluents, trace metals, ammonia, and various inorganic compounds such as sulfates and sulfides has caused effects ranging from growth inhibition to growth stimulation, reproductive alterations, toxic reactions, and changes in species diversity.

Trace elements and heavy metals may be taken up and accumulated by both aquatic and terrestrial plants, with or without injury to the plant itself. Humans and livestock have suffered toxic effects from the ingestion of selenium-bearing plants while livestock have incurred molybdenosis and fluorosis from the ingestion of plants containing (or bearing) high concentrations of molybdenum or fluorine.

Human exposure to polycyclic aromatic hydrocarbons (PAH) includes those ingested with food plants. The origin of some of these PAH has been correlated with uptake from polluted air and soil. Also, plants possibly biosynthesize these compounds, which may act as natural hormones. PAH have been observed to stimulate normal plant growth as well as the development of abnormal tissue on algae, fungi, and culture-grown higher plant structures.

The blue haze observed over some densely vegetated areas is apparently the result of volatile organics emitted by the vegetation; these emissions contain reactive hydrocarbons in a quantity six times greater than the amount produced globally by man's activities.

Certain plants, particularly the lichens and bryophytes, are used as inexpensive and sensitive indicators of pollutants. Others play a part in the cycling of materials through the ecosystem, in some cases converting the materials to other forms. The water hyacinth, for example, is capable of removing phenols from polluted wastewaters with apparent metabolic breakdown. Vegetation could be an important sink for the gaseous pollutants hydrogen fluoride, sulfur dioxide, nitrogen dioxide, ozone, chlorine gas, ammonia, and to a lesser extent, PAN. Important pollutants which plants are unable to take up effectively are carbon monoxide and nitric acid.

Although abundant descriptive data is available regarding the physiological effects of some pollutants, the primary sites of action in the plant largely remain to be defined, as do tolerance and detoxification mechanisms.

8.0 INTRODUCTION

Plants affect man in many ways: They are part of his diet, his aesthetic environment, and his economic situation; they form the basis of the environmental food chain, and they play a role in the cycling of atmospheric and aquatic pollutants. Therefore, knowledge of plant-pollutant interactions is essential for evaluating possible environmental and health effects of potential industrial effluents.

The chemical pollutants from coal conversion processes have not yet been completely characterized. Nevertheless, sufficient information about effluents from similar industries is available to allow preliminary examination of their interaction with plants.

Results of phytotoxicity research may be difficult to interpret. Even when phytotoxic effects of individual pollutants can be estimated, little is known of their effect in combination with each other or of the effects of a variety of environmental variables and their interactive effects, such as precipitation, disease, pesticides, and edaphic factors. A plant's pollutant response may be localized (acute), as in leaf-fleck injury in tobacco, or generalized (chronic), as in yield reduction. Careful observation is required to distinguish between pollutant-related effects and those resulting from other environmental stresses (Jacobson and Hill 1970). The needle tip dieback typically seen on conifers damaged by sulfur dioxide may be caused by other agents.

Positive identification of a causal agent is possible in some cases of acute injury, but the problems involved in chronic effects have not been as well defined. The term "hidden injury" has been used to describe chronic symptoms, but Feder (1973) suggests that it may be more correct to speak of chronic exposure instead of chronic symptoms because the time at which the symptoms appear after fumigation does not necessarily affect their characteristics. Although their results were not correlated with specific pollutants, Godzik and Knabe (1973) suggest that ultrastructural examination may be useful in determining the destructive effects of air pollutants. They identified certain characteristic ultrastructural changes in the chloroplasts of pine needles that showed no macroscopic signs of damage.

Recent studies involving careful measurements of crop plants have revealed subtle reductions in growth rate, plant height, root and top weight, and leaf areas, retarded floral initiation, fewer and smaller flowers, and a measurable reduction in yield as a result of long-term exposure to ambient levels of such gaseous pollutants as hydrogen fluoride, sulfur oxides, photochemical oxidants, and nitrogen oxides (Feder 1973). Table 8.1 summarizes some major pollutants and the characteristics important in inducing low-level chronic effects in vegetation (Auerbach 1975).

It is not the intent of this chapter to include every published aspect of plant interactions with the potential pollutants from a coal conversion industry. Rather, the aim has been to give some indication of the areas in which abundant material is available and to highlight areas of potential interest and concern.

Table 8.1. Major pollutants inducing chronic effects to vegetation

Pollutant	Capable of long-distance transport	Quantitatively significant or widely dispersed	Highly soluble in living tissues	Highly reactive in living tissues	Resistant to metabolic detoxification	Capable of bioaccumulation
SO_2	X	X	X	X		X
SO_4	X	X	X			X
O_3	X	X		X		
NO_x	X	X		X		
PAN	X	X		X		
HF			X		X	X
Pb	X	X	X[a]		X	X
Hg	X	X	X[a]	X[a]	X[a]	X
Cd	X				X	X
As		X		X		

[a]In organic form.

Source: Auerbach 1975, Table 1, p. 2.

8.1 CONSIDERATIONS IN PLANT STUDIES

Extensive research has been carried out on crop species because of the relative manageability
of crop systems. However, forests and other complex natural ecosystems have not lent themselves
readily to investigation and have, therefore, been less extensively studied. Heck (1973) states
that the filtered field chambers, which are presently the only feasible way to provide pollutant-
free controls, do not permit easy comparison with field-grown crop plants. However, Daines (1968)
stresses that the chambers have been indispensable in the study of pollutant effects on plants.
Yield reductions in soybeans of about 30% have been demonstrated in recent field trials with
these chambers (Howell 1975 and McLaughlin 1974, as cited in Auerbach 1975). Effects were noted
both in rural areas of the eastern United States characterized by regional scale buildup of
oxidants and near point source fossil fuel plants emitting sulfur oxides. Comparisons, however,
between chamber and greenhouse growth studies have shown that several factors must be taken into
consideration when using these devices to isolate and study the environmental factors that affect
plant growth. For instance, at comparable temperatures in the autumn at high levels of solar
radiation, the growth of tobacco, soybeans, and cotton was greater in greenhouses than in
growth chambers. At wintertime low-solar-radiation levels, growth was greater in the chambers.
Plant growth was also affected by the generally lower levels of carbon dioxide in the chambers
(Patterson and Kramer 1974).

8.1.1 Plant volatiles as pollutants

Plants can contribute to environmental contamination by emitting gaseous pollutants. Went (1960)
suggests that the blue haze present in densely vegetated areas such as the Great Smoky Mountains
is the result of vegetation-produced hydrocarbons. According to Robinson and Robbins (1968), as
cited in Rasmussen 1972, when inputs and outputs of atmospheric trace gases were calculated on a
global scale, the major sources of hydrogen sulfide, ammonia, nitrous oxide, and hydrocarbons
were found to be natural emissions, including those from plant foliage.

Rasmussen (1972) estimated world-wide terpene emissions and found that, generally, 70% of the
trees of a given forest type emitted terpenes to the atmosphere (Table 8.2). For conifers,
rates of emission ranged from 0.4 to 3.5 ppb per gram of tissue per minute in a closed atmosphere
(1 liter) at a given temperature (17 and 30 to 32°C respectively). Emission rates for undamaged
foliage varied with plant species, foliage maturity, oil gland (or resin duct) integrity, and
leaf temperature. The emission of some specific terpenes was light-dependent.

Tyson, Dement, and Mooney (1974) have contended that many of the volatile organics measured
previously may not have been terpenes and that more detailed analyses of field samples are
required to determine their true nature. They found that rates of terpenes emitted from the
highly aromatic shrub *Salvia mellifera* were sufficiently high to make a significant contribution
to the organic content of the atmosphere, but were not high enough to account for the field
levels reported by Rasmussen and Went (1965) (as cited in Tyson, Dement, and Mooney 1974).
Monoterpene (C_{10}), and isoprene (C_5) have been determined to be the major reactive hydrocarbons
emitted by trees and are reportedly responsible for a global emission rate six times greater than
that estimated for reactive hydrocarbons of anthropogenic origin (i.e., 175×10^9 kg per year
natural vs 27×10^9 kg per year anthropogenic) (Rasmussen 1972).

Table 8.2. Composition of U.S. forest-type groups by foliar terpene emissions

	Total U.S. forest area (%)	α-Pinene emitters (%)	Isoprene emitters (%)
	Eastern groups		
Softwood			
Loblolly-shortleaf pine	11	∿100	Some from oak and sweetgum associates
Longleaf-slash pine	5	∿100	Some from oak and sweetgum associates
Spruce-fir	4	∿75	25, from spruce, which also emits α-pinene
White-red-jack pine	2	∿90	10, from aspen trees
Subtotal	22	∿91	∿9
Hardwood			
Oak-hickory	23	∿10	70, diluted by hickory, maple, and black walnut
Oak-gum-cypress	7	∿50	50, from plurality of oak, cottonwood, and willow
Oak-pine	5	∿30	60, diluted by black gum and hickory associates
Maple-beech-birch	6	∿15	Terpene foliages are hemlock and white pine
Aspen-birch	5	∿20	60, diluted by birch, α-pinene-source balsam fir and balsam poplar
Elm-ash-cottonwood	4		30, from cottonwood, sycamore, willow
Subtotal	50	∿21	∿45
Total	72		
	Western groups		
Softwood			
Douglas fir	7	∿100	
Ponderosa pine	7	∿100	5, from aspen associates
Lodgepole pine	3	∿90	10, from Engelmann spruce and aspen
Fir-spruce	3	∿100	40, from spruce trees
Hemlock-Sitka spruce	2	∿100	25, from Sitka spruce
White pine	1	∿100	5, from Engelmann spruce
Larch	1	∿100	
Redwood	0.5	∿100	
Subtotal	24	∿98	∿12
Hardwood	2		∿100, from aspen trees
Total	26		

Source: Rasmussen 1972, Table III, p. 541. Reprinted by permission of the publisher.

8.1.2 Plants as bioindicators

Plants are often used as an economical means of indicating the presence of certain environmental
pollutants. For example, a test system for sulfurous pollutants based on the specificity of their
adverse effects on lichens has been suggested (Sundstrom and Hallgren 1973), and in Sweden the
geographical spread of lead, copper, nickel, zinc, and chromium has been demonstrated by analysis
of their presence in mosses (Tyler 1972). It is possible to measure fluoride levels in plants
that have been chronically exposed to low levels of the pollutant. These measurements can be
related in a quantitative manner to the intensity of symptom expression. However, pasture grasses
and small grains may be injury-free even at leaf concentrations of several hundred parts per
million of fluoride (Feder 1973). Weinstein and McCune (1970) have detailed the complexity of
the problems involved in field surveillance, vegetation sampling, and the use of plants as monitors.
A synopsis of methods using plants for the recognition and monitoring of air pollution has been
presented by Guderian and Schoenbeck (1971) and includes descriptions of the test chamber, test
plant, grass culture, and lichen exposure methods.

No high correlations between growth form of lichens and sulfur dioxide sensitivity were demonstrated by Wirth and Turk (1975), although those species having a wide ecological range, rather
than those sensitive species that are limited by environmental conditions, are recommended as bioindicators. Privet leaf analysis has been used to monitor atmospheric lead pollution in the
United Kingdom (Everett, Day, and Reynolds 1967). Goodman and Roberts (1971) have used moss
and the grass *Festuca* to analyze for zinc, lead, cadmium, copper, nickel, and magnesium. Tyler
(1972) contends that bryophytes seem to be the most sensitive organisms for measuring the
deposition of heavy metals and have been used to detect airborne mercury (Wallin 1974). Catcheside (1975) states that lichens are superior to bryophytes for mapping the extent of air pollution because they tend to occupy larger and more diverse areas and because their communities
tend to be more homogeneous. The epiphytic plant, Spanish moss, has been analyzed for lead
content and is suggested as an indicator for lead along the Gulf Coast (Martinez, Nathany, and
Dharmarajan 1971).

Rentschler (1973) has suggested that an additional factor to be taken into consideration when
using plants as indicators is the variation possible in the sensitivity of foliage as a result
of the amount of waxy deposit on leaf surfaces. He found this variation to be related to
humidity and temperature and therefore to growth location. The most resistant plants were those
in which the wax layer had aged or weathered very little and in which the wax crystals covered
the stomata. Acid rain may decrease the tolerance of a plant through deterioration of the waxy
layer (Shriner 1976).

In the aquatic ecosystem, the presence of certain algae has been widely considered an indication
of eutrophication (sometimes a result of pollution). The toxicity of heavy metals to algae may
lead to their use as bioindicators of aquatic metallic enrichment (Whitton 1970). Diatoms have
been shown to be very successful as a group in indicating the ability of a water body to support
aquatic life (Patrick 1957 and 1963, as cited in Gaufin 1974).

Table 8.3 lists indicator species and/or groups of species for some pollutants and elements,
emphasizing the wide diversity of plant types rather than attempting to present a comprehensive
list of known indicators. The factor indicating the presence of a particular pollutant may be

Table 8.3. Indicator plants (+ = accumulation)

Element or pollutant	Plant and/or plant group
Aluminum	Symplocaceae + Melastomaceae + Diapensiaceae + *Lycopodium* spp. +
Arsenic	Brown and red algae + *Artemesia tridentata*
Bismuth	*Marchantia polymorpha* +
Boron	Some bryophytes and liverworts + *Salsola nitraria*
Bromine	Brown and red algae +
Chromium	*Pimelea suteri* + *Leptospermium scoparium* +
Cobalt	*Nyssa sylvatica* + *Crotalaria cobalticola* + *Clethra barbinervis* +
Copper	*Caryophyllaceae* + *Becium homblei* + requires at least 100 ppm Labiatae + Some bryophytes and liverworts appear to require high levels Corn, beans, and squash (for excess)
Fluorine	*Gastrolobium grandiflorum* + *Dichapetalum cymosum* + *Dichapetalum toxicarium* + *Acacia georginae* +
Gallium	Low accumulations in mosses High accumulations in lichens
Iodine	Brown and red algae +
Lead	Spanish moss Low accumulations in mosses High accumulations in lichens +
Manganese	Apple, cherry, citrus, oats, beets (deficiency) Alfalfa, cabbage, cereals, pineapple, tomato (for excesses)
Mercury	*Pinus contorta* + (from very low soil levels) *Betula papyrifera* + (concentrates to highest level) Bryophytes
Molybdenum	*Brassica* spp. (for deficiency)
Nickel	*Alyssum bertolonii* + *Alyssum murale* + *Alyssum biovulatum* + *Pimelea suteri* +
Ozone	Pinto bean Tobacco variety Bel W-3

Table 8.3 (continued)

Element	Plant and/or plant group
Rare earths	Hickory trees (accumulated in the same approximate proportion as in the exchange complexes of the soil) *Clarya glabra*
Rhenium	*Astralus* spp. Myrtales Cruciferae Compositae
Selenium	*Amanita muscaria* (form low soil levels)+ *Astragalus* spp. (up to 15,000 ppm)+ Compositae Myrtales Cruciferae
Silicon	*Gramineae* + *Juncaceae* + *Cyperaceae* + *Selaginella* spp. + *Psilotum* spp. + *Equisetum* spp. +
Silver	*Eriogonium ovalifolium*
Smog	Annual bluegrass
Strontium	*Arabis stricta* + *Carex humilis* +
Sulfurous pollutants	Lichens Privet
Tin	*Silene cucubalis* Low accumulations in mosses High accumulations in lichens
Uranium	*Blechnum capense* (fern) + *Astragalus* spp. + Myrtales Compositae Cruciferae
Vanadium	*Astragalus* spp. + *Amanita muscaria* (from low soil levels) + Some bryophytes and liverworts +
Yttrium	Low accumulations in mosses High accumulations in lichens Ferns (except *Polypodium vulgare*) contain more than any other group of plants +
Zinc	*Equisetum* spp. + *Thlaspi calaminare* + Ragweed *Viola* spp. +

Source: Compiled from Peterson 1971; Wilton et al. 1972; Comanor et al. 1974; Martinez 1971; Tyler 1972; Hemphill 1972; and Oshima 1974.

the (1) foliar injury occurring on sensitive species, (2) presence or absence of a certain species, or (3) high pollutant accumulation by the plant (perhaps to levels toxic to the accumulator plant itself). Markedly different phenotypes sometimes result from high levels of the pollutant. The high concentrations may be toxic to nonaccumulators and/or other organisms, or they may be metabolized to toxic compounds by the accumulator.

8.2 ATMOSPHERIC EFFLUENTS

Numerous reviews and articles characterizing air pollutants and their effects on plants have been published (Heggestad and Heck 1971; Treshow 1971; Heck 1968a; Daines 1968; NAPCA 1970; Mudd and Kozlawski 1975; Middleton 1961; Taylor 1968; Hindawi 1970; Jacobson and Hill 1970; Thomas 1951). Well-studied air pollutants that are known to have a major impact on agriculture include ozone, other oxidants, sulfur dioxide, fluorine, and ethylene (Heck 1973), each of which may be important when considering potential pollution from the coal conversion industry (Rubin and McMichael 1975). Other possible coal conversion atmospheric pollutants, including ammonia, chlorine, volatile trace metals, phenols, other organic compounds, and particulates, may be implicated. Photochemical reactions may further result in the production of serious plant toxicants such as peroxyacyl nitrate (PAN) (Rubin and McMichael 1975).

Table 8.4 is a summary of sources, annual emissions, background concentration, and major sinks of gaseous atmospheric pollutants. Combustion of fossil fuels is the major anthropogenic source. Natural sources produce larger quantities of air pollutants than do anthropogenic sources, except for sulfur dioxide.

8.2.1 Physiological effects

Specific mechanisms may be involved in the types of effects characteristic of the pollutants listed in Table 8.5 (Auerbach 1975). However, Mudd (1973) reports that physiological observations cannot demonstrate the chemical mode of phytotoxicity of sulfur dioxide, nitric oxides, PAN, and ozone. He suggests that chemical and in vitro studies may be irrelevant at the physiological level and that the sequence of events leading to the characteristic symptoms of damage from these pollutants cannot at this point be described despite the large amount of relevant research in print. He suggests that, instead of lowering pollution levels, it might be necessary to cure symptoms with appropriate chemicals or to use resistant varieties of plants. However, Catcheside (1975) states that plants probably are unable to become more resistant without a reduction in productivity and that resistant strains are generally less fit than normal strains except in the presence of the toxin. Some research has been done regarding the prevention of pollutant injury by chemicals. In some cases, exposures to high levels of carbon dioxide prior to ozone exposure has resulted in protection from ozone damage (Heggestad and Heck 1971). Effects of ammonia in neutralizing the injurious impact of sulfur dioxide on crops is discussed in Sect. 8.2.5.3.

According to Heggestad and Heck (1971), no reliable means has been developed to protect plants from any of the commonly produced oxidants. The researchers detail the weaknesses inherent in any spray program; the frequency of spraying required for continued resistance; the cost of spraying; and reasons for the difficulty of predicting high oxidant days accurately. However, recent highly significant reductions in ozone-induced injury in pinto beans, azaleas, and bluegrass have been reported with the use of systemic fungicides (benzimidazole, benomyl, triarimol,

Table 8.4. Summary of sources, annual emission, background concentrations,
and major sinks of atmospheric gaseous pollutants

Pollutant	Major source		Estimated emission (kg/yr)		Background concentration ($\mu g/m^3$)	Major identified sinks
	Anthropogenic	Natural	Anthropogenic	Natural		
SO_2	Combustion of coal and oil	Volcanoes	130×10^9	2×10^9	1 – 4	Scavenging; chemical reactions; soil and surface water absorption; dry deposition
H_2S	Chemical processes; sewage treatment	Volcanoes; biological decay	3×10^9	100×10^9	0.3	Oxidation to SO_2
N_2O	None	Biological decay	None	590×10^9	460 – 490	Photodissociation in stratosphere; surface water and soil absorption
NO	Combustion	Bacterial action in soil; photo-dissociation of N_2O and NO_2	53×10^9	768×10^9	0.3 – 2.5	Oxidation to NO_2
NO_2	Combustion	Bacterial action in soil; oxidation of NO			2 – 2.5	Photochemical reactions; oxidation to nitrate; scavenging
NH_3	Coal burning; fertilizer; waste treatment	Biological decay	4×10^9	170×10^9	4	Reaction with SO_2; oxidation to nitrate; scavenging
CO	Auto exhaust and other combustion processes	Oxidation of methane; photo-dissociation of CO_2; forest fires; oceans	360×10^9	3000×10^9	100	Soil absorption; chemical oxidation
O_3	None	Tropospheric reactions and transport from stratosphere	None	(?)	20 – 60	Photochemical reactions; absorption by land surfaces (soil and vegetation); surface water
Nonreactive hydrocarbons	Auto exhaust; combustion of oil	Biological processes in swamps	70×10^9	300×10^9	CH = 1000; non-CH_4<1	Biological action
Reactive hydrocarbons	Auto exhaust; combustion of oil	Biological processes in forests	27×10^9	175×10^9	<1	Photochemical oxidation

Source: Rasmussen 1975, Table I, p. 34. Reprinted by permission of the publisher.

Table 8.5. Types of chronic low-level effects and probable mechanisms involved

Type of effect	Probable mechanisms	Pollutants involved[a]
Reduced growth	Decreased photosynthesis	SO_2, O_3, NO_x, HF, PAN, Pb
	Increased respiration	HF, SO_2
	Biochemical changes	
	Enzyme inhibition	SO_2, O_3, PAN, HF, As
	Electron (energy) transport	HF, O_3, PAN
	Effects on photopigments	SO_2, HF, O_3, PAN, NO_x
	Stomatal function	SO_2, O_3
	Membrane permeability	O_3, PAN, SO_2, HF, metals
	Premature senescence of leaves	SO_2, O_3, PAN, NO_2
	Physical changes	
	UV reflectivity	Acid rain
	Heat exchange	Acid rain
	Gas exchange	Acid rain
	Changes in symbiosis	O_3, acid rain
	Effects on nutrient flux	SO_2, acid rain
Reduced reproduction	Reduced carbohydrate availability for reproductive structures	SO_2, O_3, HF, PAN, Pb, NO_2
	Effects on reproductive processes	
	Flower or fruit production	Smog, HF, SO_2
	Pollen germination or tube elongation	SO_2, O_3, HF
	Mutagenesis	SO_2
	Effects on pollinating insects	HF
Increased morbidity	Inadequate carbohydrate supply to support autotrophic respiration and resist diseases; reduced growth and loss of ability to compete with other vegetation	SO_2, O_3, HF, PAN, NO_2, heavy metals

[a]Includes pollutants for which specific effects have been documented and those whose biochemical symptomology strongly implicate them as probable causal agents.

Source: Auerbach 1975, Table 2, p. 4.

and the ethyl and methyl analogues of thiophanate) and a hydrocarbon wax emulsion used as an antitranspirant (Pelissier et al. 1972; Pelissier, Lacasse, and Cole 1972*a*; Pelissier, Lacasse, and Cole 1972*b*; Seem, Cole, and Lacasse 1972; Seem, Cole, and Lacasse 1973; Moyer, Cole, and Lacasse 1974*a*; Moyer, Cole, and Lacasse 1974*b*). Degree of success varied with dose, fungicide, method of application, and species. Applications were made as soil amendments, soil drenches, or foliar sprays. Internode shortening was noted in pinto beans at higher concentrations of fungicide (Seem, Cole, and Lacasse 1972; Seem, Cole, and Lacasse 1973). The antitranspirant did not produce results as long-lasting as the benomyl (Pelissier, Lacasse, and Cole 1972*a*), and in at least one case required application on all leaf surfaces for more than 70% protection (Pelissier et al. 1972).

Hansborough (1967) has reviewed the literature from a forestry standpoint. He notes that tree mortality is high from uncontrolled emission of such pollutants as sulfur dioxide and fluorine compounds within a few miles of a major industry such as an oil refinery or a high-capacity fossil-fueled power plant. Surrounding this area is often a zone of trees which usually do not reproduce and which show obvious damage such as "burnt" foliage and/or stunted shoot growth. Hansborough estimates that thousands of acres of forest have been damaged by industrially emitted plant toxicants.

Miller and McBride (1975) describe the forest ecosystem as a receptor of air pollutants and the forest types in the United States that are particularly susceptible to injury. The details of important episodes of damage to forests in the United States, Canada, and Europe are included. For example, sulfur dioxide and ozone, acting either independently or synergistically at low concentrations for long periods of time, are believed to be responsible for the stunted growth of Eastern white pines in the northeast, central, and lake states and Scots pine in Ohio.

8.2.2 Economic effects

Wide variations in estimates of dollar losses to vegetation from air pollution are reported. Kozlowski and Mudd (1975) cite Barrett and Waddell (1973) as estimating a vegetation loss in 1968 of $120 million, of which 90% ($100 million) was traceable to effects of oxidants. On the other hand, crop losses from air pollution were believed by Benedict, Miller, and Olson (1971) to be far less than the economic threat of losses from weeds, insects, and diseases. The authors estimate the total value of all crops lost due to air pollution in the United States to be $85.5 million (0.46% of the total value of all crops). According to the U.S. Department of Agriculture (1965) (as cited in Benedict, Miller, and Olson 1971), the average annual loss of crops is $3068 million (12%) due to plant diseases, $2965 million (12%) due to insects, and $1828 million (8%) due to weeds. The figure given by the U.S. Department of Agriculture for annual crop loss due to air pollutants is $325 million, or 1.2% of the total crop value, nearly four times the amount estimated by Benedict, Miller, and Olson (1971), who believe that their figures more accurately reflect true values. The authors substantiate their belief by noting the agreement of their estimates for loss due to ozone, PAN, and NO_x with similar estimates for California and Pennsylvania, where agricultural experts in their respective states had made on-the-spot surveys. Detailed observations on a field-by-field basis are apparently the best method for obtaining reliable estimates of actual losses. For sulfur dioxide and fluorides, the authors relied on their subjective experience to judge the accuracy of their estimates because relevant data were unavailable for comparison. Losses from air pollution were found to be very high where crop production activities and heavy pollution occurred in the same general vicinity. Grape yields, for instance, were reduced by 40 to 50%, and citrus by 20%, in Los Angeles County. Naegele (1974) quotes recent surveys that indicate an approximate loss to agriculture of $100 million from air pollution. He suggests that, if estimates could be made for losses due to growth and yield (as opposed to visible injury), an estimate in billions of dollars would not be unrealistic.

The fact that crop losses due to air pollutants can vary from year to year depending on climatic, edaphic, and biotic factors should be emphasized. Estimated losses of vegetables in California amounted to more than $1.5 million in 1969 as compared with $264 thousand in 1970 (Benedict, Miller, and Olson 1971). An example of the kind of dollar loss estimates being made for an average year is presented in Table 8.6.

Table 8.6. Estimates of annual value of crops and ornamentals grown in the United States and estimated losses due to air pollution

Plant group	Value in polluted areas (thousand $)	Losses (thousand $) due to			
		Oxidants	Sulfur dioxide	Fluorides	Total
Field crops	3,357,000	17,984	3,044	2,970	24,008
Seed crops	16,000	200	10	18	228
Fruits and nuts	334,000	10,541	17	265	10,822
Vegetables	170,000	3,341	5	2	3,348
Forest and nursery	425,000	18,494	134	566	19,194
Citrus	348,000	27,429	5	388	27,823
Crop total	4,650,000	77,989	3,225	4,209	85,423
Ornamentals	1,346,000	43,407	2,979	127	46,513
Total, all plants	5,996,000	121,396	6,204	4,336	131,936

Source: Benedict, Miller, and Olson 1971, Table 17, p. 64.

8.2.3 <u>Removal by plants</u>

It has been suggested that vegetation may play an important role in cleansing the air of certain toxic pollutants (Hill 1971). An alfalfa canopy removed gases from the atmosphere in the following order: hydrogen fluoride > sulfur dioxide > chlorine > nitrogen dioxide > ozone > PAN > nitric oxide > carbon monoxide. The absorption rate for nitric oxide was low, and no absorption could be detected for carbon monoxide. Uptake increased linearly with increasing concentration of ozone and chlorine, which caused partial stomatal closure at higher concentrations (Hill 1971). Absorption rate and solubility of the pollutant in water were found to be related; in general, plant uptake rate increased with an increase in water solubility of the pollutant.

Study under highly controlled conditions by Bennett and Hill (1973) showed that hydrogen fluoride, sulfur dioxide, and nitrogen dioxide were removed efficiently by the upper portion of a highly standardized alfalfa canopy as well as by the immediate subsurface vegetation. The greatest displacement within the canopy was of hydrogen fluoride, whereas the least displacement was of nitric oxide. Ozone was primarily deposited at the canopy surfaces. Estimates of vegetal absorption of sulfur dioxide have varied from 15 to 75 x 10^9 kg of sulfur per year on a global basis (Rasmussen, Taheri, and Kabel 1975).

It has been proposed that growing plants may be a natural sink for atmospheric ammonia (Hutchinson, Millington, and Peters 1972; Porter, Viets, and Hutchinson 1972). Leaves of soybean, corn, cotton, and sunflower plants varied by species in their absorption rate for atmospheric ammonia, with large diurnal fluctuations in that rate. The nitrogen fertility level of plants within the same species caused no difference in ammonia absorption rate. Data extrapolations from experiments using single seedlings of soybean, sunflower, corn, and cotton indicated that the annual ammonia absorption by plant canopies could be 20 kg/ha (Hutchinson, Millington, and Peters 1972). Results of 24-hr exposures of corn seedlings to atmospheric ammonia showed that, at 1 ppm, 43% was absorbed, whereas at 10 and 20 ppm, 30% was absorbed (Porter, Viets, and Hutchinson 1972).

Hosker (1974) suggests that, where various effluent concentrations are at levels that would not injure the vegetation, a forested region might serve as an effective filter for particulate and gaseous emissions; Neuberger, Hosler, and Kocmond (1967), as cited in EPA (1975), note that conifers are better natural filters than deciduous trees. A computer model for the estimation of forest uptake of air pollutants has been described; using sulfur dioxide for illustrative purposes, the computer employs submodels to describe atmospheric diffusion immediately above and within a canopy and into the sink areas within or on the trees. Only minor changes may be required to use the model for any gaseous pollutant (Murphy, Sinclair, and Knoerr 1975). Estimates of the sulfur dioxide sink strength of forests derived by using the model agree with experimentally derived estimates of sulfur dioxide uptake in crops and forest trees.

8.2.4 Ozone

Ozone is one of the most widely distributed and important oxidants. Both hydrocarbons and nitrogen oxides have been implicated in the formation of ozone and PAN in smog, although it is not clear which of the two is more responsible (Benedict, Miller, and Olson 1971). However, Schuck et al. (1970) have shown a correlation between the hydrocarbon content of the atmosphere in the early morning (6 to 9 AM) and the maximum oxidant content later in the day.

8.2.4.1 Effects

It is possible that ozone causes more injury to vegetation than any other air pollutant in the United States (Taylor 1973). It displays phytotoxic reactions at distances of more than 100 miles from major ozone sources (Naegele 1974). According to Miller (1973), the forest community of the San Bernardino National Forest in California has suffered a change in composition as the result of chronic photochemical oxidant air pollution. The sensitive Ponderosa pine is being eliminated while the more resistant sugar pine and incense cedar are becoming dominant.

A marked varietal and species difference has been found in concentrations of ozone resulting in injury. Some plants are damaged by 5 to 12 pphm (parts per hundred million) for 2 to 4 hr, whereas others are resistant to 100 pphm (Hill, Heggestad, and Linzon 1970).

Even 2 pphm ozone, however, has produced leaf-fleck injury in tobacco in the presence of sulfur dioxide (Menser and Heggestad 1966). It has been shown that plants (duckweed, carnations, corn, petunias, marigolds, chrysanthemums, and turf grasses) exposed to 10 ppm ozone over a long period (to maturity) show a reduction in growth, fruit and seed set, leaf size, stem length, root weight, and flower production as well as delay in onset of floral initiation (Feder 1973). Selected species, varieties, and clones that are relatively sensitive to ozone are listed in Table 8.7.

Species of *Arabidopsis thaliana* (a member of the mustard family) grown under uniform conditions displayed consistently depressed growth and floral production when exposed to 20 pphm ozone over seven generations. Seeds from these plants produced normal plants when grown in a "clean" environment (Bruton and Anderson 1974).

Greenhouse experiments indicated that ozone was responsible for the leaf lesions and necrosis observed in five varieties of field-grown Kentucky bluegrass. In the experiment, injury scores based on the number of necrotic leaves were significantly greater when ozone was applied at 30 pphm for 4 hr than when applied at the same level for 2 hr. One cultivar demonstrated a higher tolerance for ozone, and two out of three cultivars showed a tendency to build up some tolerance to the repeated treatments (Wilton et al. 1972).

8.2.4.2 Mode of action

The extent of damage by ozone manifested as leaf-fleck injury in tobacco has been found to be related to both sugar content and stomatal width. Soluble sugar exerted an osmotic effect on stomatal movement, high levels being associated with closure. Independently of this effect, damage from ozone was prevented by increasing the sucrose content beyond the optimum for high

Table 8.7. Selected plants, certain varieties or clones of which are relatively sensitive to ozone[a]

Crops

Alfalfa
  Medicago sativo L.
Barley
  Hordeum vulgare L.
Bean
  Phaseolus vulgaris L.
Clover, red
  Trifolium pratense L.
Corn, sweet
  Zea mays L.
Grass, bent
  Agrostis palustris Huds.
Grass, brome
  Bromus inermis Leyss.

Grass, crab
  Digitaria sanguinalis L.
Grass, orchard
  Dactylis glomerata L.
Muskmelon
  Cucumis melo L.
Oat
  Avena sativa L.
Onion
  Allium cepa L.
Peanut
  Arachis hypogaea L.
Potato
  Salanum tuberosum L.

Radish
  Raphanus sativus L.
Rye
  Secale cereale L.
Spinach
  Spinecea oleracea L.
Tobacco
  Nicotiana tabacum L.
Tomato
  Lycopersicon esculentum Mill.
Wheat
  Triticum aestivum L.

Trees, shrubs, and ornamentals

Alder
  Alnus, sp.
Apple, crab
  Malus baccata Borkh.
Aspen, quaking
  Populus tremuloides Michx.
Boxelder
  Acer negundo L.
Bridalwreath
  Spiraea prunifolia Sieb.
    & Zucc.
Carnation
  Dianthus caryophyllus L.
Catalpa
  Catalpa speciosa Warder

Chrysanthemum
  Chrysanthemum sp.
Grape
  Vitis vinifera L.
Honeylocust
  Gleditsia triacanthos L.
Lilac
  Syringa vulgaris L.
Maple, silver
  Acer saccharinum L.
Oak, gambel
  Quercus gambelii
Petunia
  Petunia hybrida Vilm.

Pine, eastern white
  Pinus strobus L.
Pine, Ponderosa
  Pinus ponderosa Laws.
Privet
  Ligustrum vulgare L.
Snowberry
  Symphoricarpos albus Blake
Sycamore
  Platanus occidentalis L.
Weeping willow
  Salix babylonica L.

---

[a]Generally the crops listed are more sensitive than the trees and shrubs.

Source: Naegele 1974, Table 5.1, p. 88, from Jacobson and Hill 1970. Reprinted by permission of the publisher.

sensitivity (Lee 1965). Stomatal closure and stomatal development appear to play an important role in ozone plant responses (Rich 1964). Recently expanded leaf tissue is most susceptible to ozone (Taylor 1973).

Mudd (1973) reviews the biochemical information available regarding the effects of ozone on plants.

Much research has been conducted to clarify the mechanism of ozone injury. Results include effects on cell components (e.g., membranes and lipids) and cell processes (e.g., photosynthesis and protein metabolism) (Dugger 1974). Chlorophyll content and fresh weight of cucumbers and soybeans were reduced during early development by exposure to acute concentrations of ozone (50 pphm for 4 hr). These in vivo studies have shown that reduction of influx of amino acids into soluble pools occurred after as little as 15 min (Frick and Cherry 1974).

Color or quality and growth of leafy food and forage crops may be reduced by oxidant air pollutants, which are known to stimulate the synthesis of anthocyanin in curly dock, caffeic acid in beans, and total phenols in tobacco (Koukol and Dugger 1967; Howell 1970; and Menser and Chaplin 1969, as cited in Howell and Kremer 1973). These differences in chemical composition are associated with the appearance of pigments that result from oxidant injury and may result in a reduction in the nutritional value of the affected crops (Howell 1974). Caffeic acid is an o-diphenol and is readily oxidized to o-quinone. The total effect is that of enhancing the plant's oxidative processes. Howell (1974) demonstrated polymerization of quinones with protein components following phenol oxidation, which resulted from the release of oxidative enzymes after ozone had damaged interior membranes.

Cell wall metabolism was inhibited in oat coleoptiles by pretreatment with ozone (Ordin 1965). Soybean carbohydrate metabolism was modified by ozone, depressing the enzyme glyceraldehyde 3-phosphate dehydrogenase and activating glucose 6-phosphate dehydrogenase. Levels of ribonuclease, protease, acid phosphatase, and esterase were unaffected in a manner that might be expected if ozone enhanced senescence (Tingey, Fites, and Wickliff 1975).

According to Heath, Chimiklis, and Frederick (1974), ionic balances are basically responsible for the diverse responses to ozone. That is, by some reaction in the membrane, ozone appears to induce leakage of K^+ ions out of plant cells and abolish their uptake by inhibition of a specific device (a "K^+ pump") located on the plasmalemma. It is believed that changes in membrane fluidity may be caused by the interaction of ozone with both critical sulfhydral groups and fatty acid residues (Heath, Chimiklis, and Frederick 1974). The greater reduction in root growth (as opposed to that of shoots) that results from acute or chronic exposure to ozone has been found to be the result of certain metabolic alterations involving translocation and/or quality of photosynthate translocated to roots from soybean leaves and pine needles (Tingey 1974). Nodulation in legumes was reduced and root exudates from ozone-exposed plants inhibited root growth and nodulation in nonexposed plants (Tingey 1974).

Chloroplast polyribosomes have been shown to be very susceptible to degradation by ozone, more so than cytoplasmic polysomes (Chang 1971, as cited in Mudd 1973). In cotton, maximum susceptibility to ozone occurs when soluble amino acids are at a minimum (Ting and Mukerji 1971, as cited in Mudd 1973). An increase in soil moisture increases susceptibility to ozone, whereas moisture stress increases resistance of a plant, thus producing a combined effect (Rich 1964).

Radiomimetic aberrations were observed in root meristem cells of *Vicia fabia* seeds exposed to a relatively high concentration of ozone (0.4 wt % for 15 to 60 min) (Fetner 1958), and significant increases in mutation rate above spontaneous levels were observed in *Tradescantia* after exposure to ozone, sulfur dioxide, and nitrous oxide. Nevertheless, these pollutants are reportedly weak mutagens in these vascular species because the maximum response was only a little more than twice the spontaneous rate (Sparrow and Schairer 1974).

8.2.5 <u>Sulfur dioxide</u>

Acute damage to both agricultural and forest plants attributed to sulfur dioxide emissions has been extensively documented (Naegele 1974). A few of the many literature reviews available are Thomas and Hendricks 1956, Brandt and Heck 1968, Daines 1968, NAPCA 1970a, Jacobson and Hill 1970, and Mudd and Kozlowski 1975.

Obvious phytotoxic effects such as foliar injury occur in many sensitive plants when sulfur dioxide concentrations exceed about 0.5 ppm (1.31 mg/m^3) for 1 hr (Bennett et al. 1975), but chronic effects such as reduced growth or reproductive rates may occur at even lower concentrations over extended periods of time. According to Nash (1973), the mean annual local concentration of sulfur dioxide need only exceed 0.01 to 0.08 ppm (27 to 224 µg/m^3) for obvious injury to occur.

8.2.5.1 Acute effects

Grains such as barley, oats, wheat, and rye, which are particularly sensitive to sulfur dioxide damage, develop necrotic streaks between the veins near the leaf tip which extend toward the base as the severity of injury increases. On grasses and grains, injury usually is most severe at the bend of the leaf where the long blade curves downward (Taylor 1973). Some very resistant plants native to the U.S. desert regions (e.g., pinyon pine and juniper) can tolerate 2 ppm sulfur dioxide for 2 hr. Grama grass tolerated nine 2-hr fumigations of 6 ppm (Hill et al. 1974). Table 8.8 lists some plants that are relatively sensitive to sulfur dioxide.

Barley and corn leaves were injured by sulfur dioxide at 0.15 and 0.30 ppm (0.40 and 0.79 mg/m^3) for 7 days and less severely by 0.06 to 0.08 ppm (0.16 to 0.21 mg/m^3) of sulfur dioxide for 27 days. In both of the tested corn cultivars, injury consisted of elliptical lesions on the distal half of older leaves. Marginal and tip chlorosis occurred at the distal end of barley leaves. Beans were unaffected by these treatments (Mandl, Weinstein, and Keveny 1975).

Concentrations of sulfur dioxide from 0.3 to 1.4 ppm over 4 hr, when applied to moist pollen, caused reduction of both germination and tube elongation in several species of forest trees (Karnosky and Stairs 1974). Jensen and Kozlowski (1975) report that the relationship between species sensitivity and sulfur dioxide absorption rate changed with prefumigation treatment and that amounts of sulfur translocated from leaves to roots after eight days also varied with species. The rate of sulfur absorption was reduced in sugar maple, big tooth aspen, and yellow birch by 20 hr of prefumigation with 0.75 ppm sulfur dioxide. This did not occur in white ash, a species reported to be intermediate in sensitivity (Jensen and Kozlowski 1975).

Measurements were made on the basis of arithmetic mean values averaged over the entire experimental period for two types of open-air experiments involving spruce stands. One stand was characterized by high peak concentrations of sulfur dioxide (acute) with long emission-free intervals, whereas the other received exposure at low concentrations of sulfur dioxide for long periods of time (chronic). The results showed that the lower limits for injury lay in the same order of magnitude with both types of emission profile (Guderian 1973). Dochinger and Jensen (1975) report that foliar injury and shoot reduction in hybrid poplar clones occurred after exposure to 0.25 ppm of sulfur dioxide for six weeks; plant response was similar to that which occurred after exposures of 5 ppm of sulfur dioxide for 1-1/2, 3, and 6 hr.

8.2.5.2 Chronic effects

Chronic effects under field conditions are difficult to assess because of their subtle expression (Naegele 1974). Some investigators support the theory that injury has not occurred in the absence

Table 8.8. Selected plants which are relatively sensitive to sulfur dioxide

Crops

Alfalfa
  Medicago sativa L.
	Cotton
	  Gossypium sp. L.
		Soybean
		  Glycine max. Merr.

Barley
  Hordeum vulgare L.
	Oats
	  Avena sativa L.
		Wheat
		  Triticum sp.

Bean, field
  Phaseolus sp. L.
	Rye
	  Secale cereale L.	

Clover
  Melilotus & Trifolium sp.
	Safflower
	  Carthamus tinctorius L.	

Garden flowers

Aster
  Aster bigelovii
	Four o'clock
	  Mirabilis jalapa L.
		Verbena
		  Verbena canadensis Brit.

Bachelor's button
  Centarea cyanus L.
	Morning glory
	  Ipomoea purpurea Roth
		Violet
		  Viola sp.

Cosmos
  Cosmos bipinnatus Cau.
	Sweet pea
	  Lathyrus odoratus L.
		Zinnia
		  Zinnia elegans Lorenz

Trees

Apple
  Malus sp.
	Larch
	  Larix sp.
		Pine, Eastern white
		  Pinus strobus L.

Birch
  Betula sp.
	Mulberry
	  Morus microphylla Buckl.
		Pine, ponderosa
		  Pinus ponderosa Laws

Catalpa
  Catalpa speciosa Warder
	Pear
	  Pyrus communis L.
		Poplar, lombardy
		  Populus nigra L.

Elm, American *Ulmus americana* L.		

Garden plants

Bean
  Phaseolus vulgaris L.
	Lettuce
	  Lactuca sativa L.
		Spinach
		  Spinacea oleracea L.

Beet, table
  Beta vulgaris L.
	Okra
	  Hibiscus esculentus L.
		Squash
		  Cucurbita maxima Duchesne

Broccoli
  Brassica oleracea var.
    botrytis L.
	Pepper (bell, chili)
	  Capsicum frutescens L.
		Sweet potato
		  Ipomoea batatas Lam.

Brussel sprouts
  Brassica oleracea var.
    gemmifera L.
	Pumpkin
	  Cucurbita pepo L.
	Radish
	  Raphanus sativus L.
		Swiss chard
		  Beta vulgaris var. *cicla* L.
		Turnip
		  Brassica rapa L.

Carrot
  Daucus carota var. *sativa* L.
	Rhubard
	  Rheum rhaponticum L.	

Endive *Cichorium endivia* L.		

Weeds

Bindweed
  Convolvulus arvensis L.
	Fleabane
	  Erigeron canadensis L.
		Ragweed
		  Ambrosia artemisiifolia L.

Buckwheat
  Fagopyrum sagittatum Gilib.
	Lettuce, prickly
	  Lactuca scariola L.
		Sunflower
		  Helianthus sp.

Careless weed
  Amaranthus palmeri S. Wats
	Mallow
	  Malva parviflora
		Velvet-weed
		  Gaura parviflora Dougl.

Curly dock
  Rumex crispus L.
	Plantain
	  Plantago major L.	

Source: Naegele 1974, Table 5.2, p. 89, from Jacobson and Hill 1970. Reprinted by permission of the publisher.

of visible symptoms (Thomas 1961, Thomas, Hendricks, and Hill 1950, Katz 1949, and others, as cited in Daines 1968). Because of the absence of visible symptoms, it has been suggested that chronic indefinite exposure to 0.1 ppm (0.26 mg/m^3) or less of sulfur dioxide causes no damage in such sensitive plants as wheat and alfalfa when grown in soil culture (Daines 1968). However, research on crop plants has suggested that exposure to low levels of sulfur dioxide that are insufficient to cause visible injury may nonetheless produce reductions in growth and yield, reproductive alterations, and inhibition of photosynthesis (Feder 1973; Puckett, Nieboer, and Richardson 1973). Tingey and Reinert (1975) have reported that 0.05 ppm (0.13 mg/m^3) for 8 hr per day, 5 days per week was sufficient to cause significant reductions in tobacco leaf weight and in alfalfa foliage and root growth without producing visible symptoms of injury.

Bleasdale (1973) concluded that continuous treatment with air containing coal-smoke gases having from 0.01 to 0.2 ppm sulfur dioxide could markedly reduce the growth of ryegrass and cause earlier senescence without producing visible injury.

Linzon (1966), as cited in Linzon (1971), found a direct correlation between growth of white pines and ambient sulfur dioxide levels. Linzon (1971) estimated an income loss of $117,000 from 1953 to 1963 to owners of wood or producers of wood products in Sudbury, Ontario, from intense smelter-produced sulfur dioxide fumigations. Average daily concentrations of sulfur dioxide at 6 to 10 pphm for a 4-hr daily period were toxic to white pine needles at their most sensitive stage (Costonis 1971), producing collapse of needle tissue and finally necrosis of needle tips. Over a number of weeks of exposure, chronic symptoms, including chlorosis of both current and older foliage and premature casting of older foliage, were noted. Fluctuations in levels of ozone, which did not exceed 4 pphm, did not correlate with injury. Earlier senescence of leaves of the sessile oak in the vicinity of a smokeless fuel plant in South Wales was observed along with a marked seasonal trend in sulfur (presumably SO$_4$) levels of the leaves (Williams, Lloyd, and Ricks 1971).

8.2.5.3 Effect on growth

In considering the phenomenon of increased growth observed in plants exposed to sulfur dioxide only intermittently, Bleasdale (1973) postulated that the accelerated growth (but ultimately slightly less yield) was associated with a change in balance of partially oxidized and reduced sulfur radicals known to control cell division. Scrubbing the air improved growth of ryegrass in soils of low fertility to an extent comparable with the improvement observed when fertilizer was added to plants growing in polluted air. Bell and Clough (1973) reported depression of yield in ryegrass exposed to 0.067 ppm (191 μg/m^3) of sulfur dioxide for 182 days and 0.12 ppm (343 μg/m^3) sulfur dioxide for 63 days.

Cowling, Jones, and Lockyer (1973) report increased yields in ryegrass grown on sulfur-deficient soil for 59 days at low levels of sulfur dioxide (131 μg/m^3). On the basis of this study and others, Grennard and Ross (1974) contend that sulfur dioxide emissions from large power plants pose no problem to the health of plants, animals, or humans and therefore should not be burdened with emission restrictions. The apparent contradiction between reports of increased yield and those of decreased yield is explained by Pahlich (1975), who demonstrated that the ratio of the inorganic to organic fraction of sulfur dioxide derivatives determines whether there is a nutritionally beneficial or a damaging effect on plants. Fertilizing effects will emerge only when there is an appreciable sulfur sink. Accumulation of inorganic sulfur dioxide derivatives will develop when this sink is exceeded, resulting in damage because of uncontrolled sulfur dioxide uptake. Pahlich generalizes further to suggest that any gaseous contaminant that can be transferred to organic metabolism (sulfur dioxide, nitrogen oxides) or another sink (vacuoles) will not be as harmful as those which mainly exist as ions (hydrogen fluoride, hydrogen chloride). Varying modes of transport and deposition of accumulated ions, rather than adaptation mechanisms, may be responsible for specific responses to pollution.

8.2.5.4 Metabolism

Sulfur dioxide is rapidly absorbed by plants. One hypothesis for the mechanism of toxicity of sulfur dioxide is that the sulfite ion, formed with water in leaf mesophyll tissue after penetration, irreversibly oxidizes the chlorophyll, leading to its destruction. Chlorosis, due to chlorophyll destruction, is one symptom of sulfur dioxide damage. Also, photosynthesis can be inhibited in plants treated with sulfur dioxide (Puckett, Nieboer, and Richardson 1973). Taylor (1973) states that the sulfite ion is highly phytotoxic and is subsequently converted to the less toxic sulfate ion in leaf tissue. Cell sap buffering capacity in certain species may provide protection (Mudd 1973).

Toxicity of sulfur dioxide sometimes occurs as a result of its acidifying effects within a plant, including the formation of both sulfurous and sulfuric acids. Resistance to toxicity has been related to the plant's ability to convert sulfite to sulfate (Mudd 1973). Only a small portion of absorbed sulfur dioxide appears to remain in leaves as inorganic sulfite and sulfate ions, whereas the bulk is expelled by some plants via their root system (Feder 1973). Trees and shrubs resistant to the action of gaseous compounds of sulfur, phenols, and coal dust from a coke-oven gas factory were found to have a higher content of free and bound amino acids than did the controls (Kozyukina and Obraztsova 1973).

Susceptibility of mosses and lichens is greatest at low pH (3.2), and no damage occurs at pH 6.6 from H_2SO_3 (Mudd 1973). A threshold level for a plant's ability to regulate its internal water relations may follow from the effect of sulfur dioxide on transpiration (loss of water in the form of vapor from plant tissues) (Tamm and Aronsson 1972, as cited in Auerbach 1975). Unsworth, Biscoe, and Pinckney (1972) found that, in many crop plants in Britain, the percentage loss in stomatal resistance was approximately constant for sulfur dioxide concentrations from 5 to 50 pphm and reached a minimum value that was independent of sulfur dioxide concentration and that did not represent full stomatal opening. The slow stomatal closure that occurred when injurious concentrations (from 0.25 to 2 ppm) of sulfur dioxide were applied to geranium leaves was assumed to be a defense mechanism. The response was immediate and was preceded for a short time by an increase in opening of the stomata (Bonte and de Cornis 1973).

The injurious effects of sulfur dioxide on experimental plants (including reduced yield of bean tops, barley, and tobacco; reduced flowering in barley; retarded generative development of beans and barley; reduced yearly height increases of oak and pine shoots; and reduced leaf area in oak) were neutralized by treatment with ammonia and with the products of the reaction between sulfur dioxide and ammonia. An increase in dry weight of bean tops and in annual oak seedling growth occurred. In those plants treated only with the products of the sulfur dioxide—ammonia reaction, sulfate sulfur content was lower than after sulfur dioxide treatment, but not as low as that of the controls (Markowski, Frzesiak, and Pienkowski 1974).

8.2.5.5 <u>Mutagenic effect</u>

Sparrow and Schairer (1974) state that sulfur dioxide is a weak mutagen although it was found to enhance chromatid aberration rates in *Tradescantia* in field and in hot-house plants when pollen tube cultures were subjected to the gas 7 to 10 min after sowing the pollen. The authors suggest that the action could probably be attributed to sulfur trioxide in the cell under the experimental conditions. An average of 46 chromosome breaks occurred per 100 cells in pollen tube cultures of *Tradescantia paludosa* following treatment with a total of about 0.075 ppm sulfur dioxide over 18 hr. Aberration rates changed with season and failed to occur at high rates in cultures from GRO-LUX-lamp-preconditioned (7 to 20 days) plants (Ma et al. 1973).

It is possible that the reaction of bisulfite with aldehydes and/or the combining of sulfite with quinones and α,β-unsaturated compounds and reactions of bisulfites with pyrimidines may contribute to toxicity. Mutagenic effects of the latter occurred in *E. coli* and phage lambda at 1 *M* or higher concentrations which are probably not reached at ambient sulfur dioxide pollutant levels (Mudd 1973).

8.2.6 <u>Acid rain</u>

Acid rain, produced when sulfur dioxide is oxidized to H_2SO_4 in the presence of atmospheric moisture, is one of the more potentially serious consequences of sulfur dioxide pollution. However, internal stresses produced by sulfur dioxide differ from external plant damage resulting from acid rain. Acid rain may fall at distances of up to 1000 km from the pollutant source (Rodhe, Persson, and Akesson 1972). Acid rain is defined as any precipitation having a pH of less than 5.6, which is the normal pH of rain at prevailing carbon dioxide pressures (Likens, personal communication 1976).

8.2.6.1 <u>Contribution of sulfuric acid</u>

The increase in acidity of rain is partly a result of the oxidation of sulfur dioxide to sulfur trioxide plus water, which forms sulfuric acid, but two additional strong proton sources, nitric acid and hydrogen chloride, are contributing factors. Only negligible amounts of free hydrogen ions come from H_2CO_3, RCOOH, clay NH_4^+, Al^{3+}, and $Fe(OH)_2$ (Likens, personal communication 1976). According to Ruess (1975), the contribution to soil acidity from the oxidation of the NH_4^+ in rainfall is of a similar magnitude to direct H^+ inputs common in rainfall.

Since 1956, the hydrogen ion concentration of rain in some parts of Scandinavia has increased more than 200-fold. Indirect evidence reveals that the pH of precipitation over the northeastern United States decreased about 20 years ago to a current average annual pH of 4 with a range for individual storms of 2.1 to 5. This drop may have been associated with the increase in combustion of natural gas and use of motor fuels that occurred at that time. A decrease in forest productivity over the past 20 years in northern New England and in Scandinavia is believed to be correlated with the concurrent acidification of precipitation (Likens and Bormann 1974).

8.2.6.2 Contribution of nitric acid

Although sulfuric acid is the largest contributor to acidity, the relative contribution of nitric acid has doubled in the northeastern United States over the past 10 years and has in some areas increased four- to sixfold. A 36% increase in the input of hydrogen ions over the past two decades in a deciduous forest watershed at White Mountain, New Hampshire (the Hubbard Brook Ecosystem) is almost exclusively due to an increase in nitric acid content (Likens, personal communication 1976); the increase correlates with the increase in combustion of natural gas and use of motor fuels.

8.2.6.3 Effects

A decline in productivity concomitant with increases in acidity of precipitation in a New Hampshire watershed has been observed over the past 20 years. However, a clear cause-and-effect relationship has not been demonstrated (Whittaker et al. 1974).

Summer rain has been found to be much more acid than snow (Likens and Bormann 1974); therefore, acid rain affects vegetation at a time of heightened susceptibility due to maximum leaf area and actively growing leaf tissue. Retention by a forest canopy (as at Hubbard Brook) of 90% of the hydrogen ions results in rain of pH 4.0 being raised to about 5.6 by the time it contacts soil (Eaton, Likens, and Bormann 1973).

The problem of nitric acid deposition is complex. The deleterious effects of hydrogen ions and the possible beneficial effects of increased nitrates on plant nutrition are not yet resolved. Likens (personal communication 1976) has stated that the impact of an increase in atmospheric sulfur load on acid precipitation from increased combustion of coal and other sources is impossible to predict. Ruess (1975), however, states that the capacity of plants to use additional sulfur is often closely related to the nitrogen cycle. In the natural sulfur cycle and during soil nitrogen transformations, H^+ ions are continually being produced and consumed by various biological processes. The flux involved in the nitrogen cycle is larger than that of both the sulfur cycle and the input from acid rainfall. The oxidation of the ammonium ion in rainfall may also contribute to the input of H^+ ions. For this reason, Reuss (1975) states that there is definitely a possibility that anthropogenic sulfur inputs might result in accelerated growth in some forest ecosystems where nitrogen fixation is rapid (e.g., in alder stands). In addition, since the OH^- given off by plants in $SO_4{}^{2-}$ uptake may exert a balancing effect on sulfuric acid in rain, an increase in soil acidity to a detrimental level may not occur. In systems such as mature timber stands, where the sulfur is less likely to be limited because the nitrogen is supplied largely from internal cycling, an increased sulfur load over the system's capacity to absorb it may result in deleterious effects from acid rain. Ruess (1975) concludes that the acidity of rainfall should not be considered as an isolated factor, but that its close association with other ecosystem processes, particularly those of the nitrogen cycle, must be taken into account.

Calcium deficiency

Overrein (1972) demonstrated that in Norway the essential plant nutrient, calcium, can be leached from the soil by acid precipitation.

Leaf and plant injury

Injury from sulfuric acid has been shown to be aggravated by conditions causing leaf surfaces to be wet (i.e., fog point). Manifestation of damage is related to leaf wettability and may result in spotty or diffuse necrotic areas. Of species examined in the Los Angeles area, alfalfa, swiss chard, and beets responded most typically with these symptoms (NAPCA 1970).

Simulated acid rain (pH 1.5 to 3.5 in 0.5-unit increments), when applied to *Chenopodium quinoa* (goosefoot), *Hordeum vulgare* (barley), and *Phaseolus vulgaris* (bean), resulted in (1) failure of the plants to attain normal height, (2) necrosis and wrinkling of leaves, (3) excessive adventitious budding, and (4) premature abscission of primary leaves (Ferenbaugh 1974). At pH values below 3.0, microscopic changes observed included decreases in (1) leaf cell size, (2) intercellular space, and (3) size of starch granules within the chloroplasts. Although respiration rates increased only slightly at low pH values (2.5), apparent rates of photosynthesis increased dramatically. Carbohydrate production, root biomass, and pH of leaf cell contents were all reduced by treatment at pH 2.5. Congo red stains indicated that the leaf cell contents were reduced below pH 4.0 (Ferenbaugh 1976). The results appeared to be an acid effect since there was no indication that penetration of sulfate into the intact leaf tissue affected photosynthesis and since immersion of control leaf tissue in HCL caused a comparable increase in apparent rate of photosynthesis. A lack of sensitivity was observed in *Chenopodium quinoa* and *Hordeum vulgare* when treated similarly, indicating that plant species are differentially susceptible to the effects of acid rain. This implies that acid rain could cause changes in the species composition of plant communities, thereby altering natural food chains (Ferenbaugh 1976). The observed reduction in carbohydrate-producing capacity shows that acid rain can decrease plant productivity.

Shriner (1976) demonstrated partial destruction of surface waxes on leaves of willow oak and kidney bean after 30 to 60 days of treatment with simulated sulfuric acid rain at pH 3.2. The cutin of the kidney bean also appeared pitted. These effects could alter water repellency of the leaves, ion exchange properties, and defense mechanisms against drought, biotic pathogens, and pollutants. The sensitivity of some vascular plants (European species of tulip, turnip, dandelion) to gaseous and particulate pollutants has been observed to be related to the water-repellent characteristics of the foliage derived from the amount and position of waxy deposits (Rentschler 1973).

Shriner (1976) found that kidney beans developed halo-blight disease to a greater extent after exposure to simulated acid rain. Kidney bean nodule formation and nitrogen fixation were inhibited. Yield reductions occurred only when plants were grown in poorly buffered, unfertilized soils.

Acid rain was thought to have contributed to loss of nutrients (Ca_2^+) from tobacco leaves by formation of insoluble $CaSO_4$ (Fairfax and Lepp 1975). Decreases in growth in western pine needles inoculated with sulfuric acid at a pH of less than 4.0 have been observed (Gordon 1972, as cited in Likens 1974) as well as spot necrosis and irregular development of yellow birch leaf tissue misted with aqueous sulfuric acid at pH 3.0 (Wood and Bormann 1974). Decreased pollen germination and pollen tube growth with lower quality and production in tomato plants have also been observed (Kratky, personal communication, as cited in Likens and Bormann 1974).

8.2.6.4 Metabolism

Ferenbaugh (1974) simulated acid rain by spraying bean plants (*Phaseolus vulgaris*) with sulfuric acid solutions covering a pH range of 1.5 to 3.5. He found that the photosynthetic rate increased dramatically at low pH values, whereas respiration rates increased only slightly. Histological changes such as smaller cell size, decreased intercellular space, hypertrophied nuclei and nucleoli, and reduced size of starch granules in chloroplasts, as well as decreased root biomass were observed. These changes were thought to be due to uncoupling of photophosphorylation of adenosine diphosphate caused by hydrogen ion interference with the proton pumps associated with the electron transport chain in the light reactions of photosynthesis (Ferenbaugh 1974).

8.2.7 Nitrogen oxides

Nitrogen oxides (NO_x) are formed by numerous processes: (1) during high-temperature combustion, (2) from nitrogen in the air (Schuck 1973), (3) by soil microbial production (Rasmussen, Taheri, and Kabel 1975), (4) by the gas phase oxidation of atmospheric ammonia by OH (McConnell 1973, as cited in Rasmussen, Taheri, and Kabel 1975), and (5) by the photolysis of nitrous oxide in the stratosphere (Rasmussen, Taheri, and Kabel 1975). According to Benedict, Miller, and Olson (1971), the automobile accounts for 38% of atmospheric NO_x and their resulting estimate of $30 million crop damage. In downtown city areas, the contribution of the automobile to the atmospheric NO_x may be more than 90% (Schuck 1973). Robinson and Robbins (1970), as cited in Rasmussen, Taheri, and Kabel (1975), however, contend that natural emissions of nitrogen dioxide and nitric oxide are 15 times as high as emissions of anthropogenic origin.

8.2.7.1 Atmospheric transformations

A rapid conversion of nitric oxide to nitrogen dioxide is caused by oxidation in the atmosphere, which is followed by a photolytic reconversion to nitric oxide and oxygen atoms, giving rise to ozone, which then reacts with nitric oxide to reform nitrogen dioxide. Because of high reactivity, nitrogen oxides have relatively short residence times, probably about five days (Hidy 1973, as cited in Rasmussen, Taheri, and Kabel 1975). Because certain hydrocarbons at low reaction rates compete with oxygen for the O and with nitrous oxide for the ozone, the net result is an unbalancing of the nitrogen dioxide photolytic cycle and a steady rise in nitrogen dioxide as well as ozone, PAN, certain hydrocarbon oxidation products, and acid aerosols (Schuck 1973). Thus, the importance of nitric oxide and nitrogen dioxide arises not only from their presence, but also from their participation in photochemical reactions in the atmosphere to form other pollutants (Rasmussen, Taheri, and Kabel 1975).

8.2.7.2 Effects

Nitrogen dioxide can be phytotoxic at high levels, but exposure, dose, and response information is sparse (Naegele 1974). Likewise, very little information is available on the phytotoxicity of nitric oxide, but it appears to be less toxic than nitrogen dioxide. Higher concentrations of nitric oxide than nitrogen dioxide were required to cause a given inhibition in oats and alfalfa, although nitric oxide was more rapid in causing a response (Hill and Bennett 1970). Because atmospheric concentrations of nitric oxide or of nitrogen dioxide sufficient to cause acute plant damage rarely occur (Taylor 1973), chronic exposures may be more important than acute exposures. Taylor et al. (1975) assert that, as a single pollutant, nitrogen dioxide at many parts per million will not injure plants if exposures are for a few minutes; however, reduced performance may be expected at continuous levels of 5 to 10 pphm. This effect will be difficult to detect without careful comparison. The acute injuries which develop in many kinds of plants exposed to nitrogen dioxide resemble closely those produced by sulfur dioxide (Middleton, Darley, and Brewer 1958; NAPCA 1971).

Thompson et al. (1970) state that current levels of atmospheric nitrogen dioxide in a test area of California were causing minimal or no economic damage to vegetation, although concentrations of 0.5 and 1.0 ppm nitrogen dioxide for 35 days caused severe defoliation and chlorosis of navel orange trees. Increased leaf drop and reduced fruit yield occurred at 0.25 ppm. Random sampling of the ambient atmosphere gave levels of up to 0.24 ppm with average daytime values of 0.04 ppm.

Research with pinto beans, cucumbers, and tomatoes suggested that a 2-hr exposure to 6 ppm nitrogen dioxide will produce injury (Taylor and MacLean 1970), whereas others reportedly withstand 1 hr of 1000 ppm (Czech and Nothdurft 1952). In oats and alfalfa not displaying visible injury, both nitric oxide and nitrogen dioxide at levels below 0.6 ppm for 45 to 90 min inhibited apparent photosynthesis (Hill and Bennett 1970). Visible injury was observed in citrus plants for nitrogen dioxide doses greater than 250 ppm for 1 hr and in some ornamentals exposed to 50 ppm for 8 hr and to 150 ppm nitrogen dioxide for 4 hr. Growth suppression and leaf distortion occurred in pinto bean and tomato plants (without formation of lesions) and were caused by chronic exposures to 0.5 ppm or less of nitrogen dioxide for several weeks (Taylor and Eaton 1966; MacLean et al. 1968). The relative sensitivity of some selected plants to nitrogen dioxide is listed in Table 8.9.

8.2.7.3 Uptake and absorption

Plants readily absorb nitrogen dioxide because it is highly soluble in water. Alfalfa and oats absorbed nitrogen dioxide from the air in excess of 100×10^{-12} moles m^{-3} sec^{-1} when exposed to an atmosphere containing 460 $\mu g/m^3$ nitrogen dioxide (Tingey 1968, as cited in Rasmussen, Taheri, and Kabel 1975). Uptake of nitrogen dioxide in beans was found to be enhanced by high temperature, low carbon dioxide concentration, and high humidity (Srivastava, Jolliffe, and Runeckles 1975*a*).

8.2.7.4 Modes of toxicity

Mudd (1973) reviews some of the possible modes of toxicity for oxides of nitrogen: pH decrease in extracellular fluid, deamination reactions with amino acids and nucleic acid bases, reactions with double bonds, reactions with unsaturated compounds to form free radicals, formation of nitrosamines from reactions of nitrite with secondary amines, and diversion of the reductants required for the metabolism of nitrite and nitrate. Carbon dioxide fixation in algae and in higher plants has been observed to be inhibited by nitrogen oxides and nitrite (Mudd, Leavitt, and Kersey 1966, as cited in Mudd 1973; Hill and Bennett 1970). Experiments by Srivastava, Jolliffe, and Runeckles (1975*b*) suggest that nitrogen dioxide may cause general detrimental changes in leaf cell physiology. They found that nitrogen dioxide had only a slight effect on transpiration rate and stomatal diffusion resistance.

Table 8.9. Sensitivity of selected plants to nitrogen dioxide

Sensitive

Azalea	Hibiscus	Sunflower
Rhododendron sp.	*Hibiscus rosasinensis*	*Helianthus annuus* L.
Bean, pinto	Lettuce (head)	Tobacco
Phaseolus vulgaris L.	*Lactuca sativa* L.	*Nicotiana glutinosa* L.
Brittlewood	Mustard	
Melaleuca leucadendra	*Brassica* sp. L.	

Intermediate

Cheesewood	Dandelion	Orange
Malva parviflora L.	*Taraxacum officinal* Weber	*Citrus sinensis* Osbeck
Chickweed	Grass, annual blue	Rye
Stellaria media Cyrill	*Poa annua* L.	*Secale cereale* L.

Resistant

Asparagus	Grass, Kentucky blue	Nettle-leaf goosefoot
Asparagus officinalis L.	*Poa pratensis* L.	*Chenopodium* sp.
Bean, bush	Heath	Pigweed
Phaseolus vulgaris L.	*Erica* sp.	*Chenopodium* sp.
Carissa	Ixora	
Carissa carandas	*Ixora* sp.	
Croton	Lamb's-quarters	
Codiaeum sp.	*Chenopodium album* L.	

Source: Naegele 1974, Table 5.6, p. 93, from Jacobson and Hill 1970. Reprinted by permission of the publisher.

8.2.8 Peroxyacetyl nitrates (PAN)

Photochemical reactions of nitric oxide and nitrogen dioxide with unsaturated hydrocarbons can produce phytotoxic peroxyacetyl nitrate (PAN) ($CH_3COO_2NO_2$). Other homologs such as peroxypropionyl nitrate (PPN) and peroxybutynyl nitrate (PBN) are even more phytotoxic. Ethylene, olefins, and aromatic compounds are precursors of PAN. The toxicity of PAN increases as the alkyl chain lengthens (Taylor 1969). After exposure, PAN residues are not detectable in plant tissue; nor is it possible to detect the somewhat longer-lived oxides of nitrogen except through their visible effects. Nitrite is much more toxic than nitrate (Mudd 1973).

8.2.8.1 Effects

Severe injuries have been demonstrated in plants such as petunia and potato by a 4-hr exposure to about 14 ppb PAN (Taylor 1969). It attacks the spongy mesophyll tissue surrounding the substomatal chambers on the underside of the leaf (Taylor 1973). Typically, the leaves assume an initial glazed appearance followed by bronzing in only three or four of the rapidly expanding leaves (Taylor and MacLean 1970). Later, as the upper part of the leaf continues to grow, it cups downward and becomes wrinkled and distorted.

Acute injury is rarely observed on woody shrubs and trees, although succulent ornamentals, grasses, vegetables, and weeds may exhibit toxic symptoms when exposed for 2 hr to as little as 10 to 20 ppb (Taylor 1973). Table 8.10 lists the relative sensitivities of some selected plants to PAN.

Table 8.10. Sensitivity of selected plants to PAN

Sensitive

Bean, pinto
 Phaseolus vulgaris L.
Chard, Swiss
 Beta chilensis Hort.
Chickweed
 Stellaria media Cyrill
Dahlia
 Dahlia sp.

Grass, annual blue
 Poa annua Linn.
Lettuce
 Lactuca sativa L.
Mustard
 Brassica juncea Coss.
Nettle, little-leaf
 Urtica ureans L.

Oat
 Avena sativa L.
Petunia
 Petunia hybrida Vilm.
Tomato
 Lycopersicon esculentum Mill.

Intermediate

Alfalfa
 Medicago sativa L.
Barley
 Hordeum vulgare L.
Beet, sugar
 Beta vulgaris L.
Beet, table
 Beta vulgaris L.

Carrot
 Daucus carota L.
Cheeseweed
 Malva parviflora L.
Dock, sour
 Rumex crispus L.
Lamb's-quarter
 Chenopodium album L.

Soybean
 Glycine max Merr.
Spinach
 Spincca oleracea L.
Tobacco
 Nicotiana tabacum L.
Wheat
 Triticum sativum Lam.

Resistant

Azalea
 Rhododendron sp.
Bean, lima
 Phaseolus limensis L.
Begonia
 Begonia sp.
Broccoli
 Brassica oleracea L.
Chrysanthemum
 Chrysanthemum sp.

Corn
 Zea mays L.
Cotton
 Gossypium hirsutum L.
Cucumber
 Cucumis sativus L.
Onion
 Allium cepa L.

Periwinkle
 Vinca sp.
Radish
 Raphanus sativus L.
Sorghum
 Sorghum vulgare Pers.
Touch-me-not
 Impatiens sp.

Source: Naegele 1974, Table 5.7, p. 94, from Jacobson and Hill 1970. Reprinted by permission of the publisher.

8.2.8.2 <u>Metabolism</u>

PAN has been shown to react with the sulfhydryl groups of biological molecules such as glutathione. Both oxidating and acetylating reactions have been observed (Mudd 1973). Indoleacetic
acid (IAA), a plant growth hormone, is oxidized by PAN in vitro, but tryptophan is resistant
(Ordin and Propst 1962, as cited in Mudd 1973). It is widely believed that IAA is formed from
tryptophan by transamination (Thimann 1972).

<u>In vitro studies</u>

Some in vitro biochemical effects of PAN are listed in Table 8.11. Oxidation of IAA by PAN in
vitro produced some substances that inhibited the growth of oat coleoptiles induced by
2,4-dichlorophenoxyacetic acid (a synthetic growth hormone) (Hall, Brown, and Ordin 1971). PAN
also may react with ethylenic double bonds to form epoxides, which may affect membrane permeability
(Darnall and Pitts 1970, as cited in Mudd 1973). Under conditions of low pH, PAN has been
observed to react with bacterial nucleic acid bases (Peak and Belser 1969). Peak and Belser
observed that PAN caused a reduction in genetic transforming activity, melting temperature,
and viscosity of the DNA. Nucleic acids were susceptible to PAN attack in the following order:
thymine > guanine > uracil > cytosine > adenosine. The infectivity of bacteriophage was reduced.
The fact that high concentrations of PAN (about 1000 ppm) were used in these experiments and
that no reactions were detectable above pH 5.0 may diminish the biological importance of the
results (Mudd 1975).

<u>In vivo studies</u>

Biochemical effects of PAN in vivo are listed in Table 8.12.

In vivo, PAN lowers the sulfhydryl content of bean leaves and inhibits enzymes of cellulose
synthesis (Dugger and Ting 1968). Ultrastructural damage first occurs in the stroma of the
chloroplast (Thompson, Dugger, and Palmer 1965). At 1 ppm for 30 min, PAN has been shown to
lower the sulfhydryl content and to damage chloroplasts in bean leaves, whereas visible lesions
occurred at 0.014 ppm in 4 hr (Dugger and Ting 1968). Similar damage occurs to the alga
Chlamydomonas reinhardii at 125 ppm in 1 to 10 min (Gross and Dugger 1969). PAN-induced injury
is dependent on the physiological age of the leaf (Dugger et al. 1962). In isolated spinach
chloroplasts, PAN inhibited the electron flow in both photosynthetic systems (Photosystem I:
reduced indolphenol dye → NADP; Photosystem II: H_2O → indolphenol dye) without uncoupling
electron transport. The slow decline in Hill reaction rate, which continued after gassing (due
to the stability of PAN in solution), was eliminated but not reversed by dithiothreitol. In
aged chloroplasts, PAN inhibition was enhanced by light (Coulson and Heath 1975). Taylor et al.
(1961) found that light is required before, during, and after exposure to PAN to produce plant
damage.

8.2.9 <u>Fluorides</u>

Both the quality and quantity of an agricultural crop can be adversely affected by fluoride
pollution. More importantly, animals grazing on herbage containing high levels of fluorides
may suffer fluorosis (Less et al. 1975). Leafy food plants such as lettuce and cabbage have
been enriched by large amounts of fluorine without significant effects on yields and without

Table 8.11. Biochemical effects of peroxyacetyl nitrate in vitro

Effect	System	PAN Concentration (ppm)	Comment
Oxidation of recuced pyridine nucleotides	Pure compound in buffered solution	100 (1-5 min)	Oxidized to biologically active form
Reaction with isocitrate dehydrogenase, glucose 6-phosphate dehydrogenase, malate dehydrogenase	Pure enzyme in buffered solution	100 (1-5 min)	Enzymes can be protected by substrates and cofactors
Reaction with GSH	Pure compound in buffered solution	100 (1-5 min)	Products: disulfide and S-acetyl compound
Reaction with hemoglobin	Pure compound in buffered solution	100 (1-5 min)	No reaction with ovalbumin or RNase
Reaction with CoASH	Pure compound in buffered solution	100 (1-5 min)	Products: disulfide and higher oxidation states, but no S-acetyl. Products analogous to those obtained with H_2O_2
Oxidation of DNA, pyrimidines, and purines	Pure compound in buffered solution	1000-2000 (0-90 min)	No reaction above pH 5
Inhibition of polysaccharide synthesis	Particulate enzyme system from *Avena*	430 (3 min)	Inactivation also by IAA and *p*-hydroxymercuribenzoate
Oxidation of indoleacetic acid	Assayed by change in UV spectrum	1.3-2.6 (6 hr)	No effect on tryptophan
Inhibition of cellulose synthetase, phospho-glucomutase, UDPG pyrophosphorylase	Enzymes from *Avena* coleoptiles	100-400 (2-6 min)	UDPG pyrophosphorylase not affected in vivo

Source: Mudd 1973, Table VIII, pp. 38-39. Reprinted with permission from *Advances in Chemistry Series No. 122.*

Table 8.12. Biochemical effects of peroxyacetyl nitrate in vivo

Effect	System	PAN Concentration (ppm)	Comment
Inhibition of polysaccharide synthesis	*Avena* coleoptiles	35-50 (4 hr)	
Inhibition of phosphoglucomutase	*Avena* coleoptiles treated with PAN and enzymes assayed in subcellular fractions	35-50 (4 hr; pH 4.8)	
Chloroplast damage	Pinto beans (*Phaseolus vulgaris*)	1 (30 min)	First damage visible in EM in stroma granulation
Lesions on plants	Bean, petunia	0.014 (4 hr)	PPN and PBN[a] more toxic than PAN
Decrease in sulfhydryl content	*Phaseolus vulgaris*	1 (30 min)	Darkness lowers sulfhydryl content of control leaves
Decrease in sulfhydryl content, decrease in chlorophyll	*Chlamydomonas reinhardtii*	125 (1-10 min)	Chlorophyll *a* is more susceptible than chlorophyll *b*

[a]PPN, peroxypropionyl nitrate; PBN, peroxybutyryl nitrate.

Source: Mudd 1973, Table IX, pp. 40-41. Reprinted with permission from *Advances in Chemistry Series No. 122.*

visible injury (Guderian 1973). Ingestion of fluorine in the United States has reportedly risen by an estimated factor of 4 to 5 during the last 20 years, but is still considered a minor hazard (Guderian 1973).

8.2.9.1 Effects

Plants incorporate fluorine either from the atmosphere, generally as hydrogen fluoride, or from solution as an inorganic fluoride such as sodium fluoride. Plant damage is attributed primarily to hydrogen fluoride. Because of differences in plant susceptibility, there can be no single criterion for the effects of atmospheric fluorine on plants (McCune 1969).

Acute injury

Concentrations of hydrogen fluoride less than 0.1 ppb cause acute injury in sensitive species such as gladiolus and Chinese apricot, but most species can withstand several times that dose (Naegele 1974). Plants are regarded as sensitive when injured by a continuous exposure to 5 ppb or less fluoride for 7 to 9 days (Thomas and Hendricks 1956).

Hydrogen fluoride can penetrate open leaf stomata; it dissolves in the internal aqueous solution in contact with leaf tissues and causes acid-type burns. The cell contents leak through the damaged cell membrane, and after drying, leaf areas turn brown; necrotic tissue may drop away from the remainder of the leaf, leaving a ragged hole, but the leaf seldom separates from the plant (Taylor 1973). Rapidly growing needles of pine and fir are especially susceptible to fluoride. Necrosis starts at the tip of the needle and progresses toward the base as fluoride accumulates; the injured tissue becomes chlorotic, turns reddish brown, and dies.

Chronic effects

Fluoride injury to a plant generally occurs from gradual accumulation to a threshold concentration, but sensitivity differences between varieties and species is very marked (Feder 1973). Experimental results have suggested that development of injury is influenced by the capability of plants to recuperate between exposures (Treshow 1971).

Pasture grasses and small grains may show no injury even at leaf concentrations of several hundred parts per million of fluoride (Feder 1973). Buckwheat may absorb up to 900 ppm from soils treated with fluoride salts without exhibiting damage (Hurd-Karrer 1950, as cited in Bohn 1972). Except for gladiolus, flowers seem to be generally resistant to fluoride injury, whereas some fruits, such as peaches, are sensitive (Naegele 1974). Table 8.13 lists some plants and their level of sensitivity to fluorides.

After continuous exposure to 0.58 $\mu g/m^3$ (a level of hydrogen fluoride that may realistically be expected to occur in the vicinity of some industrial sources), tender green bean plants produced seeds of normal vigor (Pack 1971).

When the cumulative load of fluoride in a plant reaches a threshold concentration, toxic symptoms may appear, followed by death of the plant. In plants such as corn and sorghum, chlorosis without marginal necrosis of the leaf develops. Initial discoloration is marginal, starting at the leaf tip, but increasing in width, length, and intensity as time multiplied by concentration increases (Compton 1970, as cited in Feder 1973).

Table 8.13. Sensitivity of selected plants to fluorides

Sensitive

Apricot, Chinese & royal
  Prunus armeniaca L.
Boxelder
  Acer negundo L.
Blueberry
  Vaccinium sp.
Corn, sweet
  Zea mays L.
Fir, Douglas
  Pseudotsuga taxifolia Brit.
Gladiolus
  Gladiolus sp.

Grape, European
  Vitus vinifera L.
Grape, Oregon
  Mahonia repens Don.
Larch, western
  Larix occidentalis Nutt.
Peach (fruit)
  Prunus persica Sieb. & Zucc.
Pine, Eastern white, lodge-
  pole, Scotch, Mugho
  Pinus strobus L.
  Pinus contorta Dougl.

Pinus sylvestris L.
Pinus mugho
Pine, ponderosa
  Pinus ponderosa Laws.
Plum, Bradshaw
  Prunus domestica L.
Prune, Italian
  Prunus domestica L.
Spruce, blue
  Picea pungens Englm.
Tulip
  Tulipa gesneriana L.

Intermediate

Apple, delicious
  Malus sylvestris Mill.
Apricot, Moorpark & Tilton
  Prunus armeniaca L.
Arborvitae
  Thuja sp.
Ash, green
  Fraxinus pennsylvania var.
    lanceolata Borkh.
Aspen, quaking
  Populus tremuloides Michx.
Aster
  Aster sp.
Barley (young plants)
  Hordeum vulgare L.
Cherry, Bing & Royal Ann
  Prunus avium L.
Cherry, choke
  Prunus virginiana L.
Chickweed
  Cerastium sp.
Clover, yellow
  Melilotus officinalis Lam.
Citrus (lemon & tangerine)
  Citrus sp.
Geranium
  Geranium sp.
Golden rod
  Solidago sp.

Grape, Concord
  Vitis labrusca L.
Grapefruit (fruit)
  Citrus paradisi Mact.
Grass, crab
  Digitaria sanguinalis L. Scop.
Lamb's-quarters
  Chenopodium album L.
Lilac
  Syringa vulgaris L.
Linden, European
  Tilia cordata Mill.
Maple, hedge
  Acer campestre L.
Maple, silver
  Acer saccharinum L.
Mulberry, red
  Morus rubra L.
Narcissus
  Narcissus sp.
Nettle-leaf goosefoot
  Chenopodium sp.
Orange
  Citrus sinensis Osbeck
Peony
  Paeonia sp.
Poplar, Lombardy & Carolina
  Populus nigra L. and *Populus eugenei* Simon-Louis

Raspberry
  Rubus idaeus L.
Rhododendron
  Rhododendron sp.
Rose
  Rose odorata Sweet
Serviceberry
  Amelanchier alnifolia Nutt
Sorghum
  Sorghum vulgare Pers.
Spruce, white (young needles)
  Picea glauca Moench & Voss.
Sumac, smooth
  Rhus glabra L.
Sunflower
  Helianthus sp.
Violet
  Viola sp.
Walnut, black
  Juglans nigra L.
Walnut, English
  Juglans regia L.
Yew
  Taxus cuspidata Sieb & Zucc.

Resistant

Ash, European Mt.
  Sorbus aucuparia L.
Ash, Modesto
  Fraxinus velutina Torr.
Asparagus
  Asparagus sp.
Birch, cutleaf
  Betula pendula var. *gracilis*
    Roth.
Bridalwreath
  Spiraea prunifolia Sieb. & Zucc.
Burdock
  Arctium sp.
Cherry, flowering
  Prunus serrata L.
Cotton
  Gossypium hirsutum L.
Current
  Ribes sp.

Elderberry
  Sambucus sp.
Elm, American
  Ulmus americana L.
Juniper (most species)
  Juniperus sp.
Linden, American
  Tilia americana L.
Pear
  Pyrus communis L.
Pigweed
  Amaranthus retroflexus L.
Planetree
  Platanus sp.
Plum, flowering
  Prunus cerasifera Enrh.
Pyracantha
  Pyracantha sp.

Squash, summer
  Curcurbita pepo L.
Strawberry
  Fragaria sp.
Tomato
  Lycopersicon esculentum Mill.
Tree of heaven
  Ailanthus altissima L.
Virginia creeper
  Parthenocissus quinquefolia
    Planch.
Willow, several species
  Salix sp.
Wheat
  Triticum sp.

Source: Naegele 1974, Table 5.3, p. 90, from Jacobson and Hill 1970. Reprinted by permission of the publisher.

Plants chronically exposed to low fluoride levels accumulate fluoride in tissue with or without visible accompanying injury.

At concentrations of hydrogen fluoride below threshold for visible injury in tomato and corn, a positive correlation was found between frequency of chromosomal aberrations and durations of fumigation in leaf tissue. Abnormal phenotypes developed from seeds of tomato plants fumigated with hydrogen fluoride (Mohamed 1969, as cited in Pack 1971). Similar results have been obtained in bean plants (Pack 1971). Abnormalities in early leaves, resulting from continuous fumigation with relatively high levels (2.1 to 10.5 $\mu g/m^3$) of hydrogen fluoride, persisted for two generations although a rapid reversal to normality occurred in ensuing progeny.

At the 2.1-$\mu g/m^3$ level, a concentration within the range observed near some fluoride-emitting industries, bean reproduction was affected in the absence of visible foliage symptoms (Pack 1971). Reduced in vitro pollen germination occurred with long-term continuous fumigations of tomato plants at 7.9 and 13.0 $\mu g/m^3$ in low-calcium (1 mM) nutrient solution. Additions of calcium at 4 mM nullified this effect. Hydrogen fluoride treatments of even 4.2 $\mu g/m^3$ to maternal parent plants caused reductions in (1) number of pollen grains retained on the stigma, (2) pollen germination, and (3) pollen tubes reaching ovules (Sulzbach and Pack 1972). Early tomato fruit and seed development was inhibited by similar treatments.

Fluorine uptake and translocation

The form in which the fluorine is presented may affect plant uptake. Ryegrass took up gaseous hydrogen fluoride to a far greater extent than fine particulate fluorides and in an amount proportional to the average atmospheric fluorine content (Less et al. 1975). However, the absorption of hydrogen fluoride was appreciably increased by moisture on the ryegrass leaves and may thus be greater in winter as a result of frequent dews or when light rains occur frequently (Less et al. 1975).

Vegetation is known to pick up fluorine and translocate it to leaf tips, where it injures leaf tips and margins (Treshow 1971). In experiments with gladiolus, both accumulation and tip burn were found to vary considerably depending on plant variety, cultural site, and plant and leaf age (Hitchcock, Zimmerman, and Coe 1961-1962). Fluoride can be washed from foliage by rain or converted within plant tissue to compounds that do not cause active injury (Treshow and Pack 1970). The toxicity to livestock of foliage plants bearing high concentrations of fluorine depends on the degree of surface collection rather than on amount taken up (Horvath 1976).

Metabolism

Fluorine is a well-known enzyme inhibitor (McCune et al. 1964); it is either metabolized to an enzyme-fluorophosphate complex in vivo (Lovelace and Miller 1967) or complexes with enzyme cofactors (Melchior and Melchior 1956). McLaughlin and Barnes (1975) found that stimulation of dark respiration and inhibition of apparent photosynthesis occurred when foliage of pines and hardwoods was allowed to take up sodium fluoride solutions of 1.9 to 190 ppm of fluorine. This suggests that growth reduction could occur in trees accumulating low levels of fluorine. Cellulose synthesis, as well as ribosome synthesis, has been shown to be disrupted by fluoride (Treshow 1971), and low concentrations of sodium fluoride causing a disruption in oxygen metabolism in *Chlorella pyrenoidosa* (a green alga) were probably related to undisassociated hydrogen fluoride concentrations in the suspending medium (McNulty and Lords 1960).

Color changes accompanying typical hydrogen fluoride damage in sugarcane were found by Engelbrecht (1973) to be due to a decrease in the size of the internal chloroplast membranes and the destruction of chlorophyll followed by the formation of carotenoid pigments in the chloroplast. Destruction of the middle lamellae, followed by separation of the primary cell walls and loss of some cytoplasmic membranes, was responsible for necrosis and collapse of the damaged cells.

Growth stimulation

Fluoride has appeared to stimulate plant growth at concentrations characteristic of ambient levels near industrial sources (Treshow 1971). At these concentrations (an average of 2.3 $\mu g/m^3$), Treshow and Harner (1968), as cited in Treshow (1971), found that bean plants weighed significantly more than control groups although they accumulated up to 200 ppm fluoride. Hitchcock et al. (1971) (as cited in Treshow 1971) confirmed that the effect of fluorides was to cause an increase in the fresh and dry weight of the plants. The yield of some crops is not reduced by concentrations of fluoride less than 60 ppm, although plants containing more than 30 to 50 ppm may be toxic to animals (Hanson, Wiebe, and Thome 1958, as cited in Bohn 1972).

8.2.10 Ammonia

Levels of atmospheric ammonia have been found to vary widely (Ruess 1975) although they are estimated to be 36 million tons at any one time (Robinson and Robbins 1970). A number of sources, including industry, atmospheric fixation (electrical and photochemical), and volatilization from land surfaces, are widely recognized. Rasmussen, Taheri, and Kabel (1975) report that the primary source of atmospheric ammonia is from bacterial decomposition of organic material, but that combustion of coal is the chief anthropogenic source. Ammonia from coal conversion can be a marketable by-product, but some loss may occur in wastewater.

8.2.10.1 Effects

Ammonia per se is not important as an air pollutant. However, very high concentrations of ammonia that occur briefly in the atmosphere through accidental release cause leaves to appear "cooked." Several cereals and grasses have shown necrotic and chlorotic interveinal streaking at considerable distances from an accidental release (Taylor 1973). Mustard and sunflowers are among the most sensitive plants to ammonia (Waldbott 1973). Slight marginal injury to buckwheat, coleus, sunflower, and tomato foliage has occurred from a 4-hr exposure to 16.6 ppm ammonia. An exposure of 1 hr at 40 ppm also resulted in injury (Thornton and Setterstrom 1940 as cited in Heck, Daines, and Hindawi 1970).

Bioassay tests (5-min exposure) of ten plant species yielded a value of 1.0 mg/m^3 each as the maximum permissible concentration of gaseous ammonia. This amount of pollutant decreased photosynthesis by about 10% (Miroshnikova and Nikolaveskii 1974). Ammonia at 1 mM did not affect incorporation of vacuum-infiltrated nitrate or nitrite in barley, wheat, corn, and bean leaf segments; 10 mM ammonia inhibited incorporation for only 0.5 hr. Five times as much incorporation of ammonia into amino acids occurred in the light as in the dark (Canvin and Atkins 1974). Assimilation of NH_4^+ (as well as $H^{14}CO_3^-$, $^{15}NO_3^-$, and $^{15}NO_2^-$) vacuum-infiltrated into barley leaf segments was closely linked to photosynthetic electron transport (Atkins and Canvin 1975). The rate of nitrogen species assimilation was independent of the rate of carbon dioxide assimilation. It was established that dark in vivo nitrate reduction can only be analogous to the light mechanism when electron flow to oxygen is impaired (Atkins and Canvin 1975).

Blue-green algae were capable of growing when NH_4 was the only source of nitrogen (Neilson and Doudoreff 1973). Both induction and linear production of nitrite in a blue-green algal mutant were repressed by ammonia. The mutant is characterized by an impaired nitrogen metabolism resulting in an accumulation of nitrite in the growth medium. The effect of rifamycin in inhibiting the induction of nitrite production was lessened in ammonia-grown cells. The data indicate that nitrate reductase is a regulatory enzyme which can be negatively modulated by ammonia (Stevens and Van Baalen 1974). Uptake and translocation of ammonia from roots and rhizomes to stems and leaves occurred in three species of aquatic plants at highly variable rates (Toetz 1974).

Cells of the green alga *Chlorella* transferred from media containing 8 mM KNO_3 to media containing equivalent amounts of nitrogen in the form of ammonium exhibited a striking inactivation of nitrate reductase in less than 1 hr. All the enzymes of the nitrate-reducing system were fully repressed. Reactivation by removal of the ammonium ions both in vitro and in vivo was demonstrated (Losada et al. 1970). Inhibitory and toxic concentrations of some ammonia compounds for various algae and aquatic plants are listed in Table 8.14.

Table 8.14. Inhibitory effects of ammonia compounds on aquatic plants

Compound	Plant	Concentration (ppm)	Effects
Anhydrous ammonia	*Najas, Chara brasenica*	20 - 30	Disappearance over 2 to 4 weeks
Ammonia	*Navicula seminulum*	320 - 420	Inhibitory concentration
	Aphanizomenon sp.	0.4 - 0.5	Complete disappearance
	Marine diatoms	71.1	Toxicity threshold
		55	Photosynthetic ^{14}C-fixation inhibited
Benzylamine	*Scenedesmus quadricauda*	6.0	Toxicity threshold; 4 days, 24°C
Diethylamine	*Scenedesmus quadricauda*	4.0	Toxicity threshold; 4 days, 24°C
Ethylamine	*Scenedesmus quadricauda*	10.0	Toxicity threshold; 4 days, 24°C
Ethylenediamine (hydrochloride)	*Scenedesmus quadricauda*	20.0	Toxicity threshold; 4 days, 24°C
Triethylamine	*Scenedesmus quadricauda*	1.0	Toxicity threshold; 4 days, 24°C

Source: From Becker and Thatcher 1973.

8.2.10.2 <u>Mode of action</u>

When ammonium is used to fertilize plants, it is taken up directly. In the process, the plants give off H^+, which results in acidification of the soil or nutrient solution in which the plant is growing. This, in turn, exerts a toxic effect on the plant (Ruess 1975). For this reason, NH_4^+ has long been considered an unsuitable source of nitrogen for plant fertilization, but Schekel (1971), as cited in Reuss (1975), has shown that carnations thrive when NH_4^+ is used if acidification is prevented by the addition of $CaCO_3$.

8.2.11 Particulates

Atmospheric emissions from combustion or coal conversion may contain particulate matter. Fugitive emissions may also result from the receiving, handling, and storage of coal, solid wastes, or solid products (Rubin and McMichael 1975).

According to Martens (1971), plant nutrition may benefit by fly ash application to soils, but only if the differences in (1) chemical properties of the ash and its reaction with a specific soil, (2) plant tolerance to soluble salt damage and nutrient deficiencies, and (3) toxicities (especially to boron) are taken into account.

8.2.11.1 Effects

The high alkalinity and high boron concentrations that would result from the deposition of large amounts of ash fallout might reduce plant growth (Bohn 1972). Fly ash from three coal-burning power plants, when applied as fertilizer to alfalfa plants, caused decreased yield because of zinc deficiency resulting from increased soil pH. Boron availability increased with pH increase (Mulford and Martens 1971). Holliday et al. (1958) concluded that the boron content of unweathered fly ash applied as fertilizer was responsible for the toxic responses in crops observed as changes in dry weight and leaf symptoms. They have grouped agricultural and horticultural plants according to sensitivity. Chloride ions (NaCl) were found to increase boron toxicity in a southwestern forage grass even at very low concentrations. The grass was tolerant to boron during germination although percentage of germination decreased as boron levels increased from 100 to 800 ppm, especially at levels greater than 500 ppm. Individual ions (CO_3^{2-}, HCO_3^-, PO_4^{3-}, SO_4^{2-}, and NO_3^-) did not increase boron toxicity (Hyder and Yasmin 1972).

Linzon (1973) reviews the rather sparse information available concerning the effects of particulates on vegetation. Dust from cement kilns, magnesium-lime dust, soot, lead, and roadside salt spray have all proven injurious. The growth, production of shoots, and needle mass of spruce trees were substantially impaired by persistent exposure to road dust, calcium carbonate, and soot. These effects were thought to be mainly a result of reduction of light intensity, although the soot chemicals were corrosive (Rohmeder 1960). Particulate polycyclic organic matter (PPOM) may interfere with normal photosynthesis or affect other metabolic processes through attenuation of light. Its impact would depend on ambient atmospheric conditions, particle size, and the differential ability of receptors to accumulate this matter (Environmental Protection Agency 1975). It is known that vegetation generally intercepts more particles of $\leqslant$20-μm size than those of >20-μm size (Witherspoon and Taylor 1970). PPOM size ranges from a few tens of angstroms to hundreds of micrometers.

Lerman and Darley (1975) have pointed out that the little research that has been done on the effects of particulate matter on vegetation have dealt with specific kinds of dust rather than the more usually encountered conglomerate of dusts in the atmosphere.

Fluorides in particulate form will cause few, if any, effects on vegetation at fluoride particulate concentrations below 2 μg/m^3. These concentrations are rarely encountered in urban atmospheres (Lerman and Darley 1975).

Magnesium oxide deposits occurring on soils reduce plant growth although iron oxide deposits may be beneficial (Lerman and Darley 1975).

8.2.11.2 Mode of action

Particulates reportedly block the assimilation organs (stomata) of farm crops; effect is enhanced in forest trees because the pores become progressively blocked with age (Kozel and Maly 1972). In fir needles that had been heavily dusted with fly ash, gas exchange was found to be inhibited by clogging of 90% of the stomata; agricultural productivity was decreased by as much as 80% from sulfur dioxide plus fly ash pollution from a coal-fired power plant lacking filters (Guderian 1973).

Ricks and Williams (1975) report that earlier senescence of oak leaves exposed to particulate pollution in the absence of sulfur dioxide was accompanied by high levels of degradation of chlorophyll a in comparison with chlorophyll b. Relative carotenoid levels were the reverse of the carotenoid levels in control leaves that aged normally.

8.2.12 Ethylene

Ethylene, a natural plant product that acts as a growth hormone, is quite phytotoxic at relatively low concentrations in the atmosphere. It is present in high atmospheric concentrations in and near urban areas largely as a result of automotive combustion (Schuck 1973). Known as a minor gaseous by-product of combustion, it could reach the environment by accidental release during coal conversion.

8.2.12.1 Effects

Symptoms of damage in plants appear to be due to the hormonal nature of ethylene (NAS 1976). A 24-hr exposure to 1 ppb ethylene will injure the African marigold (Naegele 1974). At 0.1 ppm orchids showed symptoms of injury after 6 hr. Other sensitive species respond to 0.5 ppm after 30 hr (Heck, Daines, and Hindawi 1970). Chronic plant response is usually expressed as a distortion in growth; for example, flowers often fail to open properly. Symptoms may include epinasty, chlorosis, necrosis, leaf and bud abscission, or stimulation of lateral buds and roots (Brandt and Heck 1968).

The changes associated with the hormonal effects of ethylene are manifested as premature senescence and fruit ripening. Ethylene can cause growth inhibition, loss of geotropic sensitivity, leaf drop, and development of abnormal growths resembling tumors. Other gaseous hydrocarbons act upon plants in the same manner if they have double or triple bonds (for example, acetylene) (Howland, personal communication 1976). The relative sensitivities of some selected plants to ethylene are listed in Table 8.15.

8.2.12.2 Mode of action

Edwards and Miller (1972a) found inhibition of growth to be a specific result of inhibition in cell division. Kang and Burg (1973) showed a specific inhibition of DNA synthesis in the pea, *Pisum sativum*, by ethylene. With cessation of growth when ethylene was removed, glucose incorporation into the cell wall decreased (Apelbaum and Burg 1972a). The action of ethylene in causing mitochondrial swelling in in vitro preparations of cauliflower bud and pea seedlings

Table 8.15. Sensitivity of selected plants to ethylene

Sensitive

Bean, black valentine	Marigold, African	Privet
Phaseolus vulgaris L.	*Tagetes erecta* L.	*Ligustrum* sp.
Carnation	Orchid	Rose
Dianthus caryophyllus L.	*Cattleya* sp.	*Rosa* sp.
Cotton	Pea, cream	Sweet potato
Gossypium hirsutum L.	*Pisum sativum* L.	*Ipomoea batatas* Lam.
Cowpea	Peach	Tomato
Vigna sinensis Endl.	*Prunus persica* Sieb & Zucc.	*Lycopersicon esculentum* Mill.
Cucumber	Philodendron	
Cucumis sativus L.	*Philodendron cordatum* Kunth.	

Intermediate

Arborvitae	Gardenia	Soybean
Thuja orientalis L.	*Gardenia radicans* Thumb.	*Glycine max* Merr.
Azalea	Holly, Japanese	Squash
Rhododendron sp.	*Hex crenata* Thumb.	*Cucurbita maxima* Duchesne
Carrot		
Daucus carota L.		

Resistant

Beet	Endive	Onion
Beta vulgaris L.	*Cichorium endivia* L.	*Allium cepa* L.
Cabbage	Grass, rye	Radish
Brassica aleracea L.	*Lolium multiflorum* Lam.	*Raphanus sativus* L.
Clover	Oats	Sorghum
Trifolium sp.	*Avena sativa* L.	*Sorghum vulgare* Pers.

Source: Naegele 1974, Table 5.8, p. 95, from Jacobson and Hill 1970. Reprinted by permission of the publisher.

was obtained with six other hydrogen gases as well, indicating that mitochondria may not be the site of ethylene regulation of growth and respiration (Ku and Leopold 1970).

8.2.13 Chlorine

The amounts of chlorine in coal are small (see Table 2.2). However, chlorine is likely to be used in assorted activities, for example, during wastewater and/or cooling water treatment.

As an oxidant, chlorine may be highly phytotoxic. The white necrotic lesions occurring in plants exposed to chlorine resembles the acute injury produced by ozone (Naegele 1974) or PAN (glazing of lower surfaces) (Taylor 1973). Although chloride analysis is somewhat effective in identifying chlorine as the responsible toxicant, it is difficult because most plants can accumulate excessive chlorides through their root system. Shriner and Lacasse (1972), however, report that measurements of chloride uptake in foliage of hydrogen-chloride-fumigated plants may be useful as a diagnostic tool for the assessment of injury from atmospheric hydrogen chloride gas. For example, small chloride increases observed in the roots of tomatoes and chrysanthemums could be assumed to have come from residual soil chlorides, whereas large chloride increases in foliar portions of fumigated plants suggest that chloride metabolism in plants is similar to that of fluorides in that there is no translocation of chlorides from leaves to roots.

Chlorine was readily taken up by an alfalfa canopy at a rate changing with time, due perhaps to the tendency of chlorine to cause partial stomatal closure (Hill 1971). Complete defoliation of eucalyptus and elm trees has occurred in only a few hours following accidental chlorine spills that caused only a few minutes' exposure (Taylor 1973). Tables 8.16 and 8.17 list the relative sensitivities of selected plants to chlorine and to hydrogen chloride.

8.2.14 <u>Carbon dioxide, carbon monoxide</u>

Variations in carbon dioxide concentrations may affect a plant's sensitivity to gaseous pollutants through the possible effect of carbon dioxide on stomatal action. In some cases, exposure to high levels of carbon dioxide prior to ozone exposure has resulted in protection from ozone damage (Heggestad and Heck 1971). High levels of carbon dioxide prevented greening of etiolated wheat seedlings when exposed to light (Wolf and Kidd 1973). This was due partly to pH and partly to an interference in chlorophyll biosynthesis. Hill (1971) failed to detect uptake of carbon monoxide in an alfalfa canopy. No evidence of phytotoxic effects has been observed. Carbon dioxide is an antagonist of normal ethylene growth regulation and has been used to protect plants from ethylene damage (Smith and Parker 1966).

Carbon monoxide uptake was found to be very slow and light-dependent in barley leaves. The carbon monoxide was incorporated mainly into serine, but also into photosynthetic carbon cycle products (Krall and Tolbert 1957). There have been various reports of carbon monoxide effects: In a carbon-dioxide-free atmosphere, nasturtium leaves produced starch from carbon monoxide; tomato plants exposed to concentrations of carbon monoxide (0.1 to 1.0%) produced a profuse growth of adventitious roots within a week (Krall and Tolbert 1957). Injury has not been demonstrated on higher plants at exposures below 100 ppm for 1-3 weeks, although an exposure of 1 month to 100 ppm inhibited nitrogen fixation in clover roots (Lind and Wilson as cited in NAPCA 1970b).

Carbon monoxide has caused injury similar to that shown from exposure to ethylene and other unsaturated gases. Effects may result from the production of internal ethylene in plant tissue. Levels required for these effects are from 1000 to 10,000 times higher than those reported for ethylene (Heck, Daines, and Hindawi 1970). Total inhibition of new leaf formation in pea seedlings occurred after 18 days of exposure to 24 ppm carbon monoxide, whereas maximum leaf drop was observed at 15 days (Chakrabarti 1976).

8.2.15 <u>Other atmospheric pollutants</u>

Plants are apparently tolerant to atmospheric concentrations of hydrogen sulfide, mercaptans, amines, indole, skatole, organic acids, and some hydrocarbons to which man is sensitive (Bohn 1972). These compounds are potentially phytotoxic, however, and at high concentrations olefinic and aromatic compounds may affect plant growth. They are also precursors of PAN, PPN, and PBN (Bohn 1972). Heck (1968b) points out the possibility that "typical" symptoms ordinarily associated with pollutants such as ozone and PAN could result from the additional presence of any

Table 8.16. Sensitivity of selected plants to chlorine

Sensitive

Alfalfa[a]
Medicago sativa L.
Apple, crab
Malus baccata Borkh.
Blackberry
Rubus sp.
Boxelder
Acer negundo L.
Buckwheat
Fagopyrum esculentum Moench.
Chestnut, horse
Aesculus hippocastanum L.
Chickweed
Stellaria media Cyrill.
Coleus
Coleus sp.
Corn[b]
Zea mays L.
Cosmos
Cosmos sp.
Gomphrena[b]
Gomphrena sp.

Grass, Johnson
Holcus halepensis L.
Johnny-jump-up
Viola palmata L.
Maple, sugar
Acer saccharum Marsh.
Mustard[d]
Brassica sp.
Oak, pin
Quercus palustris L.
Onion[d]
Allium cepa L.
Pine, white
Pinus strobus L.
Primrose
Primula vulgaris Huds.
Privet
Ligustrum sp.
Radish[c]
Raphanus sativus L.
Rose, tea
Rosa odorata Sweet

Sassafras
Sassafras albidum Nutt & Nees.
Sunflower[b]
Helianthus ammuus
Sweetgum
Liquidambar styraciflua L.
Tobacco[b]
Nicotiana tabacum L.
Tree of heaven
Ailanthus altissima L.
Tulip
Tulipa sp.
Venus-looking-glass
Specularia perfoliata (L.) ADC
Virginia creeper
Parthenocissus quinquefolia
Planch.
Witch hazel
Hamamelis virginiana L.
Zinnia[b]
Zinnia sp.

Intermediate

Azalea[d]
Rhododendron sp.
Bean, Scotia[d]
Bean, pinto[d]
Phaseolus vulgaris L.
Cheeseweed
Malva rotundifolia L.
Cherry, black
Prunus serotina Ehrhe.
Cowpea[d]
Vigna sinensis Endl.
Cucumber[c]
Cucumis sativus L.
Dahlia[c]

Geranium
Geranium sp.
Grape
Vitis sp.
Grass, annual blue
Poa annua L.
Gum, black
Nyssa sylvatica Marsh.
Halesia
Halesia sp.
Nasturtium[c]
Tropaeolum sp.
Nettle-leaf goosefoot
Chenopodium murale L.

Pine, jack
Pinus banksiana Lamb.
Pine, loblolly
Pinus taeda L.
Pine, shortleaf
Pinus echinata Mill.
Pine, slash
Pinus caribaea Morelet
Rhodotypos
Rhodotypus sp.
Squash[d]
Cucurbita moschata Duchesne
Tobacco
Nicotiana tabacum L.

Table 8.16 (continued)

Dahlia sp.	Orange, mock	Tomato[c]
Dandelion[c]	*Philadelphus* sp.	*Lycopersicum esculentum* Mill.
Taraxacum officinale Weber	Peach	Wandering Jew
Fern, braken	*Prunus persica* Sieb & Zucc.	*Zebrina* sp.
Pteridium aquilinium L.	Petunia[d]	
	Petunia hybrida Vilm.	

Resistant

Begonia[e]	Lamb's-quarters[e]	Pigweed
Begonia rex Putz.	*Chenopodium album* L.	*Amaranthus retroflexus* L.
Corn, field	Oak, red	Polygonum[e]
Zea mays L.	*Quercus* sp.	*Polygonum* sp.
Eggplant	Olive, Russian	Soybean
Solanum melongena L.	*Elaeagnus angustifolia* L.	*Glycine max* Merr.
Grass, Kentucky blue	Oxalis[e]	Tobacco
Poa pratensis L.	*Oxalis* sp.	*Nicotiana tabacum* L.
Hemlock	Pepper[e]	Yew
Tsuga sp.	*Capsicum* sp.	*Taxus* sp.
Holly, Chinese		
Hex chinensis Loes.		

[a]Exposed at 0.10 ppm for 2 hr.
[b]Exposed at 0.10-0.25 ppm for 4 hr.
[c]Exposed at 0.50 ppm for 4 hr.
[d]Exposed at 0.80 ppm for 4 hr.
[e]Exposed at 1.0 ppm for 4 hr.
Source: Naegele 1974, Table 5.10, p. 97, from Jacobson and Hill 1970. Reprinted by permission of the publisher.

number of unidentified short-lived compounds such as the reaction products of ozone and unsaturated hydrocarbons. Thus, consideration must be given to other chemically active compounds which may be present in the ambient atmosphere. Atmospheric methane is assimilated and transformed by higher plants into organic acids of the Krebs cycle and amino acids (Durmishidze and Ugrekhelidze 1975). High levels of methane completely inhibited the greening of etiolated wheat seedlings upon exposure to light (Wolf and Kidd 1973).

Hydrogen sulfide has been implicated as a cause of straighthead disease and mild sulfide disease in rice (Joshi, Ibrahim, and Hollis 1975). Relative sensitivities of selected plants to hydrogen sulfide are listed in Table 8.18.

8.3 HYDROCARBONS

Hydrocarbon emission from coal conversion facilities can result from incomplete combustion at coal conversion auxiliary plant facilities; evaporation from liquid product storage areas; leakage at valves, flanges, and seals; and hydrocarbon liquid dissolved in liquid waste or cooling streams (Rubin and McMichael 1975). In addition, commonly used fuels (wood, coal, oil, natural gas) usually give off minor amounts of unburned hydrocarbon fuel elements during combustion (Schuck 1973). The automobile is a major contributor to urban atmospheric hydrocarbon pollution.

Table 8.17.   Sensitivity of selected plants to hydrogen chloride

Sensitive

Beet, sugar
  Beta vulgaris L.
Cherry
  Prunus sp.

Larch
  Larix sp.
Maple
  Acer sp.

Tomato
  Lycopersicon esculentum Mill
Viburnum
  Viburnum sp.

Intermediate

Begonia
  Begonia rex Putz.

Rose
  Rosa sp.
Rosebud
  Rosa sp.

Spruce
  Picea sp.

Resistant

Beech
  Fagus sp.
Birch
  Betula sp.
Fir
  Abies sp.

Maple
  Acer sp.
Oak
  Quercus sp.

Pear
  Pyrus sp.
Spruce
  Picea sp.

Source: Naegele 1974, Table 5.11, p. 98, from Jacobson and Hill 1970. Reprinted by permission of the publisher.

Much information on the effects of hydrocarbons on plants has been derived from accidental releases of oil in the environment. The effects of crude oil spills on flora in both the marine and the freshwater environment have been published (Mironov 1972; Cowell, Baker, and Crapp 1972; Bellamy et al. 1967; Hutchinson, Kauss, and Griffiths 1972). In fresh water, there was a decided difference in species response to crude oil in tests with planktonic algae. Three different effects were noted: (1) Growth of some species was inhibited, (2) other species were unaffected, and (3) growth of some species was stimulated. Tests with the crude oil components indicated that the benzene, toluene, xylene (BTX) fraction was most toxic to algae, with xylene having the greatest toxicity and benzene the least.

Clenndenning (1960) exposed the marine kelp *Macrocystis* to petroleum refinery wastewater diluted 50-fold with fresh seawater. The plants showed visible signs of injury within a week. In another test, there were large reductions of photosynthetic capacity after a four-day exposure to 10 ppm toluene in seawater. Concentrations of organic compounds in wastewater found to inhibit various plant species are given in Table 8.19.

Although toluene is toxic to marine plants, it appears to be tolerated by land plants. Both roots and aboveground organs of higher plants (corn, beans, maple trees, elm, ash, oleaster shrub, grape, tea) were found to assimilate and transform toluene to organic acids, amino acids, and sugars (Durmishidze, Ugrekhelidze, and Dzhikiya 1974).

Table 8.18. Sensitivity of selected plants to hydrogen sulfide

---

Sensitive

Aster
  Aster macrophyllus L.
Bean, kidney
  Phaseolus vulgaris L.
Buckwheat
  Fagopyrum esculentum Moench
Calliopsis
  Calliopsis sp.
Clover
  Trifolium sp.
Cosmos
  Cosmos bipinnatus Cau.

Cucumber
  Cucumis sativus L.
Lamb's-quarters
  Chenopodium album L.
Nettle-leaf goosefoot
  Chenopodium murale L.
Poppy
  Papaver somniferum L.
Radish
  Raphanus sativus L.

Salvia
  Salvia sp.
Soybean
  Glycine max. Merr.
Tobacco
  Nicotiana glauca Grah.
Tobacco, Turkish
  Nicotiana tabacum L.
Tomato
  Lycopersicon esculentum Mill.

Intermediate

Castor bean
  Ricinus communis L.
Chickweed
  Stellaria media Cyrill.
Cornflower
  Centaurea cyanus L.
Dandelion
  Taraxacum officinale Weber

Gladiolus
  Gladiolus sp.
Grass, Kentucky blue
  Poa pratensis L.
Nasturtium
  Tropaeolum majas L.

Pepper
  Capsicum frutescens L.
Rose
  Rosa sp.
Sunflower
  Helianthus annuus L.

Resistant

Apple
  Malus pumila Mill.
Carnation
  Dianthus caryophyllus L.
Cheeseweed
  Malva parviflora L.
Cherry
  Prunus serotina Ehrhe.
Coleus
  Coleus blumei Benth.

Fern, Boston
  Nephrolepis exaltata Schott
    var. *bostoniensis* Davenport
Grass, annual blue
  Poa annua L.
Mustard
  Brassica campestris L.
Peach
  Prunus persica Sieb & Zucc.

Pigweed
  Amaranthus retroflexus L.
Purslane
  Portulaca oleracea L.
Strawberry
  Fragaria sp.

Source: Naegele 1974, Table 5.5, p. 92, from Jacobson and Hill 1970. Reprinted by permission of the publisher.

8.3.1 Polycyclic aromatic hydrocarbons

Of the possible pollutants, the polycyclic aromatic hydrocarbons (PAH) are of special interest because of their carcinogenic potential, their apparently ubiquitous occurrence in the environment, and the potential large-scale production of coal tars during coal conversion processes.

8.3.1.1 PAH in food plants

Sources of human exposure to PAH may include foods. According to Il'nitskii and Kogan (1972), an individual may ingest 1 to 2 mg benzo[a]pyrene (BaP) and 6 to 8 mg PAH per year. Hakama and Saxen (1967) showed that significant correlations existed between high cereal consumption and stomach cancer mortality in 16 countries. The incidence of human gastric cancer and neoplastic disease has been reported to correlate with the ingestion of plants containing polycyclic organic matter (NAS 1972). Wynder et al. (1963), however, in demonstrating a strong correlation between the incidence of neoplastic disease and a vegetarian diet, emphasize the possible role of a high Intake of starchy foods with a concomitant low intake of fresh fruits and vegetables. Because plant parts (tubers, rootstocks, fruits, and grains) containing essentially stored materials (e.g., starch) yielded only 1 to 10% of the amount of PAH found in leaves and green plant parts (Graf and Diehl 1966), the relationship between gastric cancer and plant PAH content is still not clear.

Table 8.19. Inhibitory concentrations of organic compounds found in discharged wastewaters

Compound	Inhibitory concentration (mg/liter)	Test plant
Abietic acid	Not detected (most toxic of pulp mill ether-sols.)	*Scenedesmus obliquus*
Acetaldehyde	249	*Navicula seminulum*
Acetic acid	74	*Navicula seminulum*
Aniline	10	*Scenedesmus*
Benzene	>10	*Macrocytis pyrifera*
Benzyl alcohol	640	*Scenedesmus*
Benzylamine	6	*Scenedesmus*
Butyl acetate	320	*Scenedesmus*
Cresol	6 - 40	*Scenedesmus*
	5 - 10	*Macrocystis pyrifera*
Cyclohexane-carboxylic acid	28 - 29	*Navicula seminulum*
Dichloropropene	40	*Scenedesmus*
Diethylamine	4	*Scenedesmus*
Diethylphosphate	250	*Scenedesmus*
Dinitrocresol	36	*Scenedesmus*
Dinitrophenol	40	*Scenedesmus*
Ethylamine	10	*Scenedesmus*
Ethylenediamine	20	*Scenedesmus*
Ethylmercuric phosphate (lignasan)	1	General effective algicide
Formaldehyde	0.3	*Scenedesmus*
Formic acid	100	*Scenedesmus*
Hexane	10	*Macrocystis pyrifera*
Hydroquinone	4	*Scenedesmus*
Methylamine	4	*Scenedesmus*
Nitrobenzene	40	*Scenedesmus*
Nitrophenol	28 - 72	*Scenedesmus*
Phenol	1	*Platymonas*
	10	*Macrocystis pyrifera*
	40	*Scenedesmus*
	250	*Navicula seminulum*
Pyrocatechol	6	*Scenedesmus*
Pyrogallol	8	*Scenedesmus*
Quinhydrone	4	*Scenedesmus*
Quinoline	140	*Scenedesmus*
Quinone	6	*Scenedesmus*
Resorcinol	60	*Scenedesmus*
Toluene	120	*Scenedesmus*
	10	*Macrocystis pyrifera*
Toluidine	8 - 10	*Scenedesmus*
Triethylamine	1	*Scenedesmus*
Trinitrophenol	240	*Scenedesmus*

Source: North, Stephens, and North 1970, Table 4, p. 333. Reprinted by permission of the publisher.

The presence of BaP has often been used as a convenient indicator of the presence of carcinogenic PAH in the environment. Sixty-five percent of the BaP ingested by humans may come from cotton-seed, sunflower seed, maize seed, wheat, rye, and potato foodstuffs (Shabad and Kogan 1972). Levels of BaP in some food plants are given in Table 8.20. Levels of BaP were determined by fluorophotometric analysis for 17 vegetables and were found to be relatively higher in green vegetables (Table 8.21) (Shiraishi and Takabatake 1974), for example, 3.3 ppb in spinach as compared to 0.01 ppb in Japanese radish root. Representative levels of BaP in leafy green vegetables ranged from 1.2 (lettuce) to 5.0 (endive) µg/100 g dry material (Graf and Diehl 1966).

Table 8.20. Background levels of BaP in food plants

Plant	BaP	Reference
Lettuce, µg/100g	1.2	Gräf and Diehl 1966
Spinach, µg/100g	2.0	Gräf and Diehl 1966
Endive, µg/100g	5.0	Gräf and Diehl 1966
Spring wheat, µg/kg	0.29 (seed) - 27.0 (stem)	Shabad and Kogan 1972
Winter-rye seed, µg/kg	0.35 (average)	Shabad and Kogan 1972
Potato, µg/kg	0.36 (average, in peelings) 0.09 (average, in tubers)	Shabad and Kogan 1972
Vegetable oil, mg/kg	0.53 - 30.0	Il'nitskii and Kogan 1972
Grain, flour, bread, mg/kg	0.08 - 4.13	Il'nitskii and Kogan 1972
Vegetables, mg/kg	1.0 - 24.0	Il'nitskii and Kogan 1972
Coffee (roasted), mg/kg	0	Il'nitskii and Kogan 1972
Coffee (green beans), mg/kg	0 - 6.1	Il'nitskii and Kogan 1972
Dried fruit, mg/kg	0.3 - 23.9	Il'nitskii and Kogan 1972
Young wheat plants, µg/100 g dried material	4.8 - 6.6	Siddiqi and Wagner 1972
Carrots, µg/kg	1.65 - 6.0	Il'nitskii, Solenova, and Ignatova 1974
Beets, µg/kg	0.6 - 2.0	Il'nitskii, Solenova, and Ignatova 1974
Cabbage, µg/kg	0.8 - 1.2	Il'nitskii, Solenova, and Ignatova 1974
Carrots, ppb	0.08 - 0.14	Siegfried 1975
Head lettuce, ppb	0.2 - 1.3	Siegfried 1975

PAH is stable at temperatures used in cooking and baking. Flour and baked goods prepared from cereals contaminated with 0.2 to 4.1 µg/kg BaP were found to contain the same amount of contaminant (Kornreich 1975).

Tilgner (1971), as cited in Kornreich (1975), reports PAH in vegetable oil products such as oil and mayonnaise. Grigorenko et al. (1970) found that cotton seeds (largely the seed coat lipid component) used in the production of unrefined oils were the main source of BaP in the products and that the amounts of BaP were not related to production technology.

8.3.1.2 <u>Biosynthesis of PAH in plants</u>

The apparently consistent findings of PAH in a wide variety of plant materials (Borneff and Fischer 1962; Guddal 1959, Borneff 1963, and Grimmer and Hildebrandt 1966, as cited in Andelman and Suess 1970) led to experiments (Gräf and Diehl 1966) demonstrating the endogenous synthesis of PAH during growth of rye, wheat, and lentils. Levels of BaP increased from 0.2 µg/100 g in rye seeds to 3.8 µg/100 g in the seedlings (Table 8.22). The biosynthesis took place independently of photosynthesis in nutrient solutions free of contaminating PAH.

Analysis of leaves of trees (maple, oak, beech, fir, pine, etc.) from European areas remote from industrial and human activity (Rohn, Allgau, timberline of the Wetterstein Mountains, French Switzerland) yielded the consistent presence of eight specific PAH, five of which are generally considered to be highly carcinogenic (Table 8.23). The amount of BaP ranged from 10 to 20 µg/kg dry plant material and was three to five times higher in wilted or yellowed plant parts (Table 8.24). The same results were obtained when tobacco, lettuce, and kolhrabi were grown under plastic coverings (Gräf and Diehl 1966).

Confirmation of the endogenous synthesis of PAH was obtained by Borneff et al. (1968*a* and 1968*b*) in carefully structured experiments in which algae (*Chlorella vulgaris*) were grown in a medium using ^{14}C-acetate as the sole source of carbon. Seven PAH were extracted from the algae, of

Table 8.21. Amounts of BaP in vegetables

Vegetables	Sample number	Dry weight of 1 kg of fresh vegetables (g)	BaP determined (ppb, wet weight)
Potato	1	234	
	2	215	0.01
Japanese radish root	1	50	
	2	50	
Spinach	1	100	3.3
	2	105	0.38
	3	98	0.26
Carrot	1	100	0.02
	2	80	
Welsh onion greens	1	79	0.09
	2	69	0.12
Cabbage	1	67	
	2	58	
Brassica chinensis var.	1	45	0.04
	2	50	0.16
Garland chrysanthemum	1	65	1.20
Chinese cabbage	1	35	0.05
Onion	1	80	0.01
	2	80	
Lettuce	1	44	
	2	40	
Sweet pepper	1	61	
	2	60	
Cucumber	1	70	
	2	71	
Pumpkin	1	133	Trace
	2	147	
Eggplant	1	72	
	2	70	
Zingiber Mioga	1	47	
	2	39	
Sweet potato	1	305	
	2	300	

Source: Shiraishi and Takabatake 1974, Table 2, p. 19. Reprinted by permission of the publisher.

Table 8.22. Plant biosynthesis of BaP (gamma/100 g dry material)

Before growth		After growth		
Plant	BaP	Plant	BaP (with light)	BaP (without light)
Rye seeds	0.2	Rye seedlings	1.8	2.6
Rye seeds	0.05	Rye seedlings	3.8	
Lentil seeds	0.01	Lentil seedlings	2.4	
Wheat grains	0.05	Wheat seedlings	0.8	1.0
Hyacinth bulbs	0.3	Hyacinths	1.0	

Source: Gräf and Diehl 1966, Table 3, p. 56. Reprinted by permission of the publisher.

Table 8.23. PAH commonly detected in plant material

Substance	Carcinogenic activity	Structural formula
Fluoranthene	-	
Benzo[*ghi*]perylene	±	
Benzo[*b*]fluoranthene	++	
Benzo[*j*]fluoranthene	++	
Benzo[*k*]fluoranthene	±	
Indeno[1,2,3,*cd*]pyrene	+	
Benz[*a*]anthracene	+	
Benzo[*a*]pyrene	+++	

Source: Gräf and Diehl 1966, p. 52. Reprinted by permission of the publisher.

Table 8.24. Content of oak, beech, and tobacco leaves of some
PAH (gamma/500 g dry substance)

Compound	Beech		Oak		Tobacco	
	Green	Yellow	Green	Yellow	Green	Yellow
Fluoranthene	157	624	63	333	188	313
Benz[*b*]fluoranthene	31	187	31	146	16	63
Indeno[1,2,3,*cd*]pyrene	19	117	13	75	9	19
Benz[*a*]anthracene	13	83	9	50	6	22
Benzo[*ghi*]perylene	19	71	8	37	9	22
Benzo[*a*]pyrene	8	33	4	21	6	19

Source: Graf and Diehl 1966, Table 2, p. 53. Reprinted by permission of the publisher.

which BaP ranged from 0.07 to 0.08 µg per 113.5 g dried algae (Table 8.25). Control experiments precluded the possibility of exchanges of the radioactive label to other compounds. The labeled carbon atoms were incorporated into PAH molecules in the algae, indicating biosynthesis.

Table 8.25. PAH content of inactive and ^{14}C-labeled algae

Compound	PAH (µg[a]) found in	
	Inactive algae[b]	^{14}C-labeled algae[b]
Fluoranthene	5.80	6.20
Benz[*a*]anthracene	0.78	0.65
Benzo[*b*]fluoranthene	0.39	0.42
Benzo[*a*]pyrene	0.07	0.08
Benzo[*ghi*]perylene	0.22	0.23
Benzo[*k*]fluoranthene	0.14	0.15
Indeno[1,2,3,*cd*]pyrene	0.18	0.17

[a]The data refer to total extracted amount of algae disregarding losses due to chromatography and elution.
[b]113.5 g dried algae.

Source: Borneff et al. 1968*a*, Table 1, p. 26. Reprinted by permission of the publisher.

However, careful research with higher plants indicates that definitive proof of PAH biosynthesis is yet to be obtained. Benzo[*e*]pyrene, BaP, perylene, anthanthrene, benzo[*ghi*]perylene, dibenzo[*a,h*]anthracene, and coronene were not detected in various crop plants (rye, tobacco, soybeans) grown in filtered air in special rooms with air locks, whereas those grown in open air in ordinary greenhouses contained all these hydrocarbons (Grimmer and Duvel 1970). In greenhouses and open air, BaP levels were as high as 4.3 µg/kg in soybean and lettuce (Table 8.26).

Table 8.26. Amounts of PAH in test plants grown in different environments (μg/kg)

| | Lettuce seeds | Lettuce | | | Tobacco | |
		Climate-controlled chamber	Greenhouse	Open air	Climate-controlled chamber	Greenhouse
Duration, weeks		16	16	11	10	12	12
Dates		June-Oct.	Apr.-Aug.	July-Oct.	June-Sept.	Sept.-Dec.	Dec.-Mar.
Weight, g	1000	900	500	300	1000	320	400
Benzo[e]pyrene	0	0	0	4.2	4.3	0	2.5
Benzo[a]pyrene	0	0	0	3.7	4.2	0	1.8
Perylene	0	0	0	0.4	0.4	0	0.25
Anthanthrene	0	0	0	0.2	0.2	0	0
Benzo[ghi]perylene	0	0	0	2.5	2.2	0	1.1
Dibenz[a,h]anthracene	0	0	0	0.4	0.6	0	Trace
Coronene	0	0	0	0.6	0.7	0	0.1

Table 8.26. (continued)

	Rye			Soybean		
	Climate-controlled chamber	Greenhouse	Rye seeds, open air	Climate-controlled chamber		Greenhouse
Duration, weeks	16	16	40	6	6	6
Dates	Oct.-Feb.	Nov.-Mar.	Oct.-Aug.	Feb.-Mar.	Aug.-Oct.	Mar.-May
Weight, g	800	320	1000	150	250	220
Benzo[e]pyrene	0	3.4	1.20	0	0	4.3
Benzo[a]pyrene	0	1.6	0.64	0	0	3.1
Perylene	0	0.9	0.10	0	0	0.3
Anthranthrene	0	0.1	0.06	0	0	Trace
Benzo[ghi]perylene	0	0.9	0.34	0	0	1.5
Dibenz[a,h]anthracene	0	0.1	0.10	0	0	0.1
Coronene	0	0.1	0.10	0.1	0	0.4

Source: Grimmer and Duvel 1970, Tables 1 and 2, p. 1173. Reprinted by permission of the publisher.

Suess (1971) reports that, whereas endogenous PAH synthesis by flora appears to be a major source of PAH in marine waters, municipal and industrial wastewater and atmospheric precipitation may further increase their concentrations along coasts. In an examination of the origin and significance of PAH in both the marine and freshwater environments, Andelman and Snodgrass (1974) recommend monitoring and control of PAH levels in drinking water. Levels of PAH compounds determined by Borneff and Fischer (1962) in dried plankton collected from lake water (largely the diatom, *Asterionella formosa*) are reproduced in Table 8.27. Total amount of PAH was an estimated 0.7 mg/kg dried plankton.

Table 8.27. Content of PAH compounds in dried plankton (95% *Asterionella formosa*)

Compound	Amount (μg/kg)
Fluoranthene	300
Benzo[*a*]anthracene	20
Benzo[*j*]fluoranthene	50
Benzo[*b*]fluoranthene	100
Benzo[*ghi*]perylene	100
Benzo[*a*]pyrene	2
Benzo[*k*]fluoranthene	20

Source: From Borneff and Fischer 1962. Reprinted by permission of the publisher.

The green alga *Scenedesmus acutus* contained lower levels of PAH when cultured extensively in open unpolluted air in Thailand than when cultured in polluted air in Dortmund, Federal Republic of Germany. Logarithmic distributions indicated that the endogenous formation of carcinogenic PAH in the algae occurred only to a very limited extent and was completely masked by accumulation of the same substances from the environment (Payer et al. 1975).

8.3.1.3 Uptake from polluted air

Kornreich (1975) states that a strong relationship exists between air pollution and increased levels of BaP in vegetables and grain. Although some BaP was a result of dust deposition, washing was found to remove only 10% from kale (Kornreich 1975).

Fritz and Engst (1971) found that the PAH content of food plants increased conspicuously when grown in areas having increased traffic density and especially in those areas in proximity to industrial sources. Engst (1973) reports that the BaP content of fruits and vegetables was greater in heavily traveled and industrial areas; amounts ranged from 0.1 μg/kg dry substance (apples) to 150 μg/kg (lettuce) (Table 8.28).

8.3.1.4 Uptake from polluted soil

Levels of BaP in soils and vegetation near the runways of a modern airfield were found to be considerably higher than background levels. With increasing distance from the airfield, BaP levels were lower (Smirnov 1970).

Table 8.28. BaP in fruits and vegetables (µg/kg dry weight)

	City suburb	Highway	Industrial area[a]
Apple peels	0.2 - 0.5		30 - 60
Apple interior	0.1 - 0.4		5 - 6
Pear peels		9 - 10	27
Pear interior		0.5 - 0.9	3
Stone fruit	3 - 8		
Berry fruit	2 - 7		
Lettuce	23		150
Cabbage types	6 - 12		
Cucumbers	8 - 10.8		
Leeks	10.9		
Potato peels	8 - 12		
Potato interior	0		

[a]More recent investigations point to an increase of these contaminations.

Source: Engst 1973, Table 2, p. 81. Reprinted by permission of the publisher.

In another study, only 20% of the BaP found in nine fruits and vegetables in amounts proportional to the BaP content of the respective soils in which they were grown was removed by washing (Engst 1973). Cold-water washes failed to remove BaP from vegetables, grains, and fruits containing 0.1 to 24.3 µg/kg, 0.2 to 7.5 µg/kg, and 0.2 to 29.7 µg/kg BaP respectively. The amounts were inversely proportional to their distance from areas of industrial and urban activity (Kolar, Ledvina, and Hanus 1975). Aboveground plant parts contained more BaP than parts growing beneath the surface. Levels were proportional to the exposure time during the vegetative growth phase and to the extent of exposed plant surface.

8.3.1.5 Uptake from polluted wastewater

Beets, carrots, cabbage, and fodder grass grown in a turf-slight podzolic soil were not found to be enriched over ordinary levels (no more than 10 µg/kg dry weight) after irrigation with waste-waters containing high concentrations of BaP (up to 0.54 µg/liter) (Il'nitskii, Solenova, and Ignatova 1974). It was suggested that soil bacteria purified the soil by breakdown of the BaP.

8.3.1.6 Uptake from polluted compost

Siegfried (1975) determined levels of BaP in carrots and head lettuce for plants grown on garbage compost containing between 0.4 and 1.1 ppm BaP. Unwashed carrot tops contained 1.1 to 4.3 ppb, and lettuce contained 0.2 to 1.3 ppb. The carrot peels (20% of the whole) were found to contain 70 to 75% of the total root BaP. Because BaP levels did not change strongly with soil treatment, no firm conclusion was reached as to the possibility of BaP enrichment of crops by high levels in compost used as fertilizer (Siegfried 1975).

Borneff et al. (1973a; 1973b) concluded that BaP content in foods is so small as to be negligible and is unrelated to compost origin or BaP content of the compost. Different fertilizers (sewage sludge—garbage compost, manure, mineral fertilizer) containing a variety of PAH levels were applied in commonly used quantities (e.g., 30 tons of manure/ha) to soils in which crops were grown. PAH levels in soils increased in some cases, but decreased in others. Only in the outer layer of plant roots did the PAH content reflect that of the soil (with one exception). The edible portion of carrot or radish, for instance, did not contain elevated levels of PAH.

In contrast, BaP in the erosion-controlling compound nerosin (a shale-tar derivative) was found to increase BaP content in potatoes (from 0.08 to 0.22 µg/kg to as high as 29.40 µg/kg), but not in wheat and rye (Tables 8.29 through 8.31). Varietal differences were noted (Shabad and Kogan 1972a). Nerosin had no effect on the BaP content of corn, but caused an almost fivefold increase in the BaP content of sunflower seeds (Table 8.32).

Table 8.29. Content of BaP in the spring-wheat

Kind of wheat	Year of application of nerosin	Year of harvest	BaP content (µg/kg)	
			Control	Treated with nerosin
Saratovskaya-29	1969	1969	0.44	0.22
Saratovskaya-29	1969	1970		
Seed			0.29	0.38
Stem (straw)			4.52	3.60
Erythrospermum-841	1970	1970		
Seed			0.84	0.72
Stem (straw)			27.0	26.7

Source: Shabad and Kogan 1972a, Table 1, p. 239. Reprinted by permission of the publisher.

Table 8.30. BaP content in the winter-rye seed

Series	BaP content (µg/kg)
Control[a]	0.41
Control[a] + *F. nivale*	0.30
Average	0.35
Nerosin	0.28
Nerosin + *F. nivale*	0.21
Average	0.25

[a]Without nerosin.

Source: Shabad and Kogan 1972a, Table 2, p. 239. Reprinted by permission of the publisher.

8.3.1.7 Accumulation

Wagner and Siddiqi (1970) found that BaP and benzo[b]fluoranthene (benzo[e]acephenanthrylene) accumulated in summer wheat to a greater extent during the vegetative phase of growth, followed by a decline during the generative stage. The quantity of the PAH that accumulated in the wheat constituted only a small percentage of the amount found in the soil used for the experiment. Results indicate that the wheat plant does not have the ability to synthesize either of

Table 8.31. BaP content in potatoes (μg/kg)

Plot number	Control	Dose of nerosin			
		0.5 ton/ha-E	0.5 ton/ha-B	1.0 ton/ha-E	1.0 ton/ha-B
1	0.08	0.12	0.76	0.11	0.12
2	0.22	0.38	0.50	0.27	
3		0.33	23.40	29.40	
4		6.36		0.62	

Source: Shabad and Kogan 1972*a*, Table 3, p. 240. Reprinted by permission of the publisher.

Table 8.32. BaP of oil-bearing crops (μg/kg)

	Test[a]	Control
Maize seed	0.08 - 0.15	0.08 - 0.15
Maize oil	1.6 - 3.0	1.6 - 3.0
Sunflower seed	0.66 - 2.10	0.47
Sunflower oil	1.32 - 4.20	

[a]Nerosin applied to soils.

Source: From Shabad and Kogan 1972*b*. Reprinted by permission of the publisher.

these hydrocarbons. Siddiqi and Wagner (1972) found that the levels of BaP and benzo[*b*]fluoranthene were 4.8 to 6.6 and 24.5 to 76.8 mg/100 g dry matter, respectively, in young wheat plants. Levels in the growth medium were not determined.

8.3.1.8 Absorption, translocation, and metabolism

Plants have the ability to absorb PAH from water, air, and soil by roots and foliage. Roots can solubilize organic matter and absorb and translocate PAH compounds (with possible simultaneous concentration) to other parts of the plant, thus bringing these compounds into the ecological food chain (National Science Foundation 1974). There is experimental evidence that extracellular enzyme systems of plant roots are responsible for solubilizing large organic molecules in the soil environment and translocating them into the plants. There is also experimental evidence that certain plants concentrate these large organic molecules by accretion in certain parts of their anatomy, often without changes in structure (Seear, Bradfute, and McLaren 1968).

Plants appear to be capable not only of absorbing BaP but also of metabolically breaking it down for use in other compounds. The ^{14}C from labelled BaP, which was applied to corn or bean leaves, appeared in the amino acid, the organic acid, the ethanol-insoluble fraction of the plant tissues, and the carbon dioxide respired (Durmishidze et al. 1973; Durmishidze et al. 1975).

8.3.1.9 Growth stimulation — carcinogenic effect

Evidence suggests that some PAH compounds may behave as plant hormones (Gräf and Diehl 1966).
It should be noted that plant hormones act at concentrations that are insignificant in terms of
biomass. Gräf and Diehl (1966) found that the yield of rye plants was increased threefold by
addition of BaP. The clear stimulations of growth observed in both higher (tobacco, rye, radish)
and lower (algae) plants was attained with several PAH in intensities corresponding to the
intensity of their carcinogenic effects in animals (Gräf and Nowak 1966): benzo[a]pyrene >
benzo[a]anthracene > benzo[b]fluoranthene and indeno[1,2,3,cd]pyrene > 1,12-benzoperylene.

Tumor-like abnormal growths have been reported on the marine alga *Porphyra tenera* as a result of
mudwaste pollution from the coal chemical industry after a single contact (80 to 320 min) and
successive 36-day culture (Ishio, Nakagawa, and Tomiyama 1972). Two neutral-pH components were
isolated and identified as the tumor-promoting factors.

Abnormal growths on mushrooms grown in PAH-contaminated media have also been reported (NAS 1972).
Plants exposed to large amounts of hydrocarbons in the soil (primarily from oil spills and pipeline leaks) over a period of about four years have shown retarded growth. After about four
years, enhanced growth resulted from released nitrogen and the increased water-holding capacity
of the soil (Bohn 1972).

The animal carcinogen 7,12-dimethylbenz[a]anthracene has been found to stimulate callus growth
in culture-grown *Haworthia,* a succulent member of the lily family. It also induced differentiation in the absence of any other growth-promoting agents without altering chromosome number
and structure (Majumdar and Newton 1972).

BaP induced cell division in cultures of callus tissue from tobacco (*Nicotinia tabacum*), whereas
some tobacco smoke components (benzo[e]pyrene, dibenz[a,h]anthracene, and pyrene) stimulated
the formation of vegetative buds. All effects were dose-related (Sabharwal and Bhalla 1973).
Allium cepa (onion) root tips showed breakage of chromosomes and an increase in numbers of
micronuclei in a water solution of BaP 10^{-5} M when exposed to light; the BaP caused no damage
in the dark (Santamaria, Dolcher, and Calendi 1971).

8.3.2 Phenols

Whereas ordinary phenol is not found in plants in a free state, a wide range of phenolic compounds, including catechol, resorcinol, and hydroquinone, are generally present. In healthy
cells, the phenolic compounds ordinarily occur in a reduced state rather than, for instance,
as o-quinones. The latter are formed endogenously when plant tissue decomposes (Stom et al.
1973). In some cases, natural quinones apparently are closely linked with defense reactions
(Goodman, Kiraly, and Zaitlin 1967). The phenoloxidase system in normal plants evidently plays
a role in respiration, amino acid metabolism, auxin metabolism, cell wall metabolism, and the
browning reaction of plant tissue. Phenolic compounds in the protective layers of a normal plant
are known to play a role in disease resistance; for example, an accumulation of phenolic compounds
is found in the yellow- or red-pigmented bulb scales of certain varieties of onions resistant to
infective fungi. Higher levels of phenols both occurring naturally in some plants and experimentally infused in others have been correlated with greater disease resistance, although the

correlation is not universally true. It has often been observed that phenol accumulation takes place in all infected plant tissues, more rapidly in the more resistant varieties. The accumulation of aromatic compounds such as phenols in plants can also be induced by chemical and mechanical injury (Goodman, Kiraly, and Zaitlin 1967).

A concise account of the chemical nature of phenolic compounds, methods of their identification, their biosynthesis in plants, and their taxonomic and physiological importance in plants is available (Ribereau-Gayon 1972).

8.3.2.1 Uptake and toxicity

Absorption of polyphenols, benzene, and naphthalene, present in coke manufacturing wastewaters that were used to spray potato and corn plants, was not detected in the plant tissues, whereas uptake levels of absorbed monophenols and pyridines decreased 10 to 30 days after spraying (Polishchuk 1975).

In Russia, large volumes of wastewater from coking-chemical plants that could not be accommodated by reservoirs have been used to irrigate agricultural plots. The wastewaters contained high levels of various compounds — cyanides, thiocyanates, sulfides, phenol, resorcinol, benzene, pyridine, etc. (Barabanova, Polishchuk, and Stemikovskaya 1973). Corn and potatoes were grown on land irrigated with wastewaters from two by-product coking plants. Wastewaters that were diluted 1:4 with conduit water inhibited growth so strongly that further irrigation was discontinued. When the wastewaters were diluted 1:8 and were dispersed at 500 m^3/ha, both corn and potatoes accumulated nitrates, ammonia, sulfates, and chlorides after seven months' growth (Tables 8.33 and 8.34). Monosubstituted phenols were higher in potato tubers than in controls, as were nitrates; sulfides, cyanides, and thiocyanates were not present.

Hydroquinone-β-D-glucoside (arbutin) was found to be taken up actively by excised barley roots, whereas hydroquinone entered root tissues by diffusion (Glass and Bohm 1971). Uptake of the 86-rubidium ion, Rb^{86} (as an experimental substitute for potassium ions), was inhibited by 30% at 5 mM hydroquinone.

Dolgova and Kozyukina (1972) investigated the phenol-concentrating power of tree leaves continuously exposed for 14 to 16 years to atmospheric phenol (chiefly carbolic acid) and sulfur pollution in the vicinity of a coke-oven gas industry. The leaves of the trees *Ailanthus altissima, Ulmus pinnato-ramosa,* and *Populus nigra* were found to contain elevated levels of phenols as compared to leaves of the same plants grown in clean air in a botanical garden. There was a seasonal variation (Table 8.35), the highest concentrations of phenols being found in September (in *Ailanthus,* 0.490 mg/g compared with 0.025 to 0.040 mg/g in May-June).

8.3.2.2 Removal from wastewater

Vascular plants have the ability to remove organic chemicals, heavy metals, and pesticides from polluted waters (Wolverton 1975). For example, the water hyacinth (*Eichhornia crassipes*) took up 36 mg phenol per gram dry weight of plant material from distilled water, river water, and bayou water systems in 72 hr. Four-week and older plants, weighing an average 2.75 g dry wt per individual and exposed to 25, 50, and 100 ppm, showed optimum uptake at optimum growth pH.

Table 8.33. Wastewater ingredients in potatoes (mg/kg) irrigated with by-product coke plant wastewaters in experimental plots (M±m)

Compound	Conduit water	By-product coke plant sample 1			By-product coke plant sample 2		
		Wastewater dilution					
		1:8	1:4	None	1:8	1:4	None
Pyridine	0	0	0	0	0	0	0
Benzene	0	0	0	0	0	0	0
Naphthalene	0	0	0	0	0	0	0
Monoatomic phenols	0.31±0.02	0.2±0.02	0.3±0.02	0.4±0.01	0.3±0.02	0.4±0.01	0.8±0.03
Diatomic phenols	0	0	0	0	0	0	0
Cyanides	0	0	0	0	0	0	0
Thiocyanates	0	0	0	0	0	0	0
Nitrites	0.1±0.003	0.3±0.01		0.1±0.003	0.2±0.01	0.1±0.01	0.1
Nitrates	6.1±0.10	23.8±0.42	12.8±0.36	10.3±0.06	25.1±0.09	11.7±0.12	
Ammonia	8.0±0	7.8±0	12.0±0	14.0±0	9.2±0.2	23.5±0	
Sulfites	0.6±0.005	0.9±0.008	0.9±0.007	0.8±0	1.1±0.004		0.7±0.2
Sulfates	5.2±0.08	15.4±0.51		18.9±0.46	5.7±0.03		8.5±0.11
Chlorides	247.8±0	424.8±0	460.2±0	495.6±0	247.8±0	227.2±1.87	247.8±0
Sulfides	0	0	0	0	0	0	0

Source: Barbanova, Poluschuk, and Stemikovskaya 1973, Table 3, p. 78. Reprinted by permission of the publisher.

Table 8.34. Wastewater ingredients in corn cob (mg/kg) irrigated with
by-product coke plant wastewaters in experimental plots (M±m)

Compound	Conduit water	Wastewater diluted 1:8 from by-product coke	
		Sample 1	Sample 2
Pyridine	0	0	0
Benzene	0	0	0
Naphthalene	0	0	0
Monoatomic phenols	0	0	0
Diatomic phenols	0	0	0
Cyanides	0	0	0
Thiocyanites	0	0	0
Nitrites	0	0	0
Nitrates	0.3±0,03	36.3±0.05	28.3±0.05
Ammonia	14.0±0	32.0±0	24.0±0
Sulfites	0.8±0,006	1.0±0,005	0.8±0.005
Sulfates	3.0±0,09	10.0±0.18	3.8±0.09
Chlorides	106.2±0	123.9±0	123.9±0
Sulfides	0	0	0

Source: Barbanova, Poluschuk, and Stemikovskaya, Table 4, p. 78. Reprinted by permission of the publisher.

Table 8.35. Phenol content (in mg/g of dry residue) in tree leaves

Species of tree	Sampling site	May	June	July	August	September
Ailanthus altissima Sw.	Factory	0.025	0.040	0.121	0.201	0.490
	Botanical gardens	0.010	0.020	0.046	0.046	0.050
Ulmus pinnato-ramosa Dieck	Factory	0.039	0.062	0.098	0.100	0.137
	Botanical gardens	0.019	0.012	0.016	0.013	0.012
Populus nigra L.	Factory	0.044	0.040	0.420	0.483	0.518
	Botanical gardens	0.038	0.022	0.062	0.084	0.159

Source: Dolgova and Kozyukina 1972, Table 1, p. 2. Reprinted by permission of the publisher.

The larger plants displayed a faster removal rate. Evapotranspiration experiments showed that the phenol was not released to the atmosphere after uptake. Recovery of the phenol after assimilation was not accomplished, nor was a difference observed between root and leaf-floater elution components, indicating metabolic breakdown of the phenol to other compounds. Wolverton (1975) projects a removal capability of 160 kg phenol/hectare of plants/72 hr at any phenol concentration less than toxic for the plants.

The toxicity of the phenolic compounds, pyrocatechol and hydroquinone, to algae has been found to be closely associated with their ability to oxidize phenols to quinones. Increase in toxicity was associated with a drop in content of thiol group and increase in oxygen uptake (Stom et al. 1974).

Damaged plant tissue may be the direct result of interior phenolic breakdown to the highly reactive *o*-quinones initiated by a pollutant. Phenol-oxidizing enzymes probably play an important part in the processes leading to necrosis (death) of plant tissue. In some cases, naturally formed quinones apparently are closely linked with defense reactions of a plant. In diseases caused by viral, bacterial, and fungal plant pathogens, it is believed to be generally true that excessive phenol oxidation resulting from the disturbance of the normal balance between oxidative and reductive processes leads to the accumulation of toxic oxidized phenolic products. This accumulation then induces necrosis of plant tissue (Goodman, Kiraly, and Zaitlin 1967). Stom et al. (1973) demonstrated the formation of o-quinones from exogenous o-phenols in contact with plant tissue (seeds, leaves of six-day corn seedlings, leaves and stems of potatoes).

In contrast to the specific toxic action of pyrocatechol and hydroquinone on thiol group content and oxygen uptake in algae, a nonspecific action similar to the action of biological depressants (narcotics, nonelectrolytes, structurally nonspecific agents) was found for the algal toxicity of other phenolic compounds (i.e., guaicol, dimethyl ethers of hydroquinone and pyrocatechol, thymol, p- and m-creosols and β-naphthol). Data on this nonspecific action of some phenols on higher land plants are in agreement (Stom et al. 1974).

8.3.2.3 Growth inhibition and stimulation

Phenol has been implicated in plant growth inhibition; growth of wheat and corn was totally inhibited by seed germination in an atmosphere saturated with phenol or naphthalene (Chebotar et al. 1975). The growth inhibition produced by o-quinones has been demonstrated to be greater than the inhibitory action of their corresponding phenols. Benzoquinones inhibited growth of wheat coleoptile pieces (over a period of 20 hr), rooting of bean slips (over a period of six days), germination of lettuce seeds, and lengthwise growth of garden cress seedling root radicles (exposed for 2 min) to a far greater extent than did the corresponding polyphenolic dihydroxy-benzenes (Muoi et al. 1974). $Para$-benzoquinone at 120 mg/liter inhibited wheat coleoptile growth by 50%. Doses required for 50% inhibition are reproduced in Table 8.36. These data confirmed those of Stom et al. (1973) in demonstrating the greater inhibition of growth by quinones. See Table 8.19 for inhibitory concentrations of some organic compounds, including phenols, found in discharged wastewaters.

Table 8.36. Doses of polyphenols (mg/liter) depressing growth of wheat coleoptile pieces and root formation in bean slips by 50%

Substance	Growth of coleoptile pieces	Formation of roots on one radicle
Parabenzoquinone	120	500
Pyrocatechin	470	125
Resorcin	1725	1000
Hydroquinone	1825	1000

Source: Muoi et al. 1974, Table 1, p. 132. Reprinted by permission of the publisher.

Phenolic compounds have also been implicated in plant growth stimulation; root formation of Jerusalem artichoke cultivated in vitro was slightly stimulated by monophenols and highly stimulated by polyphenols. Root formation was completely inhibited by methylenedioxycinnamic acid (Rucker 1971).

Phenol stimulated growth of some algae at concentrations lower than 500 mg/liter, whereas at 500 mg/liter adaptation occurred. Many algae were able to decompose phenol at 1000 mg/liter, whereas 1500 mg/liter was lethal to most species (Maloseja, Parletic, and Munjko 1972).

Phenol at 0.5 ppm somewhat decreased the rate of carbon assimilation in phytoplankton communities, more so in August than in May and June. Concentrations at 0.1 ppm in August seemed to stimulate assimilation. At 400 ppm in May, no assimilation occurred. Above 10 ppm, dark fixation fluctuated, showing no definite trend (Niemi 1972).

8.3.2.4 <u>Tumor-promoting effect</u>

Higher plants can develop solid tumors, the formation of which is regarded as transformation of normal cells to tumor cells. Two types are recognized: crown gall disease, which is initiated by a specific bacterium, and the spontaneous neoplastic growths, which arise in interspecific hybrids, most commonly at maturity, without known cause (Braun and Wood 1969 and Tso 1972, as cited in Anderson 1973).

Growth inhibition and mutagenic effects produced in bacteria by mutagens and carcinogens have been used as measures of carcinogenic activity (Slater, Anderson, and Rosenkranz, as cited in Anderson 1973). This use is based on the assumption that, in this system, chemical carcinogens possess carcinogenic activity that is in direct proportion to mutagenic activity. An interest has arisen in using *Nicotinia* (tobacco) hybrids which are genetically tumor-prone as bioassay tests for chemical carcinogenesis. Chemical activity has been measured in these hybrids in terms of tumor counts (Buiatti 1968 and Smith 1972, as cited in Anderson 1973). Tumor growths have been reported to be stimulated in this kind of hybrid plant by stress (Smith 1962, as cited in Anderson 1973), uracil analogues (Tso and Burk 1962, as cited in Anderson 1973), plant hormone imbalance (Schaffer 1962, as cited in Anderson 1973), and radiation (Ahuja and Cameron 1963, as cited in Anderson 1973).

The phenolic compounds, pyrogallol, resorcinol, and 3-hydroxyanthranilic acid, were found to be tumor-promoting during early seedling development in one or both of two *Nicotinia* (tobacco) hybrids when applied in aqueous solution for one or two days prior to germination (Anderson 1973). Pyrogallol was the most active phenol tested, second only to 6-azauracil (not an animal carcinogen). Nitrosamines, β-propiolactone, urethane, and a water-soluble extract (containing phenolics) of cigarette-smoke condensate induced tumors in the same system.

Carbolic acid (phenol), when applied to the apical ends of stem cuttings of the poplar *Populus robusta,* was observed to cause shoot differentiation in resulting callus growths. In contrast, application of indole-butyric acid caused roots to be formed (Nanda, Kumar, and Kochhar 1975).

<u>Genetic effects</u>

Genetic effects have included chromosome breaks in *Allium cepa* (onion) root tips after treatment in solution culture with almost all of some forty substances containing phenolic OH and/or

NH_2, generally at a low level. The most active substances were pyrogallol, hydroquinone, p-phenylenediamine, hydroxyhydroquinone, pyrocatechol, p-aminophenol, and benzoquinone (Levan and Tjio 1948). The mechanism whereby creosol and nitrophenol compounds produce radiomimetic effects on chromosomes has been detailed (Sharma and Ghosh 1965). Action of some phenols and organophosphates has been classified as to antimitotic patterns (Nethery and Wilson 1966). No cytological effect was found, however, on pea seedlings exposed to phenol at up to 125 ppm in nutrient culture (Muhling et al. 1960).

Gene frequencies in the alga *Chlorella vulgaris* were unaffected by 0.01 mg/liter phenol. Control cultures exhibited a significant positive correlation between biomass and cell density, whereas correlations in phenol-treated cultures were positive but not significant. Control cultures displayed significant negative correlations between biomass, cell diameter, and cell size and density. When phenol was added, the correlation between the cell density and cell size was negative and not significant; correlations between biomass and cell diameter and between biomass and cell volume were positive, but not significant. Phenol inhibited normal development of significant correlations among growth parameters (Stein and Keller 1973).

8.3.3 Pyridines

The commercially produced compound 2-chloro-6-(trichloromethyl)-pyridine (N-Serve, Dow Chemical Co.), which selectively inhibits ammonium oxidation by *Nitrosomonas* to nitrate and is used to reduce leaching losses of nitrogen and nitrate accumulation in plant tissues, was itself phytotoxic to young bean, corn, cucumber, pea, and pumpkin plants when soil concentrations were at least 50 ppm. Up to 100 ppm failed to injure tomato plants (Mills, Barker, and Maynard 1973). The chemical frightening agent 4-aminopyridine, used to discourage blackbirds from ravishing crops, caused no visible phytotoxic effects at 0.1 to 100 ppm in corn or sorghum plants grown in nutrient solutions, but was taken up and distributed throughout the plants. Limited uptake occurred in corn grown in soils treated with the compound, indicating its unavailability for root absorption (Starr and Cunningham 1974).

8.4 TRACE ELEMENTS

During the process of combustion or coal refining, trace emissions of heavy metals, present in coal in small amounts, may occur via vaporization (Rubin and McMichael 1975). However, hazardous soil concentrations of heavy metals are probably not reached through air pollution except under conditions of heavy and protracted industrial fallout (Klein 1972). Vegetation (grasses, maple leaves, and pine needles) in the vicinity of a coal-burning power station was found to contain elevated levels of cadmium, iron, nickel, and zinc (Klein and Russell 1973). Urban soils have been reported to contain more than twice as much boron, 5 times as much copper, 17 times as much lead, and 18 times as much zinc as do rural soils (Purves 1972). Piperno (1975) states that plants are a main source of necessary trace elements for the general population. However, industrial contamination may lead to increased levels of harmful elements in food crops. Purves (1972) emphasized the long-term biochemical risks man may run if he allows contamination of the soil with toxic substances that can pass freely into plants that serve as the basis of his food supply. Although selenium rarely causes toxic responses in plants (Hemphill 1972), cases of human intoxication have resulted from the ingestion of selenized vegetables and grains (Lemley and Merryman 1941, as cited in Piperno 1975). Molybdenosis (manifested as copper deficiency) has occurred in animals that graze on plants containing naturally high levels of molybdenum to which the plants are tolerant (Hemphill 1972).

Tables 8.37 through 8.39 are compilations of the concentrations in vegetation, the effects, and the background levels in vegetation of certain trace elements and their compounds; where known, the ratio of element concentration in plants to that in soil is also given.

8.4.1 Phytotoxic effects

Reports of phytotoxic responses to heavy metals are relatively rare (Horvath 1972). However, heavy metals may alter a plant's trace element metabolism and affect its capacity for normal healthy development (Huisingh 1974). Excess exposure may cause enzymatic alterations and changes in metal requirements (vallee and Ulmer 1972). In addition, studies at Oak Ridge National Laboratory led by Van Hook and Shults (1976) indicate that terrestrial deposition of heavy metals may lead to a loss of forest productivity through a deterioration of the macronutrient cycle. Accumulation of undecomposed litter and depressions in amino sugar content and urease activity have been observed and measured in an area subjected to heavy metal fallout from lead smelters. The evidence supports previous research (Tyler 1972), which suggested that accumulations of heavy metals might have a deleterious effect on ecosystems.

Lagerwerff (1967) suggests that metal concentrations in plants and soil associated with toxicity symptoms are of little general value because they are unique for each combination of plant-soil variables. Huisingh (1974) adds that the ratios of the concentrations of various metals may be more important in determining phytotoxicity than is the absolute quantity of individual metals. For example, a high soil zinc-to-cadmium ratio reduces cadmium uptake and consequent toxicity, and an imbalance between soil copper and iron has been held responsible for iron chlorosis in citrus. Zinc-deficiency symptoms along with zinc concentrations one-fourth that of controls were found in coffee plants treated with mercury-containing sprays. The amount of lead taken up from soil by soybean plants was decreased by applications of lime, which increases the pH and calcium content of the soil, both of which may diminish the root uptake of lead (Huisingh 1974).

Whitton (1970) reports that, in general, the lack of consistency found in the literature regarding the toxic effects of heavy metals on algae precludes drawing any general conclusions.

A rise in sulfate (as well as heavy metal) levels in ponds and lakes near Sudbury, Ontario, smelters was observed to be inversely related to the numbers of aquatic plants; the pH was not responsible (Gorham and Gordon 1963). The number of species was low even in areas where the sulfuric acid was almost wholly neutralized (to pH 6.0). Table 8.37 lists effects of selected carbonate and sulfate compounds.

Cyanide is a plant respiratory inhibitor (Brook 1964) and a radiomimetic agent in the presence of oxygen (Kihlman 1957); as hydrocyanic acid, cyanide is known to influence the dark reaction in photosynthesis (Curtis and Clark 1950). Certain plants such as bird's-foot trefoil (*Lotus corniculatus*) are cyanogenic, releasing cyanide from β-glucosides upon mechanical injury or upon fungal infection (Fry and Millar 1971). The effects of cyanide and thiocyanate compounds on algae are given in Table 8.39.

8.4.1.1 Tolerance

Accumulator plants must have developed a mechanism for coping with potentially toxic concentrations. A cress (*Arabidopsis thalianum*) that commonly grows on zinc tailings has been observed

Table 8.37. Selected plant-element interactions

Compound	Organism	Concentration (ppm)[a]	Effect	Source
Arsenic				
AsO_2	Freshwater microflora	0.1	Reduces heterotrophic activity	1
	Myriophyllum spicatum (Eurasian watermilfoil)	2.9	50% inhibition of root weight	2
Sodium arsenite	*Scenedesmus*	35 - 46	Toxicity threshold (4 days)	8
	Cowpeas	1.0	Reduced yield	4
	Peas	9.0	Reduced yield	4
	Barley	2.0	Reduced yield	4
	Rice	7.0	Reduced yield	4
Sodium arsenite	Beans, strawberries	50 - 125	Detrimental to growth	4
	Apple orchards	50 - 100	Reduces growth 50%	4
	Apple orchards	>100	Very little growth	4
Lead arsenate	String beans, lima beans, various vegetable seedlings	1 - 200 (lb/acre)	Reduced germination; retarded seedling growth	4
Sodium arsenate	Apple seedlings	100 - 160	Killed	4
	Corn kernels	80 - 100	No development	4
As_2O_3	White spruce trees	1000 - 2000	Damaged foliage and buds; retarded growth	5
	Artemesia tridentata (big sagebrush)	Bathed by goldmine tailings dust	Most affected in area	6
Aluminum	*Scenedesmus* (alga)	1.5 - 2.0 (mg/liter)	Inhibitory concentration	3
Barium	*Scenedesmus*	34 (mg/liter)	Inhibitory concentration	3
			Poisonous to most plants	1
	Freshwater microflora	0.1	Reduces heterotrophic activity	
Boron	*Ulva lactuca* (alga)	0.1 - 1 (mg/liter)	Stimulatory level	3
	Higher plants	0.5 - 1 (mg/liter)	Inhibitory level	3
	Fruit trees	0.5 - 1.0	Required for growth	1
		2.0	Possibly toxic	1
	Oats, radishes, clover	>3.0	Abnormal growth	9
Cadmium	*Navicula pelliculosa* (alga)	14 - 140 (mg/liter)	Inhibitory concentration	3

Table 8.37. (continued)

Compound	Organism	Concentration (ppm)[a]	Effect	Source
Cadmium	Radishes	100 (ppb)	Growth of roots and tops reduced	4
	Rice, wheat	>10 (μg/g soil)	Decreased yields	11
	Brown algae	900[b]	No apparent injury	14
	Freshwater plants	1620[b]	No apparent injury	14
	Chlamydomonas (alga)	0.1	Critical level for growth impairment	7
	Euglena (alga)	10	Critical level for growth impairment	7
Chromium	Algae	1.39	Drastic decreases in production	8
CrO_4	Algae	0.139	Slight but significant decrease in production	8
Dichromate	Algae	0.32 - 16.0	Completely inhibiting for 56 days (15-20°C)	8
	Soybeans	0.5	Decreased concentrations and uptake of Ca, K, P, Fe and Mn in tops and of K, Mg, P, Fe, and Mn in roots (in nutrient solution)	8
Cr_2O_7	*Scenedesmus*	0.7	Inhibitory concentration	3
	Navicula seminulum	0.2	Inhibitory concentration	3
	Macrocystis pyrifera	1.0	Inhibitory concentration	3
Cobalt (as chloride)	*Scenedesmus quadricauda*	1.0	Toxic threshold; 4 days, 24°C	8
Cerium	*Scenedesmus*	0.15 - 0.20 (mg/liter)	Inhibitory concentration	3
Copper	*Chlorella* sp.	0.006	Stimulatory level	3
	Chlorella vulgaris	0.0064	Inhibitory level	3
		0.05 M $Cu_2{}^+$	Inhibition of root growth	1
	Lettuce	0.1 M $Cu_2{}^+$	Stops germination	1
		>30 ppm EDTA-extractable Cu	Toxic levels for oats, radishes, clover plants	9
Iron	*Chlorella pyrenoides* (alga)	0.0015 - 0.01 (mg/liter)	Stimulatory level	3
				3

Table 8.37. (continued)

Compound	Organism	Concentration (ppm)[a]	Effect	Source
Lanthanum	*Scenedesmus*	0.15 (mg/liter)	Inhibitory concentration	3
Lead	Microflora	0.1 (Pb_2^+)	Reduces heterotrophic activity	1
	Higher (forage) plants	150	Critical level for plant growth corresponding to 300 ppm in soils (0.5 *M* acetic acid extractable)	10
	Macrocystis pyrifera	4.1 (mg/liter) (Pb precipitated)	No inhibition	3
Manganese	*Dunaliella tertiolecta*	0.005 - 0.02 (mg/liter)	Stimulatory level	3
	Anabaena, Aphanizomenon	0.005 (mg/liter)	Inhibitory level	3
Molybdenum			Essential to higher plants	1
	Nostoc muscorum	0.00005 (NO_3 red.)	Stimulatory level	3
		0.01 (N fix.)	Stimulatory level	3
	Scenedesmus	>0.0001	Stimulatory level	3
	Anabaena cylindrica	0.001	Stimulatory level	3
	Scenedesmus	54 (mg/liter)	Inhibitory level	3
Mercury	*Scenedesmus*	0.03 (mg/liter)	Inhibitory concentration	3
	Macrocystis pyrifera (Kelp)	0.05 (mg/liter)	Inhibitory concentration	3
	Microflora	0.1	Reduces heterotrophic activity	1
Mercuric chloride	*Chlorella vulgaris*	27 (ppb, as Hg)	Inhibitory concentration	16
		13 (ppb)	Maximum tolerance level	16
	Plumaria elegans (red algae)	0.12 - 1	Growth inhibition of 20-100%	12
	Scenedesmus	30 (ppb)	Lethal concentration	13
Mercuric cyanide	*Scenedesmus*	150 (ppb)	Lethal concentration	13
Ethyl Hg	Phytoplankton, marine mixture	60 (ppb)	Lethal concentration	13
Various organic Hg compounds	Freshwater plankton	1-50 (ppb)	Drastic decrease in rate of photosynthesis	16

Table 8.37. (continued)

Compound	Organism	Concentration (ppm)[a]	Effect	Source
Nickel			A micronutrient	1
	Scenedesmus	0.09 - 1.5 (mg/liter)	Inhibitory concentration	3
	Macrocytis pyrifera	0.001	Inhibitory concentration	3
	Freshwater microflora	0.01	Reduces heterotrophic activity	1
Selenium	*Scenedesmus*	2.5 (mg/liter)	Inhibitory concentration	3
	Aquatic saprophytic microflora	2.5 (mg/liter)	Inhibits growth and biochemical oxygen demand	1
Silver	Microflora	0.0001	Reduces heterotrophic activity	1
Silicon	*Asterionella* (alga)	0.5 (mg/liter)	Stimulatory level	3
Thorium	*Scenedesmus*	0.4 - 0.8 (mg/liter)	Inhibitory concentration	3
Vanadium			Essential for green algae and stimulates higher green plants in small amounts	1
	Scenedesmus obliquus	0.1 (mg/liter)	Stimulatory level	3
	Higher plants	10-20 (mg/liter)	Inhibitory level	3
Tin	Some aerobic microbes		Toxic	1
Rubidium	*Scenedesmus*	14 (mg/liter)	Inhibitory level	3
Zinc	Phytoplankton	0.1 - 0.5	Reduced photosynthesis over 12 hr in soft water	8
	Chlorophyta (range for 7 species)	0.08 - 4.0	Maximum noninhibitory	8
		0.22 - 6.0	Minimum lethal	8
	Nitzschia closterium	0.25	Division rate reduced minimum dose	8
Zinc chloride	*N. linearis*	4.3	96-hr threshold limit mean, acute, 16 - 20°C	8
	N. linearis	4.3	5-day threshold limit mean, acute	8
Zinc sulfate	*Scenedesmus quadricauda*	1.0 - 1.4	Toxicity threshold (4 days at 24°C)	8
	Macrocystis pyrifera	1.31	No great effect on photosynthesis	8
		10.0	(4 days) 50% inactivation of lower fronds (4 days)	

Table 8.37. (continued)

Compound	Organism	Concentration (ppm)[a]	Effect	Source
Zinc (continued)				
	Microflora	0.1	Reduced heterotrophic activity	1
	Oats, radishes, clovers	200 acetic acid-extractable Zn	Abnormal growth	9
	Chlorella pyrenoidosa	0.0065 (mg/liter)	Stimulatory level	3
	Scenedesmus	0.5 - 0.7 (mg/liter)	Inhibitory level	3

[a]All concentrations in ppm except when marked otherwise.
[b]Concentration factors calculated from dividing ppm in fresh organism by ppm in water.

Sources:
1. Smith, Ferguson, and Carlson 1975.
2. Stanley 1974.
3. North, Stephens, and North 1970.
4. Ratsch 1974.
5. Rosehart and Lee 1973.
6. Comanor et al. 1974.
7. Buehler and Hirshfield 1974.
8. Becker and Thatcher 1973.
9. Purves 1972.
10. Bohn 1972.
11. Friberg 1974.
12. Boney 1971.
13. Wallace et al. 1971.
14. Fassett 1972.
15. Turner and Rust 1971.
16. D'Itri 1971.

Table 8.38. Trace elements in vegetation, soil, and coal — selected examples

Element	Background level in vegetation (ppm)	Source[a]	Levels in plants consumed as foods (ppm)	Source[b]	Ratio of plants to soil[c]	Coal[c] (ppm)	
						1.50	0.2 - 10
Arsenic	0.2	2	<0.5 (most) >1.0 (rare)	8	0.03	X	
Boron	21.0 (rural cabbage) 28.8 (urban cabbage)	4					
Aluminum			0.5 - 5.0 ppm (vegetables)	8			
Barium					0.11	X(+)	
Beryllium	0.64				0.03		X
Cadmium	0.5 - 0.6 (leaves) 0.2	2	0.14 μg/g wet wt (leafy vegetables) 0.04 μg/g wet wt (legumes) 0.08 μg/g wet wt (root vegetables) 0.07 μg/g wet wt	6	5.3		
Chromium	0.12	5	0.07 - 0.13 (various wheats) 0.1 - 0.5 (most) 1.0 (some) 0.30 - 0.38 (various wheats)	10 8 10	0.01	X	
Cobalt	None detected	5	0.6 - 0.4 (spinach) 0.2 (cabbage, lettuce) 0.01 (cornseed) 0.003 (white flour)	8			
Iodine	<0.001 - 12,500 (sea plants) <0.1 - 20 (land plants)	6	280 μg/kg wet (vegetables) (medium)	8			

Table 8.38 (continued)

Element	Background level in vegetation (ppm)	Source[a]	Levels in plants consumed as foods (ppm)	Source[b]	Ratio of plants to soil[c]	Coal[c] (ppm)	
						1.50	0.2 - 10
Copper	5 - 20	2	10 - 15 (most)	8			
	14	2	25 (leafy vegetables) (rare)				
	4.1 (rural cabbage) 3.6 (urban cabbage)	4	4.5 - 5.1 (various wheats) 5.1	10			
	10	5	20 - 40 (nuts, dried legumes, dried vine and stone fruits, cocoa)	8			
Fluoride	80	5	Tea, camellias >100	8			
Iron	80	5	43 (mean content of whole grain)	8			
Lead	25 - 150 100 0.81 (rural cabbage) 0.73 (urban cabbage)	2 2 4	<1 (most) 0.42 - 1.0 (wheat)	8 10	2.3	X	
Manganese			Concentrated in the germinal portion of grains, fruits, nuts, leafy vegetables; tea and spices are rich in Mn; root vegetables contain little. 0.30 - 1.3 mg (tea) 20 - 30 (cereals, leafy fresh vegetables)	8 8			
Mercury	0.01 - 0.10 or less	7	0.005 - 0.0035 (fruits, vegetables, and grains)[d]	8	0.05		X(-)

Table 8.38 (continued)

Element	Background level in vegetation (ppm)	Source[a]	Levels in plants consumed as foods (ppm)	Source[b]	Ratio of plants to soil[c]	Coal[c] (ppm)	
						1.50	0.2 - 10
	10 - 200 ppb (15 ppb, most plants)	2	0.04 (pome fruit)	7			
	0.015	2	0.02 (tomatoes)				
			0.01 (potatoes)				
			0.08 (wheat, barley)				
	500 - 3500 (near Hg deposits)						
	0.09	5					
	0.023 - 0.037 (marine algae)	7	0.15 (wheat, Japan)	7			
			0.15 - 0.40 (wheat, Canada)				
			0.015 (rice)				
Molybdenum	1.15 (rural cabbage)	4	0.2 - 4.7 (leguminous seeds)	8			
	1.83 (urban cabbage)		0.12 - 1.14 (cereal grains)				
			0.48 (whole grains) (mean)				
Nickel	0.85 (rural cabbage)	4	7.6 (tea)	8	0.05	X	
	0.52 (urban cabbage)		0.15 - 0.35 (fruits, tubers, grains)				
			0.15 - 0.35 (green leafy vegetables)				
			0.29 - 0.47 (various wheats)	10			
Selenium			0.007 - 1.3 (whole wheat grain)	8	1.0	X	
Strontium					0.09	X(+)	
Vanadium					0.008	X	
Titanium	None detected	5					
Zinc[e]	37	5	21.6 - 30.2 (various wheats)	10			
	20.9 (rural cabbage)	4	>1 (pome, citrus fruits)	8			
	30.2 (urban cabbage)		40 - 120 (wheat germ, bran)				

Table 8.38 (continued)

Element	Background level in vegetation (ppm)	Source[a]	Levels in plants consumed as foods (ppm)	Source[b]	Ratio of plants to soil[c]	Coal[c] (ppm)	
						1.50	0.2 - 10
	25 - 150 (normal for a variation of plants)	2	25 - 40 (whole cereal grains)	8			

[a] Refers to information given in "background level" column.
[b] Refers to information given in "plants consumed as food" column.
[c] All information from Hall, Varga, and Magee 1974.
[d] The authors consider this to be a large underestimation.
[e] Information concerning the zinc content of more than 200 foods has been compiled by the U.S. Department of Agriculture.

Sources:
1. Hall, Varga, and Magee 1974.
2. Ratsch 1974.
3. Comanor 1974.
4. Purves 1972.
5. Klein and Russell 1973.
6. Hemphill 1972.
7. Wallace et al. 1971.
8. Underwood 1973.
9. Friberg 1974.
10. Zook, Greene, and Morris 1970.

Table 8.39. Selected plant interactions with carbonates, sulfates, and cyanide compounds[a]

Compound	Plant	Concentration (ppm)	Effects
Sodium carbonate	*Nitzschia linearis* (algae)	242	5-day threshold limit mean
Sodium bicarbonate	*Nitzschia linearis*	650	5-day threshold limit mean
Sodium sulfate	*Nitzschia linearis*	1,900	5-day threshold limit mean, acute; 16 to 22°C (synthetic dilution water)
	Water plants	5000 - 10,000	Not toxic; temporarily stimulated (0.5 - 1.0%)
	Algae	<15,000	Not toxic; flourished (<1.5%)
Calcium sulfate	*Nitzschia linearis*	3,200	96-hr threshold limit mean
	Navicula seminulum	3,200	50% reduction in growth rate
Potassium cyanide	*Scenedesmus quadricauda*	0.16	Toxicity threshold, 4 days; 24°C
CN^{-}[b]	*Scenedesmus quadricauda*	26	Inhibitory concentration
Potassium ferrocyanide	*Scenedesmus quadricauda*	0.15	Toxicity threshold, 4 days; 24°C
$K_3Fe(CN)_6$[b]	*Scenedesmus quadricauda*	0.25	Inhibitory concentration
$K_4FE(CN)_6$[b]	*Scenedesmus quadricauda*	0.2	Inhibitory concentration

[a]From Becker and Thatcher 1973, unless indicated otherwise.
[b]From North, Stevens, and North 1970.

to grow on sacks of pure zinc oxide. It has been established that tolerance is genetically dominant over nontolerance in research with the lead-tolerant grass, *Festuca ovina*. In the copper-tolerant monkey flower, *Mimulus guttatus*, tolerance was dominant at low copper concentrations, intermediate at intermediate concentrations, and recessive at high concentrations (Peterson 1971). Tolerance to one element does not imply tolerance to other closely related ones. The copper-tolerant grass *Agrostis tenuis* does not tolerate nickel and zinc, both adjacent to copper in the periodic table (Peterson 1971).

Mechanisms for tolerance include

1. Exclusion of all or parts of the element from the cells (e.g., the grass *Triodia pungens* and the legume *Tephrosia* sp. n. grow on lead concentrations of up to 1000 ppm without accumulating it). The method of exclusion is unknown.

2. The element is confined to cell walls (e.g., the zinc-tolerant grass *Agrostis* spp. accumulates zinc in insoluble forms in root cell walls; in nontolerant populations, the zinc has been found in soluble form in roots, not in the cell walls). The mechanism may reside in cation binding sites in the cell wall components.

3. The element is metabolized within the cell to inactive components (e.g., selenium accumulator plants such as *Astragalus* spp. may exclude the selenium from their proteins by synthesizing it to unusual selenoamino acids; additionally, high levels of fluorine are not accumulated in species that do not metabolize fluorine), or the element binds with another compound (e.g., organic acids in plants may act as chelating agents, complexing with toxic quantities of aluminum from fly ash from coal-fired stations) (Peterson 1971).

Peterson (1971) suggests several hypothetical mechanisms for tolerance: biosynthesis of an enzyme with altered specificity towards the toxic compound; formation of an altered pathway to avoid synthesis of the toxic compound; excess accumulation of competing metabolites; and an increased activity of the normal pathway to successfully compete with the toxic compound.

Concentration and form of a trace metal may have a definite effect on algal competition by effecting shifts that may greatly reduce a system's productivity (Patrick, Bott, and Larson 1975). A dominant nuisance alga was found to have been artificially selected for by its resistance to high concentrations of trace minerals toxic to other algae studied in Raritan Bay (McLaughlin 1965, as cited in Prager 1974).

Some plants apparently develop a tolerance to normally phytotoxic levels of metals (Antonovics 1973). The mechanisms by which plants become tolerant are not clear, but they are known to be under genetic control (Huisingh 1974). Some evidence suggests that the mechanisms involve the ability of that plant to take up and accumulate the metal in cell walls because tolerant strains accumulate more than do intolerant ones in the presence of normal levels of nutrients (Catcheside 1975).

8.4.1.2 Sensitivity

There are decided species differences in uptake and accumulation and also in sensitivity. Ragweed has been observed to grow luxuriantly in high soil zinc concentrations, whereas surrounding vegetation was stunted (Hemphill 1972).

8.4.2 Uptake

The chemical form of an element is an important factor in the consideration of plant uptake. Lead oxide is readily absorbed, whereas lead sulfate (galena) is not. The presence or absence of essential nutrients and soil pH have been shown to affect lead uptake (Miller et al. 1974; John 1972). In general, heavy metal uptake by plants is greater at a soil pH of 5.5 than at 6.5 to 7.0 (Huisingh 1974).

The manner in which the element is presented to the plant, that is, by air (foliage) or water (roots), appears to affect uptake and toxicity. Red oak seedlings were planted in two soils differing in their levels of zinc and cadmium (one having from 2000 to 20,000 ppm zinc and 10 to 500 ppm cadmium and the other having 100 to 300 ppm zinc and 2 to 3 ppm cadmium), but exposed to the same very high atmospheric levels of zinc and cadmium from a smelter source. Severe stunting occurred in the seedlings grown at high soil metal levels although they had foliar levels of only 400 to 500 ppm zinc and 4 to 7 ppm cadmium (Buchauer 1973). Seedlings grown in low soil metal levels appeared healthy despite high metal foliar levels, after the leaves had been washed, of 2120 ppm zinc and 38 ppm cadmium. Possibly, the metal particles that gain entry via stomates remain biologically inert, whereas metals entering via root uptake can cause immediate phytotoxicity.

8.4.2.1 Essential trace elements

Trace elements known to be essential for plant growth are iron, copper, zinc, manganese, molybdenum, cobalt, boron, chlorine, sodium and possibly vanadium. All plants do not require all these elements, but each has been shown to be required by some plant. Current knowledge of the functions of these elements in plants has been reviewed by Nicholas (1975).

8.4.2.2 Selectivity

Plants are not indiscriminate in their uptake of elements. According to Hall, Varga, and Magee
(1974), plants seem to selectively reject (enrichment ratio below 0.03) two groups of elements
having adjacent atomic numbers: scandium, titanium, vanadium, and chromium (21 to 24); antimony
and tellurium (51,52). Plants select cadmium, lead, silicon, selenium, and lithium. Certain
plants are capable of concentrating specific elements to the extent that they are used as indicators of soil composition, mineral deposits, or pollution (Peterson 1971). It has been observed,
for instance, that plants may absorb cadmium preferentially by as much as an 18 to 1 ratio over
zinc and by a 22 to 1 ratio over lead and copper. Cadmium uptake can be reduced if the soil
cadmium-to-zinc ratio is kept at 1:1000 and pH is near neutral (Huisingh 1974).

8.4.2.3 Accumulation

The ability of plants to concentrate metal ions is minimized by a number of factors; ion inactivation, soil fixation, accumulation at the soil surface above the root zone, exclusion at the
root surface, and immobility in the root (Bohn 1972). The result is often a poor correlation
between the concentration of a metal in soil and that in the plant part. The order of reduced
plant availability (by soil fixation) of heavy metal ions roughly follows the order of increasing solubility of the hydroxyoxides: copper > lead > zinc > nickel > cadmium (Bohn 1972). The
ratio of trace element content of a soil to that in plants growing on that soil is not simple
and has been shown to vary in mature vegetables according to season by factors ranging from
1 to 30 (Warren and DelaVault 1971). Additionally, accumulation to toxic levels is species-
dependent (Ratsch 1974).

Peterson (1971) has outlined unusual accumulations of elements in higher plants. The term
accumulation refers to (1) plants that accumulate an element to levels higher than those in
the growth medium and (2) plants that accumulate to higher levels although nearby species contain normal levels (e.g., the accumulation of selenium to levels as high as 15,000 ppb on
seleniferous soil compared with less than 0.01 ppm in nearby pasture grass). With sufficiently
sensitive methods, most naturally occurring elements can be detected in plants, although some
may be confined to a certain plant part (e.g., *Lecythis ollaria* can accumulate selenium to high
levels only in the seed). The high levels accumulated may be toxic to other plants and/or
organisms or may be metabolized to a toxic compound (e.g., fluorocitrate, an organic compound
metabolized in livestock after their ingestion of fluoroacetate which is formed within the
fluorine-accumulator plant *Gastrolobium grandiflora*). Table 8.40 presents a sampling of
accumulator plants selected for their diversity.

Species that can grow on areas of concentrated element deposits (such as some ore deposits) are
often used as indicators; some species are associated only with a particular element, whereas
others are widely dispersed species that grow and predominate over an ore deposit. Occasionally,
an ecotype will develop which differs markedly in morphology from the normal. For instance, the
anemone (*Pulsatilla patens*) growing on mineral deposits in the southern Urals is abnormally
white, has no petals, and has a reduced floral envelope (Peterson 1971).

Table 8.40. Some unusual accumulations of elements by various plants[a]

Atomic Number	Element	Species	Concentration (%)
9	Fluorine	*Dichapetalum cymosum*	(0.02)
13	Aluminum	*Symplocos tinctoria*	50
14	Silicon	*Equisetum species*	40
23	Vanadium	*Astragalus confertiflorus*	0.09
23	Vanadium	*Amanita muscaria*	0.2
23	Vanadium	*Mielichhoferia macrocarpa*	0.1
24	Chromium	*Pimelea suteri*	2.6
24	Chromium	*Leptospermum scoparium*	2
27	Cobalt	*Clethra barbinervis*	0.07
27	Cobalt	*Nyssa sylvatica*	0.08[b]
27	Cobalt	*Crotalaria cobalticola*	1.8
28	Nickel	*Pimelea suteri*	2.5
28	Nickel	*Alyssum bertolonii*	10
29	Copper	*Becium homblei*	0.3[b]
29	Copper	*Mielichhoferia macrocarpa*	0.1
30	Zinc	*Thlaspi calaminare*	13
30	Zinc	*Arabidopsis thalianum*	6
30	Zinc	*Equisetum arvense*	0.9
34	Selenium	*Astragalus racemosus*	1.5[b]
38	Strontium	*Arabis stricta*	1
39	Yttrium	*Asplenium septentrionale*	0.1
50	Tin	*Silene cucubalis*	0.002
56	Barium	*Bertolettia excelsa*	0.43[b]
57+	Rare earths	*Carya glabra*	0.2[b]
74	Tungsten	*Pinus sibiricus*	0.15
75	Rhenium	*Astragalus pattersonii*	0.03
80	Mercury	*Betula papyrifera*	0.1
83	Bismuth	*Marchantia polymorpha*	0.005
92	Uranium	*Uncinia leptostachya*	3.6
92	Uranium	*Coprosma arborea*	3

[a]All values are expressed in terms of per cent ash weight unless specified otherwise; use of parentheses indicates that the units were not clearly stated; an asterisk indicates that the species belongs to the lower plant category.
[b]Dry weight.

Source: Peterson 1971, Table 1, p. 509. Reprinted by permission of the publisher.

In an aquatic environment, algal communities and leaf litter have been shown to bind large quantities of heavy metals, especially lead, in the vicinity of industrial activities (Gale, Marcellus, and Underwood 1974). Both duckweed and an aquatic fern were found to accumulate zinc and cadmium even from very low available concentrations (Hutchinson and Czyrska 1972). Windom (1973) suggests that the flux of heavy metals through salt marsh estuaries may include recycling by marsh vegetation uptake.

8.4.3 Metabolism

Lead, mercury, cadmium, and arsenic are known to form complexes with enzymes or their substrates in living systems (Auerbach 1975). Other heavy metals appear to bind specifically within the cytoplasmic membrane, causing changes in membrane permeability (Kamp-Nielson 1971, as cited in Auerbach 1975).

8.4.4 Mutagenic effects

Lead, arsenic, antimony, bismuth, barium, sodium, potassium, and magnesium are reported to have been more or less inactive in causing mutagenic changes in pea (*Pisum*) rootlets, whereas zinc, cerium, beryllium, aluminum, nickel, iron, lithium, lanthanum, calcium, manganese, strontium, and tin were somewhat more active. Very active elements (in order of decreasing activity) were tellurium, cadmium, copper, osmium, mercury, silver, titanium, tantalum, gold, platinum, chromium, and cobalt. Halogens, if in an electrically attractive form, were found to induce chromosome breakages (von Rosen 1954). Mutagenic and radiomimetic effects of heavy metal ions have been observed in pea seeds and growing pea plants (*Pisum abyssinicum*). Additive effects occurred when combinations of ions and/or individual metals in combination with radiation were applied (von Rosen 1964).

8.4.5 Elements

According to Comar (1975), trace elements of particular importance from coal combustion are lead, mercury, arsenic, vanadium, selenium, nickel, cadmium, and fluoride. These are the more volatile elements; others can be equally important as leachable residues in ash and as components of wastewater. The discussion of individual elements which follows is extremely limited. The material has been chosen merely to indicate the complexity of the interactions between plants and elements.

8.4.5.1 Arsenic

Inorganic arsenic is not readily absorbed from soils, but when absorbed, it tends to accumulate in plant roots with little translocation (Rosehart and Lee 1973; Hiltbold 1975). Surface deposits may be easily removed by washing (Underwood 1973). Some cabbage and lettuce samples near a Tacoma, Washington, copper smelter were found to exceed arsenic tolerance levels, posing a possible health hazard to humans (Ratsch 1974).

Chemical form and phytotoxicity are closely related with regard to arsenic, the compounds of which are used as aquatic weed killers with varying results (Becker and Thatcher 1973). Big sagebrush (*Artemesia tridentata*) was observed to be the plant most sensitive to the toxic effects of arsenic (1% and higher) deposited on the soils surrounding a Nevada gold mine (Comanor et al. 1974). The soil achieved a significant reduction in the phytotoxic effects by its ability to complex soluble arsenic into insoluble forms.

8.4.5.2 Aluminum

Excess exposure to aluminum has been reported to cause chromosomal aberrations and enzymatic alterations in plants (Schubert 1973). Roots accumulate excesses of aluminum from soil, which reduces their ability to translocate phosphate; the resulting phosphate starvation is not correctable by phosphate addition to soil (Russell 1954).

8.4.5.3 Boron

Dramatic increases in plant uptake can occur with small increases in soil content (Purves 1972). Although it is a plant nutrient, boron may be highly toxic to plants at high concentrations (Purves 1972). Toxic levels of boron were frequently found in bean, barley, and tomato leaves from plants irrigated with various sewage sludges (Bradford et al. 1975) (Sect. 8.2.11.1).

8.4.5.4 <u>Beryllium</u>

Beryllium is less likely to be available to plants than are the heavy metals (Bohn 1972). According to Durocher (1969), no harmful effects to vegetation at outdoor concentrations have been observed, although reduced growth rates have been seen in bush beans when beryllium is present in nutrient solutions in excess of 1 ppm. Levels of beryllium just below the reported toxicity level of 1 ppm were passively absorbed by excised roots of barley (*Hordeum vulgare*) and reduced by 44% in 1 hr with demineralized water (Holst, Schmid, and Yopp 1975). Zinc and aluminum slightly inhibited beryllium absorption, whereas magnesium had no influence and 500 μM dinitrophenol enhanced beryllium absorption. Respiration was unaffected by 1000 ppm beryllium even after 18 hr. (Holst, Schmid, and Yopp 1975).

8.4.5.5 <u>Cadmium</u>

Leaves of urban woody plants were found to accumulate excessive amounts of cadmium in solutions containing only a few tenths of a microgram per milliliter (Smith, Ferguson, and Carson 1975). Uptake is especially easy for grasses and grains (wheat, corn, rice, oats, millet) (Ratsch 1974). Tomatoes, barley, and cabbage are more tolerant to cadmium than are lettuce, radish, beans, and turnips (Smith, Ferguson, and Carson 1975).

Cadmium enrichment has been found in various plants, from bryophytes through vascular species, growing in areas subjected to pollution from industry (Tyler 1972; Klein and Russell 1973; Buchauer 1973; Hemphill and Pierce 1974) and in heavily traveled areas (Lagerwerff and Specht 1970); greater accumulations occur under these conditions than in plants growing on cadmium-enriched soils (Denaeyer-de Smet 1974). Bryophytes are a suggested indicator species (Tyler 1972). John, Chuah, and VanLaerhoven (1972) found that most absorbed cadmium remained in oat plant roots and that soil treatments affected root, but not top, content. Tomato plants took up cadmium from nutrient solutions, concentrating the largest amounts in the leaves and the lowest amounts in fruits (Steiner 1973). Linneman et al. (1973) report that absorption of cadmium was increased in wheat by lowering soil pH.

High soil cadmium levels were found to result in reduced yields for grains, radishes, and lettuce (Friberg 1974), whereas considerable concentration has been observed to occur in certain vascular water plants and algae without apparent injury (Fassett 1972). Cadmium has been shown to cause chromosomal aberrations in plants (Schubert 1973). Diffusion, when followed by sequestering, was found to be the probable mechanism by which cadmium accumulated in intact barley plants (Cutler and Rains 1974). Growth and greening of excised embryos of red kidney beans were completely inhibited by 3 mg/liter cadmium salts, which caused acute curvature in the embryonic axis (Imai and Siegel 1973). Excessive levels of cadmium were frequently found in bean, barley, and tomato plants when extracts of sewage sludges from various sources were used for irrigation (Bradford 1975). Leaves of barley contained less than 3 ppm cadmium and the grain contained less than 0.25 ppm when irrigated at different rates with liquid sewage sludge containing different concentrations of cadmium. Root content varied from 3.3 to 16.2 ppm according to the cadmium concentration and increased with decreasing frequency of irrigation (Kirkham 1975). Fulkerson and Goeller (1973) review the subject in depth.

8.4.5.6 Chromium

Chromium was not detected in vegetation around a coal-burning plant although soils were found to be chromium-enriched; nor was it detected in background level control sites (Klein and Russell 1973). Lyon et al. (1971), in reporting on the trace element content of various soils and the various plant species growing on them, found large differences between the mean chromium content of species, reflecting different relative accumulations of the element — some species excluded chromium, others accumulated it. Some plants contained up to 6% chromium in the leaf ash, indicating either a high tolerance for this metal or an unusual requirement. Bean, barley, and tomato leaves from plants irrigated with various sewage sludges frequently showed excessive levels of chromium (Bradford et al. 1975). Chromium toxicity appears to be related to its tendency to accumulate in root cells, where the toxic responses are generally elicited (Hemphill 1972). In some cases, low rates of addition may prove stimulatory; higher rates are generally thought to be toxic (Turner and Rust 1971). Soil chromium of 5 ppm or more is reported to interfere with accumulation of calcium, potassium, magnesium, phosphorus, boron, and copper by soybean tops, with little or no effect on iron, manganese, or zinc uptake (Turner and Rust 1971).

8.4.5.7 Cobalt

Cobalt uptake varies with plant species (Lyon et al. 1971). Klein and Russell (1973), however, using conventional atomic absorption analysis, failed to detect cobalt in either control plants or vascular plants growing on cobalt-enriched soils around a coal-burning plant in Michigan. Black gum is reported to absorb unusual amounts of this metal (Dorn and Phillips 1973). Excessive amounts of cobalt were frequently found in bean, barley, and tomato plants irrigated with extracts of various sewage sludges (Bradford et al. 1975). At 10 mg/liter cobalt, greening of excised red kidney bean embryos was inhibited, whereas curvature of the embryonic axis was unaffected (Imai and Siegel 1973). Cobalt is an algal micronutrient at low concentrations, but at higher levels is very toxic to algae as well as to higher plants (Becker and Thatcher 1973).

8.4.5.8 Copper

Deficiency symptoms may occur in plants at less than 5 ppm copper in soil. At soil concentrations above 20 ppm, toxic symptoms may occur (Hemphill 1972). Selective enrichment occurs in higher plants (Tyler 1972; Ratsch 1974; Buchauer 1973; Purves 1972), with many vegetables being copper-sensitive (Ratsch 1974). Death or stunting in plants is expected to occur before uptake levels toxic to animals are reached (Horvath 1972). Brams and Fiskell (1971) found that copper accumulation by citrus seedling roots occurs in the exodermis, endodermis, and pericycle cells, causing root deformation; they concluded that mechanisms of root accumulation may include reactions of copper with protein.

Excessive amounts of copper were frequently found in bean, barley, and tomato plants irrigated with extracts of various sewage sludges (Bradford et al. 1975). Distinct differences occur in copper sensitivity among algal strains, and various concentrations of copper sulfate used as an algicide are toxic or inhibitory to many algae (Stokes, Hutchinson, and Krauter 1973; Becker and Thatcher 1973). Copper sulfate is also a fungicide.

8.4.5.9 <u>Iodine</u>

Corn has been found to be sensitive to 8 ppm iodine, whereas head lettuce has tolerated 160 ppm (Piperno 1975).

8.4.5.10 <u>Iron</u>

In low concentrations, iron is an algal micronutrient (North, Stephens, and North 1970). Its uptake is favored by low pH (Horvath 1972). Excesses of manganese have caused plant iron deficiencies in plants (Piperno 1975).

8.4.5.11 <u>Lead</u>

Lead has been found in all plants, but it is not essential. Although high levels have been reported to be tolerated by many plants, others have shown retarded growth at 10 ppm in solution culture (Hemphill 1972). Lead can be accumulated by crop plants, mosses, and hardwoods when air and soil concentrations are increased by pollution (Smith 1973; Blokker 1972; Ward, Brooks, and Reeves 1974; Hampp and Holl 1974). Accumulation can occur by absorption through the roots and leaves with little translocation within the plant (Hampp and Holl 1974; Purves 1972). Ratsch (1974) reported some translocation in corn in nutrient solutions, with high concentrations accumulating in the leaves. Compared with soil concentrations, uptake was limited.

Although it has been reported that lead contamination may be largely removed by washing (Piperno 1975), foliar absorption of atmospheric lead by radishes (Rule, Hemphill, and Pierce 1974), oats, and lettuce (Rabinowitz 1972), with subsequent translocation, has been demonstrated. The highest total amount, the highest percentage translocated, and the highest final concentration in radishes were all in the edible portion of the plant. Rule, Hemphill, and Pierce (1974) suggest that, under conditions of heavy contamination, the amount of lead translocated could become significant. Translocation is evidently very dependent on growth conditions, particularly those under seasonal control (Haye et al. 1975). In many plants grown under lead pollution conditions, the edible portions have been found to contain only slightly more lead than the controls, whereas the nonedible portions contained two to three times as much lead (Bohn 1972). Research aimed at elucidating the role that cuticular and epicuticular waxes play as barriers to lead uptake led Arvik and Zimdahl (1974) to conclude that lead contamination of the fruits and leaves of various plants (tomato, pepper, onion, apple, soybean, beet, corn, radish, *Philodendron* sp.) was largely exterior to the plants and that penetration of lead in large quantities cannot occur. Even under exceptional conditions of lead solubility, pH, and exposure times, only very small amounts of lead crossed cuticular membranes. Regardless of cuticle thickness or stomatal condition, the species of plant seemed to be related to the penetration of lead (Arvik and Zimdahl 1974). On the other hand, leaf samples from bean, barley, and tomato plants that had been irrigated with extracts from various sewage sludges frequently showed excessive amounts of lead (Bradford et al. 1975). In the vicinity of lead mining and milling activities in Missouri, no biomagnification of lead was found in the grazing food chains involving aquatic vegetation heavily laden with the metal (Gale, Marcellus, and Underwood 1974). Hessler (1975) reports that $PbCl_2$ caused no appreciable increase in certain mutant types in the alga *Platymonas subcordiformias* under conditions in which ultraviolet and/or nitroguanidine produce high proportions of those mutants.

The median effective dose of Pb(NO$_3$)$_2$ for obtaining a 50% reduction in ^{14}CO$_2$ fixation in the algae *Anabaena*, *Chlamydomonas*, and *Navicula* was between 15 and 18 ppm. For the desmid *Cosmarium*, only 5 ppm was required. An unexplained increase in ^{14}CO$_2$ fixation occurred with increasing lead concentrations in the alga *Ochromonas* (Malanchuk and Gruendling 1973).

8.4.5.12 Manganese

This metal plays a key role in photosynthesis. Its uptake is favored by a low soil pH of 4 to 6 (Horvath 1972; Schroeder, Balassa, and Tipton 1966). Although manganese was concentrated in tree leaves and twigs against an environmental gradient, this was not true in some vegetable foods (Schroeder, Balassa, and Tipton 1966). Manganese is concentrated in the germinal portions of grains, fruits, and nuts, as well as in leafy vegetables; however, root vegetables contain little manganese. Tea and some spices are rich in manganese (Schroeder, Balassa, and Tipton 1966).

8.4.5.13 Molybdenum

The normal levels of molybdenum in plants have been reported to be 0.3 to 10 ppm, but levels as high as 372 ppm have been found without visible plant injury. Field observations of phytotoxicity have rarely been made (Hemphill 1972).

Usually taken up under alkaline conditions (Horvath 1972), molybdenum is an algal micronutrient that is essential to higher plants and that possibly plays a role in photosynthesis (North, Stephens, and North 1970; Smith, Ferguson, and Carlson 1975). When added in sufficient amounts to soil, molybdenum is taken up readily, resulting in large growth reductions (Battelle Northwest Laboratories 1972). Good correlation was found between the amount of molybdenum in clover and grass and the amount in the underlying soil. For the majority of plants and soils tested, however, this was not true (Briese, Runnells, and Smith 1974). Water-extractable molybdenum was much lower than total molybdenum. Excessive amounts were frequently found in bean, barley, and tomato leaves from plants irrigated with various sewage sludges (Bradford et al. 1975).

8.4.5.14 Mercury

Mercury is highly toxic to algae, fungi, and seed plants. The chemical form of the mercury, more so than the concentration, determines its toxicity in terrestrial plants. Small amounts of volatilized mercury have been found to be toxic to greenhouse roses (Ratsch 1974). Table 8.41 lists the relative sensitivities of some plants to mercury vapor.

Plant accumulation of mercury varies according to species and age of plant, soil type, type and concentration of mercury compound, and plant part; generally, the concentration is in the lower stem areas rather than in the photosynthetic upper areas (Wallace et al. 1971; Battelle Northwest Laboratories 1972). Fruits, however, were found to accumulate excessive amounts of mercury from fungicides, and seed grains absorbed 60 to 90% of the mercury from fungicides (Wallace et al. 1971). Carefully washed tomato fruit contained levels of mercury up to 50 times higher than controls when grown on alkaline soil to which sewage sludge had been added, although several other grains and vegetables showed no significant increase (Van Loon 1974).

Mercury has been widely used in fungicides and antitranspirant agents. Metallic vapors are sometimes produced which result in varying amounts of phytotoxicity. Generally, it is thought

Table 8.41. Sensitivity of selected plants to mercury vapor

Sensitive

Bean
  Phaseolus sp.
Butterflyweed
  Asclepias tuberosa L.
Cinquefoil
  Potentilla sp.
Fern, Boston
  Nephrolepis exaltata Schott.
    var. *bostoniensis* Davenport

Fern, holly
  Cyrtomium falcatum Smith
Hydrangea
  Hydrangea sp.
Mimosa
  Mimosa sp.
Oxalis
  Oxalis sp.

Privet
  Lingustrum sp.
Sunflower
  Helianthus sp.
Willow
  Salix sp.

Intermediate

Azalea
  Rhododendron sp.
Basswood (seedling)
  Tilia sp.
Begonia
  Begonia sp.
Camellia
  Camellia sp.
Columbine
  Aquilegia sp.
Cosmos
  Cosmos sp.
Cotoneaster
  Cotoneaster sp.
Forsythia
  Forsythia sp.
Fuchsia
  Fuchsia sp.

Geranium
  Geranium sp.
Holly, American
  Hex opaca Air.
Lily, Calla
  Zantedeschia aethiopica
    Spreng.
Lily, Easter
  Lilium harrisii Carr.
Maple, Japanese
  Acer palmatum Thumb.
Oak
  Quercus sp.
Peach
  Prunus persica Sieb. & Zucc.
Persimmon
  Diospyros virginiana L.

Pine, white
  Pinus strobus L.
Privet
  Ligustrum sp.
Salvia
  Salvia sp.
Saxifrage
  Saxifrage sp.
Strawberry
  Fragaria sp.
Tobacco
  Nicotiana sp.
Tomato
  Lycopersicom esculentum Mill.
Viburnum
  Viburnum sp.
Vinca
  Vinca sp.

Resistant

Aloe
  Aloe sp.
Cherry, Jerusalem
  Solanum pseudocapsicum L.

Croton
  Codiaeum sp.
Holly, Chinese
  Hex Chinensis Loes

Ivy
  Hedera sp.
Sarcococca
  Sarcococca sp.

Source: Naegele 1974, Table 5.4, p. 91, from Jacobson and Hill 1970. Reprinted by permission
of the publisher.

that preserved food plant materials, when produced under ordinary agricultural conditions, pose
no hazards to humans with regard to mercury toxicity, although the mercury is highly persistent
in fruits (Huisingh 1974).

Translocation of mercury within tomato plants and the leaves, fruits, and tubers of various
other plants has been demonstrated. Aphids and lacewing larvae placed on these plants accumulated the metal at a high rate. Beans concentrated methylmercury relative to the soil concentration by a factor of 33 and inorganic mercury by a factor of 3, with large differences in the
rate of translocation between the two forms of mercury (Huckabee and Blaylock 1973).

Maize seedlings accumulated and translocated significant amounts of mercury when germinated in
various concentrations of methylmercury hydroxide (MMH). The seedlings exhibited stunted growth,
and abnormal respiration was observed in their isolated mitochondria. In vitro photosynthesis
by chloroplasts, mitochondrial phosphate uptake, and malate and ascorbate oxidation were inhibited by the MMH; all were cysteine-reversible. $NADH_2$ and succinate oxidation were unaffected
(Lipsey 1975).

Water plants were found to concentrate mercury from water polluted by a paper mill (Johnels and
Westermark 1969 as cited in Wallace et al. 1971). The organic form of mercury is more toxic

to algae than is the inorganic (Boney 1971). Water milfoil grown in solution cultures accumulated more inorganic mercury than organic mercury, whereas in sediment cultures the reverse occurred; the amount of mercury accumulated was in direct proportion to the amount of mercury in the sediment (D'Itri 1971).

There is some indication from aquatic plant studies that organomercury resistance may depend on a plant's ability to transform the organic mercury compounds to less toxic forms. Phenylmercuric acetate, used as an antitranspirant, is rapidly decomposed in plant tissues (Huisingh 1974). In a stream receiving mercuric ion, algae did not contain methylmercury, even though their total mercury levels were high (Cox et al. 1974). In food chain studies (Huckabee and Blaylock 1973), bloodworms assimilated 60% of the inorganic mercury in green algae detritus.

Of five mercurials tested, phenylmercuric acetate (PMA) was the most toxic to algal species exposed to concentrations from 0.1 to 100.0 ppm, although species variations in sensitivity to the various compounds were found. Growth and morphology were often affected by only 0.000001 ppm of PMA or phenylmercuric nitrate (PMN). Permanent structural changes were to some extent inheritable. Some species developed increased tolerance. The green alga *Chlorella vulgaris* was able to concentrate total mercury up to 35 times the original concentration of the medium (Hudson 1973).

Boney (1971) gives LD_{50} values for sporelings of intertidal red algae for mercury as $HgCl_2$ and $n\text{-}CO_3H_7HgCl$. The data vary for the two compounds and the different species. Boney (1971) found that the effects of nonlethal but growth-inhibiting doses of $HgCl_2$ on red algae sporelings were dose-related. Growth of the alga *Coelastrum microporum* was considerably reduced at more than 0.006 ppm methylmercury chloride (Holderness, Fenwick, and Lynch 1975). In living algae, mercuric chloride (3.5 ppm) and methylmercuric chloride (2 ppm) strongly inhibited photosynthesis (Matson, Mustoe, and Chang 1972). Galactosyl transferase activity necessary for galactolipid biosynthesis in isolated chloroplasts from algal cells and spinach leaves was strongly inhibited by both mercury compounds (Matson, Mustoe, and Chang 1972).

Organomercurials have been found to cause doubling of chromosomes or mitotic abnormalities in onion roots. The fungicide Panogen 15 (methylmercury dicyandiamide) produces effects similar to colchicine in plant root tips, causing C-mitoses, multinucleated cells, and polyploidy in *Tradescantia* (Wandering Jew) and *Vicia faba* (bean) (Oak Ridge National Laboratory 1975).

8.4.5.15 Nickel

Nickel is a plant micronutrient, but it can be very phytotoxic, depending on its chemical form (Smith, Ferguson, and Carson 1975). It can be taken up in fairly significant amounts with no marked growth reduction (Battelle Northwest Laboratories 1972). Vegetation around a coal-burning plant and in polluted areas has been found to be enriched in nickel. Mosses, lichens, shrubs, and trees accumulated nickel; the roots of spruce trees showed an enrichment factor of 25 (quotient between metal concentration in the polluted site and in the control site) (Klein and Russell 1973). At 10 mg/liter, nickel inhibited greening of excised red kidney bean embryos, but did not affect curvature of the embryonic axis (Imai and Siegel 1973).

The shrub *Hybanthus floribundus* accumulates large amounts of nickel (Severne 1974). Excessive amounts of nickel were frequently found in bean, barley, and tomato leaves from plants irrigated with various sewage sludges (Bradford et al. 1975).

8.4.5.16 Selenium

Certain high-sulfur crop plants are particularly sensitive to selenium, for example, cabbage, which accumulates significant amounts from soil (Stadtman 1974; Underwood 1973). Other garden vegetables also accumulate soil sulfur, but tend to concentrate the metal in the inedible plant portion (e.g., the potato peel). Much selenium is discarded with cooking water. Forage plants containing 5 ppm selenium are considered potentially hazardous (Hemphill 1972). Selenium was found to be taken up by roots and foliage but only in limited amounts according to species; for example, barley took up 1 to 2% of the added selenite (Gissel-Nielsen 1975). Selenium, usually taken up under alkaline conditions (Horvath 1972), has been reported to stimulate growth of certain species (Hemphill 1972). *Astragalus* spp. is regarded as a selenium indicator (Hemphill 1972); levels as high as 10,000 ppm are apparently common in these species. The selenium content of timothy grass showed a positive significant correlation to soil selenium and pH on Prince Edward Island (Gupta and Winter 1975). Gutenmann et al. (1976) found that white sweet clover growing voluntarily on beds of fly ash contained more than 200 ppm (dry weight) selenium. Most mature vegetables cultivated on soil containing 10% (by weight) of these fly ashes contained up to 1 ppm, whereas cabbage absorbed selenium in direct proportion to the selenium in fly ash (up to 3.7 ppm). In the field, the addition of sulfur, boron, and molybdenum failed to affect plant tissue selenium content. The presence of selenium in food crops and its relationship to human nutrition is discussed by Allaway (1973).

8.4.5.17 Silver

Silver exerts a low level of phytotoxicity, perhaps because it is excluded from plant tops or areas of photosynthetic activity (Battelle Northwest Laboratories 1972). When applied at 10 mg/liter to excised embryos of red kidney bean, silver affected curvature of the embryonic axis without affecting greening (Imai and Siegel 1973).

8.4.5.18 Thallium

Thallium reduces the yields of crops, particularly barley; the reduction in yield is proportional to both soil concentration and increased tissue levels. Accumulation is in the lower stem areas rather than the photosynthetic areas (Battelle Northwest Laboratories 1972).

8.4.5.19 Vanadium

Vanadium in small amounts stimulates the growth of higher plants (Smith, Ferguson, and Carson 1975). It appears to concentrate in plant roots, averaging 1 ppm in measurements of plant materials (Hemphill 1972). Phytotoxicity has not been observed under field conditions (Hemphill 1972).

8.4.5.20 Titanium

Despite soil enrichment, no uptake of titanium was detected in vegetation around a coal-burning power plant, nor was it detected in vegetation at control sites (Klein and Russell 1973). However, titanium has been reported to be concentrated by the alga *Gymnodinium breve* and to be essential for green algae (Feldman 1970).

8.4.5.21 Zinc

Zinc in small quantities is required for plant growth, but at high soil concentrations, most
plants, particularly lichens, are adversely affected (Tyler 1972; Nash 1975; Dorn and Phillips
1973; Battelle Northwest Laboratories 1972). The phytotoxic threshold may be at 400 ppm (Hemphill 1972). A prolific growth of ragweed, however, may occur at these levels. Horsetail
growing on a pond, formerly used for zinc slime, contained 5400 ppm (Hemphill 1972).

Increased zinc levels in vegetation have been reported near a power plant (Klein and Russell
1973), zinc smelter (Ratsch 1974), heavily traveled roads (Lagerwerff and Specht 1970), and lead
smelters, mines, and mills (Tyler 1972). The occasionally varying levels may reflect soil geochemical differences (Ratsch 1974). In short-term experiments, a high proportion of zinc taken
into plant roots was absorbed on the cell surfaces, which, unless removed, can lead to serious
error in estimations of zinc absorption into the cell interior (Schmid, Haag, and Epstein 1965).

An increase in soil zinc content by a factor of 9.3 caused an increase in the zinc content of
bromegrass of 1.4 at pH 7.2. At pH 5.9, response was 17% greater (Lagerwerff and Specht 1970).
The rapid uptake of zinc by bean leaves and tissues was strongly dependent on external zinc concentration and pH, but not on light or temperature. Data suggest that zinc uptake by bean tissue
is primarily passive (Rathore et al. 1970). Applications of $ZnSO_4$ to sweet corn sometimes reduced forage yields, and zinc levels higher than those found in forage treated with organic
wastes were observed. Bush beans were more sensitive to zinc than was sweet corn (Giordano,
Mortvedt, and Mays 1975).

8.4.5.22 Zirconium

According to a review by Blumenthal (1973), algae collected off the south and west coasts of
Puerto Rico were found to contain concentrated zirconium, which has also been reported to be an
algicide. High levels of zirconium were reported in carrot roots, and low activity was reported
in shoots when the carrots were grown in soils containing radioactive zirconium. Levels of
0.001 to 0.1 wt % zirconium in larch, *Calmagrostis arundinacea* (a grass), and *Dactylis glomerata*
(a grass) have been reported (Blumenthal 1973). *Gymnodinium breve* has been found to concentrate
zirconium (Feldman 1970).

8.5 THERMAL EFFECTS

A temperature increase has been shown to exert an influence on many aspects of the life cycle
of aquatic plants (Coutant and Pfuderer 1974; Coutant and Talmage 1975; Coutant and Talmage
1976). A direct relationship has been shown between water temperature and biomass for all major
species of algae (Fig. 8.1). Definite shifts in species composition occurred with seasonal temperature changes; a red filamentous alga appeared in nuisance proportions at predictable temperatures (Reed et al. 1974). A review by Patrick (1963) concludes that, in general, algal growth
was increased by moderate increases in temperature toward optimal growth temperatures. Diversity
and biomass also were increased by small temperature increases with a shift in composition,
usually from communities dominated by diatoms to those dominated by blue-green algae, occurring
with temperature shifts near and beyond the upper limits of the tolerance range. Although both
long thermal exposures and great temperature rises have caused adverse effects on many algae,

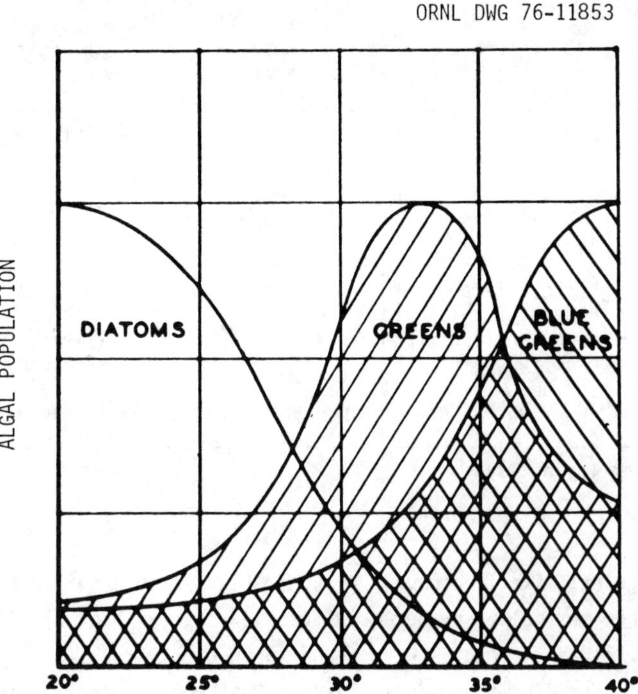

Fig. 8.1. Algal population shifts with temperature. <u>Source</u>: Cairns 1956, Fig. 4, p. 152. Reprinted by permission of the publisher.

heat shock for extremely short periods of time may have only temporary adverse effects. Gruhl (1973) states that "heated discharges have even been considered as a means of checking over-growths of algal flora at some sites" (Fig. 8.1). Photosynthesis and respiration differ in response to temperature although, in general, respiration is more dependent on temperature than is photosynthesis (Cairns, Heath, and Parker 1975).

8.6 COMBINED EFFECTS

Table 8.42 generalizes some observations regarding the effect of soil, light, stage of growth, and moisture factors on plant susceptibility to oxidant damage. According to Taylor (1974), susceptibility of plants to toxicants (and resultant injury) is a complex function of the inter-action of the plant's physiological structure with climatic factors, toxicant concentration, exposure duration, soil conditions, and the combination of these factors. Despite the repro-ducibility of injury symptoms under controlled and standardized laboratory conditions, devia-tions are observed in the extent and appearance of the injury when real-world environmental conditions are imposed.

8.6.1 Atmospheric pollutants

A well-known example of the interaction of pollutant and environment is the injury at Ducktown, Tennessee, which resulted from high atmospheric levels of sulfur dioxide combined with humid environmental conditions. The lush growth typical of the region was reduced over a relatively

Table 8.42. Factors affecting susceptibility of plants to
oxidant air pollutants

Pollutant	Enhancement of susceptibility	Reduction of susceptibility
PAN	Light before and after exposure; high relative humidity; intense light	Soil moisture deficiency
Ozone	High field temperatures; low light intensity; a long dark period (24 hr) before exposure; high relative humidities	A long photoperiod (24 hr) before exposure; soil salinity; soil moisture deficiency
Oxidants	Rapid, vigorous growth; high nutrients	Soil oxygen deficiency

Source: Compiled from Taylor 1974.

short time to a barren desert-like landscape. By contrast, in the arid Southwest, even sensitive plants develop a tolerance to smelter-produced sulfur dioxide. Hill et al. (1974) found that most of the 87 native cold-desert species were found to be highly resistant to sulfur dioxide, requiring more than 2 ppm for 2 hr to produce injury. One species, Indian ricegrass, required an exposure of more than 4 ppm sulfur dioxide for 2 hr under normal soil moisture conditions to be injured, whereas 1 ppm for 2 hr produced 2% injury under conditions simulating one year of extremely high rainfall.

8.6.1.1 Relative humidity

A plant's sensitivity to pollutant damage is influenced by every environmental factor that influences plant vigor, particularly water relations. The effect of relative humidity on the sensitivity of plants to various pollutants, including fluorine, has been documented by McLean, Schneider, and McCune (1973). They report that fluorine-induced injury increases at relative humidities (RH) between 65 and 80% and that fluorine accumulation in gladiolus increases at RH values between 50 and 65%. Four mechanisms are thought to account for the effect: (1) increased absorption of the pollutant, which is operative only during the exposure period; (2) increased translocation of the pollutant, which occurs subsequent to absorption; (3) increased cellular injury due to a lowered threshold; and (4) feedback influence of injured tissues on the first three mechanisms. Although conclusive evidence that relative humidity significantly affects sensitivity of plants to PAN (as long as stomata remain open) is not available, relative humidities of 50% or more prevail when most PAN injury occurs in the South Coastal Basin of California (Taylor 1974).

8.6.1.2 Light exposure

Light quality appears to affect the sensitivity of some plants to some pollutants. PAN injury to pinto beans is evidently dependent on the wavelength and timing of light exposure (Heggestad and Heck 1971). If a polluted air mass reaches vegetation late in the evening, PAN injury probably will not develop (Taylor 1974). Photoperiod has also been related to tobacco and blue-grass sensitivity to oxidants. Light intensity, as related to sensitivity, is a function of the toxicant involved. High light intensities tend to increase the phytotoxic effects of PAN, whereas low light intensities increase sensitivity to ozone.

Uptake of hydrogen fluoride, sulfur dioxide, chlorine, nitrogen dioxide, ozone, PAN, nitric oxide, and carbon monoxide by an alfalfa canopy was shown to be affected by wind velocity, canopy height, and light intensity (Bennett and Hill 1973).

8.6.1.3 Temperature

High irradiation, optimum temperature for apparent photosynthesis, and high humidity enhanced the inhibition of photosynthesis in bean plants (*Phaseolus vulgaris*) by nitrogen dioxide. Uptake was increased by high temperature, low carbon dioxide, and high humidity. High temperatures also increased nitrogen dioxide inhibition of dark respiration (Srivastava, Jolliffe, and Runeckles 1975*a*).

An inverse relationship between exposure temperature and degree of ozone injury of Virginia pine has been observed (Davis 1970, as cited in Taylor 1974). However, ozone injury to field-grown sweet corn, which was severe at ambient temperatures above 90°F, was absent below that temperature, even at high ozone levels (Taylor 1974).

Stomatal functioning may be implicated in sensitivity reversals due to temperature. The relationship may be inverse or positive (Heggestad and Heck 1971).

An increase in the toxicity of phenol to the alga *Selanastrum capricornutum* was observed to occur with a temperature increase according to the Arrhenius function (Reynolds, Middlebrooks, and Procella 1974).

8.6.1.4 Plant age

It appears that a plant's youngest mature cells are the most sensitive to sulfur dioxide, ammonia, chlorine, hydrogen fluoride, and nitrogen dioxide. This may be a dose effect due to the cells' differential ability to absorb a specific toxicant from the air. Severity of the ozone-induced reduction in radish root growth was modified by the stage of plant development and was not dependent on temperature. Effects of multiple exposures were found to be independent of effects of previous exposures (Tingey 1973).

8.6.1.5 Soil factors

Susceptibility to ozone damage is lessened by both inadequate and excessive soil moisture, but both conditions adversely affect plant vigor. Susceptibility is also lessened if plants are grown in heavy, compact soils such as clay as compared with growth in vermiculite or peat-perlite mixtures. Nutrient factors may also be important. Contradictory results from application of nitrogen fertilizer suggest that, although oxidant injury is enhanced by the addition of nitrogen (to correct a nitrogen deficiency), application of a luxury amount of nitrogen did not increase oxidant injury and perhaps even suppressed plant growth to the point of inducing greater tolerance to oxidants (Taylor 1974).

The effects of ozone in reducing the forage yield of alfalfa were decreased by increased salinity of irrigating nutrient solutions. Leaf diffusion resistance was unaffected by either ozone or salinity alone, but was increased by treatment with both simultaneously (Hoffman, Maas, and Rawlins 1975).

8.6.1.6 <u>Mixtures of toxicants — additive, synergistic, and antagonistic effects</u>

Foliar injury caused by combinations of gases is sometimes greater than that caused by the sum (additive effects) of the gases (Tingey and Reinert 1975). This was found to be true for foliar injury caused by exposure to mixtures of ozone and sulfur dioxide in tobacco, radish, and alfalfa. The effect was additive or less than additive for other species such as cabbage, broccoli, and tomato (Tingey et al. 1973). Tobacco of a particular variety known to be resistant to concentrations of ozone up to 5 pphm is injured by ozone at only 2 to 3 pphm when combined with 25 to 30 pphm sulfur dioxide (Menser and Heggested 1966). The mechanism involved may be the interference of ozone with precursors for normal metabolic detoxification of sulfur dioxide which allow the plant to incorporate low amounts of sulfur dioxide without injury (Ziegler 1973, as cited in Auerbach 1975).

Mandl, Weinstein, and Keveny (1975) confirm that combinations of sulfur dioxide and ozone can produce a synergistic effect, lowering the threshold for injury in other plant species as well as tobacco. Combinations of nitrogen dioxide and sulfur dioxide (Tingey et al. 1971) and nitrogen dioxide and ozone (Matsushima 1971, as cited in Mandl, Weinstein, and Keveny 1975) have also produced greater foliar injury than the additive effects of the gases alone.

In contrast to the foliar injury caused by acute exposures, chronic exposure to combinations of ozone and sulfur dioxide has caused a range of effects on growth: In tobacco, the reduction equaled the additive effects of the single gases; in alfalfa, reductions from chronic mixed treatment were less than the additive effects. Acute exposure of the radish to ozone or sulfur dioxide caused the same growth reduction as did acute exposure to a combination of the two. In contrast to reports indicating enhanced foliar injury from combinations of gases, no combination in these experiments caused growth reductions greater than the additive effects of single gases (Tingey and Reinert 1975). This is in agreement with other reports (de Koning and Jegier 1970, Tingey et al. 1971, and Keller 1973, as cited in Mandl, Weinstein, and Keveny 1975). No synergism was detected between sulfur dioxide and nitrogen dioxide at the prevailing ratio ($0.28\ NO_2:SO_2$) measured downwind from a large coal-fired power plant (Hill et al. 1974). However, concentrations of nitrogen dioxide and sulfur dioxide that were noninjurious under laboratory conditions caused injury to several crop plants when combined, but only at relatively high concentrations. The minimum dose for visible injury in the radish was 1-hr exposure to a mixture containing either 0.5 ppm sulfur dioxide plus nitrogen dioxide ($0.96\ mg/m^3\ NO_2$ plus $1.31\ mg/m^3\ SO_2$) or 0.75 ppm ($1.95\ mg/m^3$) sulfur dioxide alone. Swiss chard, oats, and sweet pea showed little incidence of enhanced phytotoxicity of the gas mixture over treatment with sulfur dioxide alone (Bennett et al. 1975). Horsman and Wellburn (1975) demonstrated a synergistic effect between sulfur dioxide and nitrogen dioxide, reflected in activity changes of the enzymes RuDPC and peroxidase at 0 to 2 ppm sulfur dioxide and 0 to 0.1 ppm nitrogen dioxide and in peroxidase at 0 to 0.2 ppm sulfur dioxide and 0 to 1 ppm nitrogen dioxide.

The effects of mixtures of hydrogen fluoride and sulfur dioxide proved to be additive rather than synergistic in causing differences in linear growth and leaf area of citrus tree seedlings (*Citrus sinensis*) and Satsuma mandarin tree foliage. No synergistic injury to the Satsuma mandarin resulted from alternate exposure to sulfur dioxide and hydrogen fluoride (Matsushima and Brewer 1972).

Corn and barley leaves were severely injured by 0.3 and 0.15 ppm sulfur dioxide, but the addition of hydrogen fluoride did not enhance symptom severity. At 0.06 to 0.08 ppm sulfur dioxide, the plants developed foliar lesions that were accentuated by hydrogen fluoride. Over 27 days, the accumulation of fluoride in corn leaves was reduced by combinations of sulfur dioxide and hydrogen fluoride as compared with reductions by hydrogen fluoride alone, perhaps because of the effects on stomates. Neither treatment caused injury to beans, nor were fresh or dry weight yields of plant tops reduced (Mandl, Weinstein, and Keveny 1975). A mixture of nitrogen dioxide and nitric oxide resulted in additive effects in depressing apparent photosynthesis in oats and alfalfa (Hill and Bennett 1970).

An optimum level of sucrose and reducing sugar was required in leaf tissue for maximum susceptibility (Lee 1965).

8.6.1.7 Genetic variability

A major factor in determining plant susceptibility to pollution is that of genetic variability. A wide variety of response has been found among species, and some effort has been expended on the selection and breeding of plants for increased tolerance (Table 8.43). Different plants react with different intensities to each pollutant and pollutant level; thus, different plant types have different threshold levels of sensitivity to different pollutants (Taylor et al. 1958 as cited in Feder 1973). Heggestad (1968) points out that the test of a single variety or selection of species can give misleading information as to the sensitivity of that species. In order to assess the impact of a pollutant on a species, knowledge of the response of a representative sample of the total population in an area is needed.

Except for some adjustment to some enzyme poisons such as heavy metals and some chemicals, Catcheside (1975) concludes that plants probably are unable to become more resistant or tolerant to pollutants. The cost of that adjustment would be a reduction in productivity, and resistant strains are generally less fit than normal strains except in the presence of the toxin.

8.6.1.8 Disease

There are many reports of changes in the incidence or severity of a disease when associated with air pollution, sometimes a lessening, as suggested by field surveys for sulfur dioxide (Skye 1968, Scheffer and Hedgcock 1955, Linzon 1958, and Pryzblski 1967, as cited in Manning 1975). McCune et al. (1973) report a decrease in severity for several fungal and bacterial vegetable diseases when fumigated with hydrogen fluoride. Recent papers detail the synergistic relationships between oxidants (ozone and sulfur dioxide) and various plant diseases (Heagle 1975; Davis and Smith 1975; Moyer and Smith 1975; Hibben and Taylor 1975; Weinstein et al. 1975). Manning (1975) reviews current knowledge about the interactions between air pollutants (ozone,

Table 8.43. Estimates of resistance and sensitivity of various
crops to different pollutants

Plant	Pollutant and sensitivity[a]					
	Ozone	Nitrogen dioxide	PAN	General oxidant	Sulfur dioxide	Hydrogen fluoride
Field crops						
Alfalfa	S		I	S	S	R
Barley	S		I		S	I
Beans (field)	S	S	S	S	S	R
Buckwheat					S	S
Clover (hay)	S			S	S	R
Corn (field)		R			R	I
Cotton					S	R
Lespedeza						
Oats	S		S	S	S	I
Peanuts	S					
Potatoes (Irish)	S				R	R
Rye	S	I			S	
Safflower					S	
Sorghum			R			I
Soybean			I		S	R
Sugar beets			I	S	I	R
Sunflower		S			S	I
Tobacco	S	S	I	S		R
Wheat	S		I		S	R
Seed crops						
Alfalfa	S		I	S	S	R
Clover	S		I	S	S	R
Alsike					S	
Crimson					S	
Ladino					S	
Red					S	
Lespedeza						
Mustard		S			I	
Citrus						
Grapefruit				S	R	I
Kumquats						
Lemons						I
Limes						
Oranges		I				I
Valencia						
Navel						
Temple						
Mandarins						
Tangerines						I
Tangelos						

Table 8.43. (continued)

Plant	Pollutant and sensitivity[a]					
	Ozone	Nitrogen dioxide	PAN	General oxidant	Sulfur dioxide	Hydrogen fluoride
Fruits and nuts						
Almonds						
Apples					S	
Apricots					I	S
Avocados						
Blackberries						
Blueberries						S
Cherries						R
Sour					R	
Sweet						
Grapes	S			S	I	S
Peaches					I	
Clingstone						
Freestone						S
Pears					S	R
Pecans						
Plums					I	S
Prunes						S
Strawberries						R
Walnuts						I
Raspberry						I
Figs						I
Vegetables						
Asparagus		R			R	R
Beans, snap		S	S		S	R
Beans, lima		S	R		S	I
Beets (table)	S		I	S	S	R
Broccoli			R		S	
Brussel sprouts			R		S	
Cabbage					I	
Carrots			I		S	
Cauliflower					I	
Celery				I	R	R
Corn (sweet)	S		R		R	S
Cucumber			R		R	
Endive and escarole				S	S	
Eggplant					I	
Lettuce (head)		S			S	R
Muskmelon	S			S	R	
Okra					S	
Onion	S		R		R	
Parsley					I	
Parsnip					I	
Peas					I	
Peppers					I	
Potato (sweet)					S	
Pumpkin					S	
Radish	S		R	I		
Romaine			S	S		
Spinach	S		I	S	S	
Squash					S	
Swiss chard			S		S	
Tomatoes	S	S	S		I	R
Turnip					S	

Table 8.43. (continued)

Plant	Pollutant and sensitivity[a]					
	Ozone	Nitrogen dioxide	PAN	General oxidant	Sulfur dioxide	Hydrogen fluoride
Flowering plants						
Aster					I	I
Azalea		S	R			
Bachelor's button					S	
Begonia			R		I	
Canna					R	
Carissa		R				
Carnation	S					
Chrysanthemum	S		R		R	
Cosmos					S	
Croton		R				
Dahlia		R				
Four-o'clock					R	
Geranium						I
Gladiolus					R	I
Goldenrod						I
Heath		R				
Hibiscus		S			R	
Hollyhock					I	
Honeysuckle					R	
Hydrangea					I	
Iris					I	
Ixora		R				
Lilac	S					I
Marigold					I	
Morning glory					S	
Narcissus						I
Nasturtium					I	
Peony						I
Periwinkle			R			
Petunia	S		S	S		
Privet	S					
Pyracantha						R
Rhododendron						I
Rose					R	I
Snowball					R	
Snowberry	S					
Spirea	S					R
Sunflower		S			I	I
Sweet pea					S	
Sweet william					I	
Touch-me-not			R			
Tulip						S
Verbena					S	
Violet				S	S	I
Virginia creeper					R	R
Wisteria					R	
Zinnia					S	

Table 8.43. (continued)

Plant	Ozone	Nitrogen dioxide	PAN	General oxidant	Sulfur dioxide	Hydrogen fluoride
Trees and native shrubs						
Alder	S					
Arbor-vitae					R	I
Ash, green						I
Aspen, quaking	S					I
Birch					S-I	R
Box elder	S				R	S
Brittlewood		S				
Catalpa	S				S-I	
Douglas fir						S
Elderberry						R
Elm (American)					S-I	R
Honeylocust	S					
Juniper (most species)					R	R
Larch					S	S
Linden, American					I	R
Linden, European						I
Maple						
Silver	S				R	I
Sugar						R
Hedge						I
Mock orange					R	
Mulberry					S	
Oak, Gambel	S					
Oak, live					R	
Oregon grape						S
Pine						
Eastern, white	S			S	S	S
Lodgepole						S
Mugho						S
Scotch						S
Western, yellow	S			S	S	S
Poplar, Carolina						I
Poplar, Lombardy					S-I	I
Serviceberry						I
Spruce, blue						S
Spruce, white						I
Sumac						I
Sycamore	S					R
Tobacco tree					S	
Tree of heaven						R
Walnut, black						I
Walnut, English						I
Willow, weeping	S					R
Yew					S	I

Table 8.43. (continued)

Plant	Ozone	Nitrogen dioxide	PAN	General oxidant	Sulfur dioxide	Hydrogen fluoride
Weeds and grasses						
Bindweed					S	
Careless weed					S	
Cheeseweed		I	I	S	S	R
Chickweed		I	S		R	I
Cocklebur					I	
Dock sour					S	
Fleabane					S	
Goosefoot nettle (leafed)		R		I	R	S
Grasses						
Annual blue			S	S	R	R
Bent	S					
Brome	S					
Crab	S					
June					S	
Kentucky blue		R			R	R
Orchard	S				I	R
Rye					S	
Salt					R	
Lamb's-quarters			I		I	I
Milkweed					R	
Mustard, black			S		I	
Mustard, yellow						
Nettle (little leaf)			S			
Nightshade					I	
Pigweed		R				
Plantain					S	
Purslane					R	
Ragweed					S	
Shepherd's purse					R	
Smartweed					I	
Sweet clover					I	

[a]S = susceptible, I = intermediate, R = resistant.

Source: Benedict, Miller, and Olson 1971, Table 11, pp. 40-46.

sulfur dioxide, hydrogen fluoride, and particulates) and fungal, bacterial, and viral plant
pathogens. Changes range from an increase in severity to no effect to a decrease in severity.
Effects of ozone and sulfur dioxide on fungal diseases have received more attention than have
effects on bacterial or viral diseases. The incidence or progress of a disease may be affected
by the pollutant-caused changes in microflora associated with plant surfaces (Manning 1975).

Interactions between plant insect infestation and fluoride fumigation are reviewed by Weinstein
et al. (1975), who report delay in growth and development in the Mexican bean beetle when cultured on hydrogen-fluoride-fumigated plants. The complexity of the air pollution impact is seen
in the mountain recreational areas of Los Angeles, where Ponderosa pine weakened by air pollution
is particularly subject to invasion by bark beetles (Heck 1973). Indications are that the relationship is a result of changes in the essential oils of the leaves (Cobb, Zavarin, and Bergot
1972).

8.6.2 Trace elements

Trace-element–plant interactions are affected by the same factors that cause synergistic effects
of plants with gaseous pollutants.

8.6.2.1 Mixtures of elements

Cadmium exhibits a synergistic relationship with lead. Noninhibitory concentrations of each
element combined to inhibit the root growth of corn (Miller et al. 1974). The addition of 5 to
50 ppm zinc to cadmium-containing soil resulted in increased cadmium concentration in soybean
shoots. The latter was primarily due to decreased plant growth. At 100 ppm zinc, a depression
occurred in cadmium concentration (Haghiri 1974).

Cadmium and zinc have been found to act synergistically in inhibiting the growth of floating
aquatic plants. Although zinc alone was stimulatory, it markedly increased the inhibitory effect
of cadmium. The presence of one also increased uptake of the other (Hutchinson and Czyrska
1972).

The synergistic relationship between copper and nickel is enhanced by low pH levels, increasing
algal toxicity, whereas molybdenum is antagonistic to copper (Becker and Thatcher 1973). Iron
chlorosis in citrus plants has resulted from an imbalance between soil copper and iron caused by
accumulated copper from fertilizers and fungicidal sprays (Huisingh 1974).

Plant deficiency of manganese can occur if amounts of available iron and/or copper are excessive
(Piperno 1975). A highly significant correlation between chromium and magnesium content was
found by Lyon et al. (1971); for the vascular plant *Hebe odora*, which was growing on serpentine
soil, the magnesium content was high. Selenium is a well-known antagonist to mercury, nullify-
ing the latter's toxic effects. The normal inhibition of wheat leaf elongation by cyanide was
increased considerably at between 0.5 to 1.0 mM cyanide when iron was present in the nutrient
solution (Israelstam 1970).

8.6.2.2 Soil factors

Cadmium uptake by radish and lettuce plants is related to various soil factors (John, VanLaerhoven, and Chuah 1972). Independent variables found to contribute significantly to plant cadmium levels included (1) the relative ability of soils to absorb cadmium, (2) acetate-soluble cadmium in the soil, (3) soil reaction, and (4) organic matter. Plant cadmium levels correlated significantly with amounts of nickel, iron, zinc, and copper in the same plant portion. Cadmium concentration in oat shoots was decreased by increasing the cation exchange capacity of the soil, whereas organic matter had no influence (Haghiri 1974). The phytotoxicity of arsenic decreases as the sodium-hydroxide-extractable soil aluminum content increases (Hemphill 1972).

8.6.2.3 Acid rain

Because acid rain leads to soil acidification (Bohn 1972) and because acid soils enhance heavy-metal uptake by plants (Huisingh 1974), an increase in uptake may be expected where acid rain and heavy metal pollution occur in concert (Auerbach 1975).

8.6.2.4 Nutrient levels

Available phosphorus appears to be related to lead uptake. During heartwood formation in certain tree trunks, a good correlation between the two was found; an estimated 70% of the lead was fixed as lead phosphate (Hampp and Holl 1974). Ratsch (1974) states that increased foliar accumulation in young corn leaves resulted when the nutrient phosphate source was limited. Lead uptake also decreases in corn with increase in soil pH and cation exchange capacity (Miller et al. 1974). Corn root elongation was inhibited by combinations of noninhibitory concentrations of cadmium and lead (Miller et al. 1974).

8.6.2.5 Temperature

Increasing the soil temperature resulted in an increase in the cadmium concentration of soybean shoots (Haghiri 1974). In bean plants, the inhibitory effect of cyanide on leaf respiration was observed to be influenced by temperature, humidity, plant and leaf age, and degree of atmospheric drought damage (Genkel and Kushnirenko 1971).

Some general points that may apply to algae have been extrapolated by Cairns, Heath, and Parker (1975) from data on the interaction between temperature and the acute toxicity of some heavy-metal ions to freshwater fish (Rehwoldt et al. 1972):

1. Effects may be direct or indirect.

2. If the temperature change remains within the tolerance range for the organism, a metabolic rate increase, accompanying numerous biochemical and physiological changes, occurs.

3. Indirect effects include solubility and diffusion rate changes.

4. "The effect of temperature on the toxicity of water-soluble toxins is difficult to predict," partly because the toxic mechanisms are poorly understood.

The blue-green alga *Pectonema boryanum* has been observed to concentrate different elements best at different temperatures (Harvey 1967, as cited in Cairns, Heath, and Parker 1975).

8.6.2.6 <u>Disease</u>

Fungicidal compounds containing various inorganic ions were observed to stimulate hatching of plant parasitic nematodes, resulting in more severe disease losses of crop plants. In both soybean and peanut, leaf applications of a number of metal-containing fungicides, when used at nonphytotoxic levels, also depressed nodulation (Huisingh 1974).

LITERATURE CITED

Allaway, W. H. 1973. Selenium in the food chain. *Cornell Veterinarian* G3(2): 151-70.

Andelman, J. B., and Suess, M. J. 1970. Polynuclear aromatic hydrocarbons in the water environment. *Bull. WHO*, 43: 479-508.

Andelman, J. B., and Snodgrass, J. E. 1974. Incidence and significance of polynuclear aromatic hydrocarbons in the water environment. *CRC Critical Rev. Environ. Control* 4(1): 69-83.

Andersen, R. A. 1973. Carcinogenicity of phenols, alkylating agents, urethan, and a cigarette-smoke fraction in *Nicotiana* seedlings. *Cancer Res.* 33: 2450-55.

Antonovics, J. 1973. The evolution of heavy metal tolerance in natural plant populations. *Environ. Health Persp. Exp.* 4: 103.

Apelbaum, A., and Burg, S. P. 1972. Effects of ethylene and 2,4-dichlorophenoxyacetic acid on cellular expansion in *"Pisum sativum."* *Plant physio.* 50: 125-31.

Arvik, J. H., and Zimdahl, R. L. 1974. Barriers to the foliar uptake of lead. *J. Environ. Qual.* 3(4): 369-73.

Atkins, C. A., Canvin, D. T. 1975. Nitrate, nitrite and ammonia assimilation by leaves: Effects of inhibitors. *Planta* 123(1): 41-51.

Auerbach, S. I. 1975. *Hearings on cost and effects of chronic low-level environmental pollution.* Testimony presented to the subcommittee on environment and atmosphere, U.S. House of Representatives, 94th Congress First Session, Nov. 12, No. 49, pp. 511-37. Washington, D.C.: U.S. Government Printing Office.

Barabanova, N. M.; Polushchuk, L. P.; and Stemikovskaya, L. A. 1973. Level of chemical substance in farm crops grown on soil irrigated by waste waters from by-product coking plants. *Voprosy Pitaniya* 5: 76-79. ORNL/tr-2957.

Barrett, L. B., and Waddell, T. E. 1973. *Cost of air pollution damage: a status report.* EPA-AP-85.

Battelle Northwest Laboratories. 1972. *Fate of heavy metals and metal complexes in soil and plants.* Richland, Wash.

Becker, C. D., and Thatcher, T. O. 1973. *Toxicity of power plant chemicals to aquatic life.* Richland, Wash.: Battelle Pacific Northwest Laboratories. WASH-1249.

Bell, J. N. B., and Clough, W. S. 1973. Depression of yield in ryegrass exposed to sulphur dioxide. *Nature* 24(1): 47-49.

Benedict, H. M.; Miller, C. J.; and Olson, R. E. 1971. *Economic impact of air pollutants on plants in the United States.* Stanford Research Institute, Calif.

Bennett, J. H., and Hill, A. C. 1973. Absorption of gaseous air pollutants by a standardized plant canopy. *J. Air Pollut. Control Assoc.* 23(3): 203-06.

Bennett. J. H.; Hill, A. C.; Soleimani, A.; and Edwards, W. H. 1975. Acute effects of combination of sulphur dioxide and nitrogen dioxide on plants. *Environ. Pollut.* 9(2): 127-32.

Bleasdale, J. K. A. 1973. Effects of coal-smoke pollution gases on the growth of ryegrass (*Lolium perenne L.*). *Environ. Pollut.* 5: 275-85.

Blokker, P. C. 1972. A literature survey on some health aspects of lead emissions from gasoline engines. *Atmos. Environ.* 6: 1-18.

Blumenthal, W. B. 1973. Zirconium in the ecology. *Amer. Ind. Hyg. Assoc. J.* 34: 128-33.

Bohn, H. L. 1972. Soil absorption of air pollutants. *J. Environ. Qual.* 1(4): 372-77.

Boney, A. D. 1971. Sub-lethal effects of mercury on marine algae. *Mar. Poll. Bull.* 2(5): 69-71.

Bonte, J., and de Cornis, L. 1973. Effects of sulfur dioxide on stomatic movements in plants. In *Proceedings of the 3rd International Clean Air Congress, Dusseldorf, Fed. Rep. Germany, 1973*, pp. A134-38. ORNL/tr-4115.

Borneff, J.; and Fischer, R. 1962. Carcinogenic substances in water and soil. *Arch. Hyg.* 146: 334-45.

Borneff, J.; Selenka, F.; Kunte, H.; and Maximos, A. 1968*a*. Experimental studies on the formation of polycyclic aromatic hydrocarbons in plants. *Envir. Res.* 2: 22-29.

Borneff, J.; Selenka, F.; Kunte, H.; and Maximos, A. 1968*b*. The synthesis of 3,4-benzopyrene and other polycyclic aromatic hydrocarbons in plants. *Arch. Hyg.* 152(3): 279-82.

Borneff, J.; Kunte, H.; Farkasdi, G.; and Glathe, H. 1973*a*. Krebs durch benzypren in naturlichem dunger. *Umschau* 73(20): 626-28.

Borneff, J.; Farkasdi, G; Glathe, H.; and Kunte, H. 1973*b*. The fate of polycyclic aromatic hydrocarbons in experiments using sewage sludge-garbage composts as fertilizers. *Zentralbl. Bakteriol. Parasitenkd. Infektionskr. Hyg. Erst Abt. Orig. Reihe B Hyg. Praev. Med.* 157(2/3): 151-64.

Bradford, G. R.; Page, A. L.; Lund, L. J.; and Olmstead, W. 1975. Trace element concentrations of sewage treatment plant effluents and sludges: Their interactions with soils and uptake by plants. *J. Environ. Qual.* 4(1): 123-27.

Brams, E. A., and Fiskell, J. G. A. 1971. Copper accumulation in citrus roots and desorption with acid. *Soil sci. Soc. Amer. Proc.* 35(5): 772-75.

Brandt, C. S., and Heck, W. W. 1968. Effect of air pollutants in vegetation. In *Air pollution*, ed. A. C. Stern, vol. I, pp. 401-43. 2nd ed. New York: Academic Press.

Briese, F. W.; Runnells, D. D.; and Smith, E. 1974. A statistical analysis and computer display of the distribution of molybdenum between alpine soils and plants. In *Proceedings of the second annual NSF-RANN trace contaminants conference, Asilomar, Pacific Grove, Calif., Aug. 29-31, 1974*, pp. 256-68. LBL-3217.

Brook, A. J. 1964. *The living plant*. Chicago: Aldine Pub. Co.

Bruton, V. C., and Anderson, C. E. 1974. The effects of ozone on the development of *Arabidopsis thaliana*. *J. Elisha Mitchell Sci. Soc.* 90(3): 97.

Buchauer, M. J. 1973. Contamination of soil and vegetation near a zinc smelter by zinc, cadmium, copper, and lead. *Environ. Sci. Technol.* 7(2): 131-35.

Buehler, K., and Hirshfield, H. I. 1974. Cadmium in an aquatic ecosystem: Effects on planktonic organisms. In *Trace contaminants in the environment. Proceedings of the second annual NSF-RANN contaminants conference, Asilomar, Pacific Grove, Calif., Aug. 29-31, 1974*, LBL-3217. Berkeley, Calif.: Lawrence Berkeley Laboratory.

Cairns, J. C. 1956. Effects of increased temperatures on aquatic organisms. *Ind. Wastes* 1(4): 150-52.

Cairns, J. C., Jr.; Heath, A. G.; and Parker, B. C. 1975. Temperature influence on chemical toxicity to aquatic organisms. *J. Water Pollut. Control Fed.* 47(2): 267-80.

Canvin, D. T., and Atkins, C. A: 1974. Nitrate, nitrite and ammonia assimilation by leaves: Effect of light, carbon dioxide and oxygen. *Planta* 116: 207-24.

Catcheside, D. G. 1975. Chemicals, radiations and heredity. *Search* 6(1-2): 23-28.

Chakrabarti, A. G. 1976. Effects of carbon monoxide and nitrogen dioxide on garden pea and string bean. *Bull. Environ. Contam. Toxicol.* 15: 214-22.

Chebotar, A. A.; Kaptar, S. G.; Suruzhiu, A. T.; and Bukhar, B. I. 1975. Chromosome and nucleoplasma changes in corn and wheat induced by the action of hexachloran, napthalene, and phenol. *Dokl. Akad. Nauk. (SSSR)* 223(1): 213-15.

Clendenning, K. A. 1960. Laboratory investigations. In *Institute of marine resources: The effects of waste discharges on kelp*, Quarterly Progress Report, Oct. 1-Dec. 31, 1959, p. 7-11. IMR-60-10.

Cobb, F. W., Jr.; Zavarin, E.; and Bergot, J. 1972. Effect of air pollution on the volatile oil from leaves of *Pinus ponderosa*. *Phytochem*. 11: 1815-18.

Comanor, P. L.; DeGuire, M. F.; Hendrix, J. L.; Vreeland, H.; and Vreeland, P. 1974. *Trans. Soc. Mining Eng.* AIME 256: 240-42.

Comar, C. L. 1975. *Conference proceedings: Workshop on health effects of fossil fuel combustion products*. PB-242-418. Electric Power Research Institute, Palo Alto, Calif.

Costonis, A. C. 1971. Effects of ambient sulfur dioxide and ozone on Eastern white pine in a rural environment. *Phytopathol*. 61: 717-20.

Coulson, C. L., and Heath, R. L. 1975. The interaction of peroxyacetyl nitrate (PAN) with the electron flow of isolated chloroplasts. *Atmos. Environ*. 9: 231-38.

Coutant, C. C., and Pfuderer, H. A. 1974. Thermal effects. *J. Water Pollut. Control Fed*. 46(6): 1476-1540.

Coutant, C. C., and Talmage, S. 1975. Thermal effects. *J. Water Pollut. Control Fed*. 47(6): 1656-1711.

Coutant, C. C., and Talmage, S. 1976. Thermal effects. *J. Water Pollut. Control Fed*. 48(6): 1486-1544.

Cowell, E. B.; Baker, J. M.; and Crapp, G. B. The biological effects of oil pollution and oil-cleaning materials on littoral communities including salt marshes. In *Marine pollution and sea life*, ed. M. Ruivo, pp. 359-64. London: Fishing News (Books) Ltd.

Cowling, D. W.; Jones, L.H.P.; and Lockyer, D. R. 1973. Increased yield through correction of sulfur deficiency in ryegrass exposed to sulfur dioxide. *Nature* 243: 479-80.

Cox, M. F.; Holm, H. W.; Kania, H. J.; and Knight, R. L. 1976. Methyl-mercury and total mercury concentrations in selected stream biota. In *Trace Substances in Environmental Health — IX. Proceedings of University of Missouri's 9th annual conference, June 10-12, 1975, Columbia, Mo.*, ed. D. D. Hemphill.

Curtis, O. F., and Clark, D. G. 1950. *An introduction to plant physiology*. New York: McGraw-Hill Book Company, Inc.

Cutler, J. M., and Rains, D. W. 1974. Characterization of cadmium uptake by plant tissue. *Plant Physiol*. 54: 67-71.

Daines, R. H. 1968. Sulfur dioxide and plant response. *J. Occup. Med*. 10(9): 516-24.

Davis, D. D., and Smith, S. H. 1975. Bean common mosaic virus reduces ozone sensitivity of pinto bean. *Environ. Pollut*. 9: 97-101.

Denaeyer-De Smet, S. 1974. Premier apercu de la distribution du cadmium dans divers ecosystemes terrestres non pollues et pollues. *Ecol. Plant* 9(2): 169-82.

D'Itri, F. M. Mercury in the aquatic ecosystem. In *Bioassay techniques and environmental chemistry*, ed. G. E. Glass, pp. 3-60. Ann Arbor, Mich.: Ann Arbor Science Publishers Inc.

Dochinger, L. S., and Jensen, K. F. 1975. Effects of chronic and acute exposure to sulphur dioxide on the growth of hybrid poplar cuttings. *Environ. Pollut*. 9(3): 219-29.

Dolgova, L. G., and Kozyukina, Z. T. 1972. On the problem of biological purification of the air under conditions of coke oven gas production. *Ukrains'kiy Botanichniy Zhurnal* 29(2): 172-75.

Dorn, C. R., and Phillips, P. E. 1973. Translocation of heavy metals in human food chains. In *Proceedings of the 77th annual meeting of the U.S. Animal Health Association, 1973*, pp. 267-81.

Dugger, M., ed. 1974. *Symposium on air pollution effects on plant growth*. ACS Symposium Series 3. Washington, D.C.: American Chemical Society.

Dugger, W. M., Jr.; Taylor, O. C., Cardiff, E.; and Thompson, C. R. 1962. Stomatal action in plants related to damage from photochemical oxidants. *Plant Physiol*. 37: 487-91.

Dugger, W. M., Jr., and Ting, I. P. 1968. The effect of peroxyacetyl nitrate on plants: Photoreductive reactions and susceptibility of bean plants to PAN. *Phytopathol.* 58: 1102-07.

Durmishidze, S. V.; Devdariani, T. V.; Kavtaradze, L. K.; and Kvartshava, L. S. 1975. Assimilation and conversion of 3,4-benzypyrene by plants under sterile conditions. *Dokl. Akad. Nauk. SSSR.* 218(1-6): 1468-71.

Durmishidze, S. V.; Devdariani, T. V.; Kavtaradze, L. K.; and Ugrekhelidze, D. S. 1973. Splitting of 3,4-benzopyrene-1,2-C_{14} by higher plants. *Soobsch Akad. Nauk Gruz. SSR* 70(2): 469-72.

Durmishidze, S. V., and Ugrekhelidze, D. S. 1975. Assimilation and transformation of methane by plants. *Sov. Plant Physiol.* 22(1): 53-55.

Durmishidze, S. V.; Ugrekhelidze, D. S.; and Dzhikiya, A. N. 1974. Toluene assimilation and transformation by higher plants. *Prikl. Biokhim. Mikrobol.* 10(5): 673-76.

Durocher, N. L. 1969. Air pollution aspects of beryllium and its compounds. PB-188-078. Bethesda, Md.: Litton Systems, Inc.

Eaton, J. S.; Likens, G. E.; and Bormann, F. H. 1973. Throughfalls and stem-flow chemistry in a northern hardwood forest. *J. Ecol.* 61: 495-508.

Edwards, M. E., and Miller, J. H. 1972. Growth regulation by ethylene in fern gametophytes. II. Inhibition of cell division. *Amer. J. Bot.* 59(5): 450-57.

Engelbrecht, A.H.P. 1973. Hydrogen fluoride injury in sugar-cane: Some ultrastructural changes. In *Proceedings of the 3rd International Clean Air Congress, Dusseldorf, Fed. Rep. Germany, 1973,* pp. A157-59.

Engst, R. 1973. Industrial contamination of food with carcinogenic hydrocarbons. *Scr. med.* 46(2): 79-85.

Environmental Protection Agency (EPA). 1975. *Scientific and technical assessment report on particulate polycyclic organic matter (PPOM).* EPA-600/675-001.

Everett, J. L.; Day, C. L.; and Reynolds, D. 1967. Comparative survey of lead at selected sites in the British Isles in relation to air pollution. *Fd. Cosmet. Toxicol.* 5: 29-35.

Fairfax, J.A.W., and Lepp, N. W. 1975. Effect of simulated "acid rain" on cation loss from leaves. *Nature* 255: 324-25.

Fassett, D. W. 1972. Cadmium. In *Metallic contaminants and human health,* ed. D.H.K. Lee, pp. 97-124. New York: Academic Press.

Feder, W. A. 1973. Cumulative effects of chronic exposure of plants to low levels of air pollutants. in *Air pollution damage to vegetation,* Advances in Chemistry Series 122, ed. J. A. Naegele, pp. 21-30. Washington, D.C.: American Chemical Society.

Feldman, M. H. 1970. *Trace materials in wastes disposed to coastal waters — fates, mechanisms, and ecological guidance and control.* PB-202 346.

Ferenbaugh, R. W. 1974. *Effects of simulated acid rain on vegetation.* Ph. D. Thesis, University of Montana. Dissertation Abstracts International, 35(5): 2053-B.

Ferenbaubh, R. W. 1976. Effects of simulated acid rain on *Phaseolus vulgaris* L. (Fabacae). *Amer. J. Bot* 63(3): 283-88.

Fetner, R. H. 1958. Chromosome breakage in *Vicia faba* by ozone. *Nature* 181: 504-05.

Friberg, L.; Piscator, M.; Nordberg, G. F.; and Kielistrom, T. 1974. Cadmium in soil and uptake by plants. In *Cadmium in the environment,* pp. 14-19. 2nd ed. Cleveland, Ohio: CRC Press.

Frick, H., and Cherry, J. H. 1974. An early site of physiological damage to soybean and cucumber seedlings following ozonation. In *Air pollution effects on plant growth,* ACS Symp. Ser. 3, ed. M. Dugger, pp. 128-47.

Fritz, W., and Engst, R. 1971. On environmental contamination of foods with carcinogenic hydrocarbons. *Z. Gesamte Hyg. Grenzgeb.* 17(4): 271-75.

Fry, W. E., and Millar, R. L. 1971. Development of cyanide tolerance in *Stemphylium loti*. *Phytopathol.* 61: 501-06.

Fulkerson, W., and Goeller, H. E., ed. 1973. *Cadmium the dissipated element*. ORNL/NSF-EP-21. Oak Ridge, Tenn.: Oak Ridge National Laboratory.

Gale, N. L.; Marcellus, P.; and Underwood, G. 1974. Life, liberty and the pursuit of lead: The impact of lead mining and milling activities on aquatic organisms. In *Proceedings of the second annual NSF-RANN trace contaminants conference, Asilomar, Pacific Grove, Calif., Aug. 29-31, 1974,* p. 295. LBL-3217.

Gaufin, A. R. 1974. Biological indices of environmental changes in aquatic habitats. In *Industrial Pollution*, ed. N. I. Sax, pp. 36-47. New York: Van Nostrand Reinhold Company.

Genkel, P. A., and Kushnirenko, S. V. 1972. Influence of cyanide on the respiration of leaves of bean plants hardened to drought before sowing. *Sov. Plant Physiol.* 18(5): 802-07.

Giordano, P.; Mortvedt, J. J.; and Mays, D. A. 1975. Effect of municipal wastes on crop yields and uptake of heavy metals. *J. Environ. Qual.* 4(3): 394-99.

Gissel-Nielsen, G. 1975. The fate of selenium added to agricultural crops. In *Symposium proceedings on nuclear techniques in comparative studies of food and environmental contamination, Otaniemi, Finland, Aug. 27-31, 1973*, pp. 333-39. International Atomic Energy Agency. IAEA-SM-175-1.

Glass, A.D.M., and Bohm, B. A. 1971. The uptake of simple phenols by barley roots. *Planta (Berl.)* 100: 93-105.

Godzik, S., and Knabe, W. 1973. Comparative electron-microscope studies of the fine structure of chloroplasts of pine varieties in industrial areas along the Ruhr and in upper Silesia. In *Proceedings of the 3rd international clean air congress, Dusseldorf, Fed. Rep. Germany, 1973*, pp. A164-69. ORNL/tr-4117.

Goodman, G. T., and Roberts, T. M. 1971. Plants and soils as indicators of metals in the air. *Nature* 231: 287-92.

Goodman, R. N.; Kiraly, Z.; and Zaitlin, M. 1967. *The biochemistry and physiology of infectious plant disease*. Princeton, N. J.: D. Van Nostrand Company, Inc.

Gorham, E., and Gordon, A. G. 1963. Some effects of smelter pollution upon aquatic vegetation near Sudbury, Ontario. *Can. J. Bot.* 41: 371-78.

Gräf, W., and Diehl, H. 1966. Concerning the naturally caused normal level of carcinogenic polycyclic aromatics and its cause. *Arch. Hyg.* 150(4): 49-59.

Gräf, W., and Nowak, W. 1966. Promotion of growth in lower and higher plants by carcinogenic polycyclic aromatics. *Arch. Hyg. Bakt.* 150: 513-28. ORNL/tr-4111.

Grennard, A., and Ross, F. 1974. Progress report on sulfur dioxide. *Combustion* 45(7): 4-9.

Grigorenko, L. T.; Dikun, P. P.; Kalinina, I. A.; Mironova, A. N.; and Rzhekhin, V. P. 1970. 3,4-Benzopyrene content in sunflower and cotton seed oils. *Prikl. Biokhim. Mikrobiol.* 6(2): 142-50.

Grimmer, G., and Duvel, D. 1970. Endogenous formation of polycyclic hydrocarbons in higher plants. 8. Carcinogenic hydrocarbons in the environment of humans. *Z. Naturforsch.* 25b: 1171-75. ORNL/tr-4118.

Gross, R. E., and Dugger, W. M., Jr. 1969. Responses of *Chlamydomonas reinhardtii* to peroxyacetyl nitrate. *Environ. Res.* 2: 256-66.

Gruhl, J. 1973. *Quantification of aquatic environmental impact of electric power generation*. PB-224 642. Massachusetts Institute of Technology.

Guderian, R. 1973. Damage to vegetation and food by emissions from conventional power plants. *P. Irahienschutz Forschung. Pravis* 12: 81-97. ORNL/tr-2986.

Guderian, R., and Schoenbeck, H. 1971. Recent results for recognition and monitoring of air pollutants with the aid of plants. In *Proceedings, second international clean air congress*, ed. H. M. Englund and W. T. Beery, pp. 167-273. New York: Academic Press.

Gupta, U. C., and Winter, K. A. 1975. Selenium content of soils and crops and the effects of lime and sulfur on plant selenium. *Can. J. Soil Sci.* 55(2): 161-66.

Gutenmann, W. H.; Bache, C. A.; Youngs, W. D.; and Lisk, D. J. Selenium in fly ash. *Science* 191: 966-67.

Haghiri, F. 1974. Plant uptake of cadmium as influenced by cation exchange capacity, organic matter, zinc, and soil temperature. *J. Environ. Qual.* 3(2): 180-83.

Hakama, M., and Saxen, E. A. 1967. Cereal consumption and gastric cancer. *Int. J. Cancer* 2: 265-68.

Hall, M. A.; Brown, R. L.; and Ordin, L. 1971. Inhibitory products of the action of peroxyacetyl nitrate upon indole-3-acetic acid. *Phytochem.* 10: 1233-38.

Hall, H. J.; Varga, G. M.; and Magee, E. M. 1974. Trace elements and potential pollutant effects in fossil fuels. In *Symposium proceedings: Environmental aspects of fuel conversion technology, St. Louis, Mo., May 1974*, pp. 35-47. EPA-650/2-74-118.

Hampp, R., and Holl, W. 1974. Radial and axial gradients of lead concentration in bark and xylem of hardwoods. *Arch. Environ. Contam. Toxicol.* 2(2): 143-51.

Hansborough, J. R. 1967. Air quality and forestry. In *Agriculture and the quality of our environment*, AAAS Pub. 85, ed. N. C. Brady, pp. 45-56. Washington, D. C.: American Association for the Advancement of Science.

Haye, S. N.; Horvath, D. J.; Bennett, O. L.; and Singh, R. 1975. A model of seasonal increase of lead in a food chain. In *Trace substances in environmental health — IX. A symposium*, ed. D. D. Hemphill, pp. 387-93. Univ. of Missouri, Columbia, Mo.

Heagle, A. S. 1975. Response of three obligate parasites to ozone. *Environ. Pollut.* 9: 91-95.

Heath, R. L.; Chimiklis, P.; and Frederick, P. 1974. The role of potassium and lipids in ozone injury to plant membranes. In *Air pollution effects on plant growth*, ACS Symp. Ser. 3, ed. M. Dugger, pp. 58-75. Washington, D. C.: American Chemical Society.

Heck, W. W. 1968*a*. Factors influencing expression of oxidant damage to plants. *Ann. Rev. Phytopathol.* 6: 165-88.

Heck, W. W. 1968*b*. Discussion of Dr. Taylor's paper. *J. Occup. Med.* 10: 496-99.

Heck, W. W. 1973. Air pollution and the future of agricultural production. In *Air pollution damamge to vegetation*, Advances in Chemistry Series 122, ed. J. A. Naegele, pp. 118-30. Washington, D.C.: American Chemical Society.

Heck, W. W.; Daines, R. H.; and Hindawi, I. J. 1970. Other phytotoxic pollutants. In *Recognition of air pollution injury to vegetation: a pictorial atlas*, ed. J. S. Jacobson and A. C. Hill, pp. F1-F10. Pittsburgh, Penn.: Air Pollution Control Association and National Air Pollution Control Administration.

Heck, W. W., and Tingey, D. T. 1971. Ozone: Time-concentration model to predict acute foliar injury. In *Proceedings of the 2nd international clean air congress, Dusseldorf, Fed. Rep. Germany*, pp. 249-55. New York: Academic Press.

Heggestad, H. E. 1968. Discussion of paper by Dr. Taylor. *J. Occup. Med.* 10: 492-96.

Heggestad, H. E., and Heck, W. W. 1971. Nature, extent, and variation of plant response to pollutants. In *Advances in agronomy*, ed. N. C. Brady, vol. 23, pp. 111-45.

Hemphill, D. D. 1972. Availability of trace elements to plants with respect to soil-plant interaction. *Ann. N. Y. Acad. Sci.* 199: 46-61.

Hemphill, D. D., and Pierce, J. O. 1974. Accumulation of lead and other heavy metals by vegetation in the vicinity of lead smelters and mines and mills in southeastern Missouri. In *Trace contaminants in the environment. Proceedings of the 2nd annual NSF-RANN trace contaminants conference, Aug. 29-31, 1974*, pp. 325-32. LBL-3217.

Hessler, A. 1975. The effects of lead on algae. II. Mutagenesis experiments on *Platymonas subcordiformis* (Chlorophyta: Volvocales). *Mutation Res.* 31: 43-47.

Hibben, C. R., and Taylor, M. P. 1975. Ozone and sulfur dioxide effects on the lilac powdery mildew fungus. *Environ. Pollut.* 9(2): 107-14.

Hill, A. C. 1971. Vegetation: A sink for atmospheric pollutants. *J. Air Pollut. Control Assoc.* 21(6): 341-46.

Hill, A. C., and Bennett, J. H. 1970. Inhibition of apparent photosynthesis by nitrogen oxides. *Atmos. Environ.* 4: 341-48.

Hill, A. C.; Heggestad, H. E.; and Linzon, S. N. 1970. Ozone. In *Recognition of air pollution injury to vegetation: a pictorial atlas*, ed. J. S. Jacobson and A. C. Hill, pp. B1-B6. Pittsburgh, Penn.: Air Pollution Control Association and National Air Pollution Control Administration.

Hill, A. C.; Hill, S.; Lamb, C.; and Barrett, T. W. 1974. Sensitivity of native desert vegetation to SO_2 and to SO_2 and NO_2 combined. *J. Air Pollut. Control Assoc.* 24(2). 153-57.

Hiltbold, A. E. 1975. Behavior of organoarsenicals in plants and soils. In *Arsenical pesticides*, ACS Symp. Ser. 7, ed. E. A. Woolson, pp. 53-69. Washington, D.C.: American Chemical Society.

Hindawi, I. J. 1970. *Air pollution injury to vegetation.* NAPCA Publication AP-71.

Hitchcock, A. E.; Zimmerman, P. W.; and Coe, R. R. 1962. Results of ten years' work (1951-1960) on the effect of fluorides on gladiolus. *Contrib. Boyce Thompson Inst.* 21: 303-44.

Hoffman, G. J.; Maas, E. V.; and Rawlins, S. L. 1975. Salinity-ozone interactive effects on alfalfa yield and water relations. *J. Environ. Qual.* 4(3): 326-30.

Holderness, J.; Fenwick, M. G.; and Lynch, D. L. 1975. The effect of methyl mercury on the growth of the green alga, *coelastrum microporum* Naeg. strain 280. *Bull. Environ. Contam. Toxicol.* 13(3): 348-50.

Holliday, R.; Hodgson, D. R.; Townsend, W. N.; and Wood, J. W. 1958. Plant growth on 'fly ash. *Nature* 181: 1079-80.

Holst, R. W.; Schmid, E.; and Yopp, J. H. 1975. Beryllium absorption by excised barley roots. *Plant Physiol.* Suppl. 56(2): 43.

Horsman, D. C., and Wellburn, A. R. 1975. Synergistic effect of SO_2 and NO_2 polluted air upon enzyme activity in pea seedlings. *Environ. Pollut.* 8(2): 123-35.

Horvath, D. J. 1972. An overview of soil/plant/animal relationships with respect to utilization of trace elements. In *Geochemical environment in relation to health and disease*, ed. H. C. Hoppe and H. L. Cannon. *Ann. N. Y. Acad. Sci.* 117: 82-94.

Horvath, D. J. 1976. Personal communication.

Hosker, R. P., Jr. 1974. *Estimates of dry deposition and plume depletion over forests and grassland.* Environmental Research Labs., 1973 annual report, pp. 231-35. Atmospheric Turbulence and Diffusion Laboratory, NOAA, Oak Ridge, Tenn. ATDL-106.

Howell, R. K. 1970. Influence of air pollution on quantities of caffeic acid isolated from leaves of *Phaseolus vulgaris* L. *Phytopathol.* 60: 1626-29.

Howell, R. K. 1974. Phenols, ozone, and their involvement in pigmentation and physiology of plant injury. In *Air pollution effects on plant growth*, ACS Symp. Ser. 3, ed. M. Dugger, pp. 94-105. Washington, D.C.: American Chemical Society.

Howell, R. K., and Kremer, D. F. 1973. The chemistry and physiology of pigmentation in leaves injured by air pollution. *J. Environ. Qual.* 2(4): 434-38.

Howland, G. P. 1976. Personal communication. Oak Ridge National Laboratory, Oak Ridge, Tenn.

Huckabee, J. D., and Blaylock, B. G. 1973. Transfer of mercury and cadmium from terrestrial to aquatic ecosystems. In *Metal ions in biological systems*, ed. S. K. Dhar, pp. 125-160. New York: Plenum Press.

Hudson, K. M. 1973. *Effects of mercury compounds on algae.* Ph.D. Dissertation, University of Southern Mississippi. MSEP-73-016.

Huisingh, D. 1974. Heavy metals: Implications for agriculture. *Ann. Phytopathol.* 12: 375-88.

Hutchinson, T. C., and Czyrska, H. 1972. Cadmium and zinc toxicity and synergism to floating aquatic plants. *Water Pollut. Res. Can.* 1972: 59-65.

Hutchinson, T. C.; Kauss, P.; and Griffiths, M. 1972. The phytotoxicity of crude oil spills in freshwater. *Water Pollut. Res. Can.* 1972: 52-58.

Hutchinson, G. L.; Millington, R. J.; and Peters, D. B. 1972. Atmospheric ammonia: Absorption by plant leaves. *Science* 175: 771-72.

Hyder, S. Z., and Yasmin, S. 1975. Boron tolerance and enhancement of boron toxicity by chloride ions in alkali sacaton during germination of *Sporobolus airoides* Torr. *Speciali* 15(4): 427-28.

Il'nitskii, A. P., and Kogan, J. P. 1972. Prophylaxis of food contamination with carcinogens. *Voprosy Onhologii* 18(5): 106-11.

Il'nitskii, A. P.; Solenova, L. G.; and Ignatova, V. V. 1974. Sanitary and oncological assessment of agricultural use of sewage containing carcinogenic hydrocarbons. *Kazanskii Meditsinskii Zh.* 2: 80-81. ORNL/tr-2959.

Imai, I., and Siegel, S. M. 1973. A specific response to toxic cadmium levels in red kidney bean embryos. *Physiol. Plant* 29: 118-20.

Ishio, S.; Nakagawa, H.; and Tomiyama, T. 1972. Algal cancer and its causes. II. Separation of carcinogenic compounds from sea bottom mud polluted by wastes of the coal chemical industry. *Bull. Jpn. Soc. Sci. Fish.* 38(6): 571-76. ORNL/tr-2848.

Israelstam, G. F. 1970. Elongation of wheat leaves in response to cyanide in the presence and absence of iron. *Can. J. Bot.* 48: 2017-19.

Jacobson, J. S., and Hill, A. C., eds. 1970. *Recognition of air pollution injury to vegetation: a pictorial atlas.* Pittsburgh, Penn.: Air Pollution Control Association.

Jensen, K. F., and Kozlowski, T. T. 1975. Absorption and translocation of sulfur dioxide by seedlings of four forest tree species. *J. Environ. Qual.* 4(3): 379-82.

John, M. K. 1972. Lead availability related to soil properties and extractable lead. *J. Environ. Qual.* 1(3): 295-98.

John, M. K.; Chuah, H. H.; and VanLaerhoven, C. J. 1972. Cadmium contamination of soil and its uptake by oats. *Environ. Sci. Technol.* 6(6): 555-57.

John, M. K.; VanLaerhoven, S. J.; and Chuah, H. H. 1972. Factors affecting plant uptake and phytotoxicity of cadmium added to soils. *Environ. Sci. Tech.* 6(12): 1005-09.

Joshi, M. M.; Ibrahim, I.K.A.; and Hollis, J. P. 1975. Hydrogen sulfide: Effects on the physiology of rice plants and relation to straight-head disease. *Phytopathol.* 65(10): 1165-70.

Kang, B. G., and Burg, S. P. 1973. Influence of ethylene on nucleic acid in etiolated *Pisum sativum. Plant Cell Physiol.* 14: 981-88.

Karnosky, D. F., and Stairs, G. R. 1974. The effects of SO_2 on *in vitro* forest tree pollen germination and tube elongation. *J. Environ. Qual.* 3(4): 406-08.

Kihlman, B. A. 1957. Experimentally induced chromosome aberrations in plants: I. The production of chromosome aberration by cyanide and other heavy metal complexing agents. *J. Biophys. Biochem. Cytol.* 3(3): 363-80.

Kirkham, M. B. 1975. Uptake of cadmium and zinc from sludge by barley grown under four different sludge irrigation regimes. *J. Environ. Qual.* 4(3): 423-26.

Klein, D. H. 1972. Mercury and other metals in urban soils. *Environ. Sci. Technol.* 6(6): 560-62.

Klein, D. H., and Russell, P. 1973. Heavy metals: Fallout around a power plant. *Environ. Sci. Technol.* 7(4): 357-58.

Kogan, Y. L. 1973. Carcinogenic hydrocarbon level in crops grown on nerosin-treated fields. *Vop. Pitan.* 3: 50-53.

Kolar, L.; Ledvina, R.; Ticha, J.; and Hanus, F. 1975. Znecisteni pud, aemedelskych plodin a zeleniny 3,4-benzpyrenem v okoli ceskych budejovic. *Cesk. Hyg.* 20(3): 135-39.

Kornreich, M. R. 1975. *A preliminary assessment of the problem of carcinogens in the atmosphere.* MTR-6874. MITRE Corporation Technical Report, McLean, Virginia.

Koukol, J., and Dugger, W. M., Jr. 1967. Anthocyanin formation as a response to ozone and smog treatment in *Rumex crispus* L. *Plant Physiol.* 42: 1023-24.

Kozel, T., and Maly, V. 1972. The effect of seven years of power plant waste on farming in neighboring regions. *Sci. Agr. Bohemsalov. 4* XXI(3): 165-90.

Kozlowski, T. T., and Mudd, J. B., eds. 1975. Introduction. In *Responses of plants to air pollution,* pp. 1-8. New York: Academic Press.

Kozyukina, Zh. T., and Obraztsova, V. I. 1973. Dynamics of free and bound amino acids in leaves of trees and shrubs under conditions of coke oven gas production. *Ukrains'kiy Botan. Zh.* 30(3): 332-39. STS Order No. 16498.

Krall, A. R., and Tolbert, N. E. 1957. A comparison of the light dependent metabolism of carbon monoxide by barley leaves with that of formaldehyde, formate and carbon dioxide. *Plant Physiol.* 32: 321-26.

Ku, H. S., and Leopold, A. C. 1970. Mitrochondrial responses to ethylene and other hydrocarbons. *Plant Physiol.* 46: 842-44.

Lagerwerff, J. V. 1967. Heavy-metal contamination of soils. In *Agriculture and the quality of our environment,* AAAS pub. 85, ed. N. C. Brady, pp. 343-66. Washington, D.C.: American Association for the Advancement of Science.

Lagerwerff, J. V., and Specht, A. W. 1970. Occurrence of environmental cadmium and zinc and their uptake by plants. In *Trace substances in environmental health. Proceedings of Univ. of Missouri's 4th annual conference on trace substance in environmental health, held at Memorial Union, University of Missouri-Columbia, Columbia, Mo., June 23-25, 1970,* ed. D. D. Hemphill, pp. 85-93.

Lee, T. T. 1965. Sugar content and stomatal width as related to ozone injury in tobacco leaves. *Can. J. Bot.* 43: 677-85.

Lerman, S. L., and Darley, E. F. 1975. Particulates. In *Responses of plants to air pollution,* ed. J. B. Mudd and T. T. Kozlowski, pp. 141-58. New York: Academic Press.

Less, L. N.; McGregor, A.; Jones, L.H.P.; Cowling, D. W.; and Leafe, E. L. 1975. Fluorine uptake by grass from aluminum smelter fume. *Int. J. Environ. Stud.* 7: 153-60.

Levan, A., and Tjio, J. H. 1948. Induction of chromosome fragmentation by phenols. *Hereditas* 34: 453-84.

Likens, G. E. 1976. Personal communication, Cornell University, Ithaca, New York.

Likens, G. E., and Bormann, F. H. 1974. Acid rain: A serious regional environmental problem. *Science* 184: 1176-79.

Linnman, L.; Andersson, A.; Nilsson, K. O.; Lind, B.; Kjellstrom, T.; and Friberg, L. 1973.
Cadmium uptake by wheat from sewage sludge used as a plant nutrient source. *Arch. Environ.
Health.* 27: 45-47.

Linzon, S. N. 1971. Economic effects of sulfur dioxide on forest growth. *J. Air Pollut.
Control Assoc.* 21(2): 81-86.

Linzon, S. N. 1973. Some effects of particulate matter on vegetation in Ontario. In
Proceedings, 3rd international clean air congress, Dusseldorf, Fed. Rep. Germany, 1973,
pp. A118-20.

Lipsey, R. L. 1975. Accumulation and physiological effects of methyl mercury hydroxide on
maize seedlings. *Environ. Pollut.* 8: 149-55.

Losada, M.; Paneque, A.; Aparicito, P. J.; Vega, J. M.; Cardenas, J.; and Herrera, J. 1970.
Inactivation and repression by ammonia of the nitrate reducing system in *Chlorella.*
Biochem. Biophys. Comm. 38(6): 1009-15.

Lovelace, C. J., and Miller, G. W. 1967. *In vitro* effects of fluoride on tricarboxylic acid
cycle dehydrogenases and oxidative phosphorylation: Part I. *J Histochem. Cytochem.*
15(4): 195-201.

Lyon, G. L.; Peterson, P. J.; Brooks, R. R.; and Butler, G. W. 1971. Calcium, magnesium and
trace elements in a New Zealand serpentine flora. *J. Ecol.* 59(2): 421-29.

Ma, T.; Isbandi, D.; Khan, S. H.; and Tseng, Y. 1973. Low level of SO_2 enhanced chromatid
aberrations in *Tradescantia* pollen tubes and seasonal variation of the aberration rates.
Mutat. Res. 21: 93-100.

MacLean, D. C.; McCune, D. C.; Weinstein, L. H.; Mandl, R. H.; and Woodruff, G. N. 1968.
Effects of acute hydrogen fluoride and nitrogen dioxide exposures on citrus and ornamental
plants of central Florida. *Environ. Sci. Technol.* 2: 444-49.

MacLean, D. C.; Schneider, R. E.; and McCune, D. C. 1973. Fluoride phytotoxicity as affected
by relative humidity. In *Proceedings, 3rd international clean air congress, Dusseldorf,
Fed. Rep. Germany, 1973,* pp. A143-45.

Maloseja, Z.; Parletic, Z.; and Munjko, I. 1972. Biological decomposition of phenol by mixed
and pure cultures of algae. *Acta. Bot. Croat.* 31: 129-38.

Majumdar, S. K., and Newton, R. S. 1972. Morphogenetic and cytogenetic effects of 7,12-
dimethylbenz(a)anthracene on *Haworthia* callus in vitro. *Experientia.* 28(12): 1497-98.

Malanchuk, J. L., and Gruendling, G. K. 1973. Toxicity of lead nitrate to algae. *Water Air
Soil Pollut.* 2: 181-90.

Mandl, R. H.; Weinstein, L. H.; and Keveny, M. 1975. Effects of hydrogen fluoride and sulphur
dioxide alone and in combination on several species of plants. *Environ. Pollut.* 9(2):
133-43.

Manning, W. J. 1975. Interactions between air pollutants and fungal, bacterial and viral
plant pathogens. *Environ. Pollut.* 9(2): 87-90.

Markowski, A.; Grzesiak, S., and Pienkowski, S. 1974. Neutralization of SO_2 with ammonia and
its effects on the vegetation of some species of plants. *Ser. Agraria.* 14(2): 59-87.

Martens, D. C. 1971. Availability of plant nutrients in fly ash. *Compost Sci.* 12(6): 15-19.

Martinez, J. D.; Nathany, M.; and Dharmarajan, V. 1971. Spanish moss, a sensor for lead.
Nature. 233: 546-65.

Matson, R. S.; Mustoe, G. E.; and Chang, S. B. 1972. Mercury inhibition on lipid biosynthesis
in freshwater algae. *Environ. Sci. Technol.* 6(2): 158-60.

Matsushima, J., and Brewer, R. F. 1972. Influence of sulfur dioxide and hydrogen fluoride
as a mix or reciprocal exposure on citrus growth and development. *J. Air Pollut. Control
Assoc.* 22(9): 710-13.

McCune, D. C. 1969. *On the establishment of air quality criteria, with reference to the effects of atmospheric fluorine on vegetation.* Air Quality Monograph 69-3. New York: American Petroleum Institute.

McCune, D. C.; Weinstein, L. H.; Jacobson, J. S.; and Hitchcock, A. E. 1964. Some effects of atmospheric fluoride on plant metabolism. *J. Air Pollut. Control Assoc.* 14: 465-68.

McCune, D. C.; Weinstein, L. H.; Macini, J. F.; and van Leuken, P. 1973. Effects of hydrogen fluoride on plant-pathogen interactions. In *Proceedings, 3rd international clean air congress, Dusseldorf, Fed. Rep. Germany, 1973,* pp. A146-49.

McLaughlin, S. B., Jr., and Barnes, R. L. 1975. Effects of fluoride on photosynthesis and respiration of some south-east American forest trees. *Environ. Pollut.* 8: 91-96.

McNulty, I. B., and Lords, J. L. 1960. Possible explanation of fluoride-induced respiration in *Chlorella pyrenoidosa. Science.* 132: 1553-54.

Melchior, N. C., and Melchoir, J. B. 1956. Inhibition of yeast hexokinase by fluoride ion. *Science.* 124: 402-03.

Menser, H. A., and Heggestad, H. E. 1966. Ozone and sulfur dioxide synergism: Injury to tobacco plants. *Science.* 153: 424-25.

Middleton, J. T. 1961. Photochemical Air Pollution Damage to Plants. Ann Rev. Plant Phys. 12: 431-448.

Middleton, J. T.; Darley, E. F.; and Brewer, R. F. 1958. Damage to vegetation from polluted atmospheres. *J. Air Pollut. Control Assoc.* 8: 9-15.

Miller, J. E.; Hassett, J.; Koeppe, D. E.; Rolfe, G. L.; and Wheeler, G. L. 1974. Effects of soil properties on Pb uptake by corn and effects of Pb and Cd on corn root elongation. In *Trace contaminants in the environment. Proceedings of the second annual NSF-RANN trace contaminant conference, Aug. 29-31,* pp. 290-94. LBL-3217.

Miller, P. L. 1973. Oxidant-induced community change in a mixed conifer forest. In *Air pollution damage to vegetation,* Advances in Chemistry Series 122, ed. J. A. Naegele, pp. 101-17. Washington, D.C.: American Chemical Society.

Miller, P. R., and McBride, J. R. 1975. Effects of air pollutants on forests. In *Responses of plants to air pollution,* ed. J. B. Mudd and T. T. Kozlowski, pp. 196-235. New York: Academic Press.

Mills, H. A.; Barker, A. V.; and Maynard, D. N. 1973. A study of the phytotoxicity of 2-chloro-6-(trichloromethyl)pyridine. *Soil Sci. Plant Anal.* 4(6): 487-94.

Mironov, O. G. 1972. Effect of oil pollution on flora and fauna of the Black Sea. In *Marine pollution and sea life,* ed. M. Ruivo, pp. 222-25. London: Fishing News (Books) Ltd.

Miroshnikova, A., and Nikolaevskii, V. 1974. Permissible standards of air pollution for plants. *Tr. Mosk. O-Va. Ispyt. Prir.* 50: 149-54.

Moyer, J. W.; Cole, H., Jr.; and Lacasse, N. L. 1974*a*. Reduction of ozone injury on *Poa annua* by benomyl and thiophanate. *Plant Disease Reporter* 58(1): 41-44.

Moyer, J. W.; Cole, H., Jr.; and Lacasse, N. L. 1974*b*. Suppression of naturally occurring oxidant injury on azalea plants by drench or foliar spray treatment with benzimidazole or oxathiin compounds. *Plant Disease Reporter* 58(2): 136-38.

Moyer, J. W., and Smith, S. H. 1975. Oxidant injury reduction on tobacco induced by tobacco etch virus infection. *Environ. Pollut.* 9: 103-06.

Mudd, J. B. 1973. Biochemical effects of some air pollutants on plants. In *Air pollution damage to vegetation,* Advances in Chemistry Series 122, ed. J. A. Naegele, pp. 31-47. Washington, D.C.: American Chemical Society.

Mudd, J. B. 1975. Peroxyacyl nitrates. In *Responses of plants to air pollution,* ed. J. B. Mudd and T. T. Kozlowski, pp. 97-119. New York: Academic Press.

Muhling, G. N.; Van't Hof, J.; Wilson, G. B.; and Grigsby, B. H. 1960. Cytological effects of herbicidal substituted phenols. *Weeds* 8: 173-81.

Mulford, F. R., and Martens, D. C. 1971. Response of alfalfa to boron in flyash. *Soil. Sci. Soc. Amer. Proc.* 35: 296-300.

Muoi, L. T.; Stom, D. I.; Kefeli, V. I.; Turetskaya, R. K.; Timofeeva, S. S.; and Vlasov, P. V. 1974. Quinones as intermediate products in oxidation of certain phenolic growth inhibitors. *Sov. Plant Physiol.* 21(1): 131-35.

Murphy, C. E.; Sinclair, T. R.; and Knoerr, K. R. 1975. A model for estimating air pollutant uptake by forests: Calculation of forest absorption of sulfur dioxide from dispersed sources. Presented at Conference on Metropolitan Physical Environment, Aug. 25-29, 1975.

Naegele, J. A. 1974. Effect of pollution on plants. In *Industrial pollution,* ed. N. I. Sax, pp. 82-100. New York: Van Nostrand Reinhold Company.

Nanda, K. K.; Kumar, P.; and Kochkar, V. K. 1975. Role of auxins, antiauxin, and phenol in the production and differentiation of callus on stem cuttings of *Populus robusta*. *N. Z. J. For. Sci.* 4(2): 338-46.

Nash, T. H., III. 1975. Influence of effluents from a zinc factory on lichens. *Ecol. Monogr.* 45: 183-98.

Nash, T. H., III. 1973. The effect of air pollution on other plants, particularly vascular plants. In *Air pollution and lichens,* ed. B. W. Ferry, M. S. Baddeley, and D. L. Hawksaworth, pp. 192-223. Toronto, Canada: Univ. of Toronto Press.

National Academy of Sciences (NAS). 1972. *Particulate polycyclic organic matter*. Washington, D.C.: National Academy of Science.

National Academy of Sciences (NAS). 1976. Interactions and effects on total environment. In *Vapor-phase organic pollutants — volatile hydrocarbons and oxidation products,* Chap. 8. Washington, D.C.: National Academy of Sciences.

National Air Pollution Control Administration (NAPCA). 1970a. *Air quality criteria for sulfur oxides*. AP-50. Washington, D.C.: National Air Pollution Control Administration.

National Air Pollution Control Administration (NAPCA). 1970b. *Air quality criteria for carbon monoxide*. AP-62. Washington, D.C.: National Air Pollution Control Administration.

National Air Pollution Control Administration (NAPCA). 1971. *Air quality criteria for nitrogen oxides*. AD-84. Washington, D.C.: National Air Pollution Control Administration.

National Science Foundation. 1974. *The disposal and environmental effects of carbonaceous solid wastes from commercial oil shale operations*. First Annual Report, N&SF GI 34282X1. Schmidt-Collerus, J. J.

Neilson, A. H., and Doudoroff, M. 1973. Ammonia assimilation in blue-green algae. *Arch. Mikrobiol.* 89: 15-22.

Nethery, A. A., and Wilson, G. B. 1966. Classification of the cytological activity of phenols and aromatic organophosphates. *Cytologia* 31: 270-75.

Nicholas, D.J.D. 1975. The functions of trace elements in plants. In *Trace elements in soil-plant-animal systems,* ed. D. J. Nicholas and A. R. Egan, pp. 181-98. New York: Academic Press, Inc.

Niemi, A. 1972. Effects of toxicants on brackish-water phytoplankton assimilation. *Commentat. Biol.* 55: 3-19.

North, W. J.; Stephens, G. C.; and North, B. B. 1970. Marine algae and their relation to pollution problems. In *Marine pollution and sea life,* ed. M. Ruivo, pp. 330-40. London: Fishing News (Books) Ltd.

Oak Ridge National Laboratory. 1975. *The environmental impact of mercury*. Oak Ridge, Tenn.: Oak Ridge National Laboratory, Biomedical Sciences Section, Information Center Complex.

Ordin, L. 1965. Effect of air pollutants on cell wall metabolism. *Arch. Environ. Health* 10: 189-94.

Oshima, R. J. 1974. A viable system of biological indicators for monitoring air pollutants. *J. Air Pollut. Control Assoc.* 24: 576-78.

Overrein, L. N. 1972. Sulphur pollution patterns observed; leaching of calcium in forest soil determined. *Ambio* 1: 145-47.

Pack, M. R. 1971. Effects of hydrogen fluoride on bean reproduction. *J. Air Pollut. Control Assoc.* 21(3): 132-38.

Pahlich, E. 1975. Effect of SO_2-pollution on cellular regulation. A general concept of the mode of action of gaseous air contamination. *Atmos. Environ.* 9(2): 261-63.

Patrick, R. 1963. The structure of diatom communities under varying ecological conditions. *Ann. N. Y. Acad. Sci.* 108: 359-65.

Patrick, R.; Bott, T.; and Larson, R. 1975. *The role of trace elements in management of nuisance growths*. Final Rept. 1, Academy of Natural Sciences. EPA-660/2-75-008.

Patterson, D. T., and Kramer, P. J. 1974. Plant growth studies in controlled environments. *J. Elisha Mitchell Sci. Soc.* 90: 97.

Payer, H. D.; Soeder, C. J.; Kunte, H.; Karuwanna, P.; Nonhof, R.; and Graf, W. 1975. Accumulation of polycyclic aromatic hydrocarbons in cultivated microalgae. *Naturwissenschaften* 62: 536-38.

Peak, M. J., and Belser, W. L. 1969. Some effects of the air pollutant, peroxyacetyl nitrate, upon deoxyribonucleic acid and upon nucleic acid bases. *Atmos. Environ.* 3: 385-97.

Pelissier, M.; Lacasse, N. L.; and Cole, H., Jr. 1972*a*. Effectiveness of benomyl and benomyl-folicote treatments in reducing ozone injury to pinto beans. *J. Air Pollut. Control Assoc.* 22(9): 722-25.

Pelissier, M.; Lacasse, N. L.; and Cole, H., Jr. 1972*b*. Effectiveness of benzimidazole, benomyl, and thiabendazole in reducing ozone injury to pinto beans. *Phytopathol.* 62(5): 580-82.

Pelissier, M.; Lacasse, N. L.; Ercegovich, C. D.; and Cole, H., Jr. 1972. Effects of hydrocarbon wax emulsion sprays in reducing visible ozone injury to *Phaseolus vulgaris* 'pinto III'. *Plant Disease Reporter* 56(1): 6-9.

Peterson, P. J. 1971. Unusual accumulations of elements by plants and animals. *Sci. Prog.* 59: 505-26.

Piperno, E. 1975. Trace elements emissions: Aspects of environmental toxicology. In *Trace elements in fuel*, ed. S. P. Babu, Advances in Chemisty Series 141, pp. 192-209 Washington, D.C.: American Chemical Society.

Polishchuk, L. R. 1975. Dynamics of the level of certain organic toxic substances in plants irrigated with waste water from by-product coke manufacture plants. *Gig. Sanit.* 4: 117-19.

Porter, L. K.; Viets, F. G., Jr.; and Hutchinson, G. L. 1972. Air containing nitrogen-15 ammonia: Foliar absorption by corn seedlings. *Science* 17: 759-61.

Prager, J. C. 1974. Factors influencing estuarine microbiota. In *Industrial pollution*, ed. N. I. Sax, pp. 48-81. New York: Van Nostrand Reinhold Company.

Puckett, K. J.; Nieboer, E.; Flora, W. P.; and Richardson, D. H. 1973. Sulphur dioxide: Its effect on photosynthetic C_{14} fixation in lichens and suggested mechanisms of phytotoxicity. *New Phytol.* 72: 141-54.

Purves, David. 1972. Consequences of trace-element contamination of soils. *Environ. Pollut.* 3: 17-24.

Rabinowitz, M. 1972. Plant uptake of soil and atmospheric lead in Southern California. *Chemosphere* 4: 175-80.

Rasmussen, R. A. 1972. What do the hydrocarbons from trees contribute to air pollution? *J. Air Pollut. Control. Assoc.* 22(7): 537-43.

Rasmussen, K. H.; Taheri, M.; and Kabel, R. I. 1975. Global emissions and natural processes for removal of gaseous pollutants. *Water Air Soil Pollut.* 4: 33-64.

Rathore, V. S.; Wittwer, S. H.; Jyung, W. H.; Bajaj, Y.P.S.; and Adams, M. W. 1970. Mechanism of zinc uptake in bean (*Phaseolus vulgaris*) tissues. *Physiol. Plant.* 23: 908-19.

Ratsch, H. C. 1971. *Heavy-metal accumulation in soil and vegetation from smelter emissions.* EPA 660/3-74-012. National Environmental Research Center.

Reed, S. E.; Davis, G. J.; Harwood, J. E.; and Jones, M. N. 1974. Seasonal dynamics of the benthic aquatic macrophytes of the Pamlico River Estuary, North Carolina: A preliminary report. *J. Elisha Mitchell Sci. Soc.* 90(3): 99.

Rehwoldt, R.; Menapace, L. W.; Nerrie, B.; and Alessandrello, D. 1972. The effect of increased temperature upon the acute toxicity of some heavy metal ions. *Bull. Environ. Contam. Toxicol.* 8(2): 91-96.

Rentshler, I. 1973. Significance of the wax structure in leaves for sensitivity of plants to air pollutants. In *Proceedings, 3rd international clean air congress, Dusseldorf, Fed. Rep. Germany, 1973,* pp. 1-5. ORNL/tr-4120.

Reuss, J. O. 1975. *Chemical/biological relationships relevant to ecological effects of acid rainfall.* EPA-660/3-75-032. National Environmental Research Center.

Reynolds, J. H.; Middlebrooks, E. J.; Asce, F.; and Procella, D. B. 1974. Temperature-toxicity model for oil refinery wastes. *J. Env. Eng. Div., Proc. Amer. Soc. Civ. Eng.* 100: 557-76.

Riberau-Gayon, P. 1968. Plant phenolics. In *University reviews in botany 3,* ed. V. H. Heywood. Edinburgh: Oliver and Boyd.

Rich, S. 1964. Ozone damage to plants. *Ann. Rev. Phytopath.* 2: 253-66.

Ricks, G. R., and Williams, R.J.H. 1975. Effects of atmospheric pollution on deciduous woodland: Part 3. Effects on photosynthetic pigments of leaves of *Quercus petrae* (Mattuschka). *Environ. Pollut.* 8: 97-106.

Robinson, E., and Robbins, R. C. 1970. Gaseous nitrogen compound pollutants from urban and natural sources. *J. Air Pollut. Control Assoc.* 20(5): 203-06.

Rodhe, H.; Persson, C.; and Akesson, O. 1972. An investigation into regional transport of soot and sulfate aerosols. *Atmos. Environ.* 6: 675-93.

Rohmeder, E. 1960. The effect of dust and soot on the growth of spruce trees. *Der Frost- und Holzwirt* 15(13): 245-48.

Rosehart, R. G., and Lee, J. Y. 1973. The effect of arsenic trioxide on the growth of white spruce seedlings. *Water Air Soil Pollut.* 2: 439-43.

Rubin, E. S., and McMichael, F. C. 1975. Impact of regulations on coal conversion plants. *Environ. Sci. Technol.* 9(2): 112-17.

Rucker, W. 1971. Synergistic and antagonistic action on phenols and indol acetic acid on plant tissue cultivated in vitro. *Monatsh. Chem.* 102: 51-57.

Rule, J.; Hemphill, D.; and Pierce, J. O. 1974. The use of ^{210}Pb and ^{109}Cd isotopes in a preliminary study of their uptake and translocation by plants. In *Trace contaminants in the environment. Proceedings of the second annual NSF-RANN trace contaminants conference,* pp. 333-39.

Russell, J. E. 1954. *Soil conditions and plant growth.* New York: Longmans, Green and Co.

Sabharwal, P. S., and Bhalla, P. R. 1973. Biological effects of tobacco smoke and its constituents on cells, tissue, and chromosomes of plants. In *Proceedings of the fourth tobacco health workshop, University of Kentucky, Lexington,* pp. 632-51.

Santamaria, L.: Dolcher, T.; and Calendi, E. 1971. Chromosome breakage in *Allium cepa* by excited 3,4-benzpyrene. *Stud. Biophys.* 29: 147-48.

Schmid, W. E.; Haag, H. P; and Epstein, E. 1965. Absorption of zinc by excised barley roots. *Physiol. Plant.* 18: 860-69.

Schroeder, H. A.; Balassa, J. J.; and Tipton, I. H. 1966. Essential trace metals in man: Manganese. *J. Chron. Dis.* 19: 545-71.

Schubert, J. 1973. Heavy metals — toxicity and environmental pollution. In *Metal ions in biological systems*, ed. S. K. Dhar, pp. 239-97. New York: Plenum Press.

Schuck, E. A. 1973. Chemical basis of the air pollution problem. In *Air pollution damage to vegetation*, Advances in Chemistry Series 122, ed. J. A. Naegele, pp. 1-8. Washington, D.C.: American Chemical Society.

Schuck, E. A.; Altshuller, A. P.; Barth, D. S.; and Morgan, G. B. 1970. Relationship of hydrocarbons to oxidants in ambient atmospheres. *J. Air Pollut. Control.* 20(5): 297-302.

Seear, J.; Bradfute, O. E.; and McLaren, A. D. 1968. Uptake of proteins by plant roots. *Physiol. Plant.* 21: 979-89.

Seem, R. C.; Cole, H., Jr.; and Lacasse, N. L. 1972. Suppression of ozone injury to *Phaseolus vulgaris* 'pinto III' with triarimol and its monochlorophenyl cyclohexyl analogue. *Plant Disease Reporter* 56(5): 386-90.

Seem, R. C.; Cole, H., Jr.; and Lacasse, N. L. 1973. Suppression of ozone injury to *Phaseolus vulgaris* L. with thiophanate ethyl and its methyl analogue. *J. Environ. Qual.* 2(2): 266-68.

Severne, B. C. 1974. Nickel accumulation by *Hybanthus floribundus*. *Nature* 248: 807-08.

Shabad, L. M., and Kogan, Y. L. 1972*a*. The contents of benzo(a)pyrene in some crops. *Arch. Geschwulstforsch* 40(3): 237-43.

Shabad, L. M., and Kogan, Y. L. 1972*b*. Level of benzo(a)pyrene in soils and vegetation after nerosin application. *Onkologiya* 3: 105-7.

Sharma, A. K., and Ghosh, S. 1965. Chemical basis of the action of cresols and nitrophenols on chromosomes. *Nucleus* 8(2): 183-90.

Shiraishi, Y., and Takabatake, E. 1974. Determination of polycyclic aromatic hydrocarbons in foods. 3. 3,4-benzopyrene in vegetables. *Shokuhin Eiseigaku Zasshi* 15(1): 18-21. ORNL/tr-4125.

Shriner, D. S. 1976. Effects of simulated rain acidified with sulfuric acid on host-parasite interactions. In *Proceedings of the first international symposium on acid precipitation and the forest ecosystem, Columbus, Ohio, May 12-15, 1975* (in press).

Shriner, D. S., and Lacasse, N. L. 1972. Rapid determination of chloride content of vegetation for assessment of air pollution injury from hydrogen chloride. *Phytopathol.* 62: 427-29.

Siddiqi, I., and Wagner, K. H. 1972. Determination of 3,4-benzpyrene and 3,4-benzofluoranthene in rain water, ground water, and wheat. *Chemosphere* 1(2): 83-88.

Siegfried, R. 1975. Effect of garbage compost on the 3,4-benzpyrene content of carrots and head lettuce. *Naturwissenschaften* 62: 300. ORNL/tr-4124.

Smirnov, G. A. 1970. The study of the benz(alpha)pyrene content in the soil and vegetation in the airfield region. *Vop. Onkol.* 16(5): 83-86.

Smith, I. C.; Ferguson, T. L.; and Carson, B. L. 1975. Metals in new and used petroleum products and by-products. Quantities and consequences. *Trace elements in petroleum*, ed. T. F. Yen, pp. 123-38. Ann Arbor, Mich.: Ann Arbor Science Publishers, Inc.

Smith, W. H. 1973. Metal contamination of urban woody plants. *Environ. Sci. Technol.* 7(7): 631-36.

Smith, W. H., and Parker, J. C. 1966. Prevention of ethylene injury to carnations by low concentrations of carbon dioxide. *Nature* 211: 100-01.

Sparrow, A. H., and Schairer, L. A. 1974. Mutagenic response of *Tradescantia* to treatment with X-rays, EMS, DBE, ozone, SO_2, N_2O and several insecticides. *Mutation Res.* 26(5): 445.

Srivastava, H. S.; Jolliffe, P. A.; and Runeckles, V. C. 1975*a*. The effects of environmental condition on the inhibition of leaf gas exchange by NO_2. *Can. J. Bot.* 53(5): 475-82.

Srivastava, H. S.; Jolliffe, P. A.; and Runeckles, V. C. 1975*b*. Inhibition of gas exchange in bean leaves by NO_2. *Can. J. Bot.* 53(5): 466-74.

Stadtman, T. C. 1974. Selenium biochemistry. *Science* 183: 915-22.

Stanley, R. A. 1974. Toxicity of heavy metals and salts to Eurasian watermilfoil. (*Miriophyllum spicatum* L.) *Arch. Environ. Contam. Toxicol.* 2(4): 331-41.

Starr, R. I., and Cunningham, D. J. 1974. Phytotoxicity, absorption, and translocation of 4-aminopyridine in corn and sorghum growing in treated nutrient cultures and soils. *J. Agr. Food Chem.* 22(3): 409-13.

Stein, A., and Keller, E. C., Jr. 1973. Gene and growth dynamics of *Chlorella* under continuous culture during phenol stress. *W. Va. Acad. Sci. Proc.* 45(2): 103-11.

Steiner, A. A. 1973. The uptake of cadmium by plants. *Landbouwkundig Tijdschrift* 85(4): 124-28.

Stevens, S. E., and Van Baalen, C. 1974. Control of nitrate reductase in a blue-green alga. *Arch. Biochem. Biophys.* 161: 146-52.

Stokes, P. M.; Hutchinson, T. C.; and Krauter, K. 1973. Heavy-metal tolerance in algae isolated from contaminated lakes near Sudbury, Ontario. *Can. J. Bot.* 51(11): 2155-68.

Stom, D. I.; Bobovskaya, L. P.; Timofeeva, S. S.; Kalmychkov, G. V.; and Kozhova, O. M. 1974. Influence of phenols and their oxidation products on aquatic plants and their content of sulfhydryl groups. *Dok. Biochem.* 216(1-6): 233-36.

Stom, D. I.; Timofeeva, S. S.; Kolmakova, E. F.; and Kalabina, A. V. 1973. Detection and inhibitory properties of ortho-quinones. *Dok. Biochem.* 205(1-6): 301-04.

Suess, M. J. 1972. Polynuclear aromatic hydrocarbon pollution of the marine environment. In *Marine pollution and sea life*, ed. M. Ruivo, pp. 568-70. London: Fishing News (Books) Ltd.

Sulzbach, C. W., and Pack, M. R. 1972. Effects of fluoride on pollen germination, pollen tube growth, and fruit development in tomato and cucumber. *Phytopath.* 62(11): 1247-53.

Sundstrom, K. R., and Hallgren, J. E. 1973. Using lichens as physiological indicators of sulfurous pollutants. *Ambio* 2(1,2): 13-21.

Taylor, O. C. 1968. Effects of oxidant air pollutants. *J. Occup. Med.* 10: 485-92.

Taylor, O. C., and Eaton, F. M. 1966. Suppression of plant growth by nitrogen dioxide. *Plant Physiol.* 41: 132-35.

Taylor, O. C.; Dugger, W. M., Jr.; Cardiff, E. A.; and Darley, E. F. 1961. Interaction of light and atmospheric photochemical products (smog) within plants. *Nature* 192: 814-16.

Taylor, O. C. 1969. Importance of peroxyacetyl nitrate (PAN) as a phytotoxic air pollutant. *J. Air Pollut. Control. Assoc.* 19(5): 347-51.

Taylor, O. C., and MacLean, D. C. 1970. Nitrogen oxides and the peroxyacyl nitrates. In *Recognition of air pollution injury to vegetation: a pictorial atlas*, ed. J. S. Jacobson and A. C. Hill, pp. E1-E5. Pittsburgh, Penn.: Air Pollution Control Association and National Air Pollution Control Administration.

Taylor, O. C. 1973. Acute responses of plants to aerial pollutants. In *Air pollution damage to vegetation*, Advances in Chemistry Series 122, ed. J. A. Naegele, pp. 9-20. Washington, D.C.: American Chemical Society.

Taylor, O. C. 1974. Air pollutant effects influenced by plant-environmental interaction. In *Air pollution effects on plant growth*, ACS Symp. Ser. 3, ed. M. Dugger, Washington, D.C.: American Chemical Society.

Taylor, O. C.; Thompson, C. R.; Tingey, D. T.; and Reinert, R. A. 1975. Oxides of nitrogen. In *Responses of plants to air pollution*, ed. J. B. Mudd and T. T. Kozlowski, pp. 122-39. New York: Academic Press.

Thimann, K. V. 1972. The natural plant horomones in plant physiology. In *Physiology of development: The horomones*, vol. VIB, ed. F. C. Steward, pp. 4-273. New York: Academic Press.

Thomas, M. D. 1951. Gas damage to plants. *Ann. Rev. Plant Physiol.* 2: 293-322.

Thomas, M. D., and Hendricks, R. H. 1956. Effects of air pollution on plants. In *Air pollution handbook*, ed. P. L. Magil et al., Sect. 9. New York: McGraw-Hill.

Thompson, C. R.; Hensel, E. G.; Kats, G.; and Taylor, O. C. 1970. Effects of continuous exposure of navel oranges to nitrogen dioxide. *Atmos. Environ.* 4: 349-55.

Thompson, W. W.; Dugger, W. M., Jr.; and Palmer, R. L. 1965. Effects of peroxyacetyl nitrate on ultrastructure of chloroplasts. *Bot. Gaz.* 126: 66-72.

Tingey, D. T.; Reinert, R. A.; Dunning, J. A.; and Heck, W. W. 1971. Vegetation injury from the interaction of nitrogen dioxide and sulfur dioxide. *Phytopathol.* 71(12): 1506-11.

Tingey, D. T. 1973. Radish root growth reduced by acute ozone exposures. *Proceedings of the 3rd international clean air congress*, pp. 154-56.

Tingey, D. T.; Reinert, R. A.; Dunning, J. A.; and Heck, W. W. 1973. Foliar injury responses of eleven plant species to ozone/sulfur dioxide mixtures. *Atmos. Environ.* 7: 201-08.

Tingey, D. T. 1974. Ozone induced alterations in the metabolite pools and enzyme activities of plants. In *Air pollution effects on plants*, ACS Symp. Ser. 3, ed. M. Dugger, pp. 40-57. Washington, D.C.: American Chemical Society.

Tingey, D. T., and Reinert, R. A. 1975. The effect of ozone and sulphur dioxide singly and in combination on plant growth. *Environ. Pollut.* 9(2): 117-26.

Tingey, D. T.; Fites, R. C.; and Wickliff, C. 1975. Activity changes in selected enzymes from soybean leaves following ozone exposure. *Physiol. Plant.* 33: 316-20.

Toetz, D. W. 1974. Uptake and translocation of ammonia by freshwater hydrophytes. *Ecology* 55: 199-201.

Treshow, M. 1971. Fluorides as air pollutants affecting plants. *Ann. Rev. Phytopathol.* 9: 21-44.

Treshow, M., and Pack, M. R. 1970. Fluoride. In *Recognition of air pollution injury to vegetation: a pictorial atlas*, ed. J. S. Jacobson and A. C. Hill, pp. D1-D6. Pittsburgh, Penn.: Air Pollution Control Association and National Air Pollution Control Administration.

Turner, M. A.; and Rust, R. H. 1971. Soil fertility and plant nutrition: Effects of chromium on growth and mineral nutrition of soybeans. *Soil Sci. Soc. Amer. Proc.* 35: 755-58.

Tyler, G. 1972. Heavy metals pollute nature, may reduce productivity. *Ambio* 4(2): 52-59.

Tyson, B. J.; Dement, W. A.; and Mooney, H. A. 1974. Volatilisation of terpenes from *Saliva mellifera*. *Nature* 252: 119-20.

Underwood, E. J. 1973. Trace elements. In *Toxicants occurring naturally in foods*. Washington, D.C.: National Academy of Sciences.

Unsworth, M. H.; Biscoe, P. V.; and Pinckney, H. R. 1972. Stomatal responses to sulphur dioxide. *Nature* 239: 458-59.

Vallee, B. L., and Ulmer, D. D. 1972. Biochemical effects of mercury, cadmium and lead. *Ann. Rev. Biochem.* 41: 91-128.

Van Hook, R. I., and Shults, W. D., eds. 1976. *Ecology and analysis of trace contaminants. Progress report, October 1974-December 1975.* ORNL/NSF/EATC-22. Oak Ridge, Tenn.: Oak Ridge National Laboratory.

Van Loon, J. C. 1974. Mercury contamination of vegetation due to the application of sewage sludge as fertilizer. *Environ. Lett.* 6: 211-18.

Von Rosen, G. 1954. Breaking of chromosomes by the action of elements of the periodical system and by some other principles. *Hereditas* 40: 258-63.

Von Rosen, G. 1964. Mutations induced by the action of metal ions in *Pisum* II. *Hereditas* 51: 89-134.

Wagner, Von K. H., and Siddiqi, I. 1970. The metabolism of 3,4-benzpyrene and 3,4-benzfluoranthene in summer wheat. *Z. Pflanzen. Bodenkundl* 127(3): 211-18.

Waldbott, G. L. 1973. *Health effects of environmental pollutants.* St. Louis: C. V. Mosby Company.

Wallace, R. A.; Fulkerson, W.; Shults, W. D.; and Lyon, W. S. 1971. *Mercury in the environment: The human element.* ORNL/NSF/EP-1.

Wallin, T. 1974. Deposition of air-borne mercury around Swedish chloralkali plants surveyed by analyses of snow and moss. *Congr. Int. Mercurio* 2: 133-39.

Ward, N. I.; Brooks, R. R.; and Reeves, R. D. 1974. Effect of lead from motor-vehicle exhausts on trees along a major thoroughfare in Palmerston North, New Zealand. *Environ. Pollut.* 6: 149-58.

Warren, H. V., and Delavault, R. E. 1971. Variations in the copper, zinc, lead, and molybdenum contents of some vegetables and their supporting soils. *Geol. Soc. Amer.* 123: 97-108.

Weinstein, L. H.; McCune, D. C.; Aluisio, A. L.; and Van Leuken, P. 1975. The effect of sulphur dioxide on the incidence and severity of bean rust and early blight of tomato. *Environ. Pollut.* 9: 145-55.

Went, F. W. 1960. Blue hazes in the atmosphere. *Nature* 187: 641-43.

Whittaker, R. H.; Bormann, F. H.; Likens, G. E.; and Siccama, T. G. 1974. The Hubbard Brook ecosystem study: Forest biomass and production. *Ecol. Monogr.* 44: 233-52.

Whitton, B. A. 1970. Toxicity of heavy metals to freshwater algae: A review. *Phykos* 9(2): 116-25.

Williams, R.J.H.; Lloyd, M. M.; and Ricks, G. R. 1971. Effects of atmospheric pollution on deciduous woodland. I. Some effects on leaves of *Quercus petraea* (Mattuschka) Leibl. *Environ. Pollut.* 2: 57-68.

Wilton, A. C.; Murray, J. J.; Heggestad, H. E.; and Juska, F. V. 1972. Tolerance and susceptibility of Kentucky bluegrass (*Poa pratensis* L.) cultivars to air pollution: In the field and in an ozone chamber. *J. Environ. Qual.* 1(1): 112-14.

Windom, H. L. 1973. Heavy metal fluxes through salt marsh estuaries. In *Proceedings of the 2nd international estuarine conference,* pp. 30-50.

Wirth, V., and Turk, R. 1975. SO_2 resistance of lichens with different forms. *Flora Bd.* 164: 133-43.

Witherspoon, J. P., and Taylor, F. G. 1970. Interception and retention of a simulated fallout by agricultural plants. *Health Phys.* 19: 493-99.

Wolf, F. T., and Kidd, G. H. 1973. Effect of various gas atmospheres upon the greening of etiolated seedlings. *Z. Pflanzenphysiol.* 70: 115-18.

Wolverton, B. C. 1975. Water hyacinths for removal of phenols from polluted waters. *NASA Technical Memorandum,* pp. 1-19.

Wood, T., and Bormann, F. H. 1974. The effects of an artificial acid mist upon the growth of Betula *alleghaniensis* Britt. *Environ. Pollut.* 7: 259-68.

Wynder, E. L.; Kmet, J.; Dungal, N.; and Segi, M. 1963. An epidemiological investigation of gastric cancer. *Cancer* 16: 1461-96.

Zook, E. G.; Greene, F. A.; and Morris, E. R. 1970. Nutrient composition of selected wheats and wheat products. VI. Distribution of manganese, copper, nickel, zinc, magnesium, lead, tin, cadmium, chromium, and selenium as determined by atomic absorption spectroscopy and colorimetry. *Cereal Chem.* 47(6): 720-38.

9. ANIMALS: BIOENVIRONMENTAL EFFECTS

D. J. Wilkes

ABSTRACT

Many organic compounds may be released into the environment from coal conversion processes.
Polycyclic aromatic hydrocarbons (PAH) and heterocyclic compounds containing either nitrogen
or sulfur have been identified as organic compounds of particular environmental concern because
of their carcinogenic potential.

Organisms may accumulate PAH and heterocyclics from air by elution in the lungs from inhaled
particles to which the compounds may be sorbed, from aquatic environments where both soluble and
particulate forms are available for absorption through surface or gill membranes, or from
ingestion of particulates or food contaminated with the compounds. Storage in tissues of high
lipid content — primarily liver, gall bladder, and hepatopancreas — is favored.

Rates and extent of metabolism are species dependent, but some organisms, such as molluscs,
apparently do not metabolize PAH and heterocyclics at all. While detoxification mechanisms vary
widely among species, some transformation products may be more toxic than the parent compound.

Although PAH and heterocyclic compounds may evoke carcinogenic, teratogenic, or mutagenic
responses in exposed organisms, no such effects have been directly correlated to the low concentrations normally found in the environment. Chronic low-level exposure, however, may affect
entire communities through modifications of population characteristics. For example, changes in
rates of birth, death, and species dispersal for a population may influence other species populations, possibly culminating in the modification of an entire community structure with far-
reaching effects throughout the ecosystem.

Many trace elements have been identified as major constituents of coals and may be released into
the environment through conversion processes; thus, an assessment of potential biological
effects is warranted.

Potential toxicity and bioaccumulation of elements is a function of concentration, chemical form,
environmental chemistry, and food web dynamics. Many trace elements are essential to sustain
life and are present at low concentrations in virtually all living organisms, but may bioaccumulate to harmful levels in organisms that either inhabit areas where anthropogenic emissions
have raised natural background levels or are situated at the upper levels of a food chain.
Although many elements are involved in natural biochemical processes that regulate accumulation
under normal conditions, some elements — notably cadmium, mercury, and lead — are cumulative
poisons; no known homeostatic mechanism exists to regulate their accumulation in animals. Such
metals may exert potent neurotoxic and nephrotoxic effects. Some metal compounds, such as nickel
carbonyl, may be highly carcinogenic.

Many elements may be acutely toxic at relatively low environmental concentrations, with chronic effects being evoked at concentrations as low as in the part-per-hundred-million range. Furthermore, environmental factors, antagonism, synergism, and variations of the toxic effect due to the action of surrounding flora and fauna may affect the toxic action of trace elements on biota.

9.0 INTRODUCTION

The purpose of this chapter is to consider the potential interaction of possible effluents of coal conversion processes with animals in their natural habitat. Natural systems are characterized by complex combinations of various components interconnected so that an intricate balance exists. Disturbing this balance reduces system stability and productivity as well as the ability of the system to regenerate renewable resources and to supply nonrenewable resources. Uncontrolled release of coal conversion effluents may disturb natural ecosystems through interaction with their biotic components.

Coal is one of the country's most abundant natural resources. It can be an environmentally unacceptable fuel, or it can be converted to clean-burning synthetic liquid and gaseous fuels (Rubin and McMichael 1975). However, coal conversion processes release complex aromatic and heterocyclic compounds (Hayatsu et al. 1975), a number of which can be hazardous to animals or to man as a higher member of the food chain. Likewise, trace metals, which are naturally abundant in coal, and inorganic compounds used as catalysts are capable of posing a biological threat (Tingey and Morrey 1973).

Potential pollutants released during coal conversion can be distributed to all environmental media; however, the bulk of available information is derived from studies of aquatic systems. Therefore, much of the data, especially for organic compounds, included in the chapter is weighted toward aquatic organisms.

9.1 POLYCYCLIC AROMATIC HYDROCARBONS

Much of the published literature about hydrocarbons deals with saturated or olefinic structures (aliphatics), but neglects aromatics, especially polycyclic aromatics (Coleman, Hirsch, and Dooley 1969). Blumer and Youngblood (1975), who also note this scarcity in the literature, attribute the lack of research on PAH toxicity to biota in their natural habitat to difficulty of analysis (Chap. 5).

However, new techniques have recently identified PAH in water (Table 6.4), in forest soil (Table 6.20), in urban air (Table 9.1), and in constituents of coal carbonization products (Table 4.54) and coal tar (Fig. 2.16).

9.1.1 Biotic availability

PAH are present throughout the environment in both biotic and abiotic sectors. A major influence on availability of any compound to biota is its concentration in the environment. Anthropogenic emissions of PAH may significantly increase biotic availability of PAH by increasing natural environmental levels (Suess 1972). Although very little information is available regarding PAH in other than aquatic environments, these compounds appear to be readily available to organisms at any level in any environmental media.

Table 9.1. Carcinogenic PAH identified in urban air

Compound	Structure	Compound	Structure
Benzo[*a*]pyrene	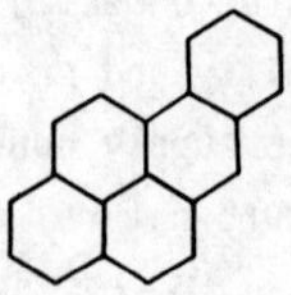	Dibenzo[*e,i*]pyrene	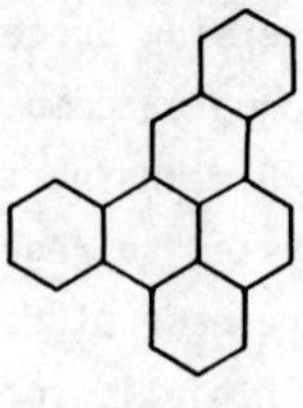
Dibenzo[*a,h*]anthracene	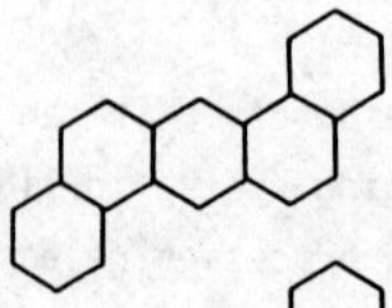	Benzo[*e*]pyrene	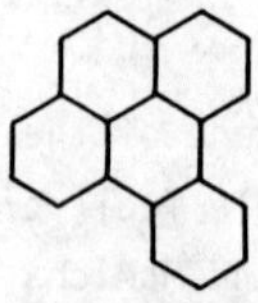
Benzo[*a*]anthracene	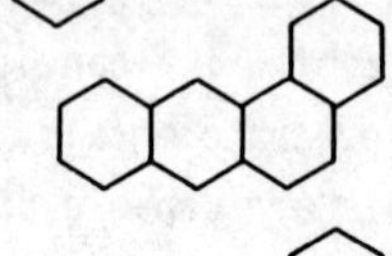	Dibenzo[*h,rst*]pentaphene	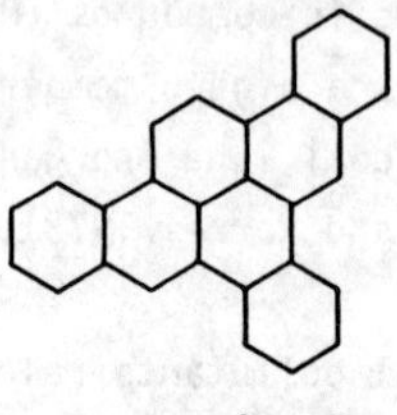
Benzo[*b*]fluoranthene	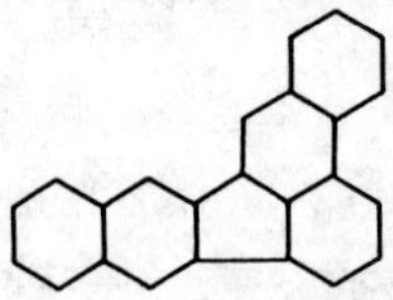	Dibenzo[*a,e*]pyrene	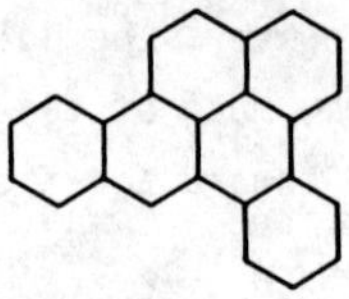
Benzo[*j*]fluoranthene	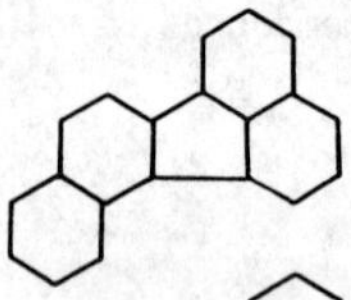	Dibenzo[*a,h*]pyrene	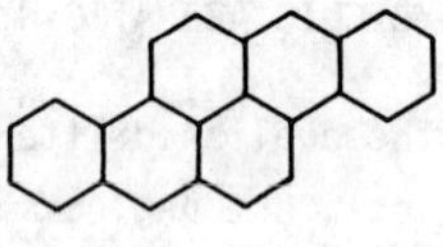
Dibenzo[*a,i*]pyrene	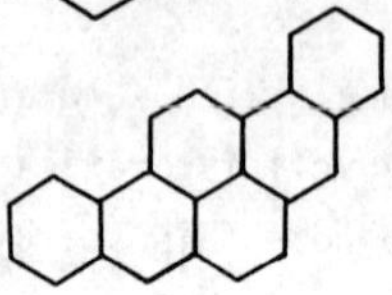	Indeno[*1,2,3-cd*]pyrene	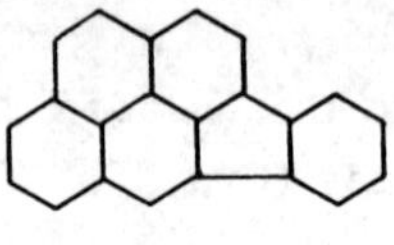

Source: Compiled from Kornreich 1975; Olsen and Haynes 1969; Fishbein 1973.

9.1.1.1 In air

PAH are found in the atmosphere primarily as compounds adsorbed on particulate matter (Olsen and Haynes 1969). The biological availability of such PAH is dependent on the fate of the particles to which they are adsorbed. PAH exist in relatively pure form on particles and are readily available to animals through ingestion, inhalation, or skin contact (EPA 1975).

Natusch (1976) has observed a number of polycyclic organic compounds, many of which are known or suspected carcinogens, on the surface of fly ash particles from coal combustion (Table 9.2). Concentrations were shown to increase with decreasing particle size, an important factor in lung uptake during respiration since smaller particles are more readily absorbed into the lungs than are large particles.

Table 9.2. Polycyclic organic compounds identified
in emitted coal fly ash

Anthracene	Fluorene
9,10-Dimethyl anthracene	**Fluoranthene**
Benzo[a]pyrene	**Perylene**
Benzo[e]pyrene	Phenanthrene
Benzo[b]phenanthrene	Pyrene
Chrysene	Triphenylenene

Source: Natusch 1976, Table III, p. 83.
Reprinted by permission of the publisher.

Airborne polycyclic organic matter may reach aquatic environments by several routes — rain, surface runoff, or direct deposition onto surface waters — thereby enriching the PAH load of aquatic environments (Neff 1975). Sawicki (1967) and Freudenthal, Lutz, and Mitchell (1975) suggest that, although many PAH are likely to be present in the atmosphere, they have not all been identified.

9.1.1.2 In natural waters

The occurrence, background concentrations, and routes of transmission of PAH in water have been reviewed (Andelman and Suess 1970; Andelman and Snodgrass 1974; Harrison, Perry, and Wellings 1975). Natural waters clearly are favored as the depositional sink for PAH, which are readily transmitted to aquatic environments from air, soil, surface runoff, industrial effluents, domestic effluents, and PAH-synthesizing microorganisms (Chap. 7) and flora (Chap. 8). Furthermore, due to recent concern over oil spills and other aquatic petroleum pollution, hydrocarbon pollution research has focused on aquatic environments, but the main concern has been with aliphatic hydrocarbons. Some oil spill studies include data on aromatics, but very few have discussed PAH.

Both soluble and particulate forms of PAH may occur in natural waters, where they remain quite stable. Ehrhardt (1972) has suggested PAH stability as a key factor in PAH exposure to a wide range of organisms throughout the food chain. Although solubilities of PAH in water are reported to be quite low (McAuliffe 1966), the presence of surfactants in water may increase their solubility from 2 to 7 times (Andelman and Snodgrass 1974). Boylan and Tripp (1971) suggest that the simpler (low-molecular-weight) PAH are sufficiently soluble in seawater to be adsorbed directly onto the surfaces of both animals and plants. More complex (higher-molecular-weight) PAH were shown to have a relatively low solubility in seawater unless substantial amounts of dissolved or suspended organic matter were present (Boehm and Quinn 1976). Hydrocarbons injected into the surface water of a marine ecosystem may be adsorbed by suspended or dissolved organic matter. Detritus then carries the hydrocarbons to the sea floor, where they are available for ingestion by filter-feeding organisms (Corner 1975).

9.1.2 Uptake by aquatic organisms

Andelman and Suess (1970) tabulated concentrations of benzo[a]pyrene (BaP) identified in marine fauna (Table 9.3) and plankton (Table 9.4). BaP concentrations ranged from 540 ppb dry weight

Table 9.3. Concentration of BaP in marine fauna

Source	Sample	BaP concentration (μg/kg of dry sample)
Greenland, west coast[a]	Codfish	15
	Mollusc	60
	Holothurian	Not detected
	Mussel	
	Shell	18
	Body	55
Italy, Bay of Naples[b]	Mussel	
	Shell	11
	Body	130 and 540
Italy, Bay of Naples[c]	Mollusc	2.4
Italy, Bay of Naples[d]	Sardine	65
Freshwater pool, Italy	Tubifex worms	50
Various French coasts[e]	Shirmp, oyster, mussel, mollusc, crab, etc.	Not detected to traces to 1.5 - 90
French Atlantic coast	Oyster	
	Shell	3.5 μg/dozen
	Body	0.4 μg/dozen
French Channel coast	Oyster	
	Lower shell	70
	Upper shell	112
France, Toulon harbor	Mussel	16 - 22[f]
Alabama	Oyster; shell	24
Virginia	Oyster	2 - 6[g]
California	Goose barnacle	Present
California	Thatched barnacle	Present

[a]Sample from depth of 40 m.
[b]Two samples.
[c]Sample from depth of 35 m.
[d]Sample from water surface.
[e]Mallet (1961) presents a detailed list of 25 samples covering 13 species.
[f]Total PAH was about 1100 - 3400 μg/kg.
[g]Total carcinogenic PAH was about 300 μg/kg and total PAH about 1200 μg/kg.

Source: Andelman and Suess 1970, Table 7, p. 488. Reprinted by permission of the publisher.

in oysters from the Bay of Naples (Italy) to 2 ppb in oysters from the Virginia coast, which contained a total PAH of 1200 ppb. Marine plankton BaP concentrations also varied widely, from undetectable concentrations to 400 ppb dry weight. Lee, Sauerheber, and Dobbs (1972) detected BaP in anchovies and smelt from San Diego Bay in amounts as high as 10 mg per fish.

Moore and Dwyer (1974) list organisms for which oil hydrocarbon uptake and tainting data had been reported: *Calanus* spp. (copepod), *Pagurus longicarpus* (hermit crab), *Aequipecten irradians*

Table 9.4. Concentration of BaP in marine plankton

Source	BaP concentration (µg/kg of dry sample)
Greenland[a]	5.5
Italy[b]	6.1 - 21.2
French Channel coast	400
French Mediterranean coast[c]	Not detected to 5
Estuary, French Channel coast[d]	100
Estuary, French Channel coast[e]	350
Diatoms serving as filters in industry	5.5

[a] One sample from depth of 30 m.
[b] Six samples collected on water surface and from a depth of 2 m.
[c] Fifteen samples.
[d] Plankton debris in foam immediately downstream of dam.
[e] Plankton debris west of outlet, downstream from dam.

Source: Andelman and Suess 1970, Table 5, p. 487.
Reprinted by permission of the publisher.

(scallop), *Modiolus dimissus* (horse mussel), *Mya arenaria* (soft-shell clam), *Asterias vulgaris* (starfish), and *Strongylocentrotus droebachienis* (sea urchin). The authors suggest that, since PAH are contained in the residual fractions of all crude oils, they may be concentrated in the same manner as are other hydrocarbons. Zitko (1975*a*) identified PAH accumulation from creosote oil in periwinkles (*Littorina littorea*), whelks (*Buccinum undatum*), mussels (*Mytilus edulis*), and clams (*Mya arenaria*). Creosote oil contains anthracene, phenanthrene, fluorene, pyrene, and other PAH. Dunn and Stich (1975) have studied the use of mussels (*Mytilus californianus* and *Mytilus edulis*) as indicators of PAH contamination in the marine environment. BaP was shown to accumulate in mussels regardless of their habitat locale (Table 9.5). However, the greatest accumulation was reported near creosoted pilings within the wharf area along the Pacific coastline of Vancouver Island, an area rich in algal growth.

Greffard and Meury (1967), as cited by Ehrhardt (1972), identified benzopyrene, benzofluoranthene, and perylene in the mussel *Mytilus edulis*.

Fazio (1971), as cited by Neff (1975), found fluoranthene and pyrene to be present at higher concentrations than other PAH in oysters from Galveston Bay. Cahnmann and Kuratsune (1957) report similar PAH analysis of oysters off the Virginia coast.

9.1.2.1 Tissue distribution

Laboratory studies by Ogata and Miyake (1973) demonstrated that benzene, toluene, and xylene rapidly infiltrated into the muscle and liver of fish (grey mullet, *Mugil japanicus*) and eels. Further research (Ogata and Miyake 1975) indicates a correlation between increase of aromatic substances in eel flesh with increased rearing time in aromatic-contaminated seawater. To investigate the sequence of flesh infiltration by the petroleum compounds, the influence of rearing

Table 9.5. Concentration of BaP in mussels on rocks in various areas

Sample location and description	n	BaP level (μg/kg wet wt)[a] Mean ± SE	Range
Open Pacific coast, no human activity within 5 km (west coast of Vancouver Island)	4[b]	0.1 ± 0.1	0.0 - 0.2
Outer harbor (Vancouver)	44	2.0 ± 0.3	0.0 - 8.3
Wharf, marina, and dock areas (Vancouver vicinity)	15	18 ± 4.9	1.5 - 60
Poorly flushed inlet, heavy marina and industrial usage (Vancouver, False Creek)	6	42 ± 6.0	27 - 63

[a]Mussels were collected from the midintertidal zone. Mussels 3 to 6 cm in size were cleaned externally and shucked into clean glass containers. Byssal threads were removed. The tissue was drained for 1 min by agitation over a screen of mesh size 1.5 mm. Samples were frozen at -10° for no more than 4 weeks before analysis.
[b]*Mytilus californianus*, all other samples are *M. edulis*.

Source: Dunn and Stich 1975, Table 1, p. 50. Reprinted by permission of the publisher.

time was charted on chromatogram peaks. The highest peak was found with toluene, and the height of peaks descended in the order of *o*-, *m*-, or *p*-xylene and benzene (Fig. 9.1).

Research by Lee, Sauerheber, and Dobbs (1972) demonstrates the ability of marine fish to readily absorb PAH from their environment. The sand goby (*Gillichthys mirabilis*), sculpin (*Oligocottus maculosus*), and sand dab (*Citharichthys stigmaeus*) accumulated naphthalene and BaP within minutes after exposure. The proposed path of uptake was through the gills, followed by accumulation in the liver, gut, and flesh; the final storage site was the gall bladder.

Scaccini et al. (1970) demonstrated that goldfish (*Carassius auratus*) incorporated BaP through ingestion of contaminated food; the compound entered through the gastroenteric tract and was absorbed by the blood. The compound was then distributed to various organs, followed by fixation by the liver, gonads, and muscle tissue. No BaP was found in the heart, bile, natatorial vesicle, brain, or scales of the fish at any time during experimentation.

Neff (1975) found that, in clams exposed to 30.5 ppb BaP for 24 hr, nearly 75% of the radioactivity was localized in the viscera (digestive system, gonads, and heart) (Table 9.6). The mantle, gills, adductor muscles, and foot each contained from 3.5 to 10% of the total radioactivity, and the viscera were shown to contain more than half of the total accumulated radioactivity at all sampling intervals. Lee, Sauerheber, and Benson (1972) suggest that the hepatopancreas is the final PAH storage site in *Mytilus edulis*. Neff and Anderson (1975) note that the hepatopancreas is contained in the visceral mass; thus, their data support the hypothesis of the hepatopancreas as the final PAH storage site. Hydrocarbon storage in tissues was postulated to be directly related to tissue lipid content. Therefore, the visceral mass of invertebrates, which is high in lipids, is favored as the ultimate hydrocarbon storage site.

Distribution of the ^{14}C label throughout the clam (*Rangia cuneata*) tissues indicated considerable individual variation (as seen by the large standard deviation in Table 9.6) in the total accumulation of BaP by the clams. This variability seems to be typical of bivalve molluscs, which are

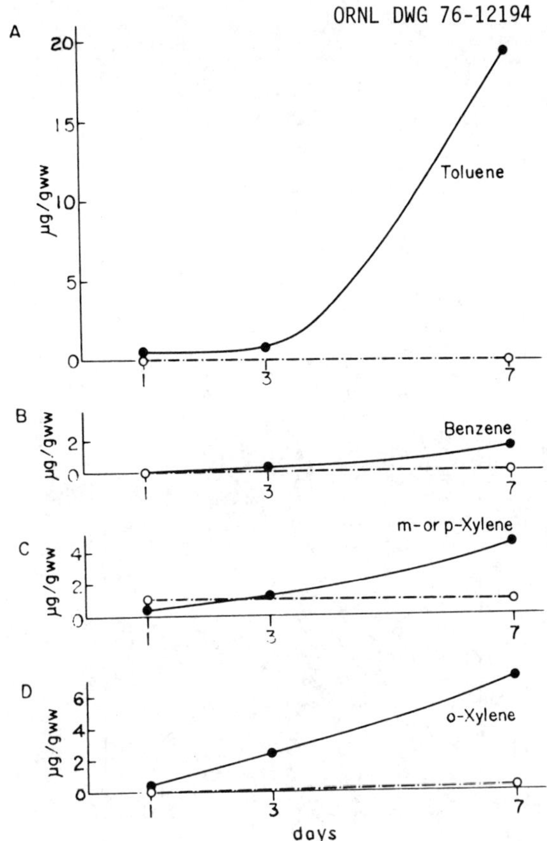

Fig. 9.1. Relation between time spent in water and concentration of petroleum products in eel flesh. (A) Toluene; (B) benzene; (C) *m*- or *p*-xylene; (D) *o*-xylene. Solid lines — flesh of eels raised in oily water; dashed lines — flesh of control eels raised in clean water. <u>Source</u>: Ogata and Miyake 1975, Fig. 3, p. 1077. Reprinted by permission of the publisher.

capable of closing their valves and remaining isolated from the external environment for variable lengths of time (Neff and Anderson 1975).

9.1.2.2 Selectivity of accumulation

Blumer, Souza, and Sass (1970) report that PAH are nonselectively accumulated from seawater by oysters (*Crassostrea virginica*) and retained indefinitely. Ehrhardt (1972) demonstrated that oysters from Galveston Bay, Texas, nonspecifically accumulate hydrocarbons from the wide range to which they are exposed. No indication of hydrocarbon avoidance or selectivity was noted, which suggests the possibility of incorporation of carcinogenic PAH. However, in bivalve molluscs, Stegeman and Teal (1973) and Anderson (1973) (as cited by Clark and Finley 1975) found that PAH were often accumulated in preference to the *n*-paraffins from petroleum.

Laboratory studies by Neff (1975) demonstrate that clams (*Rangia cuneata*), oysters (*Crassostrea virginica*), Gulf killifish (*Fundulus similis*), and brown shrimp (*Penaeus aztecus*) are all capable of rapidly accumulating PAH in their tissues from a dispersion of oil in seawater. A No. 2 fuel oil was used with a water-soluble (aromatic) fraction containing naphthalenes, biphenyls, fluorenes, and phenanthrenes. Of the four compounds, naphthalenes were shown to accumulate to the highest concentrations in all the organisms tested (Table 9.8). Naphthalenes are more water-soluble than the other aromatics tested (McAuliffe 1966) and have been characterized as major

Table 9.6. Uptake, tissue distribution, and release of BaP-C^{14} in the clam *Rangia cuneata* exposed to 0.0305 ppm BaP-C^{14} for 24 hr

Sampling time	Mean total ppm of BaP per animal ± SD[a]	Percent of 24-hr concentration remaining	Mean percentage total radioactivity per tissue fraction ± SD				
			Viscera	Mantle	Gill	Adductor	Foot
24-hr uptake (1 day)	7.2 ± 12.8		74.7 ± 15.1	10.1 ± 6.5	6.8 ± 4.3	4.9 ± 2.7	3.5 ± 2.0
144-hr depuration (6 days)	6.5 ± 5	90	61.2 ± 14.9	7.1 ± 6.2	2.9 ± 2.5	5.0 ± 3.5	3.9 ± 3.0
244-hr depuration (10 days)	2.1 ± 0.97	29	82.2 ± 12.3	5.8 ± 3.3	4.1 ± 4.4	4.5 ± 3.2	3.4 ± 1.6
312-hr depuration (13 days)	1.4 ± 0.94	19	82.7 ± 8.8	7.4 ± 3.2	2.8 ± 2.8	4.4 ± 1.8	2.8 ± 2.6
480-hr depuration (20 days)	0.10	1.4	58.8	14.3	7.1	9.4	10.5
720-hr depuration (30 days)	0.07 ± 0.07	1.2	58.5 ± 32.0	19.3 ± 16.0	12.6 ± 19.0	5.4 ± 6.0	8.7 ± 16.0
1392-hr depuration (58 days)	>0.01	>0.2					

[a]SD = standard deviation.

Source: Modified from Neff and Anderson 1975, Table 1 and 2, p. 470. Reprinted by permission of the publisher.

Table 9.7. Accumulation, release, and biomagnification factors (tissue concentration/exposure water concentration) of naphthalenes by the clam *Rangia cuneata* exposed to the water-soluble fraction of No. 2 fuel oil

Petroleum hydrocarbon	Concentration in exposure water (ppm)	Tissue con-centration, 24-hr exposure (ppm)	Biomagnification factor	Tissue con-centration, 24-hr depura-tion (ppm)	Percent released in 24 hr
Naphthalene	0.84	1.9	2.3	0.4	79
1-Methylnaphthalene	0.34	2.9	8.5	1.9	34
2-Methylnaphthalene	0.48	3.9	8.1	1.9	51
Dimethylnaphthalenes	0.24	4.1	17.1	2.8	32
Trimethylraphthalenes	0.03	0.8	26.7	0.4	50

Source: Neff 1975, Table IV.

Table 9.8. Oil-derived hydrocarbons in the tissues of oysters (*Crassostrea virginica*) following exposure to a 1% oil-in-water dispersion of No. 2 fuel oil for 4 days

Oil hydrocarbon	Tissue concentration (ppm)[a]
C_{12}-C_{25} *n*-Paraffins	3.1
Naphthalene	6.3
1-Methylnaphthalene	9.4
2-Methylnaphthalene	15.4
Dimethylnaphthalene	35.6
Trimethylnaphthalene	17.4
Subtotal	84.1
Biphenyl	0.2
Methylbiphenyl	1.0
Dimethylbiphenyl	0.7
Fluorene	1.2
Methylfluorene	1.7
Dimethylfluorene	0.5
Dibenzothiophene	0.6
Phenanthrene	1.7
Methylphenanthrene	1.7
Dimethylphenanthrene	0.2
Subtotal	9.5
Total oil-derived hydrocarbons analyzed	96.7

[a] μg/g wet wt of tissue.

Source: Neff 1975, Table I.

aromatic constituents of several crude oils and their water-soluble fractions (WSF) (Table 9.9) (Anderson et al. 1974*a*; 1974*b*); thus, they are more readily available for accumulation by marine organisms (Neff 1975).

The ability of organisms to accumulate naphthalenes from seawater has also been demonstrated by Anderson et al. (1974*b*). Sheepshead minnows (*Cyprinodon variegatus*) accumulated 205 ppm 1-methylnaphthalene or 60 ppm naphthalene in their tissues after 4 hr of exposure to a seawater solution containing either compound at a concentration of 1 ppm (Fig. 9.2). Brown shrimp (*Penaeus aztecus*) were shown to accumulate methylnaphthalenes to a greater extent than other naphthalenes from a 30% WSF of No. 2 fuel oil (1.95 ppm total hydrocarbons in solution) after a 30-min exposure (Fig. 9.3).

9.1.2.5 <u>Rate of accumulation</u>

The rate of PAH accumulation by aquatic organisms has been shown to be species-dependent. Lee, Sauerheber, and Dobbs (1972) demonstrated that fish accumulate BaP when exposed, reaching maximum levels of accumulation within an hour. Neff (1975) demonstrated that fish and shrimp accumulate naphthalenes very rapidly (Figs. 9.4 and 9.5); tissue concentrations often reach maximum levels within an hour after exposure. However, fish tissue levels usually begin to decline after the first hour of exposure.

Table 9.9. Concentrations of C_{12}-C_{24} n-paraffins and di-tri-aromatic hydrocarbons in the test oils and in water-soluble fractions (WSFs) prepared from them

Name of Compound	S. Louisiana		Kuwait		No. 2 fuel oil		Bunker C	
	Whole oil (%)[a]	WSF (ppb)[b]	Whole oil (%)[a]	WSF (ppb)[b]	Whole oil (%)[a]	WSF (ppb)[b]	Whole oil (%)[a]	WSF (ppb)[b]
n-Paraffins								
C_{14}	0.44	10.0	0.46	<0.5	0.82	5.0	0.11	0.8
C_{15}	0.48	10.0	0.41	<0.5	1.06	7.0	0.11	0.9
C_{16}	0.54	12.0	0.43	0.6	1.20	8.0	0.15	1.2
C_{17}	0.41	9.0	0.42	0.8	0.98	6.0	0.12	1.9
C_{18}	0.30	7.0	0.28	0.5	0.60	4.0	0.10	1.0
Total C_{12}-C_{24} n-paraffins	3.98	89.0	4.00	2.9	7.38	47	1.26	12
Aromatics								
Naphthalene	0.04	120	0.04	20	0.40	840	0.10	210
1-Methylnaphthalene	0.08	60	0.05	20	0.82	340	0.28	190
2-Methylnaphthalene	0.09	50	0.07	8.0	1.89	480	0.47	200
Dimethylnaphthalenes	0.36	60	0.20	20	3.11	240	1.23	200
Trimethylnaphthalenes	0.27	8.0	0.19	3.0	1.84	30	0.88	100
Biphenyls	<0.01	2.0	<0.01	1.0	0.16	28	<0.01	1.0
Fluorenes	0.02	<1.0	<0.01	<1.0	0.36	20	0.24	11.0
Phenanthrenes	0.06	4.0	0.04	3.0	0.53	20	1.11	23.0
Dibenzothiophene	0.02	1.0	0.01	<1.0	0.07	4.0	<0.01	<1.0
Total	0.94	305	0.60	75	9.18	2002	4.31	935
Total hydrocarbons measured	4.92	394	4.60	78	16.56	2049	5.57	947
Total hydrocarbons present (IR analysis)	19,800		10,400		8700		6300	

[a] g/100 ml.
[b] µg/liter in 20 % seawater.

Source: Anderson et al. 1974, Table 1, p. 288. Reprinted by permission of the publisher.

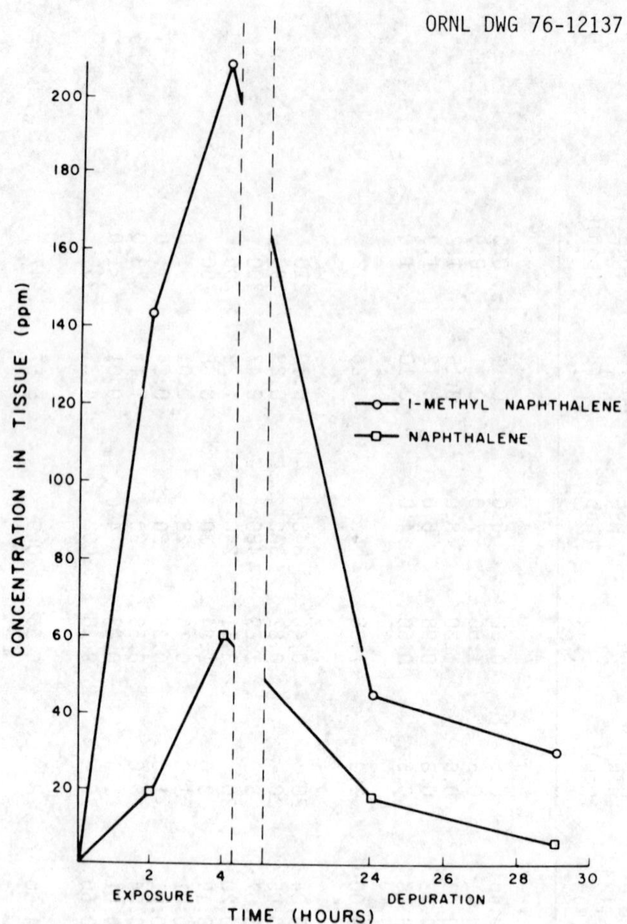

Fig. 9.2. Accumulation and release of naphthalene and 1-methyl naphthalene by *Cyprinodon variegatus*. Groups of fish were exposed for 4 hr to 1-ppm solutions of each hydrocarbon in seawater and then transferred to clean seawater. Each point represents the average of the analysis of three fish. <u>Source</u>: Anderson et al. 1974*b*, Fig. 1, p. 296. Reprinted by permission of the publisher.

Research by Anderson et al. (1974*a*; 1974*b*) has also demonstrated the rapidity of accumulation and exchange of aromatics between tissues of fish and shrimp and the environment (Sect. 9.1.5.3). Marine molluscs, on the other hand, tend to accumulate PAH more slowly over longer exposure periods (Neff 1975). Figure 9.6 shows that clams continued to incorporate naphthalenes from seawater containing an oil-in-water dispersion of No. 2 fuel oil for the duration of the exposure period — up to 20 ppm in 20 hr.

9.1.2.4 Effect of particulates

The presence of significant quantities of PAH adsorbed onto particulate material in natural waters suggests the potential for PAH uptake by benthic organisms. Herbes (1976) demonstrated that the rate and extent of PAH accumulation are influenced by particulates; *Daphnia magna*, a freshwater zooplankter, incorporated ^{14}C-anthracene quite rapidly from spring water with or without a particulate food source. Maximum levels were reached within 60 min, with bioaccumulation at equilibrium calculated to be about 200-fold. Similar results were obtained with mayfly nymphs (*Hexagenia* spp.) (Fig. 9.7). Rapid initial first-order uptake occurred with or without suspended particulate material (25 mg/liter autoclaved yeast cells). A continuous but much slower uptake corresponded to a first-order decrease in the fraction of anthracene bound to particulate material in suspension. Uptake from the soluble phase was shown to be 40 times more rapid than uptake

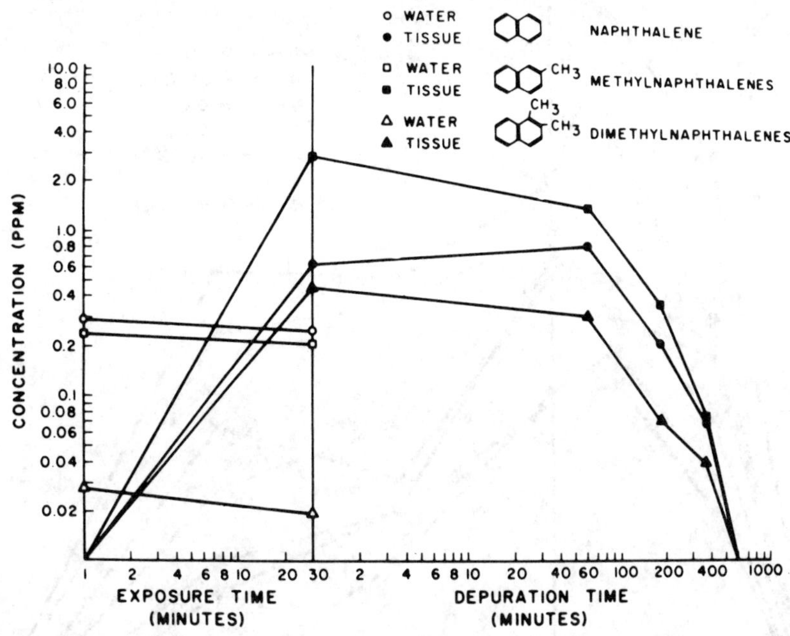

Fig. 9.3. Uptake and release of naphthalenes from a 30% WSF of No. 2 fuel oil by the brown shrimp *Penaeus aztecus*. The exposure water initially contained about 1.95 ppm total oil hydrocarbons. <u>Source</u>: Anderson et al. 1974*b*, Fig. 2, p. 297. Reprinted by permission of the publisher.

from the particulate phase. Total bioaccumulation with particulate material present, however, was twice as great as from the soluble phase alone.

9.1.2.5 <u>Absorption</u>

Whittle, Mackie, and Hardy (1974) do not feel that an ingestion mechanism accounts for the observed dynamic equilibrium of hydrocarbons between aquatic organisms and their environment. Instead, they suggest a surface adsorption effect as the route of PAH assimilation. Ames, Sims, and Grover (1972) also suggest that incorporation of PAH into the food chain is due to an adsorption mechanism, but point out that the actual adsorption process has not been determined. Ehrhardt and Heinemann (1975) report that hydrocarbons, including PAH, may be accumulated by *Mytilus edulis* through adsorption by gill tissue as well as through ingestion mechanisms.

Stegeman and Teal (1973) report that equilibration of hydrocarbons may occur between organisms and the water that passes over their exposed tissues, possibly providing the most important route of uptake for many aquatic organisms due to the large amounts of water processed during food collection and respiration. Research demonstrates that aromatics are taken up by the gill tissue of mussels and transferred to other tissues (Lee, Sauerheber, and Benson 1972).

Lee, Sauerheber, and Benson (1972) found that PAH were rapidly filtered across the gills of *Mytilus edulis* and incorporated into the body tissues. About 20 µg of tetralin (1,2,3,4-tetrahydronaphthalene) was incorporated by each mussel within 24 hr of exposure; 3 to 10 µg each of naphthalene, toluene, and BaP were taken up by the mussels. The authors propose that

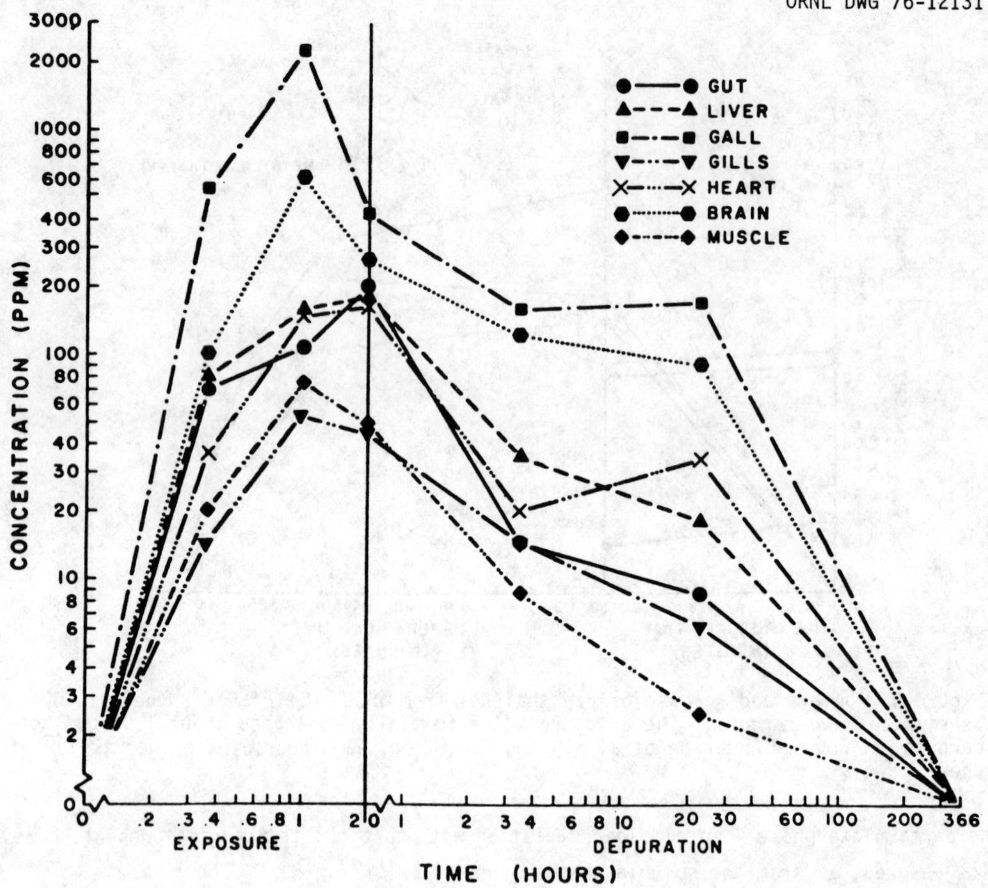

Fig. 9.4. Distribution of total naphthalenes in various tissues of the Gulf killifish *Fundulus similis* during exposure to the water-soluble fraction of No. 2 fuel oil and at different times following exposure. The water-soluble fraction contained about 2 ppm total naphthalenes. <u>Source</u>: Neff 1975, Fig. 3.

these PAH were absorbed by the micellar layer of the gill tissues, passed to the mantle, adducter muscle, and gut, and ultimately stored in the hepatopancreas. Study of hydrocarbon uptake by marine fish also demonstrated the entrance and assimilation of aromatics through the gills (Lee, Sauerheber, and Dobbs 1972).

9.1.2.6 <u>Transient uptake</u>

DiSalvo, Guard, and Hunter (1975) constructed a model to define the probable flow of hydrocarbons through the estuarine environment (Fig. 6.1). In this model, aquatic invertebrates were cited as transient storage sites for hydrocarbons. Aquatic invertebrates (*Mytilus* spp.) were shown to accumulate PAH from polluted environments and to release accumulated PAH upon transfer to relatively unpolluted environments. These experiments by DiSalvo, Guard, and Hunter (1975) demonstrated the dynamic nature of the steady-state PAH—tissue equilibrium: The PAH content of the organisms reflected the background PAH levels of their environment. Similar results have been observed in zooplankton, fish, and benthos.

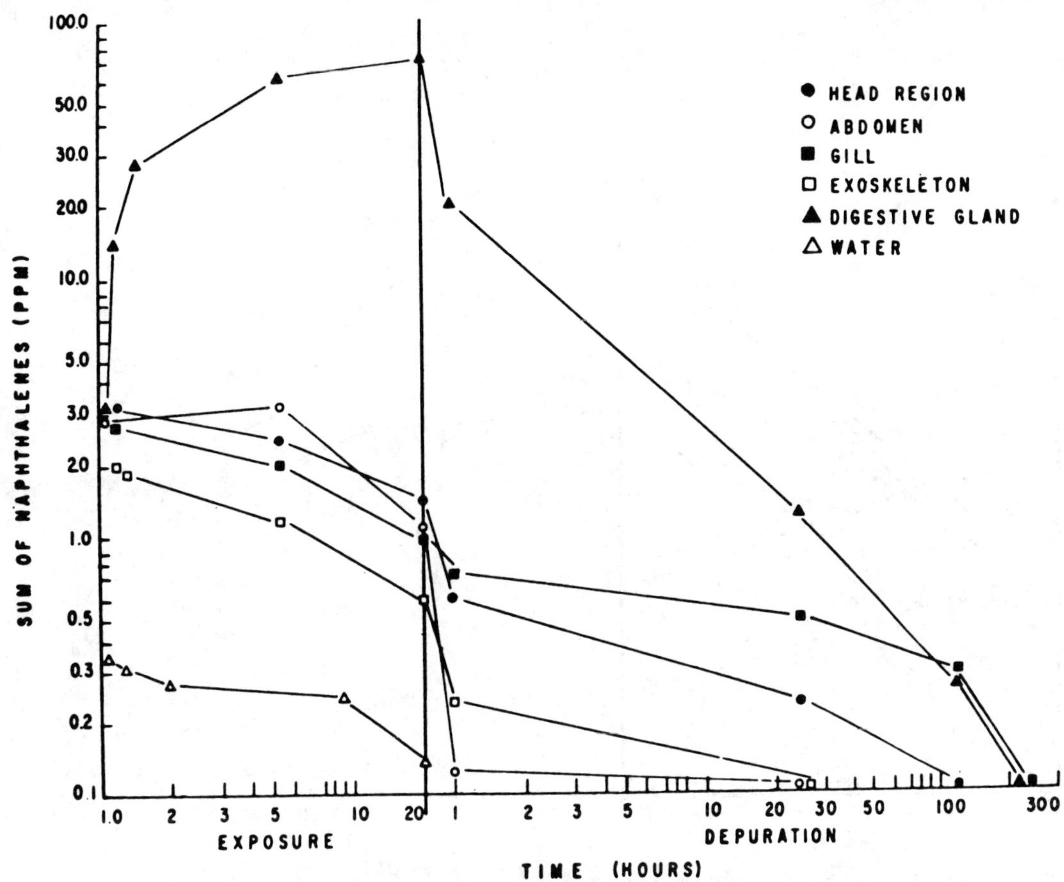

Fig. 9.5. Accumulation and release of total naphthalenes by different body regions of juvenile brown shrimp (*Penaeus aztecus*) exposed to a 20% WSF of No. 2 fuel oil. Body regions of three shrimp were pooled for each data point and background values were subtracted. <u>Source:</u> Neff 1975, Fig. 2.

9.1.3 Trophic level magnification

Food web magnification refers to increasing concentration of hydrocarbons per weight of tissue at successively higher trophic levels (NAS 1975*a*). McGinnes and Snoeyink (1974) have observed the tendency of PAH to accumulate in biota in increasing concentrations with each progressive trophic level, thus indicating consumption as a possible mechanism of PAH incorporation into the food chain. Ogata and Miyake (1973) and Neff (1975) also suggest that ingestion of PAH-contaminated seafood could transfer the PAH to higher levels of the food chain. Scaccini-Cicatelli (1966) indicates that PAH accumulation by organisms that constitute the base levels of the food chain may have far-reaching effects throughout the food web. Tubifex worms were demonstrated to absorb and concentrate BaP without being able to metabolically degrade the PAH. Without such a detoxification mechanism in the worms, the PAH could be transferred to organisms feeding on the tubifex. Scaccini-Cicatelli (1966) indicates that the PAH would continue to concentrate in progressively higher levels of the food chain, eventually being transferred to humans.

ORNL DWG 76-12130

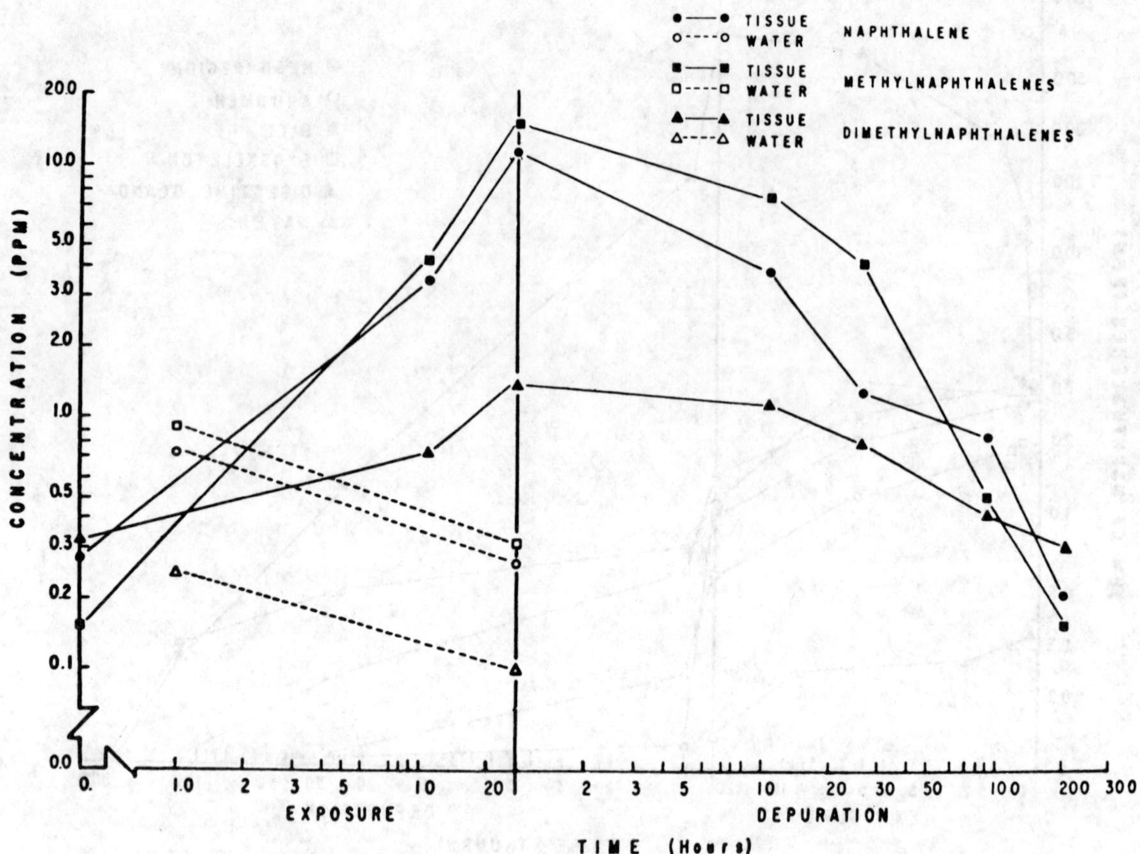

Fig. 9.6. Concentration of oil-derived naphthalene in tissues of the clam *Rangia cuneata* during exposure to a dilute oil-in-water dispersion of No. 2 fuel oil (384 ppm oil added) and following return to oil-free seawater. Each data point represents average hydrocarbon concentration in the tissues of five clams. <u>Source</u>: Neff 1975, Fig. 1. Reprinted by permission of the publisher.

Because food chain magnification may be a function of species ability to accumulate hydrocarbons from the environment rather than a function of species position in the food web, Whittle, Mackie, and Hardy (1974) attempted to define the extent to which the marine food chain incorporates PAH from the surrounding environment. Unfortunately, little information was found regarding the background levels of aromatics in either the aquatic environment or its organisms.

DiSalvo, Guard, and Hunter (1975) also indicate a need for information in the area of PAH accumulation in marine ecosystems. Hydrocarbons were shown to be present at low concentrations in sediments and middle and surface waters, but the many aromatic compounds present were not identified. An increase in PAH concentration in any sector of the aquatic ecosystem resulted in a corresponding increase throughout the ecosystem. Benthos, plankton, and fish all exhibited concentration patterns for PAH reflecting the background levels of their respective habitats. Concentrations of PAH in the livers of fish and benthos were shown to be more dependent on the background levels of PAH in the organism's immediate environment than on its trophic position. Because zooplankton forms an initial link in many aquatic food chains, routes of PAH uptake and metabolism by zooplankton may directly influence the extent of PAH bioaccumulation throughout aquatic ecosystems (Herbes 1976).

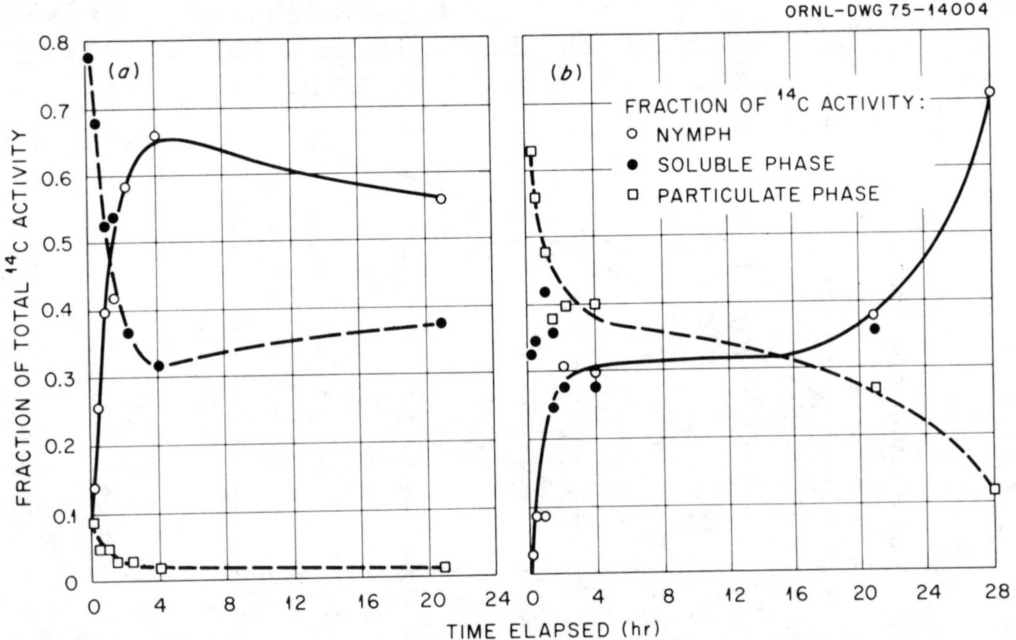

Fig. 9.7. Uptake of 9-¹⁴C-anthracene by *Hexagenia* sp. nymphs: fractions of total activity in organisms (one *Hexagenia* per 30 ml) and in soluble and suspended particulate phases. (*a*) No particulate material added to membrane-filtered spring water; (*b*) 25 mg/liter autoclaved yeast cells added. Initial anthracene concentration was about 0.02 µg/liter at 24°C. <u>Source</u>: Herbes 1976, Fig. 7.3, p. 71.

Evidence that substantial amounts of certain PAH may be released rapidly by some marine animals questions, but does not exclude, the possibility of PAH transfer via the marine food web (Corner 1975). Although animals are believed to accumulate PAH through ingestion of contaminated food, no studies were found to confirm this. Vegetation presumably provides a link in the passage of PAH through the food web, but no published data were found.

Aquatic invertebrates tend to accumulate PAH from their surrounding environment. Herbes, Southworth, and Gehrs (1976) correlated bioaccumulation factors (PAH concentration in tissues/ PAH concentration in water) for aquatic invertebrates with the molecular weight of the PAH accumulated (Fig. 9.8). A direct relationship was found between PAH of 120- to 260-unit molecular weight and increasing bioaccumulation factors (ranging from 2 to 1100); the wide range resulted from differences in lipid content and rates of excretion.

Neff and Anderson (1975) exposed the estuarine clam *Rangia cuneata* to synthetic seawater containing 0.0305 ppm BaP-¹⁴C in solution for 24 hr. The clams rapidly accumulated the PAH in their tissues, with average totals of 7.2 ppm (µg/g wet weight of tissue) BaP (Table 9.7). This represents an average magnification factor (BaP in tissues/BaP in exposure water) of 236 — well above ambient levels. Neff (1975) reports considerably lower magnification factors for naphthalenes in clams exposed to water-soluble fractions of No. 2 fuel oil for 24 hr (Table 9.7).

9.1.4 <u>Metabolism</u>

Metabolic pathways involving oxidases and other enzymes, which are important in degradation of PAH by mammalian systems, have been well studied in laboratory species. Hydroxylation is followed

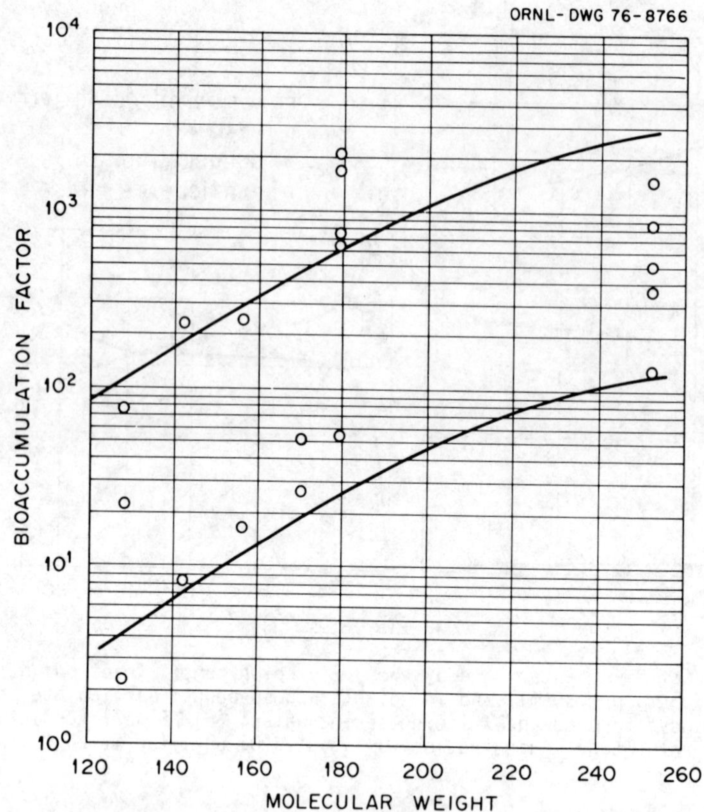

Fig. 9.8. Bioaccumulation of polycyclic aromatic hydrocarbons by aquatic invertebrates. Source: Herbes, Southworth, and Gehrs 1976, Fig. 4, p. 14.

first by conjugation with sulfate or glucose and then by excretion of the water-soluble product (Chap. 10).

Little information is available regarding metabolic rates, metabolites, or excretion products in animals other than laboratory mammals. NAS (1972) attributes this scarcity to a lack of definitive studies and incomplete characterization of the compounds involved. Metabolic rates are difficult to determine at the low concentrations usually encountered in natural environments. Generally, metabolic data at these low concentrations must be extrapolated from data obtained at higher concentrations. Even though metabolic rates and pathways may appear similar in similar species, the differences in processes make extensive extrapolation unwise. Not only do environmental conditions alter detoxification ability, but the ability to metabolize PAH and the detoxification mechanisms themselves vary significantly among species (Auerbach 1975).

9.1.4.1 Degradation and synthesis

Whereas PAH are readily metabolized by fish (Lee, Sauerheber, and Dobbs 1972), crabs (Lee, Ryan, and Neuhauser 1976; Corner, Kilvington, and O'Hara 1973), and zooplankton (*Daphnia*) (Herbes 1976), shellfish apparently are unable to degrade these compounds (Lee, Sauerheber, and Benson 1972). Therefore, PAH may accumulate to an extensive degree in the tissues of oysters, clams, and other edible shellfish (Stegeman and Teal 1973; Shimkin, Koe, and Zechmeister 1951; Blumer, Souza, and Sass 1970; Ehrhardt 1972).

Farrington and Quinn (1973) report the ability of marine clams to synthesize some low-molecular-weight hydrocarbons, but postulate that PAH (higher-molecular-weight) synthesis is too complex for these organisms. Lee, Sauerheber, and Benson (1972) note the inability of marine mussels (*Mytilus edulis*) and copepods to degrade PAH. They suggest that the hepatopancreas of these invertebrates lacks the hydroxylating enzymes necessary for PAH metabolism and that the bacteria present in these organisms lack the ability to metabolize PAH. Ehrhardt and Heinemann (1975) have shown that mussels are capable of metabolizing aliphatic hydrocarbons, but not aromatics.

9.1.4.2 Species variation

Even within individual phyla (e.g., fish), detoxification mechanisms vary widely; certain fish possess high levels of nitro-reductase and aminopyrine demethylase, whereas others contain none at all (Adamson 1967). Environmental temperature may affect detoxification ability of cold-blooded organisms; PAH metabolism in trout is most rapid at 20°C, whereas activity in toads and geckos is maximal at 30°C (activity of trout PAH detoxification is virtually zero at 30°C) (Clark and Diamond 1971). Individual strains of rodents and reptiles possess either low or high PAH-metabolizing ability (Diamond and Clark 1970) (Table 9.10).

9.1.4.3 Compound variation

Because metabolic detoxification requires that a contaminant molecule fit into a particular "slot" in a metabolizing enzyme, structural characteristics of a molecule may determine the rate at which the contaminant is detoxified. In many cases, effects of structural variation on metabolic rate are unclear or unknown. Thus, the rates of metabolism of 11 PAH compounds by the liver in rats vary over 15-fold (Sims 1970*a*); no obvious dependence on molecular size, presence or absence of substitution, solubility, or other parameter is apparent (Sims 1966; 1967; 1970*b*).

9.1.4.4 Metabolites

Although metabolism of a compound generally decreases its toxicity, the reverse can also be true. Susceptibility of tumor promotion in tissues after exposure to certain PAH compounds, some of which are potent carcinogens, has been found to be directly proportional to the rates of compound metabolism, indicating a relationship between the metabolites and tumor induction (Belitsky et al. 1970). Different species may also use different metabolic pathways, as occurs with rats and houseflies, after exposure to PAH (Terriere, Boose, and Roubal 1961).

9.1.4.5 Vertebrates

PAH have been shown to be readily metabolized by fish. Microsomes of trout (*Salmo trutta lacustris*) liver have been shown to metabolize PAH. Ahokas, Pelkonen, and Kärki (1975) identify the hydroxylating system as a typical monooxygenase system, requiring oxygen and NADPH for full activity. BaP was converted to oxidized metabolites (including 3-hydroxy-BaP, several dihydrodiol derivatives, and the 3,5-quinone of BaP) typical of similar mammalian systems.

Table 9.10. Relative efficiency of metabolic conversion of BaP to water-soluble derivatives in cell cultures derived from homeothermic and poikilothermic vertebrates[a]

| Cell | | | | Micrograms of ^{3}H-BaP degraded/10^7 cells/24 hr | | |
Culture	Common name	Species	Tissue	Low (<0.25 µg)	Intermediate (0.25 - 0.50 µg)	High (>0.50 µg)
Medium without cells				0.0026		
Aves						
CE	Domestic chicken	*Gallus domesticus*	Embryo	0.04		
Rodentia						
ME	Mouse	*Mus musculus*	Embryo			0.51
HE	Syrian hamster	*Mesocricetus auratus*	Embryo			1.12
HE 44			Spontaneously transformed HE	0.10		
HF 15/SV			SV/40-transformed hamster kidney			0.60
Reptilia						
TH-9	Box turtle	*Terrapene carolina*	Heart			1.17 1.30
TH-5			Heart	0.19		
TH-1B2			Heart	0.07		
TK			Kidney			0.59 1.58
TL-4			Lung	0.22		
TGSW	Grecian tortoise	*Testudo graeca*	Spleen	0.18		
PH-2	Side-necked turtle	*Podocnemis unifilis*	Heart	0.18		
VSW	Russell's viper	*Vipera russelli*	Spleen	0.03		
VH-2			Heart	0.18 0.05		
IgH-5	Green iguana	*Iguana iguana*	Heart			1.04
IgH-2			Heart			0.78
IgK			Kidney			0.62
IgVA			Liver		0.39	
GL-1	Tokay gecko	*Gekko gecko*	Lung			0.85 1.41 1.07
Amphibia						
BA68.1	American toad	*Bufo americanus*	Embryo			2.42
RPH67.132	Leopard frog	*Rana pipiens*	Embryo	0.19	0.35	
Pisces						
BF	Bluegill	*Lepomis microchirus*	Fry		0.45	
RTG	Rainbow trout	*Salmo gairdneri*	Embryonic gonad			0.94

[a]Culture flasks contained 5 ml medium, and input concentrations of ^{3}H-BaP ranged from 0.02 to 0.08 µg/ml. All cultures were incubated at 33°C, except RTG, which was incubated at 26°C. Cells from the same tissue, but derived from different animals, are distinguished by the numbers after the code letters.

Source: Diamond and Clark 1970, Tables 1 and 2, pp. 1006 and 1008.

Metabolite formation in the trout-liver microsome system was 5 to 10 times as fast as in the rat-liver microsome system, when measured per milligram of microsomal protein, or 15 to 30 times as fast, when measured per unit of cytochrome P-450 and NADPH-cytochrome C reductase (Table 9.11). Similar results were demonstrated with steelhead trout-liver homogenates by Pederson, Hershberger, and Juchau (1974). Metabolic enzyme activity was also studied for other tissues (Table 9.12); enzyme activity in the liver was found to exceed that in other tissues by orders of magnitude.

Table 9.11. BaP metabolites produced by trout and rat liver microsomes in a 15-min period[a]

| | BaP metabolites produced (nmoles) | | Trout-rat |
	Trout	Rat	ratio
Per mg of microsomal protein			
dihydrodiols	2.12	0.25	8.1
hydroxy	2.04	0.38	5.8
quinone	0.44	0.35	1.4
Per nmole of cytochrome P-450			
dihydrodiols	6.77	0.37	18.5
hydroxy	6.61	0.50	13.1
quinone	1.45	0.47	3.1
Per unit of cytochrome C reductase			
dihydrodiols	0.066	0.0023	28.7
hydroxy	0.065	0.0032	20.3
quinone	0.015	0.0030	5.0

[a]Metabolites were separated on TLC, and the fractions were assayed radiometrically.

Source: Ahokas, Pelkonen, and Kärki 1975, Table 3, p. 638. Reprinted by permission of the publisher.

Table 9.12. Tissue distribution of BP hydroxylase activity in fish[a]

Tissue preparation	μmoles of 8-hydroxybenzpyrene formed/g protein/min
Blood	Not detectable
Gill	Not detectable
Heart	0.006
Posterior kidney	0.026
Liver	2.404
Muscle	Not detectable

[a]Mixtures were incubated with shaking for 30 min at 28.5°C under an atmosphere of 100% oxygen.

Source: Pedersen, Hershberger, and Juchay 1974, Table 1, p. 483. Reprinted by permission of the publisher.

PAH metabolism was demonstrated in the sculpin (*Oligocottus maculosus*), mudsucker (*Gillichthys mirabilis*), and sand dab (*Citharichthys stigmaeus*) by Lee, Sauerheber, and Dobbs (1972). ^{14}C-naphthalene was metabolized to 1,2-dihydro-1,2-dihydroxynaphthalene after 24 hr of exposure. The major product of ^{3}H-BaP metabolism was identified as 7,8-dihydroxybenzopyrene. Metabolism occurred in the liver, and metabolites were stored primarily in the gall bladder. Apparently, sufficient detoxification mechanisms existed for efficient removal of accumulated PAH from body tissues.

Scaccini et al. (1970) found that BaP metabolism in goldfish (*Carassius auratus*) was slow; residues of the original PAH remained in the tissues for more than two days after initial exposure.

Knorr and Schenk (1968) refer to BaP detoxification by peroxidases in the gastrointestinal tract of fish. Although no specific data are given, degradation of BaP into relatively noncarcinogenic oxidation products was observed. BaP was introduced into the fish through esophageal probe feeding for several months. Microflora in the gastrointestinal tract were believed to be responsible for the observed detoxification (Chap. 8).

9.1.4.6 Invertebrates

Although some invertebrates apparently do not metabolize PAH, several invertebrate species have recently been found to be capable of PAH metabolism. Spider crabs (*Maia squinado*) were shown to metabolize naphthalene after introduction of the hydrocarbon into the foregut of the crabs by stomach tube (Corner, Kilvington, and O'Hara 1973). Four metabolites, including 1-naphthylglucoside and 1-naphthylsulfuric acid, were identified in the urine (Fig. 9.9). Although no quantitative data are given, chromatographic studies indicate that these compounds accounted for only small amounts of the naphthalene administered; thus, these compounds were biotransformation products.

ORNL DWG 76-6596

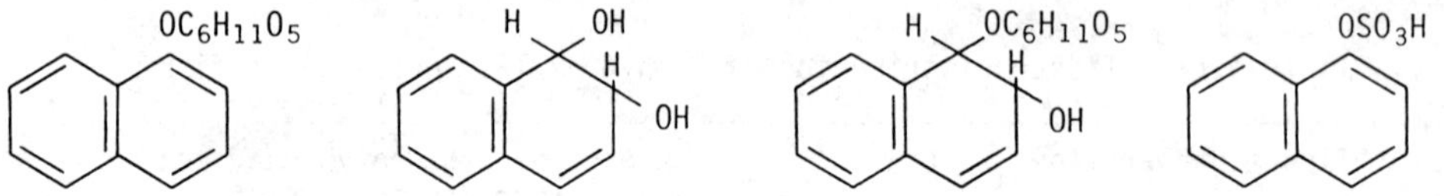

Fig. 9.9. Metabolites of naphthalene excreted by *Maia*. <u>Source</u>: Corner 1975, Fig. 14, p. 406. Reprinted by permission of the publisher.

Another species of crab (*Callinectes sapidus*) was shown by Lee, Ryan, and Neuhouser (1976) to metabolize radiolabled aromatics. Blue crabs took up benzopyrene, fluorene, naphthalene, methylnaphthalene, and methylcholanthrene from food and water, assimilated from 2 to 10%, and excreted the rest. More than half of the radioactivity assimilated by the crabs was in the hepatopancreas, suggesting this as the site of hydrocarbon metabolism. The hepatopancreas also contained polar hydrocarbon metabolites such as dihydroxy compounds and their conjugates, evidence which further supported the hepatopancreas as the site of active metabolism.

Initial studies with *Daphnia magna* indicate that these organisms are capable of excreting accumulated ^{14}C-anthracene in the form of one or possibly two water-soluble metabolites (Fig. 9.10) (Herbes 1976). Release of radioactivity by the *Daphnia*, when transferred to unlabeled water, followed sequential first-order kinetics with rate constants of 0.53 and 0.12/hr respectively. The rate constants correspond almost identically with observed rates of gut elimination and anthracene metabolite formation.

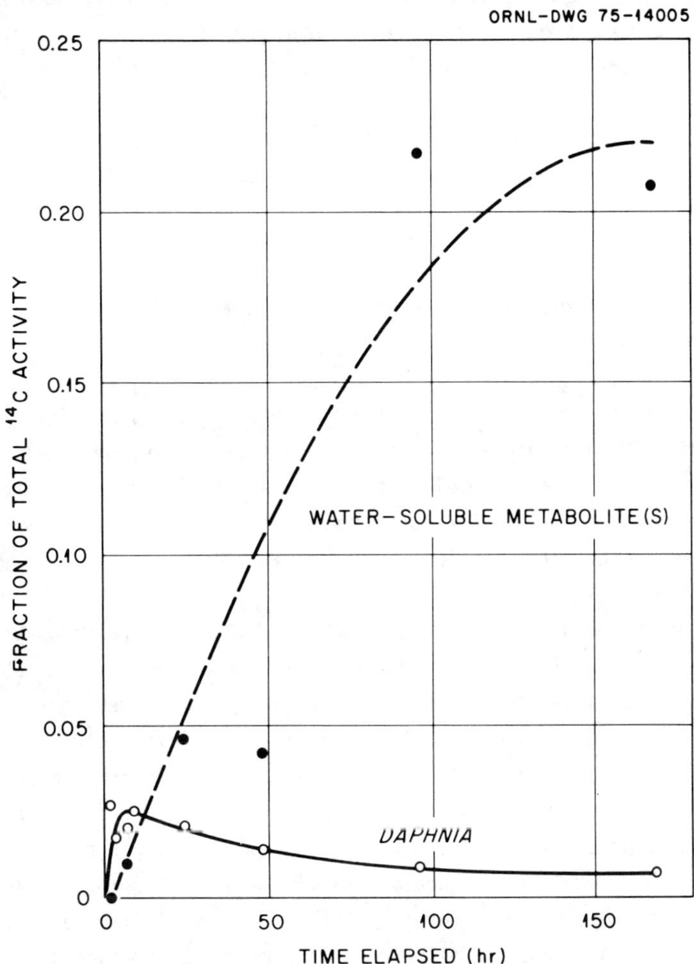

Fig. 9.10. Uptake and metabolism of 9-^{14}C-anthracene by *Daphnia magna*: fractions of total activity in organisms (one *Daphnia* per 20 ml) and in form of water-soluble metabolic products. Remainder of ^{14}C present as anthracene in solution and adsorbed onto suspended particulate matter. Initial anthracene concentration was about 0.02 μg/liter at 24°C. <u>Source</u>: Herbes 1976, Fig 7.2, p. 69.

Lee (1975) found that other species of zooplankton are also capable of PAH metabolism. Copepods, euphausiids, amphipods, crab zoea, ctenophores, and jellyfish rapidly took up ^{3}H-benzpyrene, ^{14}C-benzpyrene, ^{3}H-methylcholanthrene and ^{14}C-naphthalene from seawater solution. Crustaceans apparently metabolized these PAH to various hydroxylated and more polar metabolites, but jellyfish and ctenophores did not. Also, alkylated aromatic hydrocarbons appeared to be metabolized at a higher rate than the nonalkylated hydrocarbons.

9.1.5 <u>Elimination</u>

Interrelationships among uptake, release, retention, and a possible multicompartment system, with the added influence of environmental parameters such as temperature, play a role in determining the amount of hydrocarbon present in organisms. Stegeman and Teal (1973) speculate that such considerations (the complete exposure history of the organism), together with the current exposure level, would determine the amount of hydrocarbon observed in the organism. For example, if shellfish from polluted environments were transferred to cleaner water, the body burden of hydrocarbons would be expected to decrease, but only to a point that was still determined by equilibration. Transferral to even cleaner habitats would result in further release of hydrocarbons, down to a level which is determined largely by the bound hydrocarbon content of the stable compartment. Stegeman and Teal (1973) suggest that the organisms would not retain the hydrocarbons permanently and that even a slow turnover rate would eventually result in the ultimate release of all accumulated hydrocarbons. Considerable periods of time and virtually hydrocarbon-free seawater may be required for such a complete depuration.

9.1.5.1 Excretion

Recent work indicates that several marine species, if contaminated by certain PAH can release these compounds when transferred to clean seawater.

Anderson (1973), as cited by Clark and Finley (1975), found that oysters lost 82% of their accumulated aromatics after 13 days when exposed to a No. 2 fuel oil; no residual pollutant hydrocarbons were detected at the 0.5-ppm level (wet weight) after 52 days of depuration. Stegeman and Teal (1973) also observed a rapid loss of hydrocarbons from oysters exposed to a No. 2 fuel oil–water mixture, but the aromatic portion was not quantified.

Uptake and eventual discharge of PAH — benzene, toluene, 1,2,3,4-tetrahydronaphthalene (tetralin), naphthalene, BaP, and methylcholanthrene — by marine fish, two benthic invertebrates (mussel and lobster), and several species of zooplankton were studied by Lee, Sauerheber, and Benson (1972). Copepods and mussels were unable to degrade the aromatics into water-soluble products, but were able to discharge 80% (16 μg) of the tetralin accumulated over a 24-hr exposure period and 90% each of the remaining aromatics within 14 days after transferral to clean water. Fish were able to produce water-soluble products from the hydrocarbons and to discharge them through the urine via the gall bladder and kidney. The avenue of PAH discharge by the lobster and related invertebrates has not been determined. Possibly, bile salts or similar natural detergents were able to emulsify these hydrocarbons and allow passage through the gut and into the feces or pseudofeces.

Lee, Sauerheber, and Dobbs (1972) demonstrated the rapid release of ^{14}C-naphthalene and ^{3}H-BaP from fish tissues when the organisms were transferred to relatively PAH-free seawater. After 24 hr in clean seawater, more than 90% of the ^{14}C-naphthalene present in the fish was lost. ^{3}H-BaP in the liver, gut, gill tissue, and flesh was reduced within 24 hr by 50, 50, 90, and 20% respectively. The fish were shown to flush out naphthalene and its metabolites at a greater rate than BaP; urine was an important avenue of excretion.

Scaccini et al. (1970) found that goldfish (*Carassius auratus*) continued to incorporate BaP through ingestion of contaminated food, but that body concentrations declined if feeding was discontinued. Although mechanisms of release were not discussed, part of the compound was eliminated through the mesonephros. Elimination was observed to be slow since levels of the compound remained in the fish for several days.

9.1.5.2 Effect of means of uptake

Most experiments do not provide enough information regarding the mechanisms or pathways of petroleum hydrocarbon uptake by contaminated organisms. Clark and Finley (1975) postulate that the mechanisms of uptake and transport of hydrocarbons from the environment into organisms may have a very important effect on the degree to which subsequent depuration is reversible. For instance, hydrocarbons transported across gill membranes as dissolved particles or emulsified droplets enter the bloodstream of fishes very rapidly and can be rapidly depleted after removal of the pollutant. Hydrocarbons in food sources, however, are resorbed at a different site. For the basking shark, resorption occurs in the spiral valve, where the hydrocarbons are transferred to the liver and remain highly persistent. Whittle, Mackie, and Hardy (1974) also suggest that the mechanism of PAH acquisition by marine organisms is apparently reversible. PAH levels in fish, plankton, and benthos were observed to depurate PAH within 10 days after transfer from polluted to clean seawater.

Corner, Harris, Kilvington, and O'Hara (1976) have demonstrated in *Calanus helgolandicus* that the dietary route of ^{14}C-naphthalene entry was more important quantitatively than direct uptake from seawater solution. Furthermore, it was shown that naphthalene incorporated from food sources was depurated less rapidly than napthalene accumulated directly from seawater.

9.1.5.3 Reversibility

DiSalvo, Guard, and Hunter (1975) observed PAH accumulation by mussels inhabiting polluted environments. However, when the mussels were transferred to relatively unpolluted environments, diminution of PAH content in all tissues was observed. When the mussels were returned to their original polluted habitat, their tissues were shown to regain their original PAH content.

Table 9.13 shows the changes in hydrocarbon content of *Mytilus edulis* when transferred from polluted water conditions of San Francisco Bay (SFB) to unpolluted water conditions of Drakes Estero (DE) for 10 weeks, then returned to polluted water conditions (SFB) for 4 weeks. A well-defined loss of the aromatic fraction was apparent during the 10-week trial in clean water, and a dramatic increase of tissue hydrocarbons was observed when the organisms were returned to their original polluted water conditions.

Table 9.13. Hydrocarbon content of *Mytilus edulis* transferred from polluted water conditions (SFB) to unpolluted water conditions (DE) for 10 weeks, then returned to original polluted habitat

Station	Date	Time after transfer (weeks)	Hydrocarbon content (ppm/g dry wt)		Total
			Alkane	Aromatic	
SFB[a]	6/24/73		183	171	354
SFB[a]	7/27/73[b]		79	207	286
DE	9/4/73	6	108	75	183
DE	11/19/73[b]	10	234	61	295
SFB	12/16/73	4	169	248	417
SFB[a]	11/1/73		85	255	340

[a]Background value.
[b]Date of transfer.

Source: DiSalvo, Guard, and Hunter 1975, Table 1, p. 249. Reprinted with permission from *Environ. Sci. Technol.* Copyright by the American Chemical Society.

Table 9.14 demonstrates the hydrocarbon content of *Mytilus edulis* when transferred from the slightly polluted water conditions of San Francisco Bay for 9 weeks. This group of mussels accumulated greater amounts of hydrocarbons than those found in indigenous San Francisco Bay mussels.

Table 9.14. Hydrocarbon content of *Mytilus edulis* transferred from slightly polluted water conditions (TB) to polluted water conditions (SFB) for 9 weeks

Station	Date	Time after Transfer	Hydrocarbon content (ppm/g dry wt)		Total
			Alkane	Aromatic	
TB	11/19/73[a]		51	27	78
SFB	12/16/73	4	119	290	409
SFB	1/23/74	9	250	275	525
SFB[b]	1/25/74		325	131	456

[a]Date of transfer.
[b]Background value.

Source: DiSalvo, Guard, and Hunter 1975, Table II, p. 249. Reprinted with permission from *Environ. Sci. Technol.* Copyright by the American Chemical Society.

Further evidence is given by Anderson et al. (1974*b*). Estuarine clams (*Rangia cuneata*) and oysters (*Crassostrea virginica*) were shown to incorporate a wide variety of hydrocarbons, with naphthalenes accumulated to the greatest extent. When returned to hydrocarbon-free seawater, hydrocarbons were rapidly released from tissues with complete depuration over a 10- to 52-day period. The alkylnaphthalenes were always the last petroleum hydrocarbons to reach undetectable levels in the mollusc tissues. Marine organisms other than bivalves, such as sheepshead minnows (*Cyprinodon variegatus*) and brown shrimp (*Penaeus aztecus*), also rapidly released accumulated naphthalenes upon transfer to clean seawater (Anderson et al. 1974*b*; Neff 1975). After 29 hr in clean seawater, tissue concentrations of naphthalene and 1-methylnaphthalene in the fish had dropped to about 10 and 30 ppm respectively (Fig. 9.3). Tissue concentrations of naphthalene, methylnaphthalenes, and dimethylnaphthalenes in the shrimp generally decreased to undetectable levels (<20 ppb) in about 10 hr (Fig. 9.4).

9.1.5.4 <u>Release in eggs</u>

A mechanism of hydrocarbon release was suggested by DiSalvo, Guard, and Hunter (1975); mussels were found to purge their bodies of hydrocarbons, particularly aromatics, through release of their eggs. A decrease in the PAH content of mussels sampled during spawning (April-October) was correlated with an increase in the PAH content in mussel eggs, as compared with total body burden for populations, both in polluted waters of San Francisco Bay and in relatively unpolluted waters of Drakes Estero (Fig. 9.11). Further research to determine the viability and developmental success of gametes containing elevated levels of hydrocarbon pollutants was suggested; the hydrocarbons are presumably enriched in the gametes due to their high solubility in these fatty cells.

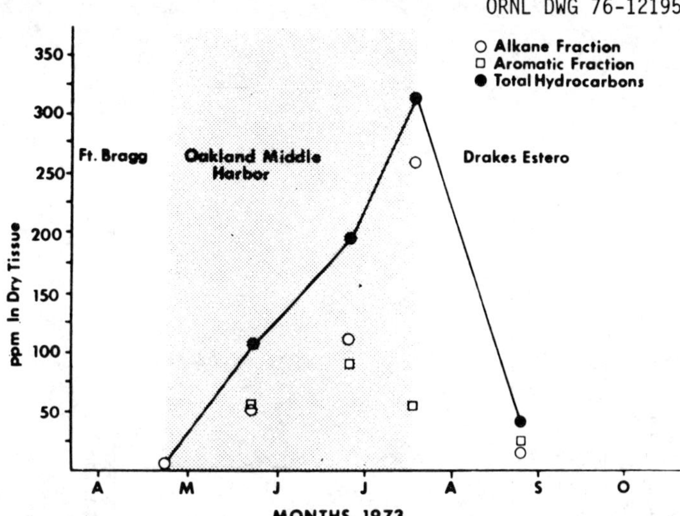

Fig. 9.11. Uptake and discharge of hydrocarbons by the California mussel *Mytilus edulis*. Source: From DiSalvo, Guard, and Hunter 1975, Fig. 3, p. 249. Reprinted with permission from *Environ. Sci. Technol.* Copyright by the American Chemical Society.

9.1.6 <u>Retention</u>

Blumer, Souza, and Sass (1970) suggest that, although molluscs discharge accumulated PAH when transferred to less polluted environments, portions of these accumulated PAH may be retained indefinitely. Stegeman and Teal (1973) found that oysters (*Crassostrea virginica*) retained a portion of accumulated PAH for at least one month and suggested that complete cleansing would require a long time.

Neff (1975), however, demonstrated that complete PAH depuration may sometimes occur; clams reached undetectable levels of PAH within 28 days after an 8-hr exposure to dispersed No. 2 fuel oil in a flow-through bioassay (Table 9.15). Oysters also were shown to be capable of depurating accumulated levels of PAH from body tissues when removed from the source of pollution (Table 9.16). Although PAH concentrations in tissues below normal environmental background levels may be insignificant, PAH were demonstrated to sometimes be partially retained, thus supporting inferences of Stegeman and Teal (1973) and Blumer, Souza, and Sass (1970).

Neely, Branson, and Blau (1974) found a linear correlation between the partition coefficient of a chemical between a polar and nonpolar solvent (water and *n*-octane) and its bioconcentration factor in fish. This correlation implies that uptake and release of a pollutant by tissue lipids is partly a function of the partition coefficient and the equilibrium concentration in fish tissue relative to that in the water. DiSalvo, Guard, and Hunter (1975) demonstrated this effect: Mussels released accumulated PAH when transferred to a habitat where the lipid-water partition coefficient for aromatics was exceeded relative to the habitat where the PAH were originally accumulated.

Stegeman and Teal (1973) suggest that partial retention of accumulated hydrocarbons after depuration reflects several possibilities:

1. Transfer of organisms from contaminated to less contaminated water may result in establishment of a new equilibrium concentration of hydrocarbons in the organism; exchange equilibrium between the animal's lipid and the surrounding water are instrumental in determining the level of a chemical's retention by animal tissues.

Table 9.15. Accumulation of petroleum hydrocarbons by the oyster *Crassostrea virginica* during exposure to dispersed No. 2 fuel oil in a flow-through system and subsequent release of hydrocarbons when the oysters were returned to oil-free seawater

Time (hr)	Petroleum hydrocarbon concentration[a] (ppm, µg/g wet wt)														
	n-p	N	1-MN	2-MN	DMN	TMN	B	MB	F	MF	DBT	P	MP	DMP	Total
							Exposure								
0		0.2	0.1	0.3	1.0	0.8									2.4
8	235	14.7	8.7	15.0	21.8	9.1	0.3	0.5	1.0	1.2	0.3	1.9	1.9	0.3	312
							Depuration								
3	156	12.0	8.4	12.0	22.7	10.8	0.3	0.4	0.7	0.7	0.3	1.3	1.3	0.2	228
6	68	7.3	5.1	7.3	13.2	5.7	0.1	0.2	0.4	0.2	0.1	0.6	0.6	0.1	109
24	18	6.5	5.7	7.6	14.8	9.5	0.2	0.2	0.5	0.7	0.2	1.2	1.3	0.3	67
120	10	8.2	4.7	6.8	13.4	4.9	0.1	0.1	0.2	0.1	0.1	0.4	0.4	0.2	54
672			0.1	0.5	0.9										1.5

[a]Abbreviations: n-p, n-paraffins; N, naphthalene; 1-MN, 1-methylnaphthalene; 2-MN, 2-methylnaphthalene; DMN, dimethylnaphthalenes; TMN, trimethylnaphthalenes; B, biphenyl; MB, methylbiphenyls; F, fluorene; MF, methylfluorenes; DBT, dibenzothiophene; P, phenanthrene; MP, methylphenanthrenes; DMP, dimethylphenanthrenes.

Source: Neff 1975, Table II. Reprinted by permission of the publisher.

Table 9.16. Accumulation of petroleum hydrocarbons by the clam *Rangia cuneata* during exposure to dispersed No. 2 fuel oil in a flow-through system and subsequent release of hydrocarbons when the clams were returned to oil-free seawater

Time (hr)	Petroleum hydrocarbon concentration[a] (ppm, μg/g wet wt)														
	n-p	N	1-MN	2-MN	DMN	TMN	B	MB	F	MF	DBT	P	MP	DMP	Total
						Exposure									
0													0.2		0.2
8	66	3.8	2.6	3.9	7.4	3.6	0.1	0.1	0.3	0.3	0.1	0.5	0.8	0.1	89
						Depuration									
3	23	2.2	1.7	2.6	4.2	1.8			0.1			0.2	0.2		36
6	1.1	2.0	1.9	2.9	4.9	1.9		0.1	0.1	0.2	0.1	0.3	0.5		16
24	1.0	0.6	0.7	1.1	2.5	1.3			0.1	0.2		0.3	0.3		8.1
120			0.1	0.1	0.6	0.6									1.4
672															0

[a]Abbreviations: n-p, n-Paraffins; N, naphthalene; 1-MN, 1-methylnaphthalene; 2-MN, 2-methylnaphthalene; DMN, dimethylnaphthalenes; TMN, trimethylnaphthalenes; B, biphenyl; MB, methylbiphenyls; F, fluorene; MF, methylfluorenes; DBT, dibenzothiophene; P, phenanthrene; MP, methylphenanthrenes; DMP, dimethylphenanthrenes.

Source: Neff 1975, Table III. Reprinted by permission of the publisher.

2. Some of the PAH retained by the animal may have entered a stable compartment characterized by a low hydrocarbon turnover rate; PAH retained in such compartments may be less available for disposal than are other accumulated PAH.

Dunn and Stitch (1976) have indicated that compartmentalization may be responsible for increased residence times of PAH in mussels. Very little is known about lipid compartmentalization in organisms. Stegeman and Teal (1973) suggest that any stable compartment responsible for long-term retention would probably not correspond to compartments affecting the pattern of uptake.

9.1.7 Acute toxicity

Biological responses evoked by toxicants are determined by the nature of the toxicant, the organism exposed, the concentration of exposure, and the mode of exposure (i.e., route of entry into the organism). Effects include acute toxicity, carcinogenicity, mutagenicity, teratogenicity, and chronic long-term effects.

9.1.7.1 Low-molecular-weight aromatic compounds

Studies to assess the toxic effect of aromatic hydrocarbons on aquatic life have been carried out, but the results are not always consistent. For example, Shelford is cited by Ryckman, Probhakara-Rao, and Buzzell (1975) as reporting toluene concentrations of between 61 and 65 mg/liter as lethal to sunfish in 1 hr, whereas Wallen et al. [also cited by Ryckman, Probhakara-Rao, and Buzzell (1975)] found the 96-hr median tolerance limit (TL_m) concentration of toluene for mosquito fish to be 1180 mg/liter.

9.1.7.2 Naphthalenes

Naphthalene toxicity curves for developing oyster embryos were drawn as in Fig. 9.12. Legore (1974) calculated the ED_{50} value (dose of toxicant added to culture system that causes abnormality in 50% of the organisms) for naphthalene as 194 mg/liter and the EMD_{50} value (dose of toxicant added to culture system that causes abnormality or mortality in 50% of the organisms) as 110 mg/liter.

The minimum lethal dose of naphthalene for minnows has been determined by LeClerc and Devlaminck (1950) as between 11 and 13 mg/liter. This concentration was noted to be high since the solubility of naphthalene in water is only 30 mg/liter at 25°C. Toxicity values of substituted naphthalene for several estuarine species are given in Table 9.17.

9.1.7.3 Anthracenes

Anthracene exhibits relatively low toxicity to animals. Legore (1974) conducted a toxicity bioassay using developing oyster embryos (*Crassostrea gigas*). He points out that embryos are much more sensitive to chemical pollutants than are adult organisms. The demonstrated toxicity of anthracene was too low to permit computation of ED_{50} or EMD_{50} values (Fig. 9.13). Legore also postulates that an increase in the dose above 5.0 g/liter would not result in an increased effect due to the low solubility of the PAH. At doses above 1.0 g/liter, substantial amounts of anthracene crystals remained in the bottom of the test containers.

At low concentrations, 7,12-dimethylbenz[*a*]anthracene (DMBA) strongly inhibits cell multiplication in normal primary cell cultures. Bourne and Jones (1973) exposed fish cells to DMBA and

ORNL DWG 76-12135

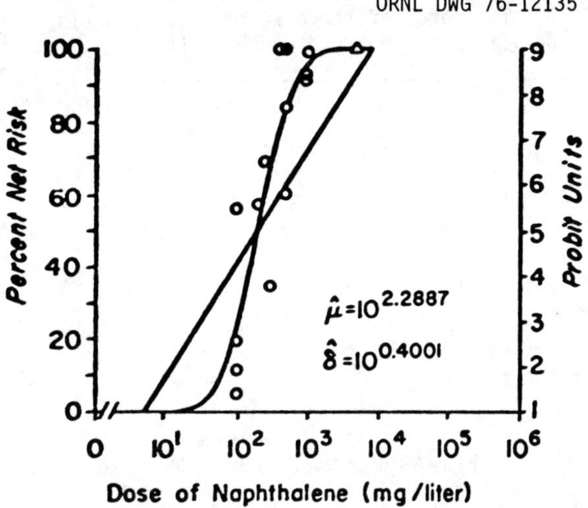

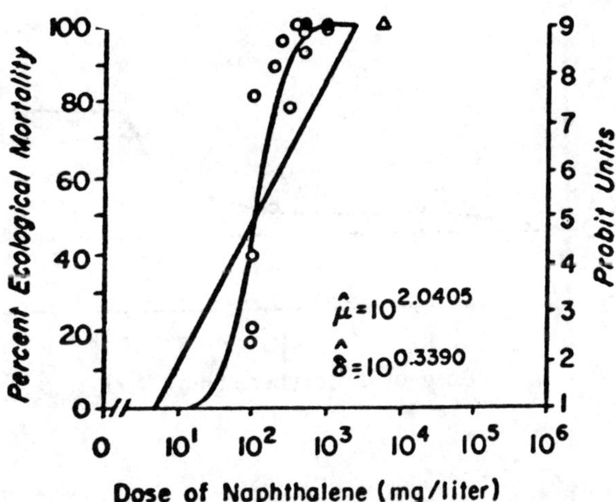

Fig. 9.12. Percent effect of naphthalene. Single data points are indicated by o; two overlapping points, ●; three overlapping points, Δ. The straight lines are prohibit lines. Percent net risk prohibit line: y = -0.721 + 2.499x; percent ecological mortality prohibit line: y = -1.020 + 2.950x. The data points do not directly relate to the prohibit lines. Source: Legore 1974, Fig. 28, p. 106. Reprinted by permission of the publisher.

found that fibroblasts from rainbow trout gonads (RTG-2) were more sensitive to inhibition of mitosis by the PAH than were epitheloid cells from the fathead minnow (FHM). This variation in sensitivity is indicated by the mitotic indices[*] as given in Table 9.18. At 1.0 ppm DMBA, mild cytotoxicity was observed in RTG-2 cells, but not in FHM cells. At concentrations ≥5.0 ppm, mitosis in all cultures ceased. Cells were hyperchromatic with vacuolated cytoplasm and highly necrotic. In the FHM cultures, many cells were extremely abnormal, showing clumped chromatin, basophilic cytoplasm, and a reduction in cytoplasm. In 10 ppm DMBA, 67.9% of the FHM cells contained hyperchromatic nuclei, as compared with 91.4% in 20 ppm. No FHM cells survived concentrations of 45 ppm DMBA, and no RTG-2 cells survived 40 ppm DMBA.

---

[*] Calculated by dividing the number of cells in mitosis, at any stage, by the total number of cells counted (but at least 1000), and then multiplying by 100.

Table 9.17. Tolerance of three species of estuarine animals
to specific petroleum hydrocarbons (naphthalenes)

Hydrocarbon	24-hr LC_{50} values (ppm)		
	Sheepshead minnow[a]	Grass shrimp[b]	Brown shrimp[c]
Naphthalene	2.4	2.6	2.5
1-Methylnaphthalene	3.4		
2-Methylnaphthalene	2.0	1.7	0.7
Dimethylnaphthalenes	5.1	0.7	0.08

[a]*Cyprinodon variegatus.*
[b]*Palaemonetes pugio.*
[c]*Penaeus aztecus.*

Source: Anderson et al. 1974*b*, Table 3, p. 299. Reprinted by permission of the publisher.

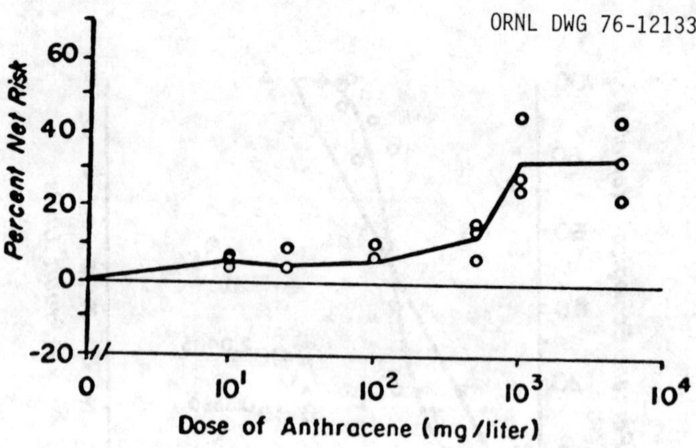

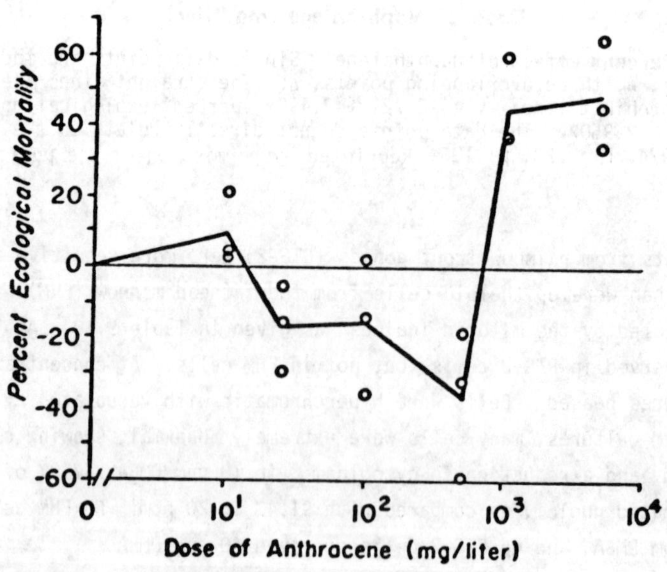

Fig. 9.13. Percent effect of anthracene. Individual data points are indicated by o, single point and ●, two overlapping points. The graph lines illustrate mean results of replicate test cultures. Source: Legore 1974, Fig. 29, p. 107. Reprinted by permission of the publisher.

Table 9.18. Effects of 7,12-dimethylbenz[a]anthracene (DMBA) to fish cells

	Control	Effect (%) due to concentration (ppm) of					
		0.1	0.5	1.0	5.0	10.0	20.0
Fathead minnow epithelioid cell-line (FHM)							
Mitotic indices	3.04	3.2	2.91	2.32	0	0	0
Range	(2.7 - 3.4)	(2.6 - 3.5)	(2.3 - 3.5)	(1.2 - 3.34)			
Significance level		a	a	a			
Multinucleation	0.52	0.92	5.53	6.69	6.1	1.56	1.23
Range	(0.28 - 1.0)	(0.4 - 1.3)	(4.20 - 7.12)	(5.30 - 8.58)	(5.0 - 8.7)	(1.0 - 2.3)	(1.0 - 1.57)
Significance level		a	0.001	0.001	0.001	0.001	0.001
Rainbow trout gonad fibroblasts cell-line (RTG-2)							
Mitotic indices	3.3	2.9	0.96	0.4	0	0	0
Range	(2.9 - 4.5)	(2.5 - 3.2)	(0.49 - 1.69)	(0.19 - 1.19)			
Significance level		a	0.001	0.001			
Multinucleation	1.16	1.57	1.15	3.3	b	b	b
Range	(1.1 - 1.9)	(1.0 - 2.0)	(0.5 - 1.9)	(2.2 - 4.2)			
Significance level		a	0.01	0.001			

[a]Not significant compared with control.
[b]Toxic, accurate determination not possible.

Source: Bourne and Jones 1973, Table 1, p. 142. Reprinted by permission of the publisher.

Jones, as cited by Bourne and Jones (1973), demonstrated that a 24-hr exposure to 0.56 ppm DMBA in the water environment of developing zebra fish (*Brachydania rerio*) embryos caused tail necrosis, tumorous growths, an enlarged pericardium, and a shortened body. Surface cells and their nuclei were enlarged and granular, and some parts of the embryos appeared to have a more rapid growth rate. Embryonic development ceased at 5.0 ppm DMBA, and death occurred after 24 hr at this exposure level.

A compound extracted from the thatched barnacle (*Tetraclita squamosa rubescens*) was demonstrated to induce tumors when injected into mice (Shimken, Koe, and Zechmeister 1951). The compound, presumed to be BaP, produced no observable response in the barnacles that had originally accumulated the PAH. Similar PAH behavior was observed in the goose barnacle (*Mitella polymerus*).

Zitko and Tibbo (1971), as cited by Ehrhardt (1972), were able to identify an intermediate coke-oven oil containing large amounts of aromatic hydrocarbons as the cause for an extensive kill of herring (*Clupea harengus*) by measuring the fluorescence emission and ultraviolet spectra of purified tissue extracts. Although not specifically characterized, PAH were known to be present in the oil.

Ryckman, Probhakara-Rao, and Buzzell (1975) report that exposure to 5 mg/liter phenanthrene was lethal to trout and bluegill within 12 hr.

9.1.7.4 Crude oil and crude oil fractions

Toxicity studies for many crude oils, oil derivatives, and various petroleum fractions are available in the literature; however, few of these studies characterize the components of the fractions or whole oil samples tested. Occurrence and nature of effects from oil hydrocarbons depend on the composition of the oil to which organisms are exposed (Sullivan 1974). Although toxicity values for various oils have been determined, as indicated in Table 9.19, characterization of the specific compound responsible for an observed biological effect from oil exposure has not been accomplished. The lower-boiling, more-soluble aromatic fractions of oil are repeatedly implicated as the most toxic (Moore and Dwyer 1974; Eisler and Kissil 1975; Anderson et al. 1974*a*; 1974*b*; Guard, Hunter and DiSalva 1975). Naphthalene aromatics, along with monocyclic aromatics, are included in this group; together they constitute a considerable percentage of the aromatic hydrocarbons in oil (Table 9.9). It is suspected that these compounds evoke more of a biological response because they are present in oil at higher concentrations than are PAH (Anderson et al. 1974*a*). The greater solubility of these lower-boiling aromatic fractions permits a greater potential for exposure in aquatic organisms, which must also be considered when comparing the relative toxicity of these lower-boiling fractions to that of PAH (Neff 1975).

Rossi, Anderson, and Ward (1976) examined the sensitivity of polychaetous annelids, *Neanthes arenaceodentata* and *Capitella capitata*, to the water-soluble fractions of two refined and two crude oils. As shown in Table 9.20, the two refined oils — No. 2 fuel oil and Bunker C residual oil — proved most toxic to both species. The authors suggest that the higher concentrations of toxic diaromatic compounds (naphthalenes) found in refined oils (Fig. 9.14) account for the greater toxicity of refined vs crude oils. Anderson et al. (1974*a*) also notes a correlation between test oil toxicity and total naphthalene (naphthalene plus methylnaphthalenes plus dimethylnaphthalenes) content.

Table 9.19. LC_{50} values, 95% confidence intervals (95% CI) and slope functions (SF)
for three species of estuarine animals and statistical analysis of the bioassay results

Hydrocarbon solution	Sheepshead minnow[a]		Grass shrimp[b]			Brown shrimp[c]		
	24 hr	96 hr	24 hr	48 hr	96 hr	24 hr	48 hr	96 hr
South Louisiana crude								
Oil-and-water dispersion								
LC_{50}	80,000[d]	29,000[d]	1700	1650	200	>1000[d]	>1000[d]	>1000[d]
95% CI			567 - 5100	589 - 4620	133 - 302			
SF			5.48	5.48	2.00			
Water-soluble fraction								
LC_{50}	>19.8[d]	>19.8[d]	>19.8[d]	>19.8[d]	>19.8[d]	>19.8[d]	>19.8[d]	>19.8[d]
95% CI								
SF								
Kuwait								
Oil-and-water dispersion								
LC_{50}	>80,000[d]	>80,000[d]	13,500	9000	6000			
95% CI			6750 - 27,000	3462 - 23,400	2400 - 15,000			
SF			3.09	6.38	4.31			
Water-soluble fraction								
LC_{50}			>10.2[d]	>10.2[d]	>10.2[d]			
95% CI								
SF								
No. 2 fuel oil								
Oil-and-water dispersion								
LC_{50}	250	93	3.8	3.4	3.0	9.4	9.4	9.4
95% CI	156 - 400	66 - 130	3.0 - 4.9	2.8 - 4.2	2.7 - 3.3	7.6 - 11.6	7.6 - 11.6	7.6 - 11.6
SF	2.55	1.78	1.51	1.42	1.18	1.26	1.26	1.26
Water-soluble fraction								
LC_{50}	>6.9[d]	>6.9[d]	4.4	4.1	3.5	5.0	5.0	4.9
95% CI			3.3 - 5.6	2.7 - 5.5	2.4 - 4.9	4.6 - 5.5	4.6 - 5.5	4.6 - 5.5
SF			1.48	1.77	1.92	1.12	1.12	1.06
Bunker C								
Water-soluble fraction								
LC_{50}	4.7[d]	3.1[d]	3.2	2.8	2.6	3.8	3.5	1.9
95% CI			2.4 - 4.3	2.1 - 3.8	2.0 - 3.3	3.3 - 4.5	2.9 - 4.1	1.0 - 3.5
SF			1.40	1.40	1.40	1.31		1.94

[a] *Cyprinodon variegatus.*
[b] *Paleamonetes pugio.*
[c] *Penaeus azteeus.*
[d] Denotes data points which could not be analyzed by the method used.

Source: Anderson et al. 1974, Table 2, p. 291. Reprinted by permission of the publisher.

Table 9.20. Statistical analysis of the results of bioassays on polychaetes[a]

Oil type	Statistical value	Neanthes arenaceodentata			Capitella capitata		
		24 h	48 h	96 h	24 h	48 h	96 h
No. 2 fuel oil WSF	TL_m	>8.7 (100)	3.2 (37)	2.7 (31)	>8.7 (100)	3.5 (42)	2.3 (56)
	95% CI		2.3 - 4.3	1.9 - 3.8		3.0 - 3.9	1.9 - 2.6
	SF		1.8	2.0		1.2	1.4
Bunker 'C' WSF	TL_m	>6.3 (100)	4.6 (73)	3.6 (57)	>6.3 (100)	1.1 (17)	0.9 (14)
	95% CI		3.7 - 5.8	4.5 - 2.9		1.0 - 1.2	0.8 - 1.0
	SF		1.8	1.5		1.9	1.7
So. La. Crude WSF	TL_m	18.0 (91)	13.9 (70)	12.5 (63)	>19.8 (100)	16.2 (82)	12.0 (61)
	95% CI	19.3 - 16.8	14.0 - 13.8	13.6 - 11.5		15.8 - 16.6	11.6 - 12.5
	SF	1.2	1.2	1.17		1.1	1.2
Kuwait Crude WSF	TL_m	>10.4 (100)	>10.4	>10.4	>10.4 (100)	>10.4	9.8 (94)
	95% CI						9.1 - 10.5
	SF						1.1

[a]95% confidence intervals (95% CI) and slope functions (SF) are given for each TL_m value; the values represent ppm of total hydro-carbons calculated from the amount present in the 100% water-soluble fraction (WSF), while those in parentheses represent the equivalent percentages.

Source: Rossi, Anderson, and Ward 1976, Table 1, p. 12. Reprinted by permission of the publisher.

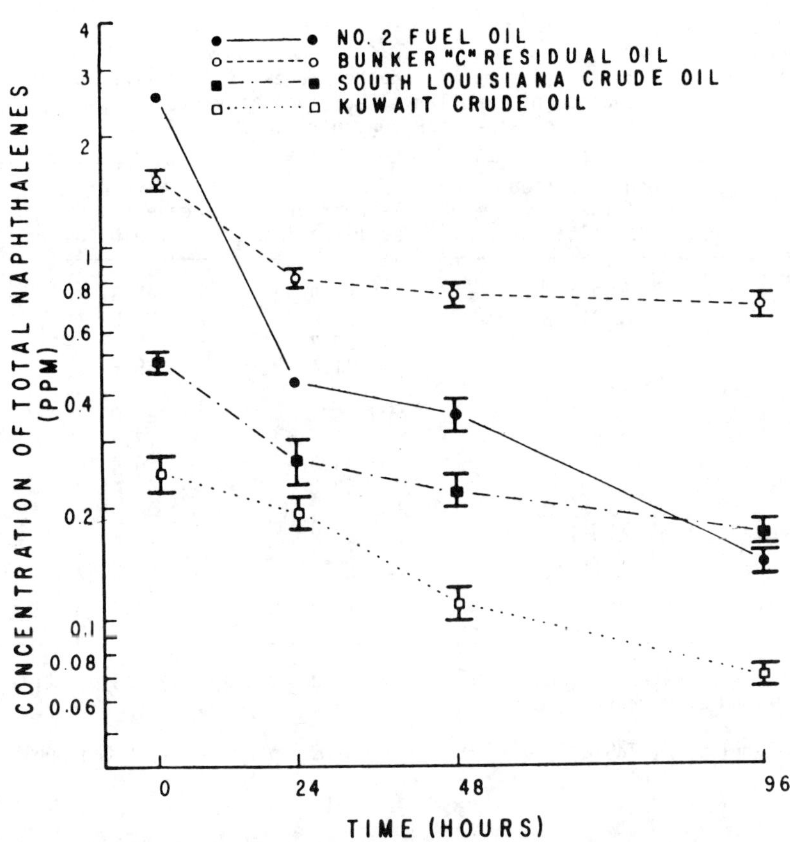

Fig. 9.14. Decrease in total naphthalenes (= naphthalene + methylnaphthalenes + dimethylnaphthalenes) concentration of 100% WSF bioassay solutions over a 96-hr time period. Mean values for four samples are given with corresponding standard deviations represented by vertical bars. Standard deviations for No. 2 fuel oil at 0 and 24 hr were <±0.020 and <±0.010 ppm respectively. <u>Source</u>: Rossi, Anderson, and Ward 1976, Fig. 2, p. 15. Reprinted by permission of the publisher.

Water-soluble fractions (WSF) of several crude oils were found to be highly toxic to bivalve gametes, embryos, and larvae (Renzoni 1975). The WSF of Alaska Prudhoe, Nigeria Bonny, and Kuwait crude oils were observed to impair fertilization processes, embryo development, and survival of larvae of *Crassostrea virginica* and *Mulinia lateralis* (Tables 9.21 and 9.22). In addition to the influence of the WSF upon the reproductive success of both species and on the larval growth of *Mulinia*, the toxicants in the WSF were found to cause abnormal embryonic development.

Table 9.21. *Crassostrea virginica*: percentages of fertilization, embryo development, and larval survival at various oil-in-water concentrations[a]

Crude oil origin	Concentration of the oil (ml/liter)	Percent of fertilization	Percent development of embryos	Percent survival of larvae
Alaska (Prudhoe)	0.001	89.9	91.5	83.0
	0.01	88.2	86.5	80.2
	0.1	83.7	81.9	75.2
	1	71.4	68.5	55.0
Nigeria (Bonny)	0.001	90.1	89.0	77.1
	0.01	80.3	77.9	72.5
	0.1	75.2	66.4	68.7
	1	63.3	50.4	43.3
Kuwait	0.001	89.9	92.8	88.3
	0.01	88.7	88.9	81.8
	0.1	85.2	85.9	77.7
	1	73.2	69.3	53.8

[a]All values are averages for duplicate cultures and are expressed as percentage of the values obtained from control cultures.

Source: Renzoni 1975, Table 1, p. 126. Reprinted by permission of the publisher.

Crude oils have been found to be toxic to fish eggs and larvae. The effects of different crude oils on hatching of cod eggs are shown in Table 9.23. Waldichuk (1974*a*) suggests that the varied effects of different crude oils on aquatic organisms are probably due to the varied concentrations of soluble aromatic constituents of the oils.

Mironov (1972) observed that fish (*Rhombus maeoticus*) died within 24 to 48 hr of exposure to seawater containing oil and soluble oil derivatives at concentrations of 0.1 to 1.0 ppm. At concentrations of 0.01 to 0.1 ppm, only 55 to 89% of the eggs hatched. Many of the larvae hatching from eggs exposed to oils were observed to exhibit developmental abnormalities.

Estimates of the soluble aromatic derivative (SAD) concentrations of petroleum have been used as a basis for comparing toxic effects induced by various oils. Estimated SAD concentrations causing toxicity are given in Table 9.24. Aromatics having greater solubilities, such as monocyclics and naphthalenes, account for the larger portion of an oil's SAD. Only a relatively small portion of total PAH would be included in the SAD for any given oil since PAH are not very soluble in water (McAuliffe 1966). Although Table 9.24 presents summary data for SAD toxicity to a variety of organisms, it should not be interpreted as a summary of PAH toxicity

Table 9.22. *Mulinia lateralis*: percentages of fertilization, embryo development, larval survival, and increase in mean length at various oil-in-water concentrations[a]

Crude oil origin	Concentration of the oil (ml/liter)	Percent of fertilization	Percent development of embryos	Percent survival of larvae	Percent increase in mean length of larvae
Alaska (Prudhoe)	0.001	93.8	91.9	84.8	82.8
	0.01	91.9	89.0	85.3	82.5
	0.1	85.1	86.9	80.6	74.5
	1	68.8	77.1	71.0	55.6
Nigeria (Bonny)	0.001	89.4	89.0	86.4	81.7
	0.01	88.5	87.0	82.6	72.6
	0.1	83.7	77.8	71.3	63.3
	1	67.8	62.8	59.2	40.3
Kuwait	0.001	96.8	95.8	88.0	81.3
	0.01	94.2	94.1	83.8	82.5
	0.1	89.5	91.1	80.9	74.5
	1	73.8	72.9	62.0	50.1

[a]All values are averages for duplicate cultures and are expressed as percentages of the values obtained from control cultures.

Source: Renzoni 1975, Table 2, p. 126. Reprinted by permission of the publisher.

Table 9.23. Percent mortality of cod eggs at various ages of transfer, subjected for 100 hr to different concentrations of oil film extracts[a] of three crude oils

Egg age at transfer (days)	Percent mortality in oil-film concentration (ppm) of			
	10^4	10^3	10^2	0 (Control)
Venezuela (Tia Juana)				
0.6	63	35	31	21
3.6	47	57	43	28
10	11	3	1	2
Iran (Agha Jari)				
1.3	42	33		8
4	34	27	14	15
10	7	10	4	
Libya (Sarir)				
0.2	3	1.5	2	3[b]
4	2	2	2	
10	5	8	7	9

[a]Oil at different concentrations (10^2, 10^3, and 10^4 ppm) was poured into 30-liter plastic containers and left for 2 days. Slow water circulation was maintained by means of small pumps to ensure a maximum saturation of soluble compounds. The clear oil extract was transferred into test jars and renewed every 2 days. Preliminary analysis showed that the amount of dissolved hydrocarbons obtained in this way from 10^4 ppm of crude oil is of the order of 10 ppm.
[b]Since the data were taken from histograms in the original paper, there is some uncertainty about percent mortality in controls of the tests for Libya crude.

Source: Waldichuk 1974a, Table 7, p. 42. Reprinted by permission of the publisher.

Table 9.24. Summary of toxicity data for
soluble aromatic derivatives of crude oils

Class of organisms	Estimated concentrations (ppm) of soluble aromatics causing toxicity
Flora	10 - 100
Finfish	5 - 50
Larvae (all species)	0.1 - 1.0
Pelagic crustaceans	1 - 10
Gastropods (snails, etc.)	1 - 100
Bivalves (oysters, clams, etc.)	5 - 50
Benthic crustaceans (lobsters, crabs, etc.)	1 - 10
Other benthic intertebrates (worms, etc.)	1 - 10

Source: Moore and Dwyer 1974, Table 4, p. 823.
Reprinted by permission of the publisher.

alone; other aromatics are contained in the SAD. More precise chemical analyses of SAD composition, and PAH content specifically, are required.

Moore and Dwyer (1974) indicate that a direct lethal response can be expected in most adult marine species from exposures of 1 to 100 ppm total SAD for periods of several hours or less. Larval stages are apparently sensitive to concentrations as low as 0.1 ppm SAD.

Morrow, Gritz, and Kirton (1975) tested the toxicity of oil compounds to young Coho salmon. Of the compounds tested, toxicity was shown to increase in the following order: cyclohexane < cyclohexene < 1,3-cyclohexadiene < toluene < xylene < benzene < ethylbenzene (Table 9.25). These results substantiate a correlation between toxicity and the number of unsaturated bonds. Survival was related to the volatility of the test substance and its structure, the aromatic ring being proposed as the toxic agent. The most toxic of the chemicals tested were the unsaturated aromatics, which were also the most volatile. Fish that withstood the first few hours of exposure, when the concentration of test substance was highest, exhibited a capacity for surviving the experiment. The volatile nature of the test substances apparently resulted in a decrease of the concentrations in solution with time. Possible mechanisms of toxicity were also suggested. The fact that most of the substances tested were lipid solvents suggests that the permeability of membranes, particularly the gills, was increased through the loss of fatty substances. With increased gill permeability, fish were unable to control the ion balance in their blood (Table 9.26). The resultant internal accumulation of carbon dioxide was a contributing factor in causing death.

9.1.7.5 Structure-activity relationship

Giger and Blumer (1974) report that, whereas low-boiling aromatics constitute the most acutely toxic fraction of petroleum, naphthalene and phenanthrene are more toxic to fish than are the smaller molecules (benzene, toluene, and xylene). Legore (1974) maintains, however, that a precise correlation between toxicity and molecular size does not exist; he cites research by Pickering and Henderson (1966), which demonstrates the toxicity of small-molecular aromatics to fathead minnows (*Pimephales promelas*), bluegills (*Lepomis macrochirus*), goldfish (*Carassius auratus*), and guppies (*Lebistes reticulatus*) in the following order: ethyl benzene < toluene <

Table 9.25. Mortalities of young Coho salmon produced by various PAH
in artificial seawater at 8°C

Test substance	Concentration (ppm)	Number of tests	Number alive after hr					Percent mortality
			0	24	48	72	96	
1,3 cyclohexadiene	0	3	9	9	9	9	9	0
	50	10	30	20	20	20	20	33.3
Toluene	0	3	30	28	28	26	26	13.3
	1	3	30	28	28	26	26	13.3
	10	3	30	30	30	29	27	10
	50	1	10	1	0	0	0	100
	100	3	30	2	0	0	0	100
Xylene	0	3	30	29	26	25	23	23.3
	1	3	30	29	26	24	23	23.3
	10	3	30	27	24	22	21	30
	100	3	30	0	0	0	0	100
Benzene	0	3	30	30	30	29	29	3.3
	1	3	30	29	29	29	29	3.3
	10	3	30	30	30	30	29	3.3
	50	2	20	9	8	8	8	60
	100	3	30	0	0	0	0	100
Ethylbenzene	0	3	30	29	28	28	28	6.7
	10	3	30	28	28	27	26	13.3
	50	3	30	0	0	0	0	100

Source: Morrow, Gritz, and Kirton 1975, Table 1, p. 327. Reprinted by permission of the publisher.

benzene < xylene. The scarcity of PAH toxicity studies prohibits correlation of physical
properties of PAH with relative toxicity.

9.1.8 Carcinogenicity

Some of the PAH that concentrate in biological material, such as BaP and benzanthracene, have
been shown to be carcinogenic (McGinnes and Snoeyink 1974). Ames, Sims, and Grover (1972) suggest that the carcinogenicity of PAH is a result of their decomposition in biological systems
and that it is the decomposition products (BA quinones, BP quinones) that are carcinogenic.
Neither the routes of PAH decomposition in biological systems nor the fate and general behavior
of PAH in aquatic systems have been determined (Attari 1973).

The potential carcinogenicity of coal conversion chemicals to organisms in their natural environments must be considered if the overall effects of coal conversion on the environment are to be
assessed. Auerbach (1975) states that such an assessment requires information, not only on the
responses produced in the component organisms but also on the ultimate fate of the compounds in
the environment. Research by Auerbach et al. (in press), as cited by Auerbach (1975), suggests
that the relationship of potential effects of compounds to their environmental fate should not
be neglected. For example, a potent carcinogen in the aquatic environment may be taken up by
the sediment, where it never encounters a target organism, whereas a weak carcinogen present
in the environment at very low levels may bioaccumulate and be transferred to higher organisms
in the food chain with devastating effects.

Table 9.26. Mean values in meq/liter of blood sodium, potassium, and chloride ions after exposure to various PAH[a]

Test Substance	Concentration (ppm)	Ion	Control	Exposure (hr)				
				1	2	3	4	24
Cyclohexene	100	Na+	175 (18)	194 (20)	173 (20)	170 (20)		
		Cl-	134 (16)	164 (12)	154 (20)	131 (14)		
Toluene	30	Na+	208 (12)	224 (4)	226 (8)	216 (10)	220 (8)	
		K+	4.1 (16)	4.8 (16)	5.8 (10)	5.1 (14)	4.9 (16)	
Xylene	30	Na+	186 (18)	249 (18)	210 (20)	197 (20)		
		K+	4.4 (16)	4.6 (16)	3.7 (16)	4.6 (16)	4.8 (8)	3.9 (16)
		Cl-	134 (19)	135 (16)	137 (10)	140 (16)		
Benzene	15	Na+	134 (16)	160 (16)	157 (16)	160 (16)	190 (14)	154 (14)
		K+	4.3 (16)	4.8 (16)	4.7 (16)	5.0 (16)	4.0 (16)	3.9 (14)

[a]Number of fish in each analysis are in parentheses.

Source: Morrow, Gritz, and Kirton 1975, Table 2, p. 328. Reprinted by permission of the publisher.

9.1.8.1 <u>Nonaquatic organisms</u>

Examination of the available literature has produced no information regarding the carcinogenic effects of PAH on domestic or commercial animals. However, Olsen and Haynes (1969) cite Pybus (1964) as reporting the occurrence of virtually every form of benign or malignant tumor among domestic animals. Pybus has found cancers of many body parts to be relatively common among older cats and dogs.

Thirty-six wild mammals and birds exposed to urban conditions have been examined for pulmonary cancer, pulmonary adenomatosis, and squamous metaplasia by Stewart (1966) at the Philadelphia Zoo. Histologic material indicated adenocarcinomas in 11 birds of the family Anatidae, a silver pheasant, a red jungle fowl, and a coypu; squamous cell carcinoma in a North American otter; undifferentiated small-cell carcinoma in a Java sparrow; alveologenic carcinoma in a striped skunk; pleural mesotheliomas in a Cape hunting dog and a clouded leopard; adenomatosis, squamous metaplasia, or both in a red fox, puma, North American otter (which also had squamous cell carcinoma of the lung), squirrel monkey, peafowl, African eared vulture, Chinese myna, two toucan barbets, and a rabbit-eared bandicoot; and metastatic renal carcinoma of the lung in a red wolf. No hypothesis was offered as to the cause of these responses other than exposure to polluted urban atmosphere.

PAH, which exist in relatively pure form on particles, can be readily absorbed by animals through ingestion, inhalation, or skin contact (EPA 1975a). Falk, Miller, and Kotin (1958) have demonstrated the carcinogenicity of PAH adsorbed onto polycyclic organic matter (POM) through inhalation studies with experimental animals.

Particle size is directly related to the biological activity of POM-containing aerosols (NAS 1972). Pierce and Katz (1975) found that the majority of PAH adsorbed on particles was associated with particles less than 3.0 μm in diameter, well within the respirable range. Many carcinogenicity studies involving atmospheric exposure of laboratory animals to PAH have been performed, but no studies are available regarding atmospheric exposure to wild animals (Kornreich 1975; EPA 1975a; NAS 1972; Olsen and Haynes 1969).

PAH have been identified in various forms of edible plant matter, where they are likely to be ingested by a variety of plant-consuming organisms. Although no studies are available to provide supporting data, the incidence of gastric cancer and neoplastic disease has been correlated with ingestion of plant material containing POM (NAS 1972).

9.1.8.2 <u>Aquatic organisms</u>

Frequency of fish tumors has been correlated with contaminated habitats (Brown et al. 1973). Study of a polluted watershed (Fox River, Illinois) allowed identity of several aromatics, including 0.01 ppm benzanthracene and 0.1 ppm each of naphthalene, toluene, and benzene. Frequency of various fish neoplasms was 4.38% in the polluted watershed as compared with 1.03% in the control waters (Canadian Watershed) (Table 9.27).

Abnormal, lethal growths have been produced in planaria (*Dugesia dorotocephala*) with PAH. Foster (1969) attributes the observed formation of papilliform tumors and other malformations to stimulated regenerative cells — totipotent neoblasts — from 3-methylcholanthrene or BaP

Table 9.27. Incidence of tumors by comparison of fish species found in two test watershed areas (1967 - 1972)

Generic name	Common	Total number of fish examined	Number of fish with tumors	Percent fish with tumors	Neoplasm most prevalent by species
		Canadian watershed			
Stizostedion vitreum vitreum	Walleye	540	3	0.55	Sarcoma
E. masquinongy	Muskellunge	99	1	0.99	Lymphosarcoma
E. lucius	Northern pike	904	15	1.65	Lymphosarcoma
Ictalurus nebulosus	Brown bullhead	101	2	1.98	Hepatoma
Salmo fario	Brown trout	403	2	0.49	Osteogenic papilloma
Hypentelium nigricans	Hog sucker	39	1	2.56	Epithelioma
Salmo gairdneri	Rainbow trout	461	3	0.65	Hepatocarcinoma
Ameiuridae	Catfish	65	1	1.54	Epidermoid carcinoma
Cyprinus carpio	Carp	97	1	1.03	Anal carcinoma
Micropterus dolomieu	Small mouth bass	427	5	1.17	Osteoma
Semotilus atromaculatus	Northern creek chub	162	1	0.61	Neurilemmoma
Lepomis gibbosus	Pumpkinseed	425	3	0.70	Epidermoid carcinoma
Perca flavescens	Yellow perch	91	2	2.19	Epidermoid carcinoma
Ichthymyzon	Lamprey	16	0	0	Unidentified tumor
Pomoxis nigromaculatus	Black crappie	36	0	0	Melanoma
Coregonus clupeaformis	Whitefish	262	1	0.38	Epidermoid carcinoma
Unidentified species	(Minnows)	511	7	1.37	Epidermoid carcinoma
Total		4639	48	1.03	
		Fox Valley watershed			
S. vitreum vitreum	Walleye	111	2	1.80	Sarcoma
E. masquinongy	Muskellunge	35	1	2.85	Lymphosarcoma
E. lucius	Northern pike	88	7	7.95	Lymphosarcoma
I. nebulosus	Brown bullhead	283	35	12.2	Hepatoma
S. fario	Brown trout	17	2	1.17	Osteogenic papilloma
H. nigricans	Hog sucker	192	8	4.17	Epithelioma
S. gairdneri	Rainbow trout	21	1	4.76	Hepatocarcinoma
Ameiuridae	Catfish	161	5	3.10	Epidermoid carcinoma
C. carpio	Carp	112	4	3.56	Anal carcinoma
M. dolomieu	Small mouth bass	93	2	2.15	Osteoma
S. atromaculatus	Northern creek chub	219	7	3.19	Neurilemmoma
L. macrochirus	Bluegill	161	2	1.24	Epidermoid carcinoma
P. flavescens	Yellow perch	148	5	3.37	Epidermoid carcinoma
Ichthymyzon	Lamprey	51	1	1.96	Unidentified tumor
P. nigromaculatus	Black crappie	92	2	2.16	Melanoma
L. chrysops	White bass	171	6	3.50	Epidermoid carcinoma
Unidentified species	(Minnows)	166	3	1.81	Epidermoid carcinoma
Total		2121	93	4.38	

Source: Brown et al. 1973, Table 3, p. 196. Reprinted by permission of the publisher.

exposure of the animals. It was acknowledged that some metaplasia of somatic cells may have occurred, but that formation of the abnormal growths was most likely derived from the neoblasts. Nine percent of the planaria exposed to 3-methylcholanthrene and seven percent of those exposed to BaP developed lethal growths.

Neukomm (1974) demonstrated that the carcinogenic potential of certain PAH can be determined through subcutaneous injection in the regenerative portions of newt (*Triturus cristatus*) epidermis. Epithelial hyperplasia was followed by infiltration of the underlying tissues. The neoplastic infiltration occurred within 4 to 5 days after injection, with activity continuing for 12 to 20 days after injection and culminating in a diffuse tumor. Newt epidermis reacted to the PAH in the following order of sensitivity: dibenzanthracene (10^{-11} to 10^{-10} mg/ml), methylcholanthrene (10^{-9} to 10^{-8} mg/ml), BaP (10^{-4} to 10^{-3} mg/ml), and 9,10-dimethyl-1,2-benzanthracene (1 to 0.1 mg/ml). Pyrene had no effect on the newt, and chrysene and 1,2-benzanthracene were weakly active. The newt's epidermis was thus shown to react to principal carcinogenic hydrocarbons in a manner similar to that of the skin of mammals, thus indicating the possible use of the newt as an experimental model in assessing carcinogenic potential of and sensitivity to low concentrations of PAH.

9.1.9 Mutagenicity

Mutagenicity in aquatic organisms is an area of research almost devoid of activity. Several studies have shown, however, that certain PAH alter normal embryonic development of some aquatic species and are thus considered by some researchers to be capable of mutagenesis.

Sea urchin (*Paracentrotus lividus*) eggs treated with BaP were demonstrated by Ceas (1974) to cleave abnormally and develop into irregular morulae. BaP was shown to interfere with normal egg development; it binds to the lipids and lipoproteins of eggs as indicated by the fluorescence pattern of the centrifuged egg. Experimentation with other sea urchin (*Arbacia lixula* and *Sphaerechinus granularis*) eggs and embryos exhibited similar membrane binding characteristics of BaP. It was suggested that BaP may possibly bind with nuclear DNA in the course of egg segmentation. It was noted that, even indirectly, linkage of BaP with the nuclear membrane might affect DNA replication.

Preliminary results of research by DeAngelis, Giordano, and Pagano (1975) indicate that sea urchin (*Paracentrotus lividus* and *Psammechinus miliaris*) embryos are capable of continued development in the presence of BaP or DMBA. Embryo development was tested through cellular dissociation and recombination. Results also indicate that, whereas BaP did not affect the morphogenesis of the developing embryos, DMBA acted specifically on the cells of the primary mesenchyme. These cells appeared disorganized and incapable of promoting normal morphogenesis.

9.1.10 Teratogenicity

Information regarding teratogenic effects (birth defects) of PAH in animals (other than laboratory species) is scarce and limited to several studies dealing with organisms capable of high regenerative potential. Induction of abnormalities by PAH can thus be observed in the rapid development of regenerative tissue. Suspected interrelationships between teratogenesis, carcinogenesis, and regeneration have prompted such studies.

A wide variety of carcinogenic substances, including methylcholanthrene, BaP, and various fractions of coal tar, were injected into healthy limbs of the adult newt (*Triturus viridescens*) by Breedis (1952). These substances were observed to lead more frequently to induction of accessory limbs than to cancer (Tables 9.28 and 9.29). BaP, however, did not stimulate regenerative growth resulting in limb duplication. In similar research with the urodele *Xenopus laevis*, Ruben and Balls (1964) found methylcholanthrene crystals to induce lymphosarcomas rather than accessory limbs, although teratogenic activity was observed in the area of active regeneration. De Lustig and Matos (1971) observed the formation of lymphomatous nodules as a result of PAH implantation into nonregenerating areas of *Bufo arenarum* tadpoles. Of the PAH tested, 20-methylcholanthrene, BaP, and DMBA also caused teratogenic effects such as accessory tail fins and accessory notochords.

Table 9.28. Incidence of accessory limbs among newts injected into the
proximal region of the forelimb with various carcinogenic or
noncarcinogenic substances or subjected to local injury[a]

Material injected[b]	Number of animals injected[c]	Animals developing accessory limbs		Time of appearance of first accessory limb (days)
		Number	Percent	
Coal tar	76	20	26	41
Coal tar + 4% MCA, BaP, AAF, and SR	50	20	40	42
Coal tar + 4% MCA	20	4	20	49
Coal tar + 4% BaP	19	2	11	87
Coal tar + 4% AAF	20	3	15	49
Coal tar + 4% SR	20	8	40	49
Olive oil + 4% MCA	4	1	25	300
Olive oil + 4% AAF	10	0	0	
Olive oil + 4% SR	10	0	0	
Vaseline + 4% BaP	23	0	0	
Vaseline	22	1	5	300
MCA crystals wet with olive oil	34	1SS[d]	3	130
Beryllium hydroxide	18	6	33	130
Rana pipiens kidney carcinoma	75	3	4	200
Triturus viridescens sarcoma	274	0	0	
Deep burn of forelimb (not injected)	21	0	0	
Fracture of humerus (not injected)	23	0	0	

[a]Animals that developed sarcoma are designated by S.
[b]MCA = 3-methylcholanthrene; BaP = benzo[*a*]pyrene; AAF = 2-acetylaminofluorene; SR = scarlet red.
[c]Animals surviving 50 or more days.
[d]Sarcoma developing in two animals; one of the tumors arose at the base of the limb.

Source: Breedis 1952, Table 1, p. 862. Reprinted by permission of the publisher.

Tail regeneration in the newt (*Triturus viridescens*) was found to be inhibited or delayed by implantation of 1,2,5,6-dibenzanthracene or 3-methylcholanthrene (Pizarello and Wolsky 1966). Matos and De Lustig (1973), however, maintain that regeneration proceeds independently of nodule development and regression. Teratogenic effects in tails of *Bufo arenarum* tadpoles not submitted to a regeneration field (normally produced via limb amputation) were induced by 20-methylcholanthrene, BaP, and DMBA.

Table 9.29. Incidence of accessory limbs among newts injected
into the proximal region of the forelimb with
various fractions of coal tar

Material injected	Number of animals injected[a]	Animals developing accessory limbs		Time of appearance of first accessory limb (days)
		Number	Percent	
Absorption fractions				
1-A	16	0	0	
1-B	16	4	25	52
1-C	16	10	63	52
Industrial tar fractions				
Tar before distillation	13	6	46	54
Middle creosote oil (287 - 343°C)	16	7	44	54
Heavy creosote oil crystals (343 - 416°C)	16	5	31	100
Heavy creosote oil mother liquor	7	5	71	54
Fuel pitch (residue in still)	16	10	63	90

[a]Animals surviving 50 or more days.

Source: Breedis 1975, Table 2, p. 863. Reprinted by permission of the publisher.

The development of nodules was modified by a regenerative field caused by amputation. The incidence and intensity of teratogenic effects (supernumerary fins and accessory notochords) were enhanced by a regenerative field. DMBA exhibited the greatest potential for inducing lymphocyte or pleomorphic cell nodules as well as the greatest teratogenicity (Tables 9.30 and 9.31).

Table 9.30. Supernumerary fins developed after
carcinogenic hydrocarbon implantation

Compound	Animals treated	Time after implantation (days)	Number of supernumerary fins
MC	38	16	1
BP	62	26	1
BP[a]	30	27	2
DMBA	50	20	5
DMBA[a]	15	17	1
Fluorene-paraffin	44	25	0

[a]Treated tadpoles submitted to a regenerative field.

Source: Matos and DeLustig 1973, Table 1, p. 169. Reprinted by permission of the publisher.

Table 9.31. Supernumerary notochords developed after
carcinogenic hydrocarbon implantation

Compound	Animals treated	Time after implantation (days)	Number of accessory notochords
MC	38	26	0
BP	62	33	0
BP[a]	30	33	0
DMBA	50	29	2
DMBA[a]	15	27	3
Fluorene-paraffin	44	25	0

[a]Treated tadpoles submitted to a regenerative field.

Source: Matos and DeLustig 1973, Table 2, p. 169. Reprinted by permission of the publisher.

In a carcinogenicity study exposing adult planaria (*Dugesia dorotocephala*) to 3-methylcholanthrene or BaP, Foster (1969) noted that offspring developed abnormal growths similar to those of the exposed adults. The offspring of the 3-methylcholanthrene-treated worms developed lethal papilliform tumors (30%) and malformations (32%); 12% of the progeny of BaP-treated animals developed similar malformations (Table 9.32).

Table 9.32. Lethal growths and developmental anomalies in the offspring
of planaria treated with carcinogens

Treatment of parents	Number of offspring	Growths in offspring	Malformations in offspring
MC, regenerated	40	12 (papilliform tumors)	6 eyespots poorly developed 4 enlarged heads and eyes 3 fused eyespots
BP, regenerated	75	2 (nodular tumors)	3 enlarged heads and eyes 2 small heads and eyes 2 fused eyespots
Acetone, regenerated	65	1 (nodular tumor)	Variation in pigmentation 2 enlarged heads and eyes
Untreated controls, whole	77	None	1 small head and eyes 1 fused eyespot
Untreated controls, regenerated	None		
MC-treated, whole	None		

Source: Foster 1969, Table 1, p. 685. Reprinted by permission of the publisher.

9.1.11 Chronic toxicity

Very limited information is available regarding the low-level chronic or sublethal effects of
PAH. Many of the available studies have focused on the long-range effects of oil spills and
other petroleum pollution, but the extent of PAH effect is not discernible from these studies.
Although contaminants at very low concentrations may cause no direct mortality to individual
organisms, entire populations and natural communities may be adversely affected (Stegeman 1974).

Chronic effects are those that occur when exposure is either continuous or repeated so often
that the biota do not have time to recover between doses (Thomas and Rice 1975). Sublethal
modifications may affect population characteristics of each species, changing the age structure,
spatial pattern, and rates of birth, death, and dispersal. Changes in the ecological communities
may occur in an affected area with far-reaching influence on other areas (NAS 1975*a*; Blumer
1971). Various sublethal effects from petroleum-associated activities or the release of
petroleum are presented in Table 9.33.

9.1.11.1 Inhibition or paralysis

Lee, Sauerheber, and Benson (1972) demonstrated that BaP, naphthalene, and toluene may inhibit
the filtering ability of mussels, but have no lethal effect, at concentrations up to 1 mg/liter.
The authors cite Purchon's (1968) proposed mechanism for the observed sublethal effect, that is,
that particles containing toxic materials in the immediate environment of the mussel could be
detected by microscopic chemoreceptors. Impulses from these organs cause changes in the muscle
fibers of the gills, which are responsible for regulating the filtration rate. Lee, Sauerheber,
and Benson (1972) also found that tetralin was highly toxic to mussels at a concentration of
100 ppm. At lower concentrations, mussels exhibited paralysis (i.e., a loss of ability to open
and close their shell), but they recover from this response when transferred to clean seawater.

DiSalvo, Guard, and Hunter (1975) suggest that, even though adult invertebrates may be highly
tolerant of hydrocarbon pollutants, including PAH, hydrocarbon insult may affect the reproductive
fitness of the organisms. No PAH studies are currently available in this area of research.

Hyperplasia of ovicells was experimentally induced in the estuarine bryozoan (*Schizoporella
unicornis*) by placing normal colonies in close proximity to coal-tar derivatives in an estuary.
Powell, Sayce, and Tufts (1970) attribute the observed uncontrolled growth of these reproductive
structures to stimulation by several PAH known to be present in coal tar; 1,2,5,6-dibenzanthra-
cene, BaP, and methylcholanthrene. It was suggested that the PAH may have inhibited the ovarian
hormone which ordinarily regulates development of the ovicell in ovicellate bryozoans. Powell
and coworkers note the scarcity of information regarding long-term effects of sublethal doses
on populations of organisms, as well as the extent to which such responses are present but
unobserved in the environment.

9.1.11.2 Behavioral effects

Concentrations of the soluble aromatic derivatives (SAD) of crude oil lower than 0.1 to 1 ppm
may cause sublethal effects. Although the percentage of PAH in the SAD of any oil is not known,
its presence may add to the biological effects elicited by the other more soluble aromatics that
are present in larger quantities — monocyclics and naphthalenes. Moore and Dwyer (1974) have

Table 9.33. Evaluation of experiments and observations of the sublethal effects on organisms both of pollution and of other associated activities of the petroleum industry

Group	Species	Reference[a]	Type of petroleum product	Concentration	Effects and evaluation
Reproduction Fecundity Crustacea	*Pollicipes polymerus*	Straughan 1971	Crude oil, Santa Barbara blowout field study No. 2 fuel oil,		Inverse relationship between the fraction of adults brooding and the amount of oil on the adults (p 0.5); heavily and moderately oiled areas had no recruitment whereas settlement was recorded from all unoiled samples.
Mollusca	*Mytilus edulis*	Blumer et al. 1971	No. 2 fuel oil, West Falmouth spill field observations		Gonads of mussels failed to develop in affected areas
Fertilization and development Polychaeta	*Capitella capitata*	Bellan et al. 1972	Detergent ("low" toxicity polyethylene-glycol fatty acid)	0.01-10 ppm	Survival of young decreased; reduction in number of females laying eggs; most complete study of sublethal effects on reproduction; laboratory strain of this hardy species probably reflects extreme tolerance conditions
Fish	*Gadus morrhua*	Kuhnhold 1970	Iranian crude extracts (paraffin based)	Aqueous extracts from 10^4, 10^3, 10^2 ppm total oil (author estimates 10^4 yields 10 ppm soluble hydrocarbons, 1 ppm may be more likely)	Eggs: "Some cases" were sublethal but embryos and larvae did not survive; apparently, 10^2 ppm does not differ from control. Larvae: "Showed typical behavior symptoms in oil extracts: increased activity was followed by a reduction of swimming activity, which finally stopped ... which slowly deepened until the 'critical point' when no responses of the larvae were obtained even by touching or prodding;" time to "critical point" varies with age of larvae and amount of oil: 10^2 ppm not different from control

Table 9.33 (continued)

Group	Species	Reference[a]	Type of petroleum product	Concentration	Effects and evaluation
					(14 - 5.5 days for 1 - 10-day-old larvae); 10^3 ppm (8.4 - 4.5); 10^4 ppm (4.2 - 0.5); herring larvae were less, and plaice larvae more resistant than cod; chemoreceptors seemed to be blocked very quickly at the first contact with oil;" insufficient quantification; no measure of uncertainty; no chemical analysis
	Pleuronectes platessa	Wilson 1970	BP1002	0 - 10 ppm	Larvae: "1/50 of 100 C_{50} (0.04 - 0.2 ppm BP 1002) disrupt photoactic and feeding behavior (the ability of the larvae to capture prey items was significantly impaired);" recovery in uncontaminated seawater in 1 - 3 days; ranges for control in photoactic response experiments not given; feeding results not documented
Lobster	*Homarus americanus*	Wells 1972	Venezuelan crude	0.1, 1, 6, 10, and 100 (emulsions)	100 ppm lethal to all larval stages; 10 ppm; stages 1 - 3 more sensitive than stage 4. Long-term experiments with newly hatched larvae: 10 ppm: 9-day mean survival time; 6 ppm; longer time to 4th, longer than at lower concentrations; concentrations at which development was prolonged are too high to be important in the field
Growth Mollusca	*Crassostrea virginica*	Mackin and Hopkins 1961	Locally produced crude Heavy loss of crude by wild well; Port Sulfur, La.; started 1/17/56, oil not evident by 3/72	Spray	Sprayed directly on oysters weekly for 6 months; no effect on growth or survival of adult oysters or spat No effect on growth, nor any difference in growth between experiment and controls for all size oysters

Table 9.33 (continued)

Group	Species	Reference[a]	Type of petroleum product	Concentration	Effects and evaluation
Metabolism Respiration Fish	*Crassostrea virginica*	Mackin and Hopkins 1961	Bleedwater		Oysters in trays and bags at varying distances from outfall; Effect: mortality at 50 - 75 ft; reduced growth and glycogen content at 75 - 150 ft; none beyond 150 ft
	Littorina littorea	Perkins 1970	BP1002	30 ppm	Significant inhibition of growth
	Cyprinodon variegatus, Lagodon thomboides, Micropogon undulatus	Steel and Copeland 1967	Petrochemical wastes	0.2 - 2.0 ppm in addition 0.4 - 4.0 phenol	Claims respiratory inhibition at low concentration, then stimulation approaching TL 48 as general pattern; only 1 of the 3 species fit this pattern; insufficient acclimation; too high concentrations
	Lagodon rhomboides	Wohlschlag and Cameron 1967	Petrochemical wastes	50% of wastewater	Depression of respiration much more severe at near-lethal temperatures than intermediate ones; water collected from Corpus Christi turning basin then diluted to 1/2; so achievable in the field; lack of confidence intervals
Molluscs	*Mytilus edulis, Modiolus demissus*	Gilfillan 1973	Extract of midcontinent sweet crudes	1% emulsion	Differential effects of salinity and oil on filtering rate and respiration; normal salinity: increasing oil produces erratic stimulation of respiration; depression of filtering rate above 1% extract; 21 ppt salinity increases respiration and increases filtering rate up to 10% extract, then a very rapid drop-off at higher concentrations of oil. 11 ppt: shutdown of activity; no additional effect of oil

Table 9.33 (continued)

Group	Species	Reference[a]	Type of petroleum product	Concentration	Effects and evaluation
Fish	Juvenile *Onchorhynchus tshanytsha* (salmon) and *Morone saxatilis* (striped bass)	Brocksen and Bailey 1973	Benzene	5 and 10 ppm changed every 48 hr.	Respiratory rate was increased during the early (24 - 48 hr) period of exposure to both 5 and 10 ppm of benzene; after longer periods, respiration decreases back to near-control levels; when tested at daily intervals after exposure, found both fish species returned to control levels
Behavior Fish	*Ictalurus natalis*	Todd 1972			Feeding unaffected; social behavior altered after 1 - 3 days and returned to normal in about 1 week; second additions after return to normal again disrupted social behavior
	Notemigonus crysoleucas				No apparent effect on social behavior and/or feeding
Crustacea	*Homarus americanus*	Atema and Stein 1972	La Rosa crude and extracts thereof		Change in feeding times (doubling of waiting time) and behavior caused by addition of 1:100,000 parts crude oil to water; soluble fractions giving same oil/water ratio had no effect; light and electron microscopy showed no change in morphology of odor receptors; response similar over 5-day period, although hydrocarbons' characteristics did change by "weathering;" experimentally good as possible, but long-term effects and recovery not considered; very difficult problem

Table 9.33 (continued)

Group	Species	Reference[a]	Type of petroleum product	Concentration	Effects and evaluation
Gastropoda	*Nassarius obsoletus*	Jacobson and Boylan 1973	Kerosene extract	1 - 4 ppb	Significant reduction in attraction to oyster extract (0.31 ppm); significant reduction in attraction to scallop extract (3 ppm); analytically and statistically sound; experimental concentrations realistic to chronically polluted harbors and spill-affected areas
·Crustacea	*Pachygrapsus crassipes*	Kittredge 1971	Crude oil (Sisquoc sand in California) dilutions (as low as 1:100) of diethyl ether extracts		Inhibition of chemically mediated feeding and 6 pheromone responses in the presence of dilutions of crude oil extracts; response recovery in 3 - 6 days; quantification less than adequate, not statistically tested; method of extract preparation of questionable validity; concentration of hydrocarbons not known
	Uca pugnax	Krebs 1973	No. 2 fuel oil, West Falmouth spill, heavily and moderately oiled marshes		Results were that male and females exhibited breeding display colors, and males exhibited threat posture even though the breeding season had been over; mortality was heavy in these areas; interpretations based on observational information only; little quantification and adequate control areas lacking
Fish	*Onchorhynchus gorbuschka* (salmon)	Rice 1973	Prudhoe Bay crude oil		Small fry are much more tolerant of oil and avoid it at much higher concentrations than large fry; large fry: TL_m 48 - 110 ppm, avoidance at 1.6 ppm. The speculation that this avoidance could disrupt migrations may be reasonable since such an effect has been reported for zinc pollution on Atlantic salmon

Table 9.33 (continued)

Group	Species	Reference[a]	Type of petroleum product	Concentration	Effects and evaluation
Histological changes Fish	*Menidia menidia*	Gardner 1972	Texas-Louisiana crude oil	1 liter oil/40 liters seawater, then separate fractions (soluble and insoluble); exposed for 168 hr.	Various types of histological abnormalities exhibited after exposure to both the soluble and insoluble fractions; largely chemoreceptor structures studies but also ventricular myocardium; no analyses of concentrations seen by fish; technique seems good, but tissue and water content of hydrocarbons needed
Clam Fish	*Mya arenaria* *Menidia menidia*	Gardner et al. 1973	Texas-Louisiana crude oil and No. 2 fuel	600 ppm	0.95 liter crude oil mixed with 37.85 liters seawater; after settling, oil on top (water-insoluble fraction) at 600 ppm used to expose *Menidia*, also water layer (water-soluble fraction) at 126 ml/liter used on other groups of *Menidia*; some histological damage to chemoreceptors of fish but concentrations seen by fish not known; *Mya arenaria* collected and sectioned 4 months after a spill of No. 2 fuel oil showed an increased incidence of gonadal tumors, compared with those collected from control area; all results indicative of possibly good technique, but neither effect nor content of hydrocarbons was well quantified; other causes for tumors possible
Mollusca	*Crassostrea virginica*	St. Amant 1970	Dredging, etc., associated with petroleum extraction in marshes of Louisiana coast		Activities have caused increased incursion of higher-salinity water into marshes that formerly had lower salinity; decreased oyster production

Table 9.33 (continued)

Group	Species	Reference[a]	Type of petroleum product	Concentration	Effects and evaluation
Benthos Zooplankton Neuston		Mackin 1971	Brine effluents from 5 different oil fields; volume and characteristics differ widely		Over a radius of about 400 ft, the effluents from Trinity Bay and Fisher's Reef fields affects the size and diversity of benthic communities; no other effect from other distances, other fields, or the zooplankton or neuston was found. A quite complete coverage of species and conducted over a 1-year period; would be good if had hydrocarbon concentration data
Benthos		Reish 1965	Refinery effluent		Bottom of L.A. harbor that received wastes from refineries was either uninhabited or inhabited only by *Capitella capitata*; after a part of region dredged, number of species increased, then rapidly declined again; correlated increase of carbon in the sediment; region studies limited to only a mile of narrow channel near the source of the effluents; in addition, this part of the harbor received wastes from many other sources; not certain that the oil was actually responsible, nor how far and to what dilutions severe damage extended.
Various	Various	Crapp 1971	Benthic Kuwait crude emulsifier BP1002		Communities, hard substrates, Milford Haven: (a) same shores compared after 7 - 9 years, only species changing were also changed elsewhere due to temperature changes over a large area; (b) field experiments crude oil had little detectable effect, but emulsifiers reduced herbivorous molluscs with the result that inner tidal alga increased; (c) stranded weathered crude, mechanical smothering of some inner tidal animals

Table 9.33 (continued)

Group	Species	Reference[a]	Type of petroleum product	Concentration	Effects and evaluation
		Crapp 1971; Baker 1973	Refinery effluents: phenol oils, etc.		Effluents poorly described, only "oil," "phenols," etc.; grazing snails reduced in numbers, particularly small ones, nearer (27 m) vs further (1960 m) from outfall; fucoid algae more abundant at nearer stations; prior to this refinery, the shore had few algae, mostly barnacles and limpets; same situation in 1970, 1972
Invertebrates		Nicholson and Cimberg 1971	Santa Barbara crude and natural seep		Coal Oil Point had a less varied invertebrate fauna than other comparable areas along the coast of Southern California; suggested that chronic exposure to the nearby natural seep might account for the low diversity by eliminating sensitive kinds; sampling along single line transects inappropriate for assessing the variety of organisms that inhibit a locality, Onuf added 21 more species to their list of 24 in 1-1/2 hr
Mussels	*Mytilus californianus*	Kanter et al, 1971	Crude	1,000, 10,000, 100,000 ppm	Coal Oil Point mussels were more resistant than ones from other areas, suggesting that chronic exposure leads to selection for tolerant forms; alternative that inherent physiological variability between populations may account for differences in oil tolerance is not eliminated and is suggested by the 10- to 100-fold difference in tolerance of mussels from 2 nonseep area samples

Table 9.33 (continued)

Group	Species	Reference[a]	Type of petroleum product	Concentration	Effects and evaluation
Fish	Gulf of Mexico sp.	Bechtel and Copeland 1970	Petrochemical wastes	Different % Houston Ship Channel water	Claims that percent polluted is a good predictor of species diversity; unfounded because confounded with salinity; to convincingly demonstrate that oil pollution responsible for decreased diversity, compare with samples covering a similar range of salinities in an unpolluted control bay
Fish Shrimp	Gulf of Mexico sp.	Spears 1971	Oil field wastes	~18 ppm	Yields of harvestable fisheries products less in every case in the one creek receiving oil field wastes than in 5 nearby relatively undisturbed creeks: 4 - 30 times fewer shrimp. 2 - 25 times fewer blue crabs, 5 - 16 times fewer game fish, 1.4 - 11 times fewer forage fish; adverse effects may extend to the bay receiving the oil-filled waters; on average, half as many stations on state tracts yielded harvestable organisms in the bay receiving oil field wastes; study useful because it does incorporate reasonable controls; drawbacks in applying to other situations: the brine is at least as toxic as the oil, control bay is not so variable in salinity, and rate of dilution in the receiving bay is undetermined

[a]As cited in NAS 1975.

Source: NAS 1975, Table 4 - 2, pp. 76 - 82.

speculated that minute concentrations of SAD may interfere with chemical communication via pheromones in many marine organisms, causing behavioral modifications. Todd et al. (1972) and Hasler (1970), both cited by Moore and Dwyer (1974), report disruptions of feeding, reproduction, and social behavior by SAD concentrations in the 10- to 100-ppb range. The implications of disruption in established behavioral patterns, the mechanisms of disruption, and the potential contribution of PAH to these observed responses are unknown.

Seawater extracts of crude oils completely inhibited feeding and pheromone-mediated sex responses in crabs (*Pachygrapsus crassipes*) (Kittredge 1971, as cited by Sutterlin 1974). A 30-min exposure depressed normal behavior for several days. Exposure to naphthalene produced a similar response.

9.1.11.3 Increased enzyme levels

Cunner fish (*Tautogolabrus adspersus*) from PAH-contaminated sites in the marine environment were shown to have elevated levels of aryl hydrocarbon hydroxylase (AHH) activity in liver and gill tissue (Payne 1976). This enzyme system, used in metabolic decomposition of PAH by fish, was suggested as a practical biological monitor for marine petroleum pollution. Considered a sublethal response to PAH contamination, metabolic activity could thus serve as an indicator of stress. Payne and Penrose (1975) also suggest the use of AHH activity as an index to monitor long-term petroleum exposure in fish, pointing out that enzymatic detoxification systems such as AHH develop as a response to specific pollutants. Thus, organisms exposed to chronic levels of a toxicant would likely show a different response to the compound than those having no prior exposure.

9.2 NITROGEN- AND SULFUR-CONTAINING HETEROCYCLIC COMPOUNDS

Heterocyclic compounds are of environmental concern for several reasons: (1) Some of the nitrogen and sulfur in coal is present in heterocyclic groups and can be released as such (Koppenaal and Manahan 1976); (2) aromatic amines (nitrogen heterocyclics) that are known carcinogens have been identified in coal liquids from conversion processes (Hueper 1956*a*; 1956*b*); (3) heterocyclic compounds appear to be quite hazardous to biota.

Shultz et al. (1972), as cited by Tingey and Morrey (1973), identify heterocyclic compounds in low-melting coal-tar pitch as substituted pyridines, nitriles, or pyroles: methyl quinoline; dimethylphenylpyridine, naphthonitrile; acridine; benzocarbazol; azapyrene; benzpyrene; dibenzacridine; azobenzoperylene; dimethylazaindole; azacarbazole; and azabenzcarbazole. Compounds of this type are listed in Table 2.37.

9.2.1 Biotic availability

Dibenzothiophene was identified in aquatic sediments by Youngblood and Blumer (1975), but no quantitative data were given. Tingey and Morrey (1973) comment on the relative stability of heterocyclic compounds and report dibenzothiophene to be the most stable of the sulfur-containing heterocyclics. It is not known whether sediments act as sinks or only as transient storage sites for heterocyclics, but the occurrence of heterocyclics in the aquatic environment suggests availability to biota, especially when the similar structural relationships to nonsubstituted hydrocarbon analogues are considered.

Nitrogen- and sulfur-containing polycyclic compounds are more water-soluble than the corresponding hydrocarbons and may be present in effluents at greater levels than are PAH. Herbes, Southworth, and Gehrs (1976) suggest that the potential for bioaccumulation and food chain exposure may be marked; the relative hazards to biota may be equivalent to or in excess of those of nonsubstituted polycyclics.

9.2.2 Metabolism

The effects of nitrogen or sulfur substitution on the metabolism of polycyclic compounds are not known. No information is available on the extent of bioaccumulation, transformation, or elimination of these compounds in animals in their natural environment.

Ogata and Miyake (1973) report that thiophene is readily incorporated into the liver and muscle of fish (grey mullet, *Mujil japanicus*) and eels from seawater. Thiophene was shown to cause an objectionable odor and taste in meats of contaminated organisms, but no effects to the organisms themselves were discussed. It is suggested that consumption of thiophene-contaminated organisms could result in thiophene accumulation by the consuming organism. No quantitative data regarding uptake, accumulation, or distribution are given.

9.2.3 Acute toxicity

Herbes, Southworth, and Gehrs (1976) observe that the increasing molecular weight of aromatic bases is indirectly related to their toxicity. Figure 9.15 indicates this negative correlation: Tenfold decreases in LC_{50} values appear to occur with 40- to 50-unit molecular weight increases. Aniline and pyridine (primary and tertiary amines respectively) appear to be similar in toxicity; the LC_{50} values range from about 100 to 400 ppm. The lower range of LC_{50} values from the figure, representing compounds of higher molecular weight, was noted to parallel the observed concentrations of arylamines in coking and other industrial effluents. Thus, Herbes and coworkers suggest that the acute toxicities of both large and small nitrogen heterocyclic compounds may be of equal concern.

McKee and Wolf (1963) note that the toxic response of fish exposed to quinoline is paralysis of the respiratory muscles resulting in death. A summary of toxicity data of several nitrogen-containing compounds to aquatic organisms is given in Table 9.34.

There is little information on acute effects of thiophenes to animals; thus, only limited generalizations can be drawn between physical properties and effects. Jones (1964), as cited by Herbes, Southworth, and Gehrs (1976), found that thiophene is considerably more toxic (33%) to sunfish than is benzene and that thiophene and 2-methylthiophene are more toxic to mammals than are the benzene analogues. It may be assumed from these correlations that higher-molecular-weight thiophene compounds may also be more toxic than are corresponding PAH compounds. A 27-mg/liter dose of thiophene was lethal to orange-spotted sunfish in 1 hr (EPA 1970).

9.2.4 Chronic toxicity

No information is available regarding carcinogenic and mutagenic effects of nitrogen and sulfur heterocyclic compounds, potential interactions between heterocyclics and other classes of compounds, and chronic effects of trace levels of heterocyclics to aquatic organisms.

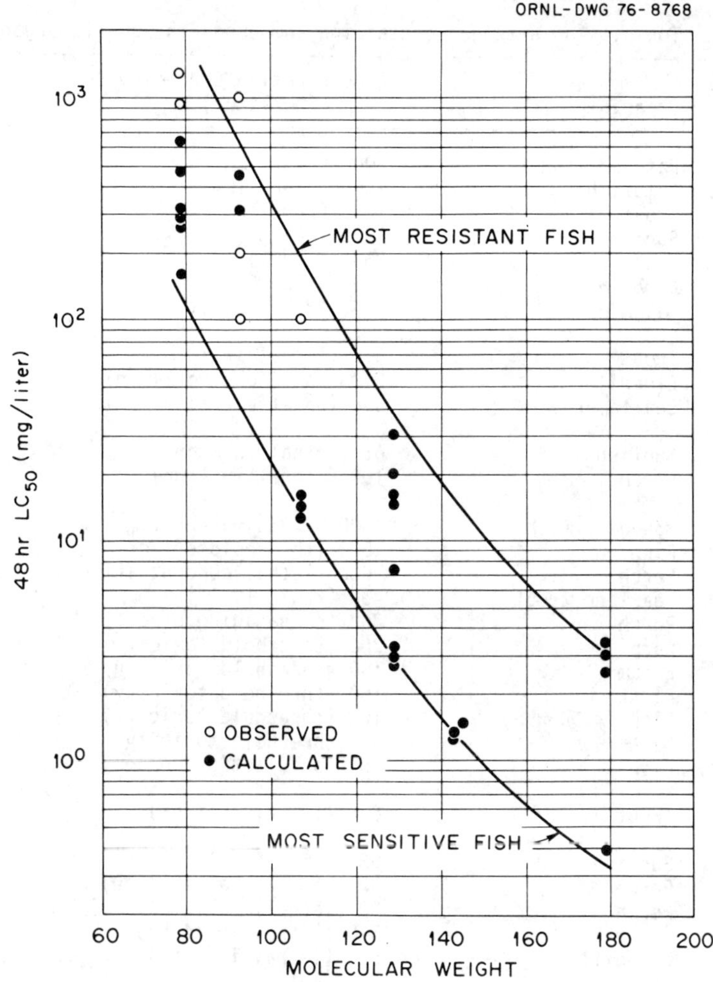

Fig. 9.15. Acute toxicities of arylamines to aquatic organisms. <u>Source</u>: Herbes, Southworth, and Gehrs 1976, Fig. 3, p. 10.

9.2.5 <u>Carcinogenicity</u>

The presence of nitrogen or sulfur heteroatoms in PAH structures has been noted to either intensify or lessen carcinogenic effects. As in the case of thiophene and its higher-ring homologues
(thionaphthene and dibenzothiophene), properties usually are similar to those of benzene, naphthalene, and anthracene, but less pronounced and more toxic. Several heterocyclic analogues of
phenanthrene have also been proven carcinogenic — acridine, carbazole, and thiophene derivatives
(Nobles and Blanton 1964).

It is generally believed that the sulfur atom is responsible for augmenting the biological
activity of thioesters. Metabolic studies of dibenzothiophene-fed rats support this suggestion.
Ambaye et al. (1961), as cited by Nobles and Blanton (1964), isolated 1-hydroxydibenzothiophene-
5,5-dioxide; the sulfur atom in thioesters, like the K-region in the corresponding hydrocarbon,
is oxidized in vivo and is apparently responsible for the carcinogenicity of these compounds.

Table 9.34. Toxicity of nitrogen-containing compounds to aquatic organisms

Compound	Organism	Concentration and response (mg/liter)	Source
Aniline	Fathead minnow	200 (96-hr TL_m)	1, 2
	Goldfish	1000 (96-hr TL_m)	1, 2
	Trout	100 (96-hr TL_m)	1, 2
	Sunfish	1020 - 1120 (lethal in 1 hr)	2
	Fish	250 (lethal)	2
	Daphnia	279 (at 25°C)	2
	Daphnia	0.4 (48-hr TL_m at 23°C)	2
Indole	Trout	5.0 (lethal in 10 hr)	2
	Bluegill	5.0 (no effect in 24 hr)	2
	Sea lamprey	5.0 (no effect in 24 hr)	2
Isoquinoline	Sunfish	65 (lethal in 1 hr)	1
	Perch	100 (lethal in 1 hr)	1
Pyridine	Mosquito fish	1300 - 1350 (96-hr TL_m)	1, 2
	Fish	1000 (threshold effect)	2
	Perch	<1000 (lethal concentrations)	2
	Yearling trout	400 (toxic limit)	2
	Perch	200 (no damaging effect)	2
	Carp	200 (threshold toxicity)	2
	Bream	180 (threshold toxicity)	2
	Bleak	160 (threshold toxicity)	2
	Fish (*Alburnus*)	100 (threshold toxicity)	2
	Daphnia	40 (threshold toxicity)	2
	Fish	15 (toxic action on nervous system)	2
Quinaldine	Trout	5.0 (lethal in 1 hr)	1, 2
Quinoline	Sunfish	52 - 56 (lethal in 1 hr)	1, 2
	Perch	30 - 50 (lethal in 1 hr)	1
	Perch	<30 (lethal)	1, 2
	Trout	5.0 (lethal in 14 hr)	1, 2
	Bluegill	5.0 (lethal in 4 hr)	1, 2
	Fish	7.5 (lethal)	2
	Trout yearlings	7.5 - 10 (lethal in 1 hr)	2
	Bleak	10 (96-hr TL_m)	2
	Bream	10 (96-hr TL_m)	2
	Carp	10 (96-hr TL_m)	2
	Daphnia	140 (at 23°C)	2
	Ciliates	750 (lethal)	2

Sources:
  1. EPA 1970.
  2. McKee and Wolf 1963.

9.3 TRACE ELEMENTS

Trace elements are defined by Baria (1975) as those elements that are present in the earth's crust at 0.1% (1000 ppm) or less. Many trace elements have been identified as natural constituents of coals (Zubovic, Stadnichenko, and Sheffey 1961; 1964; 1966; Zubovic, Sheffey, and Stadnichenko 1967; Ruch, Gluskoter, and Shimp 1974) (Chap. 2) and may be released into the environment through conversion processes (Rubin and McMichael 1975).

Additionally, metal catalysts used in coal conversion processes may ultimately reach the environment. Tingey and Morrey (1973) speculate that all the elements in the periodic table may be either present in coal or associated with conversion processes. Volatile trace metals are of major concern because they (1) may reach the aquatic environment from air (Bohn 1972;

Natusch and Wallace 1974; Natusch, Wallace, and Evans 1974; Natusch 1976) and land (Tullar and
Suffet 1975), (2) are toxic to aquatic organisms (Shaw et al. 1974), (3) may concentrate within
organisms to a level greater than that in the environment (Singer 1973), and (4) may persist
for long periods of time, gradually increasing environment concentrations (Ketchum, Zitko, and
Saward 1975).

Most of the liquid wastes from coal conversion processes find their way into the aquatic environment. Although aquatic environments appear to function as ultimate sinks for many trace elements,
chemical and biological processes may mobilize the elements, making them readily available to
organisms throughout the food chain. Furthermore, aquatic food chains concentrate certain trace
elements and pass these pollutants on to nonaquatic organisms, thus recycling the trace elements
back into nonaquatic ecosystems. Therefore, much of the data in this discussion is weighted
toward aquatic environments, emphasizing aquatic organisms.

9.3.1 Overview

A general synopsis of metabolism and potential effects of several trace elements is presented
in this section. For the most part, the elements discussed were taken from the list of trace
elements in fossil fuels identified by Comar and Nelson (1975) as being of major environmental
concern.

9.3.1.1 Arsenic

Toxicological and physiological effects of arsenic are generally related to its valence state,
organic or inorganic forms of the compounds, and individual species tolerance. Arsenic is
generally not considered an essential element of animal metabolism although recent evidence has
indicated otherwise (Anke, Grun, and Partschefeld 1976). Arsenic is commonly absorbed or ingested
into the body and is stored in liver, kidney, spleen, skin, bone, abdominal viscera, and particularly in low-level metabolic tissue such as hair and nails. Arsenic is toxic to all animals
having central nervous systems. Tolerance in some animals can be induced through gradual habitu-
ation. Trivalent arsenic compounds are generally much more toxic than the pentavalent arsenic
compounds and are carcinogenic. Elemental arsenic is considered to be nonpoisonous. The toxic
effects of arsenic generally include (1) systemic vascular poisoning, a paralytic action on
smooth and cardiac muscle, leading to hemorrhage; (2) mitotic poisoning, a blockage of chromosomal mitotic metaphase during cellular division; and (3) protoplasmic poisoning, a combination
of arsenic with certain enzymatic sulfhydryl groups that causes inhibition of various oxidative
systems requisite in tissue respiration (Bowen 1966; Smith, Ferguson, and Carson 1975).

Toxic effects of arsenic are species- and species-strain-dependent. Whereas some birds can
tolerate 4800 mg/liter lead arsenate in drinking water without harm, chickens have been noted
to suffer 58% mortality when exposed to lead arsenate concentrations in the range of 1.3 to
56.7 g per day, and no chicken survived exposure to 324 mg arsenic trioxide for 24 hr. The
lethal dose of arsenic is believed to be about 20 mg/lb body weight (Luh, Baker, and Henley
1973). The toxic dose of arsenic for various animals is given in Table 9.35.

Arsenic may be biologically concentrated through aquatic food chains; benthic algae, molluscs,
crustacea, and fish can concentrate arsenic at levels 200, 650, 400, and 700 times as great as
those in the environment respectively. Penrose, Black, and Hayward (1975) measured inorganic

Table 9.35. Toxic dose of arsenic for various animals

Animal	Toxic dose of arsenic (g)
Fowl	0.05 - 0.10
Dog	0.10 - 0.20
Swine	0.50 - 1.0
Sheep, goat, and horse	10.0 - 15.0
Cow	15.0 - 30.0

Source: Luh, Baker, and Henley 1973, Table II, p. 8.
Reprinted by permission of the publisher.

arsenic in seawater and sediments and total arsenic in sessile marine organisms to determine the effect of an abandoned stibnite mine on arsenic distributions in a small harbor in Newfoundland. The sea urchin (*Strongylocentrotus droebachiensis*) was the only animal having significantly higher concentrations near the mine. Background concentrations of arsenic from a potential lead-zinc mining area were determined by Bohn (1975). Deep-sea prawns (*Pandalus borealis*) had arsenic concentrations ranging up to 80 mg/kg dry weight, whereas planktonic copepods averaged 6.0 mg/kg dry weight.

Freshwater organisms generally contain lower levels of arsenic than do their marine counterparts (Table 9.36); concentration ratios reported for saltwater fish range from 10 to 100 times higher than those reported for freshwater fish (Woolson 1975). Lunde (1969; 1973a) determined that arsenic is present in fish as both water- and lipid-soluble arsenic-organic compounds, but that the arsenic is apparently not toxic to the organisms due to its rapid elimination by excretion (Lunde 1973b).

Table 9.36. Bioaccumulation ratio values for arsenic in aquatic organisms[a]

Fish	Arsenic in tissue (ppm)	BR values[b]
Haddock	2 - 10.8	1,000 - 5,400[c]
Kingfish	8.86	4,430
Crustacea and	1.5 - 3.1	750 - 1,550
shellfish	0.018 - 1.06	9 - 530
Assorted fish	0.076 - 2.27	38 - 1,135
Assorted fish	<1 - 6.4	<500 - 3,200[c]
Shrimp	3.6 - 48	1,836 - 64,100[c]
Mackerel	4.7 - 9.2	2,350 - 4,600[c]
Cod	24.3	12,150[c]
Assorted freshwater	0.1 - 0.2	10 - 20
fish	0.035 - 0.298	3 - 30[d]

[a]Concentration in tissue/concentration in water.
[b]A marine concentration of 2 µg arsenic/liter is used in all calculations.
[c]Dry weight basis.
[d]Water concentration assumed to be 10 µg arsenic/liter.

Source: Woolsen 1975, Table 5, p. 102. Reprinted with permission from *ACS Symposium Series No. 7*.

9.3.1.2 Boron

Boron in solution at concentrations greater than 1.5 µg/liter is toxic to plants, but its toxicity to aquatic fauna is relatively unknown; data are summarized in Table 9.37. A wide range of results is reported, the values for these fish species varying from 5 to 3100 µg/liter boron. Thurston and Russo (1976) determined LC_{50} values for fathead minnows (*Pimephales promelas*) to be 316 µg/liter for three days of exposure and 52 µg/liter for five days. At 3000 µg/liter boron, all fish died within 17 hr; at 1000 µg/liter boron, all fish died within 28 hr; at 100 µg/liter boron, one mortality occurred at 111 hr; and at 5 µg/liter boron, one mortality occurred at 115 hr.

9.3.1.3 Cadmium

Nonaquatic animals absorb cadmium, a cumulative element, through both the respiratory and digestive tracts; little cadmium penetrates the body surface. Although nutritional factors such as protein or vitamin-D deficiencies may increase cadmium absorption from the gastrointestional tract, the greatest accumulation in air-breathing animals is through inhalation; 10 to 40% of inhaled cadmium is retained (Neathery and Miller 1975). Some organisms may accumulate relatively high levels of cadmium through gastrointestinal absorption. Van Hook (1974) demonstrated that earthworms concentrate cadmium in their tissues relative to soil concentrations. Tissue levels ranged from 3.1 to 7.2 ppm (µg/g), with concentration factors (tissue Cd/Cd in top 10 cm of soil) as high as 22.5.

Blood is the transport media for absorbed cadmium, which is bound to gammaglobulin or metallothionein in erythrocytes. Cadmium rapidly disappears from blood (half-retention time of about six months) in favor of kidney or liver storage. Table 9.38 shows preferential accumulation of cadmium in the liver over other body tissues in domestic animals. Liver and kidneys together contain about 50% of the total cadmium body burden (Neathery and Miller 1975).

The protein metallothionein is involved in the transport, distribution, and excretion of cadmium. Protein binding of cadmium protects against testicular damage, but causes selective accumulation in the kidney, resulting in kidney damage or renal failure.

Chronic effects have been found to be more prevalent than acute effects. Chronic responses include pulmonary emphysema, hypertension, cardiovascular disease, interference with copper and zinc metabolism, anemia, reduced growth, testicular damage, liver and kidney damage, enlarged joints, scaly skin, and reduced milk production. Other toxicological effects have been compiled by Hise and Fulkerson (1973):

1. Amyloidosis (the accumulation in various body tissues of amyloid, an abnormal complex material, probably a glycoprotein).

2. Anemia.

3. Cancer.

4. Cirrhosis (a disease of the liver, marked by progressive destruction of liver cells) and necrosis (death of tissue) of the liver.

5. Dental changes.

6. Enteritis (inflammation of the intestine).

7. Gastritis.

Table 9.37. Boron toxicity to fishes

Fish and size	Type of bioassay	Water conditions	Toxicant used	Toxicant (mg/liter)	Boron (calculated) (mg/liter)	Results reported	Reference[a]
Mosquitofish (*Gambusia affinis*), adult females	Static	20 - 23°C, turbidity 210 - 250 ppm, pH 5.4 - 7.3	Boric acid, H_3BO_3	18,000 10,500 5,600 $\leq$1,800	3,147 1,836 979 $\leq$315	24-hr TL_m 48-hr TL_m 96-hr TL_m No mortalities in 96 hr	Wallen et al. 1957
Mosquitofish (*Gambusia affinis*), adult females	Static	22 - 26°C, turbidity 410 - 650 ppm, pH 8.6 - 9.1	Sodium tetraborate, $Na_2B_4O_7 \cdot 10\ H_2O$	12,000 8,200 3,600 1,900 $\leq$1,800	1,361 930 408 215 $\leq$204	24-hr TL_m 48-hr TL_m 96-hr TL_m 144-hr TL_m No mortalities	Wallen et al. 1957
Bluegill sunfish (*Lepomis macrochirus*), av size is 7 cm, 5 g	Static	Dilution water: 20°C, alkalinity 33.0 - 81.0 ppm, pH 6.9 - 7.5, hardness 84.0 - 163.0 ppm	Sodium tetraborate, $Na_2B_4O_7 \cdot 10\ H_2O$	15 (as B_2O_3)	4.6	24-hr TL_m	Turnbull et al. 1954
Bluegill sunfish (*Lepomis macrochirus*), av size is 7 cm, 5 g	Static	20°C, alkalinity 1750 ppm	Boron trifluoride, BF_3 (2.7% in NaOH solution)	15,000	2,392	24-hr TL_m	Turnbull et al. 1954
Rainbow trout	Static		Sodium tetraborate, $Na_2B_4O_7$	2,800	602	1-day LC_{50}	Alabaster (NAS 1973)
Rainbow trout	Static		Sodium tetraborate, $Na_2B_4O_7$	1,800	387	2-day LC_{50}	Alabaster (NAS 1973)
Rainbow trout (*Salmo gairdneri*)			Boric acid, H_3BO_3	5,000 80,000	874 13,988	Slight darkening of skin Immobilization and loss of equilibrium	Wurtz 1945

[a] As cited in Thurston and Russo 1976.

Source: Thurston and Russo 1976, Table I, p. 141. Reprinted by permission of the publisher.

Table 9.38. Cadmium-109 in tissues and total retention in goats and lactating cows after a single oral dose

Body components	Percent of dose/whole organ or section			
	Goats[a]		Lactating cows[b]	
	Average	Percent of total	Average	Percent of total
Liver	0.17	50[c]	0.24	32[c]
Kidney	0.08	23	0.08	10
Gastrointestinal tract tissue and contents	0.06	18	0.26	34
Muscle			0.07	9
Other tissues	0.03	9	0.12	15
Total body	0.34		0.75	

[a]Fourteen days after a single oral dose.
[b]Fifteen days after a single oral dose.
[c]Percent of total body retention.

Source: Neathery and Miller 1975, Table 1, p. 1769. Reprinted by permission of the publisher.

8. Hypertension.

9. Hypocalcaemia (reduction of the blood calcium below normal).

10. Life span shortening.

11. Nerve damage.

12. Ovarian changes.

13. Pancreatic atrophy.

14. Proteinuria (the presence of protein in the urine).

15. Pulmonary emphysema.

16. Renal damage.

17. Teratogenic and embryotoxic effects.

18. Testicular atrophy and lesions.

19. Toxemia of pregnancy (a series of pathologic conditions, essentially metabolic disturbances occurring during pregnancy).

20. Weight loss.

Zinc, cobalt, and selenium are antagonistic to cadmium and may act to reduce the toxic responses it evokes (Friberg, Piscator, and Nordberg 1971).

Cadmium is predominantly excreted through the feces (Doyle et al. 1972). Cadmium release from the kidney is slow, resulting in its accumulation to critical levels (200 to 400 ppm) (Anke et al. 1976). Such high levels of cadmium accumulation in the kidney causes proteinuria (the presence of protein in the urine). This condition may increase urinary cadmium by 50- to 100-fold (Hise and Fulkerson 1973). Unbound cadmium has a rapid turnover rate and relatively short biological half-life, as opposed to tissue-bound cadmium. The biological half-life for animals dosed by oral, inhalation, intraperitoneal, and intravenous routes were 206, 200, 173, and 252 days respectively. Endogenous excretion of intravenously administered radioactive cadmium was about

6% of the total dose over 14 days, most being excreted in the first 6 days (Neathery and Miller 1975).

Table 9.39 presents the data of Mathis and Cummings (as cited by Whitfield 1975), who examined cadmium levels in organisms from the Illinois River. Bottom sediments contained a mean of 2.0 ppm cadmium (range, 0.02 to 12.1 ppm), and the water contained a mean of 0.0006 ppm (range, 0.0001 to 0.002). Bottom-dwelling organisms were found to contain the highest cadmium levels; worms contained the highest concentrations, and clams contained considerably higher concentrations than fish. Ayling (1974) has found concentrations of cadmium in the Pacific oyster as high as 63 mg/kg wet weight.

Table 9.39. Cadmium levels in organisms from the Illinois River

Species	Number of samples	Cadmium level (ppm)	
		Mean	Range
Clams			
Fusonaia flava	17	0.69	0.36 - 1.17
Amblema plicata	25	0.38	0.15 - 1.41
Quadrula quadrula	20	0.56	0.31 - 1.37
Tubificid annelids	11	1.1	0.5 - 3.2
Carnivorous fish			
Esox lucius	4	0.022	0.013 - 0.031
Micropterus salmoides	7	0.022	0.004 - 0.060
Morone chrysops	2	0.024	0.004 - 0.038
Lepisoteus platostomus	2	0.030	0.004 - 0.085
Micropterus dolomieui	1	0.005	
Omnivorous fish			
Ictiobus cyprinellus	12	0.032	0.001 - 0.055
Dorosoma cepedianum	12	0.033	0.005 - 0.068
Moxostoma macrolepidotum	4	0.017	0.005 - 0.031
Carpiodes cyprins	22	0.024	0.004 - 0.046
Cyprinus carpio	14	0.035	0.011 - 0.069

Source: Whitfield 1975, Table 5.6, p. 183.

Hutcheson (1974) determined cadmium uptake by blue crabs (*Callinectes sapidus*) acclimated to a range of temperature and salinity combinations representing stressful situations in adult life. Crabs were less able to regulate cadmium influx under conditions of stress. With increasing concentrations of metal in solution, the rates of cadmium absorption by the worm *Nereis diversicolor* increased more rapidly than those of zinc absorption and were more nearly proportional to the external concentration; cadmium uptake apparently was not regulated (Bryan and Hummerstone 1973*a*). Dog whelks (*Nucella lapillus*) also exhibited more rapid accumulation of cadmium than zinc (Stenner and Nickless 1974*a*).

Kumada et al. (1972) found that a steady-state cadmium concentration was reached in 10 to 20 weeks in juvenile rainbow trout exposed to 0.0048 ppm cadmium. Maximum accumulation was up to 300 times the test water concentration. Total body content of cadmium decreased only slightly in ten weeks following transfer to cadmium-free water, suggesting stable binding. Gill levels did decrease, indicating that cadmium was absorbed mainly through the gills. Kidney levels remained high, with less decrease than other body organs (Table 9.40), indicating excretion through the kidney as the route of elimination. The rate of cadmium elimination was noted to be slow. Miettinen (1975) cites Jaakola (1972) as showing that the bulk of cadmium was rapidly eliminated from rainbow trout in a study using food labeled with $^{109}Cd^{2+}$ and counting whole-body dose of the live fish. The excretion followed a power rather than an exponential function (Fig. 9.16), showing a rather slow elimination rate; about 1% of the administered oral dose was retained in the whole body after 42 days, implying a lack of any important food chain effect. Cearley and Coleman (1974) demonstrated an equilibrium between cadmium concentrations in water and fish tissues after about a two-month exposure of largemouth bass and bluegill to sublethal concentrations of cadmium. Havre, Underdal, and Christiansen (1973) report that fish reflect the cadmium levels of their environment.

Table 9.40. Distribution of cadmium in organs (ppm) of
fish reared in cadmium solution[a]

Organ	Wet sample	Dry sample	Ash
After 30 weeks of cadmium exposure			
Muscle	0.12	0.40	5.2
Skin	0.59	1.8	18
Liver	7.3	37	350
Kidney	20	100	1100
Digestive tract			
Gill	9.5	38	240
Bone	0.85	1.1	1.2
Residue	1.6	3.1	31
10 weeks after end of cadmium exposure			
Muscle	0.04	0.14	2.2
Skin	0.32	0.84	11
Liver	4.4	19	230
Kidney	16	75	860
Gill	0.63	2.1	11
Bone	0.38	0.66	0.80
Residue	1.3	1.6	27

[a]0.0048 ppm cadmium in water.

Source: Whitfield 1975, Table 5.4, p. 180. Reprinted by permission
of the publisher.

Urinary cadmium levels have been used as an index of cadmium exposure; little cadmium is excreted by normal animals. The mechanisms of cadmium excretion in animals are not fully elucidated, and few data concerning aquatic organisms are available (Nomiyama 1973).

Fleischer et al. (1974) proposed several types of effects likely to occur from cadmium exposure: (1) direct toxicity, especially to organisms in areas of high natural cadmium levels; (2) cumulative toxicity to predatory animals that consume the kidneys (accumulating organ for cadmium) of their vertebrate prey; (3) cumulative toxicity to animals that feed regularly on molluscs (known cadmium accumulators); and (4) cumulative toxicity and adverse effects on the reproduction of fish. Table 9.41 gives acute toxicity (96-hr LC_{50}) values of cadmium to various aquatic organisms. Thorp and Lake (1974) examined the toxicity of cadmium to five freshwater invertebrate

ORNL DWG 76-12136

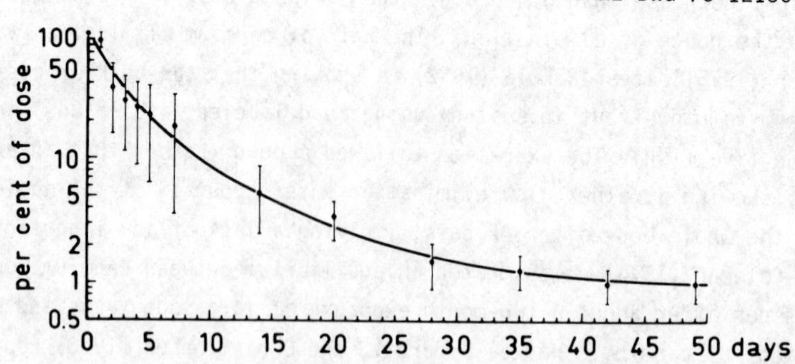

Fig. 9.16. Elimination of cadmium from rainbow trout (*Salmo gairdnerii*) determined by whole-body counting. Points indicate average retention by six fish; bars indicate one standard deviation. Source: Miettinen 1975, Fig. 13.5, p. 221. Reprinted by permission of the publisher.

Table 9.41. Acute toxic effects of cadmium

Acute dose, 96-hr LC_{50}	Species	Conditions
45 mg/liter	Orizias	$Cd(NO_3)_2 \cdot 4H_2O$
0.056 mg/liter	Guppy	$Cd(NO)_3 \cdot 4H_2O$ (Concentration as Cd)
5 ppm	*Pimephales promelas*	Static acute bioassay; hard water; $CdCl_2$
0.9 ppm	*Pimephales promelas*	Same as above, using soft water
1.05 mg/liter	*Pimephales promelas*	Static acute bioassay; soft water ($CdCl_2$ concentration as Cd)
72.6 mg/liter	*Pimephales promelas*	Same as above, using hard water
1.94 mg/liter	*Lepomis macrochirus*	Static acute bioassay; soft water ($CdCl_2$ concentration as Cd)
1.27 mg/liter	*Lebistes reticulatus*	Same as above
2.84 mg/liter	*Lepomis cyanellus*	Same as above
66.0 mg/liter	*Lepomis cyanellus*	Same as above, using hard water
0.17 ppm	*Pimephales promelas*	Cadmium cyanide complex, sodium cyanide (439 ppm CN) and cadmium sulfate (528 ppm Cd)
		Synthetic soft water; static acute bioassay (concentration as CN)
0.008 - 0.01 ppm (7 days)	*Salmo gairdneri*	Continuous flow, acute bioassay; hard water
30 mg/liter (1 day)	*Salmo gairdneri*	Same as above
30 ppm (1 day)	*Salmo gairdneri*	Continuous flow, acute bioassay
0.12 mg/liter (4 - 8 weeks)	*Crassostrea virginica*	In flowing water; 20°C, salinity 31 ppb; $(CdCl_2)2 \cdot 5H_2O$
27.0 mg/liter	*Fundulus heteroclitus*	20 to 22°C; no feeding during the 96-hr aerated water
0.2 mg/liter (8 weeks)	*Crassostrea virginica*	
0.1 mg/liter (15 weeks)	*Crassostrea virginica*	

Source: Whitfield 1975, Table 5.7, pp. 185-187.

species. Lethal concentrations (96-hr LC_{50}) varied from 0.04 mg/liter for the amphipod (*Austrochiltonia subtenuis*) to over 2000 mg/liter for a trichopteran larva of the Leptoceridae. Bioassays of the freshwater shrimp (*Paratya tasmaniensis*) indicated seasonal differences in sensitivity to cadmium. Cadmium was shown to be only moderately toxic to aquatic insects (Clubb, Gaufin, and Lords 1975*a*), but toxicity was shown to increase with dissolved oxygen (Clubb, Gaufin, and Lords 1975*b*).

9.3.1.4 Chromium

Chromium occurs in biological tissues in the trivalent (Cr^{3+}) form, where it is strongly associated with proteins, nucleic acids, and a variety of low-molecular-weight ligands. Concentrations normally range from several nanograms per gram in blood plasma to more than 1 mg/g in some liver fractions. Exposures to chromium from the environment results in accumulation of the hexavalent (Cr^{6+}) form. If soluble, Cr^{6+} will react with components of living tissue and be reduced to Cr^{3+} because of its oxidizing potential and its ability to permeate biological membranes (Taylor, Parr, and Dahlman 1975).

Routes of uptake largely consist of ingestion or inhalation. Chromium absorbed by the intestines is strongly bound to proteins. It has been demonstrated that chromium may be absorbed onto particulates of the respirable size range, and that the metal may be eluted into the lung tissue (Natusch and Wallace 1974; Natusch, Wallace, and Evans 1974; Natusch 1976). Once in the lung, chromium concentrations do not decline with age, as they do in most other organs, but tend to increase. It is believed that relatively inert forms of chromium remain in the lungs (oxides and hydroxides of Cr^{3+}), whereas the more soluble forms (chromates) penetrate into the bloodstream (NAS 1974).

Chromium is rapidly incorporated into other tissues from the blood, where it may be 10 to 100 times more concentrated than it was in the blood. Distribution in organs is dependent upon chemical form and concentration. Chromium displays a general affinity to the reticuloendothelial system, liver, spleen, and bone marrow. Chromium is excreted in the urine and feces with release of up to 88% chromium through the urinary pathway.

Trivalent chromium is poorly absorbed and thus has a low degree of toxicity. Hexavalent chromium is highly irritating and corrosive to mucous membranes, resulting in respiratory disorders such as nasal septum perforation, emphysema, and lung cancer (Smith, Ferguson, and Carson 1975).

Data from available literature regarding the toxicity of chromium to aquatic organisms were compiled by Becker and Thatcher (1973) and are given in Table 9.42. The following discussion represents their summation of chromate toxicity to aquatic organisms:

> Of the trivalent chromic salts, the chloride, nitrate and sulfate are readily soluble in water whereas the hydroxide and carbonate are quite insoluble. Of the hexavalent chromate salts, only sodium, potassium and ammonium chromates are soluble. The corresponding dichromates are quite soluble.
>
> Chromium ions occur in various forms, such as the chromous ion (Cr^{++}), the chromic ion (Cr^{+++}), the chromite ion (CrO_3^{---} or CrO_2^{-}), the chromate ion

Table 9.42. Toxicity of chromates to aquatic biota

Chemical compound	Test organism	Test conditions[a]	Concentration (ppm)	Remarks	Reference[b]
Chromium	*Cricotopus bicinctus* (midgefly)	FW, FS (river)	<25	Survived and matured; resistant species	Surber 1960
Chromium (ion)	*Daphnia magna* (cladocera)	SB, FW, LS	2.0 0.6 0.33	3-week TL_m, chronic; 18°C 50% reproduction loss in 3 weeks 16% reproduction loss in 3 weeks	Biesinger and Christensen 1972
Chromium (hexavalent)	*Lepomis macrochirus*	SB, FW, LS	110.0	96-hr TL_m, acute, soft water; aklalinity and hardness reduced toxicity	Trama and Benoit 1960
Chromium (chromate)	*Lepomis macrochirus*	SB, FW, LS	170.0		
Chromium (hexavalent)	*Salmo gairdneri*	SB, FW, LS	2.5	Tissue accumulation-elimination study, 2.5-ppm exposure level up to 24 days	Knoll and Fromm 1960
Chromium (chromate)	*Salmo gairdneri*	SB, FW, LS	5.0 10.0 - 12.5	40% kill, 15 days 80% kill, 15 days (tissue accumulation study)	Fromm and Stokes 1962
Chromium (hexavalent)	*Lepomis macrochirus*	SB, FW, LS	<50 <70	Can survive >30 days Can survive at least 7 days (hard water, pH 7.7 - 8.2)	Klassen et al. 1948
Chromium (hexavalent)	*Lepomis macrochirus*	FW	0.2	96-hr TL_m, continuous exposure	Surber 1965
Chromium (dichromate)	*Lepomis macrochirus*	SB, FW, LS	48.4 113.0 176.8	96 hr TL_m, 100% survival 96-hr TL_m, acute 96-hr, 100% kill	Cairns and Scheier 1968

Table 9.42 (continued)

Chemical compound	Test organism	Test conditions[a]	Concentration (ppm)	Remarks	Reference[b]
Chromium potassium sulfate	*Pimephales promelas*	SB, FW, LW	5.07 (S) 67.4 (H)	96-hr TL_m, acute, in hard water (H) and soft (S) water	Pickering and Henderson 1965
	Lepomis macrochirus		7.46 (S) 71.9 (H)		
	Carassius auratus		4.10 (S)		
	Lebistes reticulatus		3.33 (S)		
Chromium sulfate	*Anthocidaris* sp. (sea urchin)	SW, LS	3.2	No effects on development of eggs; 27°C	Okubo and Okubo 1962
			10.0	Effect on development of eggs; 27°C	
	Hemicentrotus sp. (sea urchin)		10.0	Effect on development of eggs; 11 - 16°C	
	Mytilus sp. (mussel)		3.2	No effect on development of embryos; 13 - 17°C	
			10.0	Effect on development of embryos; 13 - 17°C	
Potassium dichromate	*Morone saxatilus*	CB, FW, LS	150	24- and 48-hr TL_m, acute; 21.1°C, larvae	Hughes 1971
			100	72- and 96-hr TL_m, acute; larvae	
			300	24-hr TL_m, acute; 21.1°C, fingerlings	
			125	48-hr TL_m, acute; fingerlings	
			100	72-hr TL_m, acute; fingerlings	
			75	96-hr TL_m, acute; fingerlings	

Table 9.42 (continued)

Chemical compound	Test organism	Test conditions[a]	Concentration (ppm)	Remarks	Reference[b]
Potassium dichromate (Cr)	*Salmo gairdneri*	SB, FW, LS	1,000	54.6-min, mean equilibrium loss; 18°C	Grindley 1946
			500	60.6-min, mean equilibrium loss; 18°C	
			200	188-min, mean equilibrium loss; 18°C	
			20	4342-min, mean equilibrium loss; 18°C	
Potassium dichromate	*Gambusia affinis*	SB, FW, LS	370	24-hr TL_m, acute	Wallen et al. 1957
			320	48-hr TL_m, acute	
			280	96-hr TL_m, acute (turbid water, 21 - 23°C)	
Potassium dichromate	*Lepomis macrochirus*	SB, FW, LS	320	96-hr TL_m, acute (18 and 30°C), toxicity values up to 20% more in hard water	Cairns 1957
Potassium dichromate	*Pimephales promelas*	SB, FW, LS	17.6 (S) 27.3 (H)	96-hr TL_m, acute; in hard (H) and soft (S) water	Pickering and Henderson 1965
	Lepomis macrochirus		118.0 (S) 133.0 (H)		
	Carassius auratus		37.5 (S)		
	Lebistes reticulatus		30.0 (S)		
Potassium dichromate	*Lepomis macrochirus*	SB, FW, LS	320.0	96-hr TL_m, acute; both normal and low O_2 content	Cairns 1965
	Carassius carassius	SB, FW, LS	705.0	24-hr TL_m, acute; "standard reference water"	Dowden and Bennett 1965
	Lepomis macrochirus		739.0	24-hr TL_m, acute	

Table 9.42 (continued)

Chemical compound	Test organism	Test conditions[a]	Concentration (ppm)	Remarks	Reference[b]
	Daphnia magna		0.4	100-hr TL_m, acute	
Sodium dichromate	*Gambusia affinis*	SB, FW, LS	420.0	48-hr TL_m, acute; turbid water	Wallen et al. 1957
Sodium dichromate	*Daphnia magna* (cladocera)	SB, FW, LS	22.0 10.0	24-hr TL_m, acute 48-hr TL_m, acute; ("standard reference water")	Dowden and Bennett 1965
Sodium dichromate	*Daphnia magna* (cladocera)	SB, FW, LS	<0.31	Near immobilization in 48 hr; Lake Erie water, 25°C	Anderson 1946
Sodium dichromate	*Lepomis macrochirus*	SB, FW, LW	500 410 76	24-hr TL_m, acute; 20°C 48-hr TL_m, acute; 20°C "Safe concentration"	Turnbull et al. 1954
Sodium dichromate	*Salmo gairdneri*		0.013 – 0.022	Sublethal, growth inhibited; hatchery rearing	Olson and Foster 1956
Sodium dichromate	*Cyprinus carpio*	FW, FS (tank)	20	Survived but did not reproduce; 25°C	Leon 1960

[a]SB = static bioassay, CB = constant-flow bioassay, FW = freshwater, SW = sea (salt) water, LS = lab study, FS = field study.
[b]As cited in Becker and Thatcher 1973.

Source: Modified from Becker and Thatcher 1973, Table I, pp. I.3 - I.10.

(CrO_4^{--}), and as the dichromate ion ($Cr_2O_7^{--}$). In the chromic or chromite form, the chromium is trivalent whereas in the chromate and dichromate form it is hexavalent. All chromous compounds tend to be readily oxidized to the chromic state. Hexavalent chromium can be reduced to the trivalent form by heat, by organic matter, or by reducing agents.

The toxicity of chromium salts towards aquatic organisms varies widely with the species involved, as well as temperature, pH, valence of the chromium ion, and synergistic or antagonistic effects, especially due to water hardness. Fish appear to be relatively tolerant of chromium salts, while lower forms of aquatic life are more sensitive.

Under aquatic conditions, trivalent chromium often appears to be the most poisonous to fish, giving lower median tolerance limits in soft water than hexavalent chromium. Susceptibility of aquatic life to the trivalent metal may be complicated by the pH stress caused by the marked acidity of most trivalent chromium salts. In hard water, however, the solubility of trivalent chromium is decreased to an extent dependent upon the pH and mineral content, through the formation of colloidal or flocculent precipitates. Hexavalent chromium becomes somewhat more lethal to fish than trivalent chromium under these conditions.

The toxicity of hexavalent chromium towards aquatic organisms varies widely with pH, water hardness, and the individual animal. In aqueous solutions, a dichromate or chromate salt ionizes according to the following expression:

$$2CrO_4^{=} + 2H^{+} \rightleftharpoons 2HCrO_4^{-} \rightleftharpoons Cr_2O_7^{=} + H_2O \ .$$

The hydrogen ion concentration, therefore, affects the equilibrium between the chromate, hydrochromate or dichromate ions. The hydrochromate ion does not occur in significant levels at physiological pH (pH 7). The hydrochlorous ion is the most common form encountered in water at low pH (6.0-6.8) while the chromate ion predominates at higher pH (7.5-8.5). The hydrochromate is more toxic than the chromate ion, perhaps because monovalent ions tend to be more readily absorbed than divalent ions. Consequently, aqueous solutions of hexavalent chromium are more lethal to aquatic organisms when the pH of the water is below 7.0.

9.3.1.5 <u>Copper</u>

Copper is an essential element to all organisms; it is primarily a constituent of metalloenzymes concerned with oxidation and other proteins such as respiratory pigments of blood. Concentrations range from 2 to 4 ppm in nonaquatic organisms, and accumulation occurs primarily in the liver of higher organisms and in the blood of annelids and insects.

Copper toxicity is considered high for invertebrates and moderate for vertebrates. Copper can be absorbed by the lungs from airborne particulates (Natusch and Wallace 1974; Natusch, Wallace, and Evans 1974; Natusch 1976), resulting in hemochromatosis and other respiratory disturbances (Smith, Ferguson, and Carson 1975).

Ingestion of copper results in disturbances of the gastrointestinal tract. Although not a cumulative toxin, copper may cause liver and erythrocyte damage when either inhaled or ingested.

Metallic copper is insoluble in water, but many copper salts are highly soluble as cupric or cuprous ions. Copper salts occur naturally in surface waters only in trace amounts, up to about 0.05 mg/liter; higher concentrations are generally the result of anthropogenic emissions. Cupric ions introduced into natural waters at pH 7 or above will quickly precipitate as copper hydroxide or as basic copper carbonate. As illustrated by the data presented in Table 9.43,

Table 9.43. Toxicity of copper and copper salts to aquatic biota

Chemical compound	Test organism	Test conditions	Concentration (ppm)	Remarks	Reference[b]
Copper	*Nereis virens* (polychaete)	SB, SW, LS	1.5 0.5 0.1	Killed in 2 - 3 days Killed in 4 days Toxicity threshold, 21 days	Raymont and Shields 1964
	Carcinus maenas (shore crab)		1.0 - 2.0	Toxicity threshold, 11 - 12 days	
	Leander squilla (prawn)		<0.5	Toxicity threshold; copper toxicity in certain benthic organisms can be doubled by a rise of 10°C	
Copper	*Orconectes rusticus* (crayfish)	CB, FW, LS	3.0	96-hr TL_m, adults (all tests at 20°C)	Hubschman 1967
			6.0	Adults survive 24 hr, but later die even if returned to clean water	
			1.0	Total kill in 16 days, adults	
			1.0	Total kill in 6 days, juveniles	
			1.0 0.06 - 0.125	Total kill in 1 day, young Acute toxicity threshold, young	
Copper	*Gammarus pseudolimnaeus* (amphipod)	CB, FW, LS	0.020	96-hr TL_m, acute; soft water	Arthur and Leonard 1970
Copper	*Salmo gairdneri* (eggs)	SB, FW, LS	0.1	30-min exposure after fertilization did not reduce future hatching success, but increased time to hatching	Shaw and Brown 1971

Table 9.43   (continued)

Chemical compound	Test organism	Test conditions	Concentration (ppm)	Remarks	Reference[b]
Copper	*Salvelinus fontinalis*	CB, FW, LS	0.1	96-hr TL_m; 14-month-old fish	McKim and Benoit 1971
			0.1 - 0.17	Maximum acceptable toxicant dose	
			0.0034 - 0.0174	No effect on survival, growth, and reproduction of adults	
			0.0174 - 0.0325	Adverse effect on survival and growth of alevins and juveniles	
Copper (as $CuSO_4$)	*Pseudopleuronectes americanus*	CB, SW, LS	1.0 - 3.2	Histopathological effects on liver, kidney, haemopoetic tissue, and gills (fish examined just prior to death)	Baker 1969
			0.56	Histopathological effects on gills	
Copper	*Pimephales promelas*	CB, FW, LS	0.43	96-hr TL_m, chronic; hard water; sublethal concentrations affecting growth and reproduction were between 3 and 7% of the TL_m.	Mount 1968*a*
Copper	*Pimephales promelas*	FW, LS	0.75	96-hr TL_m, chronic; soft water, continuous flow test	Mount and Stephan 1969
			0.84	96-hr TL_m, chronic; soft water, static test	
Copper (Cu^{2+})	*Salmo gairdneri*	SB, FW, LS	0.4 - 0.5	48-hr TL_m, acute; lethality depends on total hardness and dissolved oxygen	Brown 1968
Copper (Cu^{2+})	*Lepomis macrochirus*	SB, FW, LS	1.25 0.63 - 0.48 1.0 - 1.97	96-hr TL_m, acute 100% survival Total kill	Cairns and Scheier 1968

Table 9.43 (continued)

Chemical compound	Test organism	Test conditions	Concentration (ppm)	Remarks	Reference[b]
Copper	*Salmo salar*	SB, FW, LS	0.032	Incipient lethal level; soft water, 17°C	Sprague and Ramsey 1965
			0.0043	Avoidance reaction	
Copper	*Daphnia magna* (cladocera)	SB, FW, LW	0.0098, 0.060	48-hr TL_m, acute; without and with food respectively	Biesinger and Christensen 1972
			0.044	3-week TL_m, chronic, with food	
			0.035	50% reproductive loss in 3 weeks (tests at 18°C, pH 7.4 - 8.2)	
Copper	*Roccus saxatilus* (larvae and juveniles)	SB, FW, LS	0.10	24-hr TL_m, acute; 21.1°C	Hughes 1969
			0.05 - 0.08	48- to 96-hr TL_m, acute; 21.1°C	
Copper	*Paracentrotus lividus* (sea urchin)	SW	0.01 - 0.02	Slowed growth of pluteus	Bougis 1965
			0.03	Growth of arms disturbed	
			0.05	Lethal	
Copper	*Acmaea scabra* (limpet)	SB, SW, LS	0.1	Lethal, 3 days	Marks 1938
	Haliotis fulgens (abalone)		0.1	100% kill, 3 days	
			0.05	Survived, 30 days	
	Ischnochiton conspicuous (chiton)		0.15	100% kill, 10 days	
			0.10	Survived 6 days	
Copper (Cu^{2+})	*Fundulus diaphanus*	SB, FW, LS	1.5	24-hr TL_m, acute (all tests at 17°C, pH 7.8 and 6.5 mg/liter dissolved oxygen	Rehwoldt et al. 1971
			0.92	48-hr TL_m, acute	
			0.86	96-hr TL_m, acute	

Table 9.43 (continued)

Chemical compound	Test organism	Test conditions	Concentration (ppm)	Remarks	Reference[b]
	Roccus saxatilus		8.3	24-hr TL_m, acute	
			6.2	48-hr TL_m, acute	
			4.3	96-hr TL_m, acute	
	Lepomis gibbosus		3.8	24-hr TL_m, acute	
			2.9	48-hr TL_m, acute	
			2.4	96-hr TL_m, acute	
	Roccus americanus		11.8	24-hr TL_m, acute	
			8.0	48-hr TL_m, acute	
			6.2	96-hr TL_m, acute	
	Anguilla rostrata		10.6	24-hr TL_m, acute	
			8.2	48-hr TL_m, acute	
			6.4	96-hr TL_m, acute	
	Cyprinus carpio		2.1	24-hr TL_m, acute	
			1.0	48-hr TL_m, acute	
			0.81	96-hr TL_m, acute	
Copper sulfate	*Micropterus salmoides* (young)	FW, FS	80.0	Median lethal dose in hard water; toxicity decreases as alkalinity increases	Warrick et al. 1943 and Warrick et al. 1948
	Flatworms		10 - 50	Killed in 24 hr	
	Isopods, damsel fly nymphs waterbeetle larvae, and caddis fly larvae			Resistant species	
	Amphipods			Sensitive species	

Table 9.43 (continued)

Chemical compound	Test organism	Test conditions	Concentration (ppm)	Remarks	Reference[b]
	Salmo trutta		25	Killed in 2.5 hr; 6 ppm alkalinity	
			2.5	Partial kill in 12.5 hr; 248 ppm alkalinity	
Copper sulfate	*Rana pipiens* (frog)	SB, FW	16.0	TL_m, chronic; 20°C (Toxic at various temperatures at doses >15 ppm)	Kaplan and Yoh 1961
Copper sulfate	*Pimephales Promelas*	SB, FW, LS	1.4	96-hr TL_m; hard water	Tarzwell and Henderson 1960
			0.05	96-hr TL_m; soft water	
	Lepomis macrochirus		10.0	96-hr TL_m; hard water	
			0.2	96-hr TL_m; soft water	
Copper sulfate	*Morone saxatilus*	CB, FW, LS	0.75	24-hr TL_m, acute; 21.1°C, larvae	Hughes 1971
			0.25	48- and 72-hr TL_m, acute; larvae	
			0.1	96-hr TL_m, acute; larvae	
			0.4	24-hr TL_m, acute; 21.1°C, fingerlings	

[a]SB = static bioassay, CB = constant-flow bioassay, FW = freshwater, SW = sea (salt) water, LS = lab study, FS = field study.
[b]As cited in Becker and Thatcher 1973.

Source: Becker and Thatcher 1973, Table M, pp. M.4 - M.21.

the toxicity of copper and copper sulfate to aquatic organisms varies considerably between individual species and also with different chemical characteristics of the water such as temperature, hardness, turbidity, and carbon dioxide content. Copper sulfate is a common salt of copper and is highly soluble in water. Copper citrate is relatively insoluble.

Pagenkopf, Russo, and Thurston (1974) examined factors affecting toxicity in waters containing copper. Equilibrium calculations on data from fish bioassays for which alkalinity, pH, hardness, and total copper concentrations had been determined indicated that Cu(II) was the chemical species most toxic to fish and that alkalinity was the factor controlling the Cu(II) concentration. Shaw and Brown (1974), however, found that the effective toxic concentrations of copper in a water body can be determined from a knowledge of the cupric-ion concentration only if it is assumed that any copper not present as Cu(II) or $CuCO_3$ has negligible toxicity. In such instances, data on alkalinity, pH, total copper, and total dissolved copper are needed to determine the concentration of dissolved Cu(II).

McLusky and Phillips (1975) studied acute toxicity and accumulation of copper in the polychaete worm *Phyllodoce maculata*. In an 80-µg/liter exposure (the lethal concentration), the rate of copper uptake was about 25 mg/kg per day.

During a 22-month exposure in soft water, adult bluegill survival was reduced, growth was retarded, and spawning was inhibited at 164 µg/liter copper (Benoit 1975). Larval fish survival was adversely affected at 40 to 162 µg/liter copper. The maximum acceptable toxicant concentration for bluegills exposed to copper in water having a hardness of 45 mg/liter (as $CaCO_3$) and a pH range of 7 to 8 was found to lie between 21 and 40 µg/liter copper, the mean 96-hr TL_{50} (threshold concentration causing 50% mortality) for larval fish to be 1100 µg/liter copper, and the application factor (mean acceptable toxicant concentration/96-hr TL_{50}) to lie between 0.002 and 0.004 µg/liter copper.

Delhaye and Cornett (1975) found that the adult *Mytilus edulis* is more susceptible to copper during its reproductive period. An important acceleration of mortality was noted during spawning, corresponding to a more rapid accumulation of metal during this period, resulting in rapid attainment of critical levels of accumulated copper. Accumulation was greatest in the gills, which was correlated to a marked decrease in respiration and oxygen consumption.

Arthur and Leonard (1970) determined the toxicity of heavy metals to three invertebrates under continuous flow conditions — an amphipod (*Gammarus* spp.) and two snails (*Campeloma* and *Physa* spp.). Responses used to measure the effects of the toxicants were survival, growth, reproduction, and feeding. The 96-hr TL_m for *Campeloma*, *Physa*, and *Gammarus* were 1.7, 0.039, and 0.020 mg/liter respectively. In extended bioassays it was determined that the total concentration of copper having no effect after a six-week exposure was between 8.0 and 14.8 ppb (µg/liter).

The gammarids reproduced in a copper concentration of 8 µg/liter or less during both six-week trials. After nine weeks of additional exposure, the newly hatched amphipods reached adult size in copper concentrations of 4.6 µg/liter or less. In their discussion, Arthur and Leonard state that the safe concentration of copper is about the same as that already existing in fresh water. This suggests that there is no safety margin for copper wastes in water supplies.

9.3.1.6 Lead

Lead is a nonessential cumulative element, which is stored primarily in the bones and kidney; however, there is some accumulation in other soft tissues. Mouw et al. (1975) determined the lead concentrations in various tissues of wild (urban and rural) rats. Data are presented in Table 9.44, indicating that lead seeks bone tissue and that the kidney functions as a secondary storage site.

Table 9.44. Tissue distribution of lead in wild urban and rural rats

| | Urban | | Rural | |
Sample source	Number of samples	Mean ± SE	Number of samples	Mean ± SE
Blood	39	0.55 ± 0.04	28	0.17 ± 0.02
Liver	21	3.34 ± 0.45	19	0.44 ± 0.09
Kidney	25	22.7 ± 2.8	20	1.14 ± 0.28
Lung	23	1.24 ± 0.20	19	0.24 ± 0.03
Brain	22	1.11 ± 0.17	19	0.21 ± 0.06
Bone	24	200.00 ± 19	18	10.3 ± 1.9
Feces	19	178.00 ± 30	19	46.6 ± 6.0

Source: Modified from Mouw et al. 1975, Table 1, p. 277. Reprinted from *Arch. Environ. Health* by permission of the publisher. Copyright 1975, American Medical Association.

Neathery and Miller (1975) point out that lead is poorly absorbed through the intestine, but that retention time is long because of the lack of effective homeostasis. Intestinal absorption of lead in sheep has been determined to be 1.3%. Inhalation of lead sorbed on particulates is another important avenue of uptake (Natusch and Wallace 1974; Natusch, Wallace, and Evans 1974; Natusch 1976). Skin absorption appears to play a minimal role in lead uptake.

Price, Rathcke, and Gentry (1974) suggest that lead is bioconcentrated from herbivore to carnivore trophic levels. It was found that insect species that suck plant juices, chew plant parts, and prey on other species of insects contained 10.3, 15.5, and 25.0 ppm lead respectively. Gish and Christensen (1973) demonstrated that earthworms accumulated up to 331.4 ppm lead, concentrations that may be lethal to earthworm-predaceous organisms; mallards were reportedly killed at dietary concentrations of 200 ppm lead.

Susceptibility to lead poisoning is affected by the type of lead compound, ruminal or intestinal acidity, animal species, and stage of lactating and/or pregnancy. As little as 6 µg lead per kilogram of body weight ingested daily for 60 days has been fatal to cattle. Young cattle have been noted to be especially susceptible to lead toxicity, partially due to its ability to cross the placental barrier (Neathery and Miller 1975).

Biochemical effects of lead include inhibition of iron utilization and heme biosynthesis, resulting in anemia; inactivation of metals from enzymes through displacement; and inhibition of enzyme synthesis.

Lead may be taken up by fish and aquatic invertebrates through the body surface, including the gills, or via food chain organisms (Adams 1975). Lead is most readily accumulated through the digestive tract in fish, as shown by the following data (Adams 1975):

Tissue	Control	1:1000 Dilution
Head	0.4	0.4
Muscles	0.2	0.3
Skin	0.4	0.6
Digestive tract	0.6	2.0
Gills	0.1	0.2

Bowen (1966) reports that accumulation is primarily in calcareous tissues and bones of aquatic organisms; these tissues were not measured by Adams.

Schulz-Baldes (1974) examined the uptake of lead by the common mussel (*Mytilus edulis*) from seawater and food. *M. edulis* accumulated a large portion of the total lead available, the percentage in the two tests being approximately equal (29% from seawater and 23.5% from food). Lead concentrations increased in all organs examined, but the rates of uptake varied markedly between organs. The highest rate of accumulation in soft tissues was in the kidney.

Distribution and bioaccumulation of lead in a river ecosystem were examined by Leland and McNurney (1974). Accumulation of lead by benthic invertebrates and fishes was shown to be a function of niche and habitat. Lead was discriminated against in food chain transfers; detrital feeders and herbivores contained higher lead concentrations than did carnivores.

Serious doubt is raised by Chow, Patterson, and Seattle (1974) about the validity of most published measurements of lead in seawater and marine organisms. Results of interlaboratory analyses of lead in seawater and tuna indicate that not one oceanographic laboratory of superior status, as of September 1973, could determine lead concentrations in seawater by atomic absorption spectrophotometry or anodic stripping voltammetry and report reliable values. Most published measurements of lead in seawater may be erroneously high by factors of 10 to 100. It was noted that similar difficulties have been experienced with measurements of lead in biological tissues. The lead concentration of tuna muscle may be four orders of magnitude lower than generally accepted levels. These findings emphasize the need for absolutely clean laboratory techniques and analytical methods employing the necessary sensitivity and accuracy. Data presented by the authors have obvious implications for investigators studying other plant and animal tissues and other trace elements of environmental concern.

9.3.1.7 Mercury

Mercury is a cumulative poison in animals; no known homeostatic mechanism for regulating concentration in tissues is known to exist. Metabolism and degree of toxicity of mercury are affected by several factors: chemical form, route of entry, length of exposure, and dietary content of interacting elements (Neathery and Miller 1975), especially selenium (Koeman et al. 1975).

Alkyl mercuries, especially methylmercury, are highly toxic to animals, presumably because of their ability to cross cell membranes. Because metal alkyls are more soluble in organic solvents than in water, they readily partition into the lipid regions of the cell (Wood 1974). This nonpolarity results in a greater solubility, allowing methylmercury to concentrate in areas where it can do the most damage.

Mercury is a potent neurotoxin, capable of causing irreversible damage to the central nervous system and resulting in death. Toxicity symptoms from different mercury sources are given in Table 9.45. An important chronic effect of mercury is egg-shell thinning in birds. Shell thickness is inversely proportional to concentrations of mercury in the diet of egg layers (Stroewsand et al. 1971).

Table 9.45. Toxicity symptoms from different mercury sources

Mercury vapor	Mercury salts	Methylmercury
Bronchitis	Anorexia	Paresthesia
Interstitial pneumonia	Weight loss	Hearing loss
Circulatory collapse	Nervous system (memory loss)	Ataxia
Renal failure	Metallic taste	Peripheral vision loss
	Gingivitis	Intellectual deterioration
	Increased saliva	Cerebral palsy-like disease in neonate

Source: Neathery and Miller 1975, Table 4, p. 1771. Reprinted by permission of the publisher.

Mercury enters the body through the respiratory system (Natusch and Wallace 1974; Natusch, Wallace, and Evans 1974; Natusch 1976), gastrointestinal tract, and skin (D'Itri 1972); 50 to 100% of inhaled mercury has been reported to be retained.

Inorganic forms of mercury are poorly absorbed through the gastrointestinal tract; absorption for several mercury species is less than 2%. In contrast, intestinal absorption of organic mercury may be as high as 95%. Ingestion of methylmercury-contaminated foods may lead to the synthesis of methylmercuric chloride in the gastrointestinal tract (Wood 1974). Thus, alkylation of mercury occurs within animals, resulting in greater absorption of mercury into the system. Alkyl mercury compounds are nonpolar and are easily transported to the blood stream; partitioning into lipids or hydrophobic regions of cells is favored. Distribution of mercury within tissues is similar after exposure to the elemental form except for the brain, where vapor readily crosses the blood-brain barrier and accumulates. The turnover rate of mercury in the brain is much slower than in most other tissues.

Mercury is accumulated in kidney and liver at greater concentrations than in other tissues, regardless of chemical form or dosing route (Table 9.46). However, relative distribution among tissues varies considerably with chemical form and route of entry. For example, the ratio of methylmercury to HgCl was 594 when given orally, as compared with only 6 when injected intravenously. Excluding the liver and kidneys, muscles accumulate more organic mercury than other body tissues. Muscles constitute the largest portion of body mass and, therefore, the largest percentage of total body organic mercury — up to 78% in lactating cows (Neathery and Miller 1975).

Feces are the predominant route of mercury excretion. The fecal-to-urinary-excretion ratio, however, may vary according to dose concentration, animal species, and chemical form. Inorganic

Table 9.46. Radioactive mercury in calves dosed orally and intravenously with
methylmercury chloride and mercuric chloride
and in cows given methylmercury chloride

	Percent of dose/kg fresh tissue						
	Calves[a]						Cows[b]
	Oral			Intravenous			
Tissues	Methylmercury chloride	Mercuric chloride	Ratio[c]	Methylmercury chloride	Mercuric chloride	Ratio[c]	Methylmercury chloride
Kidney	4.4	0.55	8	10.3	82.1	0.12	1.75
Liver	2.2	0.15	15	2.8	9.9	0.3	0.87
Brain	0.6	0.0028	214	0.6	0.15	4	0.16
Muscle	1.9	0.0032	594	1.8	0.30	6	0.50

[a]Seven days after oral and intravenous dosing.
[b]Fifteen days after a single oral methylmercury chloride dose.
[c]Ratio of methylmercury chloride to mercuric chloride.

Source: Neathery and Miller 1975, Table 6, p. 1772. Reprinted by permission of the publisher.

mercury cannot cross the placental barrier, but methylmercury may be transferred to fetuses or eggs, resulting in partial elimination of the body burden through passage into offspring.

Inorganic mercury is excreted from animals more readily than organic mercury. Organic mercury compounds such as ethyl and phenyl mercuries are biotransformed to inorganic mercury more rapidly than is methylmercury. Thus, methylmercury has a longer half-retention time (50 to 90 days) than other forms (13 to 30 days for $HgCl$).

Methylmercury is very soluble and is readily accumulated by aquatic organisms and concentrated in tissues (Schmidt-Nielsen 1974). Mercury concentration in fish muscle from relatively unpolluted areas ranges from 10 to 20 ppb. Table 9.47 indicates the mercury content of various organisms from polluted and unpolluted areas. Inorganic mercury is accumulated by organisms, but not to the extent of the organic (methylated) form; more than 90% of the mercury found in freshwater fish is in the methylated form. Total mercury values for some aquatic animals are quite high (e.g., harbor seal, total mercury at 50 ppm). This is significant since the highest level of mercury considered safe for human consumption is 0.5 ppm. Schmidt-Nielsen (1974) has also demonstrated that methylmercury is concentrated in the gills, liver, kidney, and spleen with a concentration factor between 6000 and 7000 over a period of two weeks. Mercuric chloride, although accumulated by the same organs, shows higher concentrations in the gills than in other organs and causes irreversible damage to the gill tissue.

Doi and Ui (1973) report the distribution of mercury in fish by tissues and body organs (Table 9.48). The muscle and liver appear to be the main accumulation sites. The liver and kidney have been demonstrated to be primary mercury-accumulating organs in amphibians (Table 9.49) by Byrne, Kosta, and Stegnar (1975). Accumulation levels ranged as high as 25.9 ppm fresh weight in toad livers and 24.0 ppm for toad kidneys.

Table 9.47. Total mercury vs methylmercury content in freshwater organisms as compared with marine organisms

Animal	Location	Tissue	Hg (ppm) range and average	
			Methyl	Total
Freshwater fish				
Northern pike	Clay Lake	Whole fish	7.03 - 8.71 8.00	7.03 - 8.71 8.00
Northern pike	Sweden	Muscle	1.10 - 1.88 1.36	1.10 - 1.94 1.38
Northern pike	Lake St. Clair	Muscle		0.85
Walleye pike	Lake St. Clair	Muscle		2.51
White sucker	Lake St. Clair	Muscle		1.44
Yellow perch	Lake St. Clair	Muscle		1.50
Marine animals				
Harbor seal	Boothbay Harbor	Liver	0.3 - 0.77 0.70	1.91 - 7.95 3.79
Harbor seal	Bay of Fundy	Liver	0.18 - 0.93 0.44	1.71 - 50.9 17.2
Sword fish	Open ocean	Muscle	1.17	0.48 - 2.3 1.17
Tuna fish (canned)	Open ocean	Muscle	0.43	0.34 - 0.54 0.43
Yellow fin tuna	Pacific Ocean	Muscle	0.25 - 1.00 0.48	0.24 - 1.32 0.53
Blue marlin	Pacific Ocean	Muscle	0.23 - 1.79 0.93	0.35 - 14.0 4.77
		Liver	0.16 - 1.34 0.59	0.39 - 36.0 8.03

Source: Schmidt-Nielsen 1974, Tables 1 and 2, pp. 2142 and 2143. Reprinted from *Federation Proceedings* 33: 2137-46.

Cox et al. (1975) determined CH_3Hg and total mercury concentrations in benthic invertebrates and mosquito fish from streams receiving continuous inputs of inorganic mercury; in most species, less than 1% of the total was CH_3Hg. Total mercury concentrations were highest in bottom-dwellers and carnivores. Methylmercury, although detected in all biota, was highest in mosquito fish.

Organic mercurial compounds are more toxic to aquatic organisms than are inorganic mercurials, partly because of a more rapid uptake of the organic mercurials. The uptake and retention of two mercurials by the common saltwater mullet (*Mugil cephalus*) after brief exposure (3 hr) to calculated doses in seawater were studied by Middaugh and Rose (1974). Higher total mercury concentrations were found in mullet exposed to phenylmercury than those exposed to equivalent amounts of inorganic mercury. No deaths occurred among mullets dosed with $HgCl_2$, whereas 58% of those exposed to the highest levels of $C_6H_5HgNO_3$ died within 21 hr. Mercury levels in tissues appeared to be related to the toxicant used, dose levels, and time of sampling after exposure. Peterson, Klawe, and Sharp (1973) found no significant difference in mercury contents of tuna captured 62 to 93 years ago and those caught within the last 5 years. The authors conclude that there is no evidence that mercury in tuna, swordfish, marlin, and other pelagic fishes examined can be attributed to contamination from human activities.

Table 9.48. Distribution of mercury by tissue and organ of fish

Species	No	Body length (cm)	Body weight (gm)	Total mercury concentration (ppm)				
				Muscle	Liver	Spleen	Stomach	Pyloric caeca
Sardine	1	16	70	0.04	0.07			
(*Sardinops*	2	16.5	70	*a*	0.12			
melanosticta)	3	15.5	60	*a*	0.04			
	4	17.5	77	*a*	0.09			
	5	18	83	0.12	0.12			
Saury	1	32.5	150	0.12				
(*Cololabis*	2	32.5	140	0.13				
saira)	3	33	150	0.21	0.13			
	4	33	160	0.11	0.08			
	5	33.5	144	0.09				
Grund	1	25	300	0.15	0.63			
(*Parapristipoma*	2	24	290	0.12	0.47			
trilineatum)	3	22.5	227	0.04	0.09			
	4	23	249	0.05	0.12			
	5	22.5	230	0.04	0.05			
Horse mackerel	1	18	110	0.04	0.04			
(*Trachurus*	2	18	111	0.03	0.07	0.06		
japonicus)	3	18.5	112	0.03	0.04	0.11		
	4	23.5	171	0.06	0.07			
	5	22.5	220	0.07	0.08			
Yellowtail	1		6400	0.18	0.16	0.14	0.04	0.18
(*Seriola*								
quinqueradiata)								
Common mackerel	1	31	460	0.06	0.08			
(*Scomber*	2	31	490	0.15	0.10			
japonicus)	3	35	500	0.10	0.29			
	4	33	490	0.09	0.22	0.16		
	5	32.5	486	0.09	0.19	0.13		
Skipjack tuna	1		5000	0.20	0.18	0.21	0.11	0.13
(*Katsuwonus*								
pelamis)								
Hairtail	1	60	235	0.09	0.22			
(*Trichiurus*	2	60	250	0.03	0.10			
lepturus)	3	58	255	*a*	0.04			
Black rockfish	1	19	170	0.03	0.04	0.04		
(*Sebastes*	2	19	170	0.03	0.05	0.01		
inermis)	3	19	180	0.02	0.04	0.27		
Sillago sihama	1	15.5	43	0.03				
	2	16.5	43	0.06				
	3	16	45	0.02				
	4	17	52	0.05				
	5	17.5	50	0.15				
Cuttlefish	1		410	0.04	0.04			
(*Ommastrephes*	2		350	0.03	0.13			
sloani	3		310	0.02	0.04			
pacificus)	4		397	0.03	0.09			
	5			0.07	0.03			
Shrimp	1		23	*a*				
(*Penaeus*	2		22	*a*				
orientalis)	3		17	*a*				
	4		18	0.04				
	5		20	*a*				

*a*Not detected.

Source: Doi and Ui 1973, Table 4. Reprinted by permission of the publisher.

Table 9.49. Mercury content of amphibians
(ppm, fresh weight)

Sample	Code	Liver	Kidney	Lung	Muscle	Eggs
			Background areas			
Zagorje						
Bufo bufo, male	1	0.46	0.70	0.10	0.11	
Rana dalmatina, male	2	0.42	0.29	0.078	0.25	
Reactor, Podgorica						
Bufo bufo, female	3	0.29	0.19	0.038	0.047	0.064
Ljubljana Marsh						
Bombina variegata, male	27	0.57	0.46	0.039	0.21	
Bombina variegata, female	28	0.19	0.21	0.030	0.10	0.030
Rana esculenta, male (immature)	29	0.22			0.043	
Rana esculenta, male	30	0.39			0.092	
Rana esculenta, female	30a	0.64	0.48	0.17	0.12	0.066
Rana temporaria, female	31	0.48				
Poljane						
Bufo bufo, male	19	1.51	1.24	0.17	0.17	
Bufo bufo, female	20	0.94	0.60	0.14	0.14	
Rana dalmatina, male	21	0.67	1.01	0.25	0.30	
Bombina variegata, male	22	2.07	1.42	0.11	0.37	
Bombina variegata, female	23	2.33	0.93	0.15	0.48	0.18
Rana arvalis, female	24	1.96	1.63	0.47	0.48	
Bombina variegata, male	25	1.55				
Bombina variegata, female	26	1.18				
Triturus alpestris, female	34	0.92	0.045			0.083
Triturus alpestris, male	35	1.32	0.055			0.172 (testes)
			Contaminated areas			
Podljubelj						
Salamander		9.2	0.37	0.41	0.50 (heart)	0.68 (spleen)
Idrija						
Bufo bufo, female	10	22.2	21.0	1.11	1.39	2.15
Bufo bufo, female	11	25.5	23.3	1.70	2.74	2.30
Bufo bufo, male	12	23.4	24.0	1.60	2.85	
Rana temporaria, female	13	25.9	16.2	1.54	3.44	1.25

Source: Byrne, Kosta, and Stegnar 1975, Table 2, p. 150.

Smith and Green (1975) determined mercury levels in water, sediment, and clams (family *Unionidae*) from lakes both with and without known mercury contamination. Clams concentrated mercury in the following order: methylmercuric chloride > phenylmercuric acetate > mercuric chloride. Rates of uptake increased with increased mercury in water. Temperature generally had no effect on rate of uptake or elimination.

Natural levels of trace elements in tissues of aquatic organisms are not well defined. Knowledge of these background levels are important, especially in the case of mercury because of its widespread occurrence at concentrations exceeding government standards: 1.0 mg/kg wet weight in Sweden and 0.5 mg/kg in the United States. Huckabee, Feldman, and Talmi (1974) determined total mercury concentrations in resident fishes of high-altitude streams of the Great Smoky Mountains (USA). Mean total mercury concentrations for five species ranged from 0.02 to 0.04 mg/kg. Smith and Armstrong (1975) report a mean value of 0.049 mg/kg total mercury in muscle of Arctic char (*Salvelinus alpinus*).

A mean total mercury concentration of 8.22 mg/g for American smelt (*Osmerus mordax*) from Lake Michigan is reported by Knight and Olson (1974). There were significant differences in age groups and sexes. Average total mercury concentrations were determined by Annett et al. (1975) for muscle of walleye (*Stizostedion vitreum*) and northern pike (*Esox lucius*) in a mining area of northwestern Ontario to be 3.24 and 5.55 mg/kg respectively. Scott (1974) examined total mercury concentrations in several species from a highly contaminated lake. Older and faster-growing fish were generally more contaminated; however, there was a negative correlation between mercury concentration of muscle and condition factor for fishes of the same age. Despite the small size of the lake (22 km^2), major differences (within species) between samples from different areas of the lake were found.

Olson et al. (1975) found an accumulation of mercury residues in fathead minnows (*Pimephales promelas*) exposed for 48 weeks to sublethal concentrations of methylmercury. Suzuki, Miyama, and Toyama (1973) found the highest methylmercury percentage in muscle of marine fishes to be about 15% of the total mercury. Westoo (1973) and Uthe, Atton, and Royer (1973) have found that mercury in fish muscle exists amost entirely as methylmercury and that the ratio of methylmercury to total mercury does not change with age.

Walter, Brown, and Hensley (1974) observed that mercury contents of game and nongame fishes from Lake Oahe (one of six major Missouri River storage impoundments) showed game fishes to average 0.31 mg/kg and nongame fishes to average 0.22 mg/kg. Fishes collected in an area of known mercury pollution in the Oahe Reservoir had higher average total mercury concentrations than samples taken from other reservoir locations or in the reservoir tailwaters. Twenty-four species collected from offshore regions of Lakes Erie, Michigan, and Superior were analyzed for total mercury by Thommes, Lucas, and Edgington (1972); concentrations of mercury in piscivores were higher than those in either bottom feeders or planktivores. No significant difference between lakes was noted in mercury contents of fishes from the three Great Lakes. Mercury contents of sculpins and sticklebacks ranged from 0.52 to 1.13 mg/kg and 0.96 mg/kg wet weight respectively. It was suggested that these levels result from natural geochemical sources.

Most recent studies on the distribution of mercury in fish tissues consider only the relative concentrations of total mercury or methylmercury in different organs and the persistence of these residues. A more fundamental investigation was undertaken by Reichert and Malins (1974), who demonstrated that $HgCl_2$ and CH_3HgCl readily bind with the major classes of serum lipoproteins of sockeye salmon *Onchorhynchus nerka*. Tsai, Boush, and Matsumura (1975) found inorganic mercury to be less readily translocated from water to fish in alkaline waters than in acid waters. The authors propose that the reduction in uptake rate in alkaline waters is due to unreactive forms of mercury complexes.

Luoma (1974) studied the accumulations of mercury from food and solution by a deposit-feeding shrimp (*Palaemon debilis*) and the rate of mercury loss from the animal. A mathematical description of these processes was employed. Mercury concentrations in the shrimp were never at a steady state during the study period; they varied between 200 and 5 mg/kg in response to occasional pulses of dissolved inorganic mercury entering the estuary in urban storm runoff. Neither the shrimp nor a deposit-feeding polychaete worm (*Nereis succinea*) accumulated significant quantities of mercury through ingestion of sediment.

The total mercury concentrations of chironomid larvae, sediment, and water of Sandusky Bay (Lake Erie) were examined by Skoch and Sikes (1973). No correlation was found between concentrations of mercury in sediments and in the organisms, but seasonal variation in mercury concentration of chironomids was observed.

Rasmussen and Williams (1975) surveyed mercury concentrations in intertidal organisms collected from several sites in Bellingham Bay (Washington) and found that organisms in this bay contain a higher (by an order of magnitude) concentration of mercury (0.15 vs 0.015 mg/kg wet weight) than the same species in Birch Bay, an area free of industrial pollution.

Mellinger (1973) determined the half-retention times for long-term clearances of $^{203}Hg(NO_3)_2$ and $CH_3^{203}HgCl$ from the freshwater mussel *Margaritifera margaritifera* to be 194 and 860 days respectively. Lockhart et al. (1973) confirmed that methylmercury was lost very slowly by contaminated northern pike (*Esox lucius*), the half-retention time being about two years. A year after transfer of contaminated northern pike to a clean environment, relative amounts of methylmercury in various tissues remained essentially unchanged. Organs most heavily contaminated were the kidney, liver, and eye lens. Methylmercury was shown to be taken up rapidly by blood, gills, spleen, liver, and kidney and more slowly by muscle, brain, and lens of the rainbow trout by Giblin and Massaro (1975). Skeletal muscle appeared to function as a reservoir for methylmercury and accumulated about 50% of an intragastric dose from 34 to 100 days after administration. The rate of mercury excretion from the rainbow trout appeared to be biphasic as a result of a slow elimination from skeletal muscle. A half-retention time greater than 1000 days was estimated for methylmercury loss from skeletal muscle.

9.3.1.8 Nickel

Nickel is an essential trace element at ingested concentrations of 0.3 to 0.5 mg per day. It can be very toxic, however, depending on its chemical form (Bowen 1966).

There are four routes of nickel uptake by animals: (1) oral intake; (2) inhalation; (3) percutaneous absorption; and (4) parental administration (e.g., in medications and metallic devices and prostheses). Most of the nickel that is ingested with food and water remains unabsorbed within the gastrointestinal tract and is excreted in the feces — up to 100 times more than is excreted in the urine and sweat. Absorption through the skin is of negligible quantitative significance, but may be important in the pathogenesis of nickel dermatitis. The use of nickel or nickel compounds in medications and prostheses and implantations of nickel or nickel alloys have been shown to increase nickel concentrations in parenchymal tissues and especially in tissues adjacent to implants (NAS 1975*c*).

Nickel has been shown to sorb and concentrate on particulates of the smallest respirable size, allowing deep penetration of inhaled nickel into the lungs (Natusch and Wallace 1974; Natusch, Wallace, and Evans 1974; Natusch 1976). Furthermore, most of the inhaled nickel is retained in the body; 50% is deposited in bronchial mucosa, and 25% is deposited in pulmonary parenchyma. More nickel would be expected to be retained in the body, mostly in the pulmonary parenchyma, from inhalation of gaseous nickel compounds such as nickel carbonyl. The principal route of excretion for inhaled nickel is in the urine, but it is also excreted to a minor degree in the feces. Blood serum and urine concentrations of nickel have been proposed as biologic indexes of environmental exposure to nickel compounds through inhalation and parental administration (NAS 1975*c*).

Very few data regarding distribution of nickel within the body are available. Schroeder, Balassa, and Tipton (1962), as cited by NAS (1975*c*), analyzed raw spectrographic data of nickel in tissue samples and found that most of the metal was in the small intestine or skin. Also, 65% of the samples exhibited traces of nickel in the lungs, 49% in aorta, 49% in trachea, 47% in heart, 38% in kidney, 31% in larynx, 25% in liver, and 5% in bone. Excluding the respiratory tract, the kidneys and liver appear to be the main sites of accumulation. Nickel is apparently accumulated by nucleic acids and proteins (NAS 1975*c*).

Several nickel-containing substances have been demonstrated to be carcinogenic in animals: nickel dust, nickel subsulfide, nickel oxide, nickel carbonyl, and nickel biscyclopentadiene. Effects of exposure to nickel compounds include cancers of the lung and nasal cavities, respiratory neoplasia, dermatitis, inhibition of enzyme systems, kidney damage, intestinal disorders, convulsions, and asphyxia (Bowen 1966; Smith, Ferguson, and Carson 1975).

Nickel compounds are moderately toxic to aquatic organisms; 96-hr TL_{50} values for fathead minnows range between 27 and 32 mg/liter nickel. Concentrations of 0.38 mg/liter nickel did not adversely affect survival, reproduction, or growth of the minnows, but a concentration of 0.73 mg/liter nickel caused a significant reduction in both number of eggs per spawning and the hatchability of these eggs (Pickering 1974).

9.3.1.9 <u>Selenium</u>

Selenium is an essential element to animals in low doses (0.2 mg per day), but is also highly toxic at relatively low concentrations (5 mg per day) (Diplock and Giasuddin 1976). Selenium has been shown to be carcinogenic and teratogenic, but is antagonistic to the carcinogenic and teratogenic effects of cadmium and arsenic. Selenium compounds also have a protective effect against the toxic action of mercury compounds (Koeman et al. 1975).

Nonaquatic animals have been noted to accumulate selenium to concentrations of 1.7 ppm dry weight; higher levels occur in soft tissues than in bone (Bowen 1966). Liver and kidneys are the primary accumulating organs (Blotchy et al. 1976), liver concentrations reaching 134 ppm wet weight in marine mammals and 4.6 ppm in marine birds (Koeman et al. 1975).

Selenium may enhance formation of dental caries and may affect the kidneys, liver, bone marrow, and central nervous system. "Blind staggers" and "alkali disease" of cattle have been attributed to selenium poisoning, as has "white muscle disease" in sheep. Selenium compounds are potent skin and mucous membrane irritants. Respiratory selenosis may result from inhalation of concentrations as low as 0.2 ppm. H_2Se and SeO_2 are more toxic than the sulfur analogues; LD_{50} by inhalation of 10 ppm SeO_2 results after 20 hr. Liver damage has been noted to result from selenium ingestion at 5 to 7 mg/liter of food, and liver cancer results at 10 ppm (Smith, Ferguson, and Carson 1975).

Huckabee and Blaylock (1974) found that, although selenium in the environment is bound in litter and soil, selenium may be absorbed by plant roots and slowly translocated to other portions, thus increasing selenium availability to herbivorous organisms and introducing the element into the food chain. In aquatic systems, most of the selenium is bound in sediments. Invertebrates accumulated higher levels of selenium (0.0017 to 0.05% of total dose) than did fish (0.005 to 0.009%).

Patel and Ganguly (1973) found that ^{75}Se was concentrated to levels ranging from 0.5 to 8.6 ppm in oysters and that it was accumulated in the nonlipid fraction. The biological half-retention time was determined to be about 120 days.

Cardwell et al. (1976) studied the acute toxicity of SeO_2 to six species of freshwater fish. In intermittent-flow bioassay systems, the 96-hr LC_{50} values ranged from 2.9 mg/liter SeO_2 for fathead minnow fry to 40.0 mg/liter for bluegill juveniles. The decreasing order of sensitivity was fathead minnow > flag fish > brook trout > channel catfish > goldfish > bluegill.

9.3.1.10 Silver

Bowen (1966) reports silver to be highly toxic to animals. Silver concentrations of 0.006 ppm dry weight have been identified in mammals, and 0.05 to 0.7 ppm in the shell and 0.16 to 0.8 ppm in the flesh of the freshwater mollusc *Unio mancus* have been reported. Concentrations of 3 to 11 ppm silver have been identified in aquatic organisms. Silver binds to proteins in blood plasma following absorption through the gastrointestinal tract or the lungs. Injections of colloidal silver are fatal. Silver impregnates body tissues and may reduce the immunological capacity of the body as well as cause histopathological changes in brain tissues.

Exposure of juvenile largemouth bass to silver (0.3 to 70 mg/liter) resulted in accumulation of this metal in tissues and attainment of equilibrium between concentrations in the water and tissues within two months (Coleman and Cearley 1974). Zinc concentrations of fish varied inversely with concentrations of silver in the corresponding tissues. The largemouth bass was more sensitive to silver than was the bluegill.

9.3.1.11 Thallium

Zitko (1975*b*) and Zitko, Carson, and Carson (1975) note that the pollution potential of thallium may be significant in localized areas of effluent release. Its acute toxicity to juvenile Atlantic salmon (*S. salar*) was shown to be approximately equal to that of copper and is considered to be highly toxic to all organisms.

The main storage site for thallium is the kidney, with up to 0.4 ppm thallium being accumulated by nonaquatic animals (Bowen 1966).

9.3.1.12 Vanadium

Vanadium concentrations of up to 0.15 ppm in land animals and up to 2 ppm in aquatic species have been reported. Natusch (1976) has noted vanadium sorption on respirable-size particulates. The main toxic effects in nonaquatic organisms involve the respiratory system, with damage to lung tissue resulting from concentrations as low as 0.205 µg/liter (Smith, Ferguson, and Carson 1975). Vanadium was noted to be moderately toxic to all organisms.

9.3.1.13 Zinc

Zinc is an essential element to all organisms, but is slightly toxic at relatively low concentrations (hundredths of ppm range). The main sites of accumulation in nonaquatic species are in the kidney, prostate, and eye; zinc is accumulated to up to 160 ppm.

Zinc may be sorbed on airborne particulate matter (Natusch 1976), where it may be incorporated into the lungs of air-breathing organisms (Natusch and Wallace 1974; Natusch, Wallace, and Evans 1974; Natusch 1976) as a potential cause of lung disease (Bowen 1966). Chronic effects to animals are noted by Smith, Ferguson, and Carson (1975) to include eye irritation from airborne zinc and mineral loss from bones after ingestion or accumulation through the lungs.

Zinc compounds are not particularly toxic to nonaquatic organisms unless ingested in significant quantities. Although no bioconcentration data from food sources are available, Gish and Christensen (1973) suggest that earthworms may be capable of supplying potentially lethal concentrations of zinc to predator organisms consuming the worms — birds, reptiles, amphibians, and small mammals. Earthworms have been demonstrated to accumulate up to 670 ppm zinc from soil; toxic levels in predator organisms range from 50 to 500 ppm wet weight.

Zinc has been found to accumulate in aquatic organisms to concentrations as high as 21,000 mg/kg (Ayling 1974) and to be species-dependent (Jeng and Lo 1974). Acute toxicity to aquatic life may occur within the 0.1 to 1.0 ppm range. Nehring and Goettl (1974) exposed four species of salmonids to varying dilutions of stream water containing high concentrations of zinc. The 14-day TL_m (median tolerance limit) values were as follows: rainbow trout (*Salmo gairdneri*), 0.41 mg/liter; brown trout (*S. trutta*), 0.64 mg/liter; cutthroat trout (*S. clarki*), 0.67 mg/liter; and brook trout (*S. fontinalis*), 0.96 mg/liter. Chronic and acute toxicities to aquatic biota are summarized in Table 9.50.

Bryan and Hummerstone (1973*a*) suggest that zinc uptake could be regulated by the polychaete worm *Nereis diversicolor*. Worm populations from high-zinc sediments were more resistant to zinc than worms adapted to lower environmental concentrations. The authors explain the latter adaptation by a reduced permeability to zinc and its more effective excretion by tolerant organisms.

Zinc exhibits considerable toxicity towards fish and other aquatic organisms, and it may combine with copper or nickel to produce a synergistic toxic effect (Brown and Dalton 1970). Zinc sulfate, but not zinc sulfide, is highly soluble in water. The toxicity of zinc sulfate to aquatic organisms varies according to the species and its age and the conditions and chemical characteristics of the water.

9.3.2 Potential hazard

Hildebrand, Cushman, and Carter (1976) maintain that the potential toxicity and bioaccumulation of elements released into the aquatic environment are a function of concentration, chemical form, inorganic and organic water chemistry, and food web dynamics. To identify those elements of greatest environmental concern, the dissolved element concentrations in two aqueous process streams from one experimental run of a coal conversion (COED) pilot plant were determined.

Table 9.50. Toxicity of zinc to aquatic biota

Chemical compound	Test organism	Test conditions[a]	Concentration (ppm)	Remarks	Reference[b]
Zinc (ion)	*Daphnia magna* (cladocera)	SB, FW, LS	0.1	48-hr TL_m, acute; without food	Biesinger and Christensen 1972
			0.28	48-hr TL_m, acute, with food	
			0.158	3-week TL_m, chronic, 18°C	
			0.102	50% reproduction loss in 3 weeks	
			0.07	16% reproduction loss in 3 weeks (tests at 18°C, pH pH 7.4 - 8.2)	
Zinc	*Paracentrotus lividus* (sea urchin)	SW	0.03	Growth of larvae was retarded	Bougis 1961
Zinc (nitrate)	*Balanus balanoides* (barnacle)	SB, SW, LS	30.0	90% kill of adult barnacles in 2 days	Clarke 1947
			8.0	90% kill of adult barnacles in 5 days	
Zinc ($ZnSO_4$)	*Limnaea pereger* (snail)	FW	0.2	Highest dose tolerated	Newton 1944
	Ancylastrum fluviatile (limpet)		0.2	Highest dose tolerated	
	Gammarus pulex (amphipod)		0.3	Highest dose tolerated	
	Chloeon simile (mayfly nymph)		0.2 - 0.5	Highest dose tolerated	
	Polycelis nigra (planaria)		30	Highest dose tolerated	
	Water boatman, stonefly nymph, dragonfly nymph, and caddisfly larvae		500	Highest dose tolerated	

Table 9.50. (continued)

Chemical compound	Test organism	Test conditions[a]	Concentration (ppm)	Remarks	Reference[b]
Zinc	*Roccus saxatilus*	SB, FW, LS	0.5	24-hr TL_m, acute; 21.1°C, larvae	Hughes 1969
			0.1	48- to 96-hr TL_m, acute; larvae	
			0.20	24-hr TL_m, acute; 21.1°C juveniles	
			0.10	48- to 96-hr TL_m, acute; juveniles	
Zinc sulfate (Zn)	*Anthocidaris* sp.	SW, LS	0.1	No effect on development of eggs, 27°C	Okubo and Okubo 1962
			0.32	Effect on development of eggs, 27°C	
	Crassostrea gigas (oyster)		0.1	No effect on development of eggs, 27°C	
			0.32	Effect on development of eggs, 27°C	
Zinc (Zn^{2+})	*Lepomis macrochirus*	FW	4.0 - 5.0	Lethal level in water with pH of 7.1 to 8.0 and hardness from 20 to 150 ppm	Surber 1965
			≳1.0	Safe level	
Zinc	*Lepomis macrochirus*	SB, FW, LS	20.0 - 40.0	Increase in breathing rate, death in 4 hr	Sparks et al. 1972*b*
			3.0	Significant increase in "cough" frequency, but not lethal in 24 hr	

Table 9.50. (continued)

Chemical compound	Test organism	Test conditions[a]	Concentration (ppm)	Remarks	Reference[b]
Zinc sulfate (Zn)	*Salmo gairdneri*	SB, FW, LS	2.0	800-min TL_m; soft water (12 ppm $CaCO_3$)	Lloyd 1960
			5.0	250-min TL_m; soft water (12 ppm $CaCO_3$)	
			2.0	3000-min TL_m, intermediate water (50 ppm $CaCO_3$)	
			5.0	300-min TL_m, intermediate water (50 ppm $CaCO_3$)	
			2.0	No kill in 10,000 min; hard water (320 ppm $CaCO_3$)	
			5.0	800-min TL_m; hard water (320 ppm $CaCO_3$)	
				(Above data interpolated from graph; high temperatures and low dissolved oxygen increase toxicity)	
Zinc	*Salmo salar*	CB, FW, LS	0.053	Avoidance threshold; 18.2°C, soft water	Sprague 1964*b*
Zinc sulfate	*Cyprinus carpio*	SB, FW, LS	0.05, 0.5	Improves development of embryos and viability of larvae (all tests at pH 7.8 - 8.4, 16 - 24°C);	Vladimirov 1969
			5.0	Detrimental to embryos and larvae	
Zinc	*Salmo salar* (juveniles)	FW, FS (river)	0.7 0.11	Incipient lethal concentration "Safe" level (estimated)	Sprague 1965
Zinc	*Salmo salar* (juveniles)	CB, FW, LS	0.42	Incipient lethal concentration; soft water, 17°C	Sprague and Ramsay 1965

Table 9.50. (continued)

Chemical compound	Test organism	Test conditions[a]	Concentration (ppm)	Remarks	Reference[b]
Zinc	*Fundulus heteroclitus*	SB, FW, LS	22.6	24-hr TL_m, acute (tests at 17°C, pH 7.8, and 6.5 mg/liter dissolved oxygen)	Rehwoldt et al. 1971
			20.7	48-hr TL_m, acute	
			19.1	96-hr TL_m, acute	
	Roccus saxatilus		11.2	24-hr TL_m, acute	
			10.0	48-hr TL_m, acute	
			6.7	96-hr TL_m, acute	
	Lepomis macrochirus		25.2	24-hr TL_m, acute	
			21.8	48-hr TL_m, acute	
			20.0	96-hr TL_m, acute	
	Cyprinus carpio		14.3	24-hr TL_m, acute	
			9.3	48-hr TL_m, acute	
			7.8	96-hr TL_m, acute	
Zinc	*Watersipora cucullata* (bryozoa)	SW, LS	5.0×10^{-4} moles	50% mortality, 2 hr; 19 - 24°C (molar concentrations; pH 7.6 - 8.0, all data)	Wisely and Blick 1967
	Spirorbis lamellosa (tubeworm)		7.5×10^{-5} moles	50% mortality, 2 hr; 19 - 25°C	
Zinc (chloride)	*Daphnia magna* (cladocera)	SB, FW, LS	<<0.15	Threshold concentration, immobilization in 64 hr; Lake Erie water	Anderson 1948
Zinc (chloride)	*Lepomis macrochirus*	CB, FW, LS	6.91	96-hr TL_m, acute; large fish	Cairns and Scheier 1959
			7.20	96-hr TL_m, acute; medium fish	
			7.45	96-hr TL_m, acute; small fish (all data at 19-21°C)	

Table 9.50. (continued)

Chemical compound	Test organism	Test conditions[a]	Concentration (ppm)	Remarks	Reference[b]
Zinc sulfate	*Brachydanio rerio* (Zebra fish)	SB, FW, LS	20.0	Killed in 24 hr; data are for several dose levels and O_2 uptake; toxicity to fish of different ages also considered	Skidmore 1967
Zinc sulfate (Zn)	Tubificid worms	SB, FW, LS	≳46.0	24-hr TL_m; pH 7.5, 20°C; TL_m for various pH values ranged from 5.8 - 9.7; dose more toxic at pH extremes	Whitley 1968
Zinc sulfate (Zn)	*Salmo gairdneri* (yearlings)	FW, SW, LS	20.0	No mortality, 30 hr; freshwater (all data from graphs); 24-hr TL_m; freshwater	Herbert and Wakeford 1964
	Salmo salar (smolts)		20.0 40.0	30-hr TL_m; freshwater 16-hr TL_m; freshwater	
	Salmo salar		15.0 38.0	48-hr TL_m; 18% seawater 48-hr TL_m; 36% seawater	
	Salmo gairdneri		28.0 78.0	48-hr TL_m; 19% seawater 48-hr TL_m; 36% seawater	
Zinc sulfate (Zn^{2+})	*Salmo gairdneri*	CB, FW, LS	40.0	Respiratory stress, 1 hr or more	Skidmore 1970

Table 9.50. (continued)

Chemical compound	Test organism	Test conditions[a]	Concentration (ppm)	Remarks	Reference[b]
Zinc sulfate (Zn^{2+})	*Salmo gairdneri*	CB, FW, LS	40.0	189 min, surfacing (all data pH 7.1 - 7.5, 15°C; gill damage	Skidmore and Tovell 1972
			40.0	201 min, loss of equilibrium	
			40.0	213 min, immobilization	
Zinc sulfate (Zn^{2+})	*Lepomis macrochirus*	FW, LS	5.6	No kill, 120 hr; 20°C	Morgan et al. 1971
			5.6	Some kill, 120 hr; 30°C	
			10.0	Kill increased as the rate at which thermal stress increased	
Zinc sulfate (as Zn)	*Psammechinus miliaris* (sea urchin)	SW, LS	0.16	Abnormalities in fertilization and cleavage of eggs	Cleland 1953

[a]SB = Static Bioassay, CB = Constant-flow Bioassay, FW = Freshwater, SW = Sea (Salt) Water, LS = Lab Study, FS = Field Study
[b]As cited in Becker and Thatcher 1973.

Source: Becker and Thatcher 1973, Table M, pp. M.26-M.39.

The element concentrations were used to calculate a toxicity factor (potential toxicity) to freshwater biota by dividing the concentration in the effluent by the lowest toxic concentration in the literature (Table 9.51). The toxicity factor is the highest factor anticipated on the basis of the literature; it is numerically greatest for iron.

To evaluate the threat to human health of bioaccumulation in fish used as food, the threshold bioaccumulation concentration was calculated according to the relation:

$$\text{threshold (mg/liter)} = \frac{\text{DWS (mg/liter)} \times 2 \text{ (liters water/day)}}{0.06 \text{ (kg fish/day)} \times \text{BF}} ,$$

where

    DWS = drinking water standard;
    BF = bioaccumulation factor for fish.

The calculation is based on the assumption that the only significant source of the element is 0.06 kg of fish per day in the diet. The maximum intake is established by the amount of the element ingested in drinking 2 liters of water per day. The threshold for boron at 150 mg/liter exceeds that of beryllium by one order of magnitude and that of most of the other elements by at least two orders of magnitude.

On the basis of this preliminary assessment, Hildebrand and coworkers classified the elements as having low, medium, or high potential for biological impact, depending on whether the ratio (effluent concentration/toxic concentration, or effluent concentration/threshold bioaccumulation concentration) ranged between 1 and 10, ranged between 11 and 100, or was >100 (Table 9.52).

9.3.2.1 Relative Critical Index

Ketchum (1975) proposes a quantity called a Relative Critical Index in an attempt to identify elements of present and future concern as pollutants of the marine environment. The index (Table 9.53) is obtained by dividing the annual amount of elements mobilized as a result of human activities by the concentration considered to pose minimum risk of deleterious effect. This index is numerically equal to the volume of water (in cubic kilometers) that would receive an annual increment of the element equal to the toxicity concentration shown in column D at the given rates of mobilization. Ketchum (1975) suggests that more information is needed about concentrations, distributions, speciation, and toxicity of metals before final conclusions can be reached about the hazard they present to the aquatic ecosystem.

9.3.3 Biotic availability

Apparently, metals entering the aquatic environment are fully available to biota. Metals may become insignificant at the receptor level, however, due to low emission rate, dilution in an endogenous pool in the soil or sediment, low bioaccumulation potential, or negligible toxicity.

9.3.3.1 Natural availability

Metal compounds constitute a major portion of the earth's crust and are present in all soils, rivers, oceans, and sediments (NAS 1975*b*). Many toxic metals are also essential trace elements

Table 9.51. Summary of potential toxicity and bioaccumulation for elements observed in COED aqueous effluents

Element	Highest effluent concentration (mg/liter)	Toxic concentration (mg/liter)	Bioaccumulation factor (BF) for fish	Drinking water standard (DWS) (mg/liter)	Threshold bioaccumulation concentration $\left(\frac{33\ DWS}{BF}\right)$ (mg/liter)	Effluent concentration/ toxic concentration	Effluent concentration/ threshold bioaccumulation concentration
Al	3	0.07	10	0.01	0.03	43	100
As	0.2	0.022	333	0.05	0.005	9	40
B	1.0	0.69	0.22	1.0	150	1.4	<1
Ba	0.02	5.8	4	1.0	8.25	<1	<1
Be	0.001	0.15	2	1.0	16.5	<1	<1
Bi	0.07	?	15	0.1	0.22	?	<1
Br	0.05	0.18	417	3.0	0.24	<1	<1
Cd	0.01	0.17	200	0.01	0.002	<1	5
Co	0.5	0.01	320	0.05	0.005	50	100
Cr	0.6	0.005	4000	0.05	0.0004	120	1500
Cu	5	0.8	200	0.1	0.02	6.3	250
Fe	790	0.2	100	0.3	0.10	4000	7900
Ge	0.01	?	3330	0.5	0.005	?	2
Hg	0.007	0.0002	1000	0.002	0.00007	35	100
La	0.05	0.15	25	1	1.3	<1	<1
Mg	50	17.0	50	10	6.6	2.9	8
Mn	15	0.35	660	0.05	0.003	43	5000
Mo	0.5	47.0	10	0.5	1.65	<1	<1
Ni	10	0.03	100	0.05	0.02	333	500
Pb	1	0.007	300	0.05	0.006	140	167
Rb	0.06	14.0	2000	5	0.08	<1	<1
Sb	0.007	1.3	1	0.05	1.65	<1	<1
Se	0.3	2	167	0.01	0.002	<1	150
Sn	0.1	0.35	3000	0.05	0.0006	<1	167
Ti	0.3	2.0	1000	0.1	0.003	<1	100
U	0.01	2.8	10	0.5	1.65	<1	<1
V	0.02	4.8	10	0.1	0.33	<1	<1
W	0.03	110	1200	100	2.75	<1	<1
Zn	5	0.01	8500	5	0.02	500	250
Zr	0.1	14.0	3.33	1	9.9	<1	<1

Source: Hildebrand, Cushman, and Carter 1976, Table 2, p. 10.

Table 9.52. Summary of elements detected in aqueous COED effluents that have
a potential for toxicity or bioaccumulation hazard
in aquatic systems[a]

Element	Potential toxicity	Potential bioaccumulation hazard
Al	Medium	Medium
As	Low	Medium
B	Low	
Bi	?	
Cd		Low
Co	Medium	Medium
Cr	High	High
Cu	Low	High
Fe	High	High
Ge	?	Low
Hg	Medium	Medium
Mg	Low	Low
Mn	Medium	High
Ni	High	High
Pb	High	High
Se		High
Sn		High
Ti		Medium
Zn	High	High

[a]Low potential = a calculated ratio of 1 to 10;
medium potential = a calculated ratio of 11 to 100;
high potential = a calculated ratio of >100 (see Table 9.51).

Source: Hildebrand et al. 1976, Table 3, p. 11.

necessary to sustain life. All elements are toxic at high concentrations, but some — including
the essential trace elements selenium, vanadium, copper, and zinc — are notorious poisons even
at low concentrations (Horne et al. 1972).

Frieden (1974) has compiled a list of 25 elements that are essential to living organisms (Table
9.54). The elements are classified into three categories: (1) elements that have been shown to
be essential to numerous representative animals in that group and are presumed to be required for
life and survival by virtually every member of that group; (2) elements for which a requirement
has been demonstrated for one or a few species in the group; and (3) elements for which there is
no information regarding essentiality for any member of that group. The most numerous group is
the category of essential elements, some of which are required in only trace amounts: iron,
copper, manganese, zinc, molybdenum, cobalt, vanadium, chromium, and tin. There is also growing
evidence that nickel is an essential trace element. Frieden maintains that, as new research is
undertaken, aluminum and germanium may also be found to be essential elements.

Table 9.53. Toxic elements of importance in marine pollution based on
potential supply and toxicity,
listed in order of decreasing toxicity

| Element | Rate of mobilization (10^9 g/year) | | | Toxicity D (μg/liter) | Relative Critical Index (10^{12} liters/year) | |
	Fossil fuels (man), A	River flow (natural), B	Total, C		A/D	C/D
Mercury	1.6	2.5	4.1	0.1	16,000	41,000
Cadmium	0.35	?	3.0	0.2	1,750	15,000
Silver	0.07	11	11.1	1	70	11,100
Nickel	3.7	160	164	2	1,350	82,000
Selenium	0.45	7.2	7.7	5	90	1,540
Lead	3.6	110	113.6	10	360	11,360
Copper	2.1	250	252.1	10	210	25,210
Chromium	1.5	200	201.5	10	150	20,150
Arsenic	0.7	72	72.7	10	70	7,270
Zinc	7	720	727	20	330	36,350
Manganese	7.0	250	257	20	350	12,850

Source: Ketchum 1975, Table 3.1, p. 79. Reprinted by permission of the publisher.

Table 9.54. Essentiality of the elements for living organisms[a]

Element	Bacteria	Algae	Fungi	Higher plants	Invertebrates	Vertebrates
H,C,N,O,P,S	++	++	++	++	++	++
K	++	++	++	++	++	++
Mg	++	++	++	++	++	++
Ca	+	++	+	++	++	++
Cl	++	++	-	++	++	++
Na	+	+	-	+	++	++
Fe	++	++	++	++	++	++
Cu	++	++	++	++	++	++
Zn	++	++	++	++	++	++
Mn	++	++	++	++	++	++
Mo	++	++	++	++	++	++
Co	++	+	+	+	++	++
Se	+	-	-	+	-	++
Cr	+	-	-	-	-	+
V	+	+	-	-	+	+
Sn	-	-	-	-	-	+
Ni	-	-	-	-	-	+
B	-	+	-	++	-	-
F	-	-	-	-	-	+
Si	-	+	-	+	-	+
I	-	+	-	+	+	++

[a]Key: ++, essential for several, and probably all, species in the group; +, essential
for one or a few species in the group; -, not known to be essential for any species.

Source: Frieden 1974, Table 3, p. 7. Reprinted by permission of the publisher.

Table 9.55 gives the normal, toxic, and lethal concentrations of trace elements in the diet of man and of the rat; the difference between essential and harmful levels of these metals is quite small.

9.3.3.2 Ecosystem interrelationships

Figure 9.17 illustrates the essential interrelationships that must be considered to assess the overall impact of metal pollution on the aquatic environment. The diagram (Ketchum 1975) was developed by NAS for applicability to any pollutant in the aquatic ecosystem. Entry sources into the system are outlined on the left side of the diagram, determinants of environmental quality are shown in the central portion, and exports from the system are shown on the right side. As illustrated by Fig. 9.17, a number of chemical problems are involved in natural waters, especially seawater, because they contain a complex mixture of all the elements shown. Metals added to aquatic ecosystems may form complexes, undergo chemical reactions, suffer alteration of valence states, be sorbed onto particulate matter, or be precipitated to the sediment.

9.3.3.3 Chemical interactions

Metals may be present in natural waters as dissolved species, soluble complexes, organic or inorganic particles, or dissolved species or soluble complexes adsorbed on particles (Parke et al. 1975). It is pointed out by Luoma and Jehne (1976) that very little is known about the physicochemical form of metals in the aquatic environment.

Although chemical mechanisms act to both restrict and promote solubilization of the metal, immobilization to various sinks predominates (Burrell 1974). In fact, most of the metallic elemental content of aquatic ecosystems is present in the sediments (Wolfe and Rice 1972), where it may not appear to be readily available to biota (Vaughan et al. 1975). Metals may be mobilized into interstitial water in bottom sediments, however, forming metal concentrations in solution that may be one to two orders of magnitude greater than those of the overlying water (Duinker 1975). Renfro (1973) found that polychaete worms (*Nereis diversicolor*) burrowing in the sediment caused ^{65}Zn losses from sediment to water to be three to seven times higher than ^{65}Zn losses from sediments without these organisms. Availability to benthos may thus remain high. Luoma and Jenne (1976) determined that clams (*Macoma balthica*) are able to extract silver, cadmium, cobalt, and zinc from contaminated sediments through chemical processes, thus altering the physicochemical form of these metals and accumulating them from natural sinks.

Water quality may be as significant as the concentration of the pollutant in determining the availability of metals to biota. Brown (1976) suggests that variations in water hardness, pH, dissolved oxygen, temperature, or the presence of other elements at elevated concentrations may affect the normal availability of metals to aquatic organisms.

9.3.3.4 Biological interactions

Trace metal availability to aquatic organisms is not only dependent on the influence of chemical parameters, which determine complexation forms and solubilities, but also on biological interactions. The biological activity in the water of an aquatic environment appears to be inversely proportional to trace metal content. Silvey, as cited by Andelman (1973), found that aquatic

Table 9.55. Amounts of elements in the diet of adult mammals (mg/day)[a]

Element/state	Man (*Homo sapiens*)				Rat (*Rattus norvegicus*)			
	Deficient	Normal	Toxic	Lethal	Deficient	Normal	Toxic	Lethal
Ag^+		0.06 - 0.08	60	1300				
Al^{3+}		10 - 100			0.001		200	220
As III or V		0.1 - 0.3	5 - 50	100 - 300	0.002		0.6	1.3 - 5
B(borate)		10 - 20	4000		0.0006		0.15	130 - 270
Ba^{2+} soluble		(1 - 5)	200					70 - 100
Bi^{3+}		(0.06)					1.5	
Br^-		1 - 10	3000		0.005			800
Ca^{2+}		400 - 1500			1	45 - 60		>400
Cd^{2+}		(0.6)	3				0.5	16
Cl^-	70	2400 - 4000			0.4	5 - 30		>900
Co^{2+}		0.0002	500				0.7	
Cr VI (chromate)		(0.05)	200	3000			5	
Cu^{2+}		2 - 5	250 - 500			0.05 - 0.2		20
F^-		0.5	20	2000	0.0007	0.001	0.1	30
Fe II or III		12 - 15				0.1 - 0.5		>60
Ga^{3+}		(0.02)					10	
Hg II		0.005 - 0.02		150 - 300				8
I^-	0.015	0.2	10,000			0.001 - 0.002		
In^{3+}		(0.01)						
K^+		1400 - 3700	6000				30	200 - 300
Li^+		2	200		0.3	50		>400
Mg^{2+}		220 - 400						
Mn^{2+}		3 - 9			0.1	2 - 5		
Mo VI (molybdate)		(0.7)			0.003	0.03 - 0.2		
N (organic)		8000 - 22,000			0.0005	0.0005 - 0.001	5	50
Na^+	45	1600 - 2700						
Ni^{2+}		0.3 - 0.5			0.2	5 - 50		
P (phosphate)		1200 - 2700					50	
Pb^{2+}		0.3 - 0.4		10,000		35 - 45		
Rb^+		(10)						270
S (sulfate, etc.)		420 - 3000				0.5	10	
Sb III or V		(0.1)	100					
Se IV (selenite)		(0.2)	5					11 - 75
Si (silicate)		600			0.0007		0.06	1 - 2
Sn^{2+}		17 - 45	2000					
Sr^{2+}		1.5 - 5					8	900

Table 9.55. (continued)

Ta V (tantalate)	(1)					300
Te VI (tellurate)	(0.02)	2000			0.25	1 - 9
Ti (TiO$_2$)	(1 - 10)					7.5
Tl$^+$	(0.1)	600				36
U VI (UO$_2^{2+}$)	(0.05)				0.5	1.5
V V (varadate)	(0.3)					30 - 50
W VI (tungstate)	(0.05)		0.016	0.02 - 0.04	50	150
Zn^{2+}	10 - 15					250 - 700
Zr IV	(0.1)					

[a]For comparative purposes and for order of magnitude estimates for other species of mammals, the amounts in mg per kg of body weight are more useful than the absolute amounts given above; figures given in parentheses are provisional.

Source: Bowen 1966, Table 7.4, pp. 116 and 117. Reprinted by permission of the publisher.

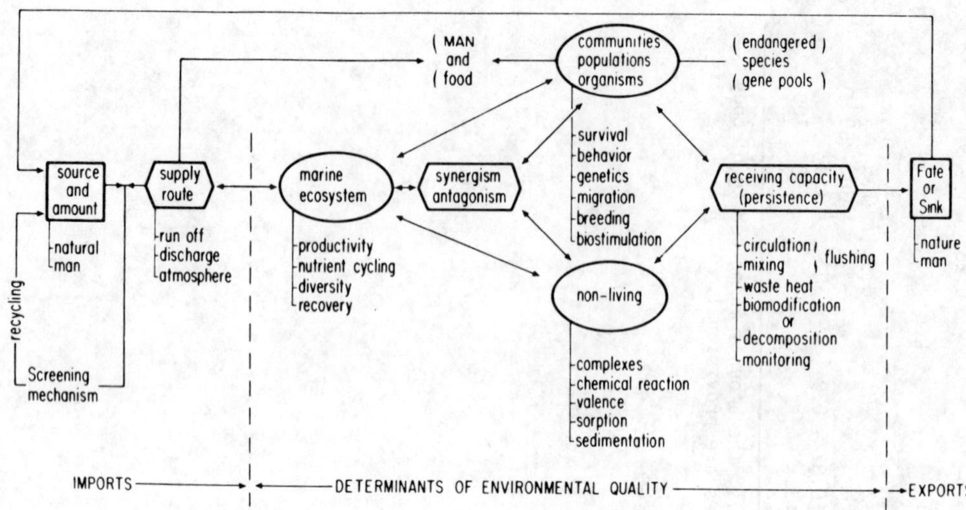

Fig. 9.17. Conceptual framework for water quality evaluation, showing the various parts of the ecosystem and processes that determine the importance of pollutants in the marine environment. <u>Source</u>: Ketchum 1975, Fig. 3.1, p. 77. Reprinted by permission of the publisher.

biota may exert the most important influence on the concentration of trace elements in aquatic ecosystems by concentrating the metals from the water environment, thus rendering them unavailable. Wolverton (1975) similarly found that water hyacinths (*Eichlornia crassipes*) were able to concentrate up to 0.67 µg cadmium and 0.50 µg nickel per gram of dry plant material when exposed for 24 hr to water polluted with 0.578 to 2.00 ppm of these metals; the water hyacinths thus function as a sink for the metals, rendering them unavailable to aquatic fauna. No considerations for possible food chain transfer of the accumulated metals were discussed in either study.

Availability of metals to aquatic organisms may depend partly on trophic relationships. Coprophagous organisms and other members of detrital food webs feeding on fecal material may potentially accumulate metals depurated through the feces of higher organisms (Frankenberg and Smith 1967, as cited by Boothe and Knauer 1972). Kelp crabs (*Pugettia producta*) have been reported to excrete arsenic, cobalt, copper, iron, lead, and zinc through their feces (Boothe and Knauer 1972), thus increasing the availability of metals to detritus feeders and decomposers.

9.3.4 <u>Uptake by invertebrates</u>

Conditions affecting uptake of metals by living organisms include the (1) availability of the metal; (2) chemical and physical properties of the metal; (3) quantity and character of ligands available for chelation, storage, and transport of the metal; and (4) nature of the organometallic complexes formed by the metal and ligands (Pringle et al. 1968). The means by which metals are taken up by aquatic biota are also summarized by Pringle and coworkers as (1) particulate ingestion of aqueous suspended matter; (2) ingestion in food material; (3) complexation by biological chelating agents; (4) incorporation into physiological systems, and (5) ion exchange and secretion on tissue or membrane surfaces.

Representative data for concentrations of metals found in biota and the environment have been compiled by Bowen (1966) (Table 6.37). All values are given in parts per million dry weight unless otherwise noted.

9.3.4.1 Rate of uptake

Environmental and biological factors affect trace element accumulation and translocation within organisms. Hutcheson (1974) found rates of cadmium uptake by the blue crab (*Callinectes sapidus*) to be greatest at low salinities and high temperatures. Physical and chemical factors affect the accumulation of elements just as they do toxicity (Sect. 9.3.7).

9.3.4.2 Effect of temperature

Uthe, Atton, and Royer (1975) suggest that mercury uptake is related to water temperature. Rainbow trout were shown to accumulate mercury more rapidly during the warm summer months than during the remainder of the year. Vernberg and O'Hara (1972) found that fiddler crabs (*Uca pagilator*) were able to transport mercury from gill tissue to the hepatopancreas more effectively at warm than at cool temperatures.

9.3.4.3 Effect of salinity and chelating agents

Concentrations of manganese were measured in a benthic annelid (*Nereis diversicolor*), interstitial water, and sediments from estuaries draining mineralized areas of southwest England by Bryan and Hummerstone (1973*b*). Manganese occurred in *N. diversicolor* in two pools, one exchanging slowly and the other more rapidly. The site of the more rapidly exchanging pool was dependent on several factors, including concentration of manganese in interstitial water and salinity. The highest concentrations of manganese were usually found in animals from the areas of lowest salinity. Huggett, Bender, and Slone (1973) found the cadmium, copper, and zinc concentrations in Chesapeake Bay oysters (*Crassostrea virginica*) to be a function of proximity to the source of pollution and the animal's position in the estuary (notably with respect to salinity).

Barica, Stainton, and Hamilton (1973) conducted parallel experiments, under both stagnant and flowing conditions, of the influence on mobilization of major cations — iron, manganese, zinc, copper, lead, and chromium. Water, sediments, and tissues of adult chironomids, crayfish, and rainbow trout were examined. Results showed that chelating agents added at concentrations of 0.2 to 5.0 mg/liter increased the concentrations of iron, manganese, lead, and zinc in water above lake sediments. Solute concentrations of copper were not affected significantly. The chelating agents had a slight effect on rates of uptake by chironomids. Metal concentrations in other organisms examined were not significantly different from controls.

9.3.4.4 Effect of seasons

Fowler and Oregioni (1976) found significant seasonal variations in metal content in mussels (*Mytilus galloprovincialis*) collected along the northwestern Mediterranean coast. Degree of seasonal variation was obtained by averaging seasonal fluctuations in metal content. Chromium showed the highest average degree of seasonal variation (by a factor of 8.8), followed by cadmium (4.5), lead (4.3), cobalt (4.0), copper (3.9), manganese (3.7), silver (3.5), nickel (3.0), iron

(2.7), and zinc (2.0). Seasonal maxima were observed in March, which was thought to be related to the reproduction state of the organisms (although not studied in detail) or, more likely, to the climatological conditions (i.e., increased runoff and subsequent metal transport resulting from the spring rainy season).

Lead concentrations of *M. galloprovincialis* from the Gulf of Trieste (Italy) were not influenced by seasonal factors, a result contrasting with observations of several macroconstituents (Favretto and Tunis 1974).

Bryan (1973) compared the concentrations of eleven trace metals in tissues of two species of scallops (*Pecten maximon* and *Chlamys opercularis*) collected from the same areas of the English Channel. Seasonal changes in the concentrations of cobalt, copper, iron, manganese, nickel, lead, and zinc were observed in both species; the changes were attributed to food supply. Distributions of zinc, copper, manganese, and lead in barnacles (*Balanus balanoides*) measured by Ireland (1974) at different times of the year at two sites in Cardigan Bay, Wales, indicated a correlation with river flow rate, tidal flow, and seasonal changes in phytoplankton densities.

9.3.4.5 Effect of bacteria

Sayler, Nelson, and Colwell (1975) found that bacteria play a significant role in mercury accumulation by the oyster (*Crassostrea virginica*). Mercury concentrations were 200 times greater in tissue fractions of oysters dosed with mercury-metabolizing bacteria (*Pseudomonas*) as compared with oysters dosed with $^{203}HgCl_2$ without the presence of the metabolizing bacteria. Both groups accumulated the metal primarily in the gills and visceral mass.

9.3.4.6 Effect of habitat

High concentrations of copper, lead, zinc, and to a lesser extent, mercury were reported for several commercially important fishes and littoral invertebrates in the Rio Tinto Estuary (Spain) by Stenner and Nickless (1975). However, animals from coastal areas near the estuary contained concentrations approaching natural levels for animals in the Atlantic Ocean.

Hirota, Fujiki, and Tajimi (1974) found mercury concentrations of plankton in the vicinity of Minamata Bay (Japan) to be high relative to offshore locations.

Navrot, Amiel, and Kronfeld (1974) determined concentrations of silver, mercury, chromium, nickel, copper, and zinc in skeletal and soft parts of the limpet (*Patella vulgata*) from various locations off the Mediterranean Coast from Tel Aviv to Dor (Israel). Metal concentrations of soft parts varied with distance from source of contamination. MacKay, Williams, Kacprzac, Kazacos, Collins, and Auty (1975) determined concentrations of copper, zinc, lead, and arsenic in oysters (*C. commercialis*) from 19 important production areas of New South Wales and found them generally low. A gradient of increasing metal concentration in oysters of a single estuary with distance upstream from the ocean was observed. Metal concentrations decreased with increasing age and weight of oysters. The shell composition of five species of barnacles from the coast of North Wales and the Irish Sea were examined by Bourget (1974). Manganese was the only trace element of those determined (manganese, silicon, lithium, barium, zinc, and aluminum) that varied significantly from one location to another. It was suggested that habitat and tidal level influence the chemical composition of balanoid shells.

9.3.4.7 Embryonic uptake

Parke et al. (1975) have stated that the rate and extent of passage of metallic compounds through the membranes of embryos of aquatic organisms are related to a variety of factors: temperature, salinity, water hardness, pH, oxygen content, presence of other metal ions, stress, previous exposure to the embryo, and particularly the age and stage of development. Boyden (1974) notes that the highest trace element concentrations in shellfish are often recorded for the smallest shellfish when values are reported on a weight-specific basis (as µg/g).

9.3.4.8 Marine plankton uptake

The uptake of heavy metals by marine plankton, which constitute the lowest trophic level in the food chain, is especially significant. Concentration factors for metal uptake from seawater by plankton are reported in Table 9.56. Ion exchange has been suggested by Leland, Shukla, and Shimp (1973) as the probable mechanism for concentration of alkali metals, and perhaps the heavier alkaline earth metals, by marine plankton. Some elements appear to be sorbed mainly by chelation since uptake increases with the ability of the element to be complexed by ligands (Goldberg, as cited by Andelman 1973). However, as indicated in Table 9.56, the order of concentration factors for divalent cations in aquatic ecosystems is not readily correlated with the order of stability of the complexes: copper > nickel > lead > cobalt > zinc > cadmium > manganese > magnesium. Bowen (1966) lists the order of concentration affinities in marine plankton as zinc > lead > copper > manganese > cobalt > nickel > cadmium. The heavier divalent metals are apparently accumulated to a greater degree than are the lighter metals (Thompson et al. 1972), presumably due to the greater polarizability of the heavier metals (Bowen 1966), although it is doubtful that any single factor can determine the observed order of concentration (Pringle et al. 1968).

9.3.4.9 Bioconcentration

All but six trace metals are considered bioaccumulative in aquatic organisms; beryllium, bismuth, molybdenum, nickel, antimony, and radium exhibit low concentration ratios (Vaughan et al. 1975). Table 6.48 compares the concentration ratios of freshwater and marine fish and shellfish. The concentration ratios are averages of numerous experimental data; however, it is not clear which variable is most appropriate for determining a concentration ratio (i.e., water free from suspended solids, water containing suspended solids, interstitial water from sediments, runoff water, diluted lake water) because metal contact is different in each variable. A ratio of 200 and above was suggested to be indicative of physiological mechanisms for conserving the metal.

The kinetics of mercury uptake in oysters (*Crassostrea virginica*) showed that mercury was accumulated 1400 and 2800 times above ambient concentrations of 100 and 10 µg/liter mercuric acetate respectively (Cunningham and Tripp 1973). Elimination of the accumulated mercury did not occur during a six-month clearance period. Kopfler (1974) continuously exposed oysters to 1 µg/liter mercury as mercuric, phenylmercuric, and methylmercuric compounds. In each instance, the oysters rapidly concentrated mercury in their tissues to levels in excess of 0.5 mg/kg wet weight.

Table 9.56. Concentration of elements from seawater by plankton and brown algae[a]

Element	State in seawater	Plankton		Brown algae	
		CF 1	CF 2	CF 1	CF 2
Ag	Anion	210		240	
Al	Particulate?	25,000		1,550	
As	Anion			2,500	200 - 6,000
Au	Anion?			270	
B	Molecule			6.6	
Ba	Cation?	120		260	
Be	Particulate?				1,500
Br	Anion			2.8	
Ca	Cation	5	10	7.2	0.5 - 10
Cd	Cation	910		890	
Cl	Anion	1		0.062	
Co	Cation?	4,600		650	450
Cr	Particulate?	17,000		6,500	300 - 10,000
Cs	Cation		1 - 5	33	1 - 100
Cu	Cation	17,000		920	100
F	Anion			0.86	
Fe	Particulate?	87,000	2,000 - 140,000	17,000	20,000 - 35,000
Ga	Particulate?	12,000		4,200	
Ge	Particulate?				15 - 200
Hg	Anion			250	
I	Anion	1,200		6,200	3,000 - 10,000
K	Cation			34	3 - 50
La	Particulate?			8,300	
Li	Cation			8?	
Mg	Cation	0.59		0.96	
Mn	Cation?	9,400	750	6,500	6,500
Mo	Anion	25		11	
N	Variable	19,000		7,500	
Na	Cation	0.14		0.78	1
Nb	Particulate?				450 - 1,000
Ni	Cation	1,700		140	500
P	Anion + organic	15,000		10,000	10,000
Pb	Cation?	41,000		70,000	
Ra	Cation	4,500	2,750	370	100
Rb	Cation			15	5 - 50
Ru	Anion?		600 - 3,000		15 - 2,000
S	Anion	1.7		3.4	10
Sb	Anion?		50		
Sc	Particulate?				1,500 - 2,600
Si	Particulate?	17,000		120	
Sn	Particulate?	2,900		92	
Sr	Cation	8	9	44	1 - 40
Th	Particulate?				10
Ti	Particulate?	20,000		3,000	
U	Anion?				10
V	Anion?	620		250	
W	Anion?			87	
Y	Particulate?				100 - 1,000
Zn	Cation	65,000	1,000	3,400	100 - 13,000
Zr	Particulate?		1,500 - 3,000		350 - 1,000

[a]CF = concentration factor = ppm in fresh organism/ppm in seawater for the element concerned; CF 1 refers to the present work, while CF 2 refers to the compilation by Mauchline and Templeton (1964) (as cited in Bowen 1966).

Source: Bowen 1966, Table 6.1, pp. 86 and 87. Reprinted by permission of the publisher.

9.3.4.10 Tissue distribution

The distribution of metals in whole bodies, shells, and various tissues of invertebrates such as
Tapes decussatus have been found by Geldiay and Uysal (1975) to vary according to several parameters: (1) the tissues and organs of the body; (2) body size of the animals; (3) habitat locale;
and (4) seasonal influence. Seasonal variations were thought to be due to changes in environmental temperature and productivity. Metal distributions in the invertebrates are given in
Table 9.57. In general, iron and zinc concentrations are higher than those of copper, manganese,
lead, cobalt, chromium, or mercury for whole body, tissues, or organs. Copper accumulated mostly
in the digestive glands, gills, and palps; manganese accumulated in gills, palps, body fluids,
and digestive glands; zinc accumulated in the body fluids, digestive glands, gills, and palps;
iron accumulated in the digestive glands, gills, and palps; lead accumulated in the body fluids,
shell, muscles, gills, and palps; cobalt accumulated in the shells and body fluids; and mercury
accumulated in the digestive glands, gills, palps, and muscles.

Accumulations of manganese, zinc, and copper in freshwater mussels (*Anodonta grandis* and *Amblema
plicata*), water, and sediments were determined by Seagle and Ehlmann (1974) for four reservoirs
in north-central Texas. The highest copper concentrations were in digestive glands; the highest
zinc and manganese values generally were found in gill tissue.

Vernberg and Vernberg (1972) found that fiddler crabs (*Uca pugilator*) are capable of rapidly
removing mercury from their surrounding aqueous media and retaining it in their tissues. Gill
tissue was the major storage site, followed by the hepatopancreas and green gland. The authors
suggest that other metals would be accumulated in similar fashion (i.e., uptake and excretion
across the gills and excess metal storage in the hepatopancreas).

9.3.4.11 Tissue binding and exchange

Gel chromatography has been used frequently to study the nature of tissue binding of trace
elements. Cadmium and zinc incorporated by the limpet (*Patella vulgata*) were shown to associate
primarily with proteins of low molecular weight (Howard and Nickless 1975). Casterline and Yip
(1975) report that cadmium in oysters (*Ostrea edulis*) is principally bound to proteins having a
molecular weight of 9,200 to 13,800. A significant amount of cadmium was also associated with
fractions having a molecular weight greater than 50,000 and less than 3,000. Coombs (1974)
separated zinc and copper in soft tissues of the oyster into soluble and cell debris-bound
components. The soluble component, which contained about 40% of the zinc and copper, was fully
exchangeable with ^{65}Zn. The soluble complexes may act as a freely available mobile reserve of
metal and ensure a constant saturation of metal-dependent enzyme in adverse environments.

The relative importance of food and water as sources of trace elements to aquatic organisms has
been a point of controversy among ecologists for many years. In a three-month model ecosystem
study, Renfro et al. (1975) found zinc pools within adult shrimp (*Lysmata seticaudata*) and crabs
(*Carcinus maenas*) to exchange very slowly, if at all, with zinc atoms available in the organism's
food or surrounding water. The authors conclude that these organisms achieve true isotopic
equilibrium only after living significant portions of their actively growing life stages in a
radioactive environment.

Table 9.57. Mean concentrations of trace metals (except mercury) in the wholy body, the tissues, the organs, and shell of *Tapes decussatus* (µg/g)[a]

Locality	Metal	Number of whole animals analyzed	Whole body (soft part)	Shell	Body fluids	Food and digestive system	Gills and palps	Mantle and siphon	Muscles	Digestive glands
Deniz Bostanli	Cu	10	6.2	2.7	2.8	5.0	7.9	2.5	2.1	30.8
	Mn	10	13.1	5.4	6.9	3.6	10.1	4.4	5.7	6.3
	Zn	10	111.4	5.5	207.0	69.8	124.5	63.9	50.1	143.0
	Fe	10	338.1	107.7	48.6	318.8	720.8	246.7	113.9	832.1
	Pb	10	20.6	15.0	18.6	6.4	10.0	5.4	13.6	9.1
	Co	10	7.6	22.8	11.9	2.8	3.0	2.2	4.2	3.0
	Cr	10								
	Hg (wet. wt.)	3	3.7	2.9	0.6	1.3	3.9	2.0	3.4	4.4
Dry weight (%)			11.7		5.0	23.6	19.6	21.5	24.5	23.5
Cakalburnu fishery	Cu	10	7.2	3.4	2.4	4.1	11.5	6.6	1.9	16.3
	Mn	10	7.3	7.5	4.0	3.1	8.5	8.6	6.0	4.7
	Zn	10	106.0	16.7	137.3	57.3	108.9	59.3	48.9	143.3
	Fe	10	267.2	27.0	46.8	209.2	607.2	178.8	79.5	582.2
	Pb	10	7.8	30.7	12.1	4.9	13.4	5.3	5.5	13.0
	Co	10	7.3	7.3	10.1	3.6	4.8	1.8	3.5	2.1
	Cr	10								
	Hg (wet. wt.)	3	3.0							
Dry weight (%)			11.7		5.5	21.8	18.1	20.8	23.0	22.3
Tuzia	Cu	10	6.3	2.5	2.7	4.4	7.3	5.6	2.4	34.3
	Mn	10	12.2	10.6	5.9	6.4	12.0	9.0	8.5	11.3
	Zn	10	211.4	5.3	389.0	85.8	133.6	90.9	75.1	159.5
	Fe	10	353.9	165.3	168.1	354.3	679.8	316.1	148.4	536.2
	Pb	10	20.7	29.0	8.5	8.7	10.2	7.8	9.9	10.2
	Co	10	10.2	12.8	7.8	2.6	4.1	2.9	4.1	5.8
	Cr	10								
	Hg (wet. wt.)	3	20.1							
Dry weight (%)			10.6		4.8	21.8	18.7	20.5	22.9	23.3

[a]Measured in terms of µg/g on a dry-weight basis. Concentrations of mercury are given on a wet-weight basis; samples consisting of one whole animal, the tissues and the organs, pooled from ten animals, were collected at different times of the year.

Source: Geldiay and Uysal 1975, Table 1, p. 72. Reprinted by permission of the publisher.

9.3.4.12 Half-retention time

Harney (1974) found the half-retention time for ^{65}Zn in the freshwater mollusc (*Anodonta* sp.) to be 103 days when the organisms were transferred to a noncontaminated environment; a half-retention time of 135 days was reported for clams left in the Columbia River (Washington). The half-retention time for inorganic mercury in the Dungeness crab (*Cancer magister*) was determined experimentally by Sloan, Thompson, and Larkin (1974) to be about 25 days. Benayoun, Fowler, and Oregioni (1974) examined the flux of cadmium through the euphausiid zooplankter, *Meganyctiphanes norvegica*. Exchange of cadmium between water and euphausiid tissue was shown to be a relatively slow process, suggesting that cadmium ingestion was probably the more important route of accumulation. Comparisons of cadmium concentrations in natural euphausiid food and the organism's fecal pellets indicated that the cadmium concentration of ingested material increased nearly fivefold during passage through the zooplankton.

Seasonal dynamics of manganese, iron, zinc, copper, and cadmium were investigated in genetically similar populations of American oysters (*Crassostrea virginica*) by Frazier (1975). Manganese and iron concentrations in soft tissues of the oyster correlated significantly with shell growth. It was suggested that zinc, copper, and cadmium concentrations were not related to shell growth, but that they may be related to gonadal development and spawning. There was a rapid loss of zinc, copper, and cadmium during late summer and early autumn, probably due to physiological readjustments involving gametogenesis.

9.3.5 Uptake in vertebrates

A general objective of studies of trace metal concentrations in fishes frequently has been to provide baseline values against which future levels of pollution in an area may be measured. Brooks and Rumsey (1974) present baseline data for cadmium, chromium, copper, iron, lead, manganese, nickel, and zinc in eight common commercial New Zealand marine fishes. Some element pairs (copper and zinc, iron and manganese) appeared to be mutually related. Significant concentrations of metals reported for other areas were lead and cadmium in selected Oregon groundfish (Childs and Gaffke 1974); mercury in fish and shrimp of the Belgian coast (De Clerck, Vanderstappen, and Vyncke 1974); cadmium, copper, mercury, lead, and zinc in fish and shellfish from Thailand coastal and inland waters (Huschenbeth and Harms 1975); and selenium, mercury, cadmium, lead, zinc, and copper in black marlin (*Makaira indica*) off Northeastern Australia (MacKay, Kazacos, Williams, and Leedow 1975). MacKay and coworkers propose a moratorium on the use of black marlin for human consumption because of its high metal content.

Trace element concentrations have been determined for a variety of marine organisms by Van As, Fourie, and Vleggaar (1975). Concentration levels (Table 9.58) determined for individual species include environmental variations because organisms were collected under a variety of conditions (e.g., pH, temperature, dissolved oxygen) and analytical uncertainties; nevertheless, a range is found in the values for similar species (e.g., line-fish and pelagic fish). Concentrations in invertebrates and seaweeds are included for comparison.

Laboratory microcosm and field studies provide information on transfer rates, sinks, and bioaccumulation of trace elements in aquatic ecosystems. Laumond et al. (1973) examined transfers of inorganic mercury in pelagic and neritic food chains in the laboratory. Concentration factors

Table 9.58. Concentration of trace elements in marine organisms (ppm)[a]

Organism	Species	Cr	Fe	Zn	Co	Mn	Sb	Cs
Molluscs	*Choromytillus meridionalis*	0.10 ± 0.06	20 ± 7	16 ± 1	0.038 ± 0.006	2.7 ± 1.1	0.20[b]	0.0072 ± 0.0027
	Donax sera	0.24 ± 0.03	59 ± 27	18 ± 4	0.038 ± 0.024	1.0 ± 0.4	0.18[b] ± 0.08	0.0072 ± 0.0054
	Haliotus midae	0.50 ± 0.25	18 ± 14	12 ± 2	0.025 ± 0.007	0.17 ± 0.07	0.083[b]	0.0010 ± 0.0011
Crustacean	*Jasus lalandii*	0.08 ± 0.08	2.7 ± 1.3	17 ± 2	0.0039 ± 0.0015	0.27 ± 0.06	0.12 ± 0.07	0.0098 ± 0.0038
Seaweeds	*Ulva spp.*	0.45 ± 0.38	39 ± 13	5.6 ± 4.1	0.038 ± 0.013	4.1 ± 2.8	0.16 ± 0.09	0.011 ± 0.001
	Ecklonia maxima	0.13 ± 0.15	3.3 ± 1.2	1.2 ± 0.6	0.0064 ± 0.0021	0.25 ± 0.09	0.082 ± 0.028	0.0095 ± 0.0025
	Porphyra capensis	0.47 ± 0.30	37 ± 22	11 ± 7	0.054 ± 0.013	6.1 ± 2.0	0.21[b] ± 0.15	0.0056 ± 0.0067
	Suhria vittata	0.87[b] ± 0.11	49 ± 4	9.4 ± 2.3	0.048 ± 0.004	3.7 ± 2.1	0.23[b]	0.0071[b] ± 0.0009
	Gigartina rachula	0.23[b]	40[b] ± 10	5.6[b] ± 1.1	0.018[b] ± 0.003	2.0 ± 0.4	0.24[b]	0.023[b]
Other	*Parechinus*	< 0.1[b]	11[b]	9.4[b]	0.006[b]			
	Pyura		230[b]	12[b]	0.075[b]			
Line-fish	*Seriola pappei*	1.0[b] ± 1.8	14[b] ± 3	6.7[b] ± 1.5	0.0080[b] ± 0.0049	0.29[b] ± 0.08	0.057[b] ± 0.042	0.040[b] ± 0.008
	Argyrozoma argyrozoma	< 0.1[b]	3.2[b]	3.2[b]	0.0026[b]	0.44[b]	0.029[b]	0.024[b]
	Johnius hololepidotus	0.72[b] ± 0.97	7.2[b] ± 5.4	3.8[b] ± 0.8	0.0022[b] ± 0.0004	0.14[b] ± 0.06	0.018[b] ± 0.006	0.031[b] ± 0.002
	Pachymetopon grande	3.0[b]	5.2[b]	3.5[b]	0.0020[b]	0.23[b]	0.19[b]	0.018[b]
	Atractoscion aequidens		7.7[b]	5.7[b]	0.0041[b]	0.26[b]	0.059[b]	0.025[b]
	Chrysoblephus gibbiceps	< 0.1[b]	5.8[b]	4.6[b]		0.16[b]	0.0020[b]	0.015[b]
	Lithognathus lithognathus	0.11[b]	5.0[b]	4.9[b]	0.0031[b]	0.21[b]	0.019[b]	0.044[b]
White fish	*Merluccius capensis*	0.26 ± 0.44	4.9 ± 3.3	3.7 ± 0.4	0.0037 ± 0.0011	0.22 ± 0.11	0.075 ± 0.062	0.038 ± 0.008
	Xiphiurus capensis	0.92 ± 1.54	4.4 ± 2.9	4.8 ± 2.0	0.0032 ± 0.0013	0.21 ± 0.10	0.037 ± 0.032	0.028 ± 0.027
	Synaptura marginata	1.3[b] ± 1.8	4.5[b] ± 3.3	4.0[b] ± 1.6	0.0031[b] ± 0.0004	0.17[b] ± 0.12	0.14[b] ± 0.09	0.0057 ± 0.0015
Pelagic fish	*Trachurus trachurus*	0.14[b]	10[b]	4.4[b]	0.0081[b]	0.11[b]	0.028[b]	0.022[b]
	Sardinop ocellata		22[b]	11[b]	0.033[b]	0.91[b]	0.26[b]	0.013[b]
	Scomber japonicus	0.65[b] ± 0.85	17[b] ± 2	7.2[b] ± 2.2	0.020 ± 0.017	0.28[b] ± 0.14	0.19[b] ± 0.17	0.037[b] ± 0.013
	Mugil richardsoni	< 0.1[b]	6.7[b]	4.1[b]		0.19[b]	0.024[b]	0.0072[b]
	Scombrops dubius		3.5[b]	3.6[b]	0.0034[b]		0.055[b]	0.025[b]
Other fish	*Lophius piscatorius*	< 0.1[b]	3.0[b]	3.7[b]	0.0062[b]	0.12[b]	0.066[b]	0.014[b]
	Triglia capensis	< 0.1[b]	4.5[b]	3.7[b]	0.011[b]	0.11[b]	0.020[b]	0.032[b]

[a] µg/g fresh weight.
[b] Single determination.

Source: Van As, Fourie, and Vleggaar 1975, Table 2, p. 152. Reprinted by permission of the publisher.

of Hg^{2+} from seawater to fish or molluscs ranged from 50 to 3000. A mercury budget, which showed mercury biomagnification in the aquatic food chain, was developed by Standiford, Potter, and Kidd (1973) for Lake Powell (Arizona-Utah).

9.3.5.1 Absorption

Very little information is available regarding intestinal absorption of metal from the gastro-intestinal tract in aquatic organisms. Most uptake studies have focused on the direct absorption from the water through the gills. Parke et al. (1975) report that the actual molecular mechanisms of intestinal absorption are not known for any species of animal, aquatic or terrestrial, but suggest several possible mechanisms: (1) carrier molecules in the mucous membranes; (2) inter-action with possible transport systems for essential trace metals (e.g., copper, iron, zinc); (3) natural or man-made chelating agents, which may be responsible for increasing or decreasing the fractional absorption of ingested metals.

9.3.5.2 Effect of size and age

A number of investigators have examined relationships between trace element concentration and fish size. Concentrations of mercury and selenium, but not of cadmium, lead, zinc, copper, or arsenic were significantly related to size of black marlin (MacKay, Kazacos, Williams, and Leedow 1975). Mercury and zinc concentrations increased with age and size of some groupers (family Serranidae) from reefs or reef banks in the Gulf of Mexico and Caribbean Sea (Taylor and Bright 1973). Arsenic showed no correlation with age or size. Interspecific differences, particularly in accumulation of mercury and arsenic, were demonstrated for certain grouper species, which may reflect differences in feeding habits or metabolism. Concentrations of mercury, copper, nickel, silver, cadmium, and lead were determined for various tissues of the northern Adriatic anchovy (*Engraulis encracicholus*) and sardine (*Sardina pilchardus*) by Gilmartin and Revelante (1975). Few interspecific differences and no unusually high concen-trations of these trace elements were reported. Small species differences were observed by Eustace (1974) in zinc and copper concentrations in muscle tissue of 39 species from the Derwent Estuary (Tasmania). The differences were attributed to feeding habits. With minor exceptions, no relationships existed between tissue concentrations of zinc or copper and length of fish.

Kelso and Frank (1974) report concentrations of mercury, copper, and cadmium for yellow perch (*Perca flavescens*), white bass (*Morone chrysops*), and smallmouth bass (*Micropterus dolomieui*) from Long Point Bay, Lake Erie. Mercury content was apparently related to weight of individuals of a particular species. Dix et al. (1975) found that mercury concentration and size of *Platycephalus bassensis*, a nonmigratory demersal fish, were highly correlated only if location of sampling was taken into account. Even then, size accounted for only 10% of the variance. In an effort to eliminate effects of sampling locations, Montgomery et al. (1975) analyzed 50 herring (*Opisthonema oglinum*) taken in one gill-net haul. Concentrations of iron, zinc, cadmium, and copper were significantly higher in small than in large fish. Lead concentration showed no detectable difference between size classes. Manganese was significantly higher in large fish.

Alexander et al. (1973) identified a significant correlation between mercury content and weight in a study of the relationship between body size and mercury content of edible flesh of bluefish

(*Pomatomus saltatrix*) and striped bass (*Morone saxatalis*) taken near Long Island, New York. Bluefish weighing less than 2.4 kg and striped bass weighing less than 3.2 kg had concentrations of mercury below 0.5 mg/kg wet weight; bluefish weighing more than 5.6 kg and striped bass weighing more than 5.7 kg had more than 0.5 mg/kg mercury in edible flesh.

Naidu (1974) found zinc and cadmium concentrations to increase with size of the Pacific hake (*Merluccius productus*), approaching an asymptotic value at maturity. Mercury concentration increased proportionately with age; the slope of the regression relating concentration and age was a function of sampling location. Zinc and cadmium concentrations were apparently regulated physiologically, whereas mercury concentration increased with age. Tong et al. (1974) examined the trace element content of lake trout (*Salvelinus namaycush*) and concluded that concentrations of most metals are apparently not related to age. However, chromium concentrations increased and molybdenum and tin concentrations decreased from ages 1 to 12 years.

Bishop and Neary (1974) found the average methylmercury-to-total-mercury ratio for 16 species from Lake St. Clair (North America) to be 88.9%. Age, weight, and length of the fishes were found to have no significant effect on the ratio.

9.3.5.3 Effect of dietary habits

Concentrations of lead, cadmium, and zinc in tissues of the flounder (*Platichthyes flesus*) from the Severn Estuary were determined by Hardisty, Kartar, and Sainsbury (1974) to examine relationships between dietary habits and concentrations of heavy metals; a positive correlation was found. A positive correlation was found between cadmium and lead concentrations of tissues and the proportion of crustaceans in the diet. No relationship was detected between diet and tissue levels of zinc. In further studies, Hardisty, Huggins, Karter, and Sainsbury (1974) observed high zinc concentrations in the flounder, and unlike the lead or cadmium levels, concentrations of zinc showed distinct seasonal and age variations.

9.3.5.4 Biomagnification

Copper, zinc, lead, and cadmium were measured in tissues of the blue shark (*Prionace glauca*) and in muscle of selected prey fishes in their diet by Stevens and Brown (1974). No biological magnification of metals from prey to the blue shark was observed; lead and cadmium concentrations were generally below the limits of detection.

High levels of several metals were detected in marine mammals by Koeman et al. (1975): 0.37 to 326 ppm mercury; 0.02 to 1.6 ppm cadmium; 0.03 to 1.9 ppm arsenic; 0.60 to 134 ppm selenium; 20 to 66 ppm zinc; and ≤0.01 ppm (the detection limit) antimony (all measurements were wet weight and determined in liver). High levels of metal concentration were presumably related to one or more reasons: (1) end position of the mammal in the food chain; (2) low rate of excretion; (3) a possible protective mechanism (e.g., by which highly toxic methylmercury is converted to less harmful compounds).

9.3.5.5 Tissue distribution

Epidermis, muscle, viscera, and bone of many species of marine fishes and some edible invertebrates were analyzed by neutron activation to determine cobalt, cesium, and zinc concentrations
by Ichikawa and Ohno (1974). Muscle contained much less cobalt and zinc than other tissues,
whereas cesium was distributed more uniformly. Tissue distribution depends primarily on the
nature of the element.

Fletcher, Watt, and King (1975) measured concentrations of copper, zinc, and total protein in
blood plasma of the sockeye salmon (*Oncorhynchus nerka*) during spawning migrations. Values for
the three constituents were 20 to 50% of those observed in fish collected in seawater. Concentrations of zinc and copper were positively correlated with plasma protein concentration. Metal-
protein ratios indicated that copper and zinc are regulated differently.

9.3.5.6 Biotransformation

Organisms use enzymes in normal biochemical processes; many metals are enzymatically essential
for maintenance of the normal physiological well-being of animals (Bowen 1966). Other metals
cause abnormal effects, interfering with normal biochemical processes.

The biotransformation of inorganic mercury to methylmercury appears to be the result of microorganisms present in the gastrointestinal tract or external mucus of the higher organisms; there
is no evidence of direct enzymatic activity in higher aquatic organisms resulting in methylation
(Jernelov 1972).

Very little is known about the metabolism of arsenic in higher organisms. Lunde (1973*b*) observed
transformations of radioactive inorganic arsenic in rainbow trout. Although no carbon-arsenic
bonds were identified, the transformation products were considered organo-arsenicals. Penrose
(1975) similarly observed such transformations of orally administered inorganic arsenic in the
gastrointestinal tract of brown trout. Injected arsenic was converted only very slowly, but
oral administration of inorganic arsenic resulted in nonanionic arsenic appearing in the gastrointestinal tract within one day. Table 9.59 gives data regarding the selective absorption of
the nonanionic arsenic, indicating that the proportion of arsenic in body tissues will shift
towards this form, regardless of the route of administration. Both organic and inorganic forms
were excreted in the bile, which is consistent with the hypothesis that conversion of arsenic
to the organic form is mediated by the intestinal flora (Chap. 7).

Although cadmium forms coordination complexes with a variety of ligands in biological systems,
no metabolic interconversions have been demonstrated for this element (Wood 1973).

9.3.5.7 Elimination

More information is available regarding mercury than for other metals in aquatic organisms.
Once mercury is accumulated by aquatic organisms, it remains for long periods of time. Biological
half-retention times of up to several years have been reported, although the organism loses a
substantial portion of mercury body burden, after transfer from a contaminated to a clean
environment, within five months (Burrows and Krenkel 1973). It is also reported that the

Table 9.59. Excretion of injected inorganic arsenic into the digestive tract,
with proportions and concentrations tabulated separately

Fish No.	Time (h)	Blood	Bile	G.I. contents	Liver
			Percent ^{74}As in nonanionic form		
1	5	0.72	5.35	-0.47	9.74
2	5	0.44	0.23	11.0	8.30
3	24	6.30	80.3	44.9	34.7
4	24	5.36	88.0	57.9	28.9
5	72	18.3	48.3	81.7	45.7
6	72	24.1	No bile	98.5	62.1
1	5	108,840	22,000	8,640	27,120
2	5	42,640	14,000	1,200	10,840
3	24	44,000	17,000	35,200	18,760
4	24	23,500	3,250	1,420	8,360
5	72	24,000	9,900	6,460	9,600
6	72	24,400	No bile	5,080	10,500

Source: Penrose 1975, Table 3, p. 2389. Reprinted by permission of the publisher.

most important factor responsible for the amount of mercury lost from an organism is the rate
of mercury removal from the environment; physiology of the organism plays a secondary role.
Losses of methylmercury from fish are very slow, although a fish removed from a contaminated
environment and released in clean water can lose a large portion of its burden in the first
few days. Burrows, Taimi, and Krenkel (1974) report a half-retention time of five months for
methylmercury in bluegill sunfish (*Lepomis macrochirus*). Other authors report half-retention
times of two to three years for different species at lower temperatures. Hasselrot and Gothberg
(1974) report that tagged northern pike (*Esox lucius*), after transfer from a lake heavily contaminated with phenylmercury, showed no evident elimination in one year. Clearance rates of
mercury residues and rates of mercury bioaccumulation from water and food indicated a preferential accumulation of methylmercury as compared with a divalent mercuric ion. Tissue distributions
of both forms of mercury approach a steady state after about two months of exposure (de Freitas,
Qadri, and Case 1974). No significant species-related differences in tissue mercury distribution
or in dynamics of mercury uptake or clearance were observed.

Giblin and Massaro (1973) administered mercury to rainbow trout intragastrically. Blood and
spleen levels were higher than other tissue levels initially, but mercury was slowly translocated to the skeletal muscle, which appeared to act as a reservoir for the methyl form of
the metal. The rate of mercury excretion appeared to be biphasic as a result of the slow
elimination from skeletal muscle relative to other tissues; the half-retention time for methylmercury was estimated to be greater than 200 days. Weisbart (1973) reports a half-retention
time of 568 hr for goldfish (*Carassius auratus*) injected with ^{210}Hg(NO$_3$)$_2$.

A summary of mercury elimination from aquatic organisms was presented by Miettinen (1975):

> The excretion rate varies greatly among the different species, but usually
> follows a biexponential equation. The biological half-time at an ambient
> water temperature of the fast component usually varies from 1 to 10 days,
> while the slow one ranges from 200 to 1,200 days. The excretion rate at
> 15°C is twice as fast as that at 4°C. Fresh, brackish, and salt water

species were found to have broadly similar rates of excretion. When the
concentration of methylmercury in fish increases, the biological half-time
of mercury decreases; in rainbow trout, dosed with Me^{203}Hg, a concentration
of 0.4 mg Hg/kg is eliminated with a half-time of 340 days, while a concentration of 4 mg Hg/kg is eliminated with a half-time of 170 days. Ionic
and protein-bound methylmercury have similar rates of elimination when
orally administered, but methylmercury absorbed by rainbow trout directly
from aquarium water via the gills is eliminated faster (half-time: 268 days)
than methylmercury administered orally as, for instance, a proteinate (half-
time: 326 days). The half-times of inorganic Hg and phenyl- and methyl-Hg
in mussels (*Pseudanodonta complanata*) were 23, 43, and (depending on age)
100-400 days, respectively, while the half-times of methyl- ethyl-, and
propyl-Hg in rainbow trout were 346, 119, and 233 days, respectively.
Thus, of all these compounds, methylmercury has the lowest rates of
elimination.

Both injected and orally administered methylmercury have approximately
the same rate of elimination in fish, while inorganic mercury, when
injected intramuscularly, is eliminated several times more slowly than
when dosed in food. Methylmercury administered orally to the seal has
a very long biological half-time (about 500 days).

The biological half-times of the slow component of ^{203}Hg elimination for ionic and protein-bound
methylmercury in three species of fish are given in Table 9.60.

Table 9.60. Biological half-time of the slow component of excretion for ionic
and protein-bound methylmercury in three species of fish (flounder, pike,
and eel) in Gulf of Finland, Tvarminne Zoological Station (salinity 0.6%)

Fish	Route of administration and form of methylmercury	$T_{\frac{1}{2}}$ biol (days)	Number of fish
Flounder	bones, proteinate	780 ± 120	9
(*Pleuronectes flesus*)	bones, ionic	700 ± 50	7
	intramuscular, ionic	1200 ± 400	10
Pike	bones, proteinate	750 ± 50	5
(*Esox lucius*)	bones, ionic	640 ± 120	2
	intramuscular, ionic	780 ± 80	3
Eel	bones, proteinate	910 ± 40	4
(*Anguilla vulgaris*)	bones, ionic	1030 ± 70	4
	intramuscular, ionic	1030 ± 80	2

Source: Miettinen 1975, Table 13.3, p. 226. Reprinted by permission of the
publisher.

9.3.6 Acute toxicity

Many factors affect the toxic action of metals in aquatic ecosystems; these include dissolved
oxygen, water hardness, pH, synergism, antagonism, temperature, sunlight, and variations of the
toxic effect due to the action of surrounding flora and fauna (Scheier and Kiry 1973; Cairns and
Scheier 1959). Thus, a multiplicity of factors must be considered in assessing the potential
impact of metals on aquatic biota (Vernberg, DeCoursay, and O'Hara 1974). The response evoked
by metal pollutants may be as much a function of general water quality as of the metal concentration (Brown 1976).

9.3.6.1 Biochemical mechanisms

The most important mechanism for toxicity action of metals, as suggested by Bowen (1966), is the inhibition of enzymes. The more electro-negative metals, notably copper, mercury, and silver, have a great affinity for amino, imino, and sulfhydryl groups, which are reactive sites on many enzymes. Metals are also readily chelated by organic molecules. Other proposed modes of toxic action are:

1. Metal compounds behave as antimetabolites (e.g., arsenate occupies sites for phosphate; borate, permanganate, antimonate, selenate, tellurate, tungstate, and beryllium may act in a similar manner).

2. Metals form stable precipitates or chelates with essential metabolites (e.g., aluminum, beryllium, scandium, titanium, yttrium, and zirconium react with phosphate, barium with sulfate, or iron with ATP).

3. Substances catalyze the decomposition of essential metabolites (e.g., La^{3+} and other lanthanide cations decompose ATP).

4. Metals combine with the cell membrane and affect its permeability (e.g., gold, cadmium, copper, mercury, and lead).

5. Metals can replace structurally or electrochemically important elements in the cell and then fail to function (e.g., lithium can replace sodium, or cesium can replace potassium).

9.3.6.2 Embryonic effects

Studies on the effects of heavy metals to marine macroinvertebrates generally emphasize toxicity to adults and forms that live in relatively stable environments. Recent reports on the toxicity of heavy metals to embryos of the American oyster (*Crassostrea virginica*) (Calabrese et al. 1973) and the hard clam (*Mercenaria mercenaria*) (Calabrese and Nelson 1974) present data on metal salt concentrations which produce 50% mortality (Table 9.61). No data were provided for effects on growth of fully developed larvae, which may be the more sensitive stage. Larvae of oysters, shrimp, crabs, and lobsters were found by Connor (1972) to be from 14 to 1000 times more susceptible than adults to mercury, copper, and zinc. The median lethal concentrations (48-hr TL_m) of each metal to the most sensitive larval stage exceeded the concentrations found in natural seawater by a factor of 100.

Servizi and Gordon (1974), as cited by Waldichuk (1974*b*), determined that certain heavy metals have an effect on the embryonic stages of some aquatic organisms which may lead to teratogenesis, mutagenesis, or carcinogenesis. Pre-eyed eggs of sockeye and pink salmon exposed to 10 ppb mercury demonstrated a teratogenic effect. Cadmium, on the other hand, was not very toxic to the eggs, but was especially toxic to the young free-swimming fry. Fishbein (1974) also notes the potential mutagenicity of mercury, cadmium, lead, and tin.

Tests of reproductive success and development during early life stages appear to be sensitive measures of pollution. A disturbance in reproduction is a common symptom of many toxic substances. Embryos at four stages of development and newly hatched larvae of zebra fish (*Brachydanio rerio*) were exposed to selenium concentrations of 0.5 to 10 µg/liter to establish relative tolerances of different life stages. Embryo mortality was negligible at all selenium concentrations. After hatching, the number of mortalities increased sharply at concentrations of 3 µg/liter or greater, irrespective of the embryonic stage of first exposure (Niimi and Laham

Table 9.61. Toxicity of heavy metals, as inorganic salts, to oyster and hard clam embryos at 26 $\pm$ 1^{0}C in synthetic sea water (25^0/oo salinity)[a]

Metals as inorganic salts	LC$_0$		LC$_{50}$		LC$_{100}$	
	Clams	Oysters	Clams	Oysters	Clams	Oysters
Mercuric chloride	0.0025	0.001	0.0048	0.0056	0.0075	0.008
Silver nitrate	0.010	0.003	0.021	0.0058	0.045	0.01
Zinc chloride	0.095	0.075	0.166	0.31	0.25	0.5
Nickel chloride	0.10	0.10	0.31	1.18	0.60	3.0
Lead nitrate	0.40	0.50	0.78	2.45	1.20	>6.0

[a]The concentrations are in parts per million of metal ion added to medium at start, producing mortality of 0, 50, and 100% (LC$_0$ and LC$_{100}$ are actual values, and LC$_{50}$ is estimated). These concentrations do not include background concentrations of heavy metals in the synthetic seawater medium. They were (in ppm): Hg — 0.00003, Ag — 0.003, Zn — 0.029, Ni — 0.0065, and Pb — 0.007.

Source: Calabrese and Nelson, Table 2, p. 96. Reprinted by permission of the publisher.

1975). In a study of effects of zinc on different developmental stages of the minnow (*Phoxinus phoxinus*), mortality of newly hatched fry was a most sensitive test, as compared with measurement by number of deposited eggs, hatchability, and mortality of under-yearlings, yearlings, and mature minnows (Bengtsson 1974).

Persoone and Uyttersprot (1975) examined the effect of several heavy metals on the reproductive rate of *Euplotes vannus*, a marine hypotrichous ciliate. Mercury was found to be the most toxic of the heavy metals tested, with total reproductive inhibition at the 1-ppm level; copper gave a 50% inhibition at 1 ppm and was lethal at 10 ppm; lead, cadmium, and zinc began to affect the rate of cell division between 1 and 10 ppm and total inhibition occurred at 100 ppm.

Maximum acceptable toxicant concentrations of lead for rainbow trout in hard water (353 mg/liter as $CaCO_3$ and alkalinity 243 mg/liter) were determined by Davies et al. (1976) to be between 18.2 and 31.7 µg/liter dissolved lead vs 120 to 360 µg/liter total lead. In soft water (28 mg/liter as $CaCO_3$) where exposure to lead was initiated at the eyed stage of egg development, the maximum acceptable toxicant concentration was determined to be between 4.1 and 7.6 µg/liter. With exposure to lead beginning after hatching of the fry, the maximum acceptable toxicant concentration was between 7.2 and 14.6 µg/liter. From these chronic toxicity bioassays, Davies and coworkers concluded that fish were more sensitive to lead when exposed as eggs. Table 9.62 indicates the observed mortalities and abnormalities of the test fish exposed at and above the determined maximum acceptable toxicant concentrations. Observed abnormalities included black tails, two types of spinal curvatures [lordosis (dorsal-ventral spinal flexures) and scoliosis (bilateral spinal flexures) described, when appearing together, as lordoscoliosis], and eroded caudal fins.

Table 9.62. Analytical, mortality, and physical abnormality results
from three chronic bioassays in hard and soft waters

Lead concentration (µg/liter)			Percent	Percent abnormalities		
Total	AAS analysis[a]	PP analysis[b]	mortality	BT[c]	LS[d]	EC[e]
Hard water (eggs not exposed to lead)						
3240	2310 ± 226[f]	64.2 ± 3.9[e]	5.7	100	100	30
1080	850 ± 75	41.2 ± 2.4	0.0	90	60	20
360	380 ± 34	31.6 ± 2.6	0.0	70	10	10
120	190 ± 20	18.2 ± 2.1	0.0	0	0	0
40	100 ± 10	10.2 ± 1.1	0.0	0	0	0
Control	0 ± 0	0.5 ± 0.3	0.0	0	0	0
Soft water (eggs exposed to lead)						
80	55.1 ± 3.7[f]		30.1	100.0	89.8	50.8
40	27.0 ± 1.7		5.0	95.0	32.2	46.2
20	13.2 ± 0.8		1.7	40.3	3.6	5.5
10	7.6 ± 0.4		0.0	4.7	0.7	0.7
5	4.1 ± 0.3		0.0	0.0	0.0	0.0
Control	0.6 ± 0.3		0.0	0.0	0.0	0.0
Soft water (eggs not exposed to lead)						
80	61.8 ± 3.1[f]		50.6	100.0	96.7	45.4
40	31.2 ± 1.7		15.0	84.5	43.8	38.2
20	14.6 ± 0.9		7.2	41.3	3.0	6.3
10	7.2 ± 0.4		0.6	0.0	0.0	0.0
5	3.6 ± 0.3		0.6	0.0	0.0	0.0
Control	0.5 ± 0.3		0.6	0.0	0.0	0.0

[a]Atomic absorption spectrophotometry.
[b]Pulse polarography.
[c]Black tail.
[d]Lordoscoliosis.
[e]Eroded caudal.
[f]95% confidence interval.

Source: Davies et al. 1976, Table 4, p. 201. Reprinted by permission of the publisher.

9.3.6.3 Neurotoxic effects

Lead and mercury have been shown to be potent neurotoxins. Lead, mercury, and cadmium have been
shown to be nephrotoxic as well as teratogenic in mammals. Waldron (1975) outlined the major
sites of toxicological action for these three metals (Table 9.63).

Table 9.63. Major sites of toxicological effects
of the heavy metals

Mercury	Lead	Cadmium
Nervous system	Red blood cells	Lung
Kidney	Nervous system	Kidney
Embryo	Kidney	Embryo
	Embryo	

Source: Waldron 1975, Table 2, p. 355.
Reprinted by permission of the publisher.

Lead is a cumulative metal causing death at 0.3 ppm in soft water and significant behavioral modification in fish at 70 ppb (Adams 1975).

Mercury compounds interfere with the ability of aquatic animals to osmoregulate, resulting in a toxic response. Schmidt-Nielsen (1974) maintains that toxic effects of mercury may also be due to its affinity for sulfhydryl groups and the consequent inhibitory effects on enzymes and ion transport systems. These result in inhibited respiration from gill damage and eventual death.

Mercury, when methylated, affects the nervous system through attack of vital brain centers. Spyker et al., as cited by Waldichuk (1974b), determined subtle behavioral effects in offspring of pregnant mothers that had been exposed to low methylmercury concentrations. Waldichuk reports that such effects have not been clearly identified in aquatic organisms, but that similar effects probably prevail. Some researchers have demonstrated that certain functions controlled by the nervous system in aquatic organisms are indeed affected by low concentrations of mercury; Lindahl and Schwanbom (1971), as cited by Waldichuk (1974a), report loss of equilibrium in fish. Little information is available regarding the neurophysiological effects of heavy metals, and none is available regarding effects on brain centers in aquatic organisms that control such functions as learning ability, temperature response, and reactions to other types of environmental stress.

9.3.6.4 Toxicity to zooplankton

Acute toxicities of various metals (including cadmium, cobalt, chromium, copper, manganese, mercury, lead, and zinc) to freshwater zooplankton (*Cyclon abyssorum*, *Epidaptomus padanus*, and *Daphnia hyalina*) were evaluated by Baudouin and Scoppa (1974). Sensitivities of the three species varied in the following order: *C. abyssorum* < *E. padanus* < *D. hyalina*.

At $CuSO_4$ doses commonly used for algal control, effects on zooplankton were shown by McIntosh and Kevern (1974) to be difficult to predict. A variety of parameters (including salinity, dissolved oxygen, pH, and temperature) affected the toxic response. Rotifers and cladocerans were susceptible to 3 mg/liter $CuSO_4$, whereas copepods and ostracods were not. The green sunfish (*Lepomis cyanellus*) shifted to alternative food sources when cladocerans disappeared from the survey area. The chronic toxicity of nitrilotriacetic acid (NTA) (tested as the trisodium salt) to *Daphnia magna* in eight natural waters studied by Biesinger, Andrew, and Arthur (1974) varied tenfold. A strong negative correlation between water hardness and NTA toxicity was demonstrated. Chelates of copper and zinc with NTA were shown to be relatively nontoxic.

9.3.6.5 Toxicity to aquatic insects

Warnick and Bell (1969) determined the 96-hr TL_m of the metals Cu^{2+}, Zn^{2+}, Cd^{2+}, Pb^{2+}, Fe^{2+}, Ni^{2+}, Co^{2+}, Cr^{2+}, and Hg^{2+} to a stone fly (*Acroneuria*), a May fly (*Ephemerella*), and a caddis fly (*Hydropsyche*). Those that did not die in 96 hr in the highest concentrations used (64 mg/liter) were retested, and the survival times at various metal concentrations were recorded. The results indicate that the aquatic insects were more tolerant to heavy metals than were fish. *Ephemerella* was the most sensitive of the three insects, and copper was the most toxic metal, followed by Fe^{2+}, Cd^{2+}, Cr^{2+}, and Hg^{2+}. Elder and Gaufin (1974) determined TL_m values for three different forms of mercury to a stone fly (*Pteronarcys californica*). The order of toxicity of the three forms was phenylmercury > methylmercury > inorganic mercury.

9.3.7 <u>Synergistic and antagonistic effects</u>

Metal toxicity to fish is affected by such variables as water alkalinity, temperature, dissolved oxygen, and pH. Reactions occurring in solution determine the form and abundance of individual chemical species and therefore the toxicity of a substance to aquatic organisms. For example, studies of the effects of additions of sewage effluent, amino acids, humic substances, and suspended organic matter on the acute toxicity of water containing $CuSO_4$ to rainbow trout showed that the addition of any of these substances reduced the toxicity of a given concentration of $CuSO_4$ (Brown, Shaw, and Shurben 1974). It was concluded that neither total concentrations of copper nor soluble copper in water could be used to determine toxicity. Stiff (1971) found between 25 and 98% of the soluble copper in river water to be in the form of nontoxic complexes and between 40 and 90% of the total copper to be associated with particulate matter.

9.3.7.1 <u>Salinity</u>

The effect of salinity on metal toxicity was considered by several investigators. Effects of mixtures of chloride salts of cadmium, copper, and zinc on survival, whole-body residues, and histopathology of mummichog (*Fundulus heteroclitus*) were investigated by Eisler and Gardner (1973). Metal residue levels in survivors exposed to mixtures of metals were different from concentrations accumulated when fish were exposed to single elements. Levels of dissolved oxygen between 4 mg/liter and saturation did not significantly affect tolerance of fish to acute cadmium poisoning, regardless of salinity and acclimation regime (Voyer 1975).

Von Westernhagen, Rosenthal, and Spedding (1974) examined effects of cadmium on embryonic development of the herring (*Clupea harengus*) at different salinities (5, 16, 25, and 32 parts per thousand). Cadmium was more toxic to developing embryos in brackish water than in seawater. At high cadmium levels (5 mg/liter), there was greater survival of embryos at high than at low salinities. Sensitive variables were viability of developing embryos, incubation time, and mean total length of newly hatched larvae.

Olson and Harrel (1973) examined the relationships between salinity and acute toxicity of mercury, copper, and chromium to *Rangia cuneata,* a brackish-water clam. Each metal was more toxic in fresh water than in brackish water, and copper and chromium were more toxic than mercury. The 96-hr LC_{50} of $HgCl_2$ to porcelain crabs (*Petrolisthes armatus*) varied from 50 to 64 µg/liter, depending upon test salinities (Roesijadi et al. 1974). Lower salinities decreased survival time. Exposure of crabs to 50 µg/liter HgCl did not alter chloride ion regulation of either acclimated crabs or individuals adjusting to new salinities.

9.3.7.2 <u>Water hardness</u>

Hardness is antagonistic towards the toxicity of zinc (Becker and Thatcher 1973). Sinley, Goettl, and Davies (1974) found juveniles to be the most sensitive life stage of rainbow trout (*Salmo gairdneri*) to zinc. The maximum acceptable toxicant concentration for rainbow trout exposed to zinc in hard water (330 mg/liter as $CaCO_3$) was determined to be between 640 and 320 µg/liter zinc, whereas for juveniles in soft waters (25 mg/liter total hardness), the value lies between 260 and 140 µg/liter zinc. Solbe and Flook (1975) determined that stone loach (fish) in hard water were more sensitive to zinc than were rainbow trout, but were much more resistant to cadmium.

An exponential correlation between beryllium sulfate tolerance in the guppy (*Lebistes reticulatus*) and water hardness was indicated by Slonim and Slonim (1973). This beryllium salt was shown to be 100 times more toxic to guppies in soft water (20 to 25 mg/liter as $CaCO_3$) than in hard water (400 mg/liter as $CaCO_3$). Slonim and Ray (1975) demonstrated that $BeSO_4$ was more toxic to salamander (*Ambystoma* spp.) larvae (by slightly greater than one order of magnitude) in soft than in hard water.

Davies et al. (1976) found the 96-hr LC_{50} values of dissolved and total lead for rainbow trout through two static bioassays in hard water (353 mg/liter as $CaCO_3$ and alkalinity 243 mg/liter) to vary considerably — 1.32 and 1.47 mg/liter for dissolved lead vs 542 and 471 mg/liter for total lead. In a flow-through bioassay in soft water (28 mg/liter as $CaCO_3$), a 96-hr LC_{50} value of 1.17 mg/liter, expressed as either total or dissolved lead, was obtained.

9.3.7.3 Temperature

Temperature is an important environmental influence on toxicity of metal species to organisms. The toxic effect may be reduced or enhanced, depending on the nature of the metal compound, the organism, and other environmental conditions (Cairns, Heath, and Parker 1975*a*; 1975*b*). Hodson (1975) suggests that water temperature modifies the acute toxicity of metals to fish, principally by shortening survival times as temperatures increase. Hodson and Sprague (1975) hypothesize that temperature stress may increase ventilation rates of fish, causing an increase in metal accumulation through the gills, resultant gill damage, and death.

Gibson, Thatcher, and Apts (1975) studied effects of heated effluents, chlorine, and copper on the coonstripe shrimp (*Pandalus danae*) by determining critical thermal maxima, the 96-hr LC_{50}, and short-term growth. Copper at 0.04 mg/liter was shown to retard growth of coonstripe shrimp in one-month tests. Times to 50% mortality (LT_{50}) of lobsters (*Homarus americanus*) exposed to copper were shown to be longer at 5 than at 13°C and were not affected by salinities in the range of 20 to 30 parts per thousand (McLeese 1974). It is often assumed that species adapted to fluctuating estuarine conditions are more capable of withstanding additional constant environment. For mercury, this assumption appears to be invalid; several species of marine and estuarine isopods showed a marked increase in susceptibility to mercury under conditions of decreased salinity and increased temperature (Jones 1973). A summary of interactions between various metals and temperature stress is presented in Table 9.64.

9.3.7.4 Metal additivity

The possibility that heavy metals may produce additive effects has received little attention to date. Waldron (1975) maintains that many toxic metals act similarly in body organs, thus indicating additive effects when present in the same organs. Examples of metal pairs having such additive effects are lead and mercury, lead and cadmium, and possibly mercury and cadmium. Eaton (1973) has demonstrated that metal toxicity is potentiated when in the presence of other metals. Fathead minnows exposed to mixtures of cadmium, copper, and zinc were compared with minnows exposed under identical conditions to the individual metals. A lethal threshold was reached when each metal in the mixture was present at a concentration of 0.4 (or less) of its individual lethal threshold (Table 9.65). Spawning and hatching success has also been shown to be impaired to a greater degree under exposure to several metals than to either metal singly

Table 9.64. Interactions of metals with temperature stress

Common name	Organism Scientific name	Stress	Condition observed	Reference
Isopods	*Idotea emarginata, I. neglecta, Faera nordmanii F. albifrons Cancer irroratus, C. borealis*	Hg	Toxicity increased with increased temperature and decreased salinity	Jones 1973[a]
Snail	*Goniobasis livescens*	Cr or Zn	Reduced mean mortality period by 12 to 60%	Carins and Messenger 1974[a]
Pacific oyster	*Crassostrea gigas*	^{65}Zn	No interactive effects	Nelson 1973[b]
Scallops	*Pectin maximus Chlamys opercularis*	Co, Cu, Fe, Mn, Ni, Pb, Zn	Highest values in fall and winter	Bryan 1973[b]
Mummichog	*Fundulus heteroclitus*	Cd	Reduced thermal resistance	McKenney and Dean 1974[a]
Freshwater shrimp	*Palaemonetes paludosus*	^{137}Cs, ^{85}Sr, ^{65}Zn	Concentration decreased by <3 between 20° and 32°C	Harvey 1973[b]
Fiddler crab	*Uca pugilator*	Hg, salinity	More toxic at 20° than at 25 and 30°C	Vernberg et al. 1973[a]
Fiddler crab	*Uca pubilator*	^{203}Hg	Increased transport from hepatopancreas to gills at higher temperatures	Vernberg and O'Hara 1972[b]
Fiddler crab	*Uca pubilator*	Cd	Accumulated maximum levels at high temperature and low salinity	O'Hara 1973a; 1973b[b]
Eel	*Anguilla rostrata*	Cu, Zn, Ni, Cd, Hg, Cr	Toxicities not different between 15 and 28°C	Rehwoldt et al 1972[b]
Goldfish	*Carassius auratus*	Cu	Copper zone more attractive at 21.5 than 21.1°C	Kleerekoper et al 1973[b]
Fathead minnow	*Pimephales promelas*	Cu	Fish selected lower temperature	Opuszynski 1971[b]
Bluegill	*Lepomis machrochirus*	Zn	Died 2.6 times faster at 30 than 20°C; 5.6 mg/liter only toxic above 20°C	Burton et al. 1972[b]
Atlantic salmon	*Salmo salar*	Zn	At 19°C, fish acclimated to 3 and 11°C had shorter survival; lethal threshold highest at 19°C	Hodson 1974[a]
Rainbow trout	*Salmo gairdneri*	HgCl	Toxicity linearly related to temperature	MacLeod and Pessah 1973[b]

Table 9.64. (continued)

| Organism | | Stress | Condition observed | Reference |
Common name	Scientific name			
Rainbow trout	*Salmo gairdneri*	Hg	Resistance dependant upon acclimation temperature survival lower at 20°C than at 10 or 15°C	Thatcher 1974[a]
Rainbow trout	*Salmo gairdneri*	Methylmercuric chloride	Accumulation increased with increasing temperature	Reinert et al. 1974[a]

[a]As cited by Coutant and Talmage 1975.
[b]As cited by Coutant and Pfuderer 1974.

Source: Modified from tables by Coutant and Pfuderer 1974 and Coutant and Talmage 1975.

Table 9.65. Results of acute toxicity determinations with a mixture of copper, cadmium, and zinc

	Measured acute test values (ppb)[a]		
	Cu	Cd	Zn
Cu, Cd, and Zn mixture lethal threshold values	145	320	2050
Cu 96-hr TLm values[b]	430		
Cd lethal threshold values		7200	
Zn lethal threshold values			5030

[a]Means of daily composited samples.
[b]As little or no mortality took place after 3 days of exposure this is considered a close approximation of the lethal threshold value.

Source: Hammons and Huff, Table 5.11, p. 199. Reprinted by permission of the publisher.

(Table 9.66). Eaton (1973) maintains that combinations of metals, as they are found in the environment, may pose more of a biological threat than laboratory studies of the individual metals indicate. A greater effect on hatchability was noted by Huckabee and Griffith (1974) in carp (*Cyprinus carpio*) eggs exposed to mixtures of mercury and selenium than when eggs were exposed to these elements singly; results indicated a synergistic effect. Becker and Thatcher (1973) note a synergistic toxic effect of zinc and copper to fish (Table 9.67).

Growth and development of the copepod *Trigriopus japonicus* were found to be inhibited to a greater degree by exposure to combinations of copper and cadmium than to either metal ion alone. D'Agostino and Finney (1974) determined a synergistic effect on this zooplankter with exposure to 0.0044 mg/liter Cd^{2+} and 0.0064 mg/liter Cu^{2+}. Gray (1974) found that the growth rate in a marine ciliate (*Cristigera* spp.) was inhibited to a greater degree (67.2% reduction) by a combination of mercury (0.005 ppm as $HgCl_2$), lead [0.3 ppm as $Pb(NO_3)_2$], and zinc (0.25 ppm as $ZnSO_4$) than for any one, or combination of two, of these metals. Singly, mercury reduced growth rate by 12%, lead by 11.7%, and zinc by 13.7%.

9.3.7.5 Detergents

The toxicity to rainbow trout of copper and mercury and three detergents (two anionics and one nonionic) was determined in 14-day exposures by Calamari and Marchetti (1973). For mixtures of anionic detergents and metals, a more-than-additive effect existed, whereas for the mixture of nonionic detergent and metal, the toxic effect was less than additive. The authors rejected the idea that surface-active agents produce the same effect in all mixtures by acting in a specific way on cellular permeability.

9.3.7.6 Insecticides

Roales and Perlmutter (1974) determined the toxicity of zinc to zebra fish (*Brachydanio rerio*) embryos singly and in conjunction with Cygon (an organophosphorus insecticide). At low doses, zinc was found to be antagonistic to the toxic action of Cygon. At higher concentrations, zinc and Cygon appeared to interact additively.

Table 9.66. Spawning and hatching success of fathead minnows at various concentrations of copper, cadmium, and zinc

Concentration (µg/liter)			Number of females	Number of spawnings	Number of spawnings/ female	Total number of embryos produced	Number of embryos/ spawning	Number of embryos/ female	Number of embryo groups hatched (50 embryos/ group)	Embryos hatching[a] (%)
Cu	Cd	Zn								
3.7	0.6 (control)	13.1	14	249	17.8	41,092	165	2935	88	79.4
5.3	3.9	27.3	12	165	13.8	24,403	147	2034	52	78.4
6.7	7.1	42.3	8	27	3.4	4,058	150	507	11	91.8
10.6	14.6	79.4	15	59	3.9	6,224	105	415	16	92.0
17.6	29.5	148.0	15	20	1.3	1,785	89	119	9	86.6
32.0	59.5	302.8	11	3	0.27	51	17	4.6	0;(6)[b]	(93.7)[c]

[a]Over 95% of mortality indicated here occurred before hatching.
[b]The six embryo groups (from three spawnings) in parentheses were transferred from the control tank on the day they were spawned.
[c]Percentage hatch of the embryos transferred from the control tank.

Source: Whitfield 1975, Table 5.12, p. 200. Reprinted by permission of the publisher.

Table 9.67. Effects of exposure to combinations of zinc and copper in aquatic organisms

Test organism	Test conditions[a]	Concentration		Remarks	Reference[b]
		Zinc	Copper		
Salmo salar (juveniles)	CB,FW,LS	0.260	0.003	12.2-hr TL_m, acute; pH 7.1 - 7.5, 15° C (all data, soft water — 20 mg/liter as $CaCO_3$)	Sprague 1964*a*
		0.055	0.003	32.4-hr TL_m, acute	
		0.046	0.003	No kill, 190 hr	
		0.002	2.260	7.2-hr TL_m, acute; pH 7.1 - 7.5, 15° C	
		0.002	0.650	24.4-hr TL_m, acute	
		0.002	0.338	No kill, 240 hr	
		0.002	4.150	11.8-hr TL_m, acute; pH 8.2 - 8.4, 17° C	
		0.002	1.920	20% kill in 217 hr; pH 8.8 - 9.2	
		0.002	1.310	No kill, 90 hr; pH 8.9 - 9.2	
		0.002	2.890	28.0-hr TL_m, acute; pH 7.1 - 7.5, 5° C	
		0.002	0.886	No kill in 209 hr; pH 7.1 - 7.5, 5° C	
Salmo salar (juveniles)	FW, FS			Avoidance threshold at about 0.35 - 0.43 "toxic unit" for Cu^{2+} and Zn^{2+}; threshold concentration for TL_m is 1.0 "toxic unit" for Cu^{2+} and Zn^{2+}	Saunders and Sprague 1967
Pimephales promelas	FW,LS	0.025	1.0	Most died within 8 hr; soft water	Doudoroff 1952
Salmo gairdneri	SB,FW,LS	1.1	3.5	3-day median survival, hard water (various ratios were tested); 17 - 18° C	Lloyd 1961
		0.044	0.56	7-day median survival, soft water; 17 - 18° C	

[a]Abbreviations: SB = static bioassay; CB = constant-flow bioassay; FW = fresh water; SW = sea (salt) water; LS = lab study; FS = field study.
[b]As cited in Becker and Thatcher 1973.

Source: Becker and Thatcher 1973, Table M, p. 21.

9.3.8 Chronic toxicity

In a review of pollution problems of coastal zones, Waldichuk (1974*a*) summarizes current types of biological studies and emphasizes the importance of examining sublethal stresses.

Preliminary results of research by the National Marine Fisheries Service (Calabrese, Collier, and Miller 1974) indicate the potential use of physiological (osmoregulation and oxygen consumption), immunological, biochemical, and histological approaches, as well as tissue assimilation, in studies of chronic cadmium toxicity to the cunner fish (*Tautogolabrus adspersus*).

A variety of approaches can be taken to examine sublethal or chronic effects of pollutants. These range from reproductive success to life span and include such factors as growth, adaptation to environmental stress, feeding behavior, respiration, blood circulation, enzyme modifications, and effects on vital organs.

Information on delayed responses that develop after relatively long exposure to toxicants can be important in assessing potential impact of a chemical in the environment. Bengtsson et al. (1975) showed that about 30% of a population of minnows (*Phoxinus phoxinus*) that survived a 70-day exposure to cadmium developed lesions in the spinal column. Thurberg, Dawson, and Collier (1973) suggest that such parameters may also be used in chronic toxicity studies with invertebrates.

Cough frequency was proposed by Drummond, Spoor, and Olson (1974) to be a good short-term indicator of the long-term effects of methylmercuric chloride. Increases in cough frequency were shown to be proportional to the concentration (from 3 to 12 µg/liter mercury) of methylmercuric and mercuric chloride.

Although aquatic organisms may concentrate cadmium from the low concentrations occurring in their environments, the nature of the effects evoked by chronic cadmium exposure are not known (EPA 1975*b*). Table 9.68 summarizes chronic toxic effects of cadmium on aquatic organisms.

Table 9.68. Chronic toxic effects of cadmium

Chronic dose	Species	Conditions
0.0026 mg/liter	*Daphnia magna*	Threshold of immobilization
0.05–0.10 mg/liter	*Australorbis glabratus*	Produced distress syndromes; distilled water
142 ppm	Sewage organisms	50% inhibition of O_2 utilization; BOD; $CdSO_1$
0.1–0.2 ppm	*Crassostrea virginica*	20-week exposure; little shell growth; lost pigmentation of mantle edge; coloration of digestive diverticulae
50 ppm	*Fundulus heteroclitus*	Pathological changes in intestinal tract, kidney, and gills; changes in essosinophil lineage
10^{-5}–10^{-2} moles	*Fundulus heteroclitus*	Concentration affecting liver enzyme activity

Source: Whitfield 1975, Table 5.8, p. 190.

9.3.8.1 Biochemical interference

Sutterlin (1974) reviews possible types of interaction between metal pollutants and chemosensory organs. He suggests that metals could interfere with the chemical-electrical transducers of aquatic organisms at the reactive sites of the organs. Sutterlin (1974) recommends caution in extending information from laboratory avoidance tests to field conditions, where other variables such as gradient steepness and the presence of other motivational factors exist.

Renfro et al. (1974) studied the effect of methyl- and inorganic mercury on the osmoregulatory mechanisms of fishes. Mercuric and methylmercuric chloride both depressed ion transport in several important osmoregulatory systems of teleosts (*Fundulus heteroclitus*, *Pseudopleuronectes americanus*). Mercuric chloride appeared to be a more effective inhibitor of sodium transport than methylmercury. At concentrations of 4.0×10^{-4} M mercuric chloride, sodium transport was totally abolished, perhaps due partly to direct interference with Na-K-ATPase activity. Methylmercury had a transient inhibitory effect on sodium transport in both gill and urinary bladder. Both mercuric and methylmercuric chloride inhibited Na-K-ATPase activity in whole bladder homogenates. Renfro and coworkers suggest that teleosts in the environment exposed to either of these mercury compounds at sublethal concentrations may experience a decrease in gill and renal sodium transport; thus, the constancy of body fluids would be altered, resulting in a weakened animal, less able to cope with environmental stress.

Glutathione peroxidase activity is proposed as a sensitive index of selenium status of animals by Hoekstra (1975).

Biochemical measurements of a zooplankter (*Daphnia magna*) exposed to various metals were made by Biesinger and Christensen (1972) (Table 9.69). Changes in body weight, total protein, and GOT activity[*] were induced by metal concentrations above those causing 16% reproductive impairment. The metal cations, except for potassium, when added in concentrations having a measurable effect on survival or reproduction, caused a decrease in animal weight. The percentage of protein per animal was notably greater under exposure to calcium, magnesium, strontium, iron, molybdenum, zinc, and cobalt, but less with arsenic or platinum exposure; other elements caused only slight change. GOT activity was stimulated appreciably by addition of sodium, magnesium, strontium, manganese, tin, lead, cobalt, and cadmium, but was inhibited by additional exposure to iron, arsenic, aluminum, nickel, copper, platinum, and mercury. The remaining elements caused only slight alterations of GOT activity. The reasons for the observed increase in protein and GOT activity were not clear, although it was suggested that changes in the molecular dynamics of the animals, such as those involving induction effects in protein-forming enzymes, may have resulted in increased protein metabolism.

9.3.8.2 Respiratory effects

Several investigators have proposed the use of measurements of respiratory distress in fishes to monitor water quality. Morgan and Kuhn (1974) describe an automatic recording device for detecting and measuring respiratory movements. Sublethal concentrations of copper and cadmium were found to markedly increase normal respiratory rates in largemouth bass (*Micropterus salmoides*).

---

[*]Metabolism as measured by glutamic oxalacetic transaminase activity.

Table 9.69. Effect of 3-week exposure to metal chlorides on body weight, total protein, and GOT activity of *Daphnia magna* (expressed as mean percentage deviation from mean control)

Metal ion	Number of experiments	Effect concentrations		Weight/animal ($\% \Delta$)	Protein/animal ($\% \Delta$)	GOT/animal ($\% \Delta$)
		($m \times 10^{-6}$)	($\mu g/liter$)			
Sodium	3	43,500	1,000,000	-18	- 7	+12
Calcium	2	10,000	400,000	-29	+18	+ 3
Magnesium	2	10,300	250,000	-22	+40	+65
Potassium	2	2,040	79,800	+27	- 5	+ 5
Strontium	2	1,140	99,900	-24	+15	+55
Barium	1	144	19,800	-12	0	+ 1
Iron	1	134	7,480	-77	+48	-13
Manganese	3	91.0	5,000	- 8	+24	+100
Arsenic	4	13.3	996	-18	-15	-18
Tin	2	25.3	3,000	-23	+ 5	+15
Chromium	4	11.9	619	-11	- 3	- 4
Aluminum	2	23.0	620	-38	+ 3	-13
Zinc	1	2.68	175	-28	+10	+ 1
Nickel	2	2.13	125	-43	- 9	-26
Lead	3	0.300	62	-12	+ 8	+15
Copper	3	0.628	40	-26	- 5	-10
Platinum	2	0.318	62	-12	-13	-20
Cobalt	1	0.423	24	-15	+12	+45
Mercury	1	0.050	10	- 5	- 5	19
Cadmium	1	0.0089	1	- 7	+ 6	+15

[a]Precision (range) = ±10% (weight); ±8% (protein); ±5% (GOT).

Source: Biesinger and Christensen 1972, Table 5, p. 1696. Reprinted by permission of the publisher.

Sellers, Heath, and Bass (1975) measured several ventilatory characteristics of rainbow trout exposed to sublethal concentrations of zinc and copper. Serial analysis of arterial oxygen tension and blood pH in fish exposed at concentrations approximating the LC_{50} showed that zinc produced a sharp decrease in both arterial oxygen tension and blood pH. Copper, however, caused little effect other than a transient increase in blood pH. These results indicate that the toxic actions of zinc and copper at low concentrations may be different.

Behavior and oxygen consumption rates of the adult mud snail (*Nassarius obsoletus*) exposed individually to arsenic, cadmium, copper, silver, and zinc and to a combination of cadmium and copper were determined by MacInnes and Thurberg (1973). Oxygen uptake of distressed and retracted gastropods was lower than for controls after exposure to all metals except cadmium.

Thurberg, Calabrese, and Dawson (1974) maintain that metal-induced changes in respiration vary according to organism species, water conditions, and the metal being tested. Silver was shown to induce elevated levels of oxygen consumption in several species of mollusc — *Crassostrea virginica*, *Mercenaria mercenaria*, *Mytilus edulis*, and *Mya arenaria*. Copper, however, decreased respiration.

Skidmore, as cited by Katz (1973), suggests that chronic exposure to zinc causes gill damage and copious mucus secretions in fish, culminating in restricted respiration and possibly death. Lloyd (1960) reports that a cytological breakdown of gill epithelium occurred in trout exposed to 20 mg/liter zinc.

Dorn (1974) showed that changes in respiration rate of *Congeria leucophaeata* can be used as a measure of sublethal exposure to inorganic mercury. Departure from normal respiratory rates appeared at concentrations above 0.01 mg/liter mercury. Alexander (1974) determined active and standard rates of oxygen consumption, fatigue velocity, and scope for activity of rainbow trout (*Salmo gairdneri*) exposed five days to sublethal concentrations of mercuric chloride. The incipient lethal concentration was 0.205 mg/liter mercury. Fatigue velocity was the only physiological measure to decrease significantly with mercury concentration in gills and muscle tissue.

9.3.8.3 Reproductive processes

The results of chronic exposure (including measurements of adverse effects on survival, growth, and reproduction) were determined through bioassay for (1) cadmium to the bluegill (*Lepomis macrochirus*) by Eaton (1974), (2) nickel to the fathead minnow (*Pimephales promelas*) by Pickering (1974), and (3) zinc to the rainbow trout by Sinley, Goettl, and Davies (1974). No effects on survival, development, or reproduction of bluegill were attributable to cadmium at 31 µg/liter. The authors state that relatively short-term exposures to embryos and larvae may possibly be substituted for long-term (greater than one generation) assays to determine maximum acceptable toxicant concentrations of cadmium to fishes. Nickel concentrations of 0.38 mg/liter and lower did not adversely affect survival, growth, and reproduction of the fathead minnow. However, a nickel concentration of 0.73 mg/liter caused a significant reduction of fecundity and hatchability of eggs. A report prepared by the European Inland Fisheries Advisory Commission (1974) on water quality criteria for freshwater fishes emphasizes the importance of field studies to validate results of laboratory bioassay data and tentatively recommends, for the maintenance of fish populations, that zinc concentrations not exceed 0.1% of the 7-day TL_m (at 15°C). It also emphasizes that allowances should be made for species variability, mixtures of toxicants, and situations in which dissolved oxygen concentrations are less than saturation.

Reproduction of various species is often affected at levels of pollution well below those that kill adults. The reproductive success of mallard ducks was reduced by half when parents received methylmercury in their diet at 3 ppm dry weight (0.6 ppm wet weight), a concentration occurring in food organisms of mercury-contaminated areas. The decrease in reproductive success resulted from a decrease in egg production, poor hatching success, and increased mortality of young (Heinz 1974).

Biesinger and Christensen (1972) studied the effects of heavy metals to *Daphnia magna*. Their work included a determination of the acute toxicity and the effects of continued exposures to metals on reproduction, growth, and metabolism. The metallic ions tested included magnesium, barium, iron, manganese, arsenic, tin, chromium, aluminum, zinc, nickel, lead, copper, platinum, cobalt, mercury, and cadmium. Table 9.70 summarizes the safe concentrations of heavy metals, as determined by testing aquatic animals [*Daphnia*, brook trout, fathead minnows, and a crustacean (*Gammarus* spp.)] through a complete generation under laboratory conditions.

Table 9.70. Safe concentrations of metal ions for aquatic animals tested through a complete generation under laboratory conditions

Metal	Safe concentration (µg/liter)	Water hardness (as $CaCO_3$) (mg/liter)	Common name	Species
Chromium				
$CrCl_3$	330	45	*Daphnia*	*Daphnia magna*
$Na_2Cr_2O_7$	400	45	Brook trout	*Salvelinus fontinalis*
Copper				
$CuCl_2$	22	45	*Daphnia*	*Daphnia magna*
$CuSO_4$	10.6	31.4	Fathead minnow	*Pimephales promelas*
$CuSO_4$	14.5	198	Fathead minnow	*Pimephales promelas*
$CuSO_4$	4.6 - 8.0	45	*Gammarus*	*Gammarus pseudolimnaeus*
$CuSO_4$	9.5	45	Brook trout	*Salvelinus fontinalis*
Zinc				
$ZnCl_2$	70	45	*Daphnia*	*Daphnia magna*
$ZnSO_4$	<180	203	Fathead minnow	*Pimephales promelas*
Cadmium				
$CdCl_2$	0.17	45	*Daphnia*	*Daphnia magna*
$CdSO_4$	37	200	Fathead minnow	*Pimephales promelas*
Nickel				
$NiCl_2$	30	45	*Daphnia*	*Daphnia magna*
$NiSO_4$	400	200	Fathead minnow	*Pimephales promelas*

Source: Biesinger and Christensen 1972, Table 7, p. 1698. Reprinted by permission of the publisher.

Acute and chronic toxicities of the metals tested on *Daphnia magna* are presented in Table 9.71. Concentrations are also given where 50 and 16% reproductive impairment was observed. The reproductive impairment for fish and crustaceans was noted to be about the same for *Daphnia* exposed to chromium, copper, and zinc.

Young and Nelson (1974) examined effects of heavy metals on motility of sea urchin spermatozoa. Concentrations of several metals greater than those of natural occurrence either accelerated or depressed swimming speed of spermatozoa in dose- or time-dependent fashion.

Sangalang and O'Halloran (1972; 1973) present evidence that cadmium can directly affect testicular steroidogenesis in the brook trout (*Salvelinus fontinalis*) and inhibit androgen synthesis. Testes of mature brook trout were exposed to 25 ppb cadmium, resulting in damage of the testicular tissues and reduction of 11-ketotestosterone, 11β-hydroxytestosterone, and testosterone levels.

Elson (1974) reports the impact of the development of base metal mining on the ecology of northwest Miramichi Atlantic salmon (*Salmo salar*). Effects attributed to metal effluents were a 25% reduction of adult stock from the spawning grounds, removal of about 25% of the rearing ground

Table 9.71. Chronic (3-week) toxicity of various metal ions to *Daphnia magna*
in Lake Superior water

Metal ion	Metal ion concentrations (µg/liter)		
	LC$_{50}$	Reproductive impairment	
		50%	16%
Sodium	1,480,000 (1,180,000-1,840,000)[a] Slope 2.22	1,020,000 (914,000-1,130,000) Slope 1.50	680,000 (596,000-775,000)
Calcium	330,000 (308,000-353,000) Slope 1.36	220,000 (193,000-251,000) Slope 1.94	116,000 (85,000-158,000)
Magnesium	190,000 (167,000-217,000) Slope 2.00	125,000 (112,000-140,000) Slope 1.63	82,000 (65,600-102,500)
Potassium	97,000 (87,000-108,000) Slope 1.42	68,000 (64,000-73,000) Slope 1,28	53,000 (48,000-58,000)
Strontium	86,000 (82,000-90,000) Slope 1.14	60,000 (53,000-68,000) Slope 1.39	42,000 (35,000-50,000)
Barium	13,500 (12,160-15,000) Slope 1.45	8,900 (8,090-9,780) Slope 1.55	5,800 (5,420-6,210)
Iron	5,900 (5,180-6,730) Slope 2.15	5,200 (4,810-5,620) Slope 1.18	4,380 (4,210-4,560)
Manganese	5,700 (5,380-6,040) Slope 1.24	5,200 (4,740-5,740) Slope 1.29	4,100 (3,870-4,350)
Arsenic	2,850 (2,520-3,220) Slope 1.63	1,400 (1,120-1,750) Slope 2.66	520 (330-820)
Tin	42,000 (23,000-76,000) Slope 5,86	1,500 (880-2,550) Slope 4.41	350 (116-1,050)
Chromium	2,000 (650-2,600) Slope 3.70	600 (468-768) Slope 1.82	330 (236-462)
Aluminum	1,400 (1,080-1,820) Slope 2.24	680 (553-836) Slope 2.14	320 (210-490)
Zinc	158 (146-170) Slope 1.40	102 (95-109) Slope 1.48	70 (64-77)
Gold	1,050 (861-1,281) Slope 1.65	180 (128-252) Slope 3.00	60 (30-100)
Nickel	130 (98-173) Slope 2.03	95 (68-132) Slope 3.16	30 (21-43)
Lead	300 (236-381) Slope 2.52	100 (77-130) Slope 3.36	30 (18-49)
Copper	44 (35-55) Slope 1.83	35 (28-43) Slope 1.60	22 (16-29)
Platinum	520 (437-619) Slope 1.51	82 (50-140) Slope 6.22	14
Cobalt	21 (14-31) Slope 1.62	12 (11-15) Slope 1.21	10 (9.6-11.0)
Mercury	13 (9-19) Slope 3.16	6.7 (5.4-8.3) Slope 1.99	3.4 (2.4-4.8)
Cadmium	5 (4.0-6.2) Slope 2.42	0.7 (0.51-0.95) Slope 4.06	0.17 (0.11-0.25)

[a]95% confidence limits.

Source: Biesinger and Christensen 1972, Table 4, pp. 1695-96. Reprinted by permission of the
publisher.

from production of salmon, and a reduction of late-run populations to a level requiring several
generations to recover.

McKim and Benoit (1974) found exposures of brook trout (*S. fontinalis*) through a reproductive
cycle (from yearling through spawning to three-month-old juveniles) to be sufficient to establish
concentrations at which no effects occur attributable to sublethal concentrations of copper. No
adverse effects were found (as determined by survival, growth, and reproductive capacity) in
trout exposed in the second generation from egg to spawning to copper concentrations (9.4, 6.1,
and 4.5 µg/liter) that had no affect on parents exposed from yearling to spawning.

9.3.8.4 Hematological responses

Courtois (1974) examined hematological responses of striped bass (*Roccus saxatilis*) to changes in
diet, salinity, and temperature and to an acute copper exposure. Freshwater fishes maintained at
10°C had higher serum protein and lower liver fat, serum potassium, and sodium than fish maintained at 17°C. Saltwater fish maintained at 10°C had low serum protein and potassium as
compared with those held at 17°C. Freshwater-acclimated adult striped bass displayed a transient
increase in plasma volume following a brief copper exposure. The magnitude of plasma expansion
correlated well with calculated plasma volume estimates based on weight change analyses of
previously tested juveniles. O'Connor and Fromm (1975) found that CH_3Hg inhibited metabolism and
plasma electrolyte regulation of the rainbow trout gill. Abel and Skidmore (1975) compared gills
of rainbow trout exposed to $ZnSO_4$ and sodium lauryl sulfate. Gill responses to pollutants appeared
to be at least partly nonspecific and inflammatory at the concentration and exposure time of the
two substances tested.

Heavy-metal toxins, in addition to producing a direct lethal effect, may also act by interfering
with other metal ions present in biological systems. Coombs (1975) found that sublethal concentrations of cadmium (2 ppm) interfered with normal tissue distributions of copper and zinc in
plaice (*Pleuronectes platessa*), resulting in a conditioned metal deficiency in blood cells, gills,
and serum. Stores of copper and zinc in the liver and kidneys were mobilized by the production
of metallothionein, a protein capable of binding cadmium, copper, mercury, and zinc. The biochemical consequences of these changes have not been studied, nor have the mechanisms by which
cadmium influences copper and zinc metabolism been determined (Webb 1975).

D'Amelio, Russo, and Ferraro (1974) found hemoglobin synthesis in *Carassius auratus* and *Eriphia
spinifrons* to be strongly inhibited by lead. Christensen et al. (1972) indicate that changes in
blood composition of brown bullhead catfish resulted from exposure to low concentrations of
copper. They suggest that chemical changes in the blood could alter the phagocytic properties
of some of the leucocytes and antibodies of the plasma, thus reducing their effectiveness to
combat disease and increasing stress susceptibility of the fish.

Serum levels of copper and zinc in the blue crab were examined in relation to concentrations of
serum protein, serum calcium, and variations in environmental temperature and salinity by
Colvocoresses and Lynch (1975). Significant positive correlations were observed between total
serum protein and serum calcium, but not between serum metal variations and variations in
environmental temperature and salinity. High ambient concentrations of heavy metals are

apparently tolerated by storage in the hepatopancreas or excretion, whereas low ambient concentrations are overcome by binding of the metal with serum proteins.

9.3.8.5 Behavioral responses

Exposure to metal compounds may significantly impair normal behavioral responses of organisms such as avoidance, thereby jeopardizing their survival in the ecosystem (Heinz 1975). Drummond, Spoor, and Olson (1973) evaluate the changes in cough frequency, locomotor activity, and feeding behavior as indicators of sublethal copper toxicity to brook trout (*Salvelinus fontinalis*). Changes were observed in each response within 2 to 24 hr at concentrations as low as 6 to 15 µg/liter copper.

Katz (1973) presents data on the lowest concentrations of arsenic, lead, mercury, and selenium that produced a significant impairment of the avoidance response in goldfish (Table 9.72). These concentrations are also given as fractions of the 50% (LC_{50}) and 1% (LC_1) lethal concentrations for these metals.

Table 9.72. Lowest concentrations of ions that significantly impaired conditioned avoidance responses and their fractions of the 50 and 1% lethal concentrations

Compound	Lowest concentrations of significant impairment				
	LC_{50} (mg/liter)	LC_1 (mg/liter)	(mg/liter)	LC_{50}(%)	LC_1(%)
As	32.0	1.5	0.1	1/320	1/15
Pb (hard water)	110.0	60.0	0.07	1/570	1/857
Pb (soft water)	6.6	1.5			
Hg	0.82	0.36	0.03	1/273	1/120
Se	12.0	1.0	0.25	1/48	1/4

Source: Katz 1973, Table 2, p. 8. Reprinted by permission of the publisher.

Westlake and Kleerekoper (1974) demonstrated that the locomotor response of goldfish to copper was dependent upon the spatial concentration gradient of the metal. Their data suggest attraction to copper-contaminated water when the exposure is to a shallow gradient. An avoidance response resulted when a similar concentration of copper was encountered in a steep gradient. Bengtsson (1974) also proposes the use of behavioral reactions in chronic toxicity testing. A school of 30 minnows (*Phoxinus phoxinus*) developed hyperactivity as a response to a gradually increased exposure to zinc.

Stirling (1975) used burrowing of the bivalve *Tellina tenuis* to measure quantitatively its behavioral response to water pollutants. Concentrations of copper greater than 0.25 mg/liter adversely affected burrowing, and the effect was more marked at higher concentrations.

Solbe and Flook (1975) observed a behavioral change — the loss of instinct to hide during daylight — in the stone loach (*Noemacheilus barbatulus*) when exposed to sublethal concentrations of zinc or cadmium.

9.3.8.6 Cellular responses

Gardner and LaRoche (1973) examined cellular changes induced by copper poisoning in the mechanoreceptors of the lateral line canals in the head of the adult mummichog (*Fundulus heteroclitus*) and Atlantic silverside (*Menidia menidia*). These manifestations, as well as lesions in the olfactory organs, were observed at copper concentrations of 0.5 to 5.0 mg/liter. These concentrations of copper also impaired the emergence of larval forms of *F. heteroclitus* from the zygote, the time required for emergence, and survival.

9.3.8.7 Growth and morphology

Morphological and growth changes in guppies were observed by Crandall and Goodnight (1962; 1963). Newborn guppies were exposed to tapwater containing 1.15 mg/liter zinc. The 79 guppies in the zinc solution grew less rapidly than the 54 controls in zinc-free water, had more mortalities, and exhibited less sexual dimorphism. After 90 days, the median weight of the test fish was 23 mg, whereas that of the control fish was 52 mg. Mortalities were 41% for the test fish and 9% for the control. Only one male of the 79 test fish developed a male reproductive organ, as opposed to 30 to 40% of the males in the control group. The authors also made histopathological examination of newborn guppies that had been exposed for long periods of time to 1.15 and 2.3 mg/liter zinc. Histological examinations of fish that had been exposed to 55 and 65 mg/liter zinc indicated that the blood vessels in the liver were poorly developed and that the mesenteries were lacking in fat. The tubules and glomeruli of the kidneys were distended, the lymphoid tissues of these organs were reduced, and the gonads were underdeveloped. After 90 days the liver contained large vacuoles, and granulocytes had accumulated in the heart muscle. The kidney tubules were even more expanded, the spleen was underdeveloped, and only one fourth of the fish were mature. The controls at 95 days were sexually mature and showed no gross or microscopic abnormalities. The authors remarked that there was no gill damage in any of the fish examined.

A delayed reaction to sublethal concentrations of copper and zinc was reported by Reish et al. (1974): Fatal abnormalities occurred in the second generation of the polychaete *Capitella capitata*.

LITERATURE CITED

Abel, P. D., and Skidmore, J. R. 1975. Toxic effects of an anionic detergent on the gills of rainbow trout. *Water Res.* 9(8): 759-65.

Adams, E. S. 1975. Effects of lead and hydrocarbons from snowmobile exhaust on brook trout *(Salvelinus fontinalis)*. *Trans. Am. Fish. Soc.* 104(2): 363-73.

Adamson, R. H. 1967. Drug metabolism in marine vertebrates. *Fed. Proc.* 26(4): 1047-55.

Ahokas, J. T.; Pelkonen, O.; and Karki, N. T. 1975. Metabolism of polycyclic hydrocarbons by a highly active aryl hydroxylase system in the liver of a trout species. *Biochem. Biophys. Res. Commun.* 63(3): 635-41.

Alexander, D. G. 1974. Mercury effects on swimming and metabolism of trout. In *Proceedings of the international conference on transport of persistent chemicals in aquatic ecosystems, Ottawa, Canada, May 1-3, 1974*, pp. III(65) — III(69). Ottawa, Canada: National Research Council of Canada.

Alexander, J. E.; Foehrenbach, J.; Fisher, S.; and Sullivan, D. 1973. Mercury in striped bass and bluefish. *N. Y. Fish Game J.* 20(2): 147-51.

Ames, B.; Sims, P.; and Grover, P. L. 1972. Epoxides of carcinogenic polycyclic hydrocarbons are frameshift mutagens. *Science* 176: 47-49.

Andelman, J. B., and Suess, M. J. 1970. Polynuclear aromatic hydrocarbons in the water environment. *Bull. W. H. O.* 43: 479-508.

Andelman, J. B. 1973. Incidence, variability and controlling factors for trace elements in natural, fresh waters. In *Trace metals and metal-organic interactions in natural waters*, ed. P. C. Singer, pp. 57-88. Ann Arbor, Mich.: Ann Arbor Science Publishers, Inc.

Andelman, J. B., and Snodgrass, J. E. 1974. Incidence and significance of polynuclear aromatic hydrocarbons in the water environment. *CRC Crit. Rev. Environ. Control* 4(1): 69-83.

Anderson, J. W.; Neff, J. M.; Cox, B. A.; Tatem, H. E.; and Hightower, G. M. 1974*a*. Characteristics of dispersions and water soluble extracts of crude and refined oils and their toxicity to estuarine crustaceans and fish. *Mar. Biol.* 27: 75-88.

Anderson, J. W.; Neff, J. M.; Cox, B. A.; Tatem, H. E.; and Hightower, G. M. 1974*b*. The effects of oil on estuarine animals: Toxicity, uptake and depuration, respiration. In *Pollution and physiology of marine organisms*, ed. F. J. Vernberg and W. B. Vernberg, pp. 285-310. New York: Academic Press.

Anke, M.; Hennig, A.; Grun, M.; Groppel, B.; and Ludke, H. 1976. Cadmium and its influence on plants, animals and man with regard to geologic and industrial conditions. Presented at 10th Annual Trace Substances in Environmental Health Conference, June 8-10, Memorial Union, Univ. of Missouri, Columbia, Mo. (to be published).

Anke, M.; Grun, M.; and Partschefeld, M. 1976. The essentiality of arsenic for animals. Presented at the 10th Annual Trace Substances in Environmental Health Conference, June 8-10, Memorial Union, Univ. of Missouri, Columbia, Mo. (to be published).

Annett, C. S.; D'Itri, F. M.; Ford, J. R.; and Prince, H. H. 1975. Mercury in fish and waterfowl from Ball Lake, Ontario. *J. Environ. Qual.* 4(2): 219-22.

Arthur, J. W., and Leonard, E. N. 1970. Effects of copper on *Gammarus pseudolimnaeus*, *Physa integra*, and *Campeloma decisum* in soft water. *J. Fish. Res. Bd. Can.* 27: 1277-83.

Attari, A. 1973. *The fate of trace constituents of coal during gasification*. Institute of Gas Technology. EPA 650/2-73-004.

Auerbach, S. I. 1975. *Testimony presented to the subcommittee on environment and the atmosphere. U.S. House of Representatives, hearings on costs and effects of chronic low-level environmental pollution, Nov. 12, 1975.*

Ayling, G. M. 1974. Uptake of cadmium, zinc, copper, lead and chromium in the Pacific oyster, *Crassostrea gigas*, grown in the Tamar River, Tasmania. *Water Res.* 8: 729-38.

Baria, D. N. 1975. A survey of trace elements in North Dakota lignite and effluent streams from combustion and gasification facilities. Engineering Experiment Station, Univ. of North Dakota, Grand Forks, N.D.

Barica, J.; Stainton, M. P.; and Hamilton, A. L. 1973. Mobilization of some metals in water and animal tissue by NTA, EDTA and TPP. *Water Res.* 7: 1791-1804.

Baudouin, M. F., and Scoppa, P. 1974. Acute toxicity of various metals to freshwater zooplankton. *Bull. Environ. Contam. Toxicol.* 12(6): 745-50.

Becker, C. D., and Thatcher, T. O., eds. 1973. Toxicity of power plant chemicals to aquatic life. Battelle Pacific Northwest Laboratories, Richland, Wash. WASH-1249; UC-11.

Belitsky, G. A.; Vasiljev, J. M.; Ivanova, O. J.; Lavrova, N. A.; Prigozhina, F. L.; Samoilina, N. L.; Stavrovskaya, A. A.; Khesina, A. Y.; and Shabad, L. M. 1970. Metabolism of benz(a)pyrene by cells of different mammals in vitro and toxic effect of polycyclic hydrocarbons on the cells. *Vopr. Onkol.* 16(2): 53-58.

Benayoun, G.; Fowler, S. W.; and Oregioni, B. 1974. Flux of cadmium through Euphausiids. *Mar. Biol.* 27: 205-12.

Bengtsson, B. E. 1974. The effects of zinc on the mortality and reproduction of the minnow, *Phoxinus phoxinus* L. *Arch. Environ. Contam. Toxicol.* 2(4): 342-355.

Bengtsson, B. E.; Carlin, C. H.; Larsson, A.; and Svanberg, O. 1975. Vertebral damage in minnows, *Phoxinus phoxinus* (L.), exposed to cadmium. *Ambio* 4(4): 166-68.

Benoit, D. A. 1975. Chronic effects of copper on survival, growth, and reproduction of the bluegill *(Lepomis Macrochirus)*. *Trans. Am. Fish. Soc.* 104(2): 353-58.

Biesinger, K. E., and Christensen, G. M. 1972. Metal effects on survival, growth, reproduction, and metabolism of *Daphnia magna*. *J. Fish. Res. Bd. Can.* 29: 1691-1700.

Biesinger, K. E.; Andrew, R. W.; and Arthus, J. W. 1974. Chronic toxicity of NTA (nitrilotriacetate) and metal-NTA complexes to *Daphnia magna*. *J. Fish. Res. Bd. Can.* 31(4): 486-90.

Bishop, J. N., and Neary, B. P. 1974. The form of mercury in freshwater fish. In *Proceedings of the international conference on transport of persistent chemicals in aquatic ecosystems, Ottawa, Canada, May 1-3, 1974,* pp. III(25) — III(29). Ottawa, Canada: National Research Council of Canada.

Blotchy, A. J.; Sullivan, J. F.; Shuman, M. S.; Woodward, G. P.; Voors, A. W.; and Johnson, W. D. 1976. Selenium levels in liver and kidney. Presented at 10th Annual Trace Substances in Environmental Health Conference, June 8-10, Memorial Union, Univ. of Missouri, Columbia, Mo. (to be published).

Blumer, M.; Souza, G.; and Sass, J. 1970. Hydrocarbon pollution of edible shellfish by an oil spill. *Mar. Biol.* 5: 195-202.

Blumer, M. 1971. Scientific aspects of the oil spill problem. *Environ. Affairs* 1: 54-73.

Blumer, M., and Youngblood, W. W. 1975. Polycyclic aromatic hydrocarbons in soils and recent sediments. *Science* 188: 53-55.

Boehm, P. D., and Quinn, J. G. 1976. The effect of dissolved organic matter in sea water on the uptake of mixed individual hydrocarbons and number 2 fuel oil by a marine filter-feeding bivalve *(Mercenaria mercenaria)*. *Estuar. Coastal Mar. Sci.* 4(1): 93-105.

Bohn, A. 1975. Arsenic in marine organisms from West Greenland. *Mar. Pollut. Bull.* 6(6): 87-89.

Bohn, H. L. 1972. Soil absorption of air pollutants. *J. Environ. Qual.* 1(4): 372-77.

Boothe, P. N., and Knauer, G. A. 1972. The possible importance of fecal material in the biological amplification of trace and heavy metals. *Limnol. Oceanogr.* 17(2): 270-74.

Bourget, E. 1974. Environmental and structural control of trace elements in barnacle shells. *Mar. Biol.* 28(1): 27-36.

Bourne, E. W., and Jones, R. W. 1973. Effects of 7,12-dimethylbenz(a)anthracene (DMBA) in fish cells in vitro. *Trans. Am. Microsc. Soc.* 92(1): 140-42.

Bowen, H.J.M. 1966. *Trace elements in biochemistry.* New York: Academic Press.

Boyden, C. R. 1974. Trace element content and body size in molluscs. *Nature* 251: 311-14.

Boylan, D. B., and Tripp, B. W. 1971. Determination of hydrocarbons in seawater extracts of crude oil and crude oil fractions. *Nature* 230: 44-47.

Breedis, C. 1952. Induction of accessory limbs and of sarcoma in the newt (*Triturus viridescens*) with carcinogenic substances. *Cancer Res.* 12(7): 861-73.

Brooks, R. R., and Rumsey, D. 1974. Heavy metals in some New Zealand commercial sea fishes. *N. Z. J. Mar. Freshwater Res.* 8(1): 155-66.

Brown, V. M., and Dalton, R. A. 1970. The acute lethal toxicity to rainbow trout of mixtures of copper, phenol, zinc and nickel. *J. Fish. Biol.* 2: 211-16.

Brown, E. R.; Hazdra, J. J.; Keith, L.; Greenspan, I.; Kwapinski, J.B.G.; and Beamer, P. 1973. Frequency of fish tumors found in a polluted watershed as compared to nonpolluted Canadian waters. *Cancer Res.* 33(2): 189-98.

Brown, V. M. 1976. Advances in testing the toxicity of substances to fish. *Chem. Ind.* 21(4): 143-49.

Brown, V. M.; Shaw, T. L.; and Shurben, D. G. 1974. Aspects of water quality and the toxicity of copper to rainbow trout. *Water Res.* 8(10): 797-803.

Bryan, G. W. 1973. The occurrence and seasonal variation of trace metals in the scallops *Pecten maximus* (L.) and *Chlamys opercularis* (L.). *J. Mar. Biol. Assoc.* 53: 145-66.

Bryan, G. W., and Hummerstone, L. G. 1973*a*. Adaptation of the polychaete *Nereis diversicolor* to estuarine sediments containing high concentrations of zinc and cadmium. *J. Mar. Biol. Assoc.* 53(1): 839-57.

Bryan, G. W., and Hummerstone, L. G. 1973*b*. Adaptation of the polychaete *Nereis diversicolor* to manganese in estuarine sediments. *J. Mar. Biol. Assoc.* 53: 859-72.

Burrell, D. C. 1974. Aqueous heavy-metal pollution. In *Atomic spectrometric analysis of heavy-metal pollutants in water*, pp. 19-45. Ann Arbor, Mich.: Ann Arbor Science Publishers, Inc.

Burrows, W. D., and Krenkel, P. A. 1973. Studies on uptake and loss of methylmercury-203 by bluegills (*Lepomis macrochirus*, Raf.). *Environ. Sci. Technol.* 7(13): 1127-30.

Burrows, W. D.; Taimi, K. L.; and Krenkel, P. A. 1974. The uptake and loss of methylmercury by freshwater fish. In *Congresso internacional del mercurio. Analysis, contamination, biological effects, applications and miscellaneous. Proceedings of a symposium at Barcelona, Spain, May 6-10, 1974*, pp. 283-91.

Byrne, A. R.; Kosta, L.; and Stegnar, P. 1975. The occurrence of mercury in amphibia. *Environ. Lett.* 8(2): 147-55.

Cahnmann, H. J., and Kuratsune, M. 1957. Determination of polycyclic aromatic hydrocarbons in oysters collected in polluted water. *Anal. Chem.* 29(9): 1312-17.

Cairns, J., and Scheier, A. 1959. The relationship of bluegill sunfish body size to tolerance for some common chemicals. *Purdue Univ. Eng. Bull., Proc. 13th Ind. Waste Conf.* 96: 243-52.

Cairns, J.; Heath, A. G.; and Parker, B. C. 1975*a*. The effects of temperature upon the toxicity of chemicals to aquatic organisms. *Hydrobiol.* 47(1): 135-71.

Cairns, J.; Heath, A. G.; and Parker, B. C. 1975*b*. Temperature influence on chemical toxicity to aquatic organisms. *J. Water Pollut. Control Fed.* 47(2): 267-80.

Calabrese, A.; Collier, R. S.; Nelson, D. A.; and MacInnes, J. R. 1973. The toxicity of heavy metals to embryos of the American oyster, *Crassostrea virginica*. *Mar. Biol.* 18(3): 162-66.

Calabrese, A., and Nelson, D. A. 1974. Inhibition of embryonic development of the hard clam, *Mercenaria mercenaria*, by heavy metals. *Bull. Environ. Contam. Toxicol.* 11(1): 92-97.

Calabrese, A.; Collier, R. S.; and Miller, J. E. 1974. *Physiological response of the cunner*, Tautogolabrus adspersus, *to cadmium. I. Introduction and experimental design.* NOAA Tech. Rept. NMFS SSRF-681.

Calamari, D., and Marchetti, R. 1973. The toxicity of mixtures of metals and surfactants to rainbow trout (*Salmo gairdneri*, Rich). *Water Res.* 7: 1453-64.

Cardwell, R. D.; Foreman, D. G.; Payne, T. R.; and Wilbur, D. J. 1976. Acute toxicity of selenium dioxide to freshwater fishes. *Arch. Environ. Contam. Toxicol.* 4(2): 129-44.

Casterline, J. L., and Yip, G. 1975. The distribution and binding of cadmium in oyster, soybean, and rat liver and kidneys. *Arch. Environ. Contam. Toxicol.* 3: 319-29.

Cearley, J. E., and Coleman, R. L. 1974. Cadmium toxicity and bioconcentration in largemouth bass and bluegill. *Bull. Environ. Contam. Toxicol.* 11(2): 146-51.

Ceas, M. P. 1974. Effects of 3-4 benzopyrene on sea urchin egg development. *Acta Embryol. Exp.* (3): 267-72.

Childs, E. A., and Gaffke, J. N. 1974. Lead and cadmium content of selected Oregon groundfish. *J. Fd. Sci.* 39: 853-54.

Chow, T. S.; Patterson, C. C.; and Settle, D. 1974. Occurrence of lead in tuna. *Nature* 251: 159-61.

Christensen, G. M.; McKim, J. M.; Brungs, W. A.; and Hunt, E. P. 1972. Changes in the blood of the brown bullhead (*Ictalurus nebulosus*, Lesueur) following short and long term exposure to copper(II). *Toxicol. Appl. Pharmacol.* 23: 417-27.

Clark, H. F., and Diamond, L. 1971. Comparative studies on the interaction of benzpyrene with cells derived from poikilothermic and homeothermic vertebrates. II. Effect of temperature on benzpyrene metabolism and cell multiplication. *J. Cell. Physiol.* 77: 385-92.

Clark, R. S., and Finley, J. S. 1975. Uptake and loss of petroleum hydrocarbons by the mussel, *Mytilus edulis*, in laboratory experiments. *Fish. Bull.* 73(3): 508-15.

Clubb, R. W.; Gaufin, A. R.; and Lords, J. L. 1975*a*. Acute cadmium toxicity studies upon nine species of aquatic insects. *Environ. Res.* 9(3): 332-41.

Clubb, R. W.; Gaufin, A. R.; and Lords, J. L. 1975*b*. Synergism between dissolved oxygen and cadmium toxicity in five species of aquatic insects. *Environ. Res.* 9(3): 285-89.

Coleman, R. L., and Cearley, J. E. 1974. Silver toxicity and accumulation in largemouth bass and bluegill. *Bull. Environ. Contam. Toxicol.* 12(1): 53-61.

Coleman, H. J.; Hirsch, D. E.; and Dooley, J. E. 1969. Separation of crude oil fractions by gel permeation chromatography. *Anal. Chem.* 6(41): 800-04.

Colvocoresses, J. A., and Lynch, M. P. 1975. Variations in serum constituents of the blue crab, *Callinectes sapidus*: Copper and zinc. *Comp. Biochem. Physiol.* 50A: 135-39.

Coombs, T. L. 1974. The nature of zinc and copper complexes in the oyster *Ostrea edulis*. *Mar. Biol.* 28:1-10.

Coombs, T. L. 1975. The significance of multielement analyses in metal pollution studies. In *Ecological toxicology research. Effects of heavy metal and organohalogen compounds. Proceedings of a NATO science committee conference,* ed. A. D. McIntyre and C. F. Mills, pp. 187-95. New York: Plenum Press.

Comar, C. L., and Nelson, M. 1975. Health effects of fossil fuel combustion products: Report of a workshop. *Environ. Health Porspect.* 12: 149-70.

Connor, P. M. 1972. Acute toxicity of heavy metals to some marine larvae. *Mar. Pollut. Bull.* 3(12): 190-93.

Corner, E.D.S. 1975. The fate of fossil fuel hydrocarbons in marine animals. *Proc. Roy. Soc. (Lond.)* 189: 391-413.

Corner, E.D.S.; Kilvington, C. C.; and O'Hara, S.C.M. 1973. Qualitative studies on the metabolism of naphthalene in *Maia squinado* (Herbst). *J. Mar. Biol. Assoc.* 53: 819-32.

Corner, E.D.S.; Harris, R. P.; Kilvington, C. C.; and O'Hara, S.C.M. 1976. Petroleum compounds in the marine food web: short-term experiments on the fate of naphthalene in *Calanus*. *J. Mar. Biol. Assoc. U.K.* 56: 121-33.

Coutant, C. C., and Pfuderer, H. A. 1974. Thermal effects. A literature review. *J. Water Pollut. Control Fed.* 46(6): 1477-1540.

Coutant, C. C., and Talmage, S. S. 1975. Thermal effects. A literature review. *J. Water Pollut. Control Fed.* 47(6): 1656-1711.

Courtois, L. A. 1974. *Physiological responses of striped bass,* Roccus saxatilus, *to changes in diet, salinity, temperature, and acute copper exposure.* Thesis, University of Calif., Davis, Calif. Ann Arbor, Mich.: Xerox University Microfilms, Order 74-29-297.

Cox, M. F.; Holm, H. W.; Kania, H. J.; and Knight, R. L. 1975. Methylmercury and total mercury concentrations in selected stream biota. Presented at the 9th Annual Conference on Trace Substances at Memorial Union, Univ. of Missouri, Columbia, Mo., June 9-12, 1975.

Crandall, C. A., and Goodnight, C. J. 1962. Effects of sublethal concentrations of several toxicants on growth of the common guppy, *Lebistes reticulatus*. *Limnol. Oceanogr.* 7: 233-39.

Crandall, C. A., and Goodnight, C. J. 1963. The effects of sublethal concentrations of several toxicants to the common guppy, *Lebistes reticulatus*. *Trans. Am. Microsc. Soc.* 82: 59-73.

Cunningham, P. A., and Tripp, M. R. 1973. Accumulation and depuration of mercury in the American oyster, *Crassostrea virginica*. *Mar. Biol.* 20(1): 14-19.

D'Agostino, A., and Finney, C. 1974. The effect of copper and cadmium on the development of *Trigriopus japonicus*. In *Pollution and physiology of marine organisms,* ed. F. J. Vernberg and W. B. Vernberg, pp. 445-63. New York: Academic Press.

D'Amelio, V.; Russo, G; and Ferraro, D. 1974. The effect of heavy metals on protein synthesis in crustaceans and fish. *Rev. Intern. Oceanogr. Med.* 33: 111-18.

Davies, P. H.; Goettl, J. P.; Sinley, J. R.; and Smith, N. F. 1976. Acute and chronic toxicity of lead to rainbow trout, *Salmo gairdneri*, in hard and soft water. *Water Res.* 10(3): 199-206.

de Angelis, E.; Giordano, G. G.; and Pagano, G. 1975. Effects of benzo(a)pyrene and 7,12-dimethylbenz(a)anthracene on sea urchin embryos. *Tumori* 61(1): 97.

DeClerck, R.; Vanderstappen, R.; and Vyncke, W. 1974. Mercury content of fish and shrimps caught off the Belgian coast. *Ocean Manage.* 2: 117-26.

de Freitas, A.S.W.; Qadri, S. U.; and Case, B. E. 1974. Origins and fate of mercury compounds in fish. In *Proceedings of the international conference on transport of persistent chemicals in aquatic ecosystems, Ottawa, Canada, May 1-3, 1974,* pp. III(31) — III(36). Ottawa, Canada: National Research Council of Canada.

De Lustig, E. S., and Matos, E. L. 1971. Teratogenic effects induced in tail of *Bufo arenarum* tadpoles following treatment with carcinogens. *Experientia* 27: 555-56.

Delhaye, W., and Cornet, D. 1975. Contribution to the study of the effect of copper on *Mytilus edulis* during reproductive period. *Comp. Biochem. Physiol.* 50A: 511-18.

Diamond, L., and Clark, H. F. 1970. Comparative studies on the interaction of benzo(a)pyrene with cells derived from poikilothermic and homeothermic vertebrates. I. Metabolism of benzo(a)pyrene. *J. Natl. Cancer Inst.* 45(5): 100-111.

D'Itri, F. M. 1972. *The environmental mercury problem.* Cleveland, Ohio: CRC Press, The Chemical Rubber Co.

Doi, R., and Ui, J. 1973. The distribution of mercury in fish and its form of occurrence. In *Heavy metals in the aquatic environment. Proceedings of a symposium at Vanderbilt University, Dec. 4-7, 1973.* Nashville, Tenn.: Vanderbilt University.

DiSalvo, L. H.; Guard, H. E.; and Hunter, L. 1975. Tissue hydrocarbon burden of mussels as potential monitor of environmental hydrocarbon insult. *Environ. Sci. Technol.* 9(3): 247-51.

Diplock, A. T., and Giasuddin, A.S.M. 1976. A tissue culture system for the study of selenium action and toxicity, and of vitamin E action. Presented at 10th Annual Trace Substances in Environmental Health Conference, June 8-10, Memorial Union, Univ. of Missouri, Columbia, Mo. (to be published).

Dix, T. G.; Martin, A.; Ayling, G. M.; Wilson, K. C.; and Ratkowsky, D. A. 1975. Sand flathead (*Platycephalus bassensis*), an indicator species for mercury pollution in Tasmanian waters. *Mar. Pollut. Bull.* 6(9): 142-44.

Dorn, P. 1974. The effects of mercuric chloride upon respiration in *Congeria leucophaeta*. *Bull. Environ. Contam. Toxicol.* 12(1): 86-91.

Doyle, J. J.; Pfander, W. H.; Grebing, S. E.; and Pierce, J. O. 1972. Effects of dietary cadmium on growth and tissue levels in sheep. In *Trace substances in environmental health. VI. Proceedings of 6th trace substances conference in environmental health, June 13, 14, 15, Memorial Union, Univ. Missouri, Columbia, Mo.*, pp. 181-86.

Drummond, R. A.; Spoor, W. A.; and Olson, G. F. 1973. Some short-term indicators of sublethal effects of copper on brook trout, *Salvelinus fontinalis*. *J. Fish. Res. Bd. Can.* 30(5): 698-701.

Duinker, J. C. 1975. Mobilization of metals in the Dutch Wadden Sea. In *Ecological toxicology research. Effects of heavy metals and organohalogen compounds. Proceedings of a NATO science committee conference*, ed. A. D. McIntyre and C. F. Mills, pp. 167-76. New York: Plenum Press.

Dunn, B. P., and Stich, H. F. 1975. The use of mussels in estimating benzo(a)pyrene contamination of the marine environment. *Proc. Soc. Exp. Biol. Med.* 150: 49-51.

Dunn, B. R., and Stich, H. F. 1976. Release of the carcinogen benzo(a)pyrene from environmentally contaminated mussels. *Bull. Environ. Contam. Toxicol.* 15(4): 398-401.

Eaton, J. G. 1973. Chronic toxicity of a copper, cadmium and zinc mixture to the fathead minnow (*Pimephales promelas*, Rafinesque). *Water Res.* 7: 1723-36.

Eaton, J. G. 1974. Chronic cadmium toxicity to the bluegill (*Lepomis macrochirus*, Rafinesque). *Trans. Am. Fish. Soc.* 103(4): 729-35.

Ehrhardt, M. 1972. Petroleum hydrocarbons in oysters from Galveston Bay. *Environ. Pollut.* 3: 257-71.

Ehrhardt, M., and Heinemann, J. 1975. Hydrocarbons in blue mussels from the Kiel Bight. *Environ. Pollut.* 9: 263-82.

Eisler, R., and Gardner, G. R. 1973. Acute toxicology to an estuarine teleost of mixtures of cadmium, copper and zinc salts. *J. Fish Biol.* 5: 131-42.

Eisler, R., and Kissil, G. W. 1975. Toxicities of crude oil and oil-dispersant mixtures to juvenile rabbitfish, *Siganus rivulatus*. *Trans. Am. Fish. Soc.* 104(3): 571-78.

Elder, J. A., and Gaufin, A. R. 1974. The toxicity of three mercurials to *Pleronarcys californica* (Newport), and some possible physiological effects which influence the toxicities. *Environ. Res.* 7: 169-75.

Elson, P. F. 1974. Impact of recent economic growth and industrial development on the ecology of northwest Miramichi Atlantic salmon (*Salmo salar*). *J. Fish. Res. Bd. Can.* 31(3): 521-44.

Environmental Protection Agency (EPA). 1970. *Water quality criteria data book. Vol. I. Organic chemical pollution of freshwater.* Cambridge, Mass.: A. D. Little, Inc.

Environmental Protection Agency (EPA). 1975*a*. *Scientific and technical assessment report on particulate polycyclic organic matter (PPOM).* EPA-600/6-75-001.

Environmental Protection Agency (EPA). 1975*b*. *Scientific and technical assessment report on cadmium.* EPA-600/6-75-003.

European Inland Fisheries Advisory Commission, Food and Agriculture Organization of the United Nations, Rome, Italy. 1974. Water quality criteria for European freshwater fish. EIFAC Tech. Paper 21. *Water Res.* 8(9): 683-84.

Eustace, I. J. 1974. Zinc, cadmium, copper and manganese in species of finfish and shellfish caught in the Derwent Estuary, Tasmania. *Aust. J. Mar. Freshwater Res.* 25: 209-20.

Falk, H. L.; Miller, A.; and Kotin, P. 1958. The elution of 3,4-benzpyrene and related hydrocarbons from soot by plasma proteins. *Science* 127: 474-75.

Fannick, N; Gonshor, L. T.; and Shockley, J. 1972. Exposure to tar pitch volatiles at coke ovens. *Am. Ind. Hyg. Assoc. J.* 33: 461.

Farrington, J. W., and Quinn, J. G. 1973. Petroleum hydrocarbons and fatty acids in wastewater effluents. *J. Water Pollut. Control Fed.* 45(4): 704-12.

Favretto, L., and Tunis, F. 1974. Typical level of lead in *Mytilus galloprovincialis* (LMK) from the Gulf of Trieste. *Rev. Interm. Oceanogr. Med.* 33: 67-74.

Fleischer, M.; Sarofim, A. F.; Fassett, D. R.; Hammond, P.; Shacklette, H. T.; Nisbet, I.C.T.; and Epstein, S. 1974. Environmental impact of cadmium: A review by the panel on hazardous trace substances. *Environ. Health Perspect.* 253-323.

Fieser, L. F., and Fieser, M. 1961. *Advanced organic chemistry*. New York: Reinhold Publishing Corporation.

Fishbein, L. 1973. Mutagens and potential mutagens in the biosphere. I. DDT and its metabolites, polychlorinated biphenyls, chlorodioxins, polycyclic aromatic hydrocarbons, haloethers. *Sci. Total Environ.* 2(4): 305-40.

Fishbein, L. 1974. Mutagens and potential mutagens in the biosphere. II. Metals: Mercury, lead, cadmium and tin. *Sci. Total Environ.* 2: 341-71.

Fletcher, G. L.; Watts, E. G.; and King, M. J. 1975. Copper, zinc, and total protein levels in plasma of sockeye salmon (*Oncorhynchus nerka*) during their spawning migration. *J. Fish. Res. Bd. Can.* 32(1): 78-82.

Foster, J. A. 1969. Malformations and lethal growths in planaria treated with carcinogens. *Natl. Cancer Inst. Monogr.* 31: 683-91.

Fowler, S. W., and Oregioni, B. 1976. Trace metals in mussels from the N. W. Mediterranean. *Mar. Pollut. Bull.* 7(2): 26-29.

Frazier, J. M. 1975. The dynamics of metals in the American oyster, *Crassostrea virginica*. I. Seasonal effects. *Chesapeake Sci.* 16(3): 162-71.

Freeman, H. C.; Horne, D. A.; McTague, B.; and McMenemy, M. 1974. Mercury in some Canadian Atlantic coast fish and shellfish. *J. Fish. Res. Bd. Can.* 31(3): 369-72.

Freudenthal, R. L.; Lutz, G. A.; and Mitchell, R. I. 1975. *Carcinogenic potential of coal and coal conversion products*. A Battelle Energy Program Report.

Friberg, L.; Piscator, M.; and Nordberg, G. 1971. *Cadmium in the environment*. Cleveland, Ohio: CRC Press, The Chemical Rubber Co.

Frieden, E. 1974. The evolution of metals as essential elements with special reference to iron and copper. In *Protein-metal interactions, advances in experimental medicine and biology,* ed. M. Friedman, pp. 1-31.

Gardner, G. R., and LaRoche, G. 1973. Copper induced lesions in estuarine teleosts. *J. Fish. Res. Bd. Can.* 30(3): 363-68.

Geldiay, R., and Uysal, H. 1975. Comparative behaviour of toxic metals in a marine ecosystem. In *Origin and fate of chemical residues in food, agriculture and fisheries. Proceedings and report of two research coordination meetings, organized by the joint FAO/IAEC division of Atomic Energy in Food and Agriculture,* pp. 69-76. Vienna: International Atomic Energy Agency.

Giblin, F. J., and Massaro, E. J. 1973. Pharmacodynamics of methylmercury in the rainbow trout (*Salmo gairdneri*): Tissue uptake, distribution and excretion. *Toxicol. Appl. Pharmacol.* 24: 81-91.

Giblin, F. J., and Massaro, E. J. 1975. The mechanism of methylmercury transport and transfer to the tissues of the rainbow trout (*Salmo gairdneri*). In *Trace substances in environmental health. Proceedings of 8th annual trace substances in environmental health conference, June 11-13, 1974, Memorial Union, Univ. of Missouri, Columbia Mo.*, pp. 349-55.

Gibson, C. I.; Thatcher, T. O.; and Apts, C. W. 1975. *Some effects of temperature, chlorine and copper on the survival and growth of the coonstripe shrimp,* Pandalus danae. BNWL-SA-5344. Battelle Pacific Northwest Laboratories, Richland, Wash.

Giger, W., and Blumer, M. 1974. Polycyclic aromatic hydrocarbons in the environment: Isolation and characterization by chromatography, visible, ultraviolet and mass spectrometry. *Anal. Chem.* 46: 1663.

Gilmartin, M., and Revelante, N. 1975. The concentration of mercury, copper, nickel, silver, cadmium, and lead in the Northern Adriatic anchovy, *Engradulis encrasicholus,* and sardine, *Sardina pilchardus. Fish. Bull.* 73(1): 193-201.

Gish, C., and Christensen, R. 1973. Cadmium, nickel, lead and zinc in earthworms from roadside soil. *Environ. Sci. Technol.* 7(11): 1060-62.

Gray, J. S. 1974. Synergistic effects of three heavy metals on growth rates of a marine ciliate protozoan. In *Pollution and physiology of marine organisms,* ed. F. J. Vernberg and W. B. Vernberg, pp. 465-85. New York: Academic Press.

Guard, H. E.; Hunter, L.; and DiSalvo, L. H. 1975. Identification and potential biological effects of the major components in the seawater extract of a bunker fuel. *Bull. Environ. Contam. Toxicol.* 14(4): 395-400.

Hardisty, M. W.; Huggins, R. J.; Karter, S.; and Sainsbury, M. 1974. Ecological implications of heavy metals in fish from Severn Estuary. *Mar. Pollut. Bull.* 5(1): 12-15.

Hardisty, M. W.; Karter, S.; and Sainsbury, M. 1974. Dietary habits and heavy metal concentrations in fish from the Severn Estuary and Bristol Channel. *Mar. Pollut. Bull.* 5(4): 61-63.

Harney, P. J. 1974. *Loss of zinc-65 and manganese-54 from the freshwater mollusc* Anodonta. RLO-1227-112-42. Master's thesis, Oregon State Univ., Corvallis, Oreg.

Harrison, R. M.; Perry, R.; and Wellings, R. A. 1975. Review paper. Polynuclear aromatic hydrocarbons in raw, potable and wastewaters. *Water Res.* 9: 331-46.

Hasselrot, T. B., and Gothberg, A. 1974. The ways of transport of mercury to fish. In *Proceedings of the international conference on transport of persistent chemicals in aquatic ecosystems, Ottawa, Canada, May 1-3, 1974,* pp. III(37) — III(47). Ottawa, Canada: National Research Council of Canada.

Havre, G. N.; Underdal, B.; and Christiansen, C. 1973. Cadmium concentrations in some fish species from a coastal area in southern Norway. *Oikos* 24(1): 155-57.

Hayatsu, R.; Scott, R. G.; Moore, L. P.; and Studier, M. H. 1975. Aromatic units in coal. *Nature* 257: 378-80.

Heinz, G. 1974. Effects of low dietary levels of methylmercury on mallard reproduction. *Bull. Environ. Contam. Toxicol.* 11(4): 386-92.

Heinz, G. 1975. Effects of methylmercury on approach and avoidance behavior of mallard ducklings. *Bull. Environ. Contam. Toxicol.* 13(5): 554-64.

Herbes, S. E. 1976. Transport and bioaccumulation of polycyclic aromatic hydrocarbons (PAH) in aquatic systems. In *Coal technology program quarterly progress report for the period ending Dec. 31, 1975,* ORNL-5120, pp. 65-73.

Herbes, S. E.; Southworth, G. R.; and Gehrs, C. W. 1976. Organic contaminants in aqueous coal conversion effluents: Environmental consequences and research priorities. Presented at 10th Annual Trace Substances in Environmental Health Conference, June 8-10, Memorial Union, Univ. of Missouri, Columbia, Mo. (to be published). Draft Pub. 880, Environmental Sciences Division, Oak Ridge National Laboratory, Oak Ridge, Tenn.

Hildebrand, S. C.; Cushman, R. M.; and Carter, J. A. 1976. The potential toxicity and bioaccumulation in aquatic systems of trace elements present in aqueous coal conversion effluents. Presented at 10th Annual Trace Substances in Environmental Health Conference, June 8-10, Memorial Union, Univ. of Missouri, Columbia, Mo. (to be published).

Hirota, R.; Fujiki, M.; and Tajima, S. 1974. Mercury contents of the plankton collected in Ariake- and Yatsushiro-kai. *Bull. Jpn. Soc. Sci. Fish.* 40(4): 393-97.

Hise, E. C., and Fulkerson, W. 1973. Environmental impact of cadmium flow. In *Cadmium, the dissipated element*, ORNL-NSF-EP-21, ed. W. Fulkerson and H. E. Goeller, pp. 203-322. Oak Ridge, Tenn.: Oak Ridge National Laboratory.

Hodson, P. V. 1975. Zinc uptake by Atlantic salmon (*Salmo salar*) exposed to lethal concentration of zinc at 3, 11, and 19°C. *J. Fish. Res. Bd. Can.* 32(12): 2552-56.

Hodson, P. V., and Sprague, J. B. 1975. Temperature-induced changes in acute toxicity of zinc to Atlantic salmon (*Salmo salar*). *J. Fish. Res. Bd. Can.* 32(1): 1-10.

Hoekstra, W. G. 1976. Glutathione peroxidase activity of animal tissues as an index of selenium status. In *Trace substances in environmental health. Proceedings, 9th annual conference of trace substances in environmental health, Memorial Union, Univ. of Missouri, Columbia — June 10, 11, 12, 1975*, pp. 331-37.

Horne, R. A.; Lockeretz, W.; Appleby, A. P.; and Dinman, B. D. 1972. Biological effects of chemical agents. *Science* 177: 1152-54.

Howard, A. G., and Nickless, G. 1975. Protein binding of cadmium, zinc and copper in environmentally insulted limpets, *Patella vulgata*. *J. Chromatogr.* 104(2): 457-59.

Huckabee, J. W.; Feldman, C.; and Talmi, Y. 1974. Mercury concentrations in fish from the Great Smoky Mountains National Park. *Anal. Chim. Acta* 70: 41-47.

Huckabee, J. W., and Griffith, N. A. 1974. Toxicity of mercury and selenium to the eggs of carp (*Cyprinus carpio*). *Trans. Am. Fish. Soc.* 103(4): 822-25.

Huckabee, J. W., and Blaylock, B. G. 1975. Microcosm studies on the transfer of mercury, cadmium, and selenium from terrestrial to aquatic ecosystems. In *Trace substances in environmental health. VIII. Proceedings of Univ. of Missouri's 8th annual conference on trace substances in environmental health, Memorial Union, June 11-13, Univ. of Missouri, Columbia, Mo.*, pp. 219-22.

Hueper, W. C. 1956*a*. Experimental carcinogenic studies on hydrogenated coal oils. I. Berguis oils. *Ind. Med.* 25: 51-55.

Hueper, W. C. 1956*b*. Experimental carcinogenic studies on hydrogenated coal oils. II. Fischer-Tropsch oils. *Ind. Med.* 25: 456-59.

Huggett, R. J.; Bender, M. E.; and Slone, H. D. 1973. Utilizing metal concentration relationships in the eastern oyster (*Crassostrea virginica*) to detect heavy metal pollution. *Water Res.* 7: 451-60.

Huschenbeth, E., and Harms, U. 1975. On the accumulation of organochlorine pesticides, PCB and certain heavy metals in fish and shellfish from Thai coastal and inland waters. *Arch. Fisch. Wiss.* 26(3): 109-22.

Hutcheson, M. S. 1974. The effect of temperature and salinity on cadmium uptake by the blue crab, *Callinectes sapidus*. *Chesapeake Sci.* 15: 237-42.

Ichikawa, R., and Ohno, S. 1974. Levels of cobalt, cesium and zinc in some marine organisms in Japan. *Bull. Jpn. Soc. Sci. Fish.* 40(5): 501-08.

Ireland, M. P. 1974. Variations in the zinc, copper, manganese and lead content of *Balanus balanoides* in Cardigan Bay, Wales. *Environ. Pollut.* 7: 65-75.

Jeng, S. S., and Lo, H. W. 1974. High zinc concentration in common carp viscera. *Bull. Jpn. Soc. Sci. Fish.* 40(5): 509.

Jernelov, A. 1972. Factors in the transformation of mercury to methylmercury. In *Environmental mercury contamination*, ed. R. Hartung and B. D. Dinman, pp. 167-78. Ann Arbor, Mich.: Ann Arbor Science Publishers, Inc.

Jones, M. B. 1973. Influence of salinity and temperature on the toxicity of mercury to marine and brackish water isopods (Crustacea). *Estuar. Coastal Mar. Sci.* 1: 425-31.

Katz, M. 1973. The effects of heavy metals on fish and aquatic organisms. In *Heavy metals in the aquatic environment. Proceedings of a symposium at Vanderbilt University, Dec. 4-7, 1973.* Nashville, Tenn.: Vanderbilt University.

Kelso, J.R.M., and Frank, R. 1974. Organochlorine residues, mercury, copper and cadmium in yellow perch, white bass and smallmouth bass, Long Point Bay, Lake Erie. *Trans. Am. Fish. Soc.* 103(3): 577-81.

Ketcham, B. H. 1975. Problems in aquatic ecosystems, with special reference to heavy metal pollution of the marine environment. In *Ecological toxicology research. Effects of heavy metal and organohalogen compounds. Proceedings of a NATO science committee conference,* ed. A. D. McIntyre and C. F. Mills, pp. 76-81. New York: Plenum Press.

Ketcham, B. H.; Zitko, V.; and Saward, D. 1975. Aspects of heavy metal and organohalogen pollution in aquatic ecosystems. In *Ecological toxicology research. Effects of heavy metal and organohalogen compounds. Proceedings of a NATO science committee conference,* ed. A. D. McIntyre and C. F. Mills, pp. 76-90. New York: Plenum Press.

Knight, H. T., and Olson, L. T. 1974. Mercury distribution in American smelt from Lake Michigan. *Am. Midl. Nat.* 91(2): 451-52.

Knorr, M., and Schenk, D. 1968. The question of the synthesis of polycyclic aromatics by bacteria. *Arch. Hyg.* 152(3): 282-85.

Koeman, J. H.; Van de Ven, W.S.M.; de Goeij, J.J.M.; Tijoe, P. S.; and Van Haaften, J. L. 1975. Mercury and selenium in marine mammals and birds. *Sci. Total Environ.* 3: 279-87.

Kopfler, F. C. 1974. The accumulation of organic and inorganic mercury compounds by the eastern oyster *(Crassostrea virginica). Bull. Environ. Contam. Toxicol.* 11(3): 275-80.

Koppenaal, D., and Manahan, S. E. 1976. Formation of organometallics and carcinogens in coal conversion processes and shale oil extraction. Presented at the 10th Annual Trace Substances in Environmental Health Conference, June 8-10, Memorial Union, Univ. of Missouri, Columbia, Mo. (to be published).

Kornreich, M. R. 1975. *A preliminary assessment of the problem of carcinogens in the atmosphere.* MITRE Corporation Tech. Rept. MTR-6874. McLean, Va.

Kumada, H.; Kimura, S.; Yokote, M.; and Matida, Y. 1972. Acute and chronic toxicity, uptake and retention of cadmium in freshwater organisms. *Bull. Freshwater Fish Res. Lab. (Tokyo)* 22(2): 157-65.

Laumond, F.; Neuburger, M.; Donnier, B.; Fourcy, A.; Bittel, R.; and Aubert, M. 1973. Experimental investigations, at laboratory, on the transfer of mercury in marine trophic chains. *Rev. Intern. Oceanogr.* 31-32: 47-53.

Leclerc, E., and Devlaminck, F. 1950. Toxicological study of some substances generally present in coking plant effluents. *Centre belge d'Etudes et de Documentation des Eaux* 8: 486-93. ORNL/tr-2973.

Lee, R. F. 1975. Fate of petroleum hydrocarbons in marine zooplankton. In *American Petroleum Institute. Proceedings 1975 conference on prevention and control of oil pollution,* pp. 549-53. Washington, D.C.

Lee, R. F.; Ryan, C.; and Neuhauser, M. L. 1976. Fate of petroleum hydrocarbons taken up from food and water by the blue crab *(Callinectes sapidus). Mar. Biol.* 37: 363-70.

Lee, R. F.; Sauerheber, R.; and Benson, A. A. 1972. Petroleum hydrocarbons: Uptake and discharge by the marine mussel *Mytilus edulis. Science* 177: 345-46.

Lee, R. F.; Sauerheber, R.; and Dobbs, G. H. 1972. Uptake, metabolism and discharge of polycyclic aromatic hydrocarbons by marine fish. *Mar. Biol.* 17: 201-08.

Legore, R. S. 1974. *The effect of Alaskan crude oil and selected hydrocarbon compounds on embryonic development of the Pacific oyster,* Crassostrea gigas. Thesis, Ph.D., University of Washington. Ann Arbor, Mich.: Xerox University Microfilms, Order No. 74-29-447.

Leland, H. V., and McNurney, J. M. 1974. Lead transport in a river ecosystem. In *Proceedings of the international conference on transport of persistent chemicals in aquatic ecosystems, Ottawa, Canada, May 1-3, 1974,* pp. III(17) — III(23). Ottawa, Canada: National Research Council of Canada.

Leland, H. V.; Shukla, S. S.; and Shimp, N. F. 1973. Factors affecting distribution of lead and other trace elements in sediment of southern Lake Michigan. In *Trace metals and metal-organic interactions in natural waters*, ed. P. C. Singer, pp. 89-129. Ann Arbor, Mich.: Ann Arbor Science Publishers, Inc.

Lloyd, R. 1960. The toxicity of zinc sulphate to rainbow trout. *Ann. Appl. Biol.* 48(1): 84-94.

Lockhart, W. L.; Uthe, J. F.; Kenney, A. R.; and Mehrle, P. M. 1973. Studies on methylmercury in northern pike (*Esox lucius*, L.). In *Trace substances in environmental health — VI. Proceedings of the sixth annual conference on trace substances in environmental health, Memorial Union, University of Missouri, Columbia, June 13-15, 1972*, ed. D. D. Hemphill, pp. 115-18. Columbia, Mo.: University of Missouri.

Luh, M. D.; Baker, R. A.; and Henley, D. 1973. Arsenic analysis and toxicity — a review. *Sci. Total Environ.* 2: 1-12.

Lunde, G. 1969. Water soluble arseno-organic compounds in marine fishes. *Nature* 224: 186-87. Also in *The analysis and characterization of trace elements, in particular, bromine, selenium and arsenic, in marine organisms*. Oslo, Norway: Central Institute for Industrial Research.

Lunde, G. 1973*a*. Separation and analysis of organic-bound and inorganic arsenic in marine organisms. *J. Sci. Fd. Agric.* 24: 1021-27. Also in *Trace elements, in particular, bromine, selenium and arsenic, in marine organisms*. Oslo, Norway: Central Institute for Industrial Research.

Lunde, G. 1973*b*. The absorption and metabolism of arsenic in fish. *Fiskeridirektoratets Skrifter. Ser. Tekn. Undersok.* 5(12): 1-16. Also in *The analysis and characterization of trace elements, in particular, bromine, selenium, and arsenic, in marine organisms*. Oslo, Norway: Central Institute for Industrial Research.

Luoma, S. N. 1974. *Mercury cycling in a small Hawaiian estuary*. Water Resources Research Center, University of Hawaii, Honolulu. Tech. Memo. Rept. 42; Sea Grant College Working Paper 5; COM 75-11170.

Luoma, S. N., and Jenne, E. A. 1976. Estimating bioavailability of sediment-bound trace metals with chemical extractants. Presented at 10th Annual Trace Substances in Environmental Health Conference, June 8-10, 1976, Memorial Union, Univ. of Missouri, Columbia, Mo. (to be published).

MacInnes, J. R., and Thurberg, F. P. 1973. Effects of metals on the behaviour and oxygen consumption of the mud snail. *Mar. Pollut. Bull.* 4(12): 185-86.

MacKay, N. J.; Kazacos, M. N.; Williams, R. J.; and Leedow, M. I. 1975. Selenium and heavy metals in black marlin. *Mar. Pollut. Bull.* 6(4): 57-61.

MacKay, N. J.; Williams, R. J.; Kacprzac, J. L.; Kazacos, N. M.; Collins, A. J.; and Auty, E. H. 1975. Heavy metals in cultivated oysters (*Crassostrea commercialis = Saccostrea cucullata*) from the estuaries of New South Wales. *Aust. J. Mar. Freshwater Res.* 26: 31-46.

Matos, E. L., and De Lustig, E. S. 1973. Teratogenic effects of carcinogen implantation in a regenerative field in *Bufo arenarium* tadpoles. *Teratology* 8: 167-74.

McAuliffe, C. 1966. Solubility in water of paraffin, cycloparaffin, olefin, acetylene, cycloolefin, and aromatic hydrocarbons. *J. Phys. Chem.* 70: 1267-75.

McGinnes, P. R., and Snoeyink, V. L. 1974. *Determination of the fate of polynuclear aromatic hydrocarbons in natural water systems*. University of Illinois at Urbana, Water Resources Center Report. UILU-WRC-74-0080.

McIntosh, A. W., and Kevern, N. R. 1974. Toxicity of copper to zooplankton. *J. Environ. Qual.* 3(2): 166-70.

McKee, J. E., and Wolf, H. W., eds. 1963. *Water quality criteria*. 2nd ed. Publication 3-A. Pasadena, Calif.: California State Water Resources Control Board.

McKim, J. M., and Benoit, D. A. 1974. Duration of toxicity tests for establishing "no effect" concentrations for copper with brook trout (*Salvelinus fontinalis*). *J. Fish. Res. Bd. Can.* 31(4): 449-52.

McLeese, D. W. 1974. Toxicity of copper at two temperatures and three salinities to the American lobster (*Homarus americanus*). *J. Fish. Res. Bd. Can.* 31(12): 1949-52.

McLusky, D. S., and Phillips, C.N.K. 1975. Some effects of copper on the polychaete *Phyllodoce maculata*. *Estuar. Coastal Sci.* 3(1): 103-08.

Mellinger, P. J. 1973. The comparative metabolism of two mercury compounds as environmental contaminants in the freshwater mussel, *Margaritifera margaritifera*. In *Trace substances in environmental health — VI. Proceedings of the sixth annual conference on trace substances in environmental health, Memorial Union, University of Missouri, Columbia, June 13-15, 1972*, ed. D. D. Hemphill, pp. 173-80.

Middaugh, D. P., and Rose, C. L. 1974. Retention of two mercurials by striped mullet, *Mugil cephalus*. *Water Res.* 8: 173-77.

Miettinen, J. K. 1975. The accumulation and excretion of heavy metals in organisms. In *Ecological toxicology research. Effects of heavy metal and organohalogen compounds. Proceedings of a NATO science committee conference*, ed. A. D. McIntyre and C. F. Mills, pp. 215-29. New York: Plenum Press.

Mironov, O. G. 1972. Effect of oil pollution on flora and fauna of the Black Sea. In *Marine pollution and sea life*, ed. M. Ruivo, pp. 222-24. London: Fishing News (Books) Ltd.

Montgomery, J. R.; Kolehmainen, S. E.; Banus, M. D.; Bendien, B. J.; Donaldson, J. L.; and Ramfrez, J. A. 1975. *Individual variation of trace metal content in fish.* CONF-741023-5.

Moore, S. F., and Dwyer, R. L. 1974. Effects of oil on marine organisms: A critical assessment of published data. *Water Res.* 8(10): 819-27.

Morgan, W.S.G., and Kuhn, P. C. 1974. A method to monitor the effects of toxicants upon breathing rate of largemouth bass (*Micropterus salmoides*, Lacepede). *Water Res.* 8: 67-77.

Morrow, J. E.; Gritz, R. L.; and Kirton, M. P. 1975. Effects of some components of crude oil on young Coho salmon. *Copeia* 2: 326-31.

Mouw, D.; Kalitis, K.; Anver, M.; Schwartz, J.; Constan, A.; Hartung, R.; Cohen, B.; and Ringler, D. 1975. Lead. Possible toxicity in urban vs. rural rats. *Arch. Environ. Health* 30: 276-80.

Naidu, J. R. 1974. *Radioactive zinc (^{65}Zn), zinc, cadmium, and mercury in the Pacific hake* Merluccius productus *(Ayres) off the west coast of the United States.* RLO-2227-T12-47. Ph.D. thesis, Oregon State University, Corvallis, Oreg.

National Academy of Sciences (NAS). 1973. *Water quality criteria.* Washington, D.C.: Environmental Studies Board, National Academy of Sciences. Also EPA-R3-73-033.

National Academy of Sciences (NAS). 1974. *Chromium*, ed. A. M. Baetjer. Washington, D.C.: National Research Council Committee on Biologic Effects of Atmospheric Pollutants, National Academy of Sciences.

National Academy of Sciences (NAS). 1972. *Particulate polycyclic organic matter.* Washington, D.C.: National Research Council Committee on Biologic Effects of Atmospheric Pollutants, National Academy of Sciences.

National Academy of Sciences (NAS). 1975a. *Petroleum in the marine environment. Workshop on inputs, fates, and effects of petroleum in the marine environment, May 21-25, 1973, Ocean Affairs Board, Airlie House, Airlie, Va.* Washington, D.C.: Ocean Affairs Board, National Academy of Sciences.

National Academy of Sciences (NAS). 1975b. *Principles for evaluating chemicals in the environment. A Report of the committee for the working conference on principles of protocols for evaluating chemicals in the environment.* Washington, D.C.: Environmental Studies Board, National Academy of Sciences-National Academy of Engineering.

National Academy of Sciences (NAS). 1975c. *Nickel.* Washington, D.C.: National Research Council Committee on Medical and Biological Effects of Environmental Pollutants, National Academy of Sciences.

Natusch, D.F.S. 1976. Characteristics of pollutants from coal combustion and conversion processes. In *Toxic effects on the aquatic biota from coal and oil shale development. Quarterly progress report October-December 1975*, pp. 73-90. Natural Resource Ecology Laboratory, Colorado State University, Fort Collins.

Natusch, D.F.S.; Wallace, J. R.; and Evans, C. A. 1974. Toxic trace elements: Preferential concentration in respirable particles. *Science* 183: 202-04.

Natusch, D.F.S., and Wallace J. R. 1974. Urban aerosol toxicity: The influence of particle size. *Science* 186: 695-99.

Navrot, J.; Amiel, A. J.; and Kronfeld, J. 1974. *Patella vulgata:* A biological monitor of coastal metal pollution — a preliminary study. *Environ. Pollut.* 7(4): 303-08.

Neathery, M. W., and Miller, W. J. 1975. Metabolism and toxicity of cadmium, mercury, and lead in animals: A review. *J. Dairy Sci.* 58(12): 1767-80.

Neely, W. B.; Branson, D. R.; and Blau, G. E. 1974. Partition coefficient to measure bioconcentration potential of organic chemicals in fish. *Environ. Sci. Technol.* 8: 1113-15.

Neff, J. M. 1975. Accumulation and release of petroleum-derived aromatic hydrocarbons by marine animals. Symposium on the chemistry, occurrence and measurement of polynuclear aromatic hydrocarbons. *ACS Petrol. Chem. Div. Prepr.* 20(4): 839-50.

Neff, J. M., and Anderson, J. W. 1975. Accumulation, release and distribution of benzo(a)-pyrene-C^{14} in the clam, *Rangia cuneata*. In *Proceedings of joint conference on prevention and control of oil pollution*, pp. 469-71. Washington, D.C.: American Petroleum Institute.

Nehring, R. B., and Goettl, J. P. 1974. Acute toxicity of a zinc-polluted stream to four species of salmonids. *Bull. Environ. Contam. Toxicol.* 12(4): 464-69.

Neukomm, S. 1974. The newt test for studying certain categories of carcinogenic substances. In *Experimental model systems in toxicology and their significance in man. Proceedings of the European Society for the study of drug toxicity, international conference*, vol. 15, pp. 228-35.

Nielsen, S. A. 1975. Cadmium in New Zealand dredge oysters: Geographic distribution. *Int. J. Environ. Anal. Chem.* 4(1): 1-7.

Niimi, A. J., and LaHam, Q. N. 1975. Selenium toxicity on the early life stages of zebrafish *(Brachydanio rerio). J. Fish. Res. Bd. Can.* 32(6): 803-06.

Nobles, L., and Blanton, C. D. 1964. Thiophene compounds of biological interest. Review article. *J. Pharm. Sci.* 53(2): 115-29.

Nomiyama, K. 1973. Toxicity and physiology. I. Toxicity of heavy metals including mercury. In *Proceedings of heavy metals in the aquatic environment, Vanderbilt University, Dec. 4-7.* Nashville, Tenn.: Vanderbilt University.

O'Conner, D. V., and Fromm, P. O. 1975. The effect of methylmercury on gill metabolism and blood parameters of rainbow trout. *Bull. Environ. Contam. Toxicol.* 13(4): 406-11.

Ogata, M., and Miyake, Y. 1973. Identification of substances in petroleum causing objectionable odour in fish. *Water Res.* 7: 1493-1504.

Ogata, M., and Miyake, Y. 1975. Compound from floating petroleum accumulating in fish. *Water Res.* 9(12): 1075-78.

Olsen, D., and Haynes, J. L. 1969. *Preliminary air pollution survey of organic carcinogens.* National Air Pollution Control Administration, Rept. Pub. APTD 69-43.

Olson, G. F.; Mount, D. I.; Snarski, V. M.; and Thorslund, T. W. 1975. Mercury residues in fathead minnows, *Pimephales promelas* (Rafinesque), chronically exposed to methylmercury in water. *Bull. Environ. Contam. Toxicol.* 14(2): 129-34.

Olson, K. R., and Harrel, R. C. 1973. Effect of salinity on acute toxicity of mercury, copper, and chromium for *Rangia cunata* (Pelecypoda, mactridae). *Contrib. Mar. Sci.* 17: 9-13.

Pagenkopf, G. K.; Russo, R. C.; and Thurston, R. V. 1974. Effect of complexation on toxicity of copper to fishes. *J. Fish. Res. Bd. Can.* 31(4): 462-65.

Parke, D. V., and Cohen, E. M., chairmen. 1975. B/C. Uptake, fate, and action of heavy metals and organohalogen compounds in living organisms. In *Ecological toxicology research. Effects of heavy metal and organohalogen compounds. Proceedings of a NATO science committee conference*, ed. A. D. McIntyre and C. F. Mills, pp. 257-83. New York: Plenum Press.

Patel, B., and Ganguly, A. K. 1973. Occurrence of ^{75}Se and ^{113}Sn in oysters. *Health Phys.* 24: 559-62.

Patrick, R.; Cairns, J.; and Scheier, A. 1968. The relative sensitivity of diatoms, snails, and fish to twenty common constituents of industrial wastes. *Prog. Fish-Cult.* 30(3): 137-40.

Payne, J. F. 1976. Field evaluation of benzopyrene hydroxylase induction as a monitor for marine petroleum pollution. *Science* 191: 945-46.

Payne, J. F., and Penrose, W. R. 1975. Induction of aryl hydrocarbon (benzo[a]pyrene) hydroxylase in fish by petroleum. *Bull. Environ. Contam. Toxicol.* 14(1): 112-16.

Pedersen, M. G.; Hershberger, W. K.; and Juchau, M. R. 1974. Metabolism of 3,4-benzpyrene in rainbow trout *(Salmo gairdneri)*. *Bull. Environ. Contam. Toxicol.* 12(4): 481-86.

Penrose, W. R. 1975. Biosynthesis of organic arsenic compounds in brown trout *(Salmo trutta)*. *J. Fish. Res. Bd. Can.* 32(12): 2385-90.

Penrose, W. R.; Black, R.; and Hayward, M. J. 1975. Limited arsenic dispersion in sea water, sediments, and biota near a continuous source. *J. Fish. Res. Bd. Can.* 32(8): 1275-81.

Pentreath, R. J. 1973. The accumulation from water of ^{65}Zn, ^{54}Mn, ^{58}Co and ^{59}Fe by the mussel, *Mytilus edulis.* *J. Mar. Biol. Assoc.* 53: 127-43.

Persoone, G., and Uyttersprot, G. 1975. The influence of inorganic and organic pollutants on the rate of reproduction of a marine hypotrichous cilate: *Euplotes vannus* (Muller). *Rev. Intern. Oceanogr. Med.* 37-38: 125-51.

Peterson, C. L.; Klawe, W. L.; and Sharp, G. D. 1973. Mercury in tunas: A review. *Fish. Bull.* 71(3): 103-13.

Pickering, Q. H. 1974. Chronic toxicity of nickel to the fathead minnow. *J. Water Pollut. Control Fed.* 46(4): 760-65.

Pickering, Q. H., and Henderson, C. 1966. Acute toxicity of some important petrochemicals to fish. *J. Water Pollut. Control Fed.* 38(9): 1419-29.

Pierce, R. C., and Katz, M. 1975. Dependency of polynuclear aromatic hydrocarbon content on size distribution of atmospheric aerosols. *Environ. Sci. Technol.* 9(4): 347-53.

Pizzarello, D. J., and Wolsky, A. 1966. Carcinogenesis and regeneration in newts. *Experientia* 22: 387-88.

Powell, N. A.; Sayce, C. S.; and Tufts, D. F. 1970. Hyperplasia in an estuarine bryozoan attributable to coal tar derivatives. *J. Fish. Res. Bd. Can.* 27(10): 2095-98.

Price, P. W.; Rathcke, B. J.; and Gentry, D. A. 1974. Lead in terrestrial arthropods: Evidence for biological concentrations. *Environ. Entomol.* 3(3): 370-72.

Pringle, B. H.; Hissong, D. E.; Katz, E. L.; and Mulawaka, S. T. 1968. Trace metal accumulation by estuarine mollusks. *J. Sanit. Engr. Div. Proc. ASCE* 94(SA3): 455-75.

Rasmussen, L. F., and Williams, D. C. 1975. The occurrence and distribution of mercury in marine organisms in Bellingham Bay. *Northwest Sci.* 49(2): 87-94.

Ratkowsky, K. A.; Thrower, S. J.; Eustace, I. J.; and Olley, J. 1974. A numerical study of concentration of some heavy metals in Tasmania oysters. *J. Fish. Res. Bd. Can.* 31(7): 1165-71.

Reichert, W. L., and Malins, D. C. 1974. Interaction of mercurials with salmon serum lipoproteins. *Nature* 247: 569-60.

Reimer, A. A., and Reimer, R. D. 1975. Total mercury in some fish and shellfish along the Mexican coast. *Bull. Environ. Contam. Toxicol.* 14(1): 105-11.

Reish, D. J.; Piltz, F.; Martin, J. M.; and Word, J. Q. 1974. Induction of abnormal polychaete larvae by heavy metals. *Mar. Pollut. Bull.* 5(8): 125-26.

Renfro, J. L.; Schmidt-Nielsen, B.; Miller, D.; Benos, D.; and Allen, J. 1974. Methylmercury and inorganic mercury: Uptake, distribution, and effect on osmoregulatory mechanisms in fishes. In *Pollution and physiology of marine organisms,* ed. F. J. Vernberg and W. B. Vernberg, pp. 101-22. New York: Academic Press.

Renfro, W. C. 1973. Transfer of ^{65}Zn from sediments of marine polychaete worms. *Mar. Biol.* 21(4): 305-16.

Renfro, W. C.; Fowler, S. W.; Heyraud, M.; and La Rosa, J. 1975. Relative importance of food and water in long-term zinc-^{65} accumulation by marine biota. *J. Fish. Res. Bd. Can.* 32(8): 1339-45.

Renzoni, A. 1975. Toxicity of three oils to bivalve gametes and larvae. *Mar. Pollut. Bull.* 6(8): 125-28.

Roales, R. R., and Perlmutter, A. 1974. Toxicity of zinc and Cygon, applied singly and jointly, to zebrafish embryos. *Bull. Environ. Contam. Toxicol.* 12(4): 475-80.

Roesijadi, G.; Petrocelli, S. R.; Anderson, J. W.; Presley, B. J.; and Sims, R. 1974. Survival and chloride ion regulation of the porcelain crab *Petrolisthes armatus* exposed to mercury. *Mar. Biol.* 27: 213-17.

Rossi, S. S.; Anderson, J. W.; and Ward, G. S. 1976. Toxicity of water-soluble fractions of four test oils for the polychaetous annelids, *Neanthes arenaceodentata* and *Capitella capitata*. *Environ. Pollut.* 10: 9-18.

Ruben, L. N., and Balls, M. 1964. The implantation of methylcholanthrene crystals into regenerating and non-regenerating forelimbs of *Xenopus laevis*. *J. Morph.* 115: 239-54.

Rubin, E. S., and McMichael, F. C. 1975. Impact of regulations on coal conversion plants. *Environ. Sci. Technol.* 9(2): 112-17.

Ruch, R. R.; Gluskoter, H. J.; and Shimp, N. F. 1974. *Occurrence and distribution of potentially volatile trace elements in coal: A final report.* Illinois State Geological Survey, Environmental Geology Notes No. 72.

Ryckman, D. W.; Prabhakara-Roa, A.V.S.; and Buzzell, J. C. 1975. *Behavior of organic chemicals in the aquatic environment. A literature critique.* Environmental and Sanitary Engineering Labs, Washington University, St. Louis. Washington, D.C.; Manufacturing Chemists Association.

Sangalang, G. B., and O'Halloran, M. J. 1972. Cadmium-induced testicular injury and alterations of androgen synthesis in brook trout. *Nature* 240: 470-71.

Sangalang, G. B., and O'Halloran, M. J. 1973. Adverse effects of cadmium on brook trout testis and on in vitro testicular androgen synthesis. *Biol. Reprod.* 9: 394-403.

Sawicki, E. 1967. Airborne carcinogens and allied compounds. *Arch. Environ. Health* 14: 46.

Sayler, G. S.; Nelson, J. D.; and Colwell, R. R. 1975. Role of bacteria in bioaccumulation of mercury in the oyster *Crassostrea virginica*. *Appl. Microbiol.* 30(1): 91-96.

Scaccini-Cicatelli, M. 1966. Sui fenomeni di accumulo del benzo-3,4-pirene nell organismo di turbifex. *Soc. Ital. Biol. Sperim. Boll.* 42(15): 957-59.

Scaccini, A.; Scaccini-Cicatelli, M.; Marani, F.; and Leonardi, V. 1970. Dosing and observing by means of spectrophotofluorimetry and fluorescence microscopy 3,4-benzopyrene on the organs and tissues of *Carassius auratus*. *Note Del Laboratorio Di Biologiamarina E Pesca-Fano, Bologna, Italy.* 3(6): 105-44.

Scheier, A., and Kiry, P. 1973. *The Delaware Estuary system, environmental impacts and socioeconomic effects. A discussion of the effects of certain potential toxicants on fish and shellfish in the upper Delaware Estuary.* Philadelphia, Penn.: Academy of Natural Sciences.

Schmidt-Nielsen, B. 1974. Osmoregulation: Effect of salinity and heavy metals. *Proc. Am. Soc. Exp. Biol.* 33(5): 2137-46.

Schulz-Baldes, M. 1974. Lead uptake from sea water and food, and lead loss in the common mussel *Mytilus edulis*. *Mar. Biol.* 25(3): 177-93.

Scott, D. P. 1974. Mercury concentration of white muscle in relation to age, growth, and condition in four species of fishes from Clay Lake, Ontario. *J. Fish. Res. Bd. Can.* 31(11): 1723-29.

Seagle, S. M., and Ehlmann, A. J. 1974. Manganese, zinc and copper in water, sediments and mussels in north central Texas reservoirs. In *Trace substances in environmental health. Proceedings of 8th trace substances in environmental Health Conference, Columbia, Missouri: University of Missouri,* pp. 101-06.

Sellers, C. M.; Heath, A. G.; and Bass, M. L. 1975. The effect of sublethal concentrations of copper and zinc on ventilatory activity, blood oxygen and pH in rainbow trout *(Salmo gairdneri). Water Res.* 9: 401-08.

Sharkey, A. G.; Shultz, J. L.; Schmidt, C. E.; and Friedel, R. A. 1975. *Mass spectrometric analysis of streams from coal gasification and liquefaction processes.* PERC/RI-75/5. Pittsburgh Energy Research Center, Pittsburg, Penn.

Shaw, T. L., and Brown, V. M. 1974. The toxicity of some forms of copper to rainbow trout. *Water Res.* 8: 377-82.

Shimkin, M. B.; Koe, B. K.; and Zechmeister, L. 1951. An instance of the occurrence of carcinogenic substances in certain barnacles. *Science* 113: 650-51.

Sims, P. 1966. The metabolism of 3-methylcholanthrene and some related compounds by rat-liver homogenates. *Biochem. J.* 98: 215-28.

Sims, P. 1967. The metabolism of benzo(a)pyrene by rat-liver homogenates. *Biochem. Pharm.* 16: 613-18.

Sims, P. 1970*a*. The metabolism of some aromatic hydrocarbons by mouse embryo cell cultures. *Biochem. Pharm.* 19: 285-97.

Sims, P. 1970*b*. Qualitative and quantitative studies on the metabolism of a series of aromatic hydrocarbons by rat-liver preparations. *Biochem. Pharm.* 19: 795-818.

Singer, P. C. 1973. *Trace metals and metal-organic interactions in natural waters.* Ann Arbor, Mich.: Science Publishers, Inc.

Sinley, J. R.; Goettl, J. P.; and Davies, P. H. 1974. The effects of zinc on rainbow trout *(Salmo gairdneri.)* in hard and soft water. *Bull. Environ. Contam. Toxicol.* 12(2): 193-201.

Skoch, E. J., and Sikes, C. S. 1973. Mercury concentrations in chironomid larvae and sediments from Sandusky Bay of Lake Erie: Evidence of seasonal cycling of mercury. In *Proceedings, 16th conference Great Lakes Research, hosted by Ohio State University and held at Sawmill Creek Hotel, Huron, Ohio, Apr. 16-18, 1973,* pp. 183-89.

Sloan, J. P.; Thompson, J.A.J.; and Larkin, P. A. 1974. The biological half-life of inorganic mercury in the Dungeness crab *(Cancer magister). J. Fish. Res. Bd. Can.* 31(10): 1571-76.

Slonim, A. R., and Ray, E. E. 1975. Acute toxicity of beryllium sulfate to salamander larvae *(Ambystoma* spp.). *Bull. Environ. Contam. Toxicol.* 13(3): 307-12.

Slonim, C. B., and Slonim, A. R. 1973. Effect of water hardness on the tolerance of the guppy to beryllium sulfate. *Bull. Environ. Contam. Toxicol.* 10(5): 295-301.

Smith, A. L., and Green, R. H. 1975. Uptake of mercury by freshwater clams (family Unionidae). *J. Fish. Res. Bd. Can.* 32(8): 1297-1303.

Smith, I. C.; Ferguson, T. L.; and Carson, B. L. 1975. Metals in new and used petroleum products and by-products: Quantities and consequences. In *The role of trace metals in petroleum,* ed. T. F. Yen, pp. 123-49. Ann Arbor, Mich.: Ann Arbor Science Publishers, Inc.

Smith, T. G., and Armstrong, F.A.J. 1975. Mercury in seals, terrestrial carnivores, and principal food items of the Inuit, from Holman, N.W.T. *J. Fish. Res. Bd. Can.* 32(6): 795-801.

Solbe, J. F. de L. G., and Flook, V. A. 1975. Studies on the toxicity of zinc sulphate and of cadmium sulphate to stone loach *Noemacheilus barbatulus* (L.) in hard water. *J. Fish. Biol.* 7(4): 631-37.

Standiford, D. R.; Potter, L. D.; and Kidd, D. E. 1973. *Mercury in the Lake Powell ecosystem.* Lake Powell Research Project Bull. 1, Department of Biology, University of New Mexico, Albuquerque. Available from J. M. Varady, Institute of Geophysics and Planetary Physics, University of California, Los Angeles 90024.

Stegeman, J. J., and Teal, J. M. 1973. Accumulation, release and retention of petroleum hydrocarbons by the oyster, *Crassostrea virginica*. *Mar. Biol.* 22: 37-44.

Stegeman, J. J. 1974. Hydrocarbons in shellfish chronically exposed to low levels of fuel oil. In *Pollution and physiology of marine organisms*, ed. F. J. Vernberg and W. B. Vernberg, pp. 329-47. New York: Academic Press.

Stenner, R. D., and Nickless, G. 1974*a*. Absorption of cadmium, copper and zinc by dog whelks in the Bristol channel. *Nature* 247: 198-99.

Stenner, R. D., and Nickless, G. 1974*b*. Distribution of some heavy metals in organisms in Hardangerfjord and Skjerstadfjord, Norway. *Water Air Soil Pollut.* 3: 279-91.

Stenner, R. D., and Nickless G. 1975. Heavy metals in organisms of the Atlantic coast of S.W. Spain and Portugal. *Mar. Pollut. Bull.* 6(6): 89-95.

Stevens, J. D., and Brown, B. E. 1974. Occurrence of heavy metals in the blue shark *Prionace glauca* and selected pelagic in the N.E. Atlantic Ocean. *Mar. Biol.* 26: 287-93.

Stewart, H. 1966. Pulmonary cancer and adenomatosis in captive wild mammals and birds from the Philadelphia Zoo. *J. Natl. Cancer Inst.* 36: 117-54.

Stiff, M. J. 1971. Copper/bicarbonate equilibria in solutions of bicarbonate ion at concentrations similar to those found in natural water. *Water Res.* 5(5): 171-76.

Stirling, E. A. 1975. Some effects of pollutants on the behaviour of the bivalve *Tellina tenuis*. *Mar. Pollut. Bull.* 6(8): 122-24.

Stoewsand, G. S.; Anderson, J. L.; Gutenmann, W. H.; Bache, C. A.; and Lisk, D. J. 1971. Eggshell thinning in Japanese quail fed mercuric chloride. *Science* 10: 1030-31.

Suess, M. J. 1972. Polynuclear aromatic hydrocarbon pollution of the marine environment. In *Marine pollution and sea life*, ed. M. Ruivo. London: Fishing News (Books) Ltd.

Sullivan, J. B. 1974. Marine pollution by carcinogenic hydrocarbons. In *Marine pollution monitoring (petroleum), proceedings of a symposium and workshop held at NBS, Gaithersburg, Maryland, May 13-17, 1974*, NBS Special Publication 409.

Sutterlin, A. M. 1974. Pollutants and the chemical senses of aquatic animals — perspective and review. *Chem. Senses Flavor* 1: 167-78.

Suzuki, T.; Miyama, T.; and Toyama, C. 1973. The chemical form and bodily distribution of mercury in marine fish. *Bull. Environ. Contam. Toxicol.* 10(6): 347-55.

Taylor, D. D., and Bright, T. J. 1973. The distribution of heavy metals in reef-dwelling groupers in the Gulf of Mexico and Bahama Islands. TAMU-SG-73-2. Department of Oceanography, Texas A&M University.

Taylor, F. G.; Parr, P. D.; and Dahlman, R. C. 1975. Distribution of chromium in vegetation and small mammals adjacent to cooling towers. Environmental Sciences Division, Oak Ridge National Laboratory, Oak Ridge, Tenn.

Terriere, L. C.; Boose, R. B.; and Roubal, W. T. 1961. The metabolism of naphthalene and 1-naphthol by houseflies and rats. *Biochem. J.* 79: 620-23.

Thomas, R. E., and Rice, S. D. 1975. Increased opercular rates of pink salmon (*Oncorhynchys gorbuscha*) fry after exposure to the water-soluble fraction of Prudhoe Bay crude oil. *J. Fish. Res. Bd. Can.* 32(11): 2221-24.

Thommes, M. M.; Lucas, H. F.; and Edgington, D. N. 1972. Mercury concentrations in fish taken from offshore areas of the Great Lakes. In *Proceedings, fifteenth conference on Great Lakes research*, pp. 192-97.

Thompson, S. E.; Burton, C. A.; Quinn, D. J.; and Ng, Y. C. 1972. *Concentration factors of chemical elements in edible aquatic organisms*. UCRL-55064 (rev. 1). Lawrence Livermore Laboratory, Bio-Medical Division, Livermore, Calif.

Thurberg, F. P.; Calabrese, A.; and Dawson, M. A. 1974. Effects of silver on oxygen consumption of bivalves at various salinities. In *Pollution and physiology of marine organisms*, ed. F. J. Vernberg and W. B. Vernberg, pp. 67-78. New York: Academic Press.

Thorp, V. J., and Lake, P. S. 1974. Toxicity bioassays of cadmium on selected freshwater
    invertebrates and the interaction of cadmium and zinc on the freshwater shrimp, *Paratya
    tasmaniensis* (Riek). *Aust. J. Mar. Freshwater Res.* 25: 97-104.

Thrower, S. J., and Eustace, I. J. 1973. Heavy metal accumulation in oysters grown in Tasmanian
    waters. *Fd. Tech. Aust.* 25: 546-53.

Thurberg, F. P.; Dawson, M. A.; and Collier, R. S. 1973. Effects of copper and cadmium on
    osmoregulation and oxygen consumption in two species of estuarine crabs. *Mar. Biol.*
    23: 171-75.

Thurston, R. V., and Russo, R. C. 1976. Fisheries bioassay laboratory field and laboratory
    studies. In *Toxic effects on the aquatic biota from coal and oil shale development.
    Research reports of environmental impacts of energy developments. Quarterly progress report
    October-December 1975*, ed. R. V. Thurston, R. K. Skogerboe, and R. C. Russo, pp. 137-54.
    Internal Project No. 5. Natural Resource Ecology Laboratory, Colorado State Univ., Fort
    Collins.

Tingey, G. L., and Morrey, J. R. 1973. *Coal structure and reactivity*. A Battelle Energy
    Program Report. Battelle Pacific Northwest Laboratories, Richland, Wash.

Tong, S.S.C.; Youngs, W. D.; Gutenmann, W. H.; and Lisk, D. J. 1974. Trace metals in Lake
    Cayuga lake trout (*Salvelinus namaycush*) in relation to age. *J. Fish. Res. Bd. Can.*
    31(2): 238-39.

Tsai, S. D.; Boush, G. M.; and Matsumura, F. 1975. Importance of water pH in accumulation of
    inorganic mercury in fish. *Bull. Environ. Contam. Toxicol.* 13(2): 188-93.

Tullar, I. V., and Suffet, I. H. 1975. The fate of vanadium in an urban air shed: the Lower
    Delaware River Valley. *J. Air Pollut. Control Assoc.* 25(3): 282-86.

Uthe, J. F.; Atton, F. M.; and Royer, L. M. 1973. Uptake of mercury by caged rainbow trout
    (*Salmo gairdneri*) in the South Saskatchewan River. *J. Fish. Res. Bd. Can.* 30(5): 643-50.

Van As, D.; Fourie, H. O.; and Vleggaar, C. M. 1975. Trace element concentrations in marine
    organisms from the Cape West Coast. *South Afr. J. Sci.* 71: 151-54.

Van Hook, R. I. 1974. Cadmium, lead, and zinc distribution between earthworms and soils:
    Potentials for biological accumulation. *Bull. Environ. Contam. Toxicol.* 12(4): 509-12.

Vaughan, B. E.; Abel, K. H.; Cataldo, D. A.; Haies, J. M.; Hane, C. E.; Rancitelli, L. A.;
    Routson, R. C.; Wildung, R. E.; and Wolf, E. G. 1975. *Review of potential impact on
    health and environmental quality from metals entering the environment as a result of coal
    utilization*. A Battelle Energy Program Report. Battelle Pacific Northwest Laboratories,
    Richland, Wash.

Vernberg, W. B.; DeCoursey, P. J.; and O'Hara, J. 1974. Multiple environmental factor effects
    on physiology and behavior of the fiddler crab, *Uca pugilator*. In *Pollution and physiology
    of marine organisms*, ed. F. J. Vernberg and W. B. Vernberg, pp. 381-425. New York:
    Academic Press.

Vernberg, W. B., and O'Hara, J. 1972. Temperature-salinity stress and mercury uptake in the
    fiddler crab, *Uca pugilator*. *J. Fish. Res. Bd. Can.* 29(10): 1491-94.

Vernberg, W. B., and Vernberg, J. 1972. The synergistic effects of temperature, salinity, and
    mercury on survival and metabolism of the adult fiddler crab, *Uca pugilator*. *Fish. Bull.*
    70(2): 415-20.

Von Westernhagen, H.; Rosenthal, H.; and Sperling, K. R. 1974. Combined effects of cadmium
    and salinity on development and survival of herring eggs. *Helgolander Wiss. Meeresuntero*
    26: 416-33.

Voyer, R. A. 1975. Effect of dissolved oxygen concentration on the acute toxicity of cadmium to
    the mummichog, *Fundulus heteroclitus* (L.), at various salinities. In *Trans. Am. Fish. Soc.*
    104(1): 129 34.

Waldichuk, M. 1974*a*. Coastal marine pollution and fish. *Ocean Manage.* 2: 1-60.

Waldichuk, M. 1974*b*. Some biological concerns in heavy metals pollution. In *Pollution and physiology of marine organisms*, ed. F. J. Vernberg and W. B. Vernberg, pp. 1-57. New York: Academic Press.

Waldron, H. A. 1975. Health standards for heavy metals. *Chem. Br.* 11(10): 354-57.

Walter, C. M.; Brown, H. G.; and Hensley, C. P. 1974. Distribution of total mercury in the fishes of Lake Oahe. *Water Res.* 8(7): 413-18.

Warnick, S. L., and Bell, H. L. 1969. The acute toxicity of some heavy metals to different species of aquatic insects. *J. Water Pollut. Control Fed.* 41(2) Part I: 280-84.

Webb, M. 1975. Metallothionein and the toxicity of cadmium. In *Ecological toxicology research. Effects of heavy metal and organohalogen compounds. Proceedings of a NATO science committee conference*, ed. A. D. McIntyre and C. F. Mills, pp. 177-86. New York: Plenum Press.

Weisbart, M. 1973. The distribution and tissue retention of mercury-203 in the goldfish (*Carassius auratus*). *Can. J. Zool.* 51: 143-50.

Westlake, G. F., and Kleerekoper, H. 1974. The locomotor response of goldfish to a steep gradient of copper ions. *Water Resour. Res.* 10(1): 103-05.

Westoo, G. 1973. Methylmercury as percentage of total mercury in flesh and viscera of salmon and sea trout of various ages. *Science* 181: 567-68.

Whitfield, B. L. 1975. Biological aspects in animals. In *Environmental impact of cadmium*, ORNL/EIS-75-76, pp. 167-216. Oak Ridge, Tenn.: Oak Ridge National Laboratory.

Whittle, K.; Mackie, P. R.; and Hardy, R. 1974. Hydrocarbons in the marine ecosystem. *South Afr. J. Sci.* 70: 141-44.

Wolfe, D. A. 1974. The cycling of zinc in the Newport River Estuary, North Carolina. In *Pollution and physiology of marine organisms*, ed. F. J. Vernberg and W. B. Vernberg, pp. 79-99. New York: Academic Press.

Wolfe, D. A., and Rice, T. R. 1972. Cycling of elements in estuaries. *Fish. Bull.* 70(3): 959-72.

Wolverton, B. C. 1975. *Water hyacinths for removal of cadmium and nickel from polluted waters.* TM-X-72721. NASA, National Space Technology Laboratories, St. Louis, Miss.

Wood, J. M. 1973. Metabolic cycles for toxic elements in the environment: A study of kinetics and mechanism. In *Heavy metals in the aquatic environment. Proceedings of a symposium at Vanderbilt University, Dec. 4-7, 1973*. Nashville, Tenn.: Vanderbilt University.

Wood, J. M. 1974. Biological cycles for elements in the environment, and the neurotoxicity of metal alkyls. In *Proceedings of 16th water quality conference on trace metals in water supplies: Occurrence, significance, and control, University of Illinois, Urbana-Champaign, Feb. 12-13, 1974*, pp. 27-38.

Woolson, E. A. 1975. Bioaccumulation of arsenicals. In *Arsenical pesticides, symposium sponsored by the Division of Pesticide Chemical Society, Atlantic City, N.J., Sept. 9, 1974*, ACS Symp. Ser. 7, ed. E. A. Woolson, pp. 97-107.

Young, L. G., and Nelson, L. 1974. The effects of heavy metal ions on the motility of sea urchin spermatozoa. *Biol. Bull.* 147: 236-46.

Youngblood, W. W., and Blumer, M. 1975. Polycyclic aromatic hydrocarbons in the environment: Homologous series in soils and recent marine sediments. *Geochim. Cosmochim. Acta* 39(9): 1303-14.

Zitko, V.; Carson, W. V.; and Carson, W. G. 1975. Thallium: Occurrence in the environment and toxicity to fish. *Bull. Environ. Contam. Toxicol.* 13(1): 23-30.

Zitko, V. 1975*a*. Aromatic hydrocarbons in aquatic fauna. *Bull. Environ. Contam. Toxicol.* 14(5): 621-31.

Zitko, V. 1975*b*. Toxicity and pollution of thallium. *Sci. Total Environ.* 4: 185-92.

Zubovic, P.; Sheffey, N. B.; and Stadnichenko, T. 1967. Distribution of minor elements in some coals in the western and southwestern regions of the Interior coal province. In *Minor elements in American coals. A study of 15 minor elements in some of the coals of Arkansas, Iowa, Missouri, Oklahoma, and Texas.* U.S. Geol. Surv. Bull. 1117-D. U.S. Dept. of the Interior, Washington, D.C.

Zubovic, P.; Stadnichenko, T.; and Sheffey, N. B. 1961. Geochemistry of minor elements in coals of the Northern Great Plains coal province. In *Minor elements in American coals. A study of 15 minor elements in some of the coals of Montana, North Dakota, and Wyoming.* U.S. Geol. Surv. Bull. 1117-A. U.S. Dept. of the Interior, Washington, D.C.

Zubovic, P.; Stadnichenko, T.; and Sheffey, N. B. 1964. Distribution of minor elements in coal beds of the Eastern Interior region. In *Minor elements in American coals. A study of 15 minor elements in coal beds of Illinois, Indiana, and Western Kentucky.* U.S. Geol. Surv. Bull. 1117-B. U.S. Dept. of the Interior, Washington, D.C.

Zubovic, P.; Stadnichenko, T.; and Sheffey, N. B. 1966. Distribution of minor elements in coals of the Appalachian Region. In *Minor elements in American coals. A study of 15 minor elements in some coals in Ohio, Pennsylvania, Maryland, Kentucky, Tennessee, Alabama, and Georgia.* U.S. Geol. Surv. Bull. 1117-C. U.S. Dept. of the Interior, Washington, D.C.

10. HUMANS: METABOLISM AND BIOLOGICAL EFFECTS

S. S. Talmage

ABSTRACT

Coal conversion processes may produce and release substances that can be hazardous to human
health; some substances, studies indicate, are potential carcinogens, cocarcinogens, mutagens, or
teratogens. Others promote chronic respiratory and/or metabolic diseases. However, rigorous
quantitative data relating human exposure with health effects are still sparse.

From the standpoint of human health, the most critical hazard probably confronts industrial
workers exposed to high levels of potentially carcinogenic polycyclic aromatic hydrocarbons
(PAH). PAH can be taken into the body by inhalation, skin contact, or ingestion, although, as
determined from laboratory animal experiments, they are poorly absorbed from the gastrointestinal
tract. Regardless of the route of entry, the hepatobiliary system and the gastrointestinal tract
are the main routes of elimination of both PAH and their metabolites.

Animal studies indicate that PAH are metabolized to epoxides, dihydrodiols, phenols, and
quinones, and exert their biological and toxic effects only after metabolism. The metabolites
of several PAH have been found to be more mutagenic, carcinogenic, and teratogenic than the
parent compounds as tested in experimental laboratory animals and human cell culture systems.
A diol-epoxide has been identified as the ultimate mutagenic metabolite of benzo[a]pyrene. PAH
metabolites have been shown to bind to DNA, RNA, and protein.

The enzyme system responsible for metabolism of PAH is aryl hydrocarbon hydroxylase (AHH), a
system of mixed function oxidases located in the microsomal fraction of cells. Recent evidence
indicates a relationship between inducibility of AHH in human lymphocytes and susceptibility to
lung cancer. Inducibility of the enzyme appears to be under genetic control in laboratory
animals and in humans.

Epidemiologic studies of workers in coal gas, coking, and coal hydrogenation industries routinely
exposed to the products of coal combustion or distillation show that they run an increased risk
of lung, skin, and other cancers.

The effect of carcinogens can be enhanced or inhibited by associated substances. For example,
particulates, sulfur dioxide, and some aliphatic hydrocarbons have been identified as possible
cocarcinogens, whereas vitamin A and several antioxidants appear to give partial protection
against the formation of cancers.

Epidemiologic studies in several U.S. cities show a relationship between long-term health effects
such as increased chronic respiratory diseases and decreased ventilatory function in children

and sulfur oxide levels as measured by sulfur dioxide, suspended particulates, and suspended
sulfates. Episodes of acute elevations of sulfur oxides and other pollutants have been associated
with increased morbidity and mortality among individuals with chronic pulmonary or cardiac
diseases and also among the very young or very old. However, studies with experimental laboratory
animals have shown little or no effect on lung function following exposure to high levels of
sulfur dioxide alone, although exposure to sulfur dioxide plus particulates produces bronchoconstriction and an irritant effect on the lungs.

An epidemiologic study has related high levels of nitrogen dioxide and particulates in the
atmosphere to decreased ventilatory performance and an excess of respiratory illness in school-
children.

Potential releases of trace elements during coal conversion processes may produce metabolic
disorders. Trace elements of particular importance are lead, mercury, arsenic, vanadium,
selenium, nickel, and possibly cadmium and fluoride. Low-level excesses of these elements can
have a deleterious effect on human health. Arsenic, beryllium, cadmium, chromate salts, and
nickel are known or suspected carcinogens.

10.0 INTRODUCTION

In assessing the risks to human health from coal processing emissions, the effects on the occupational, local, and general segments of the population must be considered. The most obvious hazard to industrial workers is exposure to potentially carcinogenic polycyclic aromatic hydrocarbons (PAH). Acute and chronic toxicity are also important considerations. The chronic effects of low levels of materials released during coal conversion will be the principal concern for the local and general population. Carcinogenicity, chronic toxicity, mutagenicity, and perhaps teratogenicity are also of concern. The same considerations should be taken into account for all three populations with regard to the PAH since the differences in their concentrations for the three populations are quantitative. Although the PAH are considered a potential health hazard and, indeed, may be the most critical hazard from coal conversion processes, less is known about them than about other potential pollutants. For this reason, they will be covered in greater detail than other pollutants in the following sections.

Although the products of coal combustion, coking, petroleum refining and coal pyrolysis are not identical to those of coal conversion, many compounds known to be carcinogenic have been found in all these products or their process streams. A study of their similarities assists in evaluating carcinogenicity in coal conversion processes. Likewise, other types of organic compounds, gases, and trace metals emitted during all these processes may be similar.

A difficult problem in all bioassays is the extrapolation of data from experimental animals to man. Marked differences in susceptibility to carcinogenic agents exist between species as well as between strains of the same species. For example, the site of origin and the histologic type of respiratory tract tumor depend on the species and strain of animal as well as the route of application and dose. Mice, which were used in early studies, have a high rate of spontaneous respiratory tract tumor of a type not found in man. On the other hand, hamsters have a low incidence of spontaneous lung tumors and tend to be resistant to respiratory tract infections, a modifying factor in tumor induction (Nettesheim and Schreiber 1975).

Animal test groups are necessarily limited in size. For statistical significance in extrapolating from animal to man for low-level concentrations of carcinogens, large groups of experimental animals are needed. For example, if a carcinogenic agent produces cancer in 1 of 10,000 persons and if the sensitivity to the carcinogen is similar in man and rodents, test groups of 10,000 rodents or more would be required to obtain one cancer. For statistical significance, 50,000 rodents would be required. Lung cancer, the most frequent cause of death from cancer, causes an annual U.S. death rate of 1 in 10,000 for females and 7 in 10,000 for males on an age-adjusted basis (NAS 1972).

In general, the absorption, distribution, and storage of drugs, and therefore PAH, are comparable in a variety of vertebrate species, although some differences in plasma protein binding are known. The metabolism of drugs also varies from species to species; different metabolites may be formed. Even when the same metabolites are formed, they may form at different rates, resulting in a difference in toxicity (Rall 1969).

Usually, dose levels in animal experiments considerably exceed the natural levels experienced by humans. Also, the artificial laboratory system does not allow for the interactions and synergisms that occur in the natural and work environment.

Despite the deficiencies of laboratory experiments, the animal model remains the best, most reliable tool for supporting epidemiological evidence and assessing the potential dangers of chemicals in the environment. In addition, in vitro mutagenic tests such as the salmonella microbioassay are useful for screening products and by-products of coal conversion processes. Laboratory-animal, biological testing, and occupational and epidemiological studies are reviewed in an attempt to present what is known of the health effects of potential coal conversion emissions.

10.1 POLYCYCLIC AROMATIC HYDROCARBONS

It has long been known that some agent, or agents, produced during the combustion or processing of coal causes skin carcinomas in man. Since the description by Pott (1775) [as cited in International Agency on Cancer Research (IARC) 1973] of skin cancer in British chimney sweeps, it has been recognized that coal soot is carcinogenic to the skin of man. Exposure of workers to coal-derived liquids and coke-oven and coal processing emissions has been correlated with increased cancers of other organ systems also. In evaluating the hazards to human health from coal processing, the PAH that may be produced and emitted during that processing are of critical importance because they are known animal carcinogens.

The carcinogenic potential of PAH has been studied in various whole animals, tissue and organ cultures, and microorganisms. The primary methods employed in whole animals include skin painting, subcutaneous injection, systemic inoculation, oral intake, local implantation (in lungs, bladder, or other organs), intratracheal inoculation, and inhalation. Exposure of animals by inhalation techniques and skin painting would appear to be most relevant in studying the problem of hazardous emissions in coal processing industries.

Benzo[a]pyrene (BaP) was identified as a chemical carcinogen following its isolation from coal tar (Kennaway 1955). β-Naphthylamine, a coal carbonization product, is a potent bladder carcinogen [National Institute for Occupational Safety and Health (NIOSH) 1973]. A variety of skin carcinogens have also been identified in coal tar.

10.1.1 Uptake, distribution, and retention

Pathways for human uptake of a contaminant are ingestion, inhalation, or absorption. In a coal conversion facility, exposed workers would probably inhale or absorb PAH as a vapor, an aerosol, or as molecules adsorbed on particulate matter in air. Although the general population is most likely to encounter PAH via ingestion of contaminated food or drinking water, inhalation also may be an important pathway for people living near an emission source.

10.1.1.1 Ingestion

Laboratory-animal studies indicate that many PAH are poorly absorbed from the gastrointestinal tract. When incorporated into the diet or when administered in a single dose to the rat, many PAH are excreted unchanged (see Table 10.1) (Chang 1943). Although there is a distribution of values, only naphthalene, phenanthrene, and acenaphthene are not excreted directly.

Rigdon and Neal (1963) and Neal and Rigdon (1964) report on the absorption and excretion of ingested BaP in several species of laboratory animals including the duck, chicken, mouse, and dog. Their studies spectrophotometrically demonstrated the presence of BaP in the blood and tissues of the animals. No quantitative data were given; nor was there an attempt to identify metabolites. Mice that were fed BaP showed a blue fluorescence of the skin, liver, kidneys, intestinal tract, abdominal fat, and gall bladder. The interval necessary for the tissues to show fluorescence was related to the amount of BaP per gram of food consumed. There was a progressive decrease of fluorescence when BaP feeding was discontinued.

Table 10.1. Fecal excretion of PAH in the rat

Hydrocarbon	Percentage of oral dose in feces	
	Single dose 200 mg in starch solution	1% Hydrocarbon in the diet
Acenaphthene	10	6
Anthracene	69	83
BaP	57	42
Chrysene	85	79
Dibenz[a,b]anthracene	97	90
3-Methylcholanthrene	69	67
Naphthalene	0	0
Phenanthrene	7	5

Source: Modified from Chang 1943. Reprinted by permission of the publisher.

In one experiment, brown mice of the CBA strain, white mice, and hairless mice were fed pellets containing 20 mg BaP per gram of food. At half-hour increments, the following results were noted: (1) the ears were faintly blue; (2) the entire body fluoresced; and (3) the kidneys, liver, gall bladder, abdominal fat, and urine had a blue fluorescence. After feeding without BaP, the kidneys, viscera, and urine showed no fluorescence after 48, 72, and 48 hr respectively. The blue color of the skin was lost slowly — some was still evident nine days later (Neal and Rigdon 1964).

Daniel, Pratt, and Pritchard (1967) studied the distribution of radioactivity in rats after radioactive benz[a,h]anthracene, 3-methylcholanthrene, and 7,12-dimethylbenz[a]anthracene (DMBA) dissolved in oil had been administered through a stomach tube. They found a prolonged retention of radioactivity in the body fat, ovaries, and adrenal glands. Hydrocarbon from the gastrointestinal tract was first absorbed into the lymphatics of the gut, from which it was carried into the bloodstream, which distributed it to the tissues. Radioactivity appeared in the lymph, bile, and urine about 1 hr after the administration of radioactive hydrocarbons. Uptake and retention were variable among tissues. At an early stage, large quantities of the hydrocarbon were held for a short period by both the kidneys and liver. From the kidneys, some of the hydrocarbon was excreted in the urine, whereas from the liver, excretion in the bile was rapid with little reabsorption.

10.1.1.2 <u>Intravenous injection</u>

Heidelberger and Weiss (1951) and Heidelberger and Jones (1948) studied the distribution of radioactivity in mice after intravenous injection of a colloidal suspension of BaP and dibenz[a,h]-anthracene (Tables 10.2 and 10.3). Almost no radioactivity appeared in the respiratory carbon dioxide. BaP and its metabolites were rapidly excreted into the bile, intestinal tract, and feces although there was some absorption into the body fat. Samples were processed into neutral,

Table 10.2. Distribution of radioactivity following intravenous injection of an aqueous colloid of 0.45 mg BaP-5-^{14}C

	90 min after injection			24 hr after injection		
	Dry weight (g)	Total counts/min	Percent of dose	Dry weight (g)	Total counts/min	Percent of dose
Dose		1,390,000			1,210,000	
Respiratory CO_2		< 400	>0.03		<100	<0.008
Feces	None			0.575	794,000	65.5
Urine		3,300	0.24		210,000	17.3
Gastrointestinal tract and contents	0.834	163,000	11.7	0.532	16,600	1.37
Carcass fat and tissue	1.746	133,000	9.6	0.982	14,200	1.17
Abdominal fat and tissue	1.778	14,000	1.01	0.971	16,100	1.33
Liver	0.371	18,900	1.36	0.270	20,900	1.72
Kidneys	0.083	4,300	0.31	0.078	4,100	0.34
Lungs	0.043	1,000	0.072	0.039	890	0.071
Heart	0.029	480	0.034	0.031	350	0.029
Stomach	0.142	2,000	0.14	0.066	3,700	0.31
Salivary glands	0.029	300	0.022	0.035	540	0.045
Plasma	0.085	3,000	0.22	0.072	1,700	0.14
Red cells	0.178	310	0.022	0.179	590	0.049
Carcass and bile and gall bladder	7.038	1,000,000	72.0	5.614	92,100	7.60
Recovery		1,344,000	97.0		1,176,000	97.2

Source: Heidelberger and Weiss, 1951, Table I, p. 887. Reprinted by permission of the publisher.

Table 10.3. Recovery of dibenz[a,h]anthracene

	Experiment 3[a]			Experiment 5[b]		
	Tissue wet weight (g)	Total counts/min	Percent of dose	Tissue wet weight (g)	Total counts/min	Percent of dose
Dose		28,300			22,800	
Respiratory CO_2	11.908	15	0.05	19.19	<100	<0.4
Urine		92	0.32		1,770	7.76
Feces		1,590	5.62		9,150	40.0
Gastrointestinal tract and contents	2.432	25,200	88.7	2.643	3,330	14.6
Liver	1.197	975	3.47	1.718	2,610	11.4
Salivary glands	0.078	73	0.26	0.053	196	0.85
Lymph nodes	0.017	0	0	0.105	96	0.41
Spleen	0.128	0	0	0.261	0	0
Seminal vescicles	0.278	0	0	0.159	0	0
Body fat	0.155	0	0	0.288	0	0
Skin and hair	2.792	0	0	3.274	0	0
Muscle	1.008	0	0	2.002	0	0
Bones	4.494	0	0	5.602	0	0
Brain	0.364	0	0	0.392	0	0
Lungs	0.101	104	0.37	0.155	87	0.38
Blood plasma	0.296	88	0.31	0.181	130	0.57
Blood cells	0.226	0	0	0.252	0	0
Kidneys	0.233	0	0	0.348	130	0.57
Heart	0.080	0	0	0.101	452	1.58
Mammary carcinoma				2.767	1,180	4.9
Total recovery		28,100	98.8		19,300	84.5

[a]Experiment conducted with a strain A male mouse weighing 17 g, sacrificed 24 hr after an intravenous dose of 0.42 mg colloid.
[b]Experiment conducted with a strain A male mouse weighing 27.6 g, sacrificed 48 hr after an intravenous dose of 0.341 mg colloid.

Source: Heidelberger and Jones 1948, Table I, p. 254. Reprinted by permission of the publisher.

acidic, loosely bound, and tightly bound protein fractions. Measurements were made of radioactivity only; no information on the chemical nature of the radioactive material was given.

Organ and organ system retention following intravenous injection of ^{14}C-BaP into rats was assessed
by Kotin, Falk, and Busser (1959). Following injection of 11 µg BaP, clearance from the blood
was extremely rapid; after 10 min, radioactivity was barely detectable in samples from the systemic
circulation. A minimal degree of localization and retention within the lung was noted. Although
the concentration in the liver at any one time was low, there was a very rapid localization of
BaP in the liver and an almost equally rapid excretion in the bile.

Organ localization and systemic clearance were influenced by the solvent used for administration
following intratracheal instillation and subcutaneous injection, but not following intravenous
injection. For example, localization persisted longer following intratracheal administration
when the solvent was water, thus forming large crystals, than when the solvent was triethylene
glycol.

10.1.1.3 Cellular uptake

Aromatic hydrocarbons are highly lipophilic and readily penetrate into cells (Wiebel and Gelboin
1974). Brunette and Katz (1975) studied the interactions of BaP with the cell surface membrane
by measuring BaP uptake into Chinese hamster ovary cells and by determining BaP fluorescence in
the presence of isolated cell surface membranes. It was found that 0.19 µg BaP was taken up by
1 million Chinese hamster ovary cells after a 30-min exposure to a solution containing 0.59 µg/ml.
Cellular BaP uptake increased over the concentration range of 0.2 to 5.0 µg/ml and showed no
evidence of approaching saturation. Experimental conditions were found to alter uptake markedly.
Low cell culture densities and increasing temperatures resulted in an increased uptake. The
uptake rate was reduced in the presence of bovine serum and with the addition of other environmental PAH such as perylene. An increased fluorescence of BaP, when added to suspensions of
isolated surface membranes, indicated that BaP may be associated with nonpolar portions of the
surface membranes. The data suggest that the mechanism of BaP uptake was probably passive
diffusion.

10.1.1.4 Skin uptake

Repeated topical application of PAH dissolved in solvent to the skin of mice and rabbits caused
systemic effects, indicating that the PAH are absorbed percutaneously (Shubik and Della Porta
1957). Pathological changes were observed in the blood, spleen, lymph nodes, and bone marrow.

Similarly, Stowell and Maas (1946) found that 3-methylcholanthrene, repeatedly painted on the
backs of mice, produced inflammation of the liver, kidney, and lungs in addition to other
systemic changes.

10.1.2 Toxicity

There is little published information on the toxicity of PAH in experimental animals other than
the mouse or rat (Gerarde 1960). The systemic acute dose LD_{50} (the lethal dose for 50% of animals tested) is usually extremely high as compared with the dose necessary to produce carcinogenic effects.

Acute toxicity is difficult to produce orally because PAH are often poorly absorbed from the gastrointestinal tract (Sect. 10.1.1.1). Most skin-painting experiments use concentrations of PAH so low (around 1%) that toxic effects are not produced in a single dose.

10.1.2.1 Systemic effects

Shubik and Della Porta (1957) describe acute effects in mice treated with large doses of several PAH. Groups of twelve 2- to 2-1/2-month-old male Swiss mice received repeated skin paintings of concentrated acetone solutions of BaP or DMBA or a benzene solution of 3-methylcholanthrene six days a week for periods of up to 19 weeks. The dose varied with the solubility of the various hydrocarbons; each dose consisted of about 0.08 ml of a 1.5% solution of BaP or 2% of 3-methylcholanthrene or DMBA. An additional group of mice received a 4% solution of 3-methylcholanthrene twice each day. Five groups of five mice each were given injections of a single dose of 1 g/kg body weight (25 mg) of DMBA, BaP, 3-methylcholanthrene, fluorene, or anthracene as either 3.3% solutions or suspensions in olive oil.

DMBA produced an acute lethal effect when applied topically to the skin and when injected intraperitoneally in mice. BaP and 3-methylcholanthrene produced a similar effect only after intraperitoneal injection of the single large dose. Intraperitoneally administered anthracene and fluorene failed to produce any toxic effects. In the skin-painting studies, all 12 mice receiving BaP developed multiple tumors. Although each animal bore at least one squamous cell carcinoma, the majority of the 71 tumors were papillomas. In the 3-methylcholanthrene group, two mice without tumors died at the second and fifth weeks; the remaining 10 mice developed at least one squamous cell carcinoma, although again most of the tumors were papillomas. All DMBA-painted mice died or were moribund by the sixteenth day, before any tumors had arisen.

In another experiment, a 2% solution of DMBA (400 mg/kg body weight) in acetone was applied to a large shaved area of skin. Two animals died and two were moribund on the seventh day after dosing; five other mice had died or were moribund by the eleventh day. The pathological pictures of tissues of the dying animals after treatment were essentially morphologically similar, but they showed quantitative variations. Myelo-, erythro-, and lymphopoiesis were markedly affected, involving the peripheral blood, spleen, lymph nodes, and bone marrow. Pathological changes were absent in other organs, except for the testes and intestinal mucosa.

In a similar, earlier study, Stowell and Maas (1946) painted the backs of Swiss mice with solutions of 1% 3-methylcholanthrene, three, six, and nine times a week. Controls were unpainted or painted with benzene. Malignant tumors developed rapidly. The mice showed systemic effects — inflammation in the liver, kidney, and lungs; hyperplasia of the bone marrow, spleen, and lymph nodes with extramedullary myelo-, lympho-, and erythropoiesis; and leukemia. Some thyroid glands had atypical acini, and some parathyroids were hypertrophic and hyperplastic.

10.1.2.2 Toxicity of metabolites

Huggins and Morii (1961) observed that a single dose of DMBA (20 to 30 mg by ingestion or 5 mg intravenously) caused selective necrosis of the adrenal cortex in the mature female rat. Massive necrosis occurred in the two inner zones of the cortex, whereas other regions of the adrenal glands were uninjured. The main products of the metabolism of DMBA by rat liver homogenates are the isomeric monohydroxymethyl derivatives, 7-hydroxymethyl-12-methylbenz[α]anthracene and 12-hydroxymethyl-7-methylbenz[a]anthracene (Boyland and Sims 1965). 7-Hydroxymethyl-12-methylbenz[a]anthracene is a more potent adrenocorticolytic agent than DMBA; 12-hydroxymethyl-7-methylbenz[a]anthracene has no adrenocorticolytic effect (Boyland, Sims, and Huggins 1965). DMBA is the only PAH known to produce an adrenal necrotic effect.

10.1.2.3 Fetal toxicity

Currie et al. (1970) investigated the effects of DMBA and its monohydroxymethyl derivatives on the pregnant rat and rat fetus. Pregnant Sprague-Dawley rats received a single intravenous injection of DMBA, 7-hydroxymethyl-12-methylbenz[a]anthracene, 12-hydroxymethyl-7-methylbenz[a]-anthracene, or 7-hydroxymethylbenz[a]anthracene at a dose of 2.5 mg/100 g body weight, from day 2 to day 18 of pregnancy. The adrenal glands of every rat treated with 7-hydroxymethyl-12-methylbenz[a]anthracene were necrotic, and the litters of the rats treated on day 17 showed extensive necrosis of the central parts of the glands. This dose of DMBA proved highly toxic to the mothers: 4 of 9 rats treated on day 8 and 7 of 11 rats treated on day 13 died by day 20. Varying the dose of 7-hydroxymethyl-12-methylbenz[a]anthracene showed a fairly close correlation between dose and the degree of damage to the maternal adrenal cortex.

10.1.2.4 Cell toxicity

Grover et al. (1971) and Huberman et al. (1972) studied the cytotoxicity of dibenz[a,h]anthracene, benz[a]anthracene (BaA), and several of their derivatives at concentrations of 2.5, 5.0, and 10 and 1.0, 2.5, and 5.0 µg/ml of medium respectively. Cytotoxicity was expressed as the percentage of colonies of plated rodent cells in PAH-treated dishes as compared with those in acetone-treated controls (cloning efficiency). Neither parent compound was toxic at the concentrations used. In contrast, the PAH derivatives, the epoxides, and cis-dihydrodiols caused increased transformations of the cells, and the K-region phenols of BaA and dibenz[a,h]anthracene were the most toxic derivatives tested (Tables 10.4 and 10.5). The results confirm the view that metabolism is necessary for hydrocarbons to produce toxicity. An increase in the concentration of the compounds was generally accompanied by an increase in toxicity to the cells (Grover et al. 1971).

Table 10.4. Transformation and cloning efficiencies of hamster embryo cells treated with K-region derivatives of BaA

Compound	Concentration (µg/ml of medium)	Total colonies counted	Cloning efficiency (%)	Transformed colonies	
				Number	Percent
Control (acetone)		818	18.2	0	0
Hydrocarbon	1.0	893	16.2	2	0.2
	2.5	1039	19.0	2	0.2
	5.0	1140	22.8	0	0
Epoxide	1.0	1082	19.6	3	0.3
	2.5	565	10.2	37	6.9
	5.0	0			
Diol	1.0	1320	22.0	3	0.2
	2.5	867	15.6	10	1.2
	5.0	596	12.0	39	6.5
Phenol	1.0	1161	21.2	3	0.3
	2.5	272	5.7	3	1.1
	5.0	0			

Source: Grover et al. 1971, Table 1, p. 1099. Reprinted by permission of the publisher.

Table 10.5. Transformation and cloning efficiencies of hamster embryo cells treated with K-region derivatives of dibenz[a,h]anthracene

Compound	Concentration (μg/ml of medium)	Total colonies counted	Cloning efficiency (%)	Transformed colonies	
				Number	Percent
Control (acetone)		1291	12.1	5	0.4
Hydrocarbon	2.5	1341	13.4	3	0.2
	5.0	1363	14.0	11	0.8
	10	1365	14.5	7	0.5
Epoxide	2.5	895	10.1	7	0.8
	5.0	866	9.3	20	2.3
	7.5	817	9.3	22	2.7
	10	707	7.7	30	4.2
Diol	2.5	739	8.5	3	0.4
	5.0	321	3.7	4	1.2
	10	2			
Phenol	2.5	729	8.0	0	0
	5.0	2			
	10	0			

Source: Grover et al. 1971, Table 2, p. 1099. Reprinted by permission of the publisher.

In another study, the K-region epoxides of 3-methylcholanthrene, chrysene, and phenanthrene were more active than the parent hydrocarbons in producing cytotoxicity (Huberman et al. 1972). Using a microfluorometric technique, Lubet, Brown, and Kouri (1973) found that the intracellular uptake of BaP alone could not account for the toxic effects produced on hamster embryo cells in culture. By either intracellular production or cellular uptake, similar amounts of 3-hydroxy-BaP resulted in similar levels of cell toxicity. Intracellular 3-hydroxy-BaP concentrations of 50 to 100 times less than those of BaP resulted in the same amount of toxicity to the cells. Toxicity probably was not due to a subsequent metabolite of 3-hydroxy-BaP since established cell lines, which have decreased abilities to metabolize these hydrocarbons, showed an exponential toxicity when exposed to pure 3-hydroxy-BaP (Gelboin, Huberman, and Sachs 1969).

10.1.3 Metabolism

Knowledge of the metabolism of PAH is crucial to an evaluation of their hazard. From the preceding data, it is clear that the metabolites are more toxic than the pure compounds. Discussions of PAH metabolism can be found in Gerarde (1960) and Williams (1959). Generally, hydrocarbons of two to five rings undergo reactions, forming (1) phenols, (2) partially saturated rings, (3) quinones, (4) mercapturic acid, and (5) glucuronide and ethereal sulfate (Williams 1959). Metabolism is carried out primarily by the drug-metabolizing enzyme system of the microsomes.

10.1.3.1 Rat liver microsomes

An example of what is known of the metabolism of a PAH is shown in Fig. 10.1. BaP added to rat liver preparations gave the following metabolites: the monophenols (3- and 6-hydroxy), the diphenols (1,6- and 3,6-dihydroxy), the corresponding quinones (3,6- and 1,6-dione), and the two dihydrodiols (1,2 and 9,10). A number of unidentified conjugated products were also present.

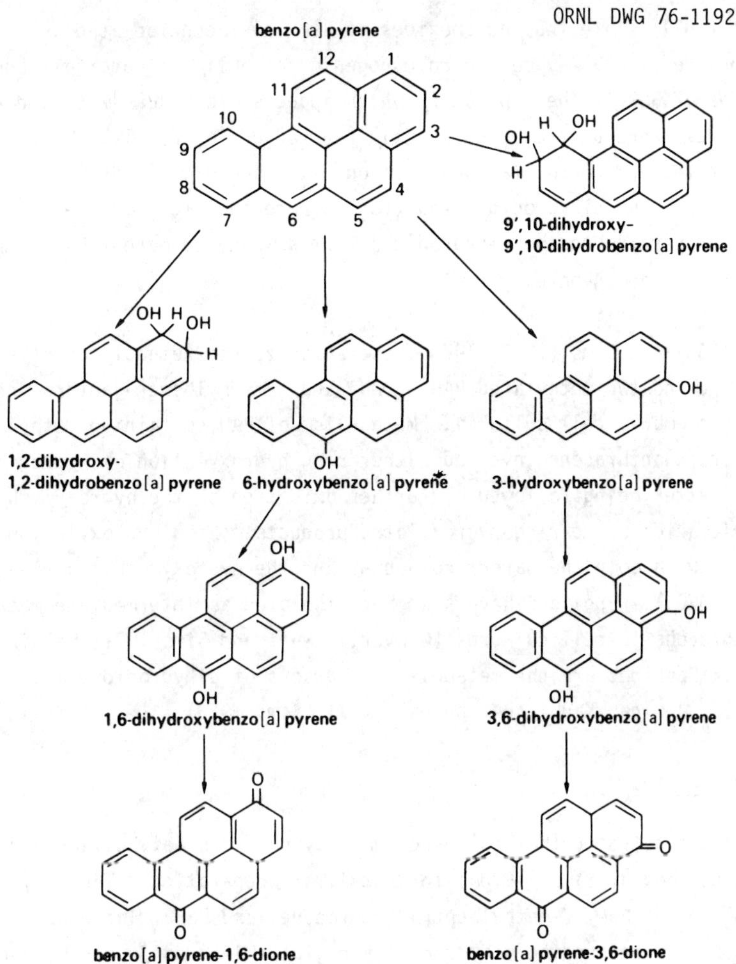

Fig. 10.1. Metabolism of BaP. <u>Source</u>: From NAS 1972, Fig. 10-1, p. 135.

The K-regions (4,5 and 11,12 bonds), postulated to be active sites, were not metabolized. None of the identified metabolites are known carcinogens. Wiebel and Gelboin (1974) listed the metabolites of BaP formed by rat liver microsomes as three dihydrodiols, two phenols, and two quinones (Table 10.6).

Table 10.6. Metabolites of BaP

7,8-dihydro-7,8-diol
9,10-dihydro-9,10-diol
4,5-dihydro-4,5-diol
3-hydroxy
9-hydroxy
1,6-quinone
3,6-quinone

Source: Wiebel and Gelboin
1974, Table 1, p. 58.

Wiebel and Gelboin (1974) state that no epoxides of BaP have been isolated as products of cell metabolism. An epoxide of BaP may be the carcinogenic metabolite. Waterfall and Sims (1972) synthesized epoxides of BaP in the laboratory which appeared to be BaP-7,8- and -9,10-oxide. These oxides (epoxides) were unstable and readily rearranged to phenols. They were converted by rat liver homogenates and microsomal preparations into phenols and dihydrodiols, but not glutathione conjugates. The dihydrodiols thus formed were identical in their chromatographic and spectrographic properties to the dihydrodiols formed when the parent BaP compound was metabolized by rat liver homogenates.

Boyland and Sims (1965) and Sims (1967; 1970a) characterize the metabolites of PAH (including BaP and benz[e]pyrene and two methylated PAH — DMBA and 7-methylbenz[a]anthracene) in rat liver homogenates and mouse embryo cell cultures. Metabolism of PAH containing methyl groups such as DMBA and 7-methylbenz[a]anthracene involved either ring hydroxylation or hydroxylation of the methyl groups, the latter being followed by further oxidation of the hydroxymethyl compounds either to carboxylic acids or to ring-hydroxylated products. Ring hydroxylation occurred mainly in the 8,9 position, with both the parent compounds and the hydroxymethyl derivatives yielding dihydrodiols (Sims 1970a). Epoxides have also been isolated as intermediate metabolites of BaA and dibenz[a,h]anthracene in cell cultures (Grover, Hewer, and Sims 1971; Selkirk, Huberman, and Heidelberger 1971). Epoxides are the metabolic precursors of dihydrodiols and phenols (Grover, Hewer, and Sims 1973; Jerina, Kaubisch, and Daly 1971; Sims 1973).

10.1.3.2 Human lymphocytes

Booth et al. (1974) demonstrated that cultured human lymphocytes metabolized PAH by pathways similar to those described by Sims (1970b) for rat liver preparation. Thus, dihydrodiols were isolated as metabolites of BaA, 7-methylbenz[a]anthracene, and BaP. The epoxide BaA-5,6-oxide was converted into the corresponding dihydrodiol and glutathione conjugate by these cells. Grover, Hewer, and Sims (1973) were first to identify an epoxide formed as a metabolite of BaA by human lung preparations as the K-region derivative BaA-5,6-oxide.

10.1.3.3 Human placental microsomes

The major metabolites of BaP produced by human placental microsomes from smokers were identified as 3-hydroxy-BaP and 7,8-dihydro-dihydroxy-BaP (7,8-diol) (Wang, Creasy, and Crocker 1974). Other metabolites included quinones, the 4,5- and 9,10-diols, and some unidentified polar metabolites.

10.1.3.4 Human fibroblast and epithelial cells

Huberman and Sachs (1973) measured the metabolism of BaP to form water- and alkali-soluble metabolites in cultured fibroblast and epithelial cells from human embryos. Fibroblasts from different organs of the same embryo metabolized similar amounts of BaP into water-soluble products. Fibroblasts derived from different embryos metabolized different amounts of BaP. The fibroblasts were divided into three groups, which metabolized an average of 350, 850, and 3400 µµmoles of water-soluble products per 1 million cells in three days. Cultures that contained more than 20% epithelial cells metabolized 3 to 25 times more BaP than did fibroblast cultures from the same embryo. Epithelial cells from different embryos also varied in the degree of BaP metabolism, and low epithelial activity was not necessarily associated with low fibroblast activity in the same embryo. Huberman and Sachs (1973) suggest that there is a genetic heterogeneity in BaP metabolism in fibroblast and epithelial cells and that the higher activity of epithelial cells might be related to the higher incidence of carcinomas rather than sarcomas in humans.

10.1.3.5 Human, rat, and hamster respiratory tract tissues

Wistar rat and Syrian hamster trachea metabolize PAH to the same products formed by human bronchial tissue (Pal, Grover, and Sims 1975). These tissues, maintained in short-term culture, metabolized BaP, BaA, and 7-methyl-BaA in a qualitatively similar manner. Preliminary results indicate that the amounts of each of the metabolites formed were also of the same order. Thus, the radioactive products formed from BaA were inseparable on thin-layer chromatograms from 5,6-dihydro-5,6-dihydroxy-BaA, 8,9-dihydro-8,9-dihydroxy-BaA, and 9-hydroxy-BaA. From 7-methyl-BaA, radioactive products having the chromatographic characteristics of the related 5,6- and 8,9-dihydrodiols were formed, together with 7-hydroxymethyl-BaA. BaP yielded products having the chromatographic characteristics of 4,5-dihydro-4,5-dihydroxy-BaP, 7,8-dihydro-7,8-dihydroxy-BaP, and 9,10-dihydro-9,10-dihydroxy-BaP; 3-hydroxy-BaP was also detected.

10.1.3.6 Mechanism of metabolism

Keysell, Booth, and Sims (1975) state that PAH are initially metabolized by microsomal oxidases into epoxides, which are then converted into dihydrodiols, phenols, and glutathione conjugates. The dihydrodiols are further metabolized by the oxidases to intermediates, which yield water-soluble products with glutathione. Their studies on the metabolism of BaA dihydrodiols showed that the 5,6-dihydrodiol is metabolized at the 8,9 bond and the 8,9-dihydrodiol at the 5,6 and 10,11 bonds. Formation of the 8,9-dihydrodiol-10,11-oxide (epoxide) of BaA by rat liver micro-somal fractions was demonstrated.

The role of the liver in the metabolism of BaP and its relation to local carcinogenesis was suggested in experiments by Kotin, Falk, and Miller (1962). An enhanced subcutaneous sarcoma response occurred in mice in which liver injury had been produced with carbon tetrachloride in comparison with mice receiving only BaP subcutaneously.

Epoxides are highly reactive. Figure 10.2 illustrates the possible pathways for an epoxide produced by mixed-function oxidative reactions. The epoxide may (1) spontaneously rearrange to a phenol, (2) be converted to a *trans*-dihydrodiol by the action of microsomal epoxide hydrases, (3) conjugate with glutathione, or (4) combine covalently with cellular nucleic acids, histones, and protein.

10.1.4 Elimination

The hepatobiliary system and the gastrointestinal tract are the main routes of elimination of PAH and their metabolites in different species, regardless of the route of administration. Because BaP and some of its metabolites are highly fluorescent in ultraviolet light, it was possible, in early studies, to follow the steps in the metabolism of intravenously injected BaP by observing color changes in the fluorescence of experimental animal tissues (Peacock 1936). Later studies used radioactive compounds.

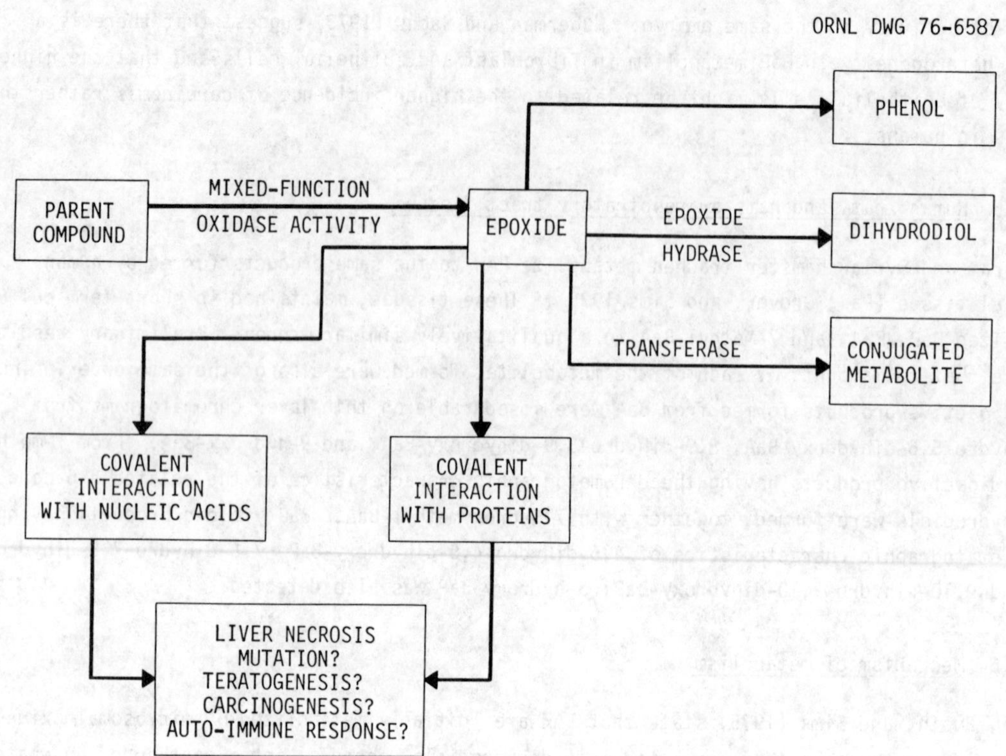

Fig. 10.2. Possible pathways in vivo for reactive epoxide intermediates of xenobiotics or endogenous substrates. Source: After Nebert 1974, Fig. 10-2, p. 219. From *The placenta: Biological and clinical aspects*. Courtesy of Charles C. Thomas, publisher, Springfield, Illinois.

10.1.4.1 Biliary excretion

Many observations have been made on the elimination of BaP and its metabolites following intravenous, subcutaneous, and intraperitoneal injections in experimental animals. Kotin, Falk, and Busser (1959), using radioactive BaP, and Heidelberger and Jones (1948), using dibenz[a,h]-anthracene, confirmed that, following those different modes of injection, the hydrocarbons are metabolized in the mouse and rat liver and excreted via the bile and intestinal tract into the feces. Radioactivity was completely cleared from the blood within 10 min.

Heidelberger and Jones (1948) observed the distribution and elimination of dibenz[a,h]anthracene labeled in the 9 and 10 atoms with ^{14}C following three different modes of administration to mice. Following intravenous injection of an aqueous colloid, an essentially quantitative recovery of the dose was found. There was an insignificant amount of radioactivity in the respiratory carbon dioxide; most of the radioactivity was found in the feces or the gastrointestinal tract. The urinary secretion of radioactivity was small compared with the amount found in the feces. Following administration by stomach tube (less than 1 mg in tricaprylin or in colloid), the radioactive dibenz[a,h]anthracene substances were eliminated almost exclusively in the feces. There was no detectable activity in the intestines or intestinal contents 48 hr after administration. There was little difference in the fate of dibenz[a,h]anthracene, whether administered as colloid or in tricaprylin, and there was no evidence that any appreciable quantity of material was absorbed in the body. Following intraperitoneal injection (1 mg dibenz[a,h]anthracene in 0.30 cm^3

tricaprylin) and subcutaneous injection, fecal elimination was again the primary route of removal of dibenz[a,h]anthracene. Dibenz[a,h]anthracene was retained for fairly long periods at the site of subcutaneous injection in fatty solvents. The distribution and rate of elimination depended on the mode of administration. Little material was absorbed into the body.

In another study, the results were the same: Radioactivity following intravenous injection of an aqueous colloid of 0.45 mg BaP-5-^{14}C into mice was recovered quantitatively, and no radio-activity appeared in the respiratory carbon dioxide. The compound and its metabolites were rapidly excreted into the bile, gastrointestinal tract, and feces (Heidelberger and Weiss 1951).

<u>Half-life</u>

The disappearance rates of radioactivity (half-lives) of three labeled hydrocarbons from the sites of subcutaneous injection in mice were 12 weeks for dibenz[a,h]anthracene; 3-1/2 weeks for 3-methylcholanthrene; and 1-3/4 weeks for BaP (Heidelberger and Weiss 1951). The relative car-cinogenicity of each compound was directly proportional to the time of retention at the injection site.

Heidelberger and Weiss (1951) also administered a single application of radioactive BaP or dibenz[a,h]anthracene in benzene to the shaved sacral region of the skin of mice. BaP or its metabolites, which had two half-lives, were eliminated much more rapidly than dibenz[a,h]-anthracene ($t_{1/2}$ = 1-2/3 and 4-1/3 d vs $t_{1/2}$ = 85 d).

<u>Metabolite excretion</u>

After intravenous injection into female Wistar rats, ^{3}H-7,12-dimethyl benz[a]anthracene (^{3}H-DMBA) was rapidly taken up by the liver, bound to particulate fractions, and metabolized to polar de-rivatives (Levine 1974). The metabolites passed into the cytosol and were rapidly excreted in the bile. Radioactivity appeared in the bile within a few minutes. Agents that induce or in-hibit microsomal drug-metabolizing enzymes increased or decreased, respectively, the biliary excretion of metabolites. Injected metabolites of ^{3}H-DMBA appeared to be taken up directly into the cytosol, and the rate of their excretion was more rapid than when the hydrocarbon itself was injected. Inducing and inhibiting agents did not affect the rate of biliary excretion of the injected metabolites. Thus, it appeared that transfer into the cytosol is the final process be-fore biliary excretion and that this step is dependent upon metabolism, which, under the condi-tions of the experiment, was probably the rate-limiting step in biliary excretion. Practically all of the injected ^{3}H-DMBA (96 to 97%) appeared in the bile, almost exclusively in the form of polar metabolites.

10.1.4.2 <u>Urinary excretion</u>

Generally, the noncarcinogenic PAH such as naphthalene, anthracene, and phenanthrene are metabolized to a variety of soluble metabolites, which are eliminated in the urine. The car-cinogenic hydrocarbons, as represented by BaA and chrysene (weak carcinogens) and dibenz[a,h]-anthracene, BaP, and DMBA, are metabolized to a variety of excretory products, which are found principally in the feces. In addition, the carcinogenic hydrocarbons are excreted principally as unconjugated phenols and quinones (Harper 1957).

10.1.4.3 Pulmonary excretion

Although pulmonary exhalation plays an important role in the elimination of volatile hydrocarbons, exhalation of PAH is negligible because their vapor pressure is low (Gerarde 1960).

10.1.5 Carcinogenicity

Formation of PAH-induced cancers in laboratory animals is well documented (Hartwell 1951; Shubik and Hartwell 1957 and 1969; Thompson & Co. 1971; Tracor/Jitco 1973a and 1973b; IARC 1973). Experimental animal studies in which the carcinogenicity of a PAH compound is clearly demonstrated, irrespective of administration route, have been reported (IARC 1973). The studies show that BaP, for example, has produced tumors in mice, rats, hamsters, guinea pigs, rabbits, ducks, and monkeys after administration by oral, skin, and intratracheal routes. BaP has both a local and systemic carcinogenic effect. In monkeys, there is some evidence of the ability of BaP to produce local sarcomas following repeated subcutaneous injections and to produce lung carcinomas following intra-tracheal instillation. BaP is also an initiator of skin carcinogenesis in mice as shown by numerous studies (Hartwell 1951; Shubik and Hartwell 1957 and 1969; Thompson & Co. 1971; Tracor/Jitco 1973a and 1973b), and it is carcinogenic in single-dose experiments and following prenatal exposure.

The amounts of different PAH necessary to produce cancer in 50% of treated animals (ED_{50}) vary greatly. In mouse skin-painting studies by Boyland (1958), the ED_{50} values for dibenz[a,h]-anthracene, 3-methylcholanthrene, and BaP were 20, 20, and 80 μg respectively.

Various polycyclic hydrocarbons that produce malignant tumors in rodents have failed to produce tumors under similar experimental conditions in rhesus, cynomolgus, and squirrel monkeys (Adamson, Cooper, and O'Gara 1970). Fibrosarcomas were produced by 3-methylcholanthrene and BaP in tree shrews and galagos after a single subcutaneous injection of 10 mg (Adamson, Cooper, and O'Gara 1970). The authors postulate that the prosimian primates, which include tree shrews, lemurs, lorises, and tarsiers, may be more closely related to the rodent than to the higher primates in their reactions to PAH carcinogens. Primates of the suborder Prosimii appear to be more susceptible to carcinogenesis by PAH and have shorter latent periods than primates of the suborder Anthropoidea, which includes monkeys, apes, and man (NAS 1972).

Only a few reports concerning the effects of a single PAH on man are available, and those describe only BaP. Cottini and Mazzone (1939) studied the effects of a 1% solution of BaP in benzol (benzene) applied daily to normal human skin and to various cutaneous lesions for periods of up to four months. This treatment resulted in definite manifestations localized to the treated areas. Erythema, pigmentation, desquamation, formation of verrucae (warts), and infiltration developed in chronological order. Pigmentary and wartlike lesions were the most commonly observed changes. After a maximum of 120 applications or when evidence of infiltration appeared, treatment was stopped and the manifestations regressed completely within two to three months. Although reversible and apparently benign, the changes were believed to represent early stages of a process that would have ultimately resulted in neoplastic proliferation.

Changes were more pronounced on unprotected skin surfaces and in older individuals. Treatment of squamous cell skin cancer with the BaP solution resulted in temporary amelioration. Similar changes and a squamous epithelioma were mentioned in two reports concerning accidental exposure to BaP (IARC 1973).

No case reports or epidemiological studies on the significance of man's exposure to other single PAH are available (IARC 1973). However, coal tar and other materials believed to be carcinogenic to man contain these hydrocarbons, and many PAH have been detected in other environmental situations.

10.1.5.1 Dose-response relationship

A dose-response relationship for atmospheric pollutants is difficult to show. Doses to experimental animals are many times greater than those received by man, either by skin contact with or inhalation of a polluted atmosphere. A rough estimation is that it would take a man living in a large urban area at least 272 years to retain 30 mg BaP in his lungs (NAS 1972). It is also difficult to evaluate factors such as (1) the effects of particles and particle size in polluted atmospheres; (2) individual breathing parameters and clearance mechanisms; (3) cocarcinogens and cigarette smoke; and (4) other less well defined exposures.

Skin tumors

In an attempt to determine a dose-response relationship and to estimate a threshold dose of BaP on the skin of mice, Schmidt, Schmal, and Misfeld (1973) painted the shaved skin of the backs of NMRI and Swiss strain mice twice weekly with BaP doses of 0.05, 0.2, 0.8, and 2.0 µg. The dose was delivered in one drop of liquid. Tumors occurred in the two groups receiving the highest doses, whereas no tumors at the site of treatment were observed in the lower-dose and control groups. Two NMRI mice receiving a dose of 0.8 µg (median total dose of 120.8 µg in 53 weeks) developed carcinomas at the site of application. Twenty-eight mice developed carcinomas after receiving applications of 2.0 µg (median total dose of 293.5 µg) twice weekly for 75.8 weeks. Values for two Swiss mice were 5 and 42 carcinomas at the same dose levels and at total doses of 106.2 µg in 57.8 weeks and 255.6 µg in 60.7 weeks. The Swiss mice were more sensitive to the carcinogen than the NMRI mice by a factor of 1.4. Since only two doses produced effects, a linear relationship of a logarithmic dose-effect curve could not be shown, but the results did suggest that there is a threshold of response.

Lung carcinomas

In a preliminary study, Hilfrich et al. (1973), using intratracheal instillations of BaP suspended in solution without a carrier dust, reported on the induction of tracheal and lung tumors in Syrian golden hamsters. At three months of age, the hamsters were intratracheally treated once a week with doses of 0.1 to 9.77 mg (increasing by a factor of 2.5) for life. Increased doses decreased survival rate. Animals receiving the highest doses (3.91 and 9.77 mg) survived three to four weeks, whereas those receiving the lowest doses (0.1 and 0.25 mg) lived more than 30 weeks. The three- to four-week survival period was not long enough for the induction of neoplasms, although hyperplastic and metaplastic changes were observed. The first neoplasm occurred after four weeks of treatment at the 3.91-mg level. With an increase in survival at the lower-dose levels, the incidence of tumors increased.

A dose-response relationship for the induction of squamous cell carcinomas was developed by
using intrabronchial pellets containing a graded series of concentrations of carcinogen in
cholesterol (Laskin, Kuschner, and Drew 1970). Pellets containing 3 mg of material ranging from
0.1 to 100% carcinogen were implanted in the bronchi of rats. Lower concentrations were produced
by dilution with cholesterol. Figure 10.3 shows a plot of the incidence of carcinoma vs
carcinogen concentration for 3-methylcholanthrene and BaP. The curves are representative of the
response observed with most drug reactions. The yield of squamous cell carcinomas with
3-methylcholanthrene ranged from 37.9% at 100% carcinogen down to 0% at the lowest level (0.1%).
The controls, which contained cholesterol or blank pellets, showed no evidence of spontaneous
carcinoma. These findings represent an overall picture without regard to dose. The squamous
cell carcinoma incidence was 20% with 3-methylcholanthrene and 11.6% with BaP. Squamous
metaplasis incidence was high in both treated and control rats. These were preliminary results
and the time length of the study was not reported.

ORNL DWG 76-11930

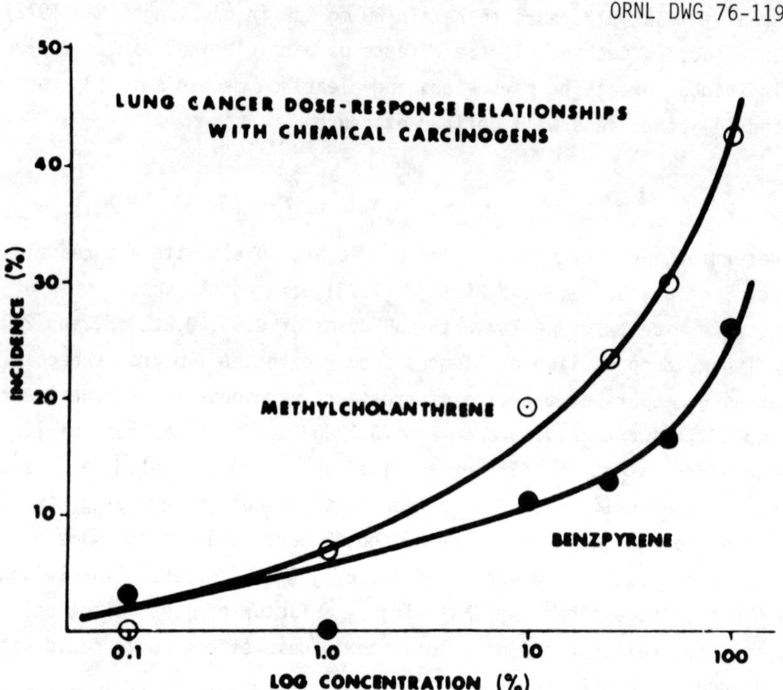

Fig. 10.3. Relationships of tumor yield to dose, using a graded series of concentrations of
carcinogen in cholesterol on an intrabronchial pellet. Source: From Laskin, Kuschner, and Drew
1970, Fig. 4, p. 326.

Subcutaneous tumors

Pfeiffer (1973) showed a dose-response relationship between BaP and tumors at the site of subcutaneous injection on the upper back of female NMRI mice. Doses ranging from 3.125 to 100 µg
BaP in 0.5 ml tricaprylin were used. The results are shown in Fig. 10.4. The spontaneous tumor
rate for this strain of mice is 0 to 2%. Similar dose-response results were found with dibenz-
[a,h]anthracene. With the lowest dose of BaP (3.125 µg), tumors developed in 5 of 100 animals
within 56 weeks. With the lowest dose of dibenz[a,h]anthracene (2.35 µg), 33 tumors developed.

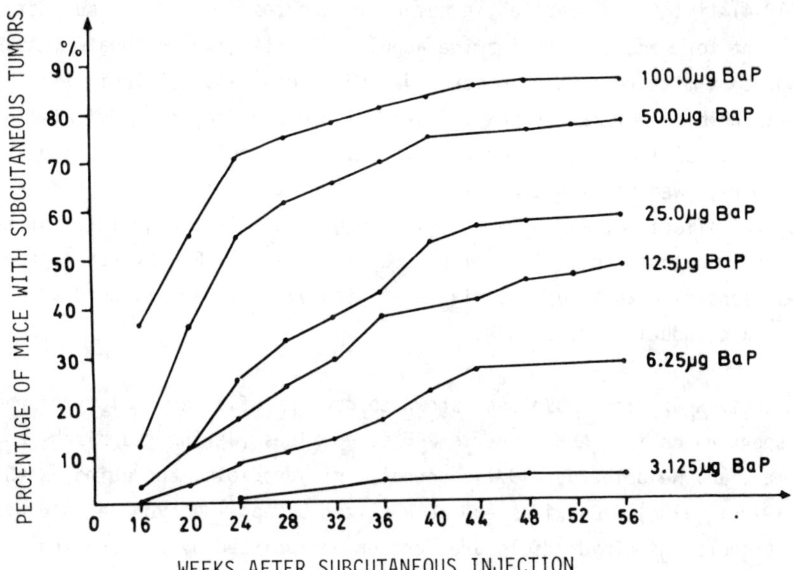

Fig. 10.4. Rates of tumors in NMRI mice after subcutaneous injection of different doses of BaP in 0.5 ml tricaprylin. Source: From Pfeiffer 1973, Fig. 1, p. 74. Reprinted by permission of the publisher.

<u>Human lung cancer</u>

Results of a long-term study of mortality among coke-oven workers (Lloyd 1971; Redmond et al. 1972) suggest a dose-response relationship between exposure to coke-oven emissions, which contain PAH, and lung cancer (Sects. 10.1.8.2 and 10.1.8.3). Increased risk of lung cancer was related to both time and degree of exposure, as indicated by area of employment. Employment at the top of the ovens, where emission concentrations were highest, was associated with an almost fivefold excess risk of lung cancer. This increased to an almost tenfold risk when only men having five or more years of experience were considered. Lloyd (1971) reports that coke-oven workers run a 7.5-fold risk of death from kidney cancers.

10.1.5.2 <u>Enzyme induction</u>

The enzyme system primarily responsible for PAH metabolism is a multicomponent, microsomal-bound system of mixed-function oxidases (oxygenases), of which aryl hydrocarbon hydroxylase (AHH) is the most important. It is well established that these enzymes may either inactivate (detoxify) carcinogenic compounds or activate them to more carcinogenic forms (Wiebel and Gelboin 1974).

Conney, Miller, and Miller (1957) describe the oxidative metabolism of PAH by enzymes located in the endoplasmic reticulum. They found that, after pretreatment with 1 mg BaP, liver homogenates from treated rats showed a tenfold greater AHH activity than did similar preparations from control rats. A doubling of enzyme activity was detected as early as 3 hr after BaP administration.

Wiebel and Gelboin (1974) list AHH properties in human lymphocytes from normal subjects. These include (1) a requirement for reduced nicotinamide adenine dinucleotide phosphate (NADPH) and oxygen, (2) inhibition by a mixture of carbon monoxide and oxygen, and (3) inhibition by 7,8-benzoflavone (a potent inhibitor of BaP hydroxylation). These requirements indicate that the AHH of human lymphocytes is similar to that of rat liver or of rodent cells in culture. Wiebel and Gelboin (1974) also reviewed studies demonstrating the relationship between the enzyme system and the biological effects of PAH. The studies suggest that the regulation of this enzyme system may be an important factor in chemical carcinogenesis. BaP hydroxylation is the basis for an extremely sensitive assay of the mixed-function oxidases and is used as an indicator for the presence and inducibility of AHH.

AHH and other mixed-function oxidases have been shown to consist of a terminal cytochrome (P450) and an electron transporting chain. AHH converts PAH to epoxides (Jerina, Kaubisch, and Daly 1971; Selkirk, Huberman, and Heidelberger 1971), phenols, dihydrodiols, and quinones (Boyland and Sims 1965; Sims 1970*a*; 1970*b*). Epoxides are metabolized by epoxide hydrase to dihydrodiols (Nebert 1974). Some phenols and dihydrodiols are further metabolized by conjugating enzymes to yield the water-soluble glucuronides or sulfates.

Aryl hydrocarbon hydroxylase inducibility

Studying the role of the AHH enzyme system in chemical carcinogenesis is particularly useful because of two properties: (1) The system is inducible (Nebert and Gelboin 1968*a*; 1969), and (2) the enzyme level and inducibility are genetically determined and vary in different species and different strains of mice (Nebert and Gelboin 1969; Wiebel and Gelboin 1974). AHH is found in 90% of the tissues of the mouse, rat, hamster, and monkey. In vivo, the prior administration of a PAH induces the hydroxylase to higher levels in the liver, lungs, gastrointestinal tract, and kidneys of the monkey, hamster, and rat. However, the specific activities of the enzyme system in all tissues of the monkey, before and after 3-methylcholanthrene administration, are considerably lower than those observed for the mouse, hamster, or rat. Generally, the monkey is rather resistant to PAH carcinogenesis; this resistance may be related to the low levels of AHH activity. AHH is present in various tissues of six strains of mice; the hepatic enzyme is inducible in the Swiss strains C-57, C3H, and A, but not in strains AKR/N or DBA. As reported by Nebert and Gelboin (1969), the magnitude of AHH induction varied greatly in the tissues and species — from no induction to more than a 100-fold increase in activity. Similar variations in AHH inducibility or activity was found to exist in the human system (Kellermann, Cantrell, and Shaw 1973).

In addition to PAH, the enzyme system is also inducible by a variety of other endogenous and exogenous compounds including steriod hormones, barbiturates, insecticides, and whole cigarette smoke (NAS 1972). Recent studies have shown that inducibility is genetically regulated (Kourie, Ratrie, and Whitmire 1973).

Transplacental induction

The enzyme system can be induced transplacentally. Treatment of the pregnant rat or hamster with PAH or phenobarbital induced the enzyme system in specific fetal tissues as well as in the placenta (Nebert and Gelboin 1969). The extent of enzyme induction was greatest in the adult,

less in neonatal tissues, and least in fetal tissues. There is a high correlation between the
placental content and the smoking habits of pregnant women (Welch et al. 1968; Nebert, Winker,
and Gelboin 1969). PAH are known constituents of cigarette smoke. The placentas from women
who smoked 20 or more cigarettes per day showed a mean AHH activity 5- to 25-fold greater than
that found in placentas from women who did not smoke. However, significant AHH activity was
found in the placentas of some women who did not smoke, and no measurable AHH activity was
found in the placentas of some women who smoked 20 or 30 cigarettes per day (Nebert, Winker,
and Gelboin 1969). No such differences were found by Welch et al. (1968).

Tissue specificity

Although earlier studies indicated that inducibility is an "all-or-none" phenomenon in all
tissues of a given mouse strain, Wiebel and Gelboin (1974) note a tissue-specific inducibility
in certain strains of mice. In four strains of mice in which hepatic AHH is either not inducible
or only slightly inducible, enzyme activity was increased about tenfold in the lungs and two- to
fivefold in the kidneys and small intestine after a single intraperitoneal BaA injection (100 mg/kg
body weight in 0.2 ml corn oil). AHH activity also increased in the skin of all the strains
16 hr after topical application of 300 µg BaA (Table 10.7).

Table 10.7. Inducibility of AHH in hepatic and extrahepatic tissues
from various mouse strains[a]

Strain	AHH activity (pmoles of phenolic product per mg of protein per 30 min)[b]									
	Liver		Lung		Small intestine		Kidney		Skin[c]	
	Control	BA	Control	BA	Control	BA	Control	BA	Control	BA
NZB	3080	2500	27	300	2.0	8.1	<1	6.2	20	79
NZW	2090	2970	37	460	4.7	23.2	3.2	13.4		
AKR	1810	2808	47	514	2.7	7.9	<1	6.1	8	72
SJL	2090	1790	35	260	2.5	17.2	4.8	10.8	25	106
DBA	920	995	56	350			2.3	8.2	36	129
C57	1550	5760	208	540			2.4	142.0	63	570

[a]BaA (100 mg/kg body weight) was injected intraperitoneally in 0.2 ml of corn oil. Control mice
received corn oil only. Animals were sacrificed 12 h later and enzyme activities were
determined in homogenates (0.5-1.2 mg protein) of the various tissues.
[b]Numbers represent the mean values from three animals.
[c]Sixteen hours after topical application of 300 µg BaA in 0.2 ml acetone or acetone alone to
the shaven backs.

Source: Wiebel and Gelboin 1974, Table 6, p. 74.

Mechanism and forms

Much is yet to be learned of the mechanisms of detoxification enzyme induction. However, the
kinetics of AHH induction at different inducer concentrations were summarized by Wiebel and
Gelboin (1974), who routinely used BaA as inducer because it is relatively nontoxic. In cell
cultures (1) the inducer was rapidly incorporated into the cell; (2) there was an initial lag
of about 30 min before enzyme activity increased; (3) induction required continuous protein

synthesis; (4) induction required RNA synthesis initially, but not later; and (5) the RNA synthesis was independent of protein synthesis. Currently, there appear to be various levels of AHH induction regulation (at the level of RNA transcription or post-transcription), and their relative importance differs for different cell types and probably for different tissues as well.

Observations by Alvares, Schilling, and Kuntzman (1968) indicate that PAH are metabolized by more than one form of microsomal hydroxylase, distinguishable by spectral and kinetic properties. The authors observed differences in the kinetics of BaP hydroxylation by hepatic hydroxylase from rats previously treated with either phenobarbital or 3-methylcholanthrene. Wiebel and Gelboin (1974) summarize the relative distribution of these two AHH forms in rats. The first form, which is stimulated by 7,8-benzoflavone, is found in newborn rats and predominates in the liver of male adult rats. This form is induced by phenobarbital. The second form, which is inhibited by 7,8-benzoflavone, is found in extrahepatic tissues, comprises a larger fraction in the liver of female adult rats, and is inducible by PAH in hepatic and extrahepatic tissues. Both forms are present in any given tissue; however, one form predominates over another. Preliminary observations indicate that there are differences in the profile of BaP metabolites formed by the constitutive and the 3-methylcholanthrene-induced form of AHH.

Rasmussen and Wang (1974) treated adult male Sprague-Dawley rats with PAH, phenobarbital, or barbital and studied the induced AHH enzymes in liver and lung in terms of the specific metabolites of BaP that were produced in an in vitro incubation system. The metabolite profile produced by phenobarbital- or barbital-induced enzymes differed from that found with the hydrocarbon-induced liver enzymes. Only the production of 4,5-dihydro-4,5-dihydroxy-BaP and 3-hydroxy-BaP was increased over the control level wher phenobarbital was used as the inducer, whereas the production of all identified metabolites was increased when the hydrocarbons were used as inducers. An exception was benzo[*e*]pyrene, which was essentially inactive as an inducer for either liver or lung enzymes. Phenobarbital and barbital did not increase enzyme activity in the lung as did BaP, 3-methylcholanthrene, and BaA. The rats received daily intraperitoneal injections of a solution (8 mg/ml) of sodium phenobarbital or sodium barbital (8 mg/100 g body weight) in NaCl for three days prior to sacrifice. Separate groups of rats received intraperitoneal injections of a solution (4 mg/ml) of BaP or other hydrocarbons (2 mg/100 g body weight) in corn oil 24 hr before sacrifice.

<u>Aryl hydrocarbon hydroxylase induction and cancer</u>

Results of studies focused on the relationship between AHH inducibility and cancer are far from definitive. Nonetheless, there is some indication in both laboratory animals and humans that greater cancer susceptibility may be associated with greater inducible AHH activity.

Kourie, Ratrie, and Whitmire (1973) found a close relationship between susceptibility to 3-methylcholanthrene-induced subcutaneous tumors and inducibility of AHH in 14 strains of mice. However, no correlation could be detected between sarcomas evoked by DMBA or BaP and the inducible AHH activity among the same inbred strains. In a genetic system where AHH inducibility is segregated as a single autosomal dominant gene, mice carrying the inducible allele were five to ten times as sensitive to 3-methylcholanthrene tumorigenesis as their noninducible litter mates. Heterozygotes seemed to have intermediate sensitivity. Of 72 mice bearing 3-methylcholanthrene-induced primary tumors, 68 were AHH inducible. In backcross experiments, 10 of 12 (83%) mice having tumors were AHH inducible. All tumors were typical fibrosarcomas and were found at the site of subcutaneous injection. In this study hepatic AHH level was assayed.

The susceptibility to tumorigenesis produced by the topical application of BaP and promotion by phorbol ester or by the intraperitoneal administration of large doses of BaP was not correlated with the AHH inducibility in litter mates of C57BL/6N, DBA/2N, and NZW/BLN strains of mice (Benedict, Considine, and Nebert 1973). It was concluded that, if this enzyme is necessary for metabolic activation of BaP to the proximal carcinogen, the basal levels of the AHH in skin, lymph nodes, or peritoneal lining are sufficient.

Kellermann, Luyten-Kellermann, and Shaw (1973*a*; 1973*b*) showed that variation in extent of AHH induction by 3-methylcholanthrene in cultured human leukocytes is under genetic control. The normal white population in the United States was divided into three distinct groups of low, intermediate, and high degrees of inducibility. The variation appeared to result from two alleles at a single locus. Phenotype frequencies were 53, 37, and 10%. A total of 161 healthy unrelated adults were sampled.

Kellermann, Shaw, and Luyten-Kellermann (1973) studied 50 patients having bronchogenic carcinomas; 4% of the patients had a low inducible AHH activity, 66% had a medium, and 30% had a high level of induction. Low, medium, and high AHH-inducibility frequencies in a control group of 85 healthy volunteers were 44.7, 45.9, and 9.4%. As an additional control group, 46 patients having other types of tumors were randomly selected to demonstrate that low or high inducibility is not simply the result of having a tumor. Percentages of AHH inducibility in this group were 43.5, 45.6, and 10.9%, which were comparable with those of the original control group. Evalua-tion of the smoking habits of the patients having lung cancer disclosed that all were heavy smokers. The data indicate that susceptibility to bronchogenic carcinoma is associated with the higher levels of inducible AHH activity. The potential usefulness of lymphocytes in screening human populations for genetic variability or possible effects of PAH was suggested.

10.1.5.3 Enhancement

In most epidemiological studies, exposure to a combination of chemicals as they appear in the environment or industry is the usual mode of contact, whereas in controlled laboratory studies, animals contact only individual chemicals or, occasionally, two chemicals. Cocarcinogens can markedly increase the effect of carcinogens, but are usually noncarcinogenic or, at most, very weakly carcinogenic.

Impairment of mucociliary transport mechanisms by chemicals could be important in determining the residence time, and thus the concentration, of damaging substances within the respiratory tract. Pulmonary clearance may be impaired by noxious gases and vapors. Reduced clearance of carcinogens has been noted when the carcinogen is attached to particulate matter (Henry and Kaufman 1973).

Particulates

There is a maximum efficiency of lung (alveolar) deposition of particulates at a size of about 1 and 2 μ; there is a minimum efficiency for a size of about 0.5 μ; the percentage of particle deposition for sizes less than 0.1 μ is just as great as for sizes of greater than 1 μ [National Air Pollution Control Administration (NAPCA) 1969]. The bulk of the particulate mass in the atmosphere ranges between 0.1 and 10 μ in size.

PAH are readily adsorbed to very small particles (i.e., those particles having a diameter of less than 0.04 μ), but are not readily eluted from them. As particle size increases, release in the presence of appropriate solvents becomes more rapid and greater in extent (NAPCA 1969). Polycyclic organic matter detected in the atmosphere is associated exclusively with particulate matter, especially soot. Most data indicate that more than 75% of the BaP associated with particles is associated with particles less than 5 μ in diameter (EPA 1975).

Within the limits of 0.01 to 100 μ (mass median diameter), the fraction of the mass of an inhaled particulate cloud deposited in the nose, tracheobronchial, and pulmonary compartments as a function of the mass median diameter is shown in Fig. 10.5.

ORNL DWG 76-11937

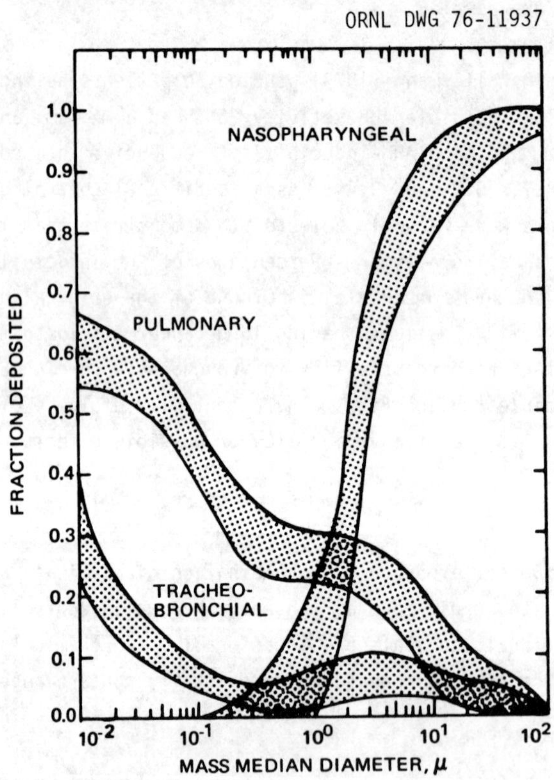

Fig. 10.5. Fraction of mass of inhaled particles deposited in the three respiratory tract compartments as a function of mass median (aerodynamic) diameter. Source: Task Group on Lung Dynamics 1966, Fig. 11, p. 181. Reprinted by permission of the publisher.

Iron oxide particulate matter appears to have properties that contribute to cancer production. Saffiotti, Cefis, and Kolb (1964) first demonstrated that BaP, when attached to iron oxide (Fe_2O_3) particles, induced an unexpectedly high incidence of bronchogenic carcinomas in hamsters after repeated intratracheal injections. Iron oxide has no extremely irritating or toxic effects. Lung cancer occurred in up to 100% of the animals in some experiments, and the lung cancers mimicked all the cell types seen in human lung cancers. Dose-response effects were indicated, as was the possibility that a single high dose could induce cancers in the system.

The influence of ferric oxide on BaP carcinogenesis was studied recently by Henry, Port, and Kaufman (1975). Three different mixtures of BaP and iron oxide with different physical properties were given intratracheally to Syrian golden hamsters. Hamsters treated with a preparation containing large aggregates of BaP and ferric oxide (95% of the particles less than 3 μ in size) showed an earlier onset and higher incidence of respiratory tract tumors than animals given a mixture containing smaller aggregates. The greatest number of tumors were present in the trachea and the predominant type was the squamous carcinoma. The third preparation, in which the carcinogen was not attached to the ferric oxide, showed a low tumor incidence similar to that present after intratracheal intubation of BaP in gelatin without a carrier particle. For this system, the physical attachment of BaP to the carrier dust was necessary for a high tumor yield, and particle size was important. The adsorption of BaP onto iron oxide particles in the atmosphere is yet to be shown (Olsen and Haynes 1969).

The enhanced tumorigenicity of BaP–iron oxide combinations may be due to a reduction in carcinogen clearance rate. Clearance of BaP-hematite particles in the hamster lung following intratracheal instillation was studied by Kennedy and Little (1974). BaP proved to be a loosely bound soluble carcinogen, which left the carrier particles and entered the upper airway epithelial cells during clearance of the particles by mucous and ciliary action. Thus, mainly epidermoid carcinomas of the major bronchi and trachea were produced.

Sulfur dioxide

Laskin, Kuschner, and Drew (1970) studied the combined effect of sulfur dioxide and BaP on lung tumor development. Rats and hamsters were exposed to atmospheres of fresh air, sulfur dioxide, and/or a BaP–sulfur dioxide mixture. No respiratory tract neoplasms or other significant pathology was found in the hamsters. The rats developed squamous cell carcinomas histologically similar to those observed in man. Of the 21 rats exposed to both sulfur dioxide and a BaP–sulfur dioxide mixture, 5 developed squamous cell carcinoma of the lung, whereas only two squamous cell carcinomas were detected in 21 rats exposed only to the BaP–sulfur dioxide. No tumors developed in the other atmospheres. The results may have been influenced by the chronic murine pneumonia observed in the rat colony. (The irritant and enhancing properties of sulfur dioxide are discussed in Sect. 10.2.2.)

Ozone

Although ozone is present in photochemical smog and causes a series of pulmonary disorders, no studies on the possible cocarcinogenic effect of ozone were found. Nettesheim and Schreiber (1975) summarize animal studies on the possible effects of ozone. Ozone has been shown to accelerate lung tumor development in mice. An important effect of ozone in relation to respiratory carcinogenesis is its suppressive effect on AHH activity in the lung. Enzyme activity in the tracheobronchial mucosa of rabbits was inhibited at ozone concentrations as low as 1.4 mg/m^3 (0.75 ppm) (Palmer, Exley, and Coffin 1972).

Nitrogen dioxide

No studies were located on the carcinogenic or cocarcinogenic effect of nitrogen dioxide that resulted in a positive relationship. Kuschner and Laskin (1974), as cited in NIOSH (1976), investigated the effects of combined nitrogen dioxide and BaP exposure in rats. Preliminary results indicated the presence of squamous cell carcinoma in the lungs of 1 of 30 rats exposed to 25 ppm nitrogen dioxide with daily 1-hr exposures to combined nitrogen dioxide and BaP (10 mg/m^3). The concentration of nitrogen dioxide used in this study was 25 times as great as the recommended ceiling level. Concentrations as high as 94.1 mg/m^3 (50 ppm) had no significant effect on AHH activity (Palmer, Exley, and Coffin 1972).

Long-chain aliphatic hydrocarbons

The accelerating properties of aliphatic and related hydrocarbons were demonstrated by Horton, Denman, and Trossett (1957) under three sets of experimental conditions: (1) as solvents for 3-methylcholanthrene or BaP in tests involving repeated application of the carcinogenic solutions to the skin of C3H mice; (2) as preconditioners (without added carcinogen) of the host by their application to the skin prior to any exposure to carcinogens; and (3) as promoting agents following a single application of DMBA. It was observed that

1. Solutions of polycyclic carcinogens in solvents having long-chain structures, such as n-dodecane and dodecylbenzene, induce skin tumors in mice at much higher rates for a given concentration of carcinogen than do solutions in hydrocarbon solvents of lower molecular weight.

2. The accelerating hydrocarbons are capable of preconditioning the skin of C3H mice, rendering it more responsive to subsequent applications of a carcinogenic oil.

3. The accelerating solvents are effective promoters of carcinogenesis initiated by a single application of a carcinogenic material.

The solvents, although primary irritants, were noncarcinogenic to the skin of mice.

In a later study, Bingham and Falk (1969), using compounds with long aliphatic chains as cocarcinogens or accelerators, compared the rates of acceleration of cutaneous tumorigenesis in mice. They found a 1000-fold increase in the enhancement of potency of low concentrations of BaP and BaA when n-dodecane, a low-temperature hydrocarbon carbonization product of coal, was the diluent. Solutions of BaP, ranging from 0.02-0.000002% dissolved in decalin or a 50:50 mixture of n-dodecane and decalin, were applied topically (50 mg by volume) to the skin of C3H/He mice three times a week. When BaP was dissolved in decalin (decahydronaphthalene), no skin tumors developed in any of the mice below the 0.02% concentration. When n-dodecane was the diluent, skin tumors were produced at 0.000002% BaP. Similar results were obtained with BaA.

In a second experiment using various concentrations of BaP and of 1-dodecanol and 1-phenyldodecane, there was a reduction in the time required to induce tumors with increasing concentrations of BaP; however, the cocarcinogen was most effective at the lower level of carcinogen (0.05% BaP in 1-dodecanol as compared with 0.20% BaP). It was impossible to distinguish between the effects of 0.05 and 0.20% BaP when 1-phenyldodecane was the solvent.

Phenols

In inhalation studies using mice, Tye and Stemmer (1967) found that coal tars with phenols removed were less potent carcinogens than coal tars containing phenols; the cocarcinogenic potential of phenols was attributed to their irritant properties.

Aldehydes

Aldehydes occur as general pollutants and are important intermediaries in many industrial processes. Several investigators have studied their carcinogenicity and mutagenicity. A 35-week pretreatment of C3H mice with gaseous formaldehyde did not enhance the response to subsequently applied coal tar at 100 mg/m^3 of air (Horton, Tye, and Stemmer 1963). However, simultaneous administration of furfural (a homolog of formaldehyde) and BaP did hasten lung tumor development in hamsters (Feron 1972).

Infections

A possible relationship between respiratory infections and susceptibility to lung cancer has been investigated and debated for years. Various respiratory infections, such as chronic bronchitis, influenza pneumonitis, interstitial pneumonia, and tuberculosis, have been thought to increase the risk of lung cancer in man. Results of numerous clinical, pathological, and epidemiological investigations have been contradictory; laboratory studies have also furnished seemingly contradictory results. Viral pneumonitis suppressed the development of pulmonary adenomas in mice, whereas chronic murine pneumonia enhanced the development of squamous cell carcinomas (Nettesheim and Schreiber 1975). Although viral pneumonitis causes extensive pulmonary scarring and consolidation of lung sections, thus reducing the potential amount of target tissue for carcinogens, other microbial effects could result in an enhanced susceptibility of respiratory tract epithelium to carcinogens. These are the prolonged stimulation of cell proliferation in the tracheobronchial epithelium, alteration of the distribution and residence time of particulates, damage to ciliated epithelium, alteration in carcinogen metabolism, and alteration of the immune response to noncross-reacting antigens (Nettesheim and Schreiber 1975).

10.1.5.4 Suppression

Two (or more) agents applied simultaneously or sequentially may interact to alter the effect of each other. This may result in a decreased biologic effect, owing to competitive reactions at the sites concerned with biologic activity or to induction or inhibition of enzymes that may detoxify or activate, respectively, chemical carcinogens. There is some evidence that excess dietary vitamin A offers partial protection to epithelial tissues against chemical carcinogens. Several compounds that inhibit AHH activity have been reported.

Vitamin A

The role of vitamin A in controlling the growth and differentiation of mucous epithelia has been known for many years. Vitamin A is necessary for normal growth and differentiation of tracheobronchial epithelium. In the absence of vitamin A, the normal ciliated and mucus-producing cells of this epithelium are replaced by squamous cells, which produce keratin. In hamster organ culture systems, addition of vitamin A or its analogs to the culture induced reversal of keratinization and growth of a new ciliated and mucus-producing epithelium (Sporn et al. 1974). Current research suggests that ample vitamin A intake imparts some protection to epithelial tissue against

chemical carcinogens. This was demonstrated in hamsters by the protection afforded by vitamin A
against lung cancer produced by instillation of hydrocarbons and other particles (NAS 1972).
However, whether vitamin A or its analogs are capable of reversing squamous metaplastic or other
preneoplastic lesions induced by chemical carcinogens remains unknown.

Saffiotti et al. (1967) investigated the effects of vitamin A on respiratory tract carcinogenesis.
Respiratory tract tumors were induced in hamsters by ten intratracheal injections of BaP (3 mg)
mixed with ferric oxide (3 mg). One group received stomach-tube feedings of vitamin-A palmitate
(5000 IV) twice a week, starting one week after the last BaP treatment. Respiratory tract tumors
(all types) developed in 17 of 53 (32%) animals receiving BaP only, but in only 5 of 46 (11%) of
animals receiving treatment with vitamin A. The incidence of squamous tumors was markedly lower
in the group receiving the vitamin treatment (only one animal in 46 as compared with 11 of 53).
Similarly, squamous metaplasia was seen in 13 cases receiving BaP treatment and in only one case
receiving BaP plus vitamin A. The number of forestomach papillomas was also significantly reduced by excess vitamin A. However, animals treated with vitamin A had a faster death rate.
The authors suggest that vitamin A has a systemic inhibitory effect on the induction of squamous
changes in the columnar mucous epithelium of the respiratory tract.

Cone and Nettesheim (1973) observed the suppression of 3-methylcholanthrene-induced squamous
metaplasias and early tumors in the respiratory tract of rats by vitamin A, which confirmed the
work of Saffiotti et al. (1967). Rats, on a diet free of vitamin A, were given normal or excess
levels of retinyl acetate (a vitamin-A compound), by stomach tube several weeks before administration of 3-methylcholanthrene. Development of metaplastic lesions and squamous cell tumors was
suppressed at all time intervals. The study suggests that sufficient vitamin A offers partial
protection of respiratory tract epithelium against high doses of 3-methylcholanthrene. Nettesheim
and Schreiber (1975) report several studies in which tumor development appeared to be enhanced
by vitamin A in experimentally prepared tumor systems.

Hill and Shih (1974) evaluated 14 vitamin-A compounds and analogs and other agents as inhibitors
of microsomal mixed-function oxidases from liver and lung tissue of mice and hamsters (Table 10.8).
The formation of BaP-X, an unidentified metabolic product of BaP that binds tightly to macromolecules, was strongly inhibited by retinol and several other vitamin-A compounds and by butylated
hydroxytoluene. Several of the tested compounds have known anticarcinogenic properties. BaP-X
accumulated in large amounts in the presence of cyclohexene oxide, an inhibitor of epoxide hydrase. The formation of tightly bound metabolites of DMBA and 3-methylcholanthrene was also
inhibited by the compounds that block production of BaP-X.

By another pathway, BaP is converted to 3-hydroxy-BaP (Fig. 10.6); this step was not inhibited
by any of the vitamin-A compounds. 3-Hydroxy-BaP is further metabolized by a microsomal oxidase
to BaP-Y. The disappearance of this compound was inhibited by retinol and some of its derivatives and by butylated hydroxytoluene. Hill and Shih (1974) proposed the scheme in Fig. 10.6
as tentative pathways for BaP metabolism. One pathway leads to BaP-X, the formation of which
probably proceeds through an epoxide intermediate. Retinol and some of its derivatives,
7,8-benzoflavone, and butylated hydroxytoluene inhibit this reaction. Cyclohexene oxide apparently inhibits epoxide hydrase to increase the yield of BaP-X.

Table 10.8. Inhibition of liver mixed-function oxidases by vitamin A compounds and analogs and other agents

Inhibitor	Cyclophosphamide (K_i)	Nicotine (N-1) (K_i)	BP (B-1) (% inhibition)[a,b]	DMBA (% inhibition)[a,c]	MC (% inhibition)[a,d]	3-Hydroxy-BP (% inhibition)[a,e]
Retinol	0.12	0.36	38 (31 - 44)	42 (40, 43)	52 (40, 61)	40 (38, 42)
13-*cis*-Retinol	0.11	0.26	32 (37 - 39)	20 (13, 27)	58 (49, 68)	32 (31, 34)
Retinyl acetate	0.12	0.40	37 (24 - 51)	40 (35, 44)	54 (49, 60)	28 (26, 30)
Retinyl palmitate	0.29	0.90	13 (0 - 22)	13 (10, 16)	-2 (-5, 0)	0 (3, -3)
Retinal	0.15	>1.0	37 (27 - 45)	37 (26, 47)	47 (34, 60)	28 (24, 33)
9-*cis*-Retinal	0.12	>2.0	45 (35 - 55)	53 (50, 56)	75 (71, 79)	46 (44, 49)
13-*cis*-Retinal	0.14	0.62	34 (30 - 36)	34 (33, 36)	56 (46, 66)	27 (26, 28)
Retinoate	>0.5	1.0	4 (0 - 17)	6 (5, 8)	-2 (-11, 7)	0 (0, 0)
13-*cis*-Retinoate	>1.0	>2.0	5 (3 - 13)	4 (3, 4)	-2 (-3, 0)	10 (10, 10)
Methyl retinoate	>1.0	>2.0	11 (0 - 23)	8 (6, 11)	17 (13, 22)	-3 (-5, -2)
Ethyl retinoate	>0.8	>2.0	10 (0 - 16)	8 (0, 16)	-3 (-18, 12)	1 (-5, 3)[f]
N-Ethylretinamide	0.25	0.64	25 (18 - 30)[f]	14 (2, 26)	30 (24, 37)	13 (5, 19)[f]
N,N-Diethylretinamide	>1.0	>2.0	26 (19 - 37)[f]	14[g]		10 (6, 13)[f]
Ro 8-7699	>0.6	>2.0	9 (0 - 22)	6 (0, 12)	26 (21, 30)	8 (3, 14)
7,8-Benzoflavore	0.1	>0.2	46 (34 - 57)[f]	39 (37, 41)	58 (57, 60)	7 (-3, -15)[f]
Butylated hydroxytoluene	0.1	>0.2	76 (67 - 83)[f]	44 (40, 49)	79 (76, 82)	90 (73, 100)[f]
Cyclohexene oxide	>2.0	>2.0	-52 (-64, -39)[h]	8 (2, 15)	3[g]	-19 (-18, -20)

[a] Inhibition of binding to reaction components produced by 50 μM test compound, except cyclohexene oxide which is 2 mM.
[b] Average value and range for 5 separate determinations; in controls, the product formed ranged from 2.3 to 3.1 nmoles.
[c] Average and individual values for 2 separate determinations; in controls, the product formed was 5.2 and 6.0 nmoles.
[d] Average and individual values for 2 separate determinations; in controls, the produced formed was 0.50 and 0.71 nmole.
[e] Average and individual values for 2 separate determinations; in controls, the substrate used was 1.2 and 1.3 nmoles.
[f] Three determinations.
[g] Single determination.
[h] Two determinations.

Source: Hill and Shih 1974, Table 1, p. 567. Reprinted by permission of the publisher.

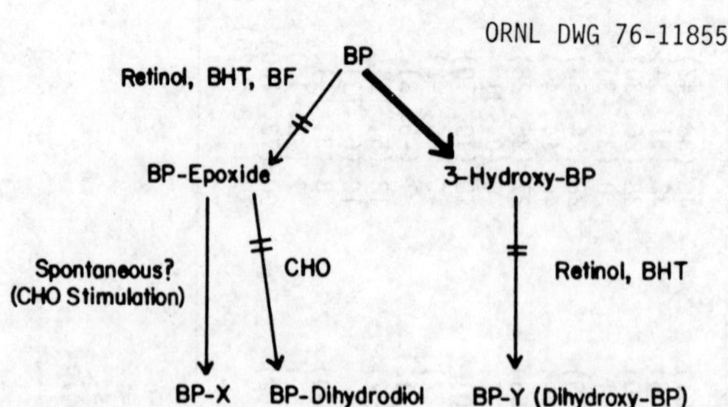

Fig. 10.6. Proposed scheme for BaP metabolism. Heavy arrow, reaction occurring both in normal liver and in lungs of mice pretreated with BaA; CHO, cyclohexene oxide; BHT, butylated hydroxytoluene; BF, 7,8-benzoflavone. <u>Source</u>: Hill and Shih 1974, Chart 3, p. 68. Reprinted by permission of the publisher.

Antioxidants

Butylated hydroxytoluene, an antioxidant that serves as a substrate for a microsomal oxidase, strongly inhibited the formation of mammary tumors induced by DMBA (Wattenberg 1972). Also, butylated hydroxyanisole, butylated hydroxytoluene, and ethoxyquin inhibited chemical carcinogenesis when administered either before or with a carcinogen, suppressing the action of several carcinogens under different experimental conditions. Inhibition occurred both when the route of administration resulted in direct contact of the carcinogen with the target tissue and when the carcinogen acted at a site remote from that of administration (Wattenberg 1974).

As an extension of the work with antioxidants, Wattenberg (1974) studied the inhibitory effects of disulfiram and several other sulfur-containing compounds in three experimental tumor systems. Disulfiram, dimethyldithiocarbamate, and benzyl thiocyanate, when added to the diet of female Sprague-Dawley rats, inhibited DMBA-induced mammary tumor formation and adrenal necrosis. A single oral dose of disulfiram, given 24 hr before the carcinogen, similarly inhibited DMBA-induced mammary tumor formation. In the mouse, disulfiram prevented the occurrence of forestomach tumors attributed to BaP in the diet, but did not affect pulmonary adenoma formation in mice given BaP by oral intubation.

7,8-Benzoflavone

7,8-Benzoflavone inhibited enzymes involved in the activation of DMBA to its carcinogenic form to a greater extent than it inhibited enzymes involved in the detoxification of DMBA (Bowden et al. 1974). However, in the case of the substrate dibenz[a,h]anthracene, 7,8-benzoflavone appeared to inhibit enzymes involved in detoxification to a greater extent than it inhibited activation of dibenz[a,h]anthracene. Data obtained with phenobarbital indicated that dibenz-[a,h]anthracene, but not DMBA, has a specific phenobarbital-inducible detoxification system in mouse-skin experiments. AHH in mouse skin epidermis was inducible by the two hydrocarbons and 5,6-benzoflavone. 7,8-Benzoflavone given at the same time as DMBA inhibited the formation of skin tumors in a two-stage system of tumorigenesis and also inhibited the formation of covalently bound complexes of DMBA with epidermal DNA, RNA, and protein. However, 7,8-benzoflavone given at the same time as dibenz[a,h]anthracene stimulated skin tumor formation, but had little or no effect on the binding of dibenz[a,h]anthracene to epidermal macromolecules. Phenobarbital had a small inhibitory effect on DMBA skin tumor initiation and significantly inhibited tumor formation initiated by dibenz[a,h]anthracene.

In a study by Diamond and Gelboin (1969), 7,8-benzoflavone inhibited AHH activity and inhibited PAH toxicity. Inhibition of AHH by 7,8-benzoflavone also inhibited tumorigenesis by DMBA in mouse skin (Gelboin, Wiebel, and Diamond 1970).

Wattenberg and Leong (1970) reported on the inhibition of lung adenoma induction by β-naphthoflavone. β-Naphthoflavone inhibited BaP-induced enzymes.

Antibiotics

The addition of the antibiotics actinomycin D, puromycin, or cycloheximide simultaneously with BaA completely prevented the induction of AHH activity in hamster fetus cell cultures (Nebert and Gelboin 1968*b*). Actinomycin D acts by inhibiting protein synthesis necessary for AHH induction.

Nicotinamide

Nicotinamide inhibited the levels of both the constitutive and induced AHH activity in liver nuclei of C3H/He and DBA/2 mice (Watanabe et al. 1975). The enzymes that catalyze the oxidation of cyclophosphamide and nicotine to hydroxy derivatives were also inhibited by the agents that block BaP metabolism in mice and hamster cell cultures (Hill and Shih 1974).

Hormones

The hormonal status of experimental animals is known to influence the appearance and rate of growth of tumors. The rate of growth of mammary tumors in rats was accelerated by pregnancy or by the injection of progesterone, and their appearance was delayed by the administration of estradiol (Huggins, Moon, and Morii 1962). Booth, Keysell, and Sims (1974) describe the effect of estradiol on the in vitro metabolism of DMBA and of its hydroxymethyl derivatives by subcellular fractions of female rat liver. The amounts of dihydrodiols, phenolic metabolites, and glutathione conjugates were greatly reduced by the presence of estradiol in such a system. The steroid probably acted by inhibiting epoxide formation rather than by inhibiting the three types of reaction involved in their further metabolism. This was supported by the finding that estradiol also reduced the total amount of each substrate metabolized. Estradiol may exert its anticarcinogenic effect by competing for binding sites on the mixed-function oxidase enzyme complex, thus interfering with the metabolism of the hydrocarbon.

Mixed hydrocarbons

Lacassagne et al. (1934), as cited in Haddow and Weinhouse (1967), applied mixtures of hydrocarbons to the skin of mice and found that the weakly carcinogenic hydrocarbons, such as chrysene and dibenz[*a,g*]fluorene, inhibited the production of skin tumors by the potent carcinogen 3-methylcholanthrene. A number of subsequent studies by the same authors and others demonstrate the inhibition, by some PAH, of carcinogenesis induced by potent PAH carcinogens.

Falk, Kotin, and Thompson (1964) studied the effects of a number of hydrocarbons found in polluted urban air and cigarette smoke on sarcoma production by BaP. Benzo[*a*]fluorene, perylene, perinaphthoxanthrene, benzo[*a*]carbazole, and chrysene were potent anticarcinogens. These compounds, when given in the ratio of 0.10 to 0.15 of anticarcinogen to 1.0 of BaP, inhibited sarcomagenesis by more than 70%. Less active were benz[k]fluoranthene, benz[m,n,o]fluoranthene, and 2-naphthol. The inhibition was probably due to a stimulation of enzyme activity capable of detoxifying the carcinogenic agent or was related to an antimetabolite type of competition of the inhibitor and the active carcinogen for a relevant receptor site in the cell.

Selenium

Selenium repressed chemical carcinogenesis in experimental animals (Shamberger 1970). Selenium (0.0005%) and vitamin E (0.13%) significantly reduced the number of skin tumors in mice that were induced by DMBA (0.125 mg in 0.25 ml acetone) and 3-methylcholanthrene (0.25 ml of 0.01% solution) in nondietary tumor production experiments. Both selenium and vitamin E reduced the total number of papillomas. Shamberger, Tytko, and Willis (1973) also presented data suggesting that selenium could be an environmental factor affecting tumor incidence in man. Selenium may exert an anticarcinogenic effect by acting as an antioxidant.

Personal hygiene

A high incidence of scrotal cancer among wax pressmen employed in the petroleum industry was attributed to contamination of their clothes with wax containing carcinogenic oil (Eckardt 1967). Of 77 men who had at least 9 years of experience as wax pressmen, 11 ultimately developed cancer of the scrotum. While removing wax from filters in the wax processing plant, the workers leaned across a supporting bar, thus heavily contaminating the crotch of their trousers with wax. At this point in the processing, the wax was grossly contaminated with oil.

An industrial hygiene program was instituted that consisted of (1) cancer detection at the time of employee quarterly examinations, (2) prompt treatment by excision of any malignancies or pre-malignant lesions, and (3) cancer prevention by protecting workers from exposure to carcinogens. All wax pressmen were issued an impervious rubber or plastic apron, a daily change of clean underwear, and double lockers for clothes changes. In addition, the men were encouraged to clean their own work clothes at least once a week, and a daily shower was enforced. After institution of the industrial hygiene program, no problems among wax pressmen were reported.

An industrial hygiene program was initiated at the Union Carbide Corporation coal hydrogenation pilot plant at Institute, West Virginia, in an effort to reduce employee exposure to potentially cancer-producing oils (Ketcham and Norton 1960). As part of the control program, a program of employee education was started. The use of protective clothing and the encouragement of personal hygiene procedures was stressed as a first line of defense against employee contamination. The program included a daily change of underwear, outer clothing, and socks, plus a daily inspection of hands, face, and neck. However, operation of the plant was curtailed before the complete clothing program was put into effect. Prior to closure, small groups of workers were tested for contamination. A skin examination using 2537 Å ultraviolet light in a dark room showed skin contamination after showering. After use of barrier clothing, consisting of pajamas buttoned at the neck and fitted at the arm and leg cuffs, in conjunction with barrier cream on the exposed parts of the body, the men became clean after several days, as determined by ultraviolet light. Dry cleaning of clothes, followed by a soap and water laundering, was successful in removing most of the pasting oil. Waterless hand soaps were most effective in removing pyrene, a noncarcinogenic material, from the skin. Double lockers and shower facilities were also provided. A drying room and inspection room, equipped with mirrors and ultraviolet lights, were planned.

Selective placement, restriction, and rotation of employees was practiced to some degree, but the burden upon the medical staff and the total number of employees made these practices difficult to accomplish (Sexton 1960*b*). In addition, medical literature does not agree on the type of person (skin type) who should be excluded from exposure to materials that contain suspected or known carcinogens.

10.1.5.5 <u>Mechanisms</u>

It is now almost generally accepted that PAH must be metabolically activated to induce cancer. However, it is apparent that the ultimate carcinogenic metabolite(s) of PAH has not been positively identified. It is possible that different PAH produce different types of carcinogenic metabolites. Although more work is needed to firmly establish the nature of the ultimate carcinogenic form of PAH, some studies have been aimed at unravelling this complex problem.

<u>Covalent binding to DNA, RNA, and protein</u>

There is appreciable evidence that derivatives of chemical carcinogens bind covalently to DNA, RNA, and protein. As reviewed by Miller (1970) and Miller and Miller (1975), the covalent binding in vivo to proteins was noted in tissues susceptible to their carcinogenic action. Subsequently, covalent bindings of derivatives of chemical carcinogens with both nucleic acids and proteins of tumor-susceptible tissues in vivo and in vitro have been observed. The amounts of these carcinogens that are bound in vivo are of the order of one carcinogen residue per 10^4 to 10^7 monomer residues in the macromolecules. Although in some cases the correlation between binding and carcinogenicity is low and some noncarcinogenic derivatives bind at appreciable levels, there are no exceptions known in which a chemical carcinogen has failed to exhibit covalent binding to macromolecules in its target tissue in vivo (Miller and Miller 1975). Chemical carcinogens have been shown to be electrophilic compounds (Miller and Miller 1975).

Brookes and Lawley (1964) painted a series of PAH on mouse skin and found a strong correlation between the reactivity of PAH with the DNA of the skin and the ability of PAH to induce cancer. This was one of the first suggestions that PAH exercise their carcinogenic activity through a reaction with DNA. Binding to RNA and protein also occurred. An earlier study by Miller (1951) had shown that derivatives of BaP bind to protein in the epidermal fraction of mouse skin in vivo.

Wiest and Heidelberger (1953), in mouse skin-painting studies, showed an irreversible binding of the carcinogen dibenz[a,h]anthracene, or its metabolites, with the nucleoproteins. This was a covalent binding to cellular constituents in vivo.

Using an in vitro system involving incubation of PAH with either DNA or protein in the presence of a microsomal hydroxylating system, Grover and Sims (1968) found that tritiated PAH reacted with the protein or DNA. Table 10.9 shows the results of incubation of a 50-mg sample of protein or DNA for 1 hr with 5 µg of hydrocarbon. When similar mixtures were incubated without enzyme cofactors (i.e., the parent hydrocarbons were not metabolized), no binding of the hydrocarbons to either DNA or protein was detected.

Table 10.9. Enzyme-catalyzed reaction of tritiated PAH
with protein and DNA in vitro

Hydrocarbon	Reaction with protein (millimoles/mole[a])	Reaction with DNA (μmoles/g-atom of DNA P)[b]
Benzo[*a*]pyrene	0.78	1.41
3-Methylcholanthrene	0.73	0.78
7,12-Dimethylbenz[*a*]anthracene	0.78	0.64
Dibenz[*a,h*]anthracene	0.95	0.44
Dibenz[*a,c*]anthracene	1.07	0.56
Benz[*a*]anthracene	0.76	0.70
Pyrene	1.15	0.31
Phenanthrene	0.72	0.05

[a]Mol wt of bovine plasma albumin taken as 64,000.
[b]Calculated on a basis of 8% phosphorus.

Source: Grover and Sims 1968, Table 1, p. 160. Reprinted by permission of the publisher.

A later study by the same authors (Grover and Sims 1970) gave results indicating that K-region epoxides may be the active intermediates involved in the binding of PAH to cellular macromolecules. At physiological temperature and pH, phenanthrene 9,10-oxide and dibenz[*a,h*]anthracene 5,6-oxide reacted with 4-(*p*-nitrobenzyl)pyridine, nucleic acids, and histone. However, neither the parent compounds (phenanthrene, dibenz[*a,h*]anthracene) nor the K-region dihydrodiols derived from these hydrocarbons reacted with *p*-nitrobenzylpyridine, nucleic acids, or histone.

Some reported data is not in agreement with the concept of an epoxide intermediate. Grilli, Rochi, and Prodi (1975) found a high level of nonenzymatic binding of PAH to nucleic acids in vitro when thymus DNA was used. The aromatic hydrocarbons were tightly bound and not extractable either by organic solvents or by warming DNA solutions above the melting point and precipitating the DNA again. Microsomal dependent binding occurred between polynucleotides and the hydrocarbons (DMBA, BaP, and 3-methylcholanthrene).

A metabolite of BaP that bound to DNA in vitro without the need for further activation was isolated and preliminarily identified on the basis of mass spectral information and hydration studies as the 4,5 epoxide (Wang, Rasmussen, and Crocker 1972).

To determine whether the K-region epoxide is the PAH derivative that binds to DNA, Baird et al. (1973) compared the DNA-bound products fromed by the reaction of the K-region epoxide of 7-methylbenz[*a*]anthracene with DNA with those of DNA isolated from cells that had been exposed to the parent hydrocarbon. These results suggest that the cellular metabolism of 7-methylbenz-[*a*]anthracene that results in the covalent binding of this hydrocarbon to DNA in cells may involve more than simple formation of K-region epoxide.

Since the carcinogenic metabolites formed in cells from methyl-substituted PAH may be different from those formed from the nonmethylated hydrocarbons (Dipple, Lawley, and Brookes 1968), the products formed by the reaction of the K-region epoxide of BaP with DNA in aqueous solution were compared with those found in DNA isolated from cells incubated with BaP in vitro (Baird, Harvey, and Brooks 1975). The hydrocarbon-deoxyribonucleoside products from enzyme digests of mouse embryo cell cultures treated with ^{3}H-BaP were not identical to those found in similar chromatograms of enzyme digests of DNA that had been reacted with BaP 4,5-epoxide in aqueous ethanol solution. This suggests that the metabolic activation of BaP that results in covalent binding to DNA in mouse embryo cells in culture may also be more complex than the simple formation of a K-region epoxide and reaction of that compound with the cellular DNA. Since tritium-labeled BaP epoxide was not available, it was not possible to determine whether the products formed in the DNA of epoxide-treated cells would be the same as those formed in aqueous solution.

On the basis of the above results, King, Thompson, and Brookes (1975) proposed a model of the mechanism of BaP metabolism and DNA binding. They suggest that BaP is metabolized by AHH to yield a series of epoxides, both K-region and non-K-region. These products are substrates for the enzyme epoxide hydrase, which is associated with the AHH complex. By epoxide hydrase action, an intermediate is produced prior to conversion to the dihydrodiols, and this intermediate is responsible for the BaP binding to DNA that occurs in vivo; for example, a diol epoxide of BaP may be the likely metabolite responsible for the observed binding to DNA. A pathway in which a dihydrodiol epoxide is produced was recently discovered by Sims et al. (1974). These investigators noted the formation of a non-K-region epoxide that was converted by epoxide hydrase to a dihydrodiol. The isolated double bond in the dihydrodiol ring was then epoxidized to yield a reactive dihydrodiol epoxide. This metabolite, 7,8-dihydro-7,8-dihydroxybenz[a]pyrene 9,10-oxide (Fig. 10.7), appeared to account for a major DNA-bound form of BaP in cultured hamster cells.

ORNL DWG 76-11854

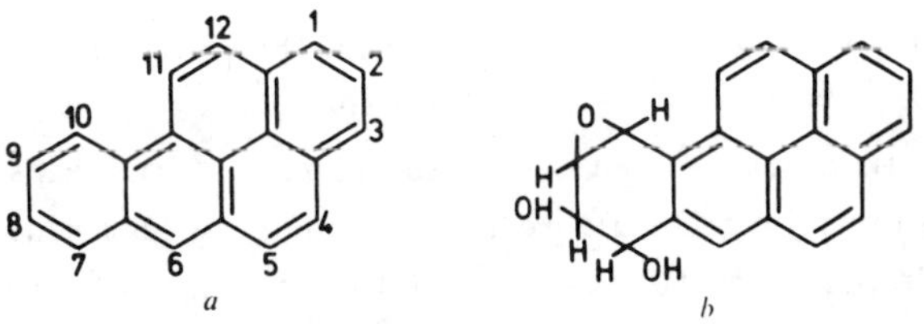

Fig. 10.7. BaP (a) and 7,8-dihydro-7,8-dihydroxybenz[a]pyrene 9,10-oxide (b).
Source: Sims et al. 1974, Fig. 1, p. 326. Reprinted by permission of the publisher.

A correlation between oncogenicity and the binding of PAH to macromolecules in man has not been studied. Harris et al. (1974) incubated bronchi from three patients with lung cancer and from one patient without lung cancer with several PAH. The bronchial epithelial cells in culture had the capability to bind and activate tritiated DMBA, BaP, 3-methylcholanthrene, and dibenz[a,h]-anthracene. Radioactivity was found in both the nuclei and cytoplasm of cells incubated with ^{3}H-DMBA. Ciliated, mucous, and basal cells contained tightly bound radioactivity. Epithelial cells in bronchial glands were overlaid with autoradiographic grains. Similar results were found with ^{3}H-BaP, ^{3}H-3-methylcholanthrene, and ^{3}H-dibenz[a,h]anthracene. Bound radioactivity was also found in mesenchymal cells, but to a lesser extent. DMBA and BaP bound to DNA more than did dibenz[a,h]anthracene. No information was given on differences of amounts of binding among patients.

DNA-binding, suppression, and enhancement

In the respiratory epithelium, vitamin-A deficiency causes basal cell hyperplasia and squamous metaplasia. Genta et al. (1974) studied the binding of tritiated BaP (^{3}H-BaP) to DNA in tracheal epithelial cells of normal and vitamin-A-deficient hamsters. When isolated tracheas were incubated in vitro in the presence of ^{3}H-BaP, substantially greater quantities of radioactivity were associated with DNA prepared from vitamin-A-deficient tracheas. When 7,8-benzoflavone, an inhibitor of AHH, was added to the system, the binding of ^{3}H-BaP was reduced. Harris et al. (unpub.), as cited in Genta et al. (1974), demonstrated that, in vitamin-A-deficient tracheas, there is a higher binding of ^{3}H-BaP to macromolecular structures of basal cells located in squamous metaplastic areas than to basal cells located in normal areas.

In hamster tracheas the binding of BaP to DNA was enhanced if the hamsters had been previously treated intratracheally with BaP plus Fe_2O_3 (Kaufman et al. 1973). There was increased binding observed in the tracheas of vitamin-A-deficient hamsters (Kaufman, Genta, and Harris 1974).

7,8-Benzoflavone given simultaneously with DMBA inhibited the formation of (1) skin tumors in mouse skin epidermis and (2) covalently bound complexes of DMBA with epidermal DNA, RNA, and protein. However, 7,8-benzoflavone given simultaneously with dibenz[*a,h*]anthracene stimulated skin tumor formation, but had little effect on the bonding of dibenz[*a,h*]anthracene to epidermal macromolecules (Bowden et al. 1974).

Metabolites

Results of studies since 1970 have provided information on the possible ultimate carcinogenic metabolite(s) of PAH in vivo. It has been demonstrated by some researchers that K-region epoxides of several PAH are more active than either the parent hydrocarbons or other metabolites in producing malignant formation in cell-culture systems. More recent studies, however, suggest that diol-epoxides may be involved in the reactions of PAH with the DNA of treated cells and thus be the ultimate carcinogenic metabolite. There is also strong evidence that the carcinogenic metabolites of methyl-substituted PAH may be different from those formed by nonmethylated hydrocarbons.

In 1950, Boyland suggested that epoxides are metabolic intermediates in the conversion of PAH to phenols and *trans*-dihydrodiols. Since that time, numerous studies have demonstrated that epoxides are intermediates in the microsomal oxidation of several PAH. Huberman et al. (1972) and Grover et al. (1971) proposed that epoxides may be the metabolically activated, and possibly the ultimate, carcinogenic form of PAH that do not contain active methyl groups. They demonstrated that K-region (5,6) epoxides of dibenz[*a,h*]anthracene and BaA were more active in producing malignant transformation of hamster embryo cells than the corresponding hydrocarbons, K-region *cis*- and *trans*-dihydrodiols and phenols. The K-region epoxides of BaA and 3-methylcholanthrene also transformed a clone of C3H-mouse ventral prostate cells, whereas the parent hydrocarbons were not active (Grover et al. 1971). These results are shown in Tables 10.4 and 10.5 (Sect. 10.1.2.4). Parent hydrocarbons gave results similar to those of the acetone controls. With either type of derivative, increased concentration of the compounds gave increases in the frequency of transformation.

Epoxides of the K-region of 3-methylcholanthrene, phenanthrene, and chrysene were also more active in producing transformation than were the corresponding hydrocarbons (Huberman et al. 1972).

Marquardt et al. (1974) tested the K-region epoxides derived from DMBA and 7-hydroxymethyl-BaA for their ability to induce malignant transformation of a clone of C3H-mouse prostate fibroblasts. In this case the K-region epoxides of both compounds were slightly less active than the parent hydrocarbons in an in vitro transformation. The corresponding hydroxylated derivatives, which were also tested, were without transforming activity. Addition of α-naphthoflavone, an inhibitor of microsomal mixed-function oxidases, prevented the toxic effects usually produced by DMBA and 7-hydroxymethyl-BaA, but did not inhibit the induction of malignant transformation in cells treated with these compounds. Thus, metabolic activation to epoxides may not be a prerequisite for the oncogenic effect of 7-methyl-substituted PAH.

However, hydroxylation of the 7-methyl group of DMBA may be a necessary first step in carcinogenesis. Liver homogenates from female Sprague-Dawley rats injected subcutaneously with 400 μg BaP were incubated with BaP or 6-methyl-BaP (Flesher and Sydnor 1973). A number of metabolites were identified, including one that was indistinguishable from 6-hydroxymethyl-BaP by either thin-layer chromatography or ultraviolet spectra. This compound was a potent carcinogen when administered by subcutaneous injection to rats. All rats injected with 100 μg BaP, 6-methyl-BaP, or 6-hydroxymethyl-BaP had sarcomas at the site of injection. No tumors at other sites were observed. These observations were in accord with the hypothesis that methylation or hydroxymethylation is one of the first steps in metabolic activation of carcinogenic PAH. Earlier studies (Flesher and Sydnor 1970; 1971) led the authors to postulate this mechanism for DMBA.

Because some K-region epoxides of DMBA do not exhibit greater carcinogenic potencies than the parent compound, Dipple, Levy, and Iype (1975) investigated the possible involvement of the ketonic derivatives, 5,6-dihydro-5-oxo-7,12-methylbenz[a]anthracene and 5,6-dihydro-6-oxo-7,12-dimethylbenz[a]anthracene. These compounds were inactive in tests for tumor-initiating activity in mouse skin and tumor production in the subcutaneous tissue of the mouse.

A diol-epoxide (Sect. 10.1.5.5 and Fig. 10.7) that binds to DNA might be the ultimate carcinogenic metabolite (Sims et al. 1974). Huberman et al. (1976) and Yang et al. (1976) demonstrated the high mutagenic activity and metabolic formation of this compound, which they identified as r-7,t-8-dihydroxy-t-9,10-oxy-7,8,9,10-tetrahydrobenzo[a]pyrene, and suggested that it may be an ultimate carcinogenic form of BaP.

10.1.6 Teratogenicity

Although only a small number of substances have been bioassayed for both teratogenicity and carcinogenicity, a relationship between the two is indicated by the large percentage of compounds that have produced both malformations and cancer. A comprehensive review by DiPaolo and Kotin (1966) shows that the majority of carcinogens exert a teratogenic effect. The rapid growth, lack of differentiation, and reactivity of the cells in the developing embryo suggest that the action of chemical or biological agents will be expressed in alterations at the level of differentiation. Somatic cell division is a prerequisite for both teratogenesis and carcinogenesis. The rapid division and differentiation of fetus cells result in a sensitivity that makes the fetus especially vulnerable to teratogens. The same compound may have different effects, depending upon the stage of development of the organism at the time of exposure. Thus, a compound found to be carcinogenic to mature cells might be teratogenic to immature, embryonic cells.

In 1973, Shepard edited a catalogue of agents that had been tested for teratogenicity. The catalogue includes a listing of agents along with a short summary of the study including animals studied, doses, route of administration, and bibliographic citation. Both positive and negative results are reported. Several carcinogens, including DMBA and 3-methylcholanthrene, produced teratogenic effects in experimental animals.

10.1.6.1 Placental transfer

Shendrikova and Aleksandrov (1974) studied the permeability of the rat placenta to DMBA, BaP, and 3-methylcholanthrene. Female rats were given a single dose (200 mg/kg body weight) of one of the three compounds (in sunflower-oil suspension) by gastric tube on the twenty-first day of pregnancy. Fetal tissue content of DMBA reached a maximum of 1.53 to 1.6 µg/g body weight 2 to 3 hr after administration (Fig. 10.8). BaP, tested 3 hr after administration, also reached high concentrations (2.77 µg/g); 3-methylcholanthrene was found only in trace quantities. The same ratio of concentrations was also found in the liver of the pregnant rats.

ORNL DWG 76-11928

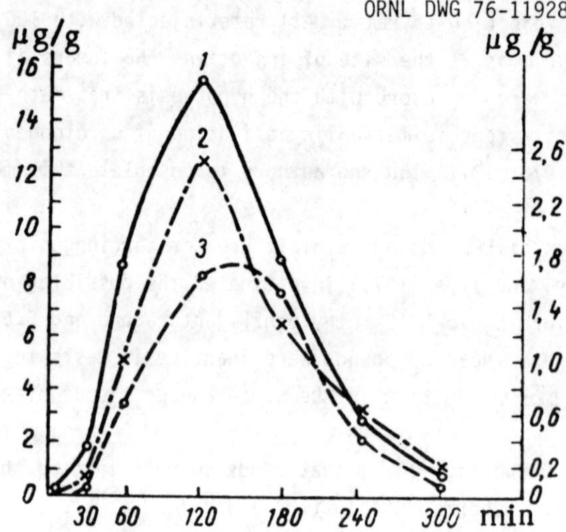

Fig. 10.8. Concentration of DMBA in liver (1) of pregnant rats, placentas (2), and fetuses (3) at various times after administration of the compound in a dose of 200 mg/kg by gastric tube on the 21st day of pregnancy. Abscissa, time after injection of DMBA (in min); ordinate, (left) DMBA concentration (in µg/g) in liver, (right) in fetuses and placentas. Source: From Shendrikova and Aleksandrov 1974, Fig. 1, p. 170. Reprinted by permission of the publisher.

When DMBA was injected intravenously into the animals (25 mg/kg body weight), maximum accumulation occurred in the fetal tissues 1 hr after injection. The period of maximum accumulation was shorter than after intragastric administration (Fig. 10.9). The fetal concentration of the compound was roughly of the same magnitude (1.5 to 2 µg/g) when administered by different routes. The similarity in fetal content following administration of doses that differed by an order of magnitude by different routes was explained by the small quantities of DMBA absorbed from the gastrointestinal tract and the consequent delay in reaching its maximal concentration.

The differences in concentrations of the different PAH in fetal tissues did not reflect their relative abilities to cross the placental barrier, but were instead due to the differential absorption of these compounds from the gastrointestinal tract (Shendrikova and Aleksandrov 1974).

10.1.6.2 <u>Enzyme activity</u>

Placental enzymes, especially those metabolizing foreign compounds, may be significant factors
in the resistance of the developing fetus to potentially toxic or teratogenic substances. The
teratogen may be inaccessible to the fetus because the placental enzyme system enzymatically
converts the teratogen to nontoxic, more polar derivatives or because the placental may metabolize
an inactive compound to an active teratogen. The functions of the placenta appear to change
during the course of gestation.

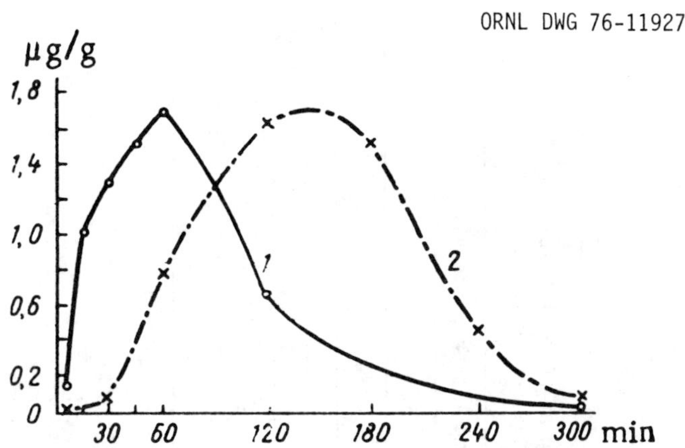

Fig. 10.9. Concentration of DMBA in fetal tissues after intravenous injection of the
compound into rats in a dose of 25 mg/kg (1) and after its administration by gastric tube in a
dose of 200 mg/kg (2) on the 21st day of pregnancy. Abscissa, time (in min) after administration
of DMBA; ordinate, DMBA concentration (in µg/g) in fetal tissues. <u>Source</u>: From Shendrikova and
Aleksandrov 1974, Fig. 2, p. 170. Reprinted by permission of the publisher.

Treatment of pregnant rats with BaP stimulated AHH activity in the entire implantation site
(embryo and decidua), decidua, placenta, fetal membranes, and fetus (Schlede and Merker 1972;
1974). Wistar rats were treated with an oral dose of BaP (40 mg/kg body weight) dissolved in
corn oil 24 hr prior to sacrifice. Maximum induction of AHH activity in the placenta, as
measured by the in vitro hydroxylation of BaP, was obtained only from day 15 of gestation,
whereas in the fetus, the degree of induction of enzyme activity increased from day 13 to day 18
of gestation (gestational period lasted 21 to 22 days). Enzyme activity in fetal membranes was
higher on days 16, 18, and 20 of gestation than on day 13. Enzyme activity was three to eight
times higher in the fetal membranes than in the placenta. In control animals little or no
enzyme activity was measured in these tissues.

In animals treated on day 8 of gestation, increased AHH was present within 6 hr in the entire
implantation site and reached a maximum activity 12 to 18 hr after treatment. By day 10, AHH
levels fell back to control levels. In the placenta, following treatment on day 12, AHH levels
were increased within 6 hr and remained at the same level until 18 hr after treatment. By
days 14 and 15, the enzyme activity had dropped to half of the highest levels. Following treatment on day 19, AHH levels dropped only slightly from day 19 to day 21. These results demonstrated that, in the rat, the fetal membranes play a more important role for metabolism of BaP
than does the placenta.

Treatment of the pregnant rat or hamster with either a PAH or phenobarbital induces the enzyme
system in specific fetal tissues as well as in the placenta (Nebert and Gelboin 1969). Table
10.10 shows the AHH activity in several tissues of the normal and 3-methylcholanthrene-treated

Table 10.10. Effect of in vivo administration of MC on AHH activity
in tissues from pregnant hamster, fetus, and newborn

| Tissue | AHH specific activity (units/mg tissue protein)[a] | | | | | |
| | Mother | | Fetus | | Newborn | |
	Control	MC	Control	MC	Control	MC
Liver	2430 - 3555 (3)	3435 - 3945 (3)	14 - 27 (3)	174 - 426 (3)	500 - 620 (2)	2505 - 2835 (2)
Lung	<6 - 11 (3)	33 - 48 (3)	<6 (3)	18 - 45 (3)	21 - 23 (2)	63 - 98 (2)
Small intestine	<6 - 18 (3)	75 - 83 (3)	<6 (3)	9 - 20 (3)	<6 - 17 (2)	134 - 155 (2)
Kidney	14 - 33 (3)	162 - 207 (3)	<6 (3)	32 - 33 (3)	12 - 15 (2)	132 - 144 (2)
Placenta	<6 (3)	15 - 42 (3)				

[a]Numbers in parentheses indicate the number of experiments.

Source: Nebert and Gelboin 1969, Table IV, p. 81. Reprinted by permission of the publisher.

hamster, hamster fetus, and newborn hamster. 3-Methylcholanthrene in corn oil (100 mg/kg body weight) was administered intraperitoneally to the pregnant hamster, and the animal was sacrificed 24 hr later, on approximately day 15 of gestation. The hydrocarbon was administered by the same route and at the same dosage on the first day of life to the newborn, which were sacrificed 24 hr later. Mothers nursing the neonates did not receive any drugs. Control levels of AHH were detectable in hamster fetal liver, and the enzyme was induced transplacentally by 10- to 20-fold after treatment of the mother with 3-methylcholanthrene. In the liver, lung, and kidney of the newborn hamster, the 3-methylcholanthrene-induced enzyme was stimulated about three- to fivefold, whereas in the small intestine it was more than nine times higher. About a 30-fold increase in AHH activity developed in less than 72 hr between the fetal and neonatal period.

Table 10.11 shows the AHH activity in several tissues of the normal and 3-methylcholanthrene-treated pregnant Sprague-Dawley rat and rat fetus. Treatment was the same as for the hamster. The transplacental induction of AHH activity in the fetal rat tissues was similar in magnitude to that found in fetal hamster tissues.

Table 10.11. Effect of in vivo administration of MC on AHH activity in tissues from pregnant rat and fetus

| | AHH specific activity (units/mg tissue protein)[a] | | | |
| | Mother | | Fetus | |
	Control	MC	Control	MC
Liver	1365 - 1530 (2)	3165 - 3525 (2)	6 - 9 (2)	201 - 311 (2)
Lung	8 (2)	159 - 191 (2)	<6 (2)	54 - 81 (2)
Small intestine	<6 (2)	605 (1)	<6 (2)	9 - 23 (2)
Kidney	6 - 17 (2)	812 - 942 (2)	<6 - 6 (2)	41 - 42 (2)
Placenta	<6 (2)	24 - 33 (2)		

[a]Numbers in parentheses indicate the number of experiments.
Source: Nebert and Gelboin 1969, Table VI, p. 82. Reprinted by permission of the publisher.

It is well established that cigarette smoke contains some of the same PAH as environmental contaminants, and several studies have linked levels of placental AHH to smoking habits in pregnant women (Welch et al. 1968; Nebert, Winker, and Gelboin 1969).

Figure 10.10 shows the distribution of AHH in human placentas assayed within 24 hr after delivery (Nebert, Winker, and Gelboin 1969). A positive relationship was observed between a relatively high level of placental AHH activity and a history of cigarette smoking. The mean AHH activity in the placentas from smoking women was 2- to 25-fold greater than that in the placentas from nonsmokers. The placentas from women who smoked 20 or more cigarettes per day showed a mean AHH activity 5- to 25-fold greater than that found in placentas from women who did not smoke. However, AHH activity was found in the placentas of some women who did not smoke, and no measurable AHH activity was found in the placentas of some women who smoked 20 to 30 cigarettes per day. The differences may reflect differences in the tar content of cigarettes smoked, method of inhalation, nutritional and medicinal status, or eating habits of the women. On the

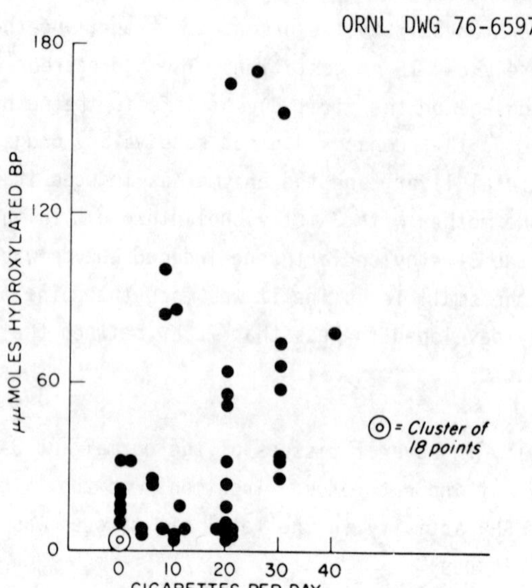

Fig. 10.10. Distribution of human placental AHH activity as a function of cigarette smoking during pregnancy (correlation coefficient r = 0.53; P = 0.001). Source: Nebert, Winker, and Gelboin, Fig. 10.5, p. 225. Reprinted by permission of the publisher.

other hand, they tend to support the studies of Kellermann, Cantrell, and Shaw (1973), which indicate that genetic differences exist in AHH inducibility by aromatic hydrocarbons in man. Of the 59 placentas assayed, 35 samples were obtained from women who smoked between 2 and 40 cigarettes daily during gestation, up to and including the day of parturition. The assay of AHH activity was the in vitro enzymatic conversion of BaP to hydroxylated metabolites. Concentration was determined spectrophotofluorometrically (Nebert and Gelboin 1968a).

10.1.6.3 Tumorigenesis

PAH have been shown to produce tumors in animal offspring of mothers treated during pregnancy (Bulay and Wattenberg 1968; Bulay 1970). Female Ha/ICR mice were injected subcutaneously on days 11, 13, and 15 of pregnancy with 0.2 ml of Carbowax-400 containing 4 mg of BaP or vehicle only. Absolute controls received no injections. On day 20 of gestation, the progeny were delivered by Cesarean section and nursed on control mothers. After weaning at 4 weeks of age, the backs of the mice were shaved, and topical administrations of croton oil (two drops of 1% croton oil in acetone) were given twice a week. (Croton oil acts as a promotor of epidermal carcinogenesis.) Mice were sacrificed at 28 weeks of age, and pulmonary adenomas and skin papillomas were noted. In both cases, the tumor incidence and the number of tumors per animal were increased in the progeny of mothers injected with BaP (Tables 10.12 and 10.13). A carcinogenic effect was not noted in other tissues (Bulay and Wattenberg 1968). Since PAH are excreted in milk (Shay et al. 1950), these experiments show the effect of prenatal exposures only.

In a more extensive study, Bulay (1970), using both BaP and DMBA as carcinogens and using different routes of administration, produced pulmonary adenomas and skin papillomas in Ha/ICR mice in utero. The effects were produced to a greater degree with DMBA, and they could be

Table 10.12. Effect of BaP administration to pregnant mice on pulmonary adenoma formation in the progeny

Experimental group	Material administered	Number of mice	Number of mice with adenomas[a]	Percent of mice with adenomas	Number of adenomas per group	Number of adenomas per mouse	Number of mice with multiple adenomas	
							3 — 6	7 — 9
Absolute controls	Nothing	82	6	7.3	7	0.09	0	0
Vehicle controls	Carbowax-400	89	10	11.2	13	0.15	1	0
BaP	BaP in carbowax-400	55	34	61.8	130	2.36	13	6

[a]Number of mice having pulmonary adenomas at 28 weeks of age.

Source: Bulay and Wattenberg 1968, Table 1, p. 85. Reprinted by permission of the publisher.

Table 10.13. Effect of BaP administration to pregnant mice on skin papilloma formation in the progeny

Experimental group	Number of mice	Number of mice with papillomas[a]	Percent of mice with papillomas	Number of papillomas per group	Number of papillomas per mouse
Absolute controls	82	6	7.3	8	0.10
Vehicle controls	89	8	9.0	9	0.10
BaP	55	13	23.6	18	0.33

[a]Number of mice having skin papillomas after 24 weeks of topical treatment twice weekly with 1% croton oil; mice were 28 weeks of age.

Source: Bulay and Wattenberg 1968, Table II, p. 85. Reprinted by permission of the publisher.

produced by either intraperitoneal or subcutaneous injections. Although adenomas and papillomas are defined as benign tumors, they may be a precarcinogenic response.

Turusov et al. (1973) investigated the effect of a low dose of 3-methylcholanthrene on the off-spring of pregnant mice. CF-1 mice received 1 mg of 3-methylcholanthrene dissolved in 0.2 ml of olive oil, by stomach tube, in late pregnancy. Delivery occurred between one and three days after treatment. Additional experimental groups, consisting of offspring of treated and un-treated mothers injected with 12.5 µg diethylnitrosamine (DENA) within 24 hr after birth, were started. For comparison, the authors included results from an earlier study in which 8.4 mg of 3-methylcholanthrene was administered by stomach tube to pregnant females. Tumor incidence is shown in Table 10.14. All mothers that had received 8.4 mg of 3-methylcholanthrene and all of their offspring died, or were killed in poor condition, within 80 weeks, and both mother and offspring bore one or more tumors.

Although tumor incidence of controls is high (82% in the females and 77% in the males), tumor incidence (nearly 100%) of treated females and offspring occurred earlier, and there were more tumors per mouse in the 3-methylcholanthrene-treated group. More treated mothers bore more than one tumor than did offspring. At the dose level of 8.4 mg, this was mainly due to the occurrence of carcinoma of the forestomach in the treated mothers.

The highest incidence of tumors was observed in the lungs and lymphoreticular tissue, both in mothers and offspring. The same tumor types also existed in untreated controls. The results also showed that prenatal treatment with 3-methylcholanthrene caused a considerable inhibition of liver carcinogenesis induced by diethylnitrosamine.

The average levels of 3-methylcholanthrene in fresh tissue per litter after the administration of 1 mg to pregnant mice were 213 ng/g and 140 ng/g, respectively, 3 and 6 hr after treatment.

Tomatis and Goodall (1969) investigated the effect of DMBA on successive generations of mice. They injected pregnant Swiss mice with 400 µg of DMBA during the last period of pregnancy. The percentages of animals having tumors in F-1, F-2, and F-3 generations were 58.6, 52.6, and 30.0 respectively. The percentage of tumors in untreated controls was 23.2. The tumors observed were lung adenomas, mammary carcinomas, granulosa cell tumors, and papillomas of the urinary bladder.

10.1.6.4 <u>Birth defects</u>

Currie et al. (1970) succeeded in inducing pronounced anomalies of the nervous system, eyes, and skeleton of rat fetuses after treatment with DMBA and its derivative, 7-hydroxymethyl-12-methylbenz[*a*]anthracene, at various stages of organogenesis.

Using the maximum dose tolerated by pregnant rats (20 mg/kg) at different stages of gestation, Alexandrov (1973) succeeded in producing embryotoxic and teratogenic effects with DMBA. An embryotoxic effect was observed after treatment on the fifth, sixth, and ninth day after coitus. Malformed fetuses were detected after treatment on days 9 and 13 post coitum. Malformations included hydrocephaly, exencephaly, spina bifida, cleft lip, and cleft maxilla; anophthalmos occurred after treatment on day 9, and ectrodactyly of the forefeet occurred after treatment on day 13.

Table 10.14. Tumor incidence in CF-1 mice treated with MC prenatally and DENA when newborn

Group	Type of mice and treatment	Number at start (sex)	Effective number[a]	Tumor-bearing animals		Number of tumors		More than one tumor		Malignant lymphoma		Lung tumor		Hepatomas		Other tumors
				Number	Percent	Total	Per mouse	Number	Percent[b]	Number	Percent[b]	Number	Percent[b]	Number	Percent[b]	Number
I	Pregnant mice treated with 8.4 mg of MC	35(F)	34	34	100	60	1.8	24	70	20	58	29	85			11
II	Offspring of pregnant mice treated with 8.4 mg of MC	54(F)	53	52	98	74	1.4	24	45	17	33	49	92			10
		45(M)	45	45	100	65	1.4	19	42	16	36	44	98	3	7	2
III	Pregnant mice treated with 1 mg of MC	38(F)	35	34	97	56	1.6	19	54	15	43	29	83	1	3	11
IV	Offspring of pregnant mice treated with 1 mg of MC	42(F)	41	41	100	54	1.3	11	27	6	15	39	95			9
		65(M)	65	64	98	81	1.2	17	26	5	8	63	97	10	15	3
V	Offspring (of pregnant mice treated with 1 mg of MC) given 12.5 µg of DENA when newborn	48(F)	45	45	100	64	1.3	17	38	12	27	41	91	7	16	4
		56(M)	54	54	100	95	1.8	39	72	6	11	53	98	33	61	3
VI	Offspring (of untreated pregnant mice) given 12.5 µg of DENA when newborn	36(F)	34	33	97	56	1.6	20	59	12	35	30	88	12	35	2
		32(M)	32	32	100	63	2.0	25	78	7	22	25	78	29	90	2
VII	Untreated controls	44(F)	38	31	82	44	1.2	10	26	20	53	13	34			11
		49(M)	47	36	77	52	1.1	13	28	19	40	24	51	3	6	6

[a]Number of animals alive at the time of appearance of the first tumor.
[b]In relation to the effective number.

Source: Turusov et al. 1973, Table 1, pp. 86-87.

10.1.7 Mutagenicity

The active forms of a large number of known chemical carcinogens are mutagens. The PAH have
been shown to bind to DNA (Sect. 10.1.5.5). This suggests that malignant transformation can
involve an alteration in the genetic constitution of treated cells. PAH carcinogens may therefore produce mutations, some of which may involve the genes that control malignancy (Huberman
and Sachs 1974).

Ames et al. (1973) propose that carcinogens that are mutagens cause cancer by somatic mutation
in which flat aromatic molecules intercalate in DNA base-pair stacks. This leads, during DNA
replication or repair, to an addition or deletion of base pairs in the DNA sequence. These
"frameshift" mutagens are more potent if they contain a side chain that covalently reacts with
DNA.

10.1.7.1 Testing methods

A variety of methods for testing mutagenicity has been developed. Submammalian tests in
bacteria, bacteriophage, *Neurospora*, plants, and *Drosophila* help to elucidate basic mechanisms
of mutagenesis. However, microorganisms do not metabolize the PAH in the same way as mammals;
thus, in vivo mammalian test systems are more relevant for extrapolating data to man. Three in
vivo mammalian tests are practical and sensitive: dominant lethal assay, host-mediated assay,
and in vivo cytogenetics (NAS 1972).

Dominant lethal assay

The induction of dominant lethal mutations in animals can be assayed with a high degree of
sensitivity and practicality after acute, subacute, or chronic administration of test materials
by several routes. After administration of a drug, male rodents are mated sequentially with
groups of untreated females over the duration of the spermatogenic cycle. Matings within three
weeks after a single drug administration represent samplings of sperm exposed during postmeiotic
stages, and matings four to eight weeks later represent samplings of sperm exposed during
premeiotic and stem cell stages. The classic form of the dominant lethal assay involves
autopsy of females about 13 days after timed matings and enumeration of corpora lutea and total
implants. Dominant lethal mutations are directly measured by enumeration of early fetal deaths
and indirectly by preimplantation losses. Results are best expressed as early fetal deaths per
pregnant female.

Host-mediated assay

In the host-mediated assay the chemical being tested is administered to the mammal, which then
receives an injection by another route of indicator microorganisms. The microorganisms are
later recovered and scored for induction of mutants. This assay determines the influence of in
vivo mamalian factors on the activation or detoxification of chemical mutagens.

In vivo cytogenic assay

Rats and Chinese hamsters are most often investigated for structural and numerical chromosomal
aberrations in in vivo cytogenetic assays. Cytogenetic effects on metaphase or anaphase

preparations of somatic or germinal cells can be studied singly or serially after recovery periods. Data on mutagenicity testing of air pollutants in in vivo mammalian systems are scarce, and there are no published data on mutagenicity testing by inhalation (NAS 1972). However, mutagenicity studies using these three systems and specific chemicals have been reported.

In vitro assay

A fourth method, the in vitro tissue or cell culture, permits quantitative analysis of genetic variations in cell populations via the plating technique, following treatment of the cells in culture with chemical agents.

Ames et al. (1973) describe a test for the detection of carcinogens as mutagens. The test consists of rat or human liver homogenate for metabolism of carcinogens and a set of *Salmonella* histidine mutants for mutagen detection. The carcinogen, liver homogenate, and bacterial tester strain are incubated together on a petri plate. Histidine revertant colonies are counted after a two-day incubation period.

10.1.7.2 Relationship between carcinogenicity and mutagenicity

Using the in vitro assay method, Chu, Bailiff, and Malling (1970) tested the mutagenicity, in Chinese hamster cells, of some chemical carcinogens and their related compounds and derivatives (Table 10.15). The genetic marker assayed in the hamster cells was the change from 8-azaguanine sensitivity to resistance. The results show a direct relation between the degree of carcinogenicity and mutagenicity. Although BaP was not mutagenic in this study, it is usually listed as a known mutagen (Hueper 1971).

Table 10.15. Relative carcinogenicity and mutagenicity
of selected compounds[a]

Test Compound	Carcinogenicity	Mutagenicity
Benzo[e]pyrene	−	−
Benzo[a]pyrene	+	−
3-Hydroxybenzo[a]pyrene	−	±
Dibenz[a,c]anthracene	−	−
Dibenz[a,h]anthracene	±	±
7,12-Dimethylbenz[a]anthracene	+++	+++

[a]Symbols: − not carcinogenic (or mutagenic); ± uncertain or weakly carcinogenic (or mutagenic); + carcinogenic (or mutagenic); +++, strongly carcinogenic (or mutagenic).

Source: Modified from EPA 1975, Table 6-1, p. 6-8.

10.1.7.3 Metabolism

A number of studies since 1969 have shown that the chemically nonreactive carcinogens, such as the PAH, must be metabolized by cellular enzymes in order to exert their biological effects (including mutagenicity). Huberman and Sachs (1974) developed a cell-mediated mutagenic assay with carcinogenic hydrocarbons in which Chinese hamster V79 cells, which are susceptible for mutagenesis but do not metabolize PAH, are cocultivated with lethally irradiated rodent cells that can metabolize these compounds. The marker used for mutagenesis in the V79 cells was the genetic change from susceptibility to resistance to 8-azaguanine. The authors found that the number of mutations was dependent on the number of metabolizing cells and that the carcinogens were not mutagenic when V79 cells were cocultivated with nonmetabolizing cells. Inhibition of the hydrocarbon metabolizing enzymes by 7,8-benzoflavone inhibited mutagenicity. Cell-mediated mutagenicity was obtained with the carcinogenic hydrocarbons (DMBA, BaP, 3-methylcholanthrene, and 7-methylbenz[a]anthracene). The most potent carcinogen, DMBA, gave the highest frequency of mutations. Pyrene and BaA, which are not carcinogenic, were also not mutagenic. The method detected mutagenicity with 1 µg/ml (Huberman and Sachs 1974; Huberman 1975).

10.1.7.4 Metabolites

In a more recent study, Huberman et al. (1976), using the same cell-mediated assay, identified the major mutagenic intermediate of BaP metabolism as a diol-epoxide. They tested the mutagenicity of BaP and 15 of its derivatives, which included the phenols; the BaP 4,5-epoxide (the K-region epoxide); dihydrodiols; two isomeric 7,8-diol 9,10-epoxides; a 6-methyl derivative; and a 6-hydroxymethyl derivative. Mutations were characterized by resistance to ouabain or 8-azaguanine. Mutagenesis was tested both in the presence and absence of normal, BaP-metabolizing, golden hamster cells. Little or no mutagenicity was shown by the phenols; 4,5-diols; *trans*-9,10-diol; and 6-methyl and 6-hydroxymethyl derivatives. The (+) 7α,8β-dihydroxy-9α,10α-epoxy-7,8,9,10-tetrahydro-BaP and K-region 4,5-epoxide exhibited similar and moderate mutagenicity in the absence of BaP-metabolizing cells, but the (+) 7α,8β-dihydroxy-9β,10β-epoxy-7,8,9,10-tetrahydro-BaP showed a 2000- and 270-fold higher mutation frequency for ouabain and 8-azaguanine resistance, respectively, than did the K-region 4,5-epoxide. The *trans*-7,8-diol was more mutagenic than BaP after metabolism in BaP-metabolizing cells. Mutagenesis by the *trans*-7,8-diol was inhibited by 7,8-benzoflavone, an inhibitor of mixed-function oxidases. The major metabolite of *trans*-7,8-diol was isolated and found to be 7α,8β-dihydroxy-9β,10β-epoxy-7,8,9,10-tetrahydro-BaP. The high mutagenic activity of this compound and the demonstration of its metabolic formation (Yang 1976) suggested that it may be an ultimate carcinogenic form of BaP.

Malaveille et al. (1975) examined the mutagenicity of a series of BaP and BaA derivatives towards *Salmonella typhimurium* strain TA 100. BaP 7,8-diol and BaA 8,9-diol were more active than the parent compounds in inducing his^+ revertant colonies. BaP 9,10-diol was less active than BaP, whereas the K-region diols (BaA 5,6-diol and BaP 4,5-diol) were inactive. None of the diols were active in the presence of AHH cofactors. The diol-epoxides (BaP 7,8-diol 9,10-oxide and BaP 7,8-diol 10,11-oxide) and the K-region epoxides (BaP 4,5-oxide and BaA 5,6-oxide) were mutagenic without further metabolism. The results support the theory that the non-K-region diol-epoxides are more mutagenic than the simple K-region epoxides. In this study, dose-response curves for mutagenicity were prepared by using varying amounts of post-mitochondrial liver supernatant. In the presence of 150 µl supernatant per plate, an increasing mutagenic response was observed with both BaP and BaA up to a concentration of 45 µ*M*, and BaP was always more active than BaA.

Induced liver enzymes from male Sprague-Dawley rats, treated prior to sacrifice with PAH or phenobarbital, were tested for their ability to activate BaP to a bacterial mutagen (Rasmussen and Wang 1974). The phenobarbital-induced enzymes were much more active than the 3-methylcholanthrene-induced enzymes when low levels (less than 1 mg/ml) of enzyme protein were present in the incubation mixtures. At higher levels of enzyme protein, the 3-methylcholanthrene-induced enzymes were more active. Addition of an epoxide hydrase inhibitor to the test mixtures with phenobarbital-induced enzymes greatly increased the mutagenic activity, but only slightly increased the activity obtained with the 3-methylcholanthrene-induced enzymes. The bacterial mutagenesis tests were carried out by using reversion of histidine auxotrophs of *Salmonella typhimurium* as indicators of mutagenesis. The authors concluded that their results favored the long-held idea that the 4,5-epoxide (K-region) metabolite is involved in mutagenesis.

10.1.7.5 Environmental pollutants

Using the dominant lethal test, Epstein and Shafner (1968) screened a wide variety of environmental pollutants for mutagenicity. In this system BaP was mutagenic, but 3-methylcholanthrene and organic extracts of atmospheric particulate pollutants and of finished drinking water displayed control values.

10.1.8 Epidemiology studies

A difficulty in relating occupational and environmental cancers to specific factors is their long latent period. Hueper (1952) estimates the latent periods for skin and lung cancer from tar and tar fumes to be 20 to 24 and 16 years respectively. Latent periods depend upon the relative potency of a particular carcinogen, its physicochemical properties, the physico-chemical and co- or anticarcinogenic properties of its vehicle or its associated agents, the route of contact, the intensity of the individual exposures, and the total duration of exposure. With increasing intensity of exposure, an increase in the incidence of cancers and a shortening of the latent period occurs.

Historically, the carcinogenicity of several coal combustion and processing chemicals was first observed in workers (occupational exposure) and later confirmed in experiments. Pott (1775), as cited by IARC (1973), observed scrotal skin cancer in British chimney sweeps and attributed it to the lodging of coal soot in the scrotal folds. Henry (1946), in a comprehensive study of the relationship between skin tumors and exposure to coal-tar products, reported an average annual scrotal cancer mortality rate of 21.1 per million in coke-oven workers during the period from 1911 to 1938, as compared with a general population rate of 4.2.

An unusual incidence of lung cancer was first reported in Japanese producer (coal) gas workers (Kuroda and Kawahata 1936). The excess lung cancer risk for gas producer workers was confirmed by British studies of death certificates in England and Wales for the years 1921 to 1932 (Kennaway and Kennaway 1936). This study also showed that other coal carbonization and by-product workers experienced greater than expected lung cancer mortality. An excess risk of bladder cancer among men employed at coal carbonization processes was first reported by Henry, Kennaway, and Kennaway (1931).

Excess mortality from cancer of other organs for coke-oven workers and other coal carbonization workers has been reported from several sources, many of them based on very limited data. Coal carbonization workers have been noted to have extremely high incidence of cancer of the skin, lungs, and urinary organs; other cancer sites, based on sometimes limited data, are the larynx, nasal sinuses, pancreas (associated with exposure to β-naphthylamine), blood-forming organs, and stomach and digestive tract (NIOSH 1973; Freudenthal, Lutz, and Miller 1975).

Accumulated evidence suggests the possibility of some difference in carcinogenic response according to the carbonization process (NIOSH 1973). A positive relationship can be seen between temperatures attained during carbonization and the lung cancer response for men employed in various areas of the industry: Excess lung cancers increased with increasing temperatures (Table 10.16). Modern coal conversion processes also carry a potential cancer hazard. Bridbord (1976), as modified from Cavanaugh et al. (1975), lists carcinogens of concern in coal liquefaction and gasification process streams (Table 10.17). An ERDA (1976) document discusses the carcinogenic potential of coal conversion processes and products.

Table 10.16. Temperature range of carbonizing chambers and excess of lung cancer reported

Carbonizing chamber	Temperature range($^\circ$C)	Percent excess of lung cancer reported
Vertical retorts	400 - 500	27
Horizontal retorts	900 - 1100	83
Coke ovens[a]	1200 - 1400	255
Japanese gas generators	1500	800

[a]The figure shown for coke oven workers is for men with five or more years experience to provide contrast with the British gas workers who had worked at least five years at the retorts.

Source: NIOSH 1973, Table VII-1, p. VII-1.

10.1.8.1 Levels in human tissue

Few investigators have found measurable amounts of BaP associated with human lung tissue. In one study, the mean content of BaP in human pulmonary tissue was measured and found to be 0.2 µg/100 g of tissue (Gräf, Eff, and Schormair 1975). The mean content differed with age, being high in infants (0.31 µg/100 g of tissue), low between ages of 1 and 15 (0.10 to 0.15 µg/100 g of tissue), and rising with increasing age (0.25 to 0.26 µg/100 g of tissue). This same curve was also noted in other tissues. Tissues having high proliferative activity, such as glands and bone marrow, averaged 0.2 µg/100 g of tissue. Adipose tissue levels were well below the average concentrations of other organ tissues. To show that the high BaP values in infants are due to exogenous factors, the same authors studied BaP levels in chicken embryos during development in the egg. During embryonic development of the chicken, BaP levels remained constant. Values rose after hatching and then dropped to trace levels in the adult chicken. Mean values in human and pig placentas were 0.09 and 0.34 µg/100 g of tissue respectively.

Tomingas, Pott, and Dehnen (1976) examined PAH content in human bronchial carcinomas obtained from surgical operations and autopsies. PAH determination was made by direct fluorescence analysis on thin-layer plates. Only 4 of 12 tested PAH were detected: BaP, benzo[b]fluoranthene, fluoranthene, and perylene. BaP was found in all carcinoma tissue. Median values of 3.5 µg BaP/g of tissue in carcinomas and 0.09 µg in the adjoining tumor-free tissue were found — a ratio of 38:1. Fluoranthene and benzo[b]fluoranthene contents of carcinoma and tumor-free tissue

Table 10.17. Carcinogens of concern in coal liquefaction
and gasification process streams

Chemical classification	Compound	Phase	Remarks
Amines	Diethylamines	Gas	Potential source of nitrosamines from reaction with nitrogen oxides
	Methylethylamines	Gas	
Heterocyclics	Pyridines	Gas/liquid	
	Pyrroles	Gas/liquid	
Hydrocarbons	Benzene	Gas/liquid	
Phenols	Cresols	Gas/liquid	May be promoters
	Alkyl cresols	Gas/liquid	
Polynuclears	Anthracene	Gas	
	BaP[a]	Gas	
	Chrysene	Gas	
	Benzo[a]anthracene	Gas	
	Benzo[a]anthrone	Gas	
	Dibenzo[a,l]pyrene	Gas	
	Dibenzo[a,n]pyrene	Gas	
	Dibenzo[a,i]pyrene	Gas	
	Indeno[1,2,3-c,d]pyrene	Gas	
	Benzoacridine	Gas	
Trace elements	Nickel	Gas	
	Beryllium	Gas	
	Cadmium[a]	Gas	
	Arsenic[a]	Liquid	
Organo-metallics	Nickel carbonyl	Gas	
Fine particulates	Sulfur particulates	Gas	May be a cocarcingoen
	Coke	Gas	
	Coal dust	Gas	

[a]Potential mutagens or teratogens, based on Teratogen/Mutagen Sulfide, *Registry of Toxic Effects of Chemical Substances*, National Institute for Occupational Safety and Health.

Source: Modified from Cavanaugh et al. (1975), as cited in Bridbord (1976), Table 2.

were not significantly different. The results demonstrated an increased concentration of BaP in tumor areas of the lung; in addition, the relative concentrations of PAH in the cancerous tissue were extremely different from those found in the air and in tobacco smoke.

Reasons for the increased concentration of PAH in cancerous tissue were discussed with respect to the deposition and elimination of inhaled particles and the metabolism of PAH.

> The deposition of inhaled particles in the area of a carcinoma may be increased by an alteration of bronchial air flow; a disturbance of the elimination of particles may not only cause a prolonged retention time of the deposited particles there but also, and more importantly, may result in a damming up of the substances moving from the terminal air passages toward the trachea. In addition to this incommodation of the "mechanical" lung clearance, a disturbance of the resorption, as well as the metabolism, of foreign substances must be taken into account. Alveolar macrophages metabolize PAH with subsequent removal of the metabolites from the lungs. Macrophages of smokers have been shown to demonstrate a higher enzyme activity than do those of nonsmokers.

Mallet and Heros (1960) measured the BaP content of urine from Paris inhabitants (4 samples) and found concentrations of 1 to 3 µg/liter.

10.1.8.2 Allegheny County coke-oven workers

In 1962 a collaborative study of chronic disease mortality among steel workers in relation to prior occupational history was initiated by the U.S. Public Health Service and the University of Pittsburgh, Graduate School of Public Health, in cooperation with three large steel firms. The results were published in a series of papers (Lloyd et al. 1970; Lloyd 1971; Redmond et al. 1972). The study included 58,828 steelworkers who were employed by three firms at seven plants in 1953. All the plants were located in Allegheny County, Pennsylvania. Of the 58,828 men employed in 1953, 3530 worked, or had previously worked, in the coke plant.

Coking

Coking refers to the heating of mixtures of different ranks of coal in slot ovens in the absence of air to form ammonia, coke-oven gas, light oil, tar, and coke. In the manufacture of steel, the primary function of the coke plant is the production of metallurgical coke for use in blast furnaces. A secondary function is the recovery of chemical by-products during the high-temperature carbonization of bituminous coal into coke. High BaP emissions have been measured in the gaseous discharge (the coal-tar pitch volatiles) of coke ovens of the steel industry. It has been roughly estimated that 1.8 g BaP are emitted per ton of coke produced (NAS 1972).

Cancer risk

Although earlier studies usually reported a high incidence of lung cancer in various occupational groups employed in the production of coal gas or coke, some studies have reported contradictory results. For example, excessive lung cancer in coke-oven workers was reported by Kennaway and Kennaway (1947), but Reid and Buck (1956) found no difference between the mortality rate for coke-oven workers and that predicted for any large industrial organization.

Lloyd et al. (1970) documented the mortality rate for coke-plant workers by work area. They report that men working in the coke plant have a risk of lung cancer two and one-half times greater than men in the total steelworkers groups. In another report, Lloyd (1971) examined the variation in the mortality rate for coke-plant workers according to calendar period, length of employment at several work stations within the coke plant, and race. The excess of respiratory cancer (again, two and one half times that predicted) was limited to men employed at the coke ovens; employment at the top of the ovens was associated with nearly a fivefold excess risk. Workers immediately above the ovens were exposed to a maximum concentration of coal volatiles.

As shown in Table 10.18, deaths among men employed at the full-time topside jobs accounted for all the excess mortality from cancer of the lung. The total mortality experience of men employed only at the side of the oven differed little from expectation, and white and nonwhite workers showed similar patterns of relative risk. As seen in Table 10.19, total mortality for men employed only at the side of the oven, or with less than five years employment at full-time topside jobs, differed little from expectation, whereas the number of deaths for men employed five or more years at full-time topside jobs was double the expected number (35 vs 17.4). A tenfold risk was observed for men employed five or more years at full-time topside jobs. Fifteen lung cancer deaths were observed among the 132 men in this group, as compared with 1.5 deaths expected.

The higher respiratory cancer rate in nonwhite coke-oven workers was accounted for by differential job distributions in which nonwhites more often encountered the high risk at the top of the oven (Lloyd et al. 1970). Interestingly, a deficit of deaths from heart disease was seen for coke-oven workers, and there appeared to be an excess risk of digestive cancer for coke-plant workers employed in the areas other than the ovens.

10.1.8.3 _Other coke-oven workers_

To more fully evaluate the nature and extent of the cancer risk among coke-oven workers, the Department of Biostatistics, University of Pittsburgh, with cooperation and funding from the U.S. Public Health Service and the American Iron and Steel Institute, designed and conducted a study of mortality patterns among coke-oven workers from various geographic areas (Redmond et al. 1972). They presented the cause-specific mortality findings for coke-oven workers at ten selected steel plants in diverse parts of the United States and Canada; compared these results with findings for the two plants previously studied (Lloyd 1971); and showed the relationships of the mortality findings for coke-oven workers to variables such as race, work area, and term of employment.

A general consistency was observed for mortality findings among the plants; therefore, the authors combined findings for all 12 plants in Tables 10.20, 10.21, and 10.22. Since the relative risks for respiratory malignancies were of the same order of magnitude for both white and nonwhite oven workers, only total (combined) mortality data is shown. Table 10.20 shows mortality for specific causes by area of employment at the time of entry to the study. Mortality for all causes was highest for topside workers. The relative risk for lung cancer was seven times as high as the expected risk for topside workers, two times as high for partial topside workers, and one and a half to two times as high for side-oven workers.

Table 10.18. Number employed; observed and expected deaths and standardized mortality ratios for selected causes by race (1953-1961): men employed in subdivisions of the coke oven in 1953 and prior years

Work area and race	Number Employed	All causes			Malignant neoplasms of lung			Other malignant neoplasms			Heart disease			All other causes		
		Observed deaths	Expected deaths	SMR	Observed deaths	Expected deaths	SMR	Observed deaths	Expected deaths	SMR	Observed deaths	Expected deaths	SMR	Observed deaths	Expected deaths	SMR
Total coke oven	2048	184	173.1	106	31	12.3	252[a]	27[d]	26.9	100	49	61.5	80	77	72.4	106
White	993	80	80.2	100	8	4.7	170	12	12.2	98	29	33.9	86	31	29.4	105
Nonwhite	1055	104	92.9	112	23	7.6	303[a]	15[d]	14.8	101	20	27.6	72	46	43.0	107
Side oven[b]	1431	106	114.6	92	10	8.0	125	13[d]	17.5	74	34	40.0	85	49	49.1	100
White	606	45	47.2	95	5	2.7	185	6	7.3	82	18	19.9	90	16	17.4	92
Nonwhite	825	61	67.3	91	5	5.4	93	7[d]	10.3	68	16	20.1	80	33	31.6	104
Partial topside[c]	315	24	26.3	91	2	1.7		6	4.0	150	5	10.9	46	11	9.7	113
White	307	22	25.1	88	2	1.6		5	3.8	132	5	10.5	48	10	9.2	109
Nonwhite	8	2	1.2		0	0.1		1	0.2		0	0.3		1	0.5	
Full topside	302	54	32.2	168[a]	19	2.6	731[a]	8	5.4	148	10	10.6	94	17	13.6	125
White	80	13	7.8	167	1	0.5		1	1.2		6	3.5	171	5	2.8	179
Nonwhite	222	41	24.4	168[a]	18	2.2	818[a]	7	4.3	163	4	7.2	56	12	10.8	111

[a]1% level.
[b]Never partial or full topside.
[c]Never full topside.
[d]Includes two respiratory cancer deaths other than lung.

Source: Lloyd 1971, Table V, p. 60. Reprinted by permission of the publisher.

Table 10.19. Number employed; observed and expected deaths and standardized mortality ratios for selected causes, (1953-1961): men employed five or more years at coke ovens and subdivisions as of January 1, 1953

Work area and race	Number employed	All causes			Malignant neoplasms of lung			Other malignant neoplasms			Heart diseases			All other causes		
		Observed deaths	Expected deaths	SMR	Observed deaths	Expected deaths	SMR	Observed deaths	Expected deaths	SMR	Observed deaths	Expected deaths	SMR	Observed deaths	Expected deaths	SMR
Side oven only	496	53	55.1	96	6	4.1	146	7	8.9	79	15	19.4	77	25	22.7	110
White	171	16	18.7	86	2	1.1		2	2.9		6	8.1	74	6	6.6	91
Nonwhite	325	37	36.4	102	4	3.0		5	6.0	83	9	11.3	80	19	16.1	118
Side and topside (less than 5 years full topside)	276	29	27.9	104	6	2.1	286[a]	9	4.3	209[b]	5	10.8	46	9	10.7	84
White	202	19	19.4	98	2	1.3		5	2.9	172	5	8.4	60	7	6.9	101
Nonwhite	74	10	8.4	119	4	0.8		4	1.4		0	2.4		2	3.8	
Full-time topside	132	35	17.4	201[a]	15	1.5	1000[a]	4	3.2		5	5.6	89	11	7.1	155
White	27	4	3.0		1	0.2		0	0.5		3	1.3		0	1.0	
Nonwhite	105	31	14.4	215[a]	14	1.3	1077[a]	4	2.8		2	4.3		11	6.1	180[b]

[a] 1% level.
[b] 5% level.

Source: Lloyd 1971, Table VII, p. 62. Reprinted by permission of the publisher.

Table 10.20. Observed and expected deaths (1951-1966) and relative risk for selected causes by coke oven subdivisions for coke oven workers

Cause of death	Full topside			Partial topside			Side oven		
	Observed deaths	Expected deaths	Rel. risk[a]	Observed deaths	Expected deaths	Rel. risk[a]	Observed deaths	Expected deaths	Rel. risk[a]
All causes	157	144.1	1.12	73	69.2	1.07	359	379.7	0.92
Malignant neoplasms — lung, bronchus, and trachea	35	10.4	7.24[b]	7	3.7	2.14	27	19.4	1.73[c]
Malignant neoplasms — genito-urinary system	4	2.5	(1.77)	2	0.8	(2.36)	15	9.2	2.02[c]
Other malignant neoplasms	15	16.8	0.86	5	9.7	0.44	35	41.4	0.79
Tuberculosis of the respiratory system	3	2.3	d	0	0.3	d	5	4.8	1.06
Other diseases of the respiratory system	7	7.0	0.99	5	3.1	1.78	18	17.8	1.01
Cardiovascular renal diseases	60	68.5	0.84	39	33.8	1.18	155	180.9	0.80[c]
Accidents	10	13.4	0.72	7	7.4	0.94	45	40.1	1.19
All other causes	23	23.1	1.00	8	10.2	0.76	59	66.2	0.85

[a]Significance of relative risk based on summary chi-square with 1 degree of freedom; numbers in parentheses indicate relative risk based on less than 5 deaths.
[b]5% level.
[c]1% level.
[d]Less than 5 deaths.

Source: Redmond et al. 1972, Table 5, p. 627. Reprinted by permission of the publisher.

Table 10.21. Observed and expected deaths (1951-1966) and relative risk for selected causes by
length of employment at entry to study for coke oven workers employed during 1951-1955
at ten nonAllegheny County plants and for coke oven workers employed during 1953
at two Allegheny County steel plants

Cause of death	Five years or more at coke oven			Less than five years at coke oven		
	Observed deaths	Expected deaths	Rel. risk[a]	Observed deaths	Expected deaths	Rel. risk[a]
All causes	383	365.9	1.08	206	235.2	0.83[b]
Malignant neoplasms — lung, bronchus, and trachea	55	28.0	3.48[c]	14	9.9	1.70
Malignant neoplasms — genito-urinary organs	17	10.6	2.01[b]	4	2.6	d
Other malignant neoplasms	42	44.0	0.93	13	26.0	0.40[c]
Tuberculosis of the respiratory system	5	5.0	1.00	3	2.9	d
Other diseases of the respiratory system	20	16.8	1.28	10	11.0	0.88
Cardiovascular renal diseases	173	186.2	0.90	81	96.4	0.79
Accidents	20	21.9	0.88	42	38.6	1.12
All other causes	51	53.4	0.94	39	47.6	0.77

[a]Significance of relative risk based on summary chi-square with 1 degree of freedom; numbers in parentheses
indicate relative risk based on less than 5 deaths.
[b]5% level.
[c]1% level.
[d]Less than 5 deaths.

Source: Redmond et al. 1972, Table 5, p. 627. Reprinted by permission of the publisher.

Table 10.22. Observed and expected deaths and relative risks for malignant neoplasms by site (1951-1966) by length of employment at the coke ovens and subdivisions at entry to study

Cause of death	Total oven			Five years or more, coke oven			Five years or more, full-time topside			Five years or more, topside & side oven exposure			Five years or more, side oven, never topside			Less than five years, coke oven		
	Observed deaths	Expected deaths	Rel. risk[a]	Observed deaths	Expected deaths	Rel. risk[a]	Observed deaths	Expected deaths	Rel. risk[a]	Observed deaths	Expected deaths	Rel. risk[a]	Observed deaths	Expected deaths	Rel. risk[a]	Observed deaths	Expected deaths	Rel. risk[a]
Malignant neoplasms, all sites	145	122.1	1.34[b]	114	82.6	1.69[b]	41	20.0	2.74[b]	27	18.9	1.51	46	35.5	1.40	31	38.6	0.74
Bronchus, lung, and trachea	69	41.5	2.85[b]	55	28.0	3.48[b]	25	7.4	6.87[b]	15	5.5	3.22[b]	15	8.7	2.10[c]	14	9.9	1.70
Other respiratory system	1	2.4	d	1	1.8	d	0	0.5	d	0	0.4	d	1	1.0	d	0	1.0	d
Buccal and pharynx	3	3.4	d	1	2.2	d	0	0.5	d	1	1.0	d	0	0.8	d	2	1.4	d
Digestive organs and peritoneum	31	37.0	0.77	25	26.0	0.95	3	5.9	0.46	4	6.3	0.62	18	14.2	1.37	6	12.2	0.42
Prostate	12	8.8	1.64	11	7.4	1.83	4	2.0	d	0	1.1	d	7	3.9	2.11	1	1.4	d
Other genital organs	0	0.4	d	0	0.3	d	0	0.1	d	0	0.1	d	0	0.2	d	0	0.2	d
Kidney	8	2.6	7.49[b]	5	1.6	5.69	0	0.1	d	3	0.4	d	2	0.7	d	3	0.7	d
Bladder and other urinary organs	1	1.5	d	1	1.3	d	0	0.3	d	0	0.3	d	1	0.8	d	0	0.4	d
Skin	1	1.5	d	1	0.8	d	0	0.1	d	0	0.3	d	1	0.3	d	0	0.8	d
Leukemias and lymphomas	9	10.8	0.76	5	6.0	0.77	2	1.4	d	3	2.1	d	0	2.3	d	4	5.3	0.68
Other and unspecified sites	10	12.1	0.69	9	7.3	1.46	7	1.7	12.37[b]	1	1.3	d	1	2.6	d	1	5.2	d

[a]Significance of relative risk based on summary chi-square with 1 degree of freedom.
[b]1% level.
[c]5% level.
[d]Less than 5 deaths.

Source: Redmond et al. 1971, Table 6, p. 628.

Reprinted by permission of the publisher.

Table 10.21 summarizes the mortality findings based on length of employment at the coke ovens upon entry to the study. Although the mortality from all causes is about the expected rate, the relative risk for lung cancer is 3.48. Malignant neoplasms of the genitourinary organs are significantly greater among coke-oven workers having five or more years of experience; there is a less pronounced excess for men having less than five years of experience.

Table 10.22 indicates a gradient in response that is related to both term and area of employment, ranging from a relative risk of 6.87 for men employed five or more years full topside at entry to the study to a relative risk of 1.70 for men having less than five years of exposure to coke ovens. This pattern suggests the existence of a dose-response relationship between some carcinogenic element or elements in the coke-oven effluents and the development of lung cancer.

Table 10.22 also indicates a risk of death from kidney cancers among all coke-oven workers seven and a half times as great as that expected. Although exposure to coal tars is known to be associated with malignancies of the skin, in only one case was skin cancer listed as the cause of death in this study.

Exposure vs risk

More recently, Mazumdar et al. (1975) studied the exposure of coke-oven workers to coal-tar pitch volatiles (CTPV). They categorized coke-oven jobs into different work areas in terms of exposure to CTPV and developed an index of cumulative exposure to investigate the dose-response relationship between exposure to CTPV and mortality from cancer, particularly lung cancer. The mortality data were based on both the long-term study of the steelworkers cited above and exposure data taken from a study conducted by the Pennsylvania Department of Health at ten coke-oven installations (Fannick, Gonsher, and Shookley 1972).

Eleven work classifications were categorized into three exposure groups (Sollecito 1970, as cited in Mazumdar et al. 1975). The exposure groups, work classifications, and average exposures to CTPV of the exposure survey are shown, by plant and job title, in Table 10.23. The table shows a considerable difference in exposure to CTPV for the various coke-oven jobs, the lidman having the highest average exposure (3.23 mg/m^3) and the pusher side-machine operator having the lowest average exposure (0.50 mg/m^3). Topside jobs had the highest exposure levels. Average levels for the occupational groups were 3.15 mg/m^3 for topside jobs, 1.99 mg/m^3 for side-oven-1 jobs, and 0.88 mg/m^3 for side-oven-2 jobs. Exposure to CTPV did not differ significantly among the ten plants. From an analysis of the data grouped into age and cumulative-exposure intervals (in which age-adjusted and age-specific death rates for all causes at all sites, by cumulative exposure and age at entry into the study, were considered), the authors concluded that both the level and the length of exposure were related to the development of cancer, particularly lung cancer. The data indicate that the time between first exposure to CTPV and death from lung cancer varies from 10 to 40 years; the average was 25 years. The authors considered the established threshold limit value of 0.2 mg/m^3 for CTPV adequate since this level of exposure for an average period of 30 years did not increase the risk of death from lung cancer.

Similar values for CTPV were reported in another study. Jackson, Warner, and Mooney (1974) surveyed a 70-oven coke manufacturing plant for concentrations of cyclohexane-soluble coal-tar pitch volatiles and BaP. Three off-battery and three on-battery sampling sites were selected, with 45 samples collected during July 1970. The on-battery locations showed measured levels

Table 10.23. Mean level of coal-tar pitch volatiles (mg/m^3) and number of samples by plant and job title

Coke plant exposure group, by job	Plant[a]										Total	Average
	A	B	C	D	E	F	G	H	I	J		
Topside												
Lidman	2.06 (3)	3.14 (6)	2.02 (6)	2.06 (9)	3.34 (3)	4.11 (5)	1.95 (5)	6.27 (10)	1.88 (5)	3.23 (7)	3.23 (59)	
Tar chaser				1.18 (2)	11.59 (2)	3.32 (3)	3.31 (7)	1.62 (3)	2.34 (6)	1.81 (4)	3.14 (27)	
Larry car operator	2.73 (1)	1.62 (4)	8.33 (2)	2.73 (4)	2.76 (3)	3.91 (7)	1.11 (4)	4.97 (3)	2.62 (5)	2.50 (8)	3.04 (41)	3.15
Side Oven 1												
Luterman			3.53 (4)		2.62 (4)		2.18 (10)				2.57 (18)	
Machine operator, coke side		2.29 (1)	1.99 (2)			0.94 (2)	3.43 (3)	3.90 (4)	1.18 (5)	3.11 (2)	2.43 (19)	
Benchman, pusher side	0.84 (8)	1.42 (2)		0.61 (1)	0.85 (1)	0.70 (5)		1.24 (4)	4.44 (8)	2.09 (2)	1.91 (31)	
Benchman, coke side	0.76 (2)	1.58 (2)		0.73 (4)		0.76 (5)			1.07 (3)	2.35 (2)	1.08 (18)	1.99
Side Oven 2												
Heater	1.12 (2)	0.61 (2)	0.65 (4)	1.39 (5)	0.61 (2)	1.38 (7)	1.40 (4)	1.11 (3)	0.78 (6)	1.05 (4)	1.07 (39)	
Miscellaneous	0.74 (2)		0.24 (2)		1.35 (1)		0.57 (3)		2.15 (1)	1.06 (9)	0.93 (18)	
Quenching car operator	0.19 (2)	0.35 (2)	0.83 (2)	0.00 (1)	0.76 (1)	0.34 (3)	0.25 (2)	0.77 (4)	0.27 (1)	2.67 (5)	0.94 (23)	
Pusher machine operator	0.56 (2)	0.94 (5)	0.15 (2)	0.57 (2)		0.46 (3)	0.37 (2)	0.22 (3)	0.53 (4)	0.31 (3)	0.50 (26)	0.88
Total	1.02 (22)	1.68 (24)	2.16 (24)	1.55 (28)	3.30 (17)	1.98 (40)	1.95 (40)	3.24 (34)	2.06 (44)	2.02 (46)	2.08 (319)	

[a]Numbers in parentheses indicate number of samples.

Source: Mazumdar et al. 1975, Table II, p. 385. Reprinted by permission of the publisher.

of 0.0313 to 1.67 mg of CTPV and 0.172 to 15.9 μg of BaP per cubic meter; the off-battery concentrations were 0.0072 to 0.322 mg of CTPV and 0.021 to 1.18 μg of BaP per cubic meter.

10.1.8.4 British gasworks workers

A selected population of 11,449 employees of a British gas works (coal carbonization) industry was monitored from 1953 to 1961 (Doll et al. 1965). Of the men aged 40 to 65 and having at least five years of employment, the heavily exposed workers had a 69% higher lung-cancer incidence (3.06 per 1000 vs 1.81 per 1000) and a 126% higher death rate from bronchitis than workers receiving no exposure (2.89 per 1000 vs 1.28 per 1000). Heavily exposed workers also had a slightly greater mortality from bladder cancer. There were no differences in the smoking habits between men experiencing heavy, intermediate, and no exposure. A final report by Doll (1972) confirmed the increased risk of death from cancer of the lung as well as bladder cancer, but raised some questions as to increased deaths from bronchitis. The absolute increase in deaths from cancer of the bladder and cancer of the scrotum and skin in the heavily exposed class of workers was 23 per 100,000 per year and 12 per 100,000 per year respectively.

Lawther, Commins, and Waller (1965) report on the concentrations of BaP and other PAH in several types of British gas works retort houses. The tarry fumes that escaped from retorts contained extremely high concentrations of PAH, but workers were exposed to these only very briefly. The mean concentration of BaP, as determined from long-period samples at sites representative of normal working conditions in three works, was 3 μg/m^3, which is more than 100 times as high as the normal level in London. Above the retorts, in an old horizontal retort house, the concentration was over 200 μg/m^3. The topman could be exposed to this concentration in the normal course of his work. No similar working areas were found in the vertical retort houses. A maximum BaP concentration of 2330 μg/m^3 was reported.

Microscopic examination of the tarry droplets present in the air showed that they varied between 0.1 and 1 μ in diam and were therefore respirable. However, smoke aggregates of the type seen in domestic pollution were not observed. Some coal-dust particles of ≥1-μ diam were present. The general atmosphere of the retort house was alkaline.

10.1.8.5 Coal-hydrogenation workers

In 1952, Union Carbide Corporation began large-scale pilot-plant production of chemicals by coal hydrogenation (liquefaction). Prior to operation of the plant, the possible existence of health hazards from carcinogenic chemicals was realized. Thus, prior to operation, all personnel were given physical examinations, and from 1951 to 1954 annual examinations of the exposed workmen were conducted. Quarterly skin inspections began in early 1955. The company instituted an animal experiment program to test the carcinogenic potential of several streams and products of the hydrogenation process. Late in 1952 and 1953, the company toxicologist reported that some of the chemicals were capable of producing cancers in experimental animals. Following these studies, an industrial hygiene program, consisting primarily of education and cleanliness, was implemented. In an effort to reduce the exposure of employees to potentially cancer-producing oils, workers were encouraged to use protective clothing and to apply effective personal hygiene procedures. During 1955, the first suspected skin cancer was detected. Although this lesion was later thought not to be a cutaneous cancer, the precautionary program was more strictly enforced (Sexton 1960*b*).

Risk

Sexton (1960*b*) reports that regular examinations of the 359 men between 1954 and 1959 revealed 42 cases of lesions, which were diagnosed as precursors of skin cancers and treated. In addition, there were 11 probable and 5 verified cancers (epitheliomas).

Including previous work, these men had been exposed from several months to 23 years. Interestingly, all significant lesions were found in workmen who had had less than 10 years of exposure. Employees having more than 10 years of exposure had worked in small experimental or laboratory-sized operations prior to employment at the pilot plant; the extent of their total exposure was not known.

Although operation of the coal hydrogenation plant at Institute was phased out in the mid-1950s, Union Carbide recently initiated a followup program of locating and medically examining workers originally associated with coal hydrogenation activities at the plant (ERDA 1976). Medical examination of 250 workers showed that, with the exception of one worker who had a precancerous horn on the upper lip, there is no evidence of cancer among the workers examined. Similar followup examination is planned for the remainder of the workers.

Animal experiments

Several streams and products of the coal-hydrogenation process were painted on the skin of mice to test their carcinogenic effect (Weil and Condra 1960). Starting at 90 days of age, fur was shaved from the backs of the mice, and one brushful of each chemical was applied three times a week to the midline of the backs of male mice of the R.A.P. and C3H strains. Additional studies included (1) the application of either of two barrier creams to the backs of the mice before painting and (2) the removal of the pasting oil with soap or soapless cleansers. In addition to the controls, a positive-control carcinogen, 3-methylcholanthrene, was also used.

Figure 10.11 shows the tumor and cancer indices of samples of 15 process materials after one year of painting. All samples, with the exception of the phenolic pitch, were tested in undiluted form. The light-oil stream and eight fractions of this stream were all without tumorigenic action. The light- and heavy-oil products were mildly tumorigenic, producing papillomas predominantly. However, the streams boiling at higher temperatures, the middle oil, light oil stream residue, pasting oil, and pitch product were all highly carcinogenic. The degree of carcinogenicity increased and the length of the median latent periods decreased as boiling points rose. The median tumor- or cancer-latent period is the time necessary to reach a 50% tumor or cancer index. Tumor induction periods were only slightly delayed by dilution of the pasting oil, application of barrier creams, or applications of various washing methods.

Products from other hydrogenation processes have been shown to be carcinogenic also. Experimental studies were performed by Hueper (1956*a*, 1956*b*) on Bergius oils and Fischer-Tropsch oils obtained from the experimental coal hydrogenation-liquefaction operation of the U.S. Bureau of Mines at Bruceton, Pennsylvania. All fractions were tested for carcinogenicity by repeated application to the skin of mice and rabbits and by intramuscular injection into the thighs of rats. Eight of nine fractions of the Bergius oil fractionation products were carcinogenic with the degree of carcinogenic potency generally increasing with increasing boiling point. Fischer-Tropsch synthesis products were less carcinogenic than Bergius products and appeared to have a narrower species and tissue susceptibility spectrum than the latter.

ORNL DWG 76-11934

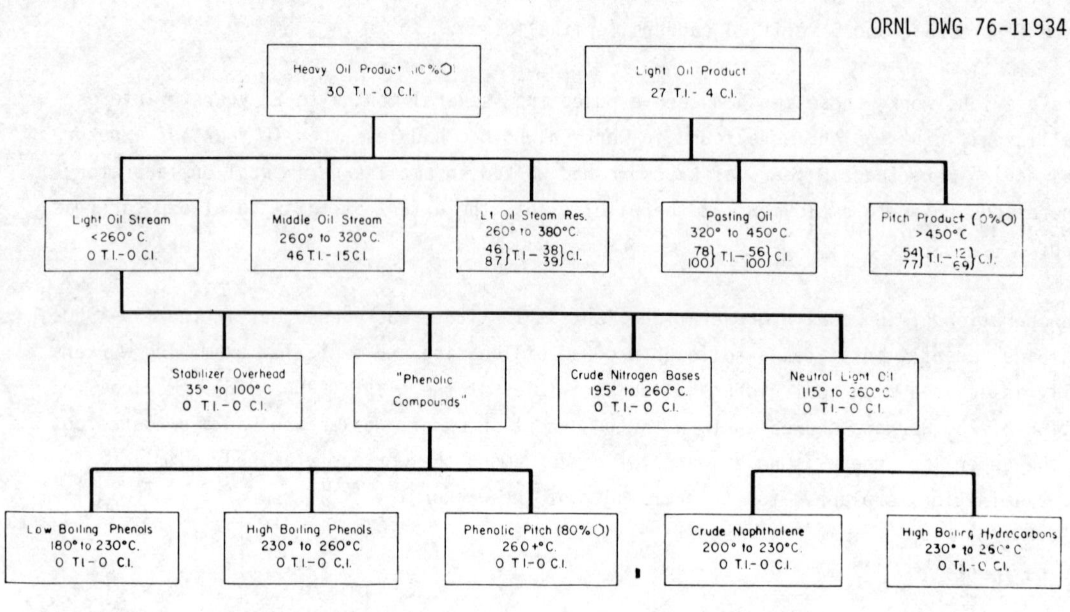

Fig. 10.11. Products, stream, and fractions of coal hydrogenation process tested (with boiling points and tumor and cancer indices indicated). The indices indicate the percentage of mice having skin tumors or cancers relative to those exposed mice that survived without tumors. Source: From Weil and Condra 1960, p. 23/189. Reprinted with permission from *Arch. Environ. Health.* Copyright 1960, American Medical Association.

Oil chemicals

The aromatic and aliphatic chemicals of the heavy and light products separation and the acidic, basic, and neutral fractions of the light products separation unit (boiling points up to 260°C) are shown in Tables 10.24 and 10.25 (Sexton 1960*a*). With further processing, over 200 individual chemicals were identified in the coal-hydrogenation process.

Air analyses

In spite of an industrial hygiene program, skin contamination with oil fractions suspected of being carcinogenic was still occurring; therefore, air-sampling studies were initiated to locate the sources and extent of fallout of airborne contaminants (Ketcham and Norton 1960). BaP was selected as a tracer material because it was known to be present in all of the middle- and heavy-oil fractions and because it was a known carcinogen. Air-sampling stations were set up around the plant and in the surrounding community. Table 10.26 shows the results of all the air samples. A wide range of BaP concentrations was found in the air samples — from 1 to almost 1900 µg per 100 m^3 of air. However, samples taken from adjacent residential areas showed concentrations comparable with those reported in the atmospheres of large cities.

Table 10.24. Aromatic and aliphatic chemicals obtained from
coal processing at the Institute, W. Va., hydro plant

Heavy products separation unit

Heavy oil stream	Middle oil stream
Anthracene	Dimethylnaphthalenes
Dihydroanthracene	Biphenyl
Phenanthrene	Acenaphthene
Pyrene	Fluorene
Chrysene	
Perylene	
Picene	

Light products separation unit — stabilizer overhead fraction

Pentane
Pentene
Cyclopentene
Hexane
Methylcyclopentane
Heptane
Heptene
Hexene
Benzene

Source: Sexton 1960*a*, Table 1, p. 184.
Reprinted by permission of the publisher.

A major source of contamination was the periodic blowing operation during pitch treatment and
solids removal. Because the materials involved in the removal of solids from hot-oil fractions
were viscous, it was necessary to blow down the equipment periodically with steam; the operation
was repeated every 15 min. Concentrations of BaP as high as 1870 µg per 100 m^3, during an
8-min sampling period, were measured during blowdown.

10.1.8.6 <u>Standards</u>

The American Conference of Governmental Industrial Hygienists (1975) has recommended a
threshold limit value for CTPV (benzene soluble fraction) of 0.2 mg/m^3 as a level that minimizes
exposure of workers to carcinogens present in coke-oven emissions. The same level has been
adopted as the Occupational Safety and Health Standard (1972) (37 CFR 202: 1910) for exposure
to coke-oven emissions.

About 10% of coal tar is composed of PAH, and at least 100 of the compounds in coal-tar pitch
are established or suspected experimental animal carcinogens (Schulte et al. 1974). BaP
constitutes about 1% of coal tar.

The National Institute for Occupational Safety and Health (NIOSH) does not recommend a safe
exposure level to coke-oven emissions due to the absence of reliable dose-response data (NIOSH
1973). Instead, it recommends more complete protection through a combination of respiratory
protection and operating procedures. NIOSH outlines five recommendations for coke-oven operating
procedures: (1) engineering controls to reduce or eliminate emissions or reduce workers' exposure,

Table 10.25. Compounds identified from three fractions
of the light products separation unit at the
Institute, W. Va., hydro plant

Acidic (phenolic)

Phenol	Indanols
Cresols	Beta-naphthol
Ethylphenols	Phenylphenols
Xylenols	Pyrocatechol
Propylphenols	Resorcinol
Methylethylphenol	Methylresorcinol

Basic (nitrogen bases)

Primary	Secondary	Tertiary
Aniline	Pyrrole	Pyridine
Toluidines	Indole	Methylpyridines
Ethylaniline	Methylindoles	Ethylpyridines
Dimethylanilines	Carbazole	Dimethylpyridines
Naphthylamines		Propylpyridine
		Methylethylpyridines
		Trimethylpyridines
		Quinoline
		Methylquinolines
		Isoquinoline
		Methylisoquinolines

Neutral

Aromatic	Aliphatic
Toluenes	Decane
Xylenes	Undecane
Ethylbenzenes	Dodecane
Propylbenzenes	Tridecane
Ethyltoluenes	Heptadecane
Trimethylbenzene	Ethylcyclopentane
Tetrahydronaphthalene	
Naphthalene	
Methylnaphthalenes	

Source: Sexton 1960*a*, Tables 2, 3, and 4, p. 185.
Reprinted by permission of the publisher.

(2) medical surveillance of all workers, (3) posted warnings of health risks in appropriate locations, (4) provision of respiratory protective devices to workers, and (5) employee education.

10.2 SULFUR OXIDES

Two major programs sponsored by the U.S. Public Health Service have provided a large body of data on atmospheric sulfur dioxide concentrations in urban areas: the Continuous Air Monitoring Project (CAMP), which monitors six large cities, and the National Air Surveillance Networks (NASN), which provides 24-hr sampling data for about 100 locations 26 times per year (NAPCA 1970). The six cities monitored by CAMP are Washington, Philadelphia, Cincinnati, Chicago, St. Louis, and Denver. Levels recorded in CAMP cities over the period 1962 to 1967 show mean annual concentrations ranging from 0.01 ppm in San Francisco to 0.18 ppm in Chicago. The NASN annual average concentrations ranged from 0.002 ppm in Kansas City, Missouri, to 0.17 ppm in New York City. Hourly average concentrations of sulfur dioxide as high as 2.9 ppm have been measured one-half mile from a coal-fired power plant, although, in general, concentrations decrease as the distance from sources increase and as the averaging time is extended (NAPCA 1970).

Table 10.26. Analyses of air samples for BaP

Sample location	Date	Description of sample[a]	Sample volume (ft^3 of air)	BaP (μg/100 m^3 of air)
1	6-29-56	Directly in fume above scraper pan of experimental north pitch treatment equipment (not a breathing zone sample).	60	240
2	6-29-56	75 ft NW from north pitch treatment equipment; gentle wind from east.	380	70
3	7-11-56	110 ft downwind from south pitch treatment equipment; in visible fumes about 30 min of total 60 min sampling time.	600	663
4	8- 1-56	10 ft downwind of paste pumps; air apparently free of fume; pumps operating.	8,384	10
4	8- 3-56	10 ft downwind of paste pumps; air apparently free of fume; pumps operating.	9,620	4
5	8- 2-56	4 ft west of Bldg. 120 at west edge of plant; gentle wind from west.	8,645	< 2
6	8- 2-56	500 ft downwind from south pitch treatment equipment; in visible fumes about 15 min of total 90-min sampling time.	1,125	25
7	8- 2-56	200 ft downwind from south pitch treatment equipment; in visible fumes about 10 min of total 60-min sampling time.	720	363
8	8- 6-56 8- 9-56	10 ft from Clock House (Bldg. FT-1) at SW corner of coal plant; wind from west.	26,105	3
9	8- 6-56	30 ft directly downwind from solids removal work equipment blown down 6 or 7 times during the 8-min sampling time.	136	1870
10	8- 7-56	60 ft east of Bldg. 111; wind variable from several directions during 2-day sampling time	22,600	2
11	8- 8-56 8-24-56	Inside shop (Bldg. 101); wind variable from several directions during 2-day sampling time.	20,490	3
11	8-20-56 8-31-56	Inside shop; wind generally from NE.	29,340	1
12	8-10-56	Near Bldg. 105, 180 ft NE of solids removal work equipment blown down about every 15 to 30 min during 400-min sampling time; wind from SW, strong and gusty.	4,340	7
13	8-13-56	In yard west of Bldg. 111, about 220 ft NE of experimental north pitch treatment equipment; wind mostly from SW; sampler in visible fume only occasionally when wind whipped to ground level.	8,720	8
13	9-13-56 9-14-56	In yard west of Bldg. 111; wind mostly from SW; coal plant shut down except distillations in Bldg. 116; new small pilot plant at Bldg. 119 operating; solids removal equipment being cleaned; paste pump vent line blown once; strong odor of natural gas noted.	25,011	8
14	8-15-56	Near scale house (Bldg. 100) at north side of coal plant; gentle wind from east; sampler in apparently clean air for entire time.	26,615	1
14	8-29-56 8-30-56	Near scale house in north side of coal plant; wind generally from SW; solids removal equipment not operating; experimental north pitch treatment work in progress most of the day.	18,670	3
15	8-16-56	On second-floor level of Bldg. 114 directly above the ground-level discharge from the solids removal equipment blowdown. During 45-min sampling time, visible fume from blowdown enveloped samplers twice.	607	1220
16	8-16-56 8-17-56	Near east fence of coal plant area; gentle wind from east during entire sampling time.	11,235	2
17	8-22-56	About 20 ft N of Bldg. 119; wind generally from NE and SE; solids removal equipment blown down about every 15 min during sampling period.	10,030	5
18	8-24-56	5 ft from east end of Bldg. 115 office; wind strong from W and NW; occasional wisps of fume from solids removal equipment blowdown whipped past the sampler.	5,360	24
18	9-18-56 9-19-56	5 ft from east end of Bldg. 115 office; coal plant not operating, except some distillations to Bldg. 116; wind variable from N, W, and S.	28,780	1

Table 10.26 (continued)

Sample location	Date	Description of sample[a]	Sample volume (ft³ of air)	BaP (μg/100) m³ of air
19	8-23-56	Immediately east of Bldg. 120; wind predominantly from the E and S; plant operation normal, including usual solids removal work; sampler may have picked up some road dust from truck traffic on nearby road	13,210	11
20	8-27-56	Between paste pumps, in visible fume from hot piston; fume mostly from lubricating oil; packing in good condition.	12,250	5
21	8-28-56	About 10 ft north of paste pump, in visible fume from hot piston; fume mostly from lubricating oil; packing in good condition.	4,750	10
22	8-28-56	At west end of pipe rack loop, across road from Bldg. 104, wind from SW; one solids separation equipment blowdown during the 350-min sampling time.	8,740	6
23	9-20-56 9-21-56	Near SW corner of Bldg. FT-2, at west end of coal plant; wind from W and S; coal plant not operating except for some distillations in Bldg. 116.	20,340	0.7
24	9- 8-56	1645 King Street, South Charleston, W. Va.; wind from west.	28,670	0.0
25	9-11-56 9-12-56	Near Fire Research Lab., NE corner of Institute Plant; wind from SW.	23,620	0.0
26	9-22-56	Residence at Cross Lanes, W. Va.; wind from W and SW.	22,040	0.0
27	9-26-56 9-27-56	In South Charleston Plant, between Bldg. 137 and back channel fence; wind from NE.	34,540	0.4
28	9-29-56	521 5th Avenue, St. Albans, W. Va.; gentle, almost non-directional wind.	24,070	0.0
29	10-13-56	In parking lot at corner of Washington and Summers Streets, Charleston, W. Va.; heavy automobile traffic; essentially no wind.	15,960	1.7

[a]All samples taken at breathing level (3-5 ft above ground or floor), unless otherwise stated; weather conditions were recorded for all samples, but described here only when particularly pertinent.

Source: Ketcham and Norton 1960, Table 4, pp. 33/199 and 34/200. Reprinted by permission of the publisher.

10.2.1 Chemical forms

Sulfur oxides in the air occur in three forms: sulfur dioxide, sulfuric acid, and inorganic sulfates. The major source of urban sulfur oxides is the combustion of fossil fuels, which forms sulfur oxides in the ratio of 40 to 80 parts sulfur dioxide to 1 part sulfur trioxide. This results in ambient urban sulfur dioxide concentrations of from a few hundredths of 1 ppm to a little over 1 ppm (Berry, Osgood, and St. John 1974). About 98% of the sulfur released to the air is in the form of sulfur dioxide, and most, if not all, of the remainder is sulfuric acid (Rall 1974).

Sulfur dioxide is a colorless gas having a pungent irritating odor, which is detectable at concentrations of about 0.5 to 1 ppm[*] (Waldbott 1973; NAPCA 1970). Sulfur dioxide is slowly oxidized by sunlight to sulfur trioxide, which combines immediately with water vapor to form sulfuric acid. Sulfuric acid and other sulfates typically account for about 5 to 29% of the total suspended particulate matter in urban air (NAPCA 1970). Studies of the particle-size distribution of suspended atmospheric sulfate showed that 80% or more of urban atmospheric sulfate is associated with particles less than 2 μ in diameter and, thus, is largely in the respirable fraction of particulate matter.

---

[*]SO_2 conversion: 1 ppm = 2860 μg/m³ (NAPCA 1970).

10.2.2 Effects

Sulfur dioxide is classified as a mild respiratory irritant. Because of its high solubility in the aqueous lining of the respiratory tract, more than 90% is absorbed in the airways above the larynx before the air reaches the bronchi or lungs. Sulfur dioxide inhalation causes bronchial narrowing, as indicated by increased airflow resistance in both experimental animals and man. This effect occurs in a matter of minutes and is readily reversible. Most people show irritation and changes in respiratory flow resistance at sulfur dioxide concentrations of 5 ppm (15,000 $\mu g/m^3$). People who are especially sensitive react to concentrations in the 1- to 2-ppm range (Rall 1974), which produces detectable changes in pulmonary function (NAPCA 1970).

Concentrations above 10 ppm produce eye, nose, and throat irritation, especially in asthma patients (Berry, Osgood, and St. John 1974). The irritation causes reflex bronchoconstriction; slowing of mucus flow; depression of clearance of inert particles from the lungs; and, in high concentrations, narrowing of airways due to inflammatory swelling of membranes (Rall 1974). However, studies on laboratory animals and human volunteers have shown that inhalation of sulfur dioxide in the absence of particulates, at concentrations 10 to 100 times as great as that commonly found in ambient air, does not affect lung function (Rall 1974).

Sulfur dioxide penetrates into the lungs only when it is absorbed onto the surface of particulate matter (5-μ-diam size or less) or converted to the sulfate form in an aerosol, such as fog. A synergistic effect is produced due to the transformation of sulfur dioxide into a variety of products, including sulfuric acid and sulfate salts, which are more highly irritant than sulfur dioxide (Rall 1974). This synergism has been demonstrated in animal studies, but there are no human exposure data on the combined effects of sulfur oxides and other commonly occurring pollutants (Rall 1974).

10.2.2.1 Enhancement

Using the guinea pig, Amdur (1957) studied the effect of simultaneous exposure to sulfur dioxide and irritant gas. Aerosols of soluble salts of ferrous iron, manganese, and vanadium produced a potentiating effect, although the concentrations used (0.7 to 1.0 mg/m^3) were considerably greater than any levels of metals reported in urban air (Amdur and Corn 1963; Amdur and Underhill 1968).

Frank, Amdur, and Whittenberger (1964) examined the response of human subjects to levels of sulfur dioxide of about 1, 5, and 15 ppm, with and without the addition of sodium chloride aerosol. No constant potentiation response was demonstrated.

In animal studies, three- to fourfold potentiation of the irritant response to sulfur dioxide was observed in the presence of particulate matter capable of oxidizing sulfur dioxide to sulfuric acid (NAPCA 1970). The degree of potentiation was related to the concentration of particulate matter.

Sulfuric acid, in combination with particulate sulfates, also produces bronchoconstriction in guinea pigs. Response is highly dependent on particle size, and the smallest particles show the greatest irritant potency. In both guinea pigs and man, sulfuric acid and irritant particles have a greater irritant potency than sulfur dioxide alone (NAPCA 1970).

Low concentrations of ozone and sulfur dioxide in combination had a greater effect on pulmonary function in humans than either gas breathed separately (Hazucha and Bates 1975). Sulfur dioxide, when present in the atmosphere at 0.37 ppm, had no significant effect on ventilatory function in 2 hr; ozone at 0.37 ppm had a barely significant effect. Combined sulfur dioxide and ozone at 0.37 ppm had an enhanced effect on pulmonary function. Midexpiratory flow rate, in intermittently exercising nonsmokers, decreased to between 60 and 70% of that of controls.

10.2.2.2 Mutagenesis

Mutagenic changes in viruses and bacteria have been attributed to the bisulfite ion, which is one of the reaction products of sulfur dioxide in water (Rall 1974).

10.2.3 Occupational exposure

A review of studies of workers occupationally exposed to high levels of sulfur dioxide showed no significant effect on health or pulmonary function at exposures up to 2 ppm for the working hours of the day over a period of several years (NAPCA 1970). No data on suspended particulates, such as are found in ambient air, were given.

10.2.4 Epidemiology studies

Epidemiologic studies of the relationship between sulfur oxide concentrations and health clearly indicate an association between health effects of varying severity. Episodes of acute elevation of sulfur oxides and other pollutants have been associated with increased mortality and increased morbidity. Those predominantly affected were individuals suffering from chronic pulmonary disease or cardiac disorders or very young or old individuals, although the general population has also been involved. An association between long-term community exposures to air pollution and respiratory disease incidence and prevalence rates is more difficult to show. However, in the absence of other explanations, the findings of increased death rates for selected causes, independent of economic status, must be considered consequential (NAPCA 1970).

Epidemiologic studies of the U.S. Environmental Protection Agency's Community Health and Environmental Surveillance System (CHESS) provided dose-response information relating short- and long-term pollutant exposures to adverse health effects in four U.S. regions (EPA 1974). Health indicators of long-term sulfur oxide effects included increased prevalence of chronic bronchitis in adults, increased acute lower respiratory infections in children, increased acute respiratory illness in families, and subtle decreases in ventilatory function of children. Health indicators for short-term effects were aggravation of cardiopulmonary symptoms and of asthma.

Threshold estimates were developed for the effects of the following pollutants: sulfur dioxide, suspended particulates, and suspended sulfates. Table 10.27 summarizes worst-case, least-case, and best-judgment estimates of the pollution levels that can be associated with adverse effects for long-term exposures. A worst-case estimate attributes an observed adverse health effect to the lowest pollution exposure suggested by the epidemiologic studies after considering only the strongest and most established covariates; and a least-case estimate attributes an observed adverse health effect to the highest pollution exposure level suggested by the epidemiologic studies after considering effects of all covariates. A best-judgment estimate is based on a synthesis of several studies; it considers interactions between pollutants and is, at times, based on special analyses that were necessary when individual studies raised questions regarding

Table 10.27. Summary of CHESS studies relating long-term pollutant exposures
to adverse effects on human health

Adverse effect	Type of estimate	Duration of exposure (years)	Annual average levels linked to adverse health effects ($\mu g/m^3$)		
			Sulfur dioxide (80)[a]	Total suspended particulates (75)[a]	Suspended sulfates (no standard)[a]
Increase in prevalence of chronic bronchitis in adults	Worst case	3	62	65	12
	Least case	10	374	179	20
	Best judgment	6	95	100	15
Increases in acute lower respiratory tract infections in children	Worst case	3	92	65	7.2
	Least case	3	177	102	15
	Best judgment	3	95	102	15
Increase in frequency or severity of acute respiratory illness in families	Worst case	1	50	104	14
	Least case	3	210	159	16
	Best judgment	3	106	151	15
Subtle decreases in childhood ventilatory function	Worst case	1	57	96	9
	Least case	9	435	200	28
	Best judgment	8-9	200	100	13

[a]National Primary Ambient Air Quality Standard; the particulate standard is a geometric mean; the equivalent arithmetic mean would be about 85 $\mu g/m^3$.

Source: EPA 1974, Table 7.1.11, p. 7-16.

interactions involving pollutants or intervening variables. In terms of the health indicators studied, these data support the existing National Primary Ambient Air Quality Standards of 80 $\mu g/m^3$ (0.03 ppm) annual average (arithmetic) for sulfur dioxide and 75 $\mu g/m^3$ annual mean (geometric) for total suspended particulates. A national standard for suspended sulfates has not been established.

Excess bronchitis in adults was associated with six-year community exposures to sulfur oxides alone, at annual levels of 95 $\mu g/m^3$ sulfur dioxide, 100 $\mu g/m^3$ total suspended particulates, and 15 $\mu g/m^3$ suspended sulfates. The studies also showed that the effects of air pollution and self-pollution, by smoking cigarettes, appeared to be additive. Personal cigarette smoking was the largest determinant of bronchitis among parents of school children, but air pollution was a significant and consistent contributing factor, leading to increased bronchitis rates in non-smokers as well as smokers from polluted communities. The relative contribution of air pollution alone ranged from one third to one seventh as strong as that of cigarette smoking as a determinant of the prevalence of chronic bronchitis in communities, with the exception of New York City. In each of the four CHESS studies, a consistent pattern of excess chronic bronchitis was found among residents of more polluted communities.

It was concluded that there is a positive association between the frequency of lower respiratory disease in children and exposure to pollution. Excess respiratory disease may reasonably be associated with community exposures for three years to about 95 $\mu g/m^3$ sulfur dioxide and 15 $\mu g/m^3$ suspended sulfates. There is no evidence that elevated levels of total particulate matter were required to produce the adverse effect.

Findings for total acute respiratory disease (combined upper and lower tract disease) among family members in the Chicago and New York City studies showed that best-judgment threshold values are 106 $\mu g/m^3$ for sulfur dioxide, 151 $\mu g/m^3$ for total suspended particulates, and 15 $\mu g/m^3$ for suspended sulfates. Parental smoking was a significant factor in the development of lower tract illness among nonsmoking young members of their households.

Ventilatory function of elementary school children, measured by the 0.75-sec forced-expiratory volume (FEV 0.75), was diminished in areas of elevated exposure to sulfur oxides. In all cases, observed decrements were subtle. It was the authors' best judgment that 8 to 9 years of exposure to about 10 to 13 $\mu g/m^3$ of suspended sulfates might reduce ventilatory function. If these exposures to suspended sulfates were accompanied by exposures to about 200 to 250 $\mu g/m^3$ sulfur dioxide and about 100 to 150 $\mu g/m^3$ total suspended particulates, further reductions in FEV 0.75 might be expected.

Least-case, worst-case, and best-judgment estimates for short-term exposures are given in Table 10.28. The data indicate that adverse effects on elderly subjects having heart and lung disease and on asthmatics are being experienced on days when the 24-hr levels of sulfur dioxide and total suspended particulate concentrations exceed the levels prescribed in the National Primary Standards. These adverse health effects, however, appeared to be associated with suspended sulfate levels rather than with sulfur dioxide and total suspended particulates, as evidenced by the consistency of the relationship between symptom aggravation and sulfate levels and by the lack of consistency of this relationship with other pollutants. It was the best judgment of the investigators that significant aggravation of cardiopulmonary symptoms could be attributed to 24-hr suspended sulfate levels as low as 8 to 10 $\mu g/m^3$ on both cool and warm days.

Table 10.28. Summary of threshold estimates for adverse effects
of short-term exposures

Adverse effect	Type of estimate	Minimum temperature (°F)	Daily average levels linked to adverse health effects[a] (μg/m³)		
			Sulfur dioxide (365)[b]	Total suspended particulate (260)[b]	Suspended sulfate (no standard)[b]
Aggravation of cardio-pulmonary symptoms in elderly					
"Well" panel	Worst case	20 - 40	81 - 365	NPE	<1
	Least case		NE	NE	NE
	Best judgment		NPE	NPE	8 - 10
	Worst case	>40	81 - 365	68	2
	Least case		NE	NE	10
	Best judgment		NPE	80 - 100	8 - 10
"Heart" panel	Worst case	>40	NPE	76 - 260	10
	Least case		NE	NE	10 - 20
	Best judgment		NPE	NPE	10
"Lung" panel	Worst case	20 - 40	NPE	76 - 260	6
	Least case		NE	NE	NE
	Best judgment		NPE	NPE	10
	Worst case	>40	NPE	76 - 260	11
	Least case		NE	NE	12
	Best judgment		NPE	NPE	12
"Heart and lung" panel	Worst case	20 - 40	181	47	9
	Least case		NE	NE	10
	Best judgment		NPE	80 - 100	10
	Worst case	>40	NPE	76	6
	Least case		NE	NE	17
	Best judgment		NPE	NPE	10
Aggravation of asthma	Worst case	30 - 50	NPE	61 - 75	8
	Least case		NE	NE	NE
	Best judgment		NPE	105	9 - 10
	Worst case	>50	23	61 - 75	<1
	Least case		NE	NE	10
	Best judgment		180 - 250[c]	70	8

[a]NE — no effect below Primary Standard, or simply no effect for suspended sulfates, for which no
Primary Standard has been established. NPE — no proven effect below Primary Standard, or simply
no proven effect for suspended sulfates.
[b]National Primary Air Quality Standard.
[c]This judgment is based on presently summarized studies and on a previously reported CHESS study
of asthma in New Cumberland, West Virginia.

Source: EPA 1974, Table 7.1.14, p. 7-21.

10.2.4.1 <u>Contribution of smoke, fog, and particulates</u>

Epidemiologic evidence of the toxic effects of acute episodes of high sulfur dioxide and smoke
or fog pollution are reviewed in several sources (Rall 1974; NAPCA 1970). These episodes have
clearly resulted in increased mortality and morbidity. Table 10.29 indicates the levels of
particulates and sulfur dioxide that may cause effects. High pollutant concentrations are often
accompanied by other environmental features, such as low temperatures, which make it difficult
to determine the extent to which pollution alone is responsible for the effects.

Table 10.29. Summary of dose-response relationships for effects
of particles and SO_2 and health

Averaging time for pollution measurements	Place	Approximate levels of pollution ($\mu g/m^3$)		Effect
		Particles	SO_2	
24 hr	London	2000	1144	Mortality
24 hr	London	750	700	Mortality
24 hr	London	300	600	Deterioration of patients
Weekly mean	London	200	400	Prevalence or incidence of respiratory illnesses
24 hr	New York	6[a]	1500	Mortality
Winter mean	Britain	100 - 200	100 - 200	Incapacity for work from bronchitis
Annual	Britain	70	90	Lower respiratory infections in children
	Britain	100	100	Upper and lower respiratory infections in children
	Britain	100	100	Brochities prevalence
	Britain	100	100	Prevalence of symptoms
	Buffalo	100	300[b]	Respiratory mortality
	Berlin, N.H.	180	731[b]	Increased respiratory symptoms Decreased pulmonary function

[a]In coefficient of haze units.
[b]As μg $SO_3/100$ cm^2/day.

Source: Rall 1974, Table 5, p. 117. Reprinted by permission of the publisher.

A study by Martin (1964) relates daily measurements of smoke and sulfur dioxide to daily deaths and illnesses in London since 1958. Mortality and morbidity for all causes and for certain respiratory diseases were correlated with both smoke and sulfur dioxide levels until 1962. Because of smoke control efforts, there has been little evidence of any effect of pollution on mortality or morbidity since 1962-1963. Smoke declined from an average annual level of about 300 $\mu g/m^3$ to less than 50 $\mu g/m^3$, and sulfur dioxide declined from a slightly higher concentration to about 200 $\mu g/m^3$.

Early acute episodes of sulfur oxide air pollution occurred in the Meuse Valley, Belgium, in 1930; in Donora, Pennsylvania, in 1948; and in London in 1952 (NAPCA 1970). Recorded deaths during these three episodes were 63, 20, and 4000 respectively. In addition, a large percentage of the populations became ill, showing symptoms of respiratory ailments. In the London incident, the maximum daily concentration of sulfur dioxide was 1.34 ppm (about 4000 $\mu g/m^3$), which appeared on the third and fourth days of the fog. Corresponding smoke levels were 4.46 mg/m^3 (about 4500 $\mu g/m^3$). (British measurements of sulfur dioxide concentrations are obtained by the hydrogen peroxide titrimetric method and may be higher than U.S. values due to the presence of other acidic gases). No fatalities that could not be explained by previous respiratory or cardiovascular lesions were found. Elderly persons and persons with preexisting pulmonary and cardiac disease were most susceptible.

The 1952 London air pollution incident and other selected episodes of acute air pollution in greater London were summarized by Brasser, Joostings, and von Zuilen (1967). The summary (Table 10.30) gives a detailed analysis of each of the episodes and points out the importance of the duration of the maximum values.

Table 10.30. Survey of selected acute air pollution
episodes in greater London

	Dec. 1952	Jan. 1956	Dec. 1962	Dec. 1957	Dec. 1956	Jan. 1955	Jan. 1959
Duration of the cumulation period in days	5	5	5	5	10	11	5
Number of days with maximum pollution	2	2	1	1	5	1 x 3[b]	1
SO_2 level preceding episode,[a] $\mu g/m^3$	500	300	400	300	300	300	300
SO_2 maximum,[a] $\mu g/m^3$	4000	1500	3300	1600	1100	1200	800
SO_2 increase per day,[a] $\mu g/m^3$	1200	500	1000	325	400	450	250
Soot level preceding episode,[a] $\mu g/m^3$	400	500	200	400	400	500	400
Soot maximum,[a] $\mu g/m^3$	4000	3250	2000	2300	1200	1750	1200
Soot increase per day,[a] $\mu g/m^3$	1200	1300	600	500	400	600	400
Number of excess deaths	3900	1000	850	800	400	240	200
Number of days with excess mortality	18	10	13	10	6	6	6
Daily mortality expected under normal circumstances	300	330	310	300	270	320	325
Average caily mortality in the period (excess mortality as a percent of normal)	170	130	120	125	125	112	110

[a]The SO_2 and soot concentrations mentioned are average values over 24 hrs.
[b]Maximum pollution values of one day's duration occurred three times.

Source: NAPCA 1970, Table 9-1, p. 125.

A major British study found an association between mortality from bronchitis and lung cancer and levels of air pollution after considering differences in age, smoking habits, social class, and occupational exposure. Average sulfur dioxide values (averaged over a year) were 116 $\mu g/m^3$ (0.040 ppm) for the polluted area and 75 $\mu g/m^3$ (0.026 ppm) for the cleaner area. Corresponding smoke values were 160 and 80 $\mu g/m^3$ respectively. The effects of smoke and sulfur dioxide could not be separated (NAPCA 1970).

Long-term exposure studies of children and housewives showed (1) increased morbidity at an annual sulfur dioxide mean of 105 $\mu g/m^3$ (0.037 ppm) in Genoa, Italy; (2) increased respiratory symptoms at a long-term sulfur dioxide value of 95 $\mu g/m^3$ (0.034 ppm), as compared with 25 $\mu g/m^3$ (0.009 ppm); and (3) increased respiratory disease frequency and severity in British school-children at long-term pollution levels $\geqslant$120 $\mu g/m^3$ (0.046 ppm) sulfur dioxide and 100 $\mu g/m^3$ smoke (NAPCA 1970).

Studies of pollution episodes in London suggest that a rise in the daily rate occurred when the concentrations of sulfur dioxide rose abruptly to about 715 $\mu g/m^3$ (0.25 ppm). A more distinct rise in deaths was noted when sulfur dioxide exceeded 1000 $\mu g/m^3$ (0.35 ppm) for one day. Daily concentrations of sulfur dioxide in excess of 1500 $\mu g/m^3$ (0.52 ppm) for one day, in conjunction with levels of suspended particles exceeding 2000 $\mu g/m^3$, appeared to have been associated with an increase in the death rate of 20% or more over base levels. At lower sulfur dioxide levels, the same effect was noted when the maximum pollution levels lasted longer.

In New York City, excess deaths were detected after a 24-hr period during which sulfur dioxide concentrations exceeded 1500 $\mu g/m^3$ (about 0.5 ppm) and suspended particulate matter was measured as a soiling index of 6 COH (coefficient of haze units) or greater. In Rotterdam, there appeared to be a positive association between total mortality and exposure, for a few days, to 24-hr mean concentrations of 500 $\mu g/m^3$ (0.19 ppm) sulfur dioxide (NAPCA 1970).

A survey of emergency clinics at four major New York City hospitals revealed a rise in visits for upper respiratory infections and cardiac disease in both children and adults during a ten-day period during which elevated sulfur dioxide levels ranged between 200 and 2460 $\mu g/m^3$ (0.07 and 0.86 ppm). In London and Chicago a one-day exposure to 600 and 715 $\mu g/m^3$ (0.20 and 0.25 ppm), respectively, caused a sharp rise in illness rates of persons having chronic respiratory disease on the following day. Particulate matter was also present (NAPCA 1970).

Effects have also been noted at lower levels of sulfur dioxide. Hospital admissions rose in Rotterdam, particularly for older people, when sulfur dioxide concentrations rose from 300 to 500 $\mu g/m^3$ (0.11 to 0.19 ppm) over a period of a few days. Absenteeism from work, particularly for those 45 years old and older, also increased. In the Ruhr area an effect on functional disturbance was noted at an estimated daily indoor mean of 270 $\mu g/m^3$ (0.10 ppm) for four days (NAPCA 1970).

10.2.5 <u>Levels vs risk</u>

The National Air Pollution Control Administration listed the following conclusions (NAPCA 1970).
> At concentrations of about 1500 $\mu g/m^3$ (0.52 ppm) of sulfur dioxide (24-hr average), and suspended particulate matter measured as a soiling index of 6 COH or greater, increased mortality may occur....At concentrations of about 715 $\mu g/m^3$ (0.25 ppm) of sulfur dioxide and higher (24-hr mean), accompanied by smoke at a concentration

of 750 µg/m^3, increased daily death rate may occur....At concentrations of about
500 µg/m^3 (0.19 ppm) of sulfur dioxide (24-hr mean), with low particulate levels,
increased mortality rates may occur....At concentrations ranging from 300 to
500 µg/m^3 (0.11 to 0.19 ppm) of sulfur dioxide (24-hr mean), with low particulate
levels, increased hospital admissions of older persons for respiratory disease may
occur; absenteeism from work, particularly with older persons, may also occur....
At concentrations of about 715 µg/m^3 (0.25 ppm) of sulfur dioxide (24-hr mean),
accompanied by particulate matter, a sharp rise in illness rates for patients over
age 54 with severe bronchitis may occur....At concentrations of about 600 µg/m^3
(about 0.21 ppm) of sulfur dioxide (24-hr mean), with smoke concentrations of about
300 µg/m^3, patients with chronic lung disease may experience accentuation of
symptoms.....At concentrations ranging from 105 to 265 µg/m^3 (0.037 to 0.092 ppm)
of sulfur dioxide (annual mean), accompanied by smoke concentrations of about
185 µg/m^3, increased frequency of respiratory symptoms and lung disease may occur....
At concentrations of about 120 µg/m^3 (0.046 ppm) of sulfur dioxide (annual mean),
accompanied by smoke concentrations of about 100 µg/m^3, increased frequency and
severity of respiratory diseases in school children may occur....At concentrations
of about 115 µg/m^3 (0.040 ppm) of sulfur dioxide (annual mean), accompanied by smoke
concentrations of about 160 µg/m^3, increases in mortality from bronchitis and from
lung cancer may occur.

10.3 NITROGEN OXIDES

The principle man-made sources of nitrogen oxides are coal, oil, natural gas, and motor vehicle
fuel combustion. In the United States, these sources produced 20.6 million tons in 1968
(Waldbott 1973). However, on a global basis the major proportion of nitrogen oxides is nitric
oxide produced by bacterial action.

10.3.1 Formation and concentrations

When nitrogen in the air is subjected to very high temperatures, as during coal conversion,
it may react with oxygen to produce nitric oxide, a gas of little significance as an air pollutant. If the gas cools slowly, nitrogen and oxygen are again formed; however, if nitric oxide
is vented into the environment and cooled rapidly, a fraction is converted to nitrogen dioxide.
Further complex reactions, in the presence of hydrocarbons and sunshine, result in formation of
ozone, nitrogen dioxide, and other oxidants commonly called photochemical smog (Waldbott 1973).

Nitrogen dioxide is the only toxic oxide of seven known nitrogen oxides. Nonurban regions
average 8 µg/m^3 (4 ppb) nitrogen dioxide and 2 µg/m^3 (2 ppb) nitric oxide (EPA 1971), whereas in
urban areas, concentrations of nitrogen oxides are 10 to 100 times higher (Waldbott 1973). Daytime urban levels depend on motor traffic and sunlight, and peak concentrations rarely exceed
0.94 mg/m^3 (0.5 ppm).

10.3.2 Effects

Nitrogen dioxide is only slightly soluble in water. In the tracheobronchial tree and lungs,
it forms nitrous acid and nitric acid. Both are very irritating and corrosive to the mucous
lining of the lungs. At high concentrations, the gas can produce effects ranging from eye
irritation and pulmonary congestion and edema to death. Animal studies show that loss of
cilia lining, changes in bronchial epithelial cells, lung changes similar to those of human
emphysema, and increased susceptibility to infection also take place (Waldbott 1973). Research
on the biochemical effects of nitrogen dioxide showed that exposure creates conditions that
stimulate anaerobic oxidation and inhibit aerobic oxidation; exposure may also alter lung
proteins. Along with structural changes in collagen and elastin, there is a disruption of cells
in the lung tissue with a concomitant release of histamine (Berry, Osgood, and St. John 1974).

The odor threshold for pure nitrogen dioxide is 0.1 to 3 ppm; this figure differs among different sources. Eye and nasal irritation become apparent at concentrations of about 13 ppm, and accidental exposure to concentrations of 150 to 200 ppm can be fatal (Berry, Osgood, and St. John 1974).

10.3.3 Acute toxicity

Acute nitrogen dioxide poisoning in humans has only been recognized during the last three decades (in farmers filling silos). Nitrogen dioxide is produced during the rapid decomposition of plant materials, as happens in silos. Death may occur up to one month after silo gas and industrial poisoning. Pathologic findings show that the effects on the lungs are typical of bronchiolitis fibrosa cystica, which includes hemorrhage; fibrous stroma replacing the terminal bronchi, alveolar ducts, and sacs; hyaline membrane formation; and hyalinization of the basement membrane (Dreisbach 1974). Levels of nitrogen dioxide during these accidental poisonings are unknown.

10.3.4 Epidemiology studies

Information on chronic nitrogen dioxide poisoning is extremely scarce because (1) no response is observed until a critical concentration is reached, (2) damage develops slowly, and (3) nitrogen dioxide is usually associated with other pollutants. According to the intensity and duration of exposure, respiratory illness ranges from slight irritation to burning and pain in the throat and chest to violent coughing and shortness of breath (Waldbott 1973).

The effects of community exposure to nitrogen dioxide were studied in four residential areas in Chattanooga, Tennessee. These studies linked high nitrogen dioxide exposure to an increase in childhood respiratory illness and to a decreased ventilatory performance of second-grade school children (Shy et al. 1970a; 1970b; Pearlman et al. 1971). Although interpretation of some of the data is questionable, the studies are reported here as they appeared.

An elaborate epidemiologic study was conducted by the U.S. Public Health Service in Chattanooga, Tennessee, on elementary school children in four areas of the city (Shy et al. 1970a). One area, in close proximity to a large TNT plant, had high nitrogen dioxide levels; another had high suspended particulate exposure; the remaining two areas served as controls. The mean range of daily nitrogen dioxide concentrations, measured over a six-month period, was between 0.062 and 0.019 ppm, or between 117 and 205 $\mu g/m^3$, in areas of high nitrogen dioxide concentrations (Table 10.31). The mean suspended nitrate level was 3.8 $\mu g/m^3$ or greater. The two control areas had nitrogen dioxide levels of 0.63 and 0.43 ppm and suspended nitrate levels of 2.6 and 1.6 $\mu g/m^3$. The ventilatory performance of second-grade school children in the area of high nitrogen dioxide exposure was significantly lower than the performance of children in the control areas. The data suggest that ventilatory performance was adversely affected only when a nitrogen dioxide threshold was exceeded, but above that threshold, no further impairment of performance could be detected.

In the second part of the study (Shy et al. 1970b), a similar temporal pattern of respiratory illness was observed in the four areas during the 24-week study period. A consistent excess of respiratory illness was reported by families residing in the two exposed areas, particularly during the A_2/Hong Kong influenza epidemic and the period between the A_2 and influenza-B outbreaks (Table 10.31). The relative excess in rates was 18.8% in the area of high nitrogen

Table 10.31. Average biweekly respiratory illness rates per 100 for each family segment according to exposure to oxides of nitrogen

| Rank of population by NO_2 exposure | | | Family segment | | | | |
Average 24-hr NO_2 (ppm)	Average 24-hr suspended nitrate ($\mu g/m^3$)	Study population	All family members	Second-graders	Siblings	Mothers	Fathers
0.109	7.2	School 1 (high-NO_2)	17.7	23.4	19.9	15.3	11.0
0.078	6.3	School 2 (high-NO_2)	17.5	23.4	18.0	14.4	12.8
0.062	3.8	School 3 (high-NO_2)	16.3	20.4	19.1	13.4	12.1
0.063	2.6	Control 1	13.9	18.0	15.6	11.8	8.8
0.043	1.6	Control 2	15.0	20.1	17.0	12.3	9.6

Source: Shy et al., 1970[b] Table 10-4, p. 589. Reprinted by permission of the publisher.

dioxide exposure and 10.4% in the area of high particulate exposure. Differences in illness rates could not be explained by differences in family composition, economic level, or prevalence of chronic conditions. Parental smoking habits did not appear to influence respiratory illness rates of second-grade children. However, concentrations of suspended nitrates and total suspended particulates were also increased in the high-exposure, as compared with the medium- and low-exposure, communities. Suspended sulfates did not differ between communities; however, concentrations of other possible contaminants such as sulfur dioxide were not reported.

In a retrospective study of three of the same Chattanooga areas (all except the high-particulate area), Pearlman et al. (1971) found that exposure to intermediate (one of the control areas) and high levels of nitrogen dioxide in ambient air was associated with a significant increase in the frequency of acute lower respiratory illness among infants exposed for three years and school children exposed for two and three years (Table 10.32). This increased frequency of acute bronchitis was observed when the mean 24-hr nitrogen dioxide concentration, measured over a six-month period, was between 0.063 and 0.083 ppm (118 and 156 $\mu g/m^3$) and the mean suspended nitrate level was 2.6 $\mu g/m^3$ or greater. The Pearlman study confirmed Shy's observation that excess acute respiratory illness can be found among children living in areas polluted with nitrogen dioxide. Pearlman et al. (1971) suggest that acute lower-respiratory illness, rather than acute general respiratory illness, may be more sensitive to pollutant effect.

Other studies suggest that high sulfur dioxide and nitrogen oxide concentrations may lead to a relative increase in resistance of blood cells to hemolysis, increase in the number of immature red blood cells, and an increase in methemoglobin in the blood of children in such areas. Occupational exposures to nitrogen oxides may produce pulmonary fibrosis and some methemoglo-binemias (EPA 1971).

Table 10.32. Distribution of children reporting one or more
episodes of bronchitis by length of exposure

6-month mean NO_2	School children exposure (year)			Infant exposure (year)		
	1	2	3	1	2	3
High-NO_2, 156 $\mu g/m^3$ (0.083 ppm)	20.9	34.7[a]	32.2[a]	33.3	37.5	46.8[a]
Intermediate-NO_2, 118 $\mu g/m^3$ (0.063 ppm)	31.6	45.5[a]	31.2[a]	26.2	29.5	50.5[a]
Low-NO_2 (control), 81 $\mu g/m^3$ (0.043 ppm)	25.1	20.3	23.2	21.1	34.0	36.3

[a]Differs significantly from low-NO_2 (control) area.

Source: EPA 1971, Table 10-6, p. 10-6. Reprinted by permission of the publisher.

The National Air Surveillance Network (NASN) measured nitrogen dioxide levels for the years 1967,
1968, and 1969 and found that yearly nitrogen dioxide averages reflect variations according to
population densities. Since nitrogen dioxide does not exhibit marked seasonal variations, there
was a direct comparison of the NASN yearly averages with the lower limit at which health effects
were noted in the Chattanooga studies (EPA 1971). A concentration of 0.06 ppm (113 $\mu g/m^3$), or
greater, exceeds the Chattanooga health-effect-related nitrogen dioxide values. Ten percent
of the cities in the United States having populations less than 50,000 and 54% of those having
populations between 50,000 and 500,000 have a yearly average nitrogen dioxide concentration
equal to or exceeding 0.06 ppm; 85% of the >500,000 population class exceed this yearly average.

10.3.5 Standards

The National Institute for Occupational Safety and Health (NIOSH 1976) recommends that
"occupational exposure to nitrogen dioxide shall be controlled so that workers are not exposed
to nitrogen dioxide at greater than a ceiling concentration of 1 ppm by volume (1.8 mg/m^3) as
determined by a sampling time of 15 minutes. Occupational exposure to nitric oxide shall be
controlled so that workers are not exposed to nitric oxide at a concentration greater than
25 ppm of air (30 mg/m^3) determined as a time weighed average (TWA) exposure for up to a 10-
hour workday, 40-hour workweek." The American Conference of Government Industrial Hygienists
(1975) recommends threshold limit values of 25 ppm or 30 mg/m^3 for nitric oxide and 5 ppm or
9 mg/m^3 for nitrogen dioxide.

The current Ambient Air Quality Standard for NO_x is 0.05 ppm or 100 $\mu g/m^3$ expressed as the
yearly arithmetic mean concentration. On the basis of basic toxicological concepts and the
most appropriate established procedures for setting safety standards for general populations,
Morrow (1975) does not consider this figure conservative.

10.4 TRACE ELEMENTS

Although there is considerable knowledge about the composition of coal, the amounts and chemical
form of trace elements that may escape as effluents or trace emissions from coal processing are
not well known. Some of the more important trace elements found in coal are lead, chromium,
manganese, arsenic, fluoride, nickel, cadmium, zinc, and copper. From the aspect of human health,

trace elements of particular importance from fossil fuel combustion are considered to be lead, mercury, arsenic, vanadium, selenium, nickel, some heavy radionuclides, and possibly cadmium and fluoride (Comar 1975). This selection was based on probable emission rates, likelihood of uptake from air, accumulation in food and water, and the known toxicity of some of the chemical forms.

10.4.1 Effects

It is difficult to determine a relationship between low-level excesses of trace elements and health effects. The emission of trace elements from the combustion of fossil fuels at stationary plants has not been shown to have health significance for the general population (Comar 1975). However, some levels appear to be rising, such as those of cadmium, zinc, lead, and chromium (NAS 1974). Baseline data on the natural and man-made occurrence of trace substances in the environment and their possible effects on human health are accumulating.

There is evidence that, at high doses, many trace elements produce toxic and/or carcinogenic effects (Patty 1967; Dreisbach 1974). Some important trace elements from coal processing, along with their threshold limit values and their effects on human health, and some gases and hydrocarbons that may be emitted are included in Table 10.33. Threshold limit values refer to airborne concentrations of substances and represent conditions under which it is believed that nearly all workers may be repeatedly exposed without adverse effects. These concentrations are determined by the American Conference of Government Industrial Hygienists annually.

10.4.2 Elements

The effects and metabolism of some specific trace elements, according to discussions by Patty (1967), Dreisbach (1974), and Berry, Osgood, and St. John (1974), are outlined in the following sections.

10.4.2.1 Lead

Lead enters the body primarily through the respiratory tract and through the walls of the digestive tract. Only about 5 to 10% of ingested lead is absorbed into the bloodstream. The average daily dietary intake is about 300 µg. Serious lead poisoning could result from prolonged breathing of air that contains 3 to 5 $\mu g/m^3$, or more, of lead (Berry, Osgood, and St. John 1974). Lead is excreted by the kidneys or stored in bone.

Lead acts on the blood, kidneys, and central nervous system. Early signs of poisoning include impairment of mental function, behavior problems, and anemia. At higher levels, lead in the system causes vomiting, cramps, and serious impairment of kidney and nervous system function.

Anemia results from the action of lead on the red blood cells. Both the life span and average number of red blood cells are reduced. Lead also interferes with the synthesis of hemoglobin, probably by interacting with enzymes containing free sulfhydryl groups. The mechanism of action on the kidneys and central nervous system is not well understood.

Table 10.33. Toxicity and threshold limit values of some potential pollutants from coal conversion processes

Pollutant	Threshold limit values		Effects
	(ppm)	(mg/m^3)	
Elements			
Antimony		0.5	Dermatitis, keratitis, conjunctivitis, and nasal septum ulceration by contact fumes or dust
Arsenic		0.5	
Arsine	0.05	0.2	Dermatitis, bronchitis, skin cancer, gastrointestinal disturbances (arsenic); inhalation of arsine produces hemolysis; combines with sulfhydral enzymes and interferes with cellular metabolism; powerful poison by inhalation or ingestion
Barium		0.5	Skin and mucous membrane irritant; toxic by inhalation or ingestion; ingestion produces convulsions by changing cell membrane permeability, thus producing indiscriminate stimulation of all muscle cells
Beryllium		0.002	Highly toxic by inhalation; produces berylliosis, dermatitis, and conjunctivitis; suspected carcinogen
Boron oxide		10	Moderately toxic via ingestion, inhalation, and skin adsorption
Cadmium (dust and soluble salts)		0.2	
Cadmium oxide		0.05	Inhalation produces pulmonary emphysema hypertension, and kidney damage; ingestion produces gastrointestinal inflammation and liver and kidney damage; interferes with Zn and Cu metabolism; cadmium oxide — possible human carcinogen
Chromium			
Chromic acid, chromates		0.1	Dermatitis and ulceration of the skin and nasal sinuses; toxic by inhalation or ingestion; carcinogenic (incidence of lung cancer increased up to 15 times normal in workers exposed to dusty chromite, chromic oxide, and ores)
Cobalt (metal fume and dust)		0.1	Dermatitis; ingestion produces goiter
Copper (fume dusts and mists)		0.2	Skin and mucous membrane irritant; inhalation causes lung and gastrointestinal disturbances
Iron oxide fume		5	Contact fumes or inhalation produces conjunctivitis, choroiditis, retinitis, and siderosis
Pentacarbonyl		0.08	
Soluble salts		1	
Lead fumes and dusts		0.15	Cumulative poison, produces behavioral disorders, brain damage, convulsions, death

Table 10.33 (continued)

Pollutant	Threshold limit values		Effects
	(ppm)	(mg/m^3)	
Manganese		5	Inhalation of dust or fumes produces a clearly characterized disease with progressive deterioration of the central nervous system
Mercury and mercuric compounds		0.01 – 0.1	Nephritis, gastrointestinal tract disturbances, nerve damage and death; depresses cellular enzymatic mechanisms by combining with (-SH) groups
Nickel (metal)		1	Dermatitis; probable nasal cavity and sinus carcinogen
Selenium (fumes and oxide fumes)		0.2	Gastrointestinal upset; respiratory tract and skin irritation; ingestion damages most organs (industrial "selenosis")
Tellurium fumes		0.1	Loss of appetite, garlic breath, liver
Thallium		0.1	Loss of hair, pains in extremities; route of uptake: ingestion or skin absorption
Tin			
Tin oxide		10	Inhalation produces benign pneumoconiosis
Organic compounds		0.1	
Inorganic compounds		2	
Vanadium (dust)		0.5	Conjunctiva and respiratory tract irritant
V$_2$O$_5$ fume		0.1	
Zinc oxide fume	5		Low toxicity; may cause dermatitis; respiratory tract irritant
Hydrocarbons			
Benzene	25	80	Inhalation produces dizziness, weakness, headache; large amounts cause central nervous system and bone marrow depression, fatty degeneration, and congestion of organs
Toluene	100	375	Produces slight bone marrow damage and central nervous system depression
Naphthalene	10	50	Dermatitis; eye irritant; hemolysis in susceptible individuals
Nitrogen and amino compounds			
Aniline	5	19	Most form methemoglobin and are skin and mucous membrane irritants; some produce kidney and liver damage and are central nervous system depressants
Toluidine	5	22	
Quinoline			
Benzidine			Skin sensitization; bladder tumors
β-Naphthamine			Bladder cancer

Table 10.33 (continued)

Pollutant	Threshold limit values		Effects
	(ppm)	(mg/m^3)	
Phenol and derivatives Creosote Resorcinol Hydroquinone	Fatal doses depend on compound		Highly toxic by ingestion, inhalation or skin absorption; skin and mucous membrane irritants; poison to all cells directly by denaturing and precipitating cellular proteins
Gases Sulfur oxides Sulfur dioxide	5 (industrial exposure)	13	Combines with water to form corrosive acid; eye, skin, and mucous membrane irritant
Nitrogen oxides Nitric oxide Nitrogen dioxide	25 5	30 9	Corrosion and irritation of skin, eyes, digestive tract or lungs following ingestion or inhalation
Hydrogen sulfide	10	15	Irritant and anoxic effects; causes damage to the central nervous system; inhalation causes pulmonary edema
Carbon disulfide	20	60	Causes damage to the central nervous system, peripheral nerves and hemopoietic system; highly toxic by ingestion, inhalation, and skin absorption
Carbon dioxide Carbon monoxide	5000 50	9000 55	Inhalation produces tissue anoxia as a result of carboxyhemoglobin production
Carbonyl sulfide			Inhalation: decomposes to carbon dioxide and hydrogen sulfide
Mercaptans	20 - 250		Inhalation causes nausea and headache; high concentrations produce cyanosis and depression of cerebral function
Cyanides (HCN) (Salts)	10	11 5	Increased respiration and cell paralysis; inhalation produces lung congestion and dizziness; poisons by inhibiting cytochrome oxidase system for oxygen utilization in cells
Thiocyanates			Ingestion produces psychotic behavior and depressed cell metabolic activity
Hydrogen fluoride	3	2	Inhalation causes inflammation and necrosis of mucous membrane; skin contact produces severe burns
Fluorine	1	2	Acts as direct cellular poison by interfering with calcium metabolism and enzyme mechanisms

Table 10.33 (continued)

Pollutant	Threshold limit values		Effects
	(ppm)	(mg/m^3)	
Hydrogen chloride	5	7	Forms corrosive acid; irritant and poison by inhalation and injection; irritant to mucous membrane of eyes and respiratory tract; produces pulmonary edema
Ammonia	25	18	Injures cells directly by alkaline caustic action; irritation of mucous membranes; highly irritant via inhalation or ingestion
Ozone	0.1	0.2	Pulmonary irritation and destruction from inhalation
Coal dust	Not established, respirable dust	2	Usually asymptomatic; produces pneumoconiosis and silicosis

Source: Compiled from Dreisbach 1974; Sax 1974; Yen 1975; American Conference of Governmental Industrial Hygienists 1975.

10.4.2.2 Mercury

The greatest risk for humans is mercury poisoning through food. Mercury and methylmercury bio-accumulate through the food chain, especially in aquatic situations. Mercury does not accumulate permanently in man; the half-life of methylmercury, which is eliminated through the feces and urine, is 70 to 74 days. Intake by inhalation is of concern only when contact with high vapor concentrations occurs regularly. Ingested metallic mercury is not absorbed.

Symptoms of methylmercury poisoning are largely neurological. Included are lack of muscular coordination, tremors, constriction of visual fields, and difficulty in swallowing. The disease may progress to more severe symptoms including deafness, blindness, paralysis, and kidney failure. Inhalation is followed within a day or two by stomatitis, salivation, metallic taste, diarrhea, pneumonitis, and kidney damage. Chronic exposure produces psychic and emotional disturbances. At Minamata, Japan, release of mercury from a vinyl chloride plant into coastal waters resulted in the deaths of 46 people between 1953 and 1960.

Mercury, like most heavy metals, interferes with the action of enzymes and proteins that function in metabolic reactions. Brain cells are particularly sensitive to methylmercury poisoning. It depresses cellular enzymatic reactions by combining with sulfhydryl groups. Active transport mechanisms involving sodium, potassium, and certain anions in the kidney are depressed. Chromosomal changes have been found.

10.4.2.3 Cadmium

Cadmium taken in by ingestion is poorly absorbed. The most common route of intake is inhalation of cadmium dust from industrial operations. Direct access to the blood is gained through the alveolar walls of the lungs. The amount absorbed from food is reduced by dietary factors, notably protein and zinc intake. Accumulation is greatest in the kidney and liver. Over 90% of cadmium in the blood is found in the red cells, probably bound to hemoglobin.

Symptoms in factory workers include tiredness, nervousness, dryness of the mouth, impaired sense of smell, shortness of breath, sore throat, chest cramps, pain in the small of the back, and poor appetite, leading to loss of weight and yellowing of teeth. Long-term effects include pulmonary emphysema, protein in the urine, and cirrhosis of the liver. Ingested cadmium is a powerful emetic. There is a strong suggestive correlation between cadmium and hypertension. Itai-Itai disease, in Japan, is characterized by degeneration of the bones and has been attributed to abnormally high levels of cadmium intake.

The active form of the metal, the charged Cd^{+2} ion, binds to the membranes of the mitochondria and inhibits enzymes required in energy-generating cellular reactions. Inhaled cadmium oxide fumes act directly on the cortical cells of the brain.

10.4.2.4 Arsenic

Compounds of arsenic may be absorbed from industrial sources by inhalation, ingestion, and/or skin contact. Arsenic is largely eliminated in the urine, but some is eliminated in the feces, hair, epithelium, nails, and possibly small amounts in the exhaled breath. Excretion is slow, requiring up to 10 days after an acute absorption and up to a year after prolonged absorption. Arsenic is retained and stored in all the tissues, bones, and especially the hair.

The element itself is nontoxic. Acute poisoning results in gastrointestinal disturbances, irritation of the nose and conjunctiva, dermatitis, and inflammation. Inhalation of arsine gas (AsH_3) causes hemolysis and may be fatal in dilutions of 50 ppm. Chronic arsenic poisoning results in muscle weakness, loss of appetite, occasional vomiting, gastrointestinal pains, irritation of nasal and oral membranes, coughing, and dermatitis. Toxicity is related to the chemical state of the element; the pentavalent form is less toxic than the trivalent form. Evidence suggesting a correlation between exposure to arsenic and skin and lung cancer is strong. Arsenic combines readily with sulfhydryl groups, thus interfering with cellular enzymatic reactions.

10.4.2.5 Vanadium

Little information is available on the absorption and metabolism of vanadium. Industrial workers may be exposed to vanadium in the form of dust and V_2O_5 fumes. Excretion is in the urine and feces. Uptake, accumulation, and excretion may depend upon the chemical form. Vanadium accumulates in the lungs following inhalation. It is generally of low toxicity when ingested. Clinical observations include irritation of the conjunctiva and nasal mucosa. Chronic exposure may result in chronic inflammatory change in the respiratory tract and bronchitis. The chief action of large doses of vanadium is on the vascular system, as manifested by peripheral constriction of the vessels of the lungs, spleen, kidneys, and intestines.

10.4.2.6 Selenium

Selenium is found in appreciable quantities in coal and efficiently passes through the combustion process, appearing on the surface of respirable particles emitted from the stack (EPRI 1976). The inhalation health hazard is virtually unexplored and it is uncertain whether the existing airborne concentration standard for occupational exposure is being met because of inadequate monitoring capability.

Selenium can be acquired from inhalation of the dust, vapors, gases, and fumes of the metal or its compounds, by ingestion, and to some extent by absorption through the skin. Fifty to eighty percent of ingested selenium is excreted through the kidneys. It is stored in the kidneys, spleen, pancreas, muscle, heart, lungs, and other tissues. The nutritional requirement lies in the range of 0.1 to 0.3 ppm. In moderately excessive amounts, selenium can be an extremely toxic element. The soluble salts are more toxic than the element.

Selenium fumes cause garlic breath, gastrointestinal upset, and nervousness. Selenium oxide causes bronchospasm, difficulty in breathing, chills, fever, and pneumonitis. Inhalation and ingestion of selenate fumes cause damage to most organs (industrial selenosis). Excess selenium may correlate with peridontal disease and dental caries in man. Animal studies on the carcinogenicity and teratogenicity of excess selenium have produced controversial results. Selenium probably acts on the cellular level by inhibiting sulfur enzymes.

10.4.2.7 Nickel

Atmospheric nickel and its compounds enter the respiratory tract by inhalation. Absorption through the skin has not been clearly demonstrated. The various salts of the metal are toxic, particularly the volatile carbonyl, $Ni(CO)_4$. Nickel carbonyl is generated and released into the atmosphere as a product of fossil fuel combustion. It is formed wherever carbon monoxide contacts

nickel and nickel alloys. Nickel compounds are widely used as catalysts in industrial processes.
Inhaled nickel carbonyl decomposes to metallic nickel, which deposits on the epithelium of the
lung. This finely divided nickel is rapidly absorbed and damages the lung and brain. Ingested
nickel metal is excreted rapidly in the feces, whereas inhaled nickel appears in the urine.

Exposure to nickel produces dermatitis, and nickel-refinery workers have severe respiratory ailments. Animal studies and circumstantial evidence in nickel refineries implicate nickel fumes
and nickel carbonyl as a nasal carcinogen. The presence of nickel in the lungs results in
significant alterations in the alveolar cells, including changes in the nucleoprotein and
increases in nuclear and cytoplasmic RNA. Following prolonged exposure, nickel can be found in
lung and liver endoplasmic reticulum, nuclei, and mitochondria.

10.4.2.8 <u>Fluorine</u>

Fluorides may be inhaled as dusts or gas and may be ingested. Ingested fluorine is stored
chiefly in the bones and teeth.

Fluorine and hydrogen fluoride are irritating and corrosive to the eyes and mucous membranes of
the respiratory tract. Neutral fluorides in 1 to 2% concentrations cause inflammation and
necrosis of mucous membranes. Chronic exposure produces bone changes, which are characterized
by abnormal density and indistinct bone structure. Fluorine and fluorides act as direct cellular
poisons by interfering with calcium metabolism and enzyme action.

LITERATURE CITED

Adamson, R. H.; Cooper, R. W.; and O'Gara, R. W. 1970. Carcinogen induced tumors in primitive primates. *J. Natl. Cancer Inst.* 45: 555-59.

Alexandrov, V. A. 1973. Embryotoxic and teratogenic effects of chemical carcinogens. In *Transplacental carcinogenesis. Proceedings of a meeting held at the Medizinsche Hochschule, Hanover, Fed. Rep. Germany, Oct. 6-7, 1971*, ed. L. Thomatis, U. Mohr, and W. Davis, pp. 112-16. Lyon: International Agency for Research on Cancer.

Alvares, A. P.; Schilling, G. R.; and Kuntzman, R. 1968. Differences in the kinetics of benzpyrene hydroxylation by hepatic drug-metabolizing enzymes from phenobarbital- and 3-methylcholanthrene-treated rats. *Biochem. Biophys. Res. Commun.* 30: 588-93.

Amdur, M. O. 1957. The influence of aerosols upon the respiratory response of guinea pigs to sulfur dioxide. *Am. Ind. Hyg. Assoc. Quart.* 18: 149-55.

Amdur, M. O., and Corn, M. 1963. The irritant potency of zinc ammonium sulfate of different particle sizes. *Am. Ind. Hyg. Assoc. J.* 24: 326-33.

Amdur, M. O., and Underhill, D. 1968. The effect of various aerosols on the response of guinea pigs to sulfur dioxide. *Arch. Environ. Health* 16: 460-68.

American Conference of Governmental Industrial Hygienists (ACGIH). 1975. *Threshold limit values for chemical substances and physical agents in the workroom environment with intended changes for 1975.* Cincinnati, Ohio: ACGIH.

Ames, B. N.; Durston, W. E.; Yamasaki, E.; and Lee, F. D. 1973. Carcinogens are mutagens: A simple test system combining liver homogenates for activation and bacteria for detection. *Proc. Natl. Acad. Sci. (USA)* 70: 2281-85.

Baird, W. M.; Dipple, A.; Grover, P. L.; Sims, P.; and Brookes, P. 1973. Studies on the formation of hydrocarbon-deoxyribonucleoside products by the binding of derivatives of 7-methylbenz(a)anthracene to DNA in aqueous solution and in mouse embryo cells in culture. *Cancer Res.* 33: 2386-92.

Baird, W. M.; Harvey, R. G.; and Brookes, P. 1975. Comparison of the cellular DNA-bound products of benzo(a)pyrene with the products formed by the reaction of benzo(a)pyrene-4, 5-oxide with DNA. *Cancer Res.* 35: 54-57.

Benedict, W. F.; Considine, N.; and Nebert, D. W. 1973. Genetic differences in aryl hydrocarbon hydroxylase induction and benzo(a)pyrene-produced tumorigenesis in the mouse. *Mol. Pharmacol.* 6: 266-77.

Berry, J. W.; Osgood, D. W.; and St. John, P. A. 1974. *Chemical villains: A biology of pollution.* St. Louis, Mo.: C. V. Mosby Company.

Bingham, E., and Falk, H. L. 1969. Environmental carcinogens: The modifying effect of cocarcinogens on the threshold response. *Arch. Environ. Health* 19: 779-83.

Booth, J.; Keysell, G. R.; Pal, K.; and Sims, P. 1974. The metabolism of polycyclic hydrocarbons by cultured human lymphocytes. *FEBS Lett.* 43(3): 341-44.

Booth, J.; Keysell, G. R.; and Sims, P. 1974. Effect of oestradiol on the *in vitro* metabolism of 7,12-dimethylbenz(a)anthracene and its hydroxymethyl derivatives. *Biochem. Pharmacol.* 23: 735-44.

Bowden, G. T.; Slaga, T. J.; Shapas, B. G.; and Boutwell, R. K. 1974. The role of aryl hydrocarbon hydroxylase in skin tumor initiation by 7,12-dimethylbenz(a)anthracene and 1,2,5,6-dibenzanthracene using DNA binding and thymidine-^{3}H incorporation into DNA as criteria. *Cancer Res.* 34: 2634-42.

Boyland, E. 1950. The biological significance of metabolism of polycyclic compounds. *Symp. Biochem. Soc.* 5: 40-54.

Boyland, E. 1958. The biological examination of carcinogenic substances. *Br. Med. Bull.* 14: 93.

Boyland, E., and Sims, P. 1965. Metabolism of polycyclic compounds: The metabolism of 7,12-dimethylbenz(a)anthracene by rat-liver homogenates. *Biochem. J.* 95(3): 780-87.

Boyland, E.; Sims, P.; and Huggins, C. 1965. Induction of adrenal damage and cancer with metabolites of 7,12-dimethylbenz(a)anthracene. *Nature* 207: 816-17.

Brasser, L. J.; Joostings, P. E.; and von Zuilen, D. 1967. *Sulfur dioxide: To what level is it acceptable?* Research Institute for Public Health Engineering, Delft, Netherlands, Report G-300.

Bridbord, Kenneth. 1976. *Carcinogenic and mutagenic risks associated with fossil fuels.* National Institute for Occupational Safety and Health, Rockville, Md.

Brookes, P., and Lawley, P. D. 1964. Evidence of the binding of polynuclear aromatic hydrocarbons to the nucleic acids of mouse skin: Relation between carcinogenic power of hydrocarbons and their binding to deoxyribonucleic acid. *Nature* 202: 781-84.

Brunette, D. M., and Katz, M. 1975. The interactions of benzo(a)pyrene with cell membranes: Uptake into Chinese hamster ovary (CHO) cells and fluorescence studies with isolated membranes. *Chem. Biol. Interactions* 11: 1-14.

Bulay, O. M. 1970. The study of development of lung and skin tumors in mice exposed in utero to polycyclic hydrocarbons. *Acta. Med. Turc.* 7: 3-38.

Bulay, O. M., and Wattenburg, L. W. 1968. Carcinogenic effects of subcutaneous administration of benzo(a)pyrene during pregnancy on the progeny. *Prof. Soc. Exp. Biol. Med.* 135: 84-86.

Chang, L. H. 1943. The fecal excretion of polycyclic hydrocarbons following their administration to the rat. *J. Biol. Chem.* 151: 93-99.

Chu, E.H.Y.; Bailiff, E. G.; and Malling, H. V. 1970. Mutagenicity of chemical carcinogenesis in mammalian cells. In *Proceedings of 10th international cancer congress*, pp. 62-63. Houston, Tex.: Medical Arts Publishing Co.

Comar, C. L. 1975. Health aspects. In *Conference proceedings: Workshop on health effects of fossil fuel combustion products*, pp. 52-55. Electric Power Research Institute, Palo Alto, Calif.

Cone, M. V., and Nettesheim, P. 1973. Effects of vitamin A on 3-methylcholanthrene-induced squamous metaplasias and early tumors in the respiratory tract of rats. *J. Natl. Cancer Inst.* 50(6): 1599-1604.

Conney, A. H.; Miller, E. G.; and Miller, J. A. 1957. Substrate-induced synthesis and other properties of benzpyrene hydroxylase in rat liver. *J. Biol. Chem.* 228: 753-66.

Cottini, G. B., and Mazzone, G. B. 1939. Effects of 3:4-benzpyrene on human skin. *Am. J. Cancer* 37: 186-95.

Currie, A. R.; Bird, C. C.; Crawford, A. M.; and Sims, P. 1970. Embryopathic effects of 7,12-dimethylbenz(a)anthracene and its hydroxymethyl derivatives in the Sprague-Dawley rat. *Nature* 226: 911-14.

Daniel, P. M.; Pratt, O. E.; and Prichard, M.M.L. 1967. Metabolism of labelled carcinogenic hydrocarbons in rats. *Nature* 215: 1142-46.

Diamond, L., and Gelboin, H. V. 1969. Alpha-naphthoflavone: An inhibitor of hydrocarbon cytotoxicity and microsomal hydroxylase. *Science* 166: 1023-25.

DiPaolo, J. A., and Kotin, P. 1966. Teratogenesis-oncogenesis: A study of possible relationships. *Arch. Path.* 81: 3-23.

Dipple, A.; Lawley, P. D.; and Brookes, P. 1968. Theory of tumour initiation by chemical carcinogens: Dependence of activity on structure of ultimate carcinogen. *Europ. J. Cancer* 4: 493-506.

Dipple, A.; Levy, L. S.; and Iype, P. T. 1975. Investigation of the possible involvement of ketonic derivatives of polycyclic hydrocarbons in hydrocarbon carcinogenesis. *Cancer Res.* 35: 652-57.

Doll, R.; Fisher, R.E.W.; Gammon, E. J.; Gunn, W.; Hughes, G. O.; Tyrer, F. H.; and Wilson, W. 1965. Mortality of gasworkers with special reference to cancers of the lung and bladder, chronic bronchitis, and pneumoconiosis. *Br. J. Ind. Med.* 22: 1-12.

Doll, R.; Vessey, M. P.; Beasley, R.W.R.; Buckley, A. R.; Fear, E. C.; Fisher, R.E.W.; Gammon, E. J.; Gunn, W.; Hughes, G. O.; Lee, K.; and Norman-Smith, B. 1972. Mortality of gasworkers — final report of a prospective study. *Br. J. Ind. Med.* 29: 394-406.

Dreisbach, R. H. 1974. *Handbook of poisoning: Diagnosis and treatment.* 8th ed. Los Altos, Calif.: Lange Medical Publications.

Eckardt, R. E. 1967. Cancer prevention in the petroleum industry. *Int. J. Cancer* 2: 656-61.

Electric Power Research Institute (EPRI). 1976. *Health effects of selenium.* EPRI 571-1.

Energy Research and Development Administration (ERDA). 1976. *Carcinogens relating to coal conversion processes.* FE-2213-1.

Environmental Protection Agency (EPA). 1971. *Air quality criteria for nitrogen oxides: Atmospheric levels of nitrogen oxides.* AP-84.

Environmental Protection Agency (EPA). 1974. *Health consequences of sulfur oxides: A report from CHESS, 1970-1971.* EPA-650/1-74-004.

Environmental Protection Agency (EPA). 1975. *Scientific and technical assessment report on particulate polycyclic organic matter (PPOM).* EPA-600/6-75-001.

Epstein, S. S., and Shafner, H. 1968. Chemical mutagens in the human environment. *Nature* 219: 385-87.

Falk, H. L.; Kotin, P.; and Thompson, S. 1964. Inhibition of carcinogenesis: The effect of polycyclic hydrocarbons and related compounds. *Arch. Environ. Health* 9: 169.

Fannick, N.; Gonsher, L. T.; and Shookley, J., Jr. 1972. Exposure to coal tar pitch volatiles at coke ovens. *Am. Ind. Hyg. Assoc. J.* 33: 461-67.

Feron, V. J. 1972. Respiratory tract tumors in hamsters after intratracheal instillations of benzo(a)pyrene alone and with furfural. *Cancer Res.* 32: 28-36.

Flesher, J. W., and Sydnor, K. L. 1970. Comparative studies on distribution of DMBA-3H and hydroxy-DMBA-3H and their carcinogenic activity. *Int. J. Cancer* 5: 253-59.

Flesher, J. W., and Sydnor, K. L. 1971. Carcinogenicity of derivatives of 7,12-dimethylbenz(a)-anthracene. *Cancer Res.* 31: 1951-54.

Flesher, J. W., and Sydnor, K. L. 1973. Possible role of 6-hydroxymethylbenzo(a)pyrene as a proximate carcinogen of benzo(a)pyrene and 6-methylbenzo(a)pyrene. *Int. J. Cancer* 11: 433-37.

Frank, N. R.; Amdur, M. O.; and Whittenberger, J. L. 1964. A comparison of the acute effects of SO_2 administered alone or in combination with NaCl particles on the respiratory mechanisms of healthy adults. *Int. J. Air Water Pollut.* 8: 125-33.

Freudenthal, R. J.; Lutz, G. A.; and Mitchell, R. I. 1975. *Carcinogenic potential of coal and coal conversion products.* Battelle Energy Program Report. Battelle Columbus Laboratories, Columbus, Ohio.

Gelboin, H. V.; Huberman, E.; and Sachs, L. 1969. Enzymatic hydroxylation of benzpyrene and its relationship to cytotoxicity. *Proc. Natl. Acad. Sci.* 64: 1188-94.

Gelboin, H. V.; Wiebel, F.; and Diamond, L. 1970. Dimethylbenzanthracene tumorigenesis and aryl hydrocarbon hydroxylase in mouse skin: Inhibition by 7,8-benzoflavone. *Science* 170: 169-71.

Genta, V. M.; Kaufman, D. G.; Harris, C. C.; Smith, J. M.; Sporn, M. B.; and Saffiotti, U. 1974. Vitamin A deficiency enhances binding of benzo(a)pyrene to tracheal epithelial DNA. *Nature* 247: 48-49.

Gerarde, H. W. 1960. *Toxicology and biochemistry of aromatic hydrocarbons.* New York: Elsevier Publishing Company.

Gräf, W.; Eff, H.; and Schormair, S. 1975. Levels of carcinogenic, polycyclic aromatic hydrocarbons in human and animal tissues. Third communication. *Zbl. Bakt. Hyg. I. Abt. Orig. B.* 161: 85-103.

Grilli, S.; Rocchi, P.; and Prodi, G. 1975. Nonenzymatic and microsome-dependent binding of polycyclic hydrocarbons to DNA and polynucleotides. *Chem. Biol. Interactions* 11: 351-63.

Grover, P. L.; Hewer, A.; and Sims, P. 1971. Epoxides as microsomal metabolites of polycyclic hydrocarbons. *FEBS Lett.* 18: 76-80.

Grover, P. L.; Hewer, A.; and Sims, P. 1973. K-region epoxides of polycyclic hydrocarbons: Formation and further metabolism of benz(a)anthracene 5,6-oxide by human lung preparations. *FEBS Lett.* 34(1): 63-68.

Grover, P. L., and Sims, P. 1968. Enzyme-catalysed reactions of polycyclic hydrocarbons with deoxyribonucleic acid and protein *in vitro*. *Biochem. J.* 110: 159-60.

Grover, P. L., and Sims, P. 1970. Interactions of the K-region epoxides of phenanthrene and dibenz(a,h)anthracene with nucleic acids and histone. *Biochem. Pharmacol.* 19: 2251-59.

Grover, P. L.; Sims, P.; Huberman, E.; Marquardt, H.; Kuroki, T.; and Heidelberger, C. 1971. *In vitro* transformation of rodent cells by K-region derivatives of polycyclic hydrocarbons. *Proc. Natl. Acad. Sci.* 68(6): 1098-1101.

Haddow, A., and Weinhouse, S., eds. 1967. *Advan. Cancer Res.* 10: 1-260.

Harper, K. H. 1957. The metabolism of pyrene. *Br. J. Cancer* 11: 499-507.

Harris, C. C.; Genta, V. M.; Frank, A. L.; Kaufman, D. G.; Barrett, L. A.; McDowell, E. M.; and Trump, B. F. 1974. Carcinogenic polynuclear hydrocarbons bind to macromolecules in cultured human bronchi. *Nature* 252: 68-69.

Hartwell, J. L. 1951. *Survey of compounds which have been tested for carcinogenic activity.* 2nd ed. Public Health Service Pub. 149. Washington, D.C.: U.S. Government Printing Office.

Hazucha, M., and Bates, D. V. 1975. Combined effect of ozone and sulphur dioxide on human pulmonary function. *Nature* 257: 50-51.

Heidelberger, C., and Jones, H. B. 1948. The distribution of radioactivity in the mouse following administration of dibenzanthracene labelled in the 9 and 10 positions with carbon[14]. *Cancer* 1: 252-60.

Heidelberger, C., and Weiss, S. M. 1951. The distribution of radioactivity in mice following administration of 3,4-benzpyrene-5-C[14] and 1,2,5,6-dibenzanthracene-9,10-C[14]. *Cancer Res.* 11: 885-91.

Henry, M. C.; Port, C. D.; and Kaufman, D. G. 1975. Importance of physical properties of benzo(a)pyrene—ferric oxide mixtures in lung tumor induction. *Cancer Res.* 35: 207-17.

Henry, M. C., and Kaufman, D. G. 1973. Clearance of benzo(a)pyrene from hamster lung after administration on coated particles. *J. Natl. Cancer Inst.* 51: 1961-64.

Henry, S. A. 1946. *Cancer of the scrotum in relation to occupation.* Oxford Medical Publications. New York: Humphrey Milford, Oxford University Press.

Henry, S. A.; Kennaway, N. M.; and Kennaway, E. L. 1931. The incidence of cancer of the bladder and prostate in certain occupations. *J. Hyg. (Camb.)* 31: 126-37.

Hilfrich, J.; Bresch, H.; Misfeld, J.; and Mohr, U. 1973. Investigation on the carcinogenic burden by air pollution in man. V. Tumors of the respiratory tract in Syrian golden hamsters after intratracheal instillation of benzo(a)pyrene. *Zbl. Bakt. Hyg. I. Abt. Orig. B.* 158: 59-61.

Hill, D. L., and Shih, T. W. 1974. Vitamin A compounds and analogs as inhibitors of mixed-function oxidases that metabolize carcinogenic polycyclic hydrocarbons and other compounds. *Cancer Res.* 34: 564-70.

Horton, A. W.; Denman, D. T.; and Trosset, R. P. 1957. Carcinogenesis of the skin. II. The accelerating properties of aliphatic and related hydrocarbons. *Cancer Res.* 17: 758-66.

Horton, A. W.; Tye, R.; and Stemmer, K. L. 1963. Experimental carcinogenesis of the lung. Inhalation of gaseous formaldehyde or an aerosol of coal tar by C3H mice. *J. Natl. Cancer Inst.* 30(1-6): 31-43.

Huberman, E. 1975. Correlation between mutagenicity and carcinogenicity: Mammalian cell transformation and cell-mediated mutagenesis by carcinogenic polycyclic hydrocarbons. *Mutat. Res.* 29(2): 285-91.

Huberman, E.; Kuroki, T.; Marquardt, H.; Selkirk, J. K.; Heidelberger, C.; Grover, P. L.; and Sims, P. 1972. Transformation of hamster embryo cells by epoxides and other derivatives of polycyclic hydrocarbons. *Cancer Res.* 32: 1391-96.

Huberman, E., and Sachs, L. 1973. Metabolism of the carcinogenic hydrocarbon benzo(a)pyrene in human fibroblast and epithelial cells. *Int. J. Cancer* 11: 412-18.

Huberman, E., and Sachs, L. 1974. Cell-mediated mutagenesis of mammalian cells with chemical carcinogens. *Int. J. Cancer* 13: 326-33.

Huberman, E.; Sachs, L.; Yang, S. K.; and Gelboin, H. V. 1976. Identification of mutagenic metabolites of benzo(a)pyrene in mammalian cells. *Proc. Natl. Acad. Sci. USA* 73(2): 607-11.

Hueper, W. C. 1952. Environmental cancers: A review. *Cancer Res.* 12(10): 691-97.

Hueper, W. C. 1956. Experimental carcinogenic studies on hydrogenated coal oils. I. Bergius oils. *Ind. Med. Surg.* 25: 51-55.

Hueper, W. C. 1956. Experimental carcinogenic studies on hydrogenated coal oils. II. Fischer-Tropsch oils. *Ind. Med. Surg.* 25: 459-62.

Hueper, W. C. 1971. Public health hazards from environmental chemical carcinogens, mutagens and teratogens. *Health Phys.* 21: 689-707.

Huggins, C.; Moon, R. C.; and Morii, S. 1962. Extinction of experimental mammary cancer. I. Estradiol-17β and progesterone. *Proc. Natl. Acad. Sci. USA* 48: 379-86.

Huggins, C., and Morii, S. 1961. Selective adrenal necrosis and apoplexy induced by 7,12-dimethylbenz(a)anthracene. *J. Exp. Med.* 114(4): 741.

International Agency for Research on Cancer (IARC). 1973. *Evaluation of the carcinogenic risk of chemicals to man: Certain polycyclic aromatic hydrocarbons and heterocyclic compounds,* vol. 3.

Jackson, J. O.; Warner, P. O.; and Mooney, T. F., Jr. 1974. Profiles of benzo[a]pyrene and coal tar pitch volatiles at and in the immediate vicinity of a coke oven battery. *Am. Ind. Hyg. Assoc. J.* 35: 276-81.

Jerina, D. M.; Kaubisch, N.; and Daly, J. W. 1971. Arene oxides as intermediates in the metabolism of aromatic substrates: Alkyl and oxygen migrations during isomerization of alkylated arene oxides. *Proc. Natl. Acad. Sci.* 68: 2545.

Kaufman, D. G.; Genta, V. M.; and Harris, C. C. 1974. Studies on carcinogen binding in vitro in isolated hamster tracheas. In *Experimental lung cancer. Carcinogenesis and bioassays,* ed. E. Karbe and J. F. Park, pp. 564-74. New York: Springer-Verlag.

Kaufman, D. G.; Genta, V. M.; Harris, C. C.; Smith, J. M.; Sporn, M. B.; and Saffiotti, U. 1973. Binding of ^{3}H-labeled benzo(a)pyrene to DNA in hamster tracheal epithelial cells. *Cancer Res.* 33: 2837-41.

Kellermann, G.; Cantrell, E.; and Shaw, C. R. 1973. Variations in extent of aryl hydrocarbon hydroxylase induction in cultured human lymphocytes. *Cancer Res.* 33: 1654-56.

Kellermann, G.; Luyten-Kellermann, M.; and Shaw, C. R. 1973a. Genetic variation in aryl hydrocarbon hydroxylase in human lymphocytes. *Am. J. Human Genet.* 25: 327-31.

Kellermann, G.; Luyten-Kellermann, M.; and Shaw, C. R. 1973b. Metabolism of polycyclic aromatic hydrocarbons in cultured human leukocytes under genetic control. *Humangenetik* 20: 257-63.

Kellermann, G.; Shaw, C. R.; and Luyten-Kellermann, M. 1973. Aryl hydrocarbon hydroxylase inducibility and bronchogenic carcinoma. *New Engl. J. Med.* 289(18): 934-36.

Kennaway, E. 1955. The identification of a carcinogenic compound in coal-tar. *Br. Med. J.* 1: 749-52.

Kenneday, N. M., and Kennaway, E. L. 1936. A study of the incidence of cancer of the lung and larynx. *J. Hyg. (Camb.)* 36: 236-67.

Kennaway, E. L., and Kennaway, N. M. 1947. A further study of the incidence of cancer of the lung and larynx. *Br. J. Cancer* 1: 260.

Kennedy, A. R., and Little, J. B. 1974. The transport and localization of benzo(a)pyrene-hematite and hematite-^{210}Po in the hamster lung following intratracheal instillation. *Cancer Res.* 34: 1344-52.

Ketcham, N. H., and Norton, W. 1960. The hazards to health in the hydrogenation of coal. III. The industrial hygiene studies. *Arch. Environ. Health* 1: 194-207.

Keysell, G. R.; Booth, J.; and Sims, P. 1975. The metabolic formation of water-soluble derivatives from dihydrodiols of polycyclic hydrocarbons. *Br. J. Cancer* 31(2): 266.

King, H.W.S.; Thompson, M. H.; and Brookes, P. 1975. The benzo(a)pyrene deoxyribonucleoside products isolated from DNA after metabolism of benzo(a)pyrene by rat liver microsomes in the presence of DNA. *Cancer Res.* 34: 1263-69.

Kotin, P.; Falk, G. L.; and Busser, R. 1959. Distribution, retention, and elimination of C^{14}-3,4-benzopyrene after administration to mice and rats. *J. Natl. Cancer Inst.* 23(3): 541-55.

Kotin, P.; Falk, H. L.; and Miller, A. 1962. Effect of carbon tetrachloride intoxication on metabolism of benzo(a)pyrene in rats and mice. *J. Natl. Cancer Inst.* 28: 725-45.

Kouri, R. E.; Ratrie, H.; and Whitmire, C. E. 1973. Evidence of a genetic relationship between susceptibility to 3-methylcholanthrene-induced subcutaneous tumors and inducibility of aryl hydrocarbon hydroxylase. *J. Natl. Cancer Inst.* 51(1): 197-200.

Kuroda, S., and Kawahata, K. 1936. Uber die gewerbliche Entstenhung des Lungenkrebses bei Generatorgasarbeitern. *Z. Krebsforsch.* 45: 36-39.

Laskin, S.; Kuschner, M.; and Drew, R. T. 1970. Studies in pulmonary carcinogenesis. In *Inhalation carcinogenesis, Proceedings of a Biology Division, Oak Ridge National Laboratory, Conference, Gatlinburg, Tenn., Oct. 8-11, 1969*, AEC Symp. Ser. 18, ed. M. G. Hanna, P. Nettesheim, and J. R. Gilbert, pp. 321-51. Conf.-691001.

Lawther, P. J.; Commins, B. T.; and Waller, R. E. 1965. A study of the concentrations of polycyclic aromatic hydrocarbons in gas works retort houses. *Br. J. Ind. Med.* 22: 13-20.

Levine, W. G. 1974. Hepatic uptake, metabolism, and biliary excretion of 7,12-dimethylbenzanthracene in the rat. *Drug Metabol. Disposition* 2(2): 169-77.

Lloyd, J. W. 1971. Long-term mortality study of steelworkers. V. Respiratory cancer in coke plant workers. *J. Occup. Med.* 13(2): 53-67.

Lloyd, J. W.; Lundin, F. E.; Redmond, C. K.; and Geiser, P. B. 1970. Long-term mortality study of steelworkers. IV. Mortality by work area. *J. Occup. Med.* 12(5): 151-56.

Lubet, R. A.; Brown, D. Q.; and Kouri, R. E. 1973. The role of 3-OH benzo(a)pyrene in mediating benzo(a)pyrene induced toxicity and transformation in cell culture. *Res. Commun. Chem. Pathol. Pharmacol.* 6(3): 929-42.

Malaveille, C.; Bartsch, H.; Grover, P. L.; and Sims, P. 1975. Mutagenicity of non-K-region diols and diol-epoxides of benz(a)anthracene and benzo(a)pyrene in *S. typhimurium* TA 100. *Biochem. Biophys. Res. Commun.* 66(2): 693-700.

Mallet, L., and Heros, M. 1960. Presence of BP in urine from man. *C. R. Acad. Sci.* 250: 943-45.

Marquardt, H.; Sondergren, J. E.; Sims, P.; and Grover, P. L. 1974. Malignant transformation *in vitro* of mouse fibroblasts by 7,12-dimethylbenz(a)anthracene and 7-hydroxymethylbenz(a)anthracene and by their K-region derivatives. *Int. J. Cancer* 13: 304-10.

Martin, A. E. 1964. Mortality and morbidity statistics and air pollution. *Roy. Soc. Med. Proc.* 57: 969-75.

Mazumdar, S.; Redmond, C.; Sollecito, W.; and Sussman, N. 1975. An epidemiological study of exposure to coal tar pitch volatiles among coke oven workers. *J. Air Pollut. Control Assoc.* 25(4): 382-89.

Miller, E. C. 1951. Studies on the formation of protein-bound derivatives of 3,4-benzpyrene in the epidermal fraction of mouse skin. *Cancer Res.* 11: 100-108.

Miller, J. A. 1970. Carcinogenesis by chemicals: An overview. *Cancer Res.* 30: 559-76.

Miller, J. A., and Miller, E. C. 1975. The metabolic activation of chemical carcinogens to reactive electrophiles. Univ. Wisconsin Medical Center, Madison, Wis. Presented at *Symposium, m-RNA: The relation of structure to function, Gatlinburg, Tenn., April 1976.*

Morrow, P. E. 1975. An evaluation of recent NO_x toxicity data and an attempt to derive an ambient air standard for NO_x by established toxicological procedures. *Environ. Res.* 10: 92-112.

National Academy of Sciences (NAS). 1972. *Particulate polycyclic organic matter.* National Research Council Committee on Biologic Effects of Atmospheric Pollutants. Washington, D.C.: NAS.

National Academy of Sciences (NAS). 1974. *The relation of selected trace elements to health and disease. Geochemistry and the environment. A report of the workshop at the Asilomar Conference Grounds, Pacific Grove, Calif., Feb. 7-12, 1972,* vol. I. Washington, D.C.: NAS.

National Air Pollution Control Administration (NAPCA). 1969. *Air quality criteria for particulate matter.* AP-49. Washington, D.C.: NAPCA.

National Air Pollution Control Administration (NAPCA). 1970. *Air quality criteria for sulfur oxides.* AP-50. Washington, D.C.: NAPCA.

National Institute for Occupational Safety and Health (NIOSH). 1973. *Criteria for a recommended standard: Occupational exposure to coke oven emissions.* Cincinnati, Ohio: NIOSH.

National Institute for Occupational Safety and Health (NIOSH). 1976. *Criteria for a recommended standard: Occupational exposure to oxides of nitrogen (nitrogen dioxide and nitric oxide).* HEW Publ. No. (NIOSH) 76-149. Cincinnati, Ohio: NIOSH.

Neal, J., and Rigdon, R. H. 1964. Absorption and excretion of benzpyrene when fed to mice. *Texas Rep. Biol. Med.* 22(Suppl.): 156-64.

Nebert, D. W. 1974. Genetic and environmental factors affecting placental and fetal metabolism of xenobiotics. In *The placenta: Biological and clinical aspects, Harold C. Mack symposium on the physiology and pathology of reproduction, 1972,* ed. S. Moghissi and E.S.E. Hafez, pp. 207-37.

Nebert, D. W., and Gelboin, H. V. 1968*a*. Substrate-inducible microsomal aryl hydroxylase in mammalian cell culture. I. Assay and properties of induced enzyme. *J. Biol. Chem.* 243(23): 6242-49.

Nebert, D. W., and Gelboin, H. V. 1968*b*. Substrate-inducible microsomal aryl hydroxylase in mammalian cell culture. II. Cellular responses during enzyme induction. *J. Biol. Chem.* 243(23): 6250-61.

Nebert, D. W., and Gelboin, H. V. 1969. The *in vivo* and *in vitro* induction of aryl hydrocarbon and hydroxylase in mammalian cells of different species, tissues, strains, and developmental and hormonal states. *Arch. Biochem. Biophys.* 134: 76-89.

Nebert, D. W.; Winker, J.; and Gelboin, H. V. 1969. Aryl hydrocarbon hydroxylase activity in human placenta from cigarette smoking and nonsmoking women. *Cancer Res.* 29: 1763-69.

Nettesheim, P., and Schreiber, H. 1975. Advances in experimental lung cancer research. In *Handbuch d. allg. Pathologie Bd: VI/7.* Heidleburg: Springer Verlag.

Olsen, D., and Haynes, J. L. 1969. *Preliminary air pollution survey of organic carcinogens. A literature review.* National Air Pollution Control Administration Pub. APTD 69-43.

Pal, K.; Grover, P. L.; and Sims, P. 1975. The metabolism of carcinogenic polycyclic hydrocarbons by tissues of the respiratory tract. *Biochem. Soc. Trans.* 3: 174-75.

Palmer, M. S.; Exley, R. W.; and Coffin, D. L. 1972. Influence of pollutant gases on benzpyrene hydroxylase activity. *Arch. Environ. Health* 25: 439-42.

Patty, F. A., ed. 1967. *Industrial hygiene and toxicology. Vol. II. Toxicology.* 2nd rev. ed. New York: Interscience Publishers, John Wiley & Sons, Inc.

Peacock, P. R. 1936. Evidence regarding the mechanism of elimination of 1:2-benzpyrene, 1:2:5:6-dibenzanthracene, and anthracene from the bloodstream of injected animals. *Br. J. Exp. Pathol.* 17: 164-72.

Pearlman, M. E.; Finklea, J. F.; Creason, J. P.; Shy, C. M.; Young, M. M.; and Horton, R.J.M. 1971. Nitrogen dioxide and lower respiratory illness. *Pediatrics* 47: 391-95.

Pfeiffer, E. H. 1973. Investigations on the carcinogenic burden by air pollution in man. VII. Studies on the oncogenetic interaction of polycyclic aromatic hydrocarbons. *Zbl. Bakt. Hyg. I. Abt. Orig. B.* 158: 69-83.

Rall, D. P. 1969. Difficulties in extrapolating the results of toxicity studies in laboratory animals to man. *Environ. Res.* 2: 360-67.

Rall, D. P. 1974. Review of the health effects of sulfur dioxide. *Environ. Health Perspect.* 8: 97-121.

Rasmussen, R. E., and Wang, I. Y. 1974. Dependence of specific metabolism of benzo(a)pyrene on the inducer of hydroxylase activity. *Cancer Res.* 34: 2290-95.

Redmond, C. K.; Ciocci, A.; Lloyd, J. W.; and Rush, H. W. 1972. Long-term mortality study of steelworkers. VI. Mortality from malignant neoplasms among coke oven workers. *J. Occup. Med.* 14(8): 621-29.

Reid, D. D., and Buck, C. 1956. Cancer in coking plant workers. *Br. J. Ind. Med.* 13: 265-69.

Rigdon, R. H., and Neal, J. 1963. Absorption and excretion of benzpyrene observations in the duck, chicken, mouse and dog. *Texas Rep. Biol. Med.* 21: 247-61.

Saffiotti, U.; Cefis, F.; and Kolb, L. H. 1964. Bronchogenic carcinoma induction by particulate carcinogens. *Proc. Am. Assoc. Cancer Res.* 5(1): 59.

Saffiotti, U.; Montesano, R.; Sellakumar, A. R.; and Borg, S. A. 1967. Experimental cancer of the lung. Inhibition by vitamin A of the induction of tracheobronchial squamous metaplasia and squamous cell tumors. *Cancer* 20(5): 857-64.

Sax, N. I., ed. 1974. *Industrial pollution*, pp. 118-35. New York: Van Nostrand Reinhold Co.

Schlede, E., and Merker, H. J. 1972. Effect of benzo(a)pyrene treatment on the benzo(a)pyrene hydroxylase activity in maternal liver, placenta, and fetus of the rat during day 13 to day 18 of gestation. *Naunyn-Schmiedeberg's Arch. Pharmacol.* 272: 80-100.

Schlede, E., and Merker, H. J. 1974. Benzo(a)pyrene hydroxylase activity in the whole implantation site, decidua, placenta and fetal membranes of the rat. *Naunyn-Schmiedeberg's Arch. Pharmacol.* 282: 59-73.

Schmidt, K. G.; Schmahl, D.; and Misfeld, J. 1973. Investigations on the carcinogenic burden by air pollution in man. VI. Experimental investigations to determine a dose-response relationship and to estimate a threshold dose of benzo(a)pyrene in the skin of two different mouse strains. *Zbl. Bakt. Hyg. I. Abt. Orig. B.* 158: 62-68.

Schulte, K. A.; Larsen, D. J.; Hornung, R. W.; and Crable, J. V. 1974. *Report on analytical methods used in coke oven effluent study. The five oven study.* U.S. Dept. of Health, Education, and Welfare, National Institute for Occupational Safety and Health, Division of Laboratories and Criteria Development, Cincinnati, Ohio.

Selkirk, J. K.; Huberman, E.; and Heidelberger, C. 1971. An epoxide is an intermediate in the microsomal metabolism of the chemical carcinogen, dibenz(a,h)anthracene. *Biochem. Biophys. Res. Commun.* 43(5): 1010-16.

Sexton, R. J. 1960a. The hazards to health in the hydrogenation of coal. I. An introductory statement on general information, process, description, and a definition of the problem. *Arch. Environ. Health* 1: 181-86.

Sexton, R. J. 1960*b*. The hazards to health in the hydrogenation of coal. IV. The control program and the clinical effects. *Arch. Environ. Health* 1: 208-31.

Shamberger, R. J. 1970. Relationship of selenium to cancer. I. Inhibitory effect of selenium on carcinogenesis. *J. Natl. Cancer Inst.* 44(4): 931-36.

Shamberger, R. J.; Tytko, S.; and Willis, C. 1973. Antioxidants and cancer. II. Selenium distribution and human cancer mortality in the United States, Canada, and New Zealand. In *Seventh annual conference on trace substances in environmental health, Memorial Union, University of Missouri-Columbia, Columbia, Mo., June 12-14, 1973*, ed. D. D. Hemphill, pp. 35-40.

Shay, H.; Friedmann, B.; Gruenstein, M.; and Weinhouse, S. 1950. Mammary excretion of 20 methylcholanthrene. *Cancer Res.* 10: 797.

Shendrikova, I. A., and Aleksandrov, V. A. 1974. Comparative penetration of polycyclic hydrocarbons through the rat placenta into the fetus. *Bull. Exp. Biol. Med.* 77(2): 169-71.

Shepard, T. H. 1973. *Catalog of teratogenic agents.* Baltimore, Md.: John Hopkins Univ. Press.

Shubik, P., and Della Porta, G. 1957. Carcinogenesis and acute intoxication with large doses of polycyclic hydrocarbons. *Arch. Pathol.* 64: 691-703.

Shubik, P., and Hartwell, J. L. 1957. *Survey of compounds which have been tested for carcinogenic activity.* Public Health Service Pub. 149, Suppl. 1. Washington, D.C.: Government Printing Office.

Shubik, P., and Hartwell, J. L. 1969. *Survey of compounds which have been tested for carcinogenic activity.* Public Health Service Pub. 149, Suppl. 2. Washington, D.C.: Government Printing Office.

Shy, C. M.; Creason, J. P.; Pearlman, M. E.; McClain, K. E.; Benson, F. B.; and Young, M. M. 1970*a*. The Chattanooga school children study: Effects of community exposure to nitrogen dioxide. I. Methods, description of pollutant exposure, and results of ventilatory function testing. *J. Air Pollut. Control Assoc.* 20(9): 539-45.

Shy, C. M.; Creason, J. P.; Pearlman, M. E.; McClain, K. E.; Benson, F. B.; and Young, M. M. 1970*b*. The Chattanooga school children study: Effects of community exposure to nitrogen dioxide. II. Incidence of acute respiratory illness. *J. Air. Pollut. Control Assoc.* 20(9): 582-88.

Sims, P. 1967. The metabolism of 7- and 12-methylbenz(a)anthracene and their derivatives. *Biochem. J.* 105: 591-98.

Sims, P. 1970*a*. The metabolism of some aromatic hydrocarbons by mouse embryo cell culture. *Biochem. Pharmacol.* 19: 285-97.

Sims, P. 1970*b*. Qualitative and quantitative studies on the metabolism of a series of aromatic hydrocarbons by rat-liver preparations. *Biochem. Pharmacol.* 19: 795-818.

Sims, P. 1973. Epoxy derivatives of aromatic polycyclic hydrocarbons. The preparation and metabolism of epoxides related to 7,12-dimethylbenz(a)anthracene. *Biochem. J.* 131: 405-13.

Sims, P.; Grover, P. L.; Swaisland, A.; Pal, K.; and Hewer, A. 1974. Metabolic activation of benzo(a)pyrene proceeds by a diol-epoxide. *Nature* 252: 326-27.

Sporn, M. B.; Clamon, G. H.; Smith, J. M.; Dunlop, N. M.; Newton, D. L.; and Saffiotti, U. 1974. The reversal of keratinized squamous metaplastic lesions of vitamin A deficiency in tracheobronchial epithelium by vitamin A and vitamin A analogs in organ culture: A model system for anticarcinogenesis studies. In *Experimental lung cancer. Carcinogenesis and bioassays,* ed. E. Karbe and J. F. Park, pp. 575-82. New York: Springer-Verlag.

Stowell, R. E., and Maas, L. C. 1946. Effects of massive doses of methylcholanthrene on epidermal carcinogenesis. *Cancer Res.* 6: 121-27.

Task Group on Lung Dynamics. 1966. Deposition and retention models for internal dosimetry of the human respiratory tract. *Health Phys.* 12: 173-207.

Thompson, J. I., and Company. 1971. *Survey of compounds which have been tested for carcinogenic activity.* Public Health Service Pub. 149. Washington, D.C.: Government Printing Office.

Tomatis, L., and Goodall, C. M. 1969. The occurrence of tumors in F-1, F-2, and F-3 descendants of pregnant mice injected with 7,12-dimethylbenz(a)anthracene. *Int. J. Cancer* 4: 219-25.

Tomingas, R.; Pott, F.; and Dehnen, W. 1976. Polycyclic aromatic hydrocarbons in human bronchial carcinoma. *Cancer Lett.* 1: 189-96.

Tracor/Jitco. 1973*a*. *Survey of compounds which have been tested for carcinogenic activity (1961-1967)*. Public Health Service Pub. 149.

Tracor/Jitco. 1973*b*. *Survey of compounds which have been tested for carcinogenic activity (1970-1971)*. Public Health Service Pub. 149.

Turusov, V.; Tomatis, L.; Guibbert, D.; Duperray, B.; and Pacheco, H. 1973. The effect of prenatal exposure of mice to methylcholanthrene combined with the neonatal administration of diethylnitrosamine. In Transplacental carcinogenesis: *Proceedings of a meeting held at the Medizinsche Hochschule, Hanover, Fed. Rep. Germany, Oct. 6-7, 1971*, ed. L. Tomatis, U. Mohr, and W. Davis, pp. 84-91. Lyon: International Agency for Research on Cancer.

Tye, R., and Stemmer, K. L. 1967. Experimental carcinogenesis of the lung. II. Influence of phenols in the production of carcinoma. *J. Natl. Cancer Inst.* 39: 175-86.

Waldbott, G. L. 1973. *Health effects of environmental pollutants*. St. Louis, Mo.: C.V. Mosby Company.

Wang, I. Y.; Creasy, R. K.; and Crocker, T. T. 1974. Effect of cigarette-smoking on metabolism of benzo(a)pyrene by human placental microsomes. *Proc. Am. Assoc. Cancer Res.* 15: 20.

Wang, I. Y.; Rasmussen, R. E.; and Crocker, T. T. 1972. Isolation and characterization of an active DNA-binding metabolite of benzo(a)pyrene from hamster liver microsomal incubation system. *Biochem. Biophys. Res. Comm.* 49(4): 1142-50.

Watanabe, M.; Ariji, F.; Takusagawa, K.; and Konno, K. 1975. Aryl hydrocarbon hydroxylase in liver nuclei of C3H/He and DBA/2 mice. *Gann* 66: 399-406.

Waterfall, J. F., and Sims, P. 1972. Epoxy derivatives of aromatic polycyclic hydrocarbons. The preparation and metabolism of epoxides related to benzo(a)pyrene and to 7,8- and 9,10-dihydrobenzo(a)pyrene. *Biochem. J.* 128: 265-77.

Wattenberg, L. W. 1972. Inhibition of carcinogenic and toxic effects of polycyclic hydrocarbons by phenolic antioxidants and ethoxyquin. *J. Natl. Cancer Inst.* 48: 1425-30.

Wattenberg, L. W. 1974. Inhibition of carcinogenic and toxic effects of polycyclic hydrocarbons by several sulfur-containing compounds. *J. Natl. Cancer Inst.* 52(5): 1583-87.

Wattenberg, L. W., and Leong, J. L. 1970. Inhibition of the carcinogenic action of benzo(a)pyrene by flavones. *Cancer Res.* 30: 1922-25.

Weil, C. S., and Condra, N. I. 1960. The hazards to health in the hydrogenation of coal. II. Carcinogenic effect of materials on the skin of mice. *Arch. Environ. Health* 1: 187-92.

Welch, R. M.; Harrison, Y. E.; Conney, A. H.; Poppers, P. J.; and Finster, M. 1968. Cigarette smoking: Stimulatory effect on metabolism of 3,4-benzpyrene by enzymes in human placenta. *Science* 160: 541-42.

Wiebel, F. J., and Gelboin, H. V. 1974. Enzyme induction and polycyclic hydrocarbon metabolism in cell culture, experimental animals and man. In *Chemical carcinogenesis essays. Proceedings of a workshop on approaches to assess the significance of experimental chemical carcinogenesis data for man, organized by IARC and the Catholic Univ. of Louvain, Brussels, Belgium, Dec. 10-12, 1973,* Pub. No. 10. Lyon: International Agency for Research on Cancer.

Wiest, W. G., and Heidelberger, C. 1953. The interaction of carcinogenic hydrocarbons with tissue constituents. II. 1:2, 5:6-dibenzanthracene-9,10-^{14}C in skin. *Cancer Res.* 13: 250.

Williams, R. T. 1959. The metabolism of aromatic hydrocarbons. In Detoxication mechanisms: *The metabolism and detoxication of drugs, toxic substances and other organic compounds,* pp. 188-236. 2nd ed. New York: John Wiley & Sons, Inc.

Yang, S. K.; McCourt, D. W.; Roller, P. P.; and Gelboin, H. V. 1976. Enzymatic conversion of benzo[*a*]pyrene leading predominantly to the diol-epoxide *r*-7,*t*-8-dihydroxy-*t*-9,10-oxy-7,8,9,10-tetrahydrobenzo[*a*]pyrene through a single enantiomer of *r*-7,*t*-8-dihydroxy-7,8-dihydrobenzo[*a*]pyrene. *Proc. Natl. Acad. Sci. (USA)* 73(8): 2594-98.

Yen, T. F. 1975. *The role of trace metals in petroleum.* Ann Arbor, Mich.: Ann Arbor Science
    Publishers, Inc.

Yen, T. F. 1975. *The role of trace metals in petroleum.* Ann Arbor, Mich.: Ann Arbor Science
    Publishers, Inc.

Appendix A

CONVERSION OF UNITS

J. T. Ensminger

This section has been prepared to aid in the conversion between units found in the literature and those used by the coal conversion researcher. Because the scientific community is moving toward worldwide acceptance of the International System of Units (SI), SI units are used as the bases for the conversion tables — both one- and two-dimensional — contained here.

Accepted SI prefixes, units of concentration and substance, and temperature conversion equations are given in Tables A.1 through A.3 respectively. A temperature conversion scale, providing fast approximate conversions between Fahrenheit, Celsius, and Kelvin units (from -40 to 2480°F), is included in Table A.3.

Table A.4 contains SI conversion factors for one-dimensional units of area, energy, force, length, mass, pressure, time, and volume; the correct SI unit is given in the center column. Table A.5 contains conversion factors for two-dimensional units of distance/time, energy/mass, energy/time, energy/volume, mass/area, mass/time, mass/volume, volume/mass, and volume/time. To determine the appropriate conversion factor in Table A.5, locate the numerator of the given unit on the horizontal axis and the denominator on the vertical axis. The number that appears at the intersection is the factor that will convert the given unit to the correct SI unit. For example, to convert 10,000 scf/ton* to the SI unit [cubic meters per kilogram (m^3/kg)], locate in Table A.5 under *Volume/mass* the number at the point where "foot (scf)" and "ton (short)" intersect: 3.121 399 E-05. Then,

$$10,000 \text{ scf/ton} \times 0.00003121399 = 0.31214 \text{ } m^3/\text{kg (rounded to 0.3)}$$

Where the conversion of given units to SI units is unnecessary, any of the given units may be converted to any of the other units by using the conversion factors given. For example, to convert 10 cal/g to Btu/lb, find the units cal/g and Btu/lb expressed as the correct SI unit in Table A.5 under *Energy/mass*:

$$1 \text{ cal/g} = 9.230313 \text{ J/kg} \text{ ;}$$
$$1 \text{ Btu/lb} = 2326.000 \text{ J/kg} \text{ .}$$

Then,

$$10 \text{ cal/g} = \frac{10 \times 9.230313}{2326.000} \text{ Btu/lb} = 0.03968320 \text{ Btu/lb (rounded to 0.04) .}$$

---

*The term standard cubic foot (scf) refers to a cubic foot of gas at atmospheric pressure and 288.71 K.

Table A.1. Accepted SI prefixes

Multiplication factors		Prefix	SI symbol
1 000 000 000 000	$= 10^{12}$	tera	T
1 000 000 000	$= 10^9$	giga	G
1 000 000	$= 10^6$	mega	M
1 000	$= 10^3$	kilo	k
100	$= 10^2$	hecto[a]	h
10	$= 10^1$	deka[a]	da
0.1	$= 10^{-1}$	deci[a]	d
0.01	$= 10^{-2}$	centi[a]	c
0.001	$= 10^{-3}$	milli	m
0.000 001	$= 10^{-6}$	micro	μ
0.000 000 001	$= 10^{-9}$	nano	n
0.000 000 000 001	$= 10^{-12}$	pico	p
0.000 000 000 000 001	$= 10^{-15}$	femto	f
0.000 000 000 000 000 001	$= 10^{-18}$	atto	a

[a]To be avoided where possible; multiple and submultiple prefixes representing steps of 1000 are recommended.

Table A.2. Units of concentration and substance

Concentration

As a general practice, units of concentration are often expressed as parts per million, parts per billion, etc. Such measurements may also be expressed appropriately in SI units.

Solids (weight per unit weight)

    1 part per million (ppm) = 1 mg/kg or 1 μg/g
    1 part per billion (ppb) = 1 μg/kg or 1 ng/g
    1 part per trillion (ppt) = 1 ng/kg or 1 pg/g

Water [weight per unit volume (1 g water $\simeq$ 1 ml)]

    1 mg/liter $\simeq$ 1 ppm (weight:weight)
    1 μg/liter $\simeq$ 1 ppb (weight:weight)
    1 ng/liter $\simeq$ 1 ppt (weight:weight)

Air (weight per unit volume)

    $1 \text{ mg/m}^3 = 10^3 \ \mu/\text{m}^3 = 10^6 \text{ ng/m}^3$

Conversion to a volume-per-unit-volume basis depends upon the molecular weight of the dispersed substance (200.6 for elemental mercury). At 25°C and 760 mm Hg pressure, the conversion formula is

$$\text{ppm, by volume} = \text{mg/m}^3 \times \frac{24.25}{\text{molecular weight in grams}},$$

where mg/m^3 is the measured concentration.

Substance

The mole is the SI unit of substance, defined by the American Society for Testing and Materials as "the amount of substance of a system which contains as many elementary entities as there are atoms in 0.012 kg of ^{12}C."[a]

[a]American Society for Testing and Materials (ASTM). 1974. Metric practice guide. E 380-74. 2nd ed. Philadelphia, Pa.: ASTM.

Table A.3. Temperature conversions

To convert from	to	Use this equation
Degree Celsius (°C)	kelvin (K)	$K = °C + 273.15$
Degree Fahrenheit (°F)	degree Celsius (°C)	$°C = (°F - 32)/1.8$
Degree Fahrenheit (°F)	kelvin (K)	$K = (°F + 459.67)/1.8$
Degree Rankine (°R)	kelvin (K)	$K = °R/1.8$
Kelvin (K)	degree Celsius (°C)	$°C = K - 273.15$

Conversion scale

°F	°C	K	°F	°C	K	°F	°C	K	°F	°C	K
−40	−40	233.15	320	160	433.15	1040	560	833.15	1760	960	1233.15
−30			340			1060			1780		
−20	−30	243.15	360	180	453.15	1080	580	853.15	1800	980	1253.15
−10			380			1100			1820		
0	−20	253.15	400	200	473.15	1120	600	873.15	1840	1000	1273.15
10	−10	263.15	420	220	493.15	1140	620	893.15	1860	1020	1293.15
20			440			1160			1880		
30	0	273.15	460	240	513.15	1180	640	913.15	1900	1040	1313.15
40			480			1200			1920		
50	10	283.15	500	260	533.15	1220	660	933.15	1940	1060	1333.15
60			520			1240			1960		
70	20	293.15	540	280	553.15	1260	680	953.15	1980	1080	1353.15
80			560			1280			2000		
90	30	303.15	580	300	573.15	1300	700	973.15	2020	1100	1373.15
100	40	313.15	600	320	593.15	1320	720	993.15	2040	1120	1393.15
110			620			1340			2060		
120	50	323.15	640	340	613.15	1360	740	1013.15	2080	1140	1413.15
130			660			1380			2100		
140	60	333.15	680	360	633.15	1400	760	1033.15	2120	1160	1433.15
150			700			1420			2140		
160	70	343.15	720	380	653.15	1440	780	1053.15	2160	1180	1453.15
170			740			1460			2180		
180	80	353.15	760	400	673.15	1480	800	1073.15	2200	1200	1473.15
190	90	363.15	780	420	693.15	1500	820	1093.15	2220	1220	1493.15
200			800			1520			2240		
210	100	373.15	820	440	713.15	1540	840	1113.15	2260	1240	1513.15
220			840			1560			2280		
230	110	383.15	860	460	733.15	1580	860	1133.15	2300	1260	1533.15
240			880			1600			2320		
250	120	393.15	900	480	753.15	1620	880	1153.15	2340	1280	1553.15
260			920			1640			2360		
270	130	403.15	940	500	773.15	1660	900	1173.15	2380	1300	1573.15
280	140	413.15	960	520	793.15	1680	920	1193.15	2400	1320	1593.15
290			980			1700			2420		
300	150	423.15	1000	540	813.15	1720	940	1213.15	2440	1340	1613.15
310			1020			1740			2460		
320	160	433.15	1040	560	833.15	1760	960	1233.15	2480	1360	1633.15

Table A.4. One-dimensional SI conversion factors

To convert from	to	Multiply by
Area		
Acre	Meter^2	4.046 873 E+03
Foot^2	Meter^2	9.290 304 E-02
Inch^2	Meter^2	6.451 600 E-04
Hectare	Meter^2	1.000 000 E+04
Mile^2 (international)	Meter^2	2.589 998 E+06
Yard^2	Meter^2	8.361 274 E-01
Energy		
Btu (international)	Joule	1.055 056 E+03
Calorie (international)	Joule	4.186 800 E+00
Erg	Joule	1.000 000 E-07
Kilocalorie (international)	Joule	4.186 800 E+03
Force		
Dyne	Newton	1.000 000 E-05
Kilogram-force	Newton	9.806 650 E+00
Pound-force	Newton	4.448 222 E+00
Length		
Angstron	Meter	1.000 000 E-10
Foot	Meter	3.048 000 E-01
Inch	Meter	2.540 000 E-02
Micron	Meter	1.000 000 E-06
Mile (international)	Meter	1.609 344 E+03
Yard	Meter	9.144 000 E-01
Mass		
Grain	Kilogram	6.479 891 E-05
Gram	Kilogram	1.000 000 E-03
Ounce	Kilogram	2.834 952 E-02
Pound	Kilogram	4.535 924 E-01
Ton (metric)	Kilogram	1.000 000 E+03
Ton (short)	Kilogram	9.071 847 E+02
Pressure		
Atmosphere (normal)	Pascal	1.013 25 E+05
Dyne/centimeter^2	Pascal	1.000 000 E-01
Kilogram-force/meter^2	Pascal	9.806 650 E+00
Pound-force/inch^2 (psi)	Pascal	6.894 757 E+03
Torr (mm Hg, 0°C)	Pascal	1.333 22 E+02
Time (mean solar units)		
Day	Second	8.640 000 E+04
Hour	Second	3.600 000 E+03
Minute	Second	6.000 000 E+01
Month (mean calendar)	Second	2.628 000 E+06
Year (calendar)	Second	3.153 600 E+07
Volume		
Barrel (oil, 42 gal)	Meter^3	1.589 873 E-01
Foot^3	Meter^3	2.831 685 E-02
Gallon (U.S. liquid)	Meter^3	3.785 412 E-03
Inch^3	Meter^3	1.638 706 E-05
Liter	Meter^3	1.000 000 E-03
Milliliter	Meter^3	1.000 000 E-06
Yard^3	Meter^3	7.645 549 E-01

Table A.5. Two-dimensional SI conversion factors

Distance/time

SI unit: meters per second (m/sec)

Time (mean solar units)	Distance					
	Meter	Centimeter	Kilometer	Inch	Foot	Mile (international)
Second	1.000 000 E+00	1.000 000 E-02	1.000 000 E+03	2.540 000 E-02	3.048 000 E-01	1.609 344 E+03
Minute	1.666 667 E-02	1.666 667 E-04	1.666 667 E+01	4.233 334 E-04	5.080 010 E-03	2.682 240 E+01
Hour	2.777 778 E-04	2.777 778 E-06	2.777 778 E-01	7.055 556 E-06	8.466 667 E-05	4.470 400 E-01
Day	1.157 407 E-05	1.157 407 E-07	1.157 407 E-02	2.939 821 E-07	3.527 786 E-06	1.862 667 E-02

Energy/mass

SI unit: joules per kilogram (J/kg)

Mass	Energy			
	Joule	Btu (international)	Calorie (international)	Kilocalorie (international)
Kilogram	1.000 000 E+00	1.055 056 E+03	4.186 800 E+00	4.186 800 E+03
Gram	1.000 000 E+03	1.055 056 E+06	4.186 800 E+03	4.186 800 E+06
Pound	2.204 622 E+00	2.326 000 E+03	9.230 313 E+00	9.230 313 E+03
Ton (short)	1.102 311 E-03	1.163 000 E+00	4.615 157 E-03	4.615 157 E+00
Ton (metric)	1.000 000 E-03	1.055 056 E+00	4.186 800 E-03	4.186 800 E+00
Grain	1.543 236 E+04	1.628 200 E+07	6.461 221 E+04	6.461 221 E+07

Energy/time

SI unit: joules per second (J/sec) = watt

Time (mean solar units)	Energy				
	Joule	Btu (international)	Calorie (international)	Kilocalorie (international)	Foot-pound force
Second	1.000 000 E+00	1.055 056 E+03	4.186 800 E+00	4.186 800 E+03	1.355 818 E+00
Minute	1.666 667 E-02	1.758 427 E+01	6.978 000 E-02	6.978 000 E+01	2.259 697 E-02
Hour	2.777 778 E-04	2.930 711 E-01	1.163 001 E-03	1.163 001 E+00	3.766 161 E-04
Day	1.157 407 E-05	1.221 130 E-02	4.845 833 E-05	4.845 833 E-02	1.569 234 E-05

Energy/volume

SI unit: joules per cubic meter (J/m^3)

Volume	Energy			
	Joule	Btu (international)	Calorie (international)	Kilocalorie (international)
Meter^3	1.000 000 E+00	1.055 056 E+03	4.186 800 E+00	4.186 800 E+03
Foot^3	3.531 466 E+01	3.725 895 E+04	1.478 554 E+02	1.478 554 E+05
Liter	1.000 000 E+03	1.055 056 E+06	4.186 800 E+03	4.186 800 E+06
Gallon (U.S. liquid)	2.641 720 E+02	2.787 163 E+05	1.106 035 E+03	1.106 035 E+06

Table A.5 (continued)

Mass/area

SI unit: kilograms per square meter (kg/m^2)

Area	Mass					
	Kilogram	Microgram	Gram	Grain	Pound	Ton (short)
Meter^2	1.000 000 E+00	1.000 000 E-09	1.000 000 E-03	6.479 891 E-05	4.535 924 E-01	9.071 847 E+02
Centimeter^2	1.000 000 E+04	1.000 000 E-05	1.000 000 E+01	6.479 891 E-01	4.535 924 E+03	9.071 847 E+06
Hectare	1.000 000 E-04	1.000 000 E-13	1.000 000 E-07	6.479 891 E-09	4.535 924 E-05	9.071 847 E-02
Acre	2.471 044 E-04	2.471 044 E-13	2.471 044 E-07	1.601 210 E-08	1.120 847 E-04	2.241 693 E-01
Foot^2	1.076 391 E+01	1.076 391 E-08	1.076 391 E-02	6.974 896 E-04	4.882 428 E+00	9.764 855 E+03
Mile^2 (international)	3.861 022 E-07	3.861 022 E-16	3.861 022 E-10	2.501 900 E-11	1.751 330 E-07	3.502 659 E-04

Mass/time

SI unit: kilograms per second (kg/sec)

Time (mean solar units)	Mass					
	Kilogram	Gram	Grain	Pound	Ton (short)	Ton (metric)
Second	1.000 000 E+00	1.000 000 E-03	6.479 891 E-05	4.535 924 E-01	9.071 847 E+02	1.000 000 E+03
Minute	1.666 667 E-02	1.666 667 E-05	1.079 982 E-06	7.559 873 E-03	1.511 974 E+01	1.666 667 E+01
Hour	2.777 778 E-04	2.777 778 E-07	1.799 997 E-08	1.259 979 E-04	2.519 958 E-01	2.777 778 E-01
Day	1.157 407 E-05	1.157 407 E-08	7.499 874 E-10	5.249 912 E-06	1.049 982 E-02	1.157 407 E-02
Year	3.170 979 E-08	3.170 979 E-11	2.054 760 E-12	1.438 332 E-08	2.876 666 E-05	3.170 979 E-05

Mass/volume

SI unit: kilograms per cubic meter (kg/m^3)

Volume	Mass						
	Kilogram	Nanogram	Microgram	Milligram	Gram	Grain	Pound
Meter^3	1.000 000 E+00	1.000 000 E-12	1.000 000 E-09	1.000 000 E-06	1.000 000 E-03	6.479 891 E-05	4.535 924 E-01
Liter	1.000 000 E+03	1.000 000 E-09	1.000 000 E-06	1.000 000 E-03	1.000 000 E+00	6.479 891 E-02	4.535 924 E+02
Milliliter, centimeter^3	1.000 000 E+06	1.000 000 E-06	1.000 000 E-03	1.000 000 E+00	1.000 000 E+03	6.479 891 E+01	4.535 924 E+05
Barrel (oil, 42 gal)	6.289 810 E+00	6.289 810 E-12	6.289 810 E-09	6.289 810 E-06	6.289 810 E-03	4.075 728 E-04	2.853 010 E+00
Gallon (U.S. liquid)	2.641 720 E+02	2.641 720 E-10	2.641 720 E-07	2.641 720 E-04	2.641 720 E-01	1.711 806 E-02	1.198 264 E+02
Foot^3	3.531 466 E+01	3.531 466 E-11	3.531 466 E-08	3.531 466 E-05	3.531 466 E-02	2.288 351 E-03	1.601 846 E+01

Volume/mass

SI unit: cubic meters per kilogram (m^3/kg)

Mass	Volume				
	Meter^3	Liter	Gallon (U.S. liquid)	Foot^3 (SCF)	Barrel (oil, 42 gal)
Kilogram	1.000 000 E+00	1.000 000 E-03	3.785 412 E-03	2.831 685 E-02	1.589 873 E-01
Pound	2.204 622 E+00	2.204 622 E-03	8.345 404 E-03	6.242 796 E-02	3.505 070 E-01
Ton (metric)	1.000 000 E-03	1.000 000 E-06	3.785 412 E-06	2.831 685 E-05	1.589 873 E-04
Ton (short)	1.102 311 E-03	1.102 311 E-06	4.172 703 E-06	3.121 399 E-05	1.752 535 E-04

Table A.5 (continued)

Volume/time

SI unit: cubic meters per second (m^3/sec)

Time (mean solar units)	Volume					
	Meter^3	Liter	Gallon (U.S. liquid)	Foot^3	Barrel (oil, 42 gal)	Yard^3
Second	1.000 000 E+00	1.000 000 E-03	3.785 412 E-03	2.831 685 E-02	1.589 873 E-01	7.645 549 E-01
Minute	1.666 667 E-02	1.666 667 E-05	6.309 021 E-05	4.719 475 E-04	2.649 788 E-03	1.274 258 E-02
Hour	2.777 778 E-04	2.777 778 E-07	1.051 504 E-06	7.865 792 E-06	4.416 314 E-05	2.123 764 E-04
Day	1.157 407 E-05	1.157 407 E-08	4.381 264 E-08	3.277 413 E-07	1.840 131 E-06	8.848 989 E-06
Year	3.170 979 E-08	3.170 979 E-11	1.200 346 E-10	8.979 214 E-10	5.041 454 E-09	2.424 402 E-08

Appendix B

GLOSSARY

R. H. Ross
T. W. Robinson

Å (angstrom) — a unit of length; 10^{-8} cm

AAS — atomic absorption spectrometry

abatement — the method of reducing the degree of pollution intensity; also, the use of such
a method

abscission — the natural separation of flowers, fruit, and leaves from plants by the development
and subsequent disorganization of the separation layer

absorption — the penetration of one substance into the inner structure of another

acclimation — the physiological and behavioral adjustments of an organism to changes in its
immediate environment

acclimatization — the acclimation or adaptation of a particular species, over several generations,
to a marked change in the environment

acfm — actual cubic feet per minute

acid gas — hydrogen sulfide (H_2S) and carbon dioxide (CO_2)

accuracy — the difference between the measured value and the true value which has been established
by an accepted reference method procedure.

acute exposure — an exposure of such intensity that a crisis rapidly develops

acute toxicity — any poisonous effect produced within a short period of time, usually within
24 to 96 hr, that results in severe biological harm or death

adaptation — a change in the structure or habit of an organism that increases the overall
compatibility of the organism with its environment

adenoma — a neoplasm of glandular epithelium

adsorption — adherence of the atoms, ions, or molecules of a gas or liquid to the surface of
another substance (the adsorbent)

adventitious buds — buds of organs or tissues that develop in an abnormal position (e.g., roots that develop from stems)

aeration — creation of intimate contact between air and a liquid by (1) spraying the liquid in the air; (2) bubbling air through the liquid or (3) agitating the liquid to promote surface absorption of air

aerobic — refers to an organism or life process that uses, or can exist only in the presence of, oxygen

aerosol — a suspension (in a body of gas) of liquid or solid particles of such size that they tend to remain suspended for an indefinite period

afterdamp — toxic mixture of gases produced by a coal mine fire, an explosion, or the partial burning of explosions used in breaking down coal

agglomerating coal — (see caking coal)

agglutinating coal — (see coking coal)

aggregative fluidized bed — a bed in such a condition of fluidization that the gas or fluid phase rises through the bed primarily in the form of bubbles

AHH — aryl hydrocarbon hydroxylase

air contaminant — any foreign material in the air; that is, material other than oxygen, nitrogen, the noble gases, water vapor, and carbon dioxide

air gasification — partial combustion; the process by which coal is burned with about one-half of the air needed for complete combustion, producing a nonfuel gas consisting mostly of carbon monoxide, a fuel species, and nitrogen; used widely by industry (in electrical power generation) and called "power gas" during the 1920s; heating value is 1/6 (per unit volume) that of natural gas

air pollution — the action of making air physically impure or unclean or contaminating air with man-made waste (The primary cause of air pollution is the burning of fossil fuels in homes, cars, factories, and power plants)

air sampling — the collection and analysis of air samples for detection or measurement of radioactive substances, particulate matter, or chemical pollutants

algae — simple plants, many microscopic, containing chlorophyll. (Freshwater algae are diverse in shape, color, size, and habitat. They are the basic link in the conversion of inorganic constituents in water into organic constituents.)

algal bloom — a proliferation of algae on the surface of lakes, streams, or ponds, which is stimulated by phosphate enrichment

alicyclic — a group of organic compounds characterized by arrangement of the carbon atoms in closed-ring structures [These compounds have properties resembling those of aliphatics. Three alicyclic subgroups are: (1) cycloparaffins (saturated), (2) cycloolefins, and (3) cycloacetylenes (cyclynes), having a triple bond.]

aliphatic — one of the major groups of organic compounds characterized by stright-chain arrangement of the constituent carbon atoms [Aliphatic hydrocarbons include three subgroups: (1) paraffins (alkanes), which are saturated and unreactive; (2) olefins (alkenes or alkadienes), which are unsaturated and quite reactive; and (3) acetylenes (alkynes), which contain a triple bond and are highly reactive. In complex structures, the chains may be branched or cross-linked.]

alkadiene — (diolefin, diene); an olefinic compound containing two double bonds, either adjacent or conjugated (see olefin)

alkene — an aliphatic hydrocarbon compound containing one double bond, such as ethylene (see olefin)

allochthonous coal — coal that originated from accumulations of plant debris that had been transported from their place of growth and deposited elsewhere (see drift theory)

ambient air quality — definition of the outdoor atmosphere as it exists around people, plants, and structures contrasted to the outdoor atmosphere in immediate proximity to emission sources

ambient air quality criteria — a scientific relationship between particular concentrations and durations of specific air contaminants and the effects they produce on persons, animals, plants, or materials ("Criteria," in this sense, has a connotation distinct from "standard;" although the dictionary lists the two as synonymous, care must be exercised not to confuse them in any consideration of air pollution control.)

ambient temperature — temperature of the surrounding cooling medium, such as gas or liquid, which comes into contact with heated parts of the apparatus in use

anadromous — type of fish that ascend rivers from the sea to spawn

anaerobic — refers to an organism or life process that does not use, or cannot exist in the presence of, oxygen

anoxic — relating to or marked by an oxygen deficiency

antagonism — opposition or contrary action, as between muscles or between medicines

anthracite (hard coal) — generally a hard, black, lustrous coal containing a high percentage of fixed carbon and a low percentage of volatile matter; a high-rank coal

anthraxylon — the vitreous-appearing components of coal, which, in thin section, are shown to be derived from the woody tissues of plants (such as stems, limbs, branches, twigs, and roots) which are changed and broken into fragments of greatly varying size through biological decomposition and weathering during the peak stage, and later are flattened and transformed into coal through the coalification process; the woody tissues still remain as definite units

anticarcinogen — a substance or agent that opposes the action of carcinogens

API gravity — scale adopted by the American Petroleum Institute for measuring the density of oils

$$°API = (141.5/sp\ gr\ 60/60°F) - 131.5$$

apparent photosynthesis — measurement of the net result of photosynthesis with no attempted correction for the gas exchange associated with respiration

aquatic biota — the sum total of living organisms of any designated aquatic area

aromatic (arene) — a major group of unsaturated cyclic hydrocarbons containing one or more rings (e.g., benzene) (These highly reactive and chemically versatile compounds have a strong, but not unpleasant, odor — thus the name aromatic.)

aromatization — the use of hydrogen in the presence of heat, pressure, and catalysts (usually platinum) to convert petroleum hydrocarbons to molecular structures that form high-octane gasoline for automobiles

aryl hydrocarbon (benzo[a]pyrene) hydroxylase (AHH) — one of the microsomally bound, mixed-function oxidases that convert a variety of lipid-soluble compounds to water-soluble forms

ash — inorganic residue remaining after ignition of combustible substances

asphaltene — any of the components of a bitumen (as asphalt) that are soluble in carbon disulfide, but not in paraffin naphtha and that constitute the solid dispersed particles of the bitumen and consist chiefly of high-molecular-weight hydrocarbons

assimilative capacity — capacity of a water body to receive, dilute, and carry away wastes without harming water quality; in the case of organic matter, also includes the capacity for natural biological oxidation, which may be expressed in pounds per day at a specific river flow rate and temperature

associated gas — raw natural gas occurring with crude oil (in the same well)

ASV — anodic stripping voltammetry; an electrochemical method of analysis

ATCC — American Type Culture Collection

atm — atmosphere

attapulgite (palygorskite) — a group of clay minerals (hydrous magnesium aluminum silicates) characterized by their distinctive rodlike shape

autochthonous bacteria — bacterial strains that originated in the locality in which they are found

autochthonous coal — coal believed to have originated from accumulations of plant debris located at the place where the plants grew (Two modes of origin can be distinguished: terrestrial and aquatic.)

autotrophs — organisms capable of using (oxidizing) simple chemical elements or compounds, such as iron, sulfur, or nitrates, to obtain energy for growth

auxiliary services — systems that may be added to coal-fired steam boilers for air or water pollution control (e.g., electrostatic precipitators; stack gas scrubbers; closed-loop, evaporative cooling-water systems; air-cooling systems; and wastewater treatment and reuse systems)

auxin — a plant hormone or growth substance

available carbon — carbon that is chemically combined with oxygen in any way and is therefore available for combustion

°B — degrees Baume

background concentration — those pollutant concentrations that are due to natural sources

bacteria — single-celled microorganisms that lack chlorophyll (Some bacteria are capable of causing human, animal or plant diseases; others are essential to pollution control because they break down organic matter in the air and water.)

banded coal — the common variety of bituminous and subbituminous coal [Banded coal consists of a sequence of irregularly alternating layers or lenses: (1) homogeneous black material having a brilliant vitreous luster; (2) grayish black, less brilliant, striated material, usually of silky luster; and (3) generally thinner bands or lenses of soft, powdery, and fibrous particles of mineral charcoal.]

BaP — benzo[a]pyrene

barrel (bbl) — a liquid measure of oil, usually crude oil, equal to 42 American gallons, or about 306 lb [One barrel equals 5.6 ft^3 or 0.159 m^3. For crude oil, 1 bbl is about 0.136 tonnes, 0.134 long tons, and 0.150 short tons. The energy values (in million Btu) of petroleum products (per bbl) are: crude petroleum, 5.6; residual fuel oil, 6.29; distillate fuel oil, 5.83; gasoline, 5.25; jet fuel (kerosene type), 5.67; jet fuel (naphtha type), 5.36; kerosene, 5.67; petroleum coke, 6.02; and asphalt, 6.64.]

bbl — barrel

bbl/day — barrels per day

Beer's Law — the absorbance of a homogeneous sample containg an absorbing substance is directly
proportional to the concentration of the absorbing substance. The absorbance, A, is given
by the expression: $A = \log_{10}\left(\dfrac{I_\circ}{I}\right)$ where $I_\circ$ is the radiant power incident on the sample
and I is the radiant power transmitted through the sample

benthic — refers to all plant or animal life inhabiting the sea bottom, lake bottom, etc.
(adjective form of benthos)

benthos — the animal and plant life that inhabit the sea bottom, lake bottom, river bottom, etc.

benzene (C_6H_6) — a colorless, liquid hydrocarbon that is made from coal tar and by catalytic
reforming of naphthenes (Benzene is used in the manufacture of phenol, styrene, nylon,
detergents, aniline, phthalic anhydride, and other compounds; as a solvent; and as a
component of high-octane gasoline.)

benzine — an old term for light petroleum distillates, covering the gasoline and naphtha range
(no longer considered good usage in current petroleum work)

Bi-Gas — a process for coal gasification being developed by the Office of Coal Research and the
American Gas Association

binary cycle — an energy recovery system that results in heat exchange between two separate
fluid-circulation systems (The purpose of a binary cycle is to obtain higher efficiencies
from the energy source.)

bioaccumulation — the ability of an organism to continue to concentrate a substance throughout
its active metabolic lifetime so that the concentration factor, if calculated, would be
continuously increasing during the lifetime of the organism

bioassay — an assay method using a change in biological activity as a qualitative or quantitative
means of analyzing a material's response to biological treatment; a method of determining
toxic effects of industrial wastes and other wastewaters by using viable organisms as test
organisms

biochemical oxygen demand (BOD) — the quantity of oxygen used in the biological oxidation of
organic matter, in a specified time and at a specified temperature, determined by its
availability to serve as food for microorganisms (BOD can be related to the oxygen resources
of a stream; for example, after dilution and mixing in a stream, one part of BOD will
consume one part of oxygen in the stream.)

bioconcentration (concentration factor) — the ability of an organism, or a population of many
organisms of the same trophic level, to concentrate a substance from an aquatic system

bioconversion — the conversion of one form of energy into another by plants or microorganisms (Synthesis of organic compounds from carbon dioxide by plants is bioconversion of solar energy into stored chemical energy. Similarly, digestion of solid wastes or sewage sludge by microorganisms to form methane is bioconversion of one form of stored chemical energy into another more useful form.)

biodegradation — the biochemical breakdown of complex, large, organic molecules into small simple molecules; decomposition by bacteria, fungi, and other microorganisms

biogeochemistry — a branch of geochemistry dealing with biologic materials and the results of biochemical research

bilogical half-life — the time required for a biological system (such as a human or animal) to eliminate, by natural processes, half the amount of a substance (such as a radioactive material) that has entered that system

biological (biochemical) oxidation — a process by which bacteria and other microorganisms decompose and feed on complex organic materials (Self-purification of waterways, as well as activated sludge and trickling-filter waste-treatment processes, depend on this principle.)

biomagnification — the existence of a substance at successively higher concentrations with increasing trophic levels in ecosystem food chains

biomass — the weight of all life in a specified unit of environment or an expression of the total mass or weight of a given population, both plant and animal.

biomonitoring — the use of living organisms to test (1) the suitability of effluents for discharge into receiving waters and (2) the quality of such waters downstream from a discharge

biosphere — the portion of the earth and its atmosphere capable of supporting life

biosynthesis — the production of substances from other, usually simpler, compounds with the aid of living organisms

biota — the sum total of the living organisms of any designated area

biotransfer — the process by which living organisms, such as bacteria, can convert a chemical compound into another

biox — biological oxidation

bitumen — a general name for various solid and semisolid hydrocarbons; a native substance of dark color that is comparatively hard and nonvolatile and is composed principally of hydrocarbon

bituminous coal (soft coal) — the most important and plentiful rank (type) of coal. [Bituminous
    coal is the chief fuel in steam-electrical plants; other important industrial uses include
    provision of coke for the steel industry, use as raw material for thousands of coke by-
    products (including gas, light oils, and chemicals), use in aluminum, cement, food, paper,
    and textile production, and heat source for homes and buildings. Large deposits of
    bituminous coal are found in many states, both east and west of the Mississippi River. The
    most important bituminous coal beds in the United States are located in a region west of the
    Appalachian Mountains, which extends southwest from Ohio and Pennsylvania to Alabama. Nearly
    three of every four tons of bituminous coal mined in the United States each year come from
    this eastern coal area.]

blackdamp — the gas found in coal mines that is responsible for most mine explosions (Due to its
    odorless and colorless characteristics, blackdamp is extremely difficult to detect; accidental
    ignition of this gas causes powerful explosions.)

blended fuel oil — a mixture of residual and distillate fuel oils

blending naphtha — a distillate used to thin heavy stocks to facilitate processing (e.g., to
    thin lubricating oil in dewaxing processes)

blending stock — any of the stocks used to make commerical gasoline (natural gasoline, straight-
    run gasoline, cracked gasoline, polymer gasoline, alkylate, and aromatics)

block coal — a peculiar kind of coal that breaks into large cubical blocks; a variety of tough
    coal

blood-brain barrier — the barrier created by semipermeable cell walls and membranes to passage
    of some molecules from the blood to the cells of the central nervous system

bloom — a visibly concentrated growth of algae and/or other aquatic plants

BOD — biochemical oxygen demand

body burden — the total amount of a specific substance (e.g., lead) in an organism, including
    the amount stored and the amount absorbed

boghead coal — a nonbanded coal having translucent attritus that consist predominantly of algae

boiling fluidized bed — a fluidized bed through which part of the fluidizing medium passes in
    the form of bubbles of a size approximately equal to the size of the solid particles, but
    very small relative to the dimensions of the containing vessel

BP — boiling point

bright coal (anthraxylon) — the constituent of banded coal that is of a jet-black pitchy appear-
    ance, is more compact than dull coal, breaks with a conchoidal fracture visible when
    viewed microscopically, and in thin section, always shows preserved cell structure of
    woody plant tissue

British thermal unit (Btu) — the quantity of heat necessary to raise the temperature of 1 lb of water 1°F (One Btu equals 252 cal/g(mean), 778 ft-lbs, 1055 J, and 0.293 Whr.)

brown coal — lignite

brown lignite — a lignite lower in rank than black lignite

bryophytes — the mosses and liverworts

bryozoan — a lophophorian coelomate animal having an exoskeleton of small calcareous cases or gelatinous masses; and/or pseudocoelomate bilateral animals distinguished by a distal circlet of ciliated tentacles, with the anus opening inside the circlet

Btu — British thermal unit

bud — a compacted stem having leaves, and sometimes flowers, that elongates to produce the mature shoot and its attached organs

butane — a colorless gas that is composed of four carbon and ten hydrogen atoms (At one time, butane was the principal component in bottled gas; but, currently, refiners are channeling it into the manufacture of petrochemicals.)

by-products (residuals) — secondary products, (possibly of commercial value) that are obtained from the processing of raw material [By-products may be the residues of the gas production process (such as coke, tar, and ammonia), or they may be the result of further processing of such residues (such as ammonium sulfate).]

by-product stream — material streams within the coal-conversion plant that are of secondary resale value as compared with the product stream (e.g., sulfur, ammonia, gypsum, phenols)

°C — degrees Celsius (centigrade)

C_1, C_2, C_3, C_4, etc. — hydrocarbons having 1, 2, 3, 4, etc., carbon atoms

caking coal — coal that, when heated, gives off volatile matter and then cakes (or cokes) into a continuous cake (All coking coals are caking coals, but not all caking coals are coking coals.)

cal — calorie

calcine — to expel volatile matter by heat (as carbon dioxide, water, or sulfur) with or without oxidation; to roast; to burn

callus — a thickened area or protuberance on the surface of a plant consisting of special wound tissue, which may be cultured in nutrient solution

calorie (large calorie) — the amount of heat required to change the temperature of 1 kg of water by 1°C.

calorific value — the heat, measured in calories or Btu, released by combustion of a unit
quantity of fuel

cancer — a malignant and invasive growth or tumor, especially one origninating in epithelium

cannel coal — a massive, noncoking, tough, clean, block coal of fine, even, compact grain, dull
luster, and commonly conchoidal cross fracture; a nonbanded coal consisting predominantly
of spores [Cannel coal, typically, has a low fuel ratio and contains a high percentage of
hydrogen. It is easily ignited; burns with a long, yellow flame; leaves a moderate
amount of ash; and is pulverulent in burning.]

carbon black — any of various black substances consisting of carbon obtained as soot from
partial combustion of hydrocarbons (Carbon black is used as a reinforcing agent in
automobile tires and is used in paint, carbon paper, printing ink, and electric resistors.
Total carbon black is produced from petroleum and natural gas.)

carbonization — in the coalification process, the progressive changes undergone by the preserved
organic matter and biochemical decomposition products between the death of the plant or
animal and the stage of essentially complete reduction to residual carbon, in situ

carbon oxides — compounds of carbon and oxygen produced when the carbon of fossil fuels combines
with oxygen during burning (The two most common carbon oxides are carbon monoxide and
carbon dioxide.)

carbureted blue (water) gas — water gas to which gaseous hydrocarbons have been added, originally
for the purpose of increasing the flame's luminosity, but now to increase the Btu content
of the water gas

carcinogen — a substance or agent that produces or incites cancerous growth

carcinogenesis — the production and development of cancer

carcinoma — an epithelial cell, new growth, or malignant tumor that is enclosed in connective
tissue and tends to infiltrate and give rise to metastases

catalyst — in a chemical reaction, an extra substance that is usually used to speed up the
reaction to produce the desired result (The catalyst usually does not appear in the
reaction product in any appreciable amount and therefore is not used up, as are the
main ingredients in the reaction; the catalyst may have to be periodically discarded
because of contamination.)

catalytic Hydrogenation process — a method for adding hydrogen to substances using a catalyst
to promote the reaction [It can be used to convert coal to a liquid and/or to remove
sulfur from residual oil, coal, crude coal liquids (such as coal tar), and coal extracts.]

catalytic cracking (Cat-Cracker) — a process of cracking in which the catalyst supplements
heat and pressure; also, the conversion of high-boiling hydrocarbons into lower-boiling
substances by means of a catalyst (Feedstocks may range from naphtha cuts to reduced
crude oils.)

cfh — cubic feet per hour

cfm — cubic feet per minute

cfs — cubic feet per second

chalcophile elements — elements that show a strong affinity for sulfur and that are readily soluble in molten iron monosulfide; elements commonly found in sulfide ores

channeling fluidized bed — an abnormality characterized by the establishment of flow paths through which disproportionately large amounts of fluid (gas) pass up the column

char — the solid, carbonaceous residue that results from incomplete combustion of organic material (Char can be burned for its energy content or, if free from large amounts of impurities, processed further for production of activated carbon for use as a filtering medium. Char produced from coal is generally called coke, whereas that produced from wood or bone is called charcoal.)

Char-Oil-Energy Development process (COED) — a process being developed by the Office of Coal Research for low-temperature distillation of coal-carbonization products (The process is designed to produce clean liquids, gases, and char for fuel, the product balance depending upon economic factors.)

chelate — the type of coordination compound in which a central atom (usually a metal) is joined by covalent bonds to two or more other atoms of one or more other molecules or ions (called ligands) so that heterocyclic rings are formed with the central (metal) atom as part of each ring

chemical oxygen demand (COD) — laboratory measurement of the amount of oxygen consumed under specific conditions in the oxidation of organic material by a strong chemical oxidant that decomposes both biodegradable (measured by biochemical oxygen demand) and nonbiodegradable organic matter

chlorite — a term used for a group of platy, hydrous silicates of aluminum, ferrous iron, and magnesium that are closely related to the micas

chloroplast — chlorophyll-containing plastid that serves as the seat of photosynthesis and starch formation

chokedamp — air that contains large proportions of carbon dioxide (Cokedamp becomes a hazard when a section of the mine is cut off from a proper air supply. A flame safety lamp is used to detect chokedamp; when present, chokedamp inhibits burning of the lamp's flame.)

chromosome — cellular elements regarded as the carriers of hereditary characteristics

chronic exposure — long-term continued exposure

clarain — an ingredient of banded coal that appears, megascopically, as thin or very thick bands that are intrinsically stratified parallel to the bedding plane, most often has as a silky luster, and has scattered or diffuse reflection markedly less intense than the specular reflection of vitrain under the same illumination

Claus unit — catalytically converts hydrogen sulfide to by-product sulfur and incinerates any residual gaseous sulfur compounds to sulfur dioxide which is removed in a subsequent tail-gas treating unit

clinker — undesirable agglomeration of ash, which is usually in the fuel bed or on the grate, but which can also be on the upper surfaces of the furnace or boiler; usually takes the form of a hard, fused mass

clone — in tissue culture, a group of cells descended from a single cell

cm — centimeter

cm^3 — cubic centimeter

c-mitosis — mitotic cell division produced by colchicine and having certain distinguishing characteristics

coal — a solid, combustible, organic material formed by the decomposition of vegetable material without free access to air (Chemically, coal is composed chiefly of condensed, aromatic ring structures of high molecular weight; thus, it has a higher ratio of carbon-to-hydrogen content than does petroleum.)

coal alkylation — a process being developed by the Bureau of Mines to convert sulfur-bearing coal into a nonpolluting fuel

coal conversion — the conversion of coal to a liquid or gas suitable for use as a fuel

coal gas — the gas that comes from retorts, mufflers, or ovens during the distillation of coal (Coal gas has a high illuminating value and is a relatively suitable engine fuel. Large quantities of coal gas are produced when coal is used to make coke, coal tar, benzoil, toluene, ammonia, and other products.)

coal gasification — the conversion of coal to a gas suitable for use as a fuel (HYGAS, CO_2-Acceptor, Bi-Gas, methanation, Lurgi ATGAS processes)

coal-gasification (high-Btu gas) — Two existing commercial processes (Lurgi, and Koppers-Totzek) that should be able to produce a synthetic pipeline gas once the methanation processes step has been proven on a large scale.

coal-gasification (low Btu gas) — production of gas from coal with air, as distinct from oxygen (The Btu content will range from 150 to 175 Btu/ft^3, which prevents any storage of the gas between producer and boiler.)

coalification — those processes involved in the genetic and metamorphic history of coal beds (The three essential processes of coalification are incorporation, vitrinization, and fusinization.)

coal liquefaction (coal hydrogenation) — the conversion of coal into liquid hydrocarbons and related compounds by hydrogenetion (Three projects of the office of Coal Research include the Consol pilot plant for low-sulfur liquid fuels, the FMC Corporation's project COED, and the Pittsburgh and Midway Mining Corporation pilot plant project for low-ash-low-sulfur solvent-refined coal.)

coal oil — oil obtained by the destructive distillation of bituminous coal; also an archaic term for kerosene made from petroleum

coal preparation — a collective term for physical and mechanical processes applied to coal to make it suitable for a particular use

coal seam — a bed of coal usually thick enough to be mined with profit

coal-slurry pipeline — a pipeline that transports pulverized coal suspended in water

coal tar — a gummy, black substnace produced as a by-product of distillation of bituminous coal

coarse particulates — suspended solids having a Svedberg coefficient greater than 20,000 S, corresponding to a diameter larger than 0.15 µ for a particle of 2.65 specific gravity

cocarcinogen — a substance or agent that enhances or potentiates the effects of a carcinogen

COED — Char-Oil-Energy Development; an Office of Coal Research project for the development of liquid fuel from coal char (see Char-Oil-Energy Development process)

coefficient of haze (COH) — a measurement of visibility interference in the atmosphere

coke — a porous, solid residue that results from the incomplete combustion of coal heated in a closed chamber, or oven, with a limited supply of air (Coke is largely carbon and is desirable fuel in certain metallurgical industries.)

coke-oven gas — the gas secured from coke ovens during the production of coke (The properties of this gas are identical to coal gas, and the two products are interchangeable. Coke is particularly useful in making iron and steel and as an industrial fuel.)

coking coal — the most important of the bituminous coals, which burns with a long yellow flame and creates an intense heat when properly attended.

coleoptile — the first leaf in germination of monocotyledons

coliform index — an index of the water purity based on a count of its coliform bacteria coliform organism — any of a number of organisms common to the intestinal tract of man and animals

colloidal particulates — suspended solids having a Svedberg value greater than 100 S and less than 20,000 S, corresponding to a diameter between 0.01 μ (100 Å) and 0.15 μ for a particle having a specific gravity of 2.65.

colloid — a substance that, when dissolved in a liquid, will not diffuse readily through vegetable or animal membranes

colorimeter — an instrument used for color measurement based on optical comparison with standard colors

complexing — in chemistry, the process of incorporation into other compounds, such as through hydration, oxygenation, halogenation, and chelation

compression — the process by which dry, purified synthetic natural gas is compressed and delivered to the pipeline with a heating value of 980 to 1000 Btu/scf

concentration ratio — the ratio of the concentration of an element or radionuclide in an aquatic organism or its tissues to the concentration in the surrounding water under equilibrium, or steady-state, conditions

condensate — liquid hydrocarbon obtained by the combustion of a vapor or gas produced from oil or gas wells and ordinarily separated at a field separator and run as crude oil

connective tissue — that tissue which is primarily concerned with supporting bodily structures and binding parts together and secondarily serving other functions such as food storage, blood formation, and body defensive mechanisms. [It includes an embryonic connective tissue (mesenchyme and mucous) and adult connective tissue. The latter is subdivided into four general groups: vascular tissues (blood, lymph); connective tissue proper (areolar, white fibrous, yellow fibrous, reticular, adipose); cartilage; and bone.]

Conradson carbon content — residue left after heating and evaporating a lubricating oil under specified conditions

convection — the transfer of heat by circulation of a liquid or gas

conventional gas — natural gas (as contrasted to synthetic gas)

conversion factors — The energy content of most fuels can vary depending on their source and composition. The following energy equivalents are among those commonly used:

<u>Coal</u>

      Anthracite = 25.4 million Btu/ton
      Bituminous = 26.2 million Btu/ton
      Subbituninous = 19.0 million Btu/ton
      Lignite = 13.4 million Btu/ton

The average heating value of bituminous coal and lignite exported and used in electricity generation and industry in 1969 in the United States was 24.7 million Btu/ton.

<u>Petroleum</u>

  Crude petroleum = 5 to 6.0 million Btu/bbl (42 gal; depends on source)
  Residual fuel oil = 6.29 million Btu/bbl
  Distillate fuel oil = 5.83 million Btu/bbl
  Gasoline (including aviation) = 5.25 million Btu/bbl
  Jet fuel (kerosene type) = 5.67 million Btu/bbl
  Jet fuel (naphtha type) = 5.36 million Btu/bbl
  Kerosene = 5.67 million Btu/bbl
  Asphalt and road oil = 6.64 million Btu/bbl
  1 gal liquid propane = 36 ft^3 of gas with 2500 Btu/ft^3 value

<u>Natural gas</u>

  Dry = 1031 Btu/ft^3 at STP
  Wet = 1103 Btu/ft^3 at STP
  Liquids (av) = 4.1 million Btu/bbl

<u>Fissionable material</u>

  74 million Btu per gram of U-235 fissioned

conversion type process — a method of removing potential pollutants from a fuel by converting it to a clean-burning fuel (e.g., elimination of sulfur from residual oil by hydrogenation; conversion of coal to a low-sulfur fuel oil)

conversion tower, dry — a unit or structure for cooling water by conduction and convection into the air, much as does the radiator of an automobile

cooling tower, wet — a unit or structure, usually built of wood, for cooling water by evaporation

coprophagous organism — one that feeds on fecal material or excrement

cracking — a refining process by which the original physical pattern of oil is disintegrated by heat and pressure, with or without a catalyst, by breaking down and rearranging the molecular structure of hydrocarbon chains so that the resulting fragments are rearranged in new combinations that differ from the originals in physical characteristic (Simple thermal cracking involves only heat treatment under high pressure whereas catalytic cracking uses a catalyst with the heat treatment. The net result of cracking is the production of considerably increased yields of gasoline of higher antiknock quality than can be produced by ordinary distillation process.)

cracking of hydrocarbons — the splitting of hydrocarbon molecules into both heavy and light molecules, all lighter than the starting molecules (Cracking results from the application of pressure, a catalyst, heat in the absence of air, or various combinations of the three.)

creosote — a colorless or yellowish oily liquid that has a burning, smoky taste, contains a mixture of phenolic compounds (as guaiacol), and is obtained by the distillation of wood tar, especially that of beechwood

crude naphtha — light distillate made in the fractionation of crude oil

crude oil — petroleum liquids as they come from the ground; also referred to, simply, as "crude"

crushed coal — usually, coal that has been crushed for use in a cyclone furnace (as opposed to pulverized coal) [Normally 90 to 97.5% passes a number 4 screen and 7 to 13% passes a number 200 screen (fine-hole U.S. standard series).

cryogenics — low-temperature operations, generally below 100°C

curtisite — a rare, PAH-contianing mineral with over 30 homologous series of CH, CHS, and CHN compounds

cyclic compound — an organic compound having a structure characterized by a closed ring [There are three major groups of cyclic compounds: (1) alicyclic, (2) aromatic (also called arene), and (3) heterocyclic.]

cyclone — a separator that depends upon centrifugal force to separate particles or droplets from the stream (used to separate solids from gases and water from steam)

cycloolefin (cycloalkene) — an alicyclic hydrocarbon having two or more double bonds

cycloparaffin (naphthene) — an alicyclic hydrocarbon in which three of more of the carbon atoms in each molecule are united in a ring structure and each of these ring carbon atoms is joined to two hydrogen atoms or alkyl groups

cytology — branch of biology concerned with the study of cells as vital units with reference to their structure, function, multiplication, pathology, and life history

daf — dry, ash-free (usually coal)

dark fixation — one of two distinct types of reactions in the conversion of raw materials to organic compounds and oxygen in plants (Dark fixation does not require light and is independent of carbon dioxide, chlorophyll, and narcotics.

decomposition — the breakdown of a substance into its constituent parts

deep mining — the exploitation of coal or mineral deposits at depths exceeding about 1000 ft (Coal is usually deep-mined at not more than 1500 ft; mineral mines are deeper.)

degradation — a type of decomposition characteristic of high-molecular-weight substances such as proteins, polymers, and branched-chain sulfonates, resulting from oxidation, heat, solvents, and bacterial action

dehydrogenation — the process by which hydrogen is removed from compounds by chemical means

deposition — the laying down or precipitation of mineral matter that may eventually form rocks or that creates secondary land forms such as deltas and sand dunes

depuration — the action of freeing from impurities

desmid — popular name of algae of the family Desmidiaceae

desulfurization — the process by which sulfur and sulfur compounds are removed from gases or liquid hydrocarbon mixtures, usually by chemical or catalytic processes

detection limit — concentration, in mg/liter, that produces a reading equal to twice the standard deviation of a series of measurements near the blank level

detergent — any substance that reduces the surface tension of water; specifically, a surface-active agent which concentrates at oil-water interfaces, exerts emulsifying action, and thus aids in removing soils (Sodium soaps are examples that are widely used.)

detritus — dead organic tissues and organisms in an ecosystem, usually including the live microorganisms engaged in decomposition of the material

devolatilization — progressive loss of volatiles by the substance undergoing the gasification process; also, treatment of coal, usually by controlled heating, to remove all or part of the volatile constituents

dfg — dry flue gas

diagenesis — the processes involving physical and chemical change in sediments that convert them to consolidated rocks

dielectric constant — the ratio of the electrical capacity of a condenser containing the material to the capacity of the same condenser with material replaced by a vacuum; the ability of a material to maintain a difference in electrical charge over any specified distance

diesel oil — the oil fraction left after petroleum and kerosene have been distilled from crude oil

diffusion, molecular — a process of spontaneous intermixing of different substances, attributable to molecular motion and tending to produce uniformity of concentration

diol — chemical compound containing two hydroxyl groups

dispersant — a chemical agent used to break up concentrations of organic material (In cleaning oil spills, dispersants are used to disperse oil from the water surface.)

dispersion — a suspension of particles in a medium; the opposite of flocculation; a scattering process

dissolved oxygen (DO) — extent to which oxygen occurs in solution in water or wastewater (usually expressed as concentration, in parts per million, or per cent of saturation)

dissolved solids — the total amount of dissolved material, organic and inorganic, contained in water or wastes (Excessive dissolved solids make water unpalatable for drinking and unsuitable for industrial uses.)

distillate fuel oil — any fuel oil, gas oil, topped crude oil, or other petroleum oils derived by refining or processing crude oil or unfinished oils, in whatever type of plant such refining or processing may occur, which has a boiling range at atmospheric pressure from 550 to 1200°F

distillation range — the difference between the initial boiling point temperature and the end point temperature

DMBA — dimethylbenzo[a]anthracene

DNA — deoxyribonucleic acid; a complex protein of high molecular weight consisting of deoxyribose, phosphoric acid, and four bases (two purines, adenine and quanine, and two pyrimidines, thymine and cytosine); a nucleic acid present in chromosomes of the nuclei of cells that is considered to be the chemical basis of heredity and the carrier of genetic information

dolomite — a mineral, $CaMg(CO_3)_2$, commonly found with iron replacing the magnesium (ankerite); a common rock-forming mineral

drift — mist or spray that is carried out with the effluent air from cooling towers

drift theory — the theory of the origin of coal which holds that the plant matter constituting coal was washed from its original place of growth and deposited in another locality where coalification occurred

dry deposition — the removal of atmospheric particles to the earth's surface by fallout

dry flue gas (dfg) — gaseous products of combustion exclusive of water vapor [Separation of the vapor from the flue gas (a practical impossibility) is a theoretical concept used in combustion calculations.]

dry gas — a gas that does not contain the heavier fractions, which may easily condense under normal atmospheric conditions (e.g., methane and ethane)

dry natural gas — a gas that does not contain any considerable amount of easily separated natural gasoline

durain — material occurring in megascopic bands in coal characterized by gray to brownish-black color, rought surface, and faintly greasy luster; consists mainly of exinite and inertite

dust — a general term used to describe solid particles that are in the micron size range (e.g., fly ash, coarse dirt, and mechanically produced particles)

dustfall jar — an open-mouthed container used to collect large particles that fall out of the
air (The particles are measured and analyzed.)

dystrophic lakes — lakes between eutrophic and swamp stages of aging (Dystrophic lakes are
shallow and have high humus content, high organic matter content, low nutrient availability,
and high BOD.)

ecology — the science dealing with the relationship of all living things with each other and
with their environment

ecosystem — a complex of the community of living things and the environment forming a functioning
whole in nature

ED_{50} — edological dose; amount of contaminant added to a culture system, regardless of whether
all of it solubilizes, behaves as a toxicant and causes 50% abnormality in a culture of
oyster larvae

efficiency, thermal — relating to heat, a percentage indicating the available Btu input that is
converted to useful purposes (generally applied to combustion equipment; $E = \dfrac{Btu\ output}{Btu\ input}$)

effluent — a discharge of pollutants into the environment, partially or completely treated or
in its natural state (generally used in regard to dishcarges into waters)

Eh — oxidation-reduction potential

electrostatic precipitator — an air pollution control device that removes particulate matter
by imparting an electrical charge to particles in a gas stream for mechanical collection
on an electrode.

EMD_{50} (ecological mortality dose) — the amount of contaminant added to a culture system, regardless of whether all of it solubilizes, that behaves as a toxicant and causes 50% mortality
and abnormality in a culture of oyster larvae

emulsion — a stable mixture of two or more immiscible liquids held in suspension by small
percentages of substances called emulsifiers (These are of two types: (1) proteins or
carbohydrate polymers and (2) long-chain alcohols and fatty acids.)

endothermic — a slow chemcial reaction that absorbs heat

energy conversion — the transformation of energy from one form to another

enrichment — the addition of nitrogen, phosphorus, and carbon compounds or other nutrients into
a lake or other waterway that greatly increases the growth potential for algae and other
aquatic plants (Most frequently, enrichment results from the inflow of sewage effluent
or from agricultural runoff.)

environment — the sum of all external conditions and influences affecting the life, development,
and ultimately, the survival of an organism

environmental fate — the result of the physical, biological, and chemical interactions of a substance released to the environment

environmental transport — the movement through the environment of a substance (chemical, trace elements, etc.), including the physical, biological, and chemical interactions undergone by the substance

epidemiology — the study of diseases as they affect populations

epigenetic — mineral deposits of later origin than the enclosing rocks; also, the formation of secondary minerals by alteration

epilimnion — in a thermally stratified lake, the turbulent layer of water that extends from the surface to the thermocline

episome — a general term for genetically active particles normally found in bacteria, but sometimes present in other organism [They can exist in either of two forms, depending on conditions: (1) noninfective and integrated on the chromosome and (2) infective and integrated on the chromosome. They may form a link between fully integrated particles (genes) and nonintegrated particles (viruses).]

epithelium — the layer of cells forming the epidermis of the skin and the surface layer of mucuous and serous membranes [Epithelium cells may be flat (squamous), cube-shaped (cuboidal), or cylindrical (columnar); regenerative ability is excellent.]

epoxide — a cyclic ether (compound containing oxygen attached to two different molecules already united)

estuaries — areas where fresh water meets salt water (e.g., bays, mouths of rivers, salt marshes, and lagoons) (Estuaries are delicate ecosystems, serving as nurseries and spawning and feeding grounds for a large group of marine life and providing shelter and food for birds and wildlife.)

etiolated — grown in absence of sunlight

ethylene — a colorless, flammable, olefinic gas (C_2H_4), having a characteristic sweet odor and taste, derived from the cracking of petroleum

eutrophication — the normally slow aging process by which a lake evolves into a bog or marsh and ultimately assumes a completely terrestrial state and disappears (During eutrophication the lake becomes so rich in nutritive compounds, especially nitrogen and phosphorus, that algae and other microscopic plant life become superabundant, thereby "choking" the lake and causing it to eventually dry up. Eutrophication may be accelerated by many human activities.)

exinite — the series of genetically related macerals forming a portion of the hydrogen-rich fraction of coal seams and derived from the waxy cell secretions represented by plant cuticles and from spore and pollen exines

exothermic — a reaction in which heat is liberated, usually rapidly and sometimes explosively

°F — degrees Fahrenheit

facultative — having the ability to live under more than one set of environmental conditions

FC — fixed carbon

feedstock — crude oil, or a fraction thereof, to be charged to any process equipment

field gas plants, gas treating plants, gas recovery plants — plants that process natural gas to recover liquids, to remove hydrogen sulfide and carbon dioxide, and to remove water

filter cake — mineral matter (ash and char) and some unreacted insoluble carbon separated from crude liquefaction oil products

fines — in general, the smallest particles of coal or mineral in any classification, process, or sample of material; especially those which are elutriated from the main body of material in the process

firedamp — a mixture of methane with air in coal mines (Explosions may result when only 5 to 15% methane is found in the air. Firedamp can be detected by a flame safety lamp, because it makes the lamp's flame burn higher.)

Fischer-Tropsch gasoline — a naphtha-like artificial gasoline produced by combining carbon monoxide and hydrogen over a cobalt-thorium oxide catalyst at 200 to 250°C (The source of the carbon monoxide and some of the hydrogen is generally coal.)

fixed bed — a bed in which the individual particles or granules of a solid are motionless and supported by contact with each other (contrast with moving bed)

fixed carbon — for coal, coke, and other bituminous materials, the solid residue, other than ash, obtained by destructive distillation and determined by prescribed methods

flocculation — the aggregation or clumping together of individual, tiny soil particles, especially fine clay, into small groups or granules

flotation — a process for separating minerals from waste rock or solids of different kinds by agitating the pulverized mixture of solids with water, oil, and special chemicals, which causes preferential wetting of solid particles of certain types by the oil (The unwetted particles are carried to the surface by the air bubbles and, thus are separated from the wetted particles. A frothing agent is also used to stablize the bubbles in the form of a froth, which can be easily separated from the body of the liquid (froth flotation).

fluidization — a condition in which the individual particles or granules of a solid are suspended by a gaseous or liquid fluid passing through them at a velocity sufficient to cause them to be disengaged somewhat from each other (The system exhibits properties of a liquid.)

fluidized bed — a bed of suitably sized solid particles through which a fluid (usually a gas) flows at a velocity high enough to buoy the particles, to overcome the influence of gravity, and to impart to them an appearance of great turbulence (Fluidized beds are used in the petroleum industry. The Office of Coal Research is developing a coal-fired fluidized-bed boiler which would permit the use of western low-sulfur coals without slagging and the use of high-sulfur coals without causing unacceptable environmental effects.)

fluidized-bed boilers — a method of power generation using the technique of moving powdered fuel (coal) into a combustion chamber in a liquidlike stream by use of air or other gases

flue gas, stack gas — synonymous terms for the gases resulting from combusion of a fuel

fly ash — the fine, solid particles of noncombustible material residue carried from a bed of solid fuel by the gaseous products of combusion

food chain — the transfer of nutrients, and hence energy, from one group of organisms to another

fossil fuel — naturally occurring substances derived from plants and animals which lived in ages past (The bodies of these long-dead organisms have become recoverable fuel which can be burned; fossil-fuels include lignite, coal, oil, and gas.)

fractional distillation — a process of separation and the fundamental process of refining [In distillation, the volatility (i.e., the ease with which a liquid vaporizes) characteristics of hydrocarbons are particularly important. Volatility depends on the boiling points of the various hydrocarbon compounds of which the crude is composed. Hydrocarbon boiling points range from over 250°F to below zero. Because different hydrocarbon compounds have different boiling points, they condense at different temperatures. Separation of crude oil into fractions by a progressive distillation process requires separating the components in the order of their boiling points: propane, butane, naphtha, gasoline, kerosene, No. 2 heating oil, lubricating oil stocks, and residuals.]

FSI — free swelling index

ft — foot

ft^3 — cubic foot

fuel — a substance used to produce heat energy, chemical energy by combustion, or nuclear energy by nuclear fission

fuel gas — synthetic gas used for heating or cooling (Fuel gas has less energy content than pipeline quality gas. The Office of Coal Research is developing a process to produce clean, low-Btu fuel gas from coal. The product could be burned in nearby power plants or used as a feed material for production of other synthetic fuels, such as high-Btu pipeline gas.)

fuel oil — any liquid or liquefiable petroleum product that is burned for the generation of heat or engine power

1. <u>residual fuel oils</u>: oil from the bottom of the crude oil barrel that cannot be further refined; viscous residua that, unless heated to temperatures of 120 to 160°F, will remain solids.

2. <u>distillate fuels</u>: distillates derived directly or indirectly from crude oil

3. <u>blended fuels</u>: (No. 3, No. 4, or No. 5) produced by blending residual and middle distillate fuel oils.

fulvic acid — a humic acid; a mixture of organic substances remaining in solution after acidification of a dilute alkali extract from the soil

fume — fine, solid particles formed by the condensation of materials that were gaseous at higher temperatures

fungi — small, often microscopic plants without chlorophyll (Some fungi infect and cause disease in plants or animals; other fungi are useful in stablizing sewage or breaking down wastes for compost.)

fusain — coal material having the appearance and structure of charcoal (friable, sooty, and generally high in ash; consists mainly of fusite)

fusinite — the series of genetically related macerals that form the charcoallike fraction of coal seams and that are produced by the rapid charring and an initially rapid alteration of cell wall substances, usually prior to or soon after incorporation into the enclosing sediment

fusinization — the process in coalification that results in the formation of fusain

fusite — coal microlithotype consisting of 95% or more of fusinite, semifusinite, and sclerotinite

g — gram

μg — microgram

gallons (gal) — a unit of measure (A U.S. gallon contains 231 in^3, 0.133 ft^3, or 3.785 liters; it is 0.83 times the Imperial gallon. One U.S. gallon of water weighs 8.3 lb.)

gas cooling — the action of cooling shifted gas again to remove additional hydrocarbon oil by-products and residual phenolic water

gas liquor (sour water) — the aqueous streams condensed from the coal conversion and processing areas by scrubbing and cooling the crude gas stream

gas manufactured — a gas obtained by destructive distillation of coal, by thermal decomposition of oil, or by reaction of steam passing through a bed or heated coal or coke [e.g., coal gases, coke oven gases, producer gas, blast furnace gas, blue (water) gas, and carbureted water gas; Btu content varies widely]

gas, natural — a naturally occurring mixture of hydrocarbon gases found in porous geologic
formations beneath the earth's surface, often in association with petroleum (The principal
constituent of natural gas is methane.)

gas scrubbing — Crude gas is scrubbed and cooled with water, which removes tar and oil by-products
and phenolic waters. The phenolic waters are subsequently processed for recovery of by-
product phenols.

gas synthesis — a method of developing liquid fuels from coal by oxidation (e.g., the Fischer-
Tropsch process) (The first step in gas synthesis is to convert pulverized coal into a gas
by exposing the coal to oxygen and superheated steam. The gas, a mixture of carbon monoxide
and hydrogen, is then passed over various solid catalysts to change it into useful products.)

gases — materials that can be condensed to liquids only by pressure or at temperatures below
ambient (such as oxygen, methane, hydrogen)

gasification — in the most commonly used sense, refers to the conversion of coal to a high-Btu
synthetic natural gas under conditions of high temperatures and pressures; in a more general
sense, conversion of coal into a usable gas

gasoline — a petroleum fraction composed primarily of small branched-chain, cyclic, and aromatic
hydrocarbons

GC-MS — Gas chromotography-mass spectrometry

geochemistry — the study of (1) the relative and absolute abundances of the elements and of the
atomic species (isotopes) in the earth and (2) the distribution and migration of the
individual elements in the various parts of the earth and in minerals and rocks

geosynthesis — geological synthesis occurring over a long period of time

geothermal energy — the heat energy available in rocks, hot water, and steam in the earth's
subsurface

germ cell — a cell set apart from the rest of the body to develop, usually after union with
another of the opposite sex, into a new individual; an egg or sperm cell

gilsonite — a solid hydrocarbon that looks like black glass but is as brittle as dried mud
(probably formed from conventional crude oil that flowed upward through subsidence fissures
where the lighter fractions evaporated)

gley — a soil horizon in which the material is bluish gray, more or less sticky, compact, and
often structureless (developed under the influence of excessive moisture)

glomerulus — in the kidney, a tuft of capillaries that is covered by epithelium, is situated at
the point of origin of each vertebrate nephron, and normally passes into the proximal
convoluted tubule

gpm — gallons per minute

grab sample — obtaining a sample of an atmosphere in a very short period of time so that this sampling time is insignificant in comparison with the duration of the operation or the period being studied

granulocyte — a cell having granule-containing cytoplasm; a polymorphonuclear leucocyte

groundwater — subsurface water which feeds wells and springs

halloysite (metahalloysite) — a common clay mineral [$Al_2Si_2O_5(OH)_4$] like kaolinite, but structurally distinct

hard coal — in U.S. usage, anthracite or semianthracite; in European usage, a coal having a calorific value greater than 10,260 Btu/lb on a moist, ash-free basis

H-Coal process — one method to convert coal to a clean (low-sulfur, low-ash) fuel by catalytic hydrogenation.

HDS (Hydrodesulfurization, a copyrighted trademark) — desulfurization by use of hydrogen and catalyst treatment

heating value — the heat released by combustion of a unit quantity of a fuel, measured in calories or Btu

heat sink — anything that absorbs heat, usually part of the environment (such as the air, a river, or outer space)

heavy hydrocarbons — higher-boiling, higher-density hydrocarbons having seven or more carbon atoms

heavy metals — metallic elements of high molecular weight, generally toxic to plant and animal life in low concentrations (Such metals are often residual in the environment and exhibit biological accumulation. Examples include mercury, chromium, cadmium, arsenic, and lead.)

heavy oils — fuel oil, heavy distillate oil, heavy furnace oil, No. 6 oil, bunker oil, and residual fuel oil, hydrocarbon mixtures of from C_{16} to C_{20+} that boil from 650 to 1000+°F and that are derived from crude oil

Hepatocellular injury — injury to the cells of the liver

heptane — any of several isomeric hydrocarbons of the alkane series; the liquid normal isomer occurring in petroleum and used especially as a solvent and in determing octane numbers.

heterocyclic — a cyclic or ring structure in which one or more of the atoms in the ring is an element other than carbon (Common heterocyclics are pyridine, pyrrole, furan, thiophene, and purine.)

heterotroph — organism that is unable to produce food from simple beginnings and, therefore, is dependent on dead or living organisms of another species for food

HHV — higher heating value

high-Btu gas — fuel gas having a higher heating value of about 1000 Btu/scf or more

High-Btu Oil-Gas process — a manufactured-gas process in which oil is converted into a fuel
    gas having a higher heating value than that of coal gas or carbureted water gas (often called
    Hi-Btu Gas Process)

higher heating value, gross heating value — the total heat released when a fuel is burned

high-sulfur coal — generally, coal that contains more than 1% sulfur by weight

HI-SUL — high sulfur content

HPLC — high-pressure liquid chromatography

hr — hour

humic acid — a gelatinous material formed as a precipitate when organic matter is treated with
    a strong base and the resulting solution is acidified

humin-ulmins — the products of decay of vegetable matter and their prototypes that occur in coal
    as a gel

hydraulic fracturing — a general term, for which there are numerous trade or service names, for
    the fracturing of rock in an oil or gas reservoir by pumping a fluid under high pressure
    into the well (The purpose is to produce artificial openings in the rock to increase
    permeability.)

hydrocarbon — an organic compound consisting exclusively of the elements carbon and hydrogen.
    The principal types are as follows:

    A. Aliphatic (straight-chain)

        1. paraffins (alkanes)
        2. olefins
            (a) alkenes
            (b) alkadienes
        3. acetylenes
        4. acyclic terpenes

    B. Cyclic (closed ring)

        1. alicyclic
            (a) cycloparaffins
            (b) cycloolefins
            (c) cycloacetylenes

2. aromatic
   (a) benzene group (ring)
   (b) naphthalene group (2 rings)
   (c) anthracene group (3 rings)
3. cyclic terpenes: monocyclic (dipentene); dicyclic (pinene)

hydrocracking — an oil-refining process in which the large molecules of crude oil are broken into smaller molecules through reaction with hydrogen (The process is used to convert heavy oil into lighter fractions such as gasoline.)

hydrodesulfurization (HDS) — desulfurization by use of hydrogen and catalyst treatment

hydrofining — a fixed-bed catalytic process to desulfurize and hydrogenate a wide range of charge stocks from gases through waxes

hydroforming — a process in which naphthas are passed over a catalyst at elevated temperatures and moderate pressures, in the presence of added hydrogen or hydrogen-containing gases, to form high-octane motor fuel or aromatics

hydrogen/synthesis gas — a process being developed by the Office of Coal Research and the American Gas Association to produce either hydrogen or synthesis gas

hydrogenation — a process of treating coal with oil and hydrogen under heat and pressure, then separating the liquid mixture into useful products. [Hydrogenation produces hydrocarbon gases (e.g., ethane, propane, and butane), low-priced pitch, and many other valuable chemicals (e.g., benzene, phenol, naphthalene, and aniline). These chemicals are used to make such common products as dyes, perfumes, paints, and plastics. The hydrogenation process also can be used to produce fuel oil and gasoline. See liquefaction of coal and gas synthesis, two alternative routes to liquid fuels from coal.]

hydrology — the science dealing with the properties, distribution, and circulation of water and snow

hydrosphere — the water on or surrounding the surface of the earth, as distinguished from the solid part of the earth (lithosphere)

hydrophilic — having a strong affinity for binding or absorbing water, which results in swelling and formation of reversible gels

hydrophobic — antagonistic to water; incapable of dissolving in water

hydrotreating — the removal of sulfur from low-octane gasoline feedstocks by replacement with hydrogen

HYGAS — a process being developed by the Office of Coal Research and the American Gas Association to produce pipeline quality gas by hydrogasification of coal (Being developed by the Institute for Gas Technology)

hyperplasia — excessive proliferation of normal cells in the normal tissue arrangement of an organ

hypha — one of the simple or branched filaments of the fungal thallus (mycelium)

hypolimnion — in a thermally stratified lake, the layer below the thermocline and extending to the bottom of the lake (water temperature is virtually uniform)

hypotrichous — refers to the hypotricha, a suborder of spirotricha comprising ciliates that have cilia only on the ventral surface and usually fused to cirri and that often have tactile bristles on the dorsum

hysteresis — retardation of the effect, as if from viscosity, when the forces acting upon a body are changing

idrialite — a rare PAH-containing mineral having over 30 homologous series of CH, CHS, and CHN compounds

impingement, dry — the process of impingement carried out so that particulate matter carried in the gas stream is retained upon the surface against which the stream is directed (The collecting surface may be treated with a film of adhesive.)

in. — inch

incorporation — a process by which material contributing to coal responds to diagenetic and metamorphic agencies of coalification and becomes a part of the coal without undergoing any material modification

individual susceptibility — the marked variability in the manner in which individuals will respond to a given exposure to a toxic agent

inducibility — a relative increase in rates of de novo synthesis or of activation of enzyme activity from preexisting moieties, or in the rates of both when compared to rate breakdown

inert gas — a gas that does not react with other substances under ordinary conditions

infrared — pertaining to the region of the electromagnetic spectrum from about 0.78 to 1000 μm (microns) (Near infrared is below 10 μm and far infrared above 10 μm.)

in. Hg — inch mercury

ion exchange — a reversible chemical reaction between a solid (ion exchanger) and a fluid (usually a water solution) by means of which ions may be interchanged from one substance to another without affecting the superficial physical structure of the solid

iron sponge — wood shavings impregnated with hydrated iron oxide

IRS — infrared spectrometry

ISE — ion selective electrode

isokinetic — a term describing a condition of sampling, in which the flow of gas into the sampling device (at the opening or face of the inlet) has the same flow rate and direction as the ambient atmosphere being sampled

K — Kelvin

kaolinite — a common clay material; a two-layer hydrous aluminum silicate having the general formula $Al_2(Si_2O_5)(OH)_4$

kcal — kilocalorie

kerogen — a resinous hydrocarbon material, of general formula $(C_6H_8O)_n$, that is the chief organic constituent of oil shale. (When heated to 450 to 600°C, kerogen releases vapors that can be converted to raw shale oil, a black, viscous mixture of hydrocarbons. Shale oil can then be converted into petroleum products by refining.)

kerosene — the petroleum fraction containing hydrocarbons that are slightly heavier than those found in gasoline and naphtha

kerosine (kerosene) — kerosine is the general name applied to the group of refined petroleum fractions, distilling after gasoline and overlapping into the light distillates and middle distillates. [Kerosine is colorless, low in sulfur, does not burn with a smoky flame, and boils over the range of 350 to 525°F. Different kerosines are called upon to burn under different conditions, so that variations in the properties can be expected. This means that, after being distilled, the kerosine is separated into different fractions, according to boiling point, and subjected to further refining and treatment to remove all undesirable constituents. Different fractions of kerosine are used for space heating (No. 1 heating oil) and blended with gas oil to make No. 2 heating oil, for tractor fuel, for jet fuel, and for solvents. Kerosine was once called coal oil because of its origin.]

Krebs cycle — a cyclic sequence of reactions occurring in the living organism and forming a phase of the metabolic function in which acetic acid or acetyl equivalent is oxidized through a series of intermediate acids to carbon dioxide and water and thus provides energy for storage in the form of energy-rich phosphate bonds (as in adenosine triphosphate) that can make it available for use in other vital processes (as muscular work)

l — liter

µl — microliter

lamellae — an organ, process, or part resembling a plate, as one of the thin plates composing the gills of a bivalve mollusc

LC — liquid chromatography

LC_{50} — median lethal concentration

LD_{50} — median lethal dose

leachate — liquid that has percolated through solid waste or other media and has extracted dissolved or suspended materials from it

leaching — the process of extracting a soluble component from a mixture by percolation of the mixture with a solvent, usually water, resulting in the solution and later separation of the soluble component

LHV — lower heating value

ligand — a molecule, ion, or atom that is attached to the central atom of a coordination compound, a chelate, or other complex

light distillates — these fractions may be used directly in the production of gasoline and are sometimes referred to as naphthas (Heavy naphtha is blended with light gas oil to make jet aircraft fuel.)

light gasoline, light naphtha — liquid C_5, C_6, C_7, C_8 derived from crude oil (about the same as condensate from raw natural gas)

light hydrocarbons, light ends — low-boiling, low-density hydrocarbons having one to six carbon atoms

lignite — a brownish-black low-rank coal in which the alteration of vegetal material has proceeded further than in peat, but not so far as in subbituminous coal

limnology — the study of the physical, chemical, meteorological, and biological aspects of fresh waters

liquefaction of coal — the conversion of coal to liquid fuels, a method for making large amounts of gasoline and heating oils (see hydrogenation and gas synthesis, two alternative routes to liquid fuels from coal.)

liquefied gases — ethane, propane, butanes, ehtylene, propylene, and butylenes; derived by refining, or other processing, of natural gas, crude oil, or unfinished oils

liquefied natural gas (LNG) — natural gas that has been changed into a liquid by cooling to about -260°F, at which point it occupies about 1/600 of its gaseous volume at normal atmospheric pressure

liquefied refinery gas (LRG) — a product of petroleum; includes butane, propane, and butane-propane mixtures

liquid petroleum gas (LPG), bottled gas — consists primarily of propanes and butanes, highly volatile gases that are extracted from refinery and natural gases. (LPG has a unique double characteristic. Under moderate pressure, LPG becomes liquid and can be easily transported by pipelines, railroad tank cars, or trucks. Released from its storage tank, LPG reverts to vapor form, burning with high heat value and a clean flame.)

lithobody — a layer in a coal seam

lithophile elements — elements enriched in a silicate crust

lithosphere — the solid part of the earth composed predominantly of rock

lithotype — refers to any of the four microscopically recognizable constituents of banded coal: vitrain, clarain, durain, fusain

low-Btu gas — fuel gas having a higher heating value of 350 to 450 Btu/scf or less

lower heating value, net heating value — the effective usable heat released when a fuel is burned (after some gross heat release is used in vaporizing the combustion product water)

low-sulfur coal and oil — generally, coal or oil that contains 1% or less of sulfur by weight

LNG — liquefied natural gas

LPG — liquefied petroleum gases, usually C_3 and C_4

LRG — liquefied refinery gas

Lurgi process — the chief commercially available process for coal gasification (Having originated in Germany, this process has limited application in the United States because of problems of scaling up the size of operations and characteristics of U.S. coal. The Office of Coal Research and the American Gas Association are jointly funding further development.)

m — meter

m, molal solution — 1 gram molecular weight per 1000 g of water

M, molar solution — 1 gram molecular weight per liter (moles per liter)

µm — micrometer

maceral — any of the elementary microscopic constituents of coal (Three groups are recognized: vitrinites, exinites, inertinites.)

macrinite — macerals of diverse origin which are white in incident light, but are not as highly reflective as fusinite, and which never display a remnant cell wall organization

maf — moisture- and ash-free

margin — the difference between the net system generating capability and system maximum load requirements, including net schedule transfers with other systems

median lethal concentration (LC_{50}) — a standard measure of toxicity; indicates the concentration of a substance that will kill 50% of a group of experimental insects or animals

median lethal dose (LD_{50}) — the dose that will kill 50% of a group of experimental animals

mercaptans (thiols) — a group of organic compounds resembling alcohols, but having the oxygen of the hydroxyl group replaced by sulfur, as ethyl mercaptan, C_2H_5SH (Mercaptans have a strong skunklike odor.)

mesentery — the membranes, or one of the membranes, that consist of a double fold of the peritoneum and enclosed tissues and that, in a vertebrate, invest the intestines and their appendages, connect them with the dorsal wall of the abdominal cavity, and serve to retain the organs in position and to support and convey to them blood vessels, nerves, and lymphatics

mesosome — any of the intracytoplasmic membraneous structures of bacteria (plasmalemmasome) that appear to originate from the cell membrane by an invagination (infolding) and pinching-off process [Mesosomes are tubelike structures and aggregations of double membranes (or lamellae) in the cytoplasm. Segregation of the bacterial DNA elements (bacterial → chromosomes) seems to be brought about by growth of the membrane. A fragment of the bacterial membrane (which may remain intact during growth and bacterial division) thus constitues the actual unit of segregation and is to be regarded as functionally analogous to the chromosome of higher organisms (eukaryotes). It provides for the coordination between DNA replication and subsequent division of the bacterial cell.]

metabolism — the sum of the processes concerned in the building up of protoplasma and its destruction incidental to life; the chemical changes in living cells by which energy is provided for the vital process and activities and new material is assimilated to repair the waste

metabolite — any product of metabolism

metallurical coal — coal with strong or moderately strong coking properties that contains no more than 8.0% ash and 1.25% sulfur, as mined or after conventional cleaning

metamorphosis — change of form, structure, or substance (e.g., the metamorphosis of peat to coal)

metaplasia — conversion of one kind of tissue into a form that is not normal for that tissue

methanation — the catalytic combination of carbon monoxide and hydrogen to produce methane and by-product water (About 60% of the end-product methane is produced in the methanation step.)

methane (CH_4) — the lightest in the paraffinic series of hydrocarbons (Methane is colorless, odorless, and flammable and forms the major portion of marsh gas and natural gas.)

methylation — replacement in a compound of one or more hydrogen atoms with the methyl group (CH_3)

mg — milligram

micelle — an electrically charged colloidal particle, usually organic in nature, composed of aggregates of large molecules (e.g., in soaps and surfactants)

micrinite — the series of genetically related macerals that form a portion of the attrital
fraction of coal seams and that are formed by the granulation and subsequent metamorphosis
of plant cell wall derived materials

microbiology — the branch of science that deals with the study of microorganisms

microflora — microscopic members of the plant kingdom (e.g., phytoplankton)

microlithotype — typical microscopic association of macerals in humic coal

microorganism — a term applied to bacteria, fungi, unicellular algae, protozoa, and the smallest
metazoa (e.g., rotifers)

microsome — one of the finer granular elements of protoplasm

middle distillates — Hydrocarbon mixtures, derived from crude oil, from C_{10} to C_{15} that boil
in the range of 350 to 650°F (e.g., kerosene, diesel oil, jet fuel, and light furnace oil)

mine-mouth plant — a steam-electric plant or coal gasification plant built close to a coal mine,
usually associated with delivery of output via transmission lines or pipelines over long
distances, as contrasted with plants located nearer load centers and at some distance from
sources of fuel supply

mist — a gaseous dispersion of liquid particles, usually less than 50 μ in diameter (When the
concentration of suspended particles is sufficient to reduce visibility, the mist becomes
a fog.)

mitochondria — small granules or rod-shaped structures seen by differential staining in the
cytoplasm of cells

ml — milliliter

mm — millimeter

mobility — the ability of a chemical element or a pollutant to move into and through the environment (e.g., the mobilization of an element from a water column to sediment)

monochromator — a device or instrument that, with an appropriate energy source, may be used to
provide a continuous calibrated series of electromagnetic energy bands of determinable
wavelength or frequency range

montmorillonite — a group of clay minerals whose formulas may be derived by substitution in the
general formula $Al_2O_{10}(OH)_2$, with deficiencies in charge in the tetrahedral and octahedral
positions balanced by the presence of cations, most commonly calcium and sodium subject to
ion exchange

morphogenesis — the various processes occurring during development by which the form of the body
and its organs is established

moving bed — a body of solids in which the particles or granules of a solid remain in mutual
contact but in which the entire bed moves in piston-like fashion with respect to the containing walls (contrast with fixed bed)

mp — melting point

MS — mass spectrometry

mutagen — an agent that tends to increase the occurrence or extent of mutation or genetic change

mutagenesis — the condition in which inheritable genetic changes are passed on from parent to
offspring

mutation — a permanent transmissible change in the characteristics of an offspring from those
of its parents

N, normal solution — 1 gram equivalent weight per liter

NAA — neutron activation analysis

naphtha — a loosely defined petroleum fraction containing primarily aliphatic (linear) hydrocarbons, with boiling points ranging from 125 to 240°C (Naphtha is intermediate between
gasoline and kerosine and contains components of both. The principal uses of naphtha are
in solvents and paint thinners and as a raw material for the production of organic chemicals,
but it is expected to be used increasingly as a raw material for the production of synthetic
natural gas.)

naphtha, gasoline, full-range naphtha — a hydrocarbon liquid mixture (C_5 to C_{11}) that boils
in the range from 100 to 400°F

naphthalene — an aromatic hydrocarbon ($C_{10}H_8$) that is the most abundant component of coal tar.
(Naphthalene is obtained by the distillation of tar and by recovery from cokeoven gas and
is used in manufacturing dyes, resins, and smokeless powder, as a moth repellent, and in
medicine.)

natural gas — a natural hydrocarbon gas composed of a variety of gases including methane, ethane,
butane, and propane (Natural gas comes from the ground with or without accompanying crude
oil and is generally much higher in heat content than manufactured gas. It is used by the
petrochemical industry as raw material for the manufacture of fertilizer and cellophane.
The energy content of natural gas is usually taken as 1032 Btu/ft^3.)

natural gas liquids (NGL) — a collective name for C_3 LPG, C_4 LPG, and C_5-C_8 condensate [The
hydrocarbon components, propane, butanes, and pentanes (also referred to as condensate),
or a combination of them are subject to recovery from raw gas liquids by processing in field
separators, scrubbers, gas processing and reprocessing plants, or cycling plants. The
propane and butane components are often referred to as liquefied petroleum gases or LPG.]

natural gasoline — a mixture of liquid hydrocarbons extracted from natural gas and stabilized to obtain a liquid product suitable for blending with refinery gasoline (This gasoline cut is used as a blending stock since it is volatile and aids in engine starting. Natural gasoline has good antiknock qualities.)

NCIB — National Collection of Industrial Bacteria

necrosis — localized or general death of plant tissue, often characterized by a brownish or black discoloration

neoblast — new growth of tissue

neoplasm — a new and abnormal formation of tissue, as a tumor or growth that serves no useful function, but grows at the expense of the healthy organism

net reserves — the recoverable quantity of an energy resonance that can be produced and delivered

ng — nanogram

nitrification — oxidative process that converts ammonium salts to nitrites and nitrites to nitrates

nitrogen oxides — compounds of nitrogen and oxygen that may be produced by the burning of fossil fuels (very harmful to health, and may be important in the formation of smogs)

nm — nanometer

node — any swelling in an elongate cylindrical structure, particularly joints on plants stems from which leaves arise

nodulation — the formation of swellings on the roots of leguminous plants by symbiotic nitrogen-fixing bacteria.

nonbanded coal — coals that do not display a striated or banded appearance on the vertical face (They consist of clarain or durain or intermediates thereof.)

nonpolar solvent — a solvent having a low dielectric constant

normal butane (total) — a molecular structure of a single unbranched chain of carbon having the chemical formula $CH_3(CH_2)_2CH_3$ (The total is the result of output from natural gas and petroleum.)

nutrient — a chemical substance (an element or an inorganic compound, e.g., nitrogen or phosphate) absorbed by a green plant and used in organic synthesis

octane — any of several isomeric liquid paraffin hydrocarbons

OES — optical emission spectrometry

oil, gas — an oil intermediate between the light distillates and the lubricating oils (Most of the gas oil produced is turned over to the cracking apparatus for the production of gasoline and is used by gas companies for the manufacture of gas.)

oil shale — a sedimentary rock containing solid organic matter from which oil can be obtained when the rock is heated to a high temperature

olefin (alkene) — a class of unsaturated aliphatic hydrocarbons having the general formula C_nH_{2n} that contain one or more double bonds and therefore are chemically reactive (Those containing one double bond are called alkenes, and those containing two are called alkadienes or diolefins.)

oligotrophic lakes — deep lakes that have a low supply of nutrients and thus contain little organic matter (Such lakes are characterized by high water transparency and high dissolved oxygen.)

oncogeny — the process of tumor formation

opacity rating — a measurement of the opacity of emissions, defined as the apparent obscuration of an observer's vision to a degree equal to the apparent obscuration of smoke of a given rating on the Ringelmann Chart

osmoregulation — the regulation of osmotic pressure, especially in the body of a living organism

ovicell — a dilation of the zooecium in many bryozoans that serves as a broad pouch

OWD — oil-in-water dispersion

oxic — relating to or marked by a sufficient supply of oxygen

oxidant — any oxygen-containing substance that reacts chemcially in the air to produce new substances (Oxidants are the primary contributors to photochemical smog.)

oxides of nitrogen — compounds formed by the fixation of nitrogen at high temperatures, as in furnaces and internal combustion engines (Primary product is nitric oxide (NO), which oxidizes slowly in air, but much more rapidly in the presence of sunlight and organic vapors, to nitrogen dioxide (NO_2). Other nitrogen oxides may have brief existence as intermediates in the atmospheric reactions involved.)

oxides of sulfur — products of the oxidation of sulfur that include both sulfurdioxide (SO_2) and sulfur trioxide (SO_3) and the acids formed by their combination with water (Of these sulfuric acid (H_2SO_4) is of principal interest.)

oxygen plant — plant at which pure oxygen is cryogenically extracted from atmospheric air

ozone — a form of oxygen (O_3) produced in the reactions of photochemical smog and in electrical discharges; a powerful oxidizing agent that is toxic to both plants and animals at relatively low concentrations

PAH — polycyclic aromatic hydrocarbons

PAN — peroxyacetylnitrate; one of the family of peroxyacylnitrates (reactive compounds formed in photochemical smog) that is highly toxic to many species of plants (The eye-irritating properties of photochemcial smog have been ascribed, at least in part, to PAN.)

papilloma — any benign epithelial tumor, such as a wart

PDU — process development unit

Paraffin (also called alkane) — a class of aliphatic hydrocarbons characterized by a straight carbon chain that have the generic formula C_nH_{2n+2} [Their physical form varies with increasing molecular weight from gases (methane) to waxy solids. They occur principally in Pennsylvania and midcontinent petroleum.]

particulate fluidized bed — a bed in such a condition of fluidization that the individual particles are discretely separated from each other and the volumetric concentration of solid particles is uniform throughout the bed

particulate matter — solid particles, such as ash, that are released in exhaust gases from the combustion process at fossil-fuel plants

particulates — small particles of solid material produced by the burning of fuels

PC — pulverized coal; also called powdered coal

pE — Negative logarithm of the activity of electrons

peat — one of the earliest stages of coal in which the remains of plants and ferns that have been preserved may be clearly seen. (Peat contains a very high percentage of water and has been used as a fuel for hundreds of years in Ireland, England, and Germany.)

pedosphere — that portion of the biosphere that contains soil

pelagic — refers to any living organism that is free-swimming in the open ocean

pendletonite — a rare PAH-containing mineral chiefly composed of coronene

pentane — any of three isomeric hydrocarbons of the methane series occurring in petroleum

percent excess air — the percentage of combustion air supplied to a burning fuel, over and above that required to combine with the fuel hydrogen and carbon

percent oxygen in flue gas — the percentage of oxygen in the flue gas as a result of using excess combustion air

peritoneum — the smooth transparent serous membrane that lines the cavity of the abdomen of a mammal, is reflected inward over the abdominal and pelvic viscera, and consists of (1) an outer layer closely adherent to the walls of the abdomen except in some places along the back where it extends forward to form (2) an inner layer that folds to invest the viscera

petrochemicals — chemicals that are made from components of crude oil and/or natural gas (The cracking process for the manufacture of gasoline produces large quantities of gaseous hydrocarbons, which were, at one time, waste products used only as illuminants and fuels in the refinery.)

petrography — the branch of science concerned with the systematic description and classification of rocks

petroleum — an oily, flammable bituminous liquid which may vary from almost colorless to black, that occurs in many places in the upper strata of the earth; a complex mixture of hydrocarbons containing small amounts of other substances that is prepared for use as gasoline, naphtha, or other products by various refining processes

petroleum coke — a solid residue; the final product of the condensation process in cracking, consisting mainly of highly polycyclic aromatic hydrocarbons very poor in hydrogen (Petroleum coke, when calcined, yields almost pure carbon or artificial graphite suitable for production of carbon or graphite electrodes, structural graphite, motor brushes, or dry cells)

petroleum naphtha — a generic term applied to refined, partially refined, or unrefined petroleum and liquid products of natural gas [The naphthas used for specific purposes (e.g., cleaning, manufacture of rubber, paints, and varnishes) are made to have more volatility than that set by the limits of this definition.]

petroleum refiner — a plant that converts crude petroleum into the many petroleum fractions (asphalt, fuel oil, gasoline, etc.) (Usually this conversion is accomplished by fractional distillation.)

petroleum tar — a viscous black or dark brown product obtained in petroleum refining that will yield a substantial quantity of solid residue when partly evaporated or fractionally distilled

pg — picogram

pH — a measure of the acidity or alkalinity of a material, liquid or solid (pH is represented on a scale of 0 to 14 with 7 representing a neutral state, 0 representing the most acid, and 14 the most alkaline.)

phenol — a class of aromatic organic compounds in which one or more hydroxy groups are attached directly to the benzene ring (e.g., phenol, cresols, xylenols, resorcinol, naphthols.)

photochemical oxidants — secondary pollutants formed by the action of sunlight on nitrogen oxides and hydrocarbons in the air; the primary contributors to photochemical smog

photochemcial smog — an eye-irritating atmospheric condition resulting from a very complex series of chemical reactions involving reactive organic substances and nitrogen oxides and initiated by ultraviolet light from the sun

photooxidation — light-initiated oxidation

photosynthesis — the manufacture of energy-containing organic compounds from raw materials by green plants when exposed to light (The process is dependent on carbon dioxide and chlorophyll and is retarded by narcotics.)

phytoplankton — the microscopic plant life found floating in lakes, oceans, etc.

phytotoxic — poisonous to plants

pilot plant — a small-scale industrial process unit operated to test the application of a chemical or other manufacturing process under conditions that will yield information useful in design and operation of full-scale manufacturing equipment

piscavore — a fish-eating organism

pitch of tar — a black or dark brown solid or semisolid residue obtained by partial evaporation or fractional distillation of tars and tar products

plankton — organisms that live in the upper portion of any body of water and that drift with the current

pleomorphism (polymorphism) — the occurrence of more than one physical shape during an organism's life span

plume — the visible emission from a flue or chimney

podsol — a highly bleached soil that is low in iron and lime and is formed under moist and cool climatic conditions

polar solvent — a solvent having a high dielectric constant

pollution — the accumulation of wastes or by-products of human activity (Pollution occurs when wastes are discharged excess of the rate at which they can be degraded, assimilated, or dispersed by natural processes. Noxious enivronmental effects not caused by human activity may also be called pollution.)

pollutant — any contaminant that, when present in the air or water, detracts from or interferes with the desired use of that air or water

polymerization — the reverse of cracking; a method of combining smaller molecules to make larger ones (Polymerization was developed during the late 1930s to utilize refinery gases, which were often wasted or burned as fuel. The product of polymerization is usually "poly" gasoline; these fractions are high in antiknock value and are used primarily to raise the antiknock value of the finished gasoline.)

polyploidy — a condition of having or being a chromosome number that is a multiple greater than two of the monoploid number

POM — polycyclic organic matter

pozzolana — burnt clay, granulated slag, certain clinkers, and burnt oil shale; natural volcanic ash and celite

ppb — parts per billion

ppm — parts per million

PPOM — particulate polycyclic organic matter

precision — in analytical chemistry, a measure of the reproducibility of a method when repeated on a homogeneous sample, regardless of whether the observed values are widely displaced from the true value as a result of systematic or constant errors present throughout the measurements. (It may be expressed in terms of the standard deviation.)

pressure gasification — the process by which coal, steam, and oxygen are caused to react under controlled conditions of temperature and pressure to produce a crude gas containing methane, hydrogen, carbon monoxide, carbon dioxide, excess steam, and various by-products and impurities (Only about 40% of the plant's end-product methane (SNG) is produced in the gasifiers. The remainder of the methane is produced in subsequent reaction steps.)

primary fuel — fuel consumed in original production of energy (as contrasted to a conversion of energy from one form to another)

probable reserves — a realistic assessment of the reserves that will be recovered from known oil or gas fields based on the estimated ultimate size and reservoir characteristics of such fields (Probable reserves include those reserves shown in the proved category.)

process stream — any material stream within the coal-conversion processing area

product stream — streams within the coal conversion plant that contain the material for which the plant was built (e.g., oil, SNG, SRC)

propane — an easily liquefiable hydrocarbon gas (C_3H_8); one of the components of raw natural gas that is also derived from petroleum refining processes

propylene — a flammable, gaseous, olefinic hydrocarbon that is usually obtained in petroleum refineries by cracking petroleum hydrocarbons and that is used chiefly in organic synthesis of compounds with which benzine is alkylated for making detergents

proved reserves — the estimated quantity of crude oil, natural gas liquids, or sulfur that analysis of geological and engineering data demonstrates with reasonable certainty to be recoverable from known oil or gas fields under existing economic and operating conditions

proximate analysis (coal) — an analysis of coal to determine moisture, volatile matter, and ash and the calculation of fixed carbon by difference

pseudovitrinite — a maceral possessing the characteristics of vitrinite, but always displaying a higher reflectance

psi — pounds per square inch

psia — pounds per square inch, absolute

psig — pounds per square inch, gauge

pulverized coal — coal that has been reduced to a fine dust (75 to 95% will pass a 200-mesh screen) by grinding mills (contrast with crushed coal)

pyrolysis — the transformation of a substance into another compound or compounds by the application of heat alone [In the context of energy, pyrolysis (also called destructive distillation) is the heating of organic materials such as coal, wood, petroleum, and solid wastes in the absence of oxygen with provisions for recovery of the desired combustible products. If heat is applied slowly, the initial products are water vapor and volatile organic compounds. Increased heat leads to recombination of the organic materials into complex hydrocarbons and water. The principal products of pyrolysis are gases, oils, and a solid residue called char. It is possible to produce great variations in the relative proportions of these products by varying the pyrolysis conditions.]

quenching — sudden cooling

°F — degrees Rankine

rainout — the removal of a pollutant within clouds

rank — those differences in the pure coal material due to geological processes designated as metamorphic, whereby the coal material changes from peat through lignite and bituminous coal to anthracite or even graphite; the degree of coal metamorphism

radiomimetic — producing especially biological effects similar to those produced by radiation

receiving waters — the bodies of water that receive effluent wastewater from treatment plants

rectisol — process in which low-temperature methanol is used to selectively absorb and remove hydrogen sulfide and carbon dioxide from the cooled gas (Precooling at the Rectisol unit entry also recovers by-product naphtha.)

refined gas — a gas similar to natural gas due to its LPG ingredients; a by-product of any of several processes used for the manufacture of cracked gasoline (The fixed or noncondensible gases are used as boiler fuel and are recycled through the process continually.)

refinery — an industrial complex for processing crude oil by distillation and chemical reactions to produce a separate petroleum product [Typical crude fractions, from top to bottom, or simple to complex, are ether, methane, and ethane (the gasolines); propane and butane; kerosene, fuel oil, and lubricants; jelly paraffin, asphalt, and tar.]

remaining reserves — those quantities of crude oil, natural gas, natural gas liquids, and sulfur as estimated under proved or probable reserves after deducting those quantities produced up to the respective date of the estimate

renal — of, relating to, or involving the kidneys; located in the region of the kidneys

repeatability — the degree of variation between repeated measurements of the same concentration

reproducibility — the degree of variation obtained when the same measurement is made with similar instruments and many operators

reserves — the amount of a fuel or other mineral resource known to exist and expected to be recoverable by existing techniques and under existing economic conditions

residence time — the period of time during which a substance resides in a designated area

residual fuel oil — petroleum oil, which is any topped crude or viscous residuum of crude or unfinished oil or one or more of petroleum oils

No. 1 fuel oil — a light distillate intended for use in burners of the vaporizing type, in which the oil is converted to a vapor by contact with a heated surface or by radiation (High volatility is necessary. Straight-run kerosine is a generally good description of the product, which is predominantly used in space heaters.)

No. 2 fuel oil — a heavier distillate than grade No. 1, intended for use in atomizing-type burners, which spray the oil into a combustion chamber where the tiny droplets burn in suspension (This grade of oil is used in most home burners that have central heating and where its ease of handling sometimes justifies its higher cost over the residual fuels.)

No. 4 fuel oil — usually a light residual, but sometimes a heavy distillate, intended for use in burners equipped with devices that atomize oils of higher viscosity than home burners can handle (In all but extremely cold weather, it requires no preheating for handling.)

No. 5 fuel oil (light) — residual fuel of intermediate viscosity for burners capable of handling fuel more viscous than grade No. 4 without preheating (Preheating may be necessary in colder climates.)

No. 5 fuel oil (heavy) — a residual fuel more viscous than grade No. 5 (light) that is intended for use in similar service (commercial, industrial, and large apartment houses) (It usually requires preheating, particularly in colder climates.)

No. 6 fuel oil — a high-viscosity oil, sometimes referred to as "Bunker C", used mostly in commercial and industrial heating (It requires preheating in the storage tank to permit pumping and additional preheating at the burner to permit atomizing. The extra equipment and maintenance required to handle this fuel usually does not permit it to be used in small installations.)

Note: Fuels No. 1, No. 2, and sometimes No. 4 are distillate fuels, sometimes called "clean fuels." Residual fuels, often referred to as "dirty fuels," are No. 6 and No. 5 (light and heavy); sometimes No. 4 is also classified as a light residual. Fuel No. 6 is diluted, or "cut back," with required amounts of No. 2 fuel to make No. 4 and both grades of No. 5.

residue — the remaining components that are not removed as distillates that can be used as fuel or residual fuel oil or for asphalt manufacture (For the latter purpose, further distillation under reduced pressure is necessary. This process yields distillates that can be used for gas oil or cracking feedstock and, in the use of certain crude oils, for lubricating oil and wax manufacture, the residue being asphalt.)

resinite — the series of genetically related macerals forming a portion of the hydrogen-rich fraction of coal seams and derived from resinous secretions and excretions of plant cells

respirable particle — particle of the size ($\leqslant 5.0$ μm) most likely to be deposited in the pulmonary portion of the respiratory tract

RF value — The R_f value of a substance A is the ratio between the distance of the respective spot from the origin and the distance between the solvent front (F) and the origin (O), i.e., $R_f = \overline{OA}/\overline{OF}$

rhizobia — eubacterial schizomycetes that are gram-negative and sparsely flagellated

Ringelmann number — a scale of values, ranging from zero through five, variously used to measure the color, or density, of a stack plume (Originally proposed to evaluate the darkness of coal smoke plumes as a shade of grey; its application to other types or colors of plumes involves an assumed "equivalence" of some related optical property, such as its opacity or its ability to obscure a target behind the plume.)

RNA — ribonucleic acid; a nucleic acid found principally in the nucleolis, microsomes, and mitochondria of cells that appears to play an important role in synthetic reactions within cells

ROM — run-of-mine coal

runoff — the portion of rainfall, melted snow, or irrigation water that flows across ground surface and eventually is returned to streams (Runoff can pick up pollutants from the air or the land and carry them to the receiving waters.)

run-of-mine — coal that is taken from the mine

sarcoma — cancer arising from underlying tissue, muscle, bone, and other connective tissue that may affect the bones, bladder, kidneys, liver, lung, parotids, and spleen

saturation pressure — the pressure at which a vapor confined above its liquid will be in stable equilibrium with it (Below saturation pressure, some of the liquid will change to vapor; above saturation pressure, some of the vapor will condense to liquid.)

saturation temperature — the boiling point of a liquid for the existing pressure

scavenging — the removal of a pollutant from the atmosphere to the ground by natural methods (rainout and washout)

scf — standard cubic foot (of gas measured at atmospheric pressure and 60°F)

scfd — standard cubic feet per day

scfh — standard cubic feet per hour

scfm — standard cubic feet per minute (60°F, 14.7 psia)

sclerotinite — maceral of the inertite group occurring as: (1) rounded bodies related to spore coats, etc., 20 to 300 μ in diameter and (2) irregular fungal filaments

scrubber — an air-pollution control device that uses a liquid spray to remove pollutants from a gas stream by absorption or chemical reaction (Scrubbers also reduce the temperature of the emission.)

secondary recovery — oil and gas obtained by the augmentation of reservoir energy, often by the injection of air, gas, or water into a production formation

sediment — soil and mineral solids of small particle size conveyed in water

seep — a small spring, pool, or other place where liquid from the ground has oozed to the earth's surface

semifusinite — a maceral that displays optical properties intermediate between vitrinite and fusinite

sensitivity — concentration, in mg/liter, that produces a reading of 1% (The term is used often in metals analysis.) (see detection limit)

septa — transverse walls in a hypha, fungal filament, or spore

settleable solids — settling solids; suspended solids that will settle out in a reasonable period of quiescence (Such a period is commonly, though arbitrarily, taken as 2 hr.)

shift conversion — the process by which excess carbon monoxide in the crude gas is "shifted" (converted) to carbon monoxide to provide the hydrogen-carbon monoxide ratio (3:1) needed for the subsequent synthesis of additional methane

shock loading — high and low surges of input contaminants to biological oxidation units capable
of causing upsets

silt — finely divided particles of soil or rock, which are often carried in cloudy suspension
in water and eventually deposited as sediment

sink — any body of water that acts as a long-term repository for pollutants

slag — ash and other residue remaining after fuel combustion, usually used to indicate a hot
molten state; solidification from this state

smog — term used broadly to mean polluted air (There are various types of smog, which are
produced from certain classes of pollutants under specific meteorologic conditions.
Photochemcial smog results from photochemical reactions involving hydrocarbons and
nitrogen oxides.)

smoke — solid and/or liquid particles formed by the imcomplete combustion of fuels and discharged
suspended in the gaseous combustion products

SNG — substitute or synthetic natural gas; a manufactured gas that contains about 97% methane,
has a higher heating value of about 1000 Btu/scf, and meet the same end-use specifications
as pipeline natural gas

solubility — the ability or tendency of one substance to blend uniformly with another

solvent extraction process — a method to convert coal and hydrogen under pressure to a clean
(low-sulfur, low-ash) fuel by dissolving most of it, then separating the undissolved coal
and mineral matter (ash) from the extract

solvent naphtha — a flammable liquid distillate containing, principally, xylenes and higher
aromatic hydrocarbons (Solvent naphtha is obtained especially from coal-tar light oils
or coke-oven-gas light oils and is used chiefly as a solvent and as a raw material for resins.)

Solvent-Refined Coal process — a process to remove ash, sulfur, and other impurities from coal
(The end-product contains about 16,000 Btu/lb, has an ash content of 0.1%, and a very low
sulfur content of about 0.5%. The product is solid at room temperature, but can be
liquefied by using relatively low heat.)

soot — agglomerations of tar-impregnated carbon particles that form when carbonaceous material
does not undergo complete combustion

sorption — a term including both adsorption and absorption (Sorption is basic to many processes
used to remove gaseous and particulate pollutants from an emission and to clean up oil
spills.)

sour gas — a gas containing hydrogen sulfide

sour water — see gas liquor

special naphthas — all finished products within the gasoline range that are specially refined to specified flash point and boiling range for use as paint thinners, cleaners, solvents, etc., but not to be marketed as motor or aviation gasoline or to be used as petrochemical feedstocks (Total includes special naphthas produced from petroleum and natural gas.)

SPG — substitute or synthetic pipeline gas; synonymous with SNG

sp gr — specific gravity

splint coal — a variety of bituminous or subbituminous coal that commonly has a dull luster, grayish-black color, and a compact structure and that often contains a few thin, irregular bonds of vitreous luster

SSMS — spark source mass spectrometry

stack gas desulfurization (scrubber) — treatment of stack gases to remove sulfur compounds

stack gases — gaseous substances emitted from power-plant smoke stacks during burning of fuel

static bed — settled bed; a bed that has no fluidizing medium passing through it (The term usually applies to bed height measurements taken under nonfluidized conditions, but also applies to a bed that has fluidizing medium passing through the bed at a rate that is insufficient to fluidize the bed

steam reforming — cracking of natural gas or other hydrogen gases, in the presence of steam, to reduce the heating value or to produce hydrogen

steroid — any of a class of compounds that are characterized by a polycyclic structure like that of the sterols and that usually include the sterols and vitamin D as well as many other naturally occurring compounds (as the bile acids and various hormones and glycosides)

steroidogenesis — the formation of steroids

stigma — the upper end of the style modified for the reception of pollen

stoma — one of the minute openings in the epidermis of leaves and other plant organs through which gaseous interchange between the atmosphere and the intercellular spaces within the leaf occurs; the opening, together with its guard and accessory cells

strip mining — the mining of coal by removing covering material (the overburden) and stripping away the entire underlying coal seam (Other forms of coal mining are (1) underground and (2) auger mining, in which coal is drilled out of seams exposed along the side of a mountain.)

style — in plants, the part of a carpel lying above the ovary (Ovary and style, capped by the stigma, make up the pistil.)

subbituminous coal — coal of intermediate rank (between lignite and bituminous); weathering and nonagglomerating coal having calorific values in the range of 8300 to 13,000 Btu, calculated on a moist, mineral-matter-free basis

sulfur oxides — compounds composed of sulfur and oxygen produced by the burning of sulfur and its compounds in coal, oil, and gas that are harmful to the health of man, plants, and animals, and that may cause damage to materials

superheat — heat in, or temperature of, steam over and above that of the saturation temperature at that pressure, usually given in degrees (°F) superheat

superheater — that part of the boiler in which vapor (steam) is further heated

surface water — all water on the surface, as distinguished from groundwater

surface mining — the mining of coal from outcroppings or by the removal of overburden from a seam of coal (as opposed to underground mining); any mining at or near the surface; also called strip mining, placer mining, open-cast or opencut mining, open-pit mining

surfactant (surface-active agent) — any compound that (1) reduces surface tension when dissolved in water or water solutions or (2) that reduces interfacial tension between two liquids or between a liquid and a solid (The three categories of surfactants are detergents, wetting agents, and emulsifiers; all have the same basic chemical mechanism and differ chiefly in the nature of the surface involved.)

susceptibility — sensitivity or predisposition to a condition or action; the state of being sensitive or predisposed to a pathogen

suspended solids — solids that either float on the surface or remain suspended in liquids; removable by filtering

Svedberg coefficient (S) — a numerical value related to the settling velocity of a spherical particle, which is a function of particle density, particle diameter, and fluid viscosity:

$$S = \frac{d^2 (SG - 1)(10^{13})}{18n} \ ,$$

where

    S = Svedberg coefficient
   SG = specific gravity of particle
    d = particle diameter (cm)
    n = fluid viscosity (poises)

Particles having the same S value will settle simultaneously. Inasmuch as S depends on both particle density and diameter, different size grains can settle together. For centrifugal work, therefore, the Svedberg coefficient is a better parameter than diameter for describing the settling characteristics of a particle in water.

synergism — the harmonious action of two agents (e.g., drugs) or organs, (e.g., muscles)
producing an effect neither could produce alone or an effect that is greater than the
total effects of each agent operating by itself

synergist — a substance that increases the effect of another substance so that the total effect
is greater than the sum of the two substances taken independently

Syngas — synthetic gas (SNG)

syngenetic — a term generally applied to mineral or ore deposits formed contemporaneously with
the enclosing rocks

synthane — a coal gasification process being developed by the Bureau of Mines to produce pipeline
quality gas

synthetic natural gas (SNG) — substitute natural gas; a manufactured gaseous fuel generally
produced from naphtha or coal that contains 95 to 98% methane and has an energy content
of 980 to 1035 Btu/scf (about the same as that of natural gas)

tail gas — the effluent from a final stage, as from a flue gas scrubber

tar sands — sedimentary rocks that contain viscous, heavy petroleum that cannot be recovered
by conventional methods of petroleum production

teleost — one of the teleostei

teleostei — a class or a subclass that contains all existing jawed fishes except the chondrichthyes
or sometimes the chondrichthyes and choanichthyes

teratogenesis — the process that produces birth defects in lower forms of life

terpene — an unsaturated hydrocarbon having the empirical formula $C_{10}H_{16}$ that occurs in most
essential oils and oleoresins of plants [The terpenes are based on the isoprene unit C_5H_8
and may be either acyclic (aliphatic) or cyclic (aromatic), with one or more benzenoid
groups. They are classified as monocyclic (dipentene), dicyclic (pinene), or acyclic
(myrcene), according to the molecular structure.]

tertiary recovery — use of heat and methods other than fluid injection to augment oil recovery
(presumably occurring after secondary recovery)

testosterone — a sex hormone that is a crystalline hydroxy steroid ketone $C_{19}H_{28}O_2$, is obtained
especially from the testes of bulls or synthetically (as from cholesterol or diosgenin),
and is used in medicine chiefly in the form of esters

thermal cracking — a process during which the less volatile heating oil fractions are subjected
to higher temperatures under increasing pressure (The heat puts a strain on the bonds
holding the larger, complex molecules together and causes them to break up into smaller
ones, including those in the gasoline range.)

thermal efficiency — the percentage of the total heating value input (including fuels) to a plant that is recovered as product and by-product heating value or equivalent energy

thermal pollution — the ejection of heated water into the environment, usually aquatic ecosystems, raising the temperature above normal limits

thermocline — in a thermally stratified lake, the layer below the epilimnion; the stratum in which there is a rapid rate of decrease in temperature with depth

thermodynamics, laws of — The first law of thermodynamics states that energy can neither be created nor destroyed. The second law of thermodynamics states that when a free exchange of heat takes place between two bodies, the heat is always transferred from the warmer to the cooler body.

thiophene (thiofuran) — CHCHCHCHS; a highly reactive cyclic organosulfur

threshold dose — the minimum dose of a given substance necessary to produce a measurable physiological or phychological effect

threshold limit value — refers to airborne concentrations of substances and represents conditions under which it is believed that nearly all workers may be repeatedly exposed for an 8-hr day, 5 days a week (expressed as parts per million (ppm) for gases and vapors and as milligrams per cubic meter (mg/m^3) for fumes, mists, and dusts)

TLC — thin-layer chromatography

TLV — threshold limit value

TNT — trinitrotoluene

tolerance — the relative capability of an organism to endure an unfavorable environmental factor

ton — a unit of weight equal to 2000 lb in the United States, Canada, and the Union of South Africa and equal to 2240 lb in Great Britain [The American ton is often called the short ton, while the British ton is called the long ton (tonne). The metric ton, or 1000 kg, equals 2204.62 lb. Depending upon specific gravity, a long, or metric, ton will equal from 6.5 to 8.5 bl of oil.]

toxicant — a substance that kills or injures an organism through chemical or physical action or by altering the organism's environment; for example, cyanides, phenols, pesticides, or heavy metals, especially used for insect control

toxicity — the quality, state, or relative degree of being toxic or poisonous; the ability of a chemical molecule or compound to produce injury when it reaches a susceptible site in the body

toxic wastes — wastes that contain substances in sufficient quantity to impinge harmfully on aquatic life irrespective of oxygen concentration

trace metals — metals found in small quantities or traces usually due to their insolubility

translocation — the transfer of metabolites, nutritive material, or other substances from one
part of a plant to another

trophic level — levels of feeding on land and in the sea and air [For example, land plants
(producers) are on one trophic level, and the primary consumers (herbivores), which eat
the plants, are on the next higher trophic level.]

troposphere — the layer of the atmosphere extending seven to ten miles above the earth (Vital
to life on earth, the trophosphere contains clouds and moisture that reach earth as rain
and snow.)

turbulent fluidized bed — a fluidized bed in a state of fluidization above the minimum fluidizing
point and in which the particles tend to mix readily

ultimate analysis (coal) — an analysis of a dried coal sample to determine carbon, hydrogen,
sulfur, nitrogen, and ash with oxygen estimated by difference

ultimate recoverable reserves — the total quantity of crude oil, natural gas, natural gas
liquids, or sulfur estimated to be ultimately producible from an oil or gas field as
determined by an analysis of current and engineering data (This includes any quantities
already produced up to the respective date of the estimate.)

ultraviolet — pertaining to the region of the electromagnetic spectrum from about 10 to 380 nm
(The term ultraviolet without further qualification usually refers to the region from
200 to 380 nm.)

underground coal gasification — the proposed process for producing synthetic gas from coal in
natural underground deposits; western coal

urban runoff — storm water from city streets and gutters that usually contains a great deal
of litter and organic and bacterial wastes

utilities section — boiler to provide steam and electric power

UVS — ultraviolet spectrometry

vapor — gaseous material that results from dilution with fixed gases but which, if pure, would
occur as a solid or liquid at the ambient temperature (such as water vapor)

vapors — gas-phase hydrocarbons that are generally derived from solvents processing

vascular plants — plants that have a specialized conducting system (xylem and phloem)

venting — release of gases or vapors under pressure to the atmosphere vermiculites — a group of platy minerals (hydrous silicates of aluminum, magnesium, and iron) closely related to the chlorites and montmorillonites

virgin naphtha, virgin oils — naphtha and oils derived from atmospheric and vacuum distillation of crude oil

vitrain — a lithotype of thin horizontal bands in coal, which are visible to the naked eye

vitrinite — the series of genetically related macerals that form the humic fraction of coal seams and that are produced by the gelification and gradual metamorphosis of cell wall substances

vitrinization — the process in coalifiction that results in the formation of vitrain

volatile matter — those products, exclusive of moisture, given off by a material as gas and vapor, determined by definite prescribed methods, which may vary according to the nature of the material

volatility — that property of a liquid that denotes its tendency to vaporize

Warburg apparatus — a complex respirometer consisting of a battery of constant-volume manometer-flask units, with a mechanically agitated constant-temperature water bath and used especially in the study of cellular respiration and metabolism or fermentation and other enzymatic reactions

washout — the removal of a pollutant by precipitation below clouds

waste stream — streams in the coal conversion plant that contain amounts of coal-conversion-derived materials that require environmentally acceptable disposal (e.g., ash, water, sludges)

water pollution — the addition of sewage, industrial wastes, or other harmful or objectionable material to water in concentration or in sufficient quantities to result in measurable degradation of water quality

watershed — land area from which water drains toward a common watercourse in a natural basin

weathering — the group of processes, such as the chemical action of air and rain water and of plants and bacteria and the mechanical action of changes of temperature, whereby rocks, on exposure to the weather, change in character, decay, and finally crumble into soil

Weber number — ratio of inertial forces to surface tension, expressed as

$$\frac{\rho L V^2}{\Sigma} \text{ , or } \frac{V}{\sqrt{\Sigma / \rho l}}$$

where

ρ = density of H_2O

L = length

$$V = \frac{\text{surface tension}}{\text{unit length}}$$

Σ = velocity

well head — oil or gas brought to the surface, ready for transportation to refinery or ship or pipeline (Well head costs usually refer to the cost to bring the oil or gas to the surface and do not include costs of transportation, refining, distribution, or profit.)

wet deposition — the removal of atmospheric particles to the earth's surface by rain or snow

wet gas — natural gas that contains distillable heavier materials (Some wet gas contains a natural gasoline.); sometimes called "casinghead gas;" a gas containing significant amounts of C_3 or heavier hydrocarbons

wetting agent — a surfactant that, when added to water, causes it to penetrate more easily into, or to spread over, the surface of another material by reducing the surface tension of the water (e.g., soaps and alcohols)

wfg — wet flue gas

WSF — water-soluble fraction

xenobiotics — substances that can only enter the environment by human activities

xylene — a commercial mixture of the three isomers, o-, m-, and p-xylene, that is derived from the fractional distillation of petroleum, coal tar, or coal gas and the catalytic re-forming of petroleum (The total includes xylene produced from coal and petroleum.)

X-RF — x-ray fluorescence

zooplankton — the microscopic animal life found floating in lakes, oceans, etc.

Appendix C

POLYCYCLIC AROMATIC HYDROCARBONS GLOSSARY

D. S. Harnden   J. T. Ensminger

Polycyclic aromatic hydrocarbons, some of which are potential carcinogens, are formed during coal conversion. This index of the most common polycyclic aromatic hydrocarbons contains structural formulas identified by the preferred IUPAC names. Because many of the polycyclic aromatic hydrocarbons were isolated and named before establishment of the 1957 IUPAC rules, many compounds have several names. The familiar synonyms included in the glossary are often but not necessarily the names used in the literature cited and thus in the text of this document. All ring structures are aromatic unless otherwise indicated.

Compound	Preferred name (synonym)
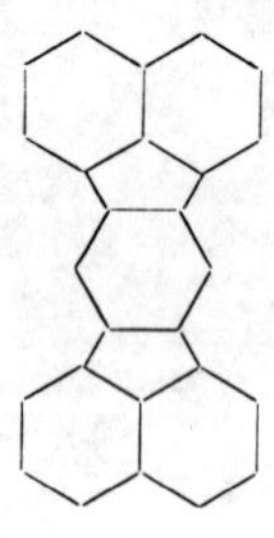	Aceanthryleno[2,1-*a*]aceanthrylene
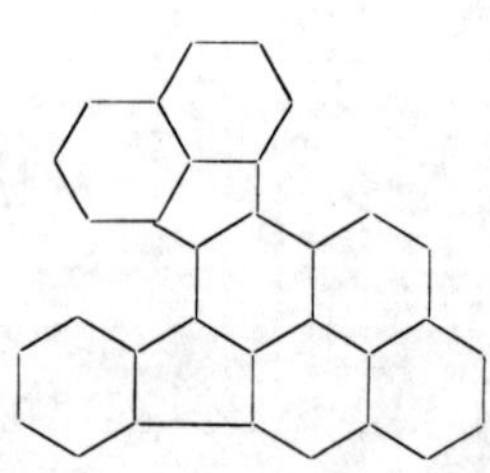	Acenaphthol[1,2-*j*]fluoranthene
	Acenaphtho[1,2-*k*]fluoranthene
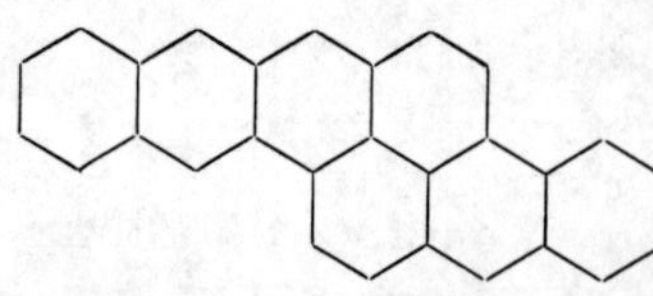	Acenaphtho[*a*]naphtho[2,1,8-*cde*]fluoranthene
	Anthra[2,1,9-*qra*]naphthacene

Compound   Preferred name (synonym)

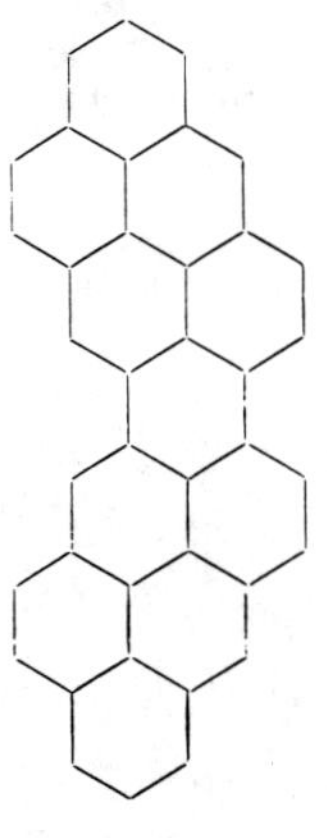

Anthra[2,1,9,8-*klmno*]naphtho[3,2,1,8,7-*vwxyz*]hexaphene

Anthracene

Anthraceno[5,4-*ab*]fluoranthene

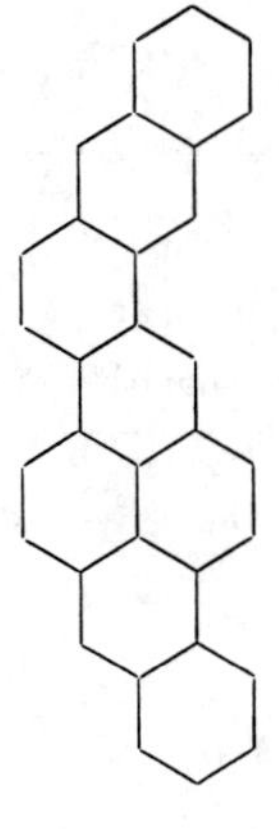

Anthraceno[5,4,3-*cde*]pentaphene

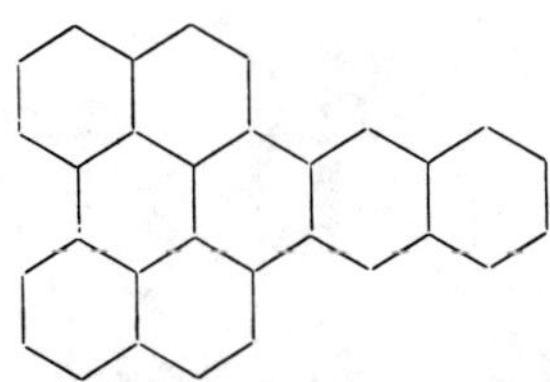

Anthraceno[4,3,2,1-*ghi*]perylene

Compound	Preferred name (synonym)
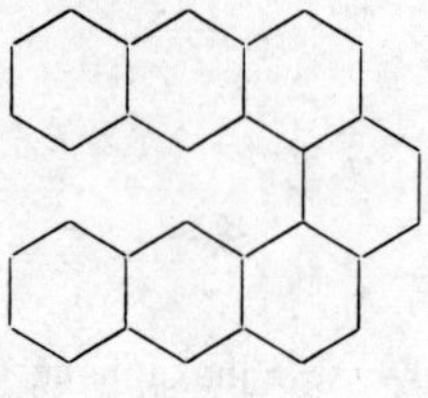	Anthraceno[3,4-*a*]tetraphene
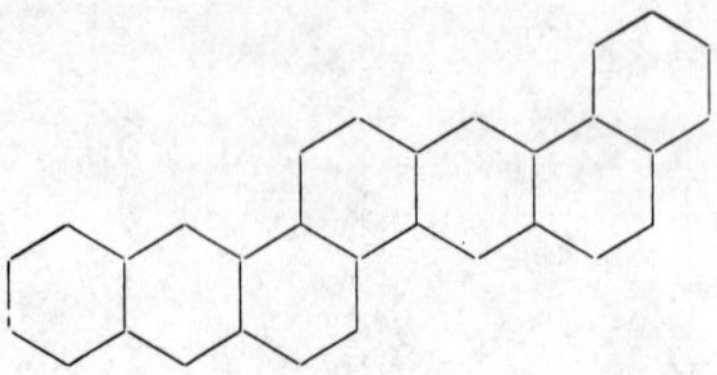	Anthraceno[1,2-*k*]tetraphene
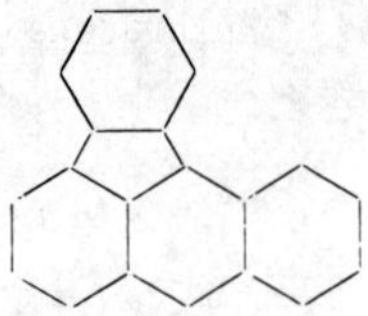	Benz[*a*]aceanthrylene (Benzo[*a*]fluoranthene)
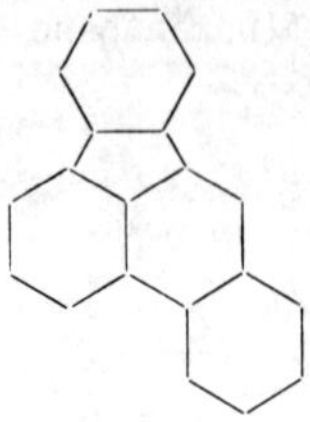	Benz[*e*]aceanthrylene (Benzo[*b*]fluoranthene)
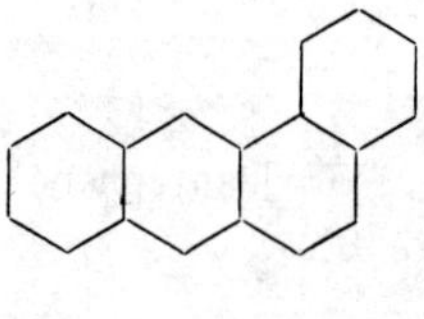	Benz[*a*]anthracene (Tetraphene)
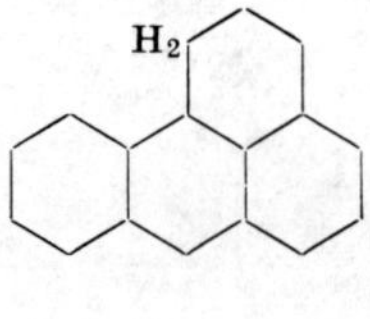	1H-Benz[*de*]anthracene

Compound	Preferred name (synonym)

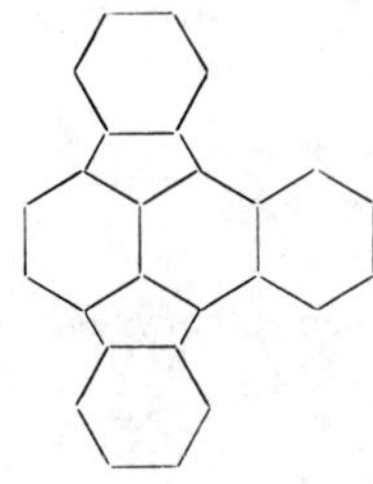

3H-Benz[*de*]anthracene

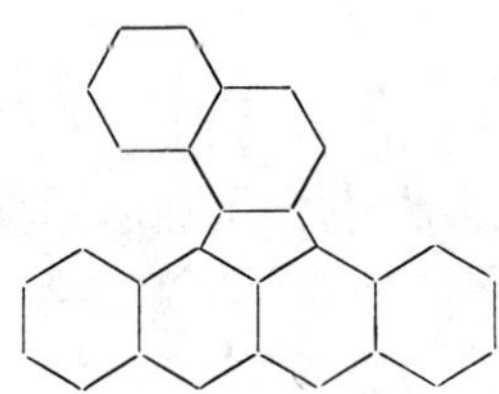

7H-Benz[*de*]anthracene

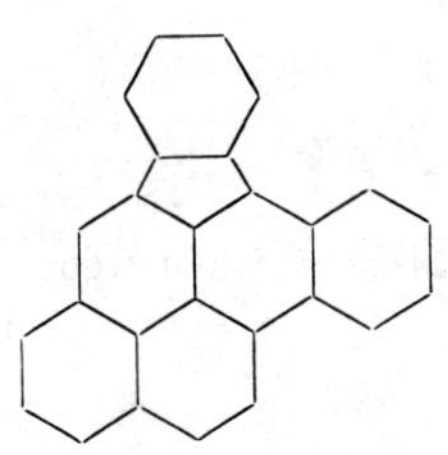

Benz[*a*]indeno[1,2,3-*fg*]aceanthrylene
(Isorubicene)

Benz[4,5]indeno[1,2,3-*fg*]naphthacene

Benz[*a*]indeno[3,2,1-*lm*]pyrene

Benzene

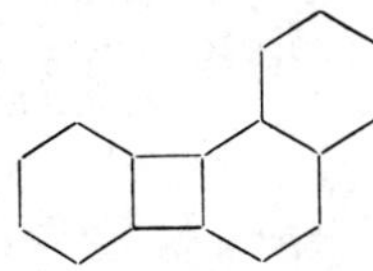

Benzo[*a*]biphenylene

Compound	Preferred name (synonym)

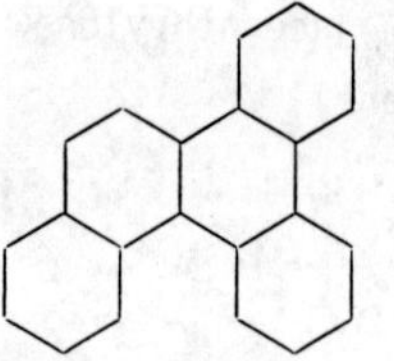

Benzo[*b*]biphenylene

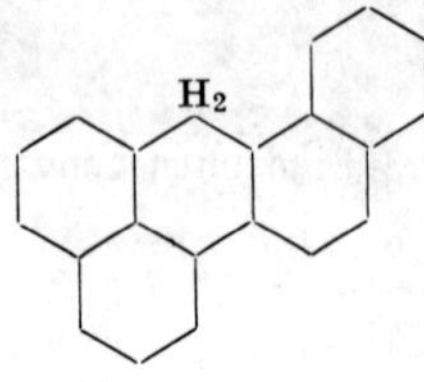

Benzo[*b*]chrysene

Benzo[*c*]chrysene

Benzo[*g*]chrysene

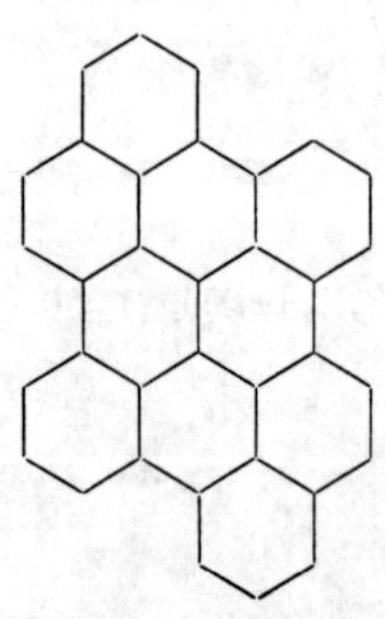

7H-Benzo[*hi*]chrysene

Benzo[*cd*]chryseno[7,6,5,4-*fghijk*]perylene

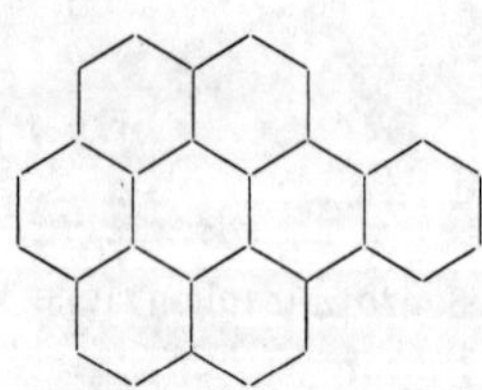

Benzo[*a*]coronene

Compound | Preferred name (synonym)

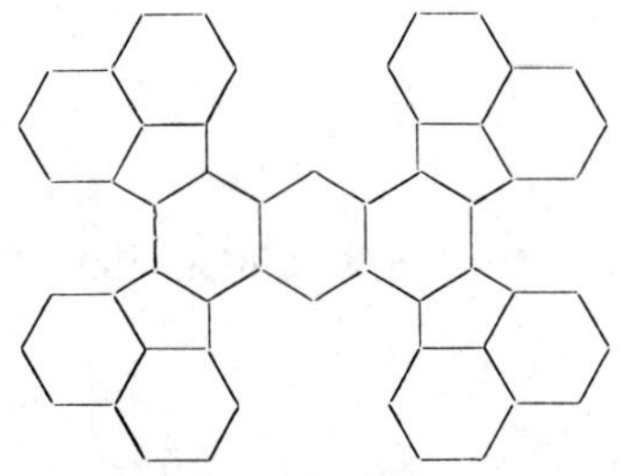

Benzo[1,2-*j*:4,5-*j'*]di(acenaphtho[1,2-*j*]fluoranthene)

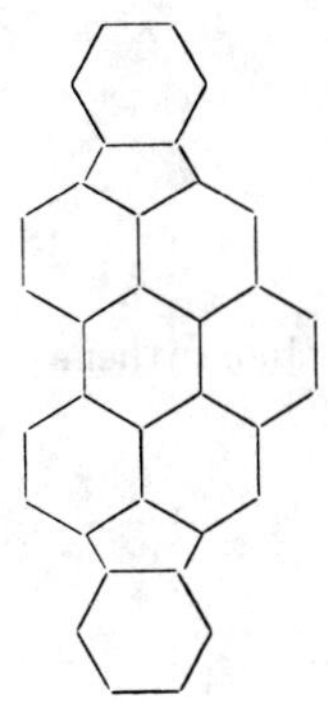

Benzo[*ghi*]diindeno[1,2,3-*cd*:1',2',2'-*lm*]perylene
(Benzo[*ghi*]periflanthene)

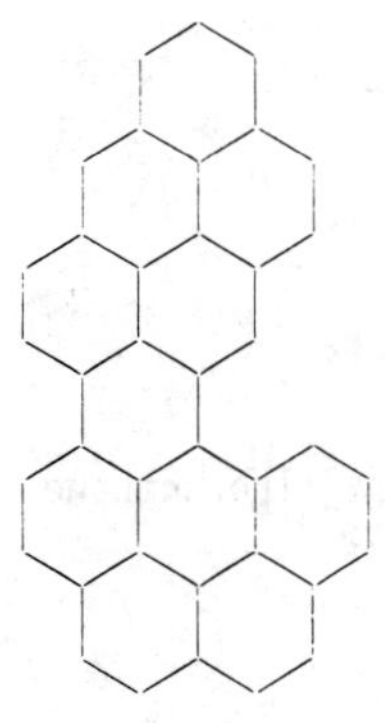

Benzo[*rs*]dinaphtho[2,1,8,7-*klmn*:3',2',1',8',7'-*vwxyz*]hexaphene

Benzo[*ghi*]dinaphtho[2,1,8-*cde*:1',8',7'-*klm*]perylene

Compound   Preferred name (synonym)

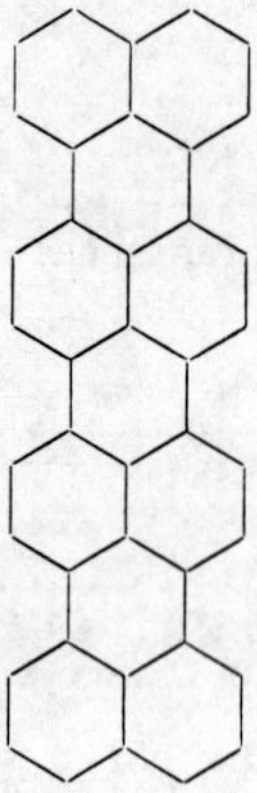

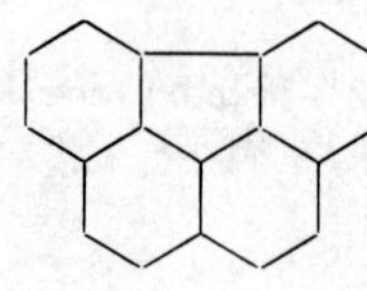

Benzo[1,2,3-*cd*:4,5,6-*c'd'*]diperylene

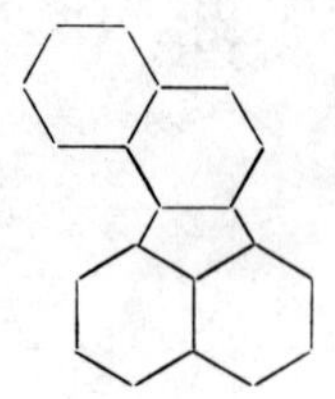

Benzo[*ghi*]fluoranthene

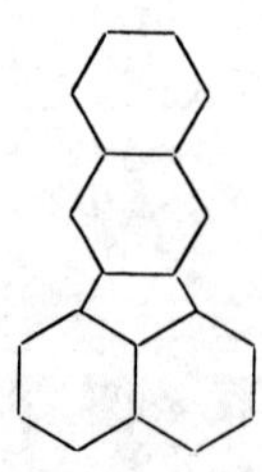

Benzo[*j*]fluoranthene

Benzo[*k*]fluoranthene

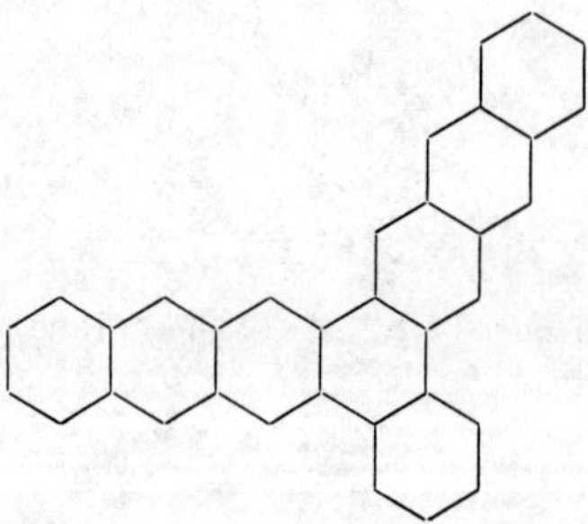

Benzo[*j*]heptaphene

Compound	Preferred name (synonym)
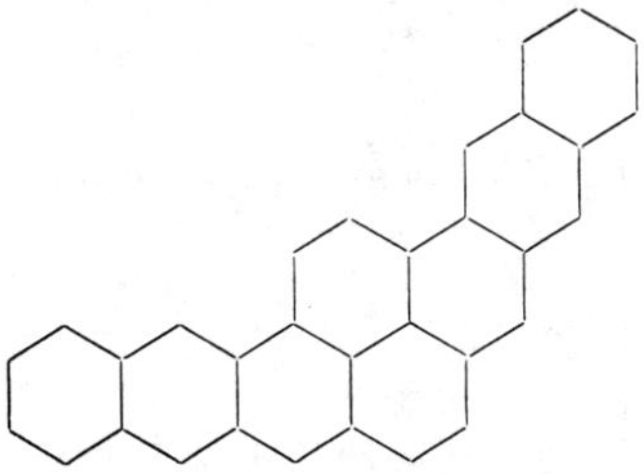	Benzo[*xyz*]heptaphene
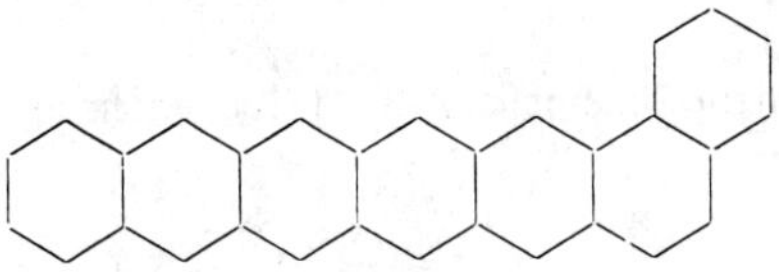	Benzo[*a*]hexacene
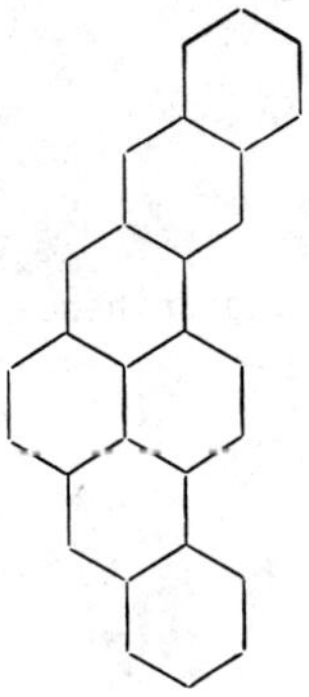	Benzo[*vwx*]hexaphene
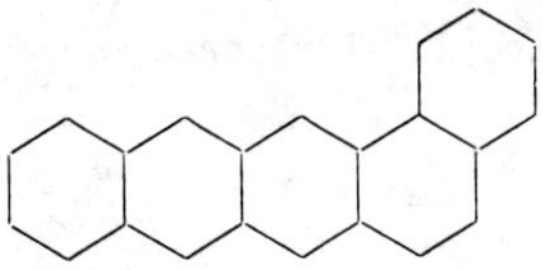	Benzo[*a*]naphthacene
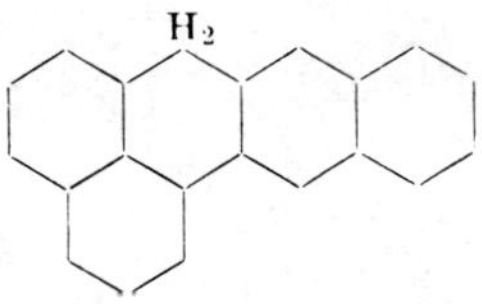	7H-Benzo[*de*]naphthacene
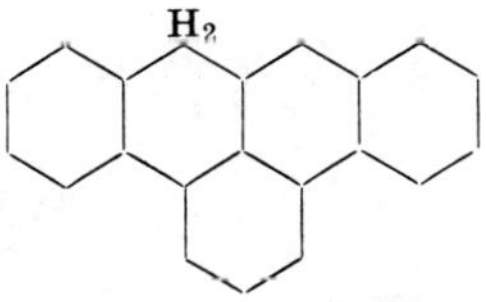	8H-Benzo[*fg*]naphthacene

Compound	Preferred name (synonym)

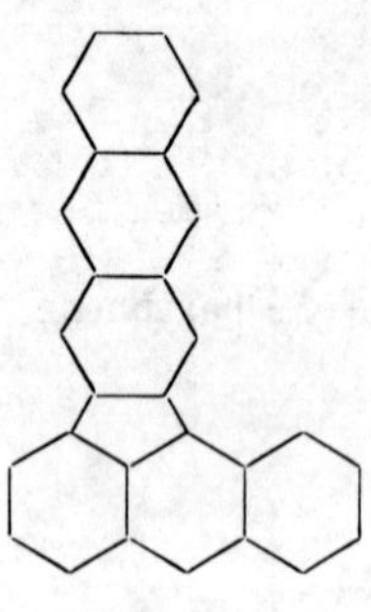

8H-Benzo[p]naphtho[3,2,1,8-$defg$]chrysene

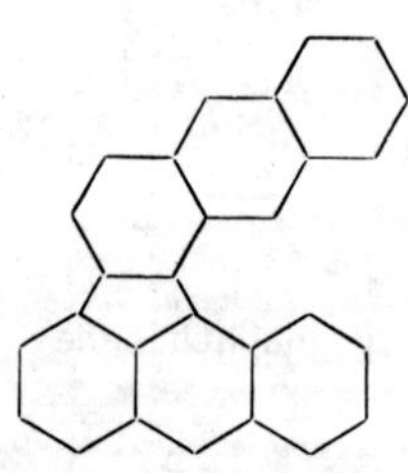

Benzo[a]naphtho[2,3-j]fluoranthene

Benzo[a]naphtho[2,3-k]fluoranthene

Benzo[a]naphtho[2,3-l]fluoranthene

Benzo[j]naphtho[2,3-b]fluoranthene

Compound   Preferred name (synonym)

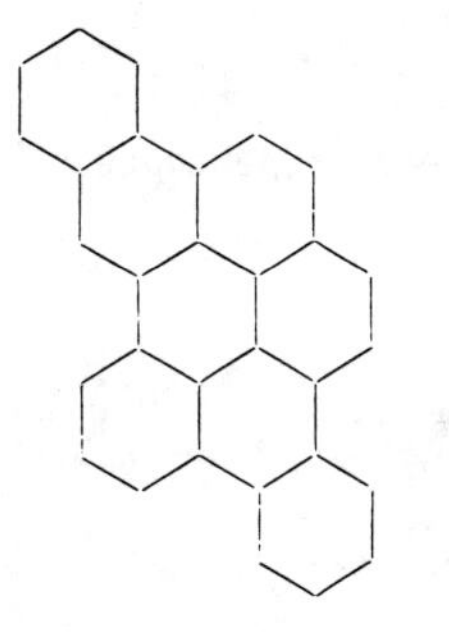

Benzo[*a*]naphtho[8,1,2-*cde*]naphthacene

Benzo[*a*]naphtho[2,1,8-*hij*]naphthacene

Benzo[*fg*]naphtho[2,1,8,7-*qrst*]pentacene

Benzo[*b*]naphtho[4,3,2,1-*pqr*]perylene

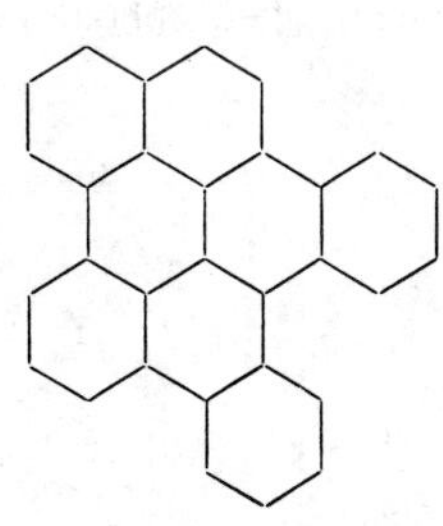

Benzo[*cd*]naphtho[1,2-*f*]phenanthro[2,3,4-*lm*]perylene

Compound	Preferred name (synonym)
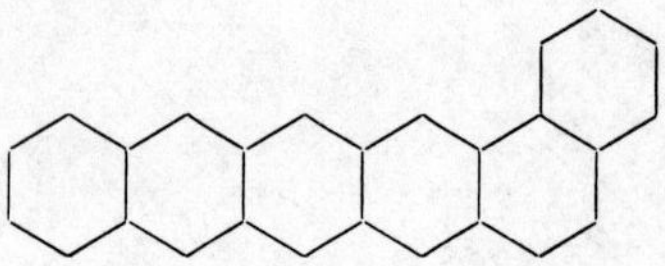	Benzo[*a*]pentacene
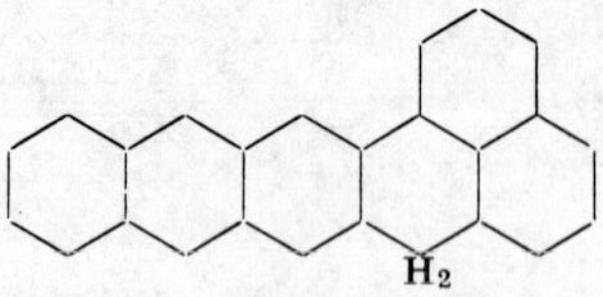	7H-Benzo[*de*]pentacene
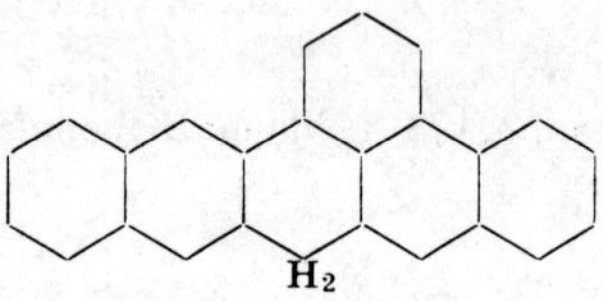	9H-Benzo[*fg*]pentacene
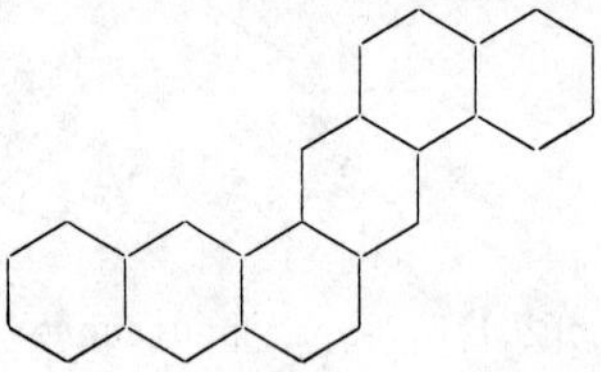	Benzo[*c*]pentaphene
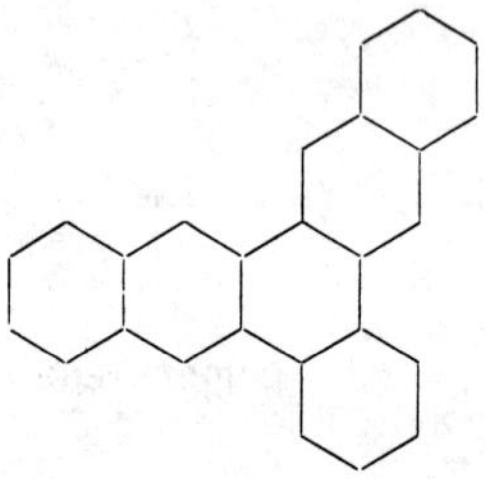	Benzo[*h*]pentaphene
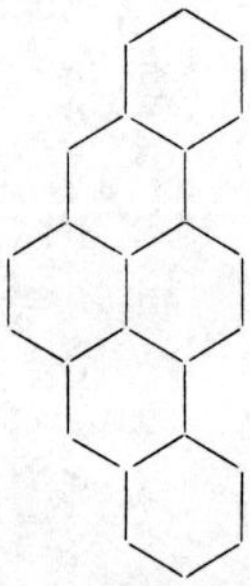	Benzo[*rst*]pentaphene
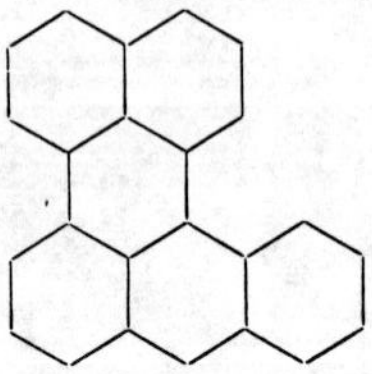	Benzo[*a*]perylene

Compound

Preferred name (synonym)

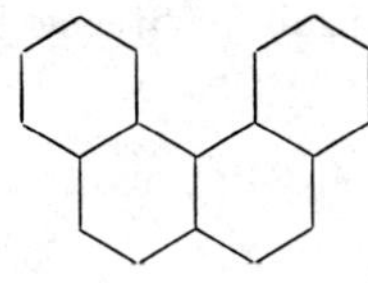

Benzo[*b*]perylene

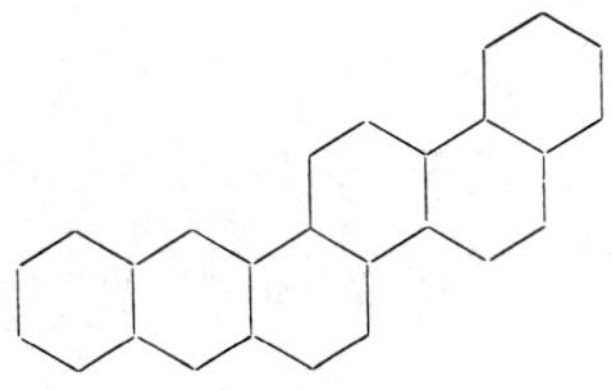

Benzo[*ghi*]perylene

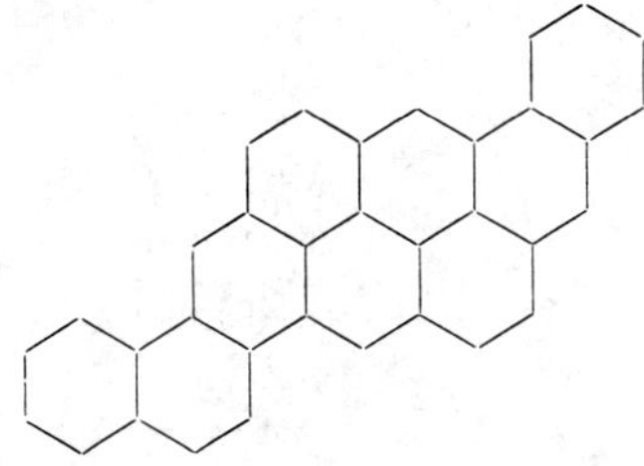

Benzo[*c*]phenanthrene

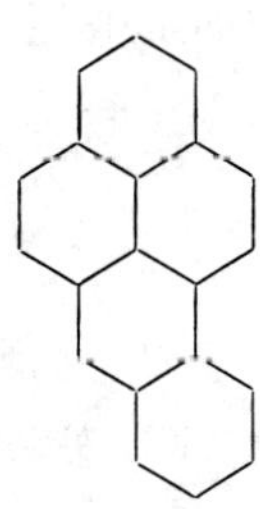

Benzo[*b*]picene

Benzo[*a*]pyranthrene

Benzo[*a*]pyrene
(3,4-Benzopyrene)

Compound	Preferred name (synonym)

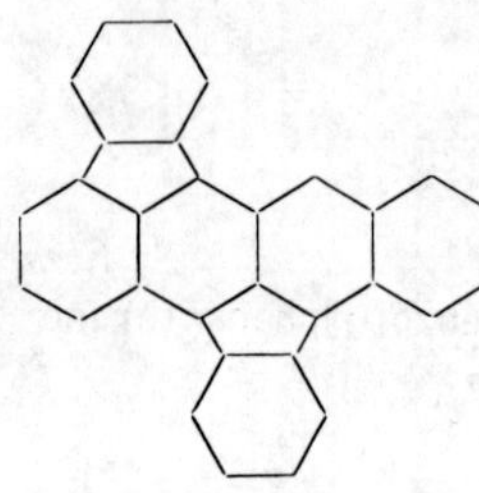

6H-Benzo[*cd*]pyrene

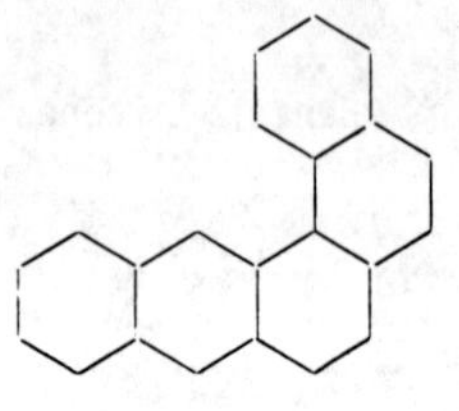

Benzo[*e*]pyrene

Benzo[*b*]rubicene

Benzo[*a*]tetraphene

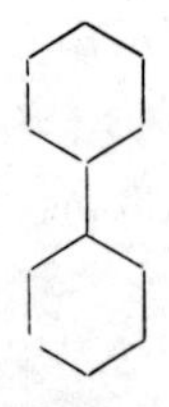

Biphenyl
1,1' Biphenyl
(Diphenyl)

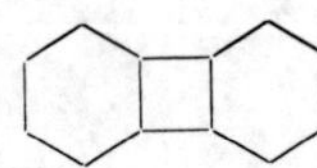

Biphenylene

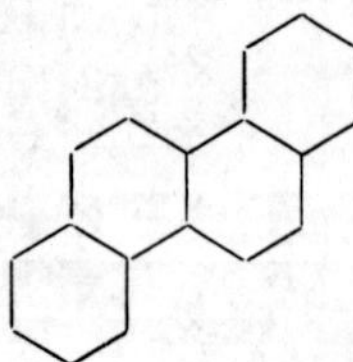

Chrysene

Compound   Preferred name (synonym)

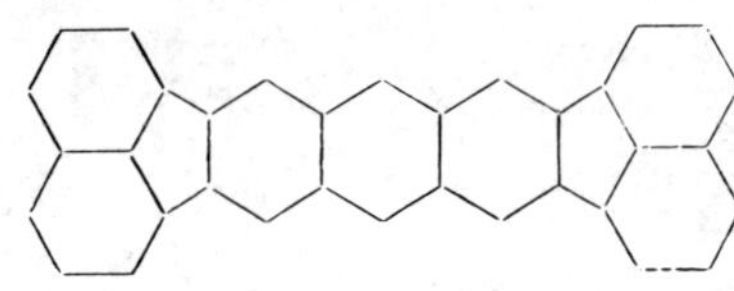

Coronene

Diacenaphth[2,1-*a*:2',1'-*c*]anthracene

Diacenaphth[1,2-*b*:1',2'-*i*]anthracene

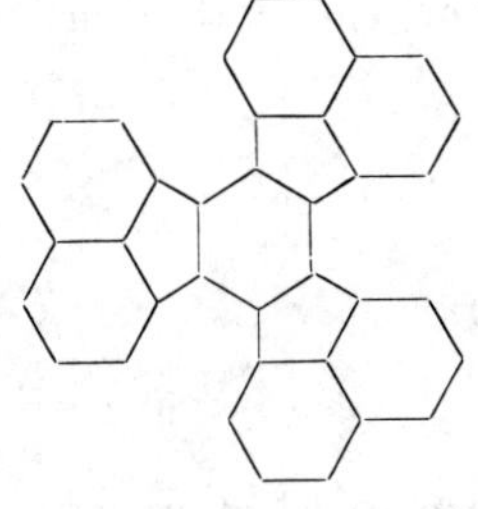

Diacenaphtho[1,2-*j*:1',2'-*l*]fluoranthene
(Decacyclene)

Diacenaphtho[1,2-*b*:1',2'-*g*]naphthalene

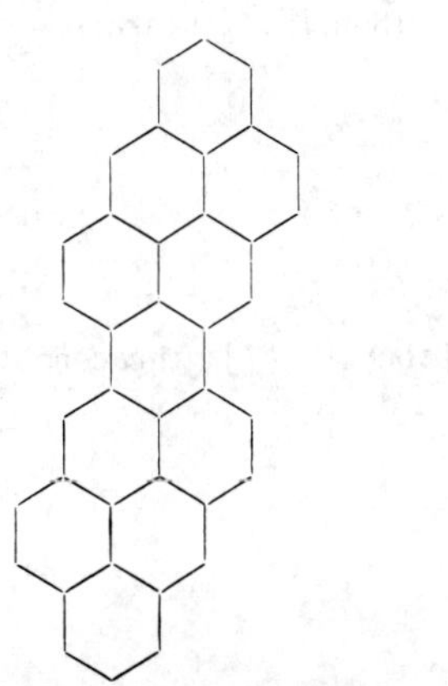

Dianthra[2,1,9,8-*stuva*:2',1',9',8'-*hijkl*]pentacene

Compound	Preferred name (synonym)
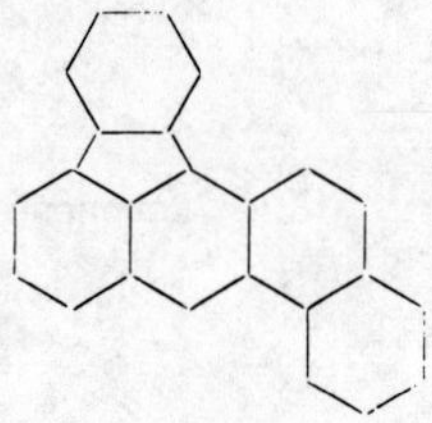	Dibenz[a,j]aceanthrylene
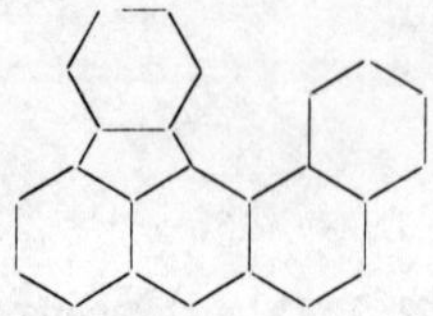	Dibenz[a,l]aceanthrylene
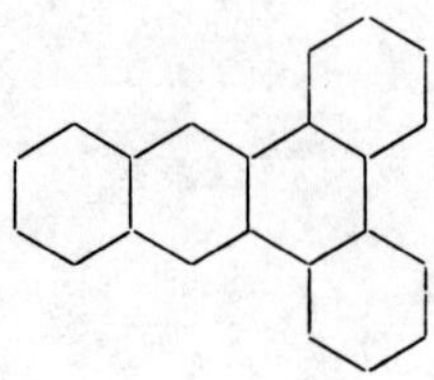	Dibenz[a,c]anthracene
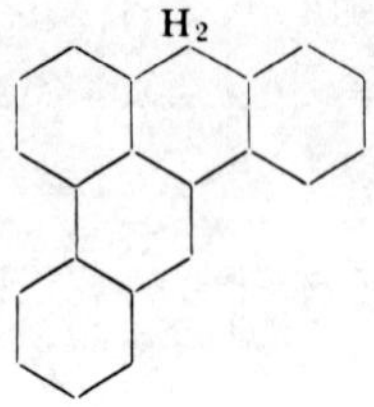	8H-Dibenz[a,de]anthracene
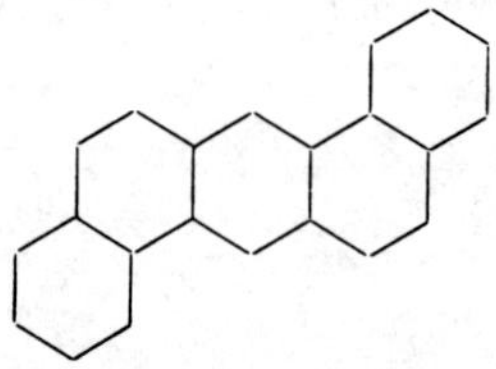	Dibenz[a,h]anthracene
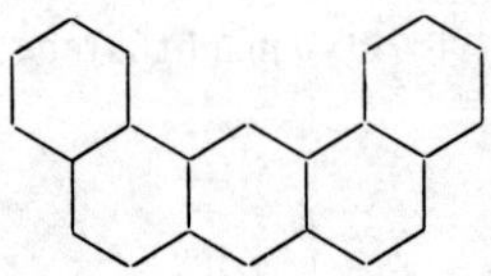	Dibenz[a,j]anthracene
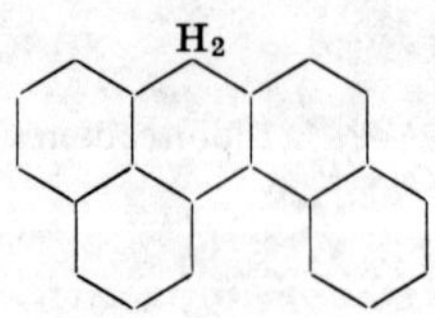	7H-Dibenz[a,kl]anthracene

Compound | Preferred name (synonym)

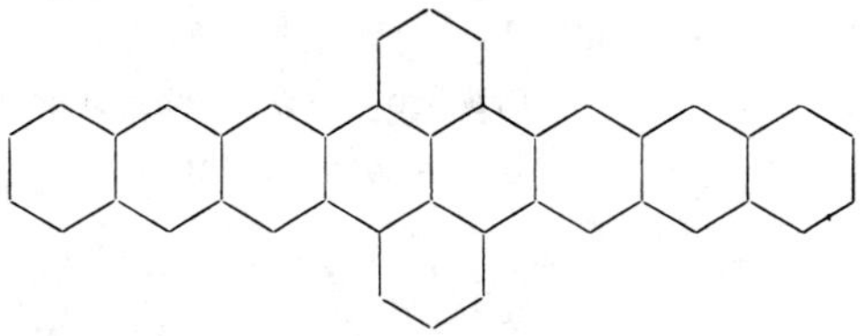

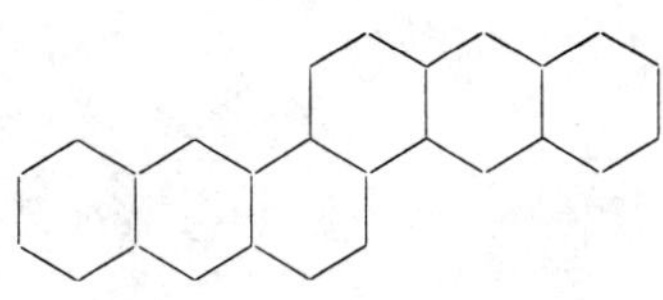

Dibenz[jk,a_1b_1]octacene

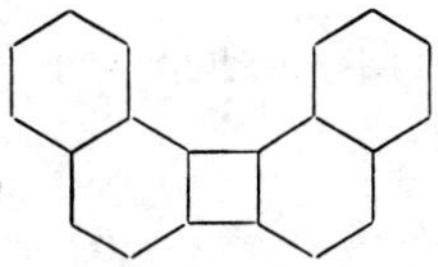

Dibenzo[a,i]biphenylene

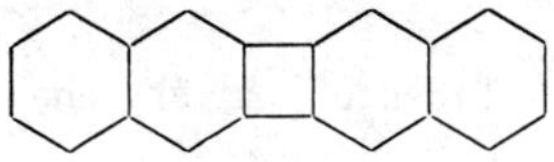

Dibenzo[b,h]biphenylene

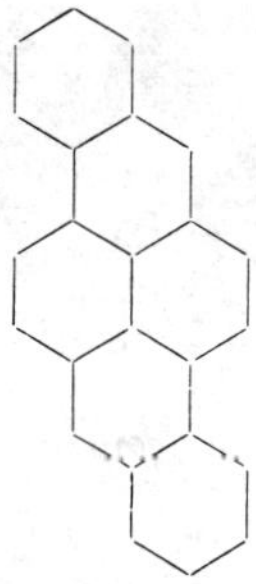

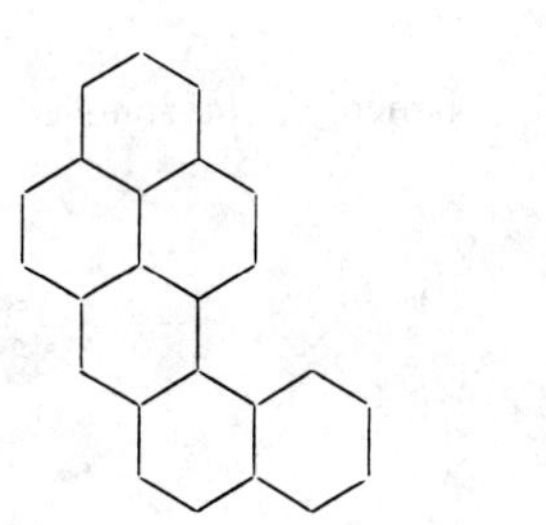

Dibenzo[b,def]chrysene

Dibenzo[b,k]chrysene

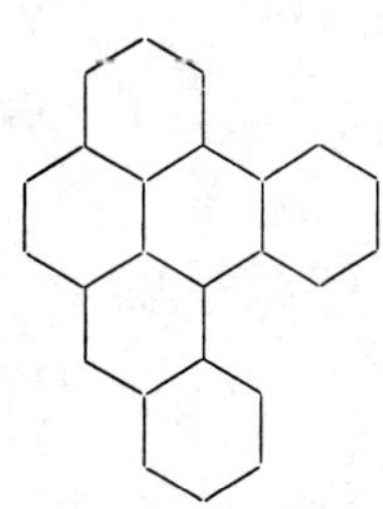

Dibenzo[c,mno]chrysene

Dibenzo[def,p]chrysene

Compound   Preferred name (synonym)

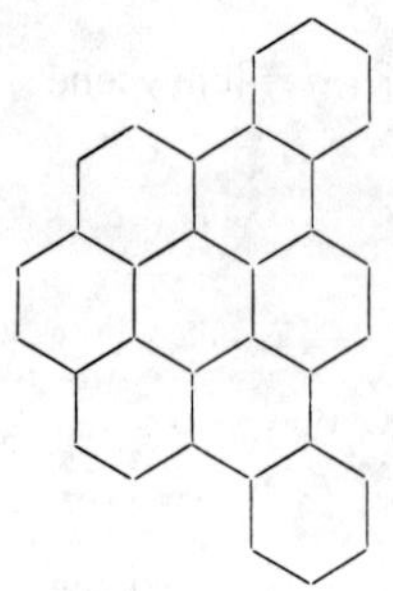

Dibenzo[*def*,*mno*]chrysene
(Anthanthrene)

8H-Dibenzo[*def*,*qr*]chrysene

Dibenzo[*q*,*p*]chrysene

Dibenzo[*a*,*g*]coronene

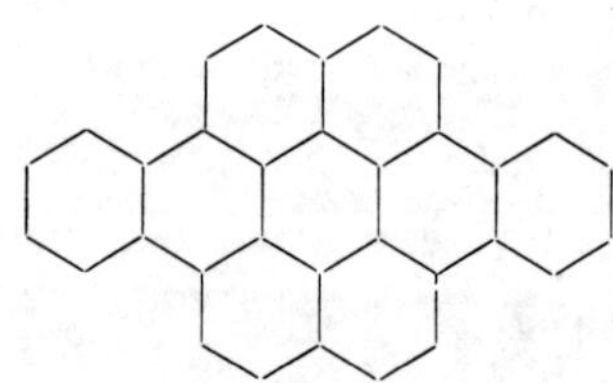

Dibenzo[*a*,*j*]coronene

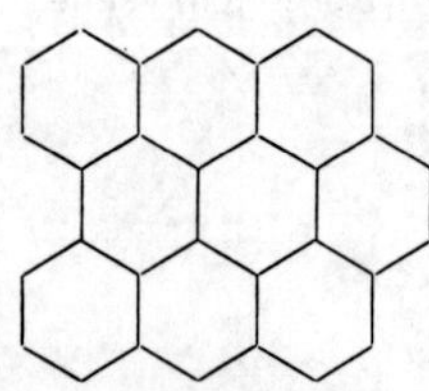

Dibenzo[*bc*,*ef*]coronene

Compound	Preferred name (synonym)

Dibenzo[*bc*,*kl*]coronene

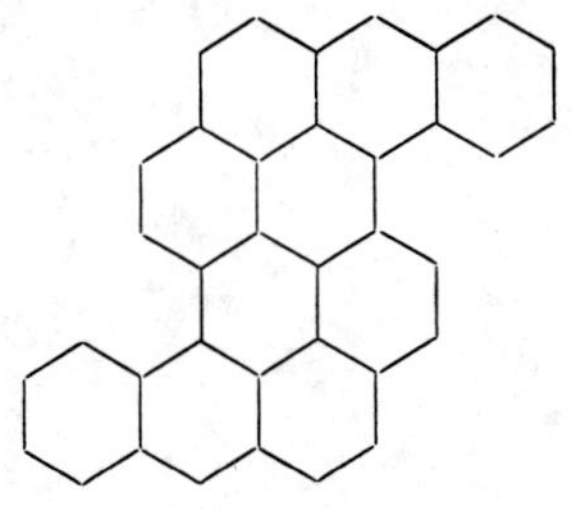

Dibenzo[*c*,*l*]dinaphtho[7,8,1,2-*fghi*:2',1',8',7'-*opqr*]naphthacene

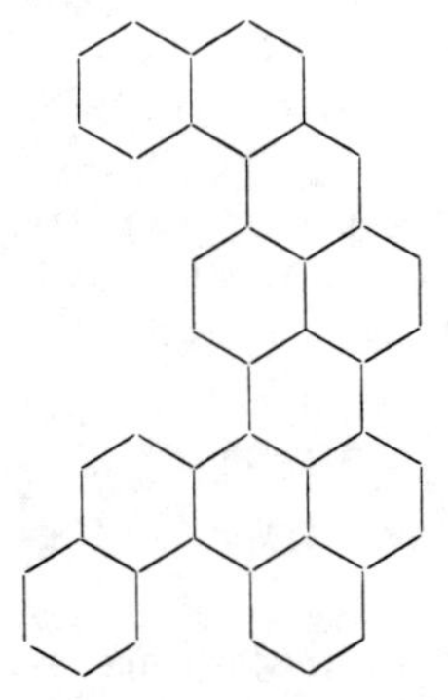

Dibenzo[*o*,*rst*]dinaphtho[1,2-*a*:8',1',2'-*cde*]pentaphene

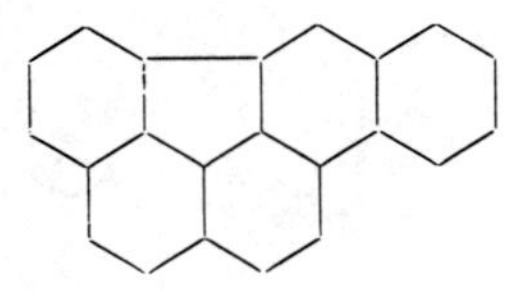

Dibenzo[*b*,*ghi*]fluoranthene

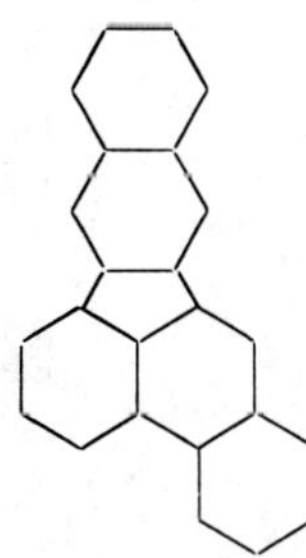

Dibenzo[*b*,*k*]fluoranthene

Compound	Preferred name (synonym)
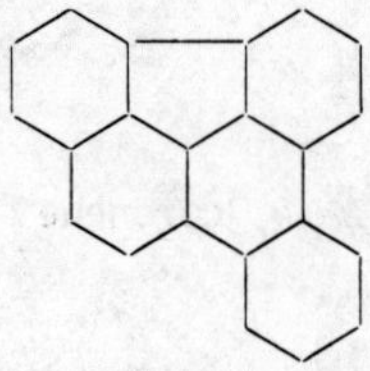	Dibenzo[e,ghi]fluoranthene
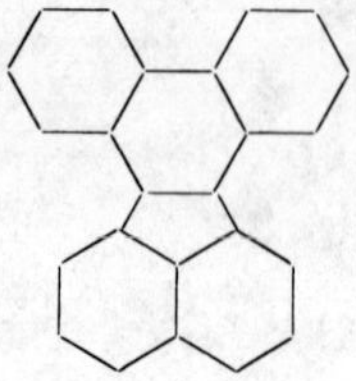	Dibenzo[j,l]fluoranthene
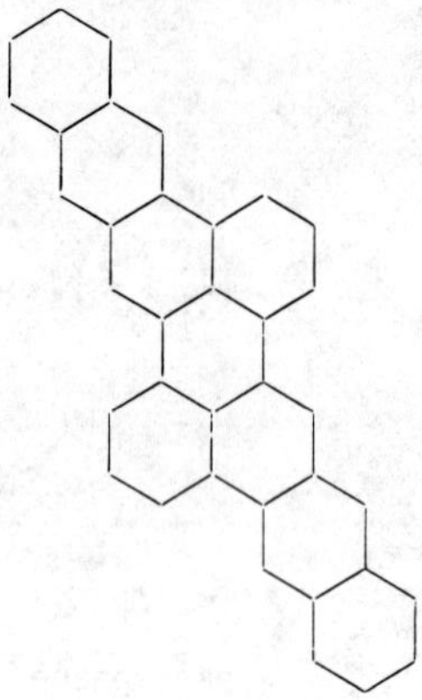	Dibenzo[hi,wx]heptacene
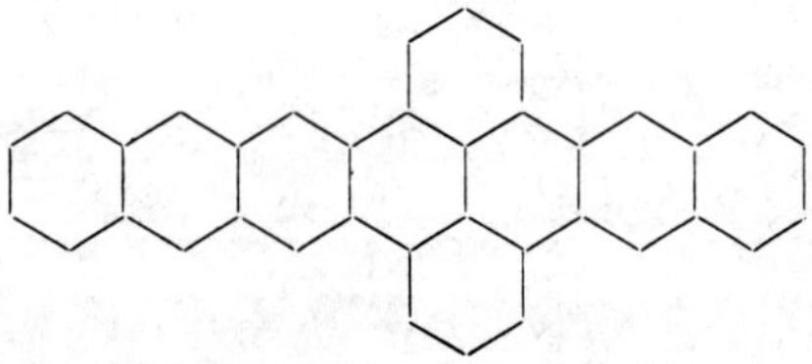	Dibenzo[hi,yz]heptacene
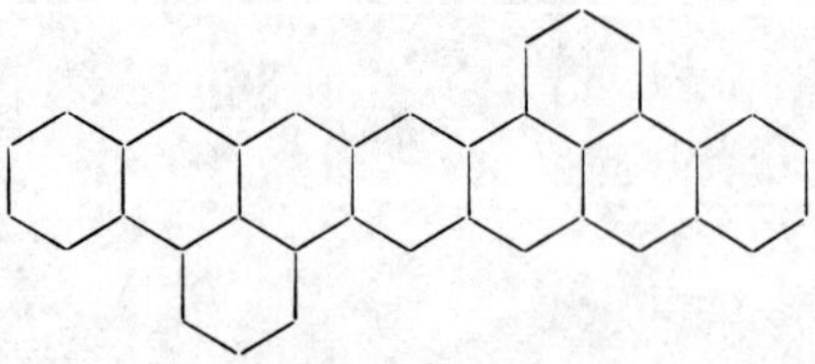	Dibenzo[lm,a_1b_1]heptacene

Compound	Preferred name (synonym)
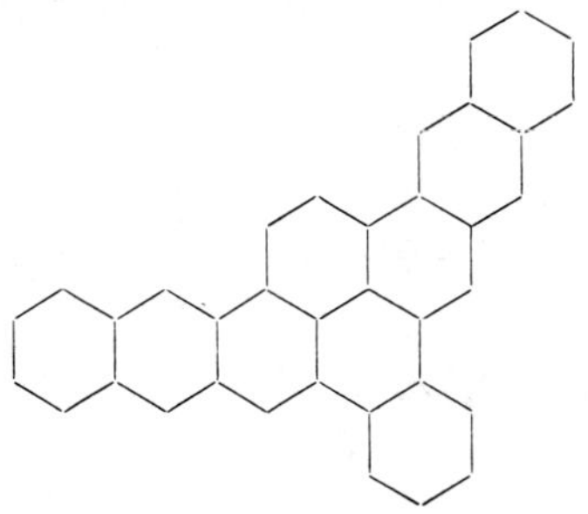	Dibenzo[*i*,*xyz*]heptaphene
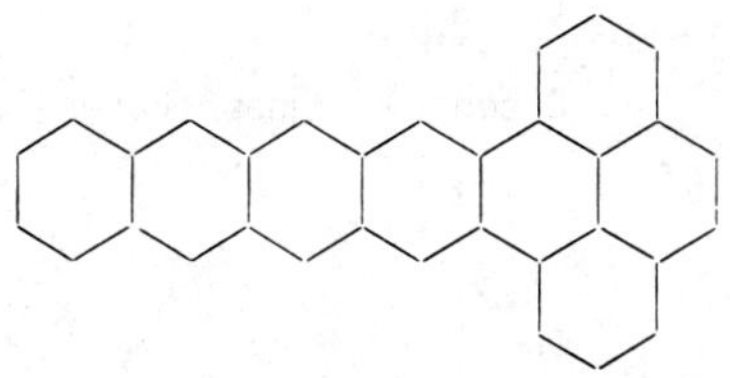	Dibenzo[*de*,*yz*]hexacene
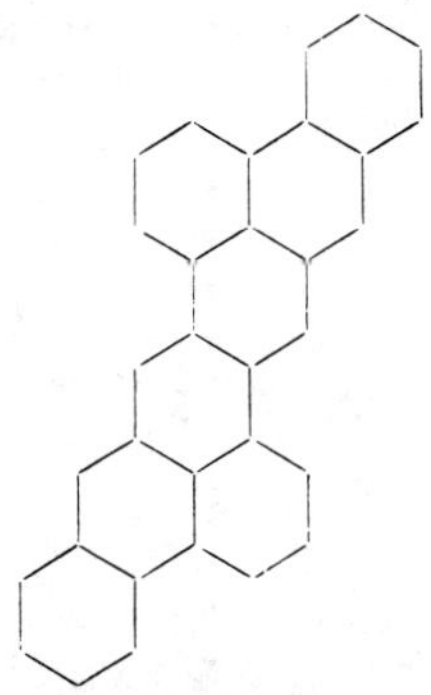	Dibenzo[*fg*,*st*]hexacene
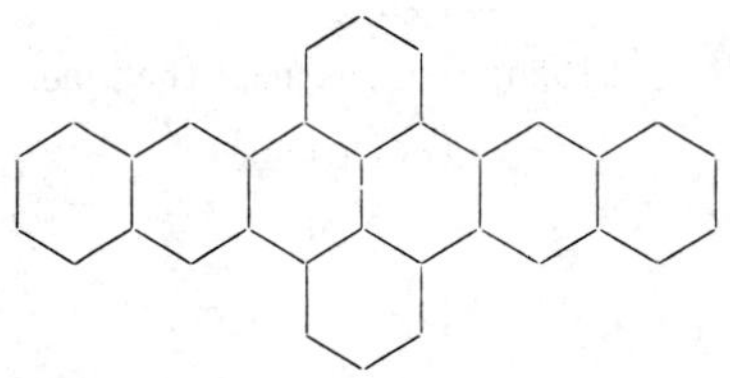	Dibenzo[*fg*,*uv*]hexacene
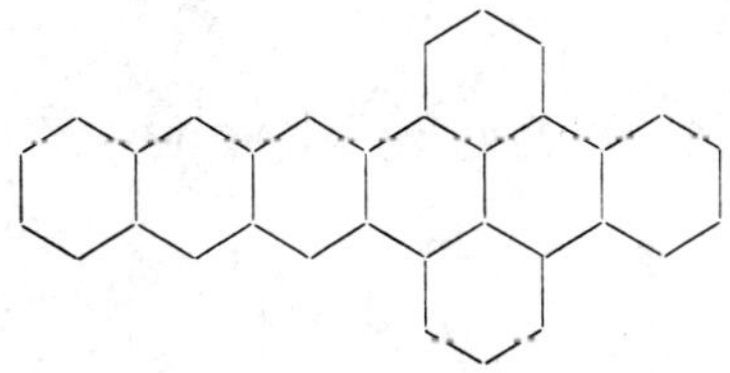	Dibenzo[*hi*,*wx*]hexacene
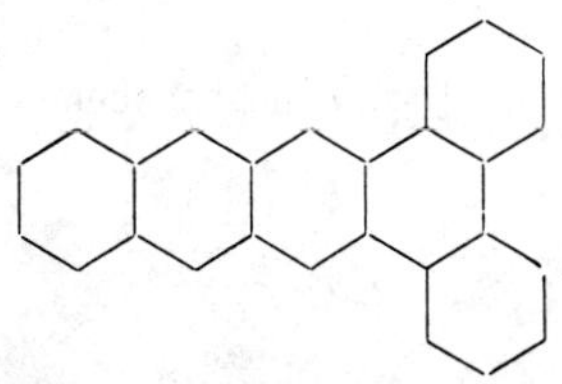	Dibenzo[*a*,*c*]naphthacene

Compound   Preferred name (synonym)

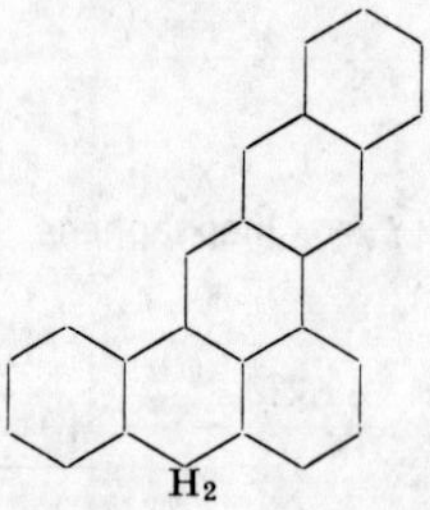

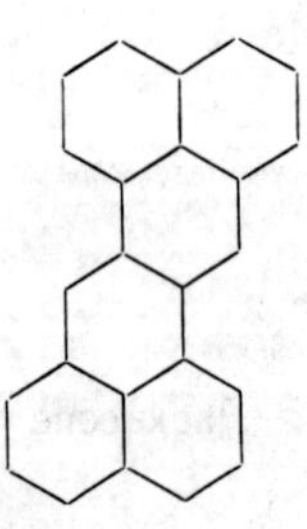

4H-Dibenzo[*a*,*de*]naphthacene

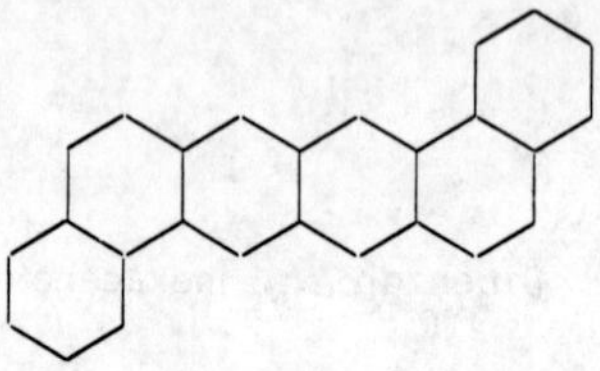

Dibenzo[*a*,*j*]naphthacene

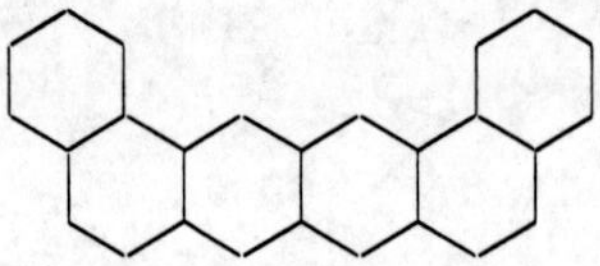

Dibenzo[*a*,*l*]naphthacene

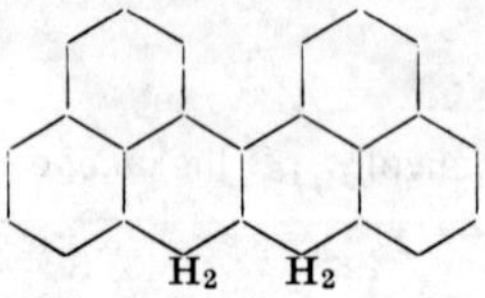

7H,8H-Dibenzo[*de*,*hi*]naphthacene

Dibenzo[*de*,*mn*]naphthacene
(Zethrene)

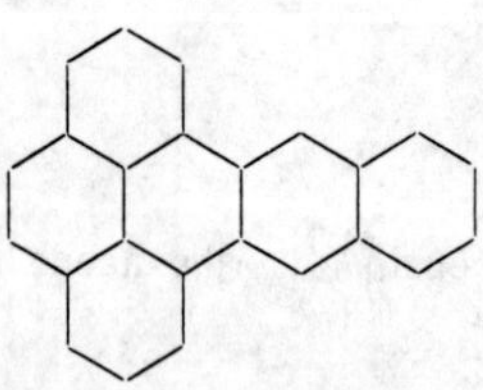

Dibenzo[*de*,*qr*]naphthacene

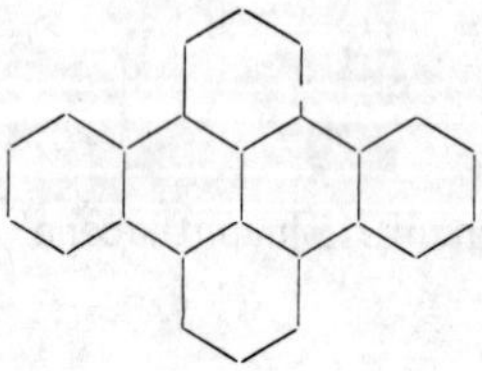

Dibenzo[*fg*,*op*]naphthacene

Compound	Preferred name (synonym)

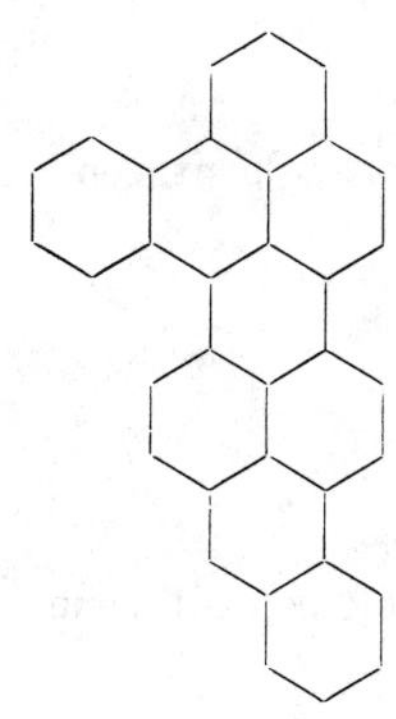

Dibenzo[*de,kl*]naphthol[1,2,3,4-*rst*]pentaphene
(Benzo[*j*]terrylene)

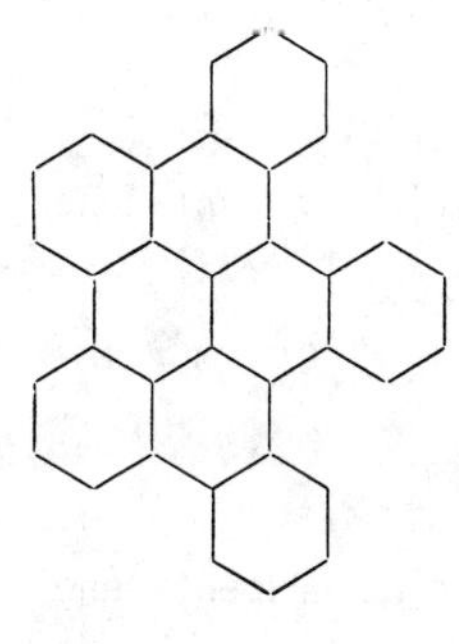

Dibenzo[*a,cd*]naphtho[4,3,2-*lm*]perylene

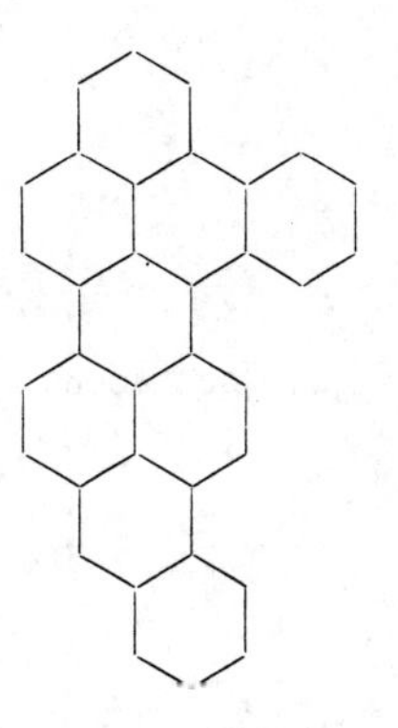

Dibenzo[*b,n*]naphtho[4,3,2,1-*pqr*]perylene

Dibenzo[*cd,f*]naphtho[4,3,2-*lm*]perylene

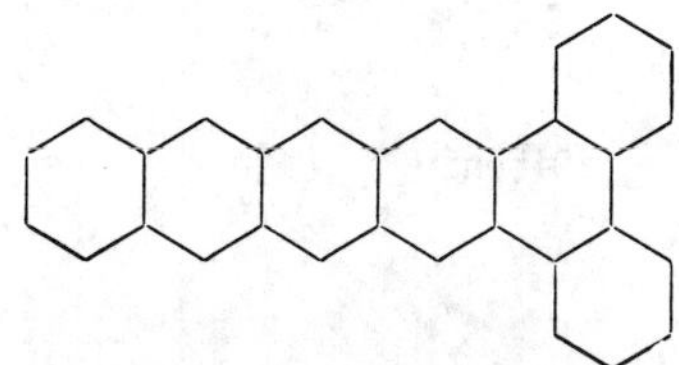

Dibenzo[*a,c*]pentacene

Compound   Preferred name (synonym)

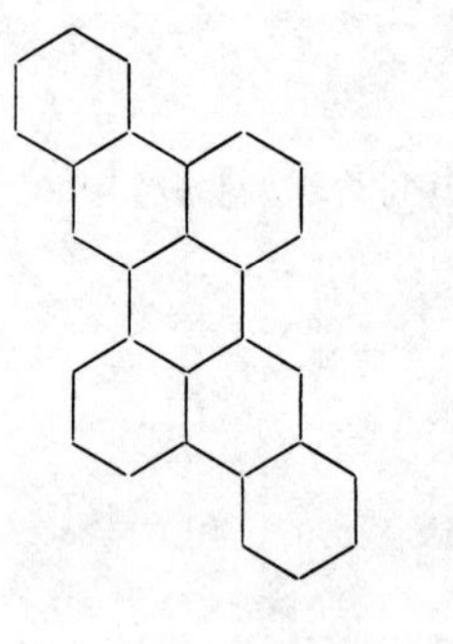

4H-Dibenzo[*a*,*de*]pentacene

Dibenzo[*a*,*l*]pentacene

Dibenzo[*a*,*n*]pentacene

7H,9H-Dibenzo[*de*,*jk*]pentacene

Dibenzo[*de*,*uv*]pentacene

Dibenzo[*fg*,*qr*]pentacene

Dibenzo[*fg*,*st*]pentacene

Compound | Preferred name (synonym)

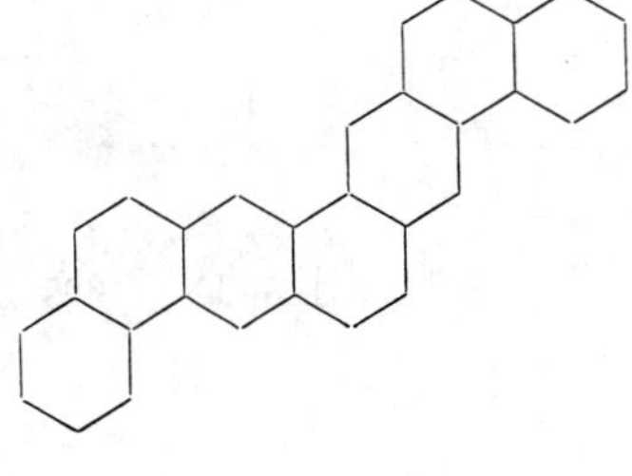

Dibenzo[*jk*,*uv*]pentacene
(Heptazethrene)

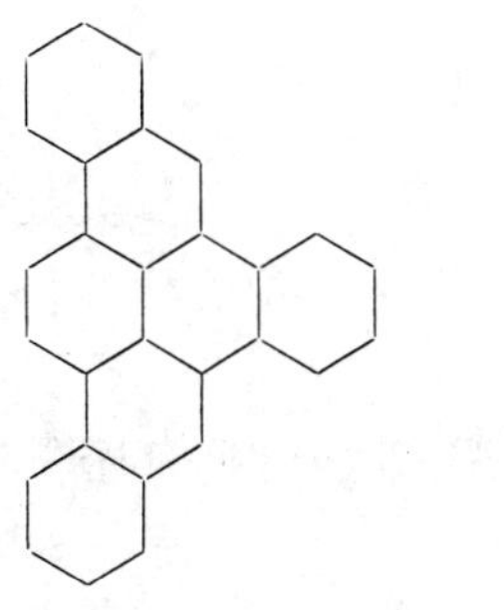

Dibenzo[*a*,*e*]pentalene

Dibenzo[*c*,*m*]pentaphene

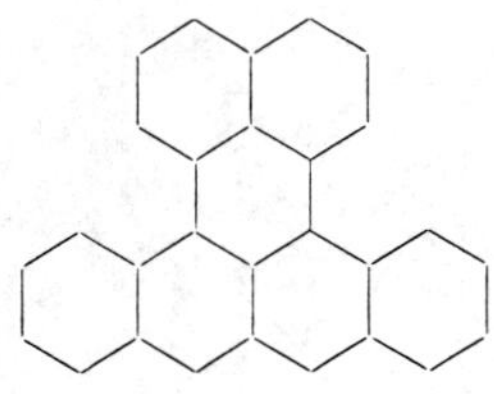

Dibenzo[*h*,*rst*]pentaphene

Dibenzo[*a*,*f*]perylene

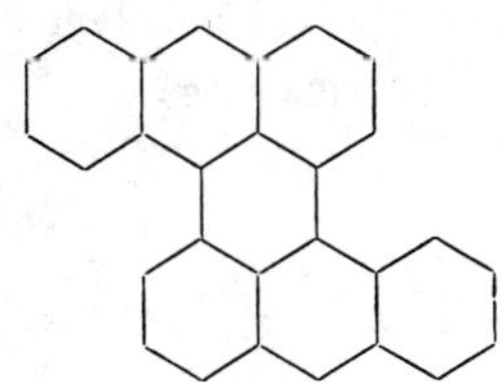

Dibenzo[*a*,*j*]perylene

Compound	Preferred name (synonym)
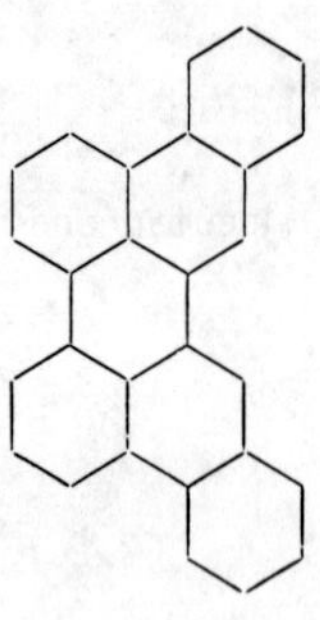	Dibenzo[*a*,*n*]perylene
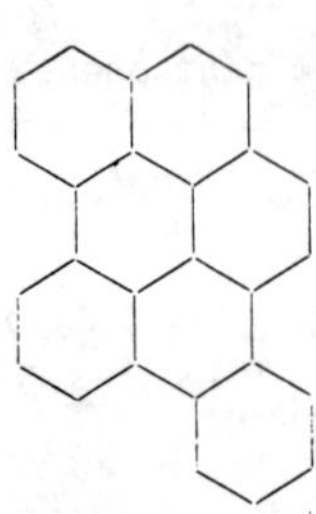	Dibenzo[*a*,*o*]perylene
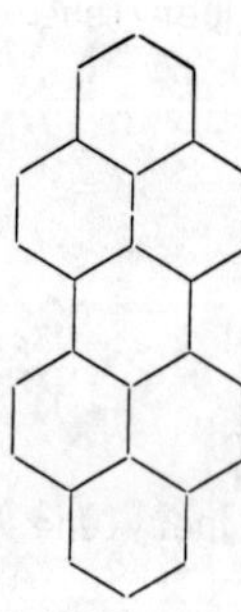	Dibenzo[*b*,*n*]perylene
	Dibenzo[*b*,*pqr*]perylene
	Dibenzo[*cd*,*lm*]perylene (Peropyrene)

Compound	Preferred name (synonym)

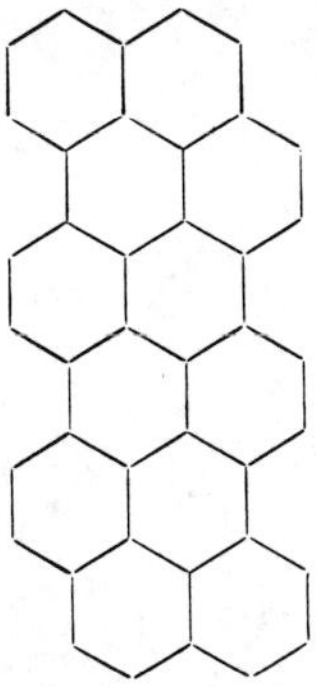

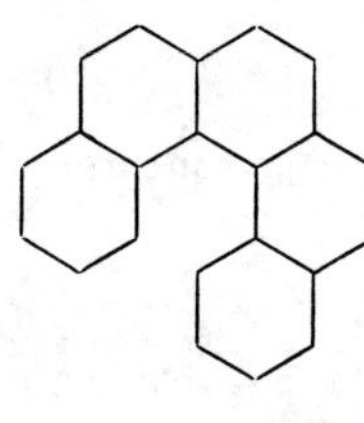

Dibenzo[6,7-15,16]perylo[5,4,3-*cde*]perylene

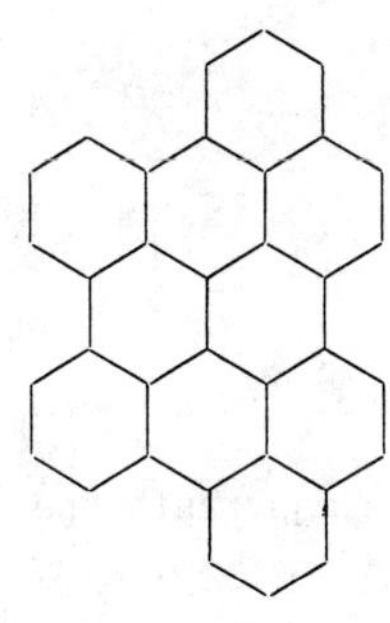

Dibenzo[*c*,*g*]phenanthrene

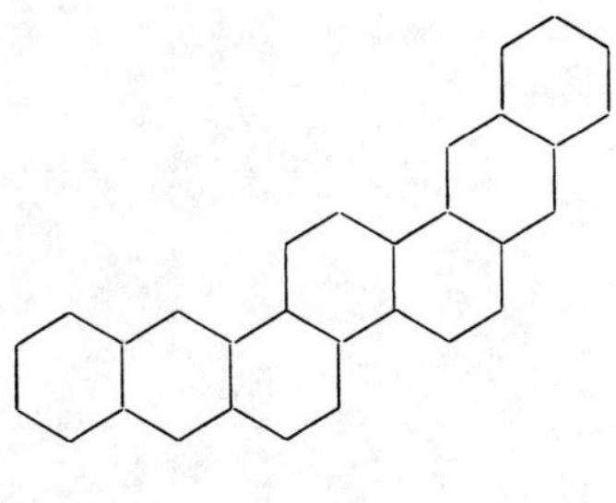

Dibenzo[*fg*,*ij*]phenanthro[7,8,9,10,1,2-*pqrstuv*]pentaphene

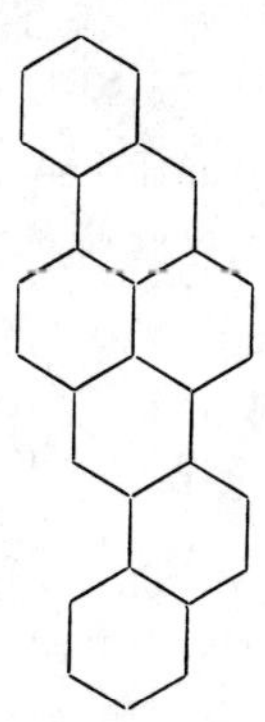

Dibenzo[*b*,*n*]picene

Dibenzo[*b*,*tuv*]picene

Compound	Preferred name (synonym)
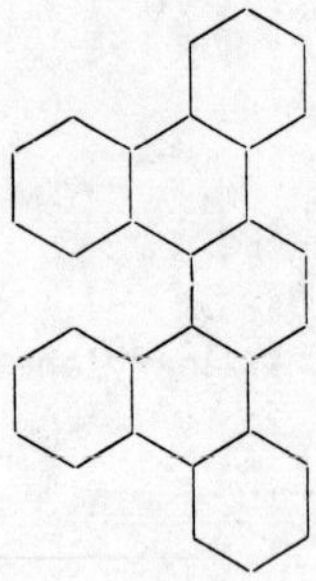	Dibenzo[*f*,*j*]picene
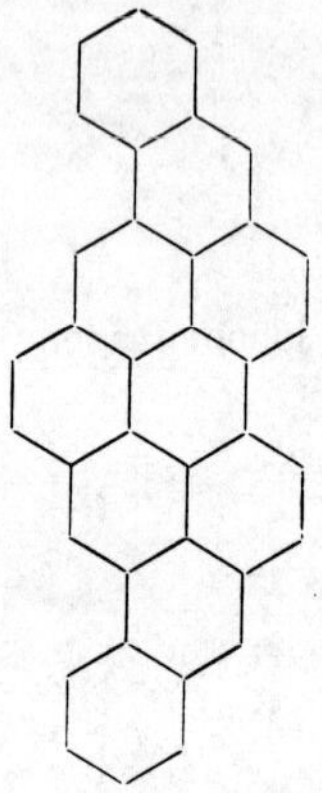	Dibenzo[*a.def*]pyranthrene
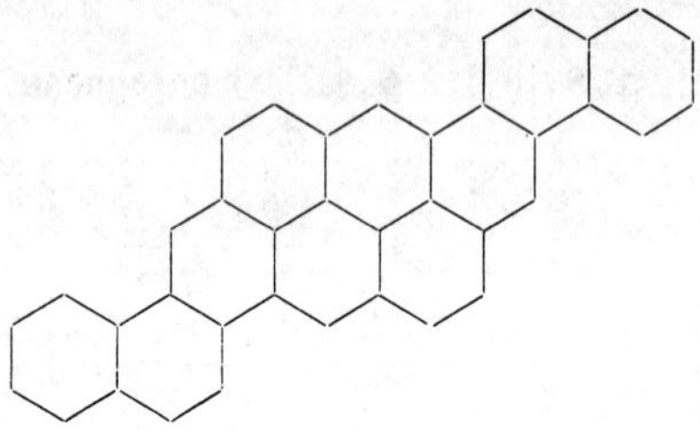	Dibenzo[*a*,*n*]pyranthrene
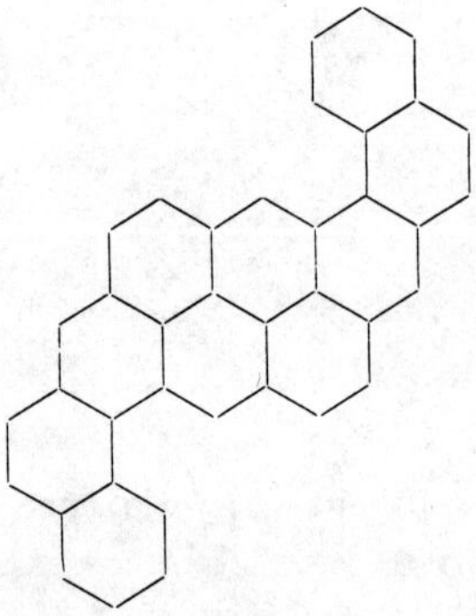	Dibenzo[*c*,*p*]pyranthrene

Compound | Preferred name (synonym)

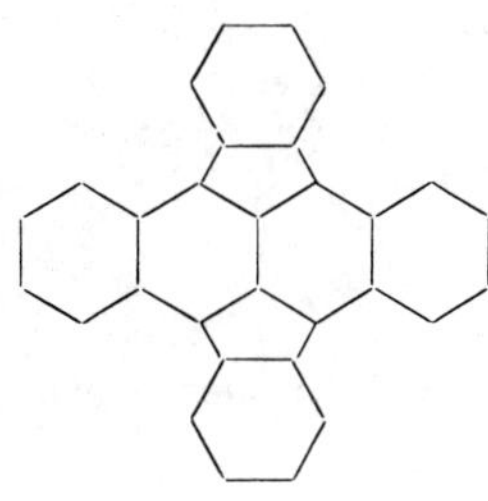

4H,8H-Dibenzo[*cd*,*mn*]pyrene

12H-Dibenzo[*gh*,*k*]tetraphene

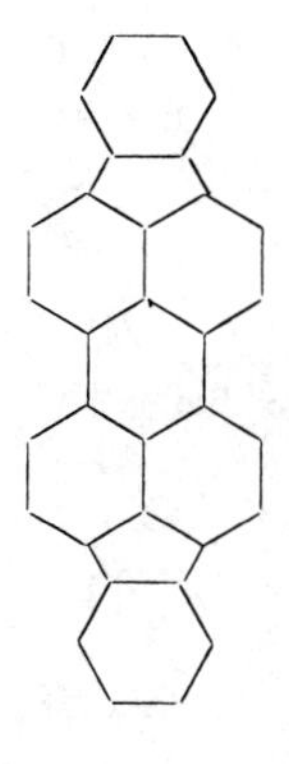

Diindeno[1,2,3-*fg*:1',2',3'-*op*]naphthacene

Diindeno[1,2,3-*cd*:1',2',3'-*lm*]perylene
(Periflanthene)

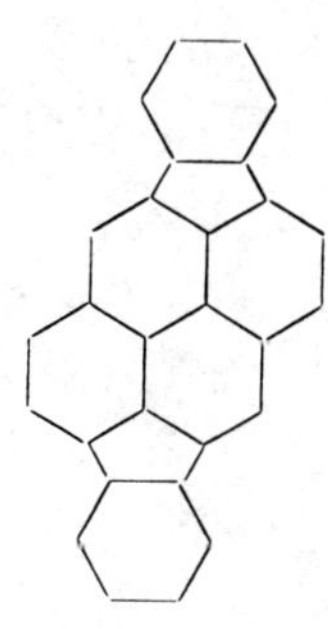

Diindeno[1,2,3-*cd*:1',2',3'-*jk*]pyrene

Compound	Preferred name (synonym)

Dinaphtho[1,2-*b*:1',2'-*k*]chrysene

Dinaphtho[8,1,2-*abc*:8',1',2'-*jkl*]coronene

Dinaphtho[2,1,8-c_1d_1a:2',1',8'-*nop*]heptacene

Dinaphtho[1,2,3-*fg*:1',2',3'-*op*]naphthacene

Dinaphtho[2,1,8,7-*defg*:2',1',8',7'-*opqr*]pentacene

Compound

Preferred name (synonym)

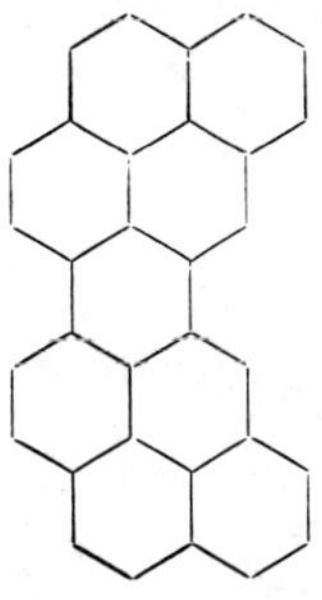

Dinaphtho[2,1,8,7-*defg*:2',1',8',7'-*ijkl*]pentaphene

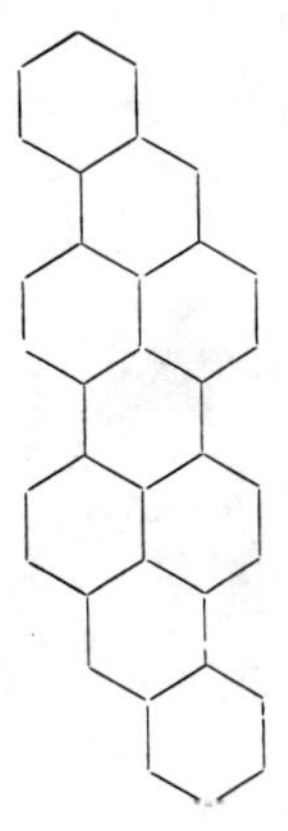

Dinaphtho[1,2,3-*cd*:1',2',3'-*lm*]perylene

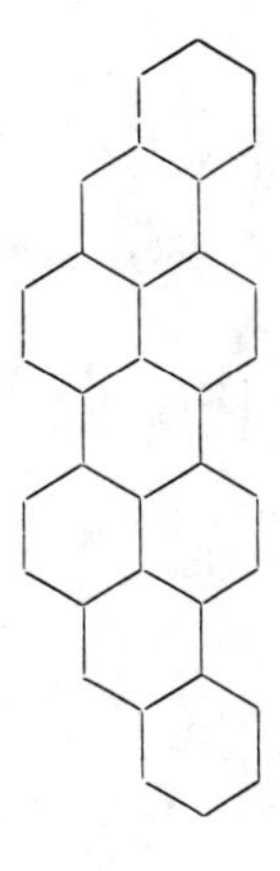

Dinaphtho[4,3,2-*cd*:3',2',1'-*lm*]perylene

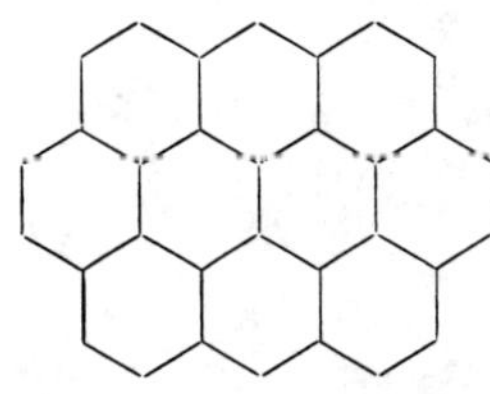

Diphenanthro[3,4,5,6-*bcdef*:3',4',5',6'-*klmno*]coronene
(Circumanthracene)

Compound   Preferred name (synonym)

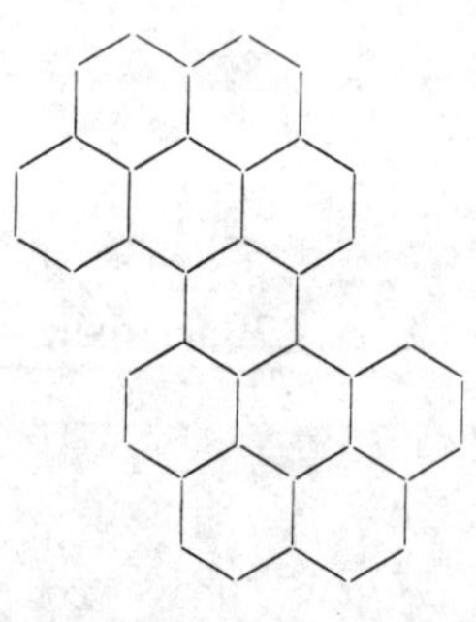

Diphenanthro[2,3,4-*cd*:2',3',4'-*lm*]perylene

Diphenanthro[2,3,4-*cd*:4',3',2'-*lm*]perylene

Diphenanthro[5,4,3-*abcd*:5',4',3'-*jklm*]perylene

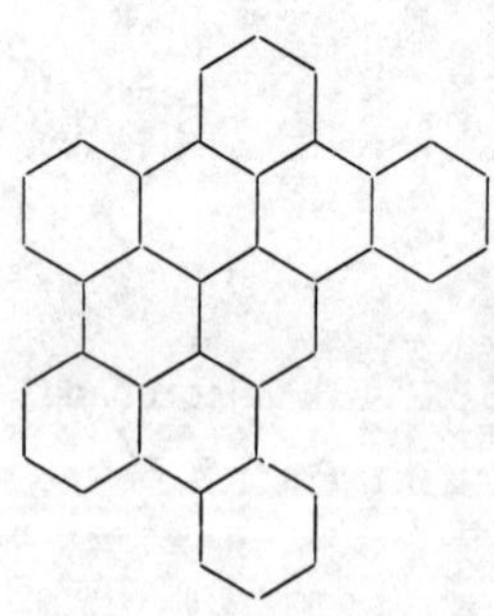

Diphenanthro[8,9,10-*efg*:9',10',1'-*jkl*]pyrene

Compound	Preferred name (synonym)

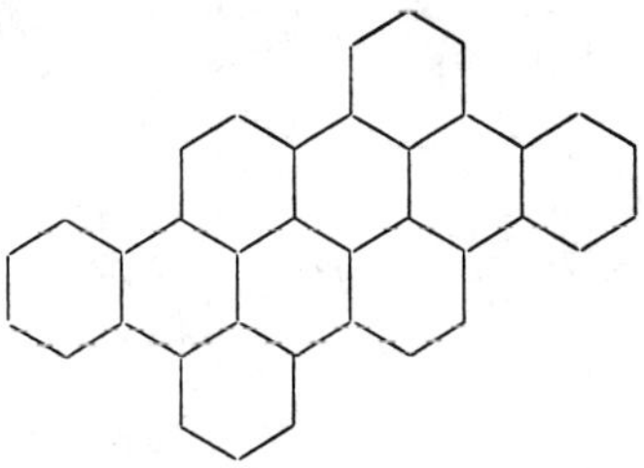

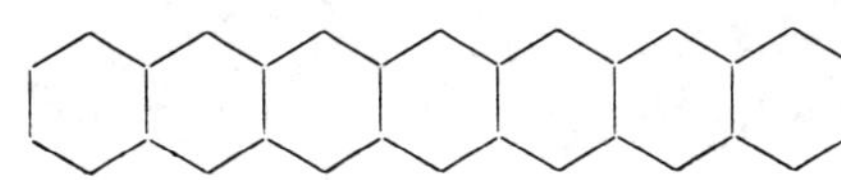

Diphenanthro[9,10,1-*cde*:9',10',1'-*jkl*]pyrene

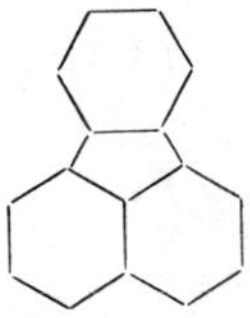

Fluoranthrene

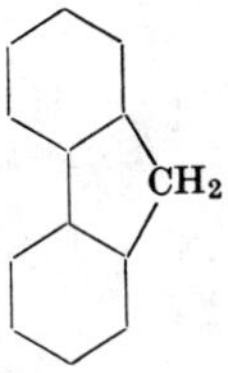

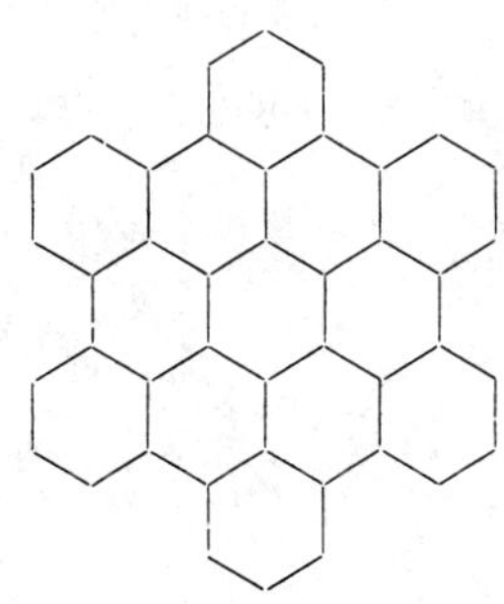

9H-Fluorene
(Fluorene)

Heptacene

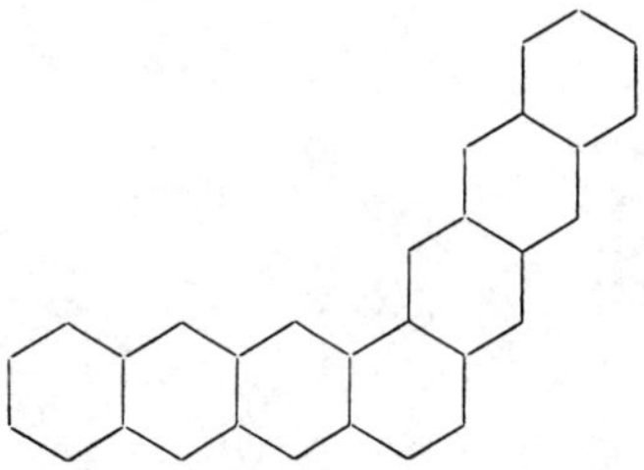

Heptaphene

Hexabenzo[*bc*,*ef*,*hi*,*kl*,*no*,*qr*]coronene

Compound	Preferred name (synonym)

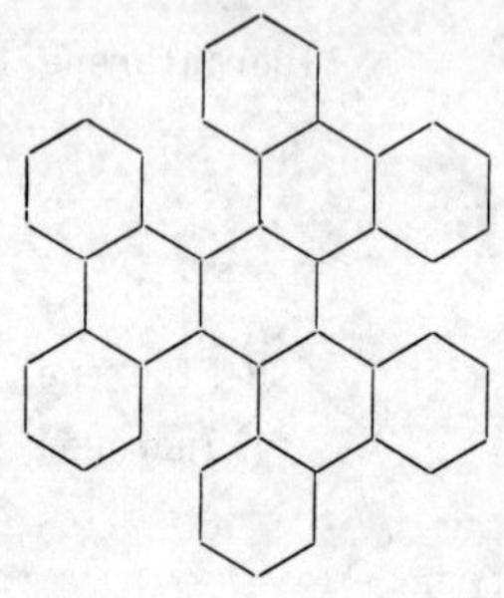

Hexabenzo[a,cd,f,j,lm,o]perylene

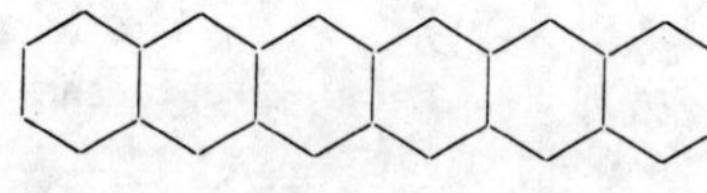

Hexabenzo[a,c,g,i,m,o]triphenylene

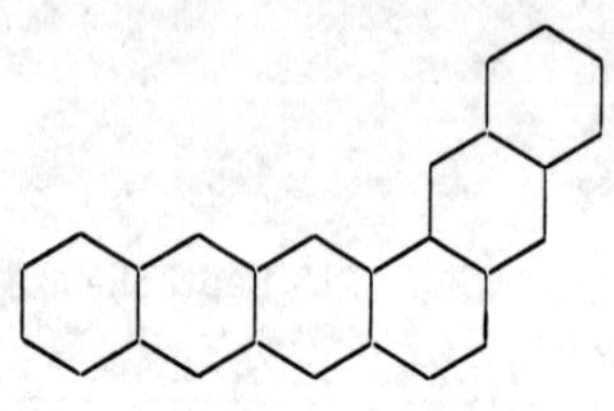

Hexacene

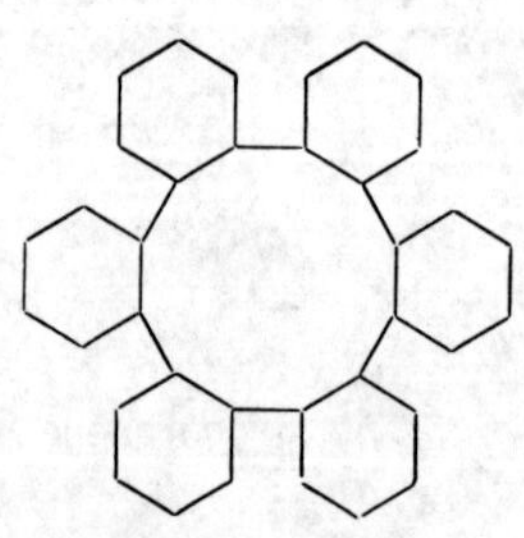

Hexaphene

Hexaphenylene

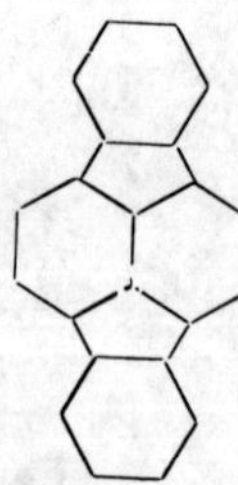

Indeno[1,2,3-cd]fluoranthene

Compound	Preferred name (synonym)
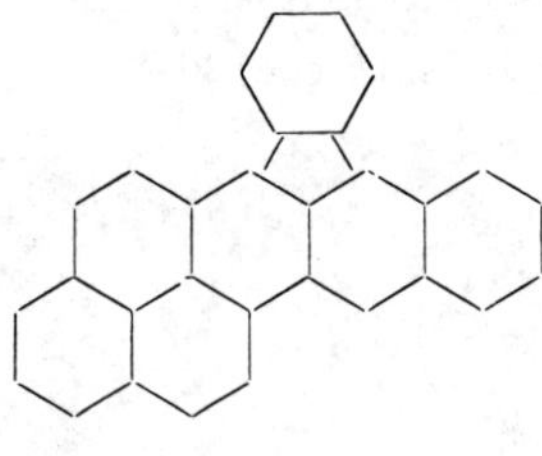	Indeno[1,2,3-*fg*]naphthacene
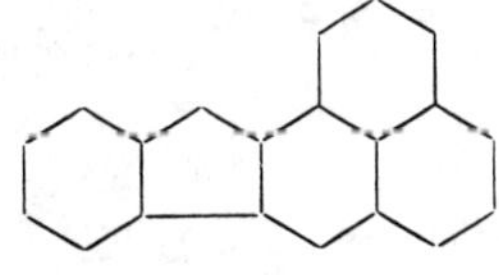	Indeno[1,2,3-*de*]naphtho[2,1,8-*qra*]naphthacene
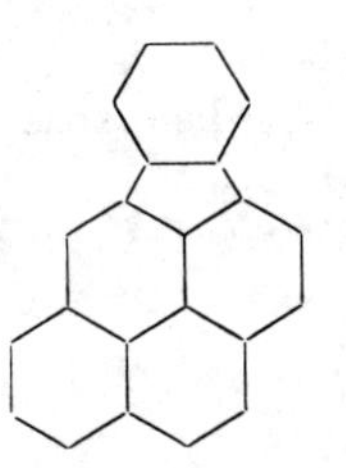	Indeno[1,2,3-*fg*]naphtho[2,1,8-*qra*]naphthacene
	Indeno[2,1-*a*]phenalene
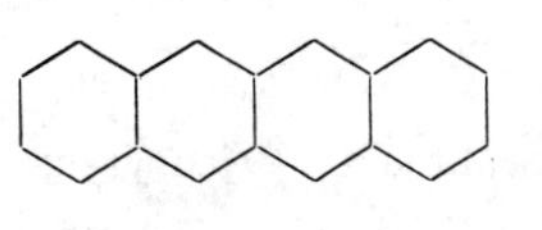	Indeno[1,2,3-*cd*]pyrene
	Naphthacene (Tetracene)
	Naphthaceno[2,1,12-*qra*]naphthacene
	Naththaceno[2,1,1],11-*opqra*]naphthacene

Compound	Preferred name (synonym)

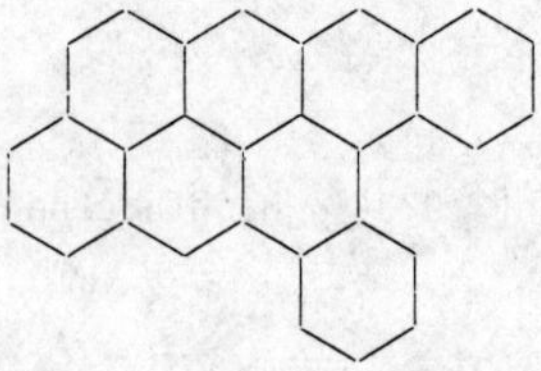

Naphthaceno[11,12,1,2-*fghi*]tetraphene

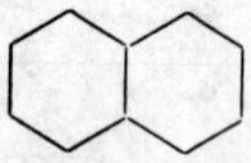

Naphthalene

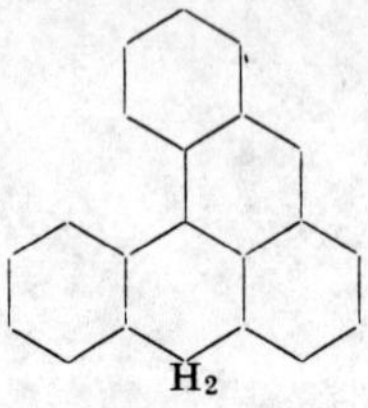

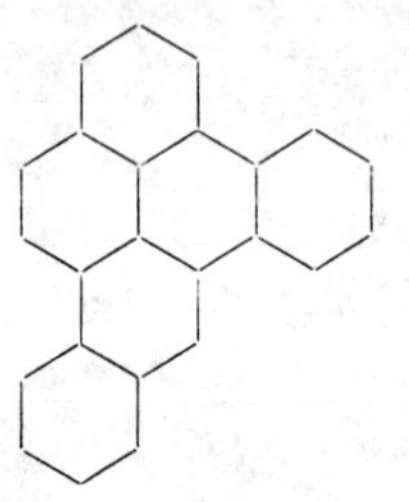

5H-Naphtho[3,2,1-*de*]anthracene
(Coeranthrene)

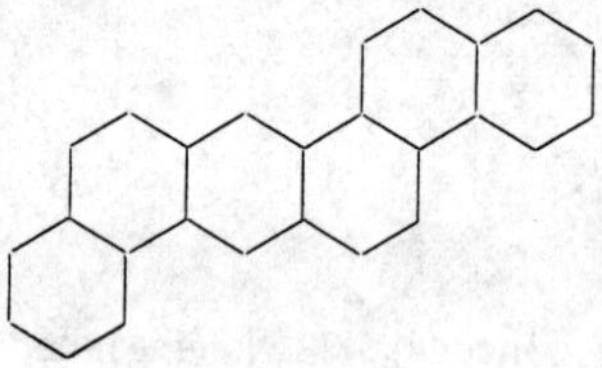

Naphtho[1,2-*b*]chrysene

Naphtho[1,2,3,4-*def*]chrysene

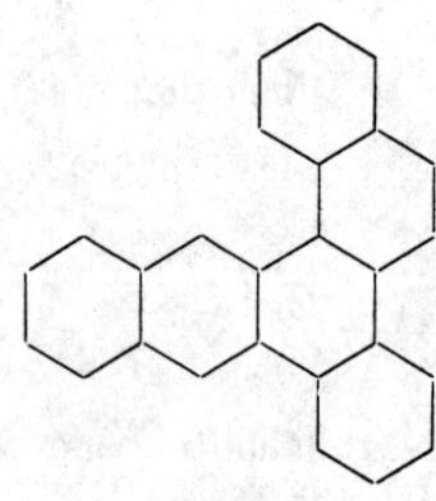

Naphtho[2,3-*g*]chrysene

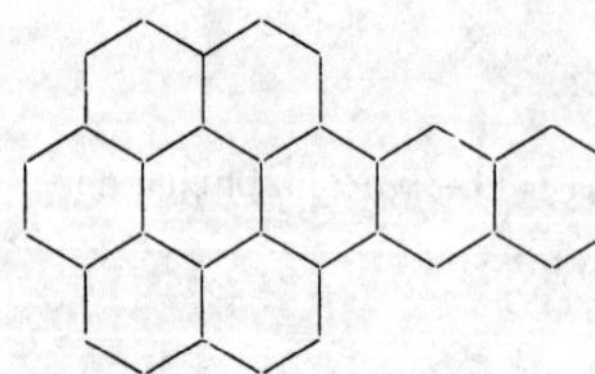

Naphtho[2,3-*a*]coronene

Compound	Preferred name (synonym)
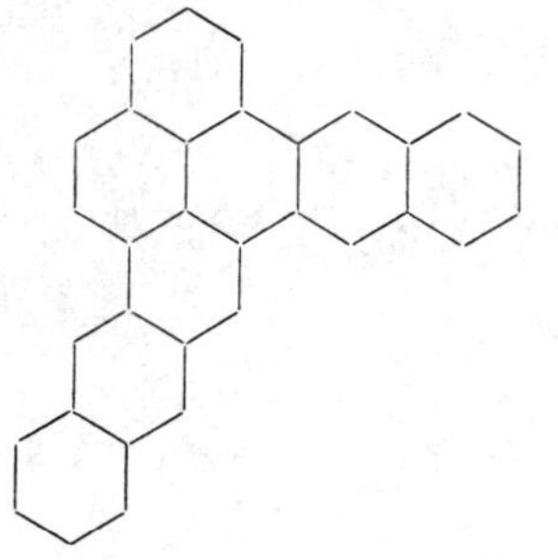	Naphtho[2,3-*b*]fluoranthene
	Naphtho[2,3-*j*]fluoranthene
	Naphtho[2,3-*k*]fluoranthene
	Naphtho[8,1,2-*azy*]hexacene
	Naphtho[2,1,8-*hij*]hexaphene
	6H-Naphtho[2,1,8,7-*defg*]naphthacene

Compound	Preferred name (synonym)
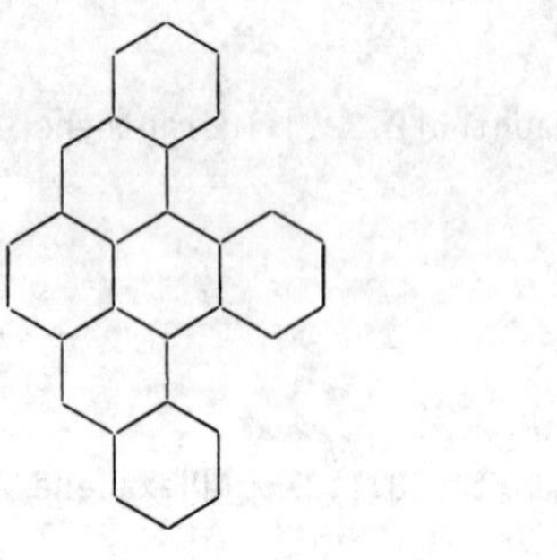	Naphtho[2,1,8-*qra*]naphthacene
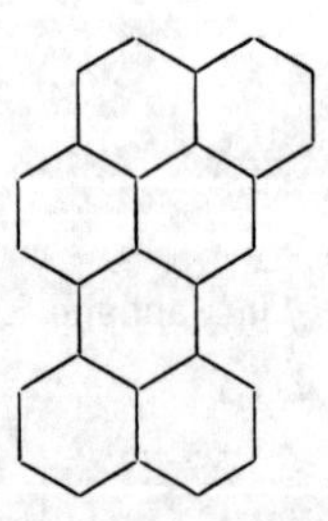	Naphtho[8,1,2-*auv*]pentacene
	Naphtho[2,3,-*c*]pentaphene
	Naphtho[4,3,2,1-*rst*]pentaphene
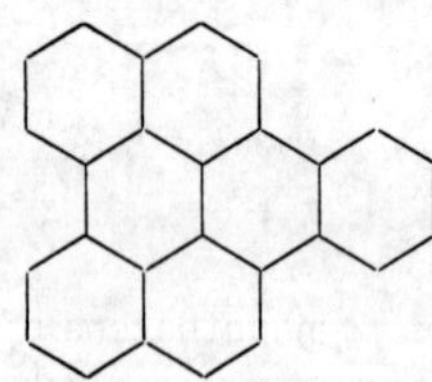	Naphtho[8,1,2-*aqr*]perylene
	Naphtho[4,3,2,1-*ghi*]perylene

Compound	Preferred name (synonym)

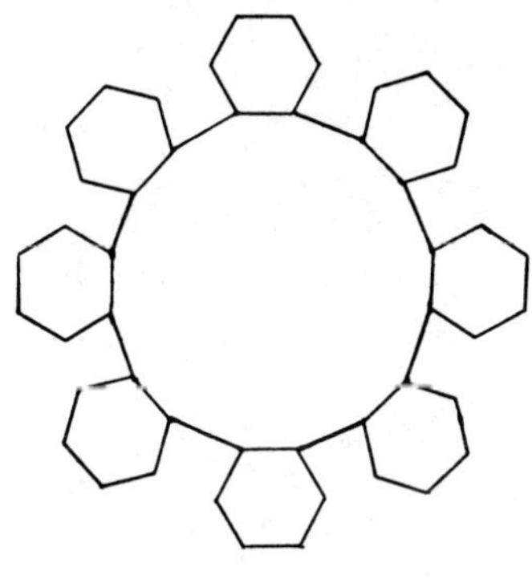

5H-Naphtho[2,3-*ghi*]tetraphene

Nonacene

Octacene

Octaphenylene

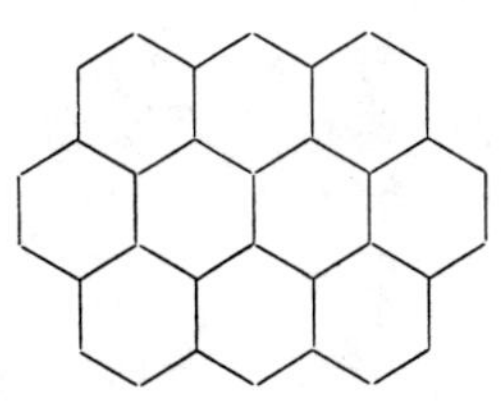

Ovalene

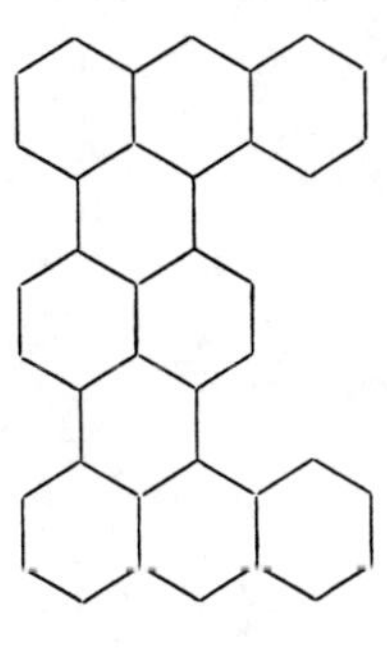

Pentabenzo[*a*,*de*,*kl*,*o*,*rst*]pentaphene

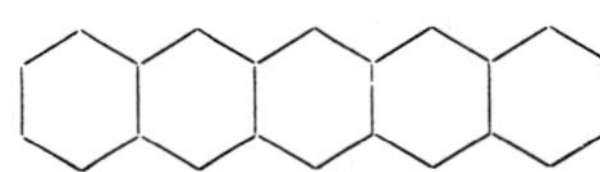

Pentacene

Compound	Preferred name (synonym)
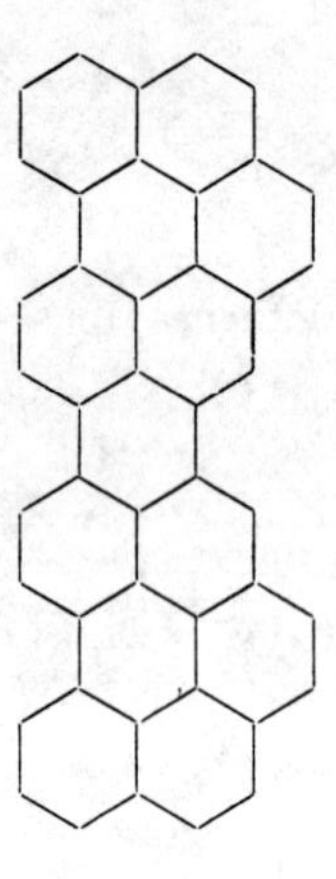	Pentaceno[2,1,14,13,12-*qrstuva*]pentacene
	Pentaphene
	Perylene
	Perylo[2,3,4-*bcd*:9,10,11-*e'd'c'*]dipyrene
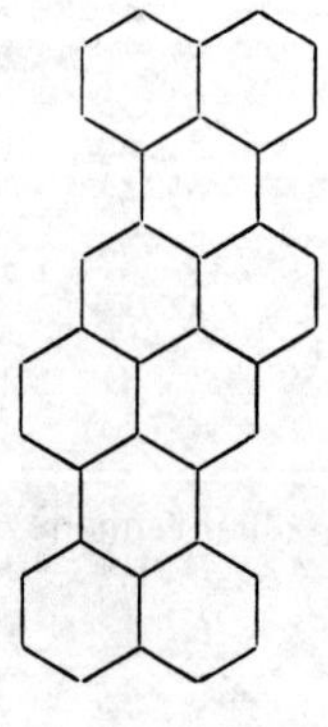	Perylo[5,4,3-*cde*]perylene
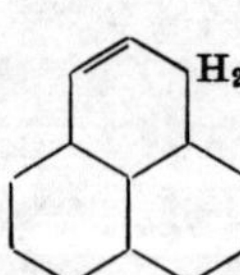	Phenalene

Compound

Preferred name (synonym)

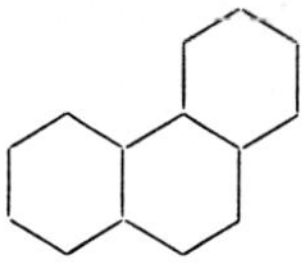

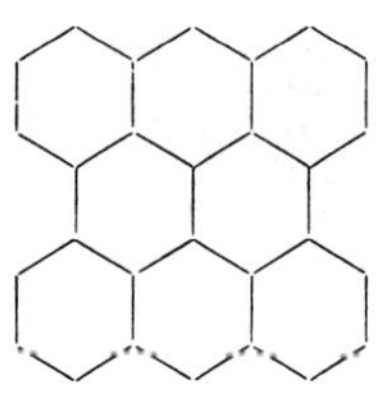

Phenanthrene

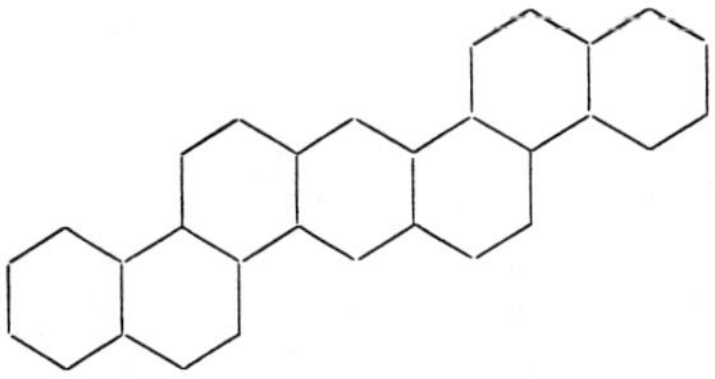

Phenanthro[1,2-b]chrysene

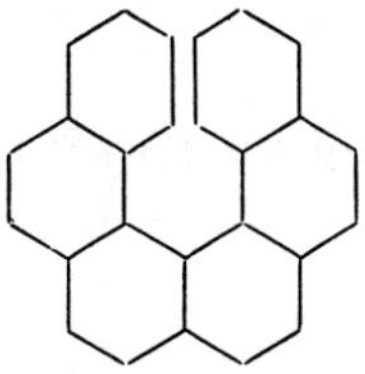

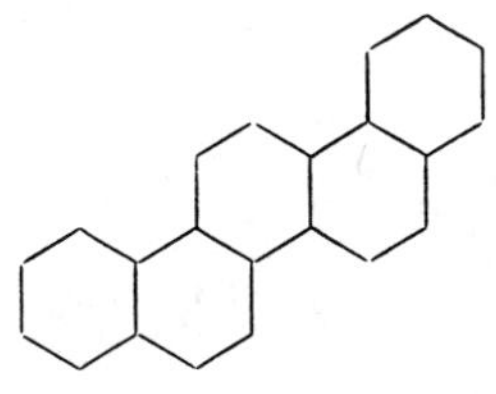

Phenanthro[3,4-c]phenanthrene
(Hexahelicene)

*a**

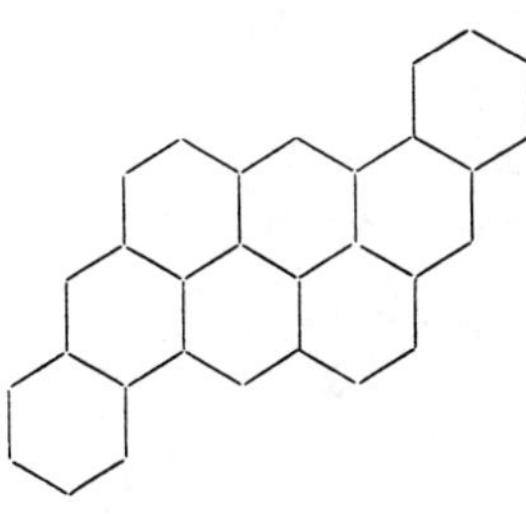

Phenanthro[1,10,9,8-$opqra$]perylene
(Bisanthene)

Picene

Pyranthrene

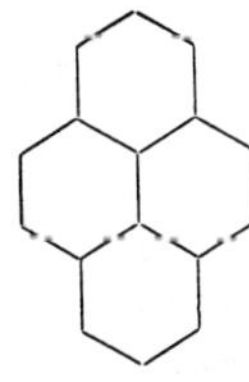

Pyrene

Compound | Preferred name (synonym)

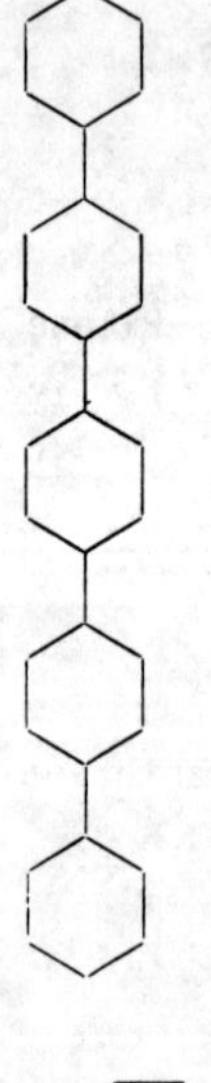

Pyreno[5,4,3-*def*]pyrene

p-Quaterphenyl

p-Quinquiphenyl

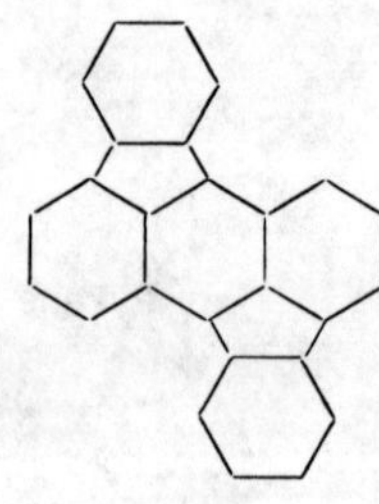

Rubicene

Compound	Preferred name (synonym)
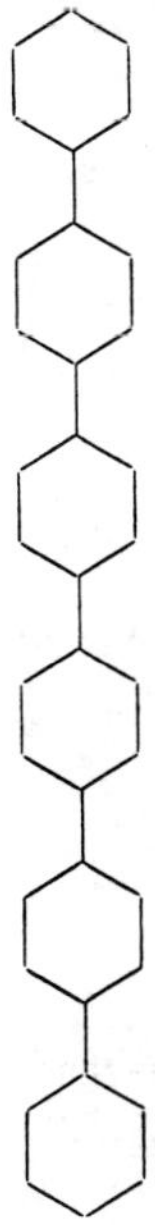	*p*-Sexiphenyl
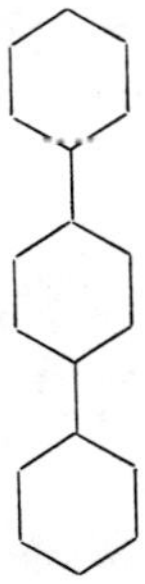	*p*-Terphenyl
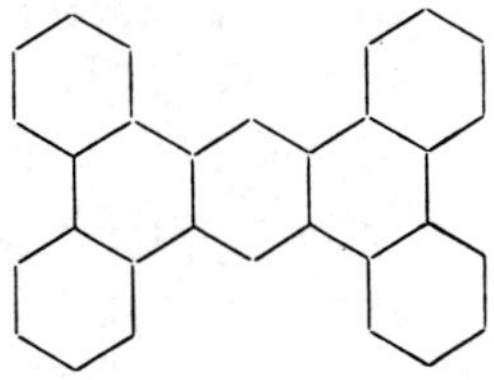	Tetrabenzo[*a*,*c*,*h*,*j*]anthracene
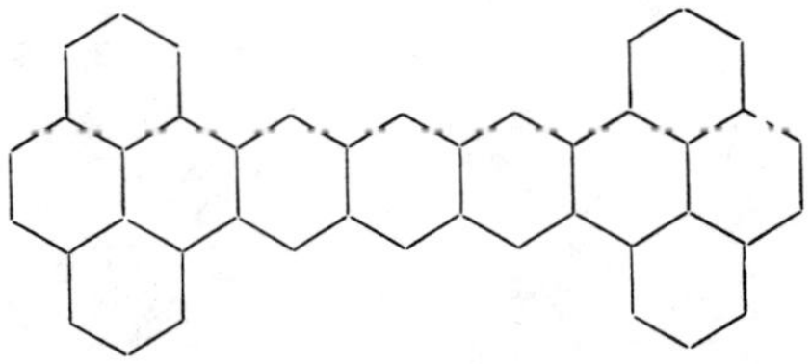	Tetrabenzo[*de*,*no*,*st*,c_1d_1]heptacene

Compound	Preferred name (synonym)
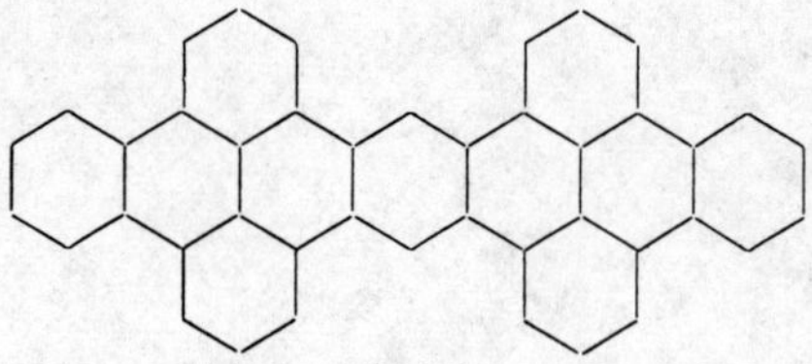	Tetrabenzo[*fg*,*lm*,*uv*,a_1b_1]heptacene
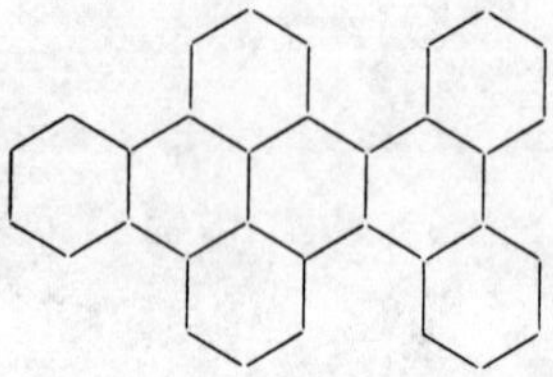	Tetrabenzo[*a*,*c*,*fg*,*op*]naphthacene
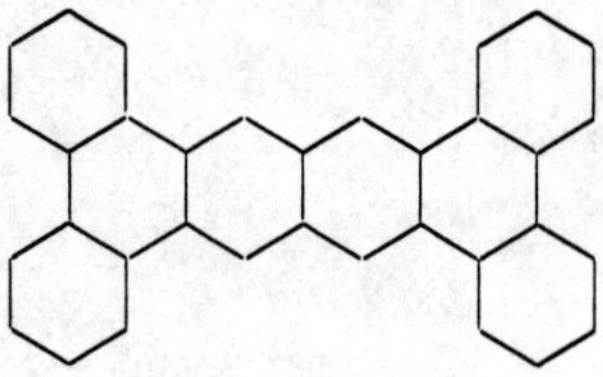	Tetrabenzo[*a*,*c*,*j*,*l*]naphthacene
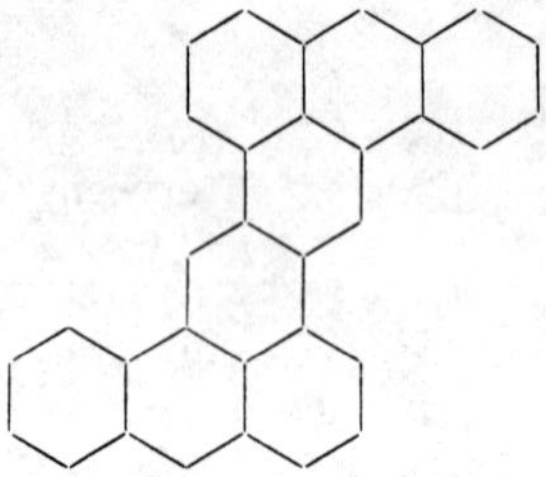	Tetrabenzo[*a*,*de*,*j*,*mn*]naphthacene
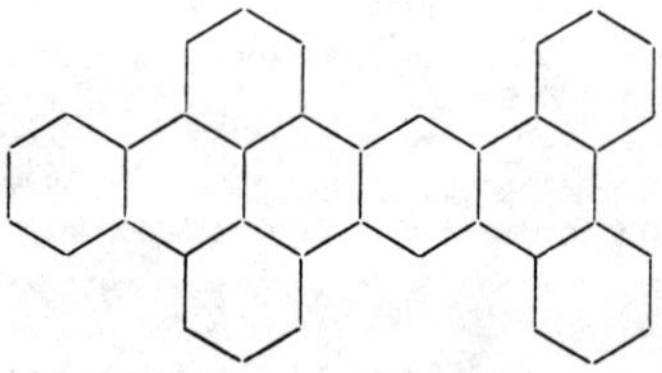	Tetrabenzo[*a*,*c*,*hi*,*qr*]pentacene
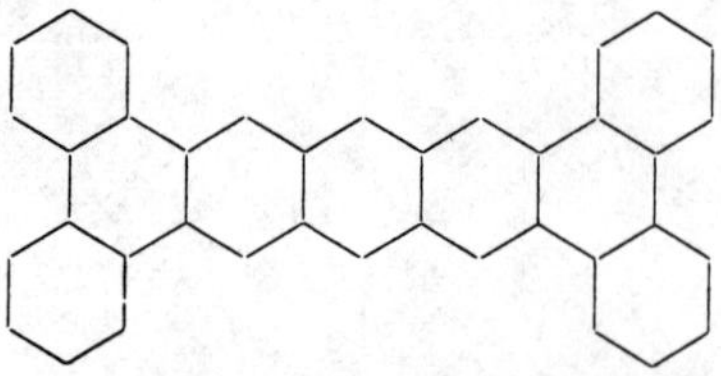	Tetrabenzo[*a*,*c*,*l*,*n*]pentacene

Compound   Preferred name (synonym)

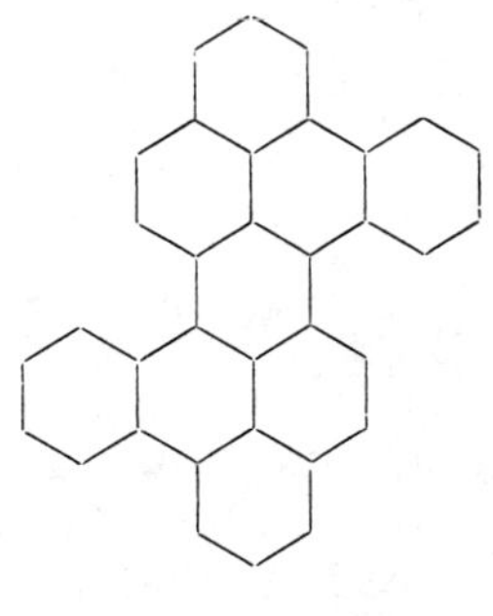

Tetrabenzo[a,de,l,op]pentacene

Tetrabenzo[de,hi,op,st]pentacene

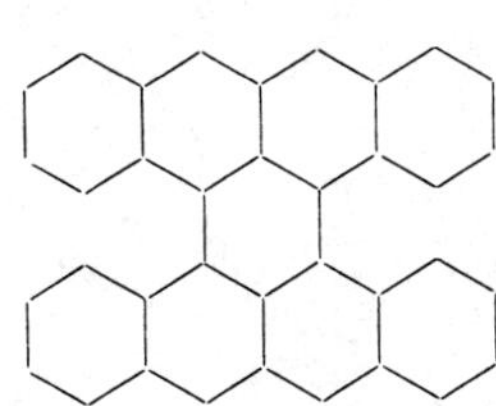

Tetrabenzo[a,cd,j,lm]perylene

Tetrabenzo[a,f,j,o]perylene

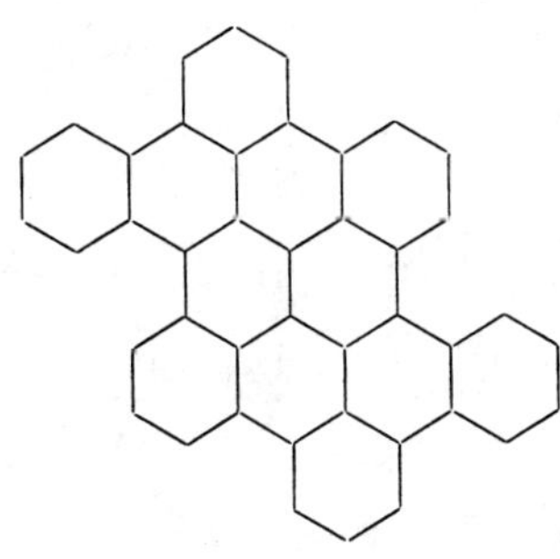

Tetrabenzo[gh,jk,tu,wx]pyranthrene

Compound	Preferred name (synonym)
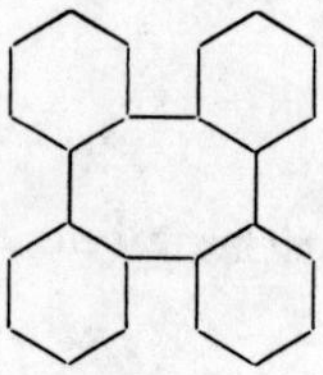	Tetraphenylene
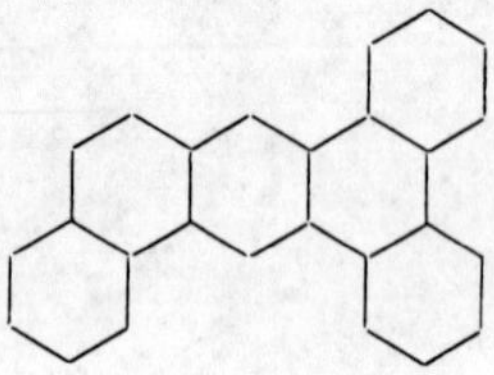	Tribenzo[*a*,*c*,*h*]anthracene
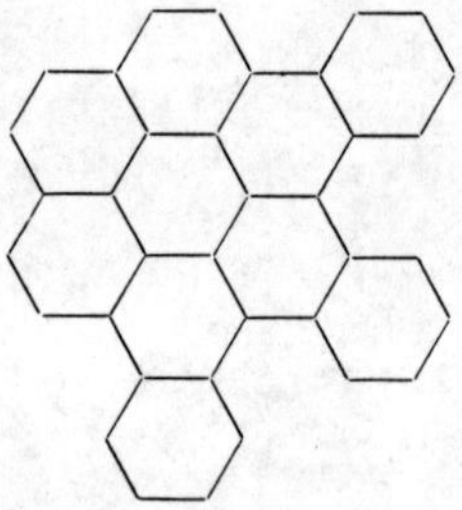	Tribenzo[*a*,*d*,*g*]coronene
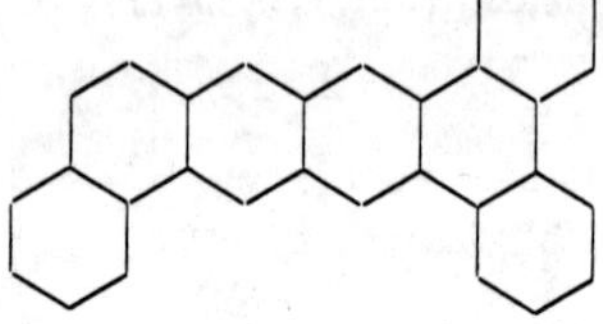	Tribenzo[*a*,*c*,*j*]naphthacene
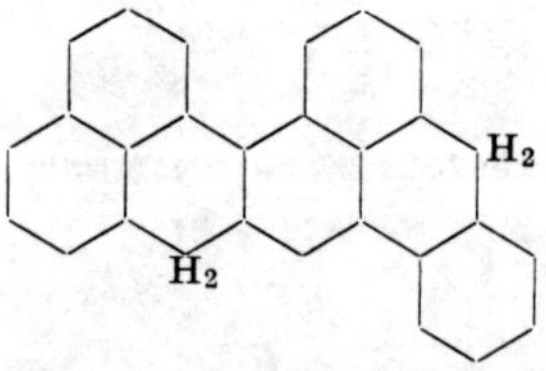	4H,10H-Tribenzo[*a*,*de*,*hi*]naphthacene
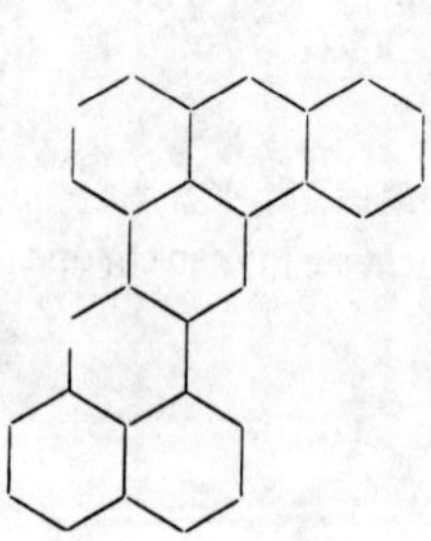	Tribenzo[*c*,*hi*,*qr*]naphthacene

Compound	Preferred name (synonym)

Tribenzo[*de*,*h*,*kl*]naphtho[1,2,3,4-*rst*]pentaphene

Tribenzo[*de*,*kl*,*rst*]pentaphene
(Terrylene)

Tribenzo[*a*,*ghi*,*o*]perylene

Tribenzo[*b*,*ghi*,*n*]perylene

Tribenzo[*b*,*n*,*pqr*]perylene

Compound	Preferred name (synonym)

Trinaphthylene

Triphenylene

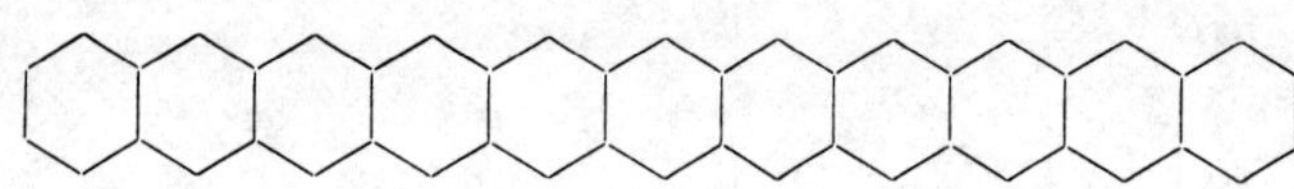

Undecacene

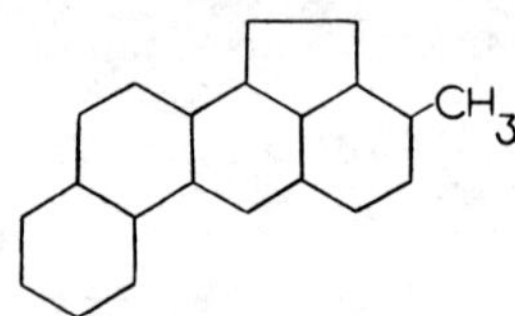

3-Methyl-benz[*j*]aceanthrylene
(3-Methyl-cholanthrene)
(20-Methyl-cholanthrene)

BIBLIOGRAPHY

Clar, E. J. 1964. *Polycyclic hydrocarbons*, vols. 1 and 2. New York: Academic Press.

International Union on Pure and Applied Chemistry. 1960. 1957 Report of the commission on the nomenclature of organic chemistry. Definitive rules for nomenclature of organic chemistry. *J. Am. Chem. Soc.* 82: 5545-84.

National Academy of Sciences. 1972. *Particulate polycyclic organic matter*. Washington, D.C.

Patterson, A. M.; Capell, L. T.; and Walker, D. F. 1960. *The ring index. A list of ring systems used in organic chemistry*. Washington, D.C.: American Chemical Society.

APPENDIX D

CHEMICAL INDEX

R. H. Ross   G. A. Dailey

abietic acid — Sect. 2.6.1.5; Table 8.19.

aceanthrene-1,2-dione — Table 5.22.

acenaphthene — Sects. 4.6.1.1, 10.1.1.1; Tables 2.38, 3.23, 4.52, 4.53, 4.54, 4.55, 4.56, 4.57,
    4.63, 5.24, 6.5, 10.1, 10.23; Figs. 2.16, 4.63.

acenaphthenols — Tables 2.38, 4.53.

acenaphtho[1,2-*b*]quinoxaline — Table 5.22.

acenaphthylenes — Sect. 4.6.1.2; Tables 2.38, 4.52, 4.57, 4.59, 4.63, 6.5.

acetaldehyde — Table 8.19; Fig. 6.7.

acetate — Sects. 7.2.2.2, 7.2.2.7, 8.3.1.2; Tables 7.16, 7.17; Fig. 7.18.

acetic acid — Tables 2.30, 8.19.

acetic anhydride — Table 2.30.

acetyl coenzyme A — Fig. 7.19.

acetone — Sects. 3.2.2.9, 10.1.2.1, 10.1.2.4; Tables 4.54, 4.55, 9.32, 10.4, 10.5; Fig. 3.43.

acetonitrile — Tables 4.54, 4.55.

acetophenone — Sect. 5.5.1; Tables 4.55, 4.71; Fig. 6.5.

2-acetylaminofluorene — Table 9.28.

acetylene — Sect. 8.2.12.1.

acetylene hydrocarbons — Sect. 6.1.3.4; Table 6.23.

acid phosphatase — Sect. 8.2.4.2.

acridine — Sects. 9.2, 9.2.5; Tables 2.37, 2.38, 4.52, 4.54, 4.55, 4.66, 4.69, 5.17, 5.18, 5.21,
    5.23, 6.5; Fig. 2.17.

acridones — Sect. 2.6.3.9; Fig. 2.15.

actinium — Table 6.34.

actinomycin D — Sect. 10.1.5.4.

adenosine — Sect. 8.2.8.2.

adenosinediphosphate (ADP) — Sect. 8.2.6.4; Fig. 7.19.

adenosinemonophosphate (AMP) — Fig. 7.19.

adenosinetriphosphate (ATP) — Sect. 9.3.6.1; Fig. 7.19.

alcohols — Sect. 6.4.4.4; Table 3.26; Fig. 3.43.

benzoitrile — Table 4.55.

benzopentacene — Table 2.37.

benzoperylene — Sect. 4.6.2.3; Tables 2.37, 2.38.

1,12-benzoperylene — Tables 6.4, 6.19.

benzo[*ghi*]perylene — Sects. 5.5.1, 5.5.3, 8.3.1.2; Tables 4.64, 4.65, 4.66, 4.67, 4.69, 4.70, 5.17, 5.18, 5.25, 6.4, 6.5, 6.11, 6.16, 6.17, 6.18, 6.20, 8.23, 8.24, 8.25, 8.26, 8.27.

benzo[*b*]phenanthrene — Table 9.2.

benzo[*c*]phenanthrene — Tables 5.17, 5.18, 5.22, 6.5.

benzophenanthrenes — Tables 4.56, 4.57.

benzophenanthridine — Table 4.69.

benzo[*lmn*]phenanthridine — Table 5.21.

benzophenone — Table 2.34.

benzopicene — Tables 2.37, 6.3.

benzopyrene — Sects. 6.1.1, 6.1.2.1, 6.1.2.2, 6.1.3.4, 6.1.3.5, 6.1.5, 7.0, 7.2.4, 7.2.4.1, 9.1.2, 9.2; Tables 6.2, 6.6, 6.13, 6.14, 6.26, 6.27, 6.28, 7.26, 7.27, 9.12, 9.30, 9.31, 9.32, 10.2, 10.8; Figs. 6.3, 10.3, 10.6.

1,2-benzopyrene — Table 4.55.

3,4-benzopyrene — Table 6.1, 6.4, 6.6, 6.19, Fig. 6.13.

4,5-benzopyrene — Table 4.55.

benzo[*a*]pyrene — Sects. 4.6.2.2, 4.6.2.3, 4.6.5, 4.6.6.2, 5.0, 5.5.1, 5.5.3, 5.5.4, 5.8.1, 6.1.1, 6.1.2.1, 6.1.3.2, 6.1.4.4, 6.1.5, 6.1.5.2, 7.0, 7.2.4, 7.2.4.1, 7.2.4.3, 7.2.4.4, 7.2.4.5, 8.3.1.1, 8.3.1.2, 8.3.1.3, 8.3.1.4, 8.3.1.5, 8.3.1.6, 8.3.1.7, 8.3.1.8, 8.3.1.9, 9.1.2, 9.1.2.1, 9.1.2.3, 9.1.2.5, 9.1.3, 9.1.4.5, 9.1.5.1, 9.1.7.3, 9.1.8, 9.1.8.2, 9.1.9, 9.1.10, 9.1.11.1, 10.1, 10.1.1.1, 10.1.1.2, 10.1.1.3, 10.1.2.1, 10.1.2.4, 10.1.3.1, 10.1.3.2, 10.1.3.3, 10.1.3.4, 10.1.3.5, 10.1.4, 10.1.4.1, 10.1.5, 10.1.5.1, 10.1.5.2, 10.1.5.3, 10.1.5.4, 10.1.5.5, 10.1.6.1, 10.1.6.2, 10.1.6.3, 10.1.7.3, 10.1.7.4, 10.1.7.5, 10.1.8.1, 10.1.8.2, 10.1.8.4, 10.1.8.5, 10.1.8.6; Tables 2.37, 4.61, 4.63, 4.64, 4.65, 4.66, 4.67, 4.68, 4.69, 4.70, 4.72, 5.17, 5.18, 5.22, 5.23, 5.25, 5.26, 5.29, 5.30, 5.33, 6.4, 6.5, 6.7, 6.8, 6.9, 6.10, 6.11, 6.15, 6.16, 6.17, 6.18, 6.20, 6.26, 6.27, 7.20, 7.21, 7.22, 7.23, 7.24, 7.25, 7.26, 7.27, 7.33, 7.34, 7.35, 7.36, 8.20, 8.21, 8.22, 8.23, 8.24, 8.25, 8.26, 8.27, 8.28, 8.29, 8.30, 8.31, 8.32, 9.1, 9.2, 9.3, 9.4, 9.5, 9.7, 9.10, 9.11, 9.28, 10.1, 10.9, 10.12, 10.13, 10.15, 10.25; Figs. 6.4, 6.13, 7.22, 7.23, 7.26, 10.1, 10.4, 10.7.

benzo[*a*]pyrene-1,6-dione — Sect. 10.1.5.2; Fig. 10.1.

benzo[*a*]pyrene-3,6-dione — Table 5.26; Fig. 10.1.

benzo[*c*]pyrene — Table 4.66.

benzo[*e*]pyrene — Sects. 4.6.2.2, 5.5.1, 5.5.3, 5.5.4, 5.8.1, 8.3.1.2, 8.3.1.9, 10.1.3.1, 10.1.5.2; Tables 2.37, 4.64, 4.65, 4.67, 4.68, 4.69, 4.70, 5.17, 5.18, 5.22, 5.25, 6.4, 6.5, 6.11, 8.26, 9.1, 9.2, 10.15; Fig. 6.13.

benzo[*ghi*]pyrene — Sect. 4.6.2.2.

benzo[*a*]pyrenequinone — Sect. 4.6.5; Table 6.5.

benzopyrene-1,6-quinone — Sect. 6.1.3.5.

benzopyrene-6,12-quinone — Sect. 6.1.3.5.

benzoquinone — Sects. 8.3.2.3, 8.3.2.4.

benzoquinoline — Tables 2.37, 4.66, 4.69, 5.17, 5.18, 5.21.

benzo[*f*]quinoline — Table 6.5.

Ra-benzo[*f*]quinoline — Table 6.5.

Rb-benzo[*f*]quinoline — Table 6.5.

benzo[*h*]quinoline — Table 6.5.

Ra-benzo[*h*]quinoline — Table 6.5.

Rb-benzo[*h*]quinoline — Table 6.5.

benzo[*c*]tetraphene — Tables 5.17, 5.18.

benzothionaphthene — Table 2.37.

benzothiophene — Sect. 2.6.3.10; Tables 2.34, 2.36, 4.17; Fig. 2.15.

benzothiophenols — Table 4.71.

benzoxanthane — Table 2.37.

benzo[*kl*]xanthene — Table 5.22.

benzoxanthanes — Sect. 2.6.3.9.

benzoxanthone — Fig. 2.15.

benzoylformic acid — Fig. 7.11.

6-benzoyloxymethylbenzo[*a*]pyrene — Table 5.26.

7-benzoyloxymethyl-12-methylbenz[*a*]anthracene — Table 5.27.

benzoyl peroxide — Sect. 6.1.5.2.

benzyl alcohol — Table 8.19.

benzylamine — Table 8.19.

benzyl thiocyanate — Sect. 10.1.5.4.

benzylthiophene — Table 2.36, 4.17.

1,2-benzypyrene — Table 6.4, Fig. 6.13.

beryllium — Sects. 2.5.2.1, 2.5.3.1, 4.3.2.1, 4.3.2.5, 5.2.2, 5.3, 5.3.1, 6.2, 6.2.2.1, 6.2.2.2, 6.2.2.3, 6.2.3.1, 8.4.4, 8.4.5.4, 9.3.4.9, 9.3.6.1; Tables 2.17, 2.18, 2.19, 2.20, 2.21, 2.22, 4.8, 4.23, 4.27, 4.31, 4.32, 4.33, 4.34, 4.36, 6.30, 6.31, 6.33, 6.34, 6.35, 6.36, 6.37, 6.38, 6.44, 6.45, 6.47, 6.48, 8.37, 8.38, 9.51, 9.56, 10.32; Tables 5.4, 5.5, 5.6, 5.7.

beryllium hydroxide — Table 9.28.

beryllium sulfate — Sect. 9.3.7.2.

bicarbonate — Sects. 8.2.11.1, 9.1.7.4; Table 8.39.

B_1B^1-binaphthyl — Tables 5.17, 5.18.

biphenols — Sect. 4.6.1.2; Table 4.71.

biphenyls — Sects. 2.6.3.1, 7.2.4.3, 9.1.2.2; Tables 4.59, 4.63, 9.8, 9.9, 9.15, 9.16; Tables 5.17, 5.18, 5.30.

bismuth — Sects. 4.3.2.1, 5.3, 6.2.3.1, 8.4.4, 9.3.4.9; Tables 2.19, 4.27, 4.32, 4.35, 5.5, 5.6, 5.7, 5.9, 6.29, 6.30, 6.34, 6.36, 6.37, 6.38, 6.45, 6.48, 8.3, 8.40, 9.51, 9.52, 9.55.

bisulfite — Sect. 8.2.5.5.

carotinoids — Table 2.27.

casamino acids — Sect. 7.2.1.6.

catechins — Table 2.27.

catechol — Sects. 4.6.5, 6.1.4.4, 7.2.1.3, 7.2.4.1, 8.3.2; Table 4.54; Figs. 4.64, 6.6, 6.7, 6.8, 6.10, 7.11, 7.21.

caustic acid — Sect. 4.4.1.

cellulose — Sects. 2.6.1.1, 2.6.1.6, 8.2.8.2, Table 2.27; Fig. 2.8.

cerium — Sects. 4.3.2.1, 8.4.4; Tables 4.27, 4.34, 5.6, 5.7, 5.9, 6.33, 6.34, 6.36, 6.37, 6.38, 6.45, 6.47, 6.61, 8.37; Figs. 6.20, 6.21.

cerotic acid — Table 2.26.

ceryl acid — Table 2.26.

cesium — Sects. 5.3.7, 9.3.5.5, 9.3.6.1; Tables 4.27, 4.34, 4.35, 5.6, 5.7, 5.9, 6.31, 6.33, 6.34, 6.35, 6.36, 6.37, 6.45, 6.60, 6.61, 9.56, 9.58, 9.64; Figs. 6.20, 6.21.

cetyl palmitate — Sect. 7.2.2.7.

char — Sects. 3.1.2.4, 3.1.4.2, 3.1.4.5, 3.1.4.8, 3.1.4.9, 3.1.4.10, 3.1.4.12, 3.2.1.1, 3.2.2.2, 3.2.2.3, 3.2.2.4, 3.2.2.10, 3.3.6.3; Tables 3.15, 3.20, 3.21, 3.25; Figs. 3.1, 3.17, 3.18, 3.19, 3.21, 3.22, 3.23, 3.24, 3.26, 3.27, 3.28, 3.31, 3.32, 3.40, 3.41, 3.42.

chinaldine — Table 4.52.

chinoline — Sects. 4.2.2, 4.6.1.2; Table 4.52.

chitin — Sect. 2.6.1.1; Fig. 2.8.

chloranil — Sect. 5.5.1.

chloride — Sects. 8.3.2.1, 9.3.1.4; Tables 8.33, 8.34.

chlorine — Sects. 8.2, 8.2.3, 8.2.13, 8.4.2.1, 8.6.1.2, 8.6.1.4, 9.3.7.3; Tables 5.2, 5.11, 8.16, 9.54, 9.55, 9.56.

chlorophyll — Sect. 8.2.4.2; Table 8.12.

z-chloro-6-(trichloromethyl)-pyridine — Sect. 8.3.3.

chloramines — Sect. 4.6.5.

chloramphenicol — Sect. 7.2.2.6.

chloride — Sect. 4.3.2.5; Tables 4.8, 4.38.

chlorine — Sects. 2.5.3.2, 3.1.4.1, 4.1.2.3, 4.3.2.5, 4.4.1, 4.4.5, 4.6.5, 4.7, 6.2.5.3, 7.1.2.1; Tables 2.22, 3.19, 3.20, 4.24, 4.27, 4.35, 4.36, 4.37, 4.72, 6.34, 6.35, 6.36, 6.27, 6.38, 6.45, 6.61; Figs. 6.20, 6.21.

chlorine dioxide — Sect. 4.4.5.

chlorite — Table 2.15.

chlorobenzene — Table 2.11.

chloroform — Sects. 6.1.3.4, 7.2.2.5, 7.2.2.7, 7.3.2.5; Table 7.15.

o-chlorophenol — Sect. 4.1.1.

cholesterol — Sect. 7.2.4.4; Table 7.36.

chromic oxide — Table 3.4.

dihydronaphthalene — Table 4.55.

dihydropyrene — Tables 5.17, 5.18.

dihydrotriphenylene — Table 5.18.

1,2-dihydroxyanthracene — Fig. 6.12.

dihydroxybenzene — Fig. 4.64.

7,8-dihydroxybenzopyrene — Sect. 9.1.4.5.

dihydroxydiphenol — Table 4.55.

1,2-dihydroxynaphthalene — Sect. 6.1.4.4; Figs. 6.8, 6.10, 6.11.

D-*trans*-1,2-dihydroxynaphthalene — Sect. 6.1.4.4.

D-*trans*-1,2-dihydro-1,2-dihydroxynaphthalene — Fig. 6.10.

2,3-dihydroxynaphthalene — Fig. 6.12.

D-*trans*-1:2-dihydronapthalenediol — Fig. 6.8.

3,4-dihydroxyphenanthrene — Sect. 6.1.4.4; Fig. 6.11.

dimeric hydrocarbons — Sect. 6.1.5.2.

4,4-bis(dimethylamino)benzophenone — Tables 7.30, 7.31, 7.32; Fig. 7.22.

dimethylaniline — Tables 3.35, 4.55, 10.24.

dimethylanthracene — Tables 2.37, 4.55, 5.18.

9,10-dimethylanthracene — Tables 7.30, 7.31, 7.32, 9.2; Fig. 7.22.

dimethylazaindole — Sect. 9.2; Table 2.38.

dimethylbenzacridine — Tables 7.30, 7.31, 7.32; Fig. 7.22.

7,12-dimethylbenz[*a*]anthracene — Sect. 9.1.7.3, 9.1.9, 9.1.10; Tables 5.17, 5.18, 9.18, 9.30, 9.31.

9,10-dimethyl-1,2-benzanthracene — Sect. 9.1.8.2; Tables 7.30, 7.31, 7.32; Fig. 7.22.

dimethylbenz[*a*]anthracene — Sect. 8.3.1.9.

7,12-dimethylbenz[*a*]anthracene — Sects. 10.1.1.1, 10.1.2.1, 10.1.2.2, 10.1.2.3, 10.1.3.1, 10.1.4.1, 10.1.4.2, 10.1.5.2, 10.1.5.4, 10.1.5.5, 10.1.6.1, 10.1.6.3, 10.1.7.3; Tables 10.8, 10.9, 10.15; Figs. 10.8, 10.9.

9,10-dimethylbenzo[*a*]anthracene — Tables 5.17, 5.18.

dimethylbenzo[*b*]fluoranthene — Table 5.18.

dimethylbenzo[*k*]fluoranthene — Table 5.18.

dimethylbenzo[*a*]pyrene — Table 5.18.

dimethylbenzo[*e*]pyrene — Table 5.18.

dimethylbenzothiophene — Tables 2.36, 4.17.

dimethylbiphenyl — Table 3.23; Fig. 4.63.

2,2-dimethylbutane — Table 6.21.

dimethylchrysene — Sect. 5.5.3; Table 5.18.

dimethylcoumarone — Table 4.55.

gallium — Sects. 4.3.2.1, 4.3.2.5, 5.3.3, 5.3.7, 6.2.4.2; Tables 2.17, 2.19, 2.20, 2.22, 4.23, 4.27, 4.35, 5.6, 5.7, 5.9, 6.29, 6.30, 6.31, 6.33, 6.34, 6.35, 6.36, 6.37, 6.38, 6.45, 6.47, 8.3, 9.55, 9.56.

gammaglobulin — Sect. 9.3.1.3.

gentisic acid — Sects. 6.1.4.4, 7.2.1.5; Fig. 7.14.

germanium — Sects. 2.5.3.2, 4.3.2.1, 5.3, 5.3.1, 8.4.4, 9.3.3.1; Tables 2.19, 2.20, 2.22, 4.23, 4.27, 4.28, 4.30, 4.32, 4.35, 5.9, 6.29, 6.30, 6.31, 6.33, 6.34, 6.36, 6.37, 6.38, 6.45, 6.48, 9.51, 9.52, 9.56.

germanium oxide — Table 2.17.

glucornide — Sects. 10.1.3, 10.1.5.2.

glucose — Sects. 7.2.2.2, 7.2.2.3, 7.2.6.4, 7.3.2.4, 8.2.12, 2; Tables 7.12, 7.44, 7.46; Fig. 7.17.

glucose 6-phosphate dehydrogenase — Sect. 8.2.4.2.

glutamate — Fig. 7.19.

glutarate — Sect. 7.2.5.1; Table 7.39.

glutarate semialdehyde — Sect. 7.2.5.1; Table 7.38.

glutarate semialdehyde dehydrogenase — Table 7.39.

glutaric dialdehyde dehydrogenase — Table 7.39.

glutathione — Sect. 8.2.8.2.

glyceraldehyde 3-phosphate dehydrogenase — Sect. 8.2.4.2.

glycerine — Sect. 2.6.1.4.

glyoxylate — Fig. 7.19.

gold — Sects. 5.3.3, 8.4.4, 9.3.6.1; Tables 4.32, 5.6, 5.7, 5.9, 6.34, 6.35, 6.36, 6.45, 6.61, 9.56, 9.71.

graphite — Sect. 2.8.2, 5.1.3.3, 5.2.2; Figs. 2.22, 2.23.

guaiacyglycerol-β-coniferylether — Sect. 2.6.1.6.

guanidines — Sect. 7.1.1.1.

guanine — Sect. 8.2.8.2.

gypsum — Sect. 4.1.2.3; Tables 2.15, 4.12; Fig. 4.21.

hafnium — Sect. 4.2.3.5; Tables 4.27, 5.9, 6.33, 6.34, 6.36, 6.37, 6.45.

heavy metals — Sect. 4.2.2.2, 4.3.2.1, 6.1.3.6, 6.2.3.1, 6.2.3.2, 6.2.3.4, 6.2.4, 6.2.4.1, 6.2.5.3, 7.0, 7.1.1.1, 7.3.2, 8.3.2.2, 8.4, 8.4.1, 8.4.2, 8.4.2.3, 8.4.3, 8.4.4, 8.4.5.4, 8.6.2.3, 9.3.1.5, 9.3.4.7, 9.3.5.3, 9.3.6.2, 9.3.7.4, 9.3.8.3, 9.3.8.4, 10.4.2.2; Tables 8.5, 9.63.

helium — Sects. 5.2.4, 5.5.3.

hematite — Table 2.15.

hemellitene — Table 4.55.

hemimellitene — Table 2.41.

heptadecane — Sect. 6.1.3.2; Table 10.24.

n-heptadecane — Table 2.37.

1,6-heptadiene — Table 6.22.

methylnaphthobenzothiophene — Table 2.36.

methylnitrocarbazole — Table 2.38.

methylnitroaniline — Table 2.38.

methyl phenanthrene — Table 2.37.

methyl pyrene — Table 2.37.

methyl pyridine — Sect. 7.2.5.1.

methylquinoline — Table 2.38.

methyltriphenylene — Table 5.18.

methyl viologen — Sect. 7.2.7; Table 7.49.

molybdenum — Sects. 2.6.3.10, 3.1.2.3, 3.2.2.7, 3.2.2.9, 4.1.2.3, 4.1.3, 4.3.2.1, 4.3.2.5, 4.7,
    5.2.2, 5.3.1, 5.3.3, 6.2, 6.2.3.1, 6.2.3.2, 6.2.3.4, 6.2.4.2, 6.2.4.3, 8.4, 8.4.2.1,
    8.4.5.13, 8.4.5.16, 8.6.2.1, 9.3.3.1, 9.3.4.9, 9.3.5.2; Tables 2.17, 2.18, 2.19, 2.20,
    2.22, 4.27, 4.30, 4.32, 4.34, 4.35, 4.73, 5.6, 5.7, 5.9, 5.11, 6.29, 6.30, 6.31, 6.32,
    6.33, 6.34, 6.35, 6.36, 6.37, 6.38, 6.44, 6.45, 6.48, 6.53, 6.54, 6.58, 6.60, 8.3, 8.37,
    8.38, 9.51, 9.54, 9.55, 9.56.

molybdenum sulfide — Table 4.7.

monobasic acids — Table 2.26.

monochloramine — Sect. 4.2.2.2.

monohexadecane — Sect. 6.1.4.4.

monohydroxy alcohols — Table 2.26.

monoterpene — Sect. 8.1.1.

montanic acid — Table 2.26.

montanone — Table 2.26.

montanyl acid — Table 2.26.

montmorillonite — Sect. 6.1.4.2; Table 2.15.

cis,cis-muconic acid — Fig. 6.6.

muscovite — Table 2.15.

myricinic acid — Table 2.26.

myricyl acid — Table 2.26.

naptha — Sects. 3.1.4.11, 3.2.2.1, 3.2.2.4, 3.2.2.5, 3.2.2.9, 3.2.2.11, 4.6, 4.6.1.2, 4.6.5,
    4.6.6.1; Tables 3.25, 4.51; Figs. 3.31, 3.35, 3.40, 3.43, 4.63.

naphthacene — Sect. 2.6.4.2, 5.5 1; Tables 2.37, 4.55, 5.22, 6.5.

naphthacenequinone — Fig. 2.15.

naphthahydroquinone sulfate — Sect. 4.1.2.1.

naphthalene — Sects. 2.6.4.1, 2.6.4.2, 4.6.1.1, 4.6.1.2, 4.6.3, 4.6.4, 4.6.5, 6.1.3.2, 6.1.3.6,
    6.1.4.4, 7.2.4.1, 7.2.4.2, 7.2.6.4, 8.3.2.1, 8.3.2.3, 9.1.2.1, 9.1.2.2, 9.1.2.3, 9.1.2.5,
    9.1.4.5, 9.1.4.6, 9.1.5.1, 9.1.5.3, 9.1.7.2, 9.1.7.4, 9.1.8.2, 9.1.11.1, 9.1.11.2, 9.2.5,
    10.1.1.1, 10.1.4.2; Tables 2.34, 3.23, 4.52, 4.53, 4.54, 4.55, 4.56, 4.57, 4.59, 4.63,
    6.5, 7.28, 7.44, 8.33, 8.34, 9.6, 9.8, 9.9, 9.15, 9.16, 9.17, 10.1, 10.24, 10.32; Figs.
    2.15, 2.16, 3.41, 4.63, 6.5, 6.8, 6.10, 7.11, 9.2, 9.3, 9.4, 9.5, 9.6, 9.9, 9.12, 9.14,
    10.11.

β-naphthamine — Table 10.32.

pentamethylbenzene — Table 2.41.

pentane — Tables 4.54, 4.55, 10.23.

n-pentane — Table 6.21.

pentene — Table 10.23.

1-pentene — Table 6.22.

2-pentene — Table 6.22.

1-pentyne — Table 6.23.

perchlorate — Sect. 6.1.4.4.

perimidylammonium sulfate — Sect. 5.4.2.3.

perinaphthoxanthrene — Sect. 10.1.5.4.

peropyrene — Table 2.38.

peroxides — Sect. 6.1.5.2.

peroxyacyl nitrate — Sects. 6.4.3.2, 8.2, 8.2.2, 8.2.3, 8.2.4, 8.2.7.1, 8.2.8, 8.2.8.1, 8.2.8.2, 8.2.15, 8.6.1.1, 8.6.1.2; Tables 8.1, 8.5, 8.10, 8.11, 8.12, 8.42, 8.43.

peroxybutynyl nitrate — Sects. 8.2.8, 8.2.15; Table 8.12.

peroxypropionly nitrate — Sects. 8.2.8, 8.2.15; Table 8.12.

perylene — Sects. 4.6.2.2, 4.6.6.2, 5.5.1, 5.5.3, 7.2.4.1, 8.3.1.2, 10.1.5.4, 10.1.8.1; Tables 2.37, 2.38, 3.23, 4.55, 4.64, 4.65, 4.66, 4.67, 4.69, 4.70, 5.17, 5.18, 5.22, 5.23, 5.25, 5.28, 6.4, 6.5, 6.11, 6.15, 7.26, 7.27, 8.26, 9.2, 10.23.

phellogenic acid — Table 2.26.

phellonic acid — Table 2.26.

phenalen-1-one — Table 6.5.

phenanthraquinone — Table 5.22.

phenanthrene — Sects. 4.6.1.1, 4.6.1.2, 4.6.2.2, 4.6.4, 5.5.1, 6.1.3.4, 6.1.4.4, 6.1.5.2, 7.2.4.1, 7.2.4.2, 7.2.4.5, 9.1.2, 9.1.2.2, 9.1.7.3, 9.2.5, 10.1.1.1, 10.1.2.4, 10.1.4.2, 10.1.5.5; Tables 2.34, 2.37, 3.23, 4.52, 4.54, 4.55, 4.56, 4.57, 4.59, 4.61, 4.63, 4.64, 4.65, 4.66, 4.67, 4.69, 4.70, 5.17, 5.18, 5.22, 5.24, 5.28, 6.4, 6.5, 7.28, 9.2, 9.8, 9.9, 9.15, 9.16, 10.1, 10.9, 10.23; Figs. 2.15, 2.16, 4.63, 6.10, 7.11, 7.21, 7.23, 7.24.

phenanthrene 9,10-oxide — Sect. 10.1.5.5.

phenanthridene — Table 4.55.

phenanthridine — Tables 2.37, 4.52, 4.66, 4.69, 5.21, 6.5.

phenanthridone — Tables 2.37, 5.22.

phenanthrol — Tables 2.37, 4.53, 4.55.

phenanthrylenemethane — Tables 2.37, 4.55.

phenobarbital — Sects. 10.1.5.2, 10.1.5.4, 10.1.6.2, 10.1.7.5.

phenol — Sects. 2.6.1.6, 2.6.4.1, 2.6.4.2, 2.8.1, 3.1.4.11, 3.2.2.11, 3.2.4.3, 4.0.1, 4.1.1, 4.2.2.2, 4.4.3.1, 4.4.5, 4.4.6, 4.6, 4.6.1.1, 4.6.1.2, 4.6.3, 4.6.5, 4.6.6, 5.5.4, 5.6, 6.1.3.6, 6.1.4.2, 7.0, 7.1.1.1, 7.2.1, 7.2.1.1, 7.2.1.3, 7.2.1.5, 7.2.1.6, 7.2.4.1, 8.2, 8.2.4.2, 8.2.5.4, 8.3.2, 8.3.2.1, 8.3.2.2, 8.3.2.3, 8.3.2.4, 8.6.1.3, 10.1.3, 10.1.3.1, 10.1.3.5, 10.1.4.2, 10.1.5.2, 10.1.5.3, 10.1.5.5, 10.1.7.4; Tables 3.23, 4.1, 4.2, 4.3, 4.38, 4.39, 4.40, 4.41, 4.48, 4.49, 4.51, 4.52, 4.53, 4.54, 4.55, 4.56, 4.57, 4.59, 4.71, 4.73, 7.50, 7.51, 8.19, 8.33, 8.34, 8.35, 8.36, 10.4, 10.5, 10.24, 10.32; Figs. 2.13, 2.18, 3.25, 3.41, 3.43, 4.58, 4.60, 4.63, 4.64, 7.11, 7.14, 10.2, 10.11.

quaiacylglycerol — Fig. 2.10.

quaiacylglycerol-β-coniferylether — Fig. 2.10.

quaiacylpyruvic acid — Fig. 2.10.

quaiaretic acid — Fig. 2.9.

quartz — Sects. 2.5.2.1, 2.5.3.2, 5.1.1.4; Table 2.15.

quaterphenyl — Table 2.41.

quinaldine — Table 9.34; Fig. 2.17.

quinhydrone — Table 8.19.

quinidines — Sect. 4.2.2.2.

quinoline — Sects. 2.6.3.10, 3.3.6.1, 4.6.1.2, 4.6.5, 9.2.3; Tables 2.36, 3.23, 3.35, 4.54, 4.55, 4.71, 5.24, 9.34, 10.24, 10.32.

quinone — Sects. 2.6.2, 6.1.5.2, 8.2.4.2, 8.2.5.5, 8.3.2.2, 10.1.3, 10.1.3.1, 10.1.3.3, 10.1.4.2, 10.1.5.2; Tables 8.19, 10.6; Fig. 2.13.

o-quinone — Sects. 8.2.4.2, 8.3.2, 8.3.2.2, 8.3.2.3.

quinone carbonyl — Sect. 2.6.3.5.

radium — Sects. 4.8, 6.2.3.1, 9.3.4.9; Tables 6.29, 6.33, 6.34, 6.35, 6.36, 6.45, 9.56.

radon — Tables 6.34, 6.35.

rectisol — Sect. 4.9.

resin — Sects. 2.6, 2.6.1.5.

resorcin — Table 8.36.

resorcinol — Sects. 4.6.5, 7.2.1.2, 8.3.2, 8.3.2.1, 8.3.2.4; Tables 4.54, 8.19, 10.24, 10.32.

retinoate — Table 10.8.

retinol — Sect. 10.1.5.4; Table 10.8; Fig. 10.6.

retinyl acetate — Table 10.8.

retinyl palmitate — Table 10.8.

rhenium — Tables 5.6, 5.7, 5.9, 6.34, 6.38, 6.45.

rhodium — Tables 5.6, 5.7, 5.9, 6.34, 6.45.

ribonuclease — Sect. 8.2.4.2.

ribonucleic acid — Sects. 10.1.5.2, 10.1.5.5, 10.4.2.7.

rosaniline — Sect. 5.4.1.

rosmerite — Table 2.16.

rozenite — Table 2.16.

rubidium — Sects. 4.3.2.1, 4.3.2.5, 5.3.7, 6.2.3.2; Tables 4.27, 4.35, 5.4, 5.5, 5.8, 5.9, 6.29, 6.31, 6.33, 6.34, 6.35, 6.36, 6.37, 6.38, 6.45, 6.60, 6.61, 8.37, 9.51, 9.55, 9.56; Figs. 6.20, 6.21.

ruthenium — Sect. 3.1.2.3; Tables 4.73, 5.6, 5.7, 5.9, 6.34, 6.45, 9.56.

rutile — Table 2.15.

saccharose — Sect. 7.2.2.2.

salicyaldehyde — Fig. 6.10.

salicylate — Fig. 6.10.

salicylic acid — Sects. 6.1.4.4, 7.2.1.3, 7.2.4.1; Table 7.12; Figs. 6.8, 7.11, 7.21.

samarium — Sect. 4.3.2.5; Tables 4.27, 4.32, 5.6, 5.7, 5.9, 6.34, 6.37, 6.38, 6.45.

sandstone — Sect. 6.2.1.1.

scandium — Sects. 4.3.2.1, 4.3.2.5, 6.2.3.2, 8.4.2.2, 9.3.6.1; Tables 2.21, 4.23, 4.27, 4.34, 4.35, 5.6, 5.7, 5.9, 6.29, 6.31, 6.33, 6.34, 6.36, 6.37, 6.38, 6.45, 6.47, 6.61, 9.56; Figs. 6.20, 6.21.

selenium — Sects. 2.5.3.1, 2.5.3.2, 4.1.2.3, 4.3.2.1, 5.3, 5.3.7, 6.2.3.4, 6.2.5.3, 8.4, 8.4.1.1, 8.4.2.2, 8.4.2.3, 8.4.5, 8.4.5.16, 8.6.2.1, 9.3.1.3, 9.3.1.9, 9.3.3.1, 9.3.5.2, 9.3.5.4, 9.3.7.4, 9.3.8.1, 9.3.8.5, 10.1.5.4, 10.4, 10.4.2.6; Tables 2.22, 2.23, 4.1, 4.8, 4.27, 4.28, 4.31, 4.32, 4.33, 4.34, 4.35, 4.36, 5.8, 5.9, 6.29, 6.31, 6.33, 6.34, 6.35, 6.36, 6.37, 6.38, 6.45, 6.48, 6.58, 6.59, 6.60, 6.61, 8.3, 8.37, 8.38, 8.40, 9.51, 9.52, 9.53, 9.54, 9.55, 9.72, 10.32; Figs. 6.20, 6.21.

selenium oxide — Sects. 4.3.2.5, 9.3.1.9.

siderite — Table 2.15.

silica — Sects. 2.6.3.10, 3.2.2.8, 3.2.2.9, 4.3.2.1, 4.3.2.5, 5.1.3.1; Tables 4.7, 4.20, 4.21, 4.22, 4.23, 4.37, 4.73.

silica gel — Sects. 4.1.2.3, 6.1.5.2.

silicates — Sects. 2.5.2.2, 2.5.3.2; Table 2.15.

silicon — Sects. 4.2.3.5, 5.1.3.3, 5.3.3, 6.2.1.1, 6.2.3.2, 8.4.2.2; Tables 2.22, 2.30, 4.27, 4.29, 4.32, 4.35, 5.11, 6.33, 6.34, 6.35, 6.36, 6.37, 6.45, 8.3, 8.37, 8.40.

silicon oxides — Sect. 2.5.2.2; Tables 5.1, 5.3.

silver — Sects. 4.3.2.1, 4.3.2.5, 5.2.2, 5.3, 5.3.1, 5.3.3, 5.3.5, 6.2.3.1, 6.2.3.2, 6.2.3.5, 8.4.4, 8.4.5.17, 9.3.1.10, 9.3.3.3, 9.3.4.4, 9.3.4.6, 9.3.5.2, 9.3.6.1, 9.3.8.2; Tables 2.19, 4.23, 4.27, 4.32, 4.34, 5.4, 5.5, 5.6, 5.7, 5.9, 6.29, 6.30, 6.31, 6.34, 6.35, 6.36, 6.37, 6.38, 6.44, 6.45, 6.47, 6.48, 6.53, 6.54, 6.57, 6.58, 6.59, 6.60, 8.3, 8.37, 9.53, 9.55, 9.56.

silver nitrate — Tables 7.50, 7.51, 9.61.

sinapin alcohol — Sect. 2.6.1.2.

skatole — Sects. 4.6.1.2, 8.2.15.

sodium — Sects. 2.5.3.2, 2.8.1, 4.1.2.3, 4.3.2.1, 4.3.2.5, 5.1.3.3, 5.3.3, 6.2, 6.2.3.3, 6.2.3.4, 6.2.5.3, 8.4.2.1, 8.4.4, 9.3.1.4, 9.3.6.1, 9.3.8.1; Tables 2.22, 4.27, 4.29, 4.32, 4.34, 4.36, 4.37, 5.4, 5.5, 5.6, 5.7, 5.9, 6.29, 6.31, 6.33, 6.34, 6.35, 6.36, 6.37, 6.38, 6.45, 6.47, 6.61, 9.54, 9.55, 9.56, 9.69, 9.71, Figs. 6.20, 6.21.

sodium arsenate — Sect. 4.1.2.1.

sodium arsenite — Sect. 4.1.2.1.

sodium bicarbonate — Sects. 4.1.2.1, 7.2.1.6.

sodium bisulfide — Sect. 4.1.2.1.

sodium bisulfite — Sect. 4.1.2.3.

sodium carbonate — Sects. 4.1.2.1, 4.1.2.3, 7.2.5.1; Tables 2.30, 4.73; Fig. 4.28.

sodium chloride — Sects. 4.2.2.2, 8.2.11.1, 10.1.5.2.

sodium dichromate — Sect. 2.6.3.9; Tables 2.34, 9.42.

sodium fluoride — Sect. 8.2.9.1.

sodium glutamate — Sect. 6.1.4.4.

sodium hydroxide — Sects. 3.1.2.3, 4.1.1, 4.1.2.3, 4.4.1, 5.4.1, 5.4.2.2, 8.6.2.2; Tables 2.30, 7.6, 7.46.

sodium hypohalite — Sect. 2.6.3.9.

sodium lauryl sulphate — Sect. 6.1.3.4.

sodium 1,4-naphthoquinone-2-sulfonate — Sect. 4.1.2.1.

sodium nitrate — Table 7.6.

sodium oleate — Table 7.36.

sodium oxides — Sects. 2.5.2.2, 4.3.2.1; Tables 4.7, 4.20, 4.22, 4.23, 5.1, 5.3.

sodium silicate — Sect. 7.2.1.6.

sodium sulfate — Sect. 4.1.2.3; Tables 4.11, 7.6, 7.46; Fig. 4.20.

sodium sulfite — Sects. 4.1.2.1, 4.1.2.3; Table 4.73; Fig. 4.28.

sodium tetrachlorometcurate — Sects. 5.4.1, 5.4.2.3.

sodium thiocyanate — Sect. 4.1.2.1.

sodium thiosulfate — Tables 7.4, 7.6.

sodium vanadate — Sect. 4.1.2.1.

stannous chloride — Sect. 3.2.2.9.

starch — Sect. 2.6.1.1; Fig. 2.8.

steroid — Sects. 2.6.1.5, 6.1.4.4.

sterols — Sects. 2.6.1.4, 7.2.4.4; Table 2.7.

stilbene — Table 2.41.

strontium — Sects. 4.3.2.1, 4.3.2.5, 4.4.5, 8.4.4, 9.3.8.1; Tables 2.18, 2.19, 4.23, 4.27, 4.28, 4.32, 4.35, 5.4, 5.5, 5.6, 5.7, 5.8, 5.9, 6.33, 6.34, 6.35, 6.36, 6.37, 6.38, 6.44, 6.45, 6.47, 8.3, 8.38, 8.40, 9.55, 9.56, 9.64, 9.69, 9.71.

styrene — Tables 2.41, 4.54, 4.55.

succinate — Sects. 7.2.4.3, 7.2.5.1; Table 7.39; Fig. 7.19.

succinic anhydride — Table 7.36.

succinic semialdehyde — Sect. 7.2.5.1.

sulfate — Sects. 2.5.2.1, 2.5.2.2, 2.5.3.4, 4.1.2.3, 4.6.6.1, 5.4.2.2, 5.4.2.3, 5.4.4, 6.3.1, 6.3.2, 6.3.3.1, 7.1.2.2, 7.1.2.4, 7.2.6.3, 8.2.5.2, 8.2.5.4, 8.2.6.3, 8.2.11.1, 8.3.2.1, 8.4.1, 9.3.1.4, 10.1.5.2, 10.2.1, 10.2.2, 10.2.4; Tables 2.15, 3.19, 4.8, 4.37, 7.7, 8.1, 8.33, 8.34, 8.39, 10.26, 10.27; Figs. 4.33, 6.22, 6.24, 7.8, 7.9.

sulfides — Sects. 2.5.2.1, 2.5.3.2, 2.6.3.1, 4.0.1, 4.1, 4.1.4.2, 4.2.2.2, 4.3.2.1, 4.3.2.5, 4.6.1, 4.6.5, 4.6.6.1, 5.1.3.1, 5.4.3, 6.2.3.4, 6.2.4.2, 7.2.7, 8.2.15, 8.3.2.1; Tables 2.15, 4.2, 4.3, 4.10, 4.16, 4.39, 8.33, 8.34; Figs. 7.6, 7.7, 7.8.

sulfite — Sects. 4.1.2.3, 4.3.2.1, 8.2.5.4, 8.2.5.5; Tables 4.7, 4.8, 4.21, 4.23, 8.33, 8.34; Fig. 4.33.

sulfonic acid — Sect. 6.1.5.2.

sulfur — Sects. 2.5.1.13, 2.5.2, 2.5.3.4, 2.6.2, 2.6.3.8, 2.6.3.10, 3.0, 3.1.2.2, 3.1.2.3, 3.1.3.3, 3.1.4.4, 3.1.4.5, 3.1.4.7, 3.1.4.8, 3.1.4.9, 3.1.4.11, 3.1.4.12, 3.2, 3.2.1.1, 3.2.2.1, 3.2.2.2, 3.2.2.3, 3.2.2.4, 3.2.2.5, 3.2.2.6, 3.2.2.7, 3.2.2.8, 3.2.2.9, 3.2.2.10, 3.2.2.11, 3.3.6.1, 3.3.6.2, 4.03, 4.1, 4.1.1, 4.1.2, 4.1.2.1, 4.1.2.2, 4.1.2.3, 4.1.3,